INSTRUCTOR
SOLUTIONS MANUAL

RANDY KRIEGER
TERRY A. KRIEGER

Rochester Community and Technical College

With contributions from Wayne Larson and typesetting assistance from Tom Krieger

A GRAPHICAL APPROACH TO PRECALCULUS WITH LIMITS:
A UNIT CIRCLE APPROACH

FOURTH EDITION

John Hornsby
University of New Orleans

Margaret L. Lial
American River College

Gary K. Rockswold
Minnesota State University, Mankato

PEARSON

Addison
Wesley

Boston San Francisco New York
London Toronto Sydney Tokyo Singapore Madrid
Mexico City Munich Paris Cape Town Hong Kong Montreal

Reproduced by Pearson Addison-Wesley from QuarkXPress® files.

Copyright © 2007 Pearson Education, Inc.
Publishing as Pearson Addison-Wesley, 75 Arlington Street, Boston, MA 02116.

ISBN   0-321-35813-9

1 2 3 4 5 6 OPM 09 08 07 06

Table of Contents

Preface

This manual contains all solutions to the exercises in *A Graphical Approach to Precalculus with Limits: A Unit Circle Approach,* 4e by Hornsby, Lial, and Rockswold. The solutions have been written in a manner that is consistent with the methods and techniques described in the text. Over 1000 graphs have been included in an effort to provide solutions that closely follow the graphical approach used in the text. Both traditional graphs and TI calculator screens have been included.

In addition to providing solutions to exercises, this manual can be helpful when selecting assignments for students. By merely browsing the solutions, an instructor can observe both the variety of exercises and the large number of application exercises available. The exercises have varying levels of difficulty. Instructors can quickly select both appropriate and meaningful exercises for their students. This manual also provides direction and guidance for explaining solutions to students in the classroom. Many of the applications included are real-data exercises that are unique to this text. This manual is consistent with the text and gives solutions that are accessible to most college algebra students.

It is this author's hope that you will find this manual helpful when teaching from *A Graphical Approach to Precalculus with Limits: A Unit Circle Approach.* Please email any comments or questions to *terry.krieger@roch.edu.* Your opinion is important. Best wishes for an enjoyable and successful precalculus or algebra and trigonometry course.

Terry A. Krieger

Rochester Community and Technical College

Chapter 1: Linear Functions, Equations, and Inequalities

1.1: Real Numbers and the Rectangular Coordinate System

1. (a) The only natural number is 10.

    (b) The whole numbers are 0 and 10.

    (c) The integers are $-6, -\dfrac{12}{4}$ (or -3), 0, 10

    (d) The rational numbers are $-6, -\dfrac{12}{4}$ (or -3), $-\dfrac{5}{8}$, 0, .31, $.\overline{3}$, and 10.

    (e) The irrational numbers are $-\sqrt{3}$, 2π, and $\sqrt{17}$.

    (f) All of the numbers listed are real numbers.

2. (a) The natural numbers are $\dfrac{6}{2}$ (or 3), 8, and $\sqrt{81}$ (or 9).

    (b) The whole numbers are $0, \dfrac{6}{2}$ (or 3), 8, and $\sqrt{81}$ (or 9).

    (c) The integers are $-8, -\dfrac{14}{7}$ (or -2), 0, $\dfrac{6}{2}$ (or 3), 8, and $\sqrt{81}$ (or 9).

    (d) The rational numbers are $-8, -\dfrac{14}{7}$ (or -2), $-.245$, 0, $\dfrac{6}{2}$ (or 3), 8, and $\sqrt{81}$ (or 9).

    (e) The only irrational number is $\sqrt{12}$.

    (f) All of the numbers listed are real numbers.

3. (a) There are no natural numbers listed.

    (b) There are no whole numbers listed.

    (c) The integers are $-\sqrt{100}$ (or -10) and -1.

    (d) The rational numbers are $-\sqrt{100}$ (or -10), $-\dfrac{13}{6}$, -1, 5.23, $9.\overline{14}$, 3.14, and $\dfrac{22}{7}$.

    (e) There are no irrational numbers listed.

    (f) All of the numbers listed are real numbers.

4. (a) The natural numbers are 3, 18, and 56.

    (b) The whole numbers are 3, 18, and 56.

    (c) The integers are $-\sqrt{49}$ (or -7), 3, 18, and 56.

    (d) The rational numbers are $-\sqrt{49}$ (or -7), $-.405$, $-.\overline{3}$, .1, 3, 18, and 56.

    (e) The only irrational number is 6π.

    (f) All of the numbers listed are real numbers.

5. The number 4,000,000,000 is a natural number, integer, rational number, and real number.

6. The number 92,000,000,000 is a natural number, integer, rational number, and real number.

7. The number -250 is an integer, rational number, and real number.

8. The number -3 is an integer, rational number, and real number.

9. The number $\frac{2}{9}$ is a rational number and real number.

10. The number -3.5 is a rational number and real number.

11. The number $5\sqrt{2}$ is a real number.

12. The number π is a real number.

13. Natural numbers would be appropriate because population is only measured in positive whole numbers.

14. Natural numbers would be appropriate because distance on road signs is only given in positive whole numbers.

15. Rational numbers would be appropriate because shoes come in fraction sizes.

16. Rational numbers would be appropriate because gas is paid for in dollars and cents, a decimal part of a dollar.

17. Integers would be appropriate because temperature is given in positive and negative whole numbers.

18. Integers would be appropriate because golf scores are given in positive and negative whole numbers.

19.

20.

21.

22.

23. A rational number can be written as a fraction, $\frac{p}{q}$, $q \neq 0$, where p and q are integers. An irrational number cannot be written in this way.

24. She should write $\sqrt{2} \approx 1.414213562$. Calculators give only approximations of irrational numbers.

25. The point $(2, 3)$ is in quadrant I. See Figure 25-34.

26. The point $(-1, 2)$ is in quadrant II. See Figure 25-34.

27. The point $(-3, -2)$ is in quadrant III. See Figure 25-34.

28. The point $(1, -4)$ is in quadrant IV. See Figure 25-34.

29. The point $(0, 5)$ is located on the y-axis, therefore is not in a quadrant. See Figure 25-34.

30. The point $(-2, -4)$ is in quadrant III. See Figure 25-34.

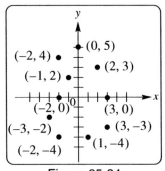

Figure 25-34

31. The point $(-2, 4)$ is in quadrant II. See Figure 25-34.

32. The point $(3, 0)$ is located on the x-axis, therefore is not in a quadrant. See Figure 25-34.

33. The point $(-2, 0)$ is located on the x-axis, therefore is not in a quadrant. See Figure 25-34.

34. The point $(3, -3)$ is in quadrant IV. See Figure 25-34.

35. If $xy > 0$, then either $x > 0$ and $y > 0 \Rightarrow$ Quadrant I, or $x < 0$ and $y < 0 \Rightarrow$ Quadrant III.

36. If $xy < 0$, then either $x > 0$ and $y < 0 \Rightarrow$ Quadrant IV, or $x < 0$ and $y > 0 \Rightarrow$ Quadrant II.

37. If $\dfrac{x}{y} < 0$, then either $x > 0$ and $y < 0 \Rightarrow$ Quadrant IV, or $x < 0$ and $y > 0 \Rightarrow$ Quadrant II.

38. If $\dfrac{x}{y} > 0$, then either $x > 0$ and $y > 0 \Rightarrow$ Quadrant I, or $x < 0$ and $y < 0 \Rightarrow$ Quadrant III.

39. Any point of the form $(0, b)$ is located on the y-axis.

40. Any point of the form $(a, 0)$ is located on the x-axis.

41. $[-5, 5]$ by $[-25, 25]$

42. $[-25, 25]$ by $[-5, 5]$

43. $[-60, 60]$ by $[-100, 100]$

44. $[-100, 100]$ by $[-60, 60]$

45. $[-500, 300]$ by $[-300, 500]$

46. $[-300, 300]$ by $[-375, 150]$

47. See Figure 47.

48. See Figure 48.

49. See Figure 49.

50. See Figure 50.

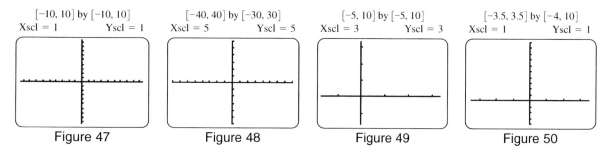

Figure 47   Figure 48   Figure 49   Figure 50

51. See Figure 51.

52. See Figure 52.

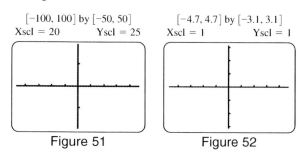

Figure 51   Figure 52

53. There are no tick marks, which is a result of setting Xscl and Yscl to 0.

54. The axes appear thicker because the tick marks are so close together. The problem can be fixed by using larger values for Xscl and Yscl such as Xscl = Yscl = 10.

55. $\sqrt{58} \approx 7.615773106 \approx 7.616$

56. $\sqrt{97} \approx 9.848857802 \approx 9.849$

57. $\sqrt[3]{33} \approx 3.20753433 \approx 3.208$

58. $\sqrt[3]{91} \approx 4.497941445 \approx 4.498$

59. $\sqrt[4]{86} \approx 3.045261646 \approx 3.045$

60. $\sqrt[4]{123} \approx 3.330245713 \approx 3.330$

61. $19^{1/2} \approx 4.358898944 \approx 4.359$

62. $29^{1/3} \approx 3.072316826 \approx 3.072$

63. $46^{1.5} \approx 311.9871792 \approx 311.987$

64. $23^{2.75} \approx 5555.863268 \approx 5555.863$

65. $(5.6 - 3.1)/(8.9 + 1.3) \approx .25$

66. $(34 + 25)/23 \approx 2.57$

67. $\sqrt{(\pi^\wedge 3 + 1)} \approx 5.66$

68. $\sqrt[3]{(2.1 - 6^2)} \approx -3.24$

69. $3(5.9)^2 - 2(5.9) + 6 = 98.63$

70. $2\pi^\wedge 3 - 5\pi^2 - 3 \approx 9.66$

71. $\sqrt{((4 - 6)^2 + (7 + 1)^2)} \approx 8.25$

72. $\sqrt{((-1 - (-3))^2 + (-5 - 3)^2)} \approx 8.25$

73. $\sqrt{(\pi - 1)}/\sqrt{(1 + \pi)} \approx .72$

74. $\sqrt[3]{(4.5E5 + 3.7E2)} \approx 76.65$

75. $2/(1 - \sqrt[3]{(5)}) \approx -2.82$

76. $1 - 4.5/(3 - \sqrt{(2)}) \approx -1.84$

77. $a^2 + b^2 = c^2 \Rightarrow 8^2 + 15^2 = c^2 \Rightarrow 64 + 225 = c^2 \Rightarrow 289 = c^2 \Rightarrow c = 17$

78. $a^2 + b^2 = c^2 \Rightarrow 7^2 + 24^2 = c^2 \Rightarrow 49 + 576 = c^2 \Rightarrow 625 = c^2 \Rightarrow c = 25$

79. $a^2 + b^2 = c^2 \Rightarrow 13^2 + b^2 = 85^2 \Rightarrow 169 + b^2 = 7225 \Rightarrow b^2 = 7056 \Rightarrow b = 84$

80. $a^2 + b^2 = c^2 \Rightarrow 14^2 + b^2 = 50^2 \Rightarrow 196 + b^2 = 2500 \Rightarrow b^2 = 2304 \Rightarrow b = 48$

81. $a^2 + b^2 = c^2 \Rightarrow 5^2 + 8^2 = c^2 \Rightarrow 25 + 64 = c^2 \Rightarrow 89 = c^2 \Rightarrow c = \sqrt{89}$

82. $a^2 + b^2 = c^2 \Rightarrow 9^2 + 10^2 = c^2 \Rightarrow 81 + 100 = c^2 \Rightarrow 181 = c^2 \Rightarrow c = \sqrt{181}$

83. $a^2 + b^2 = c^2 \Rightarrow a^2 + (\sqrt{13})^2 = (\sqrt{29})^2 \Rightarrow a^2 + 13 = 29 \Rightarrow a^2 = 16 \Rightarrow a = 4$

84. $a^2 + b^2 = c^2 \Rightarrow a^2 + (\sqrt{7})^2 = (\sqrt{11})^2 \Rightarrow a^2 + 7 = 11 \Rightarrow a^2 = 4 \Rightarrow a = 2$

85. (a) $d = \sqrt{(2 - (-4))^2 + (5 - 3)^2} = \sqrt{(6)^2 + (2)^2} = \sqrt{36 + 4} = \sqrt{40} = 2\sqrt{10}$

    (b) $M = \left(\dfrac{-4 + 2}{2}, \dfrac{3 + 5}{2}\right) = \left(\dfrac{-2}{2}, \dfrac{8}{2}\right) = (-1, 4)$

86. (a) $d = \sqrt{(2-(-3))^2 + ((-1)-4)^2} = \sqrt{(5)^2 + (-5)^2} = \sqrt{25+25} = \sqrt{50} = 5\sqrt{2}$

    (b) $M = \left(\dfrac{-3+2}{2}, \dfrac{4+(-1)}{2}\right) = \left(\dfrac{-1}{2}, \dfrac{3}{2}\right) = \left(-\dfrac{1}{2}, \dfrac{3}{2}\right)$

87. (a) $d = \sqrt{(6-(-7))^2 + (-2-4)^2} = \sqrt{(13)^2 + (-6)^2} = \sqrt{169+36} = \sqrt{205}$

    (b) $M = \left(\dfrac{-7+6}{2}, \dfrac{4+(-2)}{2}\right) = \left(\dfrac{-1}{2}, \dfrac{2}{2}\right) = \left(-\dfrac{1}{2}, 1\right)$

88. (a) $d = \sqrt{(1-(-3))^2 + (4-(-3))^2} = \sqrt{(4)^2 + (7)^2} = \sqrt{16+49} = \sqrt{65}$

    (b) $M = \left(\dfrac{-3+1}{2}, \dfrac{-3+4}{2}\right) = \left(\dfrac{-2}{2}, \dfrac{1}{2}\right) = \left(-1, \dfrac{1}{2}\right)$

89. (a) $d = \sqrt{(2-5)^2 + (11-7)^2} = \sqrt{(-3)^2 + (4)^2} = \sqrt{9+16} = \sqrt{25} = 5$

    (b) $M = \left(\dfrac{5+2}{2}, \dfrac{7+11}{2}\right) = \left(\dfrac{7}{2}, \dfrac{18}{2}\right) = \left(\dfrac{7}{2}, 9\right)$

90. (a) $d = \sqrt{(4-(-2))^2 + ((-3)-5)^2} = \sqrt{(6)^2 + (-8)^2} = \sqrt{36+64} = \sqrt{100} = 10$

    (b) $M = \left(\dfrac{-2+4}{2}, \dfrac{5+(-3)}{2}\right) = \left(\dfrac{2}{2}, \dfrac{2}{2}\right) = (1, 1)$

91. (a) $d = \sqrt{(-3-(-8))^2 + ((-5)-(-2))^2} = \sqrt{(5)^2 + (-3)^2} = \sqrt{25+9} = \sqrt{34}$

    (b) $M = \left(\dfrac{-8+(-3)}{2}, \dfrac{-2+(-5)}{2}\right) = \left(\dfrac{-11}{2}, \dfrac{-7}{2}\right) = \left(-\dfrac{11}{2}, -\dfrac{7}{2}\right)$

92. (a) $d = \sqrt{(6-(-6)^2 + (5-(-10))^2} = \sqrt{(12)^2 + (15)^2} = \sqrt{144+225} = \sqrt{369} = 3\sqrt{41}$

    (b) $M = \left(\dfrac{-6+6}{2}, \dfrac{-10+5}{2}\right) = \left(\dfrac{0}{2}, \dfrac{-5}{2}\right) = \left(0, -\dfrac{5}{2}\right)$

93. (a) $d = \sqrt{(6.2-9.2)^2 + (7.4-3.4)^2} = \sqrt{(-3)^2 + (4)^2} = \sqrt{9+16} = \sqrt{25} = 5$

    (b) $M = \left(\dfrac{9.2+6.2}{2}, \dfrac{3.4+7.4}{2}\right) = \left(\dfrac{15.4}{2}, \dfrac{10.8}{2}\right) = (7.7, 5.4)$

94. (a) $d = \sqrt{(3.9-8.9)^2 + (13.6-1.6)^2} = \sqrt{(-5)^2 + (12)^2} = \sqrt{25+144} = \sqrt{169} = 13$

    (b) $M = \left(\dfrac{8.9+3.9}{2}, \dfrac{1.6+13.6}{2}\right) = \left(\dfrac{12.8}{2}, \dfrac{15.2}{2}\right) = (6.4, 7.6)$

95. (a) $d = \sqrt{(6x-13x)^2 + (x-(-23x))^2} = \sqrt{(-7x)^2 + (24x)^2} = \sqrt{49x^2 + 576x^2} = \sqrt{625x^2} = 25x$

    (b) $M = \left(\dfrac{13x+6x}{2}, \dfrac{-23x+x}{2}\right) = \left(\dfrac{19x}{2}, \dfrac{-22x}{2}\right) = \left(\dfrac{19}{2}x, -11x\right)$

96. (a) $d = \sqrt{(20y-12y)^2 + (12y-(-3y))^2} = \sqrt{(8y)^2 + (15y)^2} = \sqrt{64y^2 + 225y^2} = \sqrt{289y^2} = 17y$

    (b) $M = \left(\dfrac{12y+20y}{2}, \dfrac{(-3y)+12y}{2}\right) = \left(\dfrac{32y}{2}, \dfrac{9y}{2}\right) = \left(16y, \dfrac{9}{2}y\right)$

97. Using the midpoint formula we get: $\left(\dfrac{7+x_2}{2}, \dfrac{-4+y_2}{2}\right) = (8, 5) \Rightarrow \dfrac{7+x_2}{2} = 8 \Rightarrow 7+x_2 = 16 \Rightarrow x_2 = 9$

    and $\dfrac{-4+y_2}{2} = 5 \Rightarrow -4+y_2 = 10 \Rightarrow y_2 = 14$. Therefore the coordinates are: $Q(9, 14)$.

98. Using the midpoint formula we get: $\left(\dfrac{13+x_2}{2}, \dfrac{5+y_2}{2}\right) = (-2, -4) \Rightarrow \dfrac{13+x_2}{2} = -2 \Rightarrow 13+x_2 = -4 \Rightarrow$

    $x_2 = -17$ and $\dfrac{5+y_2}{2} = -4 \Rightarrow 5+y_2 = -8 \Rightarrow y_2 = -13$. Therefore the coordinates are: $Q(-17, -13)$.

99. Using the midpoint formula we get: $\left(\dfrac{5.64 + x_2}{2}, \dfrac{8.21 + y_2}{2}\right) = (-4.04, 1.60) \Rightarrow \dfrac{5.64 + x_2}{2} = -4.04 \Rightarrow$

    $5.64 + x_2 = -8.08 \Rightarrow x_2 = -13.72$ and $\dfrac{8.21 + y_2}{2} = 1.60 \Rightarrow 8.21 + y_2 = 3.20 \Rightarrow y_2 = -5.01.$ Therefore

    the coordinates are: $Q(-13.72, -5.01).$

100. Using the midpoint formula we get: $\left(\dfrac{-10.32 + x_2}{2}, \dfrac{8.55 + y_2}{2}\right) = (1.55, -2.75) \Rightarrow \dfrac{-10.32 + x_2}{2} = 1.55 \Rightarrow$

    $-10.32 + x_2 = 3.10 \Rightarrow x_2 = 13.42$ and $\dfrac{8.55 + y_2}{2} = -2.75 \Rightarrow 8.55 + y_2 = -5.50 \Rightarrow y_2 = -14.05.$

    Therefore the coordinates are: $Q(13.42, -14.05).$

101. $M = \left(\dfrac{1998 + 2004}{2}, \dfrac{14{,}709 + 20{,}082}{2}\right) = \left(\dfrac{4002}{2}, \dfrac{34{,}791}{2}\right) = (2001, 17{,}395.5);$ the cost was $\approx \$17{,}396.$

102. For 2006, $M = \left(\dfrac{2005 + 2007}{2}, \dfrac{6038 + 6146}{2}\right) = \left(\dfrac{4012}{2}, \dfrac{12184}{2}\right) = (2006, 6092);$ enrollment was

    6092 thousand. For 2008, $M = \left(\dfrac{2007 + 2009}{2}, \dfrac{6146 + 6257}{2}\right) = \left(\dfrac{4016}{2}, \dfrac{12403}{2}\right) = (2008, 6201.5);$

    enrollment was ≈ 6202 thousand.

103. In 1988, $M = \left(\dfrac{1983 + 1993}{2}, \dfrac{10{,}178 + 14{,}763}{2}\right) = \left(\dfrac{3976}{2}, \dfrac{24{,}941}{2}\right) = (1988, 12{,}470.5);$ poverty level was

    $\approx \$12{,}471.$ In 1998, $M = \left(\dfrac{1993 + 2003}{2}, \dfrac{14{,}763 + 18{,}810}{2}\right) = \left(\dfrac{3996}{2}, \dfrac{33{,}573}{2}\right) = (1998, 16{,}786.5);$ it was

    $\approx \$16{,}787.$

104. (a) From $(0, 0)$ to $(3, 4)$: $d_1 = \sqrt{(3 - 0)^2 + (4 - 0)^2} = \sqrt{(3)^2 + (4)^2} = \sqrt{9 + 16} = \sqrt{25} = 5.$

    From $(3, 4)$ to $(7, 1)$: $d_2 = \sqrt{(7 - 3)^2 + (1 - 4)^2} = \sqrt{(4)^2 + (-3)^2} = \sqrt{16 + 9} = \sqrt{25} = 5.$

    From $(0, 0)$ to $(7, 1)$: $d_3 = \sqrt{(7 - 0)^2 + (1 - 0)^2} = \sqrt{(7)^2 + (1)^2} = \sqrt{49 + 1} = \sqrt{50} = 5\sqrt{2}.$

    Since $d_1 = d_2,$ the triangle is isosceles.

    (b) From $(-1, -1)$ to $(2, 3)$: $d_1 = \sqrt{(2 - (-1))^2 + (3 - (-1))^2} = \sqrt{(3)^2 + (4)^2} = \sqrt{9 + 16} = \sqrt{25} = 5.$

    From $(2, 3)$ to $(-4, 3)$: $d_2 = \sqrt{(-4 - 2)^2 + (3 - 3)^2} = \sqrt{(-6)^2 + (0)^2} = \sqrt{36 + 0} = \sqrt{36} = 6.$

    From $(-4, 3)$ to $(-1, -1)$: $d_3 = \sqrt{(-1 - (-4))^2 + (-1 - 3)^2} = \sqrt{(3)^2 + (-4)^2} = \sqrt{9 + 16} = \sqrt{25} = 5.$

    Since $d_1 \neq d_2,$ the triangle is not equilateral.

105. (a) See Figure 105.

    (b) $d = \sqrt{(50 - 0)^2 + (0 - (-40))^2} = \sqrt{(50)^2 + (40)^2} = \sqrt{2500 + 1600} = \sqrt{4100} \approx 64.0$ miles.

106. (a) See Figure 106.

    (b) $d = \sqrt{(0 - (-15t))^2 + (20t - 0)^2} = \sqrt{(15t)^2 + (20t)^2} = \sqrt{225t^2 + 400t^2} = \sqrt{625t^2} = 25t.$

    That is $d = 25t$ miles.

107. Using the area of a square produces: $(a + b)^2 = a^2 + 2ab + b^2.$ Now using the sum of the small square and

    the four right triangles produces: $c^2 + 4\left(\dfrac{1}{2}ab\right) = c^2 + 2ab.$ Therefore, $a^2 + 2ab + b^2 = c^2 + 2ab$ and

    subtracting $2ab$ from both sides produces: $a^2 + b^2 = c^2.$

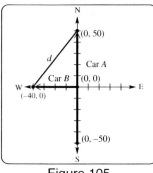

Figure 105

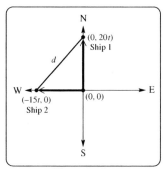

Figure 106

108.   Let d_1 represent the distance between P and M and let d_2 represent the distance between M and Q.

$$d_1 = \sqrt{\left(x_1 - \frac{x_1 + x_2}{2}\right)^2 + \left(y_1 - \frac{y_1 + y_2}{2}\right)^2} = \sqrt{\left(\frac{2x_1 - x_1 - x_2}{2}\right)^2 + \left(\frac{2y_1 - y_1 - y_2}{2}\right)^2} \Rightarrow$$

$$d_1 = \sqrt{\frac{(x_1 - x_2)^2}{4} + \frac{(y_1 - y_2)^2}{4}} = \frac{1}{2}\sqrt{(x_1 - x_2)^2 + (y_1 - y_2)^2}$$

$$d_2 = \sqrt{\left(x_2 - \frac{x_1 + x_2}{2}\right)^2 + \left(y_2 - \frac{y_1 + y_2}{2}\right)^2} = \sqrt{\left(\frac{2x_2 - x_1 - x_2}{2}\right)^2 + \left(\frac{2y_2 - y_1 - y_2}{2}\right)^2} \Rightarrow$$

$$d_2 = \sqrt{\frac{(x_2 - x_1)^2}{4} + \frac{(y_2 - y_1)^2}{4}} = \frac{1}{2}\sqrt{(x_2 - x_1)^2 + (y_2 - y_1)^2}$$

Since $(x_1 - x_2)^2 = (x_2 - x_1)^2$ and $(y_1 - y_2)^2 = (y_2 - y_1)^2$, the distances are the same.

Since $d_1 = d_2$, the sum $d_1 + d_2 = 2d_2 = 2\left(\frac{1}{2}\sqrt{(x_2 - x_1)^2 + (y_2 - y_1)^2}\right) = \sqrt{(x_2 - x_1)^2 + (y_2 - y_1)^2}$.

That is the sum is equal to the distance between P and Q.

1.2: Introduction to Relations and Functions

1.   The interval is $(-1, 4)$.

2.   The interval is $[-3, \infty)$.

3.   The interval is $(-\infty, 0)$.

4.   The interval is $(3, 8)$.

5.   The interval is $[1, 2)$.

6.   The interval is $(-5, -4]$.

7. $(-4, 3) \Rightarrow \{x \mid -4 < x < 3\}$

8. $[2, 7) \Rightarrow \{x \mid 2 \leq x < 7\}$

9. $(-\infty, -1] \Rightarrow \{x \mid x \leq -1\}$

10. $(3, \infty) \Rightarrow \{x \mid x > 3\}$

11. $\{x \mid -2 \le x < 6\}$

12. $\{x \mid 0 < x < 8\}$

13. $\{x \mid x \le -4\}$

14. $\{x \mid x > 3\}$

15. A parentheses is used if the symbol is $<$ or $>$. A square bracket is used if the symbol is $\le$ or $\ge$.

16. No real number is both greater than -7 and less than -10. Part (d) should be written $-10 < x < -7$.

17. The relation is a function. Domain: $\{5, 3, 4, 7\}$, Range: $\{1, 2, 9, 6\}$.

18. The relation is a function. Domain: $\{8, 5, 9, 3\}$, Range: $\{0, 4, 3, 8\}$.

19. The relation is a function. Domain: $\{1, 2, 3\}$, Range: $\{6\}$.

20. The relation is a function. Domain: $\{-10, -20, -30\}$, Range: $\{5\}$.

21. The relation is not a function. Domain: $\{4, 3, -2\}$, Range: $\{1, -5, 3, 7\}$.

22. The relation is not a function. Domain: $\{0, 1\}$, Range: $\{5, 3, -4\}$.

23. The relation is a function. Domain: $\{11, 12, 13, 14\}$, Range: $\{-6, -7\}$.

24. The relation is not a function. Domain: $\{1\}$, Range: $\{12, 13, 14, 15\}$.

25. The relation is a function. Domain: $\{0, 1, 2, 3, 4\}$, Range: $\{\sqrt{2}, \sqrt{3}, \sqrt{5}, \sqrt{6}, \sqrt{7}\}$.

26. The relation is a function. Domain: $\left\{1, \dfrac{1}{2}, \dfrac{1}{4}, \dfrac{1}{8}, \dfrac{1}{16}\right\}$, Range: $\{0, -1, -2, -3, -4\}$.

27. The relation is a function. Domain: $(-\infty, \infty)$, Range: $(-\infty, \infty)$.

28. The relation is a function. Domain: $(-\infty, \infty)$, Range: $(-\infty, 4]$.

29. The relation is not a function. Domain: $[-4, 4]$, Range: $[-3, 3]$.

30. The relation is a function. Domain: $[-2, 2]$, Range: $[0, 4]$.

31. The relation is a function. Domain: $[2, \infty)$, Range: $[0, \infty)$.

32. The relation is a function. Domain: $(-\infty, \infty)$, Range: $[1, \infty)$.

33. The relation is not a function. Domain: $[-9, \infty)$, Range: $(-\infty, \infty)$.

34. The relation is a function. Domain: $(-\infty, \infty)$, Range: $(-\infty, \infty)$.

35. The relation is a function. Domain: $\{-5, -2, -1, -.5, 0, 1.75, 3.5\}$, Range: $\{-1, 2, 3, 3.5, 4, 5.75, 7.5\}$.

36. The relation is a function. Domain: $\{-2, -1, 0, 5, 9, 10, 13\}$, Range: $\{5, 0, -3, 12, 60, 77, 140\}$.

37. The relation is a function. Domain: $\{2, 3, 5, 11, 17\}$, Range: $\{1, 7, 20\}$.

38. The relation is not a function. Domain: $\{1, 2, 3, 5\}$, Range: $\{10, 15, 19, 27\}$.

39. From the diagram, $f(-2) = 2$

40. From the diagram, $f(5) = 12$

41. From the diagram, $f(11) = 7$.

42. From the diagram, $f(5) = 1$.

43. $f(-2) = 3(-2) - 4 = -6 - 4 = -10$

44. $f(-5) = 5(-5) + 6 = -25 + 6 = -19$

45. $f(1) = 2(1)^2 - (1) + 3 = 2 - 1 + 3 = 4$

46. $f(2) = 3(2)^2 + 2(2) - 5 = 12 + 4 - 5 = 11$

47. $f(4) = -(4)^2 + (4) + 2 = -16 + 4 + 2 = -10$

48. $f(3) = -(3)^2 - (3) - 6 = -9 - 3 - 6 = -18$

49. $f(9) = 5$

50. $f(12) = -4$

51. $f(-2) = \sqrt{(-2)^3 + 12} = \sqrt{-8 + 12} = \sqrt{4} = 2$

52. $f(2) = \sqrt[3]{(2)^2 - (2) + 6} = \sqrt[3]{4 - 2 + 6} = \sqrt[3]{8} = 2$

53. $f(8) = |5 - 2(8)| = |-11| = 11$

54. $f(20) = |6 - \frac{1}{2}(20)| = |6 - 10| = |-4| = 4$

55. Since $f(-2) = 3$, the point $(-2, 3)$ lies on the graph of f.

56. Since $f(3) = -9.7$, the point $(3, -9.7)$ lies on the graph of f.

57. Since the point $(7, 8)$ lies on the graph of f, $f(7) = 8$.

58. Since the point $(-3, 2)$ lies on the graph of f, $f(-3) = 2$.

59. From the graph: (a) $f(-2) = 0$, (b) $f(0) = 4$, (c) $f(1) = 2$, and (d) $f(4) = 4$.

60. From the graph: (a) $f(-2) = 5$, (b) $f(0) = 0$, (c) $f(1) = 2$, and (d) $f(4) = 4$.

61. From the graph: (a) $f(-2) = -3$, (b) $f(0) = -2$, (c) $f(1) = 0$, and (d) $f(4) = 2$.

62. From the graph: (a) $f(-2) = 3$, (b) $f(0) = 3$, (c) $f(1) = 3$, and (d) $f(4) = 3$.

63. (a) – (d) Answers will vary. Refer to the definitions in the text.

64. (a) See Figure 64.

    (b) $f(2000) = 12.8$. In 2000 there were 12,800 radio stations on the air.

    (c) Domain: $\{1950, 1975, 1997, 2000, 2003\}$, Range: $\{2.8, 7.7, 12.8, 13.4\}$

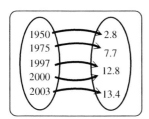

Figure 64

65. See Figure 65.

66. See Figure 66.

67. See Figure 67.

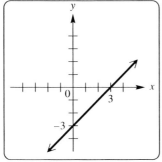

Figure 65

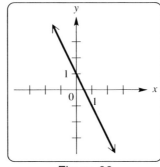

Figure 66

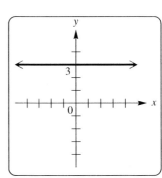

Figure 67

68. See Figure 68.

69. See Figure 69.

70. See Figure 70.

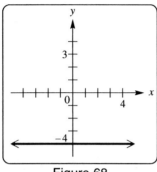

Figure 68

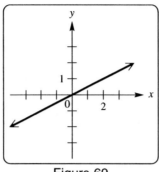

Figure 69

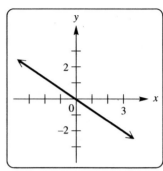

Figure 70

71. See Figure 71.

72. See Figure 72.

73. (a) $f(15) = \dfrac{15}{5} = 3$. When the delay is 15 seconds, there are 3 miles between the lightning and the observer.

    (b) See Figure 73.

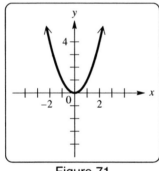

Figure 71

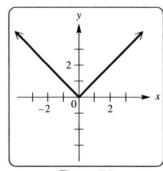

Figure 72

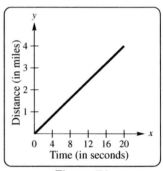

Figure 73

74. $f(3) = 80 - 5.8(3) = 62.6$. When the altitude is 3 miles, the temperature is 62.6°F.

75. The function that calculates a 7.5% tax on a sale of x dollars is $f(x) = .075x$. The tax on a purchase of \$86 is

    $f(86) = .075(86) = 6.45$ or \$6.45.

76. (a) $f = \{(N, 23{,}267), (H, 35{,}646), (B, 69{,}156), (M, 76{,}470)\}$

    (b) Domain: $\{N, H, B, M\}$, Range: $\{23{,}267, 35{,}646, 69{,}156, 76{,}470\}$

    (c) More education corresponds to a higher salary.

77. The function that calculates tuition and fees for taking x credits is $f(x) = 92x + 75$. The cost of taking 11 credits is $f(11) = 92(11) + 75 = 1087$ or \$1087.

78. The function that converts x gallons to quarts is $f(x) = 4x$. There are $f(19) = 4(19) = 76$ quarts in 19 gallons.

Reviewing Basic Concepts (Sections 1.1 and 1.2)

1. See Figure 1.

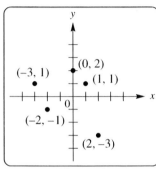

Figure 1

2. The distance is $d = \sqrt{(6 - (-4))^2 + (-2 - 5)^2} = \sqrt{100 + 49} = \sqrt{149}$.

    The midpoint is $M = \left(\dfrac{-4 + 6}{2}, \dfrac{5 - 2}{2} \right) = \left(1, \dfrac{3}{2} \right)$.

3. $\sqrt{(5 + \pi)}/(\sqrt[3]{(3)} + 1) \approx 1.168$

4. Using Pythagorean Theorem $11^2 + b^2 = 61^2 \Rightarrow b^2 = 61^2 - 11^2 \Rightarrow b^2 = 3600 \Rightarrow b = 60$ inches.

5. The set $\{x \mid -2 < x \le 5\}$ is the interval $(-2, 5]$. The set $\{x \mid x \ge 4\}$ is the interval $[4, \infty)$.

6. The relation is not a function because it does not pass the vertical line test. Domain: $[-2, 2]$, Range: $[-3, 3]$

7. $f(-5) = 3 - 4(-5) = 23$

8. From the graph $f(2) = 3$ and $f(-1) = -3$.

9. See Figure 9.

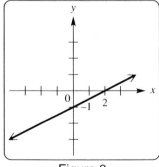

Figure 9

10. $d = \sqrt{(27 - (-3))^2 + ((-4) - 12)^2} = \sqrt{30^2 + (-16)^2} = \sqrt{900 + 256} = \sqrt{1156} = 34$

1.3: Linear Functions

1. (a) $f(-2) = (-2) + 2 = 0$ and $f(4) = (4) + 2 = 6$

    (b) $x + 2 = 0 \Rightarrow x = -2$

    (c) The x-intercept is -2 and corresponds to the zero of f. See Figure 1.

2. (a) $f(-2) = -3(-2) + 2 = 8$ and $f(4) = -3(4) + 2 = -10$

    (b) $-3x + 2 = 0 \Rightarrow -3x = -2 \Rightarrow x = \dfrac{2}{3}$

    (c) The x-intercept is $\dfrac{2}{3}$ and corresponds to the zero of f. See Figure 2.

3. (a) $f(-2) = 2 - \dfrac{1}{2}(-2) = 3$ and $f(4) = 2 - \dfrac{1}{2}(4) = 0$

    (b) $2 - \dfrac{1}{2}x = 0 \Rightarrow \dfrac{1}{2}x = 2 \Rightarrow x = 4$

    (c) The x-intercept is 4 and corresponds to the zero of f. See Figure 3.

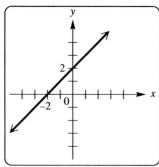

Figure 1

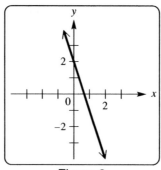

Figure 2

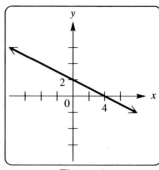

Figure 3

4. (a) $f(-2) = \dfrac{1}{4}(-2) + \dfrac{1}{2} = 0$ and $f(4) = \dfrac{1}{4}(4) + \dfrac{1}{2} = \dfrac{3}{2}$

    (b) $\dfrac{1}{4}x + \dfrac{1}{2} = 0 \Rightarrow \dfrac{1}{4}x = -\dfrac{1}{2} \Rightarrow x = -2$

    (c) The x-intercept is -2 and corresponds to the zero of f. See Figure 4.

5. (a) $f(-2) = \dfrac{1}{3}(-2) = -\dfrac{2}{3}$ and $f(4) = \dfrac{1}{3}(4) = \dfrac{4}{3}$

    (b) $\dfrac{1}{3}x = 0 \Rightarrow x = 0$

    (c) The x-intercept is 0 and corresponds to the zero of f. See Figure 5.

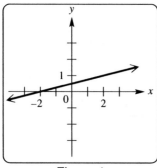

Figure 4

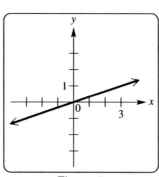

Figure 5

6. (a) $f(-2) = -3(-2) = 6$ and $f(4) = -3(4) = -12$

    (b) $-3x = 0 \Rightarrow x = 0$

    (c) The x-intercept is 0 and corresponds to the zero of f. See Figure 6.

7. (a) $f(-2) = .4(-2) + .15 = -.65$ and $f(4) = .4(4) + .15 = 1.75$

    (b) $.4x + .15 = 0 \Rightarrow .4x = -.15 \Rightarrow x = -.375$

    (c) The x-intercept is $-.375$ and corresponds to the zero of f. See Figure 7.

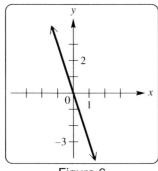

Figure 6

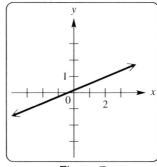

Figure 7

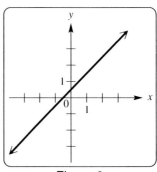

Figure 8

8. (a) $f(-2) = .5 + (-2) = -1.5$ and $f(4) = .5 + (4) = 4.5$

    (b) $.5 + x = 0 \Rightarrow x = -.5$

    (c) The x-intercept is $-.5$ and corresponds to the zero of f. See Figure 8.

9. (a) $f(-2) = \dfrac{2 - (-2)}{4} = 1$ and $f(4) = \dfrac{2 - (4)}{4} = -\dfrac{1}{2}$

    (b) $\dfrac{2 - x}{4} = 0 \Rightarrow 2 - x = 0 \Rightarrow x = 2$

    (c) The x-intercept is 2 and corresponds to the zero of f. See Figure 9.

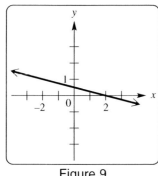

Figure 9

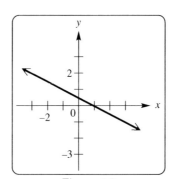

Figure 10

10. (a) $f(-2) = \dfrac{3 - 3(-2)}{6} = \dfrac{9}{6} = 1.5$ and $f(4) = \dfrac{3 - 3(4)}{6} = \dfrac{-9}{6} = -1.5$

    (b) $\dfrac{3 - 3x}{6} = 0 \Rightarrow 3 - 3x = 0 \Rightarrow -3x = -3 \Rightarrow x = 1$

    (c) The x-intercept is 1 and corresponds to the zero of f. See Figure 10

11. (a) $f(-2) = \dfrac{3(-2) + \pi}{2} = \dfrac{\pi - 6}{2}$ and $f(4) = \dfrac{3(4) + \pi}{2} = \dfrac{12 + \pi}{2}$

   (b) $\dfrac{3x + \pi}{2} = 0 \Rightarrow 3x + \pi = 0 \Rightarrow x = -\dfrac{\pi}{3}$

   (c) The *x*-intercept is $-\dfrac{\pi}{3} \approx -1.047$ and corresponds to the zero of f. See Figure 11.

12. (a) $f(-2) = \dfrac{4(-2) + \pi}{3} = \dfrac{-8 + \pi}{3}$ and $f(4) = \dfrac{4(4) + \pi}{3} = \dfrac{16 + \pi}{3}$

   (b) $\dfrac{4x + \pi}{3} = 0 \Rightarrow 4x + \pi = 0 \Rightarrow x = -\dfrac{\pi}{4}$

   (c) The *x*-intercept is $-\dfrac{\pi}{4} \approx -.7854$ and corresponds to the zero of f. See Figure 12.

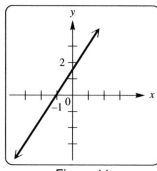
Figure 11

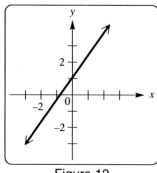
Figure 12

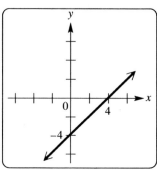
Figure 13

13. The graph is shown in Figure 13.

   (a) *x*-intercept: 4   (b) *y*-intercept: -4 (c) Domain: $(-\infty, \infty)$ (d) Range: $(-\infty, \infty)$

   (e) The equation is in slope-intercept form, therefore $m = 1$.

14. The graph is shown in Figure 14.

   (a) *x*-intercept: 4   (b) *y*-intercept: 4   (c) Domain: $(-\infty, \infty)$ (d) Range: $(-\infty, \infty)$

   (e) The equation is in slope-intercept form, therefore $m = -1$.

15. The graph is shown in Figure 15.

   (a) *x*-intercept: 2   (b) *y*-intercept: -6 (c) Domain: $(-\infty, \infty)$ (d) Range: $(-\infty, \infty)$

   (e) The equation is in slope-intercept form, therefore $m = 3$.

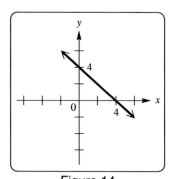
Figure 14

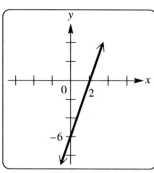
Figure 15

16. The graph is shown in Figure 16.

    (a) *x*-intercept: 3   (b) *y*-intercept: −2   (c) Domain: $(-\infty, \infty)$ (d) Range: $(-\infty, \infty)$

    (e) The equation is in slope-intercept form, therefore $m = \dfrac{2}{3}$.

17. The graph is shown in Figure 17.

    (a) *x*-intercept: 5   (b) *y*-intercept: 2   (c) Domain: $(-\infty, \infty)$ (d) Range: $(-\infty, \infty)$

    (e) The equation is in slope-intercept form, therefore $m = -\dfrac{2}{5}$.

18. The graph is shown in Figure 18.

    (a) *x*-intercept: $\dfrac{9}{4}$ (b) *y*-intercept: −3   (c) Domain: $(-\infty, \infty)$ (d) Range: $(-\infty, \infty)$

    (e) The equation is in slope-intercept form, therefore $m = \dfrac{4}{3}$.

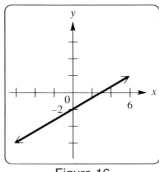

Figure 16

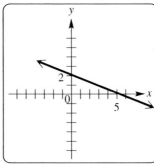

Figure 17

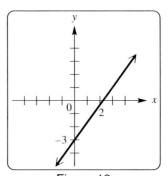

Figure 18

19. The graph is shown in Figure 19.

    (a) *x*-intercept: 0   (b) *y*-intercept: 0   (c) Domain: $(-\infty, \infty)$ (d) Range: $(-\infty, \infty)$

    (e) The equation is in slope-intercept form, therefore $m = 3$.

20. The graph is shown in Figure 20.

    (a) *x*-intercept: 0   (b) *y*-intercept: 0   (c) Domain: $(-\infty, \infty)$ (d) Range: $(-\infty, \infty)$

    (e) The equation is in slope-intercept form, therefore $m = -.5$.

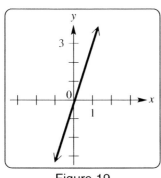

Figure 19

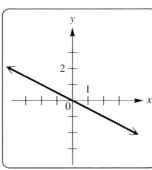

Figure 20

21. The graph of $y = ax$ always passes through $(0, 0)$.

22. Since $m = \dfrac{4}{1} = 4$, the equation of the line is $y = 4x$.

23. The graph is shown in Figure 23.

    (a) *x*-intercept: none   (b) *y*-intercept: -3 (c) Domain: $(-\infty, \infty)$ (d) Range: $\{-3\}$

    (e) The slope of all horizontal line graphs or constant functions is $m = 0$.

24. The graph is shown in Figure 24.

    (a) *x*-intercept: none   (b) *y*-intercept: 5   (c) Domain: $(-\infty, \infty)$ (d) Range: $\{5\}$

    (e) The slope of all horizontal line graphs or constant functions is $m = 0$.

25. The graph is shown in Figure 25.

    (a) *x*-intercept: -1.5 (b) *y*-intercept: none   (c) Domain: $\{-1.5\}$ (d) Range: $(-\infty, \infty)$

    (e) All vertical line graphs are not functions, therefore the slope is undefined .

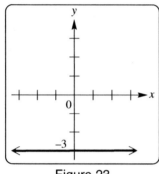
Figure 23

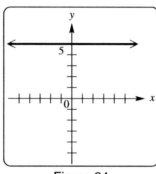
Figure 24

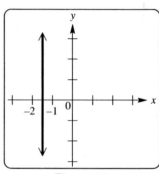
Figure 25

26. The graph is shown in Figure 26.

    (a) *x*-intercept: none   (b) *y*-intercept: $\dfrac{5}{4}$ (c) Domain: $(-\infty, \infty)$ (d) Range: $\left\{\dfrac{5}{4}\right\}$

    (e) The slope of all horizontal line graphs or constant functions is $m = 0$.

27. The graph is shown in Figure 27.

    (a) *x*-intercept: 2   (b) *y*-intercept: none   (c) Domain: $\{2\}$ (d) Range: $(-\infty, \infty)$

    (e) All vertical line graphs are not functions, therefore the slope is undefined.

28. The graph is shown in Figure 28.

    (a) *x*-intercept: -3 (b) *y*-intercept: none   (c) Domain: $\{-3\}$ (d) Range: $(-\infty, \infty)$

    (e) All vertical line graphs are not functions, therefore the slope is undefined.

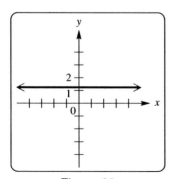
Figure 26

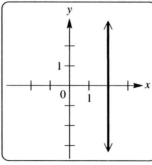
Figure 27

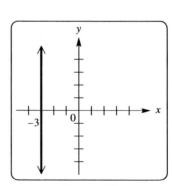
Figure 28

29. All functions in the form $f(x) = a$ are constant functions.

30. This is a vertical line graph, therefore $x = 4$.

31. This is a horizontal line graph, therefore $y = 3$.

32. This is a horizontal line graph on the x-axis, therefore $y = 0$.

33. This is a vertical line graph on the y-axis, therefore $x = 0$.

34. (a) The equation of the x-axis is $y = 0$.

    (b) The equation of the y-axis is $x = 0$.

35. Window B gives the more comprehensive graph. See Figures 35a and 35b.

36. Window A gives the more comprehensive graph. See Figures 36a and 36b.

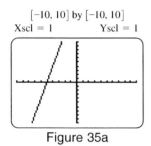

$[-10, 10]$ by $[-10, 10]$
Xscl = 1   Yscl = 1

Figure 35a

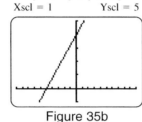

$[-10, 10]$ by $[-5, 25]$
Xscl = 1   Yscl = 5

Figure 35b

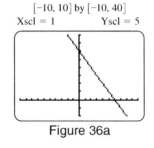

$[-10, 10]$ by $[-10, 40]$
Xscl = 1   Yscl = 5

Figure 36a

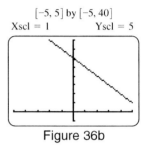

$[-5, 5]$ by $[-5, 40]$
Xscl = 1   Yscl = 5

Figure 36b

37. Window B gives the more comprehensive graph. See Figures 37a and 37b.

38. Window B gives the more comprehensive graph. See Figures 38a and 38b.

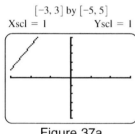

$[-3, 3]$ by $[-5, 5]$
Xscl = 1   Yscl = 1

Figure 37a

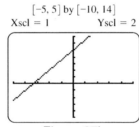

$[-5, 5]$ by $[-10, 14]$
Xscl = 1   Yscl = 2

Figure 37b

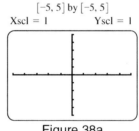

$[-5, 5]$ by $[-5, 5]$
Xscl = 1   Yscl = 1

Figure 38a

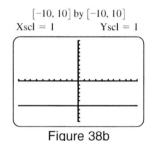

$[-10, 10]$ by $[-10, 10]$
Xscl = 1   Yscl = 1

Figure 38b

39. $m = \dfrac{6 - 1}{3 - (-2)} = \dfrac{5}{5} = 1$

40. $m = \dfrac{3 - 2}{-2 - (-1)} = \dfrac{1}{-1} = -1$

41. $m = \dfrac{4 - (-3)}{8 - (-1)} = \dfrac{7}{9}$

42. $m = \dfrac{-3 - 0}{-4 - 5} = \dfrac{-3}{-9} = \dfrac{1}{3}$

43. $m = \dfrac{5 - 3}{-11 - (-11)} = \dfrac{2}{0} \Rightarrow$ undefined slope

44. $m = \dfrac{1 - 2}{-8 - (-8)} = \dfrac{-1}{0} \Rightarrow$ undefined slope

45. $m = \dfrac{9 - 9}{\frac{1}{2} - \frac{2}{3}} = \dfrac{0}{-\frac{1}{6}} \Rightarrow 0$

46. $m = \dfrac{.36 - .36}{.18 - .12} = \dfrac{0}{.06} \Rightarrow 0$

47. Since $m = 3$ and $b = 6$, graph A most closely resembles the equation.

48. Since $m = -3$ and $b = 6$, graph D most closely resembles the equation.

49. Since $m = -3$ and $b = -6$, graph C most closely resembles the equation.

50. Since $m = 3$ and $b = -6$, graph F most closely resembles the equation.

51. Since $m = 3$ and $b = 0$, graph H most closely resembles the equation.

52. Since $m = -3$ and $b = 0$, graph G most closely resembles the equation.

53. Since $m = 0$ and $b = 3$, graph B most closely resembles the equation.

54. Since $m = 0$ and $b = -3$, graph E most closely resembles the equation.

55. (a) The graph passes through $(0, 1)$ and $(1, -1) \Rightarrow m = \dfrac{-1 - 1}{1 - 0} = \dfrac{-2}{1} = -2.$

    The y-intercept is: 1 and the x-intercept is: $\dfrac{1}{2}$.

    (b) Using the slope and y-intercept the formula is: $f(x) = -2x + 1$.

    (c) The x-intercept is the zero of $f \Rightarrow \dfrac{1}{2}$.

56. (a) The graph passes through $(0, -1)$ and $(1, 1) \Rightarrow m = \dfrac{1 - (-1)}{1 - 0} = \dfrac{2}{1} = 2.$

    The y-intercept is: -1 and the x-intercept is: $\dfrac{1}{2}$.

    (b) Using the slope and y-intercept the formula is: $f(x) = 2x - 1$.

    (c) The x-intercept is the zero of $f \Rightarrow \dfrac{1}{2}$.

57. (a) The graph passes through $(0, 2)$ and $(3, 1) \Rightarrow m = \dfrac{1 - 2}{3 - 0} = \dfrac{-1}{3} = -\dfrac{1}{3}.$

    The y-intercept is: 2 and the x-intercept is: 6.

    (b) Using the slope and y-intercept the formula is: $f(x) = -\dfrac{1}{3}x + 2$.

    (c) The x-intercept is the zero of $f \Rightarrow 6$.

58. (a) The graph passes through $(4, 0)$ and $(0, -3) \Rightarrow m = \dfrac{-3 - 0}{0 - 4} = \dfrac{-3}{-4} = \dfrac{3}{4}.$

    The y-intercept is: -3 and the x-intercept is: 4.

    (b) Using the slope and y-intercept the formula is: $f(x) = \dfrac{3}{4}x - 3$.

    (c) The x-intercept is the zero of $f \Rightarrow 4$.

59. (a) The graph passes through $(0, 300)$ and $(2, -100) \Rightarrow m = \dfrac{-100 - 300}{2 - 0} = \dfrac{-400}{2} = -200.$

    The y-intercept is: 300 and the x-intercept is: $\dfrac{3}{2}$.

    (b) Using the slope and y-intercept the formula is: $f(x) = -200x + 300$.

    (c) The x-intercept is the zero of $f \Rightarrow \dfrac{3}{2}$.

60. (a) The graph passes through $(5, 50)$ and $(0, -50) \Rightarrow m = \dfrac{-50 - 50}{0 - 5} = \dfrac{-100}{-5} = 20$.

   The y-intercept is: -50 and the x-intercept is: $\dfrac{5}{2}$.

   (b) Using the slope and y-intercept the formula is: $f(x) = 20x - 50$.

   (c) The x-intercept is the zero of $f \Rightarrow \dfrac{5}{2}$.

61. Using $(0, 2)$ and $(1, 6)$, $m = \dfrac{6 - 2}{1 - 0} = \dfrac{4}{1} = 4$. From the table the y-intercept is 2.

   Using these two answers and slope-intercept form the equation is $f(x) = 4x + 2$.

62. Using $(0, -5)$ and $(1, -2)$, $m = \dfrac{-2 - (-5)}{1 - 0} = \dfrac{3}{1} = 3$. From the table the y-intercept is -5.

   Using these two answers and slope-intercept form the equation is $f(x) = 3x - 5$.

63. Using $(0, -3.1)$ and $(.2, -3.38)$, $m = \dfrac{-3.38 - (-3.1)}{.2 - 0} = \dfrac{-.28}{.2} = -1.4$. From the table the y-intercept is -3.1.

   Using these two answers and slope-intercept form the equation is $f(x) = -1.4x - 3.1$.

64. Using $(0, -4)$ and $(50, -4)$, $m = \dfrac{-4 - (-4)}{50 - 0} = \dfrac{0}{50} = 0$. From the table the y-intercept is -4.

   Using these two answers and slope-intercept form the equation is $f(x) = -4$.

65. The graph of a constant function with positive k is a horizontal graph above the x-axis. Graph A

66. The graph of a constant function with negative k is a horizontal graph below the x-axis. Graph C

67. The graph of an equation of the form $x = k$ with $k > 0$ is a vertical line right of the y-axis. Graph D

68. The graph of an equation of the form $x = -k$ with $k > 0$ is a vertical line left of the y-axis. Graph B

69. Using $(-1, 3)$ with a rise of 3 and a run of 2 the graph also passes through $(1, 6)$. See Figure 69.

70. Using $(-2, 8)$ with a rise of -1 and a run of 1 the graph also passes through $(-1, 7)$. See Figure 70.

71. Using $(3, -4)$ with a rise of -1 and a run of 3 the graph also passes through $(6, -5)$. See Figure 71.

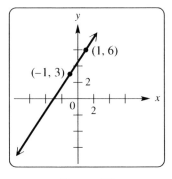

Figure 69

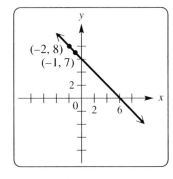

Figure 70

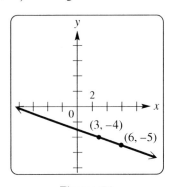

Figure 71

72. Using $(-2, -3)$ with a rise of -3 and a run of 4 the graph also passes through $(2, -6)$. See Figure 72.

73. Using $(-1, 4)$ with slope of 0 the graph is a horizontal line which also passes through $(2, 4)$. See Figure 73.

74. Using $\left(\frac{9}{4}, 2\right)$ with undefined slope the graph is a vertical line which also passes through $\left(\frac{9}{4}, -2\right)$. See Figure 74.

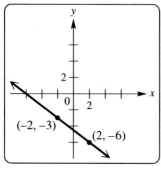

Figure 72

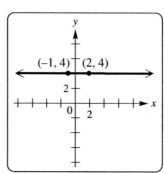

Figure 73

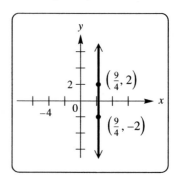

Figure 74

75. Using $(0, -4)$ with a rise of 3 and a run of 4 the graph also passes through $(4, -1)$. See Figure 75.

76. Using $(0, 5)$ with a rise of -5 and a run of 2 the graph also passes through $(2, 0)$. See Figure 76.

77. Using $(-3, 0)$ with undefined slope the graph is a vertical line which also passes through $(-3, 2)$. See Figure 77.

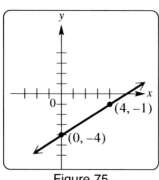

Figure 75

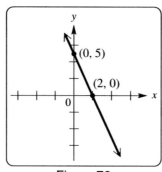

Figure 76

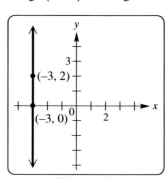

Figure 77

78. (a) Since $m = \frac{3}{4}$ and $b = -4$, the equation is $y = \frac{3}{4}x - 4$.

    (b) Since $m = -2.5$ and $b = 5$, the equation is $y = -2.5x + 5$.

79. (a) Using the points $(0, 2000)$ and $(4, 4000)$, $m = \frac{4000 - 2000}{4 - 0} = \frac{2000}{4} = 500$. The y-intercept is $b = 2000$.
    The formula is $f(x) = 500x + 2000$.

    (b) Water is entering the pool at a rate of 500 gallons per hour. The pool contains 2000 gallons initially.

    (c) From the graph $f(7) = 5500$ gallons. By evaluating, $f(7) = 500(7) + 2000 = 5500$ gallons.

80. (a) Using the points $(5, 115)$ and $(10, 230)$, $m = \frac{230 - 115}{10 - 5} = \frac{115}{5} = 23$. Using the slope-intercept form,
    $115 = 23(5) + b \Rightarrow 115 = 115 + b \Rightarrow b = 0$. Therefore $a = 23$ and $b = 0$.

    (b) The car's gas mileage is 23 miles per gallon.

    (c) Since $f(x) = ax + b$ models the data and $a = 23$, $b = 0$ the equation $f(x) = 23x$ can be used to find miles
    traveled. Therefore $f(20) = 23(20) \Rightarrow f(20) = 460$ miles traveled.

81. (a) The rain fell at a rate of $\frac{1}{4}$ inches per hour, so $m = \frac{1}{4}$. The initial amount of rain at noon was 3 inches, so $b = 3$. The equation is $f(x) = \frac{1}{4}x + 3$.

    (b) By 2:30 P.M. ($x = 2.5$), the total rainfall was $f(2.5) = \frac{1}{4}(2.5) + 3 = 3.625$ in.

82. (a) Since the rate of increase is 5.8 million people per year $m = 5.8$. Since there were 47 million cases in 1998 $b = 47$. Therefore the equation that models this is: $f(x) = 5.8x = 47$.

    (b) $x = 2006 - 1998 = 8 \Rightarrow f(8) = 5.8(8) + 47 = 93,400,000$ people will have been living with HIV in 2006.

83. (a) The birth rate is decreasing at a rate of .096 births per year, so $m = -.096$. The initial birth rate was 14.8 births per 1000 people in 1995, so $b = 14.8$. The equation is $f(x) = -.096x + 14.8$ (per thousand)

    (b) Since $x = 6$ for 2001, $f(6) = -.096(6) + 14.8 = 14.224$ births per thousand people. This favorably compares to the actual number of 14.1 births per thousand people.

84. (a) The birth rate is decreasing at a rate of .16 births per year, so $m = -.16$. The initial birth rate was 14.6 births per 1000 people in 1998, so $b = 14.6$. The equation is $f(x) = -.16x + 14.6$ (per thousand)

    (b) Since $x = 3$ for 2001, $f(3) = -.16(3) + 14.6 = 14.12$ births per thousand people. This favorably compares to the actual number of 14.1 births per thousand people.

1.4: Equations of Lines and Linear Models

1. Using Point-Slope Form yields: $y - 3 = -2(x - 1) \Rightarrow y - 3 = -2x + 2 \Rightarrow y = -2x + 5$.

2. Using Point-Slope Form yields: $y - 4 = -1(x - 2) \Rightarrow y - 4 = -x + 2 \Rightarrow y = -x + 6$.

3. Using Point-Slope Form yields: $y - 4 = 1.5(x - (-5)) \Rightarrow y - 4 = 1.5x + 7.5 \Rightarrow y = 1.5x + 11.5$.

4. Using Point-Slope Form yields: $y - 3 = .75(x - (-4)) \Rightarrow y - 3 = .75x + 3 \Rightarrow y = .75x + 6$.

5. Using Point-Slope Form yields: $y - 1 = -.5(x - (-8)) \Rightarrow y - 1 = -.5x - 4 \Rightarrow y = -.5x - 3$.

6. Using Point-Slope Form yields: $y - 9 = -.75(x - (-5)) \Rightarrow y - 9 = -.75x - 3.75 \Rightarrow y = -.75x + 5.25$.

7. Using Point-Slope Form yields: $y - (-4) = 2\left(x - \frac{1}{2}\right) \Rightarrow y + 4 = 2x - 1 \Rightarrow y = 2x - 5$.

8. Using Point-Slope Form yields: $y - \left(-\frac{1}{3}\right) = 3(x - 5) \Rightarrow y + \frac{1}{3} = 3x - 15 \Rightarrow y = 3x - \frac{46}{3}$.

9. Using Point-Slope Form yields: $y - \frac{2}{3} = \frac{1}{2}\left(x - \frac{1}{4}\right) \Rightarrow y - \frac{2}{3} = \frac{1}{2}x - \frac{1}{8} \Rightarrow y = \frac{1}{2}x + \frac{13}{24}$.

10. The slope of a line passing through $(12, 6)$ and $(12, -2)$ is: $m = \dfrac{-2 - 6}{12 - 12} = \dfrac{-8}{0}$, which is undefined. You can not write an equation in slope-intercept form with an undefined slope. The line is vertical and has the equation $x = 12$.

11. Use the points $(-4, -6)$ and $(6, 2)$ to find the slope: $m = \dfrac{2 - (-6)}{6 - (-4)} \Rightarrow m = \dfrac{4}{5}$. Now using Point-Slope Form yields: $y - 2 = \frac{4}{5}(x - 6) \Rightarrow y - 2 = \frac{4}{5}x - \frac{24}{5} \Rightarrow y = \frac{4}{5}x - \frac{14}{5}$.

12. Use the points $(6, -2)$ and $(-2, 2)$ to find the slope: $m = \dfrac{2 - (-2)}{-2 - 6} \Rightarrow m = \dfrac{4}{-8} \Rightarrow m = -\dfrac{1}{2}$. Now using

    Point-Slope Form yields: $y - 2 = -\dfrac{1}{2}(x - (-2)) \Rightarrow y - 2 = -\dfrac{1}{2}x - 1 \Rightarrow y = -\dfrac{1}{2}x + 1$.

13. Use the points $(-12, 8)$ and $(8, -12)$ to find the slope: $m = \dfrac{-12 - 8}{8 - (-12)} \Rightarrow m = \dfrac{-20}{20} \Rightarrow m = -1$. Now using

    Point-Slope Form yields: $y - 8 = -1(x + 12) \Rightarrow y - 8 = -x - 12 \Rightarrow y = -x - 4$.

14. Use the points $(12, 6)$ and $(-6, -12)$ to find the slope: $m = \dfrac{-12 - 6}{-6 - 12} \Rightarrow m = \dfrac{-18}{-18} \Rightarrow m = 1$. Now using

    Point-Slope Form yields: $y - 6 = 1(x - 12) \Rightarrow y - 6 = x - 12 \Rightarrow y = x - 6$.

15. Use the points $(4, 8)$ and $(0, 4)$ to find the slope: $m = \dfrac{4 - 8}{0 - 4} \Rightarrow m = \dfrac{-4}{-4} \Rightarrow m = 1$. Now using

    Slope-Intercept Form yields: $b = 4 \Rightarrow y = x + 4$.

16. Use the points $(3, 6)$ and $(0, 10)$ to find the slope: $m = \dfrac{10 - 6}{0 - 3} \Rightarrow m = \dfrac{4}{-3} \Rightarrow m = -\dfrac{4}{3}$. Now using

    Point-Slope Form yields: $y - 6 = -\dfrac{4}{3}(x - 3) \Rightarrow y - 6 = -\dfrac{4}{3}x + 4 \Rightarrow y = -\dfrac{4}{3}x + 10$.

17. Use the points $(3, -8)$ and $(5, -3)$ to find the slope: $m = \dfrac{-3 - (-8)}{5 - 3} \Rightarrow m = \dfrac{5}{2}$. Now using

    Point-Slope Form yields: $y - (-8) = \dfrac{5}{2}(x - 3) \Rightarrow y + 8 = \dfrac{5}{2}x - \dfrac{15}{2} \Rightarrow y = \dfrac{5}{2}x - \dfrac{31}{2}$.

18. Use the points $(-5, 4)$ and $(-3, 2)$ to find the slope: $m = \dfrac{2 - 4}{-3 - (-5)} \Rightarrow m = \dfrac{-2}{2} \Rightarrow m = -1$. Now using

    Point-Slope Form yields: $y - 4 = -1(x - (-5)) \Rightarrow y - 4 = -x - 5 \Rightarrow y = -x - 1$.

19. Use the points $(2, 3.5)$ and $(6, -2.5)$ to find the slope: $m = \dfrac{-2.5 - 3.5}{6 - 2} \Rightarrow m = \dfrac{-6}{4} \Rightarrow m = -1.5$. Now using

    Point-Slope Form yields: $y - 3.5 = -1.5(x - 2) \Rightarrow y - 3.5 = -1.5x + 3 \Rightarrow y = -1.5x + 6.5$.

20. Use the points $(-1, 6.25)$ and $(2, -4.25)$ to find the slope: $m = \dfrac{6.25 - (-4.25)}{-1 - 2} \Rightarrow m = \dfrac{10.5}{-3} \Rightarrow m = -3.5$.

    Now using Point-Slope Form yields: $y - 6.25 = -3.5(x + 1) \Rightarrow y - 6.25 = -3.5x - 3.5 \Rightarrow y = -3.5x + 2.75$.

21. Use the points $(0, 5)$ and $(10, 0)$ to find the slope: $m = \dfrac{0 - 5}{10 - 0} \Rightarrow m = \dfrac{-5}{10} \Rightarrow m = -\dfrac{1}{2}$. Now using

    Point-Slope Form yields: $y - 5 = -\dfrac{1}{2}(x - 0) \Rightarrow y - 5 = -\dfrac{1}{2}x \Rightarrow y = -\dfrac{1}{2}x + 5$.

22. Use the points $(0, -8)$ and $(4, 0)$ to find the slope: $m = \dfrac{-8 - (0)}{0 - 4} \Rightarrow m = \dfrac{-8}{-4} \Rightarrow m = 2$. Now using

    Slope-Intercept Form yields: $b = -8 \Rightarrow y = 2x - 8$.

23. Use the points $(-5, -28)$ and $(-4, -20)$ to find the slope: $m = \dfrac{-20 - (-28)}{-4 - (-5)} \Rightarrow m = \dfrac{8}{1} \Rightarrow m = 8$. Now

    using Point-Slope Form yields: $y - (-20) = 8(x - (-4)) \Rightarrow y + 20 = 8x + 32 \Rightarrow y = 8x + 12$.

24. Use the points $(-2.4, 5.2)$ and $(1.3, -24.4)$ to find the slope: $m = \dfrac{-24.4 - 5.2}{1.3 - (-2.4)} \Rightarrow m = \dfrac{-29.6}{3.7} \Rightarrow m = -8$.

    Now using Point-Slope Form yields: $y - 5.2 = -8(x - (-2.4)) \Rightarrow y - 5.2 = -8x - 19.2 \Rightarrow y = -8x - 14$.

25. To find the x-intercept set $y = 0$, then $x - 0 = 4 \Rightarrow x = 4$. Therefore $(4, 0)$ is the x-intercept. To find the

    y-intercept set $x = 0$, then $0 - y = 4 \Rightarrow y = -4$. Therefore $(0, -4)$ is the y-intercept. See Figure 25.

26. To find the x-intercept set $y = 0$, then $x + 0 = 4 \Rightarrow x = 4$. Therefore $(4, 0)$ is the x-intercept. To find the

    y-intercept set $x = 0$, then $0 + y = 4 \Rightarrow y = 4$. Therefore $(0, 4)$ is the y-intercept. See Figure 26.

27. To find the x-intercept set $y = 0$, then $3x - 0 = 6 \Rightarrow 3x = 6 \Rightarrow x = 2$. Therefore $(2, 0)$ is the x-intercept.

    To find the y-intercept set $x = 0$, then $3(0) - y = 6 \Rightarrow y = -6$. Therefore $(0, -6)$ is the y-intercept.

    See Figure 27.

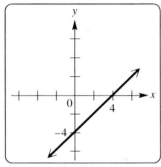
Figure 25

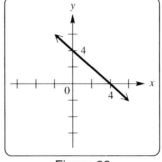
Figure 26

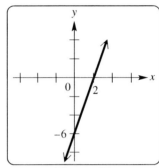
Figure 27

28. To find the x-intercept set $y = 0$, then $2x - 3(0) = 6 \Rightarrow 2x = 6 \Rightarrow x = 3$. Therefore $(3, 0)$ is the x-intercept.

    To find the y-intercept set $x = 0$, then $2(0) - 3y = 6 \Rightarrow -3y = 6 \Rightarrow y = -2$. Therefore $(0, -2)$ is the y-inter-

    cept. See Figure 28.

29. To find the x-intercept set $y = 0$, then $2x + 5(0) = 10 \Rightarrow 2x = 10 \Rightarrow x = 5$. Therefore $(5, 0)$ is the x-inter-

    cept. To find the y-intercept set $x = 0$, then $2(0) + 5y = 10 \Rightarrow 5y = 10 \Rightarrow y = 2$. Therefore $(0, 2)$ is the y-

    intercept. See Figure 29.

30. To find the x-intercept set $y = 0$, then $4x - 3(0) = 9 \Rightarrow 4x = 9 \Rightarrow x = \dfrac{9}{4}$. Therefore $\left(\dfrac{9}{4}, 0\right)$ is the x-intercept.

    To find the y-intercept set $x = 0$, then $4(0) - 3y = 9 \Rightarrow -3y = 9 \Rightarrow y = -3$. Therefore $(0, -3)$ is the y-intercept.

    See Figure 30.

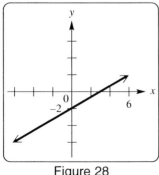
Figure 28

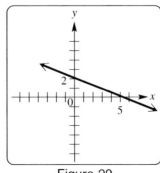
Figure 29

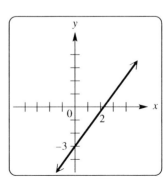
Figure 30

31. To find a second point set $x = 1$, then $y = 3(1) \Rightarrow y = 3$. A second point is $(1, 3)$. See Figure 31.

32. To find a second point set $x = 1$, then $y = -2(1) \Rightarrow y = -2$. A second point is $(1, -2)$. See Figure 32.

33. To find a second point set $x = 4$, then $y = -.75(4) \Rightarrow y = -3$. A second point is $(4, -3)$. See Figure 33.

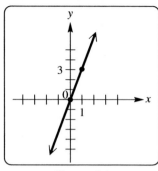

Figure 31

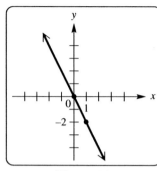

Figure 32

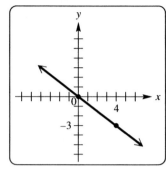

Figure 33

34. To find a second point set $x = 2$, then $y = 1.5(2) \Rightarrow y = 3$. A second point is $(2, 3)$. See Figure 34.

35. $5x + 3y = 15 \Rightarrow 3y = -5x + 15 \Rightarrow y = -\dfrac{5}{3}x + 5$. See Figure 35.

36. $6x + 5y = 9 \Rightarrow 5y = -6x + 9 \Rightarrow y = -\dfrac{6}{5}x + \dfrac{9}{5}$. See Figure 36.

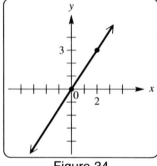

Figure 34

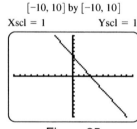

$[-10, 10]$ by $[-10, 10]$
Xscl = 1   Yscl = 1

Figure 35

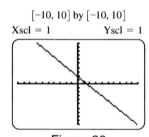

$[-10, 10]$ by $[-10, 10]$
Xscl = 1   Yscl = 1

Figure 36

37. $-2x + 7y = 4 \Rightarrow 7y = 2x + 4 \Rightarrow y = \dfrac{2}{7}x + \dfrac{4}{7}$. See Figure 37.

38. $-.23x - .46y = .82 \Rightarrow -23x - 46y = 82 \Rightarrow -46y = 23x + 82 \Rightarrow y = -\dfrac{23}{46}x + \dfrac{82}{46} \Rightarrow y = -\dfrac{1}{2}x - \dfrac{41}{23}$.
    See Figure 38.

39. $1.2x + 1.6y = 5.0 \Rightarrow 12x + 16y = 50 \Rightarrow 16y = -12x + 50 \Rightarrow y = -\dfrac{12}{16}x + \dfrac{50}{16} \Rightarrow y = -\dfrac{3}{4}x + \dfrac{25}{8}$.
    See Figure 39.

40. $2y - 5x = 0 \Rightarrow 2y = 5x + 0 \Rightarrow y = \dfrac{5}{2}x$. See Figure 40.

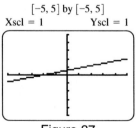

$[-5, 5]$ by $[-5, 5]$
Xscl = 1   Yscl = 1

Figure 37

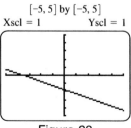

$[-5, 5]$ by $[-5, 5]$
Xscl = 1   Yscl = 1

Figure 38

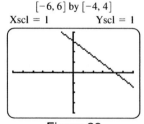

$[-6, 6]$ by $[-4, 4]$
Xscl = 1   Yscl = 1

Figure 39

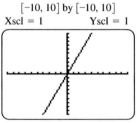

$[-10, 10]$ by $[-10, 10]$
Xscl = 1   Yscl = 1

Figure 40

41. Put into slope-intercept form to find slope: $x + 3y = 5 \Rightarrow 3y = -x + 5 \Rightarrow y = -\frac{1}{3}x + \frac{5}{3} \Rightarrow m = -\frac{1}{3}$.

    Since parallel lines have equal slopes, use $m = -\frac{1}{3}$ and $(-1, 4)$ in point-slope form to find the equation:

    $y - 4 = -\frac{1}{3}(x - (-1)) \Rightarrow y - 4 = -\frac{1}{3}x - \frac{1}{3} \Rightarrow y = -\frac{1}{3}x + \frac{11}{3}$.

42. Put into slope-intercept form to find slope: $2x - y = 5 \Rightarrow -y = -2x + 5 \Rightarrow y = 2x - 5 \Rightarrow m = 2$.

    Since parallel lines have equal slopes, use $m = 2$ and $(3, -2)$ in point-slope form to find the equation:

    $y - (-2) = 2(x - 3) \Rightarrow y + 2 = 2x - 6 \Rightarrow y = 2x - 8$.

43. Put into slope-intercept form to find slope: $3x + 5y = 1 \Rightarrow 5y = -3x + 1 \Rightarrow y = -\frac{3}{5}x + \frac{1}{5} \Rightarrow m = -\frac{3}{5}$.

    Since perpendicular lines have negative reciprocal slopes, use $m = \frac{5}{3}$ and $(1, 6)$ in point-slope form to find the

    equation: $y - 6 = \frac{5}{3}(x - 1) \Rightarrow y - 6 = \frac{5}{3}x - \frac{5}{3} \Rightarrow y = \frac{5}{3}x + \frac{13}{3}$.

44. Put into slope-intercept form to find slope: $8x - 3y = 7 \Rightarrow -3y = -8x + 7 \Rightarrow y = \frac{8}{3}x - \frac{7}{3} \Rightarrow m = \frac{8}{3}$.

    Since perpendicular lines have negative reciprocal slopes, use $m = -\frac{3}{8}$ and $(-2, 0)$ in point-slope form to find

    the equation: $y - 0 = -\frac{3}{8}(x - (-2)) \Rightarrow y = -\frac{3}{8}x - \frac{3}{4}$.

45. The equation $y = -2$ has a slope $m = 0$. A line perpendicular to this would have an undefined slope which

    would have an equation in the form $x = a$. An equation in the form $x = a$ through $(-5, 7)$ is : $x = -5$.

46. The equation $x = 4$ has an undefined slope. A line perpendicular to this would have a slope $m = 0$, which

    would have an equation in the form $y = b$. An equation in the form $y = b$ through $(1, -4)$ is : $y = -4$.

47. The equation $y = -.2x + 6$ has a slope $m = -.2$. Since parallel lines have equal slopes, use $m = -.2$ and $(-5, 8)$

    in point-slope form to find the equation: $y - 8 = -.2(x - (-5)) \Rightarrow y - 8 = -.2x - 1 \Rightarrow y = -.2x + 7$.

48. Put into slope-intercept form to find slope: $x + y = 5 \Rightarrow y = -x + 5 \Rightarrow m = -1$.

    Since parallel lines have equal slopes, use $m = -1$ and $(-4, -7)$ in point-slope form to find the equation:

    $y - (-7) = -1(x - (-4)) \Rightarrow y + 7 = -x - 4 \Rightarrow y = -x - 11$.

49. Put into slope-intercept form to find slope: $2x + y = 6 \Rightarrow y = -2x + 6 \Rightarrow m = -2$. Since perpendicular lines

    have negative reciprocal slopes, use $m = \frac{1}{2}$ and the origin $(0, 0)$ in point-slope form to find the equation:

    $y - 0 = \frac{1}{2}(x - 0) \Rightarrow y = \frac{1}{2}x$.

50. The equation $y = -3.5x + 7.4$ has a slope $m = -3.5$. Since parallel lines have equal slopes, use $m = -3.5$ and

    the origin $(0, 0)$ in point-slope form to find the equation: $y - 0 = -3.5(x - 0) \Rightarrow y = -3.5x$.

51. (a) The Pythagorean Theorem and its converse.

(b) Using the distance formula from $(0, 0)$ to (x_1, m_1x_1) yields: $d(0, P) = \sqrt{(x_1)^2 + (m_1x_1)^2}$.

(c) Using the distance formula from $(0, 0)$ to (x_2, m_2x_2) yields: $d(0, Q) = \sqrt{(x_2)^2 + (m_2x_2)^2}$.

(d) Using the distance formula from (x_1, m_1x_1) to (x_2, m_2x_2) yields: $d(P, Q) = \sqrt{(x_2 - x_1)^2 + (m_2x_2 - m_1x_1)^2}$.

(e) Using Pythagorean Theorem yields: $[d(0,P)]^2 + [d(0, Q)]^2 = [d(P, Q)]^2 \Rightarrow$

$(x_1)^2 + (m_1x_1)^2 + (x_2)^2 + (m_2x_2)^2 = (x_1 - x_2)^2 + (m_1x_1 - m_2x_2)^2 \Rightarrow$

$(x_1)^2 + (m_1x_1)^2 + (x_2)^2 + (m_2x_2)^2 = (x_1)^2 - 2x_1x_2 + (x_2)^2 + (m_1x_1)^2 - 2m_1m_2x_1x_2 + (m_2x_2)^2 \Rightarrow$

$0 = -2m_1m_2x_1x_2 - 2x_1x_2$

(f) $0 = -2x_1x_2 - 2m_1m_2x_1x_2 \Rightarrow 0 = -2x_1x_2(1 + m_1m_2)$

(g) By the zero-product property, for $-2x_1x_2(1 + m_1m_2) = 0$ either $-2x_1x_2 = 0$ or $1 + m_1m_2 = 0$. Since $x_1 \neq 0$ and $x_2 \neq 0$, $-2x_1x_2 \neq 0$, and it follows that $1 + m_1m_2 = 0 \Rightarrow m_1m_2 = -1$.

(h) The product of the slopes of two perpendicular lines, neither of which are parallel to an axis, is -1.

52. (a) To find the slope of Y_1 use $(0, -3)$ and $(1, 1)$: $m = \dfrac{-3 - 1}{0 - 1} = \dfrac{-4}{-1} = 4$. To find the slope of Y_2 use

$(0, 4)$ and $(4, 3)$: $m = \dfrac{4 - 3}{0 - 4} = \dfrac{1}{-4} = -\dfrac{1}{4}$. Since $4\left(-\dfrac{1}{4}\right) = -1$ the lines are perpendicular.

(b) To find the slope of Y_1 use $(0, -3)$ and $(1, 2)$: $m = \dfrac{-3 - 2}{0 - 1} = \dfrac{-5}{-1} = 5$. To find the slope of Y_2 use

$(0, 5)$ and $(5, 6)$: $m = \dfrac{6 - 5}{5 - 0} = \dfrac{1}{5}$. Since $5\left(\dfrac{1}{5}\right) = 1$ not -1 and they are not equal the lines are neither perpendicular or parallel.

(c) To find the slope of Y_1 use $(0, -3)$ and $(1, 2)$: $m = \dfrac{-3 - 2}{0 - 1} = \dfrac{-5}{-1} = 5$. To find the slope of Y_2 use

$(0, 12)$ and $(1, 17)$: $m = \dfrac{17 - 12}{1 - 0} = \dfrac{5}{1} = 5$. Since $5 = 5$ the lines are parallel.

(d) To find the slope of Y_1 use $(0, 2)$ and $(1, -2)$: $m = \dfrac{-2 - 2}{1 - 0} = \dfrac{-4}{1} = -4$.

To find the slope of Y_2 use $(0, -2)$ and $(1, 2)$: $m = \dfrac{2 - (-2)}{1 - 0} = \dfrac{4}{1} = 4$.

Since $4 \neq -4$ and $4(-4) \neq -1$ the lines are neither parallel or perpendicular.

53. $m = \dfrac{-6 - (-3)}{0 - 1} = \dfrac{-3}{-1} = 3$

54. $m = \dfrac{-3 - 0}{1 - 2} = \dfrac{-3}{-1} = 3$

55. If we use any two points on a line to find its slope, we find that the slope is _equal_ in all cases.

56. $d = \sqrt{(0 - 1)^2 + (-6 - (-3))^2} = \sqrt{1 + 9} = \sqrt{10}$

57. $d = \sqrt{(1 - 3)^2 + (-3 - 3)^2} = \sqrt{(-2)^2 + (-6)^2} = \sqrt{4 + 36} = 2\sqrt{10}$

58. $d = \sqrt{(0 - 3)^2 + (-6 - 3)^2} = \sqrt{(-3)^2 + (-9)^2} = \sqrt{9 + 81} = 3\sqrt{10}$

59. $\sqrt{10} + 2\sqrt{10} = 3\sqrt{10}$, this is equal to the answer in Exercise 58. The distance from A to B added to the distance from B to D is equal to the distance from A to D.

60. If points A, B, and C lies on a line in that order, then the distance between A and B added to the distance between B and C is equal to the distance between A and C.

61. $M.P. = \left(\dfrac{0 + 6}{2}, \dfrac{-6 + 12}{2}\right) = (3, 3)$. The midpoint equals the middle entry of the table.

62. Find the midpoint between $(4, 6)$ and $(5, 9)$. $M.P. = \left(\dfrac{4 + 5}{2}, \dfrac{6 + 9}{2}\right) = (4.5, 7.5)$. The y-value is 7.5.

63. (a) Use the given points to find slope, then $m = \dfrac{161 - 128}{4 - 1} = \dfrac{33}{3} \Rightarrow m = 11$. Now use point-slope form to find the equation: $y - 128 = 11(x - 1) \Rightarrow y - 128 = 11x - 11 \Rightarrow y = 11x + 117$.

    (b) From the slope the biker is traveling 11mph.

    (c) At $x = 0$, $y = 11(0) + 117 \Rightarrow y = 117$, therefore 117 miles from the highway.

    (d) Since at 1 hour and 15 minutes $x = 1.25$, then $y = 11(1.25) + 117 \Rightarrow y = 130.75$, so 130.75 miles away.

64. (a) Since the graph is falling as time increases, water is leaving the tank. 70 gallons after 3 minutes

    (b) The x-intercept: 10 and the y-intercept: 100. The tank initially held 100 gallons and is empty after 10 minutes.

    (c) Find the slope: $m = \dfrac{0 - 100}{10 - 0} = \dfrac{-100}{10} = -10$, since $b = 100$, the equation is: $y = -10x + 100$. The slope of $m = -10$, shows the rate at which the water is being drained from the tank is 10gal/min.

    (d) At $y = 50$, $x = 5 \Rightarrow (5, 50)$. The x-coordinate is: 5.

65. (a) If $x = 0$ corresponds to 1979, then the line passes through $(0, 1480)$ and $b = 1480$. The rate of increase is 280 per year, therefore $m = 280$ and $y = 280x + 1480$.

    (b) Since $1988 - 1979 = 9$, $x = 9$. At $x = 9$, $y = 280(9) + 1480 \Rightarrow y = 4000$ pollution incidents.

66. (a) First find the slope: $m = \dfrac{8.24 - 7.66}{2002 - 1990} = \dfrac{.58}{12} = .0483$, now use point-slope form to find the equation.

    $y - 7.66 = .0483(x - 1990) \Rightarrow y = .0483(x - 1990) + 7.66$

    (b) The hourly wage increased at a rate of approximately \$.05 per year between 1990 and 2002.

    (c) At $x = 1995$, $y = .0483(1995 - 1990) + 7.66 \Rightarrow y = .0483(5) + 7.66 \Rightarrow y = \$7.90$, which is higher than the actual wage and does not compare favorably.

67. (a) Since the plotted points form a line, it is a linear relation. See Figure 67 on next page.

    (b) Using the first two points find the slope: $m = \dfrac{0 - (-40)}{32 - (-40)} = \dfrac{40}{72} = \dfrac{5}{9}$, now use slope intercept form to find the function: $C(x) - 0 = \dfrac{5}{9}(x - 32) \Rightarrow C(x) = \dfrac{5}{9}(x - 32)$. The slope of $\dfrac{5}{9}$ means that the Celsius temperature changes 5° for every 9° change in Fahrenheit temperature.

    (c) $C(83) = \dfrac{5}{9}(83 - 32) = 28\dfrac{1}{3}°C$

68. (a) The slope is: $m = \dfrac{11.6 - 9.7}{2002 - 1996} = \dfrac{1.9}{6} = \dfrac{19}{60}$. Using point-slope form produces the equation:

    $y - 9.7 = \dfrac{19}{60}(x - 1996)$

    (b) At $x = 2006$, $y - 9.7 = \dfrac{19}{60}(2006 - 1996) \Rightarrow y = \dfrac{19}{60}(10) + 9.7 \Rightarrow y \approx 12.9$ million.

69. (a) The slope is: $m = \dfrac{59.6 - 51.5}{2002 - 1980} = \dfrac{8.1}{22} = \dfrac{81}{220}$. Using point-slope form produces the equation:

$$y - 51.5 = \frac{81}{220}(x - 1980)$$

(b) At $x = 1990$, $y - 51.5 = \dfrac{81}{220}(1990 - 1980) \Rightarrow y = \dfrac{81}{220}(10) + 51.5 \Rightarrow y \approx 55.2\%$.

At $x = 1995$, $y - 51.5 = \dfrac{81}{220}(1995 - 1980) \Rightarrow y = \dfrac{81}{220}(15) + 51.5 \Rightarrow y \approx 57.0\%$.

At $x = 2000$, $y - 51.5 = \dfrac{81}{220}(2000 - 1980) \Rightarrow y = \dfrac{81}{220}(20) + 51.5 \Rightarrow y \approx 58.9\%$.

The value for 2000 is close, but the other two values are too low.

70. (a) The slope is: $m = \dfrac{74.8 - 77.4}{2000 - 1980} = \dfrac{-2.6}{20} = -\dfrac{13}{100}$. Using point-slope form produces the equation:

$$y - 77.4 = -\frac{13}{100}(x - 1980).$$

(b) The slope is: $m = \dfrac{74.1 - 76.4}{2002 - 1990} = \dfrac{-2.3}{12} = -\dfrac{23}{120}$. Using point-slope form produces the equation:

$$y - 76.4 = -\frac{23}{120}(x - 1990).$$

(c) Model (a) gives: $y - 77.4 = -\dfrac{13}{100}(2003 - 1980) \Rightarrow y = -\dfrac{13}{100}(23) + 77.4 \Rightarrow y \approx 74.4$.

Model (b) gives: $y - 76.4 = -\dfrac{23}{120}(2003 - 1990) \Rightarrow y = -\dfrac{23}{120}(13) + 76.4 \Rightarrow y \approx 73.9$.

(d) The first equation (a) models the data better.

71. (a) Enter the years in L_1 and enter tuition and fees in L_2. The regression equation is:

$$y \approx 751.375x - 1,486,152.426$$

(b) See Figure 71.

(c) At $x = 1992$, $y \approx 751.375(1992) - 1,486,152.426 \Rightarrow y \approx \$10,587$, this is quite close to the actual value of $10,498.

72. (a) Enter the years in L_1 and enter tuition and fees in L_2. The regression equation is:

$$y \approx 198.493x - 393,033.872$$

(b) See Figure 72.

(c) At $x = 2006$, $y \approx 198.493(2006) - 393,033.872 \Rightarrow y \approx \$5143$.

(d) At $x = 2010$, $y \approx 198.493(2010) - 393,033.872 \Rightarrow y \approx \$5937$.

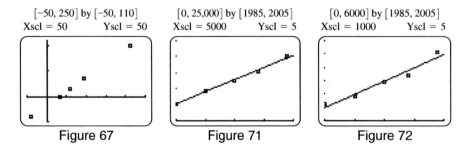

$[-50, 250]$ by $[-50, 110]$	$[0, 25{,}000]$ by $[1985, 2005]$	$[0, 6000]$ by $[1985, 2005]$
Xscl = 50 Yscl = 50	Xscl = 5000 Yscl = 5	Xscl = 1000 Yscl = 5
Figure 67	Figure 71	Figure 72

73. (a) Enter the distance in L_1 and enter velocity in L_2. The regression equation is: $y \approx 14.68x + 277.82$.

    (b) At $y = 37{,}000$, $37{,}000 \approx 14.68x + 277.82 \Rightarrow 14.68x \approx 36{,}722.18 \Rightarrow x \approx 2501.5$, or approximately 2500 light-years.

74. (a) Enter the height in L_1 and enter weight in L_2. The regression equation is: $y \approx 3.35x - 78.4$

    (b) Find y when $x = 60$: $y = 3.35(60) - 78.4 \Rightarrow y = 122.6 \Rightarrow y \approx 123$ pounds.

    (c) Find y when $x = 70$: $y = 3.35(70) - 78.4 \Rightarrow y = 156.1 \Rightarrow y \approx 156$ pounds.

    (d) The correlation coefficient is: $r = .65598 \Rightarrow r \approx .66$, there is a positive correlation, taller people are generally heavier, but it is not very strong since it is not close to 1.

75. Enter the Gestation Period in L_1 and enter Life Span in L_2. The regression equation is: $y \approx .101x + 11.6$ and the correlation coefficient is: $r \approx .909$. There is a fairly strong positive correlation, because .909 is close to 1.

76. Enter the Population in L_1 and enter Area in L_2. The regression equation is: $y \approx .01495x + 492.744$ and the correlation coefficient is: $r \approx .1583$. There is a weak positive correlation, since $r > 0$ and r is close to 0.

Reviewing Basic Concepts (Sections 1.3 and 1.4)

1. Since $m = 1.4$ and $b = -3.1$, slope-intercept form gives the function: $f(x) = 1.4x - 3.1$.

    $$f(1.3) = 1.4(1.3) - 3.1 \Rightarrow f(1.3) = -1.28$$

2. See Figure 2. x-intercept: $\dfrac{1}{2}$, y-intercept: 1, slope: -2, domain: $(-\infty, \infty)$, range: $(-\infty, \infty)$

3. $m = \dfrac{6 - 4}{5 - (-2)} = \dfrac{2}{7}$

4. Vertical line graphs are in the form $x = a$, through point $(-2, 10)$ would be $x = -2$.

    Horizontal line graphs are in the form $y = b$, through point $(-2, 10)$ would be $y = 10$.

5. See Figures 5a and 5b.

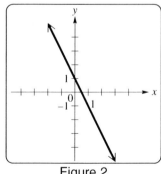

Figure 2

[-10, 10] by [-10, 10]
Xscl = 1   Yscl = 1

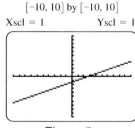

Figure 5a

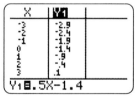

Figure 5b

6. The line of the graph rises 2 units for each 1 unit to the right, therefore the slope is: $m = \dfrac{2}{1} = 2$.

    The y-intercept is : $b = -3$. The slope-intercept form of the equation is: $y = 2x - 3$.

7. The slope is: $m = \dfrac{4 - 2}{(-2) - 5} = \dfrac{2}{-7} = -\dfrac{2}{7}$, now using point-slope form the equation is:

    $$y - 4 = -\frac{2}{7}(x + 2) \Rightarrow y - 4 = -\frac{2}{7}x - \frac{4}{7} \Rightarrow y = -\frac{2}{7}x + \frac{24}{7}.$$

8. Find the given equation in slope-intercept form: $3x - 2y = 5 \Rightarrow -2y = -3x + 5 \Rightarrow y = \frac{3}{2}x - \frac{5}{2}.$ The slope

of this equation is $m = \frac{3}{2}$, therefore the slope of a perpendicular line will be the negative reciprocal: $m = -\frac{2}{3}.$

Using point-slope form yields the equation: $y - 3 = -\frac{2}{3}(x + 1) \Rightarrow y - 3 = -\frac{2}{3}x - \frac{2}{3} \Rightarrow y = -\frac{2}{3}x + \frac{7}{3}.$

9. (a) See Figure 9.

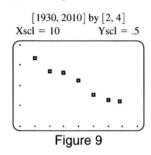

$[1930, 2010]$ by $[2, 4]$
Xscl $= 10$ Yscl $= .5$

Figure 9

(b) As x increases, y decreases, therefore a negative correlation coefficient.

(c) Enter the years in L_1 and enter people per household in L_2. The regression equation is:

$y \approx -.01904x + 40.577$ and the correlation coefficient is: $r = -.9792$.

(d) The regression equation is: $y \approx -.01904(1975) + 40.577 \Rightarrow y \approx 2.97$, which is close to the actual value 2.94.

1.5: Linear Equations and Inequalities

1. $-3x - 12 = 0 \Rightarrow -3x = 12 \Rightarrow x = -4$

2. $5x - 30 = 0 \Rightarrow 5x = 30 \Rightarrow x = 6$

3. $5x = 0 \Rightarrow x = 0$

4. $-2x = 0 \Rightarrow x = 0$

5. $2(3x - 5) + 8(4x + 7) = 0 \Rightarrow 6x - 10 + 32x + 56 = 0 \Rightarrow 38x = -46 \Rightarrow x = -\frac{46}{38} \Rightarrow x = -\frac{23}{19}$

6. $-4(2x - 3) + 8(2x + 1) = 0 \Rightarrow -8x + 12 + 16x + 8 = 0 \Rightarrow 8x = -20 \Rightarrow x = -\frac{20}{8} \Rightarrow x = -\frac{5}{2}$

7. $3x + 6(x - 4) = 0 \Rightarrow 3x + 6x - 24 = 0 \Rightarrow 9x = 24 \Rightarrow x = \frac{24}{9} \Rightarrow x = \frac{8}{3}$

8. $-8x + .5(2x + 8) = 0 \Rightarrow -8x + x + 4 = 0 \Rightarrow -7x = -4 \Rightarrow x = \frac{-4}{-7} \Rightarrow x = \frac{4}{7}$

9. $1.5x + 2(x - 3) + 5.5(x + 9) = 0 \Rightarrow 1.5x + 2x - 6 + 5.5x + 49.5 = 0 \Rightarrow 9x = -43.5 \Rightarrow$

$x = -\frac{43.5}{9} = -\frac{435}{90} \Rightarrow x = -\frac{29}{6}$

10. Since c is a zero, c is the value of x when $y = 0$, therefore the coordinate at the point the line intersects the x-axis

is: $(c, 0)$.

11. The solution to $y_1 = y_2$ is the intersection of the lines or $x = \{10\}$.

12. The solution to $y_1 = y_2$ is the intersection of the lines or $x = \{-2\}$.

13. The solution to $y_1 = y_2$ is the intersection of the lines or $x = \{1\}$.

14. When $y_1 = y_2$, $y = 0$. $y = 0$ when the graph crosses the x-axis or at the zero $x = \{-.8\}$.

15. When $y_1 = y_2$, $y = 0$. $y = 0$ when the graph crosses the x-axis or at the zero $x = \{3\}$.

16. When $y_1 = y_2$, $y = 0$. $y = 0$ when the graph crosses the x-axis or at the zero $x = \{2\}$.

17. When $x = 10$ is substituted into each function the result is 20.

18. Using the x-intercept method means using $y_1 = y_2$, which would yield: $y_1 = 2x + 3 - (4x - 12) \Rightarrow$
    $y_1 = 2x + 3 - 4x + 12$ which is not the same as graphing: $y_1 = 2x + 3 - 4x - 12$.

19. There is no real solution if $y_1 - y_2$ yields a contradiction, $y = b$, where $b \neq 0$. This equation is called a
    contradiction and the solution set is: $\varnothing$.

20. The solution set is: $x = (-\infty, \infty)$ if $y_1 - y_2$ is the line $y = 0$. This equation is called an identity.

21. $2x - 5 = x + 7 \Rightarrow x - 5 = 7 \Rightarrow x = 12$ **Check:** $2(12) - 5 = 12 + 7 \Rightarrow 19 = 19$
    The graphs of the left and right sides of the equation intersect when $x = 12$. The solution set is $\{12\}$.

22. $9x - 17 = 2x + 4 \Rightarrow 7x - 17 = 4 \Rightarrow 7x = 21 \Rightarrow x = 3$
    Check: $9(3) - 17 = 2(3) + 4 \Rightarrow 27 - 17 = 6 + 4 \Rightarrow 10 = 10$
    The graphs of the left and right sides of the equation intersect when $x = 3$. The solution set is $\{3\}$.

23. $.01x + 3.1 = 2.03x - 2.96 \Rightarrow 3.1 = 2.02x - 2.96 \Rightarrow 6.06 = 2.02x \Rightarrow x = 3$
    Check: $.01(3) + 3.1 = 2.03(3) - 2.96 \Rightarrow .03 + 3.1 = 6.09 - 2.96 \Rightarrow 3.13 = 3.13$
    The graphs of the left and right sides of the equation intersect when $x = 3$. The solution set is $\{3\}$.

24. $.04x + 2.1 = .02x + 1.92 \Rightarrow .02x + 2.1 = 1.92 \Rightarrow .02x = -.18 \Rightarrow x = -9$
    Check: $.04(-9) + 2.1 = .02(-9) + 1.92 \Rightarrow -.36 + 2.1 = -.18 + 1.92 \Rightarrow 1.74 = 1.74$
    The graphs of the left and right sides of the equation intersect when $x = -9$. The solution set is $\{-9\}$.

25. $-(x + 5) - (2 + 5x) + 8x = 3x - 5 \Rightarrow -x - 5 - 2 - 5x + 8x = 3x - 5 \Rightarrow$
    $2x - 7 = 3x - 5 \Rightarrow -2 = x$
    Check: $-(-2 + 5) - (2 + 5(-2)) + 8(-2) = 3(-2) - 5 \Rightarrow 2 - 5 - 2 + 10 - 16 = -6 - 5 \Rightarrow$
    $-11 = -11$
    The graphs of the left and right sides of the equation intersect when $x = -2$. The solution set is $\{-2\}$.

26. $-(8 + 3x) + 5 = 2x + 3 \Rightarrow -8 - 3x + 5 = 2x + 3 \Rightarrow -3 = 5x + 3 \Rightarrow -6 = 5x \Rightarrow x = -\dfrac{6}{5}$
    Check: $-\left(8 + 3\left(-\dfrac{6}{5}\right)\right) + 5 = 2\left(-\dfrac{6}{5}\right) + 3 \Rightarrow -8 + \dfrac{18}{5} + 5 = -\dfrac{12}{5} + 3 \Rightarrow \dfrac{18}{5} = -\dfrac{12}{5} + 6 \Rightarrow 6 = 6$
    The graphs of the left and right sides of the equation intersect when $x = -\dfrac{6}{5}$. The solution set is $\left\{-\dfrac{6}{5}\right\}$.

27. $\dfrac{2x + 1}{3} + \dfrac{x - 1}{4} = \dfrac{13}{2} \Rightarrow 12\left(\dfrac{2x + 1}{3} + \dfrac{x - 1}{4}\right) = 12\left(\dfrac{13}{2}\right) \Rightarrow 8x + 4 + 3x - 3 = 78 \Rightarrow$
    $11x + 1 = 78 \Rightarrow 11x = 77 \Rightarrow x = 7$ **Check:** $\dfrac{2(7) + 1}{3} + \dfrac{7 - 1}{4} = \dfrac{13}{2} \Rightarrow 5 + \dfrac{6}{4} = \dfrac{13}{2} \Rightarrow \dfrac{13}{2} = \dfrac{13}{2}$
    The graphs of the left and right sides of the equation intersect when $x = 7$. The solution set is $\{7\}$.

28. $\dfrac{x-2}{4} + \dfrac{x+1}{2} = 1 \Rightarrow 4\left[\dfrac{x-2}{4} + \dfrac{x+1}{2} = 1\right] \Rightarrow x - 2 + 2x + 2 = 4 \Rightarrow 3x = 4 \Rightarrow x = \dfrac{4}{3}$

Check: $\dfrac{\frac{4}{3}-2}{4} + \dfrac{\frac{4}{3}+1}{2} = 1 \Rightarrow \dfrac{\frac{-2}{3}}{4} + \dfrac{\frac{7}{3}}{2} = 1 \Rightarrow \dfrac{\frac{-2}{3}}{4} + \dfrac{\frac{14}{3}}{4} = 1 \Rightarrow \dfrac{\frac{12}{3}}{4} = \dfrac{4}{4} = 1$

The graphs of the left and right sides of the equation intersect when $x = \dfrac{4}{3}$. The solution set is $\left\{\dfrac{4}{3}\right\}$.

29. $\dfrac{1}{2}(x-3) = \dfrac{5}{12} + \dfrac{2}{3}(2x-5) \Rightarrow 12\left[\dfrac{1}{2}(x-3) = \dfrac{5}{12} + \dfrac{2}{3}(2x-5)\right] \Rightarrow 6(x-3) = 5 + 8(2x-5) \Rightarrow$

$6x - 18 = 5 + 16x - 40 \Rightarrow -10x = -17 \Rightarrow x = \dfrac{17}{10}$ **Check:** $\dfrac{1}{2}\left(\dfrac{17}{10} - 3\right) = \dfrac{5}{12} + \dfrac{2}{3}\left(2\left(\dfrac{17}{10}\right) - 5\right) \Rightarrow$

$\dfrac{1}{2}\left(-\dfrac{13}{10}\right) = \dfrac{5}{12} + \dfrac{2}{3}\left(-\dfrac{16}{10}\right) \Rightarrow -\dfrac{13}{20} = \dfrac{5}{12} + \left(-\dfrac{32}{30}\right) \Rightarrow -\dfrac{78}{120} = \dfrac{50}{120} + \left(-\dfrac{128}{120}\right) \Rightarrow -\dfrac{78}{120} = -\dfrac{78}{120}$

The graphs of the left and right sides of the equation intersect when $x = \dfrac{17}{10}$. The solution set is $\left\{\dfrac{17}{10}\right\}$.

30. $\dfrac{7}{3}(2x-1) = \dfrac{1}{5}x + \dfrac{2}{5}(4-3x) \Rightarrow 15\left[\dfrac{7}{3}(2x-1) = \dfrac{1}{5}x + \dfrac{2}{5}(4-3x)\right] \Rightarrow 35(2x-1) = 3x + 6(4-3x) \Rightarrow$

$70x - 35 = 3x + (24 - 18x) \Rightarrow 70x - 35 = -15x + 24 \Rightarrow 85x = 59 \Rightarrow x = \dfrac{59}{85}$

Check: $\dfrac{7}{3}\left(2\left(\dfrac{59}{85}\right) - 1\right) = \dfrac{1}{5}\left(\dfrac{59}{85}\right) + \dfrac{2}{5}\left(4 - 3\left(\dfrac{59}{85}\right)\right) \Rightarrow \dfrac{7}{3}\left(\dfrac{118}{85} - 1\right) = \dfrac{59}{425} + \dfrac{2}{5}\left(4 - \dfrac{177}{85}\right) \Rightarrow$

$\dfrac{7}{3}\left(\dfrac{33}{85}\right) = \dfrac{59}{425} + \dfrac{2}{5}\left(\dfrac{163}{85}\right) \Rightarrow \dfrac{231}{255} = \dfrac{59}{425} + \dfrac{326}{425} \Rightarrow \dfrac{231}{255} = \dfrac{385}{425} \Rightarrow \dfrac{77}{85} = \dfrac{77}{85}$

The graphs of the left and right sides of the equation intersect when $x = \dfrac{59}{85}$. The solution set is $\left\{\dfrac{59}{85}\right\}$.

31. $.1x - .05 = -.07x \Rightarrow .17x = .05 \Rightarrow 17x = 5 \Rightarrow x = \dfrac{5}{17}$

Check: $.1\left(\dfrac{5}{17}\right) - .05 = -.07\left(\dfrac{5}{17}\right) \Rightarrow 10\left(\dfrac{5}{17}\right) - 5 = -7\left(\dfrac{5}{17}\right) \Rightarrow \dfrac{50}{17} - 5 = -\dfrac{35}{17} \Rightarrow -\dfrac{35}{17} = -\dfrac{35}{17}$

The graphs of the left and right sides of the equation intersect when $x = \dfrac{5}{17}$. The solution set is $\left\{\dfrac{5}{17}\right\}$.

32. $1.1x - 2.5 = .3(x-2) \Rightarrow 11x - 25 = 3(x-2) \Rightarrow 11x - 25 = 3x - 6 \Rightarrow 8x = 19 \Rightarrow x = \dfrac{19}{8}$

Check: $1.1\left(\dfrac{19}{8}\right) - 2.5 = .3\left(\left(\dfrac{19}{8}\right) - 2\right) \Rightarrow 11\left(\dfrac{19}{8}\right) - 25 = 3\left(\left(\dfrac{19}{8}\right) - 2\right) \Rightarrow \dfrac{209}{8} - 25 = 3\left(\dfrac{3}{8}\right) \Rightarrow$

$\dfrac{9}{8} = \dfrac{9}{8}$ The graphs of the left and right sides of the equation intersect when $x = \dfrac{19}{8}$. The solution set is $\left\{\dfrac{19}{8}\right\}$.

33. $.40x + .60(100 - x) = .45(100) \Rightarrow .40x + 60 - .60x = 45 \Rightarrow -.20x = -15 \Rightarrow -20x = -1500 \Rightarrow x = 75$

Check: $.40(75) + .60(100 - 75) = .45(100) \Rightarrow 30 + 15 = 45 \Rightarrow 45 = 45$

The graphs of the left and right sides of the equation intersect when $x = 75$. The solution set is $\{75\}$.

34. $1.30x + .90(.50 - x) = 1.00(50) \Rightarrow 1.30x + .45 - .90x = 50 \Rightarrow .40x = 49.55 \Rightarrow x = 123.875$

    Check: $1.30(123.875) + .90(.50 - 123.875) = 1.00(50) \Rightarrow 161.0375 - 111.0375 = 50 \Rightarrow 50 = 50$

    The graphs of the left and right sides of the equation intersect when $x = 123.875$. The solution set is $\{123.875\}$.

35. $2[x - (4 + 2x) + 3] = 2x + 2 \Rightarrow 2[x - 4 - 2x + 3] = 2x + 2 \Rightarrow 2[-x - 1] = 2x + 2 \Rightarrow$

    $-2x - 2 = 2x + 2 \Rightarrow -4x = 4 \Rightarrow x = -1$

    Check: $2[-1 - (4 + 2(-1)) + 3] = 2(-1) + 2 \Rightarrow 2[-1 - 2 + 3] = 0 \Rightarrow 2[0] = 0 \Rightarrow 0 = 0$

    The graphs of the left and right sides of the equation intersect when $x = -1$. The solution set is $\{-1\}$.

36. $6[x - (2 - 3x) + 1] = 4x - 6 \Rightarrow 6[4x - 1] = 4x - 6 \Rightarrow 24x - 6 = 4x - 6 \Rightarrow 20x = 0 \Rightarrow x = 0$

    Check: $6[0 - (2 - 3(0)) + 1] = 4(0) - 6 \Rightarrow 6[-1] = -6 \Rightarrow -6 = -6$

    The graphs of the left and right sides of the equation intersect when $x = 0$. The solution set is $\{0\}$.

37. $\dfrac{5}{6}x - 2x + \dfrac{1}{3} = \dfrac{1}{3} \Rightarrow 6\left(\dfrac{5}{6}x - 2x + \dfrac{1}{3} = \dfrac{1}{3}\right) \Rightarrow 5x - 12x + 2 = 2 \Rightarrow -7x = 0 \Rightarrow x = 0$

    Check: $\dfrac{5}{6}(0) - 2(0) + \dfrac{1}{3} = \dfrac{1}{3} \Rightarrow \dfrac{1}{3} = \dfrac{1}{3}$

    The graphs of the left and right sides of the equation intersect when $x = 0$. The solution set is $\{0\}$.

38. $\dfrac{3}{4} + \dfrac{1}{5}x - \dfrac{1}{2} = \dfrac{4}{5}x \Rightarrow 20\left(\dfrac{3}{4} + \dfrac{1}{5}x - \dfrac{1}{2} = \dfrac{4}{5}x\right) \Rightarrow 15 + 4x - 10 = 16x \Rightarrow 5 = 12x \Rightarrow x = \dfrac{5}{12}$

    Check: $\dfrac{3}{4} + \dfrac{1}{5}\left(\dfrac{5}{12}\right) - \dfrac{1}{2} = \dfrac{4}{5}\left(\dfrac{5}{12}\right) \Rightarrow \dfrac{3}{4} + \dfrac{1}{12} - \dfrac{1}{2} = \dfrac{4}{12} \Rightarrow \dfrac{9}{12} + \dfrac{1}{12} - \dfrac{6}{12} = \dfrac{4}{12} \Rightarrow \dfrac{4}{12} = \dfrac{4}{12}$

    The graphs of the left and right sides of the equation intersect when $x = \dfrac{5}{12}$. The solution set is $\left\{\dfrac{5}{12}\right\}$.

39. $5x - (8 - x) = 2[-4 - (3 + 5x - 13)] \Rightarrow 6x - 8 = 2[-5x + 6] \Rightarrow 6x - 8 = -10x + 12 \Rightarrow$

    $16x = 20 \Rightarrow x = \dfrac{20}{16} = \dfrac{5}{4}$

    Check: $5\left(\dfrac{5}{4}\right) - \left(8 - \dfrac{5}{4}\right) = 2\left[-4 - \left(3 + 5\left(\dfrac{5}{4}\right) - 13\right)\right] \Rightarrow \dfrac{25}{4} - \dfrac{27}{4} = 2\left[-4 - \left(\dfrac{25}{4}\right) + 10\right] \Rightarrow$

    $-\dfrac{2}{4} = 2\left[6 - \dfrac{25}{4}\right] \Rightarrow -\dfrac{1}{2} = 2\left[-\dfrac{1}{4}\right] \Rightarrow -\dfrac{1}{2} = -\dfrac{1}{2}$

    The graphs of the left and right sides of the equation intersect when $x = \dfrac{5}{4}$. The solution set is $\left\{\dfrac{5}{4}\right\}$.

40. $-[x - (4x + 2)] = 2 + (2x + 7) \Rightarrow -[-3x - 2] = 2x + 9 \Rightarrow 3x + 2 = 2x + 9 \Rightarrow x = 7$

    Check: $-[7 - (4(7) + 2)] = 2 + (2(7) + 7) \Rightarrow -[7 - 30] = 2 + 21 \Rightarrow 23 = 23$

    The graphs of the left and right sides of the equation intersect when $x = 7$. The solution set is $\{7\}$.

41. When $x = 4$, both Y_1 and Y_2 have a value of 8. Therefore the solution set is $\{4\}$.

42. When $x = -3$, both Y_1 and Y_2 have a value of -17. Therefore the solution set is $\{-3\}$.

43. When $x = 1.5$, both Y_1 and Y_2 have a value of 4.5. So $Y_1 - Y_2 = 4.5 - 4.5 = 0$. The solution set is $\{1.5\}$.

44. When $x = 5.5$, both Y_1 and Y_2 have a value of 8. So $Y_1 - Y_2 = 8 - 8 = 0$. Therefore the solution set is $\{5.5\}$.

45. Graph $Y_1 = 4(.23x + \sqrt{5})$ and $Y_2 = \sqrt{2}x + 1$ as shown in Figure 45. The graphs intersect when $x \approx 16.07$. Therefore the solution set is $\{16.07\}$.

46. Graph $Y_1 = 9(-.84x + \sqrt{17})$ and $Y_2 = \sqrt{6}x - 4$ as shown in Figure 46. The graphs intersect when $x \approx 4.11$. Therefore the solution set is $\{4.11\}$.

47. Graph $Y_1 = 2\pi x + \sqrt[3]{4}$ and $Y_2 = .5\pi x - \sqrt{28}$ as shown in Figure 47. The graphs intersect when $x \approx -1.46$. Therefore the solution set is $\{-1.46\}$.

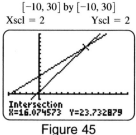

Figure 45

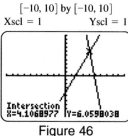

Figure 46

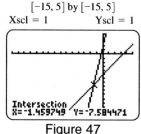

Figure 47

48. Graph $Y_1 = 3\pi x - \sqrt[4]{3}$ and $Y_2 = .75\pi x + \sqrt{19}$ as shown in Figure 48. The graphs intersect when $x \approx .80$. Therefore the solution set is: $\{.80\}$.

49. Graph $Y_1 = .23(\sqrt{3} + 4x) - .82(\pi x + 2.3)$ and $Y_2 = 5$ as shown in Figure 49. The graphs intersect when $x \approx -3.92$. Therefore the solution set is: $\{-3.92\}$.

50. Graph $Y_1 = -.15(6 + \sqrt{2}x) + 1.4(2\pi x - 6.1)$ and $Y_2 = 10$ as shown in Figure 50. The graphs intersect when $x \approx 2.26$. Therefore the solution set is: $\{2.26\}$.

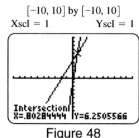

Figure 48

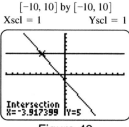

Figure 49

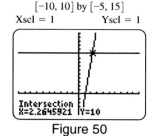

Figure 50

51. $5x + 5 = 5(x + 3) - 10 \Rightarrow 5x + 5 = 5x + 15 - 10 \Rightarrow 5x + 5 = 5x + 5 \Rightarrow 5 = 5 \Rightarrow$ Identity. The solution set is: $(-\infty, \infty)$. The table of $Y_1 = 5x + 5$ and $Y_2 = 5(x + 3) - 10$ produces all the same answers, therefore supports the Identity.

52. $5 - 4x = 5x - (9 + 9x) \Rightarrow 5 - 4x = -9 - 4x \Rightarrow 5 = -9 \Rightarrow$ Contradiction. The solution set is: $\varnothing$. The table of $Y_1 = 5 - 4x$ and $Y_2 = 5x - (9 + 9x)$ never produces the same answers, therefore supports the Contradiction.

53. $6(2x + 1) = 4x + 8\left(x + \dfrac{3}{4}\right) \Rightarrow 12x + 6 = 4x + 8x + 6 \Rightarrow 12x + 6 = 12x + 6 \Rightarrow 6 = 6 \Rightarrow$ Identity. The solution set is: $(-\infty, \infty)$. The table of $Y_1 = 6(2x + 1)$ and $Y_2 = 4x + 8\left(x + \dfrac{3}{4}\right)$ produces all the same answers, therefore supports the Identity.

54. $3(x + 2) - 5(x + 2) = -2x - 4 \Rightarrow 3x + 6 - 5x - 10 = -2x - 4 \Rightarrow -2x - 4 = -2x - 4 \Rightarrow -4 = -4$
$\Rightarrow$ Identity. The solution set is: $(-\infty, \infty)$. The table of $Y_1 = 3(x + 2) - 5(x + 2)$ and $Y_1 = -2x - 4$
produces all the same answers, therefore supports the Identity.

55. $-4[6 - (-2 + 3x)] = 21 + 12x \Rightarrow -4[8 - 3x] = 21 + 12x \Rightarrow -32 + 12x = 21 + 12x \Rightarrow -32 = 21 \Rightarrow$
Contradiction. The solution set is: $\varnothing$. The table of $Y_1 = -4[6 - (-2 + 3x)]$ and $Y_2 = 21 + 12x$ never
produces the same answers, therefore supports the Contradiction.

56. $-3[-5 - (-9 + 2x)] = 2(3x - 1) \Rightarrow -3[4 - 2x] = 6x - 2 \Rightarrow -12 + 6x = 6x - 2 \Rightarrow -12 = -2 \Rightarrow$
Contradiction. The solution set is: $\varnothing$. The table of $Y_1 = -3[-5 - (-9 + 2x)]$ and $Y_2 = 2(3x - 1)$ never
produces the same answers, therefore supports the Contradiction.

57. $\frac{1}{2}x - 2(x - 1) = 2 - \frac{3}{2}x \Rightarrow \frac{1}{2}x - 2x + 2 = 2 - \frac{3}{2}x \Rightarrow -\frac{3}{2}x + 2 = -\frac{3}{2}x + 2 \Rightarrow 2 = 2 \Rightarrow$ Identity.
The solution set is: $(-\infty, \infty)$. The table of $Y_1 = \frac{1}{2}x - 2(x - 1)$ and $Y_2 = 2 - \frac{3}{2}x$ produces all the same
answers, therefore supports the Identity.

58. $.5(x - 2) + 12 = .5x + 11 \Rightarrow 1(x - 2) + 24 = 1x + 22 \Rightarrow x - 2 + 24 = x + 22 \Rightarrow$
$x + 22 = x + 22 \Rightarrow 22 = 22 \Rightarrow$ Identity. The solution set is: $(-\infty, \infty)$. The table of
$Y_1 = .5(x - 2) + 12$ and $Y_1 = .5x + 11$ produces all the same answers, therefore supports the Identity.

59. $\frac{x - 1}{2} = \frac{3x - 2}{6} \Rightarrow 6\left[\frac{x - 1}{2} = \frac{3x - 2}{6}\right] \Rightarrow 3(x - 1) = 3x - 2 \Rightarrow 3x - 3 = 3x - 2 \Rightarrow -3 = -2 \Rightarrow$
Contradiction. The solution set is: $\varnothing$. The table of $Y_1 = \frac{x - 1}{2}$ and $Y_2 = \frac{3x - 2}{6}$ never produces the same
answers, therefore supports the Contradiction.

60. $\frac{2x - 1}{3} = \frac{2x + 1}{3} \Rightarrow 3\left[\frac{2x - 1}{3} = \frac{2x + 1}{3}\right] \Rightarrow 2x - 1 = 2x + 1 \Rightarrow -1 = 1 \Rightarrow$ Contradiction.
The solution set is: $\varnothing$. The table of $Y_1 = \frac{2x - 1}{3}$ and $Y_2 = \frac{2x + 1}{3}$ never produces the same answers, therefore
supports the Contradiction.

61. For the given functions, $f(x) = g(x)$ when the graphs intersect or when $x = 3$. The solution is: $\{3\}$

62. For the given functions, $f(x) > g(x)$ when the graph of $f(x)$ is above the graph of $g(x)$ or when $x < 3$. The
solution is: $(-\infty, 3)$.

63. For the given functions, $f(x) < g(x)$ when the graph of $f(x)$ is below the graph of $g(x)$ or when $x > 3$. The
solution is: $(3, \infty)$.

64. For the given functions, $g(x) - f(x) \geq 0 \Rightarrow g(x) \geq f(x)$ when the graph of $g(x)$ is above or intersects the graph
of $f(x)$ or when $x \geq 3$. The solution is: $[3, \infty)$.

65. For the given inequality, $y_1 - y_2 \geq 0 \Rightarrow f(x) - g(x) \geq 0 \Rightarrow f(x) \geq g(x)$ when the graph of $f(x)$ is above or
intersects the graph of $g(x)$ or when $x \leq 3$. The solution is: $(-\infty, 3]$.

66. For the given inequality, $y_2 > y_1 \Rightarrow g(x) > f(x)$ when the graph of $g(x)$ is above the graph of $f(x)$ or when
$x > 3$. The solution is: $(3, \infty)$.

67. For the given functions, $f(x) \le g(x)$ when the graph of $f(x)$ is below or intersects the graph of $g(x)$ or when $x \ge 3$. The solution is: $[3, \infty)$.

68. For the given functions, $f(x) \ge g(x)$ when the graph of $f(x)$ is above or intersects the graph of $g(x)$ or when $x \le 3$. The solution is: $(-\infty, 3]$.

69. For the given functions, $f(x) \le 2$ when the graph of $f(x)$ is below or equal to 2 or when $x \ge 3$. The solution is: $[3, \infty)$.

70. For the given functions, $g(x) \le 2$ when the graph of $g(x)$ is below or equal to 2 or when $x \le 3$. The solution is: $(-\infty, 3]$.

71. (a) The function $f(x) > 0$ when the graph is above the x-axis for the interval $(20, \infty)$.

    (b) The function $f(x) < 0$ when the graph is below the x-axis for the interval $(-\infty, 20)$.

    (c) The function $f(x) \ge 0$ when the graph intersects or is above the x-axis for the interval $[20, \infty)$.

    (d) The function $f(x) \le 0$ when the graph intersects or is below the x-axis for the interval $(-\infty, 20]$.

72. (a) The function $f(x) < 0$ when the graph is below the x-axis for the interval $(-\infty, 8)$.

    (b) The function $f(x) \le 0$ when the graph intersects or is below the x-axis for the interval $(-\infty, 8]$.

    (c) The function $f(x) \ge 0$ when the graph intersects or is above the x-axis for the interval $[8, \infty)$.

    (d) The function $f(x) > 0$ when the graph is above the x-axis for the interval $(8, \infty)$.

73. (a) If the solution set of $f(x) \ge g(x)$ is $[4, \infty)$, then $f(x) = g(x)$ at the intersection of the graphs, $x = 4$ or$\{4\}$.

    (b) If the solution set of $f(x) \ge g(x)$ is $[4, \infty)$, then $f(x) > g(x)$ is the same, but does not include the intersection of the graphs for the interval: $(4, \infty)$.

    (c) If the solution set of $f(x) \ge g(x)$ is $[4, \infty)$, then $f(x) < g(x)$ is left of the intersection of the graphs for the interval: $(-\infty, 4)$.

74. (a) If the solution set of $f(x) < g(x)$ is $(-\infty, 3)$, then $f(x) = g(x)$ at the intersection of the graphs, $x = 3$ or $\{3\}$.

    (b) If the solution set of $f(x) < g(x)$ is $(-\infty, 3)$, then $f(x) \ge g(x)$ is right of and does include the intersection of the graphs for the interval: $[3, \infty)$.

    (c) If the solution set of $f(x) < g(x)$ is $(-\infty, 3)$, then $f(x) \le g(x)$ is the same, but does include the intersection of the graphs for the interval: $(-\infty, 3]$.

75. (a) $9 - (x + 1) < 0 \Rightarrow -x + 8 < 0 \Rightarrow -x < -8 \Rightarrow x > 8 \Rightarrow$ the interval is: $(8, \infty)$

    The graph of $y_1 = 9 - (x + 1)$ is below the x-axis for the interval: $(8, \infty)$.

    (b) If $9 - (x + 1) < 0$ for: $(8, \infty)$, then $9 - (x + 1) \ge 0$ for the interval: $(-\infty, 8]$.

    The graph of $y_1 = 9 - (x + 1)$ intersects or is above the x-axis for the interval: $(-\infty, 8]$.

76. (a) $5 + 3(1 - x) \ge 0 \Rightarrow -3x + 8 \ge 0 \Rightarrow -3x \ge -8 \Rightarrow x \le \dfrac{8}{3} \Rightarrow$ the interval is: $\left(-\infty, \dfrac{8}{3}\right]$.

    The graph of $y_1 = 5 + 3(1 - x)$ intersects or is above the x-axis for the interval: $\left(-\infty, \dfrac{8}{3}\right]$.

    (b) If $5 + 3(1 - x) \ge 0$ for: $\left(-\infty, \dfrac{8}{3}\right]$, then $5 + 3(1 - x) < 0$ for the interval: $\left(\dfrac{8}{3}, \infty\right)$.

    The graph of $y_1 = 5 + 3(1 - x)$ is below the x-axis for the interval: $\left(\dfrac{8}{3}, \infty\right)$.

77. (a) $2x - 3 > x + 2 \Rightarrow x - 3 > 2 \Rightarrow x > 5 \Rightarrow$ the interval is: $(5, \infty)$

   The graph of $y_1 = 2x - 3$ is above the graph of $y_2 = x + 2$ for the interval: $(5, \infty)$.

   (b) If $2x - 3 > x + 2$ for: $(5, \infty)$, then $2x - 3 \leq x + 2$ for the interval: $(-\infty, 5]$.

   The graph $y_1 = 2x - 3$ intersects or is below the graph $y_2 = x + 2$ for the interval: $(-\infty, 5]$.

78. (a) $5 - 3x \leq 7 + x \Rightarrow 5 - 4x \leq 7 \Rightarrow -4x \leq 2 \Rightarrow x \geq -\dfrac{1}{2} \Rightarrow$ the interval is: $\left[-\dfrac{1}{2}, \infty\right)$

   The graph of $y_1 = 5 - 3x$ intersects or is below the graph of $y_2 = 7 + x$ for the interval: $\left[-\dfrac{1}{2}, \infty\right)$.

   (b) If $5 - 3x \leq 7 + x$ for: $\left[-\dfrac{1}{2}, \infty\right)$, then $5 - 3x > 7 + x$ for the interval: $\left(-\infty, -\dfrac{1}{2}\right)$.

   The graph of $y_1 = 5 - 3x$ is above the graph $y_2 = 7 + x$ for the interval: $\left(-\infty, -\dfrac{1}{2}\right)$.

79. (a) $10x + 5 - 7x \geq 8(x + 2) + 4 \Rightarrow 3x + 5 \geq 8x + 20 \Rightarrow -5x \geq 15 \Rightarrow x \leq -3 \Rightarrow$ interval is: $(-\infty, -3]$.

   The graph of $y_1 = 10x + 5 - 7x$ intersects or is above the graph of $y_2 = 8(x + 2) + 4$ for the interval:

   $(-\infty, -3]$.

   (b) If $10x + 5 - 7x \geq 8(x + 2) + 4$ for: $(-\infty, -3]$, then $10x + 5 - 7x < 8(x + 2) + 4$ for the interval:

   $(-3, \infty)$. The graph of $y_1 = 10x + 5 - 7x$ is below the graph $y_2 = 8(x + 2) + 4$ for the interval: $(-3, \infty)$.

80. (a) $6x + 2 + 10x > -2(2x + 4) + 10 \Rightarrow 16x + 2 > -4x + 2 \Rightarrow 20x > 0 \Rightarrow x > 0 \Rightarrow$ the interval is:

   $(0, \infty)$. The graph of $y_1 = 6x + 2 + 10x$ is above the graph of $y_2 = -2(2x + 4) + 10$ for the interval: $(0, \infty)$.

   (b) If $6x + 2 + 10x > -2(2x + 4) + 10$ for: $(0, \infty)$, then $6x + 2 + 10x \leq -2(2x + 4) + 10$ for the interval:

   $(-\infty, 0]$. The graph of $y_1 = 6x + 2 + 10x$ intersects or is below the graph $y_2 = -2(2x + 4) + 10$ for the

   interval: $(-\infty, 0]$.

81. (a) $x + 2(-x + 4) - 3(x + 5) < -4 \Rightarrow x - 2x + 8 - 3x - 15 < -4 \Rightarrow -4x < 3 \Rightarrow x > -\dfrac{3}{4} \Rightarrow$

   the interval is: $\left(-\dfrac{3}{4}, \infty\right)$. The graph of $y_1 = x + 2(-x + 4) - 3(x + 5)$ is below the graph of $y_2 = -4$ for

   the interval: $\left(-\dfrac{3}{4}, \infty\right)$.

   (b) If $x + 2(-x + 4) - 3(x + 5) < -4$ at: $\left(-\dfrac{3}{4}, \infty\right)$, then $x + 2(-x + 4) - 3(x + 5) \geq -4$ for the interval:

   $\left(-\infty, -\dfrac{3}{4}\right]$. The graph of $y_1 = x + 2(-x + 4) - 3(x + 5)$ intersects or is above the graph $y_2 = -4$ for the

   interval: $\left(-\infty, -\dfrac{3}{4}\right]$.

82. (a) $-11x - (6x - 4) + 5 - 3x \le 1 \Rightarrow -20x + 9 \le 1 \Rightarrow -20x \le -8 \Rightarrow x \ge \dfrac{2}{5} \Rightarrow$ the interval is: $\left[\dfrac{2}{5}, \infty\right)$.

The graph of $y_1 = -11x - (6x - 4) + 5 - 3x$ intersects or is below the graph of $y_2 = 1$ for the interval: $\left[\dfrac{2}{5}, \infty\right)$.

(b) If $-11x - (6x - 4) + 5 - 3x \le 1$ at: $\left(\dfrac{2}{5}, \infty\right)$, then $-11x - (6x - 4) + 5 - 3x > 1$ for the interval:

$\left(-\infty, \dfrac{2}{5}\right)$. The graph of $y_1 = -11x - (6x - 4) + 5 - 3x$ is below the graph $y_2 = 1$ for the interval: $\left(-\infty, \dfrac{2}{5}\right)$.

83. $\dfrac{1}{3}x - \dfrac{1}{5}x \le 2 \Rightarrow 15\left[\dfrac{1}{3}x - \dfrac{1}{5}x \le 2\right] \Rightarrow 5x - 3x \le 30 \Rightarrow 2x \le 30 \Rightarrow x \le 15 \Rightarrow (-\infty, 15]$. The graph of

$y_1 = \dfrac{1}{3}x - \dfrac{1}{5}x$ intersects or is below the graph of $y_2 = 2$ for the interval: $(-\infty, 15]$.

84. $\dfrac{3x}{2} + \dfrac{4x}{7} \ge -5 \Rightarrow 14\left[\dfrac{3x}{2} + \dfrac{4x}{7} \ge -5\right] \Rightarrow 21x + 8x \ge -70 \Rightarrow 29x \ge -70 \Rightarrow x \ge -\dfrac{70}{29} \Rightarrow \left[-\dfrac{70}{29}, \infty\right)$.

The graph of $y_1 = \dfrac{3x}{2} + \dfrac{4x}{7}$ intersects or is above the graph of $y_2 = -5$ for the interval: $\left[-\dfrac{70}{29}, \infty\right)$.

85. $\dfrac{x - 2}{2} - \dfrac{x + 6}{3} > -4 \Rightarrow 6\left[\dfrac{x - 2}{2} - \dfrac{x + 6}{3} > -4\right] \Rightarrow 3x - 6 - (2x + 12) > -24 \Rightarrow x - 18 > -24 \Rightarrow$

$x > -6 \Rightarrow (-6, \infty)$. The graph of $y_1 = \dfrac{x - 2}{2} - \dfrac{x + 6}{3}$ is above the graph of $y_2 = -4$ for the interval: $(-6, \infty)$.

86. $\dfrac{2x + 3}{5} - \dfrac{3x - 1}{2} < \dfrac{4x + 7}{2} \Rightarrow 10\left[\dfrac{2x + 3}{5} - \dfrac{3x - 1}{2} < \dfrac{4x + 7}{2}\right] \Rightarrow 4x + 6 - (15x - 5) < 20x + 35$

$\Rightarrow -11x + 11 < 20x + 35 \Rightarrow -31x < 24 \Rightarrow x > -\dfrac{24}{31} \Rightarrow \left(-\dfrac{24}{31}, \infty\right)$. The graph of

$y_1 = \dfrac{2x + 3}{5} - \dfrac{3x - 1}{2}$ is below the graph of $y_2 = \dfrac{4x + 7}{2}$ for the interval: $\left(-\dfrac{24}{31}, \infty\right)$.

87. $.6x - 2(.5x + .2) \le .4 - .3x \Rightarrow .6x - 1x - .4 \le .4 - .3x \Rightarrow 10[.6x - 1x - .4 \le .4 - .3x] \Rightarrow$

$6x - 10x - 4 \le 4 - 3x \Rightarrow -4x - 4 \le 4 - 3x \Rightarrow -x \le 8 \Rightarrow x \ge -8 \Rightarrow [-8, \infty)$. The graph of

$y_1 = .6x - 2(.5x + .2)$ intersects or is below the graph of $y_2 = .4 - .3x$ for the interval: $[-8, \infty)$.

88. $-.9x - (.5 + .1x) > -.3x - .5 \Rightarrow -.9x - .5 - .1x > -.3x - .5 \Rightarrow -x - .5 > -.3x - .5 \Rightarrow$

$10[-x - .5 > -.3x - .5] \Rightarrow -10x - 5 > -3x - 5 \Rightarrow -7x > 0 \Rightarrow x < 0 \Rightarrow (-\infty, 0)$. The graph of

$y_1 = -.9x - (.5 + .1x)$ is above the graph of $y_2 = -.3x - .5$ for the interval: $(-\infty, 0)$.

89. $-\dfrac{1}{2}x + .7x - 5 > 0 \Rightarrow 10\left[-\dfrac{1}{2}x + .7x - 5 > 0\right] \Rightarrow -5x + 7x - 50 > 0 \Rightarrow 2x > 50 \Rightarrow$

$x > 25 \Rightarrow (25, \infty)$. The graph of $y_1 = -\dfrac{1}{2}x + .7x - 5$ is above the graph of $y_2 = 0$ for the interval: $(25, \infty)$.

90. $\dfrac{3}{4}x - .2x - 6 \le 0 \Rightarrow 20\left[\dfrac{3}{4}x - .2x - 6 \le 0\right] \Rightarrow 15x - 4x - 120 \le 0 \Rightarrow 11x \le 120 \Rightarrow$

$x \le \dfrac{120}{11} \Rightarrow \left(-\infty, \dfrac{120}{11}\right]$. The graph of $y_1 = \dfrac{3}{4}x - .2x - 6$ intersects or is below the graph of $y_2 = 0$ for the

interval: $\left(-\infty, \dfrac{120}{11}\right]$.

91. $-4(3x + 2) \geq -2(6x + 1) \Rightarrow -12x - 8 \geq -12x - 2 \Rightarrow -8 \geq -2$, since this is false the solution is: $\varnothing$.

    The graph of $y_1 = -4(3x + 2)$ never intersects or is above the graph of $y_2 = -2(6x + 1)$, therefore the solution is: $\varnothing$.

92. Since $5 > 0$ is a true statement, all solutions are true for the interval: $(-\infty, \infty)$.

93. (a) As time increases, distance increases, therefore the car is moving away from Omaha.

    (b) The distance function $f(x)$ intersects the 100 mile line at 1 hour and the 200 mile line at 3 hours.

    (c) Using the answers from (b) the interval is: $[1, 3]$.

    (d) Because x hours is: $0 \leq x \leq 6$, the interval is: $(1, 6]$.

94. (a) The graph of $f(x)$ intersects the graph of $g(x)$ at: $x = 4$.

    (b) The graph of $g(x)$ intersects the graph of $h(x)$ at: $x = 2$.

    (c) $f(x) < g(x) < h(x)$ when the graph of the function $g(x)$ is between the graphs of the functions $f(x)$ and $h(x)$ for the interval: $(2, 4)$.

    (d) The function $g(x)$ is greater than $h(x)$ when it's graph is above the graph of $h(x)$ for the interval: $[0, 2)$.

    (e) The graph of $f(x)$ does not intersect the graph of $h(x)$, therefore the solution is: $\varnothing$.

    (f) The function of $f(x)$ is always graphed below the function of $h(x)$ for the the interval $[0, 6]$.

    (g) The function of $h(x)$ is greater than $g(x)$ when it's graph is above the graph of $g(x)$ for the interval: $(2, 6]$.

    (h) The function of $h(x)$ is always graphed below 500 for the interval: $[0, 6]$.

    (i) The function $g(x)$ is graphed below 200 for the interval $(4, 6]$.

    (j) The function of $f(x)$ always is equal to 200 for the interval: $[0, 6]$.

95. $4 \leq 2x + 2 \leq 10 \Rightarrow 2 \leq 2x \leq 8 \Rightarrow 1 \leq x \leq 4 \Rightarrow [1, 4]$

    The graph of $y_2 = 2x + 2$ is between the graphs of $y_1 = 4$ and $y_3 = 10$ for the interval $[1, 4]$.

96. $-4 \leq 2x - 1 \leq 5 \Rightarrow -3 \leq 2x \leq 6 \Rightarrow -\dfrac{3}{2} \leq x \leq 3 \Rightarrow \left[-\dfrac{3}{2}, 3\right]$

    The graph of $y_2 = 2x - 1$ is between the graphs of $y_1 = -4$ and $y_3 = 5$ for the interval $\left[-\dfrac{3}{2}, 3\right]$.

97. $-10 > 3x + 2 > -16 \Rightarrow -12 > 3x > -18 \Rightarrow -4 > x > -6 \Rightarrow -6 < x < -4 \Rightarrow (-6, -4)$

    The graph of $y_2 = 3x + 2$ is between the graphs of $y_1 = -10$ and $y_3 = -16$ for the interval $(-6, -4)$.

98. $4 > 6x + 5 > -1 \Rightarrow -1 > 6x > -6 \Rightarrow -\dfrac{1}{6} > x > -1 \Rightarrow -1 < x < -\dfrac{1}{6} \Rightarrow \left(-1, -\dfrac{1}{6}\right)$

    The graph of $y_2 = 6x + 5$ is between the graphs of $y_1 = 4$ and $y_3 = -1$ for the interval $\left(-1, -\dfrac{1}{6}\right)$.

99. $-3 \leq \dfrac{x - 4}{-5} < 4 \Rightarrow 15 \geq x - 4 > -20 \Rightarrow 19 \geq x > -16 \Rightarrow -16 < x \leq 19 \Rightarrow (-16, 19]$

    The graph of $y_2 = \dfrac{x - 4}{-5}$ is between the graphs of $y_1 = -3$ and $y_3 = 4$ for the interval $(-16, 19]$.

100. $1 < \dfrac{4x - 5}{-2} < 9 \Rightarrow -2 > 4x - 5 > -18 \Rightarrow 3 > 4x > -13 \Rightarrow \dfrac{3}{4} > x > -\dfrac{13}{4} \Rightarrow -\dfrac{13}{4} < x < \dfrac{3}{4} \Rightarrow$

    $\left(-\dfrac{13}{4}, \dfrac{3}{4}\right)$; The graph of $y_2 = \dfrac{4x - 5}{-2}$ is between the graphs of $y_1 = 1$ and $y_3 = 9$ for the interval $\left(-\dfrac{13}{4}, \dfrac{3}{4}\right)$.

101. $\sqrt{2} \le \dfrac{2x+1}{3} \le \sqrt{5} \Rightarrow 3\sqrt{2} \le 2x+1 \le 3\sqrt{5} \Rightarrow 3\sqrt{2} - 1 \le 2x \le 3\sqrt{5} - 1 \Rightarrow$

$\dfrac{3\sqrt{2}-1}{2} \le x \le \dfrac{3\sqrt{5}-1}{2} \Rightarrow \left[\dfrac{3\sqrt{2}-1}{2}, \dfrac{3\sqrt{5}-1}{2}\right]$; The graph of $y_2 = \dfrac{2x+1}{3}$ is between the graphs

of $y_1 = \sqrt{2}$ and $y_3 = \sqrt{5}$ for the interval $\left[\dfrac{3\sqrt{2}-1}{2}, \dfrac{3\sqrt{5}-1}{2}\right]$.

102. $\pi \le 5 - 4x < 7\pi \Rightarrow \pi - 5 \le -4x < 7\pi - 5 \Rightarrow \dfrac{\pi - 5}{-4} \ge x > \dfrac{7\pi - 5}{-4} \Rightarrow \dfrac{5 - \pi}{4} \ge x > \dfrac{5 - 7\pi}{4} \Rightarrow$

$\dfrac{5 - 7\pi}{4} < x \le \dfrac{5 - \pi}{4} \Rightarrow \left(\dfrac{5 - 7\pi}{4}, \dfrac{5 - \pi}{4}\right]$; The graph of $y_2 = 5 - 4x$ is between the graphs of

$y_1 = \pi$ and $y_3 = 7\pi$ for the interval $\left(\dfrac{5 - 7\pi}{4}, \dfrac{5 - \pi}{4}\right]$.

103. The graph of $y_1 = 3.7x - 11.1$ crosses the x-axis at $x = 3$. There is one solution to this equation. Because a linear equation can only cross the x-axis in one location, there is only one solution to any linear equation.

104. $3.7x - 11.1 < 0 \Rightarrow 3.7x < 11.1 \Rightarrow x < 3 \Rightarrow (-\infty, 3)$

$3.7x - 11.1 > 0 \Rightarrow 3.7x > 11.1 \Rightarrow x > 3 \Rightarrow (3, \infty)$

The value of $x = 3$ given by the equation represents the boundary between the sets of real numbers given by the inequality solutions $(-\infty, 3)$ and $(3, \infty)$.

105. The graph of $y_1 = -4x + 6$ crosses the x-axis at $x = 1.5$.

$-4x + 6 < 0 \Rightarrow -4x < -6 \Rightarrow x > \dfrac{-6}{-4} \Rightarrow x > 1.5 \Rightarrow (1.5, \infty)$

$-4x + 6 > 0 \Rightarrow -4x > -6 \Rightarrow x < \dfrac{-6}{-4} \Rightarrow x < 1.5 \Rightarrow (-\infty, 1.5)$

106. (a) If $a \ne 0$, then $ax + b = 0 \Rightarrow ax = -b \Rightarrow x = -\dfrac{b}{a}$

    (b) If $a > 0$, a positive slope, then $ax + b < 0 \Rightarrow ax < -b \Rightarrow x < \dfrac{-b}{a} \Rightarrow \left(-\infty, \dfrac{-b}{a}\right)$

        If $a > 0$, a positive slope, then $ax + b > 0 \Rightarrow ax > -b \Rightarrow x > \dfrac{-b}{a} \Rightarrow \left(\dfrac{-b}{a}, \infty\right)$

    (c) If $a < 0$, a negative slope, then $ax + b < 0 \Rightarrow ax < -b \Rightarrow x > \dfrac{-b}{a} \Rightarrow \left(\dfrac{-b}{a}, \infty\right)$

        If $a < 0$, a positive slope, then $ax + b > 0 \Rightarrow ax > -b \Rightarrow x < \dfrac{-b}{a} \Rightarrow \left(-\infty, \dfrac{-b}{a}\right)$

107. (a) The graph of $T(x) = 65 - 29x$ intersect the graph of $D(x) = 50 - 5.8x$ at $\approx (.6466, 46.25)$. Since the x-coordinate is altitude, the clouds will not form below .65 miles or for the interval: $[0, .65)$.

    (b) Clouds will not form when air temperature is above dew point temperature or $T(x) > D(x)$. Then

        $65 - 29x > 50 - 5.8x \Rightarrow -23.2x > -15 \Rightarrow x < \dfrac{15}{23.2}$ or for the interval: $[0, .6466)$.

108. (a) The slope of the function $f(x) = 44{,}500x + 129{,}650$ is: $m = 44{,}500$. This means that the median price of a single-family home has increased, on average, \$44,500 per year between 2001 and 2004.

   (b) The median price is in the range when the graph of the function $y_1 = 44{,}500x + 129{,}650$ is between the graphs of $y_2 = 174{,}000$ and $y_3 = 219{,}000$. From the graph this happens: $1.00 < x < 2.01$ or for $2002.00 < x < 2003.01$. The median price is in the given range approximately the years 2002-2003.

109. Since $C = 2\pi r$ and radius is in the range $1.99 \le r \le 2.01$, circumference is in the range:

   $2\pi(1.99) \le 2\pi r \le 2\pi(2.01) \Rightarrow 3.98\pi \le C \le 4.02\pi$.

110. Since $P = 4s$ and side length is in the range $9.9 \le s \le 10.1$, perimeter is in the range:

   $4(9.9) \le 4s \le 4(10.1) \Rightarrow 39.6 \le P \le 40.4$.

1.6: Applications of Linear Functions

1. $.75(40) = 30\, L$

2. If y varies directly with x, then $y = kx$ or $2 = k(4) \Rightarrow k = \dfrac{1}{2}$. Then $y = \dfrac{1}{2}(12) \Rightarrow y = 6$.

3. When combining a 26% acid solution to a 32% acid solution the result will be a solution with between 26% and 32% acid. (A) 36% is not in between these percent values and not a possible concentration.

4. (D) only discounts the original price x by \$.30 and not by 30% of the original price.

5. If x is the second number, then $6x - 3$ is the first number. The equation with the sum of these two numbers equal to 32 is: (D) $(6x - 3) + x = 32$.

6. The difference between six times a number and 9 is: $6x - 9$. This is equal to five times the sum (answer) of a number and 2 or $5(x + 2)$. The equation that does this is: (A) $6x - 9 = 5(x + 2)$.

7. If $P = 2L + 2W$ then $P = 2L + 2(19) \Rightarrow 98 = 2L + 38 \Rightarrow 60 = 2L \Rightarrow L = 30\,\text{cm}$.

8. Let x = width and $x + 3$ = length. If $P = 2W + 2L$, then $22 = 2x + 2(x + 3) \Rightarrow 22 = 2x + 2x + 6 \Rightarrow 16 = 4x \Rightarrow x = 4$. The width is: 4 feet and the length is: 7 feet.

9. Let x = width and $2x - 2.5$ = length. If $P = 2W + 2L$, then $40.6 = 2x + 2(2x - 2.5) \Rightarrow 40.6 = 6x - 5 \Rightarrow 45.6 = 6x \Rightarrow x = 7.6$. The width is: 7.6 cm.

10. Let x = width and $2x - 3$ = length. If $P = 2W + 2L$, then $54 = 2x + 2(2x - 3) \Rightarrow 54 = 6x - 6 \Rightarrow 60 = 6x \Rightarrow x = 10$. The width is: 10 cm and the length is: 17 cm.

11. Let x = the original square side length and $x + 3$ = the new square side length. If $P = 4s$, then $4(x + 3) = 2x + 40 \Rightarrow 4x + 12 = 2x + 40 \Rightarrow 2x = 28 \Rightarrow x = 14$. The original side length is: 14 cm.

12. Let x = width and $x + 120$ = length. If $P = 2W + 2L$, then $940 = 2x + 2(x + 120) \Rightarrow 940 = 4x + 240 \Rightarrow 700 = 4x \Rightarrow x = 175$. The width is: 175 inches, therefore the length is: 295 inches or 8.2 yards.

13. With an aspect ratio of 4:3, let $x =$ width and $\frac{4}{3}x =$ length. If $P = 2W + 2L$, then $98 = 2x + 2\left(\frac{4}{3}x\right) \Rightarrow$

    $98 = 2x + \frac{8}{3}x \Rightarrow 98 = \frac{14}{3}x \Rightarrow 294 = 14x \Rightarrow x = 21$. The width is: 21 inches and the length is:

    $\frac{4}{3}(21) = 28$ inches. Use the Pythagorean theorem to find the diagonal:

    $c^2 = (21)^2 + (28)^2 \Rightarrow c^2 = 441 + 784$

    $\Rightarrow c^2 = 1225 \Rightarrow c = 35$. The television is advertised as a 35 inch screen.

14. With an aspect ratio of 4:3, let $x =$ width and $\frac{4}{3}x =$ length. If $P = 2W + 2L$, then $126 = 2x + 2\left(\frac{4}{3}x\right) \Rightarrow$

    $126 = 2x + \frac{8}{3}x \Rightarrow 126 = \frac{14}{3}x \Rightarrow 378 = 14x \Rightarrow x = 27$. The width is: 27 inches and the length is:

    $\frac{4}{3}(27) = 36$ inches. Use pythagorean theorem to find the diagonal: $c^2 = (27)^2 + (36)^2 \Rightarrow c^2 = 729 + 1296$

    $\Rightarrow c^2 = 2025 \Rightarrow c = 45$. The television is advertised as a 45 inch screen.

15. Let $x =$ the short side length and $2x =$ the longer two side lengths. If $P = s + s + s$, then

    $30 = x + 2x + 2x \Rightarrow 30 = 5x \Rightarrow x = 6$. The shortest side is 6 cm long.

16. Let $x =$ the short side length, $2x - 200 =$ the longest side, and $(2x - 200) - 200 =$ the middle length side.

    If $P = s + s + s$, then $2400 = x + (2x - 200) + [(2x - 200) - 200] \Rightarrow 2400 = 5x - 600 \Rightarrow$

    $3000 = 5x \Rightarrow x = 600$. The side lengths are: 600 feet, $2(600) - 200 = 1000$ feet, and $1000 - 200 = 800$ feet.

17. Let $x =$ width and $x + 22,000 =$ length. If $P = 2W + 2L$, then $44,252 = 2x + 2(x + 22,000) \Rightarrow$

    $44,252 = 4x + 44,000 \Rightarrow 252 = 4x \Rightarrow x = 63$. If the width is: 63 inches, the length is: 22,063 inches and

    this is: 612.86 yards.

18. For the original ticket, let $x =$ width and $x + 4 =$ length. If $P = 2W + 2L$, the increased size ticket would

    produce the equation: $30 = 2(x + 1) + 2[(x + 4) + 4] \Rightarrow 30 = 4x + 18 \Rightarrow 12 = 4x \Rightarrow x = 3$.

    The width of the original ticket was: 3 mm and the length: 7 mm.

19. Let $x =$ gallons of 5% acid solution. Then $5(.10) + x(.05) = (x + 5)(.07) \Rightarrow .50 + .05x = .07x + .35 \Rightarrow$

    $.15 = .02x \Rightarrow 15 = 2x \Rightarrow x = 7.5$. Mix in 7.5 gallons of 5% acid solution.

20. Let $x =$ mL of 5% acid solution. Then $60(.20) + x(.05) = (x + 60)(.10) \Rightarrow 12 + .05x = .10x + 6 \Rightarrow$

    $6 = .05x \Rightarrow 600 = 5x \Rightarrow x = 120$. Mix in 120 mL of 5% acid solution.

21. Let $x =$ gallons of pure alcohol. Then $20(.15) + x(1.00) = (x + 20)(.25) \Rightarrow 3 + x = .25x + 5 \Rightarrow$

    $.75x = 2 \Rightarrow x = 2.67$ or $2\frac{2}{3}$. Mix in $2\frac{2}{3}$ gallons of pure alcohol.

22. Let $x =$ liters of pure alcohol. Then $7(.10) + x(1.00) = (x + 7)(.30) \Rightarrow .7 + x = .30x + 2.10 \Rightarrow$

    $.70x = 1.40 \Rightarrow 7x = 14 \Rightarrow x = 2$. Mix in 2 liters of pure alcohol.

23. Let $x =$ milliliters of water. Then $8(.06) + x(0) = (x + 8)(.04) \Rightarrow .48 = .04x + .32 \Rightarrow$

    $.16 = .04x \Rightarrow 4x = 16 \Rightarrow x = 4$. Mix in 4 milliliters of water.

24. Let $x =$ liters of water. Then $20(.18) + x(0) = (x + 20)(.15) \Rightarrow 3.6 = .15x + 3 \Rightarrow$

    $.6 = .15x \Rightarrow 15x = 60 \Rightarrow x = 4$. Mix in 4 liters of water.

25. Let x = liters of fluid to be drained and pure antifreeze added. Then $16(.80) - x(.80) + x(1.00) = 16(.90) \Rightarrow$

   $12.8 - .80x + x = 14.4 \Rightarrow .20x = 1.6 \Rightarrow 20x = 160 \Rightarrow x = 8$. Drain and add in 8 liters of antifreeze.

26. Let x = quarts of fluid to be drained and pure antifreeze added.

   Then $10(.40) - x(.40) + x(1.00) = 10(.80) \Rightarrow$

   $4 - .40x + x = 8 \Rightarrow .60x = 4 \Rightarrow 6x = 40 \Rightarrow x = 6.67$ or $6\frac{2}{3}$. Drain and add in $6\frac{2}{3}$ quarts of antifreeze.

27. Let x = gallons of 94-octane gasoline.

   Then $400(.99) + x(.94) = (x + 400)(.97) \Rightarrow 396 + .94x = .97x + 388$

   $8 = .03x \Rightarrow 3x = 800 \Rightarrow x = 266.67$ or $266\frac{2}{3}$. Mix in $266\frac{2}{3}$ gallons of 94-octane gasoline.

28. Let x = gallons of 92-octane gasoline and $120 - x$ = gallons of 98-octane gasoline. Then

   $x(.92) + (120 - x)(.98) = 120(.96) \Rightarrow .92x + 117.6 - .98x = 115.2 \Rightarrow -.06x = -2.4 \Rightarrow 6x = 240 \Rightarrow$

   $x = 40$. Mix 40 gallons of 92-octane gasoline and $120 - 40 = 80$ gallons of 98-octane gasoline.

29. (a) Since each student needs 15 ft^3/min, and there are 60 min/hr, the ventilation required by x students per hour

   is: $V(x) = 60(15x) \Rightarrow V(x) = 900x$.

   (b) The number of air exchanges needed per hour is: $A(x) = \dfrac{V(x)}{\text{ft}^3 \text{ in room}} = \dfrac{900x}{15,000} \Rightarrow A(x) = \dfrac{3}{50}x$.

   (c) $A(x) = \dfrac{3}{50}(40) \Rightarrow A(x) = 2.4$ ach are needed.

   (d) The increase is: $\dfrac{\text{smoking}}{\text{non} - \text{smoking}} = \dfrac{50 \text{ ft}^3}{15 \text{ ft}^3} = 3\frac{1}{3}$ times. Smoking areas require more than triple the ventilation.

30. (a) $k = \dfrac{.132(20)}{75} = \dfrac{2.64}{75} \Rightarrow k = .0352$

   (b) Since $R = kd$ and $k = .0352, R = .0352(.42) \Rightarrow R \approx .015$ or 1.5%.

   (c) For 5000 people 1.5% would be: $5000(.015) = 75$ people developing cancer. These 75 people divided by 72

   years would be: $\dfrac{75}{72} = 1.04$ or approximately 1 case per year.

31. (a) For one person $R = 1.5 \times 10^{-3}$ would be a $1(.0015) = .0015$ lifetime cancer risk. This divided by a life

   expectancy yields: $\dfrac{.0015}{72} = .000021$ cancer risk from second-hand tobacco smoke per year.

   (b) If one person is a .000021 chance per year, then x people would be: $C = .000021x$ per year.

   (c) $C = .000021(100,000) \Rightarrow C = 2.1$ or approximately 2 cases per year.

   (d) 26% of 260 million people is: $.26 \times 260,000,000 = 67,600,000$ smokers. For this number of smokers

   $R = .44$ would be $.44(67,600,000) = 29,744,000$ lifetime risk of death from smoking. This divided by life

   expectancy yields: $\dfrac{29,700,000}{72} = 413,111$ or approximately 413,000 deaths from smoking per year.

32. (a) The function is: $f(x) = .76x$.

   (b) If 38.7 riders do not wear helmets, then the equation is: $.76x = 38,700,000$ and the solution to the equation

   $x \approx 50,900,000$ is the total number of bike riders.

33. (a) If the fixed cost = \$200 and the variable cost = \$.02 the cost function is: $C(x) = .02x + 200$.

    (b) If she gets paid \$.04 per envelope stuffed and x = number of envelopes, the revenue function is: $R(x) = .04x$.

    (c) $R(x) = C(x)$ when $.02x + 200 = .04x \Rightarrow 200 = .02x \Rightarrow x = 10{,}000$.

    (d) Graph $C(x)$ and $R(x)$, See Figure 33. Rebecca takes a loss when stuffing less than 10,000 envelopes and makes a profit when stuffing over 10,000 envelopes.

34. (a) If the fixed cost = \$3500 and the variable cost = \$.01 the cost function is: $C(x) = .01x + 3500$.

    (b) If he gets paid \$.05 per copy and x = number of copies, the revenue function is: $R(x) = .05x$.

    (c) $R(x) = C(x)$ when $.01x + 3500 = .05x \Rightarrow 3500 = .04x \Rightarrow x = 87{,}500$.

    (d) Graph $C(x)$ and $R(x)$, See Figure 34. B.K. takes a loss when making less than 87,500 copies and makes a profit when making over 87,500 copies.

35. (a) If the fixed cost = \$2300 and the variable cost = \$3.00 the cost function is: $C(x) = 3.00x + 2300$.

    (b) If he gets paid \$5.50 per delivery and x = number of deliveries, the revenue function is: $R(x) = 5.50x$.

    (c) $R(x) = C(x)$ when $3.00x + 2300 = 5.50x \Rightarrow 2300 = 2.50x \Rightarrow x = 920$.

    (d) Graph $C(x)$ and $R(x)$, See Figure 35. Tom takes a loss when making less than 920 deliveries and makes a profit when making over 920 deliveries.

36. (a) If the fixed cost = \$40 and the variable cost = \$2.50 the cost function is: $C(x) = 2.50x + 40$.

    (b) If he gets paid \$6.50 per cake and x = number of cakes sold, the revenue function is: $R(x) = 6.50x$.

    (c) $R(x) = C(x)$ when $2.50x + 40 = 6.50x \Rightarrow 40 = 4.00x \Rightarrow x = 10$.

    (d) Graph $C(x)$ and $R(x)$, See Figure 36. Pat takes a loss when selling less than 10 cakes and makes a profit when selling over 10 cakes.

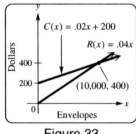

Figure 33

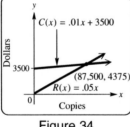

Figure 34

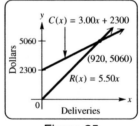

Figure 35

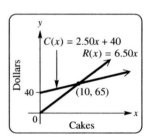

Figure 36

37. If $y = kx$, $x = 3$, and $y = 7.5$ then $7.5 = k(3) \Rightarrow k = 2.5$.

    Now, with $k = 2.5$ and $x = 8$, $y = 2.5(8) \Rightarrow y = 20$ when $x = 8$.

38. If $y = kx$, $x = 1.2$, and $y = 3.96$ then $3.96 = k(1.2) \Rightarrow k = 3.3$.

    Now, with $k = 3.3$ and $y = 23.43$, $23.43 = 3.3x \Rightarrow x = 7.1$ when $y = 23.43$.

39. If $y = kx$, $x = 25$, and $y = 1.5$ then $1.50 = k(25) \Rightarrow k = .06$.

    Now, with $k = .06$ and $y = 5.10$, $5.10 = .06x \Rightarrow x = \$85$ when $y = \$5.10$.

40. If $y = kx$, $x = 3$, and $y = 41.97$ then $41.97 = k(3) \Rightarrow k = 13.99$.

    Now, with $k = 13.99$ and $x = 5$, $y = 13.99(5) \Rightarrow y = \$69.95$ when $x = 5$.

41. Let y = pressure and x = depth for the direct proportion: $y = kx$. Then $13 = k(30) \Rightarrow k = \dfrac{13}{30}$.

    Now use $k = \dfrac{13}{30}$ and a depth of 70 feet to find the pressure: $y = \dfrac{13}{30}(70) \Rightarrow y = \dfrac{91}{3} \Rightarrow y = 30\dfrac{1}{3}$ lbs/in².

42. Let y = rate transmitted and x = diameter for the direct proportion: $y = kx$.

    Then $40 = k(6) \Rightarrow k = \dfrac{40}{6} = \dfrac{20}{3}$.

    Now use $k = \dfrac{20}{3}$ and a diameter of 8 micrometers to find the rate: $y = \dfrac{20}{3}(8) \Rightarrow y = \dfrac{160}{3} \Rightarrow y = 53\dfrac{1}{3}$ m/sec.

43. Let t = tuition and c = credits taken for the direct proportion: $t = kc$. Then $720.50 = k(11) \Rightarrow k = 65.5$.

    Now use the constant variation $k = 65.5$ and 16 credits to find the tuition: $y = 65.5(16) \Rightarrow y = \$1048$.

44. Let l = load and w = width for the direct proportion: $l = kw$. Then $250 = k(1.5) \Rightarrow k = 166.67 = 166\dfrac{2}{3}$.

    Now use the constant variation $k = 166\dfrac{2}{3}$ and a width of 3.5 inches to find the load that can be supported:

    $y = 166\dfrac{2}{3}(3.5) \Rightarrow y = 583\dfrac{1}{3}$ pounds.

45. First use proportion to find the radius of the water at a depth of 6 feet. $\dfrac{5}{11} = \dfrac{x}{6} \Rightarrow 11x = 30 \Rightarrow x \approx 2.727$

    Now use the cone volume formula to find the waters volume: $V = \dfrac{1}{3}\pi r^2 h \Rightarrow V = \dfrac{1}{3}\pi(2.727)^2(6) \approx 46.7$ ft³.

46. First use proportion to find the radius of the water at a depth of 7 feet. $\dfrac{2}{6} = \dfrac{x}{3.5} \Rightarrow 6x = 7 \Rightarrow x \approx 1.167$.

    Now use the cone volume formula to find the waters volume: $V = \dfrac{1}{3}\pi r^2 h \Rightarrow V = \dfrac{1}{3}\pi(1.167)^2(3.5) \approx 4.99$ ft³.

47. Since the triangles are similar, use a proportion to solve:

    $\dfrac{1.75}{2} = \dfrac{45}{x} \Rightarrow 1.75x = 90 \Rightarrow x \approx 51.43$ or $x = 51\dfrac{3}{7}$ feet tall.

48. Draw a picture and create similar triangles, then use a proportion to solve:

    $\dfrac{66 \text{ in.}}{84 \text{ in.}} = \dfrac{x}{15 \text{ ft.} + 84 \text{ in.}} \Rightarrow \dfrac{66}{84} = \dfrac{x}{264} \Rightarrow 84x = 17{,}424 \Rightarrow x \approx 207.43$ inches or $x \approx 17.3$ feet.

49. Let w = weight, d = distance, and the direct proportion $w = kd$ to find k. $3 = k(2.5) \Rightarrow k = 1.2$.

    Now use $k = 1.2$ and a weight of 17 pounds to find the stretch length: $17 = 1.2(d) \Rightarrow d = 14.17$ or $14\dfrac{1}{6}$ in.

50. Let w = weight, d = distance, and the direct proportion $w = kd$ to find k. $9.8 = k(.75) \Rightarrow k = \dfrac{196}{15}$.

    Now use $k = \dfrac{196}{15}$ and a stretch of 3.1 inches to find the weight: $w = \dfrac{196}{15}(3.1) \Rightarrow w = 40\dfrac{38}{75}$ or $w \approx 40.5$ lb.

51. With direct proportion $y_1 = kx_1$ and $y_2 = kx_2$, then $k = \dfrac{y_1}{x_1} = \dfrac{y_2}{x_2}$. Now let $y_1 = 250$ tagged fish,

    $y_2 = 7$ tagged fish, $x_2 = 350$ sample fish, and x_1 = total fish. Therefore $\dfrac{250}{x_1} = \dfrac{7}{350} \Rightarrow 7x_1 = 87{,}500 \Rightarrow$

    $x_1 = 12{,}500$ is the estimate for total population of fish in the lake.

52. With direct proportion $y_1 = kx_1$ and $y_2 = kx_2$, then $k = \dfrac{y_1}{x_1} = \dfrac{y_2}{x_2}$. Now let $y_1 = 4693$ tagged seal pups,

    $y_2 = 218$ tagged seal pups, $x_2 = 900$ sample seal pups, and $x_1 = $ total seal pups. Therefore $\dfrac{4963}{x_1} = \dfrac{218}{900} \Rightarrow$

    $218x_1 = 4,466,700 \Rightarrow x_1 = 20,489.45 \Rightarrow x_1 \approx 20,500$ is the estimate for total population of seal pups.

53. (a) Let $x = $ number of heaters produced and $y = $ cost, then $(10, 7500)$ and $(20, 13900)$ are two points on the

    graph of the linear function. Find the slope: $m = \dfrac{13900 - 7500}{20 - 10} = \dfrac{6400}{10} = 640$. Now use point slope

    form to find the linear function: $y - 7500 = 640(x - 10) \Rightarrow y - 7500 = 640x - 6400 \Rightarrow$

    $y = 640x + 1100$.

    (b) $y = 640(25) + 1100 \Rightarrow y = 16000 + 1100 \Rightarrow y = \$17,100$

    (c) Graph $y = 640x + 1100$ and locate the point $(25, 17,100)$ on the graph.

54. (a) Let $x = $ degrees Fahrenheit and $y = $ chirps, then $(68, 24)$ and $(40, 86)$ are two points on the

    graph of the linear function. Find the slope: $m = \dfrac{86 - 24}{40 - 68} = \dfrac{62}{-28} = -\dfrac{31}{14}$. Now use point slope form

    to find the linear function: $y - 86 = -\dfrac{31}{14}(x - 40) \Rightarrow y - 86 = -\dfrac{31}{14}x + \dfrac{620}{7} \Rightarrow y = -\dfrac{31}{14}x + \dfrac{1222}{7}$.

    (b) $y = -\dfrac{31}{14}(60) + \dfrac{1222}{7} \Rightarrow y = -\dfrac{1860}{14} + \dfrac{2444}{14} \Rightarrow y = \dfrac{584}{14} \approx 41.71 \approx 42$ times.

    (c) 40 chirps in one-half minute is 80 chirps per minute. Therefore: $80 = -\dfrac{31}{14}x + \dfrac{1222}{7} \Rightarrow$

    $1120 = -31x + 2444 \Rightarrow -1324 = -31x \Rightarrow x \approx 42.71 \approx 43$ ¡ Fahrenheit.

55. (a) Let $x = $ number of years after 1990 and $y = $ value, then $(0, 120000)$ and $(10, 146000)$ are two points on the

    graph of the linear function. Find the slope: $m = \dfrac{146000 - 120000}{10 - 0} = \dfrac{26000}{10} = 2,600$. The y-intercept is:

    120,000 therefore the linear function is: $y = 2,600x + 120,000$.

    (b) $y = 2,600(14) + 120,000 \Rightarrow y = 36,400 + 120,000 \Rightarrow y = \$156,400$ value of the house.

    (c) The value of the house increased on an average \$2,600 per year.

56. (a) Let $x = $ number of years after 1994 and $y = $ value, then $(0, 3000)$ and $(8, 600)$ are two points on the

    graph of the linear function. Find the slope: $m = \dfrac{600 - 3000}{8 - 0} = \dfrac{-2400}{8} = -300$. The y-intercept is:

    3000 therefore the linear function is: $y = -300x + 3,000$.

    (b) See Figure 56. The y-intercept $= \$3,000$, the initial value of the photocopier.

    (c) $y = -300(4) + 3000 \Rightarrow y = -1200 + 3000 \Rightarrow y = \$1,800$. Locate the point $(4, 1800)$ on the graph of

    the linear function $y = -300x + 3000$.

[0, 10] by [0, 4000]
Xscl = 1   Yscl = 1000

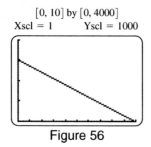

Figure 56

57. (a) The minimum amount of water pumped is 1120 gallons per minute for 12 hours per day. This results in a minimum of $1120 \times 60 \times 12 = 806{,}400$ gallons per day. Thus $A(x) = 806{,}400x$.

    (b) If $x = 30$, then $A(30) = 806{,}400(30) \Rightarrow A(30) = 24{,}192{,}000$ gallons of water.

    (c) $P(x) = \dfrac{A(x)}{20{,}000} = \dfrac{806{,}400x}{20{,}000} \Rightarrow P(x) = 40.32x$. About 40 average sized pools could be filled each day.

    (d) $1000 = 40.32x \Rightarrow x \approx 24.8$ days.

58. (a) $y = 10(76 - 65) + 50 \Rightarrow y = 10(11) + 50 \Rightarrow y = 110 + 50 \Rightarrow y = \$160$ fine.

    (b) $100 = 10(x - 65) + 50 \Rightarrow 100 = 10x - 650 + 50 \Rightarrow 700 = 10x \Rightarrow x = 70$ mph.

    (c) Because the function states $x > 65$, tickets start at 66 mph.

    (d) $200 < 10(x - 65) + 50 \Rightarrow 200 < 10x - 650 + 50 \Rightarrow 800 < 10x \Rightarrow x > 80$ mph.

59. (a) $y = \dfrac{5}{3}(27) + 455 \Rightarrow y = 45 + 455 \Rightarrow y = 500$ cm^3.

    (b) $605 = \dfrac{5}{3}x + 455 \Rightarrow 150 = \dfrac{5}{3}x \Rightarrow x = 90$ ¡ C.

    (c) $0 = \dfrac{5}{3}x + 455 \Rightarrow -455 = \dfrac{5}{3}x \Rightarrow x = -273$ ¡ C.

60. $y \approx .134(1000) - 1.18 \Rightarrow y = 134 - 1.18 \Rightarrow y = 132.82 \Rightarrow y \approx 133$ mm.

61. $I = PRT \Rightarrow \dfrac{I}{RT} = P$ or $P = \dfrac{I}{RT}$

62. $V = LWH \Rightarrow \dfrac{V}{WH} = L$ or $L = \dfrac{V}{WH}$

63. $P = 2L + 2W \Rightarrow P - 2L = 2W \Rightarrow W = \dfrac{P - 2L}{2}$ or $W = \dfrac{P}{2} - L$

64. $P = a + b + c \Rightarrow c = P - a - b$

65. $A = \dfrac{1}{2}h(b_1 + b_2) \Rightarrow 2A = h(b_1 + b_2) \Rightarrow h = \dfrac{2A}{b_1 + b_2}$

66. $A = \dfrac{1}{2}h(b_1 + b_2) \Rightarrow 2A = h(b_1 + b_2) \Rightarrow \dfrac{2A}{h} = b_1 + b_2 \Rightarrow b_1 = \dfrac{2A}{h} - b_2$ or $b_1 = \dfrac{2A - b_2h}{h}$

67. $S = 2LW + 2WH + 2HL \Rightarrow S - 2LW = H(2W + 2L) \Rightarrow \dfrac{S - 2LW}{2W + 2L}$

68. $S = 2\pi rh + 2\pi r^2 \Rightarrow S - 2\pi r^2 = 2\pi rh \Rightarrow h = \dfrac{S - 2\pi r^2}{2\pi r}$ or $h = \dfrac{S}{2\pi r} - r$

69. $V = \dfrac{1}{3}\pi r^2 h \Rightarrow 3V = \pi r^2 h \Rightarrow h = \dfrac{3V}{\pi r^2}$

70. $y = a(x - h)^2 + k \Rightarrow y - k = a(x - h)^2 \Rightarrow a = \dfrac{y - k}{(x - h)^2}$

71. $S = \dfrac{n}{2}(a_1 + a_n) \Rightarrow 2S = n(a_1 + a_n) \Rightarrow n = \dfrac{2S}{a_1 + a_n}$

72. $S = \dfrac{n}{2}[2a_1 + (n - 1)d] \Rightarrow \dfrac{2S}{n} = 2a_1 + (n - 1)d \Rightarrow \dfrac{2S}{n} - (n - 1)d = 2a_1 \Rightarrow$

    $\dfrac{2S - (n - 1)dn}{n} = 2a_1 \Rightarrow a_1 = \dfrac{2S - (n - 1)dn}{2n}$

73. $s = \dfrac{1}{2}gt^2 \Rightarrow 2s = gt^2 \Rightarrow g = \dfrac{2s}{t^2}$

74. $A = \dfrac{24f}{B(p+1)} \Rightarrow \dfrac{1}{A} = \dfrac{B(p+1)}{24f} \Rightarrow \dfrac{24f}{A} = B(p+1) \Rightarrow \dfrac{24f}{AB} = p+1 \Rightarrow p = \dfrac{24f}{AB} - 1$ or $p = \dfrac{24f - AB}{AB}$

75. Let $P =$ the amount put into the short-term note, then $240{,}000 - P =$ the amount put into the long-term note.

    With \$13,000 one year interest income, solve: $P(.06)(1) + (240{,}000 - P)(.05)(1) = 13{,}000 \Rightarrow$

    $.06P + 12{,}000 - .05P = 13{,}000 \Rightarrow .01P = 1{,}000 \Rightarrow P = 100{,}000.$

    The short-term note was \$100,000 and the long-term was \$140,000.

76. Let $P =$ the amount paid on the land that made a profit, then $120{,}000 - P =$ the amount paid on the land that

    produced a loss. With a combined profit of \$5500, solve: $.15P - .10(120{,}000 - P) = 5500$

    $\Rightarrow .15P - 12{,}000 + .10P = 5500 \Rightarrow .25P = 17{,}500 \Rightarrow P = 70{,}000.$ Therefore, \$70,000 was paid for the

    land that made a profit and \$50,000 was paid for the land that produced a loss.

77. Let $P =$ the amount deposited at 2.5% interest rate, then $2P =$ the amount deposited at 3% interest rate. With a

    one year interest income of \$850, solve: $.025P(1) + .03(2P)(1) = 850 \Rightarrow .025P + .06P = 850 \Rightarrow$

    $.085P = 850 \Rightarrow P = 10{,}000.$ Therefore, \$10,000 was deposited at 2.5% and \$20,000 was deposited at 3%.

78. Let $P =$ the amount invested at 4% interest rate, then $4P =$ the amount invested at 3.5% interest rate. With a

    one year interest income of \$3600, solve: $.04P(1) + .035(4P)(1) = 3600 \Rightarrow .04P + .14P = 3600 \Rightarrow$

    $.18P = 3600 \Rightarrow P = 20{,}000.$ Therefore, \$20,000 was invested at 4% and \$80,000 was invested at 3.5%.

79. After taxes, Marietta was able to invest 70% of the original winnings. This is $.70(200{,}000) = \$140{,}000.$

    Now let $P =$ the amount invested at 1.5%, then $140{,}000 - P =$ the amount invested at 4%. With a one year

    interest income of \$4350, solve: $.015P + .04(140{,}000 - P) = 4350 \Rightarrow .015P + 5600 - .04P = 4350 \Rightarrow$

    $-.025P = -1250 \Rightarrow P = 50{,}000.$ Therefore, \$50,000 was invested at 1.5% and \$90,000 was invested at 4%.

80. After income taxes, Latasha was able to invest 72% of the original royalties. This is $.72(48{,}000) = \$34{,}560.$

    Now let $P =$ the amount invested at 3.25%, then $34{,}560 - P =$ the amount invested at 1.75%. With a one year

    interest income of \$904.80, solve: $.0325P + .0175(34{,}560 - P) = 904.80 \Rightarrow$

    $.0325P + 604.80 - .0175P = 904.80 \Rightarrow .015P = 300 \Rightarrow P = 20{,}000.$ Therefore, \$20,000 was invested at

    3.25% and \$14,560 was invested at 1.75%.

Reviewing Basic Concepts (Sections 1.5 and 1.6)

1. $3(x - 5) + 2 = 1 - (4 + 2x) \Rightarrow 3x - 15 + 2 = 1 - 4 - 2x \Rightarrow 3x - 13 = -2x - 3 \Rightarrow 5x = 10 \Rightarrow$

    $x = 2.$ The graphs of the left and right side of the equation intersect at $x = 2$, this supports the solution set: $\{2\}$.

2. Graph $y_1 = \pi(1 - x)$ and $y_2 = .6(3x - 1)$, See Figure 2. The graphs intersect at: $x = .757$

3. $0 = \dfrac{1}{3}(4x - 2) + 1 \Rightarrow 0 = \dfrac{4}{3}x - \dfrac{2}{3} + 1 \Rightarrow -\dfrac{1}{3} = \dfrac{4}{3}x \Rightarrow x = -\dfrac{1}{4}$. Graph $y_1 = \dfrac{1}{3}(4x - 2) + 1$,

See Figure 3. The graph intersects the x-axis at $x = -\dfrac{1}{4}$.

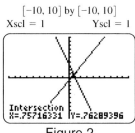

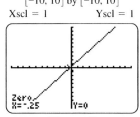

Figure 2   Figure 3

4. (a) $4x - 5 = -2(3 - 2x) + 3 \Rightarrow 4x - 5 = -6 + 4x + 3 \Rightarrow -5 = -3$. Since this is false, the equation is a

contradiction and the solution set is: $\varnothing$.

(b) $5x - 9 = 5(-2 + x) + 1 \Rightarrow 5x - 9 = -10 + 5x + 1 \Rightarrow -9 = -9$. Since this is true the equation is an

identity and the solution set is: $(-\infty, \infty)$.

(c) $5x - 4 = 3(6 - x) \Rightarrow 5x - 4 = 18 - 3x \Rightarrow 8x = 22 \Rightarrow x = \dfrac{22}{8}$ or $x = \dfrac{11}{4}$. The equation is a

conditional equation and the solution set is: $\left\{\dfrac{11}{4}\right\}$.

5. $2x + 3(x + 2) < 1 - 2x \Rightarrow 2x + 3x + 6 < 1 - 2x \Rightarrow 7x < -5 \Rightarrow x < -\dfrac{5}{7}$. The solution set is:

$\left(-\infty, -\dfrac{5}{7}\right)$. Graph $y_1 = 2x + 3(x + 2)$ and $y_2 = 1 - 2x$, the graph of y_1 is below the graph of y_2 at $x < -\dfrac{5}{7}$,

which supports the original solution.

6. $-5 \le 1 - 2x < 6 \Rightarrow -6 \le -2x < 5 \Rightarrow 3 \ge x > -\dfrac{5}{2} \Rightarrow -\dfrac{5}{2} < x \le 3 \Rightarrow \left(-\dfrac{5}{2}, 3\right]$

7. (a) The graph intersect at $x = 2 \Rightarrow \{2\}$.

(b) The graph of $f(x)$ intersects or is below the graph of $g(x)$ at $x \ge 2$ or $[2, \infty)$.

8. Since the triangles formed by the shadows are similar, use proportion to solve.

$\dfrac{x}{27} = \dfrac{6}{4} \Rightarrow 4x = 162 \Rightarrow x = 40.5$ ft.

9. (a) Since the income from each disc is \$5.50 a function R for revenue from selling x discs is: $R(x) = 5.50x$.

(b) Since the cost of producing each disc is \$1.50 and there is a one-time equipment cost of \$2500 a function C

for cost of recording x discs is: $C(x) = 1.50x + 2500$.

(c) solve $R(x) = C(x)$, $5.5x = 1.5x + 2500 \Rightarrow 4x = 2500 \Rightarrow x = 625$ discs.

10. $V = \pi r^2 h \Rightarrow h = \dfrac{V}{\pi r^2}$.

Chapter 1 Review Exercises

1. Use the distance formula: $d = \sqrt{(-1-5)^2 + (16-(-8))^2} \Rightarrow d = \sqrt{(-6)^2 + 24^2} \Rightarrow d = \sqrt{36 + 576} \Rightarrow$ $d = \sqrt{612} = 6\sqrt{17}$.

2. Use the midpoint formula: Midpoint $= \left(\dfrac{-1+5}{2}, \dfrac{16-8}{2} \right) = (2, 4)$.

3. Use the slope formula: $m = \dfrac{16-(-8)}{-1-5} \Rightarrow m = \dfrac{24}{-6} = -4$.

4. Use point-slope form and slope from ex. 3: $y - 16 = -4(x - (-1)) \Rightarrow y - 16 = -4x - 4 \Rightarrow y = -4x + 12$.

5. Change to slope intercept form: $3x + 4y = 144 \Rightarrow 4y = -3x + 144 \Rightarrow y = -\dfrac{3}{4}x + 36 \Rightarrow m = -\dfrac{3}{4}$.

6. For the *x*-intercept, $y = 0$. Therefore $3x + 4(0) = 144 \Rightarrow 3x = 144 \Rightarrow x = 48$.

7. For the *y*-intercept, $x = 0$. Therefore $3(0) + 4y = 144 \Rightarrow 4y = 144 \Rightarrow y = 36$.

8. One possible window is: $[-10, 50]$ by $[-40, 40]$.

9. Since $f(3) = 6$ and $f(-2) = 1$ then $(3, 6)$ and $(-2, 1)$ are points on the graph of the line. Using these points find slope: $m = \dfrac{6-1}{3-(-2)} = \dfrac{5}{5} = 1$. Use point-slope form to find the function: $f(x) - 6 = 1(x - 3) \Rightarrow$ $f(x) - 6 = x - 3 \Rightarrow f(x) = x + 3$. Now solve for $f(8) = 8 + 3 = 11$.

10. The slope of the given equation is -4, a line perpendicular to this will have a slope of: $\dfrac{1}{4}$. Using this and point-slope form produces: $y - 4 = \dfrac{1}{4}(x - (-2)) \Rightarrow y - 4 = \dfrac{1}{4}x + \dfrac{1}{2} \Rightarrow y = \dfrac{1}{4}x + \dfrac{9}{2}$.

11. (a) $m = \dfrac{-4-5}{2-(-1)} \Rightarrow m = \dfrac{-9}{3} = -3$

    (b) Use point-slope form: $y - 5 = -3(x - (-1)) \Rightarrow y - 5 = -3x - 3 \Rightarrow y = -3x + 2$

    (c) Midpoint $= \left(\dfrac{-1+2}{2}, \dfrac{5+(-4)}{2} \right) = \left(\dfrac{1}{2}, \dfrac{1}{2} \right)$ or $(.5, .5)$

    (d) $d = \sqrt{(2-(-1))^2 + (-4-5)^2} \Rightarrow d = \sqrt{3^2 + (-9)^2} \Rightarrow d = \sqrt{9+81} = \sqrt{90} = 3\sqrt{10}$

12. (a) $m = \dfrac{1.5-(-3.5)}{-1-(-3)} \Rightarrow m = \dfrac{5}{2} = 2.5$

    (b) Use point-slope form: $y - 1.5 = 2.5(x - (-1)) \Rightarrow y - 1.5 = 2.5x + 2.5 \Rightarrow y = 2.5x + 4$

    (c) Midpoint $= \left(\dfrac{-1+(-3)}{2}, \dfrac{1.5+(-3.5)}{2} \right) = \left(-\dfrac{4}{2}, -\dfrac{2}{2} \right) = (-2, -1)$

    (d) $d = \sqrt{(-1-(-3))^2 + (1.5-(-3.5))^2} \Rightarrow d = \sqrt{2^2 + (5)^2} \Rightarrow d = \sqrt{4+25} = \sqrt{29}$

13. *C* most closely represents: $m < 0, b < 0$.

14. *F* most closely represents: $m > 0, b < 0$.

15. *A* most closely represents: $m < 0, b > 0$.

16. *B* most closely represents: $m > 0, b > 0$.

17. *E* most closely represents: $m = 0$.

18. *D* most closely represents: $b = 0$.

19. The Domain is: $[-6, 6]$ and the Range is: $[-6, 6]$.

20. False, the slopes are different. Although the difference is small the lines are not parallel and will intersect.

21. $f(x) = g(x)$ when the graphs intersect or when $x = -3$. I $\{-3\}$ is the best match.

22. $f(x) > g(x)$ when the graph of $f(x)$ is above the graph of $g(x)$ or when $x > -3$. K $(-3, \infty)$ is the best match.

23. $f(x) < g(x)$ when the graph of $f(x)$ is below the graph of $g(x)$ or when $x < -3$. B $(-\infty, -3)$ is the best match.

24. $g(x) \geq f(x)$ when the graph of $g(x)$ intersect or is above the graph of $f(x)$ or when $x \leq -3$. A $(-\infty, -3]$ is the best match.

25. $y_2 - y_1 = 0 \Rightarrow y_2 = y_1 \Rightarrow g(x) = f(x)$ when the graphs intersect or when $x = -3$. I $\{-3\}$ is the best match.

26. $f(x) < 0$ when the graph of $f(x)$ is below the x-axis or when $x < -5$. M $(-\infty, -5)$ is the best match.

27. $g(x) > 0$ when the graph of $g(x)$ is above the x-axis or when $x < -2$. O $(-\infty, -2)$ is the best match.

28. $y_2 - y_1 < 0 \Rightarrow y_2 < y_1 \Rightarrow g(x) < f(x)$ when the graph of $g(x)$ is below the graph of $f(x)$ or when $x > -3$. K $(-3, \infty)$ is the best match.

29. $5[3 + 2(x - 6)] = 3x + 1 \Rightarrow 5[2x - 9] = 3x + 1 \Rightarrow 10x - 45 = 3x + 1 \Rightarrow 7x = 46 \Rightarrow x = \dfrac{46}{7}$.

    The graphs of $y_1 = 5[3 + 2(x - 6)]$ and $y_2 = 3x + 1$ intersect at: $\left\{\dfrac{46}{7}\right\}$ which supports the result.

30. $\dfrac{x}{4} - \dfrac{x + 4}{3} = -2 \Rightarrow 12\left[\dfrac{x}{4} - \dfrac{x + 4}{3}\right] = -2 \Rightarrow 3x - 4(x + 4) = -24 \Rightarrow -x - 16 = -24 \Rightarrow -x = -8 \Rightarrow$

    $x = 8$. The graphs of $y_1 = \dfrac{x}{4} - \dfrac{x + 4}{3}$ and $y_2 = -2$ intersect at: $\{8\}$ which supports the result.

31. $-3x - (4x + 2) = 3 \Rightarrow -7x - 2 = 3 \Rightarrow -7x = 5 \Rightarrow x = -\dfrac{5}{7}$. The graphs of $y_1 = -3x - (4x + 2)$

    and $y_2 = 3$ intersect at: $\left\{-\dfrac{5}{7}\right\}$ which supports the result.

32. $-2x + 9 + 4x = 2(x - 5) - 3 \Rightarrow 2x + 9 = 2x - 10 \Rightarrow 9 = -10$. This is false, therefore this is a

    contradiction and the solution is: $\varnothing$. The graphs of $y_1 = -2x + 9 + 4x$ and $y_2 = 2(x - 5) - 3$ are parallel and

    do not intersect, therefore no solution, $\varnothing$, which supports the result.

33. $.5x + .7(4 - 3x) = .4x \Rightarrow .5x + 2.8 - 2.1x = .4x \Rightarrow 5x + 28 - 21x = 4x \Rightarrow -20x = -28 \Rightarrow x = \dfrac{7}{5}$.

    The graphs of $y_1 = .5x + .7(4 - 3x)$ and $y_2 = .4x$ intersect at: $\left\{\dfrac{7}{5}\right\}$ which supports the results.

34. $\dfrac{x}{4} - \dfrac{5x - 3}{6} = 2 - \dfrac{7x + 18}{12} \Rightarrow 12\left[\dfrac{x}{4} - \dfrac{5x - 3}{6} = 2 - \dfrac{7x + 18}{12}\right] \Rightarrow 3x - 2(5x - 3) = 24 - (7x + 18)$

    $\Rightarrow -7x + 6 = -7x + 6 \Rightarrow 6 = 6$. This is true, therefore this is an identity and the solution is: $(-\infty, \infty)$.

    The graphs of $y_1 = \dfrac{x}{4} - \dfrac{5x - 3}{6}$ and $y_2 = 2 - \dfrac{7x + 18}{12}$ are the same line or the solution: $(-\infty, \infty)$,

    which supports the results.

35. $x - 8 < 1 - 2x \Rightarrow 3x < 9 \Rightarrow x < 3$. The graph of $y_1 = x - 8$ is below the graph of $y_2 = 1 - 2x$ for the

    interval: $(-\infty, 3)$, which supports the result.

36. $\dfrac{4x-1}{3} \geq \dfrac{x}{5} - 1 \Rightarrow 15\left[\dfrac{4x-1}{3} \geq \dfrac{x}{5} - 1\right] \Rightarrow 5(4x-1) \geq 3x - 15 \Rightarrow 20x - 5 \geq 3x - 15 \Rightarrow$

$17x \geq -10 \Rightarrow x \geq -\dfrac{10}{17}$. The graph of $y_1 = \dfrac{4x-1}{3}$ intersect or is above the graph of $y_2 = \dfrac{x}{5} - 1$ for the

interval: $\left[-\dfrac{10}{17}, \infty\right)$, which supports the result.

37. $-6 \leq \dfrac{4-3x}{7} < 2 \Rightarrow -42 \leq 4 - 3x < 14 \Rightarrow -46 \leq -3x < 10 \Rightarrow \dfrac{46}{3} \geq x > -\dfrac{10}{3} \Rightarrow$

$\dfrac{10}{3} < x \leq \dfrac{46}{3}$ or for the interval: $\left(-\dfrac{10}{3}, \dfrac{46}{3}\right]$.

38. (a) Graph $y_1 = 5\pi x + (\sqrt{3})x - 6.24(x - 8.1) + (\sqrt[3]{9})x$, See Figure 38. Find the x-intercept: $x \approx \{-3.81\}$.

    (b) $f(x) < 0$ when the graph of $f(x)$ is below the x-axis. This happens for the interval: $(-\infty, -3.81)$.

    (c) $f(x) \geq 0$ when the graph of $f(x)$ intersects or is above the x-axis. This happens for the interval: $[-3.81, \infty)$.

39. It cost \$30 to produce each CD and there is a one-time advertisement cost, therefore: $C(x) = 30x + 150$.

40. Each tape is sold for \$37.50, therefore: $R(x) = 37.50x$.

41. $C(x) = R(x)$ when $30x + 150 = 37.50x \Rightarrow 150 = 7.50x \Rightarrow x = 20$.

42. When the graph of $R(x)$ is below $C(x)$ the company is losing money, when $R(x)$ intersects $C(x)$ the company

breaks even, and when $R(x)$ is above $C(x)$ the company makes money. This happens as follows: losing money

when $x < 20$, breaking even when $x = 20$, and making money when $x > 20$.

43. $A = \dfrac{24f}{B(p+1)} \Rightarrow AB(p+1) = 24f \Rightarrow f = \dfrac{AB(p+1)}{24}$.

44. $A = \dfrac{24f}{B(p+1)} \Rightarrow AB(p+1) = 24f \Rightarrow B = \dfrac{24f}{A(p+1)}$.

45. (a) $f(x) = -3.52(5) + 58.6 \Rightarrow f(x) = -17.6 + 58.6 = 41°$ F.

    (b) $-15 = -3.52x + 58.6 \Rightarrow -73.6 = -3.52x \Rightarrow x \approx 20.9$ or about 21,000 feet.

    (c) Graph $y_1 = -3.52x + 58.6$. Find the coordinates of the point where $x = 5$ to support the answer in (a).

       Find the coordinates of the point where $y = -15$ to support the answer in (b).

46. $50 = 1.06F + 7.18 \Rightarrow 42.82 = 1.06F \Rightarrow F \approx 40.3962$ or about 40 Liters/sec.

47. (a) Enter the years in L_1 and enter wages in L_2. The regression equation is: $y \approx .1199x - 234.4228$.

    (b) Graph the table and the regression line from (a), See Figure 47. The model fits the data very well.

    (c) $y = .1199(1999) - 234.4228 \Rightarrow y = 239.6801 - 234.4228 \Rightarrow y \approx 5.26$ which is \$.11 too high.

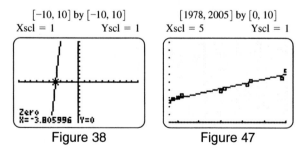

Figure 38   Figure 47

48. Enter the years in L_1 and the Pole Speed in L_2, the correlation coefficient is $r \approx .0531$; because r is so close to 0, the model would be unreliable in predicting future speed.

49. Let $x =$ bat speed and $y =$ ball travel distance, then (50, 320) and (80, 440) are two points on the graph of the function. Use the points to find slope: $m = \dfrac{440 - 320}{80 - 50} \Rightarrow m = \dfrac{120}{30} = 4$. Now use point-slope form to find the equation for the model: $y - 320 = 4(x - 50) \Rightarrow y - 320 = 4x - 200 \Rightarrow y = 4x + 120$. Because the slope is 4, the ball will travel 4 feet further for each 1 mph faster in bat speed.

50. Since surface area is: $A = 2(lw) + 2(lh) + 2(wh)$ we can solve: $496 = 2(18)(8) + 2(18)h + 2(8)h \Rightarrow 496 = 288 + 36h + 16h \Rightarrow 208 = 52h \Rightarrow h = 4$. The height of the box is 4 feet.

51. Since there are 5280 feet in a mile, there are $5280 \times 26.2 = 138{,}336$ feet in a marathon. Since there are 3.281 feet in a meter there are $100 \times 3.281 = 328.1$ feet in a 100 meter dash. Now use proportion to solve: $\dfrac{9.78}{328.1} = \dfrac{x}{138{,}336} \Rightarrow 328.1x = 1{,}352{,}926.08 \Rightarrow x \approx 4123.517464$ seconds to run a marathon. Divide by 60 to get minutes run: $4123.517464 \div 60 = 68.73$ minutes run or 1 hour 8 minutes and 44 seconds. This time is 56 minutes and 11 seconds faster.

52. $C = \dfrac{5}{9}(864 - 32) \Rightarrow C = \dfrac{5}{9}(832) \Rightarrow C = \dfrac{4160}{9} = 462\dfrac{2}{9}°\,C.$

53. Find the constant of variation: $4k = 3000 \Rightarrow k = 750$. Use this to find the pressure: $10(750) = 7500\,kg/m^2$.

54. (a) Use any two points to find slope: $m = \dfrac{1.8 - 3}{1 - 0} \Rightarrow m = \dfrac{-1.2}{1} = -1.2$. Now use point-slope form to find the equation: $y - 3 = -1.2(x - 0) \Rightarrow y = -1.2x + 3$.

 (b) $y = -1.2(-1.5) + 3 \Rightarrow y = 1.8 + 3 \Rightarrow y = 4.8$ when $x = -1.5$.

 $y = -1.2(3.5) + 3 \Rightarrow y = -4.2 + 3 \Rightarrow y = -1.2$ when $x = 3.5$.

55. (a) Enter the years in L_1 and enter test scores in L_2. The regression equation is: $y \approx 1.2x - 1886.4$.

 (b) $y = 1.2(2003) - 1886.4 \Rightarrow y = 2403.6 - 1886.4 \Rightarrow y \approx 517$.

 (c) $y = 1.2(2008) - 1886.4 \Rightarrow y = 2409.6 - 1886.4 \Rightarrow y \approx 523$.

56. (a) Enter the year 2000 in L_1 and the year 2003 in L_2. The regression equation is: $y \approx .896x + 4.067$.

 (b) $y = .896(36) + 4.067 \Rightarrow y = 32.256 + 4.067 \Rightarrow y \approx 36.3$ million passengers. This is quite close to the ACI figure of 37.4 million.

57. Let $x =$ the amount of pure alcohol to be added, then solve: $12(.10) + x(1.00) = (x + 12)(.30) \Rightarrow 1.2 + x = .30x + 3.6 \Rightarrow .70x = 2.4 \Rightarrow 70x = 240 \Rightarrow x = \dfrac{240}{70} = 3\dfrac{3}{7}$ Liters.

58. Let $x =$ the amount of 5% solution to be added, then solve: $120(.20) + x(.05) = (120 + x)(.10) \Rightarrow 24 + .05x = 12 + .10x \Rightarrow 12 = .05x \Rightarrow x = 240$ mL of 5% solution needs to be added.

59. The company will at least break even when $R(x) \geq C(x)$, therefore solve: $8x \geq 3x + 1500 \Rightarrow 5x \geq 1500 \Rightarrow x \geq 300$ or for the interval: $[300, \infty)$. 300 or more DVD's need to be sold to at least break even.

60. Let m = mental age and c = chronological age, then $IQ = \dfrac{100m}{c}$.

(a) $130 = \dfrac{100m}{7} \Rightarrow 910 = 100m \Rightarrow m = 9.1$ years.

(b) $IQ = \dfrac{100(20)}{16} \Rightarrow IQ = \dfrac{2000}{16} \Rightarrow IQ = 125.$

Chapter 1 Test

1. (a) Domain: $(-\infty, \infty)$ Range: $[2, \infty)$ x-intercept: none y-intercept: 3

   (b) Domain: $(-\infty, \infty)$ Range: $(-\infty, 0]$ x-intercept: 3 y-intercept: -3

   (c) Domain: $[-4, \infty)$ Range: $[0, \infty)$ x-intercept: -4 y-intercept: 2

2. (a) $f(x) = g(x)$ when the graph of $f(x)$ intersects the graph of $g(x)$, therefore $\{-4\}$.

   (b) $f(x) < g(x)$ when the graph of $f(x)$ is below the graph of $g(x)$, for the interval: $(-\infty, -4)$.

   (c) $f(x) \geq g(x)$ when the graph of $f(x)$ intersects or is above the graph of $g(x)$, for the interval: $[-4, \infty)$.

   (d) $y_2 - y_1 = 0 \Rightarrow y_2 = y_1 \Rightarrow g(x) = f(x)$ when the graph of $g(x)$ intersects the graph of $f(x)$, therefore $\{-4\}$.

3. (a) $y_1 = 0$ when the graph of $f(x)$ intersects the x-axis, therefore: $\{5.5\}$.

   (b) $y_1 < 0$ when the graph of $f(x)$ is below the x-axis, for the interval: $(-\infty, 5.5)$.

   (c) $y_1 > 0$ when the graph of $f(x)$ is above the x-axis, for the interval: $(5.5, \infty)$.

   (d) $y_1 \leq 0$ when the graph of $f(x)$ intersects or is below the x-axis, for the interval: $(-\infty, 5.5]$.

4. (a) $3(x - 4) - 2(x - 5) = -2(x + 1) - 3 \Rightarrow 3x - 12 - 2x + 10 = -2x - 2 - 3 \Rightarrow x - 2 = -2x - 5$

   $\Rightarrow 3x = -3 \Rightarrow x = -1.$ **Check:** $3(-1 - 4) - 2(-1 - 5) = -2(-1 + 1) - 3 \Rightarrow$

   $3(-5) - 2(-6) = -2(0) - 3 \Rightarrow -15 + 12 = 0 - 3 \Rightarrow -3 = -3$

   (b) Graph $y_1 = 3(x - 4) - 2(x - 5)$ and $y_2 = -2(x + 1) - 3$, See Figure 4. $f(x) > g(x)$ for the interval:

   $(-1, \infty)$, because the graph of $y_1 = f(x)$ is above the graph of $y_2 = g(x)$ for domain values greater than -1.

   (c) See Figure 4. $f(x) < g(x)$ for the interval: $(-\infty, -1)$, because the graph of $y_1 = f(x)$ is below the graph of

   $y_2 = g(x)$ for domain values less than -1.

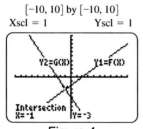

[-10, 10] by [-10, 10]
Xscl = 1   Yscl = 1

Figure 4

5. (a) $-\dfrac{1}{2}(8x + 4) + 3(x - 2) = 0 \Rightarrow -4x - 2 + 3x - 6 = 0 \Rightarrow -x - 8 = 0 \Rightarrow x = -8$ or $\{8\}$.

    (b) $-\dfrac{1}{2}(8x + 4) + 3(x - 2) \le 0 \Rightarrow -4x - 2 + 3x - 6 \le 0 \Rightarrow -x - 8 \le 0 \Rightarrow x \ge -8$

    or for the interval: $[-8, \infty)$.

    (c) Graph $y_1 = -\dfrac{1}{2}(8x + 4) + (3(x - 2)$, See Figure 5. The x-intercept is -8 supporting the result in part (a).

    The graph of the linear function lies below or on the x-axis for domain values greater than or equal to -8,

    supporting the results in part (b).

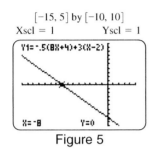

$[-15, 5]$ by $[-10, 10]$
Xscl = 1   Yscl = 1

Figure 5

6. (a) Using the points $(1982, 8.30)$ and $(2002, 34.71)$ find the midpoint:

    $$M.P. = \left(\dfrac{1982 + 2002}{2}, \dfrac{8.30 + 34.71}{2}\right) \Rightarrow M.P. = (1992, 21.51).\ \$21.51\text{ is the approximate cost in }1992.$$

    (b) Using the points $(1982, 8.30)$ and $(2002, 34.71)$ find the slope: $m = \dfrac{34.71 - 8.30}{2002 - 1982} \Rightarrow m = \dfrac{26.41}{20} \Rightarrow$

    $m = 1.3205$. This means that during the years 1982 to 2002, the monthly rate increased on the average, by

    about $\$1.32$ per year.

7. (a) Since the given line has a slope of -2 and parallel lines have equal slopes our new line has a slope of -2.

    Now use point-slope form: $y - 5 = -2(x - (-3)) \Rightarrow y - 5 = -2x - 6 \Rightarrow y = -2x - 1$.

    (b) The equation: $-2x + y = 0 \Rightarrow y = 2x$ has a slope of 2. Since perpendicular lines have slopes whose

    product equals -1, our new line has a slope of $-\dfrac{1}{2}$. Now use point-slope form: $y - 5 = -\dfrac{1}{2}(x - (-3)) \Rightarrow$

    $y - 5 = -\dfrac{1}{2}x - \dfrac{3}{2} \Rightarrow y = -\dfrac{1}{2}x + \dfrac{7}{2}$.

8. For the x-intercept $y = 0$, therefore: $3x - 4(0) = 6 \Rightarrow 3x = 6 \Rightarrow x = 2$. The x-intercept is: 2.

    For the y-intercept $x = 0$, therefore: $3(0) - 4y = 6 \Rightarrow -4y = 6 \Rightarrow y = -\dfrac{6}{4} = -\dfrac{3}{2}$. The y-intercept is: $-\dfrac{3}{2}$.

    Using the intercepts: $\left(0, -\dfrac{3}{2}\right)$ and $(2, 0)$, the slope is: $m = \dfrac{0 - (-\frac{3}{2})}{2 - 0} = \dfrac{\frac{3}{2}}{2} \Rightarrow m = \dfrac{3}{4}$.

9. The equation of the horizontal line passing through $(-3, 7)$ is: $y = 7$.

    The equation of the vertical line passing through $(-3, 7)$ is: $x = -3$.

10. (a) Enter the wind speed in L_1 and enter degrees in L_2. The regression equation is: $y \approx -.246x + 35.7$ and the

    correlation coefficient is: $r \approx -.96$.

    (b) $y \approx -.246(40) + 35.7 \Rightarrow y \approx -9.84 + 35.7 \Rightarrow y \approx 25.9°$ F.

11. (a) Use the volume of a cylinder: $V = \pi r^2 h$ and the fact that 20 feet = 240 inches.

$$V = \pi(240)^2(1) \Rightarrow V = 57,600\pi \Rightarrow V \approx 180,956 \text{ in.}^3$$

(b) Since 1 gallon = 231 cubic inches, the equation is: $g(x) = \dfrac{180,956}{231}x$.

(c) $g(x) = \dfrac{180,956}{231}(2.5) \Rightarrow g(x) \approx 1958$ gallons of water.

(d) Since the downspout can handle 400 gallons per hour, this would be 1000 gallons in 2.5 hours and one downspout would not be sufficient handle the 1958 gallons of rain water. Two downspouts handling 2000 gallons would be sufficient.

Chapter 1 Project

1. Enter the female player's height in L_1 and female player's weight in L_2. The regression equation is: $y \approx 4.512x - 154.4$ and the correlation coefficient is: $r \approx .86$. This indicates that the taller female players tend to weigh more, but the correlation is not perfect.

2. $y \approx 4.512(75) - 154.4 \Rightarrow y \approx 184$ lb.

3. Since the slope is slightly over 4.5, the increase is about 4.5 lb.

4. Enter the male player's height in L_1 and male player's weight in L_2. The regression equation is: $y \approx 4.465x - 133.3$ and the correlation coefficient is: $r \approx .90$. This indicates that the taller male players tend to weigh more, but the correlation is not perfect.

5. $y \approx 4.465(80) - 133.3 \Rightarrow y \approx 224$ lb.

Chapter 2: Analysis of Graphs of Functions

2.1: Graphs of Basic Functions and Relations; Symmetry

1. $(-\infty, \infty)$

2. $(-\infty, \infty)$; $[0, \infty)$

3. $(0, 0)$

4. $[0, \infty)$; $[0, \infty)$

5. increases

6. $(-\infty, 0]$; $[0, \infty)$

7. x-axis

8. even

9. odd

10. y-axis; origin

11. The domain can be all real numbers, therefore the function is continuous for the interval: $(-\infty, \infty)$.

12. The domain can be all real numbers, therefore the function is continuous for the interval: $(-\infty, \infty)$.

13. The domain can only be values where $x \geq 0$, therefore the function is continuous for the interval: $[0, \infty)$.

14. The domain can only be values where $x \leq 0$, therefore The function is continuous for the interval: $(-\infty, 0]$.

15. The domain can be all real numbers except -3, therefore the function is continuous for the interval:
$(-\infty, -3)$; $(-3, \infty)$.

16. The domain can be all real numbers except 1, therefore the function is continuous for the interval:
$(-\infty, 1)$; $(1, \infty)$.

17. (a) The function is increasing for the interval: $[3, \infty)$.

    (b) The function is decreasing for the interval: $(-\infty, 3]$.

    (c) The function is never constant, therefore: none

    (d) The domain can be all real numbers, therefore the interval: $(-\infty, \infty)$.

    (e) The range can only be values where $y \geq 0$, therefore the interval: $[0, \infty)$.

18. (a) The function is increasing for the interval: $[4, \infty)$.

    (b) The function is decreasing for the interval: $(-\infty, -1]$.

    (c) The function is constant for the interval: $[-1, 4]$.

    (d) The domain can be all real numbers, therefore the interval: $(-\infty, \infty)$.

    (e) The range can only be values where $y \geq 3$, therefore the interval: $[3, \infty)$.

19. (a) The function is increasing for the interval: $(-\infty, 1]$.

    (b) The function is decreasing for the interval: $[4, \infty)$.

    (c) The function is constant for the interval: $[1, 4]$.

    (d) The domain can be all real numbers, therefore the interval: $(-\infty, \infty)$.

    (e) The range can only be values where $y \leq 3$, therefore the interval: $(-\infty, 3]$.

20. (a) The function never is increasing, therefore: none

    (b) The function is always decreasing, therefore the interval: $(-\infty, \infty)$.

    (c) The function is never constant, therefore: none

    (d) The domain can be all real numbers, therefore the interval: $(-\infty, \infty)$.

    (e) The range can be all real numbers, therefore the interval: $(-\infty, \infty)$.

21. (a) The function never is increasing, therefore: none

    (b) The function is decreasing for the intervals: $(-\infty, -2]$; $[3, \infty)$.

    (c) The function is constant for the interval: $(-2, 3)$.

    (d) The domain can be all real numbers, therefore the interval: $(-\infty, \infty)$.

    (e) The range can only be values where $y \leq 1.5$ or $y \geq 2$, therefore the interval: $(-\infty, 1.5] \cup [2, \infty)$.

22. (a) The function is increasing for the interval: $(3, \infty)$.

    (b) The function is decreasing for the interval: $(-\infty, -3)$.

    (c) The function is constant for the interval: $(-3, 3]$.

    (d) The domain can be all real numbers except -3, therefore the interval: $(-\infty, -3) \cup (-3, \infty)$.

    (e) The range can only be values where $y > 1$, therefore the interval: $(1, \infty)$.

23. Graph $f(x) = x^5$, See Figure 23. As x increases for the interval: $(-\infty, \infty)$, y increases, therefore increasing.

24. Graph $f(x) = -x^3$, See Figure 24. As x increases for the interval: $(-\infty, \infty)$, y decreases, therefore decreasing.

25. Graph $f(x) = x^4$, See Figure 25. As x increases for the interval: $(-\infty, 0]$, y decreases, therefore decreasing.

26. Graph $f(x) = x^4$, See Figure 26. As x increases for the interval: $[0, \infty)$, y increases, therefore increasing.

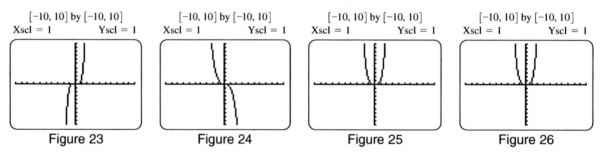

[-10, 10] by [-10, 10]	[-10, 10] by [-10, 10]	[-10, 10] by [-10, 10]	[-10, 10] by [-10, 10]
Xscl = 1 Yscl = 1	Xscl = 1 Yscl = 1	Xscl = 1 Yscl = 1	Xscl = 1 Yscl = 1
Figure 23	Figure 24	Figure 25	Figure 26

27. Graph $f(x) = -|x|$ See Figure 27. As x increases for the interval: $(-\infty, 0]$, y increases, therefore increasing.

28. Graph $f(x) = -|x|$, See Figure 28. As x increases for the interval: $[0, \infty)$, y decreases, therefore decreasing.

29. Graph $f(x) = -\sqrt[3]{x}$, See Figure 29. As x increases for the interval: $(-\infty, \infty)$, y decreases, therefore decreasing.

30. Graph $f(x) = -\sqrt{x}$, See Figure 30. As x increases for the interval: $[0, \infty)$, y decreases, therefore decreasing.

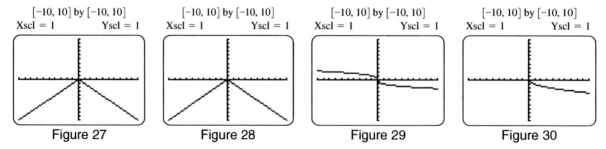

[-10, 10] by [-10, 10]	[-10, 10] by [-10, 10]	[-10, 10] by [-10, 10]	[-10, 10] by [-10, 10]
Xscl = 1 Yscl = 1	Xscl = 1 Yscl = 1	Xscl = 1 Yscl = 1	Xscl = 1 Yscl = 1
Figure 27	Figure 28	Figure 29	Figure 30

31. Graph $f(x) = 1 - x^3$, See Figure 31. As x increases for the interval: $(-\infty, \infty)$, y decreases, therefore decreasing.

32. Graph $f(x) = x^2 - 2x$, See Figure 32. As x increases for the interval: $[1, \infty)$, y increases, therefore increasing.

33. Graph $f(x) = 2 - x^2$ See Figure 33. As x increases for the interval: $(-\infty, 0]$, y increases, therefore increasing.

34. Graph $f(x) = |x + 1|$, See Figure 34. As x increases for the interval: $(-\infty, -1]$, y decreases, therefore decreasing.

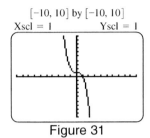

$[-10, 10]$ by $[-10, 10]$
Xscl = 1   Yscl = 1

Figure 31

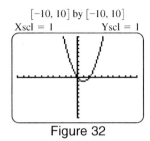

$[-10, 10]$ by $[-10, 10]$
Xscl = 1   Yscl = 1

Figure 32

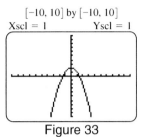

$[-10, 10]$ by $[-10, 10]$
Xscl = 1   Yscl = 1

Figure 33

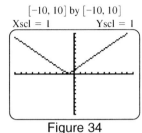

$[-10, 10]$ by $[-10, 10]$
Xscl = 1   Yscl = 1

Figure 34

35. (a) No   (b) Yes   (c) No

36. (a) Yes   (b) No   (c) No

37. (a) Yes   (b) No   (c) No

38. (a) No   (b) No   (c) Yes

39. (a) Yes   (b) Yes   (c) Yes

40. (a) Yes   (b) Yes   (c) Yes

41. (a) No   (b) No   (c) Yes

42. (a) No   (b) Yes   (c) No

43. If f is an even function then $f(-x) = f(x)$ or opposite domains have the same range. See Figure 43.

44. If g is an odd function then $g(-x) = -g(x)$ or opposite domains have the opposite range. See Figure 44.

x	$f(x)$
-3	21
-2	-12
-1	-25
1	-25
2	-12
3	21

Figure 43

x	$g(x)$
-5	13
-3	1
-2	-5
0	0
2	5
3	-1
5	-13

Figure 44

45. (a) Since $f(-x) = f(x)$, this is an even function and is symmetric with respect to the y-axis. See Figure 45a.

(b) Since $f(-x) = -f(x)$, this is an odd function and is symmetric with respect to the origin. See Figure 45b.

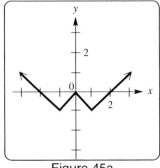

Figure 45a

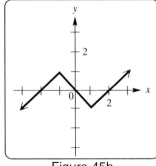

Figure 45b

46. (a) Since this is an odd function the graph is symmetric with respect to the origin. See Figure 46a.

   (b) Since this is an even function the graph is symmetric with respect to the y-axis. See Figure 46b.

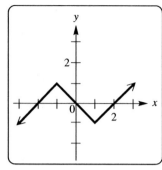
Figure 46a

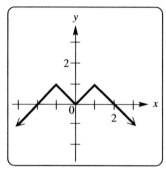
Figure 46b

47. Since $f(-x) = f(x)$, it is even.

48. Since $f(-x) = -f(x)$, it is odd.

49. If $f(x) = x^4 - 7x^2 + 6$, then $f(-x) = (-x)^4 - 7(-x)^2 + 6 \Rightarrow f(-x) = x^4 - 7x^2 + 6$. Since $f(-x) = f(x)$, the function is even.

50. If $f(x) = -2x^6 - 8x^2$, then $f(-x) = -2(-x)^6 - 8(-x)^2 \Rightarrow f(-x) = -2x^6 - 8x^2$. Since $f(-x) = f(x)$, the function is even.

51. If $f(x) = x^6 - 4x^4 + 5$, then $f(-x) = (-x)^6 - 4(-x)^4 + 5 \Rightarrow f(-x) = x^6 - 4x^4 + 5$. Since $f(-x) = f(x)$, the function is even.

52. If $f(x) = 8$, then $f(-x) = 8$. Since $f(-x) = f(x)$, the function is even.

53. If $f(x) = |5x|$, then $f(-x) = |5(-x)| \Rightarrow f(-x) = |5x|$. Since $f(-x) = f(x)$, the function is even.

54. If $f(x) = \sqrt{x^2 + 1}$, then $f(-x) = \sqrt{(-x)^2 + 1} \Rightarrow f(-x) = \sqrt{x^2 + 1}$. Since $f(-x) = f(x)$, the function is even.

55. If $f(x) = 3x^3 - x$, then $f(-x) = 3(-x)^3 - (-x) \Rightarrow f(-x) = -3x^3 + x$ and
   $-f(x) = -(3x^3 - x) \Rightarrow -f(x) = -3x^3 + x$. Since $f(-x) = -f(x)$, the function is odd.

56. If $f(x) = -x^5 + 2x^3 - 3x$, then $f(-x) = -(-x)^5 + 2(-x)^3 - 3(-x) \Rightarrow f(-x) = x^5 - 2x^3 + 3x$ and
   $-f(x) = -(-x^5 + 2x^3 - 3x) \Rightarrow -f(x) = x^5 - 2x^3 + 3x$. Since $f(-x) = -f(x)$, the function is odd.

57. If $f(x) = 3x^5 - x^3 + 7x$, then $f(-x) = 3(-x)^5 - (-x)^3 + 7(-x) \Rightarrow f(-x) = -3x^5 + x^3 - 7x$ and
   $-f(x) = -(3x^5 - x^3 + 7x) \Rightarrow -f(x) = -3x^5 + x^3 - 7x$. Since $f(-x) = -f(x)$, the function is odd.

58. If $f(x) = x^3 - 4x$, then $f(-x) = (-x)^3 - 4(-x) \Rightarrow f(-x) = -x^3 + 4x$ and
   $-f(x) = -(x^3 - 4x) \Rightarrow -f(x) = -x^3 + 4x$. Since $f(-x) = -f(x)$, the function is odd.

59. If $f(x) = \dfrac{1}{2x}$, then $f(-x) = \dfrac{1}{2(-x)} \Rightarrow f(-x) = -\dfrac{1}{2x}$ and $-f(x) = -\left(\dfrac{1}{2x}\right) \Rightarrow -f(x) = -\dfrac{1}{2x}$ Since
   $f(-x) = -f(x)$, the function is odd.

60. If $f(x) = 4x - \dfrac{1}{x}$, then $f(-x) = 4(-x) - \dfrac{1}{(-x)} \Rightarrow f(-x) = -4x + \dfrac{1}{x}$ and
   $-f(x) = -\left(4x - \dfrac{1}{x}\right) \Rightarrow -f(x) = -4x + \dfrac{1}{x}$. Since $f(-x) = -f(x)$, the function is odd.

61. If $f(x) = -x^3 + 2x$, then $f(-x) = -(-x)^3 + 2(-x) \Rightarrow f(-x) = x^3 - 2x$ and

$-f(x) = -(-x^3 + 2x) \Rightarrow -f(x) = x^3 - 2x$. Since $f(-x) = -f(x)$, the function is symmetric with respect to the origin. Graph $f(x) = -x^3 + 2x$. The graph supports symmetry with respect to the origin.

62. If $f(x) = x^5 - 2x^3$, then $f(-x) = (-x)^5 - 2(-x)^3 \Rightarrow f(-x) = -x^5 + 2x^3$ and

$-f(x) = -(x^5 - 2x^3) \Rightarrow -f(x) = -x^5 + 2x^3$. Since $f(-x) = -f(x)$, the function is symmetric with respect to the origin. Graph $f(x) = x^5 - 2x^3$. The graph supports symmetry with respect to the origin.

63. If $f(x) = .5x^4 - 2x^2 + 1$, then $f(-x) = .5(-x)^4 - 2(-x)^2 + 1 \Rightarrow f(-x) = .5x^4 - 2x^2 + 1$.   Since

$f(-x) = f(x)$, the function is symmetric with respect to the y-axis. Graph $f(x) = .5x^4 - 2x^2 + 1$. The graph supports symmetry with respect to the y-axis.

64. If $f(x) = .75x^2 + |x| + 1$, then $f(-x) = .75(-x)^2 + |(-x)| + 1 \Rightarrow f(-x) = .75x^2 + |x| + 1$. Since

$f(-x) = f(x)$, the function is symmetric with respect to the y-axis. Graph $f(x) = .75x^4 + |x| + 1$. The graph supports symmetry with respect to the y-axis.

65. If $f(x) = x^3 - x + 3$, then $f(-x) = (-x)^3 - (-x) + 3 \Rightarrow f(-x) = -x^3 + x + 3$ and $-f(x) = -(x^3 - x + 3)$

$\Rightarrow -f(x) = -x^3 + x - 3$. Since $f(x) \neq f(-x) \neq -f(x)$, the function is not symmetric with respect to the y-axis or origin. Graph $f(x) = x^3 - x + 3$. The graph supports no symmetry with respect to the y-axis or origin.

66. If $f(x) = x^4 - 5x + 2$, then $f(-x) = (-x)^4 - 5(-x) + 2 \Rightarrow f(-x) = x^4 + 5x + 2$ and $-f(x) = -(x^4 - 5x + 2)$

$\Rightarrow -f(x) = -x^4 + 5x - 2$. Since $f(x) \neq f(-x) \neq -f(x)$, the function is not symmetric with respect to the y-axis or origin. Graph $f(x) = x^4 - 5x + 2$. The graph supports no symmetry with respect to the y-axis or origin.

67. If $f(x) = x^6 - 4x^3$, then $f(-x) = (-x)^6 - 4(-x)^3 \Rightarrow f(-x) = x^6 + 4x^3$ and $-f(x) = -(x^6 - 4x^3) \Rightarrow$

$-f(x) = -x^6 + 4x^3$ Since $f(x) \neq f(-x) \neq -f(x)$, the function is not symmetric with respect to the y-axis or origin. Graph $f(x) = x^6 - 4x^3$. The graph supports no symmetry with respect to the y-axis or origin.

68. If $f(x) = x^3 - 3x$, then $f(-x) = (-x)^3 - 3(-x) \Rightarrow f(-x) = -x^3 + 3x$ and

$-f(x) = -(x^3 - 3x) \Rightarrow -f(x) = -x^3 + 3x$. Since $f(-x) = -f(x)$, the function is symmetric with respect to the origin. Graph $f(x) = x^3 - 3x$. The graph supports symmetry with respect to the origin.

69. If $f(x) = -6$, then $f(-x) = -6$.   Since $f(-x) = f(x)$, the function is symmetric with respect to the y-axis.

Graph $f(x) = -6$. The graph supports symmetry with respect to the y-axis.

70. If $f(x) = |-x| = |x|$, then $f(-x) = |-(-x)| = |x| \Rightarrow f(-x) = |x|$.   Since $f(-x) = f(x)$, the function is

symmetric with respect to the y-axis. Graph $f(x) = |-x|$. The graph supports symmetry with respect to the y-axis.

71. If $f(x) = \dfrac{1}{4x^3}$, then $f(-x) = \dfrac{1}{4(-x)^3} \Rightarrow f(-x) = -\dfrac{1}{4x^3}$ and $-f(x) = -\left(\dfrac{1}{4x^3}\right) \Rightarrow -f(x) = -\dfrac{1}{4x^3}$. Since

$f(-x) = -f(x)$, the function is symmetric with respect to the origin. Graph $f(x) = \dfrac{1}{4x^3}$. The graph supports symmetry with respect to the origin.

72. If $f(x) = \sqrt{x^2} \Rightarrow f(x) = x$, then $f(-x) = \sqrt{(-x)^2} \Rightarrow f(-x) = \sqrt{x^2} \Rightarrow f(-x) = x$.   Since $f(-x) = f(x)$, the

function is symmetric with respect to the y-axis. Graph $f(x) = \sqrt{x^2}$. The graph supports symmetry with respect to the y-axis.

73. (a) Functions where $f(-x) = f(x)$ are even, therefore exercises: 63, 64, 69, 70, and 72 are even.

    (b) Functions where $f(-x) = -f(x)$ are odd, therefore exercises: 61, 62, 68, and 71 are odd.

    (c) Functions where $f(x) \neq f(-x) \neq -f(x)$ are neither odd or even, therefore exercises: 65, 66, and 67 are neither odd or even.

74. Answers may vary. If a function f is even, then $f(x) = f(-x)$ for all x in the domain. Its graph is symmetric with respect to the y-axis. If a function f is odd, then $f(-x) = -f(x)$ for all x in the domain. Its graph is symmetric with respect to the origin.

2.2: Vertical and Horizontal Shifts of Graphs

1. The equation $y = x^2$ shifted 3 units upward is: $y = x^2 + 3$.

2. The equation $y = x^3$ shifted 2 units downward is: $y = x^3 - 2$.

3. The equation $y = \sqrt{x}$ shifted 4 units downward is: $y = \sqrt{x} - 4$.

4. The equation $y = \sqrt[3]{x}$ shifted 6 units upward is: $y = \sqrt[3]{x} + 6$.

5. The equation $y = |x|$ shifted 4 units to the right is: $y = |x - 4|$.

6. The equation $y = |x|$ shifted 3 units to the left is: $y = |x + 3|$.

7. The equation $y = x^3$ shifted 7 units to the left is: $y = (x + 7)^3$.

8. The equation $y = \sqrt{x}$ shifted 9 units to the right is: $y = \sqrt{x - 9}$.

9. Shift the graph of f 4 units upward to obtain the graph of g.

10. Shift the graph of f 4 units to the left to obtain the graph of g.

11. The equation $y = x^2 - 3$ is $y = x^2$ shifted 3 units downward, therefore graph B.

12. The equation $y = (x - 3)^2$ is $y = x^2$ shifted 3 units to the right, therefore graph C.

13. The equation $y = (x + 3)^2$ is $y = x^2$ shifted 3 units to the left, therefore graph A.

14. The equation $y = |x| + 4$ is $y = |x|$ shifted 4 units upward, therefore graph A.

15. The equation $y = |x + 4| - 3$ is $y = |x|$ shifted 4 units to the left and 3 units downward, therefore graph B.

16. The equation $y = |x - 4| - 3$ is $y = |x|$ shifted 4 units to the right and 3 units downward, therefore graph C.

17. The equation $y = (x - 3)^3$ is $y = x^3$ shifted 3 units to the right, therefore graph C.

18. The equation $y = (x - 2)^3 - 4$ is $y = x^3$ shifted 2 units to the right and 4 units downward, therefore graph A.

19. The equation $y = (x + 2)^3 - 4$ is $y = x^3$ shifted 2 units to the left and 4 units downward, therefore graph B.

20. If $y = |x - h| + k$ with $h < 0$ and $k < 0$, then the graph of $y = |x|$ is shifted to the left $-h$ units and $-k$ units downward. This would place the vertex or lowest point of the absolute value graph in the third quadrant.

21. For the equation $y = x^2$, the Domain is: $(-\infty, \infty)$ and the Range is: $[0, \infty)$. Shifting this 3 units downward gives us: (a) Domain: $(-\infty, \infty)$ (b) Range: $[-3, \infty)$

22. For the equation $y = x^2$, the Domain is: $(-\infty, \infty)$ and the Range is: $[0, \infty)$. Shifting this 3 units to the right gives us: (a) Domain: $(-\infty, \infty)$ (b) Range: $[0, \infty)$

23. For the equation $y = |x|$, the Domain is: $(-\infty, \infty)$ and the Range is: $[0, \infty)$. Shifting this 4 units to the left and 3 units downward gives us: (a) Domain: $(-\infty, \infty)$ (b) Range: $[-3, \infty)$

24. For the equation $y = |x|$, the Domain is: $(-\infty, \infty)$ and the Range is: $[0, \infty)$. Shifting this 4 units to the right and 3 units downward gives us: (a) Domain: $(-\infty, \infty)$ (b) Range: $[-3, \infty)$

25. For the equation $y = x^3$, the Domain is: $(-\infty, \infty)$ and the Range is: $(-\infty, \infty)$. Shifting this 3 units to the right gives us: (a) Domain: $(-\infty, \infty)$ (b) Range: $(-\infty, \infty)$

26. For the equation $y = x^3$, the Domain is: $(-\infty, \infty)$ and the Range is: $(-\infty, \infty)$. Shifting this 2 units to the right and 4 units downward gives us: (a) Domain: $(-\infty, \infty)$ (b) Range: $(-\infty, \infty)$

27. Using $Y_2 = Y_1 + k$ and $x = 0$, we get $19 = 15 + k \Rightarrow k = 4$.

28. Using $Y_2 = Y_1 + k$ and $x = 0$, we get $-5 = -3 + k \Rightarrow k = -2$.

29. From the graphs $(6, 2)$ is a point on Y_1 and $(6, -1)$ a point on Y_2. Using $Y_2 = Y_1 + k$ and $x = 6$, we get $-1 = 2 + k \Rightarrow k = -3$.

30. From the graphs $(-4, 3)$ is a point on Y_1 and $(-4, 8)$ a point on Y_2. Using $Y_2 = Y_1 + k$ and $x = -4$, we get $8 = 3 + k \Rightarrow k = 5$.

31. The graph of $y = (x - 1)^2$ is the graph of the equation $y = x^2$ shifted 1 unit to the right. See Figure 31.

32. The graph of $y = \sqrt{x + 2}$ is the graph of the equation $y = \sqrt{x}$ shifted 2 units to the left. See Figure 32.

33. The graph of $y = x^3 + 1$ is the graph of the equation $y = x^3$ shifted 1 unit upward. See Figure 33.

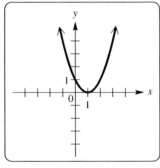
Figure 31

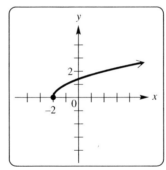
Figure 32

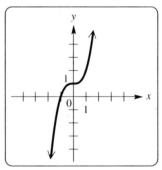
Figure 33

34. The graph of $y = |x + 2|$ is the graph of the equation $y = |x|$ shifted 2 units to the left. See Figure 34.

35. The graph of $y = (x - 1)^3$ is the graph of the equation $y = x^3$ shifted 1 unit to the right. See Figure 35.

36. The graph of $y = |x| - 3$ is the graph of the equation $y = |x|$ shifted 3 units downward. See Figure 36.

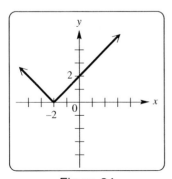
Figure 34

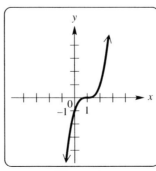
Figure 35

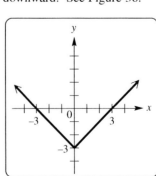
Figure 36

37. The graph of $y = \sqrt{x - 2} - 1$ is the graph of the equation $y = \sqrt{x}$ shifted 2 units to the right and 1 unit downward. See Figure 37.

38. The graph of $y = \sqrt{x + 3} - 4$ is the graph of the equation $y = \sqrt{x}$ shifted 3 units to the left and 4 units downward. See Figure 38.

39. The graph of $y = (x + 2)^2 + 3$ is the graph of the equation $y = x^2$ shifted 2 units to the left and 3 units upward. See Figure 39.

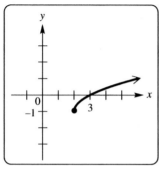
Figure 37

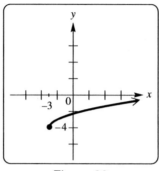
Figure 38

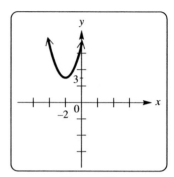
Figure 39

40. The graph of $y = (x - 4)^2 - 4$ is the graph of the equation $y = x^2$ shifted 4 units to the right and 4 units downward. See Figure 40.

41. The graph of $y = |x + 4| - 2$ is the graph of the equation $y = |x|$ shifted 4 units to the left and 2 units downward. See Figure 41.

42. The graph of $y = (x + 3)^3 - 1$ is the graph of the equation $y = x^3$ shifted 3 units to the left and 1 unit downward. See Figure 42.

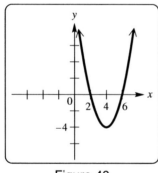
Figure 40

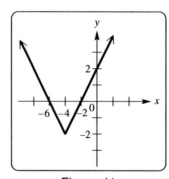
Figure 41

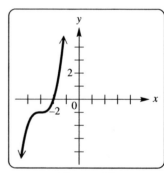
Figure 42

43. Since h and k are positive, the equation $y = (x - h)^2 - k$ is $y = x^2$ shifted to the right and down, therefore: B.

44. Since h and k are positive, the equation $y = (x + h)^2 - k$ is $y = x^2$ shifted to the left and down, therefore: D.

45. Since h and k are positive, the equation $y = (x + h)^2 + k$ is $y = x^2$ shifted to the left and up, therefore: A.

46. Since h and k are positive, the equation $y = (x - h)^2 + k$ is $y = x^2$ shifted to the right and up, therefore: C.

47. The equation $y = f(x) + 2$ is $y = f(x)$ shifted up 2 units or add 2 to the y-coordinate of each point as follows: $(-3, -2) \Rightarrow (-3, 0); (-1, 4) \Rightarrow (-1, 6); (5, 0) \Rightarrow (5, 2)$. See Figure 47.

48. The equation $y = f(x) - 2$ is $y = f(x)$ shifted down 2 units or subtract 2 from the y-coordinate of each point as follows: $(-3, -2) \Rightarrow (-3, -4); (-1, 4) \Rightarrow (-1, 2); (5, 0) \Rightarrow (5, -2)$. See Figure 48.

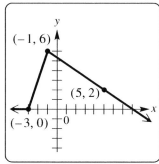

Figure 47

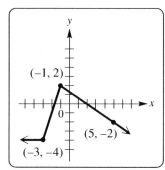

Figure 48

49. The equation $y = f(x + 2)$ is $y = f(x)$ shifted left 2 units or subtract 2 from the x-coordinate of each point as follows: $(-3, -2) \Rightarrow (-5, -2); (-1, 4) \Rightarrow (-3, 4); (5, 0) \Rightarrow (3, 0)$. See Figure 49.

50. The equation $y = f(x - 2)$ is $y = f(x)$ shifted right 2 units or add 2 to the x-coordinate of each point as follows: $(-3, -2) \Rightarrow (-1, -2); (-1, 4) \Rightarrow (1, 4); (5, 0) \Rightarrow (7, 0)$. See Figure 50.

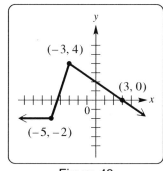

Figure 49

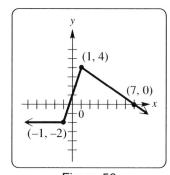

Figure 50

51. The graph is the basic function $y = x^2$ translated 4 units to the left and 3 units up, therefore the new equation is: $y = (x + 4)^2 + 3$. The equation is now increasing for the interval: (a) $[-4, \infty)$ and decreasing for the interval: (b) $(-\infty, -4]$.

52. The graph is the basic function $y = \sqrt{x}$ translated 5 units to the left, therefore the new equation is: $y = \sqrt{x + 5}$. The equation is now increasing for the interval: (a) $[-5, \infty)$ and does not decrease, therefore: (b) none.

53. The graph is the basic function $y = x^3$ translated 5 units down, therefore the new equation is: $y = x^3 - 5$. The equation is now increasing for the interval: (a) $(-\infty, \infty)$ and does not decrease, therefore: (b) none.

54. The graph is the basic function $y = |x|$ translated 10 units to the left, therefore the new equation is: $y = |x + 10|$. The equation is now increasing for the interval: (a) $[-10, \infty)$ and decreasing for the interval: (b) $(-\infty, -10]$.

55. The graph is the basic function $y = \sqrt{x}$ translated 2 units to the right and 1 unit up, therefore the new equation is: $y = \sqrt{x - 2} + 1$. The equation is now increasing for the interval: (a) $[2, \infty)$ and does not decrease, therefore: (b) none.

56. The graph is the basic function $y = x^2$ translated 2 units to the right and 3 units down, therefore the new equation is: $y = (x - 2)^2 - 3$. The equation is now increasing for the interval: (a) $[2, \infty)$ and decreasing for the interval: (b) $(-\infty, 2]$.

57. (a) $f(x) = 0$: $\{3, 4\}$

    (b) $f(x) > 0$: for the intervals $(-\infty, 3) \cup (4, \infty)$.

    (c) $f(x) < 0$: for the interval $(3, 4)$.

58. (a) $f(x) = 0$: $\{\sqrt{2}\}$

    (b) $f(x) > 0$: for the interval $(\sqrt{2}, \infty)$.

    (c) $f(x) < 0$: for the interval $(-\infty, \sqrt{2})$.

59. (a) $f(x) = 0$: $\{-4, 5\}$

    (b) $f(x) \geq 0$: for the intervals $(-\infty, -4] \cup [5, \infty)$.

    (c) $f(x) \leq 0$: for the interval $[-4, 5]$.

60. (a) $f(x) = 0$: never, therefore: $\varnothing$.

    (b) $f(x) \geq 0$: for the interval $[1, \infty)$.

    (c) $f(x) \leq 0$: never, therefore: $\varnothing$.

61. The translation is 3 units to the left and 1 unit up, therefore the new equation is: $y = |x + 3| + 1$. The form $y = |x - h| + k$ will equal $y = |x + 3| + 1$ when: $h = -3$ and $k = 1$.

62. The equation $y = x^2$ has a Domain: $(-\infty, \infty)$ and a Range: $[0, \infty)$. After the translation the Domain is still: $(-\infty, \infty)$, but now the Range is: $[38, \infty)$, a positive or upward shift of 38 units. Therefore, the horizontal shift can be any number of units, but the vertical shift is up 38. This makes h any real number and $k = 38$.

63. (a) Since 0 corresponds to 1998, our equation using exact years would be: $y = 895.5(x - 1998) + 14,709$.

    (b) $y = 895.5(2006 - 1998) + 14,709 \Rightarrow y = 7164 + 14,709 \Rightarrow y = \$21,873$

64. (a) Since 0 corresponds to 1998, our equation using exact years would be: $y = 299.8(x - 1998) + 5249.1$.

    (b) $y = 299.8(2007 - 1998) + 5249.1 \Rightarrow y = 2698.2 + 5249.1 \Rightarrow y = \$7947.3$ billion.

65. (a) Enter the year in L_1 and enter tuition and fees in L_2. The year 1991 corresponds to $x = 0$ and so on. The regression equation is: $y \approx 190x + 2071.3$.

    (b) Since $x = 0$ corresponds to 1991, the equation when the exact year is entered is:

    $y \approx 190(x - 1991) + 2071.3$

    (c) $y \approx 190(2008 - 1991) + 2071.3 \Rightarrow y \approx 3230 + 2071.3 \Rightarrow y \approx \$5300$

66. (a) Enter the year in L_1 and enter the percent of women in the workforce in L_2. The year 1965 corresponds to $x = 0$ and so on. The regression equation is: $y \approx .6162x + 40.6167$.

    (b) Since $x = 0$ corresponds to 1965, the equation when the exact year is entered is:

    $y \approx .6162(x - 1965) + 40.6167$

    (c) $y \approx .6162(2010 - 1965) + 40.6167 \Rightarrow y \approx 27.729 + 40.6167 \Rightarrow y \approx 68.3\%$

67. See Figure 67.

68. $m = \dfrac{2 - (-2)}{3 - 1} \Rightarrow m = \dfrac{4}{2} = 2$

69. Using slope-intercept form yields: $y_1 - 2 = 2(x - 3) \Rightarrow y_1 - 2 = 2x - 6 \Rightarrow y_1 = 2x - 4$

70. $(1, -2 + 6)$ and $(3, 2 + 6) \Rightarrow (1, 4)$ and $(3, 8)$

71. $m = \dfrac{8 - 4}{3 - 1} \Rightarrow m = \dfrac{4}{2} = 2$

72. Using slope-intercept form yields: $y_2 - 4 = 2(x - 1) \Rightarrow y_2 - 4 = 2x - 2 \Rightarrow y_2 = 2x + 2$

73. Graph $y_1 = 2x - 4$ and $y_2 = 2x + 2$. See Figure 73. The graph y_2 can be obtained by shifting the graph of y_1 upward 6 units. The constant, 6, comes from the 6 we added to each y-value in Exercise 70.

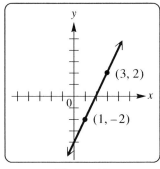

Figure 67

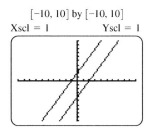

$[-10, 10]$ by $[-10, 10]$
Xscl = 1   Yscl = 1

Figure 73

74. c; c; the same as; upward (or positive vertical)

2.3: Stretching, Shrinking, and Reflecting Graphs

1. The function $y = x^2$ vertically stretched by a factor of 2 is: $y = 2x^2$.

2. The function $y = x^3$ vertically shrunk by a factor of $\frac{1}{2}$ is: $y = \frac{1}{2}x^3$.

3. The function $y = \sqrt{x}$ reflected across the y-axis is: $y = \sqrt{-x}$.

4. The function $y = \sqrt[3]{x}$ reflected across the x-axis is: $y = -\sqrt[3]{x}$.

5. The function $y = |x|$ vertically stretched by a factor of 3 and reflected across the x-axis is: $y = -3|x|$.

6. The function $y = |x|$ vertically shrunk by a factor of $\frac{1}{3}$ and reflected across the y-axis is: $y = \frac{1}{3}|-x|$.

7. The function $y = x^3$ vertically shrunk by a factor of .25 and reflected across the y-axis is:
    $y = .25(-x)^3$ or $y = -.25x^3$.

8. The function $y = \sqrt{x}$ vertically shrunk by a factor of .2 and reflected across the x-axis is: $y = -.2\sqrt{x}$.

9. Graph $y_1 = x$, $y_2 = x + 3$ (y_1 shifted up 3 units), and $y_3 = x - 3$ (y_1 shifted down 3 units). See Figure 9.

10. Graph $y_1 = x^3$, $y_2 = x^3 + 4$ (y_1 shifted up 4 units), and $y_3 = x^3 - 4$ (y_1 shifted down 4 units). See Figure 10.

11. Graph $y_1 = |x|$, $y_2 = |x - 3|$ (y_1 shifted right 3 units), and $y_3 = |x + 3|$ (y_1 shifted left 3 units). See Figure 11.

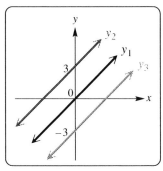

Figure 9

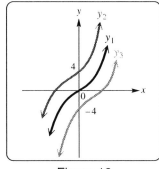

Figure 10

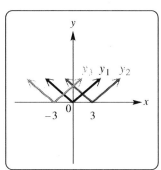

Figure 11

12. Graph $y_1 = |x|$, $y_2 = |x| - 3$ (y_1 shifted down 3 units), and $y_3 = |x| + 3$ (y_1 shifted up 3 units). See Figure 12.

13. Graph $y_1 = \sqrt{x}$, $y_2 = \sqrt{x + 6}$ (y_1 shifted left 6 units), and $y_3 = \sqrt{x - 6}$ (y_1 shifted right 6 units). See Figure 13.

14. Graph $y_1 = |x|$, $y_2 = 2|x|$ (y_1 stretched vertically by a factor of 2), and $y_3 = 2.5|x|$ (y_1 stretched vertically by a factor of 2.5). See Figure 14.

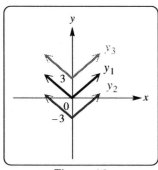
Figure 12

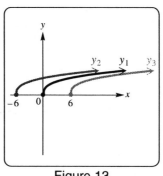
Figure 13

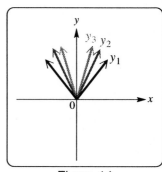
Figure 14

15. Graph $y_1 = \sqrt[3]{x}$, $y_2 = -\sqrt[3]{x}$ (y_1 reflected across the x-axis), and $y_3 = -2\sqrt[3]{x}$ (y_1 reflected across the x-axis and stretched vertically by a factor of 2). See Figure 15.

16. Graph $y_1 = x^2$, $y_2 = (x - 2)^2 + 1$ (y_1 shifted right 2 units and up 1 unit), and $y_3 = -(x + 2)^2$ (y_1 shifted left 2 units and reflected across the x-axis). See Figure 16.

17. Graph $y_1 = |x|$, $y_2 = -2|x - 1| + 1$ (y_1 reflected across the x-axis, stretched vertically by a factor of 2, shifted right 1 unit, and shifted up 1 unit,), and $y_3 = -\frac{1}{2}|x| - 4$ (y_1 reflected across the x-axis, shrunk by factor of $\frac{1}{2}$, and shifted down 4 units). See Figure 17.

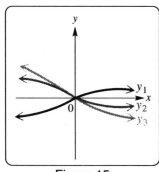
Figure 15

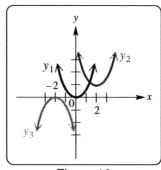
Figure 16

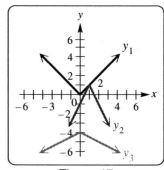
Figure 17

18. Graph $y_1 = \sqrt{x}$, $y_2 = -\sqrt{x}$ (y_1 reflected across the x-axis), and $y_3 = \sqrt{-x}$ (y_1 reflected across the y-axis). See Figure 18.

19. Graph $y_1 = x^2 - 1$ (which is $y = x^2$ shifted down 1 unit), $y_2 = \left(\frac{1}{2}x\right)^2 - 1$ (y_1 shrunk vertically by a factor of $\frac{1}{2}$), and $y_3 = (2x)^2 - 1$ (y_1 stretched vertically by a factor of 2^2 or 4). See Figure 19.

20. Graph $y_1 = 3 - |x|$ (which is $y = |x|$ reflected across the x-axis and shifted up 3 units), $y_2 = 3 - |3x|$ (y_1 stretched vertically by a factor of 3), and $y_3 = 3 - |\frac{1}{3}x|$ (y_1 shrunk vertically by a factor of $\frac{1}{3}$). See Figure 20.

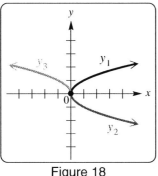

Figure 18

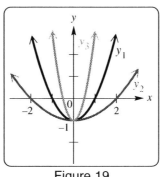

Figure 19

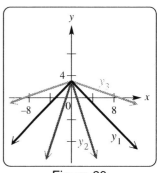

Figure 20

21. Graph $y_1 = \sqrt[3]{x}$, $y_2 = \sqrt[3]{-x}$ (y_1 reflected across the y-axis), and $y_3 = \sqrt[3]{-(x-1)}$ (y_1 reflected across the y-axis and shifted right 1 unit). See Figure 21.

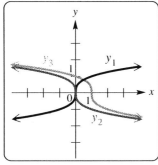

Figure 21

22. Since $y = f(x)$ is symmetric with respect to the y-axis, for every (x, y) on the graph, $(-x, y)$ is also on the graph. Reflection across the y-axis reflect onto itself and will not change the graph. It will be the same.

23. 4; x

24. 6; x

25. 2; left; $\dfrac{1}{4}$; x; 3; downward (or negative)

26. y; $\dfrac{2}{5}$; x; 6 ; upward (or positive)

27. 3; right; 6

28. 2; left; .5

29. The function $y = x^2$ is vertically shrunk by a factor of $\dfrac{1}{2}$ and shifted 7 units down, therefore: $y = \dfrac{1}{2}x^2 - 7$.

30. The function $y = x^3$ is vertically stretched by a factor of 3, reflected across the x-axis, and shifted 8 units upward, therefore: $y = -3x^3 + 8$.

31. The function $y = \sqrt{x}$ is shifted 3 units right, vertically stretched by a factor of 4.5, and shifted 6 units down, therefore: $y = 4.5\sqrt{x-3} - 6$.

32. The function $y = \sqrt[3]{x}$ is shifted 2 units left, vertically stretched by a factor of 1.5, and shifted 8 units upward, therefore: $y = 1.5\sqrt[3]{x+2} + 8$.

33. The graph $y = f(x) = x^2$ has been reflected across the x-axis, shifted 5 units to the right, and shifted 2 units downward, therefore the equation of $g(x)$ is: $g(x) = -(x - 5)^2 - 2$.

34. The graph $y = f(x) = x^3$ has been shifted 4 units to the right and shifted 3 units upward, therefore the equation of $g(x)$ is: $g(x) = (x - 4)^3 + 3$.

35. The function $f(x) = \sqrt{x - 3} + 2$ is $f(x) = \sqrt{x}$ shifted 3 units right and 2 units upward. See Figure 35.

36. The function $f(x) = |x + 2| - 3$ is $f(x) = |x|$ shifted 2 units left and 3 units downward. See Figure 36.

37. The function $f(x) = \sqrt{2x} = \sqrt{2}\sqrt{x}$ is $f(x) = \sqrt{x}$ stretched vertically by a factor of $\sqrt{2}$. See Figure 37.

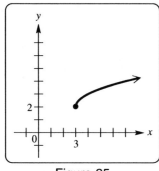
Figure 35

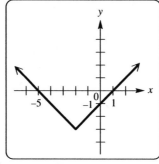
Figure 36

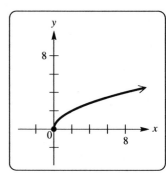
Figure 37

38. The function $f(x) = \frac{1}{2}(x + 2)^2$ is $f(x) = x^2$ shifted 2 units left and shrunk vertically by a factor of $\frac{1}{2}$. See Figure 38.

39. The function $f(x) = |2x| = 2|x|$ is $f(x) = |x|$ stretched vertically by a factor of 2. See Figure 39.

40. The function $f(x) = \frac{1}{2}|x|$ is $f(x) = |x|$ shrunk vertically by a factor of $\frac{1}{2}$. See Figure 40.

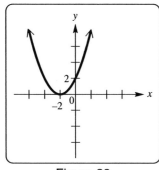
Figure 38

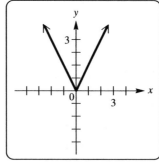
Figure 39

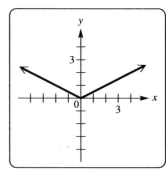
Figure 40

41. The function $f(x) = 1 - \sqrt{x}$ is $f(x) = \sqrt{x}$ reflected across the x-axis and shifted 1 unit upward. See Figure 41.

42. The function $f(x) = 2\sqrt{(x - 2)} - 1$ is $f(x) = \sqrt{x}$ shifted 2 units right, stretched vertically by a factor of 2, and shifted 1 unit downward. See Figure 42.

43. The function $f(x) = -\sqrt{1 - x} = -\sqrt{-(x - 1)}$ is $f(x) = \sqrt{x}$ reflected across both the x-axis and the y-axis and shifted 1 unit right. See Figure 43.

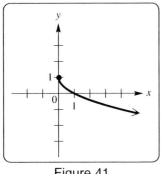

Figure 41

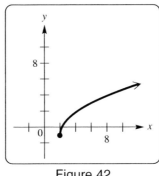

Figure 42

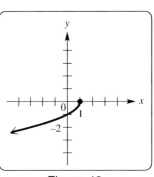

Figure 43

44. The function $f(x) = \sqrt{-x} - 1$ is $f(x) = \sqrt{x}$ reflected across the y-axis and shifted 1 unit downward. See Figure 44.

45. The function $f(x) = \sqrt{-(x + 1)}$ is $f(x) = \sqrt{x}$ reflected across the y-axis and shifted 1 unit left. See Figure 45.

46. The function $f(x) = 2 + \sqrt{-(x - 3)}$ is $f(x) = \sqrt{x}$ reflected across the y-axis, shifted 3 units right, and shifted 2 units upward. See Figure 46.

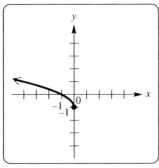

Figure 44

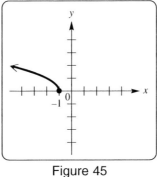

Figure 45

Figure 46

47. The function $f(x) = (x - 1)^3$ is $f(x) = x^3$ shifted 1 unit right. See Figure 47.

48. The function $f(x) = (x + 2)^3$ is $f(x) = x^3$ shifted 2 units left. See Figure 48.

49. The function $f(x) = -x^3$ is $f(x) = x^3$ reflected across the x-axis. See Figure 49.

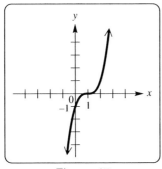

Figure 47

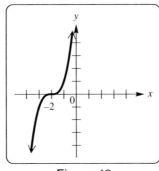

Figure 48

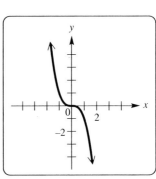

Figure 49

50. The function $f(x) = (-x)^3 + 1$ is $f(x) = x^3$ reflected across the y-axis and shifted 1 unit upward. See Figure 50.

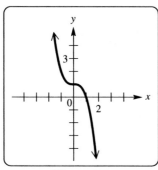

Figure 50

51. (a) The equation $y = -f(x)$ is $y = f(x)$ reflected across the x-axis. See Figure 51a.

   (b) The equation $y = f(-x)$ is $y = f(x)$ reflected across the y-axis. See Figure 51b.

   (c) The equation $y = 2f(x)$ is $y = f(x)$ stretched vertically by a factor of 2. See Figure 51c.

   (d) From the graph $f(0) = 1$.

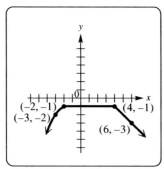

Figure 51a

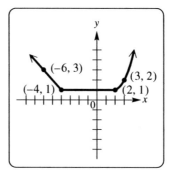

Figure 51b

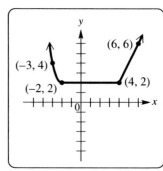

Figure 51c

52. (a) The equation $y = -f(x)$ is $y = f(x)$ reflected across the x-axis. See Figure 52a.

   (b) The equation $y = f(-x)$ is $y = f(x)$ reflected across the y-axis. See Figure 52b.

   (c) The equation $y = 3f(x)$ is $y = f(x)$ stretched vertically by a factor of 3. See Figure 52c.

   (d) From the graph $f(4) = 1$.

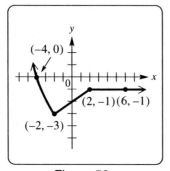

Figure 52a

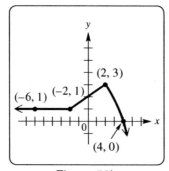

Figure 52b

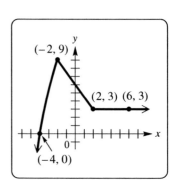

Figure 52c

53. (a) The equation $y = -f(x)$ is $y = f(x)$ reflected across the x-axis. See Figure 53a.

   (b) The equation $y = f(-x)$ is $y = f(x)$ reflected across the y-axis. See Figure 53b.

   (c) The equation $y = f(x + 1)$ is $y = f(x)$ shifted 1 unit to the left. See Figure 53c.

   (d) From the graph, there are two x-intercepts: -1 and 4.

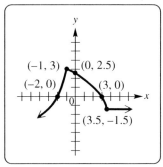

Figure 53a

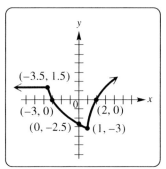

Figure 53b

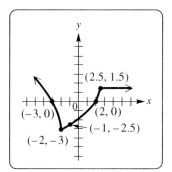

Figure 53c

54. (a) The equation $y = -f(x)$ is $y = f(x)$ reflected across the x-axis. See Figure 54a.

   (b) The equation $y = f(-x)$ is $y = f(x)$ reflected across the y-axis. See Figure 54b.

   (c) The equation $y = \frac{1}{2}f(x)$ is $y = f(x)$ shrunk vertically by a factor of $\frac{1}{2}$. See Figure 54c.

   (d) From the graph $f(x) < 0$ for the interval: $(-\infty, 0)$.

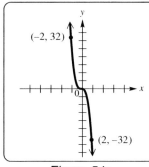

Figure 54a

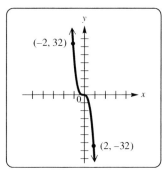

Figure 54b

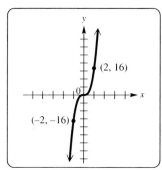

Figure 54c

55. (a) The equation $y = -f(x)$ is $y = f(x)$ reflected across the x-axis. See Figure 55a.

   (b) The equation $y = f\left(\frac{1}{3}x\right)$ is $y = f(x)$ stretched horizontally by a factor of 3. See Figure 55b.

   (c) The equation $y = .5f(x)$ is $y = f(x)$ shrunk vertically by a factor of .5. See Figure 55c.

   (d) From the graph, symmetry with respect to the origin.

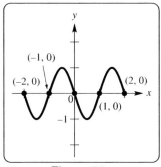

Figure 55a

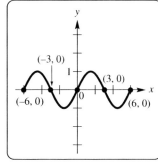

Figure 55b

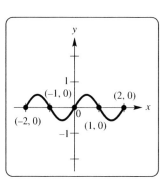

Figure 55c

56. (a) The equation $y = f(2x)$ is $y = f(x)$ stretched horizontally by a factor of $\frac{1}{2}$. See Figure 56a.

(b) The equation $y = f(-x)$ is $y = f(x)$ reflected across the y-axis. See Figure 56b.

(c) The equation $y = 3f(x)$ is $y = f(x)$ stretched vertically by a factor of 3. See Figure 56c.

(d) From the graph, symmetry with respect to the y-axis.

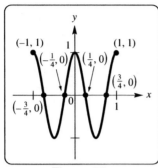

Figure 56a

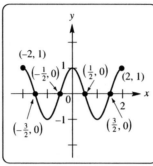

Figure 56b

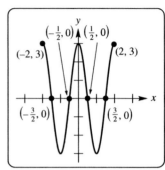

Figure 56c

57. (a) The equation $y = f(x) + 1$ is $y = f(x)$ shifted 1 unit upward. See Figure 57a.

(b) The equation $y = -f(x) - 1$ is $y = f(x)$ reflected across the x-axis and shifted 1 unit down. See Figure 57b.

(c) The equation $y = 2f\left(\frac{1}{2}x\right)$ is $y = f(x)$ stretched vertically by a factor of 2 and horizontally by a factor of 2.
See Figure 57c.

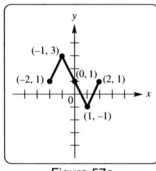

Figure 57a

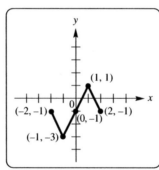

Figure 57b

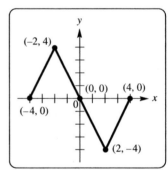

Figure 57c

58. (a) The equation $y = f(x) - 2$ is $y = f(x)$ shifted 2 units downward. See Figure 58a.

(b) The equation $y = f(x - 1) + 2$ is $y = f(x)$ shifted 1 unit right and 2 units upward. See Figure 58b.

(c) The equation $y = 2f(x)$ is $y = f(x)$ stretched vertically by a factor of 2. See Figure 58c.

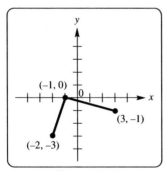

Figure 58a

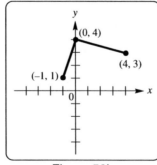

Figure 58b

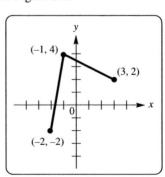

Figure 58c

59. (a) The equation $y = f(2x) + 1$ is $y = f(x)$ shrunk horizontally by a factor of $\frac{1}{2}$ and shifted 1 unit upward. See Figure 59a.

(b) The equation $y = 2f\left(\frac{1}{2}x\right) + 1$ is $y = f(x)$ stretched vertically by a factor of 2, stretched horizontally by a factor of 2, and shifted 1 unit upward. See Figure 59b.

(c) The equation $y = \frac{1}{2}f(x - 2)$ is $y = f(x)$ shrunk vertically by a factor of $\frac{1}{2}$ and shifted 2 units to the right. See Figure 59c.

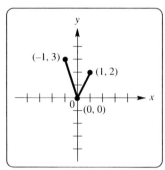 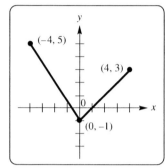 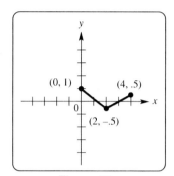

Figure 59a   Figure 59b   Figure 59c

60. (a) The equation $y = f(2x)$ is $y = f(x)$ shrunk horizontally by a factor of $\frac{1}{2}$. See Figure 60a.

(b) The equation $y = f\left(\frac{1}{2}x\right) - 1$ is $y = f(x)$ stretched horizontally by a factor of 2, and shifted 1 unit downward. See Figure 60b.

(c) The equation $y = 2f(x) - 1$ is $y = f(x)$ stretched vertically by a factor of 2 and shifted 1 unit downward. See Figure 60c.

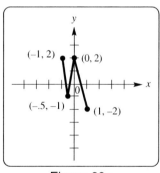 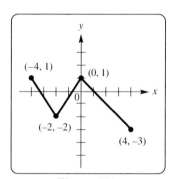 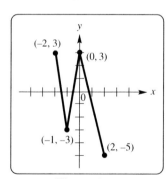

Figure 60a   Figure 60b   Figure 60c

61. (a) If r is the x-intercept of $y = f(x)$ and $y = -f(x)$ is $y = f(x)$ reflected across the x-axis, then r is also the x-intercept of $y = -f(x)$.

(b) If r is the x-intercept of $y = f(x)$ and $y = f(-x)$ is $y = f(x)$ reflected across the y-axis, then $-r$ is the x-intercept of $y = f(-x)$.

(c) If r is the x-intercept of $y = f(x)$ and $y = -f(-x)$ is $y = f(x)$ reflected across both the x-axis and y-axis, then $-r$ is the x-intercept of $y = -f(-x)$.

62. (a) If b is the y-intercept of $y = f(x)$ and $y = -f(x)$ is $y = f(x)$ reflected across the x-axis, then $-b$ is the y-intercept of $y = -f(x)$.

    (b) If b is the y-intercept of $y = f(x)$ and $y = f(-x)$ is $y = f(x)$ reflected across the y-axis, then b is also the y-intercept of $y = f(-x)$.

    (c) If b is the y-intercept of $y = f(x)$ and $y = 5f(x)$ is $y = f(x)$ stretched vertically by a factor of 5, then $5b$ is the y-intercept of $y = 5f(x)$.

    (d) If b is the y-intercept of $y = f(x)$ and $y = -3f(x)$ is $y = f(x)$ reflected across the x-axis and stretched vertically by a factor of 3, then $-3b$ is the y-intercept of of $y = -3f(x)$.

63. Since $f(x - 2)$ is $f(x)$ shifted 2 units to the right, the domain of $f(x - 2)$ is: $[-1 + 2, 2 + 2]$ or $[1, 4]$; and the range is the same: $[0, 3]$.

64. Since $5f(x + 1)$ is $f(x)$ shifted 1 unit to the left, the domain of $5f(x + 1)$ is: $[-1 - 1, 2 - 1]$ or $[-2, 1]$; and stretched vertically by a factor of 5, the range is: $[5(0), 5(3)]$ or $[0, 15]$.

65. Since $-f(x)$ is $f(x)$ reflected across the x-axis, the domain of $f(x - 2)$ is the same: $[-1, 2]$; and the range is: $[-(0), -(3)]$ or $[-3, 0]$.

66. Since $f(x - 3) + 1$ is $f(x)$ shifted 3 units to the right, the domain of $f(x - 3) + 1$ is: $[-1 + 3, 2 + 3]$ or $[2, 5]$; and shift 1 unit upward, the range is: $[0 + 1, 3 + 1]$ or $[1, 4]$.

67. Since $f(2x)$ is $f(x)$ shrunk horizontally by a factor of $\dfrac{1}{2}$, the domain of $f(2x)$ is: $\left[\dfrac{1}{2}(-1), \dfrac{1}{2}(2)\right]$ or $\left[-\dfrac{1}{2}, 1\right]$; and the range is the same: $[0, 3]$.

68. Since $2f(x - 1)$ is $f(x)$ shifted 1 unit to the right, the domain of $2f(x - 1)$ is: $[-1 + 1, 2 + 1]$ or $[0, 3]$; and stretched vertically by a factor of 2, the range is: $[2(0), 2(3)]$ or $[0, 6]$.

69. Since $3f\left(\dfrac{1}{4}x\right)$ is $f(x)$ stretched horizontally by a factor of 4, the domain of $3f\left(\dfrac{1}{4}x\right)$ is:

    $[4(-1), 4(2)]$ or $[-4, 8]$; and stretched vertically by a factor of 3, the range is: $[3(0), 3(3)]$ or $[0, 9]$.

70. Since $-2f(4x)$ is $f(x)$ shrunk horizontally by a factor of $\dfrac{1}{4}$, the domain of $-2f(4x)$ is:

    $\left[\dfrac{1}{4}(-1), \dfrac{1}{4}(2)\right]$ or $\left[-\dfrac{1}{4}, \dfrac{1}{2}\right]$; and reflected across the x-axis while being stretched vertically by a factor of 2, the range is: $[-2(0), -2(3)] = [0, -6]$ or $[-6, 0]$.

71. Since $f(-x)$ is $f(x)$ reflected across the y-axis, the domain of $f(-x)$ is: $[-(-1), -(2)] = [1, -2]$ or $[-2, 1]$; and the range is the same: $[0, 3]$.

72. Since $-2f(-x)$ is $f(x)$ reflected across the y-axis, the domain of $-2f(-x)$ is:

    $[-(-1), -(2)] = [1, -2]$ or $[-2, 1]$; and reflected across the x-axis while being stretched vertically by a factor of 2, the range is: $[-2(0), -2(3)] = [0, -6]$ or $[-6, 0]$.

73. Since $f(-3x)$ is $f(x)$ reflected across the y-axis and shrunk horizontally by a factor of $\dfrac{1}{3}$, the domain of $f(-3x)$

    is: $\left[-\dfrac{1}{3}(-1), -\dfrac{1}{3}(2)\right] = \left[\dfrac{1}{3}, -\dfrac{2}{3}\right]$ or $\left[-\dfrac{2}{3}, \dfrac{1}{3}\right]$; and the range is the same: $[0, 3]$.

74. Since $\frac{1}{3}f(x-3)$ is $f(x)$ shifted 3 units to the right, the domain of $\frac{1}{3}f(x-3)$ is: $[-1+3, 2+3]$ or $[2, 5]$; and shrunk vertically by a factor of $\frac{1}{3}$, the range is: $\left[\frac{1}{3}(0), \frac{1}{3}(3)\right]$ or $[0, 1]$.

75. Since $y = \sqrt{x}$ has an endpoint $(0, 0)$, and the graph of $y = 10\sqrt{x-20}+5$ is the graph of $y = \sqrt{x}$ shifted 20 units right, stretched vertically by a factor of 10, and shifted 5 units upward, the endpoint of $y = 10\sqrt{x-20}+5$ is: $(0+20, 10(0)+5)$ or $(20, 5)$. Therefore, the domain is: $[20, \infty)$; and the range is: $[5, \infty)$.

76. Since $y = \sqrt{x}$ has an endpoint $(0, 0)$, and the graph of $y = -2\sqrt{x+15}-18$ is the graph of $y = \sqrt{x}$ shifted 15 units left, reflected across the x-axis, stretched vertically by a factor of 2, and shifted 18 units downward, the endpoint of $y = -2\sqrt{x+15}-18$ is: $(0-15, -2(0)-18)$ or $(-15, -18)$. Therefore, the domain is: $[-15, \infty)$; and the range, because of the reflection across the x-axis, is: $(-\infty, -18]$.

77. Since $y = \sqrt{x}$ has an endpoint $(0, 0)$, and the graph of $y = -.5\sqrt{x+10}+5$ is the graph of $y = \sqrt{x}$ shifted 10 units left, reflected across the x-axis, shrunk vertically by a factor of .5, and shifted 5 units upward, the endpoint of $y = -.5\sqrt{x+10}+5$ is: $(0-10, -.5(0)+5)$ or $(-10, 5)$. Therefore, the domain is: $[-10, \infty)$; and the range, because of the reflection across the x-axis, is: $(-\infty, 5]$.

78. Using ex. 75, the domain is: $[h, \infty)$; and the range is: $[k, \infty)$.

79. The graph of $y = -f(x)$ is $y = f(x)$ reflected across the x-axis, therefore $y = -f(x)$ is decreasing for the interval: $[a, b]$.

80. The graph of $y = f(-x)$ is $y = f(x)$ reflected across the y-axis, therefore $y = f(-x)$ is decreasing for the interval: $[-b, -a]$.

81. The graph of $y = -f(-x)$ is $y = f(x)$ reflected across both the x-axis and y-axis, therefore $y = -f(-x)$ is increasing for the interval: $[-b, -a]$.

82. The graph of $y = -c \cdot f(x)$ is $y = f(x)$ reflected across the x-axis, therefore $y = -c \cdot f(x)$ is decreasing for the interval: $[a, b]$.

83. From the graph,   (a) the function is increasing for the interval: $[-1, 2]$.

    (b) the function is decreasing for the interval: $(-\infty, -1]$.

    (c) the function is constant for the interval: $[2, \infty)$.

84. From the graph,   (a) the function is increasing for the interval: $(-\infty, -1]$.

    (b) the function is decreasing for the interval: $[-1, 2]$.

    (c) the function is constant for the interval: $[2, \infty)$.

85. From the graph,   (a) the function is increasing for the interval: $[1, \infty)$.

    (b) the function is decreasing for the interval: $[-2, 1]$.

    (c) the function is constant for the interval: $(-\infty, -2]$.

86. From the graph,   (a) the function is increasing for the interval: $(-\infty, -3]$.

    (b) the function is decreasing for the interval: $[-3, \infty)$.

    (c) the function is constant for no interval: none.

87. From the graph, the point on y_2 is approximately: $(8, 10)$.

88. From the graph, the point on y_2 is approximately: $(-27, -15)$.

89. Use two points on the graph to find the slope, two points are: $(-2, -1)$ and $(-1, 1)$, therefore the slope is:

$$m = \frac{1 - (-1)}{-1 - (-2)} = \frac{2}{1} \Rightarrow m = 2.$$ The stretch factor is 2 and the graph has been shifted 2 units to the left and 1

unit down, therefore the equation is: $y = 2|x + 2| - 1$.

90. Use two points on the graph to find the slope, two points are: $(1, 2)$ and $(5, 0)$, therefore the slope is:

$$m = \frac{0 - 2}{5 - 1} = \frac{-2}{4} \Rightarrow m = -\frac{1}{2}.$$ The shrinking factor is $\frac{1}{2}$, the graph has been reflected across the x-axis,

shifted 1 unit to the right, and shifted 2 units upward, therefore the equation is: $y = -\frac{1}{2}|x - 1| + 2$.

91. Use two points on the graph to find the slope, two points are: $(0, 2)$ and $(1, -1)$, therefore the slope is:

$$m = \frac{-1 - 2}{1 - 0} = \frac{-3}{1} \Rightarrow m = -3.$$ The stretch factor is 3, the graph has been reflected across the x-axis, and

shifted 2 units upward, therefore the equation is: $y = -3|x| + 2$.

92. Use two points on the graph to find the slope, two points are: $(-1, -2)$ and $(0, 1)$, therefore the slope is:

$$m = \frac{1 - (-2)}{0 - (-1)} = \frac{3}{1} \Rightarrow m = 3.$$ The stretch factor is 3 and the graph has been shifted 1 unit to the left and 2

units down, therefore the equation is: $y = 3|x + 1| - 2$.

Reviewing Basic Concepts (Sections 2.1—2.3)

1. (a) If $y = f(x)$ is symmetric with respect to the origin, then another function value is: $f(-3) = -6$.

   (b) If $y = f(x)$ is symmetric with respect to the y-axis, then another function value is: $f(-3) = 6$.

   (c) If $f(-x) = -f(x)$, $y = f(x)$ is symmetric with respect to both the x-axis and y-axis, then another function

   value is: $f(-3) = -6$.

   (d) If $f(-x) = f(x)$, $y = f(x)$ is symmetric with respect to the y-axis, then another function value is: $f(-3) = 6$.

2. (a) The equation $y = (x - 7)^2$ is $y = x^2$ shifted 7 units to the right: B.

   (b) The equation $y = x^2 - 7$ is $y = x^2$ shifted 7 units downward: D.

   (c) The equation $y = 7x^2$ is $y = x^2$ stretched vertically by a factor of 7: E.

   (d) The equation $y = (x + 7)^2$ is $y = x^2$ shifted 7 units to the left: A.

   (e) The equation $y = \left(\frac{1}{3}x\right)^2$ is $y = x^2$ stretched horizontally by a factor of 3: C.

3. (a) The equation $y = x^2 + 2$ is $y = x^2$ shifted 2 units upward: B.

   (b) The equation $y = x^2 - 2$ is $y = x^2$ shifted 2 units downward: A.

   (c) The equation $y = (x + 2)^2$ is $y = x^2$ shifted 2 units to the left: G.

   (d) The equation $y = (x + 2)^2$ is $y = x^2$ shifted 2 units to the right: C.

   (e) The equation $y = 2x^2$ is $y = x^2$ stretched vertically by a factor of 2: F.

   (f) The equation $y = -x^2$ is $y = x^2$ reflected across the x-axis: D.

   (g) The equation $y = (x - 2)^2 + 1$ is $y = x^2$ shifted 2 units to the right and 1 unit upward: H.

   (h) The equation $y = (x + 2)^2 + 1$ is $y = x^2$ shifted 2 units to the left and 1 unit upward: E.

4. (a) The equation $y = |x| + 4$ is $y = |x|$ shifted 4 units upward. See Figure 4a.

    (b) The equation $y = |x + 4|$ is $y = |x|$ shifted 4 units to the left. See Figure 4b.

    (c) The equation $y = |x - 4|$ is $y = |x|$ shifted 4 units to the right. See Figure 4c.

    (d) The equation $y = |x + 2| - 4$ is $y = |x|$ shifted 2 units to the left and 4 units downward. See Figure 4d.

    (e) The equation $y = -|x - 2| + 4$ is $y = |x|$ reflected across the x-axis, shifted 2 units to the right, and 4 units upward. See Figure 4e.

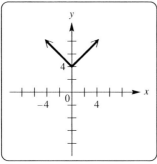

Figure 4a

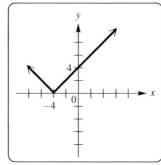

Figure 4b

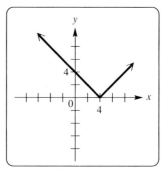

Figure 4c

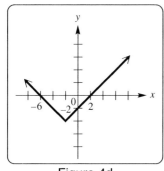

Figure 4d

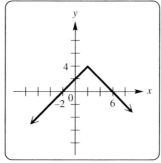

Figure 4e

5. (a) The graph is the function $f(x) = |x|$ reflected across the x-axis, shifted 1 unit left and 3 units upward. Therefore the equation is: $y = -|x + 1| + 3$.

    (b) The graph is the function $g(x) = \sqrt{x}$ reflected across the x-axis, shifted 4 units left and 2 units upward. Therefore the equation is: $y = -\sqrt{x + 4} + 2$.

    (c) The graph is the function $g(x) = \sqrt{x}$ stretched vertically by a factor of 2, shifted 4 units left and 4 units downward. Therefore the equation is: $y = 2\sqrt{x + 4} - 4$.

    (d) The graph is the function $f(x) = |x|$ shrunk vertically by a factor of $\dfrac{1}{2}$, shifted 2 units right and 1 unit downward. Therefore the equation is: $y = \dfrac{1}{2}|x - 2| - 1$.

6. (a) The graph of $g(x)$ is the graph $f(x)$ shifted 2 units upward. Therefore $c = 2$.

    (b) The graph of $g(x)$ is the graph $f(x)$ shifted 4 units to the left. Therefore $c = 4$.

7. The graph of $y = F(x + h)$ is a horizontal translation of he graph of $y = F(x)$. The graph of $y = F(x) + h$ is not the same as the graph of $y = F(x + h)$ because the graph of $y = F(x) + h$ is a vertical translation of the graph of $y = F(x)$ and $y = F(x + h)$ is a horizontal translation of the graph $y = F(x)$.

8. The effect is either a stretch or a shrink, and perhaps a reflection across the x-axis. If $c > 0$, there is a stretch or shrink by a factor of c. If $c < 0$, there is a stretch or shrink by a factor of $|c|$, and a reflection across the x-axis. If $|c| > 1$, a stretch occurs; when $|c| < 1$, a shrink occurs.

9. (a) If f is even, then $f(x) = f(-x)$. See Figure 9a.
    (b) If f is odd, then $f(-x) = -f(x)$. See Figure 9b.

10. (a) Since $x = 1$ corresponds to 1992, the equation using actual year is: $g(x) = -.279(x - 1992) + 5.532$.
    (b) $g(2006) = -.279(2006 - 1992) + 5.532 \Rightarrow g(2006) = -.279(14) + 5.532 \Rightarrow$
    $g(2006) = 1.626$ ppm.

x	$f(x)$
-3	4
-2	-6
-1	5
1	5
2	-6
3	4

Figure 9a

x	$f(x)$
-3	4
-2	-6
-1	5
1	-5
2	6
3	-4

Figure 9b

2.4: Absolute Value Functions: Graphs, Equations, Inequalities, and Applications

1. If $f(a) = -5$, then $|f(a)| = |-5| = 5$.

2. Since $f(x) = x^2$ is an even function, $f(x) = x^2$ and $f(x) = |x^2|$ are the same graph.

3. If $f(x) = -x^2$, then $y = |f(x)| \Rightarrow y = |-x^2| \Rightarrow y = x^2$. Therefore the range of $y = |f(x)|$ is: $[0, \infty)$.

4. If the range of $y = f(x)$ is $[-2, \infty)$, the range of $y = |f(x)|$ is $[0, \infty)$ since all negative values of y are reflected across the x-axis.

5. If the range of $y = f(x)$ is $(-\infty, -2]$, the range of $y = |f(x)|$ is $[2, \infty)$ since all negative values of y are reflected across the x-axis.

6. $|f(x)|$ is greater than or equal to 0 for any value of x. Since -1 is less than 0, -1 cannot be in the range of f.

7. We reflect the graph of $y = f(x)$ across the x-axis for all points for which $y < 0$. Where $y \geq 0$, the graph remains unchanged. See Figure 7.

8. We reflect the graph of $y = f(x)$ across the x-axis for all points for which $y < 0$. Where $y \geq 0$, the graph remains unchanged. See Figure 8.

9. We reflect the graph of $y = f(x)$ across the x-axis for all points for which $y < 0$. Where $y \geq 0$, the graph remains unchanged. See Figure 9.

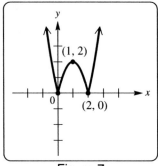

Figure 7

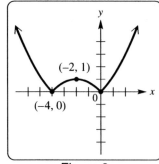

Figure 8

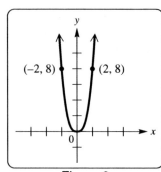

Figure 9

10. We reflect the graph of $y = f(x)$ across the x-axis for all points for which $y < 0$. Where $y \geq 0$, the graph remains unchanged. See Figure 10.

11. Since for all y, $y \geq 0$, the graph remains unchanged. That is, $y = |f(x)|$ has the same graph as $y = f(x)$.

12. We reflect the graph of $y = f(x)$ across the x-axis for all points for which $y < 0$. Where $y \geq 0$, the graph remains unchanged. See Figure 12.

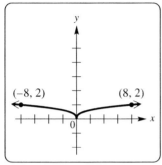
Figure 10

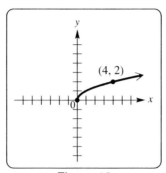
Figure 12

13. We reflect the graph of $y = f(x)$ across the x-axis for all points for which $y < 0$. Where $y \geq 0$, the graph remains unchanged. See Figure 13.

14. We reflect the graph of $y = f(x)$ across the x-axis for all points for which $y < 0$. Where $y \geq 0$, the graph remains unchanged. See Figure 14.

15. We reflect the graph of $y = f(x)$ across the x-axis for all points for which $y < 0$. Where $y \geq 0$, the graph remains unchanged. See Figure 15.

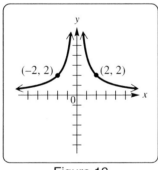
Figure 13

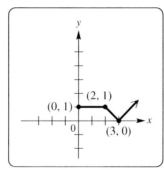
Figure 14

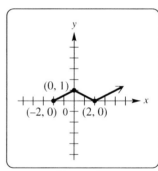
Figure 15

16. We reflect the graph of $y = f(x)$ across the x-axis for all points for which $y < 0$. Where $y \geq 0$, the graph remains unchanged.

17. From the graph of $y = (x + 1)^2 - 2$ the domain of $f(x)$ is: $(-\infty, \infty)$; and the range is: $[-2, \infty)$.
    From the graph of $y = |(x + 1)^2 - 2|$ the domain of $|f(x)|$ is: $(-\infty, \infty)$; and the range is: $[0, \infty)$.

18. From the graph of $y = 2 - \dfrac{1}{2}x$ the domain of $f(x)$ is: $(-\infty, \infty)$; and the range is: $[-\infty, \infty)$.
    From the graph of $y = \left|2 - \dfrac{1}{2}x\right|$ the domain of $|f(x)|$ is: $(-\infty, \infty)$; and the range is: $[0, \infty)$.

19. From the graph of $y = -1 - (x - 2)^2$ the domain of $f(x)$ is: $(-\infty, \infty)$; and the range is: $(-\infty, -1]$.
    From the graph of $y = |-1 - (x - 2)^2|$ the domain of $|f(x)|$ is: $(-\infty, \infty)$; and the range is: $[1, \infty)$.

20. From the graph of $y = -|x + 2| - 2$ the domain of $f(x)$ is: $(-\infty, \infty)$; and the range is: $(-\infty, -2]$.

    From the graph of $y = |(-|x + 2| - 2)|$ the domain of $|f(x)|$ is: $(-\infty, \infty)$; and the range is: $[2, \infty)$.

21. From the graph, the domain of $f(x)$ is: $[-2, 3]$; and the range is: $[-2, 3]$. For the function $y = |f(x)|$, we reflect the graph of $y = f(x)$ across the x-axis for all points for which $y < 0$ and where $y \geq 0$, the graph remains unchanged. Therefore, the domain of $y = |f(x)|$ is: $[-2, 3]$; and the range is: $[0, 3]$.

22. From the graph, the domain of $f(x)$ is: $[-3, 2]$; and the range is: $[-2, 2]$. For the function $y = |f(x)|$, we reflect the graph of $y = f(x)$ across the x-axis for all points for which $y < 0$ and where $y \geq 0$, the graph remains unchanged. Therefore, the domain of $y = |f(x)|$ is: $[-3, 2]$; and the range is: $[0, 2]$.

23. From the graph, the domain of $f(x)$ is: $[-2, 3]$; and the range is: $[-3, 1]$. For the function $y = |f(x)|$, we reflect the graph of $y = f(x)$ across the x-axis for all points for which $y < 0$ and where $y \geq 0$, the graph remains unchanged. Therefore, the domain of $y = |f(x)|$ is: $[-2, 3]$; and the range is: $[0, 3]$.

24. From the graph, the domain of $f(x)$ is: $[-3, 3]$; and the range is: $[-3, -1]$. For the function $y = |f(x)|$, we reflect the graph of $y = f(x)$ across the x-axis for all points for which $y < 0$ and where $y \geq 0$, the graph remains unchanged. Therefore, the domain of $y = |f(x)|$ is: $[-3, 3]$; and the range is: $[1, 3]$.

25. (a) The function $y = f(-x)$ is the function $y = f(x)$ reflected across the y-axis. See Figure 25a.

    (b) The function $y = -f(-x)$ is the function $y = f(x)$ reflected across both the x-axis and y-axis. See Figure 25b.

    (c) For the function $y = |-f(-x)|$ we reflect the graph of $y = -f(-x)$ (ex. b) across the x-axis for all points for which $y < 0$ and where $y \geq 0$, the graph remains unchanged. See Figure 25c.

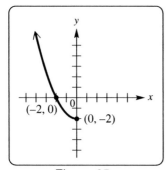

Figure 25a

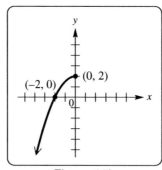

Figure 25b

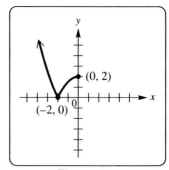

Figure 25c

26. (a) The function $y = f(-x)$ is the function $y = f(x)$ reflected across the y-axis. See Figure 26a.

    (b) The function $y = -f(-x)$ is the function $y = f(x)$ reflected across both the x-axis and y-axis. See Figure 26b.

    (c) For the function $y = |-f(-x)|$ we reflect the graph of $y = -f(-x)$ (ex. b) across the x-axis for all points for which $y < 0$ and where $y \geq 0$, the graph remains unchanged. See Figure 26c.

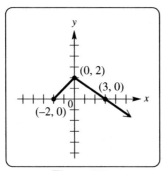

Figure 26a

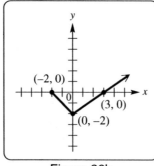

Figure 26b

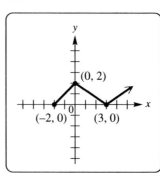

Figure 26c

27. The graph of $y = |f(x)|$ can not be below the x-axis, therefore Figure A shows the graph of $y = f(x)$, while Figure B shows the graph of $y = |f(x)|$.

28. The graph of $y = |f(x)|$ can not be below the x-axis, therefore Figure B shows the graph of $y = f(x)$, while Figure A shows the graph of $y = |f(x)|$.

29. (a) From the graph, $y_1 = y_2$ at the coordinates $(-1, 5)$ and $(6, 5)$, therefore the solution set is: $\{-1, 6\}$.

    (b) From the graph, $y_1 < y_2$ for the interval: $(-1, 6)$.

    (c) From the graph, $y_1 > y_2$ for the intervals: $(-\infty, -1) \cup (6, \infty)$.

30. (a) From the graph, $y_1 = y_2$ at the coordinates $(0, -2)$ and $(8, -2)$, therefore the solution set is: $\{0, 8\}$.

    (b) From the graph, $y_1 < y_2$ for the intervals: $(-\infty, 0) \cup (8, \infty)$.

    (c) From the graph, $y_1 > y_2$ for the intervals: $(0, 8)$.

31. (a) From the graph, $y_1 = y_2$ at the coordinate $(4, 1)$, therefore the solution set is: $\{4\}$.

    (b) From the graph, $y_1 < y_2$ never, therefore the solution set is: $\varnothing$.

    (c) From the graph, $y_1 > y_2$ for all values for x, except 4, therefore for the intervals: $(-\infty, 4) \cup (4, \infty)$.

32. (a) From the graph, $y_1 = y_2$ never, therefore the solution set is: $\varnothing$.

    (b) From the graph, $y_1 < y_2$ for all values for x, therefore for the interval: $(-\infty, \infty)$.

    (c) From the graph, $y_1 > y_2$ never, therefore the solution set is: $\varnothing$.

33. The V-shaped graph is that of $f(x) = |.5x + 6|$, since this is typical of the graphs of absolute value functions of the form $f(x) = |ax + b|$.

34. The straight line graph is that of $g(x) = 3x - 14$, which is a linear function.

35. The graph intersects at $(8, 10)$, so the solution set is: $\{8\}$.

36. From the graph, $f(x) > g(x)$ for the intervals: $(-\infty, 8)$.

37. From the graph, $f(x) < g(x)$ for the intervals: $(8, \infty)$.

38. If $|.5x + 6| - (3x - 14) = 0$ then $|.5x + 6| = 3x - 14$. Therefore the solution is the intersection of the graphs or $\{8\}$.

39. (a) $|x + 4| = 9 \Rightarrow x + 4 = 9$ or $x + 4 = -9 \Rightarrow x = 5$ or $x = -13$. The solution set is: $\{-13, 5\}$; which is supported by the graphs of $y_1 = |x + 4|$ and $y_2 = 9$.

    (b) $|x + 4| > 9 \Rightarrow x + 4 > 9$ or $x + 4 < -9 \Rightarrow x > 5$ or $x < -13$. The solution is: $(-\infty, -13) \cup (5, \infty)$; which is supported by the graphs of $y_1 = |x + 4|$ and $y_2 = 9$.

    (c) $|x + 4| < 9 \Rightarrow -9 < x + 4 < 9 \Rightarrow -13 < x < 5$. The solution is: $(-13, 5)$; which is supported by the graphs of $y_1 = |x + 4|$ and $y_2 = 9$.

40. (a) $|x - 3| = 5 \Rightarrow x - 3 = 5$ or $x - 3 = -5 \Rightarrow x = 8$ or $x = -2$. The solution set is: $\{-2, 8\}$; which is supported by the graphs of $y_1 = |x - 3|$ and $y_2 = 5$.

    (b) $|x - 3| > 5 \Rightarrow x - 3 > 5$ or $x - 3 < -5 \Rightarrow x > 8$ or $x < -2$. The solution is: $(-\infty, -2) \cup (8, \infty)$; which is supported by the graphs of $y_1 = |x - 3|$ and $y_2 = 5$.

    (c) $|x - 3| < 5 \Rightarrow -5 < x - 3 < 5 \Rightarrow -2 < x < 8$. The solution is: $(-2, 8)$; which is supported by the graphs of $y_1 = |x - 3|$ and $y_2 = 5$.

41. (a) $|-2x + 7| = 3 \Rightarrow -2x + 7 = 3$ or $-2x + 7 = -3 \Rightarrow -2x = -4$ or $-2x = -10 \Rightarrow x = 2$ or $x = 5$.

    The solution set is: $\{2, 5\}$; which is supported by the graphs of $y_1 = |-2x + 7|$ and $y_2 = 3$.

    (b) $|-2x + 7| \geq 3 \Rightarrow -2x + 7 \geq 3$ or $-2x + 7 \leq -3 \Rightarrow -2x \geq -4$ or $-2x \leq -10 \Rightarrow x \leq 2$ or $x \geq 5$.

    The solution is: $(-\infty, 2] \cup [5, \infty)$; which is supported by the graphs of $y_1 = |-2x + 7|$ and $y_2 = 3$.

    (c) $|-2x + 7| \leq 3 \Rightarrow -3 \leq -2x + 7 \leq 3 \Rightarrow -10 \leq -2x \leq -4 \Rightarrow 5 \geq x \geq 2$ or $2 \leq x \leq 5$.

    The solution is: $[2, 5]$; which is supported by the graphs of $y_1 = |-2x + 7|$ and $y_2 = 3$.

42. (a) $|-3x - 9| = 6 \Rightarrow -3x - 9 = 6$ or $-3x - 9 = -6 \Rightarrow -3x = 15$ or $-3x = 3 \Rightarrow x = -5$ or $x = -1$.

    The solution set is: $\{-5, -1\}$; which is supported by the graphs of $y_1 = |-3x - 9|$ and $y_2 = 6$.

    (b) $|-3x - 9| \geq 6 \Rightarrow -3x - 9 \geq 6$ or $-3x - 9 \leq -6 \Rightarrow -3x \geq 15$ or $-3x \leq 3 \Rightarrow x \leq -5$ or $x \geq -1$.

    The solution is: $(-\infty, -5] \cup [-1, \infty)$; which is supported by the graphs of $y_1 = |-3x - 9|$ and $y_2 = 6$.

    (c) $|-3x - 9| \leq 6 \Rightarrow -6 \leq -3x - 9 \leq 6 \Rightarrow 3 \leq -3x \leq 15 \Rightarrow -1 \geq x \geq -5$ or $-5 \leq x \leq -1$.

    The solution is: $[-5, -1]$; which is supported by the graphs of $y_1 = |-3x - 9|$ and $y_2 = 6$.

43. (a) $|2x + 1| + 3 = 5 \Rightarrow 2x + 1 = 2$ or $2x + 1 = -2 \Rightarrow 2x = 1$ or $2x = -3 \Rightarrow x = \dfrac{1}{2}$ or $x = -\dfrac{3}{2}$.

    The solution set is: $\left\{-\dfrac{3}{2}, \dfrac{1}{2}\right\}$; which is supported by the graphs of $y_1 = |2x + 1| + 3$ and $y_2 = 5$.

    (b) $|2x + 1| + 3 \leq 5 \Rightarrow -2 \leq 2x + 1 \leq 2 \Rightarrow -3 \leq 2x \leq 1 \Rightarrow -\dfrac{3}{2} \leq x \leq \dfrac{1}{2}$. The solution is:

    $\left[-\dfrac{3}{2}, \dfrac{1}{2}\right]$; which is supported by the graphs of $y_1 = |2x + 1| + 3$ and $y_2 = 5$.

    (c) $|2x + 1| + 3 \geq 5 \Rightarrow 2x + 1 \geq 2$ or $2x + 1 \leq -2 \Rightarrow 2x \geq 1$ or $2x \leq -3 \Rightarrow x \geq \dfrac{1}{2}$ or $x \leq -\dfrac{3}{2}$.

    The solution is: $\left(-\infty, -\dfrac{3}{2}\right] \cup \left[\dfrac{1}{2}, \infty\right)$; which is supported by the graphs of $y_1 = |2x + 1| + 3$ and $y_2 = 5$.

44. (a) $|4x + 7| = 0 \Rightarrow 4x + 7 = 0 \Rightarrow 4x = -7 \Rightarrow x = -\dfrac{7}{4}$. The solution set is: $\left\{-\dfrac{7}{4}\right\}$; which is supported

    by the graphs of $y_1 = |4x + 7|$ and $y_2 = 0$.

    (b) $|4x + 7| > 0 \Rightarrow 4x + 7 > 0$ or $4x + 7 < 0 \Rightarrow 4x > -7$ or $4x < -7 \Rightarrow x > -\dfrac{7}{4}$ or $x < -\dfrac{7}{4}$.

    The solution is: $\left(-\infty, -\dfrac{7}{4}\right) \cup \left(-\dfrac{7}{4}, \infty\right)$; which is supported by the graphs of $y_1 = |4x + 7|$ and $y_2 = 0$.

    (c) Absolute value is always positive, therfore the solution set is: $\varnothing$; which is supported by the graphs of

    $y_1 = |4x + 7|$ and $y_2 = 0$.

45. (a) $|7x - 5| = 0 \Rightarrow 7x - 5 = 0 \Rightarrow 7x = 5 \Rightarrow x = \dfrac{5}{7}$. The solution set is: $\left\{\dfrac{5}{7}\right\}$; which is supported

    by the graphs of $y_1 = |7x - 5|$ and $y_2 = 0$.

    (b) $|7x - 5| \geq 0 \Rightarrow 7x - 5 \geq 0$ or $7x - 5 \leq 0 \Rightarrow 7x \geq 5$ or $7x \leq 5 \Rightarrow x \geq \dfrac{5}{7}$ or $x \leq \dfrac{5}{7}$. The

    solution is: $(-\infty, \infty)$; which is supported by the graphs of $y_1 = |7x - 5|$ and $y_2 = 0$.

    (c) $|7x - 5| \leq 0 \Rightarrow 0 \leq 7x - 5 \leq 0 \Rightarrow 5 \leq 7x \leq 5 \Rightarrow \dfrac{5}{7} \leq x \leq \dfrac{5}{7}$. The solution set is: $\left\{\dfrac{5}{7}\right\}$; which is

    supported by the graphs of $y_1 = |7x - 5|$ and $y_2 = 0$.

46. (a) Absolute value is always positive, therefore the solution set is: $\varnothing$; which is supported by the graphs of

   $y_1 = |\pi x + 8|$ and $y_2 = -4$.

   (b) Absolute value is always positive and cannot be less than -4, therefore the solution set is: $\varnothing$; which is

   supported by the graphs of $y_1 = |\pi x + 8|$ and $y_2 = -4$.

   (c) Absolute value is always positive and is always greater than -4, therefore the solution is: $(-\infty, \infty)$; which

   is supported by the graphs of $y_1 = |\pi x + 8|$ and $y_2 = -4$.

47. (a) Absolute value is always positive, therefore the solution set is: $\varnothing$; which is supported by the graphs of

   $y_1 = |\sqrt{2}x - 3.6|$ and $y_2 = -1$.

   (b) Absolute value is always positive and cannot be less than or equal to -1, therefore the solution set is: $\varnothing$;

   which is supported by the graphs of $y_1 = |\sqrt{2}x - 3.6|$ and $y_2 = -1$.

   (c) Absolute value is always positive and is always greater than -1, therefore the solution is: $(-\infty, \infty)$; which

   is supported by the graphs of $y_1 = |\sqrt{2}x - 3.6|$ and $y_2 = -1$.

48. $|2x + 4| + 2 = 10 \Rightarrow |2x + 4| = 8 \Rightarrow 2x + 4 = 8$ or $2x + 4 = -8 \Rightarrow 2x = 4$ or $2x = -12 \Rightarrow$

   $x = 2$ or $x = -6$. Therefore, The solution set is: $\{-6, 2\}$.

49. $3|4 - 3x| - 4 = 8 \Rightarrow 3|4 - 3x| = 12 \Rightarrow |4 - 3x| = 4 \Rightarrow 4 - 3x = 4$ or $4 - 3x = -4 \Rightarrow$

   $-3x = 0$ or $-3x = -8 \Rightarrow x = 0$ or $x = \dfrac{8}{3}$. Therefore, the solution set is: $\left\{0, \dfrac{8}{3}\right\}$.

50. $5|x + 3| - 2 = 18 \Rightarrow 5|x + 3| = 20 \Rightarrow |x + 3| = 4 \Rightarrow x + 3 = 4$ or $x + 3 = -4 \Rightarrow$

   $x = 1$ or $x = -7$. Therefore, the solution set is: $\{-7, 1\}$.

51. $\dfrac{1}{2}\left|-2x + \dfrac{1}{2}\right| = \dfrac{3}{4} \Rightarrow \left|-2x + \dfrac{1}{2}\right| = \dfrac{3}{2} \Rightarrow -2x + \dfrac{1}{2} = \dfrac{3}{2}$ or $-2x + \dfrac{1}{2} = -\dfrac{3}{2} \Rightarrow -2x = 1$ or $-2x = -2 \Rightarrow$

   $x = -\dfrac{1}{2}$ or $x = 1$. Therefore, the solution set is: $\left\{-\dfrac{1}{2}, 1\right\}$.

52. $|3(x - 5) + 2| + 3 = 9 \Rightarrow |3(x - 5) + 2| = 6 \Rightarrow 3(x - 5) + 2 = 6$ or $3(x - 5) + 2 = -6 \Rightarrow$

   $3(x - 5) = 4$ or $3(x - 5) = -8 \Rightarrow x - 5 = \dfrac{4}{3}$ or $x - 5 = -\dfrac{8}{3} \Rightarrow x = \dfrac{19}{3}$ or $x = \dfrac{7}{3}$. Therefore, the

   solution set is: $\left\{\dfrac{7}{3}, \dfrac{19}{3}\right\}$.

53. $4.2|.5 - x| + 1 = 3.1 \Rightarrow 4.2|.5 - x| = 2.1 \Rightarrow |.5 - x| = .5 \Rightarrow .5 - x = .5$ or $.5 - x = -.5 \Rightarrow$

   $-x = 0$ or $-x = -1 \Rightarrow x = 0$ or $x = 1$. Therefore, the solution set is: $\{0, 1\}$.

54. $|3x - 1| < 8 \Rightarrow -8 < 3x - 1 < 8 \Rightarrow -7 < 3x < 9 \Rightarrow -\dfrac{7}{3} < x < 3$. Therefore, the solution is: $\left(-\dfrac{7}{3}, 3\right)$.

55. $|15 - x| < 7 \Rightarrow -7 < 15 - x < 7 \Rightarrow -22 < -x < -8 \Rightarrow 22 > x > 8$ or $8 < x < 22$. Therefore,

   the solution is: $(8, 22)$.

56. $|7 - 4x| \le 11 \Rightarrow -11 \le 7 - 4x \le 11 \Rightarrow -18 \le -4x \le 4 \Rightarrow \dfrac{9}{2} \ge x \ge -1$ or $-1 \le x \le \dfrac{9}{2}$. Therefore,

   the solution is: $\left[-1, \dfrac{9}{2}\right]$.

57. $|2x - 3| > 1 \Rightarrow 2x - 3 > 1$ or $2x - 3 < -1 \Rightarrow 2x > 4$ or $2x < 2 \Rightarrow x > 2$ or $x < 1$. Therefore,

   the solution is: $(-\infty, 1) \cup (2, \infty)$.

58. Absolute value is always positive and is always greater than -2, therefore the solution is: $(-\infty, \infty)$.

59. $|-3x + 8| \geq 3 \Rightarrow -3x + 8 \geq 3$ or $-3x + 8 \leq -3 \Rightarrow -3x \geq -5$ or $-3x \leq -11 \Rightarrow x \leq \dfrac{5}{3}$ or $x \geq \dfrac{11}{3}$.

    Therefore, the solution is: $\left(-\infty, \dfrac{5}{3}\right] \cup \left[\dfrac{11}{3}, \infty\right)$.

60. Absolute value is always positive and is always greater than -1, therefore the solution is: $(-\infty, \infty)$.

61. $\left|6 - \dfrac{1}{3}x\right| > 0 \Rightarrow 6 - \dfrac{1}{3}x > 0$ or $6 - \dfrac{1}{3}x < 0 \Rightarrow -\dfrac{1}{3}x > -6$ or $-\dfrac{1}{3}x < -6 \Rightarrow x < 18$ or $x > 18$.

    Therefore the solution is every real number except 18, the solution is: $(-\infty, 18) \cup (18, \infty)$.

62. Absolute value is always positive and cannot be less than 0, therefore the solution set is: $\varnothing$;

63. Absolute value is always positive and cannot be less than or equal to -6, therefore the solution set is: $\varnothing$;

64. Absolute value is always positive and cannot be less than -4, therefore the solution set is: $\varnothing$;

65. Absolute value is always positive and is always greater than -5, therefore the solution is: $(-\infty, \infty)$.

66. To solve such an equation, we must solve the compound equation $ax + b = cx + d$ or $ax + b = -(cx + d)$.

    The solution set consist of the union of the two individual solution sets.

67. (a) $3x + 1 = 2x - 7 \Rightarrow x + 1 = -7 \Rightarrow x = -8$ or $3x + 1 = -(2x - 7) \Rightarrow 3x + 1 = -2x + 7 \Rightarrow$

    $5x + 1 = 7 \Rightarrow 5x = 6 \Rightarrow x = \dfrac{6}{5}$. Therefore the solution set is: $\left\{-8, \dfrac{6}{5}\right\}$.

    (b) Graph $y_1 = |3x + 1|$ and $y_2 = |2x - 7|$. See Figure 67. From the graph, $|f(x)| > |g(x)|$ when

    $y_1 > y_2$ which is for the interval: $(-\infty, -8) \cup \left(\dfrac{6}{5}, \infty\right)$.

    (c) Graph $y_1 = |3x + 1|$ and $y_2 = |2x - 7|$. See Figure 67. From the graph, $|f(x)| < |g(x)|$ when

    $y_1 < y_2$ which is for the interval: $\left(-8, \dfrac{6}{5}\right)$.

68. (a) $x - 4 = 7x + 12 \Rightarrow -6x - 4 = 12 \Rightarrow -6x = 16 \Rightarrow x = -\dfrac{8}{3}$ or $x - 4 = -(7x + 12) \Rightarrow$

    $x - 4 = -7x - 12 \Rightarrow 8x - 4 = -12 \Rightarrow 8x = -8 \Rightarrow x = -1$. Therefore, the solution set is: $\left\{-\dfrac{8}{3}, -1\right\}$.

    (b) Graph $y_1 = |x - 4|$ and $y_2 = |7x + 12|$. See Figure 68. From the graph, $|f(x)| > |g(x)|$ when

    $y_1 > y_2$ which is for the interval: $\left(-\dfrac{8}{3}, -1\right)$.

    (c) Graph $y_1 = |x - 4|$ and $y_2 = |7x + 12|$. See Figure 68. From the graph, $|f(x)| < |g(x)|$ when

    $y_1 < y_2$ which is for the interval: $\left(-\infty, -\dfrac{8}{3}\right) \cup (-1, \infty)$.

[−20, 20] by [−10, 50]
   Xscl = 2   Yscl = 5

[−10, 10] by [−4, 16]
   Xscl = 1   Yscl = 1

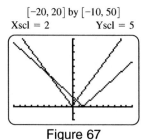
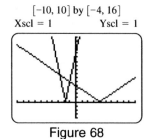

    Figure 67   Figure 68

69. (a) $-2x + 5 = x + 3 \Rightarrow -3x = -2 \Rightarrow x = \dfrac{2}{3}$ or $-2x + 5 = -(x + 3) \Rightarrow -2x + 5 = -x - 3 \Rightarrow$

$-x = -8 \Rightarrow x = 8$. Therefore, the solution set is: $\left\{\dfrac{2}{3}, 8\right\}$.

   (b) Graph $y_1 = |-2x + 5|$ and $y_2 = |x + 3|$. See Figure 69. From the graph, $|f(x)| > |g(x)|$ when

$y_1 > y_2$ which is for the interval: $\left(-\infty, \dfrac{2}{3}\right) \cup (8, \infty)$.

   (c) Graph $y_1 = |-2x + 5|$ and $y_2 = |x + 3|$. See Figure 69. From the graph, $|f(x)| < |g(x)|$ when

$y_1 < y_2$ which is for the interval: $\left(\dfrac{2}{3}, 8\right)$.

70. (a) $-5x + 1 = 3x - 4 \Rightarrow -8x = -5 \Rightarrow x = \dfrac{5}{8}$ or $-5x + 1 = -(3x - 4) \Rightarrow -5x + 1 = -3x + 4 \Rightarrow$

$-2x = 3 \Rightarrow x = -\dfrac{3}{2}$. Therefore, the solution set is: $\left\{-\dfrac{3}{2}, \dfrac{5}{8}\right\}$.

   (b) Graph $y_1 = |-5x + 1|$ and $y_2 = |3x - 4|$. See Figure 70. From the graph, $|f(x)| > |g(x)|$ when

$y_1 > y_2$ which is for the interval: $\left(-\infty, -\dfrac{3}{2}\right) \cup \left(\dfrac{5}{8}, \infty\right)$.

   (c) Graph $y_1 = |-5x + 1|$ and $y_2 = |3x - 4|$. See Figure 70. From the graph, $|f(x)| < |g(x)|$ when

$y_1 < y_2$ which is for the interval: $\left(-\dfrac{3}{2}, \dfrac{5}{8}\right)$.

[-10, 10] by [-4, 16]   Xscl = 1   Yscl = 1

[-3, 3] by [-4, 16]   Xscl = 1   Yscl = 2

[-6, 6] by [-2, 10]   Xscl = 1   Yscl = 1

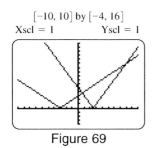

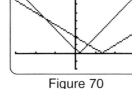

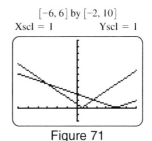

Figure 69   Figure 70   Figure 71

71. (a) $x - \dfrac{1}{2} = \dfrac{1}{2}x - 2 \Rightarrow \dfrac{1}{2}x = -\dfrac{3}{2} \Rightarrow x = -3$ or $x - \dfrac{1}{2} = -\left(\dfrac{1}{2}x - 2\right) \Rightarrow x - \dfrac{1}{2} = -\dfrac{1}{2}x + 2 \Rightarrow$

$\dfrac{3}{2}x = \dfrac{5}{2} \Rightarrow x = \dfrac{5}{3}$. Therefore, the solution set is: $\left\{-3, \dfrac{5}{3}\right\}$.

   (b) Graph $y_1 = \left|x - \dfrac{1}{2}\right|$ and $y_2 = \left|\dfrac{1}{2}x - 2\right|$. See Figure 71. From the graph, $|f(x)| > |g(x)|$ when

$y_1 > y_2$ which is for the interval: $(-\infty, -3) \cup \left(\dfrac{5}{3}, \infty\right)$.

   (c) Graph $y_1 = \left|x - \dfrac{1}{2}\right|$ and $y_2 = \left|\dfrac{1}{2}x - 2\right|$. See Figure 71. From the graph, $|f(x)| < |g(x)|$ when

$y_1 < y_2$ which is for the interval: $\left(-3, \dfrac{5}{3}\right)$.

72. (a) $x + 3 = \dfrac{1}{3}x + 8 \Rightarrow \dfrac{2}{3}x = 5 \Rightarrow x = \dfrac{15}{2}$ or $x + 3 = -\left(\dfrac{1}{3}x + 8\right) \Rightarrow x + 3 = -\dfrac{1}{3}x - 8 \Rightarrow$

$\dfrac{4}{3}x = -11 \Rightarrow x = -\dfrac{33}{4}$. Therefore, the solution set is: $\left\{-\dfrac{33}{4}, \dfrac{15}{2}\right\}$.

(b) Graph $y_1 = |x + 3|$ and $y_2 = \left|\dfrac{1}{3}x + 8\right|$. See Figure 72. From the graph, $|f(x)| > |g(x)|$ when

$y_1 > y_2$ which is for the interval: $\left(-\infty, -\dfrac{33}{4}\right) \cup \left(\dfrac{15}{2}, \infty\right)$.

(c) Graph $y_1 = |x + 3|$ and $y_2 = \left|\dfrac{1}{3}x + 8\right|$. See Figure 72. From the graph, $|f(x)| < |g(x)|$ when

$y_1 < y_2$ which is for the interval: $\left(-\dfrac{33}{4}, \dfrac{15}{2}\right)$.

73. (a) $4x + 1 = 4x + 6 \Rightarrow 1 = 6 \Rightarrow \varnothing$ or $4x + 1 = -(4x + 6) \Rightarrow 4x + 1 = -4x - 6 \Rightarrow$

$8x = -7 \Rightarrow x = -\dfrac{7}{8}$. Therefore, the solution set is: $\left\{-\dfrac{7}{8}\right\}$.

(b) Graph $y_1 = |4x + 1|$ and $y_2 = |4x + 6|$. See Figure 73. From the graph, $|f(x)| > |g(x)|$ when

$y_1 > y_2$ which is for the interval: $\left(-\infty, -\dfrac{7}{8}\right)$.

(c) Graph $y_1 = |4x + 1|$ and $y_2 = |4x + 6|$. See Figure 73. From the graph, $|f(x)| < |g(x)|$ when

$y_1 < y_2$ which is for the interval: $\left(-\dfrac{7}{8}, \infty\right)$.

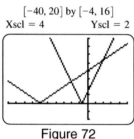

$[-40, 20]$ by $[-4, 16]$
Xscl $= 4$ Yscl $= 2$

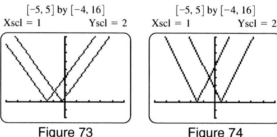

$[-5, 5]$ by $[-4, 16]$
Xscl $= 1$ Yscl $= 2$

$[-5, 5]$ by $[-4, 16]$
Xscl $= 1$ Yscl $= 2$

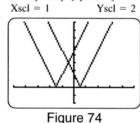

Figure 72   Figure 73   Figure 74

74. (a) $6x + 9 = 6x - 3 \Rightarrow 9 = -3 \Rightarrow \varnothing$ or $6x + 9 = -(6x - 3) \Rightarrow 6x + 9 = -6x + 3 \Rightarrow$

$12x = -6 \Rightarrow x = -\dfrac{6}{12} = -\dfrac{1}{2}$. Therefore, the solution set is: $\left\{-\dfrac{1}{2}\right\}$.

(b) Graph $y_1 = |6x + 9|$ and $y_2 = |6x - 3|$. See Figure 74. From the graph, $|f(x)| > |g(x)|$ when

$y_1 > y_2$ which is for the interval: $\left(-\dfrac{1}{2}, \infty\right)$.

(c) Graph $y_1 = |6x + 9|$ and $y_2 = |6x - 3|$. See Figure 74. From the graph, $|f(x)| < |g(x)|$ when

$y_1 < y_2$ which is for the interval: $\left(-\infty, -\dfrac{1}{2}\right)$.

75. (a) $.25x + 1 = .75x - 3 \Rightarrow -.50x = -4 \Rightarrow x = 8$ or $.25x + 1 = -(.75x - 3) \Rightarrow$

$.25x + 1 = -.75x + 3 \Rightarrow x = 2$. Therefore, the solution set is: $\{2, 8\}$.

(b) Graph $y_1 = |.25x + 1|$ and $y_2 = |.75x - 3|$. See Figure 75. From the graph, $|f(x)| > |g(x)|$ when

$y_1 > y_2$ which is for the interval: $(2, 8)$.

(c) Graph $y_1 = |.25x + 1|$ and $y_2 = |.75x - 3|$. See Figure 75. From the graph, $|f(x)| < |g(x)|$ when

$y_1 < y_2$ which is for the interval: $(-\infty, 2) \cup (8, \infty)$.

76. (a) $.40x + 2 = .60x - 5 \Rightarrow -.20x = -7 \Rightarrow x = 35$ or $.40x + 2 = -(.60x - 5) \Rightarrow$

   $.40x + 2 = -.60x + 5 \Rightarrow x = 3.$ Therefore, the solution set is: $\{3, 35\}.$

   (b) Graph $y_1 = |.40x + 2|$ and $y_2 = |.60x - 5|$. See Figure 76. From the graph, $|f(x)| > |g(x)|$ when

   $y_1 > y_2$ which is for the interval: $(3, 35)$.

   (c) Graph $y_1 = |.40x + 2|$ and $y_2 = |.60x - 5|$. See Figure 76. From the graph, $|f(x)| < |g(x)|$ when

   $y_1 < y_2$ which is for the interval: $(-\infty, 3) \cup (35, \infty)$.

$[-20, 20]$ by $[-4, 16]$
Xscl = 2   Yscl = 1

$[-30, 50]$ by $[-5, 30]$
Xscl = 5   Yscl = 5

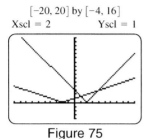
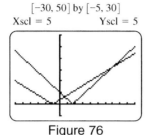

Figure 75   Figure 76

77. Graph $y_1 = |x + 1| + |x - 6|$ and $y_2 = 11$. See Figure 77. From the graph, the lines intersect at:

   $(-3, 11)$ and $(8, 11)$. Therefore the solution set is: $\{-3, 8\}$.

78. Graph $y_1 = |2x + 2| + |x + 1|$ and $y_2 = 9$. See Figure 78. From the graph, the lines intersect at:

   $(-4, 9)$ and $(2, 9)$. Therefore the solution set is: $\{-4, 2\}$.

79. Graph $y_1 = |x| + |x - 4|$ and $y_2 = 8$. See Figure 79. From the graph, the lines intersect at:

   $(-2, 8)$ and $(6, 8)$. Therefore the solution set is: $\{-2, 6\}$.

80. Graph $y_1 = |.5x + 2| + |.25x + 4|$ and $y_2 = 9$. See Figure 80. From the graph, the lines intersect at:

   $(-20, 9)$ and $(4, 9)$. Therefore the solution set is: $\{-20, 4\}$.

$[-10, 10]$ by $[-4, 16]$
Xscl = 1   Yscl = 1

$[-10, 10]$ by $[-4, 16]$
Xscl = 1   Yscl = 1

$[-10, 10]$ by $[-4, 16]$
Xscl = 1   Yscl = 1

$[-30, 10]$ by $[-4, 16]$
Xscl = 5   Yscl = 1

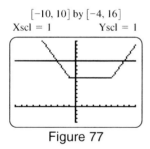
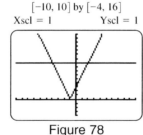
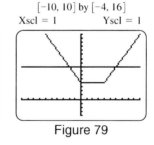
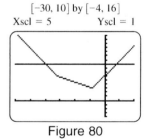

Figure 77   Figure 78   Figure 79   Figure 80

81. (a) $|T - 43| \leq 24 \Rightarrow -24 \leq T - 43 \leq 24 \Rightarrow 19 \leq T \leq 67.$

   (b) The average monthly temperatures in Marquette vary between a low of 19°F and a high of 67°F.

   The monthly averages are always within 24° of 43°F.

82. (a) $|T - 62| \leq 19 \Rightarrow -19 \leq T - 62 \leq 19 \Rightarrow 43 \leq T \leq 81.$

   (b) The average monthly temperatures in Memphis vary between a low of 43°F and a high of 81°F.

   The monthly averages are always within 19° of 62°F.

83. (a) $|T - 50| \leq 22 \Rightarrow -22 \leq T - 50 \leq 22 \Rightarrow 28 \leq T \leq 72.$

   (b) The average monthly temperatures in Boston vary between a low of 28°F and a high of 72°F.

   The monthly averages are always within 22° of 50°F.

84. (a) $|T - 10| \le 36 \Rightarrow -36 \le T - 10 \le 36 \Rightarrow -26 \le T \le 46.$

    (b) The average monthly temperatures in Chesterfield vary between a low of $-26°$F and a high of $46°$F.

    The monthly averages are always within $36°$ of $10°$F.

85. (a) $|T - 61.5| \le 12.5 \Rightarrow -12.5 \le T - 61.5 \le 12.5 \Rightarrow 49 \le T \le 74.$

    (b) The average monthly temperatures in Buenos Aires vary between a low of $49°$F (possibly in July) and a

    high of $74°$F (possibly in January). The monthly averages are always within $12.5°$ of $61.5°$F.

86. (a) $|T - 43.5| \le 8.5 \Rightarrow -8.5 \le T - 43.5 \le 8.5 \Rightarrow 35 \le T \le 52.$

    (b) The average monthly temperatures in Punta Arenas vary between a low of $35°$F and a high of $52°$F.

    The monthly averages are always within $8.5°$ of $43.5°$F.

87. $|x - 8.0| \le 1.5 \Rightarrow -1.5 \le x - 8.0 \le 1.5 \Rightarrow 6.5 \le x \le 9.5$, therefore the range is the interval: $[6.5, 9.5]$.

88. If $\dfrac{680 + 780}{2} = 730$ is the midpoint, then $680 \le F \le 780 \Rightarrow 680 - 730 \le F - 730 \le 780 - 730 \Rightarrow$

    $-50 \le F - 730 \le 50 \Rightarrow |F - 730| \le 50$ (or $|730 - F| \le 50$).

89. (a) $P_d = |116 - 125| \Rightarrow P_d = |-9| = 9.$

    (b) $17 = |P - 130| \Rightarrow P - 130 = 17$ or $P - 130 = -17 \Rightarrow P = 147$ or $P = 113.$

90. If $\dfrac{98 + 148}{2} = 123$ is the midpoint, then $98 \le x \le 148 \Rightarrow 98 - 123 \le x - 123 \le 148 - 123 \Rightarrow$

    $-25 \le x - 123 \le 25 \Rightarrow |x - 123| \le 25$ (or $|123 - x| \le 25$); and if $\dfrac{16 + 26}{2} = 21$ is the midpoint,

    then $16 \le x \le 26 \Rightarrow 16 - 21 \le x - 21 \le 26 - 21 \Rightarrow$

    $-5 \le x - 21 \le 5 \Rightarrow |x - 21| \le 5$ (or $|21 - x| \le 5$).

91. If the difference between y and 1 is less than .1, then $|y - 1| < .1 \Rightarrow |2x + 1 - 1| < .1 \Rightarrow |2x| < .1 \Rightarrow$

    $-.1 < 2x < .1 \Rightarrow -.05 < x < .05.$ The open interval of x is: $(-.05, .05)$.

92. If the difference between y and 2 is less than .01, then $|y - 2| < .1 \Rightarrow |3x - 6 - 2| < .01 \Rightarrow$

    $|3x - 8| < .01 \Rightarrow -.01 < 3x - 8 < .01 \Rightarrow 7.99 < 3x < 8.01 \Rightarrow 2.66\overline{3} < x < 2.67.$

    The open interval of x is: $(2.66\overline{3}, 2.67)$.

93. If the difference between y and 3 is less than .001, then $|y - 3| < .001 \Rightarrow |4x - 8 - 3| < .001 \Rightarrow$

    $|4x - 11| < .001 \Rightarrow -.001 < 4x - 11 < .001 \Rightarrow 10.999 < 4x < 11.001 \Rightarrow 2.74975 < x < 2.75025.$

    The open interval of x is: $(2.74975, 2.75025)$.

94. If the difference between y and 4 is less than .0001, then $|y - 4| < .0001 \Rightarrow |5x + 12 - 4| < .0001 \Rightarrow$

    $|5x + 8| < .0001 \Rightarrow -.0001 < 5x + 8 < .0001 \Rightarrow -8.0001 < 5x < -7.9999 \Rightarrow$

    $-1.60002 < x < -1.59998.$ The open interval of x is: $(-1.60002, -1.59998)$.

95. From the chart, 15 mph at $30°$ is $19°$F and 10 mph at $-10°$ is $-28°$F, Therefore, $|19 - (-28)| = |47| = 47°$F.

96. From the chart, 20 mph at $-20°$ is $-48°$F and 5 mph at $30°$ is $25°$F, Therefore, $|-48 - 25| = |-73| = 73°$F.

97. From the chart, 30 mph at $-30°$ is $-67°$F and 15 mph at $-20°$ is $-45°$F, Therefore,

    $|-67 - (-45)| = |-22| = 22°$F.

98. From the chart, 40 mph at $40°$ is $27°$F and 25 mph at $-30°$ is $-64°$F, Therefore, $|27 - (-64)| = |91| = 91°$F.

99. If $|2x + 7| = 6x - 1$ then $|2x + 7| - (6x - 1) = 0$. Graph $y_1 = |2x + 7| - (6x - 1)$,

See Figure 99. The x-intercept is: 2, therefore the solution set is: $\{2\}$.

100. If $-|3x - 12| \geq -x - 1$ then $-|3x - 12| - (-x - 1) \geq 0$. Graph $y_1 = -|3x - 12| - (-x - 1)$,

See Figure 100. The equation is ≥ 0 or the graph intersects or is above the x-axis, for the interval: $[2.75, 6.5]$.

101. If $|x - 4| > .5x - 6$ then $|x - 4| - (.5x - 6) > 0$. Graph $y_1 = |x - 4| - (.5x - 6)$,

See Figure 101. The equation is > 0 or the graph is above the x-axis, for the interval: $(-\infty, \infty)$.

102. If $2x + 8 > -|3x + 4|$ then $2x + 8 - (-|3x + 4|) > 0$. Graph $y_1 = 2x + 8 - (-|3x + 4|)$,

See Figure 102. The equation is > 0 or the graph is above the x-axis, for the interval: $(-\infty, \infty)$.

$[-10, 10]$ by $[-10, 10]$
Xscl = 1   Yscl = 1

$[-10, 10]$ by $[-10, 10]$
Xscl = 1   Yscl = 1

$[-10, 10]$ by $[-10, 10]$
Xscl = 1   Yscl = 1

$[-10, 10]$ by $[-4, 16]$
Xscl = 1   Yscl = 1

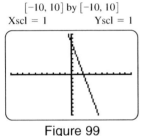
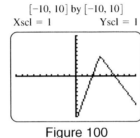
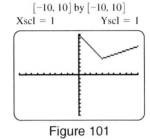
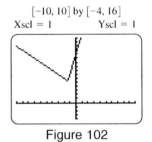

Figure 99   Figure 100   Figure 101   Figure 102

103. If $|3x + 4| < -3x - 14$ then $|3x + 4| - (-3x - 14) < 0$. Graph $y_1 = |3x + 4| - (-3x - 14)$,

See Figure 103. The equation is < 0 or the graph is below the x-axis never, therefore the solution set is: $\varnothing$.

104. If $|x - \sqrt{13}| + \sqrt{6} \leq -x - \sqrt{10}$ then $|x - \sqrt{13}| + \sqrt{6} - (-x - \sqrt{10}) \leq 0$.

Graph $y_1 = |x - \sqrt{13}| + \sqrt{6} - (-x - \sqrt{10})$, See Figure 104. The equation is ≤ 0 or the graph intersects

or is below the x-axis never, therefore the solution set is: $\varnothing$.

$[-10, 10]$ by $[-5, 30]$
Xscl = 1   Yscl = 5

$[-10, 10]$ by $[-5, 30]$
Xscl = 1   Yscl = 5

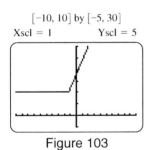
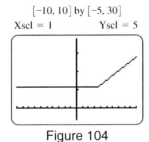

Figure 103   Figure 104

2.5: Piecewise-Defined Functions

1. (a) From the graph, the highest is 55 mph and the lowest is 30 mph.

(b) 3 stretches of 4 miles each or $3 \times 4 = 12$ miles.

(c) From the graph, $f(4) = 40$ mph; $f(12) = 30$ mph; and $f(18) = 55$ mph.

(d) From the graph, the graph is discontinuous at $x = 4, 6, 8, 12,$ and 16. The speed limit changes at each

discontinuity.

2. (a) From the graph, the initial amount was: $1000; and the final amount was: $600.

   (b) From the graph, $f(10) = \$900$; $f(50) = \$600$. The function f is not continuous.

   (c) From the graph, the discontinuity shows 3 drops or withdrawals.

   (d) From the graph, the largest drop or largest withdrawal of $300 occurred after 15 minute.

   (e) From the graph, the one increase or deposit was $200.

3. (a) From the graph, the initial amount was: 50,000 gal.; and the final amount was: 30,000 gal.

   (b) From the graph, during the first and fourth days.

   (c) From the graph, $f(2) = 45,000$ gal.; $f(4) = 40,000$ gal.

   (d) From the graph, between days 1 and 3 the water dropped: $\dfrac{50,000 - 40,000}{2} = \dfrac{10,000}{2} = 5,000$ gal./day.

4. (a) From the graph, when $x = 3$ gas was added until 20 gallons was in the tank.

   (b) When the graph is horizontal, the engine is not running; when the graph is decreasing, the engine is burning gasoline; and when the graph is increasing, gasoline is being put into the tank.

   (c) From the graph, the graph deceased the fastest or gasoline was burned the fastest between 1 and 2.9 hours.

5. (a) $f(-5) = 2(-5) = -10$ (b) $f(-1) = 2(-1) = -2$ (c) $f(0) = 0 - 1 = -1$ (d) $f(3) = 3 - 1 = 2$

6. (a) $f(-5) = -5 - 2 = -7$ (b) $f(-1) = -1 - 2 = -3$ (c) $f(0) = 0 - 2 = -2$ (d) $f(3) = 5 - 3 = 2$

7. (a) $f(-5) = 2 + (-5) = -3$ (b) $f(-1) = -(-1) = 1$ (c) $f(0) = -(0) = 0$ (d) $f(3) = 3(3) = 9$

8. (a) $f(-5) = -2(-5) = 10$ (b) $f(-1) = 3(-1) - 1 = -4$ (c) $f(0) = 3(0) - 1 = -1$

   (d) $f(3) = -4(3) = -12$

9. See Figure 9.

10. See Figure 10.

11. See Figure 11.

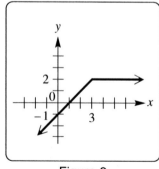

Figure 9

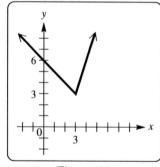

Figure 10

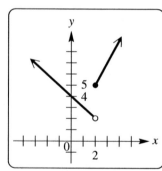

Figure 11

12 See Figure 12.

13. See Figure 13.

14. See Figure 14.

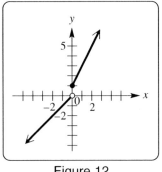

Figure 12

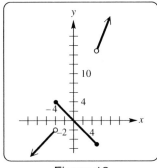

Figure 13

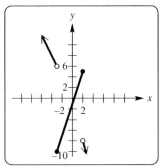

Figure 14

15. See Figure 15.

16. See Figure 16.

17. See Figure 17.

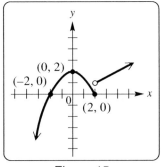

Figure 15

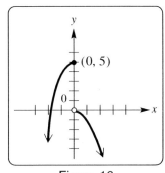

Figure 16

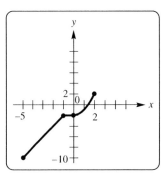

Figure 17

18. See Figure 18.

19. See Figure 19.

20. See Figure 20.

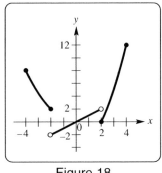

Figure 18

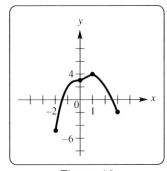

Figure 19

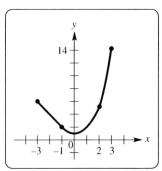

Figure 20

21. Look for a $y = x^2$ graph if $x \geq 0$; and a linear graph if $x < 0$. Therefore: B.

22. Look for a $y = |x|$ graph if $x \geq -1$; and a reflected $y = x^2$ graph if $x < -1$. Therefore: A.

23. Look for a horizontal graph above the x-axis if $x \geq 0$; and a horizontal graph below the x-axis if $x < 0$.

    Therefore D.

24. Look for a $y = \sqrt{x}$ graph if $x \geq 0$; and a reflected $y = x^2$ graph if $x < 0$. Therefore: C.

25. Graph $y_1 = (x - 1) \ast (x \leq 3) + (2) \ast (x > 3)$, See Figure 25.

26. Graph $y_1 = (6 - x) \ast (x \leq 3) + (3x - 6) \ast (x > 3)$, See Figure 26.

27. Graph $y_1 = (4 - x) \ast (x < 2) + (1 + 2x) \ast (x \geq 2)$, See Figure 27.

28. Graph $y_1 = (2x + 1) \ast (x \geq 0) + (x) \ast (x < 0)$, See Figure 28.

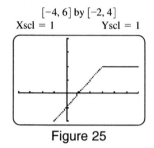
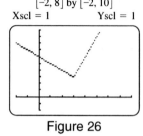
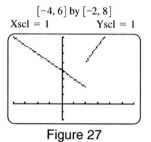
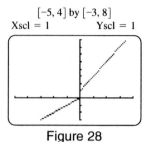

$[-4, 6]$ by $[-2, 4]$ $[-2, 8]$ by $[-2, 10]$ $[-4, 6]$ by $[-2, 8]$ $[-5, 4]$ by $[-3, 8]$
Xscl = 1   Yscl = 1   Xscl = 1   Yscl = 1   Xscl = 1   Yscl = 1   Xscl = 1   Yscl = 1

Figure 25   Figure 26   Figure 27   Figure 28

29. Graph $y_1 = (2 + x) \ast (x < -4) + (-x) \ast (-4 \leq x$ and $x \leq 5) + (3x) \ast (x > 5)$, See Figure 29.

30. Graph $y_1 = (-2x) \ast (x < -3) + (3x - 1) \ast (-3 \leq x$ and $x \leq 2) + (-4x) \ast (x > 2)$, See Figure 30.

31. Graph $y_1 = \left(-\dfrac{1}{2}x^2 + 2\right) \ast (x \leq 2) + \left(\dfrac{1}{2}x\right) \ast (x > 2)$, See Figure 31.

32. Graph $y_1 = (x^3 + 5) \ast (x \leq 0) + (-x^2) \ast (x > 0)$, See Figure 32.

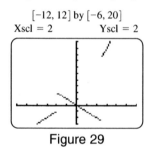
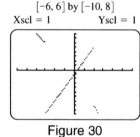
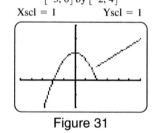
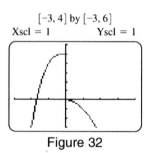

$[-12, 12]$ by $[-6, 20]$ $[-6, 6]$ by $[-10, 8]$ $[-5, 6]$ by $[-2, 4]$ $[-3, 4]$ by $[-3, 6]$
Xscl = 2   Yscl = 2   Xscl = 1   Yscl = 1   Xscl = 1   Yscl = 1   Xscl = 1   Yscl = 1

Figure 29   Figure 30   Figure 31   Figure 32

33. From the graph, the function is: $f(x) = \begin{cases} 2 & \text{if } x \leq 0 \\ -1 & \text{if } x > 1 \end{cases}$; domain: $(-\infty, 0] \cup (1, \infty)$; range: $\{-1, 2\}$.

34. From the graph, the function is: $f(x) = \begin{cases} 1 & \text{if } x \leq -1 \\ -1 & \text{if } x > 2 \end{cases}$; domain: $(-\infty, -1] \cup (2, \infty)$; range: $\{-1, 1\}$.

35. From the graph, the function is: $f(x) = \begin{cases} x & \text{if } x \leq 0 \\ 2 & \text{if } x > 0 \end{cases}$; domain: $(-\infty, \infty)$; range: $(-\infty, 0] \cup \{2\}$.

36. From the graph, the function is: $f(x) = \begin{cases} -3 & \text{if } x < 0 \\ \sqrt{x} & \text{if } x \geq 0 \end{cases}$; domain: $(-\infty, \infty)$; range: $\{-3\} \cup [0, \infty)$.

37. From the graph, the function is: $f(x) = \begin{cases} \sqrt[3]{x} & \text{if } x < 1 \\ x + 1 & \text{if } x \geq 1 \end{cases}$; domain: $(-\infty, \infty)$; range: $(-\infty, 1) \cup [2, \infty)$.

38. From the graph, the function is: $f(x) = \begin{cases} 3 & \text{if } x = 2 \\ 2x - 3 & \text{if } x \neq 2 \end{cases}$; domain: $(-\infty, \infty)$; range: $(-\infty, 1) \cup (1, \infty)$.

39. There is an overlap of intervals since the number 4 satisfies both conditions. To be a function, every x-value is used only once.

40. The value $f(4)$ cannot be found since using the first formula, $f(4) = 11$, and using the second formula, $f(4) = 16$. To have two different values violates the definition of function.

41. The graph of $y = [\![x]\!]$ is shifted 1.5 units downward.

42. The graph of $y = [\![x]\!]$ is reflected across the y-axis.

43. The graph of $y = [\![x]\!]$ is reflected across the x-axis.

44. The graph of $y = [\![x]\!]$ is shifted 2 units to the left.

45. Graph $y = [\![x]\!] - 5$, See Figure 45.

46. Graph $y = [\![-x]\!]$, See Figure 46.

47. Graph $y = -[\![x]\!]$, See Figure 47.

48. Graph $y = [\![x + 2]\!]$, See Figure 48.

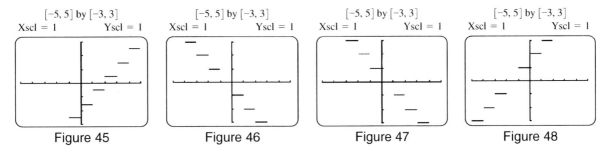

Figure 45   Figure 46   Figure 47   Figure 48

49. When $0 \leq x \leq 3$, the slope is 5, which means the inlet pipe is open and the outlet pipe is closed; when $3 < x \leq 5$, the slope is 2, which means both pipes are open; when $5 < x \leq 8$, the slope is 0, which means both pipes are closed; when $8 < x \leq 10$, the slope is -3, which means the inlet pipe is closed and the outlet pipe is open.

50. (a) Since for each year given the shoe size is 1 size smaller than his age, the formula is: $y = x - 1$.

    (b) From the table and question, graph the function. See figure 50.

51. (a) If $t = 7 - 6$ then $t = 1$. Then $40(1) + 100 = 140$.

    (b) If $t = 9 - 6$ then $t = 3$. Then $40(3) + 100 = 220$.

    (c) If $t = 10 - 6$ then $t = 4$. Then blood sugar level is: 220.

    (d) If $t = 12 - 6$ then $t = 6$. Then blood sugar level is: 220.

    (e) If $t = (12 - 6) + 2$ then $t = 8$. Then blood sugar level is: 220.

    (f) If $t = (12 - 6) + 5$ then $t = 11$. Then blood sugar level is: 60.

    (g) If $t = (12 - 6) + 12$ then $t = 18$. Then blood sugar level is: 60.

    (h) Graph the function. See Figure 51.

    (i) If f were discontinuous, the insulin level would change instantaneously from one level to a second level.

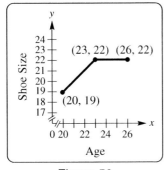

Figure 50

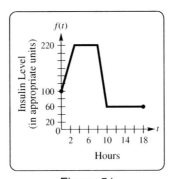

Figure 51

52. (a) Graph the function. See Figure 52.

(b) From the graph, the snow is the deepest in the fourth month: February, when the depth $= 6.5(4) = 26$ in.

(c) From the graph, the snow begins at $x = 0$, the beginning of October; and the snow ends at $x \approx 6.5$, the middle of April.

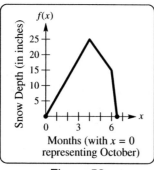

Figure 52

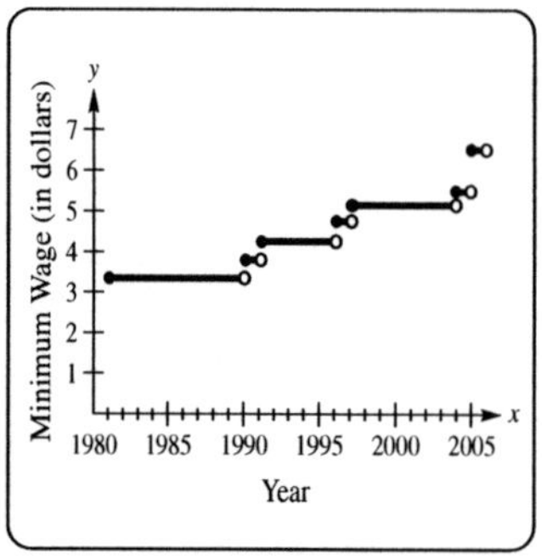

Figure 54

53. (a) From 1997 to 1999, there is a decrease of 700 cases per year. From 1999 to 2001, there is an increase of 200 cases per year.

(b) For the first line of the piecewise function, $m = -700$ (a decrease of 700 cases per year) and a starting coordinate of $(0, 8100)$, gives us the equation: $y = -700x + 8100$ for the first 2 years or $0 \leq x \leq 2$. The second line of the function has $m = 200$ (increase of 200 cases per year) and a starting coordinate of $(2, 6700)$, for an equation: $y - 6700 = 200(x - 2) \Rightarrow y - 6700 = 200x - 400 \Rightarrow y = 200x + 6300$ for years 2 through 4 or $2 < x \leq 4$. Therefore the function is: $f(x) = \begin{cases} -700x + 8100 & \text{if } 0 \leq x \leq 2 \\ 200x + 6300 & \text{if } 2 < x \leq 4 \end{cases}$.

54. From the table, graph the piecewise function. See Figure 54.

55. (a) From the table, graph the piecewise function. See Figure 55.

(b) The likelihood of being a victim peaks from age 16 up to age 20, then decreases.

56. (a) From the table, graph the piecewise function. See Figure 56.

(b) Housing starts increased and then decreased, with the maximum occurring during the 1960's.

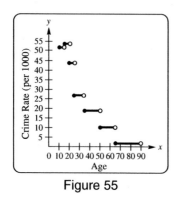

Figure 55

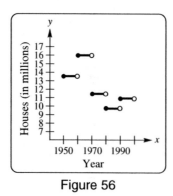

Figure 56

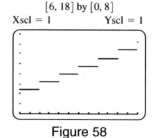

Figure 58

57. (a) A 3.5 minute call would round up to 4 minutes. A 4 minute call would cost: $.50 + 3(.25) = \$1.25$.

(b) We use a piecewise defined function where the cost increases after each whole number as follows:

$$f(x) = \begin{cases} .50 & \text{if } 0 < x \le 1 \\ .75 & \text{if } 1 < x \le 2 \\ 1.00 & \text{if } 2 < x \le 3 \\ 1.25 & \text{if } 3 < x \le 4 \\ 1.50 & \text{if } 4 < x \le 5 \end{cases}. \quad \text{Another possibility is } f(x) = \begin{cases} .50 & \text{if } 0 < x \le 1 \\ .50 - .25[\![1 - x]\!] & \text{if } 1 < x \le 5 \end{cases}.$$

58. (a) Since cost is rounded down to the nearest 2 foot interval, we can use the greatest integer function.

The function is: $f(x) = .80 \left[\!\!\left[\dfrac{x}{2} \right]\!\!\right]$ from $6 \le x \le 18$.

(b) Graph $f(x) = .80 \left[\!\!\left[\dfrac{x}{2} \right]\!\!\right]$ from $6 \le x \le 18$. See Figure 58.

(c) $f(8.5) = .80 \left[\!\!\left[\dfrac{8.5}{2} \right]\!\!\right] = .80[\![4.25]\!] = .80(4) \Rightarrow f(8.5) = \$3.20$.

$f(15.2) = .80 \left[\!\!\left[\dfrac{15.2}{2} \right]\!\!\right] = .80[\![7.6]\!] = .80(7) \Rightarrow f(8.5) = \$5.60$.

59. Sketch a piecewise function that fills a tank at a rate of 5 gallons a minute for the first 20 minutes (the time it takes to fill the 100 gallon tank) and then drains the tank at a rate of 2 gallons per minute for 50 minutes (the time it takes to drain the 100 gallon tank). See Figure 59.

60. Sketch a piecewise function that measures miles over minutes. The first piece increases at a rate of 40 mph or $\dfrac{40}{60} = \dfrac{2}{3}$ miles per minute, for 30 minutes (the time it takes to travel 20 miles at that rate). The second piece stays at a constant distance of 20 miles for the 2 hour period at the park. The third piece decreases (returns home) at a rate of 20 mph or $\dfrac{20}{60} = \dfrac{1}{3}$ miles per minute, for 60 minutes (the time it takes to travel 20 miles at that rate). See Figure 60.

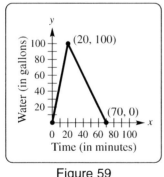

Figure 59

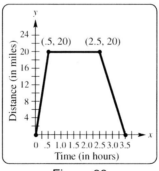

Figure 60

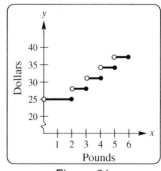

Figure 61

61. For x in the interval $(0, 2]$, $y = 25$. For x in the interval $(2, 3]$, $y = 25 + 3 = 28$. For x in the interval $(3, 4]$, $y = 28 + 3 = 31$ and so on. The graph is a step function. In this case, the first step has a different width. See Figure 61.

62. (a) For the first line of the piecewise function, $m = \dfrac{72 - 73}{6 - 3} = \dfrac{-1}{3} \Rightarrow m = -\dfrac{1}{3}$ (a decrease of $\dfrac{1}{3}$% per year)

and a starting coordinate of $(3, 73)$, gives us the equation: $y - 73 = -\dfrac{1}{3}(x - 3) \Rightarrow y - 73 = -\dfrac{1}{3}x + 1 \Rightarrow$

$y = -\dfrac{1}{3}x + 74$ for the 3 years 1993 to 1996 or $3 \le x \le 6$. The second line of the function has

$m = \dfrac{1999 - 1996}{69 - 72} = \dfrac{3}{-3} \Rightarrow m = -1$ (decrease of 1% per year) and a starting coordinate of $(6, 72)$, for

an equation: $y - 72 = -(x - 6) \Rightarrow y - 72 = -x + 6 \Rightarrow y = -x + 78$ for years 1996 through 1999

or $6 < x \le 9$.

(b) Therefore the piecewise-defined function is: $f(x) = \begin{cases} -\dfrac{1}{3}x + 74 & \text{if } 3 \le x \le 6 \\ -x + 78 & \text{if } 6 < x \le 9 \end{cases}$.

2.6: Operations and Composition

1. $x^2 + (2x - 5) = x^2 + 2x - 5 \Rightarrow$ E.

2. $x^2 - (2x - 5) = x^2 - 2x + 5 \Rightarrow$ B.

3. $x^2(2x - 5) = 2x^3 - 5x^2 \Rightarrow$ F.

4. $\dfrac{x^2}{2x - 5} \Rightarrow$ D.

5. $(2x - 5)^2 = 4x^2 - 20x + 25 \Rightarrow$ A.

6. $(2(x^2) - 5) = 2x^2 - 5 \Rightarrow$ C.

7. $4(8x + 1)^2 - 2(8x + 1) = 4(64x^2 + 16x + 1) - 16x - 2 = 256x^2 + 64x + 4 - 16x - 2 =$
    $256x^2 + 48x + 2 \Rightarrow 256(3)^2 + 48(3) + 2 = 256(9) + 144 + 2 = 2304 + 146 = 2450$

8. $8(4x^2 - 2x) + 1 = 32x^2 - 16x + 1 \Rightarrow 32(-2)^2 - 16(-2) + 1 = 32(4) + 32 + 1 = 128 + 33 = 161$

9. $4(8x + 1)^2 - 2(8x + 1) = 4(64x^2 + 16x + 1) - 16x - 2 = 256x^2 + 64x + 4 - 16x - 2 =$
    $256x^2 + 48x + 2$

10. $8(4x^2 - 2x) + 1 = 32x^2 - 16x + 1$

11. $(4x^2 - 2x) + (8x + 1) = 4x^2 + 6x + 1 = 4(3)^2 + 6(3) + 1 = 4(9) + 18 + 1 = 36 + 19 = 55$

12. $(4x^2 - 2x) + (8x + 1) = 4x^2 + 6x + 1 = 4(-5)^2 + 6(-5) + 1 = 4(25) - 30 + 1 = 100 - 30 + 1 = 71$

13. $(4x^2 - 2x)(8x + 1) = 32x^3 + 4x^2 - 16x^2 - 2x = 32x^3 - 12x^2 - 2x = 32(4)^3 - 12(4)^2 - 2(4) =$
    $32(64) - 12(16) - 8 = 2048 - 192 - 8 = 1848$

14. $(4x^2 - 2x)(8x + 1) = 32x^3 + 4x^2 - 16x^2 - 2x = 32x^3 - 12x^2 - 2x = 32(-3)^3 - 12(-3)^2 - 2(-3) =$
    $32(-27) - 12(9) + 6 = -864 - 108 + 6 = -966$

15. $\dfrac{4x^2 - 2x}{8x + 1} = \dfrac{4(-1)^2 - 2(-1)}{8(-1) + 1} = \dfrac{4(1) + 2}{-8 + 1} = \dfrac{6}{-7} = -\dfrac{6}{7}$

16. $\dfrac{4x^2 - 2x}{8x + 1} = \dfrac{4(4)^2 - 2(4)}{8(4) + 1} = \dfrac{4(16) - 8}{32 + 1} = \dfrac{64 - 8}{33} = \dfrac{56}{33}$

17. $4(8x + 1)^2 - 2(8x + 1) = 4(64x^2 + 16x + 1) - 16x - 2 = 256x^2 + 64x + 4 - 16x - 2 =$

$256x^2 + 48x + 2 \Rightarrow 256(2)^2 + 48(2) + 2 = 256(4) + 96 + 2 = 1024 + 98 = 1122$

18. $4(8x + 1)^2 - 2(8x + 1) = 4(64x^2 + 16x + 1) - 16x - 2 = 256x^2 + 64x + 4 - 16x - 2 =$

$256x^2 + 48x + 2 \Rightarrow 256(-5)^2 + 48(-5) + 2 = 256(25) - 240 + 2 = 6400 - 238 = 6162$

19. $8(4x^2 - 2x) + 1 = 32x^2 - 16x + 1 \Rightarrow 32(2)^2 - 16(2) + 1 = 32(4) - 32 + 1 = 128 - 31 = 97$

20. $8(4x^2 - 2x) + 1 = 32x^2 - 16x + 1 \Rightarrow 32(-5)^2 - 16(-5) + 1 = 32(25) + 80 + 1 = 800 + 81 = 881$

21. (a) $(f + g)(x) = (4x - 1) + (6x + 3) = 10x + 2$

$(f - g)(x) = (4x - 1) - (6x + 3) = -2x - 4$

$(fg)(x) = (4x - 1)(6x + 3) = 24x^2 + 12x - 6x - 3 = 24x^2 + 6x - 3$

(b) All values can replace x in all three equations, therefore: Domain is $(-\infty, \infty)$ in all cases.

(c) $\left(\dfrac{f}{g}\right)(x) = \dfrac{4x - 1}{6x + 3}$; all values can replace x, except $-\dfrac{1}{2}$, therefore the domain is: $\left(-\infty, -\dfrac{1}{2}\right) \cup \left(-\dfrac{1}{2}, \infty\right)$.

(d) $(f \circ g)(x) = f[g(x)] = 4(6x + 3) - 1 = 24x + 12 - 1 = 24x + 11$; all values can be input for x,

therefore the domain is: $(-\infty, \infty)$.

(e) $(g \circ f)(x) = g[f(x)] = 6(4x - 1) + 3 = 24x - 6 + 3 = 24x - 3$; all values can replace x,

therefore the domain is: $(-\infty, \infty)$.

22. (a) $(f + g)(x) = (9 - 2x) + (-5x + 2) = -7x + 11$

$(f - g)(x) = (9 - 2x) - (-5x + 2) = 3x + 7$

$(fg)(x) = (9 - 2x)(-5x + 2) = -45x + 18 + 10x^2 - 4x = 10x^2 - 49x + 18$

(b) All values can replace x in all three equations, therefore: Domain is $(-\infty, \infty)$ in all cases.

(c) $\left(\dfrac{f}{g}\right)(x) = \dfrac{9 - 2x}{-5x + 2}$; all values can replace x, except $\dfrac{2}{5}$, therefore the domain is: $\left(-\infty, \dfrac{2}{5}\right) \cup \left(\dfrac{2}{5}, \infty\right)$.

(d) $(f \circ g)(x) = f[g(x)] = 9 - 2(-5x + 2) = 9 + 10x - 4 = 10x + 5$; all values can replace x,

therefore the domain is: $(-\infty, \infty)$.

(e) $(g \circ f)(x) = g[f(x)] = -5(9 - 2x) + 2 = -45 + 10x + 2 = 10x - 43$; all values can replace x,

therefore the domain is: $(-\infty, \infty)$.

23. (a) $(f + g)(x) = |x + 3| + 2x$

$(f - g)(x) = |x + 3| - 2x$

$(fg)(x) = |x + 3|(2x)$

(b) All values can replace x in all three equations, therefore: Domain is $(-\infty, \infty)$ in all cases.

(c) $\left(\dfrac{f}{g}\right)(x) = \dfrac{|x + 3|}{2x}$; all values can replace x, except 0, therefore the domain is: $(-\infty, 0) \cup (0, \infty)$.

(d) $(f \circ g)(x) = f[g(x)] = |(2x) + 3| = |2x + 3|$ all values can replace x, therefore the domain is: $(-\infty, \infty)$.

(e) $(g \circ f)(x) = g[f(x)] = 2(|x + 3|) = 2|x + 3|$; all values can replace x, therefore the domain is: $(-\infty, \infty)$.

24. (a) $(f + g)(x) = |2x - 4| + (x + 1) = |2x - 4| + x + 1$

    $(f - g)(x) = |2x - 4| - (x + 1) = |2x - 4| - x - 1$

    $(fg)(x) = |2x - 4|(x + 1) = (x + 1)|2x - 4|$

    (b) All values can replace x in all three equations, therefore: The domain is $(-\infty, \infty)$ in all cases.

    (c) $\left(\dfrac{f}{g}\right)(x) = \dfrac{|2x - 4|}{x + 1}$; all values replace x, except -1, therefore the domain is: $(-\infty, -1) \cup (-1, \infty)$.

    (d) $(f \circ g)(x) = f[g(x)] = |2(x + 1) - 4| = |2x - 2|$; all values can replace x, so the domain is: $(-\infty, \infty)$.

    (e) $(g \circ f)(x) = g[f(x)] = |2x - 4| + 1$; all values can replace x, so the domain is: $(-\infty, \infty)$.

25. (a) $(f + g)(x) = \sqrt[3]{x + 4} + (x^3 + 5) = \sqrt[3]{x + 4} + x^3 + 5$.

    $(f - g)(x) = \sqrt[3]{x + 4} - (x^3 + 5) = \sqrt[3]{x + 4} - x^3 - 5$.

    $(fg)(x) = (\sqrt[3]{x + 4})(x^3 + 5)$.

    (b) All values can replace x in all three equations, therefore: The domain is $(-\infty, \infty)$ in all cases.

    (c) $\left(\dfrac{f}{g}\right)(x) = \dfrac{\sqrt[3]{x + 4}}{x^3 + 5}$; all values can replace x, except $\sqrt[3]{-5}$, so the domain is: $(-\infty, \sqrt[3]{-5}) \cup (\sqrt[3]{-5}, \infty)$.

    (d) $(f \circ g)(x) = f[g(x)] = \sqrt[3]{(x^3 + 5) + 4} = \sqrt[3]{x^3 + 9}$; all values can replace x, so the domain is: $(-\infty, \infty)$.

    (e) $(g \circ f)(x) = g[f(x)] = (\sqrt[3]{x + 4})^3 + 5 = x + 4 + 5 = x + 9$; all values can replace x, so the domain is: $(-\infty, \infty)$.

26. (a) $(f + g)(x) = \sqrt[3]{6 - 3x} + (2x^3 + 1) = \sqrt[3]{6 - 3x} + 2x^3 + 1$.

    $(f - g)(x) = \sqrt[3]{6 - 3x} - (2x^3 + 1) = \sqrt[3]{6 - 3x} - 2x^3 - 1$.

    $(fg)(x) = (\sqrt[3]{6 - 3x})(2x^3 + 1)$.

    (b) All values can replace x in all three equations, therefore: Domain is $(-\infty, \infty)$ in all cases.

    (c) $\left(\dfrac{f}{g}\right)(x) = \dfrac{\sqrt[3]{6 - 3x}}{2x^3 + 1}$; all values can replace x, except $\sqrt[3]{-\dfrac{1}{2}}$, so the domain is: $\left(-\infty, \sqrt[3]{-\dfrac{1}{2}}\right) \cup \left(\sqrt[3]{-\dfrac{1}{2}}, \infty\right)$.

    (d) $(f \circ g)(x) = f[g(x)] = \sqrt[3]{6 - 3(2x^3 + 1)} = \sqrt[3]{-6x^3 + 3}$; all values can replace x, so the domain is: $(-\infty, \infty)$.

    (e) $(g \circ f)(x) = g[f(x)] = 2(\sqrt[3]{6 - 3x})^3 + 1 = 2(6 - 3x) + 1 = -6x + 13$; all values can replace x, therefore the domain is: $(-\infty, \infty)$.

27. (a) $(f + g)(x) = \sqrt{x^2 + 3} + (x + 1) = \sqrt{x^2 + 3} + x + 1$.

    $(f - g)(x) = \sqrt{x^2 + 3} - (x + 1) = \sqrt{x^2 + 3} - x - 1$.

    $(fg)(x) = (\sqrt{x^2 + 3})(x + 1)$.

    (b) All values can replace x in all three equations, therefore: Domain is $(-\infty, \infty)$ in all cases.

    (c) $\left(\dfrac{f}{g}\right)(x) = \dfrac{\sqrt{x^2 + 3}}{x + 1}$; all values can replace x, except -1, therefore the domain is: $(-\infty, -1) \cup (-1, \infty)$.

    (d) $(f \circ g)(x) = f[g(x)] = \sqrt{(x + 1)^2 + 3} = \sqrt{x^2 + 2x + 1 + 3} = \sqrt{x^2 + 2x + 4}$; all values can replace x, therefore the domain is: $(-\infty, \infty)$.

    (e) $(g \circ f)(x) = g[f(x)] = (\sqrt{x^2 + 3}) + 1 = \sqrt{x^2 + 3} + 1$; all values can replace x, therefore the domain is: $(-\infty, \infty)$.

28. (a) $(f + g)(x) = \sqrt{2 + 4x^2} + (x) = \sqrt{2 + 4x^2} + x.$

    $(f - g)(x) = \sqrt{2 + 4x^2} - (x) = \sqrt{2 + 4x^2} - x.$

    $(fg)(x) = (\sqrt{2 + 4x^2})(x) = x\sqrt{2 + 4x^2}.$

    (b) All values can replace x in all three equations, therefore: Domain is $(-\infty, \infty)$ in all cases.

    (c) $\left(\dfrac{f}{g}\right)(x) = \dfrac{\sqrt{2 + 4x^2}}{x}$; all values can replace x, except 0, therefore the domain is: $(-\infty, 0) \cup (0, \infty)$.

    (d) $(f \circ g)(x) = f[g(x)] = \sqrt{2 + 4(x)^2} = \sqrt{2 + 4x^2}$; all values can replace x, therefore the domain

    is: $(-\infty, \infty)$.

    (e) $(g \circ f)(x) = g[f(x)] = (\sqrt{2 + 4x^2}) = \sqrt{2 + 4x^2}$; all values can replace x, therefore the domain

    is: $(-\infty, \infty)$.

29. (a) From the graph, $4 + (-2) = 2.$

    (b) From the graph, $1 - (-3) = 4.$

    (c) From the graph, $(0)(-4) = 0.$

    (d) From the graph, $\dfrac{1}{-3} = -\dfrac{1}{3}.$

30. (a) From the graph, $0 + 2 = 2.$

    (b) From the graph, $-2 - 1 = -3.$

    (c) From the graph, $(2)(1) = 2.$

    (d) From the graph, $\dfrac{4}{-2} = -2.$

31. (a) From the graph, $0 + 3 = 3.$

    (b) From the graph, $-1 - 4 = -5.$

    (c) From the graph, $(1)(2) = 2.$

    (d) From the graph, $\dfrac{3}{0} = $ undefined.

32. (a) From the graph, $-3 + 1 = -2.$

    (b) From the graph, $-2 - 0 = -2.$

    (c) From the graph, $(-3)(-1) = 3.$

    (d) From the graph, $\dfrac{-3}{1} = -3.$

33. (a) From the table, $7 + (-2) = 5.$

    (b) From the table, $10 - 5 = 5.$

    (c) From the table, $(0)(6) = 0.$

    (d) From the table, $\dfrac{5}{0} = $ undefined.

34. (a) From the table, $5 + 4 = 9.$

    (b) From the table, $0 - 0 = 0.$

    (c) From the table, $(-4)(2) = -8.$

    (d) From the table, $\dfrac{8}{-1} = -8.$

35. See Figure 35.

36. See Figure 36.

x	$(f+g)(x)$	$(f-g)(x)$	$(fg)(x)$	$\left(\frac{f}{g}\right)(x)$
-2	6	-6	0	0
0	5	5	0	undefined
2	5	9	-14	-3.5
4	15	5	50	2

x	$(f+g)(x)$	$(f-g)(x)$	$(fg)(x)$	$\left(\frac{f}{g}\right)(x)$
-2	-2	-6	-8	-2
0	7	9	-8	-8
2	9	1	20	1.25
4	0	0	0	undefined

Figure 35   Figure 36

37. From the graph, $G(1996) \approx 7.7$; $B(1996) \approx 11.8$; and $T(1996) \approx 19.5$.

38. From the graph, $G(1991) \approx 6.3$; $B(1991) \approx 8.2$; and $T(1991) \approx 14.5$.

39. From the graph, The slope from 1978-1991 is: $m = \dfrac{14.5 - 8}{1991 - 1978} = \dfrac{6.5}{13} \Rightarrow m = \dfrac{1}{2}$. The slope from

    1991-1996 is: $m = \dfrac{19.5 - 14.5}{1996 - 1991} = \dfrac{5}{5} \Rightarrow m = 1$. Because it's slope is higher, the period from

    1991-1996 increased more rapidly.

40. For a given x-coordinate, add the y-coordinates on the graph of $B(x)$ and $G(x)$ to obtain the corresponding

    y-coordinate on the graph of $T(x)$.

41. $(T - S)(2000) \approx 19 - 13 \approx 6$; It represents the dollars in billions spent for general science in 2000.

42. $(T - G)(2005) \approx 23 - 8 \approx 15$; It represents the dollars in billions spent for space and other technologies in 2005.

43. From the graph, the only level part of the graph is: space and other technologies from 1995-2000.

44. From the graph, space and other technologies from 2000-2005, increases the most.

45. (a) $(f \circ g)(4) = f[g(4)]$, so from the graph find $g(4) = 0$. Now find $f(0) = -4$, therefore $(f \circ g)(4) = -4$.

    (b) $(g \circ f)(3) = g[f(3)]$, so from the graph find $f(3) = 2$. Now find $g(2) = 2$, therefore $(g \circ f)(3) = 2$.

    (c) $(f \circ f)(2) = f[f(2)]$, so from the graph find $f(2) = 0$. Now find $f(0) = -4$, therefore $(f \circ f)(2) = -4$.

46. (a) $(f \circ g)(2) = f[g(2)]$, so from the graph find $g(2) = -2$. Now find $f(-2) = -4$, therefore $(f \circ g)(2) = -4$.

    (b) $(g \circ g)(0) = g[g(0)]$, so from the graph find $g(0) = 2$. Now find $g(2) = -2$, therefore $(g \circ g)(0) = -2$.

    (c) $(g \circ f)(4) = g[f(4)]$, so from the graph find $f(4) = 2$. Now find $g(2) = -2$, therefore $(g \circ f)(4) = -2$.

47. (a) $(f \circ g)(1) = f[g(1)]$, so from the graph find $g(1) = 2$. Now find $f(2) = -3$, therefore $(f \circ g)(1) = -3$.

    (b) $(g \circ f)(-2) = g[f(-2)]$, so from the graph find $f(-2) = -3$. Now find $g(-3) = -2$, therefore $(g \circ f)(-2) = -2$.

    (c) $(g \circ g)(-2) = g[g(-2)]$, so from the graph find $g(-2) = -1$. Now find $g(-1) = 0$, therefore $(g \circ g)(-1) = 0$.

48. (a) $(f \circ g)(-2) = f[g(-2)]$, so from the graph find $g(-2) = 4$. Now find $f(4) = 2$, therefore $(f \circ g)(-2) = 2$.

    (b) $(g \circ f)(1) = g[f(1)]$, so from the graph find $f(1) = 1$. Now find $g(1) = 1$, therefore $(g \circ f)(1) = 1$.

    (c) $(f \circ f)(0) = f[f(0)]$, so from the graph find $f(0) = 0$. Now find $f(0) = 0$, therefore $(f \circ f)(0) = 0$.

49. (a) $(g \circ f)(1) = g[f(1)]$, so from the table find $f(1) = 4$. Now find $g(4) = 5$, therefore $(g \circ f)(1) = 5$.

    (b) $(f \circ g)(4) = f[g(4)]$, so from the table find $g(4) = 5$. Now we find $f(5)$ is undefined, therefore $(f \circ g)(4)$

    is undefined.

    (c) $(f \circ f)(3) = f[f(3)]$, so from the table find $f(3) = 1$. Now find $f(1) = 4$, therefore $(f \circ f)(3) = 4$.

50. (a) $(g \circ f)(1) = g[f(1)]$, so from the table find $f(1) = 2$. Now find $g(2) = 4$, therefore $(g \circ f)(1) = 4$.

(b) $(f \circ g)(4) = f[g(4)]$, so from the table we find $g(4)$ is undefined, therefore $(f \circ g)(4)$ is undefined.

(c) $(f \circ f)(3) = f[f(3)]$, so from the table find $f(3) = 6$. Now find $f(6) = 7$, therefore $(f \circ f)(3) = 7$.

51. From the table, $g(3) = 4$ and $f(4) = 2$.

52. From the table, $f(6) = 7$ and $g(7) = 0$.

53. Since $Y_3 = Y_1 \circ Y_2$ and $X = -1$, we solve $Y_1[Y_2(-1)]$. First solve $Y_2 = (-1)^2 = 1$, now solve

$Y_1 = 2(1) - 5 = -3$, therefore $Y_3 = -3$.

54. Since $Y_3 = Y_1 \circ Y_2$ and $X = -2$, we solve $Y_1[Y_2(-2)]$. First solve $Y_2 = (-2)^2 = 4$, now solve

$Y_1 = 2(4) - 5 = 3$, therefore $Y_3 = 3$.

55. Since $Y_3 = Y_1 \circ Y_2$ and $X = 7$, we solve $Y_1[Y_2(7)]$. First solve $Y_2 = (7)^2 = 49$, now solve

$Y_1 = 2(49) - 5 = 93$, therefore $Y_3 = 93$.

56. Since $Y_3 = Y_1 \circ Y_2$ and $X = 8$, we solve $Y_1[Y_2(8)]$. First solve $Y_2 = (8)^2 = 64$, now solve

$Y_1 = 2(64) - 5 = 123$, therefore $Y_3 = 123$.

57. (a) $(f \circ g)(x) = f[g(x)] = (x^2 + 3x - 1)^3$; all values can be input for x, therefore the domain is: $(-\infty, \infty)$.

(b) $(g \circ f)(x) = g[f(x)] = (x^3)^2 + 3(x^3) - 1 = x^6 + 3x^3 - 1$; all values can be input for x, therefore the

domain is: $(-\infty, \infty)$.

(c) $(f \circ f)(x) = f[f(x)] = (x^3)^3 = x^9$; all values can be input for x, therefore the domain is: $(-\infty, \infty)$.

58. (a) $(f \circ g)(x) = f[g(x)] = 2 - \left(\dfrac{1}{x^2}\right) = 2 - \dfrac{1}{x^2}$; all values can be input for x, except 0, therefore the domain

is: $(-\infty, 0) \cup (0, \infty)$.

(b) $(g \circ f)(x) = g[f(x)] = \dfrac{1}{(2 - x)^2}$; all values can be input for x, except 2, therefore the domain

is: $(-\infty, 2) \cup (2, \infty)$.

(c) $(f \circ f)(x) = f[f(x)] = 2 - (2 - x) = x$; all values can be input for x, therefore the domain is: $(-\infty, \infty)$.

59. (a) $(f \circ g)(x) = f[g(x)] = (\sqrt{1 - x})^2 = 1 - x$; all values less than 1 can be input for x, therefore the domain

is: $(-\infty, 1]$.

(b) $(g \circ f)(x) = g[f(x)] = \sqrt{1 - (x^2)} = \sqrt{1 - x^2}$; only values where $x^2 \leq 1$ can be input for x, therefore the

domain is: $[-1, 1]$.

(c) $(f \circ f)(x) = f[f(x)] = (x^2)^2 = x^4$; all values can be input for x, therefore the domain is: $(-\infty, \infty)$.

60. (a) $(f \circ g)(x) = f[g(x)] = (x^4 + x^2 - 3x - 4) + 2 = x^4 + x^2 - 3x - 2$; all values can be input for x,

therefore the domain is: $(-\infty, \infty)$.

(b) $(g \circ f)(x) = g[f(x)] = (x + 2)^4 + (x + 2)^2 - 3(x + 2) - 4$; all values can be input for x, therefore the

domain is: $(-\infty, \infty)$.

(c) $(f \circ f)(x) = f[f(x)] = (x + 2) + 2 = x + 4$; all values can be input for x, therefore the domain is: $(-\infty, \infty)$.

61. (a) $(f \circ g)(x) = f[g(x)] = \dfrac{1}{(5x) + 1} = \dfrac{1}{5x + 1}$; all values can be input for x, except $-\dfrac{1}{5}$, therefore the

    domain is: $\left(-\infty, -\dfrac{1}{5}\right) \cup \left(-\dfrac{1}{5}, \infty\right)$.

    (b) $(g \circ f)(x) = g[f(x)] = 5\left(\dfrac{1}{x + 1}\right) = \dfrac{5}{x + 1}$; all values can be input for x, except -1, therefore the

    domain is: $(-\infty, -1) \cup (-1, \infty)$.

    (c) $(f \circ f)(x) = f[f(x)] = \dfrac{1}{\left(\frac{1}{x + 1}\right) + 1} = \dfrac{1}{\frac{1}{x + 1} + \frac{x + 1}{x + 1}} = \dfrac{1}{\frac{x + 2}{x + 1}} = \dfrac{x + 1}{x + 2}$; all values can be input for x,

    except those that make $\dfrac{x + 2}{x + 1} = 0$ or undefined. That would be -1 and -2, therefore the domain is:

    $(-\infty, -2) \cup (-2, -1) \cup (-1, \infty)$.

62. (a) $(f \circ g)(x) = f[g(x)] = (\sqrt{4 - x^2}) + 4 = \sqrt{4 - x^2} + 4$; only values where $x^2 \le 4$ can be input for x,

    therefore the domain is: $[-2, 2]$.

    (b) $(g \circ f)(x) = g[f(x)] = \sqrt{4 - (x + 4)^2}$; only values where $(x + 4)^2 \le 4$ can be input for x, therefore the

    domain is: $[-6, -2]$

    (c) $(f \circ f)(x) = f[f(x)] = (x + 4) + 4 = x + 8$; all values can be input for x, therefore the domain is: $(-\infty, \infty)$.

63. (a) $(f \circ g)(x) = f[g(x)] = 2(4x^3 - 5x^2) + 1 = 8x^3 - 10x^2 + 1$; all values can be input for x, therefore the

    domain is: $(-\infty, \infty)$.

    (b) $(g \circ f)(x) = g[f(x)] = 4(2x + 1)^3 - 5(2x + 1)^2 = 4(8x^3 + 12x^2 + 6x + 1) - 5(4x^2 + 4x + 1) =$

    $32x^3 + 48x^2 + 24x + 4 - (20x^2 + 20x + 5) = 32x^3 + 28x^2 + 4x - 1$; all values can be input for x,

    therefore the domain is: $(-\infty, \infty)$.

    (c) $(f \circ f)(x) = f[f(x)] = 2(2x + 1) + 1 = 4x + 3$; all values can be input for x, therefore the domain

    is: $(-\infty, \infty)$.

64. (a) $(f \circ g)(x) = f[g(x)] = \dfrac{(2x + 3) - 3}{2} = \dfrac{2x}{2} = x$; all values can be input for x, therefore the domain

    is: $(-\infty, \infty)$.

    (b) $(g \circ f)(x) = g[f(x)] = 2\left(\dfrac{x - 3}{2}\right) + 3 = (x - 3) + 3 = x$; all values can be input for x, therefore the

    domain is: $(-\infty, \infty)$.

    (c) $(f \circ f)(x) = f[f(x)] = \dfrac{\left(\frac{x - 3}{2}\right) - 3}{2} = \dfrac{\left(\frac{x - 3}{2} - \frac{6}{2}\right)}{2} = \dfrac{\left(\frac{x - 9}{2}\right)}{2} = \dfrac{x - 9}{4}$; all values can be input for x,

    therefore the domain is: $(-\infty, \infty)$.

65. $(f \circ g)(x) = f[g(x)] = 4\left(\dfrac{1}{4}(x - 2)\right) + 2 = x - 2 + 2 = x$

    $(g \circ f)(x) = g[f(x)] = \dfrac{1}{4}((4x + 2) - 2) = \dfrac{1}{4}(4x) = x$

66. $(f \circ g)(x) = f[g(x)] = -3\left(-\dfrac{1}{3}x\right) = x$

   $(g \circ f)(x) = g[f(x)] = -\dfrac{1}{3}(-3x) = x$

67. $(f \circ g)(x) = f[g(x)] = \sqrt[3]{5\left(\dfrac{1}{5}x^3 - \dfrac{4}{5}\right) + 4} = \sqrt[3]{(x^3 - 4) + 4} = \sqrt[3]{x^3} = x$

   $(g \circ f)(x) = g[f(x)] = \dfrac{1}{5}(\sqrt[3]{5x + 4})^3 - \dfrac{4}{5} = \dfrac{1}{5}(5x + 4) - \dfrac{4}{5} = x + \dfrac{4}{5} - \dfrac{4}{5} = x$

68. $(f \circ g)(x) = f[g(x)] = \sqrt[3]{(x^3 - 1) + 1} = \sqrt[3]{x^3} = x$

   $(g \circ f)(x) = g[f(x)] = (\sqrt[3]{x + 1})^3 - 1 = x + 1 - 1 = x$

69. Graph $y_1 = \sqrt[3]{x - 6}$, $y_2 = x^3 + 6$, and $y_3 = x$ in the same viewing window. See Figures 69. The graph of y_2 can be obtained by *reflecting* the graph of y_1 across the line $y_3 = x$.

70. Graph $y_1 = 5x - 3$, $y_2 = \dfrac{1}{5}(x + 3)$, and $y_3 = x$ in the same viewing window. See Figures 70. The graph of y_2 can be obtained by *reflecting* the graph of y_1 across the line $y_3 = x$.

[−10, 10] by [−10, 10]   [−10, 10] by [−10, 10]
Xscl = 1   Yscl = 1   Xscl = 1   Yscl = 1

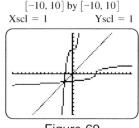   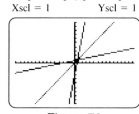

Figure 69   Figure 70

71. Using $\dfrac{f(x + h) - f(x)}{h}$ gives: $\dfrac{4(x + h) + 3 - (4x + 3)}{h} = \dfrac{4x + 4h + 3 - 4x - 3}{h} = \dfrac{4h}{h} = 4.$

72. Using $\dfrac{f(x + h) - f(x)}{h}$ gives: $\dfrac{5(x + h) - 6 - (5x - 6)}{h} = \dfrac{5x + 5h - 6 - 5x + 6}{h} = \dfrac{5h}{h} = 5.$

73. Using $\dfrac{f(x + h) - f(x)}{h}$ gives: $\dfrac{-6(x + h)^2 - (x + h) + 4 - (-6x^2 - x + 4)}{h} =$

   $\dfrac{-6(x^2 + 2xh + h^2) - x - h + 4 + 6x^2 + x - 4}{h} = \dfrac{-6x^2 - 12xh - 6h^2 - x - h + 4 + 6x^2 + x - 4}{h} =$

   $\dfrac{-12xh - 6h^2 - h}{h} = -12x - 6h - 1.$

74. Using $\dfrac{f(x + h) - f(x)}{h}$ gives: $\dfrac{\frac{1}{2}(x + h)^2 + 4(x + h) - (\frac{1}{2}x^2 + 4x)}{h} =$

   $\dfrac{\frac{1}{2}(x^2 + 2xh + h^2) + 4x + 4h - \frac{1}{2}x^2 - 4x}{h} = \dfrac{\frac{1}{2}x^2 + xh + \frac{1}{2}h^2 + 4x + 4h - \frac{1}{2}x^2 - 4x}{h} =$

   $\dfrac{xh + \frac{1}{2}h^2 + 4h}{h} = x + \dfrac{1}{2}h + 4.$

75. Using $\dfrac{f(x + h) - f(x)}{h}$ gives: $\dfrac{(x + h)^3 - x^3}{h} = \dfrac{x^3 + 3x^2h + 3xh^2 + h^3 - x^3}{h} = \dfrac{3x^2h + 3xh^2 + h^3}{h} =$

$3x^2 + 3xh + h^2$.

76. Using $\dfrac{f(x + h) - f(x)}{h}$ gives: $\dfrac{-2(x + h)^3 - (-2x^3)}{h} = \dfrac{-2(x^3 + 3x^2h + 3xh^2 + h^3) + 2x^3}{h} =$

$\dfrac{-2x^3 - 6x^2h - 6xh^2 - 2h^3 + 2x^3}{h} = \dfrac{-6x^2h - 6xh^2 - 2h^3}{h} = -6x^2 - 6xh - 2h^2$.

77. One possible solution is: $f(x) = x^2$ and $g(x) = 6x - 2$. Then $(f \circ g)(x) = f[g(x)] = (6x - 2)^2$.

78. One possible solution is: $f(x) = x^2$ and $g(x) = 11x^2 + 12x$. Then $(f \circ g)(x) = f[g(x)] = (11x^2 + 12x)^2$.

79. One possible solution is: $f(x) = \sqrt{x}$ and $g(x) = x^2 - 1$. Then $(f \circ g)(x) = f[g(x)] = \sqrt{x^2 - 1}$.

80. One possible solution is: $f(x) = x^3$ and $g(x) = 2x - 3$. Then $(f \circ g)(x) = f[g(x)] = (2x - 3)^3$.

81. One possible solution is: $f(x) = \sqrt{x} + 12$ and $g(x) = 6x$. Then $(f \circ g)(x) = f[g(x)] = \sqrt{6x} + 12$.

82. One possible solution is: $f(x) = \sqrt[3]{x} - 4$ and $g(x) = 2x + 3$. Then $(f \circ g)(x) = f[g(x)] = \sqrt[3]{2x + 3} - 4$.

83. (a) With a cost of $10 to produce each item and a fixed cost of $500, the cost function is: $C(x) = 10x + 500$.

(b) With a selling price of $35 for each item, the revenue function is: $R(x) = 35x$.

(c) The profit function is $P(x) = R(x) - C(x) \Rightarrow P(x) = 35x - (10x + 500) \Rightarrow P(x) = 25x - 500$.

(d) A profit is shown when $P(x) > 0 \Rightarrow 25x - 500 > 0 \Rightarrow 25x > 500 \Rightarrow x > 20$. Therefore, 21 items must be produced and sold to realize a profit.

(e) Graph $y_1 = 25x - 500$, the smallest whole number for which $P(x) > 0$ is 21. Use a window of $[0, 30]$ by $[-1000, 500]$, for example.

84. (a) With a cost of $11 to produce each item and a fixed cost of $180, the cost function is: $C(x) = 11x + 180$.

(b) With a selling price of $20 for each item, the revenue function is: $R(x) = 20x$.

(c) The profit function is $P(x) = R(x) - C(x) \Rightarrow P(x) = 20x - (11x + 180) \Rightarrow P(x) = 9x - 180$.

(d) A profit is shown when $P(x) > 0 \Rightarrow 9x - 180 > 0 \Rightarrow 9x > 180 \Rightarrow x > 20$. Therefore, 21 items must be produced and sold to realize a profit.

(e) Graph $y_1 = 9x - 180$, the smallest whole number for which $P(x) > 0$ is 21. Use a window of $[-5, 30]$ by $[-200, 200]$, for example.

85. (a) With a cost of $100 to produce each item and a fixed cost of $2700, the cost function is:

$C(x) = 100x + 2700$.

(b) With a selling price of $280 for each item, the revenue function is: $R(x) = 280x$.

(c) The profit function is $P(x) = R(x) - C(x) \Rightarrow P(x) = 280x - (100x + 2700) \Rightarrow P(x) = 180x - 2700$.

(d) A profit is shown when $P(x) > 0 \Rightarrow 180x - 2700 > 0 \Rightarrow 180x > 2700 \Rightarrow x > 15$. Therefore, 16 items must be produced and sold to realize a profit.

(e) Graph $y_1 = 180x - 2700$, the smallest whole number for which $P(x) > 0$ is 16. Use a window of $[0, 30]$ by $[-3000, 500]$, for example.

86. (a) With a cost of $200 to produce each item and a fixed cost of $1000, the cost function is:

$$C(x) = 200x + 1000.$$

   (b) With a selling price of $240 for each item, the revenue function is: $R(x) = 240x$.

   (c) The profit function is $P(x) = R(x) - C(x) \Rightarrow P(x) = 240x - (200x + 1000) \Rightarrow P(x) = 40x - 1000$.

   (d) A profit is shown when $P(x) > 0 \Rightarrow 40x - 1000 > 0 \Rightarrow 40x > 1000 \Rightarrow x > 25$. Therefore, 26 items must be produced and sold to realize a profit.

   (e) Graph $y_1 = 40x - 1000$, the smallest whole number for which $P(x) > 0$ is 26. Use a window of $[-5, 40]$ by $[-1200, 600]$, for example.

87. (a) If $V(r) = \frac{4}{3}\pi r^3$, then a 3 inch increase would be: $V(r) = \frac{4}{3}\pi(r + 3)^3$, and the volume gained would be:

$$V(r) = \frac{4}{3}\pi(r + 3)^3 - \frac{4}{3}\pi r^3.$$

   (b) Graph $y_1 = \frac{4}{3}\pi(x + 3)^3 - \frac{4}{3}\pi x^3$ in the window $[0, 10]$ by $[0, 1500]$. See Figure 87. Although this appears to be a portion of a parabola, it is actually a cubic function.

   (c) From the graph in exercise 87b, an input value of $x = 4$ results in a gain of: $y \approx 1168.67$.

   (d) $V(4) = \frac{4}{3}\pi(4 + 3)^3 - \frac{4}{3}\pi(4)^3 = \frac{4}{3}\pi(343) - \frac{4}{3}\pi(64) = \frac{1372}{3}\pi - \frac{256}{3}\pi = \frac{1116}{3}\pi = 372\pi \approx 1168.67$.

88. If $S(r) = 4\pi r^2$, then doubling the radius would give us a surface area gained function of:

$$S(r) = 4\pi(2r)^2 - 4\pi r^2 = 16\pi r^2 - 4\pi r^2 = 12\pi r^2.$$

89. (a) If $x = $ width, then $2x = $ length. Since the perimeter formula is: $P = 2W + 2L$ our perimeter function is:

$$P(x) = 2(x) + 2(2x) = 2x + 4x \Rightarrow P(x) = 6x. \text{ This is a linear function.}$$

   (b) Graph $P(x) = 6x$ in the window $[0, 10]$ by $[1, 100]$. See Figure 89b. From the graph when $x = 4, y = 24$. The 4 represent the width of a rectangle and 24 represents the perimeter.

   (c) If $x = 4$ is the width of a rectangle then $2x = 8$ is the length. See Figure 89c. Using the standard perimeter formula yields: $P = 2(4) + 2(8) = 24$. This compares favorably with the graph result in part b.

   (d) (Answers may vary.) If the perimeter y of a rectangle satisfying the given conditions is 36, then the width x is 6. See Figure 89d.

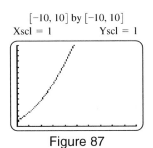

$[-10, 10]$ by $[-10, 10]$
Xscl = 1   Yscl = 1

Figure 87

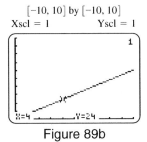

$[-10, 10]$ by $[-10, 10]$
Xscl = 1   Yscl = 1

Figure 89b

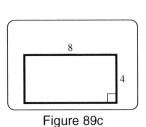

Figure 89c

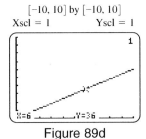

$[-10, 10]$ by $[-10, 10]$
Xscl = 1   Yscl = 1

Figure 89d

90. (a) If $x = 4s$ then $s = \dfrac{x}{4}$.

    (b) If $y = s^2$ and $s = \dfrac{x}{4}$, then $y(x) = \left(\dfrac{x}{4}\right)^2 \Rightarrow y(x) = \dfrac{x^2}{16}$.

    (c) Sine x is perimeter and $x = 6$, $y(6) = \dfrac{(6)^2}{16} = \dfrac{36}{16} = \dfrac{9}{4} = 2.25$.

    (d) Show that the point $(6, 2.25)$ is on the graph $y = \dfrac{x^2}{16}$. A square with perimeter 6 will have area 2.25 square units.

91. (a) $A(2x) = \dfrac{\sqrt{3}}{4}(2x)^2 = \dfrac{\sqrt{3}}{4}(4x^2) \Rightarrow A(2x) = \sqrt{3}\,x^2$

    (b) $A(x) = \dfrac{\sqrt{3}}{4}(16)^2 = \dfrac{\sqrt{3}}{4}(256) \Rightarrow A(x) = 64\sqrt{3}$ square units.

    (c) On the graph of $y = \dfrac{\sqrt{3}}{4}x^2$, locate the point where $x = 16$ to find $y \approx 110.85$, an approximation for $64\sqrt{3}$.

92. (a) If $A(r) = \pi r^2$ and $r(t) = 2t$ then $(A \circ r)(t) = A[r(t)] = A[2t] = \pi(2t)^2 = 4\pi t^2$.

    (b) $(A \circ r)(t)$ is a composite function that expresses the area of the circular region covered by the pollutants as a function of time t (in hours).

    (c) Since $t = 0$ is 8 A.M., noon would be $t = 4$. $(A \circ r)(4) = 4\pi(4)^2 = 64\pi$ mi^2.

    (d) Graph $y_1 = 4\pi x^2$ and show that for $x = 4$, $y \approx 201$ (an approximation for 64π).

93. (a) The function h is the addition of functions f and g.

x	1999	2000	2001	2002	2003
$h(x)$	76	82	79	89	103

    (b) The function h is the addition of functions f and g. Therefore. $h(x) = f(x) + g(x)$.

94. (a) The domain is the years. See Table. Therefore: $D = \{1998, 1999, 2000, 2001, 2002\}$.

x	1998	1999	2000	2001	2002
$h(x)$	94.2	95.3	99.3	106.4	93.2

    (b) The function h computes the value of animals produced in the U.S. and sold in billions of dollars.

95. (a) $(f + g)(1970) = 32.4 + 17.6 = 50.0$

    (b) The function $(f + g)(x)$ computes the total SO$_2$ emissions from burning coal and oil during year x.

    (c) Add functions f and g.

x	1860	1900	1940	1970	2000
$(f + g)(x)$	2.4	12.8	26.5	50.0	78.0

96. (a) The function h is the addition of functions f and g.

x	1990	2000	2010	2020	2030
$h(x)$	32	35.5	39	42.5	46

    (b) The function h is the addition of functions f and g. Therefore. $h(x) = f(x) + g(x)$.

97. (a) The function h is the subtraction of function f from g. Therefore. $h(x) = g(x) - f(x)$.

    (b) $h(1996) = g(1996) - f(1996) = 841 - 694 = 147$

    $h(2006) = g(2006) - f(2006) = 1165 - 1012 = 153$

    (c) Using the points $(1996, 147)$ and $(2006, 153)$ from part b, the slope is: $m = \dfrac{153 - 147}{2006 - 1996} = \dfrac{6}{10} = .6$.

    Now using point slope form: $y - 147 = .6(x - 1996) \Rightarrow y = .6(x - 1996) + 147$.

98. (a) Graph $h(x) = \dfrac{1900(x - 1982)^2 + 619}{3200(x - 1982)^2 + 1586}$, in the window $[1982, 1994]$ by $[0, 1]$. See Figure 98a.

    Approximately 59% of the people who contracted AIDS during this time period died.

    (b) Divide number of deaths by number of cases for each year. The results compare favorably with the graph.

    See Figure 98b.

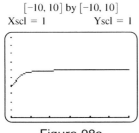

$[-10, 10]$ by $[-10, 10]$
Xscl = 1   Yscl = 1

Figure 98a

Year	1982	1984	1986	1988	1990	1992	1994
Ratio	.39	.51	.59	.58	.61	.60	.61

Figure 98b

Reviewing Basic Concepts (Sections 2.4—2.6)

1. (a) $\left| \dfrac{1}{2}x + 2 \right| = 4 \Rightarrow \dfrac{1}{2}x + 2 = 4 \Rightarrow \dfrac{1}{2}x = 2 \Rightarrow x = 4$ or $\dfrac{1}{2}x + 2 = -4 \Rightarrow \dfrac{1}{2}x = -6 \Rightarrow x = -12$.

    Therefore, the solution set is: $\{-12, 4\}$.

    (b) $\left| \dfrac{1}{2}x + 2 \right| > 4 \Rightarrow \dfrac{1}{2}x + 2 > 4 \Rightarrow \dfrac{1}{2}x > 2 \Rightarrow x > 4$ or $\dfrac{1}{2}x + 2 < -4 \Rightarrow \dfrac{1}{2}x < -6 \Rightarrow x < -12$.

    Therefore, the solution interval is: $(-\infty, -12) \cup (4, \infty)$.

    (c) $\left| \dfrac{1}{2}x + 2 \right| \le 4 \Rightarrow -4 \le \dfrac{1}{2}x + 2 \le 4 \Rightarrow -6 \le \dfrac{1}{2}x \le 2 \Rightarrow -12 \le x \le 4$.

    Therefore, the solution interval is: $[-12, 4]$.

2. For the graph of $y = |f(x)|$, we reflect the graph of $y = f(x)$ across the x-axis for all points for which $y < 0$.

    Where $y \ge 0$, the graph remains unchanged. See Figure 2.

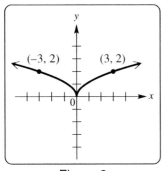

Figure 2

3. (a) The range of values for $|R_L - 26.75| \le 1.42$ is: $-1.42 \le R_L - 26.75 \le 1.42 \Rightarrow 25.33 \le R_L \le 28.17$.

    The range of values for $|R_E - 38.75| \le 2.17$ is: $-2.17 \le R_E - 38.75 \le 2.17 \Rightarrow 36.58 \le R_E \le 40.92$.

    (b) If $T_L = 225(R_L)$ then the range for T_L is: $225(25.33 \le T_L \le 28.17) = 5699.25 \le T_L \le 6338.25$.

    If $T_E = 225(R_E)$ then the range for T_L is: $225(36.58 \le T_E \le 40.92) = 8230.5 \le T_E \le 9207$.

4. (a) $f(-3) = 2(-3) + 3 = -3$ (b) $f(0) = (0)^2 + 4 = 4$ (c) $f(2) = (2)^2 + 4 = 8$

5. (a) See Figure 5a.

    (b) Graph $y_1 = (-x^2) * (x \le 0) + (x - 4) * (x > 0)$ in the window $[-10, 10]$ by $[-10, 10]$. See Figure 5b.

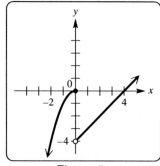

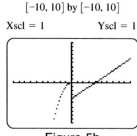

[-10, 10] by [-10, 10]
Xscl = 1   Yscl = 1

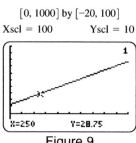

[0, 1000] by [-20, 100]
Xscl = 100   Yscl = 10

Figure 5a   Figure 5b   Figure 9

6. (a) $(f + g)(x) = (-3x - 4) + (x^2) = x^2 - 3x - 4$. Therefore, $(f + g)(1) = (1)^2 - 3(1) - 4 = -6$.

    (b) $(f - g)(x) = (-3x - 4) - (x^2) = -x^2 - 3x - 4$. Therefore, $(f - g)(3) = -(3)^2 - 3(3) - 4 = -22$.

    (c) $(fg)(x) = (-3x - 4)(x^2) = -3x^3 - 4x^2$. Therefore, $(fg)(-2) = -3(-2)^3 - 4(-2)^2 = 24 - 16 = 8$.

    (d) $\left(\dfrac{f}{g}\right)(x) = \dfrac{-3x - 4}{x^2}$. Therefore, $\left(\dfrac{f}{g}\right)(-3) = \dfrac{-3(-3) - 4}{(-3)^2} = \dfrac{5}{9}$.

    (e) $(f \circ g)(x) = f[g(x)] = -3(x^2) - 4 \Rightarrow (f \circ g)(x) = -3x^2 - 4$

    (f) $(g \circ f)(x) = g[f(x)] = (-3x - 4)^2 \Rightarrow (g \circ f)(x) = 9x^2 + 24x + 16$

7. One of many possible solutions for $(f \circ g)(x) = h(x)$ is: $f(x) = x^4$ and $g(x) = x + 2$. Then

    $(f \circ g)(x) = f[g(x)] = (x + 2)^4$.

8. $\dfrac{-2(x + h)^2 + 3(x + h) - 5 - (-2x^2 + 3x - 5)}{h} = \dfrac{-2(x^2 + 2xh + h^2) + 3x + 3h - 5 + 2x^2 - 3x + 5}{h} =$

    $\dfrac{-2x^2 - 4xh - 2h^2 + 3x + 3h - 5 + 2x^2 - 3x + 5}{h} = \dfrac{-4xh - 2h^2 + 3h}{h} = -4x - 2h + 3$.

9. (a) At 4% simple interest the equation for interest earned is: $y_1 = .04x$.

    (b) If he invested x dollars in the first account, then he invested $x + 500$ in the second account. The equation
        for the amount of interest earned on this account is: $y_2 = .025(x + 500) \Rightarrow y_2 = .025x + 12.5$.

    (c) It represents the total interest earned in both accounts for 1 year.

    (d) Graph $y_1 + y_2 = .04x + (.025x + 12.5) \Rightarrow y_1 + y_2 = .04x + .025x + 12.5$ in the window
        $[0, 1000]$ by $[0, 100]$. See Figure 9. An input value of $x = 250$, results in \$28.75 earned interest.

    (e) At $x = 250$, $y_1 + y_2 = .04(250) + .025(250) + 12.5 = 10 + 6.25 + 12.5 = \$28.75$.

10. If the radius is r, then the height is $2r$ and the equation is

    $S = \pi r \sqrt{r^2 + (2r)^2} = \pi r \sqrt{r^2 + 4r^2} = \pi r \sqrt{5r^2} \Rightarrow S = \pi r^2 \sqrt{5}$.

Chapter 2 Review Exercises

The graphs for exercises 1–10 can be found in the "Function Capsule" boxes located in section 2.1 in the text.

1. True.   Both $f(x) = x^2$ and $f(x) = |x|$ have the interval: $[0, \infty)$ as the range.

2. True.   Both $f(x) = x^2$ and $f(x) = |x|$ increase on the interval: $[0, \infty)$.

3. False.   The function $f(x) = \sqrt{x}$ has the domain: $[0, \infty)$ and $f(x) = \sqrt[3]{x}$ the domain: $(-\infty, \infty)$.

4. False.   The function $f(x) = \sqrt[3]{x}$ increases on its entire domain.

5. True.   The function $f(x) = x$ has a domain and range of: $(-\infty, \infty)$.

6. False.   The function $f(x) = \sqrt{x}$ is not defined on $(-\infty, 0)$, so certainly cannot be continuous.

7. True.   All of the functions show increases on the interval: $[0, \infty)$.

8. True.   Both $f(x) = x$ and $f(x) = x^3$ have graphs that are symmetric with respect to the origin.

9. True.   Both $f(x) = x^2$ and $f(x) = |x|$ have graphs that are symmetric with respect to the y-axis.

10. True.   No graphs are symmetric with respect to the x-axis.

11. Only values where $x \geq 0$ can be input for x, therefore the domain of $f(x) = \sqrt{x}$ is: $[0, \infty)$.

12. Only positive solutions are possible in absolute value functions, therefore the range of $f(x) = \sqrt{x}$ is: $[0, \infty)$.

13. All solutions are possible in cube root functions, therefore the range of $f(x) = \sqrt[3]{x}$ is: $(-\infty, \infty)$.

14. All values can be input for x, therefore the domain of $f(x) = x^2$ is: $(-\infty, \infty)$.

15. The function $f(x) = \sqrt[3]{x}$ increases for all inputs for x, therefore the interval is: $(-\infty, \infty)$.

16. The function $f(x) = |x|$ increases for all inputs where $x \geq 0$, therefore the interval is: $[0, \infty)$.

17. The equation $y^2 = x$ is the equation $y = \sqrt{x}$. Only values where $x \geq 0$ can be input for x, therefore the domain of $y = \sqrt{x}$ is: $[0, \infty)$.

18. The equation $y^2 = x$ is the equation $y = \sqrt{x}$. Square root functions have both positive and negative solutions and all solution are possible, therefore the range of $y = \sqrt{x}$ is: $(-\infty, \infty)$.

19. The graph of $f(x) = (x + 3) - 1$ is the graph $y = x$ shifted 3 units to the left and 1 unit downward. See Figure 19.

20. The graph of $f(x) = -\dfrac{1}{2}x + 1$ is the graph $y = x$ reflected across the x-axis, vertically shrunk by a factor of $\dfrac{1}{2}$, and shifted 1 unit upward. See Figure 20.

21. The graph of $f(x) = (x + 1)^2 - 2$ is the graph $y = x^2$ shifted 1 unit to the left and 2 units downward. See Figure 21.

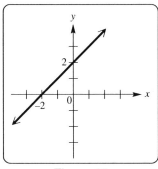

Figure 19

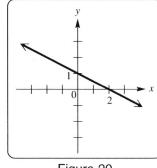

Figure 20

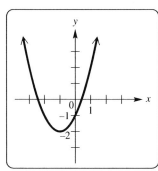

Figure 21

22. The graph of $f(x) = -2x^2 + 3$ is the graph $y = x^2$ reflected across the *x*-axis, vertically stretched by a factor of 2, and shifted 3 units upward. See Figure 22.

23. The graph of $f(x) = -x^3 + 2$ is the graph $y = x^3$ reflected across the *x*-axis and shifted 2 units upward. See Figure 23.

24. The graph of $f(x) = (x - 3)^3$ is the graph $y = x^3$ shifted 3 units to the right. See Figure 24.

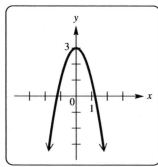

Figure 22

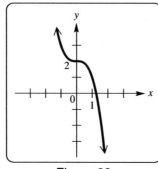

Figure 23

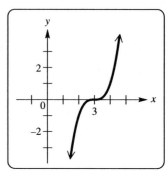

Figure 24

25. The graph of $f(x) = \sqrt{\frac{1}{2}x}$ is the graph $y = \sqrt{x}$ horizontally stretched by a factor of 2. See Figure 25.

26. The graph of $f(x) = \sqrt{x - 2} + 1$ is the graph $y = \sqrt{x}$ shifted 2 units to the right and 1 unit upward. See Figure 26.

27. The graph of $f(x) = 2\sqrt[3]{x}$ is the graph $y = \sqrt[3]{x}$ vertically stretched by a factor of 2. See Figure 27.

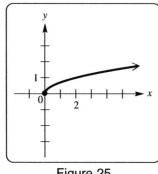

Figure 25

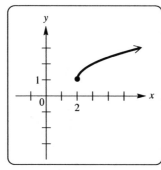

Figure 26

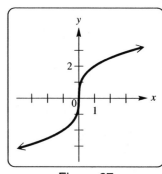

Figure 27

28. The graph of $f(x) = \sqrt[3]{x} - 2$ is the graph $y = \sqrt[3]{x}$ shifted 2 units downward. See Figure 28.

29. The graph of $f(x) = |x - 2| + 1$ is the graph $y = |x|$ shifted 2 units right and 1 unit upward. See Figure 29.

30. The graph of $f(x) = |-2x + 3|$ is the graph $y = |x|$ horizontally shrunk by a factor of $\frac{1}{2}$, shifted $\left(\frac{1}{2}\right)(3)$ or $\frac{3}{2}$ units to the left, and reflected across the *y*-axis. See Figure 30.

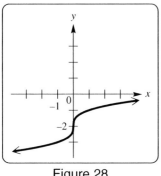

Figure 28

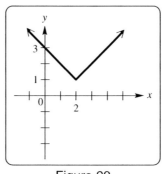

Figure 29

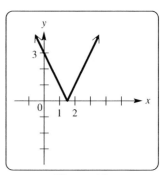

Figure 30

31. (a) From the graph, the function is continuous for the intervals: $(-\infty, -2)$, $[-2, 1]$, and $(1, \infty)$.

   (b) From the graph, the function is increasing for the interval: $[-2, 1]$.

   (c) From the graph, the function is decreasing for the interval: $(-\infty, -2)$.

   (d) From the graph, the function is constant for the interval: $(1, \infty)$.

   (e) From the graph, all values can be input for x, therefore the domain is: $(-\infty, \infty)$.

   (f) From the graph, the possible values of y or the range is: $\{-2\} \cup [-1, 1] \cup (2, \infty)$.

32. $x = y^2 - 4 \Rightarrow y^2 = x + 4 \Rightarrow y = \sqrt{x + 4}$ and $y = -\sqrt{x + 4}$

33. From the graph, the relation is symmetric with respect to the x-axis, y-axis, and origin. The relation is not a function since some inputs x have two outputs y.

34. If $F(x) = x^3 - 6$, then $F(-x) = (-x)^3 - 6 \Rightarrow F(-x) = -x^3 - 6$ and $-F(x) = -(x^3 - 6) \Rightarrow -F(x) = -x^3 + 6$. Since $F(x) \neq F(-x) \neq -F(x)$, the function has no symmetry and is neither an even nor an odd function.

35. If $f(x) = |x| + 4$, then $f(-x) = |(-x)| + 4 \Rightarrow f(-x) = |x| + 4$ and $-f(x) = -|x| - 4$. Since $f(-x) = f(x)$, the function is symmetric with respect to the y-axis and is an even function.

36. If $f(x) = \sqrt{x - 5}$, then $f(-x) = \sqrt{(-x) - 5}$ and $-f(x) = -\sqrt{x - 5}$. Since $f(x) \neq f(-x) \neq -f(x)$, the function has no symmetry and is neither an even nor an odd function.

37. If $y^2 = x - 5$ then $y = \pm\sqrt{x - 5}$. Since $f(x) = -\sqrt{x - 5}$ is the reflection of $f(x) = \sqrt{x - 5}$ across the x-axis, the relation has symmetry with respect to the x-axis. Also, one x input can produce two y outputs. The relation is not a function.

38. If $f(x) = 3x^4 + 2x^2 + 1$, then $f(-x) = 3(-x)^4 + 2(-x)^2 + 1 \Rightarrow f(-x) = 3x^4 + 2x^2 + 1$ and $-f(x) = -3x^4 - 2x^2 - 1$. Since $f(-x) = f(x)$, the function is symmetric with respect to the y-axis and is an even function.

39. True, a graph that is symmetrical with respect to the x-axis means that for every (x, y) there is also $(x, -y)$, which is not a function.

40. True, since an even function and one that is symmetric with respect to the y-axis both contain the points (x, y) and $(-x, y)$.

41. True, since an odd function and one that is symmetric with respect to the origin both contain the points (x, y) and $(-x, -y)$.

42. False, for an even function, if (a, b) is on the graph, then $(-a, b)$ is on the graph and not $(a, -b)$.

    For example, $f(x) = x^2$ is even, and $(2, 4)$ is on the graph, but $(2, -4)$ is not.

43. False, for an odd function, if (a, b) is on the graph, then $(-a, -b)$ is on the graph and not $(-a, b)$.

    For example, $f(x) = x^3$ is odd, and $(2, 8)$ is on the graph, but $(-2, 8)$ is not.

44. True, if $(x, 0)$ is on the graph of $f(x) = 0$, then $(-x, 0)$ is on the graph.

45. The graph of $y = -3(x + 4)^2 - 8$ is the graph of $y = x^2$ shifted 4 units to the left, vertically stretched by a factor of 3, reflected across the x-axis, and shifted 8 units downward.

46. The equation $y = \sqrt{x}$ reflected across the y-axis is: $y = \sqrt{-x}$, then reflected across the x-axis is: $y = -\sqrt{-x}$, now vertically shrunk by a factor of $\frac{2}{3}$ is: $y = -\frac{2}{3}\sqrt{-x}$, and finally shifted 4 units upward is: $y = -\frac{2}{3}\sqrt{-x} + 4$.

47. Shift the function f upward 3 units. See Figure 47.

48. Shift the function f to the right 2 units. See Figure 48.

49. Shift the function f to the left 3 units and downward 2 units. See Figure 49.

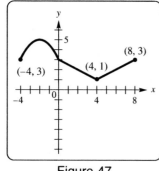

Figure 47

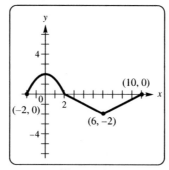

Figure 48

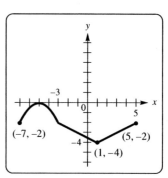

Figure 49

50. For values where $f(x) > 0$ the graph remains the same. For values where $f(x) < 0$ reflect the graph across the x-axis. See Figure 50.

51. Horizontally shrink the function f by a factor of $\frac{1}{4}$. See Figure 51.

52. Horizontally stretch the function f by a factor of 2. See Figure 52.

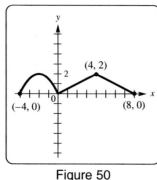

Figure 50

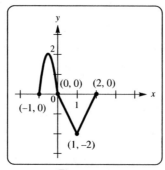

Figure 51

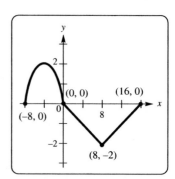

Figure 52

53. The function is shifted upward 4 units, therefore the domain remains the same: $[-3, 4]$ and the range is increased by 4 and is: $[2, 9]$.

54. The function is shifted left 10 units, therefore the domain is decreased by 10 and is: $[-13, -6]$; and the function is stretched vertically by a factor of 5, therefore the range is multiplied by 5 and is: $[-10, 25]$.

55. The function is horizontally shrunk by a factor of $\dfrac{1}{2}$, therefore the domain is divided by 2 and is: $\left[-\dfrac{3}{2}, 2\right]$; and the function is reflected across the x-axis, therefore the range is opposite of the original and is: $[-5, 2]$.

56. The function is shifted right 1 unit, therefore the domain is increased by 1 and is: $[-2, 5]$; and the function is also shifted upward 3 units, therefore the range is increased by 3 and is: $[1, 8]$.

57. We reflect the graph of $y = f(x)$ across the x-axis for all points for which $y < 0$. Where $y \geq 0$, the graph remains unchanged. See Figure 57.

58. We reflect the graph of $y = f(x)$ across the x-axis for all points for which $y < 0$. Where $y \geq 0$, the graph remains unchanged. See Figure 58.

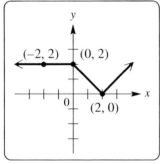
Figure 57

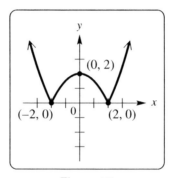
Figure 58

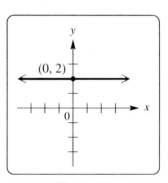
Figure 60

59. Since the range is $\{2\}$, $y \geq 0$, so the graph remains unchanged.

60. Since the range is $\{-2\}$, $y < 0$, so we reflect the graph across the x-axis. See Figure 60.

61. $|4x + 3| = 12 \Rightarrow 4x + 3 = 12 \Rightarrow 4x = 9 \Rightarrow x = \dfrac{9}{4}$ or $4x + 3 = -12 \Rightarrow 4x = -15 \Rightarrow x = -\dfrac{15}{4}$, therefore the solution set is: $\left\{-\dfrac{15}{4}, \dfrac{9}{4}\right\}$.

62. $|-2x - 6| + 4 = 1 \Rightarrow |-2x - 6| = -3$. Since an absolute value equation can not have a solution less than zero, the solution set is: $\varnothing$.

63. $|5x + 3| = |x + 11| \Rightarrow 5x + 3 = x + 11 \Rightarrow 4x = 8 \Rightarrow x = 2$ or $5x + 3 = -(x + 11) \Rightarrow 6x = -14 \Rightarrow x = -\dfrac{14}{6} = -\dfrac{7}{3}$, therefore the solution set is: $\left\{-\dfrac{7}{3}, 2\right\}$.

64. $|2x + 5| = 7 \Rightarrow 2x + 5 = 7 \Rightarrow 2x = 2 \Rightarrow x = 1$ or $2x + 5 = -7 \Rightarrow 2x = -12 \Rightarrow x = -6$, therefore the solution set is: $\{-6, 1\}$.

65. $|2x + 5| \leq 7 \Rightarrow -7 \leq 2x + 5 \leq 7 \Rightarrow -12 \leq 2x \leq 2 \Rightarrow -6 \leq x \leq 1$, therefore the interval is: $[-6, 1]$.

66. $|2x + 5| \geq 7 \Rightarrow 2x + 5 \geq 7 \Rightarrow 2x \geq 2 \Rightarrow x \geq 1$ or $2x + 5 \leq -7 \Rightarrow 2x \leq -12 \Rightarrow x \leq -6$, therefore the solution is the interval: $(-\infty, -6] \cup [1, \infty)$.

67. $|5x - 12| > 0 \Rightarrow 5x - 12 > 0 \Rightarrow 5x > 12 \Rightarrow x > \dfrac{12}{5}$ or $5x - 12 < 0 \Rightarrow 5x = 12 \Rightarrow x < \dfrac{12}{5}$,

therefore the solution is the interval: $\left(-\infty, \dfrac{12}{5}\right) \cup \left(\dfrac{12}{5}, \infty\right)$ or $\left\{x \mid x \neq \dfrac{12}{5}\right\}$.

68. Since an absolute value equation can not have a solution less than zero, the solution set is: $\varnothing$.

69. $2|3x - 1| + 1 = 21 \Rightarrow 2|3x - 1| = 20 \Rightarrow |3x - 1| = 10 \Rightarrow 3x - 1 = 10 \Rightarrow 3x = 11 \Rightarrow x = \dfrac{11}{3}$ or

$3x - 1 = -10 \Rightarrow 3x = -9 \Rightarrow x = -3$, therefore the solution set is: $\left\{-3, \dfrac{11}{3}\right\}$.

70. $|2x + 1| = |-3x + 1| \Rightarrow 2x + 1 = -3x + 1 \Rightarrow 5x = 0 \Rightarrow x = 0$ or $2x + 1 = -(-3x + 1) \Rightarrow$

$-x = -2 \Rightarrow x = 2$, therefore the solution set is: $\{0, 2\}$.

71. The x-coordinates of the points of intersection of the graphs are -6 and 1. Thus, $\{-6, 1\}$ is the solution set of

$y_1 = y_2$. The graph of y_1 lies on or below the graph of y_2 between -6 and 1, so the solution set of $y_1 \leq y_2$

is $[-6, 1]$. The graph of y_1 lies above the graph of y_2 everywhere else, so the solution set of $y_1 \geq y_2$

is $(-\infty, -6] \cup [1, \infty)$.

72. Graph $y_1 = |x + 1| + |x - 3|$ and $y_2 = 8$. See Figure 72. The intersections are $x = -3$ and $x = 5$,

therefore the solution set is: $\{-3, 5\}$.

   Check: $|(-3) + 1| + |(-3) - 3| = 8 \Rightarrow |-2| + |-6| = 8 \Rightarrow 2 + 6 = 8 \Rightarrow 8 = 8$ and

   $|(5) + 1| + |(5) - 3| = 8 \Rightarrow |6| + |2| = 8 \Rightarrow 6 + 2 = 8 \Rightarrow 8 = 8$

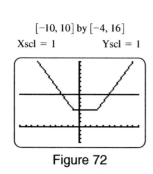

[-10, 10] by [-4, 16]
Xscl = 1   Yscl = 1

Figure 72

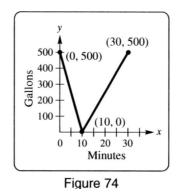

Figure 74

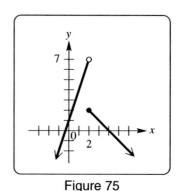

Figure 75

73. Initially, the car is at home. After traveling 30 mph for 1 hr, the car is 30 mi away from home. During the

second hour the car travels 20 mph until it is 50 mi away. During the third hour the car travels toward home at

30 mph until it is 20 mi away. During the fourth hour the car travels away from home at 40 mph until it is

60 mi away from home. During the last hour, the car travels 60 mi at 60 mph until it arrives home.

74. See Figure 74.

75. See Figure 75.

76. See Figure 76.

77. Graph $y_1 = (3x + 1) * (x < 2) + (-x + 4) * (x \geq 2)$ in the window $[-10, 10]$ by $[-10, 10]$. See Figure 77.

78. See Figure 78.

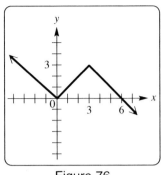
Figure 76

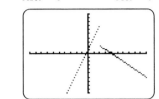
Figure 77

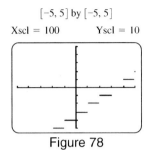
Figure 78

79. From the graphs $(f + g)(1) = 2 + 3 = 5$

80. From the graphs $(f - g)(0) = 1 - 4 = -3$

81. From the graphs $(fg)(-1) = (0)(3) = 0$

82. From the graphs $\left(\dfrac{f}{g}\right)(2) = \dfrac{3}{2}$

83. From the graphs $(f \circ g)(2) = f[g(2)] = f(2) = 3$

84. From the graphs $(g \circ f)(2) = g[f(2)] = g(3) = 2$

85. From the graphs $(g \circ f)(-4) = g[f(-4)] = g(2) = 2$

86. From the graphs $(f \circ g)(-2) = f[g(-2)] = f(2) = 3$

87. From the table $(f + g)(1) = 7 + 1 = 8$

88. From the table $(f - g)(3) = 9 - 9 = 0$

89. From the table $(fg)(-1) = (3)(-2) = -6$

90. From the table $\left(\dfrac{f}{g}\right)(0) = \dfrac{5}{0}$, which is undefined.

91. From the tables $(g \circ f)(-2) = g[f(-2)] = g(1) = 2$

92. From the graphs $(f \circ g)(3) = f[g(3)] = f(-2) = 1$

93. $\dfrac{2(x + h) + 9 - (2x + 9)}{h} = \dfrac{2x + 2h + 9 - 2x - 9}{h} = \dfrac{2h}{h} = 2$

94. $\dfrac{(x + h)^2 - 5(x + h) + 3 - (x^2 - 5x + 3)}{h} = \dfrac{x^2 + 2xh + h^2 - 5x - 5h + 3 - x^2 + 5x - 3}{h} =$

    $\dfrac{2xh + h^2 - 5h}{h} = 2x + h - 5$

95. One of many possible solutions for $(f \circ g)(x) = h(x)$ is: $f(x) = x^2$ and $g(x) = x^3 - 3x$. Then

    $(f \circ g)(x) = f[g(x)] = (x^3 - 3x)^2$.

96. One of many possible solutions for $(f \circ g)(x) = h(x)$ is: $f(x) = \dfrac{1}{x}$ and $g(x) = x - 5$. Then

    $(f \circ g)(x) = f[g(x)] = \dfrac{1}{x - 5}$.

97. If $V(r) = \dfrac{4}{3}\pi r^3$, then a 4 inch increase would be: $V(r) = \dfrac{4}{3}\pi(r + 4)^3$, and the volume gained would be:

    $V(r) = \dfrac{4}{3}\pi(r + 4)^3 - \dfrac{4}{3}\pi r^3$.

98. (a) Since $h = d, r = \dfrac{d}{2}$, and the formula for the volume of a can is: $V = \pi r^2 h$, the function is:

$$V(d) = \pi\left(\frac{d}{2}\right)^2 d \;\Rightarrow\; V(d) = \frac{\pi d^3}{4}.$$

   (b) Since $h = d, r = \dfrac{d}{2}, c = 2\pi r$, and the formula for the surface area of a can is: $A = 2\pi rh + 2\pi r^2$, the

   function is: $S(d) = 2\pi\left(\dfrac{d}{2}\right)d + 2\pi\left(\dfrac{d}{2}\right)^2 \;\Rightarrow\; S(d) = \pi d^2 + \dfrac{\pi d^2}{2} \;\Rightarrow\; S(d) = \dfrac{3\pi d^2}{2}.$

99. The function for changing yards to inches is: $f(x) = 36x$ and the function for changing miles to yards is:

   $g(x) = 1760x$. The composition of this which would change miles into inches is: $f[g(x)] = 36[1760(x)] \Rightarrow$

   $(f \circ g)(x) = 63,360x.$

100. If $x = $ width, then length $= 2x$. A formula for Perimeter can now be written as: $P = x + 2x + x + 2x$ and

   the function is: $P(x) = 6x$. This is a linear function.

Chapter 2 Test

1. (a) D, only values where $x \geq 0$ can be input into a square root function.

    (b) D, only values where $y \geq 0$ can be the range of a square root function.

    (c) C, all values can be input for x in a squaring function.

    (d) B, only values where $y \geq 3$ can be the range of $f(x) = x^2 + 3$.

    (e) C, all values can be input for x in a cube root function.

    (f) C, all values can be the range of a cube root function.

    (g) C, all values can be input for x in an absolute value function.

    (h) D, only values where $y \geq 0$ can be the range to an absolute value function.

    (i) D, if $x = y^2$ then $y = \sqrt{x}$ and only values where $x \geq 0$ can be input into a square root function.

    (j) C, all values can be the range in this function.

2. (a) This is $f(x)$ shifted 2 units upward. See Figure 2a.

    (b) This is $f(x)$ shifted 2 units to the left. See Figure 2b.

    (c) This is $f(x)$ reflected across the x-axis. See Figure 2c.

    (d) This is $f(x)$ reflected across the y-axis. See Figure 2d.

    (e) This is $f(x)$ vertically stretched by a factor of 2. See Figure 2e.

    (f) We reflect the graph of $y = f(x)$ across the x-axis for all points for which $y < 0$. Where $y \geq 0$, the graph

    remains unchanged. See Figure 2f.

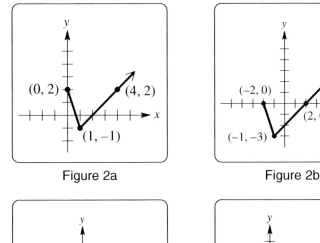

Figure 2a

Figure 2b

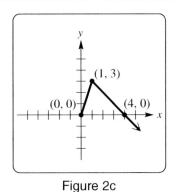

Figure 2c

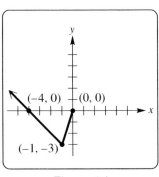

Figure 2d

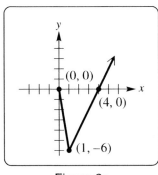

Figure 2e

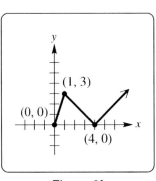

Figure 2f

3. (a) Since $y = f(2x)$ is $y = f(x)$ horizontally shrunk by a factor of $\dfrac{1}{2}$, the point $(-2, 4)$ on $y = f(x)$ becomes the point $(-1, 4)$ on the graph of $y = f(2x)$.

(b) Since $y = f\left(\dfrac{1}{2}x\right)$ is $y = f(x)$ horizontally stretched by a factor of 2, the point $(-2, 4)$ on $y = f(x)$ becomes the point $(-4, 4)$ on the graph of $y = f\left(\dfrac{1}{2}x\right)$.

4. (a) The graph of $f(x) = -(x - 2)^2 + 4$ is the basic graph $f(x) = x^2$ reflected across the *x*-axis, shifted 2 units to the right, and shifted 4 units upward. See Figure 4a.

(b) The graph of $f(x) = -2\sqrt{-x}$ is the basic graph $f(x) = \sqrt{x}$ reflected across the *y*-axis and vertically stretched by a factor of 2. See Figure 4b.

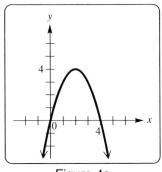

Figure 4a

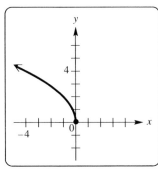

Figure 4b

5. (a) If the graph is symmetric with respect to the y-axis, then $(x, y) \Rightarrow (-x, y)$, therefore $(3, 6) \Rightarrow (-3, 6)$.

    (b) If the graph is symmetric with respect to the x-axis, then $(x, y) \Rightarrow (-x, -y)$, therefore $(3, 6) \Rightarrow (-3, -6)$.

    (c) See Figure 5. We give an actual screen here. The drawing should resemble it.

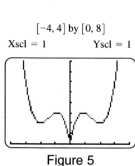

$[-4, 4]$ by $[0, 8]$

Xscl $= 1$ Yscl $= 1$

Figure 5

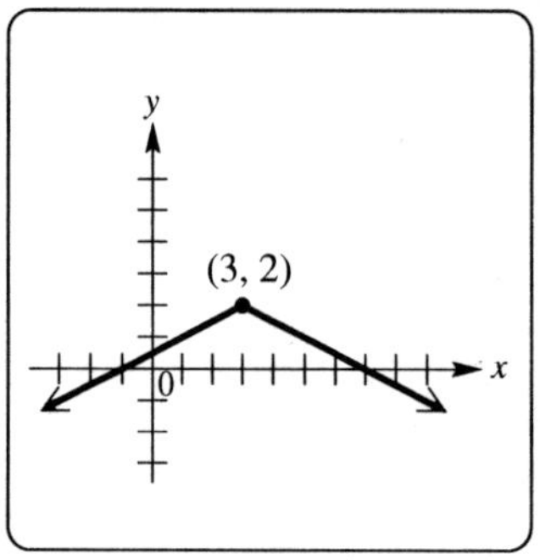

$(3, 2)$

Figure 6

6. (a) Shift the graph of $y = \sqrt[3]{x}$ to the left 2 units, vertically stretch by a factor of 4, and shift 5 units downward.

    (b) Graph $y = |x|$ reflected across the x-axis, vertically shrunk by a factor of $\dfrac{1}{2}$, shifted 3 units to the right, and shifted up 2 units. See Figure 6. From the graph the domain is: $(-\infty, \infty)$; and the range is: $(-\infty, 2]$.

7. (a) From the graph, the function is increasing for the interval: $(-\infty, -3)$.

    (b) From the graph, the function is decreasing for the interval: $(4, \infty)$.

    (c) From the graph, the function is constant for the interval: $[-3, 4]$.

    (d) From the graph, the function is continuous for the intervals: $(-\infty, -3), [-3, 4], (4, \infty)$.

    (e) From the graph, the domain is: $(-\infty, \infty)$.

    (f) From the graph, the range is: $(-\infty, 2)$.

8. (a) $|4x + 8| = 4 \Rightarrow 4x + 8 = 4 \Rightarrow 4x = -4 \Rightarrow x = -1$ or $4x + 8 = -4 \Rightarrow 4x = -12 \Rightarrow x = -3$,

    therefore the solution set is: $\{-3, -1\}$. From the graph, the x-coordinates of the points of intersection of the graphs of Y_1 and Y_2 are -3 and -1. See Figure 8.

    (b) $|4x + 8| < 4 \Rightarrow -4 < 4x + 8 < 4 \Rightarrow -12 < 4x < -4 \Rightarrow -3 < x < -1$, therefore the solution is: $(-3, -1)$. From the graph, See Figure 8, the graphs of Y_1 lies below the graph of Y_2 for x-values between -3 and -1.

    (c) $|4x + 8| > 4 \Rightarrow 4x + 8 > 4 \Rightarrow 4x > -4 \Rightarrow x > -1$ or $4x + 8 < -4 \Rightarrow 4x < -12 \Rightarrow x < -3$, therefore the solution is: $(-\infty, -3) \cup (-1, \infty)$. From the graph, See Figure 8, the graph of Y_1 lies above the graph of Y_2 for x-values less than -3 or for x-values greater than -1.

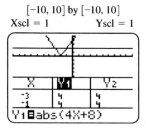

$[-10, 10]$ by $[-10, 10]$

Xscl $= 1$ Yscl $= 1$

Figure 8

9. (a) $(f - g)(x) = 2x^2 - 3x + 2 - (-2x + 1) \Rightarrow (f - g)(x) = 2x^2 - x + 1$

(b) $\left(\dfrac{f}{g}\right)(x) = \dfrac{2x^2 - 3x + 2}{-2x + 1}$

(c) The domain can be all values for x, except any that make $g(x) = 0$. Therefore $-2x + 1 \neq 0 \Rightarrow -2x \neq -1$

$\Rightarrow x \neq \dfrac{1}{2}$ or the interval: $\left(-\infty. \dfrac{1}{2}\right) \cup \left(\dfrac{1}{2}, \infty\right)$.

(d) $(f \circ g)(x) = f[g(x)] = 2(-2x + 1)^2 - 3(-2x + 1) + 2 = 2(4x^2 - 4x + 1) + 6x - 3 + 2 =$

$8x^2 - 8x + 2 + 6x - 3 + 2 = 8x^2 - 2x + 1$

(e) $\dfrac{2(x + h)^2 - 3(x + h) + 2 - (2x^2 - 3x + 2)}{h} = \dfrac{2(x^2 + 2xh + h^2) - 3x - 3h + 2 - 2x^2 + 3x - 2}{h} =$

$\dfrac{2x^2 + 4xh + 2h^2 - 3x - 3h + 2 - 2x^2 + 3x - 2}{h} = \dfrac{4xh + 2h^2 - 3h}{h} = 4x + 2h - 3$

10. (a) See Figure 10a.

(b) Graph $y_1 = (-x^2 + 3) * (x \leq 1) + (\sqrt[3]{x} + 2) * (x > 1)$ in the window $[-4.7, 4.7]$ by $[-5.1, 5.1]$.

See Figure 10b.

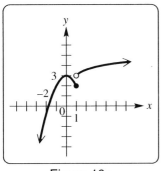

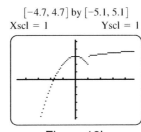
$[-4.7, 4.7]$ by $[-5.1, 5.1]$
Xscl = 1   Yscl = 1

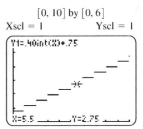
$[0, 10]$ by $[0, 6]$
Xscl = 1   Yscl = 1

Figure 10a   Figure 10b   Figure 11

11. (a) See Figure 11.

(b) Set $x = 5.5$, then $2.75 is the cost of a 5.5-minute call. See the display at the bottom of the screen.

12. (a) With an initial set-up cost of $3300 and a production cost of $4.50 the function is: $C(x) = 3300 + 4.50x$

(b) With a selling price of $10.50 the revenue function is: $R(x) = 10.50x$

(c) $P(x) = R(x) - C(x) \Rightarrow P(x) = 10.50x - (3300 + 4.50x) \Rightarrow P(x) = 6x - 3300$

(d) To make a profit $P(x) > 0$, therefore $6x - 3300 > 0 \Rightarrow 6x > 3300 \Rightarrow x > 550$.

Tyler needs to sell 551before he earns a profit.

(e) Graph $y_1 = 6x - 3300$, See Figure 12. The first integer x-value for which $P(x) > 0$ is 551.

$[0, 1000]$ by $[-4000, 4000]$
Xscl = 50   Yscl = 500

Figure 12

Chapter 2 Project

1. Since the front is moving at 40 mph for 4 hr the front moved 160 miles with each unit representing 100 miles.

    This is a shift of 1.6 units south or downward. The function $f(x) = \dfrac{1}{20}x^2$ shifted 1.6 units downward is:

    $y = \dfrac{1}{20}x^2 - 1.6.$

2. (a) Because the front has moved 250 south and 210 miles east, graph the shifted equation:

    $y = \dfrac{1}{20}(x - 2.1)^2 - 2.5.$ Plot the point $(5.5, -.8)$ for Columbus, Ohio. Here we see that the front has

    reached the city. See Figure 2a.

    (b) Because the front has moved 250 south and 210 miles east, graph the shifted equation:

    $y = \dfrac{1}{20}(x - 2.1)^2 - 2.5.$ Plot the point $(1.9, -4.3)$ for Memphis, Tennessee. Here we see that the front has

    not reached the city. See Figure 2b.

    (c) Because the front has moved 250 south and 210 miles east, graph the shifted equation:

    $y = \dfrac{1}{20}(x - 2.1)^2 - 2.5.$ Plot the point $(4.2, -2.3)$ for Louisville, Kentucky. Although the graph is

    difficult to read, repeated zooming will show that the front has not reached the city. See Figure 2c.

[−10, 10] by [−10, 10]
Xscl = 1   Yscl = 1

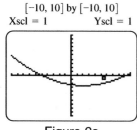

Figure 2a

[−10, 10] by [−10, 10]
Xscl = 1   Yscl = 1

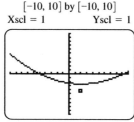

Figure 2b

[−10, 10] by [−10, 10]
Xscl = 1   Yscl = 1

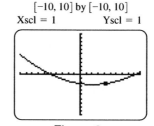

Figure 2c

Chapter 3: Polynomial Functions

3.1: Complex Numbers

1. The complex number $-9i$ can be written $0 - 9i$.

    (a) The real part is 0.   (b) The imaginary part is -9.   (c) The number is pure imaginary.

2. The complex number $3i$ can be written $0 + 3i$.

    (a) The real part is 0.   (b) The imaginary part is 3.   (c) The number is pure imaginary.

3. The complex number π can be written $\pi + 0i$.

    (a) The real part is π.   (b) The imaginary part is 0.   (c) The number is real.

4. The complex number $\sqrt{2}$ can be written $\sqrt{2} + 0i$.

    (a) The real part is $\sqrt{2}$.   (b) The imaginary part is 0.   (c) The number is real.

5. The complex number $3 + 7i$ is written in standard form.

    (a) The real part is 3.   (b) The imaginary part is 7.   (c) The number is nonreal complex.

6. The complex number $-8 + 4i$ is written in standard form.

    (a) The real part is -8.   (b) The imaginary part is 4.   (c) The number is nonreal complex.

7. The complex number $i\sqrt{7}$ can be written $0 + \sqrt{7}i$.

    (a) The real part is 0.   (b) The imaginary part is $\sqrt{7}$.   (c) The number is pure imaginary.

8. The complex number $-i\sqrt{3}$ can be written $0 - \sqrt{3}i$.

    (a) The real part is 0.   (b) The imaginary part is $-\sqrt{3}$.   (c) The number is pure imaginary.

9. The complex number $\sqrt{-7}$ can be written $0 + \sqrt{7}i$.

    (a) The real part is 0.   (b) The imaginary part is $\sqrt{7}$.   (c) The number is pure imaginary.

10. The complex number $\sqrt{-10}$ can be written $0 + \sqrt{10}i$.

    (a) The real part is 0.   (b) The imaginary part is $\sqrt{10}$.   (c) The number is pure imaginary.

11. True

12. True

13. True

14. True

15. False. *Every* real number is a complex number.

16. True

17. $\sqrt{-100} = i\sqrt{100} = 10i$

18. $\sqrt{-169} = i\sqrt{169} = 13i$

19. $-\sqrt{-400} = -i\sqrt{400} = -20i$

20. $-\sqrt{-225} = -i\sqrt{225} = -15i$

21. $-\sqrt{-39} = -i\sqrt{39}$

22. $-\sqrt{-95} = -i\sqrt{95}$

23. $5 + \sqrt{-4} = 5 + i\sqrt{4} = 5 + 2i$

24. $-7 + \sqrt{-100} = -7 + i\sqrt{100} = -7 + 10i$

25. $9 - \sqrt{-50} = 9 - i\sqrt{50} = 9 - i\sqrt{25 \cdot 2} = 9 - 5i\sqrt{2}$

26. $-11 - \sqrt{-24} = -11 - i\sqrt{24} = -11 - i\sqrt{4 \cdot 6} = -11 - 2i\sqrt{6}$

27. $i\sqrt{-9} = i^2\sqrt{9} = -3$

28. $i\sqrt{-16} = i^2\sqrt{16} = -4$

29. $\sqrt{-13} \cdot \sqrt{-13} = i\sqrt{13} \cdot i\sqrt{13} = 13i^2 = -13$

30. $\sqrt{-17} \cdot \sqrt{-17} = i\sqrt{17} \cdot i\sqrt{17} = 17i^2 = -17$

31. $\sqrt{-3} \cdot \sqrt{-8} = i\sqrt{3} \cdot i\sqrt{8} = \sqrt{24}i^2 = -2\sqrt{6}$

32. $\sqrt{-5} \cdot \sqrt{-15} = i\sqrt{5} \cdot i\sqrt{15} = \sqrt{75}i^2 = -5\sqrt{3}$

33. $\dfrac{\sqrt{-30}}{\sqrt{-10}} = \dfrac{i\sqrt{30}}{i\sqrt{10}} = \dfrac{i\sqrt{10 \cdot 3}}{i\sqrt{10}} = \dfrac{i\sqrt{10} \cdot \sqrt{3}}{i\sqrt{10}} = \sqrt{3}$

34. $\dfrac{\sqrt{-70}}{\sqrt{-7}} = \dfrac{i\sqrt{70}}{i\sqrt{7}} = \dfrac{i\sqrt{7 \cdot 10}}{i\sqrt{7}} = \dfrac{i\sqrt{7} \cdot \sqrt{10}}{i\sqrt{7}} = \sqrt{10}$

35. $\dfrac{\sqrt{-24}}{\sqrt{8}} = \dfrac{i\sqrt{24}}{\sqrt{8}} = \dfrac{i\sqrt{8 \cdot 3}}{\sqrt{8}} = \dfrac{i\sqrt{8} \cdot \sqrt{3}}{\sqrt{8}} = i\sqrt{3}$

36. $\dfrac{\sqrt{-54}}{\sqrt{27}} = \dfrac{i\sqrt{54}}{\sqrt{27}} = \dfrac{i\sqrt{27 \cdot 2}}{\sqrt{27}} = \dfrac{i\sqrt{27} \cdot \sqrt{2}}{\sqrt{27}} = i\sqrt{2}$

37. $\dfrac{\sqrt{-10}}{\sqrt{-40}} = \dfrac{i\sqrt{10}}{i\sqrt{40}} = \dfrac{i\sqrt{10}}{i\sqrt{10 \cdot 4}} = \dfrac{i\sqrt{10}}{i\sqrt{10} \cdot \sqrt{4}} = \dfrac{1}{\sqrt{4}} = \dfrac{1}{2}$

38. $\dfrac{\sqrt{-40}}{\sqrt{20}} = \dfrac{i\sqrt{40}}{\sqrt{20}} = \dfrac{i\sqrt{20 \cdot 2}}{\sqrt{20}} = \dfrac{i\sqrt{20} \cdot \sqrt{2}}{\sqrt{20}} = i\sqrt{2}$

39. $\dfrac{\sqrt{-6} \cdot \sqrt{-2}}{\sqrt{3}} = \dfrac{i\sqrt{6} \cdot i\sqrt{2}}{\sqrt{3}} = \dfrac{i^2\sqrt{3} \cdot \sqrt{2} \cdot \sqrt{2}}{\sqrt{3}} = \dfrac{i^2\sqrt{2} \cdot \sqrt{2}}{1} = -\sqrt{4} = -2$

40. $\dfrac{\sqrt{-12} \cdot \sqrt{-6}}{\sqrt{8}} = \dfrac{i\sqrt{12} \cdot i\sqrt{6}}{\sqrt{8}} = \dfrac{i^2\sqrt{12} \cdot \sqrt{6}}{\sqrt{8}} = \dfrac{i^2\sqrt{72}}{\sqrt{8}} = \dfrac{i^2\sqrt{8} \cdot \sqrt{9}}{\sqrt{8}} = \dfrac{i^2\sqrt{9}}{1} = i^2\sqrt{9} = -3$

41. $(3 + 2i) + (4 - 3i) = (3 + 4) + (2 - 3)i = 7 - i$

42. $(4 - i) + (2 + 5i) = (4 + 2) + (-1 + 5)i = 6 + 4i$

43. $(-2 + 3i) - (-4 + 3i) = (-2 - (-4)) + (3 - 3)i = 2 + 0i = 2$

44. $(-3 + 5i) - (-4 + 5i) = (-3 - (-4)) + (5 - 5)i = 1 + 0i = 1$

45. $(3 - 8i) + (2i + 4) = (3 + 4) + (-8 + 2)i = 7 - 6i$

46. $(9 - 5i) - (3i - 6) = (9 - (-6)) + (-5 - 3)i = 15 - 8i$

47. $(2 - 5i) - (3 + 4i) - (-2 + i) = (2 - 3 - (-2)) + (-5 - 4 - 1)i = 1 - 10i$

48. $(-4 - i) - (2 + 3i) + (-4 + 5i) = (-4 - 2 + (-4)) + (-1 - 3 + 5)i = -10 + i$

49. $(-6 + 5i) + (4 - 4i) + (2 - i) = (-6 + 4 + 2) + (5 + (-4) + (-1))i = 0 + 0i = 0$

50. $(7 + 9i) + (1 - 2i) + (-8 - 7i) = (7 + 1 + (-8)) + (9 + (-2) + (-7))i = 0 + 0i = 0$

51. $(2 + i)(3 - 2i) = 6 - 4i + 3i - 2i^2 = 6 - i - 2(-1) = 8 - i$

52. $(-2 + 3i)(4 - 2i) = -8 + 4i + 12i - 6i^2 = -8 + 16i - 6(-1) = -2 + 16i$

53. $(2 + 4i)(-1 + 3i) = -2 + 6i - 4i + 12i^2 = -2 + 2i + 12(-1) = -14 + 2i$

54. $(1 + 3i)(2 - 5i) = 2 - 5i + 6i - 15i^2 = 2 + i - 15(-1) = 17 + i$

55. $(-3 + 2i)^2 = (-3 + 2i)(-3 + 2i) = 9 - 6i - 6i + 4i^2 = 9 - 12i + 4(-1) = 5 - 12i$

56. $(2 + i)^2 = (2 + i)(2 + i) = 4 + 2i + 2i + i^2 = 4 + 4i + (-1) = 3 + 4i$

57. $(3 + i)(-3 - i) = -9 - 3i - 3i - i^2 = -9 - 6i - (-1) = -8 - 6i$

58. $(-5 - i)(5 + i) = -25 - 5i - 5i - i^2 = -25 - 10i - (-1) = -24 - 10i$

59. $(2 + 3i)(2 - 3i) = 4 - 9i^2 = 4 - 9(-1) = 13$

60. $(6 - 4i)(6 + 4i) = 36 - 16i^2 = 36 - 16(-1) = 52$

61. $(\sqrt{6} + i)(\sqrt{6} - i) = 6 - i^2 = 6 - (-1) = 7$

62. $(\sqrt{2} - 4i)(\sqrt{2} + 4i) = 2 - 16i^2 = 2 - 16(-1) = 18$

63. $i(3 - 4i)(3 + 4i) = i(9 - 16i^2) = i(9 - 16(-1)) = 25i$

64. $i(2 + 7i)(2 - 7i) = i(4 - 49i^2) = i(4 - 49(-1)) = 53i$

65. $3i(2 - i)^2 = 3i(4 - 4i + i^2) = 3i(4 - 4i + (-1)) = 3i(3 - 4i) = 9i - 12i^2 = 9i - 12(-1) = 12 + 9i$

66. $-5i(4 - 3i)^2 = -5i(16 - 24i + 9i^2) = -5i(16 - 24i + 9(-1)) = -5i(7 - 24i) = -35i + 120i^2 =$
    $-35i + 120(-1) = -120 - 35i$

67. $(2 + i)(2 - i)(4 + 3i) = (4 - i^2)(4 + 3i) = (4 - (-1))(4 + 3i) = 5(4 + 3i) = 20 + 15i$

68. $(3 - i)(3 + i)(2 - 6i) = (9 - i^2)(2 - 6i) = (9 - (-1))(2 - 6i) = 10(2 - 6i) = 20 - 60i$

69. $i^5 = (i^4)^1 \cdot i = 1^1 \cdot i = i$

70. $i^8 = (i^4)^2 = 1^2 = 1$

71. $i^{15} = (i^4)^3 \cdot i^3 = 1^3 \cdot (-i) = -i$

72. $i^{19} = (i^4)^4 \cdot i^3 = 1^4 \cdot (-i) = -i$

73. $i^{64} = (i^4)^{16} = 1^{16} = 1$

74. $i^{102} = (i^4)^{25} \cdot i^2 = 1^{25} \cdot (-1) = -1$

75. $i^{-6} = (i^6)^{-1} = [(i^4)^1 \cdot i^2]^{-1} = [1^1 \cdot (-1)]^{-1} = (-1)^{-1} = -1$

76. $i^{-15} = (i^{15})^{-1} = [(i^4)^3 \cdot i^3]^{-1} = [1^3 \cdot (-i)]^{-1} = (-i)^{-1} = \dfrac{1}{-i} = \dfrac{1}{-i} \cdot \dfrac{i}{i} = \dfrac{i}{-i^2} = \dfrac{i}{1} = i$

77. $\dfrac{1}{i^9} = i^{-9} = (i^9)^{-1} = [(i^4)^2 \cdot i]^{-1} = [1^2 \cdot i]^{-1} = (i)^{-1} = \dfrac{1}{i} = \dfrac{1}{i} \cdot \dfrac{i}{i} = \dfrac{i}{i^2} = \dfrac{i}{-1} = -i$

78. $\dfrac{1}{i^{12}} = i^{-12} = (i^{12})^{-1} = [(i^4)^3]^{-1} = [1^3]^{-1} = 1^{-1} = 1$

79. $\dfrac{1}{i^{-51}} = i^{51} = (i^4)^{12} \cdot i^3 = 1^{12} \cdot (-i) = -i$

80. $\dfrac{1}{i^{-46}} = i^{46} = (i^4)^{11} \cdot i^2 = 1^{11} \cdot (-1) = -1$

81. $\dfrac{-1}{-i^{12}} = \dfrac{1}{i^{12}} = \dfrac{1}{(i^4)^3} = \dfrac{1}{1^3} = \dfrac{1}{1} = 1$

82. $\dfrac{-1}{-i^{15}} = \dfrac{1}{i^{15}} = \dfrac{1}{(i^4)^3 \cdot i^3} = \dfrac{1}{1^3 \cdot (-i)} = \dfrac{1}{-i} = \dfrac{(1)-i}{i^2} = \dfrac{-i}{-1} = i$

83. $\left(\dfrac{\sqrt{2}}{2} + \dfrac{\sqrt{2}}{2}i\right)^2 = \left(\dfrac{\sqrt{2}}{2}\right)^2 + 2\left(\dfrac{\sqrt{2}}{2}\right)\left(\dfrac{\sqrt{2}}{2}i\right) + \left(\dfrac{\sqrt{2}}{2}i\right)^2 = \dfrac{2}{4} + \dfrac{2}{2}i + \dfrac{2}{4}i^2 = \dfrac{1}{2} + i - \dfrac{1}{2} = i$

84. $\left(\dfrac{\sqrt{3}}{2} + \dfrac{1}{2}i\right)^3 = \left(\dfrac{\sqrt{3}}{2} + \dfrac{1}{2}i\right)\left(\dfrac{\sqrt{3}}{2} + \dfrac{1}{2}i\right)^2 = \left(\dfrac{\sqrt{3}}{2} + \dfrac{1}{2}i\right)\left[\left(\dfrac{\sqrt{3}}{2}\right)^2 + 2\left(\dfrac{\sqrt{3}}{2}\right)\left(\dfrac{1}{2}i\right) + \left(\dfrac{1}{2}i\right)^2\right] =$
    $\left(\dfrac{\sqrt{3}}{2} + \dfrac{1}{2}i\right)\left(\dfrac{3}{4} + \dfrac{\sqrt{3}}{2}i + \dfrac{1}{4}i^2\right) = \left(\dfrac{\sqrt{3}}{2} + \dfrac{1}{2}i\right)\left(\dfrac{1}{2} + \dfrac{\sqrt{3}}{2}i\right) = \dfrac{\sqrt{3}}{4} + \dfrac{3}{4}i + \dfrac{1}{4}i + \dfrac{\sqrt{3}}{4}i^2 = \dfrac{3}{4}i + \dfrac{1}{4}i = i$

85. The conjugate of $5 - 3i$ is $5 + 3i$.

86. The conjugate of $-3 + i$ is $-3 - i$.

87. The conjugate of $-18i = 0 - 18i$ is $0 + 18i = 18i$.

88. The conjugate of $\sqrt{7} = \sqrt{7} + 0i$ is $\sqrt{7} - 0i = \sqrt{7}$.

89. $\dfrac{-19 - 9i}{4 + i} = \dfrac{-19 - 9i}{4 + i} \cdot \dfrac{4 - i}{4 - i} = \dfrac{-76 + 19i - 36i + 9i^2}{4^2 - i^2} = \dfrac{-85 - 17i}{16 - (-1)} = \dfrac{-85 - 17i}{17} = -5 - i$

90. $\dfrac{-12 - 5i}{3 - 2i} = \dfrac{-12 - 5i}{3 - 2i} \cdot \dfrac{3 + 2i}{3 + 2i} = \dfrac{-36 - 24i - 15i - 10i^2}{3^2 - (2i)^2} = \dfrac{-26 - 39i}{9 - (-4)} = \dfrac{-26 - 39i}{13} = -2 - 3i$

91. $\dfrac{1 - 3i}{1 + i} = \dfrac{1 - 3i}{1 + i} \cdot \dfrac{1 - i}{1 - i} = \dfrac{1 - i - 3i + 3i^2}{1^2 - i^2} = \dfrac{-2 - 4i}{1 - (-1)} = \dfrac{-2 - 4i}{2} = -1 - 2i$

92. $\dfrac{-3 + 4i}{2 - i} = \dfrac{-3 + 4i}{2 - i} \cdot \dfrac{2 + i}{2 + i} = \dfrac{-6 - 3i + 8i + 4i^2}{2^2 - i^2} = \dfrac{-10 + 5i}{4 - (-1)} = \dfrac{-10 + 5i}{5} = -2 + i$

93. $\dfrac{-6 + 8i}{4 + 3i} = \dfrac{-6 + 8i}{4 + 3i} \cdot \dfrac{4 - 3i}{4 - 3i} = \dfrac{-24 + 18i + 32i - 24i^2}{4^2 - (3i)^2} = \dfrac{0 + 50i}{16 - (-9)} = \dfrac{50i}{25} = 2i$

94. $\dfrac{2 - i}{2 + i} = \dfrac{2 - i}{2 + i} \cdot \dfrac{2 - i}{2 - i} = \dfrac{4 - 2i - 2i + i^2}{2^2 - i^2} = \dfrac{3 - 4i}{4 - (-1)} = \dfrac{3 - 4i}{5} = \dfrac{3}{5} - \dfrac{4}{5}i$

95. $\dfrac{4 - 3i}{4 + 3i} = \dfrac{4 - 3i}{4 + 3i} \cdot \dfrac{4 - 3i}{4 - 3i} = \dfrac{16 - 12i - 12i + 9i^2}{4^2 - (3i)^2} = \dfrac{7 - 24i}{16 - (-9)} = \dfrac{7 - 24i}{25} = \dfrac{7}{25} - \dfrac{24}{25}i$

96. $\dfrac{3}{-i} = \dfrac{3}{-i} \cdot \dfrac{i}{i} = \dfrac{3i}{-i^2} = \dfrac{3i}{-(-1)} = \dfrac{3i}{1} = 3i$

97. $\dfrac{-7}{3i} = \dfrac{-7}{3i} \cdot \dfrac{-3i}{-3i} = \dfrac{21i}{-(3i)^2} = \dfrac{21i}{-(-9)} = \dfrac{21i}{9} = \dfrac{7}{3}i$

98. $\dfrac{-10}{i} = \dfrac{-10}{i} \cdot \dfrac{-i}{-i} = \dfrac{10i}{-i^2} = \dfrac{10i}{-(-1)} = \dfrac{10i}{1} = 10i$

99. The method involves multiplying by 1, the multiplicative identity.

100. Dividing the exponent by 4 gives the total number of factors of i^4 there are. Since $(i^4)^d = 1$ for any value of d and the remainder r will be the exponent on the remaining factor i^r, the result is $1 \cdot i^r = i^r$.

3.2: Quadratic Functions and Graphs

1. Since $a > 0$, the parabola opens upward. The vertex is $(4, -3)$. The graph is shown in B.

2. Since $a < 0$, the parabola opens downward. The vertex is $(4, 3)$. The graph is shown in A.

3. Since $a > 0$, the parabola opens upward. The vertex is $(-4, -3)$. The graph is shown in D.

4. Since $a < 0$, the parabola opens downward. The vertex is $(-4, 3)$. The graph is shown in C.

5. (a) $P(x) = x^2 - 2x - 15 \Rightarrow P(x) + 15 = x^2 - 2x \Rightarrow P(x) + 15 + 1 = x^2 - 2x + 1 \Rightarrow$

    $P(x) + 16 = (x - 1)^2 \Rightarrow P(x) = (x - 1)^2 - 16$

    (b) The vertex is $(1, -16)$.

    (c) See Figure 5a and 5b.

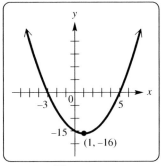

Figure 5a

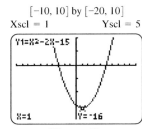

Figure 5b

6. (a) $P(x) = x^2 + 2x - 15 \Rightarrow P(x) + 15 = x^2 + 2x \Rightarrow P(x) + 15 + 1 = x^2 + 2x + 1 \Rightarrow$

$P(x) + 16 = (x + 1)^2 \Rightarrow P(x) = (x + 1)^2 - 16$

(b) The vertex is $(-1, -16)$.

(c) See Figure 6a and 6b.

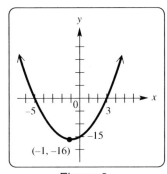

Figure 6a

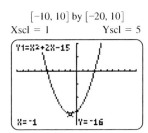

Figure 6b

7. (a) $P(x) = -x^2 - 3x + 10 \Rightarrow -P(x) = x^2 + 3x - 10 \Rightarrow -P(x) + 10 = x^2 + 3x \Rightarrow$

$-P(x) + 10 + \dfrac{9}{4} = x^2 + 3x + \dfrac{9}{4} \Rightarrow -P(x) + \dfrac{49}{4} = \left(x + \dfrac{3}{2}\right)^2 \Rightarrow -P(x) = \left(x + \dfrac{3}{2}\right)^2 - \dfrac{49}{4} \Rightarrow$

$P(x) = -\left(x + \dfrac{3}{2}\right)^2 + \dfrac{49}{4}$

(b) The vertex is $\left(-\dfrac{3}{2}, \dfrac{49}{4}\right)$ or $(-1.5, 12.25)$.

(c) See Figure 7a and 7b.

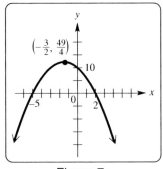

Figure 7a

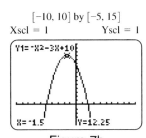

Figure 7b

8. (a) $P(x) = -x^2 + 3x + 10 \Rightarrow -P(x) = x^2 - 3x - 10 \Rightarrow -P(x) + 10 = x^2 - 3x \Rightarrow$

   $-P(x) + 10 + \dfrac{9}{4} = x^2 - 3x + \dfrac{9}{4} \Rightarrow -P(x) + \dfrac{49}{4} = \left(x - \dfrac{3}{2}\right)^2 \Rightarrow -P(x) = \left(x - \dfrac{3}{2}\right)^2 - \dfrac{49}{4} \Rightarrow$

   $P(x) = -\left(x - \dfrac{3}{2}\right)^2 + \dfrac{49}{4}$

   (b) The vertex is $\left(\dfrac{3}{2}, \dfrac{49}{4}\right)$ or $(1.5, 12.25)$.   (c) See Figure 8a and 8b.

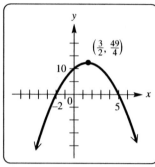

Figure 8a

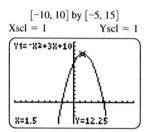

Figure 8b

9. (a) $P(x) = x^2 - 6x \Rightarrow P(x) + 9 = x^2 - 6x + 9 \Rightarrow P(x) + 9 = (x - 3)^2 \Rightarrow P(x) = (x - 3)^2 - 9$

   (b) The vertex is $(3, -9)$.   (c) See Figure 9a and 9b.

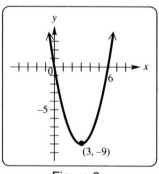

Figure 9a

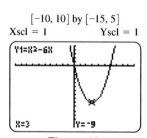

Figure 9b

10. (a) $P(x) = x^2 + 4x \Rightarrow P(x) + 4 = x^2 + 4x + 4 \Rightarrow P(x) + 4 = (x + 2)^2 \Rightarrow P(x) = (x + 2)^2 - 4$

   (b) The vertex is $(-2, -4)$.   (c) See Figure 10a and 10b.

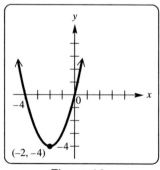

Figure 10a

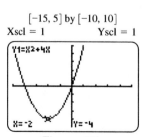

Figure 10b

11. (a) $P(x) = 2x^2 - 2x - 24 \Rightarrow \dfrac{P(x)}{2} = x^2 - x - 12 \Rightarrow \dfrac{P(x)}{2} + 12 = x^2 - x \Rightarrow$

$\dfrac{P(x)}{2} + 12 + \dfrac{1}{4} = x^2 - x + \dfrac{1}{4} \Rightarrow \dfrac{P(x)}{2} + \dfrac{49}{4} = \left(x - \dfrac{1}{2}\right)^2 \Rightarrow \dfrac{P(x)}{2} = \left(x - \dfrac{1}{2}\right)^2 - \dfrac{49}{4} \Rightarrow$

$P(x) = 2\left(x - \dfrac{1}{2}\right)^2 - \dfrac{49}{2}$

(b) The vertex is $\left(\dfrac{1}{2}, -\dfrac{49}{2}\right)$ or $(.5, -24.5)$.

(c) See Figure 11a and 11b.

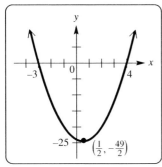

Figure 11a

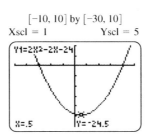

Figure 11b

12. (a) $P(x) = 3x^2 + 3x - 6 \Rightarrow \dfrac{P(x)}{3} = x^2 + x - 2 \Rightarrow \dfrac{P(x)}{3} + 2 = x^2 + x \Rightarrow$

$\dfrac{P(x)}{3} + 2 + \dfrac{1}{4} = x^2 + x + \dfrac{1}{4} \Rightarrow \dfrac{P(x)}{3} + \dfrac{9}{4} = \left(x + \dfrac{1}{2}\right)^2 \Rightarrow \dfrac{P(x)}{3} = \left(x + \dfrac{1}{2}\right)^2 - \dfrac{9}{4} \Rightarrow$

$P(x) = 3\left(x + \dfrac{1}{2}\right)^2 - \dfrac{27}{4}$

(b) The vertex is $\left(-\dfrac{1}{2}, -\dfrac{27}{4}\right)$ or $(-.5, -6.75)$.

(c) See Figure 12a and 12b.

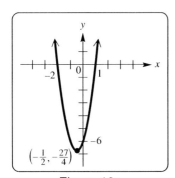

Figure 12a

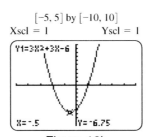

Figure 12b

13. (a) $f(x) = -2x^2 + 6x \Rightarrow \dfrac{f(x)}{-2} = x^2 - 3x \Rightarrow \dfrac{f(x)}{-2} + \dfrac{9}{4} = x^2 - 3x + \dfrac{9}{4} \Rightarrow$

$\dfrac{f(x)}{-2} + \dfrac{9}{4} = \left(x - \dfrac{3}{2}\right)^2 \Rightarrow \dfrac{f(x)}{-2} = \left(x - \dfrac{3}{2}\right)^2 - \dfrac{9}{4} \Rightarrow f(x) = -2\left(x - \dfrac{3}{2}\right)^2 + \dfrac{9}{2}$

(b) The vertex is $\left(\dfrac{3}{2}, \dfrac{9}{2}\right)$ or $(1.5, 4.5)$.

(c) See Figure 13a and 13b.

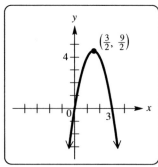

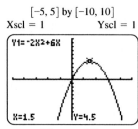

Figure 13a   Figure 13b

14. (a) $f(x) = -4x^2 + 4x \Rightarrow \dfrac{f(x)}{-4} = x^2 - x \Rightarrow \dfrac{f(x)}{-4} + \dfrac{1}{4} = x^2 - x + \dfrac{1}{4} \Rightarrow$

$\dfrac{f(x)}{-4} + \dfrac{1}{4} = \left(x - \dfrac{1}{2}\right)^2 \Rightarrow \dfrac{f(x)}{-4} = \left(x - \dfrac{1}{2}\right)^2 - \dfrac{1}{4} \Rightarrow f(x) = -4\left(x - \dfrac{1}{2}\right)^2 + 1$

(b) The vertex is $\left(\dfrac{1}{2}, 1\right)$ or $(.5, 1)$.

(c) See Figure 14a and 14b.

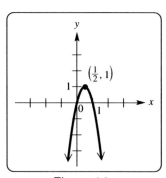

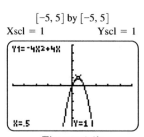

Figure 14a   Figure 14b

15. (a) $P(x) = 3x^2 + 4x - 1 \Rightarrow \dfrac{P(x)}{3} = x^2 + \dfrac{4}{3}x - \dfrac{1}{3} \Rightarrow \dfrac{P(x)}{3} + \dfrac{1}{3} = x^2 + \dfrac{4}{3}x \Rightarrow$

$\dfrac{P(x)}{3} + \dfrac{1}{3} + \dfrac{4}{9} = x^2 + \dfrac{4}{3}x + \dfrac{4}{9} \Rightarrow \dfrac{P(x)}{3} + \dfrac{7}{9} = \left(x + \dfrac{2}{3}\right)^2 \Rightarrow \dfrac{P(x)}{3} = \left(x + \dfrac{2}{3}\right)^2 - \dfrac{7}{9} \Rightarrow$

$P(x) = 3\left(x + \dfrac{2}{3}\right)^2 - \dfrac{7}{3}$

(b) The vertex is $\left(-\dfrac{2}{3}, -\dfrac{7}{3}\right)$ or $(-.67, -2.33)$.

(c) See Figure 15a and 15 b.

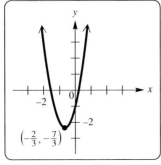

Figure 15a

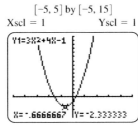

Figure 15b

16. (a) $P(x) = 4x^2 + 3x - 1 \Rightarrow \dfrac{P(x)}{4} = x^2 + \dfrac{3}{4}x - \dfrac{1}{4} \Rightarrow \dfrac{P(x)}{4} + \dfrac{1}{4} = x^2 + \dfrac{3}{4}x \Rightarrow$

$\dfrac{P(x)}{4} + \dfrac{1}{4} + \dfrac{9}{64} = x^2 + \dfrac{3}{4}x + \dfrac{9}{64} \Rightarrow \dfrac{P(x)}{4} + \dfrac{25}{64} = \left(x + \dfrac{3}{8}\right)^2 \Rightarrow \dfrac{P(x)}{4} = \left(x + \dfrac{3}{8}\right)^2 - \dfrac{25}{64} \Rightarrow$

$P(x) = 4\left(x + \dfrac{3}{8}\right)^2 - \dfrac{25}{16}$

(b) The vertex is $\left(-\dfrac{3}{8}, -\dfrac{25}{16}\right)$ or $(-.375, -1.5625)$.

(c) See Figure 16a and 16b.

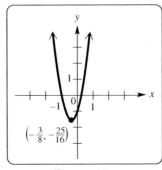

Figure 16a

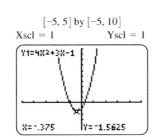

Figure 16b

17. (a) Since $a > 0$, the parabola opens upward. The vertex is $(4, -2)$. The graph is given in D.

(b) Since $a > 0$, the parabola opens upward. The vertex is $(2, -4)$. The graph is given in B.

(c) Since $a < 0$, the parabola opens downward. The vertex is $(4, -2)$. The graph is given in C.

(d) Since $a < 0$, the parabola opens downward. The vertex is $(2, -4)$. The graph is given in A.

18. (a) Since $a < 0$, the parabola opens downward. The vertex is $(-h, k)$, which is in quadrant II and will have two
    x-intercepts. The graph is given in C.

(b) Since $a > 0$, the parabola opens upward. The vertex is (h, k), which is in quadrant I and will have no
    x-intercepts. The graph is given in B.

(c) Since $a > 0$, the parabola opens upward. The vertex is $(-h, k)$, which is in quadrant II and will have no
    x-intercepts. The graph is given in D.

(d) Since $a < 0$, the parabola opens downward. The vertex is (h, k), which is in quadrant I and will have two
    x-intercepts. The graph is given in A.

19. (a) $(2, 0)$ (b) $D: (-\infty, \infty), R: [0, \infty)$ (c) $x = 2$ (d) $[2, \infty)$ (e) $(-\infty, 2]$ (f) Min.: $P(2) = 0$

20. (a) $(-4, 0)$ (b) $D: (-\infty, \infty), R: [0, \infty)$ (c) $x = -4$ (d) $[-4, \infty)$ (e) $(-\infty, -4]$ (f) Min.: $P(-4) = 0$

21. (a) $(-3, -4)$ (b) $D: (-\infty, \infty), R: [-4, \infty)$ (c) $x = -3$ (d) $[-3, \infty)$ (e) $(-\infty, -3]$ (f) Min.: $f(-3) = -4$

22. (a) $(-1, -3)$ (b) $D: (-\infty, \infty), R: (-\infty, -3]$ (c) $x = -1$ (d) $(-\infty, -1]$ (e) $[-1, \infty)$ (f) Max.: $P(-1) = -3$

23. (a) $(-3, 2)$ (b) $D: (-\infty, \infty), R: (-\infty, 2]$ (c) $x = -3$ (d) $(-\infty, -3]$ (e) $[-3, \infty)$ (f) Max.: $f(-3) = 2$

24. (a) $(2, 1)$ (b) $D: (-\infty, \infty), R: (-\infty, 1]$ (c) $x = 2$ (d) $(-\infty, 2]$ (e) $[2, \infty)$ (f) Max.: $f(2) = 1$

25. (a) $x = -\dfrac{b}{2a} = -\dfrac{-10}{2(1)} = 5; y = P(5) = (5)^2 - 10(5) + 21 = -4 \Rightarrow$ Vertex: $(5, -4)$

    (b) See Figure 25.

26. (a) $x = -\dfrac{b}{2a} = -\dfrac{-2}{2(1)} = 1; y = P(1) = (1)^2 - 2(1) + 3 = 2 \Rightarrow$ Vertex: $(1, 2)$

    (b) See Figure 26.

27. (a) $x = -\dfrac{b}{2a} = -\dfrac{4}{2(-1)} = 2; y = -(2)^2 + 4(2) - 2 = 2 \Rightarrow$ Vertex: $(2, 2)$

    (b) See Figure 27.

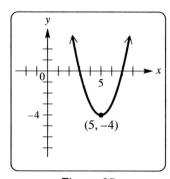

Figure 25

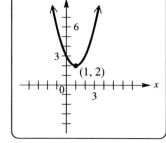

Figure 26

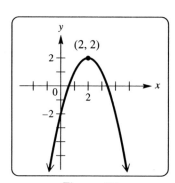

Figure 27

28. (a) $x = -\dfrac{b}{2a} = -\dfrac{2}{2(-1)} = 1; y = -(1)^2 + 2(1) + 1 = 2 \Rightarrow$ Vertex: $(1, 2)$

    (b) See Figure 28.

29. (a) $x = -\dfrac{b}{2a} = -\dfrac{(-4)}{2(2)} = 1; y = f(1) = 2(1)^2 - 4(1) + 5 = 3 \Rightarrow$ Vertex: $(1, 3)$

    (b) See Figure 29.

30. (a) $x = -\dfrac{b}{2a} = -\dfrac{(-8)}{2(2)} = 2; y = f(2) = 2(2)^2 - 8(2) + 9 = 1 \Rightarrow$ Vertex: $(2, 1)$

    (b) See Figure 30.

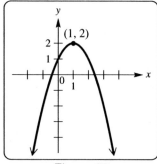

Figure 28

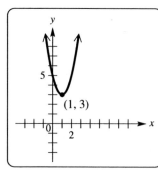

Figure 29

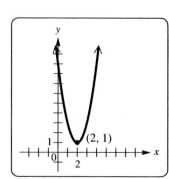

Figure 30

31. (a) $x = -\dfrac{b}{2a} = -\dfrac{24}{2(-3)} = 4$; $y = f(4) = -3(4)^2 + 24(4) - 46 = 2 \Rightarrow$ Vertex: $(4, 2)$

   (b) See Figure 31.

32. (a) $x = -\dfrac{b}{2a} = -\dfrac{-18}{2(-3)} = -3$; $y = f(-3) = -3(-3)^2 - 18(-3) - 23 = 4 \Rightarrow$ Vertex: $(-3, 4)$

   (b) See Figure 32.

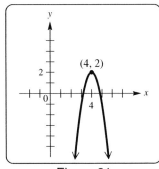
Figure 31

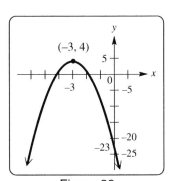
Figure 32

33. The graph is shown in Figure 33.

   (a) The vertex is approximately $(2.71, 5.20)$.

   (b) The x-intercepts are approximately -1.33 and 6.74.

34. The graph is shown in Figure 34.

   (a) The vertex is approximately $(.16, 1.43)$.

   (b) The x-intercepts are approximately $-.84$ and 1.16.

35. The graph is shown in Figure 35.

   (a) The vertex is approximately $(1.12, .56)$.

   (b) There are no x-intercepts.

36. The graph is shown in Figure 36.

   (a) The vertex is approximately $(2.92, 4.68)$.

   (b) The x-intercepts are approximately 0 and 5.84.

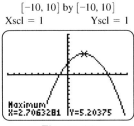
Figure 33

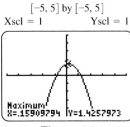
Figure 34

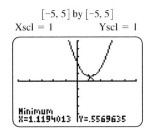
Figure 35

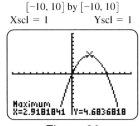
Figure 36

37. The graph is shown in Figure 37.

    (a) The vertex is approximately (.68, .57).

    (b) The *x*-intercepts are approximately 0 and 1.35.

38. The graph is shown in Figure 38.

    (a) The vertex is approximately (0, 1.31).

    (b) There are no *x*-intercepts.

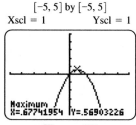

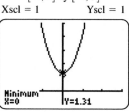

Figure 37   Figure 38

39. The minimum value of 3 is found at the vertex.

40. The value of the function is as small as possible at the vertex. Here $x = -3$.

41. The graph would not intersect the line $y = 1$. There are no solutions to $f(x) = 1$.

42. The graph would intersect the line $y = 3$ at only one point. There is one solution to $f(x) = 3$.

43. (a) From the symmetry of the *y*-values in the table, the vertex is $(4, -12)$.

    (b) Since all other *y*-values are larger than -12, the vertex is a minimum point.

    (c) The minimum value of the function is -12 and is located at the vertex.

    (d) Since the function is quadratic with a minimum value of -12, the range is $[-12, \infty)$.

44. (a) From the symmetry of the *y*-values in the table, the vertex is $(-3, 9)$.

    (b) Since all other *y*-values are larger than 9, the vertex is a minimum point.

    (c) The minimum value of the function is 9 and is located at the vertex.

    (d) Since the function is quadratic with a minimum value of 9, the range is $[9, \infty)$.

45. (a) From the symmetry of the *y*-values in the table, the vertex is $(1.5, 2)$.

    (b) Since all other *y*-values are smaller than 2, the vertex is a maximum point.

    (c) The maximum value of the function is 2 and is located at the vertex.

    (d) Since the function is quadratic with a maximum value of 2, the range is $(-\infty, 2]$.

46. (a) From the symmetry of the *y*-values in the table, the vertex is $(-2.5, -3)$.

    (b) Since all other *y*-values are smaller than -3, the vertex is a maximum point.

    (c) The maximum value of the function is -3 and is located at the vertex.

    (d) Since the function is quadratic with a maximum value of -3, the range is $(-\infty, -3]$.

47. Since the data are in the shape of a parabola that opens downward, a quadratic model with $a < 0$ is appropriate.

48. Since the data are in the shape of a line that rises from left to right, a linear model with $m > 0$ is appropriate.

49. Since the data are in the shape of a parabola that opens upward, a quadratic model with $a > 0$ is appropriate.

50. Since the data are in the shape of a parabola that opens downward, a quadratic model with $a < 0$ is appropriate.

51. Since the data are in the shape of a line that rises from left to right, a linear model with $m > 0$ is appropriate.

52. Since the data are in the shape of a parabola that opens downward, a quadratic model with $a < 0$ is appropriate.

53. (a) Since the data form a parabola that opens downward (not shown), a quadratic model with $a < 0$ (negative) is appropriate.

    (b) $x = -\dfrac{b}{2a} = -\dfrac{5.403}{2(-.3053)} \approx 8.8$; $y \approx f(8.8) = -.3053(8.8)^2 + 5.403(8.8) - 14.256 \approx 9.6$

    The vertex is approximately $(8.8, 9.6)$.

    (c) In 1999, the value of business mergers and acquisitions reached a maximum of 10 trillion dollars which is slightly higher than the value in the table.

54. (a) Since the parabolic shape suggested by the graph opens downward, the coefficient of x^2 is negative.

    (b) $x = -\dfrac{b}{2a} = -\dfrac{431.8}{2(-10.36)} \approx 20.84$; $y = f(20.84) = -10.36(20.84)^2 + 431.8(20.84) - 650 \approx 3849$

    The vertex is approximately $(20.84, 3849)$.

    (c) In 2020, Social Security assets will reach their maximum value of about \$3849 billion.

55. (a) A parabola that opens downward has a vertex with a maximum y-value.

    (b) $x = -\dfrac{b}{2a} = -\dfrac{44.43}{2(-6.15)} \approx 3.6$; $y \approx f(3.6) = -6.15(3.6)^2 + 44.43(3.6) + 400 \approx 480.2$

    The vertex is approximately $(3.6, 480.2)$. In 1998, the maximum harvest was about 480.2 million dollars of catfish.

56. (a) The year 2001 corresponds to $x = 11$. $f(11) = .156(11)^2 - 2.05(11) + 10.2 = 6.526 \approx 6.5\%$

    (b) $x = -\dfrac{b}{2a} = -\dfrac{-2.05}{2(.156)} \approx 6.57$; $y \approx f(6.57) = .156(6.57)^2 - 2.05(6.57) + 10.2 \approx 3.47$

    The vertex is approximately $(6.57, 3.47)$. In 1996, the minimum increase was about 3.5%.

57. (a) The value of t cannot be negative because t represents elapsed time after the rock is thrown.

    (b) The original height of the rock is $s_0 = 0$ which represents ground level.

    (c) Since $v_0 = 90$ and $s_0 = 0$, $s(t) = -16t^2 + v_0 t + s_0 \Rightarrow s(t) = -16t^2 + 90t$.

    (d) $s(1.5) = -16(1.5)^2 + 90(1.5) = 99$ feet

    (e) $x = -\dfrac{b}{2a} = -\dfrac{90}{2(-16)} = 2.8125$; $y \approx s(2.8125) = -16(2.8125)^2 + 90(2.8125) = 126.5625$

    The vertex is $(2.8125, 126.5625)$. The rock reaches a maximum height of 126.5625 feet after 2.8125 seconds. A graph of $y = -16x^2 + 90x$ (not shown) also gives a vertex of $(2.8125, 126.5625)$.

    (f) A graph of $y = -16x^2 + 90x$ (not shown) has x-intercepts at $(0, 0)$ and $(5.625, 0)$. The rock will hit the ground after 5.625 seconds.

58. (a) Since $v_0 = 200$ and $s_0 = 50$, $s(t) = -16t^2 + v_0 t + s_0 \Rightarrow s(t) = -16t^2 + 200t + 50$.

    (b) $x = -\dfrac{b}{2a} = -\dfrac{200}{2(-16)} = 6.25$; $y \approx s(6.25) = -16(6.25)^2 + 200(6.25) + 50 = 675$

    The vertex is $(6.25, 675)$. The rocket reaches a maximum height of 675 feet after 6.25 seconds.

    A graphical solution is found by noting that the vertex of the graph shown in Figure 58 is $(6.25, 675)$.

    (c) The graphs of $y_1 = -16x^2 + 200x + 50$ and $y_2 = 300$ (not shown). The graphs intersect when $x \approx 1.409$ and $x \approx 11.09$. The rocket is more than 300 feet high from about 1.4 to 11.1 seconds.

    (d) The x-intercept shown in Figure 58 is $(12.75, 0)$. The rocket hits the ground after 12.75 seconds.

59. (a) The graphs of $y_1 = -16x^2 + 150x$ and $y_2 = 355$ are shown in Figure 59a. The graphs do not intersect, which indicates that the ball does not reach a height of 355 feet.

    (b) The graphs of $y_1 = -16x^2 + 250x + 30$ and $y_2 = 355$ are shown in Figure 59b. The graphs intersect when $x \approx 1.43$ and $x \approx 14.19$ which indicates that the ball is 355 feet high at about 1.4 and 14.2 seconds.

$[0, 15]$ by $[0, 1000]$	$[0, 10]$ by $[300, 400]$	$[0, 20]$ by $[0, 1200]$	$[0, 15]$ by $[-30, 100]$
Xscl = 1 Yscl = 100	Xscl = 1 Yscl = 10	Xscl = 5 Yscl = 100	Xscl = 1 Yscl = 10

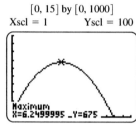
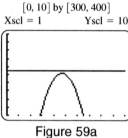
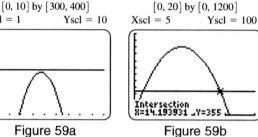
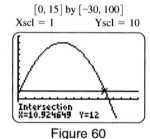

Figure 58   Figure 59a   Figure 59b   Figure 60

60. (a) The graphs of $y_1 = -2.7x^2 + 30x + 6.5$ and $y_2 = 12$ are shown in Figure 60. The graphs intersect when $x \approx .19$ and $x \approx 10.92$ which indicates that the ball is 12 feet high at about .19 and 10.92 seconds.

    (b) The x-intercept shown in Figure 60 is about $(11.32, 0)$. The ball hits the ground after about 11.32 seconds.

    (c) The vertex can be found analytically to be $(5.\overline{5}, 89.8\overline{3})$. The maximum height is only about 90 feet.

61. Let $h = -1, k = -4, x = 5$ and $P(x) = 104$ in $P(x) = a(x - h)^2 + k$ and determine the value of a.

    $104 = a(5 + 1)^2 - 4 \Rightarrow 104 = a(6)^2 - 4 \Rightarrow 108 = 36a \Rightarrow a = 3$

    $P(x) = 3(x + 1)^2 - 4 \Rightarrow P(x) = 3(x^2 + 2x + 1) - 4 \Rightarrow P(x) = 3x^2 + 6x - 1$

62. Let $h = -2, k = -3, x = 0$ and $P(x) = -19$ in $P(x) = a(x - h)^2 + k$ and determine the value of a.

    $-19 = a(0 + 2)^2 - 3 \Rightarrow -19 = a(2)^2 - 3 \Rightarrow -16 = 4a \Rightarrow a = -4$

    $P(x) = -4(x + 2)^2 - 3 \Rightarrow P(x) = -4(x^2 + 4x + 4) - 3 \Rightarrow P(x) = -4x^2 - 16x - 19$

63. Let $h = 8, k = 3, x = 10$ and $P(x) = 5$ in $P(x) = a(x - h)^2 + k$ and determine the value of a.

    $5 = a(10 - 8)^2 + 3 \Rightarrow 5 = a(2)^2 + 3 \Rightarrow 2 = 4a \Rightarrow a = .5$

    $P(x) = .5(x - 8)^2 + 3 \Rightarrow P(x) = .5(x^2 - 16x + 64) + 3 \Rightarrow P(x) = .5x^2 - 8x + 35$ or

    $P(x) = \dfrac{1}{2}x^2 - 8x + 35$

64. Let $h = -6, k = -12, x = 6$ and $P(x) = 24$ in $P(x) = a(x - h)^2 + k$ and determine the value of a.

    $24 = a(6 + 6)^2 - 12 \Rightarrow 24 = a(12)^2 - 12 \Rightarrow 36 = 144a \Rightarrow a = .25$

    $P(x) = .25(x + 6)^2 - 12 \Rightarrow P(x) = .25(x^2 + 12x + 36) - 12 \Rightarrow P(x) = .25x^2 + 3x - 3$ or

    $P(x) = \dfrac{1}{4}x^2 + 3x - 3$

65. Let $h = -4$, $k = -2$, $x = 2$ and $P(x) = -26$ in $P(x) = a(x - h)^2 + k$ and determine the value of a.

$$-26 = a(2 + 4)^2 - 2 \Rightarrow -26 = a(6)^2 - 2 \Rightarrow -24 = 36a \Rightarrow a = -\frac{2}{3}$$

$$P(x) = -\frac{2}{3}(x + 4)^2 - 2 \Rightarrow P(x) = -\frac{2}{3}(x^2 + 8x + 16) - 2 \Rightarrow P(x) = -\frac{2}{3}x^2 - \frac{16}{3}x - \frac{38}{3}$$

66. Let $h = 5$, $k = 6$, $x = 1$ and $P(x) = -6$ in $P(x) = a(x - h)^2 + k$ and determine the value of a.

$$-6 = a(1 - 5)^2 + 6 \Rightarrow -6 = a(-4)^2 + 6 \Rightarrow -12 = 16a \Rightarrow a = -\frac{3}{4}$$

$$P(x) = -\frac{3}{4}(x - 5)^2 + 6 \Rightarrow P(x) = -\frac{3}{4}(x^2 - 10x + 25) + 6 \Rightarrow P(x) = -\frac{3}{4}x^2 + \frac{15}{2}x - \frac{51}{4}$$

67. See Figure 67.

68. See Figure 68.

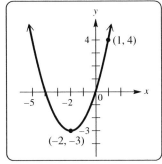
Figure 67

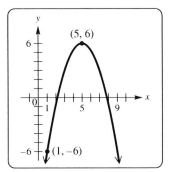
Figure 68

69. See Figure 69.

70. See Figure 70.

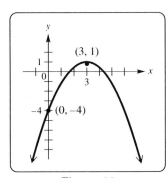
Figure 69

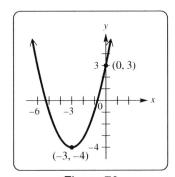
Figure 70

3.3: Quadratic Equations and Inequalities

1. $x^2 = 4 \Rightarrow \sqrt{x^2} = \sqrt{4} \Rightarrow x = \pm 2$, therefore G.

2. $x^2 = -4 \Rightarrow \sqrt{x^2} = \sqrt{-4} \Rightarrow x = \pm 2i$, therefore A.

3. $x^2 + 2 = 0 \Rightarrow x^2 = -2 \Rightarrow \sqrt{x^2} = \sqrt{-2} \Rightarrow x = \pm i\sqrt{2}$, therefore C.

4. $x^2 - 2 = 0 \Rightarrow x^2 = 2 \Rightarrow \sqrt{x^2} = \sqrt{2} \Rightarrow x = \pm\sqrt{2}$, therefore E.

5. $x^2 = -8 \Rightarrow \sqrt{x^2} = \sqrt{-8} \Rightarrow x = \pm 2i\sqrt{2}$, therefore H.

6. $x^2 = 8 \Rightarrow \sqrt{x^2} = \sqrt{8} \Rightarrow x = \pm 2\sqrt{2}$, therefore B.

7. $x - 2 = 0 \Rightarrow x = 2$, therefore D.

8. $x + 2 = 0 \Rightarrow x = -2$, therefore F.

9. Equation D, which is in the form $ab = 0$, is set up for the zero product property.

   Either $3x + 1 = 0 \Rightarrow x = -\dfrac{1}{3}$, or $x - 7 = 0 \Rightarrow x = 7$. The solution set is $\left\{ -\dfrac{1}{3}, 7 \right\}$.

10. Equation B, which is in the form $a^2 = k$, is set up for the square root property.

    $(2x + 5)^2 = 7 \Rightarrow 2x + 5 = \pm\sqrt{7} \Rightarrow 2x = -5 \pm \sqrt{7} \Rightarrow x = \dfrac{-5 \pm \sqrt{7}}{2}$

    The solution set is $\left\{ \dfrac{-5 \pm \sqrt{7}}{2} \right\}$.

11. Equation C, which has an x^2 coefficient of 1, does not require step 1 of the method of completing the square.

    $x^2 + x = 12 \Rightarrow x^2 + x + \dfrac{1}{4} = 12 + \dfrac{1}{4} \Rightarrow \left(x + \dfrac{1}{2} \right)^2 = \dfrac{49}{4} \Rightarrow x + \dfrac{1}{2} = \pm\dfrac{7}{2} \Rightarrow x = -\dfrac{1}{2} \pm \dfrac{7}{2}$

    The solution set is $\{-4, 3\}$.

12. Equation A, which is in the form $ax^2 + bx + c = 0$, is set up so that a, b, and c can be determined easily.

    $x = \dfrac{-(-17) \pm \sqrt{(-17)^2 - 4(3)(-6)}}{2(3)} = \dfrac{17 \pm \sqrt{289 + 72}}{6} = \dfrac{17 \pm \sqrt{361}}{6} = \dfrac{17 \pm 19}{6} = \dfrac{36}{6} \text{ or } \dfrac{-2}{6}$

    The solution set is $\left\{ -\dfrac{1}{3}, 6 \right\}$.

13. $x^2 = 16 \Rightarrow x = \pm\sqrt{16} \Rightarrow x = \{\pm 4\}$

    For graphical support, show that the graphs of $y_1 = x^2$ and $y_2 = 16$ intersect when $x = \{\pm 4\}$.

14. $x^2 = 144 \Rightarrow x = \pm\sqrt{144} \Rightarrow x = \{\pm 12\}$

    For graphical support, show that the graphs of $y_1 = x^2$ and $y_2 = 144$ intersect when $x = \{\pm 12\}$.

15. $2x^2 = 90 \Rightarrow x^2 = 45 \Rightarrow x = \pm\sqrt{45} \Rightarrow x = \{\pm 3\sqrt{5}\}$

    For graphical support, show that the graphs of $y_1 = 2x^2$ and $y_2 = 90$ intersect when $x = \{\pm 3\sqrt{5}\} \approx \{\pm 6.7\}$.

16. $2x^2 = 48 \Rightarrow x^2 = 24 \Rightarrow x = \pm\sqrt{24} \Rightarrow x = \{\pm 2\sqrt{6}\}$

    For graphical support, show that the graphs of $y_1 = 2x^2$ and $y_2 = 48$ intersect when $x = \{\pm 2\sqrt{6}\} \approx \{\pm 4.9\}$.

17. $x^2 = -16 \Rightarrow x = \pm i\sqrt{16} \Rightarrow x = \{\pm 4i\}$

    For graphical support, show that the graphs of $y_1 = x^2$ and $y_2 = -16$ do not intersect, therefore no real solutions.

18. $x^2 = -100 \Rightarrow x = \pm i\sqrt{100} \Rightarrow x = \{\pm 10i\}$

    For graphical support, show that the graphs of $y_1 = x^2$ and $y_2 = -100$ do not intersect, therefore no real solutions.

19. $x^2 = -18 \Rightarrow x = \pm i\sqrt{18} \Rightarrow x = \{\pm 3i\sqrt{2}\}$

    For graphical support, show that the graphs of $y_1 = x^2$ and $y_2 = -18$ do not intersect, therefore no real solutions.

20. $x^2 = -32 \Rightarrow x = \pm i\sqrt{32} \Rightarrow x = \{\pm 4i\sqrt{2}\}$

    For graphical support, show that the graphs of $y_1 = x^2$ and $y_2 = -32$ do not intersect, therefore no real solutions.

21. $(3x - 1)^2 = 12 \Rightarrow \sqrt{(3x - 1)^2} = \pm\sqrt{12} \Rightarrow 3x - 1 = \pm 2\sqrt{3} \Rightarrow 3x = \pm 2\sqrt{3} + 1 \Rightarrow$.

$x = \left\{ \dfrac{1 \pm 2\sqrt{3}}{3} \right\}$. For graphical support, show that the graphs of $y_1 = (3x - 1)^2$ and $y_2 = 12$ intersect when

$x = \left\{ \dfrac{1 \pm 2\sqrt{3}}{3} \right\} \approx \{-.8, 1.5\}$.

22. $(4x + 1)^2 = 20 \Rightarrow \sqrt{(4x + 1)^2} = \pm\sqrt{20} \Rightarrow 4x + 1 = \pm 2\sqrt{5} \Rightarrow 4x = \pm 2\sqrt{5} - 1 \Rightarrow$.

$x = \left\{ \dfrac{-1 \pm 2\sqrt{5}}{4} \right\}$. For graphical support, show that the graphs of $y_1 = (4x + 1)^2$ and $y_2 = 20$ intersect when

$x = \left\{ \dfrac{-1 \pm 2\sqrt{5}}{4} \right\} \approx \{-1.4, .9\}$.

23. $(5x - 3)^2 = -3 \Rightarrow \sqrt{(5x - 3)^2} = \pm\sqrt{-3} \Rightarrow 5x - 3 = \pm i\sqrt{3} \Rightarrow 5x = \pm i\sqrt{3} + 3 \Rightarrow$.

$x = \left\{ \dfrac{3}{5} \pm \dfrac{\sqrt{3}}{5} i \right\}$. For graphical support, show that the graphs of $y_1 = (5x - 3)^2$ and $y_2 = -3$ do not intersect,

therefore no real solutions.

24. $(-2x + 5)^2 = -8 \Rightarrow \sqrt{(-2x + 5)^2} = \pm\sqrt{-8} \Rightarrow -2x + 5 = \pm 2i\sqrt{2} \Rightarrow -2x = \pm 2i\sqrt{2} - 5 \Rightarrow$

$x = \left\{ \dfrac{5}{2} \pm i\sqrt{2} \right\}$. For graphical support, show that the graphs of $y_1 = (-2x + 5)^2$ and $y_2 = -8$ do not

intersect, therefore no real solutions.

25. $x^2 = 2x + 24 \Rightarrow x^2 - 2x - 24 = 0 \Rightarrow (x - 6)(x + 4) = 0 \Rightarrow x - 6 = 0 \text{ or } x + 4 = 0 \Rightarrow x = \{-4, 6\}$.

For graphical support, show that the graph of $y_1 = x^2 - 2x - 24 = 0$ has x-intercepts $x = \{-4, 6\}$.

26. $x^2 = 3x + 18 \Rightarrow x^2 - 3x - 18 = 0 \Rightarrow (x - 6)(x + 3) = 0 \Rightarrow x - 6 = 0 \text{ or } x + 3 = 0 \Rightarrow x = \{-3, 6\}$.

For graphical support, show that the graph of $y_1 = x^2 - 3x - 18$ has x-intercepts $x = \{-3, 6\}$.

27. $3x^2 - 2x = 0 \Rightarrow x(3x - 2) = 0 \Rightarrow x = 0 \text{ or } 3x - 2 = 0 \Rightarrow x = \left\{ 0, \dfrac{2}{3} \right\}$.

For graphical support, show that the graph of $y_1 = 3x^2 - 2x$ has x-intercepts $x = \left\{ 0, \dfrac{2}{3} \right\}$.

28. $5x^2 + 3x = 0 \Rightarrow x(5x + 3) = 0 \Rightarrow x = 0 \text{ or } 5x + 3 = 0 \Rightarrow x = \left\{ -\dfrac{3}{5}, 0 \right\}$.

For graphical support, show that the graph of $y_1 = 5x^2 + 3x$ has x-intercepts $x = \left\{ -\dfrac{3}{5}, 0 \right\}$.

29. $x(14x + 1) = 3 \Rightarrow 14x^2 + x - 3 = 0 \Rightarrow (2x + 1)(7x - 3) = 0 \Rightarrow 2x + 1 = 0 \text{ or } 7x - 3 = 0 \Rightarrow$

$x = \left\{ -\dfrac{1}{2}, \dfrac{3}{7} \right\}$. For graphical support, show that the graph of $y_1 = 14x^2 + x - 3$ has x-intercepts $x = \left\{ -\dfrac{1}{2}, \dfrac{3}{7} \right\}$.

30. $x(12x + 11) = -2 \Rightarrow 12x^2 + 11x + 2 = 0 \Rightarrow (3x + 2)(4x + 1) = 0 \Rightarrow 3x + 2 = 0 \text{ or } 4x + 1 = 0 \Rightarrow$

$x = \left\{ -\dfrac{2}{3}, -\dfrac{1}{4} \right\}$. For graphical support, show that the graph of $y_1 = 12x^2 + 11x + 2$ has x-intercepts

$x = \left\{ -\dfrac{2}{3}, -\dfrac{1}{4} \right\}$.

31. $-4 + 9x - 2x^2 = 0 \Rightarrow (-4 + x)(1 - 2x) = 0 \Rightarrow -4 + x = 0$ or $1 - 2x = 0 \Rightarrow x = \left\{\frac{1}{2}, 4\right\}$.

For graphical support, show that the graph of $y_1 = -4 + 9x - 2x^2$ has x-intercepts $x = \left\{\frac{1}{2}, 4\right\}$.

32. $-5 + 16x - 3x^2 = 0 \Rightarrow (-5 + x)(1 - 3x) = 0 \Rightarrow -5 + x = 0$ or $1 - 3x = 0 \Rightarrow x = \left\{\frac{1}{3}, 5\right\}$.

For graphical support, show that the graph of $y_1 = -5 + 16x - 3x^2$ has x-intercepts $x = \left\{\frac{1}{3}, 5\right\}$.

33. $\frac{1}{3}x^2 - \frac{1}{3}x = 24 \Rightarrow \frac{1}{3}x^2 - \frac{1}{3}x - 24 = 0 \Rightarrow x^2 - x - 72 = 0 \Rightarrow (x - 9)(x + 8) = 0 \Rightarrow$

$x - 9 = 0$ or $x + 8 = 0 \Rightarrow x = \{-8, 9\}$.

For graphical support, show that the graph of $y_1 = \frac{1}{3}x^2 - \frac{1}{3}x - 24$ has x-intercepts $x = \{-8, 9\}$.

34. $\frac{1}{6}x^2 + \frac{1}{6}x = 5 \Rightarrow \frac{1}{6}x^2 + \frac{1}{6}x - 5 = 0 \Rightarrow x^2 + x - 30 = 0 \Rightarrow (x + 6)(x - 5) = 0 \Rightarrow$

$x + 6 = 0$ or $x - 5 = 0 \Rightarrow x = \{-6, 5\}$

For graphical support, show that the graph of $y_1 = \frac{1}{6}x^2 - \frac{1}{6}x - 5$ has x-intercepts $x = \{-6, 5\}$.

35. $(x + 2)(x - 1) = 7x + 5 \Rightarrow x^2 + x - 2 = 7x + 5 \Rightarrow x^2 - 6x - 7 = 0 \Rightarrow (x - 7)(x + 1) = 0 \Rightarrow$

$x - 7 = 0$ or $x + 1 = 0 \Rightarrow x = \{-1, 7\}$.

For graphical support, show that the graph of $y_1 = x^2 - 6x - 7$ has x-intercepts $x = \{-1, 7\}$.

36. $(x + 4)(x - 1) = -5x - 4 \Rightarrow x^2 + 3x - 4 = -5x - 4 \Rightarrow x^2 + 8x = 0 \Rightarrow x(x + 8) = 0 \Rightarrow$

$x = 0$ or $x + 8 = 8 \Rightarrow x = \{-8, 0\}$.

For graphical support, show that the graph of $y_1 = x^2 + 8x$ has x-intercepts $x = \{-8, 0\}$.

37. Use the quadratic formula to solve $x^2 - 2x - 4 = 0$, therefore $a = 1, b = -2$, and $c = -4$.

$x = \frac{-(-2) \pm \sqrt{(-2)^2 - 4(1)(-4)}}{2(1)} = \frac{2 \pm \sqrt{4 + 16}}{2} = \frac{2 \pm \sqrt{20}}{2} = \frac{2 \pm 2\sqrt{5}}{2} = \frac{2(1 \pm \sqrt{5})}{2} \Rightarrow$

$x = \{1 \pm \sqrt{5}\}$. For graphical support, show that the graph of $y_1 = x^2 - 2x - 4$ has x-intercepts

$x = \{1 \pm \sqrt{5}\} \approx \{-1.24, 3.24\}$.

38. Use the quadratic formula to solve $x^2 + 8x + 13 = 0$, therefore $a = 1, b = 8$, and $c = 13$.

$x = \frac{-8 \pm \sqrt{(8)^2 - 4(1)(13)}}{2(1)} = \frac{-8 \pm \sqrt{64 - 52}}{2} = \frac{-8 \pm \sqrt{12}}{2} = \frac{-8 \pm 2\sqrt{3}}{2} = \frac{2(-4 \pm \sqrt{3})}{2} \Rightarrow$

$x = \{-4 \pm \sqrt{3}\}$. For graphical support, show that the graph of $y_1 = x^2 + 8x + 13$ has x-intercepts

$x = \{-4 \pm \sqrt{3}\} \approx \{-5.73, -2.27\}$.

39. Use the quadratic formula to solve $2x^2 + 2x = -1 \Rightarrow 2x^2 + 2x + 1 = 0$, therefore $a = 2, b = 2$, and $c = 1$.

$x = \frac{-2 \pm \sqrt{(2)^2 - 4(2)(1)}}{2(2)} = \frac{-2 \pm \sqrt{4 - 8}}{4} = \frac{-2 \pm \sqrt{-4}}{4} = \frac{-2 \pm 2i}{4} = \frac{2(-1 \pm i)}{4} \Rightarrow$

$x = \left\{-\frac{1}{2} \pm \frac{1}{2}i\right\}$. For graphical support, show that the graph of $y_1 = 2x^2 + 2x + 1$ has no x-intercepts,

therefore no real solutions.

40. Use the quadratic formula to solve $9x^2 - 12x = -8 \Rightarrow 9x^2 - 12x + 8 = 0$, therefore $a = 9$, $b = -12$, and

$c = 8$. $x = \dfrac{-(-12) \pm \sqrt{(-12)^2 - 4(9)(8)}}{2(9)} = \dfrac{12 \pm \sqrt{144 - 288}}{18} = \dfrac{12 \pm \sqrt{-144}}{18} = \dfrac{12 \pm 12i}{18} =$

$\dfrac{6(2 \pm 2i)}{18} \Rightarrow x = \left\{ \dfrac{2}{3} \pm \dfrac{2}{3}i \right\}$. For graphical support, show that the graph of $y_1 = 9x^2 - 12x + 8$ has no

x-intercepts, therefore no real solutions.

41. Use the quadratic formula to solve $x(x - 1) = 1 \Rightarrow x^2 - x - 1 = 0$, therefore $a = 1$, $b = -1$, and $c = -1$.

$x = \dfrac{-(-1) \pm \sqrt{(-1)^2 - 4(1)(-1)}}{2(1)} = \dfrac{1 \pm \sqrt{1 + 4}}{2} \Rightarrow x = \left\{ \dfrac{1 \pm \sqrt{5}}{2} \right\}$. For graphical support, show that

the graph of $y_1 = x^2 - x - 1$ has x-intercepts $x = \left\{ \dfrac{1 \pm \sqrt{5}}{2} \right\} \approx \{-.62, 1.62\}$.

42. Use the quadratic formula to solve $x(x - 3) = 2 \Rightarrow x^2 - 3x - 2 = 0$, therefore $a = 1$, $b = -3$, and $c = -2$.

$x = \dfrac{-(-3) \pm \sqrt{(-3)^2 - 4(1)(-2)}}{2(1)} = \dfrac{3 \pm \sqrt{9 + 8}}{2} \Rightarrow x = \left\{ \dfrac{3 \pm \sqrt{17}}{2} \right\}$. For graphical support, show that

the graph of $y_1 = x^2 - 3x - 2$ has x-intercepts $x = \left\{ \dfrac{3 \pm \sqrt{17}}{2} \right\} \approx \{-.56, 3.56\}$.

43. Use the quadratic formula to solve $x^2 - 5x = x - 7 \Rightarrow x^2 - 6x + 7 = 0$, therefore $a = 1$, $b = -6$, and $c = 7$.

$x = \dfrac{-(-6) \pm \sqrt{(-6)^2 - 4(1)(7)}}{2(1)} = \dfrac{6 \pm \sqrt{36 - 28}}{2} = \dfrac{6 \pm \sqrt{8}}{2} = \dfrac{6 \pm 2\sqrt{2}}{2} = \dfrac{2(3 \pm \sqrt{2})}{2} \Rightarrow$

$x = \{3 \pm \sqrt{2}\}$. For graphical support, show that the graph of $y_1 = x^2 - 6x + 7$ has x-intercepts

$x = \{3 \pm \sqrt{2}\} \approx \{1.59, 4.41\}$.

44. Use the quadratic formula to solve $11x^2 - 3x + 2 = 4x + 1 \Rightarrow 11x^2 - 7x + 1 = 0$, therefore $a = 11$,

$b = -7$, and $c = 1$. $x = \dfrac{-(-7) \pm \sqrt{(-7)^2 - 4(11)(1)}}{2(11)} = \dfrac{7 \pm \sqrt{49 - 44}}{22} \Rightarrow x = \left\{ \dfrac{7 \pm \sqrt{5}}{22} \right\}$.

For graphical support, show that the graph of $y_1 = 11x^2 - 7x + 1$ has x-intercepts

$x = \left\{ \dfrac{7 \pm \sqrt{5}}{22} \right\} \approx \{.22, .42\}$.

45. Use the quadratic formula to solve $4x^2 - 12x = -11 \Rightarrow 4x^2 - 12x + 11 = 0$, therefore $a = 4$, $b = -12$, and

$c = 11$. $x = \dfrac{-(-12) \pm \sqrt{(-12)^2 - 4(4)(11)}}{2(4)} = \dfrac{12 \pm \sqrt{144 - 176}}{8} = \dfrac{12 \pm \sqrt{-32}}{8} = \dfrac{12 \pm 4i\sqrt{2})}{8} =$

$\dfrac{4(3 \pm i\sqrt{2})}{8} \Rightarrow x = \left\{ \dfrac{3}{2} \pm \dfrac{\sqrt{2}}{2}i \right\}$. For graphical support, show that the graph of $y_1 = 4x^2 - 12x + 11$ has

no x-intercepts, therefore no real solutions.

46. Use the quadratic formula to solve $x^2 = 2x - 5 \Rightarrow x^2 - 2x + 5 = 0$, therefore $a = 1$, $b = -2$, and $c = 5$.

$x = \dfrac{-(-2) \pm \sqrt{(-2)^2 - 4(1)(5)}}{2(1)} = \dfrac{2 \pm \sqrt{4 - 20}}{2} = \dfrac{2 \pm \sqrt{-16}}{2} = \dfrac{2 \pm 4i}{2} = \dfrac{2(1 \pm 2i)}{2} \Rightarrow$

$x = \{1 \pm 2i\}$. For graphical support, show that the graph of $y_1 = x^2 - 2x + 5$ has no x-intercepts, therefore

no real solutions.

47. Use the quadratic formula to solve $\frac{1}{3}x^2 + \frac{1}{4}x - 3 = 0 \Rightarrow 12\left(\frac{1}{3}x^2 + \frac{1}{4}x - 3 = 0\right) \Rightarrow 4x^2 + 3x - 36 = 0,$

   therefore $a = 4, b = 3$, and $c = -36$. $x = \dfrac{-3 \pm \sqrt{(3)^2 - 4(4)(-36)}}{2(4)} = \dfrac{-3 \pm \sqrt{9 + 576}}{8} =$

   $\dfrac{-3 \pm \sqrt{585}}{8} \Rightarrow x = \left\{\dfrac{-3 \pm 3\sqrt{65}}{8}\right\}$. For graphical support, show that the graph of $y_1 = 4x^2 + 3x - 36$

   has x-intercepts $x = \left\{\dfrac{-3 \pm 3\sqrt{65}}{8}\right\} \approx \{-3.4, 2.6\}$.

48. Use the quadratic formula to solve $\frac{2}{3}x^2 + \frac{1}{4}x = 3 \Rightarrow \frac{2}{3}x^2 + \frac{1}{4}x - 3 = 0 \Rightarrow 12\left(\frac{2}{3}x^2 + \frac{1}{4}x - 3 = 0\right) \Rightarrow$

   $8x^2 + 3x - 36 = 0$, therefore $a = 8, b = 3$, and $c = -36$. $x = \dfrac{-3 \pm \sqrt{(3)^2 - 4(8)(-36)}}{2(8)} =$

   $\dfrac{-3 \pm \sqrt{9 + 1152}}{16} = \dfrac{-3 \pm \sqrt{1161}}{16} \Rightarrow x = \left\{\dfrac{-3 \pm 3\sqrt{129}}{16}\right\} \approx \{-2.3, 1.9\}$. For graphical support,

   show that the graph of $y_1 = 8x^2 + 3x - 36$ has x-intercepts $x = \left\{\dfrac{-3 \pm 3\sqrt{129}}{16}\right\} \approx \{-2.3, 1.9\}$.

49. Use the quadratic formula to solve $(3 - x)^2 = 25 \Rightarrow 9 - 6x + x^2 = 25 \Rightarrow x^2 - 6x - 16 = 0$, therefore

   $a = 1, b = -6$, and $c = -16$. $x = \dfrac{-(-6) \pm \sqrt{(-6)^2 - 4(1)(-16)}}{2(1)} = \dfrac{6 \pm \sqrt{36 + 64}}{2} = \dfrac{6 \pm \sqrt{100}}{2} =$

   $\dfrac{6 \pm 10}{2} \Rightarrow x = \{-2, 8\}$. For graphical support, show that the graph of $y_1 = x^2 - 6x - 16$ has

   x-intercepts $x = \{-2, 8\}$.

50. Use the quadratic formula to solve $(2 + x)^2 = 49 \Rightarrow 4 + 4x + x^2 = 49 \Rightarrow x^2 + 4x - 45 = 0$, therefore

   $a = 1, b = 4$, and $c = -45$. $x = \dfrac{-4 \pm \sqrt{(4)^2 - 4(1)(-45)}}{2(1)} = \dfrac{-4 \pm \sqrt{16 + 180}}{2} = \dfrac{-4 \pm \sqrt{196}}{2} =$

   $\dfrac{-4 \pm 14}{2} \Rightarrow x = \{-9, 5\}$. For graphical support, show that the graph of $y_1 = x^2 + 4x - 45$ has

   x-intercepts $x = \{-9, 5\}$.

51. Use the quadratic formula to solve $2x^2 - 4x = 1 \Rightarrow 2x^2 - 4x - 1 = 0$, therefore $a = 2, b = -4$, and $c = -1$.

   $x = \dfrac{-(-4) \pm \sqrt{(-4)^2 - 4(2)(-1)}}{2(2)} = \dfrac{4 \pm \sqrt{16 + 8}}{4} = \dfrac{4 \pm \sqrt{24}}{4} = \dfrac{4 \pm 2\sqrt{6}}{4} = \dfrac{2(2 \pm \sqrt{6})}{4} \Rightarrow$

   $x = \left\{\dfrac{2 \pm \sqrt{6}}{2}\right\}$. For graphical support, show that the graph of $y_1 = 2x^2 - 4x - 1$ has

   x-intercepts $x = \left\{\dfrac{2 \pm \sqrt{6}}{2}\right\} \approx \{-.2, 2.2\}$.

52. Use the quadratic formula to solve $3x^2 - 6x = 4 \Rightarrow 3x^2 - 6x - 4 = 0$, therefore $a = 3, b = -6$, and $c = -4$.

   $x = \dfrac{-(-6) \pm \sqrt{(-6)^2 - 4(3)(-4)}}{2(3)} = \dfrac{6 \pm \sqrt{36 + 48}}{6} = \dfrac{6 \pm \sqrt{84}}{6} = \dfrac{6 \pm 2\sqrt{21}}{6} = \dfrac{2(3 \pm \sqrt{21})}{6} \Rightarrow$

   $x = \left\{\dfrac{3 \pm \sqrt{21}}{3}\right\}$. For graphical support, show that the graph of $y_1 = 3x^2 - 6x - 4$ has

   x-intercepts $x = \left\{\dfrac{3 \pm \sqrt{21}}{3}\right\} \approx \{-.5, 2.5\}$.

53. Use the quadratic formula to solve $x^2 = -1 - x \Rightarrow x^2 + x + 1 = 0$, therefore $a = 1$, $b = 1$, and $c = 1$.

$$x = \frac{-1 \pm \sqrt{(1)^2 - 4(1)(1)}}{2(1)} = \frac{-1 \pm \sqrt{1 - 4}}{2} = \frac{-1 \pm \sqrt{-3}}{2} \Rightarrow x = \left\{ -\frac{1}{2} \pm \frac{\sqrt{3}}{2}i \right\}.$$ For graphical

support, show that the graph of $y_1 = x^2 + x + 1$ has no x-intercepts, therefore no real solutions.

54. Use the quadratic formula to solve $x^2 = -3 - 3x \Rightarrow x^2 + 3x + 3 = 0$, therefore $a = 1$, $b = 3$, and $c = 3$.

$$x = \frac{-3 \pm \sqrt{(3)^2 - 4(1)(3)}}{2(1)} = \frac{-3 \pm \sqrt{9 - 12}}{2} = \frac{-3 \pm \sqrt{-3}}{2} \Rightarrow x = \left\{ -\frac{3}{2} \pm \frac{\sqrt{3}}{2}i \right\}.$$ For graphical

support, show that the graph of $y_1 = x^2 + 3x + 3$ has no x-intercepts, therefore no real solutions.

55. Use the quadratic formula to solve $4x^2 - 20x + 25 = 0$, therefore $a = 4$, $b = -20$, and $c = 25$.

$$x = \frac{-(-20) \pm \sqrt{(-20)^2 - 4(4)(25)}}{2(4)} = \frac{20 \pm \sqrt{400 - 400}}{8} = \frac{20 \pm 0}{8} \Rightarrow x = \left\{ \frac{20}{8} \right\} = \left\{ \frac{5}{2} \right\}.$$

For graphical support, show that the graph of $y_1 = 3x^2 - 6x - 4$ has x-intercept $x = \left\{ \frac{5}{2} \right\} = \{2.5\}$.

56. Use the quadratic formula to solve $9x^2 + 12x + 4 = 0$, therefore $a = 9$, $b = 12$, and $c = 4$.

$$x = \frac{-12 \pm \sqrt{(12)^2 - 4(9)(4)}}{2(9)} = \frac{-12 \pm \sqrt{144 - 144}}{18} = \frac{-12 \pm 0}{18} \Rightarrow x = \left\{ \frac{-12}{18} \right\} = \left\{ -\frac{2}{3} \right\}.$$

For graphical support, show that the graph of $y_1 = 9x^2 + 12x + 4$ has x-intercept $x = \left\{ -\frac{2}{3} \right\} \approx \{-.7\}$.

57. Use the quadratic formula to solve $-3x^2 + 4x + 4 = 0$, therefore $a = -3$, $b = 4$, and $c = 4$.

$$x = \frac{-4 \pm \sqrt{(4)^2 - 4(-3)(4)}}{2(-3)} = \frac{-4 \pm \sqrt{16 + 48}}{-6} = \frac{-4 \pm \sqrt{64}}{-6} = \frac{-4 \pm 8}{-6} = \frac{-2(2 \pm 4)}{-6} = \frac{2 \pm 4}{3} \Rightarrow$$

$x = \left\{ -\frac{2}{3}, 2 \right\}$. For graphical support, show that the graph of $y_1 = -3x^2 + 4x + 4$ has x-intercepts

$x = \left\{ -\frac{2}{3}, 2 \right\} \approx \{-.7, 2\}$.

58. Use the quadratic formula to solve $-5x^2 + 28x + 12 = 0$, therefore $a = -5$, $b = 28$, and $c = 12$.

$$x = \frac{-28 \pm \sqrt{(28)^2 - 4(-5)(12)}}{2(-5)} = \frac{-28 \pm \sqrt{784 + 240}}{-10} = \frac{-28 \pm \sqrt{1024}}{-10} =$$

$$\frac{-28 \pm 32}{-10} = \frac{-2(14 \pm 16)}{-10} = \frac{14 \pm 16}{5} \Rightarrow x = \left\{ -\frac{2}{5}, 6 \right\}.$$ For graphical support, show that the graph of

$y_1 = -5x^2 + 28x + 12$ has x-intercepts $x = \left\{ -\frac{2}{5}, 6 \right\} = \{-.4, 6\}$.

59. Use the quadratic formula to solve $(x + 5)(x - 6) = (2x - 1)(x - 4) \Rightarrow x^2 - x - 30 = 2x^2 - 9x + 4 \Rightarrow$

$x^2 - 8x + 34 = 0$, therefore $a = 1$, $b = -8$, and $c = 34$. $x = \dfrac{-(-8) \pm \sqrt{(-8)^2 - 4(1)(34)}}{2(1)} =$

$$\frac{8 \pm \sqrt{64 - 136}}{2} = \frac{8 \pm \sqrt{-72}}{2} = \frac{8 \pm 6i\sqrt{2}}{2} = \frac{2(4 \pm 3i\sqrt{2})}{2} \Rightarrow x = \{4 \pm 3i\sqrt{2}\}.$$ For graphical

support, show that the graph of $y_1 = 2x^2 - 9x + 4$ has no x-intercepts, therefore no real solutions.

60. Use the quadratic formula to solve $(x + 2)(3x - 4) = (x + 5)(2x - 5) \Rightarrow 3x^2 + 2x - 8 = 2x^2 + 5x - 25 \Rightarrow$

    $x^2 - 3x + 17 = 0$, therefore $a = 1, b = -3$, and $c = 17$. $x = \dfrac{-(-3) \pm \sqrt{(-3)^2 - 4(1)(17)}}{2(1)} =$

    $\dfrac{3 \pm \sqrt{9 - 68}}{2} = \dfrac{3 \pm \sqrt{-59}}{2} = \dfrac{3 \pm i\sqrt{59}}{2} \Rightarrow x = \left\{ \dfrac{3}{2} \pm \dfrac{\sqrt{59}}{2}i \right\}$. For graphical

    support, show that the graph of $y_1 = x^2 - 3x + 17$ has no x-intercepts, therefore no real solutions.

61. For the equation $x^2 + 8x + 16 = 0$; $a = 1, b = 8$, and $c = 16$, therefore the discriminant is: $(8)^2 - 4(1)(16) =$

    $64 - 64 = 0$. Because the discriminate is 0, there is 1 real solution. Since a, b, and c are nonzero integers and

    the discriminant is a square of an integer the solution is rational.

62. For the equation $8x^2 = 14x - 3 \Rightarrow 8x^2 - 14x + 3 = 0$; $a = 8, b = -14$, and $c = 3$, therefore the discriminant

    is: $(-14)^2 - 4(8)(3) = 196 - 96 = 100$. Because the discriminate is positive, there are 2 real solutions. Since

    a, b, and c are nonzero integers and the discriminant is a square of an integer, the solutions are rational.

63. For the equation $4x^2 = 6x + 3 \Rightarrow 4x^2 - 6x - 3 = 0$; $a = 4, b = -6$, and $c = -3$, therefore the discriminant

    is: $(-6)^2 - 4(4)(-3) = 36 + 48 = 84$. Because the discriminate is positive, there are 2 real solutions. Since a,

    b, and c are nonzero integers and the discriminant is not the square of an integer, the solutions are irrational.

64. For the equation $2x^2 - 4x + 1 = 0$; $a = 2, b = -4$, and $c = 1$, therefore the discriminant is:

    $(-4)^2 - 4(2)(1) = 16 - 8 = 8$. Because the discriminate is positive, there are 2 real solutions. Since a, b, and

    c are nonzero integers and the discriminant is not the square of an integer, the solutions are irrational.

65. For the equation $9x^2 + 11x + 4 = 0$; $a = 9, b = 11$, and $c = 4$, therefore the discriminant is:

    $(11)^2 - 4(9)(4) = 121 - 144 = -23$. Because the discriminate is negative, there are no real solutions.

66. For the equation $3x^2 = 4x - 5 \Rightarrow 3x^2 - 4x + 5 = 0$; $a = 3, b = -4$, and $c = 5$, therefore the discriminant is:

    $(-4)^2 - 4(3)(5) = 16 - 60 = -44$. Because the discriminate is negative, there are no real solutions.

67. If $x = \{4, 5\}$, then $(x - 4)(x - 5) = 0 \Rightarrow x^2 - 9x + 20 = 0$ and $a = 1, b = -9$, and $c = 20$.

68. If $x = \{-3, 2\}$, then $(x + 3)(x - 2) = 0 \Rightarrow x^2 + x - 6 = 0$ and $a = 1, b = 1$, and $c = -6$.

69. If $x = \{1 + \sqrt{2}, 1 - \sqrt{2}\}$, then $(x - (1 + \sqrt{2}))(x - (1 - \sqrt{2})) = 0 \Rightarrow (x - 1 - \sqrt{2})(x - 1 + \sqrt{2}) = 0$

    $\Rightarrow x^2 - x + x\sqrt{2} - x + 1 - \sqrt{2} - x\sqrt{2} + \sqrt{2} - 2 = 0 \Rightarrow x^2 - 2x - 1 = 0$ and

    $a = 1, b = -2$, and $c = -1$.

70. If $x = \{-i, i\}$, then $(x + i)(x - i) = 0 \Rightarrow x^2 - i^2 = 0 \Rightarrow x^2 + 1 = 0$ and $a = 1, b = 0$, and $c = 1$.

71. The graph of the function $f(x) = ax^2 + bx + c$ is a parabola, $a < 0$ will make the parabola open downward,

    and $b^2 - 4ac = 0$ gives us 1 real solution or 1 x-intercept. See Figure 71.

72. The graph of the function $f(x) = ax^2 + bx + c$ is a parabola, $a > 0$ will make the parabola open upward, and

    $b^2 - 4ac < 0$ gives us no real solutions or no x-intercepts. See Figure 72.

73. The graph of the function $f(x) = ax^2 + bx + c$ is a parabola, $a < 0$ will make the parabola open downward,

    and $b^2 - 4ac < 0$ gives us no real solutions or no x-intercepts. See Figure 73.

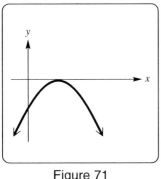

Figure 71

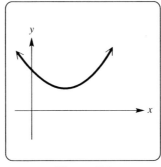

Figure 72

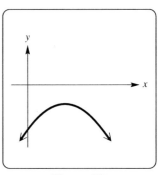

Figure 73

74. The graph of the function $f(x) = ax^2 + bx + c$ is a parabola, $a < 0$ will make the parabola open downward, and $b^2 - 4ac > 0$ gives us 2 real solutions or 2 x-intercepts. See Figure 74.

75. The graph of the function $f(x) = ax^2 + bx + c$ is a parabola, $a > 0$ will make the parabola open upward, and $b^2 - 4ac > 0$ gives us 2 real solutions or 2 x-intercepts. See Figure 75.

76. The graph of the function $f(x) = ax^2 + bx + c$ is a parabola, $a > 0$ will make the parabola open upward, and $b^2 - 4ac = 0$ gives us 1 real solution or 1 x-intercept. See Figure 76.

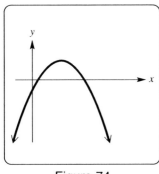

Figure 74

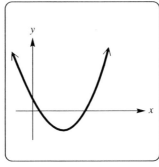

Figure 75

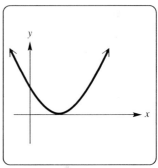

Figure 76

77. From the graph, $f(x) = 0$ when $x = \{2, 4\}$.

78. From the graph, $f(x) < 0$ for the interval: $(2, 4)$.

79. From the graph, $f(x) > 0$ for the interval: $(-\infty, 2) \cup (4, \infty)$.

80. From the graph, $g(x) = 0$ when $x = \{3\}$.

81. From the graph, $g(x) < 0$ for the interval: $(-\infty, 3) \cup (3, \infty)$.

82. From the graph, $g(x) > 0$ never or the solution is: $\varnothing$.

83. From the graph, $h(x) > 0$ always, so $(-\infty, \infty)$.

84. From the graph, $h(x) < 0$ never or the solution is: $\varnothing$.

85. From the graph, $h(x) = 0$ has no real solutions, because $h(x)$ has no x-intercepts. The graph is completely above the x-axis, therefore there are two nonreal complex solutions.

86. Since the graph of $g(x)$ has only one x-intercept or one solution, the value of the discriminate of $g(x)$ is 0.

87. The x-coordinate of the vertex of the graph of $y = f(x)$ is the midpoint of the x-intercepts or $\dfrac{2 + 4}{2} = 3$.

88. The vertex of the graph is $(3, 0)$, therefore the equation for the vertical axis of symmetry is: $x = 3$.

89. From the graph, $y = g(x)$ will have a y-intercept and it will be negative.

90. From the graph, the graph of h is always above the x-axis, therefore the minimum value is positive.

91. (a) First set $x^2 + 4x + 3 = 0$, then $(x + 1)(x + 3) = 0$ and the endpoints will be: $x = \{-3, -1\}$. The numbers -3 and -1 divide the number line into 3 intervals. The interval: $(-\infty, -3)$ has a positive product, the interval: $(-3, -1)$ has a negative product, and the interval: $(-1, \infty)$ has a positive product. Therefore $x^2 + 4x + 3 \geq 0$ for the interval: $(-\infty, -3] \cup [-1, \infty)$. The graph of $y_1 = x^2 + 4x + 3$ intersects or is above the x-axis for the interval: $(-\infty, -3] \cup [-1, \infty)$.

   (b) First set $x^2 + 4x + 3 = 0$, then $(x + 1)(x + 3) = 0$ and the endpoints will be: $x = \{-3, -1\}$. The numbers -3 and -1 divide the number line into 3 intervals. The interval: $(-\infty, -3)$ has a positive product, the interval: $(-3, -1)$ has a negative product, and the interval: $(-1, \infty)$ has a positive product. Therefore $x^2 + 4x + 3 < 0$ for the interval: $(-3, -1)$. The graph of $y_1 = x^2 + 4x + 3$ is below the x-axis for the interval: $(-3, -1)$.

92. (a) First set $x^2 + 6x + 8 = 0$, then $(x + 4)(x + 2) = 0$ and the endpoints will be: $x = \{-4, -2\}$. The numbers -4 and -2 divide the number line into 3 intervals. The interval: $(-\infty, -4)$ has a positive product, the interval: $(-4, -2)$ has a negative product, and the interval: $(-2, \infty)$ has a positive product. Therefore $x^2 + 6x + 8 < 0$ for the interval: $(-4, -2)$. The graph of $y_1 = x^2 + 6x + 8$ is below the x-axis for the interval: $(-4, -2)$.

   (b) First set $x^2 + 6x + 8 = 0$, then $(x + 4)(x + 2) = 0$ and the endpoints will be: $x = \{-4, -2\}$. The numbers -4 and -2 divide the number line into 3 intervals. The interval: $(-\infty, -4)$ has a positive product, the interval: $(-4, -2)$ has a negative product, and the interval: $(-2, \infty)$ has a positive product. Therefore $x^2 + 6x + 8 \geq 0$ for the interval: $(-\infty, -4] \cup [-2, \infty)$. The graph of $y_1 = x^2 + 6x + 8$ intersect or is above the x-axis for the interval: $(-\infty, -4] \cup [-2, \infty)$.

93. (a) First $2x^2 - 9x > -4 \Rightarrow 2x^2 - 9x + 4 > 0$, now set $2x^2 - 9x + 4 = 0$ then $(2x - 1)(x - 4) = 0$ and the endpoints will be: $x = \left\{\dfrac{1}{2}, 4\right\}$. The numbers $\dfrac{1}{2}$ and 4 divide the number line into 3 intervals. The interval: $\left(-\infty, \dfrac{1}{2}\right)$ has a positive product, the interval: $\left(\dfrac{1}{2}, 4\right)$ has a negative product, and the interval: $(4, \infty)$ has a positive product. Therefore $2x^2 - 9x > -4$ for the interval: $\left(-\infty, \dfrac{1}{2}\right) \cup (4, \infty)$. The graph of $y_1 = 2x^2 - 9x + 4$ is above the x-axis for the interval: $\left(-\infty, \dfrac{1}{2}\right) \cup (4, \infty)$.

   (b) First $2x^2 - 9x \leq -4 \Rightarrow 2x^2 - 9x + 4 \leq 0$, now set $2x^2 - 9x + 4 = 0$ then $(2x - 1)(x - 4) = 0$ and the endpoints will be: $x = \left\{\dfrac{1}{2}, 4\right\}$. The numbers $\dfrac{1}{2}$ and 4 divide the number line into 3 intervals. The interval: $\left(-\infty, \dfrac{1}{2}\right)$ has a positive product, the interval: $\left(\dfrac{1}{2}, 4\right)$ has a negative product, and the interval: $(4, \infty)$ has a positive product. Therefore $2x^2 - 9x \leq -4$ for the interval: $\left(\dfrac{1}{2}, 4\right)$. The graph of $y_1 = 2x^2 - 9x + 4$ intersects or is below the x-axis for the interval: $\left(\dfrac{1}{2}, 4\right)$.

94. (a) First set $3x^2 + 13x + 10 = 0$, then $(3x + 10)(x + 1) = 0$ and the endpoints will be: $x = \left\{ -\dfrac{10}{3}, -1 \right\}$. The

numbers $-\dfrac{10}{3}$ and -1 divide the number line into 3 intervals. The interval: $\left(-\infty, -\dfrac{10}{3} \right)$ has a positive

product, the interval: $\left(-\dfrac{10}{3}, -1 \right)$ has a negative product, and the interval: $(-1, \infty)$ has a positive product.

Therefore $3x^2 + 13x + 10 \le 0$ for the interval: $\left[-\dfrac{10}{3}, -1 \right]$. The graph of $y_1 = 3x^2 + 13x + 10$ intersects

or is below the x-axis for the interval: $\left[-\dfrac{10}{3}, -1 \right]$.

   (b) First set $3x^2 + 13x + 10 = 0$, then $(3x + 10)(x + 1) = 0$ and the endpoints will be: $x = \left\{ -\dfrac{10}{3}, -1 \right\}$. The

numbers $-\dfrac{10}{3}$ and -1 divide the number line into 3 intervals. The interval: $\left(-\infty, -\dfrac{10}{3} \right)$ has a positive

product, the interval: $\left(-\dfrac{10}{3}, -1 \right)$ has a negative product, and the interval: $(-1, \infty)$ has a positive product.

Therefore $3x^2 + 13x + 10 > 0$ for the interval: $\left(-\infty, -\dfrac{10}{3} \right) \cup (-1, \infty)$. The graph $y_1 = 3x^2 + 13x + 10$

is above the x-axis for the interval: $\left(-\infty, -\dfrac{10}{3} \right) \cup (-1, \infty)$.

95. (a) First set $-x^2 - x = 0$, then $-x(x + 1) = 0$ and the endpoints will be: $x = \{-1, 0\}$. The numbers

-1 and 0 divide the number line into 3 intervals. The interval: $(-\infty, -1)$ has a negative product, the interval:

$(-1, 0)$ has a positive product, and the interval: $(0, \infty)$ has a negative product. Therefore

$-x^2 - x \le 0$ for the interval: $(-\infty, -1] \cup [0, \infty)$. The graph of $y_1 = -x^2 - x$ intersects or is below the

x-axis for the interval: $(-\infty, -1] \cup [0, \infty)$.

   (b) First set $-x^2 - x = 0$, then $-x(x + 1) = 0$ and the endpoints will be: $x = \{-1, 0\}$. The numbers

-1 and 0 divide the number line into 3 intervals. The interval: $(-\infty, -1)$ has a negative product, the interval:

$(-1, 0)$ has a positive product, and the interval: $(0, \infty)$ has a negative product. Therefore $-x^2 - x > 0$ for

the interval: $(-1, 0)$. The graph of $y_1 = -x^2 - x$ is above the x-axis for the interval: $(-1, 0)$.

96. (a) First set $-x^2 + 2x = 0$, then $-x(x - 2) = 0$ and the endpoints will be: $x = \{0, 2\}$. The numbers

0 and 2 divide the number line into 3 intervals. The interval: $(-\infty, 0)$ has a negative product, the interval:

$(0, 2)$ has a positive product, and the interval: $(2, \infty)$ has a negative product. Therefore

$-x^2 + 2x \le 0$ for the interval: $(-\infty, 0] \cup [2, \infty)$. The graph of $y_1 = -x^2 - x$ intersects or is below the

x-axis for the interval: $(-\infty, 0] \cup [2, \infty)$.

   (b) First set $-x^2 + 2x = 0$, then $-x(x - 2) = 0$ and the endpoints will be: $x = \{0, 2\}$. The numbers

0 and 2 divide the number line into 3 intervals. The interval: $(-\infty, 0)$ has a negative product, the interval:

$(0, 2)$ has a positive product, and the interval: $(2, \infty)$ has a negative product. Therefore $-x^2 + 2x > 0$ for

the interval: $(0, 2)$. The graph of $y_1 = -x^2 + 2x$ is above the x-axis for the interval: $(0, 2)$.

97. (a) First set $x^2 - x + 1 = 0$, then by the quadratic formula $x = \dfrac{-(-1) \pm \sqrt{(-1)^2 - 4(1)(1)}}{2(1)} = \dfrac{1 \pm \sqrt{-3}}{2} \Rightarrow$

$x = \dfrac{1}{2} \pm \dfrac{\sqrt{3}}{2}i$. The inequality has no real solutions, the graph is a parabola opening upward which does

not intersect, but is completely above the x-axis. Therefore $x^2 - x + 1 < 0$ never happens and the

solution is: $\varnothing$.

(b) First set $x^2 - x + 1 = 0$, then by the quadratic formula $x = \dfrac{-(-1) \pm \sqrt{(-1)^2 - 4(1)(1)}}{2(1)} = \dfrac{1 \pm \sqrt{-3}}{2} \Rightarrow$

$x = \dfrac{1}{2} \pm \dfrac{\sqrt{3}}{2}i$. The graph of this equation is a parabola opening upward, has no real solutions, and

does not intersect, but is completely above the x-axis. Therefore $x^2 - x + 1 \geq 0$ for all values of x and the

solution is the interval: $(-\infty, \infty)$.

98. (a) First set $2x^2 - x + 3 = 0$, then by the quadratic formula $x = \dfrac{-(-1) \pm \sqrt{(-1)^2 - 4(2)(3)}}{2(2)} = \dfrac{1 \pm \sqrt{-23}}{4} \Rightarrow$

$x = \dfrac{1}{4} \pm \dfrac{\sqrt{23}}{4}i$. The inequality has no real solutions, the graph is a parabola opening upward which does

not intersect, but is completely above the x-axis. Therefore $2x^2 - x + 3 < 0$ never happens and the

solution is: $\varnothing$.

(b) First set $2x^2 - x + 3 = 0$, then by the quadratic formula $x = \dfrac{-(-1) \pm \sqrt{(-1)^2 - 4(2)(3)}}{2(2)} = \dfrac{1 \pm \sqrt{-23}}{4} \Rightarrow$

$x = \dfrac{1}{4} \pm \dfrac{\sqrt{23}}{4}i$. The graph of this equation is a parabola opening upward, has no real solutions, and

does not intersect, but is completely above the x-axis. Therefore $2x^2 - x + 3 \geq 0$ for all values of x and

the solution is the interval: $(-\infty, \infty)$.

99. (a) First $2x + 1 \geq x^2 \Rightarrow x^2 - 2x - 1 \leq 0$, now set $x^2 - 2x - 1 = 0$ then by the quadratic formula

$x = \dfrac{-(-2) \pm \sqrt{(-2)^2 - 4(1)(-1)}}{2(1)} = \dfrac{2 \pm \sqrt{8}}{2} = \dfrac{2(1 \pm \sqrt{2})}{2} \Rightarrow x = 1 \pm \sqrt{2}$ and the endpoints will

be: $x = \{1 - \sqrt{2}, 1 + \sqrt{2}\}$. The numbers $1 - \sqrt{2}$ and $1 + \sqrt{2}$ divide the number line into 3 intervals.

The interval: $(-\infty, 1 - \sqrt{2})$ has a positive product, the interval: $(1 - \sqrt{2}, 1 + \sqrt{2})$ has a negative product,

and the interval: $(1 + \sqrt{2}, \infty)$ has a positive product. Therefore $x^2 - 2x - 1 \leq 0$ for the interval:

$[1 - \sqrt{2}, 1 + \sqrt{2}]$. The graph of $y_1 = x^2 - 2x - 1$ intersects or is below the x-axis for the interval:

$[1 - \sqrt{2}, 1 + \sqrt{2}]$.

(b) First $2x + 1 < x^2 \Rightarrow x^2 - 2x - 1 > 0$, now set $x^2 - 2x - 1 = 0$ then by the quadratic formula

$x = \dfrac{-(-2) \pm \sqrt{(-2)^2 - 4(1)(-1)}}{2(1)} = \dfrac{2 \pm \sqrt{8}}{2} = \dfrac{2(1 \pm \sqrt{2})}{2} \Rightarrow x = 1 \pm \sqrt{2}$ and the endpoints will

be: $x = \{1 - \sqrt{2}, 1 + \sqrt{2}\}$. The numbers $1 - \sqrt{2}$ and $1 + \sqrt{2}$ divide the number line into 3 intervals.

The interval: $(-\infty, 1 - \sqrt{2})$ has a positive product, the interval: $(1 - \sqrt{2}, 1 + \sqrt{2})$ has a negative product,

and the interval: $(1 + \sqrt{2}, \infty)$ has a positive product. Therefore $x^2 - 2x - 1 > 0$ for the interval:

$(-\infty, 1 - \sqrt{2}) \cup (1 + \sqrt{2}, \infty)$. The graph of $y_1 = x^2 - 2x - 1$ is above the x-axis for the interval:

$(-\infty, 1 - \sqrt{2}) \cup (1 + \sqrt{2}, \infty)$.

100. (a) First $x^2 + 5x < 2 \Rightarrow x^2 + 5x - 2 < 0$, now set $x^2 + 5x - 2 = 0$ then by the quadratic formula

$$x = \frac{-5 \pm \sqrt{(5)^2 - 4(1)(-2)}}{2(1)} \Rightarrow x = \frac{-5 \pm \sqrt{33}}{2} \text{ and the endpoints will be:}$$

$x = \left\{\dfrac{-5 - \sqrt{33}}{2}, \dfrac{-5 + \sqrt{33}}{2}\right\}$. The numbers $\dfrac{-5 - \sqrt{33}}{2}$ and $\dfrac{-5 + \sqrt{33}}{2}$ divide the number line into 3

intervals. The interval: $\left(-\infty, \dfrac{-5 - \sqrt{33}}{2}\right)$ has a positive product, the interval: $\left(\dfrac{-5 - \sqrt{33}}{2}, \dfrac{-5 + \sqrt{33}}{2}\right)$

has a negative product, and the interval: $\left(\dfrac{-5 + \sqrt{33}}{2}, \infty\right)$ has a positive product. Therefore

$x^2 + 5x - 2 < 0$ for the interval: $\left(\dfrac{-5 - \sqrt{33}}{2}, \dfrac{-5 + \sqrt{33}}{2}\right)$. The graph of $y_1 = x^2 + 5x - 2$ is below

the x-axis for the interval: $\left(\dfrac{-5 - \sqrt{33}}{2}, \dfrac{-5 + \sqrt{33}}{2}\right)$.

(b) First $x^2 + 5x \geq 2 \Rightarrow x^2 + 5x - 2 \geq 0$, now set $x^2 + 5x - 2 = 0$ then by the quadratic formula

$$x = \frac{-5 \pm \sqrt{(5)^2 - 4(1)(-2)}}{2(1)} \Rightarrow x = \frac{-5 \pm \sqrt{33}}{2} \text{ and the endpoints will be:}$$

$x = \left\{\dfrac{-5 - \sqrt{33}}{2}, \dfrac{-5 + \sqrt{33}}{2}\right\}$. The numbers $\dfrac{-5 - \sqrt{33}}{2}$ and $\dfrac{-5 + \sqrt{33}}{2}$ divide the number line into 3

intervals. The interval: $\left(-\infty, \dfrac{-5 - \sqrt{33}}{2}\right)$ has a positive product, the interval: $\left(\dfrac{-5 - \sqrt{33}}{2}, \dfrac{-5 + \sqrt{33}}{2}\right)$

has a negative product, and the interval: $\left(\dfrac{-5 + \sqrt{33}}{2}, \infty\right)$ has a positive product. Therefore

$x^2 + 5x - 2 \geq 0$ for the interval: $\left(-\infty, \dfrac{-5 - \sqrt{33}}{2}\right] \cup \left[\dfrac{-5 + \sqrt{33}}{2}, \infty\right)$. The graph of

$y_1 = x^2 + 5x - 2$ intersects or is above the x-axis for the interval: $\left(-\infty, \dfrac{-5 - \sqrt{33}}{2}\right[\cup \left[\dfrac{-5 + \sqrt{33}}{2}, \infty\right)$.

101. $s = \dfrac{1}{2}gt^2 \Rightarrow \dfrac{2s}{g} = t^2 \Rightarrow t = \pm\sqrt{\dfrac{2s}{g}} \Rightarrow t = \pm\dfrac{\sqrt{2sg}}{g}$

102. $A = \pi r^2 \Rightarrow \dfrac{A}{\pi} = r^2 \Rightarrow r = \pm\sqrt{\dfrac{A}{\pi}} \Rightarrow r = \pm\dfrac{\sqrt{A\pi}}{\pi}$

103. $a^2 + b^2 = c^2 \Rightarrow a^2 = c^2 - b^2 \Rightarrow a = \pm\sqrt{c^2 - b^2}$

104. $A = s^2 \Rightarrow s = \pm\sqrt{A}$

105. $S = 4\pi r^2 \Rightarrow \dfrac{S}{4\pi} = r^2 \Rightarrow r = \pm\sqrt{\dfrac{S}{4\pi}} \Rightarrow r = \pm\dfrac{\sqrt{S\pi}}{2\pi}$

106. $V = \dfrac{1}{3}\pi r^2 h \Rightarrow \dfrac{3V}{\pi h} = r^2 \Rightarrow r = \pm\sqrt{\dfrac{3V}{\pi h}} \Rightarrow r = \pm\dfrac{\sqrt{3V\pi h}}{\pi h}$

107. $V = e^3 \Rightarrow e = \sqrt[3]{V}$

108. $V = \dfrac{4}{3}\pi r^3 \Rightarrow \dfrac{3V}{4\pi} = r^3 \Rightarrow r = \sqrt[3]{\dfrac{3V}{4\pi}} \Rightarrow r = \dfrac{\sqrt[3]{3V4\pi 4\pi}}{4\pi} \Rightarrow r = \dfrac{2\sqrt[3]{6V\pi^2}}{4\pi} \Rightarrow r = \dfrac{\sqrt[3]{6V\pi^2}}{2\pi}$

109. $F = \dfrac{kMv^4}{r} \Rightarrow \dfrac{Fr}{kM} = v^4 \Rightarrow v = \pm\sqrt[4]{\dfrac{Fr}{kM}} \Rightarrow v = \dfrac{\pm\sqrt[4]{FrkMkMkM}}{kM} \Rightarrow v = \pm\dfrac{\sqrt[4]{Frk^3 M^3}}{kM}$

110. $s = s_0 + gt^2 + k \Rightarrow s - s_0 - k = gt^2 \Rightarrow \dfrac{s - s_0 - k}{g} = t^2 \Rightarrow t = \pm\sqrt{\dfrac{s - s_0 - k}{g}} \Rightarrow$

$t = \pm\dfrac{\sqrt{(s - s_0 - k)g}}{g}$

111. $P = \dfrac{E^2R}{(r + R)^2} \Rightarrow P(r + R)^2 = E^2R \Rightarrow P(r^2 + 2rR + R^2) - E^2R = 0 \Rightarrow Pr^2 + 2PrR + PR^2 - E^2R = 0$

$\Rightarrow PR^2 + (2Pr - E^2)R + Pr^2 = 0.$ Now use the quadratic formula:

$R = \dfrac{-(2Pr - E^2) \pm \sqrt{4P^2r^2 - 4E^2Pr + E^4 - 4(P)(Pr^2)}}{2(P)} = \dfrac{E^2 - 2Pr \pm \sqrt{E^4 - 4E^2Pr}}{2P} \Rightarrow$

$R = \dfrac{E^2 - 2Pr \pm E\sqrt{E^2 - 4Pr}}{2P}$

112. $S = 2\pi rh + 2\pi r^2 \Rightarrow 2\pi r^2 + 2\pi hr - S = 0.$ Now use the quadratic formula:

$r = \dfrac{-2\pi h \pm \sqrt{4\pi^2h^2 - 4(2\pi)(-S)}}{2(2\pi)} = \dfrac{-2\pi h \pm 2\sqrt{\pi^2h^2 + 2\pi S}}{4\pi} = \dfrac{2(-\pi h \pm \sqrt{\pi^2h^2 + 2\pi S})}{4\pi} \Rightarrow$

$r = \dfrac{-\pi h \pm \sqrt{\pi^2h^2 + 2\pi S}}{2\pi}$

113. Use the quadratic formula: $x = \dfrac{-y \pm \sqrt{y^2 - 4(1)(y^2)}}{2(1)} = \dfrac{-y \pm \sqrt{-3y^2}}{2} \Rightarrow x = -\dfrac{y}{2} \pm \dfrac{\sqrt{3}}{2}yi$

For $y^2 + xy + x^2 = 0$ use the quadratic formula: $y = \dfrac{-x \pm \sqrt{x^2 - 4(1)(x^2)}}{2(1)} = \dfrac{-x \pm \sqrt{-3x^2}}{2} \Rightarrow$

$y = -\dfrac{x}{2} \pm \dfrac{\sqrt{3}}{2}xi$

114. $4x^2 - 2xy + 3y^2 = 2 \Rightarrow 4x^2 - 2yx + (3y^2 - 2) = 0.$ Now use the quadratic formula:

$x = \dfrac{2y \pm \sqrt{(-2y)^2 - 4(4)(3y^2 - 2)}}{2(4)} = \dfrac{2y \pm \sqrt{4y^2 - 48y^2 + 32}}{8} = \dfrac{2y \pm \sqrt{-44y^2 + 32}}{8} =$

$\dfrac{2y \pm \sqrt{4(-11y^2 + 8)}}{8} = \dfrac{2y \pm 2\sqrt{8 - 11y^2}}{8} \Rightarrow x = \dfrac{y \pm \sqrt{8 - 11y^2}}{4}$

$4x^2 - 2xy + 3y^2 = 2 \Rightarrow 3y^2 - 2xy + (4x^2 - 2) = 0.$ Now use the quadratic formula:

$y = \dfrac{2x \pm \sqrt{(-2x)^2 - 4(3)(4x^2 - 2)}}{2(3)} = \dfrac{2x \pm \sqrt{4x^2 - 48x^2 + 24}}{6} = \dfrac{2x \pm \sqrt{-44x^2 + 24}}{6} =$

$\dfrac{2x \pm \sqrt{4(-11x^2 + 6)}}{6} = \dfrac{2x \pm 2\sqrt{6 - 11x^2}}{6} \Rightarrow y = \dfrac{x \pm \sqrt{6 - 11x^2}}{3}$

115. $3y^2 + 4xy - 9x^2 = -1 \Rightarrow -9x^2 + 4yx + (3y^2 + 1) = 0.$ Now use the quadratic formula:

$x = \dfrac{-4y \pm \sqrt{(4y)^2 - 4(-9)(3y^2 + 1)}}{2(-9)} = \dfrac{-4y \pm \sqrt{16y^2 + 108y^2 + 36}}{-18} =$

$\dfrac{-4y \pm \sqrt{124y^2 + 36}}{-18} = \dfrac{-4y \pm \sqrt{4(31y^2 + 9)}}{-18} = \dfrac{-4y \pm 2\sqrt{31y^2 + 9}}{-18} \Rightarrow x = \dfrac{2y \pm \sqrt{31y^2 + 9}}{9}$

$3y^2 + 4xy - 9x^2 = -1 \Rightarrow 3y^2 + 4xy + (-9x^2 + 1) = 0.$ Now use the quadratic formula:

$x = \dfrac{-4x \pm \sqrt{(4x)^2 - 4(3)(-9x^2 + 1)}}{2(3)} = \dfrac{-4x \pm \sqrt{16x^2 + 108x^2 - 12}}{6} =$

$\dfrac{-4x \pm \sqrt{124x^2 - 12}}{6} = \dfrac{-4x \pm \sqrt{4(31x^2 - 3)}}{6} = \dfrac{-4x \pm 2\sqrt{31x^2 - 3}}{6} \Rightarrow y = \dfrac{-2x \pm \sqrt{31x^2 - 3}}{3}$

116. For the function $f(x) = -.2369x^2 + 1.425x + 6.905$ the vertex formula, which will give us the maximum

value is: $\dfrac{-b}{2a} = \dfrac{-1.425}{2(-.2369)} \approx 3.007598$. Therefore, $1992 + 3 = 1995$ is the year the maximum was reached.

The maximum number was: $f(3) = -.2369(3)^2 + 1.425(3) + 6.905 \Rightarrow f(3) \approx 9.05$ or about 9%.

117. For the function $f(x) = -21.15x^2 + 476.2x + 11{,}973$ the vertex formula, which will give us the maximum

value is: $\dfrac{-b}{2a} = \dfrac{-476.2}{2(-21.15)} \approx 11.25769$. Therefore, $1980 + 11 = 1991$ is the year the maximum was reached.

The maximum number was: $f(11) = -21.15(11)^2 + 476.2(11) + 11{,}973 \Rightarrow f(11) \approx 14{,}653$.

118. If $S(x) = 0$ then $-.076x^2 - .058x + 1 = 0$.   Now use the quadratic formula:

$x = \dfrac{.058 \pm \sqrt{(-.058)^2 - 4(-.076)(1)}}{2(-.076)} \Rightarrow x \approx -4.029,\ 3.266$. The value $x \approx 3.3$; means that nobody lives

beyond age 98. (There are, of course, exceptions.)

Reviewing Basic Concepts (Sections 3.1—3.3)

1. $(5 + 6i) - (2 - 4i) - 3i = 5 + 6i - 2 + 4i - 3i = 3 + 7i$

2. $i(5 + i)(5 - i) = i(25 - i^2) = i(25 - (-1)) = 26i$

3. $\dfrac{-10 - 10i}{2 + 3i} \cdot \dfrac{2 - 3i}{2 - 3i} = \dfrac{-20 + 30i - 20i + 30i^2}{4 - 9i^2} = \dfrac{-20 + 10i - 30}{4 + 9} = -\dfrac{50}{13} + \dfrac{10}{13}i$

4. See Figure 4.

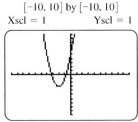

$[-10, 10]$ by $[-10, 10]$
Xscl = 1   Yscl = 1

Figure 4

5. Use the vertex formula: $\dfrac{-b}{2a} = \dfrac{-8}{2(2)} = -2$, now input $x = -2$ into the equation: $2(-2)^2 + 8(-2) + 5 = -3$.

The vertex is $(-2, -3)$ and it is a minimum because $a = 2$, which is positive, therefore the parabola opens up.

6. Since the vertex is: $(-2, -3)$ and the graph is a parabola opening up, the axis of symmetry is: $x = -2$.

7. From the graph the domain is: $(-\infty, \infty)$; and the range is: $[-3, \infty)$.

8. $9x^2 = 25 \Rightarrow x^2 = \dfrac{25}{9} \Rightarrow x = \left\{ \pm \dfrac{5}{3} \right\}$

9. $3x^2 - 5x = 2 \Rightarrow 3x^2 - 5x - 2 = 0 \Rightarrow (3x + 1)(x - 2) = 0 \Rightarrow$

$3x + 1 = 0 \text{ or } x - 2 = 0 \Rightarrow x = \left\{ -\dfrac{1}{3}, 2 \right\}$.

10. For $-x^2 + x + 3 = 0$ use the quadratic formula: $x = \dfrac{-1 \pm \sqrt{1^2 - 4(-1)(3)}}{2(-1)} = \dfrac{-1 \pm \sqrt{13}}{-2} \Rightarrow$

$x = \left\{ \dfrac{1 \pm \sqrt{13}}{2} \right\}$.

11. First set $3x^2 - 5x - 2 = 0$, then $(3x + 1)(x - 2) = 0$ and the endpoints will be: $x = \left\{-\frac{1}{3}, 2\right\}$. The numbers

$-\frac{1}{3}$ and 2 divide the number line into 3 intervals. The interval: $\left(-\infty, -\frac{1}{3}\right)$ has a positive product, the interval:

$\left(-\frac{1}{3}, 2\right)$ has a negative product, and the interval: $(2, \infty)$ has a positive product. Therefore $3x^2 - 5x - 2 \leq 0$

for the interval: $\left[-\frac{1}{3}, 2\right]$. The graph of $y_1 = 3x^2 - 5x - 2$ intersects or is below the x-axis for the interval:

$\left[-\frac{1}{3}, 2\right]$.

12. First set $x^2 - x - 3 = 0$ then by the quadratic formula $x = \dfrac{-(-1) \pm \sqrt{(-1)^2 - 4(1)(-3)}}{2(1)} \Rightarrow$

$x = \dfrac{1 \pm \sqrt{13}}{2}$. and the endpoints will be: $x = \left\{\dfrac{1 - \sqrt{13}}{2}, \dfrac{1 - \sqrt{13}}{2}\right\}$. The numbers

$\dfrac{1 - \sqrt{13}}{2}$ and $\dfrac{1 + \sqrt{13}}{2}$ divide the number line into 3 intervals. The interval: $\left(-\infty, \dfrac{1 - \sqrt{13}}{2}\right)$ has a positive

product, the interval: $\left(\dfrac{1 - \sqrt{13}}{2}, \dfrac{1 + \sqrt{13}}{2}\right)$ has a negative product, and the interval: $\left(\dfrac{1 + \sqrt{13}}{2}, \infty\right)$ has a

positive product. Therefore $x^2 - x - 3 > 0$ for the interval: $\left(-\infty, \dfrac{1 - \sqrt{13}}{2}\right) \cup \left(\dfrac{1 + \sqrt{13}}{2}, \infty\right)$. The graph

of $y_1 = x^2 - x - 3$ is above the x-axis for the interval: $\left(-\infty, \dfrac{1 - \sqrt{13}}{2}\right) \cup \left(\dfrac{1 + \sqrt{13}}{2}, \infty\right)$.

13. First $x(3x - 1) \leq -4 \Rightarrow 3x^2 - x + 4 \leq 0$, now set $3x^2 - x + 4 = 0$, then by the quadratic formula

$x = \dfrac{-(-1) \pm \sqrt{(-1)^2 - 4(3)(4)}}{2(3)} = \dfrac{1 \pm \sqrt{-47}}{6} \Rightarrow x = \dfrac{1}{6} \pm \dfrac{\sqrt{47}}{6}i$. The inequality has no real solutions,

the graph is a parabola opening upward which does not intersect, but is completely above the x-axis. Therefore

$3x^2 - x + 4 \leq 0$ never happens and the solution is: $\varnothing$.

3.4: Further Applications of Quadratic Functions and Models

1. A, The area of a rectangle is $L \cdot W = A \Rightarrow x(2x + 2) = 40,000$.

2. B, The $L = 34 - 2x$ and $W = 21 - 2x$, therefore, $L \cdot W = A \Rightarrow (34 - 2x)(21 - 2x) = 600$.

3. (a) $30 - x$ would be the other number.

    (b) Since both numbers are positive and their sum is 30, the restrictions are: $0 < x < 30$.

    (c) Multiplying $x(30 - x)$ would give the function: $P(x) = 30x - x^2 \Rightarrow P(x) = -x^2 + 30x$.

    (d) Using the Vertex formula yields: $x = \dfrac{-b}{2a} = \dfrac{-30}{2(-1)} \Rightarrow x = 15$. If $x = 15$, then $y = 30 - (15) \Rightarrow y = 15$.

    Also, if $x = 15$ the maximum product is: $P(15) = -(15)^2 + 30(15) = -225 + 450 = 225$. The graph of

    $y = -x^2 + 30x$ has a vertex point of $(15, 225)$, which supports the result.

4.   (a) $45 - x$ would be the other number.

     (b) Since both numbers are positive and their sum is 45, the restrictions are: $0 < x < 45$.

     (c) Multiplying $x(45 - x)$ would give the function: $P(x) = 45x - x^2 \Rightarrow P(x) = -x^2 + 45x$.

     (d) $504 = -x^2 + 45x \Rightarrow x^2 - 45x + 504 = 0 \Rightarrow (x - 21)(x - 24) = 0 \Rightarrow x = 21$ and 24.

     (e) Using the Vertex formula yields: $x = \dfrac{-b}{2a} = \dfrac{-45}{2(-1)} \Rightarrow x = 22.5$. If $x = 22.5$, then $y = 45 - (22.5) \Rightarrow$

        $y = 22.5$. Also, if $x = 22.5$ the maximum product is: $P(22.5) = -(22.5)^2 + 45(22.5) = -506.25 + 1012.5$

        $\Rightarrow P(x) = 506.25$. The graph of $y = -x^2 + 45x$ has a vertex point of $(22.5, 506.25)$, which supports the result.

5.   (a) $640 - 2x$ would be the other side.

     (b) Since all three numbers are positive, their sum is 640, and two of them are x, the restrictions are:

        $0 < 2x < 640 \Rightarrow 0 < x < 320$.

     (c) Multiplying $x(640 - 2x)$ would give the function: $A(x) = 640x - 2x^2 \Rightarrow A(x) = -2x^2 + 640x$.

     (d) See Figure 5. From the graph, between 57.04 ft and 85.17 ft or between 234.83 ft and 262.96 ft, will give

        and area between 30,000 and 40,000 feet.

     (e) Using the Vertex formula yields: $x = \dfrac{-b}{2a} = \dfrac{-640}{2(-2)} \Rightarrow x = 160$. If $x = 160$, then $y = 640 - 2(160) \Rightarrow$

        $y = 320$. Also, if $x = 160$ the maximum product is: $A(160) = -2(160)^2 + 640(160) = -51{,}200 + 102{,}400$

        $\Rightarrow A(x) = 51{,}200$. The graph of $y = -2x^2 + 640x$ has a vertex point of $(160, 51{,}200)$, which supports the

        result.

6.   (a) $600 - 3x$ would be the other side.

     (b) Multiplying $x(600 - 3x)$ would give the function: $A(x) = 600x - 3x^2 \Rightarrow A(x) = -3x^2 + 600x$. Since all

        four numbers are positive, their sum is 600, and three of them are x, the restrictions are:

        $0 < 3x < 600 \Rightarrow 0 < x < 200$.

     (c) $22{,}500 = -3x^2 + 600x \Rightarrow 3x^2 - 600x + 22{,}500 = 0 \Rightarrow \quad (3x - 150)(x - 150) = 0 \Rightarrow$

        $3x - 150 = 0$ or $x - 150 = 0 \Rightarrow x = 50$ or 150. Therefore the dimensions can be: 50 ft by

        $600 - 3(50) = 450$ ft or 150 ft by $600 - 3(150) = 150$ ft.

     (d) Graph the function: $A(x) = -3x^2 + 600x$. See Figure 6. From the graph the vertex is: $(100, 30.000)$ which

        gives an area of: $30{,}000$ ft^2.

$[0, 320]$ by $[0, 55{,}000]$ $[0, 200]$ by $[0, 40{,}000]$
Xscl $= 20$ Yscl $= 10{,}000$ Xscl $= 20$ Yscl $= 10{,}000$

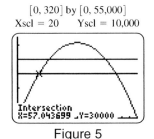 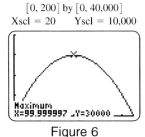

        Figure 5   Figure 6

7. (a) If the width is x units long then the length is twice the width or $2x$ units long.

    (b) The width will be $x - 2(2) = x - 4$ units long and the length will be $2x - 2(2) = 2x - 4$ units long. Since both measurements are positive and 4 inches are removed from each measurement, the restrictions are: $x > 4$.

    (c) Since volume for the box is: $V = L \cdot W \cdot H$, $V = (2x - 4)(x - 4)(2)$ and the function is:

    $V(x) = 4x^2 - 24x + 32$.

    (d) $320 = 4x^2 - 24x + 32 \Rightarrow 4x^2 - 24x - 288 = 0 \Rightarrow 4(x^2 - 6x - 72) = 0 \Rightarrow 4(x - 12)(x + 6) = 0$

    $\Rightarrow x = -6, 12$. Since length cannot be negative, $x = 12$. If $x = 12$, the dimensions are: 8 in by 20 in. From the graph of $y = 4x^2 - 24x + 32$, when $x = 12$, $y = 320$. This supports our analytical result.

    (e) From the graph of $y = 4x^2 - 24x + 32$, when $400 < y < 500$, $13.0 < x < 14.2$ in.

8. (a) If the width is x units long then the length is $2.5x$ units long.

    (b) Since both measurements are positive and 6 inches are removed from each measurement, the restriction is: $x > 6$.

    (c) Since volume for the box is: $V = L \cdot W \cdot H$, $V = (2.5x - 6)(x - 6)(3)$ and the function is:

    $V(x) = 7.5x^2 - 63x + 108$.

    (d) From the graph of $y = 7.5x^2 - 63x + 108$, when $600 < y < 800$, $13.3 < x < 14.7$ in.

9. The surface area of the can is: $V = \pi r^2 + \pi r^2 + 2\pi r(h) \Rightarrow 54.19 = 2\pi r^2 + 2(4.25)\pi r \Rightarrow$

    $6.28x^2 + 26.69x - 54.19 = 0$ now use the quadratic formula:

    $$\frac{-26.69 \pm \sqrt{(26.69)^2 - 4(6.28)(-54.19)}}{2(6.28)} = \frac{-26.69 \pm \sqrt{2073.6}}{12.56} = \frac{-26.69 \pm 45.537}{12.56} \Rightarrow$$

    $r = -5.751$ or 1.5. Since length cannot be negative, $r = 1.5$.

10. If the length of the box is x, then the width is $x - 3.1875$. Since the volume of a box is: $V = L \cdot W \cdot H$, then

    $182.742 = x(x - 3.1875)(2.3125) \Rightarrow 2.3125x^2 - 7.3711x - 182.742 = 0$. Now use the quadratic formula:

    $$x = \frac{-(-7.3711) \pm \sqrt{(-7.3711)^2 - 4(2.3125)(-182.742)}}{2(2.3125)} = \frac{7.3711 \pm \sqrt{1744.6966}}{4.625} \Rightarrow$$

    $x \approx -7.421$ or 10.625. Since the length cannot be negative, $x = 10.625$. Therefore the length is: 10.625 in and the width is: $10.625 - 3.1875 = 7.438$ in.

11. If $A = s^2$, then $800 = s^2 \Rightarrow s = \sqrt{800} = 20\sqrt{2}$. Since the lawn is square, the diagonal of the lawn would equal: $d^2 = (20\sqrt{2})^2 + (20\sqrt{2})^2 \Rightarrow d^2 = 1600 \Rightarrow d = 40$. The radius of the circular pattern is half the diagonal, therefore the radius is: $r = 20$.

12. If we use Pythagorean Theorem, we get: $x^2 + (x - 10)^2 = 50^2 \Rightarrow x^2 + x^2 - 20x + 100 = 2500 \Rightarrow$

    $2x^2 - 20x - 2400 = 0 \Rightarrow 2(x^2 - 10x - 1200) = 0 \Rightarrow 2(x - 40)(x + 30) = 0 \Rightarrow x = -30, 40$.

    Since length cannot be negative, the height of the kite is: 40 ft.

13. If we use Pythagorean Theorem, we get: $h^2 + 12^2 = (2h + 3)^2 \Rightarrow h^2 + 144 = 4h^2 + 12h + 9 \Rightarrow$

    $3h^2 + 12h - 135 = 0 \Rightarrow (3h + 27)(h - 5) = 0 \Rightarrow h = -9, 5$.

    Since length cannot be negative, the height of the dock is: 5 ft.

14. If one leg of the right triangle is x, then the other leg is $x + 700$, and the hypotenuse is $x + 800$. If we use Pythagorean Theorem, we get: $x^2 + (x + 700)^2 = (x + 800)^2 \Rightarrow x^2 + x^2 + 1400x + 490{,}000 =$ $x^2 + 1600x + 640{,}000 \Rightarrow x^2 - 200x - 150{,}000 = 0 \Rightarrow (x - 500)(x + 300) = 0 \Rightarrow x = -300, 500.$ Since length cannot be negative, the leg is 500 yds long, the second leg is 1200 yds. long, and the hypotenuse is 1300 yds long. Therefore the total length is $500 + 1200 + 1300 = 3000$ yards long.

15. If we use Pythagorean Theorem, we get: $(8 + 2)^2 + (9 + 4)^2 = x^2 \Rightarrow 100 + 169 = x^2 \Rightarrow x^2 = 269 \Rightarrow$ $x = \sqrt{269} \Rightarrow x \approx 16.4$ Since a 16 ft ladder will be too short, the ladder must be at least 17 ft.

16. If one leg of the right triangle is x, then the other leg is $2x$, and the hypotenuse is 26. If we use Pythagorean Theorem, we get: $x^2 + (2x)^2 = 26^2 \Rightarrow 5x^2 = 676 \Rightarrow x^2 = \dfrac{676}{5} \Rightarrow x = \sqrt{\dfrac{676}{5}} = 11.62755.$

    Therefore, the dimensions should be legs of: 11.6 in and 23.3 in.

17. (a) Since x equals the loss of 1 apartment for each \$20 increase, the number of apartments rented is: $80 - x$.

    (b) Since each increase is \$20 and x equals the number of increases, the rent per apartment is: $400 + 20x$.

    (c) Since revenue is the number of apartments rented times rent per apartment we multiply answers a and b.

    $(80 - x)(400 + 20x) = 32{,}000 + 1200x - 20x^2 \Rightarrow R(x) = -20x^2 + 1200x + 32{,}000.$

    (d) $37{,}500 = -20x^2 + 1200x + 32{,}000 \Rightarrow -20x^2 + 1200x - 5{,}500 = 0 \Rightarrow -20(x^2 - 60x + 275) = 0 \Rightarrow$
    $-20(x - 55)(x - 5) = 0 \;\; x = 5 \text{ or } 55.$

    (e) Use the vertex formula: $x = \dfrac{-b}{2a} = \dfrac{-1200}{-40} = 30.$ If $x = 30$, then the rent per apartment is:
    $r(30) = 400 + 20(30) = 400 + 600 = \$1000.$

18. (a) Let x be the number of \$20 decreases. If each decrease in charge nets an increase of 100 people, the number of people attending is: $1000 + 100x$. Since the charge is \$600 minus \$20 per decrease, the cost per person is: $600 - 20x$. Multiplying these yields the revenue function, therefore:
    $R(x) = (1000 + 100x)(600 - 20x) = 600{,}000 + 40{,}000x - 2000x^2 \Rightarrow$
    $R(x) = -2000x^2 + 40{,}000x + 600{,}000.$

    (b) Use the vertex formula: $x = \dfrac{-b}{2a} = \dfrac{-40{,}000}{-4000} = 10.$ If $x = 10$, then the company should charge:
    $c(10) = 600 - 20(10) = 600 - 200 = \$400.$

    (c) Since for maximum revenue $x = 10$, then $R(10) = -2000(10)^2 + 40{,}000(10) + 600{,}000 = \$800{,}000.$

19. (a) If $f(x) = 10$ and $x = 15$, then $10 = \dfrac{-16(15)^2}{.434v^2} + 1.15(15) + 8 \Rightarrow 10 = \dfrac{-3600}{.434v^2} + 25.25 \Rightarrow$
    $-15.25 = \dfrac{-3600}{.434v^2} \Rightarrow -6.6185v^2 = -3600 \Rightarrow v^2 = \dfrac{3600}{6.6185} \Rightarrow v \approx 23.32 \text{ ft/sec.}$

    (b) If $v = 23.32$, then $y = \dfrac{-16x^2}{.434(23.32)^2} + 1.15x + 8 \Rightarrow y = -.06779x^2 + 1.15x + 8.$ Graph this equation, the graph does pass through points $(0, 8)$ and $(15, 10)$.

    (c) Use the vertex formula: $x = \dfrac{-b}{2a} = \dfrac{-1.15}{2(-.06779)} \approx 8.48.$ If $x \approx 8.48$, then the maximum height is:
    $f(8.48) = -.06779(8.48)^2 + 1.15(8.48) + 8 \Rightarrow f(8.48) = 12.88 \text{ ft.}$

20. (a) If $f(x) = 10$ and $x = 15$, then $10 = \dfrac{-16(15)^2}{.117v^2} + 2.75(15) + 3 \Rightarrow 10 = \dfrac{-3600}{.117v^2} + 44.25 \Rightarrow$

$-34.25 = \dfrac{-3600}{.117v^2} \Rightarrow -4.00725v^2 = -3600 \Rightarrow v^2 = \dfrac{3600}{4.00725} \Rightarrow v \approx 29.97$ ft/sec.

(b) If $v = 29.97$, then $y = \dfrac{-16x^2}{.117(29.97)^2} + 2.75x + 3 \Rightarrow y = -.15225x^2 + 2.75x + 3$. Graph this function,

the graph does pass through points $(0, 3)$ and $(15, 10)$.

(c) Use the vertex formula: $x = \dfrac{-b}{2a} = \dfrac{-2.75}{2(-.15225)} \approx 9.03$. If $x \approx 9.03$, then the maximum height is:

$f(9.03) = -.15225(9.03)^2 + 2.75(9.03) + 3 \Rightarrow f(9.03) = 15.42$ ft.

The underhand shot produces a higher arc.

21. (a) If $x = 2$, then $h(2) = -.5(2)^2 + 1.25(2) + 3 = -2 + 2.5 + 3 \Rightarrow h(2) = 3.5$ ft.

(b) If $h(x) = 3.25$, then $3.25 = -.5x^2 + 1.25x + 3 \Rightarrow -.5x^2 + 1.25x - .25 = 0 \Rightarrow$

$4(-.5x^2 + 1.25x - .25 = 0) \Rightarrow -2x^2 + 5x - 1 = 0$. Now use the quadratic formula:

$x = \dfrac{5 \pm \sqrt{5^2 - 4(2)(1)}}{2(2)} = \dfrac{5 \pm \sqrt{17}}{4} \Rightarrow x = .219$ or 2.281. Therefore the frog is 3.25 feet above the

ground at approximately $.2$ ft and 2.3 ft.

(c) Using the vertex formula yields: $x = \dfrac{-b}{2a} = \dfrac{-1.25}{2(-.5)} \Rightarrow x = 1.25$ ft as the horizontal distance.

(d) From part (c) the maximum height was reach when the horizontal distance was $x = 1.25$, therefore:

$h(1.25) = -.5(1.25)^2 + 1.25(1.25) + 3 = -.78125 + 1.5625 + 3 \Rightarrow h(1.25) \approx 3.78$ ft high.

22. (a) Using the vertex formula yields: $x = \dfrac{-b}{2a} = \dfrac{-\frac{4}{3}}{2(-\frac{1}{3})} \Rightarrow x = 2$ ft as the horizontal distance.

(b) From part (a) the maximum height was reach when the horizontal distance was $x = 2$, therefore:

$h(2) = -\dfrac{1}{3}(2)^2 + \dfrac{4}{3}(2) + 4 = -\dfrac{4}{3} + \dfrac{8}{3} + 4 \Rightarrow h(2) = 5\dfrac{1}{3}$ ft high.

23. If $f(x) = 800$, then $800 = \dfrac{1}{10}x^2 - 3x + 22 \Rightarrow .1x^2 - 3x - 778 = 0$. Now use the quadratic formula:

$x = \dfrac{3 \pm \sqrt{9 - 4(.1)(-778)}}{2(.1)} = \dfrac{3 \pm \sqrt{320.2}}{.2} \Rightarrow x \approx 104.5$ ft/sec. This converted is:

$x = \dfrac{104.5(60)(60)}{5280} = 71.25$ mph.

24. Use the vertex formula: $x = \dfrac{-b}{2a} = \dfrac{-(-12)}{2(1.8)} = \dfrac{12}{3.6} \Rightarrow x = 3.\overline{33}$. This makes the age group between $45 - 54$,

and would make the rate: $f(3.33) = 1.8(3.33)^2 - 12(3.33) + 37.4 = 19.96 - 39.96 + 37.4 \Rightarrow x = 17.4$

fatalities per 100,000 population.

25. (a) If $x = 50$, then $T(50) = .00787(50)^2 - 1.528(50) + 75.89 = 19.675 - 76.4 + 75.89 \Rightarrow T(50) = 19.165$.

Therefore, it would take approximately 19.2 hours to start feeling the symptoms.

(b) If $T(x) = 3$, then $3 = .00787x^2 - 1.528x + 75.89 \Rightarrow .00787x^2 - 1.528x + 72.89 = 0$. Now use the

quadratic formula: $x = \dfrac{-(-1.528) \pm \sqrt{(-1.528)^2 - 4(.00787)(72.89)}}{2(.00787)} = \dfrac{1.528 \pm \sqrt{.0402068}}{.01574} \Rightarrow$

$x \approx 84.3$ or 109.817. Since $x = 109.817$ is not in the given range $50 \le x \le 100$, $x \approx 84.3$ ppm.

26. (a) If $x = 600$, then $T(600) = .0002(600)^2 - .316(600) + 127.9 = 72 - 189.6 + 127.9 \Rightarrow x = 10.3$ hours.

    (b) If $T(x) = 4$, then $4 = .0002x^2 - .316x + 127.9 \Rightarrow .0002x^2 - .316x + 123.9 = 0$. Now use the quadratic

    formula: $x = \dfrac{-(-.316) \pm \sqrt{(-.316)^2 - 4(.0002)(123.9)}}{2(.0002)} = \dfrac{.316 \pm \sqrt{.000736}}{.0004} \Rightarrow x \approx 722.2$ or 857.8.

    Since $x = 857.8$ is not in the given range $500 \le x \le 800$, $x \approx 722.2$ ppm.

27. (a) See Figure 27a.

    (b) Using the defined function and the point $(4, 50)$ yields: $f(x) = a(x - 4) + 50$. Now use $(14, 110)$ to find

    the function: $110 = a(14 - 4)^2 + 50 \Rightarrow 60 = 100a \Rightarrow a = .6$, therefore the function is:

    $f(x) = .6(x - 4)^2 + 50$.

    (c) See Figure 27c. From the graph, we see that there is a good fit.

    (d) Using the regression feature yields: $g(x) \approx .402x^2 - 1.175x + 48.343$.

    (e) For the year 2006, $x = 16$. Therefore: $f(16) = .6(16 - 4)^2 + 50 \approx 136.4$ thousand, and

    $g(16) = .402(16)^2 - 1.175(16) + 48.343 \approx 132.5$ thousand.

    For the year 2010, $x = 16$. Therefore: $f(20) = .6(20 - 4)^2 + 50 \approx 203.6$ thousand, and

    $g(20) = .402(20)^2 - 1.175(20) + 48.343 \approx 185.6$ thousand.

    For the year 2015, $x = 16$. Therefore: $f(25) = .6(25 - 4)^2 + 50 \approx 314.6$ thousand, and

    $g(25) = .402(25)^2 - 1.175(25) + 48.343 \approx 270.2$ thousand.

28. (a) See Figure 28. From the graph, we see that there is a good fit.

    (b) Using the regression feature yields: $g(x) \approx .309x^2 - 3.043x + 7.663$.

    (c) For the year 2002, $x = 12$. Therefore: $f(12) = .37(12)^2 - 4.1(12) + 12 \approx 16.1\%$ increase, and

    $g(12) = .309(12)^2 - 3.043(12) + 7.663 \approx 15.6\%$ increase.

    For the year 2003, $x = 13$. Therefore: $f(13) = .37(13)^2 - 4.1(13) + 12 \approx 21.2\%$ increase, and

    $g(13) = .309(13)^2 - 3.043(13) + 7.663 \approx 20.3\%$ increase.

    For the year 2004, $x = 14$. Therefore: $f(14) = .37(14)^2 - 4.1(14) + 12 \approx 27.1\%$ increase, and

    $g(14) = .309(14)^2 - 3.043(14) + 7.663 \approx 25.6\%$ increase.

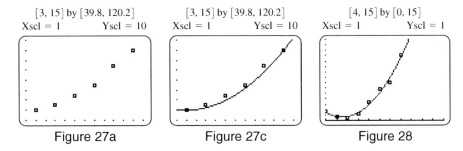

$[3, 15]$ by $[39.8, 120.2]$	$[3, 15]$ by $[39.8, 120.2]$	$[4, 15]$ by $[0, 15]$
Xscl = 1 Yscl = 10	Xscl = 1 Yscl = 10	Xscl = 1 Yscl = 1

Figure 27a   Figure 27c   Figure 28

29. (a) See Figure 29a.

   (b) $f(45) = .056057(45)^2 + 1.06657(45) = 113.515425 + 47.99565 \Rightarrow x \approx 161.5$ feet. When the speed is

   45 mph, the stopping distance is 161.5 feet.

   (c) See Figure 29c. The model is quite good, although the stopping distances are a little low for the higher speeds.

$[15, 75]$ by $[-10, 450]$
Xscl = 10   Yscl = 50

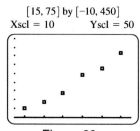

Figure 29a

$[15, 75]$ by $[-10, 450]$
Xscl = 10   Yscl = 50

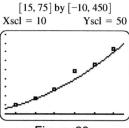

Figure 29c

30. (a) See Figure 30a.

   (b) Using the regression feature yields: $g(x) = .0074x^2 - 1.185x + 59.02$. See Figure 30b. The function

   models the data very well.

   (c) $g(70) = .0074(70)^2 - 1.185(70) + 59.02 = 36.26 - 82.95 + 59.02 \Rightarrow x \approx 12.33$ seconds.

   (d) From a graph of $g(x)$ and $y = 24$; the intersection point is: $(39.099, 24)$. The speed is about 39.1 mph.

$[25, 70]$ by $[10, 35]$
Xscl = 5   Yscl = 5

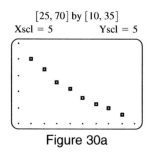

Figure 30a

$[25, 70]$ by $[10, 35]$
Xscl = 5   Yscl = 5

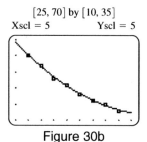

Figure 30b

3.5: Higher-Degree Polynomial Functions and Graphs

1. Shift the graph of $y = x^4$ three units to the left, stretch vertically by a factor of 2, and shift downward 7 units.

   See Figures 1a and 1b.

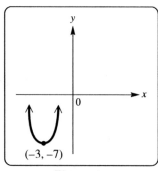

Figure 1a

$[-10, 10]$ by $[-10, 10]$
Xscl = 1   Yscl = 1

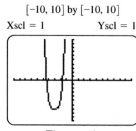

Figure 1b

2. Shift the graph of $y = x^4$ one unit to the left, stretch vertically by a factor of 3, reflect across the x-axis, and shift upward 12 units. See Figures 2a and 2b.

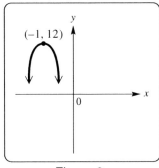

Figure 2a

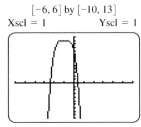

Figure 2b

3. Shift the graph of $y = x^3$ one unit to the right, stretch vertically by a factor of 3, reflect across the x-axis, and shift upward 12 units. See Figures 3a and 3b.

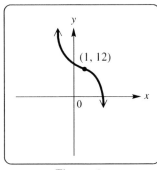

Figure 3a

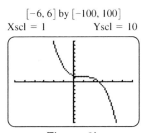

Figure 3b

4. Shift the graph of $y = x^5$ one unit to the right, stretch vertically by a factor of .5, and shift upward 13 units. See Figures 4a and 4b.

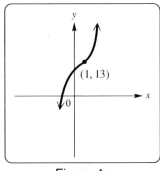

Figure 4a

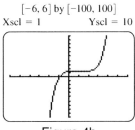

Figure 4b

5. With three extrema, the minimum degree of f is 4.

6. With four extrema, the minimum degree of g is 5.

7. The points (a, b) and (c, d) are local maxima, and (e, t) is a local minimum.

8. The points (n, p) and (j, k) are local maxima, (r, h) and (l, m) are local minima.

9. The highest point of the graph, *(a, b)* is an absolute maximum.

10. With no highest or lowest point, the function has no absolute maximum or minimum.

11. The function *f*, has local maximum values of *b* and *d*; a local minimum value of *t*; and an absolute maximum value of *b*.

12. The function *g*, has local maximum values of *p* and *k*; local minimum values of *h* and *m*; and no absolute extreme values.

13. The graph of $f(x) = x^n$ for $n \in$ {positive odd integers} will take the shape of the graph of $f(x) = x^3$, but gets steeper as *n* and *x* increase.

14. The graph of $f(x) = x^n$ for $n \in$ {positive even integers} will take the shape of the graph of $f(x) = x^2$, but the width gets narrower as *n* and *x* increase.

15. See Figure 15. As the odd exponent *n* gets larger, the graph *flattens out* in the window $[-1, 1]$ by $[-1, 1]$. The graph of $y = x^7$ will be between $y = x^5$ and the *x*-axis in this window.

16. See Figure 16. As the even exponent *n* gets larger, the graph *flattens out* in the window $[-1, 1]$ by $[-1, 1]$. The graph of $y = x^8$ will be between $y = x^6$ and the *x*-axis in this window.

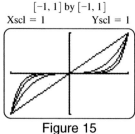

| $[-1, 1]$ by $[-1, 1]$ | | $[-1, 1]$ by $[-1, 1]$ | |
| Xscl = 1 | Yscl = 1 | Xscl = 1 | Yscl = 1 |

Figure 15   Figure 16

17. Graphing in various windows produces a local maximum of: (2, 3.67), and a local minimum of: (3, 3.5).

18. Graphing in various windows produces a local maximum of: $(-4, 9.67)$, and a local minimum of: $(-5, 9.5)$.

19. Graphing in various windows produces a local maximum of: $(-3.33, -1.85)$, and a local minimum of: $(-4, -2)$.

20. Graphing in various windows produces a local maximum of: (.5, 3.325), and a local minimum of: (.6, 3.324).

21. Graphing in various windows produces two *x*-intercepts: 2.10 and 2.15.

22. Graphing in various windows produces two *x*-intercepts: -3.55 and -3.40.

23. Graphing in various windows produces no *x*-intercept.

24. Graphing in various windows produces no *x*-intercept.

25. $P(x)$ is an positive odd-degree polynomial graph, therefore:

26. $P(x)$ is an negative odd-degree polynomial graph, therefore:

27. $P(x)$ is an negative odd-degree polynomial graph, therefore:

28. $P(x)$ is an positive odd-degree polynomial graph, therefore:

29. $P(x)$ is an positive even-degree polynomial graph, therefore:

30. $P(x)$ is an positive even-degree polynomial graph, therefore: ⌣

31. $P(x)$ is an negative even-degree polynomial graph, therefore: ⌒

32. $P(x)$ is an negative even-degree polynomial graph, therefore: ⌒

33. $P(x)$ is an positive even-degree polynomial graph, therefore: ⌣

34. $P(x)$ is an negative even-degree polynomial graph, therefore: ⌒

35. $P(x)$ is an negative odd-degree polynomial graph, therefore: ⤵

36. $P(x)$ is an positive odd-degree polynomial graph, therefore: ⤴

37. $P(x)$ is an positive odd-degree polynomial graph, therefore: D

38. $P(x)$ is an negative odd-degree polynomial graph, therefore: C

39. $P(x)$ is an negative even-degree polynomial graph, therefore: B

40. $P(x)$ is an positive even-degree polynomial graph, therefore: A

41. A. The third-degree function can have at most 2 local extrema and a positive lead coefficient will yield a right side end behavior opening up.

42. B. The fourth-degree function can have at most 3 local extrema and a positive lead coefficient will yield a right side end behavior opening up.

43. From the graph, the function graphed in C has 1 real zero.

44. C. A third-degree function can have at most 2 local extrema.

45. B and D. A third-degree function can have at most 2 local extrema.

46. From the graph, the function graphed in D has 2 *positive* real zero.

47. From the graph, the function graphed in A has 1 *negative* real zero.

48. From the graph, the absolute minimum is below the x-axis, therefore it is negative.

49. From the graph, the function graphed in B has a range of approximately $[-100, \infty)$.

50. B. From the graph, the function has a local maximum in the II quadrant.

51. False. A polynomial function of degree 3 will have at most 3 x-intercepts.

52. True. With odd-degree end behavior, it must have at least 1 x-intercept.

53. True. With a positive y-intercept and negative even-degree end behavior, it must have 2 x-intercepts.

54. True. With a positive even-degree end behavior which is shifted up 5 units, it will have no x-intercepts.

55. True. With a negative even-degree end behavior which is shifted up 5 units, it will have 2 x-intercepts.

56. True. The graph is obtained by shifting the graph of $y = x^3$ to the right 1 unit. There is one x-intercept.

57. False. With odd-degree end behavior, it will have at least 1 and at most 5 x-intercepts. Thus it may have only 1.

58. False. With a positive even-degree end behavior which is shifted up or a negative even-degree end behavior shifted down, the polynomial function will have no x-intercepts.

59. See Figure 59 for the graph of this function.

   (a) Because it is a polynomial function, its domain is: $(-\infty, \infty)$.

   (b) From the graph of P there is one local minimum which by the calculator is: $(-4.74, -27.03)$.
   Because the function is of odd-degree, it is not an absolute minimum point.

   (c) From the graph of P there is one local maximum which by the calculator is: $(.07, 84.07)$.
   Because the function is of odd-degree, it is not an absolute maximum point.

   (d) Because the polynomial function is of odd-degree, its range is: $(-\infty, \infty)$.

   (e) Using the calculator we find the x-intercepts are: $-6, -3.19$ and 2.19; the y-intercept is: 84.

   (f) From the graph and our results, the function is increasing for the interval: $[-4.74, .07]$.

   (g) From the graph and our results, the function is decreasing for the intervals: $(-\infty, -4.74]; [.07, \infty)$.

60. See Figure 60 for the graph of this function.

   (a) Because it is a polynomial function, its domain is: $(-\infty, \infty)$.

   (b) From the graph of P there is one local minimum, which by the calculator is: $(-1.52, -94.88)$.
   Because the function is of odd-degree, it is not an absolute minimum point.

   (c) From the graph of P there is one local maximum, which by the calculator is: $(2.85, 30.44)$.
   Because the function is of odd-degree, it is not an absolute maximum point.

   (d) Because the polynomial function is of odd-degree, its range is: $(-\infty, \infty)$.

   (e) Using the calculator we find the x-intercepts are: $-3.45, 1.45$ and 4; the y-intercept is: -60.

   (f) From the graph and our results, the function is increasing for the interval: $[-1.52, 2.85]$.

   (g) From the graph and our results, the function is decreasing for the intervals: $(-\infty, -1.52]; [2.85, \infty)$.

61. See Figure 61 for the graph of this function.

   (a) Because it is a polynomial function, its domain is: $(-\infty, \infty)$.

   (b) From the graph of P there are 2 local minimum points, which by the calculator are: $(-1.73, -16.39)$ and
   $(1.35, -3.49)$. Because the function is of odd-degree, neither is an absolute minimum point.

   (c) From the graph of P there are 2 local maximum points, which by the calculator are: $(-3, 0)$ and $(.17, 9.52)$.
   Because the function is of odd-degree, neither is an absolute maximum point.

   (d) Because the polynomial function is of odd-degree, its range is: $(-\infty, \infty)$.

   (e) Using the calculator we find the x-intercepts are: $-3, -.62, 1,$ and 1.62; the y-intercept is: 9.

   (f) From the graph and our results, the function is increasing for the intervals:
   $(-\infty, -3], [-1.73, .17]$ and $[1.35, \infty)$.

   (g) From the graph and our results, the function is decreasing for the intervals: $[-3, -1.73]; [.17, 1.35]$.

[-10, 10] by [-100, 100]
Xscl = 1   Yscl = 10

[-10, 10] by [-100, 100]
Xscl = 1   Yscl = 10

[-4, 4] by [-20, 20]
Xscl = 1   Yscl = 5

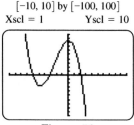

Figure 59

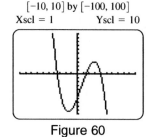

Figure 60

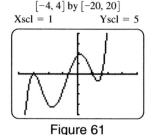

Figure 61

62. See Figure 62 for the graph of this function.

    (a) Because it is a polynomial function, its domain is: $(-\infty, \infty)$.

    (b) From the graph of P there are 2 local minimum points, which by the calculator are: $(-1.10, -.46)$ and $(1.80, -.08)$. Because the function is of odd-degree, neither is an absolute minimum point.

    (c) From the graph of P there are 2 local maximum points, which by the calculator are: $(.10, 16.20)$ and $(2, 0)$. Because the function is of odd-degree, neither is an absolute maximum point.

    (d) Because the polynomial function is of odd-degree, its range is: $(-\infty, \infty)$.

    (e) Using the calculator we find the x-intercepts are: $-1.19, -1, 1.69,$ and 2; the y-intercept is: 16.

    (f) From the graph and our results, the function is increasing for the intervals: $[-1.10, .10]$ and $[1.80, 2]$.

    (g) From the graph and our results, the function is decreasing for the intervals: $(-\infty, -1.10], [.10, 1.80]$ and $[2, \infty)$.

63. See Figure 63 for the graph of this function.

    (a) Because it is a polynomial function, its domain is: $(-\infty, \infty)$.

    (b) From the graph of P there is one absolute minimum point, which by the calculator is: $(-2.63, -132.69)$, and one local minimum point, which by the calculator is: $(1.68, -99.90)$.

    (c) From the graph of P there is one local maximum point, which by the calculator is: $(-.17, -71.48)$. Because the function is of positive even-degree, there is not an absolute maximum point.

    (d) Because the positive even-degree function has an absolute minimum, given in (b), its range is: $[-132.69, \infty)$.

    (e) Using the calculator we find the x-intercepts are: $-4,$ and 3; the y-intercept is: -72.

    (f) From the graph and our results, the function is increasing for the intervals: $[-2.63, -.17]$ and $[1.68, \infty)$.

    (g) From the graph and our results, the function is decreasing for the intervals: $(-\infty, -2.63]$ and $[-.17, 1.68]$.

64. See Figure 64 for the graph of this function.

    (a) Because it is a polynomial function, its domain is: $(-\infty, \infty)$.

    (b) From the graph of P there is two absolute minimum points, which by the calculator are: $(\pm 2.35, -36.75)$.

    (c) From the graph of P there is one local maximum point, which by the calculator is: $(0, 54)$. Because the function is of positive even-degree, there is not an absolute maximum point.

    (d) Because the positive even-degree function has an absolute minimum, given in (b), its range is: $[-36.75, \infty)$.

    (e) Using the calculator we find the x-intercepts are: $-3, -1.41, 1.41,$ and 3; the y-intercept is: 54.

    (f) From the graph and our results, the function is increasing for the intervals: $[-2.35, 0]$ and $[2.35, \infty)$.

    (g) From the graph and our results, the function is decreasing for the intervals: $(-\infty, -2.35]$ and $[0, 2.35]$.

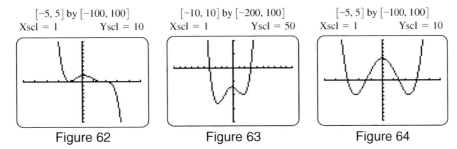

| $[-5, 5]$ by $[-100, 100]$ | $[-10, 10]$ by $[-200, 100]$ | $[-5, 5]$ by $[-100, 100]$ |
| Xscl = 1   Yscl = 10 | Xscl = 1   Yscl = 50 | Xscl = 1   Yscl = 10 |

Figure 62   Figure 63   Figure 64

65. See Figure 65 for the graph of this function.

    (a) Because it is a polynomial function, its domain is: $(-\infty, \infty)$.

    (b) From the graph of P there is two local minimum points, which by the calculator are: $(-2, 0)$ and $(2, 0)$.

    Because the function is of negative even-degree, there is not an absolute minimum point.

    (c) From the graph of P there are 3 absolute maximum points, which by the calculator are:

    $(-3.46, 256)$, $(0, 256)$, and $(3.46, 256)$.

    (d) Because the negative even-degree function has an absolute maximum, given in (c), its range is: $(-\infty, 256]$.

    (e) Using the calculator we find the x-intercepts are: $-4, -2, 2$, and 4; the y-intercept is: 256.

    (f) From the graph and our results, the function is increasing for the intervals:

    $(-\infty, -3.46]$, $[-2, 0]$, and $[2, 3.46]$.

    (g) From the graph and our results, the function is decreasing for the intervals:

    $(-3.46, -2]$, $[0, 2]$, and $[3.46, \infty)$.

66. See Figure 66 for the graph of this function.

    (a) Because it is a polynomial function, its domain is: $(-\infty, \infty)$.

    (b) From the graph of P there is one local minimum point, which by the calculator is: $(0, 4)$.

    Because the function is of negative even-degree, there is not an absolute minimum point.

    (c) From the graph of P there is one absolute maximum point, which by the calculator is: $(-1.57, 52.75)$, and

    one local maximum point, which by the calculator is: $(1.59, 29.02)$

    (d) Because the negative even-degree function has an absolute maximum, given in (c), its range is: $(-\infty, 52.75]$.

    (e) Using the calculator we find the x-intercepts are: -2 and 2; the y-intercept is: 4.

    (f) From the graph and our results, the function is increasing for the intervals: $(-\infty, -1.57]$ and $[0, 1.59]$.

    (g) From the graph and our results, the function is decreasing for the intervals: $[-1.57, 0]$, and $[1.59, \infty)$.

$[-6, 6]$ by $[-300, 300]$
Xscl $= 1$ Yscl $= 50$

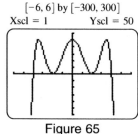

Figure 65

$[-5, 5]$ by $[-100, 100]$
Xscl $= 1$ Yscl $= 10$

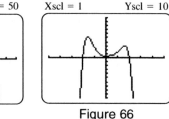

Figure 66

67. There are many possible valid windows, through experimentation one window is: $[-10, 10]$ by $[-40, 10]$.

68. There are many possible valid windows, through experimentation one window is: $[-4, 4]$ by $[-50, 15]$.

69. There are many possible valid windows, through experimentation one window is: $[-10, 20]$ by $[-1500, 500]$.

70. There are many possible valid windows, through experimentation one window is: $[-10, 10]$ by $[-500, 300]$.

71. There are many possible valid windows, through experimentation one window is: $[-10, 10]$ by $[-20, 500]$.

72. There are many possible valid windows, through experimentation one window is: $[-5, 5]$ by $[-100, 300]$.

73. Graph the data and the three equations in the same window, See Figure 73.

    All three model the data well, but (iii) models it best.

74. (a) Graph the data and the three equations in the same window, See Figure 74.

    $h(x)$ has the greatest value for R^2.

    (b) $h(14) \approx -.03(14)^3 + .698(14)^2 + 4.507(14) + 64.675 \Rightarrow h(14) \approx 182$ thousand people.

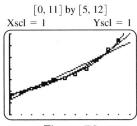

[0, 11] by [5, 12]
Xscl = 1   Yscl = 1

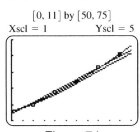

[0, 11] by [50, 75]
Xscl = 1   Yscl = 5

Figure 73   Figure 74

Reviewing Basic Concepts (Sections 3.4 and 3.5)

1. (a) The total length of the fence must equal $2L + 2x = 300 \Rightarrow 2L = 300 - 2x \Rightarrow L = 150 - x$.

   (b) Since width equals x and length equals $150 - x$, the function is: $A(x) = x(150 - x)$.

   (c) Since length and width must both be positive, he restrictions are: $0 < x < 150$.

   (d) $5000 = x(150 - x) \Rightarrow x^2 - 150x + 5000 = 0 \Rightarrow (x - 50)(x - 100) = 0 \Rightarrow x = 50$ or 100. Using

   either value will yield a garden which is 50 m by 100 m.

2. (a) See Figure 2a.

   (b) Using the defined function and the point $(51, .1)$ yields: $f(x) = a(x - 51) + .1$. Now use $(101, 4.7)$ to find

   the function: $4.7 = a(101 - 51)^2 + .1 \Rightarrow 4.6 = 2500a \Rightarrow a = .0018$, therefore the function is:

   $f(x) = .0018(x - 51)^2 + .1$.

   (c) Using the regression feature yields: $g(x) \approx .0026x^2 - .3139x + 9.426$.

   (d) Graph the data and two equations from (b) and (c), See Figure 2d. The regression function fits better

   because it is closer to or passes through more data points. Neither function would fit the data for $x < 51$.

3. $P(x)$ is an positive odd-degree polynomial graph, therefore:

4. A polynomial of degree 3 can have at most 2 extrema and at most 3 x-intercepts.

5. $P(x)$ is an positive even-degree polynomial graph, therefore:

6. See Figure 6.

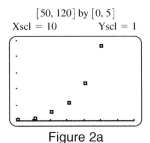

[50, 120] by [0, 5]
Xscl = 10   Yscl = 1

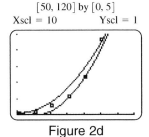

[50, 120] by [0, 5]
Xscl = 10   Yscl = 1

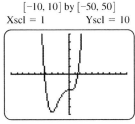

[-10, 10] by [-50, 50]
Xscl = 1   Yscl = 10

Figure 2a   Figure 2d   Figure 6

7. From the graph of P there is one extrema, an absolute minimum point, which by the calculator is: $(-3, -47)$.

8. Using the calculator we find the x-intercepts are: -4.26 and 1.53; the y-intercept is: -20.

3.6: Topics in the Theory of Polynomial Functions (I)

1. $P(1) = 3(1)^2 - 2(1) - 6 \Rightarrow P(1) = -5$ and $P(2) = 3(2)^2 - 2(2) - 6 \Rightarrow P(2) = 2$. These answers differ in sign, therefore there is a real zero between them. By graphing on the calculator the real zero is approximately: 1.79.

2. $P(2) = (2)^3 + (2)^2 - 5(2) - 5 \Rightarrow P(2) = -3$ and $P(3) = (3)^3 + (3)^2 - 5(3) - 5 \Rightarrow P(3) = 16$. These answers differ in sign, therefore there is a real zero between them. By graphing on the calculator the real zero is approximately: 2.24.

3. $P(2) = 2(2)^3 - 8(2)^2 + (2) + 16 \Rightarrow P(2) = 2$ and $P(2.5) = 2(2.5)^3 - 8(2.5)^2 + (2.5) + 16 \Rightarrow P(2.5) = -.25$. These answers differ in sign, therefore there is a real zero between them. By graphing on the calculator the real zero is approximately: 2.39.

4. $P\left(\dfrac{1}{2}\right) = 3\left(\dfrac{1}{2}\right)^3 + 7\left(\dfrac{1}{2}\right)^2 - 4 \Rightarrow P\left(\dfrac{1}{2}\right) = -1.875$ or $-1\dfrac{7}{8}$ and $P(1) = 3(1)^3 + 7(1)^2 - 4 \Rightarrow P(1) = 6$. These answers differ in sign, therefore there is a real zero between them. By graphing on the calculator the real zero is approximately: $.67$.

5. $P(1.5) = 2(1.5)^4 - 4(1.5)^2 + 3(1.5) - 6 \Rightarrow P(1.5) = -.375$ and $P(2) = 2(2)^4 - 4(2)^2 + 3(2) - 6 \Rightarrow P(2) = 16$. These answers differ in sign, therefore there is a real zero between them. By graphing on the calculator the real zero is approximately: 1.52.

6. $P(.3) = (.3)^4 - 4(.3)^3 - (.3) + 1 \Rightarrow P(.3) = .6001$ and $P(1) = (1)^4 - 4(1)^3 - (1) + 1 \Rightarrow P(1) = -3$. These answers differ in sign, therefore there is a real zero between them. By graphing on the calculator the real zero is approximately: $.52$.

7. $P(2.7) = -(2.7)^4 + 2(2.7)^3 + (2.7) + 12 \Rightarrow P(2.7) = .9219$ and $P(2.8) = -(2.8)^4 + 2(2.8)^3 + (2.8) + 12 \Rightarrow P(2.8) = -2.7616$. These answers differ in sign, therefore there is a real zero between them. By graphing on the calculator the real zero is approximately: 2.73.

8. $P(-1) = -2(-1)^4 + (-1)^3 - (-1)^2 + 3 \Rightarrow P(-1) = -1$ and $P(-.9) = -2(-.9)^4 + (-.9)^3 - (-.9)^2 + 3 \Rightarrow P(-.9) = .1488$. These answers differ in sign, therefore there is a real zero between them. By graphing on the calculator the real zero is approximately: $-.91$.

9. $P(-1.6) = (-1.6)^5 - 2(-1.6)^3 + 1 \Rightarrow P(-1.6) = -1.29376$ and $P(-1.5) = (-1.5)^5 - 2(-1.5)^3 + 1 \Rightarrow P(-1.5) = .15625$. These answers differ in sign, therefore there is a real zero between them. By graphing on the calculator the real zero is approximately: -1.51.

10. $P(1.1) = 2(1.1)^7 - (1.1)^4 + (1.1) - 4 \Rightarrow P(1.1) = -.4666658$ and $P(1.2) = 2(1.2)^7 - (1.2)^4 + (1.2) - 4 \Rightarrow P(1.2) = 2.2927616$. These answers differ in sign, therefore there is a real zero between them. By graphing on the calculator the real zero is approximately: 1.12.

11. There is at least one zero between 2 and 2.5.

12. There could be no zero or any number of zeros between these two points. For one example, see Figure 12.

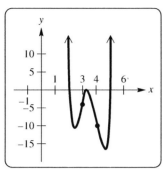

Figure 12

13.

$$
\begin{array}{r}
x^2 - 3x - 2 \\
x + 5 \overline{\smash{)}x^3 + 2x^2 - 17x - 10} \\
\underline{x^3 + 5x^2} \\
-3x^2 - 17x - 10 \\
\underline{-3x^2 - 15x} \\
-2x - 10 \\
\underline{-2x - 10} \\
0
\end{array}
$$

$\Rightarrow \dfrac{x^3 + 2x^2 - 17x - 10}{x + 5} = x^2 - 3x - 2$

14.

$$
\begin{array}{r}
x^3 + 0x^2 + 2x + 1 \\
x + 4 \overline{\smash{)}x^4 + 4x^3 + 2x^2 + 9x + 4} \\
\underline{x^4 + 4x^3} \\
2x^2 + 9x + 4 \\
\underline{2x^2 + 8x} \\
x + 4 \\
\underline{x + 4} \\
0
\end{array}
$$

$\Rightarrow \dfrac{x^4 + 4x^3 + 2x^2 + 9x + 4}{x + 4} = x^3 + 2x + 1$

15.

$$
\begin{array}{r}
3x^2 + 4x \\
x - 5 \overline{\smash{)}3x^3 - 11x^2 - 20x + 3} \\
\underline{3x^3 - 15x^2} \\
4x^2 - 20x + 3 \\
\underline{4x^2 - 20x} \\
3
\end{array}
$$

$\Rightarrow \dfrac{3x^3 - 11x^2 - 20x + 3}{x - 5} = 3x^2 + 4x + \dfrac{3}{x - 5}$

16.

$$
\begin{array}{r}
x^3 + 0x^2 - 5x - 13 \\
x - 3 \overline{\smash{)}x^4 - 3x^3 - 5x^2 + 2x - 16} \\
\underline{x^4 - 3x^3} \\
-5x^2 + 2x - 16 \\
\underline{-5x^2 + 15x} \\
-13x - 16 \\
\underline{-13x + 39} \\
-55
\end{array}
$$

$\Rightarrow \dfrac{x^4 - 3x^3 - 5x^2 + 2x - 16}{x - 3} = x^3 - 5x - 13 + \dfrac{-55}{x - 3}$

17.
$$x - 2 \overline{)\begin{array}{r} x^3 - x^2 - 6x \\ x^4 - 3x^3 - 4x^2 + 12x \end{array}} \quad \Rightarrow \quad \frac{x^4 - 3x^3 - 4x^2 + 12x}{x - 2} = x^3 - x^2 - 6x$$

$$\begin{array}{r} \underline{x^4 - 2x^3} \\ -x^3 - 4x^2 + 12x \\ \underline{-x^3 + 2x^2} \\ -6x^2 + 12x \\ \underline{-6x^2 + 12x} \\ 0 \end{array}$$

18.
$$x - 2 \overline{)\begin{array}{r} 2x^3 + 7x^2 + 9x \\ 2x^4 + 3x^3 - 5x^2 - 18x \end{array}} \quad \Rightarrow \quad \frac{2x^4 + 3x^3 - 5x^2 - 18x}{x - 2} = 2x^3 + 7x^2 + 9x$$

$$\begin{array}{r} \underline{2x^4 - 4x^3} \\ 7x^3 - 5x^2 - 18x \\ \underline{7x^3 - 14x^2} \\ 9x^2 - 18x \\ \underline{9x^2 - 18x} \\ 0 \end{array}$$

19.
$$x - 1 \overline{)\begin{array}{r} x^2 + 3x + 3 \\ x^3 + 2x^2 + 0x - 3 \end{array}} \quad \Rightarrow \quad \frac{x^3 + 2x^2 - 3}{x - 1} = x^2 + 3x + 3$$

$$\begin{array}{r} \underline{x^3 - x^2} \\ 3x^2 + 0x - 3 \\ \underline{3x^2 - 3x} \\ 3x - 3 \\ \underline{3x - 3} \\ 0 \end{array}$$

20.
$$x - 3 \overline{)\begin{array}{r} x^2 + x + 3 \\ x^3 - 2x^2 + 0x - 9 \end{array}} \quad \Rightarrow \quad \frac{x^3 - 2x^2 - 9}{x - 3} = x^2 + x + 3$$

$$\begin{array}{r} \underline{x^3 - 3x^2} \\ x^2 + 0x - 9 \\ \underline{x^2 - 3x} \\ 3x - 9 \\ \underline{3x - 9} \\ 0 \end{array}$$

21.
$$x + 1 \overline{)\begin{array}{r} -2x^2 + 2x - 3 \\ -2x^3 + 0x^2 - x - 2 \end{array}} \quad \Rightarrow \quad \frac{-2x^3 - x - 2}{x + 1} = -2x^2 + 2x - 3 + \frac{1}{x + 1}$$

$$\begin{array}{r} \underline{-2x^3 - 2x^2} \\ 2x^2 - x - 2 \\ \underline{2x^2 + 2x} \\ -3x - 2 \\ \underline{-3x - 3} \\ 1 \end{array}$$

22.

$$
\begin{array}{r}
-3x^2 + 3x - 4 \\
x + 1 \,\overline{\smash{\big)}\,-3x^3 - 0x^2 - x - 5} \\
\underline{-3x^3 - 3x^2} \\
3x^2 - x - 5 \\
\underline{3x^2 + 3x} \\
-4x - 5 \\
\underline{-4x - 4} \\
-1
\end{array}
\quad\Rightarrow\quad
\frac{-3x^3 - x - 5}{x + 1} = -3x^2 + 3x - 4 + \frac{-1}{x + 1}
$$

23.

$$
\begin{array}{r}
x^4 + x^3 + x^2 + x + 1 \\
x - 1 \,\overline{\smash{\big)}\,x^5 + 0x^4 + 0x^3 + 0x^2 + 0x - 1} \\
\underline{x^5 - x^4} \\
x^4 + 0x^3 + 0x^2 + 0x - 1 \\
\underline{x^4 - x^3} \\
x^3 + 0x^2 + 0x - 1 \\
\underline{x^3 - x^2} \\
x^2 + 0x - 1 \\
\underline{x^2 - x} \\
x - 1 \\
\underline{x - 1} \\
0
\end{array}
\quad\Rightarrow\quad
\frac{x^5 - 1}{x - 1} = x^4 + x^3 + x^2 + x + 1
$$

24.

$$
\begin{array}{r}
x^6 - x^5 + x^4 - x^3 + x^2 - x + 1 \\
x + 1 \,\overline{\smash{\big)}\,x^7 + 0x^6 + 0x^5 + 0x^4 + 0x^3 + 0x^2 + 0x + 1} \\
\underline{x^7 + x^6} \\
-x^6 + 0x^5 + 0x^4 + 0x^3 + 0x^2 + 0x + 1 \\
\underline{-x^6 - x^5} \\
x^5 + 0x^4 + 0x^3 + 0x^2 + 0x + 1 \\
\underline{x^5 + x^4} \\
-x^4 + 0x^3 + 0x^2 + 0x + 1 \\
\underline{-x^4 - x^3} \\
x^3 + 0x^2 + 0x + 1 \\
\underline{x^3 + x^2} \\
-x^2 + 0x + 1 \\
\underline{-x^2 - x} \\
x + 1 \\
\underline{x + 1} \\
0
\end{array}
\quad\Rightarrow\quad
\frac{x^7 + 1}{x + 1} = x^6 - x^5 + x^4 - x^3 + x^2 - x + 1
$$

25.

$$
\begin{array}{r}
3\,\overline{\smash{\big)}\,1 \quad -4 \quad\ 3} \\
\quad\ \ 3 \quad -3 \\
\hline
1 \quad -1 \quad\ \ 0
\end{array}
\qquad \text{Therefore, } P(3) = 0.
$$

26. $-2 \overline{)\begin{array}{rrr} 1 & 5 & 6 \end{array}}$

    $\phantom{-2 \overline{)}} \underline{\begin{array}{rr} -2 & -6 \end{array}}$

    $\phantom{-2 \overline{)}} \begin{array}{rrr} 1 & 3 & 0 \end{array}$ Therefore, $P(-2) = 0$.

27. $-2 \overline{)\begin{array}{rrrr} 5 & 2 & -1 & 5 \end{array}}$

    $\phantom{-2 \overline{)}} \underline{\begin{array}{rrr} -10 & 16 & -30 \end{array}}$

    $\phantom{-2 \overline{)}} \begin{array}{rrrr} 5 & -8 & 15 & -25 \end{array}$ Therefore, $P(-2) = -25$.

28. $2 \overline{)\begin{array}{rrrr} 2 & -3 & -5 & 4 \end{array}}$

    $\phantom{2 \overline{)}} \underline{\begin{array}{rrr} 4 & 2 & -6 \end{array}}$

    $\phantom{2 \overline{)}} \begin{array}{rrrr} 2 & 1 & -3 & -2 \end{array}$ Therefore, $P(2) = -2$.

29. $2 \overline{)\begin{array}{rrr} 1 & -5 & 1 \end{array}}$

    $\phantom{2 \overline{)}} \underline{\begin{array}{rr} 2 & -6 \end{array}}$

    $\phantom{2 \overline{)}} \begin{array}{rrr} 1 & -3 & -5 \end{array}$ Therefore, $P(2) = -5$.

30. $3 \overline{)\begin{array}{rrr} 1 & -1 & 3 \end{array}}$

    $\phantom{3 \overline{)}} \underline{\begin{array}{rr} 3 & 6 \end{array}}$

    $\phantom{3 \overline{)}} \begin{array}{rrr} 1 & 2 & 9 \end{array}$ Therefore, $P(3) = 9$.

31. $.5 \overline{)\begin{array}{rrrr} 1 & 0 & -1 & 4 \end{array}}$

    $\phantom{.5 \overline{)}} \underline{\begin{array}{rrr} .5 & .25 & -.375 \end{array}}$

    $\phantom{.5 \overline{)}} \begin{array}{rrrr} 1 & .5 & -.75 & 3.625 \end{array}$ Therefore, $P(.5) = 3.625$.

32. $1.5 \overline{)\begin{array}{rrrr} 1 & 0 & 1 & -3 \end{array}}$

    $\phantom{1.5 \overline{)}} \underline{\begin{array}{rrr} 1.5 & 2.25 & 4.875 \end{array}}$

    $\phantom{1.5 \overline{)}} \begin{array}{rrrr} 1 & 1.5 & 3.25 & 1.875 \end{array}$ Therefore, $P(1.5) = 1.875$.

33. $\sqrt{2} \overline{)\begin{array}{rrrrr} 1 & 0 & -1 & 0 & -3 \end{array}}$

    $\phantom{\sqrt{2} \overline{)}} \underline{\begin{array}{rrrr} \sqrt{2} & 2 & \sqrt{2} & 2 \end{array}}$

    $\phantom{\sqrt{2} \overline{)}} \begin{array}{rrrrr} 1 & \sqrt{2} & 1 & \sqrt{2} & -1 \end{array}$ Therefore, $P(\sqrt{2}) = -1$.

34. $\sqrt{3} \overline{)\begin{array}{rrrrr} 1 & 0 & 2 & 0 & -10 \end{array}}$

    $\phantom{\sqrt{3} \overline{)}} \underline{\begin{array}{rrrr} \sqrt{3} & 3 & 5\sqrt{3} & 15 \end{array}}$

    $\phantom{\sqrt{3} \overline{)}} \begin{array}{rrrrr} 1 & \sqrt{3} & 5 & 5\sqrt{3} & 5 \end{array}$ Therefore, $P(\sqrt{3}) = 5$.

35. $\sqrt[3]{4} \overline{)\begin{array}{rrrr} -1 & 0 & 1 & 4 \end{array}}$

    $\phantom{\sqrt[3]{4} \overline{)}} \underline{\begin{array}{rrr} -\sqrt[3]{4} & -\sqrt[3]{16} & \sqrt[3]{4} - 4 \end{array}}$

    $\phantom{\sqrt[3]{4} \overline{)}} \begin{array}{rrrr} -1 & -\sqrt[3]{4} & 1 - \sqrt[3]{16} & \sqrt[3]{4} \end{array}$ Therefore, $P(\sqrt[3]{4}) = \sqrt[3]{4}$.

36. $\sqrt[5]{3} \overline{)\begin{array}{rrrrrr} -1 & 0 & 0 & 0 & 2 & 3 \end{array}}$

    $\phantom{\sqrt[5]{3} \overline{)}} \underline{\begin{array}{rrrrr} -\sqrt[5]{3} & -\sqrt[5]{9} & -\sqrt[5]{27} & -\sqrt[5]{81} & 2\sqrt[5]{3} - 3 \end{array}}$

    $\phantom{\sqrt[5]{3} \overline{)}} \begin{array}{rrrrrr} -1 & -\sqrt[5]{3} & -\sqrt[5]{9} & -\sqrt[5]{27} & 2 - \sqrt[5]{81} & 2\sqrt[5]{3} \end{array}$ Therefore, $P(\sqrt[5]{3}) = 2\sqrt[5]{3}$.

37. $2 \overline{)\begin{array}{rrr} 1 & 2 & -8 \end{array}}$

    $\phantom{2 \overline{)}} \underline{\begin{array}{rr} 2 & 8 \end{array}}$

    $\phantom{2 \overline{)}} \begin{array}{rrr} 1 & 4 & 0 \end{array}$ Yes; since $P(2) = 0$, 2 is a zero.

38. $-1 \overline{\smash{\big)}\ 1 \quad 4 \quad -5}$
　　　　　　$\underline{\quad -1 \quad -3}$
　　　　　$1 \quad 3 \quad -8$ No; since $P(-1) \neq 0, -1$ is not a zero.

39. $4 \overline{\smash{\big)}\ 2 \quad -6 \quad -9 \quad 6}$
　　　　　　　$\underline{\quad 8 \quad 8 \quad -4}$
　　　　　$2 \quad 2 \quad -1 \quad 2$ No; since $P(4) \neq 0, 4$ is not a zero.

40. $-4 \overline{\smash{\big)}\ 9 \quad 39 \quad 12 \quad 0}$
　　　　　　　$\underline{\quad -36 \quad -12 \quad 0}$
　　　　　$9 \quad 3 \quad 0 \quad 0$ Yes; since $P(-4) = 0, -4$ is a zero.

41. $-.5 \overline{\smash{\big)}\ 4 \quad 12 \quad 7 \quad 1}$
　　　　　　　$\underline{\quad -2 \quad -5 \quad -1}$
　　　　　$4 \quad 10 \quad 2 \quad 0$ Yes; since $P(-.5) = 0, -.5$ is a zero.

42. $-.25 \overline{\smash{\big)}\ 8 \quad 6 \quad -3 \quad -1}$
　　　　　　　$\underline{\quad -2 \quad -1 \quad 1}$
　　　　　$8 \quad 4 \quad -4 \quad 0$ Yes; since $P(-.25) = 0, -.25$ is a zero.

43. $-5 \overline{\smash{\big)}\ 8 \quad 50 \quad 47 \quad 15}$
　　　　　　　$\underline{\quad -40 \quad -50 \quad 15}$
　　　　　$8 \quad 10 \quad -3 \quad 30$ No; since $P(-5) \neq 0, -5$ is not a zero.

44. $-4 \overline{\smash{\big)}\ 6 \quad 25 \quad 3 \quad -3}$
　　　　　　　$\underline{\quad -24 \quad -4 \quad 4}$
　　　　　$6 \quad 1 \quad -1 \quad 1$ No; since $P(-4) \neq 0, -4$ is not a zero.

45. $\sqrt{6} \overline{\smash{\big)}\ -2 \quad 0 \quad 5 \quad 0 \quad -3 \quad 0 \quad 270}$
　　　　　　　$\underline{\quad -2\sqrt{6} \quad -12 \quad -7\sqrt{6} \quad -42 \quad -45\sqrt{6} \quad -270}$
　　　　　$-2 \quad -2\sqrt{6} \quad -7 \quad -7\sqrt{6} \quad -45 \quad -45\sqrt{6} \quad 0$ Yes; since $P(\sqrt{6}) = 0, \sqrt{6}$ is a zero.

46. $\sqrt{7} \overline{\smash{\big)}\ -3 \quad 0 \quad 7 \quad 0 \quad -5 \quad 0 \quad 721}$
　　　　　　　$\underline{\quad -3\sqrt{7} \quad -21 \quad -14\sqrt{7} \quad -98 \quad -103\sqrt{7} \quad -721}$
　　　　　$-3 \quad -3\sqrt{7} \quad -14 \quad -14\sqrt{7} \quad -103 \quad -103\sqrt{7} \quad 0$ Yes; since $P(\sqrt{7}) = 0, \sqrt{7}$ is a zero.

47. Since the x-intercepts are: $-3, 1$, and 4, the linear factors are: $(x - (-3)), (x - 1), (x - 4)$ or

   $(x + 3), (x - 1), (x - 4)$.

48. Since the x-intercepts are: $-3, 1$, and 4, the solutions to $P(x) = 0$ are $-3, 1, 4$.

49. Since the x-intercepts are: $-3, 1$, and 4, the zeros are $-3, 1, 4$.

50. Use synthetic division and divide by 2:

   $2 \overline{\smash{\big)}\ 1 \quad -2 \quad -11 \quad 12}$
　　　　　$\underline{\quad 2 \quad 0 \quad -22}$
　　　$1 \quad 0 \quad -11 \quad -10$ The remainder is: -10, therefore $P(2) = -10$.

51. Using the x-intercepts and graph, $P(x) > 0$ for the $(-3, 1) \cup (4, \infty)$.

52. Using the x-intercepts and graph, $P(x) < 0$ for the $(-\infty, -3) \cup (1, 4)$.

53. First use synthetic division to factor out the given zero.

$$\begin{array}{r|rrrr} 3 & 1 & -2 & -5 & 6 \\ & & 3 & 3 & -6 \\ \hline & 1 & 1 & -2 & 0 \end{array}$$

Now completely factor the resulting quadratic:

$x^2 + x - 2 \Rightarrow (x - 1)(x + 2)$. The other zeros are: $-2, 1$.

54. First use synthetic division to factor out the given zero.

$$\begin{array}{r|rrrr} 3 & 1 & 2 & -11 & -12 \\ & & 3 & 15 & 12 \\ \hline & 1 & 5 & 4 & 0 \end{array}$$

Now completely factor the resulting quadratic:

$x^2 + 5x + 4 \Rightarrow (x + 4)(x + 1)$. The other zeros are: $-4, -1$.

55. First use synthetic division to factor out the given zero.

$$\begin{array}{r|rrrr} 1 & 1 & 0 & -2 & 1 \\ & & 1 & 1 & -1 \\ \hline & 1 & 1 & -1 & 0 \end{array}$$

Now solve the quadratic equation:

$x^2 + x - 1 = 0$ using the quadratic formula:

$$x = \frac{-1 \pm \sqrt{1^2 - 4(1)(-1)}}{2(1)} = \frac{-1 \pm \sqrt{5}}{2}. \text{ The other zeros are: } \frac{-1 - \sqrt{5}}{2}, \frac{-1 + \sqrt{5}}{2}.$$

56. First use synthetic division to factor out the given zero.

$$\begin{array}{r|rrrr} -5 & 2 & 8 & -11 & -5 \\ & & -10 & 10 & 5 \\ \hline & 2 & -2 & -1 & 0 \end{array}$$

Now solve the quadratic equation:

$2x^2 - 2x - 1 = 0$ using the quadratic formula:

$$x = \frac{2 \pm \sqrt{(-2)^2 - 4(2)(-1)}}{2(2)} = \frac{2 \pm \sqrt{12}}{4} = \frac{2 \pm 2\sqrt{3}}{4} = \frac{1 \pm \sqrt{3}}{2}. \text{ The other zeros are: }$$

$$\frac{1 - \sqrt{3}}{2}, \frac{1 + \sqrt{3}}{2}.$$

57. First use synthetic division to factor out the given zero.

$$\begin{array}{r|rrrr} -2 & 3 & 5 & -3 & -2 \\ & & -6 & 2 & 2 \\ \hline & 3 & -1 & -1 & 0 \end{array}$$

Now solve the quadratic equation:

$3x^2 - x - 1 = 0$ using the quadratic formula:

$$x = \frac{1 \pm \sqrt{(-1)^2 - 4(3)(-1)}}{2(3)} = \frac{1 \pm \sqrt{13}}{6}. \text{ The other zeros are: } \frac{1 - \sqrt{13}}{6}, \frac{1 + \sqrt{13}}{6}.$$

58. First use synthetic division to factor out the given zero.

$$\begin{array}{r|rrrr} 3 & 1 & -7 & 13 & -3 \\ & & 3 & -12 & 3 \\ \hline & 1 & -4 & 1 & 0 \end{array}$$

Now solve the quadratic equation:

$x^2 - 4x + 1 = 0$ using the quadratic formula:

$$x = \frac{4 \pm \sqrt{(4)^2 - 4(1)(1)}}{2(1)} = \frac{4 \pm \sqrt{12}}{2} = \frac{4 \pm 2\sqrt{3}}{2} = 2 \pm \sqrt{3}. \text{ The other zeros are: } 2 - \sqrt{3}, 2 + \sqrt{3}.$$

59. First use synthetic division to factor out the first given zero. Then use synthetic division to factor out the second given zero from the resulting expression: $x^3 - 6x^2 - 5x + 30$.

$$
\begin{array}{r|rrrrr}
-6 & 1 & 0 & -41 & 0 & 180 \\
   &   & -6 & 36 & 30 & -180 \\
\hline
   & 1 & -6 & -5 & 30 & 0
\end{array}
\qquad \Rightarrow \qquad
\begin{array}{r|rrrr}
6 & 1 & -6 & -5 & 30 \\
  &   & 6 & 0 & -30 \\
\hline
  & 1 & 0 & -5 & 0
\end{array}
$$

Finally, solve the resulting equation: $x^2 - 5 = 0 \Rightarrow x^2 = 5 \Rightarrow x = \pm\sqrt{5}$.

Therefore the other zeros are: $-\sqrt{5}, \sqrt{5}$.

60. First use synthetic division to factor out the first given zero. Then use synthetic division to factor out the second given zero from the resulting expression: $x^3 - 7x^2 - 3x + 21$.

$$
\begin{array}{r|rrrrr}
-7 & 1 & 0 & -52 & 0 & 147 \\
   &   & -7 & 49 & 21 & -147 \\
\hline
   & 1 & -7 & -3 & 21 & 0
\end{array}
\qquad \Rightarrow \qquad
\begin{array}{r|rrrr}
7 & 1 & -7 & -3 & 21 \\
  &   & 7 & 0 & -21 \\
\hline
  & 1 & 0 & -3 & 0
\end{array}
$$

Finally, solve the resulting equation: $x^2 - 3 = 0 \Rightarrow x^2 = 3 \Rightarrow x = \pm\sqrt{3}$.

Therefore the other zeros are: $-\sqrt{3}, \sqrt{3}$.

61. First use synthetic division to factor out the given zero.

$$
\begin{array}{r|rrrr}
8 & -1 & 8 & 3 & -24 \\
  &   & -8 & 0 & 24 \\
\hline
  & -1 & 0 & 3 & 0
\end{array}
$$

Now solve the resulting equation:

$-x^2 + 3 = 0 \Rightarrow -x^2 = -3 \Rightarrow x^2 = 3 \Rightarrow x = \pm\sqrt{3}$. The other zeros are: $-\sqrt{3}, \sqrt{3}$.

62. First use synthetic division to factor out the given zero.

$$
\begin{array}{r|rrrr}
4 & -1 & 4 & 7 & -28 \\
  &   & -4 & 0 & 28 \\
\hline
  & -1 & 0 & 7 & 0
\end{array}
$$

Now solve the resulting equation:

$-x^2 + 7 = 0 \Rightarrow -x^2 = -7 \Rightarrow x^2 = 7 \Rightarrow x = \pm\sqrt{7}$. The other zeros are: $-\sqrt{7}, \sqrt{7}$.

63. First use synthetic division to factor out the given zero.

$$
\begin{array}{r|rrrr}
2 & 2 & -3 & -17 & 30 \\
  &   & 4 & 2 & -30 \\
\hline
  & 2 & 1 & -15 & 0
\end{array}
$$

Now completely factor the resulting quadratic expression:

$2x^2 + x - 15 \Rightarrow (2x - 5)(x + 3)$. The linear factors are: $P(x) = (x - 2)(2x - 5)(x + 3)$.

64. First use synthetic division to factor out the given zero.

$$
\begin{array}{r|rrrr}
1 & 2 & -3 & -5 & 6 \\
  &   & 2 & -1 & -6 \\
\hline
  & 2 & -1 & -6 & 0
\end{array}
$$

Now completely factor the resulting quadratic expression:

$2x^2 - x - 6 \Rightarrow (2x + 3)(x - 2)$. The linear factors are: $P(x) = (x - 1)(2x + 3)(x - 2)$.

65. First use synthetic division to factor out the given zero.

$$\begin{array}{r|rrrr} -4 & 6 & 25 & 3 & -4 \\ & & -24 & -4 & 4 \\ \hline & 6 & 1 & -1 & 0 \end{array}$$ Now completely factor the resulting quadratic expression:

$6x^2 + x - 1 \Rightarrow (3x - 1)(2x + 1)$. Therefore the linear factors are: $P(x) = (x + 4)(3x - 1)(2x + 1)$.

66. First use synthetic division to factor out the given zero.

$$\begin{array}{r|rrrr} -5 & 8 & 50 & 47 & -15 \\ & & -40 & -50 & 15 \\ \hline & 8 & 10 & -3 & 0 \end{array}$$ Now completely factor the resulting quadratic expression:

$8x^2 + 10x - 3 \Rightarrow (4x - 1)(2x + 3)$. Therefore the linear factors are: $P(x) = (x + 5)(4x - 1)(2x + 3)$.

67. First use synthetic division to factor out the given zero.

$$\begin{array}{r|rrrr} -3 & -6 & -13 & 14 & -3 \\ & & 18 & -15 & 3 \\ \hline & -6 & 5 & -1 & 0 \end{array}$$ Now completely factor the resulting quadratic expression:

$-6x^2 + 5x - 1 \Rightarrow (-3x + 1)(2x - 1)$. The linear factors are: $P(x) = (x + 3)(-3x + 1)(2x - 1)$.

68. First use synthetic division to factor out the given zero.

$$\begin{array}{r|rrrr} -5 & -6 & -17 & 63 & -10 \\ & & 30 & -65 & 10 \\ \hline & -6 & 13 & -2 & 0 \end{array}$$ Now completely factor the resulting quadratic expression:

$-6x^2 + 13x - 2 \Rightarrow (-x + 2)(6x - 1)$. The linear factors are: $P(x) = (x + 5)(-x + 2)(6x - 1)$.

69. First use synthetic division to factor out the given zero.

$$\begin{array}{r|rrrr} -5 & 1 & 5 & -3 & -15 \\ & & -5 & 0 & 15 \\ \hline & 1 & 0 & -3 & 0 \end{array}$$ Now completely factor the resulting quadratic expression:

$x^2 - 3 = (x + \sqrt{3})(x - \sqrt{3})$. The linear factors are: $P(x) = (x + 5)(x + \sqrt{3})(x - \sqrt{3})$.

70. First use synthetic division to factor out the given zero.

$$\begin{array}{r|rrrr} -9 & 1 & 9 & -7 & -63 \\ & & -9 & 0 & 63 \\ \hline & 1 & 0 & -7 & 0 \end{array}$$ Now completely factor the resulting quadratic expression:

$x^2 - 7 = (x + \sqrt{7})(x - \sqrt{7})$. The linear factors are: $P(x) = (x + 9)(x + \sqrt{7})(x - \sqrt{7})$.

71. First use synthetic division to factor out the given zero.

$$\begin{array}{r|rrrr} -1 & 1 & -2 & -7 & -4 \\ & & -1 & 3 & 4 \\ \hline & 1 & -3 & -4 & 0 \end{array}$$ Now completely factor the resulting quadratic expression:

$x^2 - 3x - 4 \Rightarrow (x - 4)(x + 1)$. The linear factors are: $P(x) = (x + 1)(x - 4)(x + 1)$ or $(x + 1)^2(x - 4)$.

72. First use synthetic division to factor out the given zero.

$$-3 \overline{)\, \begin{array}{rrrr} 1 & 1 & -21 & -45 \end{array}}$$
$$\begin{array}{rrrr} & -3 & 6 & 45 \end{array}$$
$$\begin{array}{rrrr} \hline 1 & -2 & -15 & 0 \end{array}$$ Now completely factor the resulting quadratic expression:

$x^2 - 2x - 15 \Rightarrow (x - 5)(x + 3)$. The linear factors are: $P(x) = (x + 3)(x - 5)(x + 3)$ or $(x + 3)^2(x - 5)$.

3.7: Topics in the Theory of Polynomial Functions (II)

1. With the given zeros 4 and $2 + i$, the conjugate of $2 + i$ or $2 - i$ is also a zero. Therefore:

   $P(x) = (x - 4)(x - (2 + i))(x - (2 - i)) = (x - 4)(x - 2 - i)(x - 2 + i) = (x - 4)(x^2 - 4x + 5) =$

   $x^3 - 4x^2 + 5x - 4x^2 + 16x - 20 \Rightarrow P(x) = x^3 - 8x^2 + 21x - 20$.

2. With the given zeros -3 and $6 + 2i$, the conjugate of $6 + 2i$ or $6 - 2i$ is also a zero. Therefore:

   $P(x) = (x - (-3))(x - (6 + 2i))(x - (6 - 2i)) = (x + 3)(x - 6 - 2i)(x - 6 + 2i) =$

   $(x + 3)(x^2 - 12x + 40) = x^3 - 12x^2 + 40x + 3x^2 - 36x + 120 \Rightarrow P(x) = x^3 - 9x^2 + 4x + 120$.

3. With the given zeros 5 and i, the conjugate of i or $-i$ is also a zero. Therefore:

   $P(x) = (x - 5)(x - i)(x - (-i)) = (x - 5)(x^2 - i^2) = (x - 5)(x^2 + 1) \Rightarrow P(x) = x^3 - 5x^2 + x - 5$.

4. With the given zeros -9 and $-i$, the conjugate of $-i$ or i is also a zero. Therefore:

   $P(x) = (x - (-9))(x - (-i))(x - (i)) = (x + 9)(x + i)(x - i) = (x + 9)(x^2 - i^2) =$

   $(x + 9)(x^2 + 1) \Rightarrow P(x) = x^3 + 9x^2 + x + 9$.

5. With the given zeros 0 and $3 + i$, the conjugate of $3 + i$ or $3 - i$ is also a zero. Therefore:

   $P(x) = (x - 0)(x - (3 + i))(x - (3 - i)) = (x)(x - 3 - i)(x - 3 + i) = (x)(x^2 - 6x + 10) \Rightarrow$

   $P(x) = x^3 - 6x^2 + 10x$.

6. With the given zeros 0 and $4 - 3i$, the conjugate of $4 - 3i$ or $4 + 3i$ is also a zero. Therefore:

   $P(x) = (x)(x - (4 - 3i))(x - (4 + 3i)) = (x)(x - 4 + 3i)(x - 4 - 3i) = (x)(x^2 - 8x + 25) \Rightarrow$

   $P(x) = x^3 - 8x^2 + 25x$.

7. With the given zeros $-3, -1,$ and 4, the function $P(x)$ is: $P(x) = a(x - (-3))(x - (-1))(x - 4) =$

   $a(x + 3)(x + 1)(x - 4) = a(x + 3)(x^2 - 3x - 4) \Rightarrow P(x) = a(x^3 - 13x - 12)$. Since $P(2) = 5$, we can

   solve for a: $5 = a((2)^3 - 13(2) - 12) \Rightarrow 5 = a(8 - 26 - 12) \Rightarrow -30a = 5 \Rightarrow a = -\dfrac{5}{30} = -\dfrac{1}{6}$.

   Therefore, $P(x) = -\dfrac{1}{6}(x^3 - 13x - 12) \Rightarrow P(x) = -\dfrac{1}{6}x^3 + \dfrac{13}{6}x + 2$.

8. With the given zeros $1, -1,$ and 0, the function $P(x)$ is: $P(x) = a(x - 1)(x - (-1))(x - 0) =$

   $a(x - 1)(x + 1)(x) = ax(x^2 - 1) \Rightarrow P(x) = a(x^3 - x)$. Since $P(2) = -3$, we can

   solve for a: $-3 = a((2)^3 - (2)) \Rightarrow -3 = a(8 - 2) \Rightarrow 6a = -3 \Rightarrow a = -\dfrac{3}{6} = -\dfrac{1}{2}$.

   Therefore, $P(x) = -\dfrac{1}{2}(x^3 - x) \Rightarrow P(x) = -\dfrac{1}{2}x^3 + \dfrac{1}{2}x$.

9. With the given zeros -2, 1, and 0, the function $P(x)$ is: $P(x) = a(x - (-2))(x - 1)(x - 0) =$

    $a(x + 2)(x - 1)(x) \Rightarrow P(x) = ax(x^2 + x - 2)$. Since $P(-1) = -1$, we can solve for a:

    $-1 = a(-1)((-1)^2 + (-1) - 2) \Rightarrow -1 = -a(1 - 1 - 2) \Rightarrow -a(-2) = -1 \Rightarrow 2a = -1 \Rightarrow a = -\dfrac{1}{2}$.

    Therefore, $P(x) = -\dfrac{1}{2}x(x^2 + x - 2) \Rightarrow P(x) = -\dfrac{1}{2}x^3 - \dfrac{1}{2}x^2 + x$.

10. With the given zeros 2, 5, and -3, the function $P(x)$ is: $P(x) = a(x - 2)(x - 5)(x - (-3)) =$

    $a(x - 2)(x - 5)(x + 3) = a(x^2 - 7x + 10)(x + 3) \Rightarrow P(x) = a(x^3 - 4x^2 - 11x + 30)$. Since $P(1) = -4$,

    we can solve for a: $-4 = a((1)^3 - 4(1)^2 - 11(1) + 30) \Rightarrow -4 = a(1 - 4 - 11 + 30) \Rightarrow a(16) = -4 \Rightarrow$

    $a = -\dfrac{4}{16} \Rightarrow a = -\dfrac{1}{4}$. Therefore, $P(x) = -\dfrac{1}{4}(x^3 - 4x^2 - 11x + 30) \Rightarrow P(x) = -\dfrac{1}{4}x^3 + x^2 + \dfrac{11}{4}x - \dfrac{15}{2}$.

11. With the given zeros 4, $1 + i$, and the conjugate of $1 + i$ or $1 - i$ also a zero, the function $P(x)$ is:

    $P(x) = a(x - 4)(x - (1 + i))(x - (1 - i)) = a(x - 4)(x - 1 - i)(x - 1 + i) = a(x - 4)(x^2 - 2x + 2) =$

    $a(x^3 - 2x^2 + 2x - 4x^2 + 8x - 8) \Rightarrow P(x) = a(x^3 - 6x^2 + 10x - 8)$. Since $P(2) = 4$, we can solve for

    a: $4 = a((2)^3 - 6(2)^2 + 10(2) - 8) \Rightarrow 4 = a(8 - 24 + 20 - 8) \Rightarrow -4a = 4 \Rightarrow a = -\dfrac{1}{1} = -1$.

    Therefore, $P(x) = -1(x^3 - 6x^2 + 10x - 8) \Rightarrow P(x) = -x^3 + 6x^2 - 10x + 8$.

12. With the given zeros -7, $2 - i$, , and the conjugate of $2 - i$ or $2 + i$ also a zero, the function $P(x)$ is:

    $P(x) = a(x - (-7))(x - (2 - i))(x - (2 + i)) = a(x + 7)(x - 2 + i)(x - 2 - i) = a(x + 7)(x^2 - 4x + 5)$

    $= a(x^3 - 4x^2 + 5x + 7x^2 - 28x + 35) \Rightarrow P(x) = a(x^3 + 3x^2 - 23x + 35)$. Since $P(1) = 9$, we can solve

    for a: $9 = a((1)^3 + 3(1)^2 - 23(1) + 35) \Rightarrow 9 = a(1 + 3 - 23 + 35) \Rightarrow 16a = 9 \Rightarrow a = \dfrac{9}{16}$. Therefore,

    $P(x) = \dfrac{9}{16}(x^3 + 3x^2 - 23x + 35) \Rightarrow P(x) = \dfrac{9}{16}x^3 + \dfrac{27}{16}x^2 - \dfrac{207}{16}x + \dfrac{315}{16}$.

13. First use synthetic division to factor out the given zero.

$$
\begin{array}{r|rrrr}
3 & 1 & -1 & -4 & -6 \\
  &   & 3 & 6 & 6 \\
\hline
  & 1 & 2 & 2 & 0
\end{array}
$$

Now solve the resulting quadratic equation: $x^2 + 2x + 2 = 0$

using the quadratic formula: $x = \dfrac{-2 \pm \sqrt{(2)^2 - 4(2)(1)}}{2(1)} = \dfrac{-2 \pm \sqrt{-4}}{2} = \dfrac{-2 \pm 2i}{2} = -1 \pm i$.

Therefore the other zeros are: $-1 - i, -1 + i$.

14. First use synthetic division to factor out the given zero.

$$
\begin{array}{r|rrrr}
1 & 1 & -5 & 17 & -13 \\
  &   & 1 & -4 & 13 \\
\hline
  & 1 & -4 & 13 & 0
\end{array}
$$

Now solve the resulting quadratic equation: $x^2 - 4x + 13 = 0$

using the quadratic formula: $x = \dfrac{4 \pm \sqrt{(-4)^2 - 4(1)(13)}}{2(1)} = \dfrac{4 \pm \sqrt{-36}}{2} = \dfrac{4 \pm 6i}{2} = 2 \pm 3i$.

Therefore the other zeros are: $2 + 3i, 2 - 3i$.

15. First use synthetic division to factor out the first given zero. Then use synthetic division to factor out the second given zero from the resulting expression: $x^3 - x^2 - 7x + 3$.

$$-3 \overline{)\ \begin{array}{ccccc} 1 & 2 & -10 & -18 & 9 \\ & -3 & 3 & 21 & -9 \\ \hline 1 & -1 & -7 & 3 & 0 \end{array}} \qquad \Rightarrow \qquad 3 \overline{)\ \begin{array}{cccc} 1 & -1 & -7 & 3 \\ & 3 & 6 & -3 \\ \hline 1 & 2 & -1 & 0 \end{array}}$$

Finally, solve the resulting quadratic equation: $x^2 + 2x - 1 = 0$ using the quadratic formula:

$$x = \frac{-2 \pm \sqrt{(2)^2 - 4(1)(-1)}}{2(1)} = \frac{-2 \pm \sqrt{8}}{2} = \frac{-2 \pm 2\sqrt{2}}{2} = -1 \pm \sqrt{2}.$$

Therefore the other zeros are: $-1 + \sqrt{2}, -1 - \sqrt{2}$.

16. First use synthetic division to factor out the first given zero. Then use synthetic division to factor out the second given zero from the resulting expression: $2x^3 - 9x^2 + 9x - 20$.

$$-4 \overline{)\ \begin{array}{ccccc} 2 & -1 & -27 & 16 & -80 \\ & -8 & 36 & -36 & 80 \\ \hline 2 & -9 & 9 & -20 & 0 \end{array}} \qquad \Rightarrow \qquad 4 \overline{)\ \begin{array}{cccc} 2 & -9 & 9 & -20 \\ & 8 & -4 & 20 \\ \hline 2 & -1 & 5 & 0 \end{array}}$$

Finally, solve the resulting quadratic equation: $2x^2 - x + 5 = 0$ using the quadratic formula:

$$x = \frac{1 \pm \sqrt{(-1)^2 - 4(2)(5)}}{2(2)} = \frac{1 \pm \sqrt{-39}}{4} = \frac{1}{4} \pm \frac{\sqrt{39}}{4}i. \quad \text{Therefore the other zeros are:}$$

$$\frac{1}{4} + \frac{\sqrt{39}}{4}i, \frac{1}{4} - \frac{\sqrt{39}}{4}i.$$

17. With the given zero $3i$, , the conjugate $-3i$ is also a zero. Use synthetic division to factor out these zeros.

$$3i \overline{)\ \begin{array}{ccccc} 1 & -1 & 10 & -9 & 9 \\ & 3i & -9 - 3i & 9 + 3i & -9 \\ \hline 1 & -1 + 3i & 1 - 3i & 3i & 0 \end{array}} \quad \Rightarrow \quad -3i \overline{)\ \begin{array}{cccc} 1 & -1 + 3i & 1 - 3i & 3i \\ & & -3i & 3i & -3i \\ \hline 1 & -1 & 1 & 0 \end{array}}$$

Finally, solve the resulting quadratic equation: $x^2 - x + 1 = 0$ using the quadratic formula:

$$x = \frac{1 \pm \sqrt{(-1)^2 - 4(1)(1)}}{2(1)} = \frac{1 \pm \sqrt{-3}}{2} = \frac{1}{2} \pm \frac{\sqrt{3}}{2}i. \quad \text{Therefore the other zeros are:}$$

$$-3i, \frac{1}{2} + \frac{\sqrt{3}}{2}i, \frac{1}{2} - \frac{\sqrt{3}}{2}i.$$

18. With the given zero $-5i$, , the conjugate $5i$ is also a zero. Now use synthetic division to factor out these zeros.

$$-5i \overline{)\ \begin{array}{ccccc} 2 & -2 & 55 & -50 & 125 \\ & -10i & -50 + 10i & 50 - 25i & -125 \\ \hline 2 & -2 - 10i & 5 + 10i & -25i & 0 \end{array}} \Rightarrow \quad 5i \overline{)\ \begin{array}{cccc} 2 & -2 - 10i & 5 + 10i & -25i \\ & & 10i & -10i & 25i \\ \hline 2 & -2 & 5 & 0 \end{array}}$$

Finally, solve the resulting quadratic equation: $2x^2 - 2x + 5 = 0$ using the quadratic formula:

$$x = \frac{2 \pm \sqrt{(-2)^2 - 4(2)(5)}}{2(2)} = \frac{2 \pm \sqrt{-36}}{4} = \frac{2}{4} \pm \frac{6}{4}i = \frac{1}{2} \pm \frac{3}{2}i. \quad \text{Therefore the other zeros are:}$$

$$5i, \frac{1}{2} + \frac{3}{2}i, \frac{1}{2} - \frac{3}{2}i.$$

19. With the given zeros 5 and -4, $P(x) = a(x - 5)(x - (-4)) = a(x - 5)(x + 4) \Rightarrow P(x) = a(x^2 - x - 20)$.

   There are many possible solutions, one, when $a = 1$ is: $P(x) = x^2 - x - 20$.

20. With the given zeros 6 and -2, $P(x) = a(x - 6)(x - (-2)) = a(x - 6)(x + 2) \Rightarrow P(x) = a(x^2 - 4x - 12)$.

   There are many possible solutions, one, when $a = 1$ is: $P(x) = x^2 - 4x - 12$.

21. With the given zeros -3, 2, i, and the conjugate of i or $-i$ also a zero, the function $P(x)$ is:

   $P(x) = a(x - (-3))(x - 2)(x - i)(x - (-i)) = a(x + 3)(x - 2)(x - i)(x + i) = a(x^2 + x - 6)(x^2 + 1) \Rightarrow$
   $P(x) = a(x^4 + x^3 - 5x^2 + x - 6)$. There are many possible solutions, one, when

   $a = 1$ is: $P(x) = x^4 + x^3 - 5x^2 + x - 6$.

22. With the given zeros $1 + \sqrt{2}$, $1 - \sqrt{2}$ and 3, $P(x) = a(x - (1 + \sqrt{2}))(x - (1 - \sqrt{2}))(x - 3) =$

   $a(x - 1 - \sqrt{2})(x - 1 + \sqrt{2})(x - 3) = a(x^2 - x + \sqrt{2}x - x + 1 - \sqrt{2} - \sqrt{2}x + \sqrt{2} - 2)(x - 3) =$
   $a(x^2 - 2x - 1)(x - 3) = a(x^3 - 3x^2 - 2x^2 + 6x - x + 3) \Rightarrow P(x) = a(x^3 - 5x^2 + 5x + 3)$.

   There are many possible solutions, one, when $a = 1$ is: $P(x) = x^3 - 5x^2 + 5x + 3$.

23. With the given zeros $1 + \sqrt{3}$, $1 - \sqrt{3}$ and 1, $P(x) = a(x - (1 + \sqrt{3}))(x - (1 - \sqrt{3}))(x - 1) =$

   $a(x - 1 - \sqrt{3})(x - 1 + \sqrt{3})(x - 1) = a(x^2 - x + \sqrt{3}x - x + 1 - \sqrt{3} - \sqrt{3}x + \sqrt{3} - 3)(x - 1) =$
   $a(x^2 - 2x - 2)(x - 1) = a(x^3 - x^2 - 2x^2 + 2x - 2x + 2) \Rightarrow P(x) = a(x^3 - 3x^2 + 2)$.

   There are many possible solutions, one, when $a = 1$ is: $P(x) = x^3 - 3x^2 + 2$.

24. With the given zeros $-2 + i$, $-2 - i$, 3, and -3, $P(x) = a(x - (-2 + i))(x - (-2 - i))(x - 3)(x - (-3)) =$

   $a(x + 2 - i)(x + 2 + i)(x - 3)(x + 3) = a(x^2 + 2x + ix + 2x + 4 + 2i - ix - 2i - i^2)(x^2 - 9) =$
   $a(x^2 + 4x + 5)(x^2 - 9) = a(x^4 - 9x^2 + 4x^3 - 36x + 5x^2 - 45) \Rightarrow P(x) = a(x^4 + 4x^3 - 4x^2 - 36x - 45)$.

   There are many possible solutions, one, when $a = 1$ is: $P(x) = x^4 + 4x^3 - 4x^2 - 36x - 45$.

25. With the given zeros -1, 2, $3 + 2i$, and the conjugate of $3 + 2i$ or $3 - 2i$ also a zero, the function $P(x)$ is:

   $P(x) = a(x - (3 + 2i))(x - (3 - 2i))(x - (-1))(x - 2) = a(x - 3 - 2i)(x - 3 + 2i)(x + 1)(x - 2) =$
   $a(x^2 - 3x + 2ix - 3x + 9 - 6i - 2ix + 6i - 4i^2)(x^2 - x - 2) = a(x^2 - 6x + 13)(x^2 - x - 2) =$
   $a(x^4 - x^3 - 2x^2 - 6x^3 + 6x^2 + 12x + 13x^2 - 13x - 26) \Rightarrow P(x) = a(x^4 - 7x^3 + 17x^2 - x - 26)$.

   There are many possible solutions, one, when $a = 1$ is: $P(x) = x^4 - 7x^3 + 17x^2 - x - 26$.

26. With the given zeros 2, $3i$, and the conjugate of $3i$ or $-3i$ also a zero, the function $P(x)$ is:

   $P(x) = a(x - (3i))(x - (-3i))(x - 2) = a(x - 3i)(x + 3i)(x - 2) = a(x^2 - 9i^2)(x - 2) =$
   $a(x^2 + 9)(x - 2) \Rightarrow P(x) = a(x^3 - 2x^2 + 9x - 18)$.

   There are many possible solutions, one, when $a = 1$ is: $P(x) = x^3 - 2x^2 + 9x - 18$.

27. With the given zeros -1, $6 - 3i$, and the conjugate of $6 - 3i$ or $6 + 3i$ also a zero, the function $P(x)$ is:

   $P(x) = a(x - (6 + 3i))(x - (6 - 3i))(x - (-1)) = a(x - 6 + 3i)(x - 6 - 3i)(x + 1) =$
   $a(x^2 - 6x - 3ix - 6x + 36 + 18i + 3ix - 18i - 9i^2)(x + 1) = a(x^2 - 12x + 45)(x + 1) =$
   $a(x^3 - 12x^2 + 45x + x^2 - 12x + 45) \Rightarrow a(x^3 - 11x^2 + 33x + 45)$.

   There are many possible solutions, one, when $a = 1$ is: $P(x) = x^3 - 11x^2 + 33x + 45$.

28. With the given zeros 2 (multiplicity 2), $1 + 2i$, and the conjugate of $1 + 2i$ or $1 - 2i$ also a zero, the function

    $P(x)$ is: $P(x) = a(x - (1 + 2i))(x - (1 - 2i))(x - 2)^2 = a(x - 1 - 2i)(x - 1 + 2i)(x - 2)^2 =$

    $a(x^2 - x + 2ix - x + 1 - 2i - 2ix + 2i - 4i^2)(x^2 - 4x + 4) = a(x^2 - 2x + 5)(x^2 - 4x + 4) =$

    $a(x^4 - 4x^3 + 4x^2 - 2x^3 + 8x^2 - 8x + 5x^2 - 20x + 20) \Rightarrow a(x^4 - 6x^3 + 17x^2 - 28x + 20)$.

    There are many possible solutions, one, when $a = 1$ is: $P(x) = x^4 - 6x^3 + 17x^2 - 28x + 20$.

29. With the given zeros -3 (multiplicity 2), $2 + i$, and the conjugate of $2 + i$ or $2 - i$ also a zero, the function

    $P(x)$ is: $P(x) = a(x - (2 + i))(x - (2 - i))(x - (-3))^2 = a(x - 2 - i)(x - 2 + i)(x + 3)^2 =$

    $a(x^2 - 2x + ix - 2x + 4 - 2i - ix + 2i - i^2)(x^2 + 6x + 9) = a(x^2 - 4x + 5)(x^2 + 6x + 9) =$

    $a(x^4 + 6x^3 + 9x^2 - 4x^3 - 24x^2 - 36x + 5x^2 + 30x + 45) \Rightarrow a(x^4 + 2x^3 - 10x^2 - 6x + 45)$.

    There are many possible solutions, one, when $a = 1$ is: $P(x) = x^4 + 2x^3 - 10x^2 - 6x + 45$.

30. With the given zeros 5 (multiplicity 2), $-2i$, and the conjugate of $-2i$ or $2i$ also a zero, the function $P(x)$ is:

    $P(x) = a(x - (-2i))(x - (2i))(x - 5)^2 = a(x + 2i)(x - 2i)(x - 5)^2 =$

    $a(x^2 - 2ix + 2ix - 4i^2)(x^2 - 10x + 25) = a(x^2 + 4)(x^2 - 10x + 25) =$

    $a(x^4 - 10x^3 + 25x^2 + 4x^2 - 40x + 100) \Rightarrow a(x^4 - 10x^3 + 29x^2 - 40x + 100)$.

    There are many possible solutions, one, when $a = 1$ is: $P(x) = x^4 - 10x^3 + 29x^2 - 40x + 100$.

31. First use synthetic division to factor out the first -2. Then use synthetic division to factor out the second -2 from

    the resulting expression: $x^3 + 0x^2 - 7x - 6$.

    $$
    \begin{array}{r|rrrr}
    -2 & 1 & 2 & -7 & -20 & -12 \\
       &   & -2 & 0 & 14 & 12 \\
    \hline
       & 1 & 0 & -7 & -6 & 0
    \end{array}
    \qquad \Rightarrow \qquad
    \begin{array}{r|rrrr}
    -2 & 1 & 0 & -7 & -6 \\
       &   & -2 & 4 & 6 \\
    \hline
       & 1 & -2 & -3 & 0
    \end{array}
    $$

    Finally, completely factor the resulting quadratic expression: $x^2 - 2x - 3 = (x - 3)(x + 1)$. Therefore the

    zeros are: $-2, 3, -1$, and the factored form of $P(x)$ is: $P(x) = (x + 2)^2(x - 3)(x + 1)$.

32. First use synthetic division to factor out the first -1. Secondly use synthetic division to factor out the next -1

    from the resulting expression.

    $$
    \begin{array}{r|rrrrr}
    -1 & 1 & 9 & 33 & 55 & 42 & 12 \\
       &   & -1 & -8 & -25 & -30 & -12 \\
    \hline
       & 1 & 8 & 25 & 30 & 12 & 0
    \end{array}
    \qquad \Rightarrow \qquad
    \begin{array}{r|rrrrr}
    -1 & 1 & 8 & 25 & 30 & 12 \\
       &   & -1 & -7 & -18 & -12 \\
    \hline
       & 1 & 7 & 18 & 12 & 0
    \end{array}
    $$

    Now use synthetic division to factor out the last -1 from the resulting expression.

    $$
    \begin{array}{r|rrrr}
    -1 & 1 & 7 & 18 & 12 \\
       &   & -1 & -6 & -12 \\
    \hline
       & 1 & 6 & 12 & 0
    \end{array}
    $$

    Finally, solve the resulting quadratic equation: $x^2 + 6x + 12 = 0$ using the quadratic formula:

    $x = \dfrac{-6 \pm \sqrt{(6)^2 - 4(1)(12)}}{2(1)} = \dfrac{-6 \pm \sqrt{-12}}{2} = \dfrac{-6}{2} \pm \dfrac{2i\sqrt{3}}{2} = -3 \pm i\sqrt{3}$. Therefore the zeros are:

    $-1, -3 + i\sqrt{3}, -3 - i\sqrt{3}$, and the factored form of $P(x)$ is: $P(x) = (x + 1)^3(x + 3 + i\sqrt{3})(x + 3 - i\sqrt{3})$.

33. Since nonreal complex zeros occur in pairs, there will be 0, 2, or 4 nonreal complex zeros, leading to 5, 3, or 1 real zeros.

34. Since nonreal complex zeros occur in pairs, there will be 0, 2, or 4 nonreal complex zeros, leading to 4, 2, or 0 real zeros.

35. (a) Not Possible, with a zero of $1 + i$, the conjugate of $1 - i$ must also be a zero and this would give 4 zeros to a degree 3 function, which is not possible.

    (b) Possible, since nonreal complex zeros occur in pairs, there can be 4 nonreal complex zeros.

    (c) Not Possible, we can not have a multiplicity of 6 for a degree 5 function.

    (d) Possible, with $1 + 2i$ having a multiplicity of 2, its conjugate $1 - 2i$ must also have a multiplicity of 2. This means the function has a minimum of 4 zeros which is possible.

36. (a) Since $(x)(ax^2) = ax^3$, the degree of $P(x)$ is: 3.

    (b) Since nonreal complex zeros occur in pairs, there will be 0 or 2 nonreal complex zeros, leading to 3 or 1 real zeros. If one of 3 real zeros is of multiplicity 2, then there will be 2 real zeros, therefore there can be 1, 2, or 3 real zeros.

    (c) Since nonreal complex zeros occur in pairs, there will be 0 or 2 nonreal complex zeros.

    (d) From the given function k will be one real zero. If the discriminant $b^2 - 4ac > 0$ there will be 2 additional real zeros and $P(x)$ will have 3 real zeros. If $b^2 - 4ac = 0$ there will be 1 additional repeated real zero and $P(x)$ will have 2 real zeros, one of multiplicity 2. If $b^2 - 4ac < 0$ there will be 2 nonreal complex zeros and $P(x)$ will have 1 real zero and 2 nonreal complex conjugate zeros.

37. See Figure 37.

38. See Figure 38.

39. See Figure 39.

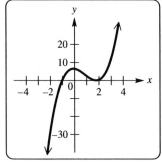

Figure 37

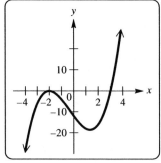

Figure 38

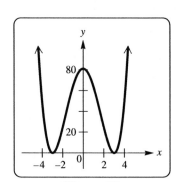

Figure 39

40. See Figure 40.

41. See Figure 41.

42. See Figure 42.

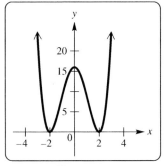

Figure 40

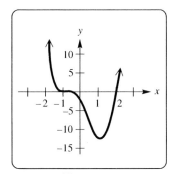

Figure 41

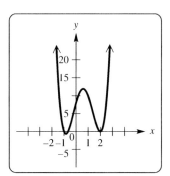

Figure 42

43. See Figure 43.

44. See Figure 44.

45. See Figure 45.

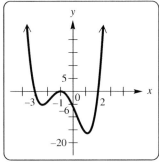

Figure 43

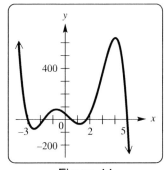

Figure 44

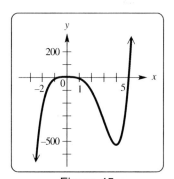

Figure 45

46. See Figure 46.

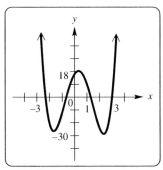

Figure 46

47. (a) Using the Rational Zeros Theorem yields: $\dfrac{p}{q} = \pm \dfrac{1, 2, 5, 10}{1} = \pm 1, \pm 2, \pm 5, \pm 10.$

    (b) A graph (not shown) would indicate that there are no zeros less than -2 or greater than 5.

    (c) First use synthetic division to factor using values discovered in steps (a) and (b).

$$-2\overline{)\begin{array}{rrrr} 1 & -2 & -13 & -10 \\ & -2 & 8 & 10 \\ \hline 1 & -4 & -5 & 0 \end{array}}$$

    Now completely factor the resulting quadratic expression: $x^2 - 4x - 5 \Rightarrow (x - 5)(x + 1)$.

    Therefore the rational zeros are: $-2, -1$, and 5.

    (d) The factored form of $P(x)$ is: $P(x) = (x + 2)(x + 1)(x - 5)$.

48. (a) Using the Rational Zeros Theorem yields: $\dfrac{p}{q} = \pm \dfrac{1, 2, 4, 8}{1} = \pm 1, \pm 2, \pm 4, \pm 8.$

    (b) A graph (not shown) would indicate that there are no zeros less than -4 or greater than 1.

    (c) First use synthetic division to factor using values discovered in steps (a) and (b).

$$-4\overline{)\begin{array}{rrrr} 1 & 5 & 2 & -8 \\ & -4 & -4 & 8 \\ \hline 1 & 1 & -2 & 0 \end{array}}$$

    Now completely factor the resulting quadratic expression: $x^2 + x - 2 \Rightarrow (x + 2)(x - 1)$.

    Therefore the rational zeros are: $-4, -2$, and 1.

    (d) The factored form of $P(x)$ is: $P(x) = (x + 4)(x + 2)(x - 1)$.

49. (a) Using the Rational Zeros Theorem yields: $\dfrac{p}{q} = \pm \dfrac{1, 2, 3, 5, 6, 10, 15, 30}{1} = \pm 1, \pm 2, \pm 3, \pm 5,$ $\pm 6, \pm 10, \pm 15, \pm 30.$

    (b) A graph (not shown) would indicate that there are no zeros less than -5 or greater than 2.

    (c) First use synthetic division to factor using values discovered in steps (a) and (b).

$$-5\overline{)\begin{array}{rrrr} 1 & 6 & -1 & -30 \\ & -5 & -5 & 30 \\ \hline 1 & 1 & -6 & 0 \end{array}}$$

    Now completely factor the resulting quadratic expression: $x^2 + x - 6 \Rightarrow (x + 3)(x - 2)$.

    Therefore the rational zeros are: $-5, -3$, and 2.

    (d) The factored form of $P(x)$ is: $P(x) = (x + 5)(x + 3)(x - 2)$.

50. (a) Using the Rational Zeros Theorem yields: $\dfrac{p}{q} = \pm \dfrac{1, 2, 4, 8}{1} = \pm 1, \pm 2, \pm 4, \pm 8.$

    (b) A graph (not shown) would indicate that there are no zeros less than -2 or greater than 4.

    (c) First use synthetic division to factor using values discovered in steps (a) and (b).

$$-2\overline{)\begin{array}{rrrr} 1 & -1 & -10 & -8 \\ & -2 & 6 & 8 \\ \hline 1 & -3 & -4 & 0 \end{array}}$$

    Now completely factor the resulting quadratic expression: $x^2 - 3x - 4 \Rightarrow (x + 1)(x - 4)$.

    Therefore the rational zeros are: $-2, -1$, and 4.

    (d) The factored form of $P(x)$ is: $P(x) = (x + 2)(x + 1)(x - 4)$.

51. (a) Using the Rational Zeros Theorem yields: $\dfrac{p}{q} = \pm\dfrac{1, 2, 3, 4, 6, 12}{1, 2, 3, 6} = \pm 1, \pm 2, \pm 3, \pm 4, \pm 6, \pm 12,$

$\pm\dfrac{1}{2}, \pm\dfrac{3}{2}, \pm\dfrac{1}{3}, \pm\dfrac{2}{3}, \pm\dfrac{4}{3}, \pm\dfrac{1}{6}.$

(b) A graph (not shown) would indicate that there are no zeros less than -4 or greater than $\dfrac{3}{2}$.

(c) First use synthetic division to factor using values discovered in steps (a) and (b).

$$
\begin{array}{r|rrrr}
-4 & 6 & 17 & -31 & -12 \\
   &   & -24 & 28 & 12 \\
\hline
   & 6 & -7 & -3 & 0
\end{array}
$$

Now completely factor the resulting quadratic expression: $6x^2 - 7x - 3 \Rightarrow (3x + 1)(2x - 3)$.

Therefore the rational zeros are: $-4, -\dfrac{1}{3}, \text{ and } \dfrac{3}{2}$.

(d) The factored form of $P(x)$ is: $P(x) = (x + 4)(3x + 1)(2x - 3)$.

52. (a) Using the Rational Zeros Theorem yields: $\dfrac{p}{q} = \pm\dfrac{1, 2, 4, 8}{1, 3, 5, 15} = \pm 1, \pm 2, \pm 4, \pm 8, \pm\dfrac{1}{3}, \pm\dfrac{2}{3},$

$\pm\dfrac{4}{3}, \pm\dfrac{8}{3}, \pm\dfrac{1}{5}, \pm\dfrac{2}{5}, \pm\dfrac{4}{5}, \pm\dfrac{8}{5}, \pm\dfrac{1}{15}, \pm\dfrac{2}{15}, \pm\dfrac{4}{15}, \pm\dfrac{8}{15}.$

(b) A graph (not shown) would indicate that there are no zeros less than -4 or greater than $\dfrac{1}{3}$.

(c) First use synthetic division to factor using values discovered in steps (a) and (b).

$$
\begin{array}{r|rrrr}
-4 & 15 & 61 & 2 & -8 \\
   &   & -60 & -4 & 8 \\
\hline
   & 15 & 1 & -2 & 0
\end{array}
$$

Now completely factor the resulting quadratic expression: $15x^2 + x - 2 \Rightarrow (5x + 2)(3x - 1)$.

Therefore the rational zeros are: $-4, -\dfrac{2}{5}, \text{ and } \dfrac{1}{3}$.

(d) The factored form of $P(x)$ is: $P(x) = (x + 4)(5x + 2)(3x - 1)$.

53. (a) Using the Rational Zeros Theorem yields: $\dfrac{p}{q} = \pm\dfrac{1, 2, 3, 6}{1, 2, 3, 4, 6, 12} = \pm 1, \pm 2, \pm 3, \pm 6, \pm\dfrac{1}{2}, \pm\dfrac{3}{2},$

$\pm\dfrac{1}{3}, \pm\dfrac{2}{3}, \pm\dfrac{1}{6}, \pm\dfrac{1}{12}, \pm\dfrac{1}{4}, \pm\dfrac{3}{4}.$

(b) A graph (not shown) would indicate that there are no zeros less than $-\dfrac{3}{2}$ or greater than $\dfrac{1}{2}$.

(c) First use synthetic division to factor using values discovered in steps (a) and (b).

$$
\begin{array}{r|rrrr}
-\frac{3}{2} & 12 & 20 & -1 & -6 \\
   &   & -18 & -3 & 6 \\
\hline
   & 12 & 2 & -4 & 0
\end{array}
$$

Now completely factor the resulting quadratic expression: $12x^2 + 2x - 4 \Rightarrow 2(2x - 1)(3x + 2)$.

Therefore the rational zeros are: $-\dfrac{3}{2}, -\dfrac{2}{3}, \text{ and } \dfrac{1}{2}$.

(d) The factored form of $P(x)$ is: $P(x) = (3x + 2)(2x + 3)(2x - 1)$.

54. (a) Using the Rational Zeros Theorem yields: $\dfrac{p}{q} = \pm\dfrac{1, 2, 3, 4, 6, 12}{1, 2, 3, 4, 6, 12} = \pm 1, \pm 2, \pm 3, \pm 4, \pm 6, \pm 12,$

$\pm\dfrac{1}{2}, \pm\dfrac{3}{2}, \pm\dfrac{1}{3}, \pm\dfrac{2}{3}, \pm\dfrac{4}{3}, \pm\dfrac{1}{4}, \pm\dfrac{3}{4}, \pm\dfrac{1}{6}, \pm\dfrac{1}{12}.$

(b) A graph (not shown) would indicate that there are no zeros less than $-\dfrac{3}{2}$ or greater than $-\dfrac{1}{2}$.

(c) First use synthetic division to factor using values discovered in steps (a) and (b).

$$-\frac{3}{2}\,\overline{)\begin{array}{rrrr} 12 & 40 & 41 & 12 \\ & -18 & -33 & -12 \\ \hline 12 & 22 & 8 & 0 \end{array}}$$

Now completely factor the resulting quadratic expression: $12x^2 + 22x + 8 \Rightarrow 2(3x + 4)(2x + 1)$.

Therefore the rational zeros are: $-\dfrac{3}{2}, -\dfrac{4}{3},$ and $-\dfrac{1}{2}$.

(d) The factored form of $P(x)$ is: $P(x) = (2x + 3)(3x + 4)(2x + 1)$.

55. (a) Using the Rational Zeros Theorem yields: $\dfrac{p}{q} = \pm\dfrac{1, 2, 3, 4, 6, 12}{1, 2, 3, 4, 6, 8, 12, 24} = \pm 1, \pm 2, \pm 3, \pm 4, \pm 6,$

$\pm 12, \pm\dfrac{1}{2}, \pm\dfrac{3}{2}, \pm\dfrac{1}{3}, \pm\dfrac{2}{3}, \pm\dfrac{4}{3}, \pm\dfrac{1}{4}, \pm\dfrac{3}{4}, \pm\dfrac{1}{6}, \pm\dfrac{1}{8}, \pm\dfrac{3}{8}, \pm\dfrac{1}{12}, \pm\dfrac{1}{24}.$

(b) A graph (not shown) would indicate that there are no zeros less than $-\dfrac{3}{2}$ or greater than $\dfrac{1}{2}$.

(c) First factor out a 2, $P(x) = 24x^3 + 40x^2 - 2x - 12 \Rightarrow P(x) = 2(12x^3 + 20x^2 - x - 6)$. Now use synthetic division to factor using values discovered in steps (a) and (b).

$$-\frac{3}{2}\,\overline{)\begin{array}{rrrr} 12 & 20 & -1 & -6 \\ & -18 & -3 & 6 \\ \hline 12 & 2 & -4 & 0 \end{array}}$$

Now completely factor the resulting quadratic expression: $12x^2 + 2x - 4 \Rightarrow 2(3x + 2)(2x - 1)$.

Therefore the rational zeros are: $-\dfrac{3}{2}, -\dfrac{2}{3},$ and $\dfrac{1}{2}$.

(d) The factored form of $P(x)$ is: $P(x) = 2(2x + 3)(3x + 2)(2x - 1)$.

56. (a) Using the Rational Zeros Theorem yields: $\dfrac{p}{q} = \pm\dfrac{1, 2, 3, 4, 6, 8, 12, 24}{1, 2, 3, 4, 6, 8, 12, 24} = \pm 1, \pm 2, \pm 3, \pm 4, \pm 6,$

$\pm 8, \pm 12, \pm 24, \pm\dfrac{1}{2}, \pm\dfrac{3}{2}, \pm\dfrac{1}{3}, \pm\dfrac{2}{3}, \pm\dfrac{4}{3}, \pm\dfrac{8}{3}, \pm\dfrac{1}{4}, \pm\dfrac{3}{4}, \pm\dfrac{1}{6}, \pm\dfrac{1}{8}, \pm\dfrac{3}{8}, \pm\dfrac{1}{12}, \pm\dfrac{1}{24}.$

(b) A graph (not shown) would indicate that there are no zeros less than $-\dfrac{3}{2}$ or greater than $-\dfrac{1}{2}$.

(c) First factor out a 2, $P(x) = 24x^3 + 80x^2 + 82x + 24 \Rightarrow P(x) = 2(12x^3 + 40x^2 + 41x + 12)$. Now use synthetic division to factor using values discovered in steps (a) and (b).

$$-\frac{3}{2}\,\overline{)\begin{array}{rrrr} 12 & 40 & 41 & 12 \\ & -18 & -33 & -12 \\ \hline 12 & 22 & 8 & 0 \end{array}}$$

Now completely factor the resulting quadratic expression: $12x^2 + 22x + 8 \Rightarrow 2(3x + 4)(2x + 1)$.

Therefore the rational zeros are: $-\dfrac{3}{2}, -\dfrac{4}{3},$ and $-\dfrac{1}{2}$.

(d) The factored form of $P(x)$ is: $P(x) = 2(2x + 3)(3x + 4)(2x + 1)$.

57. To eliminate the fractions, multiply the function by 2, therefore: $P(x) = 2\left(x^3 + \dfrac{1}{2}x^2 - \dfrac{11}{2}x - 5\right) =$

$2x^3 + x^2 - 11x - 10$. Now use the Rational Zeros Theorem to identify possible zeros: $\dfrac{p}{q} = \pm 10, \pm 5, \pm 1,$

$\pm\dfrac{1}{2}, \pm\dfrac{5}{2}$. Next, from the calculator graph of the equation, we choose to check -1 by synthetic division.

$$
\begin{array}{r|rrrr}
-1 & 2 & 1 & -11 & -10 \\
   &   & -2 & 1 & 10 \\
\hline
   & 2 & -1 & -10 & 0
\end{array}
$$

Now completely factor the resulting quadratic expression: $2x^2 - x - 10 \Rightarrow (2x - 5)(x + 2)$.

Therefore the rational zeros are: $-2, -1$, and $\dfrac{5}{2}$.

58. To eliminate the fractions, multiply the function by 7, therefore: $P(x) = 7\left(\dfrac{10}{7}x^4 - x^3 - 7x^2 + 5x - \dfrac{5}{7}\right) =$

$10x^4 - 7x^3 - 49x^2 + 35x - 5$. Now use the Rational Zeros Theorem to identify possible zeros: $\dfrac{p}{q} = \pm 1, \pm 5,$

$\pm\dfrac{1}{2}, \pm\dfrac{1}{10}, \pm\dfrac{1}{5}$. Next, from the calculator graph of the equation, we choose to check

$\dfrac{1}{2}$ and $\dfrac{1}{5}$ by synthetic division. (The second from the result of the first)

$$
\begin{array}{r|rrrrr}
\frac{1}{2} & 10 & -7 & -49 & 35 & -5 \\
            &   & 5 & -1 & -25 & 5 \\
\hline
            & 10 & -2 & -50 & 10 & 0
\end{array}
\qquad\Rightarrow\qquad
\begin{array}{r|rrrr}
\frac{1}{5} & 10 & -2 & -50 & 10 \\
            &   & 2 & 0 & -10 \\
\hline
            & 10 & 0 & -50 & 0
\end{array}
$$

Now completely factor the resulting quadratic expression: $10x^2 - 50 \Rightarrow 10(x^2 - 5) = 10(x + \sqrt{5})(x - \sqrt{5})$.

Since $\pm\sqrt{5}$ are not rational numbers, the rational zeros are: $\dfrac{1}{2}$ and $\dfrac{1}{5}$.

59. To eliminate the fractions, multiply the function by 12, therefore: $P(x) = 12\left(\dfrac{1}{6}x^4 - \dfrac{11}{12}x^3 + \dfrac{7}{6}x^2 - \dfrac{11}{12}x + 1\right) =$

$2x^4 - 11x^3 + 14x^2 - 11x + 12$. Now use the Rational Zeros Theorem to identify possible zeros: $\dfrac{p}{q} = \pm 1,$

$\pm 2, \pm 3, \pm 4, \pm 6, \pm 12, \pm\dfrac{1}{2}, \pm\dfrac{3}{2}$. Next, from the calculator graph of the equation, we choose to

check $\dfrac{3}{2}$ and 4 by synthetic division. (The second from the result of the first)

$$
\begin{array}{r|rrrrr}
\frac{3}{2} & 2 & -11 & 14 & -11 & 12 \\
            &   & 3 & -12 & 3 & -12 \\
\hline
            & 2 & -8 & 2 & -8 & 0
\end{array}
\qquad\Rightarrow\qquad
\begin{array}{r|rrrr}
4 & 2 & -8 & 2 & -8 \\
  &   & 8 & 0 & 8 \\
\hline
  & 2 & 0 & 2 & 0
\end{array}
$$

Now completely factor the resulting quadratic expression: $2x^2 + 2 \Rightarrow 2(x^2 + 1) \Rightarrow x^2 = -1 \Rightarrow x = \pm i$.

Since $\pm i$ are not real numbers, the rational zeros are: $\dfrac{3}{2}$ and 4.

60. To eliminate the fractions, multiply the function by 6, therefore: $P(x) = 6\left(x^4 - \frac{1}{6}x^3 + \frac{2}{3}x^2 - \frac{1}{6}x - \frac{1}{3}\right) =$

    $6x^4 - x^3 + 4x^2 - x - 2$. Now use the Rational Zeros Theorem to identify possible zeros: $\frac{p}{q} = \pm 1, \pm 2,$

    $\pm\frac{1}{2}, \pm\frac{1}{3}, \pm\frac{2}{3}, \pm\frac{1}{6}$. Next, from the calculator graph of the equation, we choose to check $-\frac{1}{2}$ and $\frac{2}{3}$

    by synthetic division. (The second from the result of the first)

$$
\begin{array}{r|rrrrr}
-\dfrac{1}{2} & 6 & -1 & 4 & -1 & -2 \\
& & -3 & 2 & -3 & 2 \\
\hline
& 6 & -4 & 6 & -4 & 0
\end{array}
\quad \Rightarrow \quad
\begin{array}{r|rrrr}
\dfrac{2}{3} & 6 & -4 & 6 & -4 \\
& & 4 & 0 & 4 \\
\hline
& 6 & 0 & 6 & 0
\end{array}
$$

Now completely factor the resulting quadratic expression: $6x^2 + 6 \Rightarrow 6(x^2 + 1) \Rightarrow x^2 = -1 \Rightarrow x = \pm i$.

Since $\pm i$ are not real numbers, the rational zeros are: $-\frac{1}{2}$ and $\frac{2}{3}$.

61. Because the function $P(x)$ has coefficient signs: $+\ -\ +\ +$, which is 2 sign changes, the function has 2 or 0 possible positive real zeros. Because $P(-x) = 2(-x)^3 - 4(-x)^2 + 2(-x) + 7 = -2x^3 - 4x^2 - 2x + 7$ has coefficient signs: $-\ -\ -\ +$, which is 1 sign change, the function has 1 negative real zero. From graphing the function, $P(x)$ actually has 0 positive and 1 negative real zeros.

62. Because the function $P(x)$ has coefficient signs: $+\ +\ +\ -$, which is 1 sign change, the function has 1 positive real zeros. Because $P(-x) = (-x)^3 + 2(-x)^2 + (-x) +\ - 10 = -x^3 + 2x^2 - x - 10$ has coefficient signs: $-\ +\ -\ -$, which is 2 sign changes, the function has 2 or 0 possible negative real zero. From graphing the function, $P(x)$ actually has 1 positive and 0 negative real zeros.

63. Because the function $P(x)$ has coefficient signs: $+\ +\ +\ -$, which is 1 sign change, the function has 1 positive real zeros. Because $P(-x) = 5(-x)^4 + 3(-x)^2 + 2(-x) - 9 = 5x^4 + 3x^2 - 2x - 9$ has coefficient signs: $+\ +\ -\ -$, which is 1 sign change, the function has 1 negative real zero.

64. Because the function $P(x)$ has coefficient signs: $+\ +\ -\ -\ -$, which is 1 sign change, the function has 1 positive real zero. Because $P(-x) = 3(-x)^4 + 2(-x)^3 - 8(-x)^2 - 10(-x) - 1 =$ $3x^4 - 2x^3 - 8x^2 + 10x - 1$ has coefficient signs: $+\ -\ -\ +\ -$, which is 3 sign changes, the function has 3 or 1 possible negative real zero. From graphing the function, $P(x)$ actually has 1 positive and 1 negative real zero.

65. Because the function $P(x)$ has coefficient signs: $+\ +\ -\ +\ +$, which is 2 sign changes, the function has 2 or 0 possible positive real zeros. Because $P(-x) = (-x)^5 + 3(-x)^4 - (-x)^3 + 2(-x) + 3 =$ $-x^5 + 3x^4 + x^3 - 2x + 3$ has coefficient signs: $-\ +\ +\ -\ +$, which is 3 sign changes, the function has 3 or 1 possible negative real zeros. From graphing the function, $P(x)$ actually has 0 positive and 1 negative real zero.

66. Because the function $P(x)$ has coefficient signs: $+\ -\ +\ -\ +\ +$, which is 4 sign changes, the function has 4, 2 or 0 possible positive real zeros. Because $P(-x) = 2(-x)^5 - (-x)^4 + (-x)^3 - (-x)^2 + (-x) + 5 =$ $-2x^5 - x^4 - x^3 - x^2 - x + 5$ has coefficient signs: $-\ -\ -\ -\ -\ +$, which is 1 sign change, the function has 1 negative real zero. From graphing the function, $P(x)$ actually has 0 positive and 1 negative real zero.

67. Using the Boundedness Theorem we divide synthetically by $x - 2$.

$$
\begin{array}{r|rrrrr}
2 & 1 & -1 & 3 & -8 & 8 \\
  &   & 2 & 2 & 10 & 4 \\
\hline
  & 1 & 1 & 5 & 2 & 12
\end{array}
$$

The result is all nonnegative, therefore no real zero greater than 2.

68. Using the Boundedness Theorem we divide synthetically by $x - 1$.

$$
\begin{array}{r|rrrrrr}
1 & 2 & -1 & 2 & -2 & 4 & -4 \\
  &   & 2 & 1 & 3 & 1 & 5 \\
\hline
  & 2 & 1 & 3 & 1 & 5 & 1
\end{array}
$$

The result is all nonnegative, therefore no real zero greater than 1.

69. Using the Boundedness Theorem we divide synthetically by $x - (-2)$.

$$
\begin{array}{r|rrrrr}
-2 & 1 & 1 & -1 & 0 & 3 \\
   &   & -2 & 2 & -2 & 4 \\
\hline
   & 1 & -1 & 1 & -2 & 7
\end{array}
$$

The result alternates in sign, therefore no real zero less than -2.

70. Using the Boundedness Theorem we divide synthetically by $x - (-1)$.

$$
\begin{array}{r|rrrrrr}
-1 & 1 & 0 & 2 & -2 & 5 & 5 \\
   &   & -1 & 1 & -3 & 5 & -10 \\
\hline
   & 1 & -1 & 3 & -5 & 10 & -5
\end{array}
$$

The result alternates in sign, therefore no real zero less than -1.

71. Using the Boundedness Theorem we divide synthetically by $x - 1$.

$$
\begin{array}{r|rrrrr}
1 & 3 & 2 & -4 & 1 & -1 \\
  &   & 3 & 5 & 1 & 2 \\
\hline
  & 3 & 5 & 1 & 2 & 1
\end{array}
$$

The result is all nonnegative, therefore no real zero greater than 1.

72. Using the Boundedness Theorem we divide synthetically by $x - (-2)$.

$$
\begin{array}{r|rrrrr}
-2 & 3 & 2 & -4 & 1 & -1 \\
   &   & -6 & 8 & -8 & 14 \\
\hline
   & 3 & -4 & 4 & -7 & 13
\end{array}
$$

The result alternates in sign, therefore no real zero less than -2.

73. Using the Boundedness Theorem we divide synthetically by $x - 2$.

$$
\begin{array}{r|rrrrrr}
2 & 1 & 0 & -3 & 0 & 1 & 2 \\
  &   & 2 & 4 & 2 & 4 & 10 \\
\hline
  & 1 & 2 & 1 & 2 & 5 & 12
\end{array}
$$

The result is all nonnegative, therefore no real zero greater than 2.

74. Using the Boundedness Theorem we divide synthetically by $x - (-3)$.

$$
\begin{array}{r|rrrrrr}
-3 & 1 & 0 & -3 & 0 & 1 & 2 \\
   &   & -3 & 9 & -18 & 54 & -165 \\
\hline
   & 1 & -3 & 6 & -18 & 55 & -163
\end{array}
$$

The result alternates in sign, therefore no real zero less than -3.

75. From the graph the function has zeros -6, 2, and 5, the function $P(x)$ is: $P(x) = a(x - (-6))(x - 2)(x - 5) =$

    $a(x + 6)(x - 2)(x - 5) = a(x + 6)(x^2 - 7x + 10) \Rightarrow P(x) = a(x^3 - x^2 - 32x + 60)$. Since $P(0) = 30$, we

    can solve for a: $30 = a((0)^3 - (0)^2 - 42(0) + 60) \Rightarrow 30 = a(0 - 0 - 0 + 60) \Rightarrow 60a = 30 \Rightarrow$

    $a = \dfrac{30}{60} = \dfrac{1}{2}$. Therefore, $P(x) = \dfrac{1}{2}(x^3 - x^2 - 32x + 60) \Rightarrow P(x) = \dfrac{1}{2}x^3 - \dfrac{1}{2}x^2 - 16x + 30$.

76. From the graph the function has zeros -5, and 3 (multiplicity 2, because it is tangent), the function $P(x)$ is:

$$P(x) = a(x - (-5))(x - 3)^2 = a(x + 5)(x - 3)^2 = a(x + 5)(x^2 - 6x + 9) \Rightarrow P(x) = x^3 - x^2 - 21x + 45.$$

Since $P(0) = 9$, we can solve for a: $9 = a((0)^3 - (0)^2 - 21(0) + 45) \Rightarrow 9 = a(0 - 0 - 0 + 45) \Rightarrow$

$45a = 9 \Rightarrow a = \dfrac{9}{45} = \dfrac{1}{5}$. Therefore, $P(x) = \dfrac{1}{5}(x^3 - x^2 - 21x + 45) \Rightarrow P(x) = \dfrac{1}{5}x^3 - \dfrac{1}{5}x^2 - \dfrac{21}{5}x + 9$.

77. (a) Because the function $P(x)$ has coefficient signs: $-\ -\ +\ +$, which is 1 sign change, the function has 1 positive real zero. Because $P(-x) = -2(-x)^4 - (-x)^3 + (-x) + 2 = -2x^4 + x^3 - x + 2$ has coefficient signs: $-\ +\ -\ +$, which is 3 sign changes, the function has 3 or 1 negative real zeros.

(b) By the Rational Zero Theorem the possible rational zeros are: $\dfrac{p}{q} = \pm\dfrac{1,\,2}{1,\,2} = \pm 1,\ \pm 2,\ \pm\dfrac{1}{2}$.

(c) From part (b) and a calculator graph of the equation we choose 1 and -1 to check by synthetic division.

(The second from the result of the first)

$$
\begin{array}{r|rrrrr}
1 & -2 & -1 & 0 & 1 & 2 \\
  &   & -2 & -3 & -3 & -2 \\
\hline
  & -2 & -3 & -3 & -2 & 0
\end{array}
\quad\Rightarrow\quad
\begin{array}{r|rrrr}
-1 & -2 & -3 & -3 & -2 \\
   &   & 2 & 1 & 2 \\
\hline
   & -2 & -1 & -2 & 0
\end{array}
$$

The resulting polynomial $-2x^2 - x - 2$ has no real zeros, so the rational zeros are: 1 and -1.

(d) No other real zeros, the remaining zeros are imaginary.

(e) Using the quadratic equation yields: $x = \dfrac{-(-1) \pm \sqrt{(-1)^2 - 4(-2)(-2)}}{2(-2)} = \dfrac{1 \pm \sqrt{-15}}{-4} \Rightarrow$

$x = -\dfrac{1}{4} + i\dfrac{\sqrt{15}}{4}$ and $-\dfrac{1}{4} - i\dfrac{\sqrt{15}}{4}$.

(f) The rational zeros are the x-intercepts, therefore 1 and -1.

(g) Find $P(0)$: $P(0) = -2(0)^4 - (0)^3 + (0) + 2 \Rightarrow P(0) = 2$. The y-intercept is: 2.

(h)
$$
\begin{array}{r|rrrrr}
4 & -2 & -1 & 0 & 1 & 2 \\
  &   & -8 & -36 & -144 & -572 \\
\hline
  & -2 & -9 & -36 & -143 & -570
\end{array}
$$
Therefore $f(4) = -570$; $(4, -570)$.

(i) $P(x)$ is an negative even-degree polynomial graph, therefore: $\frown$

(j) See Figure 77.

78. (a) Because the function $P(x)$ has coefficient signs: $+\ +\ +\ +\ +$, which is 0 sign changes, the function has

    0 positive real zeros. Because $P(-x) = 4(-x)^5 + 8(-x)^4 + 9(-x)^3 + 27(-x)^2 + 27(-x) =$

    $-4x^5 + 8x^4 - 9x^3 + 27x^2 - 27x$ has coefficient signs: $-\ +\ -\ +\ -$, which is 4 sign changes, the

    function has 4, 2, or 0 negative real zeros.

    (b) By using the Rational Zero Theorem after factoring out an x, the possible rational zeros are:

    $$\frac{p}{q} = \pm\frac{1, 3, 9, 27}{1, 2, 4} = 0, \pm1, \pm3, \pm9, \pm27, \pm\frac{1}{2}, \pm\frac{3}{2}, \pm\frac{9}{2}, \pm\frac{27}{2}, \pm\frac{1}{4}, \pm\frac{3}{4}, \pm\frac{9}{4}, \pm\frac{27}{4}.$$

    (c) Factoring out the x gives 0 as one of the zeros, now from part (b) and a calculator graph of the equation we

    choose $-\dfrac{3}{2}$ (multiplicity 2, because it appears tangent) to check by synthetic division. (The second from the

    result of the first)

    $$-\tfrac{3}{2}\overline{\big)\ \ 4 \quad\ \ 8 \quad\ \ 9 \quad\ 27 \quad\ 27\ }$$
    $$\phantom{-\tfrac{3}{2}\big)\ \ 4}\ \ -6\ \ -3\ \ -9\ \ -27$$
    $$\phantom{-\tfrac{3}{2}\big)}\ \ 4\quad\ \ 2\quad\ \ 6\quad\ 18\quad\ 0 \qquad\Rightarrow\qquad -\tfrac{3}{2}\overline{\big)\ \ 4 \quad\ \ 2 \quad\ \ 6 \quad\ 18\ }$$
    $$\ \ -6\ \ \ 6\ \ -18$$
    $$4\quad -4\quad\ 12\quad\ 0$$

    The resulting polynomial $4x^2 - 4x + 12$ has no real zeros, so the rational zeros are: 0 and $-\dfrac{3}{2}$ (multiplicity 2).

    (d) No other real zeros, the remaining zeros are imaginary.

    (e) Using the quadratic formula after factoring $4(x^2 - x + 3)$ yields: $x = \dfrac{-(-1) \pm \sqrt{(-1)^2 - 4(1)(3)}}{2(1)} =$

    $$\frac{1 \pm \sqrt{-11}}{2} \Rightarrow x = \frac{1}{2} + i\frac{\sqrt{11}}{2} \text{ and } \frac{1}{2} - i\frac{\sqrt{11}}{2}.$$

    (f) The rational zeros are the x-intercepts, therefore 0 and $-\dfrac{3}{2}$.

    (g) Find $P(0)$: $P(0) = 4(0)^5 + 8(0)^4 + 9(0)^3 + 27(0)^2 + 27(0) \Rightarrow P(0) = 0$. The y-intercept is: 0.

    (h)
    $$4\overline{\big)\ \ 4 \quad\ \ 8 \quad\ \ 9 \quad\ \ 27 \quad\ \ 27 \quad\ \ 0\ }$$
    $$\ \ 16\quad\ 96\quad 420\quad 1788\quad 7260$$
    $$\ \ 4\quad\ 24\quad 105\quad 447\quad 1815\quad 7260 \quad \text{Therefore } f(4) = 7260; (4, 7260).$$

    (i) $P(x)$ is an positive odd-degree polynomial graph, therefore:

    (j) See Figure 78.

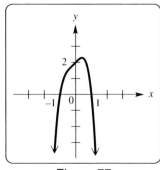

Figure 77

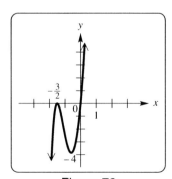

Figure 78

79. (a) Because the function $P(x)$ has coefficient signs: $+ \ - \ -$, which is 1 sign change, the function has 1 positive real zero. Because $P(-x) = 3(-x)^4 - 14(-x)^2 - 5 = 3x^4 - 14x^2 - 5$ has coefficient signs: $+ \ - \ -$, which is 1 sign change, the function has 1 negative real zero.

(b) By the Rational Zeros Theorem the possible rational zeros are: $\dfrac{p}{q} = \pm\dfrac{1,5}{1,3} = \pm 1, \pm 5, \pm\dfrac{1}{3}, \pm\dfrac{5}{3}$.

(c) Set equal to 0 and factor the function: $0 = 3x^4 - 14x^2 - 5 \Rightarrow 0 = (3x^2 + 1)(x^2 - 5)$. The factor $3x^2 + 1$ will yield imaginary zeros and $x^2 - 5$ yields irrational zeros, therefore there are no rational zeros.

(d) From part (c), the factor $x^2 - 5 = 0 \Rightarrow x^2 = 5 \Rightarrow x = \pm\sqrt{5}$, therefore $-\sqrt{5}$ and $\sqrt{5}$ are real zeros.

(e) Using the quadratic formula on the factor $3x^2 + 1 = 0$ found in part (c), yields:
$$x = \frac{-(0) \pm \sqrt{(0)^2 - 4(3)(1)}}{2(3)} = \frac{0 \pm \sqrt{-12}}{6} = \frac{\pm 2i\sqrt{3}}{6} \Rightarrow x = -i\frac{\sqrt{3}}{3} \text{ and } i\frac{\sqrt{3}}{3}.$$

(f) The real zeros are the x-intercepts, therefore $-\sqrt{5}$ and $\sqrt{5}$.

(g) Find $P(0)$: $P(0) = 3(0)^4 - 14(0)^2 - 5 \Rightarrow P(0) = -5$. The y-intercept is: -5.

(h)
```
4) 3    0   -14    0    -5
        12   48   136   544
   ─────────────────────────
     3   12   34   136   539     Therefore f(4) = 539; (4, 539).
```

(i) $P(x)$ is an positive even-degree polynomial graph, therefore: $\smile$

(j) See Figure 79.

80. (a) Because the function $P(x)$ has coefficient signs: $- \ - \ + \ + \ - \ -$, which is 2 sign changes, the function has 2 or 0 positive real zeros. Because $P(-x) = -(-x)^5 - (-x)^4 + 10(-x)^3 + 10(-x)^2 - 9(-x) - 9 = x^5 - x^4 - 10x^3 + 10x^2 + 9x - 9$ has coefficient signs: $+ \ - \ - \ + \ + \ -$, which is 3 sign changes, the function has 3 or 1 negative real zeros.

(b) By using the Rational Zeros Theorem the possible rational zeros are: $\dfrac{p}{q} = \pm\dfrac{1,3,9}{1} = \pm 1, \pm 3, \pm 9$.

(c) From part (b) and a calculator graph of the equation we choose -3, 1, and 3 to check by synthetic division.

```
-3) -1   -1   10   10   -9   -9        1) -1    2    4   -2   -3        3) -1    1    5    3
          3   -6  -12    6    9              -1    1    5    3               -3   -6   -3
   ────────────────────────────── ⇒      ──────────────────────── ⇒      ──────────────────
    -1    2    4   -2   -3    0           -1    1    5    3    0           -1   -2   -1    0
```

Factor the resulting expression to obtain the last 2 zeros. $-x^2 - 2x - 1 = -(x + 1)(x + 1) = 0 \Rightarrow$ $x = -1$ (multiplicity 2). Therefore the rational zeros are: $-3, -1$ (multiplicity 2), 1, and 3.

(d) No other real zeros, all are rational and given in part (c).

(e) No complex zeros, all are rational.

(f) The rational zeros are the x-intercepts, therefore $-3, -1$ (multiplicity 2), 1, and 3.

(g) Find $P(0)$: $P(0) = -(0)^5 - (0)^4 + 10(0)^3 + 10(0)^2 - 9(0) - 9 \Rightarrow P(0) = -9$. The y-intercept is: -9.

(h)
```
4) -1   -1    10    10    -9     -9
        -4   -20   -40  -120   -516
   ──────────────────────────────────
    -1   -5   -10   -30  -129   -525     Therefore f(4) = -525; (4, -525).
```

(i) $P(x)$ is an negative odd-degree polynomial graph, therefore: $\curvearrowright$

(j) See Figure 80.

81. (a) Because the function $P(x)$ has coefficient signs: $- \ + \ - \ + \ -$, which is 4 sign changes, the function has

    4, 2 or 0 positive real zeros. Because $P(-x) = -3(-x)^4 + 22(-x)^3 - 55(-x)^2 + 52(-x) - 12 =$

    $-3x^4 - 22x^3 - 55x^2 - 52x - 12$ has coefficient signs: $- \ - \ - \ - \ -$, which is 0 sign changes, the

    function has 0 negative real zeros.

    (b) By the Rational Zeros Theorem the possible rational zeros are: $\dfrac{p}{q} = \pm\dfrac{1, 2, 3, 4, 6, 12}{1, 3} = \pm 1, \pm 2,$

    $\pm 3, \pm 4, \pm 6, \pm 12, \pm\dfrac{1}{3}, \pm\dfrac{2}{3}, \pm\dfrac{4}{3}.$

    (c) From part (b) and a calculator graph of the equation we choose $\dfrac{1}{3}$ and 3 to check by synthetic division.

$$\begin{array}{r}
\tfrac{1}{3}\,\overline{)\ -3 \quad 22 \quad -55 \quad 52 \quad -12} \\
\underline{\quad\quad -1 \quad\ \ 7 \quad -16 \quad\ \ 12} \\
-3 \quad 21 \quad -48 \quad 36 \quad\ \ 0
\end{array}
\quad\Rightarrow\quad
\begin{array}{r}
3\,\overline{)\ -3 \quad 21 \quad -48 \quad 36} \\
\underline{\quad\quad -9 \quad\ \ 36 \quad -36} \\
-3 \quad 12 \quad -12 \quad\ \ 0
\end{array}$$

    Factor the resulting expression to obtain the last 2 zeros. $-3x^2 + 12x - 12 = -3(x-2)(x-2) = 0$

    $\Rightarrow x = 2$ (multiplicity 2). Therefore the rational zeros are: $\dfrac{1}{3}$, 2 (multiplicity 2), and 3.

    (d) No other real zeros, all are rational.

    (e) No complex zeros, all are rational.

    (f) The rational zeros are the x-intercepts, therefore $\dfrac{1}{3}$, 2 (multiplicity 2), and 3.

    (g) Find $P(0)$: $P(0) = -3(0)^4 + 22(0)^3 - 55(0)^2 + 52(0) - 12 \Rightarrow P(0) = -12$. The y-intercept is: -12.

    (h)
$$\begin{array}{r}
4\,\overline{)\ -3 \quad\ \ 22 \quad -55 \quad\ \ 52 \quad -12} \\
\underline{\quad\quad -12 \quad\ \ 40 \quad -60 \quad -32} \\
-3 \quad\ \ 10 \quad -15 \quad\ \ -8 \quad -44
\end{array}$$
    Therefore $f(4) = -44$; $(4, -44)$.

    (i) $P(x)$ is an negative even-degree polynomial graph, therefore: ⌐⌐

    (j) See Figure 81.

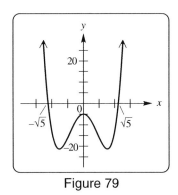

Figure 79

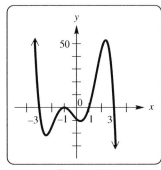

Figure 80

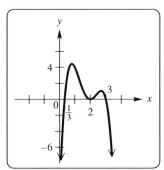

Figure 81

82. The function has 2 irrational zeros of $\pm\sqrt{5}$, which can be approximated to: ± 2.236.

3.8: Polynomial Equations and Inequalities; Further Applications and Models

1. First factor the equation: $7x^3 + x = 0 \Rightarrow x(7x^2 + 1) = 0 \Rightarrow x = 0$ and $7x^2 + 1 = 0$. Now solve:

    $7x^2 + 1 = 0 \Rightarrow 7x^2 = -1 \Rightarrow x^2 = -\dfrac{1}{7} \Rightarrow x = \pm\dfrac{\sqrt{-1}}{\sqrt{7}} \Rightarrow x = \pm\dfrac{i}{\sqrt{7}} \Rightarrow x = \pm\dfrac{i}{\sqrt{7}} \cdot \dfrac{\sqrt{7}}{\sqrt{7}} \Rightarrow$

    $x = \pm\dfrac{i\sqrt{7}}{7}$. The solution set is: $\left\{0, \pm\dfrac{i\sqrt{7}}{7}\right\}$.

2. First factor the equation: $2x^3 - 4x = 0 \Rightarrow 2x(x^2 - 2) = 0 \Rightarrow 2x = 0 \Rightarrow x = 0$ and $x^2 - 2 = 0$.

    Now solve: $x^2 - 2 = 0 \Rightarrow x^2 = 2 \Rightarrow x = \pm\sqrt{2}$. The solution set is: $\{0, \pm\sqrt{2}\}$.

3. First factor the equation: $3x^3 + 2x^2 - 3x - 2 = 0 \Rightarrow x^2(3x + 2) - 1(3x + 2) = 0 \Rightarrow (x^2 - 1)(3x + 2) = 0$

    $\Rightarrow (x + 1)(x - 1)(3x + 2) = 0$. Now solve: $x + 1 = 0 \Rightarrow x = -1; x - 1 = 0 \Rightarrow x = 1; 3x + 2 = 0 \Rightarrow$

    $x = -\dfrac{2}{3}$. The solution set is: $\left\{-1, -\dfrac{2}{3}, 1\right\}$.

4. First factor the equation: $5x^3 - x^2 + 10x - 2 = 0 \Rightarrow x^2(5x - 1) + 2(5x - 1) = 0 \Rightarrow (x^2 + 2)(5x - 1) = 0$

    $\Rightarrow 5x - 1 - 0 \Rightarrow 5x = 1 \Rightarrow x = \dfrac{1}{5}$ and $x^2 + 2 = 0$. Now solve: $x^2 + 2 = 0 \Rightarrow x^2 = -2 \Rightarrow$

    $x = \pm\sqrt{-2} \Rightarrow x = \pm i\sqrt{2}$. The solution set is: $\left\{\dfrac{1}{5}, \pm i\sqrt{2}\right\}$.

5. First factor the equation: $x^4 - 11x^2 + 10 = 0 \Rightarrow (x^2 - 1)(x^2 - 10) = 0 \Rightarrow (x + 1)(x - 1)(x^2 - 10) = 0$.

    Now solve: $x + 1 = 0 \Rightarrow x = -1; x - 1 = 0 \Rightarrow x = 1; x^2 - 10 = 0 \Rightarrow x^2 = 10 \Rightarrow x = \pm\sqrt{10}$.

    The solution set is: $\{\pm 1, \pm\sqrt{10}\}$.

6. First factor the equation: $x^4 + x^2 - 6 = 0 \Rightarrow (x^2 - 2)(x^2 + 3) = 0$. Now solve: $x^2 - 2 = 0 \Rightarrow x^2 = 2 \Rightarrow$

    $x = \pm\sqrt{2}; x^2 + 3 = 0 \Rightarrow x^2 = -3 \Rightarrow x = \pm\sqrt{-3} \Rightarrow x = \pm i\sqrt{3}$.

    The solution set is: $\{\pm\sqrt{2}, \pm i\sqrt{3}\}$.

7. First factor the equation: $4x^4 - 25x^2 + 36 = 0 \Rightarrow (4x^2 - 9)(x^2 - 4) = 0$. Now solve: $4x^2 - 9 = 0 \Rightarrow$

    $4x^2 = 9 \Rightarrow x^2 = \dfrac{9}{4} \Rightarrow x = \pm\sqrt{\dfrac{9}{4}} \Rightarrow x = \pm\dfrac{3}{2}; x^2 - 4 = 0 \Rightarrow x^2 = 4 \Rightarrow x = \pm\sqrt{4} \Rightarrow x = \pm 2$.

    The solution set is: $\left\{\pm\dfrac{3}{2}, \pm 2\right\} = \{-2, -1.5, 1.5, 2\}$. When graphing on the calculator, these values are

    supported as the x-intercepts of $y = 4x^4 - 29x^2 + 25$.

8. First factor the equation: $4x^4 - 29x^2 + 25 = 0 \Rightarrow (4x^2 - 25)(x^2 - 1) = 0$. Now solve: $4x^2 - 25 = 0 \Rightarrow$

    $4x^2 = 25 \Rightarrow x^2 = \dfrac{25}{4} \Rightarrow x = \pm\sqrt{\dfrac{25}{4}} \Rightarrow x = \pm\dfrac{5}{2}; x^2 - 1 = 0 \Rightarrow x^2 = 1 \Rightarrow x = \pm 1$.

    The solution set is: $\left\{\pm\dfrac{5}{2}, \pm 1\right\} = \{-2.5, -1, 1, 2.5\}$. When graphing on the calculator, these values are

    supported as the x-intercepts of $y = 4x^4 - 25x^2 + 36$.

9. First factor the equation: $x^4 - 15x^2 - 16 = 0 \Rightarrow (x^2 - 16)(x^2 + 1) = 0$. Now solve: $x^2 - 16 = 0 \Rightarrow$

    $x^2 = 16 \Rightarrow x = \pm\sqrt{16} \Rightarrow x = \pm 4; x^2 + 1 = 0 \Rightarrow x^2 = -1 \Rightarrow x = \pm\sqrt{-1} \Rightarrow x = \pm i$.

    The solution set is: $\{-4, 4, -i, i\}$. When graphing on the calculator, the real values -4 and 4 are

    supported as the only x-intercepts of $y = x^4 - 15x^2 - 16$.

10. First factor the equation: $9x^4 + 35x^2 - 4 = 0 \Rightarrow (9x^2 - 1)(x^2 + 4) = 0$. Now solve: $9x^2 - 1 = 0 \Rightarrow$

    $9x^2 = 1 \Rightarrow x^2 = \dfrac{1}{9} \Rightarrow x = \pm\sqrt{\dfrac{1}{9}} \Rightarrow x = \pm\dfrac{1}{3}; x^2 + 4 = 0 \Rightarrow x^2 = -4 \Rightarrow x = \pm\sqrt{-4} \Rightarrow$

    $x = \pm 2i$. The solution set is: $\left\{-\dfrac{1}{3}, \dfrac{1}{3}, -2i, 2i\right\}$. When graphing on the calculator, the real values $-\dfrac{1}{3}$ and $\dfrac{1}{3}$

    are supported as the only x-intercepts of $y = 9x^4 + 35x^2 - 4$.

11. First factor the equation: $x^3 - x^2 - 64x + 64 = 0 \Rightarrow x^2(x - 1) - 64(x - 1) = 0 \Rightarrow (x^2 - 64)(x - 1) = 0$

    $\Rightarrow (x + 8)(x - 8)(x - 1) = 0$. Now solve: $x + 8 = 0 \Rightarrow x = -8; x - 8 = 0 \Rightarrow x = 8; x - 1 = 0 \Rightarrow$

    $x = 1$. The solution set is: $\{-8, 1, 8\}$.   When graphing on the calculator, these values are supported as the

    x-intercepts of $y = x^3 - x^2 - 64x + 64$.

12. First factor the equation: $x^3 + 6x^2 - 100x - 600 = 0 \Rightarrow x^2(x + 6) - 100(x + 6) = 0 \Rightarrow$

    $(x^2 - 100)(x + 6) = 0 \Rightarrow (x + 10)(x - 10)(x + 6) = 0$. Now solve: $x + 10 = 0 \Rightarrow x = -10; x - 10 = 0$

    $x = 10; x + 6 = 0 \Rightarrow x = -6$. The solution set is: $\{-10, -6, 10\}$.   When graphing on the calculator, these

    values are supported as the x-intercepts of $y = x^3 + 6x^2 - 100x - 600$.

13. First factor the equation: $-2x^3 - x^2 + 3x = 0 \Rightarrow -x(2x^2 + x - 3) = 0 \Rightarrow -x(2x + 3)(x - 1) = 0$. Now

    solve: $-x = 0 \Rightarrow x = 0; 2x + 3 = 0 \Rightarrow x = -\dfrac{3}{2}; x - 1 = 0 \Rightarrow x = 1$. The solution set is: $\{-1.5, 0, 1\}$.

    When graphing on the calculator, these values are supported as the x-intercepts of $y = -2x^3 - x^2 + 3x$.

14. First factor the equation: $-5x^3 + 13x^2 + 6x = 0 \Rightarrow -x(5x^2 - 13x - 6) = 0 \Rightarrow -x(5x + 2)(x - 3) = 0$.

    Now solve: $-x = 0 \Rightarrow x = 0; 5x + 2 = 0 \Rightarrow x = -\dfrac{2}{5}; x - 3 = 0 \Rightarrow x = 3$. The solution set is:

    $\left\{0, -\dfrac{2}{5}, 3\right\}$.   When graphing on the calculator, these values are supported as the x-intercepts of

    $y = -5x^3 + 13x^2 + 6x$.

15. First factor the equation: $x^3 + x^2 - 7x - 7 = 0 \Rightarrow x^2(x + 1) - 7(x + 1) = 0 \Rightarrow (x^2 - 7)(x + 1) = 0$.

    Now solve: $x + 1 = 0 \Rightarrow x = -1; x^2 - 7 = 0 \Rightarrow x^2 = 7 \Rightarrow x = \pm\sqrt{7}$. The solution set is: $\{-1, \pm\sqrt{7}\}$.

    When graphing on the calculator, these values are supported as the x-intercepts of $y = x^3 + x^2 - 7x - 7$.

16. First factor the equation: $x^3 + 3x^2 - 19x - 57 = 0 \Rightarrow x^2(x + 3) - 19(x + 3) = 0 \Rightarrow (x^2 - 19)(x + 3) = 0$.

    Now solve: $x + 3 = 0 \Rightarrow x = -3; x^2 - 19 = 0 \Rightarrow x^2 = 19 \Rightarrow x = \pm\sqrt{19}$. The solution set is:

    $\{-3, \pm\sqrt{19}\}$.   When graphing on the calculator, these values are supported as the x-intercepts of

    $y = x^3 + 3x^2 - 19x - 57$.

17. First factor the equation by $-x$: $-3x^3 - x^2 + 6x = 0 \Rightarrow -x(3x^2 + x - 6) = 0$. Now one solution is $-x = 0$

    $\Rightarrow x = 0$, and we use the quadratic formula on the remaining factor to find the other solutions:

    $x = \dfrac{-1 \pm \sqrt{1^2 - 4(3)(-6)}}{2(3)} \Rightarrow x = \dfrac{-1 \pm \sqrt{73}}{6}$. The solution set is: $\left\{0, \dfrac{-1 - \sqrt{73}}{6}, \dfrac{-1 + \sqrt{73}}{6}\right\}$. When

    graphing on the calculator, these values are supported as the x-intercepts of $y = -3x^3 - x^2 + 6x$.

18. First factor the equation by $-x$: $-4x^3 - x^2 + 4x = 0 \Rightarrow -x(4x^2 + x - 4) = 0$. Now one solution is $-x = 0 \Rightarrow$

    $x = 0$, and we use the quadratic formula on the remaining factor to find the other solutions:

    $x = \dfrac{-1 \pm \sqrt{1^2 - 4(4)(-4)}}{2(4)} \Rightarrow x = \dfrac{-1 \pm \sqrt{65}}{8}$. The solution set is: $\left\{ 0, \dfrac{-1 - \sqrt{65}}{8}, \dfrac{-1 + \sqrt{65}}{8} \right\}$. When

    graphing on the calculator, these values are supported as the x-intercepts of $y = -4x^3 - x^2 + 4x$.

19. First factor the equation by $3x$: $3x^3 + 3x^2 + 3x = 0 \Rightarrow 3x(x^2 + x + 1) = 0$. Now one solution is $3x = 0 \Rightarrow$

    $x = 0$ and we use the quadratic formula on the remaining factor to find the other solutions:

    $x = \dfrac{-1 \pm \sqrt{1^2 - 4(1)(1)}}{2(1)} \Rightarrow x = \dfrac{-1 \pm i\sqrt{3}}{2}$. The solution set is: $\left\{ 0, -\dfrac{1}{2} - \dfrac{\sqrt{3}}{2}i, -\dfrac{1}{2} + \dfrac{\sqrt{3}}{2}i \right\}$.

    When graphing on the calculator, the real value 0 is the only x-intercept of $y = 3x^3 + 3x^2 + 3x$.

20. First factor the equation by $2x$: $2x^3 + 2x^2 + 12x = 0 \Rightarrow 2x(x^2 + x + 6) = 0$. Now one solution is

    $2x = 0 \Rightarrow x = 0$, and we use the quadratic formula on the remaining factor to find the other solutions:

    $x = \dfrac{-1 \pm \sqrt{1^2 - 4(1)(6)}}{2(1)} \Rightarrow x = \dfrac{-1 \pm i\sqrt{23}}{2}$. The solution set is: $\left\{ 0, -\dfrac{1}{2} - \dfrac{\sqrt{23}}{2}i, -\dfrac{1}{2} + \dfrac{\sqrt{23}}{2}i \right\}$.

    When graphing on the calculator, the real value 0 is the only x-intercept of $y = 2x^3 + 2x^2 + 12x$.

21. First factor the equation: $x^4 + 17x^2 + 16 = 0 \Rightarrow (x^2 + 16)(x^2 + 1) = 0$. Now solve: $x^2 + 16 = 0 \Rightarrow$

    $x^2 = -16 \Rightarrow x = \pm\sqrt{-16} \Rightarrow x = \pm 4i$; $x^2 + 1 = 0 \Rightarrow x^2 = -1 \Rightarrow x = \pm\sqrt{-1} \Rightarrow x = \pm i$.

    The solution set is: $\{-4i, -i, i, 4i\}$. There are no real solutions which, when graphed on the calculator, is

    supported with no x-intercepts for $y = x^4 + 17x^2 + 16$.

22. First factor the equation: $36x^4 + 85x^2 + 9 = 0 \Rightarrow (4x^2 + 9)(9x^2 + 1) = 0$. Now solve: $4x^2 + 9 = 0 \Rightarrow$

    $4x^2 = -9 \Rightarrow x^2 = -\dfrac{9}{4} \Rightarrow x = \pm\sqrt{-\dfrac{9}{4}} \Rightarrow x = \pm\dfrac{3}{2}i$; $9x^2 + 1 = 0 \Rightarrow 9x^2 = -1 \Rightarrow x^2 = -\dfrac{1}{9} \Rightarrow$

    $x = \pm\sqrt{-\dfrac{1}{9}} \Rightarrow x = \pm\dfrac{1}{3}i$. The solution set is: $\left\{ -\dfrac{3}{2}i, -\dfrac{1}{3}i, \dfrac{1}{3}i, \dfrac{3}{2}i \right\}$. There are no real solutions which,

    when graphed on the calculator, is supported with no x-intercepts for $y = 36x^4 + 85x^2 + 9$.

23. First factor the equation: $x^6 + 19x^3 - 216 = 0 \Rightarrow (x^3 - 8)(x^3 + 27) = 0 \Rightarrow$

    $(x - 2)(x^2 + 2x + 4)(x + 3)(x^2 - 3x + 9) = 0$. Now solve: $x - 2 = 0 \Rightarrow x = 2$; $x + 3 = 0 \Rightarrow x = -3$;

    and we use the quadratic formula on the remaining factors to find the other solutions:

    $x = \dfrac{-2 \pm \sqrt{2^2 - 4(1)(4)}}{2(1)} = \dfrac{-2 \pm 2i\sqrt{3}}{2} \Rightarrow -1 \pm i\sqrt{3}$; $x = \dfrac{-(-3) \pm \sqrt{(-3)^2 - 4(1)(9)}}{2(1)} =$

    $\dfrac{3 \pm 3i\sqrt{3}}{2} \Rightarrow x = \dfrac{3}{2} \pm \dfrac{3\sqrt{3}}{2}i$. The solution set is: $\left\{ -3, 2, -1 - i\sqrt{3}, -1 + i\sqrt{3}, \dfrac{3}{2} - \dfrac{3\sqrt{3}}{2}i, \dfrac{3}{2} + \dfrac{3\sqrt{3}}{2}i \right\}$.

    When graphing on the calculator, the real values -3 and 2 are supported as the only x-intercepts of

    $y = x^6 + 19x^3 - 216$.

24. First factor the equation: $8x^6 + 7x^3 - 1 = 0 \Rightarrow (8x^3 - 1)(x^3 + 1) = 0 \Rightarrow 8\left(x^3 - \frac{1}{8}\right)(x^3 + 1) = 0 \Rightarrow$

$8\left(x - \frac{1}{2}\right)\left(x^2 + \frac{1}{2}x + \frac{1}{4}\right)(x + 1)(x^2 - x + 1) = 0.$ Now solve: $x - \frac{1}{2} = 0 \Rightarrow x = \frac{1}{2}; x + 1 = 0 \Rightarrow$

$x = -1;$ and we use the quadratic formula on the remaining factors to find the other solutions:

$x = \dfrac{-\frac{1}{2} \pm \sqrt{(\frac{1}{2})^2 - 4(1)(\frac{1}{4})}}{2(1)} = \dfrac{-\frac{1}{2} \pm \sqrt{-\frac{3}{4}}}{2} = -\dfrac{1}{2} \pm \dfrac{i\frac{\sqrt{3}}{2}}{2} \Rightarrow -\dfrac{1}{4} \pm \dfrac{\sqrt{3}}{4}i.$

$x = \dfrac{-(-1) \pm \sqrt{(-1)^2 - 4(1)(1)}}{2(1)} = \dfrac{1 \pm \sqrt{-3}}{2} \Rightarrow x = \dfrac{1}{2} \pm \dfrac{\sqrt{3}}{2}i.$

The solution set is: $\left\{-1, \frac{1}{2}, -\frac{1}{4} - \frac{\sqrt{3}}{4}i, -\frac{1}{4} + \frac{\sqrt{3}}{4}i, \frac{1}{2} - \frac{\sqrt{3}}{2}i, \frac{1}{2} + \frac{\sqrt{3}}{2}i\right\}.$ When graphing on the calculator,

the real values -1 and $\dfrac{1}{2}$ are supported as the only x-intercepts of $y = 8x^6 + 7x^3 - 1.$

25. See Figure 25. From the graph the following is found:

   (a) $P(x) = 0$: $\{-2, 1, 4\}$.

   (b) $P(x) < 0$: $(-\infty, -2) \cup (1, 4)$.

   (c) $P(x) > 0$: $(-2, 1) \cup (4, \infty)$.

26. See Figure 26. From the graph the following is found:

   (a) $P(x) = 0$: $\{-5, -2, 3\}$.

   (b) $P(x) < 0$: $(-\infty, -5) \cup (-2, 3)$.

   (c) $P(x) > 0$: $(-5, -2) \cup (3, \infty)$.

27. See Figure 27. From the graph the following is found:

   (a) $P(x) = 0$: $\{-2.5, 1, 3(\text{multiplicity } 2)\}$.

   (b) $P(x) < 0$: $(-2.5, 1)$.

   (c) $P(x) > 0$: $(-\infty, -2.5) \cup (1, 3) \cup (3, \infty)$.

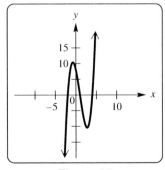

Figure 25

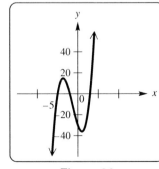

Figure 26

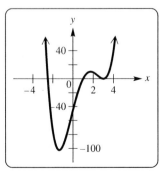

Figure 27

28. See Figure 28. From the graph the following is found:

    (a) $P(x) = 0$: $\{-5 \text{ (multiplicity 2)}, -.75, 4\}$.

    (b) $P(x) < 0$: $(-.75, 4)$.

    (c) $P(x) > 0$: $(-\infty, -5) \cup (-5, -.75) \cup (4, \infty)$.

29. See Figure 29. From the graph the following is found:

    (a) $P(x) = 0$: $\{-3 \text{ (multplicity 2)}, 0, 2\}$.

    (b) $P(x) \geq 0$: $\{-3\} \cup [0, 2]$.

    (c) $P(x) \leq 0$: $(-\infty, 0] \cup [2, \infty)$.

30. See Figure 30. From the graph the following is found:

    (a) $P(x) = 0$: $\{-2, 0 \text{ (multiplicity 2)}, 4\}$.

    (b) $P(x) \geq 0$: $[-2, 4]$.

    (c) $P(x) \leq 0$: $(-\infty, -2] \cup \{0\} \cup [4, \infty)$.

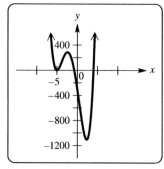

Figure 28

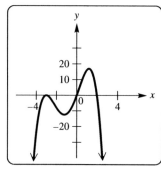

Figure 29

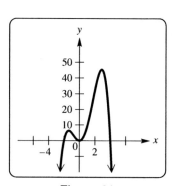

Figure 30

31. From the calculator graph of $y_1 = .86x^3 - 5.24x^2 + 3.55x + 7.84$ the solution is: $\{-.88, 2.12, 4.86\}$.

32. From the calculator graph of $y_1 = -2.47x^3 - 6.58x^2 - 3.33x + .14$ the solution is: $\{-1.96, -.74, .04\}$.

33. From the calculator graph of $y_1 = -\sqrt{7}x^3 + \sqrt{5}x^2 + \sqrt{17}$ the solution is: $\{1.52\}$.

34. From the calculator graph of $y_1 = \sqrt{10}x^3 - \sqrt{11}x - \sqrt{8}$ the solution is: $\{1.31\}$.

35. From the calculator graph of $2.45x^4 - 3.22x^3 = -.47x^2 + 6.54x + 3 \Rightarrow$

    $y_1 = 2.45x^4 - 3.22x^3 + .47x^2 - 6.54x - 3$ the solution is: $\{-.40, 2.02\}$.

36. From the calculator graph of $\sqrt{17}x^4 - \sqrt{22}x^2 = -1 \Rightarrow y_1 = \sqrt{17}x^4 - \sqrt{22}x^2 + 1$ the solution is:

    $\{\pm.92, \pm.53\}$.

37. $x^2 = -1 \Rightarrow x = \sqrt{-1} \Rightarrow x = \pm i \Rightarrow \{-i, i\}$.

38. $x^2 = -4 \Rightarrow x = \sqrt{-4} \Rightarrow x = \pm 2i \Rightarrow \{-2i, 2i\}$.

39. First factor the equation: $x^3 = -1 \Rightarrow x^3 + 1 = 0 \Rightarrow (x + 1)(x^2 - x + 1) = 0$. Now one solution is

    $x + 1 = 0 \Rightarrow x = -1$ and we use the quadratic formula on the remaining factor to find the other solutions:

    $x = \dfrac{-(-1) \pm \sqrt{(-1)^2 - 4(1)(1)}}{2(1)} \Rightarrow x = \dfrac{1 \pm i\sqrt{3}}{2}$. The solution set is: $x = \left\{-1, \dfrac{1}{2} - \dfrac{\sqrt{3}}{2}i, \dfrac{1}{2} + \dfrac{\sqrt{3}}{2}i\right\}$.

40. First factor the equation: $x^3 = -8 \Rightarrow x^3 + 8 = 0 \Rightarrow (x + 2)(x^2 - 2x + 4) = 0$. Now one solution is

    $x + 2 = 0 \Rightarrow x = -2$ and we use the quadratic formula on the remaining factor to find the other solutions:

    $x = \dfrac{-(-2) \pm \sqrt{(-2)^2 - 4(1)(4)}}{2(1)} \Rightarrow x = \dfrac{2 \pm \sqrt{-12}}{2} = \dfrac{2 \pm 2i\sqrt{3}}{2} = 1 \pm i\sqrt{3}$. The solution set is:

    $x = \{-2, 1 - \sqrt{3}i, 1 + \sqrt{3}i\}$.

41. First factor the equation: $x^3 = 27 \Rightarrow x^3 - 27 = 0 \Rightarrow (x - 3)(x^2 + 3x + 9) = 0$. Now one solution is

    $x - 3 = 0 \Rightarrow x = 3$ and we use the quadratic formula on the remaining factor to find the other solutions:

    $x = \dfrac{-3 \pm \sqrt{3^2 - 4(1)(9)}}{2(1)} = \dfrac{-3 \pm \sqrt{-27}}{2} \Rightarrow x = \dfrac{-3 \pm 3i\sqrt{3}}{2}$. The solution set is:

    $x = \left\{3, -\dfrac{3}{2} - \dfrac{3\sqrt{3}}{2}i, -\dfrac{3}{2} + \dfrac{3\sqrt{3}}{2}i\right\}$.

42. First factor the equation: $x^3 = 64 \Rightarrow x^3 - 64 = 0 \Rightarrow (x - 4)(x^2 + 4x + 16) = 0$. Now one solution is

    $x - 4 = 0 \Rightarrow x = 4$ and we use the quadratic formula on the remaining factor to find the other solutions:

    $x = \dfrac{-4 \pm \sqrt{4^2 - 4(1)(16)}}{2(1)} = \dfrac{-4 \pm \sqrt{-48}}{2} \Rightarrow x = \dfrac{-4 \pm 4i\sqrt{3}}{2} = -2 \pm 2i\sqrt{3}$. The solution set is:

    $x = \{4, -2 - 2i\sqrt{3}, -2 + 2i\sqrt{3}\}$.

43. First factor the equation: $x^4 = 16 \Rightarrow x^4 - 16 = 0 \Rightarrow (x^2 - 4)(x^2 + 4) = 0 \Rightarrow (x - 2)(x + 2)(x^2 + 4) = 0$.

    Now two solutions are $x - 2 = 0 \Rightarrow x = 2, x + 2 = 0 \Rightarrow x = -2$, and we solve the remaining factor for the

    last two solutions: $x^2 + 4 = 0 \Rightarrow x^2 = -4 \Rightarrow x = \pm\sqrt{-4} \Rightarrow x = \pm 2i$. The solution set is:

    $x = \{-2, 2, -2i, 2i\}$.

44. First factor the equation: $x^4 = 81 \Rightarrow x^4 - 81 = 0 \Rightarrow (x^2 - 9)(x^2 + 9) = 0 \Rightarrow (x - 3)(x + 3)(x^2 + 9) = 0$.

    Now two solutions are $x - 3 = 0 \Rightarrow x = 3, x + 3 = 0 \Rightarrow x = -3$, and we solve the remaining factor for the

    last two solutions: $x^2 + 9 = 0 \Rightarrow x^2 = -9 \Rightarrow x = \pm\sqrt{-9} \Rightarrow x = \pm 3i$. The solution set is:

    $x = \{-3, 3, -3i, 3i\}$.

45. First factor the equation: $x^3 = -64 \Rightarrow x^3 + 64 = 0 \Rightarrow (x + 4)(x^2 - 4x + 16) = 0$. Now one solution is

    $x + 4 = 0 \Rightarrow x = -4$ and we use the quadratic formula on the remaining factor to find the other solutions:

    $x = \dfrac{-(-4) \pm \sqrt{(-4)^2 - 4(1)(16)}}{2(1)} = \dfrac{4 \pm \sqrt{-48}}{2} \Rightarrow x = \dfrac{4 \pm 4i\sqrt{3}}{2} = 2 \pm 2i\sqrt{3}$. The solution set is:

    $x = \{-4, 2 - 2i\sqrt{3}, 2 + 2i\sqrt{3}\}$.

46. First factor the equation: $x^3 = -27 \Rightarrow x^3 + 27 = 0 \Rightarrow (x + 3)(x^2 - 3x + 9) = 0$. Now one solution is

    $x + 3 = 0 \Rightarrow x = -3$ and we use the quadratic formula on the remaining factor to find the other solutions:

    $x = \dfrac{-(-3) \pm \sqrt{(-3)^2 - 4(1)(9)}}{2(1)} = \dfrac{3 \pm \sqrt{-27}}{2} \Rightarrow x = \dfrac{3 \pm 3i\sqrt{3}}{2}$. The solution set is:

    $x = \left\{-3, \dfrac{3}{2} - \dfrac{3\sqrt{3}}{2}i, \dfrac{3}{2} + \dfrac{3\sqrt{3}}{2}i\right\}$.

47. $x^2 = -18 \Rightarrow x = \pm\sqrt{-18} \Rightarrow x = \pm 3i\sqrt{2} \Rightarrow \{-3i\sqrt{2}, 3i\sqrt{2}\}$.

48. $x^2 = -52 \Rightarrow x = \pm\sqrt{-52} \Rightarrow x = \pm 2i\sqrt{13} \Rightarrow \{-2i\sqrt{13}, 2i\sqrt{13}\}$.

49. (a) First find $f(x)$ for $d = .8$: $f(x) = \frac{\pi}{3}x^3 - 5\pi x^2 + \frac{500\pi(.8)}{3} \Rightarrow f(x) = \frac{\pi}{3}x^3 - 5\pi x^2 + \frac{400\pi}{3}$. Now graph

this equation on a graphing calculator and we find the smallest positive zero is at:

$x \approx 7.1286$ or $x \approx 7.13$ cm; the ball floats partly above the surface.

(b) First find $f(x)$ for $d = 2.7$: $f(x) = \frac{\pi}{3}x^3 - 5\pi x^2 + \frac{500\pi(2.7)}{3} \Rightarrow f(x) = \frac{\pi}{3}x^3 - 5\pi x^2 + \frac{1350\pi}{3}$. Now

graph this equation on a graphing calculator and we find the graph has no positive zero; the sphere sinks

below the surface because it is more dense than water.

(c) First find $f(x)$ for $d = 1$: $f(x) = \frac{\pi}{3}x^3 - 5\pi x^2 + \frac{500\pi(1)}{3} \Rightarrow f(x) = \frac{\pi}{3}x^3 - 5\pi x^2 + \frac{500\pi}{3}$. Now graph this

equation on a graphing calculator and we find the smallest positive zero is at: $x = 10$ cm; the balloon

floats even with the surface.

50. (a) See figure 50.

(b) By using the *maximum* function on the calculator we find that the greatest concentration is at about 17 hours.

(c) From the graph, the river is polluted from 11.4 hours to 21.2 hours.

51. First find $f(x)$ for $d = .6$: $f(x) = \frac{\pi}{3}x^3 - 10\pi x^2 + \frac{4000\pi(.6)}{3} \Rightarrow f(x) = \frac{\pi}{3}x^3 - 10\pi x^2 + 800\pi$. Now graph

this equation on a graphing calculator and we find the smallest positive zero is at: $x \approx 11.341378$ or $x \approx 11.34$ cm.

52. First find $f(x)$ for $d = .55$: $f(x) = \frac{\pi}{3}x^3 - 5\pi x^2 + \frac{500\pi(.55)}{3} \Rightarrow f(x) = \frac{\pi}{3}x^3 - 5\pi x^2 + \frac{275\pi}{3}$. Now graph this

equation on a graphing calculator and we find the smallest positive zero is at: $x \approx 5.3338$ or $x \approx 5.33$ cm.

53. (a) Since the box has sides $x > 0$ and since the box must have some width $12 - 2x > 0 \Rightarrow -2x > -12 \Rightarrow$

$x < 6$. Therefore the restrictions are: $0 < x < 6$.

(b) Since $V = lwh$, length $= 18 - 2x$, and width $= 12 - 2x$; the function is: $V(x) = (18 - 2x)(12 - 2x)(x)$

$\Rightarrow V(x) = x(4x^2 - 60x + 216) \Rightarrow V(x) = 4x^3 - 60x^2 + 216x$.

(c) By graphing the function from (b) on a graphing calculator, we can find the maximum volume at: $x \approx 2.35$,

which produces a volume of: $V(2.35) \approx 4(2.35)^3 - 60(2.35)^2 + 216(2.35) \Rightarrow V(2.35) \approx 228.16$ in³.

(d) By graphing the function from (b) on a graphing calculator, we can find the volume to be greater than

80 in³ for the interval: $.42 < x < 5$.

54. (a) Since the rain gutter has sides $x > 0$ and since the rain gutter must have some width $20 - 2x > 0 \Rightarrow$

$-2x = -20 \Rightarrow x < 10$. Therefore the restrictions are: $0 < x < 10$.

(b) Since $A = lw$, length $= 20 - 2x$, and width $= x$; the function is: $A(x) = (20 - 2x)(x)$

$\Rightarrow A(x) = x(20 - 2x) \Rightarrow A(x) = -2x^2 + 20x$.

(c) By graphing the function from (b) on a graphing calculator, we can find the maximum area at: $x = 5$,

which produces an area of: $A(5) = -2(5)^2 + 20(5) \Rightarrow A(5) = 50$ in².

(d) By graphing the function from (b) on a graphing calculator, we can find the area to be less than

40 in² for the interval: $0 < x < 2.76$ or $7.24 < x < 10$.

55. By graphing the equation on a graphing calculator, the smallest positive zero which is less than 10 is: $x \approx 2.61$ in.

56. Since $A = lw$, $y = 9 - x^2$, length $= 2x$, and width $= y$; a function for area would be: $A(x) = 2x(9 - x^2) \Rightarrow$

$A(x) = -2x^3 + 18x$. By graphing the equation on a graphing calculator, we can find the maximum area at:

$x \approx 1.73$ which is $x = \sqrt{3}$.

57. (a) A length 1 inch less than the hypotenuse (x) is: $x - 1$.

(b) Using the Pythagorean Theorem yields: $l^2 + (x - 1)^2 = x^2 \Rightarrow l^2 = x^2 - (x - 1)^2 \Rightarrow$

$l = \sqrt{x^2 - (x - 1)^2}$ or $l = \sqrt{2x - 1}$.

(c) Since $A = \dfrac{1}{2}bh = 84$, the equation is: $\dfrac{1}{2}(\sqrt{2x - 1})(x - 1) = 84 \Rightarrow (\sqrt{2x - 1})(x - 1) = 168 \Rightarrow$

$[(\sqrt{2x - 1})(x - 1)]^2 = 168^2 \Rightarrow (2x - 1)(x^2 - 2x + 1) = 28{,}224 \Rightarrow 2x^3 - 5x^2 + 4x - 28{,}225 = 0$.

(d) By graphing the equation from (c) on a graphing calculator we get an x-intercept: 25. Therefore, the

hypotenuse is: $x = 25$ in., one leg is: $x - 1 = 24$ in., and the other leg is: $\sqrt{2x - 1} = \sqrt{49} = 7$ in.

58. The volume of the tank will be $V = V_{cyl} + V_{sph} \Rightarrow V(x) = \pi x^2(12) + \dfrac{4}{3}\pi x^3$. If $\dfrac{4}{3}x^3\pi + 12x^2\pi = 144\pi$ then

$\dfrac{4}{3}x^3 + 12x^2 - 144 = 0$. By graphing this equation on the graphing calculator we find $x \approx 3$ ft.

59. (a) Since $l = 11 - 2x$, $w = 8.5 - 2x$, $h = x$, and $V = lwh$ we get an equation: $V = (11 - 2x)(8.5 - 2x)(x) \Rightarrow$

$V = (93.5 - 17x - 22x + 4x^2)(x) \Rightarrow V = 4x^3 - 39x^2 + 93.5x$. Because all sides of the box must be

positive, the restrictions on this equation will be: $8.5 - 2x > 0 \Rightarrow -2x > -8.5 \Rightarrow x < 4.25 \Rightarrow$

$0 < x < 4.25$. Now using the table feature on the graphing calculator yields a maximum value within the

restrictions of: $x \approx 1.59$ and this gives a volume of: $4(1.59)^3 - 39(1.59)^2 + 93.5(1.59) \approx 66.15$ in^3.

(b) Using the table feature on the graphing calculator for the equation: $V = 4x^3 - 39x^2 + 93x$ yields a volume

greater than 40 in^3 when x is in the following range: .54 in $< x < 2.92$ in.

60. (a) See Figure 60a.

(b) From the regression feature the best-fitting linear function is: $y \approx 33.93x + 113.4$. See Figure 60b.

(c) From the regression feature the best-fitting cubic function is: $y \approx -.0032x^3 + .4245x^2 + 16.64x + 323.1$.

See Figure 60c.

(d) When $x = 43$, the linear function yields: $y \approx 1572$ ft.; and the cubic function yields: $y \approx 1569$ ft.

(e) The cubic function is slightly better because only one data point is not on the curve.

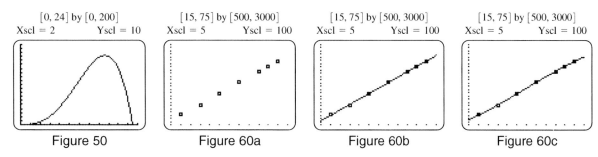

[0, 24] by [0, 200]	[15, 75] by [500, 3000]	[15, 75] by [500, 3000]	[15, 75] by [500, 3000]
Xscl = 2 Yscl = 10	Xscl = 5 Yscl = 100	Xscl = 5 Yscl = 100	Xscl = 5 Yscl = 100

Figure 50   Figure 60a   Figure 60b   Figure 60c

61. (a) See Figure 61a.

    (b) From the regression feature the best-fitting quadratic function is: $C(x) \approx .0035x^2 - .49x + 22$.

    See Figure 61b.

    (c) From the regression feature the best-fitting cubic function is: $C(x) \approx -.000068x^3 + .00987x^2 - .653x + 23$.

    See Figure 61c.

    (d) From the graphs of each, the cubic function is a slightly better fit.

    (e) Using the cubic function and graph, $C(x) > 10$ when $0 \le x < 31.92$.

62. From a calculator graph of this function the zeros are: -6.01, which has no significance, $2.15 \approx 2$ and $11.7 \approx 12$.

    The zeros 2.25 and 11.7 indicates that during February and December (or late November), the average

    temperature is 0°F.

63. From a calculator graph of $y_1 = -.006x^4 + .140x^3 - .053x^2 + 1.79x$ and $y_2 = 10$, the intersection is:

    $x \approx 3.367$ or approximately 3.4 seconds.

64. (a) See Figure 64. From the graph the two real roots are: $x = -1, 1$.

    (b) If $x^8 = 1$ then $(i)^8 = (i^4)^2 = (1)^2 = 1$.

    (c) The conjugate of i or $-i$ must also be a root. Analytically this is: $(-i)^8 = [(-i^4)]^2 = (1)^2 = 1$.

    (d) If $x = \dfrac{\sqrt{2}}{2} + \dfrac{\sqrt{2}}{2}i$, then it's conjugate $x = \dfrac{\sqrt{2}}{2} - \dfrac{\sqrt{2}}{2}i$ is a root.

    If $x = -\dfrac{\sqrt{2}}{2} + \dfrac{\sqrt{2}}{2}i$, then it's conjugate $x = -\dfrac{\sqrt{2}}{2} - \dfrac{\sqrt{2}}{2}i$ is a root.

    (e) From a − d, the eight roots are: $-1, 1, -i, i, \dfrac{\sqrt{2}}{2} + \dfrac{\sqrt{2}}{2}i, \dfrac{\sqrt{2}}{2} - \dfrac{\sqrt{2}}{2}i, -\dfrac{\sqrt{2}}{2} + \dfrac{\sqrt{2}}{2}i$, and $-\dfrac{\sqrt{2}}{2} - \dfrac{\sqrt{2}}{2}i$.

[−5, 70] by [−5, 25]	[−5, 70] by [−5, 25]	[−5, 70] by [−5, 25]	[−4.7, 4.7] by [−3.1, 3.1]
Xscl = 5 Yscl = 5	Xscl = 5 Yscl = 5	Xscl = 5 Yscl = 5	Xscl = 1 Yscl = 1

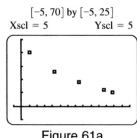
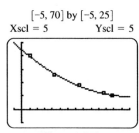
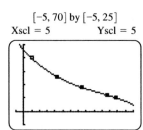
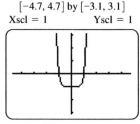

Figure 61a	Figure 61b	Figure 61c	Figure 64

Reviewing Basic Concepts (Sections 3.6—3.8)

1. $P(3) = 2(3)^4 - 7(3)^3 + 29(3) - 30 \Rightarrow P(3) = 162 - 189 + 87 - 30 \Rightarrow P(3) = 30$.

2. Since $P(2) = 2(2)^4 - 7(2)^3 + 29(2) - 30 \Rightarrow P(2) = 32 - 56 + 58 - 30 \Rightarrow P(2) = 4$, $P(2)$ is not a zero.

3. Since $2 + i$ is a zero, it's conjugate $2 - i$ is also a zero. First use Synthetic division to factor out these two zeros from the equation and the resulting equation.

$$
\begin{array}{r|rrrrr}
2+i) & 2 & -7 & 0 & 29 & -30 \\
& & 4+2i & -8+i & -17-6i & 30 \\
\hline
& 2 & -3+2i & -8+i & 12-6i & 0
\end{array}
\quad\Rightarrow\quad
\begin{array}{r|rrrr}
2-i) & 2 & -3+2i & -8+i & 12-6i \\
& & 4-2i & 2-i & -12+6i \\
\hline
& 2 & 1 & -6 & 0
\end{array}
$$

Now completely factor the resulting expression: $2x^2 + x - 6 = (2x - 3)(x + 2)$.

The linear factors are: $P(x) = (2x - 3)(x + 2)(x - 2 - i)(x - 2 + i)$.

4. Since the zeros are $\dfrac{3}{2}$, i, and the conjugate of i or $-i$, the function is: $P(x) = a\left(x - \dfrac{3}{2}\right)(x - i)(x + i) \Rightarrow$

$P(x) = a\left(x - \dfrac{3}{2}\right)(x^2 + 1) \Rightarrow P(x) = a\left(x^3 - \dfrac{3}{2}x^2 + x - \dfrac{3}{2}\right)$. Now, if $P(x) = 15$, then

$15 = a\left(3^3 - \dfrac{3}{2}(3)^2 + 3 - \dfrac{3}{2}\right) \Rightarrow 15 = a\left(27 - \dfrac{27}{2} + 3 - \dfrac{3}{2}\right) \Rightarrow 15 = a(30 - 15) \Rightarrow$

$15 = 15a \Rightarrow a = 1$. Therefore the function is: $P(x) = x^3 - \dfrac{3}{2}x^2 + x - \dfrac{3}{2}$.

5. Since the zeros are $-4, -4$, $1 + 2i$, and it's conjugate $1 - 2i$, the function is:

$P(x) = a(x + 4)^2(x - 1 - 2i)(x - 1 + 2i) \Rightarrow P(x) = a(x^2 + 8x + 16)(x^2 - 2x + 5) \Rightarrow$

$P(x) = a(x^4 + 6x^3 + 5x^2 + 8x + 80)$. If $a = 1$, then $P(x) = x^4 + 6x^3 + 5x^2 + 8x + 80$. (Answers may vary)

6. For the function the possible zeros are: $x = \pm\dfrac{1, 2, 5, 10}{1, 2} \Rightarrow x = \pm1, \pm2, \pm5, \pm10, \pm\dfrac{1}{2}, \pm\dfrac{5}{2}$.

From a calculator graph, we can choose $x = -2$ to check by synthetic division.

$$
\begin{array}{r|rrrr}
-2) & 2 & 1 & -11 & -10 \\
& & -4 & 6 & 10 \\
\hline
& 2 & -3 & -5 & 0
\end{array}
\qquad \text{Now solve the resulting quadratic equation:}
$$

$2x^2 - 3x - 5 = 0 \Rightarrow (2x - 5)(x + 1) = 0 \Rightarrow 2x - 5 = 0$ and $x + 1 = 0 \Rightarrow x = \dfrac{5}{2}, -1,$ and -2.

7. Use the quadratic formula on the equation: $x^2 = \dfrac{12 \pm \sqrt{(-12)^2 - 4(3)(1)}}{2(3)} = \dfrac{12 \pm \sqrt{132}}{6} =$

$\dfrac{12 \pm 2\sqrt{33}}{6} = \dfrac{6 \pm \sqrt{33}}{3}$. Therefore $x = \left\{\pm\sqrt{\dfrac{6 \pm \sqrt{33}}{3}}\right\}$.

8. Using the regression feature on the graphing calculator and $L_1 = \{0, 2, 4, 7\}$ and $L_2 = \{100, 137, 175, 205\}$ yields: $y \approx -1.055x^2 + 22.62x + 98.9$. Now solving for $x = 9$, yields: $P(9) \approx -1.055(9)^2 + 22.62(9) + 98.9$ $\Rightarrow P(9) \approx 217$, which means that in 2007, about 217 million debit cards will be issued.

Chapter 3 Review Exercises

1. $(17 - i) + (1 - 3i) = 18 - 4i$

2. $(17 - i) - (1 - 3i) = 17 - i - 1 + 3i = 16 + 2i$

3. $(17 - i)(1 - 3i) = 17 - 51i - i + 3i^2 = 17 - 52i - 3 = 14 - 52i$

4. $(17 - i)^2 = (17 - i)(17 - i) = 289 - 17i - 17i + i^2 = 289 - 34i - 1 = 288 - 34i$

5. $\dfrac{1}{1 - 3i} \cdot \dfrac{1 + 3i}{1 + 3i} = \dfrac{1 + 3i}{1 - 9i^2} = \dfrac{1 + 3i}{10} = \dfrac{1}{10} + \dfrac{3}{10}i$

6. $\dfrac{17 - i}{1 - 3i} \cdot \dfrac{1 + 3i}{1 + 3i} = \dfrac{17 + 51i - i - 3i^2}{1 - 9i^2} = \dfrac{20 + 50i}{10} = 2 + 5i$

7. Since all real numbers can be input for x, the domain is: $(-\infty, \infty)$.

8. By using the vertex formula we first find the x-coordinate of the vertex: $x = \dfrac{-b}{2a} = \dfrac{-(-6)}{2(2)} = \dfrac{6}{4} = \dfrac{3}{2}$.

    Now solve for y: $P\left(\dfrac{3}{2}\right) = 2\left(\dfrac{3}{2}\right)^2 - 6\left(\dfrac{3}{2}\right) - 8 \Rightarrow P\left(\dfrac{3}{2}\right) = \dfrac{9}{2} - \dfrac{18}{2} - \dfrac{16}{2} = -\dfrac{25}{2}$. The vertex is: $\left(\dfrac{3}{2}, -\dfrac{25}{2}\right)$.

9. Since the function is a positive even-degree equation, the end behavior is: $\smile$

10. For the x-intercept, $P(x) = 0$. Therefore, $2x^2 - 6x - 8 = 0 \Rightarrow x^2 - 3x - 4 = 0 \Rightarrow$

    $(x - 4)(x + 1) = 0 \Rightarrow x = -1, 4$. The x-intercepts are: $-1, 4$.

11. For the y-intercept, $x = 0$. Therefore, $P(0) = 2(0)^2 - 6(0) - 8 \Rightarrow P(0) = -8$ The y-intercept is: -8.

12. From a calculator graph of $P(x) = 2x^2 - 6x - 8$ the range is: $\left[-\dfrac{25}{2}, \infty\right)$.

13. From a calculator graph of $P(x) = 2x^2 - 6x - 8$ the graph is increasing for the interval: $\left[\dfrac{3}{2}, \infty\right)$;

    and decreasing for the interval: $\left(-\infty, \dfrac{3}{2}\right]$.

14. (a) $2x^2 - 6x - 8 = 0 \Rightarrow x^2 - 3x - 4 = 0 \Rightarrow (x - 4)(x + 1) = 0 \Rightarrow x = \{-1, 4\}$.

    (b) First set $2x^2 - 6x - 8 = 0$, then $x^2 - 3x - 4 = 0 \Rightarrow (x - 4)(x + 1) = 0$ and the solutions will be:

    $x = \{-1, 4\}$. The numbers -1 and 4 divide the number line into 3 intervals. The interval: $(-\infty, -1)$ has a

    positive product, the interval: $(-1, 4)$ has a negative product, and the interval: $(4, \infty)$ has a positive product.

    Therefore $2x^2 - 6x - 8 > 0$ for the interval: $(-\infty, -1) \cup (4, \infty)$.

    (c) First set $2x^2 - 6x - 8 = 0$, then $x^2 - 3x - 4 = 0 \Rightarrow (x - 4)(x + 1) = 0$ and the solutions will be:

    $x = \{-1, 4\}$. The numbers -1 and 4 divide the number line into 3 intervals. The interval: $(-\infty, -1)$ has a

    positive product, the interval: $(-1, 4)$ has a negative product, and the interval: $(4, \infty)$ has a positive product.

    Therefore $2x^2 - 6x - 8 \le 0$ for the interval: $[-1, 4]$.

15. The graph intersects the x-axis at -1 and 4, supporting the answer in (a). It lies above the x-axis when

    $x < -1$ or $x > 4$, supporting the answer in (b). It lies below the x-axis when x is between -1 and 4 inclusive,

    supporting the answer in (c).

16. From problem 8 the vertex of $P(x) = 2x^2 - 6x - 8$ is: $\left(\dfrac{3}{2}, -\dfrac{25}{2}\right)$. The vertical line of symmetry through this

    point has the equation: $x = \dfrac{3}{2}$.

17. The discriminate is: $b^2 - 4ac = (5.47)^2 - 4(-2.64)(3.54) = 29.9209 + 37.3824 = 67.3033$. Since the

    discriminate is greater than 0, there are two x-intercepts.

18. From the calculator graph of $P(x) = -2.64x^2 + 5.47x + 3.54$ the x-intercepts are: $x = \{-.52, 2.59\}$.

19. From the calculator graph of $P(x) = -2.64x^2 + 5.47x + 3.54$ the vertex is: $(1.04, 6.37)$.

20. Using the vertex formula yields: $x = \dfrac{-b}{2a} = \dfrac{-5.47}{2(-2.64)} \approx 1.04$. Now solve for

    $P(1.04) = -2.64(1.04)^2 + 5.47(1.04) + 3.54 \approx 6.37$. The vertex is: $(1.04, 6.37)$.

21. (a) From the graph the maximum value of $f(x)$ is: 4.

    (b) From the graph the maximum value of 4 is reached when $x = 1$.

    (c) From the graph there would be two intersections with $y = 2$, therefore two real solutions.

    (d) From the graph there would be no intersections with $y = 6$, therefore no real solutions.

22. At zero seconds the height is: $s(0) = -16(0)^2 + 800(0) + 600 = 600$. The projectile was fired from 600 feet.

23. From the calculator graph of the equation the vertex is: $(25, 10{,}600)$. The maximum height is reached after 25 sec.

24. From the calculator graph of the equation the vertex is: $(25, 10{,}600)$. The maximum height is: 10,600 feet.

25. From the calculator graph of the equation and $y = 5000$, the graph of the equation is above $y = 5000$ for the

    interval: $(6.3, 43.7)$. The projectile is above 5000 feet between 6.3 seconds and 43.7 seconds.

26. From the calculator graph of the equation the x-intercept, when the projectile hits the ground, is: $x = 50.739$.

    The projectile will be in the air for approximately 50.7 seconds.

27. (a) Let $x =$ the width and $3x =$ the length of the cardboard. The the base of the box will be

    $l = (x - 8)$, $l = (3x - 8)$, and it's height is: $h = 4$. Since $V = lwh$, we get a function:

    $V(x) = (x - 8)(3x - 8)(4) \Rightarrow V(x) = (3x^2 - 32x + 64)(4) \Rightarrow V(x) = 12x^2 - 128x + 256$.

    (b) If the volume is 2496, then $2496 = 12x^2 - 128x + 256 \Rightarrow 0 = 12x^2 - 128x - 2240 \Rightarrow$

    $0 = 3x^2 - 32x - 560$. Using the quadratic formula to solve yields: $x = \dfrac{-(-32) \pm \sqrt{(-32)^2 - 4(3)(-560)}}{2(3)}$

    $= \dfrac{32 \pm 88}{6} \Rightarrow x = 20, -\dfrac{28}{3}$. Since we can not have a negative distance we throw out

    $x = -\dfrac{28}{3}$ and $x = 20$. Therefore the original dimensions are 20 in. by 60 in.

    (c) Graphing $y_1 = V(x)$ and $y_2 = 2496$ on a graphing calculator shows that the graphs intersect at $x = 20$.

28. (a) See Figure 28.

    (b) Using the form $P(x) = a(x - h)^2 + k$ and a vertex of $(0, 353)$ yields: $P(x) = a(x - 0)^2 + 353$. Now solve

    for a using a second point of $(285, 2000)$: $2000 = a(285 - 0)^2 + 353 \Rightarrow 1647 = 81225a \Rightarrow a \approx .0203$.

    Therefore the function is: $P(x) = .0203x^2 + 353$.

    (c) Using the regression feature of the graphing calculator for the data, yields: $Q(x) = .0167x^2 + .968x + 363$.

$[-28.5, 313.5]$ by $[73.01, 2279.99]$
Xscl $= 50$ Yscl $= 100$

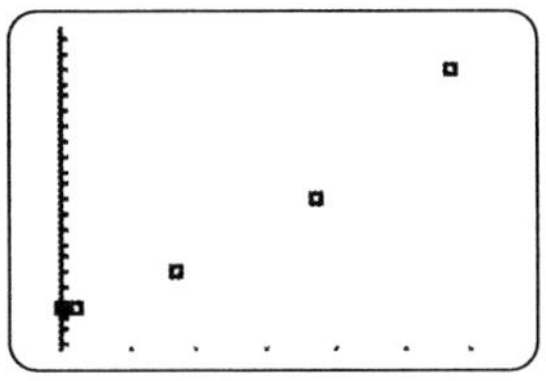

Figure 28

29. First find $P(-2) = -3(-2)^3 - (-2)^2 + 2(-2) - 4 = 24 - 4 - 4 - 4 \Rightarrow P(-2) = 12$ and

$P(-1) = -3(-1)^3 - (-1)^2 + 2(-1) - 4 = 3 - 1 - 2 - 4 \Rightarrow P(-1) = -4$. Since $P(-2) = 12$ and

$P(-1) = -4$ differ in sign, the intermediate value theorem assures us that there is a real zero between

-2 and -1.

30. (a)

$$3 \overline{)\begin{array}{rrrr} 1 & 1 & -11 & -10 \\ & 3 & 12 & 3 \\ \hline 1 & 4 & 1 & -7 \end{array}}$$ Therefore, $Q(x) = x^2 + 4x + 1; R = -7$.

(b)

$$-2 \overline{)\begin{array}{rrrr} 3 & 8 & 5 & 10 \\ & -6 & -4 & -2 \\ \hline 3 & 2 & 1 & 8 \end{array}}$$ Therefore, $Q(x) = 3x^2 + 2x + 1; R = 8$.

31.

$$2 \overline{)\begin{array}{rrrr} -1 & 5 & -7 & 1 \\ & -2 & 6 & -2 \\ \hline -1 & 3 & -1 & -1 \end{array}}$$ Therefore, $P(2) = -1$.

32.

$$2 \overline{)\begin{array}{rrrr} 2 & -3 & 7 & -12 \\ & 4 & 2 & 18 \\ \hline 2 & 1 & 9 & 6 \end{array}}$$ Therefore, $P(2) = 6$.

33.

$$2 \overline{)\begin{array}{rrrrr} 5 & 0 & -12 & 2 & -8 \\ & 10 & 20 & 16 & 36 \\ \hline 5 & 10 & 8 & 18 & 28 \end{array}}$$ Therefore, $P(2) = 28$.

34.

$$2 \overline{)\begin{array}{rrrrrr} 1 & 0 & 0 & 4 & -2 & -4 \\ & 2 & 4 & 8 & 24 & 44 \\ \hline 1 & 2 & 4 & 12 & 22 & 40 \end{array}}$$ Therefore, $P(2) = 40$.

35. The conjugate of $7 + 2i$ must also be a zero, therefore $7 - 2i$ is also a zero.

36. With the given zeros $-1, 4$, and 7, the function $P(x)$ is: $P(x) = a(x - (-1))(x - 4)(x - 7) =$

$a(x + 1)(x - 4)(x - 7) = a(x + 1)(x^2 - 11x + 28) \Rightarrow P(x) = a(x^3 - 10x^2 + 17x + 28)$. One of many

possible functions, if $a = 1$, is: $P(x) = x^3 - 10x^2 + 17x + 28$.

37. With the given zeros $8, 2$, and 3, the function $P(x)$ is: $P(x) = a(x - 8)(x - 2)(x - 3) =$

$a(x - 8)(x - 2)(x - 3) = a(x - 8)(x^2 - 5x + 6) \Rightarrow P(x) = a(x^3 - 13x^2 + 46x - 48)$. One of many

possible functions, if $a = 1$, is: $P(x) = x^3 - 13x^2 + 46x - 48$.

38. With the given zeros $\sqrt{3}, -\sqrt{3}, 2$, and 3, the function $P(x)$ is:

$P(x) = a(x - \sqrt{3})(x - (-\sqrt{3})(x - 2)(x - 3) =$

$a(x - \sqrt{3})(x + \sqrt{3})(x - 2)(x - 3) = a(x^2 - 3)(x^2 - 5x + 6) \Rightarrow a(x^4 - 5x^3 + 3x^2 + 15x - 18)$. One of

many possible functions, if $a = 1$, is: $P(x) = x^4 - 5x^3 + 3x^2 + 15x - 18$.

39. With the given zeros $-2 + \sqrt{5}, -2 - \sqrt{5}, -2$, and 1, the function $P(x)$ is:

$P(x) = a(x - (-2 + \sqrt{5}))(x - (-2 - \sqrt{5}))(x - (-2))(x - 1) =$

$a(x + 2 - \sqrt{5})(x + 2 + \sqrt{5})(x + 2)(x - 1) = a(x^2 + 4x - 1)(x^2 + x - 2) = a(x^4 + 5x^3 + x^2 - 9x + 2)$.

One of many possible functions, if $a = 1$, is: $P(x) = x^4 + 5x^3 + x^2 - 9x + 2$.

40. Use synthetic division to check if -1 is a zero.

$$
\begin{array}{r|rrrrr}
-1) & 2 & 1 & -4 & 3 & 1 \\
   &   & -2 & 1 & 3 & -6 \\
\hline
   & 2 & -1 & -3 & 6 & -5
\end{array}
$$
Since $P(-1) = -5$ and not 0, -1 is not a zero.

41. Use synthetic division to check if $x + 1$ or -1 is a zero.

$$
\begin{array}{r|rrrr}
-1) & 1 & 2 & 3 & 2 \\
   &   & -1 & -1 & -2 \\
\hline
   & 1 & 1 & 2 & 0
\end{array}
$$
Since $P(-1) = 0$, -1 or $x + 1$ is a factor.

42. With the given zeros $3, 1, -1 - 3i$, and the conjugate of $-1 - 3i$ or $-1 + 3i$ also a zero, the function $P(x)$ is:

$P(x) = a(x - 3)(x - 1)(x - (-1 - 3i))(x - (-1 + 3i)) = a(x - 3)(x - 1)(x + 1 + 3i)(x + 1 - 3i) =$

$a(x^2 - 4x + 3)(x^2 + 2x + 10) = a(x^4 - 2x^3 + 5x^2 - 34x + 30)$. Since $P(2) = -36$, we can solve for a:

$-36 = a((2)^4 - 2(2)^3 + 5(2)^2 - 34(2) + 30) \Rightarrow -36 = a(16 - 16 + 20 - 68 + 30) \Rightarrow -18a = -36 \Rightarrow$

$a = 2$. Therefore, $P(x) = 2(x^4 - 2x^3 + 5x^2 - 34x + 30) \Rightarrow P(x) = 2x^4 - 4x^3 + 10x^2 - 68x + 60$.

43. With the given zero $1 - i$, , the conjugate $1 + i$ is also a zero. Use synthetic division to factor out these zeros.

$$
\begin{array}{r|rrrrr}
1 - i) & 1 & -3 & -8 & 22 & -24 \\
      &   & 1 - i & -3 + i & -10 + 12i & 24 \\
\hline
      & 1 & -2 - i & -11 + i & 12 + 12i & 0
\end{array}
\Rightarrow
\begin{array}{r|rrrr}
1 + i) & 1 & -2 - i & -11 + i & 12 + 12i \\
      &   & 1 + i & -1 - i & -12 - 12i \\
\hline
      & 1 & -1 & -12 & 0
\end{array}
$$

Finally, solve the resulting quadratic equation: $x^2 - x - 12 = 0 \Rightarrow (x - 4)(x + 3) = 0 \Rightarrow x = -3, 4$.

The zeros are: $x = \{-3, 4, 1 - i, 1 + i\}$.

44. With the given zeros 1 and $2i$, the conjugate of $2i$ or $-2i$ is also a zero. Use synthetic division to factor out these zeros.

$$
\begin{array}{r|rrrrr}
1) & 2 & -1 & 7 & -4 & -4 \\
  &   & 2 & 1 & 8 & 4 \\
\hline
  & 2 & 1 & 8 & 4 & 0
\end{array}
\Rightarrow
\begin{array}{r|rrrr}
2i) & 2 & 1 & 8 & 4 \\
   &   & 4i & -8 + 2i & -4 \\
\hline
   & 2 & 1 + 4i & 2i & 0
\end{array}
\Rightarrow
\begin{array}{r|rrr}
-2i) & 2 & 1 + 4i & 2i \\
    &   & -4i & -2i \\
\hline
    & 2 & 1 & 0
\end{array}
$$

Finally, solve the resulting linear equation: $2x + 1 = 0 \Rightarrow 2x = -1 \Rightarrow x = -\dfrac{1}{2}$. The zeros are:

$x = \left\{ -\dfrac{1}{2}, 1, 2i, -2i \right\}$.

45. Using the Rational Zeros Theorem the possible zeros are: $x = \pm \dfrac{1, 2, 4, 8}{1, 3} = \pm 1, \pm 2, \pm 4, \pm 8, \pm \dfrac{1}{3},$

$\pm \dfrac{2}{3}, \pm \dfrac{4}{3}, \pm \dfrac{8}{3}$. From graphing the equation, we choose to check: $-2, 4, \dfrac{1}{3}$ by synthetic division.

$$
\begin{array}{r|rrrrrr}
-2) & 3 & -4 & -26 & -21 & -14 & 8 \\
   &   & -6 & 20 & 12 & 18 & -8 \\
\hline
   & 3 & -10 & -6 & -9 & 4 & 0
\end{array}
\Rightarrow
\begin{array}{r|rrrrr}
4) & 3 & -10 & -6 & -9 & 4 \\
  &   & 12 & 8 & 8 & -4 \\
\hline
  & 3 & 2 & 2 & -1 & 0
\end{array}
\Rightarrow
\begin{array}{r|rrrr}
\tfrac{1}{3}) & 3 & 2 & 2 & -1 \\
  &   & 1 & 1 & 1 \\
\hline
  & 3 & 3 & 3 & 0
\end{array}
$$

Finally, solve the resulting quadratic equation $x^2 + x + 1 = 0$ using the quadratic formula:

$x = \dfrac{-1 \pm \sqrt{1^2 - 4(1)(1)}}{2(1)} = \dfrac{-1 \pm \sqrt{-3}}{2}$, which will result in nonreal solutions, therefore the rational

solutions are: $x = \left\{ -2, \dfrac{1}{3}, 4 \right\}$.

46. Because the function $P(x)$ has coefficient signs: $+ \ + \ - \ - \ -$, which is 1 sign change, the function has 1 positive real zeros. Because $P(-x) = 3(-x)^4 + (-x)^3 - (-x)^2 - 2(-x) - 1 = 3x^4 - x^3 - x^2 + 2x - 1$ has coefficient signs: $+ \ - \ - \ + \ -$, which is 3 sign changes, the function has 3 or 1 possible negative real zero.

47. Using the Boundedness Theorem we divide synthetically by $x - 2$.

$$
\begin{array}{r|rrrrr}
2 & 2 & 3 & -5 & 8 & -10 \\
  &   & 4 & 14 & 18 & 52 \\
\hline
  & 2 & 7 & 9 & 26 & 42
\end{array}
$$

The result is all nonnegative, therefore no real zero greater than 2.

Also using the Boundedness Theorem we divide synthetically by $x - (-4)$.

$$
\begin{array}{r|rrrrr}
-4 & 2 & 3 & -5 & 8 & -10 \\
   &   & -8 & 20 & -60 & 208 \\
\hline
   & 2 & -5 & 15 & -52 & 198
\end{array}
$$

The result alternates in sign, therefore no real zero less than -4.

48. Graphing the equation on a calculator shows x-intercepts: $-1.62, 0.62, 3$. Therefore there are 3 real solutions and the integer root is: 3.

49. Using synthetic division we can factor out the root 3.

$$
\begin{array}{r|rrrr}
3 & 1 & -2 & -4 & 3 \\
  &   & 3 & 3 & -3 \\
\hline
  & 1 & 1 & -1 & 0
\end{array}
$$

Using the result of the synthetic division the factors are: $(x - 3)(x^2 + x - 1)$.

50. The remaining zeros can be found by solving $x^2 + x - 1 = 0$ using the quadratic formula:

$$x = \frac{-1 \pm \sqrt{1^2 - 4(1)(-1)}}{2(1)} = \frac{-1 \pm \sqrt{5}}{2}, \text{ the solutions are: } x = \left\{ \frac{-1 - \sqrt{5}}{2}, \frac{-1 + \sqrt{5}}{2} \right\}.$$

51. The x-intercepts from the graph are approximately equal to the solutions found in #50.

$$\frac{-1 - \sqrt{5}}{2} \approx -1,62 \text{ and } \frac{-1 + \sqrt{5}}{2} \approx 0.62$$

52. (a) From the graph and the answers of #50, $P(x) > 0$ for the interval: $\left(\frac{-1 - \sqrt{5}}{2}, \frac{-1 + \sqrt{5}}{2} \right) \cup (3, \infty)$.

    (b) From the graph and the answers of #50, $P(x) \leq 0$ for the interval: $\left(-\infty, \frac{-1 - \sqrt{5}}{2} \right] \cup \left[\frac{-1 + \sqrt{5}}{2}, 3 \right]$.

53. First factor out an x: $x^3 + 2x^2 + 5x = 0 \Rightarrow x(x^2 + 2x + 5) \Rightarrow x = 0$ is a zero. Now solve the quadratic equation using the quadratic formula: $x = \frac{-2 \pm \sqrt{2^2 - 4(1)(5)}}{2(1)} = \frac{-2 \pm \sqrt{-16}}{2} = \frac{-2 \pm 4i}{2} = -1 \pm 2i$.

Since two of the zeros are not real, the only x-intercept is: $x = 0$.

54. The x-intercepts are: $-2, 1$, and 3. The 3 has multiplicity 2, because it appears tangent.

The factored form is now: $P(x) = (x + 2)(x - 1)(x - 3)^2$.

55. Since $(-4, 0)$ is one point on $y_1 = x^3 - x^2 - 19x + 4$, one factor is $(x + 4)$ and we can factor it out by

    synthetic division.

    $$-4 \overline{)\begin{array}{rrrr} 1 & -1 & -19 & 4 \\ & -4 & 20 & -4 \\ \hline 1 & -5 & 1 & 0 \end{array}}$$

    Now solve the result using the quadratic formula:

    $x = \dfrac{-(-5) \pm \sqrt{(-5)^2 - 4(1)(1)}}{2(1)} = \dfrac{5 \pm \sqrt{21}}{2}$. Therefore, the zeros are: $-4, \dfrac{5 + \sqrt{21}}{2}, \dfrac{5 - \sqrt{21}}{2}$, and the

    linear factors are: $(x + 4)\left(x - \left(\dfrac{5 + \sqrt{21}}{2}\right)\right)\left(x - \left(\dfrac{5 - \sqrt{21}}{2}\right)\right)$.

56. From the table, there is a sign change in Y_1 from $x = .7$ to $x = .8$, therefore by the intermediate value theorem

    there exists a real zero between .7 and .8.

57. Since it's end behavior has both ends going the same direction up or down, the degree is even.

58. Since it's end behavior has one end going up and the other down the degree is odd.

59. Since it's end behavior has the right end approach positive infinity, the leading coefficient is positive.

60. Since $g(x)$ has no x-intercepts, it has no real solutions.

61. From the graph, $f(x) < 0$ for the interval: $(-\infty, a) \cup (b, c)$.

62. From the graph, $f(x) > g(x)$ for the interval: (d, h).

63. If $f(x) - g(x) = 0$ then $f(x) = g(x)$. This happens at the intersection of the graphs, which is: $\{d, h\}$.

64. If $r + pi$ is a solution, then it's conjugate $r - pi$ is also a solution.

65. Since $f(x)$ has three real zeros and a polynomial of degree 3 can have at most three zeros, there can be no other

    zeros, real or nonreal complex.

66. False. A 7th degree function can have at most 7 x-intercepts.

67. True. A 7th degree function can have at most 6 local extrema.

68. True. $f(0) = 3(0)^7 - 8(0)^6 + 9(0)^5 + 12(0)^4 - 18(0)^3 + 26(0)^2 - (0) + 500 \Rightarrow f(0) = 500$.

69. True. The end behavior of a odd-degree polynomial with a positive lead coefficient is down to the left. Because

    the function has a positive y-intercept, the graph of the equation must cross the negative x-axis at least one time.

    Therefore, it must have at least one negative x-intercept.

70. True. An even-degree polynomial with a positive lead coefficient will have end behavior of both ends going

    upward, therefore a graph with a negative y-intercept must cross the x-axis at least 2 times and will have

    at least 2 real zeros.

71. False. The conjugate must also be a zero, but the conjugate of $-\dfrac{1}{2} + i\dfrac{\sqrt{3}}{2}$ is $-\dfrac{1}{2} - i\dfrac{\sqrt{3}}{2}$ and not $\dfrac{1}{2} + i\dfrac{\sqrt{3}}{2}$.

72. From the graph there are two local maxima.

73. From the graph there is one local minimum that lies on the x-axis, it has coordinates: $(2, 0)$.

74. Use synthetic division to factor out $x - 5$.

$$
\begin{array}{r|rrrrrr}
5 & -2 & 15 & -21 & -32 & 60 & 0 \\
  &   & -10 & 25 & 20 & -60 & 0 \\
\hline
  & -2 & 5 & 4 & -12 & 0 & 0
\end{array}
$$

From this result, $Q(x) = -2x^4 + 5x^3 + 4x^2 - 12x$.

75. Because the degree of P is odd, the range is: $(-\infty, \infty)$.

76. Using the calculator functions this local minimum point has coordinates: $(-0.97, -54.15)$.

77. We first factor by grouping and then solve: $3x^3 + 2x^2 - 21x - 14 = 0 \Rightarrow x^2(3x + 2) - 7(3x + 2) = 0 \Rightarrow$

$(3x + 2)(x^2 - 7) = 0 \Rightarrow 3x + 2 = 0 \Rightarrow 3x = -2 \Rightarrow x = -\dfrac{2}{3}$ or $x^2 - 7 = 0 \Rightarrow x^2 = 7 \Rightarrow x = \pm\sqrt{7}$.

Therefore the solution set is: $\left\{-\sqrt{7}, -\dfrac{2}{3}, \sqrt{7}\right\}$. Graphing this equation shows x-intercepts of:

$x = \pm 2.65, -0.67$ which is equal to the solution set and therefore support our analytic solution.

78. An even-degree polynomial with a negative lead coefficient has an end behavior of: ⌢⌣

The function has zeros: -1 (multiplicity 2), 2, and 3, therefore x-intercepts at those values of x. The graph

will be tangent to the x-axis at $x = -1$ (multiplicity 2) and the y-intercept is $y = -6$. See Figure 78.

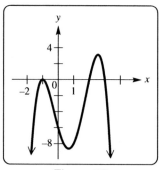

Figure 78

(a) $P(x) = 0$ at the x-intercepts, therefore: $x = \{-1, 2, 3\}$.

(b) From the graph, $P(x) > 0$ for the interval: $(2, 3)$.

(c) From the graph, $P(x) < 0$ for the interval: $(-\infty, -1) \cup (-1, 2) \cup (3, \infty)$.

79. If $x =$ cube side length and $V = s^3$, the volume of the cube after the top is cut off is: $V = (x)(x)(x - 2) = 32$

$\Rightarrow x^3 - 2x^2 - 32 = 0$. After graphing for possible roots, we use synthetic division to check 4 as a root.

$$
\begin{array}{r|rrrr}
4 & 1 & -2 & 0 & -32 \\
  &   & 4 & 8 & 32 \\
\hline
  & 1 & 2 & 8 & 0
\end{array}
$$

Now solve the resulting equation: $x^2 + 2x + 8 = 0$.

$x = \dfrac{-2 \pm \sqrt{2^2 - 4(1)(8)}}{2(1)} = \dfrac{-2 \pm \sqrt{-28}}{2}$. Both of these roots are nonreal, therefore the only real root is 4

and the dimensions of the original cube is 4 in. by 4in. by 4 in.

80. (a) Since, from a calculator graph of the equation, the zero is at approximately 9.26. Therefore, the restrictions

on t are: $0 \leq t \leq 9.26$.

(b) See Figure 80.

(c) From the graph, the object reaches its highest point at approximately $t = 4.08$ seconds.

(d) From the graph, the object reaches its highest point at approximately $s(t) = 131.63$ meters.

(e) The object reaches the ground when $s(t) = 0$, therefore the equation is: $-4.9t^2 + 40t + 50 = 0$.

We solve this using the quadratic formula: $t = \dfrac{-40 \pm \sqrt{40^2 - 4(-4.9)(50)}}{2(-4.9)} = \dfrac{-40 \pm \sqrt{2580}}{-9.8} \approx$

$-1.10, 9.26$. Since we can not have negative time, the object reaches the ground in approximately 9.26 sec.

$[0, 10]$ by $[0, 200]$

Xscl = 1   Yscl = 10

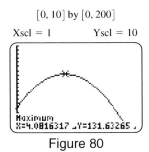

Figure 80

81. For the year 2001 $x = 11$, therefore $y = -.1885(11)^3 + 10.617(11)^2 - 157.075(11) + 2080.8 = 1386.7385$

or approximately 1387 thousand military personnel were on active duty in 2001.

82. (a) See Figure 82a.

(b) Using the capabilities of the graphing calculator the quadratic function is: $f(x) \approx -.011x^2 + .869x + 11.9$.

(c) Using the capabilities of the calculator the cubic function is: $f(x) \approx -.00087x^3 + .0456x^2 - .219x + 17.8$.

(d) See Figure 82b, for the quadratic function and See Figure 82c, for the cubic function.

(e) Both functions approximate the data well. Although the value of R^2 is closer to 1 for the cubic function, the

quadratic function is probably better for prediction because it is unlikely that the percent of out-of-pocket

spending would decrease after 2025 (as the cubic function shows) unless changes were made in Medicare law.

(f) A linear model would not really be appropriate because the data points lie in a curved nonlinear pattern.

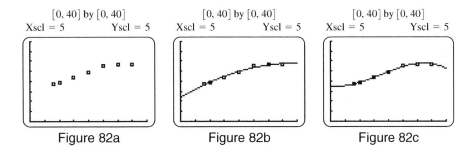

Figure 82a   Figure 82b   Figure 82c

Chapter 3 Test

1. (a) $(8 - 7i) - (-12 + 2i) = 20 - 9i$

   (b) $\dfrac{11 + 10i}{2 + 3i} \cdot \dfrac{2 - 3i}{2 - 3i} = \dfrac{22 - 33i + 20i - 30i^2}{4 - 9i^2} = \dfrac{22 - 13i + 30}{4 + 9} = \dfrac{52 - 13i}{13} = 4 - i.$

   (c) $i^{65} = (i^4)^{16}(i) = 1^{16}(i) = i.$

   (d) $2i(3 - i)^2 = 2i(9 - 6i + i^2) = 2i(9 - 6i - 1) = 2i(8 - 6i) = 16i - 12i^2 = 12 + 16i.$

2. (a) Use the vertex formula to find the x-coordinate: $x = -\dfrac{b}{2a} = -\dfrac{-4}{2(-2)} = -1.$ Now solve for $f(-1)$ to find the

   y-coordinate of the vertex: $f(-1) = -2(-1)^2 - 4(-1) + 6 = -2 + 4 + 6 = 8.$ The vertex is: $(-1, 8).$

   (b) See Figure 2b. Finding the maximum on the graph gives the vertex: $(-1, 8).$

   (c) First we find the zeros by factoring: $-2x^2 - 4x + 6 = 0 \Rightarrow -2(x^2 + 2x - 3) = 0 \Rightarrow$

   $-2(x + 3)(x - 1) = 0 \Rightarrow x = -3, 1.$ Now support this on the graphing calculator using a graph and a

   table, See Figure 2c.

   (d) For the y-intercept we let $x = 0$, therefore: $f(0) = -2(0)^2 - 4(0) + 6 \Rightarrow f(0) = 6.$

   (e) Since all values can be input for x, the Domain is: $(-\infty, \infty)$; since $(-1, 8)$ is a maximum, the Range is: $(-\infty, 8].$

   (f) From the graph, the function is increasing for the interval: $(-\infty, -1]$; and decreasing for the interval: $[-1, \infty).$

$[-9.4, 9.4]$ by $[-4.1, 8.1]$
Xscl $= 1$ Yscl $= 1$

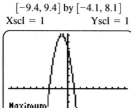

$[-9.4, 9.4]$ by $[-4.1, 8.1]$
Xscl $= 1$ Yscl $= 1$

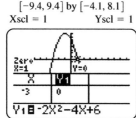

Figure 2b   Figure 2c

3. (a) We solve the equation using the quadratic formula: $x = \dfrac{-3 \pm \sqrt{3^2 - 4(3)(-2)}}{2(3)} = \dfrac{-3 \pm \sqrt{9 + 24}}{6} \Rightarrow$

   $x = \left\{ \dfrac{-3 - \sqrt{33}}{6}, \dfrac{-3 + \sqrt{33}}{6} \right\}.$

   (b) See Figure 3. From this graph, $f(x) < 0$ for the interval: $\left(\dfrac{-3 - \sqrt{33}}{6}, \dfrac{-3 + \sqrt{33}}{6} \right);$

   and (ii) $f(x) \geq 0$ for the interval: $\left(-\infty, \dfrac{-3 - \sqrt{33}}{6} \right] \cup \left[\dfrac{-3 + \sqrt{33}}{6}, \infty \right).$

$[-4.7, 4.7]$ by $[-3.1, 3.1]$
Xscl $= 1$ Yscl $= 1$

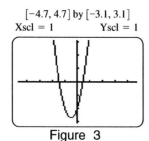

Figure 3

4. Since $V = lwh$, $720 = (11 + x)(3x)(x) \Rightarrow 720 = 3x^3 + 33x^2 \Rightarrow 0 = 3x^3 + 33x^2 - 720 \Rightarrow$

   $0 = x^3 + 11x^2 - 240$. After graphing on a calculator, we use synthetic division to check 4 as a factor.

$$\begin{array}{r|rrrr} 4 & 1 & 11 & 0 & -240 \\ & & 4 & 60 & 240 \\ \hline & 1 & 15 & 60 & 0 \end{array}$$

   Now solve the resulting equation $x^2 + 15x + 60 = 0$ using

   quadratic formula: $x = \dfrac{-15 \pm \sqrt{15^2 - 4(1)(60)}}{2(1)} = \dfrac{-15 \pm \sqrt{-15}}{2}$. These solutions will both be nonreal,

   therefore the only real solution is 4 and the dimensions are: $11 + 4$, $3(4)$, and 4 or 15 in. by 12 in. by 4 in.

5. (a) See Figure 5a.

   (b) Using the form $f(x) = a(x - h)^2 + k$ and the vertex $(0, 470)$ yields: $f(x) = a(x - 0)^2 + 470$. Now use the

   point $(10, 670)$ and this equation to find a. $670 = a(10)^2 + 470 \Rightarrow 200 = 100a \Rightarrow a = 2$.

   The function is: $f(x) = 2x^2 + 470$.

   (c) Using the regression function of the calculator the quadratic function is: $g(x) \approx .737x^2 + 13.8x + 461$.

   Now graph the data and each equation in the same window, See Figure 5c. The regression function $g(x)$

   fits the data best.

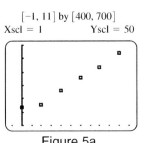

[−1, 11] by [400, 700]
Xscl = 1   Yscl = 50

Figure 5a

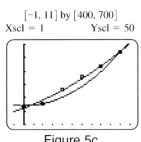

[−1, 11] by [400, 700]
Xscl = 1   Yscl = 50

Figure 5c

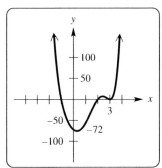

Figure 6

6. (a) Use synthetic division to factor out the given zeros.

$$\begin{array}{r|rrrrrr} 3 & 1 & -5 & 3 & 1 & 40 & -24 & -72 \\ & & 3 & -6 & -9 & -24 & 48 & 72 \\ \hline & 1 & -2 & -3 & -8 & 16 & 24 & 0 \end{array} \Rightarrow$$

$$\begin{array}{r|rrrrr} 3 & 1 & -2 & -3 & -8 & 16 & 24 \\ & & 3 & 3 & 0 & -24 & -24 \\ \hline & 1 & 1 & 0 & -8 & -8 & 0 \end{array} \Rightarrow$$

$$\begin{array}{r|rrrr} -1 & 1 & 1 & 0 & -8 & -8 \\ & & -1 & 0 & 0 & 8 \\ \hline & 1 & 0 & 0 & -8 & 0 \end{array} \Rightarrow$$

$$\begin{array}{r|rrrr} 2 & 1 & 0 & 0 & -8 \\ & & 2 & 4 & 8 \\ \hline & 1 & 2 & 4 & 0 \end{array}$$

   Now solve resulting equation $x^2 + 2x + 4 = 0$ using the quadratic formula: $x = \dfrac{-2 \pm \sqrt{2^2 - 4(1)(4)}}{2(1)} =$

   $\dfrac{-2 \pm \sqrt{-12}}{2} = \dfrac{-2 \pm 2i\sqrt{3}}{2} = -1 \pm i\sqrt{3}$. The other zeros of f are: $x = \{-1 - i\sqrt{3}, -1 + i\sqrt{3}\}$.

   (b) An even-degree polynomial with a positive lead coefficient has end-behavior: $\cup$

   See Figure 6.

7. (a) We can find the zeros by factoring: $4x^4 - 21x^2 - 25 = 0 \Rightarrow (4x^2 - 25)(x^2 + 1) = 0 \Rightarrow$

$4x^2 - 25 = 0 \Rightarrow 4x^2 = 25 \Rightarrow x^2 = \dfrac{25}{4} \Rightarrow x = \pm\sqrt{\dfrac{25}{4}} = \pm\dfrac{5}{2}$ or $x^2 + 1 = 0 \Rightarrow x^2 = -1 \Rightarrow$

$x = \pm\sqrt{-1} \Rightarrow x = \pm i.$ The zeros are: $x = \left\{-\dfrac{5}{2}, \dfrac{5}{2}, -i, i\right\}.$

(b) Use a graph and table to support the results. See Figure 7.

(c) It is symmetric with respect to the *y*-axis.

(d) From the graph and the results, (i) $f(x) \geq 0$ for the interval: $\left(-\infty, -\dfrac{5}{2}\right] \cup \left[\dfrac{5}{2}, \infty\right)$;

and (ii) $f(x) < 0$ for the interval: $\left(-\dfrac{5}{2}, \dfrac{5}{2}\right).$

$[-4, 4]$ by $[-60, 60]$
Xscl = 1   Yscl = 10

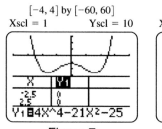

$[-1, 11]$ by $[400, 700]$
Xscl = 1   Yscl = 50

Figure 7   Figure 10

8. (a) Using the Rational Zeros Theorem the possible zeros are: $x = \pm\dfrac{1, 2, 11, 22}{1, 3} = \pm 1, \pm 2, \pm 11, \pm 22,$

$\pm\dfrac{1}{3}, \pm\dfrac{2}{3}, \pm\dfrac{11}{3},$ and $\pm\dfrac{22}{3}.$

(b) From a calculator graph of the equation, we choose to check -2 and $\dfrac{1}{3}$ by synthetic division.

$$
\begin{array}{r|rrrrr}
-2) & 3 & 5 & -35 & -55 & 22 \\
    &   & -6 & 2 & 66 & -22 \\
\hline
    & 3 & -1 & -33 & 11 & 0
\end{array}
\qquad\Rightarrow\qquad
\begin{array}{r|rrrr}
\frac{1}{3}) & 3 & -1 & -33 & 11 \\
             &   & 1 & 0 & -11 \\
\hline
             & 3 & 0 & -33 & 0
\end{array}
$$

Now solve the resulting equation $3x^2 - 33 = 0 \Rightarrow x^2 - 11 = 0 \Rightarrow x^2 = 11 \Rightarrow x = \pm\sqrt{11}.$

Since these are not rational, the rational zeros are: $-2, \dfrac{1}{3}.$

(c) Since $f(3) = 3(3)^4 + 5(3)^3 - 35(3)^2 - 55(3) + 22 = -80$ and

$f(4) = 3(4)^4 + 5(4)^3 - 35(4)^2 - 55(4) + 22 = 330$ differ in sign, the intermediate value theorem assures

us that there is a real zero between 3 and 4.

(d) Because the function $f(x)$ has coefficient signs: $+\ +\ -\ -\ +$, which is 2 sign changes, the function has

2 or 0 positive real zeros. Because $f(-x) = 3(-x)^4 + 5(-x)^3 - 35(-x)^2 - 55(-x) + 22 =$

$3x^4 - 5x^3 - 35x^2 + 55x + 22$ has coefficient signs: $+\ -\ -\ +\ +$, which is 2 sign changes, the

function has 2 or 0 possible negative real zeros.

(e) Using the Boundedness Theorem we divide synthetically by $x - (-5)$.

$$
\begin{array}{r|rrrrr}
-5 & 3 & 5 & -35 & -55 & 22 \\
   &   & -15 & 50 & -75 & 650 \\
\hline
   & 3 & -10 & 15 & -130 & 672
\end{array}
$$
The result alternates sign, therefore no real zero less than -5.

Also using the Boundedness Theorem we divide synthetically by $x - 4$.

$$
\begin{array}{r|rrrrr}
4 & 3 & 5 & -35 & -55 & 22 \\
  &   & 12 & 68 & 132 & 308 \\
\hline
  & 3 & 17 & 33 & 77 & 330
\end{array}
$$
The result is all nonnegative, therefore no real zero greater than 4.

9. (a) Using the capabilities of the calculator and graphing the equation yields real solutions of: $\{.189, 1, 3.633\}$.

   (b) A 5th degree equation has 5 solutions. Since there are 3 real solution, there must be 2 nonreal complex solutions.

10. (a) Using the regression function, the best-fitting cubic function is: $f(x) \approx -.249x^3 + 4.47x^2 + .212x + 467$.

    (b) Using the regression function, the best-fitting quartic function is:

    $g(x) \approx .0977x^4 - 2.20x^3 + 16.5x^2 - 22.1x + 470$.

    (c) See Figure 10.

    (d) The cubic function f yields: 679 million; and the quartic function g yields: 726 million. The quartic function is a better estimate because it continues to increase, while the cubic function turns downward.

Chapter 3 Project

Answers will vary depending on the social security number used. Answers are given for the illustrative social security number 539-58-0954.

1. $SSN(x) = (x - 5)(x + 3)(x - 9)(x + 5)(x - 8)(x + 0)(x - 9)(x + 5)(x - 4)$.

2. There are nine terms, so the term with the highest power will be x^9. The polynomial is of degree 9, and the dominating term is x^9.

3. A polynomial of odd-degree with a positive lead coefficient will have end behavior:

4. The zeros are: $5, -3, 9, -5, 8, 0, 9, -5$, and 4. The zeros of multiplicity one are: $5, -3, 8, 0$, and 4.

5. The zeros that are listed twice, of multiplicity two are: 9 and -5.

6. The zeros that are listed three or more times, of multiplicity three or higher are none.

7. Two graphs for our illustrative social security number are shown in the explanation of the project.

8. Using the graph, the coordinates of the local maximums are: $(-5, 0), (-1.553, 1033670.4), (4.480, 54058.1)$, and $(8.404, 37044.4)$.

9. Using the graph, the coordinates of the local minimums are: $(-3.909, -591411.9), (1.610, -916889.6)$, $(6.523, -287211.6)$, and $(9, 0)$.

10. Since all values can be input for x, the domain is: $(-\infty, \infty)$; and since the function is of odd-degree the range is: $(-\infty, \infty)$.

11. From our graph, the function is increasing for the intervals: $(-\infty, -5]$, $[-3.909, -1.553]$, $[1.610, 4.480]$, $[6.523, 8.404]$, and $[9, \infty)$.

12. From our graph, the function is decreasing for the intervals: $[-5, -3.909]$, $[-1.553, 1.610]$, $[4.480, 6.523]$, and $[8.404, 9]$.

13. In order for the polynomial graph to pass through the origin, 0 must be a zero and one factor must be $x - 0$. Therefore, the digit 0 must appear in the social security number.

Chapter 4: Rational, Power, and Root Functions

4.1: Rational Functions and Graphs

1. The only value for x that cannot be used as input is 0. The domain is $(-\infty, 0) \cup (0, \infty)$.

    It is not possible for this function to output the value 0. The range is $(-\infty, 0) \cup (0, \infty)$.

2. The only value for x that cannot be used as input is 0. The domain is $(-\infty, 0) \cup (0, \infty)$.

    The function will output only positive values. The range is $(0, \infty)$.

3. The function decreases everywhere it is defined, $(-\infty, 0) \cup (0, \infty)$. It never increases and is never constant.

4. The function increases on $(-\infty, 0)$. It decreases on $(0, \infty)$. The function is never constant.

5. Because the function is undefined when $x = 3$, the vertical asymptote has the equation $x = 3$.

    As $|x|$ increases with out bound, the graph of the function will move closer and closer to the graph of $y = 2$.

6. Because the function is undefined when $x = -2$, the vertical asymptote has the equation $x = -2$.

    As $|x|$ increases with out bound, the graph of the function will move closer and closer to the graph of $y = -4$.

7. Because $f(-x) = f(x)$, the function is even. The graph has symmetry with respect to the y-axis.

8. Because $f(-x) = -f(x)$, the function is odd. The graph has symmetry with respect to the origin.

9. Graphs A, B, and C have domain $(-\infty, 3) \cup (3, \infty)$ because each has a vertical asymptote at $x = 3$.

10. Graph B has range $(-\infty, 3) \cup (3, \infty)$ because it exists above and below the horizontal asymptote at $y = 3$.

11. Graph A has range $(-\infty, 0) \cup (0, \infty)$ because it exists above and below the horizontal asymptote at $y = 0$.

12. Graphs C and D have range $(0, \infty)$ because each exists only above the horizontal asymptote at $y = 0$.

13. The only graph that would intersect the line $y = 3$ exactly one time is graph A.

14. Because graph A exists above and below the horizontal asymptote $y = 0$, its range is $(-\infty, 0) \cup (0, \infty)$.

15. Graphs A, C, and D have the x-axis as a horizontal asymptote.

16. Noting that graph D has a hole, graph C is the only graph that is symmetric with respect to a vertical line.

17. Window C gives the most accurate depiction of the graph. See Figures 17a, 17b and 17c.

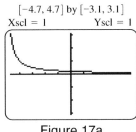
[-4.7, 4.7] by [-3.1, 3.1]
Xscl = 1   Yscl = 1

Figure 17a

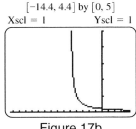
[-14.4, 4.4] by [0, 5]
Xscl = 1   Yscl = 1

Figure 17b

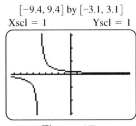
[-9.4, 9.4] by [-3.1, 3.1]
Xscl = 1   Yscl = 1

Figure 17c

18. Window A gives the most accurate depiction of the graph. See Figures 18a, 18b and 18c.

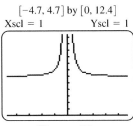
[-4.7, 4.7] by [0, 12.4]
Xscl = 1   Yscl = 1

Figure 18a

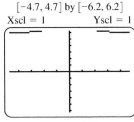
[-4.7, 4.7] by [-6.2, 6.2]
Xscl = 1   Yscl = 1

Figure 18b

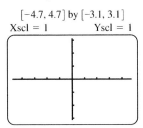
[-4.7, 4.7] by [-3.1, 3.1]
Xscl = 1   Yscl = 1

Figure 18c

19. Let $g(x) = y = \dfrac{1}{x}$, then $f(x) = 2g(x)$.

To obtain the graph of f, stretch the graph of $y = \dfrac{1}{x}$ vertically by a factor of 2. See Figures 19a and 19b.

The domain is $(-\infty, 0) \cup (0, \infty)$. The range is $(-\infty, 0) \cup (0, \infty)$.

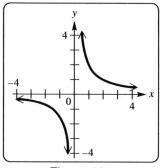

Figure 19a

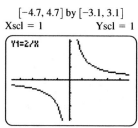

$[-4.7, 4.7]$ by $[-3.1, 3.1]$
Xscl = 1   Yscl = 1

Figure 19b

20. Let $g(x) = y = \dfrac{1}{x}$, then $f(x) = -3g(x)$.

To obtain the graph of f, stretch the graph of $y = \dfrac{1}{x}$ vertically by a factor of 3 and reflect it across the x-axis *or*

the y-axis. See Figures 20a and 20b. The domain is $(-\infty, 0) \cup (0, \infty)$. The range is $(-\infty, 0) \cup (0, \infty)$.

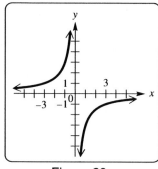

Figure 20a

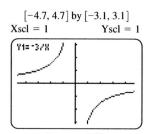

$[-4.7, 4.7]$ by $[-3.1, 3.1]$
Xscl = 1   Yscl = 1

Figure 20b

21. Let $g(x) = y = \dfrac{1}{x}$, then $f(x) = g(x + 2)$.

To obtain the graph of f, shift the graph of $y = \dfrac{1}{x}$ to the left 2 units. See Figures 21a and 21b.

The domain is $(-\infty, -2) \cup (-2, \infty)$. The range is $(-\infty, 0) \cup (0, \infty)$.

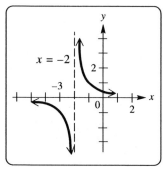

Figure 21a

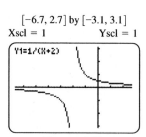

$[-6.7, 2.7]$ by $[-3.1, 3.1]$
Xscl = 1   Yscl = 1

Figure 21b

22. Let $g(x) = y = \dfrac{1}{x}$, then $f(x) = g(x - 3)$.

    To obtain the graph of f, shift the graph of $y = \dfrac{1}{x}$ to the right 3 units. See Figures 22a and 22b.

    The domain is $(-\infty, 3) \cup (3, \infty)$. The range is $(-\infty, 0) \cup (0, \infty)$.

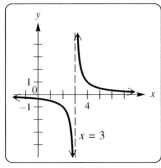

Figure 22a

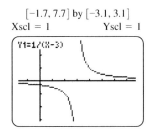

Figure 22b

23. Let $g(x) = y = \dfrac{1}{x}$, then $f(x) = g(x) + 1$.

    To obtain the graph of f, shift the graph of $y = \dfrac{1}{x}$ upward 1 unit. See Figures 23a and 23b.

    The domain is $(-\infty, 0) \cup (0, \infty)$. The range is $(-\infty, 1) \cup (1, \infty)$.

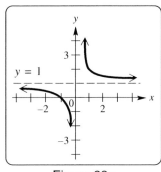

Figure 23a

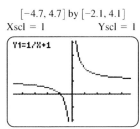

Figure 23b

24. Let $g(x) = y = \dfrac{1}{x}$, then $f(x) = g(x) - 2$.

    To obtain the graph of f, shift the graph of $y = \dfrac{1}{x}$ downward 2 units. See Figures 24a and 24b.

    The domain is $(-\infty, 0) \cup (0, \infty)$. The range is $(-\infty, -2) \cup (-2, \infty)$.

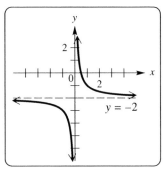

Figure 24a

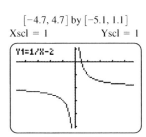

Figure 24b

25. Let $g(x) = y = \dfrac{1}{x}$, then $f(x) = g(x - 1) + 1$.

    To obtain the graph of f, shift the graph of $y = \dfrac{1}{x}$ to the right 1 unit and upward 1 unit See Figures 25a and 25b.

    The domain is $(-\infty, 1) \cup (1, \infty)$. The range is $(-\infty, 1) \cup (1, \infty)$.

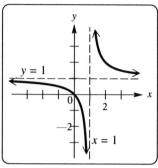

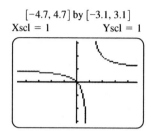

$[-4.7, 4.7]$ by $[-3.1, 3.1]$
Xscl $= 1$ Yscl $= 1$

Figure 25a   Figure 25b

26. Let $g(x) = y = \dfrac{1}{x}$, then $f(x) = 2g(x + 2) - 1$.

    To obtain the graph of f, stretch the graph of $y = \dfrac{1}{x}$ vertically by a factor of 2, shift it to the left 2 units and

    downward 1 unit. See Figures 26a and 26b. The domain is $(-\infty, -2) \cup (-2, \infty)$. The range is $(-\infty, -1) \cup (-1, \infty)$.

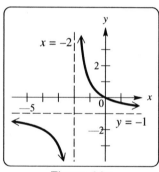

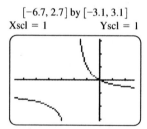

$[-6.7, 2.7]$ by $[-3.1, 3.1]$
Xscl $= 1$ Yscl $= 1$

Figure 26a   Figure 26b

27. Let $g(x) = y = \dfrac{1}{x^2}$, then $f(x) = g(x) - 2$.

    To obtain the graph of f, shift the graph of $y = \dfrac{1}{x^2}$ downward 2 units. See Figures 27a and 27b.

    The domain is $(-\infty, 0) \cup (0, \infty)$. The range is $(-2, \infty)$.

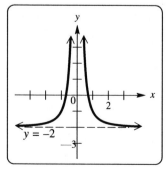

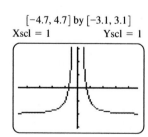

$[-4.7, 4.7]$ by $[-3.1, 3.1]$
Xscl $= 1$ Yscl $= 1$

Figure 27a   Figure 27b

28. Let $g(x) = y = \dfrac{1}{x^2}$, then $f(x) = g(x) + 3$.

To obtain the graph of f, shift the graph of $y = \dfrac{1}{x^2}$ upward 3 units. See Figures 28a and 28b.

The domain is $(-\infty, 0) \cup (0, \infty)$. The range is $(3, \infty)$.

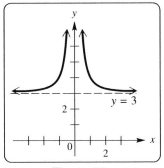

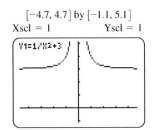

$[-4.7, 4.7]$ by $[-1.1, 5.1]$
Xscl = 1   Yscl = 1

Figure 28a   Figure 28b

29. Let $g(x) = y = \dfrac{1}{x^2}$, then $f(x) = -2g(x)$.

To obtain the graph of f, stretch the graph of $y = \dfrac{1}{x^2}$ vertically by a factor of 2 and reflect it across the x-axis.

See Figures 29a and 29b. The domain is $(-\infty, 0) \cup (0, \infty)$. The range is $(-\infty, 0)$.

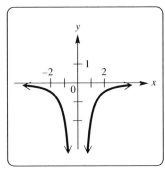

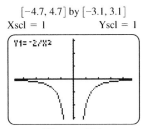

$[-4.7, 4.7]$ by $[-3.1, 3.1]$
Xscl = 1   Yscl = 1

Figure 29a   Figure 29b

30. Let $g(x) = y = \dfrac{1}{x^2}$, then $f(x) = \dfrac{1}{2}g(x)$.

To obtain the graph of f, shrink the graph of $y = \dfrac{1}{x^2}$ vertically by a factor of $\dfrac{1}{2}$. See Figures 30a and 30b.

The domain is $(-\infty, 0) \cup (0, \infty)$. The range is $(0, \infty)$.

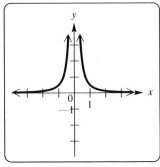

$[-4.7, 4.7]$ by $[-3.1, 3.1]$
Xscl = 1   Yscl = 1

Figure 30a   Figure 30b

31. Let $g(x) = y = \dfrac{1}{x^2}$, then $f(x) = g(x - 3)$.

To obtain the graph of f, shift the graph of $y = \dfrac{1}{x^2}$ to the right 3 units. See Figures 31a and 31b.

The domain is $(-\infty, 3) \cup (3, \infty)$. The range is $(0, \infty)$.

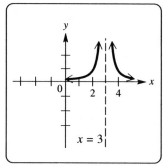

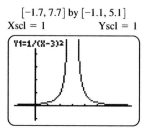

$[-1.7, 7.7]$ by $[-1.1, 5.1]$
Xscl = 1   Yscl = 1

Figure 31a   Figure 31b

32. Let $g(x) = y = \dfrac{1}{x^2}$, then $f(x) = -2g(x - 3)$.

To obtain the graph of f, stretch the graph of $y = \dfrac{1}{x^2}$ vertically by a factor of 2, reflect it across the x-axis and

shift it to the right 3 units. See Figures 32a and 32b. The domain is $(-\infty, 3) \cup (3, \infty)$. The range is $(-\infty, 0)$.

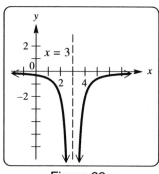

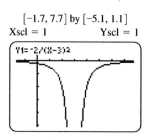

$[-1.7, 7.7]$ by $[-5.1, 1.1]$
Xscl = 1   Yscl = 1

Figure 32a   Figure 32b

33. Let $g(x) = y = \dfrac{1}{x^2}$, then $f(x) = -g(x + 2) - 3$.

To obtain the graph of f, shift the graph of $y = \dfrac{1}{x^2}$ to the left 2 units, reflect it across the x-axis and shift it

downward 3 units. See Figures 33a and 33b. The domain is $(-\infty, -2) \cup (-2, \infty)$. The range is $(-\infty, -3)$.

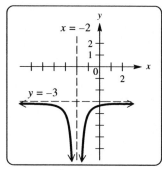

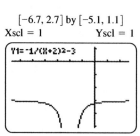

$[-6.7, 2.7]$ by $[-5.1, 1.1]$
Xscl = 1   Yscl = 1

Figure 33a   Figure 33b

34. Let $g(x) = y = \dfrac{1}{x^2}$, then $f(x) = -g(x - 4) + 2$.

    To obtain the graph of f, shift the graph of $y = \dfrac{1}{x^2}$ to the right 4 units, reflect it across the x-axis and shift it

    upward 2 units. See Figures 34a and 34b. The domain is $(-\infty, 4) \cup (4, \infty)$. The range is $(-\infty, 2)$.

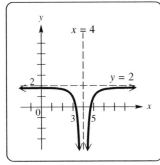

Figure 34a

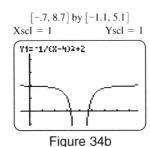

Figure 34b

35. The graph of f is obtained by shifting the graph of $y = \dfrac{1}{x^2}$ to the right 2 units. This is shown in graph C.

36. The graph of f is obtained by shifting the graph of $y = \dfrac{1}{x}$ to the right 2 units. This is shown in graph A.

37. The graph of f is obtained by shifting the graph of $y = \dfrac{1}{x}$ to the right 2 units and reflecting it across the x-axis.
    This is shown in graph B.

38. The graph of f is obtained by shifting the graph of $y = \dfrac{1}{x^2}$ to the right 2 units and reflecting it across the x-axis.
    This is shown in graph D.

39. The vertical asymptote shifts 2 units left to $x = -1$. The horizontal asymptote shifts 1 unit down to $y = 1$.
    The domain shifts 2 units left to $(-\infty, -1) \cup (-1, \infty)$. The range shifts 1 unit down to $(-\infty, 1) \cup (1, \infty)$.

40. The vertical asymptote shifts 1 unit right to $x = 2$. The horizontal asymptote shifts 3 units up to $y = 5$.
    The domain shifts 1 unit right to $(-\infty, 2) \cup (2, \infty)$. The range shifts 3 units up to $(-\infty, 5) \cup (5, \infty)$.

41. Perform long division:
$$x - 2 \overline{)\,x - 1\,} \quad \Rightarrow \quad \dfrac{x - 1}{x - 2} = 1 + \dfrac{1}{x - 2}$$
$$\dfrac{x - 2}{1}$$

    The graph of $y = 1 + \dfrac{1}{x - 2}$ is obtained by shifting the graph of $y = \dfrac{1}{x}$ to the right 2 units and 1 unit upward.

    A sketch is shown in Figure 41a. A calculator graph of $y_1 = (x - 1)/(x - 2)$ is shown in Figure 41b.

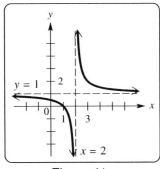

Figure 41a

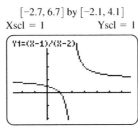

Figure 41b

42. Perform long division:

$$x - 3 \overline{\smash{\big)}\, x - 2} \quad\Rightarrow\quad \frac{x-2}{x-3} = 1 + \frac{1}{x-3}$$
$$\underline{x-3}$$
$$1$$

The graph of $y = 1 + \dfrac{1}{x-3}$ is obtained by shifting the graph of $y = \dfrac{1}{x}$ to the right 3 units and 1 unit upward.

A sketch is shown in Figure 42a. A calculator graph of $y_1 = (x-2)/(x-3)$ is shown in Figure 42b.

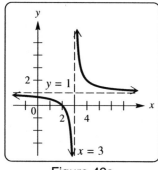

Figure 42a

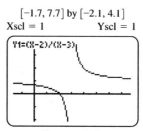

$[-1.7, 7.7]$ by $[-2.1, 4.1]$
Xscl = 1   Yscl = 1

Figure 42b

43. Perform long division:

$$x + 3 \overline{\smash{\big)}\, -2x - 5} \quad\Rightarrow\quad \frac{-2x-5}{x+3} = -2 + \frac{1}{x+3}$$
$$\underline{-2x-6}$$
$$1$$

The graph of $y = -2 + \dfrac{1}{x+3}$ is obtained by shifting the graph of $y = \dfrac{1}{x}$ to the left 3 units and 2 units downward.

A sketch is shown in Figure 43a. A calculator graph of $y_1 = (-2x-5)/(x+3)$ is shown in Figure 43b.

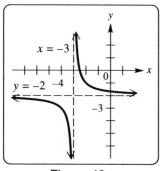

Figure 43a

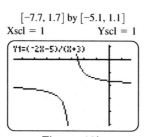

$[-7.7, 1.7]$ by $[-5.1, 1.1]$
Xscl = 1   Yscl = 1

Figure 43b

44. Perform long division:

$$x + 1 \overline{\smash{\big)}\, -2x - 1} \quad\Rightarrow\quad \frac{-2x-1}{x+1} = -2 + \frac{1}{x+1}$$
$$\underline{-2x-2}$$
$$1$$

The graph of $y = -2 + \dfrac{1}{x+1}$ is obtained by shifting the graph of $y = \dfrac{1}{x}$ to the left 1 unit and 2 units downward.

A sketch is shown in Figure 44a. A calculator graph of $y_1 = (-2x-1)/(x+1)$ is shown in Figure 44b.

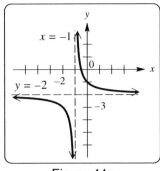

Figure 44a

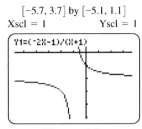

Figure 44b

45. Perform long division:

$$x - 3 \overline{)\begin{array}{r} 2 \\ 2x - 5 \\ \underline{2x - 6} \\ 1 \end{array}} \quad \Rightarrow \quad \frac{2x - 5}{x - 3} = 2 + \frac{1}{x - 3}$$

The graph of $y = 2 + \dfrac{1}{x - 3}$ is obtained by shifting the graph of $y = \dfrac{1}{x}$ to the right 3 units and 2 units upward.

A sketch is shown in Figure 45a. A calculator graph of $y_1 = (2x - 5)/(x - 3)$ is shown in Figure 45b.

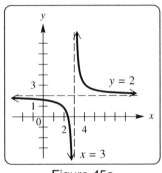

Figure 45a

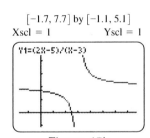

Figure 45b

46. Perform long division:

$$x - 1 \overline{)\begin{array}{r} 2 \\ 2x - 1 \\ \underline{2x - 2} \\ 1 \end{array}} \quad \Rightarrow \quad \frac{2x - 1}{x - 1} = 2 + \frac{1}{x - 1}$$

The graph of $y = 2 + \dfrac{1}{x - 1}$ is obtained by shifting the graph of $y = \dfrac{1}{x}$ to the right 1 unit and 2 units upward.

A sketch is shown in Figure 46a. A calculator graph of $y_1 = (2x - 1)/(x - 1)$ is shown in Figure 46b.

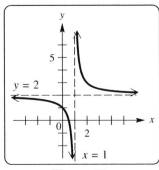

Figure 46a

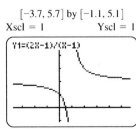

Figure 46b

4.2: More on Graphs of Rational Functions

1. Because -1 is a zero of the denominator but not of the numerator, the line $x = -1$ is a vertical asymptote: D.

2. Because $f(0) = \dfrac{0 + 10}{0 + 2} = \dfrac{10}{2} = 5$, the y-intercept is 5: B.

3. The degree of the numerator is less than the degree of the denominator, so the x-axis is a horizontal asymptote: G.

4. Because 0 is a zero of the denominator but not of the numerator, the y-axis is a vertical asymptote: H.

5. Because $f(x) = \dfrac{x^2 - 16}{x + 4} = \dfrac{(x - 4)(x + 4)}{x + 4} = x - 4$ for all $x \neq -4$, there is a hole at $x = -4$: E.

6. The degrees of the numerator and denominator are equal and the horizontal asymptote is $y = \dfrac{4}{1} = 4$: C.

7. Because $f(x) = \dfrac{x^2 + 3x + 4}{x - 5} = x + 8 + \dfrac{44}{x - 5}$, the line $y = x + 8$ is an oblique asymptote: F.

8. Because $f(-3) = \dfrac{-3 + 3}{-3 - 6} = \dfrac{0}{-9} = 0$, the x-intercept is -3: A.

9. Since 5 is a zero of the denominator but not of the numerator, the line $x = 5$ is a vertical asymptote.

   Since $f(x) \to 0$ as $|x| \to \infty$, the line $y = 0$ is a horizontal asymptote.

   Since the degree of the numerator is less than the degree of the denominator, there is no oblique asymptote.

10. Since -9 is a zero of the denominator but not of the numerator, the line $x = -9$ is a vertical asymptote.

    Since $f(x) \to 0$ as $|x| \to \infty$, the line $y = 0$ is a horizontal asymptote.

    Since the degree of the numerator is less than the degree of the denominator, there is no oblique asymptote.

11. Since $-\dfrac{1}{2}$ is a zero of the denominator but not of the numerator, the line $x = -\dfrac{1}{2}$ is a vertical asymptote.

    Since the degrees of the numerator and denominator are equal, the horizontal asymptote is $y = \dfrac{-3}{2} = -\dfrac{3}{2}$.

    Since the degree of the numerator is equal to the degree of the denominator, there is no oblique asymptote.

12. Since 4 is a zero of the denominator but not of the numerator, the line $x = 4$ is a vertical asymptote.

    Since the degrees of the numerator and denominator are equal, the horizontal asymptote is $y = \dfrac{2}{1} = 2$.

    Since the degree of the numerator is equal to the degree of the denominator, there is no oblique asymptote.

13. Since -3 is a zero of the denominator but not of the numerator, the line $x = -3$ is a vertical asymptote.

    Since the degree of the numerator is greater than the degree of the denominator, there is no horizontal asymptote.

    Since $\dfrac{x^2 - 1}{x + 3} = x - 3 + \dfrac{8}{x + 3}$, the oblique asymptote is $y = x - 3$.

14. Since 1 is a zero of the denominator but not of the numerator, the line $x = 1$ is a vertical asymptote.

    Since the degree of the numerator is greater than the degree of the denominator, there is no horizontal asymptote.

    Since $\dfrac{x^2 + 4}{x - 1} = x + 1 + \dfrac{5}{x - 1}$, the oblique asymptote is $y = x + 1$.

15. Since -2 and $\dfrac{5}{2}$ are zeros of the denominator but not of the numerator, the lines $x = -2$ and $x = \dfrac{5}{2}$ are vertical asymptotes.

    Since the degrees of the numerator and denominator are equal, the horizontal asymptote is $y = \dfrac{1}{2}$.

    Since the degree of the numerator is equal to the degree of the denominator, there is no oblique asymptote.

16. Since $\dfrac{1}{5}$ and 5 are zeros of the denominator but not of the numerator, the lines $x = \dfrac{1}{5}$ and $x = 5$ are vertical asymptotes.

    Since the degrees of the numerator and denominator are equal, the horizontal asymptote is $y = \dfrac{3}{5}$.

    Since the degree of the numerator is equal to the degree of the denominator, there is no oblique asymptote.

17. Function A, because the denominator can never be equal to 0.

18. Function C, because the degree of the numerator is greater than the degree of the denominator.

19. From the graph, the vertical asymptote is $x = 2$, the horizontal asymptote is $y = 4$, and there is no oblique asymptote.
    The function is defined for all $x \ne 2$, therefore the domain is $(-\infty, 2) \cup (2, \infty)$.

20. From the graph, the vertical asymptotes are $x = -4$ and $x = 4$, the horizontal asymptote is $y = 2$, and there is no
    oblique asymptote. The function is defined for all $x \ne \pm 4$, therefore the domain is $(-\infty, -4) \cup (-4, 4) \cup (4, \infty)$.

21. From the graph, the vertical asymptotes are $x = -2$ and $x = 2$, the horizontal asymptote is $y = -4$, and there is no
    oblique asymptote. The function is defined for all $x \ne \pm 2$, therefore the domain is $(-\infty, -2) \cup (-2, 2) \cup (2, \infty)$.

22. From the graph, the vertical asymptote is $x = -4$, the horizontal asymptote is $y = -2$, and there is no oblique asymptote.
    The function is defined for all $x \ne -4$, therefore the domain is $(-\infty, -4) \cup (-4, \infty)$.

23. From the graph, there is no vertical asymptote, the horizontal asymptote is $y = 0$, and there is no oblique asymptote.
    The function is defined for all x, therefore the domain is $(-\infty, \infty)$.

24. From the graph, the vertical asymptote is $x = 0$, the horizontal asymptote is $y = 0$, and there is no oblique asymptote.
    The function is defined for all $x \ne 0$, therefore the domain is $(-\infty, 0) \cup (0, \infty)$.

25. From the graph, the vertical asymptote is $x = -1$, there is no horizontal asymptote, and the oblique asymptote
    passes through the points $(0, -1)$ and $(1, 0)$. Thus, the equation of the oblique asymptote is $y = x - 1$.
    The function is defined for all $x \ne -1$, therefore the domain is $(-\infty, -1) \cup (-1, \infty)$.

26. From the graph, the vertical asymptote is $x = \dfrac{1}{2}$, there is no horizontal asymptote, and the oblique asymptote

    passes through the points $\left(-\dfrac{1}{2}, 0\right)$ and $(0, 1)$. Thus, the equation of the oblique asymptote is $y = 2x + 1$.

    The function is defined for all $x \ne \dfrac{1}{2}$, therefore the domain is $\left(-\infty, \dfrac{1}{2}\right) \cup \left(\dfrac{1}{2}, \infty\right)$.

27. See Figure 27.

28. See Figure 28.

29. See Figure 29.

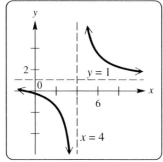

Figure 27

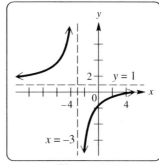

Figure 28

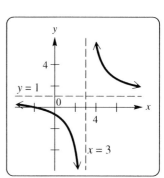

Figure 29

30. See Figure 30.

31. See Figure 31.

32. See Figure 32.

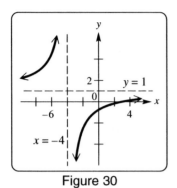

Figure 30

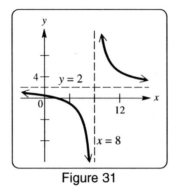

Figure 31

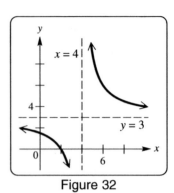

Figure 32

33. See Figure 33.

34. See Figure 34.

35. See Figure 35.

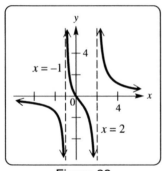

Figure 33

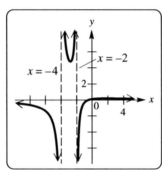

Figure 34

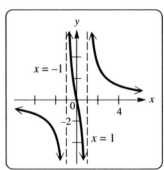

Figure 35

36. See Figure 36.

37. See Figure 37.

38. See Figure 38.

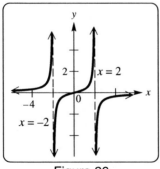

Figure 36

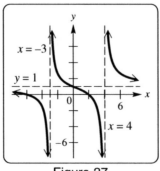

Figure 37

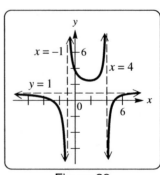

Figure 38

39. See Figure 39.

40. See Figure 40.

41. See Figure 41.

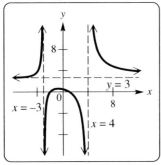

Figure 39

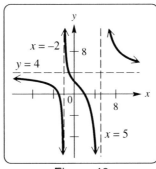

Figure 40

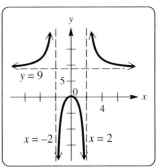

Figure 41

42. See Figure 42.

43. See Figure 43.

44. See Figure 44.

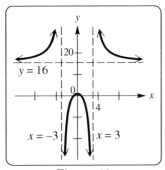

Figure 42

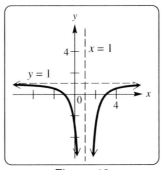

Figure 43

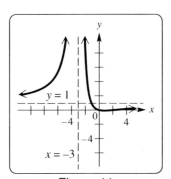

Figure 44

45. See Figure 45.

46. See Figure 46.

47. See Figure 47.

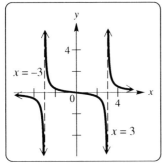

Figure 45

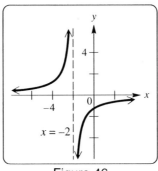

Figure 46

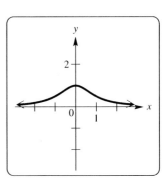

Figure 47

48. See Figure 48.

49. See Figure 49.

50. See Figure 50.

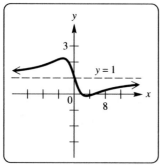

Figure 48

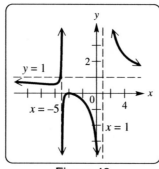

Figure 49

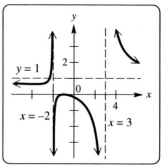

Figure 50

51. See Figure 51.

52. See Figure 52.

53. See Figure 53.

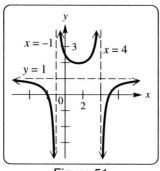

Figure 51

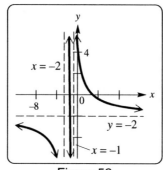

Figure 52

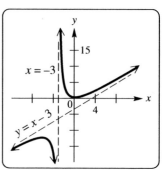

Figure 53

54. See Figure 54.

55. See Figure 55.

56. See Figure 56.

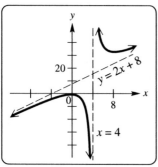

Figure 54

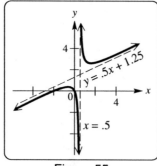

Figure 55

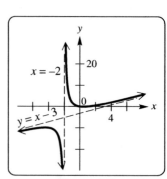

Figure 56

57. See Figure 57.

58. See Figure 58.

59. See Figure 59.

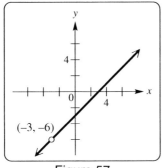

Figure 57

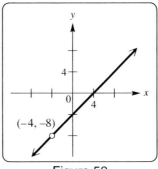

Figure 58

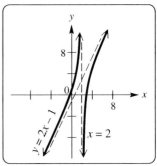

Figure 59

60. See Figure 60.

61. See Figure 61.

62. See Figure 62.

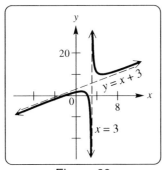

Figure 60

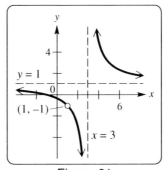

Figure 61

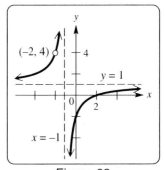

Figure 62

63. See Figure 63.

64. See Figure 64.

65. See Figure 65.

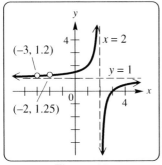

Figure 63

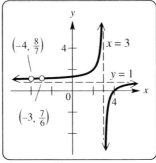

Figure 64

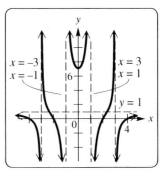

Figure 65

66. See Figure 66.

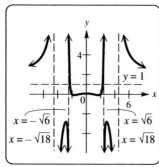

Figure 66

67. The expression is equal to 0 when $x = 4$, thus $p = 4$. The expression is undefined when $x = 2$, thus $q = 2$.

68. The expression is equal to 0 when $x = 3$, thus $p = 3$. The expression is undefined when $x = 1$, thus $q = 1$.

69. The expression is equal to 0 when $x = -2$, thus $p = -2$. The expression is undefined when $x = -1$, thus $q = -1$.

70. The expression is equal to 0 when $x = -4$, thus $p = -4$. The expression is undefined when $x = -3$, thus $q = -3$.

71. A vertical asymptote of $x = 2$ indicates that the denominator should contain the factor $(x - 2)$. A hole when $x = -2$ indicates that both the numerator and denominator should contain the factor $(x + 2)$. An x-intercept of 3 indicates that the numerator should contain the factor $(x - 3)$. One possible equation for the function is $f(x) = \dfrac{(x - 3)(x + 2)}{(x - 2)(x + 2)} = \dfrac{x^2 - x - 6}{x^2 - 4}$. Other functions are possible.

72. A vertical asymptote of $x = -2$ indicates that the denominator should contain the factor $(x + 2)$. An x-intercept of 0 indicates that the numerator should contain the factor $(x - 0)$ or x and an x-intercept of -4 indicates that the numerator should also contain the factor $(x + 4)$. Since there is a horizontal asymptote at $y = 2$, the degrees of the numerator and denominator should be equal and the leading coefficient of the numerator should be 2 times that of the denominator. Since the degree of the numerator is 2 and there are no additional vertical asymptotes, the denominator requires a second factor of $(x + 2)$ to raise its degree to 2. One possible equation for the function is $f(x) = \dfrac{2x(x + 4)}{(x + 2)^2}$. Other functions are possible.

73. Vertical asymptotes of $x = 0$ and $x = 4$ indicate that the denominator should contain the factors x and $(x - 4)$. An x-intercept of 2 indicates that the numerator should contain the factor $(x - 2)$. Since there is a horizontal asymptote at $y = 0$, the degree of the numerator should be less than the degree of the denominator. One possible equation for the function is $f(x) = \dfrac{x - 2}{x(x - 4)} = \dfrac{x - 2}{x^2 - 4x}$. Other functions are possible.

74. Since there is no vertical asymptote, the denominator must be an expression that can never equal 0, such as $x^2 + 1$. Since there is a horizontal asymptote at $y = 0$, the degree of the numerator should be less than the degree of the denominator. A y-intercept of -4 indicates that $f(0) = -4$. One possible equation for the function is $f(x) = \dfrac{-4}{x^2 + 1}$. Other functions are possible.

75. Use the quadratic formula; $a = 1, b = 1, c = 4$: $x = \dfrac{-1 \pm \sqrt{1^2 - 4(1)(4)}}{2(1)} = \dfrac{-1 \pm \sqrt{-15}}{2} = -\dfrac{1}{2} \pm \dfrac{\sqrt{15}}{2}i$

There are no real solutions, therefore, there are no vertical asymptotes.

76. The graph of $y_1 = (x + 3)/(x^2 + x + 4)$ is shown in Figure 76. Connected mode is acceptable since there are no x-values for which f is undefined. (Such values often cause problems for graphing calculators.)

77. (a) Reflect the graph across the x-axis. See Figure 77a.

    (b) Reflect the graph across the y-axis. See Figure 77b.

$[-10, 10]$ by $[-1, 3]$
Xscl = 1   Yscl = 1

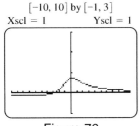

Figure 76

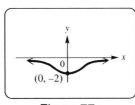

Figure 77a

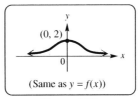

Figure 77b

78. (a) Reflect the graph across the x-axis. See Figure 78a.

    (b) Reflect the graph across the y-axis. See Figure 78b.

79. (a) Reflect the graph across the x-axis. See Figure 79a.

    (b) Reflect the graph across the y-axis. See Figure 79b.

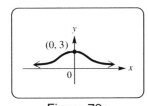

Figure 78a

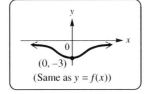

Figure 78b

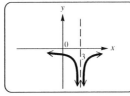

Figure 79a

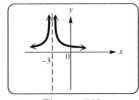

Figure 79b

80. (a) Reflect the graph across the x-axis. See Figure 80a.

    (b) Reflect the graph across the y-axis. See Figure 80b.

$[-18.8, 18.8]$ by $[-50, 25]$
Xscl = 1   Yscl = 5

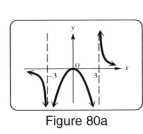

Figure 80a

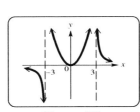

Figure 80b

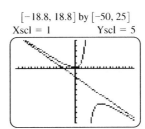

Figure 81

81. Perform long division:

$$-x + 4 \overline{\smash{\big)}\,2x^2 + 0x + 3} \quad\Rightarrow\quad \frac{2x^2 + 3}{4 - x} = -2x - 8 + \frac{35}{4 - x}$$

$$\begin{array}{r} -2x - 8 \\ -x + 4 \,\overline{\smash{\big)}\, 2x^2 + 0x + 3} \\ \underline{2x^2 - 8x} \\ 8x + 3 \\ \underline{8x - 32} \\ 35 \end{array}$$

The oblique asymptote is $y = -2x - 8$.

The graphs of $y_1 = (2x^2 + 3)/(4 - x)$ and $y_2 = -2x - 8$ are shown in Figure 81.

82. Perform long division:

$$\begin{array}{r} x - 3 \\ x + 3 \overline{\smash{\big)}\ x^2 + 0x + 9} \\ \underline{x^2 + 3x} \\ -3x + 9 \\ \underline{-3x - 9} \\ 18 \end{array}$$

$$\Rightarrow \quad \frac{x^2 + 9}{x + 3} = x - 3 + \frac{18}{x + 3}$$

The oblique asymptote is $y = x - 3$.

The graphs of $y_1 = (x^2 + 9)/(x + 3)$ and $y_2 = x - 3$ are shown in Figure 82.

83. Perform long division:

$$\begin{array}{r} -x + 3 \\ x + 2 \overline{\smash{\big)}\ -x^2 + x + 0} \\ \underline{-x^2 - 2x} \\ 3x + 0 \\ \underline{3x + 6} \\ -6 \end{array}$$

$$\Rightarrow \quad \frac{x - x^2}{x + 2} = -x + 3 + \frac{-6}{x + 2}$$

The oblique asymptote is $y = -x + 3$.

The graphs of $y_1 = (x - x^2)/(x + 2)$ and $y_2 = -x + 3$ are shown in Figure 83.

84. Perform long division:

$$\begin{array}{r} -\frac{1}{2}x - \frac{5}{4} \\ -2x + 1 \overline{\smash{\big)}\ x^2 + 2x + 0} \\ \underline{x^2 - \frac{1}{2}x} \\ \frac{5}{2}x + 0 \\ \underline{\frac{5}{2}x - \frac{5}{4}} \\ \frac{5}{4} \end{array}$$

$$\Rightarrow \quad \frac{x^2 + 2x}{1 - 2x} = -\frac{1}{2}x - \frac{5}{4} + \frac{\frac{5}{4}}{1 - 2x}$$

The oblique asymptote is $y = -\frac{1}{2}x - \frac{5}{4}$.

The graphs of $y_1 = (x^2 + 2x)/(1 - 2x)$ and $y_2 = (-1/2)x - 5/4$ are shown in Figure 84.

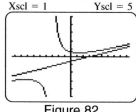

Figure 82

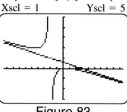

Figure 83

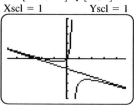

Figure 84

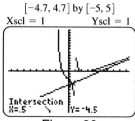

Figure 86

85. (a) Disregard the remainder to get the equation of the oblique asymptote, $y = x + 1$.

(b) Set the original function equal to the expression for the oblique asymptote and solve:

$$\frac{x^5 + x^4 + x^2 + 1}{x^4 + 1} = x + 1 \Rightarrow x^5 + x^4 + x^2 + 1 = (x^4 + 1)(x + 1) \Rightarrow$$

$$x^5 + x^4 + x^2 + 1 = x^5 + x^4 + x + 1 \Rightarrow x^2 + 1 = x + 1 \Rightarrow x^2 - x = 0 \Rightarrow x(x - 1) = 0 \Rightarrow x = 0, 1$$

The graph crosses the oblique asymptote at $x = 0$ and $x = 1$.

(c) For large values of x, $x + 1 < \dfrac{x^5 + x^4 + x^2 + 1}{x^4 + 1}$. Thus, the function approaches its asymptote from above.

86. Perform long division:

$$\begin{array}{r} x - 5 \\ x^2 + x - 2 \overline{\smash{\big)}\ x^3 - 4x^2 + x + 6} \\ \underline{x^3 + x^2 - 2x} \\ -5x^2 + 3x + 6 \\ \underline{-5x^2 - 5x + 10} \\ 8x - 4 \end{array}$$

$$\Rightarrow \quad \frac{x^3 - 4x^2 + x + 6}{x^2 + x - 2} = x - 5 + \frac{8x - 4}{x^2 + x - 2}$$

The oblique asymptote is $y = x - 5$.

$$\frac{x^3 - 4x^2 + x + 6}{x^2 + x - 2} = x - 5 \Rightarrow x^3 - 4x^2 + x + 6 = (x^2 + x - 2)(x - 5) \Rightarrow$$

$$x^3 - 4x^2 + x + 6 = x^3 - 4x^2 - 7x + 10 \Rightarrow x + 6 = -7x + 10 \Rightarrow 8x = 4 \Rightarrow x = .5$$

Noting that $f(.5) = -4.5$, the coordinates of the point of intersection are $(.5, -4.5)$. See Figure 86.

87.

$$x^2 + x - 12 \overline{\smash{\big)}\, x^4 + 0x^3 - 5x^2 + 0x + 4} \quad \begin{array}{r} x^2 - x + 8 \end{array}$$

$$\begin{array}{r} x^4 + x^3 - 12x^2 \\ \hline -x^3 + 7x^2 + 0x + 4 \\ -x^3 - x^2 + 12x \\ \hline 8x^2 - 12x + 4 \\ 8x^2 + 8x - 96 \\ \hline -20x + 100 \end{array}$$

$$\Rightarrow \quad \frac{x^4 - 5x^2 + 4}{x^2 + x - 12} = x^2 - x + 8 + \frac{-20x + 100}{x^2 + x - 12}$$

The graphs of $y_1 = (x^4 - 5x + 4)/(x^2 + x - 12)$ and $y_2 = x^2 - x + 8$ are shown in Figure 87.

In this window, the two graphs seem to overlap (coincide), suggesting that as $|x| \to \infty$, the graph of f approaches the graph of g, giving an asymptotic effect.

$[-50, 50]$ by $[0, 1000]$
Xscl = 10   Yscl = 100

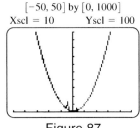

Figure 87

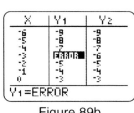

Figure 89b

$[-9.4, 9.4]$ by $[-9.3, 3.1]$
Xscl = 1   Yscl = 1

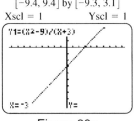

Figure 89c

88. $f(x) = \dfrac{x^2 - 25}{x + 5} = \dfrac{(x - 5)(x + 5)}{x + 5} = x - 5$, for $x \neq -5$; The graph does not have an asymptote with equation

$x = -5$, but is actually the graph of $y = x - 5$ with a "hole" at $(-5, -10)$.

89. (a) $f(x) = \dfrac{x^2 - 9}{x + 3} = \dfrac{(x - 3)(x + 3)}{x + 3} = x - 3$, for $x \neq -3$ so $g(x) = x - 3$, for $x \neq -3$

(b) The ERROR message indicates that f is undefined at $x = -3$. See Figure 89b.

(c) Since $g(-3) = -6$, the "hole" is located at $(-3, -6)$. See Figure 89c.

90. (a) $f(x) = \dfrac{x^2 - 36}{6 - x} = \dfrac{(x - 6)(x + 6)}{6 - x} = \dfrac{-1(6 - x)(x + 6)}{6 - x} = -x - 6$, for $x \neq 6$ so $g(x) = -x - 6$, for $x \neq 6$

(b) The ERROR message indicates that f is undefined at $x = 6$. See Figure 90b.

(c) Since $g(6) = -12$, the "hole" is located at $(6, -12)$. See Figure 90c.

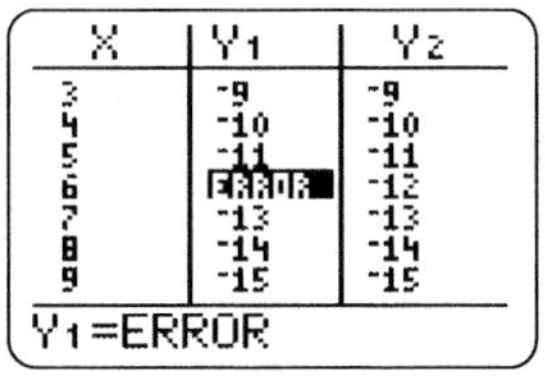

$[-9.4, 9.4]$ by $[-14.2, -1.8]$
Xscl = 1   Yscl = 1

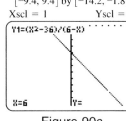

Figure 90b

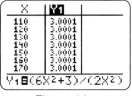

Figure 90c

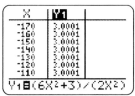

Figure 92a

Figure 92b

91. (a) The horizontal asymptote is $y = 0$ (the x-axis).

(b) The horizontal asymptote is $y = \dfrac{a}{b}$, where a and b are the leading coefficients of $p(x)$ and $q(x)$, respectively.

(c) The oblique asymptote is $y = ax + b$, where $ax + b$ is the quotient (with remainder disregarded) found by dividing $p(x)$ by $q(x)$.

92. Table $Y_1 = (6X^2 + 3)/(2X^2)$ as shown in Figures 92a and 92b. As $|X| \to \infty$, $Y_1 \to 3$.

4.3: Rational Equations, Inequalities, Applications, and Models

1. (a) The graph never intersects the x-axis. The solution set of $f(x) = 0$ is $\varnothing$.

   (b) The graph is below the x-axis on the interval $(-\infty, -2)$. Thus $f(x) < 0$ on $(-\infty, -2)$.

   (c) The graph is above the x-axis on the interval $(-2, \infty)$. Thus $f(x) > 0$ on $(-2, \infty)$.

2. (a) The graph never intersects the x-axis. The solution set of $f(x) = 0$ is $\varnothing$.

   (b) The graph is below the x-axis on the interval $(-\infty, 3)$. Thus $f(x) < 0$ on $(-\infty, 3)$.

   (c) The graph is above the x-axis on the interval $(3, \infty)$. Thus $f(x) > 0$ on $(3, \infty)$.

3. (a) The graph intersects the x-axis at $x = -1$. The solution set of $f(x) = 0$ is $\{-1\}$.

   (b) The graph is below the x-axis on the interval $(-1, 0)$. Thus $f(x) < 0$ on $(-1, 0)$.

   (c) The graph is above the x-axis on the interval $(-\infty, -1) \cup (0, \infty)$. Thus $f(x) > 0$ on $(-\infty, -1) \cup (0, \infty)$.

4. (a) The graph intersects the x-axis at $x = \dfrac{1}{2}$. The solution set of $f(x) = 0$ is $\left\{\dfrac{1}{2}\right\}$.

   (b) The graph is below the x-axis on the interval $(-\infty, 0) \cup \left(\dfrac{1}{2}, \infty\right)$. Thus $f(x) < 0$ on $(-\infty, 0) \cup \left(\dfrac{1}{2}, \infty\right)$.

   (c) The graph is above the x-axis on the interval $\left(0, \dfrac{1}{2}\right)$. Thus $f(x) > 0$ on $\left(0, \dfrac{1}{2}\right)$.

5. (a) The graph intersects the x-axis at $x = 0$. The solution set of $f(x) = 0$ is $\{0\}$.

   (b) The graph is below the x-axis on the interval $(-2, 0) \cup (2, \infty)$. Thus $f(x) < 0$ on $(-2, 0) \cup (2, \infty)$.

   (c) The graph is above the x-axis on the interval $(-\infty, -2) \cup (0, 2)$. Thus $f(x) > 0$ on $(-\infty, -2) \cup (0, 2)$.

6. (a) The graph intersects the x-axis at $x = 0$. The solution set of $f(x) = 0$ is $\{0\}$.

   (b) The graph is below the x-axis on the interval $(-\infty, -1) \cup (0, 2)$. Thus $f(x) < 0$ on $(-\infty, -1) \cup (0, 2)$.

   (c) The graph is above the x-axis on the interval $(-1, 0) \cup (2, \infty)$. Thus $f(x) > 0$ on $(-1, 0) \cup (2, \infty)$.

7. (a) The graph intersects the x-axis at $x = 0$. The solution set of $f(x) = 0$ is $\{0\}$.

   (b) The graph is below the x-axis on the interval $(-1, 0) \cup (0, 1)$. Thus $f(x) < 0$ on $(-1, 0) \cup (0, 1)$.

   (c) The graph is above the x-axis on the interval $(-\infty, -1) \cup (1, \infty)$. Thus $f(x) > 0$ on $(-\infty, -1) \cup (1, \infty)$.

8. (a) The graph intersects the x-axis at $x = -1$ and $x = 0$. The solution set of $f(x) = 0$ is $\{-1, 0\}$.

   (b) The graph is below the x-axis on the interval $(-2, -1) \cup (0, 1) \cup (1, 2)$. Thus $f(x) < 0$ on the interval $(-2, -1) \cup (0, 1) \cup (1, 2)$.

   (c) The graph is above the x-axis on the interval $(-\infty, -2) \cup (-1, 0) \cup (2, \infty)$. Thus $f(x) > 0$ on the interval $(-\infty, -2) \cup (-1, 0) \cup (2, \infty)$.

9. (a) The graph intersects the x-axis at $x = 0$. The solution set of $f(x) = 0$ is $\{0\}$.

   (b) The graph is below the x-axis on the interval $(1, 2) \cup (2, 3)$. Thus $f(x) < 0$ on $(1, 2) \cup (2, 3)$.

   (c) The graph is above the x-axis on the interval $(-\infty, -2) \cup (-2, 0) \cup (0, 1) \cup (3, \infty)$. Thus $f(x) > 0$ on the interval $(-\infty, -2) \cup (-2, 0) \cup (0, 1) \cup (3, \infty)$.

10. (a) The graph never intersects the x-axis. The solution set of $f(x) = 0$ is $\varnothing$.

    (b) The graph is below the x-axis on the interval $(-\infty, 0) \cup (0, \infty)$. Thus $f(x) < 0$ on $(-\infty, 0) \cup (0, \infty)$.

    (c) The graph is never above the x-axis. The solution set of $f(x) > 0$ is $\varnothing$.

11. (a) The graph intersects the x-axis at $x = 2.5$. The solution set of $f(x) = 0$ is $\{2.5\}$.

 (b) The graph is below the x-axis on the interval $(2.5, 3)$. Thus $f(x) < 0$ on $(2.5, 3)$.

 (c) The graph is above the x-axis on the interval $(-\infty, 2.5) \cup (3, \infty)$. Thus $f(x) > 0$ on $(-\infty, 2.5) \cup (3, \infty)$.

12. (a) The graph intersects the x-axis at $x = .75$. The solution set of $f(x) = 0$ is $\{.75\}$.

 (b) The graph is below the x-axis on the interval $(-\infty, .75) \cup (2, \infty)$. Thus $f(x) < 0$ on $(-\infty, .75) \cup (2, \infty)$.

 (c) The graph is above the x-axis on the interval $(.75, 2)$. Thus $f(x) > 0$ on $(.75, 2)$.

13. His solution is not correct. If $x + 2 < 0$, it is necessary to reverse the direction of the inequality symbol.
 He should use test values, as explained in Example 3. The correct solution set is $\left(-2, \dfrac{1}{2}\right]$.

14. Multiplying by x results in $1 < x$, which is only a partial solution because the inequality is also true for $x < 0$.
 The error occurs because x could be positive or negative. If x is negative, the inequality symbol must be reversed.

15. (a) $\dfrac{x - 3}{x + 5} = 0 \Rightarrow x - 3 = 0(x + 5) \Rightarrow x - 3 = 0 \Rightarrow x = 3$; The solution set is $\{3\}$.

 (b) Graph $y_1 = (x - 3)/(x + 5)$ as shown in Figure 15. With vertical asymptote $x = -5$, the graph is below
 or intersects the x-axis on the interval $(-5, 3]$. The solution set is $(-5, 3]$.

 (c) Graph $y_1 = (x - 3)/(x + 5)$ as shown in Figure 15. With vertical asymptote $x = -5$, the graph is above
 or intersects the x-axis on the interval $(-\infty, -5) \cup [3, \infty)$. The solution set is $(-\infty, -5) \cup [3, \infty)$.

16. (a) $\dfrac{x + 1}{x - 4} = 0 \Rightarrow x + 1 = 0(x - 4) \Rightarrow x + 1 = 0 \Rightarrow x = -1$; The solution set is $\{-1\}$.

 (b) Graph $y_1 = (x + 1)/(x - 4)$ as shown in Figure 16. With vertical asymptote $x = 4$, the graph is above or
 intersects the x-axis on the interval $(-\infty, -1] \cup (4, \infty)$. The solution set is $(-\infty, -1] \cup (4, \infty)$.

 (c) Graph $y_1 = (x + 1)/(x - 4)$ as shown in Figure 16. With vertical asymptote $x = 4$, the graph is below
 or intersects the x-axis on the interval $[-1, 4)$. The solution set is $[-1, 4)$.

17. (a) $\dfrac{x - 1}{x + 2} = 1 \Rightarrow x - 1 = 1(x + 2) \Rightarrow x - 1 = x + 2 \Rightarrow -1 = 2$ (false); The solution set is $\varnothing$.

 (b) Graph $y_1 = (x - 1)/(x + 2) - 1$ as shown in Figure 17. With vertical asymptote $x = -2$, the graph is
 above the x-axis on the interval $(-\infty, -2)$. The solution set is $(-\infty, -2)$.

 (c) Graph $y_1 = (x - 1)/(x + 2) - 1$ as shown in Figure 17. With vertical asymptote $x = -2$, the graph is
 below the x-axis on the interval $(-2, \infty)$. The solution set is $(-2, \infty)$.

18. (a) $\dfrac{x - 6}{x + 2} = -1 \Rightarrow x - 6 = -1(x + 2) \Rightarrow x - 6 = -x - 2 \Rightarrow 2x = 4 \Rightarrow x = 2$; The solution set is $\{2\}$.

 (b) Graph $y_1 = (x - 6)/(x + 2) + 1$ as shown in Figure 18. With vertical asymptote $x = -2$, the graph is
 below the x-axis on the interval $(-2, 2)$. The solution set is $(-2, 2)$.

 (c) Graph $y_1 = (x - 6)/(x + 2) + 1$ as shown in Figure 18. With vertical asymptote $x = -2$, the graph is
 above the x-axis on the interval $(-\infty, -2) \cup (2, \infty)$. The solution set is $(-\infty, -2) \cup (2, \infty)$.

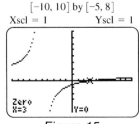
$[-10, 10]$ by $[-5, 8]$
Xscl $= 1$ Yscl $= 1$
Figure 15

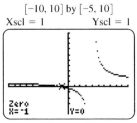
$[-10, 10]$ by $[-5, 10]$
Xscl $= 1$ Yscl $= 1$
Figure 16

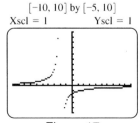
$[-10, 10]$ by $[-5, 10]$
Xscl $= 1$ Yscl $= 1$
Figure 17

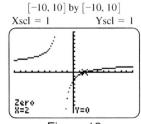
$[-10, 10]$ by $[-10, 10]$
Xscl $= 1$ Yscl $= 1$
Figure 18

19. (a) $\dfrac{1}{x-1} = \dfrac{5}{4} \Rightarrow 5(x-1) = 4(1) \Rightarrow 5x - 5 = 4 \Rightarrow 5x = 9 \Rightarrow x = \dfrac{9}{5}$; The solution set is $\left\{\dfrac{9}{5}\right\}$.

(b) Graph $y_1 = 1/(x-1) - 5/4$ as shown in Figure 19. With vertical asymptote $x = 1$, the graph is below

the x-axis on the interval $(-\infty, 1) \cup \left(\dfrac{9}{5}, \infty\right)$. The solution set is $(-\infty, 1) \cup \left(\dfrac{9}{5}, \infty\right)$.

(c) Graph $y_1 = 1/(x-1) - 5/4$ as shown in Figure 19. With vertical asymptote $x = 1$, the graph is above

the x-axis on the interval $\left(1, \dfrac{9}{5}\right)$. The solution set is $\left(1, \dfrac{9}{5}\right)$.

20. (a) $\dfrac{6}{5-3x} = 2 \Rightarrow 6 = 2(5-3x) \Rightarrow 6 = 10 - 6x \Rightarrow 6x = 4 \Rightarrow x = \dfrac{2}{3}$; The solution set is $\left\{\dfrac{2}{3}\right\}$.

(b) Graph $y_1 = 6/(5-3x) - 2$ as shown in Figure 20. With vertical asymptote $x = \dfrac{5}{3}$, the graph is below

or intersects the x-axis on the interval $\left(-\infty, \dfrac{2}{3}\right] \cup \left(\dfrac{5}{3}, \infty\right)$. The solution set is $\left(-\infty, \dfrac{2}{3}\right] \cup \left(\dfrac{5}{3}, \infty\right)$.

(c) Graph $y_1 = 6/(5-3x) - 2$ as shown in Figure 20. With vertical asymptote $x = \dfrac{5}{3}$, the graph is above

or intersects the x-axis on the interval $\left[\dfrac{2}{3}, \dfrac{5}{3}\right)$. The solution set is $\left[\dfrac{2}{3}, \dfrac{5}{3}\right)$.

21. (a) $\dfrac{4}{x-2} = \dfrac{3}{x-1} \Rightarrow 4(x-1) = 3(x-2) \Rightarrow 4x - 4 = 3x - 6 \Rightarrow x = -2$; The solution set is $\{-2\}$.

(b) Graph $y_1 = 4/(x-2) - 3/(x-1)$ as shown in Figure 21. With vertical asymptotes $x = 1$ and $x = 2$, the

graph is below or intersects the x-axis on $(-\infty, -2] \cup (1, 2)$. The solution set is $(-\infty, -2] \cup (1, 2)$.

(c) Graph $y_1 = 4/(x-2) - 3/(x-1)$ as shown in Figure 21. With vertical asymptotes $x = 1$ and $x = 2$, the

graph is above or intersects the x-axis on $[-2, 1) \cup (2, \infty)$. The solution set is $[-2, 1) \cup (2, \infty)$.

22. (a) $\dfrac{4}{x+1} = \dfrac{2}{x+3} \Rightarrow 4(x+3) = 2(x+1) \Rightarrow 4x + 12 = 2x + 2 \Rightarrow 2x = -10 \Rightarrow x = -5$;

The solution set is $\{-5\}$.

(b) Graph $y_1 = 4/(x+1) - 2/(x+3)$ as shown in Figure 22. With vertical asymptotes $x = -1$ and $x = -3$,

the graph is below the x-axis on $(-\infty, -5) \cup (-3, -1)$. The solution set is $(-\infty, -5) \cup (-3, -1)$.

(c) Graph $y_1 = 4/(x+1) - 2/(x+3)$ as shown in Figure 22. With vertical asymptotes $x = -1$ and $x = -3$,

the graph is above the x-axis on $(-5, -3) \cup (-1, \infty)$. The solution set is $(-5, -3) \cup (-1, \infty)$.

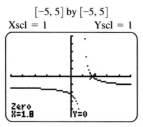

$[-5, 5]$ by $[-5, 5]$
Xscl = 1   Yscl = 1

Figure 19

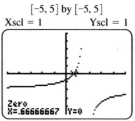

$[-5, 5]$ by $[-5, 5]$
Xscl = 1   Yscl = 1

Figure 20

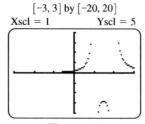

$[-3, 3]$ by $[-20, 20]$
Xscl = 1   Yscl = 5

Figure 21

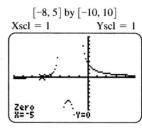

$[-8, 5]$ by $[-10, 10]$
Xscl = 1   Yscl = 1

Figure 22

23. (a) $\dfrac{1}{(x-2)^2} = 0 \Rightarrow 1 = 0(x-2)^2 \Rightarrow 1 = 0$ (false); The solution set is $\varnothing$.

    (b) Graph $y_1 = 1/(x-2)^2$ as shown in Figure 23. The graph is never below the x-axis. The solution set is $\varnothing$.

    (c) Graph $y_1 = 1/(x-2)^2$ as shown in Figure 23. With vertical asymptote $x = 2$, the graph is above

    the x-axis on the interval $(-\infty, 2) \cup (2, \infty)$. The solution set is $(-\infty, 2) \cup (2, \infty)$.

24. (a) $\dfrac{-2}{(x+3)^2} = 0 \Rightarrow -2 = 0(x+3)^2 \Rightarrow -2 = 0$ (false); The solution set is $\varnothing$.

    (b) Graph $y_1 = -2/(x+3)^2$ as shown in Figure 24. The graph is never above the x-axis. The solution set is $\varnothing$.

    (c) Graph $y_1 = -2/(x+3)^2$ as shown in Figure 24. With vertical asymptote $x = -3$, the graph is below

    the x-axis on the interval $(-\infty, -3) \cup (-3, \infty)$. The solution set is $(-\infty, -3) \cup (-3, \infty)$.

25. (a) $\dfrac{5}{x+1} = \dfrac{12}{x+1} \Rightarrow 5(x+1) = 12(x+1) \Rightarrow 5x+5 = 12x+12 \Rightarrow -7x = 7 \Rightarrow x = -1$; This solution

    is extraneous. The solution set is $\varnothing$.

    (b) Graph $y_1 = 5/(x+1) - 12/(x+1)$ as shown in Figure 25. With vertical asymptote $x = -1$, the graph is

    above the x-axis on the interval $(-\infty, -1)$. The solution set is $(-\infty, -1)$.

    (c) Graph $y_1 = 5/(x+1) - 12/(x+1)$ as shown in Figure 25. With vertical asymptote $x = -1$, the graph is

    below the x-axis on the interval $(-1, \infty)$. The solution set is $(-1, \infty)$.

26. (a) $\dfrac{7}{x+2} = \dfrac{1}{x+2} \Rightarrow 7(x+2) = 1(x+2) \Rightarrow 7x+14 = x+2 \Rightarrow 6x = -12 \Rightarrow x = -2$; This solution

    is extraneous. The solution set is $\varnothing$.

    (b) Graph $y_1 = 7/(x+2) - 1/(x+2)$ as shown in Figure 26. With vertical asymptote $x = -2$, the graph is

    above or intersects the x-axis on the interval $(-2, \infty)$. The solution set is $(-2, \infty)$.

    (c) Graph $y_1 = 7/(x+2) - 1/(x+2)$ as shown in Figure 26. With vertical asymptote $x = -2$, the graph is

    below or intersects the x-axis on the interval $(-\infty, -2)$. The solution set is $(-\infty, -2)$.

$[-5, 10]$ by $[-5, 10]$
Xscl = 1   Yscl = 1

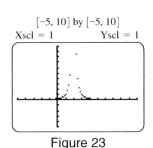

Figure 23

$[-10, 5]$ by $[-10, 5]$
Xscl = 1   Yscl = 1

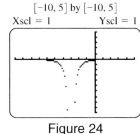

Figure 24

$[-10, 10]$ by $[-10, 10]$
Xscl = 1   Yscl = 1

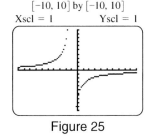

Figure 25

$[-10, 10]$ by $[-10, 10]$
Xscl = 1   Yscl = 1

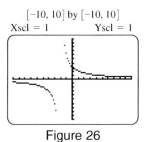

Figure 26

27. The numerator is negative and the denominator is always positive, therefore the quotient is always negative. Because the inequality requires that the rational expression be less than 0, the solution set is $(-\infty, \infty)$.

28. The numerator is positive and the denominator is always positive, therefore the quotient is always positive. Because the inequality requires that the rational expression be less than 0, the solution set is $\varnothing$.

29. The numerator is negative and the denominator is always positive, therefore the quotient is always negative. Because the inequality requires that the rational expression be greater than 0, the solution set is $\varnothing$.

30. The numerator is always positive and the denominator is negative, therefore the quotient is always negative. Because the inequality requires that the rational expression be less than or equal to 0, the solution set is $(-\infty, \infty)$.

31. The numerator is always positive and the denominator is negative, therefore the quotient is always negative. Because the inequality requires that the rational expression be greater than or equal to 0, the solution set is $\varnothing$.

32. The numerator is always positive and the denominator is always positive, therefore the quotient is always positive. Because the inequality requires that the rational expression be less than 0, the solution set is $\varnothing$.

33. The numerator is always positive and the denominator is always positive, therefore the quotient is always positive. Because the inequality requires that the rational expression be greater than 0, the solution set is $(-\infty, \infty)$.

34. The numerator is always positive or zero and the denominator is always positive, therefore the quotient is always positive or zero. Because the inequality requires that the rational expression be greater than 0, the solution set is $(-\infty, 1) \cup (1, \infty)$. Note that when $x = 1$, the rational expression equals 0 and the inequality is not satisfied.

35. The numerator is always positive or zero and the denominator is always positive, therefore the quotient is always positive or zero. Because the inequality requires that the rational expression be less than or equal to 0, the solution set is $\{1\}$. Note that when $x = 1$, the rational expression equals 0 and the inequality is satisfied.

36. (a) The rational expression equals 1 for all values of x for which it is defined. The solution set of $f(x) = 0$ is $\varnothing$.

    (b) The rational expression equals 1 for all values of x for which it is defined. The solution set of $f(x) < 0$ is $\varnothing$.

    (c) The rational expression equals 1 for all values of x for which it is defined. The solution set of $f(x) > 0$ is $(-\infty, -2) \cup (-2, 2) \cup (2, \infty)$. Note that f is undefined when $x = \pm 2$.

    (d) The rational expression equals 1 for all values of x for which it is defined. The solution set of $f(x) = 1$ is $(-\infty, -2) \cup (-2, 2) \cup (2, \infty)$. Note that f is undefined when $x = \pm 2$.

37. Multiply through by the LCD, $(x - 1)(x + 1)$.

    $$\frac{2x}{x^2 - 1} \cdot (x - 1)(x + 1) = \frac{2}{x + 1} \cdot (x - 1)(x + 1) - \frac{1}{x - 1} \cdot (x - 1)(x + 1) \Rightarrow$$

    $2x = 2(x - 1) - (x + 1) \Rightarrow 2x = 2x - 2 - x - 1 \Rightarrow 2x = x - 3 \Rightarrow x = -3$; The solution set is $\{-3\}$.

38. Multiply through by the LCD, $(2x - 1)(2x + 1)$.

    $$\frac{8x}{4x^2 - 1} \cdot (2x - 1)(2x + 1) = \frac{3}{2x + 1} \cdot (2x - 1)(2x + 1) + \frac{3}{2x - 1} \cdot (2x - 1)(2x + 1) \Rightarrow$$

    $8x = 3(2x - 1) + 3(2x + 1) \Rightarrow 8x = 6x - 3 + 6x + 3 \Rightarrow -4x = 0 \Rightarrow x = 0$; The solution set is $\{0\}$.

39. Multiply through by the LCD, $x(x - 3)(x + 3)$.

    $$\frac{4}{x^2 - 3x} \cdot x(x - 3)(x + 3) - \frac{1}{x^2 - 9} \cdot x(x - 3)(x + 3) = 0 \cdot x(x - 3)(x + 3) \Rightarrow 4(x + 3) - x = 0 \Rightarrow$$

    $4x + 12 - x = 0 \Rightarrow 3x = -12 \Rightarrow x = -4$; The solution set is $\{-4\}$.

40. Multiply through by the LCD, $x(x - 2)(x - 1)$.

    $$\frac{2}{x^2 - 2x} \cdot x(x - 2)(x - 1) - \frac{3}{x^2 - x} \cdot x(x - 2)(x - 1) = 0 \cdot x(x - 2)(x - 1) \Rightarrow$$

    $2(x - 1) - 3(x - 2) = 0 \Rightarrow 2x - 2 - 3x + 6 = 0 \Rightarrow -x = -4 \Rightarrow x = 4$; The solution set is $\{4\}$.

41. Multiply through by the LCD, x^2.

    $$1 \cdot x^2 - \frac{13}{x} \cdot x^2 + \frac{36}{x^2} \cdot x^2 = 0 \cdot x^2 \Rightarrow x^2 - 13x + 36 = 0 \Rightarrow (x - 4)(x - 9) = 0 \Rightarrow x = 4 \text{ or } 9;$$

    The solution set is $\{4, 9\}$.

42. Multiply through by the LCD, x^2.

$$1 \cdot x^2 - \frac{3}{x} \cdot x^2 - \frac{10}{x^2} \cdot x^2 = 0 \cdot x^2 \Rightarrow x^2 - 3x - 10 = 0 \Rightarrow (x + 2)(x - 5) = 0 \Rightarrow x = -2 \text{ or } 5;$$

The solution set is $\{-2, 5\}$.

43. Multiply through by the LCD, x^2.

$$1 \cdot x^2 + \frac{3}{x} \cdot x^2 = \frac{5}{x^2} \cdot x^2 \Rightarrow x^2 + 3x = 5 \Rightarrow x^2 + 3x - 5 = 0 \Rightarrow$$

$$x = \frac{-3 \pm \sqrt{3^2 - 4(1)(-5)}}{2(1)} = \frac{3 \pm \sqrt{29}}{2}; \text{ The solution set is } \left\{ \frac{3 \pm \sqrt{29}}{2} \right\}.$$

44. Multiply through by the LCD, x^2.

$$4 \cdot x^2 + \frac{7}{x} \cdot x^2 = -\frac{1}{x^2} \cdot x^2 \Rightarrow 4x^2 + 7x = -1 \Rightarrow 4x^2 + 7x + 1 = 0 \Rightarrow$$

$$x = \frac{-7 \pm \sqrt{7^2 - 4(4)(1)}}{2(4)} = \frac{-7 \pm \sqrt{33}}{8}; \text{ The solution set is } \left\{ \frac{-7 \pm \sqrt{33}}{8} \right\}.$$

45. Multiply through by the LCD, $x(2 - x)$.

$$\frac{x}{2 - x} \cdot x(2 - x) + \frac{2}{x} \cdot x(2 - x) - 5 \cdot x(2 - x) = 0 \cdot x(2 - x) \Rightarrow x^2 + 2(2 - x) - 5x(2 - x) = 0 \Rightarrow$$

$$x^2 + 4 - 2x - 10x + 5x^2 = 0 \Rightarrow 6x^2 - 12x + 4 = 0 \Rightarrow 3x^2 - 6x + 2 = 0 \Rightarrow$$

$$x = \frac{-(-6) \pm \sqrt{(-6)^2 - 4(3)(2)}}{2(3)} = \frac{6 \pm \sqrt{12}}{6} = \frac{2(3 \pm \sqrt{3})}{2(3)} = \frac{3 \pm \sqrt{3}}{3}; \text{ The solution set is } \left\{ \frac{3 \pm \sqrt{3}}{3} \right\}.$$

46. Multiply through by the LCD, $x(x - 3)$.

$$\frac{2x}{x - 3} \cdot x(x - 3) + \frac{4}{x} \cdot x(x - 3) - 6 \cdot x(x - 3) = 0 \cdot x(x - 3) \Rightarrow 2x^2 + 4(x - 3) - 6x(x - 3) = 0 \Rightarrow$$

$$2x^2 + 4x - 12 - 6x^2 + 18x = 0 \Rightarrow -4x^2 + 22x - 12 = 0 \Rightarrow -2x^2 + 11x - 6 = 0 \Rightarrow$$

$$x = \frac{-11 \pm \sqrt{11^2 - 4(-2)(-6)}}{2(-2)} = \frac{-1(11 \pm \sqrt{73})}{-1(4)} = \frac{11 \pm \sqrt{73}}{4}; \text{ The solution set is } \left\{ \frac{11 \pm \sqrt{73}}{4} \right\}.$$

47. Rewrite the equation as $\frac{1}{x^4} - \frac{3}{x^2} - 4 = 0$ and multiply through by the LCD, x^4.

$$\frac{1}{x^4} \cdot x^4 - \frac{3}{x^2} \cdot x^4 - 4 \cdot x^4 = 0 \cdot x^4 \Rightarrow 1 - 3x^2 - 4x^4 = 0 \Rightarrow 4x^4 + 3x^2 - 1 = 0 \Rightarrow$$

$$(4x^2 - 1)(x^2 + 1) = 0 \Rightarrow (2x + 1)(2x - 1)(x^2 + 1) = 0 \Rightarrow 2x + 1 = 0 \text{ or } 2x - 1 = 0 \text{ or } x^2 + 1 = 0 \Rightarrow$$

$$2x = -1 \text{ or } 2x = 1 \text{ or } x^2 = -1 \Rightarrow x = -\frac{1}{2} \text{ or } x = \frac{1}{2} \text{ or } x = \pm i; \text{ The solution set is } \left\{ \pm \frac{1}{2}, \pm i \right\}.$$

48. Rewrite the equation as $\frac{1}{x^4} - \frac{5}{x^2} - 36 = 0$ and multiply through by the LCD, x^4.

$$\frac{1}{x^4} \cdot x^4 - \frac{5}{x^2} \cdot x^4 - 36 \cdot x^4 = 0 \cdot x^4 \Rightarrow 1 - 5x^2 - 36x^4 = 0 \Rightarrow 36x^4 + 5x^2 - 1 = 0 \Rightarrow$$

$$(9x^2 - 1)(4x^2 + 1) = 0 \Rightarrow (3x + 1)(3x - 1)(4x^2 + 1) = 0 \Rightarrow 3x + 1 = 0 \text{ or } 3x - 1 = 0 \text{ or } 4x^2 + 1 = 0 \Rightarrow$$

$$3x = -1 \text{ or } 3x = 1 \text{ or } x^2 = -\frac{1}{4} \Rightarrow x = -\frac{1}{3} \text{ or } x = \frac{1}{3} \text{ or } x = \pm \frac{1}{2}i; \text{ The solution set is } \left\{ \pm \frac{1}{3}, \pm \frac{1}{2}i \right\}.$$

49. Multiply through by the LCD, $(x + 2)(x + 7)$.

$$\frac{1}{x + 2} \cdot (x + 2)(x + 7) + \frac{3}{x + 7} \cdot (x + 2)(x + 7) = \frac{5}{x^2 + 9x + 14} \cdot (x + 2)(x + 7) \Rightarrow$$

$$1(x + 7) + 3(x + 2) = 5 \Rightarrow x + 7 + 3x + 6 = 5 \Rightarrow 4x = -8 \Rightarrow x = -2; \text{ This solution is extraneous.}$$

The solution set is $\varnothing$.

50. Multiply through by the LCD, $(x + 3)(x + 5)$.

$$\frac{1}{x + 3} \cdot (x + 3)(x + 5) + \frac{4}{x + 5} \cdot (x + 3)(x + 5) = \frac{2}{x^2 + 8x + 15} \cdot (x + 3)(x + 5) \Rightarrow$$

$1(x + 5) + 4(x + 3) = 2 \Rightarrow x + 5 + 4x + 12 = 2 \Rightarrow 5x = -15 \Rightarrow x = -3$; This solution is extraneous.

The solution set is $\varnothing$.

51. Multiply through by the LCD, $(x - 3)(x + 3)$.

$$\frac{x}{x - 3} \cdot (x - 3)(x + 3) + \frac{4}{x + 3} \cdot (x - 3)(x + 3) = \frac{18}{x^2 - 9} \cdot (x - 3)(x + 3) \Rightarrow$$

$x(x + 3) + 4(x - 3) = 18 \Rightarrow x^2 + 3x + 4x - 12 = 18 \Rightarrow x^2 + 7x - 30 = 0 \Rightarrow (x + 10)(x - 3) = 0 \Rightarrow$

$x = -10$ or $x = 3$; The solution $x = 3$ is extraneous. The solution set is $\{-10\}$.

52. Rewrite the equation as $\dfrac{2x}{x - 3} + \dfrac{4}{x + 3} = \dfrac{-24}{x^2 - 9}$ and multiply through by the LCD, $(x - 3)(x + 3)$.

$$\frac{2x}{x - 3} \cdot (x - 3)(x + 3) + \frac{4}{x + 3} \cdot (x - 3)(x + 3) = \frac{-24}{x^2 - 9} \cdot (x - 3)(x + 3) \Rightarrow$$

$2x(x + 3) + 4(x - 3) = -24 \Rightarrow 2x^2 + 6x + 4x - 12 = -24 \Rightarrow 2x^2 + 10x + 12 = 0 \Rightarrow$

$x^2 + 5x + 6 = 0 \Rightarrow (x + 3)(x + 2) = 0 \Rightarrow x = -3$ or $x = -2$; The solution $x = -3$ is extraneous.

The solution set is $\{-2\}$.

53. Rewrite the equation as $\dfrac{9}{x} + \dfrac{4}{6x - 3} = \dfrac{2}{6x - 3}$ and multiply through by the LCD, $x(6x - 3)$.

$$\frac{9}{x} \cdot x(6x - 3) + \frac{4}{6x - 3} \cdot x(6x - 3) = \frac{2}{6x - 3} \cdot x(6x - 3) \Rightarrow 9(6x - 3) + 4x = 2x \Rightarrow$$

$54x - 27 + 4x = 2x \Rightarrow 56x = 27 \Rightarrow x = \dfrac{27}{56}$; The solution set is $\left\{\dfrac{27}{56}\right\}$.

54. Rewrite the equation as $\dfrac{x}{x - 2} + \dfrac{x}{x + 2} = \dfrac{8}{x^2 - 4}$ and multiply through by the LCD, $(x - 2)(x + 2)$.

$$\frac{x}{x - 2} \cdot (x - 2)(x + 2) + \frac{x}{x + 2} \cdot (x - 2)(x + 2) = \frac{8}{x^2 - 4} \cdot (x - 2)(x + 2) \Rightarrow$$

$x(x + 2) + x(x - 2) = 8 \Rightarrow x^2 + 2x + x^2 - 2x = 8 \Rightarrow 2x^2 = 8 \Rightarrow x^2 = 4 \Rightarrow x = \pm 2$;

These solution are both extraneous. The solution set is $\varnothing$.

55. The graph of $y_1 = (x + 1)^2/(x - 2)$ has a vertical asymptote at $x = 2$ and an x-intercept at $x = -1$.

The graph of y_1 intersects or is below the x-axis on the interval $(-\infty, 2)$. Note that 2 can not be included.

56. The graph of $y_1 = 2x/(x - 2)^2$ has a vertical asymptote at $x = 2$ and an x-intercept at $x = 0$.

The graph of y_1 is above the x-axis on the interval $(0, 2) \cup (2, \infty)$.

57. The graph of $y_1 = (3 - 2x)/(1 + x)$ has a vertical asymptote at $x = -1$ and an x-intercept at $x = \dfrac{3}{2}$.

The graph of y_1 is below the x-axis on the interval $(-\infty, -1) \cup \left(\dfrac{3}{2}, \infty\right)$.

58. First, rewrite the inequality:

$$\frac{x+1}{4-2x} \ge 1 \Rightarrow \frac{x+1}{4-2x} - 1 \ge 0 \Rightarrow \frac{x+1-4+2x}{4-2x} \ge 0 \Rightarrow \frac{3x-3}{4-2x} \ge 0 \Rightarrow \frac{3(x-1)}{-2(x-2)} \ge 0$$

The graph of $y_1 = 3(x-1)/(-2(x-2))$ has a vertical asymptote at $x = 2$ and an x-intercept at $x = 1$.

The graph of y_1 intersects or is above the x-axis on the interval $[1, 2)$. Note that 2 can not be included.

59. The graph of $y_1 = \dfrac{(x+1)(x-2)}{x+3}$ has a vertical asymptote of at $x = -3$ and x-intercepts at $x = -1$ and $x = 2$.

The graph of y_1 is below the x-axis on the interval $(-\infty, -3) \cup (-1, 2)$.

60. The graph of $y_1 = \dfrac{x(x-3)}{x+2}$ has a vertical asymptote of at $x = -2$ and x-intercepts at $x = 0$ and $x = 3$.

The graph of y_1 is on or above the x-axis on the interval $(-2, 0] \cup [3, \infty)$.

61. The graph of $y_1 = \dfrac{2x-5}{(x+1)(x-1)}$ has vertical asymptotes at $x = -1$ and $x = 1$ and an x-intercept at $x = \dfrac{5}{2}$.

The graph of y_1 is on or above the x-axis on the interval $(-1, 1) \cup \left[\dfrac{5}{2}, \infty\right)$.

62. The graph of $y_1 = \dfrac{5-x}{(x-2)(x+1)}$ has vertical asymptotes at $x = -1$ and $x = 2$ and an x-intercept at $x = 5$.

The graph of y_1 is below the x-axis on the interval $(-1, 2) \cup (5, \infty)$.

63. First, rewrite the inequality:

$$\frac{1}{x-3} \le \frac{5}{x-3} \Rightarrow \frac{1}{x-3} - \frac{5}{x-3} \le 0 \Rightarrow \frac{-4}{x-3} \le 0$$

The graph of $y_1 = -4/(x-3)$ has a vertical asymptote at $x = 3$ and no x-intercept.

The graph of y_1 intersects or is below the x-axis on the interval $(3, \infty)$. Note that 3 can not be included.

64. First, rewrite the inequality:

$$\frac{3}{2-x} > \frac{x}{2+x} \Rightarrow \frac{3}{2-x} - \frac{x}{2+x} > 0 \Rightarrow \frac{3(2+x) - x(2-x)}{(2-x)(2+x)} > 0 \Rightarrow \frac{x^2+x+6}{(2-x)(2+x)} > 0$$

The graph of $y_1 = (x^2+x+6)/((2-x)(2+x))$ has vertical asymptotes at $x = \pm 2$ and no x-intercept.

The graph of y_1 is above the x-axis on the interval $(-2, 2)$.

65. First, rewrite the inequality:

$$2 - \frac{5}{x} + \frac{2}{x^2} \ge 0 \Rightarrow \frac{2x^2 - 5x + 2}{x^2} \ge 0 \Rightarrow \frac{(2x-1)(x-2)}{x^2} \ge 0$$

The graph of $y_1 = (2x-1)(x-2)/x^2$ has a vertical asymptote at $x = 0$ and x-intercepts $x = \dfrac{1}{2}$ and $x = 2$.

The graph of y_1 intersects or is above the x-axis on the interval $(-\infty, 0) \cup \left(0, \dfrac{1}{2}\right] \cup [2, \infty)$.

66. First, rewrite the inequality:

$$\frac{1}{x-1} + \frac{1}{x+1} > \frac{3}{4} \Rightarrow \frac{1}{x-1} + \frac{1}{x+1} - \frac{3}{4} > 0 \Rightarrow \frac{4(x+1) + 4(x-1) - 3(x^2-1)}{4(x-1)(x+1)} > 0 \Rightarrow$$

$$\frac{8x - 3x^2 + 3}{4(x-1)(x+1)} > 0 \Rightarrow \frac{-(3x^2 - 8x - 3)}{4(x-1)(x+1)} > 0 \Rightarrow \frac{-(3x+1)(x-3)}{4(x-1)(x+1)} > 0$$

The graph of $y_1 = -(3x+1)(x-3)/(4(x-1)(x+1))$ has vertical asymptotes at $x = \pm 1$ and x-intercepts

$x = -\dfrac{1}{3}$ and $x = 3$. The graph of y_1 is above the x-axis on the interval $\left(-1, -\dfrac{1}{3}\right) \cup (1, 3)$.

67. (a) The graph of $y_1 = (\sqrt{(2)}x + 5)/(x^{\wedge}3 - \sqrt{(3)})$ is shown in Figure 67. The x-intercept is approximately -3.54. The solution set of the equation is $\{-3.54\}$.

    (b) There is a vertical asymptote when $x^3 - \sqrt{3} = 0 \Rightarrow x^3 = \sqrt{3} \Rightarrow x = \sqrt[3]{\sqrt{3}} \approx 1.20$. Thus, the graph is above the x-axis on the interval $(-\infty, -3.54) \cup (1.20, \infty)$. See Figure 67.

    (c) The graph is below the x-axis on the interval $(-3.54, 1.20)$. See Figure 67.

68. (a) The graph of $y_1 = (\sqrt[3]{(7)}x^{\wedge}3 - 1)/(x^2 + 2)$ is shown in Figure 68. The x-intercept is approximately .81. The solution set of the equation is $\{.81\}$.

    (b) The graph is above the x-axis on the interval $(.81, \infty)$. See Figure 68.

    (c) The graph is below the x-axis on the interval $(-\infty, .81)$. See Figure 68.

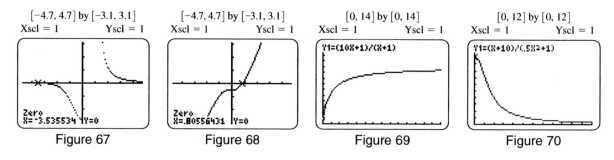

$[-4.7, 4.7]$ by $[-3.1, 3.1]$	$[-4.7, 4.7]$ by $[-3.1, 3.1]$	$[0, 14]$ by $[0, 14]$	$[0, 12]$ by $[0, 12]$
Xscl = 1 Yscl = 1	Xscl = 1 Yscl = 1	Xscl = 1 Yscl = 1	Xscl = 1 Yscl = 1
Figure 67	Figure 68	Figure 69	Figure 70

69. (a) The graph is shown in Figure 69. The equation of the horizontal asymptote is $y = \dfrac{10}{1} \Rightarrow y = 10$.

    (b) The initial insect population occurs when $x = 0$. Here $f(0) = \dfrac{10(0) + 1}{(0) + 1} = 1$ million insects.

    (c) After several months, the insect population levels off at 10 million insects.

    (d) The horizontal asymptote $y = 10$ represents the limiting population after a long time.

70. (a) The graph is shown in Figure 70. The equation of the horizontal asymptote is $y = 0$.

    (b) The initial fish population occurs when $x = 0$. Here $f(0) = \dfrac{(0) + 10}{.5(0)^2 + 1} = 10$ thousand fish.

    (c) After many years the fish population tends toward 0.

    (d) The horizontal asymptote $y = 0$ represents the limiting population after a long time. The species is dying out.

71. (a) $.25 = \dfrac{x - 5}{x^2 - 10x} \Rightarrow .25(x^2 - 10x) = x - 5 \Rightarrow .25x^2 - 2.5x = x - 5 \Rightarrow .25x^2 - 3.5x + 5 = 0 \Rightarrow$

    $x = \dfrac{-(-3.5) \pm \sqrt{(-3.5)^2 - 4(.25)(5)}}{2(.25)} = \dfrac{3.5 \pm \sqrt{7.25}}{.5} \approx 1.6$ or 12.4; Since $x > 10$, $x \approx 12.4$ cars/min.

    (b) Since 2 attendants can serve only 10 cars/min, 3 attendants are needed.

72. $3 = \dfrac{x^2}{2(1 - x)} \Rightarrow 6(1 - x) = x^2 \Rightarrow 6 - 6x = x^2 \Rightarrow x^2 + 6x - 6 = 0 \Rightarrow$

    $x = \dfrac{-6 \pm \sqrt{6^2 - 4(1)(-6)}}{2(1)} = \dfrac{-6 \pm \sqrt{60}}{2} \approx -6.9$ or $.87$; Since $x > 0$, $x \approx .87$.

73. A surface area of 280 square inches indicates that $2lw + 2lh + 2hw = 280$ or $lw + lh + hw = 140$. Solving

this equation for h gives $h = \dfrac{140 - lw}{l + w}$. Since the length is twice the width, $l = 2w$. Then by substitution,

$h = \dfrac{140 - (2w)w}{2w + w} \Rightarrow h = \dfrac{140 - 2w^2}{3w}$. Since the volume is 196 cubic inches, $lwh = 196$ and substituting for l

and h yields $(2w)w\left(\dfrac{140 - 2w^2}{3w}\right) = 196 \Rightarrow \dfrac{280w - 4w^3}{3} = 196 \Rightarrow 280w - 4w^3 = 588 \Rightarrow$

$4w^3 - 280x + 588 = 0 \Rightarrow w^3 - 70w + 147 = 0$. Graphing this equation as $y_1 = x^\wedge 3 - 70x + 147$ in

$[0, 10]$ by $[-150, 450]$ shows the possible values for the width as x-intercepts (figure not shown).

The possible values for the width are 7 inches and approximately 2.266 inches.

If $w = 7$, then $l = 2(7) = 14$ and $h = \dfrac{140 - 2(7)^2}{3(7)} = 2$. In this case, the dimensions are 7 by 14 by 2 inches.

If $w \approx 2.266$, then $l \approx 2(2.266) \approx 4.532$ and $h = \dfrac{140 - 2(2.266)^2}{3(2.266)} \approx 19.084$. In this case, the dimensions are

approximately 2.266 by 4.532 by 19.086 inches.

74. A volume of 4 cubic feet indicates that $\pi r^2 h = 4$. Solving this equation for h gives $h = \dfrac{4}{\pi r^2}$. In order to use

the least amount of material, the surface area, $S = \pi r^2 + 2\pi rh$, should be as small as possible. Substituting for

h in this equation yields $S = \pi r^2 + 2\pi r\left(\dfrac{4}{\pi r^2}\right) \Rightarrow S = \pi r^2 + \dfrac{8}{r}$. Graph this equation as $y_1 = \pi x^2 + 8/x$ in

$[0, 3]$ by $[0, 30]$. The x-coordinate of the minimum value on the graph gives the radius that will minimize the

surface area of the metal can (figure not shown). When $r \approx 1.08$ ft, $h \approx \dfrac{4}{\pi(1.08)^2} \approx 1.09$ ft, and the minimum

amount of material used is $S = \pi(1.08)^2 + \dfrac{8}{1.08} \approx 11.07$ ft^2.

75. (a) $f(400) = \dfrac{2540}{400} = 6.35$ in.; A curve designed for 60 miles per hour with a radius of 400 feet should have the

outer rail elevated 6.35 inches.

(b) See Figure 75. As the radius x of the curve increases, the elevation of the outer rail decreases.

(c) The horizontal asymptote is $y = 0$. As the radius of the curve increases without bound ($x \to \infty$), the tracks

become straight and no elevation or banking ($y \to 0$) is necessary.

(d) $12.7 = \dfrac{2540}{x} \Rightarrow 12.7x = 2540 \Rightarrow x = \dfrac{2540}{12.7} = 200$ ft

$[0, 600]$ by $[0, 50]$
Xscl $= 100$ Yscl $= 5$

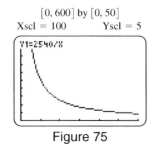

Figure 75

76. (a) See Figure 76a. A 100% participation in the recycling program is (essentially) impossible to achieve.

(b) $C(75) = \dfrac{1.2(75)}{100 - 75} = \dfrac{90}{25} = 3.6$; The cost of 75% participation is $3,600,000.

(c) Graph $y_1 = (1.2x)/(100 - x)$ as shown in Figure 76c. When $y = 5$, the participation rate is about 81%.

(d) $5 = \dfrac{1.2x}{100 - x} \Rightarrow 5(100 - x) = 1.2x \Rightarrow 500 - 5x = 1.2x \Rightarrow 500 = 6.2x \Rightarrow x = \dfrac{500}{6.2} \approx 80.6\%$

[0, 100] by [0, 10]
Xscl = 10   Yscl = 1

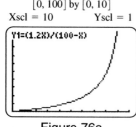

[0, 100] by [0, 10]
Xscl = 10   Yscl = 1

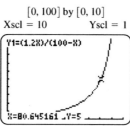

Figure 76a   Figure 76c

77. (a) $D(.05) = \dfrac{2500}{30(.3 + .05)} \approx 238$; The braking distance for a car going 50 mph on a 5% uphill grade is about 238 ft.

(b) As the uphill grade x increases, the braking distance decreases. This agrees with driving experience.

(c) $220 = \dfrac{2500}{30(.3 + x)} \Rightarrow 6600(.3 + x) = 2500 \Rightarrow .3 + x = \dfrac{2500}{6600} \Rightarrow x = \dfrac{2500}{6600} - .3 \Rightarrow x \approx .079$;

The grade associated with a braking distance of 220 feet is about 7.9% uphill.

78. (a) $D(-.1) = \dfrac{2500}{30(.3 + (-.1))} \approx 417$; The braking distance for a car going 50 mph on a 10% downhill grade is about 417 ft.

(b) As the downhill grade x increases, the braking distance increases. This agrees with driving experience.

(c) As the downhill grade x gets very close to a 30% grade, the braking distance increases without a maximum, which means stopping in time becomes impossible.

(d) $350 = \dfrac{2500}{30(.3 + x)} \Rightarrow 10{,}500(.3 + x) = 2500 \Rightarrow .3 + x = \dfrac{2500}{10{,}500} \Rightarrow x = \dfrac{2500}{10{,}500} - .3 \Rightarrow x \approx -.062$;

The grade associated with a braking distance of 350 feet is about 6.2% downhill.

79. Here $r = \dfrac{km^2}{s}$. By substitution, $12 = \dfrac{k(6)^2}{4} \Rightarrow 48 = 36k \Rightarrow k = \dfrac{48}{36} \Rightarrow k = \dfrac{4}{3}$. That is $r = \dfrac{\frac{4}{3}m^2}{s}$.

When $m = 4$ and $s = 10$, $r = \dfrac{\frac{4}{3}(4)^2}{10} = \dfrac{\frac{64}{3}}{10} = \dfrac{64}{30} = \dfrac{32}{15}$.

80. Here $p = \dfrac{kz^2}{r}$. By substitution, $\dfrac{32}{5} = \dfrac{k(4)^2}{10} \Rightarrow 320 = 80k \Rightarrow k = \dfrac{320}{80} \Rightarrow k = 4$. That is $p = \dfrac{4z^2}{r}$.

When $z = 2$ and $r = 16$, $p = \dfrac{4(2)^2}{16} = \dfrac{16}{16} = 1$.

81. Here $a = \dfrac{kmn^2}{y^3}$. By substitution, $9 = \dfrac{k(4)(9)^2}{(3)^3} \Rightarrow 243 = 324k \Rightarrow k = \dfrac{243}{324} \Rightarrow k = \dfrac{3}{4}$. That is $a = \dfrac{\frac{3}{4}mn^2}{y^3}$.

When $m = 6$, $n = 2$ and $y = 5$, $a = \dfrac{\frac{3}{4}(6)(2)^2}{(5)^3} = \dfrac{18}{125}$.

82. Here $y = \dfrac{kx}{m^2r^2}$. By substitution, $\dfrac{5}{3} = \dfrac{k(1)}{(2)^2(3)^2} \Rightarrow 180 = 3k \Rightarrow k = \dfrac{180}{3} \Rightarrow k = 60$. That is $y = \dfrac{60x}{m^2r^2}$.

When $x = 3$, $m = 1$ and $r = 8$, $y = \dfrac{60(3)}{(1)^2(8)^2} = \dfrac{180}{64} = \dfrac{45}{16}$.

83. For $k > 0$, if y varies directly as x, when x increases, y increases, and when x decreases, y decreases.

84. For $k > 0$, if y varies inversely as x, when x increases, y decreases, and when x decreases, y increases.

85. If y is inversely proportional to x, $y = \dfrac{k}{x}$. If x doubles, $y = \dfrac{k}{2x} \Rightarrow y = \dfrac{1}{2} \cdot \dfrac{k}{x}$. That is, y becomes half as much.

86. If y is inversely proportional to the second power of x, $y = \dfrac{k}{x^2}$. If x doubles, $y = \dfrac{k}{(2x)^2} \Rightarrow y = \dfrac{1}{4} \cdot \dfrac{k}{x^2}$.

    That is, y becomes one-fourth as much.

87. If y is directly proportional to the third power of x, $y = kx^3$. If x triples, $y = k(3x)^3 \Rightarrow y = 27 \cdot kx^3$.

    That is, y becomes 27 times as much.

88. If y is directly proportional to the second power of x, $y = kx^2$. If x is halved, $y = k\left(\dfrac{1}{2}x\right)^2 \Rightarrow y = \dfrac{1}{4} \cdot kx^2$.

    That is, y becomes one-fourth as much.

89. Here BMI $= \dfrac{kw}{h^2}$. By substitution, $24 = \dfrac{k(177)}{(72)^2} \Rightarrow 124{,}416 = 177k \Rightarrow k = \dfrac{124{,}416}{177} \Rightarrow k \approx 703$.

    That is BMI $\approx \dfrac{703w}{h^2}$. When $w = 130$ and $h = 66$, BMI $\approx \dfrac{703(130)}{(66)^2} \approx 21$.

90. Here $V = \dfrac{kT}{P}$. By substitution, $1.3 = \dfrac{k(300)}{18} \Rightarrow 23.4 = 300k \Rightarrow k = \dfrac{23.4}{300} \Rightarrow k \approx .078$.

    That is $V \approx \dfrac{.078T}{P}$. When $T = 340$ and $P = 24$, BMI $\approx \dfrac{.078(340)}{24} \approx 1.105$ L.

91. Here $R = \dfrac{k}{d^2}$. By substitution, $.5 = \dfrac{k}{(2)^2} \Rightarrow 2 = k$. That is $R = \dfrac{2}{d^2}$. When $d = 3$, $R = \dfrac{2}{(3)^2} = \dfrac{2}{9}$ ohm.

92. Here $R = \dfrac{kl}{r^4}$. By substitution, $25 = \dfrac{k(12)}{(.2)^4} \Rightarrow .04 = 12k \Rightarrow k = \dfrac{.04}{12} \Rightarrow k = \dfrac{1}{300}$. That is $R = \dfrac{\frac{1}{300}l}{r^4}$.

    When $l = 12$ and $r = .3$, $R = \dfrac{\frac{1}{300}(12)}{(.3)^4} \approx 4.94$ ohm.

93. Here $W = \dfrac{k}{d^2}$. By substitution, $160 = \dfrac{k}{(4000)^2} \Rightarrow 2{,}560{,}000{,}000 = k$. That is $W = \dfrac{2{,}560{,}000{,}000}{d^2}$.

    Note that 8000 miles above Earth's surface, $d = 12{,}000$. When $d = 12{,}000$, $W = \dfrac{2{,}560{,}000{,}000}{(12{,}000)^2} = 17.8$ lb

94. Here $I = \dfrac{k}{d^2}$ implies that $d^2 = \dfrac{k}{I}$ or $d = \sqrt{\dfrac{k}{I}}$. If the faintest star seen from Earth has intensity I, the Hubble

    Telescope can see a star with an intensity of $\dfrac{I}{50}$. The distance from Earth of such a star is

    $d = \sqrt{\dfrac{k}{\frac{I}{50}}} = \sqrt{\dfrac{50k}{I}} = \sqrt{50} \cdot \sqrt{\dfrac{k}{I}} \approx 7.1\sqrt{\dfrac{k}{I}}$. That is, the Hubble Telescope can see about 7.1 times farther.

95. Here $V = kr^2h$. By substitution, $300 = k(3)^2(10.62) \Rightarrow \dfrac{300}{95.58} = k \Rightarrow k \approx 3.1387$. That is $V \approx 3.1387r^2h$.

    When $r = 4$ and $h = 15.92$, $V \approx 3.1387(4)^2(15.92) \approx 799.5$. (Note that the actual value of k is π.)

96. $125 = \dfrac{25(2)^2}{200t} \Rightarrow 25{,}000t = 100 \Rightarrow t = \dfrac{100}{25{,}000} \Rightarrow t = \dfrac{1}{250}$ sec

97. Solving the equation for k gives $k = XY_1$. By substituting $X = 3$ and $Y_1 = 1.7$, $k = 1.7(3) = 5.1$.

98. Solving the equation for k gives $k = XY_1$. By substituting $X = 5$ and $Y_1 = .58$, $k = 5(.58) = 2.9$.

99. Solving the equation for k gives $k = XY_1$. By substituting $X = -2$ and $Y_1 = .7$, $k = -2(.7) = -1.4$.

100. Solving the equation for k gives $k = XY_1$. By substituting $X = -2$ and $Y_1 = 1.3$, $k = -2(1.3) = -2.6$.

Reviewing Basic Concepts (Sections 4.1—4.3)

1. A sketch of the graph is shown in Figure 1a. A Graph of $y_1 = (3x + 6)/(x - 4)$ is shown in Figure 1b.

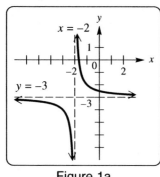

Figure 1a

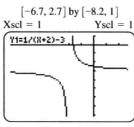

Figure 1b

2. For the function to be defined, the denominator cannot equal 0; $x^2 - 1 \neq 0 \Rightarrow x^2 \neq 1 \Rightarrow x \neq \pm 1$.
    The domain is $(-\infty, -1) \cup (-1, 1) \cup (1, \infty)$.

3. A vertical asymptote occurs when the denominator equals 0; $x - 6 = 0 \Rightarrow x = 6$.

4. Since the degrees of the numerator and the denominator are the same, the ratio of the leading coefficients of the numerator and denominator is used to find the equation of the horizontal asymptote; $y = \dfrac{1}{1} \Rightarrow y = 1$.

5. Perform long division:

$$\begin{array}{r} x - 2 \\ x + 3 \overline{)x^2 + x + 5} \\ \underline{x^2 + 3x} \\ -2x + 5 \\ \underline{-2x - 6} \\ 11 \end{array} \qquad \Rightarrow \qquad \dfrac{x^2 + x + 5}{x + 3} = x - 2 + \dfrac{11}{x + 3}$$

The oblique asymptote is $y = x - 2$.

6. A sketch of the graph is shown in Figure 6a. A Graph of $y_1 = 1/(x + 2) - 3$ is shown in Figure 6b.

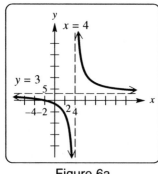

Figure 6a

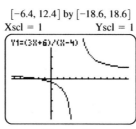

Figure 6b

7. (a) The graph intersects the x-axis at $x = -4$. The solution set is $\{-4\}$.

    (b) The graph is above the x-axis on the interval $(-\infty, -4) \cup (2, \infty)$. The solution set is $(-\infty, -4) \cup (2, \infty)$.

    (c) The graph is below the x-axis on the interval $(-4, 2)$. The solution set is $(-4, 2)$.

8. First, rewrite the inequality: $\dfrac{x + 4}{3x + 1} > 1 \Rightarrow \dfrac{x + 4}{3x + 1} - 1 > 0 \Rightarrow \dfrac{x + 4 - 3x - 1}{3x + 1} > 0 \Rightarrow \dfrac{-2x + 3}{3x + 1} > 0$

    The graph of $y_1 = (-2x + 3)/(3x + 1)$ has a vertical asymptote at $x = -\dfrac{1}{3}$ and an x-intercept at $x = \dfrac{3}{2}$.

    The graph of y_1 is above the x-axis on the interval $\left(-\dfrac{1}{3}, \dfrac{3}{2}\right)$.

9. The base of the parallelogram varies inversely as its height. The constant of variation is 24.

10. Here $v = \dfrac{k\sqrt{t}}{l}$. By substitution, $5 = \dfrac{k\sqrt{225}}{.60} \Rightarrow 3 = 15k \Rightarrow k = \dfrac{3}{15} \Rightarrow k \approx .2$. That is $v = \dfrac{.2\sqrt{t}}{l}$.

When $t = 196$ and $l = .65$, $v = \dfrac{.2\sqrt{196}}{.65} \approx 4.3$ vibrations per second.

4.4: Functions Defined by Powers and Roots

1. Since $13^2 = 169$, $\sqrt{169} = 13$.

2. Since $4^3 = 64$, $\sqrt[3]{64} = 4$ and so $-\sqrt[3]{64} = -4$.

3. Since $(-2)^5 = -32$, $\sqrt[5]{-32} = -2$.

4. Since $2^4 = 16$, $\sqrt[4]{16} = 2$.

5. $81^{3/2} = (\sqrt{81})^3 = 9^3 = 729$

6. $27^{4/3} = (\sqrt[3]{27})^4 = 3^4 = 81$

7. $125^{-2/3} = \dfrac{1}{125^{2/3}} = \dfrac{1}{(\sqrt[3]{125})^2} = \dfrac{1}{5^2} = \dfrac{1}{25}$

8. $(\sqrt[3]{-27})^2 = (-3)^2 = 9$

9. $(-1000)^{2/3} = (\sqrt[3]{-1000})^2 = (-10)^2 = 100$

10. $(-125)^{-4/3} = \dfrac{1}{(-125)^{4/3}} = \dfrac{1}{(\sqrt[3]{-125})^4} = \dfrac{1}{(-5)^4} = \dfrac{1}{625}$

11. $8^{2/3} = (8^{1/3})^2 = 2^2 = 4$

12. $-16^{3/2} = -(16^{1/2})^3 = -(\sqrt{16})^3 = -4^3 = -64$

13. $16^{-3/4} = (16^{1/4})^{-3} = (2)^{-3} = \dfrac{1}{2^3} = \dfrac{1}{8}$

14. $25^{-3/2} = (25^{1/2})^{-3} = (5)^{-3} = \dfrac{1}{5^3} = \dfrac{1}{125}$

15. $-81^{.5} = -81^{1/2} = -\sqrt{81} = -9$

16. $32^{1/5} = \sqrt[5]{32} = 2$

17. $64^{1/6} = \sqrt[6]{64} = 2$

18. $16^{-.25} = 16^{-1/4} = \dfrac{1}{\sqrt[4]{16}} = \dfrac{1}{2}$

19. $(-9^{3/4})^2 = (9^{3/4})^2 = 9^{3/2} = (\sqrt{9})^3 = 3^3 = 27$

20. $(4^{-1/2})^{-4} = 4^2 = 16$

21. $\sqrt[3]{2x} = (2x)^{1/3}$

22. $\sqrt{x+1} = (x+1)^{1/2}$

23. $\sqrt[3]{z^5} = z^{5/3}$

24. $\sqrt[5]{x^2} = x^{2/5}$

25. $(\sqrt[4]{y})^{-3} = (y^{1/4})^{-3} = y^{-3/4} = \dfrac{1}{y^{3/4}}$

26. $(\sqrt[3]{y^2})^{-5} = (y^{2/3})^{-5} = y^{-10/3} = \dfrac{1}{y^{10/3}}$

27. $\sqrt{x} \cdot \sqrt[3]{x} = x^{1/2} \cdot x^{1/3} = x^{1/2 + 1/3} = x^{5/6}$

28. $(\sqrt[5]{z})^{-3} = (z^{1/5})^{-3} = z^{-3/5} = \dfrac{1}{z^{3/5}}$

29. $\sqrt{y \cdot \sqrt{y}} = (y \cdot y^{1/2})^{1/2} = (y^{3/2})^{1/2} = y^{3/4}$

30. $\dfrac{\sqrt[3]{x}}{\sqrt{x}} = \dfrac{x^{1/3}}{x^{1/2}} = x^{1/3 - 1/2} = x^{-1/6} = \dfrac{1}{x^{1/6}}$

31. $\sqrt[3]{(-4)} \approx -1.587401052$

32. $5\sqrt[x]{(-3)} \approx -1.24573094$

33. $\sqrt[3]{(-125)} = -5$

34. $5\sqrt[x]{(-243)} = -3$

35. $\sqrt[3]{(-17)} \approx -2.571281591$

36. $5\sqrt[x]{(-8)} \approx -1.515716567$

37. $6\sqrt[x]{(\pi^2)} \approx 1.464591888$

38. $6\sqrt[x]{(\pi^{-1})} \approx .8263074871$

39. $13^{\wedge}(-1/3) \approx .4252903703$

40. $15^{\wedge}(-1/6) \approx .6367732195$

41. $32^{\wedge}.2 = 2$

42. $81^{\wedge}.25 = 3$

43. $(5/6)^{\wedge}-1.3 \approx 1.267463962$

44. $(4/7)^{\wedge}-.6 \approx 1.399016476$

45. $\pi^{\wedge}-3 \approx .0322515344$

46. $(2\pi)^{\wedge}(4/3) \approx 11.59417429$

47. $f(x) = x^{1.62} \Rightarrow f(1.2) = 1.2^{1.62} \approx 1.34$

48. $f(x) = x^{-.71} \Rightarrow f(3.8) = 3.8^{-.71} \approx .39$

49. $f(x) = x^{3/2} - x^{1/2} \Rightarrow f(50) = 50^{3/2} - 50^{1/2} \approx 346.48$

50. $f(x) = x^{5/4} - x^{-3/4} \Rightarrow f(7) = 7^{5/4} - 7^{-3/4} \approx 11.15$

51. Since $0 < a < 1$, this function is a root function. The correct graph is graph B.

52. Since $b > 1$, this function is a power function. The correct graph is graph A.

53. (a) $16^{-3/4} = \dfrac{1}{16^{3/4}} = \dfrac{1}{(\sqrt[4]{16})^3} = \dfrac{1}{(2)^3} = \dfrac{1}{8} = .125$

    (b) $16^{-3/4} = (\sqrt[4]{16})^{-3} = .125$ and $16^{-3/4} = \sqrt[4]{16^{-3}} = .125$; Other expressions are possible.

    (c) A calculator will show that $.125 = \dfrac{1}{8}$.

54. (a) $5^{\wedge}.47 \approx 2.130668582$

    (b) $5^{.47} = 5^{47/100} = (\sqrt[100]{5})^{47} \Rightarrow (100\sqrt[x]{5})^{\wedge}47 \approx 2.130668582$

55. See Figure 55.

56. See Figure 56.

57. See Figure 57.

58. See Figure 58.

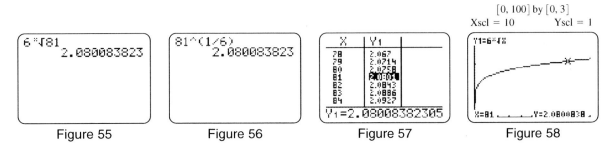

Figure 55   Figure 56   Figure 57   Figure 58

59. $S(4) = 1.27(4)^{2/3} \approx 3.2 \text{ ft}^2$

60. $L = 2.43(5.2)^{.3326} \approx 4.2 \text{ ft}$

61. $f(15) = 15^{1.5} \approx 58.1 \text{ yr}$

62. $f(30,000) = 1.22(30,000)^{.5} \approx 211 \text{ mi}$

63. (a) From the hint, $a(1)^b = 1960 \Rightarrow a = 1960$.

   (b) Since $f(3) = 525$, $1960(3)^b = 525 \Rightarrow 3^b = \dfrac{525}{1960} \Rightarrow 3^b \approx .268$. By trial and error, $b \approx -1.2$.

   (c) $f(4) = 1960(4)^{-1.2} \approx 371$; If the zinc ion concentration reaches 371 mg per L, a rainbow trout will survive, on average, 4 min.

64. (a) $N(5) = .00437(5)^{3.2} \approx .75$; $N(10) = .00437(10)^{3.2} \approx 6.93$; $N(20) = .00437(20)^{3.2} \approx 63.65$

   The likelihood of cancer increases as x increases.

   (b) No, it more than doubles. For example, $N(10) \approx 6.93$ but $N(20) \approx 63.65$.

65. $f(2) = .445(2)^{1.25} \approx 1.06 \text{ g}$

66. $f(68) = .117(68)^{1.7} \approx 153 \text{ lb}$

67. Using the Power Regression feature on a graphing calculator yields $a \approx 874.54$ and $b \approx -.49789$.

68. For the dog, $f(60) = 874.54(60)^{-.49789} \approx 114 \text{ bpm}$. For the whale, $f(4000) = 874.54(4000)^{-.49789} \approx 14 \text{ bpm}$.

69. $5 + 4x \geq 0 \Rightarrow 4x \geq -5 \Rightarrow x \geq -\dfrac{5}{4}$; The domain is $\left[-\dfrac{5}{4}, \infty \right)$.

70. $9x + 18 \geq 0 \Rightarrow 9x \geq -18 \Rightarrow x \geq -\dfrac{18}{9} \Rightarrow x \geq -2$; The domain is $[-2, \infty)$.

71. $6 - x \geq 0 \Rightarrow 6 \geq x \Rightarrow x \leq 6$; The domain is $(-\infty, 6]$.

72. $2 - .5x \geq 0 \Rightarrow 2 \geq .5x \Rightarrow 4 \geq x \Rightarrow x \leq 4$; The domain is $(-\infty, 4]$.

73. The cube root function is defined for all values of x. The domain is $(-\infty, \infty)$.

74. The fifth root function is defined for all values of x. The domain is $(-\infty, \infty)$.

75. The graph of $y = 49 - x^2 = (7 + x)(7 - x)$ is a parabola that opens downward with x-intercepts -7 and 7. This graph intersects or is above the x-axis for values of x in the interval $[-7, 7]$. That is, $49 - x^2 \geq 0$ on the interval $[-7, 7]$. The domain is $[-7, 7]$.

76. The graph of $y = 81 - x^2 = (9 + x)(9 - x)$ is a parabola that opens downward with x-intercepts -9 and 9. This graph intersects or is above the x-axis for values of x in the interval $[-9, 9]$. That is, $81 - x^2 \geq 0$ on the interval $[-9, 9]$. The domain is $[-9, 9]$.

77. The graph of $y = x^3 - x = x(x + 1)(x - 1)$ is a cubic graph with x-intercepts $-1, 0$ and 1. This graph intersects or is above the x-axis for values of x in the interval $[-1, 0] \cup [1, \infty)$. That is, $x^3 - x \geq 0$ on the interval $[-1, 0] \cup [1, \infty)$. The domain is $[-1, 0] \cup [1, \infty)$.

78. If n is even, then $ax + b$ must be greater than or equal to 0. If n is odd, then $ax + b$ may be any real number.

79. The graph is shown in Figure 79.

   (a) The range is $[0, \infty)$.

   (b) The function is increasing on the interval $\left[-\dfrac{5}{4}, \infty\right)$.

   (c) The function is never decreasing.

   (d) The graph intersects the x-axis when $x = -1.25$. The solution set of $f(x) = 0$ is $\{-1.25\}$.

80. The graph is shown in Figure 80.

   (a) The range is $[0, \infty)$.

   (b) The function is increasing on the interval $[-2, \infty)$.

   (c) The function is never decreasing.

   (d) The graph intersects the x-axis when $x = -2$. The solution set of $f(x) = 0$ is $\{-2\}$.

81. The graph is shown in Figure 81.

   (a) The range is $(-\infty, 0]$.

   (b) The function is increasing on the interval $(-\infty, 6]$.

   (c) The function is never decreasing.

   (d) The graph intersects the x-axis when $x = 6$. The solution set of $f(x) = 0$ is $\{6\}$.

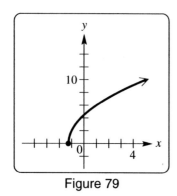
Figure 79

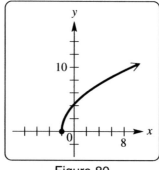
Figure 80

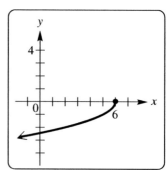
Figure 81

82. The graph is shown in Figure 82.

   (a) The range is $(-\infty, 0]$.

   (b) The function is increasing on the interval $(-\infty, 4]$.

   (c) The function is never decreasing.

   (d) The graph intersects the x-axis when $x = 4$. The solution set of $f(x) = 0$ is $\{4\}$.

83. The graph is shown in Figure 83.

    (a) The range is $(-\infty, \infty)$.

    (b) The function is increasing on the interval $(-\infty, \infty)$.

    (c) The function is never decreasing.

    (d) The graph intersects the x-axis when $x = 3$. The solution set of $f(x) = 0$ is $\{3\}$.

84. The graph is shown in Figure 84.

    (a) The range is $(-\infty, \infty)$.

    (b) The function is increasing on the interval $(-\infty, \infty)$.

    (c) The function is never decreasing.

    (d) The graph intersects the x-axis when $x = -32$. The solution set of $f(x) = 0$ is $\{-32\}$.

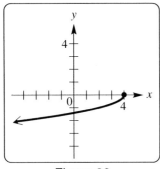

Figure 82

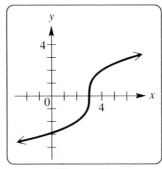

Figure 83

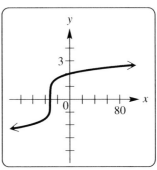

Figure 84

85. The graph is shown in Figure 85.

    (a) The range is $[0, 7]$.

    (b) The function is increasing on the interval $[-7, 0]$.

    (c) The function is decreasing on the interval $[0, 7]$.

    (d) The graph intersects the x-axis when $x = -7$ or $x = 7$. The solution set of $f(x) = 0$ is $\{-7, 7\}$.

86. The graph is shown in Figure 86.

    (a) The range is $[0, 9]$.

    (b) The function is increasing on the interval $[-9, 0]$.

    (c) The function is decreasing on the interval $[0, 9]$.

    (d) The graph intersects the x-axis when $x = -9$ or $x = 9$. The solution set of $f(x) = 0$ is $\{-9, 9\}$.

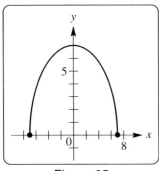

Figure 85

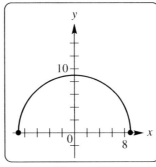

Figure 86

87. Since $y = \sqrt{9x + 27} = \sqrt{9(x + 3)} = 3\sqrt{x + 3}$, the graph can be obtained by shifting the graph of $y = \sqrt{x}$ to the left 3 units and vertically stretching by a factor of 3.

88. Since $y = \sqrt{16x + 16} = \sqrt{16(x + 1)} = 4\sqrt{x + 1}$, the graph can be obtained by shifting the graph of $y = \sqrt{x}$ to the left 1 unit and vertically stretching by a factor of 4.

89. Since $y = \sqrt{4x + 16} + 4 = \sqrt{4(x + 4)} + 4 = 2\sqrt{x + 4} + 4$, the graph can be obtained by shifting the graph of $y = \sqrt{x}$ to the left 4 units, vertically stretching by a factor of 2, and shifting upward 4 units.

90. Since $y = \sqrt{32 - 4x} - 3 = \sqrt{-4(x - 8)} - 3 = 2\sqrt{-(x - 8)} - 3$, the graph can be obtained by reflecting the graph of $y = \sqrt{x}$ across the y-axis, shifting 8 units to the right, vertically stretching by a factor of 2, and shifting downward 3 units.

91. Since $y = \sqrt[3]{27x + 54} - 5 = \sqrt[3]{27(x + 2)} - 5 = 3\sqrt[3]{x + 2} - 5$, the graph can be obtained by shifting the graph of $y = \sqrt[3]{x}$ to the left 2 units, vertically stretching by a factor of 3, and shifting downward 5 units.

92. Since $y = \sqrt[3]{8x - 8} = \sqrt[3]{8(x - 1)} = 2\sqrt[3]{x - 1}$, the graph can be obtained by shifting the graph of $y = \sqrt[3]{x}$ to the right 1 unit and vertically stretching by a factor of 2.

93. The graph is a circle. See Figure 93.
$$x^2 + y^2 = 100 \Rightarrow y^2 = 100 - x^2 \Rightarrow y = \pm\sqrt{100 - x^2}; \text{Thus } y_1 = \sqrt{100 - x^2} \text{ and } y_2 = -\sqrt{100 - x^2}.$$

94. The graph is a circle. See Figure 94.
$$x^2 + y^2 = 81 \Rightarrow y^2 = 81 - x^2 \Rightarrow y = \pm\sqrt{81 - x^2}; \text{Thus } y_1 = \sqrt{81 - x^2} \text{ and } y_2 = -\sqrt{81 - x^2}.$$

95. The graph is a (shifted) circle. See Figure 95.
$$(x - 2)^2 + y^2 = 9 \Rightarrow y^2 = 9 - (x - 2)^2 \Rightarrow y = \pm\sqrt{9 - (x - 2)^2}$$
Thus $y_1 = \sqrt{9 - (x - 2)^2}$ and $y_2 = -\sqrt{9 - (x - 2)^2}$.

96. The graph is a (shifted) circle. See Figure 96.
$$(x + 3)^2 + y^2 = 16 \Rightarrow y^2 = 16 - (x + 3)^2 \Rightarrow y = \pm\sqrt{16 - (x + 3)^2}$$
Thus $y_1 = \sqrt{16 - (x + 3)^2}$ and $y_2 = -\sqrt{16 - (x + 3)^2}$.

[−15, 15] by [−10, 10]	[−15, 15] by [−10, 10]	[−9.4, 9.4] by [−6.2, 6.2]	[−9.4, 9.4] by [−6.2, 6.2]
Xscl = 1 Yscl = 1	Xscl = 1 Yscl = 1	Xscl = 1 Yscl = 1	Xscl = 1 Yscl = 1

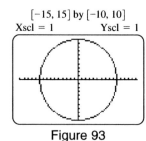

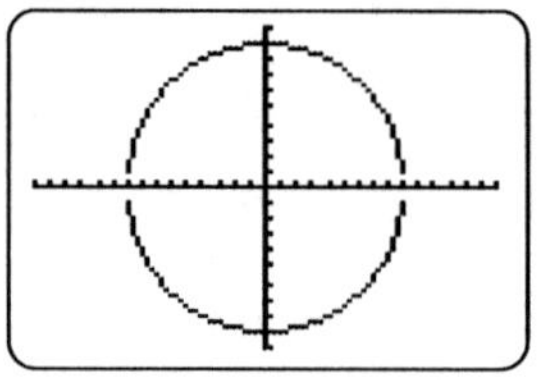

Figure 93

Figure 94

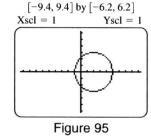

Figure 95

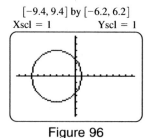

Figure 96

97. The graph is a horizontal parabola. See Figure 97.
$$x = y^2 + 6y + 9 \Rightarrow x = (y + 3)^2 \Rightarrow \pm\sqrt{x} = y + 3 \Rightarrow y = -3 \pm \sqrt{x}$$
Thus $y_1 = -3 + \sqrt{x}$ and $y_2 = -3 - \sqrt{x}$.

98. The graph is a horizontal parabola. See Figure 98.
$$x = y^2 - 8y + 16 \Rightarrow x = (y - 4)^2 \Rightarrow \pm\sqrt{x} = y - 4 \Rightarrow y = 4 \pm \sqrt{x}$$
Thus $y_1 = 4 + \sqrt{x}$ and $y_2 = 4 - \sqrt{x}$.

99. The graph is a horizontal parabola. See Figure 99.

$x = 2y^2 + 8y + 1 \Rightarrow .5x = y^2 + 4y + .5 \Rightarrow .5x = (y^2 + 4y + 4) + .5 - 4 \Rightarrow .5x = (y + 2)^2 - 3.5 \Rightarrow$

$.5x + 3.5 = (y + 2)^2 \Rightarrow \pm\sqrt{.5x + 3.5} = y + 2 \Rightarrow y = -2 \pm \sqrt{.5x + 3.5}$

Thus $y_1 = -2 + \sqrt{.5x + 3.5}$ and $y_2 = -2 - \sqrt{.5x + 3.5}$.

100. The graph is a horizontal parabola. See Figure 100.

$x = -3y^2 - 6y + 2 \Rightarrow -\dfrac{x}{3} = y^2 + 2y - \dfrac{2}{3} \Rightarrow -\dfrac{x}{3} = (y^2 + 2y + 1) - \dfrac{2}{3} - 1 \Rightarrow -\dfrac{x}{3} = (y + 1)^2 - \dfrac{5}{3} \Rightarrow$

$-\dfrac{x}{3} + \dfrac{5}{3} = (y + 1)^2 \Rightarrow \dfrac{x - 5}{-3} = (y + 1)^2 \Rightarrow \pm\sqrt{\dfrac{x - 5}{-3}} = y + 1 \Rightarrow y = -1 \pm \sqrt{\dfrac{x - 5}{-3}}$

Thus $y_1 = -1 + \sqrt{\dfrac{x - 5}{-3}}$ and $y_2 = -1 - \sqrt{\dfrac{x - 5}{-3}}$.

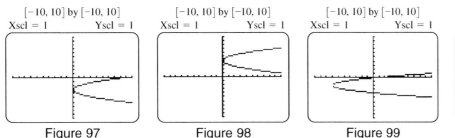
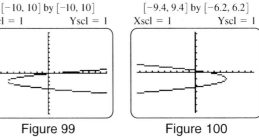

[−10, 10] by [−10, 10]   Xscl = 1   Yscl = 1

[−10, 10] by [−10, 10]   Xscl = 1   Yscl = 1

[−10, 10] by [−10, 10]   Xscl = 1   Yscl = 1

[−9.4, 9.4] by [−6.2, 6.2]   Xscl = 1   Yscl = 1

Figure 97   Figure 98   Figure 99   Figure 100

4.5: Equations, Inequalities, and Applications Involving Root Functions

1. (a) The graph of $y = \sqrt{x + 5}$ intersects the graph of $y = x - 1$ when $x = 4$. The solution set is $\{4\}$.

   (b) The graph of $y = \sqrt{x + 5}$ is below or intersects the graph of $y = x - 1$ on the interval $[4, \infty)$.
   The solution set is $[4, \infty)$.

   (c) The graph of $y = \sqrt{x + 5}$ is above or intersects the graph of $y = x - 1$ on the interval $[-5, 4]$.
   The solution set is $[-5, 4]$.

2. (a) The graph of $y = -\sqrt{x + 12}$ intersects the graph of $y = x$ when $x = -3$. The solution set is $\{-3\}$.

   (b) The graph of $y = -\sqrt{x + 12}$ is below or intersects the graph of $y = x$ on the interval $[-3, \infty)$.
   The solution set is $[-3, \infty)$.

   (c) The graph of $y = -\sqrt{x + 12}$ is above or intersects the graph of $y = x$ on the interval $[-12, -3]$.
   The solution set is $[-12, -3]$.

3. (a) The graph of $y = \sqrt[3]{2 - 2x}$ intersects the graph of $y = \sqrt[3]{2x + 14}$ when $x = -3$. The solution set is $\{-3\}$.

   (b) The graph of $y = \sqrt[3]{2 - 2x}$ is above the graph of $y = \sqrt[3]{2x + 14}$ on the interval $(-\infty, -3)$.
   The solution set is $(-\infty, -3)$.

   (c) The graph of $y = \sqrt[3]{2 - 2x}$ is below the graph of $y = \sqrt[3]{2x + 14}$ on the interval $(-3, \infty)$.
   The solution set is $(-3, \infty)$.

4. (a) The graph of $y = \sqrt[3]{3x}$ intersects the graph of $y = \sqrt[3]{7 - 4x}$ when $x = 1$. The solution set is $\{1\}$.

   (b) The graph of $y = \sqrt[3]{3x}$ is above the graph of $y = \sqrt[3]{7 - 4x}$ on the interval $(1, \infty)$.
   The solution set is $(1, \infty)$.

   (c) The graph of $y = \sqrt[3]{3x}$ is below the graph of $y = \sqrt[3]{7 - 4x}$ on the interval $(-\infty, 1)$.
   The solution set is $(-\infty, 1)$.

5. A sketch is shown in Figure 5. There is one real solution.

$$\sqrt{x} = 2x - 1 \Rightarrow (\sqrt{x})^2 = (2x - 1)^2 \Rightarrow x = 4x^2 - 4x + 1 \Rightarrow 4x^2 - 5x + 1 = 0 \Rightarrow$$

$(4x - 1)(x - 1) = 0 \Rightarrow x = \dfrac{1}{4}$ or 1; The solution set is $\{1\}$. The value $\dfrac{1}{4}$ is extraneous.

6. A sketch is shown in Figure 6. There is one real solution.

$$\sqrt{x} = x - 6 \Rightarrow (\sqrt{x})^2 = (x - 6)^2 \Rightarrow x = x^2 - 12x + 36 \Rightarrow x^2 - 13x + 36 = 0 \Rightarrow$$

$(x - 9)(x - 4) = 0 \Rightarrow x = 9$ or 4; The solution set is $\{9\}$. The value 4 is extraneous.

7. A sketch is shown in Figure 7. There is one real solution.

$$\sqrt{x} = -x + 3 \Rightarrow (\sqrt{x})^2 = (-x + 3)^2 \Rightarrow x = x^2 - 6x + 9 \Rightarrow x^2 - 7x + 9 = 0 \Rightarrow$$

$$x = \frac{-(-7) \pm \sqrt{(-7)^2 - 4(1)(9)}}{2(1)} = \frac{7 \pm \sqrt{13}}{2}; \text{ The solution set is } \left\{ \frac{7 - \sqrt{13}}{2} \right\}.$$

The value $\left\{ \dfrac{7 + \sqrt{13}}{2} \right\}$ is extraneous.

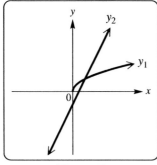

Figure 5

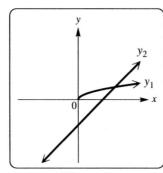

Figure 6

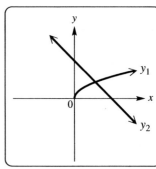

Figure 7

8. A sketch is shown in Figure 8. There are two real solutions. There are no extraneous values.

$$\sqrt{x} = 3x \Rightarrow (\sqrt{x})^2 = (3x)^2 \Rightarrow x = 9x^2 \Rightarrow 9x^2 - x = 0 \Rightarrow x(9x - 1) = 0 \Rightarrow x = 0 \text{ or } \frac{1}{9}$$

The solution set is $\left\{ 0, \dfrac{1}{9} \right\}$.

9. A sketch is shown in Figure 9. There are two real solutions. There are no extraneous values.

$$\sqrt[3]{x} = x^2 \Rightarrow (\sqrt[3]{x})^3 = (x^2)^3 \Rightarrow x = x^6 \Rightarrow x^6 - x = 0 \Rightarrow x(x^5 - 1) = 0 \Rightarrow x = 0 \text{ or } 1$$

The solution set is $\{0, 1\}$.

10. The graphs of $y_1 = \sqrt{x}$ and $y_2 = -x - 5$ do not intersect. Thus the equation $\sqrt{x} = -x - 5$ has no solution. See Figure 10.

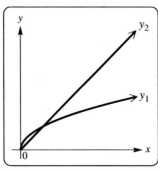

Figure 8

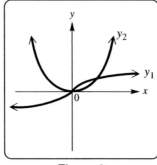

Figure 9

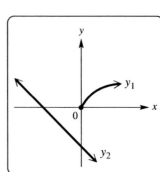

Figure 10

11. (a) $\sqrt{3x + 7} = 2 \Rightarrow (\sqrt{3x + 7})^2 = 2^2 \Rightarrow 3x + 7 = 4 \Rightarrow 3x = -3 \Rightarrow x = \dfrac{-3}{3} \Rightarrow x = -1$

    The solution set is $\{-1\}$. A graph of $y_1 = \sqrt{3x + 7}$ and $y_2 = 2$ is shown in Figure 11.

    (b) From the graph, $y_1 > y_2$ on the interval $(-1, \infty)$.

    (c) Since the graph of y_1 begins when $3x + 7 = 0 \Rightarrow x = -\dfrac{7}{3}$, $y_1 < y_2$ on the interval $\left[-\dfrac{7}{3}, -1\right)$.

12. (a) $\sqrt{2x + 13} = 3 \Rightarrow (\sqrt{2x + 13})^2 = 3^2 \Rightarrow 2x + 13 = 9 \Rightarrow 2x = -4 \Rightarrow x = \dfrac{-4}{2} \Rightarrow x = -2$

    The solution set is $\{-2\}$. A graph of $y_1 = \sqrt{2x + 13}$ and $y_2 = 3$ is shown in Figure 12.

    (b) From the graph, $y_1 > y_2$ on the interval $(-2, \infty)$.

    (c) Since the graph of y_1 begins when $2x + 13 = 0 \Rightarrow x = -\dfrac{13}{2}$, $y_1 < y_2$ on the interval $\left[-\dfrac{13}{2}, -2\right)$.

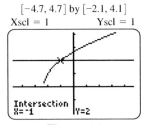

$[-4.7, 4.7]$ by $[-2.1, 4.1]$
Xscl = 1   Yscl = 1

$[-9.4, 9.4]$ by $[-4.2, 8.2]$
Xscl = 1   Yscl = 1

Figure 11   Figure 12

13. (a) $\sqrt{4x + 13} = 2x - 1 \Rightarrow (\sqrt{4x + 13})^2 = (2x - 1)^2 \Rightarrow 4x + 13 = 4x^2 - 4x + 1 \Rightarrow$

    $4x^2 - 8x - 12 = 0 \Rightarrow x^2 - 2x - 3 = 0 \Rightarrow (x + 1)(x - 3) = 0 \Rightarrow x = -1 \text{ or } 3 \; (-1 \text{ is extraneous})$

    The solution set is $\{3\}$. A graph of $y_1 = \sqrt{4x + 13}$ and $y_2 = 2x - 1$ is shown in Figure 13.

    (b) Since the graph of y_1 begins when $4x + 13 = 0 \Rightarrow x = -\dfrac{13}{4}$, $y_1 > y_2$ on the interval $\left[-\dfrac{13}{4}, 3\right)$.

    (c) From the graph, $y_1 < y_2$ on the interval $(3, \infty)$.

14. (a) $\sqrt{3x + 7} = 3x + 5 \Rightarrow (\sqrt{3x + 7})^2 = (3x + 5)^2 \Rightarrow 3x + 7 = 9x^2 + 30x + 25 \Rightarrow$

    $9x^2 + 27x + 18 = 0 \Rightarrow x^2 + 3x + 2 = 0 \Rightarrow (x + 2)(x + 1) = 0 \Rightarrow x = -2 \text{ or } -1 \; (-2 \text{ is extraneous})$

    The solution set is $\{-1\}$. A graph of $y_1 = \sqrt{3x + 7}$ and $y_2 = 3x + 5$ is shown in Figure 14.

    (b) Since the graph of y_1 begins when $3x + 7 = 0 \Rightarrow x = -\dfrac{7}{3}$, $y_1 > y_2$ on the interval $\left[-\dfrac{7}{3}, -1\right)$.

    (c) From the graph, $y_1 < y_2$ on the interval $(-1, \infty)$.

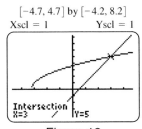
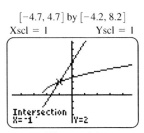

$[-4.7, 4.7]$ by $[-4.2, 8.2]$
Xscl = 1   Yscl = 1

$[-4.7, 4.7]$ by $[-4.2, 8.2]$
Xscl = 1   Yscl = 1

Figure 13   Figure 14

15. (a) $\sqrt{5x+1} + 2 = 2x \Rightarrow \sqrt{5x+1} = 2x - 2 \Rightarrow (\sqrt{5x+1})^2 = (2x-2)^2 \Rightarrow$

$5x + 1 = 4x^2 - 8x + 4 \Rightarrow 4x^2 - 13x + 3 = 0 \Rightarrow (4x-1)(x-3) = 0 \Rightarrow x = \dfrac{1}{4}$ or 3 ($\frac{1}{4}$ is extraneous)

The solution set is $\{3\}$. A graph of $y_1 = \sqrt{5x+1} + 2$ and $y_2 = 2x$ is shown in Figure 15.

(b) Since the graph of y_1 begins when $5x + 1 = 0 \Rightarrow x = -\dfrac{1}{5}$, $y_1 > y_2$ on the interval $\left[-\dfrac{1}{5}, 3\right)$.

(c) From the graph, $y_1 < y_2$ on the interval $(3, \infty)$.

16. (a) $\sqrt{3x+4} + x = 8 \Rightarrow \sqrt{3x+4} = 8 - x \Rightarrow (\sqrt{3x+4})^2 = (8-x)^2 \Rightarrow 3x + 4 = 64 - 16x + x^2 \Rightarrow$

$x^2 - 19x + 60 = 0 \Rightarrow (x-4)(x-15) = 0 \Rightarrow x = 4$ or 15 (15 is extraneous)

The solution set is $\{4\}$. A graph of $y_1 = \sqrt{3x+4} + x$ and $y_2 = 8$ is shown in Figure 16.

(b) From the graph, $y_1 > y_2$ on the interval $(4, \infty)$.

(c) Since the graph of y_1 begins when $3x + 4 = 0 \Rightarrow x = -\dfrac{4}{3}$, $y_1 < y_2$ on the interval $\left[-\dfrac{4}{3}, 4\right)$.

17. (a) $\sqrt{3x-6} + 2 = \sqrt{5x-6} \Rightarrow (\sqrt{3x-6} + 2)^2 = (\sqrt{5x-6})^2 \Rightarrow$

$3x - 6 + 4\sqrt{3x-6} + 4 = 5x - 6 \Rightarrow 4\sqrt{3x-6} = 2x - 4 \Rightarrow (4\sqrt{3x-6})^2 = (2x-4)^2 \Rightarrow$

$16(3x - 6) = 4x^2 - 16x + 16 \Rightarrow 48x - 96 = 4x^2 - 16x + 16 \Rightarrow 4x^2 - 64x + 112 = 0 \Rightarrow$

$x^2 - 16x + 28 = 0 \Rightarrow (x-2)(x-14) = 0 \Rightarrow x = 2$ or 14

The solution set is $\{2, 14\}$. A graph of $y_1 = \sqrt{3x-6} + 2$ and $y_2 = \sqrt{5x-6}$ is shown in Figure 17.

(b) From the graph, $y_1 > y_2$ on the interval $(2, 14)$.

(c) From the graph, $y_1 < y_2$ on the interval $(14, \infty)$.

18. (a) $\sqrt{2x-4} + 2 = \sqrt{3x+4} \Rightarrow (\sqrt{2x-4} + 2)^2 = (\sqrt{3x+4})^2 \Rightarrow$

$2x - 4 + 4\sqrt{2x-4} + 4 = 3x + 4 \Rightarrow 4\sqrt{2x-4} = x + 4 \Rightarrow (4\sqrt{2x-4})^2 = (x+4)^2 \Rightarrow$

$16(2x - 4) = x^2 + 8x + 16 \Rightarrow 32x - 64 = x^2 + 8x + 16 \Rightarrow x^2 - 24x + 80 = 0 \Rightarrow$

$(x-4)(x-20) = 0 \Rightarrow x = 4$ or 20

The solution set is $\{4, 20\}$. A graph of $y_1 = \sqrt{2x-4} + 2$ and $y_2 = \sqrt{3x+4}$ is shown in Figure 18.

(b) From the graph, $y_1 > y_2$ on the interval $(4, 20)$.

(c) Since the graph of y_1 begins when $2x - 4 = 0 \Rightarrow x = 2$, $y_1 < y_2$ on the interval $[2, 4) \cup (20, \infty)$.

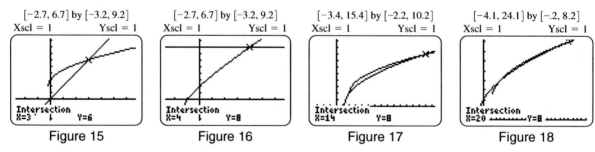

$[-2.7, 6.7]$ by $[-3.2, 9.2]$	$[-2.7, 6.7]$ by $[-3.2, 9.2]$	$[-3.4, 15.4]$ by $[-2.2, 10.2]$	$[-4.1, 24.1]$ by $[-.2, 8.2]$
Figure 15	Figure 16	Figure 17	Figure 18

19. (a) $\sqrt[3]{x^2 - 2x} = \sqrt[3]{x} \Rightarrow (\sqrt[3]{x^2 - 2x})^3 = (\sqrt[3]{x})^3 \Rightarrow x^2 - 2x = x \Rightarrow x^2 - 3x = 0 \Rightarrow x(x-3) = 0 \Rightarrow$

$x = 0$ or $x = 3$; The solution set is $\{0, 3\}$. A graph of $y_1 = \sqrt[3]{x^2 - 2x}$ and $y_2 = \sqrt[3]{x}$ is shown in Figure 19.

(b) From the graph, $y_1 > y_2$ on the interval $(-\infty, 0) \cup (3, \infty)$.

(c) From the graph, $y_1 < y_2$ on the interval $(0, 3)$.

20. (a) $\sqrt[3]{4x^2 - 4x + 1} = \sqrt[3]{x} \Rightarrow (\sqrt[3]{4x^2 - 4x + 1})^3 = (\sqrt[3]{x})^3 \Rightarrow 4x^2 - 4x + 1 = x \Rightarrow$

    $4x^2 - 5x + 1 = 0 \Rightarrow (4x - 1)(x - 1) = 0 \Rightarrow x = \dfrac{1}{4}$ or 1

    The solution set is $\left\{\dfrac{1}{4}, 1\right\}$. A graph of $y_1 = \sqrt[3]{4x^2 - 4x + 1}$ and $y_2 = \sqrt[3]{x}$ is shown in Figure 20.

    (b) From the graph, $y_1 > y_2$ on the interval $\left(-\infty, \dfrac{1}{4}\right) \cup (1, \infty)$.

    (c) From the graph, $y_1 < y_2$ on the interval $\left(\dfrac{1}{4}, 1\right)$.

21. (a) $\sqrt[4]{3x + 1} = 1 \Rightarrow (\sqrt[4]{3x + 1})^4 = 1^4 \Rightarrow 3x + 1 = 1 \Rightarrow 3x = 0 \Rightarrow x = 0$

    The solution set is $\{0\}$. A graph of $y_1 = \sqrt[4]{3x + 1}$ and $y_2 = 1$ is shown in Figure 21.

    (b) From the graph, $y_1 > y_2$ on the interval $(0, \infty)$.

    (c) Since the graph of y_1 begins when $3x + 1 = 0 \Rightarrow x = -\dfrac{1}{3}$, $y_1 < y_2$ on the interval $\left[-\dfrac{1}{3}, 0\right)$.

22. (a) $\sqrt[4]{x - 15} = 2 \Rightarrow (\sqrt[4]{x - 15})^4 = 2^4 \Rightarrow x - 15 = 16 \Rightarrow x = 31$

    The solution set is $\{31\}$. A graph of $y_1 = \sqrt[4]{x - 15}$ and $y_2 = 2$ is shown in Figure 22.

    (b) From the graph, $y_1 > y_2$ on the interval $(31, \infty)$.

    (c) Since the graph of y_1 begins when $x - 15 = 0 \Rightarrow x = 15$, $y_1 < y_2$ on the interval $[15, 31)$.

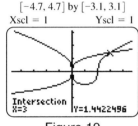

$[-4.7, 4.7]$ by $[-3.1, 3.1]$
Xscl = 1   Yscl = 1

Figure 19

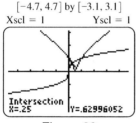

$[-4.7, 4.7]$ by $[-3.1, 3.1]$
Xscl = 1   Yscl = 1

Figure 20

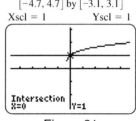

$[-4.7, 4.7]$ by $[-3.1, 3.1]$
Xscl = 1   Yscl = 1

Figure 21

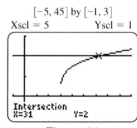

$[-5, 45]$ by $[-1, 3]$
Xscl = 5   Yscl = 1

Figure 22

23. (a) $(2x - 5)^{1/2} - 2 = (x - 2)^{1/2} \Rightarrow [(2x - 5)^{1/2} - 2]^2 = [(x - 2)^{1/2}]^2 \Rightarrow$

    $2x - 5 - 4(2x - 5)^{1/2} + 4 = x - 2 \Rightarrow x + 1 = 4(2x - 5)^{1/2} \Rightarrow (x + 1)^2 = [4(2x - 5)^{1/2}]^2 \Rightarrow$

    $x^2 + 2x + 1 = 16(2x - 5) \Rightarrow x^2 + 2x + 1 = 32x - 80 \Rightarrow x^2 - 30x + 81 = 0 \Rightarrow$

    $(x - 3)(x - 27) = 0 \Rightarrow x = 3$ or 27 (3 is extraneous)

    The solution set is $\{27\}$. A graph of $y_1 = (2x - 5)^{1/2} - 2$ and $y_2 = (x - 2)^{1/2}$ is shown in Figure 23.

    (b) From the graph, $y_1 \geq y_2$ on the interval $[27, \infty)$.

    (c) Since the graph of y_1 begins when $2x - 5 = 0 \Rightarrow x = \dfrac{5}{2}$, $y_1 \leq y_2$ on the interval $\left[\dfrac{5}{2}, 27\right]$.

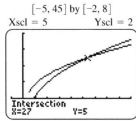

$[-5, 45]$ by $[-2, 8]$
Xscl = 5   Yscl = 2

Figure 23

24. (a) $(x + 5)^{1/2} - 2 = (x - 1)^{1/2} \Rightarrow [(x + 5)^{1/2} - 2]^2 = [(x - 1)^{1/2}]^2 \Rightarrow$

$x + 5 - 4(x + 5)^{1/2} + 4 = x - 1 \Rightarrow 10 = 4(x + 5)^{1/2} \Rightarrow (10)^2 = [4(x + 5)^{1/2}]^2 \Rightarrow$

$100 = 16(x + 5) \Rightarrow 100 = 16x + 80 \Rightarrow 16x = 20 \Rightarrow x = \dfrac{20}{16} \Rightarrow x = \dfrac{5}{4}$

The solution set is $\left\{ \dfrac{5}{4} \right\}$. A graph of $y_1 = (x + 5)^{1/2} - 2$ and $y_2 = (x - 1)^{1/2}$ is shown in Figure 24.

(b) Since the graph of y_2 begins when $x - 1 = 0 \Rightarrow x = 1$, $y_1 \geq y_2$ on the interval $\left[1, \dfrac{5}{4} \right]$.

(c) From the graph, $y_1 \leq y_2$ on the interval $\left[\dfrac{5}{4}, \infty \right)$.

25. (a) $(x^2 + 6x)^{1/4} = 2 \Rightarrow [(x^2 + 6x)^{1/4}]^4 = 2^4 \Rightarrow x^2 + 6x = 16 \Rightarrow x^2 + 6x - 16 = 0 \Rightarrow$

$(x + 8)(x - 2) = 0 \Rightarrow x = -8$ or 2

The solution set is $\{-8, 2\}$. A graph of $y_1 = (x^2 + 6x)^{1/4}$ and $y_2 = 2$ is shown in Figure 25.

(b) From the graph, $y_1 > y_2$ on the interval $(-\infty, -8) \cup (2, \infty)$.

(c) Since $x^2 + 6x = 0 \Rightarrow x(x + 6) = 0 \Rightarrow x = 0$ or -6, the graph of y_1 has endpoints when $x = -6$ and 0.

From the graph, $y_1 < y_2$ on the interval $(-8, -6] \cup [0, 2)$.

26. (a) $(x^2 + 2x)^{1/4} = 3^{1/4} \Rightarrow [(x^2 + 2x)^{1/4}]^4 = (3^{1/4})^4 \Rightarrow x^2 + 2x = 3 \Rightarrow x^2 + 2x - 3 = 0 \Rightarrow$

$(x + 3)(x - 1) = 0 \Rightarrow x = -3$ or 1

The solution set is $\{-3, 1\}$. A graph of $y_1 = (x^2 + 2x)^{1/4}$ and $y_2 = 3^{1/4}$ is shown in Figure 26.

(b) From the graph, $y_1 > y_2$ on the interval $(-\infty, -3) \cup (1, \infty)$.

(c) Since $x^2 + 2x = 0 \Rightarrow x(x + 2) = 0 \Rightarrow x = 0$ or -2, the graph of y_1 has endpoints when $x = -2$ and 0.

From the graph, $y_1 < y_2$ on the interval $(-3, -2] \cup [0, 1)$.

27. (a) $(2x - 1)^{2/3} = x^{1/3} \Rightarrow [(2x - 1)^{2/3}]^3 = (x^{1/3})^3 \Rightarrow (2x - 1)^2 = x \Rightarrow 4x^2 - 4x + 1 = x \Rightarrow$

$4x^2 - 5x + 1 = 0 \Rightarrow (4x - 1)(x - 1) = 0 \Rightarrow x = \dfrac{1}{4}$ or 1

The solution set is $\left\{ \dfrac{1}{4}, 1 \right\}$. A graph of $y_1 = (2x - 1)^{2/3}$ and $y_2 = x^{1/3}$ is shown in Figure 27.

(b) From the graph, $y_1 > y_2$ on the interval $\left(-\infty, \dfrac{1}{4} \right) \cup (1, \infty)$.

(c) From the graph, $y_1 < y_2$ on the interval $\left(\dfrac{1}{4}, 1 \right)$.

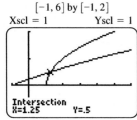

Figure 24

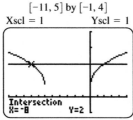

Figure 25

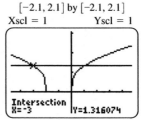

Figure 26

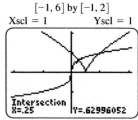

Figure 27

28. (a) $(x - 3)^{2/5} = (4x)^{1/5} \Rightarrow [(x - 3)^{2/5}]^5 = [(4x)^{1/5}]^5 \Rightarrow (x - 3)^2 = 4x \Rightarrow x^2 - 6x + 9 = 4x \Rightarrow$

$x^2 - 10x + 9 = 0 \Rightarrow (x - 1)(x - 9) = 0 \Rightarrow x = 1 \text{ or } 9$

The solution set is $\{1, 9\}$. A graph of $y_1 = (x - 3)^{2/5}$ and $y_2 = (4x)^{1/5}$ is shown in Figure 28.

(b) From the graph, $y_1 > y_2$ on the interval $(-\infty, 1) \cup (9, \infty)$.

(c) From the graph, $y_1 < y_2$ on the interval $(1, 9)$.

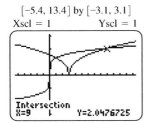

[−5.4, 13.4] by [−3.1, 3.1]
Xscl = 1   Yscl = 1

Intersection
X=9   Y=2.0476725

Figure 28

29. $\sqrt{5x - 6} = x \Rightarrow 5x - 6 = x^2 \Rightarrow x^2 - 5x + 6 = 0 \Rightarrow (x - 2)(x - 3) = 0 \Rightarrow x = 2 \text{ or } 3$

Check: $\sqrt{5(2) - 6} = 2$; $\sqrt{5(3) - 6} = 3$; The solution set is $\{2, 3\}$.

30. $x - 5 = \sqrt{5x - 1} \Rightarrow (x - 5)^2 = 5x - 1 \Rightarrow x^2 - 10x + 25 = 5x - 1 \Rightarrow x^2 - 15x + 26 = 0 \Rightarrow$

$(x - 13)(x - 2) = 0 \Rightarrow x = 13 \text{ or } 2$

Check: $13 - 5 = 8 = \sqrt{5(13) - 1}$; $2 - 5 = -3 \neq \sqrt{5(2) - 1}$ (not a solution); The solution set is $\{13\}$.

31. $\sqrt{x + 5} + 1 = x \Rightarrow \sqrt{x + 5} = x - 1 \Rightarrow x + 5 = (x - 1)^2 \Rightarrow x + 5 = x^2 - 2x + 1 \Rightarrow$

$x^2 - 3x - 4 = 0 \Rightarrow (x - 4)(x + 1) = 0 \Rightarrow x = -1 \text{ or } x = 4$

Check: $\sqrt{-1 + 5} + 1 = -1 \Rightarrow \sqrt{4} + 1 = -1 \Rightarrow 3 \neq -1$ (not a solution).

Check: $\sqrt{4 + 5} + 1 = 4 \Rightarrow \sqrt{9} + 1 = 4 \Rightarrow 4 = 4$; The solution set is $\{4\}$.

32. $\sqrt{4 - 3x} = x + 8 \Rightarrow 4 - 3x = (x + 8)^2 \Rightarrow 4 - 3x = x^2 + 16x + 64 \Rightarrow x^2 + 19x + 60 = 0 \Rightarrow$

$(x + 15)(x + 4) = 0 \Rightarrow x = -15 \text{ or } x = -4$

Check: $\sqrt{4 - 3(-15)} = -15 + 8 \Rightarrow \sqrt{49} = -7 \Rightarrow 7 \neq -7$ (not a solution).

Check: $\sqrt{4 - 3(-4)} = -4 + 8 \Rightarrow \sqrt{16} = 4 \Rightarrow 4 = 4$; The solution set is $\{-4\}$.

33. $\sqrt{2x + 3} - \sqrt{x + 1} = 1 \Rightarrow \sqrt{2x + 3} = \sqrt{x + 1} + 1 \Rightarrow 2x + 3 = (\sqrt{x + 1} + 1)^2 \Rightarrow$

$2x + 3 = x + 1 + 2\sqrt{x + 1} + 1 \Rightarrow 2x + 3 = x + 2 + 2\sqrt{x + 1} \Rightarrow x + 1 = 2\sqrt{x + 1} \Rightarrow$

$(x + 1)^2 = 4(x + 1) \Rightarrow x^2 + 2x + 1 = 4x + 4 \Rightarrow x^2 - 2x - 3 = 0 \Rightarrow (x - 3)(x + 1) = 0 \Rightarrow$

$x = -1 \text{ or } x = 3$

Check: $\sqrt{2(-1) + 3} - \sqrt{(-1) + 1} = 1 \Rightarrow \sqrt{1} - \sqrt{0} = 1 \Rightarrow 1 = 1.$

Check: $\sqrt{2(3) + 3} - \sqrt{3 + 1} = 1 \Rightarrow \sqrt{9} - \sqrt{4} = 1 \Rightarrow 3 - 2 = 1 \Rightarrow 1 = 1$; The solution set is $\{-1, 3\}$.

34. $\sqrt{3x + 4} - \sqrt{2x - 4} = -2 \Rightarrow 3x + 4 = (\sqrt{2x - 4} - 2)^2 \Rightarrow 3x + 4 = 2x - 4 - 4\sqrt{2x - 4} + 4 \Rightarrow$

$x + 4 = 4\sqrt{2x - 4} \Rightarrow (x + 4)^2 = 16(2x - 4) \Rightarrow x^2 + 8x + 16 = 32x - 64 \Rightarrow x^2 - 24x + 80 = 0 \Rightarrow$

$(x - 20)(x - 4) = 0 \Rightarrow x = 4 \text{ or } x = 20$

Check: $\sqrt{3(4) + 4} - \sqrt{2(4) - 4} = \sqrt{16} - \sqrt{4} = 4 - 2 = 2.$

Check: $\sqrt{3(20) + 4} - \sqrt{2(20) - 4} = \sqrt{64} - \sqrt{36} = 8 - 6 = 2$; The solution set is $\{4, 20\}$.

35. $\sqrt[3]{x+1} = -3 \Rightarrow x + 1 = (-3)^3 \Rightarrow x + 1 = -27 \Rightarrow x = -28$

    Check: $\sqrt[3]{-28+1} = \sqrt[3]{-27} = -3$; The solution set is $\{-28\}$.

36. $\sqrt[3]{x} + 5 = 4 \Rightarrow \sqrt[3]{x} = -1 \Rightarrow x = (-1)^3 \Rightarrow x = -1$

    Check: $\sqrt[3]{-1} + 5 = -1 + 5 = 4$; The solution set is $\{-1\}$.

37. $\sqrt[3]{x+1} = \sqrt[3]{2x-1} \Rightarrow x + 1 = 2x - 1 \Rightarrow x = 2$

    Check: $\sqrt[3]{2+1} = \sqrt[3]{3} = \sqrt[3]{2(2)-1}$; The solution set is $\{2\}$.

38. $\sqrt[3]{2x^2+1} = \sqrt[3]{1-x} \Rightarrow 2x^2 + 1 = 1 - x \Rightarrow 2x^2 + x = 0 \Rightarrow x(2x+1) = 0 \Rightarrow x = 0 \text{ or } -\dfrac{1}{2}$

    Check: $\sqrt[3]{2(0)^2+1} = \sqrt[3]{1} = \sqrt[3]{1-0}$; $\sqrt[3]{2(-\frac{1}{2})^2+1} = \sqrt[3]{\frac{1}{2}} = \sqrt[3]{1-(-\frac{1}{2})}$; The solution set is $\left\{-\dfrac{1}{2}, 0\right\}$.

39. $\sqrt[4]{x-2} + 4 = 20 \Rightarrow \sqrt[4]{x-2} = 16 \Rightarrow x - 2 = (16)^4 \Rightarrow x - 2 = 65,536 \Rightarrow x = 65,538$

    Check: $\sqrt[4]{65,538-2} + 4 = 20 \Rightarrow \sqrt[4]{65,536} + 4 = 20 \Rightarrow 20 = 20$; The solution set is $\{65,538\}$.

40. $\sqrt[4]{2x+3} = \sqrt{x+1} \Rightarrow 2x + 3 = (x+1)^2 \Rightarrow 2x + 3 = x^2 + 2x + 1 \Rightarrow x^2 = 2 \Rightarrow x = \pm\sqrt{2}$

    Check: $\sqrt[4]{2(\sqrt{2})+3} = \sqrt{\sqrt{2}+1} \Rightarrow 2\sqrt{2} + 3 = (\sqrt{2}+1)^2 \Rightarrow 2\sqrt{2} + 3 = 2 + 2\sqrt{2} + 1 \Rightarrow 3 = 3$

    Note that the value $-\sqrt{2}$ causes the right side of the equation to be undefined. The solution set is $\{\sqrt{2}\}$.

41. $x^{2/5} = 4 \Rightarrow (x^{2/5})^{5/2} = 4^{5/2} \Rightarrow x = (\sqrt{4})^5 \Rightarrow x = (\pm 2)^5 \Rightarrow x = \pm 32$

    Check: $(\pm 32)^{2/5} = (\sqrt[5]{\pm 32})^2 = (\pm 2)^2 = 4$; The solution set is $\{-32, 32\}$.

42. $x^{2/3} = 16 \Rightarrow (x^{2/3})^{3/2} = 16^{3/2} \Rightarrow x = (\sqrt{16})^3 \Rightarrow x = (\pm 4)^3 \Rightarrow x = \pm 64$

    Check: $(\pm 64)^{2/3} = (\sqrt[3]{\pm 64})^2 = (\pm 4)^2 = 16$; The solution set is $\{-64, 64\}$.

43. $2x^{1/3} - 5 = 1 \Rightarrow 2x^{1/3} = 6 \Rightarrow x^{1/3} = 3 \Rightarrow (x^{1/3})^3 = 3^3 \Rightarrow x = 27$

    Check: $2(27)^{1/3} - 5 = 2\sqrt[3]{27} - 5 = 2(3) - 5 = 6 - 5 = 1$; The solution set is $\{27\}$.

44. $4x^{3/2} + 5 = 21 \Rightarrow 4x^{3/2} = 16 \Rightarrow x^{3/2} = 4 \Rightarrow (x^{3/2})^{2/3} = 4^{2/3} \Rightarrow x = \sqrt[3]{4^2} \Rightarrow x = \sqrt[3]{16}$

    Check: $4(\sqrt[3]{16})^{3/2} + 5 = 4(16^{1/3})^{3/2} + 5 = 4(16^{1/2}) + 5 = 4(\sqrt{16}) + 5 = 4(4) + 5 = 16 + 5 = 21$

    The solution set is $\{\sqrt[3]{16}\}$.

45. $x^{-2} + 3x^{-1} + 2 = 0 \Rightarrow (x^{-1})^2 + 3(x^{-1}) + 2 = 0$, let $u = x^{-1}$, then $u^2 + 3u + 2 = 0 \Rightarrow$

    $(u+2)(u+1) = 0 \Rightarrow u = -2 \text{ or } u = -1$. Because $u = x^{-1}$, it follows that $x = u^{-1}$.

    Thus $x = (-2)^{-1} \Rightarrow x = \dfrac{1}{(-2)} \Rightarrow x = -\dfrac{1}{2}$ or $x = (-1)^{-1} \Rightarrow x = \dfrac{1}{(-1)} = -1$. Therefore, $x = -1, -\dfrac{1}{2}$.

    Check: $(-1)^{-2} + 3(-1)^{-1} + 2 = 1 - 3 + 2 = 0$; $\left(-\dfrac{1}{2}\right)^{-2} + 3\left(-\dfrac{1}{2}\right)^{-1} + 2 = 4 - 6 + 2 = 0$

    The solution set is $\left\{-1, -\dfrac{1}{2}\right\}$.

46. $2x^{-2} - x^{-1} = 3 \Rightarrow 2x^{-2} - x^{-1} - 3 = 0 \Rightarrow 2(x^{-1})^2 - (x^{-1}) - 3 = 0$, let $u = x^{-1}$, then $2u^2 - u - 3 = 0 \Rightarrow$

    $(2u-3)(u+1) = 0 \Rightarrow u = -1 \text{ or } u = \dfrac{3}{2}$. Because $u = x^{-1}$, it follows that $x = u^{-1}$.

    Thus $x = (-1)^{-1} \Rightarrow x = \dfrac{1}{(-1)} \Rightarrow x = -1$ or $x = \left(\dfrac{3}{2}\right)^{-1} \Rightarrow x = \dfrac{1}{(\frac{3}{2})} \Rightarrow x = \dfrac{2}{3}$. Therefore, $x = -1, \dfrac{2}{3}$.

    Check: $2(-1)^{-2} - (-1)^{-1} = 2 + 1 = 3$; $2\left(\dfrac{2}{3}\right)^{-2} - \left(\dfrac{2}{3}\right)^{-1} = 2\left(\dfrac{9}{4}\right) - \dfrac{3}{2} = \dfrac{9}{2} - \dfrac{3}{2} = \dfrac{6}{2} = 3$

    The solution set is $\left\{-1, \dfrac{2}{3}\right\}$.

47. $5x^{-2} + 13x^{-1} = 28 \Rightarrow 5(x^{-1})^2 + 13(x^{-1}) - 28 = 0$, let $u = x^{-1}$, then $5u^2 + 13u - 28 = 0 \Rightarrow$

$(5u - 7)(u + 4) = 0 \Rightarrow u = \dfrac{7}{5}$ or $u = -4$. Because $u = x^{-1}$, it follows that $x = u^{-1}$.

Thus $x = \left(\dfrac{7}{5}\right)^{-1} \Rightarrow x = \dfrac{5}{7}$ or $x = (-4)^{-1} \Rightarrow x = -\dfrac{1}{4}$. Therefore, $x = -\dfrac{1}{4}, \dfrac{5}{7}$.

Check: $5\left(-\dfrac{1}{4}\right)^{-2} + 13\left(-\dfrac{1}{4}\right)^{-1} = 80 - 52 = 28$; $5\left(\dfrac{5}{7}\right)^{-2} + 13\left(\dfrac{5}{7}\right)^{-1} = \dfrac{49}{5} + \dfrac{91}{5} = \dfrac{140}{5} = 28$

The solution set is $\left\{-\dfrac{1}{4}, \dfrac{5}{7}\right\}$.

48. $3x^{-2} - 19x^{-1} + 20 = 0 \Rightarrow 3(x^{-1})^2 - 19(x^{-1}) + 20 = 0$, let $u = x^{-1}$, then $3u^2 - 19u + 20 = 0 \Rightarrow$

$(3u - 4)(u - 5) = 0 \Rightarrow u = \dfrac{4}{3}$ or $u = 5$. Because $u = x^{-1}$, it follows that $x = u^{-1}$.

Thus $x = \left(\dfrac{4}{3}\right)^{-1} \Rightarrow x = \dfrac{3}{4}$ or $x = 5 \Rightarrow x = \dfrac{1}{5}$. Therefore, $x = \dfrac{1}{5}, \dfrac{3}{4}$.

Check: $3\left(\dfrac{1}{5}\right)^{-2} - 19\left(\dfrac{1}{5}\right)^{-1} + 20 = 75 - 95 + 20 = 0$; $3\left(\dfrac{3}{4}\right)^{-2} - 19\left(\dfrac{3}{4}\right)^{-1} + 20 = \dfrac{16}{3} - \dfrac{76}{3} + 20 = 0$

The solution set is $\left\{\dfrac{1}{5}, \dfrac{3}{4}\right\}$.

49. $x^{2/3} - x^{1/3} - 6 = 0 \Rightarrow (x^{1/3})^2 - x^{1/3} - 6 = 0$, let $u = x^{1/3}$, then $u^2 - u - 6 = 0 \Rightarrow$

$(u - 3)(u + 2) = 0 \Rightarrow u = -2$ or $u = 3$. Because $u = x^{1/3}$, it follows that $x = u^3$.

Thus $x = (-2)^3 \Rightarrow x = -8$ or $x = 3^3 \Rightarrow x = 27$. Therefore, $x = -8, 27$.

Check: $(-8)^{2/3} - (-8)^{1/3} - 6 = 4 + 2 - 6 = 0$; $(27)^{2/3} - (27)^{1/3} - 6 = 9 - 3 - 6 = 0$

The solution set is $\{-8, 27\}$.

50. $x^{2/3} + 9x^{1/3} + 14 = 0 \Rightarrow (x^{1/3})^2 + 9(x^{1/3}) + 14 = 0$, let $u = x^{1/3}$, then $u^2 + 9u + 14 = 0 \Rightarrow$

$(u + 7)(u + 2) = 0 \Rightarrow u = -7$ or $u = -2$. Because $u = x^{1/3}$, it follows that $x = u^3$.

Thus $x = (-7)^3 \Rightarrow x = -343$ or $x = (-2)^3 \Rightarrow x = -8$. Therefore, $x = -343, -8$.

Check: $(-343)^{2/3} + 9(-343)^{1/3} + 14 = 49 - 63 + 14 = 0$; $(-8)^{2/3} + 9(-8)^{1/3} + 14 = 4 - 18 + 14 = 0$

The solution set is $\{-343, -8\}$.

51. $x^{3/4} - x^{1/2} - x^{1/4} + 1 = 0 \Rightarrow (x^{1/4})^3 - (x^{1/4})^2 - (x^{1/4}) + 1 = 0$, let $u = x^{1/4}$, then $u^3 - u^2 - u + 1 = 0 \Rightarrow$

$(u^3 - u^2) - (u - 1) = 0 \Rightarrow u^2(u - 1) - 1(u - 1) = 0 \Rightarrow (u^2 - 1)(u - 1) = 0 \Rightarrow$

$(u + 1)(u - 1)(u - 1) = 0 \Rightarrow u = -1$ or $u = 1$. Because $u = x^{1/4}$ it follows that $x = u^4$.

Thus $x = (-1)^4 \Rightarrow x = 1$ or $x = (1)^4 \Rightarrow x = 1$. Therefore, $x = 1$.

Check: $(1)^{3/4} - (1)^{1/2} - (1)^{1/4} + 1 = 1 - 1 - 1 + 1 = 0$; The solution set is $\{1\}$.

52. $x^{3/4} - 2x^{1/2} - 4x^{1/4} + 8 = 0 \Rightarrow (x^{1/4})^3 - 2(x^{1/4})^2 - 4(x^{1/4}) + 8 = 0$, let $u = x^{1/4}$, then

$u^3 - 2u^2 - 4u + 8 = 0 \Rightarrow (u^3 - 2u^2) - (4u - 8) = 0 \Rightarrow u^2(u - 2) - 4(u - 2) = 0 \Rightarrow$

$(u^2 - 4)(u - 2) = 0 \Rightarrow (u + 2)(u - 2)(u - 2) = 0 \Rightarrow u = -2$ or $u = 2$. Because $u = x^{1/4}$ it follows that

$x = u^4$. Thus $x = (-2)^4 \Rightarrow x = 16$ or $x = (2)^4 \Rightarrow x = 16$. Therefore, $x = 16$.

Check: $(16)^{3/4} - 2(16)^{1/2} - 4(16)^{1/4} + 8 = 8 - 8 - 8 + 8 = 0$; The solution set is $\{16\}$.

53. $\sqrt[3]{4x - 4} = \sqrt{x + 1} \Rightarrow (4x - 4)^{1/3} = (x + 1)^{1/2}$

54. The LCD for $\dfrac{1}{3}$ and $\dfrac{1}{2}$ is 6.

55. $[(4x - 4)^{1/3}]^6 = [(x + 1)^{1/2}]^6 \Rightarrow (4x - 4)^2 = (x + 1)^3$

56. $(4x - 4)^2 = (x + 1)^3 \Rightarrow 16x^2 - 32x + 16 = x^3 + 3x^2 + 3x + 1 \Rightarrow x^3 - 13x^2 + 35x - 15 = 0$

57. The graph crosses the x-axis 3 times so the equation has 3 real roots. See Figure 57.

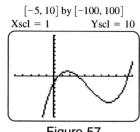

$[-5, 10]$ by $[-100, 100]$
Xscl = 1   Yscl = 10

Figure 57

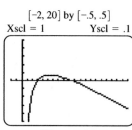

$[-2, 20]$ by $[-.5, .5]$
Xscl = 1   Yscl = .1

Figure 62

58.
$$3{\overline{\smash{\big)}\,1 \;\; -13 \;\;\; 35 \;\; -15}}$$
$$ \;\;\;\;\; 3 \;\; -30 \;\;\;\; 15$$
$$ 1 \;\; -10 \;\;\;\; 5 \;\;\;\;\;\; 0$$

The value in the remainder position is 0, therefore $P(3) = 0$.

59. From the synthetic division shown above, $P(x) = (x - 3)(x^2 - 10x + 5)$.

60. If $x^2 - 10x + 5 = 0$ then $x = \dfrac{-(-10) \pm \sqrt{(-10)^2 - 4(1)(5)}}{2(1)} = \dfrac{10 \pm \sqrt{80}}{2} = 5 \pm 2\sqrt{5} \Rightarrow \{5 \pm 2\sqrt{5}\}$

61. The three solutions are 3, $5 + 2\sqrt{5}$, and $5 - 2\sqrt{5}$.

62. The graph of $y_3 = y_1 - y_2$, where $y_1 = \sqrt[3]{4x - 4}$ and $y_2 = \sqrt{x + 1}$ is shown in Figure 62.

    There are 2 real solutions.

63. If the value $5 - 2\sqrt{5} \approx .528$ is substituted for x in the expression $\sqrt[3]{4x - 4}$, the result is negative. Since the right side of the original equation will never yield a negative value, the root $5 - 2\sqrt{5} \approx .528$ is extraneous. The solution set is $\{3, 5 + 2\sqrt{5}\}$. The calculator figure shown above supports this result.

64. The solution set of the original equation is a subset of the solution set found in the previous exercise.

    The extraneous solution was obtained when each side of the original equation was raised to the sixth power.

65. $\sqrt{\sqrt{x}} = x \Rightarrow x^{1/4} = x \Rightarrow [x^{1/4}]^4 = x^4 \Rightarrow x = x^4 \Rightarrow x - x^4 = 0 \Rightarrow x(1 - x^3) = 0 \Rightarrow x = 0, 1$

    The solution set is $\{0, 1\}$.

66. $\sqrt[3]{\sqrt[3]{x}} = x \Rightarrow x^{1/9} = x \Rightarrow [x^{1/9}]^9 = x^9 \Rightarrow x = x^9 \Rightarrow x - x^9 = 0 \Rightarrow x(1 - x^8) = 0 \Rightarrow$

    $x(1 + x^4)(1 - x^4) = 0 \Rightarrow x(1 + x^4)(1 + x^2)(1 - x^2) = 0 \Rightarrow x(1 + x^4)(1 + x^2)(1 + x)(1 - x) = 0 \Rightarrow$

    $x = 0, -1, 1$. The solution set is $\{-1, 0, 1\}$.

67. $\sqrt{\sqrt{28x + 8}} = \sqrt{3x + 2} \Rightarrow (28x + 8)^{1/4} = (3x + 2)^{1/2} \Rightarrow [(28x + 8)^{1/4}]^4 = [(3x + 2)^{1/2}]^4 \Rightarrow$

    $28x + 8 = (3x + 2)^2 \Rightarrow 28x + 8 = 9x^2 + 12x + 4 \Rightarrow 9x^2 - 16x - 4 = 0 \Rightarrow (9x + 2)(x - 2) = 0 \Rightarrow$

    $x = -\dfrac{2}{9}, 2$. The solution set is $\left\{-\dfrac{2}{9}, 2\right\}$.

68. $\sqrt{\sqrt{2x+10}} = \sqrt{2x-2} \Rightarrow (2x+10)^{1/4} = (2x-2)^{1/2} \Rightarrow [(2x+10)^{1/4}]^4 = [(2x-2)^{1/2}]^4 \Rightarrow$

    $2x+10 = (2x-2)^2 \Rightarrow 2x+10 = 4x^2 - 8x + 4 \Rightarrow 4x^2 - 10x - 6 = 0 \Rightarrow (4x+2)(x-3) = 0 \Rightarrow$

    $x = -\dfrac{1}{2}, 3$. The solution $-\dfrac{1}{2}$ is extraneous because it causes the right side of the equation to be undefined.

    The solution set is $\{3\}$.

69. $\sqrt[3]{\sqrt{32x}} = \sqrt[3]{x+6} \Rightarrow (32x)^{1/6} = (x+6)^{1/3} \Rightarrow [(32x)^{1/6}]^6 = [(x+6)^{1/3}]^6 \Rightarrow 32x = (x+6)^2 \Rightarrow$

    $32x = x^2 + 12x + 36 \Rightarrow x^2 - 20x + 36 = 0 \Rightarrow (x-2)(x-18) = 0 \Rightarrow x = 2, 18$

    The solution set is $\{2, 18\}$.

70. $\sqrt[3]{\sqrt{x+63}} = \sqrt[3]{2x+6} \Rightarrow (x+63)^{1/6} = (2x+6)^{1/3} \Rightarrow [(x+63)^{1/6}]^6 = [(2x+6)^{1/3}]^6 \Rightarrow$

    $x+63 = (2x+6)^2 \Rightarrow x+63 = 4x^2 + 24x + 36 \Rightarrow 4x^2 + 23x - 27 = 0 \Rightarrow (4x+27)(x-1) = 0 \Rightarrow$

    $x = -\dfrac{27}{4}, 1$. The solution $-\dfrac{27}{4}$ is extraneous because it causes the right side of the equation to be negative and

    the left side to be positive. The solution set is $\{1\}$.

71. $v = \dfrac{350}{\sqrt{6000}} \approx 4.5$ km per sec

72. $d = \sqrt{\dfrac{400}{14}} \approx 5.35$ ft

73. $P = 2\pi\sqrt{\dfrac{5}{32}} \approx 2.5$ sec

74. (a) $d = 1.22\sqrt{15{,}000} \approx 149$ mi

    (b) $d = 1.22\sqrt{20{,}000} \approx 173$ mi

75. $s = 30\sqrt{\dfrac{900}{97}} \approx 91$ mph

76. (a) $S = 28.6\sqrt[3]{100} \approx 133$ species

    (b) $S = 28.6\sqrt[3]{1500} \approx 327$ species

77. (a) Since the distance from C to D is 20 feet, the distance from P to C is $20 - x$.

    (b) The value must be between 0 and 20. That is, $0 < x < 20$.

    (c) $(AP)^2 = (DP)^2 + (AD)^2 \Rightarrow (AP)^2 = x^2 + 12^2 \Rightarrow AP = \sqrt{x^2 + 12^2};$

       $(BP)^2 = (CP)^2 + (BC)^2 \Rightarrow (BP)^2 = (20-x)^2 + 16^2 \Rightarrow BP = \sqrt{(20-x)^2 + 16^2}$

    (d) $f(x) = \sqrt{x^2 + 12^2} + \sqrt{(20-x)^2 + 16^2}, 0 < x < 20$

    (e) The graph of $y_1 = \sqrt{x^2 + 12^2} + \sqrt{(20-x)^2 + 16^2}$ is shown in Figure 77. Here $f(4) \approx 35.28$. When the

       stake is 4 feet from the 12-foot pole, approximately 35.28 feet of wire will be required.

    (f) Using the calculator, $f(x)$ is a minimum (about 34.41 feet) when $x \approx 8.57$ feet.

    (g) This problem examined how the total amount of wire used can be expressed in terms of the distance from

       the stake at P to the base of the 12-foot pole. We find that the amount of wire used can be minimized when

       the stake is approximately 8.57 feet from the 12-foot pole.

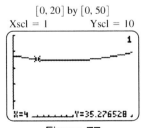

$[0, 20]$ by $[0, 50]$
$\mathrm{Xscl} = 1$ $\mathrm{Yscl} = 10$

Figure 77

78. (a) Since the distance from C to D is 16 feet, the distance from P to C is $16 - x$.

(b) The value must be between 0 and 16. That is, $0 < x < 16$.

(c) $(AP)^2 = (DP)^2 + (AD)^2 \Rightarrow (AP)^2 = x^2 + 9^2 \Rightarrow AP = \sqrt{x^2 + 9^2};$

$\quad (BP)^2 = (CP)^2 + (BC)^2 \Rightarrow (BP)^2 = (16 - x)^2 + 12^2 \Rightarrow BP = \sqrt{(16 - x)^2 + 12^2}$

(d) $f(x) = \sqrt{x^2 + 9^2} + \sqrt{(16 - x)^2 + 12^2}, 0 < x < 16$

(e) The graph of $y_1 = \sqrt{x^2 + 9^2} + \sqrt{(16 - x)^2 + 12^2}$ is shown in Figure 78. Here $f(4) \approx 26.82$. When the stake is 4 feet from the 9-foot pole, approximately 26.82 feet of wire will be required.

(f) Using the calculator, $f(x)$ is a minimum (about 26.4 feet) when $x \approx 6.86$ feet.

(g) This problem examined how the total amount of wire used can be expressed in terms of the distance from the stake at P to the base of the 9-foot pole. We find that the amount of wire used can be minimized when the stake is approximately 6.86 feet from the 9-foot pole.

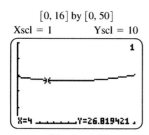

[0, 16] by [0, 50]
Xscl = 1   Yscl = 10

Figure 78

79. Using the Pythagorean theorem, the diagonal distance traveled on the river is $d = \sqrt{x^2 + 3^2}$. Noting that rate of travel on land is 5 mph, the rate of travel on the water is 2 mph and that $t = \dfrac{d}{r}$, the time needed to travel on land is $\dfrac{8 - x}{5}$ and the time needed to travel on the river is $\dfrac{\sqrt{x^2 + 9}}{2}$. The total travel time is given by the function $y_1 = \dfrac{8 - x}{5} + \dfrac{\sqrt{x^2 + 9}}{2}$. By graphing this function (not shown), we find that the minimum value for time occurs when $x \approx 1.31$. The hunter should travel $8 - 1.31 = 6.69$ miles along the river.

80. Let x represent the distance from A to P. Then the distance from B to P can be calculated by using the Pythagorean theorem. The diagonal distance is $d = \sqrt{x^2 + 1}$. The distance from P to L is given by $2 - x$. If we assume that the energy used when flying over land is 1, then the total energy used by the pigeon for its flight is $y_1 = \dfrac{4}{3}\sqrt{x^2 + 1} + (2 - x)$. By graphing this function (not shown), we find that the minimum value for energy use occurs when $x \approx 1.13$. Point P should be located 1.13 miles from point A.

81. We will refer to the original position of the *Inspiration* as the origin. At time x hours past noon, the *Celebration* is $60 - 30x$ miles south of the origin and the *Inspiration* is $20x$ miles west of the origin. Using the Pythagorean theorem, the distance between the ships is given by the function $y_1 = \sqrt{(60 - 30x)^2 + (20x)^2}$. By graphing this function (not shown), we find that the distance between the ships is a minimum when $x \approx 1.38$ hours. At the time 1.38 hours past noon (approximately 1:23 P.M.) the ships are about 33.28 miles apart.

82. Using the Pythagorean theorem, the diagonal distance traveled on the river is $d = \sqrt{x^2 + 3^2}$. Noting that rate of travel on land is 5 mph, the rate of travel on the water is 4 mph and that $t = \dfrac{d}{r}$, the time needed to travel on land is $\dfrac{6 - x}{5}$ and the time needed to travel on the river is $\dfrac{\sqrt{x^2 + 9}}{4}$. The total travel time is given by the function $y_1 = \dfrac{6 - x}{5} + \dfrac{\sqrt{x^2 + 9}}{4}$. By graphing this function (not shown), we find that the minimum value for time occurs when $x = 4$. The minimum time is 1.65 hours.

Reviewing Basic Concepts (Sections 4.4 and 4.5)

1. As the exponent increases in value, the curve rises more rapidly for $x \geq 1$. See Figure 1.

2. $S(.75) = .3(.75)^{3/4} \approx .24 \text{ m}^2$

3. Solving the equation for y yields $x^2 + y^2 = 16 \Rightarrow y^2 = 16 - x^2 \Rightarrow y = \pm\sqrt{16 - x^2}$.
   The expressions are $y_1 = \sqrt{16 - x^2}$ and $y_2 = -\sqrt{16 - x^2}$ which are graphed in a decimal window in Figure 3.

4. Solving the equation for y yields $x = y^2 + 4y + 6 \Rightarrow x - 6 + 4 = y^2 + 4y + 4 \Rightarrow x - 2 = (y + 2)^2 \Rightarrow$
   $\pm\sqrt{x - 2} = y + 2 \Rightarrow -2 \pm \sqrt{x - 2} = y \Rightarrow y = -2 \pm \sqrt{x - 2}$. The expressions are
   $y_1 = -2 + \sqrt{x - 2}$ and $y_2 = -2 - \sqrt{x - 2}$ which are graphed in a square window in Figure 4.

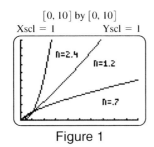
[0, 10] by [0, 10]
Xscl = 1   Yscl = 1

Figure 1

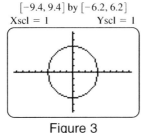
[-9.4, 9.4] by [-6.2, 6.2]
Xscl = 1   Yscl = 1

Figure 3

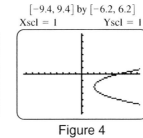
[-9.4, 9.4] by [-6.2, 6.2]
Xscl = 1   Yscl = 1

Figure 4

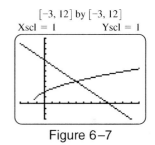
[-3, 12] by [-3, 12]
Xscl = 1   Yscl = 1

Figure 6–7

5. $\sqrt{3x + 4} = 8 - x \Rightarrow (\sqrt{3x + 4})^2 = (8 - x)^2 \Rightarrow 3x + 4 = 64 - 16x + x^2 \Rightarrow x^2 - 19x + 60 = 0 \Rightarrow$
   $(x - 4)(x - 15) = 0 \Rightarrow x = 4, 15$. The solution 15 is extraneous. The solution set is $\{4\}$.

6. The graph of $y_1 = \sqrt{3x + 4}$ is above the graph of $y_2 = 8 - x$ on the interval $(4, \infty)$. See Figure 6–7.

7. Noting that the graph of $y_1 = \sqrt{3x + 4}$ does not start until $3x + 4 = 0 \Rightarrow 3x = -4 \Rightarrow x = -\dfrac{4}{3}$,
   the graph of $y_1 = \sqrt{3x + 4}$ is below the graph of $y_2 = 8 - x$ on the interval $\left[-\dfrac{4}{3}, 4\right)$. See Figure 6–7.

8. $\sqrt{3x + 4} + \sqrt{5x + 6} = 2 \Rightarrow \sqrt{3x + 4} = 2 - \sqrt{5x + 6} \Rightarrow (\sqrt{3x + 4})^2 = (2 - \sqrt{5x + 6})^2 \Rightarrow$
   $3x + 4 = 4 - 4\sqrt{5x + 6} + 5x + 6 \Rightarrow 4\sqrt{5x + 6} = 2x + 6 \Rightarrow (4\sqrt{5x + 6})^2 = (2x + 6)^2 \Rightarrow$
   $16(5x + 6) = 4x^2 + 24x + 36 \Rightarrow 80x + 96 = 4x^2 + 24x + 36 \Rightarrow 4x^2 - 56x - 60 = 0 \Rightarrow$
   $x^2 - 14x - 15 = 0 \Rightarrow (x + 1)(x - 15) = 0 \Rightarrow x = -1, 15$. The solution 15 is an extraneous solution.
   The solution set is $\{-1\}$.

9. Any value less than 15 causes the radicand to be negative and the calculator will not return a value for the square root of a negative number in a table.

10. (a) dog: $f(24) = \dfrac{1607}{\sqrt[4]{24^3}} \approx 148$ beats per minute; person: $f(66) = \dfrac{1607}{\sqrt[4]{66^3}} \approx 69$ beats per minute

(b) $400 = \dfrac{1607}{\sqrt[4]{x^3}} \Rightarrow 400\sqrt[4]{x^3} = 1607 \Rightarrow \sqrt[4]{x^3} = \dfrac{1607}{400} \Rightarrow x^{3/4} = \dfrac{1607}{400} \Rightarrow x = \left(\dfrac{1607}{400}\right)^{4/3} \approx 6.4$ inches

Chapter 4 Review Exercises

1. (a) The graph of $y = -\dfrac{1}{x} + 6$ can be obtained by reflecting the graph of $y = \dfrac{1}{x}$ across the x-axis and shifting upward 6 units.

(b) A sketch of the graph is shown in Figure 1b.

(c) A calculator graph is shown in Figure 1c.

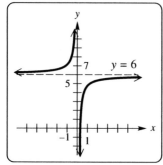

Figure 1b

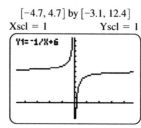

Figure 1c

2. (a) The graph of $y = \dfrac{4}{x} - 3$ can be obtained by stretching the graph of $y = \dfrac{1}{x}$ by a factor of 4 and shifting downward 3 units.

(b) A sketch of the graph is shown in Figure 2b.

(c) A calculator graph is shown in Figure 2c.

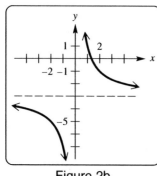

Figure 2b

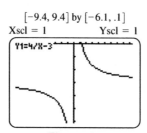

Figure 2c

3. (a) The graph of $y = -\dfrac{1}{(x-2)^2}$ can be obtained by reflecting the graph of $y = \dfrac{1}{x^2}$ across the x-axis and shifting to the right 2 units.

(b) A sketch of the graph is shown in Figure 3b.

(c) A calculator graph is shown in Figure 3c.

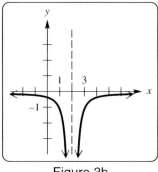

Figure 3b

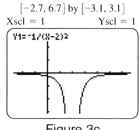

Figure 3c

4.   (a) The graph of $y = \dfrac{2}{x^2} + 1$ can be obtained by stretching the graph of $y = \dfrac{1}{x^2}$ by a factor of 2 and shifting

   upward 1 unit.

   (b) A sketch of the graph is shown in Figure 4b.

   (c) A calculator graph is shown in Figure 4c.

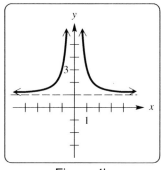

Figure 4b

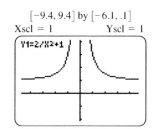

Figure 4c

5.   When the degree of the numerator is exactly 1 greater than the degree of the denominator, the graph of a

   rational function defined by an expression written in lowest terms will have an oblique asymptote.

6.   See Figure 6.

7.   See Figure 7.

8.   See Figure 8.

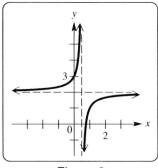

Figure 6

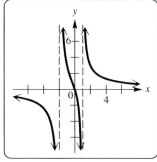

Figure 7

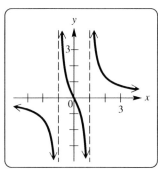

Figure 8

9. See Figure 9.

10. See Figure 10.

11. See Figure 11.

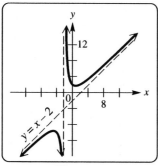
Figure 9

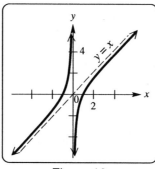
Figure 10

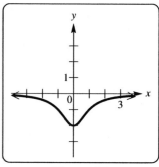
Figure 11

12. See Figure 12.

13. See Figure 13.

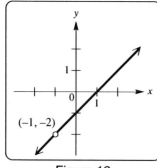
Figure 12

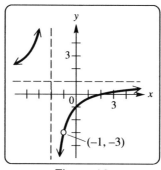
Figure 13

14. A vertical asymptote of $x = 0$ indicates that the denominator should contain the factor x. Since there are no x-intercepts, the numerator can never equal zero. Since there is a horizontal asymptote at $y = 0$, the degree of the numerator is less than the degree of the denominator. The end behavior of the graph suggests that the degree of the denominator is even. Since the graph passes through the point $\left(2, \dfrac{1}{4}\right)$, one possible equation for the function is $f(x) = \dfrac{1}{x^2}$. Other functions are possible.

15. A vertical asymptote of $x = 1$ indicates that the denominator should contain the factor $(x - 1)$. A horizontal asymptote of $y = -3$ indicates that the degrees of the numerator and denominator should be equal and the leading coefficient of the numerator should be -3 times that of the denominator. An x-intercept of 2, indicates that the numerator should contain the factor $(x - 2)$. One possible equation for the function is
$f(x) = \dfrac{-3(x - 2)}{x - 1} = \dfrac{-3x + 6}{x - 1}$. Other functions are possible.

16. An x-intercepts of -2, indicates that the numerator should contain the factor $(x + 2)$. A "hole" when $x = 2$ indicates that both the numerator and denominator should contain the factor $(x - 2)$. One possible equation for the function is $f(x) = \dfrac{(x + 2)(x - 2)}{x - 2} = \dfrac{x^2 - 4}{x - 2}$. Other functions are possible.

17. (a) $\dfrac{3x-2}{x+1} = 0 \Rightarrow 3x - 2 = 0 \Rightarrow 3x = 2 \Rightarrow x = \dfrac{2}{3}$. The solution set is $\left\{\dfrac{2}{3}\right\}$.

(b) Graph $y_1 = (3x - 2)/(x + 1)$ as shown in Figure 17. With vertical asymptote $x = -1$, the graph is below

the x-axis on the interval $\left(-1, \dfrac{2}{3}\right)$. The solution set is $\left(-1, \dfrac{2}{3}\right)$.

(c) Graph $y_1 = (3x - 2)/(x + 1)$ as shown in Figure 17. With vertical asymptote $x = -1$, the graph is above

the x-axis on the interval $(-\infty, -1) \cup \left(\dfrac{2}{3}, \infty\right)$. The solution set is $(-\infty, -1) \cup \left(\dfrac{2}{3}, \infty\right)$.

18. (a) $\dfrac{5}{2x+5} = \dfrac{3}{x+2} \Rightarrow 5(x+2) = 3(2x+5) \Rightarrow 5x + 10 = 6x + 15 \Rightarrow -5 = x \Rightarrow x = -5$;

The solution set is $\{-5\}$.

(b) Graph $y_1 = 5/(2x + 5) - 3/(x + 2)$ as shown in Figure 18. With vertical asymptotes $x = -\dfrac{5}{2}$ and $x = -2$,

the graph is below the x-axis on $\left(-5, -\dfrac{5}{2}\right) \cup (-2, \infty)$. The solution set is $\left(-5, -\dfrac{5}{2}\right) \cup (-2, \infty)$.

(c) Graph $y_1 = 5/(2x + 5) - 3/(x + 2)$ as shown in Figure 18. With vertical asymptotes $x = -\dfrac{5}{2}$ and $x = -2$,

the graph is above the x-axis on $(-\infty, -5) \cup \left(-\dfrac{5}{2}, -2\right)$. The solution set is $(-\infty, -5) \cup \left(-\dfrac{5}{2}, -2\right)$.

$[-9.4, 9.4]$ by $[-3.1, 6.2]$	$[-6.9, 2.5]$ by $[-10, 30]$	$[-9.4, 9.4]$ by $[-6.2, 6.2]$	$[-9.4, 9.4]$ by $[-6.2, 6.2]$
Xscl = 1 Yscl = 1	Xscl = 1 Yscl = 5	Xscl = 1 Yscl = 1	Xscl = 1 Yscl = 1

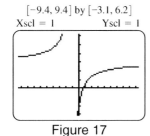 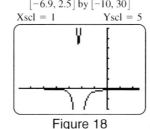 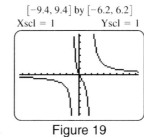 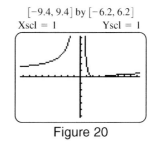

Figure 17   Figure 18   Figure 19   Figure 20

19. (a) Multiply through by the LCD, $(x - 2)(x + 1)$.

$$\dfrac{3}{x-2} \cdot (x-2)(x+1) + \dfrac{1}{x+1} \cdot (x-2)(x+1) = \dfrac{1}{x^2 - x - 2} \cdot (x-2)(x+1) \Rightarrow$$

$3(x + 1) + 1(x - 2) = 1 \Rightarrow 3x + 3 + x - 2 = 1 \Rightarrow 4x = 0 \Rightarrow x = 0$; This solution set is $\{0\}$.

(b) Graph $y_1 = 3/(x - 2) + 1/(x + 1) - 1/(x^2 - x - 2)$ as shown in Figure 19. With vertical asymptotes

$x = -1$ and $x = 2$, the graph is below or intersects the x-axis on $(-\infty, -1) \cup [0, 2)$.

The solution set is $(-\infty, -1) \cup [0, 2)$.

(c) Graph $y_1 = 3/(x - 2) + 1/(x + 1) - 1/(x^2 - x - 2)$ as shown in Figure 19. With vertical asymptotes

$x = -1$ and $x = 2$, the graph is above or intersects the x-axis on $(-1, 0] \cup (2, \infty)$.

The solution set is $(-1, 0] \cup (2, \infty)$.

20. (a) Multiply through by the LCD, x^2.

$$1 \cdot x^2 - \dfrac{5}{x} \cdot x^2 + \dfrac{6}{x^2} \cdot x^2 = 0 \cdot x^2 \Rightarrow x^2 - 5x + 6 = 0 \Rightarrow (x - 2)(x - 3) = 0 \Rightarrow x = 2, 3$$

This solution set is $\{2, 3\}$.

(b) Graph $y_1 = 1 - 5/x + 6/x^2$ as shown in Figure 20. With vertical asymptote $x = 0$, the graph is below or

intersects the x-axis on $[2, 3]$. The solution set is $[2, 3]$.

(c) Graph $y_1 = 1 - 5/x + 6/x^2$ as shown in Figure 20. With vertical asymptote $x = 0$, the graph is above or

intersects the x-axis on $(-\infty, 0) \cup (0, 2] \cup [3, \infty)$. The solution set is $(-\infty, 0) \cup (0, 2] \cup [3, \infty)$.

21. The graph intersects the x-axis when $x = -2$. The solution set is $\{-2\}$.

22. The graph is above the x-axis on the interval $(-\infty, -2) \cup (-1, \infty)$. The solution set is $(-\infty, -2) \cup (-1, \infty)$.

23. The graph is below the x-axis on the interval $(-2, -1)$. The solution set is $(-2, -1)$.

24. (a) See Figure 24.

    (b) $C(95) = \dfrac{6.7(95)}{100 - 95} = \dfrac{636.5}{5} = 127.3$ thousand dollars or $127,300

25. (a) See Figure 25.

    (b) $C(95) = \dfrac{10(95)}{49(101 - 95)} = \dfrac{950}{294} \approx 3.231$ thousand dollars or about $3231

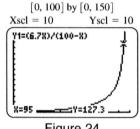
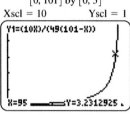

Figure 24   Figure 25

26. (a) As time increases, the value of the denominator increases. Thus concentration decreases.

    (b) $1.5 = \dfrac{5}{t^2 + 1} \Rightarrow 1.5(t^2 + 1) = 5 \Rightarrow 1.5t^2 + 1.5 = 5 \Rightarrow 1.5t^2 = 3.5 \Rightarrow t^2 = \dfrac{3.5}{1.5} \Rightarrow t = \pm\sqrt{\dfrac{3.5}{1.5}} \Rightarrow$

    $t \approx \pm 1.5$; Since the time cannot be negative, the patient should wait about 1.5 hours.

27. (a) Since $0 \le x < 40$, the denominator is positive. Multiplying by the LCD will not change the inequality.

    $\dfrac{x^2}{1600 - 40x} \le 8 \Rightarrow x^2 \le 8(1600 - 40x) \Rightarrow x^2 \le 12{,}800 - 320x \Rightarrow x^2 + 320x - 12{,}800 \le 0$

    The left side of this inequality is a quadratic polynomial whose positive zero can be obtained using the quadratic formula. The zero is approximately 35.96. The graph of this parabola is below the x-axis for values of x in the interval $[0, 36]$. Note that rounding to a whole number is appropriate since x represents a number of cars. The solution set is $[0, 36]$.

    (b) The average line length is less than or equal to 8 cars when the average arrival rate is 36 cars per hour or less.

28. (a) When the coefficient of friction becomes smaller the braking distance increases.

    (b) Since $0 < x \le 1$, the denominator is positive. Multiplying by the LCD will not change the inequality.

    $\dfrac{120}{x} \ge 400 \Rightarrow 120 \ge 400x \Rightarrow \dfrac{120}{400} \ge x \Rightarrow x \le .3$; The solution set is $(0, .3]$.

29. Here $y = \dfrac{k}{x}$. By substitution, $5 = \dfrac{k}{6} \Rightarrow 30 = k \Rightarrow k = 30$. That is $y = \dfrac{30}{x}$. When $x = 15$, $y = \dfrac{30}{15} = 2$.

30. Here $z = \dfrac{k}{t^3}$. By substitution, $.08 = \dfrac{k}{5^3} \Rightarrow .08(5^3) = k \Rightarrow k = 10$. That is $z = \dfrac{10}{t^3}$.

    When $t = 2$, $z = \dfrac{10}{2^3} = 1.25$.

31. Here $m = \dfrac{knp^2}{q}$. By substitution, $20 = \dfrac{k(5)(6^2)}{18} \Rightarrow 360 = 180k \Rightarrow k = \dfrac{360}{180} \Rightarrow k = 2$. That is $m = \dfrac{2np^2}{q}$.

    When $n = 7$, $p = 11$, and $q = 2$, $m = \dfrac{2(7)(11^2)}{2} = 847$.

32. The height of this cone varies *inversely* as the *square* of the *radius* of its base. The constant of variation is $\dfrac{300}{\pi}$.

33. Here $I = \dfrac{k}{d^2}$. By substitution, $70 = \dfrac{k}{5^2} \Rightarrow 70(5^2) = k \Rightarrow k = 1750$. That is $I = \dfrac{1750}{d^2}$.

     When $d = 12, I = \dfrac{1750}{12^2} \approx 12.15$ candela.

34. Here $R = \dfrac{k}{d^2}$. By substitution, $.4 = \dfrac{k}{.01^2} \Rightarrow .4(.01^2) = k \Rightarrow k = .00004$. That is $R = \dfrac{.00004}{d^2}$.

     When $d = .03, I = \dfrac{.00004}{.03^2} = \dfrac{.00004}{.0009} = \dfrac{4}{90} = \dfrac{2}{45}$ ohm.

35. Here $I = kPt$. By substitution, $110 = k(1000)(2) \Rightarrow \dfrac{110}{2000} = k \Rightarrow k = .055$. That is $I = .055Pt$.

     When $P = 5000$ and $t = 5, I = .055(5000)(5) = \$1375$

36. Here $F = \dfrac{kws^2}{r}$. By substitution, $3000 = \dfrac{k(2000)(30^2)}{500} \Rightarrow 1{,}500{,}000 = 1{,}800{,}000k \Rightarrow k = \dfrac{1.5}{1.8} \Rightarrow k = \dfrac{5}{6}$.

     That is $F = \dfrac{\frac{5}{6}ws^2}{r}$. When $w = 2000, s = 60,$ and $r = 800, F = \dfrac{\frac{5}{6}(2000)(60^2)}{800} = 7500$ lb.

37. Here $L = \dfrac{kd^4}{h^2}$. By substitution, $8 = \dfrac{k(1^4)}{9^2} \Rightarrow 648 = k \Rightarrow k = 648$. That is $L = \dfrac{648d^4}{h^2}$.

     When $h = 12,$ and $d = \dfrac{2}{3}, L = \dfrac{648\left(\frac{2}{3}\right)^4}{12^2} = \dfrac{128}{144} = \dfrac{8}{9}$ metric tons.

38. Here $L = \dfrac{kwh^2}{l}$. By substitution, $400 = \dfrac{k(12)(15^2)}{8} \Rightarrow 3200 = 2700k \Rightarrow k = \dfrac{3200}{2700} = \dfrac{32}{27}$.

     That is $L = \dfrac{\frac{32}{27}wh^2}{l}$. When $l = 16, w = 24$ and $h = 8, L = \dfrac{\frac{32}{27}(24)(8^2)}{16} = \dfrac{\frac{49{,}152}{27}}{16} = \dfrac{49{,}152}{432} = \dfrac{1024}{9} = 113\dfrac{7}{9}$ kg

39. Here $w = \dfrac{k}{d^2}$. By substitution, $90 = \dfrac{k}{6400^2} \Rightarrow 90(6400^2) = k \Rightarrow k = 3{,}686{,}400{,}000$.

     That is $w = \dfrac{3{,}686{,}400{,}000}{d^2}$. When $d = 7200, w = \dfrac{3{,}686{,}400{,}000}{7200^2} = \dfrac{3{,}686{,}400{,}000}{51{,}840{,}000} = \dfrac{640}{9} = 71\dfrac{1}{9}$ kg.

40. This function yields the square root graph stretched and reflected across the x-axis. See Figure 40.

41. This function yields the cube root graph shifted a units to the left. See Figure 41.

42. This function yields the cube root graph shifted a units to the right. See Figure 42.

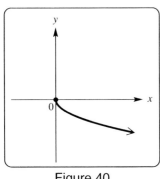

Figure 40

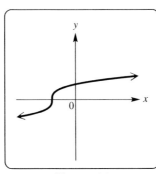

Figure 41

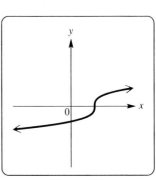

Figure 42

43. This function yields the cube root graph stretched, reflected across the *x*-axis, and shifted down *b* units. See Figure 43.

44. This function yields the square root graph shifted *a* units to the left and *b* units up. See Figure 44.

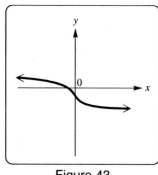

Figure 43

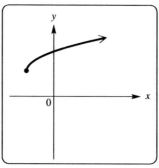

Figure 44

45. Since $12^2 = 144$, $\sqrt{144} = 12$.

46. Since $(-4)^3 = -64$, $\sqrt[3]{-64} = -4$.

47. Since $\left(\dfrac{1}{3}\right)^3 = \dfrac{1}{27}$, $\sqrt[3]{\dfrac{1}{27}} = \dfrac{1}{3}$.

48. Since $3^4 = 81$, $\sqrt[4]{81} = 3$.

49. Since $\left(-\dfrac{2}{3}\right)^5 = -\dfrac{32}{243}$, $\sqrt[5]{-\dfrac{32}{243}} = -\dfrac{2}{3}$.

50. Since $(-2)^5 = -32$, $(-32)^{1/5} = -2$ and so $-(-32)^{1/5} = -(-2) = 2$.

51. $36^{-3/2} = \dfrac{1}{36^{3/2}} = \dfrac{1}{(\sqrt{36})^3} = \dfrac{1}{6^3} = \dfrac{1}{216}$

52. $-1000^{2/3} = -(\sqrt[3]{1000})^2 = -(10^2) = -100$

53. $(-27)^{-4/3} = \dfrac{1}{(-27)^{4/3}} = \dfrac{1}{(\sqrt[3]{-27})^4} = \dfrac{1}{(-3)^4} = \dfrac{1}{81}$

54. $16^{3/4} = (\sqrt[4]{16})^3 = 2^3 = 8$

55. $\sqrt[4]{84.6} \approx 2.429260411$

56. $\sqrt[4]{\dfrac{1}{16}} = .5$

57. $\left(\dfrac{1}{8}\right)^{4/3} = .0625$

58. $12^{1/3} \approx 2.289428485$

59. $2x - 4 \geq 0 \Rightarrow 2x \geq 4 \Rightarrow x \geq \dfrac{4}{2} \Rightarrow x \geq 2$; The domain is $[2, \infty)$.

60. The graph of f is a square root graph reflected across the *x*-axis. The range is $(-\infty, 0]$.

61. (a) None. The function is never increasing.

    (b) The function is decreasing over its entire domain, $[2, \infty)$.

62. Solving the equation for *y* yields $x^2 + y^2 = 25 \Rightarrow y^2 = 25 - x^2 \Rightarrow y = \pm\sqrt{25 - x^2}$.

    The expressions are $y_1 = \sqrt{25 - x^2}$ and $y_2 = -\sqrt{25 - x^2}$ which are graphed in a square window in Figure 62.

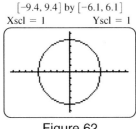

$[-9.4, 9.4]$ by $[-6.1, 6.1]$
Xscl = 1   Yscl = 1

Figure 62

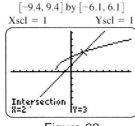

$[-9.4, 9.4]$ by $[-6.1, 6.1]$
Xscl = 1   Yscl = 1

Figure 63

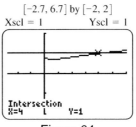

$[-2.7, 6.7]$ by $[-2, 2]$
Xscl = 1   Yscl = 1

Figure 64

63. (a) $\sqrt{5 + 2x} = x + 1 \Rightarrow (\sqrt{5 + 2x})^2 = (x + 1)^2 \Rightarrow 5 + 2x = x^2 + 2x + 1 \Rightarrow$

$x^2 - 4 = 0 \Rightarrow (x + 2)(x - 2) = 0 \Rightarrow x = -2$ or 2 (-2 is extraneous)

The solution set is $\{2\}$. A graph of $y_1 = \sqrt{5 + 2x}$ and $y_2 = x + 1$ is shown in Figure 63.

(b) Since the graph of y_1 begins when $5 + 2x = 0 \Rightarrow x = -2.5$, $y_1 > y_2$ on the interval $[-2.5, 2)$.

(c) From the graph, $y_1 < y_2$ on the interval $(2, \infty)$.

64. (a) $\sqrt{2x + 1} - \sqrt{x} = 1 \Rightarrow \sqrt{2x + 1} = \sqrt{x} + 1 \Rightarrow (\sqrt{2x + 1})^2 = (\sqrt{x} + 1)^2 \Rightarrow$

$2x + 1 = x + 2\sqrt{x} + 1 \Rightarrow x = 2\sqrt{x} \Rightarrow x^2 = (2\sqrt{x})^2 \Rightarrow x^2 = 4x \Rightarrow x^2 - 4x = 0 \Rightarrow$

$x(x - 4) = 0 \Rightarrow x = 0, 4$ The solution set is $\{0, 4\}$. A graph of $y_1 = \sqrt{2x + 1} - \sqrt{x}$ and $y_2 = 1$ is

shown in Figure 64.

(b) From the graph, $y_1 > y_2$ on the interval $(4, \infty)$.

(c) From the graph, $y_1 < y_2$ on the interval $(0, 4)$.

65. (a) $\sqrt[3]{6x + 2} = \sqrt[3]{4x} \Rightarrow (\sqrt[3]{6x + 2})^3 = (\sqrt[3]{4x})^3 \Rightarrow 6x + 2 = 4x \Rightarrow 2x = -2 \Rightarrow x = -1$

The solution set is $\{-1\}$. A graph of $y_1 = \sqrt[3]{6x + 2}$ and $y_2 = \sqrt[3]{4x}$ is shown in Figure 65.

(b) From the graph, $y_1 \geq y_2$ on the interval $[-1, \infty)$.

(c) From the graph, $y_1 \leq y_2$ on the interval $(-\infty, -1]$.

66. (a) $(x - 2)^{2/3} - x^{1/3} = 0 \Rightarrow (x - 2)^{2/3} = x^{1/3} \Rightarrow [(x - 2)^{2/3}]^3 = [x^{1/3}]^3 \Rightarrow (x - 2)^2 = x \Rightarrow$

$x^2 - 4x + 4 = x \Rightarrow x^2 - 5x + 4 = 0 \Rightarrow (x - 1)(x - 4) = 0 \Rightarrow x = 1$ or 4

The solution set is $\{1, 4\}$. A graph of $y_1 = (x - 2)^{2/3} - x^{1/3}$ and $y_2 = 0$ is shown in Figure 66.

(b) From the graph, $y_1 \geq y_2$ on the interval $(-\infty, 1] \cup [4, \infty)$.

(c) From the graph, $y_1 \leq y_2$ on the interval $[1, 4]$.

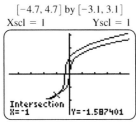

$[-4.7, 4.7]$ by $[-3.1, 3.1]$
Xscl = 1   Yscl = 1

Figure 65

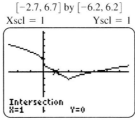

$[-2.7, 6.7]$ by $[-6.2, 6.2]$
Xscl = 1   Yscl = 1

Figure 66

67. Since the graphs do not intersect, the equation $y_1 = y_2$ has no solutions. See Figure 67.

68. Since $y_1 = y_2$ has no solutions, $y_1 - y_2$ will never equal zero. There are no x-intercepts. See Figure 68.

69. The graph of $-f(x) = y_2 - y_1$ is the reflection of the graph of $f(x) = y_1 - y_2$ across the x-axis. See Figure 69.

70. The graph of $y = f(-x)$ is the reflection of the graph of $y = f(x)$ across the y-axis. See Figure 70.

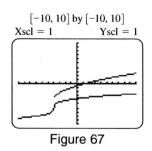
Figure 67

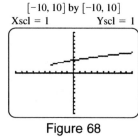
Figure 68

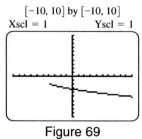
Figure 69

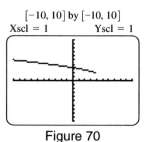
Figure 70

71. $x^5 = 1024 \Rightarrow (x^5)^{1/5} = 1024^{1/5} \Rightarrow x = 4$; **Check:** $4^5 = 1024$; The solution set is $\{4\}$.

72. $x^{1/3} = 4 \Rightarrow (x^{1/3})^3 = 4^3 \Rightarrow x = 64$; **Check:** $64^{1/3} = 4$; The solution set is $\{64\}$.

73. $\sqrt{x-2} = x - 4 \Rightarrow (\sqrt{x-2})^2 = (x-4)^2 \Rightarrow x - 2 = x^2 - 8x + 16 \Rightarrow x^2 - 9x + 18 = 0 \Rightarrow$

$(x-3)(x-6) = 0 \Rightarrow x = 3, 6$; **Check:** $\sqrt{6-2} = \sqrt{4} = 2 = 6 - 4$; $\sqrt{3-2} = \sqrt{1} = 1 \neq 3 - 4$

The solution 3 is extraneous. The solution set is $\{6\}$.

74. $x^{3/2} = 27 \Rightarrow (x^{3/2})^{2/3} = 27^{2/3} \Rightarrow x = 9$; **Check:** $9^{3/2} = 27$; The solution set is $\{9\}$.

75. $2x^{1/4} + 3 = 6 \Rightarrow 2x^{1/4} = 3 \Rightarrow x^{1/4} = \dfrac{3}{2} \Rightarrow (x^{1/4})^4 = \left(\dfrac{3}{2}\right)^4 \Rightarrow x = \dfrac{81}{16}$

Check: $2\left(\dfrac{81}{16}\right)^{1/4} + 3 = 2\left(\dfrac{3}{2}\right) + 3 = 3 + 3 = 6$; The solution set is $\left\{\dfrac{81}{16}\right\}$.

76. $\sqrt{x-2} = 14 - x \Rightarrow (\sqrt{x-2})^2 = (14-x)^2 \Rightarrow x - 2 = 196 - 28x + x^2 \Rightarrow x^2 - 29x + 198 = 0 \Rightarrow$

$(x-11)(x-18) = 0 \Rightarrow x = 11, 18$; **Check:** $\sqrt{11-2} = \sqrt{9} = 3 = 14 - 11$;

$\sqrt{18-2} = \sqrt{16} = 4 \neq 14 - 18$; The solution 18 is extraneous. The solution set is $\{11\}$.

77. $\sqrt[3]{2x-3} + 1 = 4 \Rightarrow \sqrt[3]{2x-3} = 3 \Rightarrow (\sqrt[3]{2x-3})^3 = 3^3 \Rightarrow 2x - 3 = 27 \Rightarrow 2x = 30 \Rightarrow x = 15$

Check: $\sqrt[3]{2(15)-3} + 1 = \sqrt[3]{30-3} + 1 = \sqrt[3]{27} + 1 = 3 + 1 = 4$; The solution set is $\{15\}$.

78. $x^{1/3} + 3x^{1/3} = -2 \Rightarrow 4x^{1/3} = -2 \Rightarrow x^{1/3} = -\dfrac{1}{2} \Rightarrow (x^{1/3})^3 = \left(-\dfrac{1}{2}\right)^3 \Rightarrow x = -\dfrac{1}{8}$

Check: $\left(-\dfrac{1}{8}\right)^{1/3} + 3\left(-\dfrac{1}{8}\right)^{1/3} = -\dfrac{1}{2} + 3\left(-\dfrac{1}{2}\right) = -\dfrac{1}{2} - \dfrac{3}{2} = -\dfrac{4}{2} = -2$; The solution set is $\left\{-\dfrac{1}{8}\right\}$.

79. $2x^{-2} - 5x^{-1} = 3 \Rightarrow 2(x^{-1})^2 - 5(x^{-1}) = 3$, let $u = x^{-1}$, then $2u^2 - 5u - 3 = 0 \Rightarrow$

$(2u + 1)(u - 3) = 0 \Rightarrow u = -\dfrac{1}{2}, 3$. If $u = x^{-1}$, then $x = u^{-1}$, thus $x = -2, \dfrac{1}{3}$

Check: $2(-2)^{-2} - 5(-2)^{-1} = 3 \Rightarrow 2\left(\dfrac{1}{4}\right) - 5\left(-\dfrac{1}{2}\right) = 3 \Rightarrow \dfrac{1}{2} + \dfrac{5}{2} = 3 \Rightarrow 3 = 3$;

$2\left(\dfrac{1}{3}\right)^{-2} - 5\left(\dfrac{1}{3}\right)^{-1} = 3 \Rightarrow 2(9) - 5(3) = 3 \Rightarrow 18 - 15 = 3 \Rightarrow 3 = 3$; The solution set is $\left\{-2, \dfrac{1}{3}\right\}$.

80. $x^{-3} + 2x^{-2} + x^{-1} = 0 \Rightarrow (x^{-1})^3 + 2(x^{-1})^2 + (x^{-1}) = 0$, let $u = x^{-1}$, then $u^3 + 2u^2 + u = 0 \Rightarrow$

$u(u^2 + 2u + 1) = 0 \Rightarrow u(u + 1)(u + 1) = 0 \Rightarrow u = -1, 0$. If $u = x^{-1}$, then $x = u^{-1}$, thus

$x = (-1)^{-1} \Rightarrow x = -1$ or $m = (0)^{-1} \Rightarrow x = \dfrac{1}{0}$, which is undefined.

Check: $(-1)^{-3} + 2(-1)^{-2} + (-1)^{-1} = 0 \Rightarrow -1 + 2 + (-1) = 0 \Rightarrow 0 = 0$. The solution set is $\{-1\}$.

81. $x^{2/3} - 4x^{1/3} - 5 = 0 \Rightarrow (x^{1/3})^2 - 4(x^{1/3}) - 5 = 0$, let $u = x^{1/3}$, then $u^2 - 4u - 5 = 0 \Rightarrow$

$(u - 5)(u + 1) = 0 \Rightarrow u = -1, 5$. If $u = x^{1/3}$, then $x = u^3$, thus $x = -1, 125$.

Check: $(-1)^{2/3} - 4(-1)^{1/3} - 5 = 0 \Rightarrow 1 - (-4) - 5 = 0 \Rightarrow 0 = 0$;

$(125)^{2/3} - 4(125)^{1/3} - 5 = 0 \Rightarrow 25 - 4(5) - 5 = 0 \Rightarrow 0 = 0$. The solution set is: $\{-1, 125\}$.

82. $x^{3/4} - 16x^{1/4} = 0 \Rightarrow (x^{1/4})^3 - 16(x^{1/4}) = 0$, let $u = x^{1/4}$, then $u^3 - 16u = 0 \Rightarrow u(u^2 - 16) = 0 \Rightarrow$

$u(u + 4)(u - 4) = 0 \Rightarrow u = -4, 0, 4$. If $u = x^{1/4}$, then $x = u^4$, thus $x = 0, 256$.

Check: $(0)^{3/4} - 16(0)^{1/4} = 0 \Rightarrow 0 - 0 = 0 \Rightarrow 0 = 0$.

$(256)^{3/4} - 16(256)^{1/4} = 0 \Rightarrow 64 - 16(4) = 0 \Rightarrow 0 = 0$. The solution set is $\{0, 256\}$.

83. $\sqrt{x + 1} + 1 = \sqrt{2x} \Rightarrow (\sqrt{x + 1} + 1)^2 = 2x \Rightarrow x + 1 + 2\sqrt{x + 1} + 1 = 2x \Rightarrow 2\sqrt{x + 1} = x - 2 \Rightarrow$

$4(x + 1) = x^2 - 4x + 4 \Rightarrow 4x + 4 = x^2 - 4x + 4 \Rightarrow x^2 - 8x = 0 \Rightarrow x(x - 8) = 0 \Rightarrow x = 0, 8$.

Check: $\sqrt{0 + 1} + 1 = \sqrt{2(0)} \Rightarrow 1 + 1 = 0 \Rightarrow 2 \neq 0 \Rightarrow 0$ is not a solution.

$\sqrt{8 + 1} + 1 = \sqrt{2(8)} \Rightarrow 4 = 4$. The solution set is $\{8\}$.

84. $\sqrt{x - 2} = 5 - \sqrt{x + 3} \Rightarrow (\sqrt{x - 2})^2 = (5 - \sqrt{x + 3})^2 \Rightarrow x - 2 = 25 - 10\sqrt{x + 3} + x + 3 \Rightarrow$

$-30 = -10\sqrt{x + 3} \Rightarrow 900 = 100(x + 3) \Rightarrow 9 = x + 3 \Rightarrow x = 6$.

Check: $\sqrt{6 - 2} = 5 - \sqrt{6 + 3} \Rightarrow \sqrt{4} = 5 - \sqrt{9} \Rightarrow 2 = 5 - 3 \Rightarrow 2 = 2$. The solution set is $\{6\}$.

85. (a) If the length L of the pendulum increases, so does the period of oscillation T.

   (b) There are a number of ways to find n and k. One way is to realize that $k = \dfrac{L}{T^n}$ for some integer n. The

   ratio should be the constant k for each data point when the correct n is found. Another way is to use regression.

   (c) By trial and error, $k \approx .81; n = 2$.

   (d) $5 = .81(T^2) \Rightarrow \dfrac{5}{.81} = T^2 \Rightarrow \sqrt{\dfrac{5}{.81}} = T \Rightarrow T \approx 2.48$ sec

   (e) Since $T = \sqrt{\dfrac{L}{k}}$, when length doubles T increases by a factor of $\sqrt{2} \approx 1.414$.

86. Since the volume is $\pi r^2 h = 27\pi$, solving for h yields $h = \dfrac{27}{r^2}$. Substituting this value in the formula for surface

   area yields $S = 2\pi r\left(\dfrac{27}{r^2}\right) + \pi r^2 \Rightarrow S = \dfrac{54\pi}{r} + \pi r^2$. By graphing $y_1 = \dfrac{54\pi}{r} + \pi r^2$ (not shown), the

   minimum value for surface area occurs when $r = 3$ inches.

Chapter 4 Test

1. (a) See Figure 1a.

    (b) The graph is obtained by reflecting the graph of $y = \dfrac{1}{x}$ across the x-axis or the y-axis.

    (c) See Figure 1c.

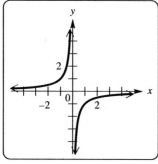

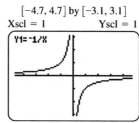

Figure 1a   Figure 1c

2. (a) See Figure 2a.

    (b) The graph is obtained by reflecting the graph of $y = \dfrac{1}{x^2}$ across the x-axis and shifting it downward 3 units.

    (c) See Figure 2c.

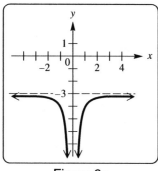

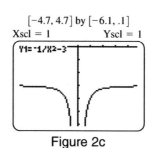

Figure 2a   Figure 2c

3. (a) The vertical asymptotes occur for values of x that make the denominator equal to zero, $x = -1$ and $x = 4$.

    (b) Since the degrees of the numerator and denominator are equal, the horizontal asymptote can be found using the ratio of the leading coefficients, $y = \dfrac{1}{1} = 1$.

    (c) The y-intercept occurs at $f(0) = \dfrac{0^2 + 0 - 6}{0^2 - 3(0) - 4} = 1.5$.

    (d) The x-intercepts occur for values of x that make the numerator equal to zero, -3 and 2.

    (e) $\dfrac{x^2 + x - 6}{x^2 - 3x - 4} = 1 \Rightarrow x^2 + x - 6 = x^2 - 3x - 4 \Rightarrow x - 6 = -3x - 4 \Rightarrow 4x = 2 \Rightarrow x = .5$

    The coordinates of the intersection point are $(.5, 1)$.

    (f) See Figure 3.

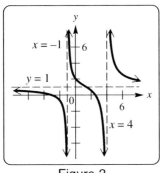

Figure 3

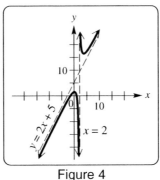

Figure 4

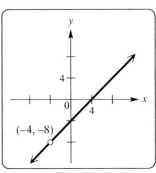

Figure 5

4. (a) Since $\dfrac{2x^2 + x - 3}{x - 2} = 2x + 5 + \dfrac{7}{x - 2}$, the oblique asymptote is $y = 2x + 5$.

(b) See Figure 4.

5. (a) Since $\dfrac{x^2 - 16}{x + 4} = \dfrac{(x + 4)(x - 4)}{x + 4} = x - 4$ for $x \neq -4$, the "hole" occurs when $x = -4$.

(b) See Figure 5.

6. (a) Multiply each side of the equation by the LCD, $(x - 2)(x + 2)$.

$$(x - 2)(x + 2) \cdot \frac{3}{x - 2} + (x - 2)(x + 2) \cdot \frac{21}{x^2 - 4} = \frac{14}{x + 2} \cdot (x - 2)(x + 2) \Rightarrow$$

$$3(x + 2) + 21 = 14(x - 2) \Rightarrow 3x + 6 + 21 = 14x - 28 \Rightarrow 11x = 55 \Rightarrow x = 5$$

The solution set is $\{5\}$.

(b) Graph $y_1 = 3/(x - 2) + 21/(x^2 - 4) - 14/(x + 2)$ as shown in Figure 6. With vertical asymptotes

$x = -2$ and $x = 2$, and x-intercept 5, the graph is above or intersects the x-axis on the interval

$(-\infty, -2) \cup (2, 5]$. The solution set is $(-\infty, -2) \cup (2, 5]$.

$[-9.4, 9.4]$ by $[-20, 10]$
Xscl = 1   Yscl = 2

$[0, 40]$ by $[-.5, 1]$
Xscl = 5   Yscl = 1

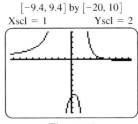

Figure 6

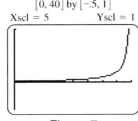

Figure 7

7. (a) $W(30) = \dfrac{1}{40 - 30} = \dfrac{1}{10}$; $W(39) = \dfrac{1}{40 - 39} = \dfrac{1}{1} = 1$; $W(39.9) = \dfrac{1}{40 - 39.9} = \dfrac{1}{.1} = 10$

When the rate is 30 vehicles per minute, the average wait time is $\dfrac{1}{10}$ minute (6 seconds). The other results

are interpreted similarly.

(b) The vertical asymptote occurs when the denominator equals zero, $x = 40$. As x approaches 40, W gets

larger and larger without bound. See Figure 7.

(c) $5 = \dfrac{1}{40 - x} \Rightarrow 5(40 - x) = 1 \Rightarrow 200 - 5x = 1 \Rightarrow 5x = 199 \Rightarrow x = \dfrac{199}{5} = 39.8$

8. Here $p = \dfrac{k\sqrt[3]{w}}{h}$. By substitution, $100 = \dfrac{k\sqrt[3]{48{,}820}}{78.7} \Rightarrow 7870 = \sqrt[3]{48{,}820}\,k \Rightarrow k = \dfrac{7870}{\sqrt[3]{48{,}820}} \Rightarrow k \approx 215.3$

    That is $p = \dfrac{215.3\sqrt[3]{w}}{h}$. When $w = 5{,}430$ and $h = 88.9$, $p = \dfrac{215.3\sqrt[3]{54{,}430}}{88.9} \approx 92$.

    The individual is undernourished.

9. Graph $y_1 = (8000 + 2\pi x^3)/x$ as shown in Figure 9. When the radius is approximately 8.6 cm, the amount of aluminum needed will be a minimum of approximately 1394.9 cm^2.

10. The graph of $y_1 = -\sqrt{5 - x}$ is shown in Figure 10.

    (a) $5 - x \geq 0 \Rightarrow 5 \geq x \Rightarrow x \leq 5$; The domain is $(-\infty, 5]$.

    (b) From the graph, the range is $(-\infty, 0]$.

    (c) This function *increases* over its entire domain.

    (d) The graph intersects the x-axis when $x = 5$. The solution set is $\{5\}$.

    (e) The graph is below the x-axis for all values in the domain except 5. The solution set is $(-\infty, 5)$.

11. (a) $\sqrt{4 - x} = x + 2 \Rightarrow (\sqrt{4 - x})^2 = (x + 2)^2 \Rightarrow 4 - x = x^2 + 4x + 4 \Rightarrow x^2 + 5x = 0 \Rightarrow$

    $x(x + 5) = 0 \Rightarrow x = 0, -5$. The solution -5 is extraneous. The solution set is $\{0\}$.

    The graph of $y_1 = \sqrt{4 - x}$ and $y_2 = x + 2$ shown in Figure 11 supports this result.

    (b) The graph of y_1 is above the graph of y_2 on the interval $(-\infty, 0)$. The solution set is $(-\infty, 0)$.

    (c) The graph of y_1 is below or intersects the graph of y_2 on the interval $[0, 4]$. The solution set is $[0, 4]$.

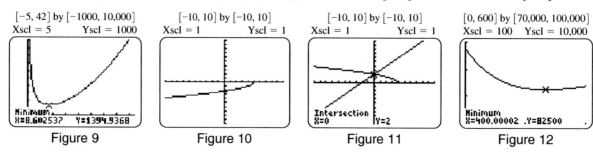

| Figure 9 | Figure 10 | Figure 11 | Figure 12 |

12. Let S be a point on land between Q and R such that the distance from Q to S is x yards. Then the distance from S to R is given by $600 - x$. By the Pythagorean theorem, the distance from P to S is $\sqrt{x^2 + 300^2}$. The total cost of the cable is given by $C = 125\sqrt{x^2 + 300^2} + 100(600 - x)$. The graph of this function is shown in Figure 12. The minimum cost is achieved when the cable is laid underwater from P to S, which is located 400 yards away from Q in the direction of R.

Chapter 4 Project

1. For a 6-inch step, $Y_1 = \dfrac{23.5}{6^{1.153}} \approx 2.98$. Thus the total distance is $2.98 \times 6 \approx 17.9$ inches.

    For a .1-inch step, $Y_1 = \dfrac{23.5}{.1^{1.153}} \approx 334.25$. Thus the total distance is $334.25 \times .1 \approx 33.4$ inches.

    For a .01-inch step, $Y_1 = \dfrac{23.5}{.01^{1.153}} \approx 4754$. Thus the total distance is $4754 \times .01 \approx 47.5$ inches.

2. Answers will vary.

Chapter 5: Inverse, Exponential, and Logarithmic Functions

5.1: Inverse Functions

1. Different x-values always produce different y-values, therefore yes, it is one-to-one.

2. Different x-values always produce different y-values, therefore yes, it is one-to-one.

3. Choosing 2 and -2 as values for x yields: $f(2) = 2^2 = 4$ and $f(-2) = (-2)^2 = 4$. Since different values of x produce the same value for $f(x)$, the function is not one-to-one.

4. Choosing 2 and -2 as values for x yields: $f(2) = -2^2 = -4$ and $f(-2) = -(-2)^2 = -4$. Since different values of x produce the same value for $f(x)$, the function is not one-to-one.

5. Choosing 6 and -6 as values for x yields: $f(6) = \sqrt{36 - 6^2} = 0$ and $f(-6) = \sqrt{36 - (-6)^2} = 0$. Since different values of x produce the same value for $f(x)$, the function is not one-to-one.

6. Choosing 10 and -10 as values for x yields: $f(6) = \sqrt{100 - 10^2} = 0$ and $f(-10) = \sqrt{100 - (-10)^2} = 0$. Since different values of x produce the same value for $f(x)$, the function is not one-to-one.

7. Every horizontal line will intersect the graph at exactly one point, therefore yes, it is one-to-one.

8. Every horizontal line will intersect the graph at exactly one point, therefore yes, it is one-to-one.

9. There are horizontal lines that will intersect the graph at more than one point, therefore it is not one-to-one.

10. A certain horizontal line intersects the whole horizontal graph (more than one point), therefore it is not one-to-one.

11. Every horizontal line will intersect the graph at exactly one point, therefore yes, it is one-to-one.

12. Every horizontal line will intersect the graph at exactly one point, therefore yes, it is one-to-one.

13. Every horizontal line will intersect the graph at exactly one point, therefore yes, it is one-to-one.

14. Every horizontal line will intersect the graph at exactly one point, therefore yes, it is one-to-one.

15. A certain horizontal line intersects the horizontal graph when $x > 0$ (more than 1 point), therefore it is not one-to-one.

16. Since different x-values greater than zero all produce the same $f(x) = 3$, it is not one-to-one.

17. Choosing 4 and 0 as values for x yields: $f(4) = (4 - 2)^2 = 4$ and $f(0) = (0 - 2)^2 = 4$. Since different values of x produce the same value for $f(x)$, the function is not one-to-one.

18. Choosing 0 and -6 as values for x yields: $f(0) = -(0 + 3)^2 - 8 = -17$ and $f(-6) = -(-6 + 3)^2 - 8 = -17$. Since different values of x produce the same value for $f(x)$, the function is not one-to-one.

19. Different x-values always produce different y-values, therefore yes, it is one-to-one.

20. Different x-values always produce different y-values, therefore yes, it is one-to-one.

21. Different x-values always produce different y-values, therefore yes, it is one-to-one.

22. Different x-values always produce different y-values, therefore yes, it is one-to-one.

23. Different x-values always produce different y-values, therefore yes, it is one-to-one.

24. All x-values produce the same $f(x) = -7$, therefore it is not one-to-one.

25. The graph fails the horizontal line test because the end behavior is either $\bigcup$ or $\bigcap$.

26. If the degree is odd and greater than or equal to 3, the graph may have extrema and would then fail the horizontal line test.

27. For a function to have an inverse, it must be one-to-one.

28. For a function f to be of the type mentioned in Exercise 27, if $a \neq b$, then $f(a) \neq f(b)$.

29. If f and g are inverses, then $(f \circ g)(x) = x$, and $(g \circ f)(x) = x$.

30. The domain of f is equal to the range of f^{-1}, and the range of f is equal to the domain of f^{-1}.

31. If the point (a, b) lies on the graph of f, and f has an inverse, then the point (b, a) lies on the graph f^{-1}.

32. If $f(x) = x$, then for any function g, $(f \circ g)(x) = (g \circ f)(x) = g(x)$.

33. If the function f has an inverse, then the graph of f^{-1} may be obtained by reflecting the graph of f across the line with equation $y = x$.

34. If a function f has an inverse and $f(-3) = 6$, then $f^{-1}(6) = -3$.

35. If $f(-4) = 16$ and $f(4) = 16$, then f does not have an inverse because it is not one-to-one.

36. If f is a function that has an inverse, and the graph of f lies completely within the second quadrant, then the graph of f^{-1} lies completely within the fourth quadrant.

37. $(f \circ g)(x) = 3\left(\dfrac{x+7}{3}\right) - 7 = x + 7 - 7 = x$; and $(g \circ f)(x) = \dfrac{(3x-7)+7}{3} = \dfrac{3x}{3} = x$.

    Since $(f \circ g)(x) = x$ and $(g \circ f)(x) = x$, the functions f and g are inverses.

38. $(f \circ g)(x) = 4\left(\dfrac{x-3}{4}\right) + 3 = x - 3 + 3 = x$; and $(g \circ f)(x) = \dfrac{(4x+3)-3}{4} = \dfrac{4x}{4} = x$.

    Since $(f \circ g)(x) = x$ and $(g \circ f)(x) = x$, the functions f and g are inverses.

39. $(f \circ g)(x) = (\sqrt[3]{x-4})^3 + 4 = x - 4 + 4 = x$; and $(g \circ f)(x) = \sqrt[3]{(x^3+4)-4} = \sqrt[3]{x^3} = x$.

    Since $(f \circ g)(x) = x$ and $(g \circ f)(x) = x$, the functions f and g are inverses.

40. $(f \circ g)(x) = (\sqrt[3]{x+7})^3 - 7 = x + 7 - 7 = x$; and $(g \circ f)(x) = \sqrt[3]{(x^3-7)+7} = \sqrt[3]{x^3} = x$.

    Since $(f \circ g)(x) = x$ and $(g \circ f)(x) = x$, the functions f and g are inverses.

41. $(f \circ g)(x) = -(-\sqrt[5]{x})^5 = -(-x) = x$; and $(g \circ f)(x) = -\sqrt[5]{(-x^5)} = -(-x) = x$.

    Since $(f \circ g)(x) = x$ and $(g \circ f)(x) = x$, the functions f and g are inverses.

42. $(f \circ g)(x) = -(-\sqrt[7]{x})^7 = -(-x) = x$; and $(g \circ f)(x) = -\sqrt[7]{(-x^7)} = -(-x) = x$.

    Since $(f \circ g)(x) = x$ and $(g \circ f)(x) = x$, the functions f and g are inverses.

43. Every x-value in f corresponds to only one y-value, and every y-value corresponds to only one x-value, so f is a one-to-one function. The inverse function is found by interchanging the x- and y-values in each ordered pair. Therefore, $f^{-1} = \{(4, 10), (5, 20), (6, 30), (7, 40)\}$.

44. Every x-value in g corresponds to only one y-value, and every y-value corresponds to only one x-value, so g is a one-to-one function. The inverse function is found by interchanging the x- and y-values in each ordered pair. Therefore, $g^{-1} = \{(12, 5), (22, 10), (32, 15), (42, 20)\}$.

45. Every x-value in f corresponds to only one y-value. However, the y-value 5 corresponds to two x-values, 1 and 3. Because some y-values corresponds to more than one x-value, f is not one-to-one.

46. Every x-value in g corresponds to only one y-value. However, the y-value 10 corresponds to two x-values, 0 and 2. Because some y-values corresponds to more than one x-value, g is not one-to-one.

47. Every x-value in f corresponds to only one y-value, and every y-value corresponds to only one x-value, so f is a one-to-one function. The inverse function is found by interchanging the x- and y-values in each ordered pair. Therefore, $f^{-1} = \{(0^2, 0), (1^2, 1), (2^2, 2), (3^2, 3), (4^2, 4)\}$.

48. Every x-value in g corresponds to only one y-value, and every y-value corresponds to only one x-value, so g is a one-to-one function. The inverse function is found by interchanging the x- and y-values in each ordered pair. Therefore, $g^{-1} = \{(0^4, 0), ((-1)^4, -1), ((-2)^4, -2), ((-3)^4, -3), ((-4)^4, -4)\}$.

49. Untying your shoelaces.

50. Pressing a car's brake.

51. Leaving a room.

52. Descending the stairs.

53. Unwrapping a package.

54. Removing a coat.

55. For the function $y = 3x - 4$, the inverse is: $x = 3y - 4 \Rightarrow x + 4 = 3y \Rightarrow y = \dfrac{x + 4}{3} \Rightarrow f^{-1}(x) = \dfrac{x + 4}{3}$.

    For the graphs, see Figure 55. Since all real numbers can be input for x in f and f^{-1}, and all real numbers can be solutions for $f(x)$ and $f^{-1}(x)$, the domain and range of both f and f^{-1} are: $(-\infty, \infty)$.

56. For the function $y = 4x - 5$, the inverse is: $x = 4y - 5 \Rightarrow x + 5 = 4y \Rightarrow y = \dfrac{x + 5}{4} \Rightarrow f^{-1}(x) = \dfrac{x + 5}{4}$.

    For the graphs, see Figure 56. Since all real numbers can be input for x in f and f^{-1}, and all real numbers can be solutions for $f(x)$ and $f^{-1}(x)$, the domain and range of both f and f^{-1} are: $(-\infty, \infty)$.

57. For the function $y = x^3 + 1$, the inverse is: $x = y^3 + 1 \Rightarrow x - 1 = y^3 \Rightarrow y = \sqrt[3]{x - 1} \Rightarrow f^{-1}(x) = \sqrt[3]{x - 1}$.

    For the graphs, see Figure 57. Since all real numbers can be input for x in f and f^{-1}, and all real numbers can be solutions for $f(x)$ and $f^{-1}(x)$, the domain and range of both f and f^{-1} are: $(-\infty, \infty)$.

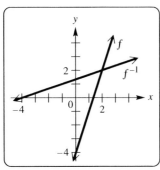

Figure 55

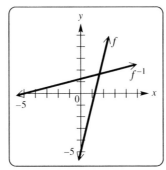

Figure 56

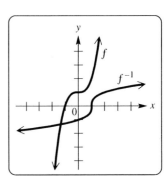

Figure 57

58. For the function $y = -x^3 - 2$, the inverse is: $x = -y^3 - 2 \Rightarrow x + 2 = -y^3 \Rightarrow -x - 2 = y^3 \Rightarrow$

    $y = \sqrt[3]{-x - 2} \Rightarrow f^{-1}(x) = \sqrt[3]{-x - 2}$. For the graphs, see Figure 58. Since all real numbers can be input

    for x in f and f^{-1}, and all real numbers can be solutions for $f(x)$ and $f^{-1}(x)$, the domain and range of both

    f and f^{-1} are: $(-\infty, \infty)$.

59. Since when $y = 4$, $x = -2$ or 2, the function is not one-to-one.

60. Since when $y = -2$, $x = -2$ or 2, the function is not one-to-one.

61. For the function $y = \dfrac{1}{x}$, the inverse is: $x = \dfrac{1}{y} \Rightarrow y = \dfrac{1}{x} \Rightarrow f^{-1}(x) = \dfrac{1}{x}$. For the graphs, see Figure 61.

    Since all real numbers except 0 can be input for x in f and f^{-1}, and all real numbers except 0 can be solutions

    for $f(x)$ and $f^{-1}(x)$, the domain and range of both f and f^{-1} are: $(-\infty, 0) \cup (0, \infty)$.

62. For the function $y = \dfrac{4}{x}$, the inverse is: $x = \dfrac{4}{y} \Rightarrow yx = 4 \Rightarrow y = \dfrac{4}{x} \Rightarrow f^{-1}(x) = \dfrac{4}{x}$. For the graphs,

    see Figure 62. Since all real numbers except 0 can be input for x in f and f^{-1}, and all real numbers

    except 0 can be solutions for $f(x)$ and $f^{-1}(x)$, the domain and range of both f and f^{-1} are: $(-\infty, 0) \cup (0, \infty)$.

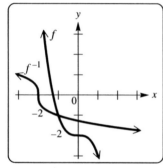

Figure 58

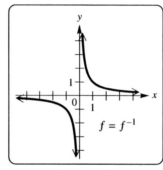

Figure 61

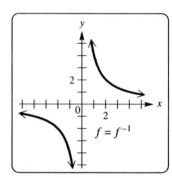

Figure 62

63. For the function $y = \dfrac{2}{x + 3}$, the inverse is: $x = \dfrac{2}{y + 3} \Rightarrow (y + 3)x = 2 \Rightarrow y + 3 = \dfrac{2}{x} \Rightarrow y = \dfrac{2}{x} - 3 \Rightarrow$

    $y = \dfrac{2 - 3x}{x} \Rightarrow f^{-1}(x) = \dfrac{2 - 3x}{x}$. For the graphs, see Figure 63. All real numbers except -3 can be input

    for x in f and all real numbers except 0 can be input for x in f^{-1}. Also all real numbers except 0 can be a

    solution for f and all real numbers except -3 can be a solution for f^{-1}. Therefore the domain of

    f is equal to the range of $f^{-1} = (-\infty, -3) \cup (-3, \infty)$; and the domain of f^{-1} is equal to the range of

    $f = (-\infty, 0) \cup (0, \infty)$.

64. For the function $y = \dfrac{3}{x - 4}$, the inverse is: $x = \dfrac{3}{y - 4} \Rightarrow (y - 4)x = 3 \Rightarrow y - 4 = \dfrac{3}{x} \Rightarrow y = \dfrac{3}{x} + 4 \Rightarrow$

    $y = \dfrac{3 + 4x}{x} \Rightarrow f^{-1}(x) = \dfrac{4x + 3}{x}$. For the graphs, see Figure 64. All real numbers except 4 can be input

    for x in f and all real numbers except 0 can be input for x in f^{-1}. Also all real numbers except 0 can be a

    solution for f and all real numbers except 4 can be a solution for f^{-1}. Therefore the domain of

    f is equal to the range of $f^{-1} = (-\infty, 4) \cup (4, \infty)$; and the domain of f^{-1} is equal to the range of

    $f = (-\infty, 0) \cup (0, \infty)$.

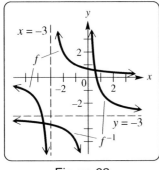

Figure 63   Figure 64

65. For the function $f(x) = \sqrt{6 + x}$, $x \geq -6$, the inverse is: $x = \sqrt{6 + y}$, $x \geq 0$ (square root can not equal a negative number) $\Rightarrow x^2 = 6 + y \Rightarrow y = x^2 - 6 \Rightarrow f^{-1}(x) = x^2 - 6$, $x \geq 0$. For the graph, see Figure 65. All real numbers $x \geq -6$ can be input for x in f and all real numbers $x \geq 0$ can be input for x in f^{-1}. Also all real numbers $y \geq 0$ can be a solution for f and all real numbers $y \geq -6$ can be a solution for f^{-1}. Therefore the domain of f is equal to the range of $f^{-1} = [-6, \infty)$; and the domain of f^{-1} is equal to the range of $f = [0, \infty)$.

66. For the function $f(x) = -\sqrt{x^2 - 16}$, $x \geq 4$, the inverse is: $x = -\sqrt{y^2 - 16}$, $x \leq 0$ (square roots can not equal a negative number, therefore a negative of a square root must be less than zero) $\Rightarrow x^2 = y^2 - 16 \Rightarrow y^2 = x^2 + 16 \Rightarrow y = \sqrt{x^2 + 16} \Rightarrow f^{-1}(x) = \sqrt{x^2 + 16}$, $x \leq 0$. For the graph, see Figure 66. All real numbers $x \geq 4$ can be input for x in f and all real numbers $x \leq 0$ can be input for x in f^{-1}. Also all real numbers $y \leq 0$ can be a solution for f and all real numbers $y \geq 4$ can be a solution for f^{-1}. Therefore the domain of f is equal to the range of $f^{-1} = [4, \infty)$; and the domain of f^{-1} is equal to the range of $f = (-\infty, 0]$.

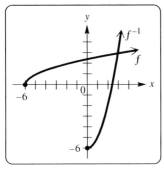 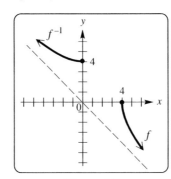

Figure 65   Figure 66

67. $f(2) = (2)^3 = 8$.

68. $f(0) = (0)^3 = 0$.

69. $f(-2) = (-2)^3 = -8$.

70. If $f(x) = x^3$, then the inverse is: $x = y^3 \Rightarrow y = \sqrt[3]{x} \Rightarrow f^{-1}(x) = \sqrt[3]{x}$, and $f^{-1}(8) = \sqrt[3]{8} = 2$.

71. If $f(x) = x^3$, then the inverse is: $x = y^3 \Rightarrow y = \sqrt[3]{x} \Rightarrow f^{-1}(x) = \sqrt[3]{x}$, and $f^{-1}(0) = \sqrt[3]{0} = 0$.

72. If $f(x) = x^3$, then the inverse is: $x = y^3 \Rightarrow y = \sqrt[3]{x} \Rightarrow f^{-1}(x) = \sqrt[3]{x}$, and $f^{-1}(-8) = \sqrt[3]{-8} = -2$.

73. For inverses, when $f^{-1}(4)$, then $f(x) = 4$. Therefore, from the graph, $x = 4$.

74. For inverses, when $f^{-1}(2)$, then $f(x) = 2$. Therefore from the graph $x = 3$.

75. For inverses, when $f^{-1}(0)$, then $f(x) = 0$. Therefore from the graph $x = 2$.

76. For inverses, when $f^{-1}(-2)$, then $f(x) = -2$. Therefore from the graph $x = 0$.

77. For inverses, when $f^{-1}(-3)$, then $f(x) = -3$. Therefore from the graph $x = -2$.

78. For inverses, when $f^{-1}(-4)$, then $f(x) = -4$. Therefore from the graph $x = -4$.

79. The graphs are reflections of each other through $x = y$, therefore the functions are inverses.

80. The graphs are reflections of each other through $x = y$, therefore the functions are inverses.

81. The graphs are not reflections of each other through $x = y$, therefore the functions are not inverses.

82. The graphs are not reflections of each other through $x = y$, therefore the functions are not inverses.

83. The graphs are reflections of each other through $x = y$, therefore the functions are inverses.

84. The graphs are not reflections of each other through $x = y$, therefore the functions are not inverses.

85. Yes, for $f(x) = -\dfrac{3}{11}x$ the inverse is: $x = -\dfrac{3}{11}y \Rightarrow -\dfrac{11}{3}x = y \Rightarrow f^{-1}(x) = -\dfrac{11}{3}x$, which is equal to $g(x)$.

86. Yes, for $f(x) = 2x + 4$ the inverse is: $x = 2y + 4 \Rightarrow x - 4 = 2y \Rightarrow \dfrac{x-4}{2} = y \Rightarrow f^{-1}(x) = \dfrac{1}{2}x - 2$, which is equal to $g(x)$.

87. Yes, for $f(x) = 5x - 5$ the inverse is: $x = 5y - 5 \Rightarrow x + 5 = 5y \Rightarrow \dfrac{x+5}{5} = y \Rightarrow f^{-1}(x) = \dfrac{1}{5}x + 1$, which is equal to $g(x)$.

88. No, for $f(x) = 8x - 7$ the inverse is: $x = 8y - 7 \Rightarrow x + 7 = 8y \Rightarrow \dfrac{x+7}{8} = y \Rightarrow f^{-1}(x) = \dfrac{x+7}{8}$, which is not equal to $g(x)$.

89. Yes, for $f(x) = \dfrac{1}{x}$ the inverse is: $x = \dfrac{1}{y} \Rightarrow xy = 1 \Rightarrow y = \dfrac{1}{x} \Rightarrow f^{-1}(x) = \dfrac{1}{x}$, which is equal to $g(x)$.

90. Yes, for $f(x) = \dfrac{2x+3}{x-1}$ the inverse is: $x = \dfrac{2y+3}{y-1} \Rightarrow xy - x = 2y + 3 \Rightarrow xy - 2y = x + 3 \Rightarrow$

    $y(x - 2) = x + 3 \Rightarrow y = \dfrac{x+3}{x-2} \Rightarrow f^{-1}(x) = \dfrac{x+3}{x-2}$, which is equal to $g(x)$.

91. Since $(x_1, y_1) = (y_2, x_2)$, the x's and y's are switched, the screen suggest that they are linear functions.

92. Since $(x_1, y_1) = (y_2, -x_2)$, the x's and y's are switched, but the x's are multiplied -1, the screen does not suggest that they are linear functions.

93. Reflect the graph across the line $y = x$. See Figure 93.

94. Reflect the graph across the line $y = x$. See Figure 94.

95. Reflect the graph across the line $y = x$. See Figure 95.

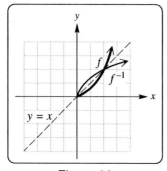

Figure 93

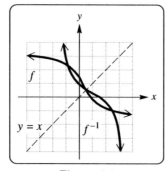

Figure 94

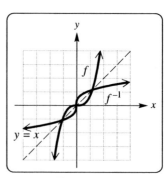

Figure 95

96. Reflect the graph across the line $y = x$. See Figure 96.

97. Reflect the graph across the line $y = x$. See Figure 97.

98. Reflect the graph across the line $y = x$. See Figure 98.

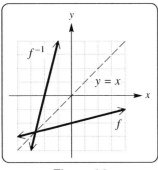
Figure 96

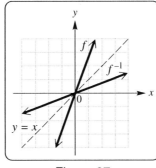
Figure 97

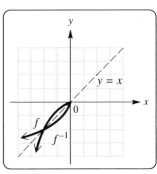
Figure 98

99. It represents the cost in dollars of building 1000 cars.

100. It represents the radius (r) of a sphere with volume 5 in.

101. With $m = a = \dfrac{a}{1}$, the inverse function will switch the x and y terms and the graph of the inverse is a reflection across the line $y = x$ which will have the slope: $\dfrac{1}{a}$.

102. Since $f(2) = 3, f^{-1}(f(2)) = f^{-1}(3)$ which because inverses switch the x and y is: $f^{-1}(3) = 2$.

103. From the graph of the function in the given window, the function fails the horizontal line test and is not one-to-one. See Figure 103.

104. From the graph of the function in the given window, the function fails the horizontal line test and is not one-to-one. See Figure 104.

$[-3, 2]$ by $[-10, 10]$
Xscl = 1   Yscl = 1

$[-3, 3]$ by $[-1, 8]$
Xscl = 1   Yscl = 1

$[-9.4, 9.4]$ by $[-6.2, 6.2]$
Xscl = 1   Yscl = 1

$[-9.4, 9.4]$ by $[-6.2, 6.2]$
Xscl = 1   Yscl = 1

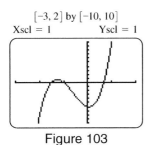
Figure 103

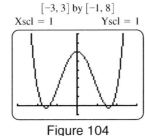
Figure 104

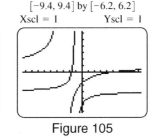
Figure 105

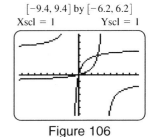
Figure 106

105. From the graph of the function in the given window, the function passes the horizontal line test and is one-to-one. The inverse is: $x = \dfrac{y - 5}{y + 3} \Rightarrow xy + 3x = y - 5 \Rightarrow xy - y = -3x - 5 \Rightarrow$

$y(x - 1) = -3x - 5 \Rightarrow y = \dfrac{-3x - 5}{x - 1} \Rightarrow f^{-1}(x) = \dfrac{-3x - 5}{x - 1}$. For both graphs, see Figure 105.

106. From the graph of the function in the given window, the function passes the horizontal line test and is one-to-one. The inverse is: $x = \dfrac{-y}{y - 4} \Rightarrow xy - 4x = -y \Rightarrow xy + y = 4x \Rightarrow$

$y(x + 1) = 4x \Rightarrow y = \dfrac{4x}{x + 1} \Rightarrow f^{-1}(x) = \dfrac{4x}{x + 1}$. For both graphs, see Figure 106.

107. From the graph on a calculator, one possible answer that passes the horizontal line test is the domain: $[0, \infty)$.

108. From the graph on a calculator, one possible answer that passes the horizontal line test is the domain: $[1, \infty)$.

109. From the graph on a calculator, one possible answer that passes the horizontal line test is the domain: $[6, \infty)$.

110. From the graph on a calculator, one possible answer that passes the horizontal line test is the domain: $[0, \infty)$.

111. From the graph on a calculator, one possible answer that passes the horizontal line test is the domain: $[0, \infty)$.

112. From the graph on a calculator, one possible answer that passes the horizontal line test is the domain: $[4, \infty)$.

113. If $f(x) = -x^2 + 4, x \geq 0$; then the inverse is: $x = -y^2 + 4, y \geq 0 \Rightarrow x - 4 = -y^2 \Rightarrow 4 - x = y^2 \Rightarrow$
     $y = \sqrt{4 - x} \Rightarrow f^{-1}(x) = \sqrt{4 - x}$.

114. If $f(x) = (x - 1)^2, x \geq 1$; then the inverse is: $x = (y - 1)^2, y \geq 1 \Rightarrow \sqrt{x} = y - 1 \Rightarrow y = \sqrt{x} + 1 \Rightarrow$
     $f^{-1}(x) = \sqrt{x} + 1$.

115. If $f(x) = |x - 6|, x \geq 6$; then the inverse is: $x = |y - 6|, y \geq 6 \Rightarrow x = y - 6 \Rightarrow x + 6 = y \Rightarrow$
     $f^{-1}(x) = x + 6, x \geq 0$.

116. If $f(x) = x^4, x \geq 0$; then the inverse is: $x = y^4, y \geq 0 \Rightarrow \sqrt[4]{x} = y \Rightarrow f^{-1}(x) = \sqrt[4]{x}$.

117. If $f(x) = 4x - 1,$; then the inverse is: $x = 4y - 1 \Rightarrow x + 1 = 4y \Rightarrow y = \dfrac{x + 1}{4} \Rightarrow f^{-1}(x) = \dfrac{x + 1}{4}$.

     Inputting the numbers given yields numbers that correspond to the letters: TREASURE HUNT IS ON.

118. If $f(x) = 3x - 3,$; then the inverse is: $x = 3y - 3 \Rightarrow x + 3 = 3y \Rightarrow y = \dfrac{x + 3}{3} \Rightarrow f^{-1}(x) = \dfrac{x + 3}{3}$.

     Inputting the numbers given yields numbers that correspond to the letters: SET SAIL TODAY.

119. Using the function $f(x) = x^3 + 1$ on the letters NO PROBLEM or the numbers: 14, 15, 16, 18, 15, 2, 12, 5, and 13, yields the numbers: 2745, 3376, 4097, 5833, 3376, 9, 1729, 126, and 2198. The inverse is:
     $x = y^3 + 1 \Rightarrow x - 1 = y^3 \Rightarrow \sqrt[3]{x - 1} = y \Rightarrow f^{-1}(x) = \sqrt[3]{x - 1}$.

120. Using the function $f(x) = (x + 2)^3$ on the letters BEGIN OPERATIONS or the numbers: 2, 5, 7, 9, 14, 15, 16, 5, 18, 1, 20, 9, 15, 14, and 19, yields the numbers: 64, 343, 729, 1331, 4096, 4913, 5832, 343, 8000, 27, 10648, 1331, 4913, 4096, and 9261. The inverse is: $x = (y + 2)^3 \Rightarrow \sqrt[3]{x} = y + 2 \Rightarrow \sqrt[3]{x} - 2 = y \Rightarrow$
     $f^{-1}(x) = \sqrt[3]{x} - 2$.

5.2: Exponential Functions

1. From the calculator, $2^{\sqrt{10}} \approx 8.952419619$

2. From the calculator, $3^{\sqrt{11}} \approx 38.23245459$

3. From the calculator, $\left(\dfrac{1}{2}\right)^{\sqrt{2}} \approx .3752142272$

4. From the calculator, $\left(\dfrac{1}{3}\right)^{\sqrt{6}} \approx .0678104116$

5. From the calculator, $4.1^{-\sqrt{3}} \approx .0868214883$

6. From the calculator, $6.4^{-\sqrt{3}} \approx .0401472058$

7. From the calculator, $\sqrt{7}^{\sqrt{7}} \approx 13.1207791$

8. From the calculator, $\sqrt{13}^{-\sqrt{13}} \approx .0098131867$

9. From the calculator, the point $(\sqrt{10}, 8.9524196)$ lies on the graph of $y = 2^x$.

10. From the calculator, the point $(\sqrt{11}, 38.232455)$ lies on the graph of $y = 3^x$.

11. From the calculator, the point $(\sqrt{2}, .37521423)$ lies on the graph of $y = \left(\dfrac{1}{2}\right)^x$.

12. From the calculator, the point $(\sqrt{6}, .06781041)$ lies on the graph of $y = \left(\dfrac{1}{3}\right)^x$.

13. Since the graph goes up to the right, $a > 1$, and since graph A is the second steepest of the graphs going up to the right, the equation has the second largest value of a. Therefore, graph A is the equation $y = 2.3^x$.

14. Since the graph goes up to the right, $a > 1$, and since graph B is the least steep of the graphs going up to the right, the equation has the lowest value of a of those where $a > 1$. Therefore, graph B is the equation $y = 1.8^x$.

15. Since the graph goes up to the left, $0 < a < 1$, and since graph C is the least steep of the graphs going up to the left, the equation has the highest value of a of those where $0 < a < 1$. Therefore, graph C is the equation $y = .75^x$.

16. Since the graph goes up to the left, $0 < a < 1$, and since graph D is the second steepest of the graphs going up to the left, the equation has the second lowest value of a of those where $0 < a < 1$. Therefore, graph D is the equation $y = .4^x$.

17. Since the graph goes up to the left, $0 < a < 1$, and since graph E is the steepest of the graphs going up to the left, the equation has the lowest value of a of those where $0 < a < 1$. Therefore, graph E is the equation $y = .31^x$.

18. Since the graph goes up to the right, $a > 1$, and since graph F is the steepest of the graphs going up to the right, the equation has the highest value of a. Therefore, graph F is the equation $y = 3.2^x$.

19. With $a > 0$, all inputs for x, yield different solutions. The function is one-to-one and an inverse function exists.

20. See Figure 20.

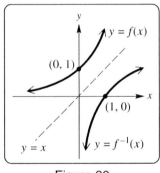   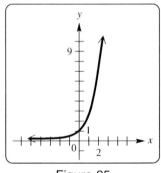   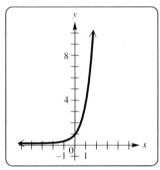

Figure 20   Figure 25   Figure 26

21. To find inverse, switch the x and y, $x = a^y$.

22. If $a = 10$, then $x = a^y$ will be $x = 10^y$.

23. If $a = e$, then $x = a^y$ will be $x = e^y$.

24. If (p, q) is on f, then switching the x and y or (q, p) will be on f^{-1}.

25. See Figure 25. All values can be input for x, therefore the domain is: $(-\infty, \infty)$. Only values where $f(x) > 0$ are possible solutions, therefore the range is: $(0, \infty)$. From the graph, the asymptote is: $y = 0$.

26. See Figure 26. All values can be input for x, therefore the domain is: $(-\infty, \infty)$. Only values where $f(x) > 0$ are possible solutions, therefore the range is: $(0, \infty)$. From the graph, the asymptote is: $y = 0$.

27. See Figure 27. All values can be input for x, therefore the domain is: $(-\infty, \infty)$. Only values where $f(x) > 0$ are possible solutions, therefore the range is: $(0, \infty)$. From the graph, the asymptote is: $y = 0$.

28. See Figure 28. All values can be input for x, therefore the domain is: $(-\infty, \infty)$. Only values where $f(x) > 0$ are possible solutions, therefore the range is: $(0, \infty)$. From the graph, the asymptote is: $y = 0$.

29. See Figure 29. All values can be input for x, therefore the domain is: $(-\infty, \infty)$. Only values where $f(x) > 0$ are possible solutions, therefore the range is: $(0, \infty)$. From the graph, the asymptote is: $y = 0$.

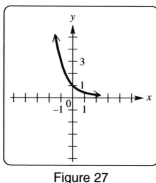
Figure 27

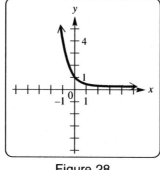
Figure 28

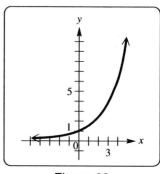
Figure 29

30. See Figure 30. All values can be input for x, therefore the domain is: $(-\infty, \infty)$. Only values where $f(x) > 0$ are possible solutions, therefore the range is: $(0, \infty)$. From the graph, the asymptote is: $y = 0$.

31. See Figure 31. All values can be input for x, therefore the domain is: $(-\infty, \infty)$. Only values where $f(x) > 0$ are possible solutions, therefore the range is: $(0, \infty)$. From the graph, the asymptote is: $y = 0$.

32. See Figure 32. All values can be input for x, therefore the domain is: $(-\infty, \infty)$. Only values where $f(x) < 0$ are possible solutions, therefore the range is: $(-\infty, 0)$. From the graph, the asymptote is: $y = 0$.

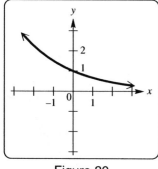
Figure 30

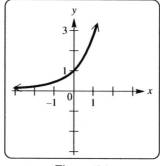
Figure 31

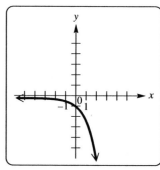
Figure 32

33. See Figure 33. All values can be input for x, therefore the domain is: $(-\infty, \infty)$. Only values where $f(x) > 0$ are possible solutions, therefore the range is: $(0, \infty)$. From the graph, the asymptote is: $y = 0$.

34. See Figure 34. All values can be input for x, therefore the domain is: $(-\infty, \infty)$. Only values where $f(x) > -1$ are possible solutions, therefore the range is: $(-1, \infty)$. From the graph, the asymptote is: $y = -1$.

35. See Figure 35. All values can be input for x, therefore the domain is: $(-\infty, \infty)$. Only values where $f(x) > 0$ are possible solutions, therefore the range is: $(0, \infty)$. From the graph, the asymptote is: $y = 0$.

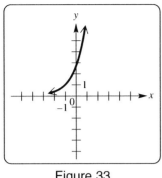

Figure 33

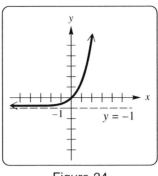

Figure 34

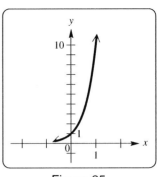

Figure 35

36. See Figure 36. All values can be input for x, therefore the domain is: $(-\infty, \infty)$. Only values where $f(x) > 0$ are possible solutions, therefore the range is: $(0, \infty)$. From the graph, the asymptote is: $y = 0$.

37. See Figure 37. All values can be input for x, therefore the domain is: $(-\infty, \infty)$. Only values where $f(x) > 0$ are possible solutions, therefore the range is: $(0, \infty)$. From the graph, the asymptote is: $y = 0$.

38. See Figure 38. All values can be input for x, therefore the domain is: $(-\infty, \infty)$. Only values where $f(x) > 0$ are possible solutions, therefore the range is: $(0, \infty)$. From the graph, the asymptote is: $y = 0$.

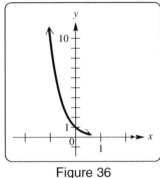

Figure 36

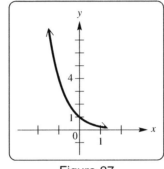

Figure 37

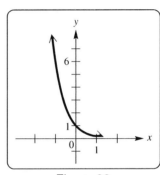

Figure 38

39. (a) Since the graph goes up to the right, $a > 1$.

(b) From the graph, the domain is: $(-\infty, \infty)$; the range is: $(0, \infty)$; and the asymptote is: $y = 0$.

(c) Reflect $f(x)$ across the x-axis. See Figure 39c.

(d) From the graph, the domain is: $(-\infty, \infty)$; the range is: $(-\infty, 0)$; and the asymptote is: $y = 0$.

(e) Reflect $f(x)$ across the y-axis. See Figure 39e.

(f) From the graph, the domain is: $(-\infty, \infty)$; the range is: $(0, \infty)$; and the asymptote is: $y = 0$.

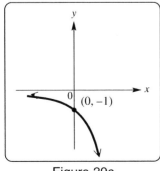

Figure 39c

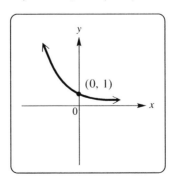

Figure 39e

40. (a) Since the graph goes up to the left, $0 < a < 1$.

  (b) From the graph, the domain is: $(-\infty, \infty)$; the range is: $(0, \infty)$; and the asymptote is: $y = 0$.

  (c) Move the graph of $f(x)$ up two units. See Figure 40c.

  (d) From the graph, the domain is: $(-\infty, \infty)$; the range is: $(2, \infty)$; and the asymptote is: $y = 2$.

  (e) Move the graph of $f(x)$ left two units. See Figure 40e.

  (f) From the graph, the domain is: $(-\infty, \infty)$; the range is: $(0, \infty)$; and the asymptote is: $y = 0$.

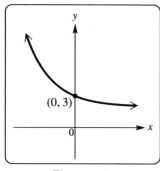

Figure 40c

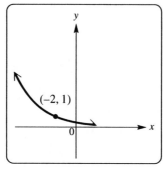

Figure 40e

41. Shift the original graph (not shown) 1 unit upward. See Figure 41.

42. Shift the original graph (not shown) 4 units downward. See Figure 42.

43. Shift the original graph (not shown) 1 unit left. See Figure 43.

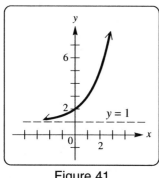

Figure 41

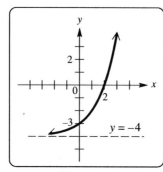

Figure 42

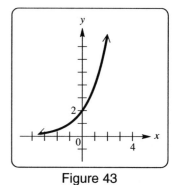

Figure 43

44. Shift the original graph (not shown) 4 units right. See Figure 44.

45. Shift the original graph (not shown) 2 units downward. See Figure 45.

46. Shift the original graph (not shown) 4 units upward. See Figure 46.

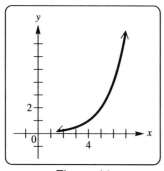

Figure 44

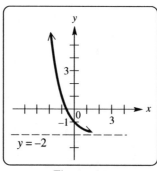

Figure 45

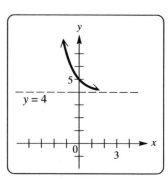

Figure 46

47. Shift the original graph (not shown) 2 units left. See Figure 47.

48. Shift the original graph (not shown) 4 units right. See Figure 48.

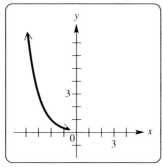

Figure 47

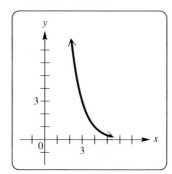

Figure 48

49. $4^x = 2 \Rightarrow (2^2)^x = 2^1 \Rightarrow 2^{2x} = 2^1 \Rightarrow 2x = 1 \Rightarrow x = \left\{\dfrac{1}{2}\right\}.$

50. $125^x = 5 \Rightarrow (5^3)^x = 5^1 \Rightarrow 5^{3x} = 5^1 \Rightarrow 3x = 1 \Rightarrow x = \left\{\dfrac{1}{3}\right\}.$

51. $\left(\dfrac{1}{2}\right)^x = 4 \Rightarrow (2^{-1})^x = 2^2 \Rightarrow 2^{-x} = 2^2 \Rightarrow -x = 2 \Rightarrow x = \{-2\}.$

52. $\left(\dfrac{2}{3}\right)^x = \dfrac{9}{4} \Rightarrow \left(\dfrac{2}{3}\right)^x = \left(\dfrac{3}{2}\right)^2 \Rightarrow \left(\dfrac{2}{3}\right)^x = \left(\dfrac{2}{3}\right)^{-2} \Rightarrow x = \{-2\}.$

53. $2^{3-x} = 8 \Rightarrow 2^{3-x} = 2^3 \Rightarrow 3 - x = 3 \Rightarrow -x = 0 \Rightarrow x = \{0\}.$

54. $5^{2x+1} = 25 \Rightarrow 5^{2x+1} = 5^2 \Rightarrow 2x + 1 = 2 \Rightarrow 2x = 1 \Rightarrow x = \left\{\dfrac{1}{2}\right\}.$

55. $12^{x-3} = 1 \Rightarrow 12^{x-3} = 12^0 \Rightarrow x - 3 = 0 \Rightarrow x = \{3\}.$

56. $(-3)^{5-x} = 1 \Rightarrow (-3)^{5-x} = (-3)^0 \Rightarrow 5 - x = 0 \Rightarrow x = \{5\}.$

57. $e^{4x-1} = (e^2)^x \Rightarrow e^{4x-1} = e^{2x} \Rightarrow 4x - 1 = 2x \Rightarrow 2x - 1 = 0 \Rightarrow 2x = 1 \Rightarrow x = \left\{\dfrac{1}{2}\right\}.$

58. $e^{3-x} = (e^3)^{-x} \Rightarrow e^{3-x} = e^{-3x} \Rightarrow 3 - x = -3x \Rightarrow 3 + 2x = 0 \Rightarrow 2x = -3 \Rightarrow x = \left\{-\dfrac{3}{2}\right\}.$

59. $27^{4x} = 9^{x+1} \Rightarrow (3^3)^{4x} = (3^2)^{x+1} \Rightarrow 3^{12x} = 3^{2x+2} \Rightarrow 12x = 2x + 2 \Rightarrow 10x = 2 \Rightarrow x = \dfrac{2}{10} \Rightarrow x = \left\{\dfrac{1}{5}\right\}.$

60. $32^x = 16^{1-x} \Rightarrow (2^5)^x = (2^4)^{1-x} \Rightarrow 2^{5x} = 2^{4-4x} \Rightarrow 5x = 4 - 4x \Rightarrow 9x = 4 \Rightarrow x = \left\{\dfrac{4}{9}\right\}.$

61. $4^{x-2} = 2^{3x+3} \Rightarrow (2^2)^{(x-2)} = 2^{3x+3} \Rightarrow 2^{2x-4} = 2^{3x+3} \Rightarrow 2x - 4 = 3x + 3 \Rightarrow -7 = x \Rightarrow x = \{-7\}.$

62. $\left(\dfrac{1}{2}\right)^{3x-6} = 8^{x+1} \Rightarrow \left(\dfrac{1}{2}\right)^{3x-6} = (2^3)^{x+1} \Rightarrow (2^{-1})^{3x-6} = 2^{3x+3} \Rightarrow 2^{-3x+6} = 2^{3x+3} \Rightarrow -3x + 6 = 3x + 3 \Rightarrow$

$3 = 6x \Rightarrow \dfrac{3}{6} = x \Rightarrow x = \left\{\dfrac{1}{2}\right\}.$

63. $(\sqrt{2})^{x+4} = 4^x \Rightarrow (2^{1/2})^{x+4} = (2^2)^x \Rightarrow 2^{1/2x+2} = 2^{2x} \Rightarrow \dfrac{1}{2}x + 2 = 2x \Rightarrow 2 = \dfrac{3}{2}x \Rightarrow x = \left\{\dfrac{4}{3}\right\}.$

64. $(\sqrt[3]{5})^{-x} = \left(\dfrac{1}{5}\right)^{x+2} \Rightarrow (5^{1/3})^{-x} = 5^{-(x+2)} \Rightarrow 5^{-1/3x} = 5^{-x-2} \Rightarrow -\dfrac{1}{3}x = -x - 2 \Rightarrow \dfrac{2}{3}x = -2 \Rightarrow x = \{-3\}.$

65. (a) $2^{x+1} = 8 \Rightarrow 2^{x+1} = 2^3 \Rightarrow x + 1 = 3 \Rightarrow x = \{2\}$.

(b) From a calculator graph of $y_1 = 2^{x+1}$ and $y_2 = 8$, $2^{x+1} > 8$ for the interval: $(2, \infty)$.

(c) From a calculator graph of $y_1 = 2^{x+1}$ and $y_2 = 8$, $2^{x+1} < 8$ for the interval: $(-\infty, 2)$.

66. (a) $3^{2-x} = 9 \Rightarrow 3^{2-x} = 3^2 \Rightarrow 2 - x = 2 \Rightarrow -x = 0 \Rightarrow x = \{0\}$.

(b) From a calculator graph of $y_1 = 3^{2-x}$ and $y_2 = 9$, $3^{2-x} > 9$ for the interval: $(-\infty, 0)$.

(c) From a calculator graph of $y_1 = 3^{2-x}$ and $y_2 = 9$, $3^{2-x} < 9$ for the interval: $(0, \infty)$.

67. (a) $27^{4x} = 9^{x+1} \Rightarrow (3^3)^{4x} = (3^2)^{x+1} \Rightarrow 3^{12x} = 3^{2x+2} \Rightarrow 12x = 2x + 2 \Rightarrow 10x = 2 \Rightarrow x = \left\{\dfrac{1}{5}\right\}$.

(b) From a calculator graph of $y_1 = 27^{4x}$ and $y_2 = 9^{x+1}$, $27^{4x} > 9^{x+1}$ for the interval: $\left(\dfrac{1}{5}, \infty\right)$.

(c) From a calculator graph of $y_1 = 27^{4x}$ and $y_2 = 9^{x+1}$, $27^{4x} < 9^{x+1}$ for the interval: $\left(-\infty, \dfrac{1}{5}\right)$.

68. (a) $32^x = 16^{1-x} \Rightarrow (2^5)^x = (2^4)^{1-x} \Rightarrow 2^{5x} = 2^{4-4x} \Rightarrow 5x = 4 - 4x \Rightarrow 9x = 4 \Rightarrow x = \left\{\dfrac{4}{9}\right\}$.

(b) From a calculator graph of $y_1 = 32^x$ and $y_2 = 16^{1-x}$, $32^x > 16^{1-x}$ for the interval: $\left(\dfrac{4}{9}, \infty\right)$.

(c) From a calculator graph of $y_1 = 32^x$ and $y_2 = 16^{1-x}$, $32^x < 16^{1-x}$ for the interval: $\left(-\infty, \dfrac{4}{9}\right)$.

69. (a) $\left(\dfrac{1}{2}\right)^{-x} = \left(\dfrac{1}{4}\right)^{x+1} \Rightarrow (2^{-1})^{-x} = (2^{-2})^{x+1} \Rightarrow 2^x = 2^{-2x-2} \Rightarrow x = -2x - 2 \Rightarrow 3x = -2 \Rightarrow x = \left\{-\dfrac{2}{3}\right\}$.

(b) From a calculator graph of $y_1 = \left(\dfrac{1}{2}\right)^{-x}$ and $y_2 = \left(\dfrac{1}{4}\right)^{x+1}$, $\left(\dfrac{1}{2}\right)^{-x} \geq \left(\dfrac{1}{4}\right)^{x+1}$ for the interval: $\left[-\dfrac{2}{3}, \infty\right)$.

(c) From a calculator graph of $y_1 = \left(\dfrac{1}{2}\right)^{-x}$ and $y_2 = \left(\dfrac{1}{4}\right)^{x+1}$, $\left(\dfrac{1}{2}\right)^{-x} \leq \left(\dfrac{1}{4}\right)^{x+1}$ for the interval: $\left(-\infty, -\dfrac{2}{3}\right]$.

70. (a) $\left(\dfrac{2}{3}\right)^{x-1} = \left(\dfrac{81}{16}\right)^{x+1} \Rightarrow \left(\dfrac{2}{3}\right)^{x-1} = \left(\left(\dfrac{3}{2}\right)^4\right)^{x+1} \Rightarrow \left(\dfrac{2}{3}\right)^{x-1} = \left(\left(\dfrac{2}{3}\right)^{-4}\right)^{x+1} \Rightarrow \left(\dfrac{2}{3}\right)^{x-1} = \left(\dfrac{2}{3}\right)^{-4x-4} \Rightarrow$

$x - 1 = -4x - 4 \Rightarrow 5x = -3 \Rightarrow x = \left\{-\dfrac{3}{5}\right\}$.

(b) From a calculator graph of $y_1 = \left(\dfrac{2}{3}\right)^{x-1}$ and $y_2 = \left(\dfrac{81}{16}\right)^{x+1}$, $\left(\dfrac{2}{3}\right)^{x-1} \leq \left(\dfrac{81}{16}\right)^{x+1}$ for the interval: $\left[-\dfrac{3}{5}, \infty\right)$.

(c) From a calculator graph of $y_1 = \left(\dfrac{2}{3}\right)^{x-1}$ and $y_2 = \left(\dfrac{81}{16}\right)^{x+1}$, $\left(\dfrac{2}{3}\right)^{x-1} \geq \left(\dfrac{81}{16}\right)^{x+1}$ for the interval: $\left(-\infty, -\dfrac{3}{5}\right]$.

71. (a) If $f(3) = 27$ then $27 = a^3 \Rightarrow \sqrt[3]{27} = a \Rightarrow a = 3$. Therefore, $f(1) = 3^1 = 3$.

(b) If $f(3) = 27$ then $27 = a^3 \Rightarrow \sqrt[3]{27} = a \Rightarrow a = 3$. Therefore, $f(-1) = 3^{-1} = \dfrac{1}{3}$.

(c) If $f(3) = 27$ then $27 = a^3 \Rightarrow \sqrt[3]{27} = a \Rightarrow a = 3$. Therefore, $f(2) = 3^2 = 9$.

(d) If $f(3) = 27$ then $27 = a^3 \Rightarrow \sqrt[3]{27} = a \Rightarrow a = 3$. Therefore, $f(0) = 3^0 = 1$.

72. If the point $(3, 8)$ in on the graph of $f(x) = a^x$, then $8 = a^3 \Rightarrow \sqrt[3]{8} = a \Rightarrow a = 2$. The equation is: $f(x) = 2^x$.

73. If the point $(-3, 64)$ in on the graph of $f(x) = a^x$, then $64 = a^{-3} \Rightarrow 64 = \dfrac{1}{a^3} \Rightarrow \dfrac{1}{64} = a^3 \Rightarrow \sqrt[3]{\dfrac{1}{64}} = a \Rightarrow$

$a = \dfrac{1}{4}$. The equation is: $f(x) = \left(\dfrac{1}{4}\right)^x$.

74. $f(t) = 3^{2t+3} = 3^{2t} \cdot 3^3 = (3^2)^t \cdot 27 \Rightarrow f(t) = (27)9^t$

75. $f(t) = \left(\dfrac{1}{3}\right)^{1-2t} = \left(\dfrac{1}{3}\right)^{1} \cdot \left(\dfrac{1}{3}\right)^{-2t} = \dfrac{1}{3} \cdot \left(\left(\dfrac{1}{3}\right)^{-2}\right)^{t} \Rightarrow f(t) = \left(\dfrac{1}{3}\right)9^{t}$

76. If $y = e^{x-3}$ then $y = e^{x} \cdot e^{-3} \Rightarrow y = e^{-3} \cdot e^{x}$. Now, since $y = Ce^{x}$ and $y = e^{-3} \cdot e^{x}$, then $C = e^{-3}$ $\left(\text{or } \dfrac{1}{e^{3}}\right)$.

77. (a) Use the formula $A = P\left(1 + \dfrac{r}{n}\right)^{nt}$ to find the amount in the account.

$A = 20{,}000\left(1 + \dfrac{.03}{1}\right)^{(1)(4)} = 20{,}000(1.03)^{4} \Rightarrow A = \$\,22{,}510.18.$

(b) Use the formula $A = P\left(1 + \dfrac{r}{n}\right)^{nt}$ to find the amount in the account.

$A = 20{,}000\left(1 + \dfrac{.03}{2}\right)^{(2)(4)} = 20{,}000(1.015)^{8} \Rightarrow A = \$\,22{,}529.85.$

78. (a) Use the formula $A = P\left(1 + \dfrac{r}{n}\right)^{nt}$ to find the amount in the account.

$A = 35{,}000\left(1 + \dfrac{.042}{1}\right)^{(1)(3)} = 35{,}000(1.042)^{3} \Rightarrow A = \$\,39{,}597.81.$

(b) Use the formula $A = P\left(1 + \dfrac{r}{n}\right)^{nt}$ to find the amount in the account.

$A = 35{,}000\left(1 + \dfrac{.042}{4}\right)^{(4)(3)} = 35{,}000(1.0105)^{12} \Rightarrow A = \$\,39{,}673.81.$

79. (a) Use the formula $A = P\left(1 + \dfrac{r}{n}\right)^{nt}$ to find the amount in the account.

$A = 27{,}500\left(1 + \dfrac{.0395}{365}\right)^{(365)(5)} = 27{,}500(1.000108219)^{1825} \Rightarrow A = \$\,33{,}504.35.$

(b) Use the formula $A = Pe^{rt}$ to find the amount in the account.

$A = 27{,}500(e^{(.0395)(5)}) = 27{,}500e^{(.1975)} \Rightarrow A = \$\,33{,}504.71.$

80. (a) Use the formula $A = P\left(1 + \dfrac{r}{n}\right)^{nt}$ to find the amount in the account.

$A = 15{,}800\left(1 + \dfrac{.046}{4}\right)^{(4)(6.5)} = 15{,}800(1.0115)^{26} \Rightarrow A = \$\,21{,}270.13.$

(b) Use the formula $A = Pe^{rt}$ to find the amount in the account.

$A = 15{,}800(e^{(.046)(6.5)}) = 15{,}800e^{(.299)} \Rightarrow A = \$\,21{,}306.45.$

81. Plan A: $A = 40{,}000\left(1 + \dfrac{.045}{4}\right)^{(4)(3)} = 40{,}000(1.01125)^{12} \Rightarrow A = \$\,45{,}746.98$

Plan B: $A = 40{,}000e^{(.044)(3)} = 40{,}000e^{(.132)} \Rightarrow A = \$\,45{,}644.33$; Plan A is better by $\$\,102.65$.

82. Plan A: $A = 50{,}000\left(1 + \dfrac{.0475}{365}\right)^{(365)(10)} = 50{,}000(1.000130137)^{3650} \Rightarrow A = \$\,80{,}398.23$

Plan B: $A = 50{,}000e^{(.047)(10)} = 50{,}000e^{(.47)} \Rightarrow A = \$\,79{,}999.71$; Plan A is better by $\$\,398.52$.

83. Set $y_{1} = 1000\left(1 + \dfrac{.05}{1}\right)^{x} \Rightarrow y_{1} = 1000(1.05)^{x}$; and $y_{2} = 1000\left(1 + \dfrac{.05}{12}\right)^{12x}$.

Graphing each on the same calculator screen shows y_{2} slight above y_{1}. Also using the table function for

y_{1}, y_{2}, and $y_{2} - y_{1}$ yields the following differences: (1 year) $- \$1.16$, (2 years) $- \$2.44$, (5 years) $- \$7.08$,

(10 years) $- \$18.11$, (20 years) $- \$59.34$, (30 years) $- \$145.80$, (40 years) $- \$318.43$.

84. Set $y_1 = 1000\left(1 + \dfrac{.075}{365}\right)^{365x}$; and $y_2 = 1000\left(1 + \dfrac{.0775}{1}\right)^x \Rightarrow y_2 = 1000(1.0775)^x$.

When graphing each on the same calculator screen it is difficult to see the difference between the graphs, but y_2 is slightly below y_1. (However, if you use the window $[35, 40]$ by $[15000, 20000]$, you will see two separate curves.) Also using the table function for y_1, y_2, and $y_3 = y_1 - y_2$ yields the following differences: (1 year) $-$ \$0.38, (2 years) $-$ \$0.81, (5 years) $-$ \$2.53, (10 years) $-$ \$7.37, (20 years) $-$ \$31.15, (30 years) $-$ \$98.73, (40 years) $-$ \$278.16. Because the lower interest rate is compounded more frequently, it yields a greater return over a long period of time.

85. (a) See Figure 85a.

(b) Because the data is graphed in a slight curve, an exponential function would fit the data better.

(c) See Figure 85c.

(d) $P(1500) = 1013e^{-.0001341(1500)} = 828.4210207 \Rightarrow P(1500) \approx 828$ mb.

 $P(11,000) = 1013e^{-.0001341(11,000)} = 231.7296764 \Rightarrow P(11,000) \approx 232$ mb.

 $P(1500)$ is slightly lower than the actual and $P(11,000)$ is slightly higher than the actual.

86. (a) $y = 5282e^{.01405(10)} = 6078.784838 \Rightarrow y \approx 6079$ million. This is a very close approximation, within 1 million.

(b) $y = 5282e^{.01405(15)} = 6521.176561 \Rightarrow y \approx 6521$ million.

(c) $y = 5282e^{.01405(20)} = 6995.763936 \Rightarrow y \approx 6996$ million.

(d) The rate of growth is affected by many factors which may have changed before 2010.

87. (a) $f(2) = 1 - e^{-.5(2)} = 1 - .367879441 = 0.632120559 \Rightarrow f(2) \approx .63$. There is a 63% chance that at least one car will enter the intersection during a 2-minute period.

(b) See Figure 87. As time progresses, the probability increases and begins to approach 1. That is, it is almost certain that at least one car will enter the intersection during a 60-minute period.

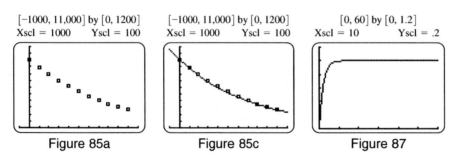

$[-1000, 11{,}000]$ by $[0, 1200]$
Xscl = 1000   Yscl = 100

$[-1000, 11{,}000]$ by $[0, 1200]$
Xscl = 1000   Yscl = 100

$[0, 60]$ by $[0, 1.2]$
Xscl = 10   Yscl = .2

Figure 85a   Figure 85c   Figure 87

88. (a) Since 99 minutes is twice 49.5 minutes, the number of bacteria will double twice or be $4(500{,}000)$ or $2{,}000{,}000$. This is verified by evaluating $N(99) = 500{,}000e^{.014(99)} = 1{,}999{,}411.364 \Rightarrow N(99) \approx 2{,}000{,}000$.

(b) Graphing the equation and using the functions of the calculator has the number reach 25 million at: 279.4 minutes or 4 hours and 39.4 minutes.

5.3: Logarithms and Their Properties

1. (a) By the definitions of logarithms, $2^4 = 16$ is equivalent to $\log_2 16 = 4$, it matches with C.

   (b) By the definitions of logarithms, $3^0 = 1$ is equivalent to $\log_3 1 = 0$, it matches with A.

   (c) By the definitions of logarithms, $10^{-1} = \dfrac{1}{10} = .1$ is equivalent to $\log_{10} .1 = -1$, it matches with E.

   (d) By the definitions of logarithms, $2^{1/2} = \sqrt{2}$ is equivalent to $\log_2 \sqrt{2} = \dfrac{1}{2}$, it matches with B.

   (e) By the definitions of logarithms, $e^{-2} = \dfrac{1}{e^2}$ is equivalent to $\log_e \dfrac{1}{e^2} = -2$, it matches with F.

   (f) By the definitions of logarithms, $\left(\dfrac{1}{2}\right)^{-3} = 8$ is equivalent to $\log_{1/2} 8 = -3$, it matches with D.

2. (a) By the definitions of logarithms, $3^4 = 81$ is equivalent to $\log_3 81 = 4$, it matches with F.

   (b) By the definitions of logarithms, $3^{-1} = \dfrac{1}{3}$ is equivalent to $\log_3 \dfrac{1}{3} = -1$, it matches with B.

   (c) By the definitions of logarithms, $10^{-2} = \dfrac{1}{100} = .01$ is equivalent to $\log_{10} .01 = -2$, it matches with A.

   (d) By the definitions of logarithms, $6^{1/2} = \sqrt{6}$ is equivalent to $\log_6 \sqrt{6} = \dfrac{1}{2}$, it matches with D.

   (e) By the definitions of logarithms, $e^0 = 1$ is equivalent to $\log_e 1 = 0$, it matches with C.

   (f) By the definitions of logarithms, $(3)^{9/2} = (3^{6/2})^{3/2} = 27^{3/2}$ is equivalent to $\log_3 27^{3/2} = \dfrac{9}{2}$, it matches with E.

3. By the definitions of logarithms, $3^4 = 81$ is equivalent to $\log_3 81 = 4$.

4. By the definitions of logarithms, $2^5 = 32$ is equivalent to $\log_2 32 = 5$.

5. By the definitions of logarithms, $\left(\dfrac{1}{2}\right)^{-4} = 16$ is equivalent to $\log_{1/2} 16 = -4$.

6. By the definitions of logarithms, $\left(\dfrac{2}{3}\right)^{-3} = \dfrac{27}{8}$ is equivalent to $\log_{2/3} \dfrac{27}{8} = -3$.

7. By the definitions of logarithms, $10^{-4} = .0001$ is equivalent to $\log_{10} .0001 = -4$ or $\log .0001 = -4$.

8. By the definitions of logarithms, $\left(\dfrac{1}{100}\right)^{-2} = 10{,}000$ is equivalent to $\log_{1/100} 10{,}000 = -2$.

9. By the definitions of logarithms and natural logarithms, $e^0 = 1$ is equivalent to $\ln 1 = 0$.

10. By the definitions of logarithms and natural logarithms, $e^{1/3} = \sqrt[3]{e}$ is equivalent to $\ln \sqrt[3]{e} = \dfrac{1}{3}$.

11. By the definitions of logarithms, $\log_6 36 = 2$ is equivalent to $6^2 = 36$.

12. By the definitions of logarithms, $\log_5 5 = 1$ is equivalent to $5^1 = 5$.

13. By the definitions of logarithms, $\log_{\sqrt{3}} 81 = 8$ is equivalent to $(\sqrt{3})^8 = 81$.

14. By the definitions of logarithms, $\log_4 \dfrac{1}{64} = -3$ is equivalent to $4^{-3} = \dfrac{1}{64}$.

15. By the definitions of logarithms, $\log_{10} .001 = -3$ is equivalent to $10^{-3} = .001$.

16. By the definitions of logarithms, $\log_3 \sqrt[3]{9} = \dfrac{2}{3}$ is equivalent to $3^{2/3} = \sqrt[3]{9}$.

17. By the definitions of logarithms, $\log \sqrt{10} = .5$ is equivalent to $10^{.5} = \sqrt{10}$.

18. By the definitions of logarithms, $\ln e^6 = 6$ is equivalent to $e^6 = e^6$.

19. By the definitions of logarithms, $\log_5 125 = x \Rightarrow 5^x = 125$, since we know $5^3 = 125$, it follows that $x = \{3\}$.

20. By the definitions of logarithms, $\log_3 81 = x \Rightarrow 3^x = 81$, since we know $3^4 = 81$, it follows that $x = \{4\}$.

21. By the definitions of logarithms, $\log_x 3^{12} = 24 \Rightarrow x^{24} = 3^{12} \Rightarrow (x^2)^{12} = 3^{12}$, since $x^2 = 3$, it follows that $x = \{\sqrt{3}\}$.

22. By the definitions of logarithms, $\log_x 25 = 6 \Rightarrow x^6 = 25 \Rightarrow (x^3)^2 = 5^2$, since $x^3 = 5$, it follows that $x = \{\sqrt[3]{5}\}$.

23. By the definitions of logarithms, $\log_6 x = -3 \Rightarrow 6^{-3} = x \Rightarrow x = \dfrac{1}{6^3} \Rightarrow x = \left\{\dfrac{1}{216}\right\}$.

24. By the definitions of logarithms, $\log_4 x = -\dfrac{1}{6} \Rightarrow 4^{-1/6} = x \Rightarrow x = \dfrac{1}{\sqrt[6]{2^2}} \Rightarrow x = \left\{\dfrac{1}{\sqrt[3]{2}}\right\} \text{or} \left\{\dfrac{1}{2^{1/3}}\right\}$.

25. By the definitions of logarithms, $\log_x 16 = \dfrac{4}{3} \Rightarrow x^{4/3} = 16 \Rightarrow (x^{4/3})^{3/4} = 16^{3/4} \Rightarrow x = \sqrt[4]{16^3} \Rightarrow x = \{8\}$.

26. By the definitions of logarithms, $\log_{16/25} x = -\dfrac{3}{2} \Rightarrow \left(\dfrac{16}{25}\right)^{-3/2} = x \Rightarrow x = \left(\dfrac{25}{16}\right)^{3/2} = \left(\dfrac{5}{4}\right)^3 \Rightarrow x = \left\{\dfrac{125}{64}\right\}$.

27. By the definitions of logarithms, $\log_x .001 = -3 \Rightarrow x^{-3} = .001$, since we know $10^{-3} = .001$, it follows that $x = \{10\}$.

28. By the definitions of logarithms, $\log_3 (x - 1) = 2 \Rightarrow 3^2 = x - 1 \Rightarrow 9 = x - 1 \Rightarrow x = \{10\}$.

29. By the definitions of logarithms, $\log_9 \dfrac{\sqrt[4]{27}}{3} = x \Rightarrow 9^x = \dfrac{\sqrt[4]{27}}{3} \Rightarrow (3^2)^x = \dfrac{3^{3/4}}{3} = 3^{-1/4}$. Since $3^{2x} = 3^{-1/4}$,
    $2x = -\dfrac{1}{4} \Rightarrow x = \left\{-\dfrac{1}{8}\right\}$.

30. By the definitions of logarithms, $\log_{1/4} \dfrac{16^2}{2^{-3}} = x \Rightarrow \left(\dfrac{1}{4}\right)^x = \dfrac{16^2}{2^{-3}} \Rightarrow \left(\dfrac{1}{2^2}\right)^x = \dfrac{2^8}{2^{-3}} \Rightarrow 2^{-2x} = 2^{11} \Rightarrow$
    $-2x = 11 \Rightarrow x = \left\{-\dfrac{11}{2}\right\}$.

31. (a) Using the properties of Logarithms, $3^{\log_3 7} = 7$

    (b) Using the properties of Logarithms, $4^{\log_4 9} = 9$

    (c) Using the properties of Logarithms, $12^{\log_{12} 4} = 4$

    (d) Using the properties of Logarithms, $a^{\log_a k} (k > 0, a > 0, a \neq 1) = k$

32. (a) Using the properties of Logarithms, $\log_3 3^{19} = 19$

    (b) Using the properties of Logarithms, $\log_4 4^{17} = 17$

    (c) Using the properties of Logarithms, $\log_{12} 12^{1/3} = \dfrac{1}{3}$

    (d) Using the properties of Logarithms, $\log_a \sqrt{a} \, (a > 0, a \neq 1) = \log_a a^{1/2} = \dfrac{1}{2}$

33. (a) Using the properties of Logarithms, $\log_3 1 = 0$

    (b) Using the properties of Logarithms, $\log_4 1 = 0$

    (c) Using the properties of Logarithms, $\log_{12} 1 = 0$

    (d) Using the properties of Logarithms, $\log_a 1 \, (a > 0, a \neq 1) = 0$

34. (a) $\text{Log}_a x$ represents the power to which a must be raised to get x.

    (b) There is no case where a power of a will lead to a negative number.

35. By properties of Logarithms, $\log 10^{1.5} = \log_{10} 10^{1.5} = 1.5$

36. By properties of Logarithms, $\log 10^{4.3} = \log_{10} 10^{4.3} = 4.3$

37. By properties of Logarithms, $\log 10^{2/5} = \log_{10} 10^{2/5} = \dfrac{2}{5}$

38. By properties of Logarithms, $\log 10^{\sqrt{3}} = \log_{10} 10^{\sqrt{3}} = \sqrt{3}$

39. By properties of Logarithms, $\ln e^{2/3} = \dfrac{2}{3}$

40. By properties of Logarithms, $\ln e^{.5} = .5$

41. By properties of Logarithms, $\ln e^{\pi} = \pi$

42. By properties of Logarithms, $\ln e^{\sqrt{6}} = \sqrt{6}$

43. By properties of Logarithms, $3 \ln e^{1.8} = 3(1.8) = 5.4$

44. By properties of Logarithms, $\sqrt{2} \ln e^{\sqrt{2}} = \sqrt{2}(\sqrt{2}) = 2$

45. From the calculator, $\log 43 \approx 1.633468456$

46. From the calculator, $\log 1247 \approx 3.095866453$

47. From the calculator, $\log .783 \approx -.1062382379$

48. From the calculator, $\log .014 \approx -1.853871964$

49. From the calculator, $\log 28^3 = 3 \log 28 = 3(1.447158031) \approx 4.341474094$

50. From the calculator, $\log (47 \times 93) = \log 47 + \log 93 \approx 1.672097858 + 1.968482949 \approx 3.640580807$

51. From the calculator, $\ln 43 \approx 3.761200116$

52. From the calculator, $\ln 1247 \approx 7.128495946$

53. From the calculator, $\ln .783 \approx -.244622583$

54. From the calculator, $\ln .014 \approx -4.268697949$

55. From the calculator, $\ln 28^3 = 3 \ln 28 = 3(3.33220451) \approx 9.996613533$

56. From the calculator, $\ln (47 \times 93) = \ln 47 + \ln 93 \approx 3.850147602 + 4.532599493 \approx 8.382747095$

57. Since $pH = -\log[H_3O^+]$ and Grapefruit has $H_3O^+ = 6.3 \times 10^{-4}$, $pH = -\log (6.3 \times 10^{-4}) \approx 3.20066 \approx 3.2$

58. Since $pH = -\log[H_3O^+]$ and Limes have $H_3O^+ = 1.6 \times 10^{-2}$, $pH = -\log (1.6 \times 10^{-2}) \approx 1.79588 \approx 1.8$

59. Since $pH = -\log[H_3O^+]$ and Crackers have $H_3O^+ = 3.9 \times 10^{-9}$, $pH = -\log (3.9 \times 10^{-9}) \approx 8.408935 \approx 8.4$

60. Since $pH = -\log[H_3O^+]$ and Lye has $H_3O^+ = 3.2 \times 10^{-14}$, $pH = -\log (3.2 \times 10^{-14}) \approx 13.49485 \approx 13.5$

61. Since $pH = -\log[H_3O^+]$ and Soda Pop has $pH = 2.7$, $2.7 = -\log [H_3O^+] \Rightarrow -2.7 = \log[H_3O^+] \Rightarrow$

    $H_3O^+ = 10^{-2.7} \Rightarrow H_3O^+ \approx .001995 \approx 2 \times 10^{-3}$

62. Since $pH = -\log[H_3O^+]$ and Wine has $pH = 3.4$, $3.4 = -\log [H_3O^+] \Rightarrow -3.4 = \log[H_3O^+] \Rightarrow$

    $H_3O^+ = 10^{-3.4} \Rightarrow H_3O^+ \approx .000398 \approx 4 \times 10^{-4}$

63. Since $pH = -\log[H_3O^+]$ and Beer has $pH = 4.8$, $4.8 = -\log [H_3O^+] \Rightarrow -4.8 = \log[H_3O^+] \Rightarrow$

    $H_3O^+ = 10^{-4.8} \Rightarrow H_3O^+ \approx .00001585 \approx 1.6 \times 10^{-5}$

64. Since $pH = -\log[H_3O^+]$ and Drinking Water has $pH = 6.5$, $6.5 = -\log [H_3O^+] \Rightarrow -6.5 = \log[H_3O^+] \Rightarrow$

    $H_3O^+ = 10^{-6.5} \Rightarrow H_3O^+ \approx .0000003162 \approx 3.2 \times 10^{-7}$

65. Since $A = Pe^{rt}$, we can find time by solving: $3000 = 2500e^{.0375t} \Rightarrow \dfrac{3000}{2500} = e^{.0375t} \Rightarrow \dfrac{6}{5} = e^{.0375t} \Rightarrow$

$\ln \dfrac{6}{5} = .0375t \Rightarrow t = \dfrac{\ln \frac{6}{5}}{.0375} \approx 4.9$ years.

66. Since $A = Pe^{rt}$, we can find time by solving: $3500 = 2500e^{.0425t} \Rightarrow \dfrac{3500}{2500} = e^{.0425t} \Rightarrow \dfrac{7}{5} = e^{.0425t} \Rightarrow$

$\ln \dfrac{7}{5} = .0425t \Rightarrow t = \dfrac{\ln \frac{7}{5}}{.0425} \approx 7.9$ years.

67. Since $A = Pe^{rt}$, we can find time by solving: $5000 = 2500e^{.05t} \Rightarrow \dfrac{5000}{2500} = e^{.05t} \Rightarrow \dfrac{2}{1} = e^{.05t} \Rightarrow$

$\ln 2 = .05t \Rightarrow t = \dfrac{\ln 2}{.05} \approx 13.9$ years.

68. Since $A = Pe^{rt}$, we can find time by solving: $5000 = 2500e^{.06t} \Rightarrow \dfrac{5000}{2500} = e^{.06t} \Rightarrow \dfrac{2}{1} = e^{.06t} \Rightarrow$

$\ln 2 = .06t \Rightarrow t = \dfrac{\ln 2}{.06} \approx 11.6$ years.

69. $\log_3 \dfrac{2}{5} = \log_3 2 - \log_3 5$

70. $\log_4 \dfrac{6}{7} = \log_4 6 - \log_4 7$

71. $\log_2 \dfrac{6x}{y} = \log_2 6 + \log_2 x - \log_2 y$

72. $\log_3 \dfrac{4p}{q} = \log_3 4 + \log_3 p - \log_3 q$

73. $\log_5 \dfrac{5\sqrt{7}}{3m} = \log_5 5 + \log_5 7^{1/2} - (\log_5 3 + \log_5 m) = 1 + \dfrac{1}{2}\log_5 7 - \log_5 3 - \log_5 m$

74. $\log_2 \dfrac{2\sqrt{3}}{5p} = \log_2 2 + \log_2 3^{1/2} - (\log_2 5 + \log_2 p) = 1 + \dfrac{1}{2}\log_2 3 - \log_2 5 - \log_2 p$

75. $\log_4 (2x + 5y)$ can not be rewritten.

76. $\log_4 (2x + 5y)$ can not be rewritten.

77. $\log_k \dfrac{pq^2}{m} = \log_k p + \log_k q^2 - \log_k m = \log_k p + 2\log_k q - \log_k m$

78. $\log_z \dfrac{x^5 y^3}{3} = \log_z x^5 + \log_z y^3 - \log_z 3 = 5\log_z x + 3\log_z y - \log_z 3$

79. $\log_m \sqrt{\dfrac{r^3}{5z^5}} = \log_m \left(\dfrac{r^3}{5z^5}\right)^{1/2} = \dfrac{1}{2}\left[\log_m r^3 - (\log_m 5 + \log_m z^5)\right] = \dfrac{1}{2}(3\log_m r - \log_m 5 - 5\log_m z)$ or

$\dfrac{3}{2}\log_m r - \dfrac{1}{2}\log_m 5 - \dfrac{5}{2}\log_m z$

80. $\log_p \sqrt[3]{\dfrac{m^5}{kt^2}} = \log_p \left(\dfrac{m^5}{kt^2}\right)^{1/3} = \dfrac{1}{3}\left[\log_p m^5 - (\log_p k + \log_p t^2)\right] = \dfrac{1}{3}(5\log_p m - \log_p k - 2\log_p t)$ or

$\dfrac{5}{3}\log_p m - \dfrac{1}{3}\log_p k - \dfrac{2}{3}\log_p t$

81. $\log_a x + \log_a y - \log_a m = \log_a \dfrac{xy}{m}$

82. $(\log_b k - \log_b m) - \log_b a = \log_b k - (\log_b m + \log_b a) = \log_b \dfrac{k}{ma}$

83. $2 \log_m a - 3 \log_m b^2 = \log_m a^2 - \log_m (b^2)^3 = \log_m \dfrac{a^2}{b^6}$

84. $\dfrac{1}{2} \log_y p^3 q^4 - \dfrac{2}{3} \log_y p^4 q^3 = \log_y p^{3/2} q^2 - \log_y p^{8/3} q^2 = \log_y \dfrac{p^{3/2} q^2}{p^{8/3} q^2} = \log_y \dfrac{p^{9/6}}{p^{16/6}} = \log_y p^{-7/6} = -\log_y p^{7/6}$

85. $2 \log_a (z - 1) + \log_a (3z + 2), z > 1 \Rightarrow \log_a (z - 1)^2 + \log_a (3z + 2) = \log_a \left[(z - 1)^2 (3z + 2)\right]$

86. $\log_b (2y + 5) - \dfrac{1}{2} \log_b (y + 3) = \log_b (2y + 5) - \log_b (y + 3)^{1/2} = \log_b \dfrac{2y + 5}{\sqrt{y + 3}}$

87. $-\dfrac{2}{3} \log_5 5m^2 + \dfrac{1}{2} \log_5 25m^2 \Rightarrow \log_5 (5m^2)^{-2/3} + \log_5 (25m^2)^{1/2} \Rightarrow \log_5 (5^{-2/3} m^{-4/3}) + \log_5 5m \Rightarrow$

$\log_5 (5^{-2/3} m^{-4/3})(5m) \Rightarrow \log_5 (5^{1/3} m^{-1/3})$ or $\log_5 \dfrac{5^{1/3}}{m^{1/3}}$

88. $-\dfrac{3}{4} \log_3 16p^4 - \dfrac{2}{3} \log_3 8p^3 \Rightarrow \log_3 (16p^4)^{-3/4} + \log_3 (8p^3)^{-2/3} \Rightarrow \log_3 2^{-3} p^{-3} + \log_3 2^{-2} p^{-2} \Rightarrow$

$\log_3 (2^{-3} p^{-3})(2^{-2} p^{-2}) \Rightarrow \log_3 (2^{-5} p^{-5}) \Rightarrow \log_3 \dfrac{1}{32p^5}$

89. $\log_5 10 = \dfrac{\log 10}{\log 5} \approx 1.430676558$

90. $\log_9 12 = \dfrac{\log 12}{\log 9} \approx 1.130929754$

91. $\log_{15} 5 = \dfrac{\log 5}{\log 15} \approx .5943161289$

92. $\log_{1/2} 3 = \dfrac{\log 3}{\log \frac{1}{2}} \approx -1.584962501$

93. $\log_{100} 83 = \dfrac{\log 83}{\log 100} \approx .9595390462$

94. $\log_{200} 175 = \dfrac{\log 175}{\log 200} \approx .9747973963$

95. $\log_{2.9} 7.5 = \dfrac{\log 7.5}{\log 2.9} \approx 1.892441722$

96. $\log_{5.8} 12.7 = \dfrac{\log 12.7}{\log 5.8} \approx 1.445851777$

97. To get the graph of $y = -3^x + 7$, reflect the graph of $y = 3^x$ across the x-axis and shift 7 units upward.

98. See Figure 98.

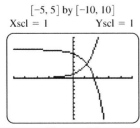

$[-5, 5]$ by $[-10, 10]$
Xscl = 1   Yscl = 1

Figure 98

99. From the calculator's capabilities the x-intercept of $y_2 = -3^x + 7$ is: $x \approx 1.7712437492$.

100. If $0 = -3^x + 7$ then $3^x = 7$. Now using base 3 logarithm yields: $\log_3 3^x = \log_3 7 \Rightarrow x = \{\log_3 7\}$.

101. $\log_3 7 = \dfrac{\log 7}{\log 3} = 1.77124374916$

102. The approximations are close enough to support the conclusion that the x-intercept is equal to $\log_3 7$.

103. With a total of 100 individuals and 50 individuals of each species, $P_1 = \dfrac{50}{100} = \dfrac{1}{2}$ and $P_2 = \dfrac{50}{100} = \dfrac{1}{2}$.

   Now using the index of diversity, $H = -\left(\dfrac{1}{2} \log_2 \left(\dfrac{1}{2} \right) + \dfrac{1}{2} \log_2 \left(\dfrac{1}{2} \right) \right) = -\log_2 \left(\dfrac{1}{2} \right) = -(-1) \Rightarrow H = 1$.

104. Using the index of diversity from problem 103,

   $H = -(.521 \log_2 .521 + .324 \log_2 .324 + .081 \log_2 .081 + .074 \log_2 .074) =$

   $-\left(\dfrac{.521 \log .521}{\log 2} + \dfrac{.324 \log .324}{\log 2} + \dfrac{.081 \log .081}{\log 2} + \dfrac{.074 \log .074}{\log 2} \right) \approx$

   $-(-.49008 - .52680 - .29370 - .27797) \approx -(-1.58855) \approx 1.59$.

105. (a) $S(100) = .36 \ln \left(1 + \dfrac{100}{.36} \right) \approx 2.0269 \approx 2$

   (b) $S(200) = .36 \ln \left(1 + \dfrac{200}{.36} \right) \approx 2.2758 \approx 2$

   (c) $S(150) = .36 \ln \left(1 + \dfrac{150}{.36} \right) \approx 2.1725 \approx 2$

   (d) $S(10) = .36 \ln \left(1 + \dfrac{10}{.36} \right) \approx 1.2095 \approx 1$

106. (a) To prove $\log_a \dfrac{x}{y} = \log_a x - \log_a y$, let $m = \log_a x$ and $n = \log_a y$, then $a^m = x$ and $a^n = y$. To prove the

   quotient rule we divide: $\dfrac{x}{y} = \dfrac{a^m}{a^n} \Rightarrow \dfrac{x}{y} = a^{m-n}$, now substituting for m and n yields:

   $\log_a \dfrac{x}{y} = \log_a x - \log_a y$.

   (b) To prove $\log_a x^r = r \log_a x$, let $m = \log_a x$, then $a^m = x \Rightarrow (a^m)^r = x^r \Rightarrow a^{mr} = x^r \Rightarrow \log_a x^r = rm$.

   To prove the power rule substitute for: $\log_a x^r = r \log_a x$.

Reviewing Basic Concepts (Sections 5.1—5.3)

1. No, because the x-value -2 and 2 both correspond to the y-value 4. In one-to-one function, each y-value must correspond to exactly one x-value (and each x-value to exactly one y-value).

2. (a) Interchange the x and y-values:

x	12	21	32	45
y	7	8	9	10

   (b) For $f(x) = \dfrac{x+5}{4}$ the inverse is: $x = \dfrac{y+5}{4} \Rightarrow 4x = y + 5 \Rightarrow y = 4x - 5 \Rightarrow f^{-1}(x) = 4x - 5$.

3. Graph $f(x) = 2x + 3$ and reflect it across the line $y = x$. See Figure 3.

4. See Figure 4.

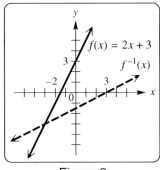

Figure 3

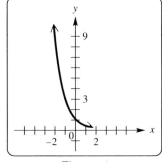

Figure 4

5. $4^{2x} = 8 \Rightarrow (2^2)^{2x} = 2^3 \Rightarrow 2^{4x} = 2^3 \Rightarrow 4x = 3 \Rightarrow x = \left\{\dfrac{3}{4}\right\}.$

6. Using $A = P\left(1 + \dfrac{r}{n}\right)^{nt}$, we get: $A = 600\left(1 + \dfrac{.04}{4}\right)^{4(3)} = 600(1.01)^{12} = 676.10.$ The interest earned is:

    $\$676.10 - \$600 = \$76.10.$

7. (a) $\log\left(\dfrac{1}{\sqrt{10}}\right) = \log_{10} 10^{-1/2} = -\dfrac{1}{2}$

    (b) $2 \ln e^{1.5} = \log_e e^{2(1.5)} = \log_e e^3 = 3$

    (c) $\log_2 4 = 2$, because we know that $2^2 = 4.$

8. From the properties of logarithms,

    $\log \dfrac{3x^2}{5y} = \log 3 + \log x^2 - (\log 5 + \log y) = \log 3 + 2 \log x - \log 5 - \log y.$

9. From the properties of logarithms, $\ln 4 + \ln x - 3 \ln 2 = \ln 4 + \ln x - \ln 2^3 = \ln \dfrac{4x}{2^3} = \ln \dfrac{4x}{8} = \ln \dfrac{x}{2}.$

10. Using $A = Pe^{rt}$, we: $1650 = 1500e^{.045t} \Rightarrow \dfrac{1650}{1500} = e^{.045t} \Rightarrow 1.1 = e^{.045t} \Rightarrow \ln 1.1 = \ln e^{.045t} \Rightarrow$

    $\ln 1.1 = .045t \Rightarrow t = \dfrac{\ln 1.1}{.045} \approx 2.1$ years.

5.4: Logarithmic Functions

1. Reflect f across the line $y = x$ and interchange the x and y-coordinates for each point. See Figure 1.

    From the graph of f^{-1}, Domain: $(0, \infty)$; Range: $(-\infty, \infty)$; the graph Increases on its domain $(0, \infty)$;

    and the Vertical Asymptote is: $x = 0.$

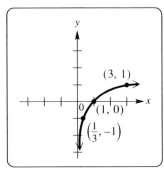

Figure 1

2. Reflect f across the line $y = x$ and interchange the x and y-coordinates for each point. See Figure 2.
    From the graph of f^{-1}, Domain: $(0, \infty)$; Range: $(-\infty, \infty)$; the graph Increases on its domain $(0, \infty)$;
    and the Vertical Asymptote is: $x = 0$.

3. Reflect f across the line $y = x$ and interchange the x and y-coordinates for each point. See Figure 3.
    From the graph of f^{-1}, Domain: $(0, \infty)$; Range: $(-\infty, \infty)$; the graph Decreases on its domain $(0, \infty)$;
    and the Vertical Asymptote is: $x = 0$.

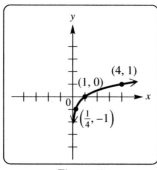

Figure 2   Figure 3

4. Reflect f across the line $y = x$ and interchange the x and y-coordinates for each point. See Figure 4.
    From the graph of f^{-1}, Domain: $(0, \infty)$; Range: $(-\infty, \infty)$; the graph Decreases on its domain $(0, \infty)$;
    and the Vertical Asymptote is: $x = 0$.

5. Reflect f across the line $y = x$ and interchange the x and y-coordinates for each point. See Figure 5.
    From the graph of f^{-1}, Domain: $(1, \infty)$; Range: $(-\infty, \infty)$; the graph Increases on its domain $(1, \infty)$;
    and the Vertical Asymptote is: $x = 1$.

6. Reflect f across the line $y = x$ and interchange the x and y-coordinates for each point. See Figure 6.
    From the graph of f^{-1}, Domain: $(-3, \infty)$; Range: $(-\infty, \infty)$; the graph Decreases on its domain $(-3, \infty)$;
    and the Vertical Asymptote is: $x = -3$.

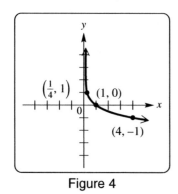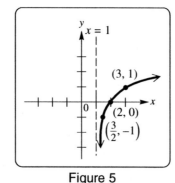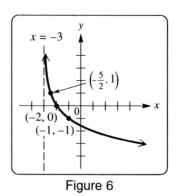

Figure 4   Figure 5   Figure 6

7. Logarithmic

8. Answers will vary. The graphs are reflections across the line $y = x$, the points on $f(x) = a^x$ can be found by
    interchanging the x and y-coordinates of the points on $f(x) = \log_a x$, both functions are increasing on the
    domain, and the asymptote of $f(x) = \log_a x$ is $x = 0$, while the asymptote of $f(x) = a^x$ is $y = 0$.

9. Since the argument of a logarithm must be positive, $2x > 0 \Rightarrow x > 0$, therefore the domain is: $(0, \infty)$.

10. Since the argument of a logarithm must be positive, $\frac{1}{2}x > 0 \Rightarrow x > 0$, therefore the domain is: $(0, \infty)$.

11. Since the argument of a logarithm must be positive, $(-x) > 0 \Rightarrow x < 0$, therefore the domain is: $(-\infty, 0)$.

12. Since the argument of a logarithm must be positive, $\left(-\frac{1}{2}x\right) > 0 \Rightarrow x < 0$, therefore the domain is: $(-\infty, 0)$.

13. Since the argument of a natural logarithm must be positive, $(x^2 + 7) > 0$, this is true for all values of x, therefore the domain is: $(-\infty, \infty)$.

14. Since the argument of a natural logarithm must be positive, $(x^4 + 8) > 0$, this is true for all values of x, therefore the domain is: $(-\infty, \infty)$.

15. Since the argument of a natural logarithm must be positive, $(-x^2 + 4) > 0$. Now set $(-x^2 + 4) = 0 \Rightarrow$ $x^2 - 4 = 0 \Rightarrow x^2 = 4 \Rightarrow x = -2, 2$ and the endpoints will be: $x = \{-2, 2\}$. The numbers -2 and 2 divide the number line into 3 intervals. The interval: $(-\infty, -2)$ has a negative solution, the interval: $(-2, 2)$ has a positive solution, and the interval: $(2, \infty)$ has a negative solution. Therefore $(-x^2 + 4) > 0$ for the interval: $(-2, 2)$ and the domain is: $(-2, 2)$.

16. Since the argument of a natural logarithm must be positive, $(-x^2 + 16) > 0$. Now set $(-x^2 + 16) = 0 \Rightarrow$ $x^2 - 16 = 0 \Rightarrow x^2 = 16 \Rightarrow x = -4, 4$ and the endpoints will be: $x = \{-4, 4\}$. The numbers -4 and 4 divide the number line into 3 intervals. The interval: $(-\infty, -4)$ has a negative solution, the interval: $(-4, 4)$ has a positive solution, and the interval: $(4, \infty)$ has a negative solution. Therefore $(-x^2 + 16) > 0$ for the interval: $(-4, 4)$ and the domain is: $(-4, 4)$.

17. Since the argument of a natural logarithm must be positive, $(x^2 - 4x - 21) > 0$. Now set $x^2 - 4x - 21 = 0 \Rightarrow (x - 7)(x + 3) = 0 \Rightarrow x = -3, 7$ and the endpoints will be: $x = \{-3, 7\}$. The numbers -3 and 7 divide the number line into 3 intervals. The interval: $(-\infty, -3)$ has a positive product, the interval: $(-3, 7)$ has a negative product, and the interval: $(7, \infty)$ has a positive product. Therefore $x^2 - 4x - 21 > 0$ for the interval: $(-\infty, -3) \cup (7, \infty)$ and the domain is: $(-\infty, -3) \cup (7, \infty)$.

18. Since the argument of a natural logarithm must be positive, $(2x^2 - 7x - 4) > 0$. Now set $2x^2 - 7x - 4 = 0 \Rightarrow (2x + 1)(x - 4) = 0 \Rightarrow x = -\frac{1}{2}, 4$ and the endpoints will be: $x = \left\{-\frac{1}{2}, 4\right\}$. The numbers $-\frac{1}{2}$ and 4 divide the number line into 3 intervals. The interval: $\left(-\infty, -\frac{1}{2}\right)$ has positive product, the interval: $\left(-\frac{1}{2}, 4\right)$ has a negative product, and the interval: $(4, \infty)$ has a positive product. Therefore $2x^2 - 7x - 4 > 0$ for the interval: $\left(-\infty, -\frac{1}{2}\right) \cup (4, \infty)$ and the domain is: $\left(-\infty, -\frac{1}{2}\right) \cup (4, \infty)$.

19. Since the argument of a natural logarithm must be positive, $(x^3 - x) > 0$. Now set $x^3 - x = 0 \Rightarrow$ $x(x^2 - 1) = 0 \Rightarrow x(x + 1)(x - 1) = 0 \Rightarrow x = -1, 0, 1$ and the endpoints will be: $x = \{-1, 0, 1\}$. The numbers $-1, 0$, and 1 divide the number line into 4 intervals. The interval: $(-\infty, -1)$ has a negative product, the interval: $(-1, 0)$ has a positive product, the interval: $(0, 1)$ has a negative product, and the interval: $(1, \infty)$ has a positive product. Therefore $x^3 - x > 0$ for the interval: $(-1, 0) \cup (1, \infty)$ and the domain is: $(-1, 0) \cup (1, \infty)$.

20. Since the argument of a natural logarithm must be positive, $(x^3 - 81x) > 0$. Now set $x^3 - 81x = 0 \Rightarrow$ $x(x^2 - 81) = 0 \Rightarrow x(x + 9)(x - 9) = 0 \Rightarrow x = -9, 0, 9$ and the endpoints will be: $x = \{-9, 0, 9\}$. The numbers $-9, 0$, and 9 divide the number line into 4 intervals. The interval: $(-\infty, -9)$ has a negative product, the interval: $(-9, 0)$ has a positive product, the interval: $(0, 9)$ has a negative product, and the interval: $(9, \infty)$ has a positive product. Therefore $x^3 - 81x > 0$ for the interval: $(-9, 0) \cup (9, \infty)$ and the domain is: $(-9, 0) \cup (9, \infty)$.

21. Since the argument of a natural logarithm must be positive, $\dfrac{x + 3}{x - 4} > 0$. Now set both $x + 3 = 0$ and $x - 4 = 0 \Rightarrow x = -3, 4$ and the endpoints will be: $x = \{-3, 4\}$. The numbers -3 and 4 divide the number line into 3 intervals. The interval: $(-\infty, -3)$ has a positive product, the interval: $(-3, 4)$ has a negative product, and the interval: $(4, \infty)$ has a positive product. Therefore $\dfrac{x + 3}{x - 4} > 0$ for the interval: $(-\infty, -3) \cup (4, \infty)$ and the domain is: $(-\infty, -3) \cup (4, \infty)$.

22. Since the argument of a natural logarithm must be positive, $\dfrac{x + 1}{x - 5} > 0$. Now set both $x + 1 = 0$ and $x - 5 = 0 \Rightarrow x = -1, 5$ and the endpoints will be: $x = \{-1, 5\}$. The numbers -1 and 5 divide the number line into 3 intervals. The interval: $(-\infty, -1)$ has a positive product, the interval: $(-1, 5)$ has a negative product, and the interval: $(5, \infty)$ has a positive product. Therefore $\dfrac{x + 1}{x - 5} > 0$ for the interval: $(-\infty, -1) \cup (5, \infty)$ and the domain is: $(-\infty, -1) \cup (5, \infty)$.

23. Since the argument of a natural logarithm must be positive, $|3x - 7| > 0$. Since this is true for all real x-values except when $3x - 7 = 0$, we solve for $3x - 7 = 0 \Rightarrow 3x = 7 \Rightarrow x = \dfrac{7}{3}$ and $|3x - 7| > 0$ for the interval: $\left(-\infty, \dfrac{7}{3}\right) \cup \left(\dfrac{7}{3}, \infty\right)$ and the domain is: $\left(-\infty, \dfrac{7}{3}\right) \cup \left(\dfrac{7}{3}, \infty\right)$.

24. Since the argument of a natural logarithm must be positive, $|6x + 6| > 0$. Since this is true for all real x-values except when $6x + 6 = 0$, we solve for $6x + 6 = 0 \Rightarrow 6x = -6 \Rightarrow x = \dfrac{-6}{6} = -1$ and $|6x + 6| > 0$ for the interval: $(-\infty, -1) \cup (-1, \infty)$ and the domain is: $(-\infty, -1) \cup (-1, \infty)$.

25. Shift the graph of $f(x) = \log_2 x$ left upward 3 units to sketch the graph of $f(x) = (\log_2 x) + 3$. See Figure 25.

26. Shift the graph of $f(x) = \log_2 x$ left 3 units to sketch the graph of $f(x) = \log_2 (x + 3)$. See Figure 26.

27. Shift the graph of $f(x) = \log_2 x$ left 3 units and reflect all negative y-values across the x-axis to sketch the graph of $f(x) = |\log_2 (x + 3)|$. See Figure 27.

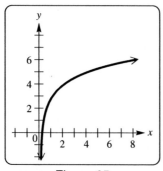

Figure 25

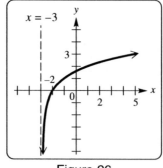

Figure 26

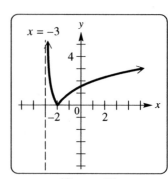

Figure 27

28. Shift the graph of $f(x) = \log_{1/2} x$ downward 2 units to sketch the graph of $f(x) = (\log_{1/2} x) - 2$. See Figure 28.

29. Shift the graph of $f(x) = \log_{1/2} x$ right 2 units to sketch the graph of $f(x) = \log_{1/2} (x - 2)$. See Figure 29.

30. Shift the graph of $f(x) = \log_{1/2} x$ right 2 units and reflect all negative y-values across the x-axis to sketch the graph of $f(x) = |\log_{1/2} (x - 2)|$. See Figure 30.

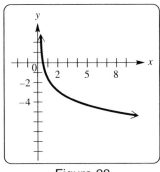

Figure 28

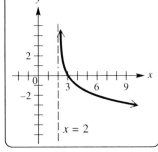

Figure 29

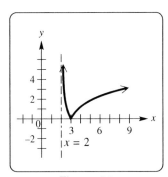

Figure 30

31. The graph of $y = e^x + 3$ is the graph of $y = e^x$ shifted 3 units upward. The correct graph is: B.

32. The graph of $y = e^x - 3$ is the graph of $y = e^x$ shifted 3 units downward. The correct graph is: G.

33. The graph of $y = e^{x+3}$ is similar to the graph of $y = e^x$ with a y-intercept of $(0, e^3)$. The correct graph is: D.

34. The graph of $y = e^{x-3}$ is similar to the graph of $y = e^x$ passing through $(3, 1)$. The correct graph is: F.

35. The graph of $y = \ln x + 3$ is the graph of $y = \ln x$ passing through $(1, 3)$. The correct graph is: A.

36. The graph of $y = \ln x - 3$ is the graph of $y = \ln x$ passing through $(1, -3)$. The correct graph is: E.

37. The graph of $y = \ln (x - 3)$ is the graph of $y = \ln x$ shifted 3 units right. The correct graph is: C.

38. The graph of $y = \ln (x + 3)$ is the graph of $y = \ln x$ shifted 3 units left. The correct graph is: H.

39. Graph a logarithmic function with base greater than 1, that has an asymptote $x = 0$, and passes through $(1, 0)$ and $(5, 1)$. See Figure 39.

40. Graph a logarithmic function with base 10, that has an asymptote $x = 0$, and passes through $(1, 0)$ and $(10, 1)$. See Figure 40.

41. Graph a logarithmic function with base between 0 and 1, that is shifted 1 unit left then reflected across the y-axis, has an asymptote $x = 1$, and passes through $(0, 0)$ and $(-1, -1)$. See Figure 41.

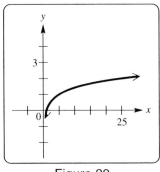

Figure 39

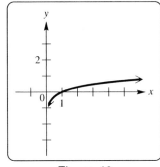

Figure 40

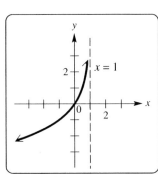

Figure 41

42. Graph a logarithmic function with base between 0 and 1, that is shifted 3 units left then reflected across the *y*-axis, has an asymptote *x* = 3, and passes through (2, 0) and (0, −1). See Figure 42.

43. Graph a logarithmic function with base greater than 1, that is shifted 1 unit right, has an asymptote *x* = 1, and passes through (2, 0) and (4, 1). See Figure 43.

44. Graph a logarithmic function with base greater than 1, that is stretched because of the squaring, has an asymptote *x* = 0, has symmetry with respect to the *y*-axis, and passes through (1, 0) and (−1, 0). See Figure 44.

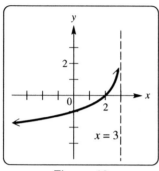

Figure 42

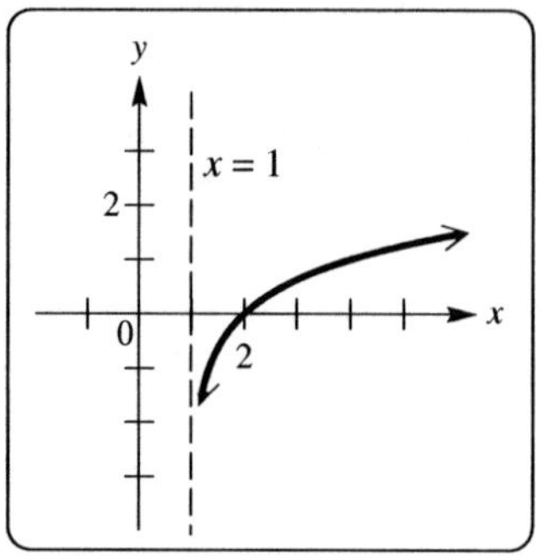

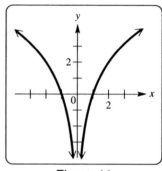

Figure 43

Figure 44

45. (a) The graph is shifted 4 units to the left.

    (b) See Figure 45.

46. (a) The graph is shifted 6 units to the right.

    (b) See Figure 46.

47. (a) The graph is stretched vertically by a factor of 3 and shifted 1 unit upward.

    (b) See Figure 47.

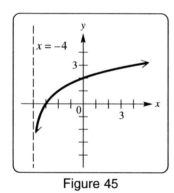

Figure 45

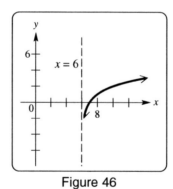

Figure 46

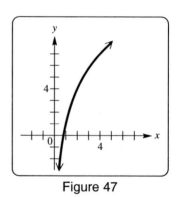

Figure 47

48. (a) The graph is stretched vertically by a factor of 4, reflected across the *x*-axis, and shifted 8 units downward.

    (b) See Figure 48.

49. (a) The graph is reflected across the *y*-axis and shifted 1 unit upward.

    (b) See Figure 49.

50. (a) The graph is reflected across the *y*-axis and reflected across the *x*-axis.

    (b) See Figure 50.

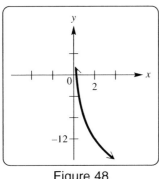

Figure 48

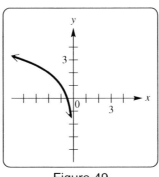

Figure 49

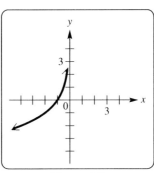

Figure 50

51. The graphs are not the same because the domain of $y = \log x^2$ is $(-\infty, 0) \cup (0, \infty)$, while the domain of

   $y = 2 \log x$ is $(0, \infty)$. The power rule does not apply if the argument is nonpositive.

52. From the graph in the given window, the domain appears to be $(-\infty, \infty)$. It is actually $(-\infty, 0) \cup (0, \infty)$, the

   argument of a logarithm can not be zero.

53. (a) If $\log_9 27 = x$ then $9^x = 27 \Rightarrow (3^2)^x = 27 \Rightarrow 3^{2x} = 3^3 \Rightarrow 2x = 3 \Rightarrow x = \left\{\dfrac{3}{2}\right\}$.

   (b) By the change of base rule, $\log_9 27 = \dfrac{\log 27}{\log 9}$, then by calculator, $\dfrac{\log 27}{\log 9} \approx \dfrac{1.43136}{.95424} = 1.5 = \dfrac{3}{2}$.

   (c) The point $\left(27, \dfrac{3}{2}\right)$ is on the calculator graph of $y = \log_9 27$, which supports the answer in part (a).

54. (a) If $\log_4 \dfrac{1}{8} = x$ then $4^x = \dfrac{1}{8} \Rightarrow (2^2)^x = \dfrac{1}{8} \Rightarrow 2^{2x} = 2^{-3} \Rightarrow 2x = -3 \Rightarrow x = \left\{-\dfrac{3}{2}\right\}$.

   (b) By the change of base rule, $\log_4 \dfrac{1}{8} = \dfrac{\log \frac{1}{8}}{\log 4}$, then by calculator, $\dfrac{\log \frac{1}{8}}{\log 4} \approx \dfrac{-.90309}{.60206} = -1.5 = -\dfrac{3}{2}$.

   (c) The point $\left(\dfrac{1}{8}, -\dfrac{3}{2}\right)$ is on the calculator graph of $y = \log_4 \dfrac{1}{8}$, which supports the answer in part (a).

55. (a) If $\log_{16} \dfrac{1}{8} = x$ then $16^x = \dfrac{1}{8} \Rightarrow (2^4)^x = \dfrac{1}{8} \Rightarrow 2^{4x} = 2^{-3} \Rightarrow 4x = -3 \Rightarrow x = \left\{-\dfrac{3}{4}\right\}$.

   (b) By the change of base rule, $\log_{16} \dfrac{1}{8} = \dfrac{\log \frac{1}{8}}{\log 16}$, then by calculator, $\dfrac{\log \frac{1}{8}}{\log 16} \approx \dfrac{-.90309}{1.20412} = -.75 = -\dfrac{3}{4}$.

   (c) The point $\left(\dfrac{1}{8}, -\dfrac{3}{4}\right)$ is on the calculator graph of $y = \log_{16} \dfrac{1}{8}$, which supports the answer in part (a).

56. (a) If $\log_2 \sqrt{8} = x$ then $2^x = \sqrt{8} \Rightarrow 2^x = (2^3)^{1/2} \Rightarrow 2^x = 2^{3/2} \Rightarrow x = \left\{\dfrac{3}{2}\right\}$.

   (b) By the change of base rule, $\log_2 \sqrt{8} = \dfrac{\log \sqrt{8}}{\log 2}$, then by calculator, $\dfrac{\log \sqrt{8}}{\log 2} \approx \dfrac{.45154}{.30103} = 1.5 = \dfrac{3}{2}$.

   (c) The point $\left(\sqrt{8}, \dfrac{3}{2}\right)$ is on the calculator graph of $y = \log_2 \sqrt{8}$, which supports the answer in part (a).

57. For the function $f(x) = 4^x - 3$, the inverse is: $x = 4^y - 3 \Rightarrow x + 3 = 4^y \Rightarrow \log_4(x + 3) = \log_4 4^y \Rightarrow$

    $\log_4(x + 3) = y \Rightarrow f^{-1}(x) = \log_4(x + 3)$. Graph f and f^{-1} in the same window. See Figure 57.

58. For the function $f(x) = \left(\dfrac{1}{2}\right)^x - 5$, the inverse is: $x = \left(\dfrac{1}{2}\right)^y - 5 \Rightarrow x + 5 = \left(\dfrac{1}{2}\right)^y \Rightarrow$

    $\log_{1/2}(x + 5) = \log_{1/2}\left(\dfrac{1}{2}\right)^y \Rightarrow \log_{1/2}(x + 5) = y \Rightarrow f^{-1}(x) = \log_{1/2}(x + 5)$. Graph f and f^{-1} in the

    same window. See Figure 58.

59. For the function $f(x) = -10^x + 4$, the inverse is: $x = -10^y + 4 \Rightarrow x - 4 = -10^y \Rightarrow -x + 4 = 10^y \Rightarrow$

    $\log(-x + 4) = \log 10^y \Rightarrow \log(-x + 4) = y \Rightarrow f^{-1}(x) = \log(4 - x)$. Graph f and f^{-1} in the same

    window. See Figure 59.

60. For the function $f(x) = -e^x + 6$, the inverse is: $x = -e^y + 6 \Rightarrow x - 6 = -e^y \Rightarrow -x + 6 = e^y \Rightarrow$

    $\ln(-x + 6) = \ln e^y \Rightarrow \ln(-x + 6) = y \Rightarrow f^{-1}(x) = \ln(6 - x)$. Graph f and f^{-1} in the same

    window. See Figure 60.

| $[-4.7, 4.7]$ by $[-3.1, 3.1]$ | $[-6.2, 6.2]$ by $[-4.4, 4.7]$ | $[-6, 6]$ by $[-4, 4]$ | $[-9.4, 9.4]$ by $[-6.2, 6.2]$ |
| Xscl = 1   Yscl = 1 | Xscl = 1   Yscl = 1 | Xscl = 1   Yscl = 1 | Xscl = 1   Yscl = 1 |

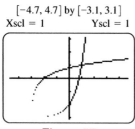 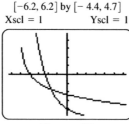 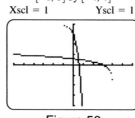 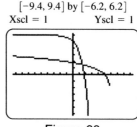

Figure 57   Figure 58   Figure 59   Figure 60

61. (a) From the graph, $\log_3 .3 \approx -1.1$.

    (b) From the graph, $\log_3 .8 \approx -.2$.

62. (a) First, find a. Using the given information, $2 = \log_a 3 \Rightarrow a^2 = 3 \Rightarrow a = \sqrt{3}$.

    Now, $f\left(\dfrac{1}{9}\right) = \log_{\sqrt{3}}\left(\dfrac{1}{9}\right) = x \Rightarrow (\sqrt{3})^x = \dfrac{1}{9} \Rightarrow (3^{1/2})^x = 3^{-2} \Rightarrow 3^{1/2x} = 3^{-2} \Rightarrow \dfrac{1}{2}x = -2 \Rightarrow x = \{-4\}$.

    (b) Since $a = \sqrt{3}$ from part a, $f(27) = \log_{\sqrt{3}} 27 = x \Rightarrow (\sqrt{3})^x = 27 \Rightarrow (3^{1/2})^x = 27 \Rightarrow 3^{1/2x} = 3^3 \Rightarrow$

    $\dfrac{1}{2}x = 3 \Rightarrow x = \{6\}$.

    (c) Since $a = \sqrt{3}$ from part a, find f^{-1}, for the function $f(x) = \log_{\sqrt{3}} x$, the inverse is:

    $x = \log_{\sqrt{3}} y \Rightarrow (\sqrt{3})^x = y \Rightarrow f^{-1}(x) = (\sqrt{3})^x$. Now, $f^{-1}(-2) = (\sqrt{3})^{-2} = \left\{\dfrac{1}{3}\right\}$.

    (d) Since $a = \sqrt{3}$ from part a, find f^{-1}, for the function $f(x) = \log_{\sqrt{3}} x$, the inverse is:

    $x = \log_{\sqrt{3}} y \Rightarrow (\sqrt{3})^x = y \Rightarrow f^{-1}(x) = (\sqrt{3})^x$. Now, $f^{-1}(0) = (\sqrt{3})^0 = \{1\}$.

63. Since logarithmic functions are in the form $y = \log_a x$, the point $(5,4)$ gives the equation: $4 = \log_a 5$.

64. Graphing $y_1 = \log_{10} x$ and $y_2 = x - 2$ in the same window yields intersection points: $x = \{.01, 2.38\}$.

65. Graphing $y_1 = 2^{-x}$ and $y_2 = \log_{10} x$ in the same window yields intersection point: $x = \{1.87\}$.

66. Graphing $y_1 = e^x$ and $y_2 = x^2$ in the same window yields intersection point: $x = \{-.70\}$.

67. (a) The left side is a reflection of the right side with respect to the axis of the tower. The graph of $f(-x)$ is the reflection of $f(x)$ with respect to the y-axis.

    (b) Since the horizontal line on the top has one-half on each side of the y-axis, the x-coordinate on the right side will be $x = \frac{1}{2}(15.7488) \Rightarrow x = 7.8744$. Using $f(7.8744) = -301 \ln\left(\frac{7.8744}{207}\right) \approx 984$ feet.

    (c) Solve for x by graphing. Graph $y_1 = -301 \ln\left(\frac{x}{207}\right)$ and $y_2 = 500$, yields the approximate intersection point: $(39.31, 500)$, therefore the height is approximately 39 feet.

68. (a) Using $L = 80$, yields: $t = -2.57 \ln\left(\frac{87-80}{63}\right) \Rightarrow t = -2.57 \ln\left(\frac{7}{63}\right) \approx 5.6$ years.

    (b) $4 = -2.57 \ln\left(\frac{87-L}{63}\right) \Rightarrow \frac{4}{-2.57} = \ln\left(\frac{87-L}{63}\right) \Rightarrow e^{-4/2.57} = \frac{87-L}{63} \Rightarrow L = 87 - 63e^{-4/2.57} \approx 73.7$ ft.

    (c) Age $t > 0$ and the expression $\frac{87-L}{63}$ is positive and in the domain of the function only if $24 < L < 87$.

69. (a) Using $x = 9$, yields: $f(9) = 27 + 1.105 \log(9+1) \Rightarrow f(9) = 27 + 1.105 \log 10 \Rightarrow f(9) \approx 28.105$ in.

    (b) It tells us that at 99 miles from the eye of a typical hurricane, the barometric pressure is 29.21 inches.

70. (a) Using $t = 9.78$, yields: $f(t) = 11.65(1 - e^{-t/1.27}) \Rightarrow f(9.78) = 11.65(1 - e^{-9.78/1.27}) \Rightarrow$

    $f(9.78) \approx 11.64$ meters per second.

    (b) Using $f(x) = 10$, yields: $10 = 11.65(1 - e^{-t/1.27}) \Rightarrow \frac{10}{11.65} = 1 - e^{-t/1.27} \Rightarrow -1 + \frac{10}{11.65} = -e^{-t/1.27} \Rightarrow$

    $1 - \frac{10}{11.65} = e^{-t/1.27} \Rightarrow \ln\left(1 - \frac{10}{11.65}\right) = \ln e^{-t/1.27} \Rightarrow \ln\left(1 - \frac{10}{11.65}\right) = \frac{-t}{1.27} \Rightarrow$

    $t = -1.27 \ln\left(1 - \frac{10}{11.65}\right) \Rightarrow t \approx 2.482$ seconds.

5.5: Exponential and Logarithmic Equations and Inequalities

1. (a) From the information of the graph, the solution for $f(x) = 0$ is: $x = \{1.4036775\}$.

    (b) From the graph and the answer to part a, $f(x) > 0$ for the interval: $(1.4036775, \infty)$.

2. (a) From the information of the graph, the solution for $f(x) = 0$ is: $x = \{1.2618595\}$.

    (b) From the graph and the answer to part a, $f(x) > 0$ for the interval: $(1.2618595, \infty)$.

3. (a) From the information of the graph, the solution for $f(x) = 0$ is: $x = \{-1\}$.

    (b) From the graph and the answer to part a, $f(x) > 0$ for the interval: $(-1, \infty)$.

4. (a) From the information of the graph, the solution for $f(x) = 0$ is: $x = \{0\}$.

    (b) From the graph and the answer to part a, $f(x) > 0$ for the interval: $(0, \infty)$.

5. (a) $3^x = 7 \Rightarrow \log 3^x = \log 7 \Rightarrow x \log 3 = \log 7 \Rightarrow x = \left\{\frac{\log 7}{\log 3}\right\}$. (Other forms of the answer are possible.)

    (b) From the calculator the solution is: $x = \{1.771\}$.

6. (a) $5^x = 13 \Rightarrow \log 5^x = \log 13 \Rightarrow x \log 5 = \log 13 \Rightarrow x = \left\{\frac{\log 13}{\log 5}\right\}$.

    (Other forms of the answer are possible.)

    (b) From the calculator the solution is: $x = \{1.594\}$.

7. (a) $\left(\dfrac{1}{2}\right)^x = 5 \Rightarrow \log\left(\dfrac{1}{2}\right)^x = \log 5 \Rightarrow x \log\left(\dfrac{1}{2}\right) = \log 5 \Rightarrow x = \left\{\dfrac{\log 5}{\log\left(\dfrac{1}{2}\right)}\right\}.$

   (Other forms of the answer are possible.)

   (b) From the calculator the solution is: $x = \{-2.322\}$.

8. (a) $\left(\dfrac{1}{3}\right)^x = 6 \Rightarrow \log\left(\dfrac{1}{3}\right)^x = \log 6 \Rightarrow x \log\left(\dfrac{1}{3}\right) = \log 6 \Rightarrow x = \left\{\dfrac{\log 6}{\log\left(\dfrac{1}{3}\right)}\right\}.$

   (Other forms of the answer are possible.)

   (b) From the calculator the solution is: $x = \{-1.631\}$.

9. (a) $.8^x = 4 \Rightarrow \log .8^x = \log 4 \Rightarrow x \log .8 = \log 4 \Rightarrow x = \left\{\dfrac{\log 4}{\log .8}\right\}.$ (Other forms of the answer are possible.)

   (b) From the calculator the solution is: $x = \{-6.213\}$.

10. (a) $.6^x = 3 \Rightarrow \log .6^x = \log 3 \Rightarrow x \log .6 = \log 3 \Rightarrow x = \left\{\dfrac{\log 3}{\log .6}\right\}.$ (Other forms of the answer are possible.)

    (b) From the calculator the solution is: $x = \{-2.151\}$.

11. (a) $4^{x-1} = 3^{2x} \Rightarrow \log 4^{x-1} = \log 3^{2x} \Rightarrow (x-1)\log 4 = (2x)\log 3 \Rightarrow x \log 4 - \log 4 = 2x \log 3 \Rightarrow$

    $x \log 4 - 2x \log 3 = \log 4 \Rightarrow x(\log 4 - 2 \log 3) = \log 4 \Rightarrow x = \left\{\dfrac{\log 4}{\log 4 - 2 \log 3}\right\}.$

    (Other forms of the answer are possible.)

    (b) From the calculator the solution is: $x = \{-1.710\}$.

12. (a) $2^{x+3} = 5^x \Rightarrow \log 2^{x+3} = \log 5^x \Rightarrow (x+3)\log 2 = x \log 5 \Rightarrow x \log 2 + 3 \log 2 = x \log 5 \Rightarrow$

    $3 \log 2 = x \log 5 - x \log 2 \Rightarrow \log 2^3 = x(\log 5 - \log 2) \Rightarrow x = \left\{\dfrac{\log 8}{\log 5 - \log 2}\right\}.$

    (Other forms of the answer are possible.)

    (b) From the calculator the solution is: $x = \{2.269\}$.

13. (a) $6^{x+1} = 4^{2x-1} \Rightarrow \log 6^{x+1} = \log 4^{2x-1} \Rightarrow (x+1)\log 6 = (2x-1)\log 4 \Rightarrow$

    $x \log 6 + \log 6 = 2x \log 4 - \log 4 \Rightarrow x \log 6 - 2x \log 4 = -\log 6 - \log 4 \Rightarrow x(\log 6 - 2 \log 4) =$

    $-\log 6 - \log 4 \Rightarrow x = \left\{\dfrac{-\log 6 - \log 4}{\log 6 - 2 \log 4}\right\}.$ (Other forms of the answer are possible.)

    (b) From the calculator the solution is: $x = \{3.240\}$.

14. (a) $3^{x-4} = 7^{2x+5} \Rightarrow \log 3^{x-4} = \log 7^{2x+5} \Rightarrow (x-4)\log 3 = (2x+5)\log 7 \Rightarrow$

    $x \log 3 - 4 \log 3 = 2x \log 7 + 5 \log 7 \Rightarrow x \log 3 - 2x \log 7 = 5 \log 7 + 4 \log 3 \Rightarrow x(\log 3 - 2 \log 7)$

    $= 5 \log 7 + 4 \log 3 \Rightarrow x = \left\{\dfrac{5 \log 7 + 4 \log 3}{\log 3 - 2 \log 7}\right\}.$ (Other forms of the answer are possible.)

    (b) From the calculator the solution is: $x = \{-5.057\}$.

15. (a) No real number value for x can produce a negative solution, therefore, the solution set is $\varnothing$.

16. (a) No real number value for x can produce a negative solution, therefore, the solution set is $\varnothing$.

17. (a) $e^{x-3} = 2^{3x} \Rightarrow \ln e^{x-3} = \ln 2^{3x} \Rightarrow (x-3)\ln e = 3x \ln 2 \Rightarrow x \ln e - 3 \ln e = 3x \ln 2 \Rightarrow$

    $x - 3x \ln 2 = 3 \Rightarrow x(1 - 3 \ln 2) = 3 \Rightarrow x = \left\{ \dfrac{3}{1 - 3 \ln 2} \right\}.$ (Other forms of the answer are possible.)

    (b) From the calculator the solution is: $x = \{-2.779\}$.

18. (a) $e^{.5x} = 3^{1-2x} \Rightarrow \ln e^{.5x} = \ln 3^{1-2x} \Rightarrow .5x \ln e = (1-2x) \ln 3 \Rightarrow .5x \ln e = \ln 3 - 2x \ln 3 \Rightarrow$

    $.5x + 2x \ln 3 = \ln 3 \Rightarrow x(.5 + 2 \ln 3) = \ln 3 \Rightarrow x = \left\{ \dfrac{\ln 3}{.5 + 2 \ln 3} \right\}.$

    (Other forms of the answer are possible.)

    (b) From the calculator the solution is: $x = \{.407\}$.

19. (a) No real number value for x can produce a negative solution, therefore $x = \varnothing$.

20. (a) No real number value for x can produce a negative solution, therefore $x = \varnothing$.

21. (a) $.05(1.15)^x = 5 \Rightarrow \log .05 + \log (1.15)^x = \log 5 \Rightarrow x \log 1.15 = \log 5 - \log .05 \Rightarrow$

    $x \log 1.15 = \log \dfrac{5}{.05} \Rightarrow x \log 1.15 = \log 100 \Rightarrow x \log 1.15 = 2 \Rightarrow x = \left\{ \dfrac{2}{\log 1.15} \right\}.$

    (Other forms of the answer are possible.)

    (b) From the calculator the solution is: $x = \{32.950\}$.

22. (a) $1.2(.9)^x = .6 \Rightarrow \log 1.2 + \log (.9)^x = \log .6 \Rightarrow x \log .9 = \log .6 - \log 1.2 \Rightarrow$

    $x \log .9 = \log \dfrac{.6}{1.2} \Rightarrow x \log .9 = \log .5 \Rightarrow x = \left\{ \dfrac{\log .5}{\log .9} \right\}.$ (Other forms of the answer are possible.)

    (b) From the calculator the solution is: $x = \{6.579\}$.

23. (a) $3(2)^{x-2} + 1 = 100 \Rightarrow 3(2)^{x-2} = 99 \Rightarrow \log 3 + \log 2^{x-2} = \log 99 \Rightarrow (x-2) \log 2 =$

    $\log 99 - \log 3 \Rightarrow x \log 2 - 2 \log 2 = \log 99 - \log 3 \Rightarrow x \log 2 = \log 99 - \log 3 + 2 \log 2 \Rightarrow$

    $x \log 2 = \log \left(\dfrac{99}{3} \right) + 2 \log 2 \Rightarrow x = \dfrac{\log 33}{\log 2} + \dfrac{2 \log 2}{\log 2} \Rightarrow x = \left\{ 2 + \dfrac{\log 33}{\log 2} \right\}.$

    (Other forms of the answer are possible.)

    (b) From the calculator the solution is: $x = \{7.044\}$.

24. (a) $5(1.2)^{3x-2} + 1 = 7 \Rightarrow 5(1.2)^{3x-2} = 6 \Rightarrow \log 5 + \log 1.2^{3x-2} = \log 6 \Rightarrow (3x-2) \log 1.2 =$

    $\log 6 - \log 5 \Rightarrow 3x \log 1.2 - 2 \log 1.2 = \log 6 - \log 5 \Rightarrow 3x \log 1.2 = \log 6 - \log 5 + 2 \log 1.2 \Rightarrow$

    $3x \log 1.2 = \log \left(\dfrac{6}{5} \right) + 2 \log 1.2 \Rightarrow 3x = \dfrac{\log 1.2}{\log 1.2} + \dfrac{2 \log 1.2}{\log 1.2} \Rightarrow 3x = 1 + 2 \Rightarrow 3x = 3 \Rightarrow x = \{1\}.$

    (Other forms of the answer are possible.)

25. (a) $2(1.05)^x + 3 = 10 \Rightarrow 2(1.05)^x = 7 \Rightarrow \log 2 + \log 1.05^x = \log 7 \Rightarrow x \log 1.05 = \log 7 - \log 2 \Rightarrow$

    $x \log 1.05 = \log \left(\dfrac{7}{2} \right) \Rightarrow x = \left\{ \dfrac{\log 3.5}{\log 1.05} \right\}.$ (Other forms of the answer are possible.)

    (b) From the calculator the solution is: $x = \{25.677\}$.

26. (a) $3(1.4)^x - 4 = 60 \Rightarrow 3(1.4)^x = 64 \Rightarrow \log 3 + \log 1.4^x = \log 64 \Rightarrow x \log 1.4 = \log 64 - \log 3 \Rightarrow$

    $x \log 1.4 = \log \left(\dfrac{64}{3} \right) \Rightarrow x = \left\{ \dfrac{\log \left(\dfrac{64}{3} \right)}{\log 1.4} \right\}.$ (Other forms of the answer are possible.)

    (b) From the calculator the solution is: $x = \{9.095\}$.

27. (a) $5(1.015)^{x-1980} = 8 \Rightarrow \log 5 + \log 1.015^{x-1980} = \log 8 \Rightarrow (x-1980)\log 1.015 = \log 8 - \log 5 \Rightarrow$

    $x\log 1.015 - 1980\log 1.015 = \log 8 - \log 5 \Rightarrow x\log 1.015 = \log 8 - \log 5 + 1980\log 1.015 \Rightarrow$

    $x\log 1.015 = \log\left(\dfrac{8}{5}\right) + 1980\log 1.015 \Rightarrow$

    $x = \dfrac{\log\left(\dfrac{8}{5}\right)}{\log 1.015} + \dfrac{1980\log 1.015}{\log 1.015} \Rightarrow x = \left\{1980 + \dfrac{\log\left(\dfrac{8}{5}\right)}{\log 1.015}\right\}.$ (Other forms of the answer are possible.)

    (b) From the calculator the solution is: $x = \{2011.568\}$.

28. (a) $30 - 3(.75)^{x-1} = 29 \Rightarrow -3(.75)^{x-1} = -1 \Rightarrow 3(.75)^{x-1} = 1 \Rightarrow \log 3 + \log(.75)^{x-1} = \log 1 \Rightarrow$

    $x\log .75 - \log .75 = 0 - \log 3 \Rightarrow x\log .75 = \log .75 - \log 3 \Rightarrow x = \left\{\dfrac{\log .75 - \log 3}{\log .75}\right\}.$

    (Other forms of the answer are possible.)

    (b) From the calculator the solution is: $x = \{4.819\}$.

29. $5\ln x = 10 \Rightarrow \ln x = \dfrac{10}{5} \Rightarrow \ln x = 2 \Rightarrow e^{\ln x} = e^2 \Rightarrow x = \{e^2\}.$

30. $3\log x = 2 \Rightarrow \log x = \dfrac{2}{3} \Rightarrow 10^{\log x} = 10^{2/3} \Rightarrow x = \sqrt[3]{10^2} \Rightarrow x = \{\sqrt[3]{100}\}.$

31. $\ln(4x) = 1.5 \Rightarrow e^{\ln 4x} = e^{1.5} \Rightarrow 4x = e^{1.5} \Rightarrow x = \left\{\dfrac{e^{1.5}}{4}\right\}.$

32. $\ln(2x) = 5 \Rightarrow e^{\ln 2x} = e^5 \Rightarrow 2x = e^5 \Rightarrow x = \left\{\dfrac{e^5}{2}\right\}.$

33. $\log(2-x) = .5 \Rightarrow 10^{\log(2-x)} = 10^{.5} \Rightarrow 2 - x = \sqrt{10} \Rightarrow -x = \sqrt{10} - 2 \Rightarrow x = \{2 - \sqrt{10}\}.$

34. $\ln(1-x) = \dfrac{1}{2} \Rightarrow e^{\ln(1-x)} = e^{1/2} \Rightarrow 1 - x = \sqrt{e} \Rightarrow -x = \sqrt{e} - 1 \Rightarrow x = \{1 - \sqrt{e}\}.$

35. $\log_6(2x+4) = 2 \Rightarrow 6^{\log_6(2x+4)} = 6^2 \Rightarrow 2x + 4 = 36 \Rightarrow 2x = 32 \Rightarrow x = \{16\}.$

36. $\log_5(8-3x) = 3 \Rightarrow 5^{\log_5(8-3x)} = 5^3 \Rightarrow 8 - 3x = 125 \Rightarrow -3x = 117 \Rightarrow x = \{-39\}.$

37. $\log_4(x^3+37) = 3 \Rightarrow 4^{\log_4(x^3+37)} = 4^3 \Rightarrow x^3 + 37 = 64 \Rightarrow x^3 = 27 \Rightarrow \sqrt[3]{x^3} = \sqrt[3]{27} \Rightarrow x = \{3\}.$

38. $\log_7(x^3+65) = 0 \Rightarrow 7^{\log_7(x^3+65)} = 7^0 \Rightarrow x^3 + 65 = 1 \Rightarrow x^3 = -64 \Rightarrow \sqrt[3]{x^3} = \sqrt[3]{-64} \Rightarrow x = \{-4\}.$

39. $\ln x + \ln x^2 = 3 \Rightarrow \ln x(x^2) = 3 \Rightarrow \ln x^3 = 3 \Rightarrow e^{\ln x^3} = e^3 \Rightarrow x^3 = e^3 \Rightarrow \sqrt[3]{x^3} = \sqrt[3]{e^3} \Rightarrow x = \{e\}.$

40. $\log x + \log x^2 = 3 \Rightarrow \log x(x^2) = 3 \Rightarrow \log x^3 = 3 \Rightarrow 10^{\log x^3} = 10^3 \Rightarrow x^3 = 10^3 \Rightarrow x = \{10\}.$

41. $\log x + \log(x-21) = 2 \Rightarrow \log x(x-21) = 2 \Rightarrow \log(x^2-21x) = 2 \Rightarrow 10^{\log(x^2-21x)} = 10^2 \Rightarrow$

    $x^2 - 21x = 100 \Rightarrow x^2 - 21x - 100 = 0 \Rightarrow (x-25)(x+4) = 0 \Rightarrow x = -4, 25.$ Since the argument of a

    logarithm can not be negative, the solution is: $x = \{25\}$.

42. $\log x + \log(3x-13) = 1 \Rightarrow \log x(3x-13) = 1 \Rightarrow \log(3x^2-13x) = 1 \Rightarrow 10^{\log(3x^2-13x)} = 10^1 \Rightarrow$

    $3x^2 - 13x = 10 \Rightarrow 3x^2 - 13x - 10 = 0 \Rightarrow (3x+2)(x-5) = 0 \Rightarrow x = -\dfrac{2}{3}, 5.$ Since the argument of a

    logarithm can not be negative, the solution is: $x = \{5\}$.

43. $\ln(4x-2) - \ln 4 = -\ln(x-2) \Rightarrow \ln(4x-2) + \ln(x-2) = \ln 4 \Rightarrow \ln(4x-2)(x-2) = \ln 4 \Rightarrow$

    $\ln(4x^2-10x+4) = \ln 4 \Rightarrow e^{\ln(4x^2-10x+4)} = e^{\ln 4} \Rightarrow 4x^2 - 10x + 4 = 4 \Rightarrow 4x^2 - 10x = 0 \Rightarrow$

    $(2x)(2x-5) = 0 \Rightarrow x = 0, \dfrac{5}{2}.$ Since the argument of a logarithm can not be zero, the solution is: $x = \{2.5\}$.

44. $\ln(5 + 4x) - \ln(3 + x) - \ln 3 = 0 \Rightarrow \ln(5 + 4x) = \ln(3 + x) + \ln 4 \Rightarrow \ln(5 + 4x) = \ln(3 + x)3 \Rightarrow$

    $\ln(5 + 4x) = \ln(9 + 3x) \Rightarrow e^{\ln(5+4x)} = e^{\ln(9+3x)} \Rightarrow 5 + 4x = 9 + 3x \Rightarrow x = \{4\}$.

45. $\log_5(x + 2) + \log_5(x - 2) = 1 \Rightarrow \log_5(x + 2)(x - 2) = 1 \Rightarrow \log_5(x^2 - 4) = 1 \Rightarrow 5^{\log_5(x^2-4)} = 5^1 \Rightarrow$

    $x^2 - 4 = 5 \Rightarrow x^2 = 9 \Rightarrow x = -3, 3$. Since the argument of a logarithm can not be negative, the solution is:

    $x = \{3\}$.

46. $\log_2(x - 7) + \log_2 x = 3 \Rightarrow \log_2(x - 7)x = 3 \Rightarrow \log_2(x^2 - 7x) = 3 \Rightarrow 2^{\log_2(x^2-7x)} = 2^3 \Rightarrow$

    $x^2 - 7x = 8 \Rightarrow x^2 - 7x - 8 = 0 \Rightarrow (x - 8)(x + 1) = 0 \Rightarrow x = -1, 8$. Since the argument of a logarithm

    can not be negative, the solution is: $x = \{8\}$.

47. $\log_7(4x) - \log_7(x + 3) = \log_7 x \Rightarrow \log_7 4x = \log_7 x + \log_7(x + 3) \Rightarrow \log_7 4x = \log_7 x(x + 3) \Rightarrow$

    $\log_7 4x = \log_7(x^2 + 3x) \Rightarrow 7^{\log_7 4x} = 7^{\log_7(x^2+3x)} \Rightarrow 4x = x^2 + 3x \Rightarrow x^2 - x = 0 \Rightarrow x(x - 1) = 0 \Rightarrow$

    $x = 0, 1$. Since the argument of a logarithm can not be zero, the solution is: $x = \{1\}$.

48. $\log_2(2x) + \log_2(x + 2) = \log_2 16 \Rightarrow \log_2 2x(x + 2) = \log_2 16 \Rightarrow \log_2(2x^2 + 4x) = \log_2 16 \Rightarrow$

    $2^{\log_2(2x^2+4x)} = 2^{\log_2 16} \Rightarrow 2x^2 + 4x = 16 \Rightarrow 2(x^2 + 2x - 8) = 0 \Rightarrow 2(x - 2)(x + 4) = 0 \Rightarrow$

    $x = -4, 2$. Since the argument of a logarithm can not be negative, the solution is: $x = \{2\}$.

49. $\ln e^x - 2\ln e = \ln e^4 \Rightarrow x - 2(1) = 4 \Rightarrow x = \{6\}$.

50. $\log_2(\log_2 x) = 1 \Rightarrow 2^{\log_2(\log_2 x)} = 2^1 \Rightarrow \log_2 x = 2 \Rightarrow 2^{\log_2 x} = 2^2 \Rightarrow x = \{4\}$.

51. $\log x = \sqrt{\log x} \Rightarrow (\log x)^2 = (\sqrt{\log x})^2 \Rightarrow (\log x)^2 = \log x \Rightarrow (\log x)^2 - \log x = 0 \Rightarrow$

    $\log x(\log x - 1) = 0$ then $\log x = 0 \Rightarrow x = 1$ or $\log x - 1 = 0 \Rightarrow \log x = 1 \Rightarrow x = 10$.

    The solution is: $x = \{1, 10\}$.

52. $\ln(\ln x) = 0 \Rightarrow e^{\ln(\ln x)} = e^0 \Rightarrow \ln x = 1 \Rightarrow e^{\ln x} = e^1 \Rightarrow x = \{e\}$.

53. (a) Graph on a calculator $y_1 = e^{-2\ln x}$ and $y_2 = \dfrac{1}{16}$. The graphs intersect at: $x = \{4\}$.

    (b) From the graph, $e^{-2\ln x} < \dfrac{1}{16}$ for the interval: $(4, \infty)$.

    (c) From the graph, $e^{-2\ln x} > \dfrac{1}{16}$ for the interval: $(0, 4)$.

54. (a) Graph on a calculator $y_1 = \ln e^{\ln x} - \ln(x - 3)$ and $y_2 = \ln 2$. The graphs intersect at: $x = \{6\}$.

    (b) From the graph, $\ln e^{\ln x} - \ln(x - 3) > \ln 2$ for the interval: $(3, 6)$.

    (c) From the graph, $\ln e^{\ln x} - \ln(x - 3) < \ln 2$ for the interval: $(6, \infty)$.

55. The statement is incorrect. We must reject any solution that is not in the domain of any logarithmic function in

    the equation.

56. Graph $y = \ln x - \ln(x + 1) - \ln 5$ on the calculator. The equation has no solution because the graph of the

    equation does not intersect the x-axis.

57. If $1.5^{\log x} = e^5$ then $1.5^{\log x} - e^5 = 0$. From a calculator graph of $y_1 = 1.5^{\log x} - e^5$, the x-intercept or

    solution is: $x \approx \{17.106\}$.

58. If $1.5^{\ln x} = 10^5$ then $1.5^{\ln x} - 10^5 = 0$. From a calculator graph of $y_1 = 1.5^{\ln x} - 10^5$, the x-intercept or

    solution is: $x \approx \{17.106\}$.

59. $r = p - k \ln t \Rightarrow r - p = -k \ln t \Rightarrow \dfrac{r - p}{-k} = \ln t \Rightarrow \ln t = \dfrac{p - r}{k} \Rightarrow e^{\ln t} = e^{(p-r)/k} \Rightarrow t = e^{(p-r)/k}$

60. $p = a + \dfrac{k}{\ln x} \Rightarrow p - a = \dfrac{k}{\ln x} \Rightarrow \ln x\,(p - a) = k \Rightarrow \ln x = \dfrac{k}{p - a} \Rightarrow e^{\ln x} = e^{k/(p-a)} \Rightarrow x = e^{k/(p-a)}$

61. $T = T_0 + (T_1 - T_0)\,10^{-kt} \Rightarrow T - T_0 = (T_1 - T_0)\,10^{-kt} \Rightarrow \dfrac{T - T_0}{T_1 - T_0} = 10^{-kt} \Rightarrow$

$\log \dfrac{T - T_1}{T_1 - T_0} = \log 10^{-kt} \Rightarrow \log \dfrac{T - T_1}{T_1 - T_0} = -kt \Rightarrow t = -\dfrac{1}{k} \log \dfrac{T - T_1}{T_1 - T_0}$

62. $A = \dfrac{Pi}{1 - (1 + i)^{-n}} \Rightarrow A\left(1 - (1 + i)^{-n}\right) = Pi \Rightarrow 1 - (1 + i)^{-n} = \dfrac{Pi}{A} \Rightarrow -(1 + i)^{-n} = \dfrac{Pi}{A} - 1 \Rightarrow$

$(1 + i)^{-n} = 1 - \dfrac{Pi}{A} \Rightarrow (1 + i)^{-n} = \dfrac{A - Pi}{A} \Rightarrow \log(1 + i)^{-n} = \log\left(\dfrac{A - Pi}{A}\right) \Rightarrow$

$-n \log(1 + i) = \log\left(\dfrac{A - Pi}{A}\right) \Rightarrow -n = \dfrac{\log\left(\dfrac{A - Pi}{A}\right)}{\log(1 + i)} \Rightarrow n = -\dfrac{\log\left(\dfrac{A - Pi}{A}\right)}{\log(1 + i)}$

63. $A = T_0 + Ce^{-kt} \Rightarrow A - T_0 = Ce^{-kt} \Rightarrow \dfrac{A - T_0}{C} = e^{-kt} \Rightarrow \ln \dfrac{A - T_0}{C} = \ln e^{-kt} \Rightarrow$

$-kt = \ln \dfrac{A - T_0}{C} \Rightarrow k = \dfrac{\ln \dfrac{A - T_0}{C}}{-t}$

64. $y = \dfrac{K}{1 + ae^{-bx}} \Rightarrow y\left(1 + ae^{-bx}\right) = K \Rightarrow y + yae^{-bx} = K \Rightarrow yae^{-bx} = K - y \Rightarrow e^{-bx} = \dfrac{K - y}{ay} \Rightarrow$

$\ln e^{-bx} = \ln\left(\dfrac{K - y}{ay}\right) \Rightarrow -bx = \ln\left(\dfrac{K - y}{ay}\right) \Rightarrow b = \dfrac{\ln\left(\dfrac{K - y}{ay}\right)}{-x}$

65. $y = A + B(1 - e^{-Cx}) \Rightarrow y = A + B - Be^{-Cx} \Rightarrow y - A - B = -Be^{-Cx} \Rightarrow \dfrac{-y + A + B}{B} = e^{-Cx} \Rightarrow$

$\ln\left(\dfrac{A + B - y}{B}\right) = \ln e^{-Cx} \Rightarrow \ln\left(\dfrac{A + B - y}{B}\right) = -Cx \Rightarrow x = \dfrac{\ln\left(\dfrac{A + B - y}{B}\right)}{-C}.$

66. $m = 6 - 2.5 \log\left(\dfrac{M}{M_0}\right) \Rightarrow m - 6 = -2.5 \log\left(\dfrac{M}{M_0}\right) \Rightarrow \dfrac{m - 6}{-2.5} = \log\left(\dfrac{M}{M_0}\right) \Rightarrow \dfrac{6 - m}{2.5} = \log\left(\dfrac{M}{M_0}\right) \Rightarrow$

$10^{(6-m)/2.5} = 10^{\log(M/M_0)} \Rightarrow \dfrac{M}{M_0} = 10^{(6-m)/2.5} \Rightarrow M = M_0\left(10^{(6-m)/2.5}\right) \text{ or } M = M_0\left(10^{(m-6)/-2.5}\right)$

67. $\log A = \log B - C \log x \Rightarrow \log A = \log B - \log x^C \Rightarrow \log A = \log \dfrac{B}{x^C} \Rightarrow 10^{\log A} = 10^{\log(B/x^C)} \Rightarrow A = \dfrac{B}{x^C}$

68. $d = 10 \log\left(\dfrac{I}{I_0}\right) \Rightarrow \dfrac{d}{10} = \log\left(\dfrac{I}{I_0}\right) = 10^{(d/10)} = 10^{\log(I/I_0)} \Rightarrow 10^{(d/10)} = \left(\dfrac{I}{I_0}\right) \Rightarrow I = I_0 \cdot 10^{d/10}$

69. $A = P\left(1 + \dfrac{r}{n}\right)^{nt} \Rightarrow \dfrac{A}{P} = \left(1 + \dfrac{r}{n}\right)^{nt} \Rightarrow \log \dfrac{A}{P} = \log\left(1 + \dfrac{r}{n}\right)^{nt} \Rightarrow$

$\log \dfrac{A}{P} = nt \log\left(1 + \dfrac{r}{n}\right) \Rightarrow t = \dfrac{\log \dfrac{A}{P}}{n \log\left(1 + \dfrac{r}{n}\right)}$

70. $D = 160 + 10 \log x \Rightarrow D - 160 = 10 \log x \Rightarrow \dfrac{D}{10} - 16 = \log x \Rightarrow 10^{\log x} = 10^{((D/10)-16)} \Rightarrow$

$x = 10^{((D/10)-16)}.$

71. It is true because $(a^m)^n = a^{mn}$, and so $(e^x)^2 = e^{x \cdot 2} = e^{2x}$.

72. $(e^x)^2 - 4e^x + 3 = 0 \Rightarrow (e^x - 3)(e^x - 1) = 0$

73. $e^x - 3 = 0 \Rightarrow e^x = 3 \Rightarrow \ln e^x = \ln 3 \Rightarrow x = \ln 3$ or $e^x - 1 = 0 \Rightarrow e^x = 1 \Rightarrow \ln e^x = \ln 1 \Rightarrow$
    $x = \ln 1 \Rightarrow x = 0$. The solution is: $x = \{0, \ln 3\}$.

74. Graph $y = e^{2x} - 4e^x + 3$ on the calculator. See Figure 74. The x-intercepts are: 0 and $\ln 3 \approx 1.099$.

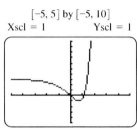

$[-5, 5]$ by $[-5, 10]$
Xscl $= 1$ Yscl $= 1$

Figure 74

75. From The graph, $e^{2x} - 4e^x + 3 > 0$ for the interval: $(-\infty, 0) \cup (\ln 3, \infty)$.

76. From The graph, $e^{2x} - 4e^x + 3 < 0$ for the interval: $(0, \ln 3)$.

77. For $x^2 = 2^x$, graph on a calculator $y_1 = x^2 - 2^x$. From the calculator the x-intercepts or solutions are:
    $\{-.767, 2, 4\}$.

78. For $x^2 - 4 = e^{x-4} + 4$, graph on a calculator $y_1 = x^2 - 4 - e^{x-4} - 4$. From the calculator the x-intercepts
    or solutions are: $\{-2.829, 2.886, 8.035\}$.

79. For $\log x = x^2 - 8x + 14$, graph on a calculator $y_1 = \log x - x^2 + 8x - 14$. From the calculator the
    x-intercepts or solutions are: $\{2.454, 5.659\}$.

80. For $\ln x = -\sqrt[3]{x + 3}$, graph on a calculator $y_1 = \ln x + \sqrt[3]{x + 3}$. From the calculator the x-intercept or
    solution is: $\{.228\}$.

81. For $e^x = \dfrac{1}{x + 2}$, graph on a calculator $y_1 = e^x - \dfrac{1}{x + 2}$. From the calculator the x-intercept or solution
    is: $\{-.443\}$.

82. For $3^{-x} = \sqrt{x + 5}$, graph on a calculator $y_1 = 3^{-x} - \sqrt{x + 5}$. From the calculator the x-intercept or solution
    is: $\{-.667\}$.

83. $\log_2 \sqrt{2x^2} - 1 = .5 \Rightarrow \log_2 \sqrt{2x^2} = 1.5 \Rightarrow 2^{\log_2 \sqrt{2x^2}} = 2^{3/2} \Rightarrow \sqrt{2x^2} = \sqrt{2^3} \Rightarrow$
    $2x^2 = 2^3 \Rightarrow 2x^2 = 8 \Rightarrow x^2 = 4 \Rightarrow x = \{-2, 2\}$.

84. $\log x^2 = (\log x)^2 \Rightarrow 2 \log x - (\log x)^2 = 0 \Rightarrow \log x (2 - \log x) = 0$, then $\log x = 0 \Rightarrow x = 1$ or
    $2 - \log x = 0 \Rightarrow -\log x = -2 \Rightarrow \log x = 2 \Rightarrow 10^{\log x} = 10^2 \Rightarrow x = 100$. The solutions are: $x = \{1, 100\}$.

85. $\ln(\ln e^{-x}) = \ln 3 \Rightarrow \ln(-x) = \ln 3 \Rightarrow -x = 3 \Rightarrow x = \{-3\}$.

86. $e^{x + \ln 3} = 4e^x \Rightarrow \ln e^{x + \ln 3} = \ln 4 + \ln e^x \Rightarrow x + \ln 3 = \ln 4 + x \Rightarrow \ln 3 = \ln 4$. Since this is always false,
    the solution is: $\varnothing$.

87. If $y = 2$, then $2 = \dfrac{2 - \log(100 - x)}{.42} \Rightarrow 2(.42) = 2 - \log(100 - x) \Rightarrow .84 = 2 - \log(100 - x) \Rightarrow$
    $-1.16 = -\log(100 - x) \Rightarrow 1.16 = \log(100 - x) \Rightarrow 10^{1.16} = 10^{\log(100 - x)} \Rightarrow 14.454398 = 100 - x \Rightarrow$
    $-85.5 = -x \Rightarrow x = 85.5\%$.

88. If $x = 75$, then $y = \dfrac{2 - \log(100 - 75)}{.42} \Rightarrow y = \dfrac{2 - \log 25}{.42} \Rightarrow y \approx 1.43$ years.

89. If $f(x) = 33$, then $33 = 31.5 + 1.1 \log(x + 1) \Rightarrow 1.5 = 1.1 \log(x + 1) \Rightarrow \dfrac{1.5}{1.1} = \log(x + 1) \Rightarrow$

    $10^{1.5/1.1} = 10^{\log(x+1)} \Rightarrow 23.1013 = x + 1 \Rightarrow x = 22.1013 \Rightarrow x \approx 22$ meters.

90. If $x = 500$, then $f(500) = 31.5 + 1.1 \log(500 + 1) \Rightarrow f(500) = 31.5 + 1.1 \log 501 \Rightarrow f(500) \approx 34.5$ g.

Reviewing Basic Concepts (Sections 5.4 and 5.5)

1. If $f(x) = 3^x$ and $g(x) = \log_3 x$, then f and g are inverse functions, and their graphs are symmetric with respect to the line with equation $y = x$. The domain of f is the range of g and visa versa.

2. See Figure 2.

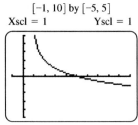

$[-1, 10]$ by $[-5, 5]$
Xscl = 1   Yscl = 1

Figure 2

3. From the graph, the asymptote is: $x = 1$; the x-intercepts is: 5; and there is no y-intercepts.

4. The graph of $f(x)$ is the same as the graph of $g(x)$, reflected across the x-axis, shifted 1 unit to the right, and shifted two units upward.

5. For $f(x) = 2 - \log_2(x - 1)$, the inverse is: $x = 2 - \log_2(y - 1) \Rightarrow x - 2 = -\log_2(y - 1) \Rightarrow$
    $2 - x = \log_2(y - 1) \Rightarrow 2^{(2-x)} = 2^{\log_2(y-1)} \Rightarrow 2^{(2-x)} = y - 1 \Rightarrow y = 1 + 2^{(2-x)} \Rightarrow f^{-1}(x) = 1 + 2^{2-x}$.

6. $3^{2x-1} = 4^x \Rightarrow \log 3^{2x-1} = \log 4^x \Rightarrow (2x - 1)\log 3 = x \log 4 \Rightarrow 2x \log 3 - \log 3 = x \log 4 \Rightarrow$

    $2x \log 3 - x \log 4 = \log 3 \Rightarrow x(2 \log 3 - \log 4) = \log 3 \Rightarrow x = \left\{\dfrac{\log 3}{2 \log 3 - \log 4}\right\}$ or $x = \left\{\dfrac{\log 3}{\log \frac{9}{4}}\right\}$.

7. $\ln 5x - \ln(x + 2) = \ln 3 \Rightarrow \ln\left(\dfrac{5x}{x + 2}\right) = \ln 3 \Rightarrow \dfrac{5x}{x + 2} = 3 \Rightarrow 3(x + 2) = 5x \Rightarrow$

    $3x + 6 = 5x \Rightarrow 6 = 2x \Rightarrow x = \{3\}$.

8. $10^{5 \log x} = 32 \Rightarrow \log 10^{5 \log x} = \log 32 \Rightarrow 5 \log x = \log 32 \Rightarrow \log x^5 = \log 32 \Rightarrow x^5 = 32 \Rightarrow x = \{2\}$.

9. $H = 1000(1 - e^{-kN}) \Rightarrow \dfrac{H}{1000} = 1 - e^{-kN} \Rightarrow 1 - \dfrac{H}{1000} = e^{-kN} \Rightarrow \ln\left(1 - \dfrac{H}{1000}\right) = \ln e^{-kN} \Rightarrow$

    $\ln\left(1 - \dfrac{H}{1000}\right) = -kN \Rightarrow N = -\dfrac{1}{k} \ln\left(1 - \dfrac{H}{1000}\right)$.

10. If $f(x) = 2300$, then $2300 = 280 \ln(x + 1) + 1925 \Rightarrow 375 = 280 \ln(x + 1) \Rightarrow \dfrac{375}{280} = \ln(x + 1) \Rightarrow$

    $e^{375/280} = e^{\ln(x+1)} \Rightarrow e^{375/280} = x + 1 \Rightarrow x = e^{375/280} - 1 \Rightarrow x \approx 2.8$ acres.

5.6: Further Applications & Modeling with Exponential & Logarithmic Functions

1. Using the given function, $\frac{1}{3}A_o = A_o e^{-.0001216t} \Rightarrow \frac{1}{3} = e^{-.0001216t} \Rightarrow \ln \frac{1}{3} = \ln e^{-.0001216t} \Rightarrow$

   $\ln \frac{1}{3} = -.0001216t \Rightarrow t = \dfrac{\ln \left(\dfrac{1}{3} \right)}{-.0001216} \Rightarrow t \approx 9034.6 \approx 9,000$ years ago.

2. Using the given function, $.6A_o = A_o e^{-.0001216t} \Rightarrow .6 = e^{-.0001216t} \Rightarrow \ln .6 = \ln e^{-.0001216t} \Rightarrow$

   $\ln .6 = -.0001216t \Rightarrow t = \dfrac{\ln (.6)}{-.0001216} \Rightarrow t \approx 4200.9 \approx 4,200$ years old.

3. Using the given function, $.15A_o = A_o e^{-.0001216t} \Rightarrow .15 = e^{-.0001216t} \Rightarrow \ln .15 = \ln e^{-.0001216t} \Rightarrow$

   $\ln .15 = -.0001216t \Rightarrow t = \dfrac{\ln (.15)}{-.0001216} \Rightarrow t \approx 15601.3 \approx 16,000$ years old.

4. Using the given function, $.2A_o = A_o e^{-.0001216t} \Rightarrow .2 = e^{-.0001216t} \Rightarrow \ln .2 = \ln e^{-.0001216t} \Rightarrow$

   $\ln .2 = -.0001216t \Rightarrow t = \dfrac{\ln (.2)}{-.0001216} \Rightarrow t \approx 13235.5 \approx 13,000$ years old.

5. (a) With a half-life of 21.7 years and using the exponential decay function, our model is: $\frac{1}{2}A_0 = A_0 e^{-21.7k} \Rightarrow$

   $.5 = e^{-21.7k} \Rightarrow \ln .5 = \ln e^{-21.7k} \Rightarrow \ln .5 = -21.7k \Rightarrow k = \dfrac{\ln .5}{-21.7} \Rightarrow k \approx .032.$

   The exponential decay model is: $A(t) = A_0 e^{-.032t}$.

   (b) $400 = 500e^{-.032t} \Rightarrow \dfrac{400}{500} = e^{-.032t} \Rightarrow \ln .8 = \ln e^{-.032t} \Rightarrow \ln .8 = -.032t \Rightarrow \dfrac{\ln .8}{-.032} = t \Rightarrow t \approx 6.97$ yr.

   (c) $A(10) = 500e^{-.032(10)} \Rightarrow A(10) = 500e^{-.32} \Rightarrow A(10) \approx 363$ grams.

6. (a) $A(50) = 100e^{-.02295(50)} \Rightarrow A(50) = 100e^{-1.1475} \Rightarrow A(50) \approx 31.7$ milligrams. Since this is less than half of

   the original amount, the half-life is less than 50 years.

   (b) Graph $y_1 = 100e^{-.02295x}$ and $y_2 = 50$ on the calculator. The intersection of the graphs is the approximate

   half-life: 30.2 years.

7. (a) Since the amount remaining is less than half of the original 2 milligrams, the half-life is less than 200 days.

   (b) Using the exponential decay function, $A_0 = 2$, and the ordered pair $(100, 1.22)$ from the table, yields the

   formula: $1.22 = 2e^{-100k} \Rightarrow .61 = e^{-100k} \Rightarrow \ln .61 = \ln e^{-100k} \Rightarrow \ln .61 = -100k \Rightarrow k = \dfrac{\ln .61}{-100} \Rightarrow$

   $k \approx .005$. The formula that models the data is: $A = 2e^{-.005t}$.

   (c) Graph $y_1 = 2e^{-.005t}$ and $y_2 = 1$ on the calculator. The intersection of the graphs is the approximate

   half-life: 140 days.

8. (a) $d = 10 \log \dfrac{3.162 \times 10^{10} I_0}{I_0} \Rightarrow d = 10 \log (3.162 \times 10^{10}) \Rightarrow d \approx 105$ decibels.

   (b) $d = 10 \log \dfrac{1.995 \times 10^9 I_0}{I_0} \Rightarrow d = 10 \log (1.995 \times 10^9) \Rightarrow d \approx 93$ decibels.

   (c) $d = 10 \log \dfrac{10^{14} I_0}{I_0} \Rightarrow d = 10 \log (10^{14}) \Rightarrow d \approx 140$ decibels.

9. (a) $7.9 = \log\dfrac{I}{I_0} \Rightarrow 10^{7.9} = 10^{\log I/I_0} \Rightarrow 10^{7.9} = \dfrac{I}{I_0} \Rightarrow I = 10^{7.9}I_0 \Rightarrow I \approx 79{,}000{,}000\,I_0.$

(b) $6.1 = \log\dfrac{I}{I_0} \Rightarrow 10^{6.1} = 10^{\log I/I_0} \Rightarrow 10^{6.1} = \dfrac{I}{I_0} \Rightarrow I = 10^{6.1}I_0 \Rightarrow I \approx 1{,}300{,}000\,I_0.$

(c) Comparing the 6.1 earthquake to a 6.0 earthquake yields: $\dfrac{10^{6.1}I_0}{10^{6.0}I_0} \approx 1.26$ times stronger.

10. (a) $9.0 = \log\dfrac{I}{I_0} \Rightarrow 10^{9} = 10^{\log I/I_0} \Rightarrow 10^{9} = \dfrac{I}{I_0} \Rightarrow I = 10^{9}I_0 \Rightarrow I \approx 1{,}000{,}000{,}000\,I_0.$

(b) $6.8 = \log\dfrac{I}{I_0} \Rightarrow 10^{6.8} = 10^{\log I/I_0} \Rightarrow 10^{6.8} = \dfrac{I}{I_0} \Rightarrow I = 10^{6.8}I_0 \Rightarrow I \approx 6{,}300{,}000\,I_0.$

(c) Comparing the 9.0 earthquake to a 6.8 earthquake yields: $\dfrac{10^{9.0}I_0}{10^{6.8}I_0} \approx 160$ times stronger.

11. $t = (1.26 \times 10^9)\left(\dfrac{\ln\left[1 + 8.33\,(.103)\right]}{\ln 2}\right) \Rightarrow t = 1.126 \times 10^9.$ The rock sample is about 1.126 billion years old.

12. Given $P(h) = 10$, find h, 10% of 14.7 is $.10 \times 14.7 = 1.47 \Rightarrow 1.47 = 14.7e^{-.0000385h} \Rightarrow \dfrac{1.47}{14.7} = e^{-.0000385h} \Rightarrow$

$\ln 0.1 = \ln e^{-.0000385h} \Rightarrow \ln 0.1 = -.0000385h \Rightarrow h = \dfrac{ln\,0.1}{-.0000385} \Rightarrow h \approx 59{,}807.41$ or about 59,800 feet.

13. Use the equation to find the intensity of a star magnitude 1: $1 = 6 - \dfrac{5}{2}\log\dfrac{I}{I_0} \Rightarrow -5 = -\dfrac{5}{2}\log\dfrac{I}{I_0} \Rightarrow$

$2 = \log\dfrac{I}{I_0} \Rightarrow 10^2 = 10^{\log I/I_0} \Rightarrow 100 = \dfrac{I}{I_0} \Rightarrow I = 100\,I_0.$

Now use the equation to find the intensity of a star magnitude 3: $3 = 6 - \dfrac{5}{2}\log\dfrac{I}{I_0} \Rightarrow -3 = -\dfrac{5}{2}\log\dfrac{I}{I_0} \Rightarrow$

$\dfrac{6}{5} = \log\dfrac{I}{I_0} \Rightarrow 10^{6/5} = 10^{\log I/I_0} \Rightarrow 10^{6/5} = \dfrac{I}{I_0} \Rightarrow I \approx 15.85\,I_0.$

Comparing intensities of the magnitude 1 star to the magnitude 3 star yields: $\dfrac{100\,I_0}{15.85\,I_0} \approx 6.31.$

The magnitude 1 star is approximately 6.3 times more intense then the magnitude 3 star.

14. First solve for k, using $f(x) = 50$, $T_0 = 0$, $t = 24$ and $C = 100$: $50 = 0 + 100e^{-k(24)} \Rightarrow 50 = 100e^{-24k} \Rightarrow$

$.5 = e^{-24k} \Rightarrow \ln .5 = \ln e^{-24k} \Rightarrow \ln .5 = -24k \Rightarrow k = \dfrac{\ln .5}{-24} \Rightarrow k \approx .02888.$ Now the equation is:

$f(t) = 100e^{-.02888t}$ and we find the temperature after 96 minutes: $f(96) = 100e^{-.02888(96)} \Rightarrow f(96) \approx 6.25^\circ C.$

15. (a) First solve for k, using $f(x) = 60$, $T_0 = 20$, $t = 1$ and $C = 100 - 20 = 80$: $60 = 20 + 80e^{-k(1)} \Rightarrow$

$40 = 80e^{-k} \Rightarrow .5 = e^{-k} \Rightarrow \ln .5 = \ln e^{-k} \Rightarrow \ln .5 = -k \Rightarrow k = -\ln .5 \Rightarrow k \approx .693.$ The equation is:

$f(t) = 20 + 80e^{-.693t}.$

(b) $f(.5) = 20 + 80e^{-.693(.5)} \Rightarrow f(.5) \approx 76.6^\circ C.$

(c) Solve for t, with $f(t) = 50$, $50 = 20 + 80e^{-.693t} \Rightarrow 30 = 80e^{-.693t} \Rightarrow \dfrac{30}{80} = e^{-.693t} \Rightarrow$

$\ln .375 = \ln e^{-.693t} \Rightarrow \ln .375 = -.693t \Rightarrow t = \dfrac{\ln .375}{-.693} \Rightarrow t \approx 1.415$ or about 1 hour 25 minutes.

16. First solve for k, using $f(x) = 175$, $T_0 = 50$, $t = 4$ and $C = 300 - 50 = 250$: $175 = 50 + 250e^{-k(4)} \Rightarrow$

$125 = 250e^{-4k} \Rightarrow .5 = e^{-4k} \Rightarrow \ln .5 = \ln e^{-4k} \Rightarrow \ln .5 = -4k \Rightarrow k = \dfrac{\ln .5}{-4} \Rightarrow k \approx .1733.$

The equation is: $f(t) = 50 + 250e^{-.1733t}.$ Now use the equation to find the temperature after 12 minutes:

$f(12) = 50 + 250e^{-.1733(12)} \Rightarrow f(12) \approx 81.25^\circ C.$

17. (a) Set $L_1 \rightarrow$ Year and $L_2 \rightarrow CFC$ 12 (ppb) and use the exponential regression capabilities of the calculator

    to find the equation: $y \approx .72(1.041)^x$. Therefore the values are: $C \approx .72$ and $a \approx 1.041$.

    (b) Use the equation from part a, to find $f(13)$: $f(13) = .72(1.041)^{13} \Rightarrow f(13) \approx 1.21$ ppb.

18. (a) Setting the year $1800 = 0$, gives us the ordered pair $(0, 2)$, now using this and the model: $R(x) = Ce^{kx}$,

    solve for C and k: $.2 = Ce^{0k} \Rightarrow C = .2$ and using the year 2000 (200, 2.4) find k: $2.4 = .2e^{200k} \Rightarrow$

    $12 = e^{200k} \Rightarrow \ln 12 = \ln e^{200k} \Rightarrow \ln 12 = 200k \Rightarrow k = \dfrac{\ln 12}{200} \Rightarrow k \approx .0124$. (Answers may vary)

    (b) Using the values from part a, gives us the equation: $R(x) = .2e^{.0124x}$. Use this and $R(x) = 3$ to solve for x.

    $3 = .2e^{.0124x} \Rightarrow 15 = e^{.0124x} \Rightarrow \ln 15 = \ln e^{.0124x} \Rightarrow \ln 15 = .0124x \Rightarrow x = \dfrac{\ln 15}{.0124} \Rightarrow x \approx 218$.

    The year is: $1800 + 218 = 2018$.

19. (a) Since $T(R) = 1.03R$, and $R(x) = .2e^{.0124x}$, then $(T \circ R)(x) = T[R(x)] = T(.2e^{.0124x}) = 1.03(.2e^{.0124x}) \Rightarrow$

    $(T \circ R)(x) = .206e^{.0124}$.

    (b) $(T \circ R)(100) = .206e^{.0124(100)} \Rightarrow (T \circ R)(x) \approx .7119$. In 1900 radiative forcing caused an approximately

    $.7 ¡ F$ increase in the average global temperature.

20. If $x = 60$, then $\ln (1 - P) = -.0034 - .0053(60) \Rightarrow \ln (1 - P) = -.3214 \Rightarrow e^{\ln (1-P)} = e^{-.3214} \Rightarrow$

    $1 - P = e^{-.3214} \Rightarrow -P = -1 + e^{-.3214} \Rightarrow P = 1 - e^{-.3214} \Rightarrow P \approx .275$. If a \$60 tax is placed on each ton of

    carbon burned, carbon dioxide emissions could decrease by 27.5%.

21. (a) Use the Compound Interest Formula $A = P\left(1 + \dfrac{r}{n}\right)^{nt}$, $5000 = 1000\left(1 + \dfrac{.035}{4}\right)^{4t} \Rightarrow 5 = 1.00875^{4t} \Rightarrow$

    $\log 5 = \log 1.00875^{4t} \Rightarrow \log 5 = 4t \log 1.00875 \Rightarrow t = \dfrac{\log 5}{4 \log 1.00875} \Rightarrow t \approx 46.2$ years.

    (b) Use the Continuous Compound Interest Formula $A = Pe^{rt}$, $5000 = 1000e^{.035t} \Rightarrow 5 = e^{.035t} \Rightarrow$

    $\ln 5 = \ln e^{.035t} \Rightarrow \ln 5 = .035t \Rightarrow t = \dfrac{\ln 5}{.035} \Rightarrow t \approx 46.0$ years.

22. (a) Use the Compound Interest Formula $A = P\left(1 + \dfrac{r}{n}\right)^{nt}$, $8400 = 5000\left(1 + \dfrac{.06}{2}\right)^{2t} \Rightarrow \dfrac{8400}{5000} = 1.03^{2t} \Rightarrow$

    $1.68 = 1.03^{2t} \Rightarrow \log 1.68 = \log 1.03^{2t} \Rightarrow \log 1.68 = 2t \log 1.03 \Rightarrow t = \dfrac{\log 1.68}{2 \log 1.03} \Rightarrow t \approx 8.78$ years.

    (b) Use the Continuous Compound Interest Formula $A = Pe^{rt}$, $8400 = 5000e^{.06t} \Rightarrow \dfrac{8400}{5000} = e^{.06t} \Rightarrow$

    $1.68 = e^{.06t} \Rightarrow \ln 1.68 = \ln e^{.06t} \Rightarrow \ln 1.68 = .06t \Rightarrow t = \dfrac{\ln 1.68}{.06} \Rightarrow t \approx 8.65$ years.

23. Use the Compound Interest Formula $A = P\left(1 + \dfrac{r}{n}\right)^{nt}$, $30,000 = 27,000\left(1 + \dfrac{.06}{4}\right)^{4t} \Rightarrow \dfrac{30,000}{27,000} = 1.015^{4t} \Rightarrow$

    $\dfrac{10}{9} = 1.015^{4t} \Rightarrow \log \dfrac{10}{9} = \log 1.015^{4t} \Rightarrow \log \dfrac{10}{9} = 4t \log 1.015 \Rightarrow t = \dfrac{\log \dfrac{10}{9}}{4 \log 1.015} \Rightarrow t \approx 1.8$ years.

24. (a) Use the Compound Interest Formula $A = P\left(1 + \dfrac{r}{n}\right)^{nt}$, $2P = P\left(1 + \dfrac{.025}{4}\right)^{4t} \Rightarrow 2 = 1.00625^{4t} \Rightarrow$

    $\log 2 = \log 1.00625^{4t} \Rightarrow \log 2 = 4t \log 1.00625 \Rightarrow t = \dfrac{\log 2}{4 \log 1.00625} \Rightarrow t \approx 27.81$ years.

    (b) Use the Continuous Compound Interest Formula $A = Pe^{rt}$, $2P = Pe^{.025t} \Rightarrow 2 = e^{.025t} \Rightarrow$

    $\ln 2 = \ln e^{.025t} \Rightarrow \ln 2 = .025t \Rightarrow t = \dfrac{\ln 2}{.025} \Rightarrow t \approx 27.73$ years.

25. Use the Compound Interest Formula $A = P\left(1 + \dfrac{r}{n}\right)^{nt}$, $A = 60{,}000\left(1 + \dfrac{.07}{4}\right)^{4(5)} \Rightarrow A = 60{,}000(1.0175)^{20} \Rightarrow$

    $A = \$84{,}886.69$. Now use the Continuous Compound Interest Formula $A = Pe^{rt}$, $A = 60{,}000e^{.0675(5)} \Rightarrow$

    $A \approx \$84{,}086.38$. The better investment is the 7% rate compounded quarterly; it will earn $800.31 more.

26. Use the Continuous Compound Interest Formula $A = Pe^{rt}$, $80{,}000 = 60{,}000e^{.0675t} \Rightarrow \dfrac{80{,}000}{60{,}000} = e^{.0675t} \Rightarrow$

    $\dfrac{4}{3} = e^{.0675t} \Rightarrow \ln \dfrac{4}{3} = \ln e^{.0675t} \Rightarrow \ln \dfrac{4}{3} = .0675t \Rightarrow t = \dfrac{\ln \frac{4}{3}}{.0675} \Rightarrow t \approx 4.3$ years.

27. $R = \left(1 + \dfrac{.06}{4}\right)^{4} - 1 \Rightarrow R \approx 1.015^{4} - 1 \Rightarrow R \approx .06136$ or $R \approx 6.14\%$.

28. $R = \left(1 + \dfrac{.045}{365}\right)^{365} - 1 \Rightarrow R \approx 1.000123288^{365} - 1 \Rightarrow R \approx .04602$ or $R \approx 4.60\%$.

29. $P = 10{,}000\left(1 + \dfrac{.12}{2}\right)^{-2(5)} \Rightarrow P = 10{,}000(1.06)^{-10} \Rightarrow P = \$5583.95$.

30. $P = 25{,}000\left(1 + \dfrac{.06}{4}\right)^{-4(2.75)} \Rightarrow P = 25{,}000(1.015)^{-11} \Rightarrow P = \$21{,}223.33$.

31. $25{,}000 = 31{,}360\left(1 + \dfrac{r}{1}\right)^{-1(2)} \Rightarrow \dfrac{25{,}000}{31{,}360} = (1 + r)^{-2} \Rightarrow \left(\dfrac{25{,}000}{31{,}360}\right)^{-1/2} = 1 + r \Rightarrow r \approx .1199776$ or about 12%.

32. $1200 = 1780\left(1 + \dfrac{r}{4}\right)^{-4(5)} \Rightarrow \dfrac{1200}{1780} = \left(1 + \dfrac{r}{4}\right)^{-20} \Rightarrow \left(\dfrac{1200}{1780}\right)^{-1/20} = \left(1 + \dfrac{r}{4}\right) \Rightarrow \left(\dfrac{1200}{1780}\right)^{-1/20} - 1 = \dfrac{r}{4} \Rightarrow$

    $r = 4\left[\left(\dfrac{1200}{1780}\right)^{-1/20} - 1\right] \Rightarrow r \approx .0796$ or $r \approx 8\%$.

33. (a) Using $R = \dfrac{P}{\dfrac{1 - (1 + i)^{-n}}{i}}$, and $n = 12(4) = 48$, $i = \dfrac{.075}{12} = .00625$ yields: $R = \dfrac{8500}{\dfrac{1 - (1 + .00625)^{-48}}{.00625}} \Rightarrow$

    $R \approx \$205.52$.

    (b) The total interest paid will be $I = nR - P \Rightarrow I = 48(205.52) - 8500 \Rightarrow I \approx \$1364.96$.

34. (a) Using $R = \dfrac{P}{\dfrac{1 - (1 + i)^{-n}}{i}}$, and $n = 12(5) = 60$, $i = \dfrac{.092}{12}$ yields: $R = \dfrac{9600}{\dfrac{1 - (1 + .092/12)^{-60}}{.092/12}} \Rightarrow$

    $R \approx \$200.21$.

    (b) The total interest paid will be $I = nR - P \Rightarrow I = 60(200.21) - 9600 \Rightarrow I \approx \$2412.60$.

35. (a) Using $R = \dfrac{P}{\dfrac{1 - (1 + i)^{-n}}{i}}$, and $n = 12(30) = 360$, $i = \dfrac{.0725}{12}$ yields: $R = \dfrac{125{,}000}{\dfrac{1 - (1 + .0725/12)^{-360}}{.0725/12}} \Rightarrow$

    $R \approx \$852.72$.

    (b) The total interest paid will be $I = nR - P \Rightarrow I = 360(852.72) - 125{,}000 \Rightarrow I \approx \$181{,}979.20$.

36. Using $y = R\left[\dfrac{1 - (1 + i)^{-(n-x)}}{i}\right]$, and $n = 360$, $x = 120$, $i = \dfrac{.0725}{12}$, $R = 852.72$ yields:

$$y = 852.72\left[\dfrac{1 - \left(1 + \dfrac{.0725}{12}\right)^{-(360-120)}}{\dfrac{.0725}{12}}\right] \Rightarrow y \approx \$107,887.89.$$

37. (a) First enter $y_1 = 1500\left(1 + \dfrac{.0575}{365}\right)^{365t}$ into the calculator, now using the table feature, the investment will

triple when $y_1 = 4500$. From the table, $t \approx 19.1078 \Rightarrow t \approx 19$ years $+ .1078(365)$ days $\Rightarrow$

$t \approx 19$ years, 39 days.

Analytically, for $y = 4500$, the solution is: $4500 = 1500\left(1 + \dfrac{.0575}{365}\right)^{365t} \Rightarrow 3 = (1.000157534)^{365t} \Rightarrow$

$\ln 3 = 365t \ln 1,000157534 \Rightarrow t = \dfrac{\ln 3}{365 \ln 1.000157534} \Rightarrow t \approx 19.1078$ or 19 years and 39 days.

(b) First enter $y_1 = 2000\left(1 + \dfrac{.08}{365}\right)^{365t}$ into the calculator using the table feature. From the table,

$t \approx 11.455 \Rightarrow t \approx 11$ years $+ .455(365)$ days $\Rightarrow t \approx 11$ years, 166 days.

38. Use the Continuous Compound Interest Formula $A = Pe^{rt}$, $2P = Pe^{rt} \Rightarrow 2 = e^{rt} \Rightarrow \ln 2 = \ln e^{rt} \Rightarrow$

$t = \dfrac{\ln 2}{r}$. Thus, t is inversely proportional to r, so when the rate is tripled, the time required for an investment

to double will be divided by 3. The time will be $\dfrac{1}{3}$ as long.

39. (a) $A(5) = 2,400,000e^{.023(5)} \Rightarrow A(5) \approx 2,692,496 \Rightarrow A(5) \approx 2,700,000$.

(b) $A(10) = 2,400,000e^{.023(10)} \Rightarrow A(10) \approx 3,020,640 \Rightarrow A(10) \approx 3,000,000$.

(c) $A(60) = 2,400,000e^{.023(60)} \Rightarrow A(60) \approx 9,539,764 \Rightarrow A(60) \approx 9,500,000$.

40. (a) Since $f(t)$ is in millions, $f(t) = 3000$. Now $3000 = 500e^{.1t} \Rightarrow 6 = e^{.1t} \Rightarrow \ln 6 = \ln e^{.1t} \Rightarrow$

$\ln 6 = .1t \Rightarrow t = \dfrac{\ln 6}{.1} \Rightarrow t \approx 17.92 \Rightarrow t \approx 18$ days.

(b) $1 + 18 = 19$, January 19th.

41. If $P(t) = 10$, then $10 = 100e^{-0.1t} \Rightarrow 0.1 = e^{-0.1t} \Rightarrow \ln 0.1 = \ln e^{-0.1t} \Rightarrow \ln 0.1 = -0.1t \Rightarrow$

$t = \dfrac{\ln 0.1}{-0.1} \Rightarrow t \approx 23$ days.

42. If $P(t) = 1$, then $1 = 100e^{-0.1t} \Rightarrow 0.01 = e^{-0.1t} \Rightarrow \ln 0.01 = \ln e^{-0.1t} \Rightarrow \ln 0.01 = -0.1t \Rightarrow$

$t = \dfrac{\ln 0.01}{-0.1} \Rightarrow t \approx 46.052 \Rightarrow t = 46$ days.

43. Drug effectiveness can be modeled by $E(t) = D_0 P^{t-1}$ where $D_0 =$ original dose, $P =$ percent of effectiveness

for each hour, and $t =$ hours since taken. Using this yields: $E(6) = 250\,(.75)^{6-1} \Rightarrow E(6) = 250(.75)^5 \Rightarrow$

$E(6) \approx 59.32617$ or $E(6) \approx 59$ mg remain.

44. Drug effectiveness can be modeled by $E(t) = D_0 P^{t-1}$ where D_0 = original dose, P = percent of effectiveness

for each hour, and t = hours since taken. Using this yields: $20 = 100\,(.8)^{t-1} \Rightarrow 0.2 = (.8)^{(t-1)} \Rightarrow$

$\log 0.2 = \log 0.8^{(t-1)} \Rightarrow \log 0.2 = (t-1)\log 0.8 \Rightarrow t - 1 = \dfrac{\log 0.2}{\log 0.8} \Rightarrow t = \dfrac{\log 0.2}{\log 0.8} + 1 \Rightarrow$

$t \approx 8.2126 \Rightarrow t \approx 8$. A new dose is needed at 8 hours.

45. (a)

x	0	15	30	45	60	75	90	125
$g(x)$	7	21	57	111	136	158	164	178

(b) Since there are 261 people in the village the equation is: $g(x) = 261 - f(x)$.

(c) Graph the data on the same calculator graph as $y_1 = \dfrac{171}{1 + 18.6e^{-.0747x}}$ and $y_2 = 18.3(1.024)^x$. From the

graph, y_1 is the better fit.

(d) Using $g(x) = 261 - f(x)$ and $g(x) = \dfrac{171}{1 + 18.6e^{-.0747x}}$ yields: $f(x) = 261 - \dfrac{171}{1 + 18.6e^{-.0747x}}$.

46. (a) Set $L_1 \rightarrow$ Age and $L_2 \rightarrow$ Death Rate and graph a scatter diagram in the window $[25, 75]$ by $[-100, 700]$.

See Figure 46.

(b) Use the capabilities of the exponential regression calculator and L_1 and L_2, to find the equation:

$f(x) \approx .3525(1.1148)^x$.

(c) $f(80) = .3525\,(1.1148)^{80} \Rightarrow f(80) \approx 2103$ deaths / 100,000 people.

47. (a) Set $L_1 \rightarrow$ Year and $L_2 \rightarrow$ Fees (in billions) and use the exponential regression capabilities of the calculator

to find the equation: $f(x) \approx 10.98(1.14)^x$.

(b) See Figure 47.

(c) $f(13) = 10.98\,(1.14)^{13} \Rightarrow f(13) \approx \$60$ billion.

48. Set $L_1 \rightarrow$ Year and $L_2 \rightarrow$ Telecommuters and use the natural logarithmic regression capabilities of the

calculator to find the equation: $f(x) = 19.962 + 18.335 \ln x$.

49. (a) For 2004, $x = 5$, then $f(5) = 9.8e^{.1345x} \Rightarrow f(x) = 9.8e^{.1345(5)} \Rightarrow f(5) \approx \$19.2$ billion.

(b) $14.7 = 9.8e^{.1345x} \Rightarrow \dfrac{14.7}{9.8} = e^{.1345x} \Rightarrow \ln 1.5 = \ln e^{.1345x} \Rightarrow \ln 1.5 = .1345x \Rightarrow x = \dfrac{\ln 1.5}{.1345} \Rightarrow x \approx 3$,

therefore the year is: $1999 + 3 = 2002$.

(c) Graph $y_1 = 9.8e^{.1345x}$. See Figure 49. The value given when $x = 5$ is about 19.2 which supports the

answer in part (a).

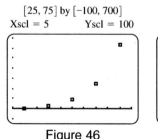

Figure 46

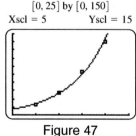

Figure 47

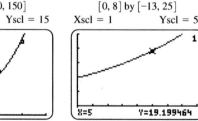

Figure 49

50. (a) Using the equation model and the ordered pairs: $(2000, 1.003)$ and $(2025, 1.362)$ solve for C and a.

Therefore, $1.003 = Ca^{2000-2000} \Rightarrow 1.003 = C(1) \Rightarrow C = 1.003$. Now using $C = 1.003$ and the second

ordered pair yields: $1.362 = 1.003a^{2025-2000} \Rightarrow \dfrac{1.362}{1.003} = a^{25} \Rightarrow \left(\dfrac{1.362}{1,003}\right)^{1/25} = a \Rightarrow a \approx 1.01231$.

(b) $f(2020) = 1.003\,(1.01231)^{2020-2000} \Rightarrow f(2020) \approx 1.281$ billion.

(c) $1.5 = 1.003\,(1.01231)^{x-2000} \Rightarrow \dfrac{1.5}{1.003} = 1.01231^{x-2000} \Rightarrow \ln\dfrac{1.5}{1.003} = \ln 1.01231^{x-2000} \Rightarrow$

$\ln\dfrac{1.5}{1.003} = (x-2000)\ln 1.01231 \Rightarrow x - 2000 = \dfrac{\ln\frac{1.5}{1.003}}{\ln 1.01231} \Rightarrow x = \dfrac{\ln\frac{1.5}{1.003}}{\ln 1.01231} + 2000 \Rightarrow$

$x \approx 2032$ or the population will hit 1.5 billion in approximately the year 2032.

51. (a) Set $L_1 \to$ Year and $L_2 \to$ Number of Midair Collisions and use the exponential regression capabilities of

the calculator to find C and a. From the regression equation, $C \approx 236.16$ and $a \approx .878$.

(b) Graph the data and the equation $f(x) = 236.16\,(.878)^{x-2000}$ in the same window. See Figure 51.

52. (a) $5.9 = 5.33e^{.0133x} \Rightarrow \dfrac{5.9}{5.33} = e^{.0133x} \Rightarrow \ln\dfrac{5.9}{5.33} = \ln e^{.0133x} \Rightarrow \ln\dfrac{5.9}{5.33} = .0133x \Rightarrow$

$x = \dfrac{\ln\dfrac{5.9}{5.33}}{.0133} \Rightarrow x \approx 8$, which is the year $1995 + 8 = 2003$.

(b) Graph $y_1 = 5.33e^{.0133x}$ and $y_2 = 5.9$ in the same window. See figure 52. The intersection is: $x \approx 8$.

(c) $6.25 = 5.33e^{.0133x} \Rightarrow \dfrac{6.25}{5.33} = e^{.0133x} \Rightarrow \ln\dfrac{6.25}{5.33} = \ln e^{.0133x} \Rightarrow \ln\dfrac{6.25}{5.33} = .0133x \Rightarrow$

$x = \dfrac{\ln\dfrac{6.25}{5.33}}{.0133} \Rightarrow x \approx 12$, which is the year $1995 + 12 = 2007$.

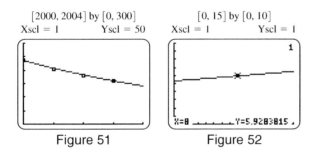

$[2000, 2004]$ by $[0, 300]$ $[0, 15]$ by $[0, 10]$
Xscl = 1   Yscl = 50   Xscl = 1   Yscl = 1

Figure 51   Figure 52

53. (a) $f(25) = \dfrac{.9}{1 + 271e^{-.122(25)}} \Rightarrow f(25) \approx .065$; and $f(65) = \dfrac{.9}{1 + 271e^{-.122(65)}} \Rightarrow f(65) \approx .82$

Among people age 25, 6.5% have some CHD, while among people age 65, 82% have some CHD.

(b) $.50 = \dfrac{.9}{1 + 271e^{-.122x}} \Rightarrow .5\left(1 + 271e^{-.122x}\right) = .9 \Rightarrow 1 + 271e^{-.122x} = \dfrac{.9}{.5} \Rightarrow$

$271e^{-.122x} = \dfrac{.9}{.5} - 1 \Rightarrow e^{-.122x} = \dfrac{\frac{.9}{.5} - 1}{271} \Rightarrow \ln e^{-.122x} = \ln\left(\dfrac{\frac{.9}{.5}-1}{271}\right) \Rightarrow -.122x = \ln\left(\dfrac{\frac{.9}{.5}-1}{271}\right) \Rightarrow$

$x = \dfrac{\ln\left(\dfrac{\frac{.9}{.5}-1}{271}\right)}{-.122} \Rightarrow x \approx 47.75$. The likelihood is 50% at about age 48.

54. (a) See Figure 54a. The maximum height appears to be 50 feet.

  (b) See Figure 54b. From the graph the asymptote is: $y = 50$. It tells us that this kind of tree can not grow taller than 50 feet.

  (c) Graphing $y_2 = 30$ in the same window gives an intersection at $x = 19.4$. The tree reaches 30 feet after about 19.4 years.

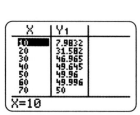

Figure 54a

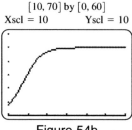

$[10, 70]$ by $[0, 60]$
Xscl $= 10$ Yscl $= 10$

Figure 54b

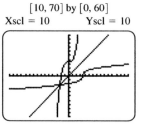

$[10, 70]$ by $[0, 60]$
Xscl $= 10$ Yscl $= 10$

Figure 9

Chapter 5 Review Exercises

1. The function is not one-to-one, the graph fails the horizontal line test.

2. The function is not one-to-one, the graph fails the horizontal line test.

3. The function is not one-to-one, the graph fails the horizontal line test.

4. The function is one-to-one, different x-values always produce different y-values.

5. All real numbers can be input for x, therefore the domain is: $(-\infty, \infty)$.

6. All real number can be solutions to $f(x)$, therefore the range is: $(-\infty, \infty)$.

7. Since f is one-to-one, it has an inverse.

8. For $f(x) = \sqrt[3]{2x - 7}$ the inverse is: $x = \sqrt[3]{2y - 7} \Rightarrow x^3 = 2y - 7 \Rightarrow x^3 + 7 = 2y \Rightarrow y = \dfrac{x^3 + 7}{2} \Rightarrow$ $f^{-1}(x) = \dfrac{x^3 + 7}{2}$.

9. See Figure 9. The graphs are reflections across the line $y = x$.

10. $(f \circ f^{-1})(x) = \sqrt[3]{2\left(\dfrac{x^3 + 7}{2}\right) - 7} = \sqrt[3]{x^3} = x$; and $(f^{-1} \circ f)(x) = \dfrac{(\sqrt[3]{2x - 7})^3 + 7}{2} = \dfrac{2x - 7 + 7}{2} = \dfrac{2x}{2} = x$.

11. $y = a^{x+2}$ is the graph of $y = a^x$ shifted 2 units left with a y-intercept of a^2. Therefore it matches: C.

12. $y = a^x + 2$ is the graph of $y = a^x$ shifted 2 units upward with a y-intercept of 3. Therefore it matches: A.

13. $y = -a^x + 2$ is the graph of $y = a^x$ reflected across the x-axis, then shifted 2 units upward, with a y-intercept of 1. Therefore it matches: D.

14. $y = a^{-x} + 2$ is the graph of $y = a^x$ reflected across the y-axis, then shifted 2 units upward, with a y-intercept of 3. Therefore it matches: B.

15. Because the graph goes up to the left, the value of a is: $0 < x < 1$.

16. All real numbers can be input for x, therefore the domain is: $(-\infty, \infty)$.

17. From the graph, only values where $f(x) > 0$, are possible solutions. Therefore, the range is: $(0, \infty)$.

18. Since every real number a raised to the zero power equals 1, $f(0) = a^0 = 1$.

19. Reflect $f(x)$ across the line $y = x$. See Figure 19.

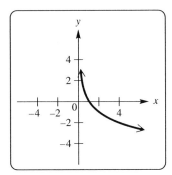

Figure 19

20. For $f(x) = a^x$, the inverse is: $x = a^y \Rightarrow y = \log_a x \Rightarrow f^{-1}(x) = \log_a x$.

21. See Figure 21. From the graph, the domain is: $(-\infty, \infty)$; and the range is: $(0, \infty)$.

22. See Figure 22. From the graph, the domain is: $(-\infty, \infty)$; and the range is: $(2, \infty)$.

23. See Figure 23. From the graph, the domain is: $(-\infty, \infty)$; and the range is: $(-\infty, 0)$.

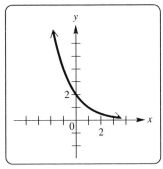   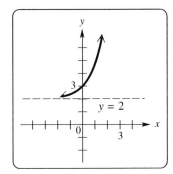   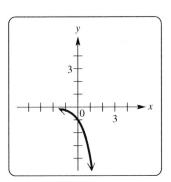

Figure 21   Figure 22   Figure 23

24. The graph of $f(x) = a^x$ with $0 < x < 1$ would have a range that decreases as the domain increases. Since the negative is on the exponent for $f(x) = a^{-x}$, the graph is reflected across the y-axis and is therefore now increasing as the domain increases. It is increasing on its domain.

25. (a) $\left(\dfrac{1}{8}\right)^{-2x} = 2^{x+3} \Rightarrow (2^{-3})^{-2x} = 2^{x+3} \Rightarrow 2^{6x} = 2^{x+3} \Rightarrow 6x = x + 3 \Rightarrow 5x = 3 \Rightarrow x = \left\{\dfrac{3}{5}\right\}$.

(b) If $\left(\dfrac{1}{8}\right)^{-2x} \geq 2^{x+3}$, then $\left(\dfrac{1}{8}\right)^{-2x} - 2^{x+3} \geq 0$. Now graph $y_1 = \left(\dfrac{1}{8}\right)^{-2x} - 2^{x+3}$ on a calculator, use the

answer from part a, and from the graph the solution to $\left(\dfrac{1}{8}\right)^{-2x} - 2^{x+3} \geq 0$ is the interval: $\left[\dfrac{3}{5}, \infty\right)$.

26. (a) $3^{-x} = \left(\dfrac{1}{27}\right)^{1-2x} \Rightarrow 3^{-x} = (3^{-3})^{1-2x} \Rightarrow 3^{-x} = 3^{-3+6x} \Rightarrow -x = -3 + 6x \Rightarrow 3 = 7x \Rightarrow x = \left\{\dfrac{3}{7}\right\}$.

(b) If $3^{-x} < \left(\dfrac{1}{27}\right)^{1-2x}$, then $3^{-x} - \left(\dfrac{1}{27}\right)^{1-2x} < 0$. Now graph $y_1 = 3^{-x} - \left(\dfrac{1}{27}\right)^{1-2x}$ on a calculator, use the

answer from part (a), and from the graph the solution to $3^{-x} - \left(\dfrac{1}{27}\right)^{1-2x} < 0$ is the interval: $\left(\dfrac{3}{7}, \infty\right)$.

27. (a) $.5^{-x} = .25^{x+1} \Rightarrow .5^{-x} = (.5^2)^{x+1} \Rightarrow .5^{-x} = .5^{2x+2} \Rightarrow -x = 2x + 2 \Rightarrow -2 = 3x \Rightarrow x = \left\{-\dfrac{2}{3}\right\}$.

    (b) If $.5^{-x} > .25^{x+1}$, then $.5^{-x} - .25^{x+1} > 0$. Now graph $y_1 = .5^{-x} - .25^{x+1}$ on a calculator, use the

    answer from part (a), and from the graph the solution to $.5^{-x} - .25^{x+1} > 0$ is the interval: $\left(-\dfrac{2}{3}, \infty\right)$.

28. (a) $.4^x = 2.5^{1-x} \Rightarrow \log .4^x = \log 2.5^{1-x} \Rightarrow x \log .4 = (1 - x) \log 2.5 \Rightarrow x \log .4 = \log 2.5 - x \log 2.5 \Rightarrow$

    $x \log .4 + x \log 2.5 = \log 2.5 \Rightarrow x (\log .4 + \log 2.5) = \log 2.5 \Rightarrow x = \dfrac{\log 2.5}{\log .4 + \log 2.5}$, using a

    calculator, this is equal to: $x = \dfrac{\log 2.5}{0}$, which is undefined. The solution is: $\varnothing$.

    (b) If $.4^x < 2.5^{1-x}$, then $.4^x - 2.5^{1-x} < 0$. Now graph $y_1 = .4^x - 2.5^{1-x}$ on a calculator, from the graph the

    solution to $.4^x - 2.5^{1-x} < 0$ is the interval: $(-\infty, \infty)$.

29. We can find the intersection points of $y = x^2$ and $y = 2^x$ by using the intersection of graphs method and graph

    the equation $y_1 = x^2 - 2^x$ to finding the x-intercepts. Using the calculator the intercepts are:

    2, 4, and $-.766664696$. The missing point in common is: $(-.766664696, .58777475603)$.

30. Using the intersection of graphs method, graph $y = 3^x - \pi$ and find the x-intercept. From the calculator, the

    x-intercepts is: $x = \{1.041978046\}$.

31. From the calculator, $\log 58.3 \approx 1.7656685547633 \approx 1.7657$.

32. From the calculator, $\log .00233 \approx -2.63264407897 \approx -2.6326$.

33. From the calculator, $\ln 58.3 \approx 4.06560209336 \approx 4.0656$.

34. From the calculator, $\log_2 .00233 = \dfrac{\log .00233}{\log 2} = -8.7455$.

35. By the definitions of logarithms, $\log_{13} 1 = x \Rightarrow 13^x = 1$, since we know $13^0 = 1$, it follows that $x = \{0\}$.

36. By the properties of logarithms, $\ln e^{\sqrt{6}} = \log_e e^{\sqrt{6}} = \sqrt{6}$.

37. By the properties of logarithms, $\log_5 5^{12} = 12$.

38. By the properties of logarithms, $7^{\log_7 13} = 13$.

39. By the properties of logarithms, $3^{\log_3 5} = 5$.

40. By the change of base rule, $\log_4 9 = \dfrac{\log 9}{\log 4} \approx 1.58496250072 \approx 1.5850$.

41. The x-intercept is 1, and the graph is increasing. The correct graph is E.

42. Since $f(x) = \log_2 (2x) = \log_2 2 + \log_2 x = 1 + \log_2 x$, the graph will be similar to that of $f(x) = \log_2 x$, but

    will have a vertical shift up 1. The x-intercept will be $\dfrac{1}{2}$, since $\log_2 2\left(\dfrac{1}{2}\right) = \log_2 1 = 0$. The correct graph is D.

43. Since $f(x) = \log_2 \left(\dfrac{1}{x}\right) = \log_2 x^{-1} = -\log_2 x$, the graph will be similar to that of $f(x) = \log_2 x$, but will be

    reflected across the x-axis. The x-intercept is 1. The correct graph is B.

44. Since $f(x) = \log_2 \left(\dfrac{x}{2}\right) = \log_2 \left(\dfrac{1}{2}x\right) = \log_2 \left(\dfrac{1}{2}\right) + \log_2 x = -1 + \log_2 x$, the graph will be similar to that of

    $f(x) = \log_2 x$, but will have a vertical shift down 1 unit. The x-intercept will be 2, since $\log_2 \left(\dfrac{2}{2}\right) = \log_2 1 = 0$.

    The correct graph is C.

45. With $f(x) = \log_2{(x-1)}$, the graph will be similar to that of $f(x) = \log_2{x}$, but will be shifted 1 unit to the right; the x-intercept will be 2. The correct graph is F.

46. With $f(x) = \log_2{(-x)}$, the graph will be similar to that of $f(x) = \log_2{x}$, but will be reflected across the y-axis. The x-intercept is -1 since $\log_2{(-(-1))} = \log_2{1} = 0$. The correct graph is A.

47. See Figure 47. From the graph, the domain is: $(0, \infty)$; and the range is: $(-\infty, \infty)$.

48. See Figure 48. From the graph, the domain is: $(2, \infty)$; and the range is: $(-\infty, \infty)$.

49. See Figure 49. From the graph, the domain is: $(-\infty, 0)$; and the range is: $(-\infty, \infty)$.

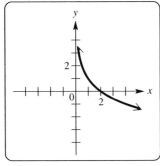

Figure 47

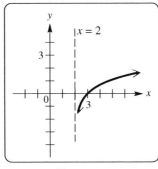

Figure 48

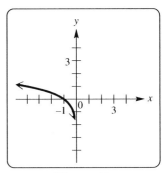

Figure 49

50. The functions in Exercises 47, 48, and 49 are the inverses, respectively, of those in Exercises 21, 22, and 23.

51. If the function has a point $(81, 4)$, then $f(81) = 4$. Therefore: $\log_a{81} = 4 \Rightarrow a^4 = 81 \Rightarrow a^4 = 3^4 \Rightarrow a = 3$. The base is: 3.

52. If the function has a point $\left(-4, \dfrac{1}{16}\right)$, then $f(-4) = \dfrac{1}{16}$. Therefore: $a^{-4} = \dfrac{1}{16} \Rightarrow a^{-4} = 2^{-4} \Rightarrow a = 2$. The base is: 2.

53. $\log_3{\dfrac{mn}{5r}} = \log_3{mn} - \log_3{5r} = \log_3{m} + \log_3{n} - (\log_3{5} + \log_3{r}) = \log_3{m} + \log_3{n} - \log_3{5} - \log_3{r}$

54. $\log_2{\dfrac{\sqrt{7}}{15}} = \log_2{\sqrt{7}} - \log_2{15} = \log_2{7^{1/2}} - \log_2{15} = \dfrac{1}{2}\log_2{7} - \log_2{15}$

55. $\log_5{(x^2 y^4 \sqrt[5]{m^3 p})} = \log_5{x^2} + \log_5{y^4} + \log_5{(m^3 p)^{1/5}} = 2\log_5{x} + 4\log_5{y} + \dfrac{1}{5}\log_5{(m^3 p)} =$

    $2\log_5{x} + 4\log_5{y} + \dfrac{1}{5}(\log_5{m^3} + \log_5{p}) = 2\log_5{x} + 4\log_5{y} + \dfrac{3}{5}\log_5{m} + \dfrac{1}{5}\log_5{p}$

56. The properties of logarithms do not apply.

57. (a) $\log{(x+3)} + \log{x} = 1 \Rightarrow \log{(x+3)(x)} = 1 \Rightarrow 10^{\log{(x+3)(x)}} = 10^1 \Rightarrow (x+3)(x) = 10 \Rightarrow$

    $x^2 + 3x - 10 = 0 \Rightarrow (x+5)(x-2) = 0 \Rightarrow x = -5, 2$. Since the argument of a logarithm can not be negative, the solution is: $x = \{2\}$.

    (b) Graph $y_1 = \log{(x+3)} + \log{x}$ and $y_2 = 1$ in the same window, from the graph $y_1 > y_2$ for the interval: $(2, \infty)$.

58. (a) $\ln{e^{\ln{x}}} - \ln{(x-4)} = \ln{3} \Rightarrow \ln{x} - \ln{(x-4)} = \ln{3} \Rightarrow \ln{\left(\dfrac{x}{x-4}\right)} = \ln{3} \Rightarrow e^{\ln{(x/x-4)}} = e^{\ln{x}} \Rightarrow$

    $\dfrac{x}{x-4} = 3 \Rightarrow x = 3x - 12 \Rightarrow -2x = -12 \Rightarrow x = \{6\}$.

    (b) Graph $y_1 = \ln{e^{\ln{x}}} - \ln{(x-4)}$ and $y_2 = \ln{3}$ in the same window, from the graph $y_1 \le y_2$ for the interval: $[6, \infty)$.

59. (a) $\ln e^{\ln 2} - \ln (x - 1) = \ln 5 \Rightarrow \ln 2 - \ln (x - 1) = \ln 5 \Rightarrow \ln \left(\dfrac{2}{x - 1}\right) = \ln 5 \Rightarrow e^{\ln (2/x - 1)} = e^{\ln 5} \Rightarrow$

$\dfrac{2}{x - 1} = 5 \Rightarrow 2 = 5x - 5 \Rightarrow 5x = 7 \Rightarrow x = \left\{\dfrac{7}{5}\right\}$ or $\{1.4\}$.

(b) Graph $y_1 = \ln e^{\ln 2} - \ln (x - 1)$ and $y_2 = \ln 5$ in the same window, from the graph $y_1 \geq y_2$ for the interval: $(1, 1.4]$.

60. (a) $8^x = 32 \Rightarrow (2^3)^x = 2^5 \Rightarrow 2^{3x} = 2^5 \Rightarrow 3x = 5 \Rightarrow x = \left\{\dfrac{5}{3}\right\}$.

61. (a) $\dfrac{8}{27} = x^{-3} \Rightarrow \left(\dfrac{2}{3}\right)^3 = x^{-3} \Rightarrow \left(\dfrac{2}{3}\right)^3 = \left(\dfrac{1}{x}\right)^3 \Rightarrow \dfrac{2}{3} = \dfrac{1}{x} \Rightarrow 2x = 3 \Rightarrow x = \left\{\dfrac{3}{2}\right\}$.

62. (a) $10^{2x - 3} = 17 \Rightarrow \log 10^{2x - 3} = \log 17 \Rightarrow (2x - 3) \log 10 = \log 17 \Rightarrow 2x - 3 = \log 17 \Rightarrow$

$2x = 3 + \log 17 \Rightarrow x = \dfrac{3 + \log 17}{2}$

(b) From the calculator the solution is: $x = \{2.115\}$.

63. (a) $e^{x + 1} = 10 \Rightarrow \ln e^{x + 1} = \ln 10 \Rightarrow (x + 1) \ln e = \ln 10 \Rightarrow x + 1 = \ln 10 \Rightarrow x = \{-1 + \ln 10\}$.

(Other forms of the answer are possible.)

(b) From the calculator the solution is: $x = \{1.303\}$.

64. (a) $\log_{64} x = \dfrac{1}{3} \Rightarrow 64^{\log_{64} x} = 64^{1/3} \Rightarrow x = 64^{1/3} \Rightarrow x = \{4\}$.

65. (a) $\ln (6x) - \ln (x + 1) = \ln 4 \Rightarrow \ln \dfrac{6x}{x + 1} = \ln 4 \Rightarrow e^{\ln 6x/x + 1} = e^{\ln 4} \Rightarrow \dfrac{6x}{x + 1} = 4 \Rightarrow 4x + 4 = 6x \Rightarrow$

$4 = 2x \Rightarrow x = \{2\}$.

66. (a) $\log_{12} (2x) + \log_{12} (x - 1) = 1 \Rightarrow \log_{12} (2x)(x - 1) = 1 \Rightarrow \log_{12} (2x^2 - 2x) = 1 \Rightarrow$

$12^{\log_{12} (2x^2 - 2x)} = 12^1 \Rightarrow 2x^2 - 2x = 12 \Rightarrow 2x^2 - 2x - 12 = 0 \Rightarrow 2(x^2 - x - 6) = 0 \Rightarrow$

$2(x - 3)(x + 2) = 0 \Rightarrow x = -2, 3$. Since the argument of a logarithm can not be negative, the solution is: $x = \{3\}$.

67. (a) $\log_{16} \sqrt{x + 1} = \dfrac{1}{4} \Rightarrow 16^{\log_{16} \sqrt{x + 1}} = 16^{1/4} \Rightarrow \sqrt{x + 1} = 16^{1/4} \Rightarrow x + 1 = (16^{1/4})^2 \Rightarrow$

$x + 1 = 16^{1/2} \Rightarrow x + 1 = 4 \Rightarrow x = \{3\}$.

68. (a) $\ln x + 3 \ln 2 = \ln \dfrac{2}{x} \Rightarrow \ln x + \ln 2^3 = \ln \dfrac{2}{x} \Rightarrow \ln [(x)(2^3)] = \ln \dfrac{2}{x} \Rightarrow \ln 8x = \ln \dfrac{2}{x} \Rightarrow$

$e^{\ln 8x} = e^{\ln 2/x} \Rightarrow 8x = \dfrac{2}{x} \Rightarrow 8x^2 = 2 \Rightarrow x^2 = \dfrac{2}{8} \Rightarrow x^2 = \dfrac{1}{4} \Rightarrow x = -\dfrac{1}{2}, \dfrac{1}{2}$. Since the argument of a

natural logarithm can not be negative, the solution is: $x = \left\{\dfrac{1}{2}\right\}$.

69. (a) $\ln [\ln (e^{-x})] = \ln 3 \Rightarrow \ln (-x) = \ln 3 \Rightarrow e^{\ln (-x)} = e^{\ln 3} \Rightarrow -x = 3 \Rightarrow x = \{-3\}$.

70. (a) $\ln e^x - \ln e^3 = \ln^e \Rightarrow x - 3 = 5 \Rightarrow x = \{8\}$.

71. No real number value for x can produce a negative solution, therefore $x = \varnothing$.

72. $N = a + b \ln \left(\dfrac{c}{d}\right) \Rightarrow N - a = b \ln \left(\dfrac{c}{d}\right) \Rightarrow \dfrac{N - a}{b} = \ln \left(\dfrac{c}{d}\right) \Rightarrow$

$e^{(N - a)/b} = e^{\ln (c/d)} \Rightarrow e^{(N - a)/b} = \dfrac{c}{d} \Rightarrow c = de^{(N - a)/b}$

73. $y = y_0 e^{-kt} \Rightarrow \dfrac{y}{y_0} = e^{-kt} \Rightarrow \ln\dfrac{y}{y_0} = \ln e^{-kt} \Rightarrow \ln\dfrac{y}{y_0} = -kt \Rightarrow t = \dfrac{\ln\dfrac{y}{y_0}}{-k}$

74. Graph $y_1 = \log_{10} x$ and $y_2 = x - 2$ in the same window. By the capabilities of the calculator the x-intercepts are: $x = \{.010, 2.376\}$.

75. Graph $y_1 = 2^{-x}$ and $y_2 = \log_{10} x$ in the same window. By the capabilities of the calculator the x-intercepts are: $x = \{1.874\}$.

76. Graph $y_1 = x^2 - 3$ and $y_2 = \log x$ in the same window. By the capabilities of the calculator the x-intercepts are: $x = \{.001, 1.805\}$.

77. Use the Compound Interest Formula $A = P\left(1 + \dfrac{r}{n}\right)^{nt}$, $A = \$8780$, $P = \$3500$, $t = 10$, and $n = 1$:

$8780 = 3500\left(1 + \dfrac{r}{1}\right)^{10(1)} \Rightarrow \dfrac{8780}{3500} = (1 + r)^{10} \Rightarrow \left(\dfrac{8780}{3500}\right)^{.1} = 1 + r \Rightarrow r = \left(\dfrac{8780}{3500}\right)^{.1} - 1 \Rightarrow$

$r \approx .096333405$. The annual interest rate to the nearest tenth, is 9.6%.

78. Use the Compound Interest Formula $A = P\left(1 + \dfrac{r}{n}\right)^{nt}$, $A = \$58344$, $P = \$48000$, $t = .05$, and $n = 2$:

$58{,}344 = 48{,}000\left(1 + \dfrac{.05}{2}\right)^{2t} \Rightarrow 58{,}344 = 48{,}000(1.025)^{2t} \Rightarrow \left(\dfrac{58{,}344}{48{,}000}\right) = 1.025^{2t} \Rightarrow 1.2155 = 1.025^{2t} \Rightarrow$

$\ln 1.2155 = \ln 1.025^{2t} \Rightarrow \ln 1.2155 = 2t \ln 1.025 \Rightarrow t = \dfrac{\ln 1.2155}{2 \ln 1.025} \Rightarrow t \approx 3.9516984$. About 4 years.

79. For the first 8 years, use the Compound Interest Formula $A = P\left(1 + \dfrac{r}{n}\right)^{nt}$, $P = \$12{,}000$, $t = 8$, $r = .05$, and

$n = 1$: $A = 12{,}000\left(1 + \dfrac{.05}{1}\right)^{8(1)} \Rightarrow A = 12{,}000(1.05)^8 \Rightarrow A \approx \$17729.47$. Now for the second 6 years,

use the Compound Interest Formula $A = P\left(1 + \dfrac{r}{n}\right)^{nt}$, $P = \$17729.47$, $t = 6$, $r = .06$, and $n = 1$:

$A = 17{,}729.47\left(1 + \dfrac{.06}{1}\right)^{6(1)} \Rightarrow A = 17{,}729.47(1.06)^6 \Rightarrow A \approx \$25{,}149.59$. At the end of the 14 year period,

about $\$25{,}149.59$ would be in the account.

80. (a) Use the Compound Interest Formula $A = P\left(1 + \dfrac{r}{n}\right)^{nt}$, $P = \$2000$, $r = .03$, $t = 5$, and $n = 4$:

$A = 2000\left(1 + \dfrac{.03}{4}\right)^{4(5)} \Rightarrow A = 2000(1.0075)^{20} \Rightarrow A \approx \$2322.37$.

(b) Use the Continuous Compound Interest Formula $A = Pe^{rt}$, $P = \$2000$, $r = .03$, and $t = 5$:

$A = 2000\, e^{(.03)(5)} \Rightarrow A \approx \$2323.67$.

(c) Use the Continuous Compound Interest Formula $A = Pe^{rt}$, $A = \$6000$, $P = \$2000$, and $r = .03$:

$6000 = 2000\, e^{.03t} \Rightarrow 3 = e^{.03t} \Rightarrow \ln 3 = \ln e^{.03t} \Rightarrow \ln 3 = .03t \Rightarrow t = \dfrac{\ln 3}{.03} \Rightarrow t \approx 36.6204$.

It would take about 36.6 years to triple.

81. (a) See Figure 81. Since L is increasing, heavier planes require longer runways.

    (b) We can find the answer by solving for $L(10)$ and $L(100)$.

      $L(10) = 3 \log 10 = 3(1) = 3$ or 3000 feet; $L(100) = 3 \log 100 = 3(2) = 6$ or 6000 feet.

      No it does not increase by a factor of 10, but rather it increases by a factor of 2 to 6000 feet.

82. $2200 = 280 \ln (x + 1) + 1925 \Rightarrow 275 = 280 \ln (x + 1) \Rightarrow \dfrac{275}{280} = \ln (x + 1) \Rightarrow e^{275/280} = e^{\ln (x+1)} \Rightarrow$

    $e^{275/280} = x + 1 \Rightarrow x = e^{275/280} - 1 \Rightarrow x \approx 1.67$ acres.

83. (a) $P(.5) = .04e^{-4(.5)} \Rightarrow P(.5) \approx .0054$ grams/liter

    (b) $P(1) = .04e^{-4(1)} \Rightarrow P(1) \approx .00073$ grams/liter

    (c) $P(2) = .04e^{-4(2)} \Rightarrow P(2) \approx .000013$ grams/liter

    (d) $.002 = .04e^{-4x} \Rightarrow .05 = e^{-4x} \Rightarrow \ln .05 = \ln e^{-4x} \Rightarrow \ln .05 = -4x \Rightarrow x = \dfrac{\ln .05}{-4} \Rightarrow x \approx .75$ miles.

84. (a) $p(2) = 250 - 120 (2.8)^{-.5(2)} \Rightarrow p(2) = 207$

    (b) $p(10) = 250 - 120 (2.8)^{-.5(10)} \Rightarrow p(10) = 249$

    (c) See Figure 84. When $x = 2$, the graph supports the answer of 207 from part (a).

85. Find t when $v(t) = 147$. $147 = 176(1 - e^{-.18t}) \Rightarrow \dfrac{147}{176} = 1 - e^{-.18t} \Rightarrow e^{-.18t} = 1 - \dfrac{147}{176} \Rightarrow e^{-.18t} = \dfrac{29}{176} \Rightarrow$

    $\ln e^{-.18t} = \ln \dfrac{29}{176} \Rightarrow -.18t = \ln \dfrac{29}{176} \Rightarrow t = \dfrac{\ln \frac{29}{176}}{-.18} \Rightarrow t \approx 10.017712.$

    It will take the skydiver about 10 seconds to attain the speed of 147 feet/second (100 mph).

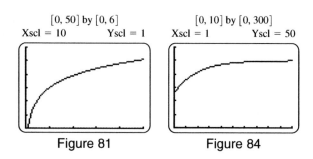

$[0, 50]$ by $[0, 6]$ Xscl $= 10$ Yscl $= 1$ — Figure 81

$[0, 10]$ by $[0, 300]$ Xscl $= 1$ Yscl $= 50$ — Figure 84

Chapter 5 Test

1. (a) This function is a decreasing graph with a y-axis asymptote. This matches with B.

    (b) This function is an increasing graph with a y-intercept of 1. This matches with A.

    (c) This function is an increasing graph with a y-axis asymptote. This matches with C.

    (d) This function is a decreasing graph with a y-intercept of 1. This matches with D.

2. (a) See Figure 2.

    (b) From the graph, the domain is: $(-\infty, \infty)$; and the range is: $(-\infty, 8)$.

    (c) From the graph, it has a horizontal asymptote with equation $y = 8$.

    (d) Set $y = 0$ to find the x-intercept. $0 = -2^{x-1} + 8 \Rightarrow 2^{x-1} = 8 \Rightarrow 2^{x-1} = 2^3 \Rightarrow x - 1 = 3 \Rightarrow x = 4$.

    The x-intercept is: $(4, 0)$. Set $x = 0$ to find the y-intercept. $y = -2^{x-1} + 8 \Rightarrow y = -\dfrac{1}{2} + 8 \Rightarrow y = 7.5$.

    The y-intercept is: $(0, 7.5)$.

    (e) To find the inverse, interchange the x and y variables and solve for y. $x = -2^{y-1} + 8 \Rightarrow$

    $x - 8 = -2^{y-1} \Rightarrow 8 - x = 2^{y-1} \Rightarrow \log(8 - x) = \log 2^{y-1} \Rightarrow \log(8 - x) = (y - 1)\log 2 \Rightarrow$

    $\log(8 - x) = y\log 2 - 1\log 2 \Rightarrow \log(8 - x) + \log 2 = y\log 2 \Rightarrow y = \dfrac{\log(8 - x) + \log 2}{\log 2}$

    or $y = \dfrac{\log(8 - x)}{\log 2} + 1$.

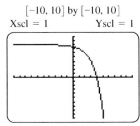

$[-10, 10]$ by $[-10, 10]$
Xscl $= 1$ Yscl $= 1$

Figure 2

3. $\left(\dfrac{1}{8}\right)^{2x+3} = 16^{x+1} \Rightarrow (2^{-3})^{2x-3} = (2^4)^{x+1} \Rightarrow 2^{-6x+9} = 2^{4x+4} \Rightarrow -6x + 9 = 4x + 4 \Rightarrow 5 = 10x \Rightarrow$

    $x = \left\{\dfrac{1}{2}\right\}$ or $x = \{.5\}$.

4. (a) Use the Compound Interest Formula $A = P\left(1 + \dfrac{r}{n}\right)^{nt}$, $P = \$10{,}000$, $r = .055$, $t = 4$, and $n = 4$:

    $A = 10{,}000\left(1 + \dfrac{.055}{4}\right)^{4(4)} \Rightarrow A \approx \$12{,}442.11$.

    (b) Use the Continuous Compound Interest Formula $A = Pe^{rt}$, $P = \$10{,}000$, $r = .055$, and $t = 4$:

    $A = 10{,}000e^{.055(4)} \Rightarrow A \approx \$12{,}460.77$.

5. The expression $\log_5 27$ is the exponent to which 5 must be raised in order to obtain 27. To find an

    approximation with a calculator, use the change-of-base rule: $\log_5 27 = \dfrac{\log 27}{\log 5}$.

6. (a) From the calculator, $\log 45.6 \approx 1.659$.

    (b) From the calculator, $\ln 470 \approx 6.153$.

    (c) From the calculator, $\log_3 769 = \dfrac{\log 769}{\log 3} \approx 6.049$.

7. $\log \dfrac{m^3 n}{\sqrt{y}} = \log m^3 + \log n - \log \sqrt{y} = \log m^3 + \log n - \log y^{1/2} \Rightarrow 3\log m + \log n - \dfrac{1}{2}\log y$.

8. $A = Pe^{rt} \Rightarrow \dfrac{A}{P} = e^{rt} \Rightarrow \ln\dfrac{A}{P} = \ln e^{rt} \Rightarrow \ln A - \ln P = rt \Rightarrow t = \dfrac{\ln A - \ln P}{r}.$

9. (a) $\log_2 x + \log_2 (x + 2) = 3; x > 0 \Rightarrow \log_2 [x(x + 2)] = 3 \Rightarrow 2^{\log_2 [x(x+2)]} = 2^3 \Rightarrow x(x + 2) = 2^3 \Rightarrow$

    $x^2 + 2x - 8 = 0 \Rightarrow (x + 4)(x - 2) = 0 \Rightarrow x = -4, 2.$ Since the argument of a logarithm can not be

    negative, the solution cannot be the extraneous solution -4, but the solution is: $x = \{2\}$.

    (b) Using the change-of-base rule on $y_1 = \log_2 x + \log_2 (x + 2) - 3$ yields: $y_1 = \dfrac{\log x}{\log 2} + \dfrac{\log (x + 2)}{\log 2} - 3.$

    Graph the equation. See Figure 9. The x-intercept is $x = \{2\}$.

    (c) From the graph, $\log_2 x + \log_2 (x + 2) > 3$ for the interval: $(2, \infty)$.

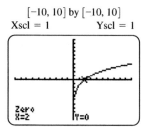

$[-10, 10]$ by $[-10, 10]$
Xscl = 1   Yscl = 1

Figure 9

10. (a) $2e^{5x+2} = 8 \Rightarrow e^{5x+2} = 4 \Rightarrow \ln e^{5x+2} = \ln 4 \Rightarrow 5x + 2 = \ln 4 \Rightarrow 5x = \ln 4 - 2 \Rightarrow x = \left\{\dfrac{\ln 4 - 2}{5}\right\}.$

    (Other forms of the answer are possible.)

    (b) From the calculator the solutions is: $x = \{-.123\}$.

11. (a) $6^{2-x} = 2^{3x+1} \Rightarrow \log 6^{2-x} = \log 2^{3x+1} \Rightarrow (2 - x)\log 6 = (3x + 1)\log 2 \Rightarrow$

    $2 \log 6 - x \log 6 = 3x \log 2 + \log 2 \Rightarrow 2 \log 6 - \log 2 = 3x \log 2 + x \log 6 \Rightarrow$

    $\log\dfrac{6^2}{2} = x (\log (2^3)(6)) \Rightarrow \log 18 = x (\log 48) \Rightarrow x = \dfrac{\log 18}{\log 48}.$ (Other forms of the answer are possible.)

    (b) From the calculator the solutions is: $x = \{.747\}$.

12. (a) $\log (\ln x) = 1 \Rightarrow 10^{\log (\ln x)} = 10^1 \Rightarrow \ln x = 10 \Rightarrow e^{\ln x} = e^{10} \Rightarrow x = \{e^{10}\}.$

    (b) From the calculator the solutions is: $x = \{22{,}026.466\}$.

13. (a) The 2 in the equation is the original population, tripling this would yield $y = 3 \cdot 2$. The match is: B.

    (b) In this function $y = 3$, therefore the match is D.

    (c) In this function $t = 3$, therefore the match is C.

    (d) Since 4 month equals $\dfrac{1}{3}$ year, in this function $t = \dfrac{1}{3}$, therefore the match is A.

14. (a) See Figure 14a.

    (b) See Figure 14b.

    (c) See Figure 14c.

    (d) See Figure 14d.

    Function (c) is the best at describing $A(t)$ because it starts at 350 and gradually decreases as time increases.

$[0, 10]$ by $[0, 500]$
Xscl = 1   Yscl = 100

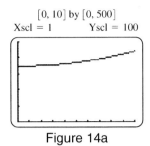

Figure 14a

$[0, 10]$ by $[0, 500]$
Xscl = 1   Yscl = 100

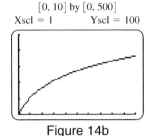

Figure 14b

$[0, 10]$ by $[0, 500]$
Xscl = 1   Yscl = 100

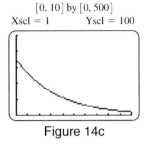

Figure 14c

$[0, 10]$ by $[0, 500]$
Xscl = 1   Yscl = 100

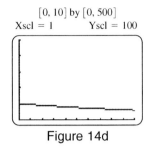

Figure 14d

15. (a) Using the exponential decay model $A(t) = A_0\, e^{-kt}$, we can solve for k with $A_0 = 2$ and $t = 1600$:

$$\frac{1}{2}(2) = 2e^{-k(1600)} \Rightarrow 1 = \; = 2e^{-k(1600)} \Rightarrow \frac{1}{2} = e^{-k(1600)} \Rightarrow \ln\frac{1}{2} = \ln e^{-k(1600)} \Rightarrow \ln\frac{1}{2} = -k(1600) \Rightarrow$$

$$k = \frac{\ln(.5)}{-1600} \Rightarrow k \approx .000433. \text{ The model is: } A(t) = 2e^{-.000433t}.$$

(b) $A(9600) = 2e^{-.000433(9600)} \Rightarrow A(9600) \approx .03$ grams.

(c) $.5 = 2e^{-.000433t} \Rightarrow .25 = e^{-.000433t} \Rightarrow \ln.25 = \ln e^{-.000433t} \Rightarrow \ln.25 = -.000433t \Rightarrow t = \dfrac{\ln.25}{-.000433} \Rightarrow$

$t \approx 3200$ years.

Chapter 5 Project

1. Using the given function model $f(x) = a(b)^x$, we know that $a = 13.1$ (Original Sales) and that $f(7) = 17.4$,

   therefore we can solve for b: $17.4 = 13.1(b)^7 \Rightarrow \dfrac{17.4}{13.1} = b^7 \Rightarrow b = \left(\dfrac{17.4}{13.1}\right)^{1/7} \Rightarrow b \approx 1.04138$.

   The model is: $f(x) = 13.1(1.04138)^x$.

2. Using the given function model $g(x) = a(b)^x$, we know that $a = 4.9$ (Original Sales) and that $f(7) = 8.7$,

   therefore we can solve for b: $8.7 = 4.9(b)^7 \Rightarrow \dfrac{8.7}{4.9} = b^7 \Rightarrow b = \left(\dfrac{8.7}{4.9}\right)^{1/7} \Rightarrow b \approx 1.08547$.

   The model is: $f(x) = 4.9(1.08547)^x$.

3. Since for new motor vehicle sales $b = 1.04138$, there is an approximately 4.1% increase in new sales.

   Since for new truck sales $b = 1.08547$, there is an approximately 8.5% increase in new sales.

4. For motor vehicle sales: $f(4) = 13.1(1.04138)^4 \Rightarrow f(4) \approx 15.4$ million new motor vehicle sales.

   For truck sales: $g(4) = 4.9(1.08547)^4 \Rightarrow g(4) \approx 6.8$ million new truck sales.

   These estimates are quite close to the actual values of 15.5 million and 6.9 million.

5. For motor vehicle sales: $f(28) = 13.1(1.04138)^{28} \Rightarrow f(28) \approx 40.769$ million new motor vehicle sales.

   For truck sales: $g(28) = 4.9(1.08547)^{28} \Rightarrow g(28) \approx 48.696$ million new truck sales.

   These results are contradictory because new motor vehicle sales include new truck sales so truck sales cannot

   exceed new motor vehicle sales.

6.   See Figure 6.

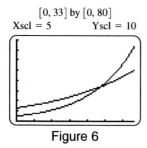

$[0, 33]$ by $[0, 80]$
Xscl = 5   Yscl = 10

Figure 6

No, because over time, truck sales begin to exceed motor vehicle sales, which is impossible.

Modeling functions generally provide better results when estimating values between data points rather than predicting values far into the future.

Chapter 6: Analytic Geometry

6.1: Circles and Parabolas

1. The equation is that of a circle with center $(3, 3)$ and radius 0. That is, the graph is the point $(3, 3)$.

2. The equation is that of a circle with center $(3, 3)$ and radius $\sqrt{-1}$. No such graph exists.

3. E. Since $x = 2y^2$ is equivalent to $y^2 = 4\left(\frac{1}{8}\right)x$, this is a parabola that opens to the right $(c > 0)$.

4. C. Since $y = 2x^2$ is equivalent to $x^2 = 4\left(\frac{1}{8}\right)y$, this is a parabola that opens upward $(c > 0)$.

5. H. Since $x^2 = -3y$ is equivalent to $x^2 = 4\left(-\frac{3}{4}\right)y$, this is a parabola that opens downward $(c < 0)$.

6. B. Since $y^2 = -3x$ is equivalent to $y^2 = 4\left(-\frac{3}{4}\right)x$, this is a parabola that opens to the left $(c < 0)$.

7. F. This is the equation of a circle centered at the origin with radius $\sqrt{5}$.

8. A. This is the equation of a circle centered at the point $(3, -4)$ with radius $\sqrt{25} = 5$.

9. D. This is the equation of a circle centered at the point $(-3, 4)$ with radius $\sqrt{25} = 5$.

10. G. This is the equation of a circle centered at the origin with radius $\sqrt{-4}$. No such graph exists.

11. Here $h = 1$, $k = 4$ and $r^2 = 3^2 = 9$. The equation is $(x - 1)^2 + (y - 4)^2 = 9$.

12. Here $h = -2$, $k = 5$ and $r^2 = 4^2 = 16$. The equation is $(x + 2)^2 + (y - 5)^2 = 16$.

13. A circle that is centered at the origin with $r^2 = 1^2 = 1$ has equation $x^2 + y^2 = 1$.

14. A circle that is centered at the origin with $r^2 = 5^2 = 25$ has equation $x^2 + y^2 = 25$.

15. Here $h = \frac{2}{3}$, $k = -\frac{4}{5}$ and $r^2 = \left(\frac{3}{7}\right)^2 = \frac{9}{49}$. The equation is $\left(x - \frac{2}{3}\right)^2 + \left(y + \frac{4}{5}\right)^2 = \frac{9}{49}$.

16. Here $h = -\frac{1}{2}$, $k = -\frac{1}{4}$ and $r^2 = \left(\frac{12}{5}\right)^2 = \frac{144}{25}$. The equation is $\left(x + \frac{1}{2}\right)^2 + \left(y + \frac{1}{4}\right)^2 = \frac{144}{25}$.

17. The radius is the distance between $(-1, 2)$ and $(2, 6)$: $r = \sqrt{(2 - (-1))^2 + (6 - 2)^2} = \sqrt{9 + 16} = 5$

    Here $h = -1$, $k = 2$ and $r^2 = 5^2 = 25$. The equation is $(x + 1)^2 + (y - 2)^2 = 25$.

18. The radius is the distance between $(2, -7)$ and $(-2, -4)$: $r = \sqrt{(2 - (-2))^2 + (-7 - (-4))^2} = \sqrt{16 + 9} = 5$

    Here $h = 2$, $k = -7$ and $r^2 = 5^2 = 25$. The equation is $(x - 2)^2 + (y + 7)^2 = 25$.

19. If the center is $(-3, -2)$, the circle must touch the x-axis at the point $(-3, 0)$. The radius is 2.

    Here $h = -3$, $k = -2$ and $r^2 = 2^2 = 4$. The equation is $(x + 3)^2 + (y + 2)^2 = 4$.

20. If the center is $(5, -1)$, the circle must touch the y-axis at the point $(0, -1)$. The radius is 5.

    Here $h = 5$, $k = -1$ and $r^2 = 5^2 = 25$. The equation is $(x - 5)^2 + (y + 1)^2 = 25$.

21. The center is the midpoint on the diameter between $(-1, 3)$ and $(5, -9)$: $M = \left(\frac{-1 + 5}{2}, \frac{3 + (-9)}{2}\right) = (2, -3)$

22. The radius is the distance from the center $(2, -3)$ to an endpoint $(-1, 3)$:
    $r = \sqrt{(2 - (-1))^2 + (-3 - 3)^2} = \sqrt{9 + 36} = \sqrt{45} = 3\sqrt{5}$

23. Here $h = 2$, $k = -3$ and $r^2 = (\sqrt{45})^2 = 45$. The equation is $(x - 2)^2 + (y + 3)^2 = 45$.

24. The center is the midpoint on the diameter between $(3, -5)$ and $(-7, 3)$: $M = \left(\dfrac{3 + (-7)}{2}, \dfrac{-5 + 3}{2} \right) = (-2, -1)$.

The radius is the distance from the center $(-2, -1)$ to an endpoint $(3, -5)$:

$r = \sqrt{(3 - (-2))^2 + (-5 - (-1))^2} = \sqrt{25 + 16} = \sqrt{41}$. Here $h = -2, k = -1$ and $r^2 = (\sqrt{41})^2 = 41$.

The equation is $(x + 2)^2 + (y + 1)^2 = 41$.

25. This is the equation of a circle centered at the origin with radius $\sqrt{4} = 2$. See Figure 25.

From the figure, the domain is $[-2, 2]$ and the range is $[-2, 2]$.

26. This is the equation of a circle centered at the origin with radius $\sqrt{36} = 6$. See Figure 26.

From the figure, the domain is $[-6, 6]$ and the range is $[-6, 6]$.

27. This is the equation of a circle centered at the origin with radius $\sqrt{0} = 0$. The graph is only the point $(0, 0)$.

See Figure 27. From the figure, the domain is $\{0\}$ and the range is $\{0\}$.

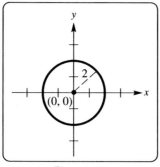

Figure 25

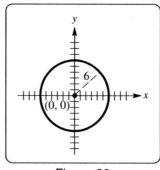

Figure 26

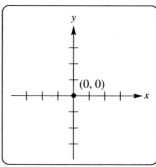

Figure 27

28. This is the equation of a circle centered at the origin with radius $\sqrt{-9}$. No such graph exists.

The domain is $\varnothing$ and the range is $\varnothing$.

29. This is the equation of a circle centered at $(2, 0)$ with radius $\sqrt{36} = 6$. See Figure 29.

From the figure, the domain is $[-4, 8]$ and the range is $[-6, 6]$.

30. This is the equation of a circle centered at $(-2, 5)$ with radius $\sqrt{16} = 4$. See Figure 30.

From the figure, the domain is $[-6, 2]$ and the range is $[1, 9]$.

31. This is the equation of a circle centered at $(5, -4)$ with radius $\sqrt{49} = 7$. See Figure 31.

From the figure, the domain is $[-2, 12]$ and the range is $[-11, 3]$.

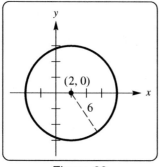

Figure 29

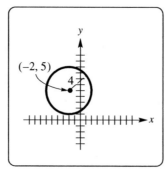

Figure 30

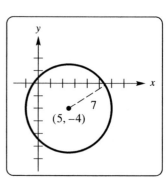

Figure 31

32. This is the equation of a circle centered at (4, 3) with radius $\sqrt{25} = 5$. See Figure 32.

    From the figure, the domain is $[-1, 9]$ and the range is $[-2, 8]$.

33. This is the equation of a circle centered at $(-3, -2)$ with radius $\sqrt{36} = 6$. See Figure 33.

    From the figure, the domain is $[-9, 3]$ and the range is $[-8, 4]$.

34. This is the equation of a circle centered at (0, 2) with radius $\sqrt{9} = 3$. See Figure 34.

    From the figure, the domain is $[-3, 3]$ and the range is $[-1, 5]$.

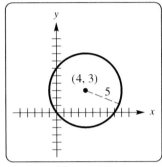

Figure 32

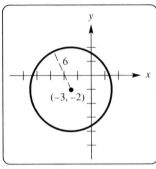

Figure 33

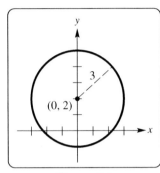

Figure 34

35. This is the equation of a circle centered at $(1, -2)$ with radius $\sqrt{16} = 4$. See Figure 35.

    From the figure, the domain is $[-3, 5]$ and the range is $[-6, 2]$.

36. Since $(x + 1)^2 + y^2 + 2 = 0$ is equivalent to $(x + 1)^2 + y^2 = -2$, this is the equation of a circle centered at $(-1, 0)$ with radius $\sqrt{-2}$. No such graph exists. The domain is $\varnothing$ and the range is $\varnothing$.

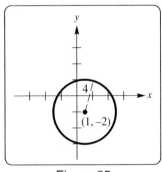

Figure 35

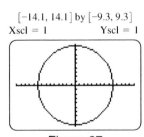

Figure 37

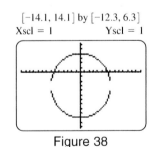

Figure 38

37. $x^2 + y^2 = 81 \Rightarrow y^2 = 81 - x^2 \Rightarrow y = \pm\sqrt{81 - x^2}$

    Graph $y_1 = -\sqrt{81 - x^2}$ and $y_2 = \sqrt{81 - x^2}$ as shown in Figure 37.

    From the figure, the domain is $[-9, 9]$ and the range is $[-9, 9]$.

38. $x^2 + (y + 3)^2 = 49 \Rightarrow (y + 3)^2 = 49 - x^2 \Rightarrow y + 3 = \pm\sqrt{49 - x^2} \Rightarrow y = -3 \pm \sqrt{49 - x^2}$

    Graph $y_1 = -3 - \sqrt{49 - x^2}$ and $y_2 = -3 + \sqrt{49 - x^2}$ as shown in Figure 38.

    From the figure, the domain is $[-7, 7]$ and the range is $[-10, 4]$.

39. $(x - 3)^2 + (y - 2)^2 = 25 \Rightarrow (y - 2)^2 = 25 - (x - 3)^2 \Rightarrow y - 2 = \pm\sqrt{25 - (x - 3)^2} \Rightarrow$
$y = 2 \pm \sqrt{25 - (x - 3)^2}$. Graph $y_1 = 2 - \sqrt{25 - (x - 3)^2}$ and $y_2 = 2 + \sqrt{25 - (x - 3)^2}$ as shown in
Figure 39. From the figure, the domain is $[-2, 8]$ and the range is $[-3, 7]$.

40. $(x + 2)^2 + (y + 3)^2 = 36 \Rightarrow (y + 3)^2 = 36 - (x + 2)^2 \Rightarrow y + 3 = \pm\sqrt{36 - (x + 2)^2} \Rightarrow$
$y = -3 \pm \sqrt{36 - (x + 2)^2}$. Graph $y_1 = -3 - \sqrt{36 - (x + 2)^2}$ and $y_2 = -3 + \sqrt{36 - (x + 2)^2}$ as shown in
Figure 40. From the figure, the domain is $[-8, 4]$ and the range is $[-9, 3]$.

$[-9.4, 9.4]$ by $[-4.2, 8.2]$
Xscl = 1   Yscl = 1

$[-9.4, 9.4]$ by $[-9.2, 3.2]$
Xscl = 1   Yscl = 1

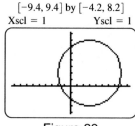

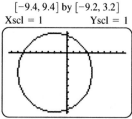

Figure 39   Figure 40

41. $x^2 + 6x + y^2 + 8y + 9 = 0 \Rightarrow (x^2 + 6x + 9) + (y^2 + 8y + 16) = -9 + 9 + 16 \Rightarrow$
$(x + 3)^2 + (y + 4)^2 = 16$. The graph is a circle with center $(-3, -4)$ and radius $r = 4$.

42. $x^2 + 8x + y^2 - 6y + 16 = 0 \Rightarrow (x^2 + 8x + 16) + (y^2 - 6y + 9) = -16 + 16 + 9 \Rightarrow$
$(x + 4)^2 + (y - 3)^2 = 9$. The graph is a circle with center $(-4, 3)$ and radius $r = 3$.

43. $x^2 - 4x + y^2 + 12y = -4 \Rightarrow (x^2 - 4x + 4) + (y^2 + 12y + 36) = -4 + 4 + 36 \Rightarrow$
$(x - 2)^2 + (y + 6)^2 = 36$. The graph is a circle with center $(2, -6)$ and radius $r = 6$.

44. $x^2 - 12x + y^2 + 10y = -25 \Rightarrow (x^2 - 12x + 36) + (y^2 + 10y + 25) = -25 + 36 + 25 \Rightarrow$
$(x - 6)^2 + (y + 5)^2 = 36$. The graph is a circle with center $(6, -5)$ and radius $r = 6$.

45. $4x^2 + 4x + 4y^2 - 16y - 19 = 0 \Rightarrow 4\left(x^2 + x + \frac{1}{4}\right) + 4(y^2 - 4y + 4) = 19 + 1 + 16 \Rightarrow$
$4\left(x + \frac{1}{2}\right)^2 + 4(y - 2)^2 = 36 \Rightarrow \left(x + \frac{1}{2}\right)^2 + (y - 2)^2 = 9$. The graph is a circle with center $\left(-\frac{1}{2}, 2\right)$ and
radius $r = 3$.

46. $9x^2 + 12x + 9y^2 - 18y - 23 = 0 \Rightarrow 9\left(x^2 + \frac{4}{3}x + \frac{4}{9}\right) + 9(y^2 - 2y + 1) = 23 + 4 + 9 \Rightarrow$
$9\left(x + \frac{2}{3}\right)^2 + 9(y - 1)^2 = 36 \Rightarrow \left(x + \frac{2}{3}\right)^2 + (y - 1)^2 = 4$. The graph is a circle with center $\left(-\frac{2}{3}, 1\right)$ and
radius $r = 2$.

47. $x^2 + 2x + y^2 - 6y + 14 = 0 \Rightarrow (x^2 + 2x + 1) + (y^2 - 6y + 9) = -14 + 1 + 9 \Rightarrow$
$(x + 1)^2 + (y - 3)^2 = -4$. The graph does not exist since the value for the radius is not a real number.

48. $x^2 + 4x + y^2 - 8y + 32 = 0 \Rightarrow (x^2 + 4x + 4) + (y^2 - 8y + 16) = -32 + 4 + 16 \Rightarrow$
$(x + 2)^2 + (y - 4)^2 = -12$. The graph does not exist since the value for the radius is not a real number.

49. $x^2 - 2x + y^2 + 4y + 9 = 0 \Rightarrow (x^2 - 2x + 1) + (y^2 + 4y + 4) = -9 + 1 + 4 \Rightarrow$
$(x - 1)^2 + (y + 2)^2 = -4$. The graph does not exist since the value for the radius is not a real number.

50. $4x^2 + 4x + 4y^2 - 4y - 3 = 0 \Rightarrow 4\left(x^2 + x + \frac{1}{4}\right) + 4\left(y^2 - y + \frac{1}{4}\right) = 3 + 1 + 1 \Rightarrow$

$4\left(x + \frac{1}{2}\right)^2 + 4\left(y - \frac{1}{2}\right)^2 = 5 \Rightarrow \left(x + \frac{1}{2}\right)^2 + \left(y - \frac{1}{2}\right)^2 = \frac{5}{4}$. The graph is a circle with center $\left(-\frac{1}{2}, \frac{1}{2}\right)$ and

radius $r = \frac{\sqrt{5}}{2}$.

51. $9x^2 + 36x + 9y^2 = -32 \Rightarrow 9(x^2 + 4x + 4) + 9y^2 = -32 + 36 \Rightarrow 9(x + 2)^2 + 9y^2 = 4 \Rightarrow$

$(x + 2)^2 + y^2 = \frac{4}{9}$. The graph is a circle with center $(-2, 0)$ and radius $r = \frac{2}{3}$.

52. $9x^2 + 9y^2 + 54y = -72 \Rightarrow 9x^2 + 9(y^2 + 6y + 9) = -72 + 81 \Rightarrow 9x^2 + 9(y + 3)^2 = 9 \Rightarrow$

$x^2 + (y + 3)^2 = 1$. The graph is a circle with center $(0, -3)$ and radius $r = 1$.

53. D. Since $(x - 4)^2 = y + 2$ is equivalent to $(x - 4)^2 = 4\left(\frac{1}{4}\right)(y + 2)$, the parabola has vertex $(4, -2)$ and it

    opens upward $(c > 0)$.

54. B. Since $(x - 2)^2 = y + 4$ is equivalent to $(x - 2)^2 = 4\left(\frac{1}{4}\right)(y + 4)$, the parabola has vertex $(2, -4)$ and it

    opens upward $(c > 0)$.

55. C. Since $y + 2 = -(x - 4)^2$ is equivalent to $(x - 4)^2 = 4\left(-\frac{1}{4}\right)(y + 2)$, the parabola has vertex $(4, -2)$ and it

    opens downward $(c < 0)$.

56. A. Since $y = -(x - 2)^2 - 4$ is equivalent to $(x - 2)^2 = 4\left(-\frac{1}{4}\right)(y + 4)$, the parabola has vertex $(2, -4)$ and it

    opens downward $(c < 0)$.

57. F. Since $(y - 4)^2 = x + 2$ is equivalent to $(y - 4)^2 = 4\left(\frac{1}{4}\right)(x + 2)$, the parabola has vertex $(-2, 4)$ and it

    opens to the right $(c > 0)$.

58. H. Since $(y - 2)^2 = x + 4$ is equivalent to $(y - 2)^2 = 4\left(\frac{1}{4}\right)(x + 4)$, the parabola has vertex $(-4, 2)$ and it

    opens to the right $(c > 0)$.

59. E. Since $x + 2 = -(y - 4)^2$ is equivalent to $(y - 4)^2 = 4\left(-\frac{1}{4}\right)(x + 2)$, the parabola has vertex $(-2, 4)$ and it

    opens to the left $(c < 0)$.

60. G. Since $x = -(y - 2)^2 - 4$ is equivalent to $(y - 2)^2 = 4\left(-\frac{1}{4}\right)(x + 4)$, the parabola has vertex $(-4, 2)$ and it

    opens to the left $(c < 0)$.

61. (a) If both coordinates of the vertex are negative, the vertex is in quadrant III.

    (b) If the first coordinate of the vertex is negative and the second is positive, the vertex is in quadrant II.

    (c) If the first coordinate of the vertex is positive and the second is negative, the vertex is in quadrant IV.

    (d) If both coordinates of the vertex are positive, the vertex is in quadrant I.

62. (a) If both coordinates of the vertex are negative, the vertex is in quadrant III.

    (b) If the first coordinate of the vertex is negative and the second is positive, the vertex is in quadrant II.

    (c) If the first coordinate of the vertex is positive and the second is negative, the vertex is in quadrant IV.

    (d) If both coordinates of the vertex are positive, the vertex is in quadrant I.

63. Since $x^2 = 16y$ is equivalent to $x^2 = 4(4)y$, the equation is in the form $x^2 = 4cy$ with $c = 4$.

    The focus is $(0, 4)$ and the equation of the directrix is $y = -4$. The axis is $x = 0$, or the y-axis.

64. Since $x^2 = 4y$ is equivalent to $x^2 = 4(1)y$, the equation is in the form $x^2 = 4cy$ with $c = 1$.

    The focus is $(0, 1)$ and the equation of the directrix is $y = -1$. The axis is $x = 0$, or the y-axis.

65. Since $x^2 = -\frac{1}{2}y$ is equivalent to $x^2 = 4\left(-\frac{1}{8}\right)y$, the equation is in the form $x^2 = 4cy$ with $c = -\frac{1}{8}$.

    The focus is $\left(0, -\frac{1}{8}\right)$ and the equation of the directrix is $y = \frac{1}{8}$. The axis is $x = 0$, or the y-axis.

66. Since $x^2 = \frac{1}{9}y$ is equivalent to $x^2 = 4\left(\frac{1}{36}\right)y$, the equation is in the form $x^2 = 4cy$ with $c = \frac{1}{36}$.

    The focus is $\left(0, \frac{1}{36}\right)$ and the equation of the directrix is $y = -\frac{1}{36}$. The axis is $x = 0$, or the y-axis.

67. Since $y^2 = \frac{1}{16}x$ is equivalent to $y^2 = 4\left(\frac{1}{64}\right)x$, the equation is in the form $y^2 = 4cx$ with $c = \frac{1}{64}$.

    The focus is $\left(\frac{1}{64}, 0\right)$ and the equation of the directrix is $x = -\frac{1}{64}$. The axis is $y = 0$, or the x-axis.

68. Since $y^2 = -\frac{1}{32}x$ is equivalent to $y^2 = 4\left(-\frac{1}{128}\right)x$, the equation is in the form $y^2 = 4cx$ with $c = -\frac{1}{128}$.

    The focus is $\left(-\frac{1}{128}, 0\right)$ and the equation of the directrix is $x = \frac{1}{128}$. The axis is $y = 0$, or the x-axis.

69. Since $y^2 = -16x$ is equivalent to $y^2 = 4(-4)x$, the equation is in the form $y^2 = 4cx$ with $c = -4$.

    The focus is $(-4, 0)$ and the equation of the directrix is $x = 4$. The axis is $y = 0$, or the x-axis.

70. Since $y^2 = -4x$ is equivalent to $y^2 = 4(-1)x$, the equation is in the form $y^2 = 4cx$ with $c = -1$.

    The focus is $(-1, 0)$ and the equation of the directrix is $x = 1$. The axis is $y = 0$, or the x-axis.

71. If the vertex is $(0, 0)$ and the focus is $(0, -2)$, then the parabola opens downward and $c = -2$.

    The equation is $x^2 = 4cy \Rightarrow x^2 = -8y$.

72. If the vertex is $(0, 0)$ and the focus is $(5, 0)$, then the parabola opens to the right and $c = 5$.

    The equation is $y^2 = 4cx \Rightarrow y^2 = 20x$.

73. If the vertex is $(0, 0)$ and the focus is $\left(-\frac{1}{2}, 0\right)$, then the parabola opens to the left and $c = -\frac{1}{2}$.

    The equation is $y^2 = 4cx \Rightarrow y^2 = -2x$.

74. If the vertex is $(0, 0)$ and the focus is $\left(0, \frac{1}{4}\right)$, then the parabola opens upward and $c = \frac{1}{4}$.

    The equation is $x^2 = 4cy \Rightarrow x^2 = y$.

75. If the vertex is $(0, 0)$ and the parabola opens to the right, the equation is in the form $y^2 = 4cx$. Find the

    value of c by using the fact that the parabola passes through $(2, -2\sqrt{2})$. Thus, $(-2\sqrt{2})^2 = 4c(2) \Rightarrow c = 1$.

    The equation is $y^2 = 4cx \Rightarrow y^2 = 4x$.

76. If the vertex is $(0, 0)$ and the parabola opens upward, the equation is in the form $x^2 = 4cy$. Find the

    value of c by using the fact that the parabola passes through $(\sqrt{3}, 3)$. Thus, $(\sqrt{3})^2 = 4c(3) \Rightarrow c = \dfrac{1}{4}$.

    The equation is $x^2 = 4cy \Rightarrow x^2 = y$.

77. If the vertex is $(0, 0)$ and the parabola opens downward, the equation is in the form $x^2 = 4cy$. Find the

    value of c by using the fact that the parabola passes through $(\sqrt{10}, -5)$. Thus, $(\sqrt{10})^2 = 4c(-5) \Rightarrow c = -\dfrac{1}{2}$.

    The equation is $x^2 = 4cy \Rightarrow x^2 = -2y$.

78. If the vertex is $(0, 0)$ and the parabola opens to the left, the equation is in the form $y^2 = 4cx$. Find the

    value of c by using the fact that the parabola passes through $(-3, 3)$. Thus, $(3)^2 = 4c(-3) \Rightarrow c = -\dfrac{3}{4}$.

    The equation is $y^2 = 4cx \Rightarrow y^2 = -3x$.

79. If the vertex is $(0, 0)$ and the parabola has y-axis symmetry, the equation is in the form $x^2 = 4cy$. Find the

    value of c by using the fact that the parabola passes through $(2, -4)$. Thus, $(2)^2 = 4c(-4) \Rightarrow c = -\dfrac{1}{4}$.

    The equation is $x^2 = 4cy \Rightarrow x^2 = -y$.

80. If the vertex is $(0, 0)$ and the parabola has x-axis symmetry, the equation is in the form $y^2 = 4cx$. Find the

    value of c by using the fact that the parabola passes through $(3, 2)$. Thus, $(2)^2 = 4c(3) \Rightarrow c = \dfrac{1}{3}$.

    The equation is $y^2 = 4cx \Rightarrow y^2 = \dfrac{4}{3}x$.

81. If the focus is $(0, 2)$ and the vertex is $(0, 1)$, the parabola opens upward and $c = 1$. Substituting in

    $(x - h)^2 = 4c(y - k)$, we get $(x - 0)^2 = 4(1)(y - 1)$ or $x^2 = 4(y - 1)$.

82. If the focus is $(-1, 2)$ and the vertex is $(3, 2)$, the parabola opens to the left and $c = -4$. Substituting in

    $(y - k)^2 = 4c(x - h)$, we get $(y - 2)^2 = -16(x - 3)$.

83. If the focus is $(0, 0)$ and the directrix has equation $x = -2$, the vertex is $(-1, 0)$ and $c = 1$. The parabola

    opens to the right. Substituting in $(y - k)^2 = 4c(x - h)$, we get $(y - 0)^2 = 4(1)(x - (-1))$ or $y^2 = 4(x + 1)$.

84. If the focus is $(2, 1)$ and the directrix has equation $x = -1$, the vertex is $\left(\dfrac{1}{2}, 1\right)$ and $c = \dfrac{3}{2}$. The parabola

    opens to the right. Substituting in $(y - k)^2 = 4c(x - h)$, we get $(y - 1)^2 = 4\left(\dfrac{3}{2}\right)\left(x - \dfrac{1}{2}\right)$ or

    $(y - 1)^2 = 6\left(x - \dfrac{1}{2}\right)$.

85. If the focus is $(-1, 3)$ and the directrix has equation $y = 7$, the vertex is $(-1, 5)$ and $c = -2$. The parabola

    opens downward. Substituting in $(x - h)^2 = 4c(y - k)$, we get $(x + 1)^2 = 4(-2)(y - 5)$ or

    $(x + 1)^2 = -8(y - 5)$.

86. If the focus is $(1, 2)$ and the directrix has equation $y = 4$, the vertex is $(1, 3)$ and $c = -1$. The parabola

    opens downward. Substituting in $(x - h)^2 = 4c(y - k)$, we get $(x - 1)^2 = 4(-1)(y - 3)$ or

    $(x - 1)^2 = -4(y - 3)$.

87. Since the parabola has a horizontal axis, the equation is in the form $(y - k)^2 = 4c(x - h)$. Find the value of c by using the fact that the parabola passes through $(-4, 0)$ and the vertex is $(-2, 3)$.

    Substituting $x = -4$, $y = 0$, $h = -2$ and $k = 3$ yields $(0 - 3)^2 = 4c(-4 - (-2)) \Rightarrow c = -\dfrac{9}{8}$.

    The equation is $(y - 3)^2 = -\dfrac{9}{2}(x + 2)$.

88. Since the parabola has a horizontal axis, the equation is in the form $(y - k)^2 = 4c(x - h)$. Find the value of c by using the fact that the parabola passes through $(2, 3)$ and the vertex is $(-1, 2)$.

    Substituting $x = 2$, $y = 3$, $h = -1$ and $k = 2$ yields $(3 - 2)^2 = 4c(2 - (-1)) \Rightarrow c = \dfrac{1}{12}$.

    The equation is $(y - 2)^2 = \dfrac{1}{3}(x + 1)$.

89. The equation $y = (x + 3)^2 - 4$ can be written $(x + 3)^2 = 4\left(\dfrac{1}{4}\right)(y + 4)$. The vertex is $(-3, -4)$.

    The vertical axis has equation $x = -3$ and the parabola opens upward. See Figure 89. From the figure, the domain is $(-\infty, \infty)$ and the range is $[-4, \infty)$.

90. The equation $y = (x - 5)^2 - 4$ can be written $(x - 5)^2 = 4\left(\dfrac{1}{4}\right)(y + 4)$. The vertex is $(5, -4)$.

    The vertical axis has equation $x = 5$ and the parabola opens upward. See Figure 90. From the figure, the domain is $(-\infty, \infty)$ and the range is $[-4, \infty)$.

91. The equation $y = -2(x + 3)^2 + 2$ can be written $(x + 3)^2 = 4\left(-\dfrac{1}{8}\right)(y - 2)$. The vertex is $(-3, 2)$.

    The vertical axis has equation $x = -3$ and the parabola opens downward. See Figure 91. From the figure, the domain is $(-\infty, \infty)$ and the range is $(-\infty, 2]$.

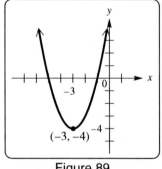

Figure 89

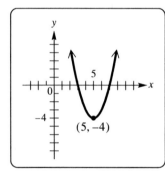

Figure 90

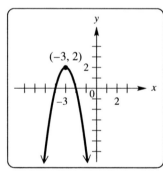

Figure 91

92. The equation $(x - 2)^2 = \dfrac{3}{2}(y + 1)$ can be written $(x - 2)^2 = 4\left(\dfrac{3}{8}\right)(y + 1)$. The vertex is $(2, -1)$.

    The vertical axis has equation $x = 2$ and the parabola opens upward. See Figure 92. From the figure, the domain is $(-\infty, \infty)$ and the range is $[-1, \infty)$.

93. Rewrite the equation: $y = x^2 - 2x + 3 \Rightarrow y - 3 + 1 = x^2 - 2x + 1 \Rightarrow y - 2 = (x - 1)^2$.

    The equation $y - 2 = (x - 1)^2$ can be written $(x - 1)^2 = 4\left(\dfrac{1}{4}\right)(y - 2)$. The vertex is $(1, 2)$.

    The vertical axis has equation $x = 1$ and the parabola opens upward. See Figure 93. From the figure, the domain is $(-\infty, \infty)$ and the range is $[2, \infty)$.

94. Rewrite the equation: $y = x^2 + 6x + 5 \Rightarrow y - 5 + 9 = x^2 + 6x + 9 \Rightarrow y + 4 = (x + 3)^2$.

The equation $y + 4 = (x + 3)^2$ can be written $(x + 3)^2 = 4\left(\dfrac{1}{4}\right)(y + 4)$. The vertex is $(-3, -4)$.

The vertical axis has equation $x = -3$ and the parabola opens upward. See Figure 94. From the figure, the domain is $(-\infty, \infty)$ and the range is $[-4, \infty)$.

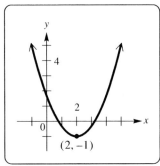

Figure 92

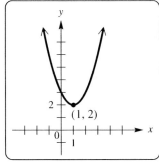

Figure 93

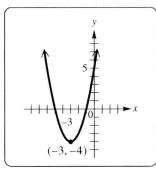

Figure 94

95. Rewrite the equation: $y = 2x^2 - 4x + 5 \Rightarrow y - 5 + 2 = 2(x^2 - 2x + 1) \Rightarrow y - 3 = 2(x - 1)^2$.

The equation $y - 3 = 2(x - 1)^2$ can be written $(x - 1)^2 = 4\left(\dfrac{1}{8}\right)(y - 3)$. The vertex is $(1, 3)$.

The vertical axis has equation $x = 1$ and the parabola opens upward. See Figure 95. From the figure, the domain is $(-\infty, \infty)$ and the range is $[3, \infty)$.

96. Rewrite the equation: $y = -3x^2 + 24x - 46 \Rightarrow y + 46 - 48 = -3(x^2 - 8x + 16) \Rightarrow y - 2 = -3(x - 4)^2$.

The equation $y - 2 = -3(x - 4)^2$ can be written $(x - 4)^2 = 4\left(-\dfrac{1}{12}\right)(y - 2)$. The vertex is $(4, 2)$.

The vertical axis has equation $x = 4$ and the parabola opens downward. See Figure 96. From the figure, the domain is $(-\infty, \infty)$ and the range is $(-\infty, 2]$.

97. The equation $x = y^2 + 2$ can be written $(y - 0)^2 = 4\left(\dfrac{1}{4}\right)(x - 2)$. The vertex is $(2, 0)$.

The horizontal axis has equation $y = 0$ and the parabola opens to the right. See Figure 97. From the figure, the domain is $[2, \infty)$ and the range is $(-\infty, \infty)$.

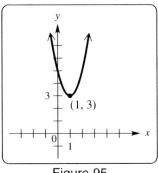

Figure 95

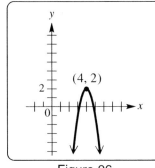

Figure 96

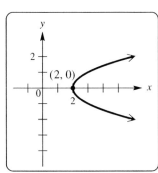

Figure 97

98. The equation $x = (y + 1)^2$ can be written $(y + 1)^2 = 4\left(\dfrac{1}{4}\right)(x - 0)$. The vertex is $(0, -1)$.

   The horizontal axis has equation $y = -1$ and the parabola opens to the right. See Figure 98. From the figure, the domain is $[0, \infty)$ and the range is $(-\infty, \infty)$.

99. The equation $x = (y - 3)^2$ can be written $(y - 3)^2 = 4\left(\dfrac{1}{4}\right)(x - 0)$. The vertex is $(0, 3)$.

   The horizontal axis has equation $y = 3$ and the parabola opens to the right. See Figure 99. From the figure, the domain is $[0, \infty)$ and the range is $(-\infty, \infty)$.

100. The equation $(y + 2)^2 = (x + 1)$ can be written $(y + 2)^2 = 4\left(\dfrac{1}{4}\right)(x + 1)$. The vertex is $(-1, -2)$.

   The horizontal axis has equation $y = -2$ and the parabola opens to the right. See Figure 100. From the figure, the domain is $[-1, \infty)$ and the range is $(-\infty, \infty)$.

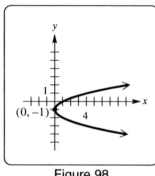

Figure 98

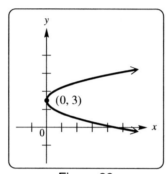

Figure 99

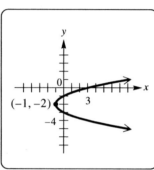

Figure 100

101. The equation $x = (y - 4)^2 + 2$ can be written $(y - 4)^2 = 4\left(\dfrac{1}{4}\right)(x - 2)$. The vertex is $(2, 4)$.

   The horizontal axis has equation $y = 4$ and the parabola opens to the right. See Figure 101. From the figure, the domain is $[2, \infty)$ and the range is $(-\infty, \infty)$.

102. The equation $x = -2(y + 3)^2$ can be written $(y + 3)^2 = 4\left(-\dfrac{1}{8}\right)(x - 0)$. The vertex is $(0, -3)$.

   The horizontal axis has equation $y = -3$ and the parabola opens to the left. See Figure 102. From the figure, the domain is $(-\infty, 0]$ and the range is $(-\infty, \infty)$.

103. Rewrite the equation: $x = \dfrac{2}{3}y^2 - 4y + 8 \Rightarrow \dfrac{3}{2}x = y^2 - 6y + 12 \Rightarrow \dfrac{3}{2}x - 12 + 9 = y^2 - 6y + 9 \Rightarrow$

   $\dfrac{3}{2}x - 3 = (y - 3)^2 \Rightarrow \dfrac{3}{2}(x - 2) = (y - 3)^2$. The equation $\dfrac{3}{2}(x - 2) = (y - 3)^2$ can be written

   $(y - 3)^2 = 4\left(\dfrac{3}{8}\right)(x - 2)$. The vertex is $(2, 3)$. The horizontal axis has equation $y = 3$ and the parabola

   opens to the right. See Figure 103. From the figure, the domain is $[2, \infty)$ and the range is $(-\infty, \infty)$.

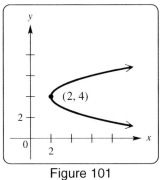

Figure 101

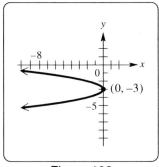

Figure 102

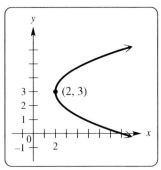

Figure 103

104. Rewrite the equation: $x = y^2 + 2y - 8 \Rightarrow x + 8 + 1 = y^2 + 2y + 1 \Rightarrow x + 9 = (y + 1)^2$

The equation $x + 9 = (y + 1)^2$ can be written $(y + 1)^2 = 4\left(\dfrac{1}{4}\right)(x + 9)$. The vertex is $(-9, -1)$.

The horizontal axis has equation $y = -1$ and the parabola opens to the right. See Figure 104.

From the figure, the domain is $[-9, \infty)$ and the range is $(-\infty, \infty)$.

105. Rewrite the equation: $x = -4y^2 - 4y - 3 \Rightarrow x + 3 - 1 = -4\left(y^2 + y + \dfrac{1}{4}\right) \Rightarrow x + 2 = -4\left(y + \dfrac{1}{2}\right)^2$

The equation $x + 2 = -4\left(y + \dfrac{1}{2}\right)^2$ can be written $\left(y + \dfrac{1}{2}\right)^2 = 4\left(-\dfrac{1}{16}\right)(x + 2)$. The vertex is $\left(-2, -\dfrac{1}{2}\right)$.

The horizontal axis has equation $y = -\dfrac{1}{2}$ and the parabola opens to the left. See Figure 105.

From the figure, the domain is $(-\infty, -2]$ and the range is $(-\infty, \infty)$.

106. Rewrite the equation: $x = 2y^2 - 4y + 6 \Rightarrow x - 6 + 2 = 2(y^2 - 2y + 1) \Rightarrow x - 4 = 2(y - 1)^2$

The equation $x - 4 = 2(y - 1)^2$ can be written $(y - 1)^2 = 4\left(\dfrac{1}{8}\right)(x - 4)$. The vertex is $(4, 1)$.

The horizontal axis has equation $y = 1$ and the parabola opens to the right. See Figure 106.

From the figure, the domain is $[4, \infty)$ and the range is $(-\infty, \infty)$.

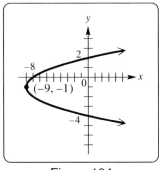

Figure 104

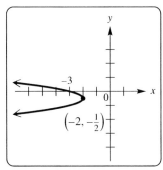

Figure 105

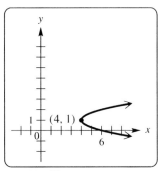

Figure 106

107. Rewrite the equation: $x = -2y^2 + 2y - 3 \Rightarrow x + 3 - \dfrac{1}{2} = -2\left(y^2 - y + \dfrac{1}{4}\right) \Rightarrow x + \dfrac{5}{2} = -2\left(y - \dfrac{1}{2}\right)^2$

The equation $x + \dfrac{5}{2} = -2\left(y - \dfrac{1}{2}\right)^2$ can be written $\left(y - \dfrac{1}{2}\right)^2 = 4\left(-\dfrac{1}{8}\right)\left(x + \dfrac{5}{2}\right)$. The vertex is $\left(-\dfrac{5}{2}, \dfrac{1}{2}\right)$.

The horizontal axis has equation $y = \dfrac{1}{2}$ and the parabola opens to the left. See Figure 107.

From the figure, the domain is $\left(-\infty, -\dfrac{5}{2}\right]$ and the range is $(-\infty, \infty)$.

108. Rewrite the equation: $2x = y^2 - 4y + 6 \Rightarrow 2x - 6 + 4 = y^2 - 4y + 4 \Rightarrow 2(x - 1) = (y - 2)^2$

The equation $2(x - 1) = (y - 2)^2$ can be written $(y - 2)^2 = 4\left(\dfrac{1}{2}\right)(x - 1)$. The vertex is $(1, 2)$.

The horizontal axis has equation $y = 2$ and the parabola opens to the right. See Figure 108.

From the figure, the domain is $[1, \infty)$ and the range is $(-\infty, \infty)$.

109. Rewrite the equation: $2x = y^2 - 2y + 9 \Rightarrow 2x - 9 + 1 = y^2 - 2y + 1 \Rightarrow 2(x - 4) = (y - 1)^2$

The equation $2(x - 4) = (y - 1)^2$ can be written $(y - 1)^2 = 4\left(\dfrac{1}{2}\right)(x - 4)$. The vertex is $(4, 1)$.

The horizontal axis has equation $y = 1$ and the parabola opens to the right. See Figure 109.

From the figure, the domain is $[4, \infty)$ and the range is $(-\infty, \infty)$.

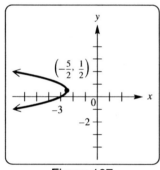
Figure 107

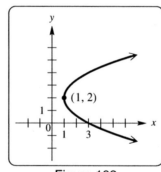
Figure 108

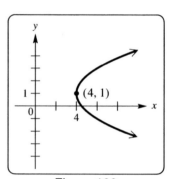
Figure 109

110. Rewrite the equation: $x = -3y^2 + 6y - 1 \Rightarrow x + 1 - 3 = -3(y^2 - 2y + 1) \Rightarrow x - 2 = -3(y - 1)^2$

The equation $x - 2 = -3(y - 1)^2$ can be written $(y - 1)^2 = 4\left(-\dfrac{1}{12}\right)(x - 2)$. The vertex is $(2, 1)$.

The horizontal axis has equation $y = 1$ and the parabola opens to the left. See Figure 110.

From the figure, the domain is $(-\infty, 2]$ and the range is $(-\infty, \infty)$.

111. Rewrite the equation: $y^2 - 4y + 4 = 4x + 4 \Rightarrow (y - 2)^2 = 4(x + 1)$

The equation $(y - 2)^2 = 4(x + 1)$ can be written $(y - 2)^2 = 4(1)(x + 1)$. The vertex is $(-1, 2)$.

The horizontal axis has equation $y = 2$ and the parabola opens to the right. See Figure 111.

From the figure, the domain is $[-1, \infty)$ and the range is $(-\infty, \infty)$.

112. Rewrite the equation: $y^2 + 2y + 1 = -2x + 4 \Rightarrow (y + 1)^2 = -2(x - 2)$

The equation $(y + 1)^2 = -2(x - 2)$ can be written $(y + 1)^2 = 4\left(-\dfrac{1}{2}\right)(x - 2)$. The vertex is $(2, -1)$.

The horizontal axis has equation $y = -1$ and the parabola opens to the left. See Figure 112.

From the figure, the domain is $(-\infty, 2]$ and the range is $(-\infty, \infty)$.

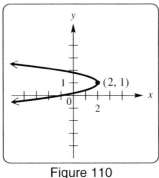

Figure 110

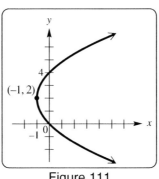

Figure 111

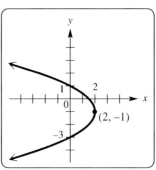

Figure 112

113. Since the directrix has equation $x = -c$, a point on the directrix has the form $(-c, y)$. Let (x, y) be a point on the parabola. By definition, the distance from the focus $(c, 0)$ to point (x, y) on the parabola must be equal to the distance from point $(-c, y)$ on the directrix to point (x, y) on the parabola. That is

$$\sqrt{(x - c)^2 + (y - 0)^2} = \sqrt{(x + c)^2 + (y - y)^2} \Rightarrow (x - c)^2 + y^2 = (x + c)^2 \Rightarrow$$

$$x^2 - 2xc + c^2 + y^2 = x^2 + 2xc + c^2 \Rightarrow -2xc + y^2 = 2xc \Rightarrow y^2 = 4xc.$$

114. (a) For Earth, $y = x - \dfrac{32.2}{1922}x^2$. For Mars, $y = x - \dfrac{12.6}{1922}x^2$.

Graph $y_1 = x - (32.2/1922)x^2$ and $y_2 = x - (12.6/1922)x^2$ as shown in Figure 114.

(b) From the graph the ball thrown on Earth hits the ground ($y = 0$) when $x \approx 153$ and the ball thrown on Mars hits the ground when $x \approx 60$. The difference in horizontal distance is about $153 - 60 = 93$ feet.

115. (a) For Mars, $y = \dfrac{19}{11}x - \dfrac{12.6}{3872}x^2$. For the moon, $y = \dfrac{19}{11}x - \dfrac{5.2}{3872}x^2$.

Graph $y_1 = (19/11)x - (12.6/3872)x^2$ and $y_2 = (19/11)x - (5.2/3872)x^2$ as shown in Figure 115.

(b) From the graph the ball thrown on Mars reaches a maximum height of $y \approx 229$ and the ball thrown on the moon reaches a maximum height of $y \approx 555$.

[0, 180] by [0, 120]
Xscl = 50   Yscl = 50

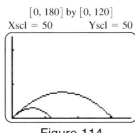

Figure 114

[0, 1500] by [0, 1000]
Xscl = 500   Yscl = 500

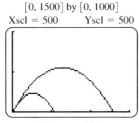

Figure 115

116. Consider a cross-section of the dish with the vertex located at $(0, 0)$ and focus located at $(0, 128.5)$. The equation for this parabola has the form $x^2 = 4cy$ and by substitution, $x^2 = 4(128.5)y \Rightarrow x^2 = 514y$. By noting that the radius of the dish is 150 and the y-coordinate of the point $(150, y)$ on the parabola corresponds to the depth of the dish, the depth can be found by substitution: $150^2 = 514y \Rightarrow y = \dfrac{150^2}{514} \approx 43.8$ feet.

117. $y = -\dfrac{5 \times 10^{-9}}{2(10^7)}(.4)^2 = -4 \times 10^{-17}$; The alpha particle is deflected 4×10^{-17} meter downward.

118. Let the vertex of the parabola be $(0, 10)$. The equation of the parabola is of the form $(x - h)^2 = 4c(y - k)$.

By substitution, the equation is $(x - 0)^2 = 4c(y - 10) \Rightarrow x^2 = 4c(y - 10)$. Since the parabola passes through the point $(200, 210)$, the value of c can be found by substitution:

$200^2 = 4c(210 - 10) \Rightarrow 200^2 = 4c(200) \Rightarrow 200 = 4c \Rightarrow c = 50$; The equation is $x^2 = 200(y - 10)$.

Noting that the x-coordinate of one of the other supports is 100, the height can be found by substitution:

$100^2 = 200(y - 10) \Rightarrow \dfrac{100^2}{200} = y - 10 \Rightarrow y = \dfrac{100^2}{200} + 10 = 60$ feet.

119. Let the vertex of the parabola be $(0, 12)$. The equation of the parabola is of the form $(x - h)^2 = 4c(y - k)$.

By substitution, the equation is $(x - 0)^2 = 4c(y - 12) \Rightarrow x^2 = 4c(y - 12)$. Since the parabola passes through the point $(6, 0)$, the value of c can be found by substitution:

$6^2 = 4c(0 - 12) \Rightarrow 36 = 4c(-12) \Rightarrow -3 = 4c \Rightarrow c = -\dfrac{3}{4}$; The equation is $x^2 = -3(y - 12)$.

Noting that the y-coordinate 9 feet up is 9, half the width can be found by substitution:

$x^2 = -3(9 - 12) \Rightarrow x^2 = 9 \Rightarrow x = 3$; The width is $2(3) = 6$ feet.

120. Let the vertex be $(0, 0)$. Substitute the point $\left(4, \dfrac{5}{2}\right)$ into $y^2 = 4cx$ and solve for c: $\left(\dfrac{5}{2}\right)^2 = 4c(4) \Rightarrow c = \dfrac{25}{64}$.

The bulb should be $\dfrac{25}{64}$ inch from the vertex.

6.2: Ellipses and Hyperbolas

1. G. This is an ellipse with $a^2 = 16$, $b^2 = 4$ and $c = \sqrt{16 - 4} = \sqrt{12} = 2\sqrt{3}$. The Foci are $(0, \pm 2\sqrt{3})$.

2. B. This is an ellipse with $a^2 = 16$, $b^2 = 4$ and $c = \sqrt{16 - 4} = \sqrt{12} = 2\sqrt{3}$. The Foci are $(\pm 2\sqrt{3}, 0)$.

3. F. This is a hyperbola centered at $(0, 0)$ with a horizontal transverse axis.

4. C. This is a a hyperbola with $a^2 = 4$, $b^2 = 16$ and $c = \sqrt{4 + 16} = \sqrt{20} = 2\sqrt{5}$. The Foci are $(0, \pm 2\sqrt{5})$.

5. E. Since $h = -2$ and $k = 4$, this is an ellipse centered at $(-2, 4)$.

6. H. Since $h = 2$ and $k = -4$, this is an ellipse centered at $(2, -4)$ with a vertical major axis.

7. D. Since $h = -2$ and $k = 4$, this is a hyperbola centered at $(-2, 4)$.

8. A. Since $h = 2$ and $k = 4$, this is a hyperbola centered at $(2, 4)$.

9. A circle can be interpreted as an ellipse whose foci have the same coordinates. The "coinciding foci" give the center of the circle.

10. The graph of the ellipse exists for values of x from -3 to 3. The domain is $[-3, 3]$. The graph of the ellipse exists for values of y from -5 to 5. The range is $[-5, 5]$.

11. $\dfrac{x^2}{9} + \dfrac{y^2}{4} = 1 \Rightarrow a = 3$ and $b = 2$. $a^2 - b^2 = 3^2 - 2^2 = 5 = c^2 \Rightarrow c = \sqrt{5}$. The foci are $(\pm \sqrt{5}, 0)$.

The endpoints of the major axis (vertices) are $(\pm 3, 0)$ so the domain is $[-3, 3]$. The endpoints of the minor axis are $(0, \pm 2)$ so the range is $[-2, 2]$. The ellipse is graphed in Figure 11.

12. $\dfrac{x^2}{16} + \dfrac{y^2}{36} = 1 \Rightarrow a = 6$ and $b = 4$. $a^2 - b^2 = 6^2 - 4^2 = 20 = c^2 \Rightarrow c = \sqrt{20}$. The foci are $(0, \pm\sqrt{20})$ or $(0, \pm 2\sqrt{5})$, the endpoints of the major axis (vertices) are $(0, \pm 6)$ so the range is $[-6, 6]$. The endpoints of the minor axis are $(\pm 4, 0)$ so the domain is $[-4, 4]$. The ellipse is graphed in Figure 12.

13. $9x^2 + 6y^2 = 54 \Rightarrow \dfrac{x^2}{6} + \dfrac{y^2}{9} = 1 \Rightarrow a = 3$ and $b = \sqrt{6}$. $a^2 - b^2 = 9 - 6 = 3 = c^2 \Rightarrow c = \sqrt{3}$. The foci are $(0, \pm\sqrt{3})$. The endpoints of the major axis (vertices) are $(0, \pm 3)$ so the range is $[-3, 3]$. The endpoints of the minor axis are $(\pm\sqrt{6}, 0)$ so the domain is $[-\sqrt{6}, \sqrt{6}]$. The ellipse is graphed in Figure 13.

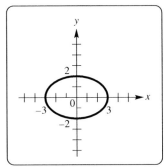

Figure 11

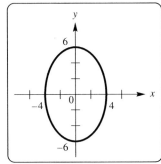

Figure 12

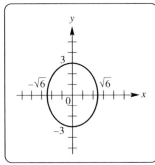

Figure 13

14. $12x^2 + 8y^2 = 96 \Rightarrow \dfrac{x^2}{8} + \dfrac{y^2}{12} = 1 \Rightarrow a = \sqrt{12}$ and $b = \sqrt{8}$. $a^2 - b^2 = 12 - 8 = 4 = c^2 \Rightarrow c = 2$. The foci are $(0, \pm 2)$. The endpoints of the major axis (vertices) are $(0, \pm 2\sqrt{3})$ so the range is $[-2\sqrt{3}, 2\sqrt{3}]$. The endpoints of the minor axis are $(\pm 2\sqrt{2}, 0)$ so the domain is $[-2\sqrt{2}, 2\sqrt{2}]$. See Figure 14.

15. $\dfrac{25y^2}{36} + \dfrac{64x^2}{9} = 1 \Rightarrow \dfrac{y^2}{\frac{36}{25}} + \dfrac{x^2}{\frac{9}{64}} = 1 \Rightarrow a = \sqrt{\dfrac{36}{25}} = \dfrac{6}{5}$ and $b = \sqrt{\dfrac{9}{64}} = \dfrac{3}{8}$

The endpoints of the major axis (vertices) are $\left(0, \pm\dfrac{6}{5}\right)$ so the range is $\left[-\dfrac{6}{5}, \dfrac{6}{5}\right]$. The endpoints of the minor axis are $\left(\pm\dfrac{3}{8}, 0\right)$ so the domain is $\left[-\dfrac{3}{8}, \dfrac{3}{8}\right]$. See Figure 15.

16. $\dfrac{16y^2}{9} + \dfrac{121x^2}{25} = 1 \Rightarrow \dfrac{y^2}{\frac{9}{16}} + \dfrac{x^2}{\frac{25}{121}} = 1 \Rightarrow a = \sqrt{\dfrac{9}{16}} = \dfrac{3}{4}$ and $b = \sqrt{\dfrac{25}{121}} = \dfrac{5}{11}$

The endpoints of the major axis (vertices) are $\left(0, \pm\dfrac{3}{4}\right)$ so the range is $\left[-\dfrac{3}{4}, \dfrac{3}{4}\right]$. The endpoints of the minor axis are $\left(\pm\dfrac{5}{11}, 0\right)$ so the domain is $\left[-\dfrac{5}{11}, \dfrac{5}{11}\right]$. See Figure 16.

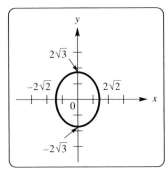

Figure 14

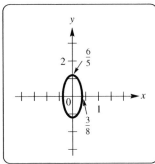

Figure 15

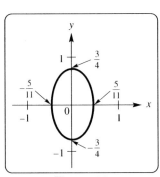

Figure 16

17. The ellipse is centered at $(1, -3)$. The major axis is vertical and has length $2a = 10$. The length of the minor axis is $2b = 6$. The graph is shown in Figure 17. The domain is $[-2, 4]$ and the range is $[-8, 2]$.

18. The ellipse is centered at $(-3, 2)$. The major axis is vertical and has length $2a = 12$. The length of the minor axis is $2b = 8$. The graph is shown in Figure 18. The domain is $[-7, 1]$ and the range is $[-4, 8]$.

19. The ellipse is centered at $(2, 1)$. The major axis is horizontal and has length $2a = 8$. The length of the minor axis is $2b = 6$. The graph is shown in Figure 19. The domain is $[-2, 6]$ and the range is $[-2, 4]$.

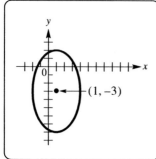

Figure 17

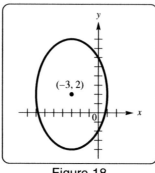

Figure 18

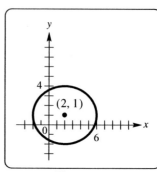

Figure 19

20. The ellipse is centered at $(-3, -2)$. The major axis is vertical and has length $2a = 12$. The length of the minor axis is $2b = 10$. The graph is shown in Figure 20. The domain is $[-8, 2]$ and the range is $[-8, 4]$.

21. The ellipse is centered at $(-1, 2)$. The major axis is horizontal and has length $2a = 16$. The length of the minor axis is $2b = 14$. The graph is shown in Figure 21. The domain is $[-9, 7]$ and the range is $[-5, 9]$.

22. The ellipse is centered at $(4, -2)$. The major axis is horizontal and has length $2a = 6$. The length of the minor axis is $2b = 4$. The graph is shown in Figure 22. The domain is $[1, 7]$ and the range is $[-4, 0]$.

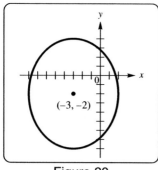

Figure 20

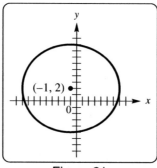

Figure 21

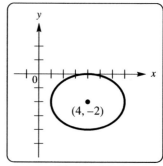

Figure 22

23. The ellipse is centered between the foci at $(0, 0)$. The major axis is horizontal with $a = 4$. Since the foci are $(\pm 2, 0)$, we know that $c = 2$. Since $c^2 = a^2 - b^2$, the value of b can be found by substitution:

$b^2 = a^2 - c^2 = 4^2 - 2^2 = 16 - 4 = 12 \Rightarrow b = \sqrt{12}$. The equation is $\dfrac{x^2}{16} + \dfrac{y^2}{12} = 1$.

24. The ellipse is centered between the foci at $(0, 0)$. The major axis is vertical with $a = 3$. Since the foci are $(0, \pm\sqrt{3})$, we know that $c = \sqrt{3}$. Since $c^2 = a^2 - b^2$, the value of b can be found by substitution:

$b^2 = a^2 - c^2 = 3^2 - (\sqrt{3})^2 = 9 - 3 = 6 \Rightarrow b = \sqrt{6}$. The equation is $\dfrac{x^2}{6} + \dfrac{y^2}{9} = 1$.

25. The ellipse is centered between the endpoint of the major axis at $(0, 0)$. The major axis is horizontal with $a = 6$.

    Since $c^2 = a^2 - b^2$, the value of b can be found by substitution:

    $b^2 = a^2 - c^2 = 6^2 - 4^2 = 36 - 16 = 20 \Rightarrow b = \sqrt{20}$. The equation is $\dfrac{x^2}{36} + \dfrac{y^2}{20} = 1$.

26. The ellipse is centered between the vertices at $(0, 0)$. The major axis is vertical with $a = 5$.

    Since $b = 2$, the equation is $\dfrac{x^2}{4} + \dfrac{y^2}{25} = 1$.

27. Since the center is $(3, -2)$, we know that $h = 3$ and $k = -2$. Since $c^2 = a^2 - b^2$, the value of b can be found by

    substitution: $b^2 = a^2 - c^2 = 5^2 - 3^2 = 25 - 9 = 16 \Rightarrow b = 4$. The major axis is vertical so the equation is

    $\dfrac{(x - 3)^2}{16} + \dfrac{(y + 2)^2}{25} = 1$.

28. Since the center is $(2, 0)$, we know that $h = 0$ and $k = 0$. Since minor axis has length 6, $b = 3$. Since the

    major axis has length 9, $a = \dfrac{9}{2}$. The major axis is horizontal so the equation is $\dfrac{(x - 2)^2}{\frac{81}{4}} + \dfrac{y^2}{9} = 1$ or

    $\dfrac{4(x - 2)^2}{81} + \dfrac{y^2}{9} = 1$.

29. The ellipse is centered between the foci at $(0, 0)$. The major axis is vertical with $a = 3$. Since the foci are

    $(0, \pm 2)$, we know that $c = 2$. Since $c^2 = a^2 - b^2$, the value of b can be found by substitution:

    $b^2 = a^2 - c^2 = 3^2 - 2^2 = 9 - 4 = 5 \Rightarrow b = \sqrt{5}$. The equation is $\dfrac{y^2}{9} + \dfrac{x^2}{5} = 1$.

30. The ellipse is centered between the foci at $(0, 0)$. The major axis is horizontal. Since the foci are $(\pm 5, 0)$, we

    know that $c = 5$. Since $c^2 = a^2 - b^2$ and $b = 2$, the value of a can be found by substitution:

    $a^2 = b^2 + c^2 = 2^2 + 5^2 = 4 + 25 = 29 \Rightarrow a = \sqrt{29}$. The equation is $\dfrac{x^2}{29} + \dfrac{y^2}{4} = 1$.

31. Since the center is $(5, 2)$, we know that $h = 5$ and $k = 2$. Since the minor axis is horizontal and has length 8,

    $b = 4$. Since $c^2 = a^2 - b^2$ and $c = 3$, the value of a can be found by substitution:

    $a^2 = b^2 + c^2 = 4^2 + 3^2 = 16 + 9 = 25 \Rightarrow a = 5$. The equation is $\dfrac{(x - 5)^2}{25} + \dfrac{(y - 2)^2}{16} = 1$.

32. Since the center is $(-3, 6)$, we know that $h = -3$ and $k = 6$. Since the major axis has length 10, $a = 5$. Since

    $c^2 = a^2 - b^2$ and $c = 2$, the value of b can be found by substitution:

    $b^2 = a^2 - c^2 = 5^2 - 2^2 = 25 - 4 = 21 \Rightarrow b = \sqrt{21}$. The major axis is vertical so the equation is

    $\dfrac{(y + 3)^2}{21} + \dfrac{(y - 6)^2}{25} = 1$.

33. The ellipse is centered between the vertices at $(4, 5)$ and $a = \dfrac{9 - 1}{2} = 4$. The major axis is vertical. Since the

    minor axis has length 6, $b = 3$. The equation is $\dfrac{(x - 4)^2}{9} + \dfrac{(y - 5)^2}{16} = 1$.

34. The ellipse is centered between the foci at $(2, -3)$ and $c = 5$. Since the point $(2, 1)$ is on the graph and it is

    located vertically above the center, it is an endpoint of the minor axis. Thus $b = 4$. Since $c^2 = a^2 - b^2$, the

    value of a can be found by substitution: $a^2 = b^2 + c^2 = 4^2 + 5^2 = 16 + 25 = 41 \Rightarrow a = \sqrt{41}$.

    The major axis is horizontal so the equation is $\dfrac{(x - 2)^2}{41} + \dfrac{(y + 3)^2}{16} = 1$.

35. $9x^2 + 18x + 4y^2 - 8y - 23 = 0 \Rightarrow 9(x^2 + 2x) + 4(y^2 - 2y) = 23 \Rightarrow$

$9(x^2 + 2x + 1) + 4(y^2 - 2y + 1) = 23 + 9 + 4 \Rightarrow 9(x + 1)^2 + 4(y - 1)^2 = 36 \Rightarrow$

$\dfrac{(x + 1)^2}{4} + \dfrac{(y - 1)^2}{9} = 1$; The center is $(-1, 1)$. The vertices are $(-1, 1 - 3), (-1, 1 + 3)$ or $(-1, -2), (-1, 4)$.

36. $9x^2 - 36x + 16y^2 - 64y - 44 = 0 \Rightarrow 9(x^2 - 4x) + 16(y^2 - 4y) = 44 \Rightarrow$

$9(x^2 - 4x + 4) + 16(y^2 - 4y + 4) = 44 + 36 + 64 \Rightarrow 9(x - 2)^2 + 16(y - 2)^2 = 144 \Rightarrow$

$\dfrac{(x - 2)^2}{16} + \dfrac{(y - 2)^2}{9} = 1$; The center is $(2, 2)$. The vertices are $(2 - 4, 2), (2 + 4, 2)$ or $(-2, 2), (6, 2)$.

37. $4x^2 + 8x + y^2 + 2y + 1 = 0 \Rightarrow 4(x^2 + 2x) + (y^2 + 2y) = -1 \Rightarrow$

$4(x^2 + 2x + 1) + (y^2 + 2y + 1) = -1 + 4 + 1 \Rightarrow 4(x + 1)^2 + (y + 1)^2 = 4 \Rightarrow$

$\dfrac{(x + 1)^2}{1} + \dfrac{(y + 1)^2}{4} = 1$; The center is $(-1, -1)$. The vertices are $(-1, -1 - 2), (-1, -1 + 2)$ or $(-1, -3), (-1, 1)$.

38. $x^2 - 6x + 9y^2 = 0 \Rightarrow (x^2 - 6x) + 9y^2 = 0 \Rightarrow (x^2 - 6x + 9) + 9y^2 = 0 + 9 \Rightarrow$

$(x - 3)^2 + 9(y - 0)^2 = 9 \Rightarrow \dfrac{(x - 3)^2}{9} + \dfrac{(y - 0)^2}{1} = 1$

The center is $(3, 0)$. The vertices are $(3 - 3, 0), (3 + 3, 0)$ or $(0, 0), (6, 0)$.

39. $4x^2 + 16x + 5y^2 - 10y + 1 = 0 \Rightarrow 4(x^2 + 4x) + 5(y^2 - 2y) = -1 \Rightarrow$

$4(x^2 + 4x + 4) + 5(y^2 - 2y + 1) = -1 + 16 + 5 \Rightarrow 4(x + 2)^2 + 5(y - 1)^2 = 20 \Rightarrow$

$\dfrac{(x + 2)^2}{5} + \dfrac{(y - 1)^2}{4} = 1$; The center is $(-2, 1)$. The vertices are $(-2 - \sqrt{5}, 1), (-2 + \sqrt{5}, 1)$.

40. $2x^2 + 4x + 3y^2 - 18y + 23 = 0 \Rightarrow 2(x^2 + 2x) + 3(y^2 - 6y) = -23 \Rightarrow$

$2(x^2 + 2x + 1) + 3(y^2 - 6y + 9) = -23 + 2 + 27 \Rightarrow 2(x + 1)^2 + 3(y - 3)^2 = 6 \Rightarrow$

$\dfrac{(x + 1)^2}{3} + \dfrac{(y - 3)^2}{2} = 1$; The center is $(-1, 3)$. The vertices are $(-1 - \sqrt{3}, 3), (-1 + \sqrt{3}, 3)$.

41. $16x^2 - 16x + 4y^2 + 12y = 51 \Rightarrow 16(x^2 - x) + 4(y^2 + 3y) = 51 \Rightarrow$

$16\left(x^2 - x + \dfrac{1}{4}\right) + 4\left(y^2 + 3y + \dfrac{9}{4}\right) = 51 + 4 + 9 \Rightarrow 16\left(x - \dfrac{1}{2}\right)^2 + 4\left(y + \dfrac{3}{2}\right)^2 = 64 \Rightarrow$

$\dfrac{(x - \frac{1}{2})^2}{4} + \dfrac{(y + \frac{3}{2})^2}{16} = 1$

The center is $\left(\dfrac{1}{2}, -\dfrac{3}{2}\right)$. The vertices are $\left(\dfrac{1}{2}, -\dfrac{3}{2} - 4\right), \left(\dfrac{1}{2}, -\dfrac{3}{2} + 4\right)$ or $\left(\dfrac{1}{2}, -\dfrac{11}{2}\right), \left(\dfrac{1}{2}, \dfrac{5}{2}\right)$.

42. $16x^2 + 48x + 4y^2 - 20y + 57 = 0 \Rightarrow 16(x^2 + 3x) + 4(y^2 - 5y) = -57 \Rightarrow$

$16\left(x^2 + 3x + \dfrac{9}{4}\right) + 4\left(y^2 - 5y + \dfrac{25}{4}\right) = -57 + 36 + 25 \Rightarrow 16\left(x + \dfrac{3}{2}\right)^2 + 4\left(y - \dfrac{5}{2}\right)^2 = 4 \Rightarrow$

$\dfrac{(x + \frac{3}{2})^2}{\frac{1}{4}} + \dfrac{(y - \frac{5}{2})^2}{1} = 1$

The center is $\left(-\dfrac{3}{2}, \dfrac{5}{2}\right)$. The vertices are $\left(-\dfrac{3}{2}, \dfrac{5}{2} - 1\right), \left(-\dfrac{3}{2}, \dfrac{5}{2} + 1\right)$ or $\left(-\dfrac{3}{2}, \dfrac{3}{2}\right), \left(-\dfrac{3}{2}, \dfrac{7}{2}\right)$.

43. The transverse axis is horizontal with $a = 4$ and $b = 3$. The asymptotes are $y = \pm\dfrac{3}{4}x$. See Figure 43.

The domain is $(-\infty, -4] \cup [4, \infty)$ and the range is $(-\infty, \infty)$.

44. The transverse axis is vertical with $a = 3$ and $b = 3$. The asymptotes are $y = \pm x$. See Figure 44.

The domain is $(-\infty, \infty)$ and the range is $(-\infty, -3] \cup [3, \infty)$.

45. $49y^2 - 36x^2 = 1764 \Rightarrow \dfrac{y^2}{36} - \dfrac{x^2}{49} = 1$. The transverse axis is vertical with $a = 6$ and $b = 7$. The asymptotes

are $y = \pm\dfrac{6}{7}x$. See Figure 45. The domain is $(-\infty, \infty)$ and the range is $(-\infty, -6] \cup [6, \infty)$.

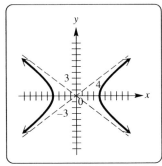
Figure 43

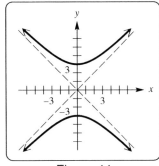
Figure 44

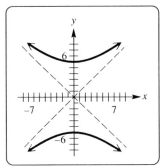
Figure 45

46. $144x^2 - 49y^2 = 7056 \Rightarrow \dfrac{x^2}{49} - \dfrac{y^2}{144} = 1$. The transverse axis is horizontal with $a = 7$ and $b = 12$. The

asymptotes are $y = \pm\dfrac{12}{7}x$. See Figure 46. The domain is $(-\infty, -7] \cup [7, \infty)$ and the range is $(-\infty, \infty)$.

47. $\dfrac{4x^2}{9} - \dfrac{25y^2}{16} = 1 \Rightarrow \dfrac{x^2}{\frac{9}{4}} - \dfrac{y^2}{\frac{16}{25}} = 1$. The transverse axis is horizontal with $a = \dfrac{3}{2}$ and $b = \dfrac{4}{5}$. The asymptotes

are $y = \pm\dfrac{8}{15}x$. See Figure 47. The domain is $\left(-\infty, -\dfrac{3}{2}\right] \cup \left[\dfrac{3}{2}, \infty\right)$ and the range is $(-\infty, \infty)$.

48. $x^2 - y^2 = 1 \Rightarrow \dfrac{x^2}{1} - \dfrac{y^2}{1} = 1$. The transverse axis is horizontal with $a = 1$ and $b = 1$. The asymptotes are

$y = \pm x$. See Figure 48. The domain is $(-\infty, -1] \cup [1, \infty)$ and the range is $(-\infty, \infty)$.

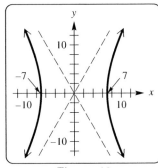
Figure 46

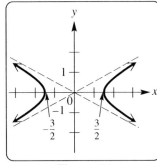
Figure 47

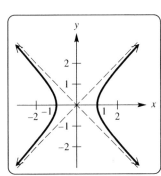
Figure 48

49. $9x^2 - 4y^2 = 1 \Rightarrow \dfrac{x^2}{\frac{1}{9}} - \dfrac{y^2}{\frac{1}{4}} = 1$. The transverse axis is horizontal with $a = \dfrac{1}{3}$ and $b = \dfrac{1}{2}$. The asymptotes are

$y = \pm\dfrac{3}{2}x$. See Figure 49. The domain is $\left(-\infty, -\dfrac{1}{3}\right] \cup \left[\dfrac{1}{3}, \infty\right)$ and the range is $(-\infty, \infty)$.

50. $25y^2 - 9x^2 = 1 \Rightarrow \dfrac{y^2}{\frac{1}{25}} - \dfrac{x^2}{\frac{1}{9}} = 1$. The transverse axis is vertical with $a = \dfrac{1}{5}$ and $b = \dfrac{1}{3}$. The asymptotes

are $y = \pm\dfrac{3}{5}x$. See Figure 50. The domain is $(-\infty, \infty)$ and the range is $\left(-\infty, -\dfrac{1}{5}\right] \cup \left[\dfrac{1}{5}, \infty\right)$.

51. The center is $(1, -3)$ and the transverse axis is horizontal with $a = 3$ and $b = 5$. See Figure 51.

The domain is $(-\infty, -2] \cup [4, \infty)$ and the range is $(-\infty, \infty)$.

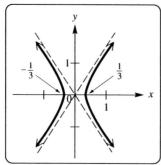

Figure 49

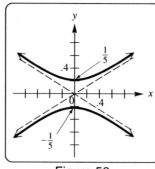

Figure 50

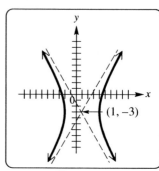

Figure 51

52. The center is $(-3, 2)$ and the transverse axis is horizontal with $a = 4$ and $b = 6$. See Figure 52.

The domain is $(-\infty, -7] \cup [1, \infty)$ and the range is $(-\infty, \infty)$.

53. The center is $(-1, 5)$ and the transverse axis is vertical with $a = 2$ and $b = 3$. See Figure 53.

The domain is $(-\infty, \infty)$ and the range is $(-\infty, 3] \cup [7, \infty)$.

54. The center is $(3, -1)$ and the transverse axis is vertical with $a = 5$ and $b = 6$. See Figure 54.

The domain is $(-\infty, \infty)$ and the range is $(-\infty, -6] \cup [4, \infty)$.

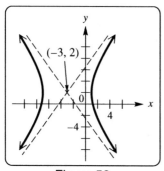

Figure 52

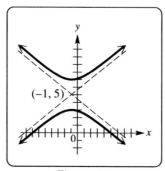

Figure 53

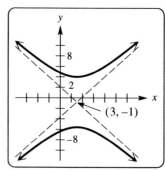

Figure 54

55. $16(x + 5)^2 - (y - 3)^2 = 1 \Rightarrow \dfrac{(x + 5)^2}{\frac{1}{16}} - \dfrac{(y - 3)^2}{1} = 1$

The center is $(-5, 3)$ and the transverse axis is horizontal with $a = \dfrac{1}{4}$ and $b = 1$. See Figure 55.

The domain is $\left(-\infty, -5 - \dfrac{1}{4}\right] \cup \left[-5 + \dfrac{1}{4}, \infty\right)$ or $\left(-\infty, -\dfrac{21}{4}\right] \cup \left[-\dfrac{19}{4}, \infty\right)$ and the range is $(-\infty, \infty)$.

56. $4(x + 9)^2 - 25(y + 6)^2 = 100 \Rightarrow \dfrac{(x + 9)^2}{25} - \dfrac{(y + 6)^2}{4} = 1$

The center is $(-9, -6)$ and the transverse axis is horizontal with $a = 5$ and $b = 2$. See Figure 56.

The domain is $(-\infty, -14] \cup [-4, \infty)$ and the range is $(-\infty, \infty)$.

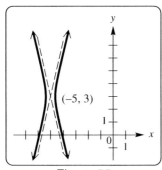

Figure 55

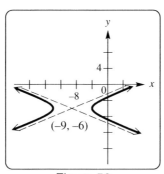

Figure 56

57. The hyperbola has a horizontal transverse axis with $c = 4$. The x-intercepts coincide with the vertices so $a = 3$.

The center is located between the foci at $(0, 0)$. Since $c^2 = a^2 + b^2$, the value of b can be found by substitution:

$b^2 = c^2 - a^2 = 4^2 - 3^2 = 16 - 9 = 7 \Rightarrow b = \sqrt{7}$. The equation is $\dfrac{x^2}{9} - \dfrac{y^2}{7} = 1$.

58. The hyperbola has a vertical transverse axis with $c = 3\sqrt{3}$. The y-intercepts coincide with the vertices so $a = 5$.

The center is located between the foci at $(0, 0)$. Since $c^2 = a^2 + b^2$, the value of b can be found by substitution:

$b^2 = c^2 - a^2 = (3\sqrt{3})^2 - 5^2 = 27 - 25 = 2 \Rightarrow b = \sqrt{2}$. The equation is $\dfrac{y^2}{25} - \dfrac{x^2}{2} = 1$.

59. The asymptotes intersect at the origin so the center is $(0, 0)$. The hyperbola has a vertical transverse axis and

the y-intercepts coincide with the vertices so $a = 3$. From the asymptotes, $\dfrac{a}{b} = \dfrac{3}{5}$ with $a = 3 \Rightarrow b = 5$.

The equation is $\dfrac{y^2}{9} - \dfrac{x^2}{25} = 1$.

60. The hyperbola has a vertical transverse axis. The y-intercept coincides with a vertex so $a = 2$.

The equation is of the form $\dfrac{y^2}{a^2} - \dfrac{x^2}{b^2} = 1$. By substitution using the point $(2, 3)$:

$\dfrac{3^2}{2^2} - \dfrac{2^2}{b^2} = 1 \Rightarrow \dfrac{9}{4} - 1 = \dfrac{4}{b^2} \Rightarrow \dfrac{5}{4} = \dfrac{4}{b^2} \Rightarrow 5b^2 = 16 \Rightarrow b^2 = \dfrac{16}{5}$

The equation is $\dfrac{y^2}{4} - \dfrac{x^2}{\frac{16}{5}} = 1$ or $\dfrac{y^2}{4} - \dfrac{5x^2}{16} = 1$.

61. The center is located between the vertices at $(0, 0)$. The hyperbola has a vertical transverse axis with $a = 6$.

From the asymptotes, $\dfrac{a}{b} = \dfrac{1}{2}$ with $a = 6 \Rightarrow b = 12$. The equation is $\dfrac{y^2}{36} - \dfrac{x^2}{144} = 1$.

62. The center is located between the vertices at $(0, 0)$. The hyperbola has a horizontal transverse axis with $a = 10$.

From the asymptotes, $\dfrac{b}{a} = 5$ with $a = 10 \Rightarrow b = 50$. The equation is $\dfrac{x^2}{100} - \dfrac{y^2}{2500} = 1$.

63. The center is located between the vertices at (0, 0). The hyperbola has a horizontal transverse axis with $a = 3$.

    The equation is of the form $\dfrac{x^2}{a^2} - \dfrac{y^2}{b^2} = 1$. By substitution using the point (6, 1):

    $\dfrac{6^2}{3^2} - \dfrac{1^2}{b^2} = 1 \Rightarrow \dfrac{36}{9} - 1 = \dfrac{1}{b^2} \Rightarrow 3 = \dfrac{1}{b^2} \Rightarrow b^2 = \dfrac{1}{3}$. The equation is $\dfrac{x^2}{3} - \dfrac{y^2}{\frac{1}{3}} = 1$ or $\dfrac{x^2}{3} - 3y^2 = 1$.

64. The center is located between the vertices at (0, 0). The hyperbola has a vertical transverse axis with $a = 5$.

    The equation is of the form $\dfrac{y^2}{a^2} - \dfrac{x^2}{b^2} = 1$. By substitution using the point (3, 10):

    $\dfrac{10^2}{5^2} - \dfrac{3^2}{b^2} = 1 \Rightarrow \dfrac{100}{25} - 1 = \dfrac{9}{b^2} \Rightarrow 3 = \dfrac{9}{b^2} \Rightarrow b^2 = 3$. The equation is $\dfrac{y^2}{25} - \dfrac{x^2}{3} = 1$.

65. The center is located between the foci at (0, 0). The hyperbola has a vertical transverse axis with $c = \sqrt{13}$.

    From the asymptotes, $\dfrac{a}{b} = 5$. Also $c^2 = a^2 + b^2 \Rightarrow a^2 + b^2 = 13$. Solving these equations simultaneously

    results in $a^2 = \dfrac{25}{2}$ and $b^2 = \dfrac{1}{2}$. The equation is $\dfrac{y^2}{\frac{25}{2}} - \dfrac{x^2}{\frac{1}{2}} = 1$ or $\dfrac{2y^2}{25} - 2x^2 = 1$.

66. The center is located between the foci at (0, 0). The hyperbola has a horizontal transverse axis with $c = 3\sqrt{5}$.

    From the asymptotes, $\dfrac{b}{a} = 2$. Also $c^2 = a^2 + b^2 \Rightarrow a^2 + b^2 = 45$. Solving these equations simultaneously

    results in $a^2 = 9$ and $b^2 = 36$. The equation is $\dfrac{x^2}{9} - \dfrac{y^2}{36} = 1$.

67. The center is located between the vertices at (4, 3). The hyperbola has a vertical transverse axis with $a = 2$.

    From the asymptotes, $\dfrac{a}{b} = 7$ with $a = 2 \Rightarrow b = \dfrac{2}{7}$.

    The equation is $\dfrac{(y-3)^2}{4} - \dfrac{(x-4)^2}{\frac{4}{49}} = 1$ or $\dfrac{(y-3)^2}{4} - \dfrac{49(x-4)^2}{4} = 1$.

68. The center is located between the vertices at (3, –2). The hyperbola has a horizontal transverse axis with $a = 2$.

    From the asymptotes, $\dfrac{b}{a} = \dfrac{3}{2}$ with $a = 2 \Rightarrow b = 3$. The equation is $\dfrac{(x-3)^2}{4} - \dfrac{(y+2)^2}{9} = 1$.

69. With center (1, –2) and vertex (3, –2), we know the hyperbola has a horizontal transverse axis with $a = 2$.

    With center (1, –2) and focus (4, –2), we know $c = 3$. Since $c^2 = a^2 + b^2$, the value of b can be found by

    substitution: $b^2 = c^2 - a^2 = 3^2 - 2^2 = 9 - 4 = 5 \Rightarrow b = \sqrt{5}$. The equation is $\dfrac{(x-1)^2}{4} - \dfrac{(y+2)^2}{5} = 1$.

70. With center (9, –7) and vertex (9, –1), we know the hyperbola has a vertical transverse axis with $a = 6$.

    With center (9, –7) and focus (9, 3), we know $c = 10$. Since $c^2 = a^2 + b^2$, the value of b can be found by

    substitution: $b^2 = c^2 - a^2 = 10^2 - 6^2 = 100 - 36 = 64 \Rightarrow b = 8$. The equation is $\dfrac{(y+7)^2}{36} - \dfrac{(x-9)^2}{64} = 1$.

71. $x^2 - 2x - y^2 + 2y = 4 \Rightarrow (x^2 - 2x + 1) - (y^2 - 2y + 1) = 4 + 1 - 1 \Rightarrow (x-1)^2 - (y-1)^2 = 4 \Rightarrow$

    $\dfrac{(x-1)^2}{4} - \dfrac{(y-1)^2}{4} = 1$. The center is (1, 1). The vertices are $(1-2, 1), (1+2, 1)$ or $(-1, 1), (3, 1)$.

72. $y^2 + 4y - x^2 + 2x = 6 \Rightarrow (y^2 + 4y + 4) - (x^2 - 2x + 1) = 6 + 4 - 1 \Rightarrow (y+2)^2 - (x-1)^2 = 9 \Rightarrow$

    $\dfrac{(y+2)^2}{9} - \dfrac{(x-1)^2}{9} = 1$. The center is (1, –2). The vertices are $(1, -2-3), (1, -2+3)$ or $(1, -5), (1, 1)$.

73. $3y^2 + 24y - 2x^2 + 12x + 24 = 0 \Rightarrow 3(y^2 + 8y) - 2(x^2 - 6x) = -24 \Rightarrow$

$3(y^2 + 8y + 16) - 2(x^2 - 6x + 9) = -24 + 48 - 18 \Rightarrow 3(y + 4)^2 - 2(x - 3)^2 = 6 \Rightarrow$

$\dfrac{(y + 4)^2}{2} - \dfrac{(x - 3)^2}{3} = 1$. The center is $(3, -4)$. The vertices are $(3, -4 - \sqrt{2}), (3, -4 + \sqrt{2})$.

74. $4x^2 + 16x - 9y^2 + 18y = 29 \Rightarrow 4(x^2 + 4x) - 9(y^2 - 2y) = 29 \Rightarrow$

$4(x^2 + 4x + 4) - 9(y^2 - 2y + 1) = 29 + 16 - 9 \Rightarrow 4(x + 2)^2 - 9(y - 1)^2 = 36 \Rightarrow$

$\dfrac{(x + 2)^2}{9} - \dfrac{(y - 1)^2}{4} = 1$. The center is $(-2, 1)$. The vertices are $(-2 - 3, 1), (-2 + 3, 1)$ or $(-5, 1), (1, 1)$.

75. $x^2 - 6x - 2y^2 + 7 = 0 \Rightarrow (x^2 - 6x + 9) - 2y^2 = -7 + 9 \Rightarrow (x - 3)^2 - 2(y - 0)^2 = 2 \Rightarrow$

$\dfrac{(x - 3)^2}{2} - \dfrac{(y - 0)^2}{1} = 1$. The center is $(3, 0)$. The vertices are $(3 - \sqrt{2}, 0), (3 + \sqrt{2}, 0)$.

76. $y^2 + 8y - 3x^2 + 13 = 0 \Rightarrow (y^2 + 8y + 16) - 3x^2 = -13 + 16 \Rightarrow (y + 4)^2 - 3(x - 0)^2 = 3 \Rightarrow$

$\dfrac{(y + 4)^2}{3} - \dfrac{(x - 0)^2}{1} = 1$. The center is $(0, -4)$. The vertices are $(0, -4 - \sqrt{3}), (0, -4 + \sqrt{3})$.

77. $4y^2 + 32y - 5x^2 - 10x + 39 = 0 \Rightarrow 4(y^2 + 8y) - 5(x^2 + 2x) = -39 \Rightarrow$

$4(y^2 + 8y + 16) - 5(x^2 + 2x + 1) = -39 + 64 - 5 \Rightarrow 4(y + 4)^2 - 5(x + 1)^2 = 20 \Rightarrow$

$\dfrac{(y + 4)^2}{5} - \dfrac{(x + 1)^2}{4} = 1$. The center is $(-1, -4)$. The vertices are $(-1, -4 - \sqrt{5}), (-1, -4 + \sqrt{5})$.

78. $5x^2 + 10x - 7y^2 + 28y = 58 \Rightarrow 5(x^2 + 2x) - 7(y^2 - 4y) = 58 \Rightarrow$

$5(x^2 + 2x + 1) - 7(y^2 - 4y + 4) = 58 + 5 - 28 \Rightarrow 5(x + 1)^2 - 7(y - 2)^2 = 35 \Rightarrow$

$\dfrac{(x + 1)^2}{7} - \dfrac{(y - 2)^2}{5} = 1$. The center is $(-1, 2)$. The vertices are $(-1 - \sqrt{7}, 2), (-1 + \sqrt{7}, 2)$.

79. $c^2 = a^2 - b^2 \Rightarrow c^2 = 16 - 12 \Rightarrow c^2 = 4 \Rightarrow c = 2$; The foci are $F_1(-2, 0)$ and $F_2(2, 0)$.

80. See Figure 80. The point $(3, 2.2912878)$ is shown. Other points include $(0, 3.4641016)$ and $(-3, -2.291288)$.

81. $\sqrt{(3 - (-2))^2 + (2.2912878 - 0)^2} + \sqrt{(3 - 2)^2 + (2.2912878 - 0)^2} \approx 7.999999937 \approx 8$

$\sqrt{(0 - (-2))^2 + (3.4641016 - 0)^2} + \sqrt{(0 - 2)^2 + (3.4641016 - 0)^2} \approx 7.999999974 \approx 8$

$\sqrt{(-3 - (-2))^2 + (-2.291288 - 0)^2} + \sqrt{(-3 - 2)^2 + (-2.291288 - 0)^2} \approx 8.000000203 \approx 8$

Note that the sums calculated here do not equal exactly 8 because of rounding.

82. $c^2 = a^2 + b^2 \Rightarrow c^2 = 4 - 12 \Rightarrow c^2 = 16 \Rightarrow c = 4$; The foci are $F_1(-4, 0)$ and $F_2(4, 0)$.

See Figure 82. The point $(-3, 3.8729833)$ is shown. Other points include $(-2, 0), (2, 0)$ and $(4, 6)$.

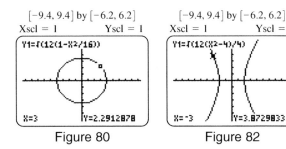

Figure 80   Figure 82

83. $\left| \sqrt{(-3-(-4))^2 + (3.8729833 - 0)^2} - \sqrt{(-3-4)^2 + (3.8729833 - 0)^2} \right| \approx 4.000000022 \approx 4$

    $\left| \sqrt{(-2-(-4))^2 + (0-0)^2} - \sqrt{(-2-4)^2 + (0-0)^2} \right| = 4$

    $\left| \sqrt{(2-(-4))^2 + (0-0)^2} - \sqrt{(2-4)^2 + (0-0)^2} \right| = 4$

    $\left| \sqrt{(4-(-4))^2 + (6-0)^2} - \sqrt{(4-4)^2 + (6-0)^2} \right| = 4$

    Note that the first difference calculated here does not equal exactly 4 because of rounding.

84. Exercise 81 demonstrates that the points on the graph satisfy the definition of an ellipse for that particular ellipse. Exercise 83 demonstrates that the points on the graph satisfy the definition of a hyperbola for that particular hyperbola.

85. The patient and the emitter are 12 units apart. These positions represent the foci of the ellipse so $c = 6$. With the minor axis measuring 16 units, $b = 8$. Since $c^2 = a^2 - b^2$, the value of a can be found by substitution: $a^2 = b^2 + c^2 = 6^2 + 8^2 = 36 + 64 = 100 \Rightarrow a = 10$. The equation is $\dfrac{x^2}{100} + \dfrac{y^2}{64} = 1$.

86. (a) The length of the major axis is $2a = 2\sqrt{5013} \approx 141.6$ million miles.

    (b) The length of the minor axis is $2b = 2\sqrt{4970} \approx 141.0$ million miles.

87. A major axis measuring 620 feet indicates that $a = 310$. A minor axis measuring 513 feet indicates that $b = 256.5$. Then $c^2 = a^2 - b^2 = 310^2 - 256.5^2 \Rightarrow c = \sqrt{310^2 - 256.5^2} \approx 174.1$. The distance between the foci is $2c \approx 2(174.1) \approx 348.2$ feet.

88. $C \approx 2\pi \sqrt{\dfrac{310^2 + 256.5^2}{2}} \approx 1788$ feet

89. Using a vertical major axis, $a = 15$. The minor axis has length 20 so $b = 10$. The equation is $\dfrac{y^2}{225} + \dfrac{x^2}{100} = 1$. Assuming the truck drives exactly in the middle of the road, we want to find y when $x = 6$.

    $\dfrac{y^2}{225} + \dfrac{6^2}{100} = 1 \Rightarrow \dfrac{y^2}{225} = 1 - \dfrac{36}{100} \Rightarrow y^2 = 225\left(1 - \dfrac{36}{100}\right) \Rightarrow y = \sqrt{225\left(1 - \dfrac{36}{100}\right)} = 12$

    The truck must be just under 12 feet high to pass through.

90. Since a difference in arrival times is being used, the curve is a hyperbola.

91. (a) Since $c = \sqrt{a^2 - b^2} = \sqrt{4465^2 - 4462^2} \approx 163.6$, one focus is located at the point $(163.6, 0)$. The graph representing Earth is a circle with radius 3960 with center $(163.6, 0)$ The equation for Earth is $(x - 163.6)^2 + y^2 = 3690^2$. To graph this equation, solve for y and graph two parts.

    $(x - 163.6)^2 + y^2 = 3690^2 \Rightarrow y^2 = 3690^2 - (x - 163.6)^2 \Rightarrow y = \pm\sqrt{3690^2 - (x - 163.6)^2}$

    To graph the ellipse, solve for y and graph two parts.

    $\dfrac{x^2}{4465^2} + \dfrac{y^2}{4462^2} = 1 \Rightarrow \dfrac{y^2}{4462^2} = 1 - \dfrac{x^2}{4465^2} \Rightarrow y^2 = 4462^2\left(1 - \dfrac{x^2}{4465^2}\right) \Rightarrow y = \pm\sqrt{4462^2\left(1 - \dfrac{x^2}{4465^2}\right)}$

    The graphs are shown in Figure 91.

    (b) The minimum distance is $4465 - (3960 + 163.6) \approx 341$ miles.

    The maximum distance is $4465 - (3960 - 163.6) \approx 669$ miles.

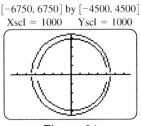

Figure 91

92. (a) The equation of the hyperbola is $400x^2 - 625y^2 = 250{,}000 \Rightarrow \dfrac{x^2}{625} - \dfrac{y^2}{400} = 1 \Rightarrow \dfrac{x^2}{25^2} - \dfrac{y^2}{20^2} = 1.$

Thus $a = 25$ and the buildings are $2(25) = 50$ meters apart at their closest point.

(b) $\dfrac{50^2}{625} - \dfrac{y^2}{400} = 1 \Rightarrow \dfrac{y^2}{400} = \dfrac{50^2}{625} - 1 \Rightarrow y^2 = 400\left(\dfrac{50^2}{625} - 1\right) \Rightarrow y = \sqrt{400\left(\dfrac{50^2}{625} - 1\right)} \approx 34.64.$

The total distance is $d \approx 2(34.64) \approx 69.3$ meters.

93. (a) Find a and b in the equation $\dfrac{x^2}{a^2} - \dfrac{y^2}{b^2} = 1.$ Because the equations of the asymptotes of a hyperbola with

horizontal transverse axis are $y = \pm\dfrac{b}{a}x$, and the given asymptotes are $y = \pm x$, it follows that

$\dfrac{b}{a} = 1$ or $a = b$. Since the line $y = x$ intersects the x-axis at a $45°$ angle, the triangle shown in the third

quadrant is a $45°$-$45°$-$90°$ right triangle and both legs must have length d. Then by the Pythagorean

theorem, $c^2 = d^2 + d^2 = 2d^2$. That is $c = d\sqrt{2}$. Also, for a hyperbola $c^2 = a^2 + b^2$, and since $a = b$,

$c^2 = a^2 + a^2 = 2a^2$. That is $c = a\sqrt{2}$. From these two equations, $a\sqrt{2} = d\sqrt{2}$ and so $a = d$. That is,

$a = b = d = 5 \times 10^{-14}$. Thus the equation of the trajectory of A, where $x > 0$, is given by

$\dfrac{x^2}{(5 \times 10^{-14})^2} - \dfrac{y^2}{(5 \times 10^{-14})^2} = 1.$ Solving for x yields

$x^2 - y^2 = (5 \times 10^{-14})^2 \Rightarrow x^2 = y^2 + 2.5 \times 10^{-27} \Rightarrow x = \sqrt{y^2 + 2.5 \times 10^{-27}}.$ This equation represents

the right half of the hyperbola, as shown in the textbook.

(b) $a = 5 \times 10^{-14}$, the distance from the origin to the vertex is 5×10^{-14}. The distance from N to the

origin can be found using the Pythagorean theorem. Let h represent this distance, then $h^2 = d^2 + d^2$. That

is, $h^2 = (5 \times 10^{-14})^2 + (5 \times 10^{-14})^2 \Rightarrow h^2 = 5 \times 10^{-27} \Rightarrow h \approx 7 \times 10^{-14}$. The minimum distance

between the centers of the alpha particle and the gold nucleus is $5 \times 10^{-14} + 7 \times 10^{-14} \approx 1.2 \times 10^{-13}.$

94. Use the formula $t = \dfrac{d}{r}$ and the distance formula to set up an equation that shows the difference in the times that it takes the sound to reach each microphone. This equation is $\dfrac{\sqrt{(x + c)^2 + y^2}}{330} - \dfrac{\sqrt{(x - c)^2 + y^2}}{330} = t$.

$\sqrt{(x + c)^2 + y^2} - \sqrt{(x - c)^2 + y^2} = 330t$	Multiply each side by 330.
$\sqrt{(x + c)^2 + y^2} = 330t + \sqrt{(x - c)^2 + y^2}$	Add $\sqrt{(x - c)^2 + y^2}$ to each side.
$(x + c)^2 + y^2 = 330^2 t^2 + 2 \cdot 330t\sqrt{(x - c)^2 + y^2} + (x - c)^2 + y^2$	Square each side.
$4cx - 330^2 t^2 = 2 \cdot 330t\sqrt{(x - c)^2 + y^2}$	Expand the binomials and simplify.
$16c^2 x^2 - 8cx \cdot 330^2 t^2 + 330^4 t^4 = 4 \cdot 330^2 t^2 [(x - c)^2 + y^2]$	Square each side.
$16c^2 x^2 + 330^4 t^4 = 4 \cdot 330^2 t^2 x^2 + 4 \cdot 330^2 c^2 t^2 + 4 \cdot 330^2 t^2 y^2$	Expand the right side and simplify.
$16c^2 x^2 - 4 \cdot 330^2 t^2 x^2 - 4 \cdot 330^2 t^2 y^2 = 4 \cdot 330^2 c^2 t^2 - 330^4 t^4$	Rewrite equation.
$4x^2(4c^2 - 330^2 t^2) - 4 \cdot 330^2 t^2 y^2 = 330^2 t^2(4c^2 - 330^2 t^2)$	Factor.
$\dfrac{x^2}{330^2 t^2} - \dfrac{y^2}{4c^2 - 330^2 t^2} = \dfrac{1}{4}$	Divide by $4 \cdot 330^2 t^2(4c^2 - 330^2 t^2)$.

95. Start with the given distance equation.

$\sqrt{(x + c)^2 + y^2} - \sqrt{(x - c)^2 + y^2} = 2a$	Given Equation
$\sqrt{(x + c)^2 + y^2} = 2a + \sqrt{(x - c)^2 + y^2}$	Add $\sqrt{(x - c)^2 + y^2}$ to each side.
$(x + c)^2 + y^2 = 4a^2 + 4a\sqrt{(x - c)^2 + y^2} + (x - c)^2 + y^2$	Square each side.
$(x + c)^2 - (x - c)^2 - 4a^2 = 4a\sqrt{(x - c)^2 + y^2}$	Subtract $(x - c)$ and $4a^2$. Simplify.
$x^2 + 2cx + c^2 - x^2 + 2cx - c^2 - 4a^2 = 4a\sqrt{(x - c)^2 + y^2}$	Expand the binomials.
$4cx - 4a^2 = 4a\sqrt{(x - c)^2 + y^2}$	Simplify.
$cx - a^2 = a\sqrt{(x - c)^2 + y^2}$	Divide each side by 4.
$c^2 x^2 - 2a^2 cx + a^4 = a^2(x^2 - 2cx + c^2 + y^2)$	Square each side.
$c^2 x^2 - 2a^2 cx + a^4 = a^2 x^2 - 2a^2 cx + a^2 c^2 + a^2 y^2$	Multiply the right side.
$c^2 x^2 - a^2 x^2 - a^2 y^2 = -a^4 + a^2 c^2$	Simplify.
$x^2(c^2 - a^2) - a^2 y^2 = a^2(c^2 - a^2)$	Factor.
$x^2 b^2 - a^2 y^2 = a^2 b^2$	Substitute b^2 for $c^2 - a^2$.
$\dfrac{x^2 b^2}{a^2 b^2} - \dfrac{a^2 y^2}{a^2 b^2} = \dfrac{a^2 b^2}{a^2 b^2}$	Divide each side by $a^2 b^2$.
$\dfrac{x^2}{a^2} - \dfrac{y^2}{b^2} = 1$	Simplify.

96. Let (x, y) be any point on the ellipse and start with the distance formula.

$\sqrt{(x + 3)^2 + y^2} + \sqrt{(x - 3)^2 + y^2} = 10$	Given Equation
$\sqrt{(x + 3)^2 + y^2} = 10 - \sqrt{(x - 3)^2 + y^2}$	Subtract $\sqrt{(x - 3)^2 + y^2}$.
$(x + 2)^2 + y^2 = 100 - 20\sqrt{(x - 3)^2 + y^2} + (x - 3)^2 + y^2$	Square each side.
$(x + 3)^2 - (x - 3)^2 - 100 = -20\sqrt{(x - 3)^2 + y^2}$	Subtract $(x - 3)^2$ and 100. Simplify.
$x^2 + 6x + 9 - x^2 + 6x - 9 - 100 = -20\sqrt{(x - 3)^2 + y^2}$	Expand the binomials.
$12x - 100 = -20\sqrt{(x - 3)^2 + y^2}$	Simplify.
$25 - 3x = 5\sqrt{(x - 3)^2 + y^2}$	Divide each side by -4.
$625 - 150x + 9x^2 = 25(x^2 - 6x + 9 + y^2)$	Square each side.
$625 - 150x + 9x^2 = 25x^2 - 150x + 225 + 25y^2$	Multiply the right side.
$-16x^2 - 25y^2 = -400$	Simplify.
$\dfrac{-16x^2}{-400} + \dfrac{-25y^2}{-400} = \dfrac{-400}{-400}$	Divide each side by -400.
$\dfrac{x^2}{25} + \dfrac{y^2}{16} = 1$	Simplify.

97. Let (x, y) be any point on the hyperbola and start with the distance formula.

$\sqrt{(x + 2)^2 + y^2} - \sqrt{(x - 2)^2 + y^2} = 2$	Given Equation
$\sqrt{(x + 2)^2 + y^2} = 2 + \sqrt{(x - 2)^2 + y^2}$	Add $\sqrt{(x - 2)^2 + y^2}$ to each side.
$(x + 2)^2 + y^2 = 4 + 4\sqrt{(x - 2)^2 + y^2} + (x - 2)^2 + y^2$	Square each side.
$(x + 2)^2 - (x - 2)^2 - 4 = 4\sqrt{(x - 2)^2 + y^2}$	Subtract $(x - 2)^2$ and 4. Simplify.
$x^2 + 4x + 4 - x^2 + 4x - 4 - 4 = 4\sqrt{(x - 2)^2 + y^2}$	Expand the binomials.
$8x - 4 = 4\sqrt{(x - 2)^2 + y^2}$	Simplify.
$2x - 1 = \sqrt{(x - 2)^2 + y^2}$	Divide each side by 4.
$4x^2 - 4x + 1 = x^2 - 4x + 4 + y^2$	Square each side.
$3x^2 - y^2 = 3$	Simplify.
$\dfrac{3x^2}{3} - \dfrac{y^2}{3} = \dfrac{3}{3}$	Divide each side by 3.
$x^2 - \dfrac{y^2}{3} = 1$	Simplify.

Reviewing Basic Concepts (Sections 6.1 and 6.2)

1. (a) The circle is defined in B.

    (b) The parabola is defined in D.

    (c) The ellipse is defined in A.

    (d) The hyperbola is defined in C.

2. $12x^2 - 4y^2 = 48 \Rightarrow \dfrac{x^2}{4} - \dfrac{y^2}{12} = 1$. The transverse axis is horizontal with $a = 2$ and $b = \sqrt{12}$. The asymptotes are $y = \pm\dfrac{\sqrt{12}}{2}x$. See Figure 2.

3. Rewrite the equation: $y = 2x^2 + 3x - 1 \Rightarrow y + 1 + \dfrac{9}{8} = 2\left(x^2 + \dfrac{3}{2}x + \dfrac{9}{16}\right) \Rightarrow y + \dfrac{17}{8} = 2\left(x + \dfrac{3}{4}\right)^2$.

    The equation $y + \dfrac{17}{8} = 2\left(x + \dfrac{3}{4}\right)^2$ can be written $y + \dfrac{17}{8} = 4\left(\dfrac{1}{2}\right)\left(x + \dfrac{3}{4}\right)^2$. The vertex is $\left(-\dfrac{3}{4}, -\dfrac{17}{8}\right)$.

    See Figure 3.

4. $x^2 + y^2 - 2x + 2y - 2 = 0 \Rightarrow (x^2 - 2x + 1) + (y^2 + 2y + 1) = 2 + 1 + 1 \Rightarrow$

    $(x - 1)^2 + (y + 1)^2 = 4$. The graph is a circle with center $(1, -1)$ and radius $r = 2$. See Figure 4.

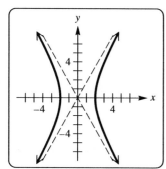

Figure 2

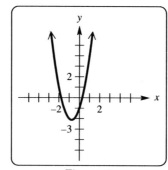

Figure 3

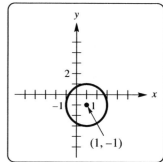

Figure 4

5. $4x^2 + 9y^2 = 36 \Rightarrow \dfrac{x^2}{9} + \dfrac{y^2}{4} = 1 \Rightarrow a = 3$ and $b = 2$. The ellipse is graphed in Figure 5.

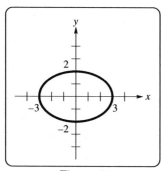

Figure 5

6. If $c < a$, it is an ellipse. If $c > a$, it is a hyperbola.

7. A circle with center $(2, -1)$ and radius 3 has equation $(x - 2)^2 + (y + 1)^2 = 9$.

8. The ellipse is centered between the foci at $(0, 0)$. The major axis is horizontal with $a = 6$. Since the foci are $(\pm 4, 0)$, we know that $c = 4$. Since $c^2 = a^2 - b^2$, the value of b can be found by substitution:

$b^2 = a^2 - c^2 = 6^2 - 4^2 = 36 - 16 = 20 \Rightarrow b = \sqrt{20}$. The equation is $\dfrac{x^2}{36} + \dfrac{y^2}{20} = 1$.

9. The hyperbola has a vertical transverse axis with $c = 4$. The vertices are $(0, \pm 2)$ so $a = 2$.

The center is located between the foci at $(0, 0)$. Since $c^2 = a^2 + b^2$, the value of b can be found by substitution:

$b^2 = c^2 - a^2 = 4^2 - 2^2 = 16 - 4 = 12 \Rightarrow b = \sqrt{12}$. The equation is $\dfrac{y^2}{4} - \dfrac{x^2}{12} = 1$.

10. If the vertex is $(0, 0)$ and the focus is $\left(0, \dfrac{1}{2}\right)$, then the parabola opens upward and $c = \dfrac{1}{2}$.

The equation is $x^2 = 4cy \Rightarrow x^2 = 2y$.

6.3: Summary of the Conic Sections

1. This is the equation of a circle with center $(0, 0)$ and radius 12.

2. This is the equation of a circle with center $(2, -3)$ and radius 5.

3. This is the equation of a parabola that opens upward.

4. This is the equation of a parabola that opens to the right.

5. This is the equation of a parabola with vertex $(1, 4)$ that opens to the left.

6. This is the equation of an ellipse centered at $(0, 0)$.

7. This is the equation of an ellipse centered at $(0, 0)$.

8. This is the equation of a hyperbola centered at $(0, 0)$ with horizontal transverse axis.

9. This is the equation of a hyperbola centered at $(0, 0)$ with horizontal transverse axis.

10. This is the equation of an ellipse centered at $(-2, 4)$.

11. This is the equation of a hyperbola centered at $(0, 0)$ with horizontal transverse axis.

12. This is the equation of a parabola with vertex $(-3, -7)$ that opens upward.

13. $\dfrac{x^2}{4} = 1 - \dfrac{y^2}{9} \Rightarrow \dfrac{x^2}{4} + \dfrac{y^2}{9} = 1$; ellipse centered at $(0, 0)$

14. $\dfrac{x^2}{4} = 1 + \dfrac{y^2}{9} \Rightarrow \dfrac{x^2}{4} - \dfrac{y^2}{9} = 1$; hyperbola centered at $(0, 0)$ with horizontal transverse axis

15. There is no graph for this equation. No real numbers satisfy this equation.

16. $x^2 = 25 + y^2 \Rightarrow x^2 - y^2 = 25 \Rightarrow \dfrac{x^2}{25} - \dfrac{y^2}{25} = 1$; hyperbola centered at $(0, 0)$ with horizontal transverse axis

17. $x^2 = 25 - y^2 \Rightarrow x^2 + y^2 = 25$; circle with center $(0, 0)$ and radius 5

18. $9x^2 + 36y^2 = 36 \Rightarrow \dfrac{x^2}{4} + \dfrac{y^2}{1} = 1$; ellipse centered at $(0, 0)$

19. $x^2 = 4y - 8 \Rightarrow (x - 0)^2 = 4(1)(y - 2)$; parabola with vertex $(0, 2)$ that opens upward

20. $\dfrac{(x + 3)^2}{16} + \dfrac{(y - 2)^2}{16} = 1 \Rightarrow (x + 3)^2 + (y - 2)^2 = 16$; circle with center $(-3, 2)$ and radius 4.

21. The only real numbers that satisfy this equation are $x = 4, y = -1$. The graph is the point $(4, -1)$.

22. $y^2 - 4y = x + 4 \Rightarrow y^2 - 4y + 4 = x + 4 + 4 \Rightarrow (y - 2)^2 = x + 8 \Rightarrow (y - 2)^2 = 4\left(\frac{1}{4}\right)(x + 8)$;

    parabola with vertex $(-8, 2)$ that opens to the right

23. There is no graph for this equation. No real numbers satisfy this equation.

24. The only real numbers that satisfy this equation are $x = 3$, $y = -4$. The graph is the point $(3, -4)$.

25. $3x^2 + 6x + 3y^2 - 12y = 12 \Rightarrow 3(x^2 + 2x + 1) + 3(y^2 - 4y + 4) = 12 + 3 + 12 \Rightarrow$

    $3(x + 1)^2 + 3(y - 2)^2 = 27 \Rightarrow (x + 1)^2 + (y - 2)^2 = 9$; circle with center $(-1, 2)$ and radius 3

26. $2x^2 - 8x + 2y^2 + 20y = 12 \Rightarrow 2(x^2 - 4x + 4) + 2(y^2 + 10y + 25) = 12 + 8 + 50 \Rightarrow$

    $2(x - 2)^2 + 2(y + 5)^2 = 70 \Rightarrow (x - 2)^2 + (y + 5)^2 = 35$; circle with center $(2, -5)$ and radius $\sqrt{35}$.

27. $x^2 - 6x + y = 0 \Rightarrow x^2 - 6x + 9 = -y + 9 \Rightarrow (x - 3)^2 = -(y - 9) \Rightarrow (x - 3)^2 = 4\left(-\frac{1}{4}\right)(y - 9)$;

    parabola with vertex $(3, 9)$ that opens upward

28. $x - 4y^2 - 8y = 0 \Rightarrow x + 4 = 4(y^2 + 2y + 1) \Rightarrow x + 4 = 4(y + 1)^2 \Rightarrow (y + 1)^2 = 4\left(\frac{1}{16}\right)(x + 4)$;

    parabola with vertex $(-4, -1)$ that opens to the right

29. $4x^2 - 8x - y^2 - 6y = 6 \Rightarrow 4(x^2 - 2x + 1) - (y^2 + 6y + 9) = 6 + 4 - 9 \Rightarrow$

    $4(x - 1)^2 - (y + 3)^2 = 1 \Rightarrow \dfrac{(x - 1)^2}{\frac{1}{4}} - \dfrac{(y + 3)^2}{1} = 1$; hyperbola centered at $(1, -3)$, horizontal transverse axis

30. $x^2 + 2x = x^2 - 4y - 2 \Rightarrow 2x = -4y - 2 \Rightarrow 2x + 2 = -4y \Rightarrow y = -\dfrac{1}{2}x - \dfrac{1}{2}$; line with slope $-\dfrac{1}{2}$

31. $4x^2 - 8x + 9y^2 + 54y = -84 \Rightarrow 4(x^2 - 2x + 1) + 9(y^2 + 6y + 9) = -84 + 4 + 81 \Rightarrow$

    $4(x - 1)^2 + 9(y + 3)^2 = 1 \Rightarrow \dfrac{(x - 1)^2}{\frac{1}{4}} + \dfrac{(y + 3)^2}{\frac{1}{9}} = 1$; ellipse centered at $(1, -3)$

32. $3x^2 + 12x + 3y^2 = -11 \Rightarrow 3(x^2 + 4x + 4) + 3y^2 = -11 + 12 \Rightarrow 3(x + 2)^2 + 3y^2 = 1 \Rightarrow$

    $(x + 2)^2 + (y - 0)^2 = \dfrac{1}{3}$; circle with center $(-2, 0)$ and radius $\dfrac{\sqrt{3}}{3}$

33. $6x^2 - 12x + 6y^2 - 18y + 25 = 0 \Rightarrow 6(x^2 - 2x + 1) + 6\left(y^2 - 3y + \dfrac{9}{4}\right) = -25 + 6 + \dfrac{27}{2} \Rightarrow$

    $6(x - 1)^2 + 6\left(y - \dfrac{3}{2}\right)^2 = -\dfrac{11}{2} \Rightarrow (x - 1)^2 + \left(y - \dfrac{3}{2}\right)^2 = -\dfrac{11}{12}$; There is no graph for this equation.

    No real numbers satisfy this equation.

34. $4x^2 - 24x + 5y^2 + 10y + 41 = 0 \Rightarrow 4(x^2 - 6x + 9) + 5(y^2 + 2y + 1) = -41 + 36 + 5 \Rightarrow$

    $4(x - 3)^2 + 5(y + 1)^2 = 0$; The only real numbers that satisfy this equation are $x = 3$, $y = -1$.

    The graph is the point $(3, -1)$.

35. The equation would be of the form $Bx + Dy + E = 0$, which is linear. The graph is a line.

36. The plane must intersect the cone along an edge of the cone to form a line.

37. Since the sum of the distances from two points (foci) is a constant, the conic section is an ellipse.

38. Since the difference of the distances from two points (foci) is a constant, the conic section is a hyperbola.

39. Since the ratio of the distance from a point to $(3, 0)$ and the distance from a point to the line $x = \dfrac{4}{3}$ is 1.5,

    the eccentricity is greater than 1. The conic section is a hyperbola.

40. Since the ratio of the distance from a point to $(2, 0)$ and the distance from a point to the line $x = 10$ is $\dfrac{1}{3}$, the eccentricity is between 0 and 1. The conic section is an ellipse.

41. $12x^2 + 9y^2 = 36 \Rightarrow \dfrac{x^2}{3} + \dfrac{y^2}{4} = 1 \Rightarrow a = 2, b = \sqrt{3},$ and $c = \sqrt{4 - 3} = 1; e = \dfrac{c}{a} = \dfrac{1}{2}$

42. $8x^2 - y^2 = 16 \Rightarrow \dfrac{x^2}{2} - \dfrac{y^2}{16} = 1 \Rightarrow a = \sqrt{2}, b = 4,$ and $c = \sqrt{2 + 16} = \sqrt{18}; e = \dfrac{c}{a} = \dfrac{\sqrt{18}}{\sqrt{2}} = 3$

43. $x^2 - y^2 = 4 \Rightarrow \dfrac{x^2}{4} - \dfrac{y^2}{4} = 1 \Rightarrow a = 2, b = 2,$ and $c = \sqrt{4 + 4} = \sqrt{8}; e = \dfrac{c}{a} = \dfrac{\sqrt{8}}{2} = \sqrt{2}$

44. $x^2 + 2y^2 = 8 \Rightarrow \dfrac{x^2}{8} + \dfrac{y^2}{4} = 1 \Rightarrow a = \sqrt{8}, b = 2,$ and $c = \sqrt{8 - 4} = 2; e = \dfrac{c}{a} = \dfrac{2}{\sqrt{8}} = \dfrac{\sqrt{2}}{2}$

45. $4x^2 + 7y^2 = 28 \Rightarrow \dfrac{x^2}{7} + \dfrac{y^2}{4} = 1 \Rightarrow a = \sqrt{7}, b = 2,$ and $c = \sqrt{7 - 4} = \sqrt{3}; e = \dfrac{c}{a} = \dfrac{\sqrt{3}}{\sqrt{7}} = \dfrac{\sqrt{21}}{7}$

46. $9x^2 - y^2 = 1 \Rightarrow \dfrac{x^2}{\frac{1}{9}} - \dfrac{y^2}{1} = 1 \Rightarrow a = \dfrac{1}{3}, b = 1,$ and $c = \sqrt{\dfrac{1}{9} + 1} = \dfrac{\sqrt{10}}{3}; e = \dfrac{c}{a} = \dfrac{\sqrt{10}}{3} \cdot \dfrac{3}{1} = \sqrt{10}$

47. $x^2 - 9y^2 = 18 \Rightarrow \dfrac{x^2}{18} - \dfrac{y^2}{2} = 1 \Rightarrow a = \sqrt{18}, b = \sqrt{2},$ and $c = \sqrt{18 + 2} = \sqrt{20}; e = \dfrac{c}{a} = \dfrac{\sqrt{20}}{\sqrt{18}} = \dfrac{\sqrt{10}}{3}$

48. $x^2 + 10y^2 = 10 \Rightarrow \dfrac{x^2}{10} + \dfrac{y^2}{1} = 1 \Rightarrow a = \sqrt{10}, b = 1,$ and $c = \sqrt{10 - 1} = 3; e = \dfrac{c}{a} = \dfrac{3}{\sqrt{10}} = \dfrac{3\sqrt{10}}{10}$

49. Since $e = 1$, the conic is a parabola. With center $(0, 0)$ and focus $(0, 8)$ the equation is $x^2 = 4cy \Rightarrow x^2 = 32y$.

50. Since $e = 1$, the conic is a parabola. With center $(0, 0)$ and focus $(-2, 0)$ the equation is $y^2 = 4cx \Rightarrow y^2 = -8x$.

51. Since $0 < e < 1$, the conic is an ellipse with $c = 3$. Now $\dfrac{c}{a} = e \Rightarrow \dfrac{3}{a} = \dfrac{1}{2} \Rightarrow a = 6$. For an ellipse,

    $b^2 = a^2 - c^2 = 36 - 9 = 27$. The equation is $\dfrac{x^2}{36} + \dfrac{y^2}{27} = 1$.

52. Since $0 < e < 1$, the conic is an ellipse with $c = 2$. Now $\dfrac{c}{a} = e \Rightarrow \dfrac{2}{a} = \dfrac{2}{3} \Rightarrow a = 3$. For an ellipse,

    $b^2 = a^2 - c^2 = 9 - 4 = 5$. The equation is $\dfrac{x^2}{5} + \dfrac{y^2}{9} = 1$.

53. Since $e > 1$, the conic is a hyperbola with $a = 6$. Now $\dfrac{c}{a} = e \Rightarrow \dfrac{c}{6} = 2 \Rightarrow c = 12$. For a hyperbola,

    $b^2 = c^2 - a^2 = 144 - 36 = 108$. The equation is $\dfrac{x^2}{36} - \dfrac{y^2}{108} = 1$.

54. Since $e > 1$, the conic is a hyperbola with $a = 4$. Now $\dfrac{c}{a} = e \Rightarrow \dfrac{c}{4} = \dfrac{5}{3} \Rightarrow c = \dfrac{20}{3}$. For a hyperbola,

    $b^2 = c^2 - a^2 = \dfrac{400}{9} - 16 = \dfrac{256}{9}$. The equation is $\dfrac{y^2}{16} - \dfrac{x^2}{\frac{256}{9}} = 1$ or $\dfrac{y^2}{16} - \dfrac{9x^2}{256} = 1$.

55. Since $e = 1$, the conic is a parabola. With center $(0, 0)$ and focus $(0, -1)$ the equation is $x^2 = 4cy \Rightarrow x^2 = -4y$.

56. Since $e > 1$, the conic is a hyperbola with $c = 2$. Now $\dfrac{c}{a} = e \Rightarrow \dfrac{2}{a} = \dfrac{6}{5} \Rightarrow a = \dfrac{5}{3}$. For a hyperbola,

    $b^2 = c^2 - a^2 = 4 - \dfrac{25}{9} = \dfrac{11}{9}$. The equation is $\dfrac{x^2}{\frac{25}{9}} - \dfrac{y^2}{\frac{11}{9}} = 1$ or $\dfrac{9x^2}{25} - \dfrac{9y^2}{11} = 1$.

57. Since $0 < e < 1$, the conic is an ellipse with $a = 3$. Now $\dfrac{c}{a} = e \Rightarrow \dfrac{c}{3} = \dfrac{4}{5} \Rightarrow c = \dfrac{12}{5}$. For an ellipse,

    $b^2 = a^2 - c^2 = 9 - \dfrac{144}{25} = \dfrac{81}{25}$. The equation is $\dfrac{x^2}{\frac{81}{25}} + \dfrac{y^2}{9} = 1$ or $\dfrac{25x^2}{81} + \dfrac{y^2}{9} = 1$.

58. Since $e > 1$, the conic is a hyperbola with $a = 4$. Now $\dfrac{c}{a} = e \Rightarrow \dfrac{c}{4} = \dfrac{7}{3} \Rightarrow c = \dfrac{28}{3}$. For a hyperbola,

$b^2 = c^2 - a^2 = \dfrac{784}{9} - 16 = \dfrac{640}{9}$. The equation is $\dfrac{y^2}{16} - \dfrac{x^2}{\frac{640}{9}} = 1$ or $\dfrac{y^2}{16} - \dfrac{9x^2}{640} = 1$.

59. For a circle, $e = 0$. For an ellipse, $0 < e < 1$. For a parabola, $e = 1$. For a hyperbola, $e > 1$. C, A, B, D

60. For an ellipse, $c = \sqrt{a^2 - b^2} = \sqrt{5013 - 4970} = \sqrt{43}$. The eccentricity is $e = \dfrac{c}{a} = \dfrac{\sqrt{43}}{\sqrt{5013}} \approx .093$.

61. (a) For Neptune: $\dfrac{c}{a} = e \Rightarrow \dfrac{c}{30.1} = .009 \Rightarrow c = .2709$ and $b = \sqrt{a^2 - c^2} = \sqrt{30.1^2 - .2709^2} \approx 30.1$.

    Since the sun is at the focus, the equation is $\dfrac{(x - .2709)^2}{30.1^2} + \dfrac{y^2}{30.1^2} = 1$.

    For Pluto: $\dfrac{c}{a} = e \Rightarrow \dfrac{c}{39.4} = .249 \Rightarrow c = 9.8106$ and $b = \sqrt{a^2 - c^2} = \sqrt{39.4^2 - 9.8106^2} \approx 38.16$.

    Since the sun is at the focus, the equation is $\dfrac{(x - 9.8106)^2}{39.4^2} + \dfrac{y^2}{38.16^2} = 1$.

    (b) Graph $y_1 = \sqrt{30.1^2 - (x - .2709)^2}$, $y_2 = -y_1$, $y_3 = 38.16\sqrt{1 - \dfrac{(x - 9.8106)^2}{39.4^2}}$ and $y_4 = -y_3$. See Figure 61.

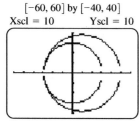

[−60, 60] by [−40, 40]
Xscl = 10   Yscl = 10

Figure 61

62. (a) Earth orbits every $365 \cdot 24 \cdot 60 \cdot 60 = 31{,}536{,}000$ seconds. Thus $P = 31{,}536{,}000$.

    The maximum velocity of Earth is $v_{max} = \dfrac{2\pi(1.496 \times 10^8)}{31{,}536{,}000}\sqrt{\dfrac{1 + .0167}{1 - .0167}} \approx 30.3$ km per sec.

    The minimum velocity of Earth is $v_{min} = \dfrac{2\pi(1.496 \times 10^8)}{31{,}536{,}000}\sqrt{\dfrac{1 - .0167}{1 + .0167}} \approx 29.3$ km per sec.

    (b) The minimum and maximum velocities are equal. Therefore, the planet's velocity is constant.

    (c) A planet is at its maximum and minimum distances from a focus when it is located at the vertices of the ellipse. Thus the minimum and maximum velocities of a planet will occur at the vertices of the elliptical orbit, which are $a + c$ for the minimum and $a - c$ for the maximum.

63. Since the eccentricity is .9673, $\dfrac{c}{a} = .9673 \Rightarrow c = .9673a$. Since the greatest distance is 3281 miles,

    $a + c = 3281$. Solving these two equations simultaneously results in $a \approx 1667.8$ and $c \approx 1613.2$.

    The minimum distance is $a - c = 1667.8 - 1613.2 \approx 55$ million miles.

64. Here $a + c = 94.6$ and $a - c = 91.4$. Solving these equations simultaneously results in $a = 93$ and $c = 1.6$.

    The eccentricity is $e = \dfrac{c}{a} = \dfrac{1.6}{93} \approx .0172$.

6.4: Parametric Equations

1. See Figure 1. From the 1st equation $x = 2t \Rightarrow t = \frac{1}{2}x$. By substitution in the 2nd equation $y = \frac{1}{2}x + 1$.

   When t is in $[-2, 3]$, the range of $x = 2t$ is x in $[-4, 6]$.

2. See Figure 2. From the 1st equation $x = t + 2 \Rightarrow t = x - 2$. By substitution in the 2nd equation $y = (x - 2)^2$.

   When t is in $[-1, 1]$, the range of $x = t + 2$ is x in $[1, 3]$.

3. See Figure 3. From the 1st equation $x = \sqrt{t} \Rightarrow t = x^2$. By substitution in the 2nd equation $y = 3x^2 - 4$.

   When t is in $[0, 4]$, the range of $x = \sqrt{t}$ is x in $[0, 2]$.

4. See Figure 4. From the 2nd equation $y = \sqrt{t} \Rightarrow t = y^2$. By substitution in the 1st equation $x = y^4$.

   When t is in $[0, 4]$, the range of $y = \sqrt{t}$ is y in $[0, 2]$.

$[-8, 8]$ by $[-8, 8]$
Xscl $= 1$ Yscl $= 1$

$[0, 4]$ by $[-2, 2]$
Xscl $= 1$ Yscl $= 1$

$[-6, 6]$ by $[-6, 10]$
Xscl $= 1$ Yscl $= 1$

$[-2, 20]$ by $[0, 4]$
Xscl $= 2$ Yscl $= 1$

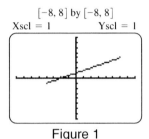
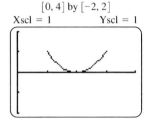
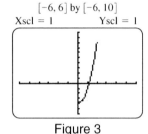
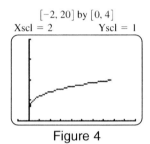

Figure 1   Figure 2   Figure 3   Figure 4

5. See Figure 5. From the 1st equation $x = t^3 + 1 \Rightarrow t^3 = x - 1$. By substitution in the 2nd equation $y = x - 2$.

   When t is in $[-3, 3]$, the range of $x = t^3 + 1$ is x in $[-26, 28]$.

6. See Figure 6. From the 1st equation $x = 2t - 1 \Rightarrow t = \frac{1}{2}(x + 1)$. By substitution in the 2nd equation

   $y = \frac{1}{4}(x + 1)^2 + 2$. When t is in $[-10, 10]$, the range of $x = 2t - 1$ is x in $[-21, 19]$.

7. See Figure 7. From the 2nd equation $y = \sqrt{3t - 1} \Rightarrow t = \frac{y^2 + 1}{3}$. By substitution in the 1st equation $x = 2^{(y^2 + 1)/3}$.

   When t is in $\left[\frac{1}{3}, 4\right]$, the range of $y = \sqrt{3t - 1}$ is y in $[0, \sqrt{11}]$.

8. See Figure 8. From the 1st equation $x = \ln(t - 1) \Rightarrow t = e^x + 1$. By substitution in the 2nd equation

   $y = 2e^x + 1$.   When t is in $(1, 10]$, the range of $x = \ln(t - 1)$ is x in $(-\infty, \ln 9]$.

$[-30, 30]$ by $[-30, 30]$
Xscl $= 2$ Yscl $= 2$

$[-20, 20]$ by $[0, 120]$
Xscl $= 5$ Yscl $= 10$

$[-2, 30]$ by $[-2, 10]$
Xscl $= 1$ Yscl $= 1$

$[-5, 5]$ by $[-2, 20]$
Xscl $= 1$ Yscl $= 1$

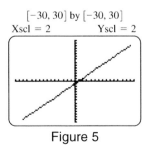
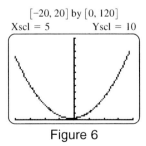
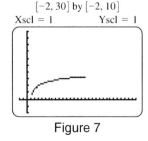
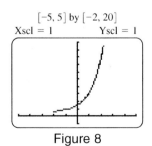

Figure 5   Figure 6   Figure 7   Figure 8

9. See Figure 9. From the 1st equation $x = t + 2 \Rightarrow t = x - 2$. By substitution in the 2nd equation

$y = -\dfrac{1}{2}\sqrt{9 - (x - 2)^2}$. When t is in $[-3, 3]$, the range of $x = t + 2$ is x in $[-1, 5]$.

10. See Figure 10. From the 1st equation $x = t \Rightarrow t = x$. By substitution in the 2nd equation $y = \sqrt{4 - x^2}$.

When t is in $[-2, 2]$, the range of $x = t$ is x in $[-2, 2]$.

11. See Figure 11. From the 1st equation $x = t \Rightarrow t = x$. By substitution in the 2nd equation $y = \dfrac{1}{x}$.

When t is in $(-\infty, 0) \cup (0, \infty)$, the range of $x = t$ is x in $(-\infty, 0) \cup (0, \infty)$.

12. See Figure 12. From the 1st equation $x = 2t - 1 \Rightarrow t = \dfrac{x + 1}{2}$. By substitution in the 2nd equation $y = \dfrac{2}{x + 1}$.

When t is in $(-\infty, 0) \cup (0, \infty)$, the range of $x = t + 3$ is x in $(-\infty, -1) \cup (-1, \infty)$.

$[-6, 6]$ by $[-4, 4]$	$[-6, 6]$ by $[-4, 4]$	$[-6, 6]$ by $[-4, 4]$	$[-6, 6]$ by $[-4, 4]$
Xscl = 1 Yscl = 1	Xscl = 1 Yscl = 1	Xscl = 1 Yscl = 1	Xscl = 1 Yscl = 1

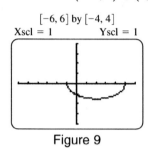 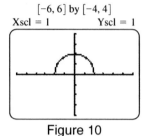 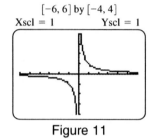 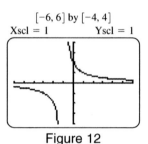

Figure 9   Figure 10   Figure 11   Figure 12

13. From the 1st equation $x = 3t \Rightarrow t = \dfrac{1}{3}x$. By substitution in the 2nd equation $y = \dfrac{1}{3}x - 1$.

When t is in $(-\infty, \infty)$, the range of $x = 3t$ is x in $(-\infty, \infty)$.

14. From the 1st equation $x = t + 3 \Rightarrow t = x - 3$. By substitution in the 2nd equation $y = 2(x - 3)$.

When t is in $(-\infty, \infty)$, the range of $x = t + 3$ is x in $(-\infty, \infty)$.

15. From the 2nd equation $y = t + 1 \Rightarrow t = y - 1$. By substitution in the 1st equation $x = 3(y - 1)^2$.

When t is in $(-\infty, \infty)$, the range of $y = t + 1$ is y in $(-\infty, \infty)$.

16. From the 1st equation $x = t - 2 \Rightarrow t = x + 2$. By substitution in the 2nd equation $y = \dfrac{1}{2}(x + 2)^2 + 1$.

When t is in $(-\infty, \infty)$, the range of $x = t - 2$ is x in $(-\infty, \infty)$.

17. From the 2nd equation $y = 4t^3 \Rightarrow t = \sqrt[3]{\dfrac{y}{4}}$. By substitution in the 1st equation $x = 3\left(\dfrac{y}{4}\right)^{2/3}$.

When t is in $(-\infty, \infty)$, the range of $y = 4t^3$ is y in $(-\infty, \infty)$.

18. From the 1st equation $x = 2t^3 \Rightarrow t = \sqrt[3]{\dfrac{x}{2}}$. By substitution in the 2nd equation $y = -\left(\dfrac{x}{2}\right)^{2/3}$.

When t is in $(-\infty, \infty)$, the range of $x = 2t^3$ is x in $(-\infty, \infty)$.

19. From the 1st equation $x = t \Rightarrow t = x$. By substitution in the 2nd equation $y = \sqrt{x^2 + 2}$.

When t is in $(-\infty, \infty)$, the range of $x = t$ is x in $(-\infty, \infty)$.

20. From the 1st equation $x = \sqrt{t} \Rightarrow t = x^2$. By substitution in the 2nd equation $y = x^4 - 1$.

When t is in $[0, \infty)$, the range of $x = \sqrt{t}$ is x in $[0, \infty)$.

21. From the 1st equation $x = e^t \Rightarrow t = \ln x$. By substitution in the 2nd equation $y = \dfrac{1}{x}$.

When t is in $(-\infty, \infty)$, the range of $x = e^t$ is x in $(0, \infty)$.

22. From the 2nd equation $y = e^t \Rightarrow t = \ln y$. By substitution in the 1st equation $x = y^2$.

    When t is in $(-\infty, \infty)$, the range of $y = e^t$ is y in $(0, \infty)$.

23. From the 1st equation $x = \dfrac{1}{\sqrt{t+2}} \Rightarrow t = \dfrac{1}{x^2} - 2$. By substitution in the 2nd equation $y = 1 - 2x^2$.

    When t is in $(-2, \infty)$, the range of $x = \dfrac{1}{\sqrt{t+2}}$ is x in $(0, \infty)$.

24. From the 1st equation $x = \dfrac{t}{t-1} \Rightarrow t = \dfrac{x}{1-x}$. By substitution in the 2nd equation $y = \sqrt{x-1}$.

    When t is in $(1, \infty)$, the range of $x = \dfrac{t}{t-1}$ is x in $(1, \infty)$.

25. From the 1st equation $x = t + 2 \Rightarrow t = x - 2$. By substitution in the 2nd equation $y = \dfrac{1}{x}$.

    When t has the restriction $t \ne -2$, the range of $x = t + 2$ has the restriction $x \ne 0$.

26. From the 1st equation $x = t - 3 \Rightarrow t = x + 3$. By substitution in the 2nd equation $y = \dfrac{2}{x}$.

    When t has the restriction $t \ne 3$, the range of $x = t - 3$ has the restriction $x \ne 0$.

27. From the 1st equation $x = t^2 \Rightarrow t = \sqrt{x}$. By substitution in the 2nd equation $y = \ln x$.

    When t is in $(0, \infty)$, the range of $x = t^2$ is x in $(0, \infty)$.

28. From the 1st equation $x = \ln t \Rightarrow t = e^x$. By substitution in the 2nd equation $y = 3x$.

    When t is in $(0, \infty)$, the range of $x = \ln t$ is x in $(-\infty, \infty)$.

29. For $x = \dfrac{1}{2}t$, $y = 2\left(\dfrac{1}{2}t\right) + 3 \Rightarrow y = t + 3$. For $x = \dfrac{t+3}{2}$, $y = 2\left(\dfrac{t+3}{2}\right) + 3 \Rightarrow y = t + 6$

30. For $x = 2t$, $y = \dfrac{3}{2}(2t) - 4 \Rightarrow y = 3t - 4$. For $x = t + 2$, $y = \dfrac{3}{2}(t+2) - 4 \Rightarrow y = \dfrac{3}{2}t - 1$

31. For $x = \dfrac{1}{3}t$, $y = \sqrt{3\left(\dfrac{1}{3}t\right) + 2} \Rightarrow y = \sqrt{t+2}$ for t in $[-2, \infty)$.

    For $x = \dfrac{t-2}{3}$, $y = \sqrt{3\left(\dfrac{t-2}{3}\right) + 2} \Rightarrow y = \sqrt{t}$ for t in $[0, \infty)$

32. For $x = t - 1$, $y = (t - 1 + 1)^2 + 1 \Rightarrow y = t^2 + 1$. For $x = 2t$, $y = (2t + 1)^2 + 1 \Rightarrow y = 4t^2 + 4t + 2$.

33. For $x = t^3 + 1$, $t^3 + 1 = y^3 + 1 \Rightarrow y = t$. For $x = t$, $t = y^3 + 1 \Rightarrow y = \sqrt[3]{t-1}$.

34. For $x = 2(t - 3)^2 - 4$, $2(t - 3)^2 - 4 = 2(y - 3)^2 - 4 \Rightarrow y = t$.

    For $x = 2t^2 - 4$, $2t^2 - 4 = 2(y - 3)^2 - 4 \Rightarrow y - 3 = t \Rightarrow y = t + 3$.

35. For $x = \sqrt{t+1}$, $\sqrt{t+1} = \sqrt{y+1} \Rightarrow y = t$ for t in $[-1, \infty)$.

    For $x = \sqrt{t}$, $\sqrt{t} = \sqrt{y+1} \Rightarrow y + 1 = t \Rightarrow y = t - 1$ for t in $[0, \infty)$.

36. For $x = \dfrac{1}{t+1}$, $\dfrac{1}{t+1} = \dfrac{1}{y+1} \Rightarrow y = t$ for $t \ne -1$.

    For $x = \dfrac{1}{t}$, $\dfrac{1}{t} = \dfrac{1}{y+1} \Rightarrow y + 1 = t \Rightarrow y = t - 1$ for $t \ne 0$.

37. (a) Find t when $y = 0$. $400 \cdot \dfrac{\sqrt{2}}{2}t - 16t^2 = 0 \Rightarrow 16t^2 - 200\sqrt{2}\,t = 0 \Rightarrow t(16t - 200\sqrt{2}) = 0 \Rightarrow$

$16t - 200\sqrt{2} = 0 \Rightarrow 16t = 200\sqrt{2} \Rightarrow t = \dfrac{200\sqrt{2}}{16} \approx 17.7$ seconds

(b) Find x when $t \approx 17.7$. $x = 400 \cdot \dfrac{\sqrt{2}}{2}(17.7) \approx 5000$ feet

(c) Find y when $x \approx 8.85$ (half the total time). $y = 400 \cdot \dfrac{\sqrt{2}}{2}(8.85) - 16(8.85)^2 \approx 1250$ feet

38. (a) Find t when $y = 0$. $\dfrac{800}{2}t - 16t^2 = 0 \Rightarrow 16t^2 - 400t = 0 \Rightarrow t(16t - 400) = 0 \Rightarrow$

$16t - 400 = 0 \Rightarrow 16t = 400 \Rightarrow t = \dfrac{400}{16} = 25$ seconds

(b) Find x when $t = 25$. $x = 800 \cdot \dfrac{\sqrt{3}}{2}(25) \approx 17{,}320.5$ feet

(c) Find y when $x = 12.5$ (half the total time). $y = \dfrac{800}{2}(12.5) - 16(12.5)^2 = 2500$ feet

39. See Figure 39. From the 1st equation $x = 60t \Rightarrow t = \dfrac{x}{60}$. By substitution in the 2nd equation

$y = 80\left(\dfrac{x}{60}\right) - 16\left(\dfrac{x}{60}\right)^2 \Rightarrow y = \dfrac{4}{3}x - \dfrac{x^2}{225}$.

40. See Figure 40. From the 1st equation $x = t^2 \Rightarrow t = \sqrt{x}$. By substitution in the 2nd equation $y = -16\sqrt{x} + 64x^{1/4}$.

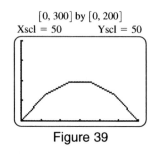

[0, 300] by [0, 200]
Xscl = 50   Yscl = 50

Figure 39

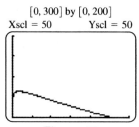

[0, 300] by [0, 200]
Xscl = 50   Yscl = 50

Figure 40

41. From the 1st equation $x = v_0\dfrac{\sqrt{2}}{2}t \Rightarrow t = \dfrac{2x}{v_0\sqrt{2}}$. By substitution in the 2nd equation

$y = v_0\dfrac{\sqrt{2}}{2}\left(\dfrac{2x}{v_0\sqrt{2}}\right) - 16\left(\dfrac{2x}{v_0\sqrt{2}}\right)^2 \Rightarrow y = x - \dfrac{32}{v_0^2}x^2$.

42. $x = -\dfrac{b}{2a} = -\dfrac{1}{2\left(-\dfrac{32}{v_0^2}\right)} = \dfrac{v_0^2}{64}$; $y = \dfrac{v_0^2}{64} - \dfrac{32}{v_0^2}\left(\dfrac{v_0^2}{64}\right)^2 = \dfrac{v_0^2}{64} - \dfrac{v_0^2}{128} \Rightarrow y = \dfrac{2v_0^2}{128} - \dfrac{v_0^2}{128} \Rightarrow y = \dfrac{v_0^2}{128}$

The vertex is $\left(\dfrac{v_0^2}{64}, \dfrac{v_0^2}{128}\right)$.

43. A line through (x_1, y_1) with slope m is given by $y - y_1 = m(x - x_1)$.

For $x = t$, $y - y_1 = m(t - x_1)$. For $t = x - x_1$, $y - y_1 = mt \Rightarrow y = mt + y_1$. Many answers are possible.

44. For $x = t$, $y = a(t - h)^2 + k$. For $x = t + h$, $y = at^2 + k$. Many answers are possible.

Reviewing Basic Concepts (Sections 6.3 and 6.4)

1. $3x^2 + y^2 - 6x + 6y = 0 \Rightarrow 3(x^2 - 2x + 1) + (y^2 + 6y + 9) = 0 + 3 + 9 \Rightarrow$

    $3(x-1)^2 + (y+3)^2 = 12 \Rightarrow \dfrac{(x-1)^2}{4} + \dfrac{(y+3)^2}{12} = 1$; ellipse centered at $(1, -3)$

2. $y^2 - 2x^2 + 8y - 8x - 4 = 0 \Rightarrow (y^2 + 8y + 16) - 2(x^2 + 4x + 4) = 4 + 16 - 8 \Rightarrow$

    $(y+4)^2 - 2(x+2)^2 = 12 \Rightarrow \dfrac{(y+4)^2}{12} - \dfrac{(x+2)^2}{6} = 1$; hyperbola centered at $(-4, -2)$

3. $3y^2 + 12y + 5x = 3 \Rightarrow 3(y^2 + 4y + 4) = -5x + 3 + 12 \Rightarrow 3(y+2)^2 = -5(x-3) \Rightarrow$

    $(y+2)^2 = -\dfrac{5}{3}(x - 3)$; parabola with center $(3, -2)$ that opens to the left

4. $x^2 + 25y^2 = 25 \Rightarrow \dfrac{x^2}{25} + \dfrac{y^2}{1} = 1 \Rightarrow a = 5, b = 1$, and $c = \sqrt{25 - 1} = \sqrt{24}$; $e = \dfrac{c}{a} = \dfrac{\sqrt{24}}{5} = \dfrac{2\sqrt{6}}{5}$

5. $8y^2 - 4x^2 = 8 \Rightarrow \dfrac{y^2}{1} - \dfrac{x^2}{2} = 1 \Rightarrow a = 1, b = \sqrt{2}$, and $c = \sqrt{1 + 2} = \sqrt{3}$; $e = \dfrac{c}{a} = \dfrac{\sqrt{3}}{1} = \sqrt{3}$

6. Since $e = 1$, the conic is a parabola. With center $(0, 0)$ and focus $(-2, 0)$ the equation is $y^2 = 4cx \Rightarrow y^2 = -8x$.

7. The ellipse is centered between the foci at $(0, 0)$. The major axis is horizontal with $a = 5$. Since the foci are

    $(\pm 3, 0)$, we know that $c = 3$. Since $c^2 = a^2 - b^2$, the value of b can be found by substitution:

    $b^2 = a^2 - c^2 = 5^2 - 3^2 = 25 - 9 = 16 \Rightarrow b = 4$. The equation is $\dfrac{x^2}{25} + \dfrac{y^2}{16} = 1$.

8. The hyperbola has a vertical transverse axis with $c = 5$. The vertices are $(0, \pm 4)$ so $a = 4$.

    The center is located between the foci at $(0, 0)$. Since $c^2 = a^2 + b^2$, the value of b can be found by substitution:

    $b^2 = c^2 - a^2 = 5^2 - 4^2 = 25 - 16 = 9 \Rightarrow b = 3$. The equation is $\dfrac{y^2}{16} - \dfrac{x^2}{9} = 1$.

9. Using a vertical minor axis, $b = 9$. The major axis has length 30 so $a = 15$. The equation is $\dfrac{x^2}{225} + \dfrac{y^2}{81} = 1$.

    When $x = 6$, $\dfrac{6^2}{225} + \dfrac{y^2}{81} = 1 \Rightarrow \dfrac{y^2}{81} = 1 - \dfrac{36}{225} \Rightarrow y^2 = 81\left(1 - \dfrac{36}{225}\right) \Rightarrow y = \sqrt{81\left(1 - \dfrac{36}{225}\right)} \approx 8.25$ feet

10. (a) See Figure 10.

    (b) From the 1st equation $x = 2t \Rightarrow t = \dfrac{x}{2}$. By substitution in the 2nd equation, $y = \sqrt{\dfrac{x^2}{4} + 1}$.

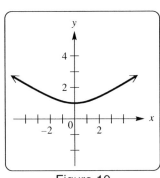

Figure 10

Chapter 6 Review Exercises

1. Here $h = -2, k = 3$ and $r^2 = 5^2 = 25$. The equation is $(x + 2)^2 + (y - 3)^2 = 25$. See Figure 1.

   From the figure, the domain is $[-7, 3]$ and the range is $[-2, 8]$.

2. Here $h = \sqrt{5}, k = -\sqrt{7}$ and $r^2 = (\sqrt{3})^2 = 3$. The equation is $(x - \sqrt{5})^2 + (y + \sqrt{7})^2 = 3$. See Figure 2.

   From the figure, the domain is $[\sqrt{5} - \sqrt{3}, \sqrt{5} + \sqrt{3}]$ and the range is $[-\sqrt{7} - \sqrt{3}, -\sqrt{7} + \sqrt{3}]$.

3. The radius is the distance between $(-8, 1)$ and $(0, 16)$: $r = \sqrt{(0 - (-8))^2 + (16 - 1)^2} = \sqrt{64 + 225} = 17$

   Here $h = -8, k = 1$ and $r^2 = 17^2 = 289$. The equation is $(x + 8)^2 + (y - 1)^2 = 289$. See Figure 3.

   From the figure, the domain is $[-25, 9]$ and the range is $[-16, 18]$.

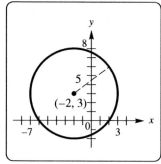 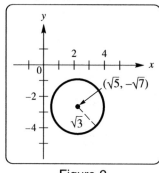 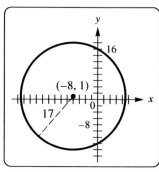

| Figure 1 | Figure 2 | Figure 3 |

4. If the center is $(3, -6)$, the circle must touch the x-axis at the point $(3, 0)$. The radius is 6.

   Here $h = 3, k = -6$ and $r^2 = 6^2 = 36$. The equation is $(x - 3)^2 + (y + 6)^2 = 36$. See Figure 4.

   From the figure, the domain is $[-3, 9]$ and the range is $[-12, 0]$.

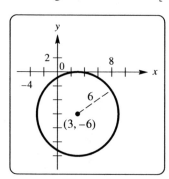

Figure 4

5. $x^2 - 4x + y^2 + 6y + 12 = 0 \Rightarrow (x^2 - 4x + 4) + (y^2 + 6y + 9) = -12 + 4 + 9 \Rightarrow$

   $(x - 2)^2 + (y + 3)^2 = 1$. The circle has center $(2, -3)$ and radius $r = 1$.

6. $x^2 - 6x + y^2 - 10y + 30 = 0 \Rightarrow (x^2 - 6x + 9) + (y^2 - 10y + 25) = -30 + 9 + 25 \Rightarrow$

   $(x - 3)^2 + (y - 5)^2 = 4$. The circle has center $(3, 5)$ and radius $r = 2$.

7. $2x^2 + 14x + 2y^2 + 6y = -2 \Rightarrow x^2 + 7x + y^2 + 3y = -1 \Rightarrow$

   $\left(x^2 + 7x + \dfrac{49}{4}\right) + \left(y^2 + 3y + \dfrac{9}{4}\right) = -1 + \dfrac{49}{4} + \dfrac{9}{4} \Rightarrow \left(x + \dfrac{7}{2}\right)^2 + \left(y + \dfrac{3}{2}\right)^2 = \dfrac{54}{4}$

   The circle has center $\left(-\dfrac{7}{2}, -\dfrac{3}{2}\right)$ and radius $r = \dfrac{\sqrt{54}}{2} = \dfrac{3\sqrt{6}}{2}$.

8. $3x^2 + 3y^2 + 33x - 15y = 0 \Rightarrow x^2 + 11x + y^2 - 5y = 0 \Rightarrow$

$$\left(x^2 + 11x + \frac{121}{4}\right) + \left(y^2 - 5y + \frac{25}{4}\right) = 0 + \frac{121}{4} + \frac{25}{4} \Rightarrow \left(x + \frac{11}{2}\right)^2 + \left(y - \frac{5}{2}\right)^2 = \frac{146}{4}$$

The circle has center $\left(-\frac{11}{2}, \frac{5}{2}\right)$ and radius $r = \frac{\sqrt{146}}{2}$.

9. The equation is that of a circle with center (4, 5) and radius 0. That is, the graph is the point (4, 5).

10. Since $y^2 = -\frac{2}{3}x$ is equivalent to $y^2 = 4\left(-\frac{1}{6}\right)x$, the equation is in the form $y^2 = 4cx$ with $c = -\frac{1}{6}$.

The focus is $\left(-\frac{1}{6}, 0\right)$ and the equation of the directrix is $x = \frac{1}{6}$. The axis is $y = 0$, or the x-axis.

The graph is shown in Figure 10. From the figure, the domain is $(-\infty, 0]$ and the range is $(-\infty, \infty)$.

11. Since $y^2 = 2x$ is equivalent to $y^2 = 4\left(\frac{1}{2}\right)x$, the equation is in the form $y^2 = 4cx$ with $c = \frac{1}{2}$.

The focus is $\left(\frac{1}{2}, 0\right)$ and the equation of the directrix is $x = -\frac{1}{2}$. The axis is $y = 0$, or the x-axis.

The graph is shown in Figure 11. From the figure, the domain is $[0, \infty)$ and the range is $(-\infty, \infty)$.

12. Since $3x^2 - y = 0$ is equivalent to $x^2 = 4\left(\frac{1}{12}\right)y$, the equation is in the form $x^2 = 4cy$ with $c = \frac{1}{12}$.

The focus is $\left(0, \frac{1}{12}\right)$ and the equation of the directrix is $y = -\frac{1}{12}$. The axis is $x = 0$, or the y-axis.

The graph is shown in Figure 12. From the figure, the domain is $(-\infty, \infty)$ and the range is $[0, \infty)$.

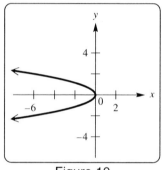   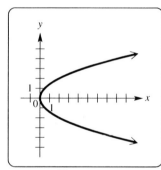   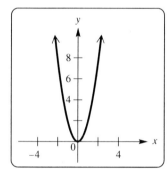

Figure 10   Figure 11   Figure 12

13. Since $x^2 + 2y = 0$ is equivalent to $x^2 = 4\left(-\frac{1}{2}\right)y$, the equation is in the form $x^2 = 4cy$ with $c = -\frac{1}{2}$.

The focus is $\left(0, -\frac{1}{2}\right)$ and the equation of the directrix is $y = \frac{1}{2}$. The axis is $x = 0$, or the y-axis.

The graph is shown in Figure 13. From the figure, the domain is $(-\infty, \infty)$ and the range is $(-\infty, 0]$.

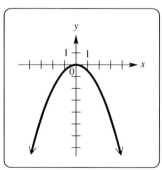

Figure 13

14. If the vertex is $(0, 0)$ and the focus is $(4, 0)$, then the parabola opens to the right and $c = 4$.

    The equation is $y^2 = 4cx \Rightarrow y^2 = 16x$.

15. If the vertex is $(0, 0)$ and the parabola opens to the right, the equation is in the form $y^2 = 4cx$. Find the

    value of c by using the fact that the parabola passes through $(2, 5)$. Thus, $(5)^2 = 4c(2) \Rightarrow c = \dfrac{25}{8}$.

    The equation is $y^2 = 4cx \Rightarrow y^2 = \dfrac{25}{2}x$.

16. If the vertex is $(0, 0)$ and the parabola opens downward, the equation is in the form $x^2 = 4cy$. Find the

    value of c by using the fact that the parabola passes through $(3, -4)$. Thus, $(3)^2 = 4c(-4) \Rightarrow c = -\dfrac{9}{16}$.

    The equation is $x^2 = 4cy \Rightarrow x^2 = -\dfrac{9}{4}y$.

17. If the focus is $(2, 6)$ and the vertex is $(-5, 6)$, the parabola opens to the right and $c = 7$. Substituting in
    $(y - k)^2 = 4c(x - h)$, we get $(y - 6)^2 = 28(x + 5)$.

18. If the focus is $(4, 5)$ and the vertex is $(4, 3)$, the parabola opens upward and $c = 2$. Substituting in
    $(x - h)^2 = 4c(y - k)$, we get $(x - 4)^2 = 8(y - 3)$.

19. $\dfrac{x^2}{5} + \dfrac{y^2}{9} = 1 \Rightarrow a = 3$ and $b = \sqrt{5}$. The endpoints of the major axis (vertices) are $(0, \pm 3)$ so the range is
    $[-3, 3]$. The endpoints of the minor axis are $(\pm\sqrt{5}, 0)$ so the domain is $[-\sqrt{5}, \sqrt{5}]$. See Figure 19.

20. $\dfrac{x^2}{16} + \dfrac{y^2}{4} = 1 \Rightarrow a = 4$ and $b = 2$. The endpoints of the major axis (vertices) are $(\pm 4, 0)$ so the domain is
    $[-4, 4]$. The endpoints of the minor axis are $(0, \pm 2)$ so the range is $[-2, 2]$. See Figure 20.

21. The transverse axis is horizontal with $a = 8$ and $b = 6$. The asymptotes are $y = \pm\dfrac{3}{4}x$. See Figure 21.

    The domain is $(-\infty, -8] \cup [8, \infty)$ and the range is $(-\infty, \infty)$. The vertices are $(-8, 0)$ and $(8, 0)$.

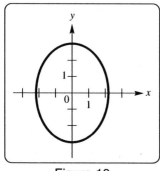

Figure 19

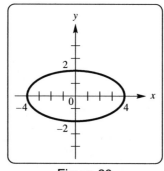

Figure 20

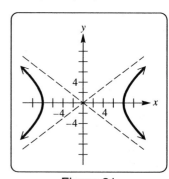

Figure 21

22. The transverse axis is vertical with $a = 5$ and $b = 3$. The asymptotes are $y = \pm\dfrac{5}{3}x$. See Figure 22.

    The domain is $(-\infty, \infty)$ and the range is $(-\infty, -5] \cup [5, \infty)$. The vertices are $(0, -5)$ and $(0, 5)$.

23. The ellipse is centered at $(3, -1)$. The major axis is horizontal and has length $2a = 4$ so the vertices are
    $(1, -1)$ and $(5, -1)$. The length of the minor axis is $2b = 2$. The graph is shown in Figure 23.
    The domain is $[1, 5]$ and the range is $[-2, 0]$.

24. The ellipse is centered at $(2, -3)$. The major axis is horizontal and has length $2a = 6$ so the vertices are
    $(-1, -3)$ and $(5, -3)$. The length of the minor axis is $2b = 4$. The graph is shown in Figure 24.
    The domain is $[-1, 5]$ and the range is $[-5, -1]$.

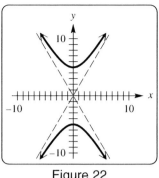

Figure 22

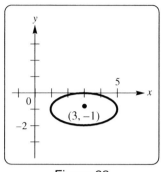

Figure 23

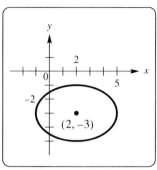

Figure 24

25. The center is $(-3, -2)$ and the transverse axis is vertical with $a = 2$ and $b = 3$. See Figure 25.

The domain is $(-\infty, \infty)$ and the range is $(-\infty, -4] \cup [0, \infty)$. The vertices are $(-3, -4)$ and $(-3, 0)$.

26. The center is $(-1, 2)$ and the transverse axis is horizontal with $a = 4$ and $b = 2$. See Figure 26.

The domain is $(-\infty, -5] \cup [3, \infty)$ and the range is $(-\infty, \infty)$. The vertices are $(-5, 2)$ and $(3, 2)$.

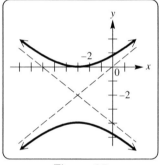

Figure 25

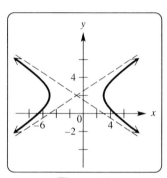

Figure 26

27. The major axis is vertical with $a = 4$. Since one focus is $(0, 2)$, we know that $c = 2$. Since $c^2 = a^2 - b^2$,

the value of b can be found by substitution: $b^2 = a^2 - c^2 = 4^2 - 2^2 = 16 - 4 = 12 \Rightarrow b = \sqrt{12}$.

The equation is $\dfrac{x^2}{12} + \dfrac{y^2}{16} = 1$.

28. The major axis is horizontal with $a = 6$. Since one focus is $(-2, 0)$, we know that $c = 2$. Since $c^2 = a^2 - b^2$,

the value of b can be found by substitution: $b^2 = a^2 - c^2 = 6^2 - 2^2 = 36 - 4 = 32 \Rightarrow b = \sqrt{32}$.

The equation is $\dfrac{x^2}{36} + \dfrac{y^2}{32} = 1$.

29. The hyperbola has a vertical transverse axis with $c = 5$. The y-intercepts coincide with the vertices so $a = 4$.

Since $c^2 = a^2 + b^2$, the value of b can be found by substitution:

$b^2 = c^2 - a^2 = 5^2 - 4^2 = 25 - 16 = 9 \Rightarrow b = 3$. The equation is $\dfrac{y^2}{16} - \dfrac{x^2}{9} = 1$.

30. The hyperbola has a vertical transverse axis. The y-intercept coincides with a vertex so $a = 2$.

The equation is of the form $\dfrac{y^2}{a^2} - \dfrac{x^2}{b^2} = 1$. By substitution using the point $(2, 3)$:

$\dfrac{3^2}{2^2} - \dfrac{2^2}{b^2} = 1 \Rightarrow \dfrac{9}{4} - 1 = \dfrac{4}{b^2} \Rightarrow \dfrac{5}{4} = \dfrac{4}{b^2} \Rightarrow 5b^2 = 16 \Rightarrow b^2 = \dfrac{16}{5}$

The equation is $\dfrac{y^2}{4} - \dfrac{x^2}{\frac{16}{5}} = 1$ or $\dfrac{y^2}{4} - \dfrac{5x^2}{16} = 1$.

31. Since $0 < e < 1$, the conic is an ellipse with $c = 3$. Now $\dfrac{c}{a} = e \Rightarrow \dfrac{3}{a} = \dfrac{2}{3} \Rightarrow a = \dfrac{9}{2}$. For an ellipse,

$b^2 = a^2 - c^2 = \dfrac{81}{4} - 9 = \dfrac{45}{4}$. The equation is $\dfrac{x^2}{\frac{45}{4}} + \dfrac{y^2}{\frac{81}{4}} = 1$ or $\dfrac{4x^2}{45} + \dfrac{4y^2}{81} = 1$.

32. Since $e > 1$, the conic is a hyperbola with $c = 5$. Now $\dfrac{c}{a} = e \Rightarrow \dfrac{5}{a} = \dfrac{5}{2} \Rightarrow a = 2$. For a hyperbola,

$b^2 = c^2 - a^2 = 25 - 4 = 21$. The equation is $\dfrac{x^2}{4} - \dfrac{y^2}{21} = 1$.

33. (a) $x^2 + y^2 + 2x + 6y - 15 = 0 \Rightarrow x^2 + 2x + 1 + y^2 + 6y + 9 = 15 + 9 + 1 \Rightarrow$

$(x + 1)^2 + (y + 3)^2 = 25$; The center is $(-1, -3)$.

(b) The radius is $r = \sqrt{25} = 5$.

(c) $(x + 1)^2 + (y + 3)^2 = 25 \Rightarrow (y + 3)^2 = 25 - (x + 1)^2 \Rightarrow y + 3 = \pm\sqrt{25 - (x + 1)^2} \Rightarrow$

$y = -3 \pm \sqrt{25 - (x + 1)^2}$; Graph $y_1 = -3 + \sqrt{25 - (x + 1)^2}$ and $y_2 = -3 - \sqrt{25 - (x + 1)^2}$.

34. D. $4x^2 + y^2 = 36 \Rightarrow \dfrac{x^2}{9} + \dfrac{y^2}{36} = 1$; This is an ellipse with major axis on the y-axis.

35. E. $x = 2y^2 + 3 \Rightarrow 4\left(\dfrac{1}{8}\right)(x - 3) = (y - 0)^2$; This is a parabola that opens to the right.

36. A. $(x - 1)^2 + (y + 2)^2 = 36$; This is a circle with center $(1, -2)$ and radius 6.

37. C. $\dfrac{x^2}{36} + \dfrac{y^2}{9} = 1$; This is an ellipse with major axis on the x-axis.

38. B. $(y - 1)^2 - (x - 2)^2 = 36 \Rightarrow \dfrac{(y - 1)^2}{36} - \dfrac{(x - 2)^2}{36} = 1$; This is a hyperbola with center $(2, 1)$.

39. F. $y^2 = 36 + 4x^2 \Rightarrow y^2 - 4x^2 = 36 \Rightarrow \dfrac{y^2}{36} - \dfrac{x^2}{9} = 1$; This is a hyperbola with transverse axis on the y-axis.

40. $4x^2 + 8x + 25y^2 - 250y = -529 \Rightarrow 4(x^2 + 2x) + 25(y^2 - 10y) = -529 \Rightarrow$

$4(x^2 + 2x + 1) + 25(y^2 - 10y + 25) = -529 + 4 + 625 \Rightarrow 4(x + 1)^2 + 25(y - 5)^2 = 100 \Rightarrow$

$\dfrac{(x + 1)^2}{25} + \dfrac{(y - 5)^2}{4} = 1$; The center is $(-1, 5)$. The vertices are $(-1 - 5, 5), (-1 + 5, 5)$ or $(-6, 5), (4, 5)$.

41. $5x^2 + 20x + 2y^2 - 8y = -18 \Rightarrow 5(x^2 + 4x) + 2(y^2 - 4y) = -18 \Rightarrow$

$5(x^2 + 4x + 4) + 2(y^2 - 4y + 4) = -18 + 20 + 8 \Rightarrow 5(x + 2)^2 + 2(y - 2)^2 = 10 \Rightarrow$

$\dfrac{(x + 2)^2}{2} + \dfrac{(y - 2)^2}{5} = 1$; The center is $(-2, 2)$. The vertices are $(-2, 2 - \sqrt{5}), (-2, 2 + \sqrt{5})$.

42. $x^2 + 4x - 4y^2 + 24y = 36 \Rightarrow (x^2 + 4x + 4) - 4(y^2 - 6y + 9) = 36 + 4 - 36 \Rightarrow$

$(x + 2)^2 - 4(y - 3)^2 = 4 \Rightarrow \dfrac{(x + 2)^2}{4} - \dfrac{(y - 3)^2}{1} = 1$.

The center is $(-2, 3)$. The vertices are $(-2 - 2, 3), (-2 + 2, 3)$ or $(-4, 3), (0, 3)$.

43. $4y^2 + 8y - 3x^2 + 6x = 11 \Rightarrow 4(y^2 + 2y + 1) - 3(x^2 - 2x + 1) = 11 + 4 - 3 \Rightarrow$

$4(y + 1)^2 - 3(x - 1)^2 = 12 \Rightarrow \dfrac{(y + 1)^2}{3} - \dfrac{(x - 1)^2}{4} = 1$.

The center is $(1, -1)$. The vertices are $(1, -1 - \sqrt{3}), (1, -1 + \sqrt{3})$.

44. $9x^2 + 25y^2 = 225 \Rightarrow \dfrac{x^2}{25} + \dfrac{y^2}{9} = 1 \Rightarrow a = 5, b = 3,$ and $c = \sqrt{25 - 9} = 4; e = \dfrac{c}{a} = \dfrac{4}{5}$

45. $4x^2 + 9y^2 = 36 \Rightarrow \dfrac{x^2}{9} + \dfrac{y^2}{4} = 1 \Rightarrow a = 3, b = 2,$ and $c = \sqrt{9 - 4} = \sqrt{5}; e = \dfrac{c}{a} = \dfrac{\sqrt{5}}{3}$

46. $9x^2 - y^2 = 9 \Rightarrow \dfrac{x^2}{1} - \dfrac{y^2}{9} = 1 \Rightarrow a = 1, b = 3,$ and $c = \sqrt{1 + 9} = \sqrt{10}; e = \dfrac{c}{a} = \dfrac{\sqrt{10}}{1} = \sqrt{10}$

47. The parabola opens to the right so the equation has the form $(y - k)^2 = 4c(x - h)$. With vertex $(-3, 2)$ and

   y-intercept $(0, 5)$, $(5 - 2)^2 = 4c(0 + 3) \Rightarrow 9 = 12c \Rightarrow c = \dfrac{3}{4}$. The equation is $(y - 2)^2 = 3(x + 3)$.

   This can also be written $x = \dfrac{1}{3}(y - 2)^2 - 3$ or $(y - 2)^2 = 3(x + 3)$.

48. The center is located between the foci at $(0, 0)$. The hyperbola has a vertical transverse axis with $c = 12$.

   From the asymptotes, $\dfrac{a}{b} = 1$. Also $c^2 = a^2 + b^2 \Rightarrow a^2 + b^2 = 144$. Solving these equations simultaneously

   results in $a^2 = 72$ and $b^2 = 72$. The equation is $\dfrac{y^2}{72} - \dfrac{x^2}{72} = 1$.

49. The foci are $(0, 0)$ and $(4, 0)$, so the center is $(2, 0)$ and $c = 2$. The sum of the distances is 8 so

   $2a = 8 \Rightarrow a = 4$. For an ellipse, $b^2 = a^2 - c^2 = 16 - 4 = 12$. The equation is $\dfrac{(x - 2)^2}{16} + \dfrac{y^2}{12} = 1$.

50. The foci are $(0, 0)$ and $(0, 4)$, so the center is $(0, 2)$ and $c = 2$. The difference of the distances is 2 so

   $2a = 2 \Rightarrow a = 1$. For a hyperbola, $b^2 = c^2 - a^2 = 4 - 1 = 3$. The equation is $\dfrac{(y - 2)^2}{1} - \dfrac{x^2}{3} = 1$.

51. See Figure 51.

52. See Figure 52.

53. See Figure 53.

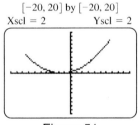

$[-20, 20]$ by $[-20, 20]$
Xscl = 2   Yscl = 2

Figure 51

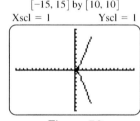

$[-15, 15]$ by $[10, 10]$
Xscl = 1   Yscl = 1

Figure 52

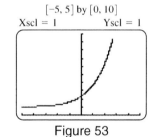

$[-5, 5]$ by $[0, 10]$
Xscl = 1   Yscl = 1

Figure 53

54. From the 1st equation $x = 3t + 2 \Rightarrow t = \dfrac{x - 2}{3}$. By substitution in the 2nd equation $y = \dfrac{x - 2}{3} - 1$.

   This is equivalent to $x - 3y = 5$. When t is in $[-5, 5]$, the range of $x = 3t + 2$ is x in $[-13, 17]$.

55. From the 1st equation $x = \sqrt{t - 1} \Rightarrow t = x^2 + 1$. By substitution in the 2nd equation $y = \sqrt{x^2 + 1}$.

   This is equivalent to $y^2 - x^2 = 1$. When t is in $[1, \infty)$, the range of $x = \sqrt{t - 1}$ is x in $[0, \infty)$.

56. From the 1st equation $x = \dfrac{1}{t + 3} \Rightarrow t = \dfrac{1}{x} - 3$. By substitution in the 2nd equation $y = \dfrac{1}{x}$ for $x \neq 0$.

57. Since the major axis has length 134.5 million miles, $2a = 134.5 \Rightarrow a = 67.25$. From the given eccentricity,

   $\dfrac{c}{a} = .006775 \Rightarrow \dfrac{c}{67.25} = .006775 \Rightarrow c = 67.25(.006775) = .4456$ million miles. The smallest distance is

   $67.25 - .4456 \approx 66.8$ million miles and the largest distance is $67.25 + .4456 \approx 67.7$ million miles.

58. Since the smallest distance between the comet and the sun is 89 million miles, $a - c = 89$. The given eccentricity provides the equation $\dfrac{c}{a} = .964$. Solving these two equations simultaneously gives $a \approx 2472.222$ and $c \approx 2383.222$. Then $b^2 = a^2 - c^2 \Rightarrow b^2 = 2472.222^2 - 2383.222^2 \Rightarrow b^2 \approx 432{,}135$.

The equation is $\dfrac{x^2}{a^2} + \dfrac{y^2}{b^2} = 1 \Rightarrow \dfrac{x^2}{6{,}111{,}883} + \dfrac{y^2}{432{,}135} = 1$.

59. The value of $\dfrac{k}{\sqrt{D}}$ is $\dfrac{2.82 \times 10^7}{\sqrt{42.5 \times 10^6}} \approx 4326$. Since $2090 < 4326$, $V < \dfrac{k}{\sqrt{D}}$. The trajectory is elliptic.

60. The velocity must be more than 4326 m per sec.. The minimum increase is $4326 - 2090 \approx 2236$ m per sec.

61. The required increase in velocity is less when D is larger.

62. $Ax^2 + Cy^2 + Dx + Ey + F = 0 \Rightarrow A\left(x^2 + \dfrac{D}{A}x\right) + C\left(y^2 + \dfrac{E}{C}y\right) = -F \Rightarrow$

$A\left(x^2 + \dfrac{D}{A}x + \dfrac{D^2}{4A^2}\right) + C\left(y^2 + \dfrac{E}{C}y + \dfrac{E^2}{4C^2}\right) = -F + \dfrac{D^2}{4A} + \dfrac{E^2}{4C} \Rightarrow$

$A\left(x + \dfrac{D}{2A}\right)^2 + C\left(y + \dfrac{E}{2C}\right)^2 = \dfrac{CD^2 + AE^2 - 4ACF}{4AC} \Rightarrow \dfrac{\left(x + \frac{D}{2A}\right)^2}{\frac{CD^2 + AE^2 - 4ACF}{4A^2C}} + \dfrac{\left(y + \frac{E}{2C}\right)^2}{\frac{CD^2 + AE^2 - 4ACF}{4AC^2}} = 1$

The center is $\left(-\dfrac{D}{2A}, -\dfrac{E}{2C}\right)$.

Chapter 6 Test

1. (a) B. This is a hyperbola with center $(-3, -2)$.

   (b) A. This is a circle with center $(3, 2)$ and radius 4.

   (c) D. This is a circle with center $(-3, 2)$ and radius 4.

   (d) E. This is a parabola that opens downward.

   (e) F. This is a parabola that opens to the right.

   (f) C. This is an ellipse with center $(-3, -2)$.

2. $y^2 = \dfrac{1}{8}x \Rightarrow (y - 0)^2 = 4\left(\dfrac{1}{32}\right)(x - 0)$; This is a parabola with vertex $(0, 0)$ that opens to the right.

   Since $c = \dfrac{1}{32}$, the focus is located at $\left(\dfrac{1}{32}, 0\right)$ and the equation of the directrix is $x = -\dfrac{1}{32}$.

3. See Figure 3. This is the graph of a function with domain $[-6, 6]$ and range $[-1, 0]$.

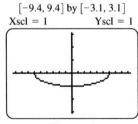

$[-9.4, 9.4]$ by $[-3.1, 3.1]$
Xscl = 1   Yscl = 1

Figure 3

4. $\dfrac{x^2}{25} - \dfrac{y^2}{49} = 1 \Rightarrow \dfrac{y^2}{49} = \dfrac{x^2}{25} - 1 \Rightarrow y^2 = 49\left(\dfrac{x^2}{25} - 1\right) \Rightarrow y = \pm\sqrt{49\left(\dfrac{x^2}{25} - 1\right)} \Rightarrow y = \pm 7\sqrt{\dfrac{x^2}{25} - 1}$

    The equations are $y_1 = 7\sqrt{\dfrac{x^2}{25} - 1}$ and $y_2 = -7\sqrt{\dfrac{x^2}{25} - 1}$.

5. This is a hyperbola with vertical transverse axis. Here $a = 2$ and $b = 3$. The asymptotes are $y = \pm\dfrac{2}{3}x$.

    See Figure 5. Since $c = \sqrt{a^2 + b^2} = \sqrt{2^2 + 3^2} = \sqrt{13}$, The foci are $(0, -\sqrt{13})$ and $(0, \sqrt{13})$.

    The center is $(0, 0)$. The vertices are $(0, -2)$ and $(0, 2)$.

6. $x^2 + 4y^2 + 2x - 16y + 17 = 0 \Rightarrow x^2 + 2x + 1 + 4(y^2 - 4y + 4) = -17 + 1 + 16 \Rightarrow$

    $(x + 1)^2 + 4(y - 2)^2 = 0$; The only point that satisfies this equation is $(-1, 2)$. See Figure 6.

7. $y^2 - 8y - 2x + 22 = 0 \Rightarrow y^2 - 8y + 16 = 2x - 22 + 16 \Rightarrow (y - 4)^2 = 2(x - 3) \Rightarrow$

    $(y - 4)^2 = 4\left(\dfrac{1}{2}\right)(x - 3)$; This is a parabola with vertex $(3, 4)$ and focus $(3.5, 4)$. See Figure 7.

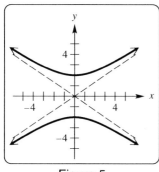
Figure 5

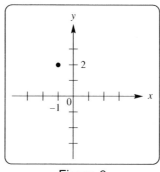
Figure 6

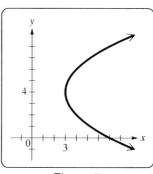
Figure 7

8. $x^2 + (y - 4)^2 = 9 \Rightarrow (x - 0)^2 + (y - 4)^2 = 9$; This is a circle with center $(0, 4)$ and radius 3. See Figure 8.

9. This is an ellipse with horizontal major axis. Here $a = 7$ and $b = 4$ and so $c = \sqrt{7^2 - 4^2} = \sqrt{33}$.

    The center is $(3, -1)$. The vertices are $(3 - 7, -1)$ and $(3 + 7, -1)$ or $(-4, -1)$ and $(10, -1)$. The foci are

    $(3 + \sqrt{33}, -1)$ and $(3 - \sqrt{33}, -1)$. See Figure 9.

10. From the 2nd equation $y = t - 1 \Rightarrow t = y + 1$. Substituting in the 1st equation yields $x = 4(y + 1)^2 - 4$.

    This equation can be written $(y + 1)^2 = 4\left(\dfrac{1}{16}\right)(x + 4)$. This is a parabola that opens to the right.

    The vertex is $(-4, -1)$ and the focus is $\left(-4 + \dfrac{1}{16}, -1\right) \Rightarrow \left(-\dfrac{63}{16}, -1\right)$. See Figure 10.

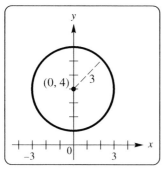
Figure 8

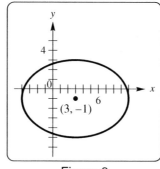
Figure 9

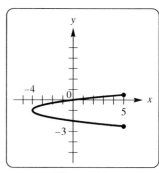
Figure 10

11. (a) Since $e = 1$, the conic is a parabola. With center $(0, 0)$ and focus $(0, -2)$, we know that $c = -2$.

The equation is $x^2 = 4cy \Rightarrow x^2 = -8y \Rightarrow y = -\dfrac{1}{8}x^2$.

(b) Since $0 < e < 1$, the conic is an ellipse with $a = 3$. Now $\dfrac{c}{a} = e \Rightarrow \dfrac{c}{3} = \dfrac{5}{6} \Rightarrow c = \dfrac{15}{6}$. For an ellipse,

$b^2 = a^2 - c^2 = 9 - \dfrac{225}{36} = \dfrac{11}{4}$. The equation is $\dfrac{x^2}{\frac{11}{4}} + \dfrac{y^2}{9} = 1$ or $\dfrac{4x^2}{11} + \dfrac{y^2}{9} = 1$.

12. Using a vertical minor axis, $b = 12$. The major axis has length 40 so $a = 20$. The equation is $\dfrac{x^2}{400} + \dfrac{y^2}{144} = 1$.

When $x = 10$, $\dfrac{10^2}{400} + \dfrac{y^2}{144} = 1 \Rightarrow \dfrac{y^2}{144} = 1 - \dfrac{1}{4} \Rightarrow y^2 = 144\left(1 - \dfrac{1}{4}\right) \Rightarrow y = 12\sqrt{\left(1 - \dfrac{1}{4}\right)} \approx 10.39$ feet.

Chapter 6 Project

1. By looking through the data, the vertices are $(.08, 2.54)$, $(.85, 1.92)$, $(1.53, 1.62)$, $(2.08, 1.42)$, and $(2.63, 1.20)$.

The times when the ball is striking the ground are .51, 1.19, 1.83, and 2.38. Thus the domains are

$[0, .51]$, $(.51, 1.19]$, $(1.19, 1.83]$, $(1.83, 2.38]$, and $(2.38, 2.76]$. The piecewise-defined function is

$$y = \begin{cases} -16(x - .08)^2 + 2.54 & \text{for } 0 \leq x \leq .51 \\ -16(x - .85)^2 + 1.92 & \text{for } .51 < x \leq 1.19 \\ -16(x - 1.53)^2 + 1.62 & \text{for } 1.19 < x \leq 1.83 \\ -16(x - 2.08)^2 + 1.42 & \text{for } 1.83 < x \leq 2.38 \\ -16(x - 2.63)^2 + 1.20 & \text{for } 2.38 < x \leq 2.76 \end{cases}$$

To graph this function let $Y_1 = (-16(X - .08)^2 + 2.54)(X \geq 0)(X \leq .51)$,

$Y_2 = (-16(X - .85)^2 + 1.92)(X > .51)(X \leq 1.19)$, $Y_3 = (-16(X - 1.53)^2 + 1.62)(X > 1.19)(X \leq 1.83)$,

$Y_4 = (-16(X - 2.08)^2 + 1.42)(X > 1.83)(X \leq 2.38)$, $Y_5 = (-16(X - 2.63)^2 + 1.20)(X > 2.38)(X \leq 2.76)$.

See Figure 1.

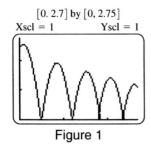

$[0. 2.7]$ by $[0, 2.75]$
Xscl = 1   Yscl = 1

Figure 1

2. Set the graph style to "Path" and graph the function in part 1.

3. Answers will vary.

Chapter 7: Systems of Equations and Inequalities; Matrices

7.1: Systems of Equations

1. The two graphs intersect at approximately (2002, 3.1). Therefore about the year 2002, both projections produce the same level of migration.

2. The two graphs intersect at approximately (2002, 3.1). Therefore , there will be about 3.1 million migrants in 2002.

3. The two graphs intersect at approximately (2002, 3.1), which is the solution to the system.

4. The graph is increasing until approximately 1998, then decreasing until approximately 2007, and constant thereafter.

5. Then t would represent time in years and y would represent the number of migrants.

6. Both graphs pass the vertical line test, that is, for each x there is only one y.

7. From the graph, the solution is $\{(2, 2)\}$. Using substitution, first solve [equation 1] for x, $x - y = 0 \Rightarrow x = y$ [equation 3]. Substitute (y) in for x, in [equation 2]: $(y) + y = 4 \Rightarrow 2y = 4 \Rightarrow y = 2$. Now substitute (2) in for y in [equation 3]: $x = y \Rightarrow x = 2$. The solution is: $\{(2, 2)\}$.

8. From the graph, the solution is $\{(-2, 1)\}$. Using substitution, first solve [equation 1] for y, $-x + y = 3 \Rightarrow y = x + 3$ [equation 3]. Substitute $(x + 3)$ in for y, in [equation 2]: $x + (x + 3) = -1 \Rightarrow 2x = -4 \Rightarrow x = -2$. Now substitute (-2) in for x in [equation 3]: $y = (-2) + 3 \Rightarrow y = 1$. The solution is: $\{(-2, 1)\}$.

9. From the graph, the solution is $\left\{\left(\dfrac{1}{2}, -2\right)\right\}$. Using substitution, first solve [equation 2] for x,

    $2x - 3y = 7 \Rightarrow 2x = 3y + 7 \Rightarrow x = \dfrac{3}{2}y + \dfrac{7}{2}$ [equation 3]. Substitute $\left(\dfrac{3}{2}y + \dfrac{7}{2}\right)$ in for x, in [equation 1]:

    $6\left(\dfrac{3}{2}y + \dfrac{7}{2}\right) + 4y = -5 \Rightarrow 9y + 21 + 4y = -5 \Rightarrow 13y = -26 \Rightarrow y = -2$. Now substitute (-2) in for y in

    [equation 3]: $x = \dfrac{3}{2}(-2) + \dfrac{7}{2} \Rightarrow x = -3 + \dfrac{7}{2} \Rightarrow x = \dfrac{1}{2}$. The solution is: $\left\{\left(\dfrac{1}{2}, -2\right)\right\}$.

10. From the graph, the solution is $\left\{\left(-1, -\dfrac{1}{2}\right)\right\}$. Using substitution, first solve [equation 2] for x, $-x + 2y = 0$

    $\Rightarrow -x = -2y \Rightarrow x = 2y$ [equation 3]. Substitute $(2y)$ in for x, in [equation 1]: $5(2y) - 2y = -4 \Rightarrow$

    $8y = -4 \Rightarrow y = -\dfrac{1}{2}$. Now substitute $\left(-\dfrac{1}{2}\right)$ in for y in [equation 3]: $x = 2\left(-\dfrac{1}{2}\right) \Rightarrow x = -1$.

    The solution is: $\left\{\left(-1, -\dfrac{1}{2}\right)\right\}$.

11. Since $y = x$ in [equation 2], we can use substitution. Substitute (x) in for y, in [equation 1]: $6x - (x) = 5 \Rightarrow$ $5x = 5 \Rightarrow x = 1$. Now substitute (1) in for x [equation 2]: $y = x \Rightarrow y = 1$. The solution is: $\{(1, 1)\}$.

12. Since $y = -3x$ in [equation 2], we can use substitution. Substitute $(-3x)$ in for y, in [equation 1]: $5x + (-3x) = 2 \Rightarrow 2x = 2 \Rightarrow x = 1$. Now substitute (1) in for x in [equation 2]: $y = -3(1) \Rightarrow y = -3$. The solution is: $\{(1, -3)\}$.

13. Using substitution, first solve [equation 1] for x, $x + 2y = -1 \Rightarrow x = -2y - 1$ [equation 3]. Substitute $(-2y - 1)$ in for x, in [equation 2]: $2(-2y - 1) + y = 4 \Rightarrow -3y - 2 = 4 \Rightarrow -3y = 6 \Rightarrow y = -2$. Now substitute (-2) in for y in [equation 3]: $x = -2(-2) - 1 \Rightarrow x = 3$.   The solution is: $\{(3, -2)\}$.

14. Using substitution, first solve [equation 2] for x, $x + 3y = -8 \Rightarrow x = -3y - 8$ [equation 3]. Substitute $(-3y - 8)$ in for x, in [equation 1]: $2(-3y - 8) + y = -11 \Rightarrow -5y - 16 = -11 \Rightarrow -5y = 5 \Rightarrow y = -1$. Now substitute (-1) in for y in [equation 3]: $x = -3(-1) - 8 \Rightarrow x = -5$.   The solution is: $\{(-5, -1)\}$.

15. Since $y = 2x + 3$ in [equation 1], we can use substitution. Substitute $(2x + 3)$ in for y, in [equation 2]: $3x + 4(2x + 3) = 78 \Rightarrow 11x + 12 = 78 \Rightarrow 11x = 66 \Rightarrow x = 6$. Now substitute (6) in for x in [equation 1]: $y = 2(6) + 3 \Rightarrow y = 15$. The solution is: $\{(6, 15)\}$.

16. Since $y = 4x - 6$ in [equation 1], we can use substitution. Substitute $(4x - 6)$ in for y, in [equation 2]: $2x + 5(4x - 6) = -8 \Rightarrow 22x - 30 = -8 \Rightarrow 22x = 22 \Rightarrow x = 1$. Now substitute (1) in for x in [equation 1]: $y = 4(1) - 6 \Rightarrow y = -2$.   The solution is: $\{(1, -2)\}$.

17. Using substitution, first solve [equation 1] for x, $3x - 2y = 12 \Rightarrow 3x = 2y + 12 \Rightarrow x = \frac{2}{3}y + 4$ [equation 3]. Substitute $\left(\frac{2}{3}y + 4\right)$ in for x, in [equation 2]: $5\left(\frac{2}{3}y + 4\right) = 4 - 2y \Rightarrow \frac{10}{3}y + 2y = 4 - 20 \Rightarrow \frac{16}{3}y = -16$ $\Rightarrow y = -3$. Now substitute (-3) in for y in [equation 3]: $x = \frac{2}{3}(-3) + 4 \Rightarrow x = 2$. The solution is: $\{(2, -3)\}$.

18. Using substitution, first solve [equation 2] for x, $5x = 17 + 6y \Rightarrow x = \frac{17 + 6y}{5}$ [equation 3]. Substitute $\left(\frac{17 + 6y}{5}\right)$ in for x, in [equation 1]: $8\left(\frac{17 + 6y}{5}\right) + 3y = 2 \Rightarrow 5\left[8\left(\frac{17 + 6y}{5}\right) + 3y = 2\right] \Rightarrow$ $8(17 + 6y) + 15y = 10 \Rightarrow 136 + 48y + 15y = 10 \Rightarrow 63y = -126 \Rightarrow y = -2$. Now substitute (-2) in for y in [equation 3]: $x = \frac{17}{5} + \frac{6}{5}(-2) \Rightarrow x = \frac{17}{5} + \left(-\frac{12}{5}\right) \Rightarrow x = 1$.   The solution is: $\{(1, -2)\}$.

19. Using substitution, first solve [equation 2] for y, $2x + y = 5 \Rightarrow y = -2x + 5$ [equation 3]. Substitute $(-2x + 5)$ in for y, in [equation 1]: $4x - 5(-2x + 5) = -11 \Rightarrow 4x + 10x - 25 = -11 \Rightarrow 14x = 14 \Rightarrow$ $x = 1$. Now substitute (1) in for y in [equation 3]: $y = -2(1) + 5 \Rightarrow y = 3$.   The solution is: $\{(1, 3)\}$.

20. Using substitution, first solve [equation 1] for y, $7x - y = -10 \Rightarrow -y = -7x - 10 \Rightarrow y = 7x + 10$ [equation 3]. Substitute $(7x + 10)$ in for y, in [equation 2]: $3(7x + 10) - x = 10 \Rightarrow 21x + 30 - x = 10 \Rightarrow 20x = -20$ $\Rightarrow x = -1$. Now substitute (-1) in for x in [equation 3]: $y = 7(-1) + 10 \Rightarrow y = 3$.   The solution is: $\{(-1, 3)\}$.

21. Using substitution, first solve [equation 2] for y, $9y = 31 + 2x \Rightarrow y = \frac{31 + 2x}{9}$ [equation 3]. Substitute $\frac{31 + 2x}{9}$ in for y, in [equation 1]: $4x + 5\left(\frac{31 + 2x}{9}\right) = 7 \Rightarrow 9\left[4x + 5\left(\frac{31 + 2x}{9}\right) = 7\right] \Rightarrow$ $36x + 5(31 + 2x) = 63 \Rightarrow 36x + 155 + 10x = 63 \Rightarrow 46x = -92 \Rightarrow x = -2$.   Now substitute (-2) in for x in [equation 3]: $y = \frac{31 + 2(-2)}{9} \Rightarrow y = 3$. The solution is: $\{(-2, 3)\}$.

22. Using substitution, first solve [equation 1] for x, $-2x = 6y + 18 \Rightarrow x = -3y - 9$ [equation 3]. Substitute

    $(-3y - 9)$ in for x, in [equation 2]: $-29 = 5y - 3(-3y - 9) \Rightarrow -29 = 5y + 9y + 27 \Rightarrow -56 = 14y \Rightarrow$

    $y = -4$. Now substitute (-4) in for y in [equation 3]: $x = -3(-4) - 9 \Rightarrow x = 3$. The solution is: $\{(3, -4)\}$.

23. Using substitution, first solve [equation 1] for x, $3x - 7y = 15 \Rightarrow 3x = 7y + 15 \Rightarrow x = \dfrac{7y + 15}{3}$ [equation 3].

    Substitute $\dfrac{7y + 15}{3}$ in for x, in [equation 2]: $3\left(\dfrac{7y + 15}{3}\right) + 7y = 15 \Rightarrow 7y + 15 + 7y = 15 \Rightarrow 14y = 0 \Rightarrow$

    $y = 0$. Now substitute (0) in for y in [equation 3]: $x = \dfrac{7(0) + 15}{3} \Rightarrow x = 5$. The solution is: $\{(5, 0)\}$.

24. Using substitution, first solve [equation 2] for y, $x + y = 2 \Rightarrow y = 2 - x$ [equation 3]. Substitute $(2 - x)$ in for

    y, in [equation 1]: $3(2 - x) = 5x + 6 \Rightarrow 6 - 3x = 5x + 6 \Rightarrow -8x = 0 \Rightarrow x = 0$. Now substitute (0) in

    for x in [equation 3]: $y = 2 - (0) \Rightarrow y = 2$. The solution is: $\{(0, 2)\}$.

25. Using substitution, first solve [equation 1] for x, $2x - 7y = 8 \Rightarrow 2x = 7y + 8 \Rightarrow x = \dfrac{7y + 8}{2}$ [equation 3].

    Substitute $\left(\dfrac{7y + 8}{2}\right)$ in for x, in [equation 2]: $-3\left(\dfrac{7y + 8}{2}\right) + \dfrac{21}{2}y = 5 \Rightarrow 2\left[-3\left(\dfrac{7y + 8}{2}\right) + \dfrac{21}{2}y = 5\right] \Rightarrow$

    $-3(7y + 8) + 21y = 10 \Rightarrow -21y - 24 + 21y = 10 \Rightarrow -24 = 10$. Since this is a false statement, the

    solution is: $\varnothing$.

26. Using substitution, first solve [equation 1] for y, $.6x - .2y = 2 \Rightarrow -.2y = -.6x + 2 \Rightarrow y = 3x + 10$ [equation 3].

    Substitute $(3x + 10)$ in for y, in [equation 2]: $-1.2x + .4(3x + 10) = 3 \Rightarrow -1.2x + 1.2x + 4 = 3 \Rightarrow 4 = 3$.

    Since this is a false statement, the solution is: $\varnothing$.

27. Solve [equation 1] for y, $4x - 5y = -11 \Rightarrow -5y = -4x - 11 \Rightarrow y = \dfrac{4}{5}x + \dfrac{11}{5}$, therefore $m = \dfrac{4}{5}$ and $b = \dfrac{11}{5}$.

    Solve [equation 2] for y, $2x + y = 5 \Rightarrow y = -2x + 5$, therefore $m = -2$ and $b = 5$.

    Screen A appears to have lines with these characteristics.

28. Since 12 is the LCD of 3 and 4 (coefficients of y), multiply [equation 1] by 3; and [equation 2] by 4.

29. Multiply the first equation by 3 and add to eliminate the y-variable.

    $$\begin{aligned} 9x - 3y &= -12 \\ x + 3y &= 12 \\ \hline 10x &= 0 \end{aligned} \Rightarrow x = 0.$$ Substitute (0) for x in the second equation: $(0) + 3y = 12 \Rightarrow y = 4$.

    The solution is: $\{(0, 4)\}$.

30. Multiply the first equation by 5, the second equation by 2, and subtract to eliminate the x-variable.

    $$\begin{aligned} 10x - 15y &= -35 \\ 10x + 8y &= 34 \\ \hline -23y &= -69 \end{aligned} \Rightarrow y = 3.$$ Substitute (3) for y in the second equation: $10x + 8(3) = 34 \Rightarrow x = 1$.

    The solution is: $\{(1, 3)\}$.

31. Multiply the second equation by 2, and subtract to eliminate the x-variable.

    $$\begin{aligned} 4x + 3y &= -1 \\ 4x + 10y &= 6 \\ \hline -7y &= -7 \end{aligned} \Rightarrow y = 1.$$ Substitute (1) for y in the first equation: $4x + 3(1) = -1 \Rightarrow 4x = -4 \Rightarrow$

    $x = -1$. The solution is: $\{(-1, 1)\}$.

32. Multiply the first equation by 2 and subtract to eliminate the x-variable.

$$
\begin{array}{rr}
10x + 14y = & 12 \\
10x - 3y = & 46 \\
\hline
17y = & -34
\end{array}
$$
$\Rightarrow y = -2$. Substitute (-2) for y in the second equation: $10x - 3(-2) = 46 \Rightarrow 10x = 40$

$\Rightarrow x = 4$. The solution is: $\{(4, -2)\}$.

33. Multiply the second equation by 4, and subtract to eliminate the x-variable.

$$
\begin{array}{rr}
12x - 5y = & 9 \\
12x - 32y = & -72 \\
\hline
27y = & 81
\end{array}
$$
$\Rightarrow y = 3$. Substitute (3) for y in the first equation: $12x - 5(3) = 9 \Rightarrow 12x = 24 \Rightarrow$

$x = 2$. The solution is: $\{(2, 3)\}$.

34. Multiply the first equation by 7, the second equation by 6, and subtract to eliminate the x-variable.

$$
\begin{array}{rr}
42x + 49y = & -14 \\
42x - 36y = & 156 \\
\hline
85y = & -170
\end{array}
$$
$\Rightarrow y = -2$. Substitute (-2) for y in the second equation: $42x - 36(-2) = 156 \Rightarrow$

$42x = 84 \Rightarrow x = 2$. The solution is: $\{(2, -2)\}$.

35. Multiply the first equation by 2 and add to eliminate both variables.

$$
\begin{array}{rr}
8x - 2y = & 18 \\
-8x + 2y = & -18 \\
\hline
0 = & 0
\end{array}
$$
$\Rightarrow$ infinite number of solutions. Therefore the solutions have the following relationship:

$4x - y = 9 \Rightarrow -y = -4x + 9 \Rightarrow y = 4x - 9$, the solution is: $\{(x, 4x - 9)\}$ or $\left\{\left(\dfrac{y + 9}{4}, y\right)\right\}$.

36. Multiply the first equation by 3 and subtract to eliminate both variables.

$$
\begin{array}{rr}
3x + 3y = & 12 \\
3x + 3y = & 12 \\
\hline
0 = & 0
\end{array}
$$
$\Rightarrow$ infinite number of solutions. Therefore the solutions have the following relationship:

$x + y = 4 \Rightarrow y = -x + 4$, the solution is: $\{(x, 4 - x)\}$ or $\{(4 - y, y)\}$.

37. Multiply the first equation by 2 and add to eliminate both variables.

$$
\begin{array}{rr}
18x - 10y = & 2 \\
-18x + 10y = & 1 \\
\hline
0 = & 3
\end{array}
$$
$\Rightarrow \varnothing$.

38. Multiply the first equation by 2 and subtract to eliminate both variables.

$$
\begin{array}{rr}
6x + 4y = & 10 \\
6x + 4y = & 8 \\
\hline
0 = & 2
\end{array}
$$
$\Rightarrow \varnothing$.

39. Multiply the first equation by 2 and subtract to eliminate both variables.

$$
\begin{array}{rr}
6x + 2y = & 12 \\
6x + 2y = & 1 \\
\hline
0 = & 11
\end{array}
$$
$\Rightarrow \varnothing$.

40. Multiply the first equation by 3 and subtract to eliminate both variables.

$$9x + 15y = -6$$
$$\underline{9x + 15y = -6}$$
$$0 = 0 \Rightarrow \text{infinite number of solutions. Therefore the solutions have the following relationship:}$$

$3x + 5y = -2 \Rightarrow 5y = -3x - 2 \Rightarrow y = \dfrac{-3x - 2}{5}$, the solution is: $\left\{\left(x, \dfrac{-3x-2}{5}\right)\right\}$ or $\left\{\left(\dfrac{-2-5y}{3}, y\right)\right\}$.

41. First multiplying both equations by 6 yields: $3x + 2y = 48$ and $4x + 9y = 102$.

   Now multiply the first equation by 4, the second equation by 3, and subtract to eliminate the x-variable.

$$12x + 8y = 192$$
$$\underline{12x + 27y = 306}$$
$$-19y = -114 \Rightarrow y = 6. \text{ Substitute } (6) \text{ for } y \text{ in the first equation (multiplied by 6): } 3x + 2(6) = 48 \Rightarrow$$
$$3x = 36 \Rightarrow x = 12. \text{ The solution is: } \{(12, 6)\}.$$

42. First multiplying both equations by 5 yields: $x + 15y = 155$ and $10x - y = 40$.

   Now multiply the first equation by 10 and subtract to eliminate the x-variable.

$$10x + 150y = 1550$$
$$\underline{10x - y = 40}$$
$$151y = 1510 \Rightarrow y = 10. \text{ Substitute } (10) \text{ for } y \text{ in the first equation (multiplied by 5): } x + 15(10) = 155$$
$$\Rightarrow x = 5. \text{ The solution is: } \{(5, 10)\}.$$

43. Multiplying the first equation by 12 and the second by 6 yields: $8x + 3y = 46$ and $x + 2y = 9$.

   Now multiply the second equation by 8, and subtract to eliminate the x-variable.

$$8x + 3y = 46$$
$$\underline{8x + 16y = 72}$$
$$-13y = -26 \Rightarrow y = 2. \text{ Substitute } (2) \text{ for } y \text{ in the second equation (multiplied by 6): }$$
$$x + 2(2) = 9 \Rightarrow x = 5. \text{ The solution is: } \{(5, 2)\}.$$

44. Multiplying the first equation by 10 and the second by 20 yields: $x + 2y = -2$ and $5x + 12y = -78$.

   Now multiply the first equation by 5, and subtract to eliminate the x-variable.

$$5x + 10y = -10$$
$$\underline{5x + 12y = -78}$$
$$-2y = 68 \Rightarrow y = -34. \text{ Substitute } (-34) \text{ for } y \text{ in the first equation (multiplied by 10): }$$
$$x + 2(-34) = -2 \Rightarrow x = 66. \text{ The solution is: } \{(66, -34)\}.$$

45. Graph each function in the same window. The intersection or solution is: $\{(.138, -4.762)\}$.

46. Graph each function in the same window. The intersection or solution is: $\{(.820, -2.508)\}$.

47. Graph each function in the same window. The intersection or solution is: $\{(-8.708, 15.668)\}$.

48. Graph each function in the same window. The intersection or solution is: $\{(-.584, 1.778)\}$.

49. An inconsistent system will conclude with no variables and a false statement, such as $0 = 1$. A system with dependent equations will conclude with no variables and a true statement, such as $0 = 0$.

50. If we multiply the first equation by (-2) yields:

$$-2x + 4y = -6$$
$$-2x + 4y = k \quad \text{From these equations, when we subtract we will get the result } 0 = 0 \text{ when } k = -6 \text{ and the}$$

   result $0 = (-6 - k)$ when $k \ne -6$. Therefore, the system will have no solution when $k \ne -6$; and system will have infinitely many solutions when $k = -6$.

51. See Figure 51.

52. See Figure 52.

53. See Figure 53.

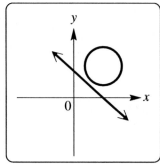

Figure 51

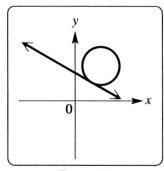

Figure 52

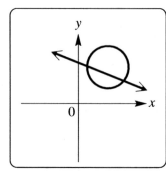

Figure 53

54. See Figure 54.

55. See Figure 55.

56. See Figure 56.

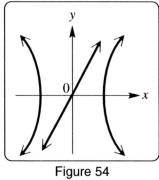

Figure 54

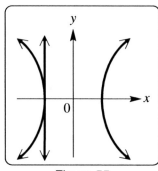

Figure 55

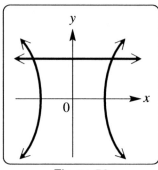

Figure 56

57. See Figure 57.

58. See Figure 58.

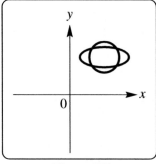

Figure 57

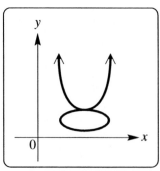

Figure 58

59. See Figure 59.

60. See Figure 60.

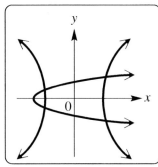

Figure 59

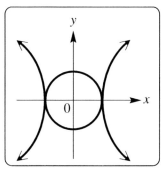

Figure 60

61. Since {equation 1} is solved for y, we substitute $(-x^2 + 2)$ in for y in [equation 2]: $x - (-x^2 + 2) = 0 \Rightarrow$

    $x^2 + x - 2 = 0 \Rightarrow (x + 2)(x - 1) = 0 \Rightarrow x = -2, 1$. Substituting these values into [equation 1] yields:

    $y = -(-2)^2 + 2 \Rightarrow y = -4 + 2 \Rightarrow y = -2$; and $y = -(1)^2 + 2 \Rightarrow y = -1 + 2 \Rightarrow y = 1$.

    The solution is: $\{(-2, -2), (1, 1)\}$.

62. Since {equation 1} is solved for y, we substitute $(x - 1)^2$ in for y in [equation 2]: $x - 3(x - 1)^2 = -1 \Rightarrow$

    $x - 3(x^2 - 2x + 1) = -1 \Rightarrow x - 3x^2 + 6x - 3 + 1 = 0 \Rightarrow -(3x^2 - 7x + 2) = 0 \Rightarrow$

    $-(x - 2)(3x - 1) = 0 \Rightarrow x = \dfrac{1}{3}, 2$. Substituting these values into [equation 1] yields:

    $y = \left(\left(\dfrac{1}{3}\right) - 1\right)^2 \Rightarrow y = \dfrac{4}{9}$; and $y = ((2) - 1)^2 \Rightarrow y = 1$. The solution is: $\left\{(2, 1), \left(\dfrac{1}{3}, \dfrac{4}{9}\right)\right\}$.

63. First solve [equation 2] for x: $x - y = -2 \Rightarrow x = y - 2$ [equation 3]. Now substitute $(y - 2)$ in for x in

    [equation 1]: $3(y - 2)^2 + 2y^2 = 5 \Rightarrow 3(y^2 - 4y + 4) + 2y^2 = 5 \Rightarrow 3y^2 - 12y + 12 + 2y^2 - 5 = 0 \Rightarrow$

    $5y^2 - 12y + 7 = 0 \Rightarrow (5y - 7)(y - 1) = 0 \Rightarrow y = 1, \dfrac{7}{5}$. Substituting these values into [equation 3] yields:

    $x = (1) - 2 \Rightarrow x = -1$; and $x = \left(\dfrac{7}{5}\right) - 2 \Rightarrow x = -\dfrac{3}{5}$. The solution is: $\left\{\left(-\dfrac{3}{5}, \dfrac{7}{5}\right), (-1, 1)\right\}$.

64. First solve [equation 2] for x: $-3x + 4y = 2 \Rightarrow -3x = 2 - 4y \Rightarrow x = \dfrac{2 - 4y}{-3}$ [equation 3]. Now substitute

    $\left(\dfrac{2 - 4y}{-3}\right)$ in for x in [equation 1]: $\left(\dfrac{2 - 4y}{-3}\right)^2 + y^2 = 5 \Rightarrow \dfrac{(2 - 4y)^2}{9} + y^2 = 5 \Rightarrow$

    $9\left[\dfrac{(2 - 4y)^2}{9} + y^2 = 5\right] \Rightarrow (2 - 4y)^2 + 9y^2 - 45 = 0 \Rightarrow 4 - 16y + 16y^2 + 9y^2 - 45 = 0 \Rightarrow$

    $25y^2 - 16y - 41 = 0 \Rightarrow (25y - 41)(y + 1) = 0 \Rightarrow y = -1, \dfrac{41}{25}$. Substituting these values into [equation 3]

    yields: $x = \dfrac{2 - 4(-1)}{-3} \Rightarrow x = \dfrac{6}{-3} \Rightarrow x = -2$; and $x = \dfrac{2 - 4\left(\dfrac{41}{25}\right)}{-3} \Rightarrow x = \dfrac{\left(\dfrac{50 - 164}{25}\right)}{-3} \Rightarrow x = \dfrac{\left(-\dfrac{114}{25}\right)}{-3}$

    $\Rightarrow x = \dfrac{38}{25}$. The solution is: $\left\{(-2, -1), \left(\dfrac{38}{25}, \dfrac{41}{25}\right)\right\}$.

65. Add the equations to eliminate the y^2.

$$x^2 + y^2 = 10$$
$$\underline{2x^2 - y^2 = 17}$$
$$3x^2 \quad\;\; = 27 \;\Rightarrow\; x^2 = 9 \;\Rightarrow\; x = \pm 3.$$

Substituting these values into [equation 1] yields: $(-3)^2 + y^2 = 10 \;\Rightarrow\; y^2 = 1 \;\Rightarrow\; y = \pm 1$; and

$(3)^2 + y^2 = 10 \;\Rightarrow\; y^2 = 1 \;\Rightarrow\; y = \pm 1$. The solution is: $\{(-3, 1), (-3, -1), (3, 1), (3, -1)\}$.

66. Multiply [equation 1] by 3 and add to eliminate the y^2.

$$3x^2 + 3y^2 = \;\;\, 12$$
$$\underline{2x^2 - 3y^2 = -12}$$
$$5x^2 \quad\;\;\; = \;\;\;\, 0 \;\Rightarrow\; 5x^2 = 0 \;\Rightarrow\; x = 0.$$

Substituting (0) into [equation 1] for x, yields: $(0)^2 + y^2 = 4 \;\Rightarrow\; y^2 = 4 \;\Rightarrow\; y = \pm 2$.

The solution is: $\{(0, 2), (0, -2)\}$.

67. Multiply [equation 1] by 3 and subtract to eliminate the x^2.

$$3x^2 + \;\;6y^2 = \;\; 27$$
$$\underline{3x^2 - \;\;4y^2 = \;\; 27}$$
$$\qquad\quad 10y^2 = \quad\;\; 0 \;\Rightarrow\; 10y^2 = 0 \;\Rightarrow\; y = 0.$$

Substituting (0) into [equation 1] for y, yields: $x^2 + 2(0)^2 = 9 \;\Rightarrow\; x^2 = 9 \;\Rightarrow\; x = \pm 3$.

The solution is: $\{(3, 0), (-3, 0)\}$.

68. Multiply [equation 1] by 4 and [equation 2] by 3 and add to eliminate the y^2.

$$8x^2 + 12y^2 = \;\; 20$$
$$\underline{9x^2 - 12y^2 = \;\; -3}$$
$$17x^2 \qquad\quad = \;\; 17 \;\Rightarrow\; x^2 = 1 \;\Rightarrow\; x = \pm 1.$$

Substituting these values into [equation 1] yields: $2(1)^2 + 3y^2 = 5 \;\Rightarrow\; 3y^2 = 3 \;\Rightarrow\; y^2 = 1 \;\Rightarrow\; y = \pm 1$; and

$2(-1)^2 + 3y^2 = 5 \;\Rightarrow\; 3y^2 = 3 \;\Rightarrow\; y^2 = 1 \;\Rightarrow\; y = \pm 1$. The solution is: $\{(1, 1), (1, -1), (-1, 1), (-1, -1)\}$.

69. Multiply [equation 1] by 3 and [equation 2] by 2 and subtract to eliminate both x^2 and y^2.

$$6x^2 + \;\;6y^2 = \;\; 60$$
$$\underline{6x^2 + \;\;6y^2 = \;\; 60}$$
$$\qquad\quad\;\; 0 \;\; = \quad\; 0 \;\Rightarrow\; \text{infinite number of solutions. The solutions have the following relationship:}$$

$2x^2 + 2y^2 = 20 \;\Rightarrow\; 2y^2 = 20 - 2x^2 \;\Rightarrow\; y^2 = 10 - x^2 \;\Rightarrow\; y = \pm\sqrt{10 - x^2}$. The solution is:

$\{(x, -\sqrt{10 - x^2}), (x, \sqrt{10 - x^2})\}$.

70. Multiply [equation 1] by 5 and subtract to eliminate both x^2 and y^2.

$$5x^2 + \;\;5y^2 = \;\; 25$$
$$\underline{5x^2 + \;\;5y^2 = \;\; 28}$$
$$\qquad\quad\;\; 0 \;\; = \;\; -3 \;\Rightarrow\; \varnothing.$$

71. If $y = |x - 1|$ and $y = x^2 - 4$ then $|x - 1| = x^2 - 4$. Solving this yields:

$x - 1 = x^2 - 4 \Rightarrow x^2 - x - 3 = 0$ or $x - 1 = -(x^2 - 4) \Rightarrow x - 1 = -x^2 + 4 \Rightarrow x^2 + x - 5 = 0$.

Solve both of these quadratic equations using the quadratic formula:

Solve for x, $x^2 - x - 3 = 0 \Rightarrow x = \dfrac{-(-1) \pm \sqrt{(-1)^2 - 4(1)(-3)}}{2(1)} \Rightarrow x = \dfrac{1 \pm \sqrt{13}}{2}$.

Solve for x, $x^2 + x - 5 = 0 \Rightarrow x = \dfrac{-1 \pm \sqrt{1^2 - 4(1)(-5)}}{2(1)} \Rightarrow x = \dfrac{-1 \pm \sqrt{21}}{2}$.

Now substitute each of these values of x into [equation 2] and solve for y.

For $x = \dfrac{1 + \sqrt{13}}{2}$, $y = \left(\dfrac{1 + \sqrt{13}}{2}\right)^2 - 4 = \dfrac{1 + 2\sqrt{13} + 13}{4} - \dfrac{16}{4} = \dfrac{2\sqrt{13} - 2}{4} \Rightarrow y = \dfrac{\sqrt{13} - 1}{2}$.

For $x = \dfrac{1 - \sqrt{13}}{2}$, $y = \left(\dfrac{1 - \sqrt{13}}{2}\right)^2 - 4 = \dfrac{1 - 2\sqrt{13} + 13}{4} - \dfrac{16}{4} = \dfrac{-2\sqrt{13} - 2}{4} \Rightarrow y = \dfrac{-\sqrt{13} - 1}{2}$.

From [equation 1] we know $y > 0$, since $y = \dfrac{-\sqrt{13} - 1}{2} < 0$ there is no solution with this value of x.

For $x = \dfrac{-1 + \sqrt{21}}{2}$, $y = \left(\dfrac{-1 + \sqrt{21}}{2}\right)^2 - 4 = \dfrac{1 - 2\sqrt{21} + 21}{4} - \dfrac{16}{4} = \dfrac{-2\sqrt{21} + 6}{4} \Rightarrow y = \dfrac{-\sqrt{21} + 3}{2}$.

From [equation 1] we know $y > 0$, since $y = \dfrac{-\sqrt{21} + 3}{2} < 0$ there is no solution with this value of x.

For $x = \dfrac{-1 - \sqrt{21}}{2}$, $y = \left(\dfrac{-1 - \sqrt{21}}{2}\right)^2 - 4 = \dfrac{1 + 2\sqrt{21} + 21}{4} - \dfrac{16}{4} = \dfrac{2\sqrt{21} + 6}{4} \Rightarrow y = \dfrac{\sqrt{21} + 3}{2}$.

The solution is: $\left\{\left(\dfrac{1 + \sqrt{13}}{2}, \dfrac{-1 + \sqrt{13}}{2}\right), \left(\dfrac{-1 - \sqrt{21}}{2}, \dfrac{3 + \sqrt{21}}{2}\right)\right\}$.

72. If $|x| = |y|$ then $x = y$ or $x = -y$. Substituting into [equation 1] when $x = y$ yields: $2(y)^2 - y^2 = 4 \Rightarrow y^2 = 4 \Rightarrow y = \pm 2$. The solutions are: $(2, 2)$ and $(-2, -2)$. Substituting into [equation 1] when $x = -y$ yields: $2(-y)^2 - y^2 = 4 \Rightarrow y^2 = 4 \Rightarrow y = \pm 2$. The solutions are: $(2, -2)$ and $(-2, 2)$.

The solution set is: $\{(-2, -2), (-2, 2), (2, -2), (2, 2)\}$.

73. Graph each function in the same window. The intersections or solutions are: $\{(-.79, .62), (.88, .77)\}$.

74. Graph each function in the same window. The intersection or solution is: $\{(.47, 2.13)\}$.

75. Graph each function in the same window. The intersection or solution is: $\{(.06, 2.88)\}$.

76. Graph each function in the same window. The intersections or solutions are: $\{(-1.68, -1.78), (2.12, -1.24)\}$.

77. (a) Let x represent the number of robberies in 2000 and let y represent the number of robberies in 2001.

The required system of equations is $x + y = 831{,}000$ and $x - y = 15{,}000$.

(b) Adding the two equations results in $2x = 846{,}000 \Rightarrow x = 423{,}000$. From the first equation,

$423{,}000 + y = 831{,}000 \Rightarrow y = 408{,}000$. The solution is: $\{(423{,}000, 408{,}000)\}$.

(c) There were 423,000 robberies in 2000 and 408,000 in 2001.

78. (a) Let x = number vaccinated in Florida and let y = number vaccinated in Texas. Then $x + y = 6767$ and

$x - y = 273$.

(b) Substitute $x - y = 273$ or $x = y + 273$ into $x + y = 6767 \Rightarrow (y + 273) + y = 6767 \Rightarrow$

$2y + 273 = 6767 \Rightarrow 2y = 6494 \Rightarrow y = 3247$. Solving for x we get: $x = 3247 + 273 \Rightarrow x = 3520$.

The solution is: $\{(3520, 3247)\}$.

(c) There were 3520 vaccinated in Florida and 3247 vaccinated in Texas.

79. Let x = hours spent on the internet in 2001. Then $0.13x = 1.3 \Rightarrow x = 10$. Therefore 10 hours were spent on the internet in 2001 and $10 + 1.3$ or 11.3 hours in 2002.

80. Let x = the length of each side of the base and let y = the height of the box. Since the volume = 576 in^3, $x^2 y = 576$, and so $y = \dfrac{576}{x^2}$. Since the surface are is 336 in^2, $x^2 + 4xy = 336$. Substituting for y in this equation yields $x^2 + 4x \cdot \dfrac{576}{x^2} = 336$. Simplifying we get: $x^3 - 336x + 2304 = 0$. By graphing the left side of the equation, the x-intercepts are approximately 9.1 and 12. See Figure 80. When $x \approx 9.1$, the dimensions are: 9.1 by 9.1 by $\dfrac{576}{(9.1)^2} \approx 6.96$ inches. When $x = 12$, the dimensions are: 12 by 12 by $\dfrac{576}{(12)^2} = 4$ inches.

$[0, 20]$ by $[-500, 1500]$
Xscl = 2   Yscl = 500

$[0, 10]$ by $[-500, 1500]$
Xscl = 1   Yscl = 500

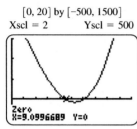

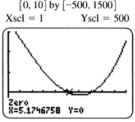

Figure 80   **Figure 81**

81. Let x = the width of the base therefore the length $2x$, and let y = the height of the box. Since the volume is 588 in^3, $2x^2 y = 588$ and so $y = \dfrac{588}{2x^2} = \dfrac{294}{x^2}$. Since the surface are is 448 in^2,

$2(2x)(x) + 2(x)(y) + 2(2x)(y) = 448 \Rightarrow 4x^2 + 6xy = 448$. Substituting for y in this equation yields

$4x^2 + 6x \cdot \dfrac{294}{x^2} = 448$. Simplifying we get: $4x^3 - 448x + 1764 = 0$. By graphing the left side of the equation, the x-intercepts are approximately 5.17 and 7. See Figure 81. When $x \approx 5.17$, the dimensions are:

5.17 by $2(5.17) \approx 10.34$ by $\dfrac{294}{(5.17)^2} \approx 11.00$ inches. When $x = 7$, the dimensions are:

7 by $2(7) = 14$ by $\dfrac{294}{(7)^2} = 6$ inches.

82. Let x = amount invested at 5% and y = amount invested at 7%, then $x + y = 5000$ and $.05x + .07y = 325$. Multiply $x + y = 5000$ by .05 and subtract from $.05x + .07y = 325$.

$$.05x + .07y = 325$$
$$\underline{.05x + .05y = 250}$$
$$.02y = \;\;\;75 \Rightarrow y = 3750$$

Substitute $y = 3750$ into $x + y = 5000 \Rightarrow x + 3750 = 5000 \Rightarrow x = 1250$. Therefore $1250 was invested at 5% and $3750 was invested at 7%.

83. First let $H = 180$ in each equation and solve for y:

$180 = .491x + .468y + 11.2 \Rightarrow .468y = 180 - 11.2 - .491x \Rightarrow y_1 = \dfrac{168.8 - .491x}{.468}$

$180 = -.981x + 1.872y + 26.4 \Rightarrow 1.872y = 180 - 26.4 + .981x \Rightarrow y_2 = \dfrac{153.6 + .981x}{1.872}$

Now graph each equation in the same window and the intersection is approximately: $(177.1, 174.9)$. This means that if an athlete's maximum heart rate is 180 beats per minute (bpm), then it will be about 177 bpm after 5 seconds and 175 bpm after 10 seconds.

84. First let $H = 195$ in each equation and solve for y:

$$195 = .491x + .468y + 11.2 \Rightarrow .468y = 195 - 11.2 - .491x \Rightarrow y_1 = \frac{183.8 - .491x}{.468}$$

$$195 = -.981x + 1.872y + 26.4 \Rightarrow 1.872y = 195 - 26.4 + .981x \Rightarrow y_2 = \frac{168.6 + .981x}{1.872}$$

Now graph each equation in the same window and the intersection is approximately: (192.4, 190.9). This

means that if an athlete's maximum heart rate is 195 beats per minute (bpm), then it will be about 194.4 bpm

after 5 seconds and 190.9 bpm after 10 seconds.

85. (a) Set the two equations equal to each other and solve for x.

$$-6.393x + 894.9 = 19.14x + 746.9 \Rightarrow -6.393x - 19.14x = 746.9 - 894.9 \Rightarrow -25.533x = -148 \Rightarrow$$

$x = \dfrac{-148}{-25.533} \Rightarrow x \approx 5.8$. Now substitute $x = 5.8$ into the first equation: $f(5.8) = -6.393(5.8) + 894.9 \Rightarrow$

$f(5.8) \approx 857.8$. The solution is: $\{(5.8, 857.8)\}$.

(b) Since $1990 + 5.8 = 1995.8$, during the year 1995, 857.8 million pounds of both canned tuna and fresh

shrimp were available.

(c) Graph the two original equations in the same window. See Figure 85. The graphs intersect at

approximately (5.8, 857.8) which supports part (a).

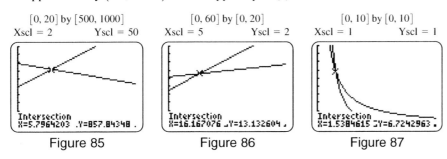

$[0, 20]$ by $[500, 1000]$	$[0, 60]$ by $[0, 20]$	$[0, 10]$ by $[0, 10]$
Xscl = 2 Yscl = 50	Xscl = 5 Yscl = 2	Xscl = 1 Yscl = 1

Figure 85	Figure 86	Figure 87

86. (a) Set the two equations equal to each other and solve for x.

$$.0515x + 12.3 = .255x + 9.01 \Rightarrow .0515x - .255x = 9.01 - 12.3 \Rightarrow -.2035x = -3.29 \Rightarrow$$

$x = \dfrac{-3.29}{-.2035} \Rightarrow x \approx 16$. Since $1990 + 16 = 2006$, during 2006 these population percents will be equal.

(b) Substitute $x = 16$ into the first equation: $y = .0515(16) + 12.3 \Rightarrow y \approx 13.1$. This means that 13.1% of the

U.S. resident population will be black and 13.1% will be Hispanic during 2006.

(c) Graph the two original equations in the same window. See Figure 86. The graphs intersect at

approximately (16, 13.1) which supports part (a).

(d) The Hispanic population is increasing most rapidly because its slope, .255, is larger than the slope for the

black population, .0515.

87. If $V = \pi r^2 h$ and $V = 50$ then $50 = \pi r^2 h$; and if $S.A. = 2\pi rh$ and $S.A. = 65$ then $65 = 2\pi rh$.

Now solve each for h: $50 = \pi r^2 h \Rightarrow h_1 = \dfrac{50}{\pi r^2}$; and $65 = 2\pi rh \Rightarrow h_2 = \dfrac{65}{2\pi r}$.

Graph each equation in the same window. See Figure 87. The intersection is approximately:

(1.538, 6.724), therefore $r \approx 1.538$ and $h \approx 6.724$.

88. If $V = \pi r^2 h$ and $V = 38$ then $38 = \pi r^2 h$; and if $S.A. = \pi r^2 + \pi r^2 + 2\pi rh$ and $S.A. = 38$ then

$38 = \pi r^2 + \pi r^2 + 2\pi rh$. Now solve each for h: $38 = \pi r^2 h \Rightarrow h_1 = \dfrac{38}{\pi r^2}$; and $38 = 2\pi r^2 + 2\pi rh \Rightarrow$

$2\pi rh = -2\pi r^2 + 38 \Rightarrow h_2 = \dfrac{-2\pi r^2 + 38}{2\pi r}$. Graph each equation in the same window. See Figure 88.

The graphs do not intersect, therefore, no such container is possible.

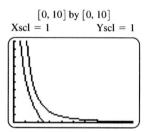

[0, 10] by [0, 10]
Xscl = 1   Yscl = 1

Figure 88

89. Add the equations to eliminate the w_2-variable.

$$w_1 + \sqrt{2}w_2 = 300$$
$$\sqrt{3}w_1 - \sqrt{2}w_2 = \ \ \ 0$$
$$(1 + \sqrt{3})w_1 = 300 \quad \Rightarrow w_1 = \frac{300}{(1 + \sqrt{3})} \approx 109.8. \text{ Now substitute (109.8) for } w_1 \text{ in the first equation:}$$

$\left(\dfrac{300}{(1 + \sqrt{3})}\right) + \sqrt{2}w_2 = 300 \Rightarrow \sqrt{2}w_2 = 300 - \dfrac{300}{(1 + \sqrt{3})} \Rightarrow \sqrt{2}w_2 = \dfrac{300(1 + \sqrt{3}) - 300}{1 + \sqrt{3}} \Rightarrow$

$\sqrt{2}w_2 = \dfrac{300\sqrt{3}}{1 + \sqrt{3}} \Rightarrow w_2 = \dfrac{300\sqrt{3}}{1 + \sqrt{3}} \times \dfrac{1}{\sqrt{2}} = \dfrac{300\sqrt{3}}{\sqrt{2} + \sqrt{6}} \Rightarrow w_2 \approx 134.5.$ The solution is: $\{(109.8, 134.5)\}$.

90. (a) Since 6 feet 11 inches is equal to 83 inches substitute (83) into each equation:

$w = 7.46(83) - 374 = 245.18 \Rightarrow w \approx 245$ pounds.

$w = 7.93(83) - 405 = 253.19 \Rightarrow w \approx 253$ pounds.

(b) Graph each equation in the same window, the graphs intersect at (65.96, 118) or the models agree when

$h \approx 65.96$ inches and $w \approx 118$ pounds.

(c) From the slope of the equations, first model: 7.46 pounds; and the second model: 7.93 pounds.

91. The total number of vehicles entering intersection A is $500 + 150 = 650$ vehicles per hour. The expression

$x + y$ represents the number of vehicles leaving intersection A each hour. Therefore, we have $x + y = 650$.

The total number of vehicles leaving intersection B is $50 + 400 = 450$ vehicles per hour. There are 100

vehicles entering intersection B from the south and y vehicles entering intersection B from the west. Thus

$y + 100 = 450 \Rightarrow y = 350$. Using substitution, substitute (350) in for y into the equation $x + y = 650$:

$x + 350 = 650 \Rightarrow x = 300$. At intersection A, a stoplight should allow for 300 vehicles per hour to travel

south and 350 vehicles per hour to continue traveling east.

92. Find where $C = R$, $20x + 10,000 = 30x - 11,000 \Rightarrow 21,000 = 10x \Rightarrow x = 2,100$. Now substitute (2,100)

in for x into the first equation: $C = 20(2,100) + 10,000 \ = 42,000 + 10,000 \Rightarrow C = 52,000$.

When $x = 2,100$; $R = C = \$52,000$.

93. Find where $C = R$, $4x + 125 = 9x - 200 \Rightarrow 325 = 5x \Rightarrow x = 65$. Now substitute (65) in for x into the first equation: $C = 4(65) + 125 = 260 + 125 \Rightarrow C = 385$. When $x = 65$; $R = C = \$385$.

94. Find where $p = p$, $80 - \dfrac{3}{5}q = \dfrac{2}{5}q \Rightarrow 5\left(80 - \dfrac{3}{5}q = \dfrac{2}{5}q\right) \Rightarrow 400 - 3q = 2q \Rightarrow 400 = 5q \Rightarrow q = 80$.

 Now substitute (80) in for q into the second equation: $p = \dfrac{2}{5}(80) \Rightarrow p = 32$. When $q = 80$; $p = \$32$.

95. Find where $p = p$, $630 - \dfrac{3}{4}q = \dfrac{3}{4}q \Rightarrow 4\left(630 - \dfrac{3}{4}q = \dfrac{3}{4}q\right) \Rightarrow 2520 - 3q = 3q \Rightarrow 2520 = 6q \Rightarrow q = 420$.

 Now substitute (420) in for q into the second equation: $p = \dfrac{3}{4}(420) \Rightarrow p = 315$. When $q = 420$; $p = \$315$.

96. (a) See Figure 96. The point of intersection is: $(36, 54)$.

 (b) Find where $S = D$, $\dfrac{3}{2}q = 81 - \dfrac{3}{4}q \Rightarrow 4\left(\dfrac{3}{2}q = 81 - \dfrac{3}{4}q\right) \Rightarrow 6q = 324 - 3q \Rightarrow 9q = 324 \Rightarrow q = 36$.

 (c) Now substitute (36) in for q into the second equation: $p = 81 - \dfrac{3}{4}(36) = 81 - 27 \Rightarrow p = \$54$.

[0, 120] by [0, 100] [0, 10] by [0, 2]
Xscl = 10 Yscl = 10 Xscl = 1 Yscl = .5

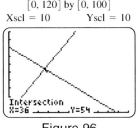
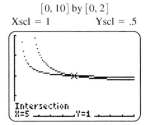

Figure 96   Figure 102

97. If $t = \dfrac{1}{x}$ and $u = \dfrac{1}{y}$, then $5\left(\dfrac{1}{x}\right) + 15\left(\dfrac{1}{y}\right) = 16 \Rightarrow 5t + 15u = 16$; and $5\left(\dfrac{1}{x}\right) + 4\left(\dfrac{1}{y}\right) = 5 \Rightarrow$

 $5t + 4u = 5$. The equations are: $5t + 15u = 16$ and $5t + 4u = 5$.

98. Subtract the two equations to eliminate the x-variable.

 $$5t + 15u = 16$$
 $$\underline{5t + 4u = 5}$$
 $$11u = 11 \quad \Rightarrow u = 1.$$ now substitute (1) in for u in the first equation:

 $5t + 15(1) = 16 \Rightarrow 5t = 1 \Rightarrow t = \dfrac{1}{5}$. Therefore $u = 1$ and $t = \dfrac{1}{5}$.

99. If $t = \dfrac{1}{x}$ and $t = \dfrac{1}{5}$, then $\dfrac{1}{x} = \dfrac{1}{5}$ and $x = 5$. If $u = \dfrac{1}{y}$ and $u = 1$, then $\dfrac{1}{y} = 1 \Rightarrow y = 1$.

100. $\dfrac{5}{x} + \dfrac{15}{y} = 16 \Rightarrow xy\left(\dfrac{5}{x} + \dfrac{15}{y} = 16\right) \Rightarrow 5y + 15x = 16xy \Rightarrow 5y - 16xy = -15x \Rightarrow$

 $y(5 - 16x) = -15x \Rightarrow y = \dfrac{-15x}{5 - 16x}$.

101. $\dfrac{5}{x} + \dfrac{4}{y} = 5 \Rightarrow xy\left(\dfrac{5}{x} + \dfrac{4}{y} = 5\right) \Rightarrow 5y + 4x = 5xy \Rightarrow 5y - 5xy = -4x \Rightarrow y(5 - 5x) = -4x \Rightarrow$

 $y = \dfrac{-4x}{5 - 5x}$.

102. See Figure 102. The graphs intersect at: $(5, 1)$ which supports the indicated exercise.

103. If we let $t = \dfrac{1}{x}$ and $u = \dfrac{1}{y}$, then our new equations are: $2t + u = \dfrac{3}{2}$ and $3t - u = 1$.

Add these to eliminate the u-variable.

$$2t + u = \dfrac{3}{2}$$
$$\underline{3t - u = 1}$$
$$5t \quad\quad = \dfrac{5}{2} \implies t = \dfrac{1}{2}. \text{ If } t = \dfrac{1}{x} \text{ and } t = \dfrac{1}{2}, \text{ then } \dfrac{1}{x} = \dfrac{1}{2} \implies x = 2$$

Now substitute $t = \dfrac{1}{2}$ into $2t + u = \dfrac{3}{2}$, which yields: $2\left(\dfrac{1}{2}\right) + u = \dfrac{3}{2} \implies u = \dfrac{1}{2}$.

If $u = \dfrac{1}{y}$ and $u = \dfrac{1}{2}$, then $\dfrac{1}{y} = \dfrac{1}{2} \implies y = 2$. The solution is: $\{(2, 2)\}$.

104. If we let $t = \dfrac{1}{x}$ and $u = \dfrac{1}{y}$, then our new equations are: $2t + u = 11$ and $3t - 5u = 10$.

Multiply the first equation by 5 and add to eliminate the u-variable.

$$10t + 5u = 55$$
$$\underline{3t - 5u = 10}$$
$$13t \quad\quad = 65 \implies t = 5. \text{ If } t = \dfrac{1}{x} \text{ and } t = 5, \text{ then } \dfrac{1}{x} = 5 \implies 5x = 1 \implies x = \dfrac{1}{5}.$$

Now substitute $t = 5$ into $3t - 5u = 10$, which yields: $3(5) - 5u = 10 \implies -5u = -5 \implies u = 1$.

If $u = \dfrac{1}{y}$ and $u = 1$, then $\dfrac{1}{y} = 1 \implies y = 1$. The solution is: $\left\{\left(\dfrac{1}{5}, 1\right)\right\}$.

105. If we let $t = \dfrac{1}{x}$ and $u = \dfrac{1}{y}$, then our new equations are: $2t + 3u = 18$ and $4t - 5u = -8$.

Multiply the first equation by 2 and subtract to eliminate the t-variable.

$$4t + 6u = 36$$
$$\underline{4t - 5u = -8}$$
$$\quad\quad 11u = 44 \implies u = 4. \text{ If } u = \dfrac{1}{y} \text{ and } u = 4, \text{ then } \dfrac{1}{y} = 4 \implies 4y = 1 \implies y = \dfrac{1}{4}.$$

Now substitute $u = 4$ into $2t + 3u = 18$, which yields: $2t + 3(4) = 18 \implies 2t = 6 \implies t = 3$.

If $t = \dfrac{1}{x}$ and $t = 3$, then $\dfrac{1}{x} = 3 \implies 3x = 1 \implies x = \dfrac{1}{3}$. The solution is: $\left\{\left(\dfrac{1}{3}, \dfrac{1}{4}\right)\right\}$.

106. If we let $t = \dfrac{1}{x}$ and $u = \dfrac{1}{y}$, then our new equations are: $t + 3u = \dfrac{16}{5}$ and $5t + 4u = 5$.

Multiply the first equation by 5 and subtract to eliminate the t-variable.

$$5t + 15u = 16$$
$$\underline{5t + 4u = 5}$$
$$\quad\quad 11u = 11 \implies u = 1. \text{ If } u = \dfrac{1}{y} \text{ and } u = 1, \text{ then } \dfrac{1}{y} = 1 \implies y = 1.$$

Now substitute $u = 1$ into $5t + 4u = 5$, which yields: $5t + 4(1) = 5 \implies 5t = 1 \implies t = \dfrac{1}{5}$.

If $t = \dfrac{1}{x}$ and $t = \dfrac{1}{5}$, then $\dfrac{1}{x} = \dfrac{1}{5} \implies x = 5$. The solution is: $\{(5, 1)\}$.

7.2: Solution of Linear Systems in Three Variables

1. (a) Multiply [equation 2] by 2 and add to [equation 1] to get $7x - z = 0$.

    (b) Multiply [equation 1] by -2 and add to [equation 3] to get $-3y + 2z = 2$.

    (c) Multiply [equation 2] by 4 and add to [equation 3] to get $14y - 3y = 2$.

2. For $2x + y - z = -1$ and $(-3, 6, 1)$: $2(-3) + 6 - 1 = -1 \Rightarrow -6 + 6 - 1 = -1 \Rightarrow -1 = -1$.

    For $x - y + 3z = -6$ and $(-3, 6, 1)$: $(-3) - 6 + 3(1) = -6 \Rightarrow -3 - 6 + 3 = -6 \Rightarrow -6 = -6$.

    For $-4x + y + z = 19$ and $(-3, 6, 1)$: $-4(-3) + 6 + 1 = 19 \Rightarrow 12 + 6 + 1 = 19 \Rightarrow 19 = 19$.

3. For $2x + 8y - 6z = -6$ and $\left(\frac{1}{2}, -\frac{3}{4}, \frac{1}{6}\right)$: $2\left(\frac{1}{2}\right) + 8\left(-\frac{3}{4}\right) - 6\left(\frac{1}{6}\right) = -6 \Rightarrow 1 - 6 - 1 = -6 \Rightarrow -6 = -6$.

    For $x + y + z = -\frac{1}{12}$ and $\left(\frac{1}{2}, -\frac{3}{4}, \frac{1}{6}\right)$: $\left(\frac{1}{2}\right) + \left(-\frac{3}{4}\right) + \left(\frac{1}{6}\right) = -\frac{1}{12} \Rightarrow$

    $\frac{6}{12} - \frac{9}{12} + \frac{2}{12} = -\frac{1}{12} \Rightarrow -\frac{1}{12} = -\frac{1}{12}$.

    For $x + 3z = 1$ and $\left(\frac{1}{2}, -\frac{3}{4}, \frac{1}{6}\right)$: $\left(\frac{1}{2}\right) + 3\left(\frac{1}{6}\right) = 1 \Rightarrow \frac{1}{2} + \frac{1}{2} = 1 \Rightarrow 1 = 1$.

4. For $5x - y + 2z = -.4$ and $(-.2, .4, .5)$: $5(-.2) - (.4) + 2(.5) = -.4 \Rightarrow -1 - .4 + 1 = .4 \Rightarrow -.4 = -.4$.

    For $x + 4z = 1.8$ and $(-.2, .4, .5)$: $(-.2) + 4(.5) = 1.8 \Rightarrow$

    $-.2 + 4(.5) = 1.8 \Rightarrow -.2 + 2.0 = 1.8 \Rightarrow 1.8 = 1.8$.

    For $-3y + z = -.7$ and $(-.2, .4, .5)$: $-3(.4) + (.5) = -.7 \Rightarrow -1.2 + .5 = -.7 \Rightarrow -.7 = -.7$.

5. First add [equation 1] and [equation 2] to eliminate z and produce [equation 4].

    Second add [equation 2] and [equation 3] to eliminate y and z and produce [equation 5].

$$\begin{array}{rrrrl} x + & y + & z = & 2 & [1] \\ 2x + & y - & z = & 5 & [2] \\ \hline 3x + & 2y & = & 7 & [4] \end{array} \qquad \begin{array}{rrrrl} 2x + & y - & z = & 5 & [2] \\ x - & y + & z = & -2 & [3] \\ \hline 3x & & = & 3 & [5] \end{array}$$

    Solve [equation 5] for x: $3x = 3 \Rightarrow x = 1$. Substitute $x = 1$ into [equation 4] to solve for y:

    $3(1) + 2y = 7 \Rightarrow 2y = 4 \Rightarrow y = 2$. Finally, substitute $x = 1$ and $y = 2$ into [equation 1] to solve for z:

    $(1) + (2) + z = 2 \Rightarrow z = -1$. The solution is: $\{(1, 2, -1)\}$.

6. First add [equation 1] and [equation 2] to eliminate y and produce [equation 4].

    Second add [equation 1] and [equation 3] to eliminate y and produce [equation 5].

$$\begin{array}{rrrrl} 2x + & y + & z = & 9 & [1] \\ -x - & y + & z = & 1 & [2] \\ \hline x & + & 2z = & 10 & [4] \end{array} \qquad \begin{array}{rrrrl} 2x + & y + & z = & 9 & [1] \\ 3x - & y + & z = & 9 & [3] \\ \hline 5x & + & 2z = & 18 & [5] \end{array}$$

    Next multiply [equation 4] by -1 and add to [equation 5] to eliminate z and produce [equation 6].

$$\begin{array}{rrrl} -x - & 2z = & -10 & -[4] \\ 5x + & 2z = & 18 & [5] \\ \hline 4x & = & 8 & [6] \end{array}$$

    Solve [equation 6] for x: $4x = 8 \Rightarrow x = 2$. Substitute $x = 2$ into [equation 4] to solve for z:

    $(2) + 2z = 10 \Rightarrow 2z = 8 \Rightarrow z = 4$. Finally, substitute $x = 2$ and $z = 4$ into [equation 1] to solve for y:

    $2(2) + y + 4 = 9 \Rightarrow y = 1$. The solution is: $\{(2, 1, 4)\}$.

7. First, add [equation 1] and [equation 2] to eliminate y and produce [equation 4].

    Second, multiply [equation 3] by 3 and add [equation 1] to eliminate y and produce [equation 5].

$$
\begin{array}{llll}
x + & 3y + & 4z = & 14 \quad [1] \\
\underline{2x - } & \underline{3y + } & \underline{2z = } & \underline{10} \quad [2] \\
3x & & + 6z = & 24 \quad [4]
\end{array}
\qquad
\begin{array}{llll}
9x - & 3y + & 3z = & 27 \quad 3[3] \\
\underline{x + } & \underline{3y + } & \underline{4z = } & \underline{14} \quad [1] \\
10x & & + 7z = & 41 \quad [5]
\end{array}
$$

Next multiply [equation 4] by 10 and add to [equation 5] multiplied by -3 to eliminate x and produce [equation 6].

$$
\begin{array}{ll}
30x + 60z = & 240 \quad 10[4] \\
\underline{-30x - 21z = } & \underline{-123} \quad -3[5] \\
39z = & 117 \quad [6]
\end{array}
$$

Solve [equation 6] for z: $39z = 117 \Rightarrow z = 3$. Substitute $z = 3$ into [equation 4] to solve for x:

$3x + 6(3) = 24 \Rightarrow 3x = 6 \Rightarrow x = 2$. Finally, substitute $x = 2$ and $z = 3$ into [equation 1] to solve for y:

$(2) + 3y + 4(3) = 14 \Rightarrow 3y = 0 \Rightarrow y = 0$. The solution is: $\{(2, 0, 3)\}$.

8. First, add [equation 1] and [equation 3] to eliminate y and produce [equation 4].

    Second, multiply [equation 3] by -5 and add [equation 2] to eliminate y and produce [equation 5].

$$
\begin{array}{llll}
4x - & y + & 3z = & -2 \quad [1] \\
\underline{-2x + } & \underline{y + } & \underline{4z = } & \underline{14} \quad [2] \\
2x & & + 7z = & 12 \quad [4]
\end{array}
\qquad
\begin{array}{llll}
10x - & 5y - & 20z = & -70 \quad -5[3] \\
\underline{3x + } & \underline{5y - } & \underline{z = } & \underline{15} \quad [2] \\
13x & & - 21z = & -55 \quad [5]
\end{array}
$$

Next multiply [equation 4] by 3 and add to [equation 5] to eliminate z and produce [equation 6].

$$
\begin{array}{ll}
6x + 21z = & 36 \quad 3[4] \\
\underline{13x - 21z = } & \underline{-55} \quad [5] \\
19x = & -19 \quad [6]
\end{array}
$$

Solve [equation 6] for x: $19x = -19 \Rightarrow x = -1$. Substitute $x = -1$ into [equation 4] to solve for z:

$2(-1) + 7z = 12 \Rightarrow 7z = 14 \Rightarrow z = 2$. Finally, substitute $x = -1$ and $z = 2$ into [equation 1] to solve for y:

$4(-1) - y + 3(2) = -2 \Rightarrow -y = -4 \Rightarrow y = 4$. The solution is: $\{(-1, 4, 2)\}$.

9. First, multiply [equation 1] by 2 and add [equation 3] to eliminate all the variables.

$$
\begin{array}{ll}
2x + 4y + 6z = & 16 \quad 2[1] \\
\underline{-2x - 4y - 6z = } & \underline{5} \quad [3] \\
0 = & 21 \qquad \text{Since this is false, the solution is: } \varnothing.
\end{array}
$$

10. First, multiply [equation 1] by -3 and add [equation 2] to eliminate all the variables.

$$
\begin{array}{ll}
-9x + 6y + 24z = & -3 \quad -3[1] \\
\underline{9x - 6y - 24z = } & \underline{-2} \quad [3] \\
0 = & -5 \qquad \text{Since this is false, the solution is: } \varnothing.
\end{array}
$$

11. First, add [equation 1] and [equation 2] to eliminate z and produce [equation 4].

    Second, multiply [equation 1] by 3 and add [equation 3] to eliminate z and produce [equation 5].

$x + 4y - z = 6$	[1]	
$2x - y + z = 3$	[2]	
$3x + 3y = 9$	[4]	

$3x + 12y - 3z = 18$	$3[1]$	
$3x + 2y + 3z = 16$	[3]	
$6x + 14y = 34$	[5]	

   Next multiply [equation 4] by -2 and add to [equation 5] to eliminate x and produce [equation 6].

$$
\begin{array}{ll}
-6x - 6y = -18 & -2[4] \\
6x + 14y = 34 & [5] \\
\hline
8y = 16 & [6]
\end{array}
$$

   Solve [equation 6] for y: $8y = 16 \Rightarrow y = 2$. Substitute $y = 2$ into [equation 4] to solve for x:

   $3x + 3(2) = 9 \Rightarrow 3x = 3 \Rightarrow x = 1$. Finally, substitute $x = 1$ and $y = 2$ into [equation 1] to solve for z:

   $(1) + 4(2) - z = 6 \Rightarrow -z = -3 \Rightarrow z = 3$. The solution is: $\{(1, 2, 3)\}$.

12. First, multiply [equation 3] by 2 and add [equation 2] to eliminate y and produce [equation 4].

    Second, multiply [equation 3] by -3 and add [equation 1] to eliminate y and produce [equation 5].

$2x - 2y + 6z = 10$	$2[3]$	
$3x + 2y - 2z = 4$	[2]	
$5x + 4z = 14$	[4]	

$-3x + 3y - 9z = -15$	$-3[3]$	
$4x - 3y + z = 9$	[1]	
$x - 8z = -6$	[5]	

   Next multiply [equation 4] by 2 and add to [equation 5] to eliminate z and produce [equation 6].

$$
\begin{array}{ll}
10x + 8z = 28 & 2[4] \\
x - 8z = -6 & [5] \\
\hline
11x = 22 & [6]
\end{array}
$$

   Solve [equation 6] for x: $11x = 22 \Rightarrow x = 2$. Substitute $x = 2$ into [equation 4] to solve for z:

   $5(2) + 4z = 14 \Rightarrow 4z = 4 \Rightarrow z = 1$. Finally, substitute $x = 2$ and $z = 1$ into [equation 1] to solve for y:

   $4(2) - 3y + (1) = 9 \Rightarrow -3y = 0 \Rightarrow y = 0$. The solution is: $\{(2, 0, 1)\}$.

13. First, multiply [equation 1] by -3 and add [equation 2] to eliminate y and produce [equation 4].

    Second, multiply [equation 1] by 2 and add [equation 3] to eliminate y and produce [equation 5].

$-15x - 3y + 9z = 18$	$-3[1]$	
$2x + 3y + z = 5$	[2]	
$-13x + 10z = 23$	[4]	

$10x + 2y - 6z = -12$	$2[1]$	
$-3x - 2y + 4z = 3$	[3]	
$7x - 2z = -9$	[5]	

   Next multiply [equation 5] by 5 and add [equation 4] to eliminate z and produce [equation 6].

$$
\begin{array}{ll}
35x - 10z = -45 & 5[5] \\
-13x + 10z = 23 & [4] \\
\hline
22x = -22 & [6]
\end{array}
$$

   Solve [equation 6] for x: $22x = -22 \Rightarrow x = -1$. Substitute $x = -1$ into [equation 4] to solve for z:

   $-13(-1) + 10z = 23 \Rightarrow 10z = 10 \Rightarrow z = 1$. Finally, substitute $x = -1$ and $z = 1$ into [equation 1] to

   solve for y: $5(-1) + y - 3(1) = -6 \Rightarrow y = 2$. The solution is: $\{(-1, 2, 1)\}$.

14. First, multiply [equation 2] by 4 and add [equation 1] to eliminate z and produce [equation 4].

Second, add [equation 2] and [equation 3] to eliminate z and produce [equation 5].

$$
\begin{array}{ll}
20x + 12y - 4z = 4 & 4[2] \\
\underline{2x - 5y + 4z = -35} & [1] \\
22x + 7y = -31 & [4]
\end{array}
\qquad
\begin{array}{ll}
5x + 3y - z = 1 & [2] \\
\underline{x + y + z = 1} & [3] \\
6x + 4y = 2 & [5]
\end{array}
$$

Next multiply [equation 4] by 4 and add [equation 5] multiplied by -7 to eliminate y and produce [equation 6].

$$
\begin{array}{ll}
88x + 28y = -124 & 4[2] \\
\underline{-42x - 28y = -14} & -7[5] \\
46x = -138 & [6]
\end{array}
$$

Solve [equation 6] for x: $46x = -138 \Rightarrow x = -3$. Substitute $x = -3$ into [equation 4] to solve for y:

$22(-3) + 7y = -31 \Rightarrow 7y = 35 \Rightarrow y = 5$. Finally, substitute $x = -3$ and $y = 5$ into [equation 1] to

solve for z: $2(-3) - 5(5) + 4z = -35 \Rightarrow 4z = -4 \Rightarrow z = -1$. The solution is: $\{(-3, 5, -1)\}$.

15. First, add [equation 1] and [equation 3] to eliminate x and produce [equation 4].

Second, multiply [equation 3] by 3 and add [equation 2] to eliminate x and produce [equation 5].

$$
\begin{array}{ll}
x - 3y - 2z = -3 & [1] \\
\underline{-x - y + 4z = 3} & [3] \\
 -4y + 2z = 0 & [4]
\end{array}
\qquad
\begin{array}{ll}
-3x - 3y + 12z = 9 & 3[3] \\
\underline{3x + 2y - z = 12} & [2] \\
 -y + 11z = 21 & [5]
\end{array}
$$

Next multiply [equation 5] by -4 and add [equation 4] to eliminate y and produce [equation 6].

$$
\begin{array}{ll}
4y - 44z = -84 & -4[5] \\
\underline{-4y + 2z = 0} & [4] \\
 -42z = -84 & [6]
\end{array}
$$

Solve [equation 6] for z: $-42z = -84 \Rightarrow z = 2$. Substitute $z = 2$ into [equation 4] to solve for y:

$-4y + 2(2) = 0 \Rightarrow -4y = -4 \Rightarrow y = 1$. Finally, substitute $y = 1$ and $z = 2$ into [equation 1] to

solve for x: $x - 3(1) - 2(2) = -3 \Rightarrow x = 4$. The solution is: $\{(4, 1, 2)\}$.

16. First, multiply [equation 3] by -1 and add [equation 1] to eliminate x, y and produce [equation 4].

$$
\begin{array}{ll}
-x - y - 3z = -11 & -1[3] \\
\underline{x + y + z = 3} & [1] \\
 -2z = -8 & [4]
\end{array}
$$

Solve [equation 4] for z: $-2z = -8 \Rightarrow z = 4$. Substituting $z = 4$ into [equation 1] and [equation 2] yields:

$x + y + 4 = 3 \Rightarrow x + y = -1$ [equation 5] and $3x - 3y - 4(4) = -1 \Rightarrow 3x - 3y = 15$ [equation 6].

Now multiply [equation 5] by 3 and add [equation 6].

$$
\begin{array}{l}
3x + 3y = -3 \\
\underline{3x - 3y = 15} \\
6x = 12 \Rightarrow x = 2.
\end{array}
$$

Finally, substitute $x = 2$ and $z = 4$ into [equation 1] to solve for y:

$(2) + y + (4) = 3 \Rightarrow y = -3$. The solution is: $\{(2, -3, 4)\}$.

17. First, multiply [equation 1] by -2 and add [equation 2] to eliminate x and produce [equation 4].

    Second, multiply [equation 1] by -3 and add [equation 3] to eliminate x and produce [equation 5].

$$
\begin{array}{rl}
-4x - 12y + 2z = -12 & -2[3] \\
\underline{4x - 3y + 5z = {-5}} & [2] \\
-15y + 7z = -17 & [4]
\end{array}
\qquad
\begin{array}{rl}
-6x - 18y + 3z = -18 & -3[1] \\
\underline{6x + 9y - 2z = 11} & [3] \\
-9y + z = -7 & [5]
\end{array}
$$

Next multiply [equation 5] by -7 and add [equation 4] to eliminate z and produce [equation 6].

$$
\begin{array}{rl}
63y - 7z = 49 & -7[5] \\
\underline{-15y + 7z = -17} & [4] \\
48y = 32 & [6]
\end{array}
$$

Solve [equation 6] for y: $48y = 32 \Rightarrow y = \dfrac{2}{3}$. Substitute $y = \dfrac{2}{3}$ into [equation 5] to solve for y:

$-9\left(\dfrac{2}{3}\right) + z = -7 \Rightarrow -6 + z = -7 \Rightarrow z = -1$. Finally, substitute $y = \dfrac{2}{3}$ and $z = -1$ into [equation 1] to

solve for x: $2x + 6\left(\dfrac{2}{3}\right) - (-1) = 6 \Rightarrow 2x + 5 = 6 \Rightarrow 2x = 1 \Rightarrow x = \dfrac{1}{2}$. The solution is: $\left\{\left(\dfrac{1}{2}, \dfrac{2}{3}, -1\right)\right\}$.

18. First, multiply [equation 1] by 3 and add [equation 2] to eliminate y and produce [equation 4], then simplify.

    Second, multiply [equation 3] by 3 and add [equation 2] to eliminate y and produce [equation 5] then simplify.

$$
\begin{array}{rl}
24x - 9y + 18z = {-6} & 3[1] \\
\underline{4x + 9y + 4z = 18} & [2] \\
28x + 22z = 12 & [4] \\
= 14x + 11z = 6 & [4a]
\end{array}
\qquad
\begin{array}{rl}
36x - 9y + 24z = {-6} & 3[2] \\
\underline{4x + 9y + 4z = 18} & [2] \\
40x + 28z = 12 & [5] \\
= 10x + 7z = 3 & [5a]
\end{array}
$$

Next multiply [equation 4a] by 5 and add [equation 5a] multiplied by -7 to eliminate x and produce [equation 6].

$$
\begin{array}{rl}
70x + 55z = 30 & 5[4a] \\
\underline{-70x - 49z = -21} & -7[5a] \\
6z = 9 & [6]
\end{array}
$$

Solve [equation 6] for z: $6z = 9 \Rightarrow z = \dfrac{3}{2}$. Substitute $z = \dfrac{3}{2}$ into [equation 5a] to solve for x:

$10x + 7\left(\dfrac{3}{2}\right) = 3 \Rightarrow 2\left[10x + 7\left(\dfrac{3}{2}\right) = 3\right] \Rightarrow 20x + 21 = 6 \Rightarrow 20x = -15 \Rightarrow x = -\dfrac{3}{4}$. Finally, substitute

$x = -\dfrac{3}{4}$ and $z = \dfrac{3}{2}$ into [equation 2] to solve for y: $4\left(-\dfrac{3}{4}\right) + 9y + 4\left(\dfrac{3}{2}\right) = 18 \Rightarrow -3 + 9y + 6 = 18 \Rightarrow$

$9y = 15 \Rightarrow y = \dfrac{15}{9} = \dfrac{5}{3}$. The solution is: $\left\{\left(-\dfrac{3}{4}, \dfrac{5}{3}, \dfrac{3}{2}\right)\right\}$.

19. First, add [equation 1] and to [equation 2] multiplied by -1 to eliminate x and produce [equation 4].

    Second, add [equation 3] and [equation 4] to eliminate y and produce [equation 5].

$$
\begin{array}{rl}
x + z = 4 & [1] \\
\underline{-x - y = -4} & -1[2] \\
z - y = 0 & [4]
\end{array}
\qquad
\begin{array}{rl}
y + z = 4 & [3] \\
\underline{-y + z = 0} & [4] \\
2z = 4 & [5]
\end{array}
$$

Solve [equation 5] for z: $2z = 4 \Rightarrow z = 2$. Substitute $z = 2$ into [equation 1] to solve for x:

$x + 2 = 4 \Rightarrow x = 2$. Finally, substitute $x = 2$ into [equation 2] to solve for y: $2 + y = 4 \Rightarrow y = 2$.

The solution is: $\{(2, 2, 2)\}$.

20. First, add [equation 1] and to [equation 2] multiplied by -1 to eliminate x and produce [equation 4].

    Second, add [equation 3] and [equation 4] to eliminate y and produce [equation 5].

    $$\begin{array}{rl} x - z = 2 & [1] \\ -x - y = 3 & -1[2] \\ \hline -y - z = 5 & [4] \end{array} \qquad \begin{array}{rl} y - z = 1 & [3] \\ -y - z = 5 & [4] \\ \hline -2z = 6 & [5] \end{array}$$

    Solve [equation 5] for z: $-2z = 6 \Rightarrow z = -3$. Substitute $z = -3$ into [equation 1] to solve for x:

    $x - (-3) = 2 \Rightarrow x = -1$. Finally, substitute $z = -3$ into [equation 3] to solve for y:

    $y - (-3) = 1 \Rightarrow y = -2$. The solution is: $\{(-1, -2, -3)\}$.

21. First, add [equation 1] and to [equation 3] multiplied by -1 to eliminate x and z.

    $$\begin{array}{rl} 2x + y - z = -4 & [1] \\ -2x + z = 4 & -1[3] \\ \hline y = 0 & \end{array}$$

    Substitute $y = 0$ to solve [equation 2] for z: $(0) + 2z = 12 \Rightarrow 2z = 12 \Rightarrow z = 6$. Finally, substitute $z = 6$

    into [equation 3] to solve for x: $2x - (6) = -4 \Rightarrow 2x = 2 \Rightarrow x = 1$. The solution is: $\{(1, 0, 6)\}$.

22. First, multiply [equation 1] by 3 and add [equation 2] multiplied by -2 to eliminate y and produce [equation 4].

    Second, add [equation 4] and [equation 3] multiplied by -9 to eliminate x.

    $$\begin{array}{rl} 9x + 6y - 3z = -3 & 3[1] \\ -6y + 2z = -24 & -2[2] \\ \hline 9x - 5z = -27 & [4] \end{array} \qquad \begin{array}{rl} 9x - 5z = -27 & [4] \\ -9x + 27z = 27 & [3] \\ \hline 22z = 0 & \Rightarrow z = 0. \end{array}$$

    Substitute $z = 0$ to solve [equation 2] for y: $3y + 0 = 12 \Rightarrow y = 4$. Finally, substitute $z = 0$ into [equation 3]

    to solve for x: $x - 3(0) = -3 \Rightarrow x = -3$. The solution is: $\{(-3, 4, 0)\}$.

23. First, multiply [equation 1] by -3 and add [equation 2] to eliminate x and produce [equation 4].

    Second, add [equation 4] and [equation 3] to eliminate y.

    $$\begin{array}{rl} -6x - 9y - 12z = -9 & -3[1] \\ 6x + 3y + 8z = 6 & -2[2] \\ \hline -6y - 4z = -3 & [4] \end{array} \qquad \begin{array}{rl} -6y - 4z = -3 & [4] \\ 6y - 4z = 1 & [3] \\ \hline -8z = -2 & \Rightarrow z = \dfrac{1}{4}. \end{array}$$

    Substitute $z = \dfrac{1}{4}$ to solve [equation 3] for y: $6y - 4\left(\dfrac{1}{4}\right) = 1 \Rightarrow 6y = 2 \Rightarrow y = \dfrac{1}{3}$. Finally, substitute $y = \dfrac{1}{3}$

    and $z = \dfrac{1}{4}$ into [equation 1] to solve for x: $2x + 3\left(\dfrac{1}{3}\right) + 4\left(\dfrac{1}{4}\right) = 3 \Rightarrow 2x = 1 \Rightarrow x = \dfrac{1}{2}$.

    The solution is: $\left\{\left(\dfrac{1}{2}, \dfrac{1}{3}, \dfrac{1}{4}\right)\right\}$.

24. First, add [equation 1] and [equation 2] multiplied by -2 to eliminate x and produce [equation 4].

Second, add [equation 4] and [equation 3] to eliminate y.

$$
\begin{array}{rrrll}
10x + & 2y - & 3z = & 0 & [1] \\
-10x - & 8y - & 12z = & 2 & -2[2] \\
\hline
 & -6y - & 15z = & 2 & [4]
\end{array}
\qquad
\begin{array}{rrll}
-6y - & 15z = & 2 & [4] \\
6y + & 3z = & 2 & [3] \\
\hline
 & -12z = & 4 & \Rightarrow z = -\dfrac{1}{3}.
\end{array}
$$

Substitute $z = -\dfrac{1}{3}$ to solve [equation 3] for y: $6y + 3\left(-\dfrac{1}{3}\right) = 2 \Rightarrow 6y = 3 \Rightarrow y = \dfrac{1}{2}.$ Finally, substitute

$y = \dfrac{1}{2}$ and $z = -\dfrac{1}{3}$ into [equation 2] to solve for x: $5x + 4\left(\dfrac{1}{2}\right) + 6\left(-\dfrac{1}{3}\right) = -1 \Rightarrow 5x = -1 \Rightarrow x = -\dfrac{1}{5}.$

The solution is: $\left\{\left(-\dfrac{1}{5}, \dfrac{1}{2}, -\dfrac{1}{3}\right)\right\}.$

25. If we let $t = \dfrac{1}{x}$, $u = \dfrac{1}{y}$, and $v = \dfrac{1}{z}$ then our new equations are: $t + u - v = \dfrac{1}{4}\,[1]$, $2t - u + 3v = \dfrac{9}{4}\,[2]$, and

$-t - 2u + 4v = 1\,[3]$. First, add [equation 1] and [equation 3] to eliminate t and produce [equation 4].

Second, add [equation 2] and [equation 3] multiplied by 2 to eliminate t and produce [equation 5].

$$
\begin{array}{rrrll}
t + & u - & v = & \dfrac{1}{4} & [1] \\[2mm]
-t - & 2u + & 4v = & 1 & [3] \\
\hline
 & -u + & 3v = & \dfrac{5}{4} & [4]
\end{array}
\qquad
\begin{array}{rrrll}
2t - & u + & 3v = & \dfrac{9}{4} & [2] \\[2mm]
-2t - & 4u + & 8v = & 2 & 2[3] \\
\hline
 & -5u + & 11v = & \dfrac{17}{4} & [5]
\end{array}
$$

Next multiply [equation 4] by -5 and add [equation 5] to eliminate u and produce [equation 6].

$$
\begin{array}{rll}
5u - 15v = & -\dfrac{25}{4} & -5[4] \\[2mm]
-5u + 11v = & \dfrac{17}{4} & [5] \\
\hline
-4v = & -\dfrac{8}{4} & [6]
\end{array}
$$

Solve [equation 6] for v: $-4v = -2 \Rightarrow v = \dfrac{1}{2}$. If $v = \dfrac{1}{2}$ and $v = \dfrac{1}{z}$ then $\dfrac{1}{z} = \dfrac{1}{2} \Rightarrow z = 2.$

Substitute $v = \dfrac{1}{2}$ into [equation 4] to solve for u: $-u + 3\left(\dfrac{1}{2}\right) = \dfrac{5}{4} \Rightarrow -u = -\dfrac{1}{4} \Rightarrow u = \dfrac{1}{4}.$

If $u = \dfrac{1}{4}$ and $u = \dfrac{1}{y}$ then $\dfrac{1}{4} = \dfrac{1}{y} \Rightarrow y = 4$. Finally, substitute $v = \dfrac{1}{2}$ and $u = \dfrac{1}{4}$ into [equation 3] to

solve for t: $-t - 2\left(\dfrac{1}{4}\right) + 4\left(\dfrac{1}{2}\right) = 1 \Rightarrow -2t - 1 + 4 = 2 \Rightarrow -2t = -1 \Rightarrow t = \dfrac{1}{2}.$

If $t = \dfrac{1}{2}$ and $= \dfrac{1}{x}$ then $\dfrac{1}{2} = \dfrac{1}{x} \Rightarrow x = 2$. The solution is: $\{(2, 4, 2)\}.$

26. If we let $t = \dfrac{1}{x}$, $u = \dfrac{1}{y}$, and $v = \dfrac{1}{z}$ then our new equations are: $3t + 2u - v = \dfrac{11}{6}$ [1], $t - u + 3v = -\dfrac{11}{12}$ [2],

and $2t + u + v = \dfrac{7}{12}$ [3]. First, add [equation 1] and [equation 2] multiplied by 2 to eliminate u and produce

[equation 4], then simplify.   Second, add [equation 2] and [equation 3] to eliminate u and produce [equation 5],

then simplify.

$$
\begin{array}{lll}
3t + 2u - v = \dfrac{11}{6} & [1] \\[2mm]
2t - 2u + 6v = -\dfrac{11}{6} & 2[2] \\[2mm]
\hline
5t \qquad + 5v = 0 \\[2mm]
= t + v = 0 & [4]
\end{array}
\qquad\qquad
\begin{array}{lll}
t - u + 3v = -\dfrac{11}{12} & [2] \\[2mm]
2t + u + v = \dfrac{7}{12} & [3] \\[2mm]
\hline
3t \qquad + 4v = -\dfrac{4}{12} \\[2mm]
= 3t + 4v = -\dfrac{1}{3} & [5]
\end{array}
$$

Next multiply [equation 4] by -3 and add [equation 5] to eliminate t.

$$
\begin{array}{ll}
-3t - 3v = 0 & -3[4] \\[2mm]
3t + 4v = -\dfrac{1}{3} & [5] \\[2mm]
\hline
v = -\dfrac{1}{3} & [6]
\end{array}
$$

If $v = -\dfrac{1}{3}$ and $v = \dfrac{1}{z}$ then $-\dfrac{1}{3} = \dfrac{1}{z} \Rightarrow z = -3$. Substitute $v = -\dfrac{1}{3}$ into [equation 4] to solve for t:

$t + \left(-\dfrac{1}{3}\right) = 0 \Rightarrow t = \dfrac{1}{3}$. If $t = \dfrac{1}{3}$ and $t = \dfrac{1}{x}$ then $\dfrac{1}{3} = \dfrac{1}{x} \Rightarrow x = 3$. Finally, substitute $v = -\dfrac{1}{3}$ and $t = \dfrac{1}{3}$

into [equation 1] to solve for u: $3\left(\dfrac{1}{3}\right) + 2u - \left(-\dfrac{1}{3}\right) = \dfrac{11}{6} \Rightarrow 1 + 2u + \dfrac{1}{3} = \dfrac{11}{6} \Rightarrow$

$6\left(1 + 2u + \dfrac{1}{3} = \dfrac{11}{6}\right) \Rightarrow 6 + 12u + 2 = 11 \Rightarrow 12u = 3 \Rightarrow u = \dfrac{1}{4}$.

If $u = \dfrac{1}{4}$ and $= \dfrac{1}{y}$ then $\dfrac{1}{4} = \dfrac{1}{y} \Rightarrow y = 4$. The solution is: $\{(3, 4, -3)\}$.

27. If we let $t = \dfrac{1}{x}$, $u = \dfrac{1}{y}$, and $v = \dfrac{1}{z}$ then our new equations are: $2t - 2u + v = -1$ [1], $4t + u - 2v = -9$ [2],

and $t + u - 3v = -9$ [3]. First, add [equation 1] and [equation 2] multiplied by 2 to eliminate u and produce

[equation 4]. Second, add [equation 1] and [equation 3] multiplied by 2 to eliminate u and produce [equation 5].

$$
\begin{array}{lr}
\begin{array}{rr}
2t - 2u + v = -1 & [1] \\
8t + 2u - 4v = -18 & 2[2] \\
\hline
10t \quad\quad - 3v = -19 & [4]
\end{array}
&
\begin{array}{rr}
2t - 2u + v = -1 & [1] \\
2t + 2u - 6v = -18 & 2[3] \\
\hline
4t \quad\quad - 5v = -19 & [5]
\end{array}
\end{array}
$$

Next multiply [equation 4] by 2 and add [equation 5] multiplied by -5 to eliminate t and produce [equation 6].

$$
\begin{array}{rr}
20t - 6v = -38 & 2[2] \\
-20t + 25v = 95 & -5[5] \\
\hline
19v = 57 & [6] \Rightarrow v = 3.
\end{array}
$$

If $v = 3$ and $v = \dfrac{1}{z}$ then $3 = \dfrac{1}{z} \Rightarrow 3z = 1 \Rightarrow z = \dfrac{1}{3}$. Substitute $v = 3$ into [equation 5] to solve for t:

$4t - 5(3) = -19 \Rightarrow 4t = -4 \Rightarrow t = -1$. If $t = -1$ and $t = \dfrac{1}{x}$ then $-1 = \dfrac{1}{x} \Rightarrow -x = 1 \Rightarrow x = -1$.

Finally, substitute $v = 3$ and $t = -1$ into [equation 1] to solve for u: $2(-1) - 2u + (3) = -1 \Rightarrow -2u = -2 \Rightarrow$

$u = 1$. If $u = 1$ and $u = \dfrac{1}{y}$, then $1 = \dfrac{1}{y} \Rightarrow y = 1$. The solution is: $\left\{\left(-1, 1, \dfrac{1}{3}\right)\right\}$.

28. If we let $t = \dfrac{1}{x}$, $u = \dfrac{1}{y}$, and $v = \dfrac{1}{z}$ then our new equations are: $5t - u - 2v = -6$ [1], $-t + 3u - 3v = -12$ [2],

and $2t - u - v = 6$ [3]. First, multiply [equation 1] by 3 and add [equation 2] to eliminate u and produce

[equation 4]. Second, add [equation 1] and [equation 3] multiplied by -1 to eliminate u and produce [equation 5].

$$
\begin{array}{lr}
\begin{array}{rr}
15t - 3u - 6v = -18 & 3[1] \\
-t + 3u - 3v = -12 & [2] \\
\hline
14t \quad\quad - 9v = -30 & [4]
\end{array}
&
\begin{array}{rr}
5t - u - 2v = -6 & [1] \\
-2t + u + v = -6 & 2[3] \\
\hline
3t \quad\quad - v = -12 & [5]
\end{array}
\end{array}
$$

Next add [equation 4] and [equation 5] multiplied by -9 to eliminate v.

$$
\begin{array}{rr}
14t - 9v = -30 & [4] \\
-27t + 9v = 108 & -9[5] \\
\hline
-13t \quad\quad = 78 & \Rightarrow t = -6.
\end{array}
$$

If $t = -6$ and $t = \dfrac{1}{x}$ then $-6 = \dfrac{1}{x} \Rightarrow -6x = 1 \Rightarrow x = -\dfrac{1}{6}$. Substitute $t = -6$ into [equation 5] to solve for v:

$3(-6) - v = -12 \Rightarrow -v = 6 \Rightarrow v = -6$. If $v = -6$ and $v = \dfrac{1}{z}$ then $-6 = \dfrac{1}{z} \Rightarrow -6z = 1 \Rightarrow z = -\dfrac{1}{6}$.

Finally, substitute $v = -6$ and $t = -6$ into [equation 3] to solve for u: $2(-6) - u - (-6) = 6 \Rightarrow -u = 12 \Rightarrow$

$u = -12$. If $u = -12$ and $u = \dfrac{1}{y}$ then $-12 = \dfrac{1}{y} \Rightarrow -12y = 1 \Rightarrow y = -\dfrac{1}{12}$.

The solution is: $\left\{\left(-\dfrac{1}{6}, -\dfrac{1}{12}, -\dfrac{1}{6}\right)\right\}$.

29. Add [equation 1] and [equation 2] to eliminate z and produce [equation 4]:

$$
\begin{array}{ll}
x - 4y + 2z = -2 & [1] \\
\underline{x + 2y - 2z = -3} & [2] \\
2x - 2y = -5 & [4]
\end{array}
$$

Add [equation 4] to [equation 3] multiplied by -2:

$$
\begin{array}{ll}
2x - 2y = -5 & [4] \\
\underline{-2x + 2y = -8} & -2[3] \\
\ 0 = -13
\end{array}
$$

Since this is a false statement, the solution is: $\varnothing$.

30. Add [equation 1] and [equation 2] to eliminate y and produce [equation 4]:

$$
\begin{array}{ll}
2x + y + 3z = 4 & [1] \\
\underline{-3x - y - 4z = 5} & [2] \\
-x \ - z = 9 & [4]
\end{array}
$$

Add [equation 2] and [equation 3] to eliminate y and produce [equation 5]:

$$
\begin{array}{ll}
-3x - y - 4z = 5 & [2] \\
\underline{x + y + 2z = 0} & [3] \\
-2x \ - 2z = 5 & [5]
\end{array}
$$

Multiply [equation 4] by -2 and add [equation 5].

$$
\begin{array}{l}
2x + 2z = -18 \\
\underline{-2x - 2z = 5} \\
\ 0 = -13
\end{array}
$$

Since this is a false statement, the solution is: $\varnothing$.

31. Subtract [equation 3] from [equation 1] to eliminate x and produce equation [4]:

$$
\begin{array}{ll}
x + y + z = 0 & [1] \\
\underline{x + 3y + 3z = 5} & [3] \\
\ -2y - 2z = -5 & [4]
\end{array}
$$

Subtract [equation 3] from [equation 2] to eliminate x and produce equation [5]:

$$
\begin{array}{ll}
x - y - z = 3 & [2] \\
\underline{x + 3y + 3z = 5} & [3] \\
\ -4y - 4z = -2 & [5]
\end{array}
$$

Multiply [equation 4] by 2 and subtract [equation 5]:

$$
\begin{array}{ll}
-4y - 4z = -10 & 2[4] \\
\underline{-4y - 4z = -2} & [5] \\
\ 0 = -8
\end{array}
$$

Since this is a false statement, the solution is: $\varnothing$.

32. There are many other examples, one is given.

    (a) $x - y - z = 0$ when added to $x + y + z = 4$ yields: $2x = 4 \Rightarrow x = 2$.

    $-x + y - z = 0$ when added to $x + y + z = 4$ yields: $2y = 4 \Rightarrow y = 2$. This one solution is: $(2, 2, 0)$.

    (b) $x - y - z = 1; -x - y - z = 0$ when added to $x + y + z = 4$ yields: $0 = 4$, which has no solution.

    (c) $x - y - z = 2; 2x + 2y + 2z = 8$ when added to $x + y + z = 4$ multiplied by -2, yields: $0 = 0$, which has infinitely many solutions.

33. Using $x + 2y + z = 4$ and $z = -y + 3$ from example 2, a solution set with x arbitrary is:

    Solving for y: $x + 2y + (-y + 3) = 4 \Rightarrow x + y = 1 \Rightarrow y = 1 - x$.

    Solving for z: $z = -(1 - x) + 3 \Rightarrow z = x + 2$. Therefore the solution: $\{(x, 1 - x, x + 2)\}$.

34. For example, two perpendicular walls meeting in a corner with the floor intersect in a point.

35. For example, three pages of a book intersect in a line.

36. First eliminate x by multiplying [equation 1] by -2 and adding [equation 2].

    $$\begin{array}{r} -2x + 4y - 6z = -12 \\ 2x - y + 2z = 5 \\ \hline 3y - 4z = -7 \end{array}$$ Now solve this for y: $3y - 4z = -7 \Rightarrow 3y = 4z - 7 \Rightarrow y = \dfrac{4z - 7}{3}$.

    Now substitute this into [equation 2]: $2x - \dfrac{4z - 7}{3} + 2z = 5 \Rightarrow 3\left(2x - \dfrac{4z - 7}{3} + 2z = 5\right) \Rightarrow$

    $6x - 4z + 7 + 6z = 15 \Rightarrow 6x + 2z = 8 \Rightarrow x = \dfrac{8 - 2z}{6} \Rightarrow x = \dfrac{4 - z}{3}$.

    We have infinitely many solutions: $\left\{\left(\dfrac{4 - z}{3}, \dfrac{4z - 7}{3}, z\right)\right\}$.

37. First eliminate x by adding [equation 1] to [equation 2] multiplied by -3:

    $$\begin{array}{r} 3x + 4y - z = 13 \\ -3x - 3y - 6z = -45 \\ \hline y - 7z = -32 \end{array}$$ Now solve this for y: $y - 7z = -32 \Rightarrow y = 7z - 32$.

    Now substitute this into [equation 2]: $x + (7z - 32) + 2z = 15 \Rightarrow x + 9z = 47 \Rightarrow x = 47 - 9z$.

    We have infinitely many solutions: $\{(47 - 9z, 7z - 32, z)\}$.

38. First eliminate y by adding {equation 1} to [equation 2] multiplied by 4:

    $$\begin{array}{r} 5x - 4y + z = 9 \\ 4x + 4y = 60 \\ \hline 9x + z = 69 \end{array}$$ Now solve this for x: $9x + z = 69 \Rightarrow 9x = 69 - z \Rightarrow x = \dfrac{69 - z}{9}$.

    Now substitute this into [equation 2]: $\dfrac{69 - z}{9} + y = 15 \Rightarrow 9\left(\dfrac{69 - z}{9} + y = 15\right) \Rightarrow 69 - z + 9y = 135 \Rightarrow$

    $-z + 9y = 66 \Rightarrow 9y = z + 66 \Rightarrow y = \dfrac{z + 66}{9}$. We have infinitely many solutions: $\left\{\left(\dfrac{69 - z}{9}, \dfrac{z + 66}{9}, z\right)\right\}$.

39. First eliminate y by adding [equation 1] to [equation 2] multiplied by 5:

    $$\begin{array}{r} 3x - 5y - 4z = -7 \\ 5y - 5z = -65 \\ \hline 3x - 9z = -72 \end{array}$$ Now solve this for x: $3x - 9z = -72 \Rightarrow x - 3y = -24 \Rightarrow x = 3z - 24$.

    Now substitute this into [equation 1]: $3(3z - 24) - 5y - 4z = -7 \Rightarrow 9z - 72 - 5y - 4z = -7 \Rightarrow$

    $5z - 5y = 65 \Rightarrow z - y = 13 \Rightarrow -y = -z + 13 \Rightarrow y = z - 13$.

    We have infinitely many solutions: $\{(3z - 24, z - 13, z)\}$.

40. First eliminate y by adding [equation 1] to [equation 2]:

$$
\begin{array}{rrrr}
x - & y + & z = & -6 \\
4x + & y + & z = & 7 \\
\hline
5x & & + 2z = & 1
\end{array}
$$

Now solve this for x: $5x + 2z = 1 \Rightarrow 5x = 1 - 2z \Rightarrow x = \dfrac{1 - 2z}{5}$.

Now substitute this into [equation 1]: $\dfrac{1 - 2z}{5} - y + z = -6 \Rightarrow 5\left(\dfrac{1 - 2z}{5} - y + z = -6\right) \Rightarrow$

$1 - 2z - 5y + 5z = -30 \Rightarrow -5y + 3z = -31 \Rightarrow -5y = -3z - 31 \Rightarrow y = \dfrac{3z + 31}{5}$.

We have infinitely many solutions: $\left\{\left(\dfrac{1 - 2z}{5}, \dfrac{3z + 31}{5}, z\right)\right\}$.

41. First eliminate y by multiplying [equation 1] by 2 and adding [equation 2]:

$$
\begin{array}{rrrr}
6x - & 4y + & 2z = & 30 \\
x + & 4y - & z = & 11 \\
\hline
7x & & + z = & 41
\end{array}
$$

Now solve this for x: $7x + z = 41 \Rightarrow 7x = 41 - z \Rightarrow x = \dfrac{41 - z}{7}$.

Now substitute this into [equation 2]: $\dfrac{41 - z}{7} + 4y - z = 11 \Rightarrow 7\left(\dfrac{41 - z}{7} + 4y - z = 11\right) \Rightarrow$

$41 - z + 28y - 7z = 77 \Rightarrow 28y - 8z = 36 \Rightarrow 28y = 8z + 36 \Rightarrow y = \dfrac{8z + 36}{28} \Rightarrow y = \dfrac{2z + 9}{7}$.

We have infinitely many solutions: $\left\{\left(\dfrac{41 - z}{7}, \dfrac{2z + 9}{7}, z\right)\right\}$.

42. Subtract the third equation from the first:

$$
\begin{array}{rrrr}
x + & 3y + & z = & 6 \\
x - & y - & z = & 0 \\
\hline
& 4y + & 2z = & 6
\end{array}
$$

Multiply the third equation by 3 and subtract it from the second equation:

$$
\begin{array}{rrrr}
3x + & y - & z = & 6 \\
3x - & 3y - & 3z = & 0 \\
\hline
& 4y + & 2z = & 6
\end{array}
$$

Subtracting these two equations we get:

$$
\begin{array}{rl}
4y + 2z = & 6 \\
4y + 2z = & 6 \\
\hline
0 = & 0
\end{array}
$$

Therefore infinitely many solutions and $4y + 2z = 6 \Rightarrow 4y = -2z + 6 \Rightarrow y = \dfrac{-z + 3}{2}$.

Adding the last two original equations we get:

$$
\begin{array}{rrrr}
3x + & y - & z = & 6 \\
x - & y - & z = & 0 \\
\hline
4x & & - 2z = & 6
\end{array}
$$

$\Rightarrow 4x = 2z + 6 \Rightarrow x = \dfrac{z + 3}{2}$. We have infinitely many solutions: $\left\{\left(\dfrac{z + 3}{2}, \dfrac{-z + 3}{2}, z\right)\right\}$.

43. Add the first two equations:

$$2x - y + 2z = 6$$
$$\underline{-x + y + z = 0}$$
$$x + 3z = 6$$

Add this equation to the third equation:

$$x + 3z = 6$$
$$\underline{-x - 3z = -6}$$
$$0 = 0$$

Therefore infinitely many solutions.

$$x + 3z = 6 \Rightarrow x = -3z + 6$$

Multiply the second equation by 2 and add to the first equation:

$$2x - y + 2z = 6$$
$$\underline{-2x + 2y + 2z = 0}$$
$$y + 4z = 6 \Rightarrow y = -4z + 6$$

We have infinitely many solutions: $\{(-3z + 6, -4z + 6, z)\}$.

44. Multiply the first equation by 3 and subtract the second equation:

$$3x + 6y + 3z = 0$$
$$\underline{3x + 2y - z = 4}$$
$$4y + 4z = -4$$

Add the first and third equations:

$$x + 2y + z = 0$$
$$\underline{-x + 2y + 3z = -4}$$
$$4y + 4z = -4$$

Subtracting the two new equations we get:

$$4y + 4z = -4$$
$$\underline{4y + 4z = -4}$$
$$0 = 0.$$

Therefore, we have infinitely many solutions.

Then, $4y + 4z = -4 \Rightarrow 4y = -4z - 4 \Rightarrow y = -z - 1.$

Subtracting the first two equations we get:

$$x + 2y + z = 0$$
$$\underline{3x + 2y - z = 4}$$
$$-2x + 2z = -4$$

Then, $-2x + 2z = -4 \Rightarrow -2x = -2z - 4 \Rightarrow x = z + 2.$

We have infinitely many solutions: $\{(z + 2, -z - 1, z)\}$.

45. Let x = number of gallons of $9 grade, y = number of gallons of $3 grade, and z = number of gallons of $4.50 grade.

    From the information the equations are: $x + y + z = 300$ [1], $9x + 3y + 4.5z = 6(300)$ [2], and

    $z = 2y$ [3]. Substitute [equation 3] into both [equation 1] and [equation 2] to produce equations [4] and [5].

    [3] into [1]: $x + y + 2y = 300 \Rightarrow x + 3y = 300$ [4].

    [3] into [2]: $9x + 3y + 4.5(2y) = 1800 \Rightarrow 9x + 12y = 1800$ [5].

    Now add {equation 5] to [equation 4] multiplied by -9:

    $$\begin{aligned} 9x + 12y &= 1800 \\ \underline{-9x - 27y} &= \underline{-2700} \\ -15y &= -900 \end{aligned}$$

    $\Rightarrow y = 60$. Substitute $y = 60$, into [3]: $z = 2(60) \Rightarrow z = 120$.

    Finally, substitute $y = 60$ and $z = 120$ into [1]: $x + 60 + 120 = 300 \Rightarrow x = 120$.

    She should use: 120 gallons of $9 water, 60 gallons of $3 water, and 120 gallons of $4.50 water.

46. Let x = number of pennies, y = number of nickels, and z = number of quarters.

    From the information the equations are: $x + y + z = 29$ [1], $z = x - 8$ [2], and

    $x + 5y + 25z = 177$ [3].

    Multiply {equation 1] by -5 and add to [equation 3] to eliminate y and produce [equation 4].

    $$\begin{aligned} -5x - 5y - 5z &= -145 \\ \underline{x + 5y + 25z} &= \underline{177} \\ -4x + 20z &= 32 \end{aligned}$$

    $\Rightarrow -x + 5z = 8$. Now substitute $z = x - 8$ into this equation:

    $-x + 5(x - 8) = 8 \Rightarrow -x + 5x - 40 = 8 \Rightarrow 4x = 48 \Rightarrow x = 12$. Substitute $x = 12$ into $z = x - 8$:

    $z = 12 - 8 \Rightarrow z = 4$. Finally, substitute $x = 12$ and $z = 4$ into [equation 1]: $12 + y + 4 = 29 \Rightarrow y = 13$.

    There were 12 pennies, 13 nickels, and 4 quarters.

47. Let x = price of Up close tickets, y = price of Middle tickets, and z = price of Farther Back tickets.

    From the information the equations are: $x = 6 + y$ [1], $y = 3 + z$ [2], and $2x = 3 + 3z$ [3].

    First substitute [equation 1] into [equation 3]: $2(6 + y) = 3 + 3z$. Now substitute [equation 2] into this new

    equation and solve for z: $2(6 + (3 + z)) = 3 + 3z \Rightarrow 2(9 + z) = 3 + 3z \Rightarrow 18 + 2z = 3 + 3z \Rightarrow$

    $z = 15$. Substitute $z = 15$ into [2]: $y = 3 + (15) \Rightarrow y = 18$. Finally, substitute $y = 18$ into [1]:

    $x = 6 + 18 \Rightarrow x = 24$. The Up close ticket was: $24; Middle ticket was: $18; and the

    Farther Back ticket was: $15.

48. Let x = number $150 barrels, y = number $190 barrels, and z = number $120 barrels.

    From the following information the equations are:

    Same number of $150 barrels as $190 barrels: $x = y$ [1].

    $120 glue is mixture of other three: $z = 150 + x + y$ [2].

    Total value of the glue: $150(100) + 150x + 190y = 120z$ [3].

    First substitute [1] into [2]: $z = 150 + x + x \Rightarrow z = 150 + 2x$ [4].

    Substitute {1] and [4] into [3]: $15,000 + 150x + 190x = 120(150 + 2x) \Rightarrow$

    $15,000 + 340x = 18,000 + 240x \Rightarrow 100x = 3000 \Rightarrow x = 30$. If $x = 30$ then by [1] $y = 30$.

    Finally, substitute $x = 30$ into [4]: $z = 150 + 2(30) \Rightarrow z = 210$.

    The company needs: 30 barrels of $150 glue and 30 barrels of $190 glue to produce 210 barrels of $120 glue.

49. Let x = measure of the largest angle, y = measure of the medium angle, and z = measure of the smallest angle.

From the information the equations are: $a + b + c = 180$ [1], $a = 2b - 55$ [2], and $c = b - 25$ [3].

Substitute both [2] and [3] into [1]: $(2b - 55) + b + (b - 25) = 180 \Rightarrow 4b - 80 = 180 \Rightarrow$

$4b = 260 \Rightarrow b = 65$. Substitute $b = 65$ into [3]: $c = 65 - 25 \Rightarrow c = 40$; and substitute $b = 65$ into [2]:

$a = 2(65) - 55 \Rightarrow a = 130 - 55 \Rightarrow a = 75$. The angles are: $75°, 65°$, and $40°$.

50. Let x = length of the longest side, y = length of the medium side, and z = length of the shortest side.

From the information the equations are: $a + b + c = 59$ [1], $a = 11 + b$ [2], and $b = 3 + c$ [3].

Substitute both [2] and [3] into [1]: $(11 + b) + (3 + c) + c = 59 \Rightarrow 11 + (3 + c) + (3 + c) + c = 59 \Rightarrow$

$17 + 3c = 59 \Rightarrow 3c = 42 \Rightarrow c = 14$. Substitute $c = 14$ into [3]: $b = 3 + 14 \Rightarrow b = 17$; and substitute

$b = 17$ into [2]: $a = 11 + 17 \Rightarrow a = 28$. The lengths are: 28 inches, 17 inches, and 14 inches.

51. Let x = the amount invested at 4%, y = the amount invested at 4.5%, and z = the amount invested at 2.5%.

From the information the equations are: $x + y + z = 10000$ [1], $.04x + .025y + .045z = 415$ [2], and

$y = 2x$ [3]. Substitute [equation 3] into [equation 1] and [equation 3] into [equation 2] multiplied by 1000,

to produce equations [4] and [5]:

[3] into [1]: $x + 2x + z = 10,000 \Rightarrow 3x + z = 10,000$ [4].

[3] into 1000[2]: $40x + 45(2x) + 25z = 415,000 \Rightarrow 130x + 25z = 415,000 \Rightarrow 26x + 5z = 83,000.$ [5].

Now multiply [equation 4] by -5 and add to [equation 5]:

$$
\begin{array}{rcr}
-15x - 5y &=& -50,000 \\
26x + 5y &=& 83,000 \\
\hline
11x &=& 33,000
\end{array}
$$

$\Rightarrow x = 3000$. Substitute $x = 3000$, into $[3]$: $y = 2(3000) \Rightarrow y = 6000$.

Finally, substitute $x = 3000$ and $y = 6000$ into $[1]$: $3000 + 6000 + z = 10,000 \Rightarrow z = 1000$.

He invested $3000 at 4%, $6000 at 4.5%, and $1000 at 2.5%.

52. Let x = the amount invested at 5%, y = the amount invested at 4.5%, and z = the amount invested at 3.75%.

From the information the equations are: $x + y + z = 100,000$ [1], $.05x + .045y + .0375z = 4450$ [2], and

$z = x + y - 20,000$ [3].

First, multiply [equation 1] by -5 and add [equation 2] multiplied by 100 to eliminate x and produce

[equation 4], then simplify.

Second, add [equation 1] and [equation 3] to eliminate x and y.

$$
\begin{array}{rcrl}
-5x - 5y - 5z &=& -500,000 & \quad -5[1] \\
5x + 4.5y + 3.75z &=& 445,000 & \quad 100[2] \\
\hline
-.5y - 1.25z &=& -55,000 & \quad [4] \\
\Rightarrow 2y + 5z &=& 220,000 & \quad [4a]
\end{array}
\qquad
\begin{array}{rcrl}
x + y + z &=& 100,000 & \quad [1] \\
-x - y + z &=& -20,000 & \quad [3] \\
\hline
2z &=& 80,000 & \quad [5] \\
\Rightarrow z &=& 40,000 &
\end{array}
$$

Now substitute $z = 40,000$ into [4a]: $2y + 5(40,000) = 220,000 \Rightarrow 2y = 20,000 \Rightarrow y = 10,000$.

Finally substitute $y = 10,000$ and $z = 40,000$ into [1]: $x + 10,000 + 40,000 = 100,000 \Rightarrow x = 50,000$.

He invested $50,000 at 5%, $10,000 at 4.5%, and $40,000 at 3.75%.

53. Let x = the number of EZ models, y = the number of Compact models, and z = the number of Commercial

    models. From the information the two equations are:

    (weight) $10x + 20y + 60z = 440$ [1] and (volume)$10x + 8y + 28z = 248$ [2].

    First, subtract [2] from [1].

    $$\begin{array}{rl} 10x + 20y + 60z = 440 & \quad [1] \\ \underline{10x + 8y + 28z = 248} & \quad [2] \\ 12y + 32z = 192 & \end{array}$$

    $\Rightarrow 12y = -32z + 192 \Rightarrow y = -\dfrac{8}{3}z + 16.$

    Now substitute into [1]: $10x + 20\left(-\dfrac{8}{3}z + 16\right) + 60z = 440 \Rightarrow 3\left[10x + 20\left(-\dfrac{8}{3}z + 16\right) + 60z = 440\right] \Rightarrow$

    $30x - 160z + 960 + 180z = 1320 \Rightarrow 30x = -20z + 360 \Rightarrow x = -\dfrac{2}{3}z + 12.$

    From the equations for x and y, z must be a multiple of 3 for x and y to be whole numbers.

    There are three possibilities:

    If $z = 0$, then $y = -\dfrac{8}{3}(0) + 16 = 16$ and $x = -\dfrac{2}{3}(0) + 12 = 12$. Therefore, 12 EZ, 16 Comp., and 0 Comm.

    If $z = 3$, then $y = -\dfrac{8}{3}(3) + 16 = 8$ and $x = -\dfrac{2}{3}(3) + 12 = 10$. Therefore, 10 EZ, 8 Comp., and 3 Comm.

    If $z = 6$, then $y = -\dfrac{8}{3}(6) + 16 = 0$ and $x = -\dfrac{2}{3}(6) + 12 = 8$. Therefore, 8 EZ, 0 Comp., and 6 Comm.

    If $z \geq 9$, then y is less than zero, therefore not possible.

54. Let b = the number of buffets, c = the number of chairs, and t = the number of tables.

    From the information the two equations are:

    (construction) $30b + 10c + 10t = 350$ [1] and (finishing)$10b + 10c + 30t = 150$ [2].

    First, subtract [2] from [1].

    $$\begin{array}{rl} 30b + 10c + 10t = 350 & \quad [1] \\ \underline{10b + 10c + 30t = 150} & \quad [2] \\ 20b - 20t = 200 & \end{array}$$

    $\Rightarrow b - t = 10 \Rightarrow b = t + 10.$

    Now substitute $b = t + 10$ into [2]: $10(t + 10) + 10c + 30t = 150 \Rightarrow 10t + 100 + 10c + 30t = 150 \Rightarrow$

    $10c = 50 - 40t \Rightarrow c = 5 - 4t.$

    The solution of the arbitrary variable t is: $(t + 10, 5 - 4t, t)$.

    There are 2 possibilities:

    If $t = 0$, then $b = 0 + 10 = 10$, and $c = 5 - 4(0) = 5$. Therefore, 10 buffets, 5 chairs, and 0 tables.

    If $t = 1$, then $b = 1 + 10 = 11$, and $c = 5 - 4(1) = 1$. Therefore, 11 buffets, 1 chairs, and 1 tables.

    If $t \geq 2$, then c is less than 0, therefore not possible.

55. (a) $a + 20b + 2c = 190$
$a + 5b + 3c = 320$
$a + 40b + c = 50$

(b) Using technology, the solution is $(30, -2, 100)$. So the equation is $P = 30 - 2A + 100S$.

(c) When $A = 10$ and $S = 2500$, $P = 30 - 2(10) + 100(2.5) = 260 \Rightarrow \$260,000$.

56. Using the form $y = ax^2 + bx + c$ and the given points yields the following equations:

For $(2, 9)$: $9 = a(2)^2 + b(2) + c \Rightarrow 4a + 2b + c = 9$.   [equation 1]

For $(-2, 1)$: $1 = a(-2)^2 + b(-2) + c \Rightarrow 4a - 2b + c = 1$.   [equation 2]

For $(-3, 4)$: $4 = a(-3)^2 + b(-3) + c \Rightarrow 9a - 3b + c = 4$.   [equation 3]

First, add [equation 1] and [equation 2] multiplied by -1 to eliminate both the a and c:

Second, add [equation 2] and [equation 3] multiplied by -1 to eliminate c and produce [equation 4].

$$
\begin{array}{lll}
4a + 2b + c = 9 & [1] \\
\underline{-4a + 2b - c = -1} & -1[2] \\
4b = 8 & \Rightarrow b = 2.
\end{array}
\qquad
\begin{array}{lll}
4a - 2b + c = 1 & [2] \\
\underline{-9a + 3b - c = -4} & -1[3] \\
-5a + b = -3 & [4]
\end{array}
$$

Substitute $b = 2$ into [4]: $-5a + (2) = -3 \Rightarrow -5a = -5 \Rightarrow a = 1$.

Finally, substitute $a = 1$ and $b = 2$ into [1]: $4(1) + 2(2) + c = 9 \Rightarrow c = 1$.

The equation is: $y = x^2 + 2x + 1$. Plot the points and equation in the same window. See Figure 56.

57. Using the form $y = ax^2 + bx + c$ and the given points yields the following equations:

For $(1.5, 6.25)$: $6.25 = a(1.5)^2 + b(1.5) + c \Rightarrow 2.25a + 1.5b + c = 6.25$.   [equation 1]

For $(0, -2)$: $-2 = a(0)^2 + b(0) + c \Rightarrow c = -2$.   [equation 2]

For $(-1.5, 3.25)$: $3.25 = a(-1.5)^2 + b(-1.5) + c \Rightarrow 2.25a - 1.5b + c = 3.25$.   [equation 3]

Substitute $c = -2$ into [1]: $2.25a + 1.5b - 2 = 6.25 \Rightarrow 2.25a + 1.5b = 8.25$.   [equation 4]

Substitute $c = -2$ into [3]: $2.25a - 1.5b - 2 = 3.25 \Rightarrow 2.25a - 1.5b = 5.25$.   [equation 5]

Now add [equation 4] and [equation 5]:

$$
\begin{array}{lll}
2.25a + 1.5b = 8.25 & [4] \\
\underline{2.25a - 1.5b = 5.25} & [5] \\
4.5a = 13.5 & \Rightarrow a = 3.
\end{array}
$$

Finally, substitute $a = 3$ and $c = -2$ into [1]: $6.25 = 2.25(3) + 1.5b + (-2) \Rightarrow b = 1$.

The equation is: $y = 3x^2 + x - 2$. Plot the points and equation in the same window. See Figure 57.

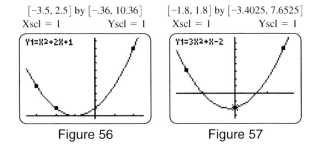

$[-3.5, 2.5]$ by $[-.36, 10.36]$ $[-1.8, 1.8]$ by $[-3.4025, 7.6525]$
Xscl = 1   Yscl = 1   Xscl = 1   Yscl = 1

Figure 56   Figure 57

58. Using the form $y = ax^2 + bx + c$ and the given points yields the following equations:

    For $(2, 14)$: $14 = a(2)^2 + b(2) + c \Rightarrow 4a + 2b + c = 14.$ [equation 1]

    For $(0, 0)$: $0 = a(0)^2 + b(0) + c \Rightarrow c = 0.$ [equation 2]

    For $(-1, -1)$: $-1 = a(-1)^2 + b(-1) + c \Rightarrow a - b + c = -1.$ [equation 3]

    Substitute $c = 0$ into [1]: $4a + 2b + 0 = 14 \Rightarrow 4a + 2b = 14.$ [equation 4]

    Substitute $c = 0$ into [3]: $a - b + 0 = -1 \Rightarrow a - b = -1.$ [equation 5]

    Now add [equation 4] and [equation 5] multiplied by 2:

    $$
    \begin{array}{rll}
    4a + 2b & = 14 & [4] \\
    \underline{2a - 2b} & \underline{= -2} & 2[5] \\
    6a & = 12 & \Rightarrow a = 2.
    \end{array}
    $$

    Finally, substitute $a = 2$ and $c = 0$ into [1]: $4(2) + 2b + (0) = 14 \Rightarrow 2b = 6 \Rightarrow b = 3.$

    The equation is: $y = 2x^2 + 3x.$ Plot the points and graph the equation in the same window. See Figure 58.

    $[-1.3, 2.3]$ by $[-3.55, 16.55]$
    Xscl = 1   Yscl = 1

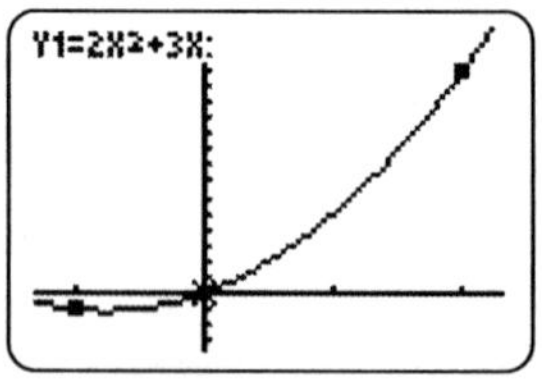

    Figure 58

59. Using the form $Y_1 = aX^2 + bX + c$ and several of the given points yields the following equations:

    For $(0, .56)$: $.56 = a(0)^2 + b(0) + c \Rightarrow c = .56.$ [equation 1]

    For $(1, .8)$: $.8 = a(1)^2 + b(1) + c \Rightarrow a + b + c = .8.$ [equation 2]

    For $(2, 1.98)$: $1.98 = a(2)^2 + b(2) + c \Rightarrow 4a + 2b + c = 1.98.$ [equation 3]

    Substitute $c = .56$ into [2]: $a + b + .56 = .8 \Rightarrow a + b = .24.$ [equation 4]

    Substitute $c = .56$ into [3]: $4a + 2b + .56 = 1.98 \Rightarrow 4a + 2b = 1.42.$ [equation 5]

    Now add [equation 5] and [equation 4] multiplied by -2:

    $$
    \begin{array}{rll}
    4a + 2b & = 1.42 & [5] \\
    \underline{-2a - 2b} & \underline{= -.48} & -2[4] \\
    2a & = .94 & \Rightarrow a = .47.
    \end{array}
    $$

    Finally, substitute $a = .47$ into [4]: $(.47) + b = .24 \Rightarrow b = -.23.$ The equation is: $Y_1 = .47X^2 - .23X + .56.$

60. Using the form $x^2 + y^2 + ax + by + c = 0$ and the given points yields the following equations:

For $(-1, 3)$: $(-1)^2 + (3)^2 + a(-1) + b(3) + c = 0 \Rightarrow 1 + 9 - a + 3b + c = 0 \Rightarrow$

$-a + 3b + c = -10$.   [equation 1]

For $(6, 2)$: $(6)^2 + (2)^2 + a(6) + b(2) + c = 0 \Rightarrow 36 + 4 + 6a + 2b + c = 0 \Rightarrow$

$6a + 2b + c = -40$.   [equation 2]

For $(-2, -4)$: $(-2)^2 + (-4)^2 + a(-2) + b(-4) + c = 0 \Rightarrow 4 + 16 - 2a - 4b + c = 0 \Rightarrow$

$-2a - 4b + c = -20$.   [equation 3]

Now, subtract [equation 1] from [equation 2] to eliminate c and produce [equation 4].

Then, subtract [equation 3] from [equation 2] to eliminate c and produce [equation 5].

$$
\begin{array}{rl}
6a + 2b + c = -40 & [2] \\
\underline{-a + 3b + c = -10} & [1] \\
7a - b = -30 & [4]
\end{array}
\qquad
\begin{array}{rl}
6a + 2b + c = -40 & [2] \\
\underline{-2a - 4b + c = -20} & [3] \\
8a + 6b = -20 & [5]
\end{array}
$$

Next multiply [equation 4] by 6 and add [equation 5] to eliminate b.

$$
\begin{array}{rl}
42a - 6b = -180 & [4] \\
\underline{8a + 6b = -20} & [5] \\
50a = -200 & \Rightarrow a = -4.
\end{array}
$$

Substitute $a = -4$ into [4]: $7(-4) - b = -30 \Rightarrow -b = -2 \Rightarrow b = 2$. Finally, substitute $a = -4$ and $b = 2$

into [1]: $-(-4) + 3(2) + c = -10 \Rightarrow c = -20$. The equation is: $x^2 + y^2 - 4x + 2y - 20 = 0$.

61. Using the form $x^2 + y^2 + ax + by + c = 0$ and the given points yields the following equations:

For $(-1, 5)$: $(-1)^2 + (5)^2 + a(-1) + b(5) + c = 0 \Rightarrow 1 + 25 - a + 5b + c = 0 \Rightarrow$

$-a + 5b + c = -26$.   [equation 1]

For $(6, 6)$: $(6)^2 + (6)^2 + a(6) + b(6) + c = 0 \Rightarrow 36 + 36 + 6a + 6b + c = 0 \Rightarrow$

$6a + 6b + c = -72$.   [equation 2]

For $(7, -1)$: $(7)^2 + (-1)^2 + a(7) + b(-1) + c = 0 \Rightarrow 49 + 1 + 7a - b + c = 0 \Rightarrow$

$7a - b + c = -50$.   [equation 3]

Now, subtract [equation 1] from [equation 2] to eliminate c and produce [equation 4].

Then, subtract [equation 3] from [equation 2] to eliminate c and produce [equation 5].

$$
\begin{array}{rl}
6a + 6b + c = -72 & [2] \\
\underline{-a + 5b + c = -26} & [1] \\
7a + b = -46 & [4]
\end{array}
\qquad
\begin{array}{rl}
6a + 6b + c = -72 & [2] \\
\underline{7a - b + c = -50} & [3] \\
-a + 7b = -22 & [5]
\end{array}
$$

Next multiply [equation 5] by 7 and add [equation 4] to eliminate a.

$$
\begin{array}{rl}
-7a + 49b = -154 & [5] \\
\underline{7a + b = -46} & [4] \\
50b = -200 & \Rightarrow b = -4.
\end{array}
$$

Substitute $b = -4$ into [4]: $7a + (-4) = -46 \Rightarrow 7a = -42 \Rightarrow a = -6$.

Finally, substitute $a = -6$ and $b = -4$ into [1]: $-(-6) + 5(-4) + c = -26 \Rightarrow c = -12$.

The equation is: $x^2 + y^2 - 6x - 4y - 12 = 0$.

62. Using the form $x^2 + y^2 + ax + by + c = 0$ and the given points yields the following equations:

For $(2, 1)$: $(2)^2 + (1)^2 + a(2) + b(1) + c = 0 \Rightarrow 4 + 1 + 2a + b + c = 0 \Rightarrow$

$2a + b + c = -5.$ [equation 1]

For $(-1, 0)$: $(-1)^2 + (0)^2 + a(-1) + b(0) + c = 0 \Rightarrow 1 + 0 - a + 0b + c = 0 \Rightarrow$

$-a + c = -1.$ [equation 2]

For $(3, 3)$: $(3)^2 + (3)^2 + a(3) + b(3) + c = 0 \Rightarrow 9 + 9 + 3a + 3b + c = 0 \Rightarrow$

$3a + 3b + c = -18.$ [equation 3]

Now, multiply [equation 1] by -3 and add [equation 3] to eliminate b and produce [equation 4].

Next multiply [equation 2] by 2 and add [equation 4] to eliminate c.

$$
\begin{array}{ll}
-6a - 3b - 3c = 15 & -3[1] \\
\underline{3a + 3b + c = -18} & [3] \\
-3a \quad\quad - 2c = -3 & [4]
\end{array}
\qquad\qquad
\begin{array}{ll}
-2a + 2c = -2 & 2[2] \\
\underline{-3a + 2c = -3} & [4] \\
-5a \quad\quad = -5 & \Rightarrow\ a = 1.
\end{array}
$$

Substitute $a = 1$ into [2]: $-(1) + c = -1 \Rightarrow c = 0$. Finally, substitute $a = 1$ and $c = 0$ into [1]:

$2(1) + b + 0 = -5 \Rightarrow b = -7$. The equation is: $x^2 + y^2 + x - 7y = 0$.

63. Using the form $x^2 + y^2 + ax + by + c = 0$ and the given points yields the following equations:

For $(0, 3)$: $(0)^2 + (3)^2 + a(0) + b(3) + c = 0 \Rightarrow 0 + 9 + 0a + 3b + c = 0 \Rightarrow$

$3b + c = -9.$ [equation 1]

For $(4, 2)$: $(4)^2 + (2)^2 + a(4) + b(2) + c = 0 \Rightarrow 16 + 4 + 4a + 2b + c = 0 \Rightarrow$

$4a + 2b + c = -20.$ [equation 2]

For $(-5, -2)$: $(-5)^2 + (-2)^2 + a(-5) + b(-2) + c = 0 \Rightarrow 25 + 4 - 5a - 2b + c = 0 \Rightarrow$

$-5a - 2b + c = -29.$ [equation 3]

Now, multiply [equation 2] by 5 and add [equation 3] multiplied by 4 to eliminate a and produce [equation 4].

Next multiply [equation 1] by -9 and add [equation 4] to eliminate c.

$$
\begin{array}{ll}
20a + 10b + 5c = -100 & 5[2] \\
\underline{-20a - 8b + 4c = -116} & 4[3] \\
2b + 9c = -216 & [4]
\end{array}
\qquad\qquad
\begin{array}{ll}
-27b - 9c = 81 & -9[1] \\
\underline{2b + 9c = -216} & [4] \\
-25b \quad\quad = -135 & \Rightarrow\ b = 5.4.
\end{array}
$$

Substitute $b = 5.4$ into [1]: $3(5.4) + c = -9 \Rightarrow c = -25.2$.

Finally, substitute $b = 5.4$ and $c = -25.2$ into [2]: $4a + 2(5.4) - 25.2 = -20 \Rightarrow 4a = -5.6 \Rightarrow a = -1.4$.

The equation is: $x^2 + y^2 - 1.4x + 5.4y - 25.2 = 0$.

64. Using the form $s(t) = at^2 + bt + c$ and the given points yields the following equations:

For $(0, 5)$: $5 = a(0)^2 + b(0) + c \Rightarrow c = 5$.

For $(1, 23)$: $23 = a(1)^2 + b(1) + c \Rightarrow a + b + c = 23$.   [equation 1]

For $(2, 37)$: $37 = a(2)^2 + b(2) + c \Rightarrow 4a + 2b + c = 37$.   [equation 2]

First, multiply [equation 1] by -2 and add [equation 2] to eliminate b and produce [equation 3]:

$$\begin{array}{ll} -2a - 2b - 2c = -46 & -2[1] \\ \underline{4a + 2b + c = 37} & [2] \\ 2a - c = -9 & [3] \end{array}$$

Substitute $c = 5$ into [4]: $2a - (5) = -9 \Rightarrow 2a = -4 \Rightarrow a = -2$.

Finally, substitute $a = -2$ and $c = 5$ into [1]: $(-2) + b + (5) = 23 \Rightarrow b = 20$.

The equation is: $s(t) = -2t^2 + 20t + 5$, and $s(8) = -2(8)^2 + 20(8) + 5 = -128 + 160 + 5 = 37$.

65. Using the form $s(t) = at^2 + bt + c$ and the given points yields the following equations:

For $(0, -10)$: $-10 = a(0)^2 + b(0) + c \Rightarrow c = -10$.

For $(1, 6)$: $6 = a(1)^2 + b(1) + c \Rightarrow a + b + c = 6$.   [equation 1]

For $(2, 30)$: $30 = a(2)^2 + b(2) + c \Rightarrow 4a + 2b + c = 30$.   [equation 2]

First, multiply [equation 1] by -2 and add [equation 2] to eliminate b and produce [equation 3]:

$$\begin{array}{ll} -2a - 2b - 2c = -12 & -2[1] \\ \underline{4a + 2b + c = 30} & [2] \\ 2a - c = 18 & [3] \end{array}$$

Substitute $c = -10$ into [4]: $2a - (-10) = 18 \Rightarrow 2a = 8 \Rightarrow a = 4$.

Finally, substitute $a = 4$ and $c = -10$ into [1]: $(4) + b + (-10) = 6 \Rightarrow b = 12$.

The equation is: $s(t) = 4t^2 + 12t - 10$, and $s(10) = 4(10)^2 + 12(10) - 10 = 400 + 120 - 10 = 510$.

7.3: Solution of Linear Systems by Row Transformations

1. For $\begin{bmatrix} 2 & 4 \\ 4 & 7 \end{bmatrix}$, $-2R_1 \rightarrow \begin{bmatrix} -4 & -8 \\ 4 & 7 \end{bmatrix}$.

2. For $\begin{bmatrix} -1 & 4 \\ 7 & 0 \end{bmatrix}$, $7R_1 \rightarrow \begin{bmatrix} -7 & 28 \\ 7 & 0 \end{bmatrix}$.

3. For $\begin{bmatrix} 1 & 5 & 6 \\ -2 & 3 & -1 \\ 4 & 7 & 0 \end{bmatrix}$, $R_1 + R_2 \rightarrow \begin{bmatrix} 1 & 5 & 6 \\ -1 & 8 & 5 \\ 4 & 7 & 0 \end{bmatrix}$.

4. For $\begin{bmatrix} 2 & 5 & 6 \\ 4 & -1 & 2 \\ 3 & 7 & 1 \end{bmatrix}$, $R_3 + R_1 \rightarrow \begin{bmatrix} 5 & 12 & 7 \\ 4 & -1 & 2 \\ 3 & 7 & 1 \end{bmatrix}$.

5. For $\begin{bmatrix} -3 & 1 & -4 \\ 2 & 1 & 3 \\ -7 & 5 & 2 \end{bmatrix}$, $-5R_2 + R_3 \rightarrow \begin{bmatrix} -3 & 1 & -4 \\ 2 & 1 & 3 \\ -17 & 0 & -13 \end{bmatrix}$.

6. For $\begin{bmatrix} 4 & 10 & -8 \\ 7 & 4 & 3 \\ -1 & 1 & 0 \end{bmatrix}$, $-4R_3 + R_2 \rightarrow \begin{bmatrix} 4 & 10 & -8 \\ 11 & 0 & 3 \\ -1 & 1 & 0 \end{bmatrix}$.

7. The augmented matrix is: $\left[\begin{array}{cc|c} 2 & 3 & 11 \\ 1 & 2 & 8 \end{array}\right]$.

8. The augmented matrix is: $\left[\begin{array}{cc|c} 3 & 5 & -13 \\ 2 & 3 & -9 \end{array}\right]$.

9. The augmented matrix is: $\left[\begin{array}{cc|c} 1 & 5 & 6 \\ 1 & 0 & 3 \end{array}\right]$.

10. The augmented matrix is: $\left[\begin{array}{cc|c} 2 & 7 & 1 \\ 5 & 0 & -15 \end{array}\right]$.

11. The augmented matrix is: $\left[\begin{array}{ccc|c} 2 & 1 & 1 & 3 \\ 3 & -4 & 2 & -7 \\ 1 & 1 & 1 & 2 \end{array}\right]$.

12. The augmented matrix is: $\left[\begin{array}{ccc|c} 4 & -2 & 3 & 4 \\ 3 & 5 & 1 & 7 \\ 5 & -1 & 4 & 7 \end{array}\right]$.

13. The augmented matrix is: $\left[\begin{array}{ccc|c} 1 & 1 & 0 & 2 \\ 0 & 2 & 1 & -4 \\ 0 & 0 & 1 & 2 \end{array}\right]$.

14. The augmented matrix is: $\left[\begin{array}{ccc|c} 1 & 0 & 0 & 6 \\ 0 & 1 & 2 & 2 \\ 1 & 0 & -3 & 6 \end{array}\right]$.

15. The system of equations is: $\begin{aligned} 2x + y &= 1 \\ 3x - 2y &= -9. \end{aligned}$

16. The system of equations is: $\begin{aligned} x - 5y &= -18 \\ 6x + 2y &= 20. \end{aligned}$

17. The system of equations is: $\begin{aligned} x &= 2 \\ y &= 3 \\ z &= -2. \end{aligned}$

18. The system of equations is: $\begin{aligned} x + z &= 4 \\ y &= 2 \\ z &= 3. \end{aligned}$

19. The system of equations is: $\begin{aligned} 3x + 2y + z &= 1 \\ 2y + 4z &= 22 \\ -x - 2y + 3z &= 15. \end{aligned}$

20. The system of equations is: $\begin{aligned} 2x + y + 3z &= 12 \\ 4x - 3y &= 10 \\ 5x - 4z &= -11. \end{aligned}$

21. The system can be written as $x + 2y = 3$ and $y = -1$. Substituting $y = -1$ into the first equation gives $x + 2(-1) = 3 \Rightarrow x = 5$. The solution is $(5, -1)$.

22. The system can be written as $x - 5y = 6$ and $0 = 0$. The solution is $\{(5y + 6, y)\}$.

23. The system can be written as $x - y = 2$ and $y = 0$. Substituting $y = 0$ into the first equation gives
    $x - 0 = 2 \Rightarrow x = 2$. The solution is $(2, 0)$.

24. The system can be written as $x + 4y = -2$ and $y = 3$. Substituting $y = 3$ into the first equation gives
    $x + 4(3) = -2 \Rightarrow x = -14$. The solution is $(-14, 3)$.

25. The system can be written as $x + y - z = 4$, $y - z = 2$, and $z = 1$. Substituting $z = 1$ into the second
    equation gives $y - (1) = 2 \Rightarrow y = 3$. Substituting $y = 3$ and $z = 1$ into the first equation gives
    $x + (3) - (1) = 4 \Rightarrow x = 2$. The solution is $(2, 3, 1)$.

26. The system can be written as $x - 2y - z = 0$, $y - 3z = 1$, and $z = 2$. Substituting $z = 2$ into the second
    equation gives $y - 3(2) = 1 \Rightarrow y = 7$. Substituting $y = 7$ and $z = 2$ into the first equation gives
    $x - 2(7) - (2) = 0 \Rightarrow x = 16$. The solution is $(16, 7, 2)$.

27. The system can be written as $x + 2y - z = 5$, $y - 2z = 1$, and $0 = 0$. Since $0 = 0$, there are an infinite
    number of solutions. The second equation gives $y = 1 + 2z$. Substituting this into the first equation gives
    $x + 2(1 + 2z) - z = 5 \Rightarrow x = 3 - 3z$. The solution can be written as
    $\{(3 - 3z, 1 + 2z, z) \mid z \text{ is a real number}\}$.

28. The system can be written as $x - y + 2z = 8$, $y - 4z = 2$, and $z = -1$. Substituting $z = -1$ into the second
    equation gives $y - 4(-1) = 2 \Rightarrow y = -2$. Substituting $y = -2$ and $z = -1$ into the first equation gives
    $x - (-2) + 2(-1) = 8 \Rightarrow x = 8$. The solution is $(8, -2, -1)$.

29. The system can be written as $x + 2y + z = -3$, $y - 3z = \dfrac{1}{2}$, and $0 = 4$. Since $0 = 4$ is false, there are no solutions.

30. The system can be written as $x - 4z = \dfrac{3}{4}$, $y + 2z = 1$, and $0 = -3$. Since $0 = -3$ is false, there are no solutions.

31. Using the Row Echelon Method, the given system of equations yields the matrix: $\begin{bmatrix} 1 & 1 & | & 5 \\ 1 & -1 & | & -1 \end{bmatrix}$, which by

    $-R_1 + R_2 \to \begin{bmatrix} 1 & 1 & | & 5 \\ 0 & -2 & | & -6 \end{bmatrix} \Rightarrow -\dfrac{1}{2}R_2 \to \begin{bmatrix} 1 & 1 & | & 5 \\ 0 & 1 & | & 3 \end{bmatrix}$. From this matrix, we have the resulting equation:

    $y = 3$. Now substitute $y = 3$ into the resulting R_1 equation: $x + (3) = 5 \Rightarrow x = 2$. The solution is: $\{(2, 3)\}$.

32. Using the Row Echelon Method, the given system of equations yields the matrix: $\begin{bmatrix} 1 & 2 & | & 5 \\ 2 & 1 & | & -2 \end{bmatrix}$, which by

    $-2R_1 + R_2 \to \begin{bmatrix} 1 & 2 & | & 5 \\ 0 & -3 & | & -12 \end{bmatrix} \Rightarrow -\dfrac{1}{3}R_2 \to \begin{bmatrix} 1 & 2 & | & 5 \\ 0 & 1 & | & 4 \end{bmatrix}$. From this matrix, we have the resulting

    equation: $y = 4$. Now substitute $y = 4$ into the resulting R_1 equation: $x + 2(4) = 5 \Rightarrow x = -3$.
    The solution is: $\{(-3, 4)\}$.

33. Using the Row Echelon Method, the given system of equations yields the matrix: $\begin{bmatrix} 1 & 1 & | & -3 \\ 2 & -5 & | & -6 \end{bmatrix}$,   which by

    $-2R_1 + R_2 \to \begin{bmatrix} 1 & 1 & | & -3 \\ 0 & -7 & | & 0 \end{bmatrix} \Rightarrow -\dfrac{1}{7}R_2 \to \begin{bmatrix} 1 & 1 & | & -3 \\ 0 & 1 & | & 0 \end{bmatrix}$. From this matrix, we have the resulting

    equation: $y = 0$. Now substitute $y = 0$ into the resulting R_1 equation: $x + (0) = -3 \Rightarrow x = -3$.
    The solution is: $\{(-3, 0)\}$.

34. Using the Row Echelon Method, the given system of equations yields the matrix: $\begin{bmatrix} 3 & -2 & | & 4 \\ 3 & 1 & | & -2 \end{bmatrix}$, which by

$-R_1 + R_2 \to \begin{bmatrix} 3 & -2 & | & 4 \\ 0 & 3 & | & -6 \end{bmatrix} \Rightarrow \left(\dfrac{1}{3}R_1\right)$ and $\left(\dfrac{1}{3}R_2\right) \to \begin{bmatrix} 1 & -\frac{2}{3} & | & \frac{4}{3} \\ 0 & 1 & | & -2 \end{bmatrix}$. From this matrix, we have the

resulting equation: $y = -2$. Now substitute $y = -2$ into the resulting R_1 equation:

$x - \dfrac{2}{3}(-2) = \dfrac{4}{3} \Rightarrow 3x + 4 = 4 \Rightarrow 3x = 0 \Rightarrow x = 0$. The solution is: $\{(0, -2)\}$.

35. Using the Row Echelon Method, the given system of equations yields the matrix: $\begin{bmatrix} 2 & -3 & | & 10 \\ 2 & 2 & | & 5 \end{bmatrix}$, which by

$-R_1 + R_2 \to \begin{bmatrix} 2 & -3 & | & 10 \\ 0 & 5 & | & -5 \end{bmatrix} \Rightarrow \left(\dfrac{1}{2}R_1\right)$ and $\left(\dfrac{1}{5}R_2\right) \to \begin{bmatrix} 1 & -\frac{3}{2} & | & 5 \\ 0 & 1 & | & -1 \end{bmatrix}$. From this matrix, we have the

resulting equation: $y = -1$. Now substitute $y = -1$ into the resulting R_1 equation:

$x - \dfrac{3}{2}(-1) = 5 \Rightarrow 2x + 3 = 10 \Rightarrow 2x = 7 \Rightarrow x = \dfrac{7}{2}$. The solution is: $\left\{\left(\dfrac{7}{2}, -1\right)\right\}$.

36. Using the Row Echelon Method, the given system of equations yields the matrix: $\begin{bmatrix} 4 & 1 & | & 5 \\ 2 & 1 & | & 3 \end{bmatrix}$, which by

$-\dfrac{1}{2}R_1 + R_2 \to \begin{bmatrix} 4 & 1 & | & 5 \\ 0 & \frac{1}{2} & | & \frac{1}{2} \end{bmatrix} \Rightarrow \left(\dfrac{1}{4}R_1\right)$ and $(2R_2) \to \begin{bmatrix} 1 & \frac{1}{4} & | & \frac{5}{4} \\ 0 & 1 & | & 1 \end{bmatrix}$. From this matrix, we have the

resulting equation: $y = 1$. Now substitute $y = 1$ into the resulting R_1 equation:

$x + \dfrac{1}{4}(1) = \dfrac{5}{4} \Rightarrow x = \dfrac{4}{4} \Rightarrow x = 1$. The solution is: $\{(1, 1)\}$.

37. Using the Row Echelon Method, the given system of equations yields the matrix: $\begin{bmatrix} 2 & -3 & | & 2 \\ 4 & -6 & | & 1 \end{bmatrix}$,   which by

$-2R_1 + R_2 \to \begin{bmatrix} 2 & -3 & | & 2 \\ 0 & 0 & | & -3 \end{bmatrix}$. Since this matrix yields the equation: $0 = -3$, which is false, therefore the

solution is: $\varnothing$.

38. Using the Row Echelon Method, the given system of equations yields the matrix: $\begin{bmatrix} 1 & 2 & | & 1 \\ 2 & 4 & | & 3 \end{bmatrix}$, which by

$-2R_1 + R_2 \to \begin{bmatrix} 1 & 2 & | & 1 \\ 0 & 0 & | & 1 \end{bmatrix}$. Since this matrix yields the equation: $0 = 1$, which is false, therefore the

solution is: $\varnothing$.

39. Using the Row Echelon Method, the given system of equations yields the matrix: $\begin{bmatrix} 6 & -3 & | & 1 \\ -12 & 6 & | & -2 \end{bmatrix}$, which by

$2R_1 + R_2 \to \begin{bmatrix} 6 & -3 & | & 1 \\ 0 & 0 & | & 0 \end{bmatrix}$. The matrix yields the equation: $0 = 0$, since this true, there are: ∞ solutions.

From the resulting R_1, the solutions have the relationship: $-3y = 1 - 6x \Rightarrow y = \dfrac{6x - 1}{3}$; $\left[\text{or } x = \dfrac{3y + 1}{6} \right]$.

The solution is: $\left\{ \left(x, \dfrac{6x - 1}{3} \right) \text{ or } \left(\dfrac{3y + 1}{6}, y \right) \right\}$.

40. Using the Row Echelon Method, the given system of equations yields the matrix: $\begin{bmatrix} 1 & -1 & | & 1 \\ -1 & 1 & | & -1 \end{bmatrix}$, which by

$R_1 + R_2 \to \begin{bmatrix} 1 & -1 & | & 1 \\ 0 & 0 & | & 0 \end{bmatrix}$. The matrix yields the equation: $0 = 0$, since this true, there are: ∞ solutions.

From the resulting R_1, the solutions have the relationship: $x = y + 1$; $[\text{or } y = x - 1]$.

The solution is: $\{(x, x - 1) \text{ or } (y + 1, y)\}$.

41. Using the Row Echelon Method, the given system of equations yields the matrix: $\begin{bmatrix} 1 & 1 & 0 & -1 \\ 0 & 1 & 1 & 4 \\ 1 & 0 & 1 & 1 \end{bmatrix}$, which by

$-R_1 + R_3 \rightarrow \begin{bmatrix} 1 & 1 & 0 & -1 \\ 0 & 1 & 1 & 4 \\ 0 & -1 & 1 & 2 \end{bmatrix} \Rightarrow R_2 + R_3 \rightarrow \begin{bmatrix} 1 & 1 & 0 & -1 \\ 0 & 1 & 1 & 4 \\ 0 & 0 & 2 & 6 \end{bmatrix} \Rightarrow \frac{1}{2}R_3 \rightarrow \begin{bmatrix} 1 & 1 & 0 & -1 \\ 0 & 1 & 1 & 4 \\ 0 & 0 & 1 & 3 \end{bmatrix}.$

From this matrix, we have the resulting equation: $z = 3$. Now use back-substitution, substituting $z = 3$ into the

resulting R_2 yields: $y + 3 = 4 \Rightarrow y = 1$; and finally substituting $y = 1$ and $z = 3$ into the resulting R_1 yields:

$x + 1 = -1 \Rightarrow x = -2$. The solution is: $\{(-2, 1, 3)\}$.

42. Using the Row Echelon Method, the given system of equations yields the matrix: $\begin{bmatrix} 1 & 0 & -1 & -3 \\ 0 & 1 & 1 & 9 \\ 1 & 0 & 1 & 7 \end{bmatrix}$, which by

$-R_1 + R_3 \rightarrow \begin{bmatrix} 1 & 0 & -1 & -3 \\ 0 & 1 & 1 & 9 \\ 0 & 0 & 2 & 10 \end{bmatrix} \Rightarrow \frac{1}{2}R_3 \rightarrow \begin{bmatrix} 1 & 0 & -1 & -3 \\ 0 & 1 & 1 & 9 \\ 0 & 0 & 1 & 5 \end{bmatrix}.$

From this matrix, we have the resulting equation: $z = 5$. Now use back-substitution, substituting $z = 5$ into the

resulting R_2 yields: $y + 5 = 9 \Rightarrow y = 4$; and finally substituting $y = 4$ and $z = 5$ into the resulting R_1 yields:

$x - 5 = -3 \Rightarrow x = 2$. The solution is: $\{(2, 4, 5)\}$.

43. Using the Row Echelon Method, the given system of equations yields the matrix: $\begin{bmatrix} 1 & 1 & -1 & 6 \\ 2 & -1 & 1 & -9 \\ 1 & -2 & 3 & 1 \end{bmatrix}$, which

by $(-2R_1 + R_2)$ and $(-R_1 + R_3) \rightarrow \begin{bmatrix} 1 & 1 & -1 & 6 \\ 0 & -3 & 3 & -21 \\ 0 & -3 & 4 & -5 \end{bmatrix} \Rightarrow -R_2 + R_3 \rightarrow \begin{bmatrix} 1 & 1 & -1 & 6 \\ 0 & -3 & 3 & -21 \\ 0 & 0 & 1 & 16 \end{bmatrix} \Rightarrow$

$-\frac{1}{3}R_2 \rightarrow \begin{bmatrix} 1 & 1 & -1 & 6 \\ 0 & 1 & -1 & 7 \\ 0 & 0 & 1 & 16 \end{bmatrix}$. From this matrix, we have the resulting equation: $z = 16$. Now use

back-substitution, substituting $z = 16$ into the resulting R_2 yields: $y - 16 = 7 \Rightarrow y = 23$; and finally substituting

$y = 23$ and $z = 16$ into the resulting R_1 yields: $x + 23 - 16 = 6 \Rightarrow x = -1$. The solution is: $\{(-1, 23, 16)\}$.

44. Using the Row Echelon Method, the given system of equations yields the matrix: $\begin{bmatrix} 1 & 3 & -6 & 7 \\ 2 & -1 & 2 & 0 \\ 1 & 1 & 2 & -1 \end{bmatrix}$, which

by $(-2R_1 + R_2)$ and $(-R_1 + R_3) \rightarrow \begin{bmatrix} 1 & 3 & -6 & 7 \\ 0 & -7 & 14 & -14 \\ 0 & -2 & 8 & -8 \end{bmatrix} \Rightarrow \left(-\frac{1}{7}R_2\right) \text{and} \left(-\frac{1}{2}R_3\right) \rightarrow \begin{bmatrix} 1 & 3 & -6 & 7 \\ 0 & 1 & -2 & 2 \\ 0 & 1 & -4 & 4 \end{bmatrix}$

$\Rightarrow -R_2 + R_3 \rightarrow \begin{bmatrix} 1 & 3 & -6 & 7 \\ 0 & 1 & -2 & 2 \\ 0 & 0 & -2 & 2 \end{bmatrix} \Rightarrow -\frac{1}{2}R_3 \rightarrow \begin{bmatrix} 1 & 3 & -6 & 7 \\ 0 & 1 & -2 & 2 \\ 0 & 0 & 1 & -1 \end{bmatrix}.$

From this matrix, we have the resulting equation: $z = -1$. Now use back-substitution: substituting $z = -1$ into

the resulting R_2 yields: $y + 2 = 2 \Rightarrow y = 0$; and finally substituting $y = 0$ and $z = -1$ into the resulting R_1

yields: $x + 0 + 6 = 7 \Rightarrow x = 1$. The solution is: $\{(1, 0, -1)\}$.

45. Using the Row Echelon Method, the given system of equations yields the matrix: $\begin{bmatrix} -1 & 1 & 0 & | & -1 \\ 0 & 1 & -1 & | & 6 \\ 1 & 0 & 1 & | & -1 \end{bmatrix}$, which

by $R_1 + R_3 \rightarrow \begin{bmatrix} -1 & 1 & 0 & | & -1 \\ 0 & 1 & -1 & | & 6 \\ 0 & 1 & 1 & | & -2 \end{bmatrix} \Rightarrow -R_2 + R_3 \rightarrow \begin{bmatrix} -1 & 1 & 0 & | & -1 \\ 0 & 1 & -1 & | & 6 \\ 0 & 0 & 2 & | & -8 \end{bmatrix} \Rightarrow (-R_1)$ and $\left(\frac{1}{2}R_3\right) \rightarrow$

$\begin{bmatrix} 1 & -1 & 0 & | & 1 \\ 0 & 1 & -1 & | & 6 \\ 0 & 0 & 1 & | & -4 \end{bmatrix}$. From this matrix, we have the resulting equation: $z = -4$. Now use back-substitution,

substituting $z = -4$ into the resulting R_2 yields: $y + 4 = 6 \Rightarrow y = 2$; and finally substituting

$y = 2$ and $z = -4$ into the resulting R_1 yields: $x - 2 = 1 \Rightarrow x = 3$. The solution is: $\{(3, 2, -4)\}$.

46. Using the Row Echelon Method, the given system of equations yields the matrix: $\begin{bmatrix} 1 & 1 & 0 & | & 1 \\ 2 & 0 & -1 & | & 0 \\ 0 & 1 & 2 & | & -2 \end{bmatrix}$, which

by $-2R_1 + R_2 \rightarrow \begin{bmatrix} 1 & 1 & 0 & | & 1 \\ 0 & -2 & -1 & | & -2 \\ 0 & 1 & 2 & | & -2 \end{bmatrix} \Rightarrow \frac{1}{2}R_2 + R_3 \rightarrow \begin{bmatrix} 1 & 1 & 0 & | & 1 \\ 0 & -2 & -1 & | & -2 \\ 0 & 0 & \frac{3}{2} & | & -3 \end{bmatrix} \Rightarrow \left(-\frac{1}{2}R_2\right)$ and $\left(\frac{2}{3}R_3\right) \rightarrow$

$\begin{bmatrix} 1 & 1 & 0 & | & 1 \\ 0 & 1 & \frac{1}{2} & | & 1 \\ 0 & 0 & 1 & | & -2 \end{bmatrix}$. From this matrix, we have the resulting equation: $z = -2$. Now use back-substitution,

substituting $z = -2$ into the resulting R_2 yields: $y + \frac{1}{2}(-2) = 1 \Rightarrow y = 2$; and finally substituting

$y = 2$ and $z = -2$ into the resulting R_1 yields: $x + 2 = 1 \Rightarrow x = -1$. The solution is: $\{(-1, 2, -2)\}$.

47. Using the Row Echelon Method, the given system of equations yields the matrix: $\begin{bmatrix} 2 & -1 & 3 & | & 0 \\ 1 & 2 & -1 & | & 5 \\ 0 & 2 & 1 & | & 1 \end{bmatrix}$, which

by $R_1 \leftrightarrow R_2 \rightarrow \begin{bmatrix} 1 & 2 & -1 & | & 5 \\ 2 & -1 & 3 & | & 0 \\ 0 & 2 & 1 & | & 1 \end{bmatrix} \Rightarrow -2R_1 + R_2 \rightarrow \begin{bmatrix} 1 & 2 & -1 & | & 5 \\ 0 & -5 & 5 & | & -10 \\ 0 & 2 & 1 & | & 1 \end{bmatrix} \Rightarrow 2R_2 + 5R_3 \rightarrow$

$\begin{bmatrix} 1 & 2 & -1 & | & 5 \\ 0 & -5 & 5 & | & -10 \\ 0 & 0 & 15 & | & -15 \end{bmatrix} \Rightarrow \left(-\frac{1}{5}R_2\right)$ and $\left(\frac{1}{15}R_3\right) \rightarrow \begin{bmatrix} 1 & 2 & -1 & | & 5 \\ 0 & 1 & -1 & | & 2 \\ 0 & 0 & 1 & | & -1 \end{bmatrix}$.

From this matrix, we have the resulting equation: $z = -1$. Now use back-substitution: substituting $z = -1$ into

the resulting R_2 yields: $y + 1 = 2 \Rightarrow y = 1$; and finally substituting $y = 1$ and $z = -1$ into the resulting R_1

yields: $x + 2 + 1 = 5 \Rightarrow x = 2$. The solution is: $\{(2, 1, -1)\}$.

48. Using the Row Echelon Method, the given system of equations yields the matrix: $\begin{bmatrix} 4 & 2 & -3 & | & 6 \\ 1 & -4 & 1 & | & -4 \\ -1 & 0 & 2 & | & 2 \end{bmatrix}$, which

by $R_1 \leftrightarrow R_2 \rightarrow \begin{bmatrix} 1 & -4 & 1 & | & -4 \\ 4 & 2 & -3 & | & 6 \\ -1 & 0 & 2 & | & 2 \end{bmatrix} \Rightarrow (-4R_1 + R_2)$ and $(R_1 + R_3) \rightarrow \begin{bmatrix} 1 & -4 & 1 & | & -4 \\ 0 & 18 & -7 & | & 22 \\ 0 & -4 & 3 & | & -2 \end{bmatrix} \Rightarrow$

$2R_2 + 9R_3 \rightarrow \begin{bmatrix} 1 & -4 & 1 & | & -4 \\ 0 & 18 & -7 & | & 22 \\ 0 & 0 & 13 & | & 26 \end{bmatrix} \Rightarrow \left(\frac{1}{18}R_2\right)$ and $\left(\frac{1}{13}R_3\right) \rightarrow \begin{bmatrix} 1 & -4 & 1 & | & -4 \\ 0 & 1 & -\frac{7}{18} & | & \frac{11}{9} \\ 0 & 0 & 1 & | & 2 \end{bmatrix}$.

From this matrix, we have the resulting equation: $z = 2$. Now use back-substitution: substituting $z = 2$ into

the resulting R_2 yields: $y - \frac{7}{18}(2) = \frac{11}{9} \Rightarrow y - \frac{7}{9} = \frac{11}{9} \Rightarrow y = \frac{18}{9} = 2$; and finally substituting

$y = 2$ and $z = 2$ into the resulting R_1 yields: $x - 4(2) + (2) = -4 \Rightarrow x - 6 = -4 \Rightarrow x = 2$.

The solution is: $\{(2, 2, 2)\}$.

49. Using the capabilities of the calculator, the Reduced Row Echelon Method of:

$\begin{bmatrix} 2.1 & .5 & 1.7 & | & 4.9 \\ -2 & 1.5 & -1.7 & | & 3.1 \\ 5.8 & -4.6 & .8 & | & 9.3 \end{bmatrix} = \begin{bmatrix} 1 & 0 & 0 & | & 5.21127 \\ 0 & 1 & 0 & | & 3.73944 \\ 0 & 0 & 1 & | & -4.65493 \end{bmatrix}$.

The solution set is approximately: $\{(5.211, 3.739, -4.655)\}$.

50. Using the capabilities of the calculator, the Reduced Row Echelon Method of:

$\begin{bmatrix} .1 & .3 & 1.7 & | & .6 \\ .6 & .1 & -3.1 & | & 6.2 \\ 0 & 2.4 & .9 & | & 3.5 \end{bmatrix} = \begin{bmatrix} 1 & 0 & 0 & | & 7.99276 \\ 0 & 1 & 0 & | & 1.60875 \\ 0 & 0 & 1 & | & -.40112 \end{bmatrix}$.

The solution set is approximately: $\{(7.993, 1.609, -.401)\}$.

51. Using the capabilities of the calculator, the Reduced Row Echelon Method of:

$\begin{bmatrix} 53 & 95 & 12 & | & 108 \\ 81 & -57 & -24 & | & -92 \\ -9 & 11 & -78 & | & 21 \end{bmatrix} = \begin{bmatrix} 1 & 0 & 0 & | & -.24997 \\ 0 & 1 & 0 & | & 1.2838 \\ 0 & 0 & 1 & | & -.05934 \end{bmatrix}$.

The solution set is approximately: $\{(-.250, 1.284, -.059)\}$.

52. Using the capabilities of the calculator, the Reduced Row Echelon Method of:

$\begin{bmatrix} 103 & -886 & 431 & | & 1200 \\ -55 & 981 & 0 & | & 1108 \\ -327 & 421 & 337 & | & 99 \end{bmatrix} = \begin{bmatrix} 1 & 0 & 0 & | & 6.07732 \\ 0 & 1 & 0 & | & 1.47019 \\ 0 & 0 & 1 & | & 4.35411 \end{bmatrix}$.

The solution set is approximately: $\{(6.077, 1.470, 4.354)\}$.

53. In both cases, we simply write the coefficients and do not write the variables. This is possible because we agree on the order in which the variables appear (descending degree).

54. The third row transformation, which eliminates a variable, is comparable to multiplying one or more equations in a system by the appropriate constants so that they are added, a variable is eliminated.

55. The given system of equations yields the matrix: $\begin{bmatrix} 1 & -3 & 2 & | & 10 \\ 2 & -1 & -1 & | & 8 \end{bmatrix}$, which by $-2R_1 + R_2 \rightarrow$

$\begin{bmatrix} 1 & -3 & 2 & | & 10 \\ 0 & 5 & -5 & | & -12 \end{bmatrix} \Rightarrow \frac{1}{5}R_2 \rightarrow \begin{bmatrix} 1 & -3 & 2 & | & 10 \\ 0 & 1 & -1 & | & -\frac{12}{5} \end{bmatrix}$. Since we cannot eliminate more than one

coefficient there are an infinite number of solutions, and they will be in the following relationship.

From the result of R_2, $y - z = -\dfrac{12}{5} \Rightarrow 5y - 5z = -12 \Rightarrow 5y = 5z - 12 \Rightarrow y = \dfrac{5z - 12}{5}$.

Now substitute this into the result of R_1: $x - 3\left(\dfrac{5z - 12}{5}\right) + 2z = 10 \Rightarrow 5x - 15z + 36 + 10z = 50 \Rightarrow$

$5x - 5z = 14 \Rightarrow 5x = 5z + 14 \Rightarrow x = \dfrac{5z + 14}{5}$. The solution is: $\left(\dfrac{5z + 14}{5}, \dfrac{5z - 12}{5}, z\right)$.

56. After switching equations 1 and 2, the given system of equations yields the matrix: $\begin{bmatrix} 1 & 2 & 1 & | & 10 \\ 3 & 1 & -1 & | & 12 \end{bmatrix}$, which by

$-3R_1 + R_2 \rightarrow \begin{bmatrix} 1 & 2 & 1 & | & 10 \\ 0 & -5 & -4 & | & -18 \end{bmatrix} \Rightarrow -\frac{1}{5}R_2 \rightarrow \begin{bmatrix} 1 & 2 & 1 & | & 10 \\ 0 & 1 & \frac{4}{5} & | & \frac{18}{5} \end{bmatrix}$. Since we cannot eliminate more than

one coefficient there are an infinite number of solutions, and they will be in the following relationship.

From the result of R_2, $y + \dfrac{4}{5}z = \dfrac{18}{5} \Rightarrow 5y + 4z = 18 \Rightarrow 5y = 18 - 4z \Rightarrow y = \dfrac{18 - 4z}{5}$.

Now substitute this into the result of R_1: $x + 2\left(\dfrac{18 - 4z}{5}\right) + z = 10 \Rightarrow 5x + 36 - 8z + 5z = 50 \Rightarrow$

$5x - 3z = 14 \Rightarrow 5x = 3z + 14 \Rightarrow x = \dfrac{3z + 14}{5}$. The solution is: $\left(\dfrac{3z + 14}{5}, \dfrac{18 - 4z}{5}, z\right)$.

57. The given system of equations yields the matrix: $\begin{bmatrix} 1 & 2 & -1 & | & 0 \\ 3 & -1 & 1 & | & 6 \\ -2 & -4 & 2 & | & 0 \end{bmatrix}$,

which by $(-3R_1 + R_2)$ and $(2R_1 + R_3) \rightarrow \begin{bmatrix} 1 & 2 & -1 & | & 0 \\ 0 & -7 & 4 & | & 6 \\ 0 & 0 & 0 & | & 0 \end{bmatrix}$. Since R_3 is $0 = 0$, there are infinitely many

solutions and the solutions are in the following relationship:

From the result of R_2, $-7y + 4z = 6 \Rightarrow -7y = 6 - 4z \Rightarrow y = \dfrac{4z - 6}{7}$.

Now substitute this into the result of R_1: $x + 2\left(\dfrac{4z - 6}{7}\right) - z = 0 \Rightarrow 7x + 8z - 12 - 7z = 0 \Rightarrow$

$7x + z = 12 \Rightarrow 7x = 12 - z \Rightarrow x = \dfrac{12 - z}{7}$. The solution is: $\left(\dfrac{12 - z}{7}, \dfrac{4z - 6}{7}, z\right)$.

58. The given system of equations yields the matrix: $\begin{bmatrix} 3 & 5 & -1 & | & 0 \\ 4 & -1 & 2 & | & 1 \\ -6 & -10 & 2 & | & 0 \end{bmatrix}$,

which by $\frac{1}{3}R_1 \rightarrow \begin{bmatrix} 1 & \frac{5}{3} & -\frac{1}{3} & | & 0 \\ 4 & -1 & 2 & | & 1 \\ -6 & -10 & 2 & | & 0 \end{bmatrix} \Rightarrow (-4R_1 + R_2)$ and $(6R_1 + R_3) \rightarrow \begin{bmatrix} 1 & \frac{5}{3} & -\frac{1}{3} & | & 0 \\ 0 & -\frac{23}{3} & \frac{10}{3} & | & 1 \\ 0 & 0 & 0 & | & 0 \end{bmatrix} \Rightarrow$

$-\frac{3}{23}R_2 \rightarrow \begin{bmatrix} 1 & \frac{5}{3} & -\frac{1}{3} & | & 0 \\ 0 & 1 & -\frac{10}{23} & | & -\frac{3}{23} \\ 0 & 0 & 0 & | & 0 \end{bmatrix}$.

Since we get an R_3 of $0 = 0$, there are infinitely many solutions and the solutions are in the following relationship:

From the result of R_2, $y - \frac{10}{23}z = -\frac{3}{23} \Rightarrow 23y - 10z = -3 \Rightarrow 23y = 10z - 3 \Rightarrow y = \frac{10z - 3}{23}$.

Now substitute this into the result of R_1: $x + \frac{5}{3}\left(\frac{10z - 3}{23}\right) - \frac{1}{3}z = 0 \Rightarrow 69x + 50z - 15 - 23z = 0 \Rightarrow$

$69x + 27z = 15 \Rightarrow 69x = 15 - 27z \Rightarrow x = \frac{15 - 27z}{69} \Rightarrow x = \frac{5 - 9z}{23}$.

The solution is: $\left(\frac{5 - 9z}{23}, \frac{10z - 3}{23}, z\right)$.

59. The given system of equations yields the matrix: $\begin{bmatrix} 1 & -2 & 1 & | & 5 \\ -2 & 4 & -2 & | & 2 \\ 2 & 1 & -1 & | & 2 \end{bmatrix}$,

$(2R_1 + R_2)$ and $(-2R_1 + R_3) \rightarrow \begin{bmatrix} 1 & -2 & 1 & | & 5 \\ 0 & 0 & 0 & | & 12 \\ 0 & 5 & -3 & | & -8 \end{bmatrix}$. Since R_2 is $0 = 12$, there is no solution or $\varnothing$.

60. After switching equations 1 and 3, the given system of equations yields the matrix: $\begin{bmatrix} 1 & 1 & -2 & | & 20 \\ -1 & -2 & 1 & | & 16 \\ 3 & 6 & -3 & | & 12 \end{bmatrix}$,

$(R_1 + R_2)$ and $(-3R_1 + R_3) \rightarrow \begin{bmatrix} 1 & 1 & -2 & | & 20 \\ 0 & -1 & -1 & | & 36 \\ 0 & 3 & 3 & | & -48 \end{bmatrix} \Rightarrow \frac{1}{3}R_3 + R_2 \rightarrow \begin{bmatrix} 1 & 1 & -2 & | & 20 \\ 0 & 0 & 0 & | & 20 \\ 0 & 3 & 3 & | & -48 \end{bmatrix}$.

Since R_2 is $0 = 12$, there is no solution or $\varnothing$.

61. The given system of equations yields the matrix: $\begin{bmatrix} 1 & 3 & -2 & -1 & | & 9 \\ 4 & 1 & 1 & 2 & | & 2 \\ -3 & -1 & 1 & -1 & | & -5 \\ 1 & -1 & -3 & -2 & | & 2 \end{bmatrix}$, which by $\left. \begin{array}{r} -4R_1 + R_2 \\ 3R_1 + R_3 \\ -R_1 + R_4 \end{array} \right\} \rightarrow$

$$\begin{bmatrix} 1 & 3 & -2 & -1 & | & 9 \\ 0 & -11 & 9 & 6 & | & -34 \\ 0 & 8 & -5 & -4 & | & 22 \\ 0 & -4 & -1 & -1 & | & -7 \end{bmatrix} \Rightarrow \left. \begin{array}{r} R_3 + 2R_4 \\ 8R_2 + 11R_3 \end{array} \right\} \rightarrow \begin{bmatrix} 1 & 3 & -2 & -1 & | & 9 \\ 0 & -11 & 9 & 6 & | & -34 \\ 0 & 0 & 17 & 4 & | & -30 \\ 0 & 0 & -7 & -6 & | & 8 \end{bmatrix} \Rightarrow 7R_3 + 17R_4 \rightarrow$$

$$\begin{bmatrix} 1 & 3 & -2 & -1 & | & 9 \\ 0 & -11 & 9 & 6 & | & -34 \\ 0 & 0 & 17 & 4 & | & -30 \\ 0 & 0 & 0 & -74 & | & -74 \end{bmatrix} \Rightarrow -\frac{1}{11}R_2, \frac{1}{17}R_3, -\frac{1}{74}R_4 \rightarrow \begin{bmatrix} 1 & 3 & -2 & -1 & | & 9 \\ 0 & 1 & -\frac{9}{11} & -\frac{6}{11} & | & \frac{34}{11} \\ 0 & 0 & 1 & \frac{4}{17} & | & -\frac{30}{17} \\ 0 & 0 & 0 & 1 & | & 1 \end{bmatrix}.$$

From this matrix, we have the resulting equation: $w = 1$. Now use back-substitution: substituting $w = 1$ into

the resulting R_3 yields: $z + \frac{4}{17}(1) = -\frac{30}{17} \Rightarrow z = -\frac{34}{17} \Rightarrow z = -2$; substituting $w = 1$ and $z = -2$ into the

resulting R_2 yields: $y - \frac{9}{11}(-2) - \frac{6}{11}(1) = \frac{34}{11} \Rightarrow y + \frac{18 - 6}{11} = \frac{34}{11} \Rightarrow y = \frac{34 - 12}{11} = \frac{22}{11} \Rightarrow y = 2$.

Finally, substitute $y = 2$, $z = -2$, and $w = 1$ into R_1: $x + 3(2) - 2(-2) - 1 = 9 \Rightarrow x + 9 = 9 \Rightarrow x = 0$.

The solution is: $\{(0, 2, -2, 1)\}$.

62. After switching equations 1 and 3, the given system of equations yields the matrix $\begin{bmatrix} 1 & 2 & -1 & 0 & | & -2 \\ 2 & 0 & 1 & 2 & | & 5 \\ 3 & 2 & 0 & -1 & | & 0 \\ 2 & -1 & 1 & 1 & | & 2 \end{bmatrix}$,

then $\left. \begin{array}{r} -2R_1 + R_2 \\ -3R_1 + R_3 \\ -2R_1 + R_4 \end{array} \right\} \rightarrow \begin{bmatrix} 1 & 2 & -1 & 0 & | & -2 \\ 0 & -4 & 3 & 2 & | & 9 \\ 0 & -4 & 3 & -1 & | & 6 \\ 0 & -5 & 3 & 1 & | & 6 \end{bmatrix} \Rightarrow \left. \begin{array}{r} -R_2 + R_3 \\ -5R_2 + 4R_4 \end{array} \right\} \rightarrow \begin{bmatrix} 1 & 2 & -1 & 0 & | & -2 \\ 0 & -4 & 3 & 2 & | & 9 \\ 0 & 0 & 0 & -3 & | & -3 \\ 0 & 0 & -3 & -6 & | & -21 \end{bmatrix} \Rightarrow$

$\left. \begin{array}{r} -\frac{1}{4}R_2 \\ -\frac{1}{3}R_3 \\ -\frac{1}{3}R_4 \end{array} \right\} \rightarrow \begin{bmatrix} 1 & 2 & -1 & 0 & | & -2 \\ 0 & 1 & -\frac{3}{4} & -\frac{1}{2} & | & -\frac{9}{4} \\ 0 & 0 & 0 & 1 & | & 1 \\ 0 & 0 & 1 & 2 & | & 7 \end{bmatrix} \Rightarrow R_3 \leftrightarrow R_4 \rightarrow \begin{bmatrix} 1 & 2 & -1 & 0 & | & -2 \\ 0 & 1 & -\frac{3}{4} & -\frac{1}{2} & | & -\frac{9}{4} \\ 0 & 0 & 1 & 2 & | & 7 \\ 0 & 0 & 0 & 1 & | & 1 \end{bmatrix}.$

From this matrix, we have the resulting equation: $w = 1$. Now use back-substitution: substituting $w = 1$ into

the resulting R_3 yields: $z + 2 = 7 \Rightarrow z = 5$, substituting $w = 1$ and $z = 5$ into the resulting R_2 yields:

$y - \frac{3}{4}(5) - \frac{1}{2}(1) = -\frac{9}{4} \Rightarrow 4y - 15 - 2 = -9 \Rightarrow 4y = 8 \Rightarrow y = 2$.

Finally, substitute $y = 2$, $z = 5$, and $w = 1$ into R_1: $x + 2(2) - 5 = -2 \Rightarrow x - 1 = -2 \Rightarrow x = -1$.

The solution is: $\{(-1, 2, 5, 1)\}$.

63. (a) Using the given equation model, the given system of equations and the capabilities of the calculator, the

Reduced Row Echelon Method of:

$$\begin{bmatrix} 1800 & 5000 & 1 & | & 1300 \\ 3200 & 12000 & 1 & | & 5300 \\ 4500 & 13000 & 1 & | & 6500 \end{bmatrix} = \begin{bmatrix} 1 & 0 & 0 & | & .5714286 \\ 0 & 1 & 0 & | & .4571429 \\ 0 & 0 & 1 & | & -2014.2857 \end{bmatrix}.$$

Therefore the equation is: $F = .5714N + .4571R - 2014$.

(b) Let $N = 3500$ and $R = 12{,}500$, then: $F = .5714(3500) + .4571(12{,}500) - 2014 \Rightarrow F = 5699.65$.

This model predicts monthly food costs of approximately \$5700.

64. (a) Using $f(x) = ax^2 + bx + c$ and the data from the table yields:

From the point $(1, 9.4)$: $9.4 = a(1)^2 + b(1) + c \Rightarrow 9.4 = a + b + c$.

From the point $(15, 18.8)$: $18.8 = a(15)^2 + b(15) + c \Rightarrow 18.8 = 225a + 15b + c$.

From the point $(30, 21.9)$: $21.9 = a(30)^2 + b(30) + c \Rightarrow 21.9 = 900a + 30b + c$.

Now using the capabilities of the calculator, the Reduced Row Echelon Method of:

$$\begin{bmatrix} 1 & 1 & 1 & | & 9.4 \\ 225 & 15 & 1 & | & 18.8 \\ 900 & 30 & 1 & | & 21.9 \end{bmatrix} = \begin{bmatrix} 1 & 0 & 0 & | & -.0160263 \\ 0 & 1 & 0 & | & .9278489 \\ 0 & 0 & 1 & | & 8.488177 \end{bmatrix}.$$

Therefore, $a \approx -.016026$, $b \approx .92785$, and $c \approx 8.4882$.

(b) Graph $f(x) = -.016026x^2 + .92785x + 8.4882$. See Figure 64.

(c) $f(3) = -.016026(3)^2 + .92785(3) + 8.4882 \Rightarrow f(3) = 11.127516$.

Approximately 11.1 days is close to the actual value.

$[-4, 32]$ by $[8, 23]$
Xscl = 5   Yscl = 1

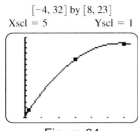

Figure 64

65. Let x represent the number of model A, and y the number of model B. Using the information yields:

$$\begin{matrix} 2x + 3y = 34 \\ 25x + 30y = 365 \end{matrix} \rightarrow \begin{bmatrix} 2 & 3 & | & 34 \\ 25 & 30 & | & 365 \end{bmatrix}, \text{ which by } \left.\begin{matrix} \frac{1}{2}R_1 \\ \frac{1}{5}R_2 \end{matrix}\right\} \rightarrow \begin{bmatrix} 1 & \frac{3}{2} & | & 17 \\ 5 & 6 & | & 73 \end{bmatrix} \Rightarrow -5R_1 + R_2 \rightarrow$$

$$\begin{bmatrix} 1 & \frac{3}{2} & | & 17 \\ 0 & -\frac{3}{2} & | & -12 \end{bmatrix} \Rightarrow -\frac{2}{3}R_2 \rightarrow \begin{bmatrix} 1 & \frac{3}{2} & | & 17 \\ 0 & 1 & | & 8 \end{bmatrix}. \text{ From this matrix, we have the resulting equation: } y = 8.$$

Now substituting $y = 8$ into R_1 yields: $x + \frac{3}{2}(8) = 7 \Rightarrow x + 12 = 17 \Rightarrow x = 5$.

The maximum number is: Model A: 5 bicycles, Model B: 8 bicycles.

66. Let x represent the number of printers, and y the number of monitors. Using the information yields:

$$\begin{aligned}3x + y &= 15 \\ x + 2y &= 15\end{aligned} \rightarrow \begin{bmatrix} 3 & 1 & | & 15 \\ 1 & 2 & | & 15 \end{bmatrix}, \text{ which by } R_1 \leftrightarrow R_2 \rightarrow \begin{bmatrix} 1 & 2 & | & 15 \\ 3 & 1 & | & 15 \end{bmatrix} \Rightarrow -3R_1 + R_2 \rightarrow$$

$$\begin{bmatrix} 1 & 2 & | & 15 \\ 0 & -5 & | & -30 \end{bmatrix} \Rightarrow -\frac{1}{5}R_2 \rightarrow \begin{bmatrix} 1 & 2 & | & 15 \\ 0 & 1 & | & 6 \end{bmatrix}. \text{ From this matrix, we have the resulting equation: } y = 6.$$

Now substituting $y = 6$ into R_1 yields: $x + 2(6) = 15 \Rightarrow x = 3$.

The maximum number is: Printers: 3 per day, Monitors: 6 per day.

67. Let x = the amount of money at 8%, y = the amount of money at 10%, and z = the amount of money at 9%.

Using the information yields: $x + y + z = 25,000$

$$y = 2000 + \frac{1}{2}x \Rightarrow -\frac{1}{2}x + y = 2000$$

$$.08x + .10y + .09z = 2220$$

This system yields: $\begin{bmatrix} 1 & 1 & 1 & | & 25,000 \\ -.5 & 1 & 0 & | & 2000 \\ .08 & .10 & .09 & | & 2220 \end{bmatrix}$, which by $\left(\frac{1}{2}R_1 + R_2\right)$ and $100R_3 \rightarrow$

$$\begin{bmatrix} 1 & 1 & 1 & | & 25,000 \\ 0 & 1.5 & .5 & | & 14,500 \\ 8 & 10 & 9 & | & 222,000 \end{bmatrix} \Rightarrow \left(\frac{2}{3}R_2\right) \text{ and } (-8R_1 + R_3) \rightarrow \begin{bmatrix} 1 & 1 & 1 & | & 25,000 \\ 0 & 1 & .\overline{33} & | & 9666.\overline{66} \\ 0 & 2 & 1 & | & 22,000 \end{bmatrix} \Rightarrow R_2 - \frac{1}{2}R_3 \rightarrow$$

$$\begin{bmatrix} 1 & 1 & 1 & | & 25,000 \\ 0 & 1 & .\overline{33} & | & 9666.\overline{66} \\ 0 & 0 & -.\overline{166} & | & -1333.\overline{33} \end{bmatrix} \Rightarrow -6R_3 \rightarrow \begin{bmatrix} 1 & 1 & 1 & | & 25,000 \\ 0 & 1 & .\overline{33} & | & 9666.\overline{66} \\ 0 & 0 & 1 & | & 8,000 \end{bmatrix}.$$

From this matrix, we have the resulting equation: $z = 8,000$. Now substituting $z = 8,000$ into R_2 yields:

$y + \frac{1}{3}(8000) = 9666\frac{2}{3} \Rightarrow 3y + 8000 = 29,000 \Rightarrow y = 7000$. Finally, substitute $y = 7000$ and $z = 8000$

into R_1: $x + 7000 + 8000 = 25,000 \Rightarrow x = 10,000$.

The loans were: $10,000 at 8%, $7000 at 10%, and $8000 at 9%.

68. Let x = the amount of money at 3%, y = the amount of money at 4%, and z = the amount of money at 4.5%.

Using the information yields: $x + y + z = 10,000$

$$y = \frac{1}{3}x \Rightarrow -\frac{1}{3}x + y = 0$$

$$.03x + .04y + .045z = 400$$

This system yields: $\begin{bmatrix} 1 & 1 & 1 & | & 10,000 \\ -\frac{1}{3} & 1 & 0 & | & 0 \\ .03 & .04 & .045 & | & 400 \end{bmatrix} \Rightarrow \text{ which by } \left.\begin{matrix} \frac{1}{3}R_1 + R_2 \\ \\ 1000R_3 \end{matrix}\right\} \rightarrow \begin{bmatrix} 1 & 1 & 1 & | & 10,000 \\ 0 & \frac{4}{3} & \frac{1}{3} & | & \frac{10,000}{3} \\ 30 & 40 & 45 & | & 400,000 \end{bmatrix}$

$$\Rightarrow \left.\begin{matrix} \frac{3}{4}R_2 \\ \\ -30R_1 + R_3 \end{matrix}\right\} \rightarrow \begin{bmatrix} 1 & 1 & 1 & | & 10,000 \\ 0 & 1 & \frac{1}{4} & | & 2500 \\ 0 & 10 & 15 & | & 100,000 \end{bmatrix} \Rightarrow \text{ which by } -10R_2 + R_3 \rightarrow \begin{bmatrix} 1 & 1 & 1 & | & 10,000 \\ 0 & 1 & \frac{1}{4} & | & 2500 \\ 0 & 0 & \frac{25}{2} & | & 75,000 \end{bmatrix} \Rightarrow$$

$$\frac{2}{25}R_3 \rightarrow \begin{bmatrix} 1 & 1 & 1 & | & 10,000 \\ 0 & 1 & \frac{1}{4} & | & 2500 \\ 0 & 0 & 1 & | & 6000 \end{bmatrix}.$$

From this matrix, we have the resulting equation: $z = 6000$. Now substituting $z = 6000$ into R_2 yields:

$y + \frac{1}{4}(6000) = 2500 \Rightarrow y + 1500 = 2500 \Rightarrow y = 1000$. Finally, substitute $y = 1000$ and $z = 6000$

into R_1: $x + 1000 + 6000 = 10,000 \Rightarrow x = 3000$.

The amounts were: $3000 at 3%, $1000 at 4%, and $6000 at 4.5%.

69. (a) Using $f(x) = ax^2 + bx + c$ and the data from the table yields:

From the point (1990, 11): $11 = a(1990)^2 + b(1990) + c \Rightarrow 1990^2a + 1990b + c = 11.$

From the point (2010, 10): $10 = a(2010)^2 + b(2010) + c \Rightarrow 2010^2a + 2010b + c = 10.$

From the point (2030, 6): $6 = a(2030)^2 + b(2030) + c \Rightarrow 2030^2a + 2030b + c = 6.$

(b) Now using the capabilities of the calculator, the Reduced Row Echelon Method of:

$$\begin{bmatrix} 1990^2 & 1990 & 1 & | & 11 \\ 2010^2 & 2010 & 1 & | & 10 \\ 2030^2 & 2030 & 1 & | & 6 \end{bmatrix} = \begin{bmatrix} 1 & 0 & 0 & | & -.00375 \\ 0 & 1 & 0 & | & 14.95 \\ 0 & 0 & 1 & | & -14{,}889.125 \end{bmatrix}.$$

Therefore the equation is: $f(x) = -.00375x^2 + 14.95x - 14{,}889.125.$

(c) See Figure 69.

(d) Answers will vary, for example in 2015 the predicted ratio is $f(2015) \approx 9.3.$

70. (a) Using $f(x) = ax^2 + bx + c$ and the data from the table yields:

From the point (1958, 315): $315 = a(1958)^2 + b(1958) + c \Rightarrow 1958^2a + 1958b + c = 315.$

From the point (1973, 325): $325 = a(1973)^2 + b(1973) + c \Rightarrow 1973^2a + 1973b + c = 325.$

From the point (2003, 376): $376 = a(2003)^2 + b(2003) + c \Rightarrow 2003^2a + 2003b + c = 376.$

(b) Now using the capabilities of the calculator, the Reduced Row Echelon Method of:

$$\begin{bmatrix} 1958^2 & 1958 & 1 & | & 315 \\ 1973^2 & 1973 & 1 & | & 325 \\ 2003^2 & 2003 & 1 & | & 376 \end{bmatrix} = \begin{bmatrix} 1 & 0 & 0 & | & .022963 \\ 0 & 1 & 0 & | & -89.6007 \\ 0 & 0 & 1 & | & 87{,}718.7 \end{bmatrix}.$$

Therefore the equation is: $f(x) \approx .022963x^2 - 89.6007x + 87{,}718.7.$

(c) See Figure 70.

(d) Answers will vary, for example in 2010 this level might be: $f(2010) \approx 394.$

[1985, 2035] by [5, 12]   [1955, 2010] by [310, 390]
Xscl = 5   Yscl = 1   Xscl = 5   Yscl = 10

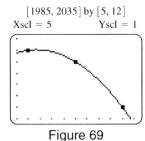

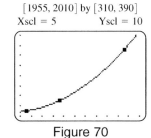

Figure 69   Figure 70

71. (a) At intersection A incoming traffic is equal to $x + 5$. The outgoing traffic is given by $y + 7$. Therefore, $x + 5 = y + 7$. The incoming traffic at intersection B is $z + 6$ and the outgoing traffic is $x + 3$, so $z + 6 = x + 3$. Finally at intersection C, the incoming flow is $y + 3$ and the outgoing flow is $z + 4$, so $y + 3 = z + 4$.

    (b) The three equations are:

    $x + 5 = y + 7 \Rightarrow x - y = 2$

    $z + 6 = x + 3 \Rightarrow x - z = 3$

    $y + 3 = z + 4 \Rightarrow y - z = 1$ which can be represented by:

    $$\begin{bmatrix} 1 & -1 & 0 & | & 2 \\ 1 & 0 & -1 & | & 3 \\ 0 & 1 & -1 & | & 1 \end{bmatrix}, \text{ which by } -R_1 + R_2 \rightarrow \begin{bmatrix} 1 & -1 & 0 & | & 2 \\ 0 & 1 & -1 & | & 1 \\ 0 & 1 & -1 & | & 1 \end{bmatrix} \Rightarrow -R_2 + R_3 \rightarrow \begin{bmatrix} 1 & -1 & 0 & | & 2 \\ 0 & 1 & -1 & | & 1 \\ 0 & 0 & 0 & | & 0 \end{bmatrix}.$$

    Since in the result of R_3 $0 = 0$, there are infinite solutions. They are: $y - z = 1 \Rightarrow y = z + 1$ and $x - z = 3 \Rightarrow x = z + 3$. The solution is: $\{(z + 3, z + 1, z), \text{ where } z \geq 0\}$.

    (c) There are infinitely many solutions since some cars could be driving around the block continually.

72. (a) At intersection A incoming traffic is equal to $x + 7$. The outgoing traffic is given by $y + 4$. Therefore, $x + 7 = y + 4$. The incoming traffic at intersection B is $4 + 5$ and the outgoing traffic is $x + z$, so $4 + 5 = x + z$. Finally at intersection C, the incoming flow is $y + 8$ and the outgoing flow is $9 + 4$, so $y + 8 = 9 + 4$.

    (b) The three equations are:

    $x + 7 = y + 4 \Rightarrow x - y = -3$

    $4 + 5 = x + z \Rightarrow x + z = 9$

    $y + 8 = 4 + 4 \Rightarrow y = 5$ which can be represented by:

    Since $y = 5$, we substitute this into [equation 1]: $x - (5) = -3 \Rightarrow x = 2$. Finally, substitute $x = 2$ into [equation 2]: $(2) + z = 9 \Rightarrow z = 7$. The solution is: $\{(2, 5, 7)\}$.

    (c) There is one solution: 2 vehicles per minute from B to A, 5 from A to C, and 7 out of B.

73. (a) Using the given model and information from the table the equations are:

    $a + 871b + 11.5c + 3d = 239$

    $a + 847b + 12.2c + 2d = 234$

    $a + 685b + 10.6c + 5d = 192$

    $a + 969b + 14.2c + 1d = 343$

    (b) The equations can be represented by: $\begin{bmatrix} 1 & 871 & 11.5 & 3 & | & 239 \\ 1 & 847 & 12.2 & 2 & | & 234 \\ 1 & 685 & 10.6 & 5 & | & 192 \\ 1 & 969 & 14.2 & 1 & | & 343 \end{bmatrix}.$

    Using the capabilities of the calculator, the Reduced Row Echelon Method produces the solutions:

    $a \approx -715.457, b \approx .34756, c \approx 48.6585, d \approx 30.71951.$

    (c) The equation is: $F = -715.457 + .34756A + 48.6585P + 30.71951W.$

    (d) Input the given data into the equation: $F = -715.457 + .34756(960) + 48.6585(12.6) + 30.71951(3) \Rightarrow$ $F \approx 323.45623$. This is approximately 323, which is close to the actual value of 320.

74. (a) Using the given model and information from the table the equations can be represented by:

$$\begin{bmatrix} 1 & 19 & 57.5 & 32 & | & 125 \\ 1 & 26 & 65 & 42 & | & 316 \\ 1 & 30 & 72 & 48 & | & 436 \\ 1 & 30.5 & 75 & 54 & | & 514 \end{bmatrix}$$

    (b) Using the capabilities of the calculator, the Reduced Row Echelon Method produces the solutions:

$a \approx -552.272, b \approx 8.733, c \approx 2.859, d \approx 10.843$

    (c) $N = 24, L = 63$ and $C = 39, W \approx -552.272 + 8.733(24) + 2.859(63) + 10.843(39) \approx 260$

       A bear with a 24-inch neck, 63-inch length and 39-inch chest weighs approximately 260 pounds.

75. The given system of equations can be represented by: $\begin{bmatrix} 1 & 1 & 1 & | & 50 \\ -2 & 0 & 1 & | & 0 \\ 15.99 & 12.99 & 10.19 & | & 618.5 \end{bmatrix}$.

    Using the capabilities of the calculator, the Reduced Row Echelon Method produces the solutions:

$a \approx 11.92$ lb, $b \approx 14.23$ lb, and $c \approx 23.85$ lb.

76. The given system of equations can be represented by: $\begin{bmatrix} 1 & 1 & 1 & | & 50 \\ -2 & 0 & 1 & | & 0 \\ 15.99 & 12.99 & 12.49 & | & 618.5 \end{bmatrix}$.

    Using the capabilities of the calculator, the Reduced Row Echelon Method produces the solutions:

$a \approx -15.5$ lb, $b \approx 96.5$ lb, and $c \approx -31$ lb.

The answer is not reasonable, because you cannot add a negative number of pounds of coffee. Note that you cannot obtain a mixture of coffee that cost \$12.37 per pound if each ingredient costs more than \$12.37 per pound.

Reviewing Basic Concepts (Sections 7.1—7.3)

1. Multiply the first equation by 5, the second equation by 2, and subtract to eliminate the x-variable.

$$\begin{array}{r} 10x - 15y = 90 \\ 10x + 4y = 14 \\ \hline -19y = 76 \end{array} \Rightarrow y = -4. \text{ Substitute } (-4) \text{ for } y \text{ in the second equation:}$$

$10x + 4(-4) = 14 \Rightarrow 10x = 30 \Rightarrow x = 3.$ The solution is: $\{(3, -4)\}$.

2. The equations are: $2x + y = -4 \Rightarrow y_1 = -2x - 4$ and $-x + 2y = 2 \Rightarrow 2y = x + 2 \Rightarrow y_2 = \frac{1}{2}x + 1$.

Graph y_1 and y_2 in the same window, the intersection or solution is: $\{(-2, 0)\}$. See Figure 2 on next page.

3. Using substitution, first solve [equation 2] for x, $x + 2y = 2 \Rightarrow x = -2y + 2$ [equation 3]. Substitute $(-2y + 2)$ in for x, in [equation 1]: $5(-2y + 2) + 10y = 10 \Rightarrow -10y + 10 + 10y = 10 \Rightarrow 10 = 10$.

Since this is true, there is an infinite number solutions. They are: $(-2y + 2, y)$ or $\left(x, \frac{2 - x}{2}\right)$.

4. Subtract [equation 2] from [equation 1] to eliminate both variables.

$$\begin{array}{r} x - y = 6 \\ x - y = 4 \\ \hline 0 = 2 \end{array} \Rightarrow \text{ Since this is false the solution is: } \varnothing.$$

5. First solve [equation 1] for y and substitute into [equation 2]: $6x + 2y = 10 \Rightarrow 2y = 10 - 6x \Rightarrow y = 5 - 3x$.

    Now substitute: $2x^2 - 3(-3x + 5) = 11 \Rightarrow 2x^2 + 9x - 15 = 11 \Rightarrow 2x^2 + 9x - 26 = 0 \Rightarrow$

    $(2x + 13)(x - 2) = 0 \Rightarrow x = -6.5$ and $x = 2$. If $x = -6.5$ then by substitution:

    $y = -3(-6.5) + 5 \Rightarrow y = 19.5 + 5 \Rightarrow y = 24.5$. If $x = 2$ then by substitution: $y = -3(2) + 5 \Rightarrow y = -1$.

    The solution is: $\{(2, -1)$ and $(-6.5, 24.5)\}$. Graphing both equations in the same window shows two

    intersection points which support our analytic results. See Figure 5.

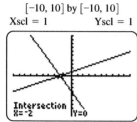

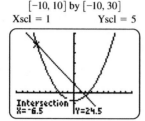

Figure 2   Figure 5

6. The equations can be represented by: $\begin{bmatrix} 1 & 1 & 1 & | & 1 \\ -1 & 1 & 1 & | & 5 \\ 0 & 1 & 2 & | & 5 \end{bmatrix}$, $R_1 + R_2 \rightarrow \begin{bmatrix} 1 & 1 & 1 & | & 1 \\ 0 & 2 & 2 & | & 6 \\ 0 & 1 & 2 & | & 5 \end{bmatrix} \Rightarrow$

    $R_2 - 2R_3 \rightarrow \begin{bmatrix} 1 & 1 & 1 & | & 1 \\ 0 & 2 & 2 & | & 6 \\ 0 & 0 & -2 & | & -4 \end{bmatrix}$.

    Now solve the resulting equation R_3: $-2z = -4 \Rightarrow z = 2$. Substitute $z = 2$ into the resulting equation R_2:

    $2y + 2(2) = 6 \Rightarrow 2y = 2 \Rightarrow y = 1$. Finally, substitute $y = 1$ and $z = 2$ into R_1:

    $x + (1) + (2) = 1 \Rightarrow x = -2$. The solution is: $\{(-2, 1, 2)\}$.

7. Using the Reduced Row Echelon method, the given system of equations yields the matrix: $\begin{bmatrix} 2 & 4 & 4 & | & 4 \\ 1 & 3 & 1 & | & 4 \\ -1 & 3 & 2 & | & -1 \end{bmatrix}$.

    Using the capabilities of the calculator, the Reduced Row Echelon Method produces the solutions:

    $x = 2$, $y = 1$, and $z = -1$.

8. Solve the augmented matrix: $\begin{bmatrix} 2 & 1 & 2 & | & 10 \\ 1 & 0 & 2 & | & 5 \\ 1 & -2 & 2 & | & 1 \end{bmatrix}$, which by $\left.\begin{matrix} R_1 - 2R_2 \\ R_1 - 2R_3 \end{matrix}\right\} \rightarrow \begin{bmatrix} 2 & 1 & 2 & | & 10 \\ 0 & 1 & -2 & | & 0 \\ 0 & 5 & -2 & | & 8 \end{bmatrix} \Rightarrow$

    $-5R_2 + R_3 \rightarrow \begin{bmatrix} 2 & 1 & 2 & | & 10 \\ 0 & 1 & -2 & | & 0 \\ 0 & 0 & 8 & | & 8 \end{bmatrix}$.

    Now solve the resulting equation R_3: $8z = 8 \Rightarrow z = 1$. Substitute $z = 1$ into the resulting equation R_2:

    $y - 2(1) = 0 \Rightarrow y = 2$. Finally, substitute $y = 2$ and $z = 1$ into R_1:

    $2x + (2) + 2(1) = 10 \Rightarrow 2x = 6 \Rightarrow x = 3$. The solution is: $\{(3, 2, 1)\}$.

9. Let x represent the number of sets with stereo sound and y represent the number of sets without stereo sound.

    Then from the information the equations are:

    $x + y = 32,000,000$ $[1]$

    $10y = 19x$ $[2]$

    First solve equation [1] for x: $x + y = 32,000,000 \Rightarrow y = 32,000,000 - x.$ $[3]$

    Substitute {3] into [2]: $10(32,000,000 - x) = 19x \Rightarrow 320,000,000 - 10x = 19x \Rightarrow 320,000,000 = 29x \Rightarrow$

    $x = 11,034,482.76.$ Finally substitute $x = 11,034,482.76$ into [3]:

    $y = 32,000,000 - 11,034,483 \Rightarrow y = 20,965,517.$

    There were approximately: 11.3 million with stereo sound; and 21million without stereo sound sold.

10. Let $x =$ amount at 8%, $y =$ amount at 11%, and $z =$ amount at 14%. Then from the information:

    $x + y + z = 5000$

    $.08x + .11y + .14z = 595$

    $z = x + y \Rightarrow x + y - z = 0$

    Using the Reduced Row Echelon method, the given system of equations yields the matrix:

    $$\begin{bmatrix} 1 & 1 & 1 & | & 5000 \\ .08 & .11 & .14 & | & 595 \\ 1 & 1 & -1 & | & 0 \end{bmatrix}.$$

    Using the capabilities of the calculator, the Reduced Row Echelon Method produces the solutions:

    $x = 1000, y = 1500,$ and $z = 2500.$

    There was $1000 invested at 8%, $1500 invested at 11%, and $2500 invested at 14%.

7.4: Matrix Properties and Operations

1. Because the number of rows equals the number of columns, this is a 2×2 square matrix.

2. Because the number of rows equals the number of columns, this is a 3×3 square matrix.

3. This is a 3×4 matrix.

4. This is a 2×4 matrix.

5. Because there is only one column, this is a 2×1 column matrix.

6. Because there is only one row, this is a 1×2 row matrix.

7. Because the number of rows equals the number of columns, there is only one row and one column,

    this is a 1×1 square, row, column matrix.

8. Because all the entries are zero, this is a 2×5 zero matrix.

9. For $\begin{bmatrix} w & x \\ y & z \end{bmatrix} = \begin{bmatrix} 3 & 2 \\ -1 & 4 \end{bmatrix}$, $w = 3, x = 2, y = -1,$ and $z = 4.$

10. For $\begin{bmatrix} 2 & 5 & 6 \\ 1 & m & n \end{bmatrix} = \begin{bmatrix} z & y & w \\ 1 & 8 & -2 \end{bmatrix}$, $z = 2, y = 5, w = 6, m = 8,$ and $n = -2.$

11. For $\begin{bmatrix} 0 & 5 & x \\ -1 & 3 & y+2 \\ 4 & 1 & z \end{bmatrix} = \begin{bmatrix} 0 & w+3 & 6 \\ -1 & 3 & 0 \\ 4 & 1 & 8 \end{bmatrix}$,

$w + 3 = 5 \Rightarrow w = 2, x = 6, y + 2 = 0 \Rightarrow y = -2,$ and $z = 8.$

12. For $\begin{bmatrix} 3+x & 4 & t \\ 5 & 8-w & y+1 \\ -4 & 3 & 2r \end{bmatrix} = \begin{bmatrix} 9 & 4 & 6 \\ z+3 & w & 9 \\ p & q & r \end{bmatrix}$,

$3 + x = 9 \Rightarrow x = 6, t = 6, z + 3 = 5 \Rightarrow z = 2, 8 - w = w \Rightarrow 8 = 2w \Rightarrow w = 4, y + 1 = 9 \Rightarrow$

$y = 8, p = -4, q = 3,$ and $2r = r \Rightarrow 3r = 0 \Rightarrow r = 0.$

13. For $\begin{bmatrix} -7+z & 4r & 8s \\ 6p & 2 & 5 \end{bmatrix} + \begin{bmatrix} -9 & 8r & 3 \\ 2 & 5 & 4 \end{bmatrix} = \begin{bmatrix} 2 & 36 & 27 \\ 20 & 7 & 12a \end{bmatrix}.$ Then:

$-7 + z - 9 = 2 \Rightarrow z = 18.$

$4r + 8r = 36 \Rightarrow r = 3.$

$8s + 3 = 27 \Rightarrow 8s = 24 \Rightarrow s = 3.$

$6p + 2 = 20 \Rightarrow 6p = 18 \Rightarrow p = 3.$

$5 + 4 = 12a \Rightarrow 9 = 12a \Rightarrow a = \dfrac{3}{4}.$

14. For $\begin{bmatrix} a+2 & 3z+1 & 5m \\ 8k & 0 & 3 \end{bmatrix} + \begin{bmatrix} 3a & 2z & 5m \\ 2k & 5 & 6 \end{bmatrix} = \begin{bmatrix} 10 & -14 & 80 \\ 10 & 5 & 9 \end{bmatrix}.$ Then:

$a + 2 + 3a = 10 \Rightarrow 4a = 8 \Rightarrow a = 2.$

$3z + 1 + 2z = -14 \Rightarrow 5z = -15 \Rightarrow z = -3.$

$5m + 5m = 80 \Rightarrow 10m = 80 \Rightarrow m = 8.$

$8k + 2k = 10 \Rightarrow 10k = 10 \Rightarrow k = 1.$

15. The two matrices must have the same dimensions. To find the sum, add the corresponding entries. The sum will be a matrix with the same dimension.

16. Multiply each entry of the matrix by the scalar to find the entries in the product. The product has the same dimensions as the original matrix.

17. $\begin{bmatrix} 6 & -9 & 2 \\ 4 & 1 & 3 \end{bmatrix} + \begin{bmatrix} -8 & 2 & 5 \\ 6 & -3 & 4 \end{bmatrix} = \begin{bmatrix} 6-8 & -9+2 & 2+5 \\ 4+6 & 1-3 & 3+4 \end{bmatrix} = \begin{bmatrix} -2 & -7 & 7 \\ 10 & -2 & 7 \end{bmatrix}.$

18. $\begin{bmatrix} 9 & 4 \\ -8 & 2 \end{bmatrix} + \begin{bmatrix} -3 & 2 \\ -4 & 7 \end{bmatrix} = \begin{bmatrix} 9-3 & 4+2 \\ -8-4 & 2+7 \end{bmatrix} = \begin{bmatrix} 6 & 6 \\ -12 & 9 \end{bmatrix}.$

19. $\begin{bmatrix} -6 & 8 \\ 0 & 0 \end{bmatrix} - \begin{bmatrix} 0 & 0 \\ -4 & -2 \end{bmatrix} = \begin{bmatrix} -6-0 & 8-0 \\ 0-(-4) & 0-(-2) \end{bmatrix} = \begin{bmatrix} -6 & 8 \\ 4 & 2 \end{bmatrix}.$

20. $\begin{bmatrix} 1 & -4 \\ 2 & -3 \\ -8 & 4 \end{bmatrix} - \begin{bmatrix} -6 & 9 \\ -2 & 5 \\ -7 & -12 \end{bmatrix} = \begin{bmatrix} 1-(-6) & -4-9 \\ 2-(-2) & -3-5 \\ -8-(-7) & 4-(-12) \end{bmatrix} = \begin{bmatrix} 7 & -13 \\ 4 & -8 \\ -1 & 16 \end{bmatrix}.$

21. $\begin{bmatrix} 6 & -2 \\ 5 & 4 \end{bmatrix} + \begin{bmatrix} -1 & 7 \\ 7 & -4 \end{bmatrix} = \begin{bmatrix} 6-1 & -2+7 \\ 5+7 & 4-4 \end{bmatrix} = \begin{bmatrix} 5 & 5 \\ 12 & 0 \end{bmatrix}.$

22. $\begin{bmatrix} 12 & -5 \\ 10 & 3 \end{bmatrix} - \begin{bmatrix} 6 & 9 \\ -2 & 0 \end{bmatrix} = \begin{bmatrix} 12-6 & -5-9 \\ 10-(-2) & 3-0 \end{bmatrix} = \begin{bmatrix} 6 & -14 \\ 12 & 3 \end{bmatrix}.$

23. $\begin{bmatrix} -8 & 4 & 0 \\ 2 & 5 & 0 \end{bmatrix} + \begin{bmatrix} 6 & 3 \\ 8 & 9 \end{bmatrix}.$ Matrices must be the same size to add or subtract. We cannot add $(2 \times 3) + (2 \times 2)$.

24. $\begin{bmatrix} 2 \\ 3 \end{bmatrix} - \begin{bmatrix} 8 & 1 \\ 9 & 4 \end{bmatrix}.$ Matrices must be the same size to add or subtract. We cannot add $(2 \times 1) + (2 \times 2)$.

25. $\begin{bmatrix} 9 & 4 & 1 & -2 \\ 5 & -6 & 3 & 4 \\ 2 & -5 & 1 & 2 \end{bmatrix} - \begin{bmatrix} -2 & 5 & 1 & 3 \\ 0 & 1 & 0 & 2 \\ -8 & 3 & 2 & 1 \end{bmatrix} + \begin{bmatrix} 2 & 4 & 0 & 3 \\ 4 & -5 & 1 & 6 \\ 2 & -3 & 0 & 8 \end{bmatrix} =$

$\begin{bmatrix} 9+2+2 & 4-5+4 & 1-1+0 & -2-3+3 \\ 5-0+4 & -6-1-5 & 3-0+1 & 4-2+6 \\ 2+8+2 & -5-3-3 & 1-2+0 & 2-1+8 \end{bmatrix} = \begin{bmatrix} 13 & 3 & 0 & -2 \\ 9 & -12 & 4 & 8 \\ 12 & -11 & -1 & 9 \end{bmatrix}.$

26. $\begin{bmatrix} 6 & -2 & 4 \\ -2 & 5 & 8 \\ 1 & 0 & 2 \end{bmatrix} + \begin{bmatrix} 3 & 0 & 8 \\ 1 & -2 & 4 \\ 6 & 9 & -2 \end{bmatrix} - \begin{bmatrix} -4 & 2 & 1 \\ 0 & 3 & -2 \\ 4 & 2 & 0 \end{bmatrix} = \begin{bmatrix} 6+3+4 & -2+0-2 & 4+8-1 \\ -2+1-0 & 5-2-3 & 8+4+2 \\ 1+6-4 & 0+9-2 & 2-2-0 \end{bmatrix} =$

$\begin{bmatrix} 13 & -4 & 11 \\ -1 & 0 & 14 \\ 3 & 7 & 0 \end{bmatrix}.$

27. $2\begin{bmatrix} 2 & -1 \\ 5 & 1 \\ 0 & 3 \end{bmatrix} + \begin{bmatrix} 5 & 0 \\ 7 & -3 \\ 1 & 1 \end{bmatrix} - \begin{bmatrix} 9 & -4 \\ 4 & 4 \\ 1 & 6 \end{bmatrix} = \begin{bmatrix} 4+5-9 & -2+0+4 \\ 10+7-4 & 2-3-4 \\ 0+1-1 & 6+1-6 \end{bmatrix} = \begin{bmatrix} 0 & 2 \\ 13 & -5 \\ 0 & 1 \end{bmatrix}.$

28. $-3\begin{bmatrix} 3 & 8 \\ -1 & -9 \end{bmatrix} + 5\begin{bmatrix} 4 & -8 \\ 1 & 6 \end{bmatrix} = \begin{bmatrix} -9+20 & -24-40 \\ 3+5 & 27+30 \end{bmatrix} = \begin{bmatrix} 11 & -64 \\ 8 & 57 \end{bmatrix}.$

29. $2\begin{bmatrix} 2 & -1 & -1 \\ -1 & 2 & -1 \\ -1 & -1 & 2 \end{bmatrix} + 3\begin{bmatrix} 1 & 2 & 3 \\ 2 & 1 & 3 \\ 2 & 3 & 1 \end{bmatrix} = \begin{bmatrix} 4+3 & -2+6 & -2+9 \\ -2+6 & 4+3 & -2+9 \\ -2+6 & -2+9 & 4+3 \end{bmatrix} = \begin{bmatrix} 7 & 4 & 7 \\ 4 & 7 & 7 \\ 4 & 7 & 7 \end{bmatrix}.$

30. $3\begin{bmatrix} 1 & 0 & 3 & -1 \\ 0 & 1 & 2 & -1 \\ 1 & 0 & -3 & 1 \end{bmatrix} - 4\begin{bmatrix} -1 & 0 & 0 & 4 \\ 0 & -1 & 3 & 2 \\ 2 & 0 & 1 & -1 \end{bmatrix} = \begin{bmatrix} 3+4 & 0-0 & 9-0 & -3-16 \\ 0-0 & 3+4 & 6-12 & -3-8 \\ 3-8 & 0-0 & -9-4 & 3+4 \end{bmatrix} =$

$\begin{bmatrix} 7 & 0 & 9 & -19 \\ 0 & 7 & -6 & -11 \\ -5 & 0 & -13 & 7 \end{bmatrix}.$

31. $3\begin{bmatrix} 6 & -1 & 4 \\ 2 & 8 & -3 \\ -4 & 5 & 6 \end{bmatrix} + 5\begin{bmatrix} -2 & -8 & -6 \\ 4 & 1 & 3 \\ 2 & -1 & 5 \end{bmatrix} = \begin{bmatrix} 18 & -3 & 12 \\ 6 & 24 & -9 \\ -12 & 15 & 18 \end{bmatrix} + \begin{bmatrix} -10 & -40 & -30 \\ 20 & 5 & 15 \\ 10 & -5 & 25 \end{bmatrix} =$

$\begin{bmatrix} 18-10 & -3-40 & 12-30 \\ 6+20 & 24+5 & -9+15 \\ -12+10 & 15-5 & 18+25 \end{bmatrix} = \begin{bmatrix} 8 & -43 & -18 \\ 26 & 29 & 6 \\ -2 & 10 & 43 \end{bmatrix}.$

32. $4\begin{bmatrix} 1 & -4 \\ 2 & -3 \\ -8 & 4 \end{bmatrix} - 3\begin{bmatrix} -6 & 9 \\ -2 & 5 \\ -7 & -12 \end{bmatrix} = \begin{bmatrix} 4 & -16 \\ 8 & -12 \\ -32 & 16 \end{bmatrix} + \begin{bmatrix} 18 & -27 \\ 6 & -15 \\ 21 & 36 \end{bmatrix} = \begin{bmatrix} 4+18 & -16-27 \\ 8+6 & -12-15 \\ -32+21 & 16+36 \end{bmatrix} =$

$\begin{bmatrix} 22 & -43 \\ 14 & -27 \\ -11 & 52 \end{bmatrix}$.

33. $2[A] = 2\begin{bmatrix} -2 & 4 \\ 0 & 3 \end{bmatrix} = \begin{bmatrix} -4 & 8 \\ 0 & 6 \end{bmatrix}$.

34. $-3[B] = -3\begin{bmatrix} -6 & 2 \\ 4 & 0 \end{bmatrix} = \begin{bmatrix} 18 & -6 \\ -12 & 0 \end{bmatrix}$.

35. $2[A] - [B] = 2\begin{bmatrix} -2 & 4 \\ 0 & 3 \end{bmatrix} - \begin{bmatrix} -6 & 2 \\ 4 & 0 \end{bmatrix} = \begin{bmatrix} -4+6 & 8-2 \\ 0-4 & 6-0 \end{bmatrix} = \begin{bmatrix} 2 & 6 \\ -4 & 6 \end{bmatrix}$.

36. $-2[A] + [B] = -2\begin{bmatrix} -2 & 4 \\ 0 & 3 \end{bmatrix} + \begin{bmatrix} -6 & 2 \\ 4 & 0 \end{bmatrix} = \begin{bmatrix} 4-6 & -8+2 \\ 0+4 & -6+0 \end{bmatrix} = \begin{bmatrix} -2 & -6 \\ 4 & -6 \end{bmatrix}$.

37. $5[A] + .5[B] = 5\begin{bmatrix} -2 & 4 \\ 0 & 3 \end{bmatrix} + .5\begin{bmatrix} -6 & 2 \\ 4 & 0 \end{bmatrix} = \begin{bmatrix} -10-3 & 20+1 \\ 0+2 & 15+0 \end{bmatrix} = \begin{bmatrix} -13 & 21 \\ 2 & 15 \end{bmatrix}$.

38. $-4[A] + 1.5[B] = -4\begin{bmatrix} -2 & 4 \\ 0 & 3 \end{bmatrix} + 1.5\begin{bmatrix} -6 & 2 \\ 4 & 0 \end{bmatrix} = \begin{bmatrix} 8-9 & -16+3 \\ 0+6 & -12+0 \end{bmatrix} = \begin{bmatrix} -1 & -13 \\ 6 & -12 \end{bmatrix}$.

39. $[A] = ([A]+[B]) - [B] = \begin{bmatrix} 6 & 12 & 0 \\ -10 & -4 & 11 \end{bmatrix} - \begin{bmatrix} 4 & 6 & -5 \\ -6 & 3 & 2 \end{bmatrix} = \begin{bmatrix} 6-4 & 12-6 & 0+5 \\ -10+6 & -4-3 & 11-2 \end{bmatrix} =$

$\begin{bmatrix} 2 & 6 & 5 \\ -4 & -7 & 9 \end{bmatrix}$.

40. $[B] = [A] - ([A]-[B]) = \begin{bmatrix} 3 & 6 & 5 \\ -2 & 1 & 4 \end{bmatrix} - \begin{bmatrix} 9 & 0 & -5 \\ -4 & 6 & -3 \end{bmatrix} = \begin{bmatrix} 3-9 & 6-0 & 5+5 \\ -2+4 & 1-6 & 4+3 \end{bmatrix} =$

$\begin{bmatrix} -6 & 6 & 10 \\ 2 & -5 & 7 \end{bmatrix}$.

41. With A: 4×2 and B: 2×4, for AB the number of columns of A is the same as the number of rows of B, therefore AB has dimensions: 4 (rows of A) $\times$ 4 (columns of B) or AB is: 4×4. For BA the number of columns of B is the same as the number of rows of A, therefore BA has dimensions:

2 (rows of B) $\times$ 2 (columns of A) or BA is: 2×2.

42. With A: 3×1 and B: 1×3, for AB the number of columns of A is the same as the number of rows of B, therefore AB has dimensions: 3 (rows of A) $\times$ 3 (columns of B) or AB is: 3×3. For BA the number of columns of B is the same as the number of rows of A, therefore BA has dimensions:

1 (rows of B) $\times$ 1 (columns of A) or BA is: 1×1.

43. With A: 3×5 and B: 5×2, for AB the number of columns of A is the same as the number of rows of B, therefore AB has dimensions: 3 (rows of A) $\times$ 2 (columns of B) or AB is: 3×2. For BA the number of columns of B is not the same as the number of rows of A, therefore BA is not defined.

44. With A: 7×3 and B: 2×7, for AB the number of columns of A is not the same as the number of rows of B, therefore AB is not defined. For BA the number of columns of B is the same as the number of rows of A, therefore BA has dimensions: 2 (rows of B) $\times$ 3 (columns of A) or BA is: 2×3.

45. With A: 4×3 and B: 2×5, for AB the number of columns of A is not the same as the number of rows of B, therefore AB is not defined. For BA the number of columns of B is not the same as the number of rows of A, therefore BA is also not defined.

46. With A: 1×6 and B: 2×4, for AB the number of columns of A is not the same as the number of rows of B, therefore AB is not defined. For BA the number of columns of B is not the same as the number of rows of A, therefore BA is also not defined.

47. . . . the number of columns of M equal the number of rows of N.

48. . . . by multiplying the first row elements in A and the second column elements in B and then adding these products.

49. $\begin{bmatrix} p & q \\ r & s \end{bmatrix} \begin{bmatrix} a & c \\ b & d \end{bmatrix} = \begin{bmatrix} pa + qb & pc + qd \\ ra + sb & rc + sd \end{bmatrix}$.

50. $\begin{bmatrix} a & b & c \\ d & e & f \\ g & h & i \end{bmatrix} \begin{bmatrix} x \\ y \\ z \end{bmatrix} = \begin{bmatrix} ax + by + cz \\ dx + ey + fz \\ gx + hy + iz \end{bmatrix}$.

51. $\begin{bmatrix} 3 & -4 & 1 \\ 5 & 0 & 2 \end{bmatrix} \begin{bmatrix} -1 \\ 4 \\ 2 \end{bmatrix} = \begin{bmatrix} 3(-1) - 4(4) + 1(2) \\ 5(-1) + 0(4) + 2(2) \end{bmatrix} = \begin{bmatrix} -3 - 16 + 2 \\ -5 + 0 + 4 \end{bmatrix} = \begin{bmatrix} -17 \\ -1 \end{bmatrix}$.

52. $\begin{bmatrix} -6 & 3 & 5 \\ 2 & 9 & 1 \end{bmatrix} \begin{bmatrix} -2 \\ 0 \\ 3 \end{bmatrix} = \begin{bmatrix} -6(-2) + 3(0) + 5(3) \\ 2(-2) + 9(0) + 1(3) \end{bmatrix} = \begin{bmatrix} 12 + 0 + 15 \\ -4 + 0 + 3 \end{bmatrix} = \begin{bmatrix} 27 \\ -1 \end{bmatrix}$.

53. $\begin{bmatrix} 5 & 2 \\ -1 & 4 \end{bmatrix} \begin{bmatrix} 3 & -2 \\ 1 & 0 \end{bmatrix} = \begin{bmatrix} 5(3) + 2(1) & 5(-2) + 2(0) \\ -1(3) + 4(1) & -1(-2) + 4(0) \end{bmatrix} = \begin{bmatrix} 15 + 2 & -10 + 0 \\ -3 + 4 & 2 + 0 \end{bmatrix} = \begin{bmatrix} 17 & -10 \\ 1 & 2 \end{bmatrix}$.

54. $\begin{bmatrix} -4 & 0 \\ 1 & 3 \end{bmatrix} \begin{bmatrix} -2 & 4 \\ 0 & 1 \end{bmatrix} = \begin{bmatrix} -4(-2) + 0(0) & -4(4) + 0(1) \\ 1(-2) + 3(0) & 1(4) + 3(1) \end{bmatrix} = \begin{bmatrix} 8 + 0 & -16 + 0 \\ -2 + 0 & 4 + 3 \end{bmatrix} = \begin{bmatrix} 8 & -16 \\ -2 & 7 \end{bmatrix}$.

55. $\begin{bmatrix} 2 & 2 & -1 \\ 3 & 0 & 1 \end{bmatrix} \begin{bmatrix} 0 & 2 \\ -1 & 4 \\ 0 & 2 \end{bmatrix} = \begin{bmatrix} 2(0) + 2(-1) + (-1)(0) & 2(2) + 2(4) + (-1)(2) \\ 3(0) + 0(-1) + 1(0) & 3(2) + 0(4) + 1(2) \end{bmatrix} =$

$\begin{bmatrix} 0 - 2 + 0 & 4 + 8 - 2 \\ 0 + 0 + 0 & 6 + 0 + 2 \end{bmatrix} = \begin{bmatrix} -2 & 10 \\ 0 & 8 \end{bmatrix}$.

56. $\begin{bmatrix} -9 & 2 & 1 \\ 3 & 0 & 0 \end{bmatrix} \begin{bmatrix} 2 \\ -1 \\ 4 \end{bmatrix} = \begin{bmatrix} -9(2) + 2(-1) + 1(4) \\ 3(2) + 0(-1) + 0(4) \end{bmatrix} = \begin{bmatrix} -18 - 2 + 4 \\ 6 + 0 + 0 \end{bmatrix} = \begin{bmatrix} -16 \\ 6 \end{bmatrix}$.

57. $\begin{bmatrix} -2 & -3 & -4 \\ 2 & -1 & 0 \\ 4 & -2 & 3 \end{bmatrix} \begin{bmatrix} 0 & 1 & 4 \\ 1 & 2 & -1 \\ 3 & 2 & -2 \end{bmatrix} =$

$\begin{bmatrix} -2(0) + (-3)(1) + (-4)(3) & -2(1) + (-3)(2) + (-4)(2) & -2(4) + (-3)(-1) + (-4)(-2) \\ 2(0) + (-1)(1) + 0(3) & 2(1) + (-1)(2) + 0(2) & 2(4) + (-1)(-1) + 0(-2) \\ 4(0) + (-2)(1) + 3(3) & 4(1) + (-2)(2) + 3(2) & 4(4) + (-2)(-1) + 3(-2) \end{bmatrix} =$

$\begin{bmatrix} 0 - 3 - 12 & -2 - 6 - 8 & -8 + 3 + 8 \\ 0 - 1 + 0 & 2 - 2 + 0 & 8 + 1 + 0 \\ 0 - 2 + 9 & 4 - 4 + 6 & 16 + 2 - 6 \end{bmatrix} = \begin{bmatrix} -15 & -16 & 3 \\ -1 & 0 & 9 \\ 7 & 6 & 12 \end{bmatrix}$.

58. $\begin{bmatrix} -1 & 2 & 0 \\ 0 & 3 & 2 \\ 0 & 1 & 4 \end{bmatrix} \begin{bmatrix} 2 & -1 & 2 \\ 0 & 2 & 1 \\ 3 & 0 & -1 \end{bmatrix} =$

$\begin{bmatrix} -1(2) + 2(0) + 0(3) & -1(-1) + 2(2) + 0(0) & -1(2) + 2(1) + 0(-1) \\ 0(2) + 3(0) + 2(3) & 0(-1) + 3(2) + 2(0) & 0(2) + 3(1) + 2(-1) \\ 0(2) + 1(0) + 4(3) & 0(-1) + 1(2) + 4(0) & 0(2) + 1(1) + 4(-1) \end{bmatrix} =$

$\begin{bmatrix} -2 + 0 + 0 & 1 + 4 + 0 & -2 + 2 + 0 \\ 0 + 0 + 6 & 0 + 6 + 0 & 0 + 3 - 2 \\ 0 + 0 + 12 & 0 + 2 + 0 & 0 + 1 - 4 \end{bmatrix} = \begin{bmatrix} -2 & 5 & 0 \\ 6 & 6 & 1 \\ 12 & 2 & -3 \end{bmatrix}$.

59. $\begin{bmatrix} -2 & 4 & 1 \end{bmatrix} \begin{bmatrix} 3 & -2 & 4 \\ 2 & 1 & 0 \\ 0 & -1 & 4 \end{bmatrix} =$

$\begin{bmatrix} -2(3) + 4(2) + 1(0) & -2(-2) + 4(1) + 1(-1) & -2(4) + 4(0) + 1(4) \end{bmatrix} =$
$\begin{bmatrix} -6 + 8 + 0 & 4 + 4 - 1 & -8 + 0 + 4 \end{bmatrix} = \begin{bmatrix} 2 & 7 & -4 \end{bmatrix}$.

60. $\begin{bmatrix} 0 & 3 & -4 \end{bmatrix} \begin{bmatrix} -2 & 6 & 3 \\ 0 & 4 & 2 \\ -1 & 1 & 4 \end{bmatrix} =$

$\begin{bmatrix} 0(-2) + 3(0) + (-4)(-1) & 0(6) + 3(4) + (-4)(1) & 0(3) + 3(2) + (-4)(4) \end{bmatrix} =$
$\begin{bmatrix} 0 + 0 + 4 & 0 + 12 - 4 & 0 + 6 - 16 \end{bmatrix} = \begin{bmatrix} 4 & 8 & -10 \end{bmatrix}$.

61. $BA = \begin{bmatrix} 5 & 1 \\ 0 & -2 \\ 3 & 7 \end{bmatrix} \begin{bmatrix} 4 & -2 \\ 3 & 1 \end{bmatrix} = \begin{bmatrix} 5(4) + 1(3) & 5(-2) + 1(1) \\ 0(4) + (-2)(3) & 0(-2) + (-2)(1) \\ 3(4) + 7(3) & 3(-2) + 7(1) \end{bmatrix} = \begin{bmatrix} 23 & -9 \\ -6 & -2 \\ 33 & 1 \end{bmatrix}$.

62. $AC = \begin{bmatrix} 4 & -2 \\ 3 & 1 \end{bmatrix} \begin{bmatrix} -5 & 4 & 1 \\ 0 & 3 & 6 \end{bmatrix} = \begin{bmatrix} 4(-5) - 2(0) & 4(4) - 2(3) & 4(1) - 2(6) \\ 3(-5) + 1(0) & 3(4) + 1(3) & 3(1) + 1(6) \end{bmatrix} = \begin{bmatrix} -20 & 10 & -8 \\ -15 & 15 & 9 \end{bmatrix}$.

63. $BC = \begin{bmatrix} 5 & 1 \\ 0 & -2 \\ 3 & 7 \end{bmatrix} \begin{bmatrix} -5 & 4 & 1 \\ 0 & 3 & 6 \end{bmatrix} = \begin{bmatrix} 5(-5) + 1(0) & 5(4) + 1(3) & 5(1) + 1(6) \\ 0(-5) - 2(0) & 0(4) - 2(3) & 0(1) - 2(6) \\ 3(-5) + 7(0) & 3(4) + 7(3) & 3(1) + 7(6) \end{bmatrix} = \begin{bmatrix} -25 & 23 & 11 \\ 0 & -6 & -12 \\ -15 & 33 & 45 \end{bmatrix}$.

64. $CB = \begin{bmatrix} -5 & 4 & 1 \\ 0 & 3 & 6 \end{bmatrix} \begin{bmatrix} 5 & 1 \\ 0 & -2 \\ 3 & 7 \end{bmatrix} = \begin{bmatrix} -5(5) + 4(0) + 1(3) & -5(1) + 4(-2) + 1(7) \\ 0(5) + 3(0) + 6(3) & 0(1) + 3(-2) + 6(7) \end{bmatrix} = \begin{bmatrix} -22 & -6 \\ 18 & 36 \end{bmatrix}$.

65. We cannot multiply AB because we cannot multiply a (2×2) by a (3×2).

66. We cannot multiply CA because we cannot multiply a (2×3) by a (2×2).

67. $A^2 = \begin{bmatrix} 4 & -2 \\ 3 & 1 \end{bmatrix} \begin{bmatrix} 4 & -2 \\ 3 & 1 \end{bmatrix} = \begin{bmatrix} 4(4) - 2(3) & 4(-2) - 2(1) \\ 3(4) + 1(3) & 3(-2) + 1(1) \end{bmatrix} = \begin{bmatrix} 10 & -10 \\ 15 & -5 \end{bmatrix}$.

68. $A^3 = A^2$(from exercise 67) $\cdot A = \begin{bmatrix} 10 & -10 \\ 15 & -5 \end{bmatrix} \begin{bmatrix} 4 & -2 \\ 3 & 1 \end{bmatrix} =$

$\begin{bmatrix} 10(4) - 10(3) & 10(-2) - 10(1) \\ 15(4) - 5(3) & 15(-2) - 5(1) \end{bmatrix} = \begin{bmatrix} 10 & -30 \\ 45 & -35 \end{bmatrix}.$

69. $AB \neq BA$; $BC \neq CB$; $AC \neq CA$.   Matrix multiplication is not commutative.

70. The number of rows of P must equal the number of columns of Q, and the number of columns of P must equal the number of rows of Q.

71. (a) From the information the matrix is: $\begin{bmatrix} 50 & 100 & 30 \\ 10 & 90 & 50 \\ 60 & 120 & 40 \end{bmatrix}.$

(b) The income matrix is: $\begin{bmatrix} 12 \\ 10 \\ 15 \end{bmatrix}.$

(c) $\begin{bmatrix} 50 & 100 & 30 \\ 10 & 90 & 50 \\ 60 & 120 & 40 \end{bmatrix} \begin{bmatrix} 12 \\ 10 \\ 15 \end{bmatrix} = \begin{bmatrix} 50(12) + 100(10) + 30(15) \\ 10(12) + 90(10) + 50(15) \\ 60(12) + 120(10) + 40(15) \end{bmatrix} = \begin{bmatrix} 2050 \\ 1770 \\ 2520 \end{bmatrix}.$

(d) $\$2050 + \$1770 + \$2520 = \$6340.$

72. (a) $\begin{bmatrix} 1 & 4 & \frac{1}{4} & \frac{1}{4} & 1 \\ 0 & 3 & 0 & \frac{1}{4} & 0 \\ 4 & 3 & 2 & 1 & 1 \\ 0 & 1 & 0 & \frac{1}{3} & 0 \end{bmatrix} \begin{bmatrix} 5 & 5 \\ 8 & 10 \\ 10 & 12 \\ 12 & 15 \\ 5 & 6 \end{bmatrix} = \begin{bmatrix} 5 + 32 + \dfrac{10}{4} + 3 + 5 & 5 + 40 + 3 + \dfrac{15}{4} + 6 \\ 0 + 24 + 0 + 3 + 0 & 0 + 30 + 0 + \dfrac{15}{4} + 0 \\ 20 + 24 + 20 + 12 + 5 & 20 + 30 + 24 + 15 + 6 \\ 0 + 8 + 0 + 4 + 0 & 0 + 10 + 0 + 5 + 0 \end{bmatrix} =$

$\begin{bmatrix} 47.5 & 57.75 \\ 27 & 33.75 \\ 81 & 95 \\ 12 & 15 \end{bmatrix}.$

(b) The order matrix is: $\begin{bmatrix} 20 & 200 & 50 & 60 \end{bmatrix}$. Now multiply this by the ingredient matrix:

$\begin{bmatrix} 20 & 200 & 50 & 60 \end{bmatrix} \begin{bmatrix} 1 & 4 & \frac{1}{4} & \frac{1}{4} & 1 \\ 0 & 3 & 0 & \frac{1}{4} & 0 \\ 4 & 3 & 2 & 1 & 1 \\ 0 & 1 & 0 & \frac{1}{3} & 0 \end{bmatrix} =$

$\begin{bmatrix} 20 + 0 + 200 + 0 & 80 + 600 + 150 + 60 & 5 + 0 + 100 + 0 & 5 + 50 + 50 + 20 & 20 + 0 + 50 + 0 \end{bmatrix}$

$= \begin{bmatrix} 220 & 890 & 105 & 125 & 70 \end{bmatrix}.$

(c) Multiply Day's Order by Cost: $\begin{bmatrix} 20 & 200 & 50 & 60 \end{bmatrix} \begin{bmatrix} 47.5 & 57.75 \\ 27 & 33.75 \\ 81 & 95 \\ 12 & 15 \end{bmatrix} =$

$\begin{bmatrix} 950 + 5400 + 4050 + 720 & 1155 + 6750 + 4750 + 900 \end{bmatrix} = \begin{bmatrix} 11,120 & 13,555 \end{bmatrix}.$

The order cost \$11,120 with the large lot; and \$13,555 with the small lot.

73. (a) From R_2 the equation is: $d_{n+1} = -.05m_n + 1.05d_n$ (hundreths). The deer population will grow at 105/year.

(b) Let $m_n = 2000$ and $d_n = 5000$ and multiply the given matrices:

$$\begin{bmatrix} m_{n+1} \\ d_{n+1} \end{bmatrix} = \begin{bmatrix} .51 & .4 \\ -.05 & 1.05 \end{bmatrix} \begin{bmatrix} 2000 \\ 5000 \end{bmatrix} = \begin{bmatrix} 1020 + 2000 \\ -100 + 5250 \end{bmatrix} = \begin{bmatrix} 3020 \\ 5150 \end{bmatrix}$$

After 1 year there will be 3020 mountain lions and 515,000 deer.

Now multiply this 1 year matrix by the given matrix to find year 2.

$$\begin{bmatrix} m_{n+2} \\ d_{n+2} \end{bmatrix} = \begin{bmatrix} .51 & .4 \\ -.05 & 1.05 \end{bmatrix} \begin{bmatrix} 3020 \\ 5150 \end{bmatrix} = \begin{bmatrix} 1540.2 + 2060 \\ -151 + 5407.5 \end{bmatrix} = \begin{bmatrix} 3600.2 \\ 5256.5 \end{bmatrix}$$

After 2 years there will be 3600 mountain lions and approximately 525,700 deer.

(c) Let $m_n = 4000$ and $d_n = 5000$ and multiply the given matrices:

$$\begin{bmatrix} m_{n+1} \\ d_{n+1} \end{bmatrix} = \begin{bmatrix} .51 & .4 \\ -.05 & 1.05 \end{bmatrix} \begin{bmatrix} 4000 \\ 5000 \end{bmatrix} = \begin{bmatrix} 2040 + 2000 \\ -200 + 5250 \end{bmatrix} = \begin{bmatrix} 4040 \\ 5050 \end{bmatrix}$$

After 1 year there will be 4040 mountain lions and 505,000 deer.

Now multiply this 1 year matrix by the given matrix to find year 2.

$$\begin{bmatrix} m_{n+2} \\ d_{n+2} \end{bmatrix} = \begin{bmatrix} .51 & .4 \\ -.05 & 1.05 \end{bmatrix} \begin{bmatrix} 4040 \\ 5050 \end{bmatrix} = \begin{bmatrix} 2060.4 + 2020 \\ -202 + 5302.5 \end{bmatrix} = \begin{bmatrix} 4080.4 \\ 5100.5 \end{bmatrix}$$

After 2 years there will be 4080 mountain lions and approximately 510,050 deer.

Mountain Lions: after 1 year $= \dfrac{4040}{4000} = 1.01$; and after 2 years $= \dfrac{4080.4}{4040} = 1.01$

Deer: after 1 year $= \dfrac{505,000}{500,000} = 1.01$; and after 2 years $= \dfrac{510,050}{505,000} = 1.01$

74. (a) Year 1: $\begin{bmatrix} 0 & 0 & .33 \\ .18 & 0 & 0 \\ 0 & .71 & .94 \end{bmatrix} \begin{bmatrix} 690 \\ 210 \\ 2100 \end{bmatrix} = \begin{bmatrix} 693 \\ 124.2 \\ 2123.1 \end{bmatrix}$; Sum ≈ 2940

Year 2: $\begin{bmatrix} 0 & 0 & .33 \\ .18 & 0 & 0 \\ 0 & .71 & .94 \end{bmatrix} \begin{bmatrix} 693 \\ 124.2 \\ 2123.1 \end{bmatrix} \approx \begin{bmatrix} 700.62 \\ 124.74 \\ 2083.9 \end{bmatrix}$; Sum ≈ 2909

Year 3: $\begin{bmatrix} 0 & 0 & .33 \\ .18 & 0 & 0 \\ 0 & .71 & .94 \end{bmatrix} \begin{bmatrix} 700.62 \\ 124.74 \\ 2083.9 \end{bmatrix} \approx \begin{bmatrix} 687.69 \\ 126.11 \\ 2047.43 \end{bmatrix}$; Sum ≈ 2861

Year 4: $\begin{bmatrix} 0 & 0 & .33 \\ .18 & 0 & 0 \\ 0 & .71 & .94 \end{bmatrix} \begin{bmatrix} 687.69 \\ 126.11 \\ 2047.43 \end{bmatrix} = \begin{bmatrix} 675.65 \\ 123.78 \\ 2014.12 \end{bmatrix}$; Sum ≈ 2814

Year 5: $\begin{bmatrix} 0 & 0 & .33 \\ .18 & 0 & 0 \\ 0 & .71 & .94 \end{bmatrix} \begin{bmatrix} 675.65 \\ 123.78 \\ 2014.12 \end{bmatrix} = \begin{bmatrix} 664.66 \\ 121.62 \\ 1981.16 \end{bmatrix}$; Sum ≈ 2767

(b) Because the scalar is less than one, the population will continue to decline and the northern spotted owl will face extinction.

(c) Reworking (a) with .18 replaced with .3 yields: Sum $\approx$ 3023, 3051, 3079, 3107, and 3135.

The population would slowly increase.

75. $A + B = B + A$: $A + B = \begin{bmatrix} a_{11} + b_{11} & a_{12} + b_{12} \\ a_{21} + b_{21} & a_{22} + b_{22} \end{bmatrix} = B + A.$

76. $A + (B + C) = (A + B) + C$:

$$A + (B + C) = \begin{bmatrix} a_{11} + (b_{11} + c_{11}) & a_{12} + (b_{12} + c_{12}) \\ a_{21} + (b_{21} + c_{21}) & a_{22} + (b_{22} + c_{22}) \end{bmatrix} =$$

$$\begin{bmatrix} (a_{11} + b_{11}) + c_{11} & (a_{12} + b_{12}) + c_{12} \\ (a_{21} + b_{21}) + c_{21} & (a_{22} + b_{22}) + c_{22} \end{bmatrix} = (A + B) + C.$$

77. $(AB)C = A(BC)$:

$$(A B)C = \begin{bmatrix} a_{11}b_{11} + a_{12}b_{21} & a_{11}b_{12} + a_{12}b_{22} \\ a_{21}b_{11} + a_{22}b_{21} & a_{21}b_{12} + a_{22}b_{22} \end{bmatrix} \begin{bmatrix} c_{11} & c_{12} \\ c_{21} & c_{22} \end{bmatrix} =$$

$$\begin{bmatrix} (a_{11}b_{11}c_{11} + a_{12}b_{21}c_{11}) + (a_{11}b_{12}c_{21} + a_{12}b_{22}c_{21}) & (a_{11}b_{11}c_{12} + a_{12}b_{21}c_{12}) + (a_{11}b_{12}c_{22} + a_{12}b_{22}c_{22}) \\ (a_{21}b_{11}c_{11} + a_{22}b_{21}c_{11}) + (a_{21}b_{12}c_{21} + a_{22}b_{22}c_{21}) & (a_{21}b_{11}c_{12} + a_{22}b_{21}c_{12}) + (a_{21}b_{12}c_{22} + a_{22}b_{22}c_{22}) \end{bmatrix}$$

$$A(B C) = \begin{bmatrix} a_{11} & a_{12} \\ a_{21} & a_{22} \end{bmatrix} \begin{bmatrix} b_{11}c_{11} + b_{12}c_{21} & b_{11}c_{12} + b_{12}c_{22} \\ b_{21}c_{11} + b_{22}c_{21} & b_{21}c_{12} + b_{22}c_{22} \end{bmatrix} =$$

$$\begin{bmatrix} (a_{11}b_{11}c_{11} + a_{11}b_{12}c_{21}) + (a_{12}b_{21}c_{11} + a_{12}b_{22}c_{21}) & (a_{11}b_{11}c_{12} + a_{11}b_{12}c_{22}) + (a_{12}b_{21}c_{12} + a_{12}b_{22}c_{22}) \\ (a_{21}b_{11}c_{11} + a_{21}b_{12}c_{21}) + (a_{22}b_{21}c_{11} + a_{22}b_{22}c_{21}) & (a_{21}b_{11}c_{12} + a_{21}b_{12}c_{22}) + (a_{22}b_{21}c_{12} + a_{22}b_{22}c_{22}) \end{bmatrix}$$

Therefore, $(A B)C = A(B C)$.

78. $A(B + C) = AB + AC$:

$$A(B + C) = \begin{bmatrix} a_{11} & a_{12} \\ a_{21} & a_{22} \end{bmatrix} \begin{bmatrix} b_{11} + c_{11} & b_{12} + c_{12} \\ b_{21} + c_{21} & b_{22} + c_{22} \end{bmatrix} =$$

$$\begin{bmatrix} (a_{11}b_{11} + a_{11}c_{11}) + (a_{12}b_{11} + a_{12}c_{21}) & (a_{11}b_{12} + a_{11}c_{12}) + (a_{12}b_{22} + a_{12}c_{22}) \\ (a_{21}b_{11} + a_{21}c_{11}) + (a_{22}b_{21} + a_{22}c_{21}) & (a_{21}b_{12} + a_{21}c_{12}) + (a_{22}b_{22} + a_{22}c_{22}) \end{bmatrix}$$

$$A B + A C = \begin{bmatrix} a_{11}b_{11} + a_{12}b_{21} & a_{11}b_{12} + a_{12}b_{22} \\ a_{21}b_{11} + a_{22}b_{21} & a_{21}b_{12} + a_{22}b_{22} \end{bmatrix} + \begin{bmatrix} a_{11}c_{11} + a_{12}c_{21} & a_{11}c_{12} + a_{12}c_{22} \\ a_{21}c_{11} + a_{22}c_{21} & a_{21}c_{12} + a_{22}c_{22} \end{bmatrix} =$$

$$\begin{bmatrix} (a_{11}b_{11} + a_{11}c_{11}) + (a_{12}b_{21} + a_{12}c_{21}) & (a_{11}b_{12} + a_{11}c_{12}) + (a_{12}b_{22} + a_{12}c_{22}) \\ (a_{21}b_{11} + a_{21}c_{11}) + (a_{22}c_{21} + a_{22}c_{21}) & (a_{21}b_{12} + a_{21}c_{12}) + (a_{22}b_{22} + a_{22}c_{22}) \end{bmatrix}$$

Therefore, $A(B + C) = AB + AC$.

79. $c(A + B) = cA + cB$: $c(A + B) = \begin{bmatrix} c(a_{11} + b_{11}) & c(a_{12} + b_{12}) \\ c(a_{21} + b_{21}) & c(a_{22} + b_{22}) \end{bmatrix} = \begin{bmatrix} ca_{11} + cb_{11} & ca_{12} + cb_{12} \\ ca_{21} + cb_{21} & ca_{22} + cb_{22} \end{bmatrix} =$

$cA + cB$.

80. $(c + d)A = cA + dA$: $(c + d)A = \begin{bmatrix} (c + d)a_{11} & (c + d)a_{12} \\ (c + d)a_{21} & (c + d)a_{22} \end{bmatrix} = \begin{bmatrix} ca_{11} + da_{11} & ca_{12} + da_{12} \\ ca_{21} + da_{21} & ca_{22} + da_{22} \end{bmatrix} = cA + dA.$

81. $(cA)d = (cd)A$: $(cA)d = \begin{bmatrix} ca_{11} & ca_{12} \\ ca_{21} & ca_{22} \end{bmatrix} \cdot d = \begin{bmatrix} cda_{11} & cda_{12} \\ cda_{21} & cda_{22} \end{bmatrix} = (cd)A.$

82. $(cd)A = c(dA)$: $(cd)A = \begin{bmatrix} cda_{11} & cda_{12} \\ cda_{21} & cda_{22} \end{bmatrix}$; $c(dA) = c\begin{bmatrix} da_{11} & da_{12} \\ da_{21} & da_{22} \end{bmatrix} = \begin{bmatrix} cda_{11} & cda_{12} \\ cda_{21} & cda_{22} \end{bmatrix}$

Therefore, $(cd)A = c(dA)$.

7.5: Determinants and Cramer's Rule

1. $\det \begin{bmatrix} -5 & 9 \\ 4 & -1 \end{bmatrix} = 5 - 36 = -31$

2. $\det \begin{bmatrix} -1 & 3 \\ -2 & 9 \end{bmatrix} = -9 - (-6) = -3$

3. $\det \begin{bmatrix} -1 & -2 \\ 5 & 3 \end{bmatrix} = -3 - (-10) = 7$

4. $\det \begin{bmatrix} 6 & -4 \\ 0 & -1 \end{bmatrix} = -6 - 0 = -6$

5. $\det \begin{bmatrix} 9 & 3 \\ -3 & -1 \end{bmatrix} = -9 - (-9) = 0$

6. $\det \begin{bmatrix} 0 & 2 \\ 1 & 5 \end{bmatrix} = 0 - 2 = -2$

7. $\det \begin{bmatrix} 3 & 4 \\ 5 & -2 \end{bmatrix} = -6 - 20 = -26$

8. $\det \begin{bmatrix} -9 & 7 \\ 2 & 6 \end{bmatrix} = -54 - 14 = -68$

9. Use $A_{ij} = (-1)^{i+j} \cdot M_{ij}$, to find the cofactor of each element a.

   For a_{21}, find $A_{21} = (-1)^{2+1} \cdot \left(\det \begin{bmatrix} 0 & 1 \\ 2 & 1 \end{bmatrix} \right) = -1(0 - 2) = 2$

   For a_{22}, find $A_{22} = (-1)^{2+2} \cdot \left(\det \begin{bmatrix} -2 & 1 \\ 4 & 1 \end{bmatrix} \right) = 1(-2 - 4) = -6$

   For a_{23}, find $A_{23} = (-1)^{2+3} \cdot \left(\det \begin{bmatrix} -2 & 0 \\ 4 & 2 \end{bmatrix} \right) = -1(-4 - 0) = 4$

10. Use $A_{ij} = (-1)^{i+j} \cdot M_{ij}$, to find the cofactor of each element a.

    For a_{21}, find $A_{21} = (-1)^{2+1} \cdot \left(\det \begin{bmatrix} -1 & 2 \\ -3 & 1 \end{bmatrix} \right) = -1(-1 + 6) = -5$

    For a_{22}, find $A_{22} = (-1)^{2+2} \cdot \left(\det \begin{bmatrix} 1 & 2 \\ 0 & 1 \end{bmatrix} \right) = 1(1 - 0) = 1$

    For a_{23}, find $A_{23} = (-1)^{2+3} \cdot \left(\det \begin{bmatrix} 1 & -1 \\ 0 & -3 \end{bmatrix} \right) = -1(-3 - 0) = 3$

11. Use $A_{ij} = (-1)^{i+j} \cdot M_{ij}$, to find the cofactor of each element a.

    For a_{21}, find $A_{21} = (-1)^{2+1} \cdot \left(\det \begin{bmatrix} 2 & -1 \\ 4 & 1 \end{bmatrix} \right) = -1(2 + 4) = -6$

    For a_{22}, find $A_{22} = (-1)^{2+2} \cdot \left(\det \begin{bmatrix} 1 & -1 \\ -1 & 1 \end{bmatrix} \right) = 1(1 - 1) = 0$

    For a_{23}, find $A_{23} = (-1)^{2+3} \cdot \left(\det \begin{bmatrix} 1 & 2 \\ -1 & 4 \end{bmatrix} \right) = -1(4 + 2) = -6$

12. Use $A_{ij} = (-1)^{i+j} \cdot M_{ij}$, to find the cofactor of each element a.

For a_{21}, find $A_{21} = (-1)^{2+1} \cdot \left(\det \begin{bmatrix} -1 & 4 \\ 1 & 4 \end{bmatrix} \right) = -1(-4 - 4) = 8$

For a_{22}, find $A_{22} = (-1)^{2+2} \cdot \left(\det \begin{bmatrix} 2 & 4 \\ -2 & 4 \end{bmatrix} \right) = 1(8 + 8) = 16$

For a_{23}, find $A_{23} = (-1)^{2+3} \cdot \left(\det \begin{bmatrix} 2 & -1 \\ -2 & 1 \end{bmatrix} \right) = -1(2 - 2) = 0$

13. Evaluate, expand by the second row. Therefore, det $= (-)(a_{21})(M_{21}) + (a_{22})(M_{22}) - (a_{23})(M_{23})$:

$-2\left(\det \begin{bmatrix} -7 & 8 \\ 3 & 0 \end{bmatrix} \right) + 1\left(\det \begin{bmatrix} 4 & 8 \\ -6 & 0 \end{bmatrix} \right) - 3\left(\det \begin{bmatrix} 4 & -7 \\ -6 & 3 \end{bmatrix} \right) = -2(-24) + 1(48) - 3(-30) =$

$48 + 48 + 90 = 186.$

14. Evaluate, expand by the second row. Therefore, det $= (-)(a_{21})(M_{21}) + (a_{22})(M_{22}) - (a_{23})(M_{23})$:

$-7\left(\det \begin{bmatrix} -2 & -4 \\ -1 & 2 \end{bmatrix} \right) + 0\left(\det \begin{bmatrix} 8 & -4 \\ 5 & 2 \end{bmatrix} \right) - 3\left(\det \begin{bmatrix} 8 & -2 \\ 5 & -1 \end{bmatrix} \right) = 7(-8) + 0(36) - 3(2) =$

$56 + 0 - 6 = 50.$

15. Evaluate, expand by the first column. Therefore, det $= (a_{11})(M_{11}) - (a_{21})(M_{21}) + (a_{31})(M_{31})$:

$1\left(\det \begin{bmatrix} 2 & -1 \\ 1 & 4 \end{bmatrix} \right) - (-1)\left(\det \begin{bmatrix} 2 & 0 \\ 1 & 4 \end{bmatrix} \right) + 0\left(\det \begin{bmatrix} 2 & 0 \\ 2 & -1 \end{bmatrix} \right) = 1(8 + 1) + 1(8 - 0) + 0 = 9 + 8 = 17.$

16. Evaluate, expand by the third column. Therefore, det $= (a_{13})(M_{13}) - (a_{23})(M_{23}) + (a_{33})(M_{33})$:

$-1\left(\det \begin{bmatrix} 4 & 7 \\ 2 & 4 \end{bmatrix} \right) - (-2)\left(\det \begin{bmatrix} 2 & 1 \\ 2 & 4 \end{bmatrix} \right) + 0\left(\det \begin{bmatrix} 2 & 1 \\ 4 & 7 \end{bmatrix} \right) = -1(16 - 14) + 2(8 - 2) + 0 =$

$-2 + 12 = 10.$

17. Evaluate, expand by the first row. Therefore, det $= (a_{11})(M_{11}) - (a_{12})(M_{12}) + (a_{13})(M_{13})$:

$10\left(\det \begin{bmatrix} 4 & 3 \\ 8 & 10 \end{bmatrix} \right) - 2\left(\det \begin{bmatrix} -1 & 3 \\ -3 & 10 \end{bmatrix} \right) + 1\left(\det \begin{bmatrix} -1 & 4 \\ -3 & 8 \end{bmatrix} \right) =$

$10(40 - 24) - 2(-10 + 9) + 1(-8 + 12) = 160 + 2 + 4 = 166.$

18. Evaluate, expand by the first column. Therefore, det $= (a_{11})(M_{11}) - (a_{21})(M_{21}) + (a_{31})(M_{31})$:

$7\left(\det \begin{bmatrix} -7 & 2 \\ 1 & 1 \end{bmatrix} \right) - 1\left(\det \begin{bmatrix} -1 & 1 \\ 1 & 1 \end{bmatrix} \right) + (-2)\left(\det \begin{bmatrix} -1 & 1 \\ -7 & 2 \end{bmatrix} \right) = 7(-7 - 2) - 1(-1 - 1) - 2(-2 + 7) =$

$-63 + 2 - 10 = -71.$

19. Evaluate, expand by the second row will give a determinant of 0.

20. Evaluate, expand by the third column will give a determinant of 0.

21. Evaluate, expand by the third column. Therefore, det $= (a_{13})(M_{13}) - (a_{23})(M_{23}) + (a_{33})(M_{33})$:

$-1\left(\det \begin{bmatrix} 2 & 6 \\ -6 & -6 \end{bmatrix} \right) - 0\left(\det \begin{bmatrix} 3 & 3 \\ -6 & -6 \end{bmatrix} \right) + 2\left(\det \begin{bmatrix} 3 & 3 \\ 2 & 6 \end{bmatrix} \right) = -1(-12 + 36) - 0 + 2(18 - 6) =$

$-24 + 24 = 0.$

22. Evaluate, expand by the third row. Therefore, det $= (a_{31})(M_{31}) - (a_{32})(M_{32}) + (a_{33})(M_{33})$:

$1\left(\det \begin{bmatrix} -3 & 2 \\ 3 & -2 \end{bmatrix} \right) - 0\left(\det \begin{bmatrix} 5 & 2 \\ -5 & -2 \end{bmatrix} \right) + 1\left(\det \begin{bmatrix} 5 & -3 \\ -5 & 3 \end{bmatrix} \right) = 1(6 - 6) - 0 + 1(15 - 15) =$

$0 - 0 + 0 = 0.$

23. Evaluate, expand by the second row. Therefore, $\det = (-)(a_{21})(M_{21}) + (a_{22})(M_{22}) - (a_{23})(M_{23})$:

$$-.3\left(\det\begin{bmatrix} -.8 & .6 \\ 4.1 & -2.8 \end{bmatrix}\right) + .9\left(\det\begin{bmatrix} .4 & .6 \\ 3.1 & -2.8 \end{bmatrix}\right) - .7\left(\det\begin{bmatrix} .4 & -.8 \\ 3.1 & 4.1 \end{bmatrix}\right) =$$

$-.3(2.24 - 2.46) + .9(-1.12 - 1.86) - .7(1.64 + 2.48) = .066 - 2.682 - 2.884 = -5.5.$

24. Evaluate, expand by the third row. Therefore, $\det = (a_{31})(M_{31}) - (a_{32})(M_{32}) + (a_{33})(M_{33})$:

$$-.1\left(\det\begin{bmatrix} -.1 & .9 \\ 4.9 & -3.2 \end{bmatrix}\right) - .4\left(\det\begin{bmatrix} -.3 & .9 \\ 2.5 & -3.2 \end{bmatrix}\right) + .8\left(\det\begin{bmatrix} -.3 & -.1 \\ 2.5 & 4.9 \end{bmatrix}\right) =$$

$-.1(.32 - 4.41) - .4(.96 - 2.25) + .8(-1.47 + .25) = .409 + .516 - .976 = -.051.$

25. If $\det\begin{bmatrix} 5 & x \\ -3 & 2 \end{bmatrix} = 6$, then $10 - (-3x) = 6 \Rightarrow 3x = -4 \Rightarrow x = -\dfrac{4}{3}$. The solution set is: $\left\{-\dfrac{4}{3}\right\}$.

26. If $\det\begin{bmatrix} -.5 & 2 \\ x & x \end{bmatrix} = 0$, then $-.5x - 2x = 0 \Rightarrow -2.5x = 0 \Rightarrow x = 0$. The solution set is: $\{0\}$.

27. If $\det\begin{bmatrix} x & 3 \\ x & x \end{bmatrix} = 4$, then $x^2 - 3x = 4 \Rightarrow x^2 - 3x - 4 = 0 \Rightarrow (x - 4)(x + 1) = 0 \Rightarrow x = 4, -1.$

The solution set is: $\{4, -1\}$.

28. If $\det\begin{bmatrix} 2x & x \\ 11 & x \end{bmatrix} = 6$, then $2x^2 - 11x = 6 \Rightarrow 2x^2 - 11x - 6 = 0 \Rightarrow (x - 6)(2x + 1) = 0 \Rightarrow x = 6, -\dfrac{1}{2}.$

The solution set is: $\left\{6, -\dfrac{1}{2}\right\}$.

29. Evaluate, expand by the first row. Therefore, $\det = (a_{11})(M_{11}) - (a_{12})(M_{12}) + (a_{13})(M_{13})$:

$$-2\left(\det\begin{bmatrix} 3 & x \\ -2 & 0 \end{bmatrix}\right) + 0 + 1\left(\det\begin{bmatrix} -1 & 3 \\ 5 & -2 \end{bmatrix}\right) = 3 \Rightarrow -2(0 + 2x) + 1(2 - 15) = 3 \Rightarrow$$

$-4x + 2 - 15 = 3 \Rightarrow -4x = 16 \Rightarrow x = -4$. The solution is: $\{-4\}$.

30. Evaluate, expand by the third column. Therefore, $\det = (a_{13})(M_{13}) - (a_{23})(M_{23}) + (a_{33})(M_{33})$:

$$0 - 1\left(\det\begin{bmatrix} 4 & 3 \\ -3 & x \end{bmatrix}\right) + (-1)\left(\det\begin{bmatrix} 4 & 3 \\ 2 & 0 \end{bmatrix}\right) = 5 \Rightarrow -1(4x + 9) - 1(0 - 6) = 5 \Rightarrow -4x - 9 + 6 = 5 \Rightarrow$$

$-4x = 8 \Rightarrow x = -2$. The solution set is: $\{-2\}$.

31. Evaluate, expand by the first column. Therefore, $\det = (a_{11})(M_{11}) - (a_{21})(M_{21}) + (a_{31})(M_{31})$:

$$5\left(\det\begin{bmatrix} 2 & -1 \\ -1 & x \end{bmatrix}\right) + 0 + 4\left(\det\begin{bmatrix} 3x & -3 \\ 2 & -1 \end{bmatrix}\right) = -7 \Rightarrow 5(2x - 1) + 4(-3x + 6) = -7 \Rightarrow$$

$10x - 5 - 12x + 24 = -7 \Rightarrow -2x = -26 \Rightarrow x = 13$. The solution set is: $\{13\}$.

32. Evaluate, expand by the second column. Therefore, $\det = (-)(a_{12})(M_{12}) + (a_{22})(M_{22}) - (a_{32})(M_{32})$:

$$-1\left(\det\begin{bmatrix} 0 & x \\ 3 & 2 \end{bmatrix}\right) + 4\left(\det\begin{bmatrix} 2x & -1 \\ 3 & 2 \end{bmatrix}\right) + 0 = x \Rightarrow -1(0 - 3x) + 4(4x + 3) = x \Rightarrow$$

$3x + 16x + 12 = x \Rightarrow 18x = -12 \Rightarrow x = -\dfrac{12}{18}$. The solution set is: $\left\{-\dfrac{2}{3}\right\}$.

33. With the given points find: $A = \dfrac{1}{2}\det\begin{bmatrix} 0 & 0 & 1 \\ 0 & 2 & 1 \\ 1 & 4 & 1 \end{bmatrix}$. Using the calculator the determinant is: -2.

Therefore, $A = \left|\dfrac{1}{2}(-2)\right| \Rightarrow A = 1$ square units.

34. With the given points find: $A = \dfrac{1}{2} \det \begin{bmatrix} 0 & 1 & 1 \\ 2 & 0 & 1 \\ 1 & 5 & 1 \end{bmatrix}$. Using the calculator the determinant is: 9.

Therefore, $A = \left| \dfrac{1}{2}(9) \right| \Rightarrow A = \dfrac{9}{2}$ square units.

35. With the given points find: $A = \dfrac{1}{2} \det \begin{bmatrix} 2 & 5 & 1 \\ -1 & 3 & 1 \\ 4 & 0 & 1 \end{bmatrix}$. Using the calculator the determinant is: 19.

Therefore, $A = \left| \dfrac{1}{2}(19) \right| \Rightarrow A = \dfrac{19}{2}$ square units.

36. With the given points find: $A = \dfrac{1}{2} \det \begin{bmatrix} 2 & -2 & 1 \\ 0 & 0 & 1 \\ -3 & -4 & 1 \end{bmatrix}$. Using the calculator the determinant is: 14.

Therefore, $A = \left| \dfrac{1}{2}(14) \right| \Rightarrow A = 7$ square units.

37. With the given points find: $A = \dfrac{1}{2} \det \begin{bmatrix} 1 & 2 & 1 \\ 4 & 3 & 1 \\ 3 & 5 & 1 \end{bmatrix}$. Using the calculator the determinant is: 7.

Therefore, $A = \left| \dfrac{1}{2}(7) \right| \Rightarrow A = \dfrac{7}{2}$ square units.

38. With the given points find: $A = \dfrac{1}{2} \det \begin{bmatrix} 101.3 & 52.7 & 1 \\ 117.2 & 253.9 & 1 \\ 313.1 & 301.6 & 1 \end{bmatrix}$. Using the calculator the determinant is:

$-38{,}656.65$. Therefore, $A = \left| \dfrac{1}{2}(-38{,}656.65) \right| \Rightarrow A \approx 19{,}328.3$ square feet.

39. If the three points form a triangle with no area ($D = 0$ using determinants), then the points must be collinear.

$D = \dfrac{1}{2} \det \begin{bmatrix} 1 & 3 & 1 \\ -3 & 11 & 1 \\ 2 & 1 & 1 \end{bmatrix} = 0$; The points are collinear.

40. If the three points form a triangle with no area ($D = 0$ using determinants), then the points must be collinear.

$D = \dfrac{1}{2} \det \begin{bmatrix} 3 & 6 & 1 \\ -1 & -6 & 1 \\ 5 & 11 & 1 \end{bmatrix} = 2 \neq 0$; The points are not collinear.

41. If the three points form a triangle with no area ($D = 0$ using determinants), then the points must be collinear.

$D = \dfrac{1}{2} \det \begin{bmatrix} -2 & -5 & 1 \\ 4 & 4 & 1 \\ 2 & 3 & 1 \end{bmatrix} = 6 \neq 0$; The points are not collinear.

42. If the three points form a triangle with no area ($D = 0$ using determinants), then the points must be collinear.

$D = \dfrac{1}{2} \det \begin{bmatrix} 4 & -5 & 1 \\ -2 & 10 & 1 \\ 6 & -10 & 1 \end{bmatrix} = 0$; The points are collinear.

43. Using the calculator, the determinant is: 298.

44. Using the calculator, the determinant is: -311.

45. Using the calculator, the determinant is: -88.

46. Using the calculator, the determinant is: 720.

47. Since the second column is all zeros, by Determinant Theorem 1, the determinant is: 0.

48. Since Row 1 $= -\dfrac{1}{4}$ Row 2, we have by Determinant Theorem 4: $\det \begin{bmatrix} -1 & 2 & 4 \\ 4 & -8 & -16 \\ 3 & 0 & 5 \end{bmatrix} =$

$-\dfrac{1}{4}\det \begin{bmatrix} 4 & -8 & -16 \\ 4 & -8 & -16 \\ 3 & 0 & 5 \end{bmatrix}$. Now by Determinant Theorem 5, the determinant equals 0. Therefore,

$\det \begin{bmatrix} -1 & 2 & 4 \\ 4 & -8 & -16 \\ 3 & 0 & 5 \end{bmatrix} = -\dfrac{1}{4}(0) = 0.$

49. Use Determinant Theorem 6, multiplying column 1 by 2 and adding the result to column 3 yields equivalent

determinant: $\det \begin{bmatrix} 6 & 8 & 0 \\ -1 & 0 & 0 \\ 4 & 0 & 0 \end{bmatrix}$. Since column 3 has all zeros the determinant is 0 (Determinant Theorem 1)

and $\det \begin{bmatrix} 6 & 8 & -12 \\ -1 & 0 & 2 \\ 4 & 0 & -8 \end{bmatrix} = 0.$

50. Use Determinant Theorem 6, multiplying column 1 by -2 and adding the result to column 2 yields equivalent

determinant: $\det \begin{bmatrix} 4 & 0 & 0 \\ -1 & 0 & 1 \\ 2 & 0 & 3 \end{bmatrix}$. Since column 2 has all zeros the determinant is 0 (Determinant Theorem 1)

and $\det \begin{bmatrix} 4 & 8 & 0 \\ -1 & -2 & 1 \\ 2 & 4 & 3 \end{bmatrix} = 0.$

51. Use Determinant Theorem 6, multiplying column 2 by 4 and adding the result to column 1 and multiplying

column 2 by -4 and add the result to column 3 yields equivalent determinant: $\det \begin{bmatrix} 0 & 1 & 0 \\ 2 & 0 & 1 \\ 8 & 2 & -4 \end{bmatrix}$. Now expand

this about row 1: $\det \begin{bmatrix} 0 & 1 & 0 \\ 2 & 0 & 1 \\ 8 & 2 & -4 \end{bmatrix} = 0(0-2) - 1(-8-8) + 0(4-0) = 16 \Rightarrow \det \begin{bmatrix} -4 & 1 & 4 \\ 2 & 0 & 1 \\ 0 & 2 & 4 \end{bmatrix} = 16.$

52. Use Determinant Theorem 6, multiplying column 1 by -2 and adding the result to column 3 yields equivalent

determinant: $\det \begin{bmatrix} 6 & 3 & -10 \\ 1 & 0 & 0 \\ 5 & 7 & -7 \end{bmatrix}$. Now expand this about row 2: $\det \begin{bmatrix} 6 & 3 & -10 \\ 1 & 0 & 0 \\ 5 & 7 & -7 \end{bmatrix} = -1(-21+70) = -49$

$\Rightarrow \det \begin{bmatrix} 6 & 3 & 2 \\ 1 & 0 & 2 \\ 5 & 7 & 3 \end{bmatrix} = -49.$

53. Expand the given determinant on column 3: $1\left(\det\begin{bmatrix} 2 & 3 \\ -1 & 4 \end{bmatrix}\right) - 1\left(\det\begin{bmatrix} x & y \\ -1 & 4 \end{bmatrix}\right) + 1\left(\det\begin{bmatrix} x & y \\ 2 & 3 \end{bmatrix}\right) =$

$1(8 + 3) - 1(4x + y) + (3x - 2y) = 11 - 4x - y + 3x - 2y = -x - 3y + 11.$ Now set equal to zero:

$-x - 3y + 11 = 0$ or $x + 3y - 11 = 0.$

54. First find the slope of the line: $m = \dfrac{4 - 3}{-1 - 2} = -\dfrac{1}{3}$, now use point-slope form: $y - 3 = -\dfrac{1}{3}(x - 2) \Rightarrow$

$3y - 9 = -x + 2 \Rightarrow x + 3y - 11 = 0.$ This is equivalent to the equation in number 53.

55. Using the slope formula and the point-slope form, the equation through (x_1, y_1) and (x_2, y_2) is:

$y - y_1 = \dfrac{y_1 - y_2}{x_1 - x_2}(x - x_1).$

56. Expanding the given equation on the third column yields:

$0 = 1\left(\det\begin{bmatrix} x_1 & y_1 \\ x_2 & y_2 \end{bmatrix}\right) - 1\left(\det\begin{bmatrix} x & y \\ x_2 & y_2 \end{bmatrix}\right) + 1\left(\det\begin{bmatrix} x & y \\ x_1 & y_1 \end{bmatrix}\right) =$

$0 = (x_1 y_2 - x_2 y_1) - (x y_2 - x_2 y) + (x y_1 - x_1 y) \Rightarrow 0 = x_1 y_2 - x_2 y_1 - x y_2 + x_2 y + x y_1 - x_1 y \Rightarrow$

$y x_2 - y x_1 - y_1 x_2 = y_2 x - y_2 x_1 - y_1 x.$ Now adding $y_1 x_1$ to both sides yields:

$y x_2 - y x_1 - y_1 x_2 + y_1 x_1 = y_2 x - y_2 x_1 - y_1 x_1 + y_1 x_1 \Rightarrow (y - y_1)(x_2 - x_1) = (y_2 - y_1)(x - x_1) \Rightarrow$

$y - y_1 = \dfrac{y_2 - y_1}{x_2 - x_1}(x - x_1).$ When the determinant is expanded, the equation is the same as the answer in 55.

57. (a) D is the determinant of the coefficients of the given system or D.

    (b) D_x is the determinant where the numbers from the answer column replaces the first column x-coefficients or A.

    (c) D_y is the determinant where the numbers from the answer column replaces the second column y-coefficients or C.

    (d) D_z is the determinant where the numbers from the answer column replaces the first column z-coefficients or B.

58. From the given information: $x = \dfrac{D_x}{D} = \dfrac{-43}{-43} = 1; y = \dfrac{D_y}{D} = \dfrac{0}{-43} = 0;$ and $z = \dfrac{D_z}{D} = \dfrac{43}{-43} = -1.$

    The solution set is: $\{(1, 0, -1)\}.$

59. Find the determinants: $D = \det\begin{bmatrix} 1 & 1 \\ 2 & -1 \end{bmatrix} = -3; D_x = \det\begin{bmatrix} 4 & 1 \\ 2 & -1 \end{bmatrix} = -6;$ and $D_y = \det\begin{bmatrix} 1 & 4 \\ 2 & 2 \end{bmatrix} = -1.$

    Then $x = \dfrac{D_x}{D} = \dfrac{-6}{-3} = 2$ and $y = \dfrac{D_y}{D} = \dfrac{-6}{-3} = 2.$ The solution is: $\{(2, 2)\}.$

60. Find the determinants: $D = \det\begin{bmatrix} 3 & 2 \\ 2 & -1 \end{bmatrix} = -7; D_x = \det\begin{bmatrix} -4 & 2 \\ -5 & -1 \end{bmatrix} = 14;$ and $D_y = \det\begin{bmatrix} 3 & -4 \\ 2 & -5 \end{bmatrix} = -7.$

    Then $x = \dfrac{D_x}{D} = \dfrac{14}{-7} = -2$ and $y = \dfrac{D_y}{D} = \dfrac{-7}{-7} = 1.$ The solution is: $\{(-2, 1)\}.$

61. Find the determinants: $D = \det\begin{bmatrix} 4 & 3 \\ 2 & 3 \end{bmatrix} = 6; D_x = \det\begin{bmatrix} -7 & 3 \\ -11 & 3 \end{bmatrix} = 12;$ and $D_y = \det\begin{bmatrix} 4 & -7 \\ 2 & -11 \end{bmatrix} = -30.$

    Then $x = \dfrac{D_x}{D} = \dfrac{12}{6} = 2$ and $y = \dfrac{D_y}{D} = \dfrac{-30}{6} = -5.$ The solution is: $\{(2, -5)\}.$

62. Find the determinants: $D = \det\begin{bmatrix} 4 & -1 \\ 2 & 3 \end{bmatrix} = 14; D_x = \det\begin{bmatrix} 0 & -1 \\ 14 & 3 \end{bmatrix} = 14;$ and $D_y = \det\begin{bmatrix} 4 & 0 \\ 2 & 14 \end{bmatrix} = 56.$

    Then $x = \dfrac{D_x}{D} = \dfrac{14}{14} = 1$ and $y = \dfrac{D_y}{D} = \dfrac{56}{14} = 4.$ The solution is: $\{(1, 4)\}.$

63. Find the determinant D: $D = \det \begin{bmatrix} 3 & 2 \\ 6 & 4 \end{bmatrix} = 0$, therefore there are infinitely many solutions or no solutions.

Now use Row Echelon Method on the augmented matrix: $\begin{bmatrix} 3 & 2 & | & 4 \\ 6 & 4 & | & 8 \end{bmatrix}$, $-2R_1 + R_2 \rightarrow \begin{bmatrix} 3 & 2 & | & 4 \\ 0 & 0 & | & 0 \end{bmatrix}$.

Since $0 = 0$, there are infinitely many solutions and using R_1, they are in the following relationship:

$3x + 2y = 4 \Rightarrow 3x = 4 - 2y \Rightarrow x = \dfrac{4 - 2y}{3}$. The solution is: $\left\{ \left(\dfrac{4 - 2y}{3}, y \right) \right\}$ or $\left\{ \left(x, \dfrac{4 - 3x}{2} \right) \right\}$.

64. Find the determinant D: $D = \det \begin{bmatrix} 1.5 & 3 \\ 2 & 4 \end{bmatrix} = 0$, therefore there are infinitely many solutions or no solutions.

Now use Row Echelon Method on the augmented matrix: $\begin{bmatrix} 1.5 & 3 & | & 5 \\ 2 & 4 & | & 3 \end{bmatrix}$, $\dfrac{2}{3}R_1 \rightarrow \begin{bmatrix} 1 & 2 & | & \frac{10}{3} \\ 2 & 4 & | & 3 \end{bmatrix}$

$-2R_1 + R_2 \rightarrow \begin{bmatrix} 1 & 2 & | & \frac{10}{3} \\ 0 & 0 & | & -\frac{11}{3} \end{bmatrix}$. Since $0 \neq -\dfrac{11}{3}$, there is no solution, the solution set is: $\varnothing$.

65. Find the determinants: $D = \det \begin{bmatrix} 2 & -3 \\ 1 & 5 \end{bmatrix} = 13$; $D_x = \det \begin{bmatrix} -5 & -3 \\ 17 & 5 \end{bmatrix} = 26$; and $D_y = \det \begin{bmatrix} 2 & -5 \\ 1 & 17 \end{bmatrix} = 39$.

Then $x = \dfrac{D_x}{D} = \dfrac{26}{13} = 2$ and $y = \dfrac{D_y}{D} = \dfrac{39}{13} = 3$. The solution is: $\{(2, 3)\}$.

66. Find the determinants: $D = \det \begin{bmatrix} 1 & 9 \\ 3 & 2 \end{bmatrix} = -25$; $D_x = \det \begin{bmatrix} -15 & 9 \\ 5 & 2 \end{bmatrix} = -75$; and $D_y = \det \begin{bmatrix} 1 & -15 \\ 3 & 5 \end{bmatrix} = 50$.

Then $x = \dfrac{D_x}{D} = \dfrac{-75}{-25} = 3$ and $y = \dfrac{D_y}{D} = \dfrac{50}{-25} = -2$. The solution is: $\{(3, -2)\}$.

67. Using your calculator, find the following determinants:

$D = \det \begin{bmatrix} 4 & -1 & 3 \\ 3 & 1 & 1 \\ 2 & -1 & 4 \end{bmatrix} = 15$; $D_x = \det \begin{bmatrix} -3 & -1 & 3 \\ 0 & 1 & 1 \\ 0 & -1 & 4 \end{bmatrix} = -15$; $D_y = \det \begin{bmatrix} 4 & -3 & 3 \\ 3 & 0 & 1 \\ 2 & 0 & 4 \end{bmatrix} = 30$

$D_z = \det \begin{bmatrix} 4 & -1 & -3 \\ 3 & 1 & 0 \\ 2 & -1 & 0 \end{bmatrix} = 15$

Then $x = \dfrac{D_x}{D} = \dfrac{-15}{15} = -1$; $y = \dfrac{D_y}{D} = \dfrac{30}{15} = 2$; and $z = \dfrac{D_z}{D} = \dfrac{15}{15} = 1$. The solution is: $\{(-1, 2, 1)\}$.

68. Using your calculator, find the following determinants:

$D = \det \begin{bmatrix} 5 & 2 & 1 \\ 2 & -1 & 1 \\ 4 & 3 & 2 \end{bmatrix} = -15$; $D_x = \det \begin{bmatrix} 15 & 2 & 1 \\ 9 & -1 & 1 \\ 13 & 3 & 2 \end{bmatrix} = -45$; $D_y = \det \begin{bmatrix} 5 & 15 & 1 \\ 2 & 9 & 1 \\ 4 & 13 & 2 \end{bmatrix} = 15$

$D_z = \det \begin{bmatrix} 5 & 2 & 15 \\ 2 & -1 & 9 \\ 4 & 3 & 13 \end{bmatrix} = -30$

Then $x = \dfrac{D_x}{D} = \dfrac{-45}{-15} = 3$; $y = \dfrac{D_y}{D} = \dfrac{15}{-15} = -1$; and $z = \dfrac{D_z}{D} = \dfrac{-30}{-15} = 2$. The solution is: $\{(3, -1, 2)\}$.

69. Using your calculator, find the following determinants:

$$D = \det \begin{bmatrix} 2 & -1 & 4 \\ 3 & 2 & -1 \\ 1 & 4 & 2 \end{bmatrix} = 63; \quad D_x = \det \begin{bmatrix} -2 & -1 & 4 \\ -3 & 2 & -1 \\ 17 & 4 & 2 \end{bmatrix} = -189; \quad D_y = \det \begin{bmatrix} 2 & -2 & 4 \\ 3 & -3 & -1 \\ 1 & 17 & 2 \end{bmatrix} = 252$$

$$D_z = \det \begin{bmatrix} 2 & -1 & -2 \\ 3 & 2 & -3 \\ 1 & 4 & 17 \end{bmatrix} = 126$$

Then $x = \dfrac{D_x}{D} = \dfrac{-189}{63} = -3$; $y = \dfrac{D_y}{D} = \dfrac{252}{63} = 4$; and $z = \dfrac{D_z}{D} = \dfrac{126}{63} = 2$. The solution is: $\{(-3, 4, 2)\}$.

70. Using your calculator, find the following determinants:

$$D = \det \begin{bmatrix} 1 & 1 & 1 \\ 2 & -1 & 3 \\ 4 & 2 & -1 \end{bmatrix} = 17; \quad D_x = \det \begin{bmatrix} 4 & 1 & 1 \\ 4 & -1 & 3 \\ -15 & 2 & -1 \end{bmatrix} = -68; \quad D_y = \det \begin{bmatrix} 1 & 4 & 1 \\ 2 & 4 & 3 \\ 4 & -15 & -1 \end{bmatrix} = 51$$

$$D_z = \det \begin{bmatrix} 1 & 1 & 4 \\ 2 & -1 & 4 \\ 4 & 2 & -15 \end{bmatrix} = 85$$

Then $x = \dfrac{D_x}{D} = \dfrac{-68}{17} = -4$; $y = \dfrac{D_y}{D} = \dfrac{51}{17} = 3$; and $z = \dfrac{D_z}{D} = \dfrac{85}{17} = 5$. The solution is: $\{(-4, 3, 5)\}$.

71. Using your calculator, find the following determinants:

$$D = \det \begin{bmatrix} 5 & -1 & 0 \\ 3 & 0 & 2 \\ 0 & 4 & 3 \end{bmatrix} = -31; \quad D_x = \det \begin{bmatrix} -4 & -1 & 0 \\ 4 & 0 & 2 \\ 22 & 4 & 3 \end{bmatrix} = 0; \quad D_y = \det \begin{bmatrix} 5 & -4 & 0 \\ 3 & 4 & 2 \\ 0 & 22 & 3 \end{bmatrix} = -124$$

$$D_z = \det \begin{bmatrix} 5 & -1 & -4 \\ 3 & 0 & 4 \\ 0 & 4 & 22 \end{bmatrix} = -62$$

Then $x = \dfrac{D_x}{D} = \dfrac{0}{-31} = 0$; $y = \dfrac{D_y}{D} = \dfrac{-124}{-31} = 4$; and $z = \dfrac{D_z}{D} = \dfrac{-62}{-31} = 2$. The solution is: $\{(0, 4, 2)\}$.

72. Using your calculator, find the following determinants:

$$D = \det \begin{bmatrix} 3 & 5 & 0 \\ 2 & 0 & 7 \\ 0 & 4 & 3 \end{bmatrix} = -114; \quad D_x = \det \begin{bmatrix} -7 & 5 & 0 \\ 2 & 0 & 7 \\ -8 & 4 & 3 \end{bmatrix} = -114; \quad D_y = \det \begin{bmatrix} 3 & -7 & 0 \\ 2 & 2 & 7 \\ 0 & -8 & 3 \end{bmatrix} = 228$$

$$D_z = \det \begin{bmatrix} 3 & 5 & -7 \\ 2 & 0 & 2 \\ 0 & 4 & -8 \end{bmatrix} = 0$$

Then $x = \dfrac{D_x}{D} = \dfrac{-114}{-114} = 1$; $y = \dfrac{D_y}{D} = \dfrac{228}{-114} = -2$; and $z = \dfrac{D_z}{D} = \dfrac{0}{-114} = 0$. The solution is: $\{(1, -2, 0)\}$.

73. Using your calculator, find the following determinant:

$$D = \det \begin{bmatrix} 2 & -1 & 3 \\ -2 & 1 & -3 \\ 5 & -1 & 1 \end{bmatrix} = 0, \text{ therefore there are infinitely many solutions or no solutions.}$$

If we add [equation 1] + [equation 2] the result is: $0 = 3$, therefore no solution or $\emptyset$.

74. Using your calculator, find the following determinant:

$$D = \det \begin{bmatrix} -2 & -2 & 3 \\ 5 & 7 & -1 \\ 2 & 2 & -3 \end{bmatrix} = 0,$$ therefore there are infinitely many solutions or no solutions.

If we add [equation 1] + [equation 3] the result is: $0 = 0$, therefore infinitely many solutions.

Now multiply [equation 1] by 5 and add to [equation 2] multiplied by 2 to eliminate x.

$$\begin{array}{r} -10x - 10y + 15z = 20 \\ 10x + 14y - 2z = 4 \\ \hline 4y + 13z = 24 \end{array} \Rightarrow 4y = 24 - 13z \Rightarrow y = \frac{24 - 13z}{4}.$$

Next substitute $y = \dfrac{24 - 13z}{4}$ into [equation 1]: $-2x - 2\left(\dfrac{24 - 13z}{4}\right) + 3z = 4 \Rightarrow$

$-4x - 24 + 13z + 6z = 8 \Rightarrow -4x + 19z = 32 \Rightarrow -4x = 32 - 19z \Rightarrow x = \dfrac{19z - 32}{4}.$

The solution is: $\left(\dfrac{19z - 32}{4}, \dfrac{24 - 13z}{4}, z\right).$ Other forms are possible.

75. Using your calculator, find the following determinants:

$$D = \det \begin{bmatrix} 3 & 2 & 0 & -1 \\ 2 & 0 & 1 & 2 \\ 1 & 2 & -1 & 0 \\ 2 & -1 & 1 & 1 \end{bmatrix} = -9; \quad D_x = \det \begin{bmatrix} 0 & 2 & 0 & -1 \\ 5 & 0 & 1 & 2 \\ -2 & 2 & -1 & 0 \\ 2 & -1 & 1 & 1 \end{bmatrix} = 9$$

$$D_y = \det \begin{bmatrix} 3 & 0 & 0 & -1 \\ 2 & 5 & 1 & 2 \\ 1 & -2 & -1 & 0 \\ 2 & 2 & 1 & 1 \end{bmatrix} = -18; \quad D_z = \det \begin{bmatrix} 3 & 2 & 0 & -1 \\ 2 & 0 & 5 & 2 \\ 1 & 2 & -2 & 0 \\ 2 & -1 & 2 & 1 \end{bmatrix} = -45; \quad D_w = \det \begin{bmatrix} 3 & 2 & 0 & 0 \\ 2 & 0 & 1 & 5 \\ 1 & 2 & -1 & -2 \\ 2 & -1 & 1 & 2 \end{bmatrix} = -9$$

Then $x = \dfrac{D_x}{D} = \dfrac{9}{-9} = -1; \; y = \dfrac{D_y}{D} = \dfrac{-18}{-9} = 2; \; z = \dfrac{D_z}{D} = \dfrac{-45}{-9} = 5; \; w = \dfrac{D_w}{D} = \dfrac{-9}{-9} = 1.$

The solution is: $\{(-1, 2, 5, 1)\}.$

76. Using your calculator, find the following determinants:

$$D = \det \begin{bmatrix} 1 & 2 & -1 & 1 \\ 2 & -1 & 0 & 2 \\ 0 & 1 & 3 & 0 \\ 1 & 0 & -1 & 0 \end{bmatrix} = 17; \quad D_x = \det \begin{bmatrix} 8 & 2 & -1 & 1 \\ 8 & -1 & 0 & 2 \\ 5 & 1 & 3 & 0 \\ 4 & 0 & -1 & 0 \end{bmatrix} = 85; \quad D_y = \det \begin{bmatrix} 1 & 8 & -1 & 1 \\ 2 & 8 & 0 & 2 \\ 0 & 5 & 3 & 0 \\ 1 & 4 & -1 & 0 \end{bmatrix} = 34$$

$$D_z = \det \begin{bmatrix} 1 & 2 & 8 & 1 \\ 2 & -1 & 8 & 2 \\ 0 & 1 & 5 & 0 \\ 1 & 0 & 4 & 0 \end{bmatrix} = 17; \quad D_w = \det \begin{bmatrix} 1 & 2 & -1 & 8 \\ 2 & -1 & 0 & 8 \\ 0 & 1 & 3 & 5 \\ 1 & 0 & -1 & 4 \end{bmatrix} = 0$$

Then $x = \dfrac{D_x}{D} = \dfrac{85}{17} = 5; \; y = \dfrac{D_y}{D} = \dfrac{34}{17} = 2; \; z = \dfrac{D_z}{D} = \dfrac{17}{17} = 1; \; w = \dfrac{D_w}{D} = \dfrac{0}{17} = 0.$

The solution is: $\{(5, 2, 1, 0)\}.$

77. If $D = 0$, Cramer's rule cannot be applied because there is no unique solution. There are either no solutions or infinitely many solutions.

78. D_x is obtained by replacing the column of x-coefficients with the corresponding constants.

D_y and D_z are obtained in a similar manner.

7.6: Solution of Linear Systems by Matrix Inverses

1. Yes, $AB = BA = \begin{bmatrix} 1 & 0 \\ 0 & 1 \end{bmatrix}$.

$$\begin{bmatrix} 5 & 7 \\ 2 & 3 \end{bmatrix}\begin{bmatrix} 3 & -7 \\ -2 & 5 \end{bmatrix} = \begin{bmatrix} 15-14 & -35+35 \\ 6-6 & -14+15 \end{bmatrix} = \begin{bmatrix} 1 & 0 \\ 0 & 1 \end{bmatrix}; \text{ and}$$

$$\begin{bmatrix} 3 & -7 \\ -2 & 5 \end{bmatrix}\begin{bmatrix} 5 & 7 \\ 2 & 3 \end{bmatrix} = \begin{bmatrix} 15-14 & 21-21 \\ -10+10 & -14+15 \end{bmatrix} = \begin{bmatrix} 1 & 0 \\ 0 & 1 \end{bmatrix}.$$

2. Yes, $AB = BA = \begin{bmatrix} 1 & 0 \\ 0 & 1 \end{bmatrix}$.

$$\begin{bmatrix} 2 & 3 \\ 1 & 1 \end{bmatrix}\begin{bmatrix} -1 & 3 \\ 1 & -2 \end{bmatrix} = \begin{bmatrix} -2+3 & 6-6 \\ -1+1 & 3-2 \end{bmatrix} = \begin{bmatrix} 1 & 0 \\ 0 & 1 \end{bmatrix}; \text{ and}$$

$$\begin{bmatrix} -1 & 3 \\ 1 & -2 \end{bmatrix}\begin{bmatrix} 2 & 3 \\ 1 & 1 \end{bmatrix} = \begin{bmatrix} -2+3 & -3+3 \\ 2-2 & 3-2 \end{bmatrix} = \begin{bmatrix} 1 & 0 \\ 0 & 1 \end{bmatrix}.$$

3. No, $AB \neq \begin{bmatrix} 1 & 0 \\ 0 & 1 \end{bmatrix}$.

$$\begin{bmatrix} -1 & 2 \\ 3 & -5 \end{bmatrix}\begin{bmatrix} -5 & -2 \\ -3 & -1 \end{bmatrix} = \begin{bmatrix} 5-6 & 2-2 \\ -15+15 & -6+5 \end{bmatrix} = \begin{bmatrix} -1 & 0 \\ 0 & -1 \end{bmatrix}.$$

4. No, $AB \neq \begin{bmatrix} 1 & 0 \\ 0 & 1 \end{bmatrix}$.

$$\begin{bmatrix} 2 & 1 \\ 3 & 2 \end{bmatrix}\begin{bmatrix} 2 & 1 \\ -3 & 2 \end{bmatrix} = \begin{bmatrix} 4-3 & 2+2 \\ 6-6 & 3+4 \end{bmatrix} = \begin{bmatrix} 1 & 4 \\ 0 & 7 \end{bmatrix}.$$

5. No, $AB \neq \begin{bmatrix} 1 & 0 & 0 \\ 0 & 1 & 0 \\ 0 & 0 & 1 \end{bmatrix}$. Use your calculator to multiply the matrices.

$$\begin{bmatrix} 0 & 1 & 0 \\ 0 & 0 & -2 \\ 1 & -1 & 0 \end{bmatrix}\begin{bmatrix} 1 & 0 & 1 \\ 1 & 0 & 0 \\ 0 & -1 & 0 \end{bmatrix} = \begin{bmatrix} 1 & 0 & 0 \\ 0 & 2 & 0 \\ 0 & 0 & 1 \end{bmatrix}.$$

6. No, $AB \neq \begin{bmatrix} 1 & 0 & 0 \\ 0 & 1 & 0 \\ 0 & 0 & 1 \end{bmatrix}$. Use the calculator to multiply the matrices.

$$\begin{bmatrix} 1 & 2 & 0 \\ 0 & 1 & 0 \\ 0 & 1 & 0 \end{bmatrix}\begin{bmatrix} 1 & -2 & 0 \\ 0 & 1 & 0 \\ 0 & -1 & 1 \end{bmatrix} = \begin{bmatrix} 1 & 0 & 0 \\ 0 & 1 & 0 \\ 0 & 1 & 0 \end{bmatrix}.$$

7. Yes. Using the calculator, multiply the matrices. $AB = \begin{bmatrix} 1 & 0 & 0 \\ 0 & 1 & 0 \\ 0 & 0 & 1 \end{bmatrix} = BA.$

8. Yes. Using the calculator, multiply the matrices. $AB = \begin{bmatrix} 1 & 0 & 0 \\ 0 & 1 & 0 \\ 0 & 0 & 1 \end{bmatrix} = BA.$

9. The inverse will not exist if its determinant is equal to 0.

10. If a column or row contains all 0s, its determinant will equal 0, and thus it will not have an inverse.

11. Using $A^{-1} = \dfrac{1}{\det A}\begin{bmatrix} d & -b \\ -c & a \end{bmatrix}$ to find the inverse yields:

$$A^{-1} = \frac{1}{15-14}\begin{bmatrix} 5 & -7 \\ -2 & 3 \end{bmatrix} = \begin{bmatrix} 5 & -7 \\ -2 & 3 \end{bmatrix}.$$

12. Using $A^{-1} = \dfrac{1}{\det A}\begin{bmatrix} d & -b \\ -c & a \end{bmatrix}$ to find the inverse yields:

$$A^{-1} = \frac{1}{-25+24}\begin{bmatrix} 5 & -3 \\ 8 & -5 \end{bmatrix} = (-1)\begin{bmatrix} 5 & -3 \\ 8 & -5 \end{bmatrix} = \begin{bmatrix} -5 & 3 \\ -8 & 5 \end{bmatrix}.$$

13. Using $A^{-1} = \dfrac{1}{\det A}\begin{bmatrix} d & -b \\ -c & a \end{bmatrix}$ to find the inverse yields:

$$A^{-1} = \frac{1}{-4+6}\begin{bmatrix} 4 & 2 \\ -3 & -1 \end{bmatrix} = \left(\frac{1}{2}\right)\begin{bmatrix} 4 & 2 \\ -3 & -1 \end{bmatrix} = \begin{bmatrix} 2 & 1 \\ -\frac{3}{2} & -\frac{1}{2} \end{bmatrix}.$$

14. Using $A^{-1} = \dfrac{1}{\det A}\begin{bmatrix} d & -b \\ -c & a \end{bmatrix}$ to find the inverse yields:

$$A^{-1} = \frac{1}{1+4}\begin{bmatrix} -1 & -2 \\ 2 & -1 \end{bmatrix} = \left(\frac{1}{5}\right)\begin{bmatrix} -1 & -2 \\ 2 & -1 \end{bmatrix} = \begin{bmatrix} -\frac{1}{5} & -\frac{2}{5} \\ \frac{2}{5} & -\frac{1}{5} \end{bmatrix}.$$

15. The $\det A = -12 - (-12) = 0$, therefore No inverse.

16. The $\det A = -30 - (-30) = 0$, therefore No inverse.

17. Using $A^{-1} = \dfrac{1}{\det A}\begin{bmatrix} d & -b \\ -c & a \end{bmatrix}$ to find the inverse yields:

$$A^{-1} = \frac{1}{.06-.1}\begin{bmatrix} .1 & -.2 \\ -.5 & .6 \end{bmatrix} = (-25)\begin{bmatrix} .1 & -.2 \\ -.5 & .6 \end{bmatrix} = \begin{bmatrix} -2.5 & 5 \\ 12.5 & -15 \end{bmatrix}.$$

18. Using $A^{-1} = \dfrac{1}{\det A}\begin{bmatrix} d & -b \\ -c & a \end{bmatrix}$ to find the inverse yields:

$$A^{-1} = \frac{1}{-.16+.15}\begin{bmatrix} -.2 & .3 \\ -.5 & .8 \end{bmatrix} = (-100)\begin{bmatrix} -.2 & .3 \\ -.5 & .8 \end{bmatrix} = \begin{bmatrix} 20 & -30 \\ 50 & -80 \end{bmatrix}.$$

19. $A \mid I_3 = \begin{bmatrix} 0 & 0 & 1 & | & 1 & 0 & 0 \\ 1 & 0 & 0 & | & 0 & 1 & 0 \\ 0 & 1 & 0 & | & 0 & 0 & 1 \end{bmatrix} \begin{matrix} R_2 \to \\ R_3 \to \\ R_1 \to \end{matrix} \begin{bmatrix} 1 & 0 & 0 & | & 0 & 1 & 0 \\ 0 & 1 & 0 & | & 0 & 0 & 1 \\ 0 & 0 & 1 & | & 1 & 0 & 0 \end{bmatrix}; \ A^{-1} = \begin{bmatrix} 0 & 1 & 0 \\ 0 & 0 & 1 \\ 1 & 0 & 0 \end{bmatrix}$

20. $A \mid I_3 = \begin{bmatrix} 1 & 0 & 0 & | & 1 & 0 & 0 \\ 1 & 1 & 0 & | & 0 & 1 & 0 \\ 0 & 1 & 1 & | & 0 & 0 & 1 \end{bmatrix} R_2 - R_1 \to \begin{bmatrix} 1 & 0 & 0 & | & 1 & 0 & 0 \\ 0 & 1 & 0 & | & -1 & 1 & 0 \\ 0 & 1 & 1 & | & 0 & 0 & 1 \end{bmatrix} R_3 - R_2 \to \begin{bmatrix} 1 & 0 & 0 & | & 1 & 0 & 0 \\ 0 & 1 & 0 & | & -1 & 1 & 0 \\ 0 & 0 & 1 & | & 1 & -1 & 1 \end{bmatrix};$

$$A^{-1} = \begin{bmatrix} 1 & 0 & 0 \\ -1 & 1 & 0 \\ 1 & -1 & 1 \end{bmatrix}$$

21. $A \mid I_3 = \begin{bmatrix} 1 & 0 & 1 & | & 1 & 0 & 0 \\ 2 & 1 & 3 & | & 0 & 1 & 0 \\ -1 & 1 & 1 & | & 0 & 0 & 1 \end{bmatrix} \begin{matrix} \\ R_2 - 2R_1 \rightarrow \\ R_3 + R_1 \rightarrow \end{matrix} \begin{bmatrix} 1 & 0 & 1 & | & 1 & 0 & 0 \\ 0 & 1 & 1 & | & -2 & 1 & 0 \\ 0 & 1 & 2 & | & 1 & 0 & 1 \end{bmatrix} \begin{matrix} \\ \\ R_3 - R_2 \rightarrow \end{matrix} \begin{bmatrix} 1 & 0 & 1 & | & 1 & 0 & 0 \\ 0 & 1 & 1 & | & -2 & 1 & 0 \\ 0 & 0 & 1 & | & 3 & -1 & 1 \end{bmatrix}$

$\begin{matrix} R_1 - R_3 \rightarrow \\ R_2 - R_3 \rightarrow \\ \\ \end{matrix} \begin{bmatrix} 1 & 0 & 0 & | & -2 & 1 & -1 \\ 0 & 1 & 0 & | & -5 & 2 & -1 \\ 0 & 0 & 1 & | & 3 & -1 & 1 \end{bmatrix} ; \quad A^{-1} = \begin{bmatrix} -2 & 1 & -1 \\ -5 & 2 & -1 \\ 3 & -1 & 1 \end{bmatrix}$

22. $A \mid I_3 = \begin{bmatrix} -2 & 1 & 0 & | & 1 & 0 & 0 \\ 1 & 0 & 1 & | & 0 & 1 & 0 \\ -1 & 1 & 0 & | & 0 & 0 & 1 \end{bmatrix} \begin{matrix} R_2 \rightarrow \\ R_1 \rightarrow \\ \\ \end{matrix} \begin{bmatrix} 1 & 0 & 1 & | & 0 & 1 & 0 \\ -2 & 1 & 0 & | & 1 & 0 & 0 \\ -1 & 1 & 0 & | & 0 & 0 & 1 \end{bmatrix} \begin{matrix} R_2 + 2R_1 \rightarrow \\ R_3 + R_1 \rightarrow \\ \\ \end{matrix} \begin{bmatrix} 1 & 0 & 1 & | & 0 & 1 & 0 \\ 0 & 1 & 2 & | & 1 & 2 & 0 \\ 0 & 1 & 1 & | & 0 & 1 & 1 \end{bmatrix}$

$\begin{matrix} \\ \\ R_3 - R_2 \rightarrow \end{matrix} \begin{bmatrix} 1 & 0 & 1 & | & 0 & 1 & 0 \\ 0 & 1 & 2 & | & 1 & 2 & 0 \\ 0 & 0 & -1 & | & -1 & -1 & 1 \end{bmatrix} \begin{matrix} \\ \\ -1R_3 \rightarrow \end{matrix} \begin{bmatrix} 1 & 0 & 1 & | & 0 & 1 & 0 \\ 0 & 1 & 2 & | & 1 & 2 & 0 \\ 0 & 0 & 1 & | & 1 & 1 & -1 \end{bmatrix} \begin{matrix} R_1 - R_3 \rightarrow \\ R_2 - 2R_3 \rightarrow \\ \\ \end{matrix} \begin{bmatrix} 1 & 0 & 0 & | & -1 & 0 & 1 \\ 0 & 1 & 0 & | & -1 & 0 & 2 \\ 0 & 0 & 1 & | & 1 & 1 & -1 \end{bmatrix} ;$

$A^{-1} = \begin{bmatrix} -1 & 0 & 1 \\ -1 & 0 & 2 \\ 1 & 1 & -1 \end{bmatrix}$

23. Solve using the calculator. Since $\det A = -1$, the inverse exists and $A^{-1} = \begin{bmatrix} 1 & 0 & 0 \\ 0 & -1 & 0 \\ -1 & 0 & 1 \end{bmatrix}$.

24. Solve using the calculator. Since $\det A = 1$, the inverse exists and $A^{-1} = \begin{bmatrix} 7 & -3 & -3 \\ -1 & 1 & 0 \\ -1 & 0 & 1 \end{bmatrix}$.

25. Solve using the calculator. Since $\det A = -1$, the inverse exists and $A^{-1} = \begin{bmatrix} 15 & 4 & -5 \\ -12 & -3 & 4 \\ -4 & -1 & 1 \end{bmatrix}$.

26. Solve using the calculator. Since $\det A = 2$, the inverse exists and $A^{-1} = \begin{bmatrix} -\frac{7}{2} & 0 & 0 \\ \frac{1}{2} & -1 & 0 \\ 2 & 0 & 1 \end{bmatrix} = \begin{bmatrix} -3.5 & 2 & -2 \\ .5 & 0 & 1 \\ 2 & -1 & 1 \end{bmatrix}$.

27. Solve using the calculator. Since $\det A = .036$, the inverse exists and $A^{-1} = \begin{bmatrix} -\frac{10}{3} & \frac{5}{9} & -\frac{10}{9} \\ \frac{20}{3} & \frac{5}{9} & \frac{80}{9} \\ -5 & \frac{5}{6} & -\frac{20}{3} \end{bmatrix}$.

28. Solve using the calculator. Since $\det A = .02$, the inverse exists and $A^{-1} = \begin{bmatrix} 0 & -5 & 3 \\ 5 & 20 & -13 \\ 0 & 0 & 2 \end{bmatrix}$.

29. Solve using the calculator. Since $\det A = 0$, there is No inverse.

30. Solve using the calculator. Since $\det A = 0$, there is No inverse.

31. Solve using the calculator. Since $\det A \approx 9.207$, the inverse exists and $A^{-1} = \begin{bmatrix} .0543058761 & -.0543058761 \\ 1.846399787 & .153600213 \end{bmatrix}$.

32. Solve using the calculator. Since $\det A \approx -14.245$, the inverse exists and $A^{-1} = \begin{bmatrix} -.1215875322 & .0491390161 \\ 1.544369078 & -.046799063 \end{bmatrix}$.

33. Solve using the calculator. Since det $A = 1.478$, the inverse exists and

$$A^{-1} = \begin{bmatrix} .9987635516 & -.252092087 & -3330564627 \\ -.5037783375 & 1.007556675 & -.2518891688 \\ -.2481013617 & -.255676976 & 1.003768868 \end{bmatrix}.$$

34. Solve using the calculator. Since det $A = -\dfrac{1}{24}$, the inverse exists and $A^{-1} = \begin{bmatrix} 2 & -2 & 0 \\ -4 & 0 & 4 \\ 3 & 3 & -3 \end{bmatrix}$.

35. Using the matrix inverse method, put the system into the proper matrix form: $\begin{bmatrix} 2 & -1 \\ 3 & 1 \end{bmatrix}\begin{bmatrix} x \\ y \end{bmatrix} = \begin{bmatrix} -8 \\ -2 \end{bmatrix} \Rightarrow$

$$\begin{bmatrix} x \\ y \end{bmatrix} = \begin{bmatrix} 2 & -1 \\ 3 & 1 \end{bmatrix}^{-1}\begin{bmatrix} -8 \\ -2 \end{bmatrix} \Rightarrow \frac{1}{5}\begin{bmatrix} 1 & 1 \\ -3 & 2 \end{bmatrix}\begin{bmatrix} -8 \\ -2 \end{bmatrix} \Rightarrow \begin{bmatrix} \frac{1}{5} & \frac{1}{5} \\ -\frac{3}{5} & \frac{2}{5} \end{bmatrix}\begin{bmatrix} -8 \\ -2 \end{bmatrix} \Rightarrow \begin{bmatrix} -\frac{8}{5} - \frac{2}{5} \\ \frac{24}{5} - \frac{4}{5} \end{bmatrix} = \begin{bmatrix} -2 \\ 4 \end{bmatrix}.$$

The solution is: $\{(-2, 4)\}$.

36. Using the matrix inverse method, put the system into the proper matrix form: $\begin{bmatrix} 1 & 3 \\ 2 & -1 \end{bmatrix}\begin{bmatrix} x \\ y \end{bmatrix} = \begin{bmatrix} -12 \\ 11 \end{bmatrix} \Rightarrow$

$$\begin{bmatrix} x \\ y \end{bmatrix} = \begin{bmatrix} 1 & 3 \\ 2 & -1 \end{bmatrix}^{-1}\begin{bmatrix} -12 \\ 11 \end{bmatrix} \Rightarrow -\frac{1}{7}\begin{bmatrix} -1 & -3 \\ -2 & 1 \end{bmatrix}\begin{bmatrix} -12 \\ 11 \end{bmatrix} \Rightarrow \begin{bmatrix} \frac{1}{7} & \frac{3}{7} \\ \frac{2}{7} & -\frac{1}{7} \end{bmatrix}\begin{bmatrix} -12 \\ 11 \end{bmatrix} \Rightarrow \begin{bmatrix} -\frac{12}{7} + \frac{33}{7} \\ -\frac{24}{7} - \frac{11}{7} \end{bmatrix} = \begin{bmatrix} 3 \\ -5 \end{bmatrix}.$$

The solution is: $\{(3, -5)\}$.

37. Using the matrix inverse method, put the system into the proper matrix form: $\begin{bmatrix} 2 & 3 \\ 3 & 4 \end{bmatrix}\begin{bmatrix} x \\ y \end{bmatrix} = \begin{bmatrix} -10 \\ -12 \end{bmatrix} \Rightarrow$

$$\begin{bmatrix} x \\ y \end{bmatrix} = \begin{bmatrix} 2 & 3 \\ 3 & 4 \end{bmatrix}^{-1}\begin{bmatrix} -10 \\ -12 \end{bmatrix} \Rightarrow -1\begin{bmatrix} 4 & -3 \\ -3 & 2 \end{bmatrix}\begin{bmatrix} -10 \\ -12 \end{bmatrix} \Rightarrow \begin{bmatrix} -4 & 3 \\ 3 & -2 \end{bmatrix}\begin{bmatrix} -10 \\ -12 \end{bmatrix} \Rightarrow \begin{bmatrix} 40 - 36 \\ -30 + 24 \end{bmatrix} = \begin{bmatrix} 4 \\ -6 \end{bmatrix}.$$

The solution is: $\{(4, -6)\}$.

38. Using the matrix inverse method, put the system into the proper matrix form: $\begin{bmatrix} 2 & -3 \\ 2 & 2 \end{bmatrix}\begin{bmatrix} x \\ y \end{bmatrix} = \begin{bmatrix} 10 \\ 5 \end{bmatrix} \Rightarrow$

$$\begin{bmatrix} x \\ y \end{bmatrix} = \begin{bmatrix} 2 & -3 \\ 2 & 2 \end{bmatrix}^{-1}\begin{bmatrix} 10 \\ 5 \end{bmatrix} \Rightarrow \frac{1}{10}\begin{bmatrix} 2 & 3 \\ -2 & 2 \end{bmatrix}\begin{bmatrix} 10 \\ 5 \end{bmatrix} \Rightarrow \begin{bmatrix} .2 & .3 \\ -.2 & .2 \end{bmatrix}\begin{bmatrix} 10 \\ 5 \end{bmatrix} \Rightarrow \begin{bmatrix} 2 + 1.5 \\ -2 + 1 \end{bmatrix} = \begin{bmatrix} 3.5 \\ -1 \end{bmatrix}.$$

The solution is: $\{(3.5, -1)\}$.

39. Using the matrix inverse method, put the system into the proper matrix form: $\begin{bmatrix} 2 & -5 \\ 4 & -5 \end{bmatrix}\begin{bmatrix} x \\ y \end{bmatrix} = \begin{bmatrix} 10 \\ 15 \end{bmatrix} \Rightarrow$

$$\begin{bmatrix} x \\ y \end{bmatrix} = \begin{bmatrix} 2 & -5 \\ 4 & -5 \end{bmatrix}^{-1}\begin{bmatrix} 10 \\ 15 \end{bmatrix} \Rightarrow \frac{1}{10}\begin{bmatrix} -5 & 5 \\ -4 & 2 \end{bmatrix}\begin{bmatrix} 10 \\ 15 \end{bmatrix} \Rightarrow \begin{bmatrix} -.5 & .5 \\ -.4 & .2 \end{bmatrix}\begin{bmatrix} 10 \\ 15 \end{bmatrix} \Rightarrow \begin{bmatrix} -5 + 7.5 \\ -4 + 3 \end{bmatrix} = \begin{bmatrix} 2.5 \\ -1 \end{bmatrix}.$$

The solution is: $\{(2.5, -1)\}$.

40. The determinant of the coefficient matrix is: $\det \begin{bmatrix} 2 & -3 \\ 4 & -6 \end{bmatrix} = 0$, therefore infinitely many solutions or no solutions.

Now use Row Echelon Method on the augmented matrix: $\begin{bmatrix} 2 & -3 & | & 2 \\ 4 & -6 & | & 1 \end{bmatrix}$, $-2R_1 + R_2 \rightarrow \begin{bmatrix} 2 & -3 & | & 0 \\ 0 & 0 & | & -3 \end{bmatrix}$.

Since $0 \neq -3$, there is no solution or the solution is: $\varnothing$.

41. Solve on the calculator using the matrix inverse method:

$$\begin{bmatrix} x \\ y \\ z \end{bmatrix} = \begin{bmatrix} 2 & 0 & 4 \\ 3 & 1 & 5 \\ -1 & 1 & -2 \end{bmatrix}^{-1} \cdot \begin{bmatrix} 14 \\ 19 \\ -7 \end{bmatrix} = \begin{bmatrix} 3 \\ 0 \\ 2 \end{bmatrix},$$ therefore the solution is: $\{(3, 0, 2)\}$.

42. Solve on the calculator using the matrix inverse method:

$$\begin{bmatrix} x \\ y \\ z \end{bmatrix} = \begin{bmatrix} 3 & 6 & 3 \\ 6 & 4 & -2 \\ 0 & 1 & -1 \end{bmatrix}^{-1} \cdot \begin{bmatrix} 12 \\ -4 \\ -3 \end{bmatrix} = \begin{bmatrix} \frac{1}{4} \\ \frac{1}{4} \\ \frac{13}{4} \end{bmatrix}, \text{ therefore the solution is: } \left\{ \left(\frac{1}{4}, \frac{1}{4}, \frac{13}{4} \right) \right\}.$$

43. Solve on the calculator using the matrix inverse method:

$$\begin{bmatrix} x \\ y \\ z \end{bmatrix} = \begin{bmatrix} 1 & 3 & 1 \\ 1 & -2 & 3 \\ 2 & -3 & -1 \end{bmatrix}^{-1} \cdot \begin{bmatrix} 2 \\ -3 \\ 34 \end{bmatrix} = \begin{bmatrix} 12 \\ -\frac{15}{11} \\ -\frac{65}{11} \end{bmatrix}, \text{ therefore the solution is: } \left\{ \left(12, -\frac{15}{11}, -\frac{65}{11} \right) \right\}.$$

44. Solve on the calculator using the matrix inverse method:

$$\begin{bmatrix} x \\ y \\ z \end{bmatrix} = \begin{bmatrix} 1 & 1 & -1 \\ 2 & -1 & 1 \\ 1 & -2 & 3 \end{bmatrix}^{-1} \cdot \begin{bmatrix} 6 \\ -9 \\ 1 \end{bmatrix} = \begin{bmatrix} -1 \\ 23 \\ 16 \end{bmatrix}, \text{ therefore the solution is: } \{(-1, 23, 16)\}.$$

45. Solve on the calculator using the matrix inverse method:

$$\begin{bmatrix} x \\ y \\ z \\ w \end{bmatrix} = \begin{bmatrix} 1 & 3 & -2 & -1 \\ 4 & 1 & 1 & 2 \\ -3 & -1 & 1 & -1 \\ 1 & -1 & -3 & -2 \end{bmatrix}^{-1} \cdot \begin{bmatrix} 9 \\ 2 \\ -5 \\ 2 \end{bmatrix} = \begin{bmatrix} 0 \\ 2 \\ -2 \\ 1 \end{bmatrix}, \text{ therefore the solution is: } \{(0, 2, -2, 1)\}.$$

46. Solve on the calculator using the matrix inverse method:

$$\begin{bmatrix} x \\ y \\ z \\ w \end{bmatrix} = \begin{bmatrix} 3 & 2 & 0 & -1 \\ 2 & 0 & 1 & 2 \\ 1 & 2 & -1 & 0 \\ 2 & -1 & 1 & 1 \end{bmatrix}^{-1} \cdot \begin{bmatrix} 0 \\ 5 \\ -2 \\ 2 \end{bmatrix} = \begin{bmatrix} -1 \\ 2 \\ 5 \\ 1 \end{bmatrix}, \text{ therefore the solution is: } \{(-1, 2, 5, 1)\}.$$

47. Solve on the calculator using the matrix inverse method:

$$\begin{bmatrix} x \\ y \end{bmatrix} = \begin{bmatrix} 1 & -\sqrt{2} \\ .75 & 1 \end{bmatrix}^{-1} \cdot \begin{bmatrix} 2.6 \\ -7 \end{bmatrix} = \begin{bmatrix} -3.542308934 \\ -4.343268299 \end{bmatrix}, \text{ therefore the solution is: }$$

$$\{(-3.542308934, -4.343268299)\}.$$

48. Solve on the calculator using the matrix inverse method:

$$\begin{bmatrix} x \\ y \end{bmatrix} = \begin{bmatrix} 2.1 & 1 \\ \sqrt{2} & -2 \end{bmatrix}^{-1} \cdot \begin{bmatrix} \sqrt{5} \\ 5 \end{bmatrix} = \begin{bmatrix} 1.68717058 \\ -1.306990242 \end{bmatrix}, \text{ therefore the solution is: } \{(1.68717058, -1.306990242)\}.$$

49. Solve on the calculator using the matrix inverse method:

$$\begin{bmatrix} x \\ y \\ z \end{bmatrix} = \begin{bmatrix} \pi & e & \sqrt{2} \\ e & \pi & \sqrt{2} \\ \sqrt{2} & e & \pi \end{bmatrix}^{-1} \cdot \begin{bmatrix} 1 \\ 2 \\ 3 \end{bmatrix} = \begin{bmatrix} -.9704156959 \\ 1.391914631 \\ .1874077432 \end{bmatrix},$$

therefore the solution is: $\{(-.9704156959, 1.391914631, .1874077432)\}$.

50. Solve on the calculator using the matrix inverse method:

$$\begin{bmatrix} x \\ y \\ z \end{bmatrix} = \begin{bmatrix} \log 2 & \ln 3 & \ln 4 \\ \ln 3 & \log 2 & \ln 8 \\ \log 12 & \ln 4 & \ln 8 \end{bmatrix}^{-1} \cdot \begin{bmatrix} 1 \\ 5 \\ 9 \end{bmatrix} = \begin{bmatrix} 13.58736702 \\ 3.929011993 \\ -5.342780076 \end{bmatrix},$$

therefore the solution is: $\{(13.58736702, 3.929011993, -5.342780076)\}$.

51. If $P(x) = ax^3 + bx^2 + cx + d$, then:

    The ordered pair: $(-1, 14)$ yields the equation: $-a + b - c + d = 14$.

    The ordered pair: $(1.5, 1.5)$ yields the equation: $3.375a + 2.25b + 1.5c + d = 1.5$.

    The ordered pair: $(2, -1)$ yields the equation: $8a + 4b + 2c + d = -1$.

    The ordered pair: $(3, -18)$ yields the equation: $27a + 9b + 3c + d = -18$.

    Now solve on the calculator using the matrix inverse method:

    $$\begin{bmatrix} a \\ b \\ c \\ d \end{bmatrix} = \begin{bmatrix} -1 & 1 & -1 & 1 \\ 3.375 & 2.25 & 1.5 & 1 \\ 8 & 4 & 2 & 1 \\ 27 & 9 & 3 & 1 \end{bmatrix}^{-1} \cdot \begin{bmatrix} 14 \\ 1.5 \\ -1 \\ -18 \end{bmatrix} = \begin{bmatrix} -2 \\ 5 \\ -4 \\ 3 \end{bmatrix}, \text{ therefore the equation is: } P(x) = -2x^3 + 5x^2 - 4x + 3.$$

52. If $P(x) = ax^3 + bx^2 + cx + d$, then:

    The ordered pair: $(-1, -4)$ yields the equation: $-a + b - c + d = -4$.

    The ordered pair: $(-2, -31)$ yields the equation: $-8a + 4b - 2c + d = -31$.

    The ordered pair: $(0, 1)$ yields the equation: $d = 1$.

    The ordered pair: $(2, 17)$ yields the equation: $8a + 4b + 2c + d = 17$.

    Now solve on the calculator using the matrix inverse method:

    $$\begin{bmatrix} a \\ b \\ c \\ d \end{bmatrix} = \begin{bmatrix} -1 & 1 & -1 & 1 \\ -8 & 4 & -2 & 1 \\ 0 & 0 & 0 & 1 \\ 8 & 4 & 2 & 1 \end{bmatrix}^{-1} \cdot \begin{bmatrix} -4 \\ -31 \\ 1 \\ 17 \end{bmatrix} = \begin{bmatrix} 3 \\ -2 \\ 0 \\ 1 \end{bmatrix}, \text{ therefore the equation is: } P(x) = 3x^3 - 2x^2 + 1.$$

53. If $P(x) = ax^4 + bx^3 + cx^2 + dx + e$, then:

    The ordered pair: $(-2, 13)$ yields the equation: $16a - 8b + 4c - 2d + e = 13$.

    The ordered pair: $(-1, 2)$ yields the equation: $a - b + c - d + e = 2$.

    The ordered pair: $(0, -1)$ yields the equation: $e = -1$.

    The ordered pair: $(1, 4)$ yields the equation: $a + b + c + d + e = 4$.

    The ordered pair: $(2, 41)$ yields the equation: $16a + 8b + 4c + 2d + e = 41$.

    Now solve on the calculator using the matrix inverse method:

    $$\begin{bmatrix} a \\ b \\ c \\ d \\ e \end{bmatrix} = \begin{bmatrix} 16 & -8 & 4 & -2 & 1 \\ 1 & -1 & 1 & -1 & 1 \\ 0 & 0 & 0 & 0 & 1 \\ 1 & 1 & 1 & 1 & 1 \\ 16 & 8 & 4 & 2 & 1 \end{bmatrix}^{-1} \cdot \begin{bmatrix} 13 \\ 2 \\ -1 \\ 4 \\ 41 \end{bmatrix} = \begin{bmatrix} 1 \\ 2 \\ 3 \\ -1 \\ -1 \end{bmatrix}, \text{ therefore the equation is: } P(x) = x^4 + 2x^3 + 3x^2 - x - 1.$$

54. If $P(x) = ax^5 + bx^4 + cx^3 + dx^2 + ex + f$, then:

The ordered pair: $(-2, -8)$ yields the equation: $-32a + 16b - 8c + 4d - 2e + f = -8$.

The ordered pair: $(-1, -1)$ yields the equation: $-a + b - c + d - e + f = -1$.

The ordered pair: $(0, -4)$ yields the equation: $f = -4$.

The ordered pair: $(1, -5)$ yields the equation: $a + b + c + d + e + f = -5$.

The ordered pair: $(2, 8)$ yields the equation: $32a + 16b + 8c + 4d + 2e + f = 8$.

The ordered pair: $(3, 167)$ yields the equation: $243a + 81b + 27c + 9d + 3e + f = 167$.

Now solve on the calculator using the matrix inverse method:

$$
\begin{bmatrix} a \\ b \\ c \\ d \\ e \\ f \end{bmatrix} =
\begin{bmatrix}
-32 & 16 & -8 & 4 & -2 & 1 \\
-1 & 1 & -1 & 1 & -1 & 1 \\
0 & 0 & 0 & 0 & 0 & 1 \\
1 & 1 & 1 & 1 & 1 & 1 \\
32 & 16 & 8 & 4 & 2 & 1 \\
243 & 81 & 27 & 9 & 3 & 1
\end{bmatrix}^{-1}
\cdot
\begin{bmatrix} -8 \\ -1 \\ -4 \\ -5 \\ 8 \\ 167 \end{bmatrix} =
\begin{bmatrix} 1 \\ 0 \\ -3 \\ 1 \\ 0 \\ -4 \end{bmatrix},
$$

therefore the equation is: $P(x) = x^5 - 3x^3 + x^2 - 4$.

55. (a) Let x be the cost of a soft drink and let y be the cost of a box of popcorn. Then:

$$\begin{aligned} 3x + 2y &= 8.5 \\ 4x + 3y &= 12 \end{aligned} \Rightarrow Ax = B: \begin{bmatrix} 3 & 2 \\ 4 & 3 \end{bmatrix} \cdot \begin{bmatrix} x \\ y \end{bmatrix} = \begin{bmatrix} 8.5 \\ 12 \end{bmatrix}.$$

Now solve on the calculator using the matrix inverse method:

$$\begin{bmatrix} x \\ y \end{bmatrix} = \begin{bmatrix} 3 & 2 \\ 4 & 3 \end{bmatrix}^{-1} \cdot \begin{bmatrix} 8.5 \\ 12 \end{bmatrix} = \begin{bmatrix} 1.5 \\ 2 \end{bmatrix}.$$ The cost of a soft drink is \$1.50 and the cost of popcorn is \$2.00.

(b) No. There are two variables. This typically requires two equations. The linear system would be dependent because the two equation would be the same. A^{-1} would not exist. If we attempt to calculate A^{-1} with a calculator, an error message tells us that the matrix is singular.

56. Let x be the number of CDs of type A, let y be the number of CDs of type B, and let z be the number of CDs of type C.

Then: $\begin{aligned} 2x + 3y + 4z &= 120.91 \\ x + 4y &= 62.95 \\ 2x + y + 3z &= 79.94 \end{aligned}$ Now solve on the calculator using the matrix inverse method:

$$\begin{bmatrix} x \\ y \\ z \end{bmatrix} = \begin{bmatrix} 2 & 3 & 4 \\ 1 & 4 & 0 \\ 2 & 1 & 3 \end{bmatrix}^{-1} \cdot \begin{bmatrix} 120.91 \\ 62.95 \\ 79.94 \end{bmatrix} = \begin{bmatrix} 10.99 \\ 12.99 \\ 14.99 \end{bmatrix}.$$

The cost of each CD is: Type A CDs \$10.99; Type B CDs \$12.99; and Type C CDs \$14.99.

57. (a) Using the model $T = aA + bI + c$ and the data from the table, the equations are:

$$113a +\ \ 308b + c = 10{,}170$$

$$133a +\ \ 622b + c = 15{,}305$$

$$155a + 1937b + c = 21{,}289$$

(b) Solve on the calculator using the matrix inverse method:

$$\begin{bmatrix} a \\ b \\ c \end{bmatrix} = \begin{bmatrix} 113 & 308 & 1 \\ 133 & 622 & 1 \\ 155 & 1937 & 1 \end{bmatrix}^{-1} \cdot \begin{bmatrix} 10{,}170 \\ 15{,}305 \\ 21{,}289 \end{bmatrix} = \begin{bmatrix} 251.3175021 \\ .3460189769 \\ -.18335.45158 \end{bmatrix}.$$

The formula is: $T \approx 251\,A + .346I - 18{,}300$.

(c) $T \approx 251(118) + .346(311) - 18{,}300 \Rightarrow T \approx 11{,}426$. This is quite close to the actual value of 11,314.

58. (a) Using the model $G = aA + bB + c$ and the data from the table, the equations are:

$$5.54a + 37.1b + c = 603$$

$$6.93a + 41.3b + c = 657$$

$$7.64a + 45.6b + c = 779$$

(b) Solve on the calculator using the matrix inverse method:

$$\begin{bmatrix} a \\ b \\ c \end{bmatrix} = \begin{bmatrix} 5.54 & 37.1 & 1 \\ 6.93 & 41.3 & 1 \\ 7.64 & 45.6 & 1 \end{bmatrix}^{-1} \cdot \begin{bmatrix} 603 \\ 657 \\ 779 \end{bmatrix} = \begin{bmatrix} -93.55592654 \\ 43.8196995 \\ -504.4110184 \end{bmatrix}.$$

The formula is: $G \approx -93.6\,A + 43.8B - 504$.

(c) $G \approx -93.61(7.75) + 43.8(47.4) - 504 \Rightarrow G \approx 847$ or $847 million.

This is quite close to the actual value of $878 million.

59. (a) Using the model $P = a + bS + cC$ and the data from the table, the equations are:

$$a + 1500b + 8c = 122$$

$$a + 2000b + 5c = 130$$

$$a + 2200b + 10c = 158$$

(b) Solve on the calculator using the matrix inverse method:

$$\begin{bmatrix} a \\ b \\ c \end{bmatrix} = \begin{bmatrix} 1 & 1500 & 8 \\ 1 & 2000 & 5 \\ 1 & 2200 & 10 \end{bmatrix}^{-1} \cdot \begin{bmatrix} 122 \\ 130 \\ 158 \end{bmatrix} = \begin{bmatrix} 30 \\ .04 \\ 4 \end{bmatrix}.$$

The formula is: $G \approx 30 + .04S + 4C$.

The selling price is: $P = 30 + .04(1800) + 4(7) \Rightarrow p = 130$ or $130,000.

60. (a) Intersection A: incoming traffic is $x_1 + 5$ and outgoing traffic is $4 + 6$, so $x_1 + 5 = 4 + 6$.

Intersection B: incoming traffic is $x_2 + 6$ and outgoing traffic is $x_1 + 3$, so $x_2 + 6 = x_1 + 3$.

Intersection C: incoming traffic is $x_3 + 4$ and outgoing traffic is $x_2 + 7$, so $x_3 + 4 = x_2 + 7$.

Intersection D: incoming traffic is $6 + 5$ and outgoing traffic is $x_3 + x_4$, so $6 + 5 = x_3 + x_4$.

(b) The equations are:

$x_1 = 5$

$x_1 - x_2 = 3$

$x_2 - x_3 = -3$

$x_3 + x_4 = 11$, then

$$AX = \begin{bmatrix} 1 & 0 & 0 & 0 \\ -1 & 1 & 0 & 0 \\ 0 & -1 & 1 & 0 \\ 0 & 0 & 1 & 1 \end{bmatrix} \begin{bmatrix} x_1 \\ x_2 \\ x_3 \\ x_4 \end{bmatrix} = \begin{bmatrix} 5 \\ -3 \\ 3 \\ 11 \end{bmatrix} = B. \text{ Now solve on the calculator using the matrix inverse method:}$$

$$\begin{bmatrix} x_1 \\ x_2 \\ x_3 \\ x_4 \end{bmatrix} = \begin{bmatrix} 1 & 0 & 0 & 0 \\ -1 & 1 & 0 & 0 \\ 0 & -1 & 1 & 0 \\ 0 & 0 & 1 & 1 \end{bmatrix}^{-1} \cdot \begin{bmatrix} 5 \\ -3 \\ 3 \\ 11 \end{bmatrix} = \begin{bmatrix} 5 \\ 2 \\ 5 \\ 6 \end{bmatrix}. \text{ Therefore } x_1 = 5, x_2 = 2, x_3 = 5, \text{ and } x_4 = 6.$$

(c) The traffic traveling west from intersection B to intersection A has a rate of $x_1 = 5$ cars per minute.

The value for x_2, x_3, and x_4 can be interpreted in a similar manner.

61. If $A = (A^{-1})^{-1}$, then $A = \begin{bmatrix} 5 & -9 \\ -1 & 2 \end{bmatrix}^{-1} = \begin{bmatrix} 2 & 9 \\ 1 & 5 \end{bmatrix}$.

62. If $A = (A^{-1})^{-1}$, then $A = \begin{bmatrix} \frac{3}{20} & \frac{1}{4} \\ -\frac{1}{20} & \frac{1}{4} \end{bmatrix}^{-1} = \begin{bmatrix} 5 & -5 \\ 1 & 3 \end{bmatrix}$.

63. If $A = (A^{-1})^{-1}$, then $A = \begin{bmatrix} \frac{2}{3} & -\frac{1}{3} & 0 \\ \frac{1}{3} & -\frac{5}{3} & 1 \\ \frac{1}{3} & \frac{1}{3} & 0 \end{bmatrix}^{-1} = \begin{bmatrix} 1 & 0 & 1 \\ -1 & 0 & 2 \\ -2 & 1 & 3 \end{bmatrix}$.

64. If $A = (A^{-1})^{-1} = \begin{bmatrix} 0 & 0 & 1 \\ 0 & 1 & 0 \\ 1 & 0 & 0 \end{bmatrix}^{-1} = \begin{bmatrix} 0 & 0 & 1 \\ 0 & 1 & 0 \\ 1 & 0 & 0 \end{bmatrix}$.

65. This is a shortened method:

First form the augmented matrix: $[A/I] = \begin{bmatrix} a & 0 & 0 & | & 1 & 0 & 0 \\ 0 & b & 0 & | & 0 & 1 & 0 \\ 0 & 0 & c & | & 0 & 0 & 1 \end{bmatrix}$.

Since a, b, c are all non-zero, $\dfrac{1}{a}, \dfrac{1}{b}, \dfrac{1}{c}$ all exist. Use these values for a, b, and c, and solve for the inverse.

$$\left.\begin{matrix} \frac{1}{a}R_1 \\ \frac{1}{b}R_2 \\ \frac{1}{c}R_3 \end{matrix}\right\} \rightarrow \begin{bmatrix} 1 & 0 & 0 & | & \frac{1}{a} & 0 & 0 \\ 0 & 1 & 0 & | & 0 & \frac{1}{b} & 0 \\ 0 & 0 & 1 & | & 0 & 0 & \frac{1}{c} \end{bmatrix}. \text{ Therefore, } A^{-1} = \begin{bmatrix} \frac{1}{a} & 0 & 0 \\ 0 & \frac{1}{b} & 0 \\ 0 & 0 & \frac{1}{c} \end{bmatrix}.$$

66. If $A = \begin{bmatrix} 1 & 0 & 0 \\ 0 & 0 & -1 \\ 0 & 1 & -1 \end{bmatrix}$, then $A^2 = \begin{bmatrix} 1 & 0 & 0 \\ 0 & 0 & -1 \\ 0 & 1 & -1 \end{bmatrix} \begin{bmatrix} 1 & 0 & 0 \\ 0 & 0 & -1 \\ 0 & 1 & -1 \end{bmatrix} = \begin{bmatrix} 1 & 0 & 0 \\ 0 & -1 & 1 \\ 0 & -1 & 0 \end{bmatrix}$.

$A^3 = AA^2 = \begin{bmatrix} 1 & 0 & 0 \\ 0 & 0 & -1 \\ 0 & 1 & -1 \end{bmatrix} \begin{bmatrix} 1 & 0 & 0 \\ 0 & -1 & 1 \\ 0 & -1 & 0 \end{bmatrix} = \begin{bmatrix} 1 & 0 & 0 \\ 0 & 1 & 0 \\ 0 & 0 & 1 \end{bmatrix} = I$, therefore $I = A^2A$.

Since $AA^2 = I$, $A^2 = A^{-1}$. Therefore $A^{-1} = \begin{bmatrix} 1 & 0 & 0 \\ 0 & -1 & 1 \\ 0 & -1 & 0 \end{bmatrix}$.

Reviewing Basic Concepts (Sections 7.4—7.6)

1. $A - B = \begin{bmatrix} -5 - 0 & 4 + 2 \\ 2 - 3 & -1 + 4 \end{bmatrix} = \begin{bmatrix} -5 & 6 \\ -1 & 3 \end{bmatrix}$.

2. $-3B = -3 \begin{bmatrix} 0 & -2 \\ 3 & -4 \end{bmatrix} = \begin{bmatrix} 0 & 6 \\ -9 & 12 \end{bmatrix}$.

3. $A^2 = \begin{bmatrix} -5 & 4 \\ 2 & -1 \end{bmatrix} \begin{bmatrix} -5 & 4 \\ 2 & -1 \end{bmatrix} = \begin{bmatrix} -5(-5) + 4(2) & -5(4) + 4(-1) \\ 2(-5) - 1(2) & 2(4) - 1(-1) \end{bmatrix} = \begin{bmatrix} 33 & -24 \\ -12 & 9 \end{bmatrix}$.

4. Using the calculator. $CD = \begin{bmatrix} 1 & 3 & -3 \\ 0 & 6 & 0 \\ 4 & 2 & 2 \end{bmatrix}$.

5. $\det A = -5(-1) - 2(4) = 5 - 8 = -3$.

6. Evaluate, expand by the first column. Therefore, $\det = (a_{11})(M_{11}) - (a_{21})(M_{21}) + (a_{31})(M_{31})$:

    $2\left(\det \begin{bmatrix} 1 & 0 \\ -1 & 4 \end{bmatrix}\right) - (-2)\left(\det \begin{bmatrix} -3 & 1 \\ -1 & 4 \end{bmatrix}\right) + 0\left(\det \begin{bmatrix} -3 & 1 \\ 1 & 0 \end{bmatrix}\right) = 2(4 + 0) + 2(-12 + 1) + 0(0 - 1) =$

    $2(4) + 2(-11) = 8 - 22 = -14$.

7. Using $A^{-1} = \dfrac{1}{\det A} \begin{bmatrix} d & -b \\ -c & a \end{bmatrix}$ to find the inverse yields: $A^{-1} = \dfrac{1}{5 - 8} \begin{bmatrix} -1 & -4 \\ -2 & -5 \end{bmatrix} = \begin{bmatrix} \frac{1}{3} & \frac{4}{3} \\ \frac{2}{3} & \frac{5}{3} \end{bmatrix}$.

8. Solve using the calculator, $C^{-1} = \begin{bmatrix} -\frac{2}{7} & -\frac{11}{14} & \frac{1}{14} \\ -\frac{4}{7} & -\frac{4}{7} & \frac{1}{7} \\ -\frac{1}{7} & -\frac{1}{7} & \frac{2}{7} \end{bmatrix}$.

9. The equations are: $\dfrac{\sqrt{3}}{2}(w_1 + w_2) = 100 \Rightarrow \dfrac{\sqrt{3}}{2}w_1 + \dfrac{\sqrt{3}}{2}w_2 = 100$ and $w_1 - w_2 = 0$. Now we

find the determinants: $D = \det \begin{bmatrix} \dfrac{\sqrt{3}}{2} & \dfrac{\sqrt{3}}{2} \\ 1 & -1 \end{bmatrix} = -\sqrt{3}; D_{w_2} = \det \begin{bmatrix} 100 & \dfrac{\sqrt{3}}{2} \\ 0 & -1 \end{bmatrix} = -100;$ and

$D_{w_2} = \det \begin{bmatrix} \dfrac{\sqrt{3}}{2} & 100 \\ 1 & 0 \end{bmatrix} = -100.$ Then $w_1 = \dfrac{D_{w_1}}{D} = \dfrac{-100}{-\sqrt{3}} = \dfrac{100}{\sqrt{3}} \cdot \dfrac{\sqrt{3}}{\sqrt{3}} = \dfrac{100\sqrt{3}}{3} \approx 57.7$ and

$w_2 = \dfrac{D_{w_2}}{D} = \dfrac{-100}{-\sqrt{3}} = \dfrac{100}{\sqrt{3}} \cdot \dfrac{\sqrt{3}}{\sqrt{3}} = \dfrac{100\sqrt{3}}{3} \approx 57.7.$ Both w_1 and w_2 are approximately 57.7 pounds.

10. $\begin{bmatrix} x \\ y \\ z \end{bmatrix} = \begin{bmatrix} 2 & 1 & 2 \\ 0 & 1 & 2 \\ 1 & -2 & 2 \end{bmatrix}^{-1} \cdot \begin{bmatrix} 10 \\ 4 \\ 1 \end{bmatrix} = \begin{bmatrix} 3 \\ 2 \\ 1 \end{bmatrix}.$ The solution is (3, 2, 1).

7.7: Systems of Inequalities and Linear Programming

1. See Figure 1.

2. See Figure 2.

3. See Figure 3.

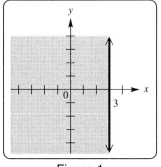

Figure 1

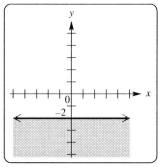

Figure 2

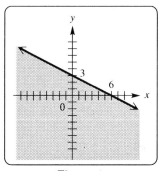

Figure 3

4. See Figure 4.

5. See Figure 5.

6. See Figure 6.

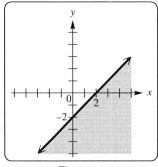

Figure 4

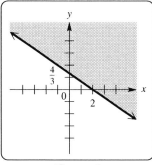

Figure 5

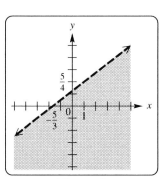

Figure 6

7. See Figure 7.

8. See Figure 8.

9. See Figure 9.

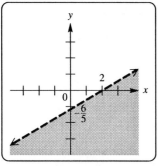

Figure 7

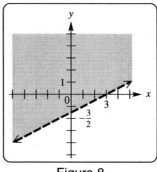

Figure 8

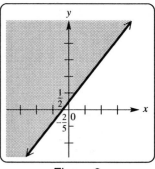

Figure 9

10. See Figure 10.

11. See Figure 11.

12. See Figure 12.

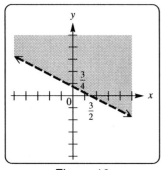

Figure 10

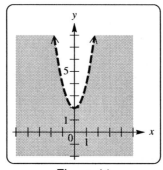

Figure 11

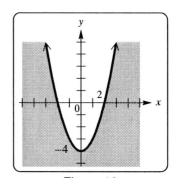

Figure 12

13. See Figure 13.

14. See Figure 14.

15. See Figure 15.

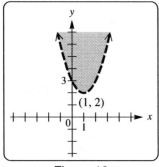

Figure 13

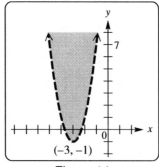

Figure 14

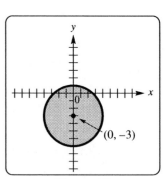

Figure 15

16. See Figure 16.

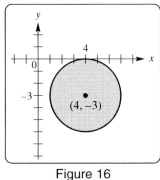

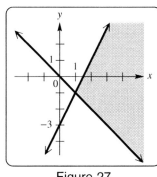

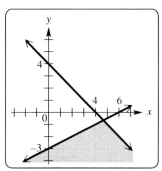

Figure 16   Figure 27   Figure 28

17. The boundary is solid if the symbol is $\geq$ or $\leq$ and dashed if the symbol is $>$ or $<$.

18. You would shade above because y must be greater then $3x - 6$.

19. Since $Ax + By \geq C \Rightarrow By \geq -Ax + C$. Now if $B > 0$, then $y \geq -\dfrac{A}{B}x + C$. Therefore shade above the line.

20. Since $Ax + By \geq C \Rightarrow By \geq -Ax + C$. Now if $B < 0$, then $y \leq -\dfrac{A}{B}x + C$. Therefore shade below the line.

21. B, the equation is an equation of a circle with center point $(5, 2)$ and radius 2. The less than symbol indicates the region inside the circle.

22. D, the equation is an equation of a ellipse centered at the origin and having x-intercepts of ± 4, y-intercepts of ± 9. The greater than symbol indicates outside the ellipse.

23. C, shaded below a line with a slope of 3.

24. B, shaded above a line with a slope of 3.

25. A, shaded below a line with a slope of -3.

26. D, shaded below a line with a slope of -3.

27. See Figure 27.

28. See Figure 28.

29. See Figure 29.

30. See Figure 30.

31. See Figure 31.

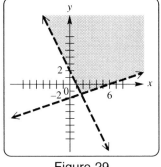

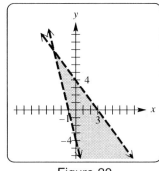

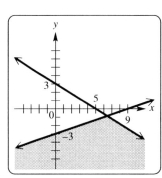

Figure 29   Figure 30   Figure 31

32. See Figure 32.

33. See Figure 33.

34. See Figure 34.

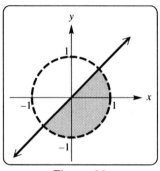

Figure 32

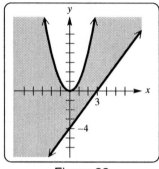

Figure 33

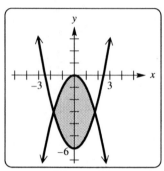

Figure 34

35. See Figure 35.

36. See Figure 36.

37. See Figure 37.

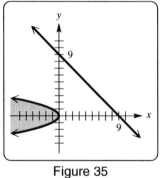

Figure 35

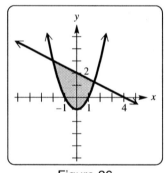

Figure 36

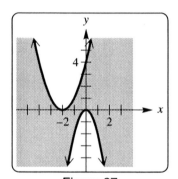

Figure 37

38. See Figure 38.

39. See Figure 39.

40. See Figure 40.

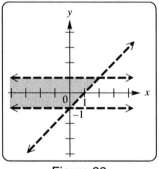

Figure 38

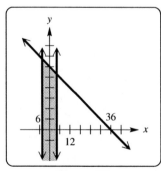

Figure 39

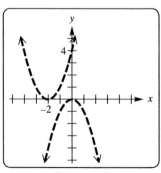

Figure 40

41. See Figure 41.

42. See Figure 42.

43. See Figure 43.

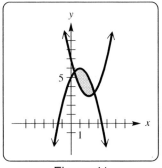

Figure 41

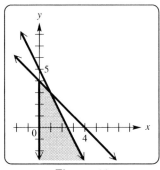

Figure 42

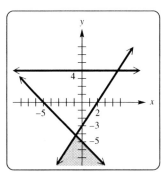

Figure 43

44. See Figure 44.

45. See Figure 45.

46. See Figure 46.

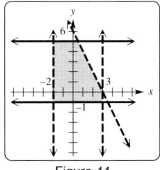

Figure 44

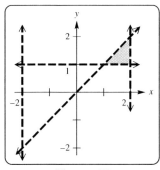

Figure 45

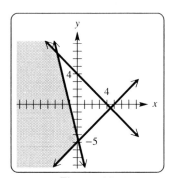

Figure 46

47. See Figure 47.

48. See Figure 48.

49. See Figure 49.

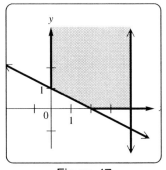

Figure 47

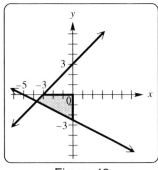

Figure 48

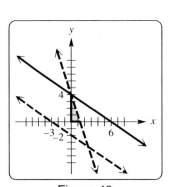

Figure 49

50. See Figure 50.

51. See Figure 51.

52. See Figure 52.

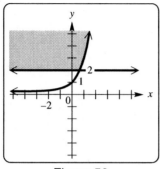
Figure 50

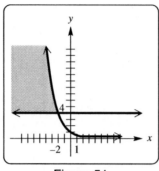
Figure 51

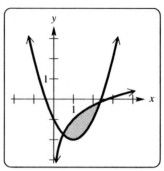
Figure 52

53. See Figure 53.

54. See Figure 54.

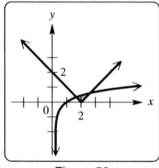
Figure 53

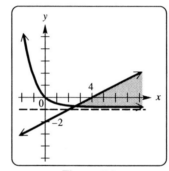
Figure 54

55. D, inside the the ellipse of [equation 1] and below the line of [equation 2].

56. The graph of the system consists of all points above the parabola $y = x^2 + 2$, inside the circle $x^2 + y^2 = 16$, and below the line $y = 7$.

57. A, the graph is of two positive slope lines. The shading is below the line with slope 2 and above the line with slope 1.

58. D, the graph has shading inside a parabola opening upward and below $y = 5$.

59. B, the graph has shading inside a circle and above $y = 0$.

60. C, the graph is of two positive slope lines. The shading is above the line with slope 2 and below the line with slope 1.

61. See Figure 61.

62. See Figure 62.

63. See Figure 63.

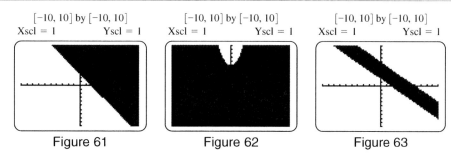

Figure 61   Figure 62   Figure 63

64. See Figure 64.

65. See Figure 65.

66. See Figure 66.

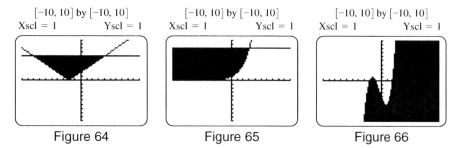

Figure 64   Figure 65   Figure 66

67. Since we are in the first quadrant, $x \geq 0$. $y \geq 0$. The lines $x + 2y - 8 = 0$ and $x + 2y = 12$ are parallel, with $x + 2y = 12$ having the greater y-intercept. Therefore, we must shade below $x + 2y = 12$ and above $x + 2y - 8 = 0$. The system is:

$x + 2y - 8 \geq 0$

$x + 2y \leq 12$

$x \geq 0, y \geq 0$

68. (a) $c = 10x + 20y \Rightarrow 20y = -10x + c \Rightarrow y = -\dfrac{10}{20}x + \dfrac{c}{20} \Rightarrow y = -\dfrac{1}{2}x + \dfrac{c}{20}$

(b) As c increases, y also increases, the line of constant cost moves up.

(c) From the graph, C gives the optimal solution.

69. Using the given expression and the ordered pairs of the vertices yields the following solutions:

$(1, 1)$: $3(1) + 5(1) = 8$; $(2, 7)$: $3(2) + 5(7) = 41$; $(5, 10)$: $3(5) + 5(10) = 65$; $(6, 3)$: $3(6) + 5(3) = 33$

Therefore, the maximum value is: 65 at $(5, 10)$; and the minimum value is: 8 at $(1, 1)$.

70. Using the given expression and the ordered pairs of the vertices yields the following solutions:

$(1, 2)$: $6(1) + (2) = 8$; $(1, 5)$: $6(1) + (5) = 11$; $(6, 8)$: $6(6) + (8) = 44$; $(9, 1)$: $6(9) + (1) = 55$

Therefore, the maximum value is: 55 at $(9, 1)$; and the minimum value is: 8 at $(1, 2)$.

71. Using the given expression and the ordered pairs of the vertices yields the following solutions:

$(1, 10)$: $3(1) + 5(10) = 53$; $(7, 9)$: $3(7) + 5(9) = 66$; $(7, 6)$: $3(7) + 5(6) = 51$; $(1, 0)$: $3(1) + 5(0) = 3$

Therefore, the maximum value is: 66 at $(7, 9)$; and the minimum value is: 3 at $(1, 0)$.

72. Using the given expression and the ordered pairs of the vertices yields the following solutions:

$(1, 10)$: $5(1) + 5(10) = 55$; $(7, 9)$: $5(7) + 5(9) = 80$; $(7, 6)$: $5(7) + 5(6) = 65$; $(1, 0)$: $5(1) + 5(0) = 5$

Therefore, the maximum value is: 80 at $(7, 9)$; and the minimum value is: 5 at $(1, 0)$.

73. Using the given expression and the ordered pairs of the vertices yields the following solutions:

    $(1, 10)$: $10(10) = 100$; $(7, 9)$: $10(9) = 90$; $(7, 6)$: $10(6) = 60$; $(1, 0)$: $10(0) = 0$

    Therefore, the maximum value is: 100 at $(1, 10)$; and the minimum value is: 0 at $(1, 0)$.

74. Using the given expression and the ordered pairs of the vertices yields the following solutions:

    $(1, 10)$: $3(1) - (10) = -7$; $(7, 9)$: $3(7) - (9) = 12$; $(7, 6)$: $3(7) - (6) = 15$; $(1, 0)$: $3(1) - (0) = 3$

    Therefore, the maximum value is: 15 at $(7, 6)$; and the minimum value is: -7 at $(1, 10)$.

75. Let x = the number of hat units and let y = the number of whistle units.

    The objective function to find the maximum number of inquiries is: $3x + 2y$.

    The constraints are: $2x + 4y \le 12$ (floor space)

    $x + y \le 5$ (total number of displays)

    $x \ge 0$, $y \ge 0$ (cannot display a negative number of displays)

    Graphing the constraints on the calculator yields four vertices. Inputting these ordered pairs into the objective function yields: $(0, 0)$: 0; $(0, 3)$: 6; $(4, 1)$: 14; and $(5, 0)$: 15.

    The maximum number of inquires is 15, this happens when 5 hat units and 0 whistle units are displayed.

76. Let x = the number of pigs and let y = the number of geese.

    The objective function to find the maximum profit is: $40x + 80y$.

    The constraints are: $x + y \le 16$ (maximum number of animals)

    $0 \le y \le 12$ (maximum and minimum number of geese)

    $50x + 20y \le 500$ (total cost of raising the animals)

    $x \ge 0$ (minimum number of pigs)

    Graphing the constraints on the calculator yields five vertices. Inputting these ordered pairs into the objective function yields: $(0, 0)$: 0; $(0, 12)$: 960; $(4, 12)$: 1120; $(6, 10)$: 1040; and $(10, 0)$: 400.

    The maximum profit is 1120, this happens when 4 pigs and 12 geese are raised and sold.

77. Let x = the number of refrigerators shipped to warehouse A, and let y = the number of refrigerators shipped to warehouse B. The objective function to find the minimum cost is: $12x + 10y$.

    The constraints are: $x + y \ge 100$ (total to be shipped)

    $0 \le x \le 75$ (maximum space available at warehouse A)

    $0 \le y \le 80$ (maximum space available at warehouse B)

    Graphing the constraints on the calculator yields three vertices. Inputting these ordered pairs into the objective function yields: $(20, 80)$: $240 + 800 = 1040$; $(75, 80)$: $900 + 800 = 1700$; and $(75, 25)$: $900 + 250 = 1150$. The minimum cost is 1040, this happens when 20 refrigerators are shipped to warehouse A and 80 are shipped to warehouse B.

78. Let x = the number of servings of product A and let y = the number of servings of product B.

    The objective function to find the minimum cost is: $.25x + .40y$.

    The constraints are: $3x + 2y \geq 15$ (grams of supplements in product A)

                     $2x + 4y \geq 15$ (grams of supplements in product B)

                     $x \geq 0, \ y \geq 0$ (cannot include negative amounts of the products)

    Graphing the constraints on the calculator yields three vertices. Inputting these ordered pairs into the objective function yields: $(0, 7.5)$: $.25(0) + .40(7.5) = 3$; $(3.75, 1.875)$: $.25(3.75) + .40(1.875) = 1.69$; $(7.5, 0)$: $.25(7.5) + .40(0) = 1.88$. The minimum cost is \$1.69, this happens when 3.75 servings of product A and 1.875 servings of B are included in the diet.

79. Let x = the number of gallons (millions) of gasoline and let y = the number of gallons (millions) of fuel oil.

    The objective function to find the maximum revenue is: $1.9x + 1.5y$.

    The constraints are: $y \leq \dfrac{1}{2}x$ (ratio requirements)

                     $y \geq 3$ (minimum daily needs of fuel oil)

                     $x \leq 6.4$ (maximum daily needs of gasoline)

    Graphing the constraints on the calculator yields three vertices. Inputting these ordered pairs into the objective function yields: $(6, 3)$: $1.9(6) + 1.5(3) = 15.9$; $(6.4, 3.2)$: $1.9(6.4) + 1.5(3.2) = 16.96$; $(6.4, 3)$: $1.9(6.4) + 1.5(3) = 16.66$. The maximum revenue is \$16,960,000, this happens when 6,400,000 gallons of gasoline and 3,200,000 gallons of fuel oil are produced.

80. Let x = the number of type A bolts and let y = the number of type B bolts.

    The objective function to find the maximum revenue is: $.10x + .12y$.

    The constraints are: $.1x + .1y \leq 240$ (machine group I)

                     $.1x + .4y \leq 720$ (machine group II)

                     $.1x + .5y \leq 160$ (machine group III)

                     $x \geq 0, \ y \geq 0$ (cannot produce negative numbers of bolts)

    Graphing the constraints on the calculator yields three vertices. Inputting these ordered pairs into the objective function yields: $(0, 0)$: 0; $(0, 320)$: 38.40; and $(1600, 0)$: 160.00.

    The maximum revenue is \$160, this happens when 1600 type A bolts and 0 type B bolts are manufactured.

81. Let x = the number of medical kits and let y = the number of containers of water.

    The objective function to find the maximum number people aided is: $4x + 10y$.

    The constraints are: $x + y \leq 6000$ (maximum space available on the plane)

                     $10x + 20y \leq 80,000$ (maximum weight plane can carry)

                     $x \geq 0, \ y \geq 0$ (cannot have negative weight or volume)

    Graphing the constraints on the calculator yields four vertices. Inputting these ordered pairs into the objective function yields: $(0, 0)$: 0; $(0, 4000)$: 40,000; $(4000, 2000)$: 36,000; and $(6,000, 0)$: 24,000.

    To maximize the number of people aided (40,000), they should take 0 medical kits and 4,000 containers of water.

82. Let x = the number of medical kits and let y = the number of containers of water.

    The objective function to find the maximum number people aided is: $6x + 10y$.

    The constraints are: $x + y \leq 6000$ (maximum space available on the plane)

    $10x + 20y \leq 80{,}000$ (maximum weight plane can carry)

    $x \geq 0,\ y \geq 0$ (cannot have negative weight or volume)

    Graphing the constraints on the calculator yields four vertices. Inputting these ordered pairs into the objective function yields: $(0, 0)$: 0; $(0, 4000)$: 40,000; $(4000, 2000)$: 44,000; and $(6{,}000, 0)$: 36,000.

    To maximize the number of people aided (44,000), they should take 4000 medical kits and 2,000 containers of water.

7.8: Partial Fractions

1. Multiply $\dfrac{5}{3x(2x + 1)} = \dfrac{A}{3x} + \dfrac{B}{2x + 1}$ by $3x(2x + 1) \Rightarrow 5 = A(2x + 1) + B(3x)$.

    Let $x = 0 \Rightarrow 5 = A(1) \Rightarrow A = 5$. Let $x = -\dfrac{1}{2} \Rightarrow 5 = B\left(-\dfrac{3}{2}\right) \Rightarrow B = -\dfrac{10}{3}$.

    The expression can be written $\dfrac{5}{3x} + \dfrac{-10}{3(2x + 1)}$.

2. Multiply $\dfrac{3x - 1}{x(x + 1)} = \dfrac{A}{x} + \dfrac{B}{x + 1}$ by $x(x + 1) \Rightarrow 3x - 1 = A(x + 1) + B(x)$.

    Let $x = 0 \Rightarrow -1 = A(1) \Rightarrow A = -1$. Let $x = -1 \Rightarrow -4 = B(-1) \Rightarrow B = 4$.

    The expression can be written $\dfrac{-1}{x} + \dfrac{4}{x + 1}$.

3. Multiply $\dfrac{4x + 2}{(x + 2)(2x - 1)} = \dfrac{A}{x + 2} + \dfrac{B}{2x - 1}$ by $(x + 2)(2x - 1) \Rightarrow 4x + 2 = A(2x - 1) + B(x + 2)$.

    Let $x = -2 \Rightarrow -6 = A(-5) \Rightarrow A = \dfrac{6}{5}$. Let $x = \dfrac{1}{2} \Rightarrow 4 = B\left(\dfrac{5}{2}\right) \Rightarrow B = \dfrac{8}{5}$.

    The expression can be written $\dfrac{6}{5(x + 2)} + \dfrac{8}{5(2x - 1)}$.

4. Multiply $\dfrac{x + 2}{(x + 1)(x - 1)} = \dfrac{A}{x + 1} + \dfrac{B}{x - 1}$ by $(x + 1)(x - 1) \Rightarrow x + 2 = A(x - 1) + B(x + 1)$.

    Let $x = -1 \Rightarrow 1 = A(-2) \Rightarrow A = -\dfrac{1}{2}$. Let $x = 1 \Rightarrow 3 = B(2) \Rightarrow B = \dfrac{3}{2}$.

    The expression can be written $\dfrac{-1}{2(x + 1)} + \dfrac{3}{2(x - 1)}$.

5. Factoring $\dfrac{x}{x^2 + 4x - 5}$ results in $\dfrac{x}{(x + 5)(x - 1)}$.

    Multiply $\dfrac{x}{(x + 5)(x - 1)} = \dfrac{A}{x + 5} + \dfrac{B}{x - 1}$ by $(x + 5)(x - 1) \Rightarrow x = A(x - 1) + B(x + 5)$.

    Let $x = -5 \Rightarrow -5 = A(-6) \Rightarrow A = \dfrac{5}{6}$. Let $x = 1 \Rightarrow 1 = B(6) \Rightarrow B = \dfrac{1}{6}$.

    The expression can be written $\dfrac{5}{6(x + 5)} + \dfrac{1}{6(x - 1)}$.

6. Multiply $\dfrac{5x - 3}{(x + 1)(x - 3)} = \dfrac{A}{x + 1} + \dfrac{B}{x - 3}$ by $(x + 1)(x - 3) \Rightarrow 5x - 3 = A(x - 3) + B(x + 1)$.

Let $x = -1 \Rightarrow -8 = A(-4) \Rightarrow A = 2$. Let $x = 3 \Rightarrow 12 = B(4) \Rightarrow B = 3$.

The expression can be written $\dfrac{2}{x + 1} + \dfrac{3}{x - 3}$.

7. Multiply $\dfrac{2x}{(x + 1)(x + 2)^2} = \dfrac{A}{x + 1} + \dfrac{B}{x + 2} + \dfrac{C}{(x + 2)^2}$ by $(x + 1)(x + 2)^2 \Rightarrow$

$2x = A(x + 2)^2 + B(x + 1)(x + 2) + C(x + 1)$. Let $x = -1 \Rightarrow -2 = A(1) \Rightarrow A = -2$.

Let $x = -2 \Rightarrow -4 = C(-1) \Rightarrow C = 4$. Let $x = 0$ with $A = -2$ and $C = 4 \Rightarrow 0 = -2(4) + B(2) + 4(1) \Rightarrow$

$4 = 2B \Rightarrow B = 2$. The expression can be written $\dfrac{-2}{x + 1} + \dfrac{2}{x + 2} + \dfrac{4}{(x + 2)^2}$.

8. Multiply $\dfrac{2}{x^2(x + 3)} = \dfrac{A}{x} + \dfrac{B}{x^2} + \dfrac{C}{x + 3}$ by $x^2(x + 3) \Rightarrow 2 = Ax(x + 3) + B(x + 3) + Cx^2$.

Let $x = 0 \Rightarrow 2 = B(3) \Rightarrow B = \dfrac{2}{3}$. Let $x = -3 \Rightarrow 2 = C(9) \Rightarrow C = \dfrac{2}{9}$.

Let $x = 1$ with $B = \dfrac{2}{3}$ and $C = \dfrac{2}{9} \Rightarrow 2 = A(4) + \dfrac{2}{3}(4) + \dfrac{2}{9}(1) \Rightarrow 18 = 36A + 24 + 2 \Rightarrow -8 = 36A \Rightarrow$

$A = -\dfrac{2}{9}$. The expression can be written $\dfrac{-2}{9x} + \dfrac{2}{3x^2} + \dfrac{2}{9(x + 3)}$.

9. Multiply $\dfrac{4}{x(1 - x)} = \dfrac{A}{x} + \dfrac{B}{1 - x}$ by $x(1 - x) \Rightarrow 4 = A(1 - x) + B(x)$.

Let $x = 0 \Rightarrow 4 = A(1) \Rightarrow A = 4$. Let $x = 1 \Rightarrow 4 = B(1) \Rightarrow B = 4$.

The expression can be written $\dfrac{4}{x} + \dfrac{4}{1 - x}$.

10. Multiply $\dfrac{x + 1}{x^2(1 - x)} = \dfrac{A}{x} + \dfrac{B}{x^2} + \dfrac{C}{1 - x}$ by $x^2(1 - x) \Rightarrow$

$x + 1 - A(x)(1 - x) + B(1 - x) + C(x)(x)$. Let $x = 0 \Rightarrow 1 = B(1) \Rightarrow B = 1$.

Let $x = 1 \Rightarrow 2 = C(1)(1) \Rightarrow C = 2$. Let $x = -1 \Rightarrow 0 = A(-1)(2) + (1)(2) + (2)(-1)(-1) \Rightarrow A = 2$.

The expression can be written $\dfrac{2}{x} + \dfrac{1}{x^2} + \dfrac{2}{1 - x}$.

11. Multiply $\dfrac{4x^2 - x - 15}{x(x + 1)(x - 1)} = \dfrac{A}{x} + \dfrac{B}{x + 1} + \dfrac{C}{x - 1}$ by $x(x + 1)(x - 1) \Rightarrow$

$4x^2 - x - 15 = A(x - 1)(x + 1) + B(x)(x - 1) + C(x)(x + 1)$. Let $x = 0 \Rightarrow -15 = A(-1) \Rightarrow A = 15$.

Let $x = 1 \Rightarrow -12 = C(1)(2) \Rightarrow C = -6$. Let $x = -1 \Rightarrow -10 = B(-1)(-2) \Rightarrow B = -5$.

The expression can be written $\dfrac{15}{x} + \dfrac{-5}{x + 1} + \dfrac{-6}{x - 1}$.

12. Multiply $\dfrac{2x + 1}{(x + 2)^3} = \dfrac{A}{x + 2} + \dfrac{B}{(x + 2)^2} + \dfrac{C}{(x + 2)^3}$ by $(x + 2)^3 \Rightarrow 2x + 1 = A(x + 2)^2 + B(x + 2) + C$.

Let $x = -2 \Rightarrow -3 = C$. Multiplying out with $C = -3$ gives $2x + 1 = Ax^2 + 4Ax + 4A + Bx + 2B - 3$.

Equate coefficients. For x^2: $0 = Ax^2 \Rightarrow A = 0$. For x: $2x = (4A + B)x \Rightarrow B = 2$.

The expression can be written $\dfrac{2}{(x + 2)^2} + \dfrac{-3}{(x + 2)^3}$.

13. By long division $\dfrac{x^2}{x^2 + 2x + 1} = 1 + \dfrac{-2x - 1}{(x + 1)^2}$.

Multiply $\dfrac{-2x - 1}{(x + 1)^2} = \dfrac{A}{x + 1} + \dfrac{B}{(x + 1)^2}$ by $(x + 1)^2 \Rightarrow -2x - 1 = A(x + 1) + B$.

Let $x = -1 \Rightarrow 1 = B$. Let $x = 0$ with $B = 1 \Rightarrow -1 = A + 1 \Rightarrow A = -2$.

The expression can be written $1 + \dfrac{-2}{x + 1} + \dfrac{1}{(x + 1)^2}$.

14. Factoring $\dfrac{3}{x^2 + 4x + 3}$ results in $\dfrac{3}{(x + 3)(x + 1)}$.

Multiply $\dfrac{3}{(x + 3)(x + 1)} = \dfrac{A}{x + 3} + \dfrac{B}{x + 1}$ by $(x + 3)(x + 1) \Rightarrow 3 = A(x + 1) + B(x + 3)$.

Let $x = -1 \Rightarrow 3 = B(2) \Rightarrow B = \dfrac{3}{2}$. Let $x = -3 \Rightarrow 3 = A(-2) \Rightarrow A = -\dfrac{3}{2}$.

The expression can be written $\dfrac{-3}{2(x + 3)} + \dfrac{3}{2(x + 1)}$.

15. By long division $\dfrac{2x^5 + 3x^4 - 3x^3 - 2x^2 + x}{2x^2 + 5x + 2} = x^3 - x^2 + \dfrac{x}{2x^2 + 5x + 2} = x^3 - x^2 + \dfrac{x}{(2x + 1)(x + 2)}$.

Multiply $\dfrac{x}{(2x + 1)(x + 2)} = \dfrac{A}{2x + 1} + \dfrac{B}{x + 2}$ by $(2x + 1)(x + 2) \Rightarrow x = A(x + 2) + B(2x + 1)$.

Let $x = -\dfrac{1}{2} \Rightarrow -\dfrac{1}{2} = A\left(\dfrac{3}{2}\right) \Rightarrow A = -\dfrac{1}{3}$. Let $x = -2 \Rightarrow -2 = B(-3) \Rightarrow B = \dfrac{2}{3}$.

The expression can be written $x^3 - x^2 + \dfrac{-1}{3(2x + 1)} + \dfrac{2}{3(x + 2)}$.

16. By long division $\dfrac{6x^5 + 7x^4 - x^2 + 2x}{3x^2 + 2x - 1} = 2x^3 + x^2 + \dfrac{2x}{3x^2 + 2x - 1} = 2x^3 + x^2 + \dfrac{2x}{(3x - 1)(x + 1)}$.

Multiply $\dfrac{2x}{(3x - 1)(x + 1)} = \dfrac{A}{3x - 1} + \dfrac{B}{x + 1}$ by $(3x - 1)(x + 1) \Rightarrow 2x = A(x + 1) + B(3x - 1)$.

Let $x = \dfrac{1}{3} \Rightarrow \dfrac{2}{3} = A\left(\dfrac{4}{3}\right) \Rightarrow A = \dfrac{1}{2}$. Let $x = -1 \Rightarrow -2 = B(-4) \Rightarrow B = \dfrac{1}{2}$.

The expression can be written $2x^3 + x^2 + \dfrac{1}{2(3x - 1)} + \dfrac{1}{2(x + 1)}$.

17. By long division $\dfrac{x^3 + 4}{9x^3 - 4x} = \dfrac{1}{9} + \dfrac{\frac{4}{9}x + 4}{9x^3 - 4x} = \dfrac{1}{9} + \dfrac{\frac{4}{9}x + 4}{x(3x + 2)(3x - 2)}$.

Multiply $\dfrac{\frac{4}{9}x + 4}{x(3x + 2)(3x - 2)} = \dfrac{A}{x} + \dfrac{B}{3x + 2} + \dfrac{C}{3x - 2}$ by $x(3x + 2)(3x - 2) \Rightarrow$

$\dfrac{4}{9}x + 4 = A(3x + 2)(3x - 2) + B(x)(3x - 2) + C(x)(3x + 2)$. Let $x = 0 \Rightarrow 4 = A(-4) \Rightarrow A = -1$.

Let $x = -\dfrac{2}{3} \Rightarrow -\dfrac{8}{27} + 4 = B\left(-\dfrac{2}{3}\right)(-4) \Rightarrow \dfrac{100}{27} = \dfrac{8}{3}B \Rightarrow B = \dfrac{25}{18}$.

Let $x = \dfrac{2}{3} \Rightarrow \dfrac{8}{27} + 4 = C\left(\dfrac{2}{3}\right)(4) \Rightarrow \dfrac{116}{27} = \dfrac{8}{3}C \Rightarrow C = \dfrac{29}{18}$.

The expression can be written $\dfrac{1}{9} + \dfrac{-1}{x} + \dfrac{25}{18(3x + 2)} + \dfrac{29}{18(3x - 2)}$.

18. By long division $\dfrac{x^3 + 2}{x^3 - 3x^2 + 2x} = 1 + \dfrac{3x^2 - 2x + 2}{x^3 - 3x^2 + 2x} = 1 + \dfrac{3x^2 - 2x + 2}{x(x - 2)(x - 1)}.$

Multiply $\dfrac{3x^2 - 2x + 2}{x(x - 2)(x - 1)} = \dfrac{A}{x} + \dfrac{B}{x - 2} + \dfrac{C}{x - 1}$ by $x(x - 2)(x - 1) \Rightarrow$

$3x^2 - 2x + 2 = A(x - 2)(x - 1) + B(x)(x - 1) + C(x)(x - 2) \Rightarrow$

$3x^2 - 2x + 2 = Ax^2 - 3Ax + 2A + Bx^2 - Bx + Cx^2 - 2Cx.$ Equate coefficients.

For x^2: $3 = A + B + C.$ For x: $-2 = -3A - B - 2C.$ For the constants: $2 = 2A \Rightarrow A = 1.$

Simultaneously solve the first two equations with $A = 1.$

$\quad\ B + \ \ C = 2$
$\underline{-B - 2C = 1}$
$\qquad -C = 3 \Rightarrow C = -3,$ then $B + C = 2 \Rightarrow B = 5$

The expression can be written $1 + \dfrac{1}{x} + \dfrac{5}{x - 2} + \dfrac{-3}{x - 1}.$

19. Multiply $\dfrac{-3}{x^2(x^2 + 5)} = \dfrac{A}{x} + \dfrac{B}{x^2} + \dfrac{Cx + D}{x^2 + 5}$ by $x^2(x^2 + 5) \Rightarrow$

$-3 = A(x)(x^2 + 5) + B(x^2 + 5) + (Cx + D)(x^2) \Rightarrow -3 = Ax^3 + 5Ax + Bx^2 + 5B + Cx^3 + Dx^2.$

Equate coefficients. For x^3: $0 = A + C.$

For x^2: $0 = B + D.$ For x: $0 = 5A \Rightarrow A = 0.$ For the constants: $-3 = 5B \Rightarrow B = -\dfrac{3}{5}.$

Substitute $A = 0$ in the first equation. $C = 0.$ Substitute $B = -\dfrac{3}{5}$ in the second equation. $D = \dfrac{3}{5}.$

The expression can be written $\dfrac{-3}{5x^2} + \dfrac{3}{5(x^2 + 5)}.$

20. Multiply $\dfrac{2x + 1}{(x + 1)(x^2 + 2)} = \dfrac{A}{x + 1} + \dfrac{Bx + C}{x^2 + 2}$ by $(x + 1)(x^2 + 2) \Rightarrow$

$2x + 1 = A(x^2 + 2) + (Bx + C)(x + 1) \Rightarrow 2x + 1 = Ax^2 + 2A + Bx^2 + Bx + Cx + C.$

Let $x = -1 \Rightarrow -1 = 3A \Rightarrow A = -\dfrac{1}{3}.$ Equate coefficients.

For x^2: $0 = A + B \Rightarrow 0 = -\dfrac{1}{3} + B \Rightarrow B = \dfrac{1}{3}.$ For x: $2 = B + C \Rightarrow 2 = \dfrac{1}{3} + C \Rightarrow C = \dfrac{5}{3}.$

The expression can be written $\dfrac{-1}{3(x + 1)} + \dfrac{x + 5}{3(x^2 + 2)}.$

21. Multiply $\dfrac{3x - 2}{(x + 4)(3x^2 + 1)} = \dfrac{A}{x + 4} + \dfrac{Bx + C}{3x^2 + 1}$ by $(x + 4)(3x^2 + 1) \Rightarrow$

$3x - 2 = A(3x^2 + 1) + (Bx + C)(x + 4) \Rightarrow 3x - 2 = 3Ax^2 + A + Bx^2 + 4Bx + Cx + 4C.$

Let $x = -4 \Rightarrow -14 = 49A \Rightarrow A = -\dfrac{2}{7}.$ Equate coefficients.

For x^2: $0 = 3A + B \Rightarrow 0 = -\dfrac{6}{7} + B \Rightarrow B = \dfrac{6}{7}.$ For x: $3 = 4B + C \Rightarrow 3 = \dfrac{24}{7} + C \Rightarrow C = -\dfrac{3}{7}.$

The expression can be written $\dfrac{-2}{7(x + 4)} + \dfrac{6x - 3}{7(3x^2 + 1)}.$

22. Multiply $\dfrac{3}{x(x+1)(x^2+1)} = \dfrac{A}{x} + \dfrac{B}{x+1} + \dfrac{Cx+D}{x^2+1}$ by $x(x+1)(x^2+1) \Rightarrow$

$3 = A(x+1)(x^2+1) + B(x)(x^2+1) + (Cx+D)(x)(x+1).$

Let $x = 0 \Rightarrow 3 = A(1) \Rightarrow A = 3$. Let $x = -1 \Rightarrow 3 = B(-1)(2) \Rightarrow B = -\dfrac{3}{2}$.

Multiply the right side out. $3 = A(x^3 + x^2 + x + 1) + Bx^3 + Bx + Cx^3 + Cx^2 + Dx^2 + Dx \Rightarrow$

$3 = Ax^3 + Ax^2 + Ax + A + Bx^3 + Bx + Cx^3 + Cx^2 + Dx^2 + Dx.$

Equate coefficients. For x^3: $0 = A + B + C \Rightarrow 0 = 3 - \dfrac{3}{2} + C \Rightarrow C = -\dfrac{3}{2}$.

For x^2: $0 = A + C + D \Rightarrow 0 = 3 - \dfrac{3}{2} + D \Rightarrow D = -\dfrac{3}{2}$.

The expression can be written $\dfrac{3}{x} + \dfrac{-3}{2(x+1)} + \dfrac{-3(x+1)}{2(x^2+1)}$.

23. Multiply $\dfrac{1}{x(2x+1)(3x^2+4)} = \dfrac{A}{x} + \dfrac{B}{2x+1} + \dfrac{Cx+D}{3x^2+4}$ by $x(2x+1)(3x^2+4) \Rightarrow$

$1 = A(2x+1)(3x^2+4) + B(x)(3x^2+4) + (Cx+D)(x)(2x+1).$

Let $x = 0 \Rightarrow 1 = A(1)(4) \Rightarrow A = \dfrac{1}{4}$. Let $x = -\dfrac{1}{2} \Rightarrow 1 = B\left(-\dfrac{1}{2}\right)\left(\dfrac{19}{4}\right) \Rightarrow B = -\dfrac{8}{19}$.

Multiply the right side out. $1 = A(6x^3 + 3x^2 + 8x + 4) + 3Bx^3 + 4Bx + 2Cx^3 + Cx^2 + 2Dx^2 + Dx \Rightarrow$

$1 = 6Ax^3 + 3Ax^2 + 8Ax + 4A + 3Bx^3 + 4Bx + 2Cx^3 + Cx^2 + 2Dx^2 + Dx.$ Equate coefficients.

For x^3: $0 = 6A + 3B + 2C \Rightarrow 0 = 6\left(\dfrac{1}{4}\right) + 3\left(-\dfrac{8}{19}\right) + 2C \Rightarrow 0 = \dfrac{9}{38} + 2C \Rightarrow C = -\dfrac{9}{76}$.

For x^2: $0 = 3A + C + 2D \Rightarrow 0 = \dfrac{3}{4} - \dfrac{9}{76} + 2D \Rightarrow 0 = \dfrac{48}{76} + 2D \Rightarrow D = -\dfrac{24}{76}$.

The expression can be written $\dfrac{1}{4x} + \dfrac{-8}{19(2x+1)} + \dfrac{-9x-24}{76(3x^2+4)}$.

24. Multiply $\dfrac{x^4+1}{x(x^2+1)^2} = \dfrac{A}{x} + \dfrac{Bx+C}{x^2+1} + \dfrac{Dx+E}{(x^2+1)^2}$ by $x(x^2+1)^2 \Rightarrow$

$x^4 + 1 = A(x^2+1)^2 + (Bx+C)(x)(x^2+1) + (Dx+E)(x).$

Let $x = 0 \Rightarrow 1 = A(1) \Rightarrow A = 1$. Multiply the right side out.

$x^4 + 1 = A(x^4 + 2x^2 + 1) + Bx^4 + Cx^3 + Bx^2 + Cx + Dx^2 + Ex \Rightarrow$

$x^4 + 1 = Ax^4 + 2Ax^2 + A + Bx^4 + Cx^3 + Bx^2 + Cx + Dx^2 + Ex.$

Equate coefficients. For x^4: $1 = A + B \Rightarrow 1 = 1 + B \Rightarrow B = 0$. For x^3: $0 = C$.

For x^2: $0 = 2A + B + D \Rightarrow 0 = 2 + 0 + D \Rightarrow D = -2$. For x: $0 = C + E \Rightarrow 0 = 0 + E \Rightarrow E = 0$

The expression can be written $\dfrac{1}{x} + \dfrac{-2x}{(x^2+1)^2}$.

25. Multiply $\dfrac{3x - 1}{x(2x^2 + 1)^2} = \dfrac{A}{x} + \dfrac{Bx + C}{2x^2 + 1} + \dfrac{Dx + E}{(2x^2 + 1)^2}$ by $x(2x^2 + 1)^2 \Rightarrow$

$3x - 1 = A(2x^2 + 1)^2 + (Bx + C)(x)(2x^2 + 1) + (Dx + E)(x)$.

Let $x = 0 \Rightarrow -1 = A(1) \Rightarrow A = -1$. Multiply the right side out.

$3x - 1 = A(4x^4 + 4x^2 + 1) + 2Bx^4 + Bx^2 + Cx + 2Cx^3 + Dx^2 + Ex \Rightarrow$

$3x - 1 = 4Ax^4 + 4Ax^2 + A + 2Bx^4 + Bx^2 + Cx + 2Cx^3 + Dx^2 + Ex$.

Equate coefficients. For x^4: $0 = 4A + 2B \Rightarrow 0 = -4 + 2B \Rightarrow B = 2$. For x^3: $0 = 2C \Rightarrow C = 0$.

For x^2: $0 = 4A + B + D \Rightarrow 0 = -4 + 2 + D \Rightarrow D = 2$. For x: $3 = C + E \Rightarrow 3 = 0 + E \Rightarrow E = 3$

The expression can be written $\dfrac{-1}{x} + \dfrac{2x}{2x^2 + 1} + \dfrac{2x + 3}{(2x^2 + 1)^2}$.

26. Multiply $\dfrac{3x^4 + x^3 + 5x^2 - x + 4}{(x - 1)(x^2 + 1)^2} = \dfrac{A}{x - 1} + \dfrac{Bx + C}{x^2 + 1} + \dfrac{Dx + E}{(x^2 + 1)^2}$ by $(x - 1)(x^2 + 1)^2 \Rightarrow$

$3x^4 + x^3 + 5x^2 - x + 4 = A(x^2 + 1)^2 + (Bx + C)(x - 1)(x^2 + 1) + (Dx + E)(x - 1)$.

Let $x = 1 \Rightarrow 12 = A(4) \Rightarrow A = 3$. Multiply the right side out.

$3x^4 + x^3 + 5x^2 - x + 4 =$

$Ax^4 + 2Ax^2 + A + Bx^4 - Bx^3 + Bx^2 - Bx + Cx^3 - Cx^2 + Cx - C + Dx^2 - Dx + Ex - E$.

Equate coefficients.

For x^4: $3 = A + B \Rightarrow 3 = 3 + B \Rightarrow B = 0$. For x^3: $1 = -B + C \Rightarrow 1 = 0 + C \Rightarrow C = 1$.

For x^2: $5 = 2A + B - C + D \Rightarrow 5 = 6 + 0 - 1 + D \Rightarrow D = 0$.

For x: $-1 = -B + C - D + E \Rightarrow -1 = 0 + 1 - 0 + E \Rightarrow E = -2$.

The expression can be written $\dfrac{3}{x - 1} + \dfrac{1}{x^2 + 1} + \dfrac{-2}{(x^2 + 1)^2}$.

27. Multiply $\dfrac{-x^4 - 8x^2 + 3x - 10}{(x + 2)(x^2 + 4)^2} = \dfrac{A}{x + 2} + \dfrac{Bx + C}{x^2 + 4} + \dfrac{Dx + E}{(x^2 + 4)^2}$ by $(x + 2)(x^2 + 4)^2 \Rightarrow$

$-x^4 - 8x^2 + 3x - 10 = A(x^2 + 4)^2 + (Bx + C)(x + 2)(x^2 + 4) + (Dx + E)(x + 2)$.

Let $x = -2 \Rightarrow -64 = A(64) \Rightarrow A = -1$. Multiply the right side out.

$-x^4 - 8x^2 + 3x - 10 =$

$Ax^4 + 8Ax^2 + 16A + Bx^4 + 2Bx^3 + 4Bx^2 + 8Bx + Cx^3 + 2Cx^2 + 4Cx + 8C + Dx^2 + 2Dx + Ex + 2E$.

Equate coefficients.

For x^4: $-1 = A + B \Rightarrow -1 = -1 + B \Rightarrow B = 0$. For x^3: $0 = 2B + C \Rightarrow 0 = 0 + C \Rightarrow C = 0$.

For x^2: $-8 = 8A + 4B + 2C + D \Rightarrow -8 = -8 + 0 + 0 + D \Rightarrow D = 0$.

For x: $3 = 8B + 4C + 2D + E \Rightarrow 3 = 0 + 0 + 0 + E \Rightarrow E = 3$.

The expression can be written $\dfrac{-1}{x + 2} + \dfrac{3}{(x^2 + 4)^2}$.

28. Factoring $\dfrac{x^2}{x^4 - 1}$ results in $\dfrac{x^2}{(x + 1)(x - 1)(x^2 + 1)}$.

Multiply $\dfrac{x^2}{(x + 1)(x - 1)(x^2 + 1)} = \dfrac{A}{x + 1} + \dfrac{B}{x - 1} + \dfrac{Cx + D}{x^2 + 1}$ by $(x + 1)(x - 1)(x^2 + 1) \Rightarrow$

$x^2 = A(x - 1)(x^2 + 1) + B(x + 1)(x^2 + 1) + (Cx + D)(x - 1)(x + 1)$

Let $x = -1 \Rightarrow 1 = A(-2)(2) \Rightarrow A = -\dfrac{1}{4}$. Let $x = 1 \Rightarrow 1 = B(2)(2) \Rightarrow B = \dfrac{1}{4}$. Multiply the right side out.

$x^2 = Ax^3 - Ax^2 + Ax - A + Bx^3 + Bx^2 + Bx + B + Cx^3 - Cx + Dx^2 - D$

Equate coefficients. For x^3: $0 = A + B + C \Rightarrow 0 = -\dfrac{1}{4} + \dfrac{1}{4} + C \Rightarrow C = 0$.

For x^2: $1 = -A + B + D \Rightarrow 1 = \dfrac{1}{4} + \dfrac{1}{4} + D \Rightarrow D = \dfrac{1}{2}$.

The expression can be written $\dfrac{-1}{4(x + 1)} + \dfrac{1}{4(x - 1)} + \dfrac{1}{2(x^2 + 1)}$.

29. By long division $\dfrac{5x^5 + 10x^4 - 15x^3 + 4x^2 + 13x - 9}{x^3 + 2x^2 - 3x} = 5x^2 + \dfrac{4x^2 + 13x - 9}{x^3 + 2x^2 - 3x} = 5x^2 + \dfrac{4x^2 + 13x - 9}{x(x + 3)(x - 1)}$.

Multiply $\dfrac{4x^2 + 13x - 9}{x(x + 3)(x - 1)} = \dfrac{A}{x} + \dfrac{B}{x + 3} + \dfrac{C}{x - 1}$ by $x(x + 3)(x - 1) \Rightarrow$

$4x^2 + 13x - 9 = A(x + 3)(x - 1) + B(x)(x - 1) + C(x)(x + 3)$ Let $x = 0 \Rightarrow -9 = A(-3) \Rightarrow A = 3$.

Let $x = -3 \Rightarrow -12 = B(-3)(-4) \Rightarrow B = -1$. Let $x = 1 \Rightarrow 8 = C(4) \Rightarrow C = 2$.

The expression can be written $5x^2 + \dfrac{3}{x} + \dfrac{-1}{x + 3} + \dfrac{2}{x - 1}$.

30. By long division $\dfrac{3x^6 + 3x^4 + 3x}{x^4 + x^2} = 3x^2 + \dfrac{3x}{x^4 + x^2} = 3x^2 + \dfrac{3x}{x^2(x^2 + 1)}$.

Multiply $\dfrac{3x}{x^2(x^2 + 1)} = \dfrac{A}{x} + \dfrac{B}{x^2} + \dfrac{Cx + D}{x^2 + 1}$ by $x^2(x^2 + 1) \Rightarrow$

$3x = A(x)(x^2 + 1) + B(x^2 + 1) + (Cx + D)(x^2) \Rightarrow 3x = Ax^3 + Ax + Bx^2 + B + Cx^3 + Dx^2$

Let $x = 0 \Rightarrow 0 = B$. Equate coefficients. For x: $3 = A$. For x^3: $0 = A + C \Rightarrow 0 = 3 + C \Rightarrow C = -3$.

For x^2: $0 = B + D \Rightarrow 0 = 0 + D \Rightarrow D = 0$. The expression can be written $3x^2 + \dfrac{3}{x} + \dfrac{-3x}{x^2 + 1}$.

31. The decomposition is correct. The graphs coincide. See Figure 31.

32. The decomposition is not correct. The graphs do not coincide. See Figure 32.

33. The decomposition is not correct. The graphs do not coincide. See Figure 33.

34. The decomposition is correct. The graphs coincide. See Figure 34.

$[-9.4, 9.4]$ by $[-6.2, 6.2]$	$[-9.4, 9.4]$ by $[-6.2, 6.2]$	$[-4.7, 4.7]$ by $[-3.1, 3.1]$	$[-4.7, 4.7]$ by $[-8.2, 4.2]$
Xscl = 1 Yscl = 1	Xscl = 1 Yscl = 1	Xscl = 1 Yscl = 1	Xscl = 1 Yscl = 1

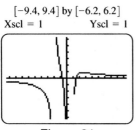

Figure 31

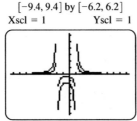

Figure 32

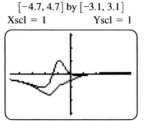

Figure 33

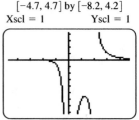

Figure 34

Reviewing Basic Concepts (Sections 7.7 and 7.8)

1. See Figure 1.

2. See Figure 2.

3. See Figure 3.

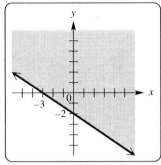

Figure 1

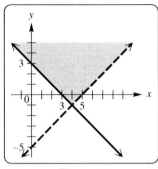

Figure 2

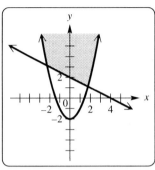

Figure 3

4. See Figure 4.

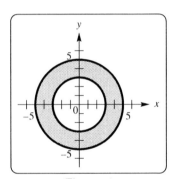

Figure 4

5. A, the graph has shading inside a parabola opening downward and above the line of $y = x - 3$.

6. The objective function is: $2x + 3y$.

   The constraints are: $x \geq 0, \ y \geq 0$

   $$x + y \geq 4$$

   $$2x + y \leq 8$$

   Graphing the constraints on the calculator yields three vertices. Inputting these ordered pairs into the

   objective function yields: $(0, 4)$: 12; $(0, 8)$: 24; and $(4, 0)$: 8.

   The minimum number is 8 at $(4,0)$.

7. Using the given expression and the ordered pairs of the vertices yields the following solutions:

   $(1, 1)$: $3(1) + 5(1) = 8$; $(2, 7)$: $3(2) + 5(7) = 41$; $(5, 10)$: $3(5) + 5(10) = 65$; $(6, 3)$: $3(6) + 5(3) = 33$

   Therefore, the maximum value is 65 at $(5, 10)$; and the minimum value is 8 at $(1, 1)$.

8. Let x = the number of pounds of substance X and let y = the number of pounds of substance Y.

The objective function to find the minimum cost is: $2x + 3y$.

The constraints are: $.2x + .5y \geq 251$ (minimum amount of ingredient A)

$.5x + .3y \geq 200$ (minimum amount of ingredient B)

$x \geq 0, \ y \geq 0$ (cannot include negative amounts of the ingredients)

Graphing the constraints on the calculator yields three vertices. Inputting these ordered pairs into the

objective function yields: $\left(0, \dfrac{2000}{3}\right)$: 2000;   (1255, 0): 2510;   (130, 450): 1610. The minimum cost of \$1610,

occurs when there are 130 pounds of substance X and 450 pounds of substance Y purchased.

9. Factoring $\dfrac{10x + 13}{x^2 - x - 20}$ results in $\dfrac{10x + 13}{(x - 5)(x + 4)}$.

Multiply $\dfrac{10x + 13}{(x - 5)(x + 4)} = \dfrac{A}{x - 5} + \dfrac{B}{x + 4}$ by $(x - 5)(x + 4) \Rightarrow 10x + 13 = A(x + 4) + B(x - 5)$.

Let $x = -4 \Rightarrow -27 = B(-9) \Rightarrow B = 3$. Let $x = 5 \Rightarrow 63 = A(9) \Rightarrow A = 7$.

The expression can be written $\dfrac{7}{x - 5} + \dfrac{3}{x + 4}$.

10. Multiply $\dfrac{2x^2 - 15x - 32}{(x - 1)(x^2 + 6x + 8)} = \dfrac{A}{x - 1} + \dfrac{B}{x + 2} + \dfrac{C}{x + 4}$ by $(x - 1)(x + 2)(x + 4) \Rightarrow$

$2x^2 - 15x - 32 = A(x + 2)(x + 4) + B(x - 1)(x + 4) + C(x - 1)(x + 2)$.

Let $x = -2 \Rightarrow 6 = B(-6) \Rightarrow B = -1$. Let $x = -4 \Rightarrow 60 = C(10) \Rightarrow C = 6$.

Let $x = 1 \Rightarrow -45 = A(15) \Rightarrow A = -3$. The expression can be written $\dfrac{-3}{x - 1} + \dfrac{-1}{x + 2} + \dfrac{6}{x + 4}$.

Chapter 7 Review Exercises

1. Using substitution, first solve [equation 1] for x, $4x - 3y = -1 \Rightarrow 4x = 3y - 1 \Rightarrow x = \dfrac{3y - 1}{4}$ [equation 3].

Substitute $\left(\dfrac{3y - 1}{4}\right)$ in for x, in [equation 2]: $3\left(\dfrac{3y - 1}{4}\right) + 5y = 50 \Rightarrow 3(3y - 1) + 20y = 200 \Rightarrow$

$9y - 3 + 20y = 200 \Rightarrow 29y = 203 \Rightarrow y = 7$. Now substitute (7) in for y in [equation 3]:

$x = \dfrac{3(7) - 1}{4} = \dfrac{20}{4} \Rightarrow x = 5$. The solution is: $\{(5, 7)\}$.

2. Using substitution, first solve [equation 1] for x, $\dfrac{x}{2} - \dfrac{y}{5} = \dfrac{11}{10} \Rightarrow 5x - 2y = 11 \Rightarrow 5x = 2y + 11 \Rightarrow$

$x = \dfrac{2y + 11}{5}$ [equation 3]. Substitute $\left(\dfrac{2y + 11}{5}\right)$ in for x, in $5 \times$ [equation 2]: $10\left(\dfrac{2y + 11}{5}\right) - 4y = 22 \Rightarrow$

$2(2y + 11) - 4y = 22 \Rightarrow 4y + 22 - 4y = 22 \Rightarrow 0 = 0$. Since this is a true statement the system is

dependent and the solution: $\left(\dfrac{2y + 11}{5}, y\right)$ or $\left(x, \dfrac{5x - 11}{2}\right)$.

3. Using substitution, first solve [equation 1] for y, $4x + 5y = 5 \Rightarrow 5y = 5 - 4x \Rightarrow y = \dfrac{5 - 4x}{5}$ [equation 3].

   Substitute $\left(\dfrac{5 - 4x}{5}\right)$ in for y, in [equation 2]: $3x + 7\left(\dfrac{5 - 4x}{5}\right) = -6 \Rightarrow 15x + 7(5 - 4x) = -30 \Rightarrow$

   $15x + 35 - 28x = -30 \Rightarrow -13x = -65 \Rightarrow x = 5$. Now substitute (5) in for x in [equation 3]:

   $y = \dfrac{5 - 4(5)}{5} = \dfrac{-15}{5} \Rightarrow y = -3$. The solution is: $\{(5, -3)\}$.

4. Since {equation 1} is solved for y, we substitute $(x^2 - 1)$ in for y in [equation 2]: $x + (x^2 - 1) = 1 \Rightarrow$

   $x^2 + x - 2 = 0 \Rightarrow (x + 2)(x - 1) = 0 \Rightarrow x = -2, 1$. Substituting these values into [equation 1] yields:

   $y = (-2)^2 - 1 \Rightarrow y = 4 - 1 \Rightarrow y = 3$; and $y = (1)^2 - 1 \Rightarrow y = 1 - 1 \Rightarrow y = 0$.

   The solution is: $\{(-2, 3), (1, 0)\}$.

5. Using substitution, first solve {equation 2} for y, $3x + y = 4 \Rightarrow y = 4 - 3x$. Now substitute $(4 - 3x)$ in for y

   in [equation 1]: $x^2 + (4 - 3x)^2 = 2 \Rightarrow x^2 + 16 - 24x + 9x^2 = 2 \Rightarrow 10x^2 - 24x + 14 = 0 \Rightarrow$

   $2(5x - 7)(x - 1) = 0 \Rightarrow x = \dfrac{7}{5}, 1$. Substituting these values into [equation 2] yields:

   $y = 4 - 3\left(\dfrac{7}{5}\right) = \dfrac{20}{5} - \dfrac{21}{5} \Rightarrow y = -\dfrac{1}{5}$; and $y = 4 - 3(1) \Rightarrow y = 1$. The solution is: $\left\{\left(\dfrac{7}{5}, -\dfrac{1}{5}\right), (1, 1)\right\}$.

6. Multiply [equation 2] by 2 and add to [equation 1] to eliminate the y^2.

   $$\begin{array}{r} x^2 + 2y^2 = 22 \\ \underline{4x^2 - 2y^2 = -2} \\ 5x^2 \quad\quad = 20 \end{array} \Rightarrow x^2 = 4 \Rightarrow x = \pm 2.$$

   Substituting -2 and 2 into [equation 2] for x, yields: $2(-2)^2 - y^2 = -1 \Rightarrow 8 - y^2 = -1 \Rightarrow 9 = y^2 \Rightarrow$

   $y = -3, 3$; and $2(2)^2 - y^2 = -1 \Rightarrow 8 - y^2 = -1 \Rightarrow 9 = y^2 \Rightarrow y = -3, 3$.

   The solution is: $\{(2, -3), (2, 3), (-2, -3), (-2, 3)\}$.

7. Multiply [equation 2] by -1 and add to [equation 1] to eliminate the x^2.

   $$\begin{array}{r} x^2 - 4y^2 = 19 \\ \underline{-x^2 - y^2 = -29} \\ -5y^2 = -10 \end{array} \Rightarrow y^2 = 2 \Rightarrow y = \pm\sqrt{2}.$$

   Substituting $-\sqrt{2}$ and $\sqrt{2}$ into [equation 2] for y, yields: $x^2 + (\sqrt{2})^2 = 29 \Rightarrow$

   $x^2 = 27 \Rightarrow x = \pm\sqrt{27} = \pm 3\sqrt{3}$; and $x^2 + (-\sqrt{2})^2 = 29 \Rightarrow x^2 = 27 \Rightarrow x = \pm\sqrt{27} = \pm 3\sqrt{3}$.

   The solution is: $\{(3\sqrt{3}, \sqrt{2}), (-3\sqrt{3}, \sqrt{2}), (3\sqrt{3}, -\sqrt{2}), (-3\sqrt{3}, -\sqrt{2})\}$.

8. Using substitution, first solve {equation 2} for x, $x - 6y = 2 \Rightarrow x = 6y + 2$. Now substitute $(6y + 2)$ in for

   x in [equation 1]: $(6y + 2)y = 4 \Rightarrow 6y^2 + 2y - 4 = 0 \Rightarrow 2(3y - 2)(y + 1) = 0 \Rightarrow y = \dfrac{2}{3}, -1$.

   Substituting these values into [equation 2] yields: $x = 6\left(\dfrac{2}{3}\right) + 2 \ x = 6$; and $x = 6(-1) + 2 \Rightarrow x = -4$.

   The solution is: $\left\{\left(6, \dfrac{2}{3}\right), (-4, -1)\right\}$.

9. Since [equation 2] is solved for x, we substitute $(y + 4)$ in for x in [equation 1]:

   $(y + 4)^2 - y^2 = 32 \Rightarrow y^2 + 8y + 16 - y^2 = 32 \Rightarrow 8y + 16 = 32 \Rightarrow 8y = 16 \Rightarrow y = 2$

   Substituting this values into [equation 2] yields: $x = 2 + 4 \Rightarrow x = 6$.

   The solution is: $\{(6, 2)\}$.

10. (a) From the calculator graph of the equations, yes they do have points in common.

    (b) From the same graph the points of intersection are approximately: $(11.8, -1.9)$ and $(-8.6, 8.3)$.

    (c) Using substitution, first solve {equation 2] for x, $x + 2y = 8 \Rightarrow x = 8 - 2y$ [equation 3].

       Substitute $(8 - 2y)$ in for x in [equation 1]:

       $(8 - 2y)^2 + y^2 = 144 \Rightarrow 64 - 32y + 4y^2 + y^2 - 144 = 0 \Rightarrow 5y^2 - 32y - 80 = 0$.

       Now use the quadratic formula to solve for y:

       $$y = \frac{-(-32) \pm \sqrt{(-32)^2 - 4(5)(-80)}}{2(5)} = \frac{32 \pm \sqrt{2624}}{10} = \frac{32 \pm 8\sqrt{41}}{10} \Rightarrow y = \frac{16 \pm 4\sqrt{41}}{5}.$$

       Finally, substitute these values into [equation 3]:

       $$x = 8 - 2\left(\frac{16 + 4\sqrt{41}}{5}\right) = \frac{40 - 32 - 8\sqrt{41}}{5} = \frac{8 - 8\sqrt{41}}{5}.$$

       $$x = 8 - 2\left(\frac{16 - 4\sqrt{41}}{5}\right) = \frac{40 - 32 + 8\sqrt{41}}{5} = \frac{8 + 8\sqrt{41}}{5}.$$

       The solution is: $\left\{\left(\dfrac{8 - 8\sqrt{41}}{5}, \dfrac{16 + 4\sqrt{41}}{5}\right), \left(\dfrac{8 + 8\sqrt{41}}{5}, \dfrac{16 - 4\sqrt{41}}{5}\right)\right\}$.

11. (a) Solve the first equation for y: $x^2 + y^2 = 2 \Rightarrow y^2 = 2 - x^2 \Rightarrow y = \pm\sqrt{2 - x^2}$.

       The two functions are: $y_1 = \sqrt{2 - x^2}$ and $y_2 = -\sqrt{2 - x^2}$.

    (b) Solve the first equation for y: $3x + y = 4 \Rightarrow y = 4 - 3x$. The function is: $y_3 = -3x + 4$.

    (c) The viewing window $[-3, 3]$ by $[-2, 2]$ should show the intersection;

       other settings are possible.

12. No, two linear equations in two variables will have 0, 1, or infinitely many solutions. Two lines cannot

    intersect in exactly two points.

13. No, a system consisting of two equations in three variables is represented by two planes in space. There will be

    no solutions or infinitely many solutions.

14. Using the Row Echelon Method, the given system of equations yields the matrix: $\begin{bmatrix} 2 & -3 & 1 & -5 \\ 1 & 4 & 2 & 13 \\ 5 & 5 & 3 & 14 \end{bmatrix}$, which

    by $(R_1 \leftrightarrow R_2) \rightarrow \begin{bmatrix} 1 & 4 & 2 & 13 \\ 2 & -3 & 1 & -5 \\ 5 & 5 & 3 & 14 \end{bmatrix} \Rightarrow (-2R_1 + R_2)$ and $(-5R_1 + R_3) \rightarrow \begin{bmatrix} 1 & 4 & 2 & 13 \\ 0 & -11 & -3 & -31 \\ 0 & -15 & -7 & -51 \end{bmatrix} \Rightarrow$

    $-\dfrac{1}{11}R_2 \rightarrow \begin{bmatrix} 1 & 4 & 2 & 13 \\ 0 & 1 & \frac{3}{11} & \frac{31}{11} \\ 0 & -15 & -7 & -51 \end{bmatrix} \Rightarrow 15R_2 + R_3 \rightarrow \begin{bmatrix} 1 & 4 & 2 & 13 \\ 0 & 1 & \frac{3}{11} & \frac{31}{11} \\ 0 & 0 & -\frac{32}{11} & -\frac{96}{11} \end{bmatrix}$.

    From this matrix, we have the resulting equation: $-\dfrac{32}{11}z = -\dfrac{96}{11} \Rightarrow z = 3$. Now use back-substitution:

    substituting $z = 3$ into the resulting R_2 yields: $y + \dfrac{3}{11}(3) = \dfrac{31}{11} \Rightarrow y = \dfrac{22}{11}; \Rightarrow y = 2$; and finally substituting

    $y = 2$ and $z = 3$ into the resulting R_1 yields: $x + 4(2) + 2(3) = 13 \Rightarrow x = -1$. The solution is: $\{(-1, 2, 3)\}$.

15. Using the Row Echelon Method, the given system of equations yields the matrix: $\begin{bmatrix} 1 & -3 & 0 & | & 12 \\ 0 & 2 & 5 & | & 1 \\ 4 & 0 & 1 & | & 25 \end{bmatrix}$, which

by $-4R_1 + R_3 \rightarrow \begin{bmatrix} 1 & -3 & 0 & | & 12 \\ 0 & 2 & 5 & | & 1 \\ 0 & 12 & 1 & | & -23 \end{bmatrix} \Rightarrow \frac{1}{2}R_2 \rightarrow \begin{bmatrix} 1 & -3 & 0 & | & 12 \\ 0 & 1 & 2.5 & | & .5 \\ 0 & 12 & 1 & | & -23 \end{bmatrix} \Rightarrow$

$-12R_2 + R_3 \rightarrow \begin{bmatrix} 1 & -3 & 0 & | & 12 \\ 0 & 1 & 2.5 & | & .5 \\ 0 & 0 & -29 & | & -29 \end{bmatrix} \Rightarrow -\frac{1}{29}R_3 \rightarrow \begin{bmatrix} 1 & -3 & 0 & | & 12 \\ 0 & 1 & 2.5 & | & .5 \\ 0 & 0 & 1 & | & 1 \end{bmatrix}$.

From this matrix, we have the resulting equation: $z = 1$. Now use back-substitution:

substituting $z = 1$ into the resulting R_2 yields: $y + 2.5(1) = .5 \Rightarrow y = -2$; and finally substituting

$y = -2$ and $z = 1$ into the resulting R_1 yields: $x - 3(-2) = 12 \Rightarrow x = 6$. The solution is: $\{(6, -2, 1)\}$.

16. Using the Row Echelon Method, the given system of equations yields the matrix: $\begin{bmatrix} 1 & 1 & -1 & | & 5 \\ 2 & 1 & 3 & | & 2 \\ 4 & -1 & 2 & | & -1 \end{bmatrix}$, which

by $\left.\begin{array}{l} -2R_1 + R_2 \\ -4R_1 + R_3 \end{array}\right\} \rightarrow \begin{bmatrix} 1 & 1 & -1 & | & 5 \\ 0 & -1 & 5 & | & -8 \\ 0 & -5 & 6 & | & -21 \end{bmatrix} \Rightarrow -R_2 \rightarrow \begin{bmatrix} 1 & 1 & -1 & | & 5 \\ 0 & 1 & -5 & | & 8 \\ 0 & -5 & 6 & | & -21 \end{bmatrix} \Rightarrow$

$5R_2 + R_3 \rightarrow \begin{bmatrix} 1 & 1 & -1 & | & 5 \\ 0 & 1 & -5 & | & 8 \\ 0 & 0 & -19 & | & 19 \end{bmatrix} \Rightarrow -\frac{1}{19}R_3 \rightarrow \begin{bmatrix} 1 & 1 & -1 & | & 5 \\ 0 & 1 & -5 & | & 8 \\ 0 & 0 & 1 & | & -1 \end{bmatrix}$.

From this matrix, we have the resulting equation: $z = -1$. Now use back-substitution:

substituting $z = -1$ into the resulting R_2 yields: $y - 5(-1) = 8 \Rightarrow y = 3$; and finally substituting

$y = 3$ and $z = -1$ into the resulting R_1 yields: $x + (3) - (-1) = 5 \Rightarrow x = 1$. The solution is: $\{(1, 3, -1)\}$.

17. Using the Row Echelon Method, the given system of equations yields the matrix: $\begin{bmatrix} 5 & -3 & 2 & | & -5 \\ 2 & 1 & -1 & | & 4 \\ -4 & -2 & 2 & | & -1 \end{bmatrix}$, which

by $\frac{1}{5}R_1 \rightarrow \begin{bmatrix} 1 & -.6 & .4 & | & -1 \\ 2 & 1 & -1 & | & 4 \\ -4 & -2 & 2 & | & -1 \end{bmatrix} \Rightarrow (-2R_1 + R_2) \text{ and } (4R_1 + R_3) \rightarrow \begin{bmatrix} 1 & -.6 & .4 & | & -1 \\ 0 & 2.2 & -1.8 & | & 6 \\ 0 & -4.4 & 3.6 & | & -5 \end{bmatrix} \Rightarrow$

$\frac{5}{11}R_2 \rightarrow \begin{bmatrix} 1 & -.6 & .4 & | & -1 \\ 0 & 1 & -\frac{9}{11} & | & \frac{30}{11} \\ 0 & -4.4 & 3.6 & | & -5 \end{bmatrix} \Rightarrow 4.4R_2 + R_3 \rightarrow \begin{bmatrix} 1 & -.6 & .4 & | & -1 \\ 0 & 1 & -\frac{9}{11} & | & \frac{30}{11} \\ 0 & 0 & 0 & | & 7 \end{bmatrix}$

Since R_3 yields: $0 = 7$, which is never true the system is inconsistent and the solution is: $\varnothing$.

18. Using the Reduced Row Echelon Method, the given system of equations yields the matrix: $\begin{bmatrix} 2 & 3 & | & 10 \\ -3 & 1 & | & 18 \end{bmatrix}$,

which by $\frac{1}{2}R_1 \rightarrow \begin{bmatrix} 1 & \frac{3}{2} & | & 5 \\ -3 & 1 & | & 18 \end{bmatrix} 3R_1 + R_2 \rightarrow \begin{bmatrix} 1 & \frac{3}{2} & | & 5 \\ 0 & \frac{11}{2} & | & 33 \end{bmatrix} \Rightarrow \frac{2}{11}R_2 \rightarrow \begin{bmatrix} 1 & \frac{3}{2} & | & 5 \\ 0 & 1 & | & 6 \end{bmatrix} \Rightarrow$

$-\frac{3}{2}R_2 + R_1 \rightarrow \begin{bmatrix} 1 & 0 & | & -4 \\ 0 & 1 & | & 6 \end{bmatrix}$.   From this Reduced Row matrix, we have the solution: $\{(-4, 6)\}$.

19. Using the Reduced Row Echelon Method, the given system of equations yields the matrix: $\begin{bmatrix} 3 & 1 & | & -7 \\ 1 & -1 & | & -5 \end{bmatrix}$,

which by $R_2 + R_1 \rightarrow \begin{bmatrix} 4 & 0 & | & -12 \\ 1 & -1 & | & -5 \end{bmatrix} \Rightarrow \frac{1}{4}R_1 \rightarrow \begin{bmatrix} 1 & 0 & | & -3 \\ 1 & -1 & | & -5 \end{bmatrix} \Rightarrow R_1 - R_2 \rightarrow \begin{bmatrix} 1 & 0 & | & -3 \\ 0 & 1 & | & 2 \end{bmatrix}$.

From this Reduced Row matrix, we have the solution: $\{(-3, 2)\}$.

20. Using the Reduced Row Echelon Method, the given system of equations yields the matrix: $\begin{bmatrix} 1 & 0 & -1 & | & -3 \\ 0 & 1 & 1 & | & 6 \\ 2 & 0 & -3 & | & -9 \end{bmatrix}$,

which by $-2R_1 + R_3 \rightarrow \begin{bmatrix} 1 & 0 & -1 & | & -3 \\ 0 & 1 & 1 & | & 6 \\ 0 & 0 & -1 & | & -3 \end{bmatrix} \Rightarrow -R_3 \rightarrow \begin{bmatrix} 1 & 0 & -1 & | & -3 \\ 0 & 1 & 1 & | & 6 \\ 0 & 0 & 1 & | & 3 \end{bmatrix} \Rightarrow$

$\left. \begin{array}{c} R_3 + R_1 \\ -R_3 + R_2 \end{array} \right\} \rightarrow \begin{bmatrix} 1 & 0 & 0 & | & 0 \\ 0 & 1 & 0 & | & 3 \\ 0 & 0 & 1 & | & 3 \end{bmatrix}$.

From this Reduced Row matrix, we have the solution: $\{(0, 3, 3)\}$.

21. Using the Reduced Row Echelon Method, the given system of equations yields the matrix: $\begin{bmatrix} 2 & -1 & 4 & | & -1 \\ -3 & 5 & -1 & | & 5 \\ 2 & 3 & 2 & | & 3 \end{bmatrix}$,

which by $\left. \begin{array}{c} \frac{3}{2}R_1 + R_2 \\ -R_1 + R_3 \end{array} \right\} \rightarrow \begin{bmatrix} 2 & -1 & 4 & | & -1 \\ 0 & \frac{7}{2} & 5 & | & \frac{7}{2} \\ 0 & 4 & -2 & | & 4 \end{bmatrix} \Rightarrow \frac{2}{7}R_2 \rightarrow \begin{bmatrix} 2 & -1 & 4 & | & -1 \\ 0 & 1 & \frac{10}{7} & | & 1 \\ 0 & 4 & -2 & | & 4 \end{bmatrix} \Rightarrow$

$-4R_2 + R_3 \rightarrow \begin{bmatrix} 2 & -1 & 4 & | & -1 \\ 0 & 1 & \frac{10}{7} & | & 1 \\ 0 & 0 & -\frac{54}{7} & | & 0 \end{bmatrix} \Rightarrow \left. \begin{array}{c} R_2 + R_1 \\ -\frac{7}{54}R_3 \end{array} \right\} \rightarrow \begin{bmatrix} 0 & 0 & \frac{38}{7} & | & 0 \\ 0 & 1 & \frac{10}{7} & | & 1 \\ 0 & 0 & 1 & | & 0 \end{bmatrix} \Rightarrow$

$\left. \begin{array}{c} -\frac{38}{7}R_3 + R_1 \\ -\frac{10}{7}R_3 + R_2 \end{array} \right\} \rightarrow \begin{bmatrix} 2 & 0 & 0 & | & 0 \\ 0 & 1 & 0 & | & 1 \\ 0 & 0 & 1 & | & 0 \end{bmatrix} \Rightarrow \frac{1}{2}R_1 \rightarrow \begin{bmatrix} 1 & 0 & 0 & | & 0 \\ 0 & 1 & 0 & | & 1 \\ 0 & 0 & 1 & | & 0 \end{bmatrix}$.

From this Reduced Row matrix, we have the solution: $\{(0, 1, 0)\}$.

22. $\begin{bmatrix} -5 & 4 & 9 \\ 2 & -1 & -2 \end{bmatrix} + \begin{bmatrix} 1 & -2 & 7 \\ 4 & -5 & -5 \end{bmatrix} = \begin{bmatrix} -5+1 & 4-2 & 9+7 \\ 2+4 & -1-5 & -2-5 \end{bmatrix} = \begin{bmatrix} -4 & 2 & 16 \\ 6 & -6 & -7 \end{bmatrix}$.

23. $\begin{bmatrix} 3 \\ 2 \\ 5 \end{bmatrix} - \begin{bmatrix} 8 \\ -4 \\ 6 \end{bmatrix} + \begin{bmatrix} 1 \\ 0 \\ 2 \end{bmatrix} = \begin{bmatrix} 3-8+1 \\ 2+4+0 \\ 5-6+2 \end{bmatrix} = \begin{bmatrix} -4 \\ 6 \\ 1 \end{bmatrix}$.

24. $\begin{bmatrix} 2 & 5 & 8 \\ 1 & 9 & 2 \end{bmatrix} - \begin{bmatrix} 3 & 4 \\ 7 & 1 \end{bmatrix}$. We cannot subtract matrices of unlike size $(2 \times 3) - (2 \times 2)$, the solution is: $\varnothing$.

25. $3\begin{bmatrix} 2 & 4 \\ -1 & 4 \end{bmatrix} - 2\begin{bmatrix} 5 & 8 \\ 2 & -2 \end{bmatrix} = \begin{bmatrix} 6 & 12 \\ -3 & 12 \end{bmatrix} - \begin{bmatrix} 10 & 16 \\ 4 & -4 \end{bmatrix} = \begin{bmatrix} 6-10 & 12-16 \\ -3-4 & 12+4 \end{bmatrix} = \begin{bmatrix} -4 & -4 \\ -7 & 16 \end{bmatrix}$.

26. $-1\begin{bmatrix} 3 & -5 & 2 \\ 1 & 7 & -4 \end{bmatrix} + 5\begin{bmatrix} 0 & 2 \\ -1 & 3 \end{bmatrix}$. We cannot add matrices of unlike size $(2 \times 3) + (2 \times 2)$, the solution is: $\varnothing$.

27. $10\begin{bmatrix} 2x+3y & 4x+y \\ x-5y & 6x+2y \end{bmatrix} + 2\begin{bmatrix} -3x-y & x+6y \\ 4x+2y & 5x-y \end{bmatrix} = \begin{bmatrix} 20x+30y-6x-2y & 40x+10y+2x+12y \\ 10x-50y+8x+4y & 60x+20y+10x-2y \end{bmatrix} =$

$\begin{bmatrix} 14x+28y & 42x+22y \\ 18x-46y & 70x+18y \end{bmatrix}$.

28. The sum of two $m \times n$ matrices A and B is founded by adding corresponding elements.

29. $\begin{bmatrix} -8 & 6 \\ 5 & 2 \end{bmatrix} \begin{bmatrix} 3 & -1 \\ 7 & 2 \end{bmatrix} = \begin{bmatrix} -8(3) + 6(7) & -8(-1) + 6(2) \\ 5(3) + 2(7) & 5(-1) + 2(2) \end{bmatrix} = \begin{bmatrix} -24 + 42 & 8 + 12 \\ 15 + 14 & -5 + 4 \end{bmatrix} = \begin{bmatrix} 18 & 20 \\ 29 & -1 \end{bmatrix}$.

30. $\begin{bmatrix} 3 & 2 & -1 \\ 4 & 0 & 6 \end{bmatrix} \begin{bmatrix} -2 & 0 \\ 0 & 2 \\ 3 & 1 \end{bmatrix} = \begin{bmatrix} 3(-2) + 2(0) - 1(3) & 3(0) + 2(2) - 1(1) \\ 4(-2) + 0(0) + 6(3) & 4(0) + 0(2) + 6(1) \end{bmatrix} = \begin{bmatrix} -9 & 3 \\ 10 & 6 \end{bmatrix}$.

31. $\begin{bmatrix} 1 & -2 & 4 & 2 \\ 0 & 1 & -1 & 8 \end{bmatrix} \begin{bmatrix} -1 \\ 2 \\ 0 \\ 1 \end{bmatrix} = \begin{bmatrix} 1(-1) - 2(2) + 4(0) + 2(1) \\ 0(-1) + 1(2) - 1(0) + 8(1) \end{bmatrix} = \begin{bmatrix} -3 \\ 10 \end{bmatrix}$.

32. $\begin{bmatrix} 1 & 2 & 5 \\ -3 & 4 & 7 \\ 0 & 2 & -1 \end{bmatrix} \begin{bmatrix} 4 & 2 & 3 \\ 10 & -5 & 6 \end{bmatrix} =$ We cannot multiply matrices of size $(3 \times 3) \times (2 \times 3)$, the solution is: $\varnothing$.

33. $\begin{bmatrix} 4 & 2 & 3 \\ 10 & -5 & 6 \end{bmatrix} \begin{bmatrix} 1 & 2 & 5 \\ -3 & 4 & 7 \\ 0 & 2 & -1 \end{bmatrix} = \begin{bmatrix} 4 - 6 + 0 & 8 + 8 + 6 & 20 + 14 - 3 \\ 10 + 15 + 0 & 20 - 20 + 12 & 50 - 35 - 6 \end{bmatrix} = \begin{bmatrix} -2 & 22 & 31 \\ 25 & 12 & 9 \end{bmatrix}$.

34. $\begin{bmatrix} 3 & -1 & 0 \end{bmatrix} \begin{bmatrix} 1 & 3 & 2 \\ 2 & -4 & 0 \\ 5 & 7 & 3 \end{bmatrix} = \begin{bmatrix} 3 - 2 + 0 & 9 + 4 + 0 & 6 + 0 + 0 \end{bmatrix} = \begin{bmatrix} 1 & 13 & 6 \end{bmatrix}$.

35. Yes; $AB = \begin{bmatrix} 3 & 2 \\ 13 & 9 \end{bmatrix} \begin{bmatrix} 9 & -2 \\ -13 & 3 \end{bmatrix} = \begin{bmatrix} 27 - 26 & -6 + 6 \\ 117 - 117 & -26 + 27 \end{bmatrix} = \begin{bmatrix} 1 & 0 \\ 0 & 1 \end{bmatrix}$.

$BA = \begin{bmatrix} 9 & -2 \\ -13 & 3 \end{bmatrix} \begin{bmatrix} 3 & 2 \\ 13 & 9 \end{bmatrix} = \begin{bmatrix} 27 - 26 & 18 - 18 \\ -39 + 39 & -26 + 27 \end{bmatrix} = \begin{bmatrix} 1 & 0 \\ 0 & 1 \end{bmatrix}$.

36. Yes; $AB = \begin{bmatrix} 1 & 0 \\ 2 & -3 \end{bmatrix} \begin{bmatrix} 1 & 0 \\ \frac{2}{3} & -\frac{1}{3} \end{bmatrix} = \begin{bmatrix} 1 + 0 & 0 + 0 \\ 2 - 2 & 0 + 1 \end{bmatrix} = \begin{bmatrix} 1 & 0 \\ 0 & 1 \end{bmatrix}$.

$BA = \begin{bmatrix} 1 & 0 \\ \frac{2}{3} & -\frac{1}{3} \end{bmatrix} \begin{bmatrix} 1 & 0 \\ 2 & -3 \end{bmatrix} = \begin{bmatrix} 1 + 0 & 0 + 0 \\ \frac{2}{3} - \frac{2}{3} & 0 + 1 \end{bmatrix} = \begin{bmatrix} 1 & 0 \\ 0 & 1 \end{bmatrix}$.

37. No; $AB = \begin{bmatrix} 2 & 0 & 6 \\ 0 & 1 & 0 \\ 1 & 0 & 1 \end{bmatrix} \begin{bmatrix} -1 & 0 & \frac{3}{2} \\ 0 & 1 & 0 \\ \frac{1}{4} & 0 & -1 \end{bmatrix} = \begin{bmatrix} -2 + 0 + \frac{3}{2} & 0 + 0 + 0 & 3 + 0 - 6 \\ 0 + 0 + 0 & 0 + 1 + 0 & 0 + 0 + 0 \\ -1 + 0 + \frac{1}{4} & 0 + 0 + 0 & \frac{3}{2} + 0 - 1 \end{bmatrix} = \begin{bmatrix} -\frac{1}{2} & 0 & -3 \\ 0 & 1 & 0 \\ -\frac{3}{4} & 0 & \frac{1}{2} \end{bmatrix}$.

38. Yes; $AB = \begin{bmatrix} 1 & 0 & 2 \\ 0 & 2 & 4 \\ 0 & 0 & 1 \end{bmatrix} \begin{bmatrix} 1 & 0 & -2 \\ 0 & \frac{1}{2} & -2 \\ 0 & 0 & 1 \end{bmatrix} = \begin{bmatrix} 1 + 0 + 0 & 0 + 0 + 0 & -2 + 0 + 2 \\ 0 + 0 + 0 & 0 + 1 + 0 & 0 - 4 + 4 \\ 0 + 0 + 0 & 0 + 0 + 0 & 0 + 0 + 1 \end{bmatrix} = \begin{bmatrix} 1 & 0 & 0 \\ 0 & 1 & 0 \\ 0 & 0 & 1 \end{bmatrix}$.

$BA = \begin{bmatrix} 1 & 0 & -2 \\ 0 & \frac{1}{2} & -2 \\ 0 & 0 & 1 \end{bmatrix} \begin{bmatrix} 1 & 0 & 2 \\ 0 & 2 & 4 \\ 0 & 0 & 1 \end{bmatrix} = \begin{bmatrix} 1 + 0 + 0 & 0 + 0 + 0 & 2 + 0 - 2 \\ 0 + 0 + 0 & 0 + 1 + 0 & 0 + 2 - 2 \\ 0 + 0 + 0 & 0 + 0 + 0 & 0 + 0 + 1 \end{bmatrix} = \begin{bmatrix} 1 & 0 & 0 \\ 0 & 1 & 0 \\ 0 & 0 & 1 \end{bmatrix}$.

39. $\det A = 30 - 30 = 0$. Since the determinant is equal to 0, A^{-1} does not exist.

40. Using $A^{-1} = \frac{1}{\det A} \begin{bmatrix} d & -b \\ -c & a \end{bmatrix}$ to find the inverse yields: $A^{-1} = \begin{bmatrix} -4 & 2 \\ 0 & 3 \end{bmatrix}^{-1} = \frac{1}{-12} \begin{bmatrix} 3 & -2 \\ 0 & -4 \end{bmatrix} = \begin{bmatrix} -\frac{1}{4} & \frac{1}{6} \\ 0 & \frac{1}{3} \end{bmatrix}$.

41. Using $A^{-1} = \frac{1}{\det A} \begin{bmatrix} d & -b \\ -c & a \end{bmatrix}$ to find the inverse yields: $A^{-1} = \begin{bmatrix} 2 & 0 \\ -1 & 5 \end{bmatrix}^{-1} = \frac{1}{10} \begin{bmatrix} 5 & 0 \\ 1 & 2 \end{bmatrix} = \begin{bmatrix} \frac{1}{2} & 0 \\ \frac{1}{10} & \frac{1}{5} \end{bmatrix}$.

42. Solve using the calculator, $A^{-1} = \begin{bmatrix} \frac{1}{4} & \frac{1}{2} & \frac{1}{2} \\ \frac{1}{4} & -\frac{1}{2} & \frac{1}{2} \\ \frac{1}{8} & -\frac{1}{4} & -\frac{1}{4} \end{bmatrix} = \begin{bmatrix} .25 & .5 & .5 \\ .25 & -.5 & .5 \\ .125 & -.25 & -.25 \end{bmatrix}$.

43. Solve using the calculator, $A^{-1} = \begin{bmatrix} \frac{2}{3} & 0 & -\frac{1}{3} \\ \frac{1}{3} & 0 & -\frac{2}{3} \\ -\frac{2}{3} & 1 & \frac{1}{3} \end{bmatrix}$.

44. Solve using the calculator, the determinant of $A = 0$, therefore there is no inverse.

45. Using the matrix inverse method, put the system into the proper matrix form: $\begin{bmatrix} 1 & 1 \\ 2 & 3 \end{bmatrix} \begin{bmatrix} x \\ y \end{bmatrix} = \begin{bmatrix} 4 \\ 10 \end{bmatrix} \Rightarrow$

$\begin{bmatrix} x \\ y \end{bmatrix} = \begin{bmatrix} 1 & 1 \\ 2 & 3 \end{bmatrix}^{-1} \begin{bmatrix} 4 \\ 10 \end{bmatrix} \Rightarrow \begin{bmatrix} 3 & -1 \\ -2 & 1 \end{bmatrix} \begin{bmatrix} 4 \\ 10 \end{bmatrix} \Rightarrow \begin{bmatrix} 12 - 10 \\ -8 + 10 \end{bmatrix} = \begin{bmatrix} 2 \\ 2 \end{bmatrix}$. The solution is: $\{(2, 2)\}$.

46. Using the matrix inverse method, put the system into the proper matrix form: $\begin{bmatrix} 5 & -3 \\ 2 & 7 \end{bmatrix} \begin{bmatrix} x \\ y \end{bmatrix} = \begin{bmatrix} -2 \\ -9 \end{bmatrix} \Rightarrow$

$\begin{bmatrix} x \\ y \end{bmatrix} = \begin{bmatrix} 5 & -3 \\ 2 & 7 \end{bmatrix}^{-1} \begin{bmatrix} -2 \\ -9 \end{bmatrix} \Rightarrow \begin{bmatrix} \frac{7}{41} & \frac{3}{41} \\ -\frac{2}{41} & \frac{5}{41} \end{bmatrix} \begin{bmatrix} -2 \\ -9 \end{bmatrix} \Rightarrow \begin{bmatrix} -\frac{14}{41} - \frac{27}{41} \\ \frac{4}{41} - \frac{45}{41} \end{bmatrix} = \begin{bmatrix} -1 \\ -1 \end{bmatrix}$. The solution is: $\{(-1, -1)\}$.

47. Using the matrix inverse method, put the system into the proper matrix form: $\begin{bmatrix} 2 & 1 \\ 3 & -2 \end{bmatrix} \begin{bmatrix} x \\ y \end{bmatrix} = \begin{bmatrix} 5 \\ 4 \end{bmatrix} \Rightarrow$

$\begin{bmatrix} x \\ y \end{bmatrix} = \begin{bmatrix} 2 & 1 \\ 3 & -2 \end{bmatrix}^{-1} \begin{bmatrix} 5 \\ 4 \end{bmatrix} \Rightarrow \begin{bmatrix} \frac{2}{7} & \frac{1}{7} \\ \frac{3}{7} & -\frac{2}{7} \end{bmatrix} \begin{bmatrix} 5 \\ 4 \end{bmatrix} \Rightarrow \begin{bmatrix} \frac{10}{7} + \frac{4}{7} \\ \frac{15}{7} - \frac{8}{7} \end{bmatrix} = \begin{bmatrix} 2 \\ 1 \end{bmatrix}$. The solution is: $\{(2, 1)\}$.

48. Using the matrix inverse method, put the system into the proper matrix form: $\begin{bmatrix} 1 & -2 \\ 3 & 1 \end{bmatrix} \begin{bmatrix} x \\ y \end{bmatrix} = \begin{bmatrix} 7 \\ 7 \end{bmatrix} \Rightarrow$

$\begin{bmatrix} x \\ y \end{bmatrix} = \begin{bmatrix} 1 & -2 \\ 3 & 1 \end{bmatrix}^{-1} \begin{bmatrix} 7 \\ 7 \end{bmatrix} \Rightarrow \begin{bmatrix} \frac{1}{7} & \frac{2}{7} \\ -\frac{3}{7} & \frac{1}{7} \end{bmatrix} \begin{bmatrix} 7 \\ 7 \end{bmatrix} \Rightarrow \begin{bmatrix} \frac{7}{7} + \frac{14}{7} \\ -\frac{21}{7} + \frac{7}{7} \end{bmatrix} = \begin{bmatrix} 3 \\ -2 \end{bmatrix}$. The solution is: $\{(3, -2)\}$.

49. Solve on the calculator using the matrix inverse method:

$\begin{bmatrix} x \\ y \\ z \end{bmatrix} = \begin{bmatrix} 1 & 2 & 0 \\ 0 & 3 & -1 \\ 1 & 2 & -1 \end{bmatrix}^{-1} \cdot \begin{bmatrix} -1 \\ -5 \\ -3 \end{bmatrix} = \begin{bmatrix} 1 \\ -1 \\ 2 \end{bmatrix}$, therefore the solution is: $\{(1, -1, 2)\}$.

50. The determinant of the coefficient matrix is: $\det \begin{bmatrix} 3 & -2 & 4 \\ 4 & 1 & -5 \\ -6 & 4 & -8 \end{bmatrix} = 0$, therefore infinitely many solutions

or no solutions. Now use Row Echelon Method on the augmented matrix: $\begin{bmatrix} 3 & -2 & 4 & | & 1 \\ 4 & 1 & -5 & | & 2 \\ -6 & 4 & -8 & | & -2 \end{bmatrix} \Rightarrow$

$2R_1 + R_3 \rightarrow \begin{bmatrix} 3 & -2 & 4 & | & 1 \\ 4 & 1 & -5 & | & 2 \\ 0 & 0 & 0 & | & 0 \end{bmatrix}$. Since $0 = 0$, there is ∞ solutions. Continue to use Row Echelon Method

to find these solutions: $R_1 + 2R_2 \rightarrow \begin{bmatrix} 3 & -2 & 4 & | & 1 \\ 11 & 0 & -6 & | & 5 \\ 0 & 0 & 0 & | & 0 \end{bmatrix}$. Now solve for R_2 for x: $11x - 6z = 5 \Rightarrow$

$11x = 6z + 5 \Rightarrow x = \dfrac{6z + 5}{11}$. Now substitute $x = \dfrac{6z + 5}{11}$ into R_1 and solve for y :

$3\left(\dfrac{6z + 5}{11}\right) - 2y + 4z = 1 \Rightarrow 18z + 15 - 22y + 44z = 11 \Rightarrow -22y = -62z - 4 \Rightarrow$

$y = \dfrac{62z + 4}{22} = \dfrac{31z + 2}{11}$. The solution is: $\left(\dfrac{6z + 5}{11}, \dfrac{31z + 2}{11}, z\right)$. Other forms are possible.

51. Solve on the calculator using the matrix inverse method:

$$\begin{bmatrix} x \\ y \\ z \end{bmatrix} = \begin{bmatrix} 1 & 1 & 1 \\ 2 & -1 & 0 \\ 0 & 3 & 1 \end{bmatrix}^{-1} \cdot \begin{bmatrix} 1 \\ -2 \\ 2 \end{bmatrix} = \begin{bmatrix} -1 \\ 0 \\ 2 \end{bmatrix}, \text{ therefore the solution is: } \{(-1, 0, 2)\}.$$

52. Solve on the calculator using the matrix inverse method:

$$\begin{bmatrix} x \\ y \\ z \end{bmatrix} = \begin{bmatrix} 1 & 0 & 0 \\ 0 & 1 & 1 \\ 2 & 0 & -3 \end{bmatrix}^{-1} \cdot \begin{bmatrix} -3 \\ 6 \\ -9 \end{bmatrix} = \begin{bmatrix} -3 \\ 5 \\ 1 \end{bmatrix}, \text{ therefore the solution is: } \{(-3, 5, 1)\}.$$

53. $\det \begin{bmatrix} -1 & 8 \\ 2 & 9 \end{bmatrix} = -9 - 16 = -25$

54. $\det \begin{bmatrix} -2 & 4 \\ 0 & 3 \end{bmatrix} = -6 - 0 = -6$

55. Evaluate, expand by the second column. Therefore, $\det = (-)(a_{12})(M_{12}) + (a_{22})(M_{22}) - (a_{32})(M_{32})$:

$$-4\left(\det \begin{bmatrix} 3 & 2 \\ -1 & 3 \end{bmatrix}\right) + 0 - 0 = -4(9 + 2) = -4(11) = -44$$

56. Evaluate, expand by the second column. Therefore, $\det = (-)(a_{12})(M_{12}) + (a_{22})(M_{22}) - (a_{32})(M_{32})$:

$$-2\left(\det \begin{bmatrix} 4 & 3 \\ 5 & 2 \end{bmatrix}\right) + 0 - (-1)\left(\det \begin{bmatrix} -1 & 3 \\ 4 & 3 \end{bmatrix}\right) = \quad -2(8 - 15) + 1(-3 - 12) = -2(-7) + 1(-15) =$$

$14 - 15 = -1.$

57. If $\det \begin{bmatrix} -3 & 2 \\ 1 & x \end{bmatrix} = 5$, then $-3x - 2 = 5 \Rightarrow -3x = 7 \Rightarrow x = -\dfrac{7}{3}$. The solution set is: $\left\{-\dfrac{7}{3}\right\}$.

58. If $\det \begin{bmatrix} 3x & 7 \\ -x & 4 \end{bmatrix} = 8$, then $12x + 7x = 8 \Rightarrow 19x = 8 \Rightarrow x = \dfrac{8}{19}$. The solution set is: $\left\{\dfrac{8}{19}\right\}$.

59. Evaluate, expand by the third column. Therefore, $\det = (a_{13})(M_{13}) - (a_{23})(M_{23}) + (a_{33})(M_{33})$:

$$0 - (-1)\left(\det \begin{bmatrix} 2 & 5 \\ 0 & 2 \end{bmatrix}\right) + 0 = 4 \Rightarrow 4 = 4. \text{ Since this is always true, all real numbers can be input for } x.$$

60. Evaluate, expand by the first row. Therefore, $\det = (a_{11})(M_{11}) - (a_{12})(M_{12}) + (a_{13})(M_{13})$:

$$6x\left(\det \begin{bmatrix} 5 & 3 \\ 2 & -1 \end{bmatrix}\right) - 2\left(\det \begin{bmatrix} 1 & 3 \\ x & -1 \end{bmatrix}\right) + 0 = 2x \Rightarrow \quad 6x(-5 - 6) - 2(-1 - 3x) = 2x \Rightarrow$$

$$-66x + 2 + 6x = 2x \Rightarrow -60x + 2 = 2x \Rightarrow -62x = -2 \Rightarrow x = \dfrac{-2}{-62}. \text{ The solution is: } \left\{\dfrac{1}{31}\right\}.$$

61. (a) $D = \det \begin{bmatrix} 3 & -1 \\ 2 & 1 \end{bmatrix} = 5$

    (b) $D_x = \det \begin{bmatrix} 28 & -1 \\ 2 & 1 \end{bmatrix} = 30$

    (c) $D_y = \det \begin{bmatrix} 3 & 28 \\ 2 & 2 \end{bmatrix} = -50$

    (d) $x = \dfrac{D_x}{D} = \dfrac{30}{5} = 6; y = \dfrac{D_y}{D} = \dfrac{-50}{5} = -10.$ The solution is: $\{(6, -10)\}$.

62. (a) A = the coefficient matrix: $A = \begin{bmatrix} 3 & -1 \\ 2 & 1 \end{bmatrix}$.

    (b) B = the answer matrix: $B = \begin{bmatrix} 28 \\ 2 \end{bmatrix}$.

    (c) To solve for x and y, multiply A^{-1} by B: $\begin{bmatrix} x \\ y \end{bmatrix} = A^{-1}B = \begin{bmatrix} 6 \\ -10 \end{bmatrix}$, therefore the solution is: $(6, -10)$.

63. If $D = 0$, there would be division by 0, which is undefined. The system will have no solutions or infinitely many solutions.

64. Find the determinants: $D = \det \begin{bmatrix} 3 & 1 \\ 5 & 4 \end{bmatrix} = 7$; $D_x = \det \begin{bmatrix} -1 & 1 \\ 10 & 4 \end{bmatrix} = -14$; and $D_y = \det \begin{bmatrix} 3 & -1 \\ 5 & 10 \end{bmatrix} = 35$.

    Then $x = \dfrac{D_x}{D} = \dfrac{-14}{7} = -2$ and $y = \dfrac{D_y}{D} = \dfrac{35}{7} = 5$. The solution is: $\{(-2, 5)\}$.

65. Find the determinants: $D = \det \begin{bmatrix} 3 & 7 \\ 5 & -1 \end{bmatrix} = -38$; $D_x = \det \begin{bmatrix} 2 & 7 \\ -22 & -1 \end{bmatrix} = 152$; and

    $D_y = \det \begin{bmatrix} 3 & 2 \\ 5 & -22 \end{bmatrix} = -76$. Then $x = \dfrac{D_x}{D} = \dfrac{152}{-38} = -4$ and $y = \dfrac{D_y}{D} = \dfrac{-76}{-38} = 2$. The solution is: $\{(-4, 2)\}$.

66. Find the determinants: $D = \det \begin{bmatrix} 2 & -5 \\ 3 & 4 \end{bmatrix} = 23$; $D_x = \det \begin{bmatrix} 8 & -5 \\ 10 & 4 \end{bmatrix} = 82$; and $D_y = \det \begin{bmatrix} 2 & 8 \\ 3 & 10 \end{bmatrix} = -4$.

    Then $x = \dfrac{D_x}{D} = \dfrac{82}{23}$ and $y = \dfrac{D_y}{D} = \dfrac{-4}{23}$. The solution is: $\left\{ \left(\dfrac{82}{23}, -\dfrac{4}{23} \right) \right\}$.

67. Using your calculator, find the following determinants:

$$D = \det \begin{bmatrix} 3 & 2 & 1 \\ 4 & -1 & 3 \\ 1 & 3 & -1 \end{bmatrix} = 3;\ D_x = \det \begin{bmatrix} 2 & 2 & 1 \\ -16 & -1 & 3 \\ 12 & 3 & -1 \end{bmatrix} = -12;\ D_y = \det \begin{bmatrix} 3 & 2 & 1 \\ 4 & -16 & 3 \\ 1 & 12 & -1 \end{bmatrix} = 18;$$

$$D_z = \det \begin{bmatrix} 3 & 2 & 2 \\ 4 & -1 & -16 \\ 1 & 3 & 12 \end{bmatrix} = 6$$

    Then $x = \dfrac{D_x}{D} = \dfrac{-12}{3} = -4$; $y = \dfrac{D_y}{D} = \dfrac{18}{3} = 6$; and $z = \dfrac{D_z}{D} = \dfrac{6}{3} = 2$. The solution is: $\{(-4, 6, 2)\}$.

68. The determinant of the coefficient matrix is: $\det \begin{bmatrix} 5 & -2 & -1 \\ -5 & 2 & 1 \\ 1 & -4 & -2 \end{bmatrix} = 0$, therefore infinitely many or no solutions.

    Now use Row Echelon Method on the augmented matrix: $\begin{matrix} R_3 \leftrightarrow R_1 \\ R_2 \leftrightarrow R_3 \end{matrix} \Bigg\} \rightarrow \left[\begin{array}{ccc|c} 1 & -4 & -2 & 0 \\ 5 & -2 & -1 & 8 \\ -5 & 2 & 1 & -8 \end{array} \right] \Rightarrow$

    Since $R_2 + R_3$ produces $0 = 0$, there are infinitely many solutions and they are dependent solutions. Continue to use

    Row Echelon Method to find these solutions: $R_1 + 2R_3 \rightarrow \left[\begin{array}{ccc|c} 1 & -4 & -2 & 0 \\ 5 & -2 & -1 & 8 \\ -9 & 0 & 0 & -16 \end{array} \right]$. Solve for

    R_3 for x: $-9x = -16 \Rightarrow x = \dfrac{16}{9}$. Now substitute $x = \dfrac{16}{9}$ into R_1 and solve for y: $\dfrac{16}{9} - 4y - 2z = 0 \Rightarrow$

    $16 - 36y - 18z = 0 \Rightarrow -36y = 18z - 16 \Rightarrow y = \dfrac{18z - 16}{-36} = \dfrac{8 - 9z}{18}$. The solution is: $\left\{ \left(\dfrac{16}{9}, \dfrac{8 - 9z}{18}, z \right) \right\}$.

    Other forms are possible.

69. Using your calculator, find the following determinants:

$$D = \det \begin{bmatrix} -1 & 3 & -4 \\ 2 & 4 & 1 \\ 3 & 0 & -1 \end{bmatrix} = 67; \; D_x = \det \begin{bmatrix} 2 & 3 & -4 \\ 3 & 4 & 1 \\ 9 & 0 & -1 \end{bmatrix} = 172; \; D_y = \det \begin{bmatrix} -1 & 2 & -4 \\ 2 & 3 & 1 \\ 3 & 9 & -1 \end{bmatrix} = -14;$$

$$D_z = \det \begin{bmatrix} -1 & 3 & 2 \\ 2 & 4 & 3 \\ 3 & 0 & 9 \end{bmatrix} = -87$$

Then $x = \dfrac{D_x}{D} = \dfrac{172}{67}; y = \dfrac{D_y}{D} = \dfrac{-14}{67};$ and $z = \dfrac{D_z}{D} = \dfrac{-87}{67}$. The solution is: $\left\{ \left(\dfrac{172}{67}, -\dfrac{14}{67}, -\dfrac{87}{67} \right) \right\}$.

70. Let x = the amount of rice in cups and y = the amount of soybeans in cups. Then from the information the

system of equations is: $15x + 22.5y = 9.5$ [equation 1] and $810x + 270y = 324$ [equation 2].

Multiply {1} by -12 and add {2} to eliminate the y:

$$-180x - 270y = -114$$
$$\underline{810x + 270y = 324}$$
$$630x = 210 \Rightarrow x = \frac{210}{630} = \frac{1}{3}.$$ Now substitute $x = \dfrac{1}{3}$ into [equation 1] and solve for y:

$$15\left(\frac{1}{3}\right) + 22.5y = 9.5 \Rightarrow 5 + 22.5y = 9.5 \Rightarrow 22.5y = 4.5 \Rightarrow y = .20 \text{ or } \frac{1}{5}.$$

The meal should include $\dfrac{1}{3}$ cup of rice and $\dfrac{1}{5}$ cup of soybeans.

71. Let x = the number of CDs and y = the number of diskettes. Then from the information the

system of equations is: $x + y = 100$ [equation 1] and $.40x + .30y = 38.00$ [equation 2].

Multiply {1} by -30 and add {2} multiplied by 100 to eliminate the y:

$$-30x - 30y = -3000$$
$$\underline{40x + 30y = 3800}$$
$$10x = 800 \Rightarrow x = 80.$$ Now substitute $x = 80$ into [equation 1] and solve for y:

$80 + y = 100 \Rightarrow y = 20$. They should send 80 CDs and 20 diskettes.

72. Let x = the number pounds of \$4.60 tea, y = the number of pounds of \$5.75 tea, and z = the number of

pounds of \$6.50 tea.. Then from the information the system of equations is: $x + y + z = 20$ [equation 1] ,

$4.6x + 5.75y + 6.5z = 20(5.25) = 105$, and $x = y + z \Rightarrow x - y - z = 0$. Now create the coefficient matrix

and solve on the calculator using the matrix inverse method:

$$A = \begin{bmatrix} 1 & 1 & 1 \\ 4.6 & 5.75 & 6.5 \\ 1 & -1 & -1 \end{bmatrix}, \text{ therefore: } \begin{bmatrix} x \\ y \\ z \end{bmatrix} = \begin{bmatrix} 1 & 1 & 1 \\ 4.6 & 5.75 & 6.5 \\ 1 & -1 & -1 \end{bmatrix}^{-1} \cdot \begin{bmatrix} 20 \\ 105 \\ 0 \end{bmatrix} = \begin{bmatrix} 10 \\ 8 \\ 2 \end{bmatrix}.$$

They should use 10 pounds of \$4.60 tea, 8 pounds of \$5.75 tea, and 2 pounds of \$6.50 tea.

73. Let $x =$ the amount of 5% solution (ml), $y =$ the amount of 15% solution (ml), and $z =$ the amount of 10% solution (ml). Then from the information the system of equations is: $x + y + z = 20$ [equation 1],

$.05x + .15y + .10z = .08(20) = 1.6$, and $x = y + z + 2 \Rightarrow x - y - z = 2$. Now create the coefficient matrix and solve on the calculator using the matrix inverse method:

$$A = \begin{bmatrix} 1 & 1 & 1 \\ 5 & 15 & 10 \\ 1 & -1 & -1 \end{bmatrix}, \text{ therefore: } \begin{bmatrix} x \\ y \\ z \end{bmatrix} = \begin{bmatrix} 1 & 1 & 1 \\ 5 & 15 & 10 \\ 1 & -1 & -1 \end{bmatrix}^{-1} \cdot \begin{bmatrix} 20 \\ 160 \\ 2 \end{bmatrix} = \begin{bmatrix} 11 \\ 3 \\ 6 \end{bmatrix}.$$

They should use 11 ml of 5% solution, 3 ml of 15% solution, and 6 ml of 10% solution.

74. (a) Using $P = a + bA + cW$ and the data from the table yields:

$a + 39b + 142c = 113$

$a + 53b + 181c = 138$

$a + 65b + 191c = 152$

Now using the capabilities of the calculator, the Reduced Row Echelon Method of:

$$\begin{bmatrix} 1 & 39 & 142 & | & 113 \\ 1 & 53 & 181 & | & 138 \\ 1 & 65 & 191 & | & 152 \end{bmatrix} = \begin{bmatrix} 1 & 0 & 0 & | & 32.780488 \\ 0 & 1 & 0 & | & .9024390 \\ 0 & 0 & 1 & | & .3170732 \end{bmatrix}.$$

Therefore the equation is: $P \approx 32.78 + .9024A + .3171W$.

(b) Using $A = 55$ and $W = 175$ yields: $P \approx 32.78 + .9024(55) + .3171(175) \Rightarrow P \approx 138$.

75. Using $f(x) = ax^2 + bx + c$ and the data from the graph yields:

From the point $(-6, 4)$: $36a - 6b + c = 4$.

From the point $(-4, -2)$: $16a - 4b + c = -2$

From the point $(2, 4)$: $4a + 2b + c = 4$

Now using your calculator, find the following determinants:

$$D = \det \begin{bmatrix} 36 & -6 & 1 \\ 16 & -4 & 1 \\ 4 & 2 & 1 \end{bmatrix} = -96; \; D_a = \det \begin{bmatrix} 4 & -6 & 1 \\ -2 & -4 & 1 \\ 4 & 2 & 1 \end{bmatrix} = -48; \; D_b = \det \begin{bmatrix} 36 & 4 & 1 \\ 16 & -2 & 1 \\ 4 & 4 & 1 \end{bmatrix} = -192;$$

$$D_c = \det \begin{bmatrix} 36 & -6 & 4 \\ 16 & -4 & -2 \\ 4 & 2 & 4 \end{bmatrix} = 192$$

Then $a = \dfrac{D_a}{D} = \dfrac{-48}{-96} = \dfrac{1}{2}; \; b = \dfrac{D_b}{D} = \dfrac{-192}{-96} = 2;$ and $c = \dfrac{D_c}{D} = \dfrac{192}{-96} = -2$.

The equation is: $P(x) = \dfrac{1}{2}x^2 + 2x - 2$.

76. Using the equation for a polynomial of degree 3 and the given points yields:

$(-2, 1)$: $-8a + 4b - 2c + d = 1$

$(-1, 6)$: $-a + b - c + d = 6$

$(2, 9)$: $8a + 4b + 2c + d = 9$

$(3, 26)$: $27a + 9b + 3c + d = 26$

The equations can be represented by: $\begin{bmatrix} -8 & 4 & -2 & 1 & | & 1 \\ -1 & 1 & -1 & 1 & | & 6 \\ 8 & 4 & 2 & 1 & | & 9 \\ 27 & 9 & 3 & 1 & | & 26 \end{bmatrix}$.

Using the capabilities of the calculator, the Reduced Row Echelon Method produces the solutions:

$a = 1, b = 0, c = -2, d = 5$. The equation is: $P(x) = x^3 - 2x + 5$.

77. See Figure 77.

78. See Figure 78.

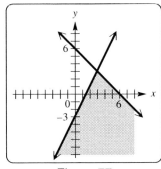

Figure 77

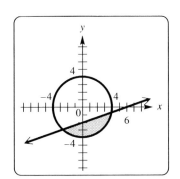

Figure 78

79. Graphing $x \geq 0, y \geq 0, 3x + 2y \leq 12$, and $5x + y \geq 5$ in the same window yields four vertices.

Using the given expression and the ordered pairs of the vertices yields the following solutions:

$(0, 5)$: $2(0) + 4(5) = 20$

$(0, 6)$: $2(0) + 4(6) = 24$

$(1, 0)$: $2(1) + 4(0) = 2$

$(4, 0)$: $2(4) + 4(0) = 8$

Therefore, the maximum value is: 24 at $(0, 6)$.

80. Graphing $x \geq 0, y \geq 0, x + y \leq 50, 2x + y \geq 20,$ and $x + 2y \geq 30$ in the same window yields five vertices.

    Using the given expression and the ordered pairs of the vertices yields the following solutions:

    $(0, 50)$: $4(0) + 2(50) = 100$

    $(0, 20)$: $4(0) + 2(20) = 40$

    $\left(\dfrac{10}{3}, \dfrac{40}{3}\right)$: $4\left(\dfrac{10}{3}\right) + 2\left(\dfrac{40}{3}\right) = 40$

    $(30, 0)$: $4(30) + 2(0) = 120$

    $(50, 0)$: $4(50) + 2(0) = 200$ Therefore, the minimum value is: 40 at $(0, 20)$ and $\left(\dfrac{10}{3}, \dfrac{40}{3}\right)$.

    Note: values on the line $2x + y = 20$ also give minimums of 40 between the two points.

81. Let $x = $ the number radios produced daily and let $y = $ the number of DVD players produced daily.

    The objective function to find the maximum profit is: $3x + 2y$.

    The constraints are: $5 \leq x \leq 25$ (radio production restrictions)

    $\qquad\qquad\qquad\quad 0 \leq y \leq 30$ (DVD player maximum production)

    $\qquad\qquad\qquad\qquad x \leq y$ (radio production less than or equal to DVD player production)

    Graphing the constraints on the calculator yields four vertices. Inputting these ordered pairs into the objective function yields: $(5, 30)$: 1125; $(5, 5)$: 250; $(25, 30)$: 1425; and $(25, 25)$: 1250.

    The maximum profit is \$1425, this happens when 25 radios and 30 DVD players are manufactured.

82. Factoring $\dfrac{5x - 2}{x^2 - 4}$ results in $\dfrac{5x - 2}{(x - 2)(x + 2)}$.

    Multiply $\dfrac{5x - 2}{(x - 2)(x + 2)} = \dfrac{A}{x - 2} + \dfrac{B}{x + 2}$ by $(x - 2)(x + 2) \Rightarrow 5x - 2 = A(x + 2) + B(x - 2)$.

    Let $x = -2 \Rightarrow -12 = B(-4) \Rightarrow B = 3$. Let $x = 2 \Rightarrow 8 = A(4) \Rightarrow A = 2$.

    The expression can be written $\dfrac{2}{x - 2} + \dfrac{3}{x + 2}$.

83. Factoring $\dfrac{x + 2}{x^3 + 2x^2 + x}$ results in $\dfrac{x + 2}{x(x + 1)^2}$.

    Multiply $\dfrac{x + 2}{x^3 - x^2 + x} = \dfrac{A}{x} + \dfrac{B}{x + 1} + \dfrac{C}{(x + 1)^2}$ by $x(x + 1)^2 \Rightarrow x + 2 = A(x + 1)^2 + Bx(x + 1) + Cx$.

    Let $x = 0 \Rightarrow 2 = A(1) \Rightarrow A = 2$. Let $x = -1 \Rightarrow 1 = C(-1) \Rightarrow C = -1$. Let $x = 1, A = 2, C = -1 \Rightarrow$

    $3 = 2(4) + B(1)(2) - 1 \Rightarrow 3 = 8 + 2B - 1 \Rightarrow B = -2$.

    The expression can be written $\dfrac{2}{x} + \dfrac{-2}{x + 1} + \dfrac{-1}{(x + 1)^2}$.

84. Factoring $\dfrac{x + 2}{x^3 - x^2 + 4x}$ results in $\dfrac{x + 2}{x(x^2 - x + 4)}$.

    Multiply $\dfrac{x + 2}{x(x^2 - x + 4)} = \dfrac{A}{x} + \dfrac{Bx + C}{x^2 - x + 4}$ by $x(x^2 - x + 4) \Rightarrow$

    $x + 2 = A(x^2 - x + 4) + (Bx + C)(x) \Rightarrow x + 2 = (A + B)x^2 + (C - A)x + 4A$. By equating coefficients

    $A + B = 0, C - A = 1$ and $4A = 2 \Rightarrow A = \dfrac{1}{2}$. Since $A + B = 0$ and $A = \dfrac{1}{2}, B = -\dfrac{1}{2}$.

    Since $C - A = 1$ and $A = \dfrac{1}{2}, C = \dfrac{3}{2}$. The expression can be written $\dfrac{\frac{1}{2}}{x} + \dfrac{-\frac{1}{2}x + \frac{3}{2}}{x^2 - x + 4} = \dfrac{1}{2x} + \dfrac{-x + 3}{2(x^2 - x + 4)}$.

Chapter 7 Test

1. (a) The first is the equation of a hyperbola; the second is the equation of a line.

    (b) A hyperbola and a line could intersect in 0, 1, or 2 points.

    (c) First solve the second equation for y: $3x + y = 1 \Rightarrow y = 1 - 3x$. Now substitute $y = 1 - 3x$ into

      equation 1: $x^2 - 4(1 - 3x)^2 = -15 \Rightarrow x^2 - 4(1 - 6x + 9x^2) = -15 \Rightarrow$

      $x^2 - 4 + 24x - 36x^2 = -15 \Rightarrow -35x^2 + 24x + 11 = 0 \Rightarrow (35x + 11)(-x + 1) = 0 \Rightarrow x = -\dfrac{11}{35}, 1$.

      Now substitute these values into $y = 1 - 3x$: $y = 1 - 3(1) \Rightarrow y = -2$ and $y = 1 - 3\left(-\dfrac{11}{35}\right) \Rightarrow y = \dfrac{68}{35}$.

      The solution is: $\left(-\dfrac{11}{35}, \dfrac{68}{35}\right), (1, -2)$.

    (d) Graph on a calculator shows two points of intersection. See Figure 1.

      $[-10, 10]$ by $[-10, 10]$
      $Xscl = 1$ $Yscl = 1$

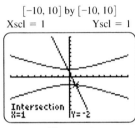

      Figure 1

2. Using the Row Echelon Method, the given system of equations yields the matrix: $\begin{bmatrix} 2 & 1 & 1 & 3 \\ 1 & 2 & -1 & 3 \\ 3 & -1 & 1 & 5 \end{bmatrix}$, which

    by $(R_1 \leftrightarrow R_2) \rightarrow \begin{bmatrix} 1 & 2 & -1 & 3 \\ 2 & 1 & 1 & 3 \\ 3 & -1 & 1 & 5 \end{bmatrix} \Rightarrow (-2R_1 + R_2) \text{ and } (-3R_1 + R_3) \rightarrow \begin{bmatrix} 1 & 2 & -1 & 3 \\ 0 & -3 & 3 & -3 \\ 0 & -7 & 4 & -4 \end{bmatrix} \Rightarrow$

    $(7R_2 - 3R_3) \rightarrow \begin{bmatrix} 1 & 2 & -1 & 3 \\ 0 & -3 & 3 & -3 \\ 0 & 0 & 9 & -9 \end{bmatrix}$. From this matrix, we have the resulting equation: $9z = -9 \Rightarrow z = -1$.

    Now use back-substitution: substituting $z = -1$ into the resulting R_2 yields:

    $-3y + 3(-1) = -3 \Rightarrow -3y = 0 \Rightarrow y = 0$; and finally substituting $y = 0$ and $z = -1$ into the resulting R_1

    yields: $x + 2(0) - (-1) = 3 \Rightarrow x = 2$. The solution is: $\{(2, 0, -1)\}$.

3. (a) $3\begin{bmatrix} 2 & 3 \\ 1 & -4 \\ 5 & 9 \end{bmatrix} - \begin{bmatrix} -2 & 6 \\ 3 & -1 \\ 0 & 8 \end{bmatrix} = \begin{bmatrix} 6 & 9 \\ 3 & -12 \\ 15 & 27 \end{bmatrix} - \begin{bmatrix} -2 & 6 \\ 3 & -1 \\ 0 & 8 \end{bmatrix} = \begin{bmatrix} 6+2 & 9-6 \\ 3-3 & -12+1 \\ 15-0 & 27-8 \end{bmatrix} = \begin{bmatrix} 8 & 3 \\ 0 & -11 \\ 15 & 19 \end{bmatrix}$.

    (b) Cannot add $(1 \times 2) + (1 \times 2) + (2 \times 2)$, therefore the solution is: $\varnothing$.

    (c) $\begin{bmatrix} 2 & 1 & -3 \\ 4 & 0 & 5 \end{bmatrix} \begin{bmatrix} 1 & 3 \\ 2 & 4 \\ 3 & -2 \end{bmatrix} = \begin{bmatrix} 2+2-9 & 6+4+6 \\ 4+0+15 & 12+0-10 \end{bmatrix} = \begin{bmatrix} -5 & 16 \\ 19 & 2 \end{bmatrix}$.

4. (a) AB can be found; it will be $n \times n$.

(b) BA can be found; it will be $n \times n$.

(c) $AB = BA$ is not necessarily true, since matrix multiplication is not commutative.

(d) Since the number of rows of the first matrix is not equal to the number of columns of the second matrix,

AC cannot be found.

Since the number of rows of the first matrix is equal to the number of columns of the second matrix,

CA can be found; it will be $m \times n$.

5. (a) $\det \begin{bmatrix} 4 & 9 \\ -5 & -11 \end{bmatrix} = -44 + 45 = 1$

(b) Evaluate, expand by the second column. Therefore, $\det = (-)(a_{12})(M_{12}) + (a_{22})(M_{22}) - (a_{32})(M_{32})$:

$0\left(\det \begin{bmatrix} -1 & 9 \\ 12 & -3 \end{bmatrix} \right) + 7\left(\det \begin{bmatrix} 2 & 8 \\ 12 & -3 \end{bmatrix} \right) - 5\left(\det \begin{bmatrix} 2 & 8 \\ -1 & 9 \end{bmatrix} \right) = 0 + 7(-6 - 96) - 5(18 + 8) \Rightarrow$

$7(-102) - 5(26) = -844.$

6. Find the determinants: $D = \det \begin{bmatrix} 2 & -3 \\ 4 & 5 \end{bmatrix} = 22;\ D_x = \det \begin{bmatrix} -33 & -3 \\ 11 & 5 \end{bmatrix} = -132;$ and

$D_y = \det \begin{bmatrix} 2 & -33 \\ 4 & 11 \end{bmatrix} = 154.$ Then $x = \dfrac{D_x}{D} = \dfrac{-132}{22} = -6$ and $y = \dfrac{D_y}{D} = \dfrac{154}{22} = 7.$ The solution is: $\{(-6, 7)\}.$

7. (a) For the system: $A = \begin{bmatrix} 1 & 1 & -1 \\ 2 & -3 & -1 \\ 1 & 2 & 2 \end{bmatrix}$, $X = \begin{bmatrix} x \\ y \\ z \end{bmatrix}$, and $B = \begin{bmatrix} -4 \\ 5 \\ 3 \end{bmatrix}$.

(b) Solve using the calculator, $A^{-1} = \begin{bmatrix} \frac{1}{4} & \frac{1}{4} & \frac{1}{4} \\ \frac{5}{16} & -\frac{3}{16} & \frac{1}{16} \\ -\frac{7}{16} & \frac{1}{16} & \frac{5}{16} \end{bmatrix}.$

(c) $\begin{bmatrix} x \\ y \\ z \end{bmatrix} = \begin{bmatrix} 1 & 1 & -1 \\ 2 & -3 & -1 \\ 1 & 2 & 2 \end{bmatrix}^{-1} \begin{bmatrix} -4 \\ 5 \\ 3 \end{bmatrix} = \begin{bmatrix} 1 \\ -2 \\ 3 \end{bmatrix}.$ The solution is: $\{(1, -2, 3)\}.$

(d) For the new system: $A = \begin{bmatrix} .5 & 1 & 1 \\ 2 & -3 & -1 \\ 1 & 2 & 2 \end{bmatrix}$ and by the calculator, the $\det A = 0.$ Since the determinant

equals zero, there is no inverse and the matrix inverse method cannot be used.

8. Using $f(x) = ax^2 + bx + c$ and the data from the table yields:

$(24, 48.9)$: $24^2 a + 24b + c = 48.9$

$(60, 62.8)$: $60^2 a + 60b + c = 62.8$

$(96, 48.8)$: $96^2 a + 96b + c = 48.8$

Now using the capabilities of the calculator, the Reduced Row Echelon Method of:

$\begin{bmatrix} 24^2 & 24 & 1 & | & 48.9 \\ 60^2 & 60 & 1 & | & 62.8 \\ 96^2 & 96 & 1 & | & 48.8 \end{bmatrix} = \begin{bmatrix} 1 & 0 & 0 & | & -.010764 \\ 0 & 1 & 0 & | & 1.2903 \\ 0 & 0 & 1 & | & 24.133 \end{bmatrix}.$

Therefore the equation is: $f(x) = -.010764x^2 + 1.2903x + 24.133.$

Graph the data and the equation in the same window. See Figure 8.

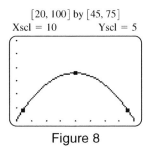

$[20, 100]$ by $[45, 75]$
Xscl = 10   Yscl = 5

Figure 8

9. B, the graph is shaded above the line, therefore: $y > 2 - x$ and outside the parabola, therefore: $y < x^2 - 5$.

10. Let $x =$ the number type X cabinets and let $y =$ the number of type Y cabinets.

    The objective function to find the maximum storage space is: $8x + 12y$.

    The constraints are: $100x + 200y \leq 1400$ (cost of cabinets)

                        $6x + 8y \leq 72$ (floor space)

                        $x \geq 0; \; y \geq 0$ (cannot have a negative number of cabinets)

    Graphing the constraints on the calculator yields four vertices. Inputting these ordered pairs into the objective function yields:

    $(0, 0)$: $8(0) + 12(0) = 0$

    $(12, 0)$: $8(12) + 12(0) = 96$

    $(8, 3)$: $8(8) + 12(3) = 100$

    (0.7): $8(0) + 12(7) = 84$

    The maximum storage is 100 cubic feet, this happens when there are 8 cabinets of type X and 3 cabinets of type Y.

11. Factoring $\dfrac{7x - 1}{x^2 - x - 6}$ results in $\dfrac{7x - 1}{(x - 3)(x + 2)}$.

    Multiply $\dfrac{7x - 1}{(x - 3)(x + 2)} = \dfrac{A}{x - 3} + \dfrac{B}{x + 2}$ by $(x - 3)(x + 2) \Rightarrow 7x - 1 = A(x + 2) + B(x - 3)$.

    Let $x = -2 \Rightarrow -15 = B(-5) \Rightarrow B = 3$. Let $x = 3 \Rightarrow 20 = A(5) \Rightarrow A = 4$.

    The expression can be written $\dfrac{4}{x - 3} + \dfrac{3}{x + 2}$.

12. The expression $\dfrac{x^2 - 11x + 6}{(x + 2)(x - 2)^2}$ is factored.

    Multiply $\dfrac{x^2 - 11x + 6}{(x + 2)(x - 2)^2} = \dfrac{A}{x + 2} + \dfrac{B}{x - 2} + \dfrac{C}{(x - 2)^2}$ by $(x + 2)(x - 2)^2 \Rightarrow$

    $x^2 - 11x + 6 = A(x - 2)^2 + B(x - 2)(x + 2) + C(x + 2)$. Let $x = -2 \Rightarrow 32 = A(16) \Rightarrow A = 2$.

    Let $x = 2$: $-12 = C(4) \Rightarrow C = -3$. Let $x = 0, A = 2$, and $C = -3$: $6 = 4(2) - 4B + 2(-3) \Rightarrow$

    $6 = 8 - 4B - 6 \Rightarrow 6 = 2 - 4B \Rightarrow 4 = -4B \Rightarrow B = -1$.

    The expression can be written $\dfrac{2}{x + 2} + \dfrac{-1}{x - 2} + \dfrac{-3}{(x - 2)^2}$.

Chapter 7 Project

1. Using, for example, 585: $(1, 5), (2, 8), (3, 5)$ and the second degree polynomial model $ax^2 + bx + c = y$ gives us the system of equations:

    $(1, 5)$: $a(1)^2 + b(1) + c = 5 \Rightarrow a^2 + b + c = 5$

    $(2, 8)$: $a(2)^2 + b(2) + c = 8 \Rightarrow 4a^2 + 2b + c = 8$

    $(3, 5)$: $a(3)^2 + b(3) + c = 5 \Rightarrow 9a^2 + 3b + c = 5$

    For this system: $A = \begin{bmatrix} 1 & 1 & 1 \\ 4 & 2 & 1 \\ 9 & 3 & 1 \end{bmatrix}$, $X = \begin{bmatrix} x \\ y \\ z \end{bmatrix}$, and $B = \begin{bmatrix} 5 \\ 8 \\ 5 \end{bmatrix}$.

    Now solve on the calculator using the matrix inverse method:

    $\begin{bmatrix} x \\ y \\ z \end{bmatrix} = \begin{bmatrix} 1 & 1 & 1 \\ 4 & 2 & 1 \\ 9 & 3 & 1 \end{bmatrix}^{-1} \begin{bmatrix} 5 \\ 8 \\ 5 \end{bmatrix} = \begin{bmatrix} -3 \\ 12 \\ -4 \end{bmatrix}$. Therefore the equation is: $-3x^2 + 12x - 4 = y$.

    Finally, follow the procedures outlined to obtain the scatterplot and the graph of the polynomial.

2. Answers will vary. Make sure students follow the steps outlined in the explanation for these activities.

3. Answers will vary. Make sure students follow the steps outlined in the explanation for these activities.

Chapter 8: Trigonometric Functions and Applications

8.1: Angles and Arcs

1. (a) $c + 30° = 90° \Rightarrow c = 60°$

    (b) $s + 30° = 180° \Rightarrow s = 150°$

2. (a) $c + 60° = 90° \Rightarrow c = 30°$

    (b) $s + 60° = 180° \Rightarrow s = 120°$

3. (a) $c + 45° = 90° \Rightarrow c = 45°$

    (b) $s + 45° = 180° \Rightarrow s = 135°$

4. (a) Since $90° = \dfrac{\pi}{2}$ radians, $c + \dfrac{\pi}{3} = \dfrac{\pi}{2} \Rightarrow c = \dfrac{3\pi}{6} - \dfrac{2\pi}{6} \Rightarrow c = \dfrac{\pi}{6}$.

    (b) Since $180° = \pi$ radians, $s + \dfrac{\pi}{3} = \pi \Rightarrow s = \dfrac{3\pi}{3} - \dfrac{\pi}{3} \Rightarrow s = \dfrac{2\pi}{3}$.

5. (a) Since $90° = \dfrac{\pi}{2}$ radians, $c + \dfrac{\pi}{4} = \dfrac{\pi}{2} \Rightarrow c = \dfrac{2\pi}{4} - \dfrac{\pi}{4} \Rightarrow c = \dfrac{\pi}{4}$.

    (b) Since $180° = \pi$ radians, $s + \dfrac{\pi}{4} = \pi \Rightarrow s = \dfrac{4\pi}{4} - \dfrac{\pi}{4} \Rightarrow s = \dfrac{3\pi}{4}$.

6. (a) Since $90° = \dfrac{\pi}{2}$ radians, $c + \dfrac{\pi}{12} = \dfrac{\pi}{2} \Rightarrow c = \dfrac{6\pi}{12} - \dfrac{\pi}{12} \Rightarrow c = \dfrac{5\pi}{12}$.

    (b) Since $180° = \pi$ radians, $s + \dfrac{\pi}{12} = \pi \Rightarrow s = \dfrac{12\pi}{12} - \dfrac{\pi}{12} \Rightarrow s = \dfrac{11\pi}{12}$.

7. (a) $c + x° = 90° \Rightarrow c = (90 - x)°$

    (b) $s + x° = 180° \Rightarrow s = (180 - x)°$

8. (a) Since $90° = \dfrac{\pi}{2}$ radians, $c + x = \dfrac{\pi}{2} \Rightarrow c = \dfrac{\pi}{2} - x$.

    (b) Since $180° = \pi$ radians, $s + x = \pi \Rightarrow s = \pi - x$.

9. Since $360° \div 12 = 30°$, every 5 minute section is $30°$. Therefore 5 o'clock equals $5 \times 30° = 150°$.

10. Since $360° \div 60 = 6°$, every 1 minute section is $6°$. Therefore 1:45 equals 20 minutes $+ \left(\dfrac{3}{4} \times 5 \text{ minutes}\right)$ or

    $23.75 \times 6° = 142.5°$.

11. $7x + 11x = 180 \Rightarrow 18x = 180 \Rightarrow x = 10$, therefore the angles are: $7(10) = 70°$ and $11(10) = 110°$.

12. $4y + 2y = 90 \Rightarrow 6y = 90 \Rightarrow y = 15$, therefore the angles are: $2(15) = 30°$ and $4(15) = 60°$.

13. $(5k + 5) + (3k + 5) = 90 \Rightarrow 8k + 10 = 90 \Rightarrow 8k = 80 \Rightarrow k = 10$, therefore the angles are:

    $5(10) + 5 = 55°$ and $3(10) + 5 = 35°$.

14. $(10m + 7) + (7m + 3) = 180 \Rightarrow 17m + 10 = 180 \Rightarrow 17m = 170 \Rightarrow m = 10$, therefore the angles are:

    $10(10) + 7 = 107°$ and $7(10) + 3 = 73°$.

15. $(6x - 4) + (8x - 12) = 180 \Rightarrow 14x - 16 = 180 \Rightarrow 14x = 196 \Rightarrow x = 14$, therefore the angles are:

    $6(14) - 4 = 80°$ and $8(14) - 12 = 100°$.

16. $(9z + 6) + (3z) = 90 \Rightarrow 12z + 6 = 90 \Rightarrow 12z = 84 \Rightarrow z = 7$, therefore the angles are:
    $9(7) + 6 = 69°$ and $3(7) = 21°$.

17. $62° \ 18' + 21° \ 41' = 83° \ 59'$

18. $75° \ 15' + 83° \ 32' = 158° \ 47'$

19. $71° \ 18' - 47° \ 29' = 70° \ 78' - 47° \ 29' = 23° \ 49'$

20. $47° \ 23' - 73° \ 48' = -73° \ 48' + 47° \ 23' = -(73° \ 48' - 47° \ 23') = -26° \ 25'$

21. $90° - 72° \ 58' \ 11'' = 89° \ 59' \ 60'' - 72° \ 58' \ 11'' = 17° \ 1' \ 49''$

22. $90° - 36° \ 18' \ 47'' = 89° \ 59' \ 60'' - 36° \ 18' \ 47'' = 53° \ 41' \ 13''$

23. $20° \ 54' = \left(20\dfrac{54}{60}\right)° = 20.9°$

24. $38° \ 42' = \left(38\dfrac{42}{60}\right)° = 38.7°$

25. $91° \ 35' \ 54'' = 91° \ 35'\dfrac{54}{60} = 91° \ 35.9' = \left(91\dfrac{35.9}{60}\right)° = 91.598°$

26. $34° \ 51' \ 35'' = 34° \ 51'\dfrac{35}{60} = 34° \ 51.583' = \left(34\dfrac{51.583}{60}\right)° = 34.860°$

27. $31.4296° = 31° + .4296(60') = 31° \ 25.776' = 31° \ 25' \ .776(60'') = 31° \ 25' \ 47''$

28. $59.0854° = 59° + .0854(60') = 59° \ 5.124' = 59° \ 5' \ .124(60'') = 59° \ 5' \ 7''$

29. $89.9004° = 89° + .9004(60') = 89° \ 54.024' = 89° \ 54' \ .024(60'') = 89° \ 54' \ 1''$

30. $102.3771° = 102° + .3771(60') = 102° \ 22.626' = 102° \ 22' \ .626(60'') = 102° \ 22' \ 38''$

31. See Figure 31. The coterminal angles are: $75° + 360° = 435°$ and $75° - 360° = -285°$, and are in quadrant I.

32. See Figure 32. The coterminal angles are: $89° + 360° = 449°$ and $89° - 360° = -271°$, and are in quadrant I.

33. See Figure 33. The coterminal angles are: $174° + 360° = 534°$ and $174° - 360° = -186°$, and are in quadrant II.

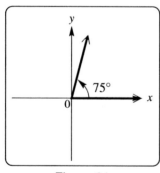

Figure 31

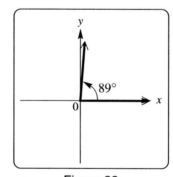

Figure 32

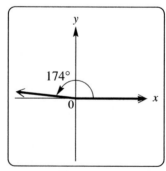

Figure 33

34. See Figure 34. The coterminal angles are: $234° + 360° = 594°$ and $234° - 360° = -126°$, and are in quadrant III.

35. See Figure 35. The coterminal angles are: $-61° + 360° = 299°$ and $-61° - 360° = -421°$, and are in quadrant IV.

36. See Figure 36. The coterminal angles are: $-159° + 360° = 201°$ and $-159° - 360° = -519°$, and are in quadrant III.

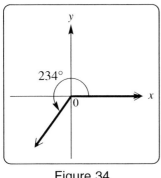

Figure 34

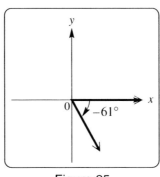

Figure 35

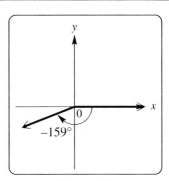

Figure 36

37. The smallest positive degree measure of the coterminal angle is: $-40° + 360° = 320°$.

38. The smallest positive degree measure of the coterminal angle is: $-98° + 360° = 262°$.

39. The smallest positive degree measure of the coterminal angle is: $450° - 360° = 90°$.

40. The smallest positive degree measure of the coterminal angle is: $539° - 360° = 179°$.

41. The smallest positive radian measure of the coterminal angle is: $-\dfrac{\pi}{4} + 2\pi = -\dfrac{\pi}{4} + \dfrac{8\pi}{4} = \dfrac{7\pi}{4}$.

42. The smallest positive radian measure of the coterminal angle is: $-\dfrac{\pi}{3} + 2\pi = -\dfrac{\pi}{3} + \dfrac{6\pi}{3} = \dfrac{5\pi}{3}$.

43. The smallest positive radian measure of the coterminal angle is: $-\dfrac{3\pi}{2} + 2\pi = -\dfrac{3\pi}{2} + \dfrac{4\pi}{2} = \dfrac{\pi}{2}$.

44. The smallest positive radian measure of the coterminal angle is: $-\pi + 2\pi = \pi$.

45. All degree measure coterminal angles will be multiples of $360°$, therefore: $30° + n \cdot 360°$.

46. All degree measure coterminal angles will be multiples of $360°$, therefore: $45° + n \cdot 360°$.

47. All degree measure coterminal angles will be multiples of $360°$, therefore: $-90° + n \cdot 360°$.

48. All degree measure coterminal angles will be multiples of $360°$, therefore: $-135° + n \cdot 360°$.

49. All radian measure coterminal angles will be multiples of 2π, therefore: $\dfrac{\pi}{4} + n \cdot 2\pi \Rightarrow \dfrac{\pi}{4} + 2n\pi$.

50. All radian measure coterminal angles will be multiples of 2π, therefore: $\dfrac{\pi}{6} + n \cdot 2\pi \Rightarrow \dfrac{\pi}{6} + 2n\pi$.

51. All radian measure coterminal angles will be multiples of 2π, therefore: $-\dfrac{3\pi}{4} + n \cdot 2\pi \Rightarrow -\dfrac{3\pi}{4} + 2n\pi$.

52. All radian measure coterminal angles will be multiples of 2π, therefore: $-\dfrac{7\pi}{6} + n \cdot 2\pi \Rightarrow -\dfrac{7\pi}{6} + 2n\pi$.

53. $60° \cdot \dfrac{\pi}{180} = \dfrac{\pi}{3}$ radians.

54. $90° \cdot \dfrac{\pi}{180} = \dfrac{\pi}{2}$ radians.

55. $150° \cdot \dfrac{\pi}{180} = \dfrac{5\pi}{6}$ radians.

56. $270° \cdot \dfrac{\pi}{180} = \dfrac{3\pi}{2}$ radians.

57. $-45° \cdot \dfrac{\pi}{180} = -\dfrac{\pi}{4}$ radians.

58. $-210° \cdot \dfrac{\pi}{180} = -\dfrac{7\pi}{6}$ radians.

59. $\dfrac{\pi}{3} \cdot \dfrac{180}{\pi} = 60°$

60. $\dfrac{8\pi}{3} \cdot \dfrac{180}{\pi} = 480°$

61. $\dfrac{7\pi}{4} \cdot \dfrac{180}{\pi} = 315°$

62. $\dfrac{2\pi}{3} \cdot \dfrac{180}{\pi} = 120°$

63. $\dfrac{11\pi}{6} \cdot \dfrac{180}{\pi} = 330°$

64. $\dfrac{15\pi}{4} \cdot \dfrac{180}{\pi} = 675°$

65. $39° \cdot \dfrac{\pi}{180} = \dfrac{39}{180} \cdot \pi = .68$

66. $74° \cdot \dfrac{\pi}{180} = \dfrac{74}{180} \cdot \pi = 1.29$

67. $139° \, 10' = \left(139 \dfrac{10}{60}\right)° = 139.167° \Rightarrow 139.167° \cdot \dfrac{\pi}{180} = \dfrac{139.167}{180} \cdot \pi = 2.43$

68. $174° \, 50' = \left(174 \dfrac{50}{60}\right)° = 174.833° \Rightarrow 174.833° \cdot \dfrac{\pi}{180} = \dfrac{174.833}{180} \cdot \pi = 3.05$

69. $64.29° \cdot \dfrac{\pi}{180} = \dfrac{64.29}{180} \cdot \pi = 1.12$

70. $122.62° \cdot \dfrac{\pi}{180} = \dfrac{122.62}{180} \cdot \pi = 2.14$

71. $2 \cdot \dfrac{180}{\pi} = \dfrac{360}{\pi} = 114.592° = 114° + .592(60') = 114° \, 35'$

72. $5 \cdot \dfrac{180}{\pi} = \dfrac{900}{\pi} = 286.479° = 286° + .479(60') = 286° \, 29'$

73. $1.74 \cdot \dfrac{180}{\pi} = \dfrac{313.2}{\pi} = 99.695° = 99° + .695(60') = 99° \, 42'$

74. $.3417 \cdot \dfrac{180}{\pi} = \dfrac{61.506}{\pi} = 19.578° = 19° + .578(60') = 19° \, 35'$

75. $-1.3 \cdot \dfrac{180}{\pi} = \dfrac{-234}{\pi} = -74.485° = -74° + .485(60') = -74° \, 29'$

76. $-4 \cdot \dfrac{180}{\pi} = \dfrac{-720}{\pi} = -229.183° = -229° + .183(60') = -229° \, 11'$

77. Going around the circle counter-clockwise and starting at $0° = 0$ radians; $30° = \dfrac{\pi}{6}$ radians; $45° = \dfrac{\pi}{4}$ radians;

    $60° = \dfrac{\pi}{3}$ radians; $90° = \dfrac{\pi}{2}$ radians; $120° = \dfrac{2\pi}{3}$ radians; $135° = \dfrac{3\pi}{4}$ radians; $150° = \dfrac{5\pi}{6}$ radians;

    $180° = \pi$ radians; $210° = \dfrac{7\pi}{6}$ radians; $225° = \dfrac{5\pi}{4}$ radians; $240° = \dfrac{4\pi}{3}$ radians; $270° = \dfrac{3\pi}{2}$ radians;

    $300° = \dfrac{5\pi}{3}$ radians; $315° = \dfrac{7\pi}{4}$ radians; and $330° = \dfrac{11\pi}{6}$ radians.

78. (a) Since grade is $\dfrac{1}{100}$ of $90°$, then grade is also $\dfrac{1}{100}$ of $\dfrac{\pi}{2}$ radians. Therefore, $\dfrac{1}{100} \cdot \dfrac{\pi}{2}$ or $\dfrac{\pi}{200}$ is the number to multiply grade to convert to radians.

   (b) The angle in degrees is: $3.5 \cdot .9° = 3.15°$.

     The angle in radians is: $3.5 \cdot \dfrac{\pi}{200} = .055$ radians.

79. Since arc length is found by the formula $s = r\theta$, the arc length is: $s = 4 \cdot \dfrac{\pi}{2} \;\Rightarrow\; s = 2\pi$.

80. Since arc length is found by the formula $s = r\theta$, the arc length is: $s = 12 \cdot \dfrac{\pi}{3} \;\Rightarrow\; s = 4\pi$.

81. Since arc length is found by the formula $s = r\theta$, the radius is: $6\pi = r \cdot \dfrac{3\pi}{4} \;\Rightarrow\; r = 6\pi \cdot \dfrac{4}{3\pi} \;\Rightarrow\; r = 8$.

82. Since arc length is found by the formula $s = r\theta$, the radius is: $3\pi = r \cdot \dfrac{\pi}{2} \;\Rightarrow\; r = 3\pi \cdot \dfrac{2}{\pi} \;\Rightarrow\; r = 6$.

83. Since arc length is found by the formula $s = r\theta$, the central angle is: $3 = 3\theta \;\Rightarrow\; \theta = \dfrac{3}{3} \;\Rightarrow\; \theta = 1$.

84. Since arc length is found by the formula $s = r\theta$, the central angle is: $6 = 4\theta \;\Rightarrow\; \theta = \dfrac{6}{4} \;\Rightarrow\; \theta = 1.5$.

85. Since arc length is found by the formula
$s = r\theta$, the arc length is: $s = 12.3\left(\dfrac{2\pi}{3}\right) \;\Rightarrow\; s = 8.2\pi \;\Rightarrow\; s \approx 25.8$ cm.

86. Since arc length is found by $s = r\theta$, the arc length is: $s = .892\left(\dfrac{11\pi}{10}\right) \;\Rightarrow\; s \approx .9812\pi \;\Rightarrow\; s \approx 3.08$ cm.

87. Since $60° = 60 \cdot \dfrac{\pi}{180} = \dfrac{\pi}{3}$ radians and arc length is found by the formula $s = r\theta$, the arc length is:
$s = 4.82\left(\dfrac{\pi}{3}\right) \;\Rightarrow\; s \approx 1.607\pi \;\Rightarrow\; s \approx 5.05$ m.

88. Since $135° = 135 \cdot \dfrac{\pi}{180} = \dfrac{3\pi}{4}$ radians and arc length is found by the formula $s = r\theta$, the arc length is:
$s = 71.9\left(\dfrac{3\pi}{4}\right) \;\Rightarrow\; s \approx 53.925\pi \;\Rightarrow\; s \approx 169$ cm.

89. Since $44° - 33° = 11°$, $11° \cdot \dfrac{\pi}{180} \approx .06111\pi$ radians, and arc length is found by the formula
$s = r\theta$, the arc length is: $s = 6400(.06111\pi) \;\Rightarrow\; s \approx 391.1\pi \;\Rightarrow\; s \approx 1200$ km.

90. Since $43° - 33° = 10°$, $10° \cdot \dfrac{\pi}{180} \approx .05556\pi$ radians, and arc length is found by the formula
$s = r\theta$, the arc length is: $s = 6400(.05556\pi) \;\Rightarrow\; s \approx 355.58\pi \;\Rightarrow\; s \approx 1100$ km.

91. Since $41° + 12° = 53°$, $53° \cdot \dfrac{\pi}{180} \approx .29444\pi$ radians, and arc length is found by the formula
$s = r\theta$, the arc length is: $s = 6400(.29444\pi) \;\Rightarrow\; s \approx 1884.42\pi \;\Rightarrow\; s \approx 5900$ km.

92. Since $45° + 34° = 79°$, $79° \cdot \dfrac{\pi}{180} \approx .43888\pi$ radians, and arc length is found by the formula
$s = r\theta$, the arc length is: $s = 6400(.43888\pi) \;\Rightarrow\; s \approx 2808.8\pi \;\Rightarrow\; s \approx 8800$ km.

93. Using $w = \dfrac{\theta}{t}$ yields: $w = \dfrac{\frac{3\pi}{4}}{8} = \dfrac{3\pi}{4} \cdot \dfrac{1}{8} \Rightarrow w = \dfrac{3\pi}{32}$ radians per second.

94. Using $w = \dfrac{\theta}{t}$ yields: $w = \dfrac{\frac{2\pi}{5}}{10} = \dfrac{2\pi}{5} \cdot \dfrac{1}{10} \Rightarrow w = \dfrac{\pi}{25}$ radians per second.

95. Using $w = \dfrac{\theta}{t}$ yields: $\dfrac{5\pi}{27} = \dfrac{\frac{2\pi}{9}}{t} \Rightarrow 5\pi t = \dfrac{27}{1} \cdot \dfrac{2\pi}{9} \Rightarrow 5\pi t = 6\pi \Rightarrow t = \dfrac{6\pi}{5\pi} \Rightarrow t = \dfrac{6}{5}$ minutes.

96. Using $w = \dfrac{\theta}{t}$ yields: $.90674 = \dfrac{\theta}{11.876} \Rightarrow \theta = (.90674)(11.876) \Rightarrow \theta \approx 10.768$ radians.

97. Using $s = rwt$ yields: $s = 6\left(\dfrac{\pi}{3}\right)(9) \Rightarrow s = 18\pi$ cm.

98. Using $s = rwt$ yields: $s = 9\left(\dfrac{2\pi}{5}\right)(12) \Rightarrow s = \dfrac{216\pi}{5}$ yd.

99. Using $s = rwt$ yields: $6\pi = 2\left(\dfrac{\pi}{4}\right)t \Rightarrow 6\pi = \dfrac{\pi}{2}t \Rightarrow 6\pi \cdot \dfrac{2}{\pi} = t \Rightarrow t = 12$ sec.

100. Using $s = rwt$ yields: $\dfrac{3\pi}{4} = 2w(4) \Rightarrow \dfrac{3\pi}{4} = 8w \Rightarrow \dfrac{3\pi}{4} \cdot \dfrac{1}{8} = w \Rightarrow w = \dfrac{3\pi}{32}$ radians per seconds.

101. (a) The weight will rise the same distance as the arc length of the rotation. Therefore since arc length is found

    by the formula $s = r\theta$, the arc length is: $s = 9.27\left(71\dfrac{50}{60} \cdot \dfrac{\pi}{180}\right) \Rightarrow s \approx 3.699\pi \Rightarrow s \approx 11.6$ in.

    (b) The weight will rise the same distance as the arc length of the rotation. Therefore since arc length is found

    by the formula $s = r\theta$, the angle is: $6 = 9.27(\theta) \Rightarrow \theta = .647$.

    Changing this to degree measure yields: $.647 \cdot \dfrac{180}{\pi} = \dfrac{116.505}{\pi} \approx 37.085° \approx 37° + .085(60') \approx 37° \, 5'$

102. The weight will rise the same distance as the arc length of the rotation. Therefore since arc length is found by

    the formula $s = r\theta$, the radius is: $11.4 = r\left(51.6 \cdot \dfrac{\pi}{180}\right) \Rightarrow 11.4 \approx .901\,r \Rightarrow r \approx 12.7$ cm.

103. First find the arc length that the smaller wheel rotates, arc length is found by the formula $s = r\theta$, the arc

    length is: $s = 5.23\left(60 \cdot \dfrac{\pi}{180}\right) \Rightarrow s \approx 1.743\pi \Rightarrow s \approx 5.477$ cm. Now the larger wheel travels the same

    arc length, therefore the angle of rotation is: $5.477 \approx 8.16\theta \Rightarrow \theta \approx .671$. Changing this to degree measure

    yields: $.671 \cdot \dfrac{180}{\pi} = \dfrac{120.78}{\pi} \approx 38.5°$

104. First find the arc length that the smaller wheel rotates, arc length is found by the formula $s = r\theta$, the arc

    length is: $s = 11.7\left(80 \cdot \dfrac{\pi}{180}\right) \Rightarrow s \approx 5.2\pi \Rightarrow s \approx 16.336$ cm. Now the larger wheel travels the same arc

    length, therefore the radius is: $16.336 = r\left(50 \cdot \dfrac{\pi}{180}\right) \Rightarrow 16.336 \approx .873\,r \Rightarrow r \approx 18.7$ cm.

105. First find the arc length that the pedal wheel rotates, arc length is found by the formula $s = r\theta$, the arc length

    is: $s = 4.72\left(180 \cdot \dfrac{\pi}{180}\right) \Rightarrow s \approx 4.72\pi \Rightarrow s \approx 14.828$ in. Now the smallest wheel travels the same arc

    length, therefore the angle of rotation is: $14.828 = 1.38\theta \Rightarrow \theta \approx 10.745$. Finally, using this radian rotation

    calculate the distance traveled with a wheel radius of 13.6 inches: $s \approx 13.6(10.745) \Rightarrow s \approx 146$ in.

106. (a) First find the distance travel in one rotation: $s = 14(2\pi) \Rightarrow s = 28\pi \Rightarrow s \approx 87.96459$ inches.

Since the car traveled 55 miles in the 1 hour, there are 5280 feet in a mile, and 12 inches in a foot, we next find the number of inches in 55 miles: $55 \cdot 5280 \cdot 12 = 3{,}484{,}800$ in. Finally divide this by the inches traveled in one rotation to find the number of rotations:

$3{,}484{,}800 \div 87.96459 \approx 39{,}615.941 \approx 39{,}616$ rotations.

(b) If the speedometer reads 55 mph, the wheel will make 39,615.941 rotations, therefore the new tires will make 39,615.941 rotations and travel: $s = (39{,}615.941)(16)(2\pi) \Rightarrow s \approx 3{,}982{,}628.8$ inches.

This is equal to: $3{,}982{,}628.8 \div 12 \div 5280 \approx 62.9$ miles. The driver does deserve a speeding ticket.

107. Using the area of a sector formula $A = \dfrac{1}{2}r^2\theta$ yields: $A = \dfrac{1}{2}(29.2)^2\left(\dfrac{5\pi}{6}\right) \Rightarrow A \approx 1120$ m^2

108. Using the area of a sector formula $A = \dfrac{1}{2}r^2\theta$ yields: $A = \dfrac{1}{2}(59.8)^2\left(\dfrac{2\pi}{3}\right) \Rightarrow A \approx 3740$ km^2

109. Using the area of a sector formula $A = \dfrac{1}{2}r^2\theta$ yields: $A = \dfrac{1}{2}(12.7)^2\left(81° \cdot \dfrac{\pi}{180}\right) \Rightarrow A \approx 114$ cm^2

110. Using the area of a sector formula $A = \dfrac{1}{2}r^2\theta$ yields: $A = \dfrac{1}{2}(18.3)^2\left(125° \cdot \dfrac{\pi}{180}\right) \Rightarrow A \approx 365$ m^2

111. First find θ using $s = r\theta$ yields: $2\pi = 6\theta \Rightarrow \theta = \dfrac{\pi}{3}$. Now using the area of a sector formula $A = \dfrac{1}{2}r^2\theta$

yields: $A = \dfrac{1}{2}(6)^2\left(\dfrac{\pi}{3}\right) \Rightarrow A \approx \dfrac{36\pi}{6} = 6\pi.$

112. First find θ using $s = r\theta$ yields: $4\pi = 8\theta \Rightarrow \theta = \dfrac{\pi}{2}$. Now using the area of a sector formula $A = \dfrac{1}{2}r^2\theta$

yields: $A = \dfrac{1}{2}(8)^2\left(\dfrac{\pi}{2}\right) \Rightarrow A \approx \dfrac{64\pi}{4} = 16\pi.$

113. Since the area scanned in 1 second would yield: $\theta = \dfrac{1}{48} \cdot 2\pi = \dfrac{\pi}{24}$ radians, we can solve for area using

$A = \dfrac{1}{2}r^2\theta$. Therefore, $A = \dfrac{1}{2}(240)^2\left(\dfrac{\pi}{24}\right) \Rightarrow A \approx 3800$ mi^2.

114. (a) Since the area of a circle is found by $A = \pi r^2$ we can find the radius:

$950{,}000 = \pi r^2 \Rightarrow r^2 = 302{,}394.39 \Rightarrow r \approx 550$ m.

(b) Using the area of a sector formula $A = \dfrac{1}{2}r^2\theta$ yields:

$950{,}000 = \dfrac{1}{2}r^2\left(35° \cdot \dfrac{\pi}{180}\right) \Rightarrow 950{,}000 \approx .30543\,r^2 \Rightarrow r^2 = 3{,}110{,}368.9 \Rightarrow r \approx 1800$ m.

115. To find the portion cleaned we will find the 10 inch radius sector and subtract the inside 3 inch radius sector.

The 10 in radius is: $A = \dfrac{1}{2}\,10^2\left(95° \cdot \dfrac{\pi}{180}\right) \Rightarrow A \approx 82.9$ in^2.

The 3 in radius is: $A = \dfrac{1}{2}\,3^2\left(95° \cdot \dfrac{\pi}{180}\right) \Rightarrow A \approx 7.5$ in^2. Subtracting these yields: $82.9 - 7.5 \approx 75.4$ in^2.

116. (a) Each sector is: $360° \div 27 = 13\frac{1}{3}°$ and $2\pi \div 27 = \dfrac{2\pi}{27}$ radians.

(b) Since $C = 2r\pi$, $C = 2(76)\pi \Rightarrow C \approx 480$ ft.

(c) Using the formula $s = r\theta$, the arc length is: $s = 76\left(\dfrac{2\pi}{27}\right) \Rightarrow s = \dfrac{152\pi}{27} \approx 17.7$ ft.

(d) Using the area of a sector formula $A = \dfrac{1}{2}r^2\theta$ yields: $A = \dfrac{1}{2}76^2\left(\dfrac{2\pi}{27}\right) \Rightarrow A \approx 672$ ft^2.

117. (a) Each sector is: $360° \div 26 = 13.85°$.

(b) Using the area of a sector formula $A = \dfrac{1}{2}r^2\theta$ and $r = \dfrac{50}{2} \Rightarrow r = 25$ yields:

$$A = \frac{1}{2}25^2\left(13.85° \cdot \frac{\pi}{180}\right) \Rightarrow A \approx 76 \text{ m}^2.$$

118. The third side of the triangle and radius of the sector is by pythagorean theorem or common right triangles is

50 yd. Now the area of the sector is: $A = \dfrac{1}{2}(50)^2\left(60° \cdot \dfrac{\pi}{180}\right) \Rightarrow A \approx 1300$ yd^2. The area of the triangle is

$A = \dfrac{1}{2}(30)(40) \Rightarrow A = 600$ yd^2. The area of the whole lot is: $1300 + 600 = 1900$ yd^2.

119. Since $A = \dfrac{1}{2}r^2(2\pi) = \pi r^2$, the area A of a circle of radius r is given by $A = \pi r^2$.

120. (a) Since one full rotation is 2π radians the speed is 2π radians per day.

(b) Since the radius (distance from the axis) of the circle of rotation at the north or south pole is $r = 0$,

the arc length of one rotation is: $s = 0(2\pi) \Rightarrow s = 0$ km, therefore the linear speed is: 0 km per hour.

(c) Since the radius (distance from the axis) of the circle of rotation at the equator is $r = 6400$, the linear

speed is: $v = \dfrac{r\theta}{t} = \dfrac{6400(2\pi)}{24} \Rightarrow v = 533\dfrac{1}{3}\pi$ km per hour.

121. If the diameter is 26, then the radius is 13, therefore 15 radians per second of rotation would have a linear

speed of: $v = rw \Rightarrow v = 13 \cdot 15 = 195$ inches per second which is equal to $195 \div 12 = 16\dfrac{1}{4}$ feet per

second. This is equal to: $(16.25 \div 5280) \cdot 60 \cdot 60 \approx 11.1$ mph.

122. First we need to find θ using the arc length formula: $56 = 12.96\theta \Rightarrow \theta = 4.321$. Now use the angular

speed formula: $w = \dfrac{4.321}{18} \Rightarrow w = .24$ radians per second.

123. First we need to find inches: 15 mph $= 15 \cdot 5280 \cdot 12 = 950{,}400$ inches. Next we need to find θ using the

arc length formula and a radius of $r = \dfrac{2.25}{2} = 1.125$: $950{,}400 = 1.125\theta \Rightarrow \theta = 844{,}800$.

Now use the angular speed formula: $w = \dfrac{844{,}800}{(60)(60)} \Rightarrow w = 234.67$ radians per second.

124. The larger pulley has an angular speed of $w = \dfrac{25(2\pi)}{36} \Rightarrow \dfrac{25\pi}{18}$.

Since the ratio of the pulleys is $\dfrac{8}{15}$ the ratio of angular speed is $\dfrac{15}{8}$, therefore the angular speed of the smaller

pulley is: $\dfrac{25\pi}{18} \cdot \dfrac{15}{8} = \dfrac{125\pi}{48}$.

125. With a radius of $r = \dfrac{10}{2} = 5$, a θ of 5000 revolutions per minute, and a time of 60 second in a minute, the

angular speed is: $w = \dfrac{\theta}{t} \implies w = \dfrac{5000(2\pi)}{60} \implies w = \dfrac{500\pi}{3}$ radians per second and the linear speed is is:

$v = rw \implies v = 5\left(\dfrac{500\pi}{3}\right) \implies v = \dfrac{2500\pi}{3}$ inches per second.

126. The linear speed with $r = 5$ and $\theta = 25$ is: $v = \dfrac{r\theta}{t} \implies v = \dfrac{5(25)}{1} \implies v = 125$ feet per second.

127. (a) The radian angle formed by 1 day is: $\theta = \dfrac{1}{365}(2\pi) \implies \theta = \dfrac{2\pi}{365}$ radians per day.

(b) With $\theta = \dfrac{2\pi}{365}$ per day and $t = 24$ hours per day, the angular speed is: $v = \dfrac{\theta}{t} \implies v = \dfrac{\frac{2\pi}{365}}{24} \implies v = \dfrac{\pi}{4380}$

radians per hour.

(c) With $r = 93{,}000{,}000$ and $\theta = \dfrac{\pi}{4380}$ radians per hour, the linear speed is: $v = rw \implies$

$v = 93{,}000{,}000\left(\dfrac{\pi}{4380}\right) \implies v = 66{,}700$ mph.

128. For 1 minute, $\theta = \dfrac{1}{360(60)} \cdot 2\pi = \dfrac{\pi}{10{,}800}$ radians. Now find a nautical mile by the arc length formula:

$s = r\theta \implies s = 3963\left(\dfrac{\pi}{10{,}800}\right) \implies s = 1.15$ miles.

129. First find the radian measure of $7°\,12'$: $\theta = \dfrac{7(60) + 12}{360(60)} \cdot 2\pi \implies \theta \approx .1257$ radians. Next use $\theta = .12566$

and the arc length formula to find the radius: $496 = r(.12566) \implies r = 3947$ miles. Now we can find the

circumference: $C = 2r\pi \implies C = 2(3947)\pi \implies C \approx 24{,}800$ miles.

130. Using $r = 238{,}900$ miles and the arc length formula yields:

$s = r\theta \implies s = 238{,}900\left(.517 \cdot \dfrac{\pi}{180}\right) \implies s \approx 2156$ miles.

8.2: The Unit Circle

1. $\dfrac{\frac{\pi}{3}}{2\pi} = \dfrac{\pi}{3} \cdot \dfrac{1}{2\pi} = \dfrac{1}{6}$

2. $\dfrac{\frac{2\pi}{3}}{2\pi} = \dfrac{2\pi}{3} \cdot \dfrac{1}{2\pi} = \dfrac{1}{3}$

3. $\dfrac{\frac{\pi}{4}}{2\pi} = \dfrac{\pi}{4} \cdot \dfrac{1}{2\pi} = \dfrac{1}{8}$

4. $\dfrac{\frac{3\pi}{4}}{2\pi} = \dfrac{3\pi}{4} \cdot \dfrac{1}{2\pi} = \dfrac{3}{8}$

5. $\dfrac{\frac{\pi}{6}}{2\pi} = \dfrac{\pi}{6} \cdot \dfrac{1}{2\pi} = \dfrac{1}{12}$

6. $\dfrac{\frac{5\pi}{6}}{2\pi} = \dfrac{5\pi}{6} \cdot \dfrac{1}{2\pi} = \dfrac{5}{12}$

7. $\dfrac{3\pi}{2\pi} = \dfrac{3}{2}$

8. $\dfrac{\frac{5\pi}{2}}{2\pi} = \dfrac{5\pi}{2} \cdot \dfrac{1}{2\pi} = \dfrac{5}{4}$

9. Since $\sin s = y$ and $\cos s = x$, $\sin s = \dfrac{3}{5}$ and $\cos s = \dfrac{4}{5}$.

10. Since $\sin s = y$ and $\cos s = x$, $\sin s = \dfrac{15}{17}$ and $\cos s = -\dfrac{8}{17}$.

11. Since $\sin s = y$ and $\cos s = x$, $\sin s = -\dfrac{5}{13}$ and $\cos s = \dfrac{12}{13}$.

12. Since $\sin s = y$ and $\cos s = x$, $\sin s = -\dfrac{24}{25}$ and $\cos s = -\dfrac{7}{25}$.

13. From the figure, the point is $(-1, 0)$. Since $\sin s = y$ and $\cos s = x$, $\sin s = 0$ and $\cos s = -1$.

    Finally, since $y = 0$, cosecant and cotangent are undefined.

14. From the figure, the point is $(-1, 0)$. Since $\sin s = y$ and $\cos s = x$, $\sin s = 0$ and $\cos s = -1$.

    Finally, since $y = 0$, cosecant and cotangent are undefined.

15. From the figure, the point is $(1, 0)$. Since $\sin s = y$ and $\cos s = x$, $\sin s = 0$ and $\cos s = 1$.

    Finally, since $y = 0$, cosecant and cotangent are undefined.

16. From the figure, the point is $(1, 0)$. Since $\sin s = y$ and $\cos s = x$, $\sin s = 0$ and $\cos s = 1$.

    Finally, since $y = 0$, cosecant and cotangent are undefined.

17. From the figure, the point is $(0, -1)$. Since $\sin s = y$ and $\cos s = x$, $\sin s = -1$ and $\cos s = 0$.

    Finally, since $x = 0$, secant and tangent are undefined.

18. From the figure, the point is $(0, 1)$. Since $\sin s = y$ and $\cos s = x$, $\sin s = 1$ and $\cos s = 0$.

    Finally, since $x = 0$, secant and tangent are undefined.

19. From the figure, the point is $(0, 1)$. Since $\sin s = y$ and $\cos s = x$, $\sin s = 1$ and $\cos s = 0$.

    Finally, since $x = 0$, secant and tangent are undefined.

20. From the figure, the point is $(0, -1)$. Since $\sin s = y$ and $\cos s = x$, $\sin s = -1$ and $\cos s = 0$.

    Finally, since $x = 0$, secant and tangent are undefined.

21. From the figure, a point on the unit circle is $(-1, 0)$. Since $x = -1$ and $y = 0$, the six functions are:

    $\cos s = x \Rightarrow \cos s = -1$ $\qquad$ $\sin s = y \Rightarrow \sin s = 0$ $\qquad$ $\tan s = \dfrac{y}{x} \Rightarrow \tan s = \dfrac{0}{-1} = 0$

    $\cot s = \dfrac{x}{y} = \dfrac{-1}{0} \Rightarrow$ undefined $\qquad$ $\sec s = \dfrac{1}{x} \Rightarrow \sec s = \dfrac{1}{-1} = -1$ $\quad$ $\csc s = \dfrac{1}{y} = \dfrac{1}{0} \Rightarrow$ undefined

22. From the figure, a point on the unit circle is $(-1, 0)$. Since $x = -1$ and $y = 0$, the six functions are:

    $\cos s = x \Rightarrow \cos s = -1$ $\qquad$ $\sin s = y \Rightarrow \sin s = 0$ $\qquad$ $\tan s = \dfrac{y}{x} \Rightarrow \tan s = \dfrac{0}{-1} = 0$

    $\cot s = \dfrac{x}{y} = \dfrac{-1}{0} \Rightarrow$ undefined $\qquad$ $\sec s = \dfrac{1}{x} \Rightarrow \sec s = \dfrac{1}{-1} = -1$ $\quad$ $\csc s = \dfrac{1}{y} = \dfrac{1}{0} \Rightarrow$ undefined

23. From the figure, a point on the unit circle is $(1, 0)$. Since $x = 1$ and $y = 0$, the six functions are:

    $\cos s = x \Rightarrow \cos s = 1$ $\qquad$ $\sin s = y \Rightarrow \sin s = 0$ $\qquad$ $\tan s = \dfrac{y}{x} \Rightarrow \tan s = \dfrac{0}{1} = 0$

    $\cot s = \dfrac{x}{y} = \dfrac{1}{0} \Rightarrow$ undefined $\qquad$ $\sec s = \dfrac{1}{x} \Rightarrow \sec s = \dfrac{1}{1} = 1$ $\qquad$ $\csc s = \dfrac{1}{y} = \dfrac{1}{0} \Rightarrow$ undefined

24. From the figure, a point on the unit circle is $(1, 0)$. Since $x = 1$ and $y = 0$, the six functions are:

$$\cos s = x \Rightarrow \cos s = 1 \qquad\qquad \sin s = y \Rightarrow \sin s = 0 \qquad\qquad \tan s = \frac{y}{x} \Rightarrow \tan s = \frac{0}{1} = 0$$

$$\cot s = \frac{x}{y} = \frac{1}{0} \Rightarrow \text{undefined} \qquad \sec s = \frac{1}{x} \Rightarrow \sec s = \frac{1}{1} = 1 \qquad \csc s = \frac{1}{y} = \frac{1}{0} \Rightarrow \text{undefined}$$

25. From the figure, a point on the unit circle is $(0, -1)$. Since $x = 0$ and $y = -1$, the six functions are:

$$\cos s = x \Rightarrow \cos s = 0 \qquad\qquad \sin s = y \Rightarrow \sin s = -1 \qquad\qquad \tan s = \frac{y}{x} = \frac{-1}{0} \Rightarrow \text{undefined}$$

$$\cot s = \frac{x}{y} \Rightarrow \cot s = \frac{0}{-1} = 0 \qquad \sec s = \frac{1}{x} = \frac{1}{0} \Rightarrow \text{undefined} \qquad \csc s = \frac{1}{y} \Rightarrow \csc s = \frac{1}{-1} = -1$$

26. From the figure, a point on the unit circle is $(0, 1)$. Since $x = 0$ and $y = 1$, the six functions are:

$$\cos s = x \Rightarrow \cos s = 0 \qquad\qquad \sin s = y \Rightarrow \sin s = 1 \qquad\qquad \tan s = \frac{y}{x} = \frac{1}{0} \Rightarrow \text{undefined}$$

$$\cot s = \frac{x}{y} \Rightarrow \cot s = \frac{0}{1} = 0 \qquad \sec s = \frac{1}{x} = \frac{1}{0} \Rightarrow \text{undefined} \qquad \csc s = \frac{1}{y} \Rightarrow \csc s = \frac{1}{1} = 1$$

27. From the figure, a point on the unit circle is $(0, 1)$. Since $x = 0$ and $y = 1$, the six functions are:

$$\cos s = x \Rightarrow \cos s = 0 \qquad\qquad \sin s = y \Rightarrow \sin s = 1 \qquad\qquad \tan s = \frac{y}{x} = \frac{1}{0} \Rightarrow \text{undefined}$$

$$\cot s = \frac{x}{y} \Rightarrow \cot s = \frac{0}{1} = 0 \qquad \sec s = \frac{1}{x} = \frac{1}{0} \Rightarrow \text{undefined} \qquad \csc s = \frac{1}{y} \Rightarrow \csc s = \frac{1}{1} = 1$$

28. From the figure, a point on the unit circle is $(0, -1)$. Since $x = 0$ and $y = -1$, the six functions are:

$$\cos s = x \Rightarrow \cos s = 0 \qquad\qquad \sin s = y \Rightarrow \sin s = -1 \qquad\qquad \tan s = \frac{y}{x} = \frac{-1}{0} \Rightarrow \text{undefined}$$

$$\cot s = \frac{x}{y} \Rightarrow \cot s = \frac{0}{-1} = 0 \qquad \sec s = \frac{1}{x} = \frac{1}{0} \Rightarrow \text{undefined} \qquad \csc s = \frac{1}{y} \Rightarrow \csc s = \frac{1}{-1} = -1$$

29. Because $0 < .75 < \dfrac{\pi}{2}$, s is found in quadrant I. Now from the calculator (radian mode):

$$\cos s \approx .7316888689 \qquad\qquad \sin s \approx .68163876 \qquad\qquad \tan s \approx .9315964599$$

$$\cot s = \frac{1}{\tan s} \approx 1.073426149 \qquad \sec s = \frac{1}{\cos s} \approx 1.366701125 \qquad \csc s = \frac{1}{\sin s} \approx 1.467052724$$

30. Because $0 < .95 < \dfrac{\pi}{2}$, s is found in quadrant I. Now from the calculator (radian mode):

$$\cos s \approx .5816830895 \qquad\qquad \sin s \approx .8134155048 \qquad\qquad \tan s \approx 1.398382589$$

$$\cot s = \frac{1}{\tan s} \approx .7151118783 \qquad \sec s = \frac{1}{\cos s} \approx 1.719149169 \qquad \csc s = \frac{1}{\sin s} \approx 1.229383991$$

31. Because $-\pi < -4.25 < -\dfrac{3\pi}{2}$, s is found in quadrant II. Now from the calculator (radian mode):

$$\cos s \approx -.4460874899 \qquad\qquad \sin s \approx .8949893582 \qquad\qquad \tan s \approx -2.006309028$$

$$\cot s = \frac{1}{\tan s} \approx -.4984277029 \qquad \sec s = \frac{1}{\cos s} \approx -2.241712719 \qquad \csc s = \frac{1}{\sin s} \approx 1.117331721$$

32. Because $-\pi < -3.75 < -\dfrac{3\pi}{2}$, s is found in quadrant II. Now from the calculator (radian mode):

$$\cos s \approx -.8205593573 \qquad\qquad \sin s \approx .5715613187 \qquad\qquad \tan s \approx -.6965508511$$

$$\cot s = \frac{1}{\tan s} \approx -1.435645364 \qquad \sec s = \frac{1}{\cos s} \approx -1.21868088 \qquad \csc s = \frac{1}{\sin s} \approx 1.749593556$$

33. Because $-\dfrac{\pi}{2} < -2.25 < -\pi$, s is found in quadrant III. Now from the calculator (radian mode):

$\cos s \approx -.6281736227$ $\sin s \approx -.77807361969$ $\tan s \approx 1.238627616$

$\cot s = \dfrac{1}{\tan s} \approx .8073451511$ $\sec s = \dfrac{1}{\cos s} \approx -1.591916572$ $\csc s = \dfrac{1}{\sin s} \approx -1.285226125$

34. Because $-\dfrac{\pi}{2} < -2.75 < -\pi$, s is found in quadrant III. Now from the calculator (radian mode):

$\cos s \approx -.9243023786$ $\sin s \approx -.3816609921$ $\tan s \approx .4129178945$

$\cot s = \dfrac{1}{\tan s} \approx 2.421788964$ $\sec s = \dfrac{1}{\cos s} \approx -1.081897032$ $\csc s = \dfrac{1}{\sin s} \approx -2.620126292$

35. Because $\dfrac{3\pi}{2} < 5.5 < 2\pi$, s is found in quadrant IV. Now from the calculator (radian mode):

$\cos s \approx .7086697743$ $\sin s \approx -.7055403256$ $\tan s \approx -.9955840522$

$\cot s = \dfrac{1}{\tan s} \approx -1.004435535$ $\sec s = \dfrac{1}{\cos s} \approx 1.411094471$ $\csc s = \dfrac{1}{\sin s} \approx -1.417353429$

36. Because $\dfrac{3\pi}{2} < 5.75 < 2\pi$, s is found in quadrant IV. Now from the calculator (radian mode):

$\cos s \approx .8611924172$ $\sin s \approx -.5082790775$ $\tan s \approx -.5902038469$

$\cot s = \dfrac{1}{\tan s} \approx -1.694329858$ $\sec s = \dfrac{1}{\cos s} \approx 1.161180684$ $\csc s = \dfrac{1}{\sin s} \approx -1.967423103$

37. For exercises 37-42 answers will vary depending on the name used. Suppose the first name is Shannon.

    Then $s = 7$, and $\cos 7 \approx .7539022543$.

38. For exercises 37-42 answers will vary depending on the name used. Suppose the last name is Mulkey.

    Then $n = 6$, and $\cos (7 + 2(6)\pi) \approx .7539022543$.

39. They are the same. The real numbers s and $s + 2n\pi$ correspond to the same point on the unit circle,

    because its circumference is 2π.

40. For exercises 37-42 answers will vary depending on the name used. Suppose the last name is Castellucio.

    Then $s = 11$, and $\sin 11 \approx -.9999902066$.

41. For exercises 37-42 answers will vary depending on the name used. Suppose the first name is Frankie.

    Then $n = 7$, and $\sin (11 + 2(7)\pi) \approx -.9999902066$.

42. They are the same. The real numbers s and $s + 2n\pi$ correspond to the same point on the unit circle,

    because its circumference is 2π.

43. The point $\left(-\dfrac{\sqrt{3}}{2}, -\dfrac{1}{2} \right)$ corresponds to $\dfrac{7\pi}{6}$ on the unit circle.

    Since the function is sine, we use the y-value of the point to obtain $\sin \dfrac{7\pi}{6} = -\dfrac{1}{2}$.

44. The point $\left(\dfrac{1}{2}, -\dfrac{\sqrt{3}}{2} \right)$ corresponds to $\dfrac{5\pi}{3}$ on the unit circle.

    Since the function is cosine, we use the x-value of the point to obtain $\cos \dfrac{5\pi}{3} = \dfrac{1}{2}$.

45. The point $\left(-\dfrac{\sqrt{2}}{2}, \dfrac{\sqrt{2}}{2}\right)$ corresponds to $\dfrac{3\pi}{4}$ on the unit circle.

Since the function is tangent and by definition $\tan s = \dfrac{y}{x}$, $\tan \dfrac{3\pi}{4} = \dfrac{\frac{\sqrt{2}}{2}}{-\frac{\sqrt{2}}{2}} = \dfrac{\sqrt{2}}{2} \cdot \left(-\dfrac{2}{\sqrt{2}}\right) = -1$

46. The point $\left(-\dfrac{\sqrt{3}}{2}, -\dfrac{1}{2}\right)$ corresponds to $\dfrac{7\pi}{6}$ on the unit circle.

Since the function is cosine, we use the x-value of the point to obtain $\cos \dfrac{7\pi}{6} = -\dfrac{\sqrt{3}}{2}$.

47. The point $\left(-\dfrac{1}{2}, \dfrac{\sqrt{3}}{2}\right)$ corresponds to $\dfrac{2\pi}{3}$ on the unit circle.

The function is secant and by definition $\sec s = \dfrac{1}{\cos s}$, therefore, $\sec \dfrac{2\pi}{3} = \dfrac{1}{\cos \frac{2\pi}{3}}$.

Since the function is cosine, we use the x-value of the point to obtain $\sec \dfrac{2\pi}{3} = \dfrac{1}{-\frac{1}{2}} = -2$.

48. The point $\left(\dfrac{\sqrt{3}}{2}, -\dfrac{1}{2}\right)$ corresponds to $\dfrac{11\pi}{6}$ on the unit circle.

The function is cosecant and by definition $\csc s = \dfrac{1}{\sin s}$, therefore, $\csc \dfrac{11\pi}{6} = \dfrac{1}{\sin \frac{11\pi}{6}}$.

Since the function is sine, we use the y-value of the point to obtain $\csc \dfrac{11\pi}{6} = \dfrac{1}{\sin \frac{11\pi}{6}} = \dfrac{1}{-\frac{1}{2}} = -2$.

49. The point $\left(-\dfrac{\sqrt{3}}{2}, \dfrac{1}{2}\right)$ corresponds to $\dfrac{5\pi}{6}$ on the unit circle.

The function is cotangent and by definition $\cot s = \dfrac{1}{\tan s}$, therefore, $\cot \dfrac{5\pi}{6} = \dfrac{1}{\tan \frac{5\pi}{6}}$.

Since the function is tangent and by definition $\tan s = \dfrac{y}{x}$,

$\cot \dfrac{5\pi}{6} = \dfrac{1}{\tan \frac{5\pi}{6}} = 1 \div \left(\dfrac{\frac{1}{2}}{-\frac{\sqrt{3}}{2}}\right) = 1 \div \left(\dfrac{1}{2} \cdot \left(-\dfrac{2}{\sqrt{3}}\right)\right) = 1 \div \left(-\dfrac{1}{\sqrt{3}}\right) = -\sqrt{3}$.

50. The point on the unit circle for $s = -\dfrac{4\pi}{3}$ is the same as that for $\dfrac{2\pi}{3}$.

The point $\left(-\dfrac{1}{2}, \dfrac{\sqrt{3}}{2}\right)$ corresponds to $\dfrac{2\pi}{3}$ on the unit circle.

Since the function is cosine, we use the x-value of the point to obtain $\cos\left(-\dfrac{4\pi}{3}\right) = \cos \dfrac{2\pi}{3} = -\dfrac{1}{2}$.

51. The point on the unit circle for $s = -\dfrac{5\pi}{6}$ is the same as that for $\dfrac{7\pi}{6}$.

The point $\left(-\dfrac{\sqrt{3}}{2}, -\dfrac{1}{2}\right)$ corresponds to $\dfrac{7\pi}{6}$ on the unit circle.

Since the function is sine, we use the y-value of the point to obtain $\sin\left(-\dfrac{5\pi}{6}\right) = \sin \dfrac{7\pi}{6} = -\dfrac{1}{2}$.

52. The point on the unit circle for $s = \dfrac{17\pi}{3}$ is the same as that for $\dfrac{5\pi}{3}$.

The point $\left(\dfrac{1}{2}, -\dfrac{\sqrt{3}}{2}\right)$ corresponds to $\dfrac{5\pi}{3}$ on the unit circle.

Since the function is tangent and by definition $\tan s = \dfrac{y}{x}$, $\tan \dfrac{17\pi}{3} = \tan \dfrac{5\pi}{3} = \dfrac{-\frac{\sqrt{3}}{2}}{\frac{1}{2}} = -\dfrac{\sqrt{3}}{2} \cdot \dfrac{2}{1} = -\sqrt{3}$

53. The point on the unit circle for $s = \dfrac{23\pi}{6}$ is the same as that for $\dfrac{11\pi}{6}$.

    The point $\left(\dfrac{\sqrt{3}}{2}, -\dfrac{1}{2}\right)$ corresponds to $\dfrac{11\pi}{6}$ on the unit circle. The function is secant and by definition

    $\sec s = \dfrac{1}{\cos s}$, therefore, $\sec \dfrac{11\pi}{6} = \dfrac{1}{\cos \frac{11\pi}{6}}$. Since the function is cosine, we use the x-value of the

    point to obtain $\sec \dfrac{23\pi}{6} = \sec \dfrac{11\pi}{6} = \dfrac{1}{\frac{\sqrt{3}}{2}} = \dfrac{2}{\sqrt{3}} = \dfrac{2\sqrt{3}}{3}$.

54. The point on the unit circle for $s = \dfrac{13\pi}{3}$ is the same as that for $\dfrac{\pi}{3}$.

    The point $\left(\dfrac{1}{2}, \dfrac{\sqrt{3}}{2}\right)$ corresponds to $\dfrac{\pi}{3}$ on the unit circle. The function is cosecant and by definition

    $\csc s = \dfrac{1}{\sin s}$, therefore, $\csc \dfrac{\pi}{3} = \dfrac{1}{\sin \frac{\pi}{3}}$. Since the function is sine, we use the y-value of the point to obtain

    $\csc \dfrac{13\pi}{3} = \csc \dfrac{\pi}{3} = \dfrac{1}{\sin \frac{\pi}{3}} = \dfrac{1}{\frac{\sqrt{3}}{2}} = \dfrac{2}{\sqrt{3}} = \dfrac{2\sqrt{3}}{3}$.

55. The point on the unit circle for $s = -\dfrac{13\pi}{6}$ is the same as that for $-\dfrac{\pi}{6}$, which is the same as that for $\dfrac{11\pi}{6}$.

    The point $\left(\dfrac{\sqrt{3}}{2}, -\dfrac{1}{2}\right)$ corresponds to $\dfrac{11\pi}{6}$ on the unit circle.

    Since the function is cosine, we use the x-value of the point to obtain $\cos\left(-\dfrac{13\pi}{6}\right) = \cos \dfrac{11\pi}{6} = \dfrac{\sqrt{3}}{2}$.

56. The point on the unit circle for $s = -\dfrac{9\pi}{4}$ is the same as that for $-\dfrac{\pi}{4}$, which is the same as that for $\dfrac{7\pi}{4}$.

    The point $\left(\dfrac{\sqrt{2}}{2}, -\dfrac{\sqrt{2}}{2}\right)$ corresponds to $\dfrac{7\pi}{4}$ on the unit circle.

    Since the function is sine, we use the y-value of the point to obtain $\sin\left(-\dfrac{9\pi}{4}\right) = \sin \dfrac{7\pi}{4} = -\dfrac{\sqrt{2}}{2}$.

57. The point on the unit circle for $s = -\dfrac{13\pi}{4}$ is the same as that for $-\dfrac{5\pi}{4}$, which is the same as that for $\dfrac{3\pi}{4}$.

    The point $\left(-\dfrac{\sqrt{2}}{2}, \dfrac{\sqrt{2}}{2}\right)$ corresponds to $\dfrac{3\pi}{4}$ on the unit circle.

    Since the function is tangent and by definition $\tan s = \dfrac{y}{x}$, $\tan -\dfrac{13\pi}{4} = \tan \dfrac{3\pi}{4} = \dfrac{\frac{\sqrt{2}}{2}}{-\frac{\sqrt{2}}{2}} = \dfrac{\sqrt{2}}{2} \cdot \left(-\dfrac{\sqrt{2}}{2}\right) = -1$

58. The statement $\sin^2 S + \cos^2 S = 1$ is true for all real numbers S. It makes no difference what value is used for S; in all cases, the result is 1.

59. Since $\cos s = x$ and $\sin s = y$, the identity $\cos^2 s + \sin^2 s = 1$ would yield:

    $\left(\dfrac{3}{5}\right)^2 + y^2 = 1 \Rightarrow y^2 = 1 - \dfrac{9}{25} \Rightarrow y^2 = \dfrac{16}{25} \Rightarrow y = \pm\dfrac{4}{5}$, but $y > 0$, therefore $y = \dfrac{4}{5}$.

    Now using $x = \dfrac{3}{5}$ and $y = \dfrac{4}{5}$, the six trigonometric functions are:

    $\cos s = x = \dfrac{3}{5}$ $\sin s = y = \dfrac{4}{5}$ $\tan s = \dfrac{y}{x} = \dfrac{4}{5} \cdot \dfrac{5}{3} = \dfrac{4}{3}$

    $\cot s = \dfrac{1}{\tan s} = \dfrac{1}{\frac{4}{3}} = \dfrac{3}{4}$ $\sec s = \dfrac{1}{\cos s} = \dfrac{1}{\frac{3}{5}} = \dfrac{5}{3}$ $\csc s = \dfrac{1}{\sin s} = \dfrac{1}{\frac{4}{5}} = \dfrac{5}{4}$

60. Since $\cos s = x$ and $\sin s = y$, the identity $\cos^2 s + \sin^2 s = 1$ would yield:

$$\left(\frac{7}{25}\right)^2 + y^2 = 1 \Rightarrow y^2 = 1 - \frac{49}{625} \Rightarrow y^2 = \frac{576}{625} \Rightarrow y = \pm\frac{24}{25}, \text{ but } y > 0, \text{ therefore } y = \frac{24}{25}.$$

Now using $x = \frac{7}{25}$ and $y = \frac{24}{25}$, the six trigonometric functions are:

$$\cos s = x = \frac{7}{25} \qquad\qquad \sin s = y = \frac{24}{25} \qquad\qquad \tan s = \frac{y}{x} = \frac{24}{25} \cdot \frac{25}{7} = \frac{24}{7}$$

$$\cot s = \frac{1}{\tan s} = \frac{1}{\frac{24}{7}} = \frac{7}{24} \qquad \sec s = \frac{1}{\cos s} = \frac{1}{\frac{7}{25}} = \frac{25}{7} \qquad \csc s = \frac{1}{\sin s} = \frac{1}{\frac{24}{25}} = \frac{25}{24}$$

61. Since $\cos s = x$ and $\sin s = y$, the identity $\cos^2 s + \sin^2 s = 1$ would yield:

$$x^2 + \left(\frac{24}{25}\right)^2 = 1 \Rightarrow x^2 = 1 - \frac{576}{625} \Rightarrow x^2 = \frac{49}{625} \Rightarrow y = \pm\frac{7}{25}, \text{ but } x < 0, \text{ therefore } x = -\frac{7}{25}.$$

Now using $x = -\frac{7}{25}$ and $y = \frac{24}{25}$, the six trigonometric functions are:

$$\cos s = x = -\frac{7}{25} \qquad\qquad \sin s = y = \frac{24}{25} \qquad\qquad \tan s = \frac{y}{x} = \frac{24}{25} \cdot \left(-\frac{25}{7}\right) = -\frac{24}{7}$$

$$\cot s = \frac{1}{\tan s} = \frac{1}{-\frac{24}{7}} = -\frac{7}{24} \qquad \sec s = \frac{1}{\cos s} = \frac{1}{-\frac{7}{25}} = -\frac{25}{7} \qquad \csc s = \frac{1}{\sin s} = \frac{1}{\frac{24}{25}} = \frac{25}{24}$$

62. Since $\cos s = x$ and $\sin s = y$, the identity $\cos^2 s + \sin^2 s = 1$ would yield:

$$x^2 + \left(\frac{8}{17}\right)^2 = 1 \Rightarrow x^2 = 1 - \frac{64}{289} \Rightarrow x^2 = \frac{225}{289} \Rightarrow y = \pm\frac{15}{17}, \text{ but } x < 0, \text{ therefore } x = -\frac{15}{17}.$$

Now using $x = -\frac{15}{17}$ and $y = \frac{8}{17}$, the six trigonometric functions are:

$$\cos s = x = -\frac{15}{17} \qquad\qquad \sin s = y = \frac{8}{17} \qquad\qquad \tan s = \frac{y}{x} = \frac{8}{17} \cdot \left(-\frac{17}{15}\right) = -\frac{8}{15}$$

$$\cot s = \frac{1}{\tan s} = \frac{1}{-\frac{8}{15}} = -\frac{15}{8} \qquad \sec s = \frac{1}{\cos s} = \frac{1}{-\frac{15}{17}} = -\frac{17}{15} \qquad \csc s = \frac{1}{\sin s} = \frac{1}{\frac{8}{17}} = \frac{17}{8}$$

63. Since $\cos s = x$ and $\sin s = y$, the identity $\cos^2 s + \sin^2 s = 1$ would yield:

$$\left(-\frac{1}{3}\right)^2 + y^2 = 1 \Rightarrow y^2 = 1 - \frac{1}{9} \Rightarrow y^2 = \frac{8}{9} \Rightarrow y = \pm\frac{2\sqrt{2}}{3}, \text{ but } y < 0, \text{ therefore } y = -\frac{2\sqrt{2}}{3}.$$

Now using $x = -\frac{1}{3}$ and $y = -\frac{2\sqrt{2}}{3}$, the six trigonometric functions are:

$$\cos s = x = -\frac{1}{3} \qquad\qquad\qquad \sin s = y = -\frac{2\sqrt{2}}{3}$$

$$\tan s = \frac{y}{x} = \left(-\frac{2\sqrt{2}}{3}\right) \cdot \left(-\frac{3}{1}\right) = 2\sqrt{2} \qquad \cot s = \frac{1}{\tan s} = \frac{1}{2\sqrt{2}} = \frac{\sqrt{2}}{4}$$

$$\sec s = \frac{1}{\cos s} = \frac{1}{-\frac{1}{3}} = -3 \qquad\qquad \csc s = \frac{1}{\sin s} = \frac{1}{-\frac{2\sqrt{2}}{3}} = -\frac{3}{2\sqrt{2}} = -\frac{3\sqrt{2}}{4}$$

64. Since $\cos s = x$ and $\sin s = y$, the identity $\cos^2 s + \sin^2 s = 1$ would yield:

$$\left(-\frac{1}{4}\right)^2 + y^2 = 1 \Rightarrow y^2 = 1 - \frac{1}{16} \Rightarrow y^2 = \frac{15}{16} \Rightarrow y = \pm\frac{\sqrt{15}}{4}, \text{ but } y < 0, \text{ therefore } y = -\frac{\sqrt{15}}{4}.$$

Now using $x = -\frac{1}{4}$ and $y = -\frac{\sqrt{15}}{4}$, the six trigonometric functions are:

$$\cos s = x = -\frac{1}{4} \qquad\qquad \sin s = y = -\frac{\sqrt{15}}{4} \qquad\qquad \tan s = \frac{y}{x} = \left(-\frac{\sqrt{15}}{4}\right) \cdot \left(-\frac{4}{1}\right) = \sqrt{15}$$

$$\cot s = \frac{1}{\tan s} = \frac{1}{\sqrt{15}} = \frac{\sqrt{15}}{15} \qquad \sec s = \frac{1}{\cos s} = \frac{1}{-\frac{1}{4}} = -4 \qquad \csc s = \frac{1}{\sin s} = \frac{1}{-\frac{\sqrt{15}}{4}} = -\frac{4}{\sqrt{15}} = -\frac{4\sqrt{15}}{15}$$

65. If $\sin s = \dfrac{1}{2}$ and $\cos s = \dfrac{\sqrt{3}}{2}$, then we can find the trigonometric functions as follows:

$$\tan s = \frac{\sin s}{\cos s} = \frac{1}{2} \cdot \frac{2}{\sqrt{3}} = \frac{1}{\sqrt{3}} = \frac{\sqrt{3}}{3} \qquad \cot s = \frac{1}{\tan s} = \frac{1}{\frac{\sqrt{3}}{3}} = \frac{3}{\sqrt{3}} = \frac{3\sqrt{3}}{3} = \sqrt{3}$$

$$\sec s = \frac{1}{\cos s} = \frac{1}{\frac{\sqrt{3}}{2}} = \frac{2}{\sqrt{3}} = \frac{2\sqrt{3}}{3} \qquad \csc s = \frac{1}{\sin s} = \frac{1}{\frac{1}{2}} = 2$$

66. If $\sin s = \dfrac{3}{4}$ and $\cos s = \dfrac{\sqrt{7}}{4}$, then we can find the trigonometric functions as follows:

$$\tan s = \frac{\sin s}{\cos s} = \frac{3}{4} \cdot \frac{4}{\sqrt{7}} = \frac{3}{\sqrt{7}} = \frac{3\sqrt{7}}{7} \qquad \cot s = \frac{1}{\tan s} = \frac{1}{\frac{3\sqrt{7}}{7}} = \frac{7}{3\sqrt{7}} = \frac{21\sqrt{7}}{63} = \frac{\sqrt{7}}{3}$$

$$\sec s = \frac{1}{\cos s} = \frac{1}{\frac{\sqrt{7}}{4}} = \frac{4}{\sqrt{7}} = \frac{4\sqrt{7}}{7} \qquad \csc s = \frac{1}{\sin s} = \frac{1}{\frac{3}{4}} = \frac{4}{3}$$

67. If $\sin s = \dfrac{4}{5}$ and $\cos s = -\dfrac{3}{5}$, then we can find the trigonometric functions as follows:

$$\tan s = \frac{\sin s}{\cos s} = \frac{4}{5} \cdot \left(-\frac{5}{3}\right) = -\frac{4}{3} \qquad \cot s = \frac{1}{\tan s} = \frac{1}{-\frac{4}{3}} = -\frac{3}{4}$$

$$\sec s = \frac{1}{\cos s} = \frac{1}{-\frac{3}{5}} = -\frac{5}{3} \qquad \csc s = \frac{1}{\sin s} = \frac{1}{\frac{4}{5}} = \frac{5}{4}$$

68. If $\sin s = -\dfrac{1}{2}$ and $\cos s = -\dfrac{\sqrt{3}}{2}$, then we can find the trigonometric functions as follows:

$$\tan s = \frac{\sin s}{\cos s} = \left(-\frac{1}{2}\right) \cdot \left(-\frac{2}{\sqrt{3}}\right) = \frac{1}{\sqrt{3}} = \frac{\sqrt{3}}{3} \qquad \cot s = \frac{1}{\tan s} = \frac{1}{\frac{\sqrt{3}}{3}} = \frac{3}{\sqrt{3}} = \frac{3\sqrt{3}}{3} = \sqrt{3}$$

$$\sec s = \frac{1}{\cos s} = \frac{1}{-\frac{\sqrt{3}}{2}} = -\frac{2}{\sqrt{3}} = -\frac{2\sqrt{3}}{3} \qquad \csc s = \frac{1}{\sin s} = \frac{1}{-\frac{1}{2}} = -2$$

69. If $\sin s = -\dfrac{\sqrt{3}}{2}$ and $\cos s = \dfrac{1}{2}$, then we can find the trigonometric functions as follows:

$$\tan s = \frac{\sin s}{\cos s} = \left(-\frac{\sqrt{3}}{2}\right) \cdot \frac{2}{1} = -\sqrt{3} \qquad \cot s = \frac{1}{\tan s} = \frac{1}{-\sqrt{3}} = -\frac{\sqrt{3}}{3}$$

$$\sec s = \frac{1}{\cos s} = \frac{1}{\frac{1}{2}} = 2 \qquad \csc s = \frac{1}{\sin s} = \frac{1}{-\frac{\sqrt{3}}{2}} = -\frac{2}{\sqrt{3}} = -\frac{2\sqrt{3}}{3}$$

70. If $\sin s = \dfrac{12}{13}$ and $\cos s = \dfrac{5}{13}$, then we can find the trigonometric functions as follows:

$$\tan s = \frac{\sin s}{\cos s} = \frac{12}{13} \cdot \frac{13}{5} = \frac{12}{5} \qquad \cot s = \frac{1}{\tan s} = \frac{1}{\frac{12}{5}} = \frac{5}{12}$$

$$\sec s = \frac{1}{\cos s} = \frac{1}{\frac{5}{13}} = \frac{13}{5} \qquad \csc s = \frac{1}{\sin s} = \frac{1}{\frac{12}{13}} = \frac{13}{12}$$

71. Since $\sin s = y$ and $\sin s > 0$, s is in either quadrant I or II.

Since $\cos s = x$ and $\cos s < 0$, s is in either quadrant II or III. The point s is found in quadrant II.

72. Since $\cos s = x$ and $\cos s > 0$, s is in either quadrant I or IV.

Since $\tan s = \dfrac{y}{x}$ and $\tan s > 0$, s is in either quadrant I or III. The point s is found in quadrant I.

73. Since $\sec s = \dfrac{1}{x}$ and $\sec s < 0$, s is in either quadrant II or III.

   Since $\csc s = \dfrac{1}{y}$ and $\csc s < 0$, s is in either quadrant III or IV. The point s is found in quadrant III.

74. Since $\tan s = \dfrac{y}{x}$ and $\tan s > 0$, s is in either quadrant I or III.

   Since $\cos s = x$ and $\cos s < 0$, s is in either quadrant II or III. The point s is found in quadrant III.

75. Since $\cos s = x$ and $\cos s > 0$, s is in either quadrant I or IV.

   Since $\sin s = y$ and $\sin s < 0$, s is in either quadrant III or IV. The point s is found in quadrant IV.

76. Since $\tan s = \dfrac{y}{x}$ and $\tan s < 0$, s is in either quadrant II or IV.

   Since $\sin s = y$ and $\sin s > 0$, s is in either quadrant I or II. The point s is found in quadrant II.

77. $1 + \tan^2 s = 1 + \dfrac{\sin^2 s}{\cos^2 s} = \dfrac{\cos^2 s + \sin^2 s}{\cos^2 s} = \dfrac{1}{\cos^2 s} = \sec^2 s$

78. $1 + \cot^2 s = 1 + \dfrac{\cos^2 s}{\sin^2 s} = \dfrac{\sin^2 s + \cos^2 s}{\sin^2 s} = \dfrac{1}{\sin^2 s} = \csc^2 s$

79. If s is a real number corresponding to point (a, b), then $\sin s = y \Rightarrow \sin s = b$ and $\cos s = x \Rightarrow \cos s = a$.

80. If s is a real number corresponding to point (a, b), then $s + 2\pi$ would be a co-terminal angle to an angle with

   measure s. Therefore $\sin s = y \Rightarrow \sin(s + 2\pi) = y \Rightarrow \sin(s + 2\pi) = b$ and

   $\cos s = x \Rightarrow \cos(s + 2\pi) = x \Rightarrow \cos(s + 2\pi) = a$.

81. If s is a real number corresponding to point (a, b), then and $s - 6\pi$ would be a co-terminal angle to an angle

   with measure s. Therefore $\sin s = y \Rightarrow \sin(s - 6\pi) = y \Rightarrow \sin(s - 6\pi) = b$ and

   $\cos s = x \Rightarrow \cos(s - 6\pi) = x \Rightarrow \cos(s - 6\pi) = a$.

82. If s is a real number corresponding to point (a, b), then $s + \pi$ would be an angle that is an additional one-half

   circle turn beyond an angle with measure s. This angle would intersect the unit circle at the point $(-a, -b)$.

   Therefore, $\sin(s + \pi) = y \Rightarrow \sin(s + \pi) = -b$ and $\cos(s + \pi) = x \Rightarrow \cos(s + \pi) = -a$.

83. If s is a real number corresponding to point (a, b), then $-s$ would be an angle that intersects the unit circle at the

   point $(a, -b)$. Therefore, $\sin -s = y \Rightarrow \sin -s = -b$ and $\cos -s = x \Rightarrow \cos -s = a$.

84. If s is a real number corresponding to point (a, b), then $-s$ would be an angle that intersects the unit circle at the

   point $(a, -b)$. Now turn one-half circle to an angle $(-s + \pi)$ with terminal side intersecting the unit circle at the

   point $(-a, b)$. Therefore, $\sin(-s + \pi) = y \Rightarrow \sin(-s + \pi) = b$ and $\cos(-s + \pi) = x \Rightarrow \cos(-s + \pi) = -a$.

85. If s is a real number corresponding to point (a, b), then $s - \dfrac{\pi}{2}$ would be an angle with an additional negative

   one-quarter turn with a terminal side intersecting the unit circle at the point $(b, -a)$. Therefore,

   $\sin\left(s - \dfrac{\pi}{2}\right) = y \Rightarrow \sin\left(s - \dfrac{\pi}{2}\right) = -a$ and $\cos\left(s - \dfrac{\pi}{2}\right) = x \Rightarrow \cos\left(s - \dfrac{\pi}{2}\right) = b$.

86. If s is a real number corresponding to point (a, b), then $s + \dfrac{\pi}{2}$ would be an angle with an additional positive

one-quarter turn with a terminal side intersecting the unit circle at the point $(-b, a)$. Therefore,

$$\sin\left(s + \dfrac{\pi}{2}\right) = y \Rightarrow \sin\left(s + \dfrac{\pi}{2}\right) = a \text{ and } \cos\left(s + \dfrac{\pi}{2}\right) = x \Rightarrow \cos\left(s + \dfrac{\pi}{2}\right) = -b.$$

Reviewing Basic Concepts (Sections 8.1 and 8.2)

1. (a) complement: $c + 35° = 90° \Rightarrow c = 55°$

    supplement: $s + 35° = 180° \Rightarrow s = 145°$

    (b) complement, since $90° = \dfrac{\pi}{2}$ radians, $c + \dfrac{\pi}{4} = \dfrac{\pi}{2} \Rightarrow c = \dfrac{2\pi}{4} - \dfrac{\pi}{4} \Rightarrow c = \dfrac{\pi}{4}.$

    supplement, since $180° = \pi$ radians, $s + \dfrac{\pi}{4} = \pi \Rightarrow s = \dfrac{4\pi}{4} - \dfrac{\pi}{4} \Rightarrow s = \dfrac{3\pi}{4}.$

2. $32.25° = 32° + .25(60)' = 32° \ 15' \ 0".$

3. $59° \ 35' \ 30" = \left(59 + \dfrac{35}{60} + \dfrac{30}{(60)(60)}\right)° = 59.591\overline{6}°.$

4. (a) The smallest positive degree measure of the coterminal angle is: $560° - 360° = 200°.$

    (b) The smallest positive radian measure of the coterminal angle is: $-\dfrac{2\pi}{3} + 2\pi = -\dfrac{2\pi}{3} + \dfrac{6\pi}{3} = \dfrac{4\pi}{3}.$

5. (a) $240° \cdot \dfrac{\pi}{180} = \dfrac{4\pi}{3}$ (b) $\dfrac{3\pi}{4} \cdot \dfrac{180}{\pi} = 135°$

6. (a) Since $120° = 120 \cdot \dfrac{\pi}{180} = \dfrac{2\pi}{3}$ and arc length is found by the formula $s = r\theta$, the arc length is:

    $s = 3\left(\dfrac{2\pi}{3}\right) \Rightarrow s = 2\pi$ cm.

    (b) Using the area of a sector formula $A = \dfrac{1}{2}r^2\theta$ yields: $A = \dfrac{1}{2}(3)^2\left(\dfrac{2\pi}{3}\right) \Rightarrow A = 3\pi$ cm^2.

7. (a) An angle of $s = -2\pi$ radians intersects the unit circle at the point $(1, 0)$.

    (b) From the properties of $45° - 45°$ right triangles, an angle of $s = \dfrac{5\pi}{4}$ radians intersects the unit circle at

    the point $\left(-\dfrac{\sqrt{2}}{2}, -\dfrac{\sqrt{2}}{2}\right).$

    (c) An angle of $s = \dfrac{5\pi}{2}$ radians intersects the unit circle at the point $(0, 1)$.

8. An angle of $s = -\dfrac{5\pi}{2}$ radians intersects the unit circle at the point $(0, -1)$. Therefore, since:

    $\sin s = y, \sin -\dfrac{5\pi}{2} = -1$ \qquad $\cos s = x, \cos -\dfrac{5\pi}{2} = 0$ \qquad $\tan s = \dfrac{y}{x}, \tan -\dfrac{5\pi}{2} = \dfrac{-1}{0} =$ undefined

    $\csc s = \dfrac{1}{y}, \csc -\dfrac{5\pi}{2} = \dfrac{1}{-1} = -1$ \quad $\sec s = \dfrac{1}{x}, \sec -\dfrac{5\pi}{2} = \dfrac{1}{0} =$ undefined \qquad $\cot s = \dfrac{x}{y}, \cot -\dfrac{5\pi}{2} = \dfrac{0}{-1} = 0$

9. (a) From the properties of $30° - 60°$ right triangles, an angle of $s = \dfrac{7\pi}{6}$ radians intersects the unit circle at the point $\left(-\dfrac{\sqrt{3}}{2}, -\dfrac{1}{2}\right)$. Therefore, since:

$$\sin s = y, \sin \frac{7\pi}{6} = -\frac{1}{2} \qquad\qquad \csc s = \frac{1}{y}, \csc \frac{7\pi}{6} = \frac{1}{-\frac{1}{2}} = -2$$

$$\cos s = x, \cos \frac{7\pi}{6} = -\frac{\sqrt{3}}{2} \qquad\qquad \sec s = \frac{1}{x}, \sec \frac{7\pi}{6} = \frac{1}{-\frac{\sqrt{3}}{2}} = -\frac{2}{\sqrt{3}} = -\frac{2\sqrt{3}}{3}$$

$$\tan s = \frac{y}{x}, \tan \frac{7\pi}{6} = \frac{-\frac{1}{2}}{-\frac{\sqrt{3}}{2}} = -\frac{1}{2} \cdot \left(-\frac{2}{\sqrt{3}}\right) = \frac{\sqrt{3}}{3} \qquad \cot s = \frac{x}{y}, \cot \frac{7\pi}{6} = \frac{-\frac{\sqrt{3}}{2}}{-\frac{1}{2}} = -\frac{\sqrt{3}}{2} \cdot \left(-\frac{2}{1}\right) = \sqrt{3}$$

(b) From the properties of $30° - 60°$ right triangles, an angle of $s = -\dfrac{2\pi}{3}$ radians intersects the unit circle at the point $\left(-\dfrac{1}{2}, -\dfrac{\sqrt{3}}{2}\right)$. Therefore, since:

$$\sin s = y, \sin -\frac{2\pi}{3} = -\frac{\sqrt{3}}{2} \qquad\qquad \csc s = \frac{1}{y}, \csc -\frac{2\pi}{3} = \frac{1}{-\frac{\sqrt{3}}{2}} = -\frac{2}{\sqrt{3}} = -\frac{2\sqrt{3}}{3}$$

$$\cos s = x, \cos -\frac{2\pi}{3} = -\frac{1}{2} \qquad\qquad \sec s = \frac{1}{x}, \sec -\frac{2\pi}{3} = \frac{1}{-\frac{1}{2}} = -2$$

$$\tan s = \frac{y}{x}, \tan -\frac{2\pi}{3} = \frac{-\frac{\sqrt{3}}{2}}{-\frac{1}{2}} = -\frac{\sqrt{3}}{2} \cdot \frac{2}{1} = \sqrt{3} \qquad \cot s = \frac{x}{y}, \cot -\frac{2\pi}{3} = \frac{-\frac{1}{2}}{-\frac{\sqrt{3}}{2}} = -\frac{1}{2} \cdot -\frac{2}{\sqrt{3}} = \frac{1}{\sqrt{3}} = \frac{\sqrt{3}}{3}$$

10. Using the calculator (radian mode), $\sin 2.25 \approx .7780731969$; $\cos 2.25 \approx -.6281736227$;

$\tan 2.25 \approx -1.238627616$; $\csc 2.25 \approx 1.285226125$; $\sec 2.25 \approx -1.591916572$; and $\cot 2.25 \approx -.8073451511$.

8.3: Graphs of the Sine and Cosine Functions

1. Since $\sin 0 = 0$, $\sin \dfrac{\pi}{2} = 1$, the amplitude is 1, and the period is $\dfrac{2\pi}{(1)} = 2\pi$, the graph is G.

2. Since $\cos 0 = 1$, $\cos \dfrac{\pi}{2} = 0$, the amplitude is 1, and the period is $\dfrac{2\pi}{(1)} = 2\pi$, the graph is A.

3. Since $-\sin 0 = 0$, $-\sin \dfrac{\pi}{2} = -1$, the amplitude is 1, and the period is $\dfrac{2\pi}{(1)} = 2\pi$, the graph is E.

4. Since $-\cos 0 = -1$, $-\cos \dfrac{\pi}{2} = 0$, the amplitude is 1, and the period is $\dfrac{2\pi}{(1)} = 2\pi$, the graph is D.

5. Since $\sin 2(0) = 0$, $\sin 2\left(\dfrac{\pi}{4}\right) = \sin \dfrac{\pi}{2} = 1$, the amplitude is 1, and the period is $\dfrac{2\pi}{(2)} = \pi$, the graph is B.

6. Since $\cos 2(0) = 1$, $\cos 2\left(\dfrac{\pi}{2}\right) = \cos \pi = -1$, the amplitude is 1, and the period is $\dfrac{2\pi}{(2)} = \pi$, the graph is H.

7. Since $2 \sin 0 = 2(0) = 0$, $2 \sin \dfrac{\pi}{2} = 2(1) = 2$, the amplitude is 2, and the period is $\dfrac{2\pi}{(1)} = 2\pi$, the graph is F.

8. Since $2 \cos 0 = 2(1) = 2$, $2 \cos \dfrac{\pi}{2} = 2(0) = 0$, the amplitude is 2, and the period is $\dfrac{2\pi}{(1)} = 2\pi$, the graph is C.

9. The graph of $y = \sin\left(x - \dfrac{\pi}{4}\right)$ is the graph of $y = \sin x$ translated $\dfrac{\pi}{4}$ units to the right, therefore the graph is D.

10. The graph of $y = \sin\left(x + \dfrac{\pi}{4}\right)$ is the graph of $y = \sin x$ translated $\dfrac{\pi}{4}$ units to the left, therefore the graph is G.

11. The graph of $y = \cos\left(x - \dfrac{\pi}{4}\right)$ is the graph of $y = \cos x$ translated $\dfrac{\pi}{4}$ units to the right, therefore the graph is H.

12. The graph of $y = \cos\left(x + \dfrac{\pi}{4}\right)$ is the graph of $y = \cos x$ translated $\dfrac{\pi}{4}$ units to the left, therefore the graph is A.

13. The graph of $y = 1 + \sin x$ is the graph of $y = \sin x$ translated 1 unit upward, therefore the graph is B.

14. The graph of $y = -1 + \sin x$ is the graph of $y = \sin x$ translated 1 unit downward, therefore the graph is E.

15. The graph of $y = 1 + \cos x$ is the graph of $y = \cos x$ translated 1 unit upward, therefore the graph is F.

16. The graph of $y = -1 + \cos x$ is the graph of $y = \cos x$ translated 1 unit downward, therefore the graph is C.

17. The equation $y = 3 \sin(2x - 4)$ has an amplitude 3; a period of $\dfrac{2\pi}{(2)} = \pi$; and a phase shift $\dfrac{4}{2} = 2$, therefore B.

18. The equation $y = 2 \sin(3x - 4)$ has an amplitude 2; a period of $\dfrac{2\pi}{(3)}$; and a phase shift $\dfrac{4}{3}$, therefore D.

19. The equation $y = 4 \sin(3x - 2)$ has an amplitude 4; a period of $\dfrac{2\pi}{(3)}$; and a phase shift $\dfrac{2}{3}$, therefore C.

20. The equation $y = 2 \sin(4x - 3)$ has an amplitude 2; a period of $\dfrac{2\pi}{(4)} = \dfrac{\pi}{2}$; and a phase shift $\dfrac{3}{4}$, therefore A.

21. With a period of 4π, solve for b: $4\pi = \dfrac{4\pi}{1} = \dfrac{2\pi}{b} \Rightarrow b = \dfrac{1}{2}$. Now since amplitude is 4, one of the possible

    answers is: $y = 4 \sin\left(\dfrac{1}{2}x\right)$.

22. With a period of 4, solve for b: $4 = \dfrac{4}{1} = \dfrac{2\pi}{b} \Rightarrow b = \dfrac{\pi}{2}$. Now since amplitude is $\dfrac{1}{4}$, one of the possible

    answers is: $y = \dfrac{1}{4} \sin\left(\dfrac{\pi}{2}x\right)$.

23. Because the maximum value of the sine function is 1 , the maximum value of $\sin\left[2\left(x - \dfrac{\pi}{2}\right)\right]$ is 1 , the

    maximum value of $3 \sin\left[2\left(x - \dfrac{\pi}{2}\right)\right]$ is 3 , and thus the maximum value of $-5 + 3 \sin\left[2\left(x - \dfrac{\pi}{2}\right)\right]$ is –2.

24. Because the minimum value of the sine function is –1 , the minimum value of $\sin\left[2\left(x - \dfrac{\pi}{2}\right)\right]$ is –1 , the

    minimum value of $3 \sin\left[2\left(x - \dfrac{\pi}{2}\right)\right]$ is –3 , and thus the minimum value of $-5 + 3 \sin\left[2\left(x - \dfrac{\pi}{2}\right)\right]$ is –8.

25. From the answers, the range of $f(x)$: $[-8, -2]$;

    The trig viewing window as defined in the text has Ymin $= -4$ and Ymax $= 4$. Because the range of f includes

    values less than -4, the local minimum points will not appear in the trig viewing window.

26. Using the range from problem 25, . . . , Ymax must be at least -2 and Ymin must be at most -8.

27. The period of f is $\dfrac{2\pi}{2} = \pi$, and the distance between -2π and 2π is 4π units. Since $\dfrac{4\pi}{\pi} = 4$, the interval

    $[-2\pi, 2\pi]$ will show 4 periods of the graph.

28. See Figure 28.

$[-2\pi, 2\pi]$ by $[-10, 5]$
Xscl $= \pi/2$ Yscl $= 1$

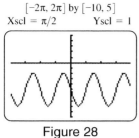

Figure 28

29. Setting the calculator at the range values would remove the borders, therefore Ymin $= -8$ and Ymax $= -2$.

30. $f(-2) = -5 + 3 \sin\left[2\left((-2) - \dfrac{\pi}{2}\right)\right] \Rightarrow f(-2) \approx -7.270407486$ and $f(-2 + \pi) \approx -7.270407486$.

    This happens because the period of f is π, $f(p) = f(p + \pi)$ for any value of p. In this case $p = -2$.

31. See Figure 31. From the equation, the amplitude is 2.

32. See Figure 32. From the equation, the amplitude is 3.

33. See Figure 33. From the equation, the amplitude is $\dfrac{2}{3}$.

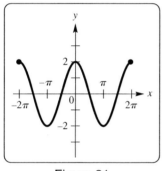

Figure 31

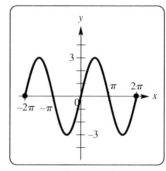

Figure 32

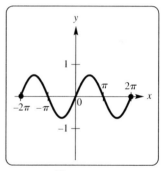

Figure 33

34. See Figure 34. From the equation, the amplitude is $\dfrac{3}{4}$.

35. See Figure 35. From the equation, the amplitude is 1.

36. See Figure 36. From the equation, the amplitude is 1.

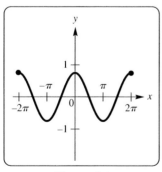

Figure 34

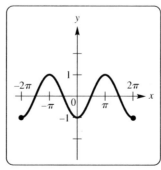

Figure 35

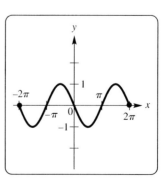

Figure 36

37. See Figure 37. From the equation, the amplitude is 2.

38. See Figure 38. From the equation, the amplitude is 3.

39. See Figure 39. From the equation, the period is 4π and the amplitude is 1.

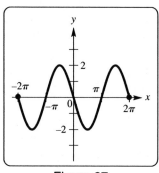

Figure 37

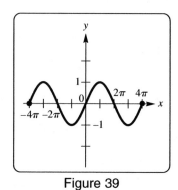

Figure 38

Figure 39

40. See Figure 40. From the equation, the period is 3π and the amplitude is 1.

41. See Figure 41. From the equation, the period is π and the amplitude is 1.

42. See Figure 42. From the equation, the period is $\dfrac{8\pi}{3}$ and the amplitude is 1.

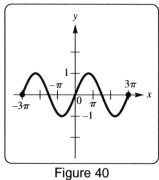

Figure 40

Figure 41

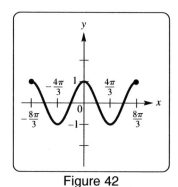

Figure 42

43. See Figure 43. From the equation, the period is 8π and the amplitude is 2.

44. See Figure 44. From the equation, the period is π and the amplitude is 3.

45. See Figure 45. From the equation, the period is $\dfrac{2\pi}{3}$ and the amplitude is 2.

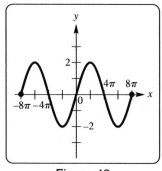

Figure 43

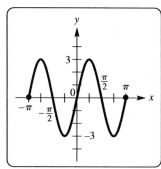

Figure 44

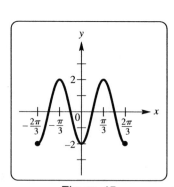

Figure 45

46. See Figure 46. From the equation, the period is π and the amplitude is 5.

47. See Figure 47. From the equation, the phase shift is $\dfrac{\pi}{4}$.

48. See Figure 48. From the equation, the phase shift is $\dfrac{\pi}{3}$.

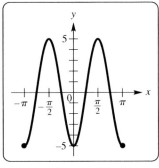

Figure 46

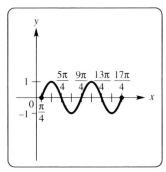

Figure 47

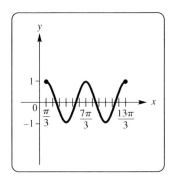

Figure 48

49. See Figure 49. From the equation, the phase shift is $\dfrac{\pi}{3}$.

50. See Figure 50. From the equation, the phase shift is $\dfrac{3\pi}{2}$.

51. (a) From the equation, the amplitude is: $|-4| = 4$.

    (b) From the equation, the period is: $\dfrac{2\pi}{(2)} = \pi$.

    (c) Since $y = -4 \sin (2x - \pi) \;\Rightarrow\; y = -4 \sin \left[2\left(x - \dfrac{\pi}{2} \right) \right]$, the phase shift is: $\dfrac{\pi}{2}$.

    (d) No vertical translation.

    (e) The range is: $[-4, 4]$. See Figure 51.

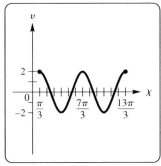

Figure 49

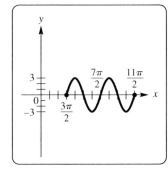

Figure 50

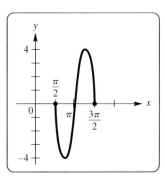

Figure 51

52. (a) From the equation, the amplitude is: 3.

    (b) From the equation, the period is: $\dfrac{2\pi}{(4)} = \dfrac{\pi}{2}$.

    (c) Since $y = 3\cos(4x + \pi) \Rightarrow y = 3\cos\left[4\left(x + \dfrac{\pi}{4}\right)\right]$, the phase shift is: $-\dfrac{\pi}{4}$.

    (d) No vertical translation.

    (e) The range is: $[-3, 3]$. See Figure 52.

53. (a) From the equation, the amplitude is: $\dfrac{1}{2}$.

    (b) From the equation, the period is: $\dfrac{2\pi}{\left(\frac{1}{2}\right)} = 4\pi$.

    (c) Since $y = \dfrac{1}{2}\cos\left(\dfrac{1}{2}x - \dfrac{\pi}{4}\right) \Rightarrow y = \dfrac{1}{2}\cos\left[\dfrac{1}{2}\left(x - \dfrac{\pi}{2}\right)\right]$, the phase shift is: $\dfrac{\pi}{2}$.

    (d) No vertical translation.

    (e) The range is: $\left[-\dfrac{1}{2}, \dfrac{1}{2}\right]$. See Figure 53.

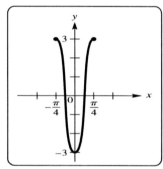

Figure 52

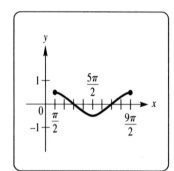

Figure 53

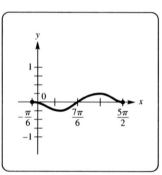

Figure 54

54. (a) From the equation, the amplitude is: $\dfrac{1}{4}$.

    (b) From the equation, the period is: $\dfrac{2\pi}{\left(\frac{3}{4}\right)} = \dfrac{8\pi}{3}$.

    (c) Since $y = -\dfrac{1}{4}\sin\left(\dfrac{3}{4}x + \dfrac{\pi}{8}\right) \Rightarrow y = -\dfrac{1}{4}\sin\left[\dfrac{3}{4}\left(x + \dfrac{\pi}{6}\right)\right]$, the phase shift is: $-\dfrac{\pi}{6}$.

    (d) No vertical translation.

    (e) The range is: $\left[-\dfrac{1}{4}, \dfrac{1}{4}\right]$. See Figure 54.

55. (a) From the equation, the amplitude is: $\left|-\dfrac{2}{3}\right| = \dfrac{2}{3}$.

    (b) From the equation, the period is: $\dfrac{2\pi}{\left(\frac{3}{4}\right)} = \dfrac{8\pi}{3}$.

    (c) From the equation, there is no phase shift.

    (d) From the equation, the vertical translation is: upward 1 unit.

    (e) The range is: $\left[-\dfrac{2}{3}, \dfrac{2}{3}\right]$ translated upward 1 unit, therefore, $\left[\dfrac{1}{3}, \dfrac{5}{3}\right]$. See Figure 55.

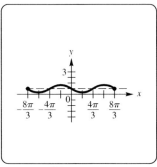

Figure 55

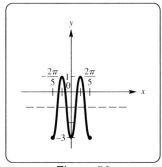

Figure 56

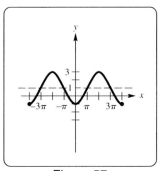

Figure 57

56. (a) From the equation, the amplitude is: $|-2| = 2$.

   (b) From the equation, the period is: $\dfrac{2\pi}{(5)}$.

   (c) From the equation, there is no phase shift.

   (d) From the equation, the vertical translation is: downward 1 unit.

   (e) The range is: $[-2, 2]$ translated downward 1 unit, therefore, $[-3, 1]$. See Figure 56.

57. (a) From the equation, the amplitude is: $|-2| = 2$.

   (b) From the equation, the period is: $\dfrac{2\pi}{\left(\frac{1}{2}\right)} = 4\pi$.

   (c) From the equation, there is no phase shift.

   (d) From the equation, the vertical translation is: upward 1 unit.

   (e) The range is: $[-2, 2]$ translated upward 1 unit, therefore, $[-1, 3]$. See Figure 57.

58. (a) From the equation, the amplitude is: $|-3| = 3$.

   (b) From the equation, the period is: $\dfrac{2\pi}{\left(\frac{1}{2}\right)} = 4\pi$.

   (c) From the equation, there is no phase shift.

   (d) From the equation, the vertical translation is: downward 3 units.

   (e) The range is: $[-3, 3]$ translated downward 3 units, therefore, $[-6, 0]$. See Figure 58.

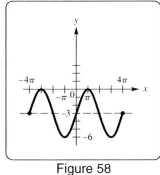

Figure 58

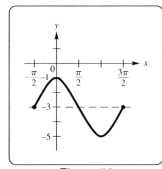

Figure 59

59. (a) From the equation, the amplitude is: 2.

   (b) From the equation, the period is: $\dfrac{2\pi}{(1)} = 2\pi$.

   (c) From the equation, the phase shift is: $-\dfrac{\pi}{2}$.

   (d) From the equation, the vertical translation is: downward 3 unit.

   (e) The range is: $[-2, 2]$ translated downward 3 units, therefore, $[-5, -1]$. See Figure 59.

60. (a) From the equation, the amplitude is: $|-3| = 3$.

    (b) From the equation, the period is: $\dfrac{2\pi}{(1)} = 2\pi$.

    (c) From the equation, the phase shift is: π.

    (d) From the equation, the vertical translation is: upward 4 units.

    (e) The range is: $[-3, 3]$ translated upward 4 units, therefore, $[1, 7]$. See Figure 60.

61. (a) From the equation, the amplitude is: 1.

    (b) From the equation, the period is: $\dfrac{2\pi}{(2)} = \pi$.

    (c) From the equation, the phase shift is: $-\dfrac{\pi}{4}$.

    (d) From the equation, the vertical translation is: upward $\dfrac{1}{2}$ unit.

    (e) The range is: $[-1, 1]$ translated upward $\dfrac{1}{2}$ unit, therefore, $\left[-\dfrac{1}{2}, \dfrac{3}{2}\right]$. See Figure 61.

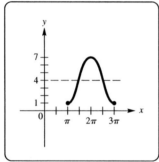

Figure 60

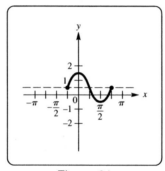

Figure 61

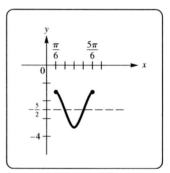

Figure 62

62. (a) From the equation, the amplitude is: 1.

    (b) From the equation, the period is: $\dfrac{2\pi}{(3)}$.

    (c) From the equation, the phase shift is: $\dfrac{\pi}{6}$.

    (d) From the equation, the vertical translation is: downward $\dfrac{5}{2}$ units.

    (e) The range is: $[-1, 1]$ translated downward $\dfrac{5}{2}$ units, therefore, $\left[-\dfrac{7}{2}, -\dfrac{3}{2}\right]$. See Figure 62.

63. (a) From the equation, the amplitude is: 2.

    (b) From the equation, the period is: $\dfrac{2\pi}{(1)} = 2\pi$.

    (c) From the equation, the phase shift is: π.

    (d) From the equation, there is no vertical translation.

    (e) The range is: $[-2, 2]$. See Figure 63.

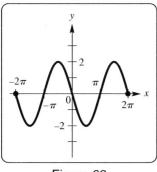

Figure 63   Figure 64

64. (a) From the equation, the amplitude is: $\frac{2}{3}$.

    (b) From the equation, the period is: $\frac{2\pi}{1} = 2\pi$.

    (c) From the equation, the phase shift is: $-\frac{\pi}{2}$.

    (d) From the equation, there is no vertical translation.

    (e) The range is: $\left[-\frac{2}{3}, \frac{2}{3}\right]$. See Figure 64.

65. (a) From the equation, the amplitude is: 4.

    (b) From the equation, the period is: $\frac{2\pi}{\left(\frac{1}{2}\right)} = 4\pi$.

    (c) Since $y = 4\cos\left(\frac{1}{2}x + \frac{\pi}{2}\right) \Rightarrow y = 4\cos\left[\frac{1}{2}(x + \pi)\right]$, the phase shift is: $-\pi$.

    (d) From the equation, there is no vertical translation.

    (e) The range is: $[-4, 4]$. See Figure 65.

66. (a) From the equation, the amplitude is: $|-1| = 1$.

    (b) From the equation, the period is: $\frac{2\pi}{\left(\frac{2}{3}\right)} = 3\pi$.

    (c) From the equation, the phase shift is: $\frac{\pi}{3}$.

    (d) From the equation, there is no vertical translation.

    (e) The range is: $[-1, 1]$. See Figure 66.

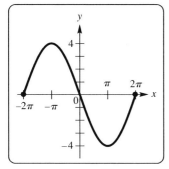

Figure 65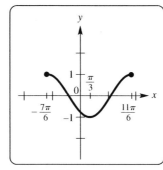

Figure 66

67. (a) From the equation, the amplitude is: $|-1| = 1$.

    (b) From the equation, the period is: $\dfrac{2\pi}{(3)}$.

    (c) Since $y = 2 - \sin\left(3x - \dfrac{\pi}{5}\right) \Rightarrow y = 2 - \sin\left[3\left(x - \dfrac{\pi}{15}\right)\right]$, the phase shift is: $\dfrac{\pi}{15}$.

    (d) From the equation, the vertical translation is: upward 2 units.

    (e) The range is: $[-1, 1]$ translated upward 2 units, therefore, $[1, 3]$. See Figure 67.

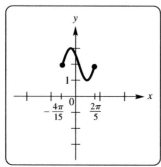   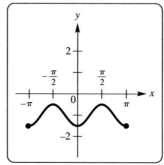

Figure 67   Figure 68

68. (a) From the equation, the amplitude is: $\dfrac{1}{2}$.

    (b) From the equation, the period is: $\dfrac{2\pi}{(2)} = \pi$.

    (c) Since $y = -1 + \dfrac{1}{2}\cos(2x - 3\pi) \Rightarrow y = -1 + \dfrac{1}{2}\cos\left[2\left(x - \dfrac{3\pi}{2}\right)\right]$, the phase shift is: $\dfrac{3\pi}{2}$.

    (d) From the equation, the vertical translation is: downward 1 unit.

    (e) The range is: $\left[-\dfrac{1}{2}, \dfrac{1}{2}\right]$ translated downward 1 unit, therefore, $\left[-\dfrac{3}{2}, -\dfrac{1}{2}\right]$. See Figure 68.

69. (a) From the graph, the maximum is 40° F and the minimum is −40° F.

    (b) From the graph, the amplitude is 40 and the period is 12. Since there is no vertical translation, the amplitude will help us find the maximum and minimum average temperature values of 40° and −40° and the period represents the 12 months of the calendar year. The monthly average temperatures vary by 80°F over a 12-month period.

    (c) The *x*-intercepts represents the two months when the average temperature is 0° F.

70. (a) Since amplitude helps us find the maximum and minimum values of 50° and −50° the new equation would be $y = 50 \cos\left[\dfrac{\pi}{6}(x - 7)\right]$.

    (b) Since the maximum and minimum temperature values of 60° and −20° are both 20° added on to the original maximum and minimum temperature values of 40° and −40°, the graph is translated up 20 units and the equation is: $y = 40 \cos\left[\dfrac{\pi}{6}(x - 7)\right] + 20$.

    (c) Since the *x* values for the months February and August ($x = 2$ and $x = 8$) are both 1 more then the *x* values for the months of January and July ($x = 1$ and $x = 7$) the minimum and maximum average temperature months, the graph is translated 1 unit to the right and the equation is: $y = 40 \cos\left[\dfrac{\pi}{6}(x - 8)\right]$.

71. (a) From the graph, the maximum is 87° F in July, and the minimum is 62° F in January or late December.

    (b) Since a minimum temperature 50° would increase the range of temperatures, the amplitude would increase.
    See Figure 71.

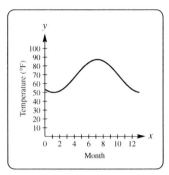

Figure 71

72. (a) Since the range of temperatures is 30°, the amplitude is $\dfrac{30°}{2} = 15°$.

    Therefore the equation is: $y = 15 \sin \left[\dfrac{\pi}{6}(x - 4.2) \right] + 75$.

    (b) Since the range of temperatures is 20°, the amplitude is $\dfrac{20°}{2} = 10°$.

    Therefore the equation is: $y = 10 \sin \left[\dfrac{\pi}{6}(x - 4.2) \right] + 75$.

73. The graph repeats each day, so the period is 24 hours.

74. The range of approximately $2.6 - .2 = 2.4$ yields a amplitude of $\dfrac{2.4}{2} = 1.2$.

75. From the graph, low tide took place at approximately 6 P.M.; at approximately .2 feet.

76. The low tide occurred at 6 P.M. + 1.19 or 7:19 P.M.; and the height was approximately $.2 - .2 = 0$ feet.

77. From the graph, high tide took place at approximately 2 A.M.; at approximately 2.6 feet.

78. The high tide occurred at 2 A.M. + 1.18 or 3:18 A.M.; and the height was approximately $2.6 - .2 = 2.4$ feet.

79. (a) From the graph, the highest temperature is 80° F and the lowest temperature is 50° F.

    (b) The amplitude is: $\dfrac{80° - 50°}{2} = 15°$.

    (c) From the graph, the period is approximately 35,000 years.

    (d) From the graph, the trend of the temperature now is downward.

80. (a) From the graph, the amplitude is: $\dfrac{120 - 80}{2} = 20$.

    (b) Since the period is .8 seconds, there are $\dfrac{1}{.8} = 1.25$ bps, and the pulse rate is $60(1.25) = 75$ bpm.

81. (a) The latest time that the animals begin their evening activity is approximately 8:00 P.M., and the earliest is
    approximately 4:00 P.M. Since there is a difference of 4 hours in these times, the amplitude is: $\dfrac{1}{2}(4) = 2$ hours.

    (b) From the graph, the length of the period is 1 year.

82. (a) From the equation, the amplitude is: 5; and the period is: $\dfrac{2\pi}{120\pi} = \dfrac{1}{60}$ second.

    (b) The number of cycles per second: $\dfrac{1}{\text{Period}} = \dfrac{1}{\frac{1}{60}} = 60$ cycles per second.

    (c) When $t = 0$, $E = 5\cos(120\pi)(0) = 5$. When $t = .03$, $E = 5\cos(120\pi)(.03) \approx 1.545$.

    When $t = .06$, $E = 5\cos(120\pi)(.06) \approx -4.045$. When $t = .09$, $E = 5\cos(120\pi)(.09) \approx -4.045$.

    When $t = .12$, $E = 5\cos(120\pi)(.12) \approx 1.545$.

    (d) See Figure 82.

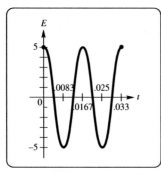

Figure 82

83. (a) From the equation, the amplitude is: 3.8; and the period is: $\dfrac{2\pi}{40\pi} = \dfrac{1}{20}$ second.

    (b) The number of cycles per second: $\dfrac{1}{\text{Period}} = \dfrac{1}{\frac{1}{20}} = 20$ cycles per second.

    (c) When $t = .02$, $E = 3.8\cos(40\pi)(.02) \approx -3.074$. When $t = .04$, $E = 3.8\cos(40\pi)(.04) \approx 1.174$.

    When $t = .08$, $E = 3.8\cos(40\pi)(.08) \approx -3.074$. When $t = .12$, $E = 3.8\cos(40\pi)(.12) \approx -3.074$.

    When $t = .14$, $E = 3.8\cos(40\pi)(.14) \approx 1.174$.

    (d) See Figure 83.

84. (a) See Figure 84.

    (b) Since $-1 \le \sin x \le 1$, the maximum occurs when $\sin(2\pi x) = 1$. Therefore,

    $$2\pi x = \frac{\pi}{2} \Rightarrow x = \frac{1}{4};\, 2\pi x = \frac{5\pi}{2} \Rightarrow x = \frac{5}{4};\text{ and } 2\pi x = \frac{9\pi}{2} \Rightarrow x = \frac{9}{4};\text{ etc.}$$

    Since $-1 \le \sin x \le 1$, the minimum occurs when $\sin(2\pi x) = -1$. Therefore,

    $$2\pi x = \frac{3\pi}{2} \Rightarrow x = \frac{3}{4};\, 2\pi x = \frac{7\pi}{2} \Rightarrow x = \frac{7}{4};\text{ and } 2\pi x = \frac{11\pi}{2} \Rightarrow x = \frac{11}{4};\text{ etc.}$$

    (c) Answers will vary. The quadratic function causes the graph to increase overall, while the sine function

    causes the graph to oscillate.

85. (a) See Figure 85.

    (b) Answers will vary. The seasons are more dramatic in Alaska.

    (c) If $x = 1970$ corresponds to 1970, then the equation is:

    $$C(x) = .04(x - 1970)^2 + .6(x - 1970) + 330 + 7.5\sin[2\pi(x - 1970)].$$

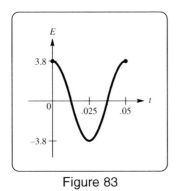

Figure 83

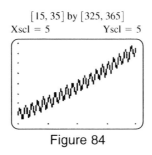

$[15, 35]$ by $[325, 365]$
Xscl = 5   Yscl = 5

Figure 84

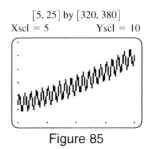

$[5, 25]$ by $[320, 380]$
Xscl = 5   Yscl = 10

Figure 85

86. (a) See Figure 86a. Yes, from the graph, the data seems to indicate a translated sine graph.

(b) See Figure 86b. The line c represents the average yearly temperature, therefore $c = 50$.

(c) The amplitude is the difference of the maximum and minimum divided by 2: $a = \dfrac{64 - 36}{2} = 14$.

The Period, since the temperature cycles every 12 months, is: 12. Which in radians is: $b = \dfrac{2\pi}{(12)} = \dfrac{\pi}{6}$.

One way to find the phase shift d, is to consider the month where the minimum temperature is found.

Since the coldest month is January, when $x = 1$, $b(x - d)$ must equal $\left(-\dfrac{\pi}{2} + 2\pi n\right)$, where n is an integer,

since the sine function is a minimum at these values. Letting $n = 0$, we can solve for d:

$\dfrac{\pi}{6}(1 - d) = -\dfrac{\pi}{2} \Rightarrow 1 - d = -3 \Rightarrow d = 4$. The table shows that temperatures are actually a little

warmer after July than before, so we choose $d = 4.2$ for a better approximation.

(d) Using a, b, c, and d from parts b and c yields: $f(x) = 14 \sin\left[\dfrac{\pi}{6}(x - 4.2)\right] + 50$.

(e) See Figure 86e. The function gives an excellent model for the given data.

(f) Set the calculator to the nearest hundredth and the regression equation is, see Figure 86f.

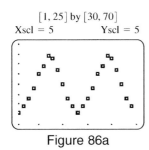

$[1, 25]$ by $[30, 70]$
Xscl = 5   Yscl = 5

Figure 86a

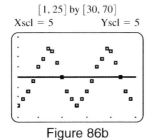

$[1, 25]$ by $[30, 70]$
Xscl = 5   Yscl = 5

Figure 86b

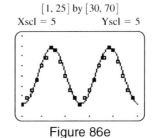

$[1, 25]$ by $[30, 70]$
Xscl = 5   Yscl = 5

Figure 86e

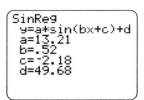

SinReg
y=a*sin(bx+c)+d
a=13.21
b=.52
c=-2.18
d=49.68

Figure 86f

87. (a) We predict the average yearly temperature by finding the mean of the average monthly temperatures:

$$\frac{51 + 55 + 63 + 67 + 77 + 86 + 90 + 90 + 84 + 71 + 59 + 52}{12} = \frac{845}{12} = 70.4° \text{ F.}$$

This is very close to the actual value of 70° F.

(b) See Figure 87b.

(c) Since amplitude $a \approx \dfrac{90 - 51}{2} \Rightarrow a \approx 19.5$; the period is 12 months or $b = \dfrac{2\pi}{(12)} = \dfrac{\pi}{6}$; the vertical

translation is $c = \dfrac{51 + 90}{2} \Rightarrow c = 70.5$; and the phase shift can be found by using the minimum

temperature value. Since the coldest month is January, when $x = 1$, $b(1 - d)$ must equal $(-\pi + 2\pi n)$,

where n is an integer, since the cosine function is a minimum at these values. Letting $n = 0$, we can solve

for d: $\dfrac{\pi}{6}(1 - d) = -\pi \Rightarrow 1 - d = -6 \Rightarrow d = 7$. The table shows that temperatures are actually a little

warmer after July than before, so we choose $d = 7.2$ for a better approximation. Now using a, b, c, and d

yields: $f(x) = 19.5 \cos\left[\dfrac{\pi}{6}(x - 7.2)\right] + 70.5$.

(d) See Figure 87d. The function gives an excellent model for the given data.

(e) Set the calculator to the nearest hundredth and the regression equation is, see Figure 87e.

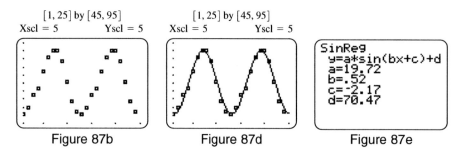

Figure 87b   Figure 87d   Figure 87e

88. Set the calculator to the nearest hundredth and the regression equation is, see Figure 88a.

See Figure 88b, for the scatter diagram and function graph.

89. Set the calculator to the nearest hundredth and the regression equation is, see Figure 89a.

See Figure 89b, for the scatter diagram and function graph.

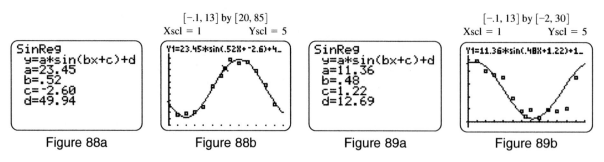

Figure 88a   Figure 88b   Figure 89a   Figure 89b

8.4 Graphs of the Other Circular Functions

1. The basic cosecant graph reflected across the x-axis, therefore graph B.

2. The basic secant graph reflected across the x-axis, therefore graph C.

3. The basic tangent graph reflected across the x-axis, therefore graph E.

4. The basic cotangent graph reflected across the x-axis, therefore graph A.

5. The basic tangent graph translated $\dfrac{\pi}{4}$ units to the right, therefore graph D.

6. The basic cotangent graph translated $\dfrac{\pi}{4}$ units to the right , therefore graph F.

7. True, $x = \dfrac{\pi}{2}$ is the smallest positive asymptote of the tangent function.

8. False, the smallest such k is π.

9. True, both the tangent and secant values will be undefined when $x = \dfrac{\pi}{2} + \pi k$.

10. False, secant values are undefined when $x = \dfrac{\pi}{2} + \pi k$, while cosecant values are undefined when $x = \pi k$.

11. False, $\tan(-x) = -\tan x$ for all x in the domain.

12. True, $\sec(-x) = \sec x$ for all x in the domain of $\sec x$.

13. (a) From the equation, the period is: $\dfrac{2\pi}{\left(\frac{1}{2}\right)} = 4\pi$.

    (b) From the equation, there is no phase shift.

    (c) From the equation, the range is $(-\infty, -2] \cup [2, \infty)$.

14. (a) From the equation, the period is: $\dfrac{2\pi}{(2)} = \pi$.

    (b) From the equation, there is no phase shift.

    (c) From the equation, the range is $(-\infty, -3] \cup [3, \infty)$.

15. (a) From the equation, the period is: $\dfrac{2\pi}{(1)} = 2\pi$.

    (b) From the equation, the phase shift is: $-\dfrac{\pi}{2}$. $\left(\dfrac{\pi}{2} \text{ units to the left}\right)$

    (c) From the equation, the range is $(-\infty, -2] \cup [2, \infty)$.

16. (a) From the equation, the period is: $\dfrac{2\pi}{(1)} = 2\pi$.

    (b) From the equation, the phase shift is: π. (π units to the right)

    (c) From the equation, the range is $\left(-\infty, -\dfrac{3}{2}\right] \cup \left[\dfrac{3}{2}, \infty\right)$.

17. (a) From the equation, the period is: $\dfrac{\pi}{\left(\frac{1}{3}\right)} = 3\pi$. (The phase shift for the basic cotangent function is π).

    (b) From the equation, the phase shift is: $\dfrac{\pi}{2}$. $\left(\dfrac{\pi}{2} \text{ units to the right}\right)$

    (c) The range of all cotangent functions is $(-\infty, \infty)$.

18. (a) From the equation, the period is: $\dfrac{\pi}{\left(\frac{1}{2}\right)} = 2\pi$. (The phase shift for the basic tangent function is π).

    (b) From the equation, the phase shift is: $-\dfrac{\pi}{4}$. $\left(\dfrac{\pi}{4} \text{ units to the left}\right)$

    (c) The range of all tangent functions is $(-\infty, \infty)$.

19. (a) From the equation, the period is: $\dfrac{2\pi}{(2)} = \pi$.

    (b) From the equation, the phase shift is: $-\dfrac{\pi}{2}$. $\left(\dfrac{\pi}{2} \text{ units to the left}\right)$

    (c) From the equation, the range is $\left(-\infty, -\dfrac{1}{2}\right] \cup \left[\dfrac{1}{2}, \infty\right)$.

20. (a) From the equation, the period is: $\dfrac{2\pi}{\left(\frac{1}{2}\right)} = 4\pi$.

    (b) Since $y = -\dfrac{1}{3}\csc\left(\dfrac{1}{2}x - \dfrac{\pi}{2}\right) \Rightarrow y = -\dfrac{1}{3}\csc\dfrac{1}{2}(x - \pi)$ and now from the this equation,

    the phase shift is: π. (π units to the right)

    (c) From the equation, the range is $\left(-\infty, -\dfrac{1}{3}\right] \cup \left[\dfrac{1}{3}, \infty\right)$.

21. (a) From the equation, the period is: $\dfrac{\pi}{(1)} = \pi$. (The phase shift for the basic tangent function is π).

    (b) From the equation, the phase shift is: $-\dfrac{\pi}{4}$. $\left(\dfrac{\pi}{4} \text{ units to the left}\right)$

    (c) The range of all tangent functions is $(-\infty, \infty)$.

22. (a) From the equation, the period is: $\dfrac{\pi}{(2)}$. (The phase shift for the basic cotangent function is π).

    (b) Since $y = 2 + \cot\left(2x - \dfrac{\pi}{3}\right) \Rightarrow y = 2 + \cot 2\left(x - \dfrac{\pi}{6}\right)$ and now from the this equation,

    the phase shift is: $\dfrac{\pi}{6}$. $\left(\dfrac{\pi}{6} \text{ units to the right}\right)$

    (c) The range of all cotangent functions is $(-\infty, \infty)$.

23. Graph a secant function with a period of $\dfrac{2\pi}{\left(\frac{1}{4}\right)} = 8\pi$, a range of $(-\infty, -3] \cup [3, \infty)$, and no phase shift.

    See Figure 23.

24. Graph a secant function with a period of $\dfrac{2\pi}{\left(\frac{1}{2}\right)} = 4\pi$, a range of $(-\infty, -2] \cup [2, \infty)$, reflected across the x-axis,

    and no phase shift. See Figure 24.

25. Graph a cosecant function with a period of $\dfrac{2\pi}{(1)} = 2\pi$, a range of $\left(-\infty, -\dfrac{1}{2}\right] \cup \left[\dfrac{1}{2}, \infty\right)$, reflected across the

    x-axis, and a phase shift $\dfrac{\pi}{2}$ units to the left. See Figure 25.

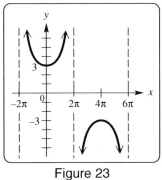

Figure 23

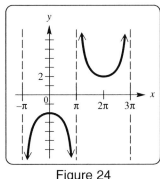

Figure 24

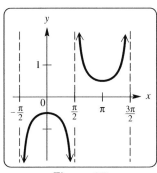

Figure 25

26. Graph a cosecant function with a period of $\dfrac{2\pi}{(1)} = 2\pi$, a range of $\left(-\infty, -\dfrac{1}{2}\right] \cup \left[\dfrac{1}{2}, \infty\right)$, and a phase shift $\dfrac{\pi}{2}$ units to the right. See Figure 26.

27. Graph a cosecant function with a period of $\dfrac{2\pi}{(1)} = 2\pi$, and a phase shift $\dfrac{\pi}{4}$ units to the right. See Figure 27.

28. Graph a secant function with a period of $\dfrac{2\pi}{(1)} = 2\pi$, and a phase shift $\dfrac{3\pi}{4}$ units to the left. See Figure 28.

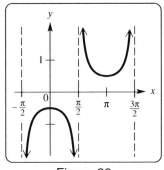

Figure 26

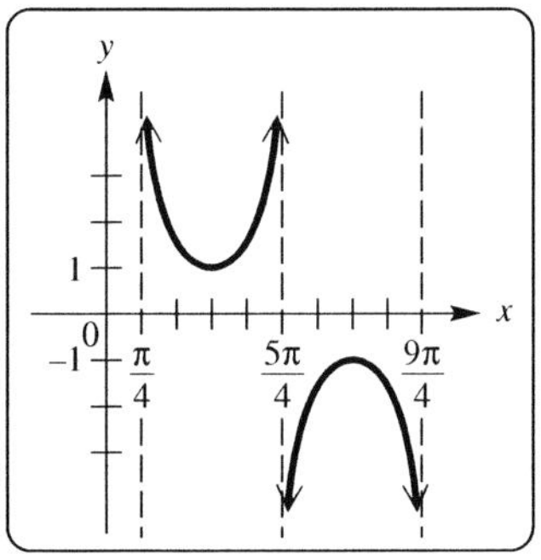

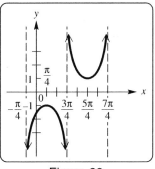

Figure 27

Figure 28

29. Graph a secant function with a period of $\dfrac{2\pi}{(1)} = 2\pi$, and a phase shift $\dfrac{\pi}{4}$ units to the left. See Figure 29.

30. Graph a cosecant function with a period of $\dfrac{2\pi}{(1)} = 2\pi$, and a phase shift $\dfrac{\pi}{3}$ units to the left. See Figure 30.

31. Since $y = \sec\left(\dfrac{1}{2}x + \dfrac{\pi}{3}\right) \Rightarrow y = \sec\dfrac{1}{2}\left(x + \dfrac{2\pi}{3}\right)$, graph a secant function with a period $\dfrac{2\pi}{\left(\frac{1}{2}\right)} = 4\pi$, and a phase shift $\dfrac{2\pi}{3}$ units to the left. See Figure 31.

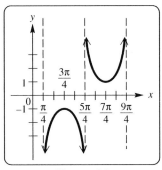

Figure 29

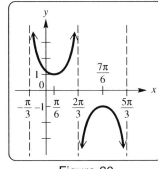

Figure 30

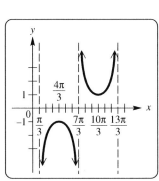

Figure 31

32. Since $y = \csc\left(\dfrac{1}{2}x - \dfrac{\pi}{4}\right) \Rightarrow y = \csc\dfrac{1}{2}\left(x - \dfrac{\pi}{2}\right)$, graph a cosecant function with a period $\dfrac{2\pi}{\left(\frac{1}{2}\right)} = 4\pi$,

and a phase shift $\dfrac{\pi}{2}$ units to the right. See Figure 32.

33. Since $y = 2 + 3\sec(2x - \pi) \Rightarrow y = 2 + 3\sec 2\left(x - \dfrac{\pi}{2}\right)$, graph a secant function with a period $\dfrac{2\pi}{(2)} = \pi$,

a vertical shift 2 units upward therefore a range of $(-\infty, -1] \cup [5, \infty)$, and a phase shift $\dfrac{\pi}{2}$ units to the right.
See Figure 33.

34. Graph a cosecant function with a period $\dfrac{2\pi}{(1)} = 2\pi$, reflected across the x-axis, a vertical shift 1 unit upward

therefore a range of $(-\infty, -1] \cup [3, \infty)$, and a phase shift $\dfrac{\pi}{2}$ units to the left. See Figure 34.

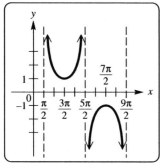

Figure 32

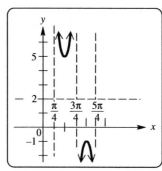

Figure 33

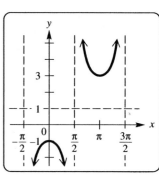

Figure 34

35. Graph a cosecant function with a period $\dfrac{2\pi}{(1)} = 2\pi$, reflected across the x-axis, a vertical shift 1 unit upward

therefore a range of $\left(-\infty, \dfrac{1}{2}\right] \cup \left[\dfrac{3}{2}, \infty\right)$, and a phase shift $\dfrac{3\pi}{4}$ units to the right. See Figure 35.

36. Since $y = 2 + \dfrac{1}{4}\sec\left(\dfrac{1}{2}x - \pi\right) \Rightarrow y = 2 + \dfrac{1}{4}\sec\dfrac{1}{2}(x - 2\pi)$, graph a secant function with a period

$\dfrac{2\pi}{\left(\frac{1}{2}\right)} = 4\pi$, a vertical shift 2 units upward therefore a range of $\left(-\infty, \dfrac{7}{4}\right] \cup \left[\dfrac{9}{4}, \infty\right)$, and a phase shift 2π units
to the right. See Figure 36.

37. Graph a tangent function with a period $\dfrac{\pi}{(4)}$. See Figure 37.

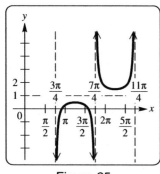

Figure 35

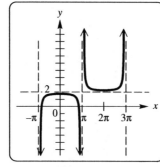

Figure 36

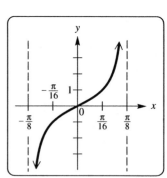

Figure 37

38. Graph a tangent function with a period $\dfrac{\pi}{\left(\frac{1}{2}\right)} = 2\pi$. See Figure 38.

39. Graph a tangent function with a period $\dfrac{\pi}{(1)} = \pi$; and passing through the points $(0, 0)$ (midpoint),

 $\left(-\dfrac{\pi}{4}, -2\right)$, and $\left(\dfrac{\pi}{4}, 2\right)$. See Figure 39.

40. Graph a cotangent function with a period $\dfrac{\pi}{(1)} = \pi$; and passing through the points $\left(\dfrac{\pi}{2}, 0\right)$ (midpoint),

 $\left(\dfrac{\pi}{4}, 2\right)$, and $\left(\dfrac{3\pi}{4}, -2\right)$. See Figure 40.

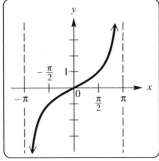

Figure 38

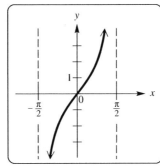

Figure 39

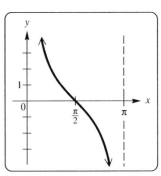

Figure 40

41. Graph a tangent function with a period $\dfrac{\pi}{\left(\frac{1}{4}\right)} = 4\pi$; and passing through the points $(0, 0)$ (midpoint),

 $(-\pi, -2)$, and $(\pi, 2)$. See Figure 41.

42. Graph a cotangent function with a period $\dfrac{\pi}{(1)} = \pi$; and passing through the points $\left(\dfrac{\pi}{2}, 0\right)$ (midpoint),

 $\left(\dfrac{\pi}{4}, \dfrac{1}{2}\right)$, and $\left(\dfrac{3\pi}{4}, -\dfrac{1}{2}\right)$. See Figure 42.

43. Graph a cotangent function with a period $\dfrac{\pi}{(3)}$; and passing through the points $\left(\dfrac{\pi}{6}, 0\right)$ (midpoint),

 $\left(\dfrac{\pi}{12}, 1\right)$, and $\left(\dfrac{\pi}{4}, -1\right)$. See Figure 43.

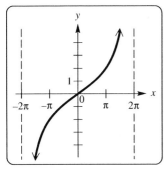

Figure 41

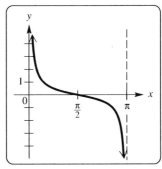

Figure 42

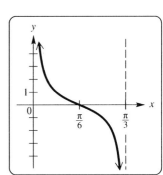

Figure 43

44. Graph a cotangent function with a period $\dfrac{\pi}{\left(\frac{1}{2}\right)} = 2\pi$, is reflected across the *x*-axis, and passes through the points

$(\pi, 0)$ (midpoint), $\left(\dfrac{\pi}{2}, -1\right)$, and $\left(\dfrac{3\pi}{2}, 1\right)$. See Figure 44.

45. Graph a tangent function with a period $\dfrac{\pi}{\left(\frac{1}{4}\right)} = 4\pi$, is reflected across the *x*-axis, and passes through the points

$(0, 0)$ (midpoint), $(-\pi, 2)$, and $(\pi, -2)$. See Figure 45.

46. Graph a tangent function with a period $\dfrac{\pi}{\left(\frac{1}{2}\right)} = 2\pi$; and passing through the points $(0, 0)$ (midpoint),

$\left(-\dfrac{\pi}{2}, -3\right)$, and $\left(\dfrac{\pi}{2}, 3\right)$. See Figure 46.

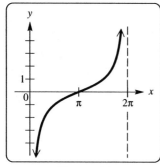

Figure 44

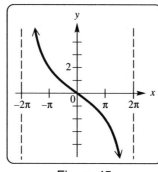

Figure 45

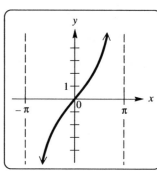

Figure 46

47. Graph a cotangent function with a period $\dfrac{\pi}{(4)}$; and passing through the points $\left(\dfrac{\pi}{8}, 0\right)$ (midpoint),

$\left(\dfrac{\pi}{16}, \dfrac{1}{2}\right)$ and $\left(\dfrac{3\pi}{16}, -\dfrac{1}{2}\right)$. See Figure 47.

48. Graph a cotangent function with a period $\dfrac{\pi}{(2)}$, is reflected across the *x*-axis, and passes through the points

$\left(\dfrac{\pi}{4}, 0\right)$ (midpoint), $\left(\dfrac{\pi}{8}, -\dfrac{1}{2}\right)$, and $\left(\dfrac{3\pi}{8}, \dfrac{1}{2}\right)$. See Figure 48.

49. Since $y = \tan(2x - \pi) \Rightarrow y = \tan 2\left(x - \dfrac{\pi}{2}\right)$, graph a tangent function, over a two period interval, with a

period $\dfrac{\pi}{(2)}$ and a phase shift of $\dfrac{\pi}{2}$ units to the right. See Figure 49.

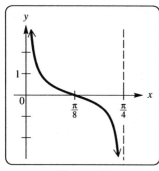

Figure 47

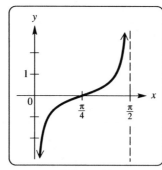

Figure 48

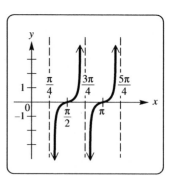

Figure 49

50. Since $y = \tan\left(\dfrac{x}{2} + \pi\right) \Rightarrow y = \tan\dfrac{1}{2}(x + 2\pi)$, graph a tangent function, over a two period interval,

    with a period $\dfrac{\pi}{\left(\frac{1}{2}\right)} = 2\pi$ and a phase shift of 2π units to the left. See Figure 50.

51. Since $y = \cot\left(3x + \dfrac{\pi}{4}\right) \Rightarrow y = \cot 3\left(x + \dfrac{\pi}{12}\right)$, graph a cotangent function, over a two period interval,

    with a period $\dfrac{\pi}{(3)}$ and a phase shift of $\dfrac{\pi}{12}$ units to the left. See Figure 51.

52. Since $y = \cot\left(2x - \dfrac{3\pi}{2}\right) \Rightarrow y = \cot 2\left(x - \dfrac{3\pi}{4}\right)$, graph a cotangent function, over a two period interval,

    with a period $\dfrac{\pi}{(2)}$ and a phase shift of $\dfrac{3\pi}{4}$ units to the right. See Figure 52.

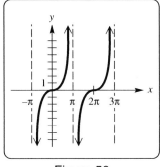   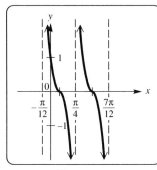   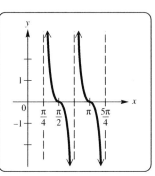

     Figure 50   Figure 51   Figure 52

53. Graph a tangent function with a period $\dfrac{\pi}{(1)} = \pi$ and vertical shift 1 unit upward. See Figure 53.

54. Graph a tangent function with a period $\dfrac{\pi}{(1)} = \pi$ and vertical shift 2 units downward. See Figure 54.

55. Graph a cotangent function with a period $\dfrac{\pi}{(1)} = \pi$, is reflected across the *x*-axis, and vertical shift 1 unit

    upward. See Figure 55.

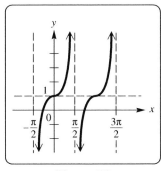   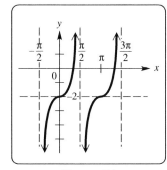   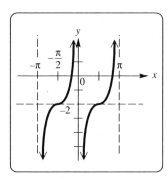

     Figure 53   Figure 54   Figure 55

56. Graph a cotangent function with a period $\dfrac{\pi}{(1)} = \pi$, is reflected across the *x*-axis, and vertical shift 2 units downward. See Figure 56.

57. Graph a tangent function with a period $\dfrac{\pi}{(1)} = \pi$, and vertical shift 1 unit downward, and passes through the points $(0, -1)$ (midpoint), $\left(-\dfrac{\pi}{4}, -3\right)$, and $\left(\dfrac{\pi}{4}, 1\right)$. See Figure 57.

58. Graph a tangent function with a period $\dfrac{\pi}{(1)} = \pi$, has a vertical shift 3 units upward, and passes through the points $(0, 3)$ (midpoint), $\left(-\dfrac{\pi}{4}, \dfrac{5}{2}\right)$, and $\left(\dfrac{\pi}{4}, \dfrac{7}{2}\right)$. See Figure 58.

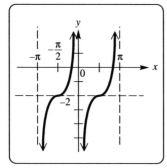

Figure 56

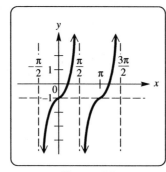

Figure 57

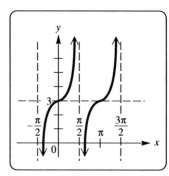

Figure 58

59. Since $y = -1 + \dfrac{1}{2}\cot(2x - 3\pi) \;\Rightarrow\; y = -1 + \dfrac{1}{2}\cot 2\left(x - \dfrac{3\pi}{2}\right)$, graph a cotangent function, over a two period interval, with a period $\dfrac{\pi}{(2)}$, a phase shift of $\dfrac{3\pi}{2}$ units to the right, has a vertical shift 1 unit downward, and passes through the points $\left(\dfrac{3\pi}{4}, -1\right)$ (midpoint), $\left(\dfrac{5\pi}{8}, -\dfrac{1}{2}\right)$, and $\left(\dfrac{7\pi}{8}, -\dfrac{3}{2}\right)$. See Figure 59.

60. Since $y = -2 + 3\tan(4x + \pi) \;\Rightarrow\; y = -2 + 3\tan 4\left(x + \dfrac{\pi}{4}\right)$, graph a tangent function, over a two period interval, with a period $\dfrac{\pi}{(4)}$, a phase shift of $\dfrac{\pi}{4}$ units to the left, has a vertical shift 2 units downward, and passes through the points $\left(-\dfrac{\pi}{4}, -2\right)$ (midpoint), $\left(-\dfrac{5\pi}{16}, -5\right)$, and $\left(-\dfrac{3\pi}{16}, 1\right)$. See Figure 60.

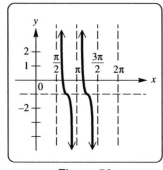

Figure 59

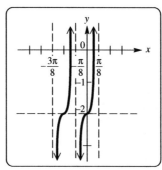

Figure 60

61. Since $c = \tan x$ has a period of $\dfrac{\pi}{(1)} = \pi$, there would be 4 of these periods in the range $(-2\pi, 2\pi)$. The graph of the function would intersect the graph of $y = c$ 4 times, for all values of c, and there are 4 solutions.

62. Since $c = \sec x$ has a range $(-\infty, -1] \cup [1, \infty)$, there would be no solutions c in $-1 < c < 1$.

63. Graph on the calculator, from this graph we can see the following: For values in the interval $-1 \le x \le 1$, x and $\tan x$ are approximately equal. As x gets closer to zero, $\tan x$ gets closer to the value of x.

64. It cannot be a parabola because it is confined between the asymptotes. A parabola with a vertical axis extends indefinitely to the left and right.

65. (a) $d = 4 \tan (2\pi)(0) \Rightarrow d = 0$ m.

    (b) $d = 4 \tan (2\pi)(.2) \Rightarrow d \approx 12.3$ m.

    (c) $d = 4 \tan (2\pi)(.8) \Rightarrow d \approx -12.3$ m.

    (d) $d = 4 \tan (2\pi)(1.2) \Rightarrow d \approx 12.3$ m.

    (e) The value $t = .25$ yields: $d = 4 \tan (2\pi)(.25) \Rightarrow d = 4 \tan \dfrac{\pi}{2}$. Since $\tan \dfrac{\pi}{2}$ is undefined .25 is a meaningless value for t.

66. (a) $a = 4 \,|\sec (2\pi)(0)| \Rightarrow a = 4$ m.

    (b) $a = 4 \,|\sec (2\pi)(.86)| \Rightarrow a \approx 6.3$ m.

    (c) $a = 4 \,|\sec (2\pi)(1.24)| \Rightarrow a = 63.7$ m.

67. From the graph, $y_1\left(\dfrac{\pi}{6}\right) + y_2\left(\dfrac{\pi}{6}\right) = \sin\left(\dfrac{\pi}{6}\right) + \sin 2\left(\dfrac{\pi}{6}\right) = .5 + \dfrac{\sqrt{3}}{2} \approx 1.366025404.$

    Also from the graph, $(y_1 + y_2)\left(\dfrac{\pi}{6}\right) \approx 1.366025404$. They both have the same results.

68. From the graph, $y_1\left(\dfrac{\pi}{6}\right) + y_2\left(\dfrac{\pi}{6}\right) = \cos\left(\dfrac{\pi}{6}\right) + \cos 2\left(\dfrac{\pi}{6}\right) = \dfrac{\sqrt{3}}{2} + .5 \approx 1.366025404.$

    Also from the graph, $(y_1 + y_2)\left(\dfrac{\pi}{6}\right) \approx 1.366025404$. They both have the same results.

69. From the graph, $y_1\left(\dfrac{\pi}{6}\right) + y_2\left(\dfrac{\pi}{6}\right) = \sin 2\left(\dfrac{\pi}{6}\right) + \cos \dfrac{1}{2}\left(\dfrac{\pi}{6}\right) = \dfrac{\sqrt{3}}{2} + .9659258263 \approx 1.83195123.$

    Also from the graph, $(y_1 + y_2)\left(\dfrac{\pi}{6}\right) \approx 1.83195123$. They both have the same results.

70. From the graph, $y_1\left(\dfrac{\pi}{6}\right) + y_2\left(\dfrac{\pi}{6}\right) = \tan\left(\dfrac{\pi}{6}\right) + \tan 2\left(\dfrac{\pi}{6}\right) \approx .57735027 + 1.7320508 \approx 2.309401077.$

    Also from the graph, $(y_1 + y_2)\left(\dfrac{\pi}{6}\right) \approx 2.309401077$. They both have the same results.

Reviewing Basic Concepts (Sections 8.3 and 8.4)

1. From the equation, the amplitude is $|-1| = 1$ and the period is $\dfrac{2\pi}{(1)} = 2\pi$. See Figure 1.

2. From the equation, the amplitude is 3, the period is $\dfrac{2\pi}{(\pi)} = 2$, and the phase shift is $-\dfrac{\pi}{\pi} = -1$. See Figure 2.

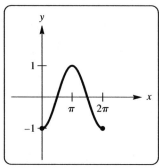

Figure 1

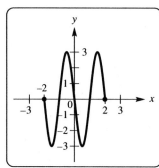

Figure 2

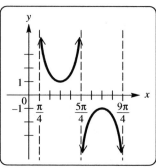

Figure 4

3. (a) Since the amplitude is 6.5 and the vertical shift is 12.4, the maximum number of daylight hours is

   $6.5 + 12.4 = 18.9$ hours and the minimum is $-6.5 + 12.4 = 5.9$ hours.

   (b) The amplitude represents half the difference in the daylight hours between the longest and shortest days.

   The period represents 12 months or one year.

4. From the equation, the period is: $\dfrac{2\pi}{(1)} = 2\pi$; the phase shift is: $\dfrac{\pi}{4}$ or $\dfrac{\pi}{4}$ units to the right; the domain is:

   $\left\{ x \mid x \neq \dfrac{\pi}{4} + n\pi, \text{ where } n \text{ is an integer} \right\}$; and the range is: $(-\infty, -1] \cup [1, \infty)$. See Figure 4.

5. From the equation, the period is: $\dfrac{2\pi}{(1)} = 2\pi$; the phase shift is: $-\dfrac{\pi}{4}$ or $\dfrac{\pi}{4}$ units to the left; the domain is:

   $\left\{ x \mid x \neq \dfrac{\pi}{4} + n\pi, \text{ where } n \text{ is an integer} \right\}$; and the range is: $(-\infty, -1] \cup [1, \infty)$. See Figure 5.

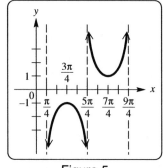

Figure 5

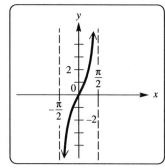

Figure 6

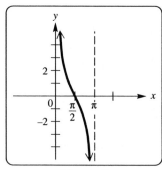

Figure 7

6. From the equation, the period is: $\dfrac{\pi}{(1)} = \pi$; the domain is: $\left\{ x \mid x \neq \dfrac{\pi}{2} + n\pi, \text{ where } n \text{ is an integer} \right\}$; and the

   range is: $(-\infty, \infty)$. See Figure 6.

7. From the equation, the period is: $\dfrac{\pi}{(1)} = \pi$; the domain is: $\{ x \mid x \neq n\pi, \text{ where } n \text{ is an integer} \}$; and the

   range is: $(-\infty, \infty)$. See Figure 7.

8.5: Trigonometric Functions and Fundamental Identities

1. See Figure 1.

2. See Figure 2.

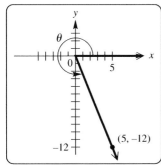

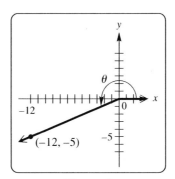

| Figure 1 | Figure 2 |

3. First find the value of r: $r = \sqrt{(-3)^2 + 4^2} \;\Rightarrow\; r = \sqrt{25} \;\Rightarrow\; r = 5$. Now each of the following is:

$$\sin\theta = \frac{y}{r} = \frac{4}{5} \qquad \cos\theta = \frac{x}{r} = -\frac{3}{5} \qquad \tan\theta = \frac{y}{x} = -\frac{4}{3}$$

$$\csc\theta = \frac{r}{y} = \frac{5}{4} \qquad \sec\theta = \frac{r}{x} = -\frac{5}{3} \qquad \cot\theta = \frac{x}{y} = -\frac{3}{4}$$

4. First find the value of r: $r = \sqrt{(-4)^2 + (-3)^2} \;\Rightarrow\; r = \sqrt{25} \;\Rightarrow\; r = 5$. Now each of the following is:

$$\sin\theta = \frac{y}{r} = -\frac{3}{5} \qquad \cos\theta = \frac{x}{r} = -\frac{4}{5} \qquad \tan\theta = \frac{y}{x} = \frac{3}{4}$$

$$\csc\theta = \frac{r}{y} = -\frac{5}{3} \qquad \sec\theta = \frac{r}{x} = -\frac{5}{4} \qquad \cot\theta = \frac{x}{y} = \frac{4}{3}$$

5. First find the value of r: $r = \sqrt{0^2 + 2^2} \;\Rightarrow\; r = \sqrt{4} \;\Rightarrow\; r = 2$. Now each of the following is:

$$\sin\theta = \frac{y}{r} = \frac{2}{2} = 1 \qquad \cos\theta = \frac{x}{r} = \frac{0}{2} = 0 \qquad \tan\theta = \frac{y}{x} = \frac{2}{0} \;\Rightarrow\; \text{undefined}$$

$$\csc\theta = \frac{r}{y} = \frac{2}{2} = 1 \qquad \sec\theta = \frac{r}{x} = \frac{2}{0} \;\Rightarrow\; \text{undefined} \qquad \cot\theta = \frac{x}{y} = \frac{0}{2} = 0$$

6. First find the value of r: $r = \sqrt{(-4)^2 + 0^2} \;\Rightarrow\; r = \sqrt{16} \;\Rightarrow\; r = 4$. Now each of the following is:

$$\sin\theta = \frac{y}{r} = \frac{0}{4} = 0 \qquad \cos\theta = \frac{x}{r} = \frac{-4}{4} = -1 \qquad \tan\theta = \frac{y}{x} = \frac{0}{-4} = 0$$

$$\csc\theta = \frac{r}{y} = \frac{4}{0} \;\Rightarrow\; \text{undefined} \qquad \sec\theta = \frac{r}{x} = \frac{4}{-4} = -1 \qquad \cot\theta = \frac{x}{y} = \frac{-4}{0} \;\Rightarrow\; \text{undefined}$$

7. First find the value of r: $r = \sqrt{1^2 + (\sqrt{3})^2} \;\Rightarrow\; r = \sqrt{4} \;\Rightarrow\; r = 2$. Now each of the following is:

$$\sin\theta = \frac{y}{r} = \frac{\sqrt{3}}{2} \qquad \cos\theta = \frac{x}{r} = \frac{1}{2} \qquad \tan\theta = \frac{y}{x} = \frac{\sqrt{3}}{1} = \sqrt{3}$$

$$\csc\theta = \frac{r}{y} = \frac{2}{\sqrt{3}} = \frac{2\sqrt{3}}{3} \qquad \sec\theta = \frac{r}{x} = \frac{2}{1} = 2 \qquad \cot\theta = \frac{x}{y} = \frac{1}{\sqrt{3}} = \frac{\sqrt{3}}{3}$$

8. First find the value of r: $r = \sqrt{(-2\sqrt{3})^2 + (-2)^2} \;\Rightarrow\; r = \sqrt{16} \;\Rightarrow\; r = 4$. Now each of the following is:

$$\sin\theta = \frac{y}{r} = \frac{-2}{4} = -\frac{1}{2} \qquad \cos\theta = \frac{x}{r} = \frac{-2\sqrt{3}}{4} = -\frac{\sqrt{3}}{2} \qquad \tan\theta = \frac{y}{x} = \frac{-2}{-2\sqrt{3}} = \frac{1}{\sqrt{3}} = \frac{\sqrt{3}}{3}$$

$$\csc\theta = \frac{r}{y} = \frac{4}{-2} = -2 \qquad \sec\theta = \frac{r}{x} = \frac{4}{-2\sqrt{3}} = \frac{4\sqrt{3}}{-6} = -\frac{2\sqrt{3}}{3} \qquad \cot\theta = \frac{x}{y} = \frac{-2\sqrt{3}}{-2} = \sqrt{3}$$

9. The sine and cosecant functions are reciprocals, and reciprocals always have the same sign, since their product is 1, positive number.

10. Since in quadrant III x is negative, y is negative, and r is always positive, $\tan \theta$ and $\cot \theta$ are $\dfrac{\text{negative}}{\text{negative}}$ and thus positive. The other four function values would be a ratio of 1 positive number and 1 negative number and thus negative.

11. In quadrant II, x is negative and r is positive, therefore the ratio is negative.

12. In quadrant III, y is negative and r is positive, therefore the ratio is negative.

13. In quadrant IV, y is negative and x is positive, therefore the ratio is negative.

14. In quadrant III, x is negative and y is negative, therefore the ratio is positive.

15. See Figure 15. Choosing $x = 1$ yields the ordered pair $(1, -2)$.

Now find the value of r: $r = \sqrt{1^2 + (-2)^2} \Rightarrow r = \sqrt{5}$. Now each of the following is:

$$\sin \theta = \frac{y}{r} = \frac{-2}{\sqrt{5}} = -\frac{2\sqrt{5}}{5} \qquad \cos \theta = \frac{x}{r} = \frac{1}{\sqrt{5}} = \frac{\sqrt{5}}{5} \qquad \tan \theta = \frac{y}{x} = \frac{-2}{1} = -2$$

$$\csc \theta = \frac{r}{y} = \frac{\sqrt{5}}{-2} = -\frac{\sqrt{5}}{2} \qquad \sec \theta = \frac{r}{x} = \frac{\sqrt{5}}{1} = \sqrt{5} \qquad \cot \theta = \frac{x}{y} = \frac{1}{-2} = -\frac{1}{2}$$

16. See Figure 16. Choosing $x = 5$ yields the ordered pair $(5, -3)$.

Now find the value of r: $r = \sqrt{5^2 + (-3)^2} \Rightarrow r = \sqrt{34}$. Now each of the following is:

$$\sin \theta = \frac{y}{r} = \frac{-3}{\sqrt{34}} = -\frac{3\sqrt{34}}{34} \qquad \cos \theta = \frac{x}{r} = \frac{5}{\sqrt{34}} = \frac{5\sqrt{34}}{34} \qquad \tan \theta = \frac{y}{x} = \frac{-3}{5} = -\frac{3}{5}$$

$$\csc \theta = \frac{r}{y} = \frac{\sqrt{34}}{-3} = -\frac{\sqrt{34}}{3} \qquad \sec \theta = \frac{r}{x} = \frac{\sqrt{34}}{5} \qquad \cot \theta = \frac{x}{y} = \frac{5}{-3} = -\frac{5}{3}$$

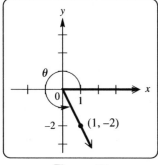

Figure 15

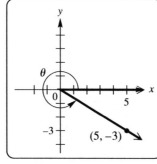

Figure 16

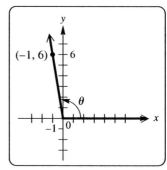

Figure 17

17. See Figure 17. Choosing $x = -1$ yields the ordered pair $(-1, 6)$.

Now find the value of r: $r = \sqrt{(-1)^2 + 6^2} \Rightarrow r = \sqrt{37}$. Now each of the following is:

$$\sin \theta = \frac{y}{r} = \frac{6}{\sqrt{37}} = \frac{6\sqrt{37}}{37} \qquad \cos \theta = \frac{x}{r} = \frac{-1}{\sqrt{37}} = -\frac{\sqrt{37}}{37} \qquad \tan \theta = \frac{y}{x} = \frac{6}{-1} = -6$$

$$\csc \theta = \frac{r}{y} = \frac{\sqrt{37}}{6} \qquad \sec \theta = \frac{r}{x} = \frac{\sqrt{37}}{-1} = -\sqrt{37} \qquad \cot \theta = \frac{x}{y} = \frac{-1}{6} = -\frac{1}{6}$$

18. See Figure 18. Choosing $x = -3$ yields the ordered pair $(-3, 5)$.

Now find the value of r: $r = \sqrt{(-3)^2 + 5^2} \Rightarrow r = \sqrt{34}$. Now each of the following is:

$$\sin \theta = \frac{y}{r} = \frac{5}{\sqrt{34}} = \frac{5\sqrt{34}}{34} \qquad \cos \theta = \frac{x}{r} = \frac{-3}{\sqrt{34}} = -\frac{3\sqrt{34}}{34} \qquad \tan \theta = \frac{y}{x} = \frac{5}{-3} = -\frac{5}{3}$$

$$\csc \theta = \frac{r}{y} = \frac{\sqrt{34}}{5} \qquad \sec \theta = \frac{r}{x} = \frac{\sqrt{34}}{-3} = -\frac{\sqrt{34}}{3} \qquad \cot \theta = \frac{x}{y} = \frac{-3}{5} = -\frac{3}{5}$$

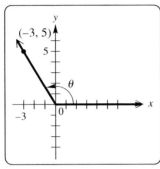

Figure 18

19. Using the degree mode on the calculator yields:

$\sin 540° = 0$; $\cos 540° = -1$; $\tan 540° = 0$; $\csc 540°$ is undefined; $\sec 540° = -1$; and $\cot 540°$ is undefined.

20. Using the degree mode on the calculator yields: $3 \sec 180° - 5 \tan 360° = 3(-1) - 5(0) = -3$

21. Using the degree mode on the calculator yields: $4 \csc 270° + 3 \cos 180° = 4(-1) + 3(-1) = -7$

22. Using the degree mode on the calculator yields: $\tan 360° + 4 \sin 180° + 5 \cos^2 180° = 0 + 4(0) + 5(-1)^2 = 5$

23. Using the degree mode on the calculator yields: $2 \sec 0° + 4 \cot^2 90° + \cos 360° = 2(1) + 4(0)^2 + 1 = 3$

24. Using the degree mode on the calculator yields: $\sin^2 180° + \cos^2 180° = (0)^2 + (-1)^2 = 1$

25. Using the degree mode on the calculator yields: $\sin^2 360° + \cos^2 360° = (0)^2 + (1)^2 = 1$

26. Odd multiples of 90° functions are equal to $-1 \times$ original function, therefore if $\cos 90° = 0$, then all odd multiples of $\cos 90° = (-1)0 = 0$.

27. Even multiples of 180° functions are equal to $-1 \times$ original function, therefore if $\tan 180° = 0$, then all even multiples of $\tan 180° = (-1)0 = 0$.

28. Odd multiples of 180° functions are equal to the original function, therefore if $\cos 180° = -1$, then all odd multiples of $\cos 180° = -1$.

29. Even multiples of 180° functions are equal to $-1 \times$ original function, therefore if $\sin 180° = 0$, then all even multiples of $\sin 180° = (-1)0 = 0$.

30. The sine of an angle is always equal to the cosine of it's complement.

31. The tangent of an angle is always equal to the cotangent of it's complement.

32. The sine of an angle and it's negative are also negative of each other. If point (x, y) is on the terminal side of θ, the point $(x, -y)$ is on the terminal side of $-\theta$, thus $\sin \theta = y$ and $\sin (-\theta) = -y$.

33. The cosine of an angle and it's negative are equal to each other. If point (x, y) is on the terminal side of θ, the point $(x, -y)$ is on the terminal side of $-\theta$, thus $\cos \theta = x$ and $\cos (-\theta) = x$.

34. From the calculator, $\cos 20° \approx .940$ and $\sin 20° \approx .342$.

35. From the calculator, $\cos T \approx .766$ when $T = 40°$.

36. From the calculator, $\sin T \approx .574$ when $T = 35°$.

37. From the calculator, $\cos T = \sin T$ when $T = 45°$.

38. From the calculator, when T increases from $0°$ to $90°$, cosine decreases and sine increases.

39. From the calculator, when T increases from $90°$ to $180°$, cosine decreases and sine decreases.

40. The positive number 1 is it's own reciprocal and $\sin \theta = \csc \theta = 1$ when $\theta = 90°$.

41. The positive number -1 is it's own reciprocal and $\cos \theta = \sec \theta = -1$ when $\theta = 180°$.

42. If $\sec \theta = -2.5$, then $\cos \theta = \dfrac{1}{\sec \theta} = \dfrac{1}{-2.5} \Rightarrow \cos \theta = -.4$.

43. If $\tan \theta = -\dfrac{1}{5}$, then $\cot \theta = \dfrac{1}{\tan \theta} = \dfrac{1}{-\frac{1}{5}} \Rightarrow \cot \theta = -5$.

44. If $\csc \theta = \sqrt{15}$, then $\sin \theta = \dfrac{1}{\csc \theta} = \dfrac{1}{\sqrt{15}} \Rightarrow \sin \theta = \dfrac{\sqrt{15}}{15}$.

45. If $\cot \theta = -\dfrac{\sqrt{5}}{3}$, then $\tan \theta = \dfrac{1}{\cot \theta} = \dfrac{1}{-\frac{\sqrt{5}}{3}} \Rightarrow \tan \theta = -\dfrac{3}{\sqrt{5}} = -\dfrac{3\sqrt{5}}{5}$.

46. If $\csc \theta = 1.42716321$, then $\sin \theta = \dfrac{1}{\csc \theta} = \dfrac{1}{1.42716321} \Rightarrow \sin \theta = .70069071$.

47. If $\sec \theta = 9.80425133$, then $\cos \theta = \dfrac{1}{\sec \theta} = \dfrac{1}{9.80425133} \Rightarrow \sin \theta = .10199657$.

48. No, the sign of the sine and the cosecant of an angle must be the same.

49. The cosine of any angle cannot be greater than 1 (or less than -1).

50. If $\cot \theta = -3$, then $\tan \theta = \dfrac{1}{\cot \theta} = \dfrac{1}{-3} \Rightarrow \tan \theta = -\dfrac{1}{3}$.

51. If $\cot \theta = \dfrac{\sqrt{3}}{3}$, then $\tan \theta = \dfrac{1}{\cot \theta} = \dfrac{1}{\frac{\sqrt{3}}{3}} \Rightarrow \tan \theta = \dfrac{3}{\sqrt{3}} = \dfrac{3\sqrt{3}}{3} = \sqrt{3}$.

52. If $\cot \theta = .4$, then $\tan \theta = \dfrac{1}{\cot \theta} = \dfrac{1}{.4} \Rightarrow \tan \theta = 2.5$.

53. If $\tan (3\theta - 4°) = \dfrac{1}{\cot (5\theta - 8°)}$, then $\tan (3\theta - 4°) = \tan (5\theta - 8°) \Rightarrow 3\theta - 4° = 5\theta - 8° \Rightarrow$
    $4° = 2\theta \Rightarrow \theta = 2°$.

54. If $\sec (2\theta + 6°) \cos (5\theta + 3°) = 1 \Rightarrow \dfrac{1}{\cos (2\theta + 6°)} \cdot \cos (5\theta + 3°) = 1 \Rightarrow \dfrac{\cos (5\theta + 3°)}{\cos (2\theta + 6°)} = \dfrac{1}{1} \Rightarrow$
    $\cos (5\theta + 3°) = \cos (2\theta + 6°) \Rightarrow 5\theta + 3° = 2\theta + 6° \Rightarrow 3\theta = 3° \Rightarrow \theta = 1°$.

55. If $\sin \theta > 0$ then $y > 0$ and if $\cos \theta < 0$ then $x < 0$, the angle is in quadrant II.

56. If $\cos \theta > 0$ then $x > 0$ and if $\tan \theta > 0$ then x and y are either both positive or both negative. Since
    $x > 0$, then $y > 0$ and the angle is in quadrant I.

57. If $\tan \theta > 0$ then x and y are either both positive or both negative and if $\cot \theta > 0$ then x and y are either both
    positive or both negative. With two signs the same the angle is in quadrant I or III.

58. If $\tan \theta < 0$ then x and y is one sign positive and one sign negative and if $\cot \theta > 0$ then x and y is one sign
    positive and one sign negative. With two opposite signs the angle is in quadrant II or IV.

59. Since $129°$ is in quadrant II, x is negative, y is positive, r is always positive, and sine functions $\frac{y}{r}$ are: (+);

    cosine functions $\frac{x}{r}$ are: (-); and tangent functions $\frac{y}{x}$ are (-).

60. Since $183°$ is in quadrant III, x is negative, y is negative, r is always positive, and sine functions $\frac{y}{r}$ are: (-);

    cosine functions $\frac{x}{r}$ are: (-); and tangent functions $\frac{y}{x}$ are (+).

61. Since $298°$ is in quadrant IV, x is positive, y is negative, r is always positive, and sine functions $\frac{y}{r}$ are: (-);

    cosine functions $\frac{x}{r}$ are: (+); and tangent functions $\frac{y}{x}$ are (-).

62. Since $412°$ is in quadrant I, x is positive, y is positive, r is always positive, and sine functions $\frac{y}{r}$ are: (+);

    cosine functions $\frac{x}{r}$ are: (+); and tangent functions $\frac{y}{x}$ are (+).

63. Since $-82°$ is in quadrant IV, x is positive, y is negative, r is always positive, and sine functions $\frac{y}{r}$ are: (-);

    cosine functions $\frac{x}{r}$ are: (+); and tangent functions $\frac{y}{x}$ are (-).

64. Since $-121°$ is in quadrant III, x is negative, y is negative, r is always positive, and sine functions $\frac{y}{r}$ are: (-);

    cosine functions $\frac{x}{r}$ are: (-); and tangent functions $\frac{y}{x}$ are (+).

65. Since sine is: $\frac{y}{r}$; tangent is: $\frac{y}{x}$; and $r > x$ because r is the hypotenuse, $\tan 30°$ is greater.

66. Since $\sin 0° = 0$ and $\sin 90° = 1$, $\sin \theta$ increases as θ increases over $0 \le \theta \le 90$. Therefore $\sin 21°$ is greater.

67. Since both sine and cosine have values between 0 and 1 for $0° < \theta < 90°$ and since $\sec \theta = \dfrac{1}{\cos \theta}$, therefore

    $\sec \theta > 1$, and $\sec 33°$ is the greater.

68. Since the range of $\sin \theta$ is: $-1 \le \sin \theta \le 1$, $\sin \theta = 2$ is impossible.

69. Since the range of $\cos \theta$ is: $-1 \le \cos \theta \le 1$, $\cos \theta = -1.001$ is impossible.

70. Since the range of $\tan \theta$ is: $-\infty \le \tan \theta \le \infty$, $\tan \theta = .92$ is possible.

71. Since the range of $\cot \theta$ is: $-\infty \le \cot \theta \le \infty$, $\cot \theta = -12.1$ is possible.

72. Since the range of $\sec \theta$ is: $\sec \theta \le -1$ or $\sec \theta \ge 1$, $\sec \theta = 1$ is possible.

73. Since the range of $\tan \theta$ is: $-\infty \le \tan \theta \le \infty$, $\tan \theta = 1$ is possible.

74. Since the range of $\sin \theta$ is: $-1 \le \sin \theta \le 1$, $\sin \theta = \dfrac{1}{2}$ is possible, and since $\csc \theta \le -1$ or $\csc \theta \ge 1$,

    $\csc \theta = 2$ is possible.

75. Since there is no single angle such that $\tan \theta = 2$ and $\cot \theta = -2$ are both satisfied, this is impossible.

76. Using the Pythagorean Identity

    $1 + \tan^2 \theta = \sec^2 \theta \Rightarrow 1 + \tan^2 \theta = 3^2 \Rightarrow \tan^2 \theta = 8 \Rightarrow \tan \theta = \pm\sqrt{8} \Rightarrow$

    $\tan \theta = \pm 2\sqrt{2}$, since θ is in quadrant IV, $\tan \theta = -2\sqrt{2}$.

77. Using the Pythagorean Identity $\sin^2 \theta + \cos^2 \theta = 1 \Rightarrow \sin^2 \theta + \left(-\dfrac{1}{4}\right)^2 = 1 \Rightarrow \sin^2 \theta = \dfrac{15}{16} \Rightarrow$

    $\sin \theta = \pm\sqrt{\dfrac{15}{16}} \Rightarrow \sin \theta = \pm\dfrac{\sqrt{15}}{4}$, since θ is in quadrant II, $\sin \theta = \dfrac{\sqrt{15}}{4}$.

78. Using the Pythagorean Identity $1 + \cot^2 \theta = \csc^2 \theta \implies 1 + \left(-\frac{1}{2}\right)^2 = \csc^2 \theta \implies \csc^2 \theta = \frac{5}{4} \implies$

    $\csc \theta = \pm\sqrt{\frac{5}{4}} \implies \csc \theta = \pm\frac{\sqrt{5}}{2}$, since θ is in quadrant IV, $\csc \theta = -\frac{\sqrt{5}}{2}$.

79. Using the Pythagorean Identity $1 + \tan^2 \theta = \sec^2 \theta \implies 1 + \left(\frac{\sqrt{7}}{3}\right)^2 = \sec^2 \theta \implies 1 + \frac{7}{9} = \sec^2 \theta \implies$

    $\sec \theta = \frac{16}{9} \implies \sec \theta = \pm\sqrt{\frac{16}{9}} \implies \sec \theta = \pm\frac{4}{3}$, since θ is in quadrant III, $\sec \theta = -\frac{4}{3}$.

80. If $\csc \theta = -4$ then $\sin \theta = \frac{1}{\csc \theta} = -\frac{1}{4}$. Using the Pythagorean Identity $\sin^2 \theta + \cos^2 \theta = 1 \implies$

    $\left(-\frac{1}{4}\right)^2 + \cos^2 \theta = 1 \implies \cos^2 \theta = \frac{15}{16} \implies \cos \theta = \pm\sqrt{\frac{15}{16}} \implies \cos \theta = \pm\frac{\sqrt{15}}{4}$,

    since θ is in quadrant III, $\cos \theta = -\frac{\sqrt{15}}{4}$.

81. If $\sec \theta = 2$ then $\cos \theta = \frac{1}{\sec \theta} = \frac{1}{2}$. Using the Pythagorean Identity $\sin^2 \theta + \cos^2 \theta = 1 \implies$

    $\sin^2 \theta + \left(\frac{1}{2}\right)^2 = 1 \implies \sin^2 \theta = \frac{3}{4} \implies \sin \theta = \pm\sqrt{\frac{3}{4}} \implies \sin \theta = \pm\frac{\sqrt{3}}{2}$, since θ is in quadrant IV,

    $\sin \theta = -\frac{\sqrt{3}}{2}$.

82. Using the Pythagorean Identity $1 + \cot^2 \theta = \csc^2 \theta \implies 1 + \cot^2 \theta = (-3.5891420)^2 \implies$

    $\cot^2 \theta = 12.8819403 - 1 \implies \cot^2 \theta = 11.8819403 \implies \cot \theta = \pm 3.447019045$, since θ is in quadrant III,

    $\cot \theta \approx 3.447019045$.

83. First, using the Pythagorean Identity $\sin^2 \theta + \cos^2 \theta = 1 \implies (.49268329)^2 + \cos^2 \theta = 1 \implies$

    $\cos^2 \theta = 1 - .24273682 \implies \cos^2 \theta = .7572631758 \implies \cos \theta = \pm.8702086967$, since θ is in quadrant II,

    $\cos \theta = -.870208697$. Now $\tan \theta = \frac{\sin \theta}{\cos \theta} = \frac{.49268329}{-.870208697} = -.56616682$.

84. Using $\sin x = .8$ in the Pythagorean Identity $\sin^2 x + \cos^2 x = 1$ yields: $(.8)^2 + \cos^2 x = 1 \implies$

    $\cos^2 x = 1 - (.64) \implies \cos^2 x = .36$.

85. Using $\tan x = 2$ in the Pythagorean Identity $1 + \tan^2 x = \sec^2 x$ yields: $1 + (2)^2 = \sec^2 x \implies \sec^2 x = 5$.

86. Yes; testing using the Pythagorean Identity $\sin^2 x + \cos^2 x = 1$ yields: $(.8)^2 + (-.6)^2 = 1 \implies .64 + .36 = 1$,

    which is true, therefore it exists.

87. Find $\sec \theta$, $1 + \tan^2 \theta = \sec^2 \theta : 1 + \left(-\frac{15}{8}\right)^2 = \sec^2 \theta \implies \sec^2 \theta = \frac{289}{64} \implies \sec \theta = \pm\frac{17}{8}$. Since θ is

    found in quadrant II, $\sec \theta = -\frac{17}{8}$. Now find $\cos \theta$, $\cos \theta = \frac{1}{\sec \theta}$, therefore $\cos \theta = -\frac{8}{17}$; find $\cot \theta$,

    $\cot \theta = \frac{1}{\tan \theta}$, therefore $\cot \theta = -\frac{8}{15}$; find $\sin \theta$, $\frac{\sin \theta}{\cos \theta} = \tan \theta \implies \sin \theta = (\tan \theta)(\cos \theta) \implies$

    $\sin \theta = \left(-\frac{15}{8}\right)\left(-\frac{8}{17}\right) \implies \sin \theta = \frac{15}{17}$; and find $\csc \theta$, $\csc \theta = \frac{1}{\sin \theta}$, therefore $\csc \theta = \frac{17}{15}$.

    The five Functions are: $\sin \theta = \frac{15}{17}$; $\cos \theta = -\frac{8}{17}$; $\cot \theta = -\frac{8}{15}$; $\sec \theta = -\frac{17}{8}$; and $\csc \theta = \frac{17}{15}$.

88. Find $\sin \theta$, $\sin^2 \theta + \cos^2 \theta = 1$: $\sin^2 \theta + \left(-\dfrac{3}{5}\right)^2 = 1 \Rightarrow \sin^2 \theta = 1 - \dfrac{9}{25} \Rightarrow \sin \theta = \sqrt{\dfrac{16}{25}} \Rightarrow$

$\sin \theta = \pm\dfrac{4}{5}$. Since θ is found in quadrant III, $\sin \theta = -\dfrac{4}{5}$. Now find $\tan \theta$, $\tan \theta = \dfrac{\sin \theta}{\cos \theta} \Rightarrow -\dfrac{4}{5} \div -\dfrac{3}{5} = \dfrac{4}{3}$,

therefore $\tan \theta = \dfrac{4}{3}$; find $\cot \theta$, $\cot \theta = \dfrac{1}{\tan \theta}$, therefore $\cot \theta = \dfrac{3}{4}$; find $\sec \theta$, $\sec \theta = \dfrac{1}{\cos \theta}$, therefore

$\sec \theta = -\dfrac{5}{3}$; and find $\csc \theta$, $\csc \theta = \dfrac{1}{\sin \theta}$, therefore $\csc \theta = -\dfrac{5}{4}$.

The five Functions are: $\sin \theta = -\dfrac{4}{5}$; $\tan \theta = \dfrac{4}{3}$; $\cot \theta = \dfrac{3}{4}$; $\sec \theta = -\dfrac{5}{3}$; and $\csc \theta = -\dfrac{5}{4}$.

89. Find $\sec \theta$, $1 + \tan^2 \theta = \sec^2 \theta$: $1 + (\sqrt{3})^2 = \sec^2 \theta \Rightarrow \sec^2 \theta = 4 \Rightarrow \sec \theta = \pm 2$. Since θ is

found in quadrant III, $\sec \theta = -2$. Now find $\cos \theta$, $\cos \theta = \dfrac{1}{\sec \theta}$, therefore $\cos \theta = -\dfrac{1}{2}$; find $\cot \theta$,

$\cot \theta = \dfrac{1}{\tan \theta} \Rightarrow \cot \theta = \dfrac{1}{\sqrt{3}}$, therefore $\cot \theta = \dfrac{\sqrt{3}}{3}$; find $\sin \theta$,

$\dfrac{\sin \theta}{\cos \theta} = \tan \theta \Rightarrow \sin \theta = (\tan \theta)(\cos \theta) \Rightarrow \sin \theta = \left(-\dfrac{1}{2}\right)(\sqrt{3}) \Rightarrow \sin \theta = -\dfrac{\sqrt{3}}{2}$; and find

$\csc \theta$, $\csc \theta = \dfrac{1}{\sin \theta} \Rightarrow \csc \theta = -\dfrac{2}{\sqrt{3}}$ therefore $\csc \theta = -\dfrac{2\sqrt{3}}{3}$.

The five Functions are: $\sin \theta = -\dfrac{\sqrt{3}}{2}$; $\cos \theta = -\dfrac{1}{2}$; $\cot \theta = \dfrac{\sqrt{3}}{3}$; $\sec \theta = -2$; and $\csc \theta = -\dfrac{2\sqrt{3}}{3}$.

90. Since $\sin \theta > 0$, θ can be in quadrant I or II and since $\tan \theta > 0$, θ can be in quadrant I or III, therefore θ must

be found in quadrant I. First find $\cos \theta$, $\sin^2 \theta + \cos^2 \theta = 1$: $\left(\dfrac{\sqrt{5}}{7}\right)^2 + \cos^2 \theta = 1 \Rightarrow$

$\cos^2 \theta = 1 - \dfrac{5}{49} \Rightarrow \cos^2 \theta = \dfrac{44}{49} \Rightarrow \cos \theta = \pm\dfrac{2\sqrt{11}}{7}$. Since θ is found in quadrant I, $\cos \theta = \dfrac{2\sqrt{11}}{7}$.

Now find $\csc \theta$, $\csc \theta = \dfrac{1}{\sin \theta} \Rightarrow \csc \theta = \dfrac{7}{\sqrt{5}}$, therefore $\csc \theta = \dfrac{7\sqrt{5}}{5}$; find $\sec \theta$,

$\sec \theta = \dfrac{1}{\cos \theta} \Rightarrow \sec \theta = \dfrac{7}{2\sqrt{11}}$,

therefore $\sec \theta = \dfrac{7\sqrt{11}}{22}$; find $\tan \theta$, $\tan \theta = \dfrac{\sin \theta}{\cos \theta} \Rightarrow \dfrac{\sqrt{5}}{7} \div \dfrac{2\sqrt{11}}{7} \Rightarrow \tan \theta = \dfrac{\sqrt{5}}{2\sqrt{11}}$, therefore

$\tan \theta = \dfrac{\sqrt{55}}{22}$; and find $\cot \theta$, $\cot \theta = \dfrac{1}{\tan \theta} \Rightarrow \cot \theta = \dfrac{22}{\sqrt{55}} = \dfrac{22\sqrt{55}}{55} \Rightarrow \cot \theta = \dfrac{2\sqrt{55}}{5}$.

The five Functions are: $\cos \theta = \dfrac{2\sqrt{11}}{7}$; $\tan \theta = \dfrac{\sqrt{55}}{22}$; $\cot \theta = \dfrac{2\sqrt{55}}{5}$; $\sec \theta = \dfrac{7\sqrt{11}}{22}$; and $\csc \theta = \dfrac{7\sqrt{5}}{5}$.

91. First find $\tan \theta$, $\tan \theta = (\cot \theta)^{-1} = (-1.49586)^{-1} \Rightarrow \tan \theta \approx -.668512$; now find $\sec \theta$, $1 + \tan^2 \theta = \sec^2 \theta$:

$1 + (-.668512)^2 = \sec^2 \theta \Rightarrow \sec^2 \theta \approx 1.446908 \Rightarrow \sec \theta \approx \pm 1.20287$. Since θ is found in quadrant IV,

$\sec \theta \approx 1.20287$. Now find $\cos \theta$, $\cos \theta = (\sec \theta)^{-1}$, therefore $\cos \theta = .831342$; find $\sin \theta$,

$\sin \theta = (\cos \theta)(\tan \theta) = (.831342)(-.668512) \Rightarrow \sin \theta \approx -.555762$; and find $\csc \theta$, $\csc \theta = (\sin \theta)^{-1} \Rightarrow$

therefore $\csc \theta = -1.79933$. The five Functions are:

$\sin \theta \approx -.555762$; $\cos \theta \approx .831342$; $\tan \theta \approx -.668512$; $\sec \theta \approx 1.20287$; and $\csc \theta \approx -1.79933$.

92. First find $\csc \theta$, $\csc \theta = (\sin \theta)^{-1} = (.164215)^{-1} \Rightarrow \csc \theta \approx 6.08958$; now find $\cos \theta$, $\sin^2 \theta + \cos^2 \theta = 1$:

    $(.164215)^2 + \cos^2 \theta = 1 \Rightarrow \cos^2 \theta = .973033 \Rightarrow \cos \theta \approx \pm .986425$. Since θ is found in quadrant II,

    $\cos \theta \approx -.986425$. Now find $\sec \theta$, $\sec \theta = (\cos \theta)^{-1}$, therefore $\sec \theta \approx -1.01376$; find $\tan \theta$,

    $\tan \theta = (\sin \theta) \div (\cos \theta) = (.164215) \div (-.986425) \Rightarrow \tan \theta \approx -.166475$; and find

    $\cot \theta$, $\cot \theta = (\tan \theta)^{-1} \Rightarrow$ therefore $\cot \theta = -6.00691$. The five Functions are:

    $\cos \theta \approx -.986425$; $\tan \theta \approx -.166475$; $\cot \theta \approx -6.00691$; $\sec \theta \approx -1.01376$; and $\csc \theta \approx 6.08958$.

93. Use the Pythagorean Identity $\sin^2 \theta + \cos^2 \theta = 1$ to write

    $\cos \theta$, $\sin^2 \theta + \cos^2 \theta = 1 \Rightarrow \cos^2 = 1 - \sin^2 \theta \Rightarrow$

    $\cos \theta = \pm \sqrt{1 - \sin^2 \theta}$. Since θ is acute, it is in quadrant I and $\cos \theta = \sqrt{1 - \sin^2 \theta}$.

94. Use a Reciprocal Identity to write $\sec \theta$, $\sec \theta = \dfrac{1}{\cos \theta}$.

95. First use the Pythagorean Identity $1 + \cot^2 \theta = \csc^2 \theta$ to write $\csc \theta$, $1 + \cot^2 \theta = \csc^2 \theta \Rightarrow$

    $\csc \theta = \pm \sqrt{1 + \cot^2 \theta}$. Since θ is in quadrant III, $\csc \theta = -\sqrt{1 + \cot^2 \theta}$.

    Substituting this into the Reciprocal Identity $\sin \theta = \dfrac{1}{\csc \theta}$ yields: $\sin \theta = -\dfrac{1}{\sqrt{1 + \cot^2 \theta}}$ or $-\dfrac{\sqrt{1 + \cot^2 \theta}}{1 + \cot^2 \theta}$.

96. Use a Reciprocal Identity to write $\sec \theta$, $\sec \theta = \dfrac{1}{\cos \theta}$. Now substituting this into the Pythagorean Identity

    $1 + \tan^2 \theta = \sec^2 \theta$ yields: $1 + \tan^2 \theta = \dfrac{1}{\cos^2 \theta} \Rightarrow \tan^2 \theta = \dfrac{1}{\cos^2 \theta} - 1 \Rightarrow \tan \theta = \pm \sqrt{\dfrac{1}{\cos^2 \theta} - 1}$.

    Since θ is in the quadrant IV, $\tan \theta = -\sqrt{\dfrac{1}{\cos^2 \theta} - 1}$ or $-\dfrac{\sqrt{1 - \cos^2 \theta}}{\cos \theta}$.

97. Use the Pythagorean Identity $\sin^2 \theta + \cos^2 \theta = 1$ to write

    $\cos \theta$, $\sin^2 \theta + \cos^2 \theta = 1 \Rightarrow \cos^2 = 1 - \sin^2 \theta \Rightarrow$

    $\cos \theta = \pm \sqrt{1 - \sin^2 \theta}$. Since θ is in quadrant I or IV, $\cos \theta = \sqrt{1 - \sin^2 \theta}$. Now substituting this

    into the Quotient Identity $\tan \theta = \dfrac{\sin \theta}{\cos \theta}$ yields: $\tan \theta = \dfrac{\sin \theta}{\sqrt{1 - \sin^2 \theta}}$ or $\dfrac{\sin \theta \sqrt{1 - \sin^2 \theta}}{1 - \sin^2 \theta}$.

98. Use a Reciprocal Identity to write $\cos \theta$, $\cos \theta = \dfrac{1}{\sec \theta}$. Now substituting this into the Pythagorean Identity

    $\sin^2 \theta + \cos^2 \theta = 1$ yields: $\sin^2 \theta = 1 - \dfrac{1}{\sec^2 \theta} \Rightarrow \sin \theta = \pm \sqrt{1 - \dfrac{1}{\sec^2 \theta}}$. Since θ is in quadrant I or II,

    $\sin \theta = \sqrt{1 - \dfrac{1}{\sec^2 \theta}}$ or $\dfrac{\sqrt{\sec^2 \theta - 1}}{|\sec \theta|}$.

99. Dividing $x^2 + y^2 = r^2$ by y^2 yields: $\dfrac{x^2}{y^2} + \dfrac{y^2}{y^2} = \dfrac{r^2}{y^2}$ which is equal to: $\cot^2 \theta + 1 = \csc^2 \theta \Rightarrow$

    $1 + \cot^2 \theta = \csc^2 \theta$.

100. $\dfrac{\cos \theta}{\sin \theta} = \dfrac{\frac{x}{r}}{\frac{y}{r}} = \dfrac{x}{r} \div \dfrac{y}{r} = \dfrac{x}{r} \cdot \dfrac{r}{y} = \dfrac{x}{y} = \cot \theta \Rightarrow \cot \theta = \dfrac{\cos \theta}{\sin \theta}$.

101. False; $\sin \theta + \cos \theta \neq 1$ for all values of θ. For example, if $\theta = 30°$ then

    $\sin 30° + \cos 30° = \dfrac{1}{2} + \dfrac{\sqrt{3}}{2} = \dfrac{1 + \sqrt{3}}{2} \neq 1$.

102. False; if $\cot \theta = \dfrac{\cos \theta}{\sin \theta}$ and $\cot \theta = \dfrac{1}{2}$, this does not imply that $\sin \theta = 2$ since $\sin \theta \leq 1$ for all θ.

103. (a) The slope of the line is .03, so when x increases by 100 feet, y increases by 3 feet. Thus the highway has a grade of 3%.

(b) Using the point from part (a), $(100, 3)$ we first solve for r and then find $\sin \theta$:

$$r = \sqrt{100^2 + 3^2} \ \Rightarrow \ r = \sqrt{10{,}009}, \text{ then } \sin \theta = \frac{y}{r} = \frac{-6}{10{,}009}.$$

Now use the formula $R = W(\sin \theta)$, to find grade resistance $R = 25{,}000\left(\dfrac{3}{\sqrt{10{,}009}}\right) \approx 750$ pounds.

104. (a) The slope of the line is $-.09$, so when x increases by 100 feet, y decreases by 9 feet. Thus the highway has a grade of -9%.

(b) Using the point from part (a), $(100, -9)$ we first solve for r and then find $\sin \theta$:

$$r = \sqrt{100^2 + (-9)^2} \ \Rightarrow \ r = \sqrt{10{,}081}, \text{ then } \sin \theta = \frac{y}{r} = \frac{-9}{\sqrt{10{,}081}}.$$

Now use the formula $R = W(\sin \theta)$, to find grade resistance $R = 3{,}800\left(\dfrac{-9}{\sqrt{10{,}081}}\right) \approx -340$ pounds.

8.6: Evaluating Trigonometric Functions

1. $\sin A = \dfrac{\text{opp}}{\text{hyp}} = \dfrac{21}{29}$ $\cos A = \dfrac{\text{adj}}{\text{hyp}} = \dfrac{20}{29}$ $\tan A = \dfrac{\text{opp}}{\text{adj}} = \dfrac{21}{20}$

   $\csc A = \dfrac{\text{hyp}}{\text{opp}} = \dfrac{29}{21}$ $\sec A = \dfrac{\text{hyp}}{\text{adj}} = \dfrac{29}{20}$ $\cot A = \dfrac{\text{adj}}{\text{opp}} = \dfrac{20}{21}$

2. $\sin A = \dfrac{\text{opp}}{\text{hyp}} = \dfrac{45}{53}$ $\cos A = \dfrac{\text{adj}}{\text{hyp}} = \dfrac{28}{53}$ $\tan A = \dfrac{\text{opp}}{\text{adj}} = \dfrac{45}{28}$

   $\csc A = \dfrac{\text{hyp}}{\text{opp}} = \dfrac{53}{45}$ $\sec A = \dfrac{\text{hyp}}{\text{adj}} = \dfrac{53}{28}$ $\cot A = \dfrac{\text{adj}}{\text{opp}} = \dfrac{28}{45}$

3. $\sin A = \dfrac{\text{opp}}{\text{hyp}} = \dfrac{n}{p}$ $\cos A = \dfrac{\text{adj}}{\text{hyp}} = \dfrac{m}{p}$ $\tan A = \dfrac{\text{opp}}{\text{adj}} = \dfrac{n}{m}$

   $\csc A = \dfrac{\text{hyp}}{\text{opp}} = \dfrac{p}{n}$ $\sec A = \dfrac{\text{hyp}}{\text{adj}} = \dfrac{p}{m}$ $\cot A = \dfrac{\text{adj}}{\text{opp}} = \dfrac{m}{n}$

4. $\sin A = \dfrac{\text{opp}}{\text{hyp}} = \dfrac{k}{z}$ $\cos A = \dfrac{\text{adj}}{\text{hyp}} = \dfrac{y}{z}$ $\tan A = \dfrac{\text{opp}}{\text{adj}} = \dfrac{k}{y}$

   $\csc A = \dfrac{\text{hyp}}{\text{opp}} = \dfrac{z}{k}$ $\sec A = \dfrac{\text{hyp}}{\text{adj}} = \dfrac{z}{y}$ $\cot A = \dfrac{\text{adj}}{\text{opp}} = \dfrac{y}{k}$

5. (a) Using the $30° - 60°$ right triangle rules, for a $30°$ angle the hypotenuse $= 2$, the opposite side $= 1$, and the adjacent side $= \sqrt{3}$, therefore $\tan 30° = \dfrac{1}{\sqrt{3}} = \dfrac{\sqrt{3}}{3}$.

   (b) Using the calculator, $\tan 30° = \dfrac{\sqrt{3}}{3} \approx .5773502692$

6. (a) Using the $30° - 60°$ right triangle rules, for a $30°$ angle the hypotenuse $= 2$, the opposite side $= 1$, and the adjacent side $= \sqrt{3}$, therefore $\cot 30° = \dfrac{\sqrt{3}}{1} = \sqrt{3}$.

   (b) Using the calculator, $\cot 30° = (\tan 30°)^{-1} = \sqrt{3} \approx 1.732050808$

7. (a) Using the $30° - 60°$ right triangle rules, for a $30°$ angle the hypotenuse $= 2$, the opposite side $= 1$, and the adjacent side $= \sqrt{3}$, therefore $\sin 30° = \dfrac{1}{2}$.

   (b) $\sin 30° = \dfrac{1}{2}$ is a rational number.

8. (a) Using the $30° - 60°$ right triangle rules, for a $30°$ angle the hypotenuse $= 2$, the opposite side $= 1$, and the adjacent side $= \sqrt{3}$, therefore $\cos 30° = \dfrac{\sqrt{3}}{2}$.

   (b) Using the calculator, $\cos 30° = \dfrac{\sqrt{3}}{2} \approx .8660254038$

9. (a) Using the $30° - 60°$ right triangle rules, for a $30°$ angle the hypotenuse $= 2$, the opposite side $= 1$, and the adjacent side $= \sqrt{3}$, therefore $\sec 30° = \dfrac{2}{\sqrt{3}} = \dfrac{2\sqrt{3}}{3}$

   (b) Using the calculator, $\sec 30° = (\cos 30°)^{-1} = \dfrac{2\sqrt{3}}{3} \approx 1.154700538$

10. (a) Using the $30° - 60°$ right triangle rules, for a $30°$ angle the hypotenuse $= 2$, the opposite side $= 1$, and the adjacent side $= \sqrt{3}$, therefore $\csc 30° = \dfrac{2}{1} = 2$

    (b) $\csc 30° = 2$ is a rational number.

11. (a) Using the $45° - 45°$ right triangle rules, for a $45°$ angle the hypotenuse $= \sqrt{2}$, the opposite side $= 1$, and the adjacent side $= 1$, therefore $\csc 45° = \dfrac{\sqrt{2}}{1} = \sqrt{2}$.

    (b) Using the calculator, $\csc 45° = (\sin 45°)^{-1} = \sqrt{2} \approx 1.414213562$

12. (a) Using the $45° - 45°$ right triangle rules, for a $45°$ angle the hypotenuse $= \sqrt{2}$, the opposite side $= 1$, and the adjacent side $= 1$, therefore $\sec 45° = \dfrac{\sqrt{2}}{1} = \sqrt{2}$.

    (b) Using the calculator, $\sec 45° = (\cos 45°)^{-1} = \sqrt{2} \approx 1.414213562$

13. (a) Using the $45° - 45°$ right triangle rules, for a $45°$ angle the hypotenuse $= \sqrt{2}$, the opposite side $= 1$, and the adjacent $= 1$, therefore $\cos 45° = \dfrac{1}{\sqrt{2}} = \dfrac{\sqrt{2}}{2}$.

    (b) Using the calculator, $\cos 45° = \dfrac{\sqrt{2}}{2} \approx .7071067812$

14. (a) Using the $45° - 45°$ right triangle rules, for a $45°$ angle the hypotenuse $= \sqrt{2}$, the opposite side $= 1$, and the adjacent $= 1$, therefore $\cot 45° = \dfrac{1}{1} = 1$.

    (b) $\csc 45° = 1$ is a rational number.

15. (a) Because $\theta = \dfrac{\pi}{3} = 60°$ we use the $30° - 60°$ right triangle rules, for a $60°$ angle the hypotenuse $= 2$, the opposite side $= \sqrt{3}$, and the adjacent side $= 1$, therefore $\sin \dfrac{\pi}{3} = \sin 60° = \dfrac{\sqrt{3}}{2}$.

    (b) $\sin \dfrac{\pi}{3} = \sin 60° = \dfrac{\sqrt{3}}{2} = .8660254038$

16. (a) Because $\theta = \dfrac{\pi}{3} = 60°$ we use the $30° - 60°$ right triangle rules, for a $60°$ angle the hypotenuse $= 2$,

the opposite side $= \sqrt{3}$, and the adjacent side $= 1$, therefore $\cos\dfrac{\pi}{3} = \sin 60° = \dfrac{1}{2}$.

(b) $\cos\dfrac{\pi}{3} = \cos 60° = \dfrac{1}{2}$ is a rational number.

17. (a) Because $\theta = \dfrac{\pi}{3} = 60°$ we use the $30° - 60°$ right triangle rules, for a $60°$ angle the hypotenuse $= 2$,

the opposite side $= \sqrt{3}$, and the adjacent side $= 1$, therefore $\tan\dfrac{\pi}{3} = \tan 60° = \dfrac{\sqrt{3}}{1} = \sqrt{3}$.

(b) $\tan\dfrac{\pi}{3} = \tan 60° = \sqrt{3} = 1.732050808$

18. (a) Because $\theta = \dfrac{\pi}{3} = 60°$ we use the $30° - 60°$ right triangle rules, for a $60°$ angle the hypotenuse $= 2$,

the opposite side $= \sqrt{3}$, and the adjacent side $= 1$, therefore $\cot\dfrac{\pi}{3} = \cot 60° = \dfrac{1}{\sqrt{3}} = \dfrac{\sqrt{3}}{3}$.

(b) $\cot\dfrac{\pi}{3} = \cot 60° = \left(\tan\dfrac{\pi}{3}\right)^{-1} = (\tan 60°)^{-1} = \dfrac{\sqrt{3}}{3} = .5773502692$

19. (a) Because $\theta = \dfrac{\pi}{3} = 60°$ we use the $30° - 60°$ right triangle rules, for a $60°$ angle the hypotenuse $= 2$,

the opposite side $= \sqrt{3}$, and the adjacent side $= 1$, therefore $\sec\dfrac{\pi}{3} = \sec 60° = \dfrac{2}{1} = 2$.

(b) $\sec\dfrac{\pi}{3} = \sec 60° = 2$ is a rational number.

20. (a) Because $\theta = \dfrac{\pi}{3} = 60°$ we use the $30° - 60°$ right triangle rules, for a $60°$ angle the hypotenuse $= 2$,

the opposite side $= \sqrt{3}$, and the adjacent side $= 1$, therefore $\csc\dfrac{\pi}{3} = \csc 60° = \dfrac{2}{\sqrt{3}} = \dfrac{2\sqrt{3}}{3}$.

(b) $\csc\dfrac{\pi}{3} = \csc 60° = \left(\sin\dfrac{\pi}{3}\right)^{-1} = (\sin 60°)^{-1} = \dfrac{2\sqrt{3}}{3} = 1.154700538$

21. The exact value of $\sin 45°$ is $\dfrac{\sqrt{2}}{2}$. The decimal value given is an approximation.

22. Since $\sin 45 \neq \sin 45°$, her calculator must be in radian mode.

23. $\cot 73° = \tan(90° - 73°) = \tan 17°$.

24. $\sec 39° = \csc(90° - 39°) = \csc 51°$.

25. $\sin 38° = \cos(90° - 38°) = \cos 52°$.

26. $\cos 19° = \sin(90° - 19°) = \sin 71°$.

27. $\tan 25° 43' = \cot(90° - 25° 43') = \cot(89° 60' - 25° 43') = \cot 64° 17'$

28. $\sin 38° 29' = \cos(90° - 38° 29') = \cos(89° 60' - 38° 29') = \cos 51° 31'$

29. $\cos\dfrac{\pi}{5} = \sin\left(\dfrac{\pi}{2} - \dfrac{\pi}{5}\right) = \sin\left(\dfrac{5\pi}{10} - \dfrac{2\pi}{10}\right) = \sin\dfrac{3\pi}{10}$

30. $\sin\dfrac{\pi}{3} = \cos\left(\dfrac{\pi}{2} - \dfrac{\pi}{3}\right) = \cos\left(\dfrac{3\pi}{6} - \dfrac{2\pi}{6}\right) = \cos\dfrac{\pi}{6}$

31. $\tan .5 = \cot\left(\dfrac{\pi}{2} - .5\right)$

32. $\csc .3 = \sec\left(\dfrac{\pi}{2} - .3\right)$

33. If $\theta = 98°$ then $\theta' = 180° - 98° = 82°$

34. If $\theta = 212°$ then $\theta' = 212° - 180° = 32°$

35. If $\theta = 230°$ then $\theta' = 230° - 180° = 50°$

36. If $\theta = 130°$ then $\theta' = 180° - 130° = 50°$

37. If $\theta = -135°$ then $\theta' = 180° + (-135°) = 45°$

38. If $\theta = -60°$ then $\theta' = 0° - (-60°) = 60°$

39. If $\theta = 750°$ then $\theta' = 750° - 720° = 30°$

40. If $\theta = 480° = 480° - 360° = 120°$ then $\theta' = 180° - 120° = 60°$

41. If $\theta = \dfrac{4\pi}{3}$ then $\theta' = \dfrac{4\pi}{3} - \pi = \dfrac{4\pi}{3} - \dfrac{3\pi}{3} = \dfrac{\pi}{3}$.

42. If $\theta = \dfrac{7\pi}{6}$ then $\theta' = \dfrac{7\pi}{6} - \pi = \dfrac{7\pi}{6} - \dfrac{6\pi}{6} = \dfrac{\pi}{6}$.

43. If $\theta = -\dfrac{4\pi}{3} = 2\pi - \dfrac{4\pi}{3} = \dfrac{6\pi}{3} - \dfrac{4\pi}{3} = \dfrac{2\pi}{3}$ then $\theta' = \pi - \dfrac{2\pi}{3} = \dfrac{3\pi}{3} - \dfrac{2\pi}{3} = \dfrac{\pi}{3}$

44. If $\theta = -\dfrac{7\pi}{6} = 2\pi - \dfrac{7\pi}{6} = \dfrac{12\pi}{6} - \dfrac{7\pi}{6} = \dfrac{5\pi}{6}$ then $\theta' = \pi - \dfrac{5\pi}{6} = \dfrac{6\pi}{6} - \dfrac{5\pi}{6} = \dfrac{\pi}{6}$

45. It is easy to find one-half of 2, which is 1. This is, then, the measure of the side opposite the 30° angle, and the ratios are easily found. Yes, any positive number could have been used.

46. For an angle in quadrant III, the trigonometric function values are found by determining the reference angle. The absolute value of the desired functions correspond to those of the reference angle. Tangent and cotangent are positive, while the other four function values are negative.

47. Using the $30° - 60°$ right triangle rules, for a 30° angle the hypotenuse = 2, the opposite side = 1, and the adjacent side = $\sqrt{3}$, therefore $\tan 30° = \dfrac{1}{\sqrt{3}} = \dfrac{\sqrt{3}}{3}$ and $\cot 30° = \dfrac{\sqrt{3}}{1} = \sqrt{3}$.

48. Using the $45° - 45°$ right triangle rules, for a 45° angle the hypotenuse = $\sqrt{2}$, the opposite side = 1, and the adjacent side = 1, therefore $\sin 45° = \dfrac{1}{\sqrt{2}} = \dfrac{\sqrt{2}}{2}$; $\cos 45° = \dfrac{1}{\sqrt{2}} = \dfrac{\sqrt{2}}{2}$; $\sec 45° = \dfrac{\sqrt{2}}{1} = \sqrt{2}$; and $\csc 45° = \dfrac{\sqrt{2}}{1} = \sqrt{2}$.

49. Using the $30° - 60°$ right triangle rules, for a 60° angle the hypotenuse = 2, the opposite side = $\sqrt{3}$, and the adjacent side = 1, therefore $\sin 60° = \dfrac{\sqrt{3}}{2}$; $\cot 60° = \dfrac{1}{\sqrt{3}} = \dfrac{\sqrt{3}}{3}$; and $\csc 60° = \dfrac{2}{\sqrt{3}} = \dfrac{2\sqrt{3}}{3}$.

50. First find the reference angle: $(180° - 120°) = 60°$. Using the $30° - 60°$ right triangle rules, for a 60° angle the hypotenuse = 2, the opposite side = $\sqrt{3}$, and the adjacent side = 1. Since $\theta = 120°$ is in quadrant II, x is negative and cosine, secant, tangent and cotangent are negative, therefore $\cos 120° = -\cos 60° = -\dfrac{1}{2}$; $\cot 120° = -\cot 60° = -\dfrac{1}{\sqrt{3}} = -\dfrac{\sqrt{3}}{3}$; and $\sec 120° = -\sec 60° = -\dfrac{2}{1} = -2$.

51. First find the reference angle: $(180° - 135°) = 45°$. Using the $45° - 45°$ right triangle rules, for a 45° angle the hypotenuse = $\sqrt{2}$, the opposite side = 1, and the adjacent = 1. Since $\theta = 135°$ is in quadrant II, x is negative and cosine, secant, tangent and cotangent are negative, therefore $\tan 135° = -\tan 45° = -\dfrac{1}{1} = -1$; and $\cot 135° = -\cot 45° = -\dfrac{1}{1} = -1$.

52. First find the reference angle: $(180° - 150°) = 30°$. Using the $30° - 60°$ right triangle rules, for a $30°$ angle the hypotenuse $= 2$, the opposite side $= 1$, and the adjacent side $= \sqrt{3}$. Since $\theta = 150°$ is in quadrant II, x is negative and cosine, secant, tangent and cotangent are negative, therefore

$$\sin 150° = \sin 30° = \frac{1}{2}; \quad \cot 150° = -\cot 30° = -\frac{\sqrt{3}}{1} = -\sqrt{3}; \quad \text{and} \quad \sec 150° = -\sec 30° = -\frac{2}{\sqrt{3}} = -\frac{2\sqrt{3}}{3}.$$

53. First find the reference angle: $(210° - 180°) = 30°$. Using the $30° - 60°$ right triangle rules, for a $30°$ angle the hypotenuse $= 2$, the opposite side $= 1$, and the adjacent side $= \sqrt{3}$. Since $\theta = 210°$ is in quadrant III, x and y are negative and sine, cosine, secant, and cosecant are negative, therefore

$$\cos 210° = -\cos 30° = -\frac{\sqrt{3}}{2}; \quad \text{and} \quad \sec 210° = -\sec 30° = -\frac{2}{\sqrt{3}} = -\frac{2\sqrt{3}}{3}.$$

54. First find the reference angle: $(240° - 180°) = 60°$. Using the $30° - 60°$ right triangle rules, for a $60°$ angle the hypotenuse $= 2$, the opposite side $= \sqrt{3}$, and the adjacent side $= 1$. Since $\theta = 240°$ is in quadrant III, x and y are negative and sine, cosine, secant, and cosecant are negative, therefore

$$\tan 240° = \tan 60° = \frac{\sqrt{3}}{1} = \sqrt{3}; \quad \text{and} \quad \cot 240° = \cot 60° = \frac{1}{\sqrt{3}} = \frac{\sqrt{3}}{3}.$$

55. First find the reference angle: $(360° - 300°) = 60°$. Using the $30° - 60°$ right triangle rules, for a $60°$ angle the hypotenuse $= 2$, the opposite side $= \sqrt{3}$, and the adjacent side $= 1$. Since $\theta = 300°$ is in quadrant IV, y is negative and sine, tangent, cotangent, and cosecant are negative, therefore:

$$\sin 300° = -\sin 60° = -\frac{\sqrt{3}}{2} \qquad\qquad \csc 300° = -\csc 60° = -\frac{2}{\sqrt{3}} = -\frac{2\sqrt{3}}{3}$$

$$\cos 300° = \cos 60° = \frac{1}{2} \qquad\qquad \sec 300° = \sec 60° = \frac{2}{1} = 2$$

$$\tan 300° = -\tan 60° = -\frac{\sqrt{3}}{1} = -\sqrt{3} \qquad\qquad \cot 300° = -\cot 60° = -\frac{1}{\sqrt{3}} = -\frac{\sqrt{3}}{3}.$$

56. First find the reference angle: $(360° - 315°) = 45°$. Using the $45° - 45°$ right triangle rules, for a $45°$ angle the hypotenuse $= \sqrt{2}$, the opposite side $= 1$, and the adjacent $= 1$. Since $\theta = 315°$ is in quadrant IV, y is negative and sine, tangent, cotangent, and cosecant are negative, therefore:

$$\sin 315° = -\sin 45° = -\frac{1}{\sqrt{2}} = -\frac{\sqrt{2}}{2} \qquad\qquad \csc 315° = -\csc 45° = -\frac{\sqrt{2}}{1} = -\sqrt{2}$$

$$\cos 315° = \cos 45° = \frac{1}{\sqrt{2}} = \frac{\sqrt{2}}{2} \qquad\qquad \sec 300° = \sec 60° = \frac{\sqrt{2}}{1} = \sqrt{2}$$

$$\tan 315° = -\tan 45° = -\frac{1}{1} = -1 \qquad\qquad \cot 315° = \cot 45° = -\frac{1}{1} = -1.$$

57. First find the reference angle: $(405° - 360°) = 45°$. Using the $45° - 45°$ right triangle rules, for a $45°$ angle the hypotenuse $= \sqrt{2}$, the opposite side $= 1$, and the adjacent $= 1$. Since $\theta = 405°$ is in quadrant I, all functions are positive, therefore:

$$\sin 405° = \sin 45° = \frac{1}{\sqrt{2}} = \frac{\sqrt{2}}{2} \qquad\qquad \csc 405° = \csc 45° = \frac{\sqrt{2}}{1} = \sqrt{2}$$

$$\cos 405° = \cos 45° = \frac{1}{\sqrt{2}} = \frac{\sqrt{2}}{2} \qquad\qquad \sec 405° = \sec 45° = \frac{\sqrt{2}}{1} = \sqrt{2}$$

$$\tan 405° = \tan 45° = \frac{1}{1} = 1 \qquad\qquad \cot 405° = \cot 45° = \frac{1}{1} = 1$$

58. First find the reference angle: $(420° - 360°) = 60°$. Using the $30° - 60°$ right triangle rules, for a $60°$ angle the hypotenuse $= 2$, the opposite side $= \sqrt{3}$, and the adjacent side $= 1$. Since $\theta = 420°$ is in quadrant I, all functions are positive, therefore:

$$\sin 420° = \sin 60° = \frac{\sqrt{3}}{2} \qquad\qquad \csc 420° = \csc 60° = \frac{2}{\sqrt{3}} = \frac{2\sqrt{3}}{3}$$

$$\cos 420° = \cos 60° = \frac{1}{2} \qquad\qquad \sec 420° = \sec 60° = \frac{2}{1} = 2$$

$$\tan 420° = \tan 60° = \frac{\sqrt{3}}{1} = \sqrt{3} \qquad\qquad \cot 420° = \cot 60° = \frac{1}{\sqrt{3}} = \frac{\sqrt{3}}{3}$$

59. First find the reference angle: $\left(2\pi - \frac{11\pi}{6}\right) = \frac{\pi}{6}$. Since $\frac{\pi}{6}$ radians is equal to $30°$ we use the $30° - 60°$ right triangle rules, for a $30°$ angle the hypotenuse $= 2$, the opposite side $= 1$, and the adjacent side $= \sqrt{3}$.

Since $\theta = \frac{11\pi}{6}$ is in quadrant IV, y is negative and sine, tangent, cotangent, and cosecant are negative, therefore:

$$\sin \frac{11\pi}{6} = -\sin \frac{\pi}{6} = -\frac{1}{2} \qquad\qquad \csc \frac{11\pi}{6} = -\csc \frac{\pi}{6} = -\frac{2}{1} = -2$$

$$\cos \frac{11\pi}{6} = \cos \frac{\pi}{6} = \frac{\sqrt{3}}{2} \qquad\qquad \sec \frac{11\pi}{6} = \sec \frac{\pi}{6} = \frac{2}{\sqrt{3}} = \frac{2\sqrt{3}}{3}$$

$$\tan \frac{11\pi}{6} = -\tan \frac{\pi}{6} = -\frac{1}{\sqrt{3}} = -\frac{\sqrt{3}}{3} \qquad\qquad \cot \frac{11\pi}{6} = -\cot \frac{\pi}{6} = -\frac{\sqrt{3}}{1} = -\sqrt{3}$$

60. First find the reference angle: $\left(2\pi - \frac{5\pi}{3}\right) = \frac{\pi}{3}$. Since $\frac{\pi}{3}$ radians is equal to $60°$ we use the $30° - 60°$ right triangle rules, for a $60°$ angle the hypotenuse $= 2$, the opposite side $= \sqrt{3}$, and the adjacent side $= 1$.

Since $\theta = \frac{5\pi}{3}$ is in quadrant IV, y is negative and sine, tangent, cotangent, and cosecant are negative, therefore:

$$\sin \frac{5\pi}{3} = -\sin \frac{\pi}{3} = -\frac{\sqrt{3}}{2} \qquad\qquad \csc \frac{5\pi}{3} = -\csc \frac{\pi}{3} = -\frac{2}{\sqrt{3}} = -\frac{2\sqrt{3}}{3}$$

$$\cos \frac{5\pi}{3} = \cos \frac{\pi}{3} = \frac{1}{2} \qquad\qquad \sec \frac{5\pi}{3} = \sec \frac{\pi}{3} = \frac{2}{1} = 2$$

$$\tan \frac{5\pi}{3} = -\tan \frac{\pi}{3} = -\frac{\sqrt{3}}{1} = -\sqrt{3} \qquad\qquad \cot \frac{5\pi}{3} = -\cot \frac{\pi}{3} = -\frac{1}{\sqrt{3}} = -\frac{\sqrt{3}}{3}$$

61. First find the reference angle: $\left(2\pi - \frac{7\pi}{4}\right) = \frac{\pi}{4}$. Since $\frac{\pi}{4}$ radians is equal to $45°$ we use the $45° - 45°$ right triangle rules, for a $45°$ angle the hypotenuse $= \sqrt{2}$, the opposite side $= 1$, and the adjacent $= 1$.

Since $\theta = -\frac{7\pi}{4}$ is in quadrant I, all functions are positive, therefore:

$$\sin -\frac{7\pi}{4} = \sin \frac{\pi}{4} = \frac{1}{\sqrt{2}} = \frac{\sqrt{2}}{2} \qquad\qquad \csc -\frac{7\pi}{4} = \csc \frac{\pi}{4} = \frac{\sqrt{2}}{1} = \sqrt{2}$$

$$\cos -\frac{7\pi}{4} = \cos \frac{\pi}{4} = \frac{1}{\sqrt{2}} = \frac{\sqrt{2}}{2} \qquad\qquad \sec -\frac{7\pi}{4} = \sec \frac{\pi}{4} = \frac{\sqrt{2}}{1} = \sqrt{2}$$

$$\tan -\frac{7\pi}{4} = \tan \frac{\pi}{4} = \frac{1}{1} = 1 \qquad\qquad \cot -\frac{7\pi}{4} = \cot \frac{\pi}{4} = \frac{1}{1} = 1$$

62. First find the reference angle: $\left(2\pi - \dfrac{4\pi}{3}\right) = \dfrac{2\pi}{3}$, then $\theta' = \pi - \dfrac{2\pi}{3} = \dfrac{\pi}{3}$. Since $\dfrac{\pi}{3}$ radians is equal to 60° we use the 30° − 60° right triangle rules, for a 60° angle the hypotenuse = 2, the opposite side = $\sqrt{3}$, and the adjacent side = 1. Since $\theta = -\dfrac{4\pi}{3}$ is in quadrant II, x is negative and cosine, secant, tangent and cotangent are negative, therefore:

$$\sin -\frac{4\pi}{3} = \sin \frac{\pi}{3} = \frac{\sqrt{3}}{2} \qquad\qquad \csc -\frac{4\pi}{3} = \csc \frac{\pi}{3} = \frac{2}{\sqrt{3}} = \frac{2\sqrt{3}}{3}$$

$$\cos -\frac{4\pi}{3} = -\cos \frac{\pi}{3} = -\frac{1}{2} \qquad\qquad \sec -\frac{4\pi}{3} = -\sec \frac{\pi}{3} = -\frac{2}{1} = -2$$

$$\tan -\frac{4\pi}{3} = -\tan \frac{\pi}{3} = -\frac{\sqrt{3}}{1} = -\sqrt{3} \qquad\qquad \cot -\frac{4\pi}{3} = -\cot \frac{\pi}{3} = -\frac{1}{\sqrt{3}} = -\frac{\sqrt{3}}{3}$$

63. Calculate in degree mode: $\tan 29° \approx .5543090515$

64. Calculate in degree mode: $\sin 38° \approx .6156614753$

65. Calculate in degree mode: $\cot 41°24' = \cot 41\dfrac{24}{60}° = \cot 41.4° = (\tan 41.4°)^{-1} \approx 1.134277349$

66. Calculate in degree mode: $\csc 145°45' = \csc 145\dfrac{45}{60}° = \csc 145.75° = (\sin 145.75°)^{-1} \approx 1.776814578$

67. Calculate in degree mode: $\sec 183°48' = \sec 145\dfrac{48}{60}° = \sec 183.8° = (\cos 183.8°)^{-1} \approx -1.002203376$

68. Calculate in degree mode: $\cos 421°30' = \cos 421\dfrac{30}{60}° = \cos 421.5° \approx .4771587603$

69. Calculate in degree mode: $\tan(-80°6') = \tan\left(-80\dfrac{6}{60}°\right) \approx \tan(-80.1°) \approx -5.729741647$

70. Calculate in degree mode: $\sin(-317°36') = \sin\left(-317\dfrac{36}{60}°\right) \approx \sin(-317.6°) \approx .6743023876$

71. Calculate in radian mode: $\sin 2.5 \approx .5984721441$

72. Calculate in radian mode: $\cos 3.8 \approx -.7909677119$

73. Calculate in radian mode: $\tan 5 \approx -3.380515006$

74. Calculate in radian mode: $\sec 10 \approx -1.191793507$

75. (a) Find the reference angle: $\theta' = \left(\dfrac{7\pi}{6} - \pi\right) = \dfrac{\pi}{6}$. Since $\theta = \dfrac{7\pi}{6}$ is in quadrant III, sine is negative and the reference angle function is: $-\sin \dfrac{\pi}{6}$.

(b) Since $\dfrac{\pi}{6}$ radians is equal to 30° we use the 30° − 60° right triangle rules, for a 30° angle the hypotenuse = 2, the opposite side = 1, and the adjacent side = $\sqrt{3}$, therefore $-\sin \dfrac{\pi}{6} = -\dfrac{1}{2}$.

(c) Calculate in radian mode: $\sin \dfrac{7\pi}{6} = -.5$, which is equal to $-\sin \dfrac{\pi}{6} = -\dfrac{1}{2}$.

76. (a) Find the reference angle: $\left(2\pi - \dfrac{5\pi}{3}\right) = \dfrac{\pi}{3}$. Since $\theta = \dfrac{5\pi}{3}$ is in quadrant IV, cosine is positive and the

    reference angle function is: $\cos \dfrac{\pi}{3}$.

    (b) Since $\dfrac{\pi}{3}$ radians is equal to $60°$ we use the $30° - 60°$ right triangle rules, for a $60°$ angle the

    hypotenuse $= 2$, the opposite side $= \sqrt{3}$, and the adjacent side $= 1$, therefore $\cos \dfrac{\pi}{3} = \dfrac{1}{2}$.

    (c) Calculate in radian mode: $\cos \dfrac{5\pi}{3} = .5$, which is equal to $\cos \dfrac{\pi}{3} = \dfrac{1}{2}$.

77. (a) Find the reference angle: $\left(\pi - \dfrac{3\pi}{4}\right) = \dfrac{\pi}{4}$. Since $\theta = \dfrac{3\pi}{4}$ is in quadrant II,

    tangent is negative and the reference angle function is: $-\tan \dfrac{\pi}{4}$.

    (b) Since $\dfrac{\pi}{4}$ radians is equal to $45°$ we use the $45° - 45°$ right triangle rules, for a $45°$ angle the

    hypotenuse $= \sqrt{2}$, the opposite side $= 1$, and the adjacent $= 1$. therefore $-\tan \dfrac{\pi}{4} = -\dfrac{1}{1} = -1$.

    (c) Calculate in radian mode: $\tan \dfrac{3\pi}{4} = -1$, which is equal to $-\tan \dfrac{\pi}{4} = -\dfrac{1}{1} = -1$.

78. (a) Find the reference angle: $\left(2\pi - \dfrac{5\pi}{3}\right) = \dfrac{\pi}{3}$. Since $\theta = \dfrac{5\pi}{3}$ is in quadrant IV, sine is negative and the

    reference angle function is: $-\sin \dfrac{\pi}{3}$.

    (b) Since $\dfrac{\pi}{3}$ radians is equal to $60°$ we use the $30° - 60°$ right triangle rules, for a $60°$ angle the

    hypotenuse $= 2$, the opposite side $= \sqrt{3}$, and the adjacent side $= 1$, therefore $-\sin \dfrac{\pi}{3} = -\dfrac{\sqrt{3}}{2}$.

    (c) Calculate in radian mode: $\sin \dfrac{5\pi}{3} \approx -.8660254038$, which is equal to $-\sin \dfrac{\pi}{3} = \dfrac{\sqrt{3}}{2}$.

79. (a) Find the reference angle: $\theta' = \left(\dfrac{7\pi}{6} - \pi\right) = \dfrac{\pi}{6}$. Since $\theta = \dfrac{7\pi}{6}$ is in quadrant III,

    cosine is negative and the reference angle function is: $-\cos \dfrac{\pi}{6}$.

    (b) Since $\dfrac{\pi}{6}$ radians is equal to $30°$ we use the $30° - 60°$ right triangle rules, for a $30°$ angle the

    hypotenuse $= 2$, the opposite side $= 1$, and the adjacent side $= \sqrt{3}$, therefore $-\cos \dfrac{\pi}{6} = -\dfrac{\sqrt{3}}{2}$.

    (c) Calculate in radian mode: $\cos \dfrac{7\pi}{6} = -.8660254038$, which is equal to $-\cos \dfrac{\pi}{6} = -\dfrac{\sqrt{3}}{2}$.

80. (a) Find the reference angle: $\theta' = \left(\dfrac{4\pi}{3} - \pi\right) = \dfrac{\pi}{3}$. Since $\theta = \dfrac{4\pi}{3}$ is in quadrant III,

    tangent is positive and the reference angle function is: $\tan \dfrac{\pi}{3}$.

    (b) Since $\dfrac{\pi}{3}$ radians is equal to $60°$ we use the $30° - 60°$ right triangle rules, for a $60°$ angle the

    hypotenuse $= 2$, the opposite side $= \sqrt{3}$, and the adjacent side $= 1$, therefore $\tan \dfrac{\pi}{3} = \dfrac{\sqrt{3}}{1} = \sqrt{3}$.

    (c) Calculate in radian mode: $\tan \dfrac{4\pi}{3} \approx 1.732050808$, which is equal to $\tan \dfrac{\pi}{3} = \sqrt{3}$.

81. Since $\sin \theta$ is positive in quadrant I and II, values for θ can only be in these quadrants.

    In quadrant I, from the $30° - 60°$ right triangle, $\sin 30° = \dfrac{1}{2}$, therefore $30°$ is one value.

    In quadrant II, use the reference angle $30°$ to find the other: $180° - 30° = 150°$. Therefore $\theta = 30°$ or $150°$.

82. Since $\cos \theta$ is positive in quadrant I and IV, values for θ can only be in these quadrants.

    In quadrant I, from the $30° - 60°$ right triangle, $\cos 30° = \dfrac{\sqrt{3}}{2}$, therefore $30°$ is one value.

    In quadrant IV, use the reference angle $30°$ to find the other: $360° - 30° = 330°$. Therefore, $\theta = 30°$ or $330°$.

83. Since $\tan \theta$ is negative in quadrant II and IV, values for θ can only be in these quadrants.

    From the $30° - 60°$ right triangle, $\tan 60° = \sqrt{3}$, therefore $60°$ is the reference angle.

    In quadrant II, use the reference angle $60°$ to find the first value: $180° - 60° = 120°$. In quadrant IV, use the reference angle $60°$ to find the other value: $360° - 60° = 300°$. Therefore, $\theta = 120°$ or $300°$.

84. Since $\sec \theta$ is negative in quadrant II and III, values for θ can only be in these quadrants.

    From the $45° - 45°$ right triangle, $\sec 45° = \sqrt{2}$, therefore $45°$ is the reference angle.

    In quadrant II, use the reference angle $45°$ to find the first value: $180° - 45° = 135°$. In quadrant III, use the reference angle $45°$ to find the other value: $180° + 45° = 225°$. Therefore, $\theta = 135°$ or $225°$.

85. Since $\cot \theta$ is negative in quadrant II and IV, values for θ can only be in these quadrants.

    From the $30° - 60°$ right triangle, $\cot 60° = \dfrac{\sqrt{3}}{3}$, therefore $60°$ is the reference angle.

    In quadrant II, use the reference angle $60°$ to find the first value: $180° - 60° = 120°$. In quadrant IV, use the reference angle $60°$ to find the other value: $360° - 60° = 300°$. Therefore, $\theta = 120°$ or $300°$.

86. Since $\cos \theta$ is positive in quadrant I and IV, values for θ can only be in these quadrants.

    In quadrant I, from the $45° - 45°$ right triangle, $\cos 45° = \dfrac{\sqrt{2}}{2}$, therefore $45°$ is one value.

    In quadrant IV, use the reference angle $45°$ to find the other: $360° - 45° = 315°$. Therefore, $\theta = 45°$ or $315°$.

87. Since $\cos \theta$ is positive in quadrant I and IV, values for θ can only be in these quadrants.

    In quadrant I, we use the inverse trigonometric function (degree mode) of our calculator to find the angle: $\cos^{-1} .68716510 \Rightarrow \theta = 46.59388121°$. In quadrant IV, use the found reference angle from quadrant I to find θ: $360° - 46.59388121° = 313.4061188°$. Therefore, $\theta \approx 46.59388121°$ or $313.4061188°$.

88. Since $\cos \theta$ is positive in quadrant I and IV, values for θ can only be in these quadrants.

    In quadrant I, we use the inverse trigonometric function (degree mode) of our calculator to find the angle: $\cos^{-1} .96476120 \Rightarrow \theta = 15.25568949°$. In quadrant IV, use the found reference angle from quadrant I to find θ: $360° - 15.7443105° = 344.7443105°$. Therefore, $\theta \approx 15.25568949°$ or $344.7443105°$.

89. Since $\sin \theta$ is positive in quadrant I and II, values for θ can only be in these quadrants.

    In quadrant I, we use the inverse trigonometric function (degree mode) of our calculator to find the angle: $\sin^{-1} .41298643 \Rightarrow \theta = 24.39257624°$. In quadrant II, use the found reference angle from quadrant I to find θ: $180° - 24.39257624° = 155.6074238°$. Therefore, $\theta \approx 24.39257624°$ or $155.6074238°$.

90. Since $\sin \theta$ is positive in quadrant I and II, values for θ can only be in these quadrants.

    In quadrant I, we use the inverse trigonometric function (degree mode) of our calculator to find the angle: $\sin^{-1} .63898531 \Rightarrow \theta = 39.71619817°$. In quadrant II, use the found reference angle from quadrant I to find θ: $180° - 39.71619817° = 140.2838018°$. Therefore, $\theta \approx 39.71619817°$ or $140.2838018°$.

91. Since $\tan \theta$ is positive in quadrant I and III, values for θ can only be in these quadrants.

    In quadrant I, we use the inverse trigonometric function (degree mode) of our calculator to find the angle: $\tan^{-1} .87692035 \Rightarrow \theta = 41.24818261°$. In quadrant III, use the found reference angle from quadrant I to find θ: $180° + 41.24818261° = 221.2481826°$. Therefore, $\theta \approx 41.24818261°$ or $221.2481826°$.

92. Since $\tan \theta$ is positive in quadrant I and III, values for θ can only be in these quadrants.

    In quadrant I, we use the inverse trigonometric function (degree mode) of our calculator to find the angle: $\tan^{-1} 1.2841996 \Rightarrow \theta = 52.09228102°$. In quadrant III, use the found reference angle from quadrant I to find θ: $180° + 52.09228102° = 232.092281°$. Therefore, $\theta \approx 52.09228102°$ or $232.092281°$.

93. Since $\tan \theta$ is positive in quadrant I and III, values for θ can only be in these quadrants.

    In quadrant I, we use the inverse trigonometric function (radian mode) of our calculator to find the angle: $\tan^{-1} .21264138 \Rightarrow \theta = .2095206607$ radians. In quadrant III, use the found reference angle from quadrant I to find θ: $\pi + .2095206607 = 3.351113314$ radians. Therefore, $\theta \approx .2095206607$ or 3.351113314 radians.

94. Since $\cos \theta$ is positive in quadrant I and IV, values for θ can only be in these quadrants.

    In quadrant I, we use the inverse trigonometric function (radian mode) of our calculator to find the angle: $\cos^{-1} .78269876 \Rightarrow \theta = .6718062049$ radians. In quadrant IV, use the found reference angle from quadrant I to find θ: $2\pi - .6718062049 = 5.611379102$ radians. Therefore, $\theta \approx .6718062049$ or 5.611379102 radians.

95. Since $\cot \theta$ is positive in quadrant I and III, values for θ can only be in these quadrants.

    In quadrant I, we use the inverse trigonometric function (radian mode) of our calculator to find the angle: Since $\cot \theta = (\tan \theta)^{-1} \Rightarrow (\tan \theta)^{-1} = .29949853 \Rightarrow \tan \theta = (.29949853)^{-1} \Rightarrow \theta = \tan^{-1} (.29949853)^{-1} \Rightarrow$ $\theta \approx 1.27979966$ radians. In quadrant III, use the found reference angle from quadrant I to find θ: $\pi + 1.27979966 = 4.421392314$ radians. Therefore, $\theta \approx 1.27979966$ or 4.421392314 radians.

96. Since $\csc \theta$ is positive in quadrant I and II, values for θ can only be in these quadrants.

    In quadrant I, we use the inverse trigonometric function (radian mode) of our calculator to find the angle: Since $\csc \theta = (\sin \theta)^{-1} \Rightarrow (\sin \theta)^{-1} = 1.0219553 \Rightarrow \sin \theta = (1.0219553)^{-1} \Rightarrow \theta = \sin^{-1} (1.0219553)^{-1} \Rightarrow$ $\theta \approx 1.363138008$ radians. In quadrant II, use the found reference angle from quadrant I to find θ: $\pi - 1.363138008 = 1.778454646$ radians. Therefore, $\theta \approx 1.363138008$ or 1.778454646 radians.

97. Using the point in the given example, $\sqrt{(x_1)^2 + (y_1)^2} = r$ would yield: $r \approx 6.9032258$. This represents the distance from the point (x_1, y_1) to the origin.

98. Using the point in the given example and degree mode yields:

    $$\tan^{-1}\left(\frac{y_1}{x_1}\right) = \tan^{-1}\left(\frac{5.9783689}{3.4516129}\right) \approx 59.9999° \approx 60°.$$

99. Using the point in the given example and degree mode yields:

$$\sin^{-1}\left(\frac{y_1}{r}\right) = \sin^{-1}\left(\frac{5.9783689}{6.9032258}\right) \approx 59.9999° \approx 60°.$$

100. Using the point in the given example and degree mode yields:

$$\cos^{-1}\left(\frac{x_1}{r}\right) = \cos^{-1}\left(\frac{3.4516129}{6.9032258}\right) \approx 60°.$$

101. It is a measure (approximately 60°, for the example) of the angle formed by the positive x-axis and the ray $y = \sqrt{3}x, x \geq 0$.

102. $\dfrac{y_1}{x_1} = \dfrac{5.9783689}{3.4516129} \approx 1.7320508.$ Now $(1.7320508)^2 \approx 2.9999 \approx 3$, therefore the exact value of $\dfrac{y_1}{x_1} = \sqrt{3}.$

103. ... make a conjecture: The <u>slope</u> of a line passing through the origin is equal to the <u>tangent</u> of the angle it forms with the positive x-axis.

104. $\left(\dfrac{x_1}{r}\right)^2 + \left(\dfrac{y_1}{r}\right)^2 \approx \left(\dfrac{3.4516129}{6.9032258}\right)^2 + \left(\dfrac{5.9783689}{6.9032258}\right)^2 \approx .25 + .75 \approx 1.$ This illustrates the identity $\cos^2\theta + \sin^2\theta = 1.$

105. $\csc 60° = \dfrac{1}{\sin 60°} = \dfrac{2\sqrt{3}}{3} \approx 1.154700538$

$\dfrac{r}{y_1} = \dfrac{6.9032258}{5.9783689} \approx 1.15470054.$ Both are approximately the same.

106. Using the graphing calculator, graph $y_1 = \sqrt{3}x$ and $y_2 = \sqrt{1 - x^2}$. From the calculator the point of intersection is: $(.5, .86602540).$

107. $\cos 60° \approx .5$; this is the x-coordinate of the point found in problem 106.

If $r = 1$, $\cos 60° = .5$ and $\cos 60° = \dfrac{x}{1}$, then $\dfrac{x}{1} = .5 \implies x = .5.$

108. $\sin 60° \approx .8660254038$; this is the y-coordinate of the point found in problem 106.

If $r = 1$, $\sin 60° = .8660254038$ and $\sin 60° = \dfrac{y}{1}$, then $\dfrac{y}{1} = .8660254038 \implies y \approx .86602540.$

109. Solve for θ using the formula and given values:

$$400 = 5000\,(\sin\theta) \implies \sin\theta = \frac{400}{5000} \implies \theta = \sin^{-1}\frac{2}{25} \approx 4.6°.$$

110. Solve for θ using the formula and given values:

$$130 = 3500\,(\sin\theta) \implies \sin\theta = \frac{130}{3500} \implies \theta = \sin^{-1}\frac{13}{350} \approx 2.1°.$$

111. Solve for c_2 using the formula and given values: $\dfrac{3 \times 10^8}{c_2} = \dfrac{\sin 46°}{\sin 31°} \implies c_2 = \dfrac{(3 \times 10^8)\sin 31°}{\sin 46°} \approx$

214,796,154 $\approx 2 \times 10^8$ min/sec.

112. Solve for c_2 using the formula and given values: $\dfrac{3 \times 10^8}{c_2} = \dfrac{\sin 39°}{\sin 28°} \implies c_2 = \dfrac{(3 \times 10^8)\sin 28°}{\sin 39°} \approx$

223,799,309 $\approx 2 \times 10^8$ min/sec.

113. Solve for θ using the formula and given values:

$$\frac{3 \times 10^8}{1.5 \times 10^8} = \frac{\sin 40°}{\sin\theta_2} \implies (3 \times 10^8)\sin\theta_2 = (1.5 \times 10^8)\sin 40° \implies \sin\theta_2 = \frac{(1.5 \times 10^8)\sin 40°}{3 \times 10^8} \implies$$

$$\theta_2 = \sin^{-1}\left(\frac{(1.5 \times 10^8)\sin 40°}{3 \times 10^8}\right) \implies \theta_2 \approx 18.747° \approx 19°.$$

114. Solve for θ using the formula and given values:

$$\frac{3 \times 10^8}{2.6 \times 10^8} = \frac{\sin 62°}{\sin \theta_2} \Rightarrow (3 \times 10^8) \sin \theta_2 = (2.6 \times 10^8) \sin 62° \Rightarrow \sin \theta_2 = \frac{(2.6 \times 10^8) \sin 62°}{3 \times 10^8} \Rightarrow$$

$$\theta_2 = \sin^{-1}\left(\frac{(2.6 \times 10^8) \sin 62°}{3 \times 10^8}\right) \Rightarrow \theta_2 \approx 49.927° \approx 50°.$$

115. Solve for θ using the formula and given values:

$$\frac{3 \times 10^8}{2.254 \times 10^8} = \frac{\sin 90°}{\sin \theta_2} \Rightarrow (3 \times 10^8) \sin \theta_2 = (2.254 \times 10^8) \sin 90° \Rightarrow$$

$$\sin \theta_2 = \frac{(2.254 \times 10^8) \sin 90°}{3 \times 10^8} \Rightarrow$$

$$\theta_2 = \sin^{-1}\left(\frac{(2.254 \times 10^8) \sin 90°}{3 \times 10^8}\right) \Rightarrow \theta_2 \approx 48.706° \approx 48.7°.$$

116. Solve for θ using the formula and given values and $\theta_1 = 90° - 29.6° = 60.4°$:

$$\frac{3 \times 10^8}{2.254 \times 10^8} = \frac{\sin 60.4°}{\sin \theta_2} \Rightarrow (3 \times 10^8) \sin \theta_2 = (2.254 \times 10^8) \sin 60.4° \Rightarrow$$

$$\sin \theta_2 = \frac{(2.254 \times 10^8) \sin 60.4°}{3 \times 10^8} \Rightarrow \theta_2 = \sin^{-1}\left(\frac{(2.254 \times 10^8) \sin 60.4°}{3 \times 10^8}\right) \Rightarrow \theta_2 \approx 40.7896° \approx 40.8°.$$

Light from the object is refracted at an angle of 40.8° from vertical. Light from the horizon is refracted at an angle of 48.7° from the vertical. Therefore, the fish thinks the object lies at an angle of 48.7° − 40.8° = 7.9° above the horizon.

117. (a) First change 55 and 30 mph to feet per second: $\dfrac{55 \text{ mi}}{1 \text{ hr}} \cdot \dfrac{5280 \text{ ft}}{1 \text{ mi}} \cdot \dfrac{1 \text{ hr}}{3600 \text{ sec}} = 80\dfrac{2}{3} \text{ ft/sec}$ and

$\dfrac{30 \text{ mi}}{1 \text{ hr}} \cdot \dfrac{5280 \text{ ft}}{1 \text{ mi}} \cdot \dfrac{1 \text{ hr}}{3600 \text{ sec}} = 44 \text{ ft/sec}$. Now solve for D using the formula, given, and found values:

$$D = \frac{1.05\left((80\tfrac{2}{3})^2 - (44)^2\right)}{64.4\left(.4 + .02 + \sin 3.5°\right)} \approx 154.9303 \Rightarrow D \approx 155 \text{ feet}.$$

   (b) Solve for D using the formula and given values:

$$D = \frac{1.05\left((80\tfrac{2}{3})^2 - (44)^2\right)}{64.4\left(.4 + .02 + \sin(-2°)\right)} \approx 193.5313 \Rightarrow D \approx 194 \text{ feet}.$$

   (c) As the grade decreases from uphill to downhill, the braking distance increases, which corresponds to driving experience.

118. First change 90 mph to feet per second: $\dfrac{90 \text{ mi}}{1 \text{ hr}} \cdot \dfrac{5280 \text{ ft}}{1 \text{ mi}} \cdot \dfrac{1 \text{ hr}}{3600 \text{ sec}} = 132 \text{ ft/sec}$.

Now, using the values from Exercise 117, determine V_2 when $D = 200$:

$$200 = \frac{1.05\left(132^2 - V_2\right)}{64.4\left(.4 + .02 + \sin(-3.5°)\right)} \Rightarrow 200 = \frac{18,295.2 - 1.05(V_2)^2}{64.4\left(.42 + \sin(-3.5°)\right)} \Rightarrow$$

$$(V_2)^2 = \frac{200(64.4)(.42 + \sin(-3.5°)) - 18,295.2}{-1.05} \Rightarrow (V_2)^2 \approx 13,020.86208 \Rightarrow V_2 \approx 114.10899$$

Thus $V_2 \approx 114$ feet per second. This is approximately: $\dfrac{114.10899 \text{ ft}}{1 \text{ sec}} \cdot \dfrac{1 \text{ mi}}{5280 \text{ ft}} \cdot \dfrac{3600 \text{ sec}}{1 \text{ hr}} \approx 77.80 \approx 78 \text{ mph}.$

8.7: Applications of Right Triangles

1. To find B: $B = 90° - 36° 20' \Rightarrow B = 89° 60' - 36° 20' \Rightarrow B = 53° 40'$.

   To find a: $\sin A = \dfrac{a}{c} \Rightarrow \sin 36° 20' = \dfrac{a}{964} \Rightarrow a = 964 \, (\sin 36° 20') \approx 571.1526 \Rightarrow a \approx 571$ m.

   To find b: $\cos A = \dfrac{b}{c} \Rightarrow \cos 36° 20' = \dfrac{b}{964} \Rightarrow b = 964 \, (\cos 36° 20') \approx 776.5827 \Rightarrow b \approx 777$ m.

2. To find B: $B = 90° - 31° 40' \Rightarrow B = 89° 60' - 31° 40' \Rightarrow B = 58° 20'$.

   To find b: $\tan A = \dfrac{a}{b} \Rightarrow b = \dfrac{a}{\tan A} = \dfrac{35.9}{\tan 31° 40'} \approx 58.2028 \Rightarrow b \approx 58.2$ km.

   To find c: $\sin A = \dfrac{a}{c} \Rightarrow c = \dfrac{a}{\sin A} = \dfrac{35.9}{\sin 31° 40'} \approx 68.384 \Rightarrow c \approx 68.4$ km.

3. To find M: $M = 90° - 51.2° \Rightarrow M = 38.8°$.

   To find n: $\tan N = \dfrac{n}{m} \Rightarrow \tan 51.2° = \dfrac{n}{124} = n = 124 \, (\tan 51.2°) \approx 154.2249 \Rightarrow n \approx 154$ m.

   To find p: $\cos N = \dfrac{m}{p} \Rightarrow \cos 51.2° = \dfrac{124}{p} \Rightarrow p = \dfrac{124}{\cos 51.2°} \approx 197.8922 \Rightarrow c \approx 198$ m.

4. To find Y: $Y = 90° - 47.8° \Rightarrow Y = 42.2°$.

   To find x: $\sin X = \dfrac{x}{z} \Rightarrow \sin 47.8° = \dfrac{x}{89.6} \Rightarrow x = 89.6 \, (\sin 47.8°) \approx 66.376 \Rightarrow x \approx 66.4$ cm.

   To find y: $\cos X = \dfrac{y}{z} \Rightarrow \cos 47.8° = \dfrac{y}{89.6} \Rightarrow y = 89.6 \, (\cos 47.8°) \approx 60.186 \Rightarrow y \approx 60.2$ cm.

5. To find A: $A = 90° - 42.0892° \Rightarrow A = 47.9108°$.

   To find a: $\tan B = \dfrac{b}{a} \Rightarrow a = \dfrac{b}{\tan B} = \dfrac{56.851}{\tan 42.0892°} \approx 62.942095 \Rightarrow a \approx 62.942$ cm.

   To find c: $\sin B = \dfrac{b}{c} \Rightarrow c = \dfrac{b}{\sin B} = \dfrac{56.851}{\sin 42.0892°} \approx 84.81594 \Rightarrow c \approx 84.816$ cm.

6. To find A: $A = 90° - 68.5142° \Rightarrow A = 21.4858°$.

   To find a: $\cos B = \dfrac{a}{c} \Rightarrow \cos 68.5142° = \dfrac{a}{3579.42} \Rightarrow a = 3579.42 \, (\cos 68.5142°) \approx 1311.0364 \Rightarrow$
   $a \approx 1311.04$ m.

   To find b: $\sin B = \dfrac{b}{c} \Rightarrow \sin 68.5142° = \dfrac{b}{3579.42} \Rightarrow b = 3579.42 \, (\sin 68.5142°) \approx 3330.6803 \Rightarrow$
   $b \approx 3330.68$ m.

7. To find A: $\tan A = \dfrac{a}{b} \Rightarrow \tan A = \dfrac{7.1}{9.7} \Rightarrow A = \tan^{-1} \dfrac{7.1}{9.7} \Rightarrow A \approx 36.2026° \Rightarrow A \approx 36°$.

   To find B: $B = 90° - 36° \Rightarrow B = 54°$.

   To find c: $\sin A = \dfrac{a}{c} \Rightarrow \sin 36° = \dfrac{7.1}{c} \Rightarrow c = \dfrac{7.1}{\sin 36°} \approx 12.07924 \Rightarrow c \approx 12$ ft.

8. To find A: $\cos A = \dfrac{b}{c} \Rightarrow \cos A = \dfrac{6.9}{12} \Rightarrow A = \cos^{-1} \dfrac{6.9}{12} \Rightarrow A \approx 54.90037° \Rightarrow A \approx 55°$.

   To find B: $B = 90° - 55° \Rightarrow B = 35°$.

   To find a: $\sin A = \dfrac{a}{c} \Rightarrow \sin 55° = \dfrac{a}{12} \Rightarrow a = 12 \, (\sin 55°) \approx 9.82982 \Rightarrow a \approx 9.8$ ft.

9. To find A: $\cos A = \dfrac{b}{c}$ $\Rightarrow$ $\cos A = \dfrac{7.3}{11}$ $\Rightarrow$ $A = \cos^{-1}\dfrac{7.3}{11}$ $\Rightarrow$ $A \approx 48.42220°$ $\Rightarrow$ $A \approx 48°$.

   To find B: $B = 90° - 48°$ $\Rightarrow$ $B = 42°$.

   To find a: $\sin A = \dfrac{a}{c}$ $\Rightarrow$ $\sin 48° = \dfrac{a}{11}$ $\Rightarrow$ a $= 11\,(\sin 48°) \approx 8.174593$ $\Rightarrow$ $a \approx 8.2$ ft.

10. To find A: $\tan A = \dfrac{a}{b}$ $\Rightarrow$ $\tan A = \dfrac{125}{85.6}$ $\Rightarrow$ $A = \tan^{-1}\dfrac{125}{85.6}$ $\Rightarrow$ $A \approx 55.59666°$ $\Rightarrow$ $A \approx 55.6°$.

   To find B: $B = 90.0° - 55.6°$ $\Rightarrow$ $B = 34.4°$.

   To find c: $\sin A = \dfrac{a}{c}$ $\Rightarrow$ $\sin 55.6° = \dfrac{125}{c}$ $\Rightarrow$ c $= \dfrac{125}{\sin 55.6°} \approx 151.149431$ $\Rightarrow$ $c \approx 152$ ft.

11. To find B: $B = 90.00° - 28.00°$ $\Rightarrow$ $B = 62.00°$.

   To find a: $\sin A = \dfrac{a}{c}$ $\Rightarrow$ $\sin 28.00° = \dfrac{a}{17.4}$ $\Rightarrow$ $a = 17.4\,(\sin 28.00°) \approx 8.1688$ $\Rightarrow$ $a \approx 8.17$ ft.

   To find b: $\cos A = \dfrac{b}{c}$ $\Rightarrow$ $\cos 28.00° = \dfrac{b}{17.4}$ $\Rightarrow$ $b = 17.4\,(\cos 28.00°) \approx 15.3633$ $\Rightarrow$ $b \approx 15.4$ ft.

12. To find A: $A = 90.00° - 46.00°$ $\Rightarrow$ $A = 44.00°$.

   To find a: $\cos B = \dfrac{a}{c}$ $\Rightarrow$ $\cos 46.00° = \dfrac{a}{29.7}$ $\Rightarrow$ $a = 29.7\,(\cos 46.00°) \approx 20.6314$ $\Rightarrow$ $a \approx 20.6$ m.

   To find b: $\sin B = \dfrac{b}{c}$ $\Rightarrow$ $\sin 46.00° = \dfrac{b}{29.7}$ $\Rightarrow$ $b = 29.7\,(\sin 46.00°) \approx 21.3644$ $\Rightarrow$ $b \approx 21.4$ m.

13. To find A: $A = 90.00° - 73.00°$ $\Rightarrow$ $A = 17.00°$.

   To find a: $\tan B = \dfrac{b}{a}$ $\Rightarrow$ $\tan 73.00° = \dfrac{128}{a}$ $\Rightarrow$ $a = \dfrac{128}{\tan 73.00°} \approx 39.1335$ $\Rightarrow$ $a \approx 39.1$ in.

   To find c: $\sin B = \dfrac{b}{c}$ $\Rightarrow$ $\sin 73.00° = \dfrac{128}{c}$ $\Rightarrow$ $c = \dfrac{128}{\sin 73.00°} \approx 133.849$ $\Rightarrow$ $c \approx 134$ in.

14. To find A: $B = 90°00' - 61°00'$ $\Rightarrow$ $B = 29°00'$.

   To find a: $\tan A = \dfrac{a}{c}$ $\Rightarrow$ $\tan 61°00' = \dfrac{a}{39.2}$ $\Rightarrow$ $a = 39.2\,(\tan 61°00') \approx 70.7187$ $\Rightarrow$ $a \approx 70.7$ cm.

   To find c: $\cos A = \dfrac{b}{c}$ $\Rightarrow$ $\cos 61°00' = \dfrac{39.2}{c}$ $\Rightarrow$ $c = \dfrac{39.2}{\cos 61°00'} \approx 80.856$ $\Rightarrow$ $c \approx 80.9$ cm.

15. To find c, use pythagorean theorem:

   $c^2 = a^2 + b^2$ $\Rightarrow$ $c^2 = (76.4)^2 + (39.3)^2$ $\Rightarrow$ $c^2 = 7381.45$ $\Rightarrow$ $c \approx 85.9154$ $\Rightarrow$ $c \approx 85.9$ yd.

   To find A: $\tan A = \dfrac{a}{b}$ $\Rightarrow$ $\tan A = \dfrac{76.4}{39.3}$ $\Rightarrow$ $A = \tan^{-1}\left(\dfrac{76.4}{39.3}\right) \approx 62.7788°$ $\Rightarrow$ $A \approx 62°46'\,44''$ $\Rightarrow$

   $A \approx 62°50'$. To find B: $B = 90°00' - 62°50'$ $\Rightarrow$ $B = 89°60' - 62°50'$ $\Rightarrow$ $B = 27°10'$.

16. To find c, use pythagorean theorem:

   $c^2 = a^2 + b^2$ $\Rightarrow$ $c^2 = (958)^2 + (489)^2$ $\Rightarrow$ $c^2 = 1{,}156{,}885$ $\Rightarrow$ $c \approx 1075.5859$ $\Rightarrow$ $c \approx 1080$ m.

   To find A: $\tan A = \dfrac{a}{b}$ $\Rightarrow$ $\tan A = \dfrac{958}{489}$ $\Rightarrow$ $A = \tan^{-1}\left(\dfrac{958}{489}\right) \approx 62.958°$ $\Rightarrow$ $A \approx 62°57.51'$ $\Rightarrow$

   $A \approx 63°00'$. To find B: $B = 90°00' - 63°00'$ $\Rightarrow$ $B = 27°00'$.

17. The other acute angle requires the least work to find, simply subtract the given angle from 90°.

18. No, because there will be infinitely many similar triangles satisfying the given conditions.

19. Because x and y are known, use tangent to find:

$$\tan \theta = \frac{y}{x} \;\Rightarrow\; \tan \theta = \frac{3.68}{4.6} \;\Rightarrow\; \theta = \tan^{-1}\!\left(\frac{3.68}{4.6}\right) \approx 38.6598°.$$

20. Because x and y are known, use tangent to find:

$$\tan \theta = \frac{y}{x} \;\Rightarrow\; \tan \theta = \frac{13.75}{5.5} \;\Rightarrow\; \theta = \tan^{-1}\!\left(\frac{13.75}{5.5}\right) \approx 68.1986°.$$

21. Because AD and BC are parallel, angle DAB is congruent to angle ABC, as they are alternate interior angles of the transversal AB. (A theorem of elementary geometry assures us of this.)

22. An angle of depression or elevation must have the horizontal line of sight as one of its sides.

23. It is measured clockwise from the north.

24. It is measured from the north (or south) in the east (or west) direction.

25. $\sin 43°50' = \dfrac{h}{13.5} \;\Rightarrow\; h = 13.5(\sin 43°50') \;\Rightarrow\; h \approx 9.34959996 \;\Rightarrow\; h \approx 9.35$ m.

26. $\cos 57.0° = \dfrac{175}{l} \;\Rightarrow\; l = \dfrac{175}{\cos 57.0°} \approx 321.31 \;\Rightarrow\; l \approx 321$ ft.

27. $\tan 23.4° = \dfrac{5.75}{x} \;\Rightarrow\; x = \dfrac{5.75}{\tan 23.4°} \approx 13.2875 \;\Rightarrow\; x \approx 13.3$ ft.

28. $\tan 34.6° = \dfrac{h}{40.6} \;\Rightarrow\; h = 40.6\,(\tan 34.6°) \approx 28.00806 \;\Rightarrow\; h \approx 28.0$ m.

29. First find the distance between the buildings, using the angle of depression and $h_1 = 30.0$ feet, since a point on the building horizontally across the street is, like the window, 30 feet from the ground.

    Find d: $\tan 20.0 = \dfrac{30.0}{d} \;\Rightarrow\; d = \dfrac{30.0}{\tan 20.0°} \approx 82.424$. Now use the found distance and the angle elevation to find h_2 (the height from the horizontal point on the building to the top of the building):

    $\tan 50° = \dfrac{h_2}{d} \;\Rightarrow\; \tan 50° = \dfrac{h_2}{82.424} \;\Rightarrow\; h_2 = 82.424\,(\tan 50°) \approx 98.229$. Finally, the height of the building is: $h_1 + h_2 = 30 + 98.229 = 128.229 \approx 128$ feet.

30. First find the distance between the buildings, using the angle of depression and $h_1 = 28.0$ meters, since a point on the tall building horizontally across the street is, like the top of the small building, 28 meters from the ground. Find d: $\tan 14°10' = \dfrac{28.0}{d} \;\Rightarrow\; d = \dfrac{28.0}{\tan 14°10'} \approx 110.9262$. Now use the found distance and the angle elevation to find h_2 (the height from the horizontal point on the building to the top of the building): $\tan 46°40' = \dfrac{h_2}{d} \;\Rightarrow\; \tan 46°40' = \dfrac{h_2}{110.9262} \;\Rightarrow\; h_2 = 110.9262\,(\tan 46°40') \approx 117.57$.

    Finally, the height of the building is: $h_1 + h_2 = 28 + 117.57 = 145.57 \approx 146$ meters.

31. First find the complementary angle to the angle of depression: $\tan C = \dfrac{12.02}{5.93} \;\Rightarrow\; C = \tan^{-1}\!\left(\dfrac{12.02}{5.93}\right) \Rightarrow$

    $C \approx 63.74°$. Therefore, the angle of depression is: $90° - 63.74° = 26.26° \approx 26.3°$ or $26°20'$.

32. First find the complementary angle to the angle of depression: $\tan F = \dfrac{51.74}{39.82} \;\Rightarrow\; F = \tan^{-1}\!\left(\dfrac{51.74}{39.82}\right) \Rightarrow$

    $C \approx 52.417°$. Therefore, the angle of depression is: $90° - 52.417° = 37.583° \approx 37.58°$ or $37°35'$.

33. To find A: $\tan A = \dfrac{a}{b} \Rightarrow \tan A = \dfrac{1.0837}{1.4923} \Rightarrow A = \tan^{-1}\dfrac{1.0837}{1.4923} \Rightarrow A \approx 35.9869° \Rightarrow$

    $A \approx 35.987°$ or $35°59'\,10''$. To find B: $B = 90.0° - 35.987° \Rightarrow B = 54.013°$ or $54°00'\,50''$.

34. Using the angle of depression and $h = 10{,}500$, find the horizontal air distance:

    $\tan 13°50' = \dfrac{10{,}500}{h} \Rightarrow h = \dfrac{10{,}500}{\tan 13°50'} \Rightarrow h \approx 42{,}641.235$, the plane flies approximately 42,600 feet.

35. Let y be the base of the smaller right triangle, then:

    $\tan 21°10' = \dfrac{x}{135 + y} \Rightarrow x = (135 + y)\tan 21°10'$ and $\tan 35°30' = \dfrac{x}{y} \Rightarrow x = y\,(\tan 35°30')$.

    Now $(135 + y)\tan 21°10' = y\,(\tan 35°30') \Rightarrow 135\,(\tan 21°10') = y\,(\tan 35°30') - y\,(\tan 21°10') \Rightarrow$

    $135\,(\tan 21°10') = y\,(\tan 35°30' - \tan 21°10') \Rightarrow y = \dfrac{135\,(\tan 21°10')}{\tan 35°30' - \tan 21°10'} \Rightarrow y \approx 160.30258$.

    Use the small triangle to solve for x: $x = y\,(\tan 35°30') \Rightarrow x = 160.30258\,(\tan 35°30') \Rightarrow x \approx 114.343$.

    The height of the pyramid is approximately: 114 feet.

36. Let y be the base of the smaller right triangle. From geometry, alternate interior angles are equal when formed

    by parallel lines cut by a transversal, so the angle of elevation from the whale's first sighting is $15°50'$, and from

    where the whale turns, it is $35°40'$. Now solve for $x + y$ and for y:

    $\tan 15°50' = \dfrac{68.7}{x + y} \Rightarrow x + y = \dfrac{68.7}{\tan 15°50'}$ and $\tan 35°40' = \dfrac{68.7}{y} \Rightarrow y = \dfrac{68.7}{\tan 35°40'}$.

    Using these equations we can use substitution for y and solve for x: $x + \dfrac{68.7}{\tan 35°40'} = \dfrac{68.7}{\tan 15°50'} \Rightarrow$

    $x = \dfrac{68.7}{\tan 15°50'} - \dfrac{68.7}{\tan 35°40'} \Rightarrow x \approx 146.519$. The distance traveled is approximately 147 meters.

37. Let h = the height of the house and x = the height of the antenna, then using the angles of elevation yields the

    following equations: $\tan 18°10' = \dfrac{h}{28.0} \Rightarrow h = 28.0\,(\tan 18°10')$ and

    $\tan 27°10' = \dfrac{h + x}{28.0} = h + x = 28.0\,(\tan 27°10')$. Using these equations we can use substitution for h and

    solve for x: $28.0\,(\tan 18°10') + x = 28.0\,(\tan 27°10') \Rightarrow x = 28.0\,(\tan 27°10') - 28.0\,(\tan 18°10') \Rightarrow$

    $x \approx 5.18157$. The height of the antenna is approximately 5.18 meters.

38. Let h = the height of Mt. Whitney and x = the base of the smaller right triangle, then using the angles of

    elevation yields the following equations: $\tan 10°50' = \dfrac{h}{7.00 + x} \Rightarrow h = (x + 7)(\tan 10°50')$ and

    $\tan 22°40' = \dfrac{h}{x} \Rightarrow h = x\,(\tan 22°40')$. Now set these equations equal to each and solve for x:

    $x\,(\tan 22°40') = (x + 7)\tan 10°50' \Rightarrow x\,(\tan 22°40') = x\,(\tan 10°50') + 7\,(\tan 10°50') \Rightarrow$

    $x\,(\tan 22°40') - x\,(\tan 10°50') = 7\,(\tan 10°50') \Rightarrow x\,(\tan 22°40' - \tan 10°50') = 7\,(\tan 10°50') \Rightarrow$

    $x = \dfrac{7\,(\tan 10°50')}{\tan 22°40' - \tan 10°50'} \Rightarrow x \approx 5.9203$. Using this value for x, solve for h:

    $h \approx 5.9203\,(\tan 22°40') \Rightarrow h \approx 2.4725$. The height of Mt. Whitney is approximately 2.47 kilometers.

39. First solve for the hypotenuse of the top triangle: $\cos 30°50' = \dfrac{198.4}{c} \Rightarrow c = \dfrac{198.4}{\cos 30°50'} \Rightarrow$

    $c \approx 231.05719$. Now use this value and the smaller angle of the bottom triangle $52°20' - 30°50' = 21°30'$ to

    solve for x: $\sin 21°30' = \dfrac{x}{c} \Rightarrow \sin 21°30' = \dfrac{x}{231.05719} \Rightarrow x = 231.05719\,(\sin 21°30') \Rightarrow x \approx 84.7$ m.

40. Let y be the hypotenuse of the smaller triangle, then solve for y: $\sin 63°40' = \dfrac{102}{y} \Rightarrow y = \dfrac{102}{\sin 63°40'} \Rightarrow$

    $y \approx 113.81026$. Now, since $x + y$ is the hypotenuse of the larger triangle, find $x + y$:

    $\cos 26°20' = \dfrac{149}{x + y} \Rightarrow x + y = \dfrac{149}{\cos 26°20'} \Rightarrow x + y \approx 166.25224$. Now substitute y from the first equa-

    tion into the second and solve for x: $x + (113.81026) \approx 166.25224 \Rightarrow x \approx 52.44198 \Rightarrow x \approx 52.4$ feet.

41. The angle between the two ships is $90°\,[180 - (28°10' + 61°50')]$; the first ship sails 96 miles (4×24),

    and the second ship sails 112 miles (4×28). Using pythagorean theorem we can find c, the distance between

    them: $c^2 = 96^2 + 112^2 \Rightarrow c^2 = 21,760 \Rightarrow c \approx 147.51$. The ships are approximately 148 miles apart.

42. A right triangle is formed, since the two acute angles at the bottom of the triangle add up to $90°$:

    $[A(90° - 36°20') + B(90° - 53°40') = 90°]$. Now using angle $A = 53°40'$, of the triangle, solve for d, the

    distance from point B to the transmitter: $\sin 53°40' = \dfrac{d}{2.50} \Rightarrow d = 2.50\,(\sin 53°40') \Rightarrow d \approx 2.014$.

    The distance is approximately 2.01 miles.

43. From exercise 42, use angle $A = 53°40'$ and solve for d, the distance from point A to the transmitter:

    $\cos 53°40' = \dfrac{d}{2.50} \Rightarrow d = 2.50\,(\cos 53°40') \Rightarrow d \approx 1.4812$. The distance is approximately 1.48 miles.

44. The angle at the top of the triangle is $90°\,(40° + (180° - 130°))$, therefore a right triangle is formed. Now

    with the first distance 165 miles (1.5×110) and the second distance 143 miles (1.3×110), use pythagorean

    theorem to solve for c, the distance from the starting point: $c^2 = 165^2 + 143^2 \Rightarrow c^2 = 47,674 \Rightarrow$

    $c \approx 218.34$. The plane is approximately 220 miles from its starting point.

45. The angle at the top of the triangle is $90°\,(27° + (180° - 117°))$, therefore a right triangle is formed.

    Now use pythagorean theorem to solve for x, the distance between starting and ending points:

    $c^2 = 50^2 + 140^2 \Rightarrow c^2 = 22,100 \Rightarrow c \approx 148.66$. The distance is approximately 150 kilometers.

46. The angle at the top of the triangle is $90°\,(130° - 40°)$, therefore a right triangle is formed. Now, with the first

    distance 27 knots (1.5×18) and the second distance 39 knots (1.5×26), use pythagorean theorem to solve for

    c, the distance between the two ships: $c^2 = 27^2 + 39^2 \Rightarrow c^2 = 2250 \Rightarrow c \approx 47.43$.

    The ships are approximately 47 nautical miles apart.

47. Let the length of RS be x: $\tan 32°10' = \dfrac{x}{53.1} \Rightarrow x = 53.1\,(\tan 32°10') \Rightarrow x \approx 33.3957 \approx 33.4$ m.

48. (a) First convert 27.0134 miles to feet: $\dfrac{27.0134 \text{ mi}}{1} \times \dfrac{5280}{1 \text{ mi}} = 142,630.752$ feet. Now let $h =$ height from

    straight line to the peak and solve for h: $\sin 5.82° = \dfrac{h}{142,630.752} \Rightarrow h = 142,630.752\,(\sin 5.82°) \Rightarrow$

    $h \approx 14,463.27$. Therefore the height of Mt. Everest is: $14,545 + 14,463.27 \approx 29,008$ feet.

    (b) It would appear shorter than it actually is.

49. First solve for h, the height of the searchlight beam: $\tan 30° = \dfrac{h}{1000} \Rightarrow h = 1000\,(\tan 30°) \Rightarrow h \approx 577.35$.

    The cloud ceiling is searchlight beam height plus observer height, therefore: $6 + 577.35 = 583.35 \approx 583$ feet.

50. Let d = error distance on the moon. Now have d and the distance to the moon form a right angle and solve

    for d: $\tan 0°0'\,30'' = \dfrac{d}{234{,}000} \Rightarrow d = 234{,}000\,(\tan 0°0'\,30'') \Rightarrow d \approx 34.0339$. The laser beam is

    approximately 34 miles from its target.

51. (a) First solve for β and then for d, using the given formula and information:

    $$\beta \approx \frac{57.3\,S}{R} \Rightarrow \beta \approx \frac{57.3\,(336)}{600} \Rightarrow \beta \approx 32.088°. \text{ Therefore: } d = 600\left(1 - \cos\left(\frac{32.088}{2}\right)°\right) \Rightarrow$$

    $d \approx 23.3702 \approx 23.4$ feet.

    (b) First solve for β and then for d, using the given formula and information:

    $$\beta \approx \frac{57.3\,S}{R} \Rightarrow \beta \approx \frac{57.3\,(485)}{600} \Rightarrow \beta \approx 46.3175°. \text{ Therefore: } d = 600\left(1 - \cos\left(\frac{46.3175}{2}\right)°\right) \Rightarrow$$

    $d \approx 48.34877 \approx 48.3$ feet.

    (c) The faster the speed, the more land that needs to be cleared on the inside of the curve.

52. (a) Using the formula and information: $r = \dfrac{66^2}{4.5 + 32.2\,(\tan 3°)} \approx 703.9965 \Rightarrow r \approx 704$ feet.

    (b) Using the formula and information: $r = \dfrac{66^2}{4.5 + 32.2\,(\tan 5°)} \approx 595.31497 \Rightarrow r \approx 595$ feet.

    (c) An increasing measure of θ produces a decreasing radius r.

    (d) Using the formula and information:

    $$1150 = \frac{v^2}{4.5 + 32.2\,(\tan 2.1°)} \Rightarrow v^2 = 1150\,(4.5 + 32.2\,(\tan 2.1°)) \Rightarrow$$

    $v^2 \approx 6{,}532.828459 \Rightarrow v \approx 80.8259 \approx 80.8$ feet/second.

53. (a) $\tan \theta = \dfrac{y}{x}$

    (b) $\tan \theta = \dfrac{y}{x} \Rightarrow x\,(\tan \theta) = y \Rightarrow x = \dfrac{y}{\tan \theta}$

54. Since $\sin \theta = \dfrac{h}{x} \Rightarrow h = x\,(\sin \theta)$, we can use this to find the area A_E of one equilateral triangle:

    $$A_E = \frac{1}{2}bh = \frac{1}{2}x\,(x \sin \theta) \Rightarrow A_E = \frac{x^2}{2}(\sin \theta). \text{ Now the area of the hexagon is:}$$

    $$A = 6A_E = 6\left(\frac{x^2}{2}(\sin \theta)\right) \Rightarrow A = 3x^2\,(\sin \theta).$$

55. Find a: $\cos 60° = \dfrac{a}{24} \Rightarrow a = 24\,(\cos 60°) = 24\left(\dfrac{1}{2}\right) = 12$

    Find b: $\sin 60° = \dfrac{b}{24} \Rightarrow b = 24\,(\sin 60°) = 24\left(\dfrac{\sqrt{3}}{2}\right) = 12\sqrt{3}$

    Using $b = 12\sqrt{3}$, find c: $\sin 45° = \dfrac{12\sqrt{3}}{c} \Rightarrow c = \dfrac{12\sqrt{3}}{\cos 45°} = \dfrac{12\sqrt{3}}{\frac{\sqrt{2}}{2}} = 12\sqrt{3} \cdot \dfrac{2}{\sqrt{2}} \cdot \dfrac{\sqrt{2}}{\sqrt{2}} = 12\sqrt{6}$

    Using $c = 12\sqrt{6}$, find d: $\sin 45° = \dfrac{d}{12\sqrt{6}} \Rightarrow d = 12\sqrt{6}\,(\sin 45°) = 12\sqrt{6}\left(\dfrac{\sqrt{2}}{2}\right) = 6\sqrt{12} =$

    $6\,(2\sqrt{3}) = 12\sqrt{3}$

56. Find x: $\cos 30° = \dfrac{x}{9} \Rightarrow x = 9(\cos 30°) = 9\left(\dfrac{\sqrt{3}}{2}\right) = \dfrac{9\sqrt{3}}{2}$

Find y: $\sin 30° = \dfrac{y}{9} \Rightarrow y = 9(\sin 30°) = 9\left(\dfrac{1}{2}\right) = \dfrac{9}{2}$

Using $y = \dfrac{9}{2}$, find z: $\tan 60° = \dfrac{\frac{9}{2}}{z} \Rightarrow z = \dfrac{\frac{9}{2}}{\tan 60°} = \dfrac{\frac{9}{2}}{\sqrt{3}} = \dfrac{9}{2} \cdot \dfrac{1}{\sqrt{3}} \cdot \dfrac{\sqrt{3}}{\sqrt{3}} = \dfrac{9\sqrt{3}}{6} = \dfrac{3\sqrt{3}}{2}$

Using $z = \dfrac{3\sqrt{3}}{2}$, find w: $\cos 60° = \dfrac{\frac{3\sqrt{3}}{2}}{w} \Rightarrow w = \dfrac{\frac{3\sqrt{3}}{2}}{\cos 60°} = \dfrac{\frac{3\sqrt{3}}{2}}{\frac{1}{2}} = \dfrac{3\sqrt{3}}{2} \cdot 2 = 3\sqrt{3}$

57. Find a: $\sin 60° = \dfrac{7}{a} \Rightarrow a = \dfrac{7}{\sin 60°} = \dfrac{7}{\frac{\sqrt{3}}{2}} = 7 \cdot \dfrac{2}{\sqrt{3}} \cdot \dfrac{\sqrt{3}}{\sqrt{3}} = \dfrac{14\sqrt{3}}{3}$

Since the triangle is a $45° - 45°$ right triangle, $a = n$ and $n = \dfrac{14\sqrt{3}}{3}$

Using $a = \dfrac{14\sqrt{3}}{3}$, find m: $\cos 60° = \dfrac{m}{\frac{14\sqrt{3}}{3}} \Rightarrow m = \dfrac{14\sqrt{3}}{3}(\cos 60°) = \dfrac{14\sqrt{3}}{3} \cdot \dfrac{1}{2} = \dfrac{7\sqrt{3}}{3}$

Using $n = \dfrac{14\sqrt{3}}{3}$, find q: $\cos 45° = \dfrac{\frac{14\sqrt{3}}{3}}{q} \Rightarrow q = \dfrac{\frac{14\sqrt{3}}{3}}{\cos 45°} = \dfrac{\frac{14\sqrt{3}}{3}}{\frac{\sqrt{2}}{2}} = \dfrac{14\sqrt{3}}{3} \cdot \dfrac{2}{\sqrt{2}} \cdot \dfrac{\sqrt{2}}{\sqrt{2}} = \dfrac{28\sqrt{6}}{6} = \dfrac{14\sqrt{6}}{3}$

58. Since p is a leg of a $45° - 45°$ right triangle, $p = 15$.

Find r: $\sin 45° = \dfrac{15}{r} \Rightarrow r = \dfrac{15}{\sin 45°} = \dfrac{15}{\frac{\sqrt{2}}{2}} = \dfrac{15}{1} \cdot \dfrac{2}{\sqrt{2}} \cdot \dfrac{\sqrt{2}}{\sqrt{2}} = \dfrac{30\sqrt{2}}{2} = 15\sqrt{2}$

Using $r = 15\sqrt{2}$, find q: $\tan 30° = \dfrac{q}{15\sqrt{2}} \Rightarrow q = 15\sqrt{2}(\tan 30°) = 15\sqrt{2}\left(\dfrac{\sqrt{3}}{3}\right) = 5\sqrt{6}$

Using $q = 5\sqrt{6}$, find t: $\sin 30° = \dfrac{5\sqrt{6}}{t} \Rightarrow t = \dfrac{5\sqrt{6}}{\sin 30°} = \dfrac{5\sqrt{6}}{\frac{1}{2}} = 5\sqrt{6} \cdot 2 = 10\sqrt{6}$

59. First, bisect the upper angle, which will also be the height (h) of the original triangle. Now solve for h using the smaller right triangle: $\sin 60° = \dfrac{h}{s} \Rightarrow h = s(\sin 60°) \Rightarrow h = \dfrac{s\sqrt{3}}{2}$. Finally find the area of the original triangle: $A = \dfrac{1}{2}bh = \dfrac{1}{2} \cdot s \cdot \dfrac{s\sqrt{3}}{2} \Rightarrow A = \dfrac{s^2\sqrt{3}}{4}$.

60. $A = \dfrac{1}{2}bh = \dfrac{1}{2} \cdot s \cdot s \Rightarrow A = \dfrac{s^2}{2}$

61. Let the length of the shadow be x or $\dfrac{x}{1}$, which is $\dfrac{\text{adj}}{\text{opp}}$, this is $\cot \theta$.

62. Let the length of the shadow be x or $\dfrac{x}{1}$, which is $\dfrac{\text{opp}}{\text{adj}}$, this is $\tan \theta$.

8.8 Harmonic Motion

1. (a) If $s(0) = 2$, then we can solve for a: $2 = a\cos(w \cdot 0) \Rightarrow 2 = a(1) \Rightarrow a = 2$. Now, since the period is .5 seconds, $\dfrac{2\pi}{w} = \dfrac{1}{2} \Rightarrow w = 4\pi$. Thus $s(t) = a\cos(wt) \Rightarrow s(t) = 2\cos(4\pi t)$.

    (b) $s(1) = 2\cos(4\pi(1)) \Rightarrow s(1) = 2$. The weight is neither moving upward nor downward. At $t = 1$ the motion of the weight is changing from up to down. The calculator graph supports this.

2. (a) If $s(0) = 5$, then we can solve for a: $5 = a \cos(w \cdot 0) \Rightarrow 5 = a(1) \Rightarrow a = 5$. Now, since the period

    is 1.5 seconds, $\dfrac{2\pi}{w} = \dfrac{3}{2} \Rightarrow 3w = 4\pi \Rightarrow w = \dfrac{4\pi}{3}$. Thus $s(t) = a \cos(wt) \Rightarrow s(t) = 5 \cos\left(\dfrac{4\pi}{3} t\right)$.

    (b) $s(1) = 5 \cos\left(\dfrac{4\pi}{3}(1)\right) = 5\left(-\dfrac{1}{2}\right) \Rightarrow s(1) = -2.5$. The spring is stretched 2.5 inches one second after

    the weight is released. The weight is moving upward. The calculator graph supports this.

3. (a) If $s(0) = -3$, then we can solve for a: $-3 = a \cos(w \cdot 0) \Rightarrow -3 = a(1) \Rightarrow a = -3$. Now, since the

    period is .8 seconds, $\dfrac{2\pi}{w} = \dfrac{8}{10} \Rightarrow 8w = 2.5\pi$. Thus $s(t) = a \cos(wt) \Rightarrow s(t) = -3 \cos(2.5\pi t)$.

    (b) $s(1) = -3 \cos(2.5\pi(1)) \Rightarrow s(1) = 0$. The spring is at its natural length one second after the weight is

    released. The weight is moving upward. The calculator graph supports this.

4. (a) If $s(0) = -4$, then we can solve for a: $-4 = a \cos(w \cdot 0) \Rightarrow -4 = a(1) \Rightarrow a = -4$. Now, since the

    period is 1.2 seconds, $\dfrac{2\pi}{w} = \dfrac{12}{10} \Rightarrow 12w = 20\pi \Rightarrow w = \dfrac{5\pi}{3}$.

    Thus $s(t) = a \cos(wt) \Rightarrow s(t) = -4 \cos\left(\dfrac{5\pi}{3} t\right)$.

    (b) $s(1) = -4 \cos\left(\dfrac{5\pi}{3}(1)\right) = -4\left(\dfrac{1}{2}\right) \Rightarrow s(1) = -2$. The spring is stretched 2 inches one second after

    the weight is released. The weight is moving downward. The calculator graph supports this.

5. Using $F = 27.5$, note that $b = w = 2\pi(27.5) = 55\pi$, therefore $s(t) = a \cos(55\pi t)$. Now since $s(0) = .21$,

    we can solve for a: $.21 = a \cos(55\pi(0)) \Rightarrow .21 = a(1) \Rightarrow a = .21$. The equation is: $s(t) = .21 \cos(55\pi t)$.

    See Figure 5.

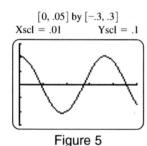
[0, .05] by [−.3, .3]
Xscl = .01   Yscl = .1

Figure 5

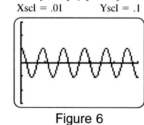
[0, .05] by [−.3, .3]
Xscl = .01   Yscl = .1

Figure 6

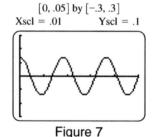
[0, .05] by [−.3, .3]
Xscl = .01   Yscl = .1

Figure 7

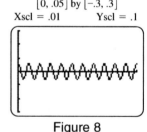
[0, .05] by [−.3, .3]
Xscl = .01   Yscl = .1

Figure 8

6. Using $F = 110$, note that $b = w = 2\pi(110) = 220\pi$, therefore $s(t) = a \cos(220\pi t)$. Now since $s(0) = .11$,

    we can solve for a: $.11 = a \cos(220\pi(0)) \Rightarrow .11 = a(1) \Rightarrow a = .11$. The equation is: $s(t) = .11 \cos(220\pi t)$.

    See Figure 6.

7. Using $F = 110$, note that $b = w = 2\pi(55) = 110\pi$, therefore $s(t) = a \cos(110\pi t)$. Now since $s(0) = .14$,

    we can solve for a: $.14 = a \cos(110\pi(0)) \Rightarrow .14 = a(1) \Rightarrow a = .14$. The equation is: $s(t) = .14 \cos(110\pi t)$.

    See Figure 7.

8. Using $F = 220$, note that $b = w = 2\pi(220) = 440\pi$, therefore $s(t) = a \cos(440\pi t)$. Now since $s(0) = .06$,

    we can solve for a: $.06 = a \cos(440\pi(0)) \Rightarrow .06 = a(1) \Rightarrow a = .06$. The equation is: $s(t) = .06 \cos(440\pi t)$.

    See Figure 8.

9. (a) Using the simple harmonic motion equation $s(t) = a \sin wt$ with $a = 2$ (radius 2) and

$w = 2$ (2 radians/sec) yields the equation: $s(t) = 2 \sin 2t$. Therefore, from this equation,

the amplitude is $|a| = |2| = 2$, the period is $\dfrac{2\pi}{2} = \pi$, and the frequency is $\dfrac{w}{2\pi} = \dfrac{2}{2\pi} = \dfrac{1}{\pi}$.

(b) Using the simple harmonic motion equation $s(t) = a \sin wt$ with $a = 2$ (radius 2) and

$w = 4$ (4 radians/sec) yields the equation: $s(t) = 2 \sin 4t$. Therefore, from this equation,

the amplitude is $|a| = |2| = 2$, the period is $\dfrac{2\pi}{4} = \dfrac{\pi}{2}$, and the frequency is $\dfrac{w}{2\pi} = \dfrac{4}{2\pi} = \dfrac{2}{\pi}$.

10. If Period $P = \dfrac{2\pi}{w}$ and $P = 2\pi\sqrt{\dfrac{L}{32}}$, then $\dfrac{2\pi}{w} = 2\pi\sqrt{\dfrac{\frac{1}{2}}{32}} \Rightarrow \dfrac{1}{w} = \sqrt{\dfrac{1}{64}} \Rightarrow w = \sqrt{64} \Rightarrow w = 8$.

If $w = 8$, then the period is $\dfrac{2\pi}{w} = \dfrac{2\pi}{8} = \dfrac{\pi}{4}$, and the frequency is $\dfrac{w}{2\pi} = \dfrac{8}{2\pi} = \dfrac{4}{\pi}$.

11. Using $P = 1$ and $P = 2\pi\sqrt{\dfrac{L}{32}}$ (from problem 10), we can solve for L: $1 = 2\pi\sqrt{\dfrac{L}{32}} \Rightarrow \dfrac{1}{2\pi} = \sqrt{\dfrac{L}{32}} \Rightarrow$

$2\pi = \sqrt{\dfrac{32}{L}} \Rightarrow (2\pi)^2 = \dfrac{32}{L} \Rightarrow 4\pi^2 = \dfrac{32}{L} \Rightarrow 4\pi^2 L = 32 \Rightarrow L = \dfrac{32}{4\pi^2} \Rightarrow L = \dfrac{8}{\pi^2}$.

12. If the period $P = 1$ and $P = \dfrac{2\pi}{w}$ then $1 = \dfrac{2\pi}{w} \Rightarrow w = 2\pi$. Now using the simple harmonic motion formula

$s(t) = a \sin wt$ and the given formula gives us $w = \sqrt{\dfrac{k}{m}}$, therefore $2\pi = \sqrt{\dfrac{4}{m}} \Rightarrow (2\pi)^2 = \dfrac{4}{m} \Rightarrow$

$4\pi^2 = \dfrac{4}{m} \Rightarrow 4\pi^2 m = 4 \Rightarrow m = \dfrac{4}{4\pi^2} \Rightarrow m = \dfrac{1}{\pi^2}$.

13. (a) Using the formula (from problem 12), $s(t) = a \sin \sqrt{\dfrac{k}{m}} \, t$ as the simple harmonic motion formula gives us

$w = \sqrt{\dfrac{k}{m}}$, therefore $w = \sqrt{\dfrac{2}{1}} \Rightarrow w = \sqrt{2}$. With a spring stretch of $\dfrac{1}{2}$ foot the amplitude is: $\dfrac{1}{2}$; the period

is: $\dfrac{2\pi}{w} = \dfrac{2\pi}{\sqrt{2}} = \dfrac{2\sqrt{2}\pi}{2} = \sqrt{2}\pi$; and the frequency is: $\dfrac{w}{2\pi} = \dfrac{\sqrt{2}}{2\pi}$.

(b) Using the answers from part (a) and the simple harmonic motion formula $s(t) = a \sin wt$ yields:

$s(t) = \dfrac{1}{2} \sin \sqrt{2}t$.

14. (a) From the equation, $a = -5$ and $w = 4\pi$, therefore the maximum height is the amplitude $|-5| = 5$ in.

(b) From the equation, $a = -5$ and $w = 4\pi$, therefore frequency $\dfrac{w}{2\pi} = \dfrac{4\pi}{2\pi} = 2$ cycles/second; and the period

is: $\dfrac{2\pi}{w} = \dfrac{2\pi}{4\pi} = \dfrac{1}{2}$ seconds.

(c) If $s(t) = 5$, solve $5 = -5\cos 4\pi t \Rightarrow -1 = \cos 4\pi t \Rightarrow 4\pi t = \pi \Rightarrow t = \dfrac{\pi}{4\pi} \Rightarrow t = \dfrac{1}{4}$.

The weight first reaches its maximum height after $\dfrac{1}{4}$ seconds.

(d) $s(1.3) = -5\cos(4\pi(1.3)) \approx 4$. After 1.3 seconds, the weight is about 4 inches above the equilibrium position.

15. (a) From the equation, $a = -4$ and $w = 10$, therefore the maximum height is the amplitude $|-4| = 4$ in.

    (b) From the equation, $a = -4$ and $w = 10$, therefore frequency $\dfrac{w}{2\pi} = \dfrac{10}{2\pi} = \dfrac{5}{\pi}$ cycles/second; and the period

    is: $\dfrac{2\pi}{w} = \dfrac{2\pi}{10} = \dfrac{\pi}{5}$ seconds.

    (c) If $s(t) = 4$, solve $4 = -4 \cos 10t \Rightarrow -1 = \cos 10t \Rightarrow 10t = \pi \Rightarrow t = \dfrac{\pi}{10}$.

    The weight first reaches its maximum height after $\dfrac{\pi}{10}$ seconds.

    (d) $s(1.466) = -4 \cos(10(1.466)) \approx 2$. After 1.466 seconds, the weight is about 2 inches above the equilibrium position.

16. (a) If the spring is pulled down 3 inches, then $a = -3$. Now use frequency to find w: $\dfrac{w}{2\pi} = \dfrac{6}{\pi} \Rightarrow$

    $w\pi = 12\pi \Rightarrow w = 12$. Therefore the equation is $s(t) = -3 \cos 12t$.

    (b) The period is: $\dfrac{2\pi}{w} = \dfrac{2\pi}{12} = \dfrac{\pi}{6}$ seconds.

17. (a) If the spring is pulled down 2 inches, then $a = -2$. Now use the given period to find w: $\dfrac{2\pi}{w} = \dfrac{1}{3} \Rightarrow$

    $w = 6\pi$. Therefore the equation is $s(t) = -2 \cos 6\pi t$.

    (b) The frequency is: $\dfrac{w}{2\pi} = \dfrac{6\pi}{2\pi} = 3$ cycles/second.

18. (a) From the graph, the x-intercepts of y_1 are 0 and π. They are the same as those of $\sin x$.

    (b) From the graph, the intersection of y_1 and y_2 is $1.570796372 \approx \dfrac{\pi}{2}$. This intersection occurs when $\sin t$ is

    at a maximum, that is, $t = \dfrac{\pi}{2}$. The graphs of y_1 and y_3 do not intersect in the window $[0, \pi]$. These graphs

    will intersect when $\sin t$ is at a minimum, but the minimum value of $\sin t$ does not occur in $[0, \pi]$.

19. (a) From the equation the amplitude is $a = 2$, the spring is compressed 2 inches.

    (b) From the equation $w = 2\pi$, therefore the frequency is $\dfrac{w}{2\pi} = \dfrac{2\pi}{2\pi} = 1$ cycle per second.

    (c) Graph the equation on a calculator, using the capabilities of the calculator $D(t) = \dfrac{1}{e}$ inches when

    $t \approx .305$ sec and .636 seconds.

20. (a) From the equation the amplitude is $a = 160$, the initial voltage is 160 volts.

    (b) From the equation $w = 120\pi$, therefore the frequency is $\dfrac{w}{2\pi} = \dfrac{120\pi}{2\pi} = 60$ cycle per second.

    (c) Graph the equation on a calculator, using the capabilities of the calculator $V(t) = 100$ volts when

    $t \approx .00228$ sec, .0152 sec, and .636 seconds.

Reviewing Basic Concepts (Sections 8.5 — 8.8)

1. First find the value of r: $r = \sqrt{(-2)^2 + 5^2} \Rightarrow r = \sqrt{29}$. Now each of the following is:

$$\sin \theta = \frac{y}{r} = \frac{5}{\sqrt{29}} = \frac{5\sqrt{29}}{29} \qquad \cos \theta = \frac{x}{r} = \frac{-2}{\sqrt{29}} = -\frac{2\sqrt{29}}{29} \qquad \tan \theta = \frac{y}{x} = \frac{5}{-2} = -\frac{5}{2}$$

$$\csc \theta = \frac{r}{y} = \frac{\sqrt{29}}{5} \qquad \sec \theta = \frac{r}{x} = \frac{\sqrt{29}}{-2} = -\frac{\sqrt{29}}{2} \qquad \cot \theta = \frac{x}{y} = \frac{-2}{5} = -\frac{2}{5}$$

2. Using a calculator: $\sin 270° = -1$ $\cos 270° = 0$ $\tan 270°$ is undefined.

    $\csc 270° = -1$ $\sec 270°$ is undefined. $\cot 270° = 0$

3. (a) Since $-1 \le \cos\theta \le 1$ for all values of θ, $\cos\theta = \dfrac{3}{2}$ is impossible.

    (b) Since the range of $\tan\theta$ is: $-\infty < \tan\theta < \infty$, $\tan\theta = 300$ is possible.

    (c) Since the range of $\csc\theta$ is: $\csc\theta \le -1$ or $\csc\theta \ge 1$, $\sec\theta = 5$ is possible.

4. Since $\sin\theta$ is in quadrant III, $\sin\theta = \dfrac{-2}{3} = -\dfrac{2}{3}$, $x = -2$ and $y = 3$. First find $\cos\theta$, $\sin^2\theta + \cos^2\theta = 1$:

    $\left(-\dfrac{2}{3}\right)^2 + \cos^2\theta = 1 \;\Rightarrow\; \cos^2\theta = 1 - \dfrac{4}{9} \;\Rightarrow\; \cos^2\theta = \dfrac{5}{9} \;\Rightarrow\; \cos\theta = \pm\dfrac{\sqrt{5}}{3}$. Since θ is found in quadrant

    III, $\cos\theta = -\dfrac{\sqrt{5}}{3}$. Find $\csc\theta$, $\csc\theta = \dfrac{1}{\sin\theta} \;\Rightarrow\; \csc\theta = -\dfrac{3}{2}$; find $\sec\theta$, $\sec\theta = \dfrac{1}{\cos\theta} \;\Rightarrow\; \sec\theta = -\dfrac{3}{\sqrt{5}}$,

    therefore $\sec\theta = -\dfrac{3\sqrt{5}}{5}$; find $\tan\theta$, $\tan\theta = \dfrac{\sin\theta}{\cos\theta} \;\Rightarrow\; -\dfrac{2}{3} \div -\dfrac{\sqrt{5}}{3} \;\Rightarrow\; \tan\theta = \dfrac{2}{\sqrt{5}}$, therefore

    $\tan\theta = \dfrac{2\sqrt{5}}{5}$; and find $\cot\theta$, $\cot\theta = \dfrac{1}{\tan\theta} \;\Rightarrow\; \cot\theta = \dfrac{5}{2\sqrt{5}} = \dfrac{5\sqrt{5}}{10} \;\Rightarrow\; \cot\theta = \dfrac{\sqrt{5}}{2}$.

    The five Functions are: $\cos\theta = -\dfrac{\sqrt{5}}{3}$; $\tan\theta = \dfrac{2\sqrt{5}}{5}$; $\cot\theta = \dfrac{\sqrt{5}}{2}$; $\sec\theta = -\dfrac{3\sqrt{5}}{5}$; and $\csc\theta = -\dfrac{3}{2}$.

5. $\sin A = \dfrac{\text{opp}}{\text{hyp}} = \dfrac{8}{17}$ $\cos A = \dfrac{\text{adj}}{\text{hyp}} = \dfrac{15}{17}$ $\tan A = \dfrac{\text{opp}}{\text{adj}} = \dfrac{8}{15}$

    $\csc A = \dfrac{\text{hyp}}{\text{opp}} = \dfrac{17}{8}$ $\sec A = \dfrac{\text{hyp}}{\text{adj}} = \dfrac{17}{15}$ $\cot A = \dfrac{\text{adj}}{\text{opp}} = \dfrac{15}{8}$

6. By the $30° - 60°$ right triangle,

    $\sin 30° = \dfrac{1}{2}$ $\cos 30° = \dfrac{\sqrt{3}}{2}$ $\tan 30° = \dfrac{\sqrt{3}}{3}$

    $\csc 30° = \dfrac{2}{1} = 2$ $\sec 30° = \dfrac{2}{\sqrt{3}} = \dfrac{2\sqrt{3}}{3}$ $\cot 30° = \dfrac{3}{\sqrt{3}} = \dfrac{3\sqrt{3}}{3} = \sqrt{3}$

    By the $45° - 45°$ right triangle,

    $\sin 45° = \dfrac{\sqrt{2}}{2}$ $\cos 45° = \dfrac{\sqrt{2}}{2}$ $\tan 45° = \dfrac{\sqrt{2}}{\sqrt{2}} = 1$

    $\csc 45° = \dfrac{2}{\sqrt{2}} = \dfrac{2\sqrt{2}}{2} = \sqrt{2}$ $\sec 45° = \dfrac{2}{\sqrt{2}} = \dfrac{2\sqrt{2}}{2} = \sqrt{2}$ $\cot 45° = \dfrac{\sqrt{2}}{\sqrt{2}} = 1$

    By the $30° - 60°$ right triangle,

    $\sin 60° = \dfrac{\sqrt{3}}{2}$ $\cos 60° = \dfrac{1}{2}$ $\tan 60° = \dfrac{\sqrt{3}}{1} = \sqrt{3}$

    $\csc 30° = \dfrac{2}{\sqrt{3}} = \dfrac{2\sqrt{3}}{3}$ $\sec 60° = \dfrac{2}{1} = 2$ $\cot 60° = \dfrac{1}{\sqrt{3}} = \dfrac{\sqrt{3}}{3}$

7. (a) $\sin 27° = \cos(90° - 27°) = \cos 63°$

    (b) $\tan\dfrac{\pi}{5} = \cot\left(\dfrac{\pi}{2} - \dfrac{\pi}{5}\right) = \cot\left(\dfrac{5\pi}{10} - \dfrac{2\pi}{10}\right) = \cot\dfrac{3\pi}{10}$

8. (a) If $\theta = 100°$ then $\theta' = 180° - 100° = 80°$

    (b) If $\theta = -365° = 720° - 365° = 355°$ then $\theta' = 360° - 355° = 5°$

    (c) If $\theta = \dfrac{8\pi}{3} - 2\pi = \dfrac{8\pi}{3} - \dfrac{6\pi}{3} = \dfrac{2\pi}{3}$ then $\theta' = \pi - \dfrac{2\pi}{3} = \dfrac{3\pi}{3} - \dfrac{2\pi}{3} = \dfrac{\pi}{3}$

9. First find the reference angle: $(360° - 315°) = 45°$. Using the $45° - 45°$ right triangle rules, for a $45°$ angle the hypotenuse $= \sqrt{2}$, the opposite side $= 1$, and the adjacent $= 1$. Since $\theta = 315°$ is in quadrant IV, y is negative and sine, tangent, cotangent, and cosecant are negative, therefore:

$$\sin 315° = -\sin 45° = -\frac{1}{\sqrt{2}} = -\frac{\sqrt{2}}{2} \qquad \csc 315° = -\csc 45° = -\frac{\sqrt{2}}{1} = -\sqrt{2}$$

$$\cos 315° = \cos 45° = \frac{1}{\sqrt{2}} = \frac{\sqrt{2}}{2} \qquad \sec 315° = \sec 45° = \frac{\sqrt{2}}{1} = \sqrt{2}$$

$$\tan 315° = -\tan 45° = -\frac{1}{1} = -1 \qquad \cot 315° = -\cot 45° = -\frac{1}{1} = -1.$$

10. (a) Using the calculator, $\sin 46°30' \approx .725374371$

    (b) Using the calculator, $\tan(-100°) \approx 5.67128182$

    (c) Using the calculator, $\csc 4 = (\sin 4)^{-1} \approx -1.321348709$

11. Since $\tan \theta$ is negative in quadrant II and IV, values for θ can only be in these quadrants.

    From the $30° - 60°$ right triangle, $\tan 30° = \frac{\sqrt{3}}{3}$, therefore $30°$ is the reference angle.

    In quadrant II, use the reference angle $30°$ to find the first value: $180° - 30° = 150°$. In quadrant IV, use the reference angle $30°$ to find the other value: $360° - 30° = 330°$. Therefore, $\theta = 150°$ or $330°$.

12. Since $\sin \theta$ is positive in quadrant I and II, values for θ can only be in these quadrants.

    In quadrant I, we use the inverse trigonometric function (radian mode) of our calculator to find the angle: $\sin^{-1} .68163876 \Rightarrow \theta = .75$. In quadrant II, use the found reference angle from quadrant I to find θ: $\pi - .75 \approx 2.391592654$. Therefore, $\theta \approx .75$ or 2.391592654.

13. Let x be the base of the smaller right triangle and h the height of Mt. Kilimanjaro, then:

    $\tan 13.7° = \dfrac{h}{x} \Rightarrow x = \dfrac{h}{\tan 13.7°}$ and $\tan 10.4° = \dfrac{h}{x + 5}$. Now substitute the second equation into the first

    equation: $\tan 10.4° = \dfrac{h}{\frac{h}{\tan 13.7°} + 5}$, now multiply by $\dfrac{\tan 13.7}{\tan 13.7}$ and solve for h:

    $\tan 10.4° = \dfrac{h}{\frac{h}{\tan 13.7°} + 5} \cdot \dfrac{\tan 13.7°}{\tan 13.7°} \Rightarrow \tan 10.4° = \dfrac{h\,(\tan 13.7°)}{h + 5\,(\tan 13.7°)} \Rightarrow$

    $\tan 10.4° \,(h + 5\,(\tan 13.7°)) = h\,(\tan 13.7°) \Rightarrow h\,(\tan 10.4°) + 5\,(\tan 13.7°)(\tan 10.4°) = h\,(\tan 13.7°) \Rightarrow$

    $h\,(\tan 10.4°) - h\,(\tan 13.7°) = -5\,(\tan 13.7°)(\tan 10.4°) \Rightarrow$

    $h\,(\tan 10.4° - \tan 13.7°) = -5\,(\tan 13.7°)(\tan 10.4°) \Rightarrow h = \dfrac{-5\,(\tan 13.7°)(\tan 10.4°)}{\tan 10.4° - \tan 13.7°} \Rightarrow$

    $h \approx 3.713588$ miles. Finally convert to feet: $h \approx \dfrac{3.713588 \text{ mi}}{1} \cdot \dfrac{5280 \text{ ft}}{1 \text{ mi}} \approx 19{,}607.7$.

    The mountain is approximately 19,600 feet high.

14. (a) From the equation, $a = -4$, therefore the amplitude is $|-4| = 4$. The maximum height is 4 inches.

    (b) Solve for t: $4 = -4\cos 8\pi t \Rightarrow -1 = \cos 8\pi t \Rightarrow 8\pi t = \pi \Rightarrow t = \dfrac{\pi}{8\pi} \Rightarrow t = \dfrac{1}{8}$ seconds.

    (c) The period is: $\dfrac{2\pi}{w} = \dfrac{2\pi}{8\pi} = \dfrac{1}{4}$; and the frequency is: $\dfrac{w}{2\pi} = \dfrac{8\pi}{2\pi} = 4$ cycles/second.

Chapter 8 Review

1. The smallest positive degree measure of the coterminal angle is: $-174° + 360° = 186°$.

2. All degree measure coterminal angles will be multiples of $360°$, therefore: $270° + n \cdot 360°$.

3. First find the radians per second: $\theta = \dfrac{320(2\pi)}{60} \Rightarrow \theta = \dfrac{32\pi}{3}$ radians per second. Now multiply θ by

   $\dfrac{2}{3}$ seconds and by $\dfrac{180}{\pi}$ to change to degrees: $\theta = \dfrac{32\pi}{3} \cdot \dfrac{2}{3} \cdot \dfrac{180}{\pi} \Rightarrow \theta = 1280°$.

4. Degrees rotated per second: $\theta = \dfrac{650(360°)}{60} \Rightarrow \theta = 3900°$. Multiply by 2.4 seconds: $3900 \cdot 2.4 = 9360°$.

5. 1 radian $= 1 \cdot \dfrac{180}{\pi} = 57.3°$, therefore 1 radian $\approx 57.3° > 1°$.

6. (a) Change to degrees and determine the quadrant: $3 \cdot \dfrac{180}{\pi} = 171.9°$, therefore quadrant II.

   (b) Change to degrees and determine the quadrant: $4 \cdot \dfrac{180}{\pi} = 229.2°$, therefore quadrant III.

   (c) Change to degrees and determine the quadrant: $-2 \cdot \dfrac{180}{\pi} = 114.6°$, therefore quadrant III.

   (d) Change to degrees and determine the quadrant: $7 \cdot \dfrac{180}{\pi} = 401.1° - 360° = 41.1°$, therefore quadrant I.

7. $120° \cdot \dfrac{\pi}{180} = \dfrac{2\pi}{3}$

8. $800° \cdot \dfrac{\pi}{180} = \dfrac{40\pi}{9}$

9. $\dfrac{5\pi}{4} \cdot \dfrac{180}{\pi} = 225°$

10. $-\dfrac{6\pi}{5} \cdot \dfrac{180}{\pi} = -216°$

11. Using the arc length formula, 1 rotation has a length: $s = r\theta \Rightarrow s = 2(2\pi) \Rightarrow s = 4\pi$ inches.

    Then 20 minutes equals: $\dfrac{20}{60} \cdot 4\pi = \dfrac{4\pi}{3}$ inches.

12. Using the arc length formula, 1 rotation has a length: $s = r\theta \Rightarrow s = 2(2\pi) \Rightarrow s = 4\pi$ inches.

    Then 3 hours equals: $3 \cdot 4\pi = 12\pi$ inches.

13. Since arc length is found by the formula $s = r\theta$, the radius is: $s = 15.2 \cdot \dfrac{3\pi}{4} \Rightarrow s = 35.8$ cm.

14. Using the area of a sector formula $A = \dfrac{1}{2}r^2\theta$ yields: $A = \dfrac{1}{2}(28.69)^2\left(\dfrac{7\pi}{4}\right) \Rightarrow A \approx 2263$ in^2

15. Because the central angle is very small, the arc length is approximately equal to the length of the inscribed

    chord. Using the arc length formula yields: $s = 2000\left(1 + \dfrac{10}{60}\right)°\left(\dfrac{\pi}{180}\right) \approx 41$ yards.

16. Since arc length is found by the formula $s = r\theta$, the central angle is: $4 = 8\theta \Rightarrow \theta = \dfrac{4}{8} \Rightarrow \theta = \dfrac{1}{2}$ radians.

    Now using the area of a sector formula $A = \dfrac{1}{2}r^2\theta$ yields: $A = \dfrac{1}{2}(8)^2\left(\dfrac{1}{2}\right) \Rightarrow A = 16$ units^2.

17. From the properties of $30° - 60°$ right triangles, an angle of $s = \dfrac{2\pi}{3}$ radians intersects the unit circle at the

    point $\left(-\dfrac{1}{2}, \dfrac{\sqrt{3}}{2}\right)$. Since $\cos s = x$, $\cos \dfrac{2\pi}{3} = -\dfrac{1}{2}$.

18. First a coterminal angle: $4\pi - \dfrac{7\pi}{3} = \dfrac{5\pi}{3} \implies \theta' = -\dfrac{\pi}{3}$, From the properties of $30° - 60°$ right triangles, an angle of $s = -\dfrac{\pi}{3}$ radians intersects the unit circle at $\left(\dfrac{1}{2}, -\dfrac{\sqrt{3}}{2}\right)$. Since $\tan s = \dfrac{y}{x}$, $\tan -\dfrac{\pi}{3} = -\dfrac{\sqrt{3}}{2} \cdot \dfrac{2}{1} = -\sqrt{3}$.

19. First find θ': $\theta' = 2\pi - \dfrac{11\pi}{6} \implies \theta' = \dfrac{\pi}{6}$. From the properties of $30° - 60°$ right triangles, an angle of $s = \dfrac{\pi}{6}$ radians intersects the unit circle at the point $\left(\dfrac{\sqrt{3}}{2}, \dfrac{1}{2}\right)$. Since $\csc s = \dfrac{1}{y}$, $\csc \dfrac{\pi}{6} = \dfrac{1}{\frac{1}{2}} = 2$

20. Using the calculator (radian mode), $\cos(-.2443) \approx .97030688 \approx .9703$

21. Using the calculator (radian mode), $\cot 3.0543 = [\tan 3.0543]^{-1} \approx -11.426605 \approx -11.4266$

22. Using the calculator (radian mode), if $\sin s = .49244294$, then $s = \sin^{-1} .49244294 \approx .51489440 \approx .5149$

23. First find θ. Since this is a unit circle, $\sin \theta = y$, therefore $\sin \theta = -.5250622$ and $\theta = \sin^{-1} -.5250622 \implies \theta \approx -.5528$. Now with $r = 1$, use the arc length formula $s = |r\theta| \implies s \approx |(1)(-.5528)| \implies s \approx .5528$. The length of the shortest arc of the circle from $(1, 0)$ to $(.85106383, -.5250622)$ is $.5528$.

24. B, from the equation, the amplitude is 4 and the period is $\dfrac{2\pi}{2} = \pi$.

25. The amplitude is 2, the period is $\dfrac{2\pi}{(1)} = 2\pi$, and there is no vertical translation or phase shift. See Figure 25.

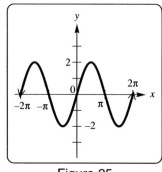

Figure 25

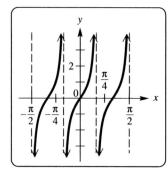

Figure 26

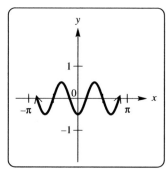

Figure 27

26. The amplitude (since a tangent function) is not applicable, the period is $\dfrac{\pi}{(3)}$, and there is no vertical translation or phase shift. See Figure 26.

27. The amplitude is $\dfrac{1}{2}$, the period is $\dfrac{2\pi}{(3)}$, and there is no vertical translation or phase shift. See Figure 27.

28. The amplitude is 2, the period is $\dfrac{2\pi}{(5)}$, and there is no vertical translation or phase shift. See Figure 28.

29. The amplitude is 2, the period is $\dfrac{2\pi}{(\frac{1}{4})} = 8\pi$, the vertical shift is 1 unit upward, and there is no phase shift. See Figure 29.

30. The amplitude is $\dfrac{1}{4}$, the period is $\dfrac{2\pi}{(\frac{2}{3})} = 3\pi$, the vertical shift is 3 units upward, and there is no phase shift. See Figure 30.

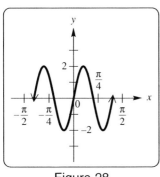

Figure 28

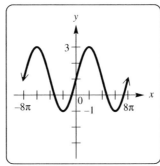

Figure 29

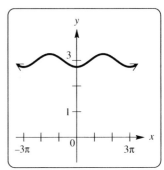

Figure 30

31. The amplitude is 3, the period is $\dfrac{2\pi}{(1)} = 2\pi$, there is no vertical shift , and the phase shift is $-\dfrac{\pi}{2}$. See Figure 31.

32. The amplitude is $|-1| = 1$, the period is $\dfrac{2\pi}{(1)} = 2\pi$, there is no vertical shift, and the phase shift is $\dfrac{3\pi}{4}$. See Figure 32.

33. Since $y = \dfrac{1}{2}\csc\left(2x - \dfrac{\pi}{4}\right) \;\Rightarrow\; y = \dfrac{1}{2}\csc 2\left(x - \dfrac{\pi}{8}\right)$ and the equation, the amplitude is not applicable

(cosecant function), the period is $\dfrac{2\pi}{(2)} = \pi$, there is no vertical shift, and the phase shift is $\dfrac{\pi}{8}$. See Figure 33.

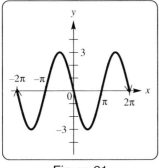

Figure 31

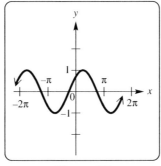

Figure 32

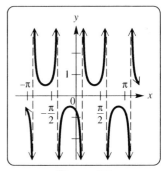

Figure 33

34. There is no amplitude, the period is $\dfrac{\pi}{\left(\frac{1}{2}\right)} = 2\pi$, and there is no vertical translation or phase shift. See Figure 34.

35. There is no amplitude, the period is $\dfrac{2\pi}{(2)} = \pi$, and there is no vertical translation or phase shift. See Figure 35.

36. Since $y = \csc(2x - \pi) \;\Rightarrow\; y = \csc 2\left(x - \dfrac{\pi}{2}\right)$ and now from the equation, the amplitude is not applicable,

(cosecant function) the period is $\dfrac{2\pi}{(2)} = \pi$, there is no vertical shift, and the phase shift is $\dfrac{\pi}{2}$. See Figure 36.

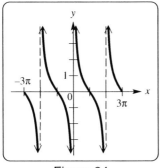

Figure 34

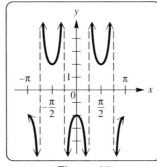

Figure 35

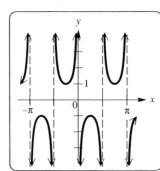

Figure 36

37. Because f has period π, $f(x) = f(x + n\pi)$ for integers n. Thus, $f\left(\dfrac{6\pi}{5}\right) = f\left(\dfrac{6\pi}{5} - 2\pi\right) = f\left(-\dfrac{4\pi}{5}\right)$,

    and it follows that $f\left(-\dfrac{4\pi}{5}\right) = 1$.

38. No. If $f(x) = \tan x$, then $f(0) = 0$ and $f(2\pi) = 0$, but the period of $\tan x$ is π.

39. Graph a cosine function with a amplitude of 3 and a period of $\dfrac{2\pi}{(2)} = \pi$. See Figure 39.

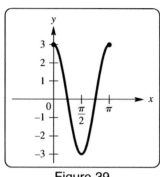
Figure 39

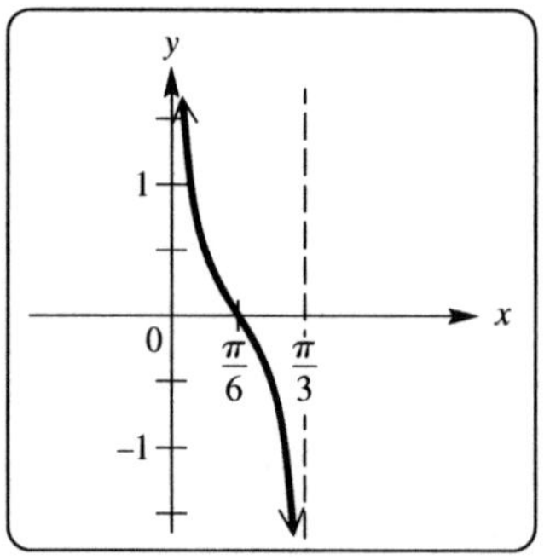

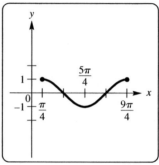
Figure 40

Figure 41

40. Graph a cotangent function with a period of $\dfrac{\pi}{(3)}$; and passing through the points $\left(\dfrac{\pi}{6}, 0\right)$ (midpoint),

    $\left(\dfrac{\pi}{12}, \dfrac{1}{2}\right)$, and $\left(\dfrac{\pi}{4}, -\dfrac{1}{2}\right)$. See Figure 40.

41. Graph a cosine function with amplitude 1, period $\dfrac{2\pi}{(1)} = 2\pi$, and phase shift $\dfrac{\pi}{4}$ units to the right. See Figure 41.

42. Graph a tangent function with a period of $\dfrac{\pi}{(1)} = \pi$, a phase shift $\dfrac{\pi}{2}$ units to the right, and passing through the

    points $\left(\dfrac{\pi}{2}, 0\right)$ (midpoint), $\left(\dfrac{\pi}{4}, -1\right)$, and $\left(\dfrac{3\pi}{4}, 1\right)$. See Figure 42.

43. Graph a cosine function with amplitude 2, period $\dfrac{2\pi}{(3)}$, and vertical translation 1 unit upward. See Figure 43.

44. Graph a sine function with a amplitude of $|-3| = 3$, a period of $\dfrac{2\pi}{(2)} = \pi$, and a vertical translation 1 unit

    downward. See Figure 44.

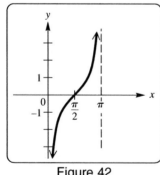
Figure 42

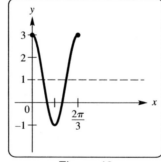
Figure 43

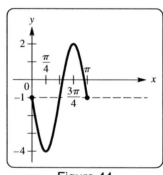
Figure 44

45. Since the period is π, it is either $\tan x$ or $\cot x$. With x-intercept $n\pi$, where n is an interger, the function is tangent.

46. Since the period is 2π, the function is not $\tan x$ or $\cot x$. With the graph passing through $(0, 0)$, the function is sine.

47. Since the period is 2π, it is not $\tan x$ or $\cot x$. The graph passes through $\left(\dfrac{\pi}{2}, 0\right)$, the function is cosine.

48. The period is 2π, so it is not $\tan x$ or $\cot x$. Since the domain is $\{x \mid x \neq n\pi$, where n is an integer$\}$ it is cosecant.

49. Since the period is π, it is either tan x or cot x. The graph decreases over the $(0, \pi)$, the function is cotangent.

50. Since the period is 2π, the function is not tan x or cot x. With the graph having asymptotes:

$x = (2n + 1)\dfrac{\pi}{2}$, where n is an integer, the function is secant.

51. Since the graph shows an amplitude of 3, a period of $\pi = \dfrac{2\pi}{(2)}$, and passes through $\left(\dfrac{\pi}{4}, 0\right)$ which means it has

a phase shift of $\dfrac{\pi}{4}$, the equation is $y = 3 \sin\left[2\left(x - \dfrac{\pi}{4}\right)\right]$. (Other answers are possible)

52. Since the graph shows an amplitude of 4, a period of $4\pi = \dfrac{2\pi}{\left(\frac{1}{2}\right)}$, and passes through $(0, 0)$ which means it has

no phase shift, the equation is $y = 4 \sin\dfrac{1}{2}x$. (Other answers are possible)

53. Since the graph shows an amplitude of $\dfrac{1}{3}$, a period of $4 = \dfrac{2\pi}{\left(\frac{\pi}{2}\right)}$, and passes through $(0, 0)$ which means it has

no phase shift, the equation is $y = \dfrac{1}{3} \sin\dfrac{\pi}{2}x$. (Other answers are possible)

54. Since the graph shows an amplitude of π, a period of $2 = \dfrac{2\pi}{(\pi)}$, and passes through $\left(\dfrac{1}{2}, 0\right)$ which means it has

a phase shift of $\dfrac{1}{2}$, the equation is $y = \pi \sin\left[\pi\left(x - \dfrac{1}{2}\right)\right]$. (Other answers are possible)

55. (a) Let January correspond to $x = 1$, February to $x = 2$, ... , and December of the 2nd year to $x = 24$.

The data appear to follow the pattern of a translated sine graph. See Figure 55a.

(b) Using the model $f(x) = a \sin\left[b(x - d)\right] + c$, we can find the constants: The maximum average monthly

temperature is 75° F, and the minimum is 25° F. Let the amplitude a be $\dfrac{75 - 25}{2} = \dfrac{50}{2}$, so $a = 25$. Since

the period is $12 = \dfrac{2\pi}{b}$, $b = \dfrac{\pi}{6}$. The data are centered vertically around the line $y = \dfrac{75 + 25}{2} = 50$,

therefore $c = 50$. The minimum temperature occurs in January. Thus, when $x = 1$, $b(x - d) = -\dfrac{\pi}{2}$ since

the sine function is minimum at $-\dfrac{\pi}{2}$. Solving for d yields: $\dfrac{\pi}{6}(1 - d) = -\dfrac{\pi}{2} \Rightarrow d = 4$. Since the months

before January are slightly colder than the months following January the value of d can be slightly adjusted

to give a better visual fit, Trying $d = 4.2$ gives a better fit. Thus: $f(x) = 25 \sin\left[\dfrac{\pi}{6}(x - 4.2)\right] + 50$.

(c) The constant values a controls the amplitude, b the period, c the vertical shift, and d the phase shift.

(d) See Figure 55d. The function gives an excellent model.

(e) The regression function gives: see Figure 55e.

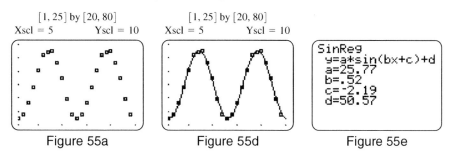

[1, 25] by [20, 80]
Xscl = 5   Yscl = 10

[1, 25] by [20, 80]
Xscl = 5   Yscl = 10

SinReg
y=a*sin(bx+c)+d
a=25.77
b=.52
c=-2.19
d=50.57

Figure 55a   Figure 55d   Figure 55e

56. (a) January 1 of the base year would be $t = 0$, which yields:

$$P(0) = 7(1 - \cos 2\pi(0))(0 + 10) + 100e^{.2(0)} = 7(1 - 1)(10) + 100(1) \implies P(0) = 100$$

(b) July 1 of the base year would be $t = .5$, which yields:

$$P(.5) = 7(1 - \cos 2\pi(.5))(.5 + 10) + 100e^{.2(.5)} = 7(1 - (-1))(10.5) + 100e^{.1} \implies P(.5) \approx 258$$

(c) January 1 of the following year would be $t = 1$, which yields:

$$P(1) = 7(1 - \cos 2\pi(1))(1 + 10) + 100e^{.2(1)} = 7(1 - 1)(11) + 100e^{.2} \implies P(1) \approx 122$$

(d) July 1 of the following year would be $t = 1.5$, which yields:

$$P(1.5) = 7(1 - \cos 2\pi(1.5))(1.5 + 10) + 100e^{.2(1.5)} = 7(1 - (-1))(11.5) + 100e^{.3} \implies P(1) \approx 296$$

57. First find the value of r: $r = \sqrt{(-3)^2 + (-3)^2} \implies r = \sqrt{18.} = 3\sqrt{2}$. Now each of the following is:

$$\sin \theta = \frac{y}{r} = \frac{-3}{3\sqrt{2}} = -\frac{\sqrt{2}}{2} \qquad \cos \theta = \frac{x}{r} = \frac{-3}{3\sqrt{2}} = -\frac{\sqrt{2}}{2} \qquad \tan \theta = \frac{y}{x} = \frac{-3}{-3} = 1$$

$$\csc \theta = \frac{r}{y} = \frac{3\sqrt{2}}{-3} = -\sqrt{2} \qquad \sec \theta = \frac{r}{x} = \frac{3\sqrt{2}}{-3} = -\sqrt{2} \qquad \cot \theta = \frac{x}{y} = \frac{-3}{-3} = 1$$

58. First find the value of r: $r = \sqrt{(1)^2 + (-\sqrt{3})^2} \implies r = \sqrt{4.} = 2$. Now each of the following is:

$$\sin \theta = \frac{y}{r} = \frac{-\sqrt{3}}{2} = -\frac{\sqrt{3}}{2} \qquad \cos \theta = \frac{x}{r} = \frac{1}{2} \qquad \tan \theta = \frac{y}{x} = \frac{-\sqrt{3}}{1} = -\sqrt{3}$$

$$\csc \theta = \frac{r}{y} = \frac{2}{-\sqrt{3}} = -\frac{2\sqrt{3}}{3} \qquad \sec \theta = \frac{r}{x} = \frac{2}{1} = 2 \qquad \cot \theta = \frac{x}{y} = \frac{1}{-\sqrt{3}} = -\frac{\sqrt{3}}{3}$$

59. Using the calculator: $\sin 180° = 0$ $\qquad \cos 180° = -1 \qquad \tan 180° = 0.$

$\qquad \csc 180°$ is undefined $\qquad \sec 180° = -1 \qquad \cot 180°$ is undefined

60. First find the value of r: $r = \sqrt{(3)^2 + (-4)^2} \implies r = \sqrt{25} = 5$. Now each of the following is:

$$\sin \theta = \frac{y}{r} = \frac{-4}{5} = -\frac{4}{5} \qquad \cos \theta = \frac{x}{r} = \frac{3}{5} \qquad \tan \theta = \frac{y}{x} = \frac{-4}{3} = -\frac{4}{3}$$

$$\csc \theta = \frac{r}{y} = \frac{5}{-4} = -\frac{5}{4} \qquad \sec \theta = \frac{r}{x} = \frac{5}{3} \qquad \cot \theta = \frac{x}{y} = \frac{3}{-4} = -\frac{3}{4}$$

61. First find the value of r: $r = \sqrt{(9)^2 + (-2)^2} \implies r = \sqrt{85}$. Now each of the following is:

$$\sin \theta = \frac{y}{r} = \frac{-2}{\sqrt{85}} = -\frac{2\sqrt{85}}{85} \qquad \cos \theta = \frac{x}{r} = \frac{9}{\sqrt{85}} = \frac{9\sqrt{85}}{85} \qquad \tan \theta = \frac{y}{x} = \frac{-2}{9} = -\frac{2}{9}$$

$$\csc \theta = \frac{r}{y} = \frac{\sqrt{85}}{-2} = -\frac{\sqrt{85}}{2} \qquad \sec \theta = \frac{r}{x} = \frac{\sqrt{85}}{9} \qquad \cot \theta = \frac{x}{y} = \frac{9}{-2} = -\frac{9}{2}$$

62. First find the value of r: $r = \sqrt{(-2\sqrt{2})^2 + (2\sqrt{2})^2} \implies r = \sqrt{16.} = 4$. Now each of the following is:

$$\sin \theta = \frac{y}{r} = \frac{2\sqrt{2}}{4} = \frac{\sqrt{2}}{2} \qquad \cos \theta = \frac{x}{r} = \frac{-2\sqrt{2}}{4} = -\frac{\sqrt{2}}{2} \qquad \tan \theta = \frac{y}{x} = \frac{2\sqrt{2}}{-2\sqrt{2}} = -1$$

$$\csc \theta = \frac{r}{y} = \frac{4}{2\sqrt{2}} = \sqrt{2} \qquad \sec \theta = \frac{r}{x} = \frac{4}{-2\sqrt{2}} = -\sqrt{2} \qquad \cot \theta = \frac{x}{y} = \frac{-2\sqrt{2}}{2\sqrt{2}} = -1$$

63. If the terminal side of a quadrantal angle lies along the y-axis, a point on the terminal side would be of the form $(0, k)$, where k is a real number, $k \neq 0$.

$$\sin \theta = \frac{y}{r} = \frac{k}{r} \qquad \cos \theta = \frac{x}{r} = \frac{0}{r} = 0 \qquad \tan \theta = \frac{y}{x} = \frac{k}{0} \text{ is undefined}$$

$$\csc \theta = \frac{r}{y} = \frac{r}{k} \qquad \sec \theta = \frac{r}{x} = \frac{r}{0} \text{ is undefined} \qquad \cot \theta = \frac{x}{y} = \frac{0}{k} = 0$$

64. Since the range of $\sec \theta$ is: $\sec \theta \leq -1$ or $\sec \theta \geq 1$, $\sec \theta = -\dfrac{2}{3}$ is impossible.

65. Since the range of $\tan \theta$ is: $-\infty \leq \tan \theta \leq \infty$, $\tan \theta = 1.4$ is possible.

66. Since $\sin \theta = \dfrac{\sqrt{3}}{5}$, $y = \sqrt{3}$ and $r = 5$. Now use pythagorean theorem to solve for x: $r^2 = x^2 + y^2 \Rightarrow$

$25 = x^2 + 3 \Rightarrow x^2 = 22 \Rightarrow x = \pm\sqrt{22}$. Since $\cos \theta < 0$, $x = -\sqrt{22}$. Therefore,

$$\sin \theta = \frac{\sqrt{3}}{5} \qquad \cos \theta = \frac{x}{r} = \frac{-\sqrt{22}}{5} = -\frac{\sqrt{22}}{5} \qquad \tan \theta = \frac{y}{x} = \frac{\sqrt{3}}{-\sqrt{22}} = -\frac{\sqrt{66}}{22}$$

$$\csc \theta = \frac{r}{y} = \frac{5}{\sqrt{3}} = \frac{5\sqrt{3}}{3} \qquad \sec \theta = \frac{r}{x} = \frac{5}{-\sqrt{22}} = -\frac{5\sqrt{22}}{22} \qquad \cot \theta = \frac{x}{y} = \frac{-\sqrt{22}}{\sqrt{3}} = -\frac{\sqrt{66}}{3}$$

67. Since $\cos \theta = \dfrac{\sqrt{3}}{5}$, $x = -5$ and $r = 8$. Now use pythagorean theorem to solve for x: $r^2 = x^2 + y^2 \Rightarrow$

$64 = 25 + y^2 \Rightarrow y^2 = 39 \Rightarrow y = \pm\sqrt{39}$. Since θ is in quadrant III, $y = -\sqrt{39}$. Therefore,

$$\sin \theta = \frac{y}{r} = \frac{-\sqrt{39}}{8} = -\frac{\sqrt{39}}{8} \qquad \cos \theta = -\frac{5}{8} \qquad \tan \theta = \frac{y}{x} = \frac{-\sqrt{39}}{-5} = \frac{\sqrt{39}}{5}$$

$$\csc \theta = \frac{r}{y} = \frac{8}{-\sqrt{39}} = -\frac{8\sqrt{39}}{39} \qquad \sec \theta = \frac{r}{x} = \frac{8}{-5} = -\frac{8}{5} \qquad \cot \theta = \frac{x}{y} = \frac{-5}{-\sqrt{39}} = \frac{5\sqrt{39}}{39}$$

68. The sine function is negative in quadrant III and IV. The cosine function is positive in quadrants I and IV.

Therefore, since $\sin \theta < 0$ and $\cos > 0$, θ must be in quadrant IV and $\tan \theta$ in quadrant IV is negative.

69. To find cotangent θ, divide 1 by tangent θ.

$$\tan \theta = 1.6778490, \quad \cot \theta = \frac{1}{\tan \theta} = \frac{1}{1.6778490} \Rightarrow \cot \theta \approx .5960.$$

70. $$\sin A = \frac{\text{opp}}{\text{hyp}} = \frac{60}{61} \qquad \cos A = \frac{\text{adj}}{\text{hyp}} = \frac{11}{61} \qquad \tan A = \frac{\text{opp}}{\text{adj}} = \frac{60}{11}$$

$$\csc A = \frac{\text{hyp}}{\text{opp}} = \frac{61}{60} \qquad \sec A = \frac{\text{hyp}}{\text{adj}} = \frac{61}{11} \qquad \cot A = \frac{\text{adj}}{\text{opp}} = \frac{11}{60}$$

71. $$\sin A = \frac{\text{opp}}{\text{hyp}} = \frac{40}{58} = \frac{20}{29} \qquad \cos A = \frac{\text{adj}}{\text{hyp}} = \frac{42}{58} = \frac{21}{29} \qquad \tan A = \frac{\text{opp}}{\text{adj}} = \frac{40}{42} = \frac{20}{21}$$

$$\csc A = \frac{\text{hyp}}{\text{opp}} = \frac{58}{40} = \frac{29}{20} \qquad \sec A = \frac{\text{hyp}}{\text{adj}} = \frac{58}{42} = \frac{29}{21} \qquad \cot A = \frac{\text{adj}}{\text{opp}} = \frac{42}{40} = \frac{21}{20}$$

72. $\cos A = \dfrac{\text{adj}}{\text{hyp}} = \dfrac{b}{c}$ and $\sin A = \dfrac{\text{opp}}{\text{hyp}} = \dfrac{b}{c}$. Since both the sine and cosine of angle $A = \dfrac{b}{c}$, $\sin A = \cos A$.

73. First find the reference angle: $(360° - 300°) = 60°$. Using the $30° - 60°$ right triangle rules, for a $60°$ angle

the hypotenuse $= 2$, the opposite side $= \sqrt{3}$, and the adjacent side $= 1$. Since $\theta = 300°$ is in quadrant IV,

y is negative and sine, tangent, cotangent, and cosecant are negative, therefore:

$$\sin 300° = -\sin 60° = -\frac{\sqrt{3}}{2} \qquad\qquad \csc 300° = -\csc 60° = -\frac{2}{\sqrt{3}} = -\frac{2\sqrt{3}}{3}$$

$$\cos 300° = \cos 60° = \frac{1}{2} \qquad\qquad \sec 300° = \sec 60° = \frac{2}{1} = 2$$

$$\tan 300° = -\tan 60° = -\frac{\sqrt{3}}{1} = -\sqrt{3} \qquad\qquad \cot 300° = -\cot 60° = -\frac{1}{\sqrt{3}} = -\frac{\sqrt{3}}{3}.$$

74. First find the reference angle: $-225°$ is coterminal with $-225° + 360° = 135°$ and the reference angle is: $360° - 135° = 45°$. Using the $45° - 45°$ right triangle rules, for a $45°$ angle the hypotenuse $= \sqrt{2}$, the opposite side $= 1$, and the adjacent $= 1$. Since $\theta = -225°$ is in quadrant II, x is negative and cosine, tangent, cotangent, and secant are negative, therefore:

$$\sin(-225°) = \sin 45° = \frac{1}{\sqrt{2}} = \frac{\sqrt{2}}{2} \qquad \csc(-225°) = \csc 45° = \frac{\sqrt{2}}{1} = \sqrt{2}$$

$$\cos(-225°) = -\cos 45° = -\frac{1}{\sqrt{2}} = -\frac{\sqrt{2}}{2} \qquad \sec(-225°) = -\sec 45° = -\frac{\sqrt{2}}{1} = -\sqrt{2}$$

$$\tan(-225°) = -\tan 45° = -\frac{1}{1} = -1 \qquad \cot(-225°) = -\cot 45° = -\frac{1}{1} = -1$$

75. First find the reference angle: $-390°$ is coterminal with $-390° + 720° = 330°$ and the reference angle is: $360° - 330° = 30°$. Using the $30° - 60°$ right triangle rules, for a $30°$ angle the hypotenuse $= 2$, the opposite side $= 1$, and the adjacent $= \sqrt{3}$. Since $\theta = -390°$ is in quadrant IV, y is negative and sine, tangent, cotangent, and cosecant are negative, therefore:

$$\sin(-390°) = -\sin 30° = -\frac{1}{2} \qquad \csc(-390°) = -\csc 30° = -\frac{2}{1} = -2$$

$$\cos(-390°) = \cos 30° = \frac{\sqrt{3}}{2} \qquad \sec(-390°) = \sec 30° = \frac{2}{\sqrt{3}} = \frac{2\sqrt{3}}{3}$$

$$\tan(-390°) = -\tan 30° = -\frac{1}{\sqrt{3}} = -\frac{\sqrt{3}}{3} \qquad \cot(-390°) = -\cot 30° = -\frac{\sqrt{3}}{1} = -\sqrt{3}$$

76. $\sin 72° 30' = \sin\left(72 + \frac{30}{60}\right)° = \sin 72.5° \approx .95371695$

77. $\sec 222°30' = \sec\left(222 + \frac{30}{60}\right)° = \sec 222.5° = (\cos 222.5°)^{-1} \approx -1.3563417$

78. $\cot 305.6° = (\tan 305.6°)^{-1} \approx -.71592968$

79. $\tan 11.7689° \approx .20834446$

80. If $\theta = 135°$, $\theta' = 45°$, if $\theta = 45°$, $\theta' = 45°$, if $\theta = 300°$, $\theta' = 60°$, and if $\theta = 140°$, $\theta' = 40°$. Of these reference angles, $40°$ is the only one which is not a special angle, so D, $\tan 140°$, is the only one which cannot be determined exactly.

81. $\theta = \sin^{-1} .82584121 \approx 55.673870° \approx 55.7°$

82. $\cot \theta = 1.1249386 \Rightarrow \frac{1}{\tan \theta} = 1.1249386 \Rightarrow \tan \theta = (1.1249386)^{-1} \Rightarrow \theta = \tan^{-1}(1.1249386)^{-1} \approx 41.635092° \approx 41.6°$

83. No, it will give an angle whose tangent equals 25.

84. For $\theta = 1997°$, $\cos 1997° \approx -.956304756$ and $\sin 1997° \approx -.292371705$. Since sine and cosine are both negative, the angle θ is in quadrant III.

85. To find B: $B = 90° - 58° 30' \Rightarrow B = 89° 60' - 58° 30' \Rightarrow B = 31° 30'$.

    To find a: $\sin A = \frac{a}{c} \Rightarrow \sin 58° 30' = \frac{a}{748} \Rightarrow a = 748(\sin 58° 30') \approx 637.7748 \Rightarrow a \approx 638$.

    To find b: $\cos A = \frac{b}{c} \Rightarrow \cos 58° 30' = \frac{b}{748} \Rightarrow b = 748(\cos 58° 30') \approx 390.8289 \Rightarrow b \approx 391$

86. To find B: $B = 90° - 39.72° \Rightarrow B = 50.28°$

To find a: $\tan A = \dfrac{a}{c} \Rightarrow \tan 39.72° = \dfrac{a}{38.97} \Rightarrow a = 38.97 \, (\tan 39.72°) \approx 32.3765 \Rightarrow a \approx 32.38$

To find c: $\cos A = \dfrac{b}{c} \Rightarrow \cos 39.72° = \dfrac{38.97}{c} \Rightarrow c = \dfrac{38.97}{\cos 39.72°} \approx 50.6646 \Rightarrow c \approx 50.66$

87. Draw a picture of these points (not shown). Since $344° - 254° = 90°$, points A, B, and C form a right triangle with angle C the right angle and angle $B = 42°$ $((254° - 180°) - 32°)$.

Now set d = the distance between point A and point B, now use the sine function to solve for d:

$\sin 42° = \dfrac{780}{d} \Rightarrow d = \dfrac{780}{\sin 42°} \Rightarrow d \approx 1165.692$. The distance from A to B is approximately 1200 meters.

88. Draw a picture of these points (not shown). A right triangle is formed since the angle where the ship turns is $90°$ $(35° + 55°$ by alternate interior angles). Solve by pythagorean:

$d^2 = 80^2 + 74^2 \Rightarrow d^2 = 11{,}876 \Rightarrow d \approx 108.97706$. The ship is approximately 110 km from the pier.

89. A right triangle is formed with a bottom angle of $36°$ $(360° - 324°)$, and the adjacent side to this $36°$ angle is 110 mph $(2 \cdot 55)$. Let x be the distance between the cars and solve for x using cosine:

$\cos 36° = \dfrac{110}{x} \Rightarrow x = \dfrac{110}{\cos 36°} \Rightarrow x \approx 135.96748$. The cars are approximately 140 miles apart.

90. Two right triangles are formed. Let x be the distance the boat travels, and y be the distance from shore at the second observation. First, solve for y using the larger right triangle:

$\angle 1 = 27°$ (by alternate interior angles) $\Rightarrow \tan 27° = \dfrac{150}{x + y} \Rightarrow x + y = \dfrac{150}{\tan 27°} \Rightarrow y = \dfrac{150}{\tan 27°} - x$.

Next, solve for y using the smaller right triangle: $\angle 2 = 39°$ (by alternate interior angles) $\Rightarrow$

$\tan 39° = \dfrac{150}{y} \Rightarrow y = \dfrac{150}{\tan 39°}$. Finally, since each equation equals y, set them equal to each other:

$\dfrac{150}{\tan 27°} - x = \dfrac{150}{\tan 39°} \Rightarrow x = \dfrac{150}{\tan 27°} - \dfrac{150}{\tan 39°} \Rightarrow x \approx 109.157$. The boat travels about 109 feet.

91. Let h = height of the tower, now use the tangent function to solve for h: $\tan 38°20' = \dfrac{h}{93.2} \Rightarrow$

$h = 93.2 \tan 38°20' \approx 73.69300534 \Rightarrow h \approx 73.7$ feet.

92. The angle at the point on the ground is $29.5°$ (by geometry, alternate interior angles are equal).

Now set h = height of the tower and use the tangent function to solve for h:

$\tan 29.5° = \dfrac{h}{36.0} \Rightarrow h = 36.0 \tan 29.5° \Rightarrow h \approx 20.36782 \Rightarrow h \approx 20.4$ meters.

93. Let x be the length of the base of the smaller right triangle. First, solve for x using the larger triangle:

$\tan 29° = \dfrac{h}{392 + x} \Rightarrow 392 + x = \dfrac{h}{\tan 29°} \Rightarrow x = \dfrac{h}{\tan 29°} - 392$. Next, solve for x using the smaller

triangle: $\tan 49° = \dfrac{h}{x} \Rightarrow x = \dfrac{h}{\tan 49°}$. Finally, since each equation is equal to x, set them equal to each other:

$\dfrac{h}{\tan 29°} - 392 = \dfrac{h}{\tan 49°} \Rightarrow \dfrac{h}{\tan 29°} - \dfrac{h}{\tan 49°} = 392 \Rightarrow$

$h \tan 49° - h \tan 29° = 392 \tan 29° \tan 49° \Rightarrow$

$h \, (\tan 49° - \tan 29°) = 392 \tan 29° \tan 49° \Rightarrow h = \dfrac{392 \tan 29° \tan 49°}{\tan 49° - \tan 29°} \Rightarrow h \approx 419.3585 \Rightarrow h \approx 419.$

94. (a) $\sin \theta = \dfrac{x_Q - x_P}{d} \Rightarrow x_Q = x_P + d \sin \theta$. Similarly, $\cos \theta = \dfrac{y_Q - y_P}{d} \Rightarrow y_Q = y_P + d \cos \theta$.

(b) Using the given information yields:

$x_Q = x_P + d \sin \theta \Rightarrow 123.62 + 193.86 \sin 17°19'22" = 123.62 + 193.86 \sin 17.3228° \Rightarrow x_Q \approx 181.34$

$y_Q = y_P + d \cos \theta \Rightarrow 337.95 + 193.86 \cos 17°19'22 = 337.95 + 193.86 \cos 17.3228° \Rightarrow y_Q \approx 523.02$

The coordinates of Q are $(181.34, 523.02)$.

95. (a) $\csc \theta = \dfrac{d}{h} \Rightarrow d = h \csc \theta$

(b) If $d = 2h$ and $\csc \theta = \dfrac{d}{h}$, then $\csc \theta = \dfrac{2h}{h} \Rightarrow \csc \theta = 2$. If $\csc \theta = 2$, then $\sin \theta = \dfrac{1}{2}$, and

$\theta = \sin^{-1}\left(\dfrac{1}{2}\right) = \dfrac{\pi}{6}$. The value of d is double that of h when the sun is 30° above the horizon.

(c) $\csc \dfrac{\pi}{2} = 1$ and $\csc \dfrac{\pi}{3} = 1.15$. When the sun is lower in the sky, $\theta = \dfrac{\pi}{3}$, sunlight is filtered by more

atmosphere. There is less ultraviolet light reaching the earth's surface, and therefore, there is less

likelihood of becoming sunburned. In the case of $\dfrac{\pi}{3}$, sunlight passes through 15% more atmosphere.

96. (a) If the shorter leg of the right triangle has length $h_2 - h_1$, then $\cot \theta = \dfrac{d}{h_2 - h_1} \Rightarrow d = (h_2 - h_1) \cot \theta$.

(b) With the given values the equation is: $d = (55 - 5) \cot \theta \Rightarrow d = 50 \cot \theta$. Since the period is π,

and the graph wanted is for the interval $0 < \theta \le \dfrac{\pi}{2}$, the graph will be the left half of a cotangent function.

The asymptote is the line $\theta = 0$ and when $\theta = \dfrac{\pi}{4}, d = 50(1) = 50$. See Figure 96.

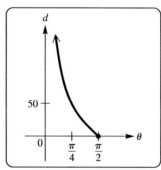

Figure 96

97. Since the coordinates of a point on the terminal side of θ are known, we will use $\tan \theta = \dfrac{y}{x}$ to solve:

$\tan \theta = \dfrac{5.1961524}{3} \Rightarrow \tan \theta \approx 1.73205$, therefore (in degree mode) $\theta = \tan^{-1} 1.73205 \Rightarrow \theta = 60°$.

This is equal to $\dfrac{60°}{1} \cdot \dfrac{\pi}{180} = \dfrac{\pi}{3}$ radians.

98. From the equation, the amplitude is 3, the period is $\dfrac{2\pi}{(2)} = \pi$, and the frequency is $\dfrac{(2)}{2\pi} = \dfrac{1}{\pi}$.

99. From the equation, the amplitude is 4, the period is $\dfrac{2\pi}{(\pi)} = 2$, and the frequency is $\dfrac{(\pi)}{2\pi} = \dfrac{1}{2}$.

100. The period is the time to complete one cycle. The amplitude is the maximum distance (on either side)

from the initial point.

101. The frequency is the number of cycles in one unit of time.

At 1.5 seconds: $s(1.5) = 4 \sin 1.5\pi = 4(-1) = -4$

At 2 seconds: $s(2) = 4 \sin 2\pi = 4(0) = 0$

At 3.25 seconds: $s(3.25) = 4 \sin 3.25\pi \approx -2.82843 = -2\sqrt{2}$

Chapter 8 Test:

1. The smallest positive degree measure of the coterminal angle is: $-157° + 360° = 203°$.

2. First find the revolutions per second: $\dfrac{450 \text{ rev.}}{1 \text{ min.}} \cdot \dfrac{1 \text{ min.}}{60 \text{ sec.}} = 7.5$ revolutions per second. Now find the degrees:

    $7.5(360°) = 2700°$. The point on the edge of the tire moves $2700°$ in one second.

3. (a) $120° \cdot \dfrac{\pi}{180} = \dfrac{2\pi}{3}$ radians.

    (b) $\dfrac{9\pi}{10} \cdot \dfrac{180}{\pi} = 162°$

4. (a) Since arc length is found by the formula

    $s = r\theta$, the central angle is: $200 = 150\theta \Rightarrow \theta = \dfrac{200}{150} \Rightarrow \theta = \dfrac{4}{3}$.

    (b) Using the area of a sector formula $A = \dfrac{1}{2}r^2\theta$ yields: $A = \dfrac{1}{2}(150)^2\left(\dfrac{4}{3}\right) \Rightarrow A = 15{,}000 \text{ km}^2$

5. Since $36° = 36 \cdot \dfrac{\pi}{180} = \dfrac{\pi}{5}$ radians and arc length is found by the formula $s = r\theta$, the arc length is:

    $s = 12\left(\dfrac{\pi}{5}\right) \Rightarrow s = \dfrac{12}{5}\pi \Rightarrow s \approx 7.54$ in.

6. From the equation, the amplitude is 2, the period is $\dfrac{2\pi}{(1)} = 2\pi$, the vertical translation is 1 unit downward, and

    the phase shift is π units to the left. See Figure 6.

7. From the equation, the amplitude is $|-1| = 1$, the period is $\dfrac{2\pi}{(2)} = \pi$, and there is no vertical translation or

    phase shift, and the graph is reflected across the x-axis. See Figure 7.

8. From the equation, the amplitude (since a tangent function) is not applicable, the period is $\dfrac{\pi}{(1)} = \pi$,

    and the phase shift is $\dfrac{\pi}{2}$ units to the right. See Figure 8.

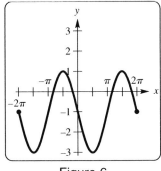
Figure 6

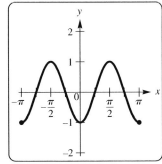
Figure 7

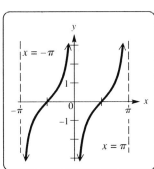
Figure 8

9. (a) See Figure 9.

    (b) From the equation, the amplitude is 17.5, the period is $\dfrac{2\pi}{\left(\frac{\pi}{6}\right)} = 12$, the phase shift is 4 units to the right, and the vertical translation is 67.5 units upward.

    (c) For December $x = 12$, therefore: $f(12) = 17.5 \sin\left[\dfrac{\pi}{6}(12 - 4)\right] + 67.5 \approx 52.34456° \Rightarrow f(12) \approx 52°$ F.

    (d) From the graph the minimum temperature takes place in January and is:

      For January $x = 1$, therefore: $f(1) = 17.5 \sin\left[\dfrac{\pi}{6}(1 - 4)\right] + 67.5 = 50° \Rightarrow f(1) = 50°$ F.

      From the graph the maximum temperature takes place in July and is:

      For July $x = 7$, therefore: $f(7) = 17.5 \sin\left[\dfrac{\pi}{6}(7 - 4)\right] + 67.5 = 85° \Rightarrow f(7) = 85°$ F.

    (e) The average yearly temperature in Austin is approximately 67.5°, the vertical translation of the equation.

[1, 25] by [45, 90]
Xscl = 5   Yscl = 5

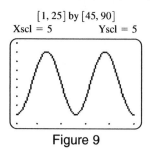

Figure 9

10. (a) The smallest positive degree measure of the coterminal angle is: $-150° + 360° = 210°$.

    (b) First find the reference angle: $(210° - 180°) = 30°$. Using the $30° - 60°$ right triangle rules, for a 30° angle the hypotenuse $= 2$, the opposite side $= 1$, and the adjacent $= \sqrt{3}$. Since $\theta = -150°$ is in quadrant III, both x and y are negative and the point is: $(-\sqrt{3}, -1)$.

    (c) $\sin(-150°) = -\sin 30° = -\dfrac{1}{2}$ $\csc(-150°) = -\csc 30° = -\dfrac{2}{1} = -2$

      $\cos(-150°) = -\cos 30° = -\dfrac{\sqrt{3}}{2}$ $\sec(-150°) = -\sec 30° = -\dfrac{2}{\sqrt{3}} = -\dfrac{2\sqrt{3}}{3}$

      $\tan(-150°) = \tan 30° = \dfrac{1}{\sqrt{3}} = \dfrac{\sqrt{3}}{3}$ $\cot(-150°) = \cot 30° = \dfrac{\sqrt{3}}{1} = \sqrt{3}$

    (d) $-150° = -\dfrac{150°}{1} \cdot \dfrac{\pi}{180°} = -\dfrac{5\pi}{6}$

11. If $\cos \theta < 0$, then θ is in quadrant II or III. If $\cot \theta > 0$, then θ is in quadrant I or III.

    Therefore, θ terminates in quadrant III.

12. With the coordinate $(2, -5)$, we will have a radius of $r = \sqrt{(2)^2 + (-5)^2} = \sqrt{4 + 25} \Rightarrow r = \sqrt{29}$.

    Therefore: $\sin \theta = \dfrac{y}{r} = \dfrac{-5}{\sqrt{29}} = -\dfrac{5\sqrt{29}}{29}$

      $\cos \theta = \dfrac{x}{r} = \dfrac{2}{\sqrt{29}} \Rightarrow \dfrac{2\sqrt{29}}{29}$

      $\tan \theta = \dfrac{y}{x} = \dfrac{-5}{2} = -\dfrac{5}{2}$

13. Since $\cos\theta = \dfrac{4}{5}$, then $x = 4$ and $r = 5$. Now use the pythagorean theorem to find y:

$5^2 = 4^2 + y^2 \;\Rightarrow\; 25 = 16 + y^2 \;\Rightarrow\; y^2 = 9 \;\Rightarrow\; y = \pm 3$. Since θ is in quadrant IV, $y = -3$.

Therefore:

$\sin\theta = \dfrac{y}{r} = \dfrac{-3}{5} = -\dfrac{3}{5}$ $\tan\theta = \dfrac{y}{x} = \dfrac{-3}{4} = -\dfrac{3}{4}$

$\csc\theta = \dfrac{r}{y} = \dfrac{5}{-3} = -\dfrac{5}{3}$ $\sec\theta = \dfrac{r}{x} = \dfrac{5}{4}$ $\cot\theta = \dfrac{x}{y} = \dfrac{4}{-3} = -\dfrac{4}{3}$

14. Use knowledge of the $(30° - 60°)$ and $(45° - 45°)$ right triangles to find each:

To find w, use $\sin 30° = \dfrac{4}{w} \;\Rightarrow\; w = \dfrac{4}{\sin 30°} = \dfrac{4}{\frac{1}{2}} = 8$

To find x, use $\tan 45° = \dfrac{4}{x} \;\Rightarrow\; x = \dfrac{4}{\tan 45°} = \dfrac{4}{1} = 4$

To find y, use $\cos 30° = \dfrac{y}{w} = \dfrac{y}{8} \;\Rightarrow\; y = 8\cos 30° = 8 \cdot \dfrac{\sqrt{3}}{2} = 4\sqrt{3}$

To find z, use $\sin 45° = \dfrac{4}{z} \;\Rightarrow\; z = \dfrac{4}{\sin 45°} = \dfrac{4}{\frac{1}{\sqrt{2}}} = 4\sqrt{2}$

15. First, $\cot(-750°)$ is coterminal with $\cot(3 \cdot 360° - 750°) = \cot 330°$. Now find the reference angle:

$\cot(360° - 330°) = \cot 30°$. Finally, since $\cot(-750°)$ is terminal in quadrant IV, we will find $-\cot 30°$ and

using the $(30° - 60°)$ right triangle: $-\cot 30° = -\sqrt{3}$.

16. (a) $\sin 78°21' = \sin\left(78 + \dfrac{21}{60}\right)° = \sin 78.35° \approx .97939940$

(b) $\tan 11.7689° \approx .20834446$

(c) $\sec 58.9041° = \dfrac{1}{\cos 58.9041°} = (\cos 58.9041°)^{-1} \approx 1.9362132$

17. If $\sin s = .82584121$, then $s = \sin^{-1} .82584121 \approx .97169234$.

18. To find B: $B = 90° - 58° 30' \;\Rightarrow\; B = 89° 60' - 58° 30' \;\Rightarrow\; B = 31° 30'$.

To find a: $\sin A = \dfrac{a}{c} \;\Rightarrow\; \sin 58° 30' = \dfrac{a}{748} \;\Rightarrow\; a = 748\,(\sin 58° 30') \approx 637.7748 \;\Rightarrow\; a \approx 638$.

To find b: $\cos A = \dfrac{b}{c} \;\Rightarrow\; \cos 58° 30' = \dfrac{b}{748} \;\Rightarrow\; b = 748\,(\cos 58° 30') \approx 390.8289 \;\Rightarrow\; b \approx 391$

19. Let x be the height of the flag pole. Now use the tangent function to solve for x:

$\tan 32°10' = \dfrac{x}{24.7} \;\Rightarrow\; x = 24.7\tan 32°10' \approx 15.5$ feet.

20. Draw a picture of the ships travels, (not shown). Angle ABX is $55°$ (Using geometry − alternate interior angles are equal) and angle ABC is $55° + 35° = 90°$, therefore triangle ABC is a right triangle and we can use pythagorean theorem to find AC: $AC^2 = 80^2 + 74^2 \;\Rightarrow\; AC = \sqrt{80^2 + 74^2} \approx 108.977 \;\Rightarrow\; AC \approx 110$ km.

21. Because the amplitude is $|-3| = 3$, the maximum height the weight rises above an equilibrium position of $y = 0$ is 3 inches.

22. If $s(t) = 3$, then $3 = -3\cos 2\pi t \;\Rightarrow\; -1 = \cos 2\pi t$. Since $\cos\theta = -1$ when $\theta = \pi$, then $2\pi t = \pi \;\Rightarrow\;$

$2t = 1 \;\Rightarrow\; t = \dfrac{1}{2}$. Therefore it reaches the maximum height at $\dfrac{1}{2}$ seconds.

Chapter 8 Project

1. Convert the times to decimals hours and enter the values in your calculator as described. Now make a scatterplot of the data on the calculator. See Figure 1.

[0, 25] by [0, 10]
Xscl = 2   Yscl = 1

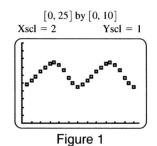

Figure 1

2. The minimum sunset time is 4:35 or 4.58; the maximum sunset time is 7:33 or 7.55. Using these values, the amplitude is $\dfrac{7.55 - 4.58}{2} = 1.48$ or 1:29. In the equation $y = a \sin b(x - d) + c$, the amplitude is a, therefore $a = 1.48$.

3. The times repeat every 12 months, so the period is $\dfrac{2\pi}{(12)} = \dfrac{\pi}{6} \approx .52$. Therefore $b \approx .52$.

4. We want the cycle to start with March 21, which corresponds to 3.68, therefore the phase shift is $d \approx 3.68$.

5. The vertical shift is the average of the maximum and minimum sunset times, so $c = \dfrac{4.58 + 7.55}{2} = 6.065$. Therefore $c \approx 6.07$.

6. Using $y = a \sin b(x - d) + c$ yields:

    $y = 1.48 \sin [.52(x - 3.68)] + 6.07 \Rightarrow y = 1.48 \sin (.52x - 1.91) + 6.07.$ See Figure 6. (Use radian mode)

[0, 25] by [0, 10]
Xscl = 2   Yscl = 1

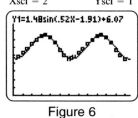
Y1=1.48sin(.52X-1.91)+6.07

Figure 6

7. The regression equation is $y = 1.42 \sin (.52x - 1.81) + 6.10$, which is very close to the equation we found above. The two forms are nearly equivalent, but not exactly.

Chapter 9: Trigonometric Identities and Equations

9.1: Trigonometric Identities

1. Since by the negative-number identities $\sin x = -\sin(-x)$, the function is : Odd.

2. Since by the negative-number identities $\cos x = \cos(-x)$, the function is : Even.

3. Since by the negative-number identities $\tan x = -\tan(-x)$, the function is : Odd.

4. Since by the negative-number identities $\cot x = -\cot(-x)$, the function is : Odd.

5. Since by the negative-number identities $\sec x = \sec(-x)$, the function is : Even.

6. Since by the negative-number identities $\csc x = -\csc(-x)$, the function is : Odd.

7. By a quotient identity, $\dfrac{\cos x}{\sin x} = \cot x$, therefore B.

8. By a quotient identity, $\tan x = \dfrac{\sin x}{\cos x}$, therefore D.

9. By a negative-number identity, $\cos(-x) = \cos x$, therefore E.

10. By a pythagorean identity, $\tan^2 x + 1 = \sec^2 x$, therefore C.

11. By a pythagorean identity, $1 = \sin^2 x + \cos^2 x$, therefore A.

12. Using a quotient identity, $-\tan x \cos x = -\dfrac{\sin x}{\cos x} \cdot \dfrac{\cos x}{1} = -\sin x = \sin(-x)$, therefore C.

13. Using a pythagorean identity and then a quotient identity, $\sec^2 x - 1 = \tan^2 x = \dfrac{\sin^2 x}{\cos^2 x}$, therefore A.

14. Using two reciprocal identities and then a quotient identity, $\dfrac{\sec x}{\csc x} = \dfrac{1}{\cos x} \div \dfrac{1}{\sin x} = \dfrac{\sin x}{\cos x} = \tan x$, therefore E.

15. Using a pythagorean identity, $1 + \sin^2 x = (\csc^2 x - \cot^2 x) + \sin^2 x$, therefore D.

16. Using a reciprocal identity, $\cos^2 x = \dfrac{1}{\sec^2 x}$, therefore B.

17. By a negative-number identity, if $\cos(-\theta) = \cos\theta$, then $\cos(-4.38) = \cos 4.38$.

18. By a negative-number identity, if $\cos(-\theta) = \cos\theta$, then $\cos(-5.46) = \cos 5.46$.

19. By a negative-number identity, if $\sin(-\theta) = -\sin\theta$, then $\sin(-.5) = -\sin .5$.

20. By a negative-number identity, if $\sin(-\theta) = -\sin\theta$, then $\sin(-2.5) = -\sin 2.5$.

21. By a negative-number identity, if $\tan(-\theta) = -\tan\theta$, then $\tan\left(-\dfrac{\pi}{7}\right) = -\tan\dfrac{\pi}{7}$.

22. By a negative-number identity, if $\cot(-\theta) = -\cot\theta$, then $\cot\left(-\dfrac{4\pi}{7}\right) = -\cot\dfrac{4\pi}{7}$.

23. The correct identity is $1 + \cot^2 x = \csc^2 x$; the function must have the argument "x" or θ, t, etc.

24. The square root of a sum does not equal the sum of the square roots:
$$\sqrt{a^2 + b^2} \neq \sqrt{a^2} + \sqrt{b^2}, \text{ so } \sqrt{\sin^2\theta + \cos^2\theta} \neq \sin\theta + \cos\theta.$$

25. $\sin\theta$ in terms of $\cot\theta$: $\sin\theta = \dfrac{1}{\csc\theta} = \dfrac{1}{\pm\sqrt{1 + \cot^2\theta}} = \pm\dfrac{\sqrt{1 + \cot^2\theta}}{1 + \cot^2\theta}$

$\sin\theta$ in terms of $\sec\theta$: $\sin\theta = \cos\theta \cdot \dfrac{\sin\theta}{\cos\theta} = \cos\theta \cdot \tan\theta = \dfrac{1}{\sec\theta} \cdot (\pm\sqrt{\sec^2\theta - 1}) = \pm\dfrac{\sqrt{\sec^2\theta - 1}}{\sec\theta}$

26. $\cos\theta$ in terms of $\sin\theta$: $\cos\theta = \pm\sqrt{1 - \sin^2\theta}$

$\cos\theta$ in terms of $\cot\theta$: $\cot\theta = \sin\theta \cdot \dfrac{\cos\theta}{\sin\theta} = \dfrac{1}{\csc\theta} \cdot \cot\theta = \dfrac{1}{\pm\sqrt{1 + \cot^2\theta}} \cdot \cot\theta = \pm\dfrac{\cot\theta\sqrt{1 + \cot^2\theta}}{1 + \cot^2\theta}$

$\cos\theta$ in terms of $\csc\theta$: $\cos\theta = \pm\sqrt{1 - \sin^2\theta} = \pm\sqrt{1 - \dfrac{1}{\csc^2\theta}} = \pm\sqrt{\dfrac{\csc^2\theta - 1}{\csc^2\theta}} = \pm\dfrac{\sqrt{\csc^2\theta - 1}}{\csc\theta}$

27. $\tan\theta$ in terms of $\sin\theta$: $\tan\theta = \dfrac{\sin\theta}{\cos\theta} = \dfrac{\sin\theta}{\pm\sqrt{1 - \sin^2\theta}} = \pm\dfrac{\sin\theta\sqrt{1 - \sin^2\theta}}{1 - \sin^2\theta}$

$\tan\theta$ in terms of $\cos\theta$: $\tan\theta = \dfrac{\sin\theta}{\cos\theta} = \pm\dfrac{\sqrt{1 - \cos^2\theta}}{\cos\theta}$

$\tan\theta$ in terms of $\sec\theta$: $\tan\theta = \pm\sqrt{\sec^2\theta - 1}$

$\tan\theta$ in terms of $\csc\theta$: $\tan\theta = \dfrac{1}{\cot\theta} = \dfrac{1}{\pm\sqrt{\csc^2\theta - 1}} = \pm\dfrac{\sqrt{\csc^2\theta - 1}}{\csc^2\theta - 1}$

28. $\cot\theta$ in terms of $\sin\theta$: $\cot\theta = \pm\sqrt{\csc^2\theta - 1} = \pm\sqrt{\dfrac{1}{\sin^2\theta} - 1} = \pm\sqrt{\dfrac{1 - \sin^2\theta}{\sin^2\theta}} = \pm\dfrac{\sqrt{1 - \sin^2\theta}}{\sin\theta}$

$\cot\theta$ in terms of $\cos\theta$: $\cot\theta = \pm\dfrac{\sqrt{1 - \sin^2\theta}}{\sin\theta} = \pm\dfrac{\cos\theta}{\sqrt{1 - \cos^2\theta}} = \pm\dfrac{\cos\theta\sqrt{1 - \cos^2\theta}}{1 - \cos^2\theta}$

$\cot\theta$ in terms of $\csc\theta$: $\cot\theta = \pm\sqrt{\csc^2\theta - 1}$

29. $\sec\theta$ in terms of $\sin\theta$: $\sec\theta = \dfrac{1}{\cos\theta} = \dfrac{1}{\pm\sqrt{1 - \sin^2\theta}} = \pm\dfrac{\sqrt{1 - \sin^2\theta}}{1 - \sin^2\theta}$

$\sec\theta$ in terms of $\tan\theta$: $\sec\theta = \pm\sqrt{\tan^2\theta + 1}$

$\sec\theta$ in terms of $\cot\theta$: $\sec\theta = \pm\sqrt{\tan^2\theta + 1} = \pm\sqrt{\dfrac{1}{\cot^2\theta} + 1} = \pm\sqrt{\dfrac{1 + \cot^2\theta}{\cot^2\theta}} = \pm\dfrac{\sqrt{1 + \cot^2\theta}}{\cot\theta}$

$\sec\theta$ in terms of $\csc\theta$: $\sec\theta = \dfrac{1}{\cos\theta} = \dfrac{1}{\pm\sqrt{1 - \sin^2\theta}} = \dfrac{1}{\pm\sqrt{1 - \dfrac{1}{\csc^2\theta}}} = \dfrac{1}{\pm\sqrt{\dfrac{\csc^2\theta - 1}{\csc^2\theta}}} =$

$\dfrac{\pm\sqrt{\csc^2\theta}}{\sqrt{\csc^2\theta - 1}} = \dfrac{\pm\csc\theta}{\sqrt{\csc^2\theta - 1}} = \pm\dfrac{\csc\theta\sqrt{\csc^2\theta - 1}}{\csc^2\theta - 1}$

30. $\csc\theta$ in terms of $\cos\theta$: $\csc\theta = \dfrac{1}{\sin\theta} = \dfrac{1}{\pm\sqrt{1 - \cos^2\theta}} = \pm\dfrac{\sqrt{1 - \cos^2\theta}}{1 - \cos^2\theta}$

$\csc\theta$ in terms of $\tan\theta$: $\csc\theta = \pm\sqrt{1 + \cot^2\theta} = \pm\sqrt{1 + \dfrac{1}{\tan^2\theta}} = \pm\sqrt{\dfrac{\tan^2\theta + 1}{\tan^2\theta}} = \pm\dfrac{\sqrt{\tan^2\theta + 1}}{\tan\theta}$

$\csc\theta$ in terms of $\cot\theta$: $\csc\theta = \pm\sqrt{1 + \cot^2\theta}$

$\csc\theta$ in terms of $\sec\theta$: $\csc\theta = \pm\dfrac{\sqrt{\tan^2\theta + 1}}{\tan\theta} = \pm\dfrac{\sec\theta}{\sqrt{\sec^2\theta - 1}} = \pm\dfrac{\sec\theta\sqrt{\sec^2\theta - 1}}{\sec^2\theta - 1}$

31. $\tan\theta\cos\theta = \dfrac{\sin\theta}{\cos\theta} \cdot \dfrac{\cos\theta}{1} = \sin\theta$

32. $\cot\alpha\sin\alpha = \dfrac{\cos\alpha}{\sin\alpha} \cdot \dfrac{\sin\alpha}{1} = \cos\alpha$

33. $\dfrac{\sin\beta\tan\beta}{\cos\beta} = \dfrac{\sin\beta}{\cos\beta} \cdot \tan\beta = \tan\beta \cdot \tan\beta = \tan^2\beta$

34. $\dfrac{\csc \theta \sec \theta}{\cot \theta} = \dfrac{1}{\sin \theta} \cdot \dfrac{1}{\cos \theta} \div \dfrac{\cos \theta}{\sin \theta} = \dfrac{1}{\sin \theta \cdot \cos \theta} \cdot \dfrac{\sin \theta}{\cos \theta} = \dfrac{1}{\cos^2 \theta} = \sec^2 \theta$

35. $\sec^2 x - 1 = (\tan^2 x + 1) - 1 = \tan^2 x$

36. $\csc^2 t - 1 = (1 + \cot^2 t) - 1 = \cot^2 t$

37. $\dfrac{\sin^2 x}{\cos^2 x} + \sin x \csc x = \tan^2 x + \dfrac{\sin x}{1} \cdot \dfrac{1}{\sin x} = \tan^2 x + 1 = \sec^2 x$

38. $\dfrac{1}{\tan^2 \alpha} + \cot \alpha \tan \alpha = \cot^2 \alpha + \dfrac{1}{\tan \alpha} \cdot \dfrac{\tan \alpha}{1} = \cot^2 \alpha + 1 = \csc^2 \alpha$

39. $\cot \theta \sin \theta = \dfrac{\cos \theta}{\sin \theta} \cdot \sin \theta = \cos \theta$

40. $\sec \theta \cot \theta \sin \theta = \dfrac{1}{\cos \theta} \cdot \dfrac{\cos \theta}{\sin \theta} \cdot \sin \theta = 1$

41. $\cos \theta \csc \theta = \cos \theta \cdot \dfrac{1}{\sin \theta} = \dfrac{\cos \theta}{\sin \theta}$, which simplified is $\cot \theta$.

42. $\cot^2 \theta (1 + \tan^2 \theta) = \dfrac{\cos^2 \theta}{\sin^2 \theta} \cdot \sec^2 \theta = \dfrac{\cos^2 \theta}{\sin^2 \theta} \cdot \dfrac{1}{\cos^2 \theta} = \dfrac{1}{\sin^2 \theta}$, which simplified is $\csc^2 \theta$.

43. $\sin^2 \theta (\csc^2 \theta - 1) = \sin^2 \theta \cdot \cot^2 \theta = \sin^2 \theta \cdot \dfrac{\cos^2 \theta}{\sin^2 \theta} = \cos^2 \theta$

44. $(\sec \theta - 1)(\sec \theta + 1) = \sec^2 \theta - 1 = \tan^2 \theta = \dfrac{\sin^2 \theta}{\cos^2 \theta}$, which simplified is $\tan^2 \theta$.

45. $(1 - \cos \theta)(1 + \sec \theta) = (1 - \cos \theta)\left(1 + \dfrac{1}{\cos}\right) = 1 + \dfrac{1}{\cos \theta} - \cos \theta - 1 = \dfrac{1}{\cos \theta} - \cos \theta$,

    which simplified is $\sec \theta - \cos \theta$.

46. It is in terms of sine and cosine, but simplified is $\dfrac{\cos \theta + \sin \theta}{\sin \theta} = \dfrac{\cos \theta}{\sin \theta} + 1 = 1 + \cot \theta$.

47. It is in terms of sine and cosine, but simplified is $\dfrac{\cos^2 \theta - \sin^2 \theta}{\sin \theta \cos \theta} = \dfrac{\cos^2 \theta}{\sin \theta \cos \theta} - \dfrac{\sin^2 \theta}{\sin \theta \cos \theta} =$

    $\dfrac{\cos \theta}{\sin \theta} - \dfrac{\sin \theta}{\cos \theta} = \cot \theta - \tan \theta$.

48. $\dfrac{1 - \sin^2 \theta}{1 + \cot^2 \theta} = \dfrac{\cos^2 \theta}{\csc^2 \theta} = \dfrac{\cos^2 \theta}{1} \div \dfrac{1}{\sin^2 \theta} = \dfrac{\cos^2 \theta}{1} \cdot \dfrac{\sin^2 \theta}{1} = \sin^2 \theta \cos^2 \theta$

49. $\sec \theta - \cos \theta = \dfrac{1}{\cos \theta} - \dfrac{\cos^2 \theta}{\cos \theta} = \dfrac{1 - \cos^2 \theta}{\cos \theta} = \dfrac{\sin^2 \theta}{\cos \theta}$

50. $\dfrac{1 + \tan^2 \theta}{1 + \cot^2 \theta} = \dfrac{\sec^2 \theta}{\csc^2 \theta} = \dfrac{1}{\cos^2 \theta} \div \dfrac{1}{\sin^2 \theta} = \dfrac{1}{\cos^2 \theta} \cdot \dfrac{\sin^2 \theta}{1} = \dfrac{\sin^2 \theta}{\cos^2 \theta}$, which simplified is $\tan^2 \theta$.

51. $\sin \theta (\csc \theta - \sin \theta) = \sin \theta \left(\dfrac{1}{\sin \theta} - \sin \theta\right) = 1 - \sin^2 \theta = \cos^2 \theta$.

52. $(\sec \theta + \csc \theta)(\cos \theta - \sin \theta) = \left(\dfrac{1}{\cos \theta} + \dfrac{1}{\sin \theta}\right)(\cos \theta - \sin \theta) = 1 - \dfrac{\sin \theta}{\cos \theta} + \dfrac{\cos \theta}{\sin \theta} - 1 =$

    $\dfrac{\cos \theta}{\sin \theta} - \dfrac{\sin \theta}{\cos \theta}$, which simplified is $\cot \theta - \tan \theta$.

53. $\cot \theta + \dfrac{1}{\cot \theta} = \dfrac{\cos \theta}{\sin \theta} + \tan \theta = \dfrac{\cos \theta}{\sin \theta} + \dfrac{\sin \theta}{\cos \theta} = \dfrac{\cos^2 \theta}{\sin \theta \cos \theta} + \dfrac{\sin^2 \theta}{\sin \theta \cos \theta} = \dfrac{\sin^2 \theta + \cos^2 \theta}{\sin \theta \cos \theta} =$

    $\dfrac{1}{\sin \theta \cos \theta}$ or $\csc \theta \sec \theta$

54. $\dfrac{\sec x}{\csc x} + \dfrac{\csc x}{\sec x} = \left(\dfrac{1}{\cos x} \div \dfrac{1}{\sin x} \right) + \left(\dfrac{1}{\sin x} \div \dfrac{1}{\cos x} \right) = \left(\dfrac{1}{\cos x} \cdot \dfrac{\sin x}{1} \right) + \left(\dfrac{1}{\sin x} \cdot \dfrac{\cos x}{1} \right) =$

$\dfrac{\sin x}{\cos x} + \dfrac{\cos x}{\sin x} = \dfrac{\sin^2 x}{\sin x \cos x} + \dfrac{\cos^2 x}{\sin x \cos x} = \dfrac{\sin^2 x + \cos^2 x}{\sin x \cos x} = \dfrac{1}{\sin x \cos x}$ or $\csc x \sec x$

55. $\tan s \left(\cot s + \csc s \right) = \tan s \left(\dfrac{1}{\tan s} + \dfrac{1}{\sin s} \right) = 1 + \dfrac{\tan s}{\sin s} = 1 + \left(\dfrac{\sin s}{\cos s} \div \sin s \right) = 1 + \left(\dfrac{\sin s}{\cos s} \cdot \dfrac{1}{\sin s} \right) =$

$1 + \dfrac{1}{\cos s} = 1 + \sec s$

56. $\cos \beta \left(\sec \beta + \csc \beta \right) = \cos \beta \left(\dfrac{1}{\cos \beta} + \dfrac{1}{\sin \beta} \right) = 1 + \dfrac{\cos \beta}{\sin \beta} = 1 + \cot \beta$

57. $\dfrac{1}{\csc^2 \theta} + \dfrac{1}{\sec^2 \theta} = \sin^2 \theta + \cos^2 \theta = 1$

58. $\dfrac{1}{\sin \alpha - 1} - \dfrac{1}{\sin \alpha + 1} = \dfrac{1(\sin \alpha + 1)}{(\sin \alpha - 1)(\sin \alpha + 1)} - \dfrac{1(\sin \alpha - 1)}{(\sin \alpha + 1)(\sin \alpha - 1)} =$

$\dfrac{2}{\sin^2 \alpha - 1} = -\dfrac{2}{\cos^2 \alpha}$ or $-2 \sec^2 \alpha$

59. $\dfrac{\cos x}{\sec x} + \dfrac{\sin x}{\csc x} = \left(\dfrac{\cos x}{1} \div \dfrac{1}{\cos x} \right) + \left(\dfrac{\sin x}{1} \div \dfrac{1}{\sin x} \right) = \left(\dfrac{\cos x}{1} \cdot \dfrac{\cos x}{1} \right) + \left(\dfrac{\sin x}{1} \cdot \dfrac{\sin x}{1} \right) =$

$\sin^2 x + \cos^2 x = 1$

60. $\dfrac{\cos \theta}{\sin \theta} + \dfrac{\sin \theta}{1 + \cos \theta} = \dfrac{\cos \theta (1 + \cos \theta)}{\sin \theta (1 + \cos \theta)} + \dfrac{\sin \theta \sin \theta}{\sin \theta (1 + \cos \theta)} = \dfrac{\cos \theta + \cos^2 \theta + \sin^2 \theta}{\sin \theta (1 + \cos \theta)} =$

$\dfrac{\cos \theta + 1}{\sin \theta (1 + \cos \theta)} = \dfrac{1}{\sin \theta} = \csc x$

61. $(1 + \sin t)^2 + \cos^2 t = 1 + 2 \sin t + \sin^2 t + \cos^2 t = 1 + 2 \sin t + 1 = 2 + 2 \sin t$

62. $(1 + \tan s)^2 - 2 \tan s = 1 + 2 \tan s + \tan^2 s - 2 \tan s = 1 + \tan^2 s = \sec^2 s$

63. $\dfrac{1}{1 + \cos x} - \dfrac{1}{1 - \cos x} = \dfrac{1(1 - \cos x)}{(1 + \cos x)(1 - \cos x)} - \dfrac{1(1 + \cos x)}{(1 + \cos x)(1 - \cos x)} = \dfrac{1 - \cos x - 1 - \cos x}{1 - \cos^2 x} =$

$\dfrac{-2 \cos x}{\sin^2 x} = -\dfrac{2 \cos x}{\sin^2 x}$ or $-2 \cot x \csc x$

64. $(\sin \alpha - \cos \alpha)^2 = \sin^2 \alpha - 2 \sin \alpha \cos \alpha + \cos^2 \alpha = \sin^2 \alpha + \cos^2 \alpha - 2 \sin \alpha \cos \alpha = 1 - 2 \sin \alpha \cos \alpha$

65. See Figure 65.

66. The graph looks like $y = \cos x$.

67. See Figure 67.

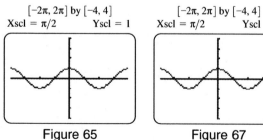

$[-2\pi, 2\pi]$ by $[-4, 4]$
Xscl = $\pi/2$ Yscl = 1

$[-2\pi, 2\pi]$ by $[-4, 4]$
Xscl = $\pi/2$ Yscl = 1

Figure 65   Figure 67

68. Yes, it suggests that $(\sec x + \tan x)(1 - \sin x) = \cos x$.

69. $(\sec x + \tan x)(1 - \sin x) = \left(\dfrac{1}{\cos x} + \dfrac{\sin x}{\cos x}\right)(1 - \sin x) = \dfrac{1 + \sin x}{\cos x} \cdot \dfrac{1 - \sin x}{1} = \dfrac{1 - \sin^2 x}{\cos x} =$

$\dfrac{\cos^2 x}{\cos x} = \cos x$

70. After graphing the function on a calculator, it appears to be the same graph as $\cot x$. Now verify this:

$\dfrac{\cos x + 1}{\sin x + \tan x} = \dfrac{\cos x + 1}{\sin x + \frac{\sin x}{\cos x}} = \dfrac{\cos x + 1}{1} \div \dfrac{\sin x \cos x + \sin x}{\cos x} = \dfrac{\cos x + 1}{1} \cdot \dfrac{\cos x}{\sin x \,(\cos x + 1)} =$

$\dfrac{\cos x}{\sin x} = \cot x$. The identity is: $\dfrac{\cos x + 1}{\sin x + \tan x} = \cot x$.

71. $\dfrac{\cot \theta}{\csc \theta} = \dfrac{\frac{\cos \theta}{\sin \theta}}{\frac{1}{\sin \theta}} = \dfrac{\cos \theta}{\sin \theta} \cdot \dfrac{\sin \theta}{1} = \dfrac{\cos \theta}{1} = \cos \theta$

72. $\dfrac{\tan \theta}{\sec \theta} = \dfrac{\frac{\sin \theta}{\cos \theta}}{\frac{1}{\cos \theta}} = \dfrac{\sin \theta}{\cos \theta} \cdot \dfrac{\cos \theta}{1} = \dfrac{\sin \theta}{1} = \sin \theta$

73. $\cos^2 \theta \,(\tan^2 \theta + 1) = \cos^2 \theta \,(\sec^2 \theta) = \cos^2 \theta \left(\dfrac{1}{\cos^2 \theta}\right) = 1$

74. $\dfrac{\cos \theta}{\sec \theta - 1} = \dfrac{\cos \theta}{\sec \theta - 1} \cdot \dfrac{\sec \theta + 1}{\sec \theta + 1} = \dfrac{\cos \theta \,(\sec \theta + 1)}{\sec^2 \theta - 1} = \dfrac{\cos \theta \left(\frac{1}{\cos \theta} + 1\right)}{\tan^2 \theta} = \dfrac{1 + \cos \theta}{\tan^2 \theta} =$

$\dfrac{1 + \cos \theta}{\tan^2 \theta} = \dfrac{\cos \theta + 1}{\tan^2 \theta}$

75. $\dfrac{\tan^2 \gamma + 1}{\sec \gamma} = \dfrac{\sec^2 \gamma}{\sec \gamma} = \sec \gamma$

76. $\sin^2 \beta \,(1 + \cot^2 \beta) = \sin^2 \beta \csc^2 \beta = \dfrac{\sin^2 \beta}{1} \cdot \dfrac{1}{\sin^2 \beta} = 1$

77. $\sin^2 \alpha + \tan^2 \alpha + \cos^2 \alpha = \sin^2 \alpha + \cos^2 \alpha + \tan^2 \alpha = 1 + \tan^2 \alpha = \sec^2 \alpha$

78. $\cot s + \tan s = \dfrac{\cos s}{\sin s} + \dfrac{\sin s}{\cos s} = \dfrac{\cos^2 s + \sin^2 s}{\sin s \cos s} = \dfrac{1}{\sin s \cos s} = \dfrac{1}{\sin s} \cdot \dfrac{1}{\cos s} = \csc s \sec s = \sec s \csc s$

79. $\dfrac{\sin^2 \gamma}{\cos \gamma} = \dfrac{1 - \cos^2 \gamma}{\cos \gamma} = \dfrac{1}{\cos \gamma} - \dfrac{\cos^2 \gamma}{\cos \gamma} = \sec \gamma - \cos \gamma$

80. First simplify the left side of the equation: $\dfrac{\cos \alpha}{\sec \alpha} + \dfrac{\sin \alpha}{\csc \alpha} = \dfrac{\cos \alpha}{\frac{1}{\cos \alpha}} + \dfrac{\sin \alpha}{\frac{1}{\sin \alpha}} = \cos^2 \alpha + \sin^2 \alpha = 1.$

Next simplify the right side of the equation: $\sec^2 \alpha - \tan^2 \alpha = (\tan^2 \alpha + 1) - \tan^2 \alpha = 1.$

Since both sides of the equation are equal to 1, $\dfrac{\cos \alpha}{\sec \alpha} + \dfrac{\sin \alpha}{\csc \alpha} = \sec^2 \alpha - \tan^2 \alpha.$

81. $\dfrac{\cos \theta}{\sin \theta \cot \theta} = \dfrac{\cos \theta}{\sin \theta \cdot \frac{\cos \theta}{\sin \theta}} = \dfrac{\cos \theta}{\cos \theta} = 1$

82. $\sin^4 \theta - \cos^4 \theta = (\sin^2 \theta - \cos^2 \theta)(\sin^2 \theta + \cos^2 \theta) = (\sin^2 \theta - (1 - \sin^2 \theta)) \cdot 1 = 2 \sin^2 \theta - 1$

83. $\tan^2 \gamma \sin^2 \gamma \equiv (\sec^2 \gamma - 1)(1 - \cos^2 \gamma) = \sec^2 \gamma - \sec^2 \gamma \cos^2 \gamma - 1 + \cos^2 \gamma =$

$\sec^2 \gamma - 1 + \cos^2 \gamma - \sec^2 \gamma \cos^2 \gamma = \tan^2 \gamma + \cos^2 \gamma - 1$

84. $(1 - \cos^2 \alpha)(1 + \cos^2 \alpha) = (\sin^2 \alpha)(1 + (1 - \sin^2 \alpha)) = (\sin^2 \alpha)(2 - \sin^2 \alpha) = 2 \sin^2 \alpha - \sin^4 \alpha$

85. $\dfrac{(\sec \theta - \tan \theta)^2 + 1}{\sec \theta \csc \theta - \tan \theta \csc \theta} = \dfrac{\sec^2 \theta - 2 \sec \theta \tan \theta + \tan^2 \theta + 1}{\csc \theta \,(\sec \theta - \tan \theta)} = \dfrac{\sec^2 \theta - 2 \sec \theta \tan \theta + \sec^2 \theta}{\csc \theta \,(\sec \theta - \tan \theta)} =$

$\dfrac{2 \sec^2 \theta - 2 \sec \theta \tan \theta}{\csc \theta \,(\sec \theta - \tan \theta)} = \dfrac{2 \sec \theta \,(\sec \theta - \tan \theta)}{\csc \theta \,(\sec \theta - \tan \theta)} = \dfrac{2 \sec \theta}{\csc \theta} = 2\left(\dfrac{1}{\cos \theta} \cdot \dfrac{\sin \theta}{1}\right) = 2 \tan \theta$

86. $\dfrac{1}{1-\sin\theta}+\dfrac{1}{1+\sin\theta}=\dfrac{(1+\sin\theta)+(1-\sin\theta)}{(1-\sin\theta)(1+\sin\theta)}=\dfrac{2}{1-\sin^2\theta}=\dfrac{2}{\cos^2\theta}=2\sec^2\theta$

87. $\dfrac{1}{\tan\alpha-\sec\alpha}+\dfrac{1}{\tan\alpha+\sec\alpha}=\dfrac{(\tan\alpha+\sec\alpha)+(\tan\alpha-\sec\alpha)}{\tan^2\alpha-\sec^2\alpha}=$

$\dfrac{2\tan\alpha}{\tan^2\alpha-(\tan^2\alpha+1)}=\dfrac{2\tan\alpha}{-1}=-2\tan\alpha$

88. $\dfrac{\csc\theta+\cot\theta}{\tan\theta+\sin\theta}=\dfrac{\frac{1}{\sin\theta}+\frac{\cos\theta}{\sin\theta}}{\frac{\sin\theta}{\cos\theta}+\sin\theta}=\dfrac{\frac{1+\cos\theta}{\sin\theta}}{\frac{\sin\theta+\sin\theta\cos\theta}{\cos\theta}}=\dfrac{1+\cos\theta}{\sin\theta}\cdot\dfrac{\cos\theta}{\sin\theta\,(1+\cos\theta)}=\dfrac{\cos\theta}{\sin^2\theta}=$

$\dfrac{\cos\theta}{\sin\theta}\cdot\dfrac{1}{\sin\theta}=\cot\theta\csc\theta$

89. While the equation is true for the particular value $\theta=\dfrac{\pi}{2}$, it is not true in general. To be an identity, the

equation must be true in all cases for which the functions involved are defined.

90. By the transitive property of equality, if $a=b$ and $b=c$, then $a=c$. Each step must be reversible.

91. (a) Using an identity yields: $I=k\left(1-\sin^2\theta\right)$.

   (b) The function reaches its maximum value of k when $\cos^2\theta=1$. This happens when $\theta=0$.

92. (a) If $P=ky^2$ and $y=4\cos(2\pi t)$, then by substitution $P=k\left(4\cos(2\pi t)\right)^2\Rightarrow P=16k\cos^2(2\pi t)$.

   (b) Using an identity yields: $P=16k[1-\sin^2(2\pi t)]$.

93. (a) See Figure 93a. The total mechanical energy E is always 2. The spring has maximum potential energy

   when it is fully stretched but not moving. The spring has maximum kinetic energy when it is not stretched

   but is moving fastest.

   (b) Let $Y_1=P(t)$, $Y_2=k(t)$, $Y_3=E(t)=2$ for all inputs, see Figure 93b. The spring is stretched the most

   (has greatest potential energy) when $t=.25, .5, .75$, ect. At these times kinetic energy is 0.

   (c) $E(t)=P(t)+k(t)=2\cos^2(4\pi t)+2\sin^2(4\pi t)=2\left(\cos^2(4\pi t)+\sin^2(4\pi t)\right)=2(1)=2$

94. (a) See Figure 94a. The sum of L and C equals 3, total energy is always 3.

   (b) Let $Y_1=L(t)$, $Y_2=C(t)$, $Y_3=3$ for all inputs, see Figure 94b. Total energy is always 3.

   (c) $E(t)=L(t)+C(t)=3\cos^2(6,000,000\,t)+3\sin^2(6,000,000\,t)=$

   $3\left[\cos^2(6,000,000\,t)+\sin^2(6,000,000)\right]=3(1)=3$

$[0, .5]$ by $[-1, 3]$
Xscl = .1   Yscl = 1

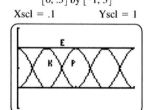

Figure 93a

Figure93b

$[0, 10^{-6}]$ by $[-1, 4]$
Xscl = 10^{-7} Yscl = 1

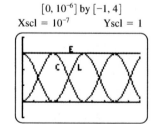

Figure 94a

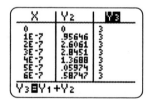

Figure 94b

9.2: Sum and Difference Identities

1. By cosine of a sum, $\cos(x+y) = \cos x \cos y - \sin x \sin y$, therefore F.

2. By cosine of a difference, $\cos(x-y) = \cos x \cos y + \sin x \sin y$, therefore A.

3. By sine of a sum, $\sin(x+y) = \sin x \cos y + \cos x \sin y$, therefore C.

4. By sine of a difference, $\sin(x-y) = \sin x \cos y - \cos x \sin y$, therefore D.

5. $\sin\dfrac{\pi}{12} = \sin\left(\dfrac{\pi}{3} - \dfrac{\pi}{4}\right) = \sin\dfrac{\pi}{3}\cos\dfrac{\pi}{4} - \cos\dfrac{\pi}{3}\sin\dfrac{\pi}{4} = \dfrac{\sqrt{3}}{2}\cdot\dfrac{\sqrt{2}}{2} - \dfrac{1}{2}\cdot\dfrac{\sqrt{2}}{2} = \dfrac{\sqrt{6}-\sqrt{2}}{4}$

6. $\tan\dfrac{\pi}{12} = \tan\left(\dfrac{\pi}{3} - \dfrac{\pi}{4}\right) = \dfrac{\tan\frac{\pi}{3} - \tan\frac{\pi}{4}}{1 + \tan\frac{\pi}{3}\tan\frac{\pi}{4}} = \dfrac{\sqrt{3}-1}{1+\sqrt{3}}\cdot\dfrac{1-\sqrt{3}}{1-\sqrt{3}} = \dfrac{-4+2\sqrt{3}}{-2} = 2-\sqrt{3}$

7. $\sin-\dfrac{5\pi}{12} = -\sin\dfrac{5\pi}{12} = -\sin\left(\dfrac{\pi}{6} + \dfrac{\pi}{4}\right) = -\left(\sin\dfrac{\pi}{6}\cos\dfrac{\pi}{4} + \cos\dfrac{\pi}{6}\sin\dfrac{\pi}{4}\right) =$

$-\left(\dfrac{1}{2}\cdot\dfrac{\sqrt{2}}{2} + \dfrac{\sqrt{3}}{2}\cdot\dfrac{\sqrt{2}}{2}\right) = \dfrac{-\sqrt{2}-\sqrt{6}}{4}$

8. $\tan\left(-\dfrac{5\pi}{12}\right) = -\tan\dfrac{5\pi}{12} = -\tan\left(\dfrac{\pi}{6} + \dfrac{\pi}{4}\right) = -\dfrac{\tan\frac{\pi}{6} + \tan\frac{\pi}{4}}{1 - \tan\frac{\pi}{6}\tan\frac{\pi}{4}} = -\dfrac{\frac{\sqrt{3}}{3}+1}{1-\frac{\sqrt{3}}{3}\cdot1}\cdot\dfrac{3}{3} = \dfrac{-3-\sqrt{3}}{3-\sqrt{3}}\cdot\dfrac{3+\sqrt{3}}{3+\sqrt{3}} =$

$\dfrac{-12-6\sqrt{3}}{6} = -2-\sqrt{3}$

9. $\sin\dfrac{13\pi}{12} = \sin\left(\dfrac{5\pi}{4} - \dfrac{\pi}{6}\right) = \sin\dfrac{5\pi}{4}\cos\dfrac{\pi}{6} - \cos\dfrac{5\pi}{4}\sin\dfrac{\pi}{6} = \dfrac{-\sqrt{2}}{2}\cdot\dfrac{\sqrt{3}}{2} - \dfrac{-\sqrt{2}}{2}\cdot\dfrac{1}{2} = \dfrac{-\sqrt{6}+\sqrt{2}}{4}$

10. $\cos\dfrac{13\pi}{12} = \cos\left(\dfrac{5\pi}{4} - \dfrac{\pi}{6}\right) = \cos\dfrac{5\pi}{4}\cos\dfrac{\pi}{6} + \sin\dfrac{5\pi}{4}\sin\dfrac{\pi}{6} = \dfrac{-\sqrt{2}}{2}\cdot\dfrac{\sqrt{3}}{2} + \dfrac{-\sqrt{2}}{2}\cdot\dfrac{1}{2} = \dfrac{-\sqrt{6}-\sqrt{2}}{4}$

11. $\cos 75° = \cos(45° + 30°) = \cos 45°\cos 30° - \sin 45°\sin 30° = \dfrac{\sqrt{2}}{2}\cdot\dfrac{\sqrt{3}}{2} - \dfrac{\sqrt{2}}{2}\cdot\dfrac{1}{2} = \dfrac{\sqrt{6}-\sqrt{2}}{4}$

12. $\sin 105° = \sin(60° + 45°) = \sin 60°\cos 45° + \cos 60°\sin 45° = \dfrac{\sqrt{3}}{2}\cdot\dfrac{\sqrt{2}}{2} + \dfrac{1}{2}\cdot\dfrac{\sqrt{2}}{2} = \dfrac{\sqrt{6}+\sqrt{2}}{4}$

13. $\tan 105° = \tan(60° + 45°) = \dfrac{\tan 60° + \tan 45°}{1 - \tan 60°\tan 45°} = \dfrac{\sqrt{3}+1}{1-\sqrt{3}}\cdot\dfrac{1+\sqrt{3}}{1+\sqrt{3}} = \dfrac{2\sqrt{3}+4}{-2} = -\sqrt{3}-2$

14. $\sin(-15°) = -\sin 15° = \sin(60° - 45°) = -[\sin 60°\cos 45° - \cos 60°\sin 45°] =$

$-\left[\dfrac{\sqrt{3}}{2}\cdot\dfrac{\sqrt{2}}{2} - \dfrac{1}{2}\cdot\dfrac{\sqrt{2}}{2}\right] = \dfrac{-\sqrt{6}+\sqrt{2}}{4}$

15. $\cos(-15°) = \cos 15° = \cos(60° - 45°) = \cos 60°\cos 45° + \sin 60°\sin 45° =$

$\dfrac{1}{2}\cdot\dfrac{\sqrt{2}}{2} + \dfrac{\sqrt{3}}{2}\cdot\dfrac{\sqrt{2}}{2} = \dfrac{\sqrt{2}+\sqrt{6}}{4}$

16. $\tan(-75°) = -\tan 75° = -\tan(45° + 30°) = -\dfrac{\tan 45° + \tan 30°}{1 - \tan 45°\tan 30°} =$

$\dfrac{-1-\frac{\sqrt{3}}{3}}{1-\frac{\sqrt{3}}{3}}\cdot\dfrac{3}{3} = \dfrac{-3-\sqrt{3}}{3-\sqrt{3}}\cdot\dfrac{3+\sqrt{3}}{3+\sqrt{3}} = \dfrac{-12-6\sqrt{3}}{6} = -2-\sqrt{3}$

17. $\cos\dfrac{\pi}{3}\cos\dfrac{2\pi}{3} - \sin\dfrac{\pi}{3}\sin\dfrac{2\pi}{3} = \cos\left(\dfrac{\pi}{3} + \dfrac{2\pi}{3}\right) = \cos\pi = -1$

18. $\cos\dfrac{7\pi}{8}\cos\dfrac{\pi}{8} + \sin\dfrac{7\pi}{8}\sin\dfrac{\pi}{8} = \cos\left(\dfrac{7\pi}{8} - \dfrac{\pi}{8}\right) = \cos\dfrac{6\pi}{8} = \cos\dfrac{3\pi}{4} = -\dfrac{\sqrt{2}}{2}$

19. $\sin 76° \cos 31° - \cos 76° \sin 31° = \sin(76° - 31°) = \sin 45° = \dfrac{\sqrt{2}}{2}$

20. $\sin 40° \cos 50° + \cos 40° \sin 50° = \sin(40° + 50°) = \sin 90° = 1$

21. $\dfrac{\tan 80° + \tan 55°}{1 - \tan 80° \tan 55°} = \tan(80° + 55°) = \tan 135° = -1$

22. $\dfrac{\tan 80° - \tan(-55°)}{1 - \tan 80° \tan(-55°)} = \tan(80° - (-55°)) = \tan 135° = -1$

23. $\sin(180° - x) = \sin 180° \cos x - \cos 180° \sin x = (0 \cdot \cos x) - (-1)\sin x = 0 - (-\sin x) = \sin x$

24. $\sin(270° + x) = \sin 270° \cos x + \cos 270° \sin x = (-1)\cos x + 0 \cdot \sin x = -\cos x$

25. $\cos(180° + x) = \cos 180° \cos x - \sin 180° \sin x = (-1)\cos x - (0 \cdot \sin x) = -\cos x$

26. $\cos(270° - x) = \cos 270° \cos x + \sin 270° \sin x = (0 \cdot \cos x) + (-1)\sin x = -\sin x$

27. $\sin(x - 90°) = \sin x \cos 90° - \cos x \sin 90° = (\sin x \cdot 0) - (\cos x \cdot 1) = -\cos x$

28. $\sin(x + 90°) = \sin x \cos 90° + \cos x \sin 90° = (\sin x \cdot 0) + (\cos x \cdot 1) = \cos x$

29. $\tan(180° - x) = \dfrac{\tan 180° - \tan x}{1 + \tan 180° \tan x} = \dfrac{0 - \tan x}{1 + (0 \cdot \tan x)} = \dfrac{-\tan x}{1} = -\tan x$

30. $\tan(360° - x) = \dfrac{\tan 360° - \tan x}{1 + \tan 360° \tan x} = \dfrac{0 - \tan x}{1 + (0 \cdot \tan x)} = \dfrac{-\tan x}{1} = -\tan x$

31. $\cos\left(\dfrac{\pi}{2} - x\right) = \cos\dfrac{\pi}{2} \cos x + \sin\dfrac{\pi}{2} \sin x = (0)(\cos x) + (1)(\sin x) = \sin x$

32. $\cos(\pi - x) = \cos \pi \cos x + \sin \pi \sin x = (-1)(\cos x) + (0)(\sin x) = -\cos x$

33. $\cos\left(\dfrac{3\pi}{2} + x\right) = \cos\dfrac{3\pi}{2} \cos x - \sin\dfrac{3\pi}{2} \sin x = (0)(\cos x) - (-1)(\sin x) = \sin x$

34. $\sin(\pi - x) = \sin \pi \cos x - \cos \pi \sin x = (0)(\cos x) - (-1)(\sin x) = \sin x$

35. $\sin(\pi + x) = \sin \pi \cos x + \cos \pi \sin x = (0)(\cos x) + (-1)(\sin x) = -\sin x$

36. $\tan(2\pi - x) = \dfrac{\tan 2\pi - \tan x}{1 + \tan 2\pi \tan x} = \dfrac{0 - \tan x}{1 + (0) \tan x} = -\tan x$

37. $\cos(135° - x) = \cos 135° \cos x + \sin 135° \sin x = -\dfrac{\sqrt{2}}{2} \cdot \cos x + \dfrac{\sqrt{2}}{2} \sin x = \dfrac{\sqrt{2}(\sin x - \cos x)}{2}$

38. $\sin(45° + x) = \sin 45° \cos x + \cos 45° \sin x = \dfrac{\sqrt{2}}{2} \cdot \cos x + \dfrac{\sqrt{2}}{2} \cdot \sin x = \dfrac{\sqrt{2}(\cos x + \sin x)}{2}$

39. $\tan(45° + x) = \dfrac{\tan 45° + \tan x}{1 - \tan 45° \tan x} = \dfrac{1 + \tan x}{1 - (1) \tan x} = \dfrac{1 + \tan x}{1 - \tan x}$

40. $\tan(\pi + x) = \dfrac{\tan \pi + \tan x}{1 - \tan \pi \tan x} = \dfrac{0 + \tan x}{1 - (0) \tan x} = \tan x$

41. $\tan(\pi - x) = \dfrac{\tan \pi - \tan x}{1 + \tan \pi \tan x} = \dfrac{0 - \tan x}{1 + (0) \tan x} = -\tan x$

42. $\sin\left(\dfrac{3\pi}{2} - x\right) = \sin\dfrac{3\pi}{2} \cos x - \cos\dfrac{3\pi}{2} \sin x = (-1)(\cos x) - (0)(\sin x) = -\cos x$

43. If $0 < A < \dfrac{\pi}{2}$ and $0 < B < \dfrac{\pi}{2}$, then A and B are both found in quadrant I.

    Next. using $\cos A = \dfrac{3}{5}$ and $\sin B = \dfrac{5}{13}$, find $\sin A$ and $\cos B$:

    $\sin^2 A = 1 - \cos^2 A = 1 - \left(\dfrac{3}{5}\right)^2 = 1 - \dfrac{9}{25} = \dfrac{16}{25} \Rightarrow \sin A = \pm\sqrt{\dfrac{16}{25}} \Rightarrow \sin A = \dfrac{4}{5};$

    $\cos^2 B = 1 - \sin^2 B = 1 - \left(\dfrac{5}{13}\right)^2 = 1 - \left(\dfrac{25}{169}\right) = \dfrac{144}{169} \Rightarrow \cos B = \pm\sqrt{\dfrac{144}{169}} \Rightarrow \cos B = \dfrac{12}{13}.$

    Now use $\cos A = \dfrac{3}{5}$, $\sin B = \dfrac{5}{13}$, $\sin A = \dfrac{4}{5}$, and $\cos B = \dfrac{12}{13}$ to find:

    (a) $\sin (A + B) = \sin A \cos B + \cos A \sin B = \dfrac{4}{5} \cdot \dfrac{12}{13} + \dfrac{3}{5} \cdot \dfrac{5}{13} = \dfrac{48 + 15}{65} \Rightarrow \sin (A + B) = \dfrac{63}{65}$

    (b) $\sin (A - B) = \sin A \cos B - \cos A \sin B = \dfrac{4}{5} \cdot \dfrac{12}{13} - \dfrac{3}{5} \cdot \dfrac{5}{13} = \dfrac{48 - 15}{65} \Rightarrow \sin (A - B) = \dfrac{33}{65}$

    (c) First find $\tan A$ and $\tan B$: $\tan A = \dfrac{\sin A}{\cos A} = \dfrac{\frac{4}{5}}{\frac{3}{5}} = \dfrac{4}{5} \cdot \dfrac{5}{3} = \dfrac{4}{3}$; $\tan B = \dfrac{\sin B}{\cos B} = \dfrac{\frac{5}{13}}{\frac{12}{13}} = \dfrac{5}{13} \cdot \dfrac{13}{12} = \dfrac{5}{12}$

    Now solve: $\tan (A + B) = \dfrac{\tan A + \tan B}{1 - \tan A \tan B} = \dfrac{\frac{4}{3} + \frac{5}{12}}{1 - \frac{4}{3} \cdot \frac{5}{12}} \cdot \dfrac{36}{36} = \dfrac{48 + 15}{36 - 20} \Rightarrow \tan (A + B) = \dfrac{63}{16}$

    (d) $\tan (A - B) = \dfrac{\tan A - \tan B}{1 + \tan A \tan B} = \dfrac{\frac{4}{3} - \frac{5}{12}}{1 + \frac{4}{3} \cdot \frac{5}{12}} \cdot \dfrac{36}{36} = \dfrac{48 - 15}{36 + 20} \Rightarrow \tan (A - B) = \dfrac{33}{56}$

    (e) Since $\sin (A + B) > 0$ and $\tan (A + B) > 0$, $A + B$ is found in quadrant I.

    (f) Since $\sin (A - B) > 0$ and $\tan (A - B) > 0$, $A - B$ is found in quadrant I.

44. If $0 < A < \dfrac{\pi}{2}$ and $\pi < B < \dfrac{3\pi}{2}$, then A is found in quadrant I and B in quadrant III.

    Next. using $\sin A = \dfrac{3}{5}$ and $\sin B = -\dfrac{12}{13}$, find $\cos A$ and $\cos B$:

    $\cos^2 A = 1 - \sin^2 A = 1 - \left(\dfrac{3}{5}\right)^2 = 1 - \dfrac{9}{25} = \dfrac{16}{25} \Rightarrow \cos A = \pm\sqrt{\dfrac{16}{25}} \Rightarrow \cos A = \dfrac{4}{5};$ (since A is in QI).

    $\cos^2 B = 1 - \sin^2 B = 1 - \left(-\dfrac{12}{13}\right)^2 = 1 - \left(\dfrac{144}{169}\right) = \dfrac{25}{169} \Rightarrow \cos B = \pm\sqrt{\dfrac{25}{169}} \Rightarrow \cos B = \pm\dfrac{5}{13}$, since B

    is in quadrant III, $\cos B = -\dfrac{5}{13}$. Now use $\sin A = \dfrac{3}{5}$, $\sin B = -\dfrac{12}{13}$, $\cos A = \dfrac{4}{5}$, and $\cos B = -\dfrac{5}{13}$ to find:

    (a) $\sin (A + B) = \sin A \cos B + \cos A \sin B = \dfrac{3}{5} \cdot \left(-\dfrac{5}{13}\right) + \dfrac{4}{5} \cdot \left(-\dfrac{12}{13}\right) = \dfrac{-15 - 48}{65} \Rightarrow \sin (A + B) = -\dfrac{63}{65}$

    (b) $\sin (A - B) = \sin A \cos B - \cos A \sin B = \dfrac{3}{5} \cdot \left(-\dfrac{5}{13}\right) - \dfrac{4}{5} \cdot \left(-\dfrac{12}{13}\right) = \dfrac{-15 + 48}{65} \Rightarrow \sin (A - B) = \dfrac{33}{65}$

    (c) First find $\tan A$ and $\tan B$: $\tan A = \dfrac{\sin A}{\cos A} = \dfrac{\frac{3}{5}}{\frac{4}{5}} = \dfrac{3}{5} \cdot \dfrac{5}{4} = \dfrac{3}{4}$; $\tan B = \dfrac{\sin B}{\cos B} = \dfrac{-\frac{12}{13}}{-\frac{5}{13}} = \left(-\dfrac{12}{13}\right)\left(-\dfrac{13}{5}\right) = \dfrac{12}{5}$

    Now solve: $\tan (A + B) = \dfrac{\tan A + \tan B}{1 - \tan A \tan B} = \dfrac{\frac{3}{4} + \frac{12}{5}}{1 - \frac{3}{4} \cdot \frac{12}{5}} \cdot \dfrac{20}{20} = \dfrac{15 + 48}{20 - 36} \Rightarrow \tan (A + B) = -\dfrac{63}{16}$

    (d) $\tan (A - B) = \dfrac{\tan A - \tan B}{1 + \tan A \tan B} = \dfrac{\frac{3}{4} - \frac{12}{5}}{1 + \frac{3}{4} \cdot \frac{12}{5}} \cdot \dfrac{20}{20} = \dfrac{15 - 48}{20 + 36} \Rightarrow \tan (A - B) = -\dfrac{33}{56}$

    (e) Since $\sin (A + B) < 0$ and $\tan (A + B) < 0$, $A + B$ is found in quadrant IV.

    (f) Since $\sin (A - B) > 0$ and $\tan (A - B) < 0$, $A - B$ is found in quadrant II.

45. If $\pi < A < \dfrac{3\pi}{2}$ and $\pi < B < \dfrac{3\pi}{2}$, then A and B are both found in quadrant III.

Next. using $\cos A = -\dfrac{8}{17}$ and $\cos B = -\dfrac{3}{5}$, find $\sin A$ and $\sin B$:

$$\sin^2 A = 1 - \cos^2 A = 1 - \left(-\dfrac{8}{17}\right)^2 = 1 - \dfrac{64}{289} = \dfrac{225}{289} \Rightarrow \sin A = \pm\sqrt{\dfrac{225}{289}} \Rightarrow \sin A = \pm\dfrac{15}{17},$$

since A is found in quadrant III, $\sin A = -\dfrac{15}{17}$.

$$\sin^2 B = 1 - \cos^2 B = 1 - \left(-\dfrac{3}{5}\right)^2 = 1 - \dfrac{9}{25} = \dfrac{16}{25} \Rightarrow \sin B = \pm\sqrt{\dfrac{16}{25}} \Rightarrow \sin B = \pm\dfrac{4}{5}, \text{ since } B$$

is in quadrant III, $\sin B = -\dfrac{4}{5}$. Now use $\cos A = -\dfrac{8}{17}, \cos B = -\dfrac{3}{5}, \sin A = -\dfrac{15}{17}$, and $\sin B = -\dfrac{4}{5}$ to find:

(a) $\sin(A + B) = \sin A \cos B + \cos A \sin B = \left(-\dfrac{15}{17}\right)\left(-\dfrac{3}{5}\right) + \left(-\dfrac{8}{17}\right)\left(-\dfrac{4}{5}\right) = \dfrac{45 + 32}{85}$

$$\Rightarrow \sin(A + B) = \dfrac{77}{85}$$

(b) $\sin(A - B) = \sin A \cos B - \cos A \sin B = \left(-\dfrac{15}{17}\right)\left(-\dfrac{3}{5}\right) - \left(-\dfrac{8}{17}\right)\left(-\dfrac{4}{5}\right) = \dfrac{45 - 32}{85}$

$$\Rightarrow \sin(A - B) = \dfrac{13}{85}$$

(c) First find $\tan A$ and $\tan B$:

$$\tan A = \dfrac{\sin A}{\cos A} = \dfrac{-\frac{15}{17}}{-\frac{8}{17}} = \left(-\dfrac{15}{17}\right)\left(-\dfrac{17}{8}\right) = \dfrac{15}{8}; \ \tan B = \dfrac{\sin B}{\cos B} = \dfrac{-\frac{4}{5}}{-\frac{3}{5}} = \left(-\dfrac{4}{5}\right)\left(-\dfrac{5}{3}\right) = \dfrac{4}{3}$$

Now solve: $\tan(A + B) = \dfrac{\tan A + \tan B}{1 - \tan A \tan B} = \dfrac{\frac{15}{8} + \frac{4}{3}}{1 - \frac{15}{8} \cdot \frac{4}{3}} \cdot \dfrac{24}{24} = \dfrac{45 + 32}{24 - 60} \Rightarrow \tan(A + B) = -\dfrac{77}{36}$

(d) $\tan(A - B) = \dfrac{\tan A - \tan B}{1 + \tan A \tan B} = \dfrac{\frac{15}{8} - \frac{4}{3}}{1 + \frac{15}{8} \cdot \frac{4}{3}} \cdot \dfrac{24}{24} = \dfrac{45 - 32}{24 + 60} \Rightarrow \tan(A - B) = \dfrac{13}{84}$

(e) Since $\sin(A + B) > 0$ and $\tan(A + B) < 0$, $A + B$ is found in quadrant II.

(f) Since $\sin(A - B) > 0$ and $\tan(A - B) > 0$, $A - B$ is found in quadrant I.

46. If $\dfrac{\pi}{2} < A < \pi$ and $0 < B < \dfrac{\pi}{2}$, then A is found in quadrant II and B in quadrant I.

Next. using $\cos A = -\dfrac{15}{17}$ and $\sin B = \dfrac{4}{5}$, find $\sin A$ and $\cos B$:

$$\sin^2 A = 1 - \cos^2 A = 1 - \left(-\dfrac{15}{17}\right)^2 = 1 - \dfrac{225}{289} = \dfrac{64}{289} \Rightarrow \sin A = \pm\sqrt{\dfrac{64}{289}} \Rightarrow \sin A = \dfrac{8}{17};$$

$$\cos^2 B = 1 - \sin^2 B = 1 - \left(\dfrac{4}{5}\right)^2 = 1 - \left(\dfrac{16}{25}\right) = \dfrac{9}{25} \Rightarrow \cos B = \pm\sqrt{\dfrac{9}{25}} \Rightarrow \cos B = \dfrac{3}{5}.$$

Now use $\cos A = -\dfrac{15}{17}, \sin B = \dfrac{4}{5}, \sin A = \dfrac{8}{17}$, and $\cos B = \dfrac{3}{5}$ to find:

(a) $\sin(A + B) = \sin A \cos B + \cos A \sin B = \dfrac{8}{17} \cdot \dfrac{3}{5} + \left(-\dfrac{15}{17}\right) \cdot \dfrac{4}{5} = \dfrac{24 - 60}{85} \Rightarrow \sin(A + B) = -\dfrac{36}{85}$

(b) $\sin(A - B) = \sin A \cos B - \cos A \sin B = \dfrac{8}{17} \cdot \dfrac{3}{5} - \left(-\dfrac{15}{17}\right) \cdot \dfrac{4}{5} = \dfrac{24 + 60}{85} \Rightarrow \sin(A - B) = \dfrac{84}{85}$

(c) First find $\tan A$ and $\tan B$: $\tan A = \dfrac{\sin A}{\cos A} = \dfrac{\frac{8}{17}}{-\frac{15}{17}} = \dfrac{8}{17}\left(-\dfrac{17}{15}\right) = -\dfrac{8}{15}$; $\tan B = \dfrac{\sin B}{\cos B} = \dfrac{\frac{4}{5}}{\frac{3}{5}} = \dfrac{4}{5} \cdot \dfrac{5}{3} = \dfrac{4}{3}$.

Now solve: $\tan(A + B) = \dfrac{\tan A + \tan B}{1 - \tan A \tan B} = \dfrac{-\frac{8}{15} + \frac{4}{3}}{1 - \left(-\frac{8}{15}\right) \cdot \frac{4}{3}} \cdot \dfrac{45}{45} = \dfrac{-24 + 60}{45 + 32} \Rightarrow \tan(A + B) = \dfrac{36}{77}$

(d) $\tan(A - B) = \dfrac{\tan A - \tan B}{1 + \tan A \tan B} = \dfrac{-\frac{8}{15} - \frac{4}{3}}{1 + \left(-\frac{8}{15}\right) \cdot \frac{4}{3}} \cdot \dfrac{45}{45} = \dfrac{-24 - 60}{45 - 32} \Rightarrow \tan(A - B) = -\dfrac{84}{13}$

(e) Since $\sin(A + B) < 0$ and $\tan(A + B) > 0$, $A + B$ is found in quadrant III.

(f) Since $\sin(A - B) > 0$ and $\tan(A - B) < 0$, $A - B$ is found in quadrant II.

47. $\sin(x + y) + \sin(x - y) = \sin x \cos y + \cos x \sin y + \sin x \cos y - \cos x \sin y = 2 \sin x \cos y$

48. $\tan(x - y) - \tan(y - x) = \dfrac{\tan x - \tan y}{1 + \tan x \tan y} - \dfrac{\tan y - \tan x}{1 + \tan y \tan x} = \dfrac{\tan x - \tan y - (\tan y - \tan x)}{1 + \tan x \tan y} =$

$\dfrac{\tan x - \tan y - \tan y + \tan x}{1 + \tan x \tan y} = \dfrac{2 \tan x - 2 \tan y}{1 + \tan x \tan y} = \dfrac{2(\tan x - \tan y)}{1 + \tan x \tan y}$

49. $\dfrac{\cos(A - B)}{\cos A \sin B} = \dfrac{\cos A \cos B + \sin A \sin B}{\cos A \sin B} = \dfrac{\cos A \cos B}{\cos A \sin B} + \dfrac{\sin A \sin B}{\cos A \sin B} = \dfrac{\cos B}{\sin B} + \dfrac{\sin A}{\cos A} =$

$\cot B + \tan A = \tan A + \cot B$

50. $\dfrac{\sin(A + B)}{\cos A \cos B} = \dfrac{\sin A \cos B + \cos A \sin B}{\cos A \cos B} = \dfrac{\sin A \cos B}{\cos A \cos B} + \dfrac{\cos A \sin B}{\cos A \cos B} = \dfrac{\sin A}{\cos A} + \dfrac{\sin B}{\cos B} = \tan A + \tan B$

51. $\dfrac{\sin(A - B)}{\sin B} + \dfrac{\cos(A - B)}{\cos B} = \dfrac{\sin A \cos B - \cos A \sin B}{\sin B} + \dfrac{\cos A \cos B + \sin A \sin B}{\cos B} =$

$\dfrac{\sin A \cos^2 B + \sin A \sin^2 B}{\sin B \cos B} = \dfrac{\sin A (\cos^2 B + \sin^2 B)}{\sin B \cos B} = \dfrac{\sin A}{\sin B \cos B}$

52. $\dfrac{\tan(A + B) - \tan B}{1 + \tan(A + B) \tan B} = \dfrac{\frac{\tan A + \tan B}{1 - \tan A \tan B} - \tan B}{1 + \left(\frac{\tan A + \tan B}{1 - \tan A \tan B}\right)(\tan B)} \cdot \dfrac{1 - \tan A \tan B}{1 - \tan A \tan B} =$

$\dfrac{\tan A + \tan B - \tan B(1 - \tan A \tan B)}{1 - \tan A \tan B + \tan B(\tan A + \tan B)} = \dfrac{\tan A + \tan B - \tan B + \tan A \tan^2 B}{1 - \tan A \tan B + \tan B \tan A + \tan^2 B} = \dfrac{\tan A + \tan A \tan^2 B}{1 + \tan^2 B} =$

$\dfrac{\tan A(1 + \tan^2 B)}{1 + \tan^2 B} = \tan A$

53. $\dfrac{\tan x - \tan y}{\tan x + \tan y} = \dfrac{\frac{\sin x}{\cos x} - \frac{\sin y}{\cos y}}{\frac{\sin x}{\cos x} + \frac{\sin y}{\cos y}} = \dfrac{\frac{\sin x \cos y - \cos x \sin y}{\cos x \cos y}}{\frac{\sin x \cos y + \cos x \sin y}{\cos x \cos y}} = \dfrac{\sin x \cos y - \cos x \sin y}{\sin x \cos y + \cos x \sin y} = \dfrac{\sin(x - y)}{\sin(x + y)}$

54. $\dfrac{1 + \cot A \cot B}{\cot A + \cot B} = \dfrac{1 + \frac{\cos A}{\sin A} \cdot \frac{\cos B}{\sin B}}{\frac{\cos A}{\sin A} + \frac{\cos B}{\sin B}} = \dfrac{1 + \frac{\cos A \cos B}{\sin A \sin B}}{\frac{\sin B \cos A + \sin A \cos B}{\sin A \sin B}} = \dfrac{\frac{\sin A \sin B + \cos A \cos B}{\sin A \sin B}}{\frac{\cos A \sin B + \sin A \cos B}{\sin A \sin B}} =$

$\dfrac{\sin A \sin B + \cos A \cos B}{\sin A \sin B} \cdot \dfrac{\sin A \sin B}{\cos A \sin B + \sin A \cos B} = \dfrac{\sin A \sin B + \cos A \cos B}{\cos A \sin B + \sin A \cos B} =$

$\dfrac{\cos A \cos B + \sin A \sin B}{\sin A \cos B + \cos A \sin B} = \dfrac{\cos(A - B)}{\sin(A + B)}$

55. Since there are 60 cycles per second, the number of cycles in .05 seconds is given by

$(.05 \text{ sec})(60 \text{ cycles/sec}) = 3$ cycles.

56. Since $V = 163 \sin wt$ and the maximum value of $\sin wt$ is 1, the maximum voltage is $163(1) = 163$ volts.

Since $V = 163 \sin wt$ and the minimum value of $\sin wt$ is -1, the maximum voltage is $163(-1) = -163$ volts.

Therefore, the voltage is not always equal to 115 volts.

57. (a) $F = \dfrac{.6(170)\sin(30° + 90°)}{\sin 12°} \approx 424.8659171 \approx 425$ pounds

    (b) Using the sine of a sum identity yields:

    $F = \dfrac{.6W\sin(\theta + 90°)}{\sin 12°} \Rightarrow F = \dfrac{.6W}{\sin 12°}(\sin\theta\cos 90° + \cos\theta\sin 90°) \approx$

    $2.8858W(\sin\theta(0) + \cos\theta(1)) \approx 2.9W\cos\theta.$

    (c) The function F is at a maximum when $\sin(\theta + 90°) = 1$: $\sin^{-1} 1 = \theta + 90° \Rightarrow 90° = \theta + 90° \Rightarrow \theta = 0°.$

58. (a) $F = \dfrac{.6(200)\sin(45° + 90°)}{\sin 12°} \approx 408.1195 \approx 408$ pounds.

    (b) Graph on a calculator: $y_1 = \dfrac{.6(200)\sin(x + 90°)}{\sin 12°}$ and $y_2 = 400$. From this graph, the intersection

    is at approximately $(46.128846, 400)$ or at approximately $46.1°$.

59. (a) Using the given values, $P = \dfrac{a}{r}\cos\left[\dfrac{2\pi r}{\lambda} - ct\right] = \dfrac{.4}{10}\cos\left[\dfrac{20\pi}{4.9} - 1026t\right]$. Graph this on

    $0 \le t \le 10$, see Figure 59a. At this distance, the pressure P is oscillating.

    (b) Using the given values, $P = \dfrac{a}{r}\cos\left[\dfrac{2\pi r}{\lambda} - ct\right] = \dfrac{3}{r}\cos\left[\dfrac{2\pi r}{4.9} - 10{,}260\right]$. Graph this on $0 \le t \le 20$,

    see Figure 59b. As the radius increases, the pressure P oscillates, and the amplitude decreases as r increases.

    (c) If $r = n\lambda$, then $P = \dfrac{a}{r}\cos\left[\dfrac{2\pi r}{\lambda} - ct\right] = \dfrac{a}{n\lambda}\cos\left[\dfrac{2\pi n\lambda}{\lambda} - ct\right] = \dfrac{a}{n\lambda}\cos\left[2\pi n - ct\right] =$

    $\dfrac{a}{n\lambda}\left[\cos(2\pi n)\cos(ct) + \sin(2\pi n)\sin(ct)\right] = \dfrac{a}{n\lambda}\left[(1)\cos(ct) + (0)\sin(ct)\right] = \dfrac{a}{n\lambda}\cos(ct).$

$[0, .05]$ by $[-.05, .05]$	$[0, 20]$ by $[-2, 2]$	$[0, .05]$ by $[-60, 60]$
Xscl = .01 Yscl = 1	Xscl = 1 Yscl = 1	Xscl = .01 Yscl = 10

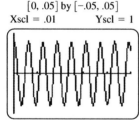 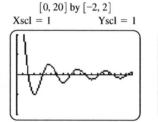 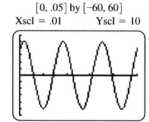

Figure 59a   Figure 59b   Figure 60

60. (a) Graph $V = 30\sin 120\pi t + 40\cos 120\pi t$ over the interval $0 \le t \le .05$, see Figure 60.

    (b) From the graph, the amplitude is 50, so let $a = 50$. Now estimate the phase shift by approximating the first

    t-intercept where the graph of V is increasing. This is located at $t \approx .0142$. Finally, find ϕ:

    $\sin(120\pi t + \phi) = 0 \Rightarrow \sin[120\pi(.0142) + \phi] = 0 \Rightarrow 120\pi(.0142) + \phi = 0 \Rightarrow$

    $\phi = -120\pi(.0142) \approx -5.353$. Therefore, $V = 50\sin(120\pi t - 5.353)$.

    (c) $50\sin(120\pi t - 5.353) = 50\left[\sin(120\pi t)(\cos 5.353) - \cos(120\pi t)(\sin 5.353)\right] =$

    $50\left[\sin(120\pi t)(.5977) - \cos(120\pi t)(-.8017)\right] = 29.885\sin(120\pi t) + 40.085\cos(120\pi t) \approx$

    $30\sin(120\pi t) + 40\cos(120\pi t)$

Reviewing Basic Concepts (Sections 9.1 and 9.2)

1. $\dfrac{\csc x}{\cot x} - \dfrac{\cot x}{\csc x} = \dfrac{1}{\sin x} \cdot \dfrac{\sin x}{\cos x} - \dfrac{\cos x}{\sin x} \cdot \dfrac{\sin x}{1} = \dfrac{1}{\cos x} - \cos x = \dfrac{1 - \cos^2 x}{\cos x} = \dfrac{\sin^2 x}{\cos x}$

2. $\tan\left(-\dfrac{\pi}{12}\right) = \tan\left(\dfrac{\pi}{4} - \dfrac{\pi}{3}\right) = \dfrac{\tan\left(\frac{\pi}{4}\right) - \tan\left(\frac{\pi}{3}\right)}{1 + \tan\left(\frac{\pi}{4}\right)\tan\left(\frac{\pi}{3}\right)} =$

   $\dfrac{1 - \sqrt{3}}{1 + 1(\sqrt{3})} = \dfrac{1 - \sqrt{3}}{1 + \sqrt{3}} \cdot \dfrac{1 - \sqrt{3}}{1 - \sqrt{3}} = \dfrac{4 - 2\sqrt{3}}{-2}$ or $\sqrt{3} - 2$

3. $\cos 18° \cos 108° + \sin 18° \sin 108° = \cos(18° - 108°) = \cos(-90°) = \cos 90° = 0$

4. $\sin\left(x - \dfrac{\pi}{4}\right) = \sin x \cos \dfrac{\pi}{4} - \cos x \sin \dfrac{\pi}{4} = \sin x\left(\dfrac{\sqrt{2}}{2}\right) - \cos x\left(\dfrac{\sqrt{2}}{2}\right) = \dfrac{\sqrt{2}}{2}(\sin x - \cos x)$

5. If A is found in quadrant II and $\sin A = \dfrac{2}{3}$, then $y = 2$, $r = 3$, and by pythagorean theorem $x^2 + 2^2 = 3^2 \Rightarrow$

   $x^2 + 4 = 9 \Rightarrow x^2 = 5 \Rightarrow x = \pm\sqrt{5}$. Because A is found in quadrant II, $x = -\sqrt{5}$, therefore

   $\sin A = \dfrac{2}{3}$, $\cos A = -\dfrac{\sqrt{5}}{3}$, and $\tan A = -\dfrac{2}{\sqrt{5}}$.

   If B in quadrant III and $\cos B = -\dfrac{1}{2}$, then $x = -1$, $r = 2$, and by pythagorean theorem $(-1)^2 + y^2 = 2^2 \Rightarrow$

   $1 + y^2 = 4 \Rightarrow y^2 = 3 \Rightarrow y = \pm\sqrt{3}$. Because B is found in quadrant III, $y = -\sqrt{3}$, therefore

   $\sin B = -\dfrac{\sqrt{3}}{2}$, $\cos B = -\dfrac{1}{2}$, and $\tan B = \dfrac{\sqrt{3}}{1} = \sqrt{3}$. Use these values to find each:

   $\sin(A + B) = \sin A \cos B + \cos A \sin B = \dfrac{2}{3}\left(-\dfrac{1}{2}\right) + \left(-\dfrac{\sqrt{5}}{3}\right)\left(-\dfrac{\sqrt{3}}{2}\right) = -\dfrac{2}{6} + \dfrac{\sqrt{15}}{6}$

   $\Rightarrow \sin(A + B) = \dfrac{-2 + \sqrt{15}}{6}$;

   $\cos(A - B) = \cos A \cos B + \sin A \sin B = \left(-\dfrac{\sqrt{5}}{3}\right)\left(-\dfrac{1}{2}\right) + \dfrac{2}{3}\left(-\dfrac{\sqrt{3}}{2}\right) = \dfrac{\sqrt{5}}{6} - \dfrac{2\sqrt{3}}{6}$

   $\Rightarrow \cos(A - B) = \dfrac{\sqrt{5} - 2\sqrt{3}}{6}$;

   $\tan(A - B) = \dfrac{\tan A - \tan B}{1 + \tan A \tan B} = \dfrac{-\frac{2}{\sqrt{5}} - \sqrt{3}}{1 + \left(-\frac{2}{\sqrt{5}}\right)\cdot(\sqrt{3})} \cdot \dfrac{\sqrt{5}}{\sqrt{5}} \Rightarrow \tan(A - B) = \dfrac{-2 - \sqrt{15}}{\sqrt{5} - 2\sqrt{3}}$.

6. $\csc^2 \theta - \cot^2 \theta = 1 + \cot^2 \theta - \cot^2 \theta = 1$

7. $\dfrac{\sin t}{1 - \cos t} = \dfrac{\sin t}{1 - \cos t} \cdot \dfrac{1 + \cos t}{1 + \cos t} = \dfrac{\sin t(1 + \cos t)}{1 - \cos^2 t} = \dfrac{\sin t(1 + \cos t)}{\sin^2 t} = \dfrac{1 + \cos t}{\sin t}$

8. $\dfrac{\cot A - \tan A}{\csc A \sec A} = \dfrac{\frac{\cos A}{\sin A} - \frac{\sin A}{\cos A}}{\frac{1}{\sin A} \cdot \frac{1}{\cos A}} \cdot \dfrac{\sin A \cos A}{\sin A \cos A} = \dfrac{\cos^2 A - \sin^2 A}{1} = \cos^2 A - \sin^2 A$

9. $\dfrac{\sin(x - y)}{\sin x \sin y} = \dfrac{\sin x \cos y - \cos x \sin y}{\sin x \sin y} = \dfrac{\sin x \cos y}{\sin x \sin y} - \dfrac{\cos x \sin y}{\sin x \sin y} = \dfrac{\cos y}{\sin y} - \dfrac{\cos x}{\sin x} = \cot y - \cot x$

10. $e = 20 \sin\left(\dfrac{\pi t}{4} - \dfrac{\pi}{2}\right) = 20\left[\sin \dfrac{\pi t}{4} \cos \dfrac{\pi}{2} - \cos \dfrac{\pi t}{4} \sin \dfrac{\pi}{2}\right] = 20\left[\sin \dfrac{\pi t}{4}(0) - \cos \dfrac{\pi t}{4}(1)\right] = -20 \cos \dfrac{\pi t}{4}$

9.3: Further Identities

1. First find $\cos\theta$: if $\sin\theta = \dfrac{2}{5}$, then $\cos^2\theta = 1 - \sin^2\theta = 1 - \left(\dfrac{2}{5}\right)^2 = 1 - \dfrac{4}{25} \Rightarrow \cos\theta = \pm\dfrac{\sqrt{21}}{5}$, since

   $\cos\theta < 0$, $\cos\theta = -\dfrac{\sqrt{21}}{5}$. Now use $\sin\theta = \dfrac{2}{5}$ and $\cos\theta = -\dfrac{\sqrt{21}}{5}$ to find:

   (a) $\sin 2\theta = 2\sin\theta\cos\theta = 2\left(\dfrac{2}{5}\right)\left(-\dfrac{\sqrt{21}}{5}\right) = -\dfrac{4\sqrt{21}}{25}$

   (b) $\cos 2\theta = \cos^2\theta - \sin^2\theta = \left(-\dfrac{\sqrt{21}}{5}\right)^2 - \left(\dfrac{2}{5}\right)^2 = \dfrac{21}{25} - \dfrac{4}{25} = \dfrac{17}{25}$

2. First find $\sin\theta$: if $\cos\theta = -\dfrac{12}{13}$, then $\sin^2\theta = 1 - \cos^2\theta = 1 - \left(-\dfrac{12}{13}\right)^2 = 1 - \dfrac{144}{169} = \dfrac{25}{169} \Rightarrow$

   $\sin\theta = \pm\dfrac{5}{13}$, since $\sin\theta > 0$, $\sin\theta = \dfrac{5}{13}$. Now use $\sin\theta = \dfrac{5}{13}$ and $\cos\theta = -\dfrac{12}{13}$ to find:

   (a) $\sin 2\theta = 2\sin\theta\cos\theta = 2\left(\dfrac{5}{13}\right)\left(-\dfrac{12}{13}\right) = -\dfrac{120}{169}$

   (b) $\cos 2\theta = \cos^2\theta - \sin^2\theta = \left(-\dfrac{12}{13}\right)^2 - \left(\dfrac{5}{13}\right)^2 = \dfrac{144}{169} - \dfrac{25}{169} = \dfrac{119}{169}$

3. First find $\sin\theta$ and $\cos\theta$: if $\tan\theta = 2$, then $\sec^2\theta = 1 + \tan^2\theta = 1 + 2^2 = 1 + 4 = 5 \Rightarrow \sec\theta = \pm\sqrt{5}$,

   since $\cos\theta > 0$, θ is in quadrant I and $\sec\theta = \sqrt{5}$. Now use this to find $\cos\theta$:

   $\cos\theta = \dfrac{1}{\sec\theta} = \dfrac{1}{\sqrt{5}} = \dfrac{\sqrt{5}}{5}$; and this to find $\sin x$: $\sin^2\theta = 1 - \cos^2\theta = 1 - \left(\dfrac{\sqrt{5}}{5}\right)^2 = 1 - \dfrac{5}{25} = \dfrac{20}{25} = \dfrac{4}{5}$

   $\Rightarrow \sin\theta = \pm\sqrt{\dfrac{4}{5}} = \pm\dfrac{2\sqrt{5}}{5}$, $\sin\theta = \dfrac{2\sqrt{5}}{5}$ (θ is in quadrant I) Use $\sin x = \dfrac{2\sqrt{5}}{5}$ and $\cos x = \dfrac{\sqrt{5}}{5}$ to find:

   (a) $\sin 2\theta = 2\sin\theta\cos\theta = 2\left(\dfrac{2\sqrt{5}}{5}\right)\left(\dfrac{\sqrt{5}}{5}\right) = \dfrac{20}{25} = \dfrac{4}{5}$

   (b) $\cos 2\theta = \cos^2\theta - \sin^2\theta = \left(\dfrac{\sqrt{5}}{5}\right)^2 - \left(\dfrac{2\sqrt{5}}{5}\right)^2 = \dfrac{5}{25} - \dfrac{20}{25} = -\dfrac{15}{25} = -\dfrac{3}{5}$

4. First find $\sin\theta$ and $\cos\theta$: if $\tan\theta = \dfrac{5}{3}$, then $\sec^2\theta = 1 + \tan^2\theta = 1 + \left(\dfrac{5}{3}\right)^2 = 1 + \dfrac{25}{9} = \dfrac{34}{9} \Rightarrow$

   $\sec\theta = \pm\dfrac{\sqrt{34}}{3}$, since $\sin\theta < 0$, θ is in quadrant III and $\sec\theta = -\dfrac{\sqrt{34}}{3}$. Now use this to find $\cos\theta$:

   $\cos\theta = \dfrac{1}{\sec\theta} = -\dfrac{3}{\sqrt{34}}$; and this to find $\sin\theta$: $\sin^2\theta = 1 - \cos^2\theta = 1 - \left(-\dfrac{3}{\sqrt{34}}\right)^2 = 1 - \dfrac{9}{34} = \dfrac{25}{34}$

   $\Rightarrow \sin\theta = \sqrt{\dfrac{25}{34}} = \dfrac{5}{\sqrt{34}}$, since θ is in quadrant III, $\sin\theta = -\dfrac{5}{\sqrt{34}}$.

   Finally, use $\sin\theta = -\dfrac{5}{\sqrt{34}}$ and $\cos\theta = -\dfrac{3}{\sqrt{34}}$ to find:

   (a) $\sin 2\theta = 2\sin\theta\cos\theta = 2\left(-\dfrac{5}{\sqrt{34}}\right)\left(-\dfrac{3}{\sqrt{34}}\right) = \dfrac{30}{34} = \dfrac{15}{17}$

   (b) $\cos 2\theta = \cos^2\theta - \sin^2\theta = \left(-\dfrac{3}{\sqrt{34}}\right)^2 - \left(-\dfrac{5}{\sqrt{34}}\right)^2 = \dfrac{9}{34} - \dfrac{25}{34} = -\dfrac{16}{34} = -\dfrac{8}{17}$

5.   First find $\cos \theta$: if $\sin \theta = -\dfrac{\sqrt{5}}{7}$, then $\cos^2 \theta = 1 - \sin^2 \theta = 1 - \left(-\dfrac{\sqrt{5}}{7}\right)^2 = 1 - \dfrac{5}{49} = \dfrac{44}{49} \Rightarrow$

$\cos \theta = \pm\sqrt{\dfrac{44}{49}} = \pm\dfrac{2\sqrt{11}}{7}$, since $\cos \theta > 0$, $\cos \theta = \dfrac{2\sqrt{11}}{7}$.   Now use $\sin \theta = -\dfrac{\sqrt{5}}{7}$ and $\cos \theta = \dfrac{2\sqrt{11}}{7}$:

(a) $\sin 2\theta = 2 \sin \theta \cos \theta = 2\left(-\dfrac{\sqrt{5}}{7}\right)\left(\dfrac{2\sqrt{11}}{7}\right) = -\dfrac{4\sqrt{55}}{49}$

(b) $\cos 2\theta = 1 - 2 \sin^2 \theta = 1 - 2\left(-\dfrac{\sqrt{5}}{7}\right)^2 = 1 - 2\left(\dfrac{5}{49}\right) = \dfrac{39}{49}$

6.   First find $\sin \theta$: if $\cos \theta = \dfrac{\sqrt{3}}{5}$, then $\sin^2 \theta = 1 - \cos^2 \theta = 1 - \left(\dfrac{\sqrt{3}}{5}\right)^2 = 1 - \dfrac{3}{25} = \dfrac{22}{25} \Rightarrow$

$\sin \theta = \pm\dfrac{\sqrt{22}}{5}$, since $\sin \theta > 0$, $\sin \theta = \dfrac{\sqrt{22}}{5}$.   Now use $\sin \theta = \dfrac{\sqrt{22}}{5}$ and $\cos \theta = \dfrac{\sqrt{3}}{5}$ to find:

(a) $\sin 2\theta = 2 \sin \theta \cos \theta = 2\left(\dfrac{\sqrt{22}}{5}\right)\left(\dfrac{\sqrt{3}}{5}\right) = \dfrac{2\sqrt{66}}{25}$

(b) $\cos 2\theta = \cos^2 \theta - \sin^2 \theta = \left(\dfrac{\sqrt{3}}{5}\right)^2 - \left(\dfrac{\sqrt{22}}{5}\right)^2 = \dfrac{3}{25} - \dfrac{22}{25} = -\dfrac{19}{25}$

7. $\cos^2 15° - \sin^2 15° = \cos[2(15°)] = \cos 30° = \dfrac{\sqrt{3}}{2}$

8. $\dfrac{2 \tan 15°}{1 - \tan^2 15°} = \tan[2(15°)] = \tan 30° = \dfrac{\sqrt{3}}{3}$

9. $1 - 2 \sin^2 15° = \cos[2(15°)] = \cos 30° = \dfrac{\sqrt{3}}{2}$

10. $1 - 2 \sin^2 22.5° = \cos[2(22.5°)] = \cos 45° = \dfrac{\sqrt{2}}{2}$

11. $2 \cos^2 67.5° - 1 = \cos[2(67.5°)] = \cos 135° = -\dfrac{\sqrt{2}}{2}$

12. $\cos^2 \dfrac{\pi}{8} - \dfrac{1}{2} = \dfrac{1}{2}\left(2 \cos^2 \dfrac{\pi}{8} - 1\right) = \dfrac{1}{2}\cos\left[2\left(\dfrac{\pi}{8}\right)\right] = \dfrac{1}{2}\cos \dfrac{\pi}{4} = \dfrac{1}{2} \cdot \dfrac{\sqrt{2}}{2} = \dfrac{\sqrt{2}}{4}$

13. $\dfrac{\tan 51°}{1 - \tan^2 51°} = \dfrac{1}{2} \cdot \dfrac{2 \tan 51°}{1 - \tan^2 51°} = \dfrac{1}{2}\tan[2 \cdot 51°] = \dfrac{1}{2}\tan 102°$

14. $\dfrac{\tan 34°}{2(1 - \tan^2 34°)} = \dfrac{1}{4} \cdot \dfrac{4 \tan 34°}{2(1 - \tan^2 34°)} = \dfrac{1}{4} \cdot \dfrac{2 \tan 34°}{1 - \tan^2 34°} = \dfrac{1}{4}\tan[2 \cdot 34°] = \dfrac{1}{4}\tan 68°$

15. $\dfrac{1}{4} - \dfrac{1}{2}\sin^2 47.1° = \dfrac{1}{4}\left[4\left(\dfrac{1}{4} - \dfrac{1}{2}\sin^2 47.1°\right)\right] = \dfrac{1}{4}[1 - 2 \sin^2 47.1°] = \dfrac{1}{4}\cos[2(47.1°)] = \dfrac{1}{4}\cos 94.2°$

16. $\dfrac{1}{8}\sin 29.5° \cos 29.5° = \dfrac{1}{16}\left[16\left(\dfrac{1}{8}\sin 29.5° \cos 29.5°\right)\right] = \dfrac{1}{16}[2 \sin 29.5° \cos 29.5°] =$

$\dfrac{1}{16}[\sin(2 \cdot 29.5°)] = \dfrac{1}{16}\sin 59°$

17. $4 \sin 15° \cos 15° = 2 \sin 2(15°) = 2 \sin 30° = 2 \cdot \dfrac{1}{2} = 1$

18. $2 - 4 \sin^2 15° = 2(1 - 2 \sin^2 15°) = 2(\cos[2(15°)]) = 2 \cos 30° = 2 \cdot \dfrac{\sqrt{3}}{2} = \sqrt{3}$

19.   When graphing the equation on the calculator, the graph looks like that of $\cos 2x$. Verify this result:

$\cos^4 x - \sin^4 x = (\cos^2 x - \sin^2 x)(\cos^2 x + \sin^2 x) = (\cos^2 x - \sin^2 x)(1) = \cos 2x$

20. When graphing the equation on the calculator, the graph looks like that of $\sin 2x$. Verify this result:

$$\frac{4 \tan x \cos^2 x - 2 \tan x}{1 - \tan^2 x} = \frac{4 \frac{\sin x}{\cos x} \cos^2 x - 2 \frac{\sin x}{\cos x}}{1 - \frac{\sin^2 x}{\cos^2 x}} = \frac{4 \sin x \cos x - 2 \frac{\sin x}{\cos x}}{\frac{\cos^2 x - \sin^2 x}{\cos^2 x}} \cdot \frac{\cos^2 x}{\cos^2 x} =$$

$$\frac{4 \sin x \cos^3 x - 2 \sin x \cos x}{\cos^2 x - \sin^2 x} = \frac{2 \sin x \cos x \left[2 \cos^2 x - 1\right]}{\cos 2x} = \frac{\sin 2x \cos 2x}{\cos 2x} = \sin 2x$$

21. $\cos 3x = \cos (2x + x) = \cos 2x \cos x - \sin 2x \sin x = (2 \cos^2 x - 1) \cos x - (2 \sin x \cos x) \sin x =$

    $2 \cos^3 x - \cos x - 2 \sin^2 x \cos x = 2 \cos^3 x - \cos x - 2 (1 - \cos^2 x) \cos x =$

    $2 \cos^3 x - \cos x - 2 \cos x + 2 \cos^3 x = 4 \cos^3 x - 3 \cos x$. The graphs support this.

22. $\tan 3x = \tan (2x + x) = \dfrac{\tan 2x + \tan x}{1 - \tan 2x \cdot \tan x} = \dfrac{\frac{2 \tan x}{1 - \tan^2 x} + \tan x}{1 - \frac{2 \tan x}{1 - \tan^2 x} \cdot \tan x} \cdot \dfrac{1 - \tan^2 x}{1 - \tan^2 x} = \dfrac{2 \tan x + \tan x - \tan^3 x}{1 - \tan^2 x - 2 \tan^2 x} =$

    $\dfrac{3 \tan x - \tan^3 x}{1 - 3 \tan^2 x}$. The graph of each equation supports that they are the equal.

23. $\tan 4x = \tan 2 (2x) = \dfrac{2 \tan 2x}{1 - \tan^2 2x} = \dfrac{2 \left(\frac{2 \tan x}{1 - \tan^2 x}\right)}{1 - \left(\frac{2 \tan x}{1 - \tan^2 x}\right)^2} \cdot \dfrac{(1 - \tan^2 x)^2}{(1 - \tan^2 x)^2} = \dfrac{4 \tan x (1 - \tan^2 x)}{(1 - \tan^2 x)^2 - 4 \tan^2 x} =$

    $\dfrac{4 \tan x - 4 \tan^3 x}{1 - 2 \tan^2 x + \tan^4 x - 4 \tan^2 x} = \dfrac{4 \tan x - 4 \tan^3 x}{1 - 6 \tan^2 x + \tan^4 x}.$

    The graph of each equation supports that they are equal.

24. $\cos 4x = \cos 2 (2x) = 2 (\cos 2x)^2 - 1 = 2 (2 \cos^2 x - 1)^2 - 1 = 2 (4 \cos^4 x - 4 \cos^2 x + 1) - 1 =$

    $8 \cos^4 x - 8 \cos^2 x + 2 - 1 = 8 \cos^4 x - 8 \cos^2 x + 1.$

    The graph of each equation supports that they are equal.

25. Since $\sin \dfrac{\pi}{12}$ is found in quadrant I, the sine value will be positive, therefore:

    $\sin \dfrac{\pi}{12} = \sin \left(\dfrac{\pi/6}{2}\right) = \sqrt{\dfrac{1 - \cos \frac{\pi}{6}}{2}} = \sqrt{\dfrac{1 - \frac{\sqrt{3}}{2}}{2} \cdot \dfrac{2}{2}} = \sqrt{\dfrac{2 - \sqrt{3}}{4}} = \dfrac{\sqrt{2 - \sqrt{3}}}{2}.$

26. Since $\cos \dfrac{\pi}{8}$ is found in quadrant I, the cosine value will be positive, therefore:

    $\cos \dfrac{\pi}{8} = \cos \left(\dfrac{\pi/4}{2}\right) = \sqrt{\dfrac{1 + \cos \frac{\pi}{4}}{2}} = \sqrt{\dfrac{1 + \frac{\sqrt{2}}{2}}{2} \cdot \dfrac{2}{2}} = \sqrt{\dfrac{2 + \sqrt{2}}{4}} = \dfrac{\sqrt{2 + \sqrt{2}}}{2}.$

27. $\tan \left(-\dfrac{\pi}{8}\right) = -\tan \left(\dfrac{\pi}{8}\right) = -\tan \left(\dfrac{\pi/4}{2}\right) = -\left(\dfrac{1 - \cos \frac{\pi}{4}}{\sin \frac{\pi}{4}}\right) = -\dfrac{1 - \frac{\sqrt{2}}{2}}{\frac{\sqrt{2}}{2}} \cdot \dfrac{2}{2} =$

    $-\dfrac{2 - \sqrt{2}}{\sqrt{2}} \cdot \dfrac{\sqrt{2}}{\sqrt{2}} = -\dfrac{2\sqrt{2} - 2}{2} = 1 - \sqrt{2}.$ An Alternate method yields: $-\sqrt{3 - 2\sqrt{2}}.$

28. Since $\cos 67.5°$ is found in quadrant I, the cosine value will be positive, therefore:

    $\cos 67.5° = \cos \left(\dfrac{135°}{2}\right) = \sqrt{\dfrac{1 + \cos 135°}{2}} = \sqrt{\dfrac{1 + \left(-\frac{\sqrt{2}}{2}\right)}{2} \cdot \dfrac{2}{2}} = \sqrt{\dfrac{2 - \sqrt{2}}{4}} = \dfrac{\sqrt{2 - \sqrt{2}}}{2}.$

29. Since $\sin 67.5°$ is found in quadrant I, the sine value will be positive, therefore:

    $\sin 67.5° = \sin \left(\dfrac{135°}{2}\right) = \sqrt{\dfrac{1 - \cos 135°}{2}} = \sqrt{\dfrac{1 - \left(-\frac{\sqrt{2}}{2}\right)}{2} \cdot \dfrac{2}{2}} = \sqrt{\dfrac{2 + \sqrt{2}}{4}} = \dfrac{\sqrt{2 + \sqrt{2}}}{2}.$

30. $\tan 195° = \tan \left(\dfrac{390°}{2}\right) = \left(\dfrac{1 - \cos 390°}{\sin 390°}\right) = \left(\dfrac{1 - \cos 30°}{\sin 30°}\right) = \dfrac{1 - \frac{\sqrt{3}}{2}}{\frac{1}{2}} \cdot \dfrac{2}{2} = 2 - \sqrt{3}.$

    An Alternate method yields: $\sqrt{7 - 4\sqrt{3}}.$

31. If $\cos x = \dfrac{1}{4}$ and $0 < x < \dfrac{\pi}{2}$, then x is found in quadrant I. With x in quadrant I the cosine value will be

positive, therefore: $\cos \dfrac{x}{2} = \sqrt{\dfrac{1 + \cos x}{2}} = \sqrt{\dfrac{1 + \frac{1}{4}}{2} \cdot \dfrac{4}{4}} = \sqrt{\dfrac{4 + 1}{8}} = \dfrac{\sqrt{5}}{\sqrt{8}} \cdot \dfrac{\sqrt{2}}{\sqrt{2}} = \dfrac{\sqrt{10}}{4}.$

32. If $\cos x = -\dfrac{5}{8}$ and $\dfrac{\pi}{2} < x < \pi$, then x is found in quadrant II. With x in quadrant II the sine value will be

positive, therefore: $\sin \dfrac{x}{2} = \sqrt{\dfrac{1 - \cos x}{2}} = \sqrt{\dfrac{1 - (-\frac{5}{8})}{2} \cdot \dfrac{8}{8}} = \sqrt{\dfrac{8 + 5}{16}} = \dfrac{\sqrt{13}}{4}.$

33. If $\sin x = \dfrac{3}{5}$ and $\dfrac{\pi}{2} < x < \pi$, then x is found in quadrant II. With x in quadrant II the cosine value will be

negative, therefore: $\cos x = -\sqrt{1 - \sin^2 x} = -\sqrt{1 - \dfrac{9}{25}} = -\sqrt{\dfrac{16}{25}} = -\dfrac{4}{5}.$

Now use $\sin x = \dfrac{3}{5}$ and $\cos x = -\dfrac{4}{5}$ to find $\tan \dfrac{x}{2}$: $\tan \dfrac{x}{2} = \dfrac{1 - \cos \theta}{\sin \theta} = \dfrac{1 - (-\frac{4}{5})}{\frac{3}{5}} \cdot \dfrac{5}{5} = \dfrac{5 + 4}{3} = 3.$

34. If $\sin x = -\dfrac{4}{5}$ and $\dfrac{3\pi}{2} < x < 2\pi$, then x is found in quadrant IV. With x in quadrant IV the cosine value will

be positive, therefore: $\cos x = \sqrt{1 - \sin^2 x} = \sqrt{1 - \dfrac{16}{25}} = \sqrt{\dfrac{9}{25}} = \dfrac{3}{5}.$

Now since $\dfrac{3\pi}{4} < \dfrac{x}{2} < \pi$ is in quadrant II the cosine value will be negative and we can use $\sin x = -\dfrac{4}{5}$

and $\cos x = \dfrac{3}{5}$ to find $\cos \dfrac{x}{2}$: $\cos \dfrac{x}{2} = -\sqrt{\dfrac{1 + \cos \theta}{2}} = -\sqrt{\dfrac{1 + \frac{3}{5}}{2} \cdot \dfrac{5}{5}} = -\sqrt{\dfrac{5 + 3}{10}} = -\dfrac{\sqrt{8}}{\sqrt{10}} \cdot \dfrac{\sqrt{10}}{\sqrt{10}} =$

$-\dfrac{4\sqrt{5}}{10} = -\dfrac{2\sqrt{5}}{5}.$

35. If $\tan x = \dfrac{\sqrt{7}}{3}$ and $\pi < x < \dfrac{3\pi}{2}$, then x is found in quadrant III. With x in quadrant III both sine and secant

values are negative, therefore: $\sec x = -\sqrt{1 + \tan^2 x} = -\sqrt{1 + \dfrac{7}{9}} = -\sqrt{\dfrac{16}{9}} = -\dfrac{4}{3}$, therefore $\cos x = -\dfrac{3}{4}$;

and $\sin x = -\sqrt{1 - \cos^2 x} = -\sqrt{1 - \dfrac{9}{16}} = -\sqrt{\dfrac{7}{16}} = -\dfrac{\sqrt{7}}{4}.$ Now use $\sin x = -\dfrac{\sqrt{7}}{4}$ and $\cos x = -\dfrac{3}{4}$

to find $\tan \dfrac{x}{2}$: $\tan \dfrac{x}{2} = \dfrac{1 - \cos x}{\sin x} = \dfrac{1 - (-\frac{3}{4})}{\frac{-\sqrt{7}}{4}} \cdot \dfrac{4}{4} = \dfrac{4 + 3}{-\sqrt{7}} \cdot \dfrac{\sqrt{7}}{\sqrt{7}} = \dfrac{7\sqrt{7}}{-7} = -\sqrt{7}.$

36. If $\tan x = -\dfrac{\sqrt{5}}{2}$ and $\dfrac{\pi}{2} < x < \pi$, then x is found in quadrant II. With x in quadrant II the sine value

is positive and the secant value is negative, therefore: $\sec x = -\sqrt{1 + \tan^2 x} = -\sqrt{1 + \dfrac{5}{4}} = -\sqrt{\dfrac{9}{4}} = -\dfrac{3}{2}$,

therefore $\cos x = -\dfrac{2}{3}$; and $\sin x = \sqrt{1 - \cos^2 x} = \sqrt{1 - \dfrac{4}{9}} = \sqrt{\dfrac{5}{9}} = \dfrac{\sqrt{5}}{3}.$ Now use $\sin x = \dfrac{\sqrt{5}}{3}$ and

$\cos x = -\dfrac{2}{3}$ to find $\tan \dfrac{x}{2}$: $\tan \dfrac{x}{2} = \dfrac{1 - \cos x}{\sin x} = \dfrac{1 - (-\frac{2}{3})}{\frac{\sqrt{5}}{3}} \cdot \dfrac{3}{3} = \dfrac{3 + 2}{\sqrt{5}} \cdot \dfrac{\sqrt{5}}{\sqrt{5}} = \dfrac{5\sqrt{5}}{5} = \sqrt{5}.$

37. $\tan \dfrac{A}{2} = \dfrac{\sin A}{1 + \cos A} = \dfrac{\sin A}{1 + \cos A} \cdot \dfrac{1 - \cos A}{1 - \cos A} = \dfrac{\sin A (1 - \cos A)}{1 - \cos^2 A} = \dfrac{\sin A (1 - \cos A)}{\sin^2 A} = \dfrac{1 - \cos A}{\sin A}$

38. Verify that $\sqrt{3 - 2\sqrt{2}} = \sqrt{2} - 1$: $(\sqrt{3 - 2\sqrt{2}})^2 = 3 - 2\sqrt{2}$, $(\sqrt{2} - 1)^2 = 2 - 2\sqrt{2} + 1 = 3 - 2\sqrt{2}$.

39. (a) The function $\tan \dfrac{\pi}{2}$ is undefined, so it cannot be used.

    (b) $\tan\left(\dfrac{\pi}{2} + x\right) = \dfrac{\sin\left(\frac{\pi}{2} + x\right)}{\cos\left(\frac{\pi}{2} + x\right)}$

    (c) $\tan\left(\dfrac{\pi}{2} + x\right) = \dfrac{\sin\frac{\pi}{2}\cos x + \cos\frac{\pi}{2}\sin x}{\cos\frac{\pi}{2}\cos x - \sin\frac{\pi}{2}\sin x} = \dfrac{1 \cdot \cos x + 0 \cdot \sin x}{0 \cdot \cos x - 1 \cdot \sin x} = \dfrac{\cos x}{-\sin x} = -\cot x$

40. (a) $\cot(A + B) = \dfrac{1}{\tan(A + B)} = \dfrac{1}{\frac{\tan A + \tan B}{1 - \tan A \tan B}} = \dfrac{1 - \tan A \tan B}{\tan A + \tan B} =$

    $\dfrac{1 - \frac{1}{\cot A} \cdot \frac{1}{\cot B}}{\frac{1}{\cot A} + \frac{1}{\cot B}} = \dfrac{1 - \frac{1}{\cot A \cot B}}{\frac{1}{\cot A} + \frac{1}{\cot B}} = \dfrac{\frac{\cot A \cot B - 1}{\cot A \cot B}}{\frac{\cot B + \cot A}{\cot A \cot B}} = \dfrac{\cot A \cot B - 1}{\cot A \cot B} \cdot \dfrac{\cot A \cot B}{\cot B + \cot A} = \dfrac{\cot A \cot B - 1}{\cot B + \cot A}$

    (b) The function $\cot \pi$ is undefined, so it cannot be used.

    (c) $\cot(\pi + x) = \dfrac{\cos(\pi + x)}{\sin(\pi + x)} = \dfrac{\cos\pi\cos x - \sin\pi\sin x}{\sin\pi\cos x + \cos\pi\sin x} = \dfrac{-1 \cdot \cos x - 0 \cdot \sin x}{0 \cdot \cos x + -1 \cdot \sin x} = \dfrac{-\cos x}{-\sin x} = \dfrac{\cos x}{\sin x} = \cot x$

41. $\sqrt{\dfrac{1 - \cos 40°}{2}} = \sin\dfrac{40°}{2} = \sin 20°$

42. $\sqrt{\dfrac{1 + \cos 76°}{2}} = \cos\dfrac{76°}{2} = \cos 38°$

43. $\sqrt{\dfrac{1 - \cos 147°}{1 + \cos 147°}} = \tan\dfrac{147°}{2} = \tan 73.5°$

44. $\sqrt{\dfrac{1 + \cos 165°}{1 - \cos 165°}} = \sqrt{\dfrac{\frac{1 + \cos 165°}{2}}{\frac{1 - \cos 165°}{2}}} = \dfrac{\cos\frac{165°}{2}}{\sin\frac{165°}{2}} = \cot\dfrac{165°}{2} = \cot 82.5°$

45. $\dfrac{1 - \cos 59.74°}{\sin 59.74°} = \tan\dfrac{59.74°}{2} = \tan 29.87°$

46. $\dfrac{\sin 158.2°}{1 + \cos 158.2°} = \tan\dfrac{158.2°}{2} = \tan 79.1°$

47. $\sin 4\alpha = 2\sin 2\alpha\cos 2\alpha = 2(2\sin\alpha\cos\alpha)\cos 2\alpha = 4\sin\alpha\cos\alpha\cos 2\alpha$

48. $\dfrac{1 + \cos 2x}{\sin 2x} = \dfrac{1 + 2\cos^2 x - 1}{2\sin x\cos x} = \dfrac{2\cos^2 x}{2\sin x\cos x} = \dfrac{\cos x}{\sin x} = \cot x$

49. $\dfrac{2\cos 2\alpha}{\sin 2\alpha} = \dfrac{2(\cos^2\alpha - \sin^2\alpha)}{2\sin\alpha\cos\alpha} = \dfrac{\cos^2\alpha}{\sin\alpha\cos\alpha} - \dfrac{\sin^2\alpha}{\sin\alpha\cos\alpha} = \dfrac{\cos\alpha}{\sin\alpha} - \dfrac{\sin\alpha}{\cos\alpha} = \cot\alpha - \tan\alpha$

50. $\sin 4\gamma = 2\sin 2\gamma\cos 2\gamma = 2(2\sin\gamma\cos\gamma)(1 - 2\sin^2\gamma) = 4\sin\gamma\cos\gamma - 8\sin^3\gamma\cos\gamma$

51. $\sin 2\alpha\cos 2\alpha = \sin 2\alpha(1 - 2\sin^2\alpha) = \sin 2\alpha - 2\sin 2\alpha\sin^2\alpha = \sin 2\alpha - 2(2\sin\alpha\cos\alpha)\sin^2\alpha =$

    $\sin 2\alpha - 4\sin^3\alpha\cos\alpha$

52. Verify backwards: $\dfrac{1 - \tan^2 x}{1 + \tan^2 x} = \dfrac{1 - \frac{\sin^2 x}{\cos^2 x}}{1 + \frac{\sin^2 x}{\cos^2 x}} = \dfrac{1 - \frac{\sin^2 x}{\cos^2 x}}{1 + \frac{\sin^2 x}{\cos^2 x}} \cdot \dfrac{\cos^2 x}{\cos^2 x} = \dfrac{\cos^2 x - \sin^2 x}{\cos^2 x + \sin^2 x} = \dfrac{\cos 2x}{1} = \cos 2x$

53. If $\tan s + \cot s = \dfrac{\sin s}{\cos s} + \dfrac{\cos s}{\sin s} = \dfrac{\sin^2 s + \cos^2 s}{\sin s\cos s} = \dfrac{1}{\sin s\cos s}$, and if $2\csc 2s = 2\left(\dfrac{1}{2\sin 2s}\right) =$

    $2\left(\dfrac{1}{2\sin s\cos s}\right) = \dfrac{1}{\sin s\cos s}$, then $\tan s + \cot s = 2\csc 2s$.

54. $\dfrac{\cot\alpha - \tan\alpha}{\cot\alpha + \tan\alpha} = \dfrac{\frac{\cos\alpha}{\sin\alpha} - \frac{\sin\alpha}{\cos\alpha}}{\frac{\cos\alpha}{\sin\alpha} + \frac{\sin\alpha}{\cos\alpha}} = \dfrac{\frac{\cos\alpha}{\sin\alpha} - \frac{\sin\alpha}{\cos\alpha}}{\frac{\cos\alpha}{\sin\alpha} + \frac{\sin\alpha}{\cos\alpha}} \cdot \dfrac{\sin\alpha\cos\alpha}{\sin\alpha\cos\alpha} = \dfrac{\cos^2\alpha - \sin^2\alpha}{\cos^2\alpha + \sin^2\alpha} = \dfrac{\cos^2\alpha - \sin^2\alpha}{1} = \cos 2\alpha$

55. $\sec^2\dfrac{x}{2} = \left(\dfrac{1}{\cos\frac{x}{2}}\right)^2 = \left(\pm\dfrac{1}{\sqrt{\frac{1+\cos x}{2}}}\right)^2 = \dfrac{1}{\frac{1+\cos x}{2}} = \dfrac{2}{1+\cos x}$

56. $\cot^2\dfrac{x}{2} = \dfrac{1}{\tan^2\frac{x}{2}} = \left(\dfrac{1}{\frac{\sin x}{1+\cos x}}\right)^2 = \dfrac{(1+\cos x)^2}{\sin^2 x}$

57. $2\sin 58° \cos 102° = 2\left[\dfrac{1}{2}\left[\sin(58° + 102°) + \sin(58° - 102°)\right]\right] = \sin 160° + \sin(-44°) =$
$\sin 160° - \sin 44°$

58. $5\cos 3x \cos 2x = 5\left[\dfrac{1}{2}\left[\cos(3x + 2x) + \cos(3x - 2x)\right]\right] = \dfrac{5}{2}(\cos 5x + \cos x) = \dfrac{5}{2}\cos 5x + \dfrac{5}{2}\cos x$

59. $2\cos 85° \sin 140° = \dfrac{1}{2}\left[2\left[\sin(85° + 140°) - \sin(85° - 140°)\right]\right] = \sin 225° - \sin(-55°) =$
$\sin 225° + \sin 55°$

60. $\sin 4x \sin 5x = \dfrac{1}{2}\left[\cos(4x - 5x) - \cos(4x + 5x)\right] = \dfrac{1}{2}(\cos(-x) - \cos 9x) = \dfrac{1}{2}\cos x - \dfrac{1}{2}\cos 9x$

61. $\cos 4x - \cos 2x = -2\sin\left(\dfrac{4x + 2x}{2}\right)\sin\left(\dfrac{4x - 2x}{2}\right) = -2\sin 3x \sin x$

62. $\cos 5t + \cos 8t = 2\cos\left(\dfrac{5t + 8t}{2}\right)\cos\left(\dfrac{5t - 8t}{2}\right) = 2\cos 6.5t \cos(-1.5t) = 2\cos 6.5t \cos 1.5t$

63. $\sin 25° + \sin(-48°) = 2\sin\left(\dfrac{25° + (-48°)}{2}\right)\cos\left(\dfrac{25° - (-48°)}{2}\right) = 2\sin(-11.5°)\cos 36.5° =$
$-2\sin 11.5° \cos 36.5°$

64. $\sin 102° - \sin 95° = 2\cos\left(\dfrac{102° + 95°}{2}\right)\sin\left(\dfrac{102° - 95°}{2}\right) = 2\cos 98.5° \sin 3.5°$

65. $\cos 4x + \cos 8x = 2\cos\left(\dfrac{4x + 8x}{2}\right)\cos\left(\dfrac{4x - 8x}{2}\right) = 2\cos 6x \cos(-2x) = 2\cos 6x \cos 2x$

66. $\sin 9B - \sin 3B = 2\cos\left(\dfrac{9B + 3B}{2}\right)\sin\left(\dfrac{9B - 3B}{2}\right) = 2\cos 6B \sin 3B$

67. (a) Since R is the radius of the circle, the dashed line has length $R - b$, so $\cos\dfrac{\theta}{2} = \dfrac{R - b}{R}$.

(b) $\tan\dfrac{\theta}{4} = \dfrac{1 - \cos\frac{\theta}{2}}{\sin\frac{\theta}{2}} = \dfrac{1 - \frac{R-b}{R}}{\frac{50}{R}} = \dfrac{R - (R - b)}{50} = \dfrac{b}{50}$

(c) If $b = 12$, then $\tan\dfrac{\theta}{4} = \dfrac{b}{50}$ yields: $\tan\dfrac{\theta}{4} = \dfrac{12}{50}$. Now solve for θ:

$\dfrac{\theta}{4} = \tan^{-1}\left(\dfrac{12}{50}\right) \Rightarrow \theta = 4\tan^{-1}\left(\dfrac{12}{50}\right) \approx 53.98° \Rightarrow \theta \approx 54°$.

68. (a) $D = \dfrac{v^2\sin\theta\cos\theta + v\cos\theta\sqrt{(v\sin\theta)^2 + 0}}{32} = \dfrac{v^2\sin\theta\cos\theta + v^2\sin\theta\cos\theta}{32} = \dfrac{2v^2\sin\theta\cos\theta}{32} =$
$\dfrac{v^2\sin 2\theta}{32}$

(b) $D = \dfrac{36^2\sin(2\cdot 30°)}{32} = \dfrac{1296 \cdot \frac{\sqrt{3}}{2}}{32} \approx 35$ feet

(c) $D = \dfrac{40^2\sin(2\cdot 32°)}{32} \approx 45$ feet

69. (a) Graph $W = VI = [163 \sin (120\pi t)][1.23 \sin (120\pi t)] = 200.49 \sin^2 (120\pi t)$. See Figure 69.

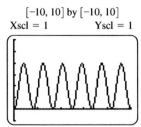

[−10, 10] by [−10, 10]
Xscl = 1   Yscl = 1

Figure 69

(b) The minimum wattage is 0 watts and occurs whenever $\sin (120\pi t) = 0$. The maximum wattage will occur when $\sin (120\pi t) = 1$. This would be $200.49 (1) = 200.49$ watts.

(c) Use the identity, $\sin^2 A = \dfrac{1}{2}(1 - \cos 2A)$: $W = 200.49 \sin^2 (120\pi t) = 200.49 \left[\dfrac{1}{2}(1 - \cos (240\pi t)) \right] = -100.245 \cos (240\pi t) + 100.245$. From this equation, $a = -100.245$, $w = 240\pi$, and $c = 100.245$.

(d) Graphing $W = 200.49 \sin^2 (120\pi t)$ and $W = -100.245 \cos (240\pi t) + 100.245$ in the same window, shows that both equations have the same graph.

(e) Graph W and $y = 100.245$ together. The cosine (or sine) graph of W appears to be vertically centered about this line. An estimate for the average wattage consumed is 100.245 watts. The light bulb would be rated at about 100 watts.

70. (a) The period is equal to $\dfrac{2\pi}{b} = \dfrac{2\pi}{2\pi w} = \dfrac{1}{w}$

(b) $W = VI \Rightarrow W = a \sin (2\pi w t) \cdot b \sin (2\pi w t) = ab \sin^2 (2\pi w t) = \dfrac{ab}{2}[1 - \cos (4\pi w t)]$.

This has a period of $\dfrac{2\pi}{4\pi w} = \dfrac{1}{2w}$, which is exactly half of $\dfrac{1}{w}$. The graph of the wattage will oscillate at twice the rate of the graph of the voltage or the amperage. For example, if the voltage oscillates at 60 cycles per second, then the wattage will oscillate at 120 cycles per seconds.

71. $\tan \theta + \cot \theta = \dfrac{\sin \theta}{\cos \theta} + \dfrac{\cos \theta}{\sin \theta} = \dfrac{\sin^2 \theta}{\sin \theta \cos \theta} + \dfrac{\cos^2 \theta}{\sin \theta \cos \theta} = \dfrac{\sin^2 \theta + \cos^2 \theta}{\sin \theta \cos \theta} = \dfrac{1}{\sin \theta \cos \theta} = \dfrac{1}{\cos \theta} \cdot \dfrac{1}{\sin \theta} = \sec \theta \csc \theta$

72. $\csc \theta \cos^2 \theta + \sin \theta = \dfrac{1}{\sin \theta} \cdot \cos^2 \theta + \sin \theta = \dfrac{\cos^2 \theta}{\sin \theta} + \sin \theta = \dfrac{\cos^2 \theta + \sin^2 \theta}{\sin \theta} = \dfrac{1}{\sin \theta} = \csc \theta$

73. $\tan \dfrac{x}{2} = \dfrac{1 - \cos x}{\sin x} = \dfrac{1}{\sin x} - \dfrac{\cos x}{\sin x} = \csc x - \cot x$

74. $\sec (\pi - x) = \dfrac{1}{\cos (\pi - x)} = \dfrac{1}{\cos \pi \cos x + \sin \pi \sin x} = \dfrac{1}{-1 \cdot \cos x + 0 \cdot \sin x} = \dfrac{1}{-\cos x} = -\sec x$

75. $\dfrac{\sin t}{1 + \cos t} = \tan \dfrac{t}{2} = \dfrac{1 - \cos t}{\sin t}$

76. $\dfrac{1 - \sin t}{\cos t} = \dfrac{1}{\cos t} - \dfrac{\sin t}{\cos t} = \sec t - \tan t = (\sec t - \tan t) \cdot \dfrac{\sec t + \tan t}{\sec t + \tan t} = \dfrac{\sec^2 t - \tan^2 t}{\sec t + \tan t} = \dfrac{1}{\sec t + \tan t}$

77. $\sin 2\theta = 2 \sin\theta \cos\theta = 2 \sin\theta \cos\theta \cdot \dfrac{\cos\theta}{\cos\theta} = \dfrac{2 \sin\theta \cos^2\theta}{\cos\theta} = 2 \cdot \dfrac{\sin\theta}{\cos\theta} \cdot \cos^2\theta = 2 \tan\theta \cos^2\theta =$

$2 \cdot \tan\theta \cdot \dfrac{1}{\sec^2\theta} = \dfrac{2 \tan\theta}{\sec^2\theta} = \dfrac{2 \tan\theta}{1 + \tan^2\theta}$

78. $\dfrac{2}{1 + \cos x} - \tan^2 \dfrac{x}{2} = \dfrac{2}{1 + \cos x} - \dfrac{\sin^2 \frac{x}{2}}{\cos^2 \frac{x}{2}} = \dfrac{2}{1 + \cos x} - \left(\sqrt{\dfrac{1 - \cos x}{1 + \cos x}}\right)^2 = \dfrac{2}{1 + \cos x} - \dfrac{1 - \cos x}{1 + \cos x} =$

$\dfrac{1 + \cos x}{1 + \cos x} = 1$

79. $\cot\theta - \tan\theta = \dfrac{\cos\theta}{\sin\theta} - \dfrac{\sin\theta}{\cos\theta} = \dfrac{\cos^2\theta}{\sin\theta \cos\theta} - \dfrac{\sin^2\theta}{\sin\theta \cos\theta} = \dfrac{\cos^2\theta - \sin^2\theta}{\sin\theta \cos\theta} =$

$\dfrac{\cos^2\theta - (1 - \cos^2\theta)}{\sin\theta \cos\theta} = \dfrac{2 \cos^2\theta - 1}{\sin\theta \cos\theta}$

80. $\dfrac{1}{\sec t - 1} + \dfrac{1}{\sec t + 1} = \dfrac{\sec t + 1}{\sec^2 t - 1} + \dfrac{\sec t - 1}{\sec^2 t - 1} = \dfrac{\sec t + 1}{\tan^2 t} + \dfrac{\sec t - 1}{\tan^2 t} = \dfrac{2 \sec t}{\tan^2 t} = 2 \sec t \cdot \dfrac{1}{\tan^2 t} =$

$2 \sec t \cdot \cot^2 t = 2 \cdot \dfrac{1}{\cos t} \cdot \dfrac{\cos^2 t}{\sin^2 t} = \dfrac{2 \cos t}{\sin^2 t} = 2 \cdot \dfrac{\cos t}{\sin t} \cdot \dfrac{1}{\sin t} = 2 \cot t \csc t$

81. $\dfrac{\sin(x + y)}{\cos(x - y)} = \dfrac{\sin x \cos y + \cos x \sin y}{\cos x \cos y + \sin x \sin y} = \dfrac{\sin x \cos y + \cos x \sin y}{\cos x \cos y + \sin x \sin y} \cdot \dfrac{\frac{1}{\sin x \sin y}}{\frac{1}{\sin x \sin y}} = \dfrac{\frac{\cos y}{\sin y} + \frac{\cos x}{\sin x}}{\frac{\cos x \cos y}{\sin x \sin y} + 1} =$

$\dfrac{\cot y + \cot x}{\cot x \cot y + 1} = \dfrac{\cot x + \cot y}{1 + \cot x \cot y}$

82. $1 - \tan^2 \dfrac{\theta}{2} = 1 - \left(\sqrt{\dfrac{1 - \cos\theta}{1 + \cos\theta}}\right)^2 = 1 - \dfrac{1 - \cos\theta}{1 + \cos\theta} = \dfrac{1 + \cos\theta}{1 + \cos\theta} - \dfrac{1 - \cos\theta}{1 + \cos\theta} = \dfrac{2 \cos\theta}{1 + \cos\theta}$

83. $\dfrac{\sin\theta + \tan\theta}{1 + \cos\theta} = \dfrac{\frac{\sin\theta}{1} + \frac{\sin\theta}{\cos\theta}}{1 + \cos\theta} = \dfrac{\frac{\cos\theta \sin\theta + \sin\theta}{\cos\theta}}{1 + \cos\theta} = \dfrac{\frac{\sin\theta(\cos\theta + 1)}{\cos\theta}}{1 + \cos\theta} = \dfrac{\sin\theta}{\cos\theta} = \tan\theta$

84. $\csc^4 x - \cot^4 x = (\csc^2\theta + \cot^2\theta)(\csc^2 - \cot^2\theta) = (\csc^2\theta + \cot^2\theta)(1 + \cot^2\theta - \cot^2\theta) =$

$(\csc^2\theta + \cot^2\theta)(1) = (\csc^2\theta + \cot^2\theta) = \dfrac{1}{\sin^2\theta} + \dfrac{\cos^2 x}{\sin^2 x} = \dfrac{1 + \cos^2 x}{\sin^2 x} = \dfrac{1 + \cos^2 x}{1 - \cos^2 x}$

85. Verify backwards: $\dfrac{1 - \tan^2 \frac{x}{2}}{1 + \tan^2 \frac{x}{2}} = \dfrac{1 - (\sqrt{\frac{1 - \cos x}{1 + \cos x}})^2}{1 + (\sqrt{\frac{1 - \cos x}{1 + \cos x}})^2} = \dfrac{1 - \frac{1 - \cos x}{1 + \cos x}}{1 + \frac{1 - \cos x}{1 + \cos x}} = \dfrac{\frac{1 + \cos x}{1 + \cos x} - \frac{1 - \cos x}{1 + \cos x}}{\frac{1 + \cos x}{1 + \cos x} + \frac{1 - \cos x}{1 + \cos x}} = \dfrac{\frac{2 \cos x}{1 + \cos x}}{\frac{2}{1 + \cos x}} =$

$\dfrac{2 \cos x}{1 + \cos x} \cdot \dfrac{1 + \cos x}{2} = \cos x$

86. $\cos 2x = 2 \cos^2 x - 1 = (2 \cos^2 x - 1) \cdot \dfrac{\frac{1}{\cos^2 x}}{\frac{1}{\cos^2 x}} = \dfrac{2 - \frac{1}{\cos^2 x}}{\frac{1}{\cos^2 x}} = \dfrac{2 - \sec^2 x}{\sec^2 x}$

87. $\dfrac{\tan^2 t + 1}{\tan t \csc^2 t} = \dfrac{\sec^2 t}{\tan t \csc^2 t} = \dfrac{\sec^2 t}{\csc^2 t} \cdot \dfrac{1}{\tan t} = (\tan^2 t) \cdot \dfrac{1}{\tan t} = \tan t$

88. $\dfrac{\sin s}{1 + \cos s} + \dfrac{1 + \cos s}{\sin s} = \dfrac{\sin^2 s}{\sin s(1 + \cos s)} + \dfrac{1 + 2\cos s + \cos^2 s}{\sin s(1 + \cos s)} = \dfrac{\sin^2 s + \cos^2 s + 2\cos s + 1}{\sin s(1 + \cos s)} =$

$\dfrac{1 + 2\cos s + 1}{\sin s(1 + \cos s)} = \dfrac{2 + 2\cos s}{\sin s(1 + \cos s)} = \dfrac{2(1 + \cos s)}{\sin s(1 + \cos s)} = \dfrac{2}{\sin s} = 2 \cdot \dfrac{1}{\sin s} = 2 \csc s$

89. $\tan 4\theta = \tan 2(2\theta) = \dfrac{2 \tan 2\theta}{1 - \tan^2 2\theta} = \dfrac{2 \tan 2\theta}{1 - (\sec^2\theta - 1)} = \dfrac{2 \tan 2\theta}{2 - \sec^2\theta}$

90. $\tan\left(\dfrac{x}{2}+\dfrac{\pi}{4}\right)=\dfrac{\tan\frac{x}{2}+\tan\frac{\pi}{4}}{1-\tan\frac{x}{2}\tan\frac{\pi}{4}}=\dfrac{\frac{\sin x}{1+\cos x}+1}{1-\frac{\sin x}{1+\cos x}(1)}=\dfrac{\frac{\sin x+1+\cos x}{1+\cos x}}{\frac{1+\cos x-\sin x}{1+\cos x}}=\dfrac{1+\cos x+\sin x}{1+\cos x-\sin x}=$

$\dfrac{1+\cos x+\sin x}{1+\cos x-\sin x}\cdot\dfrac{\cos x}{\cos x}=\dfrac{\cos x+\cos^2 x+\cos x\sin x}{(1+\cos x-\sin x)\cos x}=\dfrac{\cos x+1-\sin^2 x+\cos x\sin x}{(1+\cos x-\sin x)\cos x}=$

$\dfrac{\cos x+(1+\sin x)(1-\sin x)+\cos x\sin x}{(1+\cos x-\sin x)\cos x}=\dfrac{\cos x(1+\sin x)+(1+\sin x)(1-\sin x)}{(1+\cos x-\sin x)\cos x}=$

$\dfrac{(1+\sin x)[\cos x+(1-\sin x)]}{(1+\cos x-\sin x)\cos x}=\dfrac{1+\sin x}{\cos x}=\dfrac{1}{\cos x}+\dfrac{\sin x}{\cos x}=\sec x+\tan x$

9.4: The Inverse Circular Functions

1. (a) The domain of $y=\sin^{-1}x$ is the range of $y=\sin x$, therefore the domain is: $[-1, 1]$.

   (b) The range of $y=\sin^{-1}x$ is the domain of $y=\sin x$, therefore the range is: $\left[-\dfrac{\pi}{2},\dfrac{\pi}{2}\right]$.

   (c) For this function, as x increases, y increases. Therefore, it in an <u>increasing</u> function.

   (d) Arcsin (-2) is not defined since -2 is not in the domain.

2. (a) The domain of $y=\cos^{-1}x$ is the range of $y=\cos x$, therefore the domain is: $[-1, 1]$.

   (b) The range of $y=\cos^{-1}x$ is the domain of $y=\cos x$, therefore the range is: $[0, \pi]$.

   (c) For this function, as x increases, y decreases. Therefore, it in an <u>decreasing</u> function.

   (d) The value $-\dfrac{4\pi}{3}$ is not in the range.

3. (a) The domain of $y=\tan^{-1}x$ is the range of $y=\tan x$, therefore the domain is: $(-\infty, \infty)$.

   (b) The range of $y=\tan^{-1}x$ is the domain of $y=\tan x$, therefore the range is: $\left(-\dfrac{\pi}{2},\dfrac{\pi}{2}\right)$.

   (c) For this function, as x increases, y increases. Therefore, it in an <u>increasing</u> function.

   (d) No, since the domain is $(-\infty, \infty)$, arctan x is defined for all numbers.

4. (a) The domain of $y=\csc^{-1}x$ is: $(-\infty, -1]\cup[1, \infty)$.

   The range of $y=\csc^{-1}x$ is: $\left[-\dfrac{\pi}{2},0\right)\cup\left(0,\dfrac{\pi}{2}\right]$.

   (b) The domain of $y=\sec^{-1}x$ is: $(-\infty, -1]\cup[1, \infty)$.

   The range of $y=\sec^{-1}x$ is: $\left[0,\dfrac{\pi}{2}\right)\cup\left(\dfrac{\pi}{2},\pi\right]$.

   (c) The domain of $y=\cot^{-1}x$ is: $(-\infty, \infty)$.

   The range of $y=\cot^{-1}x$ is: $(0, \pi)$.

5. Since $\sec^{-1}a=y\Rightarrow a=\sec y\Rightarrow a=\dfrac{1}{\cos y}\Rightarrow\dfrac{1}{a}=\cos y$, therefore $y=\cos^{-1}\left(\dfrac{1}{a}\right)$.

6. For $x<0$, $\tan^{-1}x$ is in quadrant IV, but $\cot^{-1}x$ is found in quadrant II. Therefore $\cot^{-1}a=\tan^{-1}\left(\dfrac{1}{a}\right)+\pi$

7. $y=\tan^{-1}1\Rightarrow1=\tan y\Rightarrow y=\dfrac{\pi}{4}$ $\left(\text{range:}\left(-\dfrac{\pi}{2},\dfrac{\pi}{2}\right)\right)$

8. $y=\sin^{-1}0\Rightarrow0=\sin y\Rightarrow y=0$ $\left(\text{range:}\left[-\dfrac{\pi}{2},\dfrac{\pi}{2}\right]\right)$

9. $y=\cos^{-1}(-1)\Rightarrow-1=\cos y\Rightarrow y=\pi$ $(\text{range:}[0, \pi])$

10. $y = \arctan(-1) \Rightarrow -1 = \tan y \Rightarrow y = -\dfrac{\pi}{4}$ $\left(\text{range: } \left(-\dfrac{\pi}{2}, \dfrac{\pi}{2}\right)\right)$

11. $y = \sin^{-1}(-1) \Rightarrow -1 = \sin y \Rightarrow y = -\dfrac{\pi}{2}$ $\left(\text{range: } \left[-\dfrac{\pi}{2}, \dfrac{\pi}{2}\right]\right)$

12. $y = \cos^{-1}\left(\dfrac{1}{2}\right) \Rightarrow \dfrac{1}{2} = \cos y \Rightarrow y = \dfrac{\pi}{3}$ (range: $[0, \pi]$)

13. $y = \arctan 0 \Rightarrow 0 = \tan y \Rightarrow y = 0$ $\left(\text{range: } \left(-\dfrac{\pi}{2}, \dfrac{\pi}{2}\right)\right)$

14. $y = \arcsin\left(-\dfrac{\sqrt{3}}{2}\right) \Rightarrow -\dfrac{\sqrt{3}}{2} = \sin y \Rightarrow y = -\dfrac{\pi}{3}$ $\left(\text{range: } \left[-\dfrac{\pi}{2}, \dfrac{\pi}{2}\right]\right)$

15. $y = \arccos 0 \Rightarrow 0 = \cos y \Rightarrow y = \dfrac{\pi}{2}$ (range: $[0, \pi]$)

16. $y = \tan^{-1}(-1) \Rightarrow -1 = \tan y \Rightarrow y = -\dfrac{\pi}{4}$ $\left(\text{range: } \left(-\dfrac{\pi}{2}, \dfrac{\pi}{2}\right)\right)$

17. $y = \sin^{-1}\left(\dfrac{\sqrt{2}}{2}\right) \Rightarrow \dfrac{\sqrt{2}}{2} = \sin y \Rightarrow y = \dfrac{\pi}{4}$ $\left(\text{range: } \left[-\dfrac{\pi}{2}, \dfrac{\pi}{2}\right]\right)$

18. $y = \arccos\left(-\dfrac{1}{2}\right) \Rightarrow -\dfrac{1}{2} = \cos y \Rightarrow y = \dfrac{2\pi}{3}$ (range: $[0, \pi]$)

19. $y = \arccos\left(-\dfrac{\sqrt{3}}{2}\right) \Rightarrow -\dfrac{\sqrt{3}}{2} = \cos y \Rightarrow y = \dfrac{5\pi}{6}$ (range: $[0, \pi]$)

20. $y = \arcsin\left(-\dfrac{\sqrt{2}}{2}\right) \Rightarrow -\dfrac{\sqrt{2}}{2} = \sin y \Rightarrow y = -\dfrac{\pi}{4}$ $\left(\text{range: } \left[-\dfrac{\pi}{2}, \dfrac{\pi}{2}\right]\right)$

21. $y = \cot^{-1}(-1) \Rightarrow y = \tan^{-1}\left(\dfrac{1}{-1}\right) + \pi \Rightarrow y = \tan^{-1}(-1) + \pi \Rightarrow (-1 = \tan y) + \pi \Rightarrow$

     $y = -\dfrac{\pi}{4} + \pi \Rightarrow y = \dfrac{3\pi}{4}$ (range: $(0, \pi)$)

22. $y = \sec^{-1}(-\sqrt{2}) \Rightarrow y = \cos^{-1}\left(\dfrac{1}{-\sqrt{2}}\right) \Rightarrow y = \cos^{-1}\left(-\dfrac{\sqrt{2}}{2}\right) \Rightarrow -\dfrac{\sqrt{2}}{2} = \cos y \Rightarrow y = \dfrac{3\pi}{4}$

     $\left(\text{range: } \left[0, \dfrac{\pi}{2}\right) \cup \left(\dfrac{\pi}{2}, \pi\right]\right)$

23. $y = \csc^{-1}(-2) \Rightarrow y = \sin^{-1}\left(\dfrac{1}{-2}\right) \Rightarrow y = \sin^{-1}\left(-\dfrac{1}{2}\right) \Rightarrow -\dfrac{1}{2} = \sin y \Rightarrow y = -\dfrac{\pi}{6}$

     $\left(\text{range: } \left[-\dfrac{\pi}{2}, 0\right) \cup \left(0, \dfrac{\pi}{2}\right]\right)$

24. $y = \operatorname{arccot}(-\sqrt{3}) \Rightarrow y = \tan^{-1}\left(\dfrac{1}{-\sqrt{3}}\right) + \pi \Rightarrow y = \tan^{-1}\left(-\dfrac{\sqrt{3}}{3}\right) + \pi \Rightarrow \left(-\dfrac{\sqrt{3}}{3} = \tan y\right) + \pi \Rightarrow$

     $y = -\dfrac{\pi}{6} + \pi \Rightarrow y = \dfrac{5\pi}{6}$ (range: $(0, \pi)$)

25. $y = \operatorname{arcsec}\dfrac{2\sqrt{3}}{3} \Rightarrow y = \cos^{-1}\left(\dfrac{1}{\frac{2\sqrt{3}}{3}}\right) \Rightarrow y = \cos^{-1}\left(\dfrac{3}{2\sqrt{3}}\right) \Rightarrow y = \cos^{-1}\dfrac{\sqrt{3}}{2} \Rightarrow \dfrac{\sqrt{3}}{2} = \cos y \Rightarrow y = \dfrac{\pi}{6}$

     $\left(\text{range: } \left[0, \dfrac{\pi}{2}\right) \cup \left(\dfrac{\pi}{2}, \pi\right]\right)$

26. $y = \csc^{-1}\sqrt{2} \Rightarrow y = \sin^{-1}\left(\dfrac{1}{\sqrt{2}}\right) \Rightarrow y = \sin^{-1}\left(\dfrac{\sqrt{2}}{2}\right) \Rightarrow \dfrac{\sqrt{2}}{2} = \sin y \Rightarrow y = \dfrac{\pi}{4}$

     $\left(\text{range: } \left[-\dfrac{\pi}{2}, 0\right) \cup \left(0, \dfrac{\pi}{2}\right]\right)$

27. $\theta = \arctan(-1) \Rightarrow -1 = \tan\theta \Rightarrow \theta = -45°$ (range: $(-90°, 90°)$)

28. $\theta = \arccos\left(-\dfrac{1}{2}\right) \Rightarrow -\dfrac{1}{2} = \cos\theta \Rightarrow \theta = 120°$ (range: $[0°, 180°]$)

29. $\theta = \arcsin\left(-\dfrac{\sqrt{3}}{2}\right) \Rightarrow -\dfrac{\sqrt{3}}{2} = \sin\theta \Rightarrow \theta = -60°$ (range: $[-90°, 90°]$)

30. $\theta = \arcsin\left(-\dfrac{\sqrt{2}}{2}\right) \Rightarrow -\dfrac{\sqrt{2}}{2} = \sin\theta \Rightarrow \theta = -45°$ (range: $[-90°, 90°]$)

31. $\theta = \cot^{-1}\left(-\dfrac{\sqrt{3}}{3}\right) \Rightarrow \theta = \tan^{-1}\left(\dfrac{1}{-\frac{\sqrt{3}}{3}}\right) + 180° \Rightarrow \theta = \tan^{-1}\left(-\dfrac{3}{\sqrt{3}}\right) + 180° \Rightarrow \theta = \tan^{-1}(-\sqrt{3}) + 180°$

    $\Rightarrow (-\sqrt{3} = \tan\theta) + 180° \Rightarrow \theta = -60° + 180° \Rightarrow \theta = 120°$ (range: $(0°, 180°)$)

32. $\theta = \sec^{-1}(-2) \Rightarrow \theta = \cos^{-1}\left(\dfrac{1}{-2}\right) \Rightarrow \theta = \cos^{-1}\left(-\dfrac{1}{2}\right) \Rightarrow \left(-\dfrac{1}{2} = \cos\theta\right) \Rightarrow \theta = 120°$

    (range: $[0°, 90°) \cup (90°, 180°]$)

33. $\theta = \csc^{-1}(-2) \Rightarrow \theta = \sin^{-1}\left(\dfrac{1}{-2}\right) \Rightarrow \theta = \sin^{-1}\left(-\dfrac{1}{2}\right) \Rightarrow \left(-\dfrac{1}{2} = \sin\theta\right) \Rightarrow \theta = -30°$

    (range: $[-90°, 0°) \cup (0°, 90°]$)

34. $\theta = \csc^{-1}(-1) \Rightarrow \theta = \sin^{-1}\left(\dfrac{1}{-1}\right) \Rightarrow \theta = \sin^{-1}(-1) \Rightarrow (-1 = \sin\theta) \Rightarrow \theta = -90°$

    (range: $[-90°, 0°) \cup (0°, 90°]$)

35. $\theta = \sin^{-1}(-.13349122) \Rightarrow \theta \approx -7.6713835°$

36. $\theta = \cos^{-1}(-.13348816) \Rightarrow \theta \approx 97.671207°$

37. $\theta = \arccos(-.39876459) \Rightarrow \theta \approx 113.500970°$

38. $\theta = \arcsin(.77900016) \Rightarrow \theta \approx 51.1691219°$

39. $\theta = \csc^{-1} 1.9422833 \Rightarrow \theta = \sin^{-1}\dfrac{1}{1.9422833} \Rightarrow \theta = \sin^{-1}(1.9422833)^{-1} \Rightarrow \theta \approx 30.987961°$

40. $\theta = \cot^{-1} 1.7670492 \Rightarrow \theta = \tan^{-1}\dfrac{1}{1.7670492} \Rightarrow \theta = \tan^{-1}(1.7670492)^{-1} \Rightarrow \theta \approx 29.50618°$

41. $y = \arctan 1.1111111 \Rightarrow y \approx .83798122$

42. $y = \arcsin .81926439 \Rightarrow y \approx .96012698$

43. $y = \cot^{-1}(-.92170128) \Rightarrow y = \left(\tan^{-1}\dfrac{1}{-.92170128}\right) + \pi \Rightarrow y = (\tan^{-1}(-.92170128)^{-1}) + \pi \Rightarrow$

    $y \approx -.8261201193 + \pi \Rightarrow y \approx 2.315472534$

44. $y = \sec^{-1}(-1.2871684) \Rightarrow y = \cos^{-1}\dfrac{1}{-1.2871684} \Rightarrow y = \cos^{-1}(-1.2871684)^{-1} \Rightarrow y \approx 2.4605221$

45. $y = \arcsin .92837781 \Rightarrow y \approx 1.1900238$

46. $y = \arccos .44624593 \Rightarrow y \approx 1.1082303$

47. Graph $y = \cot^{-1} x$, see Figure 47.

48. Graph $y = \csc^{-1} x$, see Figure 48.

49. Graph $y = \sec^{-1} x$, see Figure 49.

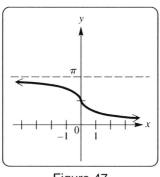

Figure 47

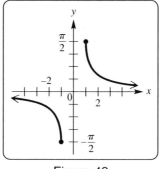

Figure 48

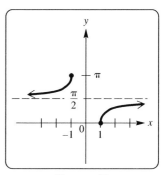

Figure 49

50. Graph $y = \text{arccsc } 2x$ $\left(\text{like the graph of } y = \csc^{-1} x, \text{ with a domain } \left(-\infty, -\dfrac{1}{2}\right] \cup \left[\dfrac{1}{2}, \infty\right)\right)$. See Figure 50.

51. Graph $y = \text{arcsec } \dfrac{1}{2}x$ (like the graph of $y = \sec^{-1} x$, with a domain $(-\infty, -2] \cup [2, \infty)$). See Figure 51.

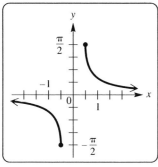

Figure 50

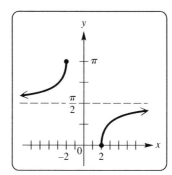

Figure 51

52. The error message occurs because 1.003 is not in the domain of $y = \sin^{-1} x$.

53. The domain of $y = \tan^{-1} x$, is $(-\infty, \infty)$. Therefore 1.003 is in this domain and $\tan^{-1} 1.003$ can be found.

54. $f[f^{-1}(x)] = 3\left(\dfrac{x + 2}{3}\right) - 2 = \dfrac{3x + 6}{3} - 2 = x + 2 - 2 = x;$

   $f^{-1}[f(x)] = \dfrac{(3x - 2) + 2}{3} = \dfrac{3x}{3} = x.$ In both cases, the result is x.

   The graph in each case is a straight line bisecting quadrants I and III (i.e. the line $y = x$).

55. See Figure 55. It is the same graph as in exercise 54, it is the graph of $y = x$.

56. See Figure 56. It does not agree because the range of the inverse tangent function is $\left(-\dfrac{\pi}{2}, \dfrac{\pi}{2}\right)$, not $(-\infty, \infty)$, as was the case in exercise 55.

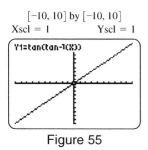

Figure 55

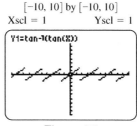

Figure 56

57. Since $\sin(\sin^{-1} x) = x$ when $-1 < x < 1$, $\sin\left(\sin^{-1}\dfrac{1}{2}\right) = \dfrac{1}{2}$.

58. Since $\cos(\cos^{-1} x) = x$ when $-1 < x < 1$, $\cos\left(\cos^{-1}\dfrac{\sqrt{3}}{2}\right) = \dfrac{\sqrt{3}}{2}$.

59. First find $\sin\dfrac{4\pi}{3} = \dfrac{-\sqrt{3}}{2}$, now solve for $\sin^{-1}\left(\dfrac{-\sqrt{3}}{2}\right) \Rightarrow \dfrac{-\sqrt{3}}{2} = \sin\theta \Rightarrow \theta = -\dfrac{\pi}{3}$.

60. First find $\cos-\dfrac{\pi}{6} = \dfrac{\sqrt{3}}{2}$, now solve for $\cos^{-1}\left(\dfrac{\sqrt{3}}{2}\right) \Rightarrow \dfrac{\sqrt{3}}{2} = \cos\theta \Rightarrow \theta = \dfrac{\pi}{6}$.

61. First find $\cos\dfrac{3\pi}{2} = 0$, now solve for $\cos^{-1} 0 \Rightarrow 0 = \cos\theta \Rightarrow \theta = \dfrac{\pi}{2}$.

62. First find $\sin\dfrac{3\pi}{2} = -1$, now solve for $\sin^{-1}(-1) \Rightarrow -1 = \sin\theta \Rightarrow \theta = -\dfrac{\pi}{2}$.

63. Since $\tan(\tan^{-1} x) = x$ for all values of x, $\tan(\tan^{-1} 5) = 5$.

64. Since $\tan(\tan^{-1} x) = x$ for all values of x, $\tan(\tan^{-1}(-4)) = -4$.

65. Since $\sec(\sec^{-1} x) = x$ when $x \le -1$ or $x \ge 1$, $\sec(\sec^{-1} 2) = 2$.

66. Since $\csc(\csc^{-1} x) = x$ when $x \le -1$ or $x \ge 1$, $\csc(\csc^{-1} 3) = 3$.

67. First find $\tan\dfrac{5\pi}{6} = \dfrac{1}{-\sqrt{3}} = \dfrac{-\sqrt{3}}{3}$, now solve for $\tan^{-1}\left(\dfrac{-\sqrt{3}}{3}\right) \Rightarrow \dfrac{-\sqrt{3}}{3} = \tan\theta \Rightarrow \theta = -\dfrac{\pi}{6}$.

68. First find $\tan\dfrac{3\pi}{4} = -1$, now solve for $\tan^{-1}(-1) \Rightarrow -1 = \tan\theta \Rightarrow \theta = -\dfrac{\pi}{4}$.

69. Since $\cos\theta = \dfrac{3}{4}$, then θ is in quadrant I. Sketch and label a triangle in quadrant I, with $x = 3$ and $r = 4$.

    Now use pythagorean theorem to find y: $3^2 + y^2 = 4^2 \Rightarrow y^2 = 16 - 9 \Rightarrow y^2 = 7 \Rightarrow y = \sqrt{7}$.

    Therefore, $\tan\theta = \dfrac{\sqrt{7}}{3}$ and $\tan\left(\arccos\dfrac{3}{4}\right) = \dfrac{\sqrt{7}}{3}$.

70. Since $\cos\theta = \dfrac{1}{4}$, then θ is in quadrant I. Sketch and label a triangle in quadrant I, with $x = 1$ and $r = 4$.

    Now use pythagorean theorem to find y: $1^2 + y^2 = 4^2 \Rightarrow y^2 = 16 - 1 \Rightarrow y^2 = 15 \Rightarrow y = \sqrt{15}$.

    Therefore, $\sin\theta = \dfrac{\sqrt{15}}{4}$ and $\sin\left(\arccos\dfrac{1}{4}\right) = \dfrac{\sqrt{15}}{4}$.

71. Since $\tan\theta = -2$, then θ is in quadrant IV. Sketch and label a triangle in quadrant IV, with $x = 1$ and $y = -2$.

    Now use pythagorean theorem to find r: $1^2 + (-2)^2 = r^2 \Rightarrow r^2 = 1 + 4 \Rightarrow r^2 = 5 \Rightarrow r = \sqrt{5}$.

    Therefore, $\cos\theta = \dfrac{1}{\sqrt{5}} = \dfrac{\sqrt{5}}{5}$ and $\cos(\tan^{-1}(-2)) = \dfrac{\sqrt{5}}{5}$.

72. Since $\sin\theta = -\dfrac{1}{5}$, then θ is in quadrant IV. Sketch and label a triangle in quadrant IV, with $y = -1$ and $r = 5$.

    Now use pythagorean theorem to find x: $x^2 + (-1)^2 = 5^2 \Rightarrow x^2 = 25 - 1 \Rightarrow x^2 = 24 \Rightarrow x = \sqrt{24} = 2\sqrt{6}$.

    Therefore, $\sec\theta = \dfrac{5}{2\sqrt{6}} = \dfrac{5\sqrt{6}}{12}$ and $\sec\left(\sin^{-1}\left(-\dfrac{1}{5}\right)\right) = \dfrac{5\sqrt{6}}{12}$.

73. Since $\tan\theta = \dfrac{12}{5}$, then θ is in quadrant I. Sketch and label a triangle in quadrant I, with $x = 5$ and $y = 12$.

    Use pythagorean theorem to find r: $5^2 + 12^2 = r^2 \Rightarrow r^2 = 25 + 144 \Rightarrow r^2 = 169 \Rightarrow r = 13$.

    Now $\sin 2\theta = 2\sin\theta\cos\theta$, therefore, $\sin 2\theta = 2\left(\dfrac{12}{13}\right)\left(\dfrac{5}{13}\right) = \dfrac{120}{169}$ and $\sin\left(2\tan^{-1}\dfrac{12}{5}\right) = \dfrac{120}{169}$.

74. Since $\sin \theta = \dfrac{1}{4}$, then θ is in quadrant I. Sketch and label a triangle in quadrant I, with $y = 1$ and $r = 4$.

Use pythagorean theorem to find x: $x^2 + 1^2 = 4^2 \Rightarrow x^2 = 16 - 1 \Rightarrow x^2 = 15 \Rightarrow x = \sqrt{15}$.

Now $\cos 2\theta = \cos^2 \theta - \sin^2 \theta$, therefore, $\cos 2\theta = \left(\dfrac{\sqrt{15}}{4}\right)^2 - \left(\dfrac{1}{4}\right)^2 = \dfrac{14}{16} = \dfrac{7}{8}$ and $\cos\left(2 \sin^{-1} \dfrac{1}{4}\right) = \dfrac{7}{8}$.

75. Since $\tan \theta = \dfrac{4}{3}$, then θ is in quadrant I. Sketch and label a triangle in quadrant I, with $x = 3$ and $y = 4$.

Use pythagorean theorem to find r: $3^2 + 4^2 = r^2 \Rightarrow r^2 = 9 + 16 \Rightarrow r^2 = 25 \Rightarrow r = 5$.

Now $\cos 2\theta = \cos^2 \theta - \sin^2 \theta$, therefore, $\cos 2\theta = \left(\dfrac{3}{5}\right)^2 - \left(\dfrac{4}{5}\right)^2 = -\dfrac{7}{25}$ and $\cos\left(2 \arctan \dfrac{4}{3}\right) = -\dfrac{7}{25}$.

76. Since $\cos \theta = \dfrac{1}{4}$, then θ is in quadrant I. Sketch and label a triangle in quadrant I, with $x = 1$ and $r = 4$.

Use pythagorean theorem to find y: $1^2 + y^2 = 4^2 \Rightarrow y^2 = 16 - 1 \Rightarrow y^2 = 15 \Rightarrow y = \sqrt{15}$.

Now $\tan 2\theta = \dfrac{2 \tan \theta}{1 - \tan^2 \theta}$ therefore, $\tan 2\theta = \dfrac{2\left(\frac{\sqrt{15}}{1}\right)}{1 - \left(\frac{\sqrt{15}}{1}\right)^2} = \dfrac{2\sqrt{15}}{-14} = -\dfrac{\sqrt{15}}{7}$ and $\tan\left(2 \cos^{-1} \dfrac{1}{4}\right) = -\dfrac{\sqrt{15}}{7}$.

77. Since $\cos \theta = \dfrac{1}{5}$, then θ is in quadrant I. Sketch and label a triangle in quadrant I, with $x = 1$ and $r = 5$.

Use pythagorean theorem to find y: $1^2 + y^2 = 5^2 \Rightarrow y^2 = 25 - 1 \Rightarrow y^2 = 24 \Rightarrow y = \sqrt{24} = 2\sqrt{6}$.

Now $\sin 2\theta = 2 \sin \theta \cos \theta$, therefore, $\sin 2\theta = 2\left(\dfrac{2\sqrt{6}}{5}\right)\left(\dfrac{1}{5}\right) = \dfrac{4\sqrt{6}}{25}$ and $\sin\left(2 \cos^{-1} \dfrac{1}{5}\right) = \dfrac{4\sqrt{6}}{25}$.

78. Since $\tan \theta = (-2)$, then θ is in quadrant IV. Sketch and label a triangle in quadrant IV, with $x = 1$ and $y = -2$.

Use pythagorean theorem to find r: $1^2 + (-2)^2 = r^2 \Rightarrow r^2 = 1 + 4 \Rightarrow r^2 = 5 \Rightarrow r = \sqrt{5}$.

Now $\cos 2\theta = \cos^2 \theta - \sin^2 \theta$, therefore, $\cos 2\theta = \left(\dfrac{1}{\sqrt{5}}\right)^2 - \left(\dfrac{-2}{\sqrt{5}}\right)^2 = -\dfrac{3}{5}$ and $\cos\left(2 \tan^{-1}(-2)\right) = -\dfrac{3}{5}$.

79. Since $\sin \theta = \dfrac{3}{5}$, then θ is in quadrant I. Sketch and label a triangle in quadrant I, with $y = 3$ and $r = 5$.

Use pythagorean theorem to find x: $x^2 + 3^2 = 5^2 \Rightarrow x^2 = 25 - 9 \Rightarrow x^2 = 16 \Rightarrow x = 4$.

Since $\cos \beta = \dfrac{12}{13}$, then β is in quadrant I. Sketch and label a triangle in quadrant I, see Figure 79b.

Use pythagorean theorem to find y: $12^2 + y^2 = 13^2 \Rightarrow y^2 = 169 - 144 \Rightarrow y^2 = 25 \Rightarrow y = 5$.

Now $\cos(\theta - \beta) = \cos \theta \cos \beta + \sin \theta \sin \beta$, therefore, $\cos(\theta + \beta) = \dfrac{4}{5} \cdot \dfrac{12}{13} + \dfrac{3}{5} \cdot \dfrac{5}{13} = \dfrac{48}{65} + \dfrac{15}{65} = \dfrac{63}{65}$

and $\cos\left(\sin^{-1} \dfrac{3}{5} - \cos^{-1} \dfrac{12}{13}\right) = \dfrac{63}{65}$

80. Since $\cos \theta = \dfrac{3}{5}$, then θ is in quadrant I. Sketch and label a triangle in quadrant I, with $x = 3$ and $r = 5$.

Use pythagorean theorem to find y: $3^2 + y^2 = 5^2 \Rightarrow y^2 = 25 - 9 \Rightarrow y^2 = 16 \Rightarrow y = 4$.

Since $\sin \beta = \dfrac{5}{13}$, then β is in quadrant I. Sketch and label a triangle in quadrant I, with $y = 5$ and $r = 13$.

Use pythagorean theorem to find x: $x^2 + 5^2 = 13^2 \Rightarrow x^2 = 169 - 25 \Rightarrow x^2 = 144 \Rightarrow x = 12$.

Now $\cos(\theta + \beta) = \cos \theta \cos \beta - \sin \theta \sin \beta$, therefore, $\cos(\theta + \beta) = \dfrac{3}{5} \cdot \dfrac{12}{13} - \dfrac{4}{5} \cdot \dfrac{5}{13} = \dfrac{36}{65} - \dfrac{20}{65} = \dfrac{16}{65}$

and $\cos\left(\cos^{-1} \dfrac{3}{5} + \sin^{-1} \dfrac{5}{13}\right) = \dfrac{16}{65}$

81. Use $\tan \theta = \dfrac{3}{4}$ and $\tan \beta = \dfrac{12}{5}$ in $\tan (\theta + \beta) = \dfrac{\tan \theta + \tan \beta}{1 - \tan \theta \tan \beta}$, therefore, $\dfrac{\frac{3}{4} + \frac{12}{5}}{1 - (\frac{3}{4} \cdot \frac{12}{5})} = \dfrac{\frac{15}{20} + \frac{48}{20}}{1 - \frac{36}{20}} =$

    $\dfrac{\frac{63}{20}}{-\frac{16}{20}} = \dfrac{63}{20} \cdot \left(-\dfrac{20}{16}\right) = -\dfrac{63}{16}$ and $\tan \left(\tan^{-1} \dfrac{3}{4} + \tan^{-1} \dfrac{12}{5}\right) = -\dfrac{63}{16}$

82. Since $\sin \theta = \dfrac{8}{17}$, then θ is in quadrant I. Sketch and label a triangle in quadrant I, with $y = 8$ and $r = 17$.

    Use pythagorean theorem to find x: $x^2 + 8^2 = 17^2 \Rightarrow x^2 = 289 - 64 \Rightarrow x^2 = 225 \Rightarrow x = 15$.

    Now use $\tan \theta = \dfrac{8}{15}$ and $\tan \beta = \dfrac{4}{3}$ in $\tan (\theta + \beta) = \dfrac{\tan \theta + \tan \beta}{1 - \tan \theta \tan \beta}$, therefore, $\dfrac{\frac{8}{15} + \frac{4}{3}}{1 - (\frac{8}{15} \cdot \frac{4}{3})} = \dfrac{\frac{8}{15} + \frac{20}{15}}{1 - \frac{32}{45}} =$

    $\dfrac{\frac{28}{15}}{\frac{13}{45}} = \dfrac{28}{15} \cdot \dfrac{45}{13} = \dfrac{84}{13}$ and $\tan \left(\sin^{-1} \dfrac{8}{17} + \tan^{-1} \dfrac{4}{3}\right) = \dfrac{84}{13}$

83. Since $\tan \theta = \dfrac{5}{12}$, then θ is in quadrant I. Sketch and label a triangle in quadrant I, with $x = 12$ and $y = 5$.

    Use pythagorean theorem to find r: $12^2 + 5^2 = r^2 \Rightarrow r^2 = 144 + 25 \Rightarrow r^2 = 169 \Rightarrow r = 13$.

    Since $\tan \beta = \dfrac{3}{4}$, then β is in quadrant I. Sketch and label a triangle in quadrant I, with $x = 4$ and $y = 3$.

    Use pythagorean theorem to find y: $4^2 + 3^2 = r^2 \Rightarrow r^2 = 16 + 9 \Rightarrow r^2 = 25 \Rightarrow r = 5$.

    Now $\cos (\theta - \beta) = \cos \theta \cos \beta + \sin \theta \sin \beta$, therefore, $\cos (\theta + \beta) = \dfrac{12}{13} \cdot \dfrac{4}{5} + \dfrac{5}{13} \cdot \dfrac{3}{5} = \dfrac{48}{65} + \dfrac{15}{65} = \dfrac{63}{65}$

    and $\cos \left(\tan^{-1} \dfrac{5}{12} - \tan^{-1} \dfrac{3}{4}\right) = \dfrac{63}{65}$

84. Since $\sin \theta = \dfrac{3}{5}$, then θ is in quadrant I. Sketch and label a triangle in quadrant I, with $y = 3$ and $r = 5$.

    Use pythagorean theorem to find x: $x^2 + 3^2 = 5^2 \Rightarrow x^2 = 25 - 9 \Rightarrow x^2 = 16 \Rightarrow x = 4$.

    Since $\cos \beta = \dfrac{5}{13}$, then β is in quadrant I. Sketch and label a triangle in quadrant I, with $x = 5$ and $r = 13$.

    Use pythagorean theorem to find y: $5^2 + y^2 = 13^2 \Rightarrow y^2 = 169 - 25 \Rightarrow y^2 = 144 \Rightarrow y = 12$.

    Now $\cos (\theta + \beta) = \cos \theta \cos \beta - \sin \theta \sin \beta$, therefore, $\cos (\theta + \beta) = \dfrac{4}{5} \cdot \dfrac{5}{13} - \dfrac{3}{5} \cdot \dfrac{12}{13} = \dfrac{20}{65} - \dfrac{36}{65} = -\dfrac{16}{65}$

    and $\cos \left(\sin^{-1} \dfrac{3}{5} + \cos^{-1} \dfrac{5}{13}\right) = -\dfrac{16}{65}$

85. Since $\sin \theta = \dfrac{1}{2}$, then θ is in quadrant I. Sketch and label a triangle in quadrant I, with $y = 1$ and $r = 2$.

    Use pythagorean theorem to find x: $x^2 + 1^2 = 2^2 \Rightarrow x^2 = 4 - 1 \Rightarrow x^2 = 3 \Rightarrow x = \sqrt{3}$.

    Since $\tan \beta = (-3)$, then β is in quadrant IV. Sketch and label a triangle in quadrant IV, with $x = 1$ and $y = -3$

    Use pythagorean theorem to find r: $1^2 + (-3)^2 = r^2 \Rightarrow r^2 = 1 + 9 \Rightarrow r^2 = 10 \Rightarrow y = \sqrt{10}$.

    Now $\sin (\theta + \beta) = \sin \theta \cos \beta + \cos \theta \sin \beta$, therefore, $\sin (\theta + \beta) = \dfrac{1}{2} \cdot \dfrac{1}{\sqrt{10}} + \dfrac{\sqrt{3}}{2} \cdot \dfrac{-3}{\sqrt{10}} =$

    $\dfrac{1}{2\sqrt{10}} + \dfrac{-3\sqrt{3}}{2\sqrt{10}} = \dfrac{1 - 3\sqrt{3}}{2\sqrt{10}} = \dfrac{\sqrt{10} - 3\sqrt{30}}{20}$ and $\sin \left(\sin^{-1} \dfrac{1}{2} - \tan^{-1} (-3)\right) = \dfrac{\sqrt{10} - 3\sqrt{30}}{20}$

86. Since $\cos\theta = \dfrac{\sqrt{3}}{2}$, then θ is in quadrant I. Sketch and label a triangle in quadrant I, with $x = \sqrt{3}$ and $r = 2$

    Use pythagorean theorem to find y: $(\sqrt{3})^2 + y^2 = 2^2 \Rightarrow y^2 = 4 - 3 \Rightarrow y^2 = 1 \Rightarrow y = 1.$

    Since $\sin\beta = -\dfrac{3}{5}$, then β is in quadrant IV. Sketch and label a triangle in quadrant IV, with $y = -3$ and $r = 5.$

    Use pythagorean theorem to find x: $x^2 + (-3)^2 = 5^2 \Rightarrow x^2 = 25 - 9 \Rightarrow x^2 = 16 \Rightarrow x = 4.$

    Now use $\tan(\theta - \beta) = \dfrac{\tan\theta - \tan\beta}{1 + \tan\theta\tan\beta}$, therefore, $\dfrac{\frac{1}{\sqrt{3}} - \left(-\frac{3}{4}\right)}{1 + \left(\frac{1}{\sqrt{3}}\right)\left(-\frac{3}{4}\right)} = \dfrac{\frac{1}{\sqrt{3}} - \left(-\frac{3}{4}\right)}{1 + \left(\frac{1}{\sqrt{3}}\right)\left(-\frac{3}{4}\right)} \cdot \dfrac{4\sqrt{3}}{4\sqrt{3}} = \dfrac{4 + 3\sqrt{3}}{4\sqrt{3} - 3} =$

    $\dfrac{4 + 3\sqrt{3}}{4\sqrt{3} - 3} \cdot \dfrac{4\sqrt{3} + 3}{4\sqrt{3} + 3} = \dfrac{48 + 25\sqrt{3}}{39}$ and $\tan\left(\cos^{-1}\dfrac{\sqrt{3}}{2} - \sin^{-1}\left(-\dfrac{3}{5}\right)\right) = \dfrac{48 + 25\sqrt{3}}{39}$

87. $\cos\left(\tan^{-1}.5\right) \approx .894427191$

88. $\sin\left(\cos^{-1}.25\right) \approx .9682458366$

89. $\tan(\arcsin .12251014) \approx .1234399811$

90. $\cot(\arccos .58236841) = \dfrac{1}{\tan(\arccos .58236841)} \approx .716386406$

91. $\sec\left(\cos^{-1}u\right) = \dfrac{1}{\cos\left(\cos^{-1}u\right)} = \dfrac{1}{u}$

92. $\cot\left(\tan^{-1}u\right) = \dfrac{1}{\tan\left(\tan^{-1}u\right)} = \dfrac{1}{u}$

93. Since $\cos\theta = u$, then θ is in quadrant I. Sketch and label a triangle in quadrant I, with $x = u$ and $r = 1.$

    Use pythagorean theorem to find y: $u^2 + y^2 = 1^2 \Rightarrow y^2 = 1 - u^2 \Rightarrow y = \sqrt{1 - u^2}.$

    Now $\sin\theta = \dfrac{\sqrt{1 - u^2}}{1} = \sqrt{1 - u^2}$ and $\sin(\arccos u) = \sqrt{1 - u^2}.$

94. Since $\cos\theta = u$, then θ is in quadrant I. Sketch and label a triangle in quadrant I, with $x = u$ and $r = 1.$

    Use pythagorean theorem to find y: $u^2 + y^2 = 1^2 \Rightarrow y^2 = 1 - u^2 \Rightarrow y = \sqrt{1 - u^2}.$

    Now $\tan\theta = \dfrac{\sqrt{1 - u^2}}{u}$ and $\tan(\arccos u) = \dfrac{\sqrt{1 - u^2}}{u}.$

95. Since $\sin\theta = u$, then θ is in quadrant I. Sketch and label a triangle in quadrant I, with $y = u$ and $r = 1.$

    Use pythagorean theorem to find x: $x^2 + u^2 = 1^2 \Rightarrow x^2 = 1 - u^2 \Rightarrow x = \sqrt{1 - u^2}.$

    Now $\cot\theta = \dfrac{\sqrt{1 - u^2}}{u}$ and $\cot(\arcsin u) = \dfrac{\sqrt{1 - u^2}}{u}.$

96. Since $\sin\theta = u$, then θ is in quadrant I. Sketch and label a triangle in quadrant I, $y = u$ and $r = 1.$

    Use pythagorean theorem to find x: $x^2 + u^2 = 1^2 \Rightarrow x^2 = 1 - u^2 \Rightarrow x = \sqrt{1 - u^2}.$

    Now $\cos\theta = \dfrac{\sqrt{1 - u^2}}{1} = \sqrt{1 - u^2}$ and $\cos(\arcsin u) = \sqrt{1 - u^2}.$

97. Since $\sec\theta = \dfrac{u}{2}$, then θ is in quadrant I. Sketch and label a triangle in quadrant I, $x = 2$ and $r = u.$

    Use pythagorean theorem to find y: $2^2 + y^2 = u^2 \Rightarrow y^2 = u^2 - 4 \Rightarrow y = \sqrt{u^2 - 4}.$

    Now $\sin\theta = \dfrac{\sqrt{u^2 - 1}}{u}$ and $\sin\left(\sec^{-1}\dfrac{u}{2}\right) = \dfrac{\sqrt{u^2 - 4}}{u}.$

98. Since $\tan \theta = \dfrac{3}{u}$, then θ is in quadrant I. Sketch and label a triangle in quadrant I, $x = u$ and $y = 3$.

    Use pythagorean theorem to find r: $u^2 + 3^2 = r^2 \Rightarrow r^2 = u^2 + 9 \Rightarrow r = \sqrt{u^2 + 9}$.

    Now $\cos \theta = \dfrac{u}{\sqrt{u^2 + 9}} = \dfrac{u\sqrt{u^2 + 9}}{u^2 + 9}$ and $\cos\left(\tan^{-1}\dfrac{3}{u}\right) = \dfrac{u\sqrt{u^2 + 9}}{u^2 + 9}$.

99. Since $\sin \theta = \dfrac{u}{\sqrt{u^2 + 2}}$, then θ is in quadrant I. Sketch and label a triangle in quadrant I.

    Use $r = \sqrt{u^2 + 2}$ and pythagorean theorem to find x: $x^2 + u^2 = (\sqrt{u^2 + 2})^2 \Rightarrow$

    $x^2 = u^2 + 2 - u^2 \Rightarrow x^2 = 2 \Rightarrow x = \sqrt{2}$. Now $\tan \theta = \dfrac{u}{\sqrt{2}} = \dfrac{u\sqrt{2}}{2}$ and $\tan\left(\sin^{-1}\dfrac{u}{\sqrt{u^2 + 2}}\right) = \dfrac{u\sqrt{2}}{2}$.

100. Since $\cos \theta = \dfrac{u}{\sqrt{u^2 + 5}}$, then θ is in quadrant I. Sketch and label a triangle in quadrant I.

    Use $r = \sqrt{u^2 + 5}$ to find $\sec \theta = \dfrac{\sqrt{u^2 + 5}}{u}$ and $\sec\left(\cos^{-1}\dfrac{u}{\sqrt{u^2 + 5}}\right) = \dfrac{\sqrt{u^2 + 5}}{u}$.

    Note that $\sec(\cos^{-1}x) = \dfrac{1}{x}$, this would yield: $\dfrac{1}{\frac{u}{\sqrt{u^2 + 5}}} = \dfrac{\sqrt{u^2 + 5}}{u}$, which is the same answer.

101. Since $\cot \theta = \dfrac{\sqrt{4 - u^2}}{u}$, then θ is in quadrant I. Sketch and label a triangle in quadrant I.

    Use $x = \sqrt{4 - u^2}$ and pythagorean theorem to find r: $(\sqrt{4 - u^2})^2 + u^2 = r^2 \Rightarrow r^2 = 4 - u^2 + u^2 \Rightarrow$

    $r^2 = 4 \Rightarrow r = 2$. Now $\sec \theta = \dfrac{2}{\sqrt{4 - u^2}} = \dfrac{2\sqrt{4 - u^2}}{4 - u^2}$ and $\sec\left(\text{arccot}\dfrac{\sqrt{4 - u^2}}{u}\right) = \dfrac{2\sqrt{4 - u^2}}{4 - u^2}$.

102. Since $\tan \theta = \dfrac{\sqrt{9 - u^2}}{u}$, then θ is in quadrant I. Sketch and label a triangle in quadrant I.

    Use $y = \sqrt{9 - u^2}$ and pythagorean theorem to find r: $u^2 + (\sqrt{9 - u^2})^2 = r^2 \Rightarrow r^2 = u^2 + 9 - u^2 \Rightarrow$

    $r^2 = 9 \Rightarrow r = 3$. Now $\csc \theta = \dfrac{3}{\sqrt{9 - u^2}} = \dfrac{3\sqrt{9 - u^2}}{9 - u^2}$ and $\csc\left(\arctan\dfrac{\sqrt{9 - u^2}}{u}\right) = \dfrac{3\sqrt{9 - u^2}}{9 - u^2}$.

103. (a) $\theta = \arcsin\left(\sqrt{\dfrac{v^2}{2v^2 + 64(0)}}\right) = \arcsin\left(\sqrt{\dfrac{v^2}{2v^2}}\right) = \arcsin\sqrt{\dfrac{1}{2}} = \arcsin\dfrac{1}{\sqrt{2}} = 45°$

    (b) $\theta = \arcsin\left(\sqrt{\dfrac{v^2}{2v^2 + 64(6)}}\right) = \arcsin\left(\sqrt{\dfrac{v^2}{2v^2 + 384}}\right)$. As $v \to \infty$, $\sqrt{\dfrac{v^2}{2v^2 + 384}} \to \sqrt{\dfrac{1}{2}}$,

    therefore $\theta = \arcsin\sqrt{\dfrac{1}{2}} = \arcsin\dfrac{1}{\sqrt{2}} = 45°$. The equation of the asymptote is

    $\theta = 45°$ or since $\sin 45° = \cos 45°$, $x = y$.

104. (a) Use a calculator, for $m = 1.2$: $\alpha = 2\arcsin\dfrac{1}{1.2} \approx 113°$.

    (b) Use a calculator, for $m = 1.5$: $\alpha = 2\arcsin\dfrac{1}{1.5} \approx 84°$.

    (c) Use a calculator, for $m = 2$: $\alpha = 2\arcsin\dfrac{1}{2} \approx 60°$.

    (d) Use a calculator, for $m = 2.5$: $\alpha = 2\arcsin\dfrac{1}{2.5} \approx 47°$.

105. (a) For $x = 3$: $\theta = \tan^{-1}\left(\dfrac{3(3)}{(3)^2 + 4}\right) = \tan^{-1}\dfrac{9}{13} \approx 34.70 \approx 35°$

(b) For $x = 6$: $\theta = \tan^{-1}\left(\dfrac{3(6)}{(6)^2 + 4}\right) = \tan^{-1}\dfrac{18}{40} = \tan^{-1}\dfrac{9}{20} \approx 24.23 \approx 24°$

(c) For $x = 9$: $\theta = \tan^{-1}\left(\dfrac{3(9)}{(9)^2 + 4}\right) = \tan^{-1}\dfrac{27}{85} \approx 17.62 \approx 18°$

(d) Sketch and label the two triangles formed above the sight line.

From this sketch $\tan(\theta + \alpha) = \dfrac{3 + 1}{x} = \dfrac{4}{x}$ and $\tan \alpha = \dfrac{1}{x}$.

Now use these values in the tangent of a sum identity $\tan(\theta + \alpha) = \dfrac{\tan\theta + \tan\alpha}{1 - \tan\theta\tan\alpha}$:

$\dfrac{4}{x} = \dfrac{\tan\theta + \frac{1}{x}}{1 - (\tan\theta)\left(\frac{1}{x}\right)} \Rightarrow \dfrac{4}{x} = \dfrac{x\tan\theta + 1}{x - \tan\theta} \Rightarrow 4(x - \tan\theta) = x(x\tan\theta + 1) =$

$4x - 4\tan\theta = x^2\tan\theta + x \Rightarrow 4x - x = x^2\tan\theta + 4\tan\theta \Rightarrow 3x = \tan\theta(x^2 + 4) \Rightarrow$

$\tan\theta = \dfrac{3x}{x^2 + 4} \Rightarrow \theta = \tan^{-1}\left(\dfrac{3x}{x^2 + 4}\right).$

(e) Graph $y_1 = \tan^{-1}\left(\dfrac{3x}{x^2 + 4}\right)$, see Figure 105. The maximum value occurs when $x \approx 2$ feet.

$[-4, 4]$ by $[-1, 1]$
Xscl $= 1$ Yscl $= 1$

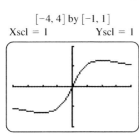

Figure 105

106. Since the diameter of the earth is 7927 miles at the equator, the radius of the earth is 3963.5 miles.

Now from the picture, $\cos\theta = \dfrac{3963.5}{20{,}000 + 3963.5} = \dfrac{3963.5}{23{,}963.5} \Rightarrow \theta = \cos^{-1}\dfrac{3963.5}{23{,}963.5} \approx 80.48°.$

Finally, the percent of the equator that can be seen by the satellite is: $\dfrac{2\theta}{360} \cdot 100 = \dfrac{2(80.48)}{360} \cdot 100 \approx 44.7\%.$

107. Let α be the angle to the right of θ, and let β be the angle to the left of θ. Then $\beta + \theta + \alpha = \pi$ and

$\theta = \pi - \alpha - \beta$. Now from the picture $\tan\alpha = \dfrac{150}{x}$, therefore $\alpha = \tan^{-1}\left(\dfrac{150}{x}\right)$; and $\tan\beta = \dfrac{75}{100 - x}$,

therefore $\beta = \tan^{-1}\left(\dfrac{75}{100 - x}\right)$. As a result: $\theta = \pi - \tan^{-1}\left(\dfrac{150}{x}\right) - \tan^{-1}\left(\dfrac{75}{100 - x}\right) =$

$\pi - \arctan\left(\dfrac{150}{x}\right) - \arctan\left(\dfrac{75}{100 - x}\right).$

Reviewing Basic Concepts (Sections 9.3 and 9.4)

1. Since $\dfrac{\pi}{2} < x < \pi$ and $\cos 2x = -\dfrac{5}{12}$, x is found in quadrant II.

    Now solve for x: $\cos 2x = 1 - 2 \sin^2 x \Rightarrow -\dfrac{5}{12} = = 1 - 2 \sin^2 x \Rightarrow 2 \sin^2 x = \dfrac{17}{12} \Rightarrow \sin^2 x = \dfrac{17}{24} \Rightarrow$

    $\sin x = \dfrac{\sqrt{17}}{\sqrt{24}}$ (positive because in quadrant II) $\Rightarrow \sin x = \dfrac{\sqrt{17}}{2\sqrt{6}} \cdot \dfrac{\sqrt{6}}{\sqrt{6}} \Rightarrow \sin x = \dfrac{\sqrt{102}}{12}$.

    Sketch and label a triangle in quadrant II, with $y = \sqrt{102}$ and $r = 12$.

    Now use pythagorean theorem to find x: $x^2 + (\sqrt{102})^2 = 12^2 \Rightarrow x^2 = 144 - 102 \Rightarrow x^2 = 42 \Rightarrow x = \sqrt{42}$.

    Therefore, since x is in quadrant II, $\tan x = -\dfrac{\sqrt{102}}{\sqrt{42}} = -\dfrac{\sqrt{102}}{\sqrt{42}} \cdot \dfrac{\sqrt{42}}{\sqrt{42}} = -\dfrac{\sqrt{4284}}{42} = -\dfrac{6\sqrt{119}}{42} \Rightarrow$

    $\tan x = -\dfrac{\sqrt{119}}{7}$.

2. Since $\sin \theta = -\dfrac{1}{3}$ and θ is in quadrant III. Sketch and label a triangle in quadrant III, with $y = -1$ and $r = 3$.

    Now use pythagorean theorem to find x: $x^2 + (-1)^2 = 3^2 \Rightarrow x^2 = 9 - 1 \Rightarrow x^2 = 8 \Rightarrow x = \sqrt{8} = 2\sqrt{2}$,

    since x is in quadrant III, $x = -2\sqrt{2}$. Therefore, $\cos \theta = -\dfrac{2\sqrt{2}}{3}$. Now find:

    $\sin 2\theta = 2 \sin \theta \cos \theta = 2\left(-\dfrac{1}{2}\right)\left(-\dfrac{2\sqrt{2}}{3}\right) = \dfrac{4\sqrt{2}}{9}$;

    $\cos 2\theta = 2 \cos^2 \theta - 1 = 2\left(-\dfrac{2\sqrt{2}}{3}\right)^2 - 1 = 2\left(\dfrac{8}{9}\right) - 1 = \dfrac{16}{9} - \dfrac{9}{9} = \dfrac{7}{9}$;

    $\tan 2\theta = \dfrac{\sin 2\theta}{\cos 2\theta} = \dfrac{\frac{4\sqrt{2}}{9}}{\frac{7}{9}} = \dfrac{4\sqrt{2}}{9} \cdot \dfrac{9}{7} = \dfrac{4\sqrt{2}}{7}$.

3. $\sin 75° = \sin \dfrac{150°}{2} = \sqrt{\dfrac{1 - \cos 150°}{2}} = \sqrt{\dfrac{1 - (-\frac{\sqrt{3}}{2})}{2} \cdot \dfrac{2}{2}} = \dfrac{\sqrt{2 + \sqrt{3}}}{2}$

    or: $\sin 75° = \sin (30° + 45°) = \sin 30° \cos 45° + \cos 30° \sin 45° = \dfrac{1}{2} \cdot \dfrac{\sqrt{2}}{2} + \dfrac{\sqrt{3}}{2} \cdot \dfrac{\sqrt{2}}{2} =$

    $\dfrac{\sqrt{2}}{4} + \dfrac{\sqrt{6}}{4} = \dfrac{\sqrt{2} + \sqrt{6}}{4}$.

4. $2 \sin 25° \cos 150° = 2\left[\dfrac{1}{2}\left[\sin (25° + 150°) + \sin (25° - 150°)\right]\right] = \sin 175° + \sin (-125°) =$

    $\sin 175° - \sin 125°$

5. (a) If the left side equals: $\sin^2 \dfrac{x}{2} = \left(\pm\sqrt{\dfrac{1 - \cos x}{2}}\right)^2 = \dfrac{1 - \cos x}{2}$, and the right side equals:

    $\dfrac{\tan x - \sin x}{2 \tan x} = \dfrac{\frac{\sin x}{\cos x} - \sin x}{2 \frac{\sin x}{\cos x}} = \dfrac{\frac{\sin x}{\cos x} - \sin x}{2 \frac{\sin x}{\cos x}} \cdot \dfrac{\cos x}{\cos x} = \dfrac{\sin x - \sin x \cos x}{2 \sin x} = \dfrac{\sin x (1 - \cos x)}{2 \sin x} =$

    $\dfrac{1 - \cos x}{2}$, then $\sin^2 \dfrac{x}{2} = \dfrac{\tan x - \sin x}{2 \tan x}$.

    (b) If the left side equals: $\dfrac{\sin 2x}{2 \sin x} = \dfrac{2 \sin x \cos x}{2 \sin x} = \cos x$, and the right side equals:

    $\cos^2 \dfrac{x}{2} - \sin^2 \dfrac{x}{2} = \left(\pm\sqrt{\dfrac{1 + \cos x}{2}}\right)^2 - \left(\pm\sqrt{\dfrac{1 - \cos x}{2}}\right)^2 = \dfrac{1 + \cos x}{2} - \dfrac{1 - \cos x}{2} =$

    $\dfrac{1 + \cos x - (1 - \cos x)}{2} = \dfrac{2 \cos x}{2} = \cos x$, then $\dfrac{\sin 2x}{2 \sin x} = \cos^2 \dfrac{x}{2} - \sin^2 \dfrac{x}{2}$.

6. (a) $y = \arccos \dfrac{\sqrt{3}}{2} \Rightarrow \dfrac{\sqrt{3}}{2} = \cos y \Rightarrow y = \dfrac{\pi}{6}$

   (b) $y = \sin^{-1}\left(-\dfrac{\sqrt{2}}{2}\right) \Rightarrow -\dfrac{\sqrt{2}}{2} = \sin y \Rightarrow y = -\dfrac{\pi}{4}$

7. (a) $\theta = \arccos .5 \Rightarrow .5 = \cos \theta \Rightarrow \theta = 60°$

   (b) $\theta = \cot^{-1}(-1) \Rightarrow -1 = \cot \theta \Rightarrow \theta = 135°$

8. Graph $y = 2\csc^{-1} x$ (like the graph of $y = \csc^{-1} x$, with a range $2\left(-\dfrac{\pi}{2}, \dfrac{\pi}{2}\right) = (-\pi, \pi)$), see Figure 8.

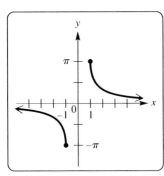

Figure 8

9. (a) Since $\sin \theta = -\dfrac{2}{3}$, then θ is in quadrant IV. Sketch and label a triangle in quadrant IV, with $y = -2$ and

   $r = 3$. Now use pythagorean theorem to find x: $x^2 + (-2)^2 = 3^2 \Rightarrow x^2 = 9 - 4 \Rightarrow x^2 = 5 \Rightarrow x = \sqrt{5}$.

   Therefore, $\cot \theta = \dfrac{\sqrt{5}}{-2} = -\dfrac{\sqrt{5}}{2}$ and $\cot\left(\arcsin\left(-\dfrac{2}{3}\right)\right) = -\dfrac{\sqrt{5}}{2}$.

   (b) Since $\tan \theta = \dfrac{5}{12}$, then θ is in quadrant I. Sketch and label a triangle in quadrant I, with $x = 12$ and $y = 5$.

   Use pythagorean theorem to find r: $12^2 + 5^2 = r^2 \Rightarrow r^2 = 144 + 25 \Rightarrow r^2 = 169 \Rightarrow r = 13$.

   Since $\sin \beta = \dfrac{3}{5}$, then β is in quadrant I. Sketch and label a triangle in quadrant I, with $y = 3$ and $r = 5$.

   Use pythagorean theorem to find x: $x^2 + 3^2 = 5^2 \Rightarrow r^2 = 25 - 9 \Rightarrow x^2 = 16 \Rightarrow x = 4$.

   Now $\cos(\theta - \beta) = \cos\theta\cos\beta + \sin\theta\sin\beta$, therefore, $\cos(\theta + \beta) = \dfrac{12}{13}\cdot\dfrac{4}{5} + \dfrac{5}{13}\cdot\dfrac{3}{5} = \dfrac{48}{65} + \dfrac{15}{65} = \dfrac{63}{65}$

   and $\cos\left(\tan^{-1}\dfrac{5}{12} - \sin^{-1}\dfrac{3}{5}\right) = \dfrac{63}{65}$.

10. Since $\cot \theta = u$, then θ is in quadrant I. Sketch and label a triangle in quadrant I, with $x = u$ and $y = 1$.

    Use pythagorean theorem to find r: $u^2 + 1^2 = r^2 \Rightarrow r^2 = u^2 + 1 \Rightarrow r = \sqrt{u^2 + 1}$.

    Now $\sin \theta = \dfrac{1}{\sqrt{u^2 + 1}} = \dfrac{1}{\sqrt{u^2 + 1}}\cdot\dfrac{\sqrt{u^2 + 1}}{\sqrt{u^2 + 1}} = \dfrac{\sqrt{u^2 + 1}}{u^2 + 1}$ and $\sin(\text{arccot } u) = \dfrac{\sqrt{u^2 + 1}}{u^2 + 1}$.

9.5: Trigonometric Equations and Inequalities (I)

1. $2\cos x + 1 = 0 \Rightarrow 2\cos x = -1 \Rightarrow \cos x = -\dfrac{1}{2} \Rightarrow x = \dfrac{2\pi}{3}, \dfrac{4\pi}{3}$. The solution set is: $\left\{\dfrac{2\pi}{3}, \dfrac{4\pi}{3}\right\}$.

2. $2\sin x + 1 = 0 \Rightarrow 2\sin x = -1 \Rightarrow \sin x = -\dfrac{1}{2} \Rightarrow x = \dfrac{7\pi}{6}, \dfrac{11\pi}{6}$. The solution set is: $\left\{\dfrac{7\pi}{6}, \dfrac{11\pi}{6}\right\}$.

3. $5\sin x - 5 = 0 \Rightarrow 5\sin x = 5 \Rightarrow \sin x = 1 \Rightarrow x = \dfrac{\pi}{2}$. The solution set is: $\left\{\dfrac{\pi}{2}\right\}$.

4. $3\cos x + 3 = 0 \Rightarrow 3\cos x = -3 \Rightarrow \cos x = -1 \Rightarrow x = \pi$. The solution set is: $\{\pi\}$.

5. $2\tan x + 1 = -1 \Rightarrow 2\tan x = -2 \Rightarrow \tan x = -1 \Rightarrow x = \dfrac{3\pi}{4}, \dfrac{7\pi}{4}$. The solution set is: $\left\{\dfrac{3\pi}{4}, \dfrac{7\pi}{4}\right\}$.

6. $2\cot x + 1 = -1 \Rightarrow 2\cot x = -2 \Rightarrow \cot x = -1 \Rightarrow x = \dfrac{3\pi}{4}, \dfrac{7\pi}{4}$. The solution set is: $\left\{\dfrac{3\pi}{4}, \dfrac{7\pi}{4}\right\}$.

7. $2\cos x + 5 = 6 \Rightarrow 2\cos x = 1 \Rightarrow \cos x = \dfrac{1}{2} \Rightarrow x = \dfrac{\pi}{3}, \dfrac{5\pi}{3}$. The solution set is: $\left\{\dfrac{\pi}{3}, \dfrac{5\pi}{3}\right\}$.

8. $2\sin x + 3 = 4 \Rightarrow 2\sin x = 1 \Rightarrow \sin x = \dfrac{1}{2} \Rightarrow x = \dfrac{\pi}{6}, \dfrac{5\pi}{6}$. The solution set is: $\left\{\dfrac{\pi}{6}, \dfrac{5\pi}{6}\right\}$.

9. $2\csc x + 4 = \csc x + 6 \Rightarrow \csc x = 2 \Rightarrow \sin x = \dfrac{1}{2} \Rightarrow x = \dfrac{\pi}{6}, \dfrac{5\pi}{6}$. The solution set is: $\left\{\dfrac{\pi}{6}, \dfrac{5\pi}{6}\right\}$.

10. $2\sec x + 1 = \sec x + 3 \Rightarrow \sec x = 2 \Rightarrow x = \dfrac{\pi}{3}, \dfrac{5\pi}{3}$. The solution set is: $\left\{\dfrac{\pi}{3}, \dfrac{5\pi}{3}\right\}$.

11. From the factored equation $(\cot x - 1)(\sqrt{3}\cot x + 1) = 0$, either $\cot x - 1 = 0 \Rightarrow \cot x = 1 \Rightarrow x = \dfrac{\pi}{4}, \dfrac{5\pi}{4}$;

    or $\sqrt{3}\cot x + 1 = 0 \Rightarrow \sqrt{3}\cot x = -1 \Rightarrow \cot x = -\dfrac{1}{\sqrt{3}} \Rightarrow \cot x = -\dfrac{\sqrt{3}}{3} \Rightarrow x = \dfrac{2\pi}{3}, \dfrac{5\pi}{3}$.

    The solution set is: $\left\{\dfrac{\pi}{4}, \dfrac{2\pi}{3}, \dfrac{5\pi}{4}, \dfrac{5\pi}{3}\right\}$.

12. From the factored equation $(\csc x + 2)(\csc x - \sqrt{2}) = 0$, either

    $\csc x + 2 = 0 \Rightarrow \csc x = -2 \Rightarrow x = \dfrac{7\pi}{6}, \dfrac{11\pi}{6}$; or $\csc x - \sqrt{2} = 0 \Rightarrow \csc x = \sqrt{2} \Rightarrow x = \dfrac{\pi}{4}, \dfrac{3\pi}{4}$.

    The solution set is: $\left\{\dfrac{\pi}{4}, \dfrac{3\pi}{4}, \dfrac{7\pi}{6}, \dfrac{11\pi}{6}\right\}$.

13. First set the equation equal to zero and factor: $\cos x \cot x = \cos x \Rightarrow \cos x \cot x - \cos x = 0 \Rightarrow$

    $\cos x (\cot x - 1) = 0$. Now from this factored equation either $\cos x = 0 \Rightarrow x = \dfrac{\pi}{2}, \dfrac{3\pi}{2}$;

    or $\cot x - 1 = 0 \Rightarrow \cot x = 1 \Rightarrow x = \dfrac{\pi}{4}, \dfrac{5\pi}{4}$. The solution set is: $\left\{\dfrac{\pi}{4}, \dfrac{\pi}{2}, \dfrac{5\pi}{4}, \dfrac{3\pi}{2}\right\}$.

14. First set the equation equal to zero and factor: $\sin x \cot x = \sin x \Rightarrow \sin x \cot x - \sin x = 0 \Rightarrow$

    $\sin x (\cot x - 1) = 0$. Now from this factored equation, either $\sin x = 0 \Rightarrow x = 0, \pi$;

    or $\cot x - 1 = 0 \Rightarrow \cot x = 1 \Rightarrow x = \dfrac{\pi}{4}, \dfrac{5\pi}{4}$. The solution set is: $\left\{0, \dfrac{\pi}{4}, \pi, \dfrac{5\pi}{4}\right\}$.

15. First set the equation equal to zero and factor: $\sin^2 x - 2\sin x + 1 = 0 \Rightarrow (\sin x - 1)^2 = 0$. Now from this

    factored equation, $\sin x - 1 = 0 \Rightarrow \sin x = 1 \Rightarrow x = \dfrac{\pi}{2}$. The solution set is: $\left\{\dfrac{\pi}{2}\right\}$.

16. First factor: $\cos^2 x + 2\cos x + 1 = 0 \Rightarrow (\cos x + 1)^2 = 0$. For this factored equation,

    $\cos x + 1 = 0 \Rightarrow \cos x = -1 \Rightarrow x = \pi$. The solution set is: $\{\pi\}$.

17. If $4(1 + \sin x)(1 - \sin x) = 3$, then $1 - \sin^2 x = \dfrac{3}{4} \Rightarrow \cos^2 x = \dfrac{3}{4} \Rightarrow \cos x = \pm\dfrac{\sqrt{3}}{2} \Rightarrow$

    $x = \left\{\dfrac{\pi}{6}, \dfrac{5\pi}{6}, \dfrac{7\pi}{6}, \dfrac{11\pi}{6}\right\}$. The solution set is: $\left\{\dfrac{\pi}{6}, \dfrac{5\pi}{6}, \dfrac{7\pi}{6}, \dfrac{11\pi}{6}\right\}$.

18. From the factored equation $(\cot x - \sqrt{3})(2 \sin x + \sqrt{3}) = 0$, either $\cot x - \sqrt{3} = 0 \Rightarrow \cot x = \sqrt{3} \Rightarrow$

    $x = \dfrac{\pi}{6}, \dfrac{7\pi}{6}$; or $2 \sin x + \sqrt{3} = 0 \Rightarrow 2 \sin x = -\sqrt{3} \Rightarrow \sin x = -\dfrac{\sqrt{3}}{2} \Rightarrow x = \dfrac{4\pi}{3}, \dfrac{5\pi}{3}$.

    The solution set is: $\left\{\dfrac{\pi}{6}, \dfrac{7\pi}{6}, \dfrac{4\pi}{3}, \dfrac{5\pi}{3}\right\}$.

19. First set the equation equal to zero and factor: $\tan x + 1 = \sqrt{3} + \sqrt{3} \cot x \Rightarrow \tan x + 1 - \sqrt{3} - \dfrac{\sqrt{3}}{\tan x} = 0$

    $\Rightarrow \tan^2 x + \tan x (1 - \sqrt{3}) - \sqrt{3} = 0 \Rightarrow (\tan x + 1)(\tan x - \sqrt{3}) = 0$. Now from this factored equation,

    either, $\tan x + 1 = 0 \Rightarrow \tan x = -1 \Rightarrow x = \dfrac{3\pi}{4}, \dfrac{7\pi}{4}$; or $\tan x - \sqrt{3} = 0 \Rightarrow \tan x = \sqrt{3} \Rightarrow x = \dfrac{\pi}{3}, \dfrac{4\pi}{3}$.

    The solution set is: $\left\{\dfrac{\pi}{3}, \dfrac{3\pi}{4}, \dfrac{4\pi}{3}, \dfrac{7\pi}{4}\right\}$.

20. $\tan x - \cot x = 0 \Rightarrow \tan x - \dfrac{1}{\tan x} = 0 \Rightarrow \tan^2 x - 1 = 0 \Rightarrow \tan^2 x = 1 \Rightarrow \tan x = \pm 1 \Rightarrow \dfrac{\pi}{4}, \dfrac{3\pi}{4}, \dfrac{5\pi}{4}, \dfrac{7\pi}{4}$.

    The solution set is: $\left\{\dfrac{\pi}{4}, \dfrac{3\pi}{4}, \dfrac{5\pi}{4}, \dfrac{7\pi}{4}\right\}$.

21. First set the equation equal to zero and factor: $2 \sin x - 1 = \csc x \Rightarrow 2 \sin x - 1 - \dfrac{1}{\sin x} = 0 \Rightarrow$

    $2 \sin^2 x - \sin x - 1 = 0 \Rightarrow (2 \sin x + 1)(\sin x - 1) = 0$.

    From this factored equation either, $2 \sin x + 1 = 0 \Rightarrow 2 \sin x = -1 \Rightarrow \sin x = -\dfrac{1}{2} \Rightarrow x = \dfrac{7\pi}{6}, \dfrac{11\pi}{6}$;

    or $\sin x - 1 = 0 \Rightarrow \sin x = 1 \Rightarrow x = \dfrac{\pi}{2}$. The solution set is: $\left\{\dfrac{\pi}{2}, \dfrac{7\pi}{6}, \dfrac{11\pi}{6}\right\}$.

22. $\cos^2 x = \sin^2 x \Rightarrow \dfrac{\cos^2 x}{\sin^2 x} = 1 \Rightarrow \tan^2 x = 1 \Rightarrow \tan x = \pm 1 \Rightarrow x = \dfrac{\pi}{4}, \dfrac{3\pi}{4}, \dfrac{5\pi}{4}, \dfrac{7\pi}{4}$.

    The solution set is: $\left\{\dfrac{\pi}{4}, \dfrac{3\pi}{4}, \dfrac{5\pi}{4}, \dfrac{7\pi}{4}\right\}$

23. $\cos^2 x - \sin^2 x = 1 \Rightarrow 1 - \sin^2 x - \sin^2 x = 1 \Rightarrow 1 - 2 \sin^2 x = 1 \Rightarrow$

    $-2 \sin^2 x = 0 \Rightarrow \sin^2 x = 0 \Rightarrow \sin x = 0 \Rightarrow x = 0, \pi$. The solution set is: $\{0, \pi\}$.

24. $\csc^2 x = 2 \cot x \Rightarrow \csc^2 x - 2 \cot x = 0 \Rightarrow 1 + \cot^2 x - 2 \cot x = 0 \Rightarrow \cot^2 x - 2 \cot x + 1 = 0 \Rightarrow$

    $(\cot x - 1)^2 = 0 \Rightarrow \cot x = 1 \Rightarrow x = \dfrac{\pi}{4}, \dfrac{5\pi}{4}$. The solution set is: $\left\{\dfrac{\pi}{4}, \dfrac{5\pi}{4}\right\}$.

25. (a) First simplify: $-2 \cos x + 1 = 0 \Rightarrow -2 \cos x = -1 \Rightarrow \cos x = \dfrac{1}{2}$.

    Now the reference angle is $\dfrac{\pi}{3}$ and x is found in quadrant I and IV, therefore $f(x) = 0$ when $x = \left\{\dfrac{\pi}{3}, \dfrac{5\pi}{3}\right\}$.

    (b) Graph $y_1 = -2 \cos x + 1$ on the calculator. From this graph, $f(x) > 0$ when the graph is above the x-axis,

    therefore for the interval: $\left(\dfrac{\pi}{3}, \dfrac{5\pi}{3}\right)$.

    (c) Graph $y_1 = -2 \cos x + 1$ on the calculator. From this graph, $f(x) < 0$ when the graph is below the x-axis,

    therefore for the interval: $\left[0, \dfrac{\pi}{3}\right) \cup \left(\dfrac{5\pi}{3}, 2\pi\right)$.

26. (a) First simplify: $2 \sin x + 1 \Rightarrow 2 \sin x = -1 \Rightarrow \sin x = -\dfrac{1}{2}$.

    Now the reference angle is $\dfrac{\pi}{6}$ and x is found in quadrant III and IV, therefore $f(x) = 0$ when $x = \left\{\dfrac{7\pi}{6}, \dfrac{11\pi}{6}\right\}$.

    (b) Graph $y_1 = 2 \sin x + 1$ on the calculator. From this graph, $f(x) > 0$ when the graph is above the x-axis,

    therefore for the interval: $\left[0, \dfrac{7\pi}{6}\right) \cup \left(\dfrac{11\pi}{6}, 2\pi\right)$.

    (c) Graph $y_1 = 2 \sin x + 1$ on the calculator. From this graph, $f(x) < 0$ when the graph is below the x-axis,

    therefore for the interval: $\left(\dfrac{7\pi}{6}, \dfrac{11\pi}{6}\right)$.

27. (a) First simplify: $\tan^2 x - 3 = 0 \Rightarrow \tan^2 x = 3 \Rightarrow \tan x = \pm\sqrt{3}$. Now the reference angle is $\dfrac{\pi}{3}$ and x is

    found in all quadrants, therefore $f(x) = 0$ when $x = \left\{\dfrac{\pi}{3}, \dfrac{2\pi}{3}, \dfrac{4\pi}{3}, \dfrac{5\pi}{3}\right\}$.

    (b) Graph $y_1 = \tan^2 x - 3$ on the calculator. From this graph, the asymptotes are $\dfrac{\pi}{2}$ and $\dfrac{3\pi}{2}$, and $f(x) > 0$

    when the graph is above the x-axis, therefore for the interval:

    $\left(\dfrac{\pi}{3}, \dfrac{\pi}{2}\right) \cup \left(\dfrac{\pi}{2}, \dfrac{2\pi}{3}\right) \cup \left(\dfrac{4\pi}{3}, \dfrac{3\pi}{2}\right) \cup \left(\dfrac{3\pi}{2}, \dfrac{5\pi}{3}\right)$.

    (c) Graph $y_1 = \tan^2 x - 3$ on the calculator. From this graph, $f(x) < 0$ when the graph is below the x-axis,

    therefore for the interval: $\left[0, \dfrac{\pi}{3}\right) \cup \left(\dfrac{2\pi}{3}, \dfrac{4\pi}{3}\right) \cup \left(\dfrac{5\pi}{3}, 2\pi\right)$.

28. (a) First simplify: $\sec^2 x - 1 = 0 \Rightarrow \sec^2 x = 1 \Rightarrow \sec x = \pm 1$.

    Now the reference angles are 0 and π, therefore $f(x) = 0$ when $x = \{0, \pi\}$.

    (b) Graph $y_1 = \sec^2 x - 1$ on the calculator. From this graph, the asymptotes are $\dfrac{\pi}{2}$ and $\dfrac{3\pi}{2}$, and $f(x) > 0$

    when the graph is above the x-axis, therefore for the interval: $\left(0, \dfrac{\pi}{2}\right) \cup \left(\dfrac{\pi}{2}, \pi\right) \cup \left(\pi, \dfrac{3\pi}{2}\right) \cup \left(\dfrac{3\pi}{2}, 2\pi\right)$.

    (c) Graph $y_1 = \sec^2 x - 1$ on the calculator. From this graph, $f(x) < 0$ when the graph is below the x-axis,

    therefore the solution is: $\varnothing$.

29. (a) First set the equation equal to zero and factor: $2 \cos^2 x - \sqrt{3} \cos x = 0 \Rightarrow \cos x (2 \cos x - \sqrt{3}) = 0$.

    Now from this factored equation, either, $\cos x = 0 \Rightarrow x = \dfrac{\pi}{2}, \dfrac{3\pi}{2}$; or $2 \cos x - \sqrt{3} = 0 \Rightarrow$

    $2 \cos x = \sqrt{3} \Rightarrow \cos x = \dfrac{\sqrt{3}}{2}$, therefore the reference angle is $\dfrac{\pi}{6}$, x is found in quadrant I and IV, and

    $x = \dfrac{\pi}{6}, \dfrac{11\pi}{6}$. The solution for $f(x) = 0$ is when $x = \left\{\dfrac{\pi}{6}, \dfrac{\pi}{2}, \dfrac{3\pi}{2}, \dfrac{11\pi}{6}\right\}$.

    (b) Graph $y_1 = 2 \cos^2 x - \sqrt{3} \cos x$ on the calculator. From this graph, $f(x) > 0$ when the graph is above

    the x-axis, therefore for the interval: $\left(0, \dfrac{\pi}{6}\right) \cup \left(\dfrac{\pi}{2}, \dfrac{3\pi}{2}\right) \cup \left(\dfrac{11\pi}{6}, 2\pi\right)$.

    (c) Graph $y_1 = 2 \cos^2 x - \sqrt{3} \cos x$ on the calculator. From this graph, $f(x) < 0$ when the graph is below

    the x-axis, therefore for the interval: $\left(\dfrac{\pi}{6}, \dfrac{\pi}{2}\right) \cup \left(\dfrac{3\pi}{2}, \dfrac{11\pi}{6}\right)$.

30. (a) First set the equation equal to zero and factor: $2\sin^2 x + 3\sin x + 1 = 0 \Rightarrow (2\sin x + 1)(\sin x + 1) = 0$.

    Now from this factored equation, either, $2\sin x + 1 = 0 \Rightarrow 2\sin x = -1 \Rightarrow \sin x = -\dfrac{1}{2}$,

    therefore the reference angle is $\dfrac{\pi}{6}$, x is found in quadrant III and IV, and $x = \dfrac{7\pi}{6}, \dfrac{11\pi}{6}$; or

    $\sin x + 1 = 0 \Rightarrow \sin x = -1 \Rightarrow x = \dfrac{3\pi}{2}$. The solution for $f(x) = 0$ is when $x = \left\{\dfrac{7\pi}{6}, \dfrac{3\pi}{2}, \dfrac{11\pi}{6}\right\}$.

    (b) Graph $y_1 = 2\sin^2 x + 3\sin x + 1$ on the calculator. From this graph, $f(x) > 0$ when the graph is above

    the x-axis, therefore for the interval: $\left[0, \dfrac{7\pi}{6}\right) \cup \left(\dfrac{11\pi}{6}, 2\pi\right)$.

    (c) Graph $y_1 = 2\sin^2 x + 3\sin x + 1$ on the calculator. From this graph, $f(x) < 0$ when the graph is below

    the x-axis, therefore for the interval: $\left(\dfrac{7\pi}{6}, \dfrac{3\pi}{2}\right) \cup \left(\dfrac{3\pi}{2}, \dfrac{11\pi}{6}\right)$.

31. (a) First set the equation equal to zero and factor: $\sin^2 x \cos x - \cos x = 0 \Rightarrow \cos x (\sin^2 x - 1) = 0$.

    Now from this factored equation, either, $\cos x = 0 \Rightarrow x = \dfrac{x}{2}, \dfrac{3\pi}{2}$; or $\sin^2 x - 1 = 0 \Rightarrow \sin^2 x = 1 \Rightarrow$

    $\sin x = \pm 1 \Rightarrow x = \dfrac{\pi}{2}, \dfrac{3\pi}{2}$, therefore $f(x) = 0$ when $x = \left\{\dfrac{\pi}{2}, \dfrac{3\pi}{2}\right\}$.

    (b) Graph $y_1 = \sin^2 x \cos x - \cos x$ on the calculator. From this graph, $f(x) > 0$ when the graph is above

    the x-axis, therefore for the interval: $\left(\dfrac{\pi}{2}, \dfrac{3\pi}{2}\right)$.

    (c) Graph $y_1 = \sin^2 x \cos x - \cos x$ on the calculator. From this graph, $f(x) < 0$ when the graph is below

    the x-axis, therefore for the interval: $\left[0, \dfrac{\pi}{2}\right) \cup \left(\dfrac{3\pi}{2}, 2\pi\right)$.

32. (a) First set the equation equal to zero and factor: $\sin^2 x \cos^2 x = 0 \Rightarrow (1 - \cos^2 x)(\cos^2 x) = 0$.

    Now from this factored equation, either, $1 - \cos^2 x \Rightarrow \cos^2 x = 1 \Rightarrow \cos x = \pm 1 \Rightarrow x = 0, \pi$;

    or $\cos^2 x = 0 \Rightarrow \cos x = 0 \Rightarrow x = \dfrac{\pi}{2}, \dfrac{3\pi}{2}$, therefore $f(x) = 0$ when $x = \left\{0, \dfrac{\pi}{2}, \pi, \dfrac{3\pi}{2}\right\}$.

    (b) Graph $y_1 = \sin^2 x \cos^2 x$ on the calculator. From this graph, $f(x) > 0$ when the graph is above

    the x-axis, therefore for the interval: $\left(0, \dfrac{\pi}{2}\right) \cup \left(\dfrac{\pi}{2}, \pi\right) \cup \left(\pi, \dfrac{3\pi}{2}\right) \cup \left(\dfrac{3\pi}{2}, 2\pi\right)$.

    (c) Graph $y_1 = \sin^2 x \cos^2 x$ on the calculator. From this graph, $f(x) < 0$ when the graph is below

    the x-axis, therefore the solution is: $\varnothing$.

33. (a) First set the equation equal to zero and factor: $2\tan^2 x \sin x - \tan^2 x = 0 \Rightarrow \tan^2 x (2\sin x - 1) = 0$.

    Now from this factored equation, either, $\tan^2 x = 0 \Rightarrow \tan x = 0 \Rightarrow x = 0, \pi$; or $2\sin x - 1 = 0 \Rightarrow$

    $2\sin x = 1 \Rightarrow \sin x = \dfrac{1}{2} \Rightarrow x = \dfrac{\pi}{6}, \dfrac{5\pi}{6}$, therefore $f(x) = 0$ when $x = \left\{0, \dfrac{\pi}{6}, \dfrac{5\pi}{6}, \pi\right\}$.

    (b) Graph $y_1 = 2\tan^2 x \sin x - \tan^2 x$ on the calculator. From this graph, the asymptotes are $\dfrac{\pi}{2}$ and $\dfrac{3\pi}{2}$,

    and $f(x) > 0$ when the graph is above the x-axis, therefore for the interval: $\left(\dfrac{\pi}{6}, \dfrac{\pi}{2}\right) \cup \left(\dfrac{\pi}{2}, \dfrac{5\pi}{6}\right)$.

    (c) Graph $y_1 = 2\tan^2 x \sin x - \tan^2 x$ on the calculator. From this graph, the asymptotes are $\dfrac{\pi}{2}$ and $\dfrac{3\pi}{2}$,

    and $f(x) < 0$ when the graph is below the x-axis, therefore for the interval:

    $\left(0, \dfrac{\pi}{6}\right) \cup \left(\dfrac{5\pi}{6}, \pi\right) \cup \left(\pi, \dfrac{3\pi}{2}\right) \cup \left(\dfrac{3\pi}{2}, 2\pi\right)$.

34. $\tan^3 x = 3 \tan x \Rightarrow \tan^3 - 3 \tan x = 0$

35. First factor: $\tan^3 x - 3 \tan x = 0 \Rightarrow \tan x \left(\tan^2 x - 3\right) = 0$.

    Now from this factored equation, either $\tan x = 0 \Rightarrow x = 0, \pi$; or $\tan^2 x - 3 = 0 \Rightarrow \tan^2 x = 3 \Rightarrow$

    or $\tan x = \pm\sqrt{3} \Rightarrow x = \dfrac{\pi}{3}, \dfrac{2\pi}{3}, \dfrac{4\pi}{3}, \dfrac{5\pi}{3}$, therefore the solution set is: $\left\{0, \dfrac{\pi}{3}, \dfrac{2\pi}{3}, \pi, \dfrac{4\pi}{3}, \dfrac{5\pi}{3}\right\}$.

36. $x = 0: (\tan 0)^3 - 3 \tan 0 = 0 \Rightarrow 0 - 0 = 0 \Rightarrow 0 = 0$

    $x = \dfrac{\pi}{3}: \left(\tan \dfrac{\pi}{3}\right)^3 - 3 \tan \dfrac{\pi}{3} = 0 \Rightarrow (\sqrt{3})^3 - 3(\sqrt{3}) = 3\sqrt{3} - 3\sqrt{3} = 0 \Rightarrow 0 = 0$

    $x = \pi: (\tan \pi)^3 - 3 \tan \pi = 0 \Rightarrow 0 - 0 = 0 \Rightarrow 0 = 0$

    $x = \dfrac{4\pi}{3}: \left(\tan \dfrac{4\pi}{3}\right)^3 - 3 \tan \dfrac{4\pi}{3} = 0 \Rightarrow (\sqrt{3})^3 - 3\sqrt{3} = 3\sqrt{3} - 3\sqrt{3} = 0 \Rightarrow 0 = 0$

    $x = \dfrac{5\pi}{3}: \left(\tan \dfrac{5\pi}{3}\right)^3 - 3 \tan \dfrac{5\pi}{3} = 0 \Rightarrow (-\sqrt{3})^3 - 3(-\sqrt{3}) = -3\sqrt{3} + 3\sqrt{3} = 0 \Rightarrow 0 = 0$

37. $\dfrac{\tan^3 x}{\tan x} = \dfrac{3 \tan x}{\tan x} \Rightarrow \tan^2 x = 3 \Rightarrow \tan x = \pm\sqrt{3} \Rightarrow x = \dfrac{\pi}{3}, \dfrac{2\pi}{3}, \dfrac{4\pi}{3}, \dfrac{5\pi}{3}$

38. The answers do not agree. The solutions 0 and π were lost when dividing by $\tan x$.

39. First put the equation in quadratic form and factor:

    $3 \sin^2 x - \sin x = 2 \Rightarrow 3 \sin^2 x - \sin x - 2 = 0 \Rightarrow (3 \sin x + 2)(\sin x - 1) = 0$.

    From this factored form, either $3 \sin x + 2 = 0 \Rightarrow 3 \sin x = -2 \Rightarrow \sin x = -\dfrac{2}{3}$; or

    $\sin x - 1 = 0 \Rightarrow \sin x = 1$.

    Now find x:

    (i) For $\sin x = 1: x = \sin^{-1}(1) \Rightarrow 1 = \sin x \Rightarrow x = \dfrac{\pi}{2}$

    (ii) For $\sin x = -\dfrac{2}{3}$: find the reference angle (calculate in radian mode): $\sin^{-1}\left(\dfrac{2}{3}\right) \approx .7297$.

      Since x is in quadrants III and IV, $x = \pi + .7297 \approx 3.87$ and $x = 2\pi - .7297 \approx 5.55$.

    Therefore $f(x) = 0$, when $x = \left\{\dfrac{\pi}{2}, 3.87, 5.55\right\}$.

40. First put the equation in quadratic form and factor by the quadratic formula:

    $\sin x = \dfrac{6 \pm \sqrt{(-6)^2 - 4(9)(-1)}}{2(9)} = \dfrac{6 \pm \sqrt{72}}{18} = \dfrac{6 \pm 6\sqrt{2}}{18} \Rightarrow \sin x = \dfrac{1 \pm \sqrt{2}}{3}$.

    Now find x:

    (i) For $\sin x = \dfrac{1 + \sqrt{2}}{3}$: find the reference angle (calculate in radian mode): $\sin^{-1}\left(\dfrac{1 + \sqrt{2}}{3}\right) \approx .9352$.

      Since x is found in quadrants I and II, $x \approx .935$ and $x = \pi - .9352 \approx 2.206$.

    (ii) For $\sin x = \dfrac{1 - \sqrt{2}}{3}$: find the reference angle (calculate in radian mode): $\sin^{-1}\left(\dfrac{1 - \sqrt{2}}{3}\right) \approx .1385$.

      Since x is found in quadrants III and IV, $x = \pi + .1385 \approx 3.280$ and $x = 2\pi - .1385 \approx 6.1447$.

    Therefore $f(x) = 0$, when $x = \{.94, 2.21, 3.28, 6.14\}$.

41. First put the equation in quadratic form and factor by the quadratic formula:

    $$\tan x = \frac{-4 \pm \sqrt{(4)^2 - 4(1)(2)}}{2(1)} = \frac{-4 \pm \sqrt{8}}{2} = \frac{-4 \pm 2\sqrt{2}}{2} \Rightarrow \tan x = -2 \pm \sqrt{2}.$$

    Now find x:

    (i) For $\tan x = -2 + \sqrt{2}$: find the reference angle (calculate in radian mode): $\tan^{-1}(-2 + \sqrt{2}) \approx .5299$.

    Since x is found in quadrants II and IV, $x = \pi - .5299 \approx 2.61$ and $x = 2\pi - .5299 \approx 5.75$.

    (ii) For $\tan x = -2 - \sqrt{2}$: find the reference angle (calculate in radian mode) $\tan^{-1}(-2 - \sqrt{2}) \approx 1.2859$.

    Since x is found in quadrants II and IV, $x = \pi - 1.2859 \approx 1.86$ and $x = 2\pi - 1.2859 \approx 5.00$.

    Therefore $f(x) = 0$, when $x = \{1.86, 2.61, 5.00, 5.75\}$.

42. First set the equation equal to zero and factor by the quadratic formula:

    $$3 \cot^2 x - 3 \cot x = 1 \Rightarrow 3 \cot^2 x - 3 \cot x - 1 = 0 \Rightarrow \cot x = \frac{3 \pm \sqrt{(-3)^2 - 4(3)(-1)}}{2(3)} = \frac{3 \pm \sqrt{21}}{6} \Rightarrow$$

    $$\tan x = \frac{6}{3 \pm \sqrt{21}}$$

    Now find x:

    (i) For $\tan x = \dfrac{6}{3 + \sqrt{21}}$: find the reference angle (calculate in radian mode): $\tan^{-1}\left(\dfrac{6}{3 + \sqrt{21}}\right) \approx .6694$.

    Since x is found in quadrants I and III, $x \approx .67$ and $x = \pi + .6694 = 3.81$.

    (ii) For $\tan x = \dfrac{6}{3 - \sqrt{21}}$: find the reference angle (calculate in radian mode) $\tan^{-1}\left(\dfrac{6}{3 - \sqrt{21}}\right) \approx 1.313$.

    Since x is found in quadrants II and IV, $x = \pi - 1.313 \approx 1.83$ and $x = 2\pi - 1.313 \approx 4.97$.

    Therefore $f(x) = 0$, when $x = \{.67, 1.83, 3.81, 4.97\}$.

43. First set the equation equal to zero and factor by the quadratic formula:

    $$2 \cos^2 x + 2 \cos x = 1 \Rightarrow 2 \cos^2 x + 2 \cos x - 1 = 0 \Rightarrow$$

    $$\cos x = \frac{-2 \pm \sqrt{(2)^2 - 4(2)(-1)}}{2(2)} = \frac{-2 \pm \sqrt{12}}{4} = \frac{-2 \pm 2\sqrt{3}}{4} \Rightarrow \cos x = \frac{-1 \pm \sqrt{3}}{2}$$

    Now find x:

    (i) For $\cos x = \dfrac{-1 + \sqrt{3}}{2}$: find the reference angle (calculate in radian mode) $\cos^{-1}\left(\dfrac{-1 + \sqrt{3}}{2}\right) \approx 1.196$.

    Since x is found in quadrants I and IV, $x \approx 1.20$ and $x = 2\pi - 1.196 = 5.09$.

    (ii) For $\cos x = \dfrac{-1 - \sqrt{3}}{2}$: find the reference angle (calculate in radian mode) $\cos^{-1}\left(\dfrac{-1 + \sqrt{3}}{2}\right) = \varnothing$,

    because $\dfrac{-1 - \sqrt{3}}{2}$ is outside the range of $\cos x$.

    Therefore $f(x) = 0$, when $x = \{1.20, 5.09\}$.

44. First put the equation in quadratic form and factor by the quadratic formula:

    $$\sin x = \frac{2 \pm \sqrt{(-2)^2 - 4(1)(3)}}{2(1)} = \frac{2 \pm \sqrt{-8}}{2} = \frac{2 \pm 2i\sqrt{2}}{2} \Rightarrow \sin x = 1 \pm i\sqrt{2}.$$

    There is no angle that produces this imaginary solution, therefore: $\varnothing$.

45. First set the equation equal to zero, then use identities to put it into the quadratic form and factor:

$\sec^2 \theta = 2 \tan \theta + 4 \Rightarrow \tan^2 \theta + 1 - 2 \tan \theta - 4 = 0 \Rightarrow \tan^2 \theta - 2 \tan \theta - 3 = 0 \Rightarrow$

$(\tan \theta - 3)(\tan \theta + 1) = 0.$ From this factored form, either $\tan \theta - 3 = 0 \Rightarrow \tan \theta = 3$; or

$\tan \theta + 1 = 0 \Rightarrow \tan \theta = -1$, therefore $\tan x = -1, 3$.

Now find x:

(i) For $\tan x = 3$: find the reference angle (calculate in degree mode) $\cos^{-1} 3 \approx 71.57°$.

Since x is found in quadrants I and III, $x \approx 71.57°$ and $x = 180° + 71.57° = 251.57°$.

(ii) For $\tan x = -1$: find the reference angle (calculate in degree mode) $\tan^{-1}(-1) = 45°$.

Since x is found in quadrants II and IV, $x = 180° - 45° = 135°$ and $x = 360° - 45° = 315°$.

Therefore $f(x) = 0$, when $x = \{71.6°, 135°, 251.6°, 315°\}$.

46. Use identities to get the equation in quadratic form and factor by the quadratic formula:

$\cot \theta + 2 \csc \theta = 3 \Rightarrow \dfrac{\cos \theta}{\sin \theta} + \dfrac{2}{\sin \theta} = 3 \Rightarrow \cos \theta + 2 = 3 \sin \theta \Rightarrow (\cos \theta + 2)^2 = (3 \sin \theta)^2 \Rightarrow$

$\cos^2 \theta + 4 \cos x + 4 = 9 \sin^2 \theta \Rightarrow \cos^2 \theta + 4 \cos \theta + 4 = 9(1 - \cos^2 \theta) \Rightarrow$

$\cos^2 \theta + 4 \cos \theta + 4 = 9 - 9 \cos^2 \theta \Rightarrow 10 \cos^2 \theta + 4 \cos \theta - 5 = 0 \Rightarrow$

$\cos \theta = \dfrac{-4 \pm \sqrt{(4)^2 - 4(10)(-5)}}{2(10)} = \dfrac{-4 \pm \sqrt{216}}{20} = \dfrac{-4 \pm 6\sqrt{6}}{20} \Rightarrow \cos \theta = \dfrac{-2 \pm 3\sqrt{6}}{10}.$

Now find x:

(i) For $\cos x = \dfrac{-2 + 3\sqrt{6}}{10}$: find the reference angle (calculate in degree mode): $\cos \theta = \dfrac{-2 + 3\sqrt{6}}{10} \Rightarrow$

$\cos \theta \approx .53484692 \Rightarrow \theta = \cos^{-1} .53484692 \approx 57.67°$.

Since x is found in quadrants I and IV, $x \approx 57.67°$ and $x = 360° - 57.67 = 302.33°$.

(ii) For $\cos \theta = \dfrac{-2 - 3\sqrt{6}}{10}$: find the reference angle (calculate in degree mode): $\cos \theta = \dfrac{-2 - 3\sqrt{6}}{10} \Rightarrow$

$\cos \theta \approx -.9348692 \Rightarrow \theta = \cos^{-1}(-.9348692) \approx 20.8°$.

Since x is found in quadrants II and III, $x = 180° - 20.8° \approx 159.20°$ and $x = 180° + 20.8° \approx 200.80°$.

Finally, since the equation was square, we will need to check each solution:

Check $57.67°$: $\cos 57.67° + 2 \approx 2.53$ and $3 \sin 57.67° \approx 2.53$, this is a solution.

Check $302.33°$: $\cos 302.33° + 2 \approx 2.53$ and $3 \sin 302.33° \approx -2.53$, this is not a solution.

Check $159.20°$: $\cos 159.20° + 2 \approx 1.07$ and $3 \sin 159.20° \approx 1.07$, this is a solution.

Check $57.67°$: $\cos 200.80° + 2 \approx 1.07$ and $3 \sin 200.80° \approx -1.07$, this is not a solution.

Therefore $f(x) = 0$, when $x = \{57.7°, 159.2°\}$.

47. Use identities to get the equation in quadratic form and factor by the quadratic formula:

$2 \sin \theta = 1 - 2 \cos \theta \Rightarrow (2 \sin \theta)^2 = (1 - 2 \cos \theta)^2 \Rightarrow 4 \sin^2 \theta = 1 - 4 \cos \theta + 4 \cos^2 \theta \Rightarrow$

$4(1 - \cos^2 \theta) = 1 - 4 \cos \theta + 4 \cos^2 \theta \Rightarrow 4 - 4 \cos^2 \theta = 1 - 4 \cos \theta + 4 \cos^2 \theta \Rightarrow$

$0 = 8 \cos^2 \theta - 4 \cos \theta - 3 \Rightarrow \cos \theta = \dfrac{4 \pm \sqrt{(-4)^2 - 4(8)(-3)}}{2(8)} = \dfrac{4 \pm \sqrt{112}}{16} = \dfrac{4 \pm 4\sqrt{7}}{16} \Rightarrow$

$\cos \theta = \dfrac{1 \pm \sqrt{7}}{4}.$

Now find x:

(i) For $\cos x = \dfrac{1 + \sqrt{7}}{4}$: find the reference angle (calculate in degree mode): $\cos \theta = \dfrac{1 + \sqrt{7}}{4} \Rightarrow$

$\cos \theta \approx .91143783 \Rightarrow \theta = \cos^{-1} .91143783 \approx 24.30°$.

Since x is found in quadrants I and IV, $x \approx 24.30°$ and $x = 360° - 24.30 = 335.70°$.

(ii) For $\cos \theta = \dfrac{1 - \sqrt{7}}{4}$: find the reference angle (calculate in degree mode): $\cos \theta = \dfrac{1 - \sqrt{7}}{4} \Rightarrow$

$\cos \theta \approx -.41143783 \Rightarrow \theta = \cos^{-1}(-.41143783) \approx 65.70°$.

Since x is found in quadrants II and III, $x = 180° - 65.70° \approx 114.30°$ and $x = 180° + 65.70° \approx 245.70°$.

Finally, since the equation was square, we will need to check each solution:

Check $24.30°$: $2 \sin 24.30° \approx .82$ and $1 - 2 \cos 24.30° \approx -.82$, this is not a solution.

Check $114.30°$: $2 \sin 114.30° \approx 1.82$ and $1 - 2 \cos 114.30° \approx 1.82$, this is a solution.

Check $245.70°$: $2 \sin 245.70° \approx -1.82$ and $1 - 2 \cos 245.70° \approx 1.82$, this is not a solution.

Check $335.70°$: $2 \sin 335.70° \approx -.82$ and $1 - 2 \cos 335.70° \approx -.82$, this is a solution.

Therefore $f(x) = 0$, when $x = \{114.3°, 335.7°\}$.

48. Graph $y = (\tan x)^{-1} + 2(\sin x)^{-1} - 3$, (not shown). The x-intercepts are: $\{1.01, 2.78\}$

49. Graph $y = 2 \sin x - 1 + 2 \cos x$, (not shown). The x-intercepts are: $\{1.99, 5.86\}$

50. Graph $y = \sin^3 x + \sin x - 1$, (not shown). The x-intercepts are: $\{0.75, 2.39\}$

51. Graph $y = 2 \cos^3 x + \sin x + 1$, (not shown). The x-intercepts are: $\{2.68, 4.46, 4.71\}$

52. Graph $y = e^x - \sin x - 3$, (not shown). The x-intercept is: $\{1.38\}$

53. Graph $y = \ln x - \cos x$, (not shown). The x-intercept is: $\{1.30\}$

54. Dividing by $\sin x$ in the second step causes two solutions to be lost. It should have been factored:

$\sin^2 x - \sin x = 0 \Rightarrow \sin x(\sin x - 1) = 0$, therefore either $\sin x = 0 \Rightarrow x = 0, \pi$; or

$\sin x - 1 = 0 \Rightarrow \sin x = 1 \Rightarrow x = \dfrac{\pi}{2}$.

55. (a) $14 = \dfrac{35}{3} + \dfrac{7}{3} \sin \dfrac{2\pi x}{365} \Rightarrow \dfrac{7}{3} = \dfrac{7}{3} \sin \dfrac{2\pi x}{365} \Rightarrow 1 = \sin \dfrac{2\pi x}{365} \Rightarrow \dfrac{2\pi x}{365} = \dfrac{\pi}{2} \Rightarrow 4\pi x = 365\pi \Rightarrow x \approx 91.25$

Fourteen hours of daylight will occur approximately 91.3 days after March 21, or June 20.

(b) The minimum will occur when $\sin \dfrac{2\pi x}{365} = -1$, therefore $\dfrac{2\pi x}{365} = \dfrac{3\pi}{2} \Rightarrow 4\pi x = 1095\pi \Rightarrow x \approx 273.75$.

The minimum hours of daylight will occur approximately 273.8 days after March 21, or on December 19.

(c) $10 = \dfrac{35}{3} + \dfrac{7}{3} \sin \dfrac{2\pi x}{365} \Rightarrow 30 = 35 + 7 \sin \dfrac{2\pi x}{365} \Rightarrow -\dfrac{5}{7} = \sin \dfrac{2\pi x}{365}$.

Now for $\sin \dfrac{2\pi x}{365}$, find the reference angle (calculate in radian mode): $\dfrac{2\pi x}{365} = \sin^{-1} \dfrac{5}{7} \Rightarrow \dfrac{2\pi x}{365} \approx .7956$.

Since $\sin \dfrac{2\pi x}{365} = -\dfrac{5}{7}, \dfrac{2\pi x}{365}$ is found in quadrants III and IV, and $\dfrac{2\pi x}{365} \approx \pi + .7956 \Rightarrow \dfrac{2\pi x}{365} \approx 3.9372 \Rightarrow$

$x \approx 228.7; or \dfrac{2\pi x}{365} \approx 2\pi - .7956 \Rightarrow \dfrac{2\pi x}{365} \approx 5.4876 \Rightarrow x \approx 318.8$.

There will be about 10 hours of daylight twice, and they are approximately:

228.7 days after March 21, or Nov 4; and 318.8 days after March 21, or Feb 2.

56. Substituting 1.5 for m in the equation and solving yields the following:

$$\sin\frac{\alpha}{2} = \frac{1}{1.5} \Rightarrow \sin\frac{\alpha}{2} = \frac{2}{3} \Rightarrow \frac{\alpha}{2} = \sin^{-1}\left(\frac{2}{3}\right) \Rightarrow \frac{\alpha}{2} \approx 41.8° \Rightarrow \alpha \approx 83.6°$$

57. First put the equation in quadratic form and then factor by the quadratic formula:

$$.342\,(80)\cos\theta + 2\cos^2\theta = \frac{16\,(80)^2}{(60)^2} \Rightarrow 27.36\cos\theta + 2\cos^2\theta = \frac{256}{9} \Rightarrow$$

$$2\cos^2\theta + 27.36\cos\theta - \frac{256}{9} = 0 \Rightarrow \cos\theta = \frac{-27.36 \pm \sqrt{748.5696 - 4(2)\left(-\frac{256}{9}\right)}}{2\,(2)} = \frac{-27.36 \pm 31.243}{4}.$$

Therefore $\cos\theta = \dfrac{-27.36 + 31.243}{4} = .97075$ or $\cos\theta = \dfrac{-27.36 - 31.243}{4} = -14.65.$

Since the range of $\cos\theta$ is $-1 < \cos\theta < 1$, the only solution is: $\theta = \cos^{-1}.97075 \approx 13.892° \approx 14°.$

58. First find $\tan x$: $\dfrac{\tan x + .4}{1 - .4\tan x} = 2.4 \Rightarrow \tan x + .4 = 2.4 - .96\tan x \Rightarrow 1.96\tan x = 2 \Rightarrow$

$\tan x = \dfrac{2}{1.96}$. Now find x: $x = \tan^{-1}\left(\dfrac{2}{1.96}\right) \Rightarrow x \approx 45.5787° \approx 46°.$

59. Let $s\,(t) = \dfrac{2 + \sqrt{3}}{2}$ and find $\sin t$: $\sin t + 2\cos t = \dfrac{2 + \sqrt{3}}{2} \Rightarrow 2\sin t + 4\cos t = 2 + \sqrt{3} \Rightarrow$

$4\cos t = (2 + \sqrt{3}) - 2\sin t \Rightarrow (4\cos t)^2 = ((2 + \sqrt{3}) - 2\sin t)^2 \Rightarrow$

$16\cos^2 t = (7 + 4\sqrt{3}) - 4\,(2 + \sqrt{3})\sin t + 4\sin^2 t \Rightarrow$

$16\,(1 - \sin^2 t) = (7 + 4\sqrt{3}) - 4(2 + \sqrt{3})\sin t + 4\sin^2 t \Rightarrow$

$0 = 20\sin^2 t - 4\,(2 + \sqrt{3})\sin t + (-9 + 4\sqrt{3}) \Rightarrow 0 = (2\sin t - \sqrt{3})(10\sin t - 4 + 3\sqrt{3}).$

Now when $2\sin t - \sqrt{3} = 0 \Rightarrow 2\sin t = \sqrt{3} \Rightarrow \sin t = \dfrac{\sqrt{3}}{2}$; and when

$10\sin t - 4 + 3\sqrt{3} = 0 \Rightarrow 10\sin t = 4 - 3\sqrt{3} \Rightarrow \sin t = \dfrac{4 - 3\sqrt{3}}{10}.$

To find a solution choose $\sin t = \dfrac{\sqrt{3}}{2}$ which yields: $t = \dfrac{\pi}{3}.$

60. Let $s\,(t) = \dfrac{3\sqrt{2}}{2}$ and find $\sin t$: $\sin t + 2\cos t = \dfrac{3\sqrt{2}}{2} \Rightarrow 2\sin t + 4\cos t = 3\sqrt{2} \Rightarrow$

$4\cos t = 3\sqrt{2} - 2\sin t \Rightarrow (4\cos t)^2 = (3\sqrt{2} - 2\sin t)^2 \Rightarrow 16\cos^2 t = 18 - 12\sqrt{2}\sin t + 4\sin^2 t \Rightarrow$

$16\,(1 - \sin^2 t) = 18 - 12\sqrt{2}\sin t + 4\sin^2 t \Rightarrow 0 = 20\sin^2 t - 12\sqrt{2}\sin t + 2 \Rightarrow$

$0 = 10\sin^2 t - 6\sqrt{2}\sin t + 1 \Rightarrow 0 = (5\sqrt{2}\sin t - 1)(\sqrt{2}\sin t - 1).$

Now when $5\sqrt{2}\sin t - 1 = 0 \Rightarrow 5\sqrt{2}\sin t = 1 \Rightarrow \sin t = \dfrac{1}{5\sqrt{2}}$; and when

$\sqrt{2}\sin t - 1 = 0 \Rightarrow \sqrt{2}\sin t = 1 \Rightarrow \sin t = \dfrac{1}{\sqrt{2}}.$

To find a solution choose $\sin t = \dfrac{1}{\sqrt{2}}$ which yields: $t = \dfrac{\pi}{4}.$

9.6: Trigonometric Equations and Inequalities (II)

1. Dividing each by 2 yields: $x = \dfrac{\pi}{3}, \pi, \dfrac{4\pi}{3}$.

2. Multiplying each by 3 yields: $\theta = 135°, 180°, 225°, 270°$.

3. (a) If the required interval for x is: $0 \le x < 2\pi$, then $0 \le 2x < 4\pi$.

    Therefore, if $\cos 2x = \dfrac{\sqrt{3}}{2}$, then $2x = \dfrac{\pi}{6}, \dfrac{11\pi}{6}, \dfrac{13\pi}{6}, \dfrac{23\pi}{6}$.

    Now dividing each solution by 2 $\left(\text{or multipling by } \dfrac{1}{2}\right)$ yields: $x = \left\{\dfrac{\pi}{12}, \dfrac{11\pi}{12}, \dfrac{13\pi}{12}, \dfrac{23\pi}{12}\right\}$.

    (b) Using a calculator, the graph of $\cos 2x = \dfrac{\sqrt{3}}{2}$ is above the graph of $y = \dfrac{\sqrt{3}}{2}$, for the interval:

    $\left[0, \dfrac{\pi}{12}\right) \cup \left(\dfrac{11\pi}{12}, \dfrac{13\pi}{12}\right) \cup \left(\dfrac{23\pi}{12}, 2\pi\right)$.

4. (a) If the required interval for x is: $0 \le x < 2\pi$, then $0 \le 2x < 4\pi$.

    Therefore, if $\cos 2x = -\dfrac{1}{2}$, then $2x = \dfrac{2\pi}{3}, \dfrac{4\pi}{3}, \dfrac{8\pi}{3}, \dfrac{10\pi}{3}$.

    Now dividing each solution by 2 $\left(\text{or multipling by } \dfrac{1}{2}\right)$ yields: $x = \left\{\dfrac{\pi}{3}, \dfrac{2\pi}{3}, \dfrac{4\pi}{3}, \dfrac{5\pi}{3}\right\}$.

    (b) Using a calculator, the graph of $\cos 2x = -\dfrac{1}{2}$ is above the graph of $y = -\dfrac{1}{2}$, for the interval:

    $\left[0, \dfrac{\pi}{3}\right) \cup \left(\dfrac{2\pi}{3}, \dfrac{4\pi}{3}\right) \cup \left(\dfrac{5\pi}{3}, 2\pi\right)$.

5. (a) If the required interval for x is: $0 \le x < 2\pi$, then $0 \le 3x < 6\pi$.

    Therefore, if $\sin 3x = -1$, then $3x = \dfrac{3\pi}{2}, \dfrac{7\pi}{2}, \dfrac{11\pi}{2}$.

    Now dividing each solution by 3 $\left(\text{or multipling by } \dfrac{1}{3}\right)$ yields: $x = \left\{\dfrac{\pi}{2}, \dfrac{7\pi}{6}, \dfrac{11\pi}{6}\right\}$.

    (b) Using a calculator, the graph of $\sin 3x = -1$ is never below the graph of $y = -1$, therefore $\varnothing$.

6. (a) If the required interval for x is: $0 \le x < 2\pi$, then $0 \le 3x < 6\pi$.

    Therefore, if $\sin 3x = 0$, then $3x = 0, \pi, 2\pi, 3\pi, 4\pi, 5\pi$.

    Now dividing each solution by 3 $\left(\text{or multipling by } \dfrac{1}{3}\right)$ yields: $x = \left\{0, \dfrac{\pi}{3}, \dfrac{2\pi}{3}, \pi, \dfrac{4\pi}{3}, \dfrac{5\pi}{3}\right\}$.

    (b) Using a calculator, the graph of $\cos 3x = 0$ is below the graph of $y = 0$, for the interval:

    $\left(\dfrac{\pi}{3}, \dfrac{2\pi}{3}\right) \cup \left(\pi, \dfrac{4\pi}{3}\right) \cup \left(\dfrac{5\pi}{3}, 2\pi\right)$.

7. (a) First solve for $\cos 2x$: $\sqrt{2}\cos 2x = -1 \Rightarrow \cos 2x = -\dfrac{1}{\sqrt{2}}$.

    If the required interval for x is: $0 \le x < 2\pi$, then $0 \le 2x < 4\pi$.

    Therefore, if $\cos 2x = -\dfrac{1}{\sqrt{2}}$, then $2x = \dfrac{3\pi}{4}, \dfrac{5\pi}{4}, \dfrac{11\pi}{4}, \dfrac{13\pi}{4}$.

    Now dividing each solution by 2 $\left(\text{or multipling by } \dfrac{1}{2}\right)$ yields: $x = \left\{\dfrac{3\pi}{8}, \dfrac{5\pi}{8}, \dfrac{11\pi}{8}, \dfrac{13\pi}{8}\right\}$.

    (b) Using a calculator, the graph of $\sqrt{2}\cos 2x = -1$ intersects or is below the graph of $y = -1$, for the interval:

    $\left[\dfrac{3\pi}{8}, \dfrac{5\pi}{8}\right] \cup \left[\dfrac{11\pi}{8}, \dfrac{13\pi}{8}\right]$.

8. (a) First solve for $\sin 2x$: $2\sqrt{3} \sin 2x = \sqrt{3} \Rightarrow \sin 2x = \dfrac{\sqrt{3}}{2\sqrt{3}} \Rightarrow \sin 2x = \dfrac{1}{2}$.

   If the required interval for x is: $0 \le x < 2\pi$, then $0 \le 2x < 4\pi$.

   Therefore, if $\sin 2x = \dfrac{1}{2}$, then $2x = \dfrac{\pi}{6}, \dfrac{5\pi}{6}, \dfrac{13\pi}{6}, \dfrac{17\pi}{6}$.

   Now dividing each solution by 2 $\left(\text{or multipling by } \dfrac{1}{2}\right)$ yields: $x = \left\{\dfrac{\pi}{12}, \dfrac{5\pi}{12}, \dfrac{13\pi}{12}, \dfrac{17\pi}{12}\right\}$.

   (b) Using a calculator, the graph of $2\sqrt{3} \sin 2x = \sqrt{3}$ intersects or is below the graph of $y = \sqrt{3}$,

   for the interval: $\left[0, \dfrac{\pi}{12}\right] \cup \left[\dfrac{5\pi}{12}, \dfrac{13\pi}{12}\right] \cup \left[\dfrac{17\pi}{12}, 2\pi\right)$.

9. (a) First solve for $\sin \dfrac{x}{2}$: $\sin \dfrac{x}{2} = \sqrt{2} - \sin \dfrac{x}{2} \Rightarrow 2 \sin \dfrac{x}{2} = \sqrt{2} \Rightarrow \sin \dfrac{x}{2} = \dfrac{\sqrt{2}}{2}$.

   If the required interval for x is: $0 \le x < 2\pi$, then $0 \le \dfrac{x}{2} < \pi$.

   Therefore, if $\sin \dfrac{x}{2} = \dfrac{\sqrt{2}}{2}$, then $\dfrac{x}{2} = \dfrac{\pi}{4}, \dfrac{3\pi}{4}$.

   Now multiplying each solution by 2 yields: $x = \left\{\dfrac{\pi}{2}, \dfrac{3\pi}{2}\right\}$.

   (b) Using a calculator, the graph of $y = \sin \dfrac{x}{2}$ is above the graph of $y = \sqrt{2} - \sin \dfrac{x}{2}$,

   for the interval: $\left(\dfrac{\pi}{2}, \dfrac{3\pi}{2}\right)$.

10. (a) First solve for $\sin x$ by setting the equation equal to zero and factoring: $\sin x = \sin 2x \Rightarrow$

   $\sin x - \sin 2x = 0 \Rightarrow \sin x - 2 \sin x \cos x = 0 \Rightarrow \sin x(1 - 2 \cos x) = 0$. Now either $\sin x = 0 \Rightarrow$

   $x = 0, \pi$; or $1 - 2 \cos x = 0 \Rightarrow 2 \cos x = 1 \Rightarrow \cos x = \dfrac{1}{2} \Rightarrow x = \dfrac{\pi}{3}, \dfrac{5\pi}{3}$.

   Therefore $x = \left\{0, \dfrac{\pi}{3}, \pi, \dfrac{5\pi}{3}\right\}$.

   (b) Using a calculator, the graph of $y = \sin x$ is above the graph of $y = \sin 2x$,

   for the interval: $\left(\dfrac{\pi}{3}, \pi\right) \cup \left(\dfrac{5\pi}{3}, 2\pi\right)$.

11. Use identities to solve for x: $\sin \dfrac{x}{2} = \cos \dfrac{x}{2} \Rightarrow \left(\sin \dfrac{x}{2}\right)^2 = \left(\cos \dfrac{x}{2}\right)^2 \Rightarrow \dfrac{1 - \cos x}{2} = \dfrac{1 + \cos x}{2} =$

   $1 - \cos x = 1 + \cos x \Rightarrow -2 \cos x = 0 \Rightarrow x = \dfrac{\pi}{2}, \dfrac{3\pi}{2}$. Since both sides of the equation have been squared,

   we must check each solution: Check $x = \dfrac{\pi}{2}$: $\sin \dfrac{\left(\frac{\pi}{2}\right)}{2} = \sin \dfrac{\pi}{4} = \dfrac{\sqrt{2}}{2}$ and $\cos \dfrac{\left(\frac{\pi}{2}\right)}{2} = \cos \dfrac{\pi}{4} = \dfrac{\sqrt{2}}{2}$,

   therefore $x = \dfrac{\pi}{2}$ is a solution. Check $x = \dfrac{3\pi}{2}$: $\sin \dfrac{\left(\frac{3\pi}{2}\right)}{2} = \sin \dfrac{3\pi}{4} = \dfrac{\sqrt{2}}{2}$ and $\cos \dfrac{\left(\frac{3\pi}{2}\right)}{2} = \cos \dfrac{3\pi}{4} = -\dfrac{\sqrt{2}}{2}$,

   therefore $x = \dfrac{3\pi}{2}$ is not a solution. The solution is: $x = \left\{\dfrac{\pi}{2}\right\}$.

12. Use identities to solve for x: $\sec \dfrac{x}{2} = \cos \dfrac{x}{2} \Rightarrow \dfrac{1}{\cos \frac{x}{2}} = \cos \dfrac{x}{2} \Rightarrow 1 = \left(\cos \dfrac{x}{2}\right)^2 \Rightarrow 1 = \dfrac{1 + \cos x}{2} \Rightarrow$

   $2 = 1 + \cos x \Rightarrow \cos x = 1 \Rightarrow x = \{0\}$.

13. Use identities to solve for x: $\sin^2\left(\dfrac{x}{2}\right) - 1 = 0 \Rightarrow \left(\sqrt{\dfrac{1-\cos x}{2}}\right)^2 - 1 = 0 \Rightarrow \dfrac{1-\cos x}{2} - 1 = 0 \Rightarrow$

    $1 - \cos x - 2 = 0 \Rightarrow -\cos x = 1 \Rightarrow \cos x = -1 \Rightarrow x = \{\pi\}$.

14. Use identities to solve for x: $\sin x \cos x = \dfrac{1}{4} \Rightarrow 2\sin x \cos x = \dfrac{1}{2} \Rightarrow \sin 2x = \dfrac{1}{2}$.

    If the required interval for x is: $0 \le x < 2\pi$, then $0 \le 2x < 4\pi$.

    Therefore, if $\sin 2x = \dfrac{1}{2}$, then $2x = \dfrac{\pi}{6}, \dfrac{5\pi}{6}, \dfrac{13\pi}{6}, \dfrac{17\pi}{6}$.

    Now dividing each solution by 2 $\left(\text{or multipling by } \dfrac{1}{2}\right)$ yields: $x = \left\{\dfrac{\pi}{12}, \dfrac{5\pi}{12}, \dfrac{13\pi}{12}, \dfrac{17\pi}{12}\right\}$.

15. Use identities to solve for x: $\sin 2x = 2\cos^2 x \Rightarrow 2\sin x \cos x = 2\cos^2 x \Rightarrow 2\sin x \cos x - 2\cos^2 x = 0 \Rightarrow$

    $2\cos x (\sin x - \cos x) = 0$. Now either $2\cos x = 0 \Rightarrow \cos x = 0 \Rightarrow x = \dfrac{\pi}{2}, \dfrac{3\pi}{2}$;

    or $\sin x - \cos x = 0 \Rightarrow \sin x = \cos x \Rightarrow x = \dfrac{\pi}{4}, \dfrac{5\pi}{4}$. Therefore $x = \left\{\dfrac{\pi}{4}, \dfrac{\pi}{2}, \dfrac{5\pi}{4}, \dfrac{3\pi}{2}\right\}$.

16. Use identities to solve for x: $\csc^2\left(\dfrac{x}{2}\right) = 2\sec x \Rightarrow \dfrac{1}{(\sin\frac{x}{2})^2} = \dfrac{2}{\cos x} \Rightarrow \cos x = 2\left(\sin\dfrac{x}{2}\right)^2 \Rightarrow$

    $\cos x = \dfrac{2(1-\cos x)}{2} \Rightarrow \cos x = 1 - \cos x \Rightarrow 2\cos x = 1 \Rightarrow \cos x = \dfrac{1}{2} \Rightarrow x = \left\{\dfrac{\pi}{3}, \dfrac{5\pi}{3}\right\}$.

17. Use identities to solve for x: $\cos x - 1 = \cos 2x \Rightarrow \cos x - 1 = 2\cos^2 x - 1 \Rightarrow \cos x - 2\cos^2 x = 0 \Rightarrow$

    $(\cos x)(1 - 2\cos x) = 0$. Now either $\cos x = 0 \Rightarrow x = \dfrac{\pi}{2}, \dfrac{3\pi}{2}$; or $1 - 2\cos x = 0 \Rightarrow$

    $-2\cos x = -1 \Rightarrow \cos x = \dfrac{1}{2} \Rightarrow x = \dfrac{\pi}{3}, \dfrac{5\pi}{3}$. Therefore $x = \left\{\dfrac{\pi}{3}, \dfrac{\pi}{2}, \dfrac{3\pi}{2}, \dfrac{5\pi}{3}\right\}$.

18. Use identities to solve for x: $1 - \sin x = \cos 2x \Rightarrow 1 - \sin x = 1 - 2\sin^2 x \Rightarrow 2\sin^2 x - \sin x = 0 \Rightarrow$

    $(\sin x)(2\sin x - 1) = 0$. Now either $\sin x = 0 \Rightarrow x = 0, \pi$; or $2\sin x - 1 = 0 \Rightarrow$

    $2\sin x = 1 \Rightarrow \sin x = \dfrac{1}{2} \Rightarrow x = \dfrac{\pi}{6}, \dfrac{5\pi}{6}$. Therefore $x = \left\{0, \dfrac{\pi}{6}, \dfrac{5\pi}{6}, \pi\right\}$.

19. First solve for $\sin 3\theta$: $\sqrt{2}\sin 3\theta - 1 = 0 \Rightarrow \sqrt{2}\sin 3\theta = 1 \Rightarrow \sin 3\theta = \dfrac{1}{\sqrt{2}}$.

    If the required interval for θ is: $0° \le \theta < 360°$, then $0° \le 3\theta < 1080°$.

    Therefore, if $\sin 3\theta = \dfrac{1}{\sqrt{2}}$, then $3\theta = 45°, 135°, 405°, 495°, 765°, 855°$.

    Now dividing each solution by 3 $\left(\text{or multiplying by } \dfrac{1}{3}\right)$ yields: $\theta = \{15°, 45°, 135°, 165°, 255°, 285°\}$.

20. First solve for $\cos 2\theta$: $-2\cos 2\theta = \sqrt{3} \Rightarrow \cos 2\theta = -\dfrac{\sqrt{3}}{2}$.

    If the required interval for θ is: $0° \le \theta < 360°$, then $0° \le 2\theta < 720°$.

    Therefore, if $\cos 2\theta = -\dfrac{\sqrt{3}}{2}$, then $2\theta = 150°, 210°, 510°, 570°$.

    Now dividing each solution by 2 $\left(\text{or multiplying by } \dfrac{1}{2}\right)$ yields: $\theta = \{75°, 105°, 255°, 285°\}$.

21. If the required interval for θ is: $0° \leq \theta < 360°$, then $0° \leq \dfrac{\theta}{2} < 180°$.

    Therefore, if $\cos \dfrac{\theta}{2} = 1$, then $\dfrac{\theta}{2} = 0°$, and multiplying this by 2 gives a solution of $\theta = \{0°\}$.

22. If the required interval for θ is: $0° \leq \theta < 360°$, then $0° \leq \dfrac{\theta}{2} < 180°$.

    Therefore, if $\sin \dfrac{\theta}{2} = 1$, then $\dfrac{\theta}{2} = 90°$, and multiplying this by 2 gives a solution of $\theta = \{180°\}$.

23. First solve for $\sin \dfrac{\theta}{2}$: $2\sqrt{3} \sin \dfrac{\theta}{2} = 3 \Rightarrow \sin \dfrac{\theta}{2} = \dfrac{3}{2\sqrt{3}} = \dfrac{\sqrt{3}}{2}$.

    If the required interval for θ is: $0° \leq \theta < 360°$, then $0° \leq \dfrac{\theta}{2} < 180°$.

    Therefore, if $\sin \dfrac{\theta}{2} = \dfrac{\sqrt{3}}{2}$, then $\dfrac{\theta}{2} = 60°, 120°$. Now multiplying each solution by 2 yields: $\theta = \{120°, 240°\}$.

24. First solve for $\cos \dfrac{\theta}{2}$: $2\sqrt{3} \cos \dfrac{\theta}{2} = -3 \Rightarrow \cos \dfrac{\theta}{2} = \dfrac{-3}{2\sqrt{3}} = -\dfrac{\sqrt{3}}{2}$.

    If the required interval for θ is: $0° \leq \theta < 360°$, then $0° \leq \dfrac{\theta}{2} < 180°$.

    Therefore, if $\cos \dfrac{\theta}{2} = -\dfrac{\sqrt{3}}{2}$, then $\dfrac{\theta}{2} = 150°$. Now multiplying each solution by 2 yields: $\theta = \{300°\}$.

25. Use identities to solve for θ: $2 \sin \theta = 2 \cos 2\theta \Rightarrow 2 \sin \theta = 2(1 - 2\sin^2 \theta) \Rightarrow 2 \sin \theta = 2 - 4 \sin^2 \theta \Rightarrow$
    $4 \sin^2 \theta + 2 \sin \theta - 2 = 0 \Rightarrow 2(2 \sin^2 \theta + \sin \theta - 1) = 0 \Rightarrow 2(2 \sin \theta - 1)(\sin \theta + 1) = 0$.

    Now either $2 \sin \theta - 1 = 0 \Rightarrow 2 \sin \theta = 1 \Rightarrow \sin \theta = \dfrac{1}{2} \Rightarrow \theta = 30°, 150°$; or $\sin \theta + 1 = 0 \Rightarrow$
    $\sin \theta = -1 \Rightarrow \theta = 270°$. Therefore $\theta = \{30°, 150°, 270°\}$.

26. Use identities to solve for θ: $\cos \theta = \left(\pm \sqrt{\dfrac{1 - \cos \theta}{2}} \right)^2 \Rightarrow \cos \theta = \dfrac{1 - \cos \theta}{2} \Rightarrow 2 \cos \theta = 1 - \cos \theta \Rightarrow$
    $3 \cos \theta = 1 \Rightarrow \cos \theta = \dfrac{1}{3}$. For $\cos \theta = \dfrac{1}{3}$: find the reference angle (calculate in degree mode):

    $\cos^{-1}\left(\dfrac{1}{3} \right) \approx 70.5°$. Since θ is in quadrants I and IV, $\theta \approx 70.5°$ and $\theta = 360° - 70.5° \approx 289.5°$.
    Therefore, $\theta = \{70.5°, 289.5°\}$.

27. Use identities to solve for $\sin 2\theta$: $2 - \sin 2\theta = 4 \sin 2\theta \Rightarrow 2 = 5 \sin 2\theta \Rightarrow \sin 2\theta = \dfrac{2}{5}$.

    If the required interval for θ is: $0° \leq \theta < 360°$, then $0° \leq 2\theta < 720°$.

    For $\sin 2\theta = \dfrac{2}{5}$: find the reference angle (calculate in degree mode): $\sin^{-1} \dfrac{2}{5}\theta \approx 23.6°$.

    Since θ is in quadrants I and II, $2\theta \approx 23.6°$, $2\theta = 180° - 23.6° \approx 156.4°$, $2\theta = 360° + 23.6 \approx 383.6°$,
    and $2\theta = 540° - 23.6° \approx 516.4°$ Therefore, $2\theta = \{23.6°, 156.4°, 383.6°, 516.4°\}$.

    Now dividing each solution by 2 $\left(\text{or multipling by } \dfrac{1}{2}\right)$ yields: $\theta = \{11.8°, 78.2°, 191.8°, 258.2°\}$.

28. Use identities to solve for $\tan 2\theta$: $4 \cos 2\theta = 8 \sin \theta \cos \theta \Rightarrow 4 \cos 2\theta = 4(2 \sin \theta \cos \theta) \Rightarrow$
    $4 \cos 2\theta = 4(\sin 2\theta) \Rightarrow \dfrac{\cos 2\theta}{\sin 2\theta} = 1 \Rightarrow \tan 2\theta = 1$.

    If the required interval for θ is: $0° \leq \theta < 360°$, then $0° \leq 2\theta < 720°$.

    Therefore, if $\tan 2\theta = 1$, then $2\theta = 45°, 225°, 405°, 585°$.

    Now dividing each solution by 2 $\left(\text{or multipling by } \dfrac{1}{2}\right)$ yields: $\theta = \{22.5°, 112.5°, 202.5°, 292.5°\}$.

29. Use identities to solve for $\cos 2\theta$: $2\cos^2 2\theta = 1 - \cos 2\theta \Rightarrow 2\cos^2 2\theta + \cos 2\theta - 1 = 0 \Rightarrow$

    $(2\cos 2\theta - 1)(\cos 2\theta + 1) = 0$. Now either $2\cos 2\theta - 1 = 0 \Rightarrow 2\cos 2\theta = 1 \Rightarrow \cos 2\theta = \dfrac{1}{2}$;

    or $\cos 2\theta + 1 = 0 \Rightarrow \cos 2\theta = -1$. If the required interval for θ is: $0° \le \theta < 360°$, then $0° \le 2\theta < 720°$.

    Therefore, if $\cos 2\theta = \dfrac{1}{2}$, then $2\theta = 60°, 300°, 420°, 660°$; or if $\cos 2\theta = -1$, then $2\theta = 180°, 540°$.

    Now dividing each solution by 2 $\left(\text{or multipling by } \dfrac{1}{2}\right)$ yields: $\theta = \{30°, 90°, 150°, 210°, 270°, 330°\}$.

30. Use identities to solve for $\sin \theta$ and $\cos \theta$: $\sin \theta - \sin 2\theta = 0 \Rightarrow \sin \theta - 2\sin \theta \cos \theta = 0 \Rightarrow$

    $\sin \theta (1 - 2\cos \theta) = 0$. Now either $\sin \theta = 0$; or $1 - 2\cos \theta = 0 \Rightarrow -2\cos \theta = -1 \Rightarrow \cos \theta = \dfrac{1}{2}$.

    If $\sin \theta = 0$, then $\theta = 0°, 180°$; or if $\cos \theta = \dfrac{1}{2}$, then $\theta = 60°, 300°$. Therefore, $\theta = \{0°, 60°, 180°, 300°\}$.

31. Graph $y = \sin x + \sin 3x - \cos x$, from the capabilities of the calculator the x-intercepts are:

    $\{.262, 1.309, 1.571, 3.403, 4.451, 4.712\}$.

32. Graph $y = \sin 3x - \sin x$, from the capabilities of the calculator the x-intercepts are:

    $\{0, .785, 2.356, 3.142, 3.927, 5.498\}$.

33. Graph $y = \cos 2x + \cos x$, from the capabilities of the calculator the x-intercepts are:

    $\{1.047, 3.142, 5.236\}$.

34. Graph $y = \sin 4x + \sin 2x - 2\cos x$, from the capabilities of the calculator the x-intercepts are:

    $\{.524, 1.571, 2.618, 4.712\}$.

35. Graph $y = \cos \dfrac{x}{2} - 2\sin 2x$, from the capabilities of the calculator the x-intercepts are:

    $\{.259, 1.372, 3.142, 4.911, 6.024\}$.

36. Graph $y = \sin \dfrac{x}{2} + \cos 3x$, from the capabilities of the calculator the x-intercepts are:

    $\{.628, 1.346, 3.142, 4.937, 5.655\}$.

37. The functions, $\dfrac{\tan 2\theta}{2} \ne \tan \theta$, because the 2 in 2θ is not a factor of the numerator.

    It is a factor in the argument of the tangent function.

38. If there are no solutions over the interval $[0, 2\pi)$, the graph does not intersect the x-axis on that interval.

39. (a) If $e = 0$, then $0 = 20 \sin\left(\dfrac{\pi t}{4} - \dfrac{\pi}{2}\right)$. Since $\arcsin 0 = 0$, we can solve the following equation for t:

    $\dfrac{\pi t}{4} - \dfrac{\pi}{2} = 0 \Rightarrow \dfrac{\pi t}{4} = \dfrac{\pi}{2} \Rightarrow 2\pi t = 4\pi \Rightarrow t = 2$ seconds.

    (b) If $e = 10\sqrt{3}$, then $10\sqrt{3} = 20 \sin\left(\dfrac{\pi t}{4} - \dfrac{\pi}{2}\right) \Rightarrow \dfrac{\sqrt{3}}{2} = \sin\left(\dfrac{\pi t}{4} - \dfrac{\pi}{2}\right)$.

    Since $\arcsin \dfrac{\sqrt{3}}{2} = \dfrac{\pi}{3}$, we can solve the following equation for t:

    $\dfrac{\pi t}{4} - \dfrac{\pi}{2} = \dfrac{\pi}{3} \Rightarrow \dfrac{\pi t}{4} = \dfrac{\pi}{3} + \dfrac{\pi}{2} \Rightarrow \dfrac{\pi t}{4} = \dfrac{10\pi}{12} \Rightarrow 12\pi t = 40\pi \Rightarrow t = \dfrac{10}{3}$ seconds.

40. (a) First solve for $\cos 2\theta$: $6074 = 6077 - 31 \cos 2\theta \Rightarrow -3 = -31 \cos 2\theta \Rightarrow \cos 2\theta = \dfrac{3}{31}$.

   If the required interval for θ is: $0° \le \theta < 90°$, then $0° \le 2\theta < 180°$.

   For $\cos 2\theta = \dfrac{3}{31}$: find the reference angle (calculate in degree mode): $\cos^{-1}\left(\dfrac{3}{31}\right) \approx 84.447°$.

   Since θ is in quadrants I, $2\theta \approx 84.447°$. Now dividing this by 2 yields: $\theta \approx 42.2°$.

   (b) First solve for $\cos 2\theta$: $6108 = 6077 - 31 \cos 2\theta \Rightarrow 31 = -31 \cos 2\theta \Rightarrow \cos 2\theta = -1$.

   If the required interval for θ is: $0° \le \theta < 180°$, then $0° \le 2\theta < 360°$.

   For $\cos 2\theta = -1$: find the reference angle in degree mode: $\cos^{-1}(-1) = 180°$.

   Now dividing this by 2 yields: $\theta \approx 90°$.

   (c) First solve for $\cos 2\theta$: $6080.2 = 6077 - 31 \cos 2\theta \Rightarrow 3.2 = -31 \cos 2\theta \Rightarrow \cos 2\theta = -.1032258$.

   If the required interval for θ is: $0° \le \theta < 90°$, then $0° \le 2\theta < 180°$.

   For $\cos 2\theta = -.1032258$: find the reference angle (calculate in degree mode): $\cos^{-1}(-.1032258) \approx 95.925°$.

   Since θ is in quadrants II, $2\theta \approx 95.925°$. Now dividing this by 2 yields: $\theta \approx 48.0°$.

41. (a) For $x = t$, graph $P(t) = .004 \sin\left(2\pi(261.63)t + \dfrac{\pi}{7}\right)$, see Figure 41.

   (b) $0 = .004 \sin\left[2\pi(261.63)t + \dfrac{\pi}{7}\right] \Rightarrow 0 = \sin(1643.87t + .45)$.

   Since $\sin(1643.87t + .45) = 0$, $1643.87t + .45 = n\pi$, n is an integer, and $t = \dfrac{n\pi - .45}{1643.87}$.

   If $n = 0$, then $t \approx -.000274$. If $n = 1$, then $t \approx .00164$. If $n = 2$, then $t \approx .00355$.

   If $n = 3$, then $t \approx .00546$

   The only solutions for $t \in [0, .005]$ are $t = .00164, .00355$. A calculator supports this.

   (c) Graph $y = \sin(1643.87t + .45)$ on the calculator, the x-intercepts are $t \approx .00164$ and $.00355$,

   therefore from the graph, $P < 0$ for the interval: $(.00164, .00355)$.

   (d) The inequality $P < 0$ implies that there is a decrease in pressure so an eardrum would be vibrating outward.

42. (a) Graph $P(t) = .004 \sin 220\pi t + \dfrac{.003}{3} \sin 660\pi t + \dfrac{.003}{5} \sin 110\pi t + \dfrac{.003}{7} \sin 1540\pi t$, see Figure 42.

   (b) The graph is periodic, and the wave has "jagged square" tops and bottoms.

   (c) From the capabilities of the calculator, the x-intercepts are: $.004545, .00909, .01364, .01818, .02273, .02727$.

   The eardrum will be moving outward when the graph of the equation is below the x-axis,

   therefore for the intervals: $(.0045, .0091) \cup (.0136, .0182) \cup (.0227, .0273)$.

$[0, .005]$ by $[-.005, .005]$
Xscl $= .001$ Yscl $= 1$

$[0, .03]$ by $[-.005, .005]$
Xscl $= .01$ Yscl $= 1$

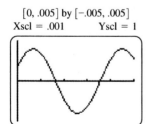

Figure 41   Figure 42

43. (a) For $x = t$, graph $P(t) = .005 \sin 440\pi t + .005 \sin 446\pi t$, see Figure 43a. There are 3 beats per second.

(b) For $x = t$, graph $P(t) = .005 \sin 440\pi t + .005 \sin 432\pi t$, see Figure 43b. There are 4 beats per second.

(c) The number of beats is equal to the absolute value of the difference in the frequencies of the two tones.

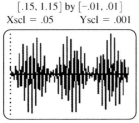

$[.15, 1.15]$ by $[-.01, .01]$
Xscl = .05   Yscl = .001

Figure 43a

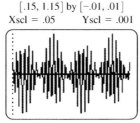

$[.15, 1.15]$ by $[-.01, .01]$
Xscl = .05   Yscl = .001

Figure 43b

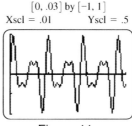

$[0, .03]$ by $[-1, 1]$
Xscl = .01   Yscl = .5

Figure 44a

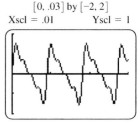

$[0, .03]$ by $[-2, 2]$
Xscl = .01   Yscl = 1

Figure 44d

44. (a) For $x = t$, graph $P(t) = \dfrac{1}{2} \sin [2\pi(220)t] + \dfrac{1}{2} \sin [2\pi(330)t] + \dfrac{1}{2} \sin [2\pi(440)t]$, see Figure 44a.

(b) P is a maximum at approximately: .0007576, .009847, .01894, .02803.

(c) A person also hears 110 Hz.

(d) For $x = t$, graph $P(t) = \dfrac{1}{2} \sin [2\pi(110)t] + \dfrac{1}{2} \sin [2\pi(220)t] + \dfrac{1}{2} \sin [2\pi(330)t] + \dfrac{1}{2} \sin [2\pi(440)t]$,

see Figure 44d.

Reviewing Basic Concepts (Sections 9.5 and 9.6)

1. If the required interval for x is: $0 \le x < 2\pi$, then $0 \le 2x < 4\pi$.

Therefore, if $\cos 2x = \dfrac{\sqrt{3}}{2}$, then $2x = \dfrac{\pi}{6}, \dfrac{11\pi}{6}, \dfrac{13\pi}{6}, \dfrac{23\pi}{6}$.

Now dividing each solution by 2 $\left(\text{or multipling by } \dfrac{1}{2}\right)$ yields: $x = \left\{\dfrac{\pi}{12}, \dfrac{11\pi}{12}, \dfrac{13\pi}{12}, \dfrac{23\pi}{12}\right\}$.

2. First solve for $\sin x$: $2 \sin x + 1 \Rightarrow 2 \sin x = -1 \Rightarrow \sin x = -\dfrac{1}{2}$. If $\sin x = -\dfrac{1}{2}$ then $x = \left\{\dfrac{7\pi}{6}, \dfrac{11\pi}{6}\right\}$.

3. Using the factored equation, either $\tan x - 1 = 0 \Rightarrow \tan x = 1 \Rightarrow x = \dfrac{\pi}{4}, \dfrac{5\pi}{4}$;

   or $\cos x - 1 = 0 \Rightarrow \cos x = 1 \Rightarrow x = 0$. Therefore, $x = \left\{0, \dfrac{\pi}{4}, \dfrac{5\pi}{4}\right\}$.

4. Solve for x: $2 \cos^2 x = \sqrt{3} \cos x \Rightarrow 2 \cos^2 x - \sqrt{3} \cos x = 0 \Rightarrow \cos x(2 \cos x - \sqrt{3}) = 0$.

   Now either $\cos x = 0 \Rightarrow x = \dfrac{\pi}{2}, \dfrac{3\pi}{2}$; or $2 \cos x - \sqrt{3} = 0 \Rightarrow 2 \cos x = \sqrt{3} \Rightarrow \cos x = \dfrac{\sqrt{3}}{2} \Rightarrow$

   $x = \dfrac{\pi}{6}, \dfrac{11\pi}{6}$. Therefore $x = \left\{\dfrac{\pi}{6}, \dfrac{\pi}{2}, \dfrac{3\pi}{2}, \dfrac{11\pi}{6}\right\}$.

5. First set the equation equal to zero and factor by the quadratic formula: $3 \cot^2 \theta - 3 \cot \theta = 1 \Rightarrow$

$3 \cot^2 \theta - 3 \cot \theta - 1 = 0 \Rightarrow \cot \theta = \dfrac{3 \pm \sqrt{(-3)^2 - 4(3)(-1)}}{2(3)} = \dfrac{3 \pm \sqrt{21}}{6}$.

If $\cot \theta = \dfrac{3 \pm \sqrt{21}}{6}$, then $\tan \theta = \dfrac{6}{3 \pm \sqrt{21}}$.

Now find θ:

(i) For $\tan \theta = \dfrac{6}{3 + \sqrt{21}}$: find the reference angle (calculate in degree mode): $\tan \theta = \dfrac{6}{3 + \sqrt{21}} \Rightarrow$

$\tan \theta \approx .79128785 \Rightarrow \theta = \tan^{-1} .79128785 \approx 38.4°$.

Since θ is found in quadrants I and III, $\theta \approx 38.4°$ and $\theta = 180° + 38.4 = 218.4°$.

(ii) For $\tan \theta = \dfrac{6}{3 - \sqrt{21}}$: find the reference angle (calculate in degree mode): $\tan \theta = \dfrac{6}{3 - \sqrt{21}}$:

$\tan \theta \approx -3.79128785 \Rightarrow \theta = \tan^{-1} (-3.79128785) \approx 75.2°$.

Since θ is found in quadrants II and IV, $\theta = 180° - 75.2° \approx 104.8°$ and $\theta = 360° - 75.2° \approx 284.8°$.

Therefore, $\theta = \{38.4°, 104.8°, 218.4°, 284.8°\}$.

6. First factor by the quadratic formula: $4 \cos^2 \theta + 4 \cos \theta - 1 = 0 \Rightarrow$

$\cos \theta = \dfrac{-4 \pm \sqrt{(4)^2 - 4(4)(-1)}}{2(4)} = \dfrac{-4 \pm \sqrt{32}}{8} = \dfrac{-4 \pm 4\sqrt{2}}{8} \Rightarrow \cos \theta = \dfrac{-1 \pm \sqrt{2}}{2}$

Now find x:

(i) For $\cos \theta = \dfrac{-1 + \sqrt{2}}{2}$: find the reference angle (calculate in degree mode): $\cos \theta = \dfrac{-1 + \sqrt{2}}{2}$:

$\cos \theta \approx .20710678 \Rightarrow \theta = \cos^{-1} (.20710678) \approx 78.0°$

Since θ is found in quadrants I and IV, $\theta \approx 78.0°$ and $\theta = 360° - 78.8 = 282.0°$.

(ii) For $\cos \theta = \dfrac{-1 - \sqrt{2}}{2}$: find the reference angle (calculate in degree mode): $\cos \theta = \dfrac{-1 - \sqrt{2}}{2}$:

$\cos \theta \approx -1.207$, the is outside the range of $\cos \theta$, therefore $\varnothing$.

Therefore, $\theta = \{78.0°, 282.0°\}$.

7. Use identities to solve for θ: $2 \sin \theta - 1 = \csc \theta \Rightarrow 2 \sin \theta - 1 = \dfrac{1}{\sin \theta} \Rightarrow 2 \sin^2 \theta - \sin \theta = 1 \Rightarrow$

$2 \sin^2 \theta - \sin \theta - 1 = 0 \Rightarrow (2 \sin \theta + 1)(\sin \theta - 1) = 0$.

Now either $2 \sin \theta + 1 = 0 \Rightarrow 2 \sin \theta = -1 \Rightarrow \sin \theta = -\dfrac{1}{2} \Rightarrow \theta = 210°, 330°$; or $\sin \theta - 1 = 0 \Rightarrow$

$\sin \theta = 1 \Rightarrow \theta = 90°$. Therefore $\theta = \{90°, 210°, 330°\}$.

8. If $\sec^2 \dfrac{\theta}{2} = 2$, then $\cos^2 \dfrac{\theta}{2} = \dfrac{1}{2}$. If the required interval for θ is: $0 \le \theta < 360°$, then $0 \le \dfrac{\theta}{2} < 180°$.

If $\cos \dfrac{\theta}{2} = \pm\dfrac{1}{\sqrt{2}} = \pm\dfrac{\sqrt{2}}{2}$, then $\dfrac{\theta}{2} = 45°, 135°$. Multiplying these solutions by 2 yields: $\theta = \{90°, 270°\}$.

9. Graph $y = x^2 + \sin x - x^3 - \cos x$, by the capabilities of the calculator the x-intercepts are:

$x \approx \{.68058878, 1.4158828\}$.

10. Graph $y_1 = x^3 - \cos^2 x$ and $y_2 = \dfrac{1}{2}x - 1$ in the same screen, by the capabilities of the calculator the

intersection of these two graphs is: $x \approx \{0, .37600772\}$.

Chapter 9 Review Exercises

1. If $f(-x) = -f(x)$, then the function is an odd function and (x, y) yields $(-x, -y)$. The following functions are odd: sine, tangent, cotangent, and cosecant.

2. If $f(-x) = f(x)$, then the function is an even function and (x, y) yields $(-x, y)$. The following functions are even: cosine and secant.

3. Since cosine is an even function, $\cos(-x) = \cos x$, and $\cos(-3) = \cos 3$.

4. Since sine is an odd function, $\sin(-x) = -\sin x$, and $\sin(-3) = -\sin 3$.

5. Since tangent is an odd function, $\tan(-x) = -\tan x$, and $\tan(-3) = -\tan 3$.

6. Since secant is an even function, $\sec(-x) = \sec x$, and $\sec(-3) = \sec 3$.

7. Since cosecant is an odd function, $\csc(-x) = -\csc x$, and $\csc(-3) = -\csc 3$.

8. Since cotangent is an odd function, $\cot(-x) = -\cot x$, and $\cot(-3) = -\cot 3$.

9. By reciprocal identity, $\sec x = \dfrac{1}{\cos x}$, therefore B.

10. By quotient identity, $\tan x = \dfrac{\sin x}{\cos x}$, therefore C.

11. By quotient identity, $\cot x = \dfrac{\cos x}{\sin x}$, therefore F.

12. By pythagorean identity, $\tan^2 x + 1 = \sec^2 x = \dfrac{1}{\cos^2 x}$, therefore E.

13. By reciprocal identity, $\tan^2 x = \dfrac{1}{\cot^2 x}$, therefore D.

14. By reciprocal identity, $\csc x = \dfrac{1}{\sin x}$, therefore A.

15. $\dfrac{\cot\theta}{\sec\theta} = \dfrac{\frac{\cos\theta}{\sin\theta}}{\frac{1}{\cos\theta}} = \dfrac{\cos\theta}{\sin\theta} \cdot \dfrac{\cos\theta}{1} = \dfrac{\cos^2\theta}{\sin\theta}$

16. $\tan^2\theta(1 + \cot^2\theta) = \dfrac{\sin^2\theta}{\cos^2\theta}(\csc^2\theta) = \dfrac{\sin^2\theta}{\cos^2\theta}\left(\dfrac{1}{\sin^2\theta}\right) = \dfrac{1}{\cos^2\theta}$

17. $\csc\theta + \cot\theta = \dfrac{1}{\sin\theta} + \dfrac{\cos\theta}{\sin\theta} = \dfrac{1 + \cos\theta}{\sin\theta}$

18. Since $\cos x = \dfrac{3}{5}$ and x is in quadrant IV, $\sin x = -\sqrt{1 - \cos^2 x}$,

therefore $\sin x = -\sqrt{1 - \left(\dfrac{3}{5}\right)^2} = -\sqrt{1 - \dfrac{9}{25}} = -\sqrt{\dfrac{16}{25}} = -\dfrac{4}{5}$. Now use these values to solve for:

$$\tan x = \dfrac{\sin x}{\cos x} = \dfrac{-\frac{4}{5}}{\frac{3}{5}} = -\dfrac{4}{3} \qquad\qquad \sec x = \dfrac{1}{\cos x} = \dfrac{5}{3}$$

$$\csc x = \dfrac{1}{\sin x} = -\dfrac{5}{4} \qquad\qquad \cot x = -\dfrac{3}{4}$$

19. Since $\tan x = -\dfrac{5}{4}$ and, because $\dfrac{\pi}{2} < x < \pi$, x is in quadrant II, $\sec x = -\sqrt{1 + \tan^2 x} = -\sqrt{1 + \dfrac{25}{16}} =$

    $-\sqrt{\dfrac{41}{16}} = -\dfrac{\sqrt{41}}{4}$. Now use these to find the other functions: $\cos x = \dfrac{1}{\sec x} = -\dfrac{4}{\sqrt{41}} = -\dfrac{4\sqrt{41}}{41}$;

    $\cot x = \dfrac{1}{\tan x} = -\dfrac{4}{5}$; $\sin x = \sqrt{1 - \cos^2 x} = \sqrt{1 - \left(-\dfrac{4\sqrt{41}}{41}\right)^2} = \sqrt{1 - \dfrac{656}{1681}} = \sqrt{\dfrac{1025}{1681}} = \dfrac{5\sqrt{41}}{41}$;

    and $\csc x = \dfrac{1}{\sin x} = \dfrac{41}{5\sqrt{41}} \cdot \dfrac{\sqrt{41}}{\sqrt{41}} = \dfrac{41\sqrt{41}}{5 \cdot 41} = \dfrac{\sqrt{41}}{5}$.

20. E, by a sum identity, $\cos 210° = \cos(150° + 60°) = \cos 150° \cos 60° - \sin 150° \sin 60°$.

21. B, by a cofunctional identity, $\sin 35° = \cos(90° - 35°) = \cos 55°$.

22. J, by a cofunctional identity, $\tan 35° = \tan(90° - (-35°)) = \cot 125°$.

23. A, by a negative number identity, $-\sin 35° = \sin(-35°)$.

24. I, by a negative number identity, $\cos 35° = \cos(-35°)$.

25. C, by a half-number identity, $\cos 75° = \cos \dfrac{150°}{2} = \sqrt{\dfrac{1 + \cos 150°}{2}}$.

26. H, by a sum identity, $\sin 75° = \sin 15° \cos 60° + \cos 15° \sin 60°$.

27. D, by a double number identity, $\sin 300° = 2 \sin 150° \cos 150°$.

28. G, by a double number identity, $\cos 300° = \cos(2 \cdot 150°) = \cos^2 150° - \sin^2 150°$.

29. F, by a cofunctional identity, $\tan(-55°) = \cot(-35°)$.

30. $\sin^2 x - \sin^2 y = (1 - \cos^2 x) - (1 - \cos^2 y) = -\cos^2 x + \cos^2 y = \cos^2 y - \cos^2 x$

31. $2 \cos^3 x - \cos x = \cos x (2 \cos^2 x - 1) = \dfrac{1}{\sec x}(\cos 2x) = \dfrac{\cos^2 x - \sin^2 x}{\sec x}$

32. The right side of the equation equals: $\dfrac{\sin 2x + \sin x}{\cos 2x - \cos x} = \dfrac{2 \sin x \cos x + \sin x}{2 \cos^2 x - 1 - \cos x} = \dfrac{\sin x (2 \cos x + 1)}{(2 \cos x + 1)(\cos x - 1)} =$

    $\dfrac{\sin x}{\cos x - 1}$, the left side of the equation equals: $-\cot \dfrac{x}{2} = \dfrac{1}{\tan \frac{x}{2}} = -\dfrac{1}{\frac{1 - \cos x}{\sin x}} = \dfrac{-\sin x}{1 - \cos x} = \dfrac{\sin x}{\cos x - 1}$.

    Since both sides of the equation equal the same expression, $-\cot \dfrac{x}{2} = \dfrac{\sin 2x + \sin x}{\cos 2x - \cos x}$.

33. The left side of the equation equals: $\dfrac{\sin^2 x}{2 - 2 \cos x} = \dfrac{1 - \cos^2 x}{2(1 - \cos x)} = \dfrac{(1 - \cos x)(1 + \cos x)}{2(1 - \cos x)} = \dfrac{1 + \cos x}{2}$.

    The right side of the equation equals: $\cos^2 \dfrac{x}{2} = \left(\pm\sqrt{\dfrac{1 + \cos x}{2}}\right)^2 = \dfrac{1 + \cos x}{2}$.

    Since both sides of the equation equal the same expression, $\dfrac{\sin^2 x}{2 - 2 \cos x} = \cos^2 \dfrac{x}{2}$.

34. $\dfrac{\sin 2x}{\sin x} = \dfrac{2 \sin x \cos x}{\sin x} = 2 \cos x = \dfrac{2}{\sec x}$

35. The right side of the equation equals: $\cos A - \dfrac{\tan A}{\csc A} = \cos A - \dfrac{\frac{\sin A}{\cos A}}{\frac{1}{\sin A}} = \cos A - \dfrac{\sin^2 A}{\cos A} = \dfrac{\cos^2 A - \sin^2 A}{\cos A} =$

    $\dfrac{\cos 2A}{\cos A}$. The left side of the equation equals: $2 \cos A - \sec A = 2 \cos A - \dfrac{1}{\cos A} = \dfrac{2 \cos^2 A - 1}{\cos A} = \dfrac{\cos 2A}{\cos A}$.

    Since both sides of the equation equal the same expression, $2 \cos A - \sec A = \cos A - \dfrac{\tan A}{\csc A}$.

36. $\dfrac{2\tan B}{\sin 2B} = \dfrac{\frac{2\sin B}{\cos B}}{2\sin B\cos B} = \dfrac{2\sin B}{\cos B}\cdot\dfrac{1}{2\sin B\cos B} = \dfrac{1}{\cos^2 B} = \sec^2 B$

37. $2\tan\alpha\csc 2\alpha = \dfrac{2\sin\alpha}{\cos\alpha}\cdot\dfrac{1}{\sin 2\alpha} = \dfrac{2\sin\alpha}{\cos\alpha\,(2\sin\alpha\cos\alpha)} = \dfrac{1}{\cos^2\alpha} = \sec^2\alpha = 1 + \tan^2\alpha$

38. $\cot\dfrac{t}{2} = \dfrac{1}{\tan\frac{t}{2}} = \dfrac{1}{\frac{1-\cos t}{\sin t}} = \dfrac{\sin t}{1-\cos t}$

39. $\dfrac{2\cot x}{\tan 2x} = \dfrac{2\cos x}{\sin x}\div\dfrac{\sin 2x}{\cos 2x} = \dfrac{2\cos x}{\sin x}\cdot\dfrac{1-2\sin^2 x}{2\sin x\cos x} = \dfrac{1-2\sin^2 x}{\sin^2 x} = \dfrac{1}{\sin^2 x} - \dfrac{2\sin^2 x}{\sin^2 x} = \csc^2 x - 2$

40. $\tan\theta\sin 2\theta = \dfrac{\sin\theta}{\cos\theta}(2\sin\theta\cos\theta) = 2\sin^2\theta = 2(1-\cos^2\theta) = 2 - 2\cos^2\theta$

41. $2\tan x\csc 2x - \tan^2 x = \tan x\,(2\csc 2x - \tan x) = \tan x\left(\dfrac{2}{\sin 2x} - \dfrac{\sin x}{\cos x}\right) = \tan x\left(\dfrac{2}{2\sin x\cos x} - \dfrac{\sin x}{\cos x}\right) =$

$\tan x\left(\dfrac{2 - 2\sin^2 x}{2\sin x\cos x}\right) = \tan x\left(\dfrac{2(1-\sin^2 x)}{2\sin x\cos x}\right) = \dfrac{\sin x}{\cos x}\cdot\left(\dfrac{\cos^2 x}{\sin x\cos x}\right) = \dfrac{\sin x}{\cos x}\cdot\dfrac{\cos x}{\sin x} = 1$

42. $y = \sin^{-1}\dfrac{\sqrt{2}}{2} \Rightarrow \dfrac{\sqrt{2}}{2} = \sin y \Rightarrow y = \dfrac{\pi}{4}$ $\left(\text{range: }\left[-\dfrac{\pi}{2},\dfrac{\pi}{2}\right]\right)$

43. $y = \arccos\left(-\dfrac{1}{2}\right) \Rightarrow -\dfrac{1}{2} = \cos y \Rightarrow y = \dfrac{2\pi}{3}$ (range: $[0,\pi]$)

44. $y = \arctan\dfrac{\sqrt{3}}{3} \Rightarrow \dfrac{\sqrt{3}}{3} = \tan y \Rightarrow y = \dfrac{\pi}{6}$ $\left(\text{range: }\left[-\dfrac{\pi}{2},\dfrac{\pi}{2}\right]\right)$

45. $y = \sec^{-1}(-2) \Rightarrow y = \cos^{-1}\left(-\dfrac{1}{2}\right) \Rightarrow -\dfrac{1}{2} = \cos y \Rightarrow y = \dfrac{2\pi}{3}$ $\left(\text{range: }\left[0,\dfrac{\pi}{2}\right)\cup\left(\dfrac{\pi}{2},\pi\right]\right)$

46. $y = \text{arccsc}\dfrac{2\sqrt{3}}{3} \Rightarrow y = \arcsin\dfrac{3}{2\sqrt{3}} \Rightarrow y = \arcsin\dfrac{\sqrt{3}}{2} \Rightarrow \left(\dfrac{\sqrt{3}}{2} = \sin y\right) \Rightarrow y = \dfrac{\pi}{3}$

$\left(\text{range: }\left[-\dfrac{\pi}{2},0\right)\cup\left(0,\dfrac{\pi}{2}\right]\right)$

47. $y = \cot^{-1}(-1) \Rightarrow y = \tan^{-1}\left(\dfrac{1}{-1}\right) + \pi \Rightarrow y = \tan^{-1}(-1) + \pi \Rightarrow (-1 = \tan y) + \pi \Rightarrow$

$y = -\dfrac{\pi}{4} + \pi \Rightarrow y = \dfrac{3\pi}{4}$ (range: $(0,\pi)$)

48. $\theta = \arccos\dfrac{1}{2} \Rightarrow \dfrac{1}{2} = \cos\theta \Rightarrow \theta = 60°$ (range: $[0°, 180°]$)

49. $\theta = \arcsin\left(-\dfrac{\sqrt{3}}{2}\right) \Rightarrow -\dfrac{\sqrt{3}}{2} = \sin\theta \Rightarrow \theta = -60°$ (range: $[-90°, 90°]$)

50. $\theta = \tan^{-1}(0) \Rightarrow 0 = \tan\theta \Rightarrow \theta = 0°$ (range: $[-90°, 90°]$)

51. $\theta = \arcsin(-.656059029) \Rightarrow \theta \approx -41°$

52. $\theta = \arccos(.7095707365) \Rightarrow \theta \approx 44.8°$

53. $\theta = \arctan(-.1227845609) \Rightarrow \theta \approx -7°$

54. $\theta = \cot^{-1}(4.704630109) \Rightarrow \theta = \tan^{-1}\dfrac{1}{4.704630109} \Rightarrow \theta = \tan^{-1}(4.704630109)^{-1} \Rightarrow \theta \approx 12°$

55. $\theta = \sec^{-1}(28.65370835) \Rightarrow \theta = \cos^{-1}\dfrac{1}{28.65370835} \Rightarrow \theta = \cos^{-1}(28.65370835)^{-1} \Rightarrow \theta \approx 88°$

56. $\theta = \csc^{-1}19.10732261 \Rightarrow \theta = \sin^{-1}\dfrac{1}{19.10732261} \Rightarrow \theta = \sin^{-1}(19.10732261)^{-1} \Rightarrow \theta \approx 3°$

57. The value is -3 is in the domain of the inverse tangent function but not in the domain of the inverse sine function.

58. The function $\cos x$ is defined for every real number, but $\arccos x$ is defined only on the interval $[-1, 1]$.

The function $\arccos (\cos x) = x$ only for x in the interval $[0, \pi]$.

59. Since $\sin (\sin^{-1} x) = x$ when $-1 < x < 1$, $\sin \left(\sin^{-1} \dfrac{1}{2} \right) = \dfrac{1}{2}$.

60. Since $\cos \theta = \dfrac{3}{4}$, then θ is in quadrant I. Sketch and label a triangle in quadrant I, with $x = 3$ and $r = 4$.

Now use pythagorean theorem to find y: $3^2 + y^2 = 4^2 \Rightarrow y^2 = 16 - 9 \Rightarrow y^2 = 7 \Rightarrow y = \sqrt{7}$.

Therefore, $\sin \theta = \dfrac{\sqrt{7}}{4}$ and $\sin \left(\cos^{-1} \dfrac{3}{4} \right) = \dfrac{\sqrt{7}}{4}$.

61. Since $\tan \theta = \dfrac{3}{1}$, then θ is in quadrant I. Sketch and label a triangle in quadrant I, with $x = 1$ and $y = 3$.

Use pythagorean theorem to find r: $1^2 + 3^2 = r^2 \Rightarrow r^2 = 1 + 9 \Rightarrow r^2 = 10 \Rightarrow r = \sqrt{10}$.

Therefore, $\cos \theta = \dfrac{1}{\sqrt{10}} = \dfrac{\sqrt{10}}{10}$ and $\cos (\arctan 3) = \dfrac{\sqrt{10}}{10}$.

62. Since $\sin \theta = -\dfrac{1}{3}$, then θ is in quadrant IV. Sketch and label a triangle in quadrant I, with $y = -1$ and $r = 3$.

Use pythagorean theorem to find x: $x^2 + (-1)^2 = 3^2 \Rightarrow x^2 = 9 - 1 \Rightarrow x^2 = 8 \Rightarrow x = \sqrt{8} = 2\sqrt{2}$.

Now $\cos 2\theta = \cos^2 \theta - \sin^2 \theta$, thus, $\cos 2\theta = \left(\dfrac{2\sqrt{2}}{3} \right)^2 - \left(-\dfrac{1}{3} \right)^2 = \dfrac{8}{9} - \dfrac{1}{9} = \dfrac{7}{9}$ and $\cos \left(2 \sin^{-1} \dfrac{1}{3} \right) = \dfrac{7}{9}$.

Finally, if $\cos \left(2 \sin^{-1} \left(-\dfrac{1}{3} \right) \right) = \dfrac{7}{9}$, then $\sec \left(2 \sin^{-1} \left(-\dfrac{1}{3} \right) \right) = \dfrac{9}{7}$.

63. Since the $\cos \dfrac{3\pi}{2} = 0$, we can solve $\cos^{-1} (0) \Rightarrow 0 = \cos y \Rightarrow y = \dfrac{\pi}{2}$. Therefore, $\cos^{-1} \left(\cos \dfrac{3\pi}{2} \right) = \dfrac{\pi}{2}$.

64. Since $\sin \theta = \dfrac{3}{5}$, then θ is in quadrant I. Sketch and label a triangle in quadrant I, with $y = 3$ and $r = 5$.

Use pythagorean theorem to find x: $x^2 + 3^2 = 5^2 \Rightarrow x^2 = 25 - 9 \Rightarrow x^2 = 16 \Rightarrow x = 4$.

Since $\cos \beta = \dfrac{5}{7}$, then β is in quadrant I. Sketch and label a triangle in quadrant I, with $x = 5$ and $r = 7$.

Use pythagorean theorem to find y: $5^2 + y^2 = 7^2 \Rightarrow y^2 = 49 - 25 \Rightarrow y^2 = 24 \Rightarrow y = \sqrt{24} = 2\sqrt{6}$.

Now use $\tan \theta = \dfrac{3}{4}$ and $\tan \beta = \dfrac{2\sqrt{6}}{5}$ in $\tan (\theta + \beta) = \dfrac{\tan \theta + \tan \beta}{1 - \tan \theta \tan \beta}$, therefore, $\dfrac{\dfrac{3}{4} + \dfrac{2\sqrt{6}}{5}}{1 - \left(\dfrac{3}{4} \cdot \dfrac{2\sqrt{6}}{5} \right)} =$

$\dfrac{\dfrac{3}{4} + \dfrac{2\sqrt{6}}{5}}{1 - \left(\dfrac{3}{4} \cdot \dfrac{2\sqrt{6}}{5} \right)} \cdot \dfrac{20}{20} = \dfrac{15 + 8\sqrt{6}}{20 - 6\sqrt{6}} = \dfrac{15 + 8\sqrt{6}}{20 - 6\sqrt{6}} \cdot \dfrac{20 + 6\sqrt{6}}{20 + 6\sqrt{6}} = \dfrac{588 + 250\sqrt{6}}{184} = \dfrac{294 + 125\sqrt{6}}{92}$.

Therefore, $\tan \left(\sin^{-1} \dfrac{3}{5} + \cos^{-1} \dfrac{5}{7} \right) = \dfrac{294 + 125\sqrt{6}}{92}$.

65. Since $\tan \theta = u$, then θ is in quadrant I. Sketch and label a triangle in quadrant I, with $x = 1$ and $y = u$.

Use pythagorean theorem to find y: $1^2 + u^2 = r^2 \Rightarrow r^2 = 1 + u^2 \Rightarrow r = \sqrt{u^2 + 1}$.

Now $\sin \theta = \dfrac{u}{\sqrt{u^2 + 1}} = \dfrac{u\sqrt{u^2 + 1}}{u^2 + 1}$ and $\sin (\tan^{-1} u) = \dfrac{u\sqrt{u^2 + 1}}{u^2 + 1}$.

66. Since $\tan \theta = \dfrac{u}{\sqrt{1 - u^2}}$, then θ is in quadrant I. Sketch and label a triangle in quadrant I, $x = \sqrt{1 - u^2}$

and $y = u$. Use pythagorean theorem to find r: $(\sqrt{1 - u^2})^2 + u^2 = r^2 \Rightarrow r^2 = 1 - u^2 + u^2 \Rightarrow r = 1$.

Now $\cos \theta = \dfrac{\sqrt{1 - u^2}}{1} = \sqrt{1 - u^2}$ and $\cos\left(\arctan \dfrac{u}{\sqrt{1 - u^2}}\right) = \sqrt{1 - u^2}$.

67. Since $\cos \theta = \dfrac{u}{\sqrt{u^2 + 1}}$, then θ is in quadrant I. Sketch and label a triangle in quadrant I, with $x = u$ and

$r = \sqrt{u^2 + 1}$. Use pythagorean theorem to find y: $u^2 + y^2 = (\sqrt{u^2 + 1})^2 \Rightarrow y^2 = u^2 + 1 - u^2 \Rightarrow y = 1$.

Now $\tan \theta = \dfrac{1}{u}$ and $\tan\left(\arccos u\right) = \dfrac{1}{u}$.

68. $\sin^2 x = 1 \Rightarrow \sin x = \pm 1 \Rightarrow x = \dfrac{\pi}{2}, \dfrac{3\pi}{2}$. The solution set is: $x = \left\{\dfrac{\pi}{2}, \dfrac{3\pi}{2}\right\}$.

69. $2 \tan x - 1 = 0 \Rightarrow 2 \tan x = 1 \Rightarrow \tan x = \dfrac{1}{2}$. Now for $\tan x = \dfrac{1}{2}$, find the reference angle

(calculate in radian mode) $x = \tan^{-1}\left(\dfrac{1}{2}\right) \approx .463647609$. Since x is found in quadrants I and III,

$x \approx .463647609$ and $x = \pi + .463647609 \approx 3.605240263$.

The solution set is: $x = \{.463647609, 3.605240263\}$.

70. First factor and solve for $\sin x$: $3 \sin^2 x - 5 \sin x + 2 = 0 \Rightarrow (3 \sin x - 2)(\sin x - 1) = 0$.

Now either $3 \sin x - 2 = 0 \Rightarrow 3 \sin x = 2 \Rightarrow \sin x = \dfrac{2}{3}$; or $\sin x - 1 = 0 \Rightarrow \sin x = 1$.

If $\sin x = \dfrac{2}{3}$, find the reference angle (calculate in radian mode) $x = \sin^{-1}\left(\dfrac{2}{3}\right) \approx .729726562$.

Since x is found in quadrants I and II, $x \approx .729726562$ and $x = \pi - .729726562 \approx 2.411864997$.

If $\sin x = 1$, then $x = \dfrac{\pi}{2}$. The solution is $x = \left\{.729726562, \dfrac{\pi}{2}, 2.411864997\right\}$.

71. Use identities to find $\tan x$: $\tan x = \cot x \Rightarrow \tan x = \dfrac{1}{\tan x} \Rightarrow \tan^2 x = 1 \Rightarrow \tan x = \pm 1$.

If $\tan x = \pm 1$, then the solution is $x = \left\{\dfrac{\pi}{4}, \dfrac{3\pi}{4}, \dfrac{5\pi}{4}, \dfrac{7\pi}{4}\right\}$.

72. First set equal to zero, factor, and solve for $\cot x$: $5 \cot^2 x + 3 \cot x = 2 \Rightarrow 5 \cot^2 x + 3 \cot x - 2 = 0 \Rightarrow$

$(5 \cot x - 2)(\cot x + 1) = 0$. Now either $5 \cot x - 2 = 0 \Rightarrow 5 \cot x = 2 \Rightarrow \cot x = \dfrac{2}{5}$;

or $\cot x + 1 = 0 \Rightarrow \cot x = -1$.

If $\cot x = \dfrac{2}{5}$, then $\tan x = \dfrac{5}{2}$. Now find the reference angle in radian mode: $x = \tan^{-1}\left(\dfrac{5}{2}\right) \approx 1.19028995$

Since x is found in quadrants I and III, $x \approx 1.19028995$ and $x = \pi + 1.19028995 \approx 4.331882603$.

If $\cot x = -1$, then $x = \dfrac{3\pi}{4}, \dfrac{7\pi}{4}$. The solution is $x \approx \left\{1.19028995, \dfrac{3\pi}{4}, 4.331882603, \dfrac{7\pi}{4}\right\}$.

73. Use identities to find $\cos x$: $\sec \dfrac{x}{2} = \cos \dfrac{x}{2} \Rightarrow \dfrac{1}{\cos \frac{x}{2}} = \cos \dfrac{x}{2} \Rightarrow 1 = \cos^2 \dfrac{x}{2} \Rightarrow 1 = \left(\pm\sqrt{\dfrac{1 + \cos x}{2}}\right)^2 \Rightarrow$

$1 = \dfrac{1 + \cos x}{2} \Rightarrow 2 = 1 + \cos x \Rightarrow \cos x = 1$. If $\cos x = 1$, then the solution is $x = \{0\}$.

74. Use identities and factor: $\sin 2x = \cos 2x + 1 \Rightarrow \sin 2x - \cos 2x - 1 = 0 \Rightarrow$

$2 \sin x \cos x - (2 \cos^2 x - 1) - 1 = 0 \Rightarrow 2 \sin x \cos x - 2 \cos^2 x = 0 \Rightarrow 2 \cos x (\sin x - \cos x) = 0.$

Now either $2 \cos x = 0 \Rightarrow \cos x = 0 \Rightarrow x = \dfrac{\pi}{2}, \dfrac{3\pi}{2}$; or $\sin x - \cos x = 0 \Rightarrow \sin x = \cos x \Rightarrow$

$\cos^2 x = \sin^2 x \Rightarrow \cos^2 x = 1 - \cos^2 x \Rightarrow 2 \cos^2 x = 1 \Rightarrow \cos^2 x = \dfrac{1}{2} \Rightarrow \cos x = \pm \dfrac{1}{\sqrt{2}} \Rightarrow$

$x = \dfrac{\pi}{4}, \dfrac{3\pi}{4}, \dfrac{5\pi}{4}, \dfrac{7\pi}{4}$. Since we squared to solve we need to check the second set of answers. The functions

$\cos x$ and $\sin x$ have different signs in quadrants II and IV, so only $\dfrac{\pi}{4}$ and $\dfrac{5\pi}{4}$ will check.

The solution is $x = \left\{ \dfrac{\pi}{4}, \dfrac{\pi}{2}, \dfrac{5\pi}{4}, \dfrac{3\pi}{2} \right\}$.

75. First solve for $\sin 2x$: $2 \sin 2x = 1 \Rightarrow \sin 2x = \dfrac{1}{2}$.

If the required interval for x is: $0 \le x < 2\pi$, then $0 \le 2x < 4\pi$.

Therefore, if $\sin 2x = \dfrac{1}{2}$, then $2x = \dfrac{\pi}{6}, \dfrac{5\pi}{6}, \dfrac{13\pi}{6}, \dfrac{17\pi}{6}$.

Now dividing each solution by 2 $\left(\text{or multipling by } \dfrac{1}{2}\right)$ yields: $x = \left\{ \dfrac{\pi}{12}, \dfrac{5\pi}{12}, \dfrac{13\pi}{12}, \dfrac{17\pi}{12} \right\}$.

76. Use identities and factor: $\sin 2x + \sin 4x = 0 \Rightarrow \sin 2x + 2 \sin 2x \cos 2x = 0 \Rightarrow \sin 2x (1 + 2 \cos 2x) = 0.$

If the required interval for x is: $0 \le x < 2\pi$, then $0 \le 2x < 4\pi$.

Now either $\sin 2x = 0 \Rightarrow 2x = 0, \pi, 2\pi, 3\pi$; or $1 + 2 \cos 2x = 0 \Rightarrow 2 \cos 2x = -1 \Rightarrow \cos 2x = -\dfrac{1}{2} \Rightarrow$

$2x = \dfrac{2\pi}{3}, \dfrac{4\pi}{3}, \dfrac{8\pi}{3}, \dfrac{10\pi}{3}$. Divide each solution by 2 $\left(\text{multiply by } \dfrac{1}{2}\right)$ to get: $x = \left\{ 0, \dfrac{\pi}{3}, \dfrac{\pi}{2}, \dfrac{2\pi}{3}, \pi, \dfrac{4\pi}{3}, \dfrac{3\pi}{2}, \dfrac{5\pi}{3} \right\}$.

77. Use identities and solve for x: $\cos x - \cos 2x = 2 \cos x \Rightarrow -\cos x - (2 \cos^2 x - 1) = 0 \Rightarrow$

$-\cos x - 2 \cos^2 x + 1 = 0 \Rightarrow 2 \cos^2 x + \cos x - 1 = 0 \Rightarrow (2 \cos x - 1)(\cos x + 1) = 0.$

Now either $2 \cos x - 1 = 0 \Rightarrow 2 \cos x = 1 \Rightarrow \cos x = \dfrac{1}{2} \Rightarrow x = \dfrac{\pi}{3}, \dfrac{5\pi}{3}$; or $\cos x + 1 = 0 \Rightarrow$

$\cos x = -1 \Rightarrow x = \pi$. The solution is $x = \left\{ \dfrac{\pi}{3}, \pi, \dfrac{5\pi}{3} \right\}$.

78. If the required interval for x is: $0 \le x < 2\pi$, then $0 \le 2x < 4\pi$.

Therefore, if $\tan 2x = \sqrt{3}$, then $2x = \dfrac{\pi}{3}, \dfrac{4\pi}{3}, \dfrac{7\pi}{3}, \dfrac{10\pi}{3}$.

Now dividing each solution by 2 $\left(\text{or multipling by } \dfrac{1}{2}\right)$ yields: $x = \left\{ \dfrac{\pi}{6}, \dfrac{2\pi}{3}, \dfrac{7\pi}{6}, \dfrac{5\pi}{3} \right\}$.

79. Factor to solve for $\cos \dfrac{x}{2}$: $\cos^2 \dfrac{x}{2} - 2 \cos \dfrac{x}{2} + 1 = 0 \Rightarrow \left(\cos \dfrac{x}{2} - 1 \right)\left(\cos \dfrac{x}{2} - 1 \right) = 0.$

If the required interval for x is: $0 \le x < 2\pi$, then $0 \le \dfrac{x}{2} < \pi$.

If $\cos \dfrac{x}{2} - 1 = 0 \Rightarrow \cos \dfrac{x}{2} = 1 \Rightarrow \dfrac{x}{2} = 0$, multiplying this by 2 yields the solution $x = \{0\}$.

80. Graph $y = \sin 2x - \cos 2x - 1$, from the calculator, the x-intercepts are $\dfrac{\pi}{4}, \dfrac{\pi}{2}, \dfrac{5\pi}{4}, \dfrac{3\pi}{2}$.

(a) For $\sin 2x > \cos 2x + 1$, the graph is above the x-axis for the interval: $\left(\dfrac{\pi}{4}, \dfrac{\pi}{2} \right) \cup \left(\dfrac{5\pi}{4}, \dfrac{3\pi}{2} \right)$.

(b) For $\sin 2x < \cos 2x + 1$, the graph is below the x-axis for the interval: $\left[0, \dfrac{\pi}{4} \right) \cup \left(\dfrac{\pi}{2}, \dfrac{5\pi}{4} \right) \cup \left(\dfrac{3\pi}{2}, 2\pi \right)$.

81. (a) Let α be the angle to the left of θ, then $\tan \alpha = \dfrac{5}{x}$ and $\alpha = \arctan \dfrac{5}{x}$.

   Now $\tan(\alpha + \theta) = \dfrac{5 + 10}{x} \Rightarrow \alpha + \theta = \arctan \dfrac{15}{x} \Rightarrow \theta = \arctan \dfrac{15}{x} - \alpha \Rightarrow \theta = \arctan \dfrac{15}{x} - \arctan \dfrac{5}{x}$.

   (b) Graph $f(x) = \arctan\left(\dfrac{15}{x}\right) - \arctan\left(\dfrac{5}{x}\right)$, see Figure 81. The maximum occurs at $x \approx 8.6602567$.

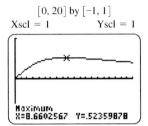

$[0, 20]$ by $[-1, 1]$
Xscl = 1   Yscl = 1

Figure 81

82. (a) $3y = \sin \dfrac{4\pi t}{3} \Rightarrow \dfrac{4\pi t}{3} = \sin^{-1} 3y \Rightarrow t = \dfrac{3 \sin^{-1} 3y}{4\pi}$

   (b) Calculating $t = \dfrac{3 \sin^{-1} 3(.3)}{4\pi}$, yields: $t \approx .267325 \approx .27$ seconds.

83. $40 = 100 \sin(2\pi \cdot 60)t \Rightarrow .4 = \sin 120\pi t \Rightarrow 120\pi t = \sin^{-1} .4 \Rightarrow$

   $t = \dfrac{\sin^{-1} .4}{120\pi} \Rightarrow t \approx .00109 \approx .001$ seconds.

84. $50 = 100 \sin(2\pi \cdot 120)t \Rightarrow .5 = \sin 240\pi t \Rightarrow 240\pi t = \sin^{-1} .5 \Rightarrow$

   $t = \dfrac{\sin^{-1} .5}{240\pi} \Rightarrow t \approx .000694 \approx .0007$ seconds.

85. For $.752 = \dfrac{\sin \theta_1}{\sin \theta_2}$, if $\theta_2 = 90°$, then $.752 = \dfrac{\sin \theta_1}{\sin 90°} \Rightarrow .752 = \dfrac{\sin \theta_1}{1} \Rightarrow .752 = \sin \theta_1 \Rightarrow$

   $\theta_1 = \sin^{-1} .752 \Rightarrow \theta_1 \approx 48.8°$.

86. If $\theta_1 > 48.8°$, then $\theta_2 > 90°$ and the light beam will stay completely under the water. The beam will be reflected at the surface of the water.

Chapter 9 Test

1. If $\sin y = -\dfrac{2}{3}$ and $\pi < y < \dfrac{3\pi}{2}$, then y is in quadrant III.

   Sketch and label a triangle in quadrant III, with $y = -2$ and $r = 3$.

   Use pythagorean theorem to find x: $x^2 + (-2)^2 = 3^2 \Rightarrow x^2 = 9 - 4 \Rightarrow x^2 = 5 \Rightarrow x = \sqrt{5}$.

   Because x is found in quadrant III, $x = -\sqrt{5}$. Therefore, $\cos y = -\dfrac{\sqrt{5}}{3}$.

   If $\cos x = -\dfrac{1}{5}$ and $\dfrac{\pi}{2} < x < \pi$, then x is in quadrant II.

   Sketch and label a triangle in quadrant II, with $x = -1$ and $r = 5$.

   Use pythagorean theorem to find x: $(-1)^2 + y^2 = 5^2 \Rightarrow y^2 = 25 - 1 \Rightarrow y^2 = 24 \Rightarrow y = \sqrt{24} = 2\sqrt{6}$.

   Therefore, $\sin x = \dfrac{2\sqrt{6}}{5}$. Using these values, $\sin(x + y) = \sin x \cos y + \cos x \sin y \Rightarrow$

   $\left(\dfrac{2\sqrt{6}}{5}\right)\left(-\dfrac{\sqrt{5}}{3}\right) + \left(-\dfrac{1}{5}\right)\left(-\dfrac{2}{3}\right) = \dfrac{-2\sqrt{30}}{15} + \dfrac{2}{15} = \dfrac{2 - 2\sqrt{30}}{15}$.

2. Use the values found in exercise 1 to solve, $\cos(x - y) = \cos x \cos y + \sin x \sin y \Rightarrow$

$$\left(-\frac{1}{5}\right)\left(-\frac{\sqrt{5}}{3}\right) + \left(\frac{2\sqrt{6}}{5}\right)\left(-\frac{2}{3}\right) = \frac{\sqrt{5}}{15} - \frac{4\sqrt{6}}{15} = \frac{\sqrt{5} - 4\sqrt{6}}{15}.$$

3. Use the values found in exercise 1 to solve, $\tan\dfrac{y}{2} = \dfrac{1 - \cos y}{\sin y} = \dfrac{1 - (-\frac{\sqrt{5}}{3})}{-\frac{2}{3}} = \dfrac{1 - (-\frac{\sqrt{5}}{3})}{-\frac{2}{3}} \cdot \dfrac{3}{3} =$

$$\frac{3 + \sqrt{5}}{-2} = \frac{-3 - \sqrt{5}}{2}.$$

4. Use the values found in exercise 1 to solve, $\cos 2x = 2\cos^2 x - 1 = 2\left(-\dfrac{1}{5}\right)^2 - 1 = \dfrac{2}{25} - \dfrac{25}{25} = -\dfrac{23}{25}.$

5. $\tan^2 x - \sec^2 x = \dfrac{\sin^2 x}{\cos^2 x} - \dfrac{1}{\cos^2 x} = \dfrac{\sin^2 x - 1}{\cos^2 x} = -\dfrac{\cos^2 x}{\cos^2 x} = -1$

6. Graph $y = \sec x - \sin x \tan x$ on the calculator. The graph looks like $\cos x$, so the identity is

   $\sec x - \sin x \tan x = \cos x$. Verify this:

   $$\sec x - \sin x \tan x = \frac{1}{\cos x} - \frac{\sin x}{1} \cdot \frac{\sin x}{\cos x} = \frac{1 - \sin^2 x}{\cos x} = \frac{\cos^2 x}{\cos x} = \cos x \text{ Therefore,}$$

   $\sec x - \sin x \tan x = \cos x$

7. $\sec^2 B = \dfrac{1}{\cos^2 B} = \dfrac{1}{1 - \sin^2 B}$

8. $\dfrac{\cot A - \tan A}{\csc A \sec A} = \dfrac{\frac{\cos A}{\sin A} - \frac{\sin A}{\cos A}}{\frac{1}{\sin A} \cdot \frac{1}{\cos A}} = \dfrac{\cos^2 A - \sin^2 A}{\sin A \cos A} \cdot \dfrac{\sin A \cos A}{1} = \cos^2 A - \sin^2 A = \cos 2A$

9. $\cos(270° - \theta) = \cos 270° \cos \theta + \sin 270° \sin \theta = 0 \cdot \cos \theta + (-1) \sin \theta = -\sin \theta$

10. $\sin(\pi + \theta) = \sin \pi \cos \theta + \cos \pi \sin \theta = 0 \cdot \cos \theta + (-1) \sin \theta = -\sin \theta$

11. (a) See Figure 11.

    (b) The domain is $[-1, 1]$, the same as $f(x) = \sin^{-1} x$.

       The range is that of $f(x) = \sin^{-1} x$, multiplied by -2, therefore $[-\pi, \pi]$.

    (c) The domain of $f(x) = -2\sin^{-1} x$ is $[-1, 1]$, and 2 is not in this interval. (No number has sine value 2.)

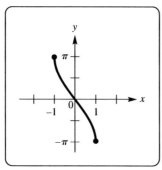

Figure 11

12. $y = \arccos\left(-\dfrac{1}{2}\right) \Rightarrow -\dfrac{1}{2} = \cos y \Rightarrow y = \dfrac{2\pi}{3}$

13. $y = \tan^{-1}(0) \Rightarrow 0 = \tan y \Rightarrow y = 0$

14. $y = \csc^{-1}\dfrac{2\sqrt{3}}{3} \Rightarrow \dfrac{2\sqrt{3}}{3} = \csc y \Rightarrow \sin y = \dfrac{3}{2\sqrt{3}} \Rightarrow \sin y = \dfrac{\sqrt{3}}{2} \Rightarrow y = \dfrac{\pi}{3}.$

15. Since $\sin\left(\dfrac{5\pi}{6}\right) = \dfrac{1}{2}$, then $y = \sin^{-1}\left(\dfrac{1}{2}\right) \Rightarrow \dfrac{1}{2} = \sin y \Rightarrow y = \dfrac{\pi}{6}$.

16. Since $\sin\theta = \dfrac{2}{3}$, then θ is in quadrant I. Sketch and label a triangle in quadrant I, with $y = 2$ and $r = 3$.

Use pythagorean theorem to find x: $x^2 + 2^2 = 3^2 \Rightarrow x^2 = 9 - 4 \Rightarrow x^2 = 5 \Rightarrow x = \pm\sqrt{5}$.

Therefore $\cos\theta = \dfrac{\sqrt{5}}{3}$ and $\cos\left(\arcsin\dfrac{2}{3}\right) = \dfrac{\sqrt{5}}{3}$.

17. Since $\cos\theta = \dfrac{1}{3}$, then θ is in quadrant I. Sketch and label a triangle in quadrant I, with $x = 1$ and $r = 3$.

Use pythagorean theorem to find y: $1^2 + y^2 = 3^2 \Rightarrow y^2 = 9 - 1 \Rightarrow y^2 = 8 \Rightarrow y = \pm\sqrt{8} = \pm2\sqrt{2}$.

Now $\sin 2\theta = 2\sin\theta\cos\theta$, therefore, $\sin 2\theta = 2\left(\dfrac{2\sqrt{2}}{3}\right)\left(\dfrac{1}{3}\right) = \dfrac{4\sqrt{2}}{9}$ and $\sin\left(2\cos^{-1}\dfrac{1}{3}\right) = \dfrac{4\sqrt{2}}{9}$.

18. $\sec\left(\cos^{-1} u\right) = \dfrac{1}{\cos\left(\cos^{-1} u\right)} = \dfrac{1}{u}$

19. Since $\sin\theta = \dfrac{u}{1}$, then θ is in quadrant I. Sketch and label a triangle in quadrant I.

Use $r = 1$, $y = u$ and pythagorean theorem to find x: $x^2 + u^2 = 1^2 \Rightarrow x^2 = 1 - u^2 \Rightarrow x = \sqrt{1 - u^2}$.

Now $\tan\theta = \dfrac{u}{\sqrt{1 - u^2}} = \dfrac{u\sqrt{1 - u^2}}{1 - u^2}$ and $\tan\left(\arcsin u\right) = \dfrac{u\sqrt{1 - u^2}}{1 - u^2}$.

20. Use identities to solve for θ: $\sin^2\theta = \cos^2\theta + 1 \Rightarrow 1 - \cos^2\theta = \cos^2\theta + 1 \Rightarrow 0 = 2\cos^2\theta \Rightarrow$

$\cos\theta = 0 \Rightarrow \theta = \left\{\dfrac{\pi}{2}, \dfrac{3\pi}{2}\right\}$.

21. Use identities and factor: $\csc^2\theta - 2\cot\theta = 4 \Rightarrow 1 + \cot^2\theta - 2\cot\theta = 4 \Rightarrow$

$\cot^2\theta - 2\cot\theta - 3 = 0 \Rightarrow (\cot\theta - 3)(\cot x + 1) = 0$.

Now either $\cot\theta - 3 = 0 \Rightarrow \cot\theta = 3$; or $\cot\theta + 1 = 0 \Rightarrow \cot\theta = -1$.

If $\cot\theta = 3$, then $\tan\theta = \dfrac{1}{3} \Rightarrow \theta = \tan^{-1}\left(\dfrac{1}{3}\right)$ and we find the reference angle in degree mode:

$\tan^{-1}\left(\dfrac{1}{3}\right) \approx 18.4°$. Since θ is in quadrants I and III, $\theta \approx 18.4°$ and $\theta \approx 180° + 18.4 \approx 198.4°$.

If $\cot\theta = -1 \Rightarrow \theta = 135°, 315°$. Therefore the solution set is $\theta \approx \{18.4°, 135°, 198.4°, 315°\}$.

22. Use identities and factor: $\cos x = \cos 2x \Rightarrow \cos x = 2\cos^2 x - 1 \Rightarrow 0 = 2\cos^2 x - \cos x - 1 \Rightarrow$

$0 = (2\cos x + 1)(\cos x - 1)$.

Now either $2\cos x + 1 = 0 \Rightarrow 2\cos x = -1 \Rightarrow \cos x = -\dfrac{1}{2} \Rightarrow x = \dfrac{2\pi}{3}, \dfrac{4\pi}{3}$;

or $\cos x - 1 = 0 \Rightarrow \cos x = 1 \Rightarrow x = 0$. Therefore the solution set is $x = \left\{0, \dfrac{2\pi}{3}, \dfrac{4\pi}{3}\right\}$.

23. If the required interval for θ is: $0° \le \theta < 360°$, then $0° \le \dfrac{\theta}{2} < 180°$.

Now solve for $\dfrac{\theta}{2}$: $2\sqrt{3}\sin\dfrac{\theta}{2} = 3 \Rightarrow \sin\dfrac{\theta}{2} = \dfrac{3}{2\sqrt{3}} \Rightarrow \sin\dfrac{\theta}{2} = \dfrac{\sqrt{3}}{2} \Rightarrow \dfrac{\theta}{2} = 60°, 120°$.

Finally, multiplying the solutions by 2 yields the solution set: $\theta = \{120°, 240°\}$.

24. Graph $y = 2\sin x - 1$, from the capabilities of the calculator the x-intercepts are $\dfrac{\pi}{6}$ and $\dfrac{5\pi}{6}$. From the graph,

$2\sin x - 1 \le 0$ or the graph of the equation is below the x-axis for the interval: $\left[0, \dfrac{\pi}{6}\right] \cup \left[\dfrac{5\pi}{6}, 2\pi\right)$.

25. (a) Since t will repeat every 12 months, the domain for 1 year will be $[0,11]$. The function $T(x)$ is limited

by the range of cosine, at the minimum range (-1), $50 + 50 \cos(-1) = 0$ and at the maximum range (1),

$50 + 50 \cos(1) = 100$, therefore the range of the function is $[0, 100]$ in hundreds.

(b) See Figure 25.

(c) From the graph, the maximum of 10,000 animals occurs at 0 or 12 months (July). The minimum

of 0 animals occurs at 6 months (January).

(d) $T(3) = 50 + 50 \cos\left(\dfrac{(3)\pi}{6}\right) = 50 + 50\,(0) = 50$ hundreds $= 5000$.

There are 5000 animals at 3 months and at 9 months. This is supported by the graph.

(e) $75 = 50 + 50 \cos\left(\dfrac{\pi}{6}t\right) = 25 \Rightarrow 50\cos\left(\dfrac{\pi}{6}t\right) \Rightarrow \dfrac{1}{2} = \cos\left(\dfrac{\pi}{6}t\right) \Rightarrow \dfrac{\pi}{6}t = \dfrac{\pi}{3}$ or $\dfrac{\pi}{6}t = \dfrac{5\pi}{3} \Rightarrow t = 2, 10$.

At 2 months (September) and at 10 months (May), there will be 7500 animals.

(f) $T = 50 + 50 \cos\left(\dfrac{\pi}{6}t\right) \Rightarrow T - 50 = 50\cos\left(\dfrac{\pi}{6}t\right) \Rightarrow \dfrac{T - 50}{50} = \cos\left(\dfrac{\pi}{6}t\right) \Rightarrow$

$\dfrac{\pi}{6}t = \arccos\left(\dfrac{T - 50}{50}\right) \Rightarrow t = \dfrac{6}{\pi}\,\text{acrcos}\left(\dfrac{T - 50}{50}\right)$

$[0, 12]$ by $[0, 110]$	$[0, 10]$ by $[-1, 5]$	$[0, 10]$ by $[-1, 5]$
Xscl $= 1$ Yscl $= 10$	Xscl $= 1$ Yscl $= 1$	Xscl $= 1$ Yscl $= 1$

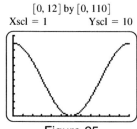   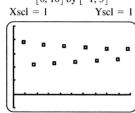   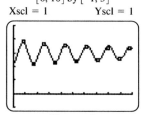

Figure 25   Figure 1   Figure 3

Chapter 9 Project

1. See Figure 1.

2. Find a: 3.81 is the greatest height and 2.17 is the least height, so the amplitude is $a = \dfrac{3.81 - 2.17}{2} = .82$.

Find b: The time difference between successive maximums (or minimums) is 1.8 seconds, so the period is

$b = \dfrac{2\pi}{1.8} \approx 3.49066 \Rightarrow b \approx 3.49$.

Find d: The smallest time corresponding to the mean height can be found by taking half the difference of the time

for the first maximum and the first minimum and subtracting that from the time at the first maximum.

$\dfrac{1.7 - .8}{2} = .45$, and so $d = .8 - .45 \Rightarrow d = .35$.

Find r: The amplitude diminishing rate can be found by dividing the second maximum by the first maximum.

$r = \dfrac{3.61}{3.81} \approx .9475066 \Rightarrow r \approx .95$.

Find c: To find c, the vertical shift, find the mean height by averaging the maximum and minimum heights:

$c = \dfrac{3.81 + 2.17}{2} \Rightarrow c = 2.99$.

With these values, the function $y = r^x a \sin\left[b\,(x - d)\right] + c$ will be $y = .95^x\,(.82)\sin\left[3.49\,(x - .35)\right] + 2.99$.

3. Graph the new equation, see Figure 3. The function is a reasonably good fit for the data.

Chapter 10: Applications of Trigonometry; Vectors

10.1: The Law of Sines

1. The proportion shown in C is not valid because it cannot be written in the form of the Law of Sines.

2. Situation C is SAS and situation D is SSS. Neither of these can be solved using the Law of Sines.

3. The given information is AAS which does not result in the ambiguous case.

4. The given information is AAS which does not result in the ambiguous case.

5. The given information is SSS which does not result in the ambiguous case.

6. The given information is SSA which results in the ambiguous case.

7. The given information is SSA which results in the ambiguous case.

8. The given information is SAS which does not result in the ambiguous case.

9. The given information is SSA which results in the ambiguous case.

10. The given information is AAS which does not result in the ambiguous case.

11. Angle C has measure $180° - 60° - 75° = 45°$. Now, using the Law of Sines,

$$\frac{a}{\sin 60°} = \frac{\sqrt{2}}{\sin 45°} \Rightarrow a = \frac{\sqrt{2}\sin 60°}{\sin 45°} \Rightarrow a = \frac{\sqrt{2} \cdot \frac{\sqrt{3}}{2}}{\frac{\sqrt{2}}{2}} \Rightarrow a = \frac{\frac{\sqrt{6}}{2}}{\frac{\sqrt{2}}{2}} \Rightarrow a = \frac{\sqrt{6}}{2} \cdot \frac{2}{\sqrt{2}} = \sqrt{3}.$$

12. Angle B has measure $180° - 105° - 45° = 30°$. Now, using the Law of Sines,

$$\frac{a}{\sin 45°} = \frac{10}{\sin 30°} \Rightarrow a = \frac{10\sin 45°}{\sin 30°} \Rightarrow a = \frac{10 \cdot \frac{\sqrt{2}}{2}}{\frac{1}{2}} \Rightarrow a = \frac{5\sqrt{2}}{\frac{1}{2}} \Rightarrow a = 10\sqrt{2}.$$

13. Angle C has measure $180° - 37° - 48° = 95°$. Now, using the Law of Sines,

$$\frac{a}{\sin 37°} = \frac{18}{\sin 95°} \Rightarrow a = \frac{18\sin 37°}{\sin 95°} \Rightarrow a \approx 10.87; \quad \frac{b}{\sin 48°} = \frac{18}{\sin 95°} \Rightarrow b = \frac{18\sin 48°}{\sin 95°} \Rightarrow b \approx 13.43.$$

The solutions are $C = 95°$, $b \approx 13$ m, $a \approx 11$ m.

14. Angle A has measure $180° - 52° - 29° = 99°$. Now, using the Law of Sines,

$$\frac{b}{\sin 52°} = \frac{43}{\sin 99°} \Rightarrow b = \frac{43\sin 52°}{\sin 99°} \Rightarrow b \approx 34.31; \quad \frac{c}{\sin 29°} = \frac{43}{\sin 99°} \Rightarrow c = \frac{43\sin 29°}{\sin 99°} \Rightarrow c \approx 21.11.$$

The solutions are $A = 99°$, $b \approx 34$ cm, $c \approx 21$ cm.

15. Angle B has measure $180° - 115.5° - 27.2° = 37.3°$. Now, using the Law of Sines,

$$\frac{a}{\sin 27.2°} = \frac{76.0}{\sin 115.5°} \Rightarrow a = \frac{76.0\sin 27.2°}{\sin 115.5°} \Rightarrow a \approx 38.49;$$

$$\frac{b}{\sin 37.3°} = \frac{76.0}{\sin 115.5°} \Rightarrow b = \frac{76.0\sin 37.3°}{\sin 115.5°} \Rightarrow b \approx 51.03.$$

The solutions are $B = 37.3°$, $a \approx 38.5$ ft, $b \approx 51.0$ ft.

16. Angle A has measure $180° - 124.1° - 18.7° = 37.2°$. Now, using the Law of Sines,

$$\frac{a}{\sin 37.2°} = \frac{94.6}{\sin 18.7°} \Rightarrow a = \frac{94.6 \sin 37.2°}{\sin 18.7°} \Rightarrow a \approx 178.39;$$

$$\frac{c}{\sin 124.1°} = \frac{94.6}{\sin 18.7°} \Rightarrow c = \frac{94.6 \sin 124.1°}{\sin 18.7°} \Rightarrow c \approx 244.33.$$

The solutions are $A = 37.2°$, $a \approx 178$ m, $c \approx 244$ m.

17. Angle B has measure $180° - 37° - 95° = 48°$. Now, using the Law of Sines,

$$\frac{a}{\sin 37°} = \frac{18}{\sin 95°} \Rightarrow a = \frac{18 \sin 37°}{\sin 95°} \Rightarrow a \approx 10.87; \frac{b}{\sin 48°} = \frac{18}{\sin 95°} \Rightarrow b = \frac{18 \sin 48°}{\sin 95°} \Rightarrow b \approx 13.43.$$

The solutions are $B = 48°$, $a \approx 11$ m, $b \approx 13$ m.

18. Angle A has measure $180° - 52° - 29° = 99°$. Now, using the Law of Sines,

$$\frac{b}{\sin 52°} = \frac{43}{\sin 99°} \Rightarrow b = \frac{43 \sin 52°}{\sin 99°} \Rightarrow b \approx 34.31; \frac{c}{\sin 29°} = \frac{43}{\sin 99°} \Rightarrow c = \frac{43 \sin 29°}{\sin 99°} \Rightarrow c \approx 21.11.$$

The solutions are $A = 99°$, $b \approx 34$ cm, $c \approx 21$ cm.

19. Angle A has measure $180° - 74.08° - 69.38° = 36.54°$. Now, using the Law of Sines,

$$\frac{a}{\sin 36.54°} = \frac{45.38}{\sin 74.08°} \Rightarrow a = \frac{45.38 \sin 36.54°}{\sin 74.08°} \Rightarrow a \approx 28.096;$$

$$\frac{b}{\sin 69.38°} = \frac{45.38}{\sin 74.08°} \Rightarrow b = \frac{45.38 \sin 69.38°}{\sin 74.08°} \Rightarrow b \approx 44.167.$$

The solutions are $A = 36.54°$, $a \approx 28.10$ m, $b \approx 44.17$ m.

20. Angle B has measure $180° - 87.2° - 74.3° = 18.5°$. Now, using the Law of Sines,

$$\frac{a}{\sin 87.2°} = \frac{75.9}{\sin 18.5°} \Rightarrow a = \frac{75.9 \sin 87.2°}{\sin 18.5°} \Rightarrow a \approx 238.92;$$

$$\frac{c}{\sin 74.3°} = \frac{75.9}{\sin 18.5°} \Rightarrow c = \frac{75.9 \sin 74.3°}{\sin 18.5°} \Rightarrow c \approx 230.28.$$

The solutions are $B = 18.5°$, $a \approx 239$ yd, $c \approx 230$ yd.

21. Angle A has measure $180° - 38° 40' - 91° 40' = 49° 40'$. Now, using the Law of Sines,

$$\frac{b}{\sin 38° 40'} = \frac{19.7}{\sin 49° 40'} \Rightarrow b = \frac{19.7 \sin 38° 40'}{\sin 49° 40'} \Rightarrow b \approx 16.146;$$

$$\frac{c}{\sin 91° 40'} = \frac{19.7}{\sin 49° 40'} \Rightarrow c = \frac{19.7 \sin 91° 40'}{\sin 49° 40'} \Rightarrow c \approx 25.832.$$

The solutions are $A = 49° 40'$, $b \approx 16.1$ cm, $c \approx 25.8$ cm.

22. Angle A has measure $180° - 20° 50' - 103° 10' = 56° 00'$. Now, using the Law of Sines,

$$\frac{a}{\sin 56° 00'} = \frac{132}{\sin 20° 50'} \Rightarrow a = \frac{132 \sin 56° 00'}{\sin 20° 50'} \Rightarrow a \approx 307.698;$$

$$\frac{c}{\sin 103° 10'} = \frac{132}{\sin 20° 50'} \Rightarrow c = \frac{132 \sin 103° 10'}{\sin 20° 50'} \Rightarrow c \approx 361.394.$$

The solutions are $A = 56° 00'$, $a \approx 308$ ft, $c \approx 361$ ft.

23. Angle C has measure $180° - 35.3° - 52.8° = 91.9°$. Now, using the Law of Sines,

$$\frac{a}{\sin 35.3°} = \frac{675}{\sin 52.8°} \Rightarrow a = \frac{675 \sin 35.3°}{\sin 52.8°} \Rightarrow a \approx 489.69;$$

$$\frac{c}{\sin 91.9°} = \frac{675}{\sin 52.8°} \Rightarrow c = \frac{675 \sin 91.9°}{\sin 52.8°} \Rightarrow c \approx 846.96.$$

The solutions are $C = 91.9°$, $a \approx 490$ ft, $c \approx 847$ ft.

24. Angle C has measure $180° - 68.41° - 54.23° = 57.36°$. Now, using the Law of Sines,

$$\frac{b}{\sin 54.23°} = \frac{12.75}{\sin 68.41°} \Rightarrow b = \frac{12.75 \sin 54.23°}{\sin 68.41°} \Rightarrow b \approx 11.126;$$

$$\frac{c}{\sin 57.36°} = \frac{12.75}{\sin 68.41°} \Rightarrow c = \frac{12.75 \sin 57.36°}{\sin 68.41°} \Rightarrow c \approx 11.547.$$

The solutions are $C = 57.36°$, $b \approx 11.13$ ft, $c \approx 11.55$ ft.

25. Angle B has measure $180° - 39.70° - 30.35° = 109.95° \approx 110.0°$. Now, using the Law of Sines,

$$\frac{a}{\sin 39.70°} = \frac{39.74}{\sin 110.0°} \Rightarrow a = \frac{39.74 \sin 39.70°}{\sin 110.0°} \Rightarrow a \approx 27.01;$$

$$\frac{c}{\sin 30.35°} = \frac{39.74}{\sin 110.0°} \Rightarrow c = \frac{39.74 \sin 30.35°}{\sin 110.0°} \Rightarrow c \approx 21.37.$$

The solutions are $B = 110.0°$, $a \approx 27.01$ m, $c \approx 21.37$ m.

26. Angle A has measure $180° - 71.83° - 42.57° = 65.60°$. Now, using the Law of Sines,

$$\frac{b}{\sin 42.57°} = \frac{2.614}{\sin 65.60°} \Rightarrow b = \frac{2.614 \sin 42.57°}{\sin 65.60°} \Rightarrow b \approx 1.942;$$

$$\frac{c}{\sin 71.83°} = \frac{2.614}{\sin 65.60°} \Rightarrow c = \frac{2.614 \sin 71.83°}{\sin 65.60°} \Rightarrow c \approx 2.727.$$

The solutions are $A = 65.60°$, $b \approx 1.942$ cm, $c \approx 2.727$ cm.

27. Angle A has measure $180° - 42.88° - 102.40° = 34.72°$. Now, using the Law of Sines,

$$\frac{a}{\sin 34.72°} = \frac{3974}{\sin 42.88°} \Rightarrow a = \frac{3974 \sin 34.72°}{\sin 42.88°} \Rightarrow a \approx 3326;$$

$$\frac{c}{\sin 102.40°} = \frac{3974}{\sin 42.88°} \Rightarrow c = \frac{3974 \sin 102.40°}{\sin 42.88°} \Rightarrow c \approx 5704.$$

The solutions are $A = 34.72°$, $a \approx 3326$ ft, $c \approx 5704$ ft.

28. Angle C has measure $180° - 18.75° - 51.53° = 109.72°$. Now, using the Law of Sines,

$$\frac{a}{\sin 18.75°} = \frac{2798}{\sin 109.72°} \Rightarrow a = \frac{2798 \sin 18.75°}{\sin 109.72°} \Rightarrow a \approx 955.4;$$

$$\frac{b}{\sin 51.53°} = \frac{2798}{\sin 109.72°} \Rightarrow b = \frac{2798 \sin 51.53°}{\sin 109.72°} \Rightarrow b \approx 2327.$$

The solutions are $C = 109.72°$, $a \approx 955.4$ yd, $b \approx 2327$ yd.

29. Three given angles do not determine a unique triangle. Data set A does not determine a unique triangle.

30. Since three given angles do not determine a unique triangle we may eliminate data sets A and C. In data set B, the two shorter sides do not sum to at least the length of the longest side. The only data set that determines a unique triangle is data set D.

644 CHAPTER 10 Applications of Trigonometry; Vectors

31. Using the Law of Sines,

$$\frac{\sin B}{41.5} = \frac{\sin 29.7°}{27.2} \Rightarrow \sin B = \frac{41.5 \sin 29.7°}{27.2} \Rightarrow B = \sin^{-1}\left(\frac{41.5 \sin 29.7°}{27.2}\right) \approx 49.1° \text{ or } 180 - 49.1 = 130.9°.$$

If $B = 49.1, C = 180° - 29.7° - 49.1° = 101.2°$. If $B = 130.9°, C = 180° - 29.7° - 130.9° = 19.4°$.

The solutions are $B_1 = 49.1°, C_1 = 101.2°$ or $B_2 = 130.9°, C_2 = 19.4°$.

32. Using the Law of Sines,

$$\frac{\sin A}{890} = \frac{\sin 48.2°}{697} \Rightarrow \sin A = \frac{890 \sin 48.2°}{697} \Rightarrow A = \sin^{-1}\left(\frac{890 \sin 48.2°}{697}\right) \approx 72.2° \text{ or } 180 - 72.2 = 107.8°.$$

If $A = 72.2°, C = 180° - 48.2° - 72.2° = 59.6°$. If $A = 107.8°, C = 180° - 48.2° - 107.8° = 24.0°$.

The solutions are $A_1 = 72.2°, C_1 = 59.6°$ or $A_2 = 107.8°, C_2 = 24.0°$.

33. Using the Law of Sines,

$$\frac{\sin A}{859} = \frac{\sin 74.3°}{783} \Rightarrow \sin A = \frac{859 \sin 74.3°}{783} \Rightarrow A = \sin^{-1}\left(\frac{859 \sin 74.3°}{783}\right) \Rightarrow A \approx \sin^{-1}(1.056).$$

Since 1.056 is not in the domain of the inverse sine function, no such angle exists. Thus no such triangle exists.

34. Using the Law of Sines,

$$\frac{\sin A}{10.9} = \frac{\sin 82.2°}{7.62} \Rightarrow \sin A = \frac{10.9 \sin 82.2°}{7.62} \Rightarrow A = \sin^{-1}\left(\frac{10.9 \sin 82.2°}{7.62}\right) \Rightarrow A \approx \sin^{-1}(1.417).$$

Since 1.417 is not in the domain of the inverse sine function, no such angle exists. Thus no such triangle exists.

35. Using the Law of Sines,

$$\frac{\sin B}{5.432} = \frac{\sin 142.13°}{7.297} \Rightarrow \sin B = \frac{5.432 \sin 142.13°}{7.297} \Rightarrow B = \sin^{-1}\left(\frac{5.432 \sin 142.13°}{7.297}\right) \approx 27.19°.$$

Since $180° - 27.19° = 152.81°$ and $142.13° + 152.81° > 180°$, there is only one triangle.

$C = 180° - 142.13° - 27.19° = 10.68°$. The solutions are $B = 27.19°, C = 10.68°$.

36. Using the Law of Sines,

$$\frac{\sin A}{189.6} = \frac{\sin 113.72°}{243.8} \Rightarrow \sin A = \frac{189.6 \sin 113.72°}{243.8} \Rightarrow A = \sin^{-1}\left(\frac{189.6 \sin 113.72°}{243.8}\right) \approx 45.40°.$$

Since $180° - 45.40° = 134.6°$ and $113.72° + 134.6° > 180°$, there is only one triangle.

$C = 180° - 113.72° - 45.40° = 20.88°$. The solutions are $A = 45.40°, C = 20.88°$.

37. (a) Consider the point $(3, 0)$ which is on the x-axis and 4 units below the given point. If the length of the line segment is $4 < h < 5$, two triangles can be drawn. One triangle would intersect the x-axis to the left of $(3, 0)$ and to the right of $(0, 0)$, and the other would intersect the x-axis to the right of $(3, 0)$.

(b) A line segment with length $h = 4$ would form a single right triangle. Also, if $h \geq 5$, a single triangle would be formed with the line segment intersecting the positive x-axis to the right of $(3, 0)$.

(c) If $h < 4$, no triangle could be drawn because the line segment would not intersect the positive x-axis.

38. (a) Two triangles cannot be formed using a line segment that intersects the positive x-axis.

(b) If $h > 5$, a single triangle would be formed with the line segment intersecting the positive x-axis to the right of the point $(0, 0)$.

(c) If $h \leq 5$, no triangle could be drawn because the line segment would not intersect the positive x-axis.

39. Using the Law of Sines, $\dfrac{\sin A}{31} = \dfrac{\sin 48°}{26} \Rightarrow \sin A = \dfrac{31 \sin 48°}{26} \Rightarrow A = \sin^{-1}\left(\dfrac{31 \sin 48°}{26}\right) \approx 62.4°.$

Since $180° - 62.4° = 117.6°$ and $48° + 117.6° < 180°$, there are two triangles.

40. Using the Law of Sines, $\dfrac{\sin B}{30} = \dfrac{\sin 40°}{35} \Rightarrow \sin B = \dfrac{30 \sin 40°}{35} \Rightarrow B = \sin^{-1}\left(\dfrac{30 \sin 40°}{35}\right) \approx 33.4°.$

Since $180° - 33.4° = 146.6°$ and $40° + 146.6° > 180°$, there is only one triangle.

41. Using the Law of Sines,

$\dfrac{\sin B}{61} = \dfrac{\sin 58°}{50} \Rightarrow \sin B = \dfrac{61 \sin 58°}{50} \Rightarrow B = \sin^{-1}\left(\dfrac{61 \sin 58°}{50}\right) \Rightarrow B \approx \sin^{-1}(1.03).$

Since 1.03 is not in the domain of the inverse sine function, no such angle exists. Thus no such triangle exists.

42. Using the Law of Sines, $\dfrac{\sin C}{28} = \dfrac{\sin 54°}{23} \Rightarrow \sin C = \dfrac{28 \sin 54°}{23} \Rightarrow C = \sin^{-1}\left(\dfrac{28 \sin 54°}{23}\right) \approx 80.0°.$

Since $180° - 80.0° = 100.0°$ and $54° + 100.0° < 180°$, there are two triangles.

43. Using the Law of Sines, $\dfrac{\sin B}{8.14} = \dfrac{\sin 42.5°}{15.6} \Rightarrow \sin B = \dfrac{8.14 \sin 42.5°}{15.6} \Rightarrow B = \sin^{-1}\left(\dfrac{8.14 \sin 42.5°}{15.6}\right) \approx 20.6°.$

Since $180° - 20.6° = 159.4°$ and $42.5° + 159.4° > 180°$, there is one triangle.

$C = 180° - 42.5° - 20.6° = 116.9°.$ $\dfrac{c}{\sin 116.9°} = \dfrac{15.6}{\sin 42.5°} \Rightarrow c = \dfrac{15.6 \sin 116.9°}{\sin 42.5°} \approx 20.6.$

The solutions are $B = 20.6°$, $C = 116.9°$, $c = 20.6$ ft.

44. Using the Law of Sines, $\dfrac{\sin A}{32.5} = \dfrac{\sin 52.3°}{59.8} \Rightarrow \sin A = \dfrac{32.5 \sin 52.3°}{59.8} \Rightarrow A = \sin^{-1}\left(\dfrac{32.5 \sin 52.3°}{59.8}\right) \approx 25.5°.$

Since $180° - 25.5° = 154.5°$ and $52.3° + 154.5° > 180°$, there is one triangle.

$B = 180° - 52.3° - 25.5° = 102.2°.$ $\dfrac{b}{\sin 102.2°} = \dfrac{59.8}{\sin 52.3°} \Rightarrow b = \dfrac{59.8 \sin 102.2°}{\sin 52.3°} \approx 73.9.$

The solutions are $A = 25.5°$, $B = 102.2°$, $b = 73.9$ yd.

45. Using the Law of Sines,

$\dfrac{\sin C}{145} = \dfrac{\sin 72.2°}{78.3} \Rightarrow \sin C = \dfrac{145 \sin 72.2°}{78.3} \Rightarrow C = \sin^{-1}\left(\dfrac{145 \sin 72.2°}{78.3}\right) \Rightarrow C \approx \sin^{-1}(1.76).$

Since 1.76 is not in the domain of the inverse sine function, no such angle exists. Thus no such triangle exists.

46. Using the Law of Sines,

$\dfrac{\sin B}{386} = \dfrac{\sin 68.5°}{258} \Rightarrow \sin B = \dfrac{386 \sin 68.5°}{258} \Rightarrow B = \sin^{-1}\left(\dfrac{386 \sin 68.5°}{258}\right) \Rightarrow B \approx \sin^{-1}(1.39).$

Since 1.39 is not in the domain of the inverse sine function, no such angle exists. Thus no such triangle exists.

47. Using the Law of Sines,

$\dfrac{\sin B}{11.8} = \dfrac{\sin 38° 40'}{9.72} \Rightarrow \sin B = \dfrac{11.8 \sin 38° 40'}{9.72} \Rightarrow B = \sin^{-1}\left(\dfrac{11.8 \sin 38° 40'}{9.72}\right) \approx 49° 20'.$

There are two angles: $B_1 = 49° 20'$ and $B_2 = 180 - 49° 20' = 130° 40'$ (since $38° 40' + 130° 40' < 180°$).

$C_1 = 180° - 38° 40' - 49° 20' = 92° 00'$ and $C_2 = 180° - 38° 40' - 130° 40' = 10° 40'$

$\dfrac{c_1}{\sin 92° 00'} = \dfrac{9.72}{\sin 38° 40'} \Rightarrow c_1 = \dfrac{9.72 \sin 92° 00'}{\sin 38° 40'} \approx 15.5$

$\dfrac{c_2}{\sin 10° 40'} = \dfrac{9.72}{\sin 38° 40'} \Rightarrow c_2 = \dfrac{9.72 \sin 10° 40'}{\sin 38° 40'} \approx 2.88$

The solutions are $B_1 = 49° 20'$, $C_1 = 92° 00'$, $c_1 = 15.5$ km or $B_2 = 130° 40'$, $C_2 = 10° 40'$, $c_2 = 2.88$ km.

48. Using the Law of Sines,

$$\frac{\sin A}{8.61} = \frac{\sin 29° 50'}{5.21} \Rightarrow \sin A = \frac{8.61 \sin 29° 50'}{5.21} \Rightarrow A = \sin^{-1}\left(\frac{8.61 \sin 29° 50'}{5.21}\right) \approx 55° 20'.$$

There are two angles: $A_1 = 55° 20'$ and $A_2 = 180 - 55° 20' = 124° 40'$ (since $29° 50' + 124° 40' < 180°$).

$B_1 = 180° - 29° 50' - 55° 20' = 94° 50'$ and $B_2 = 180° - 29° 50' - 124° 40' = 25° 30'$

$$\frac{b_1}{\sin 94° 50'} = \frac{5.21}{\sin 29° 50'} \Rightarrow b_1 = \frac{5.21 \sin 94° 50'}{\sin 29° 50'} \approx 10.4$$

$$\frac{b_2}{\sin 25° 30'} = \frac{5.21}{\sin 29° 50'} \Rightarrow b_2 = \frac{5.21 \sin 25° 30'}{\sin 29° 50'} \approx 4.51$$

The solutions are $A_1 = 55° 20'$, $B_1 = 94° 50'$, $b_1 = 10.4$ m or $A_2 = 124° 40'$, $B_2 = 25° 30'$, $b_2 = 4.51$ m.

49. Using the Law of Sines,

$$\frac{\sin A}{7540} = \frac{\sin 32° 50'}{5180} \Rightarrow \sin A = \frac{7540 \sin 32° 50'}{5180} \Rightarrow A = \sin^{-1}\left(\frac{7540 \sin 32° 50'}{5180}\right) \approx 52° 10'.$$

There are two angles: $A_1 = 52° 10'$ and $A_2 = 180 - 52° 10' = 127° 50'$ (since $32° 50' + 127° 50' < 180°$).

$C_1 = 180° - 32° 50' - 52° 10' = 95° 00'$ and $C_2 = 180° - 32° 50' - 127° 50' = 19° 20'$

$$\frac{c_1}{\sin 95° 00'} = \frac{5180}{\sin 32° 50'} \Rightarrow c_1 = \frac{5180 \sin 95° 00'}{\sin 32° 50'} \approx 9520$$

$$\frac{c_2}{\sin 19° 20'} = \frac{5180}{\sin 32° 50'} \Rightarrow c_2 = \frac{5180 \sin 19° 20'}{\sin 32° 50'} \approx 3160$$

The solutions are $A_1 = 52° 10'$, $C_1 = 95° 00'$, $c_1 = 9520$ cm or $A_2 = 127° 50'$, $C_2 = 19° 20'$, $c_2 = 3160$ cm.

50. Using the Law of Sines,

$$\frac{\sin B}{159} = \frac{\sin 22° 50'}{132} \Rightarrow \sin B = \frac{159 \sin 22° 50'}{132} \Rightarrow B = \sin^{-1}\left(\frac{159 \sin 22° 50'}{132}\right) \approx 27° 50'.$$

There are two angles: $B_1 = 27° 50'$ and $B_2 = 180 - 27° 50' = 152° 10'$ (since $22° 50' + 152° 10' < 180°$).

$A_1 = 180° - 22° 50' - 27° 50' = 129° 20'$ and $A_2 = 180° - 22° 50' - 152° 10' = 5° 00'$

$$\frac{a_1}{\sin 129° 20'} = \frac{132}{\sin 22° 50'} \Rightarrow a_1 = \frac{132 \sin 129° 20'}{\sin 22° 50'} \approx 263$$

$$\frac{a_2}{\sin 5° 00'} = \frac{132}{\sin 22° 50'} \Rightarrow a_2 = \frac{132 \sin 5° 00'}{\sin 22° 50'} \approx 29.6$$

The solutions are $A_1 = 129° 20'$, $B_1 = 27° 50'$, $a_1 = 263$ mm or $A_2 = 5° 00'$, $B_2 = 152° 10'$, $a_2 = 29.6$ mm.

51. The Pythagorean theorem only applies to right triangles.

52. It is a valid statement. If we are given two angles, the third angle can be found. Then the ASA congruence theorem assures us of a unique triangle.

53. If we are given only three sides, then any equation from the law of sines will contain two unknowns. At least one angle must be given.

54. No. If $a = 2b$ then $\frac{\sin A}{a} = \frac{\sin B}{b} \Rightarrow \sin A = \frac{a \sin B}{b} \Rightarrow \sin A = \frac{2b \sin B}{b} \Rightarrow \sin A = 2 \sin B$.

55. Angle A has measure $180° - 112° 10' - 15° 20' = 52° 30'$. Now, using the Law of Sines,

$$\frac{AB}{\sin 15° 20'} = \frac{354}{\sin 52° 30'} \Rightarrow AB = \frac{354 \sin 15° 20'}{\sin 52° 30'} \Rightarrow AB \approx 118 \text{ m}.$$

56. Angle S has measure $180° - 102°\,20' - 32°\,50' = 44°\,50'$. Now, using the Law of Sines,

$$\frac{RS}{\sin 32°\,50'} = \frac{582}{\sin 44°\,50'} \Rightarrow RS = \frac{582 \sin 32°\,50'}{\sin 44°\,50'} \Rightarrow RS \approx 448 \text{ yd.}$$

57. Triangle ABC has the following measurements:

$A = 90° - 47.7° = 42.3°$, $B = 302.5° - 270° = 32.5°$, $C = 180° - 42.3° - 32.5° = 105.2°$, and $c = 3.46$.

Now, using the Law of Sines, $\dfrac{b}{\sin 32.5°} = \dfrac{3.46}{\sin 105.2°} \Rightarrow b = \dfrac{3.46 \sin 32.5°}{\sin 105.2°} \Rightarrow b \approx 1.93 \text{ mi.}$

58. Let A, B, and C be the ship's initial position, new position and position of the lighthouse, respectively.

Triangle ABC has the following measurements:

$A = 38.8°$, $B = 44.2°$, $C = 180° - 38.8° - 44.2° = 97°$, and $b = 12.5$.

Now, using the Law of Sines, $\dfrac{c}{\sin 97°} = \dfrac{12.5}{\sin 44.2°} \Rightarrow c = \dfrac{12.5 \sin 97°}{\sin 44.2°} \Rightarrow c \approx 17.8 \text{ mi.}$

59. Using the Law of Sines, $\dfrac{x}{\sin 54.8°} = \dfrac{12}{\sin 70.4°} \Rightarrow x = \dfrac{12 \sin 54.8°}{\sin 70.4°} \Rightarrow x \approx 10.4 \text{ in.}$

60. Let M, L, and T be the positions of Mark, Lisa, and the tree, respectively.

Triangle MLT has the following measurements:

$M = 180° - 115.45° = 64.55°$, $L = 45.47°$, $T = 180° - 64.55° - 45.47° = 69.98°$, and $t = 428.3$.

Now, using the Law of Sines, $\dfrac{m}{\sin 64.55°} = \dfrac{428.3}{\sin 69.98°} \Rightarrow m = \dfrac{428.3 \sin 64.55°}{\sin 69.98°} \Rightarrow m \approx 411.6 \text{ m.}$

If a line segment joining T to the opposite side of the river intersects the river bank at D, we can use right

triangle LTD to find the distance d across the river: $\sin 45.47° = \dfrac{d}{411.6} \Rightarrow d = 411.6 \sin 45.47° \approx 293.4 \text{ m.}$

61. Note that the distance between the centers of the small gear and middle-size gear is $2.7 + 1.6 = 4.3$ and the

distance between the centers of the small gear and large gear is $3.6 + 1.6 = 5.2$. Use the Law of Sines to find

the angle α on the large gear. $\dfrac{\sin \alpha}{4.3} = \dfrac{\sin 38°}{5.2} \Rightarrow \sin \alpha = \dfrac{4.3 \sin 38°}{5.2} \Rightarrow \alpha \approx \sin^{-1}\left(\dfrac{4.3 \sin 38°}{5.2}\right) \approx 30.6°.$

It follows that $\theta = 180° - 38° - 30.6° = 111.4° \approx 111°.$

62. Note that $AB = 3.0 + 4.5 = 7.5$ and $BC = 2.0 + 3.0 = 5.0$. Use the Law of Sines to find angle C.

$\dfrac{\sin C}{7.5} = \dfrac{\sin 18°}{5.0} \Rightarrow \sin C = \dfrac{7.5 \sin 18°}{5.0} \Rightarrow C \approx \sin^{-1}\left(\dfrac{7.5 \sin 18°}{5.0}\right) \approx 27.6°.$

It follows that $B = 180° - 18° - 27.6° = 134.4°$. Now use the Law of Sines to find AC.

$\dfrac{AC}{\sin 134.4°} = \dfrac{5.0}{\sin 18°} \Rightarrow AC = \dfrac{5.0 \sin 134.4°}{\sin 18°} \Rightarrow AC \approx 11.6 \text{ units.}$

63. Let A, B, and C be the ship's initial position, new position and position of the lighthouse, respectively.

Triangle ABC has the following measurements:

$A = 180° - 37° = 143°$, $B = 25°$, $C = 180° - 143° - 25° = 12°$, and $c = 2.5$.

Now, using the Law of Sines, $\dfrac{a}{\sin 143°} = \dfrac{2.5}{\sin 12°} \Rightarrow a = \dfrac{2.5 \sin 143°}{\sin 12°} \Rightarrow a \approx 7.2 \text{ mi}$ and

$\dfrac{b}{\sin 25°} = \dfrac{2.5}{\sin 12°} \Rightarrow a = \dfrac{2.5 \sin 25°}{\sin 12°} \Rightarrow a \approx 5.1 \text{ mi.}$

64. Let A, B, and C be the positions of the town on the left, the balloon, and the town on the right, respectively.

Triangle ABC has the following measurements:

$A = 35°$, $B = 180° - 35° - 31° = 114°$, $C = 31°$, and $b = 1.5$.

Now, using the Law of Sines, $\dfrac{c}{\sin 31°} = \dfrac{1.5}{\sin 114°} \Rightarrow c = \dfrac{1.5 \sin 31°}{\sin 114°} \Rightarrow c \approx .85$ mi.

If a perpendicular line segment joining B to the the ground intersects the ground at D, we can use right triangle ABD to find the distance d to the balloon: $\sin 35° = \dfrac{d}{.85} \Rightarrow d = .85 \sin 35° \approx .49$ mi.

65. Triangle ABC has the following measurements:

$A = 90° - 22.4° + 45° = 112.6°$, $B = 45° - 10.6° = 34.4°$, $C = 180° - 112.6° - 34.4° = 33°$, $c = 25.5$

Now, using the Law of Sines, $\dfrac{b}{\sin 34.4°} = \dfrac{25.5}{\sin 33°} \Rightarrow b = \dfrac{25.5 \sin 34.4°}{\sin 33°} \Rightarrow b \approx 26.5$ km.

66. Let A, B, and C be the positions of the observer on the left, the helicopter, and the observer on the right, respectively.

Triangle ABC has the following measurements:

$A = 20.5°$, $C = 180° - 27.8° = 152.2°$, $B = 180° - 20.5° - 152.2° = 7.3°$, and $b = 3$.

Now, using the Law of Sines, $\dfrac{a}{\sin 20.5°} = \dfrac{3}{\sin 7.3°} \Rightarrow a = \dfrac{3 \sin 20.5°}{\sin 7.3°} \Rightarrow a \approx 8.27$ mi.

If the perpendicular line segment joining B to the the ground intersects the ground at D, we can use right triangle CBD to find the distance d to the helicopter: $\sin 27.8° = \dfrac{d}{8.27} \Rightarrow d = 8.27 \sin 27.8° \approx 3.86$ mi.

67. Let R be the position of the rocket. Triangle T_1RT_2 has the following measurements:

$T_1 = 28.1°$, $T_2 = 180° - 79.5° = 100.5°$, $R = 180° - 28.1° - 100.5° = 51.4°$, and $r = 1.73$.

Now, using the Law of Sines, $\dfrac{t_2}{\sin 100.5°} = \dfrac{1.73}{\sin 51.4°} \Rightarrow t_2 = \dfrac{1.73 \sin 100.5°}{\sin 51.4°} \Rightarrow t_2 \approx 2.18$ km.

68. The height to the top of the clock tower is $\tan 45.6° = \dfrac{h}{48.0} \Rightarrow h = 48.0 \tan 45.6° \approx 49.0$ m.

The height to the top of the building is $\tan 37.4° = \dfrac{h}{48.0} \Rightarrow h = 48.0 \tan 37.4° \approx 36.7$ m.

The height of the clock tower is the difference $49.0 - 36.7 = 12.3$ m.

69. Use the Law of Sines to find angle W.

$\dfrac{\sin W}{11.2} = \dfrac{\sin 25.5°}{28.6} \Rightarrow \sin W = \dfrac{11.2 \sin 25.5°}{28.6} \Rightarrow W = \sin^{-1}\left(\dfrac{11.2 \sin 25.5°}{28.6}\right) \approx 9.7°$.

Thus, $P = 180° - 25.5° - 9.7° = 144.8°$. Now use the Law of Sines to find the required distance.

$\dfrac{p}{\sin 144.8°} = \dfrac{11.2}{\sin 9.7°} \Rightarrow p = \dfrac{11.2 \sin 144.8°}{\sin 9.7°} \Rightarrow p \approx 38.3$ cm.

70. Let A, B, and C be the positions of the tracking station, the satellite, and the center of Earth, respectively.

Triangle ABC has the following measurements:

$A = 90° + 30° = 120°$, $a = 6400 + 1600 = 8000$, and $b = 6400$. Now, using the Law of Sines,

$\dfrac{\sin B}{6400} = \dfrac{\sin 120°}{8000} \Rightarrow \sin B = \dfrac{6400 \sin 120°}{8000} \Rightarrow B = \sin^{-1}\left(\dfrac{6400 \sin 120°}{8000}\right) \approx 43.85°$.

Thus, $C = 180° - 120° - 43.85° = 16.15°$. If the satellite travels one orbit in 2 hours, it is traveling at a rate of $360 \div 120 = 3°$ per minute. The satellite gets to the tracking beam $16.15 \div 3 \approx 5.4$ minutes before noon.

The time would be approximately 11:55 A.M.

71. $\dfrac{\sin C}{2\sqrt{5}} = \dfrac{\sin 30°}{\sqrt{5}} \Rightarrow \sin C = \dfrac{2\sqrt{5}\,\sin 30°}{\sqrt{5}} = 1$. Thus $C = 90°$. This is a right triangle.

72. If $\sin A > 1$, then no triangle exists.

73. The longest side must be opposite the largest angle. Thus, B must be larger than A, which is impossible because A and B cannot both be obtuse.

74. $\dfrac{\sin B}{31.3} = \dfrac{\sin 104°}{26.8} \Rightarrow \sin B = \dfrac{31.3\,\sin 104°}{26.8} \approx 1.13$. Since 1.13 is not in the domain of the inverse sine function, no such angle exists. The calculator gives an error because $\sin B > 1$.

75. $\dfrac{\sin A}{78.3} = \dfrac{\sin 38° \; 50'}{21.9} \Rightarrow \sin A = \dfrac{78.3\,\sin 38° \; 50'}{21.9} \approx 2.24$. Since 2.24 is not in the domain of the inverse sine function, no such angle exists. The property could not exist.

76. $\dfrac{\sin A}{26.5} = \dfrac{\sin 28° \; 10'}{21.2} \Rightarrow \sin A = \dfrac{26.5\,\sin 28° \; 10'}{21.2} \Rightarrow A = \sin^{-1}\left(\dfrac{26.5\,\sin 28° \; 10'}{21.2}\right) \approx 36° \; 10'$. A second triangle could exist. The angle opposite the side with length 26.5 could be either $36° \; 10'$ or $180° - 36° \; 10' = 143° \; 50'$.

77. Let B, M, and D be the positions of Bochum, the moon, and Donaueschingen, respectively.

    Triangle BMD has the following measurements:

    $B = 52.6997°$, $D = 180° - 52.7430° = 127.2570°$, $M = 180° - 52.6997° - 127.2570° = .0433°$, $m = 398$.

    Now, using the Law of Sines, $\dfrac{d}{\sin 127.2570°} = \dfrac{398}{\sin .0433°} \Rightarrow a = \dfrac{398\,\sin 127.2570°}{\sin .0433°} \Rightarrow a \approx 419{,}000$ km.

    This compares favorably to the actual value.

78. Since the camera is tilted $35°$, the slanted dashed line represents the center of the angular coverage. That is, it divides the $60°$ angle into two $30°$ angles. Thus, the acute angles of the *smallest* right triangle are $5°$ and $85°$ and the hypotenuse of the *smallest* right triangle is $\cos 5° = \dfrac{5000}{h} \Rightarrow h = \dfrac{5000}{\cos 5°} \approx 5019$ ft. The angle adjacent to the $85°$ angle has measure $180° - 85° = 95°$. This is an angle of the large non-dashed triangle.

    The third angle of this triangle is $180° - 60° - 95° = 25°$. To find d, use the Law of Sines on the large non-dashed triangle: $\dfrac{d}{\sin 60°} = \dfrac{5019}{\sin 25°} \Rightarrow d = \dfrac{5019\,\sin 60°}{\sin 25°} \Rightarrow d \approx 10{,}285$ ft or about 1.95 mi.

79. (a) Note: this solution utilizes the area formula introduced in section 10.2.

    First note that $\dfrac{R}{\sin C} = \dfrac{r}{\sin A} \Rightarrow r = \dfrac{R\sin A}{\sin C} \Rightarrow r = \dfrac{R\sin A}{\sin (A + B)}$. Substituting in the area formula gives

    $A = 10\left(\dfrac{1}{2}Rr\sin B\right) = 5R\left(\dfrac{R\sin A}{\sin (A + B)}\right)\sin B = \left[5\dfrac{\sin A \sin B}{\sin (A + B)}R^2\right]$.

    (b) $A = \left[5\dfrac{\sin 18° \sin 36°}{\sin (18° + 36°)}R^2\right] \approx 1.12257R^2$.

    (c) (i) The stripe has height $10 \div 13 \approx .76923$ in. and area of $11.4(.76923) \approx 8.77$ in.^{2}.

    (ii) The 50 stars have area $A = 50(1.12257)(.308)^2 \approx 5.32$ in.^{2}.

    (iii) Red occupies the greatest area on the flag.

80. On the given interval, the graph is increasing.

81. If $B < A$, then $\sin B < \sin A$ because $y = \sin x$ is increasing on $\left[0, \dfrac{\pi}{2}\right]$.

82. $\dfrac{b}{\sin B} = \dfrac{a}{\sin A} \Rightarrow b = \dfrac{a\,\sin B}{\sin A}$

83. $b = \dfrac{a\,\sin B}{\sin A} = a \cdot \dfrac{\sin B}{\sin A}$. Since $\dfrac{\sin B}{\sin A} < 1$, $b = a \cdot \dfrac{\sin B}{\sin A} < a \cdot 1 = a$, so $b < a$.

84. If $B < A$, then $b < a$, but $b > a$ in triangle ABC.

10.2: The Law of Cosines and Area Formulas

1. (a) Two sides and the included angle are given. This is form SAS.

    (b) SAS should be solved using the Law of Cosines.

2. (a) Two angles and a non-included side are given. This is form SAA.

    (b) SAA should be solved using the Law of Sines.

3. (a) Two sides and a non-included angle are given. This is form SSA.

    (b) SSA should be solved using the Law of Sines.

4. (a) Three sides are given. This is form SSS.

    (b) SSS should be solved using the Law of Cosines.

5. (a) Two angles and the included side are given. This is form ASA.

    (b) ASA should be solved using the Law of Sines.

6. (a) Two sides and a non-included angle are given. This is form SSA.

    (b) SSA should be solved using the Law of Sines.

7. (a) Two angles and the included side are given. This is form ASA.

    (b) ASA should be solved using the Law of Sines.

8. (a) Two sides and the included angle are given. This is form SAS.

    (b) SAS should be solved using the Law of Cosines.

9. $c^2 = 3^2 + 8^2 - 2(3)(8)\cos 60^\circ \Rightarrow c^2 = 73 - 48(.5) \Rightarrow c^2 = 73 - 24 \Rightarrow c^2 = 49 \Rightarrow c = 7$

10. $c^2 = 1^2 + (4\sqrt{2})^2 - 2(1)(4\sqrt{2})\cos 45^\circ \Rightarrow c^2 = 33 - 8\sqrt{2}\left(\dfrac{1}{\sqrt{2}}\right) \Rightarrow c^2 = 33 - 8 \Rightarrow c^2 = 25 \Rightarrow c = 5$

11. $1^2 = 1^2 + (\sqrt{3})^2 - 2(1)(\sqrt{3})\cos\theta \Rightarrow 1 = 4 - 2\sqrt{3}\cos\theta \Rightarrow \cos\theta = \dfrac{-3}{-2\sqrt{3}} \Rightarrow \cos\theta = \dfrac{\sqrt{3}}{2} \Rightarrow \theta = 30^\circ$

12. $7^2 = 3^2 + 5^2 - 2(3)(5)\cos\theta \Rightarrow 49 = 34 - 30\cos\theta \Rightarrow \cos\theta = \dfrac{15}{-30} \Rightarrow \cos\theta = -\dfrac{1}{2} \Rightarrow \theta = 120^\circ$

13. Use the Law of Cosines to find a: $a^2 = 4^2 + 6^2 - 2(4)(6)\cos 61^\circ \Rightarrow a = \sqrt{52 - 48\cos 61^\circ} \approx 5.36$

    Use the Law of Sines to find B: $\dfrac{\sin B}{4} = \dfrac{\sin 61^\circ}{5.36} \Rightarrow \sin B = \dfrac{4\sin 61^\circ}{5.36} \Rightarrow B = \sin^{-1}\left(\dfrac{4\sin 61^\circ}{5.36}\right) \approx 40.7^\circ$.

    Then $C = 180^\circ - 61^\circ - 40.7^\circ = 78.3^\circ$. The solutions are $a \approx 5.4$, $B \approx 40.7^\circ$, $C \approx 78.3^\circ$.

14. Use the Law of Cosines to find a: $a^2 = 3^2 + 5^2 - 2(3)(5)\cos 121^\circ \Rightarrow a = \sqrt{34 - 30\cos 121^\circ} \approx 7.03$

    Use the Law of Sines to find B: $\dfrac{\sin B}{5} = \dfrac{\sin 121^\circ}{7.03} \Rightarrow \sin B = \dfrac{5\sin 121^\circ}{7.03} \Rightarrow B = \sin^{-1}\left(\dfrac{5\sin 121^\circ}{7.03}\right) \approx 37.6^\circ$.

    Then $C = 180^\circ - 121^\circ - 37.6^\circ = 21.4^\circ$. The solutions are $a \approx 7.0$, $B \approx 37.6^\circ$, $C \approx 21.4^\circ$.

15. Use the Law of Cosines to find A: $4^2 = 8^2 + 10^2 - 2(8)(10)\cos A \Rightarrow A = \cos^{-1}\left(\dfrac{16 - 164}{-160}\right) \approx 22.33°$.

    Use the Law of Sines to find B:

    $\dfrac{\sin B}{10} = \dfrac{\sin 22.33°}{4} \Rightarrow \sin B = \dfrac{10 \sin 22.33°}{4} \Rightarrow B = \sin^{-1}\left(\dfrac{10 \sin 22.33°}{4}\right) \approx 71.8°$.

    Since angle B is obtuse, $B = 180° - 71.8° = 108.2°$. The value of C is $180° - 22.3° - 108.2° = 49.5°$.

    The solutions are $A \approx 22.3°$, $B \approx 108.2°$, $C \approx 49.5°$.

16. Use the Law of Cosines to find A: $12^2 = 10^2 + 10^2 - 2(10)(10)\cos A \Rightarrow A = \cos^{-1}\left(\dfrac{144 - 200}{-200}\right) \approx 73.7°$.

    Use the Law of Sines to find B:

    $\dfrac{\sin B}{10} = \dfrac{\sin 73.7°}{12} \Rightarrow \sin B = \dfrac{10 \sin 73.7°}{12} \Rightarrow B = \sin^{-1}\left(\dfrac{10 \sin 73.7°}{12}\right) \approx 53.1°$.

    Since the triangle is isosceles, $C = B \approx 53.1$. The solutions are $A \approx 73.7°$, $B \approx 53.1°$, $C \approx 53.1°$.

17. Use the Law of Cosines to find A: $5^2 = 7^2 + 9^2 - 2(7)(9)\cos A \Rightarrow A = \cos^{-1}\left(\dfrac{25 - 130}{-126}\right) \approx 33.55°$.

    Use the Law of Sines to find B:

    $\dfrac{\sin B}{7} = \dfrac{\sin 33.55°}{5} \Rightarrow \sin B = \dfrac{7 \sin 33.55°}{5} \Rightarrow B = \sin^{-1}\left(\dfrac{7 \sin 33.55°}{5}\right) \approx 50.7°$.

    The value of C is $180° - 33.6° - 50.7° = 95.7°$. The solutions are $A \approx 33.6°$, $B \approx 50.7°$, $C \approx 95.7°$.

18. Use the Law of Cosines to find b:

    $b^2 = 100^2 + 90^2 - 2(100)(90)\cos 55° \Rightarrow b = \sqrt{18,100 - 18,000\cos 55°} \approx 88.2$

    Use the Law of Sines to find A: $\dfrac{\sin A}{90} = \dfrac{\sin 55°}{88.2} \Rightarrow \sin A = \dfrac{90 \sin 55°}{88.2} \Rightarrow A = \sin^{-1}\left(\dfrac{90 \sin 55°}{88.2}\right) \approx 56.7°$.

    Then $C = 180° - 55° - 56.7° = 68.3°$. The solutions are $b \approx 88.2$, $A \approx 56.7°$, $C \approx 68.3°$.

19. Use the Law of Cosines to find c:

    $c^2 = 5.71^2 + 4.21^2 - 2(5.71)(4.21)\cos 28.3° \Rightarrow c = \sqrt{50.3282 - 48.0782\cos 28.3°} \approx 2.83$

    Use the Law of Sines to find A:

    $\dfrac{\sin A}{4.21} = \dfrac{\sin 28.3°}{2.83} \Rightarrow \sin A = \dfrac{4.21 \sin 28.3°}{2.83} \Rightarrow A = \sin^{-1}\left(\dfrac{4.21 \sin 28.3°}{2.83}\right) \approx 44.9°$.

    Then $B = 180° - 28.3° - 44.9° = 106.8°$. The solutions are $c \approx 2.83$ in., $A \approx 44.9°$, $B \approx 106.8°$.

20. Use the Law of Cosines to find a:

    $a^2 = 2.78^2 + 3.92^2 - 2(2.78)(3.92)\cos 41.4° \Rightarrow a = \sqrt{23.0948 - 21.7952\cos 41.4°} \approx 2.597$

    Use the Law of Sines to find B:

    $\dfrac{\sin B}{2.78} = \dfrac{\sin 41.4°}{2.597} \Rightarrow \sin B = \dfrac{2.78 \sin 41.4°}{2.597} \Rightarrow B = \sin^{-1}\left(\dfrac{2.78 \sin 41.4°}{2.597}\right) \approx 45.1°$.

    Then $C = 180° - 41.4° - 45.1° = 93.5°$. The solutions are $a \approx 2.60$ yd, $B \approx 45.1°$, $C \approx 93.5°$.

21. Use the Law of Cosines to find c:

    $c^2 = 8.94^2 + 7.23^2 - 2(8.94)(7.23)\cos 45.6° \Rightarrow c = \sqrt{132.1965 - 129.2724\cos 45.6°} \approx 6.46$

    Use the Law of Sines to find A:

    $\dfrac{\sin A}{7.23} = \dfrac{\sin 45.6°}{6.46} \Rightarrow \sin A = \dfrac{7.23 \sin 45.6°}{6.46} \Rightarrow A = \sin^{-1}\left(\dfrac{7.23 \sin 45.6°}{6.46}\right) \approx 53.1°$.

    Then $B = 180° - 45.6° - 53.1° = 81.3°$. The solutions are $c \approx 6.46$ m, $A \approx 53.1°$, $B \approx 81.3°$.

22. Use the Law of Cosines to find a:

$$a^2 = 37.9^2 + 40.8^2 - 2(37.9)(40.8)\cos 67.3° \Rightarrow a = \sqrt{3101.05 - 3092.64\cos 67.3°} \approx 43.68$$

Use the Law of Sines to find B:

$$\frac{\sin B}{37.9} = \frac{\sin 67.3°}{43.68} \Rightarrow \sin B = \frac{37.9 \sin 67.3°}{43.68} \Rightarrow B = \sin^{-1}\left(\frac{37.9 \sin 67.3°}{43.68}\right) \approx 53.2°.$$

Then $C = 180° - 67.3° - 53.2° = 59.5°$. The solutions are $a \approx 43.7$ km, $B \approx 53.2°$, $C \approx 59.5°$.

23. Use the Law of Cosines to find A:

$$9.3^2 = 5.7^2 + 8.2^2 - 2(5.7)(8.2)\cos A \Rightarrow A = \cos^{-1}\left(\frac{86.49 - 99.73}{-93.48}\right) \approx 81.86°.$$

Use the Law of Sines to find B:

$$\frac{\sin B}{5.7} = \frac{\sin 81.86°}{9.3} \Rightarrow \sin B = \frac{5.7 \sin 81.86°}{9.3} \Rightarrow B = \sin^{-1}\left(\frac{5.7 \sin 81.86°}{9.3}\right) \approx 37.35°.$$

The value of C is $180° - 82° - 37° = 61°$. The solutions are $A \approx 82°$, $B \approx 37°$, $C \approx 61°$.

24. Use the Law of Cosines to find A:

$$28^2 = 47^2 + 58^2 - 2(47)(58)\cos A \Rightarrow A = \cos^{-1}\left(\frac{784 - 5573}{-5452}\right) \approx 28.55°.$$

Use the Law of Sines to find B:

$$\frac{\sin B}{47} = \frac{\sin 28.55°}{28} \Rightarrow \sin B = \frac{47 \sin 28.55°}{28} \Rightarrow B = \sin^{-1}\left(\frac{47 \sin 28.55°}{28}\right) \approx 53.34°.$$

The value of C is $180° - 29° - 53° = 98°$. The solutions are $A \approx 29°$, $B \approx 53°$, $C \approx 98°$.

25. Use the Law of Cosines to find A:

$$42.9^2 = 37.6^2 + 62.7^2 - 2(37.6)(62.7)\cos A \Rightarrow A = \cos^{-1}\left(\frac{1840.41 - 5345.05}{-4715.04}\right) \approx 41.99°.$$

Use the Law of Sines to find B:

$$\frac{\sin B}{37.6} = \frac{\sin 41.99°}{42.9} \Rightarrow \sin B = \frac{37.6 \sin 41.99°}{42.9} \Rightarrow B = \sin^{-1}\left(\frac{37.6 \sin 41.99°}{42.9}\right) \approx 35.90°.$$

$C \approx 180° - 42° 00' - 35° 50' = 102° 10'$. The solutions are $A \approx 42° 00'$, $B \approx 35° 50'$, $C \approx 102° 10'$.

26. Use the Law of Cosines to find A:

$$189^2 = 214^2 + 325^2 - 2(214)(325)\cos A \Rightarrow A = \cos^{-1}\left(\frac{35,721 - 151,421}{-139,100}\right) \approx 33.72°.$$

Use the Law of Sines to find B:

$$\frac{\sin B}{214} = \frac{\sin 33.72°}{189} \Rightarrow \sin B = \frac{214 \sin 33.72°}{189} \Rightarrow B = \sin^{-1}\left(\frac{214 \sin 33.72°}{189}\right) \approx 38.94°.$$

$C \approx 180° - 33° 40' - 39° 00' = 107° 20'$. The solutions are $A \approx 33° 40'$, $B \approx 39° 00'$, $C \approx 107° 20'$.

27. Use the Law of Cosines to find A:

$$965^2 = 1240^2 + 876^2 - 2(1240)(876)\cos A \Rightarrow A = \cos^{-1}\left(\frac{931,225 - 2,304,976}{-2,172,480}\right) \approx 50.78°.$$

Use the Law of Sines to find B:

$$\frac{\sin B}{876} = \frac{\sin 50.78°}{965} \Rightarrow \sin B = \frac{876 \sin 50.78°}{965} \Rightarrow B = \sin^{-1}\left(\frac{876 \sin 50.78°}{965}\right) \approx 44.69°.$$

$C \approx 180° - 50° 50' - 44° 40' = 84° 30'$. The solutions are $A \approx 50° 50'$, $B \approx 44° 40'$, $C \approx 84° 30'$.

28. Use the Law of Cosines to find A:

$$324^2 = 298^2 + 421^2 - 2(298)(421)\cos A \Rightarrow A = \cos^{-1}\left(\frac{104{,}976 - 266{,}045}{-250{,}916}\right) \approx 50.06°.$$

Use the Law of Sines to find B:

$$\frac{\sin B}{421} = \frac{\sin 50.06°}{324} \Rightarrow \sin B = \frac{421\sin 50.06°}{324} \Rightarrow B = \sin^{-1}\left(\frac{421\sin 50.06°}{324}\right) \approx 85.09°.$$

$C \approx 180° - 50°\,00' - 85°\,10' = 44°\,50'$. The solutions are $A \approx 50°\,00'$, $B \approx 85°\,10'$, $C \approx 44°\,50'$.

29. Use the Law of Cosines to find a:

$$a^2 = 143^2 + 89.6^2 - 2(143)(89.6)\cos 80°\,40' \Rightarrow a = \sqrt{28{,}477.16 - 25{,}625.6\cos 80°\,40'} \approx 155.95$$

Use the Law of Sines to find B:

$$\frac{\sin B}{143} = \frac{\sin 80°\,40'}{155.95} \Rightarrow \sin B = \frac{143\sin 80°\,40'}{155.95} \Rightarrow B = \sin^{-1}\left(\frac{143\sin 80°\,40'}{155.95}\right) \approx 64°\,50'.$$

Then $C = 180° - 80°\,40' - 64°\,50' = 34°\,30'$. The solutions are $a \approx 156$ cm, $B \approx 64°\,50'$, $C \approx 34°\,30'$.

30. Use the Law of Cosines to find c:

$$c^2 = 327^2 + 251^2 - 2(327)(251)\cos 72°\,40' \Rightarrow c = \sqrt{169{,}930 - 164{,}154\cos 72°\,40'} \approx 347.89$$

Use the Law of Sines to find A:

$$\frac{\sin A}{327} = \frac{\sin 72°\,40'}{347.89} \Rightarrow \sin A = \frac{327\sin 72°\,40'}{347.89} \Rightarrow A = \sin^{-1}\left(\frac{327\sin 72°\,40'}{347.89}\right) \approx 63°\,50'.$$

Then $B = 180° - 72°\,40' - 63°\,50' = 43°\,30'$. The solutions are $c \approx 348$ ft, $A \approx 63°\,50'$, $B \approx 43°\,30'$.

31. Use the Law of Cosines to find b:

$$b^2 = 8.919^2 + 6.427^2 - 2(8.919)(6.427)\cos 74.80° \Rightarrow b = \sqrt{120.85489 - 114.644826\cos 74.80°} \approx 9.529$$

Use the Law of Sines to find A:

$$\frac{\sin A}{8.919} = \frac{\sin 74.80°}{9.529} \Rightarrow \sin A = \frac{8.919\sin 74.80°}{9.529} \Rightarrow A = \sin^{-1}\left(\frac{8.919\sin 74.80°}{9.529}\right) \approx 64.59°.$$

Then $C = 180° - 74.80° - 64.59° = 40.61°$. The solutions are $b \approx 9.529$ in., $A \approx 64.59°$, $C \approx 40.61°$.

32. Use the Law of Cosines to find c:

$$c^2 = 3.725^2 + 4.698^2 - 2(3.725)(4.698)\cos 59.70° \Rightarrow c = \sqrt{35.946829 - 35.0001\cos 59.70°} \approx 4.276$$

Use the Law of Sines to find A:

$$\frac{\sin A}{3.725} = \frac{\sin 59.70°}{4.276} \Rightarrow \sin A = \frac{3.725\sin 59.70°}{4.276} \Rightarrow A = \sin^{-1}\left(\frac{3.725\sin 59.70°}{4.276}\right) \approx 48.77°.$$

Then $B = 180° - 59.70° - 48.77° = 71.53°$. The solutions are $c \approx 4.276$ mi, $A \approx 48.77°$, $B \approx 71.53°$.

33. Use the Law of Cosines to find a:

$$a^2 = 6.28^2 + 12.2^2 - 2(6.28)(12.2)\cos 112.8° \Rightarrow a = \sqrt{188.2784 - 153.232\cos 112.8°} \approx 15.7$$

Use the Law of Sines to find B:

$$\frac{\sin B}{6.28} = \frac{\sin 112.8°}{15.7} \Rightarrow \sin B = \frac{6.28\sin 112.8°}{15.7} \Rightarrow B = \sin^{-1}\left(\frac{6.28\sin 112.8°}{15.7}\right) \approx 21.6°.$$

Then $C = 180° - 112.8° - 21.6° = 45.6°$. The solutions are $a \approx 15.7$ m, $B \approx 21.6°$, $C \approx 45.6°$.

34. Use the Law of Cosines to find b:

$b^2 = 15.1^2 + 19.2^2 - 2(15.1)(19.2)\cos 168.2° \Rightarrow b = \sqrt{596.65 - 579.84\cos 168.2°} \approx 34.1$

Use the Law of Sines to find A:

$\dfrac{\sin A}{15.1} = \dfrac{\sin 168.2°}{34.1} \Rightarrow \sin A = \dfrac{15.1\sin 168.2°}{34.1} \Rightarrow A = \sin^{-1}\left(\dfrac{15.1\sin 168.2°}{34.1}\right) \approx 5.2°.$

Then $C = 180° - 168.2° - 5.2° = 6.6°$. The solutions are $b \approx 34.1$ cm, $A \approx 5.2°$, $C \approx 6.6°$.

35. Use the Law of Cosines to find A:

$3.0^2 = 5.0^2 + 6.0^2 - 2(5.0)(6.0)\cos A \Rightarrow A = \cos^{-1}\left(\dfrac{9 - 61}{-60}\right) \approx 30°.$

Use the Law of Sines to find B:

$\dfrac{\sin B}{5.0} = \dfrac{\sin 30°}{3.0} \Rightarrow \sin B = \dfrac{5.0\sin 30°}{3.0} \Rightarrow B = \sin^{-1}\left(\dfrac{5.0\sin 30°}{3.0}\right) \approx 56°.$

The value of C is $180° - 30° - 56° = 94°$. The solutions are $A \approx 30°$, $B \approx 56°$, $C \approx 94°$.

36. Use the Law of Cosines to find A:

$4.0^2 = 5.0^2 + 8.0^2 - 2(5.0)(8.0)\cos A \Rightarrow A = \cos^{-1}\left(\dfrac{16 - 89}{-80}\right) \approx 24°.$

Use the Law of Sines to find B:

$\dfrac{\sin B}{5.0} = \dfrac{\sin 24°}{4.0} \Rightarrow \sin B = \dfrac{5.0\sin 24°}{4.0} \Rightarrow B = \sin^{-1}\left(\dfrac{5.0\sin 24°}{4.0}\right) \approx 31°.$

The value of C is $180° - 24° - 31° = 125°$. The solutions are $A \approx 24°$, $B \approx 31°$, $C \approx 125°$.

37. The absolute value of $\cos\theta$ will be greater than 1. The calculator will give an error message (or a complex number) when using the inverse cosine function.

38. Answers will vary. The shortest distance between two vertices of a triangle is the side joining the vertices. The distance from one vertex to another, when passing through the third vertex, must be greater than this side.

39. Use the Law of Cosines to find c:

$c^2 = 350^2 + 286^2 - 2(350)(286)\cos 46.3° \Rightarrow c = \sqrt{204{,}296 - 200{,}200\cos 46.3°} \approx 257$ m

40. Use the Law of Cosines to find each diagonal d:

$d^2 = 4.0^2 + 6.0^2 - 2(4.0)(6.0)\cos 122° \Rightarrow d = \sqrt{52 - 48\cos 122°} \approx 8.8$ cm

$d^2 = 4.0^2 + 6.0^2 - 2(4.0)(6.0)\cos 58° \Rightarrow d = \sqrt{52 - 48\cos 58°} \approx 5.2$ cm

41. Let the vertical line passing through airport C intersect the line from A to B at the point D. Now, using the right triangle CDB we can find the measure of angle B. The angle of this right triangle at position C is given by $180° - 128° 40' = 51° 20'$. Thus $B = 180° - 90° - 51° 20' = 38° 40'$. Use the Law of Cosines on ABC to find b: $b^2 = 450^2 + 359^2 - 2(450)(359)\cos 38° 40' \Rightarrow d = \sqrt{331{,}381 - 323{,}100\cos 38° 40'} \approx 281$ km.

42. Use the Law of Cosines to find the distance d:

$d^2 = 402^2 + 402^2 - 2(402)(402)\cos 135° 40' \Rightarrow d = \sqrt{323{,}208 - 323{,}208\cos 135° 40'} \approx 745$ mi

43. Let A, B, and C be the ship's initial position, new position and position of the rock, respectively. Triangle ABC has the following measurements: $A = 90° - 45° 20' = 44° 40'$,

$B = 308° 40' - 270° = 38° 40'$, $C = 180° - 44° 40' - 38° 40' = 96° 40'^{\circ}$, and $c = 15.2$

Now, using the Law of Sines, $\dfrac{a}{\sin 44° 40'} = \dfrac{15.2}{\sin 96° 40'} \Rightarrow a = \dfrac{15.2\sin 44° 40'}{\sin 96° 40'} \Rightarrow a \approx 10.8$ mi.

44. Let A, B, and C be the positions of the airplane, battleship and submarine, respectively.

    Triangle ABC has the following measurements: $A = 24°\ 10' - 17°\ 30' = 6°\ 40'$,

    $B = 17°\ 30'$, $C = 180° - 6°\ 40' - 17°\ 30' = 155°\ 50'°$, and $c = 5120$

    Now, using the Law of Sines, $\dfrac{a}{\sin 6°\ 40'} = \dfrac{5120}{\sin 155°\ 50'} \Rightarrow a = \dfrac{5120 \sin 6°\ 40'}{\sin 155°\ 50'} \Rightarrow a \approx 1450$ ft.

45. After 3 hours the ships have traveled $36.2(3) = 108.6$ miles and $45.6(3) = 136.8$ miles.

    Use the Law of Cosines to find the distance d:

    $d^2 = 108.6^2 + 136.8^2 - 2(108.6)(136.8)\cos 54°\ 10' \Rightarrow d = \sqrt{30{,}508.2 - 29{,}712.96 \cos 54°\ 10'} \approx 115$ km

46. Use the Law of Cosines to find θ:

    $13^2 = 16^2 + 20^2 - 2(16)(20)\cos\theta \Rightarrow \theta = \cos^{-1}\left(\dfrac{169 - 656}{-640}\right) \approx 40.5°$.

47. Use the Law of Cosines to find the distance d:

    $d^2 = 10^2 + 10^2 - 2(10)(10)\cos 128° \Rightarrow d = \sqrt{200 - 200 \cos 128°} \approx 18$ ft

48. Use the Law of Cosines to find each angle θ:

    $60^2 = 45^2 + 90^2 - 2(45)(90)\cos\theta \Rightarrow \theta = \cos^{-1}\left(\dfrac{3600 - 10{,}125}{-8100}\right) \approx 36.3°$.

    $45^2 = 60^2 + 90^2 - 2(60)(90)\cos\theta \Rightarrow \theta = \cos^{-1}\left(\dfrac{2025 - 11{,}700}{-10{,}800}\right) \approx 26.4°$.

49. Use the Law of Cosines to find the distance c:

    $c^2 = 3800^2 + 2900^2 - 2(3800)(2900)\cos 110° \Rightarrow c = \sqrt{22{,}850{,}000 - 22{,}040{,}000 \cos 110°} \approx 5500$ m

50. Use the Law of Cosines to find the distance c from the pitcher's rubber to first base (third base is the same):

    $c^2 = 60.5^2 + 90^2 - 2(60.5)(90)\cos 45° \Rightarrow c = \sqrt{11{,}760.25 - 10{,}890 \cos 45°} \approx 63.7$ ft

    Find the diagonal of the square: $d^2 = 90^2 + 90^2 \Rightarrow d^2 = 16{,}200 \Rightarrow d = \sqrt{16{,}200} \approx 127.28$.

    The distance from the pitcher's rubber to second base is $127.28 - 60.5 \approx 66.8$ ft.

51. Use the Law of Cosines to find the distance d:

    $d^2 = 246.75^2 + 246.75^2 - 2(246.75)(246.75)\cos 125°\ 12' \Rightarrow$

    $d = \sqrt{121{,}771.125 - 121{,}771.125 \cos 125°\ 12'} \approx 438.14$ ft

52. Let A, B, and C be the ship's initial position, turning position and final position, respectively.

    Triangle ABC has the following measurements:

    $B = 360° - [317° - (189° - 180°)] = 52°$, $a = 47.8$, and $c = 18.5$

    Now, $b^2 = 18.5^2 + 47.8^2 - 2(18.5)(47.8)\cos 52° \Rightarrow b = \sqrt{2627.09 - 1768.6 \cos 52°} \approx 39.2$ km

53. Use the Law of Cosines to find the angle of the triangle located at position A:

    $9^2 = 17^2 + 21^2 - 2(17)(21)\cos A \Rightarrow A = \cos^{-1}\left(\dfrac{81 - 730}{-714}\right) \approx 25°$. The bearing is $325° + 25° = 350°$.

54. The distances traveled by the sound are $3(344) = 1032$ m and $6(344) = 2064$ m.

    Use the Law of Cosines to find the distance d:

    $d^2 = 1032^2 + 2064^2 - 2(1032)(2064)\cos 42.2° \Rightarrow d = \sqrt{5{,}325{,}120 - 4{,}260{,}096 \cos 42.2°} \approx 1473$ m

55. If the satellite travels one orbit in 2 hours, it is traveling at a rate of $360 \div 120 = 3°$ per minute. At 12:03 P.M.

The angle formed at the center of Earth is $9°$. Noting that the sides of the tringle shown are 6400 km and

$6400 + 1600 = 8000$ km, use the Law of Cosines to find the distance d:

$$d^2 = 8000^2 + 6400^2 - 2(8000)(6400)\cos 9° \Rightarrow d = \sqrt{104,960,000 - 102,400,000\cos 9°} \approx 2000 \text{ km}$$

56. Let A, B, and C be the ship's initial position, turning position and final position, respectively.

Triangle ABC has the following measurements: $A = 90° - 62° = 28°$, $B = 62° + (180° - 115°) = 127°$,

$C = 180° - 28° - 127° = 25°$, and $b = 50$.   Use the Law of Sines to find a and c:

$$\frac{a}{\sin 28°} = \frac{50}{\sin 127°} \Rightarrow a = \frac{50\sin 28°}{\sin 127°} \approx 29.4; \frac{c}{\sin 25°} = \frac{50}{\sin 127°} \Rightarrow a = \frac{50\sin 25°}{\sin 127°} \approx 26.5$$

The ship traveled $26.5 + 29.4 = 55.9$ or 5.9 extra miles.

57. Use the Law of Cosines to find angle θ:

$$57.8^2 = 25.9^2 + 32.5^2 - 2(25.9)(32.5)\cos \theta \Rightarrow \theta = \cos^{-1}\left(\frac{3340.84 - 1727.06}{-1683.5}\right) \approx 163.5°.$$

58. Let A, B, and C be the plane's initial position, new position and position of the mountain, respectively.

Triangle ABC has the following measurements: $A = 24.1°$, $B = 180° - 32.7° = 147.3°$, and $b = 7.92$

Use the Law of Sines to find a: $\dfrac{a}{\sin 24.1°} = \dfrac{7.92}{\sin 147.3°} \Rightarrow a = \dfrac{7.92\sin 24.1°}{\sin 147.3°} \approx 5.99$ km

59. Use the Law of Cosines to find x:

$$x^2 = 25^2 + 25^2 - 2(25)(25)\cos 52° \Rightarrow x = \sqrt{1250 - 1250\cos 52°} \approx 22 \text{ ft}$$

60. Use the Law of Cosines to find d:

$$d^2 = 3428^2 + 5631^2 - 2(3428)(5631)\cos 43.33° \Rightarrow d = \sqrt{43,459,345 - 38,606,136\cos 43.33°} \approx 3921 \text{ m}$$

61. Since A is obtuse, $90° < A < 180°$. The cosine of a quadrant II angle is negative.

62. In $a^2 = b^2 + c^2 - 2ab\cos A$, $\cos A$ is negative, so $a^2 = b^2 + c^2 +$ (a positive quantity). Thus, $a^2 > b^2 + c^2$.

63. $b^2 + c^2 > b^2$ and $b^2 + c^2 > c^2$. If $a^2 > b^2 + c^2$, then $a^2 > b^2$ and $a^2 > c^2$ from which $a > b$ and $a > c$

because a, b, and c are nonnegative.

64. Because A is obtuse, it is the largest angle, so the longest side should be a, not c.

65. $A = \dfrac{1}{2}(13.6)(10.1)\sin 42.5° \approx 46.4 \text{ m}^2$

66. $A = \dfrac{1}{2}(30.4)(28.4)\sin 124.5° \approx 356 \text{ cm}^2$

67. $s = \dfrac{1}{2}(12 + 16 + 25) = 26.5; A = \sqrt{26.5(26.5 - 12)(26.5 - 16)(26.5 - 25)} \approx 78 \text{ m}^2$

68. $s = \dfrac{1}{2}(154 + 179 + 183) = 258; A = \sqrt{258(258 - 154)(258 - 179)(258 - 183)} \approx 12,600 \text{ cm}^2$

69. $s = \dfrac{1}{2}(76.3 + 109 + 98.8) = 142.05;$

$A = \sqrt{142.05(142.05 - 76.3)(142.05 - 109)(142.05 - 98.8)} \approx 3650 \text{ ft}^2$

70. $s = \dfrac{1}{2}(22 + 45 + 31) = 49; A = \sqrt{49(49 - 22)(49 - 45)(49 - 31)} \approx 310 \text{ in.}^2$

71. $s = \dfrac{1}{2}(25.4 + 38.2 + 19.8) = 41.7; A = \sqrt{41.7(41.7 - 25.4)(41.7 - 38.2)(41.7 - 19.8)} \approx 228 \text{ yd}^2$

72. $s = \dfrac{1}{2}(15.89 + 21.74 + 10.92) = 24.275;$

    $A = \sqrt{24.275(24.275 - 15.89)(24.275 - 21.74)(24.275 - 10.92)} \approx 83.01 \text{ in.}^2$

73. $s = \dfrac{1}{2}(75 + 68 + 85) = 114; A = \sqrt{114(114 - 75)(114 - 68)(114 - 85)} \approx 2435 \text{ m}^2$

    The number of cans needed is $2435 \div 75 \approx 32.5$ or 33 cans.

74. $A = \dfrac{1}{2}(52.1)(21.3)\sin 42.2° \approx 373 \text{ m}^2$

75. $A = \dfrac{1}{2}(15.2)(16.1)\sin 125° \approx 100 \text{ m}^2$

76. $s = \dfrac{1}{2}(850 + 925 + 1300) = 1537.5;$

    $A = \sqrt{1537.5(1537.5 - 850)(1537.5 - 925)(1537.5 - 1300)} \approx 392{,}000 \text{ mi}^2$

77. $s = \dfrac{1}{2}(9 + 10 + 17) = 18; A = \sqrt{18(18 - 9)(18 - 10)(18 - 17)} = 36 \text{ m}^2$ and $9 + 10 + 17 = 36$

78. (a) $s = \dfrac{1}{2}(11 + 13 + 20) = 22; A = \sqrt{22(22 - 11)(22 - 13)(22 - 20)} = 66,$ and 66 is an integer.

    (b) $s = \dfrac{1}{2}(13 + 14 + 15) = 21; A = \sqrt{21(21 - 13)(21 - 14)(21 - 15)} = 84,$ and 84 is an integer.

    (c) $s = \dfrac{1}{2}(7 + 15 + 20) = 21; A = \sqrt{21(21 - 7)(21 - 15)(21 - 20)} = 42,$ and 42 is an integer.

10.3: Vectors and Their Applications

1. The vector pairs that have the same direction and length are **m** & **p** and **n** & **r**.

2. The vector pairs that have the same length but opposite direction are **m** & **q**, **p** & **q**, **r** & **r**, and **n** & **s**.

3. $\mathbf{m} = 2\mathbf{t}, \mathbf{p} = 2\mathbf{t},$ or $\mathbf{t} = \dfrac{1}{2}\mathbf{m}$ and $\mathbf{t} = \dfrac{1}{2}\mathbf{p}$; also $\mathbf{m} = 1\mathbf{p}$ and $\mathbf{n} = 1\mathbf{r}$ or $\mathbf{p} = 1\mathbf{m}$ and $\mathbf{r} = 1\mathbf{n}$

4. $\mathbf{m} = -1\mathbf{q}$ and $\mathbf{p} = -1\mathbf{q},$ or $\mathbf{q} = -1\mathbf{m}$ and $\mathbf{q} = -1\mathbf{p}$; also $\mathbf{r} = -1\mathbf{s}$ and $\mathbf{n} = -1\mathbf{s}$ or $\mathbf{s} = -1\mathbf{r}$ and $\mathbf{s} = -1\mathbf{n}$;

    also $\mathbf{q} = -2\mathbf{t}$ or $\mathbf{t} = -\dfrac{1}{2}\mathbf{q}$

5. See Figure 5.

6. See Figure 6.

7. See Figure 7.

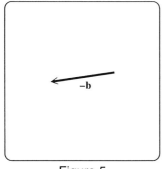

Figure 5

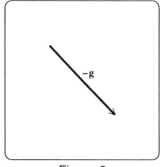

Figure 6

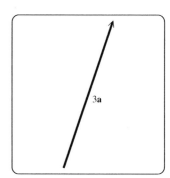

Figure 7

8. See Figure 8.

9. See Figure 9.

10. See Figure 10.

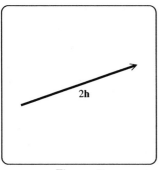

Figure 8

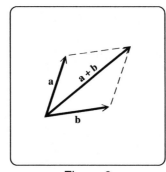

Figure 9

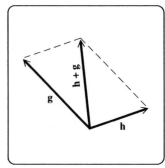

Figure 10

11. See Figure 11.

12. See Figure 12.

13. See Figure 13.

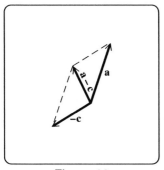

Figure 11

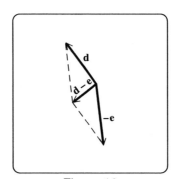

Figure 12

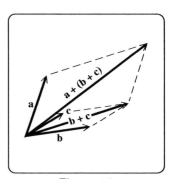

Figure 13

14. See Figure 14.

15. See Figure 15.

16. See Figure 16.

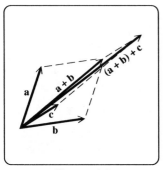

Figure 14

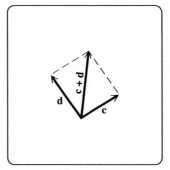

Figure 15

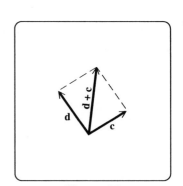

Figure 16

17. (a) $\mathbf{a} + \mathbf{b} = \langle -8, 8 \rangle + \langle 4, 8 \rangle = \langle -8 + 4, 8 + 8 \rangle = \langle -4, 16 \rangle$

  (b) $\mathbf{a} - \mathbf{b} = \langle -8, 8 \rangle - \langle 4, 8 \rangle = \langle -8 - 4, 8 - 8 \rangle = \langle -12, 0 \rangle$

  (c) $-\mathbf{a} = -\langle -8, 8 \rangle = \langle 8, -8 \rangle$

18. (a) $\mathbf{a} + \mathbf{b} = \langle 4, -4 \rangle + \langle -8, -4 \rangle = \langle 4 + (-8), -4 + (-4) \rangle = \langle -4, -8 \rangle$

  (b) $\mathbf{a} - \mathbf{b} = \langle 4, -4 \rangle - \langle -8, -4 \rangle = \langle 4 - (-8), -4 - (-4) \rangle = \langle 12, 0 \rangle$

  (c) $-\mathbf{a} = -\langle 4, -4 \rangle = \langle -4, 4 \rangle$

19. (a) $\mathbf{a} + \mathbf{b} = \langle 4, 8 \rangle + \langle 4, -8 \rangle = \langle 4 + 4, 8 + (-8) \rangle = \langle 8, 0 \rangle$

  (b) $\mathbf{a} - \mathbf{b} = \langle 4, 8 \rangle - \langle 4, -8 \rangle = \langle 4 - 4, 8 - (-8) \rangle = \langle 0, 16 \rangle$

  (c) $-\mathbf{a} = -\langle 4, 8 \rangle = \langle -4, -8 \rangle$

20. (a) $\mathbf{a} + \mathbf{b} = \langle -4, -4 \rangle + \langle 8, 4 \rangle = \langle -4 + 8, -4 + 4 \rangle = \langle 4, 0 \rangle$

  (b) $\mathbf{a} - \mathbf{b} = \langle -4, -4 \rangle - \langle 8, 4 \rangle = \langle -4 - 8, -4 - 4 \rangle = \langle -12, -8 \rangle$

  (c) $-\mathbf{a} = -\langle -4, -4 \rangle = \langle 4, 4 \rangle$

21. (a) $\mathbf{a} + \mathbf{b} = \langle -8, 4 \rangle + \langle 8, 8 \rangle = \langle -8 + 8, 4 + 8 \rangle = \langle 0, 12 \rangle$

  (b) $\mathbf{a} - \mathbf{b} = \langle -8, 4 \rangle - \langle 8, 8 \rangle = \langle -8 - 8, 4 - 8 \rangle = \langle -16, -4 \rangle$

  (c) $-\mathbf{a} = -\langle -8, 4 \rangle = \langle 8, -4 \rangle$

22. (a) $\mathbf{a} + \mathbf{b} = \langle 8, -4 \rangle + \langle -4, 8 \rangle = \langle 8 + (-4), -4 + 8 \rangle = \langle 4, 4 \rangle$

  (b) $\mathbf{a} - \mathbf{b} = \langle 8, -4 \rangle - \langle -4, 8 \rangle = \langle 8 - (-4), -4 - 8 \rangle = \langle 12, -12 \rangle$

  (c) $-\mathbf{a} = -\langle 8, -4 \rangle = \langle -8, 4 \rangle$

23. (a) $2\mathbf{a} = 2(2\mathbf{i}) = 4\mathbf{i}$

  (b) $2\mathbf{a} + 3\mathbf{b} = 2(2\mathbf{i}) + 3(\mathbf{i} + \mathbf{j}) = 4\mathbf{i} + 3\mathbf{i} + 3\mathbf{j} = 7\mathbf{i} + 3\mathbf{j}$

  (c) $\mathbf{b} - 3\mathbf{a} = \mathbf{i} + \mathbf{j} - 3(2\mathbf{i}) = \mathbf{i} + \mathbf{j} - 6\mathbf{i} = -5\mathbf{i} + \mathbf{j}$

24. (a) $2\mathbf{a} = 2(-\mathbf{i} + 2\mathbf{j}) = -2\mathbf{i} + 4\mathbf{j}$

  (b) $2\mathbf{a} + 3\mathbf{b} = 2(-\mathbf{i} + 2\mathbf{j}) + 3(\mathbf{i} - \mathbf{j}) = -2\mathbf{i} + 4\mathbf{j} + 3\mathbf{i} - 3\mathbf{j} = \mathbf{i} + \mathbf{j}$

  (c) $\mathbf{b} - 3\mathbf{a} = \mathbf{i} - \mathbf{j} - 3(-\mathbf{i} + 2\mathbf{j}) = \mathbf{i} - \mathbf{j} + 3\mathbf{i} - 6\mathbf{j} = 4\mathbf{i} - 7\mathbf{j}$

25. (a) $2\mathbf{a} = 2\langle -1, 2 \rangle = \langle -2, 4 \rangle$

  (b) $2\mathbf{a} + 3\mathbf{b} = 2\langle -1, 2 \rangle + 3\langle 3, 0 \rangle = \langle -2, 4 \rangle + \langle 9, 0 \rangle = \langle -2 + 9, 4 + 0 \rangle = \langle 7, 4 \rangle$

  (c) $\mathbf{b} - 3\mathbf{a} = \langle 3, 0 \rangle - 3\langle -1, 2 \rangle = \langle 3, 0 \rangle + \langle 3, -6 \rangle = \langle 3 + 3, 0 + (-6) \rangle = \langle 6, -6 \rangle$

26. (a) $2\mathbf{a} = 2\langle -2, -1 \rangle = \langle -4, -2 \rangle$

  (b) $2\mathbf{a} + 3\mathbf{b} = 2\langle -2, -1 \rangle + 3\langle -3, 2 \rangle = \langle -4, -2 \rangle + \langle -9, 6 \rangle = \langle -4 + (-9), -2 + 6 \rangle = \langle -13, 4 \rangle$

  (c) $\mathbf{b} - 3\mathbf{a} = \langle -3, 2 \rangle - 3\langle -2, -1 \rangle = \langle -3, 2 \rangle + \langle 6, 3 \rangle = \langle -3 + 6, 2 + 3 \rangle = \langle 3, 5 \rangle$

27. $\mathbf{u} + \mathbf{v} = \langle -2, 5 \rangle + \langle 4, 3 \rangle = \langle -2 + 4, 5 + 3 \rangle = \langle 2, 8 \rangle$

28. $\mathbf{u} - \mathbf{v} = \langle -2, 5 \rangle - \langle 4, 3 \rangle = \langle -2 - 4, 5 - 3 \rangle = \langle -6, 2 \rangle$

29. $\mathbf{v} - \mathbf{u} = \langle 4, 3 \rangle - \langle -2, 5 \rangle = \langle 4 - (-2), 3 - 5 \rangle = \langle 6, -2 \rangle$

30. $5\mathbf{v} = 5\langle 4, 3 \rangle = \langle 20, 15 \rangle$

31. $-5\mathbf{v} = -5\langle 4, 3 \rangle = \langle -20, -15 \rangle$

32. $3\mathbf{u} + 6\mathbf{v} = 3\langle -2, 5 \rangle + 6\langle 4, 3 \rangle = \langle -6, 15 \rangle + \langle 24, 18 \rangle = \langle -6 + 24, 15 + 18 \rangle = \langle 18, 33 \rangle$

33. $\mathbf{v} = \langle 6\cos 30°, 6\sin 30° \rangle = \left\langle 6 \cdot \dfrac{\sqrt{3}}{2}, 6 \cdot \dfrac{1}{2} \right\rangle = \langle 3\sqrt{3}, 3 \rangle$

34. $\mathbf{v} = \langle 7\cos 120°, 7\sin 120° \rangle = \left\langle 7 \cdot \left(-\dfrac{1}{2}\right), 7 \cdot \dfrac{\sqrt{3}}{2} \right\rangle = \left\langle -\dfrac{7}{2}, \dfrac{7\sqrt{3}}{2} \right\rangle$

35. $\mathbf{v} = \langle 9\cos 225°, 9\sin 225° \rangle = \left\langle 9 \cdot \left(-\dfrac{\sqrt{2}}{2}\right), 9 \cdot \left(-\dfrac{\sqrt{2}}{2}\right) \right\rangle = \left\langle -\dfrac{9\sqrt{2}}{2}, -\dfrac{9\sqrt{2}}{2} \right\rangle$

36. $\mathbf{v} = \langle 4\cos(-60°), 4\sin(-60°) \rangle = \left\langle 4 \cdot \dfrac{1}{2}, 4 \cdot \left(-\dfrac{\sqrt{3}}{2}\right) \right\rangle = \langle 2, -2\sqrt{3} \rangle$

37. $\mathbf{v} = \langle 4\cos 40°, 4\sin 40° \rangle \approx \langle 3.06, 2.57 \rangle$

38. $\mathbf{v} = \langle 3\cos 130°, 3\sin 130° \rangle \approx \langle -1.93, 2.30 \rangle$

39. $\mathbf{v} = \langle 5\cos(-35°), 5\sin(-35°) \rangle \approx \langle 4.10, -2.87 \rangle$

40. $\mathbf{v} = \langle 2\cos 220°, 2\sin 220° \rangle \approx \langle -1.53, -1.29 \rangle$

41. The adjacent angle in the parallelogram is supplementary to 40°, thus it has measure 140°.

   $|\mathbf{v}|^2 = 40^2 + 60^2 - 2(40)(60)\cos 140° \approx 8877.0133 \Rightarrow |\mathbf{v}| = \sqrt{8877.0133} \approx 94.2 \text{ lb}$

42. The adjacent angle in the parallelogram is supplementary to 65°, thus it has measure 115°.

   $|\mathbf{v}|^2 = 85^2 + 102^2 - 2(85)(102)\cos 115° \approx 24{,}957.2007 \Rightarrow |\mathbf{v}| = \sqrt{24{,}957.2007} \approx 158.0 \text{ lb}$

43. The adjacent angle in the parallelogram is supplementary to 110°, thus it has measure 70°.

   $|\mathbf{v}|^2 = 15^2 + 25^2 - 2(15)(25)\cos 70° \approx 593.4849 \Rightarrow |\mathbf{v}| = \sqrt{593.4849} \approx 24.4 \text{ lb}$

44. The adjacent angle in the parallelogram is supplementary to 140°, thus it has measure 40°.

   $|\mathbf{v}|^2 = 1500^2 + 2000^2 - 2(1500)(2000)\cos 40° \approx 1{,}653{,}733.341 \Rightarrow |\mathbf{v}| = \sqrt{1{,}653{,}733.341} \approx 1286.0 \text{ lb}$

45. $\langle -5, 8 \rangle = -5\mathbf{i} + 8\mathbf{j}$

46. $\langle 6, -3 \rangle = 6\mathbf{i} - 3\mathbf{j}$

47. $\langle 2, 0 \rangle = 2\mathbf{i} + 0\mathbf{j} = 2\mathbf{i}$

48. $\langle 0, -4 \rangle = 0\mathbf{i} - 4\mathbf{j} = -4\mathbf{j}$

49. $(8\cos 45°)\mathbf{i} + (8\sin 45°)\mathbf{j} = \left(8 \cdot \dfrac{\sqrt{2}}{2}\right)\mathbf{i} + \left(8 \cdot \dfrac{\sqrt{2}}{2}\right)\mathbf{j} = 4\sqrt{2}\,\mathbf{i} + 4\sqrt{2}\,\mathbf{j}$

50. $(3\cos 210°)\mathbf{i} + (3\sin 210°)\mathbf{j} = \left(3 \cdot \left(-\dfrac{\sqrt{3}}{2}\right)\right)\mathbf{i} + \left(3 \cdot \left(-\dfrac{1}{2}\right)\right)\mathbf{j} = -\dfrac{3\sqrt{3}}{2}\mathbf{i} - \dfrac{3}{2}\mathbf{j}$

51. $(.6\cos 115°)\mathbf{i} + (.6\sin 115°)\mathbf{j} = -.25\mathbf{i} + .54\mathbf{j}$

52. $(.9\cos 208°)\mathbf{i} + (.9\sin 208°)\mathbf{j} = -.79\mathbf{i} - .42\mathbf{j}$

53. $|\langle 1, 1 \rangle| = \sqrt{1^2 + 1^2} = \sqrt{1+1} = \sqrt{2}$; direction angle: $\tan^{-1}\left(\dfrac{1}{1}\right) = 45°$

54. $|\langle -4, 4\sqrt{3} \rangle| = \sqrt{(-4)^2 + (4\sqrt{3})^2} = \sqrt{16+48} = \sqrt{64} = 8$; direction angle: $\tan^{-1}\left(\dfrac{4\sqrt{3}}{-4}\right) = 120°$

55. $|\langle 8\sqrt{2}, -8\sqrt{2} \rangle| = \sqrt{(8\sqrt{2})^2 + (-8\sqrt{2})^2} = \sqrt{128+128} = \sqrt{256} = 16$;

   direction angle: $\tan^{-1}\left(\dfrac{-8\sqrt{2}}{8\sqrt{2}}\right) = 315°$

56. $|\langle \sqrt{3}, -1 \rangle| = \sqrt{(\sqrt{3})^2 + (-1)^2} = \sqrt{3+1} = \sqrt{4} = 2$; direction angle: $\tan^{-1}\left(\dfrac{-1}{\sqrt{3}}\right) = 330°$

57. $|\langle 15, -8 \rangle| = \sqrt{15^2 + (-8)^2} = \sqrt{225+64} = \sqrt{289} = 17$; direction angle: $\tan^{-1}\left(\dfrac{-8}{15}\right) = 331.9°$

58. $|\langle-7, 24\rangle| = \sqrt{(-7)^2 + 24^2} = \sqrt{49 + 576} = \sqrt{625} = 25$; direction angle: $\tan^{-1}\left(\dfrac{24}{-7}\right) = 106.3°$

59. $|\langle-6, 0\rangle| = \sqrt{(-6)^2 + 0^2} = \sqrt{36 + 0} = \sqrt{36} = 6$; direction angle: $\tan^{-1}\left(\dfrac{0}{-6}\right) = 180°$

60. $|\langle 0, -12\rangle| = \sqrt{0^2 + (-12)^2} = \sqrt{0 + 144} = \sqrt{144} = 12$; direction angle: $\tan^{-1}\left(\dfrac{-12}{0}\right) \Rightarrow$ undefined $\Rightarrow 270°$

61. $\langle 6, -1\rangle \cdot \langle 2, 5\rangle = (6)(2) + (-1)(5) = 12 - 5 = 7$

62. $\langle-3, 8\rangle \cdot \langle 7, -5\rangle = (-3)(7) + (8)(-5) = -21 - 40 = -61$

63. $\langle 2, -3\rangle \cdot \langle 6, 5\rangle = (2)(6) + (-3)(5) = 12 - 15 = -3$

64. $\langle 1, 2\rangle \cdot \langle 3, -1\rangle = (1)(3) + (2)(-1) = 3 - 2 = 1$

65. $4\mathbf{i} \cdot (5\mathbf{i} - 9\mathbf{j}) = (4)(5) + (0)(-9) = 20 + 0 = 20$

66. $(2\mathbf{i} + 4\mathbf{j}) \cdot (-\mathbf{j}) = (2)(0) + (4)(-1) = 0 - 4 = -4$

67. $\cos\theta = \dfrac{\mathbf{u} \cdot \mathbf{v}}{|\mathbf{u}|\,|\mathbf{v}|} = \dfrac{\langle 2, 1\rangle \cdot \langle-3, 1\rangle}{|\langle 2, 1\rangle|\,|\langle-3, 1\rangle|} = \dfrac{(2)(-3) + (1)(1)}{\sqrt{4 + 1}\,\sqrt{9 + 1}} = \dfrac{-5}{\sqrt{5}\sqrt{10}} = \dfrac{-5}{\sqrt{50}} = \dfrac{-5}{5\sqrt{2}} = -\dfrac{1}{\sqrt{2}} \Rightarrow$

$\theta = \cos^{-1}\left(-\dfrac{1}{\sqrt{2}}\right) = 135°$

68. $\cos\theta = \dfrac{\mathbf{u} \cdot \mathbf{v}}{|\mathbf{u}|\,|\mathbf{v}|} = \dfrac{\langle 1, 7\rangle \cdot \langle 1, 1\rangle}{|\langle 1, 7\rangle|\,|\langle 1, 1\rangle|} = \dfrac{(1)(1) + (7)(1)}{\sqrt{1 + 49}\,\sqrt{1 + 1}} = \dfrac{8}{\sqrt{50}\sqrt{2}} = \dfrac{8}{\sqrt{100}} = \dfrac{8}{10} = \dfrac{4}{5} \Rightarrow$

$\theta = \cos^{-1}\left(\dfrac{4}{5}\right) \approx 36.87°$

69. $\cos\theta = \dfrac{\mathbf{u} \cdot \mathbf{v}}{|\mathbf{u}|\,|\mathbf{v}|} = \dfrac{\langle 1, 2\rangle \cdot \langle-6, 3\rangle}{|\langle 1, 2\rangle|\,|\langle-6, 3\rangle|} = \dfrac{(1)(-6) + (2)(3)}{\sqrt{1 + 4}\,\sqrt{36 + 9}} = \dfrac{0}{\sqrt{5}\sqrt{45}} = 0 \Rightarrow \theta = \cos^{-1}0 = 90°$

70. $\cos\theta = \dfrac{\mathbf{u} \cdot \mathbf{v}}{|\mathbf{u}|\,|\mathbf{v}|} = \dfrac{\langle 4, 0\rangle \cdot \langle 2, 2\rangle}{|\langle 4, 0\rangle|\,|\langle 2, 2\rangle|} = \dfrac{(4)(2) + (0)(2)}{\sqrt{16 + 0}\,\sqrt{4 + 4}} = \dfrac{8}{\sqrt{16}\sqrt{8}} = \dfrac{8}{8\sqrt{2}} = \dfrac{1}{\sqrt{2}} \Rightarrow$

$\theta = \cos^{-1}\left(\dfrac{1}{\sqrt{2}}\right) = 45°$

71. $\cos\theta = \dfrac{\mathbf{u} \cdot \mathbf{v}}{|\mathbf{u}|\,|\mathbf{v}|} = \dfrac{(3\mathbf{i} + 4\mathbf{j}) \cdot \mathbf{j}}{|3\mathbf{i} + 4\mathbf{j}|\,|\mathbf{j}|} = \dfrac{(3)(0) + (4)(1)}{\sqrt{9 + 16}\,\sqrt{0 + 1}} = \dfrac{4}{\sqrt{25}\sqrt{1}} = \dfrac{4}{5} \Rightarrow \theta = \cos^{-1}\left(\dfrac{4}{5}\right) \approx 36.87°$

72. $\cos\theta = \dfrac{\mathbf{u} \cdot \mathbf{v}}{|\mathbf{u}|\,|\mathbf{v}|} = \dfrac{(-5\mathbf{i} + 12\mathbf{j}) \cdot (3\mathbf{i} + 2\mathbf{j})}{|-5\mathbf{i} + 12\mathbf{j}|\,|3\mathbf{i} + 2\mathbf{j}|} = \dfrac{(-5)(3) + (12)(2)}{\sqrt{25 + 144}\,\sqrt{9 + 4}} = \dfrac{9}{\sqrt{169}\sqrt{13}} = \dfrac{9}{13\sqrt{13}} = \dfrac{9\sqrt{13}}{169} \Rightarrow$

$\theta = \cos^{-1}\left(\dfrac{9\sqrt{13}}{169}\right) \approx 78.93°$

73. $(3\mathbf{u}) \cdot \mathbf{v} = 3\langle-2, 1\rangle \cdot \langle 3, 4\rangle = \langle-6, 3\rangle \cdot \langle 3, 4\rangle = (-6)(3) + (3)(4) = -18 + 12 = -6$

74. $\mathbf{u} \cdot (\mathbf{v} - \mathbf{w}) = \langle-2, 1\rangle \cdot (\langle 3, 4\rangle - \langle-5, 12\rangle) = \langle-2, 1\rangle \cdot \langle 8, -8\rangle = (-2)(8) + (1)(-8) = -16 - 8 = -24$

75. $\mathbf{u} \cdot \mathbf{v} - \mathbf{u} \cdot \mathbf{w} = \langle-2, 1\rangle \cdot \langle 3, 4\rangle - \langle-2, 1\rangle \cdot \langle-5, 12\rangle = [(-2)(3) + (1)(4)] - [(-2)(-5) + (1)(12)] = -24$

76. $\mathbf{u} \cdot (3\mathbf{v}) = \langle-2, 1\rangle \cdot 3\langle 3, 4\rangle = \langle-2, 1\rangle \cdot \langle 9, 12\rangle = (-2)(9) + (1)(12) = -18 + 12 = -6$

77. $\langle 1, 2\rangle \cdot \langle-6, 3\rangle = (1)(-6) + (2)(3) = -6 + 6 = 0 \Rightarrow$ orthogonal

78. $\langle 3, 4\rangle \cdot \langle 6, 8\rangle = (3)(6) + (4)(8) = 18 + 32 = 50 \Rightarrow$ not orthogonal

79. $\langle 1, 0\rangle \cdot \langle\sqrt{2}, 0\rangle = (1)(\sqrt{2}) + (0)(0) = \sqrt{2} + 0 = \sqrt{2} \Rightarrow$ not orthogonal

80. $\langle 1, 1 \rangle \cdot \langle 1, -1 \rangle = (1)(1) + (1)(-1) = 1 - 1 = 0 \Rightarrow$ orthogonal

81. $(\sqrt{5}\mathbf{i} - 2\mathbf{j}) \cdot (-5\mathbf{i} + 2\sqrt{5}\mathbf{j}) = (\sqrt{5})(-5) + (-2)(2\sqrt{5}) = -5\sqrt{5} - 4\sqrt{5} = -9\sqrt{5} \Rightarrow$ not orthogonal

82. $(-4\mathbf{i} + 3\mathbf{j}) \cdot (8\mathbf{i} - 6\mathbf{j}) = (-4)(8) + (3)(-6) = -32 - 18 = -50 \Rightarrow$ not orthogonal

83. By completing the parallelogram with the resultant vector as the diagonal, the triangle formed by two sides of the parallelogram and the resultant has a side of measure 176 and angles with the following measures:

    $41° \, 10'$, $78° \, 50' - 41° \, 10' = 37° \, 40'$, and $180° - 41° \, 10' - 37° \, 40' = 101° \, 10'$. Using the Law of Sines,

    $$\frac{F_2}{\sin 41° \, 10'} = \frac{176}{\sin 37° \, 40'} \Rightarrow F_2 = \frac{176 \sin 41° \, 10'}{\sin 37° \, 40'} \approx 190 \text{ lb}$$

    $$\frac{R}{\sin 101° \, 10'} = \frac{176}{\sin 37° \, 40'} \Rightarrow F_2 = \frac{176 \sin 101° \, 10'}{\sin 37° \, 40'} \approx 283 \text{ lb}$$

84. By completing the parallelogram with the resultant vector as the diagonal, the triangle formed by two sides of the parallelogram and the resultant has a side of measure 28.7 and angles with the following measures:

    $32° \, 40'$, $42° \, 10' - 32° \, 40' = 9° \, 30'$, and $180° - 32° \, 40' - 9° \, 30' = 137° \, 50'$. Using the Law of Sines,

    $$\frac{F_2}{\sin 32° \, 40'} = \frac{28.7}{\sin 9° \, 30'} \Rightarrow F_2 = \frac{28.7 \sin 32° \, 40'}{\sin 9° \, 30'} \approx 93.9 \text{ lb}$$

    $$\frac{R}{\sin 137° \, 50'} = \frac{28.7}{\sin 9° \, 30'} \Rightarrow R = \frac{28.7 \sin 137° \, 50'}{\sin 9° \, 30'} \approx 117 \text{ lb}$$

85. $\theta = \sin^{-1}\left(\dfrac{25}{80}\right) \approx 18°$

86. $\sin 15° = \dfrac{F}{3000} \Rightarrow F = 3000 \sin 15° \approx 780 \text{ lb}$

87. $\sin 2.3° = \dfrac{F}{60} \Rightarrow F = 60 \sin 2.3° \approx 2.4 \text{ tons}$

88. Since the forces are in equilibrium, the vector sum must be zero. Arrange the forces in a triangle.

    Using the Law of Cosines and supplementary angles, the angles are found as follows:

    $$980^2 = 760^2 + 1220^2 - 2(760)(1220)\cos\theta \Rightarrow \theta = \cos^{-1}\left(\frac{960,400 - 2,066,000}{-1,854,400}\right) \approx 53.4°.$$

    $$1220^2 = 760^2 + 980^2 - 2(760)(980)\cos\theta \Rightarrow \theta = \cos^{-1}\left(\frac{1,488,400 - 1,538,000}{-1,489,600}\right) \approx 88.1°.$$

    $$760^2 = 1220^2 + 980^2 - 2(1220)(980)\cos\theta \Rightarrow \theta = \cos^{-1}\left(\frac{577,600 - 2,448,800}{-2,391,200}\right) \approx 38.5°.$$

    The angles between the forces are $180° - 53.4° = 126.6°$, $180° - 88.1° = 91.9°$, and $180° - 38.5° = 141.5°$.

89. $\theta = \sin^{-1}\left(\dfrac{18}{60}\right) \approx 17.5°$

90. $\theta = \sin^{-1}\left(\dfrac{30}{80}\right) \approx 22.0°$

91. The weight of the crate is $\sin 46° \, 20' = \dfrac{F}{89.6} \Rightarrow F = 89.6 \sin 46° \, 20' \approx 64.8 \text{ lb}.$

    The tension on the other rope is $\cos 46° \, 20' = \dfrac{F}{89.6} \Rightarrow F = 89.6 \cos 46° \, 20' \approx 61.9 \text{ lb}.$

92. $\sin 62.4° = \dfrac{F_1}{150} \Rightarrow F_1 = 150 \sin 62.4° \approx 132.9 \text{ lb}$; $\sin 54.9° = \dfrac{F_2}{114} \Rightarrow F_2 = 114 \sin 54.9° \approx 93.3 \text{ lb}$

    The weight of the box is the sum of the two vertical forces, $132.9 + 93.3 = 226.2 \approx 226 \text{ lb}.$

93. Let A, B, and C be the ship's initial position, turning position and final position, respectively.

    Triangle ABC has the following measurements: $B = 90° + 34.0° = 124.0°$, $a = 4.6$, and $c = 10.4$

    Now, $b^2 = 10.4^2 + 4.6^2 - 2(10.4)(4.6)\cos 124.0° \Rightarrow b = \sqrt{129.32 - 95.68\cos 124.0°} \approx 13.5$ mi

    $\dfrac{\sin A}{4.6} = \dfrac{\sin 124.0°}{13.5} \Rightarrow \sin A = \dfrac{4.6\sin 124.0°}{13.5} \Rightarrow A = \sin^{-1}\left(\dfrac{4.6\sin 124.0°}{13.5}\right) \approx 16.4°.$

    The ship is 13.5 miles from its starting point on a bearing of $34.0° + 16.4° = 50.4°$.

94. Let A, B, and C be the ship's initial position, turning position and final position, respectively.

    Triangle ABC has the following measurements: $B = 110.0° - 90° = 20.0°$, $a = 2.4$, and $c = 8.8$

    Now, $b^2 = 8.8^2 + 2.4^2 - 2(8.8)(2.4)\cos 20° \Rightarrow b = \sqrt{83.2 - 42.24\cos 20°} \approx 6.6$ mi

    $\dfrac{\sin A}{2.4} = \dfrac{\sin 20.0°}{6.6} \Rightarrow \sin A = \dfrac{2.4\sin 20.0°}{6.6} \Rightarrow A = \sin^{-1}\left(\dfrac{2.4\sin 20.0°}{6.6}\right) \approx 7.1°.$

    The ship is 6.6 miles from its starting point on a bearing of $110.0° + 7.1° = 117.1°$.

95. Let A, B, and C be the ship's initial position, turning position and final position, respectively.

    Triangle ABC has the following measurements:

    $B = 360° - [317° - (189° - 180°)] = 52°$, $a = 47.8$, and $c = 18.5$

    Now, $b^2 = 47.8^2 + 18.5^2 - 2(47.8)(18.5)\cos 52° \Rightarrow b = \sqrt{2627.09 - 1768.6\cos 52°} \approx 39.2$ km

96. Let X, B, and C be the ship's initial position, turning position and final position, respectively.

    Triangle XBC has the following measurements:

    $B = 360° - [320° - (200° - 180°)] = 60°$, $x = 2.4$, and $c = 15.5$

    Now, $b^2 = 15.5^2 + 2.4^2 - 2(15.5)(2.4)\cos 60° \Rightarrow b = \sqrt{246.01 - 74.4\cos 60°} \approx 14.5$ km

97. The speed of the boat in still water, 20, is the hypotenuse of a right triangle with the shorter leg representing the speed of the current and the longer leg representing the actual speed of the boat. The smallest angle of the right triangle is $90° - 80° = 10°$. The actual speed of the boat is $20\cos 10° \approx 19.7$ mph. The speed of the current is $20\sin 10° \approx 3.5$ mph.

98. Let x represent the time, in hours, until the plane turns. Since the plane will travel for 2.6 hours, the length of the side of the triangle representing the distance traveled by the ship is $2.6(32) = 83.2$. The length of the side of the triangle representing the distance traveled by the plane before the turn is $520x$. The length of the side of the triangle representing the distance traveled after the plane turns is $520(2.6 - x) = 1352 - 520x$. The angle at the bottom of the triangle is $30° + (360° - 338°) = 52°$. Use the Law of Cosines to find x.

    $(1352 - 520x)^2 = (520x)^2 + 83.2^2 - 2(520x)(83.2)\cos 52° \Rightarrow$

    $1{,}827{,}904 - 1{,}406{,}080 + (520x)^2 = (520x)^2 + 6922.24 - 53{,}272x \Rightarrow 1{,}820{,}981.76 = 1{,}352{,}808x \Rightarrow$

    $x \approx 1.35$. The plane turns about 1.35 hours after it leaves the ship, or about 3:21 P.M.

99. If the wind is from the direction of $114°$, the bearing of the wind is $114° + 180° = 294°$. The angle between the wind and the flight of the jet is $(360° - 294°) + [180° - (360° - 233°)] = 119°$. If b is the missing side of the triangle, $b^2 = 450^2 + 39^2 - 2(450)(39)\cos 119° \Rightarrow b = \sqrt{204{,}021 - 35{,}100\cos 119°} \approx 470$ mph.

    Then $\dfrac{\sin A}{39} = \dfrac{\sin 119°}{470} \Rightarrow \sin A = \dfrac{39\sin 119°}{470} \Rightarrow A = \sin^{-1}\left(\dfrac{39\sin 119°}{470}\right) \approx 4.2°.$

    The ground speed of the jet is 470 mph and the bearing is $233° + 4.2° = 237°$.

100. (a) Using the Pythagorean theorem, $7^2 = 3^2 + b^2 \Rightarrow b^2 = 49 - 9 \Rightarrow b^2 = 40 \Rightarrow b = \sqrt{40} \approx 6.3246$.

   The speed of the boat relative to the banks is about 6.32 mph.

   (b) Noting that 6.3246 mph $\approx$ 9.2761 ft/sec, the boat takes $132 \div 9.2761 \approx 14.23$ seconds to cross the river.

   (c) $\sin \theta = \dfrac{3}{7} \Rightarrow \theta = \sin^{-1} \dfrac{3}{7} \approx 25.38°$

101. The triangle formed is a right triangle with one leg representing the wind speed of 42 mph, the hypotenuse representing airspeed and the other leg representing the groundspeed. The angle between the airspeed and groundspeed is $90° - 74.9° = 15.1°$. Let g represent the hypotenuse and let a represent the unknown leg. The groundspeed is given by $\sin 15.1° = \dfrac{42}{g} \Rightarrow g = \dfrac{42}{\sin 15.1°} \approx 161$ mph.

   The airspeed is given by $\tan 15.1° = \dfrac{42}{a} \Rightarrow a = \dfrac{42}{\tan 15.1°} \approx 156$ mph.

102. If the wind is from the direction of $266.6°$, the bearing of the wind is $266.6° - 180° = 86.6°$. The angle between the wind and the flight of the plane is $86.6° + (180° - 175.3°) = 91.3°$. If b is the missing side of the triangle, $b^2 = 650^2 + 25^2 - 2(650)(25)\cos 91.3° \Rightarrow b = \sqrt{423{,}125 - 32{,}500 \cos 91.3°} \approx 651$ mph.

   Then $\dfrac{\sin A}{25} = \dfrac{\sin 91.3°}{651} \Rightarrow \sin A = \dfrac{25 \sin 91.3°}{651} \Rightarrow A = \sin^{-1}\left(\dfrac{25 \sin 91.3°}{651}\right) \approx 2.2°$.

   The bearing is $175.3° - 2.2° = 173.1°$.

103. Let A, B, and C represent angle between the flight path and the desired bearing, the angle between the wind vector and the original flight path, and the angle between the wind vector and the desired bearing. Triangle ABC has the following measurements: $C = 64° 30'$, $a = 35$, and $c = 190$. Using the Law of Sines,

   $\dfrac{\sin A}{35} = \dfrac{\sin 64° 30'}{190} \Rightarrow \sin A = \dfrac{35 \sin 64° 30'}{190} \Rightarrow A = \sin^{-1}\left(\dfrac{35 \sin 64° 30'}{190}\right) \approx 9° 30'$

   So $B = 180° - 9° 30' - 64° 30' = 106°$. Use the Law of Sines to find b.

   $\dfrac{b}{\sin 106°} = \dfrac{190}{\sin 64° 30'} \Rightarrow b = \dfrac{190 \sin 106°}{\sin 64° 30'} \approx 202$ mph

   The bearing is $64° 30' + 9° 30' = 74°$ and the ground speed is 202 mph

104. Let A, B, and C represent angle between the flight path and the desired bearing, the angle between the wind vector and the original flight path, and the angle between the wind vector and the desired bearing, respectively. Triangle ABC has the following measurements: $C = 57° 40'$, $a = 27.1$, and $c = 168$. Using the Law of Sines,

   $\dfrac{\sin A}{27.1} = \dfrac{\sin 57° 40'}{168} \Rightarrow \sin A = \dfrac{27.1 \sin 57° 40'}{168} \Rightarrow A = \sin^{-1}\left(\dfrac{27.1 \sin 57° 40'}{168}\right) \approx 7° 50'$

   So $B = 180° - 7° 50' - 57° 40' = 114° 30'$. Use the Law of Sines to find b.

   $\dfrac{b}{\sin 114° 30'} = \dfrac{168}{\sin 57° 40'} \Rightarrow b = \dfrac{168 \sin 114° 30'}{\sin 57° 40'} \approx 181$ mph

   The bearing is $57° 40' + 7° 50' = 65° 30'$ and the ground speed is 181 mph

105. Let A, B, and C represent angle between the required flight path and due north, the angle between the wind vector and due north, and the angle between the wind vector and the required flight path, respectively.

Triangle ABC has the following measurements: $C = 360° - 328° = 32°$, $a = 11$, and $c = 400 \div 2.5 = 160$.

Using the Law of Sines, $\dfrac{\sin A}{11} = \dfrac{\sin 32°}{160} \Rightarrow \sin A = \dfrac{11 \sin 32°}{160} \Rightarrow A = \sin^{-1}\left(\dfrac{11 \sin 32°}{160}\right) \approx 2°$

So $B = 180° - 2° - 32° = 146°$. Use the Law of Sines to find b.

$\dfrac{b}{\sin 146°} = \dfrac{160}{\sin 32°} \Rightarrow b = \dfrac{160 \sin 146°}{\sin 32°} \approx 170$ mph

The bearing is $360° - 2° = 358°$ and the ground speed is 170 mph.

106. If the wind is from the direction of $78°$, the bearing of the wind is $78° + 180° = 258°$. The angle between the wind and the due south flight path is $360° - 258° = 102°$. If b is the missing side of the triangle,

$b^2 = 192^2 + 23^2 - 2(192)(23)\cos 102° \Rightarrow b = \sqrt{37{,}393 - 8832 \cos 102°} \approx 198$ mph.

Then $\dfrac{\sin A}{23} = \dfrac{\sin 102°}{198} \Rightarrow \sin A = \dfrac{23 \sin 102°}{198} \Rightarrow A = \sin^{-1}\left(\dfrac{23 \sin 102°}{198}\right) \approx 6.5°$.

The groundspeed is 198 mph and the bearing is $180° + 6.5° = 186.5°$.

107. If the wind is from the direction of $245°$, the bearing of the wind is $245° - 180° = 65°$. The angle between the wind and the flight of the plane is $65° + (180° - 174°) = 71°$. If b is the missing side of the triangle,

$b^2 = 240^2 + 30^2 - 2(240)(30)\cos 71° \Rightarrow b = \sqrt{58{,}500 - 14{,}400 \cos 71°} \approx 230$ km per hr.

Then $\dfrac{\sin A}{30} = \dfrac{\sin 71°}{230} \Rightarrow \sin A = \dfrac{30 \sin 71°}{230} \Rightarrow A = \sin^{-1}\left(\dfrac{30 \sin 71°}{230}\right) \approx 7°$.

The groundspeed is 230 km per hr and the bearing is $174° - 7° = 167°$.

108. (a) First convert 10.34" to radians: $\dfrac{10.34''}{1\text{yr}} \cdot \dfrac{1'}{60''} \cdot \dfrac{1°}{60'} \cdot \dfrac{\pi}{180°} \approx 5.012973463 \times 10^{-5}$ rad/yr.

Now using the arc length formula with a radius of 35 trillion miles yields

$\mathbf{v}_t = (35 \times 10^{12})(5.012973463 \times 10^{-5}) \approx 1{,}754{,}540{,}712$ mi/yr. Convert this to miles per second.

$\mathbf{v}_t = \dfrac{1{,}754{,}540{,}712 \text{ mi}}{1\text{yr}} \cdot \dfrac{1\text{yr}}{356 \text{ days}} \cdot \dfrac{1 \text{ day}}{24 \text{ hr}} \cdot \dfrac{1 \text{ hr}}{60 \text{ min}} \cdot \dfrac{1 \text{ min}}{60 \text{ sec}} \approx 56$ miles per second.

(b) Using the Pythagorean theorem, $\mathbf{v}^2 = 67^2 + 56^2 \Rightarrow \mathbf{v}^2 = 7625 \Rightarrow \mathbf{v}^2 = \sqrt{7625} \approx 87$

109. (a) $|\mathbf{R}| = \sqrt{1^2 + (-2)^2} = \sqrt{5} \approx 2.2$; $|\mathbf{A}| = \sqrt{.5^2 + 1^2} = \sqrt{1.25} \approx 1.1$

About 2.2 inches of rain fell. The area of the opening is about 1.1 in^2.

(b) $V = |\mathbf{R} \cdot \mathbf{A}| = |(1)(.5) + (-2)(1)| = |-1.5| = 1.5$; The volume of rain was 1.5 in^3.

(b) To collect the maximum amount of rain, $\mathbf{R}$ and $\mathbf{A}$ should be parallel and point in opposite directions.

110. $\|\mathbf{a} - \mathbf{b}\|^2 = \|\mathbf{a}\|^2 + \|\mathbf{b}\|^2 - 2\|\mathbf{a}\|\|\mathbf{b}\|\cos\theta \Rightarrow \|\mathbf{a} - \mathbf{b}\|^2 - \|\mathbf{a}\|^2 - \|\mathbf{b}\|^2 = -2\|\mathbf{a}\|\|\mathbf{b}\|\cos\theta$

$\Rightarrow \left(\sqrt{(a_1 - b_1)^2 + (a_2 - b_2)^2}\right)^2 - \left(\sqrt{a_1^2 + a_2^2}\right)^2 - \left(\sqrt{b_1^2 + b_2^2}\right)^2 = -2\|\mathbf{a}\|\|\mathbf{b}\|\cos\theta$

$\Rightarrow (a_1 - b_1)^2 + (a_2 - b_2)^2 - (a_1^2 + a_2^2) - (b_1^2 + b_2^2) = -2\|\mathbf{a}\|\|\mathbf{b}\|\cos\theta$

$\Rightarrow a_1^2 - 2a_1 b_1 + b_1^2 + a_2^2 - 2a_2 b_2 + b_2^2 - a_1^2 - a_2^2 - b_1^2 - b_2^2 = -2\|\mathbf{a}\|\|\mathbf{b}\|\cos\theta$

$\Rightarrow -2a_1 b_1 - 2a_2 b_2 = -2\|\mathbf{a}\|\|\mathbf{b}\|\cos\theta$

$\Rightarrow a_1 b_1 + a_2 b_2 = \|\mathbf{a}\|\|\mathbf{b}\|\cos\theta$

$\Rightarrow \mathbf{a} \cdot \mathbf{b} = \|\mathbf{a}\|\|\mathbf{b}\|\cos\theta$

Reviewing Basic Concepts (Sections 10.1—10.3)

1. Angle B has measure $180° - 44° - 62° = 74°$. Now, using the Law of Sines,

$$\frac{c}{\sin 62°} = \frac{12}{\sin 44°} \Rightarrow c = \frac{12 \sin 62°}{\sin 44°} \Rightarrow c \approx 15.3; \quad \frac{b}{\sin 74°} = \frac{12}{\sin 44°} \Rightarrow b = \frac{12 \sin 74°}{\sin 44°} \Rightarrow b \approx 16.6.$$

The solutions are $B = 74°, b \approx 16.6, c \approx 15.3$.

2. There are two solutions. Using the Law of Sines,

$$\frac{\sin B}{8} = \frac{\sin 32°}{6} \Rightarrow \sin B = \frac{8 \sin 32°}{6} \Rightarrow B = \sin^{-1}\left(\frac{8 \sin 32°}{6}\right) \approx 45.0°.$$

There are two angles: $B_1 = 45.0°$ and $B_2 = 180° - 45.0° = 135.0°$ (since $32° + 135.0° < 180°$).

$C_1 = 180° - 32° - 45.0° = 103°$ and $C_2 = 180° - 32° - 135° = 13°$

$$\frac{c_1}{\sin 103°} = \frac{6}{\sin 32°} \Rightarrow c_1 = \frac{6 \sin 103°}{\sin 32°} \approx 11.0$$

$$\frac{c_2}{\sin 13°} = \frac{6}{\sin 32°} \Rightarrow c_2 = \frac{6 \sin 13°}{\sin 32°} \approx 2.5$$

The solutions are $B_1 = 45.0°, C_1 = 103°, c_1 = 11.0$ or $B_2 = 135.0°, C_2 = 13°, c_2 = 2.5$.

3. Using the Law of Sines, $\dfrac{\sin A}{7} = \dfrac{\sin 41°}{12} \Rightarrow \sin A = \dfrac{7 \sin 41°}{12} \Rightarrow A = \sin^{-1}\left(\dfrac{7 \sin 41°}{12}\right) \approx 22.5°$.

Since $180° - 22.5° = 157.5°$ and $41° + 157.5° > 180°$, there is one triangle.

$B = 180° - 41° - 22.5° = 116.5°$. $\dfrac{b}{\sin 116.5°} = \dfrac{12}{\sin 41°} \Rightarrow b = \dfrac{12 \sin 116.5°}{\sin 41°} \approx 16.4$.

The solutions are $A = 22.5°, B = 116.5°, b = 16.4$.

4. (a) Use the Law of Cosines to find b:

$$b^2 = 8.1^2 + 8.3^2 - 2(8.1)(8.3)\cos 51° \Rightarrow b = \sqrt{134.5 - 134.46 \cos 51°} \approx 7.063$$

Use the Law of Sines: $\dfrac{\sin A}{8.1} = \dfrac{\sin 51°}{7.063} \Rightarrow \sin A = \dfrac{8.1 \sin 51°}{7.063} \Rightarrow A = \sin^{-1}\left(\dfrac{8.1 \sin 51°}{7.063}\right) \approx 63.0°$.

Then $C = 180° - 51° - 63.0° = 66.0°$. The solutions are $b \approx 7.1, A \approx 63.0°, C \approx 66.0°$.

(b) Use the Law of Cosines to find A: $14^2 = 8^2 + 9^2 - 2(8)(9)\cos A \Rightarrow A = \cos^{-1}\left(\dfrac{196 - 145}{-144}\right) \approx 110.7°$.

Use the Law of Sines to find B:

$$\frac{\sin B}{9} = \frac{\sin 110.7°}{14} \Rightarrow \sin B = \frac{9 \sin 110.7°}{14} \Rightarrow B = \sin^{-1}\left(\frac{9 \sin 110.7°}{14}\right) \approx 37.0°.$$

The value of C is $180° - 110.7° - 37.0° = 32.3°$. The solutions are $A \approx 110.7°, B \approx 37.0°, C \approx 32.3°$.

5. $A = \dfrac{1}{2}(4.5)(5.2)\sin 55° \approx 9.6$

6. $s = \dfrac{1}{2}(6 + 7 + 9) = 11; A = \sqrt{11(11 - 6)(11 - 7)(11 - 9)} \approx 21$

7. (a) $2\mathbf{v} + \mathbf{u} = 2(2\mathbf{i} - \mathbf{j}) + (-3\mathbf{i} + 2\mathbf{j}) = 4\mathbf{i} - 2\mathbf{j} + (-3\mathbf{i} + 2\mathbf{j}) = (4\mathbf{i} - 3\mathbf{i}) + (-2\mathbf{j} + 2\mathbf{j}) = \mathbf{i}$

(b) $2\mathbf{v} = 2(2\mathbf{i} - \mathbf{j}) = 4\mathbf{i} - 2\mathbf{j}$

(c) $\mathbf{v} - 3\mathbf{u} = (2\mathbf{i} - \mathbf{j}) - 3(-3\mathbf{i} + 2\mathbf{j}) = 2\mathbf{i} - \mathbf{j} + 9\mathbf{i} - 6\mathbf{j} = (2\mathbf{i} + 9\mathbf{i}) + (-\mathbf{j} - 6\mathbf{j}) = 11\mathbf{i} - 7\mathbf{j}$

8. $\langle 3, -2 \rangle \cdot \langle -1, 3 \rangle = (3)(-1) + (-2)(3) = -3 - 6 = -9$

$$\cos \theta = \frac{\mathbf{a} \cdot \mathbf{b}}{|\mathbf{a}| \, |\mathbf{b}|} = \frac{-9}{\sqrt{9 + 4}\sqrt{1 + 9}} = \frac{-9}{\sqrt{13}\sqrt{10}} = \frac{-9}{\sqrt{130}} \Rightarrow \theta = \cos^{-1}\left(-\frac{9}{\sqrt{130}}\right) \approx 142.1°$$

9. The adjacent angle in the parallelogram is supplementary to $52°$, thus it has measure $128°$.

    $|\mathbf{v}|^2 = 100^2 + 130^2 - 2(100)(130)\cos 128° \approx 42{,}907.19836 \Rightarrow |\mathbf{v}| = \sqrt{42{,}907.19836} \approx 207 \text{ lb}$

10. Angle B is supplementary to $52°$ and has measure $180° - 57° = 123°$. The other angle of the triangle, C, has

    measure $180° - 52° - 123° = 5°$. Using the Law of Sines:

    $\dfrac{a}{\sin 52°} = \dfrac{950}{\sin 5°} \Rightarrow a = \dfrac{950 \sin 52°}{\sin 5°} \Rightarrow a \approx 8589.$ If a vertical line from the plane to the ground intersects

    the ground at D, then triangle CBD is a right triangle. Thus, $\sin 57° = \dfrac{h}{8589} \Rightarrow h = 8589 \sin 57° \approx 7200 \text{ ft.}$

10.4: Trigonometric (Polar) Form of Complex Numbers

1. The modulus of a complex number represents the magnitude (length) of the vector representing it in the complex plane.

2. The geometric interpretation is the angle formed by the vector and the positive horizontal axis.

3. See Figure 3.

4. See Figure 4.

5. See Figure 5.

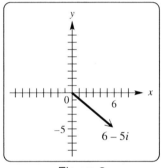
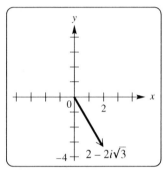
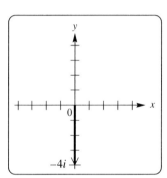

Figure 3   Figure 4   Figure 5

6. See Figure 6.

7. See Figure 7.

8. See Figure 8.

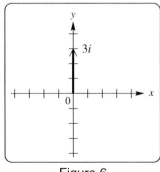
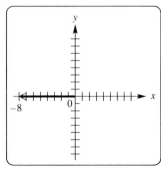
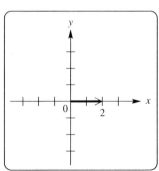

Figure 6   Figure 7   Figure 8

9. Since the vector has terminal point $(1, -4)$, the rectangular form is $1 - 4i$.

10. Since the vector has terminal point $(-4, -1)$, the rectangular form is $-4 - i$.

11. The imaginary part must be 0.

12. The vector will lie on the x-axis.

13. $a = 0$

14. No, $a + bi$ represents a complex number and $a\mathbf{i} + b\mathbf{j}$ represents a vector.

15. $(4 - 3i) + (-1 + 2i) = (4 - 1) + (-3 + 2)i = 3 - i$

16. $(2 + 3i) + (-4 - i) = (2 - 4) + (3 - 1)i = -2 + 2i$

17. $(-3 + 0i) + (0 + 3i) = (-3 + 0) + (0 + 3)i = -3 + 3i$

18. $(6 + 0i) + (0 - 2i) = (6 + 0) + (0 - 2)i = 6 - 2i$

19. $(2 + 6i) + (0 - 2i) = (2 + 0) + (6 - 2)i = 2 + 4i$

20. $(-5 - 8i) + (-1 + 0i) = (-5 - 1) + (-8 + 0)i = -6 - 8i$

21. $r = \sqrt{1^2 + 1^2} = \sqrt{2}$

22. $r = \sqrt{3^2 + (-4)^2} = \sqrt{25} = 5$

23. $r = \sqrt{12^2 + (-5)^2} = \sqrt{169} = 13$

24. $r = \sqrt{(-24)^2 + 7^2} = \sqrt{625} = 25$

25. $r = \sqrt{(-6)^2 + 0^2} = \sqrt{36} = 6$

26. $r = \sqrt{0^2 + 15^2} = \sqrt{225} = 15$

27. $r = \sqrt{2^2 + (-3)^2} = \sqrt{13}$

28. $r = \sqrt{11^2 + 60^2} = \sqrt{3721} = 61$

29. $2(\cos 45° + i\sin 45°) = 2\left(\dfrac{\sqrt{2}}{2} + i\dfrac{\sqrt{2}}{2}\right) = \sqrt{2} + i\sqrt{2}$

30. $4(\cos 60° + i\sin 60°) = 4\left(\dfrac{1}{2} + i\dfrac{\sqrt{3}}{2}\right) = 2 + 2i\sqrt{3}$

31. $10 \operatorname{cis} 90° = 10(\cos 90° + i\sin 90°) = 10(0 + 1i) = 10i$

32. $8 \operatorname{cis} 270° = 8(\cos 270° + i\sin 270°) = 8(0 - 1i) = -8i$

33. $4(\cos 240° + i\sin 240°) = 4\left(-\dfrac{1}{2} - i\dfrac{\sqrt{3}}{2}\right) = -2 - 2i\sqrt{3}$

34. $2(\cos 330° + i\sin 330°) = 2\left(\dfrac{\sqrt{3}}{2} - i\dfrac{1}{2}\right) = \sqrt{3} - i$

35. $\left(\cos \dfrac{\pi}{6} + i\sin \dfrac{\pi}{6}\right) = \left(\dfrac{\sqrt{3}}{2} + i\dfrac{1}{2}\right) = \dfrac{\sqrt{3}}{2} + \dfrac{1}{2}i$

36. $3\left(\cos \dfrac{5\pi}{6} + i\sin \dfrac{5\pi}{6}\right) = 3\left(-\dfrac{\sqrt{3}}{2} + i\dfrac{1}{2}\right) = -\dfrac{3\sqrt{3}}{2} + \dfrac{3}{2}i$

37. $5 \operatorname{cis}\left(-\dfrac{\pi}{6}\right) = 5\left[\cos\left(-\dfrac{\pi}{6}\right) + i\sin\left(-\dfrac{\pi}{6}\right)\right] = 5\left(\dfrac{\sqrt{3}}{2} - i\dfrac{1}{2}\right) = \dfrac{5\sqrt{3}}{2} - \dfrac{5}{2}i$

38. $6 \operatorname{cis}\dfrac{3\pi}{4} = 6\left(\cos \dfrac{3\pi}{4} + i\sin \dfrac{3\pi}{4}\right) = 6\left(-\dfrac{\sqrt{2}}{2} + i\dfrac{\sqrt{2}}{2}\right) = -3\sqrt{2} + 3i\sqrt{2}$

39. $\sqrt{2} \operatorname{cis} \pi = \sqrt{2}(\cos \pi + i\sin \pi) = \sqrt{2}(-1 - 0i) = -\sqrt{2}$

40. $\sqrt{3} \operatorname{cis}\dfrac{3\pi}{2} = \sqrt{3}\left(\cos \dfrac{3\pi}{2} + i\sin \dfrac{3\pi}{2}\right) = \sqrt{3}(0 - 1i) = -i\sqrt{3}$

41. $r = \sqrt{3^2 + (-3)^2} = \sqrt{18} = 3\sqrt{2}$; $\tan\theta = \dfrac{y}{x} \Rightarrow \tan\theta = \dfrac{-3}{3} \Rightarrow \tan\theta = -1 \Rightarrow \theta = \tan^{-1}(-1) = -45°$

    Since θ is in quadrant IV, $\theta = -45°$. Therefore $3 - 3i = 3\sqrt{2}[\cos(-45°) + i\sin(-45°)]$.

42. $r = \sqrt{(-2)^2 + (2\sqrt{3})^2} = \sqrt{16} = 4$;

    $\tan\theta = \dfrac{y}{x} \Rightarrow \tan\theta = \dfrac{2\sqrt{3}}{-2} \Rightarrow \tan\theta = -\sqrt{3} \Rightarrow \theta = \tan^{-1}(-\sqrt{3}) = -60°$

    Since θ is in quadrant II, $\theta = -60° + 180° = 120°$. Therefore $-2 + 2i\sqrt{3} = 4(\cos 120° + i\sin 120°)$.

43. $r = \sqrt{1^2 + (\sqrt{3})^2} = \sqrt{4} = 2$; $\tan\theta = \dfrac{y}{x} \Rightarrow \tan\theta = \dfrac{\sqrt{3}}{1} \Rightarrow \tan\theta = \sqrt{3} \Rightarrow \theta = \tan^{-1}(\sqrt{3}) = 60°$

    Since θ is in quadrant I, $\theta = 60°$. Therefore $1 + i\sqrt{3} = 2(\cos 60° + i\sin 60°)$.

44. $r = \sqrt{(-3)^2 + (-3\sqrt{3})^2} = \sqrt{36} = 6$;

    $\tan\theta = \dfrac{y}{x} \Rightarrow \tan\theta = \dfrac{-3\sqrt{3}}{-3} \Rightarrow \tan\theta = \sqrt{3} \Rightarrow \theta = \tan^{-1}(\sqrt{3}) = 60°$

    Since θ is in quadrant III, $\theta = 60° - 180° = -120°$. Therefore $-3 - 3i\sqrt{3} = 6[\cos(-120°) + i\sin(-120°)]$.

45. $r = \sqrt{0^2 + (-2)^2} = \sqrt{4} = 2$; $\tan\theta = \dfrac{y}{x} \Rightarrow \tan\theta = \dfrac{-2}{0} \Rightarrow \tan\theta$ is undefined $\Rightarrow \theta = 90°$

    Since θ is on the negative y-axis, $\theta = 90° - 180° = -90°$. Therefore $-2i = 2[\cos(-90°) + i\sin(-90°)]$.

46. $r = \sqrt{7^2 + 0^2} = \sqrt{49} = 7$; $\tan\theta = \dfrac{y}{x} \Rightarrow \tan\theta = \dfrac{0}{7} \Rightarrow \tan\theta = 0 \Rightarrow \theta = \tan^{-1}(0) = 0°$

    Since θ is on the positive x-axis, $\theta = 0°$. Therefore $7 = 7(\cos 0° + i\sin 0°)$.

47. $r = \sqrt{(4\sqrt{3})^2 + 4^2} = \sqrt{64} = 8$; $\tan\theta = \dfrac{y}{x} \Rightarrow \tan\theta = \dfrac{4}{4\sqrt{3}} \Rightarrow \tan\theta = \dfrac{1}{\sqrt{3}} \Rightarrow \theta = \tan^{-1}\left(\dfrac{1}{\sqrt{3}}\right) = \dfrac{\pi}{6}$

    Since θ is in quadrant I, $\theta = \dfrac{\pi}{6}$. Therefore $4\sqrt{3} + 4i = 8\left(\cos\dfrac{\pi}{6} + i\sin\dfrac{\pi}{6}\right)$.

48. $r = \sqrt{(\sqrt{3})^2 + (-1)^2} = \sqrt{4} = 2$;

    $\tan\theta = \dfrac{y}{x} \Rightarrow \tan\theta = \dfrac{-1}{\sqrt{3}} \Rightarrow \tan\theta = -\dfrac{1}{\sqrt{3}} \Rightarrow \theta = \tan^{-1}\left(-\dfrac{1}{\sqrt{3}}\right) = -\dfrac{\pi}{6}$

    Since θ is in quadrant IV, $\theta = -\dfrac{\pi}{6}$. Therefore $\sqrt{3} - i = 2\left[\cos\left(-\dfrac{\pi}{6}\right) + i\sin\left(-\dfrac{\pi}{6}\right)\right]$.

49. $r = \sqrt{(-\sqrt{2})^2 + (\sqrt{2})^2} = \sqrt{4} = 2$; $\tan\theta = \dfrac{y}{x} \Rightarrow \tan\theta = \dfrac{\sqrt{2}}{-\sqrt{2}} \Rightarrow \tan\theta = -1 \Rightarrow \theta = \tan^{-1}(-1) = -\dfrac{\pi}{4}$

    Since θ is in quadrant II, $\theta = -\dfrac{\pi}{4} + \pi = \dfrac{3\pi}{4}$. Therefore $-\sqrt{2} + i\sqrt{2} = 2\left(\cos\dfrac{3\pi}{4} + i\sin\dfrac{3\pi}{4}\right)$.

50. $r = \sqrt{(-5)^2 + (-5)^2} = \sqrt{50} = 5\sqrt{2}$; $\tan\theta = \dfrac{y}{x} \Rightarrow \tan\theta = \dfrac{-5}{-5} \Rightarrow \tan\theta = 1 \Rightarrow \theta = \tan^{-1}(1) = \dfrac{\pi}{4}$

    Since θ is in quadrant III, $\theta = \dfrac{\pi}{4} - \pi = -\dfrac{3\pi}{4}$. Therefore $-5 - 5i = 5\sqrt{2}\left[\cos\left(-\dfrac{3\pi}{4}\right) + i\sin\left(-\dfrac{3\pi}{4}\right)\right]$.

51. $r = \sqrt{(-4)^2 + 0^2} = \sqrt{16} = 4$; $\tan\theta = \dfrac{y}{x} \Rightarrow \tan\theta = \dfrac{0}{-4} \Rightarrow \tan\theta = 0 \Rightarrow \theta = \tan^{-1}(0) = 0$

    Since θ is on the negative x-axis, $\theta = 0 + \pi = \pi$. Therefore $-4 = 4(\cos\pi + i\sin\pi)$.

52. $r = \sqrt{0^2 + 5^2} = \sqrt{25} = 5$; $\tan\theta = \dfrac{y}{x} \Rightarrow \tan\theta = \dfrac{5}{0} \Rightarrow \tan\theta$ is undefined $\Rightarrow \theta = \dfrac{\pi}{2}$

    Since θ is on the positive y-axis, $\theta = \dfrac{\pi}{2}$. Therefore $5i = 5\left(\cos\dfrac{\pi}{2} + i\sin\dfrac{\pi}{2}\right)$.

53. $[3(\cos 60° + i \sin 60°)][2(\cos 90° + i \sin 90°)] = (3 \cdot 2)[\cos (60° + 90°) + i \sin (60° + 90°)] =$

$6(\cos 150° + i \sin 150°) = 6\left(-\dfrac{\sqrt{3}}{2} + \dfrac{1}{2}i\right) = -3\sqrt{3} + 3i$

54. $[4(\cos 30° + i \sin 30°)][5(\cos 120° + i \sin 120°)] = (4 \cdot 5)[\cos (30° + 120°) + i \sin (30° + 120°)] =$

$20(\cos 150° + i \sin 150°) = 20\left(-\dfrac{\sqrt{3}}{2} + \dfrac{1}{2}i\right) = -10\sqrt{3} + 10i$

55. $[2(\cos 45° + i \sin 45°)][2(\cos 225° + i \sin 225°)] = (2 \cdot 2)[\cos (45° + 225°) + i \sin (45° + 225°)] =$

$4(\cos 270° + i \sin 270°) = 4(0 - 1i) = -4i$

56. $[8(\cos 300° + i \sin 300°)][5(\cos 120° + i \sin 120°)] = (8 \cdot 5)[\cos (300° + 120°) + i \sin (300° + 120°)] =$

$40(\cos 420° + i \sin 420°) = 40\left(\dfrac{1}{2} + \dfrac{\sqrt{3}}{2}i\right) = 20 + 20i\sqrt{3}$

57. $\left[5 \operatorname{cis} \dfrac{\pi}{2}\right]\left[3 \operatorname{cis} \dfrac{\pi}{4}\right] = \left[5\left(\cos \dfrac{\pi}{2} + i \sin \dfrac{\pi}{2}\right)\right]\left[3\left(\cos \dfrac{\pi}{4} + i \sin \dfrac{\pi}{4}\right)\right] =$

$(5 \cdot 3)\left[\cos\left(\dfrac{\pi}{2} + \dfrac{\pi}{4}\right) + i \sin\left(\dfrac{\pi}{2} + \dfrac{\pi}{4}\right)\right] =$

$15\left(\cos \dfrac{3\pi}{4} + i \sin \dfrac{3\pi}{4}\right) = 15\left(-\dfrac{\sqrt{2}}{2} + \dfrac{\sqrt{2}}{2}i\right) = -\dfrac{15\sqrt{2}}{2} + \dfrac{15\sqrt{2}}{2}i$

58. $\left[6 \operatorname{cis} \dfrac{2\pi}{3}\right]\left[5 \operatorname{cis}\left(-\dfrac{\pi}{6}\right)\right] = \left[6\left(\cos \dfrac{2\pi}{3} + i \sin \dfrac{2\pi}{3}\right)\right]\left[5\left(\cos\left(-\dfrac{\pi}{6}\right) + i \sin\left(-\dfrac{\pi}{6}\right)\right)\right] =$

$(6 \cdot 5)\left[\cos\left(\dfrac{2\pi}{3} - \dfrac{\pi}{6}\right) + i \sin\left(\dfrac{2\pi}{3} - \dfrac{\pi}{6}\right)\right] = 30\left(\cos \dfrac{\pi}{2} + i \sin \dfrac{\pi}{2}\right) = 30(0 + 1i) = 30i$

59. $\left[\sqrt{3} \operatorname{cis} \dfrac{\pi}{4}\right]\left[\sqrt{3} \operatorname{cis} \dfrac{5\pi}{4}\right] = \left[\sqrt{3}\left(\cos \dfrac{\pi}{4} + i \sin \dfrac{\pi}{4}\right)\right]\left[\sqrt{3}\left(\cos \dfrac{5\pi}{4} + i \sin \dfrac{5\pi}{4}\right)\right] =$

$(\sqrt{3} \cdot \sqrt{3})\left[\cos\left(\dfrac{\pi}{4} + \dfrac{5\pi}{4}\right) + i \sin\left(\dfrac{\pi}{4} + \dfrac{5\pi}{4}\right)\right] = 3\left(\cos \dfrac{3\pi}{2} + i \sin \dfrac{3\pi}{2}\right) = 3(0 - 1i) = -3i$

60. $\left[\sqrt{2} \operatorname{cis} \dfrac{5\pi}{6}\right]\left[\sqrt{2} \operatorname{cis} \dfrac{3\pi}{2}\right] = \left[\sqrt{2}\left(\cos \dfrac{5\pi}{6} + i \sin \dfrac{5\pi}{6}\right)\right]\left[\sqrt{2}\left(\cos \dfrac{3\pi}{2} + i \sin \dfrac{3\pi}{2}\right)\right] =$

$(\sqrt{2} \cdot \sqrt{2})\left[\cos\left(\dfrac{5\pi}{6} + \dfrac{3\pi}{2}\right) + i \sin\left(\dfrac{5\pi}{6} + \dfrac{3\pi}{2}\right)\right] = 2\left(\cos \dfrac{7\pi}{3} + i \sin \dfrac{7\pi}{3}\right) = 2\left(\dfrac{1}{2} + \dfrac{\sqrt{3}}{2}i\right) = 1 + i\sqrt{3}$

61. $\dfrac{4(\cos 120° + i \sin 120°)}{2(\cos 150° + i \sin 150°)} = \dfrac{4}{2}[\cos (120° - 150°) + i \sin (120° - 150°)] =$

$2[\cos (-30°) + i \sin (-30°)] = 2\left(\dfrac{\sqrt{3}}{2} - \dfrac{1}{2}i\right) = \sqrt{3} - i$

62. $\dfrac{16(\cos 300° + i \sin 300°)}{8(\cos 60° + i \sin 60°)} = \dfrac{16}{8}[\cos (300° - 60°) + i \sin (300° - 60°)] =$

$2(\cos 240° + i \sin 240°) = 2\left(-\dfrac{1}{2} - \dfrac{\sqrt{3}}{2}i\right) = -1 - i\sqrt{3}$

63. $\dfrac{10(\cos \frac{5\pi}{4} + i \sin \frac{5\pi}{4})}{5(\cos \frac{\pi}{4} + i \sin \frac{\pi}{4})} = \dfrac{10}{5}[\cos (\frac{5\pi}{4} - \frac{\pi}{4}) + i \sin (\frac{5\pi}{4} - \frac{\pi}{4})] = 2(\cos \pi + i \sin \pi) = 2(-1 - 0i) = -2$

64. $\dfrac{24(\cos \frac{5\pi}{6} + i \sin \frac{5\pi}{6})}{2(\cos \frac{\pi}{6} + i \sin \frac{\pi}{6})} = \dfrac{24}{2}[\cos (\frac{5\pi}{6} - \frac{\pi}{6}) + i \sin (\frac{5\pi}{6} - \frac{\pi}{6})] =$

$12(\cos \frac{2\pi}{3} + i \sin \frac{2\pi}{3}) = 12\left(-\dfrac{1}{2} + \dfrac{\sqrt{3}}{2}i\right) = -6 + 6i\sqrt{3}$

65. $\dfrac{3 \operatorname{cis} \frac{61\pi}{36}}{9 \operatorname{cis} \frac{13\pi}{36}} = \dfrac{3\left(\cos \frac{61\pi}{36} + i\sin \frac{61\pi}{36}\right)}{9\left(\cos \frac{13\pi}{36} + i\sin \frac{13\pi}{36}\right)} = \dfrac{3}{9}\left[\cos\left(\frac{61\pi}{36} - \frac{13\pi}{36}\right) + i\sin\left(\frac{61\pi}{36} - \frac{13\pi}{36}\right)\right] =$

$\dfrac{1}{3}\left(\cos \frac{4\pi}{3} + i\sin \frac{4\pi}{3}\right) = \dfrac{1}{3}\left(-\dfrac{1}{2} - \dfrac{\sqrt{3}}{2}i\right) = -\dfrac{1}{6} - \dfrac{\sqrt{3}}{6}i$

66. $\dfrac{12 \operatorname{cis} 293°}{6 \operatorname{cis} 23°} = \dfrac{12(\cos 293° + i\sin 293°)}{6(\cos 23° + i\sin 23°)} = \dfrac{12}{6}\left[\cos(293° - 23°) + i\sin(293° - 23°)\right] =$

$2(\cos 270° + i\sin 270°) = 2(0 - 1i) = -2i$

67. First find the trigonometric form of $Z = 6 + 3i$: $r = \sqrt{6^2 + 3^2} = \sqrt{45} \approx 6.71$;

$\tan\theta = \dfrac{3}{6} \Rightarrow \tan\theta = \dfrac{1}{2} \Rightarrow \theta = \tan^{-1}\dfrac{1}{2} \Rightarrow \theta \approx 26.6°.$ Therefore $Z \approx 6.71(\cos 26.6° + i\sin 26.6°).$

$I = \dfrac{E}{Z} = \dfrac{8(\cos 20° + i\sin 20°)}{6.71(\cos 26.57° + i\sin 26.57°)} = \dfrac{8}{6.71}\left[\cos(20° - 26.57°) + i\sin(20° - 26.57°)\right] \approx 1.18 - .14i$

68. First find the trigonometric form of $R + (X_L - X_c)i = 3 - 2i$: $r = \sqrt{3^2 + (-2)^2} = \sqrt{13} \approx 3.61$;

$\tan\theta = \dfrac{-2}{3} \Rightarrow \Rightarrow \theta = \tan^{-1}\left(-\dfrac{2}{3}\right) \Rightarrow \theta \approx -33.7°.$ The value is $3.61[\cos(-33.7°) + i\sin(-33.7°)].$

$I = \dfrac{12(\cos 25° + i\sin 25°)}{3.61[\cos(-33.7°) + i\sin(-33.7°)]} = \dfrac{12}{3.61}\left[\cos(25° + 33.7°) + i\sin(20° + 33.7°)\right] \approx 1.7 + 2.8i$

69. Noting the $Z = \dfrac{1}{\dfrac{1}{Z_1} + \dfrac{1}{Z_2}} = \dfrac{1}{\dfrac{Z_2 + Z_1}{Z_1 Z_2}} = \dfrac{Z_1 Z_2}{Z_1 + Z_2}$, we find the following.

First find the trigonometric form of $Z_1 = 50 + 25i$: $r = \sqrt{50^2 + 25^2} = \sqrt{3125}$;

$\tan\theta = \dfrac{25}{50} \Rightarrow \Rightarrow \theta = \tan^{-1}\dfrac{1}{2} \Rightarrow \theta \approx 26.565°.$ The value is $\sqrt{3125}(\cos 26.565° + i\sin 26.565°).$

Next find the trigonometric form of $Z_2 = 60 + 20i$: $r = \sqrt{60^2 + 20^2} = \sqrt{4000}$;

$\tan\theta = \dfrac{20}{60} \Rightarrow \Rightarrow \theta = \tan^{-1}\dfrac{1}{3} \Rightarrow \theta \approx 18.435°.$ The value is $\sqrt{4000}(\cos 18.435° + i\sin 18.435°).$

Finally find the trigonometric form of $Z_1 + Z_2 = 110 + 45i$: $r = \sqrt{110^2 + 45^2} = \sqrt{14{,}125}$;

$\tan\theta = \dfrac{45}{110} \Rightarrow \Rightarrow \theta = \tan^{-1}\dfrac{9}{22} \Rightarrow \theta \approx 22.249°.$ The value is $\sqrt{14{,}125}(\cos 22.249° + i\sin 22.249°).$

$Z = \dfrac{[\sqrt{3125}(\cos 26.565° + i\sin 26.565°)][\sqrt{4000}(\cos 18.435° + i\sin 18.435°)]}{\sqrt{14{,}125}(\cos 22.249° + i\sin 22.249°)} =$

$\dfrac{\sqrt{3125} \cdot \sqrt{4000}}{\sqrt{14{,}125}}\left[(\cos(26.565° + 18.435° - 22.249°) + i\sin(26.565° + 18.435° - 22.249°)\right] \approx$

$27.43 + 11.5i$

70. $z = -.2i \Rightarrow z^2 - 1 = .04i^2 - 1 = -.04 - 1 = -1.04 \Rightarrow (z^2 - 1)^2 - 1 = (-1.04)^2 - 1 = .0816$

Continuing in this way, the absolute values of the moduli will never exceed 2. This number is in the Julia set.

71. (a) $|a + bi| = \sqrt{a^2 + b^2}$ and $|a - bi| = \sqrt{a^2 + (-b^2)} = \sqrt{a^2 + b^2}$

 (b) $z_1^2 - 1 = (a + bi)^2 - 1 = a^2 + 2abi - b^2 - 1 = a^2 - b^2 - 1 + 2abi$

 $z_2^2 - 1 = (a - bi)^2 - 1 = a^2 - 2abi - b^2 - 1 = a^2 - b^2 - 1 - 2abi$

 (c) If $z_1 = a + bi$ and $z_2 = a - bi$, then $z_1^2 - 1$ and $z_2^2 - 1$ are also conjugates with the same modulus.

 Therefore, if z_1 is in the Julia set, so is z_2. Thus (a, b) in the Julia set implies $(a, -b)$ is also in the set.

 (d) Yes, see part (e) below.

 (e) $z_1^2 - 1 = (a + bi)^2 - 1 = a^2 + 2abi - b^2 - 1 = a^2 - b^2 - 1 + 2abi$

 $z_2^2 - 1 = (-a + bi)^2 - 1 = a^2 - 2abi - b^2 - 1 = a^2 - b^2 - 1 - 2abi$

 If $z_1 = a + bi$ and $z_2 = -a + bi$, then $z_1^2 - 1$ and $z_2^2 - 1$ are also conjugates with the same modulus.

 Therefore, if z_1 is in the Julia set, so is z_2. Thus (a, b) in the Julia set implies $(-a, b)$ is also in the set.

72. The angles $45°$ and $-315°$ are coterminal angles so their trigonometric functions values are equal.

 The same is true for the angles $90°$ and $-270°$.

73. To square a complex number in trigonometric form, square the modulus r and double the argument θ.

74. If $b = d$, the imaginary part would equal 0 in the difference.

75. (a) If the real part equals 0, then $\cos \theta = 0 \Rightarrow \theta = \dfrac{\pi}{2}$.

 (b) If the imaginary part equals 0, then $\sin \theta = 0 \Rightarrow \theta = \pi$.

76. $r = \sqrt{3^2 + 5^2} = \sqrt{34} \approx 5.83$; $\tan \theta = \dfrac{y}{x} \Rightarrow \tan \theta = \dfrac{5}{3} \Rightarrow \Rightarrow \theta = \tan^{-1} \dfrac{5}{3} \approx 1.03$

 Therefore $3 + 5i \approx 5.83(\cos 1.03 + i \sin 1.03)$ or 5.83 cis 1.03.

10.5: Powers and Roots of Complex Numbers

1. $[3(\cos 30° + i \sin 30°)]^3 = 3^3[\cos(3 \cdot 30°) + i \sin(3 \cdot 30°)] = 27(\cos 90° + i \sin 90°) =$
 $27(0 + 1i) = 27i$

2. $[2(\cos 135° + i \sin 135°)]^4 = 2^4[\cos(4 \cdot 135°) + i \sin(4 \cdot 135°)] = 16(\cos 540° + i \sin 540°) =$
 $16(-1 + 0i) = -16$

3. $\left(\cos \dfrac{\pi}{4} + i \sin \dfrac{\pi}{4}\right)^8 = \cos\left(8 \cdot \dfrac{\pi}{4}\right) + i \sin\left(8 \cdot \dfrac{\pi}{4}\right) = \cos 2\pi + i \sin 2\pi = 1 + 0i = 1$

4. $\left(\cos \dfrac{\pi}{5} + i \sin \dfrac{\pi}{5}\right)^5 = \cos\left(5 \cdot \dfrac{\pi}{5}\right) + i \sin\left(5 \cdot \dfrac{\pi}{5}\right) = \cos \pi + i \sin \pi = -1 + 0i = -1$

5. $\left[2\left(\cos \dfrac{2\pi}{3} + i \sin \dfrac{2\pi}{3}\right)\right]^3 = 2^3\left[\cos\left(3 \cdot \dfrac{2\pi}{3}\right) + i \sin\left(3 \cdot \dfrac{2\pi}{3}\right)\right] = 8(\cos 2\pi + i \sin 2\pi) = 8(1 + 0i) = 8$

6. $\left[3\left(\cos \dfrac{3\pi}{4} + i \sin \dfrac{3\pi}{4}\right)\right]^4 = 3^4\left[\cos\left(4 \cdot \dfrac{3\pi}{4}\right) + i \sin\left(4 \cdot \dfrac{3\pi}{4}\right)\right] = 81(\cos 3\pi + i \sin 3\pi) = 8(-1 + 0i) = -81$

7. $[3 \text{ cis } 100°]^3 = [3(\cos 100° + i \sin 100°)]^3 = 3^3[\cos(3 \cdot 100°) + i \sin(3 \cdot 100°)] =$
 $27(\cos 300° + i \sin 300°) = 27\left(\dfrac{1}{2} - \dfrac{\sqrt{3}}{2}i\right) = \dfrac{27}{2} - \dfrac{27\sqrt{3}}{2}i$

8. $[3 \operatorname{cis} 40°]^3 = [3(\cos 40° + i\sin 40°)]^3 = 3^3[\cos (3 \cdot 40°) + i\sin (3 \cdot 40°)] =$

$27(\cos 120° + i\sin 120°) = 27\left(-\dfrac{1}{2} + \dfrac{\sqrt{3}}{2}i\right) = -\dfrac{27}{2} + \dfrac{27\sqrt{3}}{2}i$

9. First find the trigonometric form of $\sqrt{3} + i$; $r = \sqrt{(\sqrt{3})^2 + 1^2} = \sqrt{4} = 2$;

$\tan \theta = \dfrac{y}{x} \Rightarrow \tan \theta = \dfrac{1}{\sqrt{3}} \Rightarrow \Rightarrow \theta = \tan^{-1} \dfrac{1}{\sqrt{3}} = 30°$. Therefore $\sqrt{3} + i = 2(\cos 30° + i\sin 30°)$.

$(\sqrt{3} + i)^3 = [2(\cos 30° + i\sin 30°)]^3 = 2^3[\cos (3 \cdot 30°) + i\sin (3 \cdot 30°)] =$

$8(\cos 90° + i\sin 90°) = 8(0 + 1i) = 8i$

10. First find the trigonometric form of $1 + i\sqrt{3}$; $r = \sqrt{1^2 + (\sqrt{3})^2} = \sqrt{4} = 2$;

$\tan \theta = \dfrac{y}{x} \Rightarrow \tan \theta = \dfrac{\sqrt{3}}{1} \Rightarrow \Rightarrow \theta = \tan^{-1} \sqrt{3} = 60°$. Therefore $1 + i\sqrt{3} = 2(\cos 60° + i\sin 60°)$.

$(1 + i\sqrt{3})^4 = [2(\cos 60° + i\sin 60°)]^4 = 2^4[\cos (4 \cdot 60°) + i\sin (4 \cdot 60°)] =$

$16(\cos 240° + i\sin 240°) = 16\left(-\dfrac{1}{2} - \dfrac{\sqrt{3}}{2}i\right) = -8 - 8i\sqrt{3}$

11. First find the trigonometric form of $2\sqrt{2} - 2i\sqrt{2}$; $r = \sqrt{(2\sqrt{2})^2 + (-2\sqrt{2})^2} = \sqrt{16} = 4$;

$\tan \theta = \dfrac{y}{x} \Rightarrow \tan \theta = \dfrac{-2\sqrt{2}}{2\sqrt{2}} \Rightarrow \Rightarrow \theta = \tan^{-1} (-1) = -45°$.

Therefore $2\sqrt{2} - 2i\sqrt{2} = 4[\cos (-45°) + i\sin (-45°)]$.

$(2\sqrt{2} - 2i\sqrt{2})^6 = [4(\cos (-45°) + i\sin (-45°))]^6 = 4^6[\cos (6 \cdot (-45°)) + i\sin (6 \cdot (-45°))] =$

$4096[\cos (-270°) + i\sin (-270°)] = 4096(0 + 1i) = 4096i$

12. First find the trigonometric form of $2 - 2i\sqrt{3}$; $r = \sqrt{2^2 + (-2\sqrt{3})^2} = \sqrt{16} = 4$;

$\tan \theta = \dfrac{y}{x} \Rightarrow \tan \theta = \dfrac{-2\sqrt{3}}{2} \Rightarrow \Rightarrow \theta = \tan^{-1} (-\sqrt{3}) = -60°$.

Therefore $2 - 2i\sqrt{3} = 4[\cos (-60°) + i\sin (-60°)]$.

$(2 - 2i\sqrt{3})^4 = [4(\cos (-60°) + i\sin (-60°))]^4 = 4^4[\cos (4 \cdot (-60°)) + i\sin (4 \cdot (-60°))] =$

$256[\cos (-240°) + i\sin (-240°)] = 256\left(-\dfrac{1}{2} + \dfrac{\sqrt{3}}{2}i\right) = -128 + 128i\sqrt{3}$

13. First find the trigonometric form of $-\dfrac{\sqrt{2}}{2} + \dfrac{\sqrt{2}}{2}i$; $r = \sqrt{\left(-\dfrac{\sqrt{2}}{2}\right)^2 + \left(\dfrac{\sqrt{2}}{2}\right)^2} = \sqrt{1} = 1$;

$\tan \theta = \dfrac{y}{x} \Rightarrow \tan \theta = \left(\dfrac{\sqrt{2}}{2} \div \left(-\dfrac{\sqrt{2}}{2}\right)\right) \Rightarrow \tan \theta = -1 \Rightarrow \theta = \tan^{-1} 1 = -45°$. Since θ is in quadrant II,

$\theta = -45° + 180° = 135°$. Therefore $-\dfrac{\sqrt{2}}{2} + \dfrac{\sqrt{2}}{2}i = \cos 135° + i\sin 135°$.

$\left(-\dfrac{\sqrt{2}}{2} + \dfrac{\sqrt{2}}{2}i\right)^4 = (\cos 135° + i\sin 135°)^4 = \cos (4 \cdot 135°) + i\sin (4 \cdot 135°) = \cos 540° + i\sin 540° = -1$

14. First find the trigonometric form of $\dfrac{\sqrt{2}}{2} - \dfrac{\sqrt{2}}{2}i$; $r = \sqrt{\left(\dfrac{\sqrt{2}}{2}\right)^2 + \left(-\dfrac{\sqrt{2}}{2}\right)^2} = \sqrt{1} = 1$;

$\tan \theta = \dfrac{y}{x} \Rightarrow \tan \theta = \left(-\dfrac{\sqrt{2}}{2} \div \dfrac{\sqrt{2}}{2}\right) \Rightarrow \tan \theta = -1 \Rightarrow \theta = \tan^{-1} (-1) = -45°$.

Therefore $\dfrac{\sqrt{2}}{2} - \dfrac{\sqrt{2}}{2}i = \cos (-45°) + i\sin (-45°)$. Now $\left(\dfrac{\sqrt{2}}{2} - \dfrac{\sqrt{2}}{2}i\right)^8 = [\cos (-45°) + i\sin (-45°)]^8 =$

$\cos [8 \cdot (-45°)] + i\sin [8 \cdot (-45°)] = \cos (-360°) + i\sin (-360°) = 1$

15. First find the trigonometric form of $1 - i$; $r = \sqrt{1^2 + (-1)^2} = \sqrt{2}$;

$\tan \theta = \dfrac{y}{x} \Rightarrow \tan \theta = \dfrac{-1}{1} \Rightarrow \Rightarrow \theta = \tan^{-1}(-1) = -45°$. Therefore $1 - i = \sqrt{2}[\cos(-45°) + i\sin(-45°)]$.

$(1 - i)^6 = [\sqrt{2}(\cos(-45°) + i\sin(-45°))]^6 = (\sqrt{2})^6[\cos(6 \cdot (-45°)) + i\sin(6 \cdot (-45°))] =$

$8[\cos(-270°) + i\sin(-270°)] = 8(0 + 1i) = 8i$

16. First find the trigonometric form of $-1 + i$; $r = \sqrt{(-1)^2 + 1^2} = \sqrt{2}$;

$\tan \theta = \dfrac{y}{x} \Rightarrow \tan \theta = \dfrac{1}{-1} \Rightarrow \Rightarrow \theta = \tan^{-1}(-1) = 135°$. Therefore $-1 + i = \sqrt{2}(\cos 135° + i\sin 135°)$.

$(-1 + i)^7 = [\sqrt{2}(\cos 135° + i\sin 135°)]^7 = (\sqrt{2})^7[\cos(7 \cdot 135°) + i\sin(7 \cdot 135°)] =$

$8\sqrt{2}(\cos 945° + i\sin 945°) = 8\sqrt{2}\left(-\dfrac{\sqrt{2}}{2} - \dfrac{\sqrt{2}}{2}i\right) = -8 - 8i$

17. First find the trigonometric form of $-2 - 2i$; $r = \sqrt{(-2)^2 + (-2)^2} = \sqrt{8} = 2\sqrt{2}$;

$\tan \theta = \dfrac{y}{x} \Rightarrow \tan \theta = \dfrac{-2}{-2} \Rightarrow \Rightarrow \theta = \tan^{-1}1 = 225°$. Therefore $-2 - 2i = 2\sqrt{2}(\cos 225° + i\sin 225°)$.

$(-2 - 2i)^5 = [2\sqrt{2}(\cos 225° + i\sin 225°)]^5 = (2\sqrt{2})^5[\cos(5 \cdot 225°) + i\sin(5 \cdot 225°)] =$

$128\sqrt{2}(\cos 1125° + i\sin 1125°) = 128\sqrt{2}\left(\dfrac{\sqrt{2}}{2} + \dfrac{\sqrt{2}}{2}i\right) = 128 + 128i$

18. First find the trigonometric form of $-3 - 3i$; $r = \sqrt{(-3)^2 + (-3)^2} = \sqrt{18} = 3\sqrt{2}$;

$\tan \theta = \dfrac{y}{x} \Rightarrow \tan \theta = \dfrac{-3}{-3} \Rightarrow \Rightarrow \theta = \tan^{-1}1 = 225°$. Therefore $-3 - 3i = 3\sqrt{2}(\cos 225° + i\sin 225°)$.

$(-3 - 3i)^3 = [3\sqrt{2}(\cos 225° + i\sin 225°)]^3 = (3\sqrt{2})^3[\cos(3 \cdot 225°) + i\sin(3 \cdot 225°)] =$

$54\sqrt{2}(\cos 675° + i\sin 675°) = 54\sqrt{2}\left(\dfrac{\sqrt{2}}{2} - \dfrac{\sqrt{2}}{2}i\right) = 54 - 54i$

19. Note that $1 = (\cos 0° + i\sin 0°)$. Here $r = 1$ so $\sqrt[3]{r} = 1$, $n = 3$ and $k = 0, 1,$ or 2.

$\alpha = \dfrac{0° + 360° \cdot k}{3}$: For $k = 0$, $\alpha = 0°$. For $k = 1$, $\alpha = 120°$. For $k = 2$, $\alpha = 240°$. The cube roots are

$\cos 0° + i\sin 0°$, $\cos 120° + i\sin 120°$, and $\cos 240° + i\sin 240°$. See Figure 19.

20. Note that $i = (\cos 90° + i\sin 90°)$. Here $r = 1$ so $\sqrt[3]{r} = 1$, $n = 3$ and $k = 0, 1,$ or 2.

$\alpha = \dfrac{90° + 360° \cdot k}{3}$: For $k = 0$, $\alpha = 30°$. For $k = 1$, $\alpha = 150°$. For $k = 2$, $\alpha = 270°$. The cube roots are

$\cos 30° + i\sin 30°$, $\cos 150° + i\sin 150°$, and $\cos 270° + i\sin 270°$. See Figure 20.

21. The complex number is $8(\cos 60° + i\sin 60°)$. Here $r = 8$ so $\sqrt[3]{r} = 2$, $n = 3$ and $k = 0, 1,$ or 2.

$\alpha = \dfrac{60° + 360° \cdot k}{3}$: For $k = 0$, $\alpha = 20°$. For $k = 1$, $\alpha = 140°$. For $k = 2$, $\alpha = 260°$. The cube roots are

$2(\cos 20° + i\sin 20°)$, $2(\cos 140° + i\sin 140°)$, and $2(\cos 260° + i\sin 260°)$. See Figure 21.

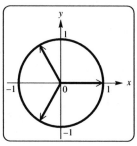

Figure 19

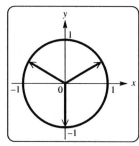

Figure 20

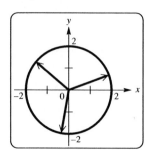

Figure 21

22. The complex number is $27(\cos 300° + i\sin 300°)$. Here $r = 27$ so $\sqrt[3]{r} = 3$, $n = 3$ and $k = 0, 1,$ or 2.

$\alpha = \dfrac{300° + 360° \cdot k}{3}$: For $k = 0$, $\alpha = 100°$. For $k = 1$, $\alpha = 220°$. For $k = 2$, $\alpha = 340°$. The cube roots are

$3(\cos 100° + i\sin 100°)$, $3(\cos 220° + i\sin 220°)$, and $3(\cos 340° + i\sin 340°)$. See Figure 22.

23. Note that $-8i = 8(\cos 270° + i\sin 270°)$. Here $r = 8$ so $\sqrt[3]{r} = 2$, $n = 3$ and $k = 0, 1,$ or 2.

$\alpha = \dfrac{270° + 360° \cdot k}{3}$: For $k = 0$, $\alpha = 90°$. For $k = 1$, $\alpha = 210°$. For $k = 2$, $\alpha = 330°$. The cube roots are

$2(\cos 90° + i\sin 90°)$, $2(\cos 210° + i\sin 210°)$, and $2(\cos 330° + i\sin 330°)$. See Figure 23.

24. Note that $27i = 27(\cos 90° + i\sin 90°)$. Here $r = 27$ so $\sqrt[3]{r} = 3$, $n = 3$ and $k = 0, 1,$ or 2.

$\alpha = \dfrac{90° + 360° \cdot k}{3}$: For $k = 0$, $\alpha = 30°$. For $k = 1$, $\alpha = 150°$. For $k = 2$, $\alpha = 270°$. The cube roots are

$3(\cos 30° + i\sin 30°)$, $3(\cos 150° + i\sin 150°)$, and $3(\cos 270° + i\sin 270°)$. See Figure 24.

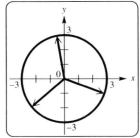
Figure 22

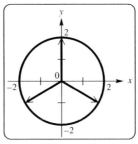
Figure 23

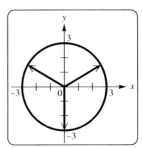
Figure 24

25. Note that $-64 = 64(\cos 180° + i\sin 180°)$. Here $r = 64$ so $\sqrt[3]{r} = 4$, $n = 3$ and $k = 0, 1,$ or 2.

$\alpha = \dfrac{180° + 360° \cdot k}{3}$: For $k = 0$, $\alpha = 60°$. For $k = 1$, $\alpha = 180°$. For $k = 2$, $\alpha = 300°$. The cube roots are

$4(\cos 60° + i\sin 60°)$, $4(\cos 180° + i\sin 180°)$, and $4(\cos 300° + i\sin 300°)$. See Figure 25.

26. Note that $27 = 27(\cos 0° + i\sin 0°)$. Here $r = 27$ so $\sqrt[3]{r} = 3$, $n = 3$ and $k = 0, 1,$ or 2.

$\alpha = \dfrac{0° + 360° \cdot k}{3}$: For $k = 0$, $\alpha = 0°$. For $k = 1$, $\alpha = 120°$. For $k = 2$, $\alpha = 240°$. The cube roots are

$3(\cos 0° + i\sin 0°)$, $3(\cos 120° + i\sin 120°)$, and $3(\cos 240° + i\sin 240°)$. See Figure 26.

27. Note that $1 + i\sqrt{3} = 2(\cos 60° + i\sin 60°)$. Here $r = 2$ so $\sqrt[3]{r} = \sqrt[3]{2}$, $n = 3$ and $k = 0, 1,$ or 2.

$\alpha = \dfrac{60° + 360° \cdot k}{3}$: For $k = 0$, $\alpha = 20°$. For $k = 1$, $\alpha = 140°$. For $k = 2$, $\alpha = 260°$. The cube roots are

$\sqrt[3]{2}(\cos 20° + i\sin 20°)$, $\sqrt[3]{2}(\cos 140° + i\sin 140°)$, and $\sqrt[3]{2}(\cos 260° + i\sin 260°)$. See Figure 27.

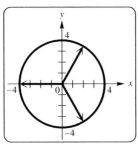
Figure 25

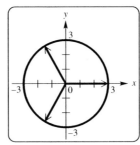
Figure 26

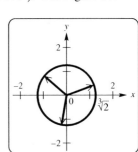
Figure 27

28. Note that $2 - 2i\sqrt{3} = 4(\cos 300° + i\sin 300°)$. Here $r = 4$ so $\sqrt[3]{r} = \sqrt[3]{4}$, $n = 3$ and $k = 0$, 1, or 2.

$\alpha = \dfrac{300° + 360° \cdot k}{3}$: For $k = 0$, $\alpha = 100°$. For $k = 1$, $\alpha = 220°$. For $k = 2$, $\alpha = 340°$. The cube roots are

$\sqrt[3]{4}(\cos 100° + i\sin 100°)$, $\sqrt[3]{4}(\cos 220° + i\sin 220°)$, and $\sqrt[3]{4}(\cos 340° + i\sin 340°)$. See Figure 28.

29. Note that $-2\sqrt{3} + 2i = 4(\cos 150° + i\sin 150°)$. Here $r = 4$ so $\sqrt[3]{r} = \sqrt[3]{4}$, $n = 3$ and $k = 0$, 1, or 2.

$\alpha = \dfrac{150° + 360° \cdot k}{3}$: For $k = 0$, $\alpha = 50°$. For $k = 1$, $\alpha = 170°$. For $k = 2$, $\alpha = 290°$. The cube roots are

$\sqrt[3]{4}(\cos 50° + i\sin 50°)$, $\sqrt[3]{4}(\cos 170° + i\sin 170°)$, and $\sqrt[3]{4}(\cos 290° + i\sin 290°)$. See Figure 29.

30. Note that $\sqrt{3} - i = 2(\cos 330° + i\sin 330°)$. Here $r = 2$ so $\sqrt[3]{r} = \sqrt[3]{2}$, $n = 3$ and $k = 0$, 1, or 2.

$\alpha = \dfrac{330° + 360° \cdot k}{3}$: For $k = 0$, $\alpha = 110°$. For $k = 1$, $\alpha = 230°$. For $k = 2$, $\alpha = 350°$. The cube roots are

$\sqrt[3]{2}(\cos 110° + i\sin 110°)$, $\sqrt[3]{2}(\cos 230° + i\sin 230°)$, and $\sqrt[3]{2}(\cos 350° + i\sin 350°)$. See Figure 30.

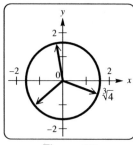

Figure 28

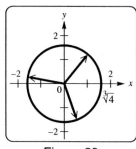

Figure 29

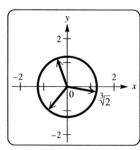

Figure 30

31. The complex number is $4(\cos 120° + i\sin 120°)$. Here $r = 4$ so $\sqrt{r} = 2$, $n = 2$ and $k = 0$ or 1.

$\alpha = \dfrac{120° + 360° \cdot k}{2}$: For $k = 0$, $\alpha = 60°$. For $k = 1$, $\alpha = 240°$. The square roots are

$2(\cos 60° + i\sin 60°) = 1 + i\sqrt{3}$ and $2(\cos 240° + i\sin 240°) = -1 - i\sqrt{3}$.

32. The complex number is $27(\cos 180° + i\sin 180°)$. Here $r = 27$ so $\sqrt[3]{r} = 3$, $n = 3$ and $k = 0$, 1, or 2.

$\alpha = \dfrac{180° + 360° \cdot k}{3}$: For $k = 0$, $\alpha = 60°$. For $k = 1$, $\alpha = 180°$. For $k = 2$, $\alpha = 300°$. The cube roots are

$3(\cos 60° + i\sin 60°) = \dfrac{3}{2} + \dfrac{3\sqrt{3}}{2}i$, $3(\cos 180° + i\sin 180°) = -3$, and

$3(\cos 300° + i\sin 300°) = \dfrac{3}{2} - \dfrac{3\sqrt{3}}{2}i$.

33. The complex number is $\cos 180° + i\sin 180°$. Here $r = 1$ so $\sqrt[3]{r} = 1$, $n = 3$ and $k = 0$, 1, or 2.

$\alpha = \dfrac{180° + 360° \cdot k}{3}$: For $k = 0$, $\alpha = 60°$. For $k = 1$, $\alpha = 180°$. For $k = 2$, $\alpha = 300°$. The cube roots are

$\cos 60° + i\sin 60° = \dfrac{1}{2} + \dfrac{\sqrt{3}}{2}i$, $\cos 180° + i\sin 180° = -1$, and $\cos 300° + i\sin 300° = \dfrac{1}{2} - \dfrac{\sqrt{3}}{2}i$.

34. The complex number is $16(\cos 240° + i \sin 240°)$. Here $r = 16$ so $\sqrt[4]{r} = 2$, $n = 4$ and $k = 0, 1, 2,$ or 3.

    $\alpha = \dfrac{240° + 360° \cdot k}{4}$: For $k = 0$, $\alpha = 60°$. For $k = 1$, $\alpha = 150°$. For $k = 2$, $\alpha = 240°$. For $k = 3$, $\alpha = 330°$.

    The fourth roots are $2(\cos 60° + i \sin 60°) = 1 + i\sqrt{3}$, $2(\cos 150° + i \sin 150°) = -\sqrt{3} + i$,

    $2(\cos 240° + i \sin 240°) = -1 - i\sqrt{3}$, and $2(\cos 330° + i \sin 330°) = \sqrt{3} - i$.

35. Note that $i = \cos 90° + i \sin 90°$. Here $r = 1$ so $\sqrt{r} = 1$, $n = 2$ and $k = 0$ or 1.

    $\alpha = \dfrac{90° + 360° \cdot k}{2}$: For $k = 0$, $\alpha = 45°$. For $k = 1$, $\alpha = 225°$. The square roots are

    $\cos 45° + i \sin 45° = \dfrac{\sqrt{2}}{2} + \dfrac{\sqrt{2}}{2} i$ and $\cos 225° + i \sin 225° = -\dfrac{\sqrt{2}}{2} - \dfrac{\sqrt{2}}{2} i$.

36. Note that $-4i = 4(\cos 270° + i \sin 270°)$. Here $r = 4$ so $\sqrt{r} = 2$, $n = 2$ and $k = 0$ or 1.

    $\alpha = \dfrac{270° + 360° \cdot k}{2}$: For $k = 0$, $\alpha = 135°$. For $k = 1$, $\alpha = 315°$. The square roots are

    $2(\cos 135° + i \sin 135°) = -\sqrt{2} + i\sqrt{2}$ and $2(\cos 315° + i \sin 315°) = \sqrt{2} - i\sqrt{2}$.

37. Note that $64i = 64(\cos 90° + i \sin 90°)$. Here $r = 64$ so $\sqrt[3]{r} = 4$, $n = 3$ and $k = 0, 1,$ or 2.

    $\alpha = \dfrac{90° + 360° \cdot k}{3}$: For $k = 0$, $\alpha = 30°$. For $k = 1$, $\alpha = 150°$. For $k = 2$, $\alpha = 270°$.

    The cube roots are $4(\cos 30° + i \sin 30°) = 2\sqrt{3} + 2i$, $4(\cos 150° + i \sin 150°) = -2\sqrt{3} + 2i$, and

    $4(\cos 270° + i \sin 270°) = -4i$.

38. Note that $-1 = \cos 180° + i \sin 180°$. Here $r = 1$ so $\sqrt[4]{r} = 1$, $n = 4$ and $k = 0, 1, 2,$ or 3.

    $\alpha = \dfrac{180° + 360° \cdot k}{4}$: For $k = 0$, $\alpha = 45°$. For $k = 1$, $\alpha = 135°$. For $k = 2$, $\alpha = 225°$. For $k = 3$, $\alpha = 315°$.

    The fourth roots are $\cos 45° + i \sin 45° = \dfrac{\sqrt{2}}{2} + \dfrac{\sqrt{2}}{2} i$, $\cos 135° + i \sin 135° = -\dfrac{\sqrt{2}}{2} + \dfrac{\sqrt{2}}{2} i$,

    $\cos 225° + i \sin 225° = -\dfrac{\sqrt{2}}{2} - \dfrac{\sqrt{2}}{2} i$, and $\cos 315° + i \sin 315° = \dfrac{\sqrt{2}}{2} - \dfrac{\sqrt{2}}{2} i$.

39. Note that $81 = 81(\cos 0° + i \sin 0°)$. Here $r = 81$ so $\sqrt[4]{r} = 3$, $n = 4$ and $k = 0, 1, 2,$ or 3.

    $\alpha = \dfrac{0° + 360° \cdot k}{4}$: For $k = 0$, $\alpha = 0°$. For $k = 1$, $\alpha = 90°$. For $k = 2$, $\alpha = 180°$. For $k = 3$, $\alpha = 270°$.

    The fourth roots are $3(\cos 0° + i \sin 0°) = 3$, $3(\cos 90° + i \sin 90°) = 3i$,

    $3(\cos 180° + i \sin 180°) = -3$, and $3(\cos 270° + i \sin 270°) = -3i$.

40. Note that $-1 + i\sqrt{3} = 2(\cos 120° + i \sin 120°)$. Here $r = 2$ so $\sqrt{r} = \sqrt{2}$, $n = 2$ and $k = 0$ or 1.

    $\alpha = \dfrac{120° + 360° \cdot k}{2}$: For $k = 0$, $\alpha = 60°$. For $k = 1$, $\alpha = 240°$. The square roots are

    $\sqrt{2}(\cos 60° + i \sin 60°) = \dfrac{\sqrt{2}}{2} + \dfrac{\sqrt{6}}{2} i$ and $\sqrt{2}(\cos 240° + i \sin 240°) = -\dfrac{\sqrt{2}}{2} - \dfrac{\sqrt{6}}{2} i$.

41. (a) Note that $1 = \cos 0° + i\sin 0°$. Here $r = 1$ so $\sqrt[4]{r} = 1$, $n = 4$ and $k = 0, 1, 2,$ or 3.

$\alpha = \dfrac{0° + 360° \cdot k}{4}$: For $k = 0, \alpha = 0°$. For $k = 1, \alpha = 90°$. For $k = 2, \alpha = 180°$. For $k = 3, \alpha = 270°$.

The fourth roots are $\cos 0° + i\sin 0° = 1$, $\cos 90° + i\sin 90° = i$, $\cos 180° + i\sin 180° = -1$, and

$\cos 270° + i\sin 270° = -i$. See Figure 41a.

(b) Note that $1 = \cos 0° + i\sin 0°$. Here $r = 1$ so $\sqrt[6]{r} = 1$, $n = 6$ and $k = 0, 1, 2, 3, 4,$ or 5.

$\alpha = \dfrac{0° + 360° \cdot k}{6}$: For $k = 0, \alpha = 0°$. For $k = 1, \alpha = 60°$. For $k = 2, \alpha = 120°$. For $k = 3, \alpha = 180°$.

For $k = 4, \alpha = 240°$. For $k = 5, \alpha = 300°$. The sixth roots are $\cos 0° + i\sin 0° = 1$,

$\cos 60° + i\sin 60° = \dfrac{1}{2} + \dfrac{\sqrt{3}}{2}i$, $\cos 120° + i\sin 120° = -\dfrac{1}{2} + \dfrac{\sqrt{3}}{2}i$, $\cos 180° + i\sin 180° = -1$

$\cos 240° + i\sin 240° = -\dfrac{1}{2} - \dfrac{\sqrt{3}}{2}i$, $\cos 300° + i\sin 300° = \dfrac{1}{2} - \dfrac{\sqrt{3}}{2}i$. See Figure 41b.

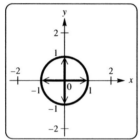

Figure 41a

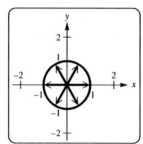

Figure 41b

42. (a) Note that $i = \cos 90° + i\sin 90°$. Here $r = 1$ so $\sqrt{r} = 1$, $n = 2$ and $k = 0$ or 1.

$\alpha = \dfrac{90° + 360° \cdot k}{2}$: For $k = 0, \alpha = 45°$. For $k = 1, \alpha = 225°$. The square roots are

$\cos 45° + i\sin 45°$ and $\cos 225° + i\sin 225°$. See Figure 42a.

(b) Note that $i = \cos 90° + i\sin 90°$. Here $r = 1$ so $\sqrt[4]{r} = 1$, $n = 4$ and $k = 0, 1, 2$ or 3.

$\alpha = \dfrac{90° + 360° \cdot k}{4}$: For $k = 0, \alpha = 22.5°$. For $k = 1, \alpha = 112.5°$. For $k = 2, \alpha = 202.5°$.

For $k = 3, \alpha = 292.5°$. The fourth roots are $\cos 22.5° + i\sin 22.5°$, $\cos 112.5° + i\sin 112.5°$.

$\cos 202.5° + i\sin 202.5°$ and $\cos 292.5° + i\sin 292.5°$. See Figure 42b.

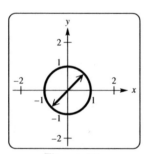

Figure 42a

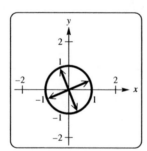

Figure 42b

43. The argument for a positive real number is $\theta = 0°$. By the nth root theorem with $k = 0$, $\alpha = 0°$. Thus, one nth root has an argument of $0°$ and it must be real.

44. (a) False. The square roots of negative numbers are complex.

    (b) False. Two of the cube roots of any real number must be complex.

45. $x^3 + 8 = (x + 2)(x^2 - 2x + 4)$

46. $x + 2 = 0 \Rightarrow x = -2$

47. $x^2 - 2x + 4 = 0 \Rightarrow x^2 - 2x + 1 = -4 + 1 \Rightarrow (x - 1)^2 = -3 \Rightarrow x - 1 = \pm\sqrt{-3} \Rightarrow x = 1 \pm i\sqrt{3}$

48. Note that $-8 = 8(\cos 180° + i\sin 180°)$. Here $r = 8$ so $\sqrt[3]{r} = 2$, $n = 3$ and $k = 0, 1$, or 2.

    $\alpha = \dfrac{180° + 360° \cdot k}{3}$: For $k = 0$, $\alpha = 60°$. For $k = 1$, $\alpha = 180°$. For $k = 2$, $\alpha = 300°$. The cube roots are

    $2(\cos 60° + i\sin 60°)$, $2(\cos 180° + i\sin 180°)$, and $2(\cos 300° + i\sin 300°)$.

49. $2(\cos 60° + i\sin 60°) = 1 + i\sqrt{3}$, $2(\cos 180° + i\sin 180°) = -2$, and $2(\cos 300° + i\sin 300°) = 1 - i\sqrt{3}$.

50. The results are the same.

51. $x^4 + 1 = 0 \Rightarrow x^4 = -1$. Find the fourth roots of -1. Note that $-1 = \cos 180° + i\sin 180°$.

    Here $r = 1$ so $\sqrt[4]{r} = 1$, $n = 4$ and $k = 0, 1, 2$, or 3.

    $\alpha = \dfrac{180° + 360° \cdot k}{4}$: For $k = 0$, $\alpha = 45°$. For $k = 1$, $\alpha = 135°$. For $k = 2$, $\alpha = 225°$. For $k = 3$, $\alpha = 315°$.

    The fourth roots are $\cos 45° + i\sin 45°$, $\cos 135° + i\sin 135°$, $\cos 225° + i\sin 225°$, and $\cos 315° + i\sin 315°$.

52. $x^4 + 16 = 0 \Rightarrow x^4 = -16$. Find the fourth roots of -16. Note that $-16 = 16(\cos 180° + i\sin 180°)$.

    Here $r = 16$ so $\sqrt[4]{r} = 2$, $n = 4$ and $k = 0, 1, 2$, or 3. $\alpha = \dfrac{180° + 360° \cdot k}{4}$: For $k = 0$, $\alpha = 45°$.

    For $k = 1$, $\alpha = 135°$. For $k = 2$, $\alpha = 225°$. For $k = 3$, $\alpha = 315°$. The fourth roots are

    $2(\cos 45° + i\sin 45°)$, $2(\cos 135° + i\sin 135°)$, $2(\cos 225° + i\sin 225°)$, and $2(\cos 315° + i\sin 315°)$.

53. $x^5 - i = 0 \Rightarrow x^5 = i$. Find the fifth roots of i. Note that $i = \cos 90° + i\sin 90°$.

    Here $r = 1$ so $\sqrt[5]{r} = 1$, $n = 5$ and $k = 0, 1, 2, 3$, or 4. $\alpha = \dfrac{90° + 360° \cdot k}{5}$: For $k = 0$, $\alpha = 18°$.

    For $k = 1$, $\alpha = 90°$. For $k = 2$, $\alpha = 162°$. For $k = 3$, $\alpha = 234°$. For $k = 4$, $\alpha = 306°$. The fifth roots are

    $\cos 18° + i\sin 18°$, $\cos 90° + i\sin 90°$, $\cos 162° + i\sin 162°$, $\cos 234° + i\sin 234°$, and $\cos 306° + i\sin 306°$.

54. $x^4 - i = 0 \Rightarrow x^4 = i$. Find the fourth roots of i. Note that $i = \cos 90° + i\sin 90°$.

    Here $r = 1$ so $\sqrt[4]{r} = 1$, $n = 4$ and $k = 0, 1, 2$ or 3. $\alpha = \dfrac{90° + 360° \cdot k}{4}$: For $k = 0$, $\alpha = 22.5°$.

    For $k = 1$, $\alpha = 112.5°$. For $k = 2$, $\alpha = 202.5°$. For $k = 3$, $\alpha = 292.5°$. The fourth roots are

    $\cos 22.5° + i\sin 22.5°$, $\cos 112.5° + i\sin 112.5°$, $\cos 202.5° + i\sin 202.5°$ and $\cos 292.5° + i\sin 292.5°$.

55. $x^3 + 1 = 0 \Rightarrow x^3 = -1$. Find the cube roots of -1. Note that $-1 = \cos 180° + i\sin 180°$.

    Here $r = 1$ so $\sqrt[3]{r} = 1$, $n = 3$ and $k = 0, 1$, or 2.

    $\alpha = \dfrac{180° + 360° \cdot k}{3}$: For $k = 0$, $\alpha = 60°$. For $k = 1$, $\alpha = 180°$. For $k = 2$, $\alpha = 300°$. The cube roots are

    $\cos 60° + i\sin 60°$, $\cos 180° + i\sin 180°$, and $\cos 300° + i\sin 300°$.

56. $x^3 + i = 0 \Rightarrow x^3 = -i$. Find the cube roots of $-i$. Note that $-i = \cos 270° + i \sin 270°$.

Here $r = 1$ so $\sqrt[3]{r} = 1$, $n = 3$ and $k = 0, 1,$ or 2.

$\alpha = \dfrac{270° + 360° \cdot k}{3}$: For $k = 0$, $\alpha = 90°$. For $k = 1$, $\alpha = 210°$. For $k = 2$, $\alpha = 330°$. The cube roots are

$\cos 90° + i \sin 90°$, $\cos 210° + i \sin 210°$, and $\cos 330° + i \sin 330°$.

57. $x^3 - 8 = 0 \Rightarrow x^3 = 8$. Find the cube roots of 8. Note that $8 = 8(\cos 0° + i \sin 0°)$. Here

$r = 8$ so $\sqrt[3]{r} = 2$, $n = 3$ and $k = 0, 1,$ or 2.

$\alpha = \dfrac{0° + 360° \cdot k}{3}$: For $k = 0$, $\alpha = 0°$. For $k = 1$, $\alpha = 120°$. For $k = 2$, $\alpha = 240°$. The cube roots are

$2(\cos 0° + i \sin 0°)$, $2(\cos 120° + i \sin 120°)$, and $2(\cos 240° + i \sin 240°)$.

58. $x^5 + 1 = 0 \Rightarrow x^5 = -1$. Find the fifth roots of -1. Note that $-1 = \cos 180° + i \sin 180°$.

Here $r = 1$ so $\sqrt[5]{r} = 1$, $n = 5$ and $k = 0, 1, 2, 3,$ or 4. $\alpha = \dfrac{180° + 360° \cdot k}{5}$: For $k = 0$, $\alpha = 36°$.

For $k = 1$, $\alpha = 108°$. For $k = 2$, $\alpha = 180°$. For $k = 3$, $\alpha = 252°$. For $k = 4$, $\alpha = 324°$.

The fifth roots are $\cos 36° + i \sin 36°$, $\cos 108° + i \sin 108°$, $\cos 180° + i \sin 180°$, $\cos 252° + i \sin 252°$, and

$\cos 324° + i \sin 324°$.

59. (a) Using a calculator yields: $f(z_1) = \dfrac{2(i)^3 + 1}{3(i)^2} = -\dfrac{1}{3} + \dfrac{2}{3}i$; $f(z_2) = \dfrac{2(-\frac{1}{3} + \frac{2}{3}i)^3 + 1}{3(-\frac{1}{3} + \frac{2}{3}i)^2} = -\dfrac{131}{225} + \dfrac{208}{225}i$

The value is approaching w_2 and the pixel should be colored blue.

(b) Using a calculator yields: $f(z_1) = \dfrac{2(2 + i)^3 + 1}{3(2 + i)^2} = \dfrac{103}{75} + \dfrac{46}{75}i$; $f(z_2) = \dfrac{2(\frac{103}{75} + \frac{46}{75}i)^3 + 1}{3(\frac{103}{75} + \frac{46}{75}i)^2} \approx 1.01 + .23i$

The value is approaching w_1 and the pixel should be colored red.

(c) Using a calculator yields: $f(z_1) = \dfrac{2(-1 - i)^3 + 1}{3(-1 - i)^2} = -\dfrac{2}{3} - \dfrac{5}{6}i$; $f(z_2) = \dfrac{2(-\frac{2}{3} - \frac{5}{6}i)^3 + 1}{3(-\frac{2}{3} + \frac{5}{6}i)^2} \approx -.51 - .84i$

The value is approaching w_3 and the pixel should be colored yellow.

60. (a) Using the technique described the roots are

$1, .30901699 + .95105652i, -.809017 + .58778525i, -.809017 - .58778525i, .30901699 - .95105652i$.

(b) Using the technique described the roots are $1, .80901699 + .58778525i, .30901699 + .95105652i$.

Reviewing Basic Concepts (Sections 10.4 and 10.5)

1. $2(\cos 60° + i \sin 60°) = 2\left(\dfrac{1}{2} + \dfrac{\sqrt{3}}{2}\right) = 1 + i\sqrt{3}$.

2. $|3 - 4i| = \sqrt{3^2 + (-4)^2} = \sqrt{25} = 5$

3. $r = \sqrt{(-\sqrt{2})^2 + (\sqrt{2})^2} = \sqrt{4} = 2$; $\tan \theta = \dfrac{y}{x} \Rightarrow \tan \theta = \dfrac{\sqrt{2}}{-\sqrt{2}} \Rightarrow \tan \theta = -1 \Rightarrow \theta = \tan^{-1}(-1) = -45°$

Since θ is in quadrant II, $\theta = -45° + 180° = 135°$. Therefore $-\sqrt{2} + i\sqrt{2} = 2(\cos 135° + i \sin 135°)$.

4. $[4(\cos 135° + i\sin 135°)][2(\cos 45° + i\sin 45°)] = (4 \cdot 2)[\cos(135° + 45°) + i\sin(135° + 45°)] =$

    $8(\cos 180° + i\sin 180°) = 8(-1 + 0i) = -8$

5. $\dfrac{4(\cos 135° + i\sin 135°)}{2(\cos 45° + i\sin 45°)} = \dfrac{4}{2}[\cos(135° - 45°) + i\sin(135° - 45°)] =$

    $2(\cos 90° + i\sin 90°) = 2(0 + i) = 2i$

6. $[4 \operatorname{cis} 17°]^3 = [4(\cos 17° + i\sin 17°)]^3 = 4^3[\cos(3 \cdot 17°) + i\sin(3 \cdot 17°)] = 64(\cos 51° + i\sin 51°)$

7. First find the trigonometric form of $2 - 2i$; $r = \sqrt{2^2 + 2^2} = \sqrt{8} = 2\sqrt{2}$;

    $\tan\theta = \dfrac{y}{x} \Rightarrow \tan\theta = \dfrac{-2}{2} \Rightarrow \Rightarrow \theta = \tan^{-1}(-1) = -45°$. Therefore $2 - 2i = 2\sqrt{2}[\cos(-45°) + i\sin(-45°)]$.

    $(2 - 2i)^4 = [2\sqrt{2}(\cos(-45°) + i\sin(-45°))]^4 = (2\sqrt{2})^4[\cos(4 \cdot (-45°)) + i\sin(4 \cdot (-45°))] =$

    $64[\cos(-180°) + i\sin(-180°)] = 64(-1 + 0i) = -64$

8. Note that $-64 = 64(\cos 180° + i\sin 180°)$. Here $r = 64$ so $\sqrt[3]{r} = 4$, $n = 3$ and $k = 0, 1,$ or 2.

    $\alpha = \dfrac{180° + 360° \cdot k}{3}$: For $k = 0$, $\alpha = 60°$. For $k = 1$, $\alpha = 180°$. For $k = 2$, $\alpha = 300°$. The cube roots are

    $4(\cos 60° + i\sin 60°) = 2 + 2i\sqrt{3}$, $4(\cos 180° + i\sin 180°) = -4$, and $4(\cos 300° + i\sin 300°) = 2 - 2i\sqrt{3}$.

9. Note that $2i = 2(\cos 90° + i\sin 90°)$. Here $r = 2$ so $\sqrt{r} = \sqrt{2}$, $n = 2$ and $k = 0$ or 1.

    $\alpha = \dfrac{90° + 360° \cdot k}{2}$: For $k = 0$, $\alpha = 45°$. For $k = 1$, $\alpha = 225°$. The square roots are

    $\sqrt{2}(\cos 45° + i\sin 45°) = 1 + i$ and $\sqrt{2}(\cos 225° + i\sin 225°) = -1 - i$.

10. Find the cube roots of -1. Note that $-1 = \cos 180° + i\sin 180°$.

    Here $r = 1$ so $\sqrt[3]{r} = 1$, $n = 3$ and $k = 0, 1,$ or 2.

    $\alpha = \dfrac{180° + 360° \cdot k}{3}$: For $k = 0$, $\alpha = 60°$. For $k = 1$, $\alpha = 180°$. For $k = 2$, $\alpha = 300°$. The cube roots are

    $\cos 60° + i\sin 60°$, $\cos 180° + i\sin 180°$, and $\cos 300° + i\sin 300°$.

10.6: Polar Equations and Graphs

1. (a) The point is in quadrant II.

    (b) The point is in quadrant I.

    (c) The point is in quadrant IV.

    (d) The point is in quadrant III.

2. (a) The point lies on the positive *x*-axis.

    (b) The point lies on the negative *x*-axis.

    (c) The point lies on the negative *y*-axis.

    (d) The point lies on the positive *y*-axis.

3. See Figure 3-11. Two other forms are $(1, 45° + 360°) = (1, 405°)$ or $(-1, 45° + 180°) = (-1, 225°)$.

4. See Figure 4-12. Two other forms are $(3, 120° + 360°) = (3, 480°)$ or $(-3, 120° + 180°) = (-3, 300°)$.

5. See Figure 3-11. Two other forms are $(-2, 135° + 360°) = (-2, 495°)$ or $(2, 135° + 180°) = (2, 315°)$.

6. See Figure 4-12. Two other forms are $(-4, 27° + 360°) = (-4, 387°)$ or $(4, 27° + 180°) = (4, 207°)$.

7. See Figure 3-11. Two other forms are $(5, -60° + 360°) = (5, 300°)$ or $(-5, -60° + 180°) = (-5, 120°)$.

8. See Figure 4-12. Two other forms are $(2, -45° + 360°) = (2, 315°)$ or $(-2, -45° + 180°) = (-2, 135°)$.

9. See Figure 3-11. Two other forms are $(-3, -210° + 360°) = (-3, 150°)$ or $(3, -210° + 180°) = (3, -30°)$.

10. See Figure 4-12. Two other forms are $(-1, -120° + 360°) = (-1, 240°)$ or $(1, -120° + 180°) = (1, 60°)$.

11. See Figure 3-11. Two other forms are $(3, 300° + 360°) = (3, 660°)$ or $(-3, 300° - 180°) = (-3, 120°)$.

12. See Figure 4-12. Two other forms are $(4, 270° - 360°) = (4, -90°)$ or $(-4, 270° - 180°) = (-4, 90°)$.

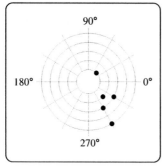

Figure 3-11

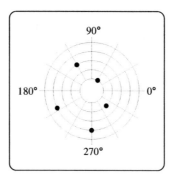

Figure 4-12

13. See Figure 13-21. $r = \sqrt{(-1)^2 + 1^2} = \sqrt{2}$; $\theta = \tan^{-1}\left(\dfrac{1}{-1}\right) = -45°$. Since the point is in QII, $\theta = 135°$.

    Two forms are $(\sqrt{2}, 135°)$ and $(-\sqrt{2}, 315°)$.

14. See Figure 14-22. $r = \sqrt{1^2 + 1^2} = \sqrt{2}$; $\theta = \tan^{-1}\left(\dfrac{1}{1}\right) = 45°$.

    Two forms are $(\sqrt{2}, 45°)$ and $(-\sqrt{2}, 225°)$.

15. See Figure 13-21. $r = \sqrt{0^2 + 3^2} = 3$; $\theta = \tan^{-1}\left(\dfrac{3}{0}\right) \Rightarrow$ undefined $\Rightarrow \theta = 90°$ (positive y-axis).

    Two forms are $(3, 90°)$ and $(-3, 270°)$.

16. See Figure 14-22. $r = \sqrt{0^2 + (-3)^2} = 3$; $\theta = \tan^{-1}\left(\dfrac{-3}{0}\right) \Rightarrow$ undefined $\Rightarrow \theta = 270°$ (negative y-axis).

    Two forms are $(3, 270°)$ and $(-3, 90°)$.

17. See Figure 13-21. $r = \sqrt{(\sqrt{2})^2 + (\sqrt{2})^2} = 2$; $\theta = \tan^{-1}\left(\dfrac{\sqrt{2}}{\sqrt{2}}\right) = 45°$.

    Two forms are $(2, 45°)$ and $(-2, 225°)$.

18. See Figure 14-22. $r = \sqrt{(-\sqrt{2})^2 + (\sqrt{2})^2} = 2$; $\theta = \tan^{-1}\left(\dfrac{\sqrt{2}}{-\sqrt{2}}\right) = -45°$. The point is in QII, so $\theta = 135°$.

    Two forms are $(2, 135°)$ and $(-2, 315°)$.

19. See Figure 13-21. $r = \sqrt{\left(\dfrac{\sqrt{3}}{2}\right)^2 + \left(\dfrac{3}{2}\right)^2} = \sqrt{3}; \theta = \tan^{-1}\left(\dfrac{3}{2} \div \dfrac{\sqrt{3}}{2}\right) = 60°.$

 Two forms are $(\sqrt{3}, 60°)$ and $(-\sqrt{3}, 240°)$.

20. See Figure 14-22. $r = \sqrt{\left(-\dfrac{\sqrt{3}}{2}\right)^2 + \left(-\dfrac{1}{2}\right)^2} = 1; \theta = \tan^{-1}\left[-\dfrac{1}{2} \div \left(-\dfrac{\sqrt{3}}{2}\right)\right] = 30°.$

 Since the point is in QIII, $\theta = 210°$. Two forms are $(1, 210°)$ and $(-1, 30°)$.

21. See Figure 13-21. $r = \sqrt{3^2 + 0^2} = 3; \theta = \tan^{-1}\left(\dfrac{0}{3}\right) = 0°$ (positive x-axis).

 Two forms are $(3, 0°)$ and $(-3, 180°)$.

22. See Figure 14-22. $r = \sqrt{(-2)^2 + 0^2} = 2; \theta = \tan^{-1}\left(\dfrac{0}{-2}\right) = 0° \Rightarrow \theta = 180°$ (negative x-axis).

 Two forms are $(2, 180°)$ and $(-2, 0°)$.

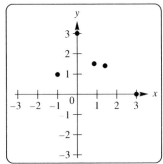   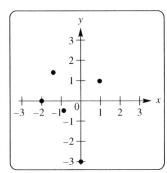   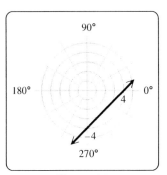

Figure 13-21   Figure 14-22   Figure 23

23. $x - y = 4 \Rightarrow r\cos\theta - r\sin\theta = 4 \Rightarrow r(\cos\theta - \sin\theta) = 4 \Rightarrow r = \dfrac{4}{\cos\theta - \sin\theta}.$ See Figure 23.

24. $x + y = -7 \Rightarrow r\cos\theta + r\sin\theta = -7 \Rightarrow r(\cos\theta + \sin\theta) = -7 \Rightarrow r = \dfrac{-7}{\cos\theta + \sin\theta}.$ See Figure 24.

25. $x^2 + y^2 = 16 \Rightarrow (r\cos\theta)^2 + (r\sin\theta)^2 = 16 \Rightarrow r^2(\cos^2\theta + \sin^2\theta) = 16 \Rightarrow r^2 = 16 \Rightarrow r = 4$ or $r = -4$

 See Figure 25.

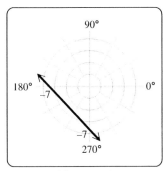   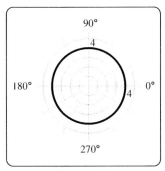   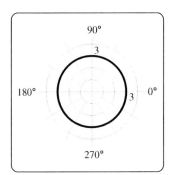

Figure 24   Figure 25   Figure 26

26. $x^2 + y^2 = 9 \Rightarrow (r\cos\theta)^2 + (r\sin\theta)^2 = 9 \Rightarrow r^2(\cos^2\theta + \sin^2\theta) = 9 \Rightarrow r^2 = 9 \Rightarrow r = 3$ or $r = -3$

 See Figure 26.

27. $2x + y = 5 \Rightarrow 2r\cos\theta + r\sin\theta = 5 \Rightarrow r(2\cos\theta + \sin\theta) = 5 \Rightarrow r = \dfrac{5}{2\cos\theta + \sin\theta}$. See Figure 27.

28. $3x - 2y = 6 \Rightarrow 3r\cos\theta - 2r\sin\theta = 6 \Rightarrow r(3\cos\theta - 2\sin\theta) = 6 \Rightarrow r = \dfrac{6}{3\cos\theta - 2\sin\theta}$. See Figure 28.

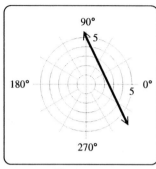

Figure 27

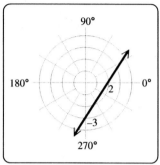

Figure 28

29. This is a circle of radius 3. Graph C.

30. This is a rose with three petals. Graph D.

31. This is a rose with four petals. Graph A.

32. This is a straight line. Graph B.

33. The graph is a cardioid. See Figure 33.

34. The graph is a limaçon. See Figure 34.

35. The graph is a limaçon. See Figure 35.

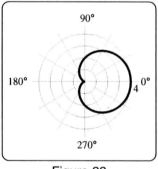

Figure 33

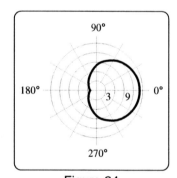

Figure 34

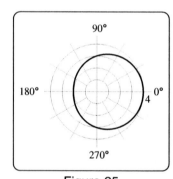

Figure 35

36. The graph is a limaçon. See Figure 36.

37. The graph is a four-leaved rose. See Figure 37.

38. The graph is a five-leaved rose. See Figure 38.

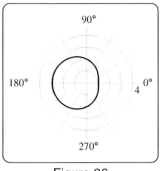

Figure 36

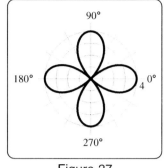

Figure 37

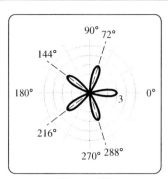

Figure 38

39. The graph is a lemniscate. See Figure 39.

40. The graph is a lemniscate. See Figure 40.

41. The graph is a cardioid. See Figure 41.

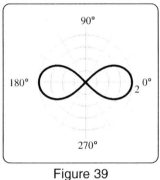

Figure 39

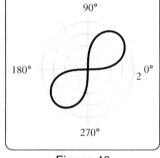

Figure 40

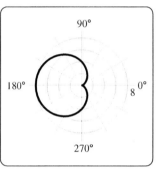

Figure 41

42. The graph is a cardioid. See Figure 42.

43. The graph is a cissoid. See Figure 43.

44. The graph is a cissoid with a loop. See Figure 44.

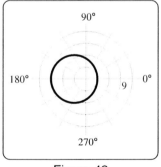

Figure 42

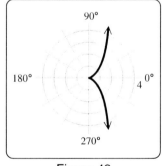

Figure 43

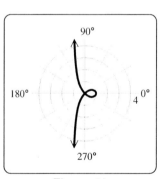

Figure 44

45. (a) The missing ordered pair is $(r, -\theta)$.

(b) The missing ordered pairs are $(r, \pi - \theta)$ or $(-r, -\theta)$.

(c) The missing ordered pairs are $(r, \pi + \theta)$ or $(-r, \theta)$.

46. (a) $-\theta$ (b) $\pi - \theta$ (c) $-r; -\theta$ (d) $-r$ (e) $\pi + \theta$ (f) the polar axis   (g) the line $\theta = \dfrac{\pi}{2}$

47. To graph (r, θ), $r < 0$, you could locate θ, add 180° to it, and move $|r|$ units along the terminal ray of $\theta + 180°$ in standard position.

48. $(-r, \theta + 180°) = (-r\cos(\theta + 180°), -r\sin(\theta + 180°)) = (-r(-\cos\theta), -r(-\sin\theta)) = (r\cos\theta, r\cos\theta) = (r, \theta)$

49. The angle must be quadrantal. That is, it must be coterminal with 0°, 90°, 180°, or 270°.

50. The graph of $r = k, k > 0$ is a circle with radius k.

51. The graph would be reflected across the line $\theta = \dfrac{\pi}{2}$ (the y-axis).

52. The graph would be rotated 90° counterclockwise.

53. The value of a determines the length of the petals and the value of n determines the number of petals. If n is odd, there are n petals. If n is even, there are $2n$ petals.

54. If $n = 1$, the graph will be a circle. The value of a determines the diameter of the circle.

55. $r = 2\sin\theta \Rightarrow r^2 = 2r\sin\theta \Rightarrow x^2 + y^2 = 2y \Rightarrow x^2 + y^2 - 2y + 1 = 1 \Rightarrow x^2 + (y - 1)^2 = 1$
    See Figure 55.

56. $r = 2\cos\theta \Rightarrow r^2 = 2r\cos\theta \Rightarrow x^2 + y^2 = 2x \Rightarrow x^2 - 2x + 1 + y^2 = 1 \Rightarrow (x - 1)^2 + y^2 = 1$
    See Figure 56.

57. $r = \dfrac{2}{1 - \cos\theta} \Rightarrow r - r\cos\theta = 2 \Rightarrow \sqrt{x^2 + y^2} - x = 2 \Rightarrow \sqrt{x^2 + y^2} = x + 2 \Rightarrow x^2 + y^2 = (x + 2)^2$
    $\Rightarrow x^2 + y^2 = x^2 + 4x + 4 \Rightarrow y^2 = 4x + 4 \Rightarrow y^2 = 4(x + 1)$. See Figure 57.

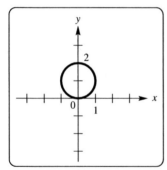

Figure 55

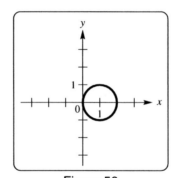

Figure 56

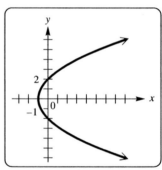

Figure 57

58. $r = \dfrac{3}{1 - \sin\theta} \Rightarrow r - r\sin\theta = 3 \Rightarrow \sqrt{x^2 + y^2} - y = 3 \Rightarrow \sqrt{x^2 + y^2} = y + 3 \Rightarrow x^2 + y^2 = (y + 3)^2$

$\Rightarrow x^2 + y^2 = y^2 + 6y + 9 \Rightarrow x^2 = 6y + 9 \Rightarrow x^2 = 6\left(y + \dfrac{3}{2}\right)$. See Figure 58.

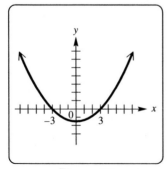

Figure 58

59. $r = -2\cos\theta - 2\sin\theta \Rightarrow r^2 = -2r\cos\theta - 2r\sin\theta \Rightarrow x^2 + y^2 = -2x - 2y \Rightarrow x^2 + y^2 + 2x + 2y = 0 \Rightarrow$

   $x^2 + 2x + 1 + y^2 + 2y + 1 = 2 \Rightarrow (x + 1)^2 + (y + 1)^2 = 2.$ See Figure 59.

60. $r = \dfrac{3}{4\cos\theta - \sin\theta} \Rightarrow 4r\cos\theta - r\sin\theta = 3 \Rightarrow 4x - y = 3.$ See Figure 60.

61. $r = 2\sec\theta \Rightarrow r = \dfrac{2}{\cos\theta} \Rightarrow r\cos\theta = 2 \Rightarrow x = 2.$ See Figure 61.

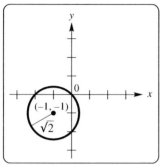

Figure 59

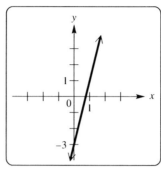

Figure 60

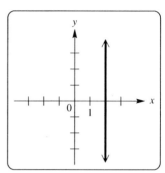

Figure 61

62. $r = -5\csc\theta \Rightarrow r = \dfrac{-5}{\sin\theta} \Rightarrow r\sin\theta = -5 \Rightarrow y = -5.$ See Figure 62.

63. $r = \dfrac{2}{\cos\theta + \sin\theta} \Rightarrow r\cos\theta + r\sin\theta = 2 \Rightarrow x + y = 2.$ See Figure 63.

64. $r = \dfrac{2}{2\cos\theta + \sin\theta} \Rightarrow 2r\cos\theta + r\sin\theta = 2 \Rightarrow 2x + y = 2.$ See Figure 64.

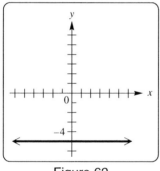

Figure 62

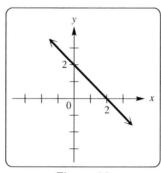

Figure 63

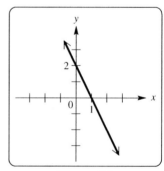

Figure 64

65. See Figure 65.

66. See Figure 66.

67. See Figure 67.

68. See Figure 68.

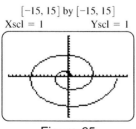

Figure 65

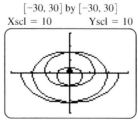

Figure 66

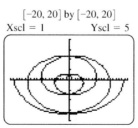

Figure 67

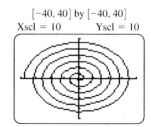

Figure 68

69. $4\sin\theta = 1 + 2\sin\theta \Rightarrow 2\sin\theta = 1 \Rightarrow \sin\theta = \dfrac{1}{2} \Rightarrow \theta = \sin^{-1}\dfrac{1}{2} \Rightarrow \theta = \dfrac{\pi}{6} \text{ or } \dfrac{5\pi}{6}$

Then $r = 4\sin\theta \Rightarrow r = 4\sin\dfrac{\pi}{6} \Rightarrow r = 4\left(\dfrac{1}{2}\right) \Rightarrow r = 2$. The intersection points are $\left(2, \dfrac{\pi}{6}\right)$ and $\left(2, \dfrac{5\pi}{6}\right)$.

70. $3 = 2 + 2\cos\theta \Rightarrow 2\cos\theta = 1 \Rightarrow \cos\theta = \dfrac{1}{2} \Rightarrow \theta = \cos^{-1}\dfrac{1}{2} \Rightarrow \theta = 60° \text{ or } 300°$

The intersection points are $(3, 60°)$ and $(3, 300°)$.

71. $2 + \sin\theta = 2 + \cos\theta \Rightarrow \sin\theta = \cos\theta \Rightarrow \theta = \dfrac{\pi}{4} \text{ or } \dfrac{5\pi}{4}$

When $\theta = \dfrac{\pi}{4}$, $r = 2 + \sin\theta \Rightarrow r = 2 + \sin\dfrac{\pi}{4} \Rightarrow r = 2 + \left(\dfrac{\sqrt{2}}{2}\right) \Rightarrow r = \dfrac{4 + \sqrt{2}}{2}$.

When $\theta = \dfrac{5\pi}{4}$, $r = 2 + \sin\theta \Rightarrow r = 2 + \sin\dfrac{5\pi}{4} \Rightarrow r = 2 + \left(-\dfrac{\sqrt{2}}{2}\right) \Rightarrow r = \dfrac{4 - \sqrt{2}}{2}$.

The intersection points are $\left(\dfrac{4 + \sqrt{2}}{2}, \dfrac{\pi}{4}\right)$ and $\left(\dfrac{4 - \sqrt{2}}{2}, \dfrac{5\pi}{4}\right)$.

72. $\sin 2\theta = \sqrt{2}\cos\theta \Rightarrow 2\sin\theta\cos\theta = \sqrt{2}\cos\theta \Rightarrow 2\sin\theta\cos\theta - \sqrt{2}\cos\theta = 0 \Rightarrow$

$\cos\theta(2\sin\theta - \sqrt{2}) = 0 \Rightarrow \cos\theta = 0 \text{ or } \sin\theta = \dfrac{\sqrt{2}}{2} \Rightarrow \theta = \cos^{-1}0 \text{ or } \theta = \sin^{-1}\dfrac{\sqrt{2}}{2} \Rightarrow \theta = \dfrac{\pi}{2}, \dfrac{\pi}{4}, \dfrac{3\pi}{4}$

When $\theta = \dfrac{\pi}{4}$, $r = \sin 2\theta \Rightarrow r = \sin\left(2 \cdot \dfrac{\pi}{4}\right) \Rightarrow r = \sin\dfrac{\pi}{2} \Rightarrow r = 1$.

When $\theta = \dfrac{\pi}{2}$, $r = \sin 2\theta \Rightarrow r = \sin\left(2 \cdot \dfrac{\pi}{2}\right) \Rightarrow r = \sin\pi \Rightarrow r = 0$.

When $\theta = \dfrac{3\pi}{4}$, $r = \sin 2\theta \Rightarrow r = \sin\left(2 \cdot \dfrac{3\pi}{4}\right) \Rightarrow r = \sin\dfrac{3\pi}{2} \Rightarrow r = -1$.

The intersection points are $\left(1, \dfrac{\pi}{4}\right), \left(0, \dfrac{\pi}{2}\right)$ and $\left(-1, \dfrac{3\pi}{4}\right)$.

73. (a) See Figure 73a.

(b) See Figure 73b. Earth is closest to the sun.

(c) By graphing the orbits of Neptune and Pluto (not shown), we see that Pluto is not always farthest from the sun.

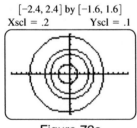

[-2.4, 2.4] by [-1.6, 1.6]
Xscl = .2   Yscl = .1

Figure 73a

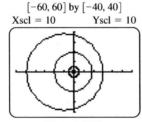

[-60, 60] by [-40, 40]
Xscl = 10   Yscl = 10

Figure 73b

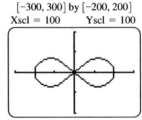

[-300, 300] by [-200, 200]
Xscl = 100   Yscl = 100

Figure 74a

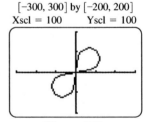

[-300, 300] by [-200, 200]
Xscl = 100   Yscl = 100

Figure 74b

74. (a) See Figure 74a. Inside the "figure eight" the radio signal can be received. This region is generally in an east-west direction from the two radio towers with a maximum distance of 200 miles.

(b) See Figure 74b. The radio signal can be received inside the "figure eight". This region is generally in an southwest-northeast direction from the two radio towers with a maximum distance of 150 miles.

10.7: More Parametric Equations

1. Note that $x = 3\sin t \Rightarrow x^2 = 9\sin^2 t$ and $y = 3\cos t \Rightarrow y^2 = 9\cos^2 t$.

    Thus $x^2 + y^2 = 9\sin^2 t + 9\cos^2 t \Rightarrow x^2 + y^2 = 9(\sin^2 t + \cos^2 t) \Rightarrow x^2 + y^2 = 9$.

    This is a circle centered at $(0, 0)$ with radius 3.

2. Note that $x = 2\sin t \Rightarrow x^2 = 4\sin^2 t$ and $y = 2\cos t \Rightarrow y^2 = 4\cos^2 t$.

    Thus $x^2 + y^2 = 4\sin^2 t + 4\cos^2 t \Rightarrow x^2 + y^2 = 4(\sin^2 t + \cos^2 t) \Rightarrow x^2 + y^2 = 4$.

    This is a circle centered at $(0, 0)$ with radius 2.

3. Here $y + x = 2\sin^2 t + 2\cos^2 t \Rightarrow y + x = 2(\sin^2 t + \cos^2 t) \Rightarrow y + x = 2 \Rightarrow y = 2 - x$.

    This is a line segment connecting the points $(2, 0)$ and $(0, 2)$.

4. Note that $x = \sqrt{5}\sin t \Rightarrow x^2 = 5\sin^2 t$ and $y = \sqrt{3}\cos t \Rightarrow y^2 = 3\cos^2 t$.

    Thus $3x^2 + 5y^2 = 15\sin^2 t + 15\cos^2 t \Rightarrow \dfrac{3x^2}{15} + \dfrac{5y^2}{15} = \sin^2 t + \cos^2 t \Rightarrow \dfrac{x^2}{5} + \dfrac{y^2}{3} = 1$. This is a ellipse.

5. Note that $x = 3\tan t \Rightarrow x^2 = 9\tan^2 t$ and $y = 2\sec t \Rightarrow y^2 = 4\sec^2 t$.

    Thus $9y^2 - 4x^2 = 36\sec^2 t - 36\tan^2 t \Rightarrow \dfrac{9y^2}{36} - \dfrac{4x^2}{36} = \sec^2 t - \tan^2 t \Rightarrow \dfrac{y^2}{4} - \dfrac{x^2}{9} = 1$.

    This is the upper branch of a hyperbola.

6. Note that $x = \cot t \Rightarrow x^2 = \cot^2 t$ and $y = \csc t \Rightarrow y^2 = \csc^2 t$.

    Thus $y^2 - x^2 = \csc^2 t - \cot^2 t \Rightarrow y^2 - x^2 = 1$. This is the upper branch of a hyperbola.

7. (a) The graph traces a circle of radius 3 once. See Figure 7a.

    (b) The graph traces a circle of radius 3 twice. See Figure 7b.

8. (a) The graph traces a circle of radius 2 once counterclockwise starting at $(2, 0)$. See Figure 8a.

    (b) The graph traces a circle of radius 2 once clockwise starting at $(2, 0)$. See Figure 8b.

$[-4.7, 4.7]$ by $[-3.1, 3.1]$	$[-4.7, 4.7]$ by $[-3.1, 3.1]$	$[-4.7, 4.7]$ by $[-3.1, 3.1]$	$[-4.7, 4.7]$ by $[-3.1, 3.1]$
Xscl = 1 Yscl = 1	Xscl = 1 Yscl = 1	Xscl = 1 Yscl = 1	Xscl = 1 Yscl = 1

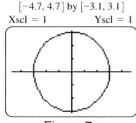 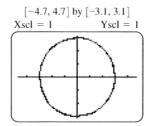 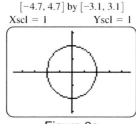 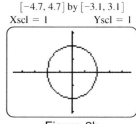

Figure 7a	Figure 7b	Figure 8a	Figure 8b

9. (a) The graph traces a circle of radius 3 once counterclockwise starting at $(3, 0)$. See Figure 9a.

    (b) The graph traces a circle of radius 3 once clockwise starting at $(3, 0)$. See Figure 9b.

10. (a) The graph traces a circle of radius 1 centered at $(-1, 2)$. See Figure 10a.

    (b) The graph traces a circle of radius 1 centered at $(1, 2)$. See Figure 10b.

$[-4.7, 4.7]$ by $[-3.1, 3.1]$	$[-4.7, 4.7]$ by $[-3.1, 3.1]$	$[-4.7, 4.7]$ by $[-3.1, 3.1]$	$[-4.7, 4.7]$ by $[-3.1, 3.1]$
Xscl = 1 Yscl = 1	Xscl = 1 Yscl = 1	Xscl = 1 Yscl = 1	Xscl = 1 Yscl = 1

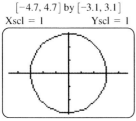 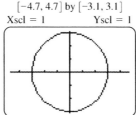 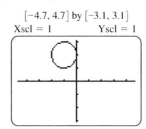 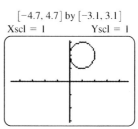

Figure 9a	Figure 9b	Figure 10a	Figure 10b

11. $xy = \sin t \csc t \Rightarrow xy = 1 \Rightarrow y = \dfrac{1}{x}$. For t in the interval $(0, \pi)$, $x = \sin t$ takes on the values in $(0, 1)$.

    See Figure 11.

12. $xy = \tan t \cot t \Rightarrow xy = 1 \Rightarrow y = \dfrac{1}{x}$. For t in the interval $\left(0, \dfrac{\pi}{2}\right)$, $x = \tan t$ takes on the values in $(0, \infty)$.

    See Figure 12.

13. $x = 2 + \sin t \Rightarrow x - 2 = \sin t \Rightarrow (x - 2)^2 = \sin^2 t$ and $y = 1 + \cos t \Rightarrow y - 1 = \cos t \Rightarrow$

    $(y - 1)^2 = \cos^2 t$. Thus $(x - 2)^2 + (y - 1)^2 = \sin^2 t + \cos^2 t \Rightarrow (x - 2)^2 + (y - 1)^2 = 1$.

    For t in the interval $[0, 2\pi]$, $x = 2 + \sin t$ takes on the values in $[1, 3]$. See Figure 13.

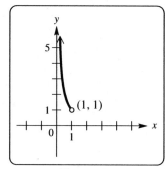

Figure 11

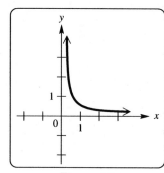

Figure 12

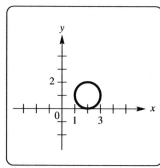

Figure 13

14. $x = 1 + 2\sin t \Rightarrow x - 1 = 2\sin t \Rightarrow (x - 1)^2 = 4\sin^2 t$ and $y = 2 + 3\cos t \Rightarrow y - 2 = 3\cos t \Rightarrow$

    $(y - 2)^2 = 9\cos^2 t$. Thus $9(x - 1)^2 + 4(y - 2)^2 = 36\sin^2 t + 36\cos^2 t \Rightarrow 9(x - 1)^2 + 4(y - 2)^2 = 36 \Rightarrow.$

    $\dfrac{(x - 1)^2}{4} + \dfrac{(y - 2)^2}{9} = 1$. For t in the interval $[0, 2\pi]$, $x = 1 + 2\sin t$ takes on values in $[-1, 3]$. See Figure 14.

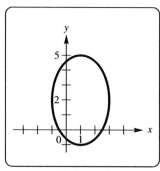

Figure 14

$[-4.7, 4.7]$ by $[-3.1, 3.1]$
Xscl $= 1$ Yscl $= 1$

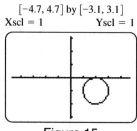

Figure 15

$[-4.7, 4.7]$ by $[-3.1, 3.1]$
Xscl $= 1$ Yscl $= 1$

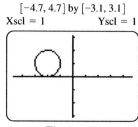

Figure 16

15. See Figure 15.

16. See Figure 16.

17. See Figure 17.

18. See Figure 18.

19. See Figure 19.

20. See Figure 20.

[−1.5, 1.5] by [−1, 1]
Xscl = .5   Yscl = .5

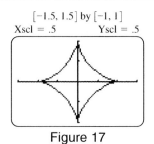

Figure 17

[−1.5, 1.5] by [−1, 1]
Xscl = .5   Yscl = .5

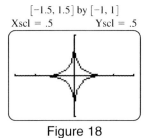

Figure 18

[−4.7, 4.7] by [−3.1, 3.1]
Xscl = 1   Yscl = 1

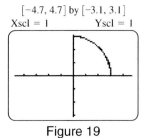

Figure 19

[−4.7, 4.7] by [−3.1, 3.1]
Xscl = 1   Yscl = 1

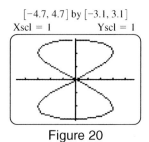

Figure 20

21. See Figure 21.

22. See Figure 22.

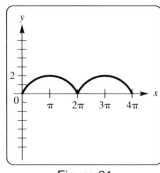

Figure 21

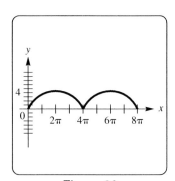

Figure 22

23. The letter graphed is F. See Figure 23.

24. The letter graphed is E. See Figure 24.

25. The letter graphed is D. See Figure 25.

26. The letter graphed is e. See Figure 26.

[0, 6] by [0, 4]
Xscl = 1   Yscl = 1

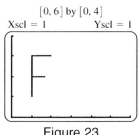

Figure 23

[0, 6] by [0, 4]
Xscl = 1   Yscl = 1

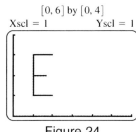

Figure 24

[0, 6] by [0, 4]
Xscl = 1   Yscl = 1

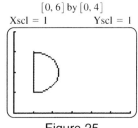

Figure 25

[0, 6] by [0, 4]
Xscl = 1   Yscl = 1

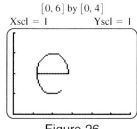

Figure 26

27. Answers may vary. One possibility is: $x_1 = 0$, $y_1 = 2t$, $x_2 = t$, $y_2 = 0$ for $0 \le t \le 1$.

28. Answers may vary. One possibility is: $x_1 = 2 \cos t$, $y_1 = 2 \sin t$ for $.8 \le t \le 5.5$.

29. Answers may vary. One possibility is: $x_1 = \sin t$, $y_1 = \cos t$, $x_2 = 0$, $y_2 = t - 2$ for $0 \le t \le \pi$.

30. Answers may vary. One possibility is: $x_1 = 0$, $y_1 = t$, $x_2 = 2$, $y_2 = t$, $x_3 = t$, $y_3 = 2 - t$ for $0 \le t \le 2$.

31. Answers may vary.

32. Answers may vary. One way to add a nose to the face is to graph $x_5 = .2 \cos t$, $y_5 = -.1 + .2 \sin t$ as part of the solution to the previous exercise.

33. See Figure 33.

34. See Figure 34.

35. See Figure 35.

36. See Figure 36.

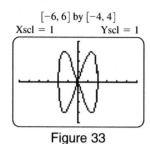
Figure 33

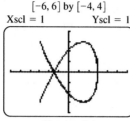
Figure 34

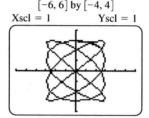
Figure 35

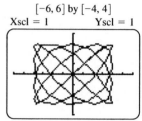
Figure 36

37. (a) $x = (48\cos 60°)t \Rightarrow x = 48\left(\dfrac{1}{2}\right)t \Rightarrow x = 24t$

$y = (48\sin 60°)t - 16t^2 + 0 \Rightarrow y = 48\left(\dfrac{\sqrt{3}}{2}\right)t - 16t^2 \Rightarrow y = -16t^2 + 24\sqrt{3}\,t$

(b) $x = 24t \Rightarrow t = \dfrac{x}{24}$ thus $y = -16t^2 + 24\sqrt{3}\,t \Rightarrow y = -16\left(\dfrac{x}{24}\right)^2 + 24\sqrt{3}\left(\dfrac{x}{24}\right) \Rightarrow y = -\dfrac{1}{36}x^2 + \sqrt{3}\,x$

(c) Solve the equation $-16t^2 + 24\sqrt{3}\,t = 0$ to determine when the rocket is at ground level. The solutions are 0 and 2.6 seconds. Therefore the rocket is in flight for about 2.6 seconds. The horizontal distance traveled is $x = 24t \Rightarrow x = 24(2.6) \approx 62$ feet.

38. (a) $x = (150\cos 60°)t \Rightarrow x \approx 150\left(\dfrac{1}{2}\right)t \Rightarrow x = 75t$

$y = (150\sin 60°)t - 16t^2 + 0 \Rightarrow y = 150\left(\dfrac{\sqrt{3}}{2}\right)t - 16t^2 \Rightarrow y = -16t^2 + 75\sqrt{3}\,t$

(b) $x = 75t \Rightarrow t = \dfrac{x}{75}$ thus $y = -16t^2 + 75\sqrt{3}\,t \Rightarrow y = -16\left(\dfrac{x}{75}\right)^2 + 75\sqrt{3}\left(\dfrac{x}{75}\right) \Rightarrow$

$y = -\dfrac{16}{5625}x^2 + \sqrt{3}\,x$

(c) Solve the equation $-16t^2 + 75\sqrt{3}\,t = 0$ to determine when the ball is at ground level. The solutions are 0 and 8.12 seconds. Therefore the ball is in flight for about 8.1 seconds. The horizontal distance traveled is $x = 75t \Rightarrow x = 75(8.12) \approx 609$ feet.

39. (a) $x = (88\cos 20°)t$

$y = (88\sin 20°)t - 16t^2 + 2 \Rightarrow y = 2 - 16t^2 + (88\sin 20°)t$

(b) $x = (88\cos 20°)t \Rightarrow t = \dfrac{x}{88\cos 20°}$ thus $y = 2 - 16\left(\dfrac{x}{88\cos 20°}\right)^2 + (88\sin 20°)\left(\dfrac{x}{88\cos 20°}\right) \Rightarrow$

$y = 2 - \dfrac{x^2}{484\cos^2 20°} + (\tan 20°)x$

(c) Use the quadratic formula to solve the equation $2 - 16t^2 + (88\sin 20°)t = 0$ to determine when the ball is at ground level. The solutions are $-.064$ and 1.945 seconds. Therefore the ball is in flight for about 1.9 seconds. The horizontal distance traveled is $x = (88\cos 20°)(1.945) \approx 161$ feet.

40. (a) $x = (136\cos 29°)t$ and $y = (136\sin 29°)t - 16t^2 + 2.5 \Rightarrow y = 2.5 - 16t^2 + (136\sin 29°)t$

(b) $x = (136\cos 29°)t \Rightarrow t = \dfrac{x}{136\cos 29°}$ thus $y = 2.5 - 16\left(\dfrac{x}{136\cos 29°}\right)^2 + (136\sin 29°)\left(\dfrac{x}{136\cos 29°}\right) \Rightarrow$

$y = 2.5 - \dfrac{x^2}{1156\cos^2 29°} + (\tan 29°)x$

(c) Use the quadratic formula to solve the equation $2.5 - 16t^2 + (136\sin 29°)t = 0$ to determine when the ball

is at ground level. The solutions are $-.038$ and 4.158 seconds. Therefore the ball is in flight for about 4.2

seconds. The horizontal distance traveled is $x = (136\cos 29°)(4.158) \approx 495$ feet.

41. $x = (88\cos 45°)t \Rightarrow x = 88\left(\dfrac{\sqrt{2}}{2}\right)t \Rightarrow x = 44\sqrt{2}\,t$

$y = (88\sin 45°)t - 2.66t^2 + 0 \Rightarrow y = 88\left(\dfrac{\sqrt{2}}{2}\right)t - 2.66t^2 \Rightarrow y = -2.66t^2 + 44\sqrt{2}\,t$

Solve the equation $-2.66t^2 + 44\sqrt{2}\,t = 0$ to determine when the ball is at ground level. The solutions are

0 and 23.393 seconds. The horizontal distance traveled is $x = 44\sqrt{2}\,t \Rightarrow x = 24\sqrt{2}(23.393) \approx 1456$ feet.

42. (a) $x = (64\cos 60°)t \Rightarrow x = 64\left(\dfrac{1}{2}\right)t \Rightarrow x = 32t$

$y = (64\sin 60°)t - 16t^2 + 3 \Rightarrow y = 64\left(\dfrac{\sqrt{3}}{2}\right)t - 16t^2 + 3 \Rightarrow y = 3 - 16t^2 + 32\sqrt{3}\,t$

(b) Use the quadratic formula to solve the equation $3 - 16t^2 + 32\sqrt{3}\,t = 0$ to determine when the ball

is at ground level. The solutions are $-.053$ and 3.51741 seconds. The horizontal distance traveled is

$x = 32(3.51741) \approx 112.6$ feet.

(c) By graphing the parametric equations (not shown), the maximum height can be determined to be

approximately 51 feet. At that time the ball has traveled about 55.4 feet horizontally.

(d) The ball is 100 feet from the batter when $t = \dfrac{100}{32} = 3.125$ seconds. At that moment the ball is

$y = 3 - 16(3.125)^2 + 32\sqrt{3}(3.125) \approx 20$ feet high. The ball will clear the fence.

43. (a) See Figure 43.

(b) $(88\cos\theta)t = 82.69265063t \Rightarrow \cos\theta = \dfrac{82.69265063}{88} \Rightarrow \theta = \cos^{-1}\left(\dfrac{82.69265063}{88}\right) \approx 20.0°$

(c) $x = (88\cos 20.0°)t$ and $y = -16t^2 + (88\sin 20.0°)t$

44. (a) See Figure 44.

(b) $(88\cos\theta)t = 56.56530965t \Rightarrow \cos\theta = \dfrac{56.56530965}{88} \Rightarrow \theta = \cos^{-1}\left(\dfrac{56.56530965}{88}\right) \approx 50.0°$

(c) $x = (88\cos 50.0°)t$ and $y = -16t^2 + (88\sin 50.0°)t$

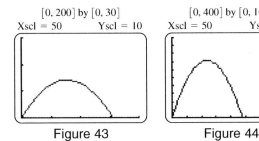

$[0, 200]$ by $[0, 30]$ Xscl $= 50$ Yscl $= 10$

$[0, 400]$ by $[0, 100]$ Xscl $= 50$ Yscl $= 10$

Figure 43   Figure 44

45. Since $x = r\cos\theta$ and $y = r\sin\theta$, by substitution, $x = a\theta\cos\theta$ and $y = a\theta\sin\theta$ for θ in $(-\infty, \infty)$.

46. Since $r = \dfrac{a}{\theta}$, $x = r\cos\theta$ and $y = r\sin\theta$, by substitution, $x = \dfrac{a}{\theta}\cos\theta$ and $y = \dfrac{a}{\theta}\sin\theta$.

    That is $x = \dfrac{a\cos\theta}{\theta}$ and $y = \dfrac{a\sin\theta}{\theta}$ for θ in $(-\infty, 0)\cup(0, \infty)$.

Reviewing Basic Concepts (Sections 10.6 and 10.7)

1. The point is in quadrant IV.

2. $r = \sqrt{(-2)^2 + 2^2} = \sqrt{8} = 2\sqrt{2}; \theta = \tan^{-1}\left(\dfrac{2}{-2}\right) = -45°$. Since the point is in QII, $\theta = 135°$.

    Two forms are $(2\sqrt{2}, 135°)$ and $(-2\sqrt{2}, -45°)$.

3. The graph is a cardioid. See Figure 3.

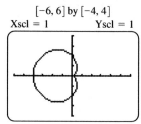

   [-6, 6] by [-4, 4]   [-3, 3] by [-2, 2]
   Xscl = 1   Yscl = 1   Xscl = 1   Yscl = 1

   Figure 3   Figure 7

4. $r = 2\cos\theta \Rightarrow r^2 = 2r\cos\theta \Rightarrow x^2 + y^2 = 2x \Rightarrow x^2 - 2x + 1 + y^2 = 1 \Rightarrow (x - 1)^2 + y^2 = 1$

5. $x + y = 6 \Rightarrow r\cos\theta + r\sin\theta = 6 \Rightarrow r(\cos\theta + \sin\theta) = 6 \Rightarrow r = \dfrac{6}{\cos\theta + \sin\theta}$

6. Note that $x = 2\cos t \Rightarrow x^2 = 4\cos^2 t$ and $y = 4\sin t \Rightarrow y^2 = 16\sin^2 t$.

    Thus $4x^2 + y^2 = 16\cos^2 t + 16\sin^2 t \Rightarrow 4x^2 + y^2 = 16(\cos^2 t + \sin^2 t) \Rightarrow 4x^2 + y^2 = 16 \Rightarrow \dfrac{x^2}{4} + \dfrac{y^2}{16} = 1$.

7. See Figure 7.

8. $x = (88\cos 45°)t \Rightarrow x = 88\left(\dfrac{\sqrt{2}}{2}\right)t \Rightarrow x = 44\sqrt{2}\,t$

    $y = (88\sin 45°)t - 16t^2 + 50 \Rightarrow y = 88\left(\dfrac{\sqrt{2}}{2}\right)t - 16t^2 + 50 \Rightarrow y = 50 - 16t^2 + 88\sqrt{2}\,t$

    Use the quadratic formula to solve the equation $50 - 16t^2 + 44\sqrt{2}\,t = 0$ to determine when the ball is at ground level. The solutions are $-.683$ and 4.5725 seconds. The horizontal distance traveled is $x = 44\sqrt{2}(4.5725) \approx 285$ feet.

Chapter 10 Review Exercises

1. Using the Law of Sines, $\dfrac{b}{\sin 39° 30'} = \dfrac{96.3}{\sin 74° 10'} \Rightarrow b = \dfrac{96.3 \sin 39° 30'}{\sin 74° 10'} \Rightarrow b \approx 63.7$ m.

2. Using the Law of Sines, $\dfrac{b}{\sin 25.0°} = \dfrac{165}{\sin 100.2°} \Rightarrow b = \dfrac{165 \sin 25.0°}{\sin 100.2°} \Rightarrow b \approx 70.9$ m.

3. Using the Law of Cosines,

$$86.14^2 = 253.2^2 + 241.9^2 - 2(253.2)(241.9)\cos A \Rightarrow A = \cos^{-1}\left(\frac{7420.0996 - 122,625.85}{-122,498.16}\right) \approx 19.87°.$$

4. Using the Law of Cosines,

$$19.7^2 = 14.8^2 + 31.8^2 - 2(14.8)(31.8)\cos B \Rightarrow B = \cos^{-1}\left(\frac{388.09 - 1230.28}{-941.28}\right) \approx 26.5°.$$

5. Using the Law of Sines,

$$\frac{\sin B}{69.8} = \frac{\sin 129° \, 40'}{127} \Rightarrow \sin B = \frac{69.8 \sin 129° \, 40'}{127} \Rightarrow B = \sin^{-1}\left(\frac{69.8 \sin 129° \, 40'}{127}\right) \approx 25° \, 00'.$$

6. Using the Law of Sines,

$$\frac{\sin A}{340} = \frac{\sin 39° \, 50'}{268} \Rightarrow \sin A = \frac{340 \sin 39° \, 50'}{268} \Rightarrow A = \sin^{-1}\left(\frac{340 \sin 39° \, 50'}{268}\right) \approx 54° \, 20'.$$

   Since $180° - 54° \, 20' = 125° \, 40'$ and $39° \, 50' + 125° \, 40' < 180°$, the other possible angle is $125° \, 40'$.

7. Using the Law of Cosines,

$$b^2 = 127^2 + 69.8^2 - 2(127)(69.8)\cos 120.7° \Rightarrow b = \sqrt{21,001.04 - 17,729.2 \cos 120.7°} \approx 173 \text{ ft}$$

8. Using the Law of Cosines,

$$a^2 = 184^2 + 192^2 - 2(184)(192)\cos 46.2° \Rightarrow a = \sqrt{70,720 - 70,656 \cos 46.2°} \approx 148 \text{ cm}$$

9. $A = \dfrac{1}{2}(840.6)(715.9)\sin 149.3° \approx 153,600 \text{ m}^2$

10. $A = \dfrac{1}{2}(6.90)(10.2)\sin 35° \, 10' \approx 20.3 \text{ ft}^2$

11. $s = \dfrac{1}{2}(.913 + .816 + .582) = 1.1555;$

$$A = \sqrt{1.1555(1.1555 - .913)(1.1555 - .816)(1.1555 - .582)} \approx .234 \text{ km}^2$$

12. $s = \dfrac{1}{2}(43 + 32 + 51) = 63; \; A = \sqrt{63(63 - 43)(63 - 32)(63 - 51)} \approx 680 \text{ m}^2$

13. First note that $C = 180° - 47° \, 20' - 24° \, 50' = 107° \, 50'$.

   Now, using the Law of Sines, $\dfrac{a}{\sin 24° \, 50'} = \dfrac{8.4}{\sin 107° \, 50'} \Rightarrow a = \dfrac{8.4 \sin 24° \, 50'}{\sin 107° \, 50'} \Rightarrow a \approx 3.71 \text{ mi.}$

   If a perpendicular line segment joining C to the the ground intersects the ground at D, we can use right

   triangle BDC to find the distance d to the balloon: $\sin 47° \, 20' = \dfrac{d}{3.71} \Rightarrow d = 3.71 \sin 47° \, 20' \approx 2.7 \text{ mi.}$

14. Triangle XYZ has the following measurements: $X = 48°, Z = 36°, Y = 180° - 48° - 36° = 96°$, and $y = 10$.

   Using the Law of Sines, $\dfrac{x}{\sin 48°} = \dfrac{10}{\sin 96°} \Rightarrow x = \dfrac{10 \sin 48°}{\sin 96°} \Rightarrow x \approx 7 \text{ km.}$

15. Let C represent the position of the illegal transmitter. Triangle ABC has the following measurements:

   $A = 90° - 48° = 42°, B = 302° - 270° = 32°, C = 180° - 42° - 32° = 106°$, and $c = 3.46$.

   Using the Law of Sines, $\dfrac{b}{\sin 32°} = \dfrac{3.46}{\sin 106°} \Rightarrow b = \dfrac{3.46 \sin 32°}{\sin 106°} \Rightarrow b \approx 1.91 \text{ mi.}$

16. First note that $A = 180° - 58.4° - 27.9° = 93.7°$.

   Now, using the Law of Sines, $\dfrac{c}{\sin 27.9°} = \dfrac{125}{\sin 93.7°} \Rightarrow c = \dfrac{125 \sin 27.9°}{\sin 93.7°} \Rightarrow c \approx 58.6 \text{ ft.}$

17. Let A, B, and C represent the lower end of the brace, the lower end of the pole, and the upper end of the pole, respectively. Then $A = 180° - 115° - 22° = 43°$, $B = 115°$, $C = 22°$, and $a = 8.0$.

Using the Law of Sines, $\dfrac{b}{\sin 115°} = \dfrac{8.0}{\sin 43°} \Rightarrow b = \dfrac{8.0 \sin 115°}{\sin 43°} \Rightarrow b \approx 11$ ft.

18. Using the Law of Cosines,
$$c^2 = 15^2 + 12.2^2 - 2(15)(12.2)\cos 70.3° \Rightarrow c = \sqrt{373.84 - 366\cos 70.3°} \approx 15.8 \text{ ft}$$

19. Using the Law of Cosines,
$$150^2 = 102^2 + 135^2 - 2(102)(135)\cos C \Rightarrow C = \cos^{-1}\left(\frac{22{,}500 - 28{,}629}{-27{,}540}\right) \approx 77.1°.$$

20. This is not the ambiguous case because we are not given SSA.

21. This triangle cannot exist because $a + b = c$.

22. (a) If A and B lie on a horizontal line with C positioned above the line, then C lies $10\sin 30° = 5$ units above the line. Thus there will be one solution if $b = 5$ or if $b \geq 10$.

(b) There will be two possible solutions if $5 < b < 10$.

(c) There will be no solution if $b < 5$.

23. If $C = 90°$, the value of $2ab\cos\theta$ is 0. The Law of Cosines becomes the Pythagorean theorem.

24. If the cosine value is positive, the angle is acute. If the cosine value is negative, the angle is obtuse.

25. See Figure 25.

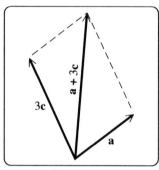

Figure 25

26. $|\langle 21, -20 \rangle| = \sqrt{21^2 + (-20)^2} = \sqrt{441 + 400} = \sqrt{841} = 29$; direction angle: $\tan^{-1}\left(\dfrac{-20}{21}\right) \approx 316.4°$

27. $|\langle -9, 12 \rangle| = \sqrt{(-9)^2 + 12^2} = \sqrt{81 + 144} = \sqrt{225} = 15$; direction angle: $\tan^{-1}\left(\dfrac{12}{-9}\right) \approx 126.9°$

28. $\langle 50\cos 45°, 50\sin 45° \rangle = \left\langle 50 \cdot \dfrac{\sqrt{2}}{2}, 50 \cdot \dfrac{\sqrt{2}}{2} \right\rangle = \langle 25\sqrt{2}, 25\sqrt{2} \rangle$

29. $\langle 69.2\cos 75°, 69.2\sin 75° \rangle \approx \langle 17.9, 66.8 \rangle$

30. $\langle 964\cos 154° \, 20', 964\sin 154° \, 20' \rangle \approx \langle -869, 418 \rangle$

31. (a) $\langle 6, 2 \rangle \cdot \langle 3, -2 \rangle = (6)(3) + (2)(-2) = 18 - 4 = 14$

(b) $\cos\theta = \dfrac{\mathbf{u} \cdot \mathbf{v}}{|\mathbf{u}| \, |\mathbf{v}|} = \dfrac{\langle 6, 2 \rangle \cdot \langle 3, -2 \rangle}{|\langle 6, 2 \rangle| \, |\langle 3, -2 \rangle|} = \dfrac{(6)(3) + (2)(-2)}{\sqrt{36 + 4} \, \sqrt{9 + 4}} = \dfrac{14}{\sqrt{40}\sqrt{13}} = \dfrac{14}{\sqrt{520}} \Rightarrow$

$\theta = \cos^{-1}\left(\dfrac{14}{\sqrt{520}}\right) = 52.13°$

32. (a) $\langle 2\sqrt{3}, 2\rangle \cdot \langle 5, 5\sqrt{3}\rangle = (2\sqrt{3})(5) + (2)(5\sqrt{3}) = 10\sqrt{3} + 10\sqrt{3} = 20\sqrt{3}$

    (b) $\cos\theta = \dfrac{\mathbf{u} \cdot \mathbf{v}}{|\mathbf{u}|\,|\mathbf{v}|} = \dfrac{\langle 2\sqrt{3}, 2\rangle \cdot \langle 5, 5\sqrt{3}\rangle}{|\langle 2\sqrt{3}, 2\rangle|\,|\langle 5, 5\sqrt{3}\rangle|} = \dfrac{(2\sqrt{3})(5) + (2)(5\sqrt{3})}{\sqrt{12 + 4}\,\sqrt{25 + 75}} = \dfrac{20\sqrt{3}}{\sqrt{16}\sqrt{100}} = \dfrac{20\sqrt{3}}{40} = \dfrac{\sqrt{3}}{2} \Rightarrow$

    $\theta = \cos^{-1}\left(\dfrac{\sqrt{3}}{2}\right) = 30°$

33. The vectors are orthogonal because $\mathbf{u} \cdot \mathbf{v} = \langle 5, -1\rangle \cdot \langle -2, -10\rangle = (5)(-2) + (-1)(-10) = -10 + 10 = 0$.

34. The adjacent angle in the parallelogram is supplementary to $10° + 15° = 25°$, thus it has measure $155°$.

    $|\mathbf{v}|^2 = 12^2 + 18^2 - 2(12)(18)\cos 155° \approx 859.524964 \Rightarrow |\mathbf{v}| = \sqrt{859.524964} \approx 29 \text{ lb}$

35. The adjacent angle in the parallelogram is supplementary to $45°$, thus it has measure $135°$.

    $|\mathbf{v}|^2 = 1000^2 + 2000^2 - 2(1000)(2000)\cos 135° \approx 7{,}828{,}427.125 \Rightarrow |\mathbf{v}| = \sqrt{7{,}828{,}427.125} \approx 2798$

    The resultant is about 2800 newtons. The required angle can be found using the Law of Cosines:

    $2000^2 = 1000^2 + 2798^2 - 2(1000)(2798)\cos A \Rightarrow A = \cos^{-1}\left(\dfrac{4{,}000{,}000 - 8{,}828{,}804}{-5{,}596{,}000}\right) \approx 30.4°.$

36. Let A represent the plane's initial position. Let B represent the point at which the vector representing the required flight path intersects the vector representing the wind. Let C represent the point at which the vector representing the desired flight path (310° bearing) intersects the vector representing the wind. Triangle ABC has the following measurements: $C = (360° - 310°) + (212° - 180°) = 82°$, $a = 37$, and $c = 520$.

    Using the Law of Sines, $\dfrac{\sin A}{37} = \dfrac{\sin 82°}{520} \Rightarrow \sin A = \dfrac{37\sin 82°}{520} \Rightarrow A = \sin^{-1}\left(\dfrac{37\sin 82°}{520}\right) \approx 4°.$

    The bearing the plane should take is $310° - 4° = 306°$. Noting that $B = 180° - 82° - 4° = 94°$,

    $\dfrac{b}{\sin 94°} = \dfrac{520}{\sin 82°} \Rightarrow b = \dfrac{520\sin 94°}{\sin 82°} \Rightarrow b \approx 524.$ The plane's actual speed is about 524 mph.

37. The angle between the current and the due north swim path is $180° - 12° = 168°$. If b is the missing side of the triangle, $b^2 = 3.2^2 + 5.1^2 - 2(3.2)(5.1)\cos 168° \Rightarrow b = \sqrt{36.25 - 32.64\cos 168°} \approx 8.3$ mph.

    Then $\dfrac{\sin A}{5.1} = \dfrac{\sin 168°}{8.3} \Rightarrow \sin A = \dfrac{5.1\sin 168°}{8.3} \Rightarrow A = \sin^{-1}\left(\dfrac{5.1\sin 168°}{8.3}\right) \approx 7°\,20'.$

    The swimmer's actual speed is 8.3 mph and the resulting bearing is $7°\,20'$.

38. (a) The speed of the wind is $|\mathbf{v}| = \sqrt{6^2 + 8^2} = \sqrt{36 + 64} = \sqrt{100} = 10$ mph.

    (b) $3\mathbf{v} = 3(6\mathbf{i} + 8\mathbf{j}) = 18\mathbf{i} + 24\mathbf{j}$; This represents a 30 mph wind in the direction of $\mathbf{v}$.

    (c) $\mathbf{u}$ represents a southeast wind with speed $|\mathbf{u}| = \sqrt{(-8)^2 + 8^2} = \sqrt{64 + 64} = \sqrt{128} \approx 11.3$ mph.

39. By placing vector $\mathbf{v}$ at the end of vector $\mathbf{u}$, the angle between them is $(110° - 90°) + (270° - 260°) = 30°$.

    $|\mathbf{u} + \mathbf{v}|^2 = 3^2 + 12^2 - 2(3)(12)\cos 30° \approx 90.64617093 \Rightarrow |\mathbf{u} + \mathbf{v}| = \sqrt{90.64617093} \approx 9.52082827$

    Let A represent the angle between $\mathbf{u}$ and $\mathbf{u} + \mathbf{v}$.

    $\dfrac{\sin A}{3} = \dfrac{\sin 30°}{9.52082827} \Rightarrow \sin A = \dfrac{3\sin 30°}{9.52082827} \Rightarrow A = \sin^{-1}\left(\dfrac{3\sin 30°}{9.52082827}\right) \approx 9.0646784°.$

    The direction angle is $110° + 9.0646784° = 119.0646784°.$

40. $\mathbf{u} = \langle 12\cos 110°, 12\sin 110°\rangle = \langle -4.10424172, 11.27631145\rangle$

41. $\mathbf{u} = \langle 3\cos 260°, 3\sin 260°\rangle = \langle -.520944533, -2.954423259\rangle$

42. $\mathbf{u} + \mathbf{v} = \langle -4.10424172 + (-.520944533), 11.27631145 + (-2.954423259)\rangle = \langle -4.625186253, 8.321888191\rangle$

43. $|\mathbf{u} + \mathbf{v}| = \sqrt{(-4.625186253)^2 + 8.321888191^2} \approx 9.52082827$

$\theta = \tan^{-1}\left(\dfrac{8.321888191}{-4.625186253}\right) \approx -60.93532161°$; The direction angle is $-60.93532161° + 180° = 119.0646784°$.

44. The answers are the same. Preference of method is an individual choice.

45. See Figure 45.

46. See Figure 46.

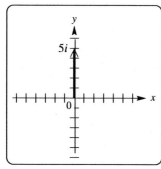

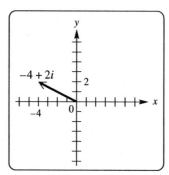

Figure 45   Figure 46

47. $(7 + 3i) + (-2 + i) = (7 + (-2)) + (3 + 1)i = 5 + 4i$

48. $(2 - 4i) + (5 + i) = (2 + 5) + (-4 + 1)i = 7 - 3i$

49. $r = \sqrt{(-2)^2 + 2^2} = \sqrt{8} = 2\sqrt{2}$; $\tan \theta = \dfrac{2}{-2} \Rightarrow \tan \theta = -1 \Rightarrow \theta = \tan^{-1}(-1) = -45°$

   Since θ is in quadrant II, $\theta = -45° + 180° = 135°$. Therefore $-2 + 2i = 2\sqrt{2} \text{ cis } 135°$.

50. $3(\cos 90° + i \sin 90°) = 3(0 + 1i) = 3i$

51. $2 \text{ cis } 225° = 2(\cos 225° + i \sin 225°) = 2\left(-\dfrac{\sqrt{2}}{2} - \dfrac{\sqrt{2}}{2}i\right) = -\sqrt{2} - \sqrt{2}i$

52. $r = \sqrt{(-4)^2 + (4\sqrt{3})^2} = \sqrt{64} = 8$; $\tan \theta = \dfrac{4\sqrt{3}}{-4} \Rightarrow \tan \theta = -\sqrt{3} \Rightarrow \theta = \tan^{-1}(-\sqrt{3}) = -60°$

   Since θ is in quadrant II, $\theta = -60° + 180° = 120°$. Therefore $-4 + 4i\sqrt{3} = 8 \text{ cis } 120°$.

53. $[5(\cos 90° + i \sin 90°)][6(\cos 180° + i \sin 180°)] = (5 \cdot 6)[\cos(90° + 180°) + i \sin(90° + 180°)] =$

   $30(\cos 270° + i \sin 270°) = 30(0 - i) = -30i$

54. $[3 \text{ cis } 135°][2 \text{ cis } 105°][3(\cos 135° + i \sin 135°)][2(\cos 105° + i \sin 105°)] =$

   $(3 \cdot 2)[\cos(135° + 105°) + i \sin(135° + 105°)] =$

   $6(\cos 240° + i \sin 240°) = 6\left(-\dfrac{1}{2} - \dfrac{\sqrt{3}}{2}i\right) = -3 - 3\sqrt{3}i$

55. $\dfrac{2(\cos 60° + i \sin 60°)}{8(\cos 300° + i \sin 300°)} = \dfrac{2}{8}[\cos(60° - 300°) + i \sin(60° - 300°)] =$

   $\dfrac{1}{4}[\cos(-240°) + i \sin(-240°)] = \dfrac{1}{4}\left(-\dfrac{1}{2} + \dfrac{\sqrt{3}}{2}i\right) = -\dfrac{1}{8} + \dfrac{\sqrt{3}}{8}i$

56. $\dfrac{4 \operatorname{cis} 270°}{2 \operatorname{cis} 90°} = \dfrac{4(\cos 270° + i \sin 270°)}{2(\cos 90° + i \sin 90°)} = \dfrac{4}{2}[\cos (270° - 90°) + i \sin (270° - 90°)] =$

$2(\cos 180° + i \sin 180°) = 2(-1 + 0i) = -2$

57. First find the trigonometric form of $\sqrt{3} + i$; $r = \sqrt{(\sqrt{3})^2 + 1^2} = \sqrt{4} = 2$;

$\tan \theta = \dfrac{y}{x} \Rightarrow \tan \theta = \dfrac{1}{\sqrt{3}} \Rightarrow \Rightarrow \theta = \tan^{-1} \dfrac{1}{\sqrt{3}} = 30°$. Therefore $\sqrt{3} + i = 2(\cos 30° + i \sin 30°)$.

$(\sqrt{3} + i)^3 = [2(\cos 30° + i \sin 30°)]^3 = 2^3[\cos (3 \cdot 30°) + i \sin (3 \cdot 30°)] =$

$8(\cos 90° + i \sin 90°) = 8(0 + 1i) = 8i$

58. First find the trigonometric form of $2 - 2i$; $r = \sqrt{2^2 + (-2)^2} = \sqrt{8} = 2\sqrt{2}$;

$\tan \theta = \dfrac{y}{x} \Rightarrow \tan \theta = \dfrac{-2}{2} \Rightarrow \Rightarrow \theta = \tan^{-1}(-1) = 315°$. Therefore $2 - 2i = 2\sqrt{2}(\cos 315° + i \sin 315°)$.

$(2 - 2i)^5 = [2\sqrt{2}(\cos 315° + i \sin 315°)]^5 = (2\sqrt{2})^5[\cos (5 \cdot 315°) + i \sin (5 \cdot 315°)] =$

$128\sqrt{2}(\cos 1575° + i \sin 1575°) = 128\sqrt{2}\left(-\dfrac{\sqrt{2}}{2} + \dfrac{\sqrt{2}}{2}i\right) = -128 + 128i$

59. $(\cos 100° + i \sin 100°)^6 = \cos (6 \cdot 100°) + i \sin (6 \cdot 100°) = \cos 600° + i \sin 600° = -\dfrac{1}{2} - \dfrac{\sqrt{3}}{2}i$

60. $(\operatorname{cis} 20°)^3 = (\cos 20° + i \sin 20°)^3 = \cos (3 \cdot 20°) + i \sin (3 \cdot 20°) = \cos 60° + i \sin 60° = \dfrac{1}{2} + \dfrac{\sqrt{3}}{2}i$

61. Note that $-27i = 27(\cos 270° + i \sin 270°)$. Here $r = 27$ so $\sqrt[3]{r} = 3$, $n = 3$ and $k = 0, 1,$ or 2.

$\alpha = \dfrac{270° + 360° \cdot k}{3}$: For $k = 0$, $\alpha = 90°$. For $k = 1$, $\alpha = 210°$. For $k = 2$, $\alpha = 330°$.

The cube roots are $3 \operatorname{cis} 90°$, $3 \operatorname{cis} 210°$, and $3 \operatorname{cis} 330°$. See Figure 61.

62. Note that $16i = 16(\cos 90° + i \sin 90°)$. Here $r = 16$ so $\sqrt[4]{r} = 4$, $n = 4$ and $k = 0, 1, 2,$ or 3.

$\alpha = \dfrac{90° + 360° \cdot k}{4}$: For $k = 0$, $\alpha = 22.5°$. For $k = 1$, $\alpha = 112.5°$. For $k = 2$, $\alpha = 202.5°$. For

$k = 3$, $\alpha = 292.5°$. The fourth roots are $4 \operatorname{cis} 22.5°$, $4 \operatorname{cis} 112.5°$, $4 \operatorname{cis} 202.5°$, and $4 \operatorname{cis} 292.5°$. See Figure 62.

63. Note that $32 = 32(\cos 0° + i \sin 0°)$.

Here $r = 32$ so $\sqrt[5]{r} = 2$, $n = 5$ and $k = 0, 1, 2, 3,$ or 4. $\alpha = \dfrac{0° + 360° \cdot k}{5}$: For $k = 0$, $\alpha = 0°$.

For $k = 1$, $\alpha = 72°$. For $k = 2$, $\alpha = 144°$. For $k = 3$, $\alpha = 216°$. For $k = 4$, $\alpha = 288°$.

The fifth roots are $2 \operatorname{cis} 0°$, $2 \operatorname{cis} 72°$, $2 \operatorname{cis} 144°$, $2 \operatorname{cis} 216°$, and $2 \operatorname{cis} 288°$. See Figure 63.

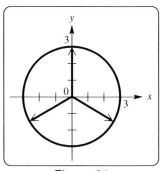

Figure 61

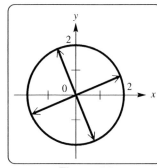

Figure 62

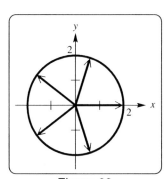

Figure 63

64. $x^4 + i = 0 \Rightarrow x^4 = -i$. Find the fourth roots of $-i$. Note that $-i = \cos 270° + i \sin 270°$.

   Here $r = 1$ so $\sqrt[4]{r} = 1$, $n = 4$ and $k = 0, 1, 2$ or 3. $\alpha = \dfrac{270° + 360° \cdot k}{4}$: For $k = 0$, $\alpha = 67.5°$.

   For $k = 1$, $\alpha = 157.5°$. For $k = 2$, $\alpha = 247.5°$. For $k = 3$, $\alpha = 337.5°$. The fourth roots are

   cis 67.5°, cis 157.5°, cis 247.5°, and cis 337.5°.

65. $x = r \cos \theta = 12 \cos 225° = 12\left(-\dfrac{\sqrt{2}}{2}\right) = -6\sqrt{2}$; $y = r \sin \theta = 12 \sin 225° = 12\left(-\dfrac{\sqrt{2}}{2}\right) = -6\sqrt{2}$.

   The rectangular coordinates are $(-6\sqrt{2}, -6\sqrt{2})$.

66. $x = r \cos \theta = -8 \cos\left(-\dfrac{\pi}{3}\right) = -8\left(\dfrac{1}{2}\right) = -4$; $y = r \sin \theta = -8 \sin\left(-\dfrac{\pi}{3}\right) = -8\left(-\dfrac{\sqrt{3}}{2}\right) = 4\sqrt{3}$.

   The rectangular coordinates are $(-4, 4\sqrt{3})$.

67. $r = \sqrt{(-6)^2 + 6^2} = \sqrt{72} = 6\sqrt{2}$; $\theta = \tan^{-1}\left(\dfrac{6}{-6}\right) = -45°$. Since the point is in QII, $\theta = 135°$.

   One possible form is $(6\sqrt{2}, 135°)$.

68. $r = \sqrt{0^2 + (-5)^2} = 5$; $\theta = \tan^{-1}\left(\dfrac{-5}{0}\right) \Rightarrow$ undefined $\Rightarrow \theta = 270° \Rightarrow \theta = -90°$ (negative y-axis).

   One possible form is $(5, -90°)$.

69. See Figure 69.

70. See Figure 70.

71. See Figure 71.

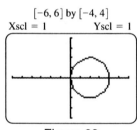

Figure 69

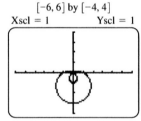

Figure 70

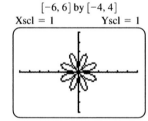

Figure 71

72. See Figure 72.

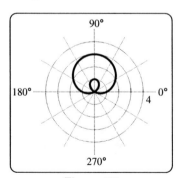

Figure 72

73. $r = \dfrac{3}{1 + \cos\theta} \Rightarrow r + r\cos\theta = 3 \Rightarrow \sqrt{x^2 + y^2} + x = 3 \Rightarrow \sqrt{x^2 + y^2} = 3 - x \Rightarrow x^2 + y^2 = (3 - x)^2$

     $\Rightarrow x^2 + y^2 = x^2 - 6x + 9 \Rightarrow y^2 = -6x + 9 \Rightarrow y^2 + 6x - 9 = 0.$

74. $r = \dfrac{4}{2\sin\theta - \cos\theta} \Rightarrow 2r\sin\theta - r\cos\theta = 4 \Rightarrow 2y - x = 4.$

75. $r = \sin\theta + \cos\theta \Rightarrow r^2 = r\sin\theta + r\cos\theta \Rightarrow x^2 + y^2 = y + x \Rightarrow x^2 - x + y^2 - y = 0$

76. $r = 2 \Rightarrow r^2 = 4 \Rightarrow x^2 + y^2 = 4$

77. $x = -3 \Rightarrow r\cos\theta = -3 \Rightarrow r = \dfrac{-3}{\cos\theta} \Rightarrow r = -\dfrac{3}{\cos\theta}$

78. $y = x \Rightarrow r\sin\theta = r\cos\theta \Rightarrow \dfrac{r\sin\theta}{r\cos\theta} = 1 \Rightarrow \tan\theta = 1$

79. $y = x^2 \Rightarrow r\sin\theta = r^2\cos^2\theta \Rightarrow \dfrac{r\sin\theta}{r\cos^2\theta} = r \Rightarrow r = \tan\theta \cdot \dfrac{1}{\cos\theta} \Rightarrow r = \dfrac{\tan\theta}{\cos\theta}$

80. $x = y^2 \Rightarrow r\cos\theta = r^2\sin^2\theta \Rightarrow \dfrac{r\cos\theta}{r\sin^2\theta} = r \Rightarrow r = \cot\theta \cdot \dfrac{1}{\sin\theta} \Rightarrow r = \dfrac{\cot\theta}{\sin\theta}$

81. Note that $x = \cos 2t \Rightarrow x = \cos^2 t - \sin^2 t$ and $y = \sin t \Rightarrow y^2 = \sin^2 t.$

     Thus $x + 2y^2 = \cos^2 t - \sin^2 t + 2\sin^2 t \Rightarrow x + 2y^2 = \cos^2 t + \sin^2 t \Rightarrow x + 2y^2 = 1 \Rightarrow 2y^2 + x - 1 = 0.$

     This can also be written $y^2 = -\dfrac{1}{2}(x - 1).$ For t in the interval $(-\pi, \pi)$, $x = \cos 2t$ takes on the values in $[-1, 1].$

82. Note that $x = 5\tan t \Rightarrow x^2 = 25\tan^2 t$ and $y = 3\sec t \Rightarrow y^2 = 9\sec^2 t.$

     Thus $25y^2 - 9x^2 = 225\sec^2 t - 225\tan^2 t \Rightarrow \dfrac{25y^2}{225} - \dfrac{9x^2}{225} = \sec^2 t - \tan^2 t \Rightarrow \dfrac{y^2}{9} - \dfrac{x^2}{25} = 1.$ This can also

     be written $y = 3\sqrt{1 + \dfrac{x^2}{25}}.$ For t in the interval $\left(-\dfrac{\pi}{2}, \dfrac{\pi}{2}\right)$, $x = 5\tan t$ takes on the values in $(-\infty, \infty).$

83. See Figure 83.

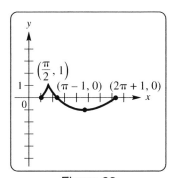

Figure 83

84. $x = (118\cos 27°)t;\ y = (118\sin 27°)t - 16t^2 + 3.2 \Rightarrow y = 3.2 - 16t^2 + (118\sin 27°)t$

     Use the quadratic formula to solve the equation $y = 3.2 - 16t^2 + (118\sin 27°)t = 0$ to determine when the

     ball is at ground level. The solutions are $-.059$ and 3.4069 seconds. The horizontal distance traveled is

     $x = (118\cos 27°)(3.4069) \approx 360$ feet.

Chapter 10 Test

1. (a) By the Law of Sines, $\dfrac{\sin B}{4.82} = \dfrac{\sin 25.2°}{6.92} \Rightarrow \sin B = \dfrac{4.82 \sin 25.2°}{6.92} \Rightarrow B = \sin^{-1}\left(\dfrac{4.82 \sin 25.2°}{6.92}\right) \approx 17.3°.$

   $C = 180° - 25.2° - 17.3° = 137.5°$

   (b) By the Law of Cosines,

   $c^2 = 132^2 + 75.1^2 - 2(132)(75.1)\cos 118° \Rightarrow c = \sqrt{23,064.01 - 19,826.4\cos 118°} \approx 180$ km

   (c) By the Law of Cosines,

   $22.6^2 = 17.3^2 + 29.8^2 - 2(17.3)(29.8)\cos B \Rightarrow B = \cos^{-1}\left(\dfrac{510.76 - 1187.33}{-1031.08}\right) \approx 49.0°.$

2. (a) Since angle B is obtuse, there will be one solution if $b > 10$.

   (b) Since angle B is obtuse, it is not possible to have two solutions.

   (c) There will be no solution if $b \le 10$.

3. The sum of the two smaller numbers must be greater than the largest number.

4. (a) After 3 hours the ships have traveled $36.2(3) = 108.6$ miles and $45.6(3) = 136.8$ miles.

   Use the Law of Cosines to find the distance d:

   $d^2 = 108.6^2 + 136.8^2 - 2(108.6)(136.8)\cos 54.2° \Rightarrow d = \sqrt{30,508.2 - 29,712.96\cos 54.2°} \approx 115$ km.

   (b) If H represents home plate, then angle MHT is $45°$ (half of $90°$). Use the Law of Cosines to find the

   distance d: $d^2 = 46^2 + 60^2 - 2(46)(60)\cos 45° \Rightarrow d = \sqrt{5716 - 5520\cos 45°} \approx 43$ ft.

   (c) The horizontal component is $569\cos 127.5° \approx -346$. The vertical component is $569\sin 127.5° \approx 451$.

   (d) The adjacent angle in the parallelogram is supplementary to $78.2°$, thus it has measure $101.8°$.

   $|\mathbf{v}|^2 = 475^2 + 586^2 - 2(475)(586)\cos 101.8° \approx 682,863.9521 \Rightarrow |\mathbf{v}| = \sqrt{682,863.9521} \approx 826$ lb

5. (a) $\mathbf{u} + \mathbf{v} = \langle -1, 3\rangle + \langle 2, -6\rangle = \langle -1 + 2, 3 + (-6)\rangle = \langle 1, -3\rangle$

   (b) $-3\mathbf{v} = -3\langle 2, -6\rangle = \langle -6, 18\rangle$

   (c) $\mathbf{u} \cdot \mathbf{v} = \langle -1, 3\rangle \cdot \langle 2, -6\rangle = (-1)(2) + (3)(-6) = -20$

   (d) $\cos\theta = \dfrac{\mathbf{u} \cdot \mathbf{v}}{|\mathbf{u}|\,|\mathbf{v}|} = \dfrac{\langle -1, 3\rangle \cdot \langle 2, -6\rangle}{|\langle -1, 3\rangle|\,|\langle 2, -6\rangle|} = \dfrac{(-1)(2) + (3)(-6)}{\sqrt{1 + 9}\,\sqrt{4 + 36}} = \dfrac{-20}{\sqrt{10}\sqrt{40}} = -1 \Rightarrow \theta = \cos^{-1}(-1) = 180°$

6. (a) $4\operatorname{cis} 240° = 4(\cos 240° + i\sin 240°) = 4\left(-\dfrac{1}{2} - i\dfrac{\sqrt{3}}{2}\right) = -2 - 2i\sqrt{3}$

   (b) $r = \sqrt{(-4)^2 + (4\sqrt{3})^2} = \sqrt{64} = 8;$

   $\tan\theta = \dfrac{y}{x} \Rightarrow \tan\theta = \dfrac{4\sqrt{3}}{-4} \Rightarrow \tan\theta = -\sqrt{3} \Rightarrow \theta = \tan^{-1}(-\sqrt{3}) = -60°$

   Since θ is in quadrant II, $\theta = -60° + 180° = 120°$. Therefore $-4 + 4i\sqrt{3} = 8\operatorname{cis} 120°$.

   (c) $(-2 - 2i\sqrt{3}) + (-4 + 4i\sqrt{3}) = -6 + 2i\sqrt{3}.$

7. (a) $[3(\cos 30° + i\sin 30°)][5(\cos 90° + i\sin 90°)] = (3 \cdot 5)[\cos (30° + 90°) + i\sin (30° + 90°)] =$

$15(\cos 120° + i\sin 120°) = 15\left(-\dfrac{1}{2} + \dfrac{\sqrt{3}}{2}i\right) = -\dfrac{15}{2} + \dfrac{15\sqrt{3}}{2}i$

(b) $\dfrac{2 \operatorname{cis} 315°}{4 \operatorname{cis} 45°} = \dfrac{2(\cos 315° + i\sin 315°)}{4(\cos 45° + i\sin 45°)} = \dfrac{2}{4}[\cos (315° - 45°) + i\sin (315° - 45°)] =$

$\dfrac{1}{2}(\cos 270° + i\sin 270°) = \dfrac{1}{2}(0 - 1i) = -\dfrac{1}{2}i$

(c) First find the trigonometric form of $1 - i\sqrt{3}$; $r = \sqrt{1^2 + (-\sqrt{3})^2} = \sqrt{4} = 2$;

$\tan \theta = \dfrac{y}{x} \Rightarrow \tan \theta = \dfrac{-\sqrt{3}}{1} \Rightarrow \Rightarrow \theta = \tan^{-1}(-\sqrt{3}) = -60°$. Therefore $1 - i\sqrt{3} = 2 \operatorname{cis}(-60°)$.

$(1 - i\sqrt{3})^5 = [2(\cos (-60°) + i\sin (-60°))]^5 = 2^5[\cos (5 \cdot (-60°)) + i\sin (5 \cdot (-60°))] =$

$32[\cos (-300°) + i\sin (-300°)] = 32\left(\dfrac{1}{2} + \dfrac{\sqrt{3}}{2}i\right) = 16 + 16i\sqrt{3}$

(d) Note that $\sqrt{3} + i = 2(\cos 30° + i\sin 30°)$. Here $r = 2$ so $\sqrt[4]{r} = \sqrt[4]{2}$, $n = 4$ and $k = 0, 1, 2,$ or 3.

$\alpha = \dfrac{30° + 360° \cdot k}{4}$: For $k = 0$, $\alpha = 7.5°$. For $k = 1$, $\alpha = 97.5°$. For $k = 2$, $\alpha = 187.5°$.

For $k = 3$, $\alpha = 277.5°$. The fourth roots are $\sqrt[4]{2} \operatorname{cis} 7.5°$, $\sqrt[4]{2} \operatorname{cis} 97.5°$, $\sqrt[4]{2} \operatorname{cis} 187.5°$, and $\sqrt[4]{2} \operatorname{cis} 277.5°$.

8. (a) See Figure 8.

(b) $r = 4 \cos \theta \Rightarrow r^2 = 4r \cos \theta \Rightarrow x^2 + y^2 = 4x \Rightarrow x^2 - 4x + y^2 = 0$

(c) Yes, by rewriting the equation, we see that the graph is a circle with center $(2, 0)$ and radius 2.

$x^2 - 4x + y^2 = 0 \Rightarrow x^2 - 4x + 4 + y^2 = 4 \Rightarrow (x - 2)^2 + y^2 = 4$

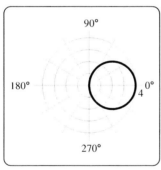

Figure 8

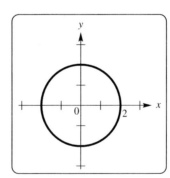

Figure 10

9. $-x + 2y = 4 \Rightarrow -r\cos \theta + 2r\sin \theta = 4 \Rightarrow r(2\sin \theta - \cos \theta) = 4 \Rightarrow r = \dfrac{4}{2\sin \theta - \cos \theta}$

10. See Figure 10.

704   CHAPTER 10   Applications of Trigonometry; Vectors

Chapter 10 Project

1. To make a 360-gon, θ-step $= \dfrac{360}{360} = 1°$.

2. To make a square, θ-step $= \dfrac{360}{4} = 90$. To make a pentagon, θ-step $= \dfrac{360}{5} = 72°$.

    To make a hexagon, θ-step $= \dfrac{360}{6} = 60$. To make a heptagon, θ-step $= \dfrac{360}{7} \approx 51.43°$.

    To make an octagon, θ-step $= \dfrac{360}{8} = 45$. To make a nonagon, θ-step $= \dfrac{360}{9} = 40°$.

    To make a decagon, θ-step $= \dfrac{360}{10} = 36°$.

3. To make a 5-pointed star, θ-step $= \dfrac{360}{5} \cdot 2 = 144$; θ-max $= 360 \cdot 2 = 720°$.

    To make a 9-pointed star, θ-step $= \dfrac{360}{9} \cdot 4 = 160$; θ-max $= 360 \cdot 4 = 1440°$.

    To make a 11-pointed star, θ-step $= \dfrac{360}{11} \cdot 5 \approx 163.64$; θ-max $= 360 \cdot 5 = 1800°$.

Chapter 11: Further Topics in Algebra

11.1: Sequences and Series

1. $a_1 = 4(1) + 10 = 14$; $a_2 = 4(2) + 10 = 18$; $a_3 = 4(3) + 10 = 22$; $a_4 = 4(4) + 10 = 26$;

    $a_5 = 4(5) + 10 = 30$. The first five terms of the sequence are 14, 18, 22, 26, 30.

2. $a_1 = 6(1) - 3 = 3$; $a_2 = 6(2) - 3 = 9$; $a_3 = 6(3) - 3 = 15$; $a_4 = 6(4) - 3 = 21$; $a_5 = 6(5) - 3 = 27$.

    The first five terms of the sequence are 3, 9, 15, 21, 27.

3. $a_1 = 2^{1-1} = 2^0 = 1$; $a_2 = 2^{2-1} = 2^1 = 2$; $a_3 = 2^{3-1} = 2^2 = 4$; $a_4 = 2^{4-1} = 2^3 = 8$; $a_5 = 2^{5-1} = 2^4 = 16$.

    The first five terms of the sequence are 1, 2, 4, 8, 16.

4. $a_1 = -3^1 = -3$; $a_2 = -3^2 = -9$; $a_3 = -3^3 = -27$; $a_4 = -3^4 = -81$; $a_5 = -3^5 = -243$.

    The first five terms of the sequence are $-3, -9, -27, -81, -243$.

5. $a_1 = \left(\dfrac{1}{3}\right)^1 (1 - 1) = \left(\dfrac{1}{3}\right)(0) = 0$; $a_2 = \left(\dfrac{1}{3}\right)^2 (2 - 1) = \left(\dfrac{1}{9}\right)(1) = \dfrac{1}{9}$; $a_1 = \left(\dfrac{1}{3}\right)^3 (3 - 1) = \left(\dfrac{1}{27}\right)(2) = \dfrac{2}{27}$;

    $a_4 = \left(\dfrac{1}{3}\right)^4 (4 - 1) = \left(\dfrac{1}{81}\right)(3) = \dfrac{1}{27}$; $a_5 = \left(\dfrac{1}{3}\right)^5 (5 - 1) = \left(\dfrac{1}{243}\right)(4) = \dfrac{4}{243}$.

    The first five terms of the sequence are $0, \dfrac{1}{9}, \dfrac{2}{27}, \dfrac{1}{27}, \dfrac{4}{243}$.

6. $a_1 = (-2)^1 (1) = -2$; $a_2 = (-2)^2 (2) = 8$; $a_3 = (-2)^3 (3) = -24$; $a_4 = (-2)^4 (4) = 64$; $a_5 = (-2)^5 (5) = -160$.

    The first five terms of the sequence are $-2, 8, -24, 64, -160$.

7. $a_1 = (-1)^1 [(2)(1)] = -2$; $a_1 = (-1)^2 [(2)(2)] = 4$; $a_3 = (-1)^3 [(2)(3)] = -6$; $a_4 = (-1)^4 [(2)(4)] = 8$;

    $a_5 = (-1)^5 [(2)(5)] = -10$. The first five terms of the sequence are $-2, 4, -6, 8, -10$.

8. $a_1 = (-1)^{1-1} (1 + 1) = (-1)^0 (2) = 2$; $a_2 = (-1)^{2-1} (2 + 1) = (-1)^1 (3) = -3$;

    $a_3 = (-1)^{3-1} (3 + 1) = (-1)^2 (4) = 4$; $a_4 = (-1)^{4-1} (4 + 1) = (-1)^3 (5) = -5$;

    $a_5 = (-1)^{5-1} (5 + 1) = (-1)^4 (6) = 6$. The first five terms of the sequence are $2, -3, 4, -5, 6$.

9. $a_1 = \dfrac{4(1) - 1}{1^2 + 2} = \dfrac{3}{3} = 1$; $a_2 = \dfrac{4(2) - 1}{2^2 + 2} = \dfrac{7}{6}$; $a_3 = \dfrac{4(3) - 1}{3^2 + 2} = \dfrac{11}{11} = 1$; $a_4 = \dfrac{4(4) - 1}{4^2 + 2} = \dfrac{15}{18} = \dfrac{5}{6}$;

    $a_5 = \dfrac{4(5) - 1}{5^2 + 2} = \dfrac{19}{27}$. The first five terms of the sequence are $1, \dfrac{7}{6}, 1, \dfrac{5}{6}, \dfrac{19}{27}$.

10. $a_1 = \dfrac{1^2 - 1}{1^2 + 1} = \dfrac{0}{2} = 2$; $a_2 = \dfrac{2^2 - 1}{2^2 + 1} = \dfrac{3}{5}$; $a_3 = \dfrac{3^2 - 1}{3^2 + 1} = \dfrac{8}{10} = \dfrac{4}{5}$; $a_4 = \dfrac{4^2 - 1}{4^2 + 1} = \dfrac{15}{17}$;

    $a_5 = \dfrac{5^2 - 1}{5^2 + 1} = \dfrac{24}{26} = \dfrac{12}{13}$. The first five terms of the sequence are $0, \dfrac{3}{5}, \dfrac{4}{5}, \dfrac{15}{17}, \dfrac{12}{13}$.

11. The terms of a sequence are numbered according to their position in the sequence; a_1 is the first term, a_2 is the second term, etc. So a_n will be the nth term in the sequence for any positive integer value of n.

12. A sequence is a function which is defined on the positive integers. That is, the positive integers are the domain of the function and the terms of the sequence constitute the range of the function.

13. The sequence has 7 terms; it is finite.

14. The sequence has 30 terms; it is finite.

15. The sequence has 4 terms; it is finite.

16. The sequence has 4 terms; it is finite.

17. The sequence has no last term; it is infinite.

18. The sequence has no last term; it is infinite.

19. The sequence is defined for integer values of n from 1 through 10; it is finite.

20. The sequence is defined for all positive integer values of n; it is infinite.

21. $a_1 = -2; a_2 = a_1 + 3 = -2 + 3 = 1; a_3 = a_2 + 3 = 1 + 3 = 4; a_4 = a_3 + 3 = 4 + 3 = 7.$

    The first four terms of the sequence are $-2, 1, 4, 7$.

22. $a_1 = -1; a_2 = a_1 - 4 = -1 - 4 = -5; a_3 = a_2 - 4 = -5 - 4 = -9; a_4 = a_3 - 4 = -9 - 4 = -13.$

    The first four terms of the sequence are $-1, -5, -9, -13$.

23. $a_1 = 1; a_2 = 1; a_3 = a_2 + a_1 = 1 + 1 = 2; a_4 = a_3 + a_2 = 2 + 1 = 3.$

    The first four terms of the sequence are $1, 1, 2, 3$.

24. $a_1 = 2; a_2 = 2a_1 = 2(2) = 4; a_3 = 3a_2 = 3(4) = 12; a_4 = 4a_3 = 4(12) = 48.$

    The first four terms of the sequence are $2, 4, 12, 48$.

25. $\displaystyle\sum_{i=1}^{5} (2i + 1) = 3 + 5 + 7 + 9 + 11 = 35$

26. $\displaystyle\sum_{i=1}^{6} (3i - 2) = 1 + 4 + 7 + 10 + 13 + 16 = 51$

27. $\displaystyle\sum_{j=1}^{4} \frac{1}{j} = \frac{1}{1} + \frac{1}{2} + \frac{1}{3} + \frac{1}{4} = \frac{25}{12}$

28. $\displaystyle\sum_{i=1}^{5} (i + 1)^{-1} = \frac{1}{2} + \frac{1}{3} + \frac{1}{4} + \frac{1}{5} + \frac{1}{6} = \frac{29}{20}$

29. $\displaystyle\sum_{i=1}^{4} i^i = 1^1 + 2^2 + 3^3 + 4^4 = 288$

30. $\displaystyle\sum_{k=1}^{4} (k + 1)^2 = 2^2 + 3^2 + 4^2 + 5^2 = 54$

31. $\displaystyle\sum_{k=1}^{6} (-1)^k \cdot k = (-1)^1(1) + (-1)^2(2) + (-1)^3(3) + (-1)^4(4) + (-1)^5(5) + (-1)^6(6) = 3$

32. $\displaystyle\sum_{i=1}^{7} (-1)^{i+1} \cdot i^2 = (-1)^2(1)^2 + (-1)^3(2)^2 + (-1)^4(3)^2 + (-1)^5(4)^2 + (-1)^6(5)^2 + (-1)^7(6)^2 + (-1)^8(7)^2 = 28$

33. $\displaystyle\sum_{i=2}^{5} (6 - 3i) = [6 - 3(2)] + [6 - 3(3)] + [6 - 3(4)] + [6 - 3(5)] = -18$

34. $\displaystyle\sum_{i=3}^{7} (5i + 2) = [5(3) + 2] + [5(4) + 2] + [5(5) + 2] + [5(6) + 2] + [5(7) + 2] = 135$

35. $\displaystyle\sum_{i=-2}^{3} 2(3)^i = 2(3)^{-2} + 2(3)^{-1} + 2(3)^0 + 2(3)^1 + 2(3)^2 + 2(3)^3 = \frac{728}{9}$

36. $\displaystyle\sum_{i=-1}^{2} 5(2)^i = 5(2)^{-1} + 5(2)^0 + 5(2)^1 + 5(2)^2 = \frac{75}{2} = 37.5$

37. $\displaystyle\sum_{i=-1}^{5} (i^2 - 2i) = [(-1)^2 - 2(-1)] + [0^2 - 2(0)] + [1^2 - 2(1)] + [2^2 - 2(2)] + [3^2 - 2(3)] +$

    $[4^2 - 2(4)] + [5^2 - 2(5)] = 28$

38. $\displaystyle\sum_{i=3}^{6}(2i^2 + 1) = [2(3)^2 + 1] + [2(4)^2 + 1] + [2(5)^2 + 1] + [2(6)^2 + 1] = 176$

39. $\displaystyle\sum_{i=1}^{5}(3^i - 4) = (3^1 - 4) + (3^2 - 4) + (3^3 - 4) + (3^4 - 4) + (3^5 - 4) = 343$

40. $\displaystyle\sum_{i=1}^{4}[(-2)^i - 3] = [(-2)^1 - 3] + [(-2)^2 - 3] + [(-2)^3 - 3] + [(-2)^4 - 3] = -2$

41. $\displaystyle\sum_{i=1}^{5}x_i = (-2) + (-1) + (0) + (1) + (2) = -2 - 1 + 0 + 1 + 2$

42. $\displaystyle\sum_{i=1}^{5}-x_i = -(-2) - (-1) - (0) - (1) - (2) = 2 + 1 - 0 - 1 - 2$

43. $\displaystyle\sum_{i=1}^{5}(2x_i + 3) = [2(-2) + 3] + [2(-1) + 3] + [2(0) + 3] + [2(1) + 3] + [2(2) + 3] =$

     $-1 + 1 + 3 + 5 + 7$

44. $\displaystyle\sum_{i=1}^{4}x_i^2 = (-2)^2 + (-1)^2 + (0)^2 + (1)^2 = 4 + 1 + 0 + 1$

45. $\displaystyle\sum_{i=1}^{3}(3x_i - x_i^2) = [3(-2) - (-2)^2] + [3(-1) - (-1)^2] + [3(0) - (0)^2] = -10 - 4 + 0$

46. $\displaystyle\sum_{i=1}^{3}(x_i^2 + 1) = [(-2)^2 + 1] + [(-1)^2 + 1] + [(0)^2 + 1] = 5 + 2 + 1$

47. $\displaystyle\sum_{i=2}^{5}\frac{x_i + 1}{x_i + 2} = \frac{-1 + 1}{-1 + 2} + \frac{0 + 1}{0 + 2} + \frac{1 + 1}{1 + 2} + \frac{2 + 1}{2 + 2} = 0 + \frac{1}{2} + \frac{2}{3} + \frac{3}{4}$

48. $\displaystyle\sum_{i=1}^{5}\frac{x_i}{x_i + 3} = \frac{-2}{-2 + 3} + \frac{-1}{-1 + 3} + \frac{0}{0 + 3} + \frac{1}{1 + 3} + \frac{2}{2 + 3} = -2 - \frac{1}{2} + 0 + \frac{1}{4} + \frac{2}{5}$

49. $\displaystyle\sum_{i=1}^{4}f(x_i)\Delta x = [4(0) - 7](.5) + [4(2) - 7](.5) + [4(4) - 7](.5) + [4(6) - 7](.5) = -3.5 + .5 + 4.5 + 8.5$

50. $\displaystyle\sum_{i=1}^{4}f(x_i)\Delta x = [6 + 2(0)](.5) + [6 + 2(2)](.5) + [6 + 2(4)](.5) + [6 + 2(6)](.5) = 3 + 5 + 7 + 9$

51. $\displaystyle\sum_{i=1}^{4}f(x_i)\Delta x = [2(0)^2](.5) + [2(2)^2](.5) + [2(4)^2](.5) + [2(6)^2](.5) = 0 + 4 + 16 + 36$

52. $\displaystyle\sum_{i=1}^{4}f(x_i)\Delta x = [(0)^2 - 1](.5) + [(2)^2 - 1](.5) + [(4)^2 - 1](.5) + [(6)^2 - 1](.5) = -.5 + 1.5 + 7.5 + 17.5$

53. $\displaystyle\sum_{i=1}^{4}f(x_i)\Delta x = \left(\frac{-2}{0 + 1}\right)(.5) + \left(\frac{-2}{2 + 1}\right)(.5) + \left(\frac{-2}{4 + 1}\right)(.5) + \left(\frac{-2}{6 + 1}\right)(.5) = -1 - \frac{1}{3} - \frac{1}{5} - \frac{1}{7}$

54. $\displaystyle\sum_{i=1}^{4}f(x_i)\Delta x = \left(\frac{5}{2(0) - 1}\right)(.5) + \left(\frac{5}{2(2) - 1}\right)(.5) + \left(\frac{5}{2(4) - 1}\right)(.5) + \left(\frac{5}{2(6) - 1}\right)(.5) = -\frac{5}{2} + \frac{5}{6} + \frac{5}{14} + \frac{5}{22}$

55. $\displaystyle\sum_{i=1}^{100}6 = (100)(6) = 600$

56. $\displaystyle\sum_{i=1}^{20}5i = 5\left[\frac{(20)(20 + 1)}{2}\right] = 1050$

57. $\displaystyle\sum_{i=1}^{15} i^2 = \frac{(15)(15+1)[2(15)+1]}{6} = 1240$

58. $\displaystyle\sum_{i=1}^{50} 2i^3 = 2\left[\frac{(50)^2(50+1)^2}{4}\right] = 3,251,250$

59. $\displaystyle\sum_{i=1}^{5} (5i+3) = 5\left[\frac{(5)(5+1)}{2}\right] + 5(3) = 90$

60. $\displaystyle\sum_{i=1}^{5} (8i-1) = 8\left[\frac{(5)(5+1)}{2}\right] + 5(-1) = 115$

61. $\displaystyle\sum_{i=1}^{5} (4i^2 - 2i + 6) = 4\left[\frac{(5)(5+1)[2(5)+1]}{6}\right] - 2\left[\frac{5(5+1)}{2}\right] + 5(6) = 220$

62. $\displaystyle\sum_{i=1}^{6} (2+i-i^2) = 6(2) + \left[\frac{(6)(6+1)}{2}\right] - \left[\frac{6(6+1)[2(6)+1]}{6}\right] = -58$

63. $\displaystyle\sum_{i=1}^{4} (3i^3 + 2i - 4) = 3\left[\frac{4^2(4+1)^2}{4}\right] + 2\left[\frac{(4)(4+1)}{2}\right] + 4(-4) = 304$

64. $\displaystyle\sum_{i=1}^{6} (i^2 + 2i^3) = \frac{(6)(6+1)[2(6)+1]}{6} + 2\left[\frac{6^2(6+1)^2}{4}\right] = 973$

65. The graphing calculator graph of the first ten terms of this sequence is shown in Figure 65. From this graph, the terms of the sequence appear to converge to the value $\dfrac{1}{2}$ which, in fact, they do.

66. The graphing calculator graph of the first ten terms of this sequence is shown in Figure 66. From this graph, the terms of the sequence appear to converge to the value 2 which, in fact, they do.

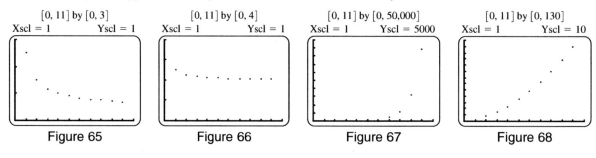

[0, 11] by [0, 3]	[0, 11] by [0, 4]	[0, 11] by [0, 50,000]	[0, 11] by [0, 130]
Xscl = 1 Yscl = 1	Xscl = 1 Yscl = 1	Xscl = 1 Yscl = 5000	Xscl = 1 Yscl = 10
Figure 65	Figure 66	Figure 67	Figure 68

67. The graphing calculator graph of the first ten terms of this sequence is shown in Figure 67. From this graph, the terms of the sequence appear to diverge and, in fact, they do.

68. The graphing calculator graph of the first ten terms of this sequence is shown in Figure 68. From this graph, the terms of the sequence appear to diverge and, in fact, they do.

69. The graphing calculator graph of the first ten terms of this sequence is shown in Figure 69. From this graph, the terms of the sequence appear to converge to the value somewhat greater than 2.5. In fact, they converge to $e \approx 2.71828$.

70. The graphing calculator graph of the first ten terms of this sequence is shown in Figure 70. From this graph, the terms of the sequence appear to converge to the value 5 which, in fact, they do.

71. The sum of the first six terms of the series is: $\dfrac{1}{1^4} + \dfrac{1}{2^4} + \dfrac{1}{3^4} + \dfrac{1}{4^4} + \dfrac{1}{5^4} + \dfrac{1}{6^4} =$

$1 + \dfrac{1}{16} + \dfrac{1}{81} + \dfrac{1}{256} + \dfrac{1}{625} + \dfrac{1}{1296} \approx 1.081123534 \Rightarrow \dfrac{\pi^4}{90} \approx 1.081123534 \Rightarrow \pi^4 \approx 97.30111806 \Rightarrow$

$\pi \approx 3.140721718$. This approximation of π is accurate to three decimal places when rounded to 3.141.

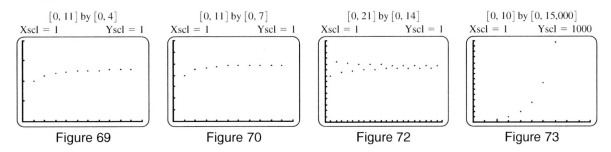

$[0, 11]$ by $[0, 4]$
Xscl = 1   Yscl = 1

$[0, 11]$ by $[0, 7]$
Xscl = 1   Yscl = 1

$[0, 21]$ by $[0, 14]$
Xscl = 1   Yscl = 1

$[0, 10]$ by $[0, 15{,}000]$
Xscl = 1   Yscl = 1000

Figure 69   Figure 70   Figure 72   Figure 73

72. (a) $a_1 = 8$ thousand per acre. $a_2 = 2.9(8) - .2(8)^2 = 10.4$ thousand per acre.

$a_3 = 2.9(10.4) - .2(10.4)^2 = 8.528$ thousand per acre.

(b) The graph for the first twenty terms of this sequence is shown in Figure 72. From this graph, the population density can be seen to oscillate above and below a value of approximately 9.5 thousand per acre.

73. (a) The number of bacteria doubles every 40 minutes, so $N_{j+1} = 2N_j$ *for* $j \geq 1$.

(b) Two hours = 120 minutes $\Rightarrow$ $120 = 40(j - 1) \Rightarrow 3 = j - 1 \Rightarrow j = 4$.

$N_1 = 230; N_2 = 2(230) = 460; N_3 = 2(460) = 920; N_4 = 2(920) = 1840$. If there are initially 230 bacteria, then there will be 1840 bacteria after two hours.

(c) The graph of the first seven terms of this sequence is shown in Figure 73.

(d) From the graph, the growth rate is seen to be very rapid. Doubling the number of bacteria at equal intervals of time produces an exponential growth rate.

74. (a) $N_{j+1} = \left[\dfrac{2}{1 + \frac{N_j}{K}}\right] N_j$; $N_2 = \left[\dfrac{2}{1 + \frac{N_1}{K}}\right] N_1 = \left[\dfrac{2}{1 + \frac{230}{5000}}\right] 230 \approx 439.77$. Similarly,

$N_3 = \left[\dfrac{2}{1 + \frac{N_2}{K}}\right] N_2 = \left[\dfrac{2}{1 + \frac{439.77}{5000}}\right] 439.77 \approx 808.45$. Repeating this process, generate the table shown in Figure 74a.

(b) Graph the sequence $N_{j+1} = \left[\dfrac{2}{1 + \frac{N_j}{K}}\right] N_j$ for $j = 1, 2, 3, \ldots, 20$ if $N_1 = 230, K = 5000$. See Figure 74b.

(c) The growth is rapid at first but slows and seems to approach 5000 asymptotically.

(d) Since the final number of bacteria is 5000, which is equal to K in this exercise, it is reasonable to conjecture that the carrying capacity of the medium is K. That is, K gives the maximum number of bacteria that can be supported on the given medium. When $N_j = K$, the medium becomes saturated. Changing K in the formula demonstrates that this conjecture is correct.

j	N_j	j	N_j
1	230	11	4901
2	440	12	4950
3	808	13	4975
4	1392	14	4987
5	2178	15	4994
6	3034	16	4997
7	3776	17	4998
8	4303	18	4999
9	4625	19	5000
10	4805	20	5000

Figure 74a

$[0, 20]$ by $[0, 6000]$
Xscl = 1   Yscl = 1000

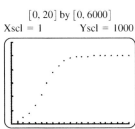

Figure 74b

75. (a) $1 + 1 + \dfrac{1^2}{2!} + \dfrac{1^3}{3!} + \dfrac{1^4}{4!} + \dfrac{1^5}{5!} + \dfrac{1^6}{6!} + \dfrac{1^7}{7!} \approx 2.718254;\ e \approx 2.718282.$ Eight terms of the series gives an

estimate that matches to four decimal places.

(b) $1 + (-1) + \dfrac{(-1)^2}{2!} + \dfrac{(-1)^3}{3!} + \dfrac{(-1)^4}{4!} + \dfrac{(-1)^5}{5!} + \dfrac{(-1)^6}{6!} + \dfrac{(-1)^7}{7!} \approx .367857;\ e^{-1} \approx .367879.$ Eight terms of

the series gives an estimate that matches to four decimal places.

(c) $1 + \dfrac{1}{2} + \dfrac{\left(\frac{1}{2}\right)^2}{2!} + \dfrac{\left(\frac{1}{2}\right)^3}{3!} + \dfrac{\left(\frac{1}{2}\right)^4}{4!} + \dfrac{\left(\frac{1}{2}\right)^5}{5!} + \dfrac{\left(\frac{1}{2}\right)^6}{6!} + \dfrac{\left(\frac{1}{2}\right)^7}{7!} \approx 1.648721;\ \sqrt{e} = e^{1/2} \approx 1.648721.$ Eight terms of the

series gives an estimate that matches to six decimal places.

76. (a) $a_1 = 2;\ a_2 = \dfrac{1}{2}\left(2 + \dfrac{2}{2}\right) = 1.5;\ a_3 = \dfrac{1}{2}\left(1.5 + \dfrac{2}{1.5}\right) \approx 1.416666667;$

$a_4 = \dfrac{1}{2}\left(1.416666667 + \dfrac{2}{1.416666667}\right) \approx 1.414245686;$

$a_5 = \dfrac{1}{2}\left(1.414215686 + \dfrac{2}{1.414215686}\right) \approx 1.414213562;$

$a_6 = \dfrac{1}{2}\left(1.414213562 + \dfrac{2}{1.414213562}\right) \approx 1.414213562;$

$\sqrt{2} \approx 1.414213562.$ The sixth term of the sequence gives an estimate that is accurate to nine decimal places.

(b) $a_1 = 8;\ a_2 = \dfrac{1}{2}\left(8 + \dfrac{8}{8}\right) = 4.5;\ a_3 = \dfrac{1}{2}\left(4.5 + \dfrac{8}{4.5}\right) \approx 3.138888889;$

$a_4 = \dfrac{1}{2}\left(3.138888889 + \dfrac{8}{3.138888889}\right) \approx 2.843780728;$

$a_5 = \dfrac{1}{2}\left(2.843780728 + \dfrac{8}{2.843780728}\right) \approx 2.828468572;$

$a_6 = \dfrac{1}{2}\left(2.828468572 + \dfrac{8}{2.828468572}\right) \approx 2.828427125;$

$\sqrt{8} \approx 2.828427128.$ The sixth term of the sequence gives an estimate that is accurate to eight decimal places.

(c) $a_1 = 11;\ a_2 = \dfrac{1}{2}\left(11 + \dfrac{11}{11}\right) = 6;\ a_3 = \dfrac{1}{2}\left(6 + \dfrac{11}{6}\right) \approx 3.916666667;$

$a_4 = \dfrac{1}{2}\left(3.916666667 + \dfrac{11}{3.916666667}\right) \approx 3.362588652;$

$a_5 = \dfrac{1}{2}\left(3.362588652 + \dfrac{11}{3.362588652}\right) \approx 3.316938935;$

$a_6 = \dfrac{1}{2}\left(3.316938935 + \dfrac{11}{3.316938935}\right) \approx 3.316624805;$

$\sqrt{11} \approx 3.31662479.$ The sixth term of the sequence gives an estimate that is accurate to six decimal places.

11.2: Arithmetic Sequences and Series

1. $5 - 2 = 3; 8 - 5 = 3; 11 - 8 = 3$. The common difference for this arithmetic sequence is 3.

2. $10 - 4 = 6; 16 - 10 = 6; 22 - 16 = 6$. The common difference for this arithmetic sequence is 6.

3. $-2 - 3 = -5; -7 - (-2) = -5; -12 - (-7) = -5$. The common difference for this arithmetic sequence is -5.

4. $-12 - (-8) = -4; -16 - (-12) = -4; -20 - (-16) = -4$.

   The common difference for this arithmetic sequence is -4.

5. $(2x + 5y) - (x + 3y) = x + 2y; (3x + 7y) - (2x + 5y) = x + 2y$. The common difference for this arithmetic

   sequence is $x + 2y$.

6. $(-4t^2 + 2q) - (t^2 + q) = -5t^2 + q; (-9t^2 + 3q) - (-4t^2 + 2q) = -5t^2 + q$. The common difference for this

   arithmetic sequence is $-5t^2 + q$.

7. $8; 8 + 6 = 14; 14 + 6 = 20; 20 + 6 = 26; 26 + 6 = 32$. The first five terms of this arithmetic sequence

   are 8, 14, 20, 26, 32.

8. $-2; -2 + 12 = 10; 10 + 12 = 22; 22 + 12 = 34; 34 + 12 = 46$. The first five terms of this arithmetic

   sequence $-2, 10, 22, 34, 46$.

9. $5; 5 + (-2) = 3; 3 + (-2) = 1; 1 + (-2) = -1; (-1) + (-2) = -3$. The first five terms of this arithmetic

   sequence $5, 3, 1, -1, -3$.

10. $4; 4 + 3 = 7; 7 + 3 = 10; 10 + 3 = 13; 13 + 3 = 16$. The first five terms of this arithmetic sequence

    are 4, 7, 10, 13, 16.

11. $a_3 = 10; a_1 = 10 - 2(-2) = 14; a_2 = 10 - (-2) = 12; a_4 = 10 + (-2) = 8; a_5 = 8 + (-2) = 6$. The first five

    terms of this arithmetic sequence are 14, 12, 10, 8, 6.

12. $a_1 = 3 - \sqrt{2}; a_2 = 3; d = 3 - (3 - \sqrt{2}) = \sqrt{2}; a_3 = 3 + \sqrt{2}; a_4 = 3 + \sqrt{2} + \sqrt{2} = 3 + 2\sqrt{2};$

    $a_5 = 3 + 2\sqrt{2} + \sqrt{2} = 3 + 3\sqrt{2}$. The first five terms of this arithmetic sequence

    are $3 - \sqrt{2}, 3, 3 + \sqrt{2}, 3 + 2\sqrt{2}, 3 + 3\sqrt{2}$.

13. $a_8 = 5 + (8 - 1)(2) = 19; a_n = 5 + (n - 1)(2) = 2n + 3$. The eighth term of the sequence is 19. The nth

    term of the sequence is $2n + 3$.

14. $a_8 = -3 + (8 - 1)(-4) = -31; a_n = -3 + (n - 1)(-4) = -4n + 1$. The eighth term of the sequence is -31.

    The nth term of the sequence is $-4n + 1$.

15. $a_3 = 2; d = 1 \Rightarrow a_1 = 2 - 2(1) = 0; a_8 = 0 + (8 - 1)(1) = 7; a_n = 0 + (n - 1)(1)$. The eighth term of the

    sequence is 7. The nth term of the sequence is $n - 1$.

16. $a_4 = 5; d = -2 \Rightarrow a_1 = 5 - 3(-2) = 11; a_8 = 11 + (8 - 1)(-2) = -3; a_n = 11 + (n - 1)(-2)$. The eighth

    term of the sequence is -3. The nth term of the sequence is $-2n + 13$.

17. $a_1 = 8; a_2 = 6 \Rightarrow d = 6 - 8 = -2; a_8 = 8 + (8 - 1)(-2) = -6; a_n = 8 + (n - 1)(-2) = -2n + 10$. The

    eighth term of the sequence is -6. The nth term of the sequence is $-2n + 10$.

18. $a_1 = 6; a_2 = 3 \Rightarrow d = 3 - 6 = -3; a_8 = 6 + (8 - 1)(-3) = -15; a_n = 6 + (n - 1)(-3) = -3n + 9$. The

    eighth term of the sequence is -15. The nth term of the sequence is $-3n + 9$.

19. $a_{10} = 6; a_{12} = 15 \Rightarrow 2d = 15 - 6 = 9; d = 4.5; a_1 = 6 - (9)(4.5) = -34.5;$

$a_8 = -34.5 + (8 - 1)(4.5) = -3; a_n = -34.5 + (n - 1)(4.5) = 4.5n - 39.$

The eighth term of the sequence is -3. The nth term of the sequence is $4.5n - 39$.

20. $a_{15} = 8; a_{17} = 2 \Rightarrow 2d = 2 - 8 = -6; d = -3; a_1 = 8 - (14)(-3) = 50;$

$a_8 = 50 + (8 - 1)(-3) = 29; a_n = 50 + (n - 1)(-3) = -3n + 53.$

The eighth term of the sequence is 29. The nth term of the sequence is $-3n + 53$.

21. $a_1 = x; a_2 = x + 3 \Rightarrow d = (x + 3) - x = 3; a_8 = x + (8 - 1)(3) = x + 21;$

$a_n = x + (n - 1)(3) = x + 3n - 3.$ The eighth term of the sequence is $x + 21$.

The nth term of the sequence is $x + 3n - 3$.

22. $a_2 = y + 1; d = -3 \Rightarrow a_1 = (y + 1) - (-3) = y + 4; a_8 = (y + 4) + (8 - 1)(-3) = y - 17;$

$a_n = (y + 4) + (n - 1)(-3) = y - 3n + 7.$ The eighth term of the sequence is $y - 17$.

The nth term of the sequence is $y - 3n + 7$.

23. $a_5 = 27; a_{15} = 87 \Rightarrow 87 = 27 + 10d \Rightarrow 10d = 60 \Rightarrow d = 6; 27 = a_1 + (5 - 1)(6) \Rightarrow$

$a_1 = 27 - 24 = 3.$ The first term of the sequence is 3.

24. $a_{12} = 60; a_{20} = 84 \Rightarrow 84 = 60 + 8d \Rightarrow 8d = 24 \Rightarrow d = 3; 60 = a_1 + (12 - 1)(3) \Rightarrow$

$a_1 = 60 - 33 = 27.$ The first term of the sequence is 27.

25. $S_{16} = -160; a_{16} = -25 \Rightarrow -160 = \dfrac{16}{2}[a_1 + (-25)] \Rightarrow 8a_1 = -160 + 200 \Rightarrow a_1 = 5.$

The first term of the sequence is 5.

26. $S_{28} = 2926; a_{28} = 199 \Rightarrow 2926 = \dfrac{28}{2}(a_1 + 199) \Rightarrow 14a_1 = 2926 - 2786 \Rightarrow a_1 = 10.$

The first term of the sequence is 10.

27. $a_8 = -160; d = 3 \Rightarrow a_{10} = 8 + (10 - 1)(3) = 35 \Rightarrow S_{10} = \dfrac{10}{2}(8 + 35) = 215.$

The sum of the first ten terms of the sequence is 215.

28. $a_1 = -9; d = 4 \Rightarrow a_{10} = -9 + (10 - 1)(4) = 27 \Rightarrow S_{10} = \dfrac{10}{2}(-9 + 27) = 90.$

The sum of the first ten terms of the sequence is 90.

29. $a_3 = 5; a_4 = 8 \Rightarrow d = 8 - 5 = 3 \Rightarrow 5 = a_1 + (3 - 1)(3) \Rightarrow a_1 = -1 \Rightarrow$

$a_{10} = -1 + (10 - 1)(3) = 26 \Rightarrow S_{10} = \dfrac{10}{2}(-1 + 26) = 125.$

The sum of the first ten terms of the sequence is 125.

30. $a_2 = 9; a_4 = 13 \Rightarrow 2d = 13 - 9 = 4 \Rightarrow d = 2 \Rightarrow 9 = a_1 + 2 \Rightarrow a_1 = 7 \Rightarrow$

$a_{10} = 7 + (10 - 1)(2) = 25 \Rightarrow S_{10} = \dfrac{10}{2}(7 + 25) = 160.$

The sum of the first ten terms of the sequence is 160.

31. $a_1 = 5; d = 9 - 5 = 4 \Rightarrow a_{10} = 5 + (10 - 1)(4) = 41 \Rightarrow S_{10} = \dfrac{10}{2}(5 + 41) = 230.$

The sum of the first ten terms of the sequence is 230.

32. $a_1 = 8; d = 6 - 8 = -2 \Rightarrow a_{10} = 8 + (10 - 1)(-2) = -10 \Rightarrow S_{10} = \dfrac{10}{2}[8 + (-10)] = -10.$

The sum of the first ten terms of the sequence is -10.

33. $a_1 = 10; a_{10} = 5.5 \Rightarrow S_{10} = \dfrac{10}{2}(10 + 5.5) = 77.5.$ The sum of the first ten terms of the sequence is 77.5.

34. $a_1 = -8; a_{10} = -1.25 \Rightarrow S_{10} = \dfrac{10}{2}[(-8) + (-1.25)] = -46.25.$ The sum of the first ten terms of the sequence is -46.25.

35. $S_{20} = 1090; a_{20} = 102 \Rightarrow 1090 = \dfrac{20}{2}(a_1 + 102) \Rightarrow 109 = a_1 + 102 \Rightarrow a_1 = 7 \Rightarrow$

    $102 = 7 + (20 - 1)d \Rightarrow 19d = 95 \Rightarrow d = 5.$ The first term of the sequence is 7 and the common difference is 5.

36. $S_{31} = 5580; a_{31} = 360 \Rightarrow 5580 = \dfrac{31}{2}(a_1 + 360) \Rightarrow \dfrac{31}{2}a_1 = 5580 - 5580 = 0 \Rightarrow$

    $a_1 = 0 \Rightarrow 360 = 0 + (31 - 1)d \Rightarrow 30d = 360 \Rightarrow d = 12.$ The first term of the sequence is 0 and the common difference is 12.

37. $S_{12} = -108; a_{12} = -19 \Rightarrow -108 = \dfrac{12}{2}[a_1 + (-19)] \Rightarrow 6a_1 = -108 + 114 = 6 \Rightarrow$

    $a_1 = 1 \Rightarrow -19 = 1 + (12 - 1)d \Rightarrow 11d = -20 \Rightarrow d = -\dfrac{20}{11}.$ The first term of the sequence is 1 and the common difference is $-\dfrac{20}{11}$.

38. $S_{25} = 650; a_{25} = 62 \Rightarrow 650 = \dfrac{25}{2}(a_1 + 62) \Rightarrow \dfrac{25}{2}a_1 = 650 - 775 = -125 \Rightarrow$

    $a_1 = -10 \Rightarrow 62 = -10 + (25 - 1)d \Rightarrow 24d = 72 \Rightarrow d = 3.$ The first term of the sequence is -10 and the common difference is 3.

39. From the graph, $a_1 = -2, a_2 = -1 \Rightarrow d = -1 - (-2) = 1 \Rightarrow a_n = -2 + (n - 1) = n - 3.$ Also from the graph, $D: \{1, 2, 3, 4, 5, 6\}; R: \{-2, -1, 0, 1, 2, 3\}.$ The nth term of the sequence is $n - 3$. The domain of the sequence is the set $\{1, 2, 3, 4, 5, 6\}$ and the range of the sequence is set $\{-2, -1, 0, 1, 2, 3\}$.

40. From the graph, $a_1 = 1, a_2 = 0 \Rightarrow d = 0 - 1 = -1 \Rightarrow a_n = 1 + (n - 1)(-1) = -n + 2.$ Also from the graph, $D: \{1, 2, 3, 4, 5\}; R: \{-3, -2, -1, 0, 1\}.$ The nth term of the sequence is $-n + 2$. The domain of the sequence is the set $\{1, 2, 3, 4, 5\}$ and the range of the sequence is set $\{-3, -2, -1, 0, 1\}$.

41. From the graph, $a_1 = 2.5, a_2 = 2 \Rightarrow d = 2 - 2.5 = -.5 \Rightarrow a_n = 2.5 + (n - 1)(-.5) = -.5n + 3.$ Also from the graph, $D: \{1, 2, 3, 4, 5, 6\}; R: \{0, .5, 1, 1.5, 2, 2.5\}.$ The nth term of the sequence is $-.5n + 3$. The domain of the sequence is the set $\{1, 2, 3, 4, 5, 6\}$ and the range of the sequence is set $\{0, .5, 1, 1.5, 2, 2.5\}$.

42. From the graph, $a_1 = -5, a_2 = 0 \Rightarrow d = 0 - (-5) = 5 \Rightarrow a_n = -5 + (n - 1)(5) = 5n - 10.$ Also from the graph, $D: \{1, 2, 3, 4, 5\}; R: \{-5, 0, 5, 10, 15\}.$ The nth term of the sequence is $5n - 10$. The domain of the sequence is the set $\{1, 2, 3, 4, 5\}$ and the range of the sequence is set $\{-5, 0, 5, 10, 15\}$.

43. From the graph, $a_1 = 10, a_2 = -10 \Rightarrow d = -10 - 10 = -20 \Rightarrow a_n = 10 + (n - 1)(-20) = -20n + 30.$ Also from the graph, $D: \{1, 2, 3, 4, 5\}; R: \{-70, -50, -30, -10, 10\}.$ The nth term of the sequence is $-20n + 30$. The domain of the sequence is the set $\{1, 2, 3, 4, 5\}$ and the range of the sequence is set $\{-70, -50, -30, -10, 10\}$.

44. From the graph, $a_1 = -2, a_2 = 0 \Rightarrow d = 0 - (-2) = 2 \Rightarrow a_n = -2 + (n - 1)(2) = 2n - 4.$ Also from the graph, $D: \{1, 2, 3, 4, 5\}; R: \{-2, 0, 2, 4, 6\}.$ The nth term of the sequence is $2n - 4$. The domain of the sequence is the set $\{1, 2, 3, 4, 5\}$ and the range of the sequence is set $\{-2, 0, 2, 4, 6\}$.

45. $a_1 = 3, a_8 = 17; 3 + 5 + 7 + 9 + 11 + 13 + 15 + 17 = S_8 = \dfrac{8}{2}(3 + 17) = 80$. The sum of the series is 80.

46. $a_1 = 7.5, a_7 = -1.5; 7.5 + 6 + 4.5 + 3 + 1.5 + 0 + (-1.5) = S_7 = \dfrac{7}{2}[(7.5 + (-1.5)] = 21$. The sum of the series is 21.

47. $a_1 = 1, a_{50} = 50; 1 + 2 + 3 + 4 + \cdots + 50 = S_{50} = \dfrac{50}{2}(1 + 50) = 1275$. The sum of the series is 1275.

48. $a_1 = 1; d = 2; 97 = 1 + (n - 1)(2) \Rightarrow 2n = 98 \Rightarrow n = 49 \Rightarrow a_{49} = 97;$

    $1 + 3 + 5 + 7 + \cdots + 97 = S_{49} = \dfrac{49}{2}(1 + 97) = 2401$. The sum of the series is 2401.

49. $a_1 = -7; d = -4 - (-7) = 3; 101 = -7 + (n - 1)(3) \Rightarrow 3n = 111 \Rightarrow n = 37;$

    $-7 + (-4) + (-1) + 2 + 5 + \cdots + 98 + 101 = S_{37} = \dfrac{37}{2}(-7 + 101) = 1739$. The sum of the series is 1739.

50. $a_1 = 89; d = 84 - 89 = -5; 4 = 89 + (n - 1)(-5) \Rightarrow -5n = -90 \Rightarrow n = 18;$

    $89 + 84 + 79 + 74 + \cdots + 9 + 4 = S_{18} = \dfrac{18}{2}(89 + 4) = 837$. The sum of the series is 837.

51. $a_1 = 5(1) = 5, a_{40} = 5(40) = 200; S_{40} = \dfrac{40}{2}(5 + 200) = 4100$. The sum of the series is 4100.

52. $a_1 = 1 - 3(1) = -2, a_{50} = 1 - 3(50) = -149; S_{50} = \dfrac{50}{2}[(-2) + (-149)] = -3775.$

    The sum of the series is -3775.

53. $a_1 = 1 + 4 = 5; a_3 = 3 + 4 = 7; \displaystyle\sum_{i=1}^{3}(i + 4) = S_3 = \dfrac{3}{2}(5 + 7) = 18$. The sum is 18.

54. $a_1 = 1 - 8 = -7; a_5 = 5 - 8 = -3; \displaystyle\sum_{i=1}^{5}(i - 8) = S_5 = \dfrac{5}{2}[(-7) + (-3)] = -25$. The sum is -25.

55. $a_1 = 2(1) + 3 = 5; a_{10} = 2(10) + 3 = 23; \displaystyle\sum_{j=1}^{10}(2j + 3) = S_{10} = \dfrac{10}{2}(5 + 23) = 140$. The sum is 140.

56. $a_1 = 5(1) - 9 = -4; a_{15} = 5(15) - 9 = 66; \displaystyle\sum_{j=1}^{15}(5j - 9) = S_{15} = \dfrac{15}{2}(-4 + 66) = 465$. The sum is 465.

57. $a_1 = -5 - 8(1) = -13; a_{12} = -5 - 8(12) = -101; \displaystyle\sum_{i=1}^{12}(-5 - 8i) = S_{12} = \dfrac{12}{2}[-13 + (-101)] = -684.$

    The sum is -684.

58. $a_1 = -3 - 4(1) = -7; a_{19} = -3 - 4(19) = -79; \displaystyle\sum_{k=1}^{19}(-3 - 4k) = S_{19} = \dfrac{19}{2}[-7 + (-79)] = -817.$

    The sum is -817.

59. $a_1 = 1; a_{1000} = 1000; \displaystyle\sum_{i=1}^{1000}i = S_{1000} = \dfrac{1000}{2}(1 + 1000) = 500,500$. The sum is 500,500.

60. $a_1 = 1; a_{2000} = 2000; \displaystyle\sum_{k=1}^{2000}k = S_{2000} = \dfrac{2000}{2}(1 + 2000) = 2,001,000$. The sum is 2,001,000.

61. $f(1) = m(1) + b = m + b; f(2) = m(2) + b = 2m + b; f(3) = m(3) + b = 3m + b$

62. $a_1 = f(1) = m + b; a_2 = f(2) = 2m + b; a_3 = f(3) = 3m + b; \ldots$. Yes, the sequence $f(1), f(2), f(3), \ldots$ is an arithmetic sequence.

63. $a_2 - a_1 = (2m + b) - (m + b) = m; a_3 - a_2 = (3m + b) - (2m + b) = m$. The common difference for this arithmetic sequence is m.

64. $a_n = a_1 + (n - 1)d = (m + b) + (n - 1)m = mn + b$. The nth term of the sequence is $mn + b$.

65. $a_n = 4.2n + 9.73$. Using the sequence feature of a graphing calculator, we obtain $S_{10} = 328.3$.

66. $a_n = 84.2n + 36.18$. Using the sequence feature of a graphing calculator, we obtain $S_{10} = 824.9$.

67. $a_n = \sqrt{8}n + \sqrt{3}$. Using the sequence feature of a graphing calculator, we obtain $S_{10} = 172.884$.

68. $a_n = -\sqrt[3]{4n} + \sqrt{7}$. Using the sequence feature of a graphing calculator, we obtain $S_{10} = -60.850$.

69. $a_1 = 51; d = 1; 71 = 51 + (n - 1)(1) \Rightarrow n = 71 - 50 = 21; \sum_{i=51}^{71} i = S_{21} = \frac{21}{2}(51 + 71) = 1281$.

    The sum of all the integers from 51 to 71 is 1281.

70. $a_1 = -8; d = 1; 30 = -8 + (n - 1)(1) \Rightarrow n = 30 + 9 = 39; \sum_{i=-8}^{30} i = S_{39} = \frac{39}{2}(-8 + 30) = 429$.

    The sum of all the integers from -8 to 30 is 429.

71. $a_1 = 1; a_{12} = 12; S_{12} = \frac{12}{2}(1 + 12) = 78$; chimes per 24 hours $= 2(78) = 156$.

    Chimes per 30 days $= (30)(156) = 4680$. The clock will chime 4680 times in a month of 30 days.

72. $a_1 = 30; a_{30} = 1; S_{30} = \frac{30}{2}(30 + 1) = 465$. There will be 465 poles in the stack.

73. $a_1 = 49,000, d = 580, n = 11; a_{11} = 49,000 + (11 - 1)(580) = 54,800$. The population five years from now will be 54,800.

74. $a_1 = 2; a_{20} = 15; S_{20} = \frac{20}{2}(2 + 15) = 170$. The total length of all the supports will be 170 meters.

75. $a_1 = 18; a_{31} = 28; S_{31} = \frac{31}{2}(18 + 28) = 713$. A total of 713 inches of material will be needed.

76. (a) $a_1 = 98.2, a_3 = 109.8 \Rightarrow 109.8 = 98.2 + (3 - 1)d \Rightarrow 2d = 11.6 \Rightarrow d = 5.8$. The common difference of the arithmetic sequence describing the child's height would be 5.8 centimeters.

    (b) $a_6 = 98.2 + (6 - 1)(5.8) = 127.2$. We would expect the child's height to be 127.2 centimeters at age 8.

77. $a_1, a_2, a_3, \ldots$ is an arithmetic sequence $\Rightarrow a_2 = a_1 + d, a_3 = a_2 + d$. For $a_1^2, a_2^2, a_3^2, \ldots$ to also be an arithmetic sequence, $a_2^2 - a_1^2 = a_3^2 - a_2^2 \Rightarrow (a_1 + d)^2 - a_1^2 = (a_2 + d)^2 - a_2^2 \Rightarrow$

    $a_1^2 + 2a_1d + d^2 - a_1^2 = a_2^2 + 2a_2d + d^2 - a_2^2 \Rightarrow 2a_1d + d^2 = 2a_2d + d^2 \Rightarrow a_1d = a_2d \Rightarrow$

    $a_1 = a_2 \Rightarrow d = 0$. For both $a_1, a_2, a_3, \ldots$ and $a_1^2, a_2^2, a_3^2, \ldots$ to be arithmetic sequences, the terms of each are constant. That is, $a_1, a_2, a_3, \ldots = k, k, k, \ldots$ and $a_1^2, a_2^2, a_3^2, \ldots = k^2, k^2, k^2, \ldots$.

78. $d_{n+1} - d_n = (a_{n+1} + c \cdot b_{n+1}) - (a_n + c \cdot b_n) = (a_{n+1} - a_n) + c(b_{n+1} - b_n) = j + ck$, where j is the common difference for the sequence $a_1, a_2, a_3, \ldots$ and k is the common difference for the sequence $b_1, b_2, b_3, \ldots$.

    Thus, since $d_{n+1} - d_n$ is a constant, $d_1, d_2, d_3, \ldots$ is also an arithmetic sequence.

79. $a_1, a_2, a_3, a_4, a_5, \ldots$ is an arithmetic sequence with a common difference of d. Thus, $a_1, a_3, a_5, \ldots$ will be an arithmetic sequence with a common difference of $2d$.

80. $a_1 = \log 2; a_2 = \log 4 = 2\log 2; a_3 = \log 8 = 3\log 2; a_4 = \log 16 = 4\log 2; \ldots$

    $a_2 - a_1 = 2\log 2 - \log 2 = \log 2; a_3 - a_2 = 3\log 2 - 2\log 2 = \log 2;$

    $a_4 - a_3 = 4\log 2 - 3\log 2 = \log 2; \ldots$. Thus, $\log 2, \log 4, \log 8, \log 16, \ldots$ is an arithmetic sequence with common difference $\log 2$.

11.3: Geometric Sequences and Series

1. $a_1 = \dfrac{5}{3}; a_2 = \left(\dfrac{5}{3}\right)(3) = 5; a_3 = (5)(3) = 15; a_4 = (15)(3) = 45.$ The first four terms of the sequence

    are $\dfrac{5}{3}, 5, 15, 45.$

2. $a_1 = -\dfrac{3}{4}; a_2 = \left(-\dfrac{3}{4}\right)\left(\dfrac{2}{3}\right) = -\dfrac{1}{2}; a_3 = \left(-\dfrac{1}{2}\right)\left(\dfrac{2}{3}\right) = -\dfrac{1}{3}; a_4 = \left(-\dfrac{1}{3}\right)\left(\dfrac{2}{3}\right) = -\dfrac{2}{9}.$ The first four terms of the

    sequence are $-\dfrac{3}{4}, -\dfrac{1}{2}, -\dfrac{1}{3}, -\dfrac{2}{9}.$

3. $a_4 = 5; a_5 = 10 \Rightarrow r = \dfrac{10}{5} = 2; a_3 = \dfrac{5}{2}; a_2 = \dfrac{\frac{5}{2}}{2} = \dfrac{5}{4}; a_1 = \dfrac{\frac{5}{4}}{2} = \dfrac{5}{8}.$ The first five terms of the

    sequence are $\dfrac{5}{8}, \dfrac{5}{4}, \dfrac{5}{2}, 5, 10.$

4. $a_3 = 16; a_4 = 8 \Rightarrow r = \dfrac{8}{16} = \dfrac{1}{2}; a_2 = \dfrac{16}{\frac{1}{2}} = 32; a_1 = \dfrac{32}{\frac{1}{2}} = 64; a_5 = (8)\left(\dfrac{1}{2}\right) = 4.$ The first five terms of the

    sequence are 64, 32, 16, 8, 4.

5. $a_5 = 5(-2)^{5-1} = (5)(16) = 80; a_n = 5(-2)^{n-1}.$ The fifth term of the sequence is 80 and the nth term of the

    sequence is $5(-2)^{n-1}.$

6. $a_5 = 8(-5)^{5-1} = 5000; a_n = 8(-5)^{n-1}.$ The fifth term of the sequence is 5000 and the nth term of the

    sequence is $8(-5)^{n-1}.$

7. $a_2 = -4, r = 3 \Rightarrow a_1 = -\dfrac{4}{3}; a_5 = \left(-\dfrac{4}{3}\right)(3)^{5-1} = -108; a_n = \left(-\dfrac{4}{3}\right)(3)^{n-1}.$ The fifth term of the sequence

    is -108 and the nth term of the sequence is $\left(-\dfrac{4}{3}\right)(3)^{n-1}.$

8. $a_3 = -2, r = 4 \Rightarrow a_1 = \dfrac{-2}{(4)^{3-1}} = -\dfrac{1}{8}; a_5 = \left(-\dfrac{1}{8}\right)(4)^{5-1} = -32; a_n = \left(-\dfrac{1}{8}\right)(4)^{n-1}.$ The fifth term of the

    sequence is -32 and the nth term of the sequence is $\left(-\dfrac{1}{8}\right)(4)^{n-1}.$

9. $a_4 = 243, r = -3 \Rightarrow a_1 = \dfrac{243}{(-3)^{4-1}} = -9; a_5 = (-9)(-3)^{5-1} = -729; a_n = (-9)(-3)^{n-1}.$ The fifth term of the

    sequence is -729 and the nth term of the sequence is $(-9)(-3)^{n-1}.$

10. $a_4 = 18, r = 2 \Rightarrow a_1 = \dfrac{18}{(2)^{4-1}} = \dfrac{9}{4}; a_5 = \left(\dfrac{9}{4}\right)(2)^{5-1} = 36; a_n = \left(\dfrac{9}{4}\right)(2)^{n-1}.$ The fifth term of the sequence

    is 36 and the nth term of the sequence is $\left(\dfrac{9}{4}\right)(2)^{n-1}.$

11. $r = \dfrac{-12}{-4} = \dfrac{-36}{-12} = \dfrac{-108}{-36} = 3; a_5 = (-4)(3)^{5-1} = -324; a_n = (-4)(3)^{n-1}.$ The fifth term of the sequence

    is -324 and the nth term of the sequence is $(-4)(3)^{n-1}.$

12. $r = \dfrac{6}{-2} = \dfrac{-18}{6} = \dfrac{54}{-18} = -3; a_5 = (-2)(-3)^{5-1} = -162; a_n = (-2)(-3)^{n-1}.$ The fifth term of the sequence

    is -162 and the nth term of the sequence is $(-2)(-3)^{n-1}.$

13. $r = \dfrac{2}{\frac{4}{5}} = \dfrac{5}{2} = \dfrac{\frac{25}{2}}{5} = \dfrac{5}{2}$; $a_5 = \left(\dfrac{4}{5}\right)\left(\dfrac{5}{2}\right)^{5-1} = \dfrac{125}{4}$; $a_n = \left(\dfrac{4}{5}\right)\left(\dfrac{5}{2}\right)^{n-1}$. The fifth term of the sequence is $\dfrac{125}{4}$ and

the nth term of the sequence is $\left(\dfrac{4}{5}\right)\left(\dfrac{5}{2}\right)^{n-1}$.

14. $r = \dfrac{\frac{2}{3}}{\frac{1}{2}} = \dfrac{\frac{8}{9}}{\frac{2}{3}} = \dfrac{\frac{32}{27}}{\frac{8}{9}} = \dfrac{4}{3}$; $a_5 = \left(\dfrac{1}{2}\right)\left(\dfrac{4}{3}\right)^{5-1} = \dfrac{128}{81}$; $a_n = \left(\dfrac{1}{2}\right)\left(\dfrac{4}{3}\right)^{n-1}$. The fifth term of the sequence is $\dfrac{128}{81}$ and

the nth term of the sequence is $\left(\dfrac{1}{2}\right)\left(\dfrac{4}{3}\right)^{n-1}$.

15. $r = \dfrac{-5}{10} = \dfrac{\frac{5}{2}}{-5} = \dfrac{-\frac{5}{4}}{\frac{5}{2}} = -\dfrac{1}{2}$; $a_5 = (10)\left(-\dfrac{1}{2}\right)^{5-1} = \dfrac{5}{8}$; $a_n = 10\left(-\dfrac{1}{2}\right)^{n-1}$. The fifth term of the sequence is $\dfrac{5}{8}$ and

the nth term of the sequence is $10\left(-\dfrac{1}{2}\right)^{n-1}$.

16. $r = \dfrac{-\frac{9}{4}}{3} = \dfrac{\frac{27}{16}}{-\frac{9}{4}} = \dfrac{-\frac{81}{64}}{\frac{27}{16}} = -\dfrac{3}{4}$; $a_5 = 3\left(-\dfrac{3}{4}\right)^{5-1} = \dfrac{243}{256}$; $a_n = 3\left(-\dfrac{3}{4}\right)^{n-1}$. The fifth term of the sequence is $\dfrac{243}{256}$

and the nth term of the sequence is $3\left(-\dfrac{3}{4}\right)^{n-1}$.

17. $a_3 = 5, a_8 = \dfrac{1}{625} \Rightarrow r^{7-2} = \dfrac{\frac{1}{625}}{5} = \dfrac{1}{3125} \Rightarrow r = \dfrac{1}{5}$; $a_1 = \dfrac{5}{\left(\frac{1}{5}\right)^2} = 125$. The first term of the sequence is 125

and the common ratio is $\dfrac{1}{5}$.

18. $a_2 = -6, a_7 = -192 \Rightarrow r^{6-1} = \dfrac{-192}{-6} = 32 \Rightarrow r = 2$; $a_1 = \dfrac{-6}{2} = -3$. The first term of the sequence is -3

and the common ratio is 2.

19. $a_4 = -\dfrac{1}{4}, a_9 = -\dfrac{1}{128} \Rightarrow r^{8-3} = \dfrac{-\frac{1}{128}}{-\frac{1}{4}} = \dfrac{1}{32} \Rightarrow r = \dfrac{1}{2}$; $a_1 = \dfrac{-\frac{1}{4}}{\left(\frac{1}{2}\right)^3} = -2$. The first term of the sequence is -2

and the common ratio is $\dfrac{1}{2}$.

20. $a_3 = 50, a_7 = .005 \Rightarrow r^{6-2} = \dfrac{.005}{50} = .0001 \Rightarrow r = \pm.1$; $a_1 = \dfrac{50}{(\pm.1)^2} = 5000$. The first term of the

sequence is 5000 and the common ratio is either .1 or $-.1$.

21. $r = \dfrac{8}{2} = \dfrac{32}{8} = \dfrac{128}{32} = 4$; $S_5 = \dfrac{2(1 - 4^5)}{1 - 4} = 682$. The sum of the first five terms of the sequence is 682.

22. $r = \dfrac{16}{4} = \dfrac{64}{16} = \dfrac{256}{64} = 4$; $S_5 = \dfrac{4(1 - 4^5)}{1 - 4} = 1364$. The sum of the first five terms of the sequence is 1364.

23. $r = \dfrac{-9}{18} = \dfrac{\frac{9}{2}}{-9} = \dfrac{-\frac{9}{4}}{\frac{9}{2}} = -\dfrac{1}{2}$; $S_5 = \dfrac{18\left[1 - \left(-\frac{1}{2}\right)^5\right]}{1 - \left(-\frac{1}{2}\right)} = \dfrac{99}{8}$. The sum of the first five terms of the sequence is $\dfrac{99}{8}$.

24. $r = \dfrac{-4}{12} = \dfrac{\frac{4}{3}}{-4} = \dfrac{-\frac{4}{9}}{\frac{4}{3}} = -\dfrac{1}{3}$; $S_5 = \dfrac{12\left[1 - \left(-\frac{1}{3}\right)^5\right]}{1 - \left(-\frac{1}{3}\right)} = \dfrac{244}{27}$. The sum of the first five terms of the sequence is $\dfrac{244}{27}$.

25. $S_5 = \dfrac{(8.423)\left[1 - (2.859)^5\right]}{1 - 2.859} \approx 860.95$. Rounded to the nearest hundredth, the sum of the first five terms of

the sequence is 860.95.

26. $S_5 = \dfrac{(-3.772)[1 - (-1.553)^5]}{1 - (-1.553)} = -14.82$. Rounded to the nearest hundredth, the sum of the first five terms of

the sequence is -14.82.

27. $a_1 = 3; r = 3; S_5 = \dfrac{3(1 - 3^5)}{1 - 3} = 363$. The sum is 363.

28. $a_1 = -2; r = -2; S_4 = \dfrac{(-2)[1 - (-2)^4]}{1 - (-2)} = 10$. The sum is 10.

29. $a_1 = (48)\left(\dfrac{1}{2}\right) = 24; r = \dfrac{1}{2}; S_6 = \dfrac{24[1 - (\frac{1}{2})^6]}{1 - \frac{1}{2}} = \dfrac{189}{4}$. The sum is $\dfrac{189}{4}$.

30. $a_1 = (243)\left(\dfrac{2}{3}\right) = 162; r = \dfrac{2}{3}; S_5 = \dfrac{162[1 - (\frac{2}{3})^5]}{1 - \frac{2}{3}} = 422$. The sum is 422.

31. $a_1 = 2^4 = 16; r = 2; n = 10 - 3 = 7; S_7 = \dfrac{16(1 - 2^7)}{1 - 2} = 2032$. The sum is 2032.

32. $a_1 = (-3)^3 = -27; r = -3; n = 9 - 2 = 7; S_7 = \dfrac{(-27)[1 - (-3)^7]}{1 - (-3)} = -14{,}469$. The sum is $-14{,}769$.

33. The sum of the terms of an infinite geometric sequence exists if the absolute value of the common ratio is less

than 1.

34. $S_\infty = \dfrac{.9}{1 - .1} = \dfrac{.9}{.9} = 1$. The sum of the terms of this infinite sequence is 1. This may seem counter intuitive to

many people.

35. $a_1 = .8, r = .1 \Rightarrow S_\infty = \dfrac{.8}{1 - .1} = \dfrac{8}{9}$. The sum of the geometric series is $\dfrac{8}{9}$.

36. $a_1 = .7, r = .1 \Rightarrow S_\infty = \dfrac{.7}{1 - .1} = \dfrac{7}{9}$. The sum of the geometric series is $\dfrac{7}{9}$.

37. $a_1 = .45, r = .01 \Rightarrow S_\infty = \dfrac{.45}{1 - .01} = \dfrac{5}{11}$. The sum of the geometric series is $\dfrac{5}{11}$.

38. $a_1 = .36, r = .01 \Rightarrow S_\infty = \dfrac{.36}{1 - .01} = \dfrac{4}{11}$. The sum of the geometric series is $\dfrac{4}{11}$.

39. $r = \dfrac{24}{12} = \dfrac{48}{24} = \dfrac{96}{48} = 2; |r| > 1 \Rightarrow$ the sum will not converge. Because the common ratio, 2, is greater

than 1, the sum of the terms of the geometric sequence does not converge.

40. $r = \dfrac{125}{625} = \dfrac{25}{125} = \dfrac{5}{25} = \dfrac{1}{5}; |r| < 1 \Rightarrow$ the sum would converge. The common ratio is $\dfrac{1}{5}$ and the sum of the

terms of the geometric sequence would converge.

41. $r = \dfrac{-24}{-48} = \dfrac{-12}{-24} = \dfrac{-6}{-12} = \dfrac{1}{2}; |r| < 1 \Rightarrow$ the sum would converge. The common ratio is $\dfrac{1}{2}$ and the sum of

the terms of the geometric sequence would converge.

42. $r = \dfrac{-10}{2} = \dfrac{50}{-10} = \dfrac{-250}{50} = -5; |r| > 1 \Rightarrow$ the sum will not converge. Because the common ratio, -5, has

absolute value greater than 1, the sum of the terms of the geometric sequence does not converge.

43. $r = \dfrac{2}{16} = \dfrac{\frac{1}{4}}{2} = \dfrac{\frac{1}{32}}{\frac{1}{4}} = \dfrac{1}{8}; S_\infty = \dfrac{16}{1 - \frac{1}{8}} = \dfrac{128}{7}$. The sum is $\dfrac{128}{7}$.

44. $r = \dfrac{6}{18} = \dfrac{2}{6} = \dfrac{\frac{2}{3}}{2} = \dfrac{1}{3}; S_\infty = \dfrac{18}{1 - \frac{1}{3}} = 27.$ The sum is 27.

45. $r = \dfrac{10}{100} = \dfrac{1}{10}; S_\infty = \dfrac{100}{1 - \frac{1}{10}} = \dfrac{1000}{9}.$ The sum is $\dfrac{1000}{9}$.

46. $r = \dfrac{64}{128} = \dfrac{32}{64} = \dfrac{1}{2}; S_\infty = \dfrac{128}{1 - \frac{1}{2}} = 256.$ The sum is 256.

47. $r = \dfrac{\frac{2}{3}}{\frac{4}{3}} = \dfrac{\frac{1}{3}}{\frac{2}{3}} = \dfrac{1}{2}; S_\infty = \dfrac{\frac{4}{3}}{1 - \frac{1}{2}} = \dfrac{8}{3}.$ The sum is $\dfrac{8}{3}$.

48. $r = \dfrac{-\frac{1}{6}}{\frac{1}{4}} = \dfrac{\frac{1}{9}}{-\frac{1}{6}} = \dfrac{-\frac{2}{27}}{\frac{1}{9}} = -\dfrac{2}{3}; S_\infty = \dfrac{\frac{1}{4}}{1 - (-\frac{2}{3})} = \dfrac{3}{20}.$ The sum is $\dfrac{3}{20}$.

49. $a_1 = 3, r = \dfrac{1}{4} \Rightarrow S_\infty = \dfrac{3}{1 - \frac{1}{4}} = 4.$ The sum is 4.

50. $a_1 = 5, r = -\dfrac{1}{4} \Rightarrow S_\infty = \dfrac{5}{1 - (-\frac{1}{4})} = 4.$ The sum is 4.

51. $a_1 = .3, r = .3 \Rightarrow S_\infty = \dfrac{.3}{1 - .3} = \dfrac{3}{7}.$ The sum is $\dfrac{3}{7}$.

52. $a_1 = \dfrac{1}{10}, r = \dfrac{1}{10} \Rightarrow S_\infty = \dfrac{\frac{1}{10}}{1 - \frac{1}{10}} = \dfrac{1}{9}.$ The sum is $\dfrac{1}{9}$.

53. $g(1) = ab; g(2) = ab^2; g(3) = ab^3.$ The value of the function for $x = \{1, 2, 3\}$ is $\{ab, ab^2, ab^3\}$.

54. $\{g(1), g(2), g(3), \dots\} = (ab, ab^2, ab^3, \dots)$ is a geometric sequence with a common ratio of b.

55. $a_n = g(n) = ab^n.$ The nth term of the sequence is ab^n.

56. In both geometric sequences and exponential functions, the independent variable is in the exponent.

57. See Figure 57. Rounded to the nearest thousandth, the sum is 97.739.

58. See Figure 58. Rounded to the nearest thousandth, the sum is −3012.622.

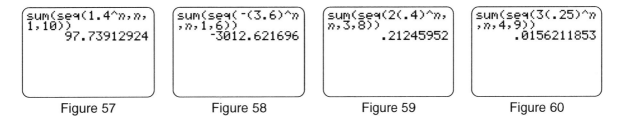

Figure 57   Figure 58   Figure 59   Figure 60

59. See Figure 59. Rounded to the nearest thousandth, the sum is .212.

60. See Figure 60. Rounded to the nearest thousandth, the sum is .016.

61. $S = 1000\left[\dfrac{(1 + .04)^9 - 1}{.04}\right] \approx 10{,}582.80.$ Rounded to the nearest penny, the future value of the annuity is \$10,582.80.

62. $S = 800\left[\dfrac{(1 + .03)^{12} - 1}{.03}\right] \approx 11{,}353.62.$ Rounded to the nearest penny, the future value of the annuity is \$11,353.62.

63. $S = 2430\left[\dfrac{(1 + .025)^{10} - 1}{.025}\right] \approx 27{,}224.22.$ Rounded to the nearest penny, the future value of the annuity

is \$27,224.22.

64. $S = 1500\left[\dfrac{(1 + .015)^6 - 1}{.015}\right] \approx 9344.33.$ Rounded to the nearest penny, the future value of the annuity

is \$9,344.33.

65. (a) $a_1 = 1276(.916)^1 \approx 1169; r = .916.$ Rounded to the nearest whole number, the first term of the sequence

is 1169. The common ratio is .916.

(b) $a_{10} = 1276(.916)^{10} \approx 531; a_{20} = 1276(.916)^{20} \approx 221.$ Rounded to the nearest whole number, the tenth

term of the sequence is 531 and the twentieth term of the sequence is 221. This means that a person 10

years from retirement should have saved 531% of his or her annual income and a person 20 years from

retirement should have saved 221% of his or her annual income.

66. (a) $a_1 = 1278(.935)^1 \approx 1195; r = .935.$ Rounded to the nearest whole number, the first term of the sequence

is 1195. The common ratio is .935.

(b) $a_{10} = 1278(.935)^{10} \approx 653; a_{20} = 1278(.935)^{20} \approx 333.$ Rounded to the nearest whole number, the tenth

term of the sequence is 653 and the twentieth term of the sequence is 333. This means that a person 10

years from retirement should have saved 653% of his or her annual income and a person 20 years from

retirement should have saved 333% of his or her annual income.

(c) Using a conservative investment strategy, typically a lower annual return will be realized in exchange for

lower risk. With lower annual return, the individual needs to have saved a larger percentage of his or her

ultimate goal during each year of employment.

67. (a) $a_2 = 2a_1; a_3 = 2a_2 = 4a_1; \ldots a_n = a_1(2)^{n-1}.$ The nth term of the geometric series is $a_1(2)^{n-1}$.

(b) $a_1 = 100; a_n > 1{,}000{,}000 \Rightarrow (2)^{n-1} > 10{,}000 \Rightarrow n - 1 > \dfrac{4}{\log{(2)}} \Rightarrow n > 14.25 = 15.$

(c) $n = 15 \Rightarrow \text{time} = (15 - 1)40 \text{ minutes} = 560 \text{ minutes} = 9 \text{ hours } 20 \text{ minutes}.$ After 9 hours and

20 minutes, the number of bacteria will exceed one million.

68. $a_1 = 1 - .98 = .02; r = 1 - .98 = .02 \Rightarrow a_4 = .02(.02)^{4-1} = .00000016 = .000016\%.$ After one hour of

washing, .000016% of the original fixer will remain.

69. $a_1 = .8, r = .8 \Rightarrow a_9 = (.8)(.8)^{9-1} \approx .134 = 13.4\%.$ After nine drainings and replacements with water, the

strength of the mixture will be approximately 13.4%.

70. Amount remaining $= (10^{15})\left(\dfrac{1}{2}\right)^{15/3} = 3.125 \times 10^{13}$ molecules. After 15 years, 5 half lives, 3.125×10^{13}

molecules of the substance will remain.

71. $a_1 = (100{,}000)(1 - .2) = 80{,}000; r = .8 \Rightarrow a_6 = 80{,}000(.8)^{6-1} = 26{,}214.40..$ The value of the machine at

the end of six years will be \$26,214.40.

72. $a_1 = 1000, r = .1 \Rightarrow S_\infty = \dfrac{1000}{1 - .1} = \dfrac{10{,}000}{9} = 1111\dfrac{1}{9}.$ To produce 1000 units of sugar, he should have

ordered $1111\dfrac{1}{9}$ units.

73. $a_1 = 40, r = .8 \Rightarrow S_\infty = \dfrac{40}{1 - .8} = 200$. The total length of arc through which the pendulum will swing is

200 centimeters.

74. Traveling down, $a_1 = 10, r = \dfrac{3}{4} \Rightarrow S_\infty = \dfrac{10}{1 - \frac{3}{4}} = 40$.

Traveling up, $a_1 = (10)\left(\dfrac{3}{4}\right) = \dfrac{30}{4}, r = \dfrac{3}{4} \Rightarrow S_\infty = \dfrac{\frac{30}{4}}{1 - \frac{3}{4}} = 30$.

S_∞ down $+ S_\infty$ up $= 70$. The total distance the ball will fall is 70 meters.

75. $a_1 = 2, r = 2 \Rightarrow S_5 = \dfrac{2(1 - 2^5)}{1 - 2} = 62; S_{10} = \dfrac{2(1 - 2^{10})}{1 - 2} = 2046$. Going back 5 generations, a person has 62

ancestors and going back 10 generations, a person has 2046 ancestors.

76. (a) On the first day, 2mg is administered and at the end of that day $2(.6)$mg remains. Then, on the second day,

another 2mg dose is administered, so at the end of that day, $(2 + 2(.6)(.6)$mg remains. After n days,

$2(.6)^n + 2(.6)^{n-1} + 2(.6)^{n-2} + \cdots + 2(.6)$mg of the drug are present.

That is, the amount present $= \displaystyle\sum_{i=1}^{n} 2(.6)^i$ milligrams.

(b) Amount $= 2\displaystyle\sum_{i=1}^{\infty}(.6)^i \Rightarrow a_1 = .6, r = .6 \Rightarrow S_\infty = \dfrac{.6}{1 - .6} = 1.5 \Rightarrow$ Amount $= 2(1.5) = 3$ milligrams.

After a long period of treatments, approximately 3 milligrams of the drug will be present.

77. Option 1: Arithmetic sequences with $a_1 = 5000$ and $d = 10,000 \Rightarrow$

$S_{30} = \dfrac{30}{2}[2(5000) + (30 - 1)(10,000)] = 4,500,000$.

Option 2: Geometric sequences with $a_1 = .01$ and $r = 2 \Rightarrow S_{30} = \dfrac{.01(1 - 2^{30})}{1 - 2} = 10,737,418.23$.

Option 1 pays a total of \$4,500,000 while Option 2 pays a total of \$10,737,418.23. So Option 2 should be selected.

78. $a_1 = 2, r = 2 \Rightarrow S_{12} = \dfrac{2(1 - 2^{12})}{1 - 2} = 8190$. Up to as many as 8190 ancestors might be found going back

12 generations.

79. $a_1 = 2, r = \dfrac{1}{2} \Rightarrow a_8 = 2\left(\dfrac{1}{2}\right)^{8-1} = \dfrac{1}{64}$. The eighth such triangle will have sides of length $\dfrac{1}{64}$ meter.

80. (a) $a_1 = 6, r = \dfrac{1}{2} \Rightarrow S_\infty = \dfrac{6}{1 - \frac{1}{2}} = 12$. The total length of the perimeters of all the triangles would be 12 meters.

(b) $a_1 = \dfrac{1}{2}(2)(\sqrt{3}) = \sqrt{3}; a_2 = \dfrac{1}{2}(1)\left(\dfrac{\sqrt{3}}{2}\right) = \dfrac{\sqrt{3}}{4} \Rightarrow r = \dfrac{1}{4} \Rightarrow S_\infty = \dfrac{\sqrt{3}}{1 - \frac{1}{4}} = \dfrac{4\sqrt{3}}{3}$. The total area of

the triangles would be $\dfrac{4\sqrt{3}}{3}$ square meters.

81. Let geometric series $a_1, a_2, a_3, \ldots$ have common ratio r and let geometric series $b_1, b_2, b_3, \ldots$ have common ratio s.

$d_n = c \cdot a_n \cdot b_n \Rightarrow d_{n+1} = c \cdot a_{n+1} \cdot b_{n+1} \Rightarrow \dfrac{d_{n+1}}{d_n} = \dfrac{c \cdot a_{n+1} \cdot b_{n+1}}{c \cdot a_n \cdot b_n} = r \cdot s$ for any n. Thus, $d_1, d_2, d_3, \ldots$ is

a geometric series with common ratio $r \cdot s$.

82. $a_1 = \log 6, a_2 = \log 36 = 2\log 6, a_3 = \log 1296 = 4\log 6, a_4 = \log 1,679,616 = 8\log 6, \ldots$. The sequence is

a geometric sequence with a common ratio 2.

Reviewing Basic Concepts (Sections 11.1—11.3)

1. $a_1 = (-1)^{1-1}(4 \cdot 1) = 4; a_2 = (-1)^{2-1}(4 \cdot 2) = -8; a_3 = (-1)^{3-1}(4 \cdot 3) = 12; a_4 = (-1)^{4-1}(4 \cdot 4) = -16;$
    $a_5 = (-1)^{5-1}(4 \cdot 5) = 20.$ The first five terms of the sequence are $4, -8, 12, -16, 20$.

2. $a_1 = 3 \cdot 1 + 1 = 4; a_5 = 3 \cdot 5 + 1 = 16; S_5 = \dfrac{5}{2}(4 + 16) = 50.$ The series sum is 50.

3. From Figure 3, it can be seen that as n increases, the terms of the sequence converge to 1.

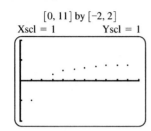

$[0, 11]$ by $[-2, 2]$

Xscl = 1   Yscl = 1

Figure 3

4. $a_1 = 8, d = -2 \Rightarrow a_2 = 8 - 2 = 6, a_3 = 6 - 2 = 4, a_4 = 4 - 2 = 2, a_5 = 2 - 2 = 0.$ The first five terms
    of the sequence are $8, 6, 4, 2, 0$.

5. $a_5 = 5, a_8 = 17 \Rightarrow 3d = 17 - 5 \Rightarrow d = 4 \Rightarrow a_1 = 5 - 4 \cdot 4 = -11.$ The first term of the sequence is -11.

6. $a_{10} = 2 + 5(10 - 1) = 47; S_{10} = \dfrac{10}{2}(2 + 47) = 245.$ The sum of the first ten terms of the sequence is 245.

7. $a_3 = (-2)(-3)^{3-1} = -18; a_n = (-2)(-3)^{n-1}.$ The third term of the sequence is -18 and the nth term of the
    sequence is $(-2)(-3)^{n-1}.$

8. $r = \dfrac{3}{5} = \dfrac{\frac{9}{5}}{3} = \dfrac{3}{5}; \dfrac{243}{625} = 5\left(\dfrac{3}{5}\right)^{n-1} \Rightarrow \left(\dfrac{3}{5}\right)^{n-1} = \dfrac{243}{3125} \Rightarrow n - 1 = 5 \Rightarrow n = 6. \ S_6 = \dfrac{5[1 - (\frac{3}{5})^6]}{1 - \frac{3}{5}} = \dfrac{7448}{625}.$

    The series is geometric with common ratio $\dfrac{3}{5}$. The series sum is $\dfrac{7448}{625}$.

9. $a_1 = 3\left(\dfrac{2}{3}\right) = 2. \ S_\infty = \dfrac{2}{1 - \frac{2}{3}} = 6.$ The series sum is 6.

10. $S = 500\left[\dfrac{(1 + .035)^{13} - 1}{.035}\right] \approx 8056.52.$ Rounded to the nearest penny, the future value of the annuity is \$8056.52.

11.4: The Binomial Theorem

1. $\dfrac{6!}{3!3!} = 6 \, _nC_r \, 3 = 20.$ The expression is equal to 20.

2. $\dfrac{5!}{2!3!} = 5 \, _nC_r \, 2 = 10.$ The expression is equal to 10.

3. $\dfrac{7!}{3!4!} = 7 \, _nC_r \, 3 = 35.$ The expression is equal to 35.

4. $\dfrac{8!}{5!3!} = 8 \, _nC_r \, 5 = 56.$ The expression is equal to 56.

5. $\dbinom{8}{3} = 8 \, _nC_r \, 3 = 56.$ The expression is equal to 56.

6. $\dbinom{7}{4} = 7\,{}_nC_r\,4 = 35$. The expression is equal to 35.

7. $\dbinom{10}{8} = 10\,{}_nC_r\,8 = 45$. The expression is equal to 45.

8. $\dbinom{9}{6} = 9\,{}_nC_r\,6 = 84$. The expression is equal to 84.

9. $\dbinom{13}{13} = 13\,{}_nC_r\,13 = 1$. The expression is equal to 1.

10. $\dbinom{12}{12} = 12\,{}_nC_r\,12 = 1$. The expression is equal to 1.

11. $\dbinom{n}{n-1} = \dfrac{n!}{(n-1)!(n-(n-1))!} = \dfrac{n!}{(n-1)!(1)!} = \dfrac{n(n-1)!}{(n-1)!} = n$. The expression is equal to n.

12. $\dbinom{n}{n-2} = \dfrac{n!}{(n-2)!(n-(n-2))!} = \dfrac{n!}{(n-2)!(2)!} = \dfrac{n(n-1)(n-2)!}{(n-2)!(2)} = \dfrac{n(n-1)}{2}$. The expression is equal to $\dfrac{n(n-1)}{2}$.

13. ${}_8C_3 = 8\,{}_nC_r\,3 = 56$. The expression is equal to 56.

14. ${}_9C_7 = 9\,{}_nC_r\,7 = 36$. The expression is equal to 36.

15. ${}_{100}C_2 = 100\,{}_nC_r\,2 = 4950$. The expression is equal to 4950.

16. ${}_{20}C_{15} = 20\,{}_nC_r\,15 = 15{,}504$. The expression is equal to 15,504.

17. ${}_5C_0 = 5\,{}_nC_r\,0 = 1$. The expression is equal to 1.

18. ${}_6C_0 = 6\,{}_nC_r\,0 = 1$. The expression is equal to 1.

19. The expansion of $(x+y)^8$ has a term where x is raised to each of the following powers: 8, 7, 6, 5, 4, 3, 2, 1, 0. Thus, there are 9 terms.

20. The expansion of $(x+y)^{10}$ has a term where x is raised to each of the following powers:

    10, 9, 8, 7, 6, 5, 4, 3, 2, 1, 0. Thus, there are 11 terms.

21. $(2x)^4 = 16x^4;\ (3y)^4 = 81y^4$. In the expansion of $(2x+3y)^4$, $16x^4$ is the first term and $81y^4$ is the last term.

22. $\dbinom{8}{4}x^{8-4}y^4 = 70x^4y^4$. The coefficient in the fifth term in the expansion of $(x+y)^8$ would be $\dbinom{8}{4} = 70$. The exponent of x would be $8 - 4 = 4$ and the exponent of y would be 4.

23. The binomial expansion for $(x+y)^6$ is given by:

$$(x+y)^6 = \binom{6}{0}x^6 + \binom{6}{1}x^5y + \binom{6}{2}x^4y^2 + \binom{6}{3}x^3y^3 + \binom{6}{4}x^2y^4 + \binom{6}{5}xy^5 + \binom{6}{6}y^6 =$$

$$x^6 + 6x^5y + 15x^4y^2 + 20x^3y^3 + 15x^2y^4 + 6xy^5 + y^6$$

24. The binomial expansion for $(m+n)^4$ is given by:

$$(m+n)^4 = \binom{4}{0}m^4 + \binom{4}{1}m^3n + \binom{4}{2}m^2n^2 + \binom{4}{3}mn^3 + \binom{4}{4}n^4 = m^4 + 4m^3n + 6m^2n^2 + 4mn^3 + n^4$$

25. The binomial expansion for $(p - q)^5$ is given by:

$$(p - q)^5 = \binom{5}{0}p^5 + \binom{5}{1}p^4(-q) + \binom{5}{2}p^3(-q)^2 + \binom{5}{3}p^2(-q)^3 + \binom{5}{4}p(-q)^4 + \binom{5}{5}(-q)^5 =$$

$$p^5 - 5p^4q + 10p^3q^2 - 10p^2q^3 + 5pq^4 - q^5$$

26. The binomial expansion for $(a - b)^7$ is given by:

$$(a - b)^7 = \binom{7}{0}a^7 + \binom{7}{1}a^6(-b) + \binom{7}{2}a^5(-b)^2 + \binom{7}{3}a^4(-b)^3 + \binom{7}{4}a^3(-b)^4 + \binom{7}{5}a^2(-b)^5 + \binom{7}{6}a(-b)^6 +$$

$$\binom{7}{7}(-b)^7 = a^7 - 7a^6b + 21a^5b^2 - 35a^4b^3 + 35a^3b^4 - 21a^2b^5 + 7ab^6 - b^7$$

27. The binomial expansion for $(r^2 + s)^5$ is given by:

$$(r^2 + s)^5 = \binom{5}{0}(r^2)^5 + \binom{5}{1}(r^2)^4s + \binom{5}{2}(r^2)^3s^2 + \binom{5}{3}(r^2)^2s^3 + \binom{5}{4}(r^2)s^4 + \binom{5}{5}s^5 =$$

$$r^{10} + 5r^8s + 10r^6s^2 + 10r^4s^3 + 5r^2s^4 + s^5$$

28. The binomial expansion for $(m + n^2)^4$ is given by:

$$(m + n^2)^4 = \binom{4}{0}m^4 + \binom{4}{1}m^3(n^2) + \binom{4}{2}m^2(n^2)^2 + \binom{4}{3}m(n^2)^3 + \binom{4}{4}(n^2)^4 =$$

$$m^4 + 4m^3n^2 + 6m^2n^4 + 4mn^6 + n^8$$

29. The binomial expansion for $(p + 2q)^4$ is given by:

$$(p + 2q)^4 = \binom{4}{0}p^4 + \binom{4}{1}p^3(2q) + \binom{4}{2}p^2(2q)^2 + \binom{4}{3}p(2q)^3 + \binom{4}{4}(2q)^4 =$$

$$p^4 + 8p^3q + 24p^2q^2 + 32pq^3 + 16q^4$$

30. The binomial expansion for $(3r - s)^6$ is given by:

$$(3r - s)^6 = \binom{6}{0}(3r)^6 + \binom{6}{1}(3r)^5(-s) + \binom{6}{2}(3r)^3(-s)^2 + \binom{6}{3}(3r)^3(-s)^3 + \binom{6}{4}(3r)^2(-s)^4 +$$

$$\binom{6}{5}(3r)(-s)^5 + \binom{6}{5}(-s)^6 = 729r^6 - 1458r^5s + 1215r^4s^2 - 540r^3s^3 + 135r^2s^4 - 18rs^5 + s^6$$

31. The binomial expansion for $(7p + 2q)^4$ is given by:

$$(7p + 2q)^4 = \binom{4}{0}(7p)^4 + \binom{4}{1}(7p)^3(2q) + \binom{4}{2}(7p)^2(2q)^2 + \binom{4}{3}(7p)(2q)^3 + \binom{4}{4}(2q)^4 =$$

$$2401p^4 + 2744p^3q + 1176p^2q^2 + 224pq^3 + 16q^4$$

32. The binomial expansion for $(4a - 5b)^5$ is given by:

$$(4a - 5b)^5 = \binom{5}{0}(4a)^5 + \binom{5}{1}(4a)^4(-5b) + \binom{5}{2}(4a)^3(-5b)^2 + \binom{5}{3}(4a)^2(-5b)^3 + \binom{5}{4}(4a)(-5b)^4 +$$

$$\binom{5}{5}(-5b)^5 = 1024a^5 - 6400a^4b + 16,000a^3b^2 - 20,000a^2b^3 + 12,500ab^4 - 3125b^5$$

33. The binomial expansion for $(3x - 2y)^6$ is given by:

$$(3x - 2y)^6 = \binom{6}{0}(3x)^6 + \binom{6}{1}(3x)^5(-2y) + \binom{6}{2}(3x)^4(-2y)^2 + \binom{6}{3}(3x)^3(-2y)^3 + \binom{6}{4}(3x)^2(-2y)^4 +$$

$$\binom{6}{5}(3x)(-2y)^5 + \binom{6}{6}(-2y)^6 = 729x^6 - 2916x^5y + 4860x^4y^2 - 4320x^3y^3 + 2160x^2y^4 - 576xy^5 + 64y^6$$

34. The binomial expansion for $(7k - 9j)^4$ is given by:

$$(7k - 9j)^4 = \binom{4}{0}(7k)^4 + \binom{4}{1}(7k)^3(-9j) + \binom{4}{2}(7k)^2(-9j)^2 + \binom{4}{3}(7k)(-9j)^3 + \binom{4}{4}(-9j)^4 =$$

$$2401k^4 - 12{,}348k^3j + 23{,}814k^2j^2 - 20{,}412kj^3 + 6561j^4$$

35. The binomial expansion for $\left(\dfrac{m}{2} - 1\right)^6$ is given by:

$$\left(\frac{m}{2} - 1\right)^6 = \binom{6}{0}\left(\frac{m}{2}\right)^6 + \binom{6}{1}\left(\frac{m}{2}\right)^5(-1) + \binom{6}{2}\left(\frac{m}{2}\right)^4(-1)^2 + \binom{6}{3}\left(\frac{m}{2}\right)^3(-1)^3 + \binom{6}{4}\left(\frac{m}{2}\right)^2(-1)^4 +$$

$$\binom{6}{5}\left(\frac{m}{2}\right)(-1)^5 + \binom{6}{6}(-1)^6 = \frac{m^6}{64} - \frac{3m^5}{16} + \frac{15m^4}{16} - \frac{5m^3}{2} + \frac{15m^2}{4} - 3m + 1$$

36. The binomial expansion for $\left(3 + \dfrac{y}{3}\right)^5$ is given by:

$$\left(3 + \frac{y}{3}\right)^5 = \binom{5}{0}(3)^5 + \binom{5}{1}(3)^4\left(\frac{y}{3}\right) + \binom{5}{2}(3)^3\left(\frac{y}{3}\right)^2 + \binom{5}{3}(3)^2\left(\frac{y}{3}\right)^3 + \binom{5}{4}(3)\left(\frac{y}{3}\right)^4 + \binom{5}{5}\left(\frac{y}{3}\right)^5 =$$

$$243 + 135y + 30y^2 + \frac{10}{3}y^3 + \frac{5}{27}y^4 + \frac{y^5}{243}$$

37. The binomial expansion for $\left(\sqrt{2}r + \dfrac{1}{m}\right)^4$ is given by:

$$\left(\sqrt{2}r + \frac{1}{m}\right)^4 = \binom{4}{0}(\sqrt{2}r)^4 + \binom{4}{1}(\sqrt{2}r)^3\left(\frac{1}{m}\right) + \binom{4}{2}(\sqrt{2}r)^2\left(\frac{1}{m}\right)^2 + \binom{4}{3}(\sqrt{2}r)\left(\frac{1}{m}\right)^3 + \binom{4}{4}\left(\frac{1}{m}\right)^4 =$$

$$4r^4 + \frac{8\sqrt{2}r^3}{m} + \frac{12r^2}{m^2} + \frac{4\sqrt{2}r}{m^3} + \frac{1}{m^4}$$

38. The binomial expansion for $\left(\dfrac{1}{k} - \sqrt{3}p\right)^3$ is given by:

$$\left(\frac{1}{k} - \sqrt{3}p\right)^3 = \binom{3}{0}\left(\frac{1}{k}\right)^3 + \binom{3}{1}\left(\frac{1}{k}\right)^2(-\sqrt{3}p) + \binom{3}{2}\left(\frac{1}{k}\right)(-\sqrt{3}p)^2 + \binom{3}{3}(-\sqrt{3}p)^3 =$$

$$\frac{1}{k^3} - \frac{3\sqrt{3}p}{k^2} + \frac{9p^2}{k} - 3\sqrt{3}p^3$$

39. $\binom{8}{5}(4h)^3(-j)^5 = -3584h^3j^5$. The sixth term of the binomial expansion of $(4h - j)^8$ is $-3584n^3j^5$.

40. $\binom{14}{7}(2c)^7(-3d)^7 = -960{,}740{,}352c^7d^7$. The eighth term of the binomial expansion of

$(2c - 3d)^{14}$ is $-960{,}740{,}352c^7d^7$.

41. $\binom{22}{14}(a^2)^8(b)^{14} = 319{,}770a^{16}b^{14}$. The fifteenth term of the binomial expansion of $(a^2 + b)^{22}$ is $319{,}770a^{16}b^{14}$.

42. $\binom{16}{11}(2x)^5(y^2)^{11} = 139{,}776x^5y^{22}$. The twelfth term of the binomial expansion of $(2x + y^2)^{16}$ is $139{,}776x^5y^{22}$.

43. $\binom{20}{14}(x)^6(-y^3)^{14} = 38{,}760x^6y^{42}$. The fifteenth term of the binomial expansion of $(x - y^3)^{20}$ is $38{,}760x^6y^{42}$.

44. $\binom{11}{9}(a^3)^2(3b)^9 = 1{,}082{,}565a^6b^9$. The tenth term of the binomial expansion of $(a^3 + 3b)^{11}$ is $1{,}082{,}565a^6b^9$.

45. $\binom{8}{4}(3x^7)^4(2y^3)^4 = 90{,}720x^{28}y^{12}$. The middle term of the binomial expansion of $(3x^7 + 2y^3)^8$ is the fifth term which is $90{,}720x^{28}y^{12}$.

46. $\binom{11}{5}(-2m^{-1})^6(3n^{-2})^5 = 7{,}185{,}024m^{-6}n^{-10}$; $\binom{11}{6}(-2m^{-1})^5(3n^{-2})^6 = -10{,}777{,}536m^{-5}n^{-12}$

    The two middle terms of the binomial expansion of $(-2m^{-1} + 3n^{-2})^{11}$ are the sixth and seventh terms which are, respectively, $7{,}185{,}024m^{-6}n^{-10}$ and $-10{,}777{,}536m^{-5}n^{-12}$.

47. $n - 4 = 7$ and $n - 7 = 4 \Rightarrow n = 11$. The coefficients of the fifth and eighth terms in the expansion of $(x + y)^n$ are the same for $n = 11$.

48. $(\sqrt{x})^k = x^4$ for $k = 8$; $\binom{11}{8}(3)^3(\sqrt{x})^8 = 4455x^4$. The term in the expansion of $(3 + \sqrt{x})^{11}$ that contains x^4 is $4455x^4$.

49. $10! = 3{,}628{,}800$; $\sqrt{2\pi(10)}(10^{10})(e^{-10}) \approx 3{,}598{,}695.619$. The exact value of $10!$ is $3{,}628{,}800$ and the Stirling's formula approximation of $10!$ is about $3{,}598{,}695.619$.

50. $\dfrac{3{,}628{,}800 - 3{,}598{,}695.619}{3{,}628{,}800} \approx .00830 = .830\%$. The percentage error in the Stirling's formula approximation of $10!$ is about $.830\%$.

51. $12! = 479{,}001{,}600$; $\sqrt{2\pi(12)}(12^{12})(e^{-12}) \approx 475{,}687{,}486.5$;

    $\dfrac{479{,}001{,}600 - 475{,}687{,}486.5}{479{,}001{,}600} \approx .00692 = .692\%$. The exact value of $12!$ is $479{,}001{,}600$ and the Stirling's

    formula approximation of $12!$ is about $475{,}687{,}486.5$ which has a percent error of about $.692\%$.

52. $13! = 6{,}227{,}020{,}800$; $\sqrt{2\pi(13)}(13^{13})(e^{-13}) \approx 6{,}187{,}239{,}475$;

    $\dfrac{6{,}227{,}020{,}800 - 6{,}187{,}239{,}475}{6{,}227{,}020{,}800} \approx .00639 = .639\%$. The exact value of $13!$ is $6{,}227{,}020{,}800$ and the Stirling's

    formula approximation of $13!$ is about $6{,}187{,}239{,}475$ which has a percent error of about $.639\%$. Based on this series of exercises, it appears that the percent error in the Stirling's formula approximation of $n!$ decreases as n increases.

53. $(1.02)^{-3} = (1 + .02)^{-3} \approx 1 + (-3)(.02) + \dfrac{(-3)(-3 - 1)}{2!}(.02)^2 + \dfrac{(-3)(-3 - 1)(-3 - 2)}{3!}(.02)^3 \approx .942$.

    To the nearest thousandth, $(1.02)^{-3}$ is $.942$.

54. $\dfrac{1}{1.04^5} = (1 + .04)^{-5} \approx 1 + (-5)(.04) + \dfrac{(-5)(-5 - 1)}{2!}(.04)^2 + \dfrac{(-5)(-5 - 1)(-5 - 2)}{3!}(.04)^3 \approx .822$.

    To the nearest thousandth, $\dfrac{1}{1.04^5}$ is $.822$.

55. $(1.01)^{3/2} = (1 + .01)^{3/2} \approx 1 + \left(\dfrac{3}{2}\right)(.01) + \dfrac{(\frac{3}{2})(\frac{3}{2} - 1)}{2!}(.01)^2 + \dfrac{(\frac{3}{2})(\frac{3}{2} - 1)(\frac{3}{2} - 2)}{3!}(.01)^3 \approx 1.015$.

    To the nearest thousandth, $(1.01)^{3/2}$ is 1.015.

56. $(1.03)^{.2} = (1 + .03)^{.2} \approx 1 + (.2)(.03) + \dfrac{(.2)(.2 - 1)}{2!}(.03)^2 + \dfrac{(.2)(.2 - 1)(.2 - 2)}{3!}(.03)^3 \approx 1.006$.

    To the nearest thousandth, $(1.03)^{.2}$ is 1.006.

11.5: Mathematical Induction

1. The domain of the variable must be all positive integers. (natural numbers)

2. No, both steps must be true.

3. Since $2^1 = 2(1)$ and $2^2 = 2(2)$, the statement is not true for $n = 1$ and $n = 2$.

4. $S_1: 2 = 1(1 + 1); \ S_2: 2 + 4 = 2(2 + 1); \ S_3: 2 + 4 + 6 = 3(3 + 1); \ S_4: 2 + 4 + 6 + 8 = 4(4 + 1);$

    $S_5: 2 + 4 + 6 + 8 + 10 = 5(5 + 1)$

5. $3 + 6 + 9 + \cdots + 3n = \dfrac{3n(n + 1)}{2}$

    (i) Show that the statement is true for $n = 1$: $3(1) = \dfrac{3(1)(2)}{2} \Rightarrow 3 = 3$

    (ii) Assume that S_k is true: $3 + 6 + 9 + \cdots + 3k = \dfrac{3k(k + 1)}{2}$

    Show that S_{k+1} is true: $3 + 6 + \cdots + 3(k + 1) = \dfrac{3(k + 1)(k + 2)}{2}$

    Add $3(k + 1)$ to each side of S_k: $3 + 6 + 9 + \cdots + 3k + 3(k + 1) = \dfrac{3k(k + 1)}{2} + 3(k + 1) =$

    $\dfrac{3k(k + 1) + 6(k + 1)}{2} = \dfrac{(k + 1)(3k + 6)}{2} = \dfrac{3(k + 1)(k + 2)}{2}$

    Since S_k implies S_{k+1}, the statement is true for every positive integer n.

6. $1 + 3 + 5 + \cdots + (2n - 1) = n^2$

    (i) Show that the statement is true for $n = 1$: $2(1) - 1 = 1^2 \Rightarrow 1 = 1$

    (ii) Assume that S_k is true: $1 + 3 + 5 + \cdots + (2k - 1) = k^2$

    Show that S_{k+1} is true: $1 + 3 + \cdots + (2(k + 1) - 1) = (k + 1)^2$

    Add $2k + 1$ to each side of S_k: $1 + 3 + 5 + \cdots + (2k - 1) + (2k + 1) = k^2 + 2k + 1 = (k + 1)^2$

    Since S_k implies S_{k+1}, the statement is true for every positive integer n.

7. $5 + 10 + 15 + \cdots + 5n = \dfrac{5n(n + 1)}{2}$

    (i) Show that the statement is true for $n = 1$: $5(1) = \dfrac{5(1)(2)}{2} \Rightarrow 5 = 5$

    (ii) Assume that S_k is true: $5 + 10 + 15 + \cdots + 5k = \dfrac{5k(k + 1)}{2}$

    Show that S_{k+1} is true: $5 + 10 + \cdots + 5(k + 1) = \dfrac{5(k + 1)(k + 2)}{2}$

    Add $5(k + 1)$ to each side of S_k: $5 + 10 + 15 + \cdots + 5k + 5(k + 1) = \dfrac{5k(k + 1)}{2} + 5(k + 1) =$

    $\dfrac{5k(k + 1) + 10(k + 1)}{2} = \dfrac{(k + 1)(5k + 10)}{2} = \dfrac{5(k + 1)(k + 2)}{2}$

    Since S_k implies S_{k+1}, the statement is true for every positive integer n.

8. $4 + 7 + 10 + \cdots + (3n + 1) = \dfrac{n(3n + 5)}{2}$

(i) Show that the statement is true for $n = 1$: $3(1) + 1 = \dfrac{1(3(1) + 5)}{2} \Rightarrow 4 = 4$

(ii) Assume that S_k is true: $4 + 7 + 10 + \cdots + (3k + 1) = \dfrac{k(3k + 5)}{2}$

Show that S_{k+1} is true: $4 + 7 + 10 + \cdots + (3(k + 1) + 1) = \dfrac{(k + 1)(3(k + 1) + 5)}{2}$

Add $3(k + 1) + 1$ to each side of S_k: $4 + 7 + \cdots + (3k + 1) + 3(k + 1) + 1$

$= \dfrac{k(3k + 5)}{2} + 3(k + 1) + 1 = \dfrac{3k^2 + 5k}{2} + 3k + 4 = \dfrac{3k^2 + 5k + 6k + 8}{2} = \dfrac{3k^2 + 11k + 8}{2}$

$= \dfrac{(k + 1)(3k + 8)}{2} = \dfrac{(k + 1)(3k + 3 + 5)}{2} = \dfrac{(k + 1)(3(k + 1) + 5)}{2}$

Since S_k implies S_{k+1}, the statement is true for every positive integer n.

9. $3 + 3^2 + 3^3 + \cdots + 3^n = \dfrac{3(3^n - 1)}{2}$

(i) Show that the statement is true for $n = 1$: $3^1 = \dfrac{3(3^1 - 1)}{2} \Rightarrow 3 = 3$

(ii) Assume that S_k is true: $3 + 3^2 + 3^3 + \cdots + 3^k = \dfrac{3(3^k - 1)}{2}$

Show that S_{k+1} is true: $3 + 3^2 + 3^3 + \cdots + 3^{k+1} = \dfrac{3(3^{k+1} - 1)}{2}$

Add 3^{k+1} to each side of S_k: $3 + 3^2 + 3^3 + \cdots + 3^k + 3^{k+1} = \dfrac{3(3^k - 1)}{2} + 3^{k+1} = \dfrac{3(3^k - 1) + 2(3^{k+1})}{2}$

$= \dfrac{3^{k+1} - 3 + 2(3^{k+1})}{2} = \dfrac{3(3^{k+1}) - 3}{2} = \dfrac{3(3^{k+1} - 1)}{2}$

Since S_k implies S_{k+1}, the statement is true for every positive integer n.

10. $1^2 + 2^2 + 3^2 + \cdots + n^2 = \dfrac{n(n + 1)(2n + 1)}{6}$

(i) Show that the statement is true for $n = 1$: $1^2 = \dfrac{1(1 + 1)(2(1) + 1)}{6} \Rightarrow 1 = 1$

(ii) Assume that S_k is true: $1^2 + 2^2 + 3^2 + \cdots + k^2 = \dfrac{k(k + 1)(2k + 1)}{6}$

Show that S_{k+1} is true: $1^2 + 2^2 + 3^2 + \cdots + k^2 + (k + 1)^2 = \dfrac{(k + 1)(k + 2)(2(k + 1) + 1)}{6}$

Add $(k + 1)^2$ to each side of S_k: $1^2 + 2^2 + 3^2 + \cdots + k^2 + (k + 1)^2 = \dfrac{k(k + 1)(2k + 1)}{6} + (k + 1)^2$

$= \dfrac{k(k + 1)(2k + 1) + 6(k + 1)^2}{6} = \dfrac{(k + 1)[k(2k + 1) + 6(k + 1)]}{6} = \dfrac{(k + 1)(2k^2 + 7k + 6)}{6}$

$= \dfrac{(k + 1)(k + 2)(2k + 3)}{6} = \dfrac{(k + 1)(k + 2)(2k + 2 + 1)}{6} = \dfrac{(k + 1)(k + 2)(2(k + 1) + 1)}{6}$

Since S_k implies S_{k+1}, the statement is true for every positive integer n.

11. $1^3 + 2^3 + 3^3 + \cdots + n^3 = \dfrac{n^2(n + 1)^2}{4}$

  (i) Show that the statement is true for $n = 1$: $1^3 = \dfrac{1^2(1 + 1)^2}{4} \Rightarrow 1 = 1$

  (ii) Assume that S_k is true: $1^3 + 2^3 + 3^3 + \cdots + k^3 = \dfrac{k^2(k + 1)^2}{4}$

   Show that S_{k+1} is true: $1^3 + 2^3 + \cdots + (k + 1)^3 = \dfrac{(k + 1)^2(k + 2)^2}{4}$

   Add $(k + 1)^3$ to each side of S_k: $1^3 + 2^3 + 3^3 + \cdots + k^3 + (k + 1)^3 = \dfrac{k^2(k + 1)^2}{4} + (k + 1)^3$

   $= \dfrac{k^2(k + 1)^2 + 4(k + 1)^3}{4} = \dfrac{(k + 1)^2(k^2 + 4k + 4)}{4} = \dfrac{(k + 1)^2(k + 2)^2}{4}$

   Since S_k implies S_{k+1}, the statement is true for every positive integer n.

12. $5 \cdot 6 + 5 \cdot 6^2 + \cdots + 5 \cdot 6^n = 6(6^n - 1)$

  (i) Show that the statement is true for $n = 1$: $5 \cdot 6^1 = 6(6^1 - 1) \Rightarrow 30 = 30$

  (ii) Assume that S_k is true: $5 \cdot 6 + 5 \cdot 6^2 + \cdots + 5 \cdot 6^k = 6(6^k - 1)$

   Show that S_{k+1} is true: $5 \cdot 6 + 5 \cdot 6^2 + \cdots + 5 \cdot 6^{k+1} = 6(6^{k+1} - 1)$

   Add $5 \cdot 6^{k+1}$ to each side of S_k: $5 \cdot 6 + 5 \cdot 6^2 + \cdots + 5 \cdot 6^k + 5 \cdot 6^{k+1} = 6(6^k - 1) + 5 \cdot 6^{k+1}$

   $= 6^{k+1} - 6 + 5 \cdot 6^{k+1} = 6 \cdot 6^{k+1} - 6 = 6(6^{k+1} - 1)$

   Since S_k implies S_{k+1}, the statement is true for every positive integer n.

13. $\dfrac{1}{1 \cdot 2} + \dfrac{1}{2 \cdot 3} + \cdots + \dfrac{1}{n(n + 1)} = \dfrac{n}{n + 1}$

  (i) Show that the statement is true for $n = 1$: $\dfrac{1}{1(1 + 1)} = \dfrac{1}{1 + 1} \Rightarrow \dfrac{1}{2} = \dfrac{1}{2}$

  (ii) Assume that S_k is true: $\dfrac{1}{1 \cdot 2} + \dfrac{1}{2 \cdot 3} + \cdots + \dfrac{1}{k(k + 1)} = \dfrac{k}{k + 1}$

   Show that S_{k+1} is true: $\dfrac{1}{1 \cdot 2} + \dfrac{1}{2 \cdot 3} + \cdots + \dfrac{1}{(k + 1)(k + 2)} = \dfrac{k + 1}{k + 2}$

   Add $\dfrac{1}{(k + 1)(k + 2)}$ to each side of S_k: $\dfrac{1}{1 \cdot 2} + \dfrac{1}{2 \cdot 3} + \cdots + \dfrac{1}{(k + 1)(k + 2)} = \dfrac{k}{k + 1} + \dfrac{1}{(k + 1)(k + 2)}$

   $= \dfrac{k(k + 2) + 1}{(k + 1)(k + 2)} = \dfrac{k^2 + 2k + 1}{(k + 1)(k + 2)} = \dfrac{(k + 1)(k + 1)}{(k + 1)(k + 2)} = \dfrac{k + 1}{k + 2}$

   Since S_k implies S_{k+1}, the statement is true for every positive integer n.

14. $7 \cdot 8 + 7 \cdot 8^2 + \cdots + 7 \cdot 8^n = 8(8^n - 1)$

  (i) Show that the statement is true for $n = 1$: $7 \cdot 8^1 = 8(8^1 - 1) \Rightarrow 56 = 56$

  (ii) Assume that S_k is true: $7 \cdot 8 + 7 \cdot 8^2 + \cdots + 7 \cdot 8^k = 8(8^k - 1)$

   Show that S_{k+1} is true: $7 \cdot 8 + 7 \cdot 8^2 + \cdots + 7 \cdot 8^{k+1} = 8(8^{k+1} - 1)$

   Add $7 \cdot 8^{k+1}$ to each side of S_k: $7 \cdot 8 + 7 \cdot 8^2 + \cdots + 7 \cdot 8^k + 7 \cdot 8^{k+1} = 8(8^k - 1) + 7 \cdot 8^{k+1}$

   $= 8^{k+1} - 8 + 7 \cdot 8^{k+1} = 8 \cdot 8^{k+1} - 8 = 8(8^{k+1} - 1)$

   Since S_k implies S_{k+1}, the statement is true for every positive integer n.

15. $\dfrac{4}{5} + \dfrac{4}{5^2} + \dfrac{4}{5^3} + \cdots + \dfrac{4}{5^n} = 1 - \dfrac{1}{5^n}$

(i) Show that the statement is true for $n = 1$: $\dfrac{4}{5^1} = 1 - \dfrac{1}{5^1} \Rightarrow \dfrac{4}{5} = \dfrac{4}{5}$

(ii) Assume that S_k is true: $\dfrac{4}{5} + \dfrac{4}{5^2} + \dfrac{4}{5^3} + \cdots + \dfrac{4}{5^k} = 1 - \dfrac{1}{5^k}$

Show that S_{k+1} is true: $\dfrac{4}{5} + \dfrac{4}{5^2} + \cdots + \dfrac{4}{5^{k+1}} = 1 - \dfrac{1}{5^{k+1}}$

Add $\dfrac{4}{5^{k+1}}$ to each side of S_k: $\dfrac{4}{5} + \dfrac{4}{5^2} + \dfrac{4}{5^3} + \cdots + \dfrac{4}{5^k} + \dfrac{4}{5^{k+1}} = 1 - \dfrac{1}{5^k} + \dfrac{4}{5^{k+1}}$

$= 1 - \dfrac{1}{5^k} \cdot \dfrac{5}{5} + \dfrac{4}{5^{k+1}} = 1 - \dfrac{5}{5^{k+1}} + \dfrac{4}{5^{k+1}} = 1 - \dfrac{1}{5^{k+1}}$

Since S_k implies S_{k+1}, the statement is true for every positive integer n.

16. $\dfrac{1}{2} + \dfrac{1}{2^2} + \dfrac{1}{2^3} + \cdots + \dfrac{1}{2^n} = 1 - \dfrac{1}{2^n}$

(i) Show that the statement is true for $n = 1$: $\dfrac{1}{2^1} = 1 - \dfrac{1}{2^1} \Rightarrow \dfrac{1}{2} = \dfrac{1}{2}$

(ii) Assume that S_k is true: $\dfrac{1}{2} + \dfrac{1}{2^2} + \dfrac{1}{2^3} + \cdots + \dfrac{1}{2^k} = 1 - \dfrac{1}{2^k}$

Show that S_{k+1} is true: $\dfrac{1}{2} + \dfrac{1}{2^2} + \cdots + \dfrac{1}{2^{k+1}} = 1 - \dfrac{1}{2^{k+1}}$

Add $\dfrac{1}{2^{k+1}}$ to each side of S_k: $\dfrac{1}{2} + \dfrac{1}{2^2} + \dfrac{1}{2^3} + \cdots + \dfrac{1}{2^k} + \dfrac{1}{2^{k+1}} = 1 - \dfrac{1}{2^k} + \dfrac{1}{2^{k+1}} = 1 - \dfrac{1}{2^k} \cdot \dfrac{2}{2} + \dfrac{1}{2^{k+1}}$

$= 1 - \dfrac{2}{2^{k+1}} + \dfrac{1}{2^{k+1}} = 1 - \dfrac{1}{2^{k+1}}$

Since S_k implies S_{k+1}, the statement is true for every positive integer n.

17. $\dfrac{1}{1 \cdot 4} + \dfrac{1}{4 \cdot 7} + \cdots + \dfrac{1}{(3n-2)(3n+1)} = \dfrac{n}{3n+1}$

(i) Show that the statement is true for $n = 1$: $\dfrac{1}{1 \cdot 4} = \dfrac{1}{3(1)+1} \Rightarrow \dfrac{1}{4} = \dfrac{1}{4}$

(ii) Assume that S_k is true: $\dfrac{1}{1 \cdot 4} + \cdots + \dfrac{1}{(3k-2)(3k+1)} = \dfrac{k}{3k+1}$

Show that S_{k+1} is true: $\dfrac{1}{1 \cdot 4} + \cdots + \dfrac{1}{[3(k+1)-2][3(k+1)+1]} = \dfrac{k+1}{3(k+1)+1}$

Add $\dfrac{1}{[3(k+1)-2][3(k+1)+1]}$ to each side of S_k: $\dfrac{1}{1 \cdot 4} + \cdots + \dfrac{1}{[3(k+1)-2][3(k+1)+1]}$

$= \dfrac{k}{3k+1} + \dfrac{1}{[3(k+1)-2][3(k+1)+1]} = \dfrac{k}{3k+1} + \dfrac{1}{(3k+1)(3k+4)} = \dfrac{k(3k+4)+1}{(3k+1)(3k+4)}$

$= \dfrac{3k^2+4k+1}{(3k+1)(3k+4)} = \dfrac{(3k+1)(k+1)}{(3k+1)(3k+4)} = \dfrac{k+1}{3k+4} = \dfrac{k+1}{3(k+1)+1}$

Since S_k implies S_{k+1}, the statement is true for every positive integer n.

18. $x^{2n} + x^{2n-1}y + \cdots + xy^{2n-1} + y^{2n} = \dfrac{x^{2n+1} - y^{2n+1}}{x - y}$

    (i) Show that the statement is true for $n = 1$: $x^2 + xy + y^2 = \dfrac{x^3 - y^3}{x - y} \Rightarrow$

        $x^2 + xy + y^2 = \dfrac{(x - y)(x^2 + xy + y^2)}{x - y} \Rightarrow x^2 + xy + y^2 = x^2 + xy + y^2$

    (ii) Assume that S_k is true: $x^{2k} + x^{2k-1}y + \cdots + xy^{2k-1} + y^{2k} = \dfrac{x^{2k+1} - y^{2k+1}}{x - y}$

        Show that S_{k+1} is true: $x^{2(k+1)} + x^{2(k+1)-1}y + \cdots + xy^{2(k+1)-1} + y^{2(k+1)} = \dfrac{x^{2(k+1)+1} - y^{2(k+1)+1}}{x - y}$

        Multiply each side of S_k by x^2 and then add $xy^{2k+1} + y^{2k+2}$ to each side.

        The left side is $x^2(x^{2k} + x^{2k-1}y + \cdots + xy^{2k-1} + y^{2k}) + xy^{2k+1} + y^{2k+2}$

        $= x^{2k+2} + x^{2k+1}y + \cdots + xy^{2k-1} + y^{2k} + xy^{2k+1} + y^{2k+2} = x^{2(k+1)} + x^{2(k+1)-1}y + \cdots + xy^{2(k+1)-1} + y^{2(k+1)}$

        $= S_{k+1}$. The right side is $x^2 \cdot \dfrac{x^{2k+1} - y^{2k+1}}{x - y} + xy^{2k+1} + y^{2k+2}$.

        Multiply, write the expression using the LCD, and then simplify to obtain

        $\dfrac{x^{2k+3} - y^{2k+3}}{x - y} = \dfrac{x^{2(k+1)+1} - y^{2(k+1)+1}}{x - y}$.

    Since S_k implies S_{k+1}, the statement is true for every positive integer n.

19. When $n = 1$, $3^1 < 6(1) \Rightarrow 3 < 6$. When $n = 2$, $3^2 < 6(2) \Rightarrow 9 < 12$.

    When $n = 3$, $3^3 < 6(3) \Rightarrow 27 > 18$. For all $n \geq 3$, $3^n > 6n$. The only values are 1 and 2.

20. When $n = 1$, $3^1 = 2(1) + 1 \Rightarrow 3 = 3$. When $n = 2$, $3^2 > 2(2) + 1 \Rightarrow 9 > 5$.

    When $n = 3$, $3^3 > 2(3) + 1 \Rightarrow 27 > 7$. For all $n \geq 2$, $3^n > 2n + 1$. The only value is 1.

21. When $n = 1$, $2^1 > 1^2 \Rightarrow 2 > 1$. When $n = 2$, $2^2 = 2^2 \Rightarrow 4 = 4$. When $n = 3$, $2^3 < 3^2 \Rightarrow 8 < 9$.

    When $n = 4$, $2^4 = 4^2 \Rightarrow 16 = 16$. For all $n \geq 5$, $2^n > n^2$. The only values are 2, 3, and 4.

22. When $n = 1$, $1! < 2(1) \Rightarrow 1 < 2$. When $n = 2$, $2! < 2(2) \Rightarrow 2 < 4$. When $n = 3$, $3! = 2(3) \Rightarrow 6 = 6$.

    When $n = 4$, $4! > 2(4) \Rightarrow 24 > 8$. For all $n \geq 4$, $n! > 2n$. The only values are 1, 2, and 3.

23. $(a^m)^n = a^{mn}$

    (i) Show that the statement is true for $n = 1$: $(a^m)^1 = a^{m \cdot 1} \Rightarrow a^m = a^m$

    (ii) Assume that S_k is true: $(a^m)^k = a^{mk}$

        Show that S_{k+1} is true: $(a^m)^{k+1} = a^{m(k+1)}$

        Multiply each side of S_k by a^m: $(a^m)^k \cdot (a^m)^1 = a^{mk} \cdot a^m \Rightarrow (a^m)^{k+1} = a^{mk+m} \Rightarrow (a^m)^{k+1} = a^{m(k+1)}$

    Since S_k implies S_{k+1}, the statement is true for every positive integer n.

24. $(ab)^n = a^n b^n$

    (i) Show that the statement is true for $n = 1$: $(ab)^1 = a^1 b^1 \Rightarrow ab = ab$

    (ii) Assume that S_k is true: $(ab)^k = a^k b^k$

        Show that S_{k+1} is true: $(ab)^{k+1} = a^{k+1} b^{k+1}$

        Multiply each side of S_k by ab: $(ab)^k \cdot (ab)^1 = a^k b^k \cdot a^1 b^1 \Rightarrow (ab)^{k+1} = (a^k \cdot a^1)(b^k \cdot b^1) \Rightarrow$

        $(ab)^{k+1} = a^{k+1} b^{k+1}$

    Since S_k implies S_{k+1}, the statement is true for every positive integer n.

25. $2^n > 2n$, if $n \geq 3$

    (i) Show that the statement is true for $n = 3$: $2^3 > 2(3) \Rightarrow 8 > 6$

    (ii) Assume that S_k is true: $2^k > 2k$

    Show that S_{k+1} is true: $2^{k+1} > 2(k + 1)$

    Multiply each side of S_k by 2: $2^k \cdot 2 > 2k \cdot 2 \Rightarrow 2^{k+1} > 2k \Rightarrow 2^{k+1} > 2(k + 1)$, for all $k > 1$.

    Since S_k implies S_{k+1}, the statement is true for every positive integer $n \geq 3$.

26. $3^n > 2n + 1$, if $n \geq 2$

    (i) Show that the statement is true for $n = 2$: $3^2 > 2(2) + 1 \Rightarrow 9 > 5$

    (ii) Assume that S_k is true: $3^k > 2k + 1$

    Show that S_{k+1} is true: $3^{k+1} > 2(k + 1) + 1$

    Multiply each side of S_k by 3: $3^k \cdot 3 > (2k + 1) \cdot 3 \Rightarrow 3^{k+1} > 6k + 3 \Rightarrow 3^{k+1} > 6k + 2 + 1 \Rightarrow$

    $3^{k+1} > 2(3k + 1) + 1$
    Because $3k + 1 > k + 1$ for all $k \geq 2$, we may substitute $k + 1$ for $3k + 1$ in the expression.

    That is $3^{k+1} > 2(k + 1) + 1$.

    Since S_k implies S_{k+1}, the statement is true for every positive integer $n \geq 2$.

27. $a^n > 1$, if $a > 1$

    (i) Show that the statement is true for $n = 1$: $a^1 > 1 \Rightarrow a > 1$, which is true by the given restriction.

    (ii) Assume that S_k is true: $a^k > 1$

    Show that S_{k+1} is true: $a^{k+1} > 1$

    Multiply each side of S_k by a: $a^k \cdot a > 1 \cdot a \Rightarrow a^{k+1} > a$

    Because $a > 1$, we may substitute 1 for a in the expression. That is $a^{k+1} > 1$

    Since S_k implies S_{k+1}, the statement is true for every positive integer n.

28. $a^n > a^{n-1}$, if $a > 1$

    (i) Show that the statement is true for $n = 1$: $a^1 > a^0 \Rightarrow a > 1$, which is true by the given restriction.

    (ii) Assume that S_k is true: $a^k > a^{k-1}$

    Show that S_{k+1} is true: $a^{k+1} > a^k$

    Multiply each side of S_k by a: $a^k \cdot a > a^{k-1} \cdot a \Rightarrow a^{k+1} > a^k$

    Since S_k implies S_{k+1}, the statement is true for every positive integer n.

29. $a^n < a^{n-1}$, if $0 < a < 1$

    (i) Show that the statement is true for $n = 1$: $a^1 < a^0 \Rightarrow a < 1$, which is true by the given restriction.

    (ii) Assume that S_k is true: $a^k < a^{k-1}$

    Show that S_{k+1} is true: $a^{k+1} < a^k$

    Multiply each side of S_k by a: $a^k \cdot a < a^{k-1} \cdot a \Rightarrow a^{k+1} < a^k$

    Since S_k implies S_{k+1}, the statement is true for every positive integer n.

30. $2^n > n^2$, if $n > 4$

    (i) Show that the statement is true for $n = 5$: $2^5 > 5^2 \Rightarrow 32 > 25$

    (ii) Assume that S_k is true: $2^k > k^2$

    Show that S_{k+1} is true: $2^{k+1} > (k+1)^2$

    Multiply each side of S_k by 2: $2^k \cdot 2 > k^2 \cdot 2 \Rightarrow 2^{k+1} > 2k^2$

    Because $2k^2 > (k+1)^2$ for all $k \geq 5$, we may substitute $(k+1)^2$ for $2k^2$ in the expression.

    That is $2^{k+1} > (k+1)^2$.

    Since S_k implies S_{k+1}, the statement is true for every positive integer $n > 4$.

31. $n! > 2^n$, if $n \geq 4$

    (i) Show that the statement is true for $n = 4$: $4! > 2^4 \Rightarrow 24 > 16$

    (ii) Assume that S_k is true: $k! > 2^k$

    Show that S_{k+1} is true: $(k+1)! > 2^{k+1}$

    Multiply each side of S_k by $k + 1$: $(k+1)k! > 2^k(k+1) \Rightarrow (k+1)! > 2^k(k+1)$

    Because $(k+1) > 2$ for all $k \geq 4$, we may substitute 2 for $(k+1)$ in the expression.

    That is $(k+1)! > 2^k(2)$ or $(k+1)! > 2^{k+1}$.

    Since S_k implies S_{k+1}, the statement is true for every positive integer $n \geq 4$.

32. $4^n > n^4$, if $n \geq 5$

    (i) Show that the statement is true for $n = 5$: $4^5 > 5^4 \Rightarrow 1024 > 625$

    (ii) Assume that S_k is true: $4^k > k^4$

    Show that S_{k+1} is true: $4^{k+1} > (k+1)^4$

    Multiply each side of S_k by 4: $4^k \cdot 4 > k^4 \cdot 4 \Rightarrow 4^{k+1} > 4k^4$

    Because $4k^4 > (k+1)^4$ for all $k \geq 5$, we may substitute $(k+1)^4$ for $4k^4$ in the expression.

    That is $4^{k+1} > (k+1)^4$.

    Since S_k implies S_{k+1}, the statement is true for every positive integer $n \geq 5$.

33. The number of handshakes is $\dfrac{n^2 - n}{2}$ if $n \geq 2$.

    (i) Show that the statement is true for $n = 2$: The number of handshakes for 2 people is $\dfrac{2^2 - 2}{2} = \dfrac{2}{2} = 1$,

    which is true.

    (ii) Assume that S_k is true: The number of handshakes for k people is $\dfrac{k^2 - k}{2}$.

    Show that S_{k+1} is true: The number of handshakes for $k + 1$ people is

    $$\frac{(k+1)^2 - (k+1)}{2} = \frac{k^2 + 2k + 1 - k - 1}{2} = \frac{k^2 + k}{2}.$$

    When a person joins a group of k people, each person must shake hands with the new person.

    Since there are a total of k people that will shake hands with the new person, the total number of handshakes

    for $k + 1$ people is $\dfrac{k^2 - k}{2} + k = \dfrac{k^2 - k + 2k}{2} = \dfrac{k^2 + k}{2}$.

    Since S_k implies S_{k+1}, the statement is true for every positive integer $n \geq 2$.

34. The number of sides is $3(4)^{n-1}$.

    (i) Show that the statement is true for $n = 1$: The number of sides in the first figure is

    $3(4)^{1-1} = 3(4)^0 = 3(1) = 3$, which is true.

    (ii) Assume that S_k is true: The number of sides in figure k is $3(4)^{k-1}$.

    Show that S_{k+1} is true: The number of sides in figure $k + 1$ is $3(4)^k$.

    When a new figure is made, each side of the previous figure was broken into 4 smaller sides.

    That is, the total number of sides in figure $k + 1$ is 4 times the number of sides in figure k.

    This is given by $4 \cdot 3(4)^{k-1} = 3(4)^k$.

    Since S_k implies S_{k+1}, the statement is true for every positive integer n.

35. The first figure has perimeter $P = 3$. When a new figure is generated, each side if the previous figure increases

    in length by a factor of $\dfrac{4}{3}$. Thus, the second figure has perimeter $P = 3\left(\dfrac{4}{3}\right)$, the third figure has perimeter

    $P = 3\left(\dfrac{4}{3}\right)^2$, and so on. In general, the nth figure has perimeter $P = 3\left(\dfrac{4}{3}\right)^{n-1}$.

36. The area of an equilateral triangle is $A = \dfrac{\sqrt{3}}{4}s^2$. In the first figure, $s = 1$ so $A_1 = \dfrac{\sqrt{3}}{4} \cdot 1^2 = \dfrac{\sqrt{3}}{4}$.

    The second figure has three smaller equilateral triangles with $s = \dfrac{1}{3}$. Each of these triangles has area

    $A = \dfrac{\sqrt{3}}{4} \cdot \left(\dfrac{1}{3}\right)^2 = \dfrac{\sqrt{3}}{36}$ and $3A = \dfrac{\sqrt{3}}{12}$. The total area of the second figure is $A_1 + 3A$. The third figure has

    twelve smaller equilateral triangles with $s = \dfrac{1}{9}$. Each of these triangles has area

    $A = \dfrac{\sqrt{3}}{4} \cdot \left(\dfrac{1}{9}\right)^2 = \dfrac{\sqrt{3}}{324}$ and $12A = \dfrac{\sqrt{3}}{27}$. The total area of the third figure is $A_2 + 12A$. Likewise the fourth

    figure has an additional area of $\dfrac{4\sqrt{3}}{243}$. The sequence of *additional* areas is given by $\dfrac{\sqrt{3}}{4}, \dfrac{\sqrt{3}}{12}, \dfrac{\sqrt{3}}{27}, \dfrac{4\sqrt{3}}{243}, \ldots$.

    The sum of the first n numbers in this sequence gives the area of the nth figure. Starting with the second

    number listed, the sequence is geometric with $a_1 = \dfrac{\sqrt{3}}{12}$ and $r = \dfrac{4}{9}$.

    $$A = \frac{a_1(1 - r^{n-1})}{1 - r} + \frac{\sqrt{3}}{4} = \frac{\frac{\sqrt{3}}{12}\left(1 - \left(\frac{4}{9}\right)^{n-1}\right)}{1 - \frac{4}{9}} + \frac{\sqrt{3}}{4} = \frac{\frac{\sqrt{3}}{12}\left(1 - \left(\frac{4}{9}\right)^{n-1}\right)}{\frac{5}{9}} \cdot \frac{9}{9} + \frac{\sqrt{3}}{4} = \frac{\frac{3\sqrt{3}}{4}\left(1 - \left(\frac{4}{9}\right)^{n-1}\right)}{5} + \frac{\sqrt{3}}{4}$$

    $$= \frac{3\sqrt{3}\left(1 - \left(\frac{4}{9}\right)^{n-1}\right) + 5\sqrt{3}}{20} = \frac{3\sqrt{3} - 3\sqrt{3}\left(\frac{4}{9}\right)^{n-1} + 5\sqrt{3}}{20} = \frac{8\sqrt{3} - 3\sqrt{3}\left(\frac{4}{9}\right)^{n-1}}{20} = \sqrt{3}\left[\frac{8 - 3\left(\frac{4}{9}\right)^{n-1}}{20}\right]$$

    $$= \sqrt{3}\left[\frac{2}{5} - \frac{3}{20}\left(\frac{4}{9}\right)^{n-1}\right]$$

37. With 1 ring, 1 move is required. With 2 rings, 3 moves are required. Note that $3 = 2 + 1$. With 3 rings, 7 moves are required. Note that $7 = 2^2 + 2 + 1$.

   With n rings $2^{n-1} + 2^{n-2} + \cdots + 2^1 + 1 = 2^n - 1$ moves are required.

   (i) Show that the statement is true for $n = 1$: The number of moves for 1 ring is $2^1 - 1 = 1$, which is true.

   (ii) Assume that S_k is true: The number of moves for k rings is $2^k - 1$.

   Show that S_{k+1} is true: The number of moves for $k + 1$ rings is $2^{k+1} - 1$.

   Assume $k + 1$ rings are on the first peg. Since S_k is true, the top k rings can be moved to the second peg in $2^k - 1$ moves. Now move the bottom ring to the third peg. Since S_k is true, move the k rings from the second peg on top of the ring on the third peg in $2^k - 1$ moves. The total number of moves is

   $$(2^k - 1) + 1 + (2^k - 1) = 2 \cdot 2^k - 1 = 2^{k+1} - 1$$

   Since S_k implies S_{k+1}, the statement is true for every positive integer n.

Reviewing Basic Concepts (Sections 11.4 and 11.5)

1. $_5C_3 = 5\ _nC_r\ 3 = 10$. The value of $_5C_3$ is 10.

2. $\binom{6}{2} = 6\ _nC_r\ 2 = 15$. The value of $\binom{6}{2}$ is 15.

3. The expansion of $(x + y)^n$ has a term involving each of these powers of x: $\{n, n - 1, n - 2, \ldots, 2, 1, 0\}$. Thus, the expansion of $(x + y)^n$ has $n + 1$ terms.

4. The binomial expansion for $(a + 2b)^4$ is given by:

   $$(a + 2b)^4 = \binom{4}{0}a^4 + \binom{4}{1}a^3(2b) + \binom{4}{2}a^2(2b)^2 + \binom{4}{3}a(2b)^3 + \binom{4}{4}(2b)^4 =$$

   $$a^4 + 8a^3b + 24a^2b^2 + 32ab^3 + 16b^4$$

5. The binomial expansion for $\left(\dfrac{1}{m} - n^2\right)^3$ is given by:

   $$\left(\frac{1}{m} - n^2\right)^3 = \binom{3}{0}\left(\frac{1}{m}\right)^3 + \binom{3}{1}\left(\frac{1}{m}\right)^2(-n^2) + \binom{3}{2}\left(\frac{1}{m}\right)(-n^2)^2 + \binom{3}{3}(-n^2)^3 = \frac{1}{m^3} - \frac{3n^2}{m^2} + \frac{3n^4}{m} - n^6$$

6. $\binom{6}{2}(x)^4(-2y)^2 = 60x^4y^2$. The third term of the expansion of $(x - 2y)^6$ is $60x^4y^2$.

7. (i) $4 + 8 = 12$; $2(2)(2 + 1) = (4)(3) = 12$; Thus, $4 + 8 + \cdots + 4n = 2n(n + 1)$ for $n = 2$.

   (ii) Assume $4 + 8 + \cdots + 4n = 2n(n + 1)$ for $n = k$, that is $4 + 8 + \cdots + 4k = 2k(k + 1)$

   Then $4 + 8 + \cdots + 4k + 4(k + 1) = 2k(k + 1) + 4(k + 1) = (2k + 4)(k + 1) = 2(k + 2)(k + 1) = 2(k + 1)(k + 2)$. So $4 + 8 + \cdots + 4n = 2n(n + 1)$ for $n = k + 1$ also. Thus, by induction, $4 + 8 + \cdots + 4n = 2n(n + 1)$ for all $n \geq 2$.

8. (i) $4^2 = 16$; $2^4 = 16 \Rightarrow n^2 = 2^n$ for $n = 4$

    $5^2 = 25$; $2^5 = 32 \Rightarrow n^2 < 2^n$ for $n = 5$

  (ii) Assume $n^2 \le 2^n$ for $n = k \ge 4$, that is $k^2 \le 2^k$, now, $(k + 1)^2 = k^2 + 2k + 1$.

    Since $k \ge 4$, $2k \ge 8 > 1$, so $(k + 1)^2 \le k^2 + 2k + 2k = k^2 + 4k$.

    Since $k \ge 4$, $k^2 \ge 4k$, so $(k + 1)^2 \le k^2 + k^2 = 2k^2 \le 2(2^k) = 2^{k+1}$, so $n^2 \le 2^n$ for $n = k + 1$ also.

11.6: Counting Theory

1. $P(12, 8) = \dfrac{12!}{(12 - 8)!} = \dfrac{12!}{4!} = 12 \times 11 \times 10 \times 9 \times 8 \times 7 \times 6 \times 5 = 19{,}958{,}400.$

   The value of $P(12, 8)$ is 19,958,400.

2. $P(5, 5) = \dfrac{5!}{(5 - 5)!} = \dfrac{5!}{0!} = 5! = 120.$ The value of $P(5, 5)$ is 120.

3. $P(9, 2) = \dfrac{9!}{(9 - 2)!} = \dfrac{9!}{7!} = 9 \times 8 = 72.$ The value of $P(9, 2)$ is 72.

4. $P(10, 9) = \dfrac{10!}{(10 - 9)!} = \dfrac{10!}{1!} = 10! = 3{,}628{,}800.$ The value of $P(10, 9)$ is 3,628,800.

5. $P(5, 1) = \dfrac{5!}{(5 - 1)!} = \dfrac{5!}{4!} = 5.$ The value of $P(5, 4)$ is 5.

6. $P(6, 0) = \dfrac{6!}{(6 - 0)!} = \dfrac{6!}{6!} = 1.$ The value of $P(6, 0)$ is 1.

7. $C(4, 2) = \dfrac{4!}{(4 - 2)!2!} = \dfrac{4!}{2!2!} = \dfrac{4 \times 3}{2 \times 1} = 6.$ The value of $C(4, 2)$ is 6.

8. $C(9, 3) = \dfrac{9!}{(9 - 3)!3!} = \dfrac{9!}{6!3!} = \dfrac{9 \times 8 \times 7}{3 \times 2 \times 1} = 84.$ The value of $C(9, 3)$ is 84.

9. $C(6, 0) = \dfrac{6!}{(6 - 0)!0!} = \dfrac{6!}{6!} = 1.$ The value of $C(6, 0)$ is 1.

10. $C(8, 1) = \dfrac{8!}{(8 - 1)!1!} = \dfrac{8!}{7!} = 8.$ The value of $C(8, 1)$ is 8.

11. $\dbinom{12}{4} = \dfrac{12!}{(12 - 4)!4!} = \dfrac{12!}{8!4!} = \dfrac{12 \times 11 \times 10 \times 9}{4 \times 3 \times 2 \times 1} = 495.$ The value of $\dbinom{12}{4}$ is 495.

12. $\dbinom{16}{3} = \dfrac{16!}{(16 - 3)!3!} = \dfrac{16!}{13!3!} = \dfrac{16 \times 15 \times 14}{3 \times 2 \times 1} = 560.$ The value of $\dbinom{16}{3}$ is 560.

13. $_{20}P_5 = 20\ _nP_r\ 5 = 1{,}860{,}480.$ The value of $_{20}P_5$ is 1,860,480.

14. $_{100}P_5 = 100\ _nP_r\ 5 = 9{,}034{,}502{,}400.$ The value of $_{100}P_5$ is 9,034,502,400.

15. $_{15}P_8 = 15\ _nP_r\ 8 = 259{,}459{,}200.$ The value of $_{15}P_8$ is 259,459,200.

16. $_{32}P_4 = 32\ _nP_r\ 4 = 863{,}040.$ The value of $_{32}P_4$ is 863,040.

17. $_{20}C_5 = 20\ _nC_r\ 5 = 15{,}504.$ The value of $_{20}C_5$ is 15,504.

18. $_{100}C_5 = 100\ _nC_r\ 5 = 75{,}287{,}520.$ The value of $_{100}C_5$ is 75,287,520.

19. $\binom{15}{8} = 15 \,_nC_r\, 8 = 6435$. The value of $\binom{15}{8}$ is 6435.

20. $\binom{32}{4} = 32 \,_nC_r\, 4 = 35{,}960$. The value of $\binom{32}{4}$ is 35,960.

21. (a) A telephone number involves a permutation of digits because order matters.

    (b) A social security number involves a permutation of digits because order matters.

    (c) A hand of cards in poker involves a combination of cards because order does not matter.

    (d) A committee of politicians involves a combination of persons because order does not matter.

    (e) The "combination" on a combination lock involves a permutation of numbers because order matters.

    (f) A lottery choice of six numbers where the order does not matter involves a combination of numbers.

    (g) An automobile license plate involves a permutation of characters because order matters.

22. A permutation is an arrangement of objects in a particular order, while a combination is a grouping of objects where order is unimportant. You should look for the clue words in the table preceeding Example 9 in this section.

23. $5 \times 3 \times 2 = 30$. 30 different types of homes are available.

24. $7 \times 6 \times 4 \times 5 = 840$. 840 varieties of automobiles are available.

25. (a) $2 \times 25 \times 24 \times 23 = 27{,}600$. 27,600 radio station calls can be made.

    (b) $2 \times 26 \times 26 \times 26 = 35{,}152$. 35,152 radio station calls can be made.

    (c) $2 \times 24 \times 23 \times 1 = 1104$. 1104 radio station calls can be made.

26. $3 \times 8 \times 5 = 120$. 120 three course meals are possible.

27. $3 \times 5 = 15$. 15 first and middle name combinations are possible.

28. $5 \,_nP_r\, 5 = 120$. The program can be arranged in 120 ways.

29. (a) $26 \times 26 \times 26 \times 10 \times 10 \times 10 = 17{,}576{,}000$. 17,576,000 different license plates are possible.

    (b) $10 \times 10 \times 10 \times 26 \times 26 \times 26 = 17{,}576{,}000$. 17,576,000 additional plates are possible.

    (c) $26 \times 10 \times 10 \times 10 \times 26 \times 26 \times 26 = 456{,}976{,}000$. the new scheme provides 456,976,000 plates.

30. (a) $5 \times 5 \times 5 \times 5 \times 5 \times 5 \times 5 = 78{,}125$. 78,125 telephone numbers are possible.

    (b) $9 \times 10 \times 10 \times 10 \times 10 \times 10 \times 1 = 900{,}000$. 900,000 telephone numbers are possible.

    (c) $9 \times 10 \times 10 \times 10 \times 10 \times 1 \times 1 = 90{,}000$. 90,000 telephone numbers are possible.

    (d) $1 \times 1 \times 1 \times 10 \times 10 \times 10 \times 10 = 10{,}000$. 10,000 telephone numbers are possible.

    (e) $9 \times 9 \times 8 \times 7 \times 6 \times 5 \times 4 = 544{,}320$. 544,320 telephone numbers are possible.

31. $6 \,_nP_r\, 6 = 720$. 720 arrangements of the people are possible.

32. $10 \,_nP_r\, 7 = 604{,}800$. 604,800 arrangements of the monkeys are possible.

33. $6 \,_nP_r\, 3 = 120$. 120 course schedules are possible.

34. $380 \,_nP_r\, 4 = 20{,}523{,}714{,}120$. 20,523,714,120 course schedules are possible.

35. $15 \,_nP_r\, 3 = 2730$. 2730 slates of 3 officers are possible.

36. $20 \,_nP_r\, 9 = 60{,}949{,}324{,}800$. 60,949,324,800 batting orders are possible.

37. $5 \,_nP_r\, 5 = 120$; $10 \,_nP_r\, 5 = 30{,}240$. With 5 players, 120 assignments are possible and with 10 players, 30,240 assignments are possible.

38. $5\,_nP_r\,5 = 120$. The letters can be arrange in 120 ways.

39. $30\,_nP_r\,4 = 27{,}405$. 27,405 different groups are possible.

40. $25\,_nC_r\,3 = 2300$. 2300 different samples are possible.

41. $6\,_nC_r\,3 = 20$. 20 different garnished hamburgers are possible.

42. $12\,_nC_r\,3 = 220$; $12\,_nC_r\,9 = 220$. 220 groups of participants are possible. Not surprisingly, 220 groups of non-participants are also possible.

43. $5\,_nC_r\,2 = 10$. 10 different hands are possible.

44. $15\,_nC_r\,2 = 105$; $15\,_nC_r\,4 = 1365$. 105 samples of 2 may be drawn and 1365 samples of 4 may be drawn.

45. $8\,_nC_r\,2 = 28$. 28 samples of 2 may be drawn where both marbles are blue.

46. (a) $5\,_nC_r\,3 = 10$. 10 samples may be drawn in which all 3 apples are rotten.

    (b) $(5\,_nC_r\,1)(20\,_nC_r\,2) = (5)(190) = 950$. 950 samples may be drawn with 1 rotten apple and 2 good apples.

47. (a) $9\,_nC_r\,3 = 84$. 84 delegations are possible.

    (b) $5\,_nC_r\,3 = 10$. 10 delegations could have all liberals.

    (c) $(5\,_nC_r\,2)(4\,_nC_r\,1) = (10)(4) = 40$. 40 delegations could have 2 liberals and 1 conservative.

    (d) $8\,_nC_r\,2 = 28$. 28 delegations are possible that include the mayor.

48. (a) $7\,_nC_r\,2 = 21$. 21 delegations are possible.

    (b) $6\,_nC_r\,1 = 6$. 6 delegations are possible that include a particular emplyee.

    (c) $(2\,_nC_r\,1)(5\,_nC_r\,1) + (2\,_nC_r\,2)(5\,_nC_r\,0) = (2)(5) + (1)(1) = 11$. 11 delegations are possible that include at least 1 woman.

49. $8\,_nP_r\,4 = 1680$. 1680 course schedules are possible.

50. $12\,_nC_r\,3 = 220$. 220 samples are possible.

51. $6\,_nC_r\,4 = 15$. 15 different soups can be made.

52. $7\,_nP_r\,3 = 210$. 210 different assignments can be made.

53. $12\,_nP_r\,11 = 479{,}001{,}600$. 479,001,600 different seatings are possible.

54. (a) $11\,_nC_r\,4 = 330$. 330 different selections are possible.

    (b) $(6\,_nC_r\,2)(5\,_nC_r\,2) = (15)(10) = 150$. 150 selections are possible that include exactly 2 wheat plants.

55. (a) $8\,_nC_r\,5 = 56$. 56 committees of all men may be chosen.

    (b) $11\,_nC_r\,5 = 462$. 462 committees of all women may be chosen.

    (c) $(8\,_nC_r\,3)(11\,_nC_r\,2) = (56)(55) = 3080$. 3080 committees of 3 men and 2 women may be chosen.

    (d) $(8\,_nC_r\,5)(11\,_nC_r\,0) + (8\,_nC_r\,4)(11\,_nC_r\,1) + (8\,_nC_r\,3)(11\,_nC_r\,2) + (8\,_nC_r\,2)(11\,_nC_r\,3) =$
    $(56)(1) + (70)(11) + (56)(55) + (28)(165) = 8526$. 8526 committees with no more than 3 women are possible.

56. $10\,_nP_r\,4 = 5040$. 5040 committees can be formed.

57. $2^{12} = 4096$. 4096 codes are possible.

58. $40 \times 40 \times 40 = 64{,}000$. 64,000 different combinations are possible.

59. $(10 \times 10 \times 10)(10 \times 10 \times 10) = 1{,}000{,}000$. 1,000,000 different combinations are possible.

60. $10 \times 10 \times 10 = 1000$. 1000 different plays are possible.

61. Circular ring $\Rightarrow$ 1 key to start; then, $3\,{}_nP_r\,3 = 6$ ways to add the remaining 3. The keys can be put on in 6 distinguishable ways.

62. Round table $\Rightarrow$ 1 person to start; then, $6\,{}_nP_r\,6 = 720$ ways to add the remaining 6. Since clockwise and counterclockwise are considered to be the same, we divide by 2. The people can sit in 360 distinguishable patterns.

63. $P(n, n-1) = \dfrac{n!}{[n-(n-1)]!} = \dfrac{n!}{1!} = n!;\;\; P(n, n) = \dfrac{n!}{(n-n)!} = \dfrac{n!}{0!} = n!$. Thus, $P(n, n-1) = P(n, n)$.

64. $P(n, 1) = \dfrac{n!}{(n-1)!} = \dfrac{n(n-1)!}{(n-1)!} = n$. Thus, $P(n, 1) = n$.

65. $P(n, 0) = \dfrac{n!}{(n-0)!} = \dfrac{n!}{n!} = 1$. Thus, $P(n, 0) = 1$.

66. $\dbinom{n}{n} = C(n, n) = \dfrac{n!}{(n-n)!n!} = \dfrac{n!}{0!n!} = \dfrac{n!}{n!} = 1$. Thus, $\dbinom{n}{n} = 1$.

67. $\dbinom{n}{0} = C(n, 0) = \dfrac{n!}{(n-0)!0!} = \dfrac{n!}{n!} = 1$. Thus, $\dbinom{n}{0} = 1$.

68. $\dbinom{n}{1} = \dfrac{n!}{(n-1)!\,1!} = \dfrac{n(n-1)!}{(n-1)!} = \dfrac{n}{1} = n$

69. $\dbinom{n}{n-1} = C(n, n-1) = \dfrac{n!}{[n-(n-1)]!(n-1)!} = \dfrac{n!}{1!(n-1)!} = \dfrac{n(n-1)!}{(n-1)!} = n$. Thus, $\dbinom{n}{n-1} = n$.

70. $\dbinom{n}{n-r} = C(n, n-r) = \dfrac{n!}{[n-(n-r)]!(n-r)!} = \dfrac{n!}{r!(n-r)!} = C(n, r) = \dbinom{n}{r}$. Thus, $\dbinom{n}{n-r} = \dbinom{n}{r}$.

71. Let $a = (n-r)$ is not defined for negative values of a. Thus, if $r > n$, the factor $(n-r)!$ in the denominator of the formula for $P(n, r)$ would be undefined.

72. If $P(n, r)$ is entered into a graphing calculator with non-integer values for n and/or r, then an error is indicated.

11.7: Probability

1. Since the coin has a head on each side, the sample space is $S = \{H\}$.

2. Since each coin can be either heads or tails, the sample space is $S = \{(H, H), (H, T), (T, H), (T, T)\}$.

3. Since each of the three coins can be either heads or tails, the sample space is
   $S = \{(H, H, H), (H, H, T), (H, T, H), (H, T, T), (T, H, H), (T, H, T), (T, T, H), (T, T, T)\}$

4. There are $5\,{}_nC_r\,2 = 10$ possible outcomes. The sample space is
   $S = \{(1, 2), (1, 3), (1, 4), (1, 5), (2, 3), (2, 4), (2, 5), (3, 4), (3, 5), (4, 5)\}$

5. On each spin, the spinner may land on 1, 2, or 3. The sample space is
   $S = \{(1, 1), (1, 2), (1, 3), (2, 1), (2, 2), (2, 3), (3, 1), (3, 2), (3, 3)\}$

6. The die may come up 1, 2, 3, 4, 5, or 6 and the coin may come up H or T. The sample space is
   $S = \{(1, H), (2, H), (3, H), (4, H), (5, H), (6, H), (1, T), (2, T), (3, T), (4, T), (5, T), (6, T)\}$

7. (a) The event $E_1 = \{H\}$. The probability of the event $P(E_1) = 1$.

(b) The event $E_2 = \emptyset$. The probability of the event $P(E_2) = 0$.

8. (a) The event $E_1 = \{(H, H), (T, T)\}$. The probability of the event $P(E_1) = \frac{2}{4} = \frac{1}{2}$.

(b) The event $E_2 = \{(H, H), (H, T), (T, H)\}$. The probability of the event $P(E_2) = \frac{3}{4}$.

9. (a) The event $E_1 = \{(1, 1), (2, 2), (3, 3)\}$. The probability of the event $P(E_1) = \frac{3}{9} = \frac{1}{3}$.

(b) The event $E_2 = \{(1, 1), (1, 3), (2, 1), (2, 3), (3, 1), (3, 3)\}$. The probability of the event $P(E_2) = \frac{6}{9} = \frac{2}{3}$.

(c) The event $E_3 = \{(2, 1), (2, 3)\}$. The probability of the event $P(E_3) = \frac{2}{9}$.

10. (a) The event $E_1 = \{(2, 4)\}$. The probability of the event $P(E_1) = \frac{1}{10}$.

(b) The event $E_2 = \{(1, 3), (1, 5), (3, 5)\}$. The probability of the event $P(E_2) = \frac{3}{10}$.

(c) The event $E_3 = \emptyset$. The probability of the event $P(E_3) = 0$.

(d) The event $E_4 = \{(1, 2), (1, 4), (2, 3), (2, 5), (3, 4), (4, 5)\}$. The probability of the event $P(E_4) = \frac{6}{10} = \frac{3}{5}$.

11. All probability values must be greater than or equal to 0 and less than or equal to 1. Since $\frac{6}{5} > 1$, it cannot be a probability.

12. If the probability of an event $P(E) = .857$, then the probability that the event will not occur is
$P(E') = 1 - .857 = .143$.

13. (a) The probability of drawing a yellow marble is $P(\text{yellow}) = \frac{3}{15} = \frac{1}{5}$.

(b) The probability of drawing a black marble is $P(\text{black}) = \frac{0}{15} = 0$.

(c) The probability of drawing a yellow or white marble is $P(\text{yellow} \cup \text{white}) = \frac{3 + 4}{15} = \frac{7}{15}$.

(d) $P(\text{yellow}) = \frac{1}{5}$; $P(\text{not yellow}) = \frac{4}{5}$. The odds in favor of drawing a yellow marble are $\frac{\frac{1}{5}}{\frac{4}{5}} = \frac{1}{4}$ or 1 to 4.

(e) $P(\text{blue}) = \frac{8}{15}$; $P(\text{not blue}) = \frac{7}{15}$. The odds against drawing a blue marble are $\frac{\frac{7}{15}}{\frac{8}{15}} = \frac{7}{8}$ or 7 to 8.

14. $P(\text{hit}) = .3$; $P(\text{no hit}) = .7$. The odds in favor of getting a hit are $\frac{.3}{.7} = \frac{3}{7}$ or 3 to 7.

15. $P(\text{sum is 5}) = \frac{2}{10} = \frac{1}{5}$; $P(\text{sum is not 5}) = \frac{4}{5}$. The odds in favor of the sum being 5 are $\frac{\frac{1}{5}}{\frac{4}{5}} = \frac{1}{4}$ or 1 to 4.

16. Odds in favor of rain are 4 to 5 $\Rightarrow P(\text{rain}) = \frac{4}{4 + 5} = \frac{4}{9}$. The probability of rain is $\frac{4}{9}$.

17. Odds in favor of candidate are 3 to 2 $\Rightarrow P(\text{lose}) = \frac{2}{3 + 2} = \frac{2}{5}$. The probability the candidate will lose is $\frac{2}{5}$.

18. (a) The probability of drawing a 9 is $P(9) = \dfrac{4}{52} = \dfrac{1}{13}$.

   (b) The probability of drawing a black card is $P(\text{black}) = \dfrac{26}{52} = \dfrac{1}{2}$.

   (c) The probability of drawing a black 9 is $P(\text{black } 9) = \dfrac{2}{52} = \dfrac{1}{26}$.

   (d) The probability of drawing a heart is $P(\text{heart}) = \dfrac{13}{52} = \dfrac{1}{4}$.

   (e) The probability of a face card is $P(\text{face}) = \dfrac{12}{52} = \dfrac{3}{13}$.

   (f) The probability of red or 3 is $P(\text{red} \cup 3) = P(\text{red}) + P(3) - P(\text{red} \cap 3) = \dfrac{26}{52} + \dfrac{4}{52} - \dfrac{2}{52} = \dfrac{28}{52} = \dfrac{7}{13}$.

   (g) The probability of a card less than 4 is $P(1 \cup 2 \cup 3) = \dfrac{12}{52} = \dfrac{3}{13}$.

19. (a) The probability of an uncle or brother arriving first is $P(\text{uncle} \cup \text{brother}) = \dfrac{2+3}{10} = \dfrac{5}{10} = \dfrac{1}{2}$.

   (b) The probability of a brother or cousin arriving first is $P(\text{brother} \cup \text{cousin}) = \dfrac{3+4}{10} = \dfrac{7}{10}$.

   (c) The probability of a brother or her mother arriving first is $P(\text{brother} \cup \text{mother}) = \dfrac{3+1}{10} = \dfrac{4}{10} = \dfrac{2}{5}$.

20. (a) The probability of rolling at least 10 is $P(10 \cup 11 \cup 12) = \dfrac{3+2+1}{36} = \dfrac{6}{36} = \dfrac{1}{6}$.

   (b) The probability of rolling 7 or least 10 is $P(7 \cup 10 \cup 11 \cup 12) = \dfrac{6+3+2+1}{36} = \dfrac{12}{36} = \dfrac{1}{3}$.

   (c) The sum being 2 is one of the cases of both showing the same number. So the probability of rolling 2 or both the same is equal to the probability of both the same which is

   $P(\text{two 1's} \cup \text{two 2's} \cup \text{two 3's} \cup \text{two 4's} \cup \text{two 5's} \cup \text{two 6's}) = \dfrac{6}{36} = \dfrac{1}{6}$.

21. (a) $P(E) = -.1$ matches with statement F, the event is impossible, because a probability value cannot be negative.

   (b) $P(E) = .01$ matches with statement D, the event is very unlikely to occur, because the probability value is relatively low.

   (c) $P(E) = 1$ matches with statement A, the event is certain to occur.

   (d) $P(E) = 2$ matches with statement F, the probability cannot occur, because a probability value cannot be greater than 1.

   (e) $P(E) = .99$ matches with statement C, the event is very likely to occur, because the probability value is relatively high.

   (f) $P(E) = 0$ matches with statement B, the event is impossible.

   (g) $P(E) = .5$ matches with statement E, the event is just as likely to occur as not occur.

22. $P(\text{making a loan}) = .002$; $P(\text{not making a loan}) = .998$. The odds against such a bank making a small business loan are $\dfrac{.998}{.002} = \dfrac{998}{2} = \dfrac{499}{1}$ or 499 to 1.

23. (a) The probability that a randomly selected patient will need a kidney or a heart transplant is

   $P(\text{kidney} \cup \text{heart}) = \dfrac{35{,}025 + 3774}{51{,}277} \approx .76$.

   (b) The probability that a randomly selected patient will need neither a kidney nor a heart transplant is

   $P(\text{not kidney} \cup \text{not heart}) \approx 1 - .76 = .24$.

24. From the table, the total population (in thousands) will be 335,050.

    (a) The probability that a randomly selected resident will be Hispanic is $P(\text{Hispanic}) = \dfrac{58,930}{335,050} \approx .176$.

    (b) The probability that a randomly selected resident will not be White is

    $$P(\text{not White}) = 1 - (\text{White}) = 1 - \frac{209,117}{335,050} \approx .376.$$

    (c) The probability that a randomly selected resident will will be Native American or Black is

    $$P(\text{Native American} \cup \text{Black}) = \frac{2744 + 43,511}{335,050} \approx .138.$$

    (d) $P(\text{Asian}) = \dfrac{20,748}{335,050} \approx .062$; $P(\text{not Asian}) \approx 1 - .062 = .938$. The odds in favor of a randomly selected

    resident being Asian are approximately $\dfrac{.062}{.938} \approx .066$ or about 1 to 15.

25. (a) $P(\text{less than } \$20) = .25 + .37 = .62$. The probability of a purchase that is less then $20 is .62.

    (b) $P(\$40 \text{ or more}) = .09 + .07 + .08 + .03 = .27$. The probability of a purchase that is $40 or more is .27.

    (c) $P(\text{more than } \$99.99) = .08 + .03 = .11$. The probability of a purchase that is more than $99.99 is .11.

    (d) $P(\text{less than } \$100) = 1 - .11 = .89$. The probability of a purchase that is less than $100 is .89.

26. Number of possible picks $= 13 \times 13 \times 13 \times 13 = 28,561$. $P(\text{winning pick}) = \dfrac{1}{28,561} \approx .000035$.

    The probability of getting the winning lottery pick is $\dfrac{1}{28,561}$ or approximately .000035.

27. Number of picks with all cards correct except the heart $= 12$.

    Number of picks with all cards correct except the club $= 12$.

    Number of picks with all cards correct except the diamond $= 12$.

    Number of picks with all cards correct except the spade $= 12$.

    Total number of picks with all cards correct but one $= 48$. $P(\text{all cards correct but one}) = \dfrac{48}{28,561} \approx .001681$.

    The probability of getting three of the four selections correct is $\dfrac{48}{28,561}$ or approximately .001681.

28. (a) Total number of possible selections $= 5\,{}_nC_r\,2 = 10$.

    Total number of possible selections which include Chinn $= 4\,{}_nC_r\,1 = 4$. $P(\text{Chinn is included}) = \dfrac{4}{10} = \dfrac{2}{5}$.

    The probability that Chinn is included is $\dfrac{2}{5}$.

    (b) The number of selections which include Alam and Dickinson $= 3\,{}_nC_r\,0 = 1$.

    $P(\text{Alam and Dickinson are selected}) = \dfrac{1}{10}$. The probability that Alam and Dickinson are selected is $\dfrac{1}{10}$.

    (c) $1 \times 3 = 3$ selections include Alam but not Bartolini. $1 \times 3 = 3$ selections include Bartolini but not Alam.

    $1 \times 1 = 1$ selection includes both Alam and Bartolini. Therefore, 7 selections include at least one senior

    partner. $P(\text{at least one senior partner}) = \dfrac{7}{10}$. The probability that at least one senior partner is included is $\dfrac{7}{10}$.

29. (a) $P(\text{male selected}) = 1 - .28 = .72$. The probability that a male worker is selected is .72.

    (b) $P(5 \text{ years or less}) = 1 - .3 = .7$. The probability that a worker is selected who has worked for the company 5 years or less is .7.

    (c) $P(\text{contribute} \cup \text{female}) = P(\text{contribute}) + P(\text{female}) - P(\text{contribute} \cap \text{female}) = .65 + .28 - \dfrac{.28}{2} = .79$.

    The probability that a worker is selected who contributes to the retirement plan or is female is .79.

30. The number of outcomes which constitute event E is zero or greater and is less than or equal to the number of outcomes which constitute the sample space S. So the probability of event E, which is the ratio of the number of outcomes which constitute E to the number of outcomes which constitute S, must be a number between 0 and 1, inclusively.

31. There are $5^2 = 32$ possible outcomes, each with probability $\dfrac{1}{32}$.

    $\dbinom{5}{2} = 10$ outcomes with 2 girls (and 3 boys) $\Rightarrow P(2 \text{ girls and 3 boys}) = 10\left(\dfrac{1}{32}\right) = \dfrac{5}{16} = .3125$.

    The probability of having exactly 2 girls and 3 boys is $\dfrac{5}{16} = .3125$.

32. There are $5^2 = 32$ possible outcomes, each with probability $\dfrac{1}{32}$.

    $\dbinom{5}{3} = 10$ outcomes with 3 girls (and 2 boys) $\Rightarrow P(3 \text{ girls and 2 boys}) = 10\left(\dfrac{1}{32}\right) = \dfrac{5}{16} = .3125$.

    The probability of having exactly 2 girls and 3 boys is $\dfrac{5}{16} = .3125$.

33. $\dbinom{5}{0} = 1$ outcome has no girls $\Rightarrow P(\text{no girls}) = 1\left(\dfrac{1}{32}\right) = \dfrac{1}{32} = .03125$.

    The probability of having no girls is $\dfrac{1}{32} = .03125$.

34. $\dbinom{5}{5} = 1$ outcome has no boys $\Rightarrow P(\text{no boys}) = 1\left(\dfrac{1}{32}\right) = \dfrac{1}{32} = .03125$.

    The probability of having no boys is $\dfrac{1}{32} = .03125$.

35. $\dbinom{5}{5} + \dbinom{5}{4} + \dbinom{5}{3} = 1 + 5 + 10 = 16$ outcomes have at least 3 boys $\Rightarrow$

    $P(\text{at least 3 boys}) = 16\left(\dfrac{1}{32}\right) = \dfrac{1}{2} = .5$. The probability of having at least 3 boys is $\dfrac{1}{2} = .5$.

36. $P(\text{no more than 4 girls}) = 1 - P(5 \text{ girls}) = 1 - \dfrac{1}{32} = \dfrac{31}{32} = .96875$. The probability of having no more than 4 girls is $\dfrac{31}{32} = .96875$.

37. $P(1 \text{ student smokes less than 10 per day}) = .45 + .24 = .69$.

    $P(4 \text{ of 10 smoke less than 10 per day}) = \dbinom{10}{4}(.69)^4(1 - .69)^6 \approx .042246$. The probability that 4 of 10 students selected at random smoked less than 10 cigarettes per day is approximately .042246.

38. $P(5 \text{ of } 10 \text{ smoke a pack or more per day}) = \binom{10}{5}(.11)^5(1 - .11)^5 \approx .002266$. The probability that 5 of 10

    students selected at random smoke a pack or more per day is approximately .002266.

39. $P(1 \text{ student smokes between } 1 \text{ and } 19 \text{ per day}) = .24 + .20 = .44$.

    $P(\text{fewer than 2 smoke between 1 and 19 per day}) =$

    $P(0 \text{ smoke between 1 and 19 per day}) + P(1 \text{ smokes between 1 and 19 per day}) =$

    $\binom{10}{0}(.44)^0(1 - .44)^{10} + \binom{10}{1}(.44)^1(1 - .44)^9 \approx .026864$. The probability that fewer than 2 of 10 students

    selected at random smoked between 1 and 19 cigarettes per day is approximately .026864.

40. $P(\text{no more than 3 smoked less than 1 per day}) =$

    $P(0 \text{ smoked less than 1 per day}) + P(1 \text{ smoked less than 1 per day}) +$

    $P(2 \text{ smoked less than 1 per day}) + P(3 \text{ smoked less than 1 per day}) =$

    $\binom{10}{0}(.45)^0(1 - .45)^{10} + \binom{10}{1}(.45)^1(1 - .45)^9 + \binom{10}{2}(.45)^2(1 - .45)^8 + \binom{10}{3}(.45)^3(1 - .45)^7 \approx .266038$.

    The probability that no more than 3 of 10 students selected at random smoked less than 1 cigarette per day is

    approximately .266038.

41. $P(\text{exactly 12 ones}) = \binom{12}{12}\left(\frac{1}{6}\right)^{12}\left(\frac{5}{6}\right)^0 = \frac{1}{6^{12}} \approx 4.6 \times 10^{-10}$. The probability of rolling exactly 12 ones is

    $\frac{1}{6^{12}} \approx 4.6 \times 10^{-10}$.

42. $P(\text{exactly 6 ones}) = \binom{12}{6}\left(\frac{1}{6}\right)^6\left(\frac{5}{6}\right)^6 \approx .0066$. The probability of rolling exactly 6 ones is approximately .0066.

43. $P(\text{no more than 3 ones}) = P(0 \text{ ones}) + P(1 \text{ one}) + P(2 \text{ ones}) + P(3 \text{ ones}) =$

    $\binom{12}{0}\left(\frac{1}{6}\right)^0\left(\frac{5}{6}\right)^{12} + \binom{12}{1}\left(\frac{1}{6}\right)^1\left(\frac{5}{6}\right)^{11} + \binom{12}{2}\left(\frac{1}{6}\right)^2\left(\frac{5}{6}\right)^{10} + \binom{12}{3}\left(\frac{1}{6}\right)^3\left(\frac{5}{6}\right)^9 \approx .875$.

    The probability of rolling no more than 3 ones is approximately .875.

44. $P(\text{no more than 1 one}) = P(0 \text{ ones}) + P(1 \text{ one}) = \binom{12}{0}\left(\frac{1}{6}\right)^0\left(\frac{5}{6}\right)^{12} + \binom{12}{1}\left(\frac{1}{6}\right)^1\left(\frac{5}{6}\right)^{11} \approx .381$.

    The probability of rolling no more than 1 one is approximately .381.

45. $P(\text{fewer than 4}) = P(1) + P(2 \text{ or } 3) = .2 + .29 = .49$. The probability that a randomly selected student

    applied to fewer than 4 colleges is .49.

46. $P(\text{at least 2 colleges}) = 1 - P(1) = 1 - .2 = .8$. The probability that a randomly selected student

    applied to at least 2 colleges is .8.

47. $P(\text{more than 3}) = P(4 - 6) + P(7 \text{ or more}) = .37 + .14 = .51$. The probability that a randomly selected

    student applied to more than 3 colleges is .51.

48. $P(0) = 1 - (P(1) + P(2 \text{ or } 3) + P(4 - 6) + P(7 \text{ or more})) = 0$. The probability that a randomly selected

    student applied to colleges is 0. Note that the subjects in this survey were all college freshmen.

49. (a) $P(5 \text{ of } 53 \text{ are color blind}) = \binom{53}{5}(.042)^5(1 - .042)^{48} \approx .047822$. The probability that exactly 5 of 53 men

    are color blind is approximately .047822.

    (b) $P(\text{no more than } 5) = P(0) + P(1) + P(2) + P(3) + P(4) + P(5) =$

    $\binom{53}{0}(.042)^0(1 - .042)^{53} + \binom{53}{1}(.042)^1(1 - .042)^{52} + \binom{53}{2}(.042)^2(1 - .042)^{51} +$

    $\binom{53}{3}(.042)^3(1 - .042)^{50} + \binom{53}{4}(.042)^4(1 - .042)^{49} + \binom{53}{5}(.042)^5(1 - .042)^{48} \approx .976710$.

    The probability that no more than 5 of 53 men are color blind is approximately .976710.

    (c) $P(\text{none are color blind}) = \binom{53}{0}(.042)^0(1 - .042)^{53} \approx .102890$.

    $P(\text{at least 1 color blind}) = 1 - P(\text{none are color blind}) \approx .897110$.

    The probability that at least 1 of 53 men is color blind is approximately .897110.

50. The zeros in the table appear because $_nC_r = \dfrac{n!}{(n - r)!r!}$ involves the factorial of a negative number when $r > n$.
    This is not defined and zero results.

    (a) From Figure 50a-1 and figure 50a-2, the probabilities are seen to be .125, .375, .375 and .125, in order.

    (b) From Figure 50b-1 and figure 50b-2, the probabilities are seen to be .015625, .09375, .234375, .3125, .234375,
    .09375, and .015625, in order.

| Figure 50a-1 | Figure 50a-2 | Figure 50b-1 | Figure 50b-2 |

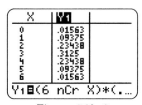

51. (a) $I = 2, S = 4, p = .1 \Rightarrow q = (1 - p)^I = (1 - .1)^2 = .81$.

    $P(3 \text{ not becoming infected}) = \binom{4}{3}(.81)^3(1 - .81)^{4-3} \approx .404 = 40.4\%$. The probability of 3 family

    members not becoming infected is approximately 40.4%.

    (b) $I = 2, S = 4, p = .5 \Rightarrow q = (1 - p)^I = (1 - .5)^2 = .25$.

    $P(3 \text{ not becoming infected}) = \binom{4}{3}(.25)^3(1 - .25)^{4-3} \approx .047 = 4.7\%$. The probability of 3 family

    members not becoming infected is approximately 4.7%.

    (c) $I = 1, S = 9, p = .5 \Rightarrow q = (1 - p)^I = (1 - .5)^1 = .5$.

    $P(0 \text{ not becoming infected}) = \binom{9}{0}(.5)^0(1 - .5)^{9-0} \approx .002 = .2\%$. The probability that all of the other

    family members becoming sick is approximately .2%. Thus, it is unlikely that everyone in a large family

    would become sick even though the disease is likely infections.

52. $I = 2 \Rightarrow q = (1 - p)^I = (1 - p)^2$; $s = 4, k = 2 \Rightarrow$

$$P = \binom{s}{k}q^k(1 - q)^{s-k} = \binom{4}{2}((1 - p)^2)^2(1 - (1 - p)^2)^{4-2} \Rightarrow .25 = 6(1 - p)^4(1 - (1 - p)^2)^2.$$

From Figure 52, the possible values of P are seen to be approximately .1549 or approximately .4654.

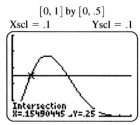

[0, 1] by [0, .5]
Xscl = .1   Yscl = .1

Intersection
X=.15490445 Y=.25

Figure 52

Reviewing Basic Concepts (Sections 11.6 and 11.7)

1. $4\,_nP_r\,4 = 24$. The number of arrangements is 24.

2. $P(7, 3) = 7\,_nP_r\,3 = 210$. The value of $P(7, 3)$ is 210.

3. $\binom{6}{2}\binom{5}{2}\binom{3}{1} = 15 \times 10 \times 3 = 450$. The number of selections possible is 450.

4. $\binom{10}{4} = 10\,_nC_r\,4 = 210$. The value of $\binom{10}{4}$ is 210.

5. $9 \times 4 \times 2 = 72$. The number of different homes is 72.

6. The sample space S for tossing a coin twice is $S = \{(H, H), (H, T), (T, H), (T, T)\}$.

7. 36 possible outcomes, each with probability $\dfrac{1}{36}$. 2 outcomes yield 11, $\{(5, 6), (6, 5)\}$.

    Therefore $P(11) = 2\left(\dfrac{1}{36}\right) = \dfrac{1}{18}$. The probability of rolling a sum of 11 is $\dfrac{1}{18}$.

8. $\binom{4}{4} = 1$ way of drawing 4 aces; $\binom{4}{1} = 4$ way of drawing 1 queen;

    $\binom{52}{5} = 2{,}598{,}960$ ways of drawing 5 cards. $P(4 \text{ aces and } 1 \text{ queen}) = \dfrac{1 \times 4}{2{,}598{,}960} \approx .0000015.$

    The probability of drawing 4 aces and 1 queen is approximately .0000015.

9. $P(\text{rain}) = \dfrac{3}{3 + 7} = \dfrac{3}{10}$. The probability of rain is $\dfrac{3}{10}$.

10. $P(\text{female}) = \dfrac{2.81 - 1.45}{2.81} \approx .484$. The probability that a randomly selected graduate is female is approximately .484.

Chapter 11 Review Exercises

1. $a_1 = \dfrac{1}{1 + 1} = \dfrac{1}{2}$; $a_2 = \dfrac{2}{2 + 1} = \dfrac{2}{3}$; $a_3 = \dfrac{3}{3 + 1} = \dfrac{3}{4}$; $a_4 = \dfrac{4}{4 + 1} = \dfrac{4}{5}$; $a_5 = \dfrac{5}{5 + 1} = \dfrac{5}{6}$. The first five

    terms of the sequence are $\dfrac{1}{2}, \dfrac{2}{3}, \dfrac{3}{4}, \dfrac{4}{5}$, and $\dfrac{5}{6}$. The sequence is neither arithmetic nor geometric.

2. $a_1 = (-2)^1 = -2$; $a_2 = (-2)^2 = 4$; $a_3 = (-2)^3 = -8$; $a_4 = (-2)^4 = 16$; $a_5 = (-2)^5 = -32$. The first five

    terms of the sequence are $-2, 4, -8, 16$, and -32. The sequence is geometric with common ratio -2.

3. $a_1 = 2(1 + 3) = 8$; $a_2 = 2(2 + 3) = 10$; $a_3 = 2(3 + 3) = 12$; $a_4 = 2(4 + 3) = 14$; $a_5 = 2(5 + 3) = 16$.

    The first five terms of the sequence are $8, 10, 12, 14$, and 16. The sequence is arithmetic with common

    difference 2.

4. $a_1 = 1(1 + 1) = 2$; $a_2 = 2(2 + 1) = 6$; $a_3 = 3(3 + 1) = 12$; $a_4 = 4(4 + 1) = 20$; $a_5 = 5(5 + 1) = 30$.

    The first five terms of the sequence are $2, 6, 12, 20$, and 30. The sequence is neither arithmetic nor geometric.

5. $a_1 = 5$; $a_2 = 5 - 3 = 2$; $a_3 = 2 - 3 = -1$; $a_4 = -1 - 3 = -4$; $a_5 = -4 - 3 = -7$.

    The first five terms of the sequence are $5, 2, -1, -4$, and -7. The sequence is arithmetic with common

    difference -3.

6. $a_2 = 10, d = -2 \Rightarrow a_1 = 10 + 2 = 12$; $a_3 = 10 - 2 = 8$; $a_4 = 8 - 2 = 6$; $a_5 = 6 - 2 = 4$.

    The first five terms of the sequence are $12, 10, 8, 6$, and 4.

7. $a_3 = \pi, a_4 = 1 \Rightarrow d = 1 - \pi \Rightarrow a_2 = \pi - (1 - \pi) = 2\pi - 1$; $a_1 = 2\pi - 1 - (1 - \pi) = 3\pi - 2$;

    $a_5 = 1 + (1 - \pi) = -\pi + 2$. The first five terms of the sequence are $3\pi - 2, 2\pi - 1, \pi, 1$, and $-\pi + 2$.

8. $a_1 = 6, r = 2 \Rightarrow a_2 = 6 \cdot 2 = 12$; $a_3 = 12 \cdot 2 = 24$; $a_4 = 24 \cdot 2 = 48$; $a_5 = 48 \cdot 2 = 96$.

    The first five terms of the sequence are $6, 12, 24, 48$, and 96.

9. $a_1 = -5, a_2 = -1 \Rightarrow r = \dfrac{1}{5} \Rightarrow a_3 = (-1)\left(\dfrac{1}{5}\right) = -\dfrac{1}{5}$; $a_4 = \left(-\dfrac{1}{5}\right)\left(\dfrac{1}{5}\right) = -\dfrac{1}{25}$; $a_5 = \left(-\dfrac{1}{25}\right)\left(\dfrac{1}{5}\right) = -\dfrac{1}{125}$.

    The first five terms of the sequence are $-5, -1, -\dfrac{1}{5}, -\dfrac{1}{25}, -\dfrac{1}{125}$.

10. $a_5 = -3, a_{15} = 17 \Rightarrow 17 = -3 + 10d \Rightarrow d = 2 \Rightarrow -3 = a_1 + 4 \cdot 2 \Rightarrow a_1 = -11 \Rightarrow$

    $a_n = -11 + 2(n - 1) = 2n - 13$. The first term of the sequence is -11 and the nth term of the sequence is

    given by $a_n = 2n - 13$.

11. $a_1 = -8, a_7 = -\dfrac{1}{8} \Rightarrow -\dfrac{1}{8} = (-8)r^6 \Rightarrow r^6 = \dfrac{1}{64} \Rightarrow r = \pm\dfrac{1}{2}$. Therefore,

    $a_4 = (-8)\left(\dfrac{1}{2}\right)^3 = -1$ and $a_n = (-8)\left(\dfrac{1}{2}\right)^{n-1} = -\left(\dfrac{1}{2}\right)^{n-4}$ or

    $a_4 = (-8)\left(-\dfrac{1}{2}\right)^3 = 1$ and $a_n = (-8)\left(-\dfrac{1}{2}\right)^{n-1} = \left(-\dfrac{1}{2}\right)^{n-4}$. Either the common ratio is $\dfrac{1}{2}$ and the fourth term of

    the sequence is -1 and the nth term of the sequence is given by $a_n = (-8)\left(\dfrac{1}{2}\right)^{n-1} = -\left(\dfrac{1}{2}\right)^{n-4}$ or the common

    ratio is $-\dfrac{1}{2}$ and the fourth term of the sequence is 1 and the nth term of the sequence is given by

    $a_n = (-8)\left(-\dfrac{1}{2}\right)^{n-1} = \left(-\dfrac{1}{2}\right)^{n-4}$.

12. $a_1 = 6, d = 2 \Rightarrow a_8 = 6 + (8 - 1)(2) = 20$. The eighth term of the sequence is 20.

13. $a_1 = 6x - 9, a_2 = 5x + 1 \Rightarrow d = (5x + 1) - (6x - 9) = -x + 10 \Rightarrow$

    $a_8 = (6x - 9) + (8 - 1)(-x + 10) = -x + 61$. The eighth term of the sequence is $-x + 61$.

14. $a_1 = 2, d = 3 \Rightarrow a_{12} = 2 + (12 - 1)(3) = 35 \Rightarrow S_{12} = \dfrac{12}{2}(2 + 35) = 222.$ The sum of the first twelve

terms of the sequence is 222.

15. $a_2 = 6, d = 10 \Rightarrow a_1 = 6 - 10 = -4 \Rightarrow a_{12} = -4 + (12 - 1)(10) = 106 \Rightarrow S_{12} = \dfrac{12}{2}(-4 + 106) = 612.$

The sum of the first twelve terms of the sequence is 612.

16. $a_1 = -2, r = 3 \Rightarrow a_5 = (-2)(3)^{5-1} = -162.$ The fifth term of the sequence is -162.

17. $a_3 = 4, r = \dfrac{1}{5} \Rightarrow a_1 = \dfrac{4}{\left(\frac{1}{5}\right)^2} = 100 \Rightarrow a_5 = (100)\left(\dfrac{1}{5}\right)^{5-1} = \dfrac{100}{625} = \dfrac{4}{25}.$ The fifth term of the sequence is $\dfrac{4}{25}$.

18. $a_1 = 3, r = 2 \Rightarrow S_4 = \dfrac{3(1 - 2^4)}{1 - 2} = 45.$ The sum of the first four terms of the sequence is 45.

19. $a_1 = \dfrac{3}{4}, a_2 = -\dfrac{1}{2}, a_3 = \dfrac{1}{3} \Rightarrow r = -\dfrac{2}{3} \Rightarrow S_4 = \dfrac{\frac{3}{4}\left(1 - (-\frac{2}{3})^4\right)}{1 - (-\frac{2}{3})} = \dfrac{13}{36}.$ The sum of the first four terms is $\dfrac{13}{36}$.

20. $S = 2000\left[\dfrac{(1 + .03)^5 - 1}{.03}\right] = 10{,}618.27.$ The future value of the annuity is \$10,618.27.

21. $a_1 = (-1)^{1-1} = 1;\; r = -1;\; \displaystyle\sum_{i=1}^{7}(-1)^{i-1} = S_7 = \dfrac{1[1 - (-1)^7]}{1 - (-1)} = \dfrac{1(2)}{2} = 1.$ The value of the sum is 1.

22. $\displaystyle\sum_{i=1}^{5}(i^2 + i) = \sum_{i=1}^{5}i^2 + \sum_{i=1}^{5}i = \dfrac{5(5 + 1)(2 \cdot 5 + 1)}{6} + \dfrac{5(5 + 1)}{2} = 55 + 15 = 70.$ The value of the sum is 70.

23. $\displaystyle\sum_{i=1}^{4}\dfrac{i + 1}{i} = \dfrac{2}{1} + \dfrac{3}{2} + \dfrac{4}{3} + \dfrac{5}{4} = \dfrac{73}{12}.$ The value of the sum is $\dfrac{73}{12}$.

24. $a_1 = 3(1) - 4 = -1;\; a_{10} = 3(10) - 4 = 26;\; \displaystyle\sum_{j=1}^{10}(3j - 4) = S_{10} = \dfrac{10}{2}(-1 + 26) = 125.$

The value of the sum is 125.

25. $\displaystyle\sum_{j=1}^{2500}j = \dfrac{2500(2500 + 1)}{2} = 3{,}126{,}250.$ The value of the sum is 3,126,250.

26. $a_1 = 4 \cdot 2^1 = 8;\; r = 2;\; \displaystyle\sum_{i=1}^{5}4 \cdot 2^i = S_5 = \dfrac{8(1 - 2^5)}{1 - 2} = 248.$ The value of the sum is 248.

27. $a_1 = \left(\dfrac{4}{7}\right)^1 = \dfrac{4}{7};\; r = \dfrac{4}{7};\; \displaystyle\sum_{i=1}^{\infty}\left(\dfrac{4}{7}\right)^i = S_\infty = \dfrac{\frac{4}{7}}{1 - \frac{4}{7}} = \dfrac{4}{3}.$ The value of the sum is $\dfrac{4}{3}$.

28. $r = \dfrac{6}{5} \Rightarrow |r| \geq 1 \Rightarrow \displaystyle\sum_{i=1}^{\infty}-2\left(\dfrac{6}{5}\right)^i$ does not exist.

29. $a_1 = 24, a_2 = 8, a_3 = \dfrac{8}{3}, a_4 = \dfrac{8}{9} \Rightarrow r = \dfrac{1}{3} \Rightarrow S_\infty = \dfrac{24}{1 - \frac{1}{3}} = 36.$ The value of the sum is 36.

30. $a_1 = -\dfrac{3}{4}, a_2 = \dfrac{1}{2}, a_3 = -\dfrac{1}{3}, a_4 = \dfrac{2}{9} \Rightarrow r = -\dfrac{2}{3} \Rightarrow S_\infty = \dfrac{-\frac{3}{4}}{1 - (-\frac{2}{3})} = -\dfrac{9}{20}.$ The value of the sum is $-\dfrac{9}{20}$.

31. $a_1 = \dfrac{1}{12}, a_2 = \dfrac{1}{6}, a_3 = \dfrac{1}{3}, a_4 = \dfrac{2}{3} \Rightarrow r = 2 \Rightarrow |r| \geq 1 \Rightarrow S_\infty$ diverges. The value of the sum diverges and

does not exist.

32. $a_1 = .9, a_2 = .09, a_3 = .009, a_4 = .0009 \Rightarrow r = \dfrac{1}{10} \Rightarrow S_\infty = \dfrac{.9}{1 - \frac{1}{10}} = 1.$ The value of the sum is 1.

33. $\displaystyle\sum_{i=1}^{4}(x_i^2 - 6) = \sum_{i=1}^{4}x_i^2 - \sum_{i=1}^{4}6 = \sum_{i=1}^{3}(x_i + 1)^2 - 24 = \frac{3(3+1)(2\cdot 3+1)}{6} - 24 = -10.$

The value of the sum is -10.

34. $\displaystyle\sum_{i=1}^{6}f(x_i)\Delta x = (0-2)^3(.1) + (1-2)^3(.1) + (2-2)^3(.1) + (3-2)^3(.1) + (4-2)^3(.1) +$

$(5-2)^3(.1) = 2.7.$ The value of the sum is 2.7.

35. $a_1 = 4, a_2 = -1, a_3 = -6 \Rightarrow d = -5 \Rightarrow a_n = 4 - 5(n-1) = -5n + 9.$

$a_n = -66 = -5n + 9 \Rightarrow n = 15 \Rightarrow 4 - 1 - 6 - \ldots - 66 = \displaystyle\sum_{i=1}^{15}(-5i + 9).$

The sum may be written as $\displaystyle\sum_{i=1}^{15}(-5i + 9)$.

36. $a_1 = 10, a_2 = 14, a_3 = 18 \Rightarrow d = 4 \Rightarrow a_n = 10 + 4(n-1) = 4n + 6.$

$a_n = 86 = 4n + 6 \Rightarrow n = 20 \Rightarrow 10 + 14 + 18 + \ldots + 86 = \displaystyle\sum_{i=1}^{20}(4i + 6).$

The sum may be written as $\displaystyle\sum_{i=1}^{20}(4i + 6)$.

37. $a_1 = 4, a_2 = 12, a_3 = 36 \Rightarrow r = 3 \Rightarrow a_n = 4(3)^{n-1}.$

$a_n = 972 = 4(3)^{n-1} \Rightarrow n - 1 = 5 \Rightarrow n = 6 \Rightarrow 4 + 12 + 36 + \ldots + 972 = \displaystyle\sum_{i=1}^{6}4(3)^{n-1}.$

The sum may be written as $\displaystyle\sum_{i=1}^{6}4(3)^{n-1}$.

38. $a_n = \dfrac{n}{n+1}$ for $5 \le n \le 12$; $\dfrac{5}{6} + \dfrac{6}{7} + \dfrac{7}{8} + \cdots + \dfrac{12}{13} = \displaystyle\sum_{i=5}^{12}\dfrac{i}{i+1}.$ The sum may be written as $\displaystyle\sum_{i=5}^{12}\dfrac{i}{i+1}.$

39. $(x + 2y)^4 = \dbinom{4}{0}x^4(2y)^0 + \dbinom{4}{1}x^3(2y)^1 + \dbinom{4}{2}x^2(2y)^2 + \dbinom{4}{3}x^1(2y)^3 + \dbinom{4}{4}x^0(2y)^4 =$

$x^4 + 8x^3y + 24x^2y^2 + 32xy^3 + 16y^4.$

The binomial expansion of $(x + 2y)^4$ is $x^4 + 8x^3y + 24x^2y^2 + 32xy^3 + 16y^4.$

40. $(3z - 5w)^3 = \dbinom{3}{0}(3z)^3(-5w)^0 + \dbinom{3}{1}(3z)^2(-5w)^1 + \dbinom{3}{2}(3z)^1(-5w)^2 + \dbinom{3}{3}(3z)^0(-5w)^3 =$

$27z^3 - 135z^2w + 225zw^2 - 125w^3.$

The binomial expansion of $(3z - 5w)^3$ is $27z^3 - 135z^2w + 225zw^2 - 125w^3.$

41. $\left(3\sqrt{x} - \dfrac{1}{\sqrt{x}}\right)^5 = \dbinom{5}{0}(3x^{1/2})^5(-x^{-1/2})^0 + \dbinom{5}{1}(3x^{1/2})^4(-x^{-1/2})^1 + \dbinom{5}{2}(3x^{1/2})^3(-x^{-1/2})^2 + \dbinom{5}{3}(3x^{1/2})^2(-x^{-1/2})^3$

$+ \dbinom{5}{4}(3x^{1/2})^1(-x^{-1/2})^4 + \dbinom{5}{5}(3x^{1/2})^0(-x^{-1/2})^5 = 243x^{5/2} - 405x^{3/2} + 270x^{1/2} - 90x^{-1/2} + 15x^{-3/2} - x^{-5/2}.$

The binomial expansion of $\left(3\sqrt{x} - \dfrac{1}{\sqrt{x}}\right)^5$ is $243x^{5/2} - 405x^{3/2} + 270x^{1/2} - 90x^{-1/2} + 15x^{-3/2} - x^{-5/2}.$

42. $(m^3 - m^{-2})^4 = \dbinom{4}{0}(m^3)^4(-m^{-2})^0 + \dbinom{4}{1}(m^3)^3(-m^{-2})^1 + \dbinom{4}{2}(m^3)^2(-m^{-2})^2 + \dbinom{4}{3}(m^3)^1(-m^{-2})^3 +$

$\dbinom{4}{4}(m^3)^0(-m^{-2})^4 = m^{12} - 4m^7 + 6m^2 - 4m^{-3} + m^{-8}.$

The binomial expansion of $(m^3 - m^{-2})^4$ is $m^{12} - 4m^7 + 6m^2 - 4m^{-3} + m^{-8}.$

43. $\dbinom{8}{5}(4x)^3(-y)^5 = -3584x^3y^5.$ The sixth term of $(4x - y)^8$ is $-3584x^3y^5.$

44. $\binom{14}{6}(m)^8(-3n)^6 = 2,189,187m^8n^6$. The seventh term of $(m - 3n)^{14}$ is $2,189,187m^8n^6$.

45. $(x + 2)^{12} = \binom{12}{0}(x)^{12}(2)^0 + \binom{12}{1}(x)^{11}(2)^1 + \binom{12}{2}(x)^{10}(2)^2 + \binom{12}{3}(x)^9(2)^3 + \cdots =$

   $x^{12} + 24x^{11} + 264x^{10} + 1760x^9 + \cdots$. The first four terms of $(x + 2)^{12}$ are $x^{12} + 24x^{11} + 264x^{10} + 1760x^9$.

46. $(2a + 5b)^{16} = \cdots \binom{16}{14}(2a)^2(5b)^{14} + \binom{16}{15}(2a)^1(5b)^{15} + \binom{16}{16}(2a)^0(5b)^{16} =$

   $\cdots + 480 \cdot 5^{14}a^2b^{14} + 32 \cdot 5^{15}ab^{15} + 5^{16}b^{16}$.

   The last three terms of $(2a + 5b)^{16}$ are $480 \cdot 5^{14}a^2b^{14} + 32 \cdot 5^{15}ab^{15} + 5^{16}b^{16}$.

47. Statements which are defined on the natural numbers, that is, statements which have the natural numbers as

   their domain, are proved by mathematical induction. An example is $\displaystyle\sum_{i=1}^{n} i = \frac{n(n + 1)}{2}$.

48. A proof by mathematical induction consists of two steps. First, show that the statement is true for $n = 1$.

   Then show that if the statement is true for $n = k$, it must follow that the statement is also true for $n = k + 1$.

49. Prove $1 + 3 + 5 + 7 + \cdots + (2n - 1) = n^2$.

   Step 1: $1 = 1^2 \Rightarrow 1 + 3 + 5 + 7 + \cdots + (2n - 1) = n^2$ for $n = 1$.

   Step 2: Assume $1 + 3 + 5 + 7 + \cdots + (2k - 1) = k^2$, then

   $\qquad 1 + 3 + 5 + 7 + \cdots + (2k - 1) + (2k + 1) = k^2 + 2k + 1 \Rightarrow$

   $\qquad 1 + 3 + 5 + 7 + \cdots + (2k - 1) + (2(k + 1) - 1) = (k + 1)^2$.

   $\qquad$ Thus if $1 + 3 + 5 + 7 + \cdots + (2n - 1) = n^2$ for $n = k$, then it must follow that

   $\qquad 1 + 3 + 5 + 7 + \cdots + (2n - 1) = n^2$ for $n = k + 1$ also.

50. Prove $2 + 6 + 10 + 14 + \cdots + (4n - 2) = 2n^2$.

   Step 1: $2 = 2(1)^2 \Rightarrow 2 + 6 + 10 + 14 + \cdots + (4n - 2) = 2n^2$ for $n = 1$.

   Step 2: Assume $2 + 6 + 10 + 14 + \cdots + (4k - 2) = 2k^2$, then

   $\qquad 2 + 6 + 10 + 14 + \cdots + (4k - 2) + (4k + 2) = 2k^2 + 4k + 2 \Rightarrow$

   $\qquad 2 + 6 + 10 + 14 + \cdots + (4k - 2) + (4(k + 1) - 2) = 2(k + 1)^2$.

   $\qquad$ Thus if $2 + 6 + 10 + 14 + \cdots + (4n - 2) = 2n^2$ for $n = k$, then it must follow that

   $\qquad 2 + 6 + 10 + 14 + \cdots + (4n - 2) = 2n^2$ for $n = k + 1$ also.

51. Prove $2 + 2^2 + 2^3 + \cdots + 2^n = 2(2^n - 1)$.

   Step 1: $2 = 2(2^1 - 1) \Rightarrow 2 + 2^2 + 2^3 + \cdots + 2^n = 2(2^n - 1)$ for $n = 1$.

   Step 2: Assume $2 + 2^2 + 2^3 + \cdots + 2^k = 2(2^k - 1)$, then

   $\qquad 2 + 2^2 + 2^3 + \cdots + 2^k + 2^{k+1} = 2(2^k - 1) + 2^{k+1} = 2^{k+1} - 2 + 2^{k+1} = 2(2^{k+1} - 1)$

   $\qquad$ Thus if $2 + 2^2 + 2^3 + \cdots + 2^n = 2(2^n - 1)$ for $n = k$, then it must follow that

   $\qquad 2 + 2^2 + 2^3 + \cdots + 2^n = 2(2^n - 1)$ for $n = k + 1$ also.

52. Prove $1^3 + 3^3 + 5^3 + \cdots + (2n - 1)^3 = n^2(2n^2 - 1)$.

    Step 1: $1^3 = 1$; $1^2(2(1)^2 - 1) = 1(2 - 1) = 1$; so $1^3 + 3^3 + 5^3 + \cdots + (2n - 1)^3 = n^2(2n^2 - 1)$ for $n = 1$.

    Step 2: Assume $1^3 + 3^3 + 5^3 + \cdots + (2k - 1)^3 = k^2(2k^2 - 1)$ then,

        $1^3 + 3^3 + 5^3 + \cdots + (2k - 1)^3 + (2k + 1)^3 = k^2(2k^2 - 1) + (2k + 1)^3$ then

        $1^3 + 3^3 + 5^3 + \cdots + (2k - 1)^3 + (2(k + 1) - 1)^3 = 2k^4 + 8k^3 + 11k^2 + 6k + 1 =$

        $(k^2 + 2k + 1)(2k^2 + 4k + 1) = (k + 1)^2(2(k + 1)^2 - 1)$

        Thus if $1^3 + 3^3 + 5^3 + \cdots + (2n - 1)^3 = n^2(2n^2 - 1)$ for $n = k$, then it must follow that

        $1^3 + 3^3 + 5^3 + \cdots + (2n - 1)^3 = n^2(2n^2 - 1)$ for $n = k + 1$ also.

53. Order is significant in the digits of a student identification number. Thus, such a number is an example of a permutation.

54. $P(9, 2) = 9\,_nP_r\,2 = 72$. The value of $P(9, 2)$ is 72.

55. $P(6, 0) = 6\,_nP_r\,0 = 1$. The value of $P(6, 0)$ is 1.

56. $\binom{8}{3} = 8\,_nC_r\,3 = 56$. The value of $\binom{8}{3}$ is 56.

57. $9! = 362{,}880$. The value of $9!$ is 362,880.

58. $C(10, 5) = 10\,_nC_r\,5 = 252$. The value of $C(10, 5)$ is 252.

59. $2 \times 4 \times 3 \times 2 = 48$. The number of wedding arrangements possible is 48.

60. $5 \times 3 \times 6 = 90$. The number of available couches is 90.

61. $4\,_nP_r\,4 = 24$. The jobs can be assigned in 24 different ways.

62. (a) $\binom{6}{3} = 20$. The number of different delegations possible is 20.

    (b) $\binom{5}{2} = 10$. If the president must attend, 10 different delegations are possible.

63. $9\,_nP_r\,3 = 504$. There are 504 different ways the winners could be determined.

64. $26 \times 10 \times 10 \times 10 \times 26 \times 26 \times 26 = 456{,}976{,}000$. The number of possible different license plates is 456,976,000. The number of license plates that have no repeats is

    $26 \times 10 \times 9 \times 8 \times 25 \times 24 \times 23 = 258{,}336{,}000$.

65. (a) $P(\text{green}) = \dfrac{4}{15}$. The probability of drawing a green marble is $\dfrac{4}{15}$.

    (b) $P(\text{not black}) = \dfrac{4 + 6}{15} = \dfrac{10}{15} = \dfrac{2}{3}$. The probability of drawing a marble that is not black is $\dfrac{2}{3}$.

    (c) $P(\text{blue}) = \dfrac{0}{15} = 0$. The probability of drawing a green blue is 0.

66. (a) $P(\text{green}) = \dfrac{4}{15} \Rightarrow \dfrac{4}{15 - 4} = \dfrac{4}{11}$. The odds in favor of drawing a green marble are 4 to 11.

    (b) $P(\text{not white}) = \dfrac{4 + 5}{15} = \dfrac{9}{15} = \dfrac{3}{5} \Rightarrow \dfrac{3}{5 - 3} = \dfrac{3}{2}$. The odds against drawing a white marble are 3 to 2.

    (c) From (b), the odds in favor of drawing a marble that is not white are 3 to 2.

67. $P(\text{black king}) = \dfrac{2}{52} = \dfrac{1}{26}$. The probability of drawing a black king is $\dfrac{1}{26}$.

68. $P(\text{face or ace}) = \dfrac{16}{52} = \dfrac{4}{13}$. The probability of drawing a face card or an ace is $\dfrac{4}{13}$.

69. $P(\text{ace} \cup \text{diamond}) = P(\text{ace}) + P(\text{diamond}) - P(\text{ace} \cap \text{diamond}) = \dfrac{4}{52} + \dfrac{13}{52} - \dfrac{1}{52} = \dfrac{16}{52} = \dfrac{4}{13}$.

    The probability of drawing an ace or a diamond is $\dfrac{4}{13}$.

70. $P(\text{not a diamond}) = 1 - P(\text{diamond}) = 1 - \dfrac{1}{4} = \dfrac{3}{4}$. The probability of drawing a card that is not a diamond is $\dfrac{3}{4}$.

71. $P(\text{no more than 3}) = 1 - [P(4) + P(5)] = 1 - (.08 + .06) = .86$. The probability that no more than 3 filters are defective is .86.

72. $P(\text{at least 2}) = 1 - [P(0) + P(1)] = 1 - (.31 + .25) = .44$. The probability that at least 2 filters are defective is .44.

73. $P(\text{more than 5}) = 0$. Only 5 filters were selected, so it is not possible for more than 5 to be defective. Note that, from the probabilities in the table, $P(0) + P(1) + P(2) + P(3) + P(4) + P(5) = 1$.

74. $\dbinom{12}{2}\left(\dfrac{1}{6}\right)^2\left(\dfrac{5}{6}\right)^{10} \approx .296$. The probability that exactly 2 of 12 rolls results in a 5 is approximately .296.

75. $\dbinom{10}{4}\left(\dfrac{1}{2}\right)^4\left(\dfrac{1}{2}\right)^6 \approx .205$. The probability that exactly 4 of 10 coins tossed result in a tail is approximately .205.

76. (a) $P(\text{conservation}) = \dfrac{56.51}{282.2} \approx .2002$. The probability that a randomly selected student is in the conservative group is approximately .2002.

    (b) $P(\text{far left or far right}) = \dfrac{7.06 + 3.673}{282.2} \approx .0380$. The probability that a randomly selected student is on the far left or the far right is approximately .0380.

    (c) $P(\text{not middle of road}) = 1 - P(\text{middle of road}) = 1 - \dfrac{143.5}{282.2} \approx .4915$. The probability that a randomly selected student is not middle of the road is approximately .2002.

Chapter 11 Test

1. (a) $a_1 = (-1)^1(1 + 2) = -3$; $a_2 = (-1)^2(2 + 2) = 4$; $a_3 = (-1)^3(3 + 2) = -5$; $a_4 = (-1)^4(4 + 2) = 6$;
    $a^5 = (-1)^5(5 + 2) = -7$. The first five terms of the sequence are $-3, 4, -5, 6,$ and -7. The sequence is neither arithmetic nor geometric.

    (b) $a_1 = (-3)\left(\dfrac{1}{2}\right)^1 = -\dfrac{3}{2}$; $a_2 = (-3)\left(\dfrac{1}{2}\right)^2 = -\dfrac{3}{4}$; $a_3 = (-3)\left(\dfrac{1}{2}\right)^3 = -\dfrac{3}{8}$; $a_4 = (-3)\left(\dfrac{1}{2}\right)^4 = -\dfrac{3}{16}$;
    $a^5 = (-3)\left(\dfrac{1}{2}\right)^5 = -\dfrac{3}{32}$. The first five terms of the sequence are $-\dfrac{3}{2}, -\dfrac{3}{4}, -\dfrac{3}{8}, -\dfrac{3}{16},$ and $-\dfrac{3}{32}$. The sequence is geometric with common ratio $\dfrac{1}{2}$.

    (c) $a_1 = 2$; $a_2 = 3$; $a_3 = 3 + 2(2) = 7$; $a_4 = 7 + 2(3) = 13$; $a_5 = 13 + 2(7) = 27$. The first five terms of the sequence are 2, 3, 7, 13, and 27. The sequence is neither arithmetic nor geometric.

2. (a) $a_1 = 1, a_3 = 25 \Rightarrow 25 = 1 + (3 - 1)d \Rightarrow 2d = 24 \Rightarrow d = 12 \Rightarrow a_5 = 1 + (5 - 1)(12) = 49$.

   The fifth term of the sequence is 49.

   (b) $a_1 = 81, r = -\dfrac{2}{3} \Rightarrow a_5 = 81\left(-\dfrac{2}{3}\right)^{5-1} = 16$. The fifth term of the sequence is 16.

3. (a) $a_1 = -43, d = 12 \Rightarrow a_{10} = -43 + 12(10 - 1) = 65 \Rightarrow S_{10} = \dfrac{10}{2}(-43 + 65) = 110$. The sum of the

   first ten terms of the sequence is 110.

   (b) $a_1 = 5, r = -2 \Rightarrow S_{10} = \dfrac{5(1 - (-2)^{10})}{1 - (-2)} = -1705$. The sum of the first ten terms of the sequence is -1705.

4. (a) $a_1 = 5(1) + 2 = 7; a_{30} = 5(30) + 2 = 152; d = 5 \Rightarrow \displaystyle\sum_{i=1}^{30}(5i + 2) = S_{30} = \dfrac{30}{2}(7 + 152) = 2385$.

   The value of the sum is 2385.

   (b) $a_1 = (-3)(2)^1 = -6; r = 2 \Rightarrow \displaystyle\sum_{i=1}^{5}(-3 \cdot 2^i) = \dfrac{-6(1 - (2)^5)}{1 - 2} = -186$. The value of the sum is -186.

   (c) $r = 2 \Rightarrow |r| \geq 1 \Rightarrow \displaystyle\sum_{i=1}^{\infty}(2^i) \cdot 4$ does not exist. The value of the sum diverges and does not exist.

   (d) $a_1 = 54\left(\dfrac{2}{9}\right)^1 = 12; r = \dfrac{2}{9} \Rightarrow \displaystyle\sum_{i=1}^{\infty}54\left(\dfrac{2}{9}\right)^i = S_\infty = \dfrac{12}{1 - \frac{2}{9}} = \dfrac{108}{7}$. The value of the sum is $\dfrac{108}{7}$.

5. (a) $(2x - 3y)^4 = \dbinom{4}{0}(2x)^4(-3y)^0 + \dbinom{4}{1}(2x)^3(-3y)^1 + \dbinom{4}{2}(2x)^2(-3y)^2 + \dbinom{4}{3}(2x)^1(-3y)^3 + \dbinom{4}{4}(2x)^0(-3y)^4 =$

   $16x^4 - 96x^3y + 216x^2y^2 - 216xy^3 + 81y^4$.

   The binomial expansion of $(2x - 3y)^4$ is $16x^4 - 96x^3y + 216x^2y^2 - 216xy^3 + 81y^4$.

   (b) $\dbinom{6}{2}(w)^4(-2y)^2 = 60w^4y^2$. The third term of the expansion of $(w - 2y)^6 = 60w^4y^2$.

6. (a) $10\,{_nC_r}\,2 = 45$. The value of $10\,{_nC_r}\,2 = 45$.

   (b) $\dbinom{7}{3} = 7\,{_nC_r}\,3 = 35$. The value of $\dbinom{7}{3}$ is 35.

   (c) $7! = 5040$. The value of $7!$ is 5040.

   (d) $P(11, 3) = 11\,{_nP_r}\,3 = 990$. The value of $P(11, 3)$ is 990.

7. Prove $8 + 14 + 20 + 26 + \cdots + (6n + 2) = 3n^2 + 5n$.

   Step 1: $3(1)^2 + 5(1) = 8 \Rightarrow 8 + 14 + 20 + 26 + \cdots + (6n + 2) = 3n^2 + 5n$ for $n = 1$.

   Step 2: Assume $8 + 14 + 20 + 26 + \cdots + (6k + 2) = 3k^2 + 5k$, then

   $\qquad 8 + 14 + 20 + 26 + \cdots + (6k + 2) + (6k + 8) = 3k^2 + 5k + 6k + 8 \Rightarrow$

   $\qquad 8 + 14 + 20 + 26 + \cdots + (6k + 2) + [6(k + 1) + 2] = 3k^2 + 11k + 8 =$

   $\qquad 3k^2 + 6k + 3 + 5k + 5 = 3(k + 1)^2 + 5(k + 1)$

   $\qquad$ Thus if $8 + 14 + 20 + 26 + \cdots + (6n + 2) = 3n^2 + 5n$ for $n = k$, then it must follow that

   $\qquad 8 + 14 + 20 + 26 + \cdots + (6n + 2) = 3n^2 + 5n$ for $n = k + 1$ also.

8. $4 \times 3 \times 2 = 24$. 24 different types of shoes can be made.

9. $20\,{_nP_r}\,3 = 6840$. The three offices can be filled in 6840 different ways.

10. $\dbinom{8}{2}\dbinom{12}{3} = 28 \cdot 220 = 6160$. Two men and three women can be chosen in 6160 different ways.

11. (a) $P(\text{red }3) = \dfrac{2}{52} = \dfrac{1}{26}$. The probability of drawing a red three is $\dfrac{1}{26}$.

   (b) $4 \times 3 = 12$ cards are face cards $\Rightarrow 52 - 12 = 40$ cards are not face cards $\Rightarrow$

   $P(\text{not a face card}) = \dfrac{40}{52} = \dfrac{10}{13}$. The probability of drawing a card that is not a face card is $\dfrac{10}{13}$.

   (c) $P(\text{king} \cup \text{spade}) = P(\text{king}) + P(\text{spade}) - P(\text{king} \cap \text{spade}) = \dfrac{4}{52} + \dfrac{13}{52} - \dfrac{1}{52} = \dfrac{16}{52} = \dfrac{4}{13}$.

   (d) $P(\text{face card}) = \dfrac{12}{52} = \dfrac{3}{13} \Rightarrow \dfrac{3}{13 - 3} = \dfrac{3}{10}$. The odds in favor of drawing a face card are 3 to 10.

12. (a) $\dbinom{8}{3}\left(\dfrac{1}{6}\right)^3\left(\dfrac{5}{6}\right)^5 \approx .104$. The probability that exactly 3 of 8 rolls will result in a 4 is approximately .104.

   (b) $\dbinom{8}{8}\left(\dfrac{1}{6}\right)^8\left(\dfrac{5}{6}\right)^0 \approx .000000595$. The probability that all 8 of 8 rolls will result in a 6 is

   approximately .000000595.

Chapter 11 Project

1. The results of the experiment will vary.

   (a) The sample space for tossing two coins is $S = \{(H, H). (H, T), (T, H), (T, T)\}$. The probabilities for these

   outcomes are $P(H, H) = \dfrac{1}{4}$, $P(H, T) = \dfrac{1}{4}$, $P(T, H) = \dfrac{1}{4}$, and $P(T, T) = \dfrac{1}{4}$. Therefore, the theoretical

   probabilities of the events listed in the table are $P(2 \text{ heads}) = \dfrac{1}{4} = .25$, $P(1 \text{ head, 1 tail}) = \dfrac{2}{4} = .5$,

   and $P(2 \text{ tails}) = \dfrac{1}{4} = .25$.

   (b) While there is only one way to obtain two heads and only one way to obtain two tails, there are two ways to

   obtain one head and one tail.

   (c) Answers will vary. Better results could be obtained by flipping the coin more times.

2. The results of the experiment will vary.

   (a) The sample space for tossing four coins contains 16 possible outcomes as listed here.

   $S = \{(H, H, H, H). (H, H, H, T), (H, H, T, H), (H, H, T, T), (H, T, H, H), (H, T, H, T), (H, T, T, H),$

   $(H, T, T, T), (T, H, H, H), (T, H, H, T), (T, H, T, H), (T, H, T, T), (T, T, H, H), (T, T, H, T), (T, T, T, H)$

   $(T, T, T, T)\}$

   Since these outcomes are equally likely, the probability of each is $\dfrac{1}{4}$. The theoretical probabilities of the

   events listed in the table are $P(4 \text{ heads}) = \dfrac{1}{16} = .0625$, $P(3 \text{ heads, 1 tail}) = \dfrac{4}{16} = .25$,

   $P(2 \text{ heads, 2 tails}) = \dfrac{6}{16} = .375$, $P(1 \text{ head, 3 tails}) = \dfrac{4}{16} = .25$, $P(4 \text{ tails}) = \dfrac{1}{16} = .0625$.

   (b) Answers will vary.

   (c) Answers will vary. Better results could be obtained by flipping the coin more times.

3. The results of the experiment will vary, however the experimental probabilities should become closer to the

   theoretical probabilities as the number of times the experiment is run increases.

Chapter 12: Limits, Derivatives, and Definite Integrals

12.1: An Introduction to Limits

1. The statement is false. Consider the following function: $f(x) = \begin{cases} x \text{ if } x \neq 0 \\ 2 \text{ if } x = 0 \end{cases}$.

    For this function, $\lim_{x \to 0} f(x) = 0 \neq 2 = f(0)$.

2. The statement is false. The function $f(x) = \dfrac{1}{x^2}$ used in example 6 had no limits as x approached 0 because $f(x)$ became arbitrarily large for x near 0. Also, the function $f(x) = \sin \dfrac{1}{x}$ used in example 7 had no limits as x approached 0 because $f(x)$ oscillated wildly for x near 0.

3. The statement is false. Consider the following function: $f(x) = \begin{cases} x + 4 \text{ for } x \neq 1 \\ \text{undefined for } x = 1 \end{cases}$.

    For this function, $\lim_{x \to 1} f(x) = 5$, but $x = 1$ is not in the domain of $f(x)$. For $\lim_{x \to a} f(x)$ to exist, values of x arbitrarily near a must be in the domain of $f(x)$, but $x = a$ need not be in the domain of $f(x)$.

4. The statement is false. Consider the following function: $f(x) = \begin{cases} x + 4 \text{ for } x \neq 1 \\ 10 \text{ for } x = 1 \end{cases}$.

    For this function, $\lim_{x \to 1} f(x) = 5$, but $f(x)$ never takes on the value 5.

5. The statement is true. If $\lim_{x \to a} f(x) = -5$ then there must be some value of x near a where

    $|f(x) - (-5)| \leq .001 \Rightarrow -5.001 \leq f(x) \leq -4.999$.

6. The statement is true. If $\lim_{x \to a} f(x) = b$ then there must be some value of x near a where $|f(x) - b| < .0001$.

7. From the graph, $\lim_{x \to 3} f(x) = 2.5$.

8. From the graph, $\lim_{x \to 2} f(x) = 4$.

9. From the graph, as x approaches 2, $|f(x)|$ becomes arbitrarily large. Therefore, $\lim_{x \to 2} f(x)$ does not exist.

10. From the graph, as x approaches 3, $|f(x)|$ becomes arbitrarily large. Therefore, $\lim_{x \to 3} f(x)$ does not exist.

11. From the graph, $\lim_{x \to 0} f(x) = 0$.

12. From the graph, $\lim_{x \to 1} h(x) = 1$.

13. From the graph, as x approaches 1 from below, $f(x)$ approaches 1.5, but as x approaches 1 from above, $f(x)$ approaches 3. Thus, $\lim_{x \to 1} f(x)$ does not exist.

14. From the graph, as x approaches 2 from above, $f(x)$ approaches 2, but as x approaches 2 from below, $f(x)$ becomes arbitrarily large. Thus, $\lim_{x \to 2} f(x)$ does not exist.

15. From the graph, $\lim_{x \to 3} g(x) = 2$.

16. From the graph, as x approaches -2 from below, $g(x)$ approaches -1, but as x approaches -2 from above, $g(x)$ approaches $-.5$. Thus, $\lim_{x \to -2} g(x)$ does not exist.

17. From the graph, as x approaches .5, $h(x)$ has infinitely many oscillations between the values of $y = 0$ and $y = 2$. Thus, $\lim_{x \to .5} h(x)$ does not exist.

18. From the graph, as x approaches 2, $f(x)$ has infinitely many oscillations. However, the minimum and maximum values of these oscillations converge to $y = 1.5$. Thus, $\lim_{x \to 2} f(x) = 1.5$.

19. From the table of values, it appears that $\lim_{x \to 1} f(x) = 2$.

20. From the table of values, it appears that $\lim_{x \to 2} f(x) = -1$.

21. The completed table is shown in Figure 21. From the table, $\lim_{x \to 1} f(x) = 1$.

x	.9	.99	.999	1.001	1.01	1.1
$f(x)$	1.02	1.0002	1.000002	1.000002	1.0002	1.02

Figure 21

22. The completed table is shown in Figure 22. From the table, $\lim_{x \to 2} k(x) = 10$.

x	1.9	1.99	1.999	2.001	2.01	2.1
$k(x)$	9.41	9.9401	9.994	10.006	10.0601	10.61

Figure 22

23. The completed table is shown in Figure 23. From the table, $\lim_{x \to -1} f(x) = -4$.

x	-1.1	-1.01	-1.001	$-.999$	$-.99$	$-.9$
$f(x)$	-3.68	-3.9698	-3.997	-4.003	-4.0298	-4.28

Figure 23

24. The completed table is shown in Figure 24. From the table, as x approaches 1 from below, $h(x)$ goes to large positive values, but as x approaches 1 from above, $h(x)$ goes to large negative values. Thus, $\lim_{x \to 1} h(x)$ does not exist.

x	.9	.99	.999	1.001	1.01	1.1
$h(x)$	10.5132	100.501	1000.5	-999.5	-99.5012	-9.51191

Figure 24

25. The completed table is shown in Figure 25. From the table, as x approaches 3 from below, $f(x)$ goes to large positive values, but as x approaches 3 from above, $f(x)$ goes to large negative values. Thus, $\lim_{x \to 3} f(x)$ does not exist.

x	2.9	2.99	2.999	3.001	3.01	3.1
$f(x)$	12.9706	127.084	1268.24	-1267.66	-126.506	-12.3932

Figure 25

26. The completed table is shown in Figure 26. From the table, $\lim_{x \to 3} f(x) = 10$.

x	-3.1	-3.01	-3.001	-2.999	-2.99	-2.9
$f(x)$	10.61	10.0601	10.006	9.994	9.9401	9.41

Figure 26

27. The completed table is shown in Figure 27. From the table, $\lim_{x \to 0} f(x) = 2$.

x	$-.1$	$-.01$	$-.001$	$.001$	$.01$	$.1$
$f(x)$	1.987	1.99987	1.9999987	1.9999987	1.99987	1.987

Figure 27

28. The completed table is shown in Figure 28. From the table, $\lim_{x \to 0} f(x) = 2.5$.

x	$-.1$	$-.01$	$-.001$	$.001$	$.01$	$.1$
$f(x)$	2.39713	2.49896	2.49999	2.49999	2.49896	2.39713

Figure 28

29. Graphing $y_1 = \text{abs}(2x - 4)$ in the window $0 \le x \le 10, 0 \le y \le 10$ (not shown), it can be seen that $\lim_{x \to 5} |2x - 4| = 6$.

30. Graphing $y_1 = \sqrt{5 + 3x}$ and $y_2 = \sqrt{2}$ in the window $-2 \le x \le 0, 0 \le y \le 5$ (not shown), it can be seen that $\lim_{x \to -1} \sqrt{5 + 3x} = \sqrt{2}$.

31. Graphing $y_1 = (x^2 - 3x - 10)/(x - 5)$ in the window $0 \le x \le 10, 0 \le y \le 10$ (not shown), it can be seen that even though $\dfrac{x^2 - 3x - 10}{x - 5}$ does not exist at $x = 5$, $\lim_{x \to 5} \dfrac{x^2 - 3x - 10}{x - 5} = 7$.

32. Graphing $y_1 = (x + 1)/(x - 3)^2$ in the window $0 \le x \le 6, 0 \le y \le 100$ (not shown), it can be seen that as x approaches 3, $\dfrac{x + 1}{(x - 3)^2}$ grows without bounds. Thus $\lim_{x \to 3} \dfrac{x + 1}{(x - 3)^2}$ does not exist.

33. Graphing $y_1 = (x^2 + 2)/(x + 2)$ in the window $-4 \le x \le 0, -100 \le y \le 100$ (not shown), it can be seen that $\dfrac{x^2 + 2}{x + 2}$ goes to unbounded negative values as x approaches -2 from below and unbounded positive values as x approaches -2 from above. Thus $\lim_{x \to -2} \dfrac{x^2 + 2}{x + 2}$ does not exist.

34. Graphing $y_1 = (x^2 + 5x + 4)/(x + 4)$ in the window $-8 \le x \le 0, -6 \le y \le 0$ (not shown), it can be seen that even though $\dfrac{x^2 + 5x + 4}{x + 4}$ does not exist at $x = -4$, $\lim_{x \to -4} \dfrac{x^2 + 5x + 4}{x + 4} = -3$.

35. Graphing $y_1 = (x^2 - x - 2)/(x - 2)$ in the window $0 \le x \le 4, 0 \le y \le 6$ (not shown), it can be seen that even though $\dfrac{x^2 - x - 2}{x - 2}$ does not exist at $x = 2$, $\lim_{x \to 2} \dfrac{x^2 - x - 2}{x - 2} = 3$.

36. Graphing $y_1 = (4^x - x - 1)/(x - 1)$ in the window $0 \le x \le 2, -100 \le y \le 100$ (not shown), it can be seen that $\dfrac{4^x - x - 1}{x - 1}$ goes to unbounded negative values as x approaches 1 from below and unbounded positive values as x approaches 1 from above. Thus $\lim_{x \to 1} \dfrac{4^x - x - 1}{x - 1}$ does not exist.

37. Graphing $y_1 = (x + 7)/(x \le 3)$ and $y_2 = (5x - 5)/(x > 3)$ in the window $0 \le x \le 6, 0 \le y \le 20$ (not shown), it can be seen that $\lim_{x \to 3} f(x) = 10$.

38. Graphing $y_1 = (3x - 5)/(x \le 1)$ and $y_2 = (6 - 2x)/(x > 1)$ in the window $0 \le x \le 2, -5 \le y \le 5$ (not shown), it can be seen that $f(x)$ goes to -2 as x approaches 1 from below and $f(x)$ approaches 4 as x approaches 1 from above. Thus $\lim_{x \to 1} f(x)$ does not exist.

39. Graphing $y_1 = (x^2 - 3)/(x < 2)$ and $y_2 = (5 - x^2)/(x > 2)$ in the window $1 \le x \le 3, .5 \le y \le 1.5$

   (not shown), it can be seen that even though $f(x)$ does not exist at $x = 2$, $\lim_{x \to 2} f(x) = 1$.

40. Graphing $y_1 = \sqrt{(3 - x)}/(x < -1)$ and $y_2 = (4/(1 - x))/(x > -1)$ in the window $-2 \le x \le 0, 0 \le y \le 4$

   (not shown), it can be seen that even though $f(x)$ does not exist at $x = -1$, $\lim_{x \to -1} f(x) = 2$.

41. Graphing $y_1 = (e^x)/(x \le 1)$, $y_2 = (\sqrt{x})/(x > 1)$ and $y_3 = e^1$ in the window $0 \le x \le 2, 0 \le y \le 4$

   (not shown), it can be seen that $f(x)$ goes to e as x approaches 1 from below and $f(x)$ approaches 1 as x

   approaches 1 from above. Thus $\lim_{x \to 1} f(x)$ does not exist.

42. Graphing $y_1 = (\sqrt{x} - 1)/(x - 1)$ in the window $0 \le x \le 2, 0 \le y \le 1$ (not shown), it can be seen that even

   though $\dfrac{\sqrt{x} - 1}{x - 1}$ is not defined at $x = 1$, $\lim_{x \to 1} \dfrac{\sqrt{x} - 1}{x - 1} = .5$.

43. Graphing $y_1 = (x^3)/(x - \sin x)$ in the window $-1 \le x \le 1, 5 \le y \le 7$ (not shown), it can be seen that even

   though $\dfrac{x^3}{x - \sin x}$ does not exist at $x = 0$, $\lim_{x \to 0} \dfrac{x^3}{x - \sin x} = 6$.

44. Graphing $y_1 = (\sin x)/(\sin (2x))$ in the window $-1 \le x \le 1, 0 \le y \le 1$ (not shown), it can be seen that even

   though $\dfrac{\sin x}{\sin 2x}$ does not exist at $x = 0$, $\lim_{x \to 0} \dfrac{\sin x}{\sin 2x} = .5$.

45. Graphing $y_1 = (\cos x - 1)/x$ in the window $-1 \le x \le 1, -1 \le y \le 1$ (not shown), it can be seen that even

   though $\dfrac{\cos x - 1}{x}$ does not exist at $x = 0$, $\lim_{x \to 0} \dfrac{\cos x - 1}{x} = 0$.

46. Graphing $y_1 = (\tan x)^2/(1 + (1/\cos x))$ in the window $0 \le x \le 2\pi, -10 \le y \le 10$ (not shown), it can be seen

   that even though $\dfrac{\tan^2 x}{1 + \sec x}$ does not exist at $x = \pi$, $\lim_{x \to \pi} \dfrac{\tan^2 x}{1 + \sec x} = -2$.

47. Graphing $y_1 = (e^{2x} - 1)/(e^x - 1)$ in the window $-1 \le x \le 1, 0 \le y \le 4$ (not shown), it can be seen that even

   though $\dfrac{e^{2x} - 1}{e^x - 1}$ does not exist at $x = 0$, $\lim_{x \to 0} \dfrac{e^{2x} - 1}{e^x - 1} = 2$.

48. Graphing $y_1 = (\ln x)/(x - 1)$ in the window $0 \le x \le 2, 0 \le y \le 2$ (not shown), it can be seen that even

   though $\dfrac{\ln x}{x - 1}$ does not exist at $x = 1$, $\lim_{x \to 1} \dfrac{\ln x}{x - 1} = 1$.

49. Graphing $y_1 = (\ln x^2)/(\ln x)$ in the window $0 \le x \le 2, 0 \le y \le 4$ (not shown), it can be seen that even

   though $\dfrac{\ln x^2}{\ln x}$ does not exist at $x = 1$, $\lim_{x \to 1} \dfrac{\ln x^2}{\ln x} = 2$.

50. Graphing $y_1 = (e^{-x} - 1)/x$ in the window $-1 \le x \le 1, -2 \le y \le 2$ (not shown), it can be seen that even

   though $\dfrac{e^{-x} - 1}{x}$ does not exist at $x = 0$, $\lim_{x \to 0} \dfrac{e^{-x} - 1}{x} = -1$.

51. Graphing $y_1 = x \sin x$ in the window $-1 \le x \le 1, -1 \le y \le 1$ (not shown), it can be seen that $\lim_{x \to 0} x \sin x = 0$.

52. Graphing $y_1 = \cos (1/x)$ in the window $-1 \le x \le 1, -2 \le y \le 2$ (not shown), it can be seen that $\cos \dfrac{1}{x}$

   oscillates infinitely many times between $y = -1$ and $y = 1$ as x approaches 0. Thus $\lim_{x \to 0} \cos \dfrac{1}{x}$ does not exist.

53. There are cases where the limit of a function $f(x)$ at the point $x = a$ cannot be evaluated as $f(a)$. An example

   is $\lim_{x \to 2} \dfrac{x^2 - 4}{x - 2}$. This limit is 4, but $\dfrac{x^2 - 4}{x - 2}$ cannot be evaluated at $x = 2$ because 2 is not in it's domain.

12.2: Techniques for Calculating Limits

1. $\lim_{x \to 4} [f(x) - g(x)] = \lim_{x \to 4} f(x) - \lim_{x \to 4} g(x) = 16 - 8 = 8$. Thus, the limit is 8.

2. $\lim_{x \to 4} [g(x) \cdot f(x)] = [\lim_{x \to 4} g(x)] \cdot [\lim_{x \to 4} f(x)] = 8 \cdot 16 = 128$. Thus, the limit is 128.

3. Since $\lim_{x \to 4} g(x) = 8 \neq 0$, $\lim_{x \to 4} \dfrac{f(x)}{g(x)} = \dfrac{\lim_{x \to 4} f(x)}{\lim_{x \to 4} g(x)} = \dfrac{16}{8} = 2$. Thus, the limit is 2.

4. Since $\lim_{x \to 4} f(x) = 16 > 0$, $\lim_{x \to 4} [\log_2 f(x)] = \log_2 [\lim_{x \to 4} f(x)] = \log_2 (16) = 4$. Thus, the limit is 4.

5. Since $\lim_{x \to 4} f(x) = 16 \geq 0$, $\lim_{x \to 4} \sqrt{f(x)} = [\lim_{x \to 4} f(x)]^{1/2} = (16)^{1/2} = 4$. Thus, the limit is 4.

6. $\lim_{x \to 4} [1 + f(x)]^2 = [1 + \lim_{x \to 4} f(x)]^2 = (17)^2 = 289$. Thus, the limit is 289.

7. $\lim_{x \to 4} 2^{g(x)} = 2^{\lim_{x \to 4} g(x)} = 2^8 = 256$. Thus, the limit is 256.

8. $\lim_{x \to 4} \sqrt[3]{g(x)} = [\lim_{x \to 4} g(x)]^{1/3} = (8)^{1/3} = 2$. Thus, the limit is 2.

9. Since $\lim_{x \to 4} g(x) = 8 \neq 0$, $\lim_{x \to 4} \dfrac{f(x) + g(x)}{2g(x)} = \dfrac{\lim_{x \to 4} f(x) + \lim_{x \to 4} g(x)}{2 \cdot \lim_{x \to 4} g(x)} = \dfrac{16 + 8}{2 \cdot 8} = \dfrac{3}{2}$. Thus, the limit is $\dfrac{3}{2}$.

10. Since $\lim_{x \to 4} f(x) = 16 \neq 1$, $\lim_{x \to 4} \dfrac{5g(x) + 2}{1 - f(x)} = \dfrac{5 \cdot \lim_{x \to 4} g(x) + 2}{1 - \lim_{x \to 4} f(x)} = \dfrac{5 \cdot 8 + 2}{1 - 16} = -\dfrac{14}{5}$. Thus, the limit is $-\dfrac{14}{5}$.

11. $\lim_{x \to -3} 7 = 7$. The limit is 7.

12. $\lim_{x \to 6} (-5) = -5$. The limit is -5.

13. $\lim_{x \to \pi} x = \pi$. The limit is π.

14. $\lim_{x \to -\sqrt{2}} x = -\sqrt{2}$. The limit is $-\sqrt{2}$.

15. $\lim_{x \to 3} 4x^2 = 4(3)^2 = 36$. The limit is 36.

16. $\lim_{x \to -2} (-3x^5) = -3(-2)^5 = 96$. The limit is 96.

17. $\lim_{x \to -1} 4x^3 = 4(-1)^3 = -4$. The limit is -4.

18. $\lim_{x \to 1} (5x^8 - 3x^2 + 2) = 5(1)^8 - 3(1)^2 + 2 = 4$. The limit is 4.

19. $\lim_{x \to 2} (x^3 + 4x^2 - 5) = (2)^3 + 4(2)^2 - 5 = 19$. The limit is 19.

20. $\lim_{x \to 3} \dfrac{x^3 - 1}{x^2 + 1} = \dfrac{(3)^3 - 1}{(3)^2 + 1} = \dfrac{13}{5}$. The limit is $\dfrac{13}{5}$.

21. $\lim_{x \to -1} \dfrac{2x + 3}{3x + 4} = \dfrac{2(-1) + 3}{3(-1) + 4} = 1$. The limit is 1.

22. $\lim_{x \to 0} \dfrac{x^2 + 2x}{x} = \lim_{x \to 0} \dfrac{x(x + 2)}{x} = \lim_{x \to 0} (x + 2) = 2$. The limit is 2.

23. $\lim_{x \to 3} \dfrac{x^2 - 9}{x - 3} = \lim_{x \to 3} \dfrac{(x + 3)(x - 3)}{x - 3} = \lim_{x \to 3} (x + 3) = 6$. The limit is 6.

24. $\lim_{x \to -2} \dfrac{x^2 - 4}{x + 2} = \lim_{x \to -2} \dfrac{(x + 2)(x - 2)}{x + 2} = \lim_{x \to -2} (x - 2) = -4$. The limit is -4.

25. $\lim_{x \to -2} \dfrac{x^2 - x - 6}{x + 2} = \lim_{x \to -2} \dfrac{(x - 3)(x + 2)}{x + 2} = \lim_{x \to -2} (x - 3) = -5$. The limit is -5.

26. $\lim\limits_{x \to 5} \dfrac{x^2 - 3x - 10}{x - 5} = \lim\limits_{x \to 5} \dfrac{(x - 5)(x + 2)}{x - 5} = \lim\limits_{x \to 5} (x + 2) = 7.$ The limit is 7.

27. $\lim\limits_{x \to 1} \dfrac{x^2 + x - 2}{x - 1} = \lim\limits_{x \to 1} \dfrac{(x + 2)(x - 1)}{x - 1} = \lim\limits_{x \to 1} (x + 2) = 3.$ The limit is 3.

28. $\lim\limits_{x \to 5} \dfrac{x^2 - 7x + 10}{x^2 - 25} = \lim\limits_{x \to 5} \dfrac{(x - 5)(x - 2)}{(x - 5)(x + 5)} = \lim\limits_{x \to 5} \dfrac{x - 2}{x + 5} = \dfrac{3}{10}.$ The limit is $\dfrac{3}{10}.$

29. $\lim\limits_{x \to 3} \sqrt{6x - 2} = \sqrt{6(3) - 2}. = 4.$ The limit is 4.

30. $\lim\limits_{x \to -4} (1 - 6x)^{3/2} = (1 - 6(-4))^{3/2} = 125.$ The limit is 125.

31. $\lim\limits_{x \to 1} 9^{1/(x+1)} = 9^{1/(1+1)} = 3.$ The limit is 3.

32. $\lim\limits_{x \to 3} 5^{\sqrt{x+1}} = 5^{\sqrt{3+1}} = 25.$ The limit is 25.

33. $\lim\limits_{x \to 5} [\log_3 (2x - 1)] = \log_3 (2(5) - 1) = 2.$ The limit is 2.

34. $\lim\limits_{x \to 4} [\log_2 (14 + \sqrt{x})] = \log_2 (14 + \sqrt{4}) = 4.$ The limit is 4.

35. $\lim\limits_{x \to 1} [\sqrt{x}(1 + x)] = \sqrt{1}(1 + 1) = 2.$ The limit is 2.

36. $\lim\limits_{x \to 0} [2^{3x} - \ln (x + 1)] = 2^{3(0)} - \ln (0 + 1) = 1.$ The limit is 1.

37. $\lim\limits_{x \to 3} \dfrac{\sqrt{x + 1}}{\log_2 (5x + 1)} = \dfrac{\sqrt{3 + 1}}{\log_2 (5(3) + 1)} = \dfrac{2}{4} = \dfrac{1}{2}.$ The limit is $\dfrac{1}{2}.$

38. $\sqrt{-1}$ does not exist and $\sqrt{x}$ does not exist for values of x near -1. Thus, $\lim\limits_{x \to -1} \sqrt{x}$ does not exist.

39. $\lim\limits_{x \to 1} \sqrt{3 - x} = \sqrt{3 - 1} = \sqrt{2}.$ The limit is $\sqrt{2}.$

40. $\sqrt{x}$ goes to 0 as x approaches 0 from above. However, $\sqrt{x}$ does not exist as x approaches 0 from below. Thus, $\lim\limits_{x \to 0} \sqrt{x}$ does not exist.

41. $\lim\limits_{x \to 0} \dfrac{\sin x - 3x}{x} = \lim\limits_{x \to 0} \dfrac{\sin x}{x} - 3 \lim\limits_{x \to 0} \dfrac{x}{x} = 1 - 3(1) = -2.$ The limit is $-2.$

42. $\lim\limits_{x \to 0} \dfrac{\sin x}{5x} = \dfrac{1}{5} \lim\limits_{x \to 0} \dfrac{\sin x}{x} = \dfrac{1}{5}(1) = \dfrac{1}{5}.$ The limit is $\dfrac{1}{5}.$

43. $\lim\limits_{x \to 0} (x \cot x) = \lim\limits_{x \to 0} \left(x \dfrac{\cos x}{\sin x} \right) = \dfrac{1}{\lim\limits_{x \to 0} \frac{\sin x}{x}} \cdot \lim\limits_{x \to 0} \cos x = 1 \cdot 1 = 1.$ The limit is 1.

44. $\lim\limits_{x \to 0} \dfrac{\sin^2 x}{x^2} = \left(\lim\limits_{x \to 0} \dfrac{\sin x}{x} \right)^2 = (1)^2 = 1.$ The limit is 1.

45. $\dfrac{\cos x - 1}{3x} = \dfrac{\cos x - 1}{3x} \cdot \dfrac{\cos x + 1}{\cos x + 1} = \dfrac{\cos^2 x - 1}{3x(\cos x + 1)} = \dfrac{-\sin^2 x}{3x(\cos x + 1)} = -\dfrac{1}{3}(\sin x)\left(\dfrac{\sin x}{x} \right)\left(\dfrac{1}{\cos x + 1} \right).$

Thus, $\lim\limits_{x \to 0} \dfrac{\cos x - 1}{3x} = \lim\limits_{x \to 0} \left[-\dfrac{1}{3}(\sin x)\left(\dfrac{\sin x}{x} \right)\left(\dfrac{1}{\cos x + 1} \right) \right] = -\dfrac{1}{3}(0)(1)\left(\dfrac{1}{2} \right) = 0.$ The limit is 0.

46. $\dfrac{\cos x + 2 \sin x - 1}{3x} = \dfrac{2}{3}\left(\dfrac{\sin x}{x} \right) + \dfrac{\cos x - 1}{3x}.$

Thus, $\lim\limits_{x \to 0} \dfrac{\cos x + 2 \sin x - 1}{3x} = \lim\limits_{x \to 0} \left[\dfrac{2}{3}\left(\dfrac{\sin x}{x} \right) + \dfrac{1}{3}\left(\dfrac{\cos x - 1}{x} \right) \right] = \dfrac{2}{3}(1) + \dfrac{1}{3}(0) = \dfrac{2}{3}.$ The limit is $\dfrac{2}{3}.$

47. As the angle x approaches 0, the length of the line segment AB approaches 0 and the length of the line segment OB approaches 1 (the radius of the unit circle). Thus, $\lim\limits_{x \to 0} \sin x = 0$ and $\lim\limits_{x \to 0} \cos x = 1.$

48. The area of triangle $OAB = \dfrac{1}{2}(AB)(OB) = \dfrac{1}{2}\sin x \cos x$.

49. The area of triangle $OCD = \dfrac{1}{2}(1)(CD) = \dfrac{1}{2}\tan x$.

50. The area of sector $OAD = \pi(1)^2\left(\dfrac{x}{2\pi}\right) = \dfrac{x}{2}$. From the figure, we see that area of triangle OAB < area of

    sector OAD < area of triangle OCD. Thus, $\dfrac{1}{2}\sin x \cos x < \dfrac{1}{2}x < \dfrac{1}{2}\dfrac{\sin x}{\cos x} = \cos x < \dfrac{x}{\sin x} < \dfrac{1}{\cos x}$.

    Thus, $\dfrac{1}{\cos x} > \dfrac{\sin x}{x} > \cos x$ or $\cos x < \dfrac{\sin x}{x} < \dfrac{1}{\cos x}$.

51. From the inequality $\cos x < \dfrac{\sin x}{x} < \dfrac{1}{\cos x}$ we see that, as x approaches 0, the values of $\dfrac{\sin x}{x}$ are between the

    values of two other functions, both of which go to 1. Thus, $\displaystyle\lim_{x \to 0} \dfrac{\sin x}{x} = 1$.

52. $\dfrac{1 - \cos x}{x} = \dfrac{1 - \cos x}{x} \cdot \dfrac{1 + \cos x}{1 + \cos x} = \dfrac{1 - \cos^2 x}{x(1 + \cos x)} = \dfrac{\sin^2 x}{x(1 + \cos x)} = (\sin x)\left(\dfrac{\sin x}{x}\right)\left(\dfrac{1}{1 + \cos x}\right)$.

    Thus, $\displaystyle\lim_{x \to 0} \dfrac{1 - \cos x}{x} = \lim_{x \to 0}\left[(\sin x)\left(\dfrac{\sin x}{x}\right)\left(\dfrac{1}{1 + \cos x}\right)\right] = (0)(1)\left(\dfrac{1}{2}\right) = 0$. The limit is 0.

12.3: One-Sided Limits; Limits Involving Infinity

1. (a) Since $f(x) = 4$ for $x > 2$, $\displaystyle\lim_{x \to 2^+} f(x) = 4$.

   (b) Since $f(x) = x$ for $x > 2$, $\displaystyle\lim_{x \to 2^-} f(x) = 2$.

2. (a) Since $f(x) = \sqrt{x}$ for $x > 4$, $\displaystyle\lim_{x \to 4^+} f(x) = 2$.

   (b) Since $f(x) = \dfrac{1}{2}x^2$ for $x < 4$, $\displaystyle\lim_{x \to 4^-} f(x) = 8$.

3. (a) As shown in the graph of $f(x) = \dfrac{x}{5(3 - x)^3}$, $\displaystyle\lim_{x \to 3^+} f(x) = -\infty$.

   (b) As shown in the same graph, $\displaystyle\lim_{x \to 3} f(x) = \infty$.

4. (a) As shown in the graph of $f(x) = \dfrac{x^2}{3(x + 2)}$, $\displaystyle\lim_{x \to -2^+} f(x) = \infty$.

   (b) As shown in the same graph, $\displaystyle\lim_{x \to -2^-} f(x) = -\infty$.

5. (a) As shown in the graph of $f(x) = \dfrac{x}{(x + 1)^2}$, $\displaystyle\lim_{x \to -1^+} f(x) = -\infty$.

   (b) As shown in the same graph, $\displaystyle\lim_{x \to -1} f(x) = -\infty$.

6. (a) As shown in the graph of $f(x) = \dfrac{2x}{(x - 1)^2}$, $\displaystyle\lim_{x \to 1^+} f(x) = \infty$.

   (b) As shown in the same graph, $\displaystyle\lim_{x \to 1} f(x) = \infty$.

7. Since $3x - 5$ is continuous at $x = 5$, $\displaystyle\lim_{x \to 5^+} (3x - 5) = 3(5) - 5 = 10$.

8. Since x^3 is continuous at $x = -4$, $\displaystyle\lim_{x \to -4} (x^3) = (-4)^3 = -64$.

9. Since 100 is continuous at $x = 7$, $\displaystyle\lim_{x \to 7^-} 100 = 100$.

10. Since $\sqrt{x - 1}$ is defined for $x > 1$, $\displaystyle\lim_{x \to 1^+} \sqrt{x - 1} = \sqrt{1 - 1} = 0$.

11. Since $\sqrt{2-x}$ is defined at $x < 2$, $\lim\limits_{x \to 2^-} \sqrt{2-x} = \sqrt{2-2} = 0$.

12. Since $\sqrt{x+3}$ does not exist for $x < -3$, $\lim\limits_{x \to -3^-} \sqrt{x+3}$ does not exist.

13. Since $\dfrac{|x|}{x} = -1$ for $x < 0$, $\lim\limits_{x \to 0^-} \dfrac{|x|}{x} = -1$.

14. Since $\dfrac{|x|}{x} = 1$ for $x > 0$, $\lim\limits_{x \to 0^+} \dfrac{|x|}{x} = 1$.

15. Since $\dfrac{|x+3|}{x+3} = -1$ for $x < -3$, $\lim\limits_{x \to -3^-} \dfrac{|x+3|}{x+3} = -1$.

16. As shown in the graph of $\dfrac{6x^2+1}{2x^2+3}$, $\lim\limits_{x \to -\infty} \dfrac{6x^2+1}{2x^2+3} = 3$.

17. As shown in the graph of $2 + e^{-x}$, $\lim\limits_{x \to \infty} (2 + e^{-x}) = 2$.

18. As shown in the graph of $x \sin x$, $\lim\limits_{x \to \infty} (x \sin x)$ does not exist.

19. As shown in the graph of $x + \dfrac{1}{x}$, $\lim\limits_{x \to \infty} \left(x + \dfrac{1}{x} \right) = \infty$.

20. Graphing $y_1 = x \cdot \sin(5/x)$ in the window $0 \le x \le 100, 0 \le y \le 10$ (not shown), it can be seen that

$$\lim_{x \to \infty} \left(x \sin \frac{5}{x} \right) = 5.$$

21. Graphing $y_1 = x \cdot \sin(1/x^2)$ in the window $0 \le x \le 100, -1 \le y \le 1$ (not shown), it can be seen that

$$\lim_{x \to \infty} \left(x \sin \frac{1}{x^2} \right) = 0.$$

22. Graphing $y_1 = \sqrt{(x^2+x)} - x$ in the window $0 \le x \le 100, 0 \le y \le 1$ (not shown), it can be seen that

$$\lim_{x \to \infty} (\sqrt{x^2+x} - x) = .5.$$

23. Graphing $y_1 = x - \sqrt{(x^2+5)}$ in the window $0 \le x \le 100, -1 \le y \le 1$ (not shown), it can be seen that

$$\lim_{x \to \infty} (x - \sqrt{x^2+5}) = 0.$$

24. Graphing $y_1 = e^x/(e^x - 1)$ in the window $-10 \le x \le 10, -2 \le y \le 2$ (not shown), it can be seen that $x = 0$ is a vertical asymptote and $y = 1$ is a horizontal asymptote as $x \to \infty$ and $y = 0$ is a horizontal asymptote as $x \to -\infty$.

25. Graphing $y_1 = \tan^{-1}(x/(x-1))$ in the window $-100 \le x \le 100, -\dfrac{\pi}{2} \le y \le \dfrac{\pi}{2}$ (not shown), it can be seen that $y = \dfrac{\pi}{4}$ is a horizontal asymptote both when $x \to \infty$ and when $x \to -\infty$. Changing the window to $0 \le x \le 2, -\pi \le y \le \pi$, shows that $x = 1$ is not a vertical asymptote.

26. Graphing $y_1 = (x - \cos x)/(x + \sin x)$ in the window $-1 \le x \le 1, -10 \le y \le 10$ (not shown), it can be seen that $x = 0$ is a vertical asymptote. Changing the window to $-1000 \le x \le 1000, 0 \le y \le 2$, shows that even though $\dfrac{x - \cos x}{x + \sin x}$ oscillates back and forth above and below $y = 1$ for large $|x|$, $y = 1$ still is a horizontal asymptote.

27. Graphing $y_1 = 5 - e^{-x}$ in the window $-10 \le x \le 10, -10 \le y \le 10$ (not shown), it can be seen that $y = 5$ is a horizontal asymptote as $x \to \infty$.

28. (a) Since $f(x) = x^2$ if $x > 1$, $\lim\limits_{x \to 1^+} f(x) = (1)^2 = 1$.

   (b) Since $f(x) = 2x + 3$ if $x < 1$, $\lim\limits_{x \to 1^-} f(x) = 2(1) + 3 = 5$.

29. (a) Since $f(x) = x - 1$ if $x > 2$, $\lim\limits_{x \to 2^+} f(x) = 2 - 1 = 1$.

   (b) Since $f(x) = 7x$ if $x \le 2$, $\lim\limits_{x \to 2^-} f(x) = 7(2) = 14$.

30. (a) Graphing $y_1 = 1/(1 + x)^3$ in the window $-3 \le x \le 1, -50 \le y \le 50$ (not shown), shows that

   $$\lim\limits_{x \to -1^+} f(x) = \infty.$$

   (b) The same graph shows that $\lim\limits_{x \to -1^-} f(x) = -\infty$.

31. (a) Graphing $y_1 = x/(4 - x)^3$ in the window $0 \le x \le 8, -50 \le y \le 50$ (not shown), shows that

   $$\lim\limits_{x \to 4^+} f(x) = -\infty.$$

   (b) The same graph shows that $\lim\limits_{x \to 4^-} f(x) = \infty$.

32. (a) Graphing $y_1 = 1/(x - 3)^2$ in the window $0 \le x \le 6, -50 \le y \le 50$ (not shown), shows that

   $$\lim\limits_{x \to 3^+} f(x) = \infty.$$

   (b) The same graph shows that $\lim\limits_{x \to 3^-} f(x) = \infty$.

33. (a) Graphing $y_1 = x/(x + 3)^3$ in the window $-6 \le x \le 0, -50 \le y \le 50$ (not shown), shows that

   $$\lim\limits_{x \to -3^+} f(x) = -\infty.$$

   (b) The same graph shows that $\lim\limits_{x \to -3^-} f(x) = \infty$.

34. $\lim\limits_{x \to \infty} \dfrac{3x}{5x - 1} = \lim\limits_{x \to \infty} \dfrac{3}{5 - \frac{1}{x}} = \dfrac{3}{5}$. The limit is $\dfrac{3}{5}$.

35. $\lim\limits_{x \to \infty} \dfrac{5x}{3x - 1} = \lim\limits_{x \to \infty} \dfrac{5}{3 - \frac{1}{x}} = \dfrac{5}{3}$. The limit is $\dfrac{5}{3}$.

36. $\lim\limits_{x \to -\infty} \dfrac{2x + 3}{4x - 7} = \lim\limits_{x \to -\infty} \dfrac{2 + \frac{3}{x}}{4 - \frac{7}{x}} = \dfrac{2}{4} = \dfrac{1}{2}$. The limit is $\dfrac{1}{2}$.

37. $\lim\limits_{x \to -\infty} \dfrac{8x + 2}{2x - 5} = \lim\limits_{x \to -\infty} \dfrac{8 + \frac{2}{x}}{2 - \frac{5}{x}} = \dfrac{8}{2} = 4$. The limit is 4.

38. $\lim\limits_{x \to \infty} \dfrac{x^2 + 2x}{2x^3 - 2x + 1} = \lim\limits_{x \to \infty} \dfrac{\frac{1}{x} + \frac{2}{x^2}}{2 - \frac{2}{x^2} + \frac{1}{x^3}} = \dfrac{0}{2} = 0$. The limit is 0.

39. $\lim\limits_{x \to \infty} \dfrac{x^2 + 2x - 5}{3x^2 + 2} = \lim\limits_{x \to \infty} \dfrac{1 + \frac{2}{x} - \frac{5}{x^2}}{3 + \frac{2}{x^2}} = \dfrac{1}{3}$. The limit is $\dfrac{1}{3}$.

40. $\lim\limits_{x \to \infty} \dfrac{3x^3 + 2x - 1}{2x^4 - 3x^3 - 2} = \lim\limits_{x \to \infty} \dfrac{\frac{3}{x} + \frac{2}{x^3} - \frac{1}{x^4}}{2 - \frac{3}{x} - \frac{2}{x^4}} = \dfrac{0}{2} = 0$. The limit is 0.

41. $\lim\limits_{x \to \infty} \dfrac{2x^2 - 1}{3x^4 + 2} = \lim\limits_{x \to \infty} \dfrac{\frac{2}{x^2} - \frac{1}{x^4}}{3 + \frac{2}{x^4}} = \dfrac{0}{3} = 0$. The limit is 0.

42. $\lim\limits_{x \to \infty} \dfrac{2x^3 - x - 3}{6x^2 - x - 1} = \lim\limits_{x \to \infty} \dfrac{2x - \frac{1}{x} - \frac{3}{x^2}}{6 - \frac{1}{x} - \frac{1}{x^2}} = \infty$. The limit is ∞.

43. $\lim\limits_{x \to \infty} \dfrac{x^4 - x^3 - 3x}{7x^2 + 9} = \lim\limits_{x \to \infty} \dfrac{x^2 - x - \frac{3}{x}}{7 + \frac{9}{x^2}} = \infty$. The limit is ∞.

44. $\lim\limits_{x \to -\infty} \dfrac{-x^3 - 3x + 1}{4x^3 + 5x^2 - x} = \lim\limits_{x \to -\infty} \dfrac{-1 - \frac{3}{x^2} + \frac{1}{x^3}}{4 + \frac{5}{x} - \frac{1}{x^2}} = \dfrac{-1}{4} = -\dfrac{1}{4}$. The limit is $-\dfrac{1}{4}$.

45. The function $f(x) = \dfrac{1}{x - 5}$ has the properties that $\lim\limits_{x \to 5^+} f(x) = \infty$ and $\lim\limits_{x \to 5^-} f(x) = -\infty$.

46. The function $f(x) = \dfrac{x^3}{x^4 + 1}$ has the properties that $f(x)$ is a quotient of two polynomials of degree greater than

    2 and $\lim\limits_{x \to \infty} f(x) = 0$.

47. The function $f(x) = -x^3$ has the properties that $f(x)$ is a polynomial and $\lim\limits_{x \to \infty} f(x) = -\infty$.

48. The function $f(x) = \dfrac{1}{(x - 5)^2}$ has the properties that $\lim\limits_{x \to 5^+} f(x) = \infty$ and $\lim\limits_{x \to 5^-} f(x) = \infty$.

49. The functions $f(x) = 2x$ and $g(x) = x$ have the properties that $\lim\limits_{x \to \infty} f(x) = \infty$, $\lim\limits_{x \to \infty} g(x) = \infty$, and

    $\lim\limits_{x \to \infty} [f(x) - g(x)] = \infty$.

50. The functions $f(x) = 2 + x$ and $g(x) = x$ have the properties that $\lim\limits_{x \to \infty} f(x) = \infty$, $\lim\limits_{x \to \infty} g(x) = \infty$, and

    $\lim\limits_{x \to \infty} [f(x) - g(x)] = 2$.

51. The functions $f(x) = \dfrac{1}{x}$ and $g(x) = x^2$ have the properties that $\lim\limits_{x \to \infty} f(x) = 0$, $\lim\limits_{x \to \infty} g(x) = \infty$, and

    $\lim\limits_{x \to \infty} [f(x) \cdot g(x)] = \infty$.

52. The functions $f(x) = \dfrac{1}{x^2}$ and $g(x) = x$ have the properties that $\lim\limits_{x \to \infty} f(x) = 0$, $\lim\limits_{x \to \infty} g(x) = \infty$, and

    $\lim\limits_{x \to \infty} [f(x) \cdot g(x)] = 0$.

53. If the graph of $y = f(x)$ has oblique asymptote $y = \dfrac{1}{2}x + 3$, then $\lim\limits_{x \to \infty} f(x) = \infty$.

54. If the graph of $y = f(x)$ has oblique asymptote $y = -2x + 3$, then $\lim\limits_{x \to \infty} f(x) = -\infty$.

55. (a) Graphing $y_1 = xe^{-x}$ in the window $0 \le x \le 10, -.1 \le y \le .5$ (not shown), shows that $\lim\limits_{x \to \infty} (xe^{-x}) = 0$.

    (b) Graphing $y_1 = x^2 e^{-x}$ in the window $0 \le x \le 20, -.1 \le y \le .7$ (not shown), shows that $\lim\limits_{x \to \infty} (x^2 e^{-x}) = 0$.

    (c) Graphing $y_1 = x^{10} e^{-x}$ in the window $0 \le x \le 25, 0 \le y \le 500{,}000$ (not shown), shows that

    $\lim\limits_{x \to \infty} (x^{10} e^{-x}) = 0$. It would seem that $\lim\limits_{x \to \infty} (x^n e^{-x}) = 0$ for any positive integer n.

56. (a) Graphing $y_1 = (\ln x)/x$ in the window $0 \le x \le 1000, -.1 \le y \le .5$ (not shown), shows that $\lim\limits_{x \to \infty} \dfrac{\ln x}{x} = 0$.

    (b) Graphing $y_1 = (\ln x)^2/x$ in the window $0 \le x \le 10{,}000, -.1 \le y \le .5$ (not shown), shows that

    $\lim\limits_{x \to \infty} \dfrac{(\ln x)^2}{x} = 0$.

    (c) Graphing $y_1 = ((\ln x)^{10})/x$ in the window $0 \le x \le 10^{17}, -1 \le y \le 5$ (not shown), shows that

    $\lim\limits_{x \to \infty} \dfrac{(\ln x)^{10}}{x} = 0$. It would seem that $\lim\limits_{x \to \infty} \dfrac{(\ln x)^n}{x} = 0$ for any positive integer n.

57. From the graph in the figure $\lim\limits_{t \to 1^-} f(t)$ is the quantity of the drug in the body just prior to the second injection

    and $\lim\limits_{t \to 1^+} f(t)$ is the quantity of the drug in the body immediately after the second injection.

58. $\lim\limits_{t \to \infty} f(t) = 176$. The limit is 176 feet per second. This limit represents the terminal speed of a skydiver in a

    free fall.

59. $\lim\limits_{t\to\infty} p(t) = 12$. The limit is \$12. This limit represents the long term ultimate price for the commodity.

60. As $x \to \infty$, $\sin x$ continuously oscillates between -1 and 1 without approaching a particular value. Therefore, $\lim\limits_{t\to\infty} \sin x$ does not exist.

61. $f(x) = a_n x^n + a_{n-1} x^{n-1} + \cdots + a_1 x + a_0 = a_n x^n \left(1 + \dfrac{a_{n-1}}{a_n} \cdot \dfrac{1}{x} + \dfrac{a_{n-2}}{a_n} \cdot \dfrac{1}{x^2} + \cdots + \dfrac{a_1}{a_n} \cdot \dfrac{1}{x^{n-1}} + \dfrac{a_0}{a_n} \cdot \dfrac{1}{x^n} \right)$

62. $\lim\limits_{x\to\infty} f(x) = \left[\lim\limits_{x\to\infty} (a_n x^n) \right] \cdot \left[\lim\limits_{x\to\infty} \left(1 + \dfrac{a_{n-1}}{a_n} \cdot \dfrac{1}{x} + \cdots + \dfrac{a_0}{a_n} \cdot \dfrac{1}{x^n} \right) \right] = \left[a_n \lim\limits_{x\to\infty} x^n \right] \cdot [1] = a_n \lim\limits_{x\to\infty} x^n$

$\lim\limits_{x\to-\infty} f(x) = \left[\lim\limits_{x\to-\infty} (a_n x^n) \right] \cdot \left[\lim\limits_{x\to-\infty} \left(1 + \dfrac{a_{n-1}}{a_n} \cdot \dfrac{1}{x} + \cdots + \dfrac{a_0}{a_n} \cdot \dfrac{1}{x^n} \right) \right] = \left[a_n \lim\limits_{x\to-\infty} x^n \right] \cdot [1] = a_n \lim\limits_{x\to-\infty} x^n$

63. Since $\lim\limits_{x\to\infty} f(x) = a_n \lim\limits_{x\to\infty} x^n$ and $\lim\limits_{x\to-\infty} f(x) = a_n \lim\limits_{x\to-\infty} x^n$,

    (a) if a_n is positive and n is even, then $\lim\limits_{x\to\infty} f(x) = \infty$ and $\lim\limits_{x\to-\infty} f(x) = \infty$.

    (b) if a_n is negative and n is even, then $\lim\limits_{x\to\infty} f(x) = -\infty$ and $\lim\limits_{x\to-\infty} f(x) = -\infty$.

    (c) if a_n is positive and n is odd, then $\lim\limits_{x\to\infty} f(x) = \infty$ and $\lim\limits_{x\to-\infty} f(x) = -\infty$.

    (d) if a_n is negative and n is odd, then $\lim\limits_{x\to\infty} f(x) = -\infty$ and $\lim\limits_{x\to-\infty} f(x) = \infty$.

64. (a) ∪   (b) ∩   (c) ⤴   (d) ⤵

Reviewing Basic Concepts (Sections 12.1—12.3)

1. (a) Based on the intercept shown in the graph, $f(x) = \dfrac{2}{3}x + 1$, $\lim\limits_{x\to3} f(x) = \dfrac{2}{3}(3) + 1 = 3$. The limit is 3.

    (b) Even though $f(x)$ is not defined at $x = 2$, the graph shows that $\lim\limits_{x\to2} f(x) = 4$. The limit is 4.

    (c) Even though $f(x)$ is not defined at $x = 0$, the graph shows that $\lim\limits_{x\to0} f(x) = 0$. The limit is 0.

    (d) The graph shows that $\lim\limits_{x\to3} g(x) = 2$. The limit is 2.

2. (a) The graph shows that $\lim\limits_{x\to-2^-} f(x) = -1$ and that $\lim\limits_{x\to-2^+} f(x) = -\dfrac{1}{2}$ and that $\lim\limits_{x\to-2} f(x)$ does not exist.

    (b) The graph shows that $\lim\limits_{x\to-1^-} f(x) = -\dfrac{1}{2}$ and that $\lim\limits_{x\to-1^+} f(x) = -\dfrac{1}{2}$ and that $\lim\limits_{x\to-1} f(x) = -\dfrac{1}{2}$.

3. (a) The graph shows that $\lim\limits_{x\to1^-} f(x) = 1$ and that $\lim\limits_{x\to1^+} f(x) = 1$ and, even though $f(1) = 2$, $\lim\limits_{x\to1} f(x) = 1$.

    (b) The graph shows that $\lim\limits_{x\to2^-} f(x) = 0$ and that $\lim\limits_{x\to2^+} f(x) = 0$ and that $\lim\limits_{x\to2} f(x) = 0$.

4. (a) The graph shows that $\lim\limits_{x\to\infty} f(x) = 3$. The limit is 3.

    (b) The graph shows that $\lim\limits_{x\to-\infty} g(x) = \infty$. The limit is ∞ (or does not exist).

5. The completed table is shown in Figure 5. From the table of values, $\lim\limits_{x\to4} \dfrac{\sqrt{x} - 2}{x - 4} = \dfrac{1}{4}$.

x	3.9	3.99	3.999	4.001	4.01	4.1
$f(x)$	.251582	.250156	.250015	.249984	.249844	.248457

Figure 5

6. (a) $\lim\limits_{x\to 8}\left[f(x) - g(x)\right] = \lim\limits_{x\to 8} f(x) - \lim\limits_{x\to 8} g(x) = 32 - 4 = 28$. The limit is 28.

   (b) $\lim\limits_{x\to 8}\left[g(x)\cdot f(x)\right] = \left[\lim\limits_{x\to 8} g(x)\right]\cdot\left[\lim\limits_{x\to 8} f(x)\right] = 4\cdot 32 = 128$. The limit is 128.

   (c) $\lim\limits_{x\to 8}\dfrac{f(x)}{g(x)} = \dfrac{\lim\limits_{x\to 8} f(x)}{\lim\limits_{x\to 8} g(x)} = \dfrac{32}{4} = 8$. The limit is 8.

   (d) $\lim\limits_{x\to 8}\left[\log_2 f(x)\right] = \log_2\left[\lim\limits_{x\to 8} f(x)\right] = \log_2(32) = 5$. The limit is 5.

   (e) $\lim\limits_{x\to 8}\sqrt{f(x)} = \sqrt{\lim\limits_{x\to 8} f(x)} = \sqrt{32} = 4\sqrt{2}$. The limit is $4\sqrt{2}$.

   (f) $\lim\limits_{x\to 8}\sqrt[3]{g(x)} = \sqrt[3]{\lim\limits_{x\to 8} g(x)} = \sqrt[3]{4}$. The limit is $\sqrt[3]{4}$.

   (g) $\lim\limits_{x\to 8} 2^{g(x)} = 2^{\lim\limits_{x\to 8} g(x)} = 2^4 = 16$. Thus, the limit is 16.

   (h) $\lim\limits_{x\to 8}\left[1 + f(x)\right]^2 = \left[1 + \lim\limits_{x\to 8} f(x)\right]^2 = [1 + 32]^2 = 1089$. The limit is 1089.

   (i) $\lim\limits_{x\to 8}\dfrac{f(x) - g(x)}{4g(x)} = \dfrac{\lim\limits_{x\to 8} f(x) - \lim\limits_{x\to 8} g(x)}{4\lim\limits_{x\to 8} g(x)} = \dfrac{32 - 4}{4(4)} = \dfrac{7}{4}$. The limit is $\dfrac{7}{4}$.

   (j) $\lim\limits_{x\to 8}\dfrac{2g(x) + 3}{1 + f(x)} = \dfrac{2\lim\limits_{x\to 8} g(x) + 3}{1 + \lim\limits_{x\to 8} f(x)} = \dfrac{2(4) + 3}{1 + 32} = \dfrac{1}{3}$. The limit is $\dfrac{1}{3}$.

7. $\lim\limits_{x\to 3}\dfrac{x^2 - 9}{x + 3} = \lim\limits_{x\to 3}\dfrac{(x + 3)(x - 3)}{x - 3} = 6$. Even though $\dfrac{x^2 - 9}{x - 3}$ does not exist at $x = 3$, the limit is 6.

8. $\lim\limits_{x\to\infty}\dfrac{2x^2 - 1}{3x^4 + 5} = \lim\limits_{x\to\infty}\dfrac{\frac{2}{x^2} - \frac{1}{x^4}}{3 + \frac{5}{x^4}} = \dfrac{0}{3} = 0$. The limit is 0.

9. $\lim\limits_{x\to -\infty}\dfrac{2x^3 - x + 3}{6x^3 + 4x - 9} = \lim\limits_{x\to -\infty}\dfrac{2 - \frac{1}{x^2} + \frac{3}{x^3}}{6 + \frac{4}{x^2} - \frac{9}{x^3}} = \dfrac{2}{6} = \dfrac{1}{3}$. The limit is $\dfrac{1}{3}$.

10. $\lim\limits_{x\to 4^-} f(x) = \lim\limits_{x\to 4^-}(x + 2) = 6$ and $\lim\limits_{x\to 4^+} f(x) = \lim\limits_{x\to 4^+}(x^2 - 6) = 10$. Therefore $\lim\limits_{x\to 4} f(x)$ does not exist.

12.4: Tangent Lines and Derivatives

1. The tangent line passes through the points (5, 3) and (6, 5) and has slope $m = \dfrac{5 - 3}{6 - 5} = 2$.

2. The tangent line passes through the points (2, 2) and (0, 4) and has slope $m = \dfrac{4 - 2}{0 - 2} = -1$.

3. The tangent line passes through the points (−2, 2) and (3, 3) and has slope $m = \dfrac{3 - 2}{3 - (-2)} = \dfrac{1}{5}$.

4. The tangent line is a horizontal line and has slope $m = 0$.

5. $m = \lim\limits_{x\to 4}\dfrac{x^2 - 4^2}{x - 4} = \lim\limits_{x\to 4}\dfrac{(x + 4)(x - 4)}{x - 4} = 8$. The slope of the tanget line is 8.

6. $m = \lim\limits_{x\to 2}\dfrac{(x^2 + 5) - (2^2 + 5)}{x - 2} = \lim\limits_{x\to 2}\dfrac{(x + 2)(x - 2)}{x - 2} = 4$. The slope of the tanget line is 4.

7. $m = \lim\limits_{x\to -2}\dfrac{(-4x^2 + 11x) - [-4(-2)^2 + 11(-2)]}{x - (-2)} = \lim\limits_{x\to -2}\dfrac{-4x^2 + 11x + 38}{x + 2} = \lim\limits_{x\to -2}\dfrac{(-4x + 19)(x + 2)}{x + 2} = 27$.

The slope of the tangent line is 27.

8. $m = \lim\limits_{x\to -1} \dfrac{(6x^2 - 4x) - [6(-1)^2 - 4(-1)]}{x - (-1)} = \lim\limits_{x\to -1} \dfrac{6x^2 - 4x - 10}{x + 1} = \lim\limits_{x\to -1} \dfrac{(6x - 10)(x + 1)}{x + 1} = -16.$

The slope of the tangent line is -16.

9. $m = \lim\limits_{x\to 4} \dfrac{\left(-\frac{2}{x}\right) - \left(-\frac{2}{4}\right)}{x - 4} = \lim\limits_{x\to 4} \dfrac{-\frac{2}{x} + \frac{1}{2}}{x - 4} = \lim\limits_{x\to 4} \dfrac{\frac{x-4}{2x}}{x - 4} = \dfrac{1}{8}.$ The slope of the tangent line is $\dfrac{1}{8}$.

10. $m = \lim\limits_{x\to -1} \dfrac{\left(\frac{6}{x}\right) - \left(-\frac{6}{1}\right)}{x - (-1)} = \lim\limits_{x\to -1} \dfrac{\frac{6(x+1)}{x}}{x + 1} = = -6.$ The slope of the tangent line is -6.

11. $m = \lim\limits_{x\to 1} \dfrac{-3\sqrt{x} + 3\sqrt{1}}{x - 1} = \lim\limits_{x\to 1} \dfrac{-3(\sqrt{x} - 1)}{(\sqrt{x} - 1)(\sqrt{x} + 1)} = -\dfrac{3}{2}.$ The slope of the tangent line is $-\dfrac{3}{2}$.

12. $m = \lim\limits_{x\to 1} \dfrac{(x^3 - 1^3)}{x - 1} = \lim\limits_{x\to 1} \dfrac{(x - 1)(x^2 + x + 1)}{x - 1} = 3.$ The slope of the tanget line is 3.

13. $m = \lim\limits_{x\to 3} \dfrac{(x^2 + 2x) - [3^2 + 2(3)]}{x - 3} = \lim\limits_{x\to 3} \dfrac{x^2 + 2x - 15}{x - 3} = \lim\limits_{x\to 3} \dfrac{(x + 5)(x - 3)}{x - 3} = 8.$

$f(3) = 15;\ y - 15 = 8(x - 3) = 8x - 24.$ The tangent line is $y = 8x - 9.$ Graphing $y_1 = x^2 + 2x$ and

$y_2 = 8x - 9$ in the window $0 \le x \le 6, 0 \le y \le 50$ (not shown), shows this result.

14. $m = \lim\limits_{x\to -1} \dfrac{(6 - x^2) - [6 - (-1)^2]}{x - (-1)} = \lim\limits_{x\to -1} \dfrac{1 - x^2}{x + 1} = \lim\limits_{x\to -1} \dfrac{(1 + x)(1 - x)}{1 + x} = 2.$

$f(-1) = 5;\ y - 5 = 2(x - (-1)) = 2x + 2.$ The tangent line is $y = 2x + 7.$ Graphing $y_1 = 6 - x^2$ and

$y_2 = 2x + 7$ in the window $-2 \le x \le 0, 0 \le y \le 10$ (not shown), shows this result.

15. $m = \lim\limits_{x\to 2} \dfrac{\frac{5}{x} - \frac{5}{2}}{x - 2} = \lim\limits_{x\to 2} \dfrac{\frac{-5(x-2)}{2x}}{x - 2} = -\dfrac{5}{4}.$ $f(2) = \dfrac{5}{2};\ y - \dfrac{5}{2} = -\dfrac{5}{4}(x - 2) = -\dfrac{5}{4}x + \dfrac{5}{2}.$

The tangent line is $y = -\dfrac{5}{4}x + 5.$ Graphing $y_1 = 5/x$ and $y_2 = -(5/4)x + 5$ in the window

$0 \le x \le 4, 0 \le y \le 10$ (not shown), shows this result.

16. $m = \lim\limits_{x\to 1} \dfrac{\left(\frac{-3}{x+1}\right) - \left(\frac{-3}{1+1}\right)}{x - 1} = \lim\limits_{x\to 1} \dfrac{\frac{3(x-1)}{2(x+1)}}{x - 1} = \dfrac{3}{4}.$ $f(1) = -\dfrac{3}{2};\ y + \dfrac{3}{2} = \dfrac{3}{4}(x - 1) = \dfrac{3}{4}x - \dfrac{3}{4}.$

The tangent line is $y = \dfrac{3}{4}x - \dfrac{9}{4}.$ Graphing $y_1 = -3/(x + 1)$ and $y_2 = (3/4)x - 9/4$ in the window

$0 \le x \le 2, -5 \le y \le 0$ (not shown), shows this result.

17. $m = \lim\limits_{x\to 9} \dfrac{4\sqrt{x} - 4\sqrt{9}}{x - 9} = \lim\limits_{x\to 9} \dfrac{4(\sqrt{x} - 3)}{(\sqrt{x} - 3)(\sqrt{x} + 3)} = \dfrac{2}{3}.$ $f(9) = 12;\ y - 12 = \dfrac{2}{3}(x - 9) = \dfrac{2}{3}x - 6.$

The tangent line is $y = \dfrac{2}{3}x + 6.$ Graphing $y_1 = 4\sqrt{x}$ and $y_2 = (2/3)x + 6$ in the window

$0 \le x \le 18, 0 \le y \le 20$ (not shown), shows this result.

18. $m = \lim\limits_{x\to 25} \dfrac{\sqrt{x} - \sqrt{25}}{x - 25} = \lim\limits_{x\to 25} \dfrac{\sqrt{x} - 5}{(\sqrt{x} - 5)(\sqrt{x} + 5)} = \dfrac{1}{10}.$ $f(25) = 5;\ y - 5 = \dfrac{1}{10}(x - 25) = \dfrac{1}{10}x - \dfrac{5}{2}.$

The tangent line is $y = \dfrac{1}{10}x + \dfrac{5}{2}.$ Graphing $y_1 = \sqrt{x}$ and $y_2 = (1/10)x + 5/2$ in the window

$0 \le x \le 50, 0 \le y \le 10$ (not shown), shows this result.

19. Graphing $y_1 = 5$, we see that the graph of $f(x)$ is a horizontal line which never changes. Thus, the rate of

change of $f(x)$ is 0 everywhere and $f'(2) = 0.$

20. Graphing $y_1 = x$, we see that the graph of $f(x)$ is a straight line with slope $m = 1$. Thus, the rate of change of $f(x)$ is 1 everywhere and $f'(2) = 1$.

21. Graphing $y_1 = -x$, we see that the graph of $f(x)$ is a straight line with slope $m = -1$. Thus, the rate of change of $f(x)$ is -1 everywhere and $f'(2) = -1$.

22. Graphing $y_1 = 3x + 4$, we see that the graph of $f(x)$ is a straight line with slope $m = 3$. Thus, the rate of change of $f(x)$ is 3 everywhere and $f'(2) = 3$.

23. On a TI-86, nDer$(e^x, x, 0) = 1.00000016667$. Thus, we conclude that $f'(0)$ is 1.

24. On a TI-86, nDer$(\sin x, x, 0) = .999999833333$. Thus, we conclude that $f'(0)$ is 1.

25. On a TI-86, nDer$(10x/(1 + .25x^2), x, 2) = 6.25\,E^{-7}$. Thus, we conclude that $f'(2)$ is 0.

26. On a TI-86, nDer$(1/(1 + x^2), x, 0) = 0$. Thus, we conclude that $f'(0)$ is 0.

27. On a TI-86, nDer$(x \cos x, x, \pi/4) = .151746152925$. Thus, we conclude that $f'\left(\dfrac{\pi}{4}\right)$ is approximately .1517.

28. On a TI-86, nDer$(xe^x, x, 1) = 5.4365654691$. Thus, we conclude that $f'(1)$ is approximately 5.4366.

29. From the figure, the tangent line at $(0, 1)$ has slope $m = \dfrac{1}{3}$. Thus, $f'(0)$ is $\dfrac{1}{3}$.

30. From the figure, the tangent line at $(1, 0)$ has slope $m = 1$. Thus, $f'(1)$ is 1.

31. Using the first two secant line equations, we get $f(a) = 2.03a - .53$ and $f(a) = 2.02a - .52$. Solving this simultaneously, we get $a = 1$ and $f(a) = 1.5$. The slope of the tangent line is the limit of the slopes of the secant lines as $x \to a$ and, from the figure, this limit appears to be 2. Thus $f'(a)$ is 2.

32. The slope of the secant line through $(1, .8)$ and $(1.6, 1.4)$ is $m_1 = 1$. The slope of the secant line through $(1, .8)$ and $(1.4, 1.3)$ is $m_2 = 1.25$. The slope of the secant line through $(1, .8)$ and $(1.2, 1.1)$ is $m_3 = 1.5$. The slope of the tangent line through $(1, .8)$ is the limit of the slopes of the secant lines as $x \to 1$ and we would conclude that it is a value slightly larger than 1.5.

33. $s'(3) = \lim\limits_{t \to 3} \dfrac{9t^2 - 9(3)^2}{t - 3} = \lim\limits_{t \to 3} \dfrac{9(t + 3)(t - 3)}{t - 3} = 9(6) = 54$. The cars velocity when $t = 3$ seconds is 54 feet/second.

34. $s'(3) = \lim\limits_{t \to 3} \dfrac{50\sqrt{t} - 50\sqrt{3}}{t - 3} = \lim\limits_{t \to 3} \dfrac{50(\sqrt{t} - \sqrt{3})}{(\sqrt{t} - \sqrt{3})(\sqrt{t} + \sqrt{3})} = \dfrac{50}{2\sqrt{3}} = \dfrac{25}{\sqrt{3}} \approx 14.4$. The cars velocity when $t = 3$ seconds is approximately 14.4 feet/second.

35. $s'(3) = \lim\limits_{t \to 3} \dfrac{(3t^3 - t^2) - [3(3)^3 - (3)^2]}{t - 3} = \lim\limits_{t \to 3} \dfrac{3t^3 - t^2 - 72}{t - 3} = \lim\limits_{t \to 3} \dfrac{(t - 3)(3t^2 + 8t + 24)}{t - 3} = $ $3(3)^2 + 8(3) + 24 = 75$. The cars velocity when $t = 3$ seconds is approximately 75 feet/second.

36. $s'(3) = \lim\limits_{t \to 3} \dfrac{(4t^2 + 5t + 1) - [4(3)^2 + 5(3) + 1]}{t - 3} = \lim\limits_{t \to 3} \dfrac{4t^2 + 5t - 51}{t - 3} = \lim\limits_{t \to 3} \dfrac{(t - 3)(4t + 17)}{t - 3} = $ $4(3) + 17 = 29$. The cars velocity when $t = 3$ seconds is approximately 29 feet/second.

37. The tangent line when $t = 10$ years passes through the points $(12, 10)$ and $(7, 6)$ and, therefore, has slope $m = \dfrac{10 - 6}{12 - 7} = \dfrac{4}{5}$. The interest rates were rising at a rate of $\dfrac{4}{5}\%$ per year on January 1, 1989.

38. The tangent line when $t = 14$ years passes through the points (6, 120) and (24, 300) and, therefore, has slope
$$m = \frac{300 - 120}{24 - 6} = 10.$$ The balance is growing at a rate of \$10 per year when $t = 14$ years.

39. $W'(4) = \lim_{t \to 4} \frac{.1t^2 - .1(4)^2}{t - 4} = \lim_{t \to 4} \frac{.1(t - 4)(t + 4)}{t - 4} = .8.$ The tumor is growing at a rate of .8 grams/week when $t = 4$ weeks.

40. $S(t) = -16t^2 + 128t + 5;$

$$S'(2) = \lim_{t \to 2} \frac{(-16t^2 + 128t + 5) - [-16(2)^2 + 128(2) + 5]}{t - 2} = \lim_{t \to 2} \frac{-16t^2 + 128t - 192}{t - 2} =$$

$\lim_{t \to 2} \frac{(t - 2)(-16t + 96)}{t - 2} = 64.$ The velocity of the ball when $t = 2$ seconds is 64 feet/second.

41. (a) $20 = t^2 + t \Rightarrow t^2 + t - 20 = 0 \Rightarrow (t + 5)(t - 4) = 0 \Rightarrow t = -5$ or $t = 4.$ The helicopter will reach a height of 20 feet in 4 seconds.

   (b) $s'(4) = \lim_{t \to 4} \frac{(t^2 + t) - [4^2 + 4]}{t - 4} = \lim_{t \to 4} \frac{t^2 + t - 20}{t - 4} = \lim_{t \to 4} \frac{(t + 5)(t - 4)}{t - 4} = 9.$ The velocity of the helicopter when it is 20 feet above the ground will be 9 feet/second.

42. $P'(2) = \lim_{x \to 2} \frac{(2x^2 - 5x + 6) - [2(2)^2 - 5(2) + 6]}{x - 2} = \lim_{x \to 2} \frac{2x^2 - 5x + 2}{x - 2} = \lim_{x \to 2} \frac{(x - 2)(2x - 1)}{x - 2} = 3.$
   The profit margin at $x = 2$ is \$300 per item.

43. $R'(1000) = \lim_{x \to 1000} \frac{(10x - .002x^2) - [10(1000) - .002(1000)^2]}{x - 1000} = \lim_{x \to 1000} \frac{-.002x^2 + 10x - 8000}{x - 1000} =$

$\lim_{x \to 1000} \frac{(x - 1000)(-.002x + 8)}{x - 1000} = 6.$ The marginal revenue at $x = 1000$ is \$6000 per unit.

44. $p'(3) = \lim_{t \to 3} \frac{(t^2 + 2t) - [3^2 + 2(3)]}{t - 3} = \lim_{t \to 3} \frac{t^2 + 2t - 15}{t - 3} = \lim_{t \to 3} \frac{(t - 3)(t + 5)}{t - 3} = 8.$

The instantaneous rate of change of the percent of the population infected at $t = 3$ days is 8% per day.

45. Graph $y_1 = 100,000/(1 + 9.134(.8)^x)$ in the window $0 \le x \le 32, 0 \le y \le 100,000,$ to obtain the graph shown in Figure 45. nDer$(y_1, x, 8) = 5331.9986515.$ Thus, we conclude that the rumor is spreading by about 5332 people per day when $t = 8$ days.

[0, 32] by [0, 100,000]   [0, 16] by [0, 70]
Xscl = 5   Yscl = 10,000   Xscl = 1   Yscl = 10

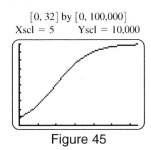
Figure 45

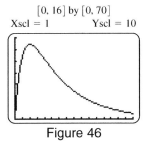
Figure 46

46. Graph $y_1 = 120(e^{-2x} - e^{-x})$ in the window $0 \le x \le 16, 0 \le y \le 70,$ to obtain the graph shown in Figure 46. Then, using the table with $x = 6,$ we see that approximately 35.85 units of the drug are in the bloodstream when $t = 6$ hours. Finally nDer$(y_1, x, 6) = -6.9312108235$ and we conclude that the level of the drug in the bloodstream is decreasing by approximately 6.93 units per hour when $t = 6$ hours.

47. $f(8) = \dfrac{200{,}000(\frac{8}{1200})(1 + \frac{8}{1200})^{360}}{(1 + \frac{8}{1200})^{360} - 1} \approx 1467.53.$ The monthly payment is \$1467.53.

48. $\text{nDer}(200{,}000(x/1200)(1 + x/1200)^{360}/((1 + x/1200)^{360-1}), x, 8) \approx 139.42.$

49. $1467.53 = 1349.42(8) + b \Rightarrow b = 352.17.$ The equation of the tangent line to $f(x)$ is $y = 139.42x + 352.17.$

50. (a) $f(8.25) - f(8) \approx (139.42)(.25) = 34.855.$ If the interest rate rises to 8.25%, the monthly payment will

    increase by approximately \$34.86.

    (b) $f(7.5) - f(8) \approx (139.42)(-.5) = -69.710.$ If the interest rate falls to 7.5%, the monthly payment will

    decrease by approximately \$69.71.

12.5: Area and the Definite Integral

1. (a) $f(x_1) = 2(0) + 1 = 1;\ f(x_2) = 2(2) + 1 = 5;\ f(x_3) = 2(4) + 1 = 9;\ f(x_4) = 2(6) + 1 = 13;$

    $\displaystyle\sum_{i=1}^{4} f(x_i)\Delta x = 1(2) + 5(2) + 9(2) + 13(2) = 56.$ The sum is 56.

    (b) $\displaystyle\int_{0}^{8} (2x + 1)\,dx$ is the definite integral approximated by the sum in part (a).

2. The intervals will be $0 - 1, 1 - 2, 2 - 3,$ and $3 - 4$ with midpoints $x_1 = .5, x_2 = 1.5, x_3 = 2.5,$ and $x_4 = 3.5.$

    $f(x_1) \approx 3.97;\ f(x_2) \approx 3.71;\ f(x_3) \approx 3.12;\ f(x_4) \approx 1.94;$

    $\displaystyle\sum_{i=1}^{4} f(x_i)\Delta x = 3.97(1) + 3.71(1) + 3.12(1) + 1.94(1) = 12.74.$ The area above the x-axis bounded by the

    graph of $f(x) = \sqrt{16 - x^2}$ is 12.74.

3. (a) $x_1 = 1;\ x_2 = 2;\ x_3 = 3;\ x_4 = 4;$

    $f(x_1) = 3(1) + 2 = 5;\ f(x_2) = 3(2) + 2 = 8;\ f(x_3) = 3(3) + 2 = 11;\ f(x_4) = 3(4) + 2 = 14;$

    $\displaystyle\sum_{i=1}^{4} f(x_i)\Delta x = 5(1) + 8(1) + 11(1) + 14(1) = 38.$ Using left endpoints, the area is approximately 38.

    (b) $x_1 = 2;\ x_2 = 3;\ x_3 = 4;\ x_4 = 5;$

    $f(x_1) = 3(2) + 2 = 8;\ f(x_2) = 3(3) + 2 = 11;\ f(x_3) = 3(4) + 2 = 14;\ f(x_4) = 3(5) + 2 = 17;$

    $\displaystyle\sum_{i=1}^{4} f(x_i)\Delta x = 8(1) + 11(1) + 14(1) + 17(1) = 50.$ Using right endpoints, the area is approximately 50.

    (c) $\dfrac{38 + 50}{2} = 44.$ Averaging the answers to (a) and (b), the area is approximately 44.

    (d) $x_1 = 1.5;\ x_2 = 2.5;\ x_3 = 3.5;\ x_4 = 4.5;\ f(x_1) = 3(1.5) + 2 = 6.5;\ f(x_2) = 3(2.5) + 2 = 9.5;$

    $f(x_3) = 3(3.5) + 2 = 12.5;\ f(x_4) = 3(4.5) + 2 = 15.5;$

    $\displaystyle\sum_{i=1}^{4} f(x_i)\Delta x = 6.5(1) + 9.5(1) + 12.5(1) + 15.5(1) = 44.$ Using midpoints, the area is approximately 44.

4. (a) $x_1 = 2$; $x_2 = 2.5$; $x_3 = 3$; $x_4 = 3.5$;

$f(x_1) = 2 + 5 = 7$; $f(x_2) = 2.5 + 5 = 7.5$; $f(x_3) = 3 + 5 = 8$; $f(x_4) = 3.5 + 5 = 8.5$;

$\displaystyle\sum_{i=1}^{4} f(x_i)\Delta x = 7(.5) + 7.5(.5) + 8(.5) + 8.5(.5) = 15.5$. Using left endpoints, the area is approximately 15.5.

(b) $x_1 = 2.5$; $x_2 = 3$; $x_3 = 3.5$; $x_4 = 4$;

$f(x_1) = 2.5 + 5 = 7.5$; $f(x_2) = 3 + 5 = 8$; $f(x_3) = 3.5 + 5 = 8.5$; $f(x_4) = 4 + 5 = 9$;

$\displaystyle\sum_{i=1}^{4} f(x_i)\Delta x = 7.5(.5) + 8(.5) + 8.5(.5) + 9(.5) = 16.5$. Using right endpoints, the area is approximately 16.5.

(c) $\dfrac{15.5 + 16.5}{2} = 16$. Averaging the answers to (a) and (b), the area is approximately 16.

(d) $x_1 = 2.25$; $x_2 = 2.75$; $x_3 = 3.25$; $x_4 = 3.75$; $f(x_1) = 2.25 + 5 = 7.25$; $f(x_2) = 2.75 + 5 = 7.75$;

$f(x_3) = 3.25 + 5 = 8.25$; $f(x_4) = 3.75 + 5 = 8.75$;

$\displaystyle\sum_{i=1}^{4} f(x_i)\Delta x = 7.25(.5) + 7.75(.5) + 8.25(.5) + 8.75(.5) = 16$. Using midpoints, the area is approximately 16.

5. (a) $x_1 = 0$; $x_2 = 1$; $x_3 = 2$; $x_4 = 3$;

$f(x_1) = 0 + 2 = 2$; $f(x_2) = 1 + 2 = 3$; $f(x_3) = 2 + 2 = 4$; $f(x_4) = 3 + 2 = 5$;

$\displaystyle\sum_{i=1}^{4} f(x_i)\Delta x = 2(1) + 3(1) + 4(1) + 5(1) = 14$. Using left endpoints, the area is approximately 14.

(b) $x_1 = 1$; $x_2 = 2$; $x_3 = 3$; $x_4 = 4$;

$f(x_1) = 1 + 2 = 3$; $f(x_2) = 2 + 2 = 4$; $f(x_3) = 3 + 2 = 5$; $f(x_4) = 4 + 2 = 6$;

$\displaystyle\sum_{i=1}^{4} f(x_i)\Delta x = 3(1) + 4(1) + 5(1) + 6(1) = 18$. Using right endpoints, the area is approximately 18.

(c) $\dfrac{14 + 18}{2} = 16$. Averaging the answers to (a) and (b), the area is approximately 16.

(d) $x_1 = .5$; $x_2 = 1.5$; $x_3 = 2.5$; $x_4 = 3.5$; $f(x_1) = .5 + 2 = 2.5$; $f(x_2) = 1.5 + 2 = 3.5$;

$f(x_3) = 2.5 + 2 = 4.5$; $f(x_4) = 3.5 + 2 = 5.5$;

$\displaystyle\sum_{i=1}^{4} f(x_i)\Delta x = 2.5(1) + 3.5(1) + 4.5(1) + 5.5(1) = 16$. Using midpoints, the area is approximately 16.

6. (a) $x_1 = 1$; $x_2 = 1.5$; $x_3 = 2$; $x_4 = 2.5$;

$f(x_1) = 3 + 1 = 4$; $f(x_2) = 3 + 1.5 = 4.5$; $f(x_3) = 3 + 2 = 5$; $f(x_4) = 3 + 2.5 = 5.5$;

$\displaystyle\sum_{i=1}^{4} f(x_i)\Delta x = 4(.5) + 4.5(.5) + 5(.5) + 5.5(.5) = 9.5$. Using left endpoints, the area is approximately 9.5.

(b) $x_1 = 1.5$; $x_2 = 2$; $x_3 = 2.5$; $x_4 = 3$;

$f(x_1) = 3 + 1.5 = 4.5$; $f(x_2) = 3 + 2 = 5$; $f(x_3) = 3 + 2.5 = 5.5$; $f(x_4) = 3 + 3 = 6$;

$\displaystyle\sum_{i=1}^{4} f(x_i)\Delta x = 4.5(.5) + 5(.5) + 5.5(.5) + 6(.5) = 10.5$. Using right endpoints, the area is approximately 10.5.

(c) $\dfrac{9.5 + 10.5}{2} = 10$. Averaging the answers to (a) and (b), the area is approximately 10.

(d) $x_1 = 1.25$; $x_2 = 1.75$; $x_3 = 2.25$; $x_4 = 2.75$; $f(x_1) = 3 + 1.25 = 4.25$; $f(x_2) = 3 + 1.75 = 4.75$;

$f(x_3) = 3. + 2.25 = 5.25$; $f(x_4) = 3 + 2.75 = 5.75$;

$\displaystyle\sum_{i=1}^{4} f(x_i)\Delta x = 4.25(.5) + 4.75(.5) + 5.25(.5) + 5.75(.5) = 10$. Using midpoints, the area is approximately 10.

7. (a) $x_1 = 1$; $x_2 = 2$; $x_3 = 3$; $x_4 = 4$; $f(x_1) = 1^2 = 1$; $f(x_2) = 2^2 = 4$; $f(x_3) = 3^2 = 9$; $f(x_4) = 4^2 = 16$;

$\sum_{i=1}^{4} f(x_i)\Delta x = 1(1) + 4(1) + 9(1) + 16(1) = 30$. Using left endpoints, the area is approximately 30.

(b) $x_1 = 2$; $x_2 = 3$; $x_3 = 4$; $x_4 = 5$; $f(x_1) = 2^2 = 4$; $f(x_2) = 3^2 = 9$; $f(x_3) = 4^2 = 16$; $f(x_4) = 5^2 = 25$;

$\sum_{i=1}^{4} f(x_i)\Delta x = 4(1) + 9(1) + 16(1) + 25(1) = 54$. Using right endpoints, the area is approximately 54.

(c) $\dfrac{30 + 54}{2} = 42$. Averaging the answers to (a) and (b), the area is approximately 42.

(d) $x_1 = 1.5$; $x_2 = 2.5$; $x_3 = 3.5$; $x_4 = 4.5$; $f(x_1) = (1.5)^2 = 2.25$; $f(x_2) = (2.5)^2 = 6.25$;

$f(x_3) = (3.5)^2 = 12.25$; $f(x_4) = (4.5)^2 = 20.25$;

$\sum_{i=1}^{4} f(x_i)\Delta x = 2.25(1) + 6.25(1) + 12.25(1) + 20.25(1) = 41$. Using midpoints, the area is approximately 41.

8. (a) $x_1 = -2$; $x_2 = -1$; $x_3 = 0$; $x_4 = 1$;

$f(x_1) = -(-2)^2 + 4 = 0$; $f(x_2) = -(-1)^2 + 4 = 3$; $f(x_3) = -(0)^2 + 4 = 4$; $f(x_4) = -(1)^2 + 4 = 3$;

$\sum_{i=1}^{4} f(x_i)\Delta x = 0(1) + 3(1) + 4(1) + 3(1) = 10$. Using left endpoints, the area is approximately 10.

(b) $x_1 = -1$; $x_2 = 0$; $x_3 = 1$; $x_4 = 2$;

$f(x_1) = -(-1)^2 + 4 = 3$; $f(x_2) = -(0)^2 + 4 = 4$; $f(x_3) = -(1)^2 + 4 = 3$; $f(x_4) = -(2)^2 + 4 = 0$;

$\sum_{i=1}^{4} f(x_i)\Delta x = 3(1) + 4(1) + 3(1) + 0(1) = 10$. Using right endpoints, the area is approximately 10.

(c) $\dfrac{10 + 10}{2} = 10$. Averaging the answers to (a) and (b), the area is approximately 10.

(d) $x_1 = -1.5$; $x_2 = -.5$; $x_3 = .5$; $x_4 = 1.5$; $f(x_1) = -(-1.5)^2 + 4 = 1.75$; $f(x_2) = -(-.5)^2 + 4 = 3.75$;

$f(x_3) = -(.5)^2 + 4 = 3.75$; $f(x_4) = -(1.5)^2 + 4 = 1.75$;

$\sum_{i=1}^{4} f(x_i)\Delta x = 1.75(1) + 3.75(1) + 3.75(1) + 1.75(1) = 11$. Using midpoints, the area is approximately 11.

9. (a) $x_1 = 0$; $x_2 = 1$; $x_3 = 2$; $x_4 = 3$;

$f(x_1) = e^0 - 1 = 0$; $f(x_2) = e^1 - 1 \approx 1.718$; $f(x_3) = e^2 - 1 \approx 6.389$; $f(x_4) = e^3 - 1 \approx 19.086$;

$\sum_{i=1}^{4} f(x_i)\Delta x = 0(1) + 1.718(1) + 6.389(1) + 19.086(1) \approx 27.19$.

Using left endpoints, the area is approximately 27.19.

(b) $x_1 = 1$; $x_2 = 2$; $x_3 = 3$; $x_4 = 4$;

$f(x_1) = e^1 - 1 \approx 1.718$; $f(x_2) = e^2 - 1 \approx 6.389$; $f(x_3) = e^3 - 1 \approx 19.086$; $f(x_4) = e^4 - 1 \approx 53.598$;

$\sum_{i=1}^{4} f(x_i)\Delta x = 1.718(1) + 6.389(1) + 19.086(1) + 53.598(1) \approx 80.79$.

Using right endpoints, the area is approximately 80.79.

(c) $\dfrac{27.19 + 80.79}{2} = 53.99$. Averaging the answers to (a) and (b), the area is approximately 53.99.

(d) $x_1 = .5$; $x_2 = 1.5$; $x_3 = 2.5$; $x_4 = 3.5$; $f(x_1) = e^{.5} - 1 \approx .649$; $f(x_2) = e^{1.5} - 1 \approx 3.482$;

$f(x_3) = e^{2.5} - 1 \approx 11.182$; $f(x_4) = e^{3.5} - 1 \approx 32.115$;

$\sum_{i=1}^{4} f(x_i)\Delta x = .649(1) + 3.482(1) + 11.182(1) + 32.115(1) \approx 47.43$.

Using midpoints, the area is approximately 47.43.

10. (a) $x_1 = -2$; $x_2 = -1$; $x_3 = 0$; $x_4 = 1$;

$f(x_1) = e^{-2} + 1 \approx 1.135$; $f(x_2) = e^{-1} + 1 \approx 1.368$; $f(x_3) = e^0 + 1 = 2$; $f(x_4) = e^1 + 1 \approx 3.718$;

$\sum_{i=1}^{4} f(x_i)\Delta x = 1.135(1) + 1.368(1) + 2(1) + 3.718(1) \approx 8.22$.

Using left endpoints, the area is approximately 8.22.

(b) $x_1 = -1$; $x_2 = 0$; $x_3 = 1$; $x_4 = 2$;

$f(x_1) = e^{-1} + 1 \approx 1.368$; $f(x_2) = e^0 + 1 \approx 2$; $f(x_3) = e^1 + 1 \approx 3.718$; $f(x_4) = e^2 + 1 \approx 8.389$;

$\sum_{i=1}^{4} f(x_i)\Delta x = 1.368(1) + 2(1) + 3.718(1) + 8.389(1) \approx 15.48$.

Using right endpoints, the area is approximately 15.48.

(c) $\dfrac{8.22 + 15.48}{2} = 11.85$. Averaging the answers to (a) and (b), the area is approximately 11.85.

(d) $x_1 = -1.5$; $x_2 = -.5$; $x_3 = .5$; $x_4 = 1.5$; $f(x_1) = e^{-1.5} + 1 \approx 1.223$; $f(x_2) = e^{-.5} + 1 \approx 1.607$;

$f(x_3) = e^{.5} + 1 \approx 2.649$; $f(x_4) = e^{1.5} + 1 \approx 5.482$;

$\sum_{i=1}^{4} f(x_i)\Delta x = 1.223(1) + 1.607(1) + 2.649(1) + 5.482(1) \approx 10.96$.

Using midpoints, the area is approximately 10.96.

11. (a) $x_1 = 1$; $x_2 = 2$; $x_3 = 3$; $x_4 = 4$; $f(x_1) = \dfrac{1}{1} = 1$; $f(x_2) = \dfrac{1}{2}$; $f(x_3) = \dfrac{1}{3}$; $f(x_4) = \dfrac{1}{4}$;

$\sum_{i=1}^{4} f(x_i)\Delta x = (1)(1) + \left(\dfrac{1}{2}\right)(1) + \left(\dfrac{1}{3}\right)(1) + \left(\dfrac{1}{4}\right)(1) \approx \dfrac{25}{12}$.

Using left endpoints, the area is approximately $\dfrac{25}{12}$.

(b) $x_1 = 2$; $x_2 = 3$; $x_3 = 4$; $x_4 = 5$; $f(x_1) = \dfrac{1}{2}$; $f(x_2) = \dfrac{1}{3}$; $f(x_3) = \dfrac{1}{4}$; $f(x_4) = \dfrac{1}{5}$;

$\sum_{i=1}^{4} f(x_i)\Delta x = \left(\dfrac{1}{2}\right)(1) + \left(\dfrac{1}{3}\right)(1) + \left(\dfrac{1}{4}\right)(1) + \left(\dfrac{1}{2}\right)(1) \approx \dfrac{77}{60}$.

Using right endpoints, the area is approximately $\dfrac{77}{60}$.

(c) $\dfrac{\frac{25}{12} + \frac{77}{60}}{2} = \dfrac{101}{60}$. Averaging the answers to (a) and (b), the area is approximately $\dfrac{101}{60}$.

(d) $x_1 = 1.5$; $x_2 = 2.5$; $x_3 = 3.5$; $x_4 = 4.5$; $f(x_1) = \dfrac{1}{1.5} = \dfrac{2}{3}$; $f(x_2) = \dfrac{1}{2.5} = \dfrac{2}{5}$;

$f(x_3) = \dfrac{1}{3.5} = \dfrac{2}{7}$; $f(x_4) = \dfrac{1}{4.5} = \dfrac{2}{9}$; $\sum_{i=1}^{4} f(x_i)\Delta x = \left(\dfrac{2}{3}\right)(1) + \left(\dfrac{2}{5}\right)(1) + \left(\dfrac{2}{7}\right)(1) + \left(\dfrac{2}{9}\right)(1) \approx \dfrac{496}{315}$.

Using midpoints, the area is approximately $\dfrac{496}{315}$.

12. (a) $x_1 = 1$; $x_2 = 3$; $x_3 = 5$; $x_4 = 7$; $f(x_1) = \dfrac{2}{1} = 2$; $f(x_2) = \dfrac{2}{3}$; $f(x_3) = \dfrac{2}{5}$; $f(x_4) = \dfrac{2}{7}$;

$$\sum_{i=1}^{4} f(x_i)\Delta x = (2)(2) + \left(\frac{2}{3}\right)(2) + \left(\frac{2}{5}\right)(2) + \left(\frac{2}{7}\right)(2) \approx \frac{704}{105} \approx 6.70.$$

Using left endpoints, the area is approximately 6.70.

(b) $x_1 = 3$; $x_2 = 5$; $x_3 = 7$; $x_4 = 9$; $f(x_1) = \dfrac{2}{3}$; $f(x_2) = \dfrac{2}{5}$; $f(x_3) = \dfrac{2}{7}$; $f(x_4) = \dfrac{2}{9}$;

$$\sum_{i=1}^{4} f(x_i)\Delta x = \left(\frac{2}{3}\right)(2) + \left(\frac{2}{5}\right)(2) + \left(\frac{2}{7}\right)(2) + \left(\frac{2}{9}\right)(2) \approx \frac{992}{315} \approx 3.15.$$

Using right endpoints, the area is approximately 3.15.

(c) $\dfrac{\frac{704}{105} + \frac{992}{315}}{2} = \dfrac{1552}{315} \approx 4.93$ Averaging the answers to (a) and (b), the area is approximately 4.93.

(d) $x_1 = 2$; $x_2 = 4$; $x_3 = 6$; $x_4 = 8$; $f(x_1) = \dfrac{2}{2} = 1$; $f(x_2) = \dfrac{2}{4} = \dfrac{1}{2}$; $f(x_3) = \dfrac{2}{6} = \dfrac{1}{3}$; $f(x_4) = \dfrac{2}{8} = \dfrac{1}{4}$;

$$\sum_{i=1}^{4} f(x_i)\Delta x = (1)(2) + \left(\frac{1}{2}\right)(2) + \left(\frac{1}{3}\right)(2) + \left(\frac{1}{4}\right)(2) \approx \frac{25}{6} \approx 4.17.$$

Using midpoints, the area is approximately 4.17.

13. (a) $x_1 = .5$; $x_2 = 1.5$; $x_3 = 2.5$; $x_4 = 3.5$; $f(x_1) = \dfrac{.5}{2} = \dfrac{1}{4}$; $f(x_2) = \dfrac{1.5}{2} = \dfrac{3}{4}$;

$f(x_3) = \dfrac{2.5}{2} = \dfrac{5}{4}$; $f(x_4) = \dfrac{3.5}{2} = \dfrac{7}{4}$; $\sum_{i=1}^{4} f(x_i)\Delta x = \left(\dfrac{1}{4}\right)(1) + \left(\dfrac{3}{4}\right)(1) + \left(\dfrac{5}{4}\right)(1) + \left(\dfrac{7}{4}\right)(1) = \dfrac{16}{4} = 4.$

Using midpoints, the area is approximately 4.

(b) Triangle base $= 4 - 0 = 4$; Triangle height $= \dfrac{4}{2} = 2$; Triangle area $= \dfrac{1}{2}(4)(2) = 4.$

Using the formula for the area of a triangle, the area is 4.

14. Triangle base $= 5 - 0 = 5$; Triangle height $= 5$; Triangle area $= \dfrac{1}{2}(5)(5) = 12.5.$

Using the formula for the area of a triangle, the area is 12.5.

15. (a) Triangle base $= 4$; Triangle height $= 2$; Triangle area $= \dfrac{1}{2}(4)(2) = 4.$

Using the formula for the area of a triangle, the area is 4.

(b) First triangle base $= 3$; First triangle height $= 3$; First triangle area $= \dfrac{1}{2}(3)(3) = \dfrac{9}{2}.$

Second triangle base $= 1$; Second triangle height $= 1$; Second triangle area $= \dfrac{1}{2}(1)(1) = \dfrac{1}{2}.$

Using the formula for the area of a triangle, the area is $\dfrac{9}{2} + \dfrac{1}{2} = 5.$

16. $\sqrt{9 - x^2}$ for $-3 \le x \le 3$ is a semicircle with center $(0, 0)$ and radius 3.

Area of semicircle $= \left(\dfrac{1}{2}\right)(\pi)(3)^2 = \dfrac{9\pi}{2}$. The area is $\dfrac{9\pi}{2}$.

17. $\sqrt{16 - x^2}$ for $-4 \le x \le 0$ is a quarter circle with center $(0, 0)$ and radius 4.

Area of quarter circle $= \left(\dfrac{1}{4}\right)(\pi)(4)^2 = 4\pi$. The area is 4π.

18. $(5 - x)$ for $1 \le x \le 3$ is a (sideways) trapezoid with height $= 2$, first base $= 4$, second base $= 2$.

Area of trapezoid $= \left(\dfrac{1}{2}\right)(4 + 2)(2) = 6$. The area is 6.

19. $(1 + 2x)$ for $2 \le x \le 5$ is a (sideways) trapezoid with height $= 3$, first base $= 11$, second base $= 5$.

Area of trapezoid $= \left(\dfrac{1}{2}\right)(11 + 5)(3) = 24$. The area is 24.

20. (a) $x_1 = 2.5$; $x_2 = 7.5$; $x_3 = 12.5$; $x_4 = 17$; From the figure, $f(x_1) \approx 30$; $f(x_2) \approx 63$;

$f(x_3) \approx 85$; $f(x_4) \approx 95$; $\displaystyle\sum_{i=1}^{4} f(x_i)\Delta x = (30)(5) + (63)(5) + (85)(5) + (95)(4) = 1270$.

The area is approximately $\left(1270 \dfrac{\text{mile seconds}}{\text{hour}}\right)\left(\dfrac{1}{3600} \dfrac{\text{hour}}{\text{second}}\right)\left(5280 \dfrac{\text{feet}}{\text{mile}}\right) \approx 1900$ feet.

(b) $x_1 = 2.5$; $x_2 = 7.5$; $x_3 = 12.5$; $x_4 = 17.5$; $x_5 = 22.5$; $x_6 = 26.5$;

From the graph, $f(x_1) \approx 25$; $f(x_2) \approx 52$; $f(x_3) \approx 70$; $f(x_4) \approx 83$; $f(x_5) \approx 93$; $f(x_6) \approx 98$;

$\displaystyle\sum_{i=1}^{6} f(x_i)\Delta x = (25)(5) + (52)(5) + (70)(5) + (83)(5) + (93)(5) + (98)(3) = 1909$.

The area is approximately $\left(1909 \dfrac{\text{mile seconds}}{\text{hour}}\right)\left(\dfrac{1}{3600} \dfrac{\text{hour}}{\text{second}}\right)\left(5280 \dfrac{\text{feet}}{\text{mile}}\right) \approx 2800$ feet.

Reviewing Basic Concepts (Sections 12.4 and 12.5)

1. The tangent line passes through the points $(-3, 4)$ and $(0, 1.5)$ and has slope $m = \dfrac{4 - 1.5}{-3 - 0} = -\dfrac{5}{6}$.

2. The tangent line is a horizontal line and has slope $m = 0$.

3. The tangent line is a vertical line and its slope does not exist.

4. Since $f(x) = k$, the value of $f(x)$ never changes. Therefore, its rate of change $= 0$ for all x. Thus, $f'(x) = 0$.

5. $\displaystyle\lim_{x \to 1800} R(x) = \lim_{x \to 1800} \dfrac{(-.0012x^2 + 3x) - [-.0012(1800)^2 + 3(1800)]}{x - 1800} = \lim_{x \to 1800} \dfrac{-.0012x^2 + 3x - 1512}{x - 1800} =$

$\displaystyle\lim_{x \to 1800} \dfrac{(x - 1800)(-.0012x + .84)}{x - 1800} = -1.32$. The marginal revenue when producing 1800 units is $-\$1.32$ per unit.

6. $\displaystyle\lim_{t \to -1} f(x) = \lim_{t \to -1} \dfrac{(6 - x^2) - (6 - (-1)^2)}{x - -1} = \lim_{t \to -1} \dfrac{1 - x^2}{x + 1} = \lim_{t \to -1} \dfrac{(1 - x)(1 + x)}{x + 1} = 2$. The slope of the tangent

line is $m = 2$. $f(-1) = 6 - (-1)^2 = 5 \Rightarrow$ the tangent line passes through $(-1, 5)$.

$y - 5 = 2(x - (-1)) = 2x + 2$. The equation of the tangent line is $y = 2x + 7$.

7. $f'(9) = \displaystyle\lim_{x \to 9} \dfrac{(\sqrt{x} + 2) - (\sqrt{9} + 2)}{x - 9} = \lim_{x \to 9} \dfrac{\sqrt{x} - 3}{(\sqrt{x} - 3)(\sqrt{x} + 3)} = \dfrac{1}{6}$. $f'(9)$ is $\dfrac{1}{6}$.

8. $x_1 = .5$; $x_2 = 1.5$; $x_3 = 2.5$; $x_4 = 3.5$; $x_5 = 4.5$; $f(x_1) = (.5)^2 = .25$; $f(x_2) = (1.5)^2 = 2.25$;

$f(x_3) = (2.5)^2 = 6.25$; $f(x_4) = (3.5)^2 = 12.25$; $f(x_5) = (4.5)^2 = 20.25$;

$\displaystyle\sum_{i=1}^{5} f(x_i)\Delta x = (.25)(1) + (2.25)(1) + (6.25)(1) + (12.25)(1) + (20.25)(1) = 41.25$.

The area is approximately 41.25.

9. Triangle base = 8; Triangle height = 8. Triangle area = $\frac{1}{2}(8)(8) = 32$.

Using the formula for the area of a triangle, the area is 32.

10. $\sqrt{9 - x^2}$ for $0 \leq x \leq 3$ is a quarter circle with center (0, 0) and radius 3.

Area of quarter circle $= \left(\frac{1}{4}\right)(\pi)(3)^2 = \frac{9\pi}{4}$. The area is $\frac{9\pi}{4}$.

Chapter 12 Review Exercises

1. (a) From the graph, $\lim\limits_{x \to 1^-} f(x) = 2$.

(b) From the graph, $\lim\limits_{x \to 1^+} f(x) = 2$.

(c) From the graph, $\lim\limits_{x \to 1} f(x) = 2$.

2. (a) From the graph, $\lim\limits_{x \to -1^-} f(x) = -2$.

(b) From the graph, $\lim\limits_{x \to -1^+} f(x) = 2$.

(c) From the graph, $\lim\limits_{x \to -1} f(x)$ does not exist.

3. (a) From the graph, $\lim\limits_{x \to 4^-} f(x) = \infty$.

(b) From the graph, $\lim\limits_{x \to 4^+} f(x) = -\infty$.

(c) From the graph, $\lim\limits_{x \to 4} f(x)$ does not exist.

4. (a) From the graph, $\lim\limits_{x \to \infty} f(x) = -3$.

(b) From the graph, $\lim\limits_{x \to 0^-} f(x) = -3$.

(c) From the graph, $\lim\limits_{x \to 0^+} f(x) = -3$.

5. $\lim\limits_{x \to 1} (2x^2 - 3x) = 2(1)^2 - 3(1) = -1$. The limit is -1.

6. $\lim\limits_{x \to -1} (x - x^3) = (-1) - (-1)^3 = 0$. The limit is 0.

7. $\lim\limits_{x \to 2} \dfrac{3x + 4}{x + 3} = \dfrac{3(2) + 4}{2 + 3} = 2$. The limit is 2.

8. $\lim\limits_{x \to 4} \dfrac{x - 4}{x^3 + 5x} = \dfrac{4 - 4}{4^3 + 5(4)} = \dfrac{0}{84} = 0$. The limit is 0.

9. $\lim\limits_{x \to -1} \sqrt{5x + 21} = \sqrt{5(-1) + 21} = 4$. The limit is 4.

10. $\lim\limits_{x \to 2} 27^{(3 - x)/(1 + x)} = 27^{(3 - 2)/(1 + 2)} = 3$. The limit is 3.

11. $\lim\limits_{x \to 1} \left[\log_2 (5x + 3)\right] = \log_2 (5(1) + 3) = 3$. The limit is 3.

12. $\lim\limits_{x \to -3} (5 - 9x)^{2/5} = (5 - 9(-3))^{2/5} = 4$. The limit is 4.

13. $\lim\limits_{x \to 5} \dfrac{2x - 10}{5 - x} = \lim\limits_{x \to 5} \dfrac{-2(5 - x)}{5 - x} = -2$. The limit is -2.

14. $\lim\limits_{x \to 6} \dfrac{x^2 - 36}{x - 6} = \lim\limits_{x \to 6} \dfrac{(x - 6)(x + 6)}{x - 6} = 12$. The limit is 12.

15. $\lim\limits_{x \to -3} \dfrac{x^2 + 2x - 3}{x + 3} = \lim\limits_{x \to -3} \dfrac{(x + 3)(x - 1)}{x + 3} = -4$. The limit is -4.

16. $\lim\limits_{x\to 1} \dfrac{x^2 - 2x + 1}{x - 1} = \lim\limits_{x\to 1} \dfrac{(x - 1)(x - 1)}{x - 1} = 0$. The limit is 0.

17. $\lim\limits_{x\to 2} \dfrac{x^2 - x - 2}{x^2 - 5x + 6} = \lim\limits_{x\to 2} \dfrac{(x - 2)(x + 1)}{(x - 2)(x - 3)} = \dfrac{3}{-1} = -3$. The limit is -3.

18. $\lim\limits_{x\to -2} \dfrac{x^3 + 3x^2 + 2x}{x^2 + 2x} = \lim\limits_{x\to -2} \dfrac{x(x + 2)(x + 1)}{x(x + 2)} = -1$. The limit is -1.

19. $\lim\limits_{x\to 1} \dfrac{x^2 + x}{x - 1} = \lim\limits_{x\to 1} \dfrac{x(x + 1)}{x - 1} = \infty$. The limit does not exist.

20. $\lim\limits_{x\to 0} \dfrac{x^2 + 1}{x} = \infty$. The limit does not exist.

21. $\lim\limits_{x\to 0} \dfrac{\sin x}{3x} = \dfrac{1}{3} \lim\limits_{x\to 0} \dfrac{\sin x}{x} = \dfrac{1}{3}(1) = \dfrac{1}{3}$. The limit is $\dfrac{1}{3}$.

22. Graphing $y_1 = (\sin(1/x))/x$ in the window $-.5 \le x \le .5, -20 \le y \le 20$ (not shown), it can be seen

    that $\dfrac{\sin\frac{1}{x}}{x}$ oscillates wildly as $x \to 0$. Thus, $\lim\limits_{x\to 0} \dfrac{\sin\frac{1}{x}}{x}$ does not exist.

23. Graphing $y_1 = (x\cos x - 1)/x^2$ in the window $-.5 \le x \le .5, -100 \le y \le 0$ (not shown), it can be seen

    that $\lim\limits_{x\to 0} \dfrac{x\cos x - 1}{x^2} = -\infty$. The limit is $-\infty$.

24. $\lim\limits_{x\to 0} \dfrac{\tan x}{2x} = \lim\limits_{x\to 0} \left(\dfrac{\sin x}{\cos x}\right)\left(\dfrac{1}{2x}\right) = \dfrac{1}{2} \lim\limits_{x\to 0} \left(\dfrac{\sin x}{x}\right)(\cos x) = \left(\dfrac{1}{2}\right)(1)(1) = \dfrac{1}{2}$. The limit is $\dfrac{1}{2}$.

25. $\lim\limits_{x\to 2^-} f(x) = \lim\limits_{x\to 2^-} (3x - 1) = 5$; $\lim\limits_{x\to 2^+} f(x) = \lim\limits_{x\to 2^+} (x + 3) = 5$; Thus, $\lim\limits_{x\to 2} f(x) = 5$. The limit is 5.

26. $\lim\limits_{x\to 1^-} f(x) = \lim\limits_{x\to 1^-} x^2 = 1$; $\lim\limits_{x\to 1^+} f(x) = \lim\limits_{x\to 1^+} (3x - 2) = 1$; Thus, even though $f(1) = 5$, $\lim\limits_{x\to 1} f(x) = 1$.

    The limit is 1.

27. $\lim\limits_{x\to 2} f(x) = \lim\limits_{x\to 2} (x^2 - 1) = 3$. The limit is 3.

28. $\lim\limits_{x\to 0^+} f(x) = \lim\limits_{x\to 0^+} \sqrt{x} = 0$. The limit is 0.

29. $\lim\limits_{x\to 0^-} f(x) = \lim\limits_{x\to 0^-} \ln(-x) = \lim\limits_{x\to 0^+} \ln(x) = -\infty$. The limit is $-\infty$.

30. $\lim\limits_{x\to 0} f(x) = \lim\limits_{x\to 0} \left(\dfrac{\sin x}{x}\right) = 1$. The limit is 1.

31. $\lim\limits_{x\to 3^+} (\pi - 1) = \pi - 1$. The limit is $\pi - 1$.

32. $\lim\limits_{x\to -2} f(x) = -5x^3 = -5(-2)^3 = 40$. The limit is 40.

33. Graphing $y_1 = 1/(x - 7)$ in the window $6.5 \le x \le 7.5, -100 \le y \le 100$ (not shown), shows that

    $\lim\limits_{x\to 7} \dfrac{1}{x - 7} = -\infty$. The limit is $-\infty$.

34. $x \to -2^+ \Rightarrow x \ge -2 \Rightarrow x + 2 \ge 0 \Rightarrow \lim\limits_{x\to -2^+} \sqrt{x + 2} = \sqrt{-2 + 2} = 0$. The limit is 0.

35. $x \to 1^+ \Rightarrow x \ge 1 \Rightarrow 1 - x \le 0 \Rightarrow \lim\limits_{x\to 1^+} \sqrt{1 - x}$ does not exist. The limit does not exist.

36. $\lim\limits_{x\to 2} (3x - 1) = 3(2) - 1 = 5$. The limit is 5.

37. $\lim\limits_{x\to 1^-} \dfrac{|x - 1|}{x - 1} = \lim\limits_{x\to 1^-} (-1) = -1$. The limit is -1.

38. $\displaystyle\lim_{x\to-2^+}\frac{x+2}{|x+2|}=\lim_{x\to-2^+}(1)=1.$ The limit is 1.

39. Graphing $y_1=x/(x-4)^2$ in the window $3.5\le x\le 4.5, 0\le y\le 1000$ (not shown), shows that

$$\lim_{x\to4^+}\frac{x}{(x-4)^2}=\infty.\text{ The limit is }\infty.$$

40. Graphing $y_1=x/(x-4)^2$ in the window $3.5\le x\le 4.5, 0\le y\le 1000$ (not shown), shows that

$$\lim_{x\to4^-}\frac{x}{(x-4)^2}=\infty.\text{ The limit is }\infty.$$

41. Graphing $y_1=x^2/(x-1)^3$ in the window $.75\le x\le 1.25, -10{,}000\le y\le 10{,}000$ (not shown), shows that

$$\lim_{x\to1^+}\frac{x^2}{(x-1)^3}=\infty.\text{ The limit is }\infty.$$

42. Graphing $y_1=x^2/(x-1)^3$ in the window $.75\le x\le 1.25, -10{,}000\le y\le 10{,}000$ (not shown), shows that

$$\lim_{x\to1^-}\frac{x^2}{(x-1)^3}=-\infty.\text{ The limit is }-\infty.$$

43. $\displaystyle\lim_{x\to\infty}\frac{5x+1}{2x-7}=\lim_{x\to\infty}\frac{5+\frac{1}{x}}{2-\frac{7}{x}}=\frac{5}{2}.$ The limit is $\dfrac{5}{2}$.

44. $\displaystyle\lim_{x\to\infty}\frac{4x^2-5x}{2x^2}=\lim_{x\to\infty}\frac{4-\frac{5}{x}}{2}=\frac{4}{2}=2.$ The limit is 2.

45. $\displaystyle\lim_{x\to-\infty}\frac{x^3+1}{x^2-1}=\lim_{x\to-\infty}\frac{x+\frac{1}{x^2}}{1-\frac{1}{x^2}}=-\infty.$ The limit is $-\infty$.

46. $\displaystyle\lim_{x\to-\infty}\frac{x^4+x+1}{x^5-2}=\lim_{x\to-\infty}\frac{\frac{1}{x}+\frac{1}{x^4}+\frac{1}{x^5}}{1-\frac{2}{x^5}}=\frac{0}{1}=0.$ The limit is 0.

47. $\displaystyle\lim_{x\to\infty}\left(5+\frac{x}{1+x^2}\right)=5+\lim_{x\to\infty}\left(\frac{\frac{1}{x}}{\frac{1}{x^2}+1}\right)=5+0=5.$ The limit is 5.

48. $\displaystyle\lim_{x\to\infty}(e^{-2x}+7)=\lim_{x\to\infty}e^{-2x}+7=0+7=7.$ The limit is 7.

49. $\displaystyle\lim_{x\to\infty}\left(2-\frac{3}{1-e^{-x}}\right)=2-\frac{3}{1-\lim_{x\to\infty}e^{-x}}=2-3=-1.$ The limit is -1.

50. $\displaystyle\lim_{x\to\infty}\left(\frac{5}{1+2^{-x}}-3\right)=\frac{5}{1+\lim_{x\to\infty}2^{-x}}-3=5-3=2.$ The limit is 2.

51. The functions $f(x)=\dfrac{1}{x-2}$ and $g(x)=-\dfrac{1}{x-2}$ have the properties that neither $\displaystyle\lim_{x\to2}f(x)$ nor $\displaystyle\lim_{x\to2}g(x)$ exists, but $\displaystyle\lim_{x\to2}[f(x)+g(x)]=0$ does exists.

52. The functions $f(x)=x^2$ and $g(x)=x$ have the properties that $\displaystyle\lim_{x\to\infty}f(x)=\infty,\lim_{x\to\infty}g(x)=\infty,$ and

$$\lim_{x\to\infty}\frac{f(x)}{g(x)}=\infty.$$

53. $\displaystyle\lim_{x\to\infty}f(t)=70+110\lim_{x\to\infty}e^{-25t}=70.$ In time, the coffee ultimately cools to 70°.

54. $m=\displaystyle\lim_{x\to3}\frac{(1+x^2)-(1+3^2)}{x-3}=\lim_{x\to3}\frac{x^2-9}{x-3}=\lim_{x\to3}\frac{(x-3)(x+3)}{x-3}=6.$ The slope of the tangent line to $f(x)$ at $x=3$ is 6.

55. $m = \lim\limits_{x \to 1} \dfrac{\frac{3}{x} - \frac{3}{1}}{x - 1} = \lim\limits_{x \to 1} \dfrac{\frac{3 - 3x}{x}}{x - 1} = \lim\limits_{x \to 1} \dfrac{\frac{-3(x - 1)}{x}}{x - 1} = -3.$ The slope of the tangent line to $f(x)$ at $x = 1$ is -3.

56. $m = \lim\limits_{x \to 2} \dfrac{\frac{4}{x} - \frac{4}{2}}{x - 2} = \lim\limits_{x \to 2} \dfrac{\frac{-2(x - 2)}{x}}{x - 2} = -1.$ $f(2) = \dfrac{4}{2} = 2;$ $y - 2 = -1(x - 2) = -x + 2.$ The equation of the

tangent line is $y = -x + 4$. Graphing $y_1 = 4/x$ and $y_2 = -x + 4$ in the window $0 \le x \le 4, 0 \le y \le 10$

(not shown), shows this to be so.

57. $m = \lim\limits_{x \to 2} \dfrac{(x^2 - x) - (2^2 - 2)}{x - 2} = \lim\limits_{x \to 2} \dfrac{x^2 - x - 2}{x - 2} = \lim\limits_{x \to 2} \dfrac{(x - 2)(x + 1)}{x - 2} = 3.$

$f(2) = 2^2 - 2 = 2;$ $y - 2 = 3(x - 2) = 3x - 6.$ The equation of the tangent line is $y = 3x - 4.$ Graphing

$y_1 = x^2 - x$ and $y_2 = 3x - 4$ in the window $0 \le x \le 4, -5 \le y \le 10$ (not shown), shows this to be so.

58. $\lim\limits_{x \to 4} \dfrac{\sqrt{x} - 2}{x - 4}$ will be the slope of the tangent line to $f(x) = \sqrt{x}$ at the point $(4, 2)$. From the figure, this slope

is .25. Thus, $\lim\limits_{x \to 4} \dfrac{\sqrt{x} - 2}{x - 4} = .25.$

59. $f'\left(\dfrac{\pi}{2}\right)$ will be the slope of the tangent line to $f(x) = \sin x$ at the point $\left(\dfrac{\pi}{2}, 1\right)$. From the figure, this slope

is 0. Thus, $f'\left(\dfrac{\pi}{2}\right) = 0.$

60. $\text{nDer}(x \sin x, x, \pi) = -3.14159212999 \approx -\pi.$ The value of $f'(\pi)$ is $-\pi$.

61. $\text{nDer}(e^x, x, \pi) = 1.00000016667 \approx 1.$ The value of $f'(0)$ is 1.

62. The tangent line passes through the points $(40, 150)$ and $(80, 400)$. Thus, the slope of the tangent line is

$m = \dfrac{400 - 150}{80 - 40} = 6.25.$ The average farm size was increasing by 6.25 acres per year on January 1, 1950.

63. $S(t) = -16t^2 + 100t + 4;$

$S'(3) = \lim\limits_{t \to 3} \dfrac{(-16t^2 + 100t + 4) - [-16(3)^2 + 100(3) + 4]}{t - 3} = \lim\limits_{t \to 3} \dfrac{-16t^2 + 100t - 156}{t - 3} =$

$\lim\limits_{t \to 3} \dfrac{(t - 3)(-16t + 52)}{t - 3} = 4.$ The velocity of the ball when $t = 3$ seconds is 4 feet per second.

64. $\text{nDer}(.006x^3 - .7x^2 + 32x + 250, x, 72) = 24.512.$ The marginal cost at $x = 72$ units is \$2451.20 per unit.

65. $f(x_1) = 3(-1) + 1 = -2;$ $f(x_2) = 3(0) + 1 = 1;$ $f(x_3) = 3(1) + 1 = 4;$ $f(x_4) = 3(2) + 1 = 7;$

$f(x_5) = 3(3) + 1 = 10;$ $\sum\limits_{i=1}^{5} f(x_i)\Delta x = (-2)(1) + (1)(1) + (4)(1) + (7)(1) + (10)(1) = 20.$ The sum is 20.

66. $\displaystyle\int_0^3 f(x)\,dx = $ area of a trapezoid. Height of the trapezoid $= 3$, first base of the trapezoid $= 2$, second base of the

trapezoid $= 1$, area of the trapezoid $= \left(\dfrac{1}{2}\right)(3)(2 + 1) = \dfrac{9}{2}.$ Thus, the value of $\displaystyle\int_0^3 f(x)\,dx$ is $\dfrac{9}{2} = 4.5.$

67. $x_1 = 0;$ $x_2 = 1;$ $x_3 = 2;$ $x_4 = 3;$ $f(x_1) = 2(0) + 3 = 3;$ $f(x_2) = 2(1) + 3 = 5;$

$f(x_3) = 2(2) + 3 = 7;$ $f(x_4) = 2(3) + 3 = 9;$ $\sum\limits_{i=1}^{4} f(x_i)\Delta x = 3(1) + 5(1) + 7(1) + 9(1) = 24.$

The area is approximately 24.

68. Height of the trapezoid = 4, first base of the trapezoid = $2(4) + 3 = 11$, second base of the

    trapezoid = $2(0) + 3 = 3$, area of the trapezoid = $\left(\frac{1}{2}\right)(4)(11 + 3) = 28$. The area is 28, which is 4 greater

    than the approximation in exercise 67. We also conclude that the value of $\int_0^4 (2x + 3)\,dx$ is 28.

69. Height of the triangle = 5, and the base of the triangle = 5. Thus, the area of the

    triangle = $\left(\frac{1}{2}\right)(5)(5) = \frac{25}{2} = 12.5$. Thus, we conclude that the value of $\int_0^5 (5 - x)\,dx$ is 12.5.

70. $\sqrt{1 - x^2}$ for $0 \le x \le 1$ is a quarter of a circle centered at $(0, 0)$ with radius 1. The area of the quarter circle is

    $\left(\frac{1}{4}\right)(\pi)(1)^2 = \frac{\pi}{4}$. Thus, we conclude that the value of $\int_0^1 \sqrt{1 - x^2}\,dx$ is $\frac{\pi}{4}$.

Chapter 12 Test

1. From the graph, we see that $\lim_{x \to 4^-} f(x)$ is 5.

2. From the graph, we see that $\lim_{x \to \infty} f(x)$ is 0.

3. (a) From the graph, we see that $\lim_{x \to 3^+} f(x)$ is ∞.

    (b) From the graph, we see that $\lim_{x \to 3^-} f(x)$ is 4.

4. $\lim_{x \to 1} \dfrac{x^2 + x + 1}{x^2 + 1} = \dfrac{1^2 + 1 + 1}{1^2 + 1} = \dfrac{3}{2}$. The limit is $\dfrac{3}{2}$.

5. $\lim_{x \to -2} \dfrac{x^2 + 2x}{x + 2} = \lim_{x \to -2} \dfrac{x(x + 2)}{x + 2} = -2$. The limit is -2.

6. $\lim_{x \to 3} \dfrac{x^2 - 6x + 9}{x - 3} = \lim_{x \to 3} \dfrac{(x - 3)(x - 3)}{x - 3} = 0$. The limit is 0.

7. $\lim_{x \to 2} \dfrac{x^2 + x - 6}{x^2 - 4} = \lim_{x \to 2} \dfrac{(x - 2)(x + 3)}{(x - 2)(x + 2)} = \dfrac{5}{4}$. The limit is $\dfrac{5}{4}$.

8. $\lim_{x \to 3} \sqrt{x^2 + 7} = \sqrt{3^2 + 7} = 4$. The limit is 4.

9. $\lim_{x \to \infty} \dfrac{2x^2 - 3}{5x^2 + x + 1} = \lim_{x \to \infty} \dfrac{2 - \frac{3}{x^2}}{5 + \frac{1}{x} + \frac{1}{x^2}} = \dfrac{2}{5}$. The limit is $\dfrac{2}{5}$.

10. $\lim_{x \to -\infty} \dfrac{3x - 4}{4x - 3} = \lim_{x \to -\infty} \dfrac{3 - \frac{4}{x}}{4 - \frac{3}{x}} = \dfrac{3}{4}$. The limit is $\dfrac{3}{4}$.

11. $\lim_{x \to 0} \dfrac{\sin x}{x \cos x} = \left(\lim_{x \to 0} \dfrac{\sin x}{x}\right)\left(\lim_{x \to 0} \dfrac{1}{\cos x}\right) = (1)(1) = 1$. The limit is 1.

12. $\lim_{x \to 0} \dfrac{x^2 - 10}{x^3 + 1} = \dfrac{0^2 - 10}{0^3 + 1} = -10$. The limit is -10.

13. $\lim_{x \to 1^+} f(x) = \lim_{x \to 1^+} (1 - 2x^2) = 1 - 2(1)^2 = -1$. The limit is -1.

14. $\lim_{x \to 4^+} 55 = 55$. The limit is 55.

15. $\lim_{x \to 3} (4x^2 - 3x) = 4(3)^2 - 3(3) = 27$. The limit is 27.

16. For $x > 2, 2 - x < 0$. Thus, $\lim_{x \to 2^+} \sqrt{2 - x}$ does not exist. The limit is 55.

17. $\lim_{x \to 3} \dfrac{|3 - x|}{3 - x} = 1$. The limit is 1.

18. For $x > -1, x + 1 > 0$. Thus, $\lim_{x \to -1^+} \sqrt{x + 1} = \sqrt{-1 + 1} = 0$. The limit is 0.

19. $m = \lim_{x \to 1} \dfrac{(2x^2 - 1) - (2(1)^2 - 1)}{x - 1} = \lim_{x \to 1} \dfrac{2x^2 - 2}{x - 1} = \lim_{x \to 1} \dfrac{2(x - 1)(x + 1)}{x - 1} = 4$. The slope is 4.

20. $m = \lim_{x \to 1} \dfrac{\frac{-3}{x} - \left(\frac{-3}{1}\right)}{x - 1} = \lim_{x \to 1} \dfrac{\frac{3(x - 1)}{x}}{x - 1} = 3$. $y + 3 = 3(x - 1) = 3x - 3$.

The equation of the tangent line is $y = 3x - 6$.

21. nDer$((1 + e^x)/x, x, 4) = 10.174654336$. $f'(4)$ is approximately 10.1747.

22. The tangent line passes through the points (14, .80 and (22, 0). Thus, the slope of the tangent line is

$m = \dfrac{0 - .8}{22 - 14} = -.1$. At $t = 14$ years, the cobalt is disintegrating at a rate of $\dfrac{1}{10}$ gram per year.

23. The graph of $y_1 = 40,000(1 - e^{-.25x})$ in the window $0 \le x \le 32, 0 \le y \le 45,000$ is shown in Figure 23.

nDer$(40,000(1 - e^{-.25x}), x, 3) \approx 4723.67$. When $t = 3$ days, the information is spreading at a rate of

approximately 4724 persons per day.

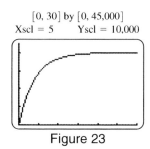

$[0, 30]$ by $[0, 45,000]$
Xscl = 5   Yscl = 10,000

Figure 23

24. $x_1 = .5$; $x_2 = 1.5$; $x_3 = 2.5$; $x_4 = 3.5$; $f(x_1) = (.5)^2 = .25$; $f(x_2) = (1.5)^2 = 2.25$; $f(x_3) = (2.5)^2 = 6.25$;

$f(x_4) = (3.5)^2 = 12.25$; $\displaystyle\sum_{i=1}^{4} f(x_i)\Delta x = (.25)(1) + (2.25)(1) + (6.25)(1) + (12.25)(1) = 21$.

The area is approximately 21.

25. Height of the triangle $= 6$, and the base of the triangle $= 6$. Thus, the area of the

triangle $= \left(\dfrac{1}{2}\right)(6)(6) = \dfrac{36}{2} = 18$. Thus, we conclude that the value of $\displaystyle\int_0^6 (6 - x)dx$ is 18.

Chapter 12 Project

1. Graph the function $y_1 = .0004985x^2 + .04527x + 4.054$ in the window $0 \le x \le 98, -1 \le y \le 25$ (not shown). For a person born in 1949, for example, $x = 49$ and $\frac{dy}{dx} \approx .094$ or 9.4%.

2. **Cubic:** Graph $y_1 = -9.364x^3 + 287.07x^2 + 252.8x + 808$ in the window $0 \le x \le 9.4, -1000 \le y \le 30,000$ (not shown). In 1998, $x = 3$ and $\frac{dy}{dx} \approx 1722$.

    Quartic: Graph $y_1 = 3.831x^4 - 85.99x^3 + 757.9x^2 - 624.7x + 829$ in the window $0 \le x \le 9.4, -1000 \le y \le 30,000$ (not shown). In 1998, $x = 3$ and $\frac{dy}{dx} \approx 2015$.

3. Graph $y_1 = 364(1.005)^x$ in the window $0 \le x \le 98, 300 \le y \le 1000$ (not shown). For a person born in 1949, for example, 100 years after birth is 2049 or $x = 49$. The instantaneous rate of change for $x = 49$ is $\frac{dy}{dx} \approx 2.32$.

4. Graph $y_1 = 19.962 + 18.335 \ln(x)$ in the window $0 \le x \le 9.4, 0 \le y \le 100$ (not shown). If the current year is 2007, for example, $x = 7$ and $\frac{dy}{dx} \approx 2.62$.

Chapter R: Basic Algebraic Concepts

R.1: Exponents and Polynomials

1. $(-4)^3 \cdot (-4)^2 = (-4)^{3+2} = (-4)^5$

2. $(-5)^2 \cdot (-5)^6 = (-5)^{2+6} = (-5)^8$

3. $2^0 = 1$

4. $-2^0 = -(1) = -1$

5. $(5m)^0 = 1$, if $m \neq 0$

6. $(-4z)^0 = 1$, if $z \neq 0$

7. $(2^2)^5 = 2^{2 \cdot 5} = 2^{10}$

8. $(6^4)^3 = 6^{4 \cdot 3} = 6^{12}$

9. $(2x^5y^4)^3 = 2^3(x^5)^3(y^4)^3 = 2^3x^{15}y^{12}$ or $8x^{15}y^{12}$

10. $(-4m^3n^9)^2 = (-4)^2(m^3)^2(n^9)^2 = 4^2m^6n^{18}$ or $16m^6n^{18}$

11. $-\left(\dfrac{p^4}{q}\right)^2 = -\left(\dfrac{p^{4 \cdot 2}}{q^2}\right) = -\dfrac{p^8}{q^2}$

12. $\left(\dfrac{r^8}{s^2}\right)^3 = \dfrac{r^{8 \cdot 3}}{s^{2 \cdot 3}} = \dfrac{r^{24}}{s^6}$

13. $-5x^{11}$ is a polynomial. It is a monomial since it has one term. It has degree 11 since 11 is the highest exponent.

14. $9y^{12} + y^2$ is a polynomial. It is a binomial since it has two terms. It has degree 12 since 12 is the highest exponent.

15. $18p^5q + 6pq$ is a polynomial. It is a binomial since it has two terms. It has degree 6 since 6 is the sum of the exponents in the term $18p^5q$. (The term $6pq$ has degree 2.)

16. $2a^6 + 5a^2 + 4a$ is a polynomial. It is a trinomial since it has three terms. It has degree 6 since 6 is the highest exponent.

17. $\sqrt{2}x^2 + \sqrt{3}x^6$ is a polynomial. It is a binomial since it has two terms. It has degree 6 since 6 is the highest exponent.

18. $-\sqrt{7}m^5n^2 + 2\sqrt{3}m^3n^2$ is a polynomial. It is a binomial since it has two terms. It has degree 7 since 7 is the sum of the exponents in the term $-\sqrt{7}m^5n^2$. (The other term has degree 5.)

19. $\dfrac{1}{3}r^2s^2 - \dfrac{3}{5}r^4s^2 + rs^3$ is a polynomial. It is a trinomial since it has three terms. It has degree 6 since 6 is the sum of the exponents in the term $-\dfrac{3}{5}r^4s^2$. (The other terms have degree 4.)

20. $\dfrac{5}{p} + \dfrac{2}{p^2} + \dfrac{5}{p^3}$ is not a polynomial since positive exponents in the denominator are equivalent to negative exponents in the numerator.

21. $-5\sqrt{z} + 2\sqrt{z^3} - 5\sqrt{z^5} = -5z^{1/2} + 2z^{3/2} - 5z^{5/2}$ is not a polynomial since the exponents are not integers.

22. $(3x^2 - 4x + 5) + (-2x^2 + 3x - 2) = (3x^2 - 2x^2) + (-4x + 3x) + (5 - 2) = x^2 - x + 3$

23. $(4m^3 - 3m^2 + 5) + (-3m^3 - m^2 + 5) = (4m^3 - 3m^3) + (-3m^2 - m^2) + (5 + 5) = m^3 - 4m^2 + 10$

24. $(12y^2 - 8y + 6) - (3y^2 - 4y + 2) = 12y^2 - 8y + 6 - 3y^2 + 4y - 2 = 9y^2 - 4y + 4$

25. $(8p^2 - 5p) - (3p^2 - 2p + 4) = 8p^2 - 5p - 3p^2 + 2p - 4 = 5p^2 - 3p - 4$

26. $(6m^4 - 3m^2 + m) - (2m^3 + 5m^2 + 4m) + (m^2 - m) = 6m^4 - 3m^2 + m - 2m^3 - 5m^2 - 4m + m^2 - m = $

    $6m^4 - 2m^3 - 7m^2 - 4m$

27. $-(8x^3 + x - 3) + (2x^3 + x^2) - (4x^2 + 3x - 1) = -8x^3 - x + 3 + 2x^3 + x^2 - 4x^2 - 3x + 1 = $

    $-6x^3 - 3x^2 - 4x + 4$

28. $(4r - 1)(7r + 2) = 4r \cdot 7r + 4r \cdot 2 - 1 \cdot 7r - 1 \cdot 2 = 28r^2 + 8r - 7r - 2 = 28r^2 + r - 2$

29. $(5m - 6)(3m + 4) = 5m \cdot 3m + 5m \cdot 4 - 6 \cdot 3m - 6 \cdot 4 = 15m^2 + 20m - 18m - 24 = 15m^2 + 2m - 24$

30. $\left(3x - \dfrac{2}{3}\right)\left(5x + \dfrac{1}{3}\right) = 3x \cdot 5x + 3x \cdot \dfrac{1}{3} - \dfrac{2}{3} \cdot 5x - \dfrac{2}{3} \cdot \dfrac{1}{3} = 15x^2 + x - \dfrac{10}{3}x - \dfrac{2}{9} = 15x^2 - \dfrac{7}{3}x - \dfrac{2}{9}$

31. $\left(2m - \dfrac{1}{4}\right)\left(3m + \dfrac{1}{2}\right) = 2m \cdot 3m + 2m \cdot \dfrac{1}{2} - \dfrac{1}{4} \cdot 3m - \dfrac{1}{4} \cdot \dfrac{1}{2} = 6m^2 + m - \dfrac{3}{4}m - \dfrac{1}{8} = 6m^2 + \dfrac{1}{4}m - \dfrac{1}{8}$

32. $4x^2(3x^3 + 2x^2 - 5x + 1) = 4x^2 \cdot 3x^3 + 4x^2 \cdot 2x^2 + 4x^2 \cdot (-5x) + 4x^2 \cdot 1 = 12x^5 + 8x^4 - 20x^3 + 4x^2$

33. $2b^3(b^2 - 4b + 3) = 2b^3 \cdot b^2 + 2b^3 \cdot (-4b) + 2b^3 \cdot 3 = 2b^5 - 8b^4 + 6b^3$

34. $(2z - 1)(-z^2 + 3z - 4) = 2z(-z^2) + 2z(3z) + 2z(-4) - 1(-z^2) - 1(3z) - 1(-4) = $

    $-2z^3 + 6z^2 - 8z + z^2 - 3z + 4 = -2z^3 + 7z^2 - 11z + 4$

35. $(m - n + k)(m + 2n - 3k) = m^2 + 2mn - 3km - mn - 2n^2 + 3kn + km + 2kn - 3k^2 = $

    $m^2 + mn - 2km - 2n^2 + 5kn - 3k^2$

36. $(r - 3s + t)(2r - s + t) = 2r^2 - rs + rt - 6rs + 3s^2 - 3st + 2rt - st + t^2 = $

    $2r^2 - 7rs + 3rt + 3s^2 - 4st + t^2$

37. To find the square of a binomial, find the sum of the square of the first term, twice the product of the two terms, and the square of the last term.

38. To find the product of the sum and difference of two terms, find the difference between the square of the first term and the square of the last term.

39. $(2m + 3)(2m - 3) = (2m)^2 - (3)^2 = 4m^2 - 9$

40. $(8s - 3t)(8s + 3t) = (8s)^2 - (3t)^2 = 64s^2 - 9t^2$

41. $(4m + 2n)^2 = (4m)^2 + 2(4m)(2n) + (2n)^2 = 16m^2 + 16mn + 4n^2$

42. $(a - 6b)^2 = (a)^2 - 2(a)(6b) + (6b)^2 = a^2 - 12ab + 36b^2$

43. $(5r + 3t^2)^2 = (5r)^2 + 2(5r)(3t^2) + (3t^2)^2 = 25r^2 + 30rt^2 + 9t^4$

44. $(2z^4 - 3y)^2 = (2z^4)^2 - 2(2z^4)(3y) + (3y)^2 = 4z^8 - 12z^4y + 9y^2$

45. $[(2p - 3) + q]^2 = (2p - 3)^2 + 2(2p - 3)(q) + (q)^2 = [(2p)^2 - 2(2p)(3) + 3^2] + 4pq - 6q + q^2 = $

    $4p^2 - 12p + 9 + 4pq - 6q + q^2$

46. $[(4y - 1) + z]^2 = (4y - 1)^2 + 2(4y - 1)(z) + (z)^2 = [(4y)^2 - 2(4y)(1) + 1^2] + 8yz - 2z + z^2 = $

    $16y^2 - 8y + 1 + 8yz - 2z + z^2$

47. $[(3q + 5) - p][(3q + 5) + p] = (3q + 5)^2 - (p)^2 = [(3q)^2 + 2(3q)(5) + (5)^2] - p^2 = $

    $9q^2 + 30q + 25 - p^2$

48. $[(9r - s) + 2][(9r - s) - 2] = (9r - s)^2 - (2)^2 = [(9r)^2 - 2(9r)(s) + (s)^2] - 4 = 81r^2 - 18rs + s^2 - 4$

49. $[(3a + b) - 1]^2 = (3a + b)^2 - 2(3a + b)(1) + (-1)^2 = [(3a)^2 + 2(3a)(b) + (b)^2] - 6a - 2b + 1 =$

    $9a^2 + 6ab + b^2 - 6a - 2b + 1$

50. $[(2m + 7) - n]^2 = (2m + 7)^2 - 2(2m + 7)(n) + (n)^2 = [(2m)^2 + 2(2m)(7) + (7)^2] - 4mn - 14n + n^2 =$

    $4m^2 + 28m + 49 - 4mn - 14n + n^2$

51. $(6p + 5q)(3p - 7q) = (6p)(3p) + (6p)(-7q) + (5q)(3p) + (5q)(-7q) = 18p^2 - 42pq + 15pq - 35q^2 =$

    $18p^2 - 27pq - 35q^2$

52. $(2p - 1)(3p^2 - 4p + 5) = (2p)(3p^2) + (2p)(-4p) + (2p)(5) - 1(3p^2) - 1(-4p) - 1(5) =$

    $6p^3 - 8p^2 + 10p - 3p^2 + 4p - 5 = 6p^3 - 11p^2 + 14p - 5$

53. $(p^3 - 4p^2 + p) - (3p^2 + 2p + 7) = p^3 - 4p^2 + p - 3p^2 - 2p - 7 = p^3 - 7p^2 - p - 7$

54. $(6k - 3)^2 = (6k)^2 - 2(6k)(3) + (3)^2 = 36k^2 - 36k + 9$

55. $y(4x + 3y)(4x - 3y) = y[(4x)^2 - (3y)^2] = y[16x^2 - 9y^2] = 16x^2y - 9y^3$

56. $(r^5 - r^3 + r) + (3r^5 - 4r^4 + r^3 + 2r) = 4r^5 - 4r^4 + 3r$

57. $(2z + y)(3z - 4y) = (2z)(3z) + (2z)(-4y) + (y)(3z) + (y)(-4y) = 6z^2 - 8yz + 3yz - 4y^2 =$

    $6x^2 - 5yz - 4y^2$

58. $(7m + 2n)(7m - 2n) = (7m)^2 - (2n)^2 = 49m^2 - 4n^2$

59. $(3p + 5)^2 = (3p)^2 + 2(3p)(5) + 5^2 = 9p^2 + 30p + 25$

60. $2(3r^2 + 4r + 2) - 3(-r^2 + 4r - 5) = 6r^2 + 8r + 4 + 3r^2 - 12r + 15 = 9r^2 - 4r + 19$

61. $p(4p - 6) + 2(3p - 8) = 4p^2 - 6p + 6p - 16 = 4p^2 - 16$

62. $m(5m - 2) + 9(5 - m) = 5m^2 - 2m + 45 - 9m = 5m^2 - 11m + 45$

63. $-y(y^2 - 4) + 6y^2(2y - 3) = -y^3 + 4y + 12y^3 - 18y^2 = 11y^3 - 18y^2 + 4y$

64. $-z^3(9 - z) + 4z(2 + 3z) = -9z^3 + z^4 + 8z + 12z^2 = z^4 - 9z^3 + 12z^2 + 8z$

R.2: Factoring

1. (a) $(x + 5y)^2 = x^2 + 10xy + 25y^2$; B

    (b) $(x - 5y)^2 = x^2 - 10xy + 25y^2$; C

    (c) $(x + 5y)(x - 5y) = x^2 - 25y^2$; A

    (d) $(5y + x)(5y - x) = 25y^2 - x^2$; D

2. (a) $(2x - 3)(4x^2 + 6x + 9) = 8x^3 - 27$; B

    (b) $(2x + 3)(4x^2 - 6x + 9) = 8x^3 + 27$; C

    (c) $(3 - 2x)(9 + 6x + 4x^2) = 27 - 8x^3$; A

3. $4k^2m^3 + 8k^4m^3 - 12k^2m^4 = 4k^2m^3(1 + 2k^2 - 3m)$

4. $28r^4s^2 + 7r^3s - 35r^4s^3 = 7r^3s(4rs + 1 - 5rs^2)$

5. $2(a + b) + 4m(a + b) = (a + b)(2 + 4m) = 2(a + b)(1 + 2m)$

6. $4(y - 2)^2 + 3(y - 2) = (y - 2)(4(y - 2) + 3) = (y - 2)(4y - 5)$

7. $(2y - 3)(y + 2) + (y + 5)(y + 2) = (y + 2)(2y - 3 + y + 5) = (y + 2)(3y + 2)$

8. $(6a - 1)(a + 2) + (6a - 1)(3a - 1) = (6a - 1)(a + 2 + 3a - 1) = (6a - 1)(4a + 1)$

9. $(5r - 6)(r + 3) - (2r - 1)(r + 3) = (r + 3)(5r - 6 - (2r - 1)) = (r + 3)(3r - 5)$

10. $(3z + 2)(z + 4) - (z + 6)(z + 4) = (z + 4)(3z + 2 - (z + 6)) = (z + 4)(2z - 4) = 2(z + 4)(z - 2)$

11. $2(m - 1) - 3(m - 1)^2 + 2(m - 1)^3 = (m - 1)(2 - 3(m - 1) + 2(m - 1)^2) =$

    $(m - 1)(2 - 3m + 3 + 2(m^2 - 2m + 1)) = (m - 1)(5 - 3m + 2m^2 - 4m + 2) = (m - 1)(2m^2 - 7m + 7)$

12. $5(a + 3)^3 - 2(a + 3) + (a + 3)^2 = (a + 3)(5(a + 3)^2 - 2 + (a + 3)) =$

    $(a + 3)(5(a^2 + 6a + 9) + 1 + a) = (a + 3)(5a^2 + 30a + 45 + 1 + a) = (a + 3)(5a^2 + 31a + 46)$

13. $6st + 9t - 10s - 15 = 3t(2s + 3) - 5(2s + 3) = (2s + 3)(3t - 5)$

14. $10ab - 6b + 35a - 21 = 2b(5a - 3) + 7(5a - 3) = (5a - 3)(2b + 7)$

15. $10x^2 - 12y + 15x - 8xy = 10x^2 + 15x - 8xy - 12y = 5x(2x + 3) - 4y(2x + 3) = (2x + 3)(5x - 4y)$

16. $2m^4 + 6 - am^4 - 3a = 2(m^4 + 3) - a(m^4 + 3) = (m^4 + 3)(2 - a)$

17. $t^3 + 2t^2 - 3t - 6 = t^2(t + 2) - 3(t + 2) = (t + 2)(t^2 - 3)$

18. $x^3 + 3x^2 - 5x - 15 = x^2(x + 3) - 5(x + 3) = (x + 3)(x^2 - 5)$

19. $(8a - 3)(2a - 5) = [-1(-8a + 3)][-1(-2a + 5)] = (3 - 8a)(5 - 2a);$ Both are correct.

20. $6a^2 - 48a - 120 = 6(a^2 - 8a - 20) = 6(a - 10)(a + 2)$

21. $8h^2 - 24h - 320 = 8(h^2 - 3h - 40) = 8(h - 8)(h + 5)$

22. $3m^3 + 12m^2 + 9m = 3m(m^2 + 4m + 3) = 3m(m + 3)(m + 1)$

23. $9y^4 - 54y^3 + 45y^2 = 9y^2(y^2 - 6y + 5) = 9y^2(y - 1)(y - 5)$

24. $6k^2 + 5kp - 6p^2 = (3k - 2p)(2k + 3p)$

25. $14m^2 + 11mr - 15r^2 = (7m - 5r)(2m + 3r)$

26. $5a^2 - 7ab - 6b^2 = (5a + 3b)(a - 2b)$

27. $12s^2 + 11st - 5t^2 = (3s - t)(4s + 5t)$

28. $9x^2 - 6x^3 + x^4 = x^2(9 - 6x + x^2) = x^2(3 - x)^2$

29. $30a^2 + am - m^2 = (5a + m)(6a - m)$

30. $24a^4 + 10a^3b - 4a^2b^2 = 2a^2(12a^2 + 5ab - 2b^2) = 2a^2(3a + 2b)(4a - b)$

31. $18x^5 + 15x^4z - 75x^3z^2 = 3x^3(6x^2 + 5xz - 25z^2) = 3x^3(2x + 5z)(3x - 5z)$

32. $9m^2 - 12m + 4 = (3m - 2)^2$

33. $16p^2 - 40p + 25 = (4p - 5)^2$

34. $32a^2 - 48ab + 18b^2 = 2(16a^2 - 24ab + 9b^2) = 2(4a - 3b)^2$

35. $20p^2 - 100pq + 125q^2 = 5(4p^2 - 20pq + 25q^2) = 5(2p - 5q)^2$

36. $4x^2y^2 + 28xy + 49 = (2xy + 7)^2$

37. $9m^2n^2 - 12mn + 4 = (3mn - 2)^2$

38. $(a - 3b)^2 - 6(a - 3b) + 9 = ((a - 3b) - 3)((a - 3b) - 3) = (a - 3b - 3)^2$

39. $(2p + q)^2 - 10(2p + q) + 25 = ((2p + q) - 5)((2p + q) - 5) = (2p + q - 5)^2$

40. $(5r + 2s)^2 + 6(5r + 2s) + 9 = ((5r + 2s) + 3)((5r + 2s) + 3) = (5r + 2s + 3)(5r + 2s + 3) =$

    $(5r + 2s + 3)^2$

41. $9a^2 - 16 = (3a + 4)(3a - 4)$

42. $16q^2 - 25 = (4q + 5)(4q - 5)$

43. $25s^4 - 9t^2 = (5s^2 + 3t)(5s^2 - 3t)$

44. $36z^2 - 81y^4 = 9(4z^2 - 9y^4) = 9(2z + 3y^2)(2z - 3y^2)$

45. $(a + b)^2 - 16 = (a + b + 4)(a + b - 4)$

46. $(p - 2q)^2 - 100 = (p - 2q + 10)(p - 2q - 10)$

47. $p^4 - 625 = (p^2 + 25)(p^2 - 25) = (p^2 + 25)(p + 5)(p - 5)$

48. $m^4 - 81 = (m^2 + 9)(m^2 - 9) = (m^2 + 9)(m + 3)(m - 3)$

49. $x^4 - 1 = (x^2 + 1)(x^2 - 1) = (x^2 + 1)(x + 1)(x - 1);$ b

50. $x^3 + 8 = (x + 2)(x^2 - 2x + 4);$ c

51. $8 - a^3 = 2^3 - a^3 = (2 - a)(4 + 2a + a^2)$

52. $r^3 + 27 = r^3 + 3^3 = (r + 3)(r^2 - 3r + 9)$

53. $125x^3 - 27 = (5x)^3 - 3^3 = (5x - 3)(25x^2 + 15x + 9)$

54. $8m^3 - 27n^3 = (2m)^3 - (3n)^3 = (2m - 3n)(4m^2 + 6mn + 9n^2)$

55. $27y^9 + 125z^6 = (3y^3)^3 + (5z^2)^3 = (3y^3 + 5z^2)(9y^6 - 15y^3z^2 + 25z^4)$

56. $27z^3 + 729y^3 = 27(z^3 + 27y^3) = 27(z + 3y)(z^2 - 3yz + 9y^2)$

57. $(r + 6)^3 - 216 = (r + 6)^3 - 6^3 = (r + 6 - 6)((r + 6)^2 + 6(r + 6) + 36) =$

    $r(r^2 + 12r + 36 + 6r + 36 + 36) = r(r^2 + 18r + 108)$

58. $(b + 3)^3 - 27 = (b + 3)^3 - 3^3 = (b + 3 - 3)((b + 3)^2 + 3(b + 3) + 9) =$

    $b(b^2 + 6b + 9 + 3b + 9 + 9) = b(b^2 + 9b + 27)$

59. $27 - (m + 2n)^3 = 3^3 - (m + 2n)^3 = (3 - (m + 2n))(9 + 3(m + 2n) + (m + 2n)^2) =$

    $(3 - m - 2n)(9 + 3m + 6n + m^2 + 4mn + 4n^2)$

60. $125 - (4a - b)^3 = 5^3 - (4a - b)^3 = (5 - (4a - b))(25 + 5(4a - b) + (4a - b)^2) =$

    $(5 - 4a + b)(25 + 20a - 5b + 16a^2 - 8ab + b^2)$

61. $3a^4 + 14a^2 - 5;$ Let $u = a^2 \Rightarrow 3u^2 + 14u - 5 = (3u - 1)(u + 5)$

    Then $a^2 = u \Rightarrow (3a^2 - 1)(a^2 + 5).$ a^2 was not substituted back in for u.

62. $m^4 - 3m^2 - 10;$ Let $u = m^2 \Rightarrow u^2 - 3u - 10 = (u + 2)(u - 5)$ Then $m^2 = u \Rightarrow (m^2 + 2)(m^2 - 5)$

63. $a^4 - 2a^2 - 48;$ Let $u = a^2 \Rightarrow u^2 - 2u - 48 = (u - 8)(u + 6)$ Then $a^2 = u \Rightarrow (a^2 - 8)(a^2 + 6)$

64. $7(3k - 1)^2 + 26(3k - 1) - 8;$ Let $u = 3k - 1 \Rightarrow 7u^2 + 26u - 8 = (7u - 2)(u + 4)$

    Then $3k - 1 = u \Rightarrow (7(3k - 1) - 2)(3k - 1 + 4) = (21k - 9)(3k + 3) = 3(7k - 3)(3)(k + 1) =$

    $9(7k - 3)(k + 1)$

65. $6(4z - 3)^2 + 7(4z - 3) - 3$; Let $u = 4z - 3 \Rightarrow 6u^2 + 7u - 3 = (3u - 1)(2u + 3)$

   Then $4z - 3 = u \Rightarrow (3(4z - 3) - 1)(2(4z - 3) + 3) = (12z - 10)(8z - 3) = 2(6z - 5)(8z - 3)$

66. $9(a - 4)^2 + 30(a - 4) + 25$; Let $u = a - 4 \Rightarrow 9u^2 + 30u + 25 = (3u + 5)^2$

   Then $a - 4 = u \Rightarrow (3(a - 4) + 5)^2 = (3a - 7)^2$

67. $20(4 - p)^2 - 3(4 - p) - 2$; Let $u = 4 - p \Rightarrow 20u^2 - 3u - 2 = (5u - 2)(4u + 1)$

   Then $4 - p = u \Rightarrow (5(4 - p) - 2)(4(4 - p) + 1) = (18 - 5p)(17 - 4p)$

68. $a^3(r + s) + b^2(r + s) = (r + s)(a^3 + b^2)$

69. $4b^2 + 4bc + c^2 - 16 = (4b^2 + 4b + c^2) - 16 = (2b + c)^2 - 16 = (2b + c + 4)(2b + c - 4)$

70. $(2y - 1)^2 - 4(2y - 1) + 4$; Let $u = 2y - 1 \Rightarrow u^2 - 4u + 4 = (u - 2)^2$

   Then $2y - 1 = u \Rightarrow (2y - 1 - 2)^2 = (2y - 3)^2$

71. $x^2 + xy - 5x - 5y = x(x + y) - 5(x + y) = (x + y)(x - 5)$

72. $8r^2 - 3rs + 10s^2$ does not factor, it is prime.

73. $p^4(m - 2n) + q(m - 2n) = (m - 2n)(p^4 + q)$

74. $36a^2 + 60a + 25 = (6a + 5)^2$

75. $4z^2 + 28z + 49 = (2z + 7)^2$

76. $6p^4 + 7p^2 - 3 = (3p^2 - 1)(2p^2 + 3)$

77. $1000x^3 + 343y^3 = (10x)^3 + (7y)^3 = (10x + 7y)(100x^2 - 70xy + 49y^2)$

78. $b^2 + 8b + 16 - a^2 = (b^2 + 8b + 16) - a^2 = (b + 4)^2 - a^2 = (b + 4 + a)(b + 4 - a)$

79. $125m^6 - 216 = (5m^2)^3 - 6^3 = (5m^2 - 6)(25m^4 + 30m^2 + 36)$

80. $q^2 + 6p + 9 - p^2 = (q^2 + 6p + 9) - p^2 = (q + 3)^2 - p^2 = (q + 3 + p)(q + 3 - p)$

81. $12m^2 + 16mn - 35n^2 = (6m - 7n)(2m + 5n)$

82. $216p^3 + 125q^3 = (6p)^3 + (5q)^3 = (6p + 5q)(36p^2 - 30pq + 25q^2)$

83. $4p^2 + 3p - 1 = (4p - 1)(p + 1)$

84. $100r^2 - 169s^2 = (10r + 13s)(10r - 13s)$

85. $144z^2 + 121$ does not factor, it is prime.

86. $(3a + 5)^2 - 18(3a + 5) + 81$; Let $u = 3a + 5 \Rightarrow u^2 - 18u + 81 = (u - 9)^2$

   Then $3a + 5 = u \Rightarrow (3a + 5 - 9)^2 = (3a - 4)^2$

87. $(x + y)^2 - (x - y)^2 = (x + y + (x - y))(x + y - (x - y)) = (2x)(2y) = 4xy$

88. $4z^4 - 7z^2 - 15 = (4z^2 + 5)(z^2 - 3)$

R.3: Rational Expressions

1. $\dfrac{x - 2}{x + 6}$; $x + 6 = 0, \{x \mid x \neq -6\}$

2. $\dfrac{x + 5}{x - 3}$; $x - 3 = 0, \{x \mid x \neq 3\}$

3. $\dfrac{2x}{5x-3}$; $5x-3=0$, $\left\{x \mid x \neq \dfrac{3}{5}\right\}$

4. $\dfrac{6x}{2x-1}$; $2x-1=0$, $\left\{x \mid x \neq \dfrac{1}{2}\right\}$

5. $\dfrac{-8}{x^2+1}$; No restrictions since $x^2+1>0$. Domain: $(-\infty, \infty)$

6. $\dfrac{3x}{3x^2+7}$; No restrictions since $3x^2+7>0$. Domain: $(-\infty, \infty)$

7. $\dfrac{3x+7}{(4x+2)(x-1)}$; $4x+2=0 \Rightarrow 4x=-2 \Rightarrow x=-\dfrac{2}{4}=-\dfrac{1}{2}$ and $x-1=0 \Rightarrow x=1$; $\left\{x \mid x \neq -\dfrac{1}{2}, 1\right\}$

8. $\dfrac{9x+12}{(2x+3)(x-5)}$; $2x+3=0 \Rightarrow 2x=-3 \Rightarrow x=-\dfrac{3}{2}$ and $x-5=0 \Rightarrow x=5$; $\left\{x \mid x \neq -\dfrac{3}{2}, 5\right\}$

9. $\dfrac{25p^3}{10p^2}=\dfrac{5p^2(5p)}{5p^2(2)}=\dfrac{5p}{2}$

10. $\dfrac{14z^3}{6z^2}=\dfrac{2z^2(7z)}{2z^2(3)}=\dfrac{7z}{3}$

11. $\dfrac{8k+16}{9k+18}=\dfrac{8(k+2)}{9(k+2)}=\dfrac{8}{9}$

12. $\dfrac{20r+10}{30r+15}=\dfrac{10(2r+1)}{15(2r+1)}=\dfrac{2}{3}$

13. $\dfrac{3(t+5)}{(t+5)(t-3)}=\dfrac{3}{t-3}$

14. $\dfrac{-8(y-4)}{(y+2)(y-4)}=\dfrac{-8}{y+2}$

15. $\dfrac{8x^2+16x}{4x^2}=\dfrac{8x(x+2)}{4x^2}=\dfrac{2(x+2)}{x}=\dfrac{2x+4}{x}$

16. $\dfrac{36y^2+72y}{9y}=\dfrac{36y(y+2)}{9y}=4(y+2)=4y+8$

17. $\dfrac{m^2-4m+4}{m^2+m-6}=\dfrac{(m-2)^2}{(m-2)(m+3)}=\dfrac{m-2}{m+3}$

18. $\dfrac{r^2-r-6}{r^2+r-12}=\dfrac{(r-3)(r+2)}{(r-3)(r+4)}=\dfrac{r+2}{r+4}$

19. $\dfrac{8m^2+6m-9}{16m^2-9}=\dfrac{(4m-3)(2m+3)}{(4m-3)(4m+3)}=\dfrac{2m+3}{4m+3}$

20. $\dfrac{6y^2+11y+4}{3y^2+7y+4}=\dfrac{(3y+4)(2y+1)}{(3y+4)(y+1)}=\dfrac{2y+1}{y+1}$

21. $\dfrac{15p^3}{9p^2} \div \dfrac{6p}{10p^2}=\dfrac{15p^3}{9p^2} \times \dfrac{10p^2}{6p}=\dfrac{25p^2}{9}$

22. $\dfrac{3r^2}{9r^3} \div \dfrac{8r^3}{6r}=\dfrac{3r^2}{9r^3} \times \dfrac{6r}{8r^3}=\dfrac{3}{12r^3}=\dfrac{1}{4r^3}$

23. $\dfrac{2k + 8}{6} \div \dfrac{3k + 12}{2} = \dfrac{2(k + 4)}{6} \times \dfrac{2}{3(k + 4)} = \dfrac{2}{9}$

24. $\dfrac{5m + 25}{10} \cdot \dfrac{12}{6m + 30} = \dfrac{5(m + 5)}{10} \cdot \dfrac{12}{6(m + 5)} = 1$

25. $\dfrac{x^2 + x}{5} \cdot \dfrac{25}{xy + y} = \dfrac{x(x + 1)}{5} \cdot \dfrac{25}{y(x + 1)} = \dfrac{5x}{y}$

26. $\dfrac{3m - 15}{4m - 20} \cdot \dfrac{m^2 - 10m + 25}{12m - 60} = \dfrac{3(m - 5)}{4(m - 5)} \cdot \dfrac{(m - 5)(m - 5)}{12(m - 5)} = \dfrac{m - 5}{16}$

27. $\dfrac{4a + 12}{2a - 10} \div \dfrac{a^2 - 9}{a^2 - a - 20} = \dfrac{4(a + 3)}{2(a - 5)} \cdot \dfrac{(a + 4)(a - 5)}{(a - 3)(a + 3)} = \dfrac{2(a + 4)}{a - 3}$

28. $\dfrac{6r - 18}{9r^2 + 6r - 24} \cdot \dfrac{12r - 16}{4r - 12} = \dfrac{6(r - 3)}{3(3r - 4)(r + 2)} \cdot \dfrac{4(3r - 4)}{4(r - 3)} = \dfrac{2}{r + 2}$

29. $\dfrac{p^2 - p - 12}{p^2 - 2p - 15} \cdot \dfrac{p^2 - 9p + 20}{p^2 - 8p + 16} = \dfrac{(p - 4)(p + 3)}{(p + 3)(p - 5)} \cdot \dfrac{(p - 5)(p - 4)}{(p - 4)(p - 4)} = 1$

30. $\dfrac{x^2 + 2x - 15}{x^2 + 11x + 30} \cdot \dfrac{x^2 + 2x - 24}{x^2 - 8x + 15} = \dfrac{(x + 5)(x - 3)}{(x + 5)(x + 6)} \cdot \dfrac{(x + 6)(x - 4)}{(x - 3)(x - 5)} = \dfrac{x - 4}{x - 5}$

31. $\dfrac{m^2 + 3m + 2}{m^2 + 5m + 4} \div \dfrac{m^2 + 5m + 6}{m^2 + 10m + 24} = \dfrac{(m + 2)(m + 1)}{(m + 4)(m + 1)} \cdot \dfrac{(m + 4)(m + 6)}{(m + 3)(m + 2)} = \dfrac{m + 6}{m + 3}$

32. $\dfrac{y^2 + y - 2}{y^2 + 3y - 4} \div \dfrac{y^2 + 3y + 2}{y^2 + 4y + 3} = \dfrac{(y + 2)(y - 1)}{(y + 4)(y - 1)} \cdot \dfrac{(y + 3)(y + 1)}{(y + 1)(y + 2)} = \dfrac{y + 3}{y + 4}$

33. $\dfrac{2m^2 - 5m - 12}{m^2 - 10m + 24} \div \dfrac{4m^2 - 9}{m^2 - 9m + 18} = \dfrac{(2m + 3)(m - 4)}{(m - 4)(m - 6)} \cdot \dfrac{(m - 3)(m - 6)}{(2m + 3)(2m - 3)} = \dfrac{m - 3}{2m - 3}$

34. $\dfrac{6n^2 - 5n - 6}{6n^2 + 5n - 6} \cdot \dfrac{12n^2 - 17n + 6}{12n^2 - n - 6} = \dfrac{(3n + 2)(2n - 3)}{(2n + 3)(3n - 2)} \cdot \dfrac{(3n - 2)(4n - 3)}{(3n + 2)(4n - 3)} = \dfrac{2n - 3}{2n + 3}$

35. $\dfrac{x^3 + y^3}{x^2 - y^2} \cdot \dfrac{x + y}{x^2 - xy + y^2} = \dfrac{(x + y)(x^2 - xy + y^2)}{(x + y)(x - y)} \cdot \dfrac{x + y}{x^2 - xy + y^2} = \dfrac{x + y}{x - y}$

36. $\dfrac{8y^3 - 125}{4y^2 - 20y + 25} \cdot \dfrac{2y - 5}{y} = \dfrac{(2y - 5)(4y^2 + 10y + 25)}{(2y - 5)(2y - 5)} \cdot \dfrac{2y - 5}{y} = \dfrac{4y^2 + 10y + 25}{y}$

37. $\dfrac{x^3 + y^3}{x^3 - y^3} \cdot \dfrac{x^2 - y^2}{x^2 + 2xy + y^2} = \dfrac{(x + y)(x^2 - xy + y^2)}{(x - y)(x^2 + xy + y^2)} \cdot \dfrac{(x - y)(x + y)}{(x + y)(x + y)} = \dfrac{x^2 - xy + y^2}{x^2 + xy + y^2}$

38. $\dfrac{x^2 - y^2}{(x - y)^2} \cdot \dfrac{x^2 - xy + y^2}{x^2 - 2xy + y^2} \div \dfrac{x^3 + y^3}{(x - y)^4} = \dfrac{(x + y)(x - y)}{(x - y)^2} \cdot \dfrac{x^2 - xy + y^2}{(x - y)^2} \cdot \dfrac{(x - y)^4}{(x + y)(x^2 - xy + y^2)} = x - y$

39. A: $\dfrac{x - 4}{x + 4} \neq -1$, B: $\dfrac{-x - 4}{x + 4} = \dfrac{-1(x + 4)}{x + 4} = -1$, C: $\dfrac{x - 4}{4 - x} = \dfrac{x - 4}{-1(x - 4)} = -1$, D: $\dfrac{x - 4}{-x - 4} = \dfrac{x - 4}{-1(x + 4)} \neq -1$

40. To find the lowest common denominator for two fractions, factor each denominator; find the product of all of the prime factors that appear in either denominator. Choose the largest exponent of those prime factors that appears in that factor.

41. $\dfrac{3}{2k} + \dfrac{5}{3k} = \dfrac{9}{6k} + \dfrac{10}{6k} = \dfrac{19}{6k}$

42. $\dfrac{8}{5p} + \dfrac{3}{4p} = \dfrac{32}{20p} + \dfrac{15}{20p} = \dfrac{47}{20p}$

43. $\dfrac{a+1}{2} - \dfrac{a-1}{2} = \dfrac{a+1-(a-1)}{2} = \dfrac{2}{2} = 1$

44. $\dfrac{y+6}{5} - \dfrac{y-6}{5} = \dfrac{y+6-(y-6)}{5} = \dfrac{12}{5}$

45. $\dfrac{3}{p} + \dfrac{1}{2} = \dfrac{6}{2p} + \dfrac{p}{2p} = \dfrac{6+p}{2p}$

46. $\dfrac{9}{r} - \dfrac{2}{3} = \dfrac{27}{3r} - \dfrac{2r}{3r} = \dfrac{27-2r}{3r}$

47. $\dfrac{1}{6m} + \dfrac{2}{5m} + \dfrac{4}{m} = \dfrac{5(1)+6(2)+30(4)}{30m} = \dfrac{5+12+120}{30m} = \dfrac{137}{30m}$

48. $\dfrac{8}{3p} + \dfrac{5}{4p} + \dfrac{9}{2p} = \dfrac{4(8)+3(5)+6(9)}{12p} = \dfrac{32+15+54}{12p} = \dfrac{101}{12p}$

49. $\dfrac{1}{a+1} - \dfrac{1}{a-1} = \dfrac{a-1-(a+1)}{(a+1)(a-1)} = \dfrac{-2}{(a+1)(a-1)}$

50. $\dfrac{1}{x+z} + \dfrac{1}{x-z} = \dfrac{x-z+(x+z)}{(x+z)(x-z)} = \dfrac{2x}{(x+z)(x-z)}$

51. $\dfrac{m+1}{m-1} + \dfrac{m-1}{m+1} = \dfrac{(m+1)^2+(m-1)^2}{(m-1)(m+1)} = \dfrac{m^2+2m+1+m^2-2m+1}{(m-1)(m+1)} = \dfrac{2m^2+2}{(m-1)(m+1)}$

52. $\dfrac{2}{x-1} + \dfrac{1}{1-x} = \dfrac{2}{x-1} + \dfrac{-1}{x-1} = \dfrac{1}{x-1}$ or $\dfrac{-1}{1-x}$

53. $\dfrac{3}{a-2} - \dfrac{1}{2-a} = \dfrac{3}{a-2} - \dfrac{-1}{a-2} = \dfrac{3+1}{a-2} = \dfrac{4}{a-2}$ or $\dfrac{-4}{2-a}$

54. $\dfrac{q}{p-q} - \dfrac{q}{q-p} = \dfrac{q}{p-q} - \dfrac{-q}{p-q} = \dfrac{q+q}{p-q} = \dfrac{2q}{p-q}$ or $\dfrac{-2q}{q-p}$

55. $\dfrac{x+y}{2x-y} - \dfrac{2x}{y-2x} = \dfrac{x+y}{2x-y} - \dfrac{-2x}{2x-y} = \dfrac{x+y+2x}{2x-y} = \dfrac{3x+y}{2x-y}$ or $\dfrac{-3x-y}{y-2x}$

56. $\dfrac{m-4}{3m-4} + \dfrac{3m+2}{4-3m} = \dfrac{m-4}{3m-4} + \dfrac{-(3m+2)}{3m-4} = \dfrac{m-4-3m-2}{3m-4} = \dfrac{-2m-6}{3m-4}$ or $\dfrac{2m+6}{4-3m}$

57. $\dfrac{1}{a^2-5a+6} - \dfrac{1}{a^2-4} = \dfrac{1}{(a-2)(a-3)} - \dfrac{1}{(a-2)(a+2)} = \dfrac{1(a+2)-1(a-3)}{(a-2)(a-3)(a+2)} =$
$\dfrac{a+2-a+3}{(a-2)(a-3)(a+2)} = \dfrac{5}{(a-2)(a-3)(a+2)}$

58. $\dfrac{-3}{m^2-m-2} - \dfrac{1}{m^2+3m+2} = \dfrac{-3}{(m-2)(m+1)} - \dfrac{1}{(m+2)(m+1)} = \dfrac{-3(m+2)-1(m-2)}{(m-2)(m+1)(m+2)} =$
$\dfrac{-3m-6-m+2}{(m-2)(m+1)(m+2)} = \dfrac{-4m-4}{(m-2)(m+1)(m+2)} = \dfrac{-4(m+1)}{(m-2)(m+1)(m+2)} = \dfrac{-4}{(m-2)(m+2)}$

59. $\dfrac{1}{x^2+x-12} - \dfrac{1}{x^2-7x+12} + \dfrac{1}{x^2-16} = \dfrac{1}{(x+4)(x-3)} - \dfrac{1}{(x-3)(x-4)} + \dfrac{1}{(x-4)(x+4)} =$
$\dfrac{1(x-4)-1(x+4)+1(x-3)}{(x+4)(x-3)(x-4)} = \dfrac{x-4-x-4+x-3}{(x+4)(x-3)(x-4)} = \dfrac{x-11}{(x+4)(x-3)(x-4)}$

60. $\dfrac{2}{2p^2 - 9p - 5} + \dfrac{p}{3p^2 - 17p + 10} - \dfrac{2p}{6p^2 - p - 2} =$

$\dfrac{2}{(2p + 1)(p - 5)} + \dfrac{p}{(3p - 2)(p - 5)} - \dfrac{2p}{(3p - 2)(2p + 1)} = \dfrac{2(3p - 2) + p(2p + 1) - 2p(p - 5)}{(2p + 1)(p - 5)(3p - 2)} =$

$\dfrac{6p - 4 + 2p^2 + p - 2p^2 + 10p}{(2p + 1)(p - 5)(3p - 2)} = \dfrac{17p - 4}{(2p + 1)(p - 5)(3p - 2)}$

61. $\dfrac{3a}{a^2 + 5a - 6} - \dfrac{2a}{a^2 + 7a + 6} = \dfrac{3a}{(a + 6)(a - 1)} - \dfrac{2a}{(a + 6)(a + 1)} = \dfrac{3a(a + 1) - 2a(a - 1)}{(a + 6)(a - 1)(a + 1)} =$

$\dfrac{3a^2 + 3a - 2a^2 + 2a}{(a + 6)(a - 1)(a + 1)} = \dfrac{a^2 + 5a}{(a + 6)(a - 1)(a + 1)}$

62. $\dfrac{2k}{k^2 + 4k + 3} + \dfrac{3k}{k^2 + 5k + 6} = \dfrac{2k}{(k + 3)(k + 1)} + \dfrac{3k}{(k + 2)(k + 3)} = \dfrac{2k(k + 2) + 3k(k + 1)}{(k + 3)(k + 1)(k + 2)} =$

$\dfrac{2k^2 + 4k + 3k^2 + 3k}{(k + 3)(k + 1)(k + 2)} = \dfrac{5k^2 + 7k}{(k + 3)(k + 1)(k + 2)}$

63. $\dfrac{1 + \frac{1}{x}}{1 - \frac{1}{x}} = \dfrac{x(1 + \frac{1}{x})}{x(1 - \frac{1}{x})} = \dfrac{x + 1}{x - 1}$

64. $\dfrac{2 - \frac{2}{y}}{2 + \frac{2}{y}} = \dfrac{y(2 - \frac{2}{y})}{y(2 + \frac{2}{y})} = \dfrac{2y - 2}{2y + 2} = \dfrac{2(y - 1)}{2(y + 1)} = \dfrac{y - 1}{y + 1}$

65. $\dfrac{\frac{1}{x + 1} - \frac{1}{x}}{\frac{1}{x}} = \dfrac{x(x + 1)(\frac{1}{x + 1} - \frac{1}{x})}{x(x + 1)(\frac{1}{x})} = \dfrac{x - (x + 1)}{x + 1} = \dfrac{-1}{x + 1}$

66. $\dfrac{\frac{1}{y + 3} - \frac{1}{y}}{\frac{1}{y}} = \dfrac{y(y + 3)(\frac{1}{y + 3} - \frac{1}{y})}{y(y + 3)(\frac{1}{y})} = \dfrac{y - (y + 3)}{y + 3} = \dfrac{-3}{y + 3}$

67. $\dfrac{1 + \frac{1}{1 - b}}{1 - \frac{1}{1 + b}} = \dfrac{(1 - b)(1 + b)(1 + \frac{1}{1 - b})}{(1 - b)(1 + b)(1 - \frac{1}{1 + b})} = \dfrac{(1 - b)(1 + b) + (1 + b)}{(1 - b)(1 + b) - (1 - b)} = \dfrac{1 - b^2 + 1 + b}{1 - b^2 - 1 + b} =$

$\dfrac{-b^2 + b + 2}{-b^2 + b} = \dfrac{(2 - b)(1 + b)}{b(1 - b)}$

68. $m - \dfrac{m}{m + \frac{1}{2}} = m - \dfrac{2(m)}{2(m + \frac{1}{2})} = m - \dfrac{2m}{2m + 1} = \dfrac{m(2m + 1) - 2m}{2m + 1} = \dfrac{2m^2 + m - 2m}{2m + 1} =$

$\dfrac{2m^2 - m}{2m + 1} = \dfrac{m(2m - 1)}{2m + 1}$

69. $\dfrac{m - \frac{1}{m^2 - 4}}{\frac{1}{m + 2}} = \dfrac{(m^2 - 4)(m - \frac{1}{m^2 - 4})}{(m^2 - 4)(\frac{1}{m + 2})} = \dfrac{m(m^2 - 4) - 1}{m - 2} = \dfrac{m^3 - 4m - 1}{m - 2}$

70. $\dfrac{\frac{3}{p^2 - 16} + p}{\frac{1}{p - 4}} = \dfrac{(p^2 - 16)(\frac{3}{p^2 - 16} + p)}{(p^2 - 16)(\frac{1}{p - 4})} = \dfrac{3 + p(p^2 - 16)}{p + 4} = \dfrac{3 + p^3 - 16p}{p + 4} = \dfrac{p^3 - 16p + 3}{p + 4}$

R.4: Negative and Rational Exponents

1. $\left(\dfrac{4}{9}\right)^{3/2} = \left[\left(\dfrac{4}{9}\right)^{1/2}\right]^3 = \left(\dfrac{2}{3}\right)^3 = \dfrac{2^3}{3^3} = \dfrac{8}{27}$; E

2. $\left(\dfrac{4}{9}\right)^{-3/2} = \left(\dfrac{9}{4}\right)^{3/2} = \left[\left(\dfrac{9}{4}\right)^{1/2}\right]^3 = \left(\dfrac{3}{2}\right)^3 = \dfrac{3^3}{2^3} = \dfrac{27}{8}$; G

3. $-\left(\dfrac{9}{4}\right)^{3/2} = -\left[\left(\dfrac{9}{4}\right)^{1/2}\right]^3 = -\left(\dfrac{3}{2}\right)^3 = -\dfrac{3^3}{2^3} = -\dfrac{27}{8}$; F

4. $-\left(\dfrac{4}{9}\right)^{-3/2} = -\left(\dfrac{9}{4}\right)^{3/2} = -\left[\left(\dfrac{9}{4}\right)^{1/2}\right]^3 = -\left(\dfrac{3}{2}\right)^3 = -\dfrac{3^3}{2^3} = -\dfrac{27}{8}$; F

5. $\left(\dfrac{8}{27}\right)^{2/3} = \left[\left(\dfrac{8}{27}\right)^{1/3}\right]^2 = \left(\dfrac{2}{3}\right)^2 = \dfrac{2^2}{3^2} = \dfrac{4}{9}$; D

6. $\left(\dfrac{8}{27}\right)^{-2/3} = \left(\dfrac{27}{8}\right)^{2/3} = \left[\left(\dfrac{27}{8}\right)^{1/3}\right]^2 = \left(\dfrac{3}{2}\right)^2 = \dfrac{3^2}{2^2} = \dfrac{9}{4}$; A

7. $-\left(\dfrac{27}{8}\right)^{2/3} = -\left[\left(\dfrac{27}{8}\right)^{1/3}\right]^2 = -\left(\dfrac{3}{2}\right)^2 = -\dfrac{3^2}{2^2} = -\dfrac{9}{4}$; B

8. $-\left(\dfrac{27}{8}\right)^{-2/3} = -\left(\dfrac{8}{27}\right)^{2/3} = -\left[\left(\dfrac{8}{27}\right)^{1/3}\right]^2 = -\left(\dfrac{2}{3}\right)^2 = -\dfrac{2^2}{3^2} = -\dfrac{4}{9}$; C

9. $(-4)^{-3} = \left(-\dfrac{1}{4}\right)^3 = -\dfrac{1}{64}$

10. $(-5)^{-2} = \left(-\dfrac{1}{5}\right)^2 = \dfrac{1}{25}$

11. $\left(\dfrac{1}{2}\right)^{-3} = 2^3 = 8$

12. $\left(\dfrac{2}{3}\right)^{-2} = \left(\dfrac{3}{2}\right)^2 = \dfrac{9}{4}$

13. $-4^{1/2} = -(4^{1/2}) = -2$

14. $25^{1/2} = 5$

15. $8^{2/3} = (8^{1/3})^2 = 2^2 = 4$

16. $-81^{3/4} = -(81^{1/4})^3 = -(3)^3 = -27$

17. $27^{-2/3} = \left[\left(\dfrac{1}{27}\right)^{1/3}\right]^2 = \left(\dfrac{1}{3}\right)^2 = \dfrac{1}{9}$

18. $(-32)^{-4/5} = \left[\left(-\dfrac{1}{32}\right)^{1/5}\right]^4 = \left(-\dfrac{1}{2}\right)^4 = \dfrac{1}{16}$

19. $\left(\dfrac{27}{64}\right)^{-4/3} = \left[\left(\dfrac{64}{27}\right)^{1/3}\right]^4 = \left(\dfrac{4}{3}\right)^4 = \dfrac{256}{81}$

20. $\left(\dfrac{121}{100}\right)^{-3/2} = \left[\left(\dfrac{100}{121}\right)^{1/2}\right]^3 = \left(\dfrac{10}{11}\right)^3 = \dfrac{1000}{1331}$

21. $(16p^4)^{1/2} = 4p^2$

22. $(36r^6)^{1/2} = 6r^3$

23. $(27x^6)^{2/3} = [(27x^6)^{1/3}]^2 = (3x^2)^2 = 9x^4$

24. $(64a^{12})^{5/6} = [(64a^{12})^{1/6}]^5 = (2a^2)^5 = 32a^{10}$

25. $2^{-3} \cdot 2^{-4} = 2^{-7} = \dfrac{1}{2^7}$

26. $5^{-2} \cdot 5^{-6} = 5^{-8} = \dfrac{1}{5^8}$

27. $27^{-2} \cdot 27^{-1} = 27^{-3} = \dfrac{1}{27^3}$

28. $9^{-4} \cdot 9^{-1} = 9^{-5} = \dfrac{1}{9^5}$

29. $\dfrac{4^{-2} \cdot 4^{-1}}{4^{-3}} = 4^{(-2+(-1)-(-3))} = 4^0 = 1$

30. $\dfrac{3^{-1} \cdot 3^{-4}}{3^2 \cdot 3^{-2}} = 3^{(-1+(-4)-2-(-2))} = 3^{-5} = \dfrac{1}{3^5}$

31. $(m^{2/3})(m^{5/3}) = m^{(2/3+5/3)} = m^{7/3}$

32. $(x^{4/5})(x^{2/5}) = x^{(4/5+2/5)} = x^{6/5}$

33. $(1+n)^{1/2}(1+n)^{3/4} = (1+n)^{(2/4+3/4)} = (1+n)^{5/4}$

34. $(m+7)^{-1/6}(m+7)^{-2/3} = (m+7)^{(-1/6-4/6)} = (m+7)^{-5/6} = \dfrac{1}{(m+7)^{5/6}}$

35. $(2y^{3/4}z)(3y^{-2}z^{-1/3}) = 6y^{(3/4-8/4)}z^{(1-1/3)} = 6y^{-5/4}z^{2/3} = \dfrac{6z^{2/3}}{y^{5/4}}$

36. $(4a^{-1}b^{2/3})(a^{3/2}b^{-3}) = 4a^{(-2/2+3/2)}b^{(2/3-9/3)} = 4a^{1/2}b^{-7/3} = \dfrac{4a^{1/2}}{b^{7/3}}$

37. $(4a^{-2}b^7)^{1/2} \cdot (2a^{1/4}b^3)^5 = (2a^{-1}b^{7/2})(2^5a^{5/4}b^{15}) = 2^6a^{(-1+5/4)}b^{(7/2+30/2)} = 2^6a^{1/4}b^{37/2}$

38. $(x^{-2}y^{1/3})^5 \cdot (8x^2y^{-2})^{-1/3} = (x^{-10}y^{5/3})(8^{-1/3}x^{-2/3}y^{2/3}) = \dfrac{1}{2}x^{(-30/3-2/3)}y^{(5/3+2/3)} = \dfrac{1}{2}x^{-32/3}y^{7/3} = \dfrac{y^{7/3}}{2x^{32/3}}$

39. $\left(\dfrac{r^{-2}}{s^{-5}}\right)^{-3} = \dfrac{r^6}{s^{15}}$

40. $\left(\dfrac{p^{-1}}{q^{-5}}\right)^{-2} = \dfrac{p^2}{q^{10}}$

41. $\left(\dfrac{-a}{b^{-3}}\right)^{-1} = \left(\dfrac{b^{-3}}{-a}\right)^1 = \dfrac{-1}{ab^3}$

42. $\dfrac{7^{-1/3} \cdot 7r^{-3}}{7^{2/3}r^{-2}} = 7^{(-1/3+1-2/3)}r^{(-3-(-2))} = 7^0r^{-1} = \dfrac{1}{r}$

43. $\dfrac{12^{5/4} \cdot y^{-2}}{12^{-1}y^{-3}} = 12^{(5/4-(-1))}y^{(-2-(-3))} = 12^{9/4}y$

44. $\dfrac{6k^{-4}(3k^{-1})^{-2}}{2^3k^{1/2}} = \dfrac{6k^{-4}3^{-2}k^2}{2^3k^{1/2}} = 6 \cdot 3^{-2} \cdot 2^{-3}k^{(-4+2-1/2)} = \dfrac{6}{9 \cdot 8}k^{-5/2} = \dfrac{1}{12k^{5/2}}$

45. $\dfrac{8p^{-3}(4p^2)^{-2}}{p^{-5}} = \dfrac{8p^5}{p^3(4p^2)^2} = \dfrac{8p^5}{16p^7} = \dfrac{1}{2p^2}$

46. $\dfrac{k^{-3/5}h^{-1/3}t^{2/5}}{k^{-1/5}h^{-2/3}t^{1/5}} = k^{(-3/5+1/5)}h^{(-1/3+2/3)}t^{(2/5-1/5)} = k^{-2/5}h^{1/3}t^{1/5} = \dfrac{h^{1/3}t^{1/5}}{k^{2/5}}$

47. $\dfrac{m^{7/3}n^{-2/5}p^{3/8}}{m^{-2/3}n^{3/5}p^{-5/8}} = m^{(7/3+2/3)}n^{(-2/5-3/5)}p^{(3/8+5/8)} = m^3n^{-1}p^1 = \dfrac{m^3p}{n}$

48. $\dfrac{m^{2/5}m^{3/5}m^{-4/5}}{m^{1/5}m^{-6/5}} = m^{(2/5+3/5-4/5-1/5+6/5)} = m^{6/5}$

49. $\dfrac{-4a^{-1}a^{2/3}}{a^{-2}} = -4a^{(-3/3+2/3+6/3)} = -4a^{5/3}$

50. $\dfrac{8y^{2/3}y^{-1}}{2^{-1}y^{3/4}y^{-1/6}} = 8(2^1)y^{(2/3-1-3/4+1/6)} = 16y^{(8/12-12/12-9/12+2/12)} = 16y^{-11/12} = \dfrac{16}{y^{11/12}}$

51. $\dfrac{(k+5)^{1/2}(k+5)^{-1/4}}{(k+5)^{3/4}} = (k+5)^{(1/2-1/4-3/4)} = (k+5)^{-1/2} = \dfrac{1}{(k+5)^{1/2}}$

52. $\dfrac{(x+y)^{-5/8}(x+y)^{3/8}}{(x+y)^{1/8}(x+y)^{-1/8}} = (x+y)^{(-5/8+3/8-1/8+1/8)} = (x+y)^{-2/8} = \dfrac{1}{(x+y)^{1/4}}$

53. $y^{5/8}(y^{3/8} - 10y^{11/8}) = y^{(5/8+3/8)} - 10y^{(5/8+11/8)} = y - 10y^2$

54. $p^{11/5}(3p^{4/5} + 9p^{19/5}) = 3p^{(11/5+4/5)} + 9p^{(11/5+19/5)} = 3p^3 + 9p^6$

55. $-4k(k^{7/3} - 6k^{1/3}) = -4k^{(1+7/3)} + 24k^{(1+1/3)} = -4k^{10/3} + 24k^{4/3}$

56. $-5y(3y^{9/10} + 4y^{3/10}) = -15y^{(1+9/10)} - 20y^{(1+3/10)} = -15y^{19/10} - 20y^{13/10}$

57. $(x + x^{1/2})(x - x^{1/2}) = x^2 - x^{(1+1/2)} + x^{(1+1/2)} - x^{(1/2+1/2)} = x^2 - x$

58. $(2z^{1/2} + z)(z^{1/2} - z) = 2z^{(1/2+1/2)} - 2z^{(1/2+1)} + z^{(1+1/2)} - z^{(1+1)} = 2z - z^{3/2} - z^2$

59. $(r^{1/2} - r^{-1/2})^2 = (r^{1/2})^2 - 2(r^{1/2})(r^{-1/2}) + (r^{-1/2})^2 = r - 2r^0 + r^{-1} = r - 2 + \dfrac{1}{r}$

60. $(p^{1/2} - p^{-1/2})(p^{1/2} + p^{-1/2}) = (p^{1/2})^2 - (p^{-1/2})^2 = p^1 - p^{-1} = p - \dfrac{1}{p}$

61. $4k^{-1} + k^{-2} = 4k^1 \cdot k^{-2} + k^{-2} = k^{-2}(4k + 1)$

62. $y^{-5} - 3y^{-3} = y^{-5} - 3y^2 \cdot y^{-5} = y^{-5}(1 - 3y^2)$

63. $9z^{-1/2} + 2z^{1/2} = 9z^{-1/2} + 2z^{-1/2} \cdot z^{2/2} = z^{-1/2}(9 + 2z)$

64. $3m^{2/3} - 4m^{-1/3} = 3m^{-1/3} \cdot m^{3/3} - 4m^{-1/3} = m^{-1/3}(3m - 4)$

65. $p^{-3/4} - 2p^{-7/4} = p^{-7/4} \cdot p^{4/4} - 2p^{-7/4} = p^{-7/4}(p - 2)$

66. $6r^{-2/3} - 5r^{-5/3} = 6r^{-5/3} \cdot r^{3/3} - 5r^{-5/3} = r^{-5/3}(6r - 5)$

67. $(p+4)^{-3/2} + (p+4)^{-1/2} + (p+4)^{1/2} = (p+4)^{-3/2}(1 + (p+4) + (p+4)^2) =$

$(p+4)^{-3/2}(1 + p + 4 + p^2 + 8p + 16) = (p+4)^{-3/2}(p^2 + 9p + 21)$

68. $(3r+1)^{-2/2} + (3r+1)^{1/3} + (3r+1)^{4/3} = (3r+1)^{-2/3}(1 + (3r+1) + (3r+1)^2) =$

$(3r+1)^{-2/3}(1 + 3r + 1 + 9r^2 + 6r + 1) = (3r+1)^{-2/3}(9r^2 + 9r + 3) = 3(3r+1)^{-2/3}(3r^2 + 3r + 1)$

R.5: Radicals

1. $(-3x)^{1/3} = \sqrt[3]{-3x}$; F

2. $-3x^{1/3} = -3\sqrt[3]{x}$; B

3. $(-3x)^{-1/3} = \dfrac{1}{(-3x)^{1/3}} = \dfrac{1}{\sqrt[3]{-3x}}$; H

4. $-3x^{-1/3} = \dfrac{-3}{x^{1/3}} = \dfrac{-3}{\sqrt[3]{x}};$ D

5. $(3x)^{1/3} = \sqrt[3]{3x};$ G

6. $3x^{-1/3} = \dfrac{3}{x^{1/3}} = \dfrac{3}{\sqrt[3]{x}};$ A

7. $(3x)^{-1/3} = \dfrac{1}{(3x)^{1/3}} = \dfrac{1}{\sqrt[3]{3x}};$ C

8. $3x^{1/3} = 3\sqrt[3]{x};$ E

9. $(-m)^{2/3} = \sqrt[3]{(-m)^2}$ or $(\sqrt[3]{(-m)})^2$

10. $p^{5/4} = \sqrt[4]{p^5}$ or $(\sqrt[4]{p})^5$

11. $(2m + p)^{2/3} = \sqrt[3]{(2m + p)^2}$ or $(\sqrt[3]{(2m + p)})^2$

12. $(5r + 3t)^{4/7} = \sqrt[7]{(5r + 3t)^4}$ or $(\sqrt[7]{5r + 3t})^4$

13. $\sqrt[5]{k^2} = k^{2/5}$

14. $-\sqrt[4]{z^5} = -z^{5/4}$

15. $-3\sqrt{5p^3} = -3(5p^3)^{1/2} = -3 \cdot 5^{1/2}p^{3/2}$

16. $m\sqrt{2y^5} = m(2y^5)^{1/2} = m \cdot 2^{1/2}y^{5/2}$

17. A: $\sqrt{ab} = \sqrt{a} \cdot \sqrt{b}$ is true for $a > 0, b > 0.$

18. $3, 5, 7, \dots$ (odd positive integers greater than or equal to 3)

19. $\sqrt{9ax^2} = 3x\sqrt{a}$ is true for all $x \geq 0.$

20. The expression D is not simplified because the radicand is a fraction. The expression may be simplified by using the rule $\sqrt[n]{\dfrac{a}{b}} = \dfrac{\sqrt[n]{a}}{\sqrt[n]{b}};\ \sqrt{\dfrac{3}{4}} = \dfrac{\sqrt{3}}{\sqrt{4}} = \dfrac{\sqrt{3}}{2}.$

21. $\sqrt[3]{125} = \sqrt[3]{5^3} = 5$

22. $\sqrt[4]{81} = \sqrt[4]{3^4} = 3$

23. $\sqrt[5]{-3125} = \sqrt[5]{(-5)^5} = -5$

24. $\sqrt[3]{343} = \sqrt[3]{7^3} = 7$

25. $\sqrt{50} = \sqrt{25 \cdot 2} = 5\sqrt{2}$

26. $\sqrt{45} = \sqrt{9 \cdot 5} = 3\sqrt{5}$

27. $\sqrt[3]{81} = \sqrt[3]{3^4} = 3\sqrt[3]{3}$

28. $\sqrt[3]{250} = \sqrt[3]{5^3 \cdot 2} = 5\sqrt[3]{2}$

29. $-\sqrt[4]{32} = -\sqrt[4]{2^5} = -2\sqrt[4]{2}$

30. $-\sqrt[4]{243} = -\sqrt[4]{3^5} = -3\sqrt[4]{3}$

31. $-\sqrt{\dfrac{9}{5}} = -\dfrac{\sqrt{9}}{\sqrt{5}} \cdot \dfrac{\sqrt{5}}{\sqrt{5}} = -\dfrac{3\sqrt{5}}{5}$

32. $-\sqrt[3]{\dfrac{3}{2}} = -\dfrac{\sqrt[3]{3}}{\sqrt[3]{2}} \cdot \dfrac{\sqrt[3]{2^2}}{\sqrt[3]{2^2}} = -\dfrac{\sqrt[3]{12}}{2}$

33. $-\sqrt[3]{\dfrac{4}{5}} = -\dfrac{\sqrt[3]{4}}{\sqrt[3]{5}} \cdot \dfrac{\sqrt[3]{5^2}}{\sqrt[3]{5^2}} = -\dfrac{\sqrt[3]{100}}{5}$

34. $\sqrt[4]{\dfrac{3}{2}} = \dfrac{\sqrt[4]{3}}{\sqrt[4]{2}} \cdot \dfrac{\sqrt[4]{2^3}}{\sqrt[4]{2^3}} = \dfrac{\sqrt[4]{24}}{2}$

35. $\sqrt[3]{16(-2)^4(2)^8} = \sqrt[3]{(2)^4(-2)^4(2)^8} = \sqrt[3]{(2)^{12}(-2)^4} = 2^4(-2)\sqrt[3]{-2} = -32\sqrt[3]{-2}$ or $32\sqrt[3]{2}$

36. $\sqrt[3]{25(3)^4(5)^3} = \sqrt[3]{(5)^2(3)^4(5)^3} = \sqrt[3]{(5)^5(3)^4} = 5 \cdot 3\sqrt[3]{5^2 \cdot 3} = 15\sqrt[3]{75}$

37. $\sqrt{8x^5z^8} = \sqrt{2^3x^5z^8} = 2x^2z^4\sqrt{2x}$

38. $\sqrt{24m^6n^5} = \sqrt{2^3 \cdot 3m^6n^5} = 2m^3n^2\sqrt{6n}$

39. $\sqrt[3]{16z^5x^8y^4} = \sqrt[3]{2^4z^5x^8y^4} = 2zx^2y\sqrt[3]{2z^2x^2y}$

40. $-\sqrt[6]{64a^{12}b^8} = -\sqrt[6]{2^6a^{12}b^8} = -2a^2b\sqrt[6]{b^2} = -2a^2b\sqrt[3]{b}$

41. $\sqrt[4]{m^2n^7p^8} = np^2\sqrt[4]{m^2n^3}$

42. $\sqrt[4]{x^8y^7z^9} = x^2yz^2\sqrt[4]{y^3z}$

43. $\sqrt[4]{x^4 + y^4}$ cannot be simplified.

44. $\sqrt[3]{27 + a^3}$ cannot be simplified.

45. $\sqrt{\dfrac{2}{3x}} = \dfrac{\sqrt{2}}{\sqrt{3x}} \cdot \dfrac{\sqrt{3x}}{\sqrt{3x}} = \dfrac{\sqrt{6x}}{3x}$

46. $\sqrt{\dfrac{5}{3p}} = \dfrac{\sqrt{5}}{\sqrt{3p}} \cdot \dfrac{\sqrt{3p}}{\sqrt{3p}} = \dfrac{\sqrt{15p}}{3p}$

47. $\sqrt{\dfrac{x^5y^3}{z^2}} = \dfrac{\sqrt{x^5y^3}}{\sqrt{z^2}} = \dfrac{x^2y\sqrt{xy}}{z}$

48. $\sqrt{\dfrac{g^3h^5}{r^3}} = \dfrac{\sqrt{g^3h^5}}{\sqrt{r^3}} = \dfrac{gh^2\sqrt{gh}}{r\sqrt{r}} \cdot \dfrac{\sqrt{r}}{\sqrt{r}} = \dfrac{gh^2\sqrt{ghr}}{r^2}$

49. $\sqrt[3]{\dfrac{8}{x^2}} = \dfrac{\sqrt[3]{2^3}}{\sqrt[3]{x^2}} \cdot \dfrac{\sqrt[3]{x}}{\sqrt[3]{x}} = \dfrac{2\sqrt[3]{x}}{x}$

50. $\sqrt[3]{\dfrac{9}{16p^4}} = \dfrac{\sqrt[3]{3^2}}{\sqrt[3]{2^4p^4}} = \dfrac{\sqrt[3]{3^2}}{2p\sqrt[3]{2p}} \cdot \dfrac{\sqrt[3]{2^2p^2}}{\sqrt[3]{2^2p^2}} = \dfrac{\sqrt[3]{36p^2}}{4p^2}$

51. $\sqrt[4]{\dfrac{g^3h^5}{9r^6}} = \dfrac{\sqrt[4]{g^3h^5}}{\sqrt[4]{3^2r^6}} = \dfrac{h\sqrt[4]{g^3h}}{r\sqrt[4]{3^2r^2}} \cdot \dfrac{\sqrt[4]{3^2r^2}}{\sqrt[4]{3^2r^2}} = \dfrac{h\sqrt[4]{9g^3hr^2}}{3r^2}$

52. $\sqrt[4]{\dfrac{32x^5}{y^5}} = \dfrac{\sqrt[4]{2^5x^5}}{\sqrt[4]{y^5}} = \dfrac{2x\sqrt[4]{2x}}{y\sqrt[4]{y}} \cdot \dfrac{\sqrt[4]{y^3}}{\sqrt[4]{y^3}} = \dfrac{2x\sqrt[4]{2xy^3}}{y^2}$

53. $\dfrac{\sqrt[3]{mn} \cdot \sqrt[3]{m^2}}{\sqrt[3]{n^2}} \cdot \dfrac{\sqrt[3]{n}}{\sqrt[3]{n}} = \dfrac{\sqrt[3]{m^3n^2}}{n} = \dfrac{m\sqrt[3]{n^2}}{n}$

54. $\dfrac{\sqrt[3]{8m^2n^3} \cdot \sqrt[3]{2m^2}}{\sqrt[3]{32m^4n^3}} = \dfrac{\sqrt[3]{2^4m^4n^3}}{\sqrt[3]{2^5m^4n^3}} = \dfrac{2mn\sqrt[3]{2m}}{2mn\sqrt[3]{2^2m}} \cdot \dfrac{\sqrt[3]{2m^2}}{\sqrt[3]{2m^2}} = \dfrac{2mn\sqrt[3]{4m^3}}{4m^2n} = \dfrac{2m^2n\sqrt[3]{4}}{4m^2n} = \dfrac{\sqrt[3]{4}}{2}$

55. $\dfrac{\sqrt[4]{32x^5y} \cdot \sqrt[4]{2xy^4}}{\sqrt[4]{4x^3y^2}} = \dfrac{\sqrt[4]{2^6x^6y^5}}{\sqrt[4]{2^2x^3y^2}} \cdot \dfrac{\sqrt[4]{2^2xy^2}}{\sqrt[4]{2^2xy^2}} = \dfrac{\sqrt[4]{2^8x^7y^7}}{2xy} = \dfrac{4xy\sqrt[4]{x^3y^3}}{2xy} = 2\sqrt[4]{x^3y^3}$

56. $\dfrac{\sqrt[4]{rs^2t^3} \cdot \sqrt[4]{r^3s^2t}}{\sqrt[4]{r^2t^3}} = \dfrac{\sqrt[4]{r^4s^4t^4}}{\sqrt[4]{r^2t^3}} \cdot \dfrac{\sqrt[4]{r^2t}}{\sqrt[4]{r^2t}} = \dfrac{rst\sqrt[4]{r^2t}}{rt} = s\sqrt[4]{r^2t}$

57. $\sqrt[3]{\sqrt{4}} = \sqrt[3]{2}$

58. $\sqrt[4]{\sqrt[3]{2}} = (2^{1/3})^{1/4} = 2^{1/12} = \sqrt[12]{2}$

59. $\sqrt[6]{\sqrt[3]{x}} = (x^{1/3})^{1/6} = x^{1/18} = \sqrt[18]{x}$

60. $\sqrt[8]{\sqrt[4]{y}} = (y^{1/4})^{1/8} = y^{1/32} = \sqrt[32]{y}$

61. $4\sqrt{3} - 5\sqrt{12} + 3\sqrt{75} = 4\sqrt{3} - 5(2\sqrt{3}) + 3(5\sqrt{3}) = 4\sqrt{3} - 10\sqrt{3} + 15\sqrt{3} = 9\sqrt{3}$

62. $2\sqrt{5} - 3\sqrt{20} + 2\sqrt{45} = 2\sqrt{5} - 3(2\sqrt{5}) + 2(3\sqrt{5}) = 2\sqrt{5} - 6\sqrt{5} + 6\sqrt{5} = 2\sqrt{5}$

63. $3\sqrt{28p} - 4\sqrt{63p} + \sqrt{112p} = 3(2\sqrt{7p}) - 4(3\sqrt{7p}) + 4\sqrt{7p} = 6\sqrt{7p} - 12\sqrt{7p} + 4\sqrt{7p} = -2\sqrt{7p}$

64. $9\sqrt{8k} + 3\sqrt{18k} - \sqrt{32k} = 9(2\sqrt{2k}) + 3(3\sqrt{2k}) - 4\sqrt{2k} = 18\sqrt{2k} + 9\sqrt{2k} - 4\sqrt{2k} = 23\sqrt{2k}$

65. $2\sqrt[3]{3} + 4\sqrt[3]{24} - \sqrt[3]{81} = 2\sqrt[3]{3} + 4(2\sqrt[3]{3}) - 3\sqrt[3]{3} = 2\sqrt[3]{3} + 8\sqrt[3]{3} - 3\sqrt[3]{3} = 7\sqrt[3]{3}$

66. $\sqrt[3]{32} - 5\sqrt[3]{4} + 2\sqrt[3]{108} = 2\sqrt[3]{4} - 5\sqrt[3]{4} + 2(3\sqrt[3]{4}) = 2\sqrt[3]{4} - 5\sqrt[3]{4} + 6\sqrt[3]{4} = 3\sqrt[3]{4}$

67. $\dfrac{1}{\sqrt{3}} - \dfrac{2}{\sqrt{12}} + 2\sqrt{3} = \dfrac{1}{\sqrt{3}} \cdot \dfrac{\sqrt{3}}{\sqrt{3}} - \dfrac{2}{2\sqrt{3}} \cdot \dfrac{\sqrt{3}}{\sqrt{3}} + 2\sqrt{3} = \dfrac{\sqrt{3}}{3} - \dfrac{2\sqrt{3}}{6} + 2\sqrt{3} =$

$\dfrac{2\sqrt{3} - 2\sqrt{3} + 12\sqrt{3}}{6} = \dfrac{12\sqrt{3}}{6} = 2\sqrt{3}$

68. $\dfrac{1}{\sqrt{2}} + \dfrac{3}{\sqrt{8}} + \dfrac{1}{\sqrt{32}} = \dfrac{1}{\sqrt{2}} \cdot \dfrac{\sqrt{2}}{\sqrt{2}} + \dfrac{3}{2\sqrt{2}} \cdot \dfrac{\sqrt{2}}{\sqrt{2}} + \dfrac{1}{4\sqrt{2}} \cdot \dfrac{\sqrt{2}}{\sqrt{2}} = \dfrac{\sqrt{2}}{2} + \dfrac{3\sqrt{2}}{4} + \dfrac{\sqrt{2}}{8} =$

$\dfrac{4\sqrt{2} + 6\sqrt{2} + \sqrt{2}}{8} = \dfrac{11\sqrt{2}}{8}$

69. $\dfrac{5}{\sqrt[3]{2}} - \dfrac{2}{\sqrt[3]{16}} + \dfrac{1}{\sqrt[3]{54}} = \dfrac{5}{\sqrt[3]{2}} \cdot \dfrac{\sqrt[3]{4}}{\sqrt[3]{4}} - \dfrac{2}{\sqrt[3]{16}} \cdot \dfrac{\sqrt[3]{4}}{\sqrt[3]{4}} + \dfrac{1}{\sqrt[3]{54}} \cdot \dfrac{\sqrt[3]{4}}{\sqrt[3]{4}} = \dfrac{5\sqrt[3]{4}}{2} - \dfrac{\sqrt[3]{4}}{2} + \dfrac{\sqrt[3]{4}}{6} =$

$\dfrac{15\sqrt[3]{4} - 3\sqrt[3]{4} + \sqrt[3]{4}}{6} = \dfrac{13\sqrt[3]{4}}{6}$

70. $\dfrac{-4}{\sqrt[3]{3}} + \dfrac{1}{\sqrt[3]{24}} - \dfrac{2}{\sqrt[3]{81}} = \dfrac{-4}{\sqrt[3]{3}} \cdot \dfrac{\sqrt[3]{9}}{\sqrt[3]{9}} + \dfrac{1}{2\sqrt[3]{3}} \cdot \dfrac{\sqrt[3]{9}}{\sqrt[3]{9}} - \dfrac{2}{3\sqrt[3]{3}} \cdot \dfrac{\sqrt[3]{9}}{\sqrt[3]{9}} = \dfrac{-4\sqrt[3]{9}}{3} + \dfrac{\sqrt[3]{9}}{6} - \dfrac{2\sqrt[3]{9}}{9} =$

$\dfrac{-24\sqrt[3]{9} + 3\sqrt[3]{9} - 4\sqrt[3]{9}}{18} = \dfrac{-25\sqrt[3]{9}}{18}$

71. $(\sqrt{2} + 3)(\sqrt{2} - 3) = (\sqrt{2})^2 - 3^2 = 2 - 9 = -7$

72. $(\sqrt{5} + \sqrt{2})(\sqrt{5} - \sqrt{2}) = (\sqrt{5})^2 - (\sqrt{2})^2 = 5 - 2 = 3$

73. $(\sqrt[3]{11} - 1)(\sqrt[3]{11^2} + \sqrt[3]{11} + 1) = \sqrt[3]{11} \cdot \sqrt[3]{11^2} + \sqrt[3]{11} \cdot \sqrt[3]{11} + \sqrt[3]{11} \cdot 1 - \sqrt[3]{11^2} - \sqrt[3]{11} - 1 =$

$11 + \sqrt[3]{11^2} + \sqrt[3]{11} - \sqrt[3]{11^2} - \sqrt[3]{11} - 1 = 11 - 1 = 10$

74. $(\sqrt[3]{7} + 3)(\sqrt[3]{7^2} - 3\sqrt[3]{7} + 9) = \sqrt[3]{7} \cdot \sqrt[3]{7^2} - 3\sqrt[3]{7} \cdot \sqrt[3]{7} + 9\sqrt[3]{7} + 3\sqrt[3]{7^2} - 9\sqrt[3]{7} + 27 =$

$7 - 3\sqrt[3]{7^2} + 9\sqrt[3]{7} + 3\sqrt[3]{7^2} - 9\sqrt[3]{7} + 27 = 7 + 27 = 34$

75. $(\sqrt{3} + \sqrt{8})^2 = (\sqrt{3})^2 + 2\sqrt{3} \cdot \sqrt{8} + (\sqrt{8})^2 = 3 + 2\sqrt{24} + 8 = 11 + 2(2\sqrt{6}) = 11 + 4\sqrt{6}$

76. $(\sqrt{2} - 1)^2 = (\sqrt{2})^2 - 2\sqrt{2} \cdot 1 + (-1)^2 = 2 - 2\sqrt{2} + 1 = 3 - 2\sqrt{2})$

77. $(3\sqrt{2} + \sqrt{3})(2\sqrt{3} - \sqrt{2}) = 3\sqrt{2} \cdot 2\sqrt{3} + 3\sqrt{2}(-\sqrt{2}) + \sqrt{3} \cdot 2\sqrt{3} + \sqrt{3}(-\sqrt{2}) =$

$6\sqrt{6} - 6 + 6 - \sqrt{6} = 5\sqrt{6}$

78. $(4\sqrt{5} - 1)(3\sqrt{5} + 2) = 4\sqrt{5} \cdot 3\sqrt{5} + 4\sqrt{5} \cdot 2 - 1 \cdot 3\sqrt{5} - 2 =$

$60 + 8\sqrt{5} - 3\sqrt{5} - 2 = 58 + 5\sqrt{5}$

79. $\dfrac{8}{\sqrt{5}} = \dfrac{8}{\sqrt{5}} \cdot \dfrac{\sqrt{5}}{\sqrt{5}} = \dfrac{8\sqrt{5}}{5}$

80. $\dfrac{3}{\sqrt{10}} = \dfrac{3}{\sqrt{10}} \cdot \dfrac{\sqrt{10}}{\sqrt{10}} = \dfrac{3\sqrt{10}}{10}$

81. $\dfrac{6}{\sqrt[3]{x^2}} = \dfrac{6}{\sqrt[3]{x^2}} \cdot \dfrac{\sqrt[3]{x}}{\sqrt[3]{x}} = \dfrac{6\sqrt[3]{x}}{x}$

82. $\dfrac{4}{\sqrt[3]{a^2}} = \dfrac{4}{\sqrt[3]{a^2}} \cdot \dfrac{\sqrt[3]{a}}{\sqrt[3]{a}} = \dfrac{4\sqrt[3]{a}}{a}$

83. $\dfrac{\sqrt{3}}{\sqrt{5} + \sqrt{3}} \cdot \dfrac{\sqrt{5} - \sqrt{3}}{\sqrt{5} - \sqrt{3}} = \dfrac{\sqrt{3} \cdot \sqrt{5} - \sqrt{3} \cdot \sqrt{3}}{(\sqrt{5})^2 - (\sqrt{3})^2} = \dfrac{\sqrt{15} - 3}{5 - 3} = \dfrac{\sqrt{15} - 3}{2}$

84. $\dfrac{\sqrt{7}}{\sqrt{3} - \sqrt{7}} \cdot \dfrac{\sqrt{3} + \sqrt{7}}{\sqrt{3} + \sqrt{7}} = \dfrac{\sqrt{7} \cdot \sqrt{3} + \sqrt{7} \cdot \sqrt{7}}{(\sqrt{3})^2 - (\sqrt{7})^2} = \dfrac{\sqrt{21} + 7}{3 - 7} = -\dfrac{\sqrt{21} + 7}{4}$

85. $\dfrac{1 + \sqrt{3}}{3\sqrt{5} + 2\sqrt{3}} \cdot \dfrac{3\sqrt{5} - 2\sqrt{3}}{3\sqrt{5} - 2\sqrt{3}} = \dfrac{3\sqrt{5} - 2\sqrt{3} + \sqrt{3} \cdot 3\sqrt{5} - 2(\sqrt{3})^2}{(3\sqrt{5})^2 - (2\sqrt{3})^2} = \dfrac{3\sqrt{5} - 2\sqrt{3} + 3\sqrt{15} - 6}{45 - 12} =$

$\dfrac{3\sqrt{5} - 2\sqrt{3} + 3\sqrt{15} - 6}{33}$

86. $\dfrac{\sqrt{7} - 1}{2\sqrt{7} + 4\sqrt{2}} \cdot \dfrac{2\sqrt{7} - 4\sqrt{2}}{2\sqrt{7} - 4\sqrt{2}} = \dfrac{\sqrt{7} \cdot 2\sqrt{7} + \sqrt{7}(-4\sqrt{2}) - 2\sqrt{7} + 4\sqrt{2}}{(2\sqrt{7})^2 - (4\sqrt{2})^2} =$

$\dfrac{14 - 4\sqrt{14} - 2\sqrt{7} + 4\sqrt{2}}{28 - 32} = \dfrac{2(7 - 2\sqrt{14} - \sqrt{7} + 2\sqrt{2})}{-4} = \dfrac{-7 + 2\sqrt{14} + \sqrt{7} - 2\sqrt{2}}{2}$

87. $\dfrac{p}{\sqrt{p} + 2} \cdot \dfrac{\sqrt{p} - 2}{\sqrt{p} - 2} = \dfrac{p\sqrt{p} - 2p}{(\sqrt{p})^2 - (2)^2} = \dfrac{p(\sqrt{p} - 2)}{p - 4}$

88. $\dfrac{\sqrt{r}}{3 - \sqrt{r}} \cdot \dfrac{3 + \sqrt{r}}{3 + \sqrt{r}} = \dfrac{3\sqrt{r} + (\sqrt{r})^2}{(3)^2 - (\sqrt{r})^2} = \dfrac{3\sqrt{r} + r}{9 - r}$

89. $\dfrac{a}{\sqrt{a + b} - 1} \cdot \dfrac{\sqrt{a + b} + 1}{\sqrt{a + b} + 1} = \dfrac{a(\sqrt{a + b} + 1)}{(\sqrt{a + b})^2 - (1)^2} = \dfrac{a(\sqrt{a + b} + 1)}{a + b - 1}$

90. $\dfrac{3m}{2 + \sqrt{m + n}} \cdot \dfrac{2 - \sqrt{m + n}}{2 - \sqrt{m + n}} = \dfrac{3m(2 - \sqrt{m + n})}{(2)^2 - (\sqrt{m + n})^2} = \dfrac{3m(2 - \sqrt{m + n})}{4 - m - n}$

Appendix B: Deciding Which Model Best Fits a Set of Data

1. Enter the years in L_1 starting with year 3 and ending with year 12. Enter the number of deaths in L_2.

   The LinReg($ax + b$) feature of the graphing calculator yields the linear model $y = -.129091x + 15.418182$.

   Enter this equation in Y_1 before utilizing the SSE program. The SSE program yields SSE = .430182.

   The ExpReg feature of the graphing calculator yields the exponential model $y = 15.434482(.991194)^x$.

   Enter this equation in Y_1 before utilizing the SSE program. The SSE program yields SSE = .418686.

   The exponential model is the better of these two models since its SSE value is smaller.

2. Enter the years in L_1 starting with year 3 and ending with year 13. Enter the average hourly earnings in L_2.

   The LinReg($ax + b$) feature of the graphing calculator yields the linear model $y = .451909x + 9.458364$.

   Enter this equation in Y_1 before utilizing the SSE program. The SSE program yields SSE = .116654. The

   LnReg feature of the graphing calculator yields the natural logarithmic model $y = 6.980077 + 3.066430 \ln x$.

   Enter this equation in Y_1 before utilizing the SSE program. The SSE program yields SSE = 1.560233.

   The linear model is the better of these two models since its SSE value is smaller.

3. Enter the years in L_1 starting with year 4 and ending with year 9. Enter the number of lab cleanups in L_2.

   The LinReg($ax + b$) feature of the graphing calculator yields the linear model $y = 99.971429x - 373.314286$.

   Enter this equation in Y_1 before utilizing the SSE program. The SSE program yields SSE = 7533.485714.

   The QuadReg feature of the graphing calculator yields the quadratic model

   $y = -6.428571x^2 + 183.542857x - 626.171429$. Enter this equation in Y_1 before utilizing the SSE program.

   The SSE program yields SSE = 5990.628571. The quadratic model is the better of these two models since its

   SSE value is smaller.

4. Enter the years in L_1 starting with year 1 and ending with year 5. Enter the number of viewers in L_2.

   The LinReg($ax + b$) feature of the graphing calculator yields the linear model $y = -11.6x + 196$.

   Enter this equation in Y_1 before utilizing the SSE program. The SSE program yields SSE = 87.2. The LnReg

   feature of the graphing calculator yields the natural logarithmic model $y = 189.512044 - 29.568765 \ln x$.

   Enter this equation in Y_1 before utilizing the SSE program. The SSE program yields SSE = 20.358801.

   The natural logarithmic model is the better of these two models since its SSE value is smaller.

5. Enter the years in L_1 starting with year 3 and ending with year 33. Enter the percent of women in L_2.

   The QuadReg feature of the graphing calculator yields the quadratic model

   $y = -.008459x^2 + .688792x + 1.161391$. Enter this equation in Y_1 before utilizing the SSE program.

   The SSE program yields SSE = 2.870134. The LnReg feature of the graphing calculator yields the natural

   logarithmic model $y = -3.454802 + 5.093917 \ln x$. Enter this equation in Y_1 before utilizing the SSE program.

   The SSE program yields SSE = 2.261093. The natural logarithmic model is the better of these two models

   since its SSE value is smaller.

Appendix C: Vectors in Space

1. The plane determined by the x-axis and the z-axis is called the xz-plane.

2. A position vector has its initial point at the origin.

3. The component form of the position vector with terminal point $(5, 3, -2)$ is $\langle 5, 3, -2 \rangle$.

4. The **i, j, k** form of the position vector with terminal point $(6, -1, -3)$ is $6\mathbf{i} - \mathbf{j} - 3\mathbf{k}$.

5. $d = \sqrt{(2-0)^2 + (-2-0)^2 + (5-0)^2} = \sqrt{4 + 4 + 25} = \sqrt{33}$

6. $d = \sqrt{(7-0)^2 + (4-0)^2 + (-1-0)^2} = \sqrt{49 + 16 + 1} = \sqrt{66}$

7. $d = \sqrt{(8-10)^2 + (3-15)^2 + (-4-9)^2} = \sqrt{4 + 144 + 169} = \sqrt{317}$

8. $d = \sqrt{(3-5)^2 + (7-4)^2 + (2-(-4))^2} = \sqrt{4 + 9 + 36} = \sqrt{49} = 7$

9. $d = \sqrt{(5-20)^2 + (5-25)^2 + (6-16)^2} = \sqrt{225 + 400 + 100} = \sqrt{725} = 5\sqrt{29}$

10. $d = \sqrt{(-2-14)^2 + (4-10)^2 + (9-18)^2} = \sqrt{256 + 36 + 81} = \sqrt{373}$

11. $\mathbf{PQ} = \langle (2-0), (-2-0), (5-0) \rangle = \langle 2, -2, 5 \rangle;\ 2\mathbf{i} - 2\mathbf{j} + 5\mathbf{k}$

12. $\mathbf{PQ} = \langle (7-0), (4-0), (-1-0) \rangle = \langle 7, 4, -1 \rangle;\ 7\mathbf{i} + 4\mathbf{j} - \mathbf{k}$

13. $\mathbf{PQ} = \langle (8-10), (3-15), (-4-0) \rangle = \langle -2, -12, -4 \rangle;\ -2\mathbf{i} - 12\mathbf{j} - 4\mathbf{k}$

14. $\mathbf{PQ} = \langle (3-0), (7-4), (2-(-4)) \rangle = \langle 3, 3, 6 \rangle;\ 3\mathbf{i} + 3\mathbf{j} + 6\mathbf{k}$

15. $\mathbf{PQ} = \langle (5-20), (5-25), (16-6) \rangle = \langle -15, -20, 10 \rangle;\ -15\mathbf{i} - 20\mathbf{j} + 10\mathbf{k}$

16. $\mathbf{PQ} = \langle (-2-14), (4-10), (9-18) \rangle = \langle -16, -6, -9 \rangle;\ -16\mathbf{i} - 6\mathbf{j} - 9\mathbf{k}$

17. $\mathbf{QP} = \langle (0-2), (0-(-2)), (0-5) \rangle = \langle -2, 2, -5 \rangle;\ \mathbf{QP}$ is the opposite of $\mathbf{PQ}$. That is $\mathbf{QP} = -\mathbf{PQ}$.

18. The distance between points P and Q and the magnitude of $\mathbf{PQ}$ are the same number.

19. $\mathbf{u} - \mathbf{w} = (2-4)\mathbf{i} + (4-(-3))\mathbf{j} + (7-(-6))\mathbf{k} = -2\mathbf{i} + 7\mathbf{j} + 13\mathbf{k}$

20. $\mathbf{v} + \mathbf{w} = (-3+4)\mathbf{i} + (5+(-3))\mathbf{j} + (2+(-6))\mathbf{k} = \mathbf{i} + 2\mathbf{j} - 4\mathbf{k}$

21. $4\mathbf{u} + 5\mathbf{v} = [4(2) + 5(-3)]\mathbf{i} + [4(4) + 5(5)]\mathbf{j} + [4(7) + 5(2)]\mathbf{k} = -7\mathbf{i} + 41\mathbf{j} + 38\mathbf{k}$

22. $-\mathbf{v} + 3\mathbf{u} = [-1(-3) + 3(2)]\mathbf{i} + [-1(5) + 3(4)]\mathbf{j} + [-1(2) + 3(7)]\mathbf{k} = 9\mathbf{i} + 7\mathbf{j} + 19\mathbf{k}$

23. $|\mathbf{u}| = \sqrt{2^2 + 4^2 + 7^2} = \sqrt{69}$

24. $|\mathbf{w}| = \sqrt{4^2 + (-3)^2 + (-6)^2} = \sqrt{61}$

25. $\mathbf{w} + \mathbf{u} = (4+2)\mathbf{i} + ((-3)+4)\mathbf{j} + ((-6)+7)\mathbf{k} = 6\mathbf{i} + \mathbf{j} + \mathbf{k} \Rightarrow |\mathbf{w} + \mathbf{u}| = \sqrt{6^2 + 1^2 + 1^2} = \sqrt{38}$

26. $2\mathbf{v} = 2(-3)\mathbf{i} + 2(5)\mathbf{j} + 2(2)\mathbf{k} = -6\mathbf{i} + 10\mathbf{j} + 4\mathbf{k} \Rightarrow |2\mathbf{v}| = \sqrt{(-6)^2 + 10^2 + 4^2} = \sqrt{152} = 2\sqrt{38}$

27. $\mathbf{v} \cdot \mathbf{w} = (-3)(4) + (5)(-3) + (2)(-6) = -39$

28. $\mathbf{u} \cdot \mathbf{w} = (2)(4) + (4)(-3) + (7)(-6) = -46$

29. $\mathbf{v} \cdot \mathbf{v} = (-3)(-3) + (5)(5) + (2)(2) = 38$

30. $\mathbf{u} \cdot \mathbf{u} = (2)(2) + (4)(4) + (7)(7) = 69$

31. Let $\mathbf{v} = \langle 2, -2, 0 \rangle$ and $\mathbf{w} = \langle 5, -2, -1 \rangle$. The dot product is $\mathbf{v} \cdot \mathbf{w} = (2)(5) + (-2)(-2) + (0)(-1) = 14$.

The required magnitudes are $|\mathbf{v}| = \sqrt{2^2 + (-2)^2 + 0^2} = \sqrt{8}$ and $|\mathbf{w}| = \sqrt{5^2 + (-2)^2 + (-1)^2} = \sqrt{30}$.

$\cos\theta = \dfrac{\mathbf{v} \cdot \mathbf{w}}{|\mathbf{v}||\mathbf{w}|} \Rightarrow \cos\theta = \dfrac{14}{\sqrt{8}\sqrt{30}} \Rightarrow \theta = \cos^{-1}\left(\dfrac{14}{\sqrt{8}\sqrt{30}}\right) \approx 25.4°$

32. Let $\mathbf{v} = \langle 4, 0, 0 \rangle$ and $\mathbf{w} = \langle 5, 3, -2 \rangle$. The dot product is $\mathbf{v} \cdot \mathbf{w} = (4)(5) + (0)(3) + (0)(-2) = 20$.

The required magnitudes are $|\mathbf{v}| = \sqrt{4^2 + 0^2 + 0^2} = 4$ and $|\mathbf{w}| = \sqrt{5^2 + 3^2 + (-2)^2} = \sqrt{38}$.

$\cos\theta = \dfrac{\mathbf{v} \cdot \mathbf{w}}{|\mathbf{v}||\mathbf{w}|} \Rightarrow \cos\theta = \dfrac{20}{4\sqrt{38}} \Rightarrow \theta = \cos^{-1}\left(\dfrac{20}{4\sqrt{38}}\right) \approx 35.8°$

33. Let $\mathbf{v} = \langle 6, 0, 0 \rangle$ and $\mathbf{w} = \langle 8, 3, -4 \rangle$. The dot product is $\mathbf{v} \cdot \mathbf{w} = (6)(8) + (0)(3) + (0)(-4) = 48$.

The required magnitudes are $|\mathbf{v}| = \sqrt{6^2 + 0^2 + 0^2} = 6$ and $|\mathbf{w}| = \sqrt{8^2 + 3^2 + (-4)^2} = \sqrt{89}$.

$\cos\theta = \dfrac{\mathbf{v} \cdot \mathbf{w}}{|\mathbf{v}||\mathbf{w}|} \Rightarrow \cos\theta = \dfrac{48}{6\sqrt{89}} \Rightarrow \theta = \cos^{-1}\left(\dfrac{48}{6\sqrt{89}}\right) \approx 32.0°$

34. Let $\mathbf{v} = \langle -1, 2, -3 \rangle$ and $\mathbf{w} = \langle 0, -2, 1 \rangle$. The dot product is $\mathbf{v} \cdot \mathbf{w} = (-1)(0) + (2)(-2) + (-3)(1) = -7$.

The required magnitudes are $|\mathbf{v}| = \sqrt{(-1)^2 + 2^2 + (-3)^2} = \sqrt{14}$ and $|\mathbf{w}| = \sqrt{0^2 + (-2)^2 + 1^2} = \sqrt{5}$.

$\cos\theta = \dfrac{\mathbf{v} \cdot \mathbf{w}}{|\mathbf{v}||\mathbf{w}|} \Rightarrow \cos\theta = \dfrac{-7}{\sqrt{14}\sqrt{5}} \Rightarrow \theta = \cos^{-1}\left(\dfrac{-7}{\sqrt{14}\sqrt{5}}\right) \approx 146.8°$

35. Let $\mathbf{v} = \langle 1, 0, 0 \rangle$ and $\mathbf{w} = \langle 0, 1, 0 \rangle$. The dot product is $\mathbf{v} \cdot \mathbf{w} = (1)(0) + (0)(1) + (0)(0) = 0$.

The required magnitudes are $|\mathbf{v}| = \sqrt{1^2 + 0^2 + 0^2} = 1$ and $|\mathbf{w}| = \sqrt{0^2 + 1^2 + 0^2} = 1$.

$\cos\theta = \dfrac{\mathbf{v} \cdot \mathbf{w}}{|\mathbf{v}||\mathbf{w}|} \Rightarrow \cos\theta = \dfrac{0}{1(1)} \Rightarrow \theta = \cos^{-1} 0 = 90°$

36. Let $\mathbf{v} = \langle 0, 0, 1 \rangle$ and $\mathbf{w} = \langle 0, 1, 0 \rangle$. The dot product is $\mathbf{v} \cdot \mathbf{w} = (0)(0) + (0)(1) + (1)(0) = 0$.

The required magnitudes are $|\mathbf{v}| = \sqrt{0^2 + 0^2 + 1^2} = 1$ and $|\mathbf{w}| = \sqrt{0^2 + 1^2 + 0^2} = 1$.

$\cos\theta = \dfrac{\mathbf{v} \cdot \mathbf{w}}{|\mathbf{v}||\mathbf{w}|} \Rightarrow \cos\theta = \dfrac{0}{1(1)} \Rightarrow \theta = \cos^{-1} 0 = 90°$

37. $|\mathbf{u}| = \sqrt{2^2 + 4^2 + 7^2} = \sqrt{69}; \cos\alpha = \dfrac{a}{|\mathbf{u}|} = \dfrac{2}{\sqrt{69}} \Rightarrow \alpha = \cos^{-1}\left(\dfrac{2}{\sqrt{69}}\right) \approx 76.1°$

$\cos\beta = \dfrac{b}{|\mathbf{u}|} = \dfrac{4}{\sqrt{69}} \Rightarrow \beta = \cos^{-1}\left(\dfrac{4}{\sqrt{69}}\right) \approx 61.2°; \cos\gamma = \dfrac{c}{|\mathbf{u}|} = \dfrac{7}{\sqrt{69}} \Rightarrow \gamma = \cos^{-1}\left(\dfrac{7}{\sqrt{69}}\right) \approx 32.6°$

38. $|\mathbf{v}| = \sqrt{(-3)^2 + 5^2 + 2^2} = \sqrt{38}; \cos\alpha = \dfrac{a}{|\mathbf{v}|} = \dfrac{-3}{\sqrt{38}} \Rightarrow \alpha = \cos^{-1}\left(\dfrac{-3}{\sqrt{38}}\right) \approx 119.1°$

$\cos\beta = \dfrac{b}{|\mathbf{v}|} = \dfrac{5}{\sqrt{38}} \Rightarrow \beta = \cos^{-1}\left(\dfrac{5}{\sqrt{38}}\right) \approx 35.8°; \cos\gamma = \dfrac{c}{|\mathbf{v}|} = \dfrac{2}{\sqrt{38}} \Rightarrow \gamma = \cos^{-1}\left(\dfrac{2}{\sqrt{38}}\right) \approx 71.1°$

39. $|\mathbf{w}| = \sqrt{4^2 + (-3)^2 + (-6)^2} = \sqrt{61}; \cos\alpha = \dfrac{a}{|\mathbf{w}|} = \dfrac{4}{\sqrt{61}} \Rightarrow \alpha = \cos^{-1}\left(\dfrac{4}{\sqrt{61}}\right) \approx 59.2°$

$\cos\beta = \dfrac{b}{|\mathbf{w}|} = \dfrac{-3}{\sqrt{61}} \Rightarrow \beta = \cos^{-1}\left(\dfrac{-3}{\sqrt{61}}\right) \approx 112.6°; \cos\gamma = \dfrac{c}{|\mathbf{w}|} = \dfrac{-6}{\sqrt{61}} \Rightarrow \gamma = \cos^{-1}\left(\dfrac{-6}{\sqrt{61}}\right) \approx 140.2°$

40. $|\mathbf{y}| = \sqrt{2^2 + (-3)^2 + 4^2} = \sqrt{29}; \cos\alpha = \dfrac{a}{|\mathbf{y}|} = \dfrac{2}{\sqrt{29}} \Rightarrow \alpha = \cos^{-1}\left(\dfrac{2}{\sqrt{29}}\right) \approx 68.2°$

$\cos\beta = \dfrac{b}{|\mathbf{y}|} = \dfrac{-3}{\sqrt{29}} \Rightarrow \beta = \cos^{-1}\left(\dfrac{-3}{\sqrt{29}}\right) \approx 123.9°; \cos\gamma = \dfrac{c}{|\mathbf{y}|} = \dfrac{4}{\sqrt{29}} \Rightarrow \gamma = \cos^{-1}\left(\dfrac{4}{\sqrt{29}}\right) \approx 42.0°$

41. Here $\cos \alpha = \cos 45° = \dfrac{\sqrt{2}}{2}$ and $\cos \beta = \cos 120° = -\dfrac{1}{2}$. Since $\cos^2 \alpha + \cos^2 \beta + \cos^2 \gamma = 1$,

$$\cos^2 \gamma = 1 - \cos^2 \alpha - \cos^2 \beta = 1 - \frac{1}{2} - \frac{1}{4} = \frac{1}{4} \Rightarrow \cos \gamma = \frac{1}{2} \Rightarrow \gamma = \cos^{-1} \frac{1}{2} = 60°$$

42. Here $\cos \beta = \cos 135° = -\dfrac{\sqrt{2}}{2}$ and $\cos \gamma = \cos 90° = 0$. Since $\cos^2 \alpha + \cos^2 \beta + \cos^2 \gamma = 1$,

$$\cos^2 \alpha = 1 - \cos^2 \beta - \cos^2 \gamma = 1 - \frac{1}{2} - 0 = \frac{1}{2} \Rightarrow \cos \alpha = \frac{\sqrt{2}}{2} \Rightarrow \alpha = \cos^{-1} \frac{\sqrt{2}}{2} = 45°$$

43. Two vectors are parallel if the position vector for one is a scalar multiple of the position vector of the other.

44. The two vectors are parallel because $\langle -12, -20, 4 \rangle = -4\langle 3, 5, -1 \rangle$.

45. $\mathbf{PQ} = \langle (1 - 0), (3 - 0), (2 - 0) \rangle = \langle 1, 3, 2 \rangle$; $\mathbf{F} \cdot \mathbf{PQ} = (2)(1) + (0)(3) + (5)(2) = 12$ work units

46. $\mathbf{PQ} = \langle (4 - 0), (-2 - 0), (5 - 0) \rangle = \langle 4, -2, 5 \rangle$; $\mathbf{F} \cdot \mathbf{PQ} = (3)(4) + (2)(-2) + (0)(5) = 8$ work units

47. $\mathbf{PQ} = \langle (5 - 2), (7 - (-1)), (8 - 2) \rangle = \langle 3, 8, 6 \rangle$; $\mathbf{F} \cdot \mathbf{PQ} = (1)(3) + (2)(8) + (-1)(6) = 13$ work units

48. $\mathbf{PQ} = \langle (10 - 4), (15 - 7), (12 - 6) \rangle = \langle 6, 8, 6 \rangle$; $\mathbf{F} \cdot \mathbf{PQ} = (3)(6) + (2)(8) + (0)(6) = 34$ work units

49. $\cos^2 \alpha + \cos^2 \beta + \cos^2 \gamma = \left(\dfrac{a}{|\mathbf{v}|}\right)^2 + \left(\dfrac{b}{|\mathbf{v}|}\right)^2 + \left(\dfrac{c}{|\mathbf{v}|}\right)^2 = \dfrac{a^2}{|\mathbf{v}|^2} + \dfrac{b^2}{|\mathbf{v}|^2} + \dfrac{c^2}{|\mathbf{v}|^2} =$

$$\frac{a^2}{(\sqrt{a^2 + b^2 + c^2})^2} + \frac{b^2}{(\sqrt{a^2 + b^2 + c^2})^2} + \frac{c^2}{(\sqrt{a^2 + b^2 + c^2})^2} = \frac{a^2 + b^2 + c^2}{a^2 + b^2 + c^2} = 1$$

Appendix D: Polar Form of Conic Sections

1. Only the calculator graph is shown here. See Figure 1.
2. Only the calculator graph is shown here. See Figure 2.
3. Only the calculator graph is shown here. See Figure 3.
4. Only the calculator graph is shown here. See Figure 4.

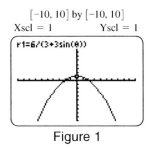
Figure 1

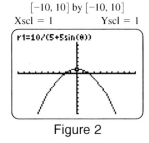
Figure 2

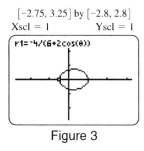
Figure 3

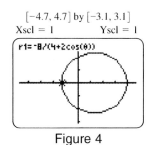
Figure 4

5. Only the calculator graph is shown here. See Figure 5.
6. Only the calculator graph is shown here. See Figure 6.
7. Only the calculator graph is shown here. See Figure 7.
8. Only the calculator graph is shown here. See Figure 8.

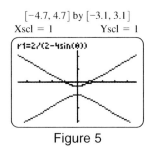
Figure 5

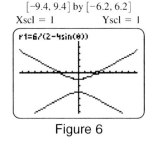
Figure 6

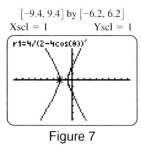
Figure 7

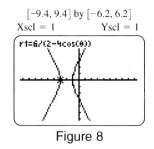
Figure 8

9. Only the calculator graph is shown here. See Figure 9.
10. Only the calculator graph is shown here. See Figure 10.
11. Only the calculator graph is shown here. See Figure 11.
12. Only the calculator graph is shown here. See Figure 12.

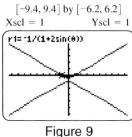
Figure 9

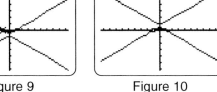
Figure 10

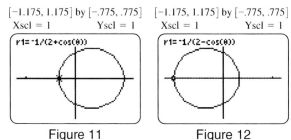
Figure 11

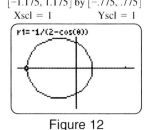
Figure 12

13. The conic is a parabola, so $e = 1$. Since the vertical directrix is 3 units to the right of the pole, the equation is

    of the form $r = \dfrac{ep}{1 + e\cos\theta}$ with $p = 3$. The equation for this parabola is $r = \dfrac{3}{1 + \cos\theta}$.

14. The conic is a parabola, so $e = 1$. Since the vertical directrix is 4 units to the left of the pole, the equation is

    of the form $r = \dfrac{ep}{1 - e\cos\theta}$ with $p = 4$. The equation for this parabola is $r = \dfrac{4}{1 - \cos\theta}$.

15. The conic is a parabola, so $e = 1$. Since the horizontal directrix is 5 units below the pole, the equation is of the

    form $r = \dfrac{ep}{1 - e\sin\theta}$ with $p = 5$. The equation for this parabola is $r = \dfrac{5}{1 - \sin\theta}$.

16. The conic is a parabola, so $e = 1$. Since the horizontal directrix is 6 units above the pole, the equation is of the

    form $r = \dfrac{ep}{1 + e\sin\theta}$ with $p = 6$. The equation for this parabola is $r = \dfrac{6}{1 + \sin\theta}$.

17. Since the vertical directrix is 5 units to the right of the pole, the equation is of the form $r = \dfrac{ep}{1 + e\cos\theta}$ with

    $p = 5$. Since $e = \dfrac{2}{3} < 1$, the conic is an ellipse with equation $r = \dfrac{(\frac{4}{5})5}{1 + \frac{4}{5}\cos\theta} \Rightarrow r = \dfrac{4}{1 + \frac{4}{5}\cos\theta} \Rightarrow$

    $r = \dfrac{4}{1 + \frac{4}{5}\cos\theta} \cdot \dfrac{5}{5} \Rightarrow r = \dfrac{20}{5 + 4\cos\theta}$.

18. Since the vertical directrix is 6 units to the left of the pole, the equation is of the form $r = \dfrac{ep}{1 - e\cos\theta}$ with

    $p = 6$. Since $e = \dfrac{2}{3} < 1$, the conic is an ellipse with equation $r = \dfrac{(\frac{2}{3})6}{1 - \frac{2}{3}\cos\theta} \Rightarrow r = \dfrac{4}{1 - \frac{2}{3}\cos\theta} \Rightarrow$

    $r = \dfrac{4}{1 - \frac{2}{3}\cos\theta} \cdot \dfrac{3}{3} \Rightarrow r = \dfrac{12}{3 - 2\cos\theta}$.

19. Since the horizontal directrix is 8 units below the pole, the equation is of the form $r = \dfrac{ep}{1 - e\sin\theta}$ with

    $p = 8$. Since $e = \dfrac{5}{4} > 1$, the conic is a hyperbola with equation $r = \dfrac{(\frac{5}{4})8}{1 - \frac{5}{4}\sin\theta} \Rightarrow r = \dfrac{10}{1 - \frac{5}{4}\sin\theta} \Rightarrow$

    $r = \dfrac{10}{1 - \frac{5}{4}\sin\theta} \cdot \dfrac{4}{4} \Rightarrow r = \dfrac{40}{4 - 5\sin\theta}$.

20. Since the horizontal directrix is 4 units above the pole, the equation is of the form $r = \dfrac{ep}{1 + e\sin\theta}$ with

    $p = 4$. Since $e = \dfrac{3}{2} > 1$, the conic is a hyperbola with equation $r = \dfrac{(\frac{3}{2})4}{1 + \frac{3}{2}\sin\theta} \Rightarrow r = \dfrac{6}{1 + \frac{3}{2}\sin\theta} \Rightarrow$

    $r = \dfrac{6}{1 + \frac{3}{2}\sin\theta} \cdot \dfrac{2}{2} \Rightarrow r = \dfrac{12}{2 + 3\sin\theta}$.

21. $r = \dfrac{6}{3 - \cos\theta} \cdot \dfrac{\frac{1}{3}}{\frac{1}{3}} = \dfrac{2}{1 - \frac{1}{3}\cos\theta}$; Since $e = \dfrac{1}{3} < 1$, this equation represents an ellipse.

    $r = \dfrac{6}{3 - \cos\theta} \Rightarrow r(3 - \cos\theta) = 6 \Rightarrow 3r - r\cos\theta = 6 \Rightarrow 3r = r\cos\theta + 6 \Rightarrow (3r)^2 = (r\cos\theta + 6)^2 \Rightarrow$

    $(3r)^2 = (x + 6)^2 \Rightarrow 9r^2 = x^2 + 12x + 36 \Rightarrow 9(x^2 + y^2) = x^2 + 12x + 36 \Rightarrow 8x^2 + 9y^2 - 12x - 36 = 0$

    The rectangular form of the equation is $8x^2 + 9y^2 - 12x - 36 = 0$.

22. $r = \dfrac{8}{4 - \cos\theta} \cdot \dfrac{\frac{1}{4}}{\frac{1}{4}} = \dfrac{2}{1 - \frac{1}{4}\cos\theta}$; Since $e = \dfrac{1}{4} < 1$, this equation represents an ellipse.

$r = \dfrac{8}{4 - \cos\theta} \Rightarrow r(4 - \cos\theta) = 8 \Rightarrow 4r - r\cos\theta = 8 \Rightarrow 4r = r\cos\theta + 8 \Rightarrow (4r)^2 = (r\cos\theta + 8)^2 \Rightarrow$

$(4r)^2 = (x + 8)^2 \Rightarrow 16r^2 = x^2 + 16x + 64 \Rightarrow 16(x^2 + y^2) = x^2 + 16x + 64 \Rightarrow$

$15x^2 + 16y^2 - 16x - 64 = 0$; The rectangular form of the equation is $15x^2 + 16y^2 - 16x - 64 = 0$.

23. $r = \dfrac{-2}{1 + 2\cos\theta}$; Since $e = 2 > 1$, this equation represents a hyperbola.

$r = \dfrac{-2}{1 + 2\cos\theta} \Rightarrow r(1 + 2\cos\theta) = -2 \Rightarrow r + 2r\cos\theta = -2 \Rightarrow r = -2r\cos\theta - 2 \Rightarrow$

$r^2 = (-2r\cos\theta - 2)^2 \Rightarrow r^2 = (-2x - 2)^2 \Rightarrow r^2 = 4x^2 + 8x + 4 \Rightarrow x^2 + y^2 = 4x^2 + 8x + 4 \Rightarrow$

$3x^2 - y^2 + 8x + 4 = 0$; The rectangular form of the equation is $3x^2 - y^2 + 8x + 4 = 0$.

24. $r = \dfrac{-3}{1 + 3\cos\theta}$; Since $e = 3 > 1$, this equation represents a hyperbola.

$r = \dfrac{-3}{1 + 3\cos\theta} \Rightarrow r(1 + 3\cos\theta) = -3 \Rightarrow r + 3r\cos\theta = -3 \Rightarrow r = -3r\cos\theta - 3 \Rightarrow$

$r^2 = (-3r\cos\theta - 3)^2 \Rightarrow r^2 = (-3x - 3)^2 \Rightarrow r^2 = 9x^2 + 18x + 9 \Rightarrow x^2 + y^2 = 9x^2 + 18x + 9 \Rightarrow$

$8x^2 - y^2 + 18x + 9 = 0$; The rectangular form of the equation is $8x^2 - y^2 + 18x + 9 = 0$.

25. $r = \dfrac{-6}{4 + 2\sin\theta} \cdot \dfrac{\frac{1}{4}}{\frac{1}{4}} = \dfrac{-\frac{3}{2}}{1 + \frac{1}{2}\sin\theta}$; Since $e = \dfrac{1}{2} < 1$, this equation represents an ellipse.

$r = \dfrac{-6}{4 + 2\sin\theta} \Rightarrow r(4 + 2\sin\theta) = -6 \Rightarrow 4r + 2r\sin\theta = -6 \Rightarrow 4r = -2r\sin\theta - 6 \Rightarrow$

$(4r)^2 = (-2r\sin\theta - 6)^2 \Rightarrow (4r)^2 = (-2y - 6)^2 \Rightarrow 16r^2 = 4y^2 + 24y + 36 \Rightarrow$

$16(x^2 + y^2) = 4y^2 + 24y + 36 = 0 \Rightarrow 16x^2 + 12y^2 - 24y - 36 = 0 \Rightarrow 4x^2 + 3y^2 - 6y - 9 = 0$

The rectangular form of the equation is $4x^2 + 3y^2 - 6y - 9 = 0$.

26. $r = \dfrac{-12}{6 + 3\sin\theta} \cdot \dfrac{\frac{1}{6}}{\frac{1}{6}} = \dfrac{-2}{1 + \frac{1}{2}\sin\theta}$; Since $e = \dfrac{1}{2} < 1$, this equation represents an ellipse.

$r = \dfrac{-12}{6 + 3\sin\theta} \Rightarrow r(6 + 3\sin\theta) = -12 \Rightarrow 6r + 3r\sin\theta = -12 \Rightarrow 6r = -3r\sin\theta - 12 \Rightarrow$

$(6r)^2 = (-3r\sin\theta - 12)^2 \Rightarrow (6r)^2 = (-3y - 12)^2 \Rightarrow 36r^2 = 9y^2 + 72y + 144 \Rightarrow$

$36(x^2 + y^2) = 9y^2 + 72y + 144 = 0 \Rightarrow 36x^2 + 27y^2 - 72y - 144 = 0 \Rightarrow 4x^2 + 3y^2 - 8y - 16 = 0$

The rectangular form of the equation is $4x^2 + 3y^2 - 8y - 16 = 0$.

27. $r = \dfrac{10}{2 - 2\sin\theta} \cdot \dfrac{\frac{1}{2}}{\frac{1}{2}} = \dfrac{5}{1 - \sin\theta}$; Since $e = 1$, this equation represents a parabola.

$r = \dfrac{10}{2 - 2\sin\theta} \Rightarrow r(2 - 2\sin\theta) = 10 \Rightarrow 2r - 2r\sin\theta = 10 \Rightarrow 2r = 2r\sin\theta + 10 \Rightarrow$

$(2r)^2 = (2r\sin\theta + 10)^2 \Rightarrow (2r)^2 = (2y + 10)^2 \Rightarrow 4r^2 = 4y^2 + 40y + 100 \Rightarrow$

$4(x^2 + y^2) = 4y^2 + 40y + 100 = 0 \Rightarrow 4x^2 - 40y - 100 = 0 \Rightarrow x^2 - 10y - 25 = 0$

The rectangular form of the equation is $x^2 - 10y - 25 = 0$.

28. $r = \dfrac{12}{4 - 4\sin\theta} \cdot \dfrac{\frac{1}{4}}{\frac{1}{4}} = \dfrac{3}{1 - \sin\theta}$; Since $e = 1$, this equation represents a parabola.

$r = \dfrac{12}{4 - 4\sin\theta} \Rightarrow r(4 - 4\sin\theta) = 12 \Rightarrow 4r - 4r\sin\theta = 12 \Rightarrow 4r = 4r\sin\theta + 12 \Rightarrow$

$(4r)^2 = (4r\sin\theta + 12)^2 \Rightarrow (4r)^2 = (4y + 12)^2 \Rightarrow 16r^2 = 16y^2 + 96y + 144 \Rightarrow$

$16(x^2 + y^2) = 16y^2 + 96y + 144 = 0 \Rightarrow 16x^2 - 96y - 144 = 0 \Rightarrow x^2 - 6y - 9 = 0$

The rectangular form of the equation is $x^2 - 6y - 9 = 0$.

Appendix E: Rotation of Axes

1. $4x^2 + 3y^2 + 2xy - 5x = 8 \Rightarrow B^2 - 4AC = 2^2 - 4(4)(3) = 4 - 48 < 0.$ Circle, ellipse, or a point.

2. $x^2 + 2xy - 3y^2 + 2y = 12 \Rightarrow B^2 - 4AC = 2^2 - 4(1)(-3) = 4 + 12 > 0.$ Hyperbola or 2 intersecting lines.

3. $2x^2 + 3xy - 4y^2 = 0 \Rightarrow B^2 - 4AC = 3^2 - 4(2)(-4) = 9 + 32 > 0.$ Hyperbola or 2 intersecting lines.

4. $x^2 - 2xy + y^2 + 4x - 8y = 0 \Rightarrow B^2 - 4AC = (-2)^2 - 4(1)(1) = 4 - 4 = 0.$ Parabola, one line, or 2 parallel lines.

5. $4x^2 + 4xy + y^2 + 15 = 0 \Rightarrow B^2 - 4AC = 4^2 - 4(4)(1) = 16 - 16 = 0.$ Parabola, one line, or 2 parallel lines.

6. $-x^2 + 2xy - y^2 + 16 = 0 \Rightarrow B^2 - 4AC = 2^2 - 4(-1)(-1) = 4 - 4 = 0.$ Parabola, one line, or 2 parallel lines.

7. $2x^2 + \sqrt{3}xy + y^2 + x = 5 \Rightarrow \cot 2\theta = \dfrac{A - C}{B} = \dfrac{2 - 1}{\sqrt{3}} \Rightarrow \cot 2\theta = \dfrac{1}{\sqrt{3}} \Rightarrow 2\theta = 60° \Rightarrow \theta = 30°$

8. $4\sqrt{3}x^2 + xy + 3\sqrt{3}y^2 = 10 \Rightarrow \cot 2\theta = \dfrac{A - C}{B} = \dfrac{4\sqrt{3} - 3\sqrt{3}}{1} \Rightarrow \cot 2\theta = \sqrt{3} \Rightarrow 2\theta = 30° \Rightarrow$
   $\theta = 15°$

9. $3x^2 + \sqrt{3}xy + 4y^2 + 2x - 3y = 12 \Rightarrow \cot 2\theta = \dfrac{A - C}{B} = \dfrac{3 - 4}{\sqrt{3}} \Rightarrow \cot 2\theta = -\dfrac{1}{\sqrt{3}} \Rightarrow 2\theta = 120° \Rightarrow$
   $\theta = 60°$

10. $4x^2 + 2xy + 2y^2 + x - 7 = 0 \Rightarrow \cot 2\theta = \dfrac{A - C}{B} = \dfrac{4 - 2}{2} = 1 \Rightarrow \cot 2\theta = 1 \Rightarrow 2\theta = 45° \Rightarrow$
    $\theta = 22.5°$

11. $x^2 - 4xy + 5y^2 = 18 \Rightarrow \cot 2\theta = \dfrac{A - C}{B} = \dfrac{1 - 5}{-4} = 1 \Rightarrow \cot 2\theta = 1 \Rightarrow 2\theta = 45° \Rightarrow \theta = 22.5°$

12. $3\sqrt{3}x^2 - 2xy + \sqrt{3}y^2 = 25 \Rightarrow \cot 2\theta = \dfrac{A - C}{B} = \dfrac{3\sqrt{3} - \sqrt{3}}{-2} = \dfrac{-2\sqrt{3}}{2} \Rightarrow \cot 2\theta = -\sqrt{3} \Rightarrow$
    $2\theta = 150° \Rightarrow \theta = 75°$

13. $x^2 - xy + y^2 = 6$ [1]; $\theta = 45°$; $x = x' \cos\theta - y' \sin\theta = \dfrac{\sqrt{2}}{2}x' - \dfrac{\sqrt{2}}{2}y'$ [2];

    $y = x' \sin\theta + y' \cos\theta = \dfrac{\sqrt{2}}{2}x' + \dfrac{\sqrt{2}}{2}y'$ [3]. Substitute [2] and [3] in [1].

    $\left(\dfrac{\sqrt{2}}{2}x' - \dfrac{\sqrt{2}}{2}y'\right)^2 - \left(\dfrac{\sqrt{2}}{2}x' - \dfrac{\sqrt{2}}{2}y'\right)\left(\dfrac{\sqrt{2}}{2}x' + \dfrac{\sqrt{2}}{2}y'\right) + \left(\dfrac{\sqrt{2}}{2}x' + \dfrac{\sqrt{2}}{2}y'\right)^2 = 6 \Rightarrow$

    $\dfrac{1}{2}x'^2 - x'y' + \dfrac{1}{2}y'^2 - \dfrac{1}{2}x'^2 + \dfrac{1}{2}y'^2 + \dfrac{1}{2}x'^2 + x'y' + \dfrac{1}{2}y'^2 = 6 \Rightarrow \dfrac{1}{2}x'^2 + \dfrac{3}{2}y'^2 = 6 \Rightarrow \dfrac{x'^2}{12} + \dfrac{y'^2}{4} = 1.$

    See Figure 13.

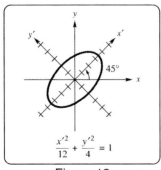

Figure 13

14. $2x^2 - xy + 2y^2 = 25$ [1]; $\theta = 45°$; $x = x' \cos \theta - y' \sin \theta = \dfrac{\sqrt{2}}{2}x' - \dfrac{\sqrt{2}}{2}y'$ [2];

$y = x' \sin \theta + y' \cos \theta = \dfrac{\sqrt{2}}{2}x' + \dfrac{\sqrt{2}}{2}y'$ [3]. Substitute [2] and [3] in [1].

$2\left(\dfrac{\sqrt{2}}{2}x' - \dfrac{\sqrt{2}}{2}y'\right)^2 - \left(\dfrac{\sqrt{2}}{2}x' - \dfrac{\sqrt{2}}{2}y'\right)\left(\dfrac{\sqrt{2}}{2}x' + \dfrac{\sqrt{2}}{2}y'\right) + 2\left(\dfrac{\sqrt{2}}{2}x' + \dfrac{\sqrt{2}}{2}y'\right)^2 = 25 \Rightarrow$

$2\left(\dfrac{1}{2}x'^2 - x'y' + \dfrac{1}{2}y'^2\right) - \left(\dfrac{1}{2}x'^2 - \dfrac{1}{2}y'^2\right) + 2\left(\dfrac{1}{2}x'^2 + x'y' + \dfrac{1}{2}y'^2\right) = 25 \Rightarrow$

$x'^2 - 2x'y' + y'^2 - \dfrac{1}{2}x'^2 + \dfrac{1}{2}y'^2 + x'^2 + 2x'y' + y'^2 = 25 \Rightarrow \dfrac{3}{2}x'^2 + \dfrac{5}{2}y'^2 = 25 \Rightarrow \dfrac{3x'^2}{50} + \dfrac{5y'^2}{50} = 1 \Rightarrow$

$\dfrac{3x'^2}{50} + \dfrac{y'^2}{10} = 1$. See Figure 14.

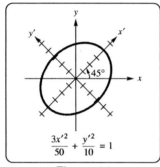

$\dfrac{3x'^2}{50} + \dfrac{y'^2}{10} = 1$

Figure 14

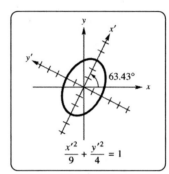

$\dfrac{x'^2}{9} + \dfrac{y'^2}{4} = 1$

Figure 15

15. $8x^2 - 4xy + 5y^2 = 36$ [1]; $\sin \theta = \dfrac{2}{\sqrt{5}}, y = 2, r = \sqrt{5}, x = \sqrt{5-4} = 1 \Rightarrow \cos \theta = \dfrac{1}{\sqrt{5}}$;

$x = x' \cos \theta - y' \sin \theta = \dfrac{1}{\sqrt{5}}x' - \dfrac{2}{\sqrt{5}}y'$ [2]; $y = x' \sin \theta + y' \cos \theta = \dfrac{2}{\sqrt{5}}x' + \dfrac{1}{\sqrt{5}}y'$ [3].

Substitute [2] and [3] in [1].

$8\left(\dfrac{1}{\sqrt{5}}x' - \dfrac{2}{\sqrt{5}}y'\right)^2 - 4\left(\dfrac{1}{\sqrt{5}}x' - \dfrac{2}{\sqrt{5}}y'\right)\left(\dfrac{2}{\sqrt{5}}x' + \dfrac{1}{\sqrt{5}}y'\right) + 5\left(\dfrac{2}{\sqrt{5}}x' + \dfrac{1}{\sqrt{5}}y'\right)^2 = 36 \Rightarrow$

$8\left(\dfrac{1}{5}x'^2 - \dfrac{4}{5}x'y' + \dfrac{4}{5}y'^2\right) - 4\left(\dfrac{2}{5}x'^2 - \dfrac{3}{5}x'y' - \dfrac{2}{5}y'^2\right) + 5\left(\dfrac{4}{5}x'^2 + \dfrac{4}{5}x'y' + \dfrac{1}{5}y'^2\right) = 36 \Rightarrow$

$\dfrac{8}{5}x'^2 - \dfrac{32}{5}x'y' + \dfrac{32}{5}y'^2 - \dfrac{8}{5}x'^2 + \dfrac{12}{5}x'y' + \dfrac{8}{5}y'^2 + 4x'^2 + 4x'y' + y'^2 = 36 \Rightarrow 4x'^2 + 9y'^2 = 36 \Rightarrow$

$\dfrac{x'^2}{9} + \dfrac{y'^2}{4} = 1$. See Figure 15.

16. $5y^2 + 12xy = 10$ [1]; $\sin\theta = \dfrac{3}{\sqrt{13}}$, $y = 3$, $r = \sqrt{13}$, $x = \sqrt{13 - 9} = 2 \Rightarrow \cos\theta = \dfrac{2}{\sqrt{13}}$;

$x = x'\cos\theta - y'\sin\theta = \dfrac{2}{\sqrt{13}}x' - \dfrac{3}{\sqrt{13}}y'$ [2]; $y = x'\sin\theta + y'\cos\theta = \dfrac{3}{\sqrt{13}}x' + \dfrac{2}{\sqrt{13}}y'$ [3].

Substitute [2] and [3] in [1]. $5\left(\dfrac{3}{\sqrt{13}}x' + \dfrac{2}{\sqrt{13}}y'\right)^2 + 12\left(\dfrac{2}{\sqrt{13}}x' - \dfrac{3}{\sqrt{13}}y'\right)\left(\dfrac{3}{\sqrt{13}}x' + \dfrac{2}{\sqrt{13}}y'\right) = 10 \Rightarrow$

$5\left(\dfrac{9}{13}x'^2 + \dfrac{12}{13}x'y' + \dfrac{4}{13}y'^2\right) + 12\left(\dfrac{6}{13}x'^2 - \dfrac{5}{13}x'y' - \dfrac{6}{13}y'^2\right) = 10 \Rightarrow$

$\dfrac{45}{13}x'^2 + \dfrac{60}{13}x'y' + \dfrac{20}{13}y'^2 + \dfrac{72}{13}x'^2 - \dfrac{60}{13}x'y' - \dfrac{72}{13}y'^2 = 10 \Rightarrow \dfrac{117}{13}x'^2 - \dfrac{52}{13}y'^2 = 10 \Rightarrow$

$9x'^2 - 4y'^2 = 10 \Rightarrow \dfrac{9}{10}x'^2 - \dfrac{2}{5}y'^2 = 1$. See Figure 16.

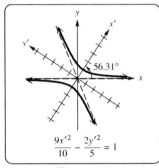

$\dfrac{9x'^2}{10} - \dfrac{2y'^2}{5} = 1$

Figure 16

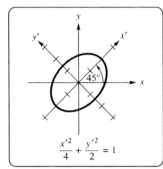

$\dfrac{x'^2}{4} + \dfrac{y'^2}{2} = 1$

Figure 17

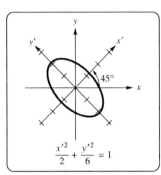

$\dfrac{x'^2}{2} + \dfrac{y'^2}{6} = 1$

Figure 18

17. $3x^2 - 2xy + 3y^2 = 8$ [1]; $\cot 2\theta = \dfrac{A - C}{B} = \dfrac{3 - 3}{-2} = 0 \Rightarrow 2\theta = 90° \Rightarrow \theta = 45°$;

$x = x'\cos\theta - y'\sin\theta = \dfrac{\sqrt{2}}{2}x' - \dfrac{\sqrt{2}}{2}y'$ [2]; $y = x'\sin\theta + y'\cos\theta = \dfrac{\sqrt{2}}{2}x' + \dfrac{\sqrt{2}}{2}y'$ [3].

Substitute [2] and [3] in [1].

$3\left(\dfrac{\sqrt{2}}{2}x' - \dfrac{\sqrt{2}}{2}y'\right)^2 - 2\left(\dfrac{\sqrt{2}}{2}x' - \dfrac{\sqrt{2}}{2}y'\right)\left(\dfrac{\sqrt{2}}{2}x' + \dfrac{\sqrt{2}}{2}y'\right) + 3\left(\dfrac{\sqrt{2}}{2}x' + \dfrac{\sqrt{2}}{2}y'\right)^2 = 8 \Rightarrow$

$3\left(\dfrac{1}{2}x'^2 - x'y' + \dfrac{1}{2}y'^2\right) - 2\left(\dfrac{1}{2}x'^2 - \dfrac{1}{2}y'^2\right) + 3\left(\dfrac{1}{2}x'^2 + x'y' + \dfrac{1}{2}y'^2\right) = 8 \Rightarrow$

$\dfrac{3}{2}x'^2 - 3x'y' + \dfrac{3}{2}y'^2 - x'^2 + y'^2 + \dfrac{3}{2}x'^2 + 3x'y' + \dfrac{3}{2}y'^2 = 8 \Rightarrow 2x'^2 + 4y'^2 = 8 \Rightarrow \dfrac{x'^2}{4} + \dfrac{y'^2}{2} = 1$

See Figure 17.

18. $x^2 + xy + y^2 = 3$ [1]; $\cot 2\theta = \dfrac{A - C}{B} = \dfrac{1 - 1}{1} = 0 \Rightarrow 2\theta = 90° \Rightarrow \theta = 45°$;

$x = x'\cos\theta - y'\sin\theta = \dfrac{\sqrt{2}}{2}x' - \dfrac{\sqrt{2}}{2}y'$ [2]; $y = x'\sin\theta + y'\cos\theta = \dfrac{\sqrt{2}}{2}x' + \dfrac{\sqrt{2}}{2}y'$ [3].

Substitute [2] and [3] in [1].

$\left(\dfrac{\sqrt{2}}{2}x' - \dfrac{\sqrt{2}}{2}y'\right)^2 + \left(\dfrac{\sqrt{2}}{2}x' - \dfrac{\sqrt{2}}{2}y'\right)\left(\dfrac{\sqrt{2}}{2}x' + \dfrac{\sqrt{2}}{2}y'\right) + \left(\dfrac{\sqrt{2}}{2}x' + \dfrac{\sqrt{2}}{2}y'\right)^2 = 3 \Rightarrow$

$\dfrac{1}{2}x'^2 - x'y' + \dfrac{1}{2}y'^2 + \dfrac{1}{2}x'^2 - \dfrac{1}{2}y'^2 + \dfrac{1}{2}x'^2 + x'y' + \dfrac{1}{2}y'^2 = 3 \Rightarrow \dfrac{3}{2}x'^2 + \dfrac{1}{2}y'^2 = 3 \Rightarrow \dfrac{x'^2}{2} + \dfrac{y'^2}{6} = 1$

See Figure 18.

19. $x^2 - 4xy + y^2 = -5$ [1]; $\cot 2\theta = \dfrac{A - C}{B} = \dfrac{1 - 1}{-4} = 0 \Rightarrow 2\theta = 90° \Rightarrow \theta = 45°$;

$x = x' \cos \theta - y' \sin \theta = \dfrac{\sqrt{2}}{2}x' - \dfrac{\sqrt{2}}{2}y'$ [2]; $y = x' \sin \theta + y' \cos \theta = \dfrac{\sqrt{2}}{2}x' + \dfrac{\sqrt{2}}{2}y'$ [3].

Substitute [2] and [3] in [1].

$\left(\dfrac{\sqrt{2}}{2}x' - \dfrac{\sqrt{2}}{2}y'\right)^2 - 4\left(\dfrac{\sqrt{2}}{2}x' - \dfrac{\sqrt{2}}{2}y'\right)\left(\dfrac{\sqrt{2}}{2}x' + \dfrac{\sqrt{2}}{2}y'\right) + \left(\dfrac{\sqrt{2}}{2}x' + \dfrac{\sqrt{2}}{2}y'\right)^2 = -5 \Rightarrow$

$\dfrac{1}{2}x'^2 - x'y' + \dfrac{1}{2}y'^2 - 4\left(\dfrac{1}{2}x'^2 - \dfrac{1}{2}y'^2\right) + \dfrac{1}{2}x'^2 + x'y' + \dfrac{1}{2}y'^2 = -5 \Rightarrow$

$\dfrac{1}{2}x'^2 - x'y' + \dfrac{1}{2}y'^2 - 2x'^2 + 2y'^2 + \dfrac{1}{2}x'^2 + x'y' + \dfrac{1}{2}y'^2 = -5 \Rightarrow -x'^2 + 3y'^2 = -5 \Rightarrow \dfrac{x'^2}{5} - \dfrac{3y'^2}{5} = 1.$

See Figure 19.

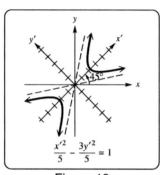

$\dfrac{x'^2}{5} - \dfrac{3y'^2}{5} = 1$

Figure 19

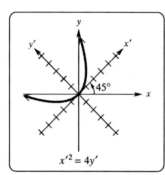

$x'^2 = 4y'$

Figure 20

20. $x^2 + 2xy + y^2 + 4\sqrt{2}x - 4\sqrt{2}y = 0$ [1]; $\cot 2\theta = \dfrac{A - C}{B} = \dfrac{1 - 1}{2} = 0 \Rightarrow 2\theta = 90° \Rightarrow \theta = 45°$;

$x = x' \cos \theta - y' \sin \theta = \dfrac{\sqrt{2}}{2}x' - \dfrac{\sqrt{2}}{2}y'$ [2]; $y = x' \sin \theta + y' \cos \theta = \dfrac{\sqrt{2}}{2}x' + \dfrac{\sqrt{2}}{2}y'$ [3].

Substitute [2] and [3] in [1]. $\left(\dfrac{\sqrt{2}}{2}x' - \dfrac{\sqrt{2}}{2}y'\right)^2 + 2\left(\dfrac{\sqrt{2}}{2}x' - \dfrac{\sqrt{2}}{2}y'\right)\left(\dfrac{\sqrt{2}}{2}x' + \dfrac{\sqrt{2}}{2}y'\right) +$

$\left(\dfrac{\sqrt{2}}{2}x' + \dfrac{\sqrt{2}}{2}y'\right)^2 + 4\sqrt{2}\left(\dfrac{\sqrt{2}}{2}x' - \dfrac{\sqrt{2}}{2}y'\right) - 4\sqrt{2}\left(\dfrac{\sqrt{2}}{2}x' + \dfrac{\sqrt{2}}{2}y'\right) = 0 \Rightarrow$

$\dfrac{1}{2}x'^2 - x'y' + \dfrac{1}{2}y'^2 + 2\left(\dfrac{1}{2}x'^2 - \dfrac{1}{2}y'^2\right) + \dfrac{1}{2}x'^2 + x'y' + \dfrac{1}{2}y'^2 + 4x' - 4y' - 4x' - 4y' = 0 \Rightarrow$

$\dfrac{1}{2}x'^2 - x'y' + \dfrac{1}{2}y'^2 + x'^2 - y'^2 + \dfrac{1}{2}x'^2 + x'y' + \dfrac{1}{2}y'^2 + 4x' - 4y' - 4x' - 4y' = 0 \Rightarrow$

$2x'^2 - 8y' = 0 \Rightarrow x'^2 = 4y'.$ See Figure 20.

21. $7x^2 + 6\sqrt{3}xy + 13y^2 = 64$ [1]; $\cot 2\theta = \dfrac{A - C}{B} = \dfrac{7 - 13}{6\sqrt{3}} = \dfrac{-6}{6\sqrt{3}} = -\dfrac{1}{\sqrt{3}} \Rightarrow 2\theta = 120° \Rightarrow \theta = 60°;$

$x = x' \cos\theta - y' \sin\theta = \dfrac{1}{2}x' - \dfrac{\sqrt{3}}{2}y'$ [2]; $y = x' \sin\theta + y' \cos\theta = \dfrac{\sqrt{3}}{2}x' + \dfrac{1}{2}y'$ [3].

Substitute [2] and [3] in [1].

$$7\left(\dfrac{1}{2}x' - \dfrac{\sqrt{3}}{2}y'\right)^2 + 6\sqrt{3}\left(\dfrac{1}{2}x' - \dfrac{\sqrt{3}}{2}y'\right)\left(\dfrac{\sqrt{3}}{2}x' + \dfrac{1}{2}y'\right) + 13\left(\dfrac{\sqrt{3}}{2}x' + \dfrac{1}{2}y'\right)^2 = 64 \Rightarrow$$

$$7\left(\dfrac{1}{4}x'^2 - \dfrac{\sqrt{3}}{2}x'y' + \dfrac{3}{4}y'^2\right) + 6\sqrt{3}\left(\dfrac{\sqrt{3}}{4}x'^2 - \dfrac{1}{2}x'y' - \dfrac{\sqrt{3}}{4}y'^2\right) + 13\left(\dfrac{3}{4}x'^2 + \dfrac{\sqrt{3}}{2}x'y' + \dfrac{1}{4}y'^2\right) = 64 \Rightarrow$$

$$\dfrac{7}{4}x'^2 - \dfrac{7\sqrt{3}}{2}x'y' + \dfrac{21}{4}y'^2 + \dfrac{18}{4}x'^2 - \dfrac{6\sqrt{3}}{2}x'y' - \dfrac{18}{4}y'^2 + \dfrac{39}{4}x'^2 + \dfrac{13\sqrt{3}}{2}x'y' + \dfrac{13}{4}y'^2 = 64 \Rightarrow$$

$$16x'^2 + 4y'^2 = 64 \Rightarrow \dfrac{x'^2}{4} + \dfrac{y'^2}{16} = 1. \text{ See Figure 21.}$$

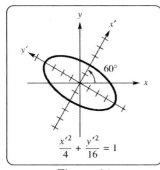

$$\dfrac{x'^2}{4} + \dfrac{y'^2}{16} = 1$$

Figure 21

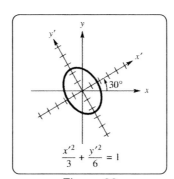

$$\dfrac{x'^2}{3} + \dfrac{y'^2}{6} = 1$$

Figure 22

22. $7x^2 + 2\sqrt{3}xy + 5y^2 = 24$ [1]; $\cot 2\theta = \dfrac{A - C}{B} = \dfrac{7 - 5}{2\sqrt{3}} = \dfrac{1}{\sqrt{3}} \Rightarrow 2\theta = 60° \Rightarrow \theta = 30°;$

$x = x' \cos\theta - y' \sin\theta = \dfrac{\sqrt{3}}{2}x' - \dfrac{1}{2}y'$ [2]; $y = x' \sin\theta + y' \cos\theta = \dfrac{1}{2}x' + \dfrac{\sqrt{3}}{2}y'$ [3].

Substitute [2] and [3] in [1].

$$7\left(\dfrac{\sqrt{3}}{2}x' - \dfrac{1}{2}y'\right)^2 + 2\sqrt{3}\left(\dfrac{\sqrt{3}}{2}x' - \dfrac{1}{2}y'\right)\left(\dfrac{1}{2}x' + \dfrac{\sqrt{3}}{2}y'\right) + 5\left(\dfrac{1}{2}x' + \dfrac{\sqrt{3}}{2}y'\right)^2 = 24 \Rightarrow$$

$$7\left(\dfrac{3}{4}x'^2 - \dfrac{\sqrt{3}}{2}x'y' + \dfrac{1}{4}y'^2\right) + 2\sqrt{3}\left(\dfrac{\sqrt{3}}{4}x'^2 + \dfrac{1}{2}x'y' - \dfrac{\sqrt{3}}{4}y'^2\right) + 5\left(\dfrac{1}{4}x'^2 + \dfrac{\sqrt{3}}{2}x'y' + \dfrac{3}{4}y'^2\right) = 24 \Rightarrow$$

$$\dfrac{21}{4}x'^2 - \dfrac{7\sqrt{3}}{2}x'y' + \dfrac{7}{4}y'^2 + \dfrac{6}{4}x'^2 + \sqrt{3}x'y' - \dfrac{6}{4}y'^2 + \dfrac{5}{4}x'^2 + \dfrac{5\sqrt{3}}{2}x'y' + \dfrac{15}{4}y'^2 = 24 \Rightarrow$$

$$8x'^2 + 4y'^2 = 24 \Rightarrow \dfrac{x'^2}{3} + \dfrac{y'^2}{6} = 1. \text{ See Figure 22.}$$

23. $3x^2 - 2\sqrt{3}xy + y^2 - 2x - 2\sqrt{3}y = 0$ [1]; $\cot 2\theta = \dfrac{A - C}{B} = \dfrac{3 - 1}{-2\sqrt{3}} = -\dfrac{1}{\sqrt{3}} \Rightarrow 2\theta = 120° \Rightarrow$

$\theta = 60°$; $x = x' \cos\theta - y' \sin\theta = \dfrac{1}{2}x' - \dfrac{\sqrt{3}}{2}y'$ [2]; $y = x' \sin\theta + y' \cos\theta = \dfrac{\sqrt{3}}{2}x' + \dfrac{1}{2}y'$ [3].

Substitute [2] and [3] in [1]. $3\left(\dfrac{1}{2}x' - \dfrac{\sqrt{3}}{2}y'\right)^2 - 2\sqrt{3}\left(\dfrac{1}{2}x' - \dfrac{\sqrt{3}}{2}y'\right)\left(\dfrac{\sqrt{3}}{2}x' + \dfrac{1}{2}y'\right) +$

$\left(\dfrac{\sqrt{3}}{2}x' + \dfrac{1}{2}y'\right)^2 - 2\left(\dfrac{1}{2}x' - \dfrac{\sqrt{3}}{2}y'\right) - 2\sqrt{3}\left(\dfrac{\sqrt{3}}{2}x' + \dfrac{1}{2}y'\right) = 0 \Rightarrow 3\left(\dfrac{1}{4}x'^2 - \dfrac{\sqrt{3}}{2}x'y' + \dfrac{3}{4}y'^2\right) -$

$2\sqrt{3}\left(\dfrac{\sqrt{3}}{4}x'^2 - \dfrac{1}{2}x'y' - \dfrac{\sqrt{3}}{4}y'^2\right) + \left(\dfrac{3}{4}x'^2 + \dfrac{\sqrt{3}}{2}x'y' + \dfrac{1}{4}y'^2\right) - x' + \sqrt{3}y' - 3x' - \sqrt{3}y' = 0 \Rightarrow$

$\dfrac{3}{4}x'^2 - \dfrac{3\sqrt{3}}{2}x'y' + \dfrac{9}{4}y'^2 - \dfrac{3}{2}x'^2 + \sqrt{3}x'y' + \dfrac{3}{2}y'^2 + \dfrac{3}{4}x'^2 + \dfrac{\sqrt{3}}{2}x'y' + \dfrac{1}{4}y'^2 - 4x' = 0 \Rightarrow$

$4y'^2 - 4x' = 0 \Rightarrow 4y'^2 = 4x' \Rightarrow y'^2 = x'$. See Figure 23.

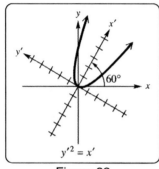

$y'^2 = x'$

Figure 23

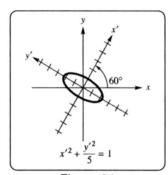

$x'^2 + \dfrac{y'^2}{5} = 1$

Figure 24

24. $2x^2 + 2\sqrt{3}xy + 4y^2 = 5$ [1]; $\cot 2\theta = \dfrac{A - C}{B} = \dfrac{2 - 4}{2\sqrt{3}} = -\dfrac{1}{\sqrt{3}} \Rightarrow 2\theta = 120° \Rightarrow \theta = 60°$;

$x = x' \cos\theta - y' \sin\theta = \dfrac{1}{2}x' - \dfrac{\sqrt{3}}{2}y'$ [2]; $y = x' \sin\theta + y' \cos\theta = \dfrac{\sqrt{3}}{2}x' + \dfrac{1}{2}y'$ [3].

Substitute [2] and [3] in [1].

$2\left(\dfrac{1}{2}x' - \dfrac{\sqrt{3}}{2}y'\right)^2 + 2\sqrt{3}\left(\dfrac{1}{2}x' - \dfrac{\sqrt{3}}{2}y'\right)\left(\dfrac{\sqrt{3}}{2}x' + \dfrac{1}{2}y'\right) + 4\left(\dfrac{\sqrt{3}}{2}x' + \dfrac{1}{2}y'\right)^2 = 5 \Rightarrow$

$2\left(\dfrac{1}{4}x'^2 - \dfrac{\sqrt{3}}{2}x'y' + \dfrac{3}{4}y'^2\right) + 2\sqrt{3}\left(\dfrac{\sqrt{3}}{4}x'^2 - \dfrac{1}{2}x'y' - \dfrac{\sqrt{3}}{4}y'^2\right) + 4\left(\dfrac{3}{4}x'^2 + \dfrac{\sqrt{3}}{2}x'y' + \dfrac{1}{4}y'^2\right) = 5 \Rightarrow$

$\dfrac{1}{2}x'^2 - \sqrt{3}x'y' + \dfrac{3}{2}y'^2 + \dfrac{3}{2}x'^2 - \sqrt{3}x'y' - \dfrac{3}{2}y'^2 + 3x'^2 + 2\sqrt{3}x'y' + y'^2 = 5 \Rightarrow 5x'^2 + y'^2 = 5 \Rightarrow$

$x'^2 + \dfrac{y'^2}{5} = 1$. See Figure 24.

25. $x^2 + 3xy + y^2 - 5\sqrt{2}y = 15$ [1]; $\cot 2\theta = \dfrac{A - C}{B} = \dfrac{1 - 1}{3} = 0 \Rightarrow 2\theta = 90° \Rightarrow \theta = 45°$;

$x = x' \cos\theta - y' \sin\theta = \dfrac{\sqrt{2}}{2}x' - \dfrac{\sqrt{2}}{2}y'$ [2]; $y = x' \sin\theta + y' \cos\theta = \dfrac{\sqrt{2}}{2}x' + \dfrac{\sqrt{2}}{2}y'$ [3].

Substitute [2] and [3] in [1]. $\left(\dfrac{\sqrt{2}}{2}x' - \dfrac{\sqrt{2}}{2}y'\right)^2 + 3\left(\dfrac{\sqrt{2}}{2}x' - \dfrac{\sqrt{2}}{2}y'\right)\left(\dfrac{\sqrt{2}}{2}x' + \dfrac{\sqrt{2}}{2}y'\right) +$

$\left(\dfrac{\sqrt{2}}{2}x' + \dfrac{\sqrt{2}}{2}y'\right)^2 - 5\sqrt{2}\left(\dfrac{\sqrt{2}}{2}x' + \dfrac{\sqrt{2}}{2}y'\right) = 15 \Rightarrow$

$\dfrac{1}{2}x'^2 - x'y' + \dfrac{1}{2}y'^2 + 3\left(\dfrac{1}{2}x'^2 - \dfrac{1}{2}y'^2\right) + \dfrac{1}{2}x'^2 + x'y' + \dfrac{1}{2}y'^2 - 5x' - 5y' = 15 \Rightarrow$

$\dfrac{1}{2}x'^2 - x'y' + \dfrac{1}{2}y'^2 + \dfrac{3}{2}x'^2 - \dfrac{3}{2}y'^2 + \dfrac{1}{2}x'^2 + x'y' + \dfrac{1}{2}y'^2 - 5x' - 5y' = 15 \Rightarrow$

$\dfrac{5}{2}x'^2 - \dfrac{1}{2}y'^2 - 5x' - 5y' = 15 \Rightarrow 5x'^2 - 10x' - y'^2 - 10y' = 30 \Rightarrow$

$5(x'^2 - 2x' + 1) - (y'^2 + 10y' + 25) = 30 + 5 - 25 \Rightarrow 5(x' - 1)^2 - (y' + 5)^2 = 10 \Rightarrow$

$\dfrac{(x' - 1)^2}{2} - \dfrac{(y' + 5)^2}{10} = 1$. See Figure 25.

The graph of the equation is a hyperbola with its center at $(1, -5)$. By translating the axes of the $x'y'$-system

down 5 units and right 1 unit, we get an $x''y''$-coordinate system, in which the hyperbola is centered at the

origin. Thus $\dfrac{x''^2}{2} - \dfrac{y''^2}{10} = 1$.

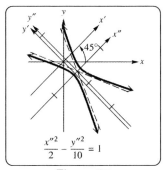

Figure 25

26. $x^2 - \sqrt{3}xy + 2\sqrt{3}x - 3y - 3 = 0$ [1]; $\cot 2\theta = \dfrac{A - C}{B} = \dfrac{1 - 0}{-\sqrt{3}} = -\dfrac{1}{\sqrt{3}} \Rightarrow 2\theta = 120° \Rightarrow \theta = 60°$;

$x = x' \cos\theta - y' \sin\theta = \dfrac{1}{2}x' - \dfrac{\sqrt{3}}{2}y'$ [2]; $y = x' \sin\theta + y' \cos\theta = \dfrac{\sqrt{3}}{2}x' + \dfrac{1}{2}y'$ [3].

Substitute [2] and [3] in [1]. $\left(\dfrac{1}{2}x' - \dfrac{\sqrt{3}}{2}y'\right)^2 - \sqrt{3}\left(\dfrac{1}{2}x' - \dfrac{\sqrt{3}}{2}y'\right)\left(\dfrac{\sqrt{3}}{2}x' + \dfrac{1}{2}y'\right) +$

$2\sqrt{3}\left(\dfrac{1}{2}x' - \dfrac{\sqrt{3}}{2}y'\right) - 3\left(\dfrac{\sqrt{3}}{2}x' + \dfrac{1}{2}y'\right) = 3 \Rightarrow$

$\dfrac{1}{4}x'^2 - \dfrac{\sqrt{3}}{2}x'y' + \dfrac{3}{4}y'^2 - \sqrt{3}\left(\dfrac{\sqrt{3}}{4}x'^2 - \dfrac{1}{2}x'y' - \dfrac{\sqrt{3}}{4}y'^2\right) + \sqrt{3}x' - 3y' - \dfrac{3\sqrt{3}}{2}x' - \dfrac{3}{2}y' = 3 \Rightarrow$

$\dfrac{1}{4}x'^2 - \dfrac{\sqrt{3}}{2}x'y' + \dfrac{3}{4}y'^2 - \dfrac{3}{4}x'^2 + \dfrac{\sqrt{3}}{2}x'y' + \dfrac{3}{4}y'^2 + \sqrt{3}x' - 3y' - \dfrac{3\sqrt{3}}{2}x' - \dfrac{3}{2}y' = 3 \Rightarrow$

$-\dfrac{1}{2}x'^2 + \dfrac{3}{2}y'^2 - \dfrac{\sqrt{3}}{2}x' - \dfrac{9}{2}y' = 3 \Rightarrow -x'^2 + 3y'^2 - \sqrt{3}x' - 9y' = 6 \Rightarrow$

$-\left(x'^2 + \sqrt{3}x' + \dfrac{3}{4}\right) + 3\left(y'^2 - 3y' + \dfrac{9}{4}\right) = 6 - \dfrac{3}{4} + \dfrac{27}{4} \Rightarrow -\left(x' + \dfrac{\sqrt{3}}{2}\right)^2 + 3\left(y' - \dfrac{3}{2}\right)^2 = 12 \Rightarrow$

$\dfrac{\left(y' - \frac{3}{2}\right)^2}{4} - \dfrac{\left(x' + \frac{\sqrt{3}}{2}\right)^2}{12} = 1$. See Figure 26.

The graph of the equation is a hyperbola with its center at $\left(-\dfrac{\sqrt{3}}{2}, \dfrac{3}{2}\right)$. By translating the axes of the

$x'y'$-system up $\dfrac{3}{2}$ units and left $\dfrac{\sqrt{3}}{2}$ units, we get an $x''y''$-coordinate system, in which the hyperbola is centered

at the origin. Thus $\dfrac{y''^2}{4} - \dfrac{x''^2}{12} = 1$.

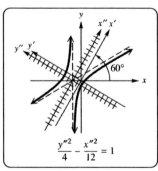

Figure 26

27. $4x^2 + 4xy + y^2 - 24x + 38y - 19 = 0$ [1]; $\cot 2\theta = \dfrac{A - C}{B} = \dfrac{4 - 1}{4} = \dfrac{3}{4} \Rightarrow 2\theta \approx 53.13° \Rightarrow$

$\theta \approx 26.57°$. For 2θ; $x = 3, y = 4, r = \sqrt{3^2 + 4^2} = 5 \Rightarrow \cos 2\theta = \dfrac{3}{5}$.

$\sin\theta = \sqrt{\dfrac{1 - \cos 2\theta}{2}} = \sqrt{\dfrac{1 - \frac{3}{5}}{2}} = \sqrt{\dfrac{2}{10}} = \dfrac{\sqrt{5}}{5}$; $\cos\theta = \sqrt{\dfrac{1 + \cos 2\theta}{2}} = \sqrt{\dfrac{1 + \frac{3}{5}}{2}} = \sqrt{\dfrac{8}{10}} = \dfrac{2\sqrt{5}}{5}$;

$x = x' \cos\theta - y' \sin\theta = \dfrac{2\sqrt{5}}{5}x' - \dfrac{\sqrt{5}}{5}y'$ [2]; $y = x' \sin\theta + y' \cos\theta = \dfrac{\sqrt{5}}{5}x' + \dfrac{2\sqrt{5}}{5}y'$ [3].

Substitute [2] and [3] in [1]. $4\left(\dfrac{2\sqrt{5}}{5}x' - \dfrac{\sqrt{5}}{5}y'\right)^2 + 4\left(\dfrac{2\sqrt{5}}{5}x' - \dfrac{\sqrt{5}}{5}y'\right)\left(\dfrac{\sqrt{5}}{5}x' + \dfrac{2\sqrt{5}}{5}y'\right) +$

$\left(\dfrac{\sqrt{5}}{5}x' + \dfrac{2\sqrt{5}}{5}y'\right)^2 - 24\left(\dfrac{2\sqrt{5}}{5}x' - \dfrac{\sqrt{5}}{5}y'\right) + 38\left(\dfrac{\sqrt{5}}{5}x' + \dfrac{2\sqrt{5}}{5}y'\right) = 19 \Rightarrow$

$4\left(\dfrac{4}{5}x'^2 - \dfrac{4}{5}x'y' + \dfrac{1}{5}y'^2\right) + 4\left(\dfrac{2}{5}x'^2 + \dfrac{3}{5}x'y' - \dfrac{2}{5}y'^2\right) + \left(\dfrac{1}{5}x'^2 + \dfrac{4}{5}x'y' + \dfrac{4}{5}y'^2\right) - \dfrac{48\sqrt{5}}{5}x' + \dfrac{25\sqrt{5}}{5}y' +$

$\dfrac{38\sqrt{5}}{5}x' + \dfrac{76\sqrt{5}}{5}y' = 19 \Rightarrow$

$\dfrac{16}{5}x'^2 - \dfrac{16}{5}x'y' + \dfrac{4}{5}y'^2 + \dfrac{8}{5}x'^2 + \dfrac{12}{5}x'y' - \dfrac{8}{5}y'^2 + \dfrac{1}{5}x'^2 + \dfrac{4}{5}x'y' + \dfrac{4}{5}y'^2 - \dfrac{48\sqrt{5}}{5}x' + \dfrac{24\sqrt{5}}{5}y' +$

$\dfrac{38\sqrt{5}}{5}x' + \dfrac{76\sqrt{5}}{5}y' = 19 \Rightarrow 5x'^2 - 2\sqrt{5}x' + 20\sqrt{5}y' = 19 \Rightarrow$

$5\left(x'^2 - \dfrac{2\sqrt{5}}{5}x' + \dfrac{1}{5}\right) + 20\sqrt{5}y' = 19 + 1 \Rightarrow 5\left(x' - \dfrac{\sqrt{5}}{5}\right)^2 = 20 - 20\sqrt{5}y \Rightarrow$

$\left(x' - \dfrac{\sqrt{5}}{5}\right)^2 = 4 - 4\sqrt{5}y' \Rightarrow \left(x' - \dfrac{\sqrt{5}}{5}\right)^2 = -4\sqrt{5}\left(y' - \dfrac{\sqrt{5}}{5}\right)$. See Figure 27.

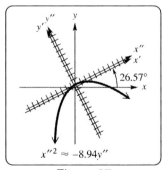

Figure 27

28. $12x^2 + 24xy + 19y^2 - 12x - 40y + 31 = 0$ [1]; $\cot 2\theta = \dfrac{A-C}{B} = \dfrac{12-19}{24} = -\dfrac{7}{24} \Rightarrow$

$2\theta \approx 106.26° \Rightarrow \theta \approx 53.13°$. For 2θ; $x = -7, y = 24, r = \sqrt{(-7)^2 + 24^2} = 25 \Rightarrow \cos 2\theta = -\dfrac{7}{25}$.

$\sin \theta = \sqrt{\dfrac{1 - \cos 2\theta}{2}} = \sqrt{\dfrac{1 + \frac{7}{25}}{2}} = \sqrt{\dfrac{32}{50}} = \dfrac{4}{5}$; $\cos \theta = \sqrt{\dfrac{1 + \cos 2\theta}{2}} = \sqrt{\dfrac{1 - \frac{7}{25}}{2}} = \sqrt{\dfrac{18}{50}} = \dfrac{3}{5}$;

$x = x' \cos \theta - y' \sin \theta = \dfrac{3}{5}x' - \dfrac{4}{5}y'$ [2]; $y = x' \sin \theta + y' \cos \theta = \dfrac{4}{5}x' + \dfrac{3}{5}y'$ [3].

Substitute [2] and [3] in [1]. $12\left(\dfrac{3}{5}x' - \dfrac{4}{5}y'\right)^2 + 24\left(\dfrac{3}{5}x' - \dfrac{4}{5}y'\right)\left(\dfrac{4}{5}x' + \dfrac{3}{5}y'\right) +$

$19\left(\dfrac{4}{5}x' + \dfrac{3}{5}y'\right)^2 - 12\left(\dfrac{3}{5}x' - \dfrac{4}{5}y'\right) - 40\left(\dfrac{4}{5}x' + \dfrac{3}{5}y'\right) = -31 \Rightarrow$

$12\left(\dfrac{9}{25}x'^2 - \dfrac{24}{25}x'y' + \dfrac{16}{25}y'^2\right) + 24\left(\dfrac{12}{25}x'^2 - \dfrac{7}{25}x'y' - \dfrac{12}{25}y'^2\right) + 19\left(\dfrac{16}{25}x'^2 + \dfrac{24}{25}x'y' + \dfrac{9}{25}y'^2\right) -$

$\dfrac{36}{5}x' + \dfrac{48}{5}y' - \dfrac{160}{5}x' - \dfrac{120}{5}y' = -31 \Rightarrow$

$\dfrac{108}{25}x'^2 - \dfrac{288}{25}x'y' + \dfrac{192}{25}y'^2 + \dfrac{288}{25}x'^2 - \dfrac{168}{25}x'y' - \dfrac{288}{25}y'^2 + \dfrac{304}{25}x'^2 + \dfrac{456}{25}x'y' + \dfrac{171}{25}y'^2 - \dfrac{196}{5}x' -$

$\dfrac{72}{5}y' = -31 \Rightarrow 28x'^2 + 3y'^2 - \dfrac{196}{5}x' - \dfrac{72}{5}y' = -31 \Rightarrow 140x'^2 + 15y'^2 - 196x' - 72y' = -155 \Rightarrow$

$140\left(x'^2 - \dfrac{7}{5}x' + \dfrac{49}{100}\right) + 15\left(y'^2 - \dfrac{72}{15}y' + \dfrac{1296}{225}\right) = -155 + 68.6 + 86.4 \Rightarrow$

$140(x' - 0.7)^2 + 15(y' - 2.4)^2 = 0$. The graph of the equation is the point $(0.7, 2.4)$.

29. $16x^2 + 24xy + 9y^2 - 130x + 90y = 0$ [1]; $\cot 2\theta = \dfrac{A - C}{B} = \dfrac{16 - 9}{24} = \dfrac{7}{24} \Rightarrow$

$2\theta \approx 73.74° \Rightarrow \theta \approx 36.87°$. For 2θ; $x = 7$, $y = 24$, $r = \sqrt{7^2 + 24^2} = 25 \Rightarrow \cos 2\theta = \dfrac{7}{25}$.

$\sin\theta = \sqrt{\dfrac{1 - \cos 2\theta}{2}} = \sqrt{\dfrac{1 - \frac{7}{25}}{2}} = \sqrt{\dfrac{18}{50}} = \dfrac{3}{5}$; $\cos\theta = \sqrt{\dfrac{1 + \cos 2\theta}{2}} = \sqrt{\dfrac{1 + \frac{7}{25}}{2}} = \sqrt{\dfrac{32}{50}} = \dfrac{4}{5}$;

$x = x'\cos\theta - y'\sin\theta = \dfrac{4}{5}x' - \dfrac{3}{5}y'$ [2]; $y = x'\sin\theta + y'\cos\theta = \dfrac{3}{5}x' + \dfrac{4}{5}y'$ [3].

Substitute [2] and [3] in [1]. $16\left(\dfrac{4}{5}x' - \dfrac{3}{5}y'\right)^2 + 24\left(\dfrac{4}{5}x' - \dfrac{3}{5}y'\right)\left(\dfrac{3}{5}x' + \dfrac{4}{5}y'\right) +$

$9\left(\dfrac{3}{5}x' + \dfrac{4}{5}y'\right)^2 - 130\left(\dfrac{4}{5}x' - \dfrac{3}{5}y'\right) + 90\left(\dfrac{3}{5}x' + \dfrac{4}{5}y'\right) = 0 \Rightarrow$

$16\left(\dfrac{16}{25}x'^2 - \dfrac{24}{25}x'y' + \dfrac{9}{25}y'^2\right) + 24\left(\dfrac{12}{25}x'^2 + \dfrac{7}{25}x'y' - \dfrac{12}{25}y'^2\right) + 9\left(\dfrac{9}{25}x'^2 + \dfrac{24}{25}x'y' + \dfrac{16}{25}y'^2\right) -$

$104x' + 78y' + 54x' + 72y' = 0 \Rightarrow \dfrac{256}{25}x'^2 - \dfrac{384}{25}x'y' + \dfrac{144}{25}y'^2 + \dfrac{288}{25}x'^2 + \dfrac{168}{25}x'y' - \dfrac{288}{25}y'^2 +$

$\dfrac{81}{25}x'^2 + \dfrac{216}{25}x'y' + \dfrac{144}{25}y'^2 - 50x' + 150y' = 0 \Rightarrow 25x'^2 - 50x' + 150y' = 0 \Rightarrow$

$25(x'^2 - 2x' + 1) = -150y' + 25 \Rightarrow 25(x' - 1)^2 = -150\left(y' - \dfrac{1}{6}\right) \Rightarrow (x' - 1)^2 = -6\left(y' - \dfrac{1}{6}\right)$.

See Figure 29. The graph of the equation is a parabola with its vertex at $\left(1, \dfrac{1}{6}\right)$. By translating the axes of the

$x'y'$-system up $\dfrac{1}{6}$ units and right 1 unit, we get an $x''y''$-coordinate system, in which the parabola is centered at

the origin. Thus $x''^2 = -6y''$.

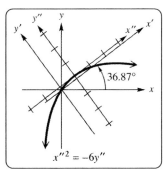

Figure 29

30. $9x^2 - 6xy + y^2 - 12\sqrt{10}x - 36\sqrt{10}y = 0$ [1]; $\cot 2\theta = \dfrac{A - C}{B} = \dfrac{9 - 1}{-6} = -\dfrac{4}{3} \Rightarrow$

$2\theta \approx 143.13° \Rightarrow \theta \approx 71.57°$. For 2θ; $x = -4, y = 3, r = \sqrt{(-4)^2 + 3^2} = 5 \Rightarrow \cos 2\theta = -\dfrac{4}{5}$.

$\sin \theta = \sqrt{\dfrac{1 - \cos 2\theta}{2}} = \sqrt{\dfrac{1 + \frac{4}{5}}{2}} = \sqrt{\dfrac{9}{10}} = \dfrac{3\sqrt{10}}{10}$; $\cos \theta = \sqrt{\dfrac{1 + \cos 2\theta}{2}} = \sqrt{\dfrac{1 - \frac{4}{5}}{2}} = \sqrt{\dfrac{1}{10}} = \dfrac{\sqrt{10}}{10}$;

$x = x' \cos \theta - y' \sin \theta = \dfrac{\sqrt{10}}{10}x' - \dfrac{3\sqrt{10}}{10}y'$ [2]; $y = x' \sin \theta + y' \cos \theta = \dfrac{3\sqrt{10}}{10}x' + \dfrac{\sqrt{10}}{10}y'$ [3].

Substitute [2] and [3] in [1]. $9\left(\dfrac{\sqrt{10}}{10}x' - \dfrac{3\sqrt{10}}{10}y'\right)^2 - 6\left(\dfrac{\sqrt{10}}{10}x' - \dfrac{3\sqrt{10}}{10}y'\right)\left(\dfrac{3\sqrt{10}}{10}x' + \dfrac{\sqrt{10}}{10}y'\right) +$

$\left(\dfrac{3\sqrt{10}}{10}x' + \dfrac{\sqrt{10}}{10}y'\right)^2 - 12\sqrt{10}\left(\dfrac{\sqrt{10}}{10}x' - \dfrac{3\sqrt{10}}{10}y'\right) - 36\sqrt{10}\left(\dfrac{3\sqrt{10}}{10}x' + \dfrac{\sqrt{10}}{10}y'\right) = 0 \Rightarrow$

$9\left(\dfrac{1}{10}x'^2 - \dfrac{3}{5}x'y' + \dfrac{9}{10}y'^2\right) - 6\left(\dfrac{3}{10}x'^2 - \dfrac{4}{5}x'y' - \dfrac{3}{10}y'^2\right) + \dfrac{9}{10}x'^2 + \dfrac{3}{5}x'y' + \dfrac{1}{10}y'^2 -$

$12x' + 36y' - 108x' - 36y' = 0 \Rightarrow \dfrac{9}{10}x'^2 - \dfrac{27}{5}x'y' + \dfrac{81}{10}y'^2 - \dfrac{18}{10}x'^2 + \dfrac{24}{5}x'y' + \dfrac{18}{10}y'^2 +$

$\dfrac{9}{10}x'^2 + \dfrac{3}{5}x'y' + \dfrac{1}{10}y'^2 - 120x' = 0 \Rightarrow 10y'^2 - 120x' = 0 \Rightarrow 10y'^2 = 120x' \Rightarrow y'^2 = 12x'$.

See Figure 30. The graph of the equation is a parabola with its vertex at $(0, 0)$.

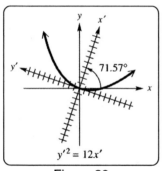

$y'^2 = 12x'$

Figure 30